Biology of Microorganisms

Fifth Edition

Thomas D. Brock
University of Wisconsin

Michael T. Madigan
Southern Illinois University

PRENTICE HALL
Englewood Cliffs, New Jersey 07632

Library of Congress Cataloging-in-Publication Data

Brock, Thomas D.
Biology of microorganisms.

Bibliography: p.
Includes index.
1. Microbiology. I. Madigan, Michael T., (date).
II. Title.
QR41.2.B77 1988 576 87-25811
ISBN 0-13-076829-4

Book Design and Page Layout: Thomas D. Brock
Line Art: Science Tech Publishers
Supervision: Ed Phillips and Thomas D. Brock
Typesetting and Production: Science Tech Publishers/Impressions, Inc.
Editorial/Production Supervision:
Prentice Hall: Eleanor Henshaw Hiatt
Science Tech: Thomas D. Brock
Cover Design: Jules Perlmutter, Off-Broadway Graphics
Manufacturing Buyer: Paula Benevento
Cover Photograph: Ben B. Bohlool, University of Hawaii

Cover photo shows an infection thread formed by ***Rhizobium trifolii*** on white clover, *Trifolium repens*, examined by fluorescence microscopy. A number of bacteria can be seen attached to the root hair. The infection thread consists of a cellulose deposit down which the bacterial cells move to the root.

(Credits and acknowledgments appear on pp. 807-808, which constitute a continuation of the copyright page.)

A Division of Simon & Schuster
Englewood Cliffs, New Jersey 07632

Printed in the United States of America
10 9 8 7 6 5 4 3 2 1

ISBN 0-13-076829-4 01

Prentice-Hall International (UK) Limited, *London*
Prentice-Hall of Australia Pty. Limited, *Sydney*
Prentice-Hall Canada Inc., *Toronto*
Prentice-Hall Hispanoamericana, S.A., *Mexico*
Prentice-Hall of India Private Limited, *New Delhi*
Prentice-Hall of Japan, Inc., *Tokyo*
Simon & Schuster Asia Pte. Ltd., *Singapore*
Editora Prentice-Hall do Brasil, Ltda., *Rio de Janeiro*

Contents

4 Nutrition, Metabolism, and Biosynthesis 113

5 Macromolecules and Molecular Genetics 150

6 Viruses 203

15 Major Microbial Diseases 513

16 Metabolic Diversity among the Microorganisms 552

17 Microbial Ecology 598

Preface

We are pleased to present this fifth edition of *Biology of Microorganisms* to students and instructors of microbiology. Although each edition of this textbook has represented a marked advance over earlier editions, this present edition represents the most complete revision that has ever been accomplished. A quick glance will tell the reader why: The book has been produced for the first time completely in full color. The extensive use of color made it necessary to redesign the whole book. Most of the line drawings are completely new, and even those that were retained from earlier editions have, to a great extent, been fully redrawn. The number of full-color photographs has also been greatly increased. Microbiology is a colorful subject. For the first time in a college textbook, this colorful subject is revealed to the student.

Although the new full-color design is the most obvious change from earlier editions, the content of the book has also been greatly altered. The science of microbiology continues to undergo rapid changes through the impact of new developments in cell and molecular biology, and through major advances in medical and environmental microbiology. The revolution in microbial genetics, first reflected in the previous edition, is now fully presented in the fifth edition. Gene cloning and genetic engineering are having profound impact on the teaching and practice of microbiology. Microbes continue to make important research tools for the study of many fundamental biological processes. Yet microbes are more than research tools—they are of considerable importance and interest in themselves. Therefore, the second major focus of this book is on what microbes are doing in the world at large. Basic research in such areas as ecology and evolution is advancing rapidly because of our increased understanding of fundamental microbial processes. At the same time, practical developments in biotechnology, food processing, and agriculture arise from applications of microbiological principles. And whole areas, such as virology and immunology, are moving so rapidly that they must be treated in completely new ways. However, the basic approach which has made this book so popular in earlier editions has not been changed.

So many new things have been added to this edition that it is more efficient to list them in outline form. Details can be found in the table of contents and in the chapters themselves.

1. Chapter 1, greatly enlarged in this new edition, is now subtitled "An Overview of Microbiology and Cell Biology." It not only includes a brief, well-illustrated introduction to the nature of microbial life but also an extensive section on microbiological methods, material requested by many instructors.

2. Chapter 2, "Cell Chemistry," is a completely new chapter, added to provide a compact source of information on the important chemicals of living organisms. Chemistry is, to a great extent, the language of biology, and we have felt that by providing this overview early in the book, students will have a convenient reference source for review of the basic chemistry of microbial life.
3. Chapter 3, "Cell Biology," incorporates much of the material found in Chapters 2 and 3 of the previous edition, but greatly enhanced and presented in an integrated fashion. Most of the figures in this chapter are completely new, and have been greatly enhanced by the full-color treatment.
4. Chapter 4, "Nutrition, Metabolism, and Biosynthesis," contains a more compact treatment of microbial metabolism than that found in Chapters 4 and 5 of the previous edition. The present Chapter 4 presents only the key concepts of energetics and heterotrophic metabolism, and the specialized material has been moved to the completely new Chapter 16.
5. Chapter 5, "Macromolecules and Molecular Genetics," is of course almost completely new, reflecting the phenomenal advances that have been made in this field. We have also placed this chapter earlier in the book, reflecting our conviction that genetics is so important and pervasive in modern microbiology that it must be covered as early as possible. Chapter 5 also provides an introduction to the following molecularly oriented chapters.
6. The field of virology has changed so remarkably that the virology chapter in this book has been completely rewritten, and the material on animal viruses greatly expanded. The chapter has been organized in such a way that instructors can choose those kinds of materials that will be most useful for their specific courses.
7. In Chapter 7, "Microbial Genetics," the material on transposons and plasmids has been considerably enhanced, and a whole new section on genetics of yeast has been added, reflecting the current exciting research on these important eucaryotic microorganisms.
8. The chapter "Gene Manipulation and Genetic Engineering," so popular in the previous edition, has been updated and enlarged, and new material on the uses of genetic engineering added.
9. The growth process can be viewed either as a process of macromolecular biosynthesis or as a process with enormous practical implications. Chapter 9, "Growth and Its Control," has been moved so that it comes *after* rather than before the molecular material, thus making it possible to discuss microbial growth in macromolecular terms. But at the same time, the practical material on the *control* of microbial growth has been considerably increased. Because microbial growth in foods is one of the most important practical considerations, a whole new section on *food microbiology* has been added.
10. Continuing with the innovations in this new edition, a whole new chapter, "Microbial Biotechnology," has been added. A number of important industrial microbial processes predate the era of genetic engineering so that this chapter attempts to provide a complete overview of economically important microbial processes. For added student interest, beer- and wine-making are given extensive treatment for the first time in this book (including a brief section on home brewing).
11. Beginning with Chapter 11, a series of chapters, either new or greatly rewritten, provides extensive coverage of the field of medical microbiology and immunology. Chapter 11 begins this part of the book with a presentation of the principles of *host-parasite relationships*.
12. Probably no field in biology has advanced so rapidly, and has become so important both theoretically and practically, as immunology. "Immunology and Immunity," a chapter so popular in the last edition, has been considerably enhanced and has benefited from extensive new illustrations.
13. Chapter 13, "Clinical and Diagnostic Microbiology", is completely new in this edition. We have been impressed not only by the great sophistication of modern diagnostic methods in microbiology but by the way in which genetics is revolutionizing this field. Building on the genetics background of Chapters 5–9, the present chapter has an extensive treatment of the new *monoclonal antibody* and *nucleic acid probe* technology for diagnostic microbiology.
14. Continuing the expanded coverage of medical microbiology, Chapter 14, "Epidemiology and Public Health Microbiology," is a brief but almost completely new treatment of this subject.
15. Responding to the requests of many instructors, we have added a separate, rather lengthy, chapter "Major Microbial Diseases." The diseases in this chapter are arranged, as much as is possible, by *organ systems* affected or by *mechanisms of transmission*. As such, this chapter will complement Chapter 19, which provides the conventional taxonomic treatment of pathogens along with the groups of other bacteria.
16. Probably no new chapter has been written so enthusiastically by your two authors as has Chapter

16, "Metabolic Diversity among the Microorganisms." In the previous editions, some of this material was included in the chapter on metabolism or in separate chapters on biosynthesis and autotrophy, but it came too early in the book to warrant the detailed treatment that it deserves. In the present chapter, an integrated picture of metabolic diversity is presented for the first time. Note that this chapter also provides an overview of material needed for the definitive discussion of ecology, evolution, and systematics in the next several chapters.

17. Chapter 17, "Microbial Ecology," while retaining much of the material from the previous edition, has been considerably expanded by the addition of material on *enrichment culture*, *measurement of microbial activity in nature*, and *plant-microbe interactions* (with emphasis on *Rhizobium* and *Agrobacterium*).

18. Chapter 18, "Molecular Systematics and Microbial Evolution," is completely new and reflects the current excitement in the use of molecular techniques to classify microorganisms. In this chapter, the new classification of microbes which emphasizes the vast differences between *eubacteria* and *archaebacteria* has been given in considerable detail.

19. Finally, Chapter 19, "The Bacteria," provides that detailed and authoritative reference to procaryotic diversity that instructors have come to expect from this book.

The instructor or student reading the above list of innovations can well imagine why we are so excited and enthusiastic about this new edition of "Biology of Microorganisms." New chapters, extensive revisions, new design, and especially, full-color art! However, this is not the end of the innovations. A whole array of new supplements have been developed by the publisher to make the book more accessible to the student and more useful to the instructor. These supplements include a Study Guide, Test Item File, Computerized Test Item File, Instructor's Manual, and Color Transparencies. In addition, Study Questions have been included at the end of each chapter in the book itself.

This edition has been greatly benefited from the computer-assisted publication skills of Science Tech Publishers, the company responsible for complete editorial and production matters. The use of the computer has made it possible to revise and insert important new information almost up until the end of the book production cycle. The following individuals working at or under the auspices of Science Tech deserve special recognition: Irene Slater for careful keyboarding of a lengthy and extremely complex manuscript; Susan Fariss, Elizabeth McBride, and Kathy Rosner for editorial assistance; Ed Phillips, Lisa Buckley, Cathy Schmaeng, Julie Whitty, Randy Kampen, Lucy Taylor, and Tony Dunn for art; Ruth Siegel for photo research, Barbara Littlewood for the index, and Kathie Brock for consultation. Thomas D. Brock was responsible for the design and for supervision of all editorial, art, typesetting, and production.

Many individuals have reviewed the manuscript or portions of it, or made special efforts to provide color photographs and we are extremely grateful for their efforts. These include:

- Jan-Hendrik Becking, Wageningen University
- Ben B. Bohlool, University of Hawaii
- Willy Burgdorfer, Rocky Mountain Laboratory
- Thomas R. Corner, Michigan State University
- Robert Gallo, National Institutes of Health
- Patricia Grilione, San Jose State University
- Holger Jannasch, Woods Hole Oceanographic Institute
- James Hu, Massachusetts Institute of Technology
- Ronald E. Hurlbert, Washington State University
- Allan Konopka, Purdue University
- Donald Lehman, Wright State University
- Alan Liss, State University of New York at Binghamton
- John Martinko, Southern Illinois University
- William Matthai, Tarrant County Junior College
- William McClain, University of Wisconsin-Madison
- James E. Miller, Delaware Valley College
- Kenneth Nealson, University of Wisconsin-Milwaukee
- Norbert Pfennig, University of Konstanz
- Francis Steiner, Hillsdale College
- Howard Temin, University of Wisconsin-Madison
- Carl Woese, University of Illinois-Urbana
- Stephen Zinder, Cornell University

In addition, the Credits section at the end of the book lists the many more who have provided photographs.

The authors would also like to pay special thanks to David Smith of the University of Delaware, a coauthor on the fourth edition, who was unable because of circumstances to participate in the present revision. His advice and consultation during the early planning stages for the present edition were extremely helpful. MTM wishes to thank his wife, Nancy, for her patience and understanding during this very time-consuming revision. TDB wishes to thank MTM for his efforts above and beyond the call of duty, Robert Sickles of Prentice-Hall Editorial for his support in bringing out this innovative edition, and Eleanor Hiatt of Prentice-Hall Production for her careful supervision and unfailing good humor.

Introduction: An Overview of Microbiology and Cell Biology

Microbiology is the study of microorganisms, a large and diverse group of organisms that exist as single cells or cell clusters. Microbial cells are thus distinct from the cells of animals and plants, which are unable to live alone in nature but can exist only as parts of multicellular organisms (Figure 1.1). A single microbial cell is generally able to carry out its life processes of growth, energy generation, and reproduction independently of other cells, either of the same kind or of different kinds.

What, then, is microbiology all about? We can list several aspects of microbiology here: 1) It is about living cells and how they work. 2) It is about microorganisms, an important class of cells capable of free-living (independent) existence. 3) It is about microbial diversity and evolution, about how different kinds of microbes arise, and why. 4) It is about what microbes do, in the world at large, in human society, in our own bodies, in the bodies of animals and plants. 5) It is about the central role which microbiology plays as a basic biological science and how an understanding of microbiology helps in the understanding of the biology of higher organisms, even including humans.

Why study microbiology? Microbiology, one of the most important of the biological sciences, is studied for two major reasons:

1. As a *basic biological science*, microbiology provides some of the most accessible research tools for probing the nature of life processes. Our most sophisticated understanding of the chemical and physical principles behind living processes has arisen from studies using microorganisms.
2. As an *applied biological science*, microbiology deals with many important practical problems in medicine, agriculture, and industry. Some of the most important diseases of humans, animals, and plants are caused by microorganisms. Microorganisms play major roles in soil fertility and animal production. Many large-scale industrial processes are microbially based, and are now classified under the heading of *biotechnology*.

In this book, both the basic and applied aspects of microbiology are covered in an integrated fashion. We discuss the experimental basis of microbiology, the general principles of cell structure and function, the classification and diversity of microorganisms, biochemical processes in cells, and the genetic basis of microbial growth and evolution. From an applied viewpoint, we discuss disease processes in humans that are caused by microorganisms, the roles of microorganisms in food and agriculture, and industrial (biotechnological) processes employing microorganisms.

The material in this textbook serves as a foundation for advanced work in microbiology. It also serves as a basis for further studies in cell biology,

(a)

(b)

(c)

Figure 1.1 Living organisms are composed of cells. Plants (a) and animals (b) are composed of many cells; they are called *multicellular*. A single plant or animal cell cannot have an independent existence; each of its cells is dependent on the other. Microorganisms (c) are free-living cells. A single microbial cell can have an independent existence.

biochemistry, molecular biology, and genetics. Although a student may begin to study microbiology primarily because of its applied problems, the basic concepts learned will serve as a foundation for advanced study in many areas of contemporary biology. A firm grasp of microbiological principles will also serve as a basis for understanding processes in higher organisms, including humans.

1.1 Microorganisms as Cells

The **cell** is the fundamental unit of all living matter. A single cell is an entity, isolated from other cells by a cell membrane (and perhaps a cell wall) and containing within it a variety of materials and subcellular structures (Figure 1.2). The **cell membrane** is the *barrier* which separates the inside of the cell from the outside. Inside the cell membrane are the various structures and chemicals which make it possible for the cell to function. Key structures are the **nucleus** or nuclear region, where the *information* needed to make more cells is stored, and the **cytoplasm**, where the *machinery* for cell growth and function is present.

All cells contain certain types of chemical components: proteins, nucleic acids, lipids, and polysaccharides. Because these chemical components are common throughout the living world, it is thought that all cells have descended from a single common ancestor, the *universal ancestor*. Through billions of years of evolution, the tremendous diversity of cell types that exist today has arisen.

Although each kind of cell has a definite structure and size, a cell is a dynamic unit, constantly undergoing change and replacing its parts. Even when it is not growing, a cell is continually taking materials from its environment and working them into its own

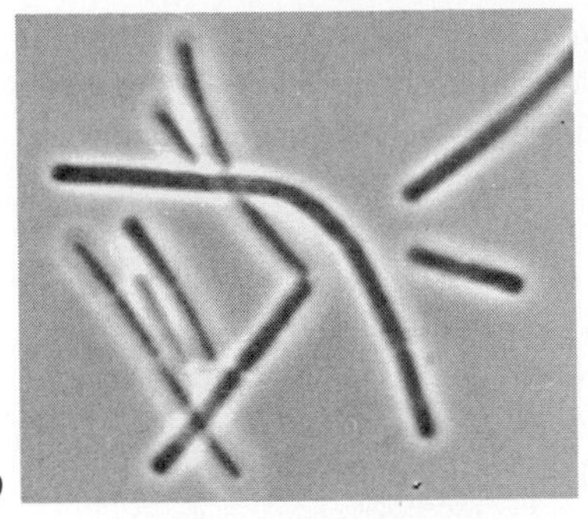

(a)

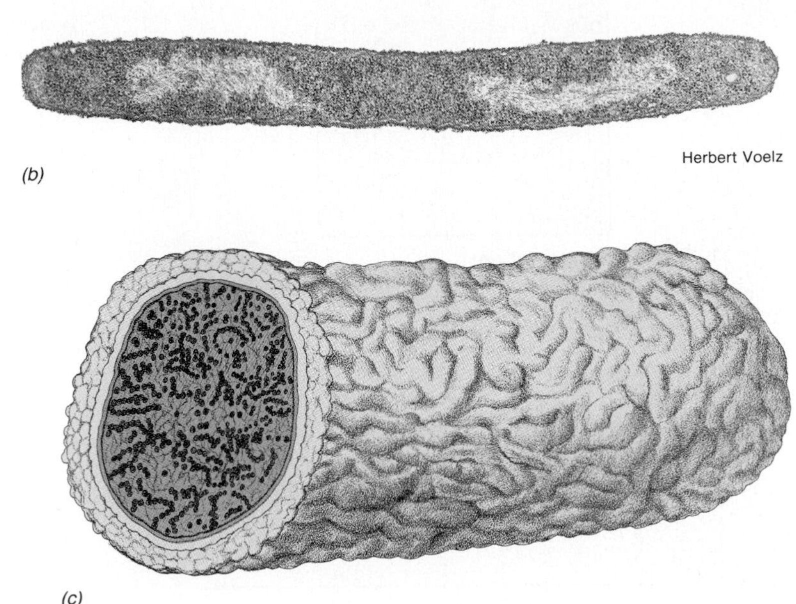

(b)

(c)

Figure 1.2 Cells. (a) Photomicrograph of microbial cells, as seen in the light microscope. (b) Cross section through a bacterial cell, as viewed with an electron microscope. The two lighter areas represent the nuclear region. (c) An artist's interpretation in three dimensions of a bacterial cell, showing the characteristic wrinkled outer structure.

fabric. At the same time, it perpetually discards into its environment cellular materials and waste products. A cell is thus an open system, forever changing, yet generally remaining the same.

The hallmarks of a cell A living cell is a complex chemical system. What are the characteristics that set living cells apart from nonliving chemical systems? We list four major characteristics here, discuss them briefly in this chapter, and in more detail in later chapters.

1. **Self-feeding or nutrition**. Cells take up chemicals from the environment, transform these chemicals from one form to another, release energy, and eliminate waste products (Figure 1.3).
2. **Self-replication or growth**. Cells are capable of directing their own synthesis. A cell grows and divides, forming two cells, each nearly identical to the original cell.
3. **Differentiation**. Most kinds of cells also can undergo changes in form or function. When a cell differentiates, certain substances or structures which were not formed previously are now formed, or substances or structures which had been formed previously are no longer formed. Cell differentiation is often a part of the cellular life cycle, in which cells form spores or other structures involved in sexual reproduction, dispersal, or survival of unfavorable conditions.
4. **Chemical signalling**. Another attribute of cells is that they often *interact* or *communicate* with other cells, generally by means of **chemical signals**. Multicellular organisms, such as plants and animals, are composed of many different cell types that have arisen as a result of differentiation from single cells. In multicellular organisms, complex interactions between these different cell types lead to the behavior and function of these cells. One of the striking things about the cells of multicellular organisms is that they are incapable of independent existence in nature, but only exist as part of a whole plant or animal. This interdependence of the cells of higher organisms is one of the hallmarks of multicellular life. Even in the microbial world chemical communication occurs, although it is less highly developed.

The improbable cell A common view of physics is that the universe is moving toward a condition where everything is in great disorder, with molecules and atoms and elementary particles arranged in the most random manner. If the universe is random, life then seems like a miracle, since the living cell is anything but random, being a highly ordered, exceedingly improbable (nonrandom) structure. More careful analysis of life, however, convinces one that there is no divergence between biology and physics. How can this be? First, we note that the idea of randomness implies that the physical system is in *equilibrium* with its surroundings. A living cell, on the other hand, is definitely *not* in equilibrium with its surroundings. We say that the living cell is a *nonequilibrium* system. How is it possible for a cell to

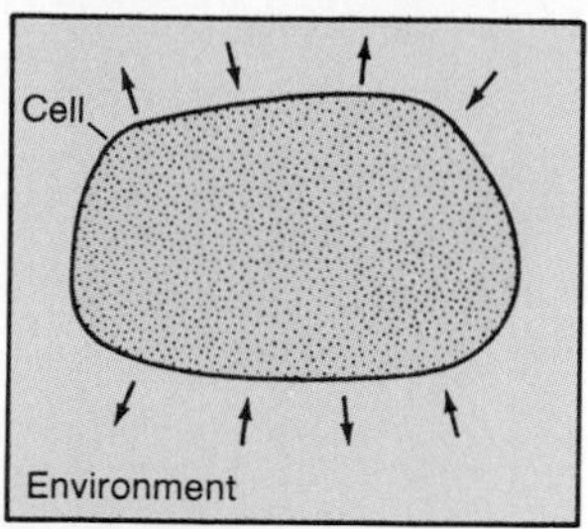

1. Self-feeding (nutrition)
Uptake of chemicals from the environment and elimination of wastes into the environment.

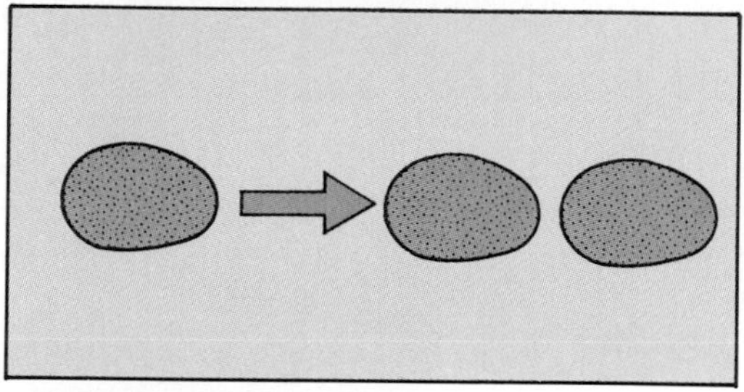

2. Self-replication (growth)
Chemicals from the environment are turned into new cells under the direction of preexisting cells.

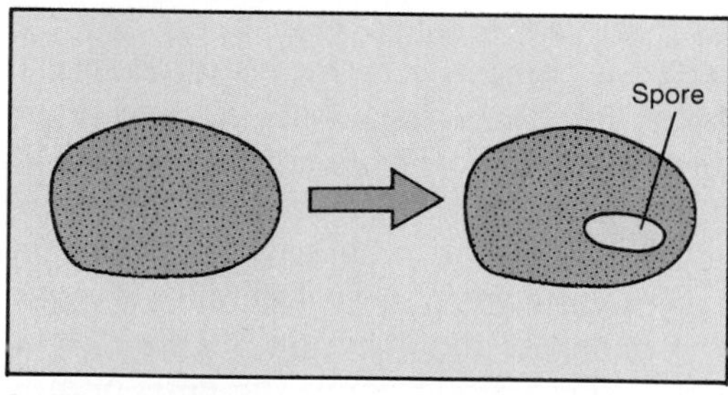

3. Differentiation
Formation of a new cell structure, usually as part of a cellular *life cycle*

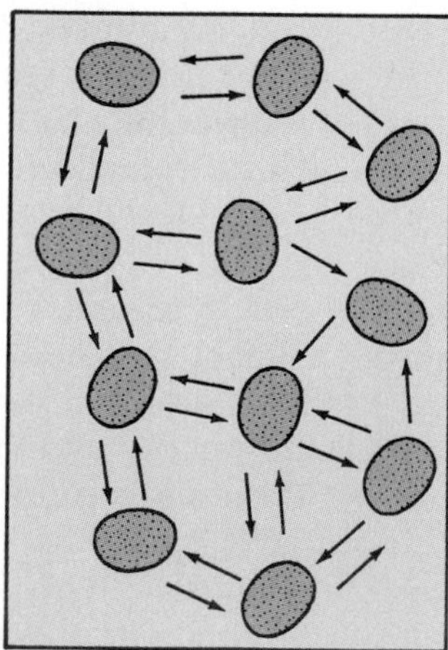

4. Chemical signalling
Cells *communicate* or *interact* primarily by means of chemicals which are released or taken up.

Figure 1.3 The hallmarks of cellular life.

maintain this nonequilibrium condition? A living cell is actually an *open system*, a system in which energy is taken in from the surroundings and used to maintain cell structure.

Viewed in this way, a living cell can be considered as a *chemical system that works*. A cell carries out energy transformations, and some of this energy is used to maintain the structure of the cell itself. What happens if the cell runs out of energy? It deteriorates and eventually dies.

For a cell to function as a cell, its structure must be maintained. If the structure of a cell is destroyed, then the cell usually dies. Biological information, the information needed to produce a new cell, is structural. However, because of the highly improbable nature of a cell, the cell is constantly tending toward degradation, and energy is needed to restore structures as they are being degraded.

Thus, the basis of cell function is cell structure. The importance of structure as a foundation of life is further emphasized when we note that cells are *self-replicating systems*. Cell reproduces cell. Therefore, making a living cell is a matter of making the right structure. This principle, which is the basis of biology, was enunciated many years ago by the famous German cellular pathologist Rudolph Virchow: "Omnis cellula e cellula." "*Every cell from a cell.*"

If every cell comes from a preexisting cell, where did the first cell come from? This question is an example of the classic "chicken and egg" conundrum. Which came first, the chicken or the egg? All those trained in genetics know that it must be the egg. In the same way, the first cell must have come from a noncell, something before the cell, a procellular structure. We will discuss the origin of life and the evolution of cells in some detail in Chapter 18.

The nonrandom nature of a living cell is shown most dramatically by an analysis of its chemical composition and a comparison of that chemical composition with the chemical composition of the earth. The average chemical composition of a living cell is quite different from the average chemical composition of the earth. The cell therefore is not a random assortment of chemical elements found on earth (Figure 1.4). This further emphasizes the special or nonrandom nature of a living cell.

However, because a cell is a nonrandom structure it does not necessarily follow that life is a miracle. Life is indeed an improbable phenomenon, and probably arose only once as a result of a series of highly improbable events that occurred on the primitive earth. However, once a cell arose, subsequent events appear highly probable, a result of the inherent chemical reactions which cells are able to carry out.

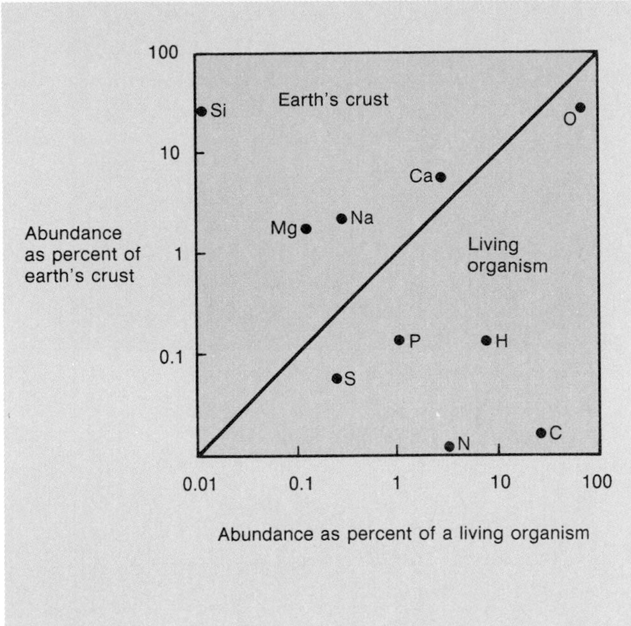

Figure 1.4 Chemical differences between a living organism and the earth. Note that the key elements C, H, O, N, P, and S are much more abundant in living organisms than in nonliving matter. Thus, living organisms *concentrate* these elements from the environment.

Astronomers tell us that there are probably vast numbers of planets in the universe with earth-like conditions on which life might arise or might have arisen. If life is an almost inevitable event when the proper physical and chemical conditions are available, then we can anticipate that there are other planets in the universe with living organisms similar to those found on earth. The precise assemblage of organisms would be different, of course, because so much of evolution depends upon accidents of history, and the time stage and evolution on these other planets would likely be earlier or later than those on earth. But if we were able to sample some of these other planets, it is likely we would not find any surprises.

Two views of a cell Cells can be studied as *machines* which carry out chemical transformations. In this view, the cell is a chemical machine which converts energy from one form to another, breaking molecules down into smaller units, building up larger molecules from smaller ones and carrying out many other kinds of chemical transformations. The term **metabolism** is used to refer to the collective series of chemical processes which occur in living organisms. When we speak of metabolic reactions, we mean *chemical* reactions occurring in living organisms.

Cells can also be studied as *coding devices*, analogous to computers, possessors of information which is either passed on to offspring or is translated. Depending upon our interest, and upon the problem at hand, we may study cells either as chemical machines or as coding devices. In the following we discuss these two attributes briefly.

If a cell is a chemical machine, what are the components which make it function? The components of the cell's chemical machine are **enzymes**, protein molecules capable of catalyzing specific chemical reactions. Whether or not a cell can carry out a particular chemical reaction will depend in the first instance on the presence in the cell of an enzyme that catalyzes the reaction. Specificity of enzymes is frequently very high, so that even very closely related chemical reactions are catalyzed by separate enzymes. The specificity of an enzyme is determined primarily by its structure. Enzymes are proteins. As proteins, enzymes consist of long chains of amino acids (20 different amino acids) which are connected in specific and highly precise ways. It is the amino acid *sequence* of an enzyme that determines its structure as well as its catalytic specificity. Proteins frequently have 100 or more amino acid residues, composed of the 20 types of amino acids. The long chain of amino acids becomes folded into a specific configuration, leading to the formation of various regions or domains which play specific roles in the function of the protein (Figure 1.5). The machine function of a cell (metabolism) is ultimately determined by the amounts and types of the various enzymes of which it is composed.

We now turn to the idea of cells as coding devices. It is when we consider how the amino acid sequence of a protein is determined that we consider cells as coding devices. How is the cell able to arrange amino acids into precise sequences, each leading to the production of a separate kind of protein? To understand this, we must consider the cell as a device for storing information and converting it into the ap-

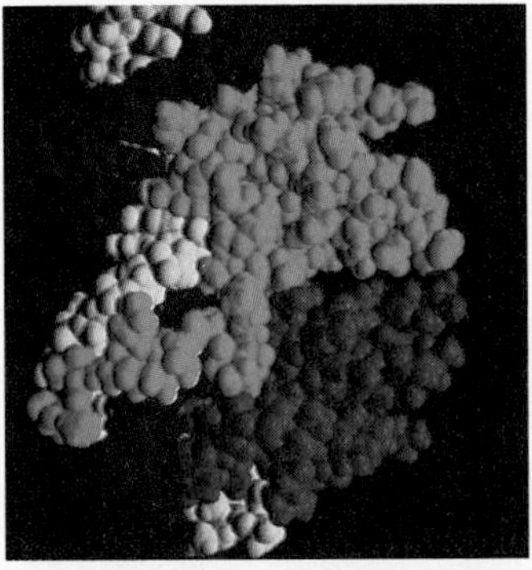

Figure 1.5 The structure of a protein as shown in a computer-generated model. The different domains of the protein are shown in different colors.

propriate form. In this context, the cell can be viewed as a repository of protein sequence information stored in a coded manner.

This code, called the *genetic code*, is stored in a sequence of bases in the hereditary molecule, *deoxyribonucleic acid*, DNA (Figure 1.6). DNA is present in the cell in two molecules that are intertwined to form a helix, the famous DNA double helix. Each molecule is chemically *complementary* to the other, the two molecules being held together by highly specific pairing of the bases of which the molecules are constructed. The genetic code is contained in the *sequence* of bases in the molecule. Each amino acid in a protein is coded by a sequence of three bases, and the genetic code is thus said to be a *triplet* code.

The genetic code has two functions. First, the code must be copied or *replicated* into the two offspring which result when a cell divides. Replication involves copying the information from the DNA molecule into new molecules. As each strand of the DNA double helix opens, a new complement is made by the complementary base pairing just mentioned (Figure 1.6*b*).

Second, the code must be *translated*: the DNA base sequence must be converted into the amino acid sequence of the protein. This translation process is carried out by a special and highly complex translation machinery in the cell. The translation apparatus is at the very core of cell function.

The translation system In human language, it is primarily a convenience if a text is translated from one language to another. In the language of the cell, however, translation is essential if the cell is to function. This is so because the language of the DNA molecule is incapable of being interpreted directly, the protein only being made via the translation system as an intermediate. The translation apparatus is thus the central and most basic attribute of the cell.

Because of the central role of the translation system, it is difficult to change it. In fact, the translation system, although differing in certain details, is essentially the same in all kinds of cells. It is very likely that in the origin of life the translation system was one of the first things that arose, and once formed was retained in essentially the same form throughout the long history of life on earth.

Although we will describe the translation process in some detail in Chapter 5, at the moment we note that translation consists of two processes (Figure 1.7):

1. **Transcription**, the formation of messenger RNA molecules, which contain a copy of the information of the DNA.
2. **Translation**, a process occurring on ribosomes, structures composed of ribonucleic acid (RNA)

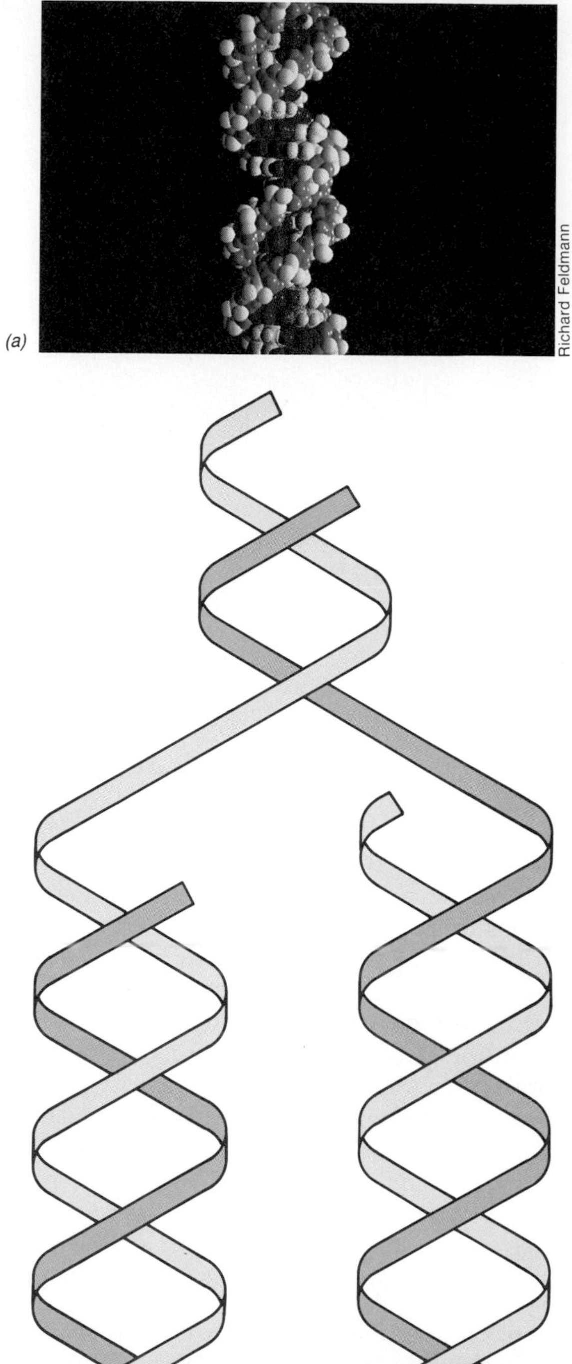

Figure 1.6 The DNA molecule: deoxyribonucleic acid. (a) A computer model of DNA. The blue colored atoms represent the backbone of the double helix. The bases which hold the two chains together are shown in red and yellow. Only a small part of the DNA double helix is shown. (b) The DNA replication process. Each strand of the DNA double helix is copied into a complementary strand.

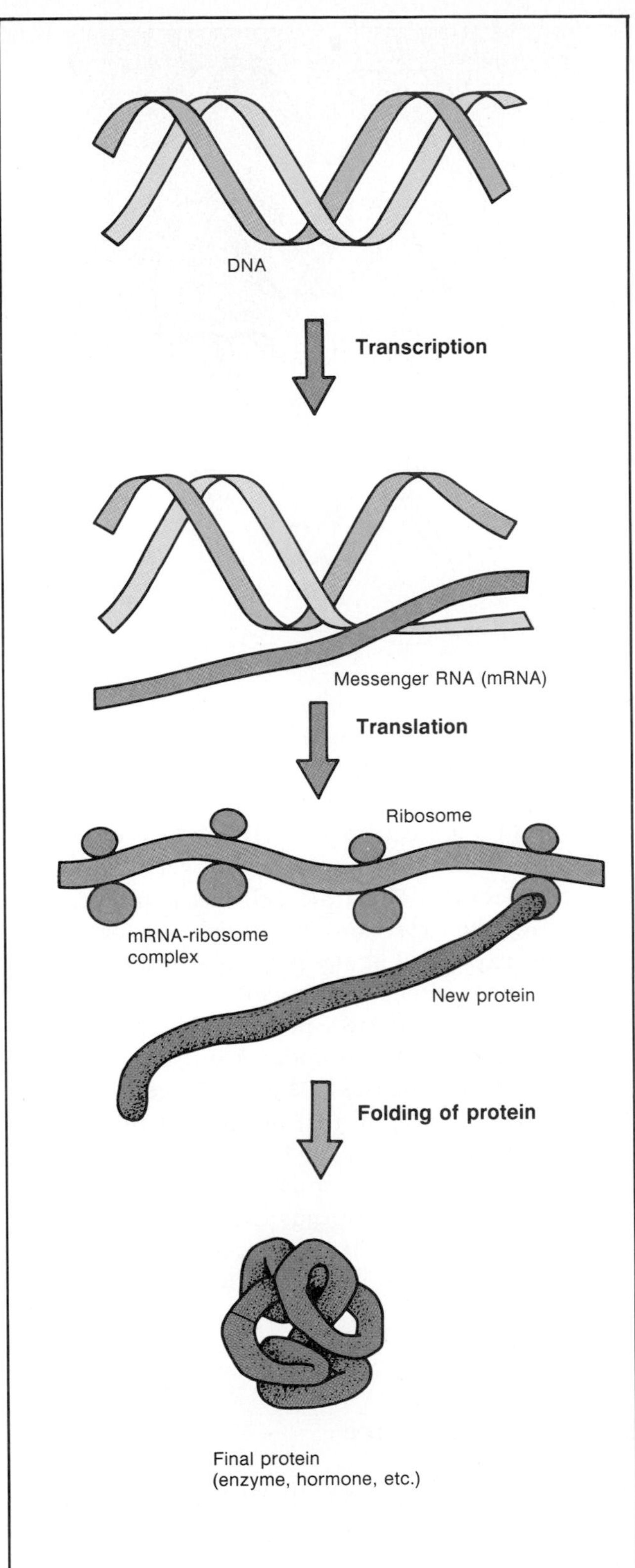

Figure 1.7 The two processes that lead to protein synthesis, transcription and translation.

and protein, on which the translation process occurs.

In the translation process, other RNA molecules called **transfer RNA** act as the critical adaptor molecules of the translation process, recognizing both the amino acid and the genetic code. The messenger RNA combines with ribosomes, forming a complex, and the amino acid-carrying transfer RNA molecules move up to the messenger RNA one at a time and the amino acid being carried is transferred onto the growing protein chain. For each protein, there are three parts of the genetic code, the **start site**, which is the triplet coding unit that indicates where the translation process should begin, the **gene**, a long series of triplet codes that specify the amino acid sequences, and the **stop site**, the triplet coding unit that indicates where the translation process should stop.

The growth of the cell The connection between the two attributes of a cell, its machine function and its coding function, is expressed through the process of *cell growth*. A cell grows in size, then divides and forms two cells. In the orderly growth process that results in two cells being formed from a single cell, all of the constitutents of the cell are doubled in amount. Growth in size requires the functioning of the chemical machinery of the cell, such as energy transformation, biosynthesis, and metabolism. But each of the two cells must contain *all* of the information necessary for the formation of more cells, so that during cell growth and division, there must be a duplication of the genetic code. Thus, it is not just that the DNA content doubles when one cell turns into two, but that the *precise* sequence of bases in the DNA must be copied (see Figure 1.6). The fidelity of this copying function is very high, so that the two progeny of a single cell are identical in DNA base sequence to the parent. We say that the two progeny are *genetically identical* to the parent (Figure 1.8).

However, mistakes in copying DNA do occur occasionally, so that one of the two offspring cells may not always be identical to the parent (Figure 1.8). The mistakes that are made in copying DNA are called **mutations**. A mutation is a change in the DNA molecule which is passed on to one or more offspring. In most cases, the mutation results in the formation of a malfunctioning enzyme (or no enzyme at all), so that the cell with the mutation is defective, either dying or deteriorating. Such cells eventually disappear from the population and are no longer seen. Thus, mutations are generally harmful. However, in rare cases the mutation may result in the formation of an improved enzyme, an enzyme better able to

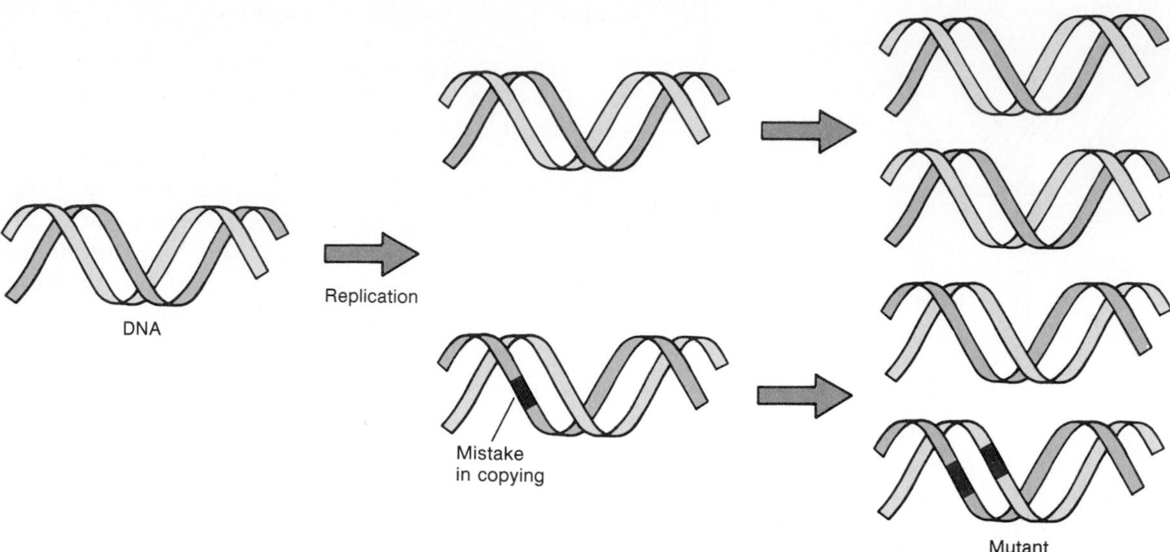

Figure 1.8 Fidelity of DNA replication and the phenomenon of mutation.

function than that of the parent cell. The cell containing this mutant enzyme thus has a *selective* advantage and will probably eventually replace, as a result of further cell divisions, the parent type. The survival value of a mutation usually depends on the environment in which the organism is living. In some environments, a mutation may be advantageous, and in other cases it may be harmful.

The process which we have just discussed here is called **natural selection**, a phenomenon which is at the basis of the process of **evolution**. Although Charles Darwin first discovered the process of evolution from observations of higher organisms, Darwinian evolution is demonstrated most dramatically and effectively at the microbial level. Not only has microbial evolution provided some of the most critical support for the theory of evolution, it has led to some very important practical consequences. For instance, disease-causing microbes have arisen which are resistant to antibiotics, and therefore the diseases caused by these microbes cannot be treated with these antibiotics. We see this as a dramatic and very practical result of microbial evolution. Many other important consequences of natural selection will be discussed later in this book. We note here most importantly that the vast diversity of microorganisms, as well as the diversity of higher organisms, is due to the process of natural selection.

As we have noted, the DNA coding sequence which specifies the amino acid sequence of a specific protein molecule is called a *gene*. How many proteins are there in a cell? A simple bacterial cell, the organism *Escherichia coli*, has about 1000 different kinds of proteins and contains about 2.4 million separate protein molecules. Some proteins in the *E. coli* cell are present in very large numbers, even greater than 100,000 molecules per cell, whereas other proteins are present in the cell only in a very small number of molecules. Thus, the cell has some means of controlling the *expression* of its genes so that not all genes are expressed at any particular time.

How many genes does a cell have? If we determine the amount of DNA per cell we find that our simple bacterial cell has about 4 million base pairs of DNA (4000 kilobases). If a single gene is about 1.1 kilobases in length, that means that the cell has about 3500 genes, if all the DNA coded for protein. However, we know that some of the DNA does not code for protein and we thus estimate that the bacterial cell has around 3000 or fewer total genes, of which only about 1000 appear to be expressed at any particular time.

1.2 Cell Structure

What is the structure of a cell? All cells have a barrier separating inside from outside which is called the **cell membrane** (Figure 1.9). It is through the cell membrane that all food materials and other substances of vital importance to the cell pass in, and it is through this membrane that waste materials and other cell products pass out. If the membrane is damaged, the interior contents of the cell leak out, and the cell usually dies. As we shall see, some drugs and

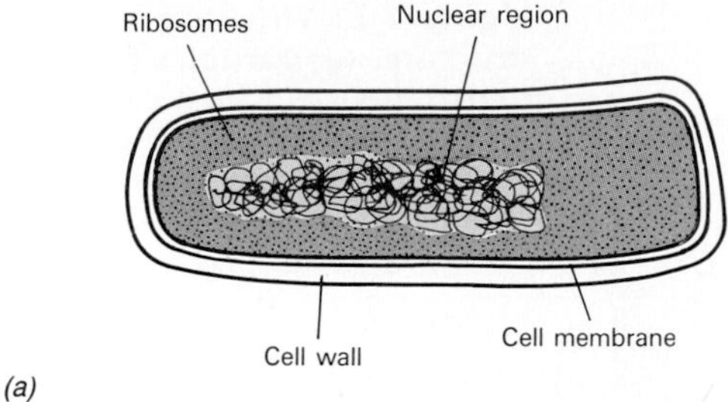

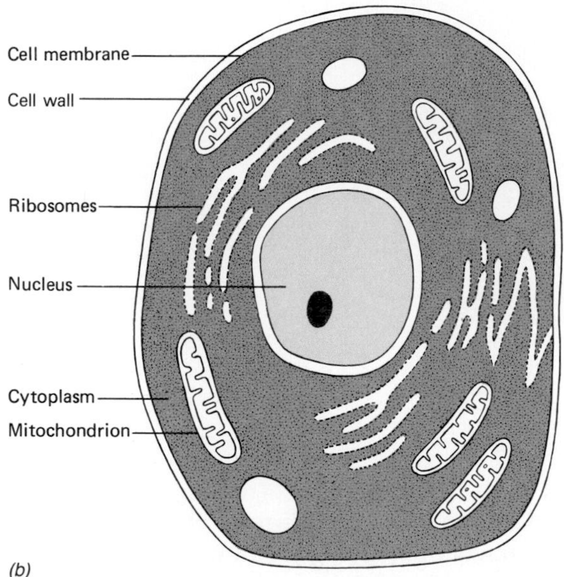

Figure 1.9 Structure of microbial cells. (a) Procaryote (bacterium). (b) Eucaryote.

other chemical agents damage the cell membrane and, in this way, bring about the destruction of cells.

The cell membrane is a very thin, highly flexible layer and is structurally weak. By itself, it usually cannot hold the cell together, and an additional stronger layer, called the **cell wall**, is usually necessary (Figure 1.9). The wall is a relatively rigid layer which is present outside the membrane and protects the membrane and strengthens the cell. Plant cells and most microorganisms have such rigid cell walls. Animal cells, however, do not have walls; these cells have developed other means of support and protection.

Within a cell is a complicated mixture of substances and structures called the **cytoplasm**. These materials and structures, bathed in water, carry out the functions of the cell. The word **metabolism** is often used to refer to the collection of chemical processes that cells carry out. Microbial metabolism will be discussed in Chapter 4.

Upon careful study of the structures of cells, two basic types have been recognized which differ greatly in structure. These are called **procaryote** and **eucaryote** (Figure 1.9*a* and *b*). These two types of cells are so different that the separation of procaryotic from eucaryotic cell lines is thought to have been a major event in biological evolution. **Bacteria** are the only procaryotes. There are several groups of eucaryotic microorganisms, including **algae**, **fungi**, and **protozoa**. In addition, all higher life forms (plants and animals) are constructed of eucaryotic cells. Figure 1.10 summarizes the major types of cellular life forms.

Microorganisms are very small. A bacterium of typical size is about 1 to 2 μm long, and thus completely invisible to the naked eye. To illustrate how minute a bacterium is, consider that 1,000 bacteria could be placed end to end across the period at the end of this sentence.

Viruses are not cells. Viruses lack many of the attributes of cells, of which the most important is that they are not dynamic open systems. A single virus particle is a static structure, quite stable and unable to change or replace its parts. Only when it is associated with a cell does a virus acquire attributes of a living system. Thus, unlike cells, viruses have no metabolism of their own. Although viruses have their own genetic elements (either DNA or RNA), viruses lack the translation apparatus and use the cell machinery for protein synthesis. Are viruses living? The answer to this question depends on how we define life and will be discussed later in this book. Viruses are known which infect all kinds of living organisms: animals, plants, algae, fungi, protozoa, and bacteria. Some size comparisons of viruses and cells are shown in Figure 1.11. Although many viruses cause disease in the organisms they infect, virus infection does not always lead to disease; many viruses cause inapparent infections.

Microorganisms are so diverse that it is useful to give them names, and to do this we must have ways of telling them apart. After close study of the structure, composition, and behavior of a microorganism, we can usually recognize a group of characteristics

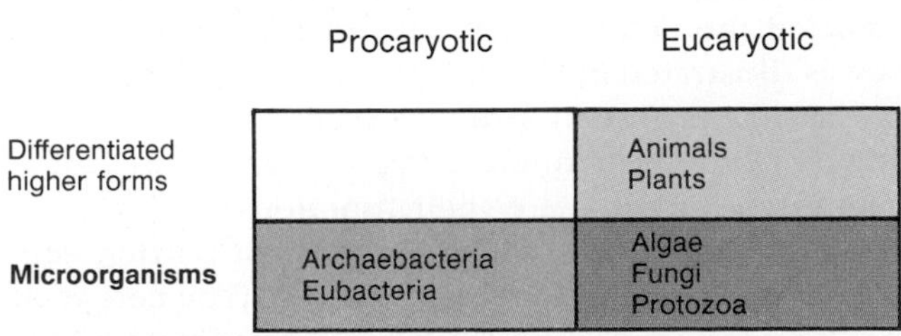

	Procaryotic	Eucaryotic
Differentiated higher forms		Animals Plants
Microorganisms	Archaebacteria Eubacteria	Algae Fungi Protozoa

Figure 1.10 The major types of cellular life forms. All multicellular organisms are eucaryotic.

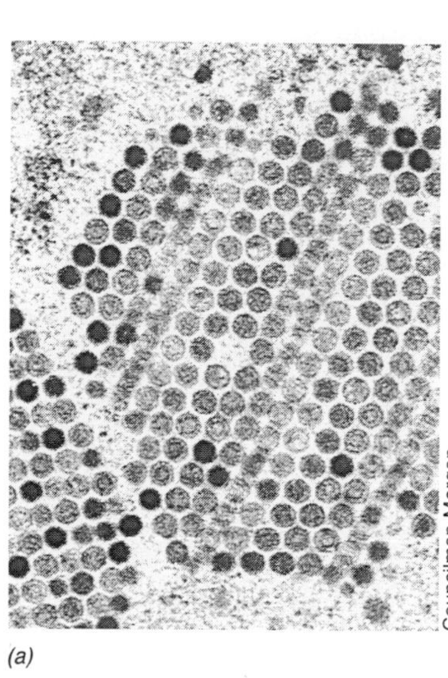

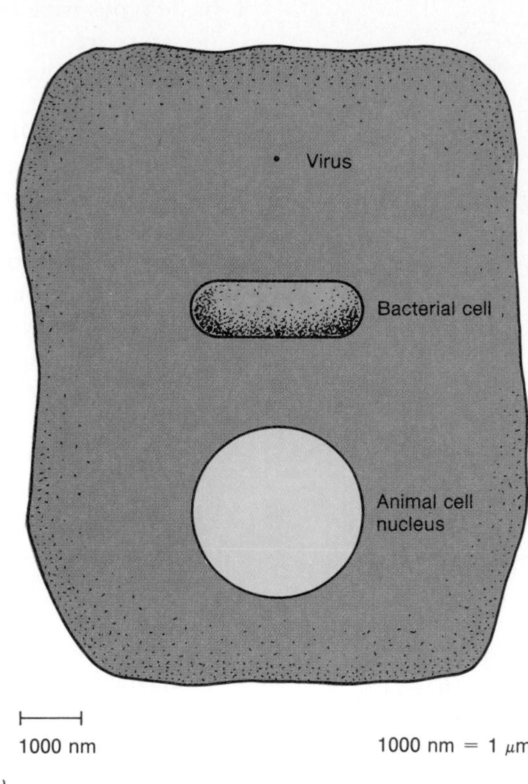

Figure 1.11 Virus size and structure. (a) Particles of *adenovirus*, a virus which causes respiratory infections in humans. (b) The size of a virus in comparison to a bacterial and animal cell.

unique to a certain organism. Once an organism has been defined by such a set of characteristics, it can be given a name. Microbiologists use the binomial system of nomenclature first developed by Linnaeus for plants and animals. The *genus* name is applied to a number of related organisms; each different type of organism within the genus has a *species* name. Genus and species names are always used together to describe a specific type of organism, whether it be a single cell or a group of such cells. (In writing, the genus and species names are either underlined or printed in italics.)

1.3 The Procaryotic Cell

Careful study with light and electron microscopes has revealed the detailed structure of procaryotic cells. As was illustrated in Figure 1.9*a*, a typical procaryotic cell has the following structures: cell wall, cell membrane, ribosomes, nuclear region. What are these structures and what are their functions?

The **cell membrane** is the critical barrier, separating inside from outside of the cell. The **cell wall** is a rigid structure outside the cell membrane which provides support and protection. The **ribosomes** are small particles composed of **protein** and **ribonucleic acid** (RNA), which are just barely visible in the electron microscope. A single cell may have as many as 10,000 ribosomes. Ribosomes are part of the translation apparatus, and synthesis of cell proteins takes place upon these structures. We discuss ribosomes, proteins, and RNA in Chapter 5.

As we have noted, the **nuclear region** of the procaryotic cell is primitive, in contrast to that of the eucaryotic cell to be discussed in the next section. Procaryotic cells do not possess a true nucleus, the functions of the nucleus being carried out by a single long strand of **deoxyribonucleic acid** (DNA) (Figure 1.12). This nucleic acid is present in a more or less free state within the cell. The procaryote is thus said to have a nuclear region (the place where the DNA is present) rather than a nucleus. An analogy to the eucaryote, the DNA molecule of the procaryote is called a **chromosome**. In addition to the major genetic material of the procaryote in the nuclear region, procaryotes have accessory genetic material in small circular elements called **plasmids** (Figure 1.12).

Many, but not all, procaryotic microorganisms are able to move. Movement of a procaryotic cell is usually by means of a structure called a **flagellum** (plural, **flagella**). Each flagellum (Figure 1.13) consists of a single coiled tube of protein. The *rotation* of

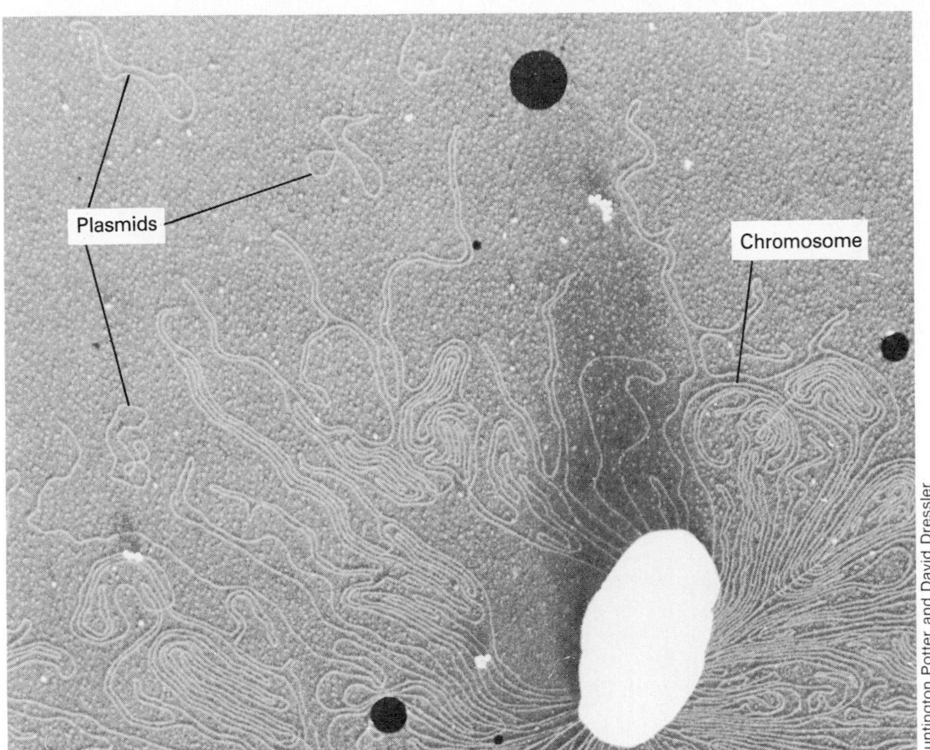

Figure 1.12 The bacterial chromosome and bacterial plasmids, as shown in the electron microscope. The plasmids are the circular structures, much smaller than the main chromosomal DNA. The cell (large white structure) was broken gently so that the DNA would remain intact.

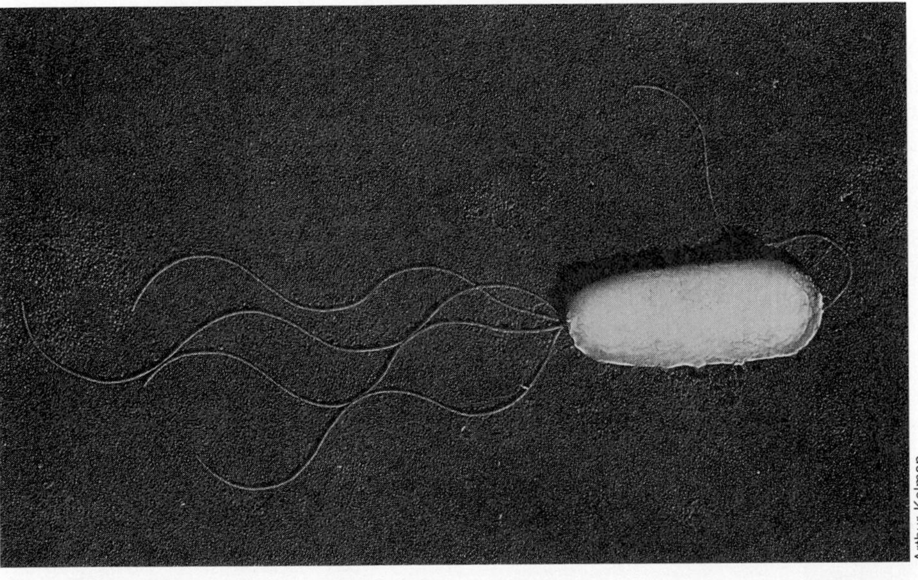

Figure 1.13 Electron micrograph of a bacterial cell, showing flagella. Magnification, 17,000×.

flagella propels the cell through the water. Flagella are too small to be seen in the light microscope without special staining techniques, but are readily visible in the electron microscope.

Size and shape of procaryotes Bacteria vary in size from cells as small as 0.1 μm in width to those more than 5 μm in diameter (Figure 1.14). Several distinct shapes of bacteria can be recognized, which have been given names. Examples of some of these various bacterial shapes are given in Figure 1.15. A bacterium which is spherical or egg-shaped is called a **coccus** (plural, **cocci**). A bacterium with a cylindrical shape is called, a **bacillus**, or simply, a **rod**. Some rods are curved, frequently forming spiral-shaped patterns, and are called **spirilla**.

In many microorganisms, the cells remain together in groups or clusters and the arrangements in these groups are often characteristic of different or-

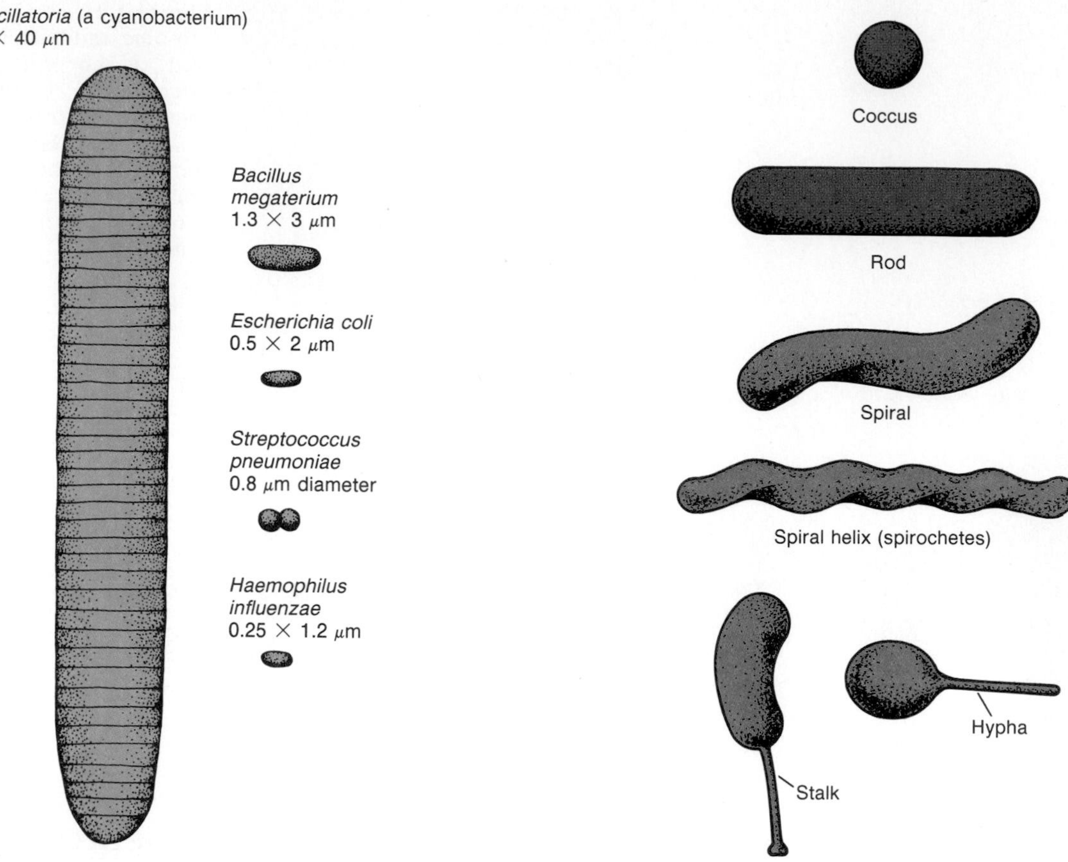

Figure 1.14 Comparison of sizes of a variety of procaryotic microorganisms.

Figure 1.15 Representative cell shapes in procaryotes.

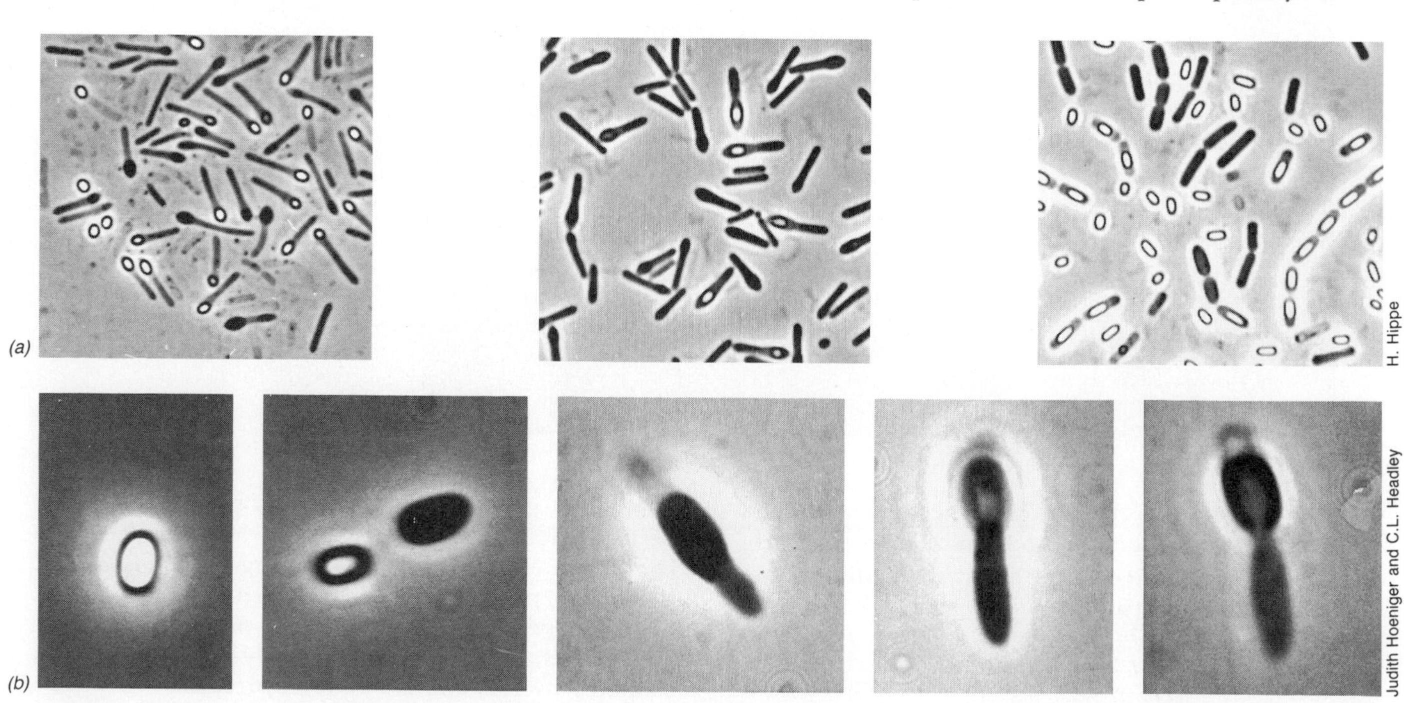

Figure 1.16 The bacterial endospore. (a) Light photomicrographs illustrating several types of endospore morphologies. (b) Photomicrographs showing the sequence of events during endospore germination.

ganisms. For instance, cocci or rods may occur in long chains (as shown for cocci in Figure 1.15). Some cocci form thin sheets of cells, while others occur in three-dimensional cubes or irregular grapelike clusters. Bacteria with unusual shapes include coiled bacteria, the **spirochetes** (Figure 1.15), and bacteria which possess extensions of their cells as long tubes or stalks, the **appendaged bacteria** (Figure 1.15).

Bacterial endospores Bacteria of another group, the spore-forming bacteria, produce special structures called **endospores** within their cells (Figure 1.16*a*). Endospores are very resistant to heat and cannot be killed easily, even by boiling. When spore-forming bacteria are present, heat sterilization of foods and other perishable products is difficult. A knowledge of the nature and properties of spores is of considerable importance in applied microbiology. Spore-forming bacteria are found most commonly in the soil, and virtually any sample of soil will have some bacterial spores present. Since soil particles contaminate nearly all materials, it must be assumed that spores are present in any item to be sterilized. Bacteria that are actively growing do not form spores, but when growth ceases due to starvation or some other cause, spore formation may be initiated. Spores are more resistant than normal cells to drying, radiation, and drugs, as well as to heat. Spores are able to remain alive but dormant and inactive for many years; however, they can convert back to normal cells in a matter of minutes, given proper conditions. This process, called **spore germination**, involves outgrowth of a new cell from within the spore (Figure 1.16*b*). As soon as germination begins, resistance to heat and to other harmful agents is lost.

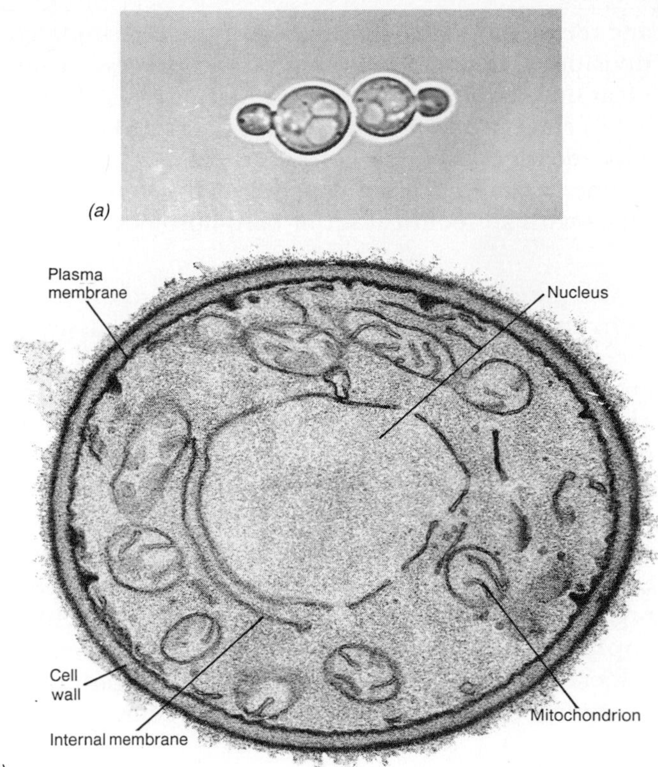

Figure 1.17 The yeast cell, a typical eucaryotic microorganism. (a) Photomicrograph of a yeast cell in the process of budding. (b) Electron micrograph of a thin section through a yeast cell, showing various structures.

1.4 The Eucaryotic Cell

Eucaryotic cells are larger and more complex in structure than procaryotic cells. The key difference is, of course, that eucaryotes contain **true nuclei**. Eucaryotic cells also contain within them distinct structures called **organelles** within which important cellular functions occur (Figure 1.17). These structures are lacking in procaryotes, in which similar functions, if they occur, are not restricted to special organelles.

Nucleus The most obvious eucaryotic organelle is the **nucleus**, a special membrane-surrounded structure within which the genetic material of the cell, the DNA, is located. The DNA in the nucleus is organized into **chromosomes** (Figure 1.18), which are invisible except at the time of cell division. Before cell division occurs, the chromosomes are duplicated

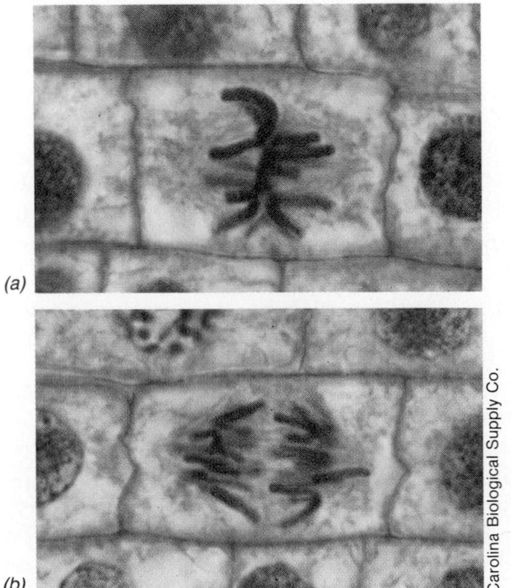

Figure 1.18 Nuclei and chromosomes. Light photomicrographs of onion root tip cells stained to reveal nuclei and chromosomes and the process of *mitosis*. (a) Chromosomes are paired in the center of the cell, a stage called *metaphase*. (b) The chromosomes separating, a stage called *anaphase* which occurs just before cell division. After mitosis, each of the two cells formed has a complete chromosome complement.

and then condense, become thicker and then undergo division as the nucleus divides. The process of nuclear division in eucaryotes is called **mitosis** (Figure 1.18) and is a complex but highly organized process. Two identical daughter cells result from the division of one parent cell. Each daughter cell receives a nucleus with an identical set of chromosomes.

Organelles As we have noted, some of the functions in eucaryotes are carried out in special structures called *organelles*. One organelle found in most eucaryotes is the **mitochondrion** (plural, **mitochondria**) (Figure 1.17*b*). Mitochondria are the organelles within which the energy-generating functions of the cell occur. The energy produced in the mitochondria is then used throughout the cell.

The **algae** are eucaryotic microorganisms which carry out the process of **photosynthesis**. In these organisms, an additional organelle is found: the **chloroplast**. The chloroplast, the site where chlorophyll is localized, is green, and is where the light-gathering functions involved in photosynthesis occur.

Motility in eucaryotes Many eucaryotic cells are motile, and two types of organelles of motility are recognized: flagella and cilia (Figure 1.19). **Flagella** are long filamentous structures which move in a whiplike manner. It should be emphasized that the flagella of eucaryotes are quite different in structure from those of procaryotes, even though the same name is used for both. The flagellum of a eucaryote is a complex structure composed of two central protein fibers surrounded by nine outer fibers, with each of the latter composed of two subfibrils. Eucaryotic flagella are large enough to be seen with the light microscope, and their movements can be easily followed.

Cilia (singular, **cilium**) are similar to eucaryotic flagella in structure but are shorter and more numerous. A single cell may have over 10,000 cilia (see Figure 3.46). These organelles operate like oars and propel the cell through the water. Protozoa are the only microorganisms which possess cilia, but ciliated cells are also common in higher organisms.

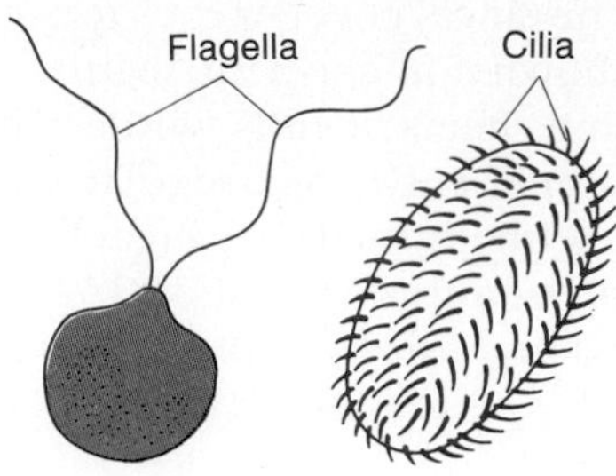

Figure 1.19 Cilia and flagella, organs of motility in eucaryotic cells.

To emphasize further the vast structural differences between procaryotes and eucaryotes, a comparison between the cells of these two kinds of organisms is given in Table 1.1. Although procaryotes and eucaryotes differ greatly in structure, it should be emphasized that they are quite similar in chemical composition and metabolism. Evidence is strong that procaryotes and eucaryotes have had a common origin. We now discuss a few eucaryotic microorganisms in a little more detail.

1.5 Algae

Algae contain **chlorophyll**, a green pigment that serves as a light-gathering substance and makes it possible for them to use light as a source of energy. Because of chlorophyll, most algae are green, and they can be recognized first by their green color. However, a few kinds of common algae are not green but appear brown or red because in addition to chlorophyll other pigments are present that mask the green color.

Table 1.1 Characteristics of procaryotic and eucaryotic cells

Characteristic	Procaryote	Eucaryote
Size of cell	Small, 0.5–2.0 μm	Larger, 2–200 μm
Nuclear body	No nuclear membrane; no mitosis	True nucleus; nuclear membrane; mitosis
DNA	Single molecule; not in chromosomes	Several or many chromosomes
Organelles	None	Mitochondria, chloroplasts, vacuoles, others
Cell wall	Relatively thin; usually peptidoglycan	Thick or absent; chemically different
Manner of movement	Flagella of submicroscopic size; single protein fiber	Flagella or cilia of microscopic size; complex pattern of fibers

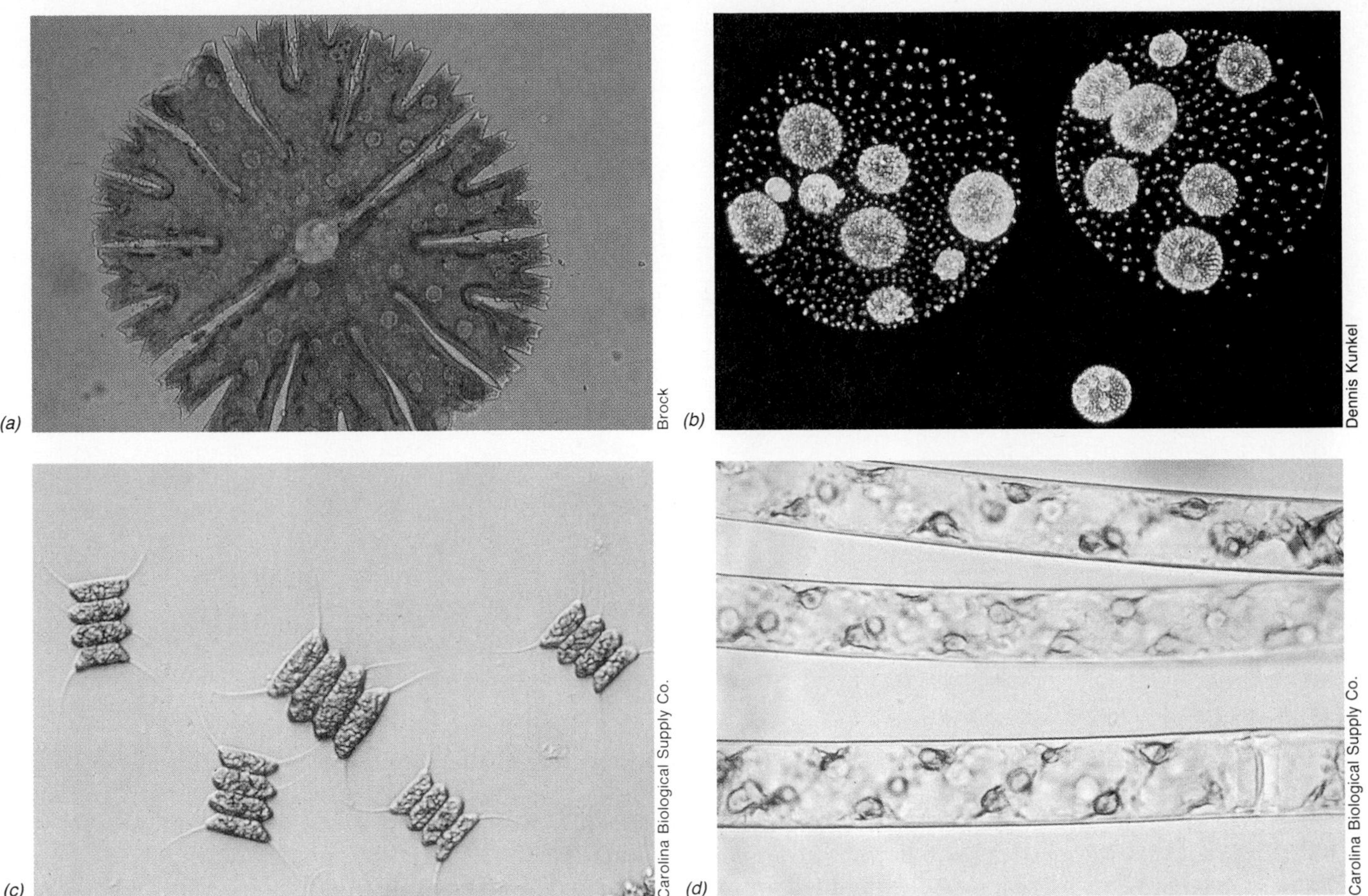

Figure 1.20 Representative algae. (a) *Micrasterias*. A single cell. Magnification, 250×. (b) *Volvox* colony, with a large number of cells. Magnification, 1000×. (c) *Scenedesmus*, a packet of four cells. Magnification, 100×. (d) *Spirogyra*, a filamentous alga. Note the spiral-shaped chloroplasts. Magnification, 350×.

A characteristic structure of algae is the **chloroplast**, the site of the chlorophyll pigments. Chloroplasts can often be recognized microscopically within algal cells by their green color.

Algae can be either **unicellular** (Figure 1.20) or **colonial**, occurring as aggregates of cells. When the cells are arranged end-to-end, the alga is said to be **filamentous**. Among the filamentous forms, both unbranched filaments and more complicated branched filaments occur. In many cases, the filamentous algae occur in such large masses that they can be seen easily without a microscope. Small ponds or streams or even fish tanks frequently show such massive accumulations of algae.

1.6 Fungi

Fungi can be distinguished from algae because the fungi do not have chlorophyll. Fungi can be differentiated from bacteria by the fact that fungal cells are much larger, and vacuoles, nuclei, and other intracellular organelles typical of eucaryotic cells can usually be seen inside, even with the ordinary light microscope. Although the fungi are a large and rather diverse group of eucaryotic microorganisms, three groups of fungi are of major practical importance: the molds, the yeasts, and the mushrooms.

Molds The molds are filamentous fungi (Figure 1.21). They are widespread in nature and are commonly seen on stale bread, cheese, or fruit. Each filament grows mainly at the tip, by extension of the terminal cell. A single filament is called a **hypha** (plural, **hyphae**). Hyphae usually grow together across a surface and form rather compact tufts, collectively called a **mycelium**, which can be easily seen without a microscope. The mycelium arises because the individual hyphae form branches as they grow, and these branches intertwine, resulting in a compact mat, or felt. From the mycelium, other branches may

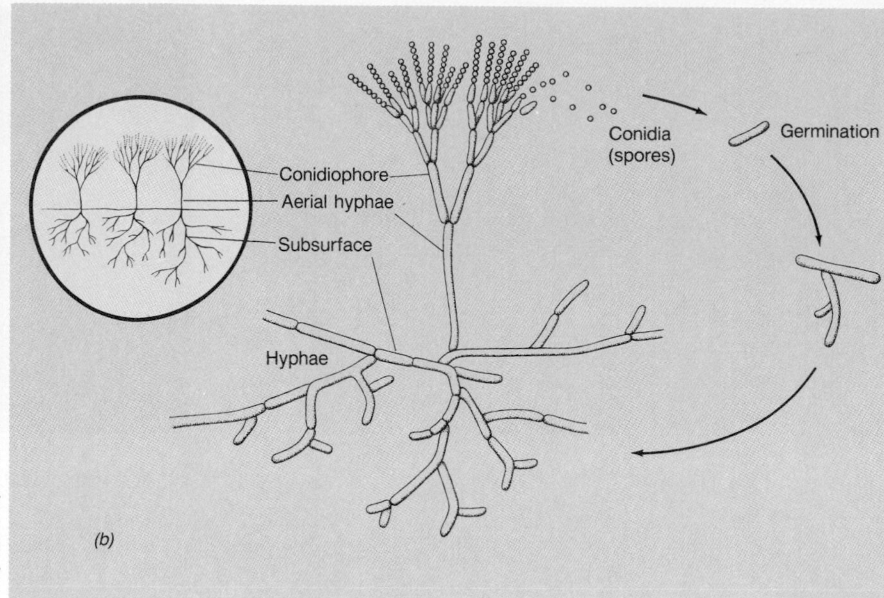

Figure 1.21 Mold structure and growth. (a) Photomicrograph of a typical mold. (b) Diagram of a mold life cycle.

reach up into the air above the surface, and on these aerial branches spores are formed. **Conidia** are round structures, often highly pigmented, that are resistant to drying, are very lightweight, and permit the fungus to be dispersed to new habitats. When conidia form, the white color of the mat changes, taking on the color of the conidia, which may be black, blue-green, red, or brown. The presence of these spores gives the mat a rather dusty appearance. The spores just described are called **asexual spores**, since no sexual reproduction is involved in their formation. Because these spores are so numerous and spread so easily through the air, molds are common laboratory contaminants. Airborne mold spores are also often responsible for allergies.

Some molds also produce **sexual spores**, formed as a result of sexual reproduction. These sexual spores are usually resistant to drying, heat, freezing, and some chemical agents. However, these sexual spores are not as resistant to heat as bacterial endospores. Either an asexual or sexual spore of a fungus can germinate and develop into a new hypha and mycelium.

Yeasts The yeasts are unicellular fungi. Yeast cells are usually spherical, oval, or cylindrical, and cell division generally takes place by **budding** (Figure 1.22). In the budding process, a new cell forms as a small outgrowth of the old cell; the bud gradually enlarges, and then separates. Yeasts do not usually form filaments or a mycelium, and the population of yeast cells remains a collection of single cells. Yeast cells are considerably larger than bacterial cells and can be distinguished from bacteria by their size and by the obvious presence of internal cell structures. Some yeasts also exhibit sexual reproduction by a process called **mating**, in which two yeast cells fuse. Within the fused cell, called a **zygote**, spores form. For the most part, yeasts spread from place to place as ordinary vegetative cells rather than as spores. We discuss the life cycle of yeast in some detail in Chapter 7.

Mushrooms The mushrooms are a group of filamentous fungi that form large complicated structures called **fruiting bodies** (Figure 1.23). The fruiting body is commonly called the "mushroom". The fruiting body is formed through the association of a large number of individual hyphae (Figure 1.24). If the mushroom fruiting body is cut and examined under the microscope, the individual hyphae can be seen pressed tightly together.

During most of its existence, the mushroom fungus lives as a simple mycelium, growing within soil, leaf litter, or decaying logs (Figure 1.23). When environmental conditions are favorable, the fruiting body develops, beginning first as a small button-shaped structure underground and then expanding into the full-grown fruiting body that we see above ground. The nutrients for growth come from organic matter in the soil and are taken up by the hyphal filaments which, like the roots of a plant, feed the growing fruiting body. Sexual spores, called **basidiospores**,

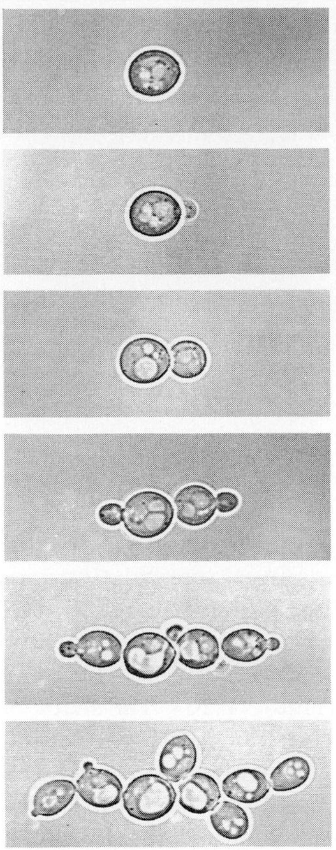

Figure 1.22 Yeast: development of buds from a single cell.

Figure 1.23 A typical mushroom growing on a decaying log. This is the Japanese forest mushroom, shiitake, *Lentinus edulis*, a widespread commercial mushroom.

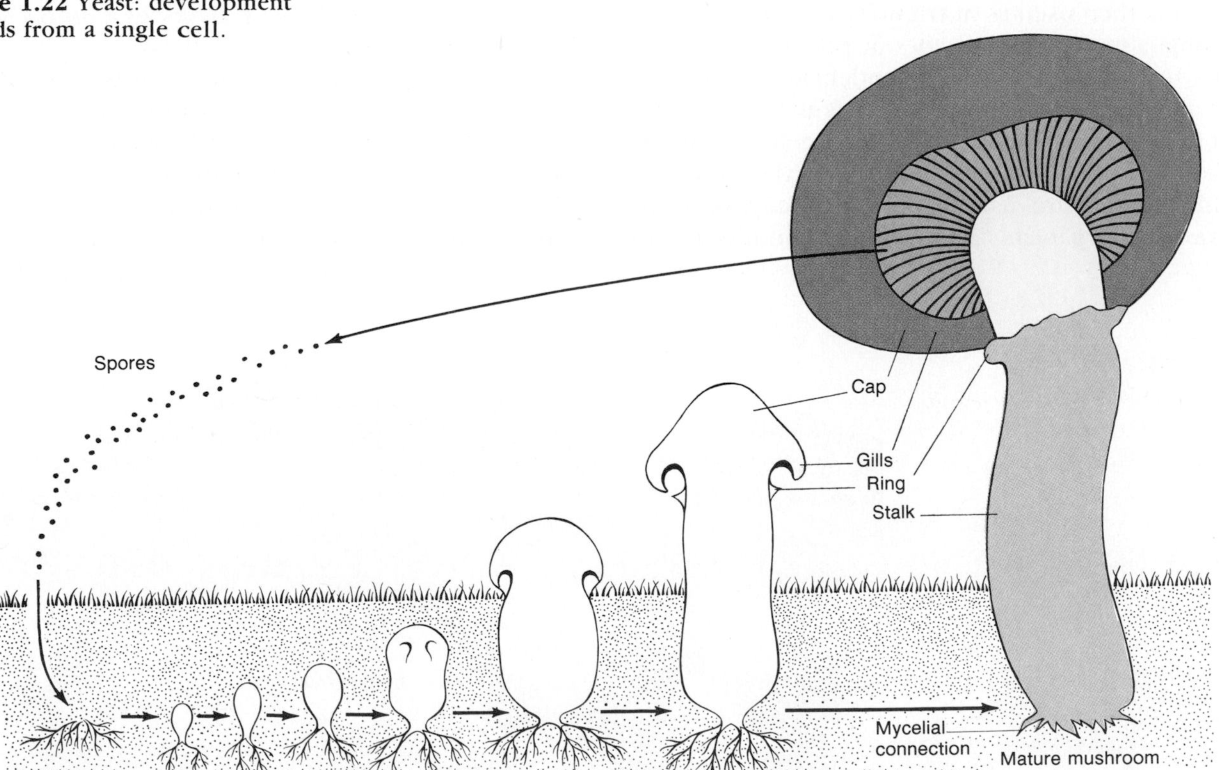

Figure 1.24 Mushroom life cycle, showing how the fruiting body develops from underground hyphae.

are formed, borne on the underside of the fruiting body, either on flat plates called **gills** or within deep **pores**. The spores are often colored and impart a color to the underside of the fruiting body cap. Some mushrooms, called **puff balls**, produce their spores within spherical fruiting bodies instead of on gills or pores, the spores puffing out through cracks or holes that develop in the fruiting body as it dries. The spore is the agent of dispersal of mushrooms and is carried away by the wind. If it alights in a favorable place, the spore will germinate and initiate the growth of new hyphae, mycelium, and fruiting body.

Figure 1.25 A lichen growing on a branch of a dead tree.

1.7 Lichens

Lichens are leafy or encrusting growths that are widespread in nature and are often found growing on bare rocks, tree trunks, house roofs, and on the surfaces of bare soils (Figure 1.25). The lichen is one of the best examples of **mutualism**, a trait commonly found in the biological world. Mutualism refers to two organisms living together for their own benefit. The lichen plant, usually called a **thallus**, consists of two organisms, a fungus and an alga, always found living together in nature. The alga is photosynthetic and is able to produce organic matter from carbon dioxide of the air. Some of the organic matter produced by the alga is then used as nutrient by the fungus. Since the fungus is unable to carry out photosynthesis, its ability to live in nature is dependent on the activity of its algal partner. Lichens are extremely interesting associations because they demonstrate so clearly the value and importance of cooperation in the biological world. Lichens are usually found in environments where other organisms do not grow, and it is almost certain that their success in colonizing such extreme environments is due to the mutual interrelationships between the alga and fungus partners.

The lichen thallus usually consists of a tight association of many fungus hyphae, within which the algal cells are embedded (Figure 1.26). The shape of the lichen thallus is determined primarily by the fungal partner, and a wide variety of fungi are able to form lichen associations. The diversity of algal types is much smaller, and many different kinds of lichens may have the same algal component. The algae are usually present in defined layers or clumps within the lichen thallus.

The fungus clearly benefits from associating with the alga, but how does the alga benefit? The fungus provides a firm anchor within which the alga can grow protected from erosion by rain or wind. In addition, the fungus absorbs from the rock or other substrate upon which the lichen is living the inorganic nu-

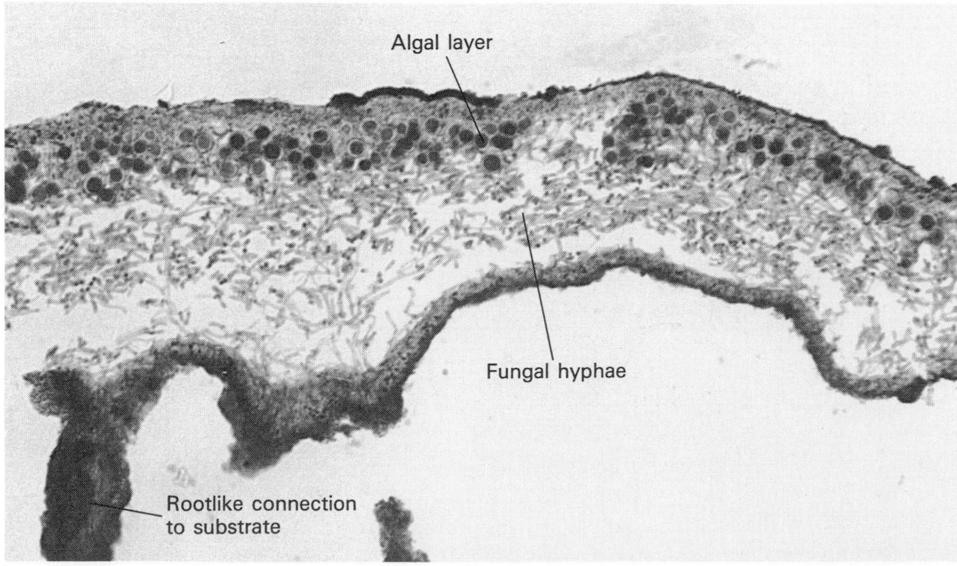

Figure 1.26 Photomicrograph of a cross section through a lichen.

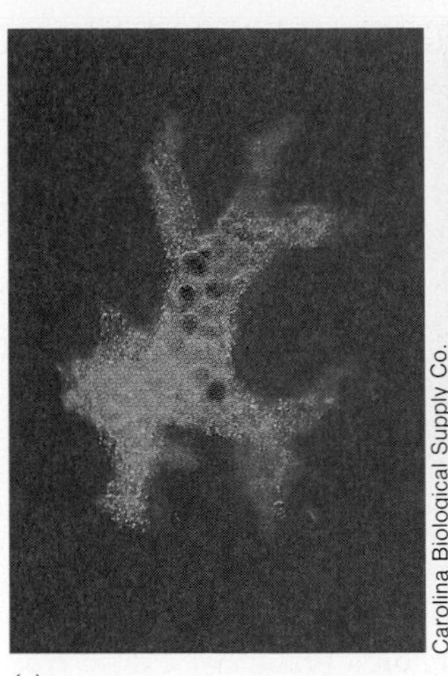

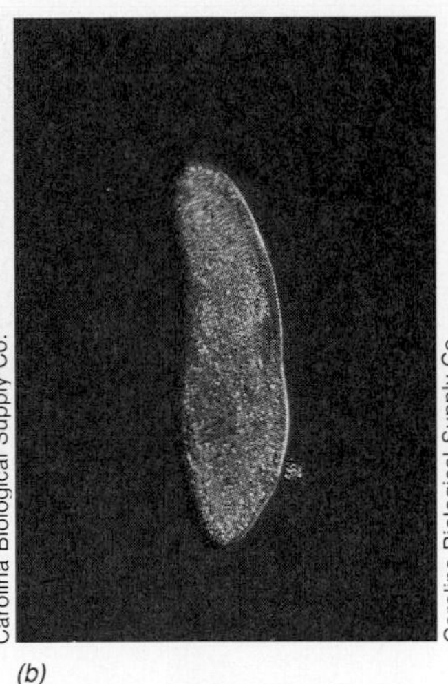

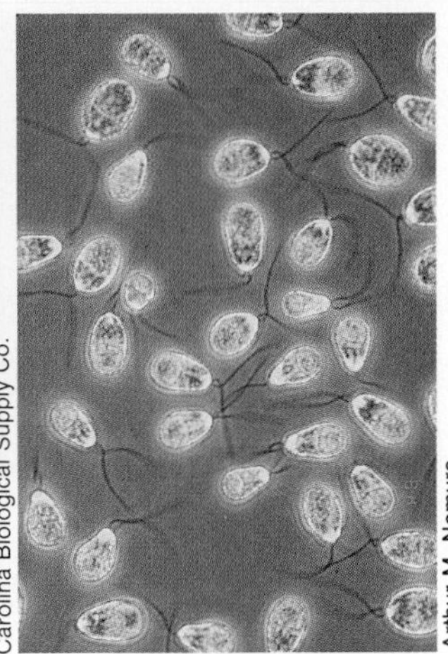

Figure 1.27 Typical protozoa. (a) Amoeba. (b) A typical ciliate, *Paramecium*. (c) A flagellate, *Dunaliella*. Although this flagellate has chloroplasts, it is frequently classified with the protozoa because of its animal-like nature. Many similar flagellates lack chloroplasts.

trients essential for the growth of the alga. Another role of the fungus is to protect the alga from drying; most of the habitats in which lichens live are dry (rock, bare soil, roof tops), and fungi are in general much better able to tolerate dry conditions than are algae.

Although lichens live in nature under rather harsh conditions, they are extremely sensitive to air pollution and quickly disappear from large cities when heavy air pollution occurs. One reason for the great sensitivity of lichens to air pollution is that they absorb and concentrate materials from rainwater and air and have no means for excreting them, so that lethal concentrations are reached. Lichens have even been used as indicators of air pollution and might provide some early warnings of impending chronic air pollution problems. Studies made on the richness of the lichen flora in cities and their surroundings show that the number of species of lichens found on various habitats decreases as one moves from the countryside to the center of the city.

1.8 Protozoa

Protozoa are unicellular, generally colorless and motile, eucaryotic microorganisms that lack cell walls (Figure 1.27). They are distinguished from bacteria by their greater size and eucaryotic nature, from algae by their lack of chlorophyll, and from yeasts and other fungi by their motility and absence of the cell wall. Protozoa usually obtain food by ingesting other organisms or organic particles. They do this by surrounding the food particle with a portion of their flexible cell membrane and engulfing the particle, or by swallowing the particle through a special structure called the *gullet*.

Depending on the group, the type of protozoal movement varies. The **amoebae** move by what is called *amoeboid locomotion*; the cytoplasm of the cell flows forward in a lobe of the cell, called a *pseudopodium* (false foot), and the rest of the cell flows toward this lobe. Amoeboid motion requires a solid substrate and is rather slow. A few amoebas are disease-causing agents, but most are harmless and live in soil or water.

The **flagellates** move by use of flagella, usually by a single long flagellum attached at one end of the cell. The **trypanosomes**, a subgroup of flagellates, cause African sleeping sickness and a number of other diseases of humans and animals.

The **ciliates**, a third group of protozoa, move by the action of a large number of smaller appendages called **cilia**. The ciliates are the most complicated protozoa structurally, and in addition they have complicated sexual reproduction mechanisms as well as the regular cell division process.

A fourth group of protozoa, the **sporozoans**, contains parasites of man or animals; the agent that causes malaria, *Plasmodium vivax*, is a member of this group. These nonmotile organisms do not engulf food particles but absorb dissolved food materials directly through their membranes. Despite their name, sporozoans do not form true spores as do bacteria, algae, and fungi.

1.9 The Slime Molds*

The slime molds are organisms that have at times been classified by some workers as fungi and by others as protozoa. The confusion arises because the vegetative

*This section is dedicated to Kenneth B. Raper, the discoverer of *Dictyostelium discoideum*, who died January 15, 1987.

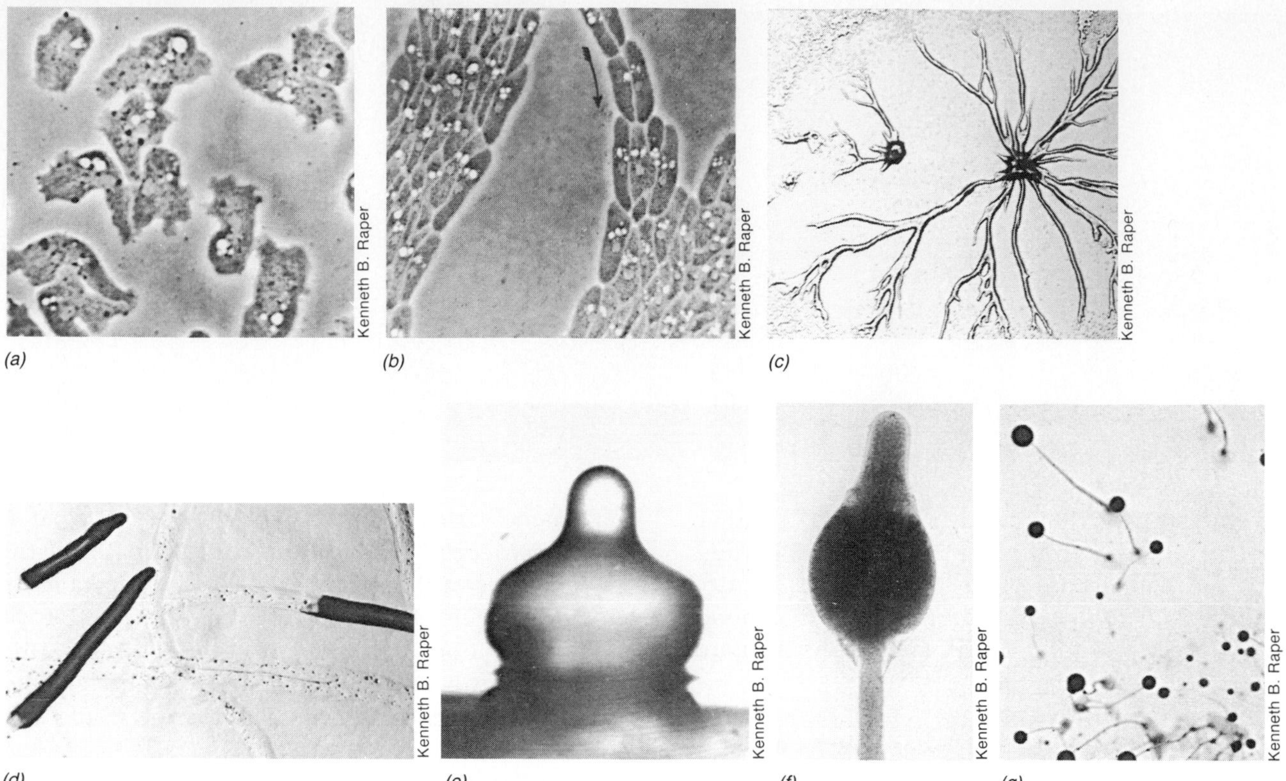

Figure 1.28 Photomicrographs of various stages in the life cycle of the cellular slime mold *Dictyostelium discoideum.* (a) Amoebas in preaggregation stage. Note irregular shape, lack of orientation. Magnification, 330×. (b) Aggregating amoebas. Notice the regular shape and orientation. The cells are moving in streams in one direction. Magnification, 330×. (c) Low-power view of aggregating amoebas. Magnification, 11×. (d) Migrating pseudoplasmodia (slugs) moving on an agar surface and leaving trails of slime in their wake. Magnification, 13×. (e) Early stage of fruiting-body formation. Magnification, 53×. (f) A late stage of a developing fruiting body. Magnification, 87×. (g) Mature fruiting bodies. Magnification, 6×.

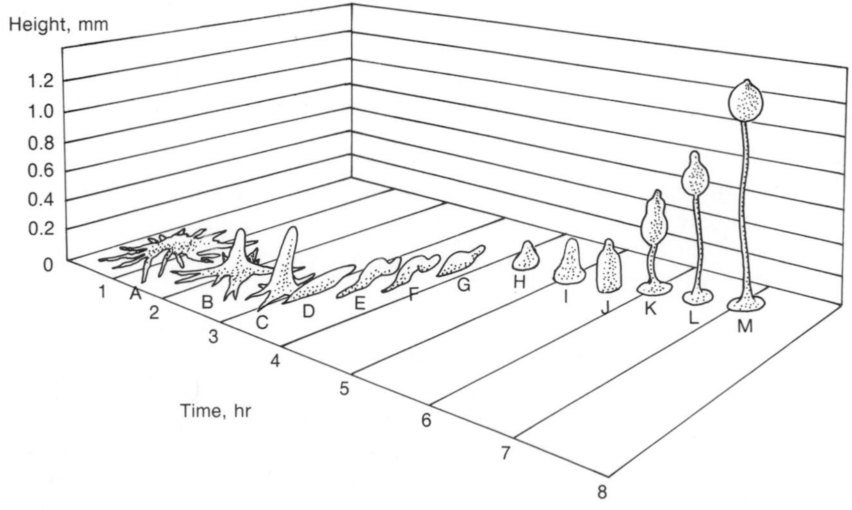

Figure 1.29 Stages in fruiting-body formation in the cellular slime mold *D. discoideum*: (A–C) aggregation of amoebas; (D–G) migration of the slug formed from aggregated amoeba; (H–L) culmination and formation of the fruiting body; (M) mature fruiting body.

structure of a slime mold is a protozoanlike amoeboid mass, whereas its fruiting body resembles the fruiting body of a fungus and produces spores with cell walls. Because the slime mold fruiting body was first studied by mycologists, the slime molds have traditionally been discussed in mycology rather than protozoology texts, even though the affinities of these organisms to protozoa are just as close as they are to fungi.

The slime molds can be divided into two groups, the **cellular slime molds**, whose vegetative forms are composed of single amoeba-like cells, and the **acellular slime molds**, whose vegetative forms are acellular masses of protoplasm of indefinite size and shape (called *plasmodia*). Both types live primarily on decaying plant matter, such as leaf litter, logs, and soil. Their food consists mainly of other microorganisms, especially bacteria, which they engulf and digest. One of the easiest ways of detecting and isolating slime molds is to bring into the laboratory small pieces of rotting logs and place them in moist chambers; the amoebas or plasmodia will proliferate, migrate over the surface of the wood, and eventually form fruiting bodies, which then can be observed under a dissecting microscope. The fruiting bodies of slime molds are often very ornate and colorful. Only the cellular slime molds will be discussed here.

Cellular slime molds Few of the described species of cellular slime molds have been studied in detail. An exception is the genus *Dictyostelium*, for which the life cycle has been worked out and for which a considerable amount of information is known about the physiology and biochemistry of fruiting-body formation. The vegetative cell of *Dictyostelium discoideum* is an amoeboid cell (Figure 1.28*a*) that feeds on living or dead bacteria and grows and divides indefinitely. When the food supply has been exhausted, a number of amoebas aggregate to form a pseudoplasmodium, a structure in which the cells lose some of their individuality but do not fuse (Figure 1.28*b* and *c*). This aggregation is triggered by the production of *acrasin*, a substance identified as cyclic adenosine monophosphate (Section 5.12), which attracts other amoebas in a chemotactic fashion. Those cells which are the first to produce acrasin serve as centers for the attraction of other amoebas, and since each newcomer also produces acrasin, the centers increase rapidly in size. The swarm patterns around such aggregating masses are shown in Figure 1.28*b* and *c*.

Because the pseudoplasmodium formed resembles a slimy, shell-less snail in appearance and movement, it is called a "slug." Around the outside is secreted a mucoid sheath that probably protects the periphery of the slug from drying. The pseudoplasmodial slug migrates as a unit across the substratum as a result of the collective action of the amoebas (Figure 1.28*d*). Presumably there is some mechanism that coordinates the movements of the individual amoebas so that the slug maintains its unity, but the nature of this mechanism is unknown. The pseudoplasmodium usually migrates for a period of hours, and as it is positively phototactic, it will migrate toward the light. In nature the pseudoplasmodium is presumably formed within the dim recesses of soil or decaying bark and migrates in response to light to the surface, where fruiting-body formation takes place. Thus the formation of the pseudoplasmodium and its behavioral responses are essential features to ensure that the spores formed will be borne in the air and effectively dispersed.

Fruiting-body formation begins when the slug ceases to migrate and becomes vertically oriented (Figures 1.28*e–g* and 1.29). The fruiting body is differentiated into a stalk and a head; cells in the forward end of the slug become stalk cells and those from the posterior end become spores. The amoebas that form the stalk cells begin to secrete cellulose, which provides the rigidity of the stalk. Other amoebas, from the rear of the slug, swarm up the stalk to the tip and form the head; most of these become differentiated into spores that are usually embedded in slime. Upon maturation of the head, the spores are released and dispersed. Each spore germinates and forms a single amoeba, which then initiates another round of vegetative growth.

The cycle of fruiting-body and spore formation described above is asexual. There is some evidence for a sexual cycle as well, but not all of the details are known. In cellular slime molds, the sexual cycle apparently involves formation of a reproductive structure, the **macrocyst**. Macrocysts are structures of multicellular origin that develop from simple aggregates of amoebas, which, when mature, are enclosed in a thick cellulose wall. The occurrence of a sexual cycle may enable the slime mold to survive cold, drought, and other adverse conditions, to which the macrocyst is quite resistant. In comparison, the spores formed in the asexual cycle are not very resistant, and function only to disperse the organism to new locations.

1.10 Classification of Living Organisms

The classification of microorganisms is a complex problem that we will discuss in some detail later. Traditionally, living organisms were separated into plants and animals, but when biologists began to study

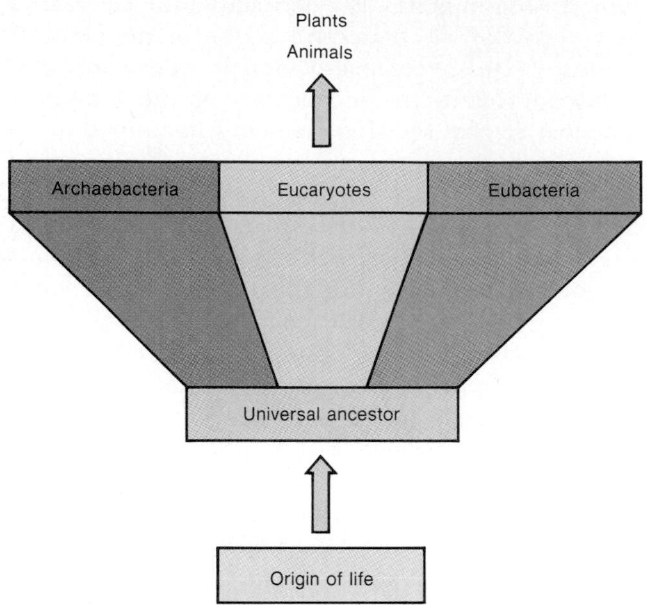

Figure 1.30 Evolution of the living world, based on the structure of RNA of the ribosomes.

microorganisms they discerned that many microrganisms could be classified as *neither* plant nor animal. Subsequently, it was discovered that all living organisms could be divided into two basic cellular types, *procaryotes* and *eucaryotes*, based on fundamental differences in cell structure. As we have noted, all animals and plants are eucaryotes, as are three major groupings of microorganisms: algae, fungi, and protozoa. There are two major groups of procaryotes, the true bacteria, called *eubacteria*, and the *archaebacteria*. These two groups of procaryotes differ in many aspects of their translation apparatus as well as in many other important ways and are thought to have diverged early in the history of life on earth. Because the cells of higher animals and plants are all eucaryotic, it has often been thought that eucaryotic microorganisms were the forerunners of higher organisms, whereas procaryotes represent a branch which never evolved past the microbial stage.

Which came first, procaryote or eucaryote? We do not know, but studies on genetic differences between eubacteria, archaebacteria, and eucaryotes suggest that all three groups diverged early in earth history from a common ancestor organism, the "original ancestor." This hypothesis is summarized in Figure 1.30.

1.11 Populations, Communities, and Ecosystems

Up to now, we have been discussing cells as if they lived in isolation in their environments. Nothing could be further from the truth. Cells live in nature in association with other cells, in assemblages which we call **populations**. Such populations are composed of groups of related cells, generally derived by successive cell divisions from a single parent cell. The location in the environment where a population lives is called the **habitat** of that population. **Ecology** is the discipline which deals with the study of living organisms in their natural environments.

In nature, populations of cells rarely live alone. Rather, populations of cells live in association with other populations of cells, in assemblages which are usually called **communities** (Figure 1.31). A community is an assemblage of populations which live together in the same habitat. Frequently, the populations of communities interact, either in beneficial ways or in harmful ways. If two populations interact in a beneficial way, these populations will then maintain themselves better when together than when separate. In such cases we speak of the cooperative nature of the populations.

In other cases, two populations living in the same habitat may interact in a way which is harmful to one of the populations. If such harmful interaction occurs, the population which is harmed will be reduced in number, or even replaced. If the effect is severe enough, the population may be completely eliminated.

The living organisms in a habitat also interact with the physical and chemical environment of that habitat. Habitats differ markedly in their physical and chemical characteristics, and a habitat which is favorable for the growth of one organism may be harmful for another organism. Thus, the community which we see in any given habitat will be determined to a great extent by the physical and chemical characteristics of that environment.

In addition, the organisms of the habitat modify the physical and chemical properties of the environment. Organisms carrying out metabolic processes remove chemical consituents from the environment and use these constituents as energy or nutrient sources. At the same time, organisms excrete into the environment waste products of their metabolism. Therefore, as time progresses the environment is gradually changed through life processes. We speak collectively of the living organisms together with the physical and chemical consituents of the environment as an **ecosystem**. We can also think of the earth as a whole as an ecosystem; this global ecosystem is sometimes called the **biosphere** (Figure 1.32). The biosphere interacts with two major regions of the earth, the **atmosphere**, and the **lithosphere**.

Since single microbial cells are too small to be seen with the unaided eye, our knowledge of microorganisms in nature begins with studies using the mi-

(a)

(b)

Figure 1.31 Microbial communities. (a) The microbial community of a nutrient-rich lake. (b) Microbial communities at the basis of the food chain in deep-sea hydrothermal vent animals. Giant tube worms and mussels abound near hydrothermal vents on the ocean floor near the Galapagos Islands and elsewhere. Sulfur-oxidizing bacteria exist in the worm and mussel tissue and oxidize hydrogen sulfide emitted from the vents as an energy source. Organic carbon compounds, synthesized from CO_2 by these bacteria are excreted and used as the primary food source of the animals in this permanently dark habitat where photosynthesis cannot occur.

Figure 1.32 The global biosphere. A view of the earth taken from a satellite orbiting the moon.

croscope. Examination of natural materials such as soil, mud, water, spoiling food, human bodily excretions, and other living and dead material under the microscope reveals that such materials teem with vast numbers of very tiny cells. Although such tiny cells seem harmless, we can state that microbes are *small but powerful*. Even though a single cell can never, by itself, cause a harmful effect that we can perceive, this single cell is often capable of multiplying rapidly and producing a massive number of progeny which may be capable of harm. One of the most important and impressive attributes of microorganisms is *rapid growth*, so that single cells do not remain single very long. Thus, although microorganisms may seem to occupy inconsequential niches in nature, they are extremely important components of ecosystems. Microorganisms also have very serious effects on higher plants and animals. In later chapters, after we have learned some of the details of microbial structure and function, we will discuss again the ways in which microorganisms affect animals, plants, and the whole global ecosystem.

1.12 The Laboratory Study of Microorganisms

In order to study the structure and function of microorganisms, it is necessary to have each type of microbe separate from all other types. This requires that a homogeneous population of an individual microorganism be obtained. Microbial culture procedures are thus the basis of the science of microbiology. The general approach to the laboratory study of microorganisms begins with microscopic examination of samples of natural material, and proceeds quickly to the cultivation of microorganisms in the pure state. We discuss these laboratory procedures briefly here.

1.13 Microscopy

Microscopic examination of microorganisms makes use of either the **light microscope** or the **electron microscope**. For most routine use, the light microscope is used, whereas for research purposes, especially in studies on cell structure, the electron microscope is used.

The light microscope The light microscope has been of crucial importance for the development of microbiology as a science and remains a basic tool of routine microbiological research (Figure 1.33). Several types of light microscopes are commonly used in microbiology: bright-field, phase-contrast, and fluorescence. The **bright-field microscope** is most commonly used in elementary biology and microbiology courses (Figure 1.33*b* and *d*). With this microscope, objects are visualized because of the contrast differences that exist between them and the surrounding medium. Contrast differences arise because cells absorb or scatter light in varying degrees. Most bacterial cells are difficult to see well with the bright-field microscope because of their lack of contrast (Figure 1.33*d*) with the surrounding medium. Staining, which will soon be discussed, is commonly used to increase contrast. But because staining either kills or greatly modifies cells, it is desirable to avoid its use for many purposes.

The **phase-contrast microscope** was developed in order to make it possible to see small cells easily, even without staining. It is based on the principle that cells differ in refractive index from their surrounding medium and hence bend some of the light rays that pass through them. This difference can be used to create an image of much higher degree of contrast (Figure 1.33*e*) than can be obtained in the bright-field microscope.

A fluorescent substance emits light of one color when light of another color shines upon it. The **fluorescence microscope** is used to visualize specimens that fluoresce, either because they contain natural fluorescent substances (for example, chlorophyll fluoresces brilliant red), or because they have been treated with fluorescent dyes (Figure 1.34). Fluorescence microscopy is widely used in medical microbiology and in immunology, as discussed in Sections 12.8 and 13.6).

Staining Dyes can be used to stain cells and increase their contrast so that they can be more easily seen in the bright-field microscope. Dyes are organic compounds, and each dye used has an affinity for specific cellular materials. For example, many commonly used dyes are positively charged (cationic) and combine strongly with negatively charged cellular constituents such as nucleic acids and acidic polysaccharides. Examples of cationic dyes are *methylene blue, crystal violet*, and *safranin*. Since cell surfaces are generally negatively charged, these dyes combine with structures on the surfaces of cells and hence are excellent general stains. Other dyes (for example, *eosin, acid fuchsin, Congo red*) are negatively charged (anionic) and combine with positively charged cellular constituents, such as many proteins.

The simplest staining procedure is done with dried preparations (Figure 1.35). The slide containing dried and washed organisms is flooded for a minute or two with a dilute solution of a dye, then rinsed several times in water, and blotted dry. It is usual to observe dried stained preparations of bacteria with a high-power (oil-immersion) lens.

Differential stains are so named because they are procedures which do not stain all kinds of cells equally. For example, there are a number of staining techniques which reveal spores by making the spore and the cell body different colors. Cells with and without spores can therefore be differentiated. Another important differential staining procedure widely used in bacteriology, is the **Gram stain** (Figure 1.36).

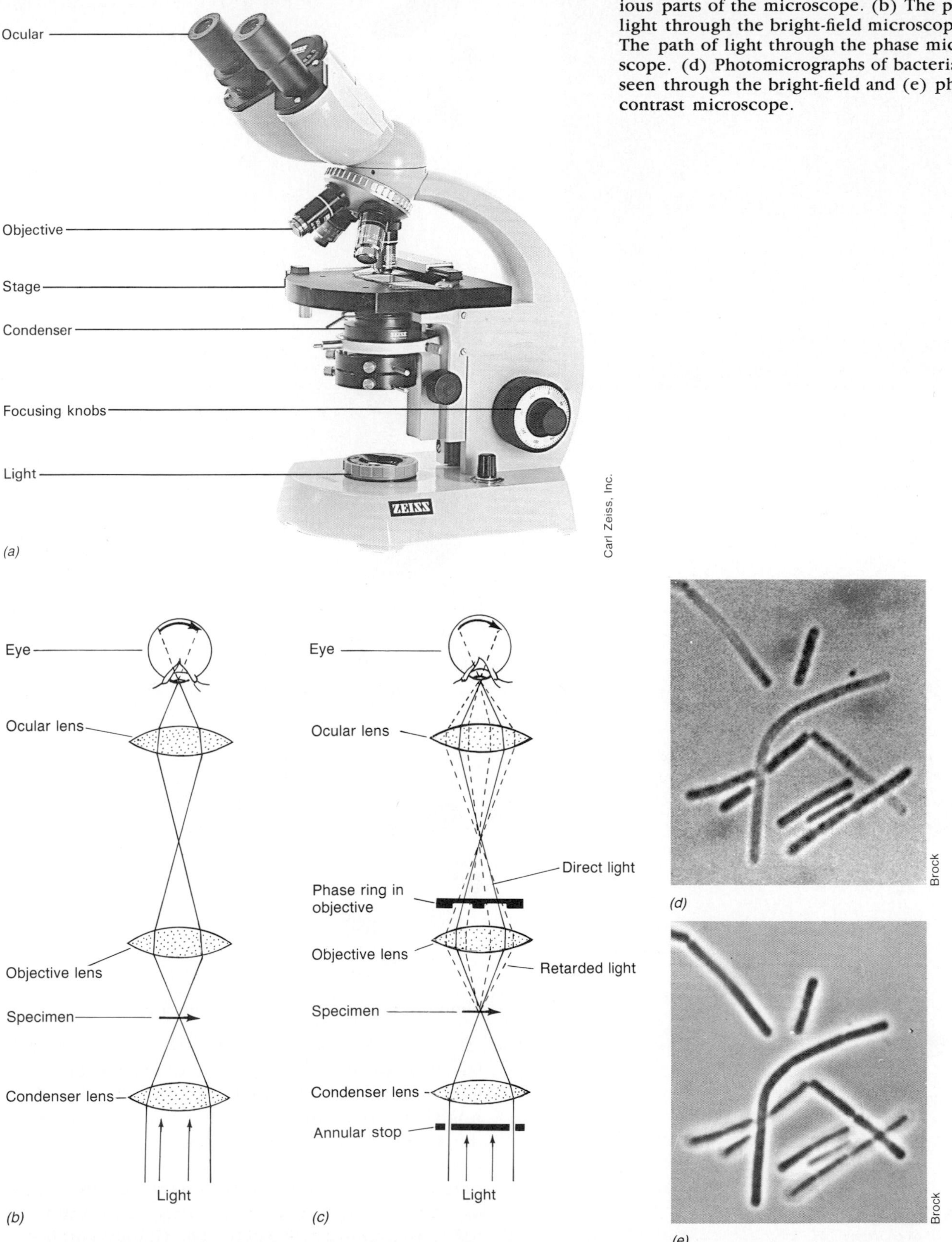

Figure 1.33 The light microscope. (a) The various parts of the microscope. (b) The path of light through the bright-field microscope. (c) The path of light through the phase microscope. (d) Photomicrographs of bacteria as seen through the bright-field and (e) phase-contrast microscope.

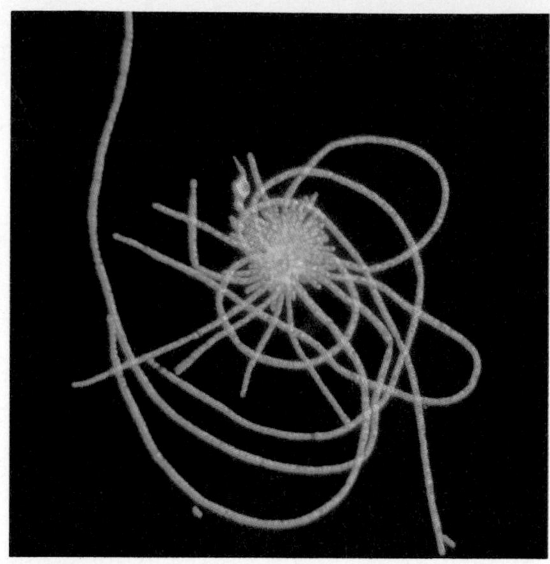

Figure 1.34 Photomicrograph of microorganisms as visualized by fluorescence microscopy.

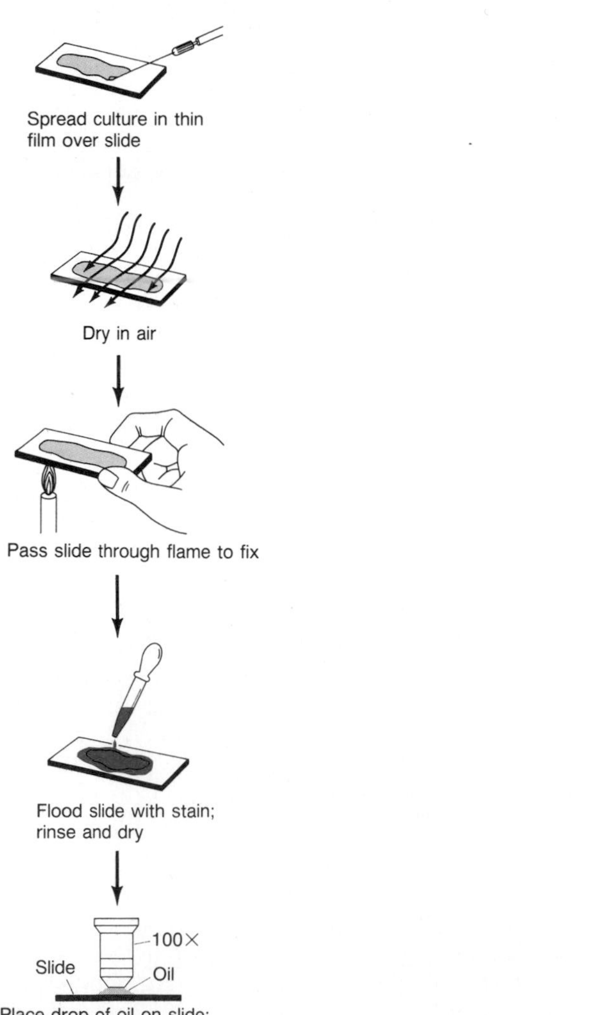
Spread culture in thin film over slide
Dry in air
Pass slide through flame to fix
Flood slide with stain; rinse and dry
100×
Slide
Oil
Place drop of oil on slide; examine with 100× objective

Figure 1.35 Staining cells for microscopic observation.

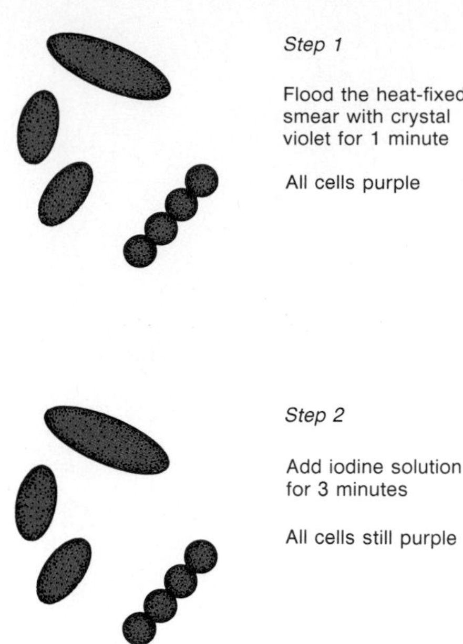
Step 1
Flood the heat-fixed smear with crystal violet for 1 minute
All cells purple
Step 2
Add iodine solution for 3 minutes
All cells still purple

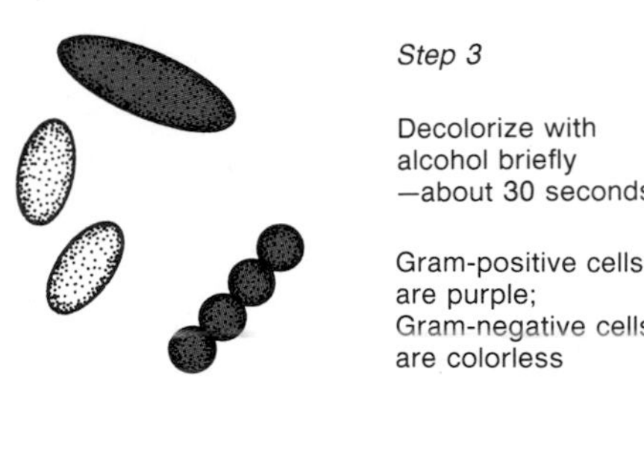
Step 3
Decolorize with alcohol briefly —about 30 seconds
Gram-positive cells are purple; Gram-negative cells are colorless

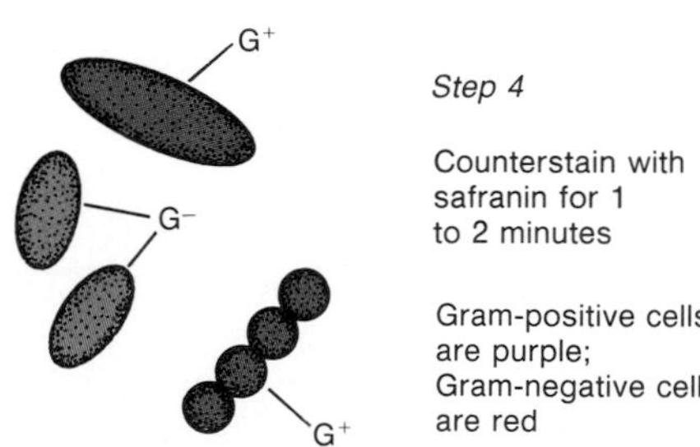
G+
G−
G+
Step 4
Counterstain with safranin for 1 to 2 minutes
Gram-positive cells are purple; Gram-negative cells are red

Figure 1.36 Steps in the Gram stain procedure.

On the basis of their reaction to the Gram stain, bacteria can be divided into two major groups, **Gram-positive** and **Gram-negative**. After Gram staining, the Gram-positive bacteria appear purple, and the Gram-negative bacteria appear red (Figure 1.37). This difference in reaction to the Gram stain arises because of differences in the cell wall structure of Gram-positive and Gram-negative cells. The Gram-positive cell has a single thick wall layer, through which the decolorizing solvent (alcohol) does not readily pene-

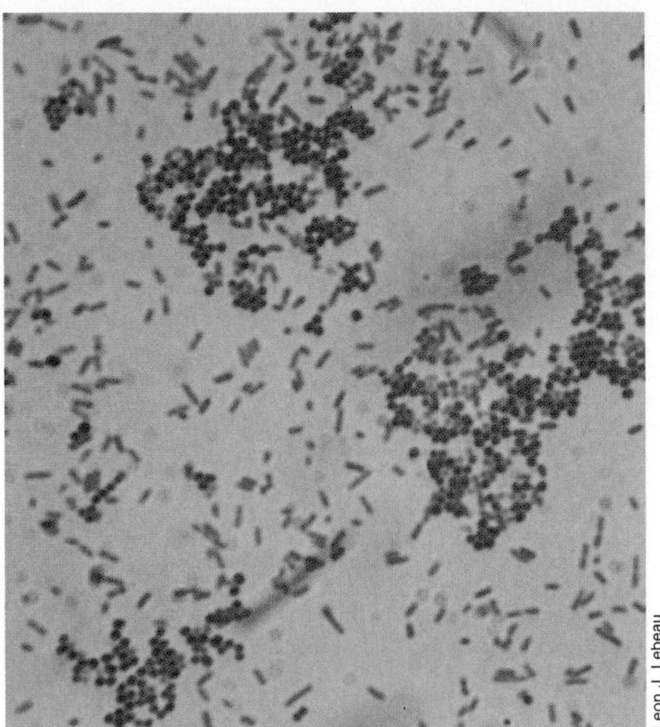

Figure 1.37 Photomicrographs of bacteria which are Gram-positive (blue) and Gram-negative (red). The species are *Staphylococcus aureus* and *Escherichia coli*, respectively.

trate, whereas the Gram-negative cell has a wall with several thinner layers through which the decolorizer readily penetrates. The Gram stain is one of the most useful staining procedures in the bacteriological laboratory; it is almost essential in identifying an unknown bacterium to know first whether it is Gram-positive or Gram-negative.

The electron microscope Electron microscopes are widely used for studying the structures of cells (Figure 1.38). To study the internal structure of procaryotes, a **transmission electron microscope** (TEM) is essential. With the TEM, electrons are used instead of light rays, and electromagnets function as lenses, the whole system operating in a high vacuum. The **resolving power**, or ability to distinguish two objects as distinct and separate, of the electron microscope is much greater than that of the light microscope. Whereas with the light or phase-contrast microscope the smallest structure that can be seen is about 0.2 μm in length, with the electron microscope objects of 0.001 μm can readily be seen. The electron microscope enables us to see many substances of molecular size. However, electron beams do not penetrate very well and if one is interested in seeing internal structure, even a single cell is too

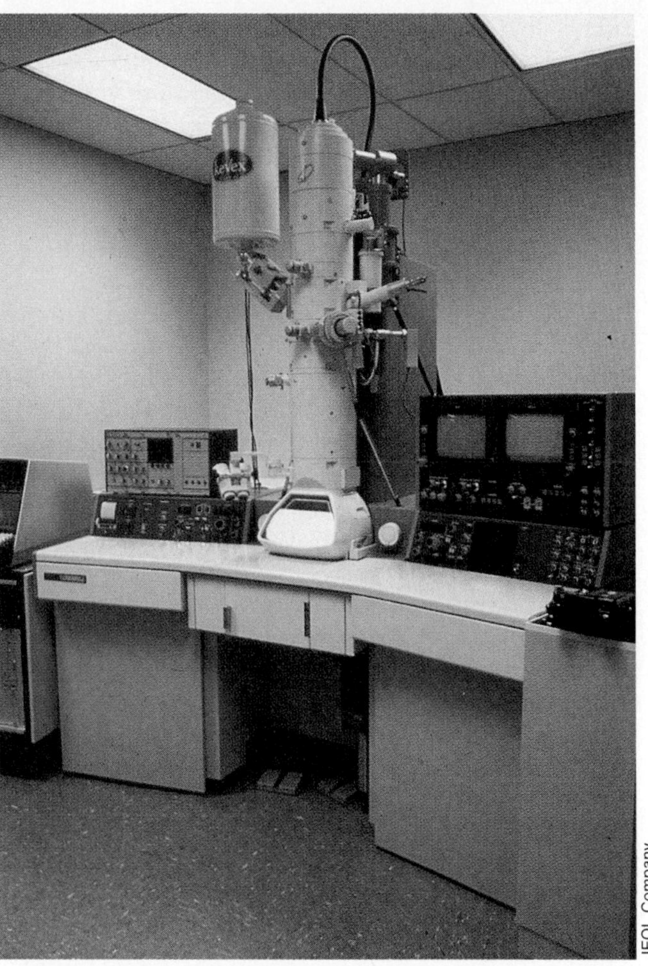

Figure 1.38 A modern electron microscope. This instrument encompasses both transmission and scanning electron microscope functions.

thick to be viewed directly. Consequently, special techniques of *thin sectioning* are needed in order to prepare specimens for the electron microscope.For sectioning, cells must first be fixed, that is, treated with chemicals that prevent tissue distortion. Then the cells are dehydrated by immersing them in an organic solvent. After dehydration the specimen is embedded in plastic. Thin sections are cut from this plastic with a special machine called an ultramicrotome, usually equipped with a diamond knife. A single bacterial cell, for instance, may be cut into five or more very thin slices, which are then examined individually with the electron microscope (see Figure 1.2*b*). To obtain sufficient contrast, the preparations are treated with special electron-microscope stains, such as osmic acid, permanganate, uranium, lanthanum, or lead. Because these materials are composed of atoms of high atomic weight, they scatter electrons well.

Fixation, embedding, and staining are potentially damaging treatments and may greatly alter cell structures. For this reason, great care must be taken in interpretation of the electron-microscope images obtained; artifacts are much more common and serious than in light microscopy. Procedures that work well for some organisms may fail with others; in general, different methods are used for electron microscopy of procaryotic and eucaryotic organisms.

If only the outlines of an organism need be observed, thin sections are not necessary, and whole cells can be examined directly. To increase contrast of whole cells, *shadowing* is usually done. This involves coating the specimen with a thin layer of a metal such as palladium, chromium, or gold. The metal is deposited upon the specimen from one side so that a shadow is created, and the object is thus seen to have thickness and shape. Shadowing is often used to observe flagellar and other surface appendages of bacteria (see Figure 1.13).

Another way of examining whole organisms and their surface structures is by the use of the **scanning electron microscope** (SEM) procedure (see Figure 1.38). The material to be studied is coated with a thin film of a metal such as gold. The electron beam is directed down on the specimen and scans back and forth across it. Electrons scattered by the metal are collected and activate a viewing screen to produce an image. With the scanning electron microscope even fairly large specimens can be observed, and the depth of field is very good. A wide range of magnifications can be obtained, from as low as 15× up to about 100,000×, but only the surface of an object can be visualized. An example of the image obtained in this manner is given in Figure 3.4.

1.14 The Culture of Microorganisms

Although we can get an idea of what microorganisms look like from a microscopic study, we can only get an idea of what microorganisms do by studying their activities in pure cultures. A **pure culture** is a culture consisting of only one type of microorganism. When a microorganism is cultured, it multiplies, and the cell number increases. This process is called **growth**. The term *clone* is often used in microbiology as synonomous with pure culture. A **clone** is a collection of cells which are all derived from a single cell. Because of the rapid growth of microorganisms, it is very easy to obtain in a short period of time a pure culture when starting with only a single cell. In order to obtain a pure culture, we must be able to obtain growth of the organism in the laboratory. Laboratory study requires that we provide the organism with the proper nutrients and environmental conditions so that it can grow. It is also essential that we be able to keep other organisms from entering the pure culture. Such unwanted organisms, called **contaminants**, are ubiquitous, and microbiological technique revolves around the avoidance of contaminants. Once we have isolated a pure culture, we can then proceed to a study of the characteristics of the organism and a determination of its capabilities.

The laboratory study of microorganisms is not difficult to understand. In practice, however, great care must be taken to ensure that pure cultures remain pure and that appropriate nutrients and environmental conditions are provided. When disease-causing (*pathogenic*) microorganisms are to be studied, special precautions must also be observed to prevent infection of people nearby. The term **aseptic technique** refers to the manipulation of cultures in such a way that contamination does not occur.

What are the conditions required for microbial growth? Microorganisms are cultured in water to which appropriate nutrients have been added. The aqueous solution containing such necessary nutrients is called a **culture medium**. The nutrients present in the culture medium provide the microbial cell with those ingredients required for the cell to produce more cells like itself. We divide the nutrients of a culture medium into three major kinds: energy sources, cell structural components, and growth factors.

Energy sources All organisms require energy to live and grow. We can divide energy sources into three groups: organic chemicals, inorganic chemicals, and light. An organic compound is a substance containing carbon (C). **Organic energy sources** include sugars, starch, proteins, fats, and a wide variety of other organic materials. These are, in fact, the kinds of energy sources that humans also use, and they are the energy sources used by many microorganisms. Organic energy sources are used by all fungi and protozoa and by the majority of bacteria. Organisms using organic energy sources are frequently called *heterotrophs*. Thus fungi, protozoa, and most bacteria are heterotrophs, as are humans.

A number of **inorganic energy sources** are used by a few special groups of bacteria. The inorganic energy sources used include ammonium (NH_4^+), nitrite (NO_2^-), ferrous iron (Fe^{2+}), hydrogen sulfide (H_2S), sulfur (S^0), and hydrogen gas (H_2). As for organic energy sources, the utilization of an inorganic energy source also requires that it be oxidized; the end products of such oxidations are compounds that are more oxidized than the starting materials. The organisms that use inorganic energy sources are called *lithotrophs* (literally, "rock eaters").

Light is used as an energy source by photosynthetic organisms. Photosynthetic microorganisms include the algae, which are eucaryotes, and the photosynthetic bacteria, which include the cyanobacteria and the purple and green bacteria (Figure 1.39). All photosynthetic organisms contain at least one form of *chlorophyll*, the crucial light-gathering pigment of photosynthesis. An organism that uses light as an energy source is called a **phototroph**.

Organisms such as the lithotrophs and phototrophs are usually able to grow in completely inorganic media, using carbon dioxide (CO_2) as their sole source of carbon. Organisms using inorganic carbon are called **autotrophs** (literally, self-feeding organisms) and are to be contrasted with the **heterotrophs**, discussed above, which need organic carbon.

Cell structural components As we have noted earlier in this chapter, cells are constructed of a number of chemical elements. The major elements in cells are carbon, hydrogen, oxygen, nitrogen, sulfur, phosphorous, potassium, magnesium, calcium, iron, and sodium. In addition to these major elements, there are several elements present in very small amounts, which are called **trace elements**. The most important trace elements are cobalt, zinc, molybdenum, copper, and manganese. In order to obtain microbial growth, all required elements must be provided in the culture medium in the proper form. The study of how organisms acquire the elements they need for growth is called *nutrition*, and is discussed in Chapter 4.

Other environmental factors In addition to appropriate nutrients, it is essential that other environmental factors be adjusted appropriately for each organism to be cultured. One of the most important environmental factors that must be adjusted is **temperature**. Each microorganism has a defined temperature range over which it is capable of growing and if a temperature too low or too high for this organism is used, satisfactory growth will not be obtained (Figure 1.40). The **optimum** temperature is the temperature at which the organism grows the fast-

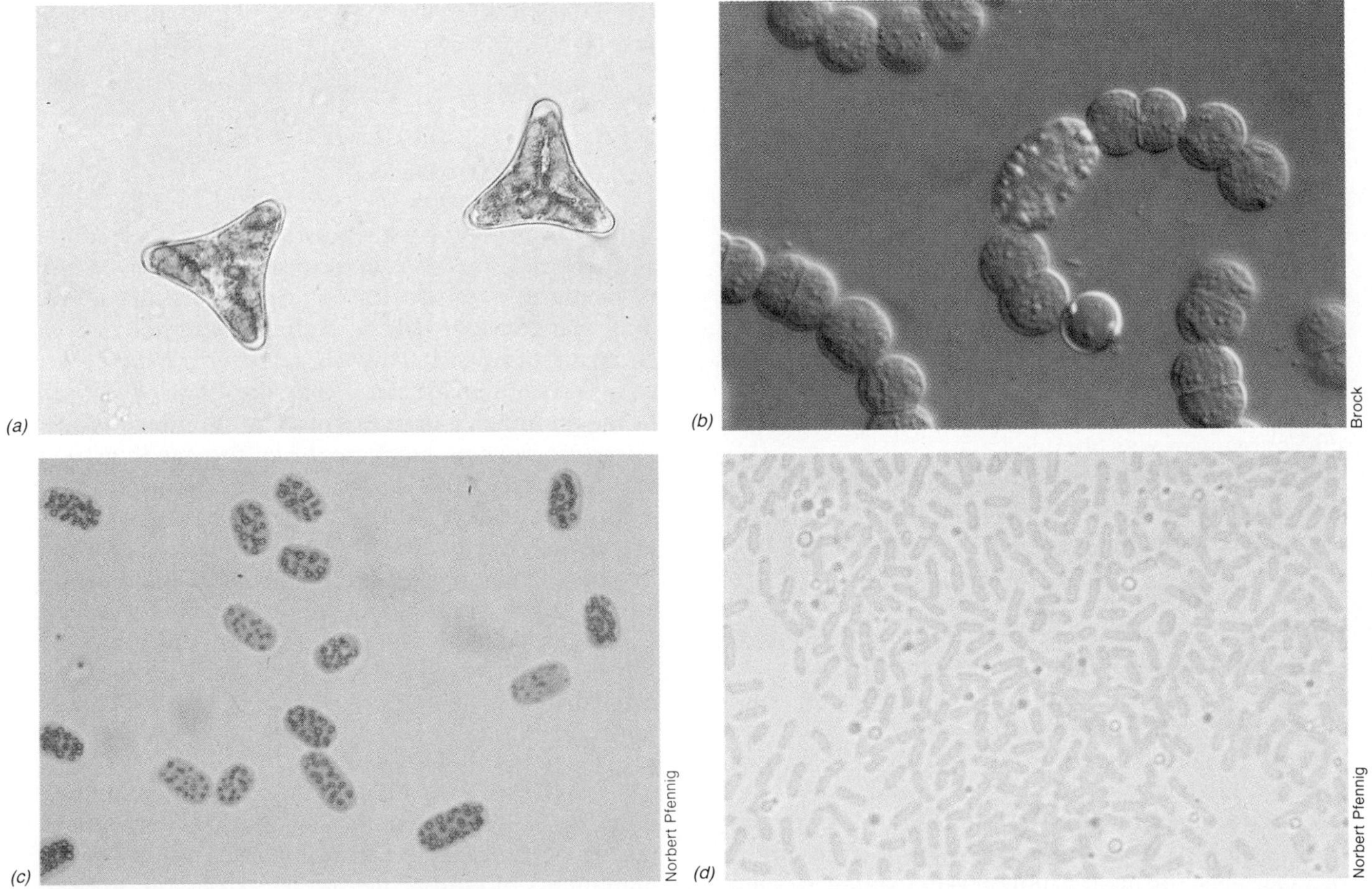

Figure 1.39 Photosynthetic microorganisms. (a) A green alga (eucaryote). (b) Cyanobacterium. (c) Purple bacterium. (d) Green bacterium.

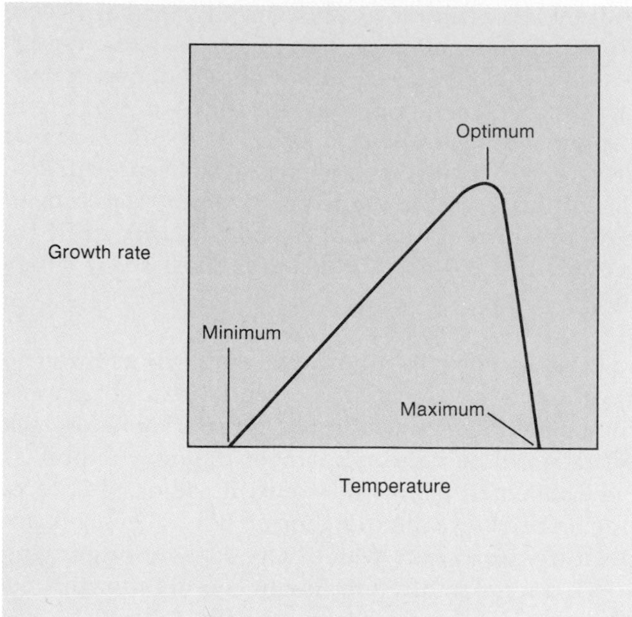

Figure 1.40 The typical growth response of a microorganism to temperature, showing the minimum, maximum, and optimum growth temperatures.

est. Some microorganisms have optimum temperatures as low as 5°C, and others are known with optima as high as 105°C. However, no one organism spans more than a small part of its temperature range. Microorganisms capable of causing disease in warm-blooded animals grow best near body temperature, which is about 37°C. Microorganisms found in soil and water usually grow best at temperatures around 25°C, whereas microorganisms from excessively heated environments require higher temperatures. As a general rule, it can be stated that the most appropriate temperature for culture of a microorganism is near the temperature of the habitat from which that microorganism was obtained.

A *proper acidity* or *alkalinity* must be provided for each organism. Acidity or alkalinity of a solution is expressed by its *pH value* on a scale in which neutrality is pH 7. Those pH values that are less than 7 are acidic and those greater than 7 are alkaline. We shall discuss acids, bases, and pH in greater detail in Section 9.12.

Microorganisms vary in their need for, or tolerance of, **oxygen**. Microorganisms can be divided into several groups, depending on the effect of oxygen. Microorganisms able to use oxygen are called **aerobes**. Some organisms are unable to use oxygen; such organisms are called **anaerobes**. For the culture of anaerobes, the main problem is to exclude oxygen. Because oxygen is ubiquitous in the air, it is not a trivial task to culture microorganisms under anaerobic conditions. Bottles or tubes filled completely to the top with culture medium and provided with tightly fitting stoppers will provide anaerobic conditions for organisms that are not too sensitive to small amounts of oxygen. It is also possible to add a chemical to the culture medium which reacts with oxygen and therefore removes it from the culture medium. Such a substance is called a *reducing agent*. A number of organisms, called **facultative aerobes**, can grow either in the presence or absence of oxygen.

Culture media We have described briefly the basic nutritional components that must be provided for organisms in order to obtain growth. The aqueous solution containing such necessary nutrients is called a **culture medium** (plural, **media**). We discuss culture media in some detail in Chapter 4. Culture media can be prepared for use either in a liquid state, or in a gel (semisolid) state. If a semisolid culture medium is used, a liquid culture medium is converted into a solid form by adding a gelling agent. **Agar** is the most commonly used gelling agent. Agar is manufactured from certain seaweeds and is not a nutrient for most microorganisms. Culture media containing agar are often dispensed in flat dishes called Petri plates (Figure 1.41).

1.15 Sterilization and Aseptic Technique

We have considered what must be put into a medium and how to prepare it. Before proceeding to the use of media to obtain cultures of microorganisms, we must first consider how to exclude unwanted or contaminant organisms. Microbes are everywhere. Because of their small sizes, they are easily dispersed in the air and on surfaces (such as the human skin). Therefore, we must **sterilize** the culture medium soon after its preparation to eliminate microorganisms already contaminating it. It is equally important to take precautions during the subsequent handling of a sterile culture medium, to exclude from the medium all but the organisms desired. Thus, other materials which come into contact with culture media must also be sterilized. It is also necessary to sterilize contaminated materials in order to control the spread of harmful bacteria.

Sterilization of media The most common way of sterilizing culture media is by heat. However, since heat may also cause harmful changes in many of the ingredients of culture media, it is desirable to keep the heating time as short as possible. The best procedure is to heat under pressure, since at pressures

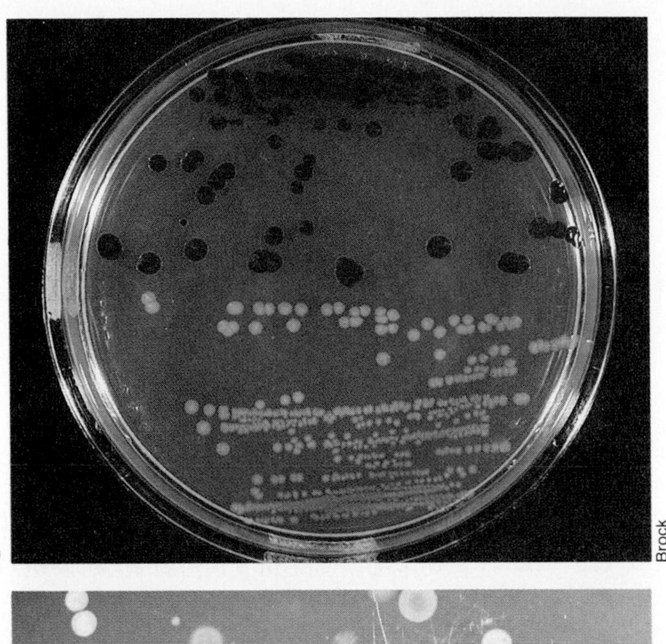

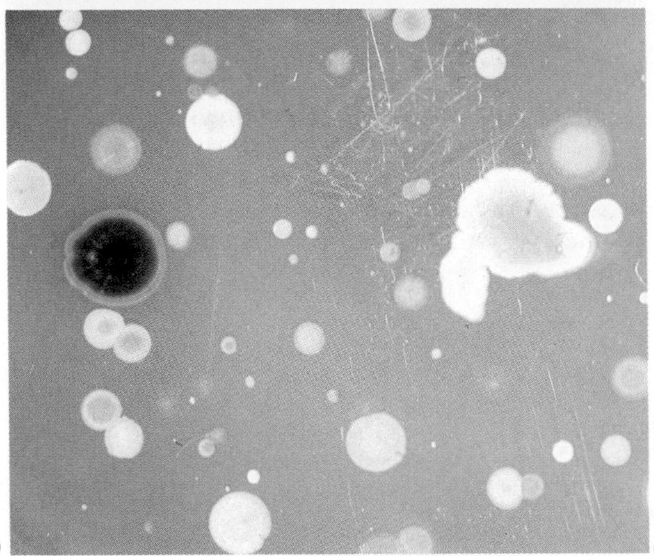

Figure 1.41 An agar-containing culture medium in a Petri plate. Bacterial colonies are visible on the surface of the agar. Note the distinct and characteristic colors. (a) Red and white colonies. (b) Purple, yellow, and white colonies.

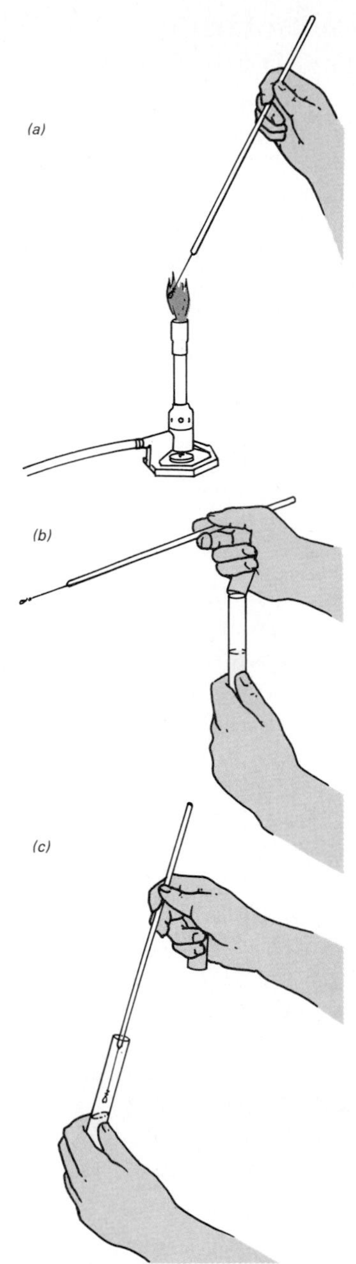

Figure 1.42 Aseptic transfer. (a) Loop is heated until red-hot. (b) Tube is uncapped and loop is cooled in air briefly. (c) Sample is removed and tube is recapped. Sample is transferred to a sterile tube. Loop is reheated before being taken out of service.

above atmospheric the temperature can be increased above 100°C, thus decreasing the time necessary to make the medium sterile. Devices for heating under pressure are called **autoclaves** (see Figure 9.30). The conditions used to sterilize in an autoclave are generally 15 minutes at 121°C.

The technique used in the prevention of contamination during manipulations of cultures and sterile culture media is called **aseptic technique**. Its mastery is required for success in the microbiology laboratory. Air-borne contaminants are the most common problem, since air always contains dust particles that generally have a population of microorganisms. When containers are opened, they must be handled in such a way that contaminant-laden air does not enter. This is best done by keeping the containers at an angle so that most of the opening is not exposed to the air. Manipulations should preferably be carried out in a dust-free room in which air currents are absent. Aseptic transfer of a culture from one tube of medium to another is usually accomplished with an inoculating loop or needle which has been sterilized by incineration in a flame (Figure 1.42).

1.16 Procedures for Isolating Microorganisms from Nature

In nature, microorganisms almost always live in mixtures. However, before most characteristics of a particular microorganism can be determined, the organism must first be isolated in pure culture. How does one go about isolating a pure culture of an organism? The general procedure is to use a culture medium and incubation conditions which encourage the growth of the desired organism and discourage the growth of other organisms. Once the desired organism or organisms have been increased in amount relative to the others, it is purified so that the culture contains only the single type of microorganism desired.

Enrichment culture Organisms vary widely in their nutritional and environmental requirements. It is possible, by choosing the appropriate growth conditions, to select from a mixture an organism of interest. The procedure of adjusting culture conditions to select a particular organism is called **enrichment culture** and is a procedure of considerable importance in microbiology.

The usual practice in enrichment culture procedures is to prepare a culture medium that will favor the growth of the organism of interest. The medium is then inoculated with a sample of material thought to contain the organism, such as soil, water, blood, or tissue. The appropriate conditions of aeration, temperature, and pH are provided and the inoculated medium is then incubated. After an appropriate period of time, usually several days, the enrichment culture is examined visually and microscopically for evidence of microbial growth. Cultures in which growth has taken place are then used further for isolation of pure cultures by streaking on medium solidified with agar (Figure 1.43).

Some examples of enrichment cultures:

1. Photosynthetic organisms can be enriched by using media devoid of any organic or inorganic energy sources (but with CO_2 as a source of carbon) and incubating the cultures in the light. Inocula from rivers, lakes, oceans, or the soil may be used.
2. Lithotrophic bacteria able to utilize ammonium can be enriched by preparing medium with NH_4^+ as sole energy source and incubating in the dark. Inocula can be from soil, sewage, or mud.
3. Organisms using specific organic compounds as energy sources can be enriched by preparing cul-

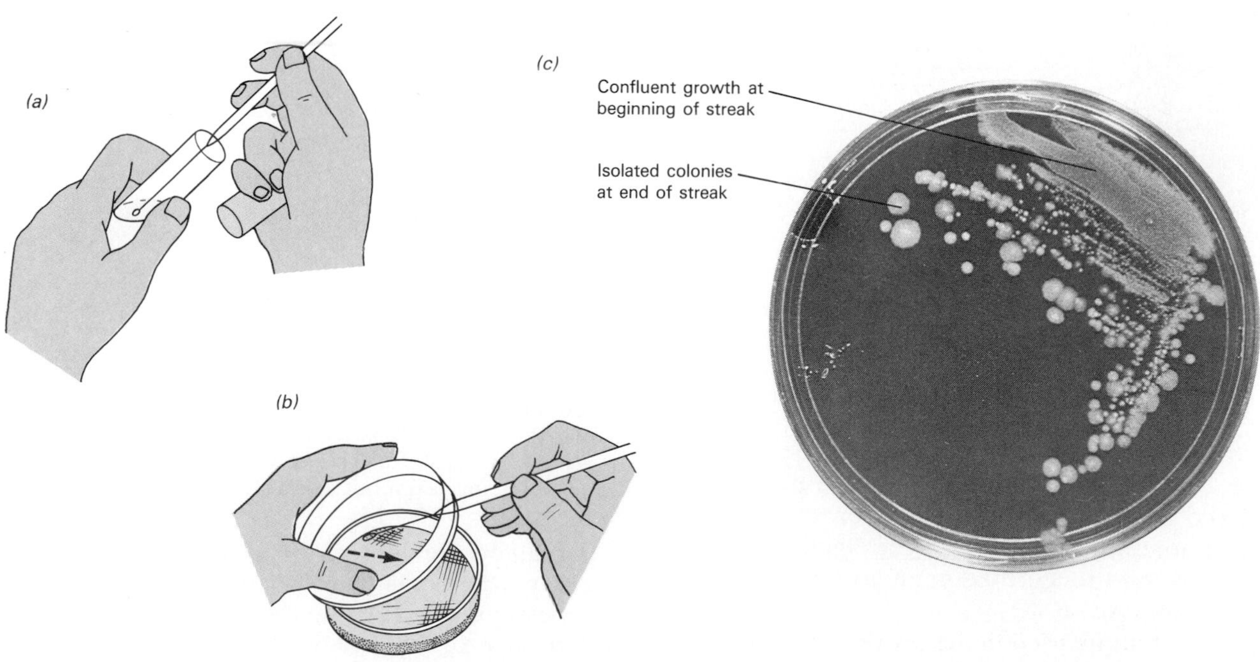

Figure 1.43 Method of making a streak plate to obtain pure cultures. (a) Loopful of inoculum is removed from tube. (b) Streak is made over a sterile agar plate, spreading out the organisms. (c) Appearance of the streaked plate after incubation. Note the presence of isolated colonies. It is from such well-isolated colonies that pure cultures usually can be obtained.

ture media in which the organic compound in question is the sole energy and carbon source. As an example, if we were interested in an organism that can utilize the herbicide 2,4-D (2,4-dichlorophenoxyacetic acid), we could prepare a culture medium in which 2,4-D is the sole carbon and energy source.

4. If organisms capable of growth at high temperatures are desired, enrichment cultures can be prepared using high temperatures for incubation.
5. If anaerobic organisms are desired, enrichment cultures devoid of O_2 can be prepared.

These examples are just a few of many that might have been given. Using appropriate enrichment culture procedures, thousands of cultures have been obtained and most of our knowledge of microbial diversity has come from a study of pure cultures derived by enrichment procedures. We discuss details of enrichment culture in Section 17.3.

Maintaining pure cultures Once a pure culture is obtained, it must be kept pure. One of the most frequent ways in which erroneous results and conclusions are obtained in microbiology is by the use of contaminated cultures. Cultures of organisms of interest that are maintained in the laboratory for study and reference are called *stock cultures*. The stock culture must be maintained so that it is free from contamination, retains viability, and remains genetically homogeneous. Cultures which are infrequently used can be purchased from a commercial culture collection (see Section 10.1). For long-term storage of cultures, they may be frozen in *liquid nitrogen*, at which low temperature viability is effectively maintained. Many cultures can also be dried by a *freeze-drying* process (*lyophilization*) and preserved almost indefinitely in the dried state.

1.17 The Impact of Microorganisms on Human Affairs

We have already suggested that the actions of microorganisms have important consequences for human affairs. A goal of the microbiologist is to understand how microorganisms work, and through this understanding to devise ways that benefits may be increased and damages curtailed. Microbiologists have been eminently successful in achieving these goals, and microbiology has played a major role in the advancement of human health and welfare. One measure of the microbiologist's success is shown by the statistics in Figure 1.44, which compare the present causes of death in the United States to those at the beginning of this century. At the beginning of the century, the

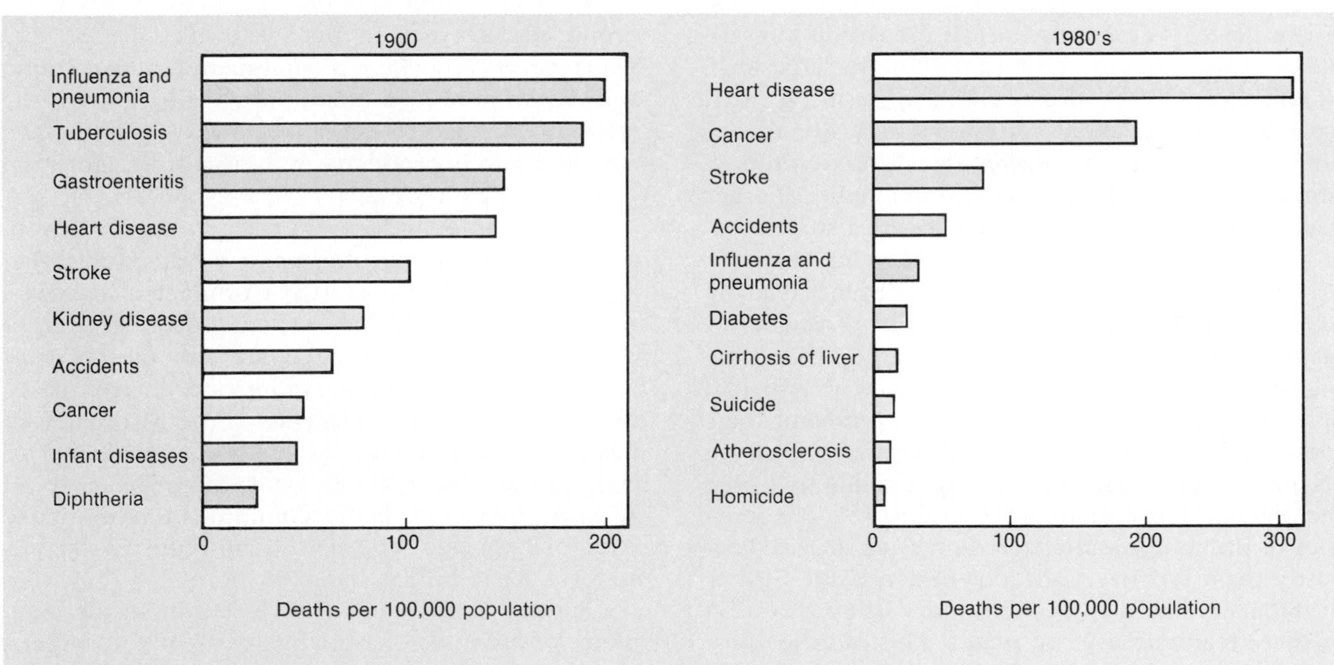

Figure 1.44 Death rates for the 10 leading causes of death: 1900 and 1980. Infectious diseases were the leading causes of death in 1900, whereas today they are much less important. Microbial diseases are shown in color. Data from the National Center for Health Statistics.

major causes of death were infectious diseases; currently, such diseases are of only minor importance. Control of infectious disease has come as a result of our comprehensive understanding of disease processes. As we will see later in this chapter, microbiology had its beginnings in these studies of disease.

However, although we now live in a world where microbes are mostly under control, for the individual dying slowly of acquired immunodeficiency syndrome (AIDS), or the cancer patient whose immune system has been devastated as a result of treatment with an anti-cancer drug, microbes are still the major threat to survival. Although such tragic situations barely appear in our health statistics, they are of no less concern. Further, microbial diseases still constitute the major causes of death in many of the less developed countries of the world. Although eradication of smallpox from the world has been a stunning triumph for medical science, millions still die yearly in developing countries from such pervasive illnesses as malaria, cholera, African sleeping sickness, and severe diarrheal diseases.

Thus, microorganisms are still serious threats to human existence. But we must emphasize that most microbes are *not* harmful to humans. In fact, most microorganisms cause no harm at all. More interestingly, a large number of microorganisms display *beneficial* activities, carrying out processes that are of immense value to human society. Even in the healthcare industry, microorganisms play beneficial roles. For instance, the pharmaceutical industry is a multibillion dollar industry built in part on the large-scale production of antibiotics by microorganisms. We will have much to say about antibiotics in this book, but note here that current worldwide antibiotic production is over *100,000 tons* per year! A number of other major pharmaceutical product lines are also derived, at least in part, from the activities of microbes.

Our whole system of *agriculture* depends in many important ways on microbial activities. A number of major crops are members of a plant group called the **legumes**, which live in close association with special bacteria which form structures called *nodules* on their roots. In these root nodules, atmospheric nitrogen (N_2) is converted into fixed nitrogen compounds that the plants can use for growth. In this way, the activities of the root nodule bacteria reduce the need for costly plant fertilizer. Millions of acres and billions of dollars are annually committed to the cultivation of these leguminous crop plants. Also of major agricultural importance are the microorganisms that are essential for the digestive process in cattle and sheep. These important farm animals have a special digestive organ called the **rumen** in which microorganisms carry out the digestive process. Many billions of dollars of meat and milk production are linked to these rumen microbes. Without these microbes, cattle and sheep production would be virtually impossible. In addition to benefits to agriculture, microorganisms also cause harmful effects. Animal and plant diseases due to microorganisms have major economic impact.

Once food crops are produced, they must be delivered in wholesome form to consumers. Microbes play important roles in the *food industry*. We note first that food spoilage wastes vast amounts of money every year. The canning, frozen-food, and dried-food industries exist to prepare foods in such ways that they will not spoil. Food preservation is an almost $30 billion industry today.

However, not all microbes have harmful effects on foods. Dairy products manufactured, at least in part, via microbial activity include cheese, yogurt, and buttermilk, all products of major economic value. Sauerkraut, pickles, and some sausages also owe their existence to microbial activity. Baked goods are made using yeast. Even more pervasive in our society are the alcoholic beverages, also based on the activities of yeast, which build a $60 billion industry.

All of these applications of microbes in food and agriculture are of ancient origin, but microbiology has not rested on the past. Consider, for instance, the microbe's contribution to a carbonated soft drink. The major sugar in many soft drinks is *fructose*, produced from corn starch via microbial activity (see Section 10.9). Over 16 billion pounds of corn sweeteners are produced each year! In diet soft drinks, the artificial sweetener *aspartame* is a combination of two amino acids, both produced microbiologically. Finally, the *citric acid* added to many soft drinks to give them tang and bite is produced in a large-scale industrial process using a fungus.

Our complex industrial society is energy-driven, and here also microorganisms play major roles. Natural gas, methane, is mostly a product of bacterial action. It is harvested in vast amounts. A few other mineral and energy products are also the result of microbial activity, but of even greater interest is the relationship of microorganisms to the petroleum industry. Crude oil is subject to vigorous microbial attack, and drilling, recovery, and storage of crude oil all have to be done under conditions that minimize microbial damage. The petroleum industry today is almost a $100 billion industry.

Further, human activity will result in the complete consumption of available fossil fuels during the next few decades, so that we must seek new ways to supply the energy needs of society. In the future, microorganisms may provide alternative energy sources. Photosynthetic microorganisms can harvest light energy for the production of **biomass**, energy stored in living organisms. Microbial biomass and existing waste materials such as domestic refuse, surplus grain, and animal wastes, can be converted to "biofuels," such as methane and ethanol, by other microorganisms. Other microbial products may be used as "chemical feedstocks," the chemicals from which synthetic materials are manufactured, and which are now also commonly derived from petroleum. Human activity is also decreasing the supplies of other substances, such as metals, and microorganisms are being used increasingly in metal recovery from low grade ores.

But the above recital of the benefits of microbiology is only the beginning. One of the most exciting new areas of microbiology at present is that called **biotechnology**. In the broad sense, biotechnology entails the use of microorganisms in large-scale industrial processes, but by "biotechnology" today we usually mean the application of novel genetic procedures, generally involving microorganisms. Biotechnology is frequently equated with **genetic engineering**, the discipline that concerns the artificial manipulation of genes and their products.

Genes from human sources, for instance, can be broken into pieces, modified, and added to or subtracted from, using microbes and their enzymes as precise and sophisticated molecular tools. It is even possible to make completely artificial genes using genetic engineering techniques. And once the desired gene has been selected or created, it can be inserted into a microbe where it can be made to reproduce and make the desired gene product. For instance, human insulin, a hormone found in abnormally low amounts in people with the disease diabetes, has now been produced microbiologically with the human insulin gene engineered into a microbe. We discuss genetic engineering and biotechnology in detail in Chapters 8 and 10.

The overwhelming influence of microorganisms in human society is clear. We have many reasons to be aware of microorganisms and their activities. As the eminent French scientist Louis Pasteur, one of the founders of microbiology, expressed it: "The role of the infinitely small is infinitely large." Therefore, before we begin a detailed study of microbiology, let us consider briefly the contributions which Pasteur and others made to the origins of our science.

1.18 A Brief History of Microbiology

Although the existence of creatures too small to be seen with the eye had long been suspected, their discovery was linked to the invention of the microscope. Robert Hooke described the fruiting structures of molds in 1664 (Figure 1.45), but the first person to see microorganisms in any detail was the Dutch amateur microscope builder Anton van Leeuwenhoek, who used simple microscopes of his own construction (Figure 1.46). Leeuwenhoek's microscopes were extremely crude by today's standards, but by careful manipulation and focusing he was able to see organisms as small as bacteria. He reported his observations in a series of letters to the Royal Society of London, which published them in English translation. Drawings of some of Leeuwenhoek's "wee animalcules" are shown in Figure 1.46*b*. His observations were confirmed by other workers, but progress in understanding the nature of these tiny organisms came slowly. Only in the nineteenth century did improved microscopes become available and widely distributed. During its history, the science of microbiology has taken the greatest steps forward when better microscopes have been developed, for these have enabled scientists to penetrate ever deeper into the mysteries of the cell.

Microbiology as a science did not develop until the latter part of the nineteenth century. This long delay occurred because, in addition to microscopy, certain basic techniques for the study of microorganisms needed to be devised. In the nineteenth century, investigation of two perplexing questions led to the development of these techniques and laid the foundation of microbiological science: (1) Does spontaneous generation occur? (2) What is the nature of contagious disease? By the end of the century both questions were answered, and the science of microbiology was firmly established as a distinct and growing field.

Spontaneous generation The basic idea of spontaneous generation can easily be understood. If food is allowed to stand for some time, it putrefies. When the putrefied material is examined microscopically, it is found to be teeming with bacteria. Where do these bacteria come from, since they are not seen in fresh food? Some people said they developed from seeds or germs that had entered the food from the air, whereas others said that they arose spontaneously.

Spontaneous generation would mean that life could arise from something nonliving, and many people could not imagine something so complex as a living cell arising spontaneously from dead materials. The most powerful opponent of spontaneous generation was the French chemist Louis Pasteur, whose work on this problem was the most exacting and convincing. Pasteur first showed that structures were present in air that resembled closely the microorganisms seen in putrefying materials. He did this by passing air through guncotton filters, the fibers of

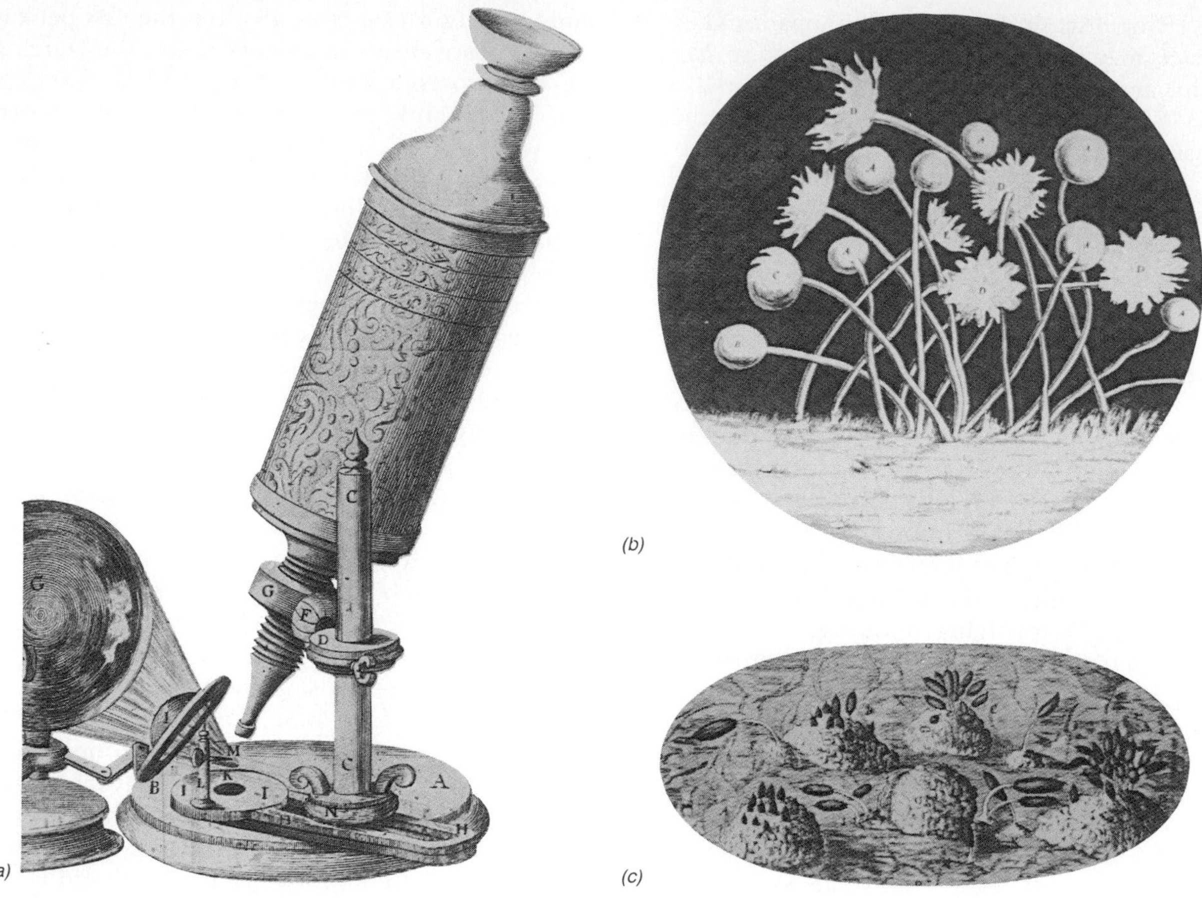

Figure 1.45 (a) The microscope used by Robert Hooke. (b) Two drawings by Robert Hooke that represent one of the first microscopic descriptions of microorganisms. Top: a blue mold growing on the surface of leather; the round structures contain spores of the mold. Bottom: a mold growing on the surface of an aging and deteriorating rose leaf.

which stopped solid particles. After the guncotton was dissolved in a mixture of alcohol and ether, the particles that it had trapped fell to the bottom of the liquid and were examined on a microscope slide. Pasteur found that in ordinary air there exists constantly a variety of solid structures ranging in size from 0.01 mm to more than 1.0 mm. Many of these structures resemble the spores of common molds, the cysts of protozoa, and various other microbial cells. As many as 20 to 30 of them were found in 15 liters of ordinary air, and they could not be distinguished from the organisms found in much larger numbers in putrefying materials. Pasteur concluded that the organisms found in putrefying materials originated from the organized bodies present in the air. He postulated that these bodies are constantly being deposited on all objects. If this conclusion was correct, it would mean that, if food were treated to destroy all the living organisms contaminating it, then it should not putrefy.

Pasteur used heat to eliminate contaminants, since it had already been established that heat effectively kills living organisms. In fact, many workers had shown that if a nutrient solution was sealed in a glass flask and heated to boiling, it never putrefied. The proponents of spontaneous generation criticized such experiments by declaring that fresh air was necessary for spontaneous generation and that the air itself inside the sealed flask was affected in some way by heating so that it would no longer support spontaneous generation. Pasteur skirted this objection simply and brilliantly by constructing a swan-necked flask, now called the "Pasteur flask" (Figure 1.47). In such a flask putrefying materials can be heated to boiling; after the flask is cooled, air can reenter but the bends in the neck prevent particulate matter, bacteria, or other microorganisms from getting in the flask. Material sterilized in such a flask did not putrefy, and no microorganisms ever appeared as long as the neck

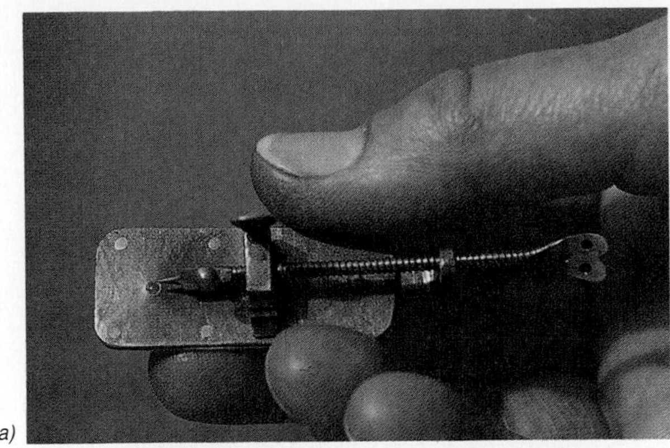

(a)

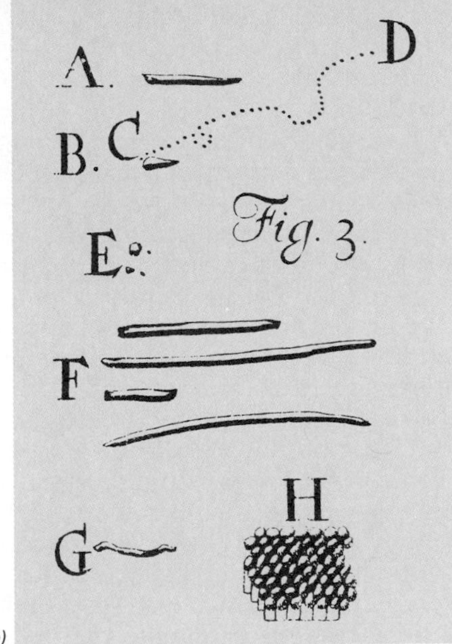

(b)

Figure 1.46 (a) Photograph of a replica of Leeuwenhoek's microscope. (b) Leeuwenhoek's drawings of bacteria, published in 1684. Even from these crude drawings we can recognize several kinds of common bacteria. Those lettered A, C, F, and G are rod-shaped; E, spherical or coccus-shaped.

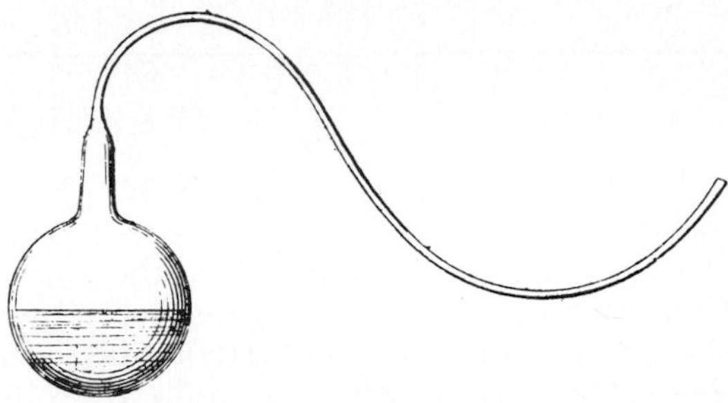

Figure 1.47 Pasteur's drawing of a swan-necked flask. In his own words: "In a glass flask I placed one of the following liquids which are extremely alterable through contact with ordinary air: yeast water, sugared yeast water, urine, sugar beet juice, pepper water. Then I drew out the neck of the flask under a flame, so that a number of curves were produced in it. . . . I then boiled the liquid for several minutes until steam issued freely through the extremity of the neck. This end remained open without any other precautions. The flasks were then allowed to cool. Any one who is familiar with the delicacy of experiments concerning the so-called 'spontaneous' generation will be astounded to observe that the liquid treated in this casual manner remains indefinitely without alteration. The flasks can be handled in any manner, can be transported from one place to another, can be allowed to undergo all the variations in temperature of the different seasons, [yet] the liquid does not undergo the slightest alteration. . . . After one or more months in the incubator, if the neck of the flask is removed by a stroke of a file, without otherwise touching the flask, molds and infusoria begin to appear after 24, 36, or 48 hours, just as usual, or as if dust from the air had been inoculated into the flask. . . . At this moment I have in my laboratory many alterable liquids which have remained unchanged for 18 months in open vessels with curved or inclined necks."

of the flask remained intact. If the neck was broken, however, putrefaction occurred and the liquid soon teemed with living organisms. This simple experiment served to effectively settle the controversy surrounding the theory of spontaneous generation.

Killing all the bacteria or other microorganisms in or on objects is a process we now call **sterilization**, and the procedures that Pasteur and others used were eventually carried over into microbiological research. Disproving the theory of spontaneous generation thus led to the development of effective sterilization procedures, without which microbiology as a science could not have developed (see Section 1.15).

It was later shown that flasks and other vessels could be protected from contamination by cotton stoppers, which still permit the exchange of air. The principles of aseptic technique, developed so effectively by Pasteur, are the first procedures learned by the novice microbiologist. Food science also owes a debt to Pasteur, as his principles are applied in the canning and preservation of many foods.

Although Pasteur was successful in sterilizing materials with simple boiling, some workers found that boiling was insufficient. We now know that the failure resulted from the presence in these materials of bacteria which formed unusually heat-resistant structures called *endospores* (see Figure 1.16). Initial work on the endospore was carried out by two men: John Tyndall in England, and Ferdinand Cohn in Germany. Both men observed that some preparations, such as the fruit juice solutions used by Pasteur, were relatively easy to sterilize, requiring only 5 minutes of

The Origin of Pasteurization

Pasteur's name is forever linked in the public mind with the process of pasteurization. The background of the development of the pasteurization process has been nicely discussed by René Dubos in his little book on Pasteur's life, *Pasteur and Modern Science*. We quote from this book here:

"The demonstration that microbes do not generate spontaneously encouraged the development of techniques to destroy them and to prevent or minimize subsequent contamination. Immediately these advances brought about profound technological changes in the preparation and preservation of food products and subsequently in other industrial processes as well . . .

"It was soon discovered that the introduction of microorganisms in biological products can be minimized by an intelligent and rigorous control of the technological operations, but cannot be prevented entirely. The problem therefore was to inhibit the further development of these organisms after they had been introduced into the product. To this end, Pasteur first tried to add a variety of antiseptics, but the results were mediocre and, after much hesitation, he considered the possibility of using heat as a sterilizing agent.

"Pasteur's first studies of heat as a preserving agent were carried out with wine. Pasteur had grown up in one of the best wine districts in France, and, as a connoisseur of the beverage, was much disturbed at the thought that heating might alter its flavor and bouquet. He therefore proceeded with very great caution and eventually convinced himself that heating at 55°C would not alter appreciably the bouquet of wine . . . These considerations led to the process of partial sterilization, which soon became known the world over under the name of 'pasteurization,' and which was found applicable to wine, beer, cider, vinegar, milk, and countless other perishable beverages, foods, and organic products.

"It was characteristic of Pasteur that he did not remain satisfied with formulating the theoretical basis of heat sterilization, but took an active interest in designing industrial equipment adapted to the heating of fluids in large volumes and at low cost. His treatises on vinegar, wine, and beer are illustrated with drawings and photographs of this type of equipment, and describe in detail the operations involved in the process. The word "pasteurization" is, indeed, a symbol of his scientific life; it recalls the part he played in establishing the theoretical basis of the germ theory, and the phenomenal effort that he devoted to making it useful to his fellow humans. It reminds us also of his well-known statement: 'There are no such things as pure and applied science—there are only science, and the applications of science.'

"While it is of no scientific interest, it might be worth mentioning an incident that bears on a point of ethics of increasing relevance to the behavior of university scientists in modern industrial societies. Pasteur was not a wealthy man, and there is no doubt that his family responsibilities often weighed on his mind. It is known that after he had developed techniques for the preservation of vinegar, wine, and beer with the use of heat, he took patents to protect the rights to his discovery. That there were discussions within his family concerning the possible financial exploitation of these patents is revealed in one of his letters: 'My wife . . . who worries concerning the future of our children, gives me good reasons for overcoming my scruples.' Nevertheless, he decided to release his patents to the public, and he did not derive financial profit even from the development or sale of large-scale industrial equipment devised for pasteurization.

"For American readers, it will be of interest to learn that pasteurization was immediately adopted in the United States, even in far-away and as yet undeveloped California. Pasteur took great pride in this recognition of his work so far away from France, and he stated in one of his articles: 'It is inspiring to hear from the citizen of a country where the grapevine did not exist twenty years ago, that, to credit a French discovery, he has experimented at one stroke on 100,000 liters of wine. These men go forward with giant steps, while we timidly place one foot in front of the other . . . "

boiling, whereas others were not sterilized by much longer periods of boiling, sometimes even hours. Notably difficult to sterilize were hay infusions. In addition, once hay had been brought into the laboratory, even the sugar solutions could no longer be reliably sterilized by hours of boiling. Cohn performed detailed microscopic observations and discovered endospores inside cells of old cultures of species of *Bacillus*. Cohn and another German scientist, Robert Koch, applied this observation to the study of disease, as discussed below.

The germ theory of disease Proof that microorganisms could cause disease provided the greatest impetus for the development of the science of microbiology. Indeed, even in the sixteenth century it was thought that something could be transmitted from a diseased person to a well person to induce in the latter the disease of the former. Many diseases seemed to spread through populations and were called *contagious*; the unknown thing that did the spreading was called the *contagion*. After the discovery of microorganisms, it was more or less widely held that these organisms might be responsible for contagious diseases, but proof was lacking. Discoveries by Ignaz Semmelweis and Joseph Lister provided some evidence for the importance of microorganisms in causing human diseases, but it was not until the work of Robert Koch, a physician, that the germ theory of disease was placed on a firm footing.

In his early work, published in 1876, Koch studied *anthrax*, a disease of cattle, which sometimes also occurs in humans. Anthrax is caused by a spore-forming bacterium now called *Bacillus anthracis*, and the blood of an animal infected with anthrax teems with cells of this large bacterium. Koch established by careful microscopy that the bacteria were always present in the blood of an animal that had the disease. However, the mere association of the bacterium with the disease did not prove that it actually *caused* the disease; it might instead be a *result* of the disease. Therefore, Koch demonstrated that it was possible to take a small amount of blood from a diseased animal and inject it into another animal, which in turn became diseased and died. He could then take blood from this second animal, inject it into another, and again obtain the characteristic disease symptoms. By repeating this process as often as 20 times, successively transferring small amounts of blood containing bacteria from one animal to another, he proved that the bacteria did indeed cause anthrax: the twentieth animal died just as rapidly as the first; and in each case Koch could demonstrate by microscopy that the blood of the dying animal contained large numbers of the bacterium.

Koch carried this experiment further. He found that the bacteria could also be cultivated in nutrient fluids outside the animal body, and even after many transfers in culture the bacteria could still cause the disease when reinoculated into an animal. Bacteria from a diseased animal and bacteria in culture both induced the same disease symptoms upon injection. On the basis of these and other experiments Koch formulated the following criteria, now called **Koch's postulates**, for proving that a specific type of bacterium causes a specific disease:

1. The organism should always be found in animals suffering from the disease and should not be present in healthy individuals.
2. The organism must be cultivated in pure culture away from the animal body.
3. Such a culture, when inoculated into susceptible animals, should initiate the characteristic disease symptoms.
4. The organism should be reisolated from these experimental animals and cultured again in the laboratory, after which it should still be the same as the original organism.

Koch's postulates not only supplied a means of demonstrating that specific organisms cause specific diseases but also provided a tremendous spur for the development of the science of microbiology by stressing the use of laboratory culture.

As we have noted, in order to successfully study the activities of a microorganism, such as a microorganism which causes a disease, one must be sure that it alone is present in culture. That is, the culture must be *pure*. With objects as small as microorganisms, ascertaining purity is not easy, for even a very tiny sample of blood or animal fluid may contain several kinds of organisms that may all grow together in culture. Koch realized the importance of pure cultures. He developed several ingenious methods of obtaining them, of which the most useful is that involving the isolation of single *colonies*. Koch observed that when a solid nutrient surface, such as a potato slice, was exposed to air and then incubated, bacterial colonies developed, each having a characteristic shape and color. He inferred that each colony had arisen from a single bacterial cell that fell on the surface, found suitable nutrients, and began to multiply. Because the solid surface prevented the bacteria from moving around, all of the offspring of the initial cell remained together, and when a large enough number of organisms were present, the mass of cells became visible to the naked eye. He assumed that colonies with different shapes and colors were derived from different kinds of organisms. When the cells of a single colony were spread out on a fresh surface, many colonies developed, each with the same shape and color as the original.

Koch realized that this discovery provided a simple way of obtaining pure cultures: he found that if mixed cultures were spread on solid nutrient surfaces, the individual cells were so far apart that the colonies they produced did not mingle. Many organisms could not grow on potato slices; so he devised semisolid media, in which gelatin was added to a nutrient fluid such as blood serum in order to solidify it. When the gelatin-containing fluid was warmed, it liquefied and could be poured out on glass plates; upon cooling, the solidified medium could be inoculated. Later *agar* (a material derived from seaweed) was found to be a better solidifying agent than gelatin, and this substance is widely used today (Figure 1.41).

In the 20 years following the formulation of Koch's postulates, the causal agents of a wide variety of contagious diseases were isolated. These discoveries led to the development of successful treatments for the prevention and cure of contagious diseases and contributed to the development of modern medical practice. The impact of Koch's work has been felt throughout the world.

It is important to realize that Koch's postulates have relevance beyond identifying organisms which cause specific diseases. The essential general conclusion to be gathered is that specific organisms have specific effects. This principle that different organisms have unique activities was important in establishing microbiology as an independent biological science. By the beginning of the twentieth century, the disciplines of bacteriology and microbiology were on a firm footing.

Study Questions

1. What is a pure culture, and why was a knowledge of how to obtain one important for the development of microbiology?
2. Explain why the invention of solid media was of great importance in the development of microbiology.
3. In unicellular forms, such as bacteria and yeasts, the cell and the organism are one and the same. Most multicellular organisms, such as human beings, consist of an indefinite but very large number of cells. From a single yeast cell, by division, billions of cells can arise, and we can get a yeast cake. From a fertilized human egg, billions of cells can arise and we can get a baby. It is undoubtedly clear to you that the billions of cells making up a yeast cake differ in a fairly fundamental way from the billions of cells making up a baby. Describe briefly this fundamental difference and discuss its significance.
4. Why is the phase-contrast microscope now almost universally used in microbiological research laboratories instead of the bright-field microscope?
5. If the bright-field microscope must be used in visualizing bacteria, what special problem must be overcome?
6. For what purpose is staining most commonly used?
7. Write the steps that are performed to carry out a Gram stain and indicate the colors of both Gram-positive and Gram-negative cells at each stage of the process.
8. What would be the result if the alcohol decolorization step were erroneously left out in a Gram staining procedure and the organism was Gram positive? If the organism was Gram negative?
9. Make a list of five different places in which you might expect to get specimens for your microscope that would reveal microorganisms, and explain why each place would be a likely source.
10. Wine has been made since ancient times and by peoples with no formal scientific knowledge or culture, who had no idea of the real process involved. This is an example of how practical advances may come before the scientific knowledge is available. Imagine you were living in biblical times and were acquainted with the winemaking process. (a) What clues would you have which might indicate that conversion of grape juice to wine required the action of something living? (b) What experiment might you do which would help in deciding if a living agency was involved?
11. Pasteur's experiments on spontaneous generation were of enormous importance for the advance of microbiology. They had impact on the following areas: (a) methodology of microbiology; (b) ideas on the origin of life; (c) preservation of food. Explain briefly how the impact of his experiments was felt in each of these areas.
12. Why was the development of sterilization methods of crucial importance for the science of microbiology?
13. If spontaneous generation was an everyday occurrence, would sterilization be possible?
14. Suppose you were Leeuwenhoek and had told a friend of your discovery of "little animalcules." Your friend says "So what?" How might you indicate to him the possible importance of your discovery?
15. If microscopes had never been invented, explain how the existence of microorganisms

might still have been suspected.

16. Explain the principle behind the use of the Pasteur flask in studies on spontaneous generation.
17. If you were an opponent of Pasteur, how might you attempt to refute his experiments on spontaneous generation?
18. List four key properties associated with the living state. Can you think of inanimate systems that might exhibit any of these properties? Explain.
19. Cells can be thought of either as machines or as coding devices. Explain how these two attributes of a cell differ. Can you imagine a cell existing that lacks one of these two attributes?
20. Compare and contrast the procaryotic and the eucaryotic cell. List three properties that are common to both cell types. List three properties that are different in these two cell types.
21. List three ways in which viruses differ from cells.
22. What is a bacterial *endospore*? Of what practical significance is a knowledge of endospores?
23. In what ways are the *flagella* of procaryotes and eucaryotes different? List one property they have in common.
24. Compare and contrast *flagella* and *cilia*.
25. List three major groups of eucaryotic microorganisms and indicate how they differ. Can you think of any property that they all have in common?
26. Molds, yeasts, and mushrooms are all fungi. In what ways do these groups differ?
27. Define *clone, contaminant, aseptic, culture medium*.
28. Define *heterotroph, lithotroph, phototroph*, and *autotroph*.
29. Define *sterilization*. How does *sterilization* differ from *pasteurization*? Present a commonly used means of sterilizing culture media.
30. Explain the principle behind the enrichment culture procedure. Describe in general terms an enrichment culture procedure you might use for isolating an organism capable of growing at refrigerator temperature.

Supplementary Readings

Arms, K., and **P. S. Camp**. 1987. *Biology*. Third Edition. Saunders College Publishing, Philadelphia. A widely used biology textbook with useful background on cell and molecular biology.

Brock, T. D. 1975. *Milestones in Microbiology*. American Society for Microbiology, Washington, D. C. The key papers of Pasteur, Koch, and others are translated, edited, and annotated for the beginning student.

Bold, H. C., and **M. J. Wynne**. 1978. *Introduction to the algae: structure and reproduction*. Prentice-Hall, Englewood Cliffs, N. J.

Bulloch, W. 1935. *The History of Bacteriology*. Oxford University Press, London. The standard history of bacteriology; emphasis on medical aspects.

Dobell, C., ed. and trans. *Antoni van Leeuwenhoek and his "little animals."* Constable and Co., London (1960, Dover Publications, New York). An introduction to Leeuwenhoek's life, his work, and his times.

Kendrick, Bryce. 1985. *The Fifth Kingdom*. Mycologie Publications, Waterloo, Canada. A fully illustrated guide to the fungi.

2

Cell Chemistry

The science of microbiology has matured through the years to the point where our understanding of microorganisms depends more and more on our understanding of the *chemical* processes taking place within cells. As pointed out in Chapter 1, the chemistry of living cells is governed by the same chemical and physical principles that dictate the properties of nonliving matter. However, as self-replicating entities, cells contain a variety of molecules not usually found in inanimate structures. The chemical nature of these molecules is the subject of this chapter. We begin by introducing chemical principles of atoms, molecules, and atomic bonding, and proceed through a discussion of the biochemistry of macromolecules, the building blocks of the cell. The student may find it useful to refer to material in this chapter from time to time while mastering the principles of cellular metabolism (Chapter 4) and genetics and molecular biology (Chapter 5).

2.1 Atoms, Molecules, and Chemical Bonding

Atoms are the basic structural units of matter and can be defined as the smallest particles of an element that can enter into chemical combinations. The atoms of different elements have characteristic chemical and physical properties that are determined by the number of constituent particles of which they are composed. Atoms consist of particles called **protons, electrons**, and **neutrons**. Protons are positively charged and electrons are negatively charged; neutrons, as their name implies, are uncharged particles. About twenty different elements are found in cells. However only six elements: hydrogen, carbon, nitrogen, oxygen, phosphorus, and sulfur, play major roles in living systems. The atomic structure of these six important elements is shown in Table 2.1.

The **nucleus** of an atom contains protons and neutrons and constitutes the majority of the atomic mass. Each proton carries an electronic charge of +1, and the number of protons in the atoms of an element correspond to the **atomic number** of that element. The **atomic weight** of an element is the total mass of protons and neutrons in the atoms of that element. Hence the atomic number of carbon is 6 and its atomic weight is 12 (Table 2.1). Although the number of protons in the atoms of a given element are constant, the number of neutrons may vary, giving rise to **isotopes** of a given element. Thus sulfur atoms always contain 16 protons, but exist in nature in three isotopic forms. Sulfur^{32} (S^{32}), the most abundant form, contains 16 protons and 16 neutrons, sulfur^{34} (S^{34}), a less abundant sulfur isotope, contains 16 protons and 18 neutrons, and sulfur^{35} (S^{35}) is a radioactive isotope or **radioisotope**, and contains 16 protons and 19 neutrons. Unlike S^{32} and S^{34}, the atomic nucleus of S^{35} is highly unstable and spontaneously disintegrates forming P^{33} and releasing an electron (beta particle). All of these forms of sulfur are chemically alike because they have the same number of protons and the same number of electrons in their outer shell (Table 2.1). We will learn elsewhere in this book of the value of stable isotopes, such as S^{32} and S^{34}, for analyzing biogeochemical transformations (see Section 17.5).

Electrons are negatively charged particles of negligible mass and are present in atoms in equal number to protons. Electrons orbit the nucleus in "shells," zones of electron density at increasing distance from the nucleus of the atom. The chemical properties of elements are for the most part determined by the number of electrons in the *outermost* shell of their atoms (see Table 2.1). The outermost shell of most elements of atomic number 8 or greater contains up to 8 electrons. The potential of an element for combination with other elements to form molecules is to a major degree governed by whether or not the outer electron shell is full or partially depleted. For the biologically relevant elements, the most stable electron configuration in the outer shell of atoms is *eight*, except for the first shell which holds only two electrons. Table 2.1 shows that the outer shells of atoms of the major bioelements, hydrogen, carbon, nitrogen, oxygen, phosphorus, and sulfur, are all incomplete; thus these elements readily combine to form molecules in which all atoms achieve stable outer electron shells.

2.2 Chemical Bonds

Several types of chemical bonds are known including ionic, covalent, and hydrogen bonds. **Ionic bonds** are bonds in which electrons are not equally shared between two atoms, resulting in charged chemical species. For example, in NaCl, the atom of chlorine is much more electronegative (electron attracting) than that of sodium, and formally removes the electron from the outer shell of sodium to form, in effect, Na^+Cl^-. The electrical attraction between the two atoms holds the molecule together. Ionic bonds are important in the binding of magnesium ions to macromolecules such as the nucleic acids because of the strong attraction of positive Mg^{2+} by negative phosphate, PO_4^{3-}. Ionic contacts are also important in nucleic acid/protein interactions. However, since ionic bonds are not significant in the bonding of the important bioelements listed in Table 2.1, they will not be further discussed here. As discussed below, biologically relevant molecules are held together by covalent and hydrogen bonds.

Covalent bonds Bonds in which electrons are more or less shared equally between two atoms are referred to as **covalent bonds**. Consider the element hydrogen. A hydrogen atom contains a single outer shell electron and readily combines with itself to form hydrogen gas, H_2 (Figure 2.1*a*). The H_2 molecule consists of two hydrogen atoms, each containing two shared electrons, the maximum allowed in the initial shell surrounding any nucleus. The atoms in the H_2

Table 2.1 The atomic structure of the major biologically relevant elements

Element	Atomic number	Protons	Neutrons	Electrons	Electrons in outer shell	Simplified structure
Hydrogen	1	1	0	1	1	H•
Carbon	6	6	6	6	4	•C• (4 dots)
Nitrogen	7	7	7	7	5	•N• (5 dots)
Oxygen	8	8	8	8	6	:O: (6 dots)
Phosphorus	15	15	16	15	5	•P• (5 dots)
Sulfur	16	16	16	16	6	:S: (6 dots)

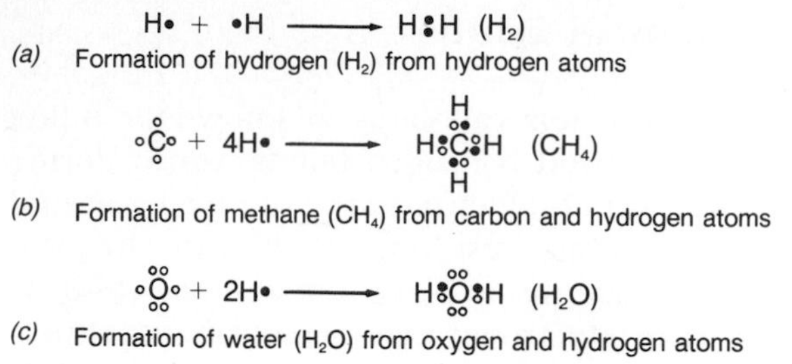

H• + •H ⟶ H:H (H_2)

(a) Formation of hydrogen (H_2) from hydrogen atoms

C + 4H• ⟶ H:C:H (CH_4)

(b) Formation of methane (CH_4) from carbon and hydrogen atoms

O + 2H• ⟶ H:O:H (H_2O)

(c) Formation of water (H_2O) from oxygen and hydrogen atoms

Figure 2.1 Formation of typical single covalently-bonded molecules.

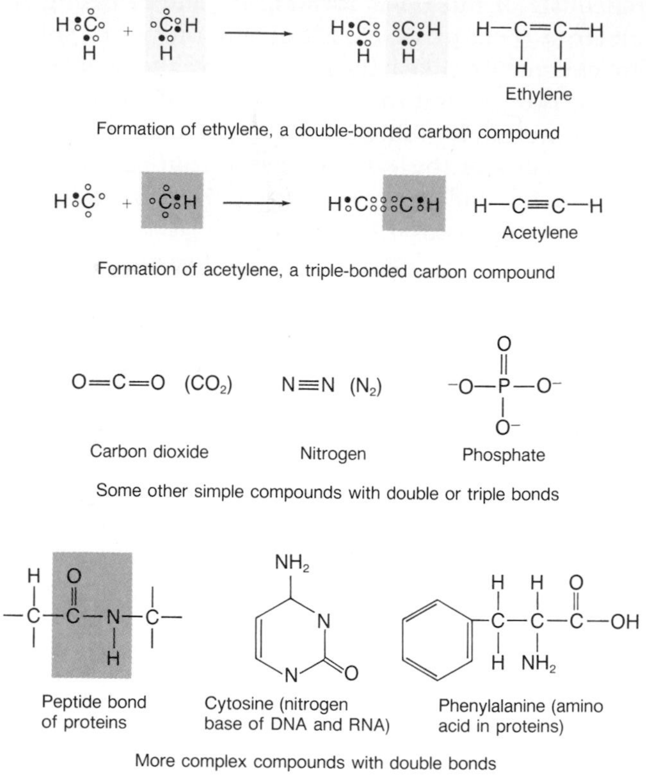

molecule are held together by covalent bonds. By comparison, carbon has an outer shell containing four electrons and is completed by addition of four more electrons. Hence, a carbon atom readily combines with four hydrogen atoms, leading to the formation of methane (Figure 2.1*b*) a simple organic (carbon-containing) compound. By sharing electrons in this fashion, the carbon atom and the hydrogen atoms achieve stable outer electron shells: eight electrons in the outer shell for carbon, and two each for the hydrogens. Likewise, when water forms from the elements hydrogen and oxygen (Figure 2.1*c*) the outer electron shells in both elements reach the stable configuration.

If more than one electron pair is shared between two atoms, double or even triple bonds are formed (Figure 2.2). Double bonds occur most frequently as C=O or C=C bonds and are found in a variety of important biological molecules. For triple bonds, three electron pairs are shared between two atoms. Triple bonds are rare in biological molecules, but microorganisms occasionally metabolize triply bonded compounds, the best example being N_2 (Figure 2.2), which is reduced in nitrogen fixation (see Section 16.24).

Hydrogen bonds A second type of bond that is extremely important in biological systems is the **hydrogen bond**. Hydrogen bonds form between hydrogen atoms and relatively electronegative elements like oxygen and nitrogen. Hydrogen bonds are very weak bonds. However, when many hydrogen bonds are formed within and between macromolecules, the overall stability of the molecule is greatly increased.

Water molecules readily undergo hydrogen bonding (Figure 2.3). Since oxygen is relatively electronegative and hydrogen is not, the covalent bond between oxygen and hydrogen is one in which the shared electrons in the outer shells orbit nearer the oxygen nucleus than the hydrogen nucleus. This creates a *slight* electrical charge separation, oxygen slightly negative, hydrogen slightly positive. The latter attracts the oxygen of a second water molecule near

H:C: + :C:H ⟶ H:C::C:H H—C=C—H

Ethylene

Formation of ethylene, a double-bonded carbon compound

H:C: + :C:H ⟶ H:C:::C:H H—C≡C—H

Acetylene

Formation of acetylene, a triple-bonded carbon compound

O=C=O (CO_2) N≡N (N_2) $^-$O—P(=O)(O$^-$)—O$^-$

Carbon dioxide Nitrogen Phosphate

Some other simple compounds with double or triple bonds

Peptide bond of proteins

Cytosine (nitrogen base of DNA and RNA)

Phenylalanine (amino acid in proteins)

More complex compounds with double bonds

Figure 2.2 Bonding of some biologically important molecules containing double or triple bonds.

the positive charge, in effect creating a positively charged "bridge" between the two electronegative oxygens of adjacent water molecules. The most common hydrogen bonds in macromolecules are those which involve O—H$^+$—O$^-$, O$^-$—H$^+$—$^-$N, and N$^-$—H$^+$—$^-$N interactions in proteins and nucleic acids (Figure 2.3).

Other atomic interactions Molecules benefit from other types of atomic interactions not usually referred to as bonds. **van der Waals forces** are nonspecific **attractive** forces which occur when the distance between two atoms is reduced to the range of 3–4Å. van der Waals forces occur because of momentary charge asymmetries around atoms due to electron movement. If atoms become closer than 3–4Å, however, overlap of the atom's electron shells causes repulsive forces to occur. van der Waals forces can play a significant role in the binding of substrates to enzymes (see Section 4.3) and in protein/nucleic acid interactions.

Hydrophobic interactions can also be important in the properties of biologically relevant molecules. Hydrophobic interactions occur because nonpolar molecules tend to cluster together in an aqueous

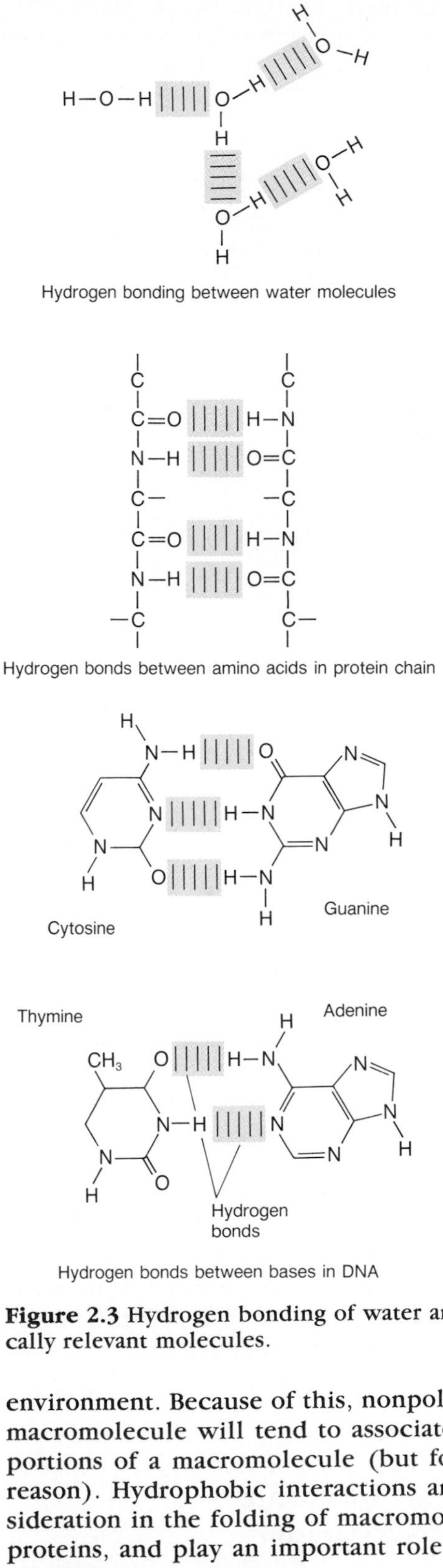

Hydrogen bonding between water molecules

Hydrogen bonds between amino acids in protein chain

Cytosine Guanine

Thymine Adenine

Hydrogen bonds

Hydrogen bonds between bases in DNA

Figure 2.3 Hydrogen bonding of water and some biologically relevant molecules.

environment. Because of this, nonpolar portions of a macromolecule will tend to associate, as will polar portions of a macromolecule (but for the opposite reason). Hydrophobic interactions are a major consideration in the folding of macromolecules such as proteins, and play an important role in the binding of substrates to enzymes.

Bond strengths and the importance of carbon The relative strength of chemical bonds can be measured by the energy (heat) required to dissociate them. Table 2.2 lists the bond energies of a number of biologically relevant chemical bonds. Note that double and triple bonds are much stronger than single bonds and that hydrogen bonds are by comparison very weak. Hydrogen bonds will form and dissociate in the cell spontaneously and rapidly, whereas covalent bonds will only be formed and broken by specific chemical reactions brought about by enzymes (see Section 4.3). Noncovalent interactions such as van der Waals forces and hydrophobic bonds are also very weak. However, like hydrogen bonds, these interactions become significant when large numbers of them develop either within or between various molecules.

All cellular structures contain an abundance of carbon in chemical combination with other elements. Carbon is a unique element in that it is able to combine not only with many other elements but also with itself, thus forming larger chemical structures of considerable diversity and complexity. An enormous number of different carbon compounds have been identified in various biological systems. For example, many biological compounds contain a carbon atom bonded to an oxygen atom (Table 2.3). Differences in the carbon-oxygen bond (single or double) and the nature of the atoms surrounding the carbon-oxygen bond will dictate the chemical properties of the

Table 2.2 Energies of some covalent bonds and noncovalent interactions*

Covalent bonds	Bond energy
Single bonds	
H—H	104
C—H	98
C—O	88
C—N	70
C—S	62
Double bonds	
C=C	147
C=O	168
O=O	96
Triple bonds	
C≡C	192
N≡N	228
Noncovalent interactions	**Energy**
Hydrogen bonds	1–2
Hydrophobic interactions	1–2
van der Waal's attractions	1–2

**The bond energies are given as the amount of heat, in kilocalories per mole, needed to break the bonds.*

Table 2.3 Carbon-oxygen compounds of importance to biochemistry

Chemical species	Structure	Biological importance
Carboxylic acid	$-C(=O)-OH$	Organic, amino, and fatty acids
Aldehyde	$-C(=O)-H$	Functional group of reducing sugars such as glucose
Alcohol	$-CH_2-OH$	Lipids
Keto	$-C(=O)-$	Pyruvate, citric acid cycle intermediates
Ester	$-CH_2-O-C(=O)-$	Lipids, amino acid attachment to tRNAs
Ether	$-CH_2-O-CH_2-$	Archaebacterial lipids, sphingolipids

molecule. Hence, carboxylic acids are chemically distinct from alcohols and each plays specific roles in cellular biochemistry (Table 2.3).

2.3 Small Molecules: Monomers

The main chemical components of cells are structures called **macromolecules**. Macromolecules are built up of individual "building blocks" which are connected in specific ways. A single building block is called a **monomer**, and the macromolecule is called a **polymer**. There are only a few types of polymers important in cell biochemistry, and each is made of a characteristic set of monomers.

Monomers are organic compounds containing up to about 30 carbon atoms and are grouped in classes according to their chemical properties. There are four classes of monomers to be considered here: *sugars*, the monomeric constituents of polysaccharides; *fatty acids*, the monomeric units of lipids; *nucleotides*, the basic units of the nucleic acids (DNA and RNA); and *amino acids*, the monomeric constituents of the proteins. The four classes of macromolecules can be subdivided into "informational" and "non-informational" types. Nucleic acids and proteins are considered informational macromolecules because the *sequence* of monomeric units within them is highly specific and carries biological information and the means to process this information. Lipids and polysaccharides, on the other hand, are not informational, because the sequence of monomers in these polymers is frequently highly repetitive and the sequence itself is generally of no functional importance. Approximately 500 chemically different monomeric units have been identified in various macromolecules, some of which differ in only subtle ways. Informational macromolecules contain a relatively small number of different monomeric units—it is the *sequence* of these units that is important. However, it should be appreciated that both the nature *and* the sequence of the monomeric units of any macromolecule is important in distinguishing it chemically from related macromolecules.

2.4 From Carbohydrate to Polysaccharide

Carbohydrates (sugars) are organic compounds containing carbon, hydrogen, and oxygen in the ratio of 1:2:1. The structural formula for glucose, the most

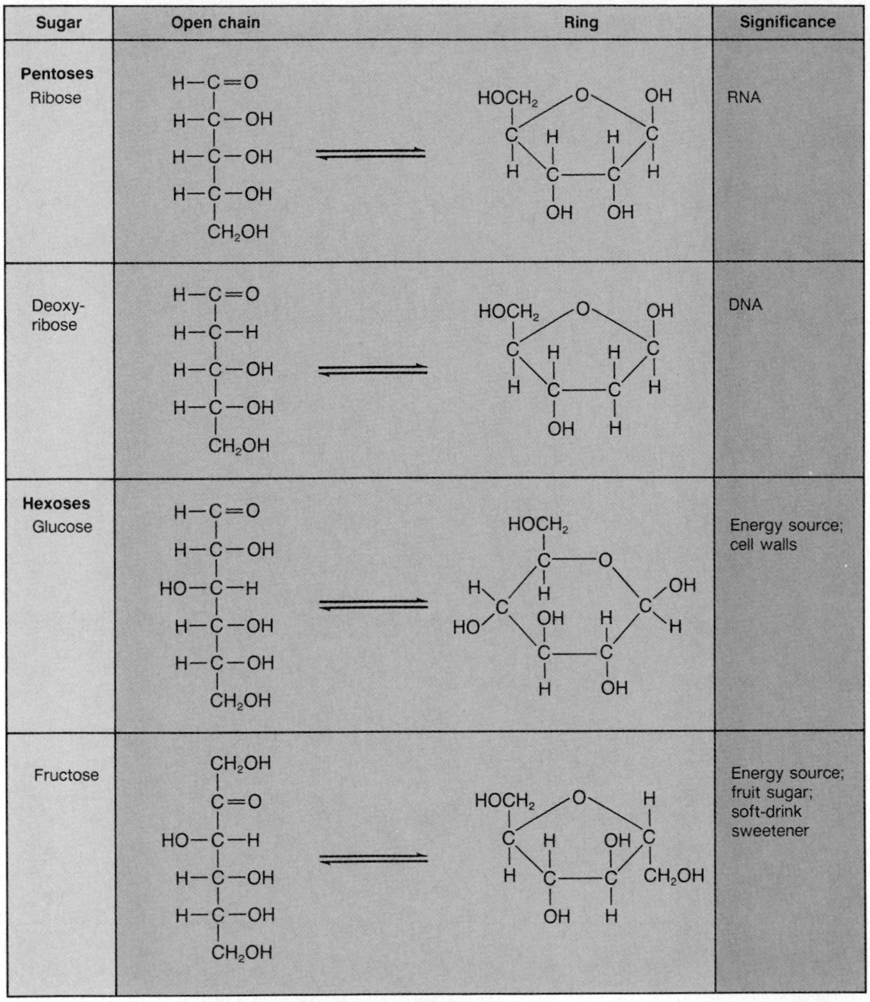

Sugar	Open chain	Ring	Significance
Pentoses Ribose	H—C=O H—C—OH H—C—OH H—C—OH CH_2OH		RNA
Deoxy-ribose	H—C=O H—C—H H—C—OH H—C—OH CH_2OH		DNA
Hexoses Glucose	H—C=O H—C—OH HO—C—H H—C—OH H—C—OH CH_2OH		Energy source; cell walls
Fructose	CH_2OH C=O HO—C—H H—C—OH H—C—OH CH_2OH		Energy source; fruit sugar; soft-drink sweetener

Figure 2.4 Structural formulas of a few common sugars. The formulas can be represented in two alternate ways, open chain and ring. The open chain is easier to visualize but the ring form is the commonly used structure.

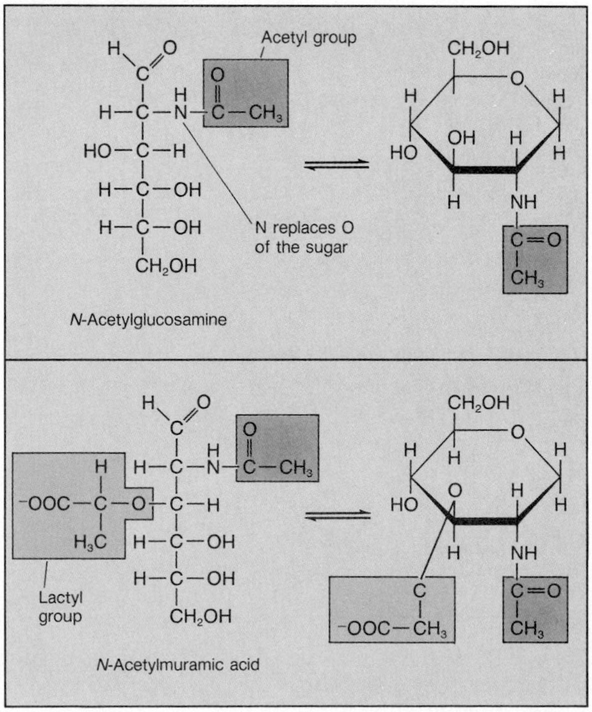

Figure 2.5 Sugar derivatives found in the cell walls of most bacteria. Note that the parent structure is glucose in both cases.

abundant sugar, is $C_6H_{12}O_6$ (Figure 2.4). The most biologically relevant carbohydrates are those containing 4, 5, 6, and 7 carbon atoms (C_4, C_5, C_6, and C_7). C_5 sugars (pentoses) and C_6 sugars (hexoses) are of special significance because of their roles as the structural backbones of nucleic acids. Likewise, C_6 sugars (hexoses) are the monomeric constituents of cell wall polymers and energy reserves. Figure 2.4 shows the structural formulas of a few common sugars. In addition to the simple sugars, derivatives of carbohydrates can be formed by replacing one or more of the hydroxyl groups with other chemical species. For example, the important bacterial cell wall polymer **peptidoglycan** (see Section 3.5) contains the glucose derivatives N-acetylglucosamine and muramic acid (Figure 2.5). Besides sugar derivatives, sugars having the same structural formula can still differ in their stereoisomeric properties (see Section 2.8). Hence, a large number of different sugars are potentially available for the construction of polysaccharides.

Removal of water

α-D-Glucose + α-D-Glucose → H_2O + α-1,4-Glycosidic bond

β-D-Glucose + β-D-Glucose → H_2O + β-1,4-Glycosidic bond

α-1,6-Glycosidic bond

(a)

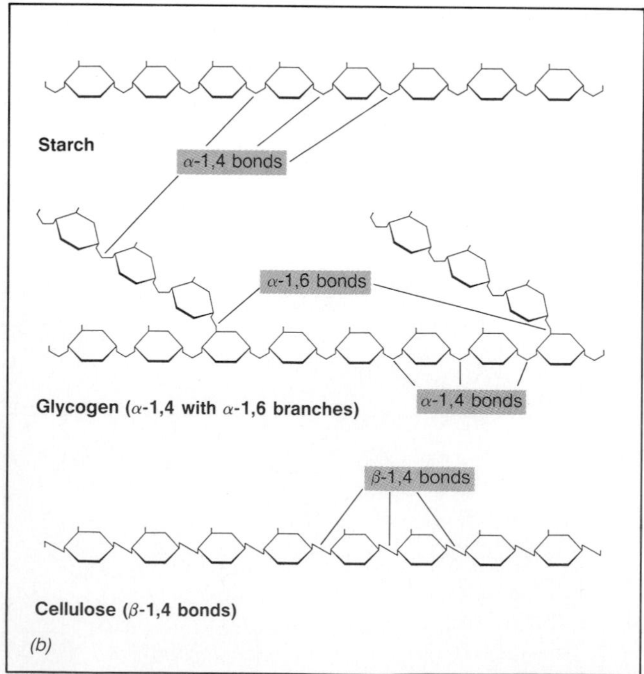

Figure 2.6 Polysaccharides. (a) Formation of polysaccharides from monosaccharides. (b) General structures of some common polysaccharides.

Polysaccharides are defined as high-molecular-weight carbohydrates containing many (hundreds or thousands) of monomeric units connected to one another by a type of covalent bond referred to as a **glycosidic bond** (Figure 2.6). If two sugar units (monosaccharides) are joined by glycosidic linkage the resulting molecule is called a **disaccharide**. The addition of one more monosaccharide would yield a **trisaccharide**, several more an **oligosaccharide**, and an extremely long chain of monosaccharides in glycosidic linkage is called a **polysaccharide**.

The glycosidic bond can exist in two different orientations, referred to as *alpha* (α) and *beta* (β) (Figure 2.6*a*). Polysaccharides with a repeating structure composed of glucose units linked between carbons 1 and 4 in the *alpha* orientation (glycogen and starch, Figure 2.6*b*) function as important carbon and energy reserves in bacteria, plants and animals. Alternatively, glucose units joined by *beta* 1-4 linkages are present in cellulose (Figure 2.6*b*), a plant and algal cell wall component functionally unrelated to glycogen or starch. Polysaccharides can also combine with other classes of macromolecules, such as protein or lipid, to form glycoproteins or glycolipids. These compounds play important roles in cell membranes

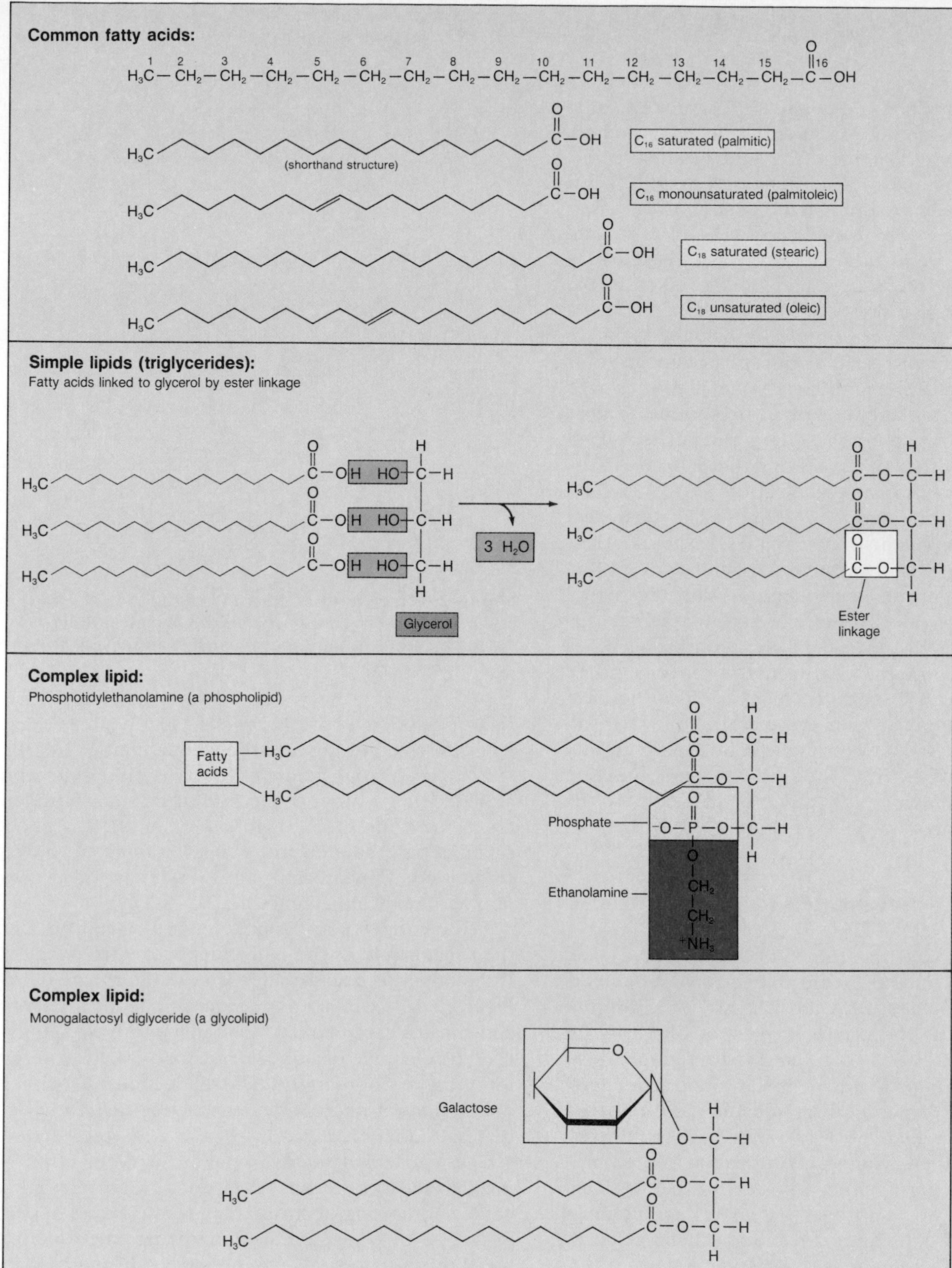

Figure 2.7 Fatty acids, simple lipids (fats) and complex lipids.

as cell surface receptor molecules. The compounds reside on the external surfaces of the membrane where they are in contact with the environment. Glycolipids also constitute a major portion of the cell wall of Gram-negative bacteria, and as such impart a number of unique surface properties on these organisms (see Section 3.5).

2.5 From Fatty Acids to Lipids

Fatty acids are the main constituents of **lipids**. Fatty acids have interesting chemical properties because they contain both highly *hydrophobic* (water repelling) and highly *hydrophilic* (water soluble) regions. Palmitate, for example (Figure 2.7), is a 16 carbon fatty acid composed of a chain of 15 saturated (fully hydrogenated) carbon atoms and a single carboxylic acid group. Other common fatty acids are stearic (C_{18} saturated) and oleic (C_{18} monounsaturated) acids (Figure 2.7). Simple lipids (fats) consist of fatty acids bonded to the C_3 alcohol glycerol (Figure 2.7). Simple lipids are frequently referred to as **triglycerides** because of the three fatty acids being linked to the glycerol molecule.

Complex lipids are simple lipids which contain additional elements such as phosphate, nitrogen, or sulfur, or small hydrophilic carbon compounds such as sugars, ethanolamine, serine, or choline (Figure 2.7). Phospholipids are a very important class of complex lipids, as they play a major structural role in the cytoplasmic membrane (see Section 3.3).

The chemical properties of lipids make them ideal structural components of membranes. Because of their dual properties of hydrophobicity and hydrophilicity, lipids aggregate in membranes with the hydrophilic portions toward the external or internal (cytoplasmic) environment, while maintaining their hydrophobic portions away from the aqueous milieu (see Section 3.3). Such structures are ideal permeability barriers because of the inability of water-soluble substances to flow through the hydrophobic fatty acid portion of the lipids. Indeed, the major function of the cell membrane is to serve as the barrier to the diffusion of substances into or out of the cell.

2.6 From Nucleotides to Nucleic Acids

The nucleic acids, DNA (deoxyribonucleic acid), and RNA (ribonucleic acid), are polymers of monomers called **nucleotides**. DNA and RNA are thus both **polynucleotides**. DNA carries the genetic blueprint for the cell and RNA acts as an intermediary molecule to convert the blueprint into defined protein sequences. Despite these important functions, nucleic acids are composed of a relatively few simple building blocks. Each nucleotide is composed of three separate units: a five carbon sugar, either ribose (in RNA) or deoxyribose (in DNA), a nitrogen base, and a molecule of phosphate, PO_4^{3-}. Figure 2.8 shows schematic drawings of single nucleotides of DNA and RNA.

The nitrogen bases of nucleic acids belong to either of two chemical classes. Purine bases, **adenine** and **guanine**, contain two fused carbon-nitrogen rings while the pyrimidine bases, **thymine, cytosine** and **uracil**, contain single six-membered carbon-nitrogen ring (Figure 2.9). Guanine and cytosine are found in both DNA and RNA; thymine is present (with minor exception) only in DNA, uracil is present only in RNA.

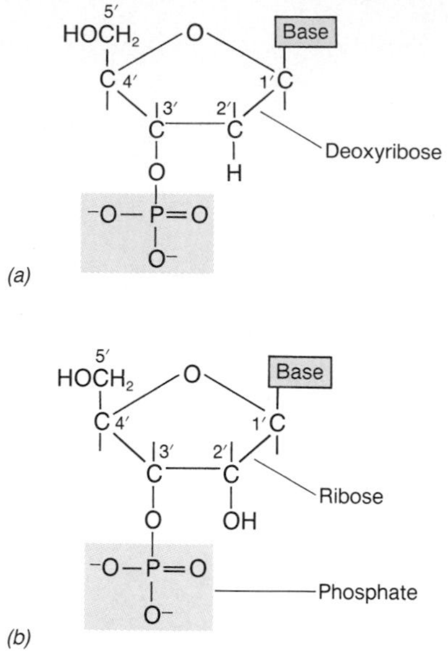

Figure 2.8 Nucleotide of DNA (a) and RNA (b). Note the numbering system employed and the chemical differences on carbon atom 2′ between DNA and RNA.

In a nucleotide a base is attached to a pentose sugar by glycosidic linkage to carbon atom number 1 of the sugar and a nitrogen atom of the base, either nitrogen atom labeled as atom 1 (pyrimidine base) or nitrogen atom 9 (purine base) (Figure 2.9). Without a phosphate, a base bonded to its sugar is referred to as a **nucleoside**. Nucleotides are thus nucleosides containing phosphate (Figure 2.10).

Nucleotides play other roles in the cell besides their major role as the constituents of nucleic acids. Nucleotides, especially adenosine triphosphate (ATP, Figure 2.10), can act as carriers of chemical energy and can release sufficient free energy during the hydrolytic cleavage of a phosphate bond to drive energy requiring reactions in the cell (see Section 4.6). Other nucleotides or nucleotide derivatives function in oxidation-reduction reactions in the cell (see Section 4.5) as carriers of sugars in the biosynthesis of polysaccharides (see Section 4.17), and as regulatory molecules, inhibiting or stimulating the activities of certain enzymes or even controlling the overall metabolism of various compounds as in the case of cyclic adenosine monophosphate, "cyclic AMP," (see Section 5.12). However, the central role of a nucleotide is as a building block of nucleic acid.

Nucleic acids are long polymers in which nucleotides are covalently bonded to one another in a *defined sequence*, forming structures called **polynucleotides**. The backbone of the nucleic acid is a polymer in which sugar and phosphate moieties al-

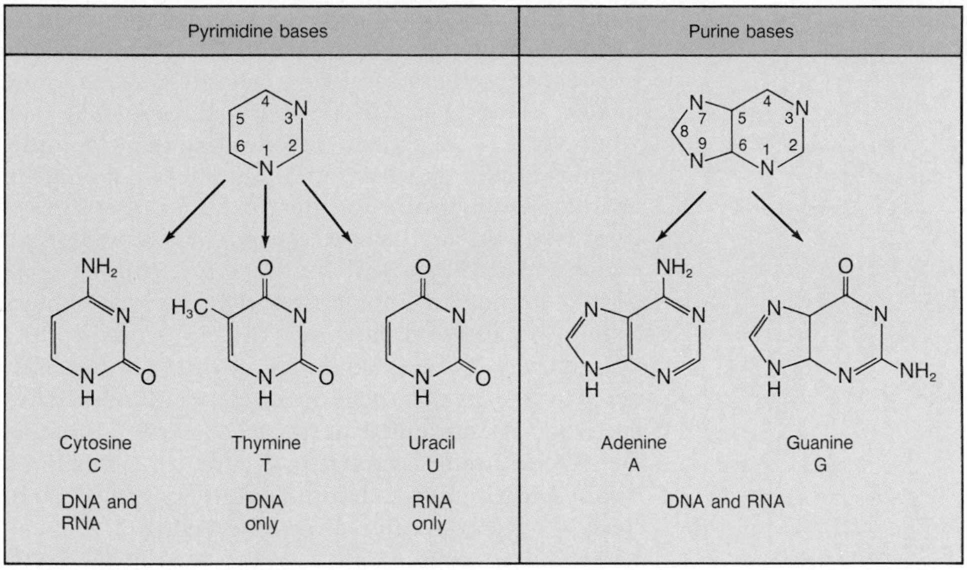

Figure 2.9 Structure of bases of DNA and RNA. The letters C, T, U, A, and G, are often used to refer to the individual bases.

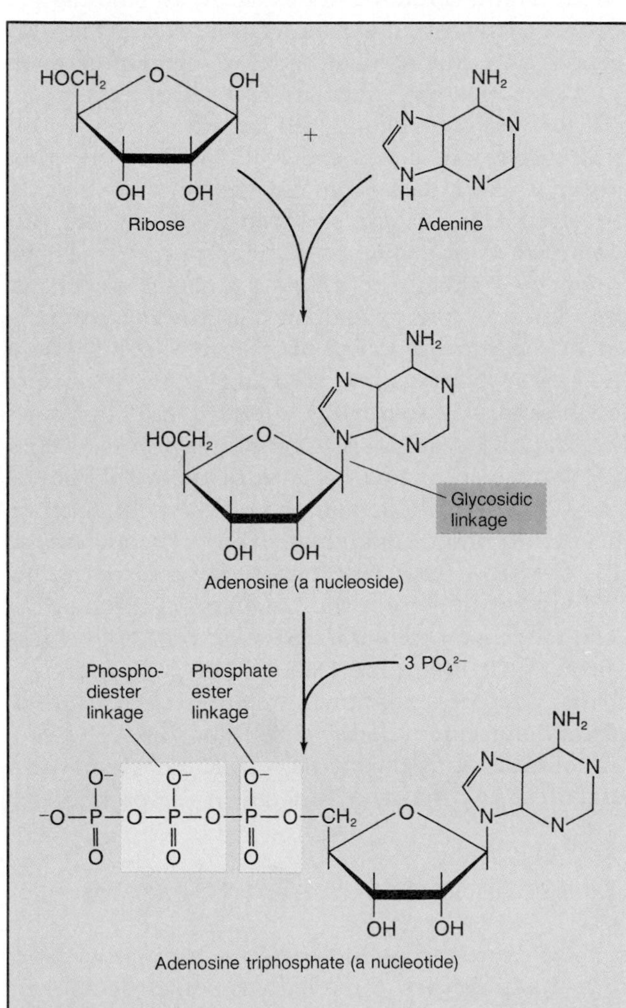

Figure 2.10 Components of the important nucleotide triphosphate, adenosine triphosphate (ATP).

ternate (Figure 2.11). Hence, when we refer to a specific *sequence* of nucleotides in a nucleic acid, we are really referring to the variable portions of the nucleotide, the bases; the sugar-phosphate backbone is always present. The sequence of bases in a DNA or RNA molecule is *informational*, representing the genetic information necessary to reproduce an identical copy of the organism. We will see later that the replication of DNA and the production of RNA are highly complex processes (see Chapter 5) and that a virtually error-free mechanism is necessary to insure the faithful transfer of genetic traits from one generation to another.

In biochemical terms nucleic acids are composed of nucleotides covalently attached via phosphate residues from carbon 3 (referred to as the 3′ [3 prime] carbon) of one sugar to carbon 5 (5′) of the adjacent sugar (Figure 2.11). The phosphate linkage is chemically a **phosphodiester**, since a single phosphate is connected by ester linkage to two separate sugars.

DNA Cells contain two strands of DNA, each strand containing several million nucleotides linked by phosphodiester bonds. The strands themselves associate with one another by hydrogen bonds which form between the nucleotides of one strand and the nucleotides of the other. When positioned adjacent to one another, purine and pyrimidine bases can undergo hydrogen bonding (see Figure 2.3). Chemically, the most stable hydrogen bonding configuration occurs when guanine (G) forms hydrogen bonds with cytosine (C), and adenine (A) forms hydrogen bonds with thymine (T) (see Figure 2.3*c*). It will be recalled that hydrogen bonds, although individually very weak, collectively serve to stabilize macromo-

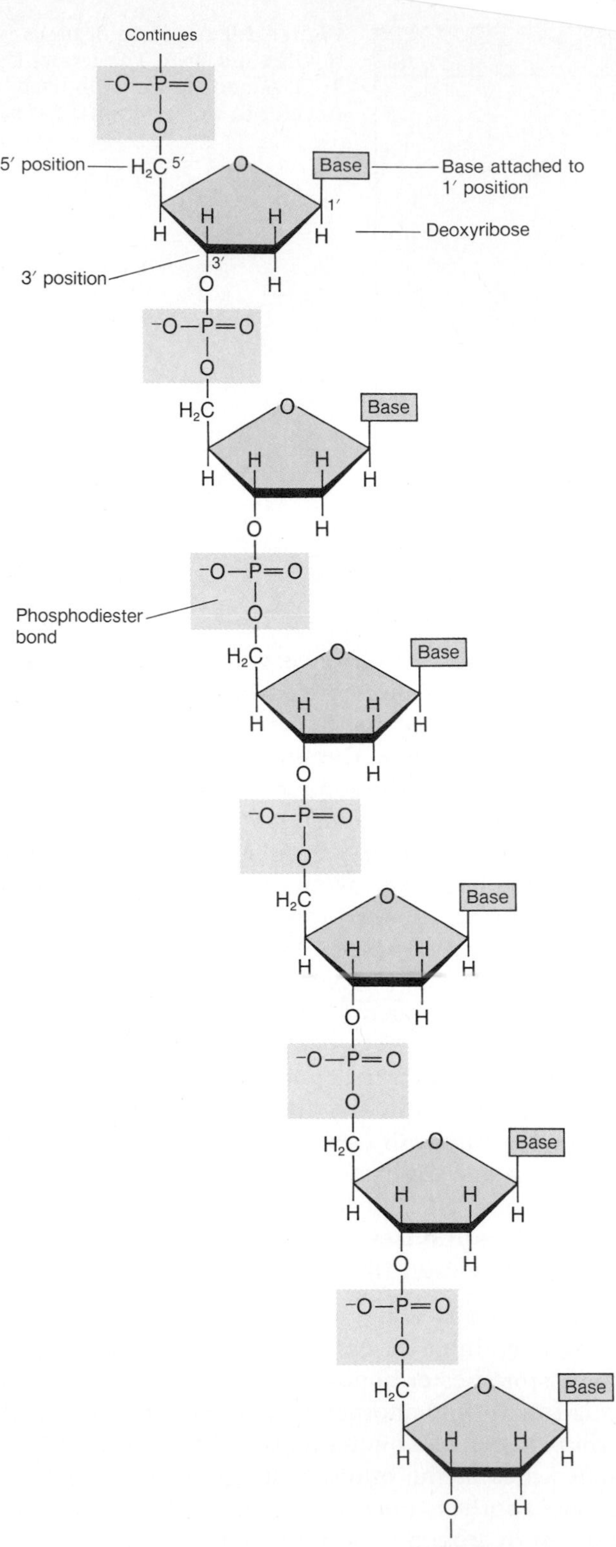

Figure 2.11 Structure of part of a DNA chain.

lecules. Specific base pairing, A with T and G with C, means that the two strands of DNA will be *complementary* in base sequence; wherever a G is found in one strand, a C will be found in the other, and wherever a T is present in one strand its complementary strand will have an A. The molar amounts of guanine and cytosine are therefore *identical* in DNA from any source; likewise the molar amount of adenine and thymine will be the same. The two opposing strands are thus arranged in complementary fashion. In addition, however, the two strands of the DNA molecule are arranged in an *antiparallel* manner. This means that the individual strands of the DNA molecule are arranged in a "head to toe" arrangement—one strand (the left in Figure 2.12) runs in a 5′→3′ direction top to bottom, while its complement runs 3′→5′ in the same direction (Figure 2.12).

RNA With the exception of certain viruses which contain double-stranded helical RNA, all ribonucleic acids are *single*-stranded molecules. However, RNA molecules frequently fold back upon themselves in regions where complementary base pairing can occur to form a variety of highly folded structures.

RNA plays three crucial roles in the cell. **Messenger RNA** (mRNA) contains the genetic information of DNA in a single stranded molecule complementary in base sequence to a portion of the base sequence of DNA. The mRNA sequence directs incorporation of amino acids into a growing polypeptide in the process called translation (see Sections 5.8–5.10). These events take place on the surface of the ribosome. **Transfer-RNA** (tRNA) molecules are the "adaptor" molecules in protein synthesis. Transfer RNAs are so named because of their dual specificities, one for a specific three-base sequence on mRNA (the codon) and the other for a specific amino acid. The tRNA molecule effectively adapts the genetic information from the language of nucleotides to the language of amino acids, the building blocks of proteins. **Ribosomal RNA**, (rRNA), of which several distinct types are known, are important structural and catalytic components of the ribosome, the protein-synthesizing system of the cell. These various RNA molecules are discussed in detail in Chapter 5.

2.7 From Amino Acids to Proteins

Amino acids are the monomeric units of proteins. Most amino acids consist only of carbon, hydrogen, oxygen, and nitrogen, but two of the twenty common amino acids found in cells also contain sulfur atoms as well. All amino acids contain two important func-

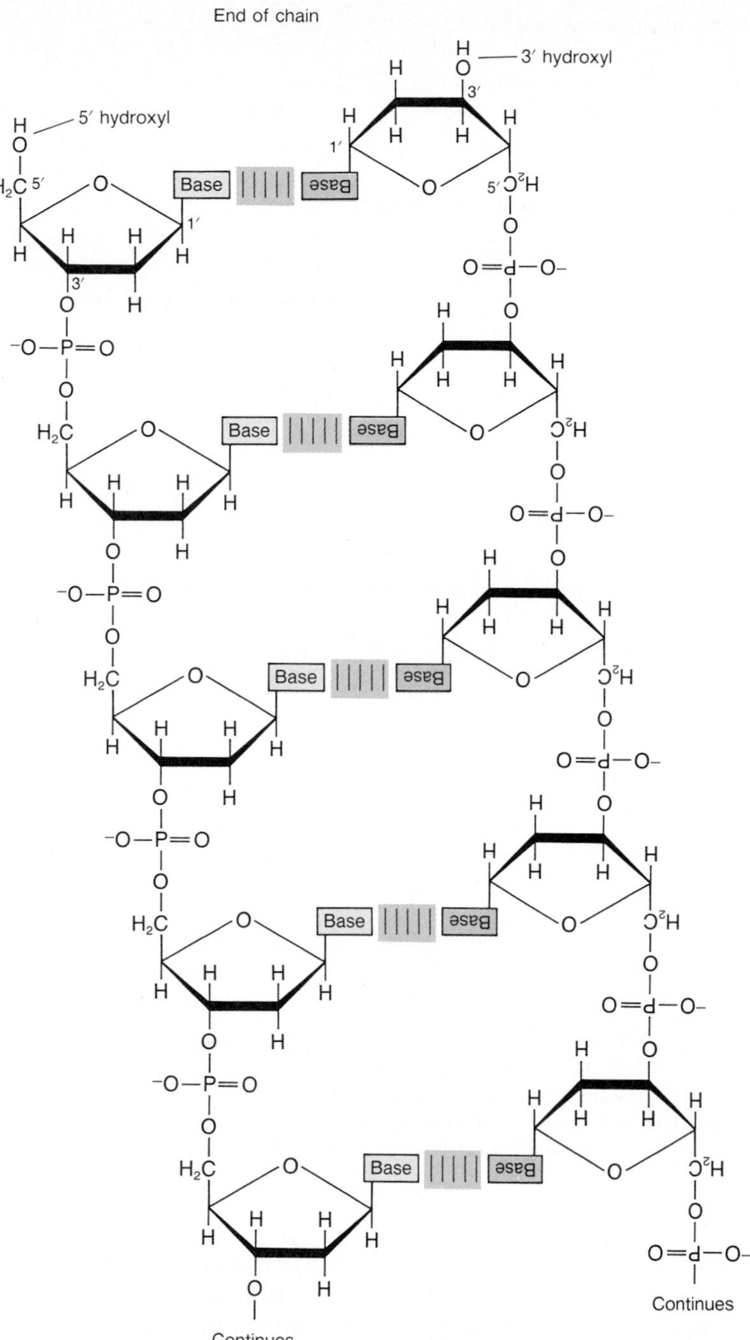

Figure 2.12 Complementary and antiparallel nature of DNA. One end of the double helix is shown. Note that one chain ends in a 5′ hydroxyl group whereas the other ends in a 3′ hydroxyl.

tional groups, a *carboxylic acid* group (—COOH) and an *amino* group (—NH_2). These groups are functionally important because covalent bonds between the carbon of the carboxyl group and the nitrogen of the amino group (with elimination of a molecule of water), forms the **peptide bond**, a type of covalent bond characteristic of proteins. All amino acids conform to the general structure shown in Figure 2.13*a*. Amino acids differ in the nature of the side group attached to the alpha carbon. The alpha carbon is the carbon atom immediately adjacent to the carbon atom of the carboxylic acid group. The side chains on the alpha carbon vary considerably, from as simple as a hydrogen atom in the amino acid glycine, to complex ringed structures in amino acids such as phenylalanine (Figure 2.13). The chemical properties of an amino acid are to a major degree governed by the nature of the side chain, and biochemists tend to group together amino acids which show similar chemical properties into amino acid "families," as shown in Figure 2.13. For example, the side chain may itself contain a carboxylic acid group, such as in aspartate

General structure of an amino acid

$$R-\underset{NH_2}{\overset{H}{C}}-\overset{O}{\overset{\|}{C}}-OH$$

"R" indicates the side chain — see below

Detailed structures of the amino acid "R" groups

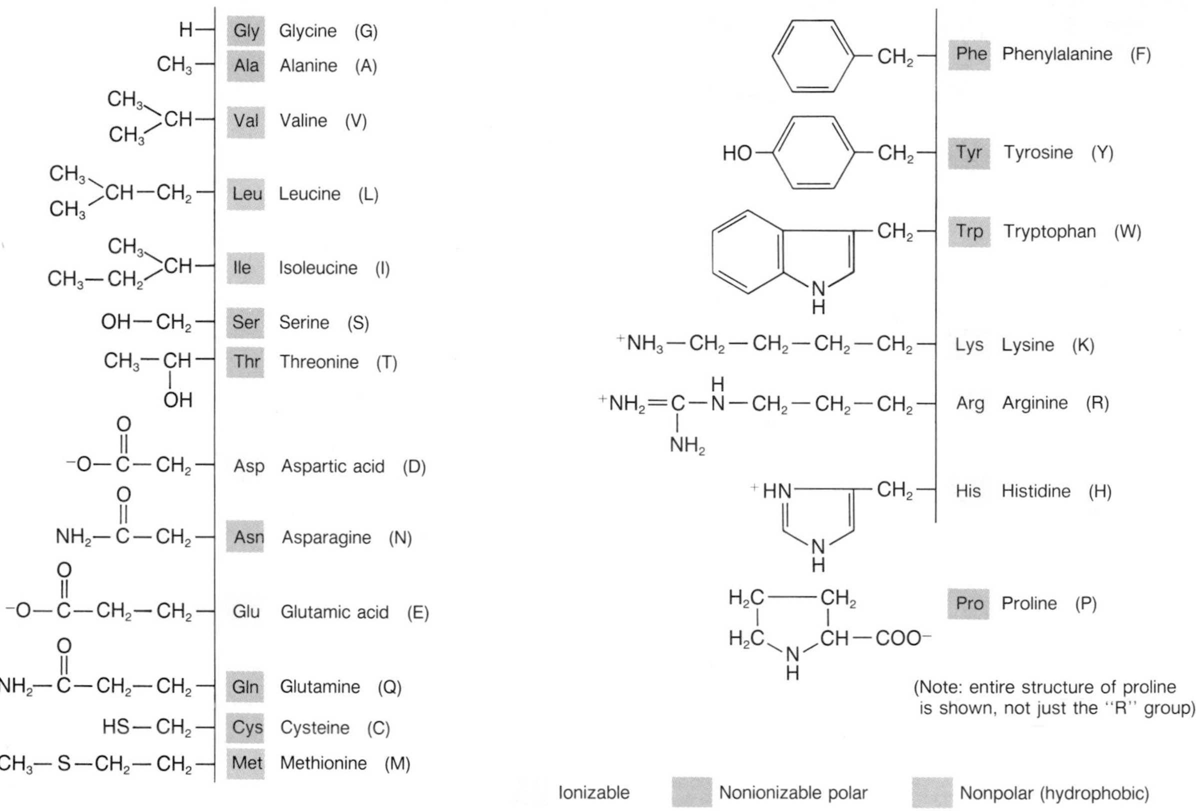

Figure 2.13 Structure of the twenty common amino acids. The three-letter codes for the amino acids are within the boxes, the one-letter codes are in parentheses to the right of the names Note that *proline* is an *imino* rather than an *amino* acids and lacks the NH_2 group.

or glutamate, rendering the amino acid acidic. Alternatively, several amino acids contain nonpolar hydrophobic side chains and are grouped together as nonpolar amino acids. The amino acid cysteine contains a sulfhydryl group (—SH) which is frequently important in connecting one chain of amino acids to another by disulfide linkage (R—S—S—R).

The chemical variety inherent in the side chains of the twenty common amino acids is sufficiently great to allow a compound as structurally simple as an amino acid to possess the chemical diversity necessary to make proteins with widely different chemical properties. For example, proteins that are in direct contact with highly hydrophobic regions of the cell, such as proteins embedded in the lipid-rich cytoplasmic membrane, generally contain higher overall proportions of hydrophobic amino acids (or contain regions extremely rich in hydrophobic amino acids) than proteins which function in the aqueous environment of the cytoplasm.

Structure of proteins Proteins play key roles in cell function. Two kinds of proteins are recognized in cells: *catalytic* proteins (enzymes) and *structural* proteins. The proteins called enzymes serve as catalysts for the wide variety of chemical reactions which occur in cells. We will discuss enzymes as catalysts extensively in Chapter 4. Structural proteins are those which become integral parts of the structures of cells, in membranes, walls, and cytoplasmic components. In essence, a cell is what it is because of the kinds of proteins which it contains. Therefore, an understanding of protein structure is essential for an understanding of cell function.

Proteins are comprised of linear polymers of various lengths containing defined sequences of amino acids covalently bonded by peptide linkages. The combination of two amino acids head to tail (carboxyl group to amino group) along with the elimination of water generates the peptide bond (Figure 2.14). Two amino acids connected together constitute a dipeptide, three amino acids a tripeptide, and so on. Many amino acids covalently linked via peptide bonds constitute a **polypeptide**. Proteins consist of one or more polypeptides. The number of amino acids varies from one protein to another. Proteins with as few as 20 and as many as 10,000 amino acids are known. Since proteins may vary in their composition, sequence, and number of amino acids, it is easy to see that enormous variation in protein types is possible.

All proteins are folded and show complex arrangements of structure. The linear array of amino acids is referred to as the **primary structure** of the polypeptide. In many ways the primary structure of a polypeptide is the most important, because a given primary structure will allow only certain types of folding to occur. The juxtaposition of α-carbon side groups dictated by the primary structure forces the polypeptide to twist and fold in a specific way. This twisting and coiling of the polypeptide molecule leads to the formation of the **secondary structure** of the protein. Hydrogen bonds, the weak non-covalent linkages discussed earlier (see Section 2.2), play important roles in the secondary structure of a protein. A typical secondary structure for many proteins is that called the α-helix. If a linear polypeptide were wound around a cylinder, oxygen and nitrogen atoms from different amino acids would become positioned close enough together in such a structure to allow hydrogen bonding to occur. Such is the case in the polypeptide α-helix (Figure 2.15*a*).

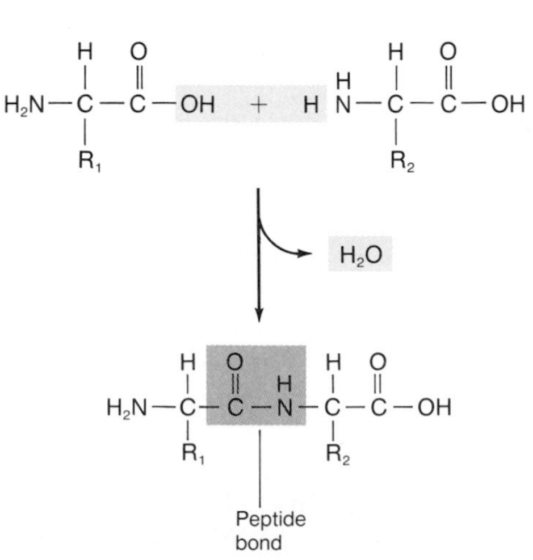

Many polypeptides conform to a different type of secondary structure referred to as the β-sheet. In the β-sheet (sometimes referred to as the pleated sheet) the chain of amino acids in the polypeptide folds back and forth upon itself, exposing hydrogen atoms which can undergo extensive hydrogen bonding (Figure 2.15*b*). Some polypeptides contain both regions of α-helix and regions of β-sheet secondary structure, the type of folding being determined by the available hydrogen-bonding opportunities (which in turn is set by the primary structure of the polypeptides). Since β-sheet secondary structure generally yields a rather rigid structure while α-helical secondary structures are usually more flexible, the secondary structure of a given polypeptide will to some degree dictate a functional role for the protein in the cell.

Once a polypeptide has achieved a given secondary structure it folds back upon itself to form an even more stable molecule. This folding leads to the formation of the **tertiary structure** of the protein. The tertiary structure of a protein is determined by primary structure, but tertiary structure is also governed to some extent by the secondary structure of the molecule. As a result of the formation of the primary and secondary structures, the side chain of each amino acid is positioned in a specific way. If additional hydrogen bonds, covalent bonds, or other atomic interactions can form due to this positioning of amino acids, the molecule will fold to attain a unique three-dimensional shape (Figure 2.16). Frequently the polypeptide will fold in such a way that adjacent sulfhydryl (—SH) groups of cysteine residues are exposed (Figure 2.16*b*). These free —SH groups can join covalently to form a disulfide (—S—S—) bridge between the two amino acids. If the two cysteine residues are located in different chains, for example, the disulfide linkage would strengthen the alignment of the rows of amino acids. By contrast, a linear polypeptide with α-helix secondary structure could fold back upon itself extensively if cysteine residues were positioned near the ends of the molecule. If highly nonpolar regions are brought close together by specific tertiary folds, the polypeptide will gain additional stability due to hydrophobic interactions. The tertiary folding of the polypeptide ultimately forms exposed regions or grooves in the molecule which may be of importance in binding other molecules.

If a protein consists of more than one polypeptide, and many proteins do, the arrangement of polypeptide subunits to form the final molecule is referred to as the **quaternary structure** of the protein

Figure 2.14 Peptide bond formation.

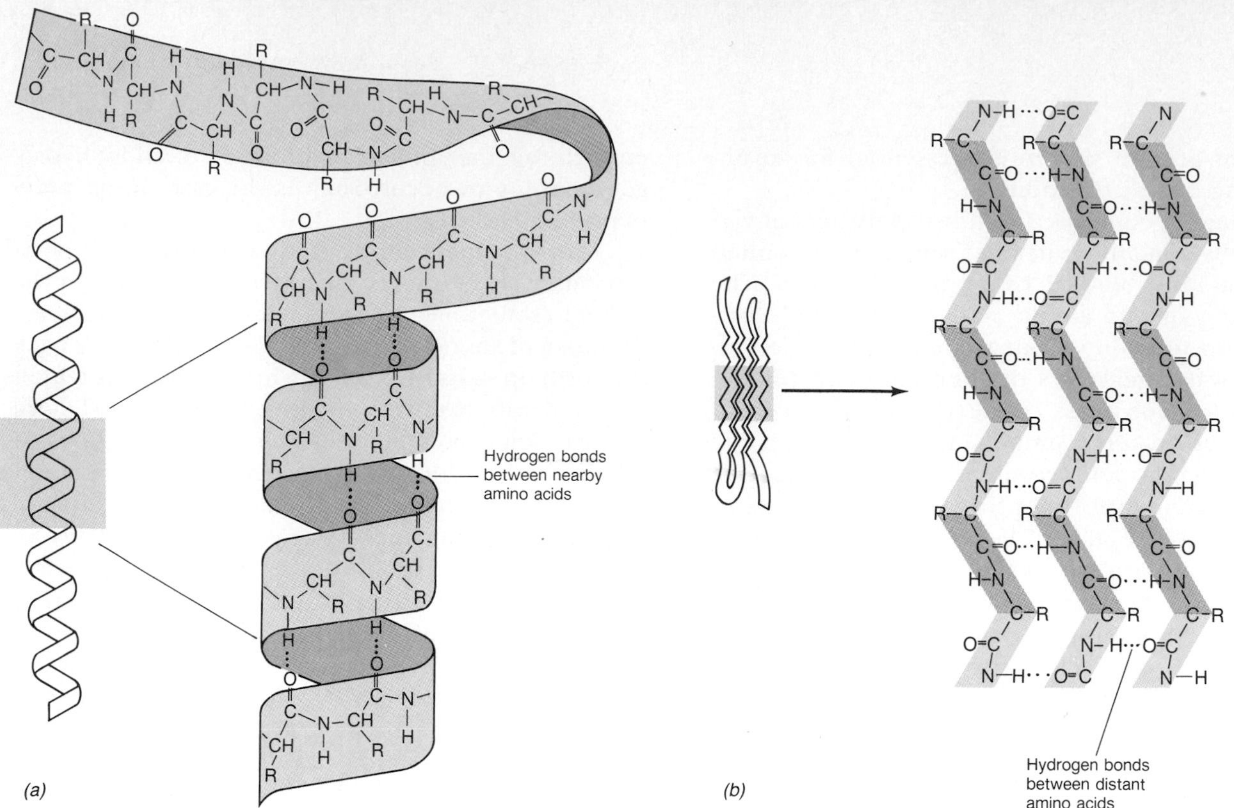

Figure 2.15 Secondary structure of polypeptides. (a) Alpha-helix secondary structure. Note that hydrogen bonding does not involve the R groups, but the peptide bonds. (b) Beta-(pleated) sheet secondary structure.

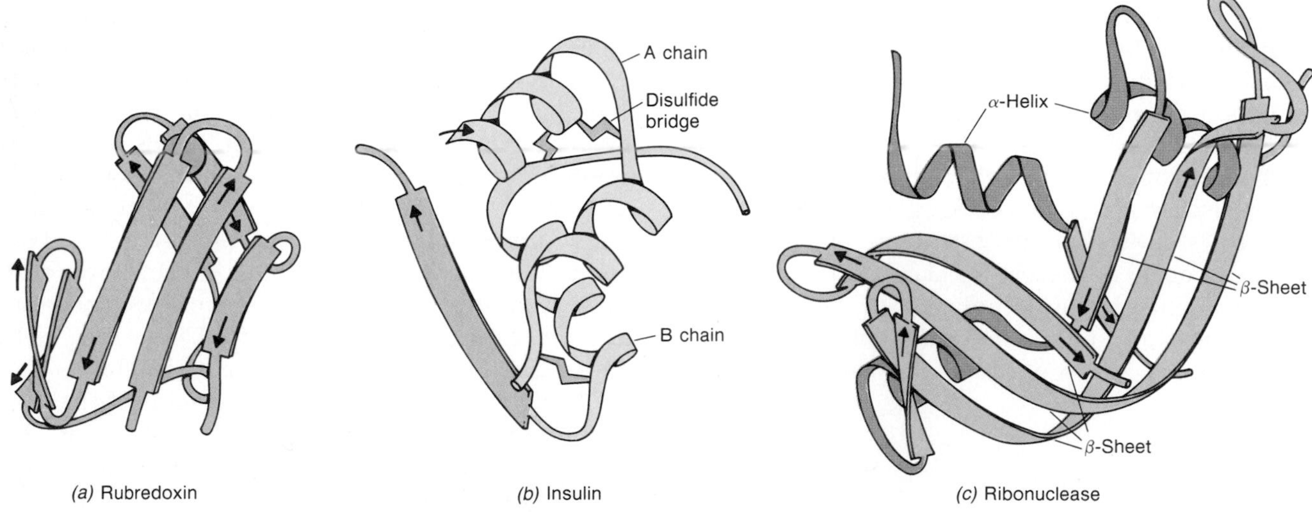

Figure 2.16 Tertiary structure of polypeptides showing where regions of alpha-helix or beta-sheet secondary structure might be located. (a) Rubredoxin. (b) Insulin, a protein containing two polypeptide chains. Note how disulfide linkages (—SS—) may help in dictating folding patterns. (c) Ribonuclease, a large protein with several regions of alpha helix and beta sheet.

(Figure 2.17). It should be remembered that in proteins showing quaternary structure each subunit of the final protein will contain primary, secondary, and tertiary structure. Some proteins displaying quaternary structure contain many identical subunits; others contain several nonidentical subunits, while still others may contain more than one identical subunit and a second nonidentical subunit in the final molecule. The subunits of multisubunit proteins are held together either by noncovalent interactions (hydrogen bonding, van der Waals forces or hydrophobic interactions) or by covalent linkages, generally intersubunit disulfide bonds.

Denaturation of proteins When proteins are exposed to extremes of heat or pH, or to certain chemicals which affect their folding properties, they are said to undergo **denaturation** (Figure 2.18). In

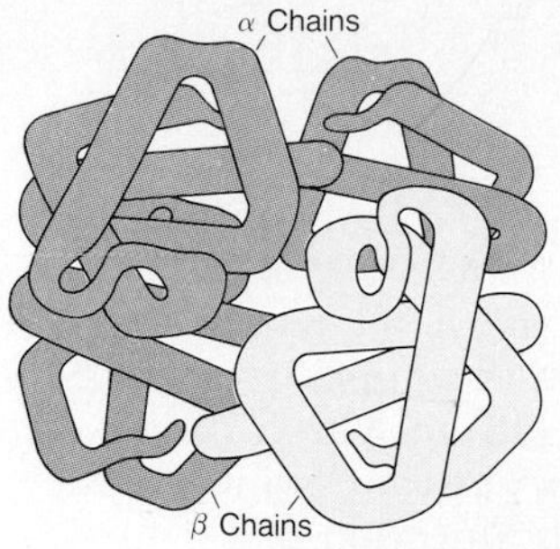

Figure 2.17 Quaternary structure of hemoglobin, a protein containing four polypeptide subunits. There are two kinds of polypeptide in hemoglobin, α chains (top) and β chains bottom. Separate colors are used to distinguish the four chains.

general, the biological properties of a protein are lost when it is denatured. When proteins are denatured, peptide bonds are generally unaffected, and the sequence of amino acids (primary structure) in the polypeptide remains unchanged. However, denaturation causes the polypeptide chain to unfold, destroying higher order structure of the molecule, in particular hydrogen bonds. The denatured polypeptide retains its primary structure because it is held together by covalent bonds. Depending upon the severity of the denaturing conditions, refolding of the polypeptide may reappear after removing the denaturant (Figure 2.18). However, the fact that dena-

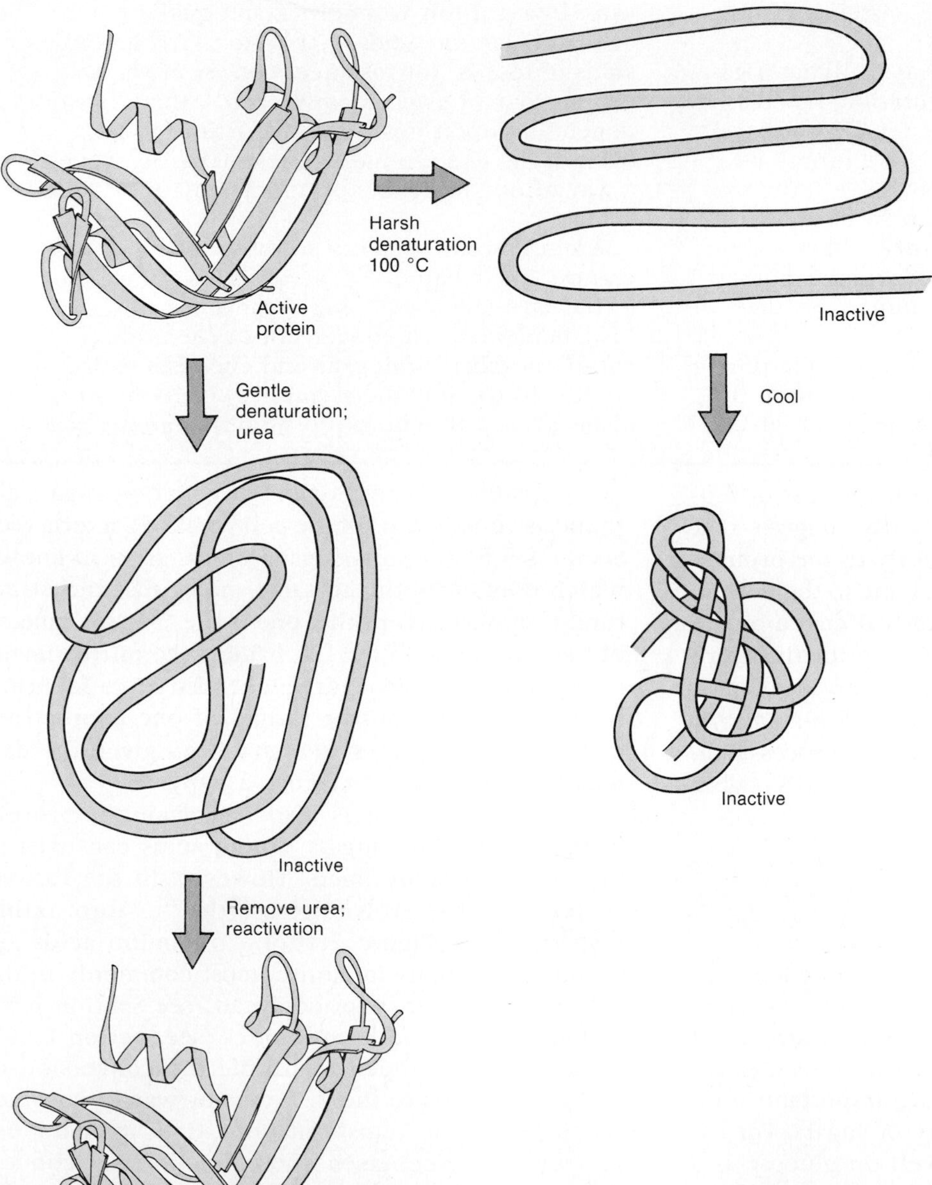

Figure 2.18 Denaturation of a protein.

Chemical Isomers and Living Organisms

Louis Pasteur was trained as a chemist and his first research work was on the chemistry of stereoisomers. He was one of the first people to recognize the significance of stereoisomers in living organisms, and his interest in this problem led him into biology, a field that would occupy him for the rest of his life. For Pasteur, the fact that living organisms discriminated between stereoisomers was of profound philosophical significance. He saw life processes as inherently asymmetric, whereas nonliving chemical processes were not asymmetric. Pasteur's vision of the asymmetry of life has been well confirmed over the more than 100 years since Pasteur did his work.

Why is it that the amino acids of living organisms are of only the L configuration, whereas when the same amino acids are made chemically they consist of mixtures of D and L forms? We know that this is because chemical reactions of living organisms are carried out by asymmetric catalysts, the **enzymes**. Asymmetry begats asymmetry, and since enzymes are asymmetric, they introduce asymmetry into the molecules they produce.

The above statement begs the question, of course, of how the *first* asymmetry arose in the living world. Was it by chance or by design? Is there somewhere in the universe a planet with living organisms whose proteins are comprised of D amino acids? A planet, perhaps, where everything got started along the opposite symmetry to that of life on earth, and in which life is now trapped forever in this opposite symmetry? We will return to the whole question of evolution and the origin of life in Chapter 18.

Isomers such as the L amino acids will gradually, with time, racemize spontaneously, resulting in a mixture of D and L forms. Organic material derived from dead organisms, occurring as part of fossils, gradually becomes a racemic mixture. In a million years or so, an equal amount of D and L amino acids will be left. The amino acids of fossils, for instance, consist of variable proportions of L and D amino acids, the L/D ratio depending upon how long the organism has been dead. The L/D ratio can actually be used as a way of obtaining an approximate date of a fossil.

When Pasteur began to study spontaneous generation (see Chapter 1), he used his knowledge of the asymmetry of living organisms to bolster his confidence that spontaneous generation could not exist. Biologists and chemists have continued to view the asymmetry of living organisms as of profound philosophical significance.

turation is generally associated with the loss of biological activity of the protein, clearly suggests that biological activity is not due directly to the primary structure of proteins, but instead is due to the unique folding of the molecule as ultimately dictated by the primary structure. Folding of a polypeptide therefore accomplishes two things: (1) the polypeptide obtains a unique shape which is compatible with a *specific* biological function, and (2) the folding process converts the molecule into its most chemically stable form.

2.8 Stereoisomerism

It was mentioned in reference to carbohydrates in Section 2.4 that two molecules may have the same molecular formula but exist in different structural forms. These related, but not identical molecules are referred to as **isomers**. Isomers are important in biology, especially in the chemistry of sugars. For example, *Escherichia coli* grows well on glucose as a carbon and energy source, but will not grow on the closely related sugar allose (Figure 2.19*a*), presumably because it cannot incorporate or metabolize this hexose isomer. Many isomers of common sugars are found as constituents of the cell walls of bacteria (see Sections 3.5 and 3.6). Sugar isomers are also known which contain both the same molecular and structural formulas except that one is the "mirror image" of the other, just as the left hand is the mirror image of the right. In carbohydrate chemistry two identical sugars which are mirror images of one another are called **stereoisomers** and have been given the designation "D" and "L" (Figure 2.19*b*).

Stereoisomerism is also important in protein chemistry. As with sugars, amino acids can exist as "D" or "L" stereoisomers. However, in the case of protein, life has evolved to use the "L" form rather than the "D" (Figure 2.19*c*). "D" amino acids are found occasionally in nature, most commonly in the cell wall polymer, peptidoglycan (see Section 3.5), and in certain peptide antibiotics (see Section 10.6). Bacteria are equipped to handle the conversion of "D" amino acids to the "L" form by way of enzymes specific for this transformation. Cells contain enzymes called **racemases** whose function is to convert the unusual form (L-sugar or D-amino acid) into the readily metabolizable form, the D-sugar or L-amino acid.

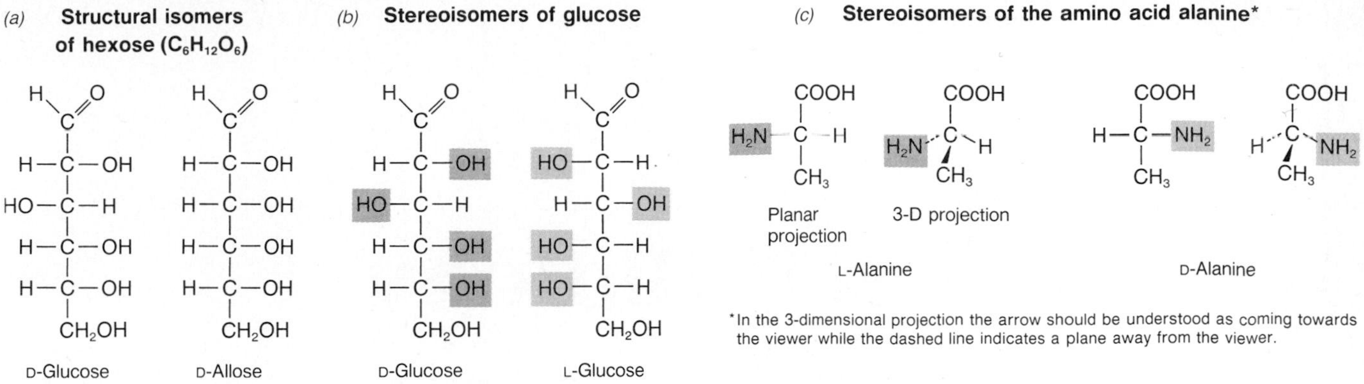

(a) **Structural isomers of hexose ($C_6H_{12}O_6$)**

D-Glucose D-Allose

(b) **Stereoisomers of glucose**

D-Glucose L-Glucose

(c) **Stereoisomers of the amino acid alanine***

Planar projection 3-D projection

L-Alanine D-Alanine

*In the 3-dimensional projection the arrow should be understood as coming towards the viewer while the dashed line indicates a plane away from the viewer.

Figure 2.19 Stereoisomers and structural isomers.

Study Questions

1. Compare and contrast covalent and noncovalent bonds. Give three examples of the noncovalent type.
2. Why are the weak noncovalent bonds of special importance for the molecules of living organisms?
3. Compare and contrast *monomer* and *polymer*. Give three examples of biologically important polymers and list the key constituents of each.
4. Polymers are sometimes called macromolecules. Why? Contrast *informational* and *noninformational* macromolecules. Give two examples of each kind.
5. RNA and DNA are similar types of macromolecules but they show distinct differences. List three ways in which RNA differs chemically from DNA.
6. *Complementarity* is an important feature of the two polynucleotides of the DNA double helix. What is complementarity and how is it expressed chemically in DNA? Are the two DNA molecules held together by covalent or noncovalent bonds?
7. Write a general structure for an *amino acid* and explain how the R groups of the amino acid differ. Show how two amino acids can be connected in the formation of a *peptide bond*. Does the peptide bond involve covalent or noncovalent interactions?
8. Define each of the following as it relates to protein structure: primary, secondary, tertiary, quaternary.
9. Describe briefly the effects of heat on protein structure. Heat sensitivity of proteins is one reason why most living organisms do not live at high temperatures, yet a few bacteria not only live but thrive at temperatures up to boiling. Can you offer an explanation, based on your knowledge of protein structure? (You should be able to refine your answer after covering the material in Chapter 9.)

Supplementary Readings

Alberts, B., D. Bray, J. Lewis, M. Raff, K. Roberts, and **J. D. Watson**. 1983. *Molecular Biology of the Cell.* Garland Publishing, Inc., New York. An excellent textbook of cell biology with emphasis on the molecular aspects of cells. Chapters 2 and 3 give a detailed account of cell chemistry and macromolecular structure. Highly recommended for advanced reading.

Edwards, N. A., and **K. A. Hassall**. 1980. *Biochemistry and Physiology of the Cell—an Introductory Text.* Second edition. McGraw-Hill Book Company, New York. A nice treatment of cellular macromolecules with emphasis on microbial cells.

Stryer, L. 1981. *Biochemistry.* Second edition. W. H. Freeman and Company, San Francisco. One of the best general biochemistry texts with excellent art work making the material come alive. Highly recommended.

Watson, J. D., N. H. Hopkins, J. W. Roberts, J. Argetsinger Steitz, and **A. M. Weiner**. 1987. *Molecular Biology of the Gene.* Fourth edition. Benjamin/Cummings Publishing Co., Inc., Menlo Park, CA. A two-volume treatise of molecular biology. Chapters 4 and 5 of volume one consider aspects of cell chemistry, in particular nucleic acid structure.

3

Cell Biology

All cells, whether procaryotic or eucaryotic, share a number of properties in common. For example, all cells employ a lipid bilayer membrane as the permeability barrier between the cytoplasm and the "outside world," and ribosomes are the universal protein synthesis "factories" of all cells. On the other hand, procaryotes and eucaryotes differ in many of their solutions to the common problems of cellular life. Structures uniformly present in one group (a membrane-bounded nucleus in eucaryotes, for example), may be totally absent in the other. Structures which serve the same function in both groups (flagella, for example) may differ considerably at the subcellular and molecular level. Presumably these differences reflect the great evolutionary divergence between eucaryotes and procaryotes.

This chapter on cell biology is intended to give the student an overview of structure and function in microbial cells. We emphasize the biology of the procaryotic cell, but will compare and contrast procaryotes with eucaryotes at several points, and discuss in detail a few key eucaryotic cell structures.

3.1 The Importance of Being Small

Microorganisms are small, as the name implies, and being small has several physiological advantages. The rate at which nutrients and waste products pass into or out of the cell is in general inversely proportional to cell size; transport rates in turn affect an organism's metabolic rate and growth rate. Thus, the smaller the cell, the faster its potential growth rate.

The accumulation of nutrients and elimination of waste products by a cell involves the cell surface, especially the cell membrane. The cytoplasm of the cell, where many essential metabolic activities take place, communicates with the external environment through the cell membrane, and the rate of internal reactions are to a large degree controlled by the amount of membrane surface available to transport materials in and out of the cell. That is, a relation exists between cell volume and cell surface area, the latter of which is a good measure of the amount of available membrane. However, the relation between volume and surface is not constant. This point may be seen most easily in the case of a sphere, in which the volume is a function of the cube of the radius ($V=\frac{4}{3}\pi r^3$) and the surface area is a function of the square of the radius ($A=4\pi r^2$). The surface/volume ratio of a sphere can therefore be expressed as $3/r$ (Figure 3.1). Thus a smaller sphere (smaller r value) has a *higher* ratio of surface area to volume than a larger sphere. To return to a biological example, a small cell should therefore have more efficient exchange with its surroundings than a large cell. This pressure for

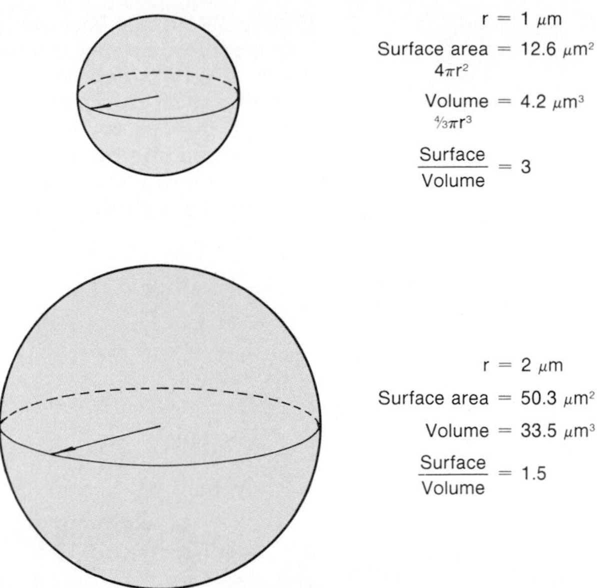

Figure 3.1 As a cell increases in size, its surface/volume ratio decreases.

a cell to be small has limits, however, as a certain minimum volume is necessary for a cell to contain all the genetic information and biochemical apparatus such as enzymes and ribosomes.

Although most procaryotic cells are small, there is a wide variation in size among different organisms (Figure 3.2). Most bacteria have distinctive cell shapes, which remain more or less constant, although shape is influenced to some extent by the environment. The shape of a cell definitely affects its behavior and ecology. Cocci, for instance, being round, become less distorted upon drying and thus can usually survive more severe desiccation than can rods or spirals. Rods, on the other hand, have more surface exposed per unit volume than cocci, and thus can more readily take up nutrients from dilute solutions. Spiral forms are always motile and move by a corkscrew motion, which means that they meet with less resistance from the surrounding medium than do motile rods, in the same way that a screw moves into hardwood more easily than a nail. Even square bacteria are known. These unusual organisms are quite distinctive in their straight sides and right-angle corners (Figure 3.3). To date square bacteria have been found only in extremely salty environments, such as brines used for commercial production of salt (see Section 19.33). It is thought that their unusual morphology is related to the stresses they must deal with in their environment because of the high salt concentration (see Section 9.13).

When cells divide they often remain attached to each other, and the manner of attachment is frequently characteristic of both the organism and the type of division the cell has undergone. Thus many cocci form chains, *Streptococcus* being the best known example, by always dividing along the same axis (Figure 3.4*a*). The length of such chains may be either short (2–4 cells) or long (over 20 cells). Some cocci divide along two axes at right angles to each other, forming sheets of cells. If there is no pattern to the orientation of successive divisions an irregular clump will be formed; such a random cluster is characteristic of *Staphylococcus* (Figure 3.4*b*). Rods always divide in only one plane (Figure 3.5*a*) and hence may form chains, such as in many *Bacillus* species (Figure 3.5*b*), but not more complicated arrangements. Spirally shaped organisms also divide in only one plane, but they usually separate immediately and do not form chains.

3.2 Comparative Structure and Function of Procaryotic and Eucaryotic Cells

The electron microscope has made possible a careful study of cellular architecture, and through electron microscopic studies we have become aware of the profound differences between procaryotic and eu-

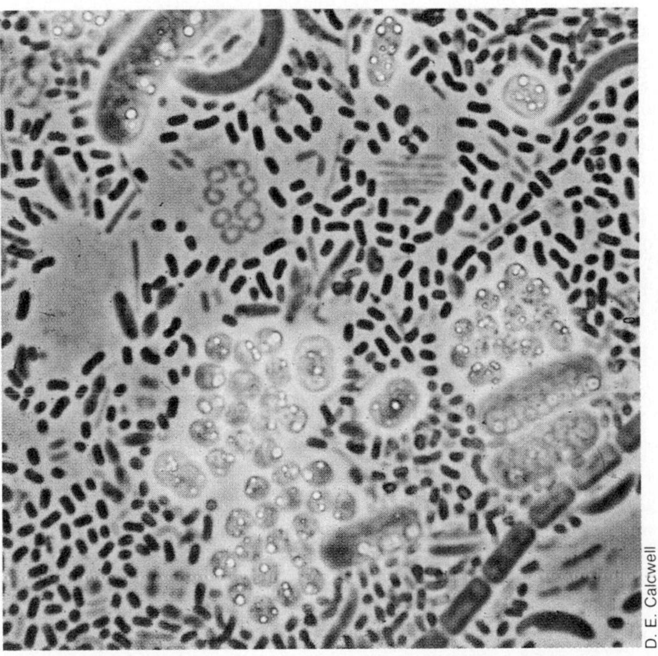

Figure 3.2 Photomicrograph of a bacterial community that developed in the depths of a small lake in Michigan, showing procaryotic cells of various sizes. The large purple-red cells measure approximately 4 μm × 12 μm, the spherical purple cells containing bright refractile droplets measure approximately 2 μm in diameter, the small dark green cells measure approximately 0.8 μm × 1.6 μm, and the small ringed-shaped cells (center) measure approximately 0.5 μm in diameter. The volume of the cells shown vary by a factor of over 200 and the surface/volume ratios by a factor of 12.

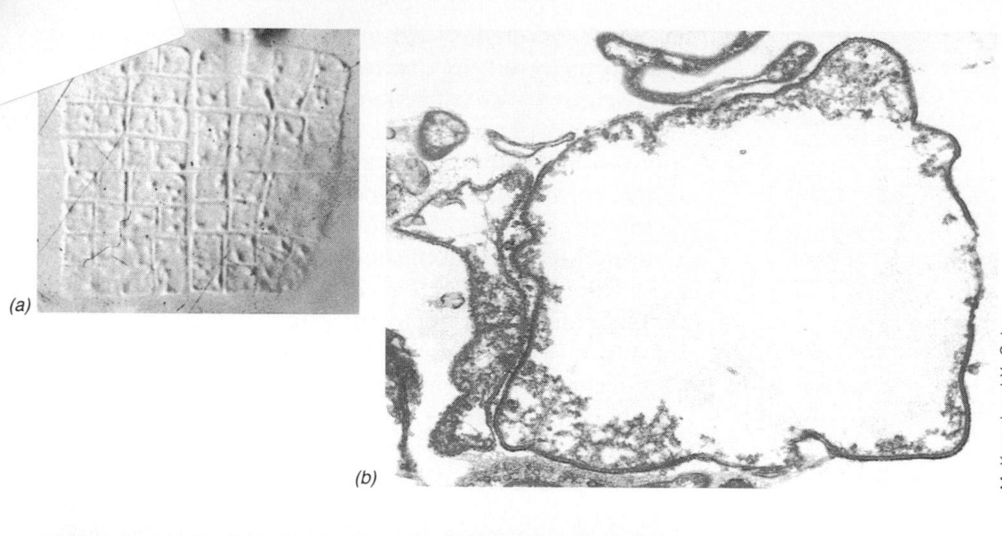

Figure 3.3 Square bacteria. (a) Nomarski interference micrograph of cells which have formed a sheet by remaining associated through six divisions. (b) Electron micrograph of thin section of lysed cells showing that cell shape stays constant.

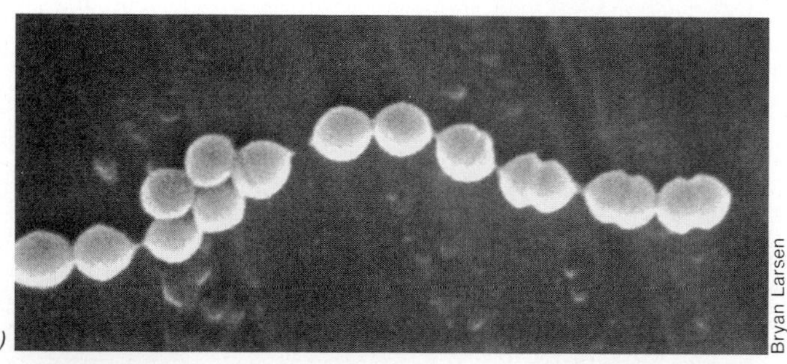

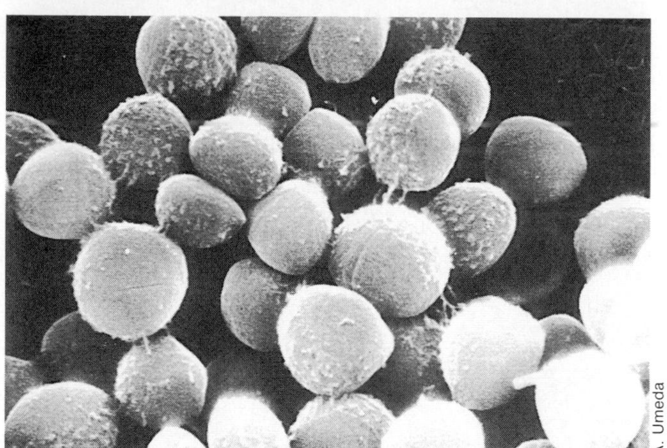

Figure 3.4 Spherical bacteria which associate in characteristically different ways. (a) *Streptococcus*, a chain-forming organism which divides only in one plane. Scanning electron micrograph. (b) *Staphylococcus*, a coccus which divides in more than one plane. Scanning electron micrograph.

(a) L. A. Mangels et al.

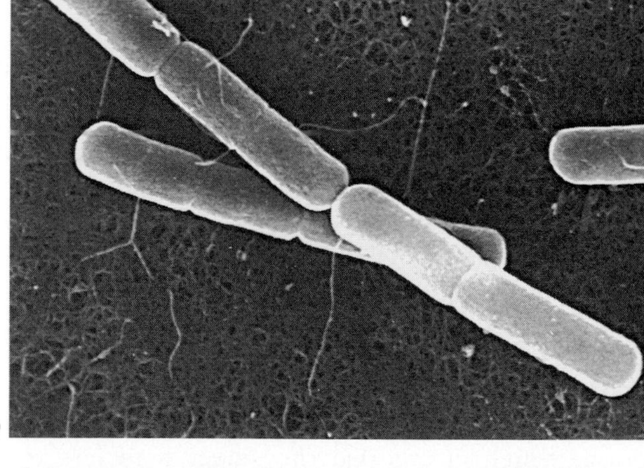

Figure 3.5 Examples of rod-shaped cells. (a) Transmission electron micrograph of thin section of dividing cells of the marine photosynthetic bacterium, *Rhodopseudomonas marina*. (b) Scanning electron micrograph of chains of *Bacillus*.

caryotic cells. In the following discussion it should be emphasized that although we contrast procaryotes and eucaryotes, with the procaryotes we are mainly discussing the most common bacteria, the eubacteria. What is true for eubacteria may not necessarily be true for archaebacteria, even though both groups are procaryotic.

Cellular components can be divided into two groups, *invariant structures*, found in all organisms, and *variant structures*, found in some cells but not all. Invariant structures include the cell membrane, ribosomes, and the genomic (nuclear) region. Variant structures include the cell wall, flagella, cilia, pili, and several additional structures unique to procaryotic cells. We begin our discussion with a consideration of the plasma membrane, a key structure in all living organisms.

3.3 Cell Membrane: Structure

The cell membrane, usually called the **plasma membrane**, is a thin structure that completely surrounds the cell. Only 4–5 nm thick, this vital structure is the critical barrier separating the inside of the cell from its environment. If the membrane is broken, the integrity of the cell is destroyed, the internal contents leak into the environment, and the cell dies. The plasma membrane is also a highly selective barrier, enabling a cell to concentrate specific metabolites and excrete waste materials.

Chemical composition of membranes The general structure of all biological membranes is the same. Despite significant differences in chemical composition, the solution to the basic problem of serving a permeability function seems restricted to some variation on a **lipid bilayer** theme (Figure 3.6*a*). Lipids are chemically unique molecules. As discussed in Section 2.5, lipids contain both highly hydrophobic (fatty acid) and highly hydrophilic (glycerol) moieties, and can exist in many different chemical forms as a result of variation in the nature of the fatty acids or phosphate-containing groups attached to the glycerol backbone. As lipids aggregate in an aqueous solution, they tend to form lipid bilayer structures spontaneously—the fatty acids point inward toward each other in a hydrophobic environment, while the glycerol molecules remain exposed to the aqueous external environment (Figure 3.6*a*).

In experimental studies of membrane function, **membrane vesicles** have been used extensively as models of the cytoplasmic membrane. Membrane vesicles are miniature versions of the plasma membrane, except that they do not surround all the cytoplasmic components of the living cell, but instead trap only a small amount of cytoplasm when they form from a larger membrane. Membrane vesicles form spontaneously when a bacterial cell is broken under carefully controlled conditions. The membrane fragments generated during cell breakage become associated together in spherical structures via the hydrophobic/hydrophilic bonding of phospholipids (Figure 3.6*b*). Therefore the lipid bilayer character of membranes probably represents the most stable arrangement of lipid molecules in an aqueous environment.

Thin sections of the plasma membrane can be visualized with the electron microscope; a representative example is seen in Figure 3.7*a*. To prepare the plasma membrane for electron microscopy, the cells must first be treated with osmic acid or some other electron-dense material that combines with hydrophilic components of the membrane (Figure 3.7*b*). By careful high-resolution electron microscopy, the plasma membrane appears as two thin lines separated by a lighter area (Figure 3.7*a*). This main component of the unit membrane is the lipid bilayer, composed of phospholipid (see Figure 2.7), and proteins, the phospholipids forming the basic lipid bilayer. The major proteins of the membrane associate with either side of or become embedded in the phospholipid matrix (Figure 3.8). Those proteins that are embedded in the membrane generally have very hydrophobic external surfaces which make intimate association with the highly nonpolar fatty acid chains possible. Membrane surface proteins may or may not be highly hydrophobic, depending largely on their degree of association with nonpolar regions of the membrane. Protein molecules are also bound to the ionic groups of the phospholipid, and water molecules congregate in a more or less ordered structure around the outside of the bimolecular leaflet. The structure of the plasma membrane is stabilized by hydrogen bonds and hydrophobic interactions. However, cations such as Mg^{2+} and Ca^{2+} also combine by ionic bonding with negative charges of the phospholipids and help stabilize the membrane structure.

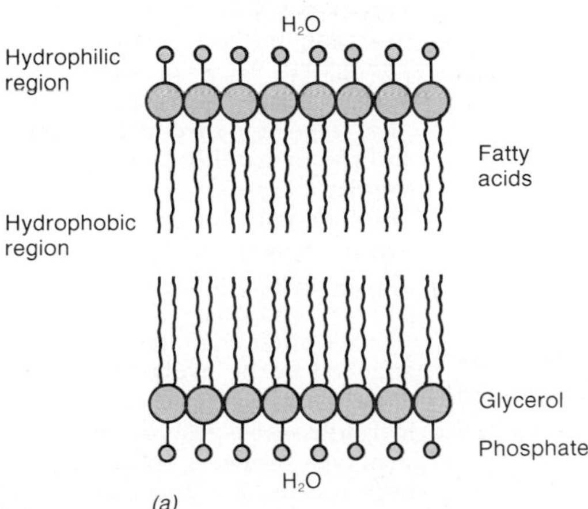

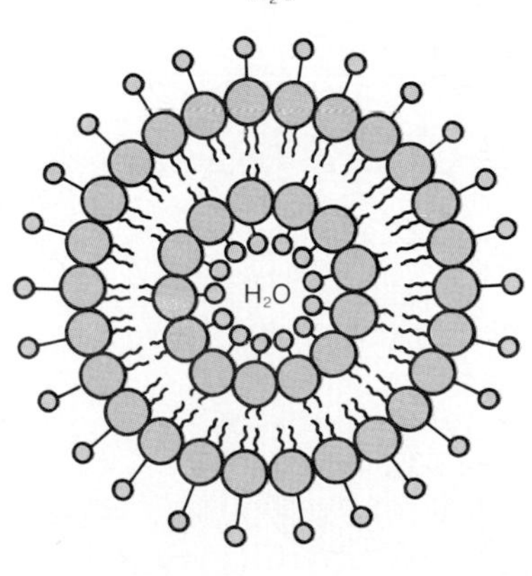

Figure 3.6 Fundamental structure of a phospholipid bilayer. (a) Phospholipid bilayer. (b) Phospholipid membrane vesicle—a stable arrangement of lipids in water.

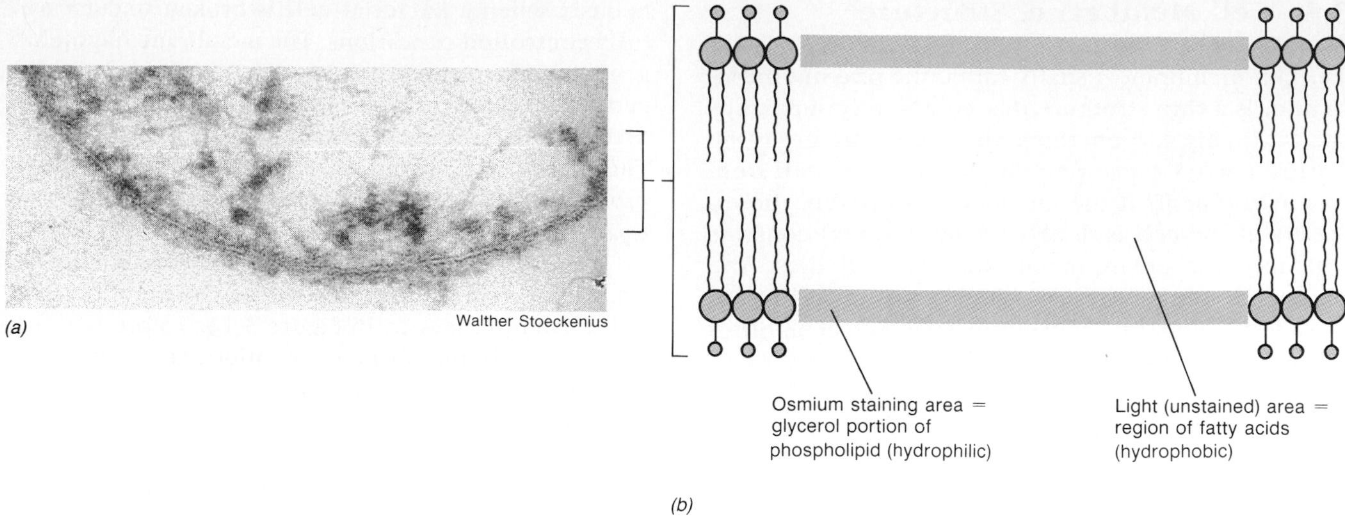

Figure 3.7 (a) Electron micrograph of a thin section of a plasma membrane. Note the distinct double track. Magnification, 221,000×. (b) Enlarged schematic view of membrane shown in (a).

Eucaryotic cell membranes One major difference in chemical composition of membranes between eucaryotic and procaryotic cells is that the eucaryotes have **sterols** in their membranes (Figure 3.9). Sterols are absent from the membranes of virtually all procaryotes. Sterols are rigid, planar molecules, whereas fatty acids are flexible. The association of sterols with the membrane serves to stabilize its structure and make it less flexible. It has been shown that when artificial lipid bilayers are supplemented with sterols, they are much less leaky than when they are composed of pure phospholipid. Why membrane rigidity is necessary in eucaryotes is not known. One possibility, however, is that the eucaryotic cell, because it is much larger than the procaryotic cell, must endure greater physical stresses on the membrane, necessitating a more rigid membrane structure in order to keep the cell stable and functional. One group of antibiotics, the polyenes (filipin, nystatin, and candicidin, for example) react with sterols and destabilize the membrane. It is of interest that these antibiotics are active against eucaryotes but generally do not affect procaryotes, probably because the latter lack sterols in their membranes. However, some bacteria of the mycoplasma group (Section 19.27), which lack a cell wall, require sterol for growth. The sterol becomes incorporated into the plasma membrane and probably stabilizes the mem-

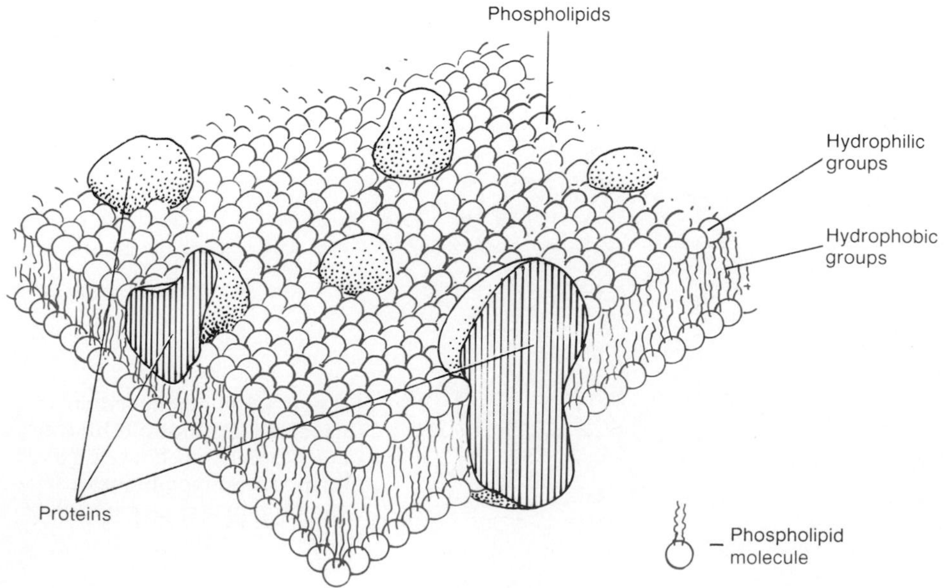

Figure 3.8 Diagram of the structure of a cell membrane. The matrix is composed of phospholipids, with the hydrophobic groups directed inward and the hydrophilic groups toward the outside, where they associate with water. Embedded in the matrix are hydrophobic proteins. Hydrophilic proteins and other charged substances, such as metal ions, are attached to the hydrophilic surfaces. Although there are chemical differences, the overall structure shown is similar in both procaryotes and eucaryotes.

(a)

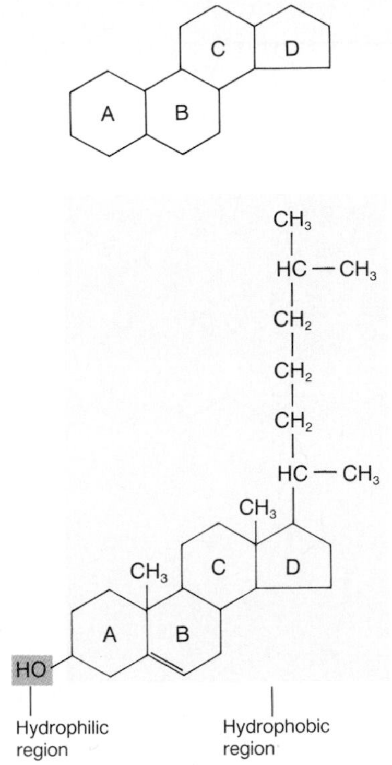

(b)

Figure 3.9 (a) The general structure of a steroid. All steroids contain the same four rings, labeled A, B, C, and D. (b) The structure of cholesterol.

brane structure. Mycoplasmas are inhibited by polyene antibiotics, in contrast to the insensitivity to these antibiotics displayed by all other procaryotes.

Archaebacterial membranes The lipids of archaebacteria are chemically unique. Unlike the lipids of eubacteria and eucaryotes, where ester linkages bond the fatty acids to the glycerol molecule (see Section 2.5), archaebacterial lipids are of two basic types, neither of which contains fatty acids. Instead, glycerol *diethers* and diglycerol *tetraethers* (Figure 3.10) are the major lipids present (see Figure 3.10*d* for the chemical differences between *ester* and *ether* linkages). The major polar lipids of archaebacteria include phospholipids, sulfolipids (lipids containing sulfur), and glycolipids (lipids containing sugars).

In addition, certain archaebacteria contain substantial amounts of nonpolar lipids—most of which are derivatives of the C_{30} isoprenoid, squalene (Figure 3.11). The methanogens and sulfur-dependent archaebacteria produce a variety of isoprenoid compounds ranging from C_{15} up to C_{30}. Literally dozens of other isoprenoid derivatives, including the carotenoids α and β carotene (Figure 3.11) and geranylgeraniol, have been identified in lipid fractions of one or another archaebacterium.

By comparison to eucaryotic and eubacterial membranes, the membranes of archaebacteria are chemically unique but structurally familiar. Archaebacterial membranes consist of polar inner and outer surfaces (due to glycerol molecules) and contain a highly hydrophobic interior. This is the same pattern found in all other cell membranes. However, the glycerol molecule contains either phosphate, sulfate, or carbohydrate on the third carbon atom. Thus, unlike eubacterial or eucaryotic membranes, archaebacterial membranes are not necessarily composed of phospholipid.

Glycerol diethers (Figure 3.10*a*) would be expected to form a true lipid bilayer typical of all plasma membranes. However, some archaebacteria have little or no diether content, but instead produce glycerol *tetra*ethers (Figure 3.10*b*). If one considers how tetraethers must be arranged in archaebacterial membranes, it becomes apparent that tetraether membranes are really **lipid monolayers** and not lipid bilayers (Figure 3.12). Tetraethers create the chemical equivalent of a lipid bilayer, with the major difference being that the two layers are covalently linked in the middle. Good evidence exists for this lipid monolayer in the membranes of certain methanogens and in sulfur-dependent archaebacteria.

Since the length of the hydrocarbon moiety of the di- and tetraethers is fixed at 20 carbons or 40 carbons, respectively, the ability of archaebacteria to adjust hydrocarbon chain length to varying physiological conditions seems limited. However, the linear structure of archaebacterial glycerol tetraethers can be modified by the production of 5-membered (pentacyclic) rings while still maintaining a C_{40} structure (see Figure 3.10). The introduction or removal of pentacyclic rings results in alteration in the chain length, and this in turn could affect the width and perhaps functional properties of the archaebacterial membrane.

The production of ether-linked lipids is a hallmark of archaebacteria—no bona fide archaebacterium has been shown to lack them—and these unique structures have come to be used as biomarkers for identifying archaebacteria (see Section 18.8). The chemical differences between archaebacterial and eubacterial lipids points out once again the huge evolutionary distance between these two primary kingdoms, despite the fact that both are procaryotic.

Sidedness of the plasma membrane It is now well established that the two sides of the plasma membrane are different. Certain proteins embedded in the membrane project toward the inside and others toward the outside; of those that span the membrane, the part of the protein facing outward is different from that facing inward (see Figure 3.8). This sidedness can be more easily visualized by freeze-fracturing electron microscopy of membranes. This tech-

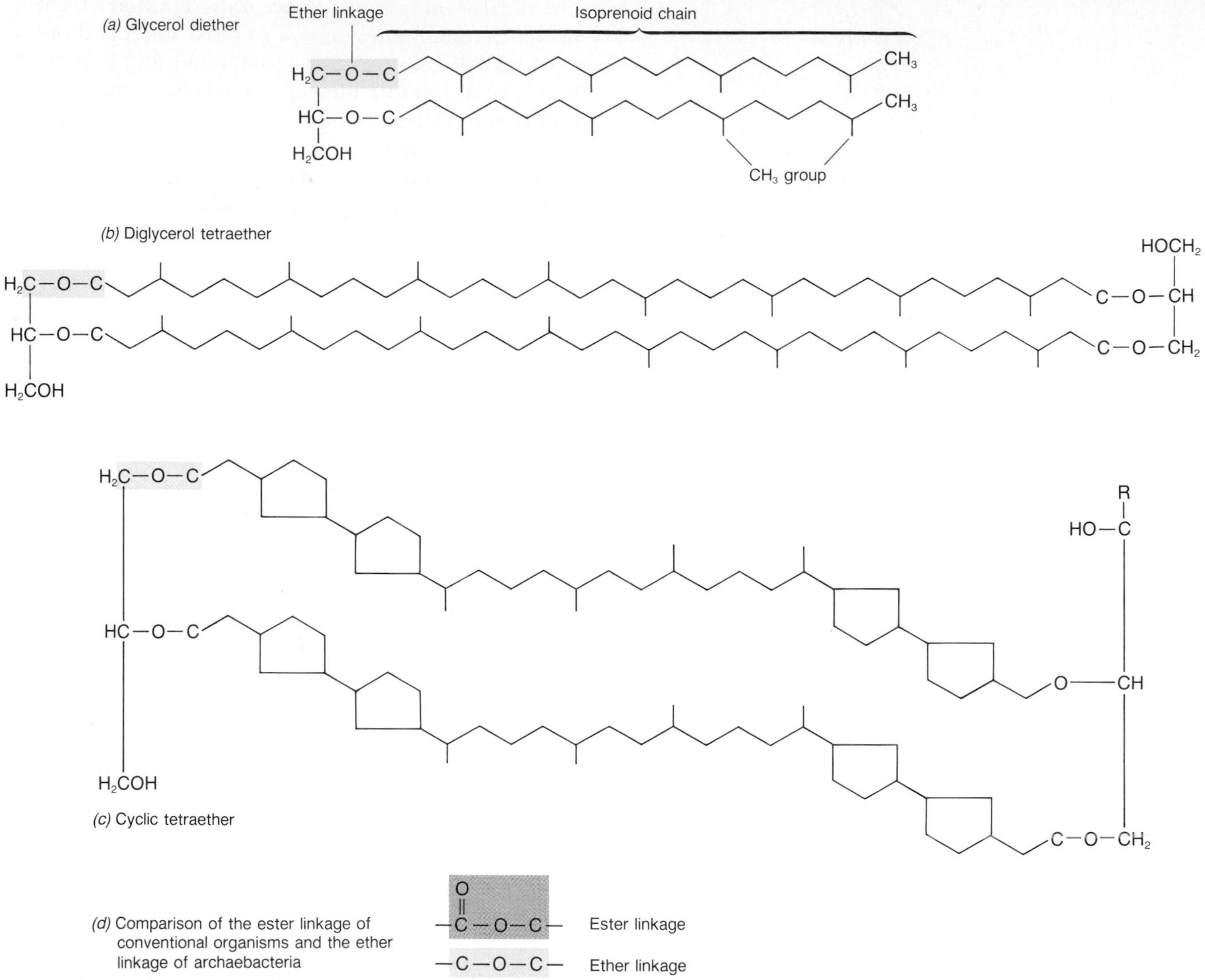

(a) Glycerol diether

Ether linkage

Isoprenoid chain

H_2C-O-C

$HC-O-C$

H_2COH

CH_3

CH_3

CH_3 group

(b) Diglycerol tetraether

H_2C-O-C

$HC-O-C$

H_2COH

$HOCH_2$

$C-O-CH$

$C-O-CH_2$

H_2C-O-C

$HC-O-C$

H_2COH

(c) Cyclic tetraether

R

$HO-C$

$O-CH$

$C-O-CH_2$

(d) Comparison of the ester linkage of conventional organisms and the ether linkage of archaebacteria

$-\overset{O}{\overset{\|}{C}}-O-C-$ Ester linkage

$-C-O-C-$ Ether linkage

Figure 3.10 Major lipids of archaebacteria. (a) Glycerol diethers. (b) Diglycerol tetraethers. (c) Tetrapentacyclic diglycerol tetraethers. Note that in all cases the hydrocarbon is attached to the glycerol by *ether* linkages. Hydrocarbon in (a) phytanyl; (b) and (c) dibiphytanyl. Note also how the introduction of rings in (c) shortens the length of the hydrocarbon chain. (d) Comparison of the ester linkage of conventional organisms and the ether linkage of the archaebacteria.

nique involves freezing the membranes rapidly at liquid nitrogen temperatures and then fracturing the membrane with a sharp knife. The membrane peels apart along the hydrophobic fatty acid region to yield two membrane halves (Figure 3.13). Typically, proteins are observed to stick with one side of the membrane or the other, but transmembrane proteins are readily apparent. Thus the plasma membrane has distinctly different inner and outer surface characteristics, and this sidedness is of considerable importance in membrane function.

Even membrane vesicles (Figure 3.5) possess sidedness. When vesicles are formed under appropriate conditions the sidedness of the membrane is not altered; the surface of the membrane which was the inner face in the cell becomes the inner face of the membrane vesicle. It is possible to incorporate experimentally materials (sugars, salts, enzymes, and so on) inside the membrane vesicles at the time they form, and then study the behavior of these incorporated solutes during membrane function. This has provided an important experimental tool in determining how membranes function in the uptake of nutrients (transport, see Section 3.4).

Internal membranes In addition to the plasma membrane, many procaryotes possess internal membranes, which can be seen in electron micrographs. These internal membranes may be simply extensions or invaginations of the cell membrane, or they may

H_3C ... CH_3 ... CH_3 group

(a) Squalene

H_3C ... C—OH

(b) Geranylgeraniol

(c) β-Carotene

Figure 3.11 Structure of (a) squalene and major squalene derivatives, (b) geranylgeraniol, and (c) β-carotene, found in the lipids of halophilic archaebacteria.

be much more complicated. In photosynthetic procaryotes, the photosynthetic membranes often form an extensive internal membrane system (Figure 3.14; see also Section 19.1), and the nitrifying bacteria (Section 19.4) and methane-oxidizing bacteria (Section 19.7) also frequently have elaborate internal membranes (see Section 16.2 for a discussion of photosynthetic membrane systems).

3.4 Cell Membrane: Function

The plasma membrane is not just the barrier separating inside from outside. It has a major role in cell function. It is through the cell membrane that all nutrients must pass, and it is also through the cell membrane that all waste products leave the cell. The term **permeability** is used to refer to the property of membranes that permits movement of molecules back and forth through the cell membrane. However, most molecules that move through the membrane do not move passively; the cell membrane is *selectively permeable*. Further, the cell itself plays an active role in the movement of molecules across the membrane; molecules are said to be **transported** through the membrane. In the present section we discuss the permeability and transport properties of membranes.

The plasma membrane as a permeability barrier Despite its thinness, the plasma membrane functions as a tight barrier, so that the passive movement of polar solute molecules does not readily occur. Some small nonpolar and fat-soluble substances, such as fatty acids, alcohols, and benzene, may penetrate cell membranes readily by becoming dissolved in the lipid phase of the membrane, but movement of most other molecules occurs only by means of specific transport systems. Charged molecules, such as organic acids, amino acids, and inorganic salts, which are hydrophilic, do not readily pass the membrane

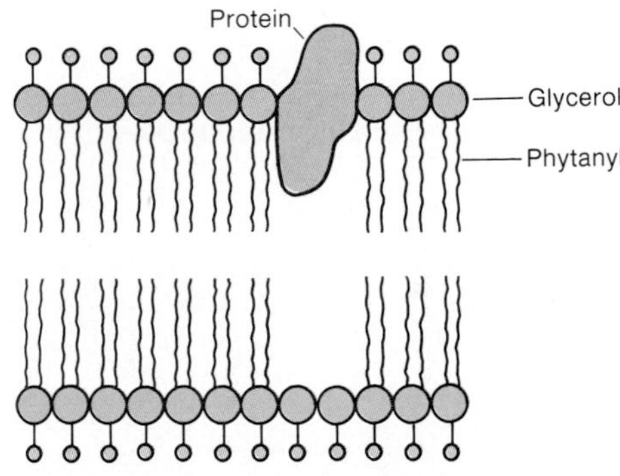

(a) Lipid bilayer

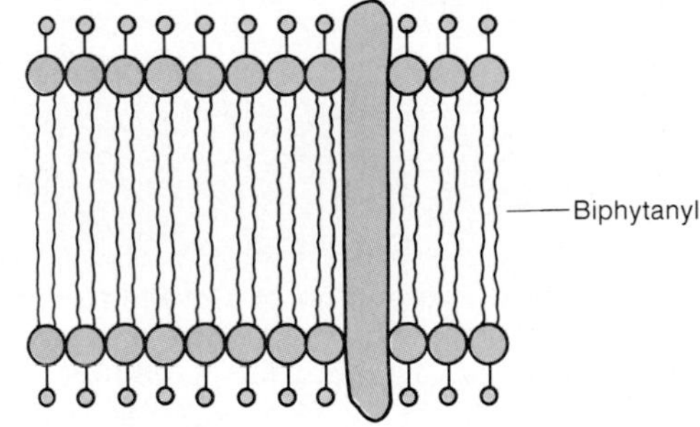

(b) Lipid monolayer

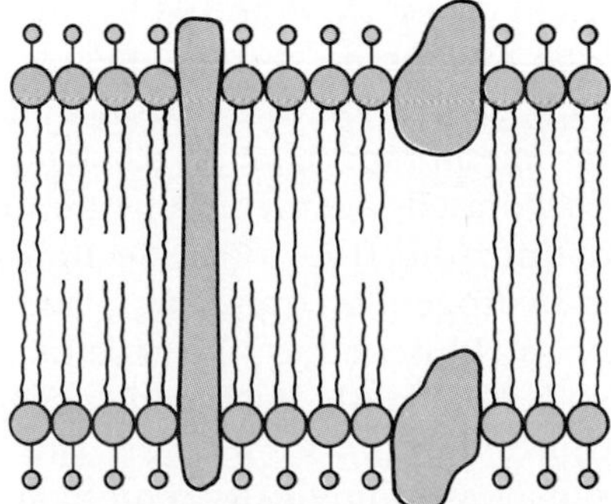

(c) Lipid monolayer/bilayer mixtures

Figure 3.12 (a) Glycerol diether lipid bilayer membrane structure. (b) Diglycerol tetraether lipid monolayer membrane. (c) Mixed membrane type.

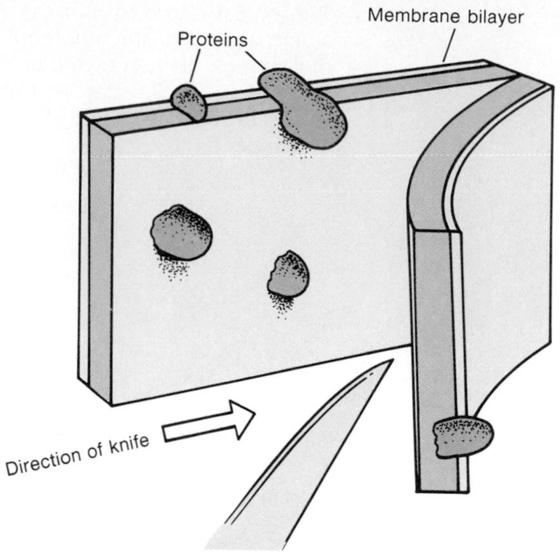

Figure 3.13 Schematic view of a freeze-fractured plasma membrane showing the distinct sidedness of the membrane.

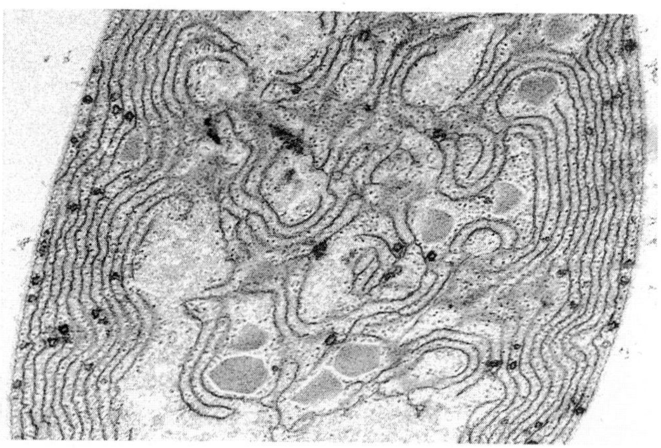

Figure 3.14 Electron micrograph of a cyanobacterial cell (*Anabaena azollae*), showing the extensive array of photosynthetic membranes. Magnification, 14,800×.

barrier, because of the highly hydrophobic nature of the inner portion of the membrane and also because (in the case of anions) they are repelled by the electrical charge on the membrane surface. Even a substance as small as a hydrogen ion, H^+, does not readily breach the plasma membrane barrier passively because it is always hydrated, occurring in solution as the charged hydronium ion (H_3O^+) (see Section 4.12 for a discussion of hydrogen ion transport). One molecule which does freely penetrate the membrane is water itself, which is sufficiently small and uncharged to pass between phospholipid molecules. The relative permeability of a few biologically important substances is shown in Table 3.1.

The interior of the cell (called the *cytoplasm*) consists of an aqueous solution of salts, sugars, amino acids, vitamins, coenzymes, and a wide variety of other soluble materials. If the membrane is damaged, most of these materials are able to leak out, and only substances too large to pass through the cell-wall pores are retained. Components enter the cytoplasm either as nutrients taken up from the environment or as materials synthesized from other constituents in the cell. Compounds called **ionophores** destroy the selective permeability properties of membranes. Ionophores are small hydrophobic molecules that dissolve in lipid bilayers and allow the passive diffusion of ionized substances into or out of the cell. Certain antibiotics are effective ionophores. *Valinomycin*, for example, acts as a potassium channel, transporting potassium down the electrochemical gradient until the concentration is equal on both sides of the membrane. *Tyrocidin*, another ionophore (Figure 3.15), is a small hydrophobic peptide that can form a transmembrane channel and allow massive leakage of monovalent cations. Since the formation of concentration gradients are essential in cell function, ionophores are obviously lethal agents to cells.

Transport across biological membranes Although intact cell membranes do not allow the free diffusion of various polar molecules such as sugars, amino acids, ions, and the like, these substances do pass through the membrane by other means and can be concentrated to over 1000 times that of the external environment through the action of **membrane transport proteins**. Three classes of transport proteins have been identified. *Uniporters* are proteins that transport a substance from one side of the membrane to the other (Figure 3.16). The other transport proteins move the substance of interest across the membrane along with a second substance required for transport of the first. They are thus cotransport proteins. *Symporters* are membrane proteins that carry

Table 3.1 Relative permeability rates of various biologically important molecules or ions through artificial lipid bilayers

Substance	Rate of permeability*
Water	100
Glycerol	0.1
Tryptophan	0.001
Glucose	0.001
Chloride ion (Cl^-)	0.000001
Potassium ion (K^+)	0.0000001
Sodium ion (Na^+)	0.00000001

**Relative scale—permeability expressed in respect to permeability of water, expressed as 100.*

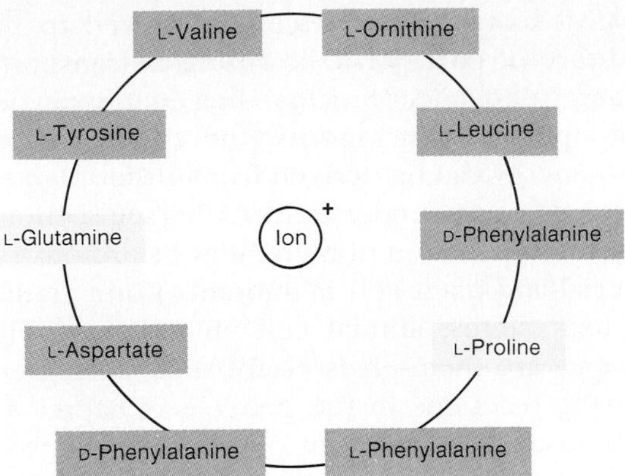

Figure 3.15 Structure of the peptide ionophore (and antibiotic), tyrocidin. Note the predominance of nonpolar and aromatic amino acids. The ion is carried in the center of the molecule.

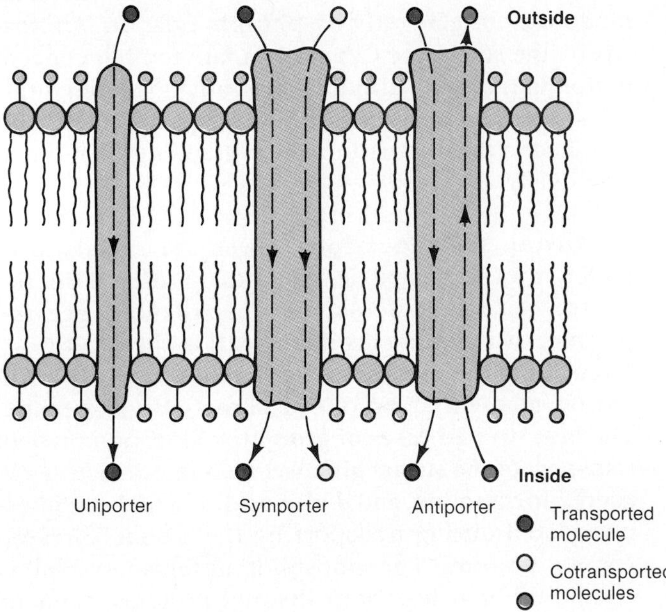

Figure 3.16 Schematic drawing showing the operation of transport proteins.

both substances across the membrane in the same direction (Figure 3.16). *Antiporters* transport one substance across the membrane in one direction while transporting the second substance in the opposite direction (Figure 3.16).

The necessity for carrier-mediated transport mechanisms in microorganisms can readily be appreciated. Most microbes live in environments whose concentrations of salts and other nutrients are many times lower than concentrations within the cells. If diffusion were the only type of transport mechanism available, cells would not be able to acquire the proper concentrations of solutes. In diffusion-limited uptake, both the rate of uptake and the intracellular level are proportional to the external concentration. Carrier-mediated transport mechanisms overcome this problem by enabling the cell to accumulate solutes *against* a concentration gradient. As shown in Figure 3.17, carrier-mediated transport shows a saturation effect: if the concentration of substrate in the medium is high enough to saturate the carrier, which is likely to be the case even at quite low substrate concentrations, the rate of uptake (and often the internal level as well) becomes maximal. The value of carrier-mediated transport to organisms living in dilute environments is thus obvious.

One characteristic of carrier-mediated transport processes is the highly specific nature of the transport event. The binding and carrying of a substance across the membrane resembles an enzyme reaction (see Section 4.3). Certain carrier proteins react with only a single molecule, but many show affinities for a chemical class of molecules. For instance, there is a carrier which transports the aromatic amino acids but has no affinity for other amino acids. This can lead to competition for uptake; for instance, L-alanine, L-serine and L-glycine, all structurally related amino acids, compete with one another for uptake, as do L-valine, L-leucine, and L-isoleucine. It is thus clear that transport proteins involved in the uptake of amino acids recognize a chemical *class* of molecules, each of which can be transported (probably with different affinities) at a rate sufficient to accumulate each to satisfactory levels. Similar competition can be shown among related cations, such as Na^+, K^+, and Rb^+.

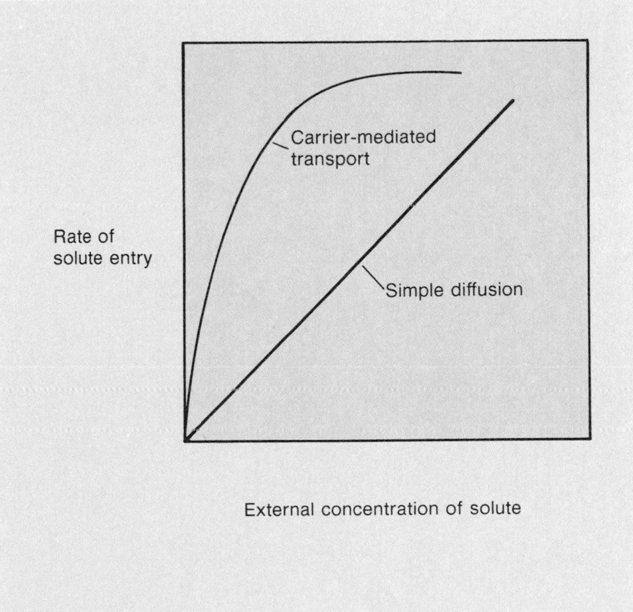

Figure 3.17 Relationship between uptake rate and external concentration in passive uptake and active transport. Note that in the carrier-mediated process the uptake rate shows saturation at low external concentrations.

The action of membrane transport proteins Although the molecular details of the transport process by membrane carriers are not clear, it appears that most membrane transport proteins are **transmembrane proteins** (see Figures 3.8 and 3.13). Transmembrane proteins span the membrane, with portions of the protein being exposed to both the cytoplasm and to the external environment. Arranged in this fashion it is possible for solutes bound on the external surface of the cell to be carried through the membrane by a conformational change in the transport protein (Figure 3.18). The conformational change would then serve as a "gate," releasing the substance on the inside of the cell. However, even if the substance is carried in by a specific transport protein, the law of mass action will prevent any significant concentration gradient from forming and the concentration of the solute inside the cell will be about the same as that outside the cell. This type of carrier-mediated diffusion is known as **facilitated diffusion**.

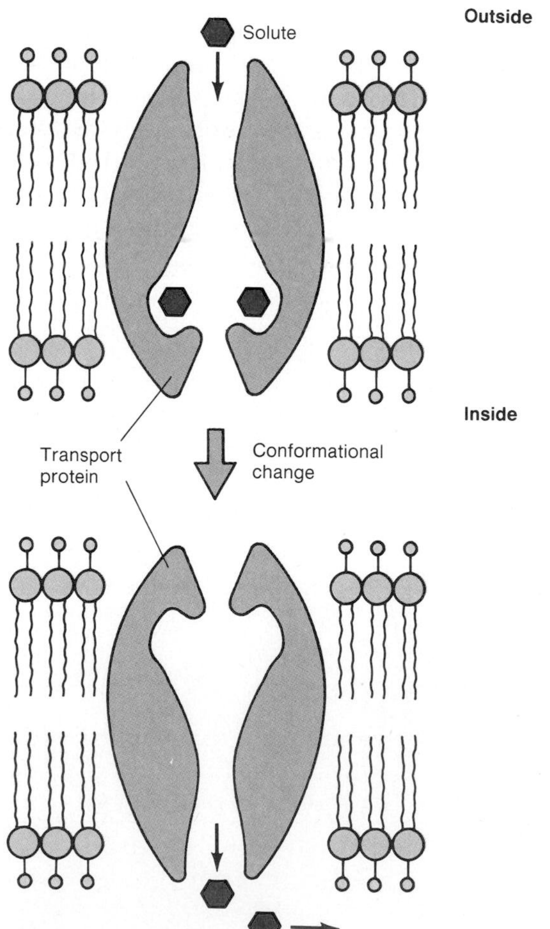

Figure 3.18 Schematic diagram showing how a conformational change in a transmembrane protein could serve to drive the transport event.

Most transport processes are linked to the expenditure of energy. If the solute is transported by an energy-dependent process, then energy can be used to pump the solute *against* the concentration gradient. Energy can be derived from either high energy phosphate compounds, such as ATP (see Section 4.6), or by the dissipation of a gradient of protons (across bacterial and plant cell membranes) or a gradient of Na^+ ions (across animal cell membranes). The ion gradients are themselves established during energy-releasing reactions in the cell (see Chapter 4) and can be used as a source of potential energy to drive the uptake of solutes against the concentration gradient (see below).

Two major mechanisms of energy-linked transport are known. **Group translocation** is the process wherein a substance is transported while simultaneously being converted into a derivative by chemical modification, generally by phosphorylation. Alternatively, the substance can accumulate to high concentration in the cytoplasm in chemically unaltered form. This process of **active transport** requires energy and is linked to the dissipation of ion gradients mentioned previously.

Group translocation These are processes in which the substance is chemically altered in the course of passage across the membrane. Since the product that appears inside the cell is chemically different from the external substrate, no actual concentration gradient is produced across the membrane. The best studied cases of group translocation involve transport of the sugars glucose, mannose, fructose, *N*-acetylglucosamine, and β-glucosides, which are phosphorylated during transport by the **phosphotransferase system**. The phosphotransferase system is composed of at least four distinct proteins, some of which are specific for the sugars (Figure 3.19). One of the protein components of the system is a small heat-stable protein designated HPr, which acts as a carrier of high-energy phosphate:

$$\text{Phosphoenolpyruvate} + \text{HPr} \rightarrow \text{HPr—P} + \text{pyruvate}$$

$$\text{HPr—P} + \text{sugar} \rightarrow \text{sugar—P} + \text{HPr}$$

The phosphorylation of HPr by phosphoenolpyruvate is carried out by an enzyme present in the cytoplasm, and HPr-phosphate then moves to the membrane. The sugar being transported is then phosphorylated by HPr-phosphate, the reaction being catalyzed by an enzyme within the membrane specific for that sugar. Finally, the phosphorylated sugar moves into the cell, where it can be metabolized by normal catabolic pathways. (This process is illustrated for glucose transport in Figure 3.19.)

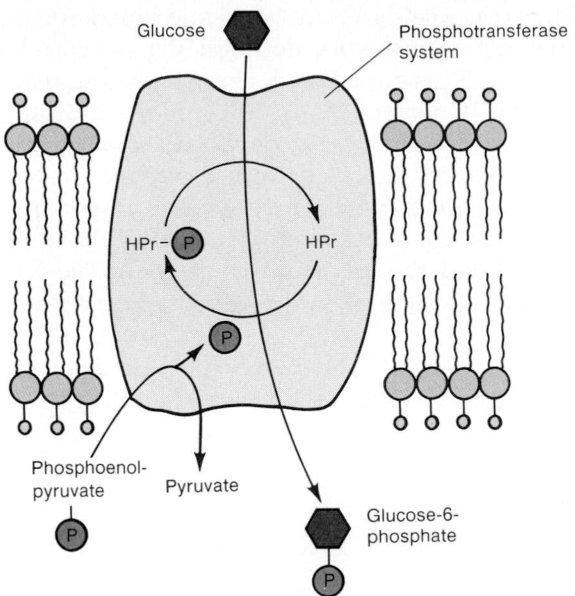

Figure 3.19 Mechanism of the phosphotransferase system of *Escherichia coli*. The system consists of four proteins, one of which is the direct phosphate donor, HPr. See text for details.

One of the best pieces of evidence for the involvement of the phosphotransferase system in bacterial transport comes from studies on genetically altered (mutant) strains unable to utilize sugars. Two kinds of mutants can be isolated, those unable to utilize specific sugars, due to a block in synthesis of a specific membrane-bound enzyme, and those unable to utilize all of the sugars transported by the phosphotransferase system, due to a block in synthesis of either HPr or the enzyme involved in its phosphorylation. One ingenious experiment showed that phosphorylation occurred only *during* transport and not after the sugar had entered the cytoplasm. This involved the study of transport by membrane vesicles (Figure 3.6*b*). Vesicles will carry out transport reactions and have the virtue that substances can be placed inside the vesicles at the time of closure, so that movement in either direction can be studied. If a sugar is placed inside the vesicles along with the enzymes of the phosphotransferase system, it is not phosphorylated. However, if a sugar is applied externally, it becomes phosphorylated when it is transported into the vesicles. This experiment shows that phosphorylation occurs during the actual transport event.

Other substances thought to be transported by group translocation include purines, pyrimidines, and fatty acids. However, many substances are not taken up by group translocation, including even several sugars which seem to be accumulated instead by the process of active transport discussed below.

Active transport This is an energy-dependent transport process in which the substance being transported combines with a membrane-bound carrier which then releases the *chemically unchanged* substance inside the cell. Since the substance is not altered during the transport process, if it is not consumed in cell reactions, its concentration inside may reach many times the external concentration. Substances transported by active transport include some sugars, amino acids, organic acids, and a number of inorganic ions such as sulfate, phosphate, and potassium. Glucose is taken up by active transport processes in some bacteria and by the phosphotransferase system in others (Table 3.2).

As in any other pump, active transport requires that work be performed. In bacteria, the energy for driving the pump is the separation of hydrogen ions (protons) across the membrane, called the *proton motive force* (see Section 4.12). Energy released from the breakdown of organic or inorganic compounds, or the energy of light, is used to establish a separation of protons across the membrane, with the proton concentration highest outside the cell and lowest inside (see Section 4.12). It is the electrochemical potential residing in this separation that drives the uptake of nutrients by active transport (Figure 3.20). Each membrane carrier has specific sites for both its substrate (for example, glucose or potassium) and a proton (or protons). As the substrate is taken up, protons move across the membrane and the concentration difference decreases (Figure 3.20).

There are at least four independent mechanisms by which bacteria generate proton separation and these will be discussed in detail in Chapter 4. Each mechanism results in an "energized" membrane containing a proton separation as depicted in Figure 3.20. The difference in pH and in electrical potential across the membrane, however established, is then used by the cell to transport both ions and uncharged molecules.

Let us recall that the transport process involves protein carriers specific for each substrate; the H^+

Table 3.2 Mechanism of glucose uptake by various bacteria

Phosphotransferase system	Active transport
Escherichia coli	*Pseudomonas aeruginosa*
Bacillus subtilis	*Azotobacter vinelandii*
Clostridium pasteurianum	*Micrococcus luteus*
Staphylococcus aureus	*Mycobacterium smegmatis*

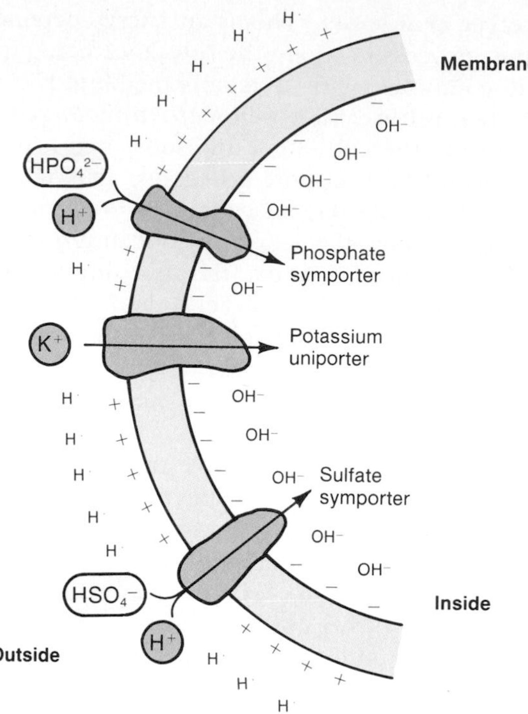

Figure 3.20 Use of ion separation, in this case the separation of protons from hydroxyl ions across the membrane, to transport inorganic ions by specific transport proteins. Note that there is both a separation of protons and of electrical charge.

concentration difference serves as the energy link between these membrane carriers and the metabolic machinery, making it possible for the carriers to "pump" nutrients inward. Cations, such as K^+, may be transported into the cytoplasm by uniporters in response to the electrical charge separation, since the interior of the cell is negative (Figure 3.20). Uptake of anions probably occurs together with protons by symporters, so that it is effectively the undissociated acid that enters the cell (Figure 3.20). Transport of uncharged molecules such as sugars or amino acids can also be linked to the electrical charge differences: the symporter transports both the substrate and one or more protons. Proton pumps linked to transport are key constituents of all eubacterial (and probably archaebacterial) membranes, and in the inner membranes of mitochondria and chloroplasts.

Ion gradients in eucaryotes In the plasma membrane of animal cells, a second type of ion pump is largely responsible for the energy required to transport nutrients by active transport. As we will see in Chapter 4, adenosine triphosphate (ATP) is a key energy carrier in cell function. An enzyme, the Na^+/K^+ ATPase pumps Na^+ out of the cell while pumping K^+ in. A Na^+/K^+ concentration difference is thus established that, analogous to the proton separation of bacteria, represents work done on the system (Figure 3.21). ATP produced from energy-yielding reactions in the cell powers the Na^+/K^+ pump and then the transport of solutes, glucose for example, is driven by the energy released during passage of Na^+/K^+ across the membrane. The glucose transport protein in this case could be a symporter which binds glucose and Na^+ and carries them into the cell. The Na^+/K^+ pump also functions in the uptake of amino acids in animal cells.

Plant and algal cell membranes also contain Na^+/K^+ pumps, however solute transport in plant cells and in algae is primarily driven by a proton pump.

3.5 Cell Wall: Eubacteria

One of the most important structural features of the procaryotic cell is the cell wall, which confers rigidity and shape. The procaryotic cell wall is chemically quite different from that of any eucaryotic cell, and this is one of the features distinguishing procaryotic from eucaryotic organisms. The cell wall is difficult to visualize well with the light microscope, but can be readily seen in thin sections of cells with the electron microscope.

Bacteria can be divided into two major groups, called **Gram-positive** and **Gram-negative**. The original distinction between Gram-positive and Gram-negative was based on a special staining procedure, the Gram stain (see Section 1.13), but differences in cell wall structure are at the base of these differences in Gram staining reaction. Gram-positive and Gram-negative cells differ considerably in the appearance

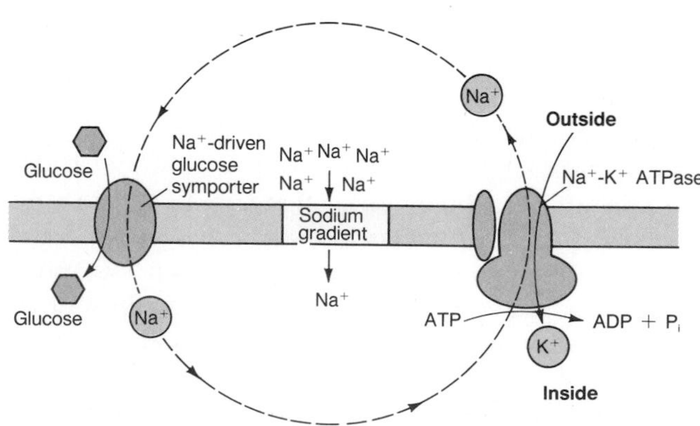

Figure 3.21 Transport of metabolites linked to the Na^+/K^+ pump in the membranes of animal cells. The ATPase pumps Na^+ out of the cell while K^+ is pumped in, and is driven by the energy released from the reaction $ATP \rightarrow ADP + P_i$.

Plasma membrane

Peptidoglycan

(a)

Outer wall layer

Plasma membrane

Peptidoglycan

(b)

A. Umeda and K. Amako

A. Umeda and K. Amako

(c)

(d)

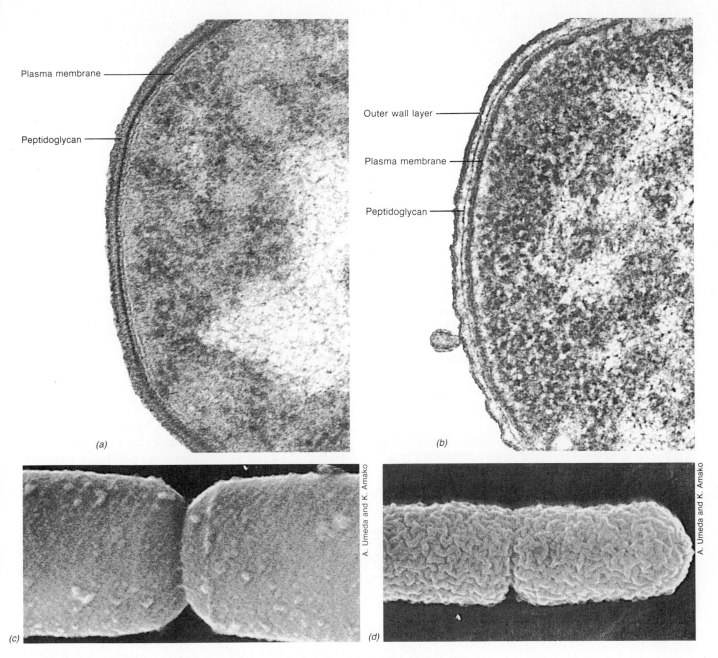

Figure 3.22 Electron micrographs of bacterial cell walls: (a) Gram-positive, *Arthrobacter crystallopoietes*. Magnification, 126,000×. (b) Gram-negative, *Leucothrix mucor*. Magnification, 165,000×. (c) and (d) scanning electron micrographs of Gram-positive (*Bacillus subtilis*) and Gram-negative (*Escherichia coli*) bacteria. Magnification *c*, 78,625×; *d*, 56,700×.

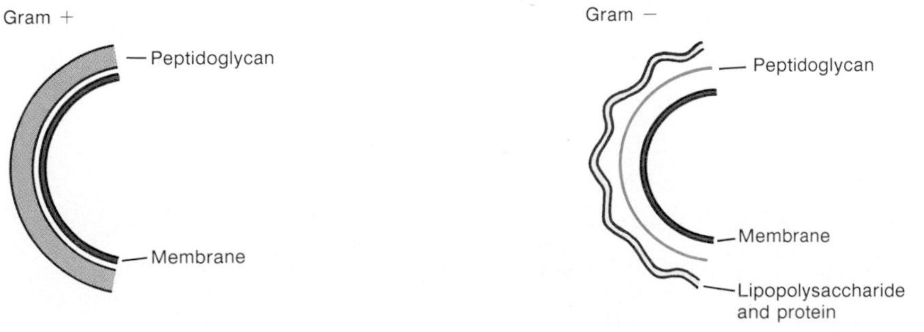

Figure 3.23 Comparison of the cell walls of Gram-positive and Gram-negative bacteria.

(a)

COOH
|
H_2N-CH
|
CH_2
|
CH_2
|
CH_2
|
H_2N-CH
|
COOH

(b)

COOH
|
H_2N-CH
|
CH_2
|
CH_2
|
CH_2
|
CH_2
|
NH_2

Figure 3.24 (a) Diaminopimelic acid. (b) Lysine.

of their cell walls, as is shown in Figure 3.22. The Gram-negative cell wall is a multilayered structure and quite complex, while the Gram-positive cell wall consists of a single layer and is often much thicker. Close examination of Figure 3.22 shows that there is also a significant textural difference between the surfaces of Gram-positive and Gram-negative bacteria, as revealed by negative staining procedures. A schematic representation of Gram-positive and Gram-negative cell walls is given in Figure 3.23.

The rigid layer In the cell walls of eubacteria, there is one layer, the *rigid layer*, immediately adjacent to the plasma membrane, which is primarily responsible for the strength of the wall. In most bacteria, additional layers are present outside the rigid layer. The rigid layer of both Gram-negative and Gram-positive bacteria is very similar in chemical composition. Called **peptidoglycan**, this layer is a thin sheet composed of two sugar derivatives; *N*-acetylglucosamine and *N*-acetylmuramic acid, and a small group of amino acids, consisting of L-alanine, D-alanine, D-glutamic acid, and either lysine or diaminopimelic acid (DAP) (Figure 3.24). These constituents are connected to form a repeating structure, the *glycan tetrapeptide* (Figure 3.25).

The basic structure is in reality a thin sheet in which the *glycan* chains formed by the sugars are connected by *peptide* cross-links formed by the amino acids. The glycosidic bonds connecting the sugars in the glycan chains are very strong, but these chains alone cannot provide rigidity in all directions. The full strength of the peptidoglycan structure is obtained when these chains are joined by peptide cross-links. This cross-linking occurs to characteristically different extents in different bacteria, with greater rigidity coming from more complete cross-linking. In Gram-negative eubacteria, cross-linkage usually occurs by direct peptide linkage of the amino group of diaminopimelic acid to the carboxyl group of the terminal D-alanine (Figure 3.26*a*). In Gram-positive eubacteria, cross-linkage is usually by a peptide interbridge, the kinds and numbers of cross-linking amino acids varying from organism to organism. In *Staphylococcus aureus*, the best studied Gram-positive or-

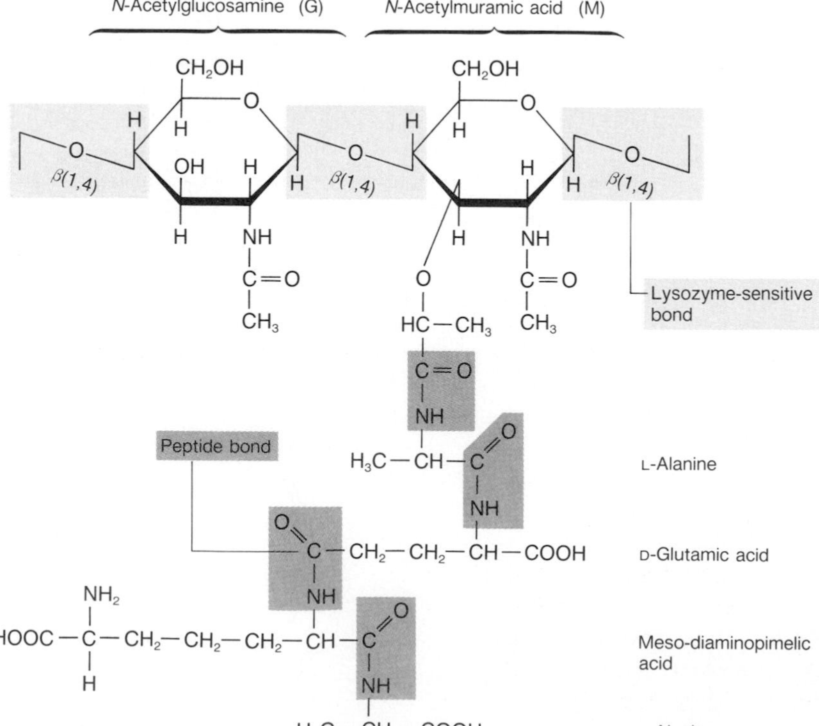

Figure 3.25 Structure of one of the repeating units of the peptidoglycan cell-wall structure, the glycan tetrapeptide. The structure given is that found in *Escherichia coli* and most other Gram-negative bacteria. In some bacteria, other amino acids are found.

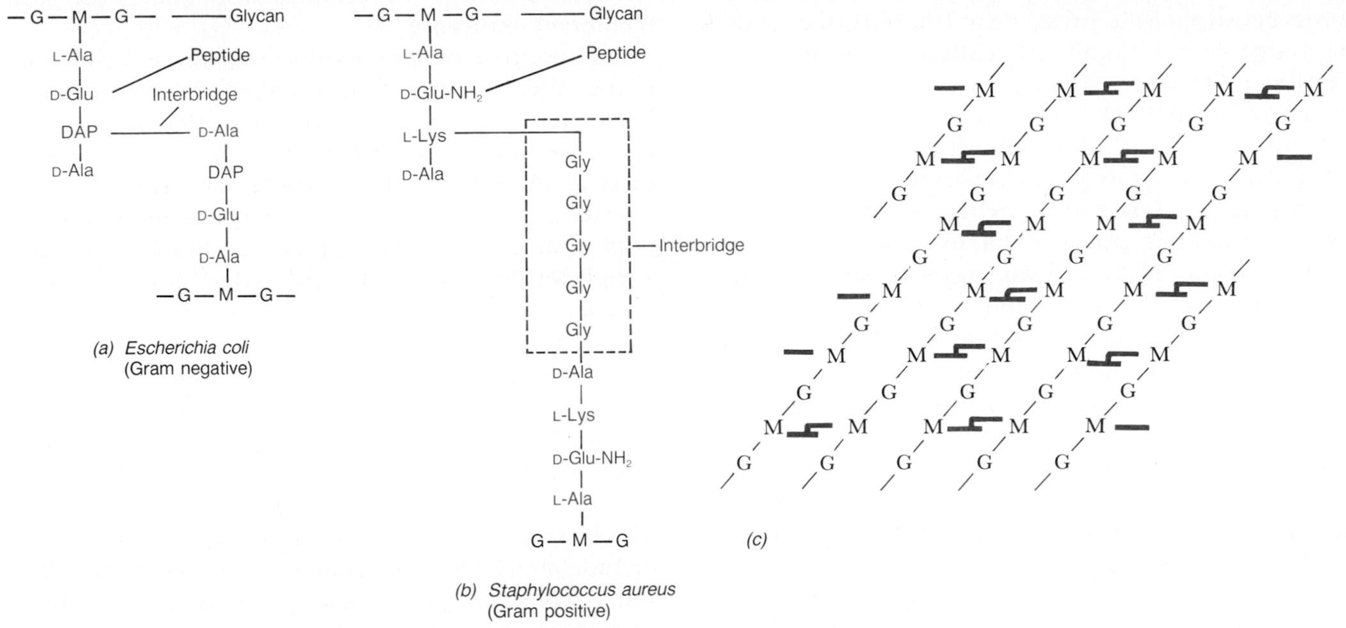

(a) Escherichia coli (Gram negative)

(b) Staphylococcus aureus (Gram positive)

(c)

Figure 3.26 Manner in which the peptide and glycan units are connected in the formation of the peptidoglycan sheet. (a) Direct interbridge in Gram-negative bacteria. (b) Glycine interbridge in *Staphylococcus aureus* (Gram-positive). (c) Overall structure of the peptidoglycan. G, N-acetylglucosamine; M, N-acetylmuramic acid; heavy lines in c, peptide cross-links.

ganism, each interbridge peptide consists of five molecules of the amino acid glycine connected by peptide bonds (Figure 3.26*b*). The overall structure of a peptidoglycan is shown in Figure 3.26*c*. The shape of a cell is determined by the lengths of the peptidoglycan chains and by the manner and extent of cross-linking of the chains.

The peptidoglycan structure is present only in procaryotes, and is found in the wall of virtually all species of eubacteria. The sugar *N*-acetylmuramic acid and the amino acid diaminopimelic acid (DAP) are never found in eucaryotic walls. However, not all procaryotic organisms have DAP in their peptidoglycan. This amino acid is present in all Gram-negative bacteria and in some Gram-positive species, but most Gram-positive cocci have lysine instead of DAP, and a few other Gram-positive bacteria have other amino acids. Another unusual feature of the procaryotic cell wall is the presence of two amino acids that have the D configuration, D-alanine and D-glutamic acid. As we have seen in Chapter 2, in proteins amino acids are always in the L configuration.

Several generalizations regarding peptidoglycan structure can be made. The glycan portion is uniform throughout the eubacterial kingdom, with only the sugars *N*-acetylglucosamine and *N*-acetylmuramic acid being present, and these sugars are always connected in β-1,4 linkage. The tetrapeptide of the repeating unit shows variation only in one amino acid, the lysine-diaminopimelic acid alternation. However, the D-glutamic acid at position 2 can be hydroxylated in some organisms.

The greatest variation in peptidoglycans occurs in the interbridge. Any of the amino acids present in the tetrapeptide can also occur in the interbridge, but in addition, a number of other amino acids can be found in the interbridge, such as glycine, threonine, serine, and aspartic acid. However, certain amino acids are never found in the interbridge: branched-chain amino acids, aromatic amino acids, sulfur-containing amino acids, and histidine, arginine, and proline.

In Gram-positive bacteria, as much as 90 percent of the wall consists of peptidoglycan, although another kind of constituent, teichoic acid (discussed below), is usually present in small amounts. In Gram-negative bacteria, only 5 to 20 percent of the wall is peptidoglycan, the remainder being present in an outer wall layer outside the peptidoglycan layer as discussed below.

Outer-wall (lipopolysaccharide) layer In Gram-negative bacteria, the outer wall layer is a membrane-like structure. However, the membrane of the outer wall is not constructed solely of phospholipid, as is the plasma membrane, but also contains additional lipid plus polysaccharide and protein. The lipid and polysaccharide are intimately linked in the outer layer to form specific **lipopolysaccharide** structures. Because of the presence of lipopolysaccharide, the outer layer is frequently called the lipopolysaccharide or LPS layer. Although complex, the chemical structures of some LPS layers are now understood. As seen in Figure 3.27, the polysaccharide consists of two portions, the core polysaccharide and the O-polysaccharide. In *Salmonella,* where it has been best studied, the **core polysaccharide** consists of ketodeoxyoctonate, seven-carbon sugars (heptoses), glucose, galactose, and *N*-acetylglucosamine. Connected to the core is the **O-polysaccharide**, which usually contains galactose, glucose, rhamnose, and mannose (all six-carbon sugars) as well as one or more unusual dideoxy sugars such as abequose, colitose, paratose, or tyvelose. These sugars are connected in four- or five-sugar sequences, which often are branched. When the sugar sequences are repeated, the long O-polysaccharide is formed.

The relationship of the O-polysaccharide to the rest of the LPS layer is shown in Figure 3.28. The lipid portion of the lipopolysaccharide, referred to as *lipid A,* is not a glycerol lipid, but instead the fatty acids are connected by ester linkage to *N*-acetylglucosamine. Fatty acids frequently found in the lipid include β-hydroxymyristic, lauric, myristic, and palmitic acids. In the outer membrane, the LPS associates with phospholipids to form the *outer* half of the unit membrane structure.

A lipoprotein complex is found on the *inner* side of the outer membrane of a number of Gram-negative bacteria (Figure 3.28). This lipoprotein is a small (~7200 molecular weight) protein that appears to serve as an *anchor* between the outer membrane and peptidoglycan. The terminal amino acid of the lipoprotein is a cysteine residue, modified to contain a diglyceride instead of a sulfhydryl moiety, and a fatty acid bonded by amide linkage to the amino group of the amino acid. Presumably, this hydrophobic-rich end of the lipoprotein is the end which associates with the outer membrane phospholipids. One important biological property of the outer membrane layer of many Gram-negative bacteria is that it is frequently *toxic* to animals. Gram-negative bacteria which are pathogenic for humans and other mammals include members of the genera *Salmonella, Shigella,* and *Escherichia.* The toxic property of the outer membrane layer of these bacteria is responsible for some of the symptoms of infection which these bacteria bring about. The toxic properties are associated with part of the lipopolysaccharide (LPS) layer of these organisms. The term *endotoxin* is frequently used to refer to this toxic component of LPS, as we discuss in Section 11.12.

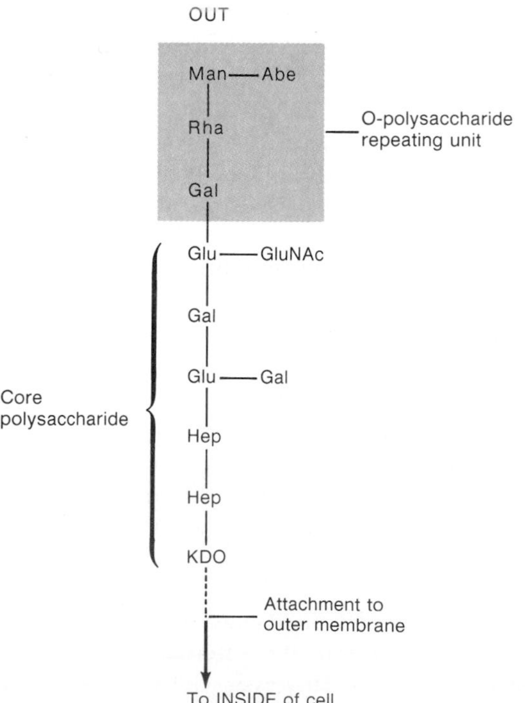

Figure 3.27 Structure of the lipopolysaccharide of *Salmonella.* KDO, ketodeoxyoctonate; Hep, heptose; Glu, glucose; Gal, galactose; GluNac, N-acetylglucosamine; Rha, rhamnose; Man, mannose; Abe, abequose.

Permeability in the Gram-negative cell Unlike the plasma membrane, the outer membrane of Gram-negative eubacteria is relatively permeable to small molecules, even though it is a true lipid bilayer. How then, can molecules move so readily across the outer membrane? It is shown that proteins called **porins** are present in the lipopolysaccharide of Gram-negative bacteria (Figure 3.28). Porins serve as membrane channels for the entrance and exit of low-molecular-weight substances. Several porins have now been identified, and most appear to be rather nonspecific. For example, one porin of *Escherichia coli* allows phosphate, sulfate, and certain other anionic compounds to pass. Another porin of *E. coli* is specific only for small peptides. The number of different porins that exist is not known, but it is likely to be of the order of 30 or more. Structural studies have shown that each porin is a protein containing two, or more commonly three polypeptides. Porins are transmembrane proteins (Figure 3.28) which form small membrane holes about 1 nm in diameter. Apparently a mechanism exists for opening and closing the pores, because resistance to certain antibiotics is related to porin structure. Presumably, such porins can be closed to prevent antibiotic uptake.

Thus, the outer layer serves as a barrier through which materials must penetrate if they are to reach the cell. And, although the outer layer is permeable

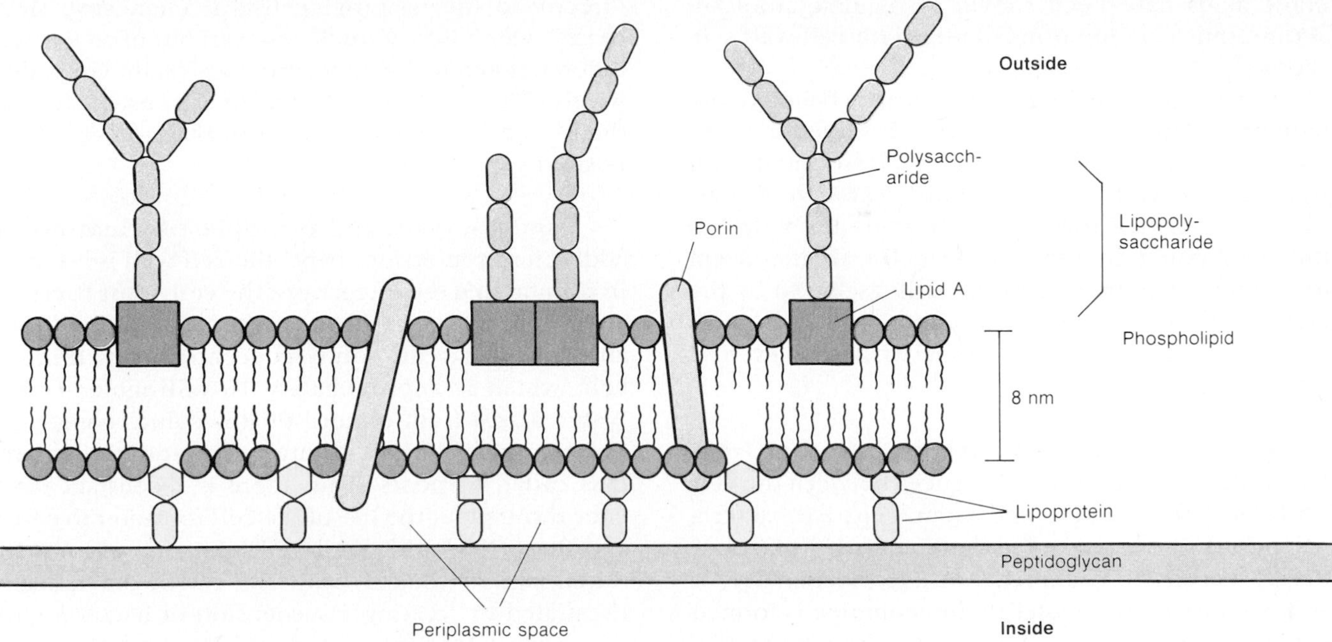

Figure 3.28 Arrangement of lipopolysaccharide, lipid A, phospholipid, porins, and lipoprotein in the Gram-negative bacterium *Salmonella*.

to small molecules, it is not permeable to enzymes or other large molecules. In fact, one of the main functions of the outer layer may be its ability to keep certain enzymes, which are present outside the peptidoglycan, from leaving the cell. These enzymes are present in a region called the **periplasmic space** (see Figure 3.28). In addition to enzymes, the periplasmic space is home for a number of specific proteins involved in transport. These proteins are called **periplasmic binding proteins** because, although not enzymes, they bind specifically the substances that they transport. These binding proteins are readily released from cells by treatments that disrupt the outer lipopolysaccharide layer, and because they leave the cell so readily they are not thought to be an integral part of the plasma membrane, as are the carrier proteins described in Section 3.4. Periplasmic binding proteins function in the initial stages of transport, binding the substance and bringing it to the membrane-bound carrier; these processes are probably not linked to the H^+ gradient but may instead use ATP as an energy source. Binding proteins seem to be absent in Gram-positive bacteria, which also lack a lipopolysaccharide layer and a defined periplasmic space.

Teichoic acids Although Gram-positive bacteria do not have a lipopolysaccharide outer layer, they generally do have acidic polysaccharides attached to their cell wall called **teichoic acids** (from the Greek word *teichos*, meaning "wall"). Teichoic acids contain repeating units of either glycerol or ribitol (both polyols). The *polyol units* are connected by phosphate esters and usually have other sugars and D-alanine attached (Figure 3.29). Because they are negatively charged, teichoic acids are partially responsible for the negative charge of the cell surface as a whole. Another function of certain teichoic acids is the part they play in the regulation of cell wall enlargement during growth and cell division. As discussed below, cell-wall growth involves enzymes, called *autolysins*, which open up the existing cell wall to make room for new subunits. It is crucial for the cell that autolysin activity be regulated, because if breakdown of existing cell walls were to occur before synthesis of new cell walls, large holes would develop in the cell wall and lysis (cell rupture) would occur. Tei-

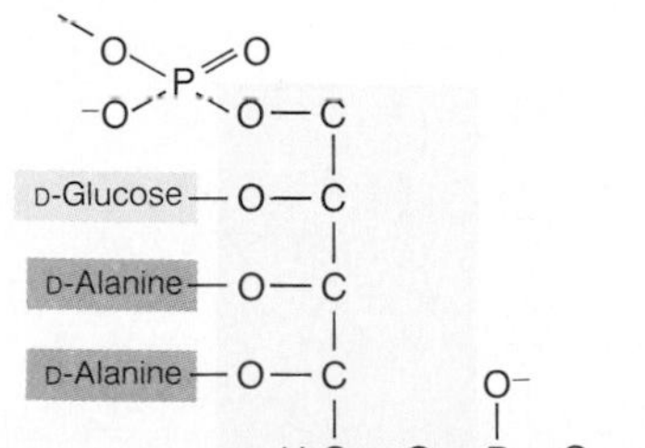

D-Glucose — O — C
D-Alanine — O — C
D-Alanine — O — C
H_2C — O — P — O
Ribitol

Figure 3.29 Structure of the ribitol teichoic acid of *Bacillus subtilis*. The teichoic acid is a polymer of the repeating ribitol units shown above.

choic acids have been shown to regulate autolysin action, thus keeping it in balance with cell-wall synthesis.

Certain glycerol-containing acids are bound to the membrane lipid of Gram-positive bacteria; because these teichoic acids are intimately associated with lipid, they have been called *lipoteichoic acids*. Despite the fact that these teichoic acids are bound to the membrane rather than to the cell wall, they seem to extend to the outside of the cell, as shown by the fact that agents (such as antibodies) which bind to these teichoic acids can combine with them even in whole cells.

Relation of cell-wall structure to the Gram stain Are the structural differences between the cell walls of Gram-positive and Gram-negative bacteria responsible in any way for the Gram-stain reaction? Recall that in the Gram reaction (see Section 1.13), an insoluble crystal violet-iodine complex is formed inside the cell and that this complex is extracted by alcohol from Gram-negative but not from Gram-positive bacteria. Gram-positive bacteria have very thick cell walls, which become dehydrated by the alcohol. This causes the pores in the walls to close, preventing the insoluble crystal violet-iodine complex from escaping. In Gram-negative bacteria, the solvent readily dissolves in and penetrates the outer layer, and the thin peptidoglycan layer also does not prevent solvent passage. However, a Gram reaction is not related directly to the bacterial cell-wall chemistry, since yeasts, which have a thick cell wall but of an entirely different chemical composition, also stain Gram-positively. Thus it is not the chemical constituents but the physical structure of the wall that confers Gram-positivity.

Osmosis, lysis, and protoplast formation In addition to conferring shape, the cell wall is essential in maintaining the integrity of the cell. Most bacterial environments have solute concentrations considerably lower than the solute concentration within the cell, which is approximately 10 millimolar (mM). Water passes from regions of low solute concentration to regions of high solute concentration in a process called **osmosis**. Thus there is a constant tendency throughout the life of the cell for water to enter, and the cell would swell and burst were it not for the strength of the cell wall. This can be dramatically illustrated by treating a suspension of bacteria with the enzyme **lysozyme**. Lysozyme is found in tears, saliva, other body fluids, and egg white. It hydrolyzes the cell-wall polysaccharide (Figure 3.25), thereby weakening the wall. Water then enters, the cell swells, and bursts, a process called **lysis** (Figure 3.30).

If the proper concentration of a solute which does not penetrate the cell, such as sucrose, is added to the medium, the solute concentration outside the cell balances that inside. Under these conditions, lysozyme still digests the cell wall, but lysis does not

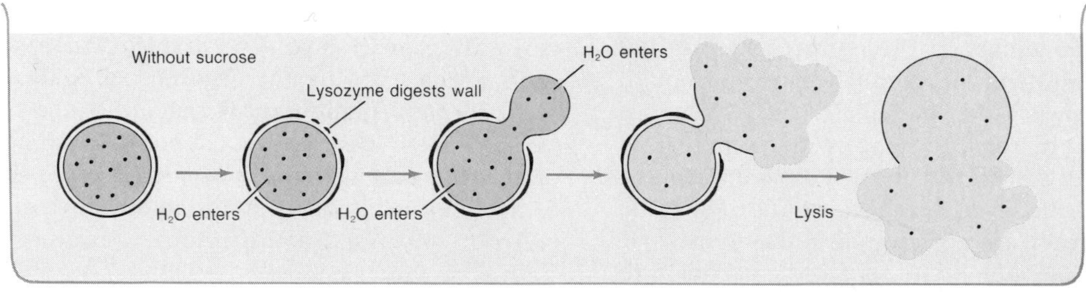

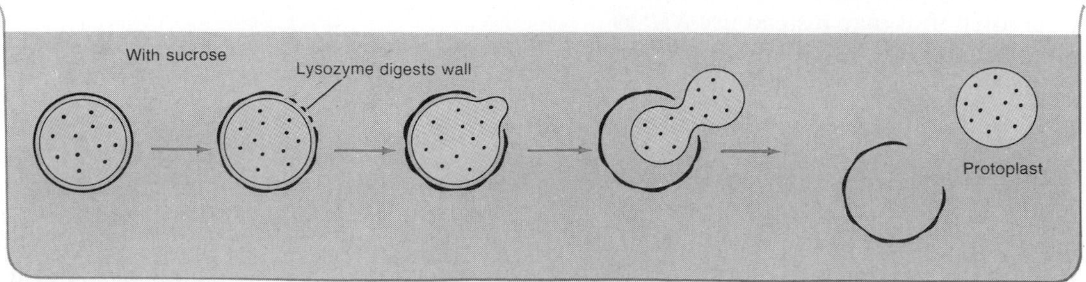

Figure 3.30 Lysis of protoplast in dilute solution and stabilization by sucrose.

occur, and an intact **protoplast** is formed (Figure 3.30). If such sucrose-stabilized protoplasts are placed in water, lysis occurs immediately.

Osmotically stabilized protoplasts are always spherical in liquid medium, even if derived from rod-shaped organisms, further emphasizing that the shape of the intact cell is conferred by the cell wall. However, the wall is not a completely inelastic structure, and is able to stretch and contract to some extent. Thus it might be viewed as somewhat like the skin of a football.

If the solute concentration is higher in the medium than in the cell, water flows out, the cell becomes dehydrated, and the protoplast collapses, a process called **plasmolysis**. This is one reason that foods can be protected from bacterial spoilage by curing them with strong salt or sugar solutions.

Although most bacteria cannot survive without their cell walls, a few organisms are able to do so; these are the mycoplasmas, a group of organisms that cause certain infectious diseases. Mycoplasmas are essentially free-living protoplasts, and are probably able to survive without cell walls either because they have unusually tough membranes or because they live in osmotically protected habitats, such as the animal body. These organisms will be discussed in Section 19.27.

Cell-wall synthesis and cell division When a cell enlarges during the division process, new cell-wall synthesis must take place, and this new wall material must be added in some way to the preexisting wall. This process can occur in different ways, as shown in Figure 3.31*a*. Small openings in the macromolecular structure of the wall are created by enzymes called **autolysins**, similar to lysozyme, that are produced within the cell. New wall material is then added across the openings. The junction between new and old peptidoglycan forms a ridge on the cell surface of Gram-positive bacteria (Figure 3.31*b*) analogous to a scar. Figure 3.31*b* also gives another example of incomplete separation of cells following division. In nongrowing cells, spontaneous lysis can occur, a result of cell enzymes hydrolyzing the cell wall without concomitant new cell-wall synthesis—a process called **autolysis**. As discussed below, the antibiotic *penicillin* inhibits the synthesis of new cell-wall material in growing cells and induces cell lysis.

Biosynthesis of peptidoglycan Although complex, the biosynthesis of peptidoglycan is of great interest and practical importance. Most information is known about the synthesis of the cell wall of *Staphylococcus aureus*, and the discussion here will be restricted to this organism.

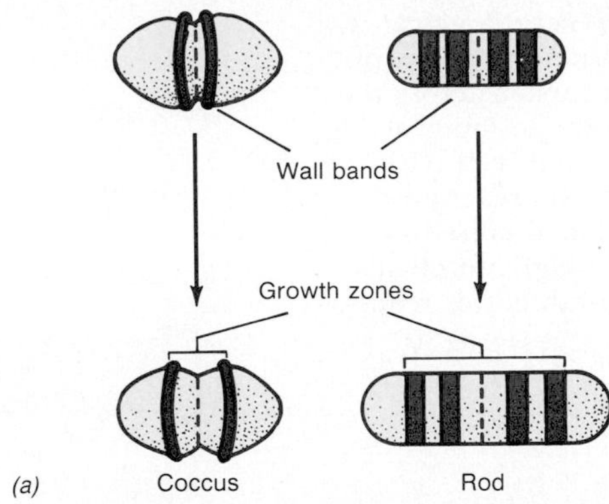

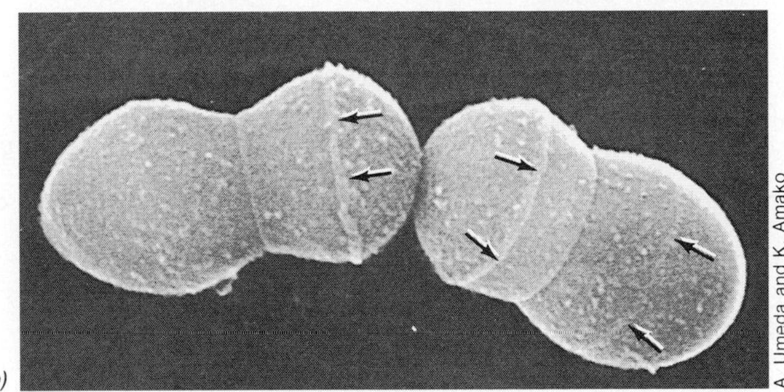

Figure 3.31 (a) Localization of new cell-wall synthesis during cell division. In cocci, new cell-wall synthesis is localized at only one point, whereas in rod-shaped bacteria it occurs at several locations along the cell wall. (b) Scanning electron micrograph of *Streptococcus hemolyticus* showing division bridges (arrows). Magnification, 29,800×.

Two carrier molecules participate in bacterial cell-wall synthesis, uridine diphosphate and a lipid carrier. The lipid carrier, called **bactoprenol** is a C_{55} isoprenoid alcohol which is connected via phosphodiester linkage to *N*-acetylmuramic acid (Figure 3.32). The second amino sugar of peptidoglycan, *N*-acetylglucosamine, is then added followed by addition of the pentaglycine bridge.

The assembly of polymers such as peptidoglycan outside the cell membrane presents special problems

$H_3C{-}C(CH_3){=}CHCH_2(CH_2C(CH_3){=}CHCH_2)_9CH_2C(CH_3){=}CHCH_2{-}O{-}P(O)(O^-){-}O{-}P(O)(O^-){-}O{-}$[*N*-Acetylmuramic acid]

Figure 3.32 Bactoprenol, the lipid carrier of the cell-wall peptidoglycan.

Figure 3.33 Peptidoglycan synthesis. (a) Transport of peptidoglycan precursors across the cell membrane to the growing point of the cell wall. (b) The transpeptidation reaction that leads to the final cross-linking of two peptidoglycan chains. Penicillin inhibits this reaction.

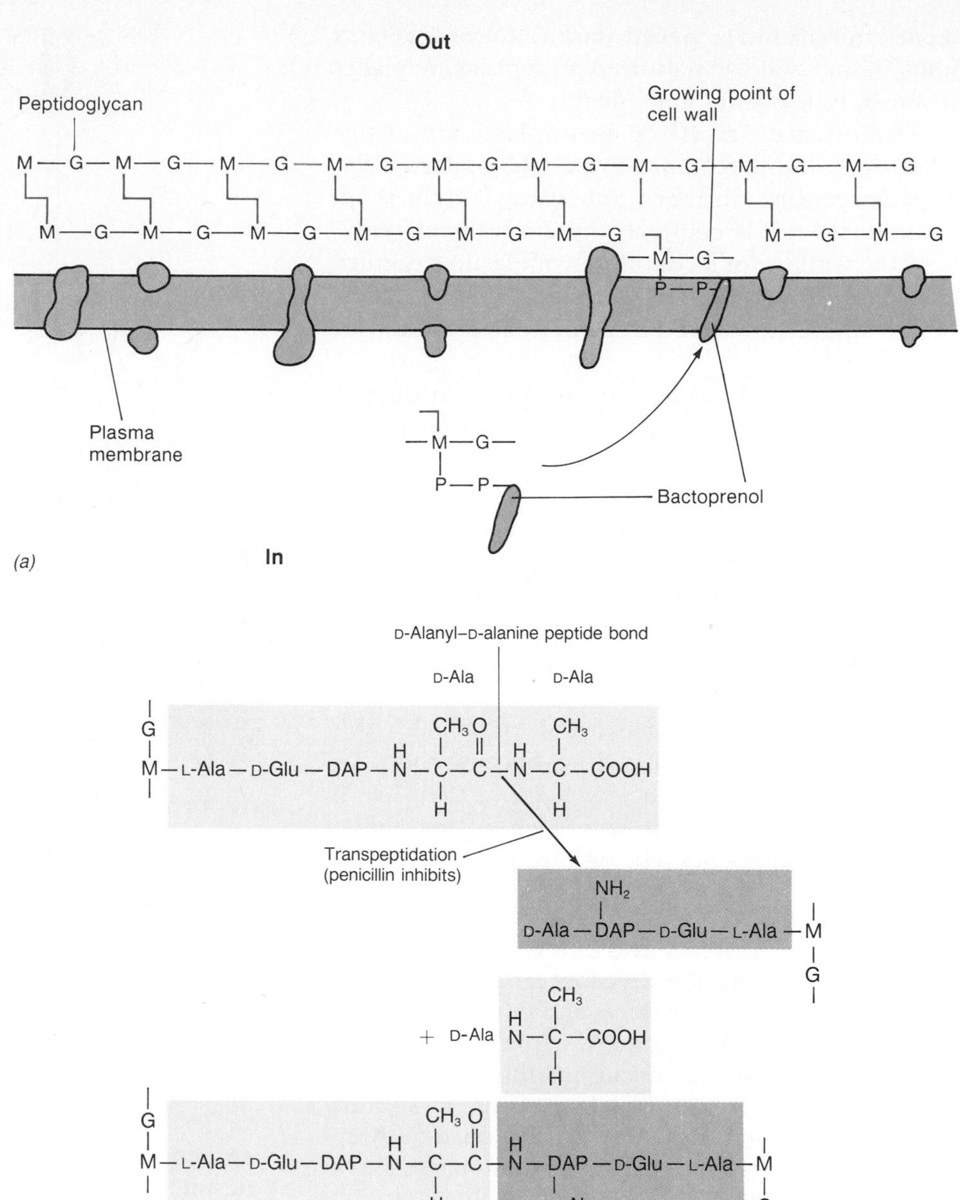

of transport and control. Bactoprenol is involved in transport of the peptidoglycan building blocks across the cell membrane, where the peptidoglycan is then inserted into a growing point of the cell wall (Figure 3.33). The function of bactoprenol is to render sugar intermediates sufficiently hydrophobic so that they will pass through the hydrophobic cell membrane. The lipid carrier inserts the disaccharide into the glycan backbone and then moves back inside the cell to pick up another peptidoglycan unit (Figure 3.33).

The final step in cell-wall synthesis is the formation of the peptide cross-links between adjacent glycan chains. The formation of the peptide cross-links involves an unusual type of peptide bond formation, called **transpeptidation**, which is also noteworthy because it is inhibited by the antibiotic *penicillin*. This cross-linking reaction involves peptide formation with one of several different amino acids, depending on the bacterium involved. In Gram-negative bacteria, such as *Escherichia coli*, the cross-linking is between diaminopimelic acid (DAP) on one peptide and D-alanine on an adjacent peptide (Figure 3.33*b*). Initially, there are *two* D-alanine groups at the end of the peptidoglycan precursor, but one D-alanine group is split off during the transpeptidation reaction. The peptide bond between the two molecules of D-alanine serves to *activate* the subterminal D-alanine, thereby favoring its reaction with the DAP. This reaction occurs outside the cell membrane, where energy is not available, and the transpeptidation reaction replaces the requirement for energy input. In *Staphylococcus aureus* the transpeptidation reaction occurs via the pentaglycine bridge.

Inhibition of transpeptidation by penicillin thus leads to the formation of peptidoglycan which lacks strength. The further damage to the cell, resulting in lysis and death, occurs through the action of **autolysins** that are involved in the opening up of the peptidoglycan structure as growth occurs. These enzymes continue to act, but because new peptidoglycan cross-links cannot occur, the cell wall becomes progressively weaker and osmotic lysis occurs. As with lysozyme, lysis by penicillin can be prevented by adding an osmotic stabilizing agent such as sucrose. Under such conditions, continued growth in the presence of penicillin leads to the formation of structures lacking rigid cell walls, called *protoplasts* or *spheroplasts*. Penicillin-induced lysis only occurs with growing cells. In nongrowing cells, action of autolysins does not occur, so that breakdown of the cell-wall peptidoglycan is prevented. It is fascinating to consider that one of the key developments in human medicine, the discovery of penicillin, is linked to a specific biochemical reaction involved in cell-wall synthesis, transpeptidation.

3.6 Cell Wall: Archaebacteria

We have already emphasized the profound differences between archaebacteria and eubacteria in the structure and composition of the cell membrane. Major differences also exist in cell wall structure. All eubacteria are either Gram-positive or Gram-negative. Although various species of archaebacteria stain Gram-positively or Gram-negatively, no archaebacterium has been shown to contain muramic acid and D-amino acids, the "signature molecules" of peptidoglycan. Instead, several chemically distinct cell wall types are present in various archaebacteria. Among methanogenic bacteria, a large and diverse group of strictly anaerobic procaryotes which make methane as a major product of cell metabolism (see Section 19.34), at least four wall types have been described. *Methanobacterium* species contain a peptidoglycan-like material referred to as *pseudopeptidoglycan*. Pseudopeptidoglycan is composed of alternating repeats of two amino sugars, *N*-acetylglucosamine and *N*-acetyltalosaminuronic acid (Figure 3.34). The latter compound is unique to these archaebacteria while the former makes up one-half of the backbone of eubacterial peptidoglycan (see Figure 3.25).

The acetyltalosaminuronic acid residues of pseudopeptidoglycan are cross-linked by amino acids (as is peptidoglycan), but the amino acids are all of the L form instead of the D-amino acids found in peptidoglycan. The linkage between amino sugars in pseudopeptidoglycan is chemically a (1→3) bond rather than the (1→4) bond between acetylglucosamine and muramic acid of peptidoglycan; since 1→4 intersugar linkages are the ones broken by the enzyme lysozyme, pseudopeptidoglycan-containing methanogens are resistant to the action of lysozyme (all other archaebacteria are lysozyme resistant as well because they lack true peptidoglycan).

Methanosarcina species contain a thick Gram-positive wall consisting exclusively of polysaccharide (see Section 19.34). The major sugars present are galactosamine, glucuronic acid, and glucose; a considerable amount of acetate is also present. Phosphate and sulfate are absent. Although the polysaccharide wall of *Methanosarcina* is complex, the major structural elements, galactosamine and glucuronic acid, suggest a relationship to chondroitin, a major component of connecting tissue of animal cells. Unlike chondroitin, however, the polysaccharide wall of *Methanosarcina* is not sulfated.

The cell-wall of *Halococcus*, an extremely halophilic archaebacterium, is a sulfated polysaccharide which otherwise resembles that of *Methanosarcina*.

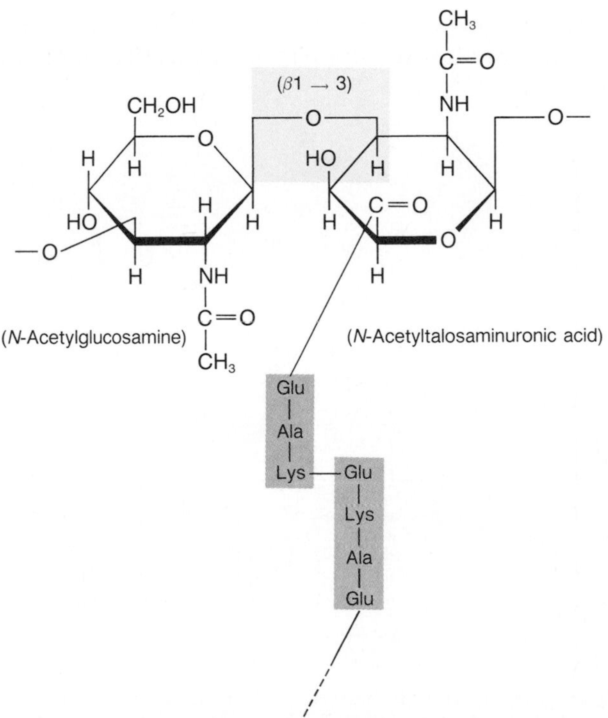

Figure 3.34 Structure of pseudopeptidoglycan, the cell wall polymer of *Methanobacterium* species. Note the resemblance to the structure of peptidoglycan in Figure 3.25.

The chemical composition of *Halococcus* cell walls is complex. Glucose, mannose, and galactose are present both as neutral sugars and in amino form. Uronic acids and acetate are abundant and most sugar residues are sulfated. A proposed structure for the *Halococcus* cell wall is shown in Figure 3.35.

Certain methanogens, the extreme halophile *Halobacterium*, and the sulfur-dependent archaebacteria produce glycoprotein cell walls. The carbohydrates present include hexoses such as glucose, glucosamine, galactose, and mannose, and pentoses such as ribose and arabinose. In *Halobacterium*, there is a great excess of acidic (negatively-charged) amino acids, especially aspartate, in the cell walls, and these probably serve to balance the abundance of positive charges contributed by the high concentrations of sodium in the organism's environment (see Section 19.13). Cells of *Halobacterium* require 20–25 percent NaCl to maintain an intact cell wall. As the salt concentration is reduced to about 15 percent NaCl, *Halobacterium* cells (which are normally rod-shaped) begin to form spherical structures, presumably due to unstable cell wall components. At lower NaCl concentrations the cells actually lyse.

A few methanogens lack carbohydrate in their cell walls and instead have walls of pure protein. For example, *Methanococcus* and *Methanomicrobium* cell walls consist of several distinct proteins, whereas *Methanospirillum* (see Section 19.34) contains a protein subunit wall which forms a sheath around the organism.

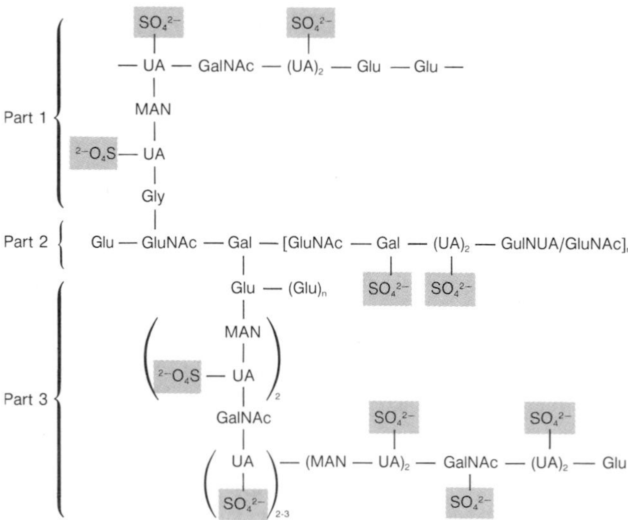

Figure 3.35 Cell-wall structure of *Halococcus*, an extremely halophilic archaebacterium. The wall basically consists of a repeating three-part pattern. Note the sulfate groups. Abbreviations: UA, uronic acid; Glu, glucose; Gal, galactose; GluNAc, N-acetylglucosamine; GalNAc, N-acetylgalactosamine; Gly, glycine; GulNUA, N-acetylgulosaminuronic acid; MAN, mannose.

The thermoacidophilic archaebacterium *Thermoplasma* contains no cell wall at all, but does contain unique tetraether membrane lipids (see Section 3.6 and 18.6) and glycoprotein in its cell membrane. The membrane of this organism must be quite stable in order to withstand not only osmotic forces but also the hot acidic conditions of its environment; *Thermoplasma* grows optimally at 60°C and pH 2 (see Section 19.36).

3.7 Cell Wall: Eucaryotes

Animal cells do not contain cell walls but plant and algal cells contain structurally rigid cell walls of variable chemical composition. Plant and algal cell walls consist of an interwoven network of cellulose fibers forming a strong amorphous matrix that surrounds the entire cell (Figure 3.36). **Cellulose** is a linear polymer of glucose connected by $\beta(1{\rightarrow}4)$ linkages (see Section 2.4). The wall matrix also contains fibers of two other polysaccharides—hemicellulose and pectin. **Hemicellulose** is a branched polysaccharide containing glucose and other sugars. Hemicellulose fibrils form hydrogen bonds with cellulose fibers in the cell wall matrix. **Pectins** are highly hydrated heterogeneous polysaccharides containing galacturonic acid as the major carbohydrate and the hexose rhamnose in lesser amounts. Pectins associate with hem-

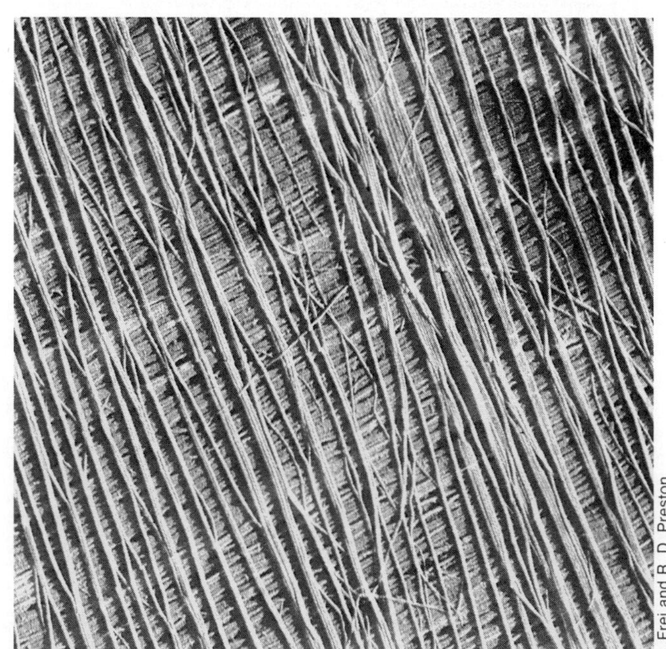

Figure 3.36 Electron micrograph of the cellulosic cell wall of the filamentous green alga *Chaetomorpha melagonium*. Note the regular arrangement of the cellulose microfibrils in alternating layers.

icelluloses and cellulose and, because of their high water content, tend to contribute a gel-like consistency to the rigid cellulose/hemicellulose matrix.

Algae show considerable diversity in the structure and chemistry of their cell walls. In many cases the cell wall is composed primarily of cellulose (Figure 3.36; Figure 3.37*a*), but it is usually modified by the addition of other polysaccharides such as pectin (see above), xylans, mannans, alginic acids, fucinic acid, and so on. In some eucaryotic microorganisms the wall is strengthened by the deposition of calcium carbonate; these are often called calcareous or coralline ("coral-like") algae (Figure 3.37*b*). Sometimes chitin is also present in the cell wall.

In diatoms the cell wall is composed of silica (Figure 3.37*c*), to which protein and polysaccharide are added. Even after the diatom dies and the organic materials have disappeared, the external structure (frustule) remains, showing that the siliceous component is indeed responsible for the rigidity of the cell. Because of the extreme resistance of diatom frustules to decay, they remain intact for long periods of time and constitute some of the best algal fossils. Diatomaceous earth, an industrial filter aid material, is composed of diatom fossil cells.

Plant and algal cell walls are freely permeable to low molecular weight constituents such as water, ions, gases and other nutrients. Their cell walls are essentially impermeable, on the other hand, to larger molecules or macromolecules. Plant cell walls contain pores 3–5 nm wide which are sufficiently small to only pass molecules of a molecular weight of less than about 15,000. Thus, although animal cells can eat particles, a process called **phagocytosis**, phagocytic activities are impossible in plant cells; particles large enough to be phagocytized never reach the plasma membrane because they are unable to penetrate the cell wall. Thus, although providing immense structural strength, the plant cell wall does dictate certain aspects of plant physiology: plants must subsist on molecules of low molecular weight, and any molecules used as intercellular communicators must also be small. Indeed, it has been shown that most plant hormones—auxins, gibberellins, and cytokinins, for example—are all low molecular weight molecules.

Fungal cells also contain rigid cell walls. Fungal cell walls resemble plant cell walls architecturally, but not chemically. Although cellulose is present in the walls of certain fungi, many fungi have noncellulosic walls. **Chitin**, a polymer of *N*-acetylglucosamine, is a common constituent of fungal cell walls. It is laid down in microfibrillar bundles as for cellulose; other glucans such as mannans, galactosans, and chitosans replace chitin in some fungal cell walls. Fungal cell walls are generally 80–90 percent polysaccharide, with proteins, lipids, polyphosphates and inorganic ions making up the wall-cementing matrix. An understanding of fungal cell wall chemistry is important because of the extensive biotechnological uses of fungi (see Chapter 10) and because the chemical nature of the fungal cell wall has been useful in classifying fungi for research and industrial purposes.

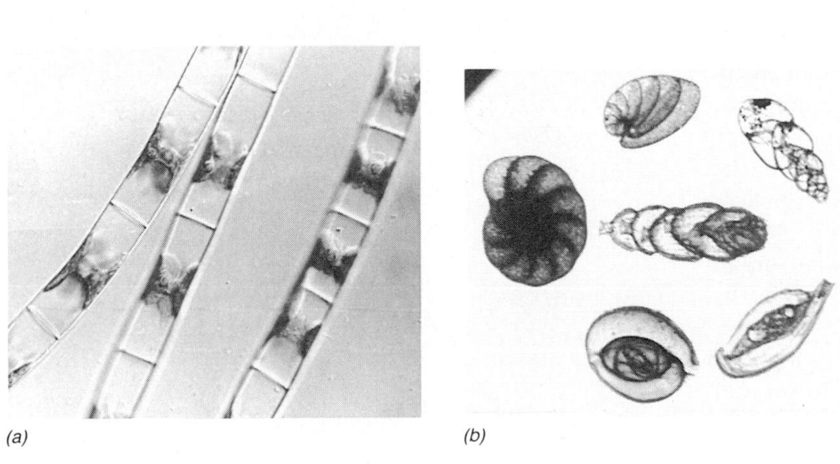

(a) (b)

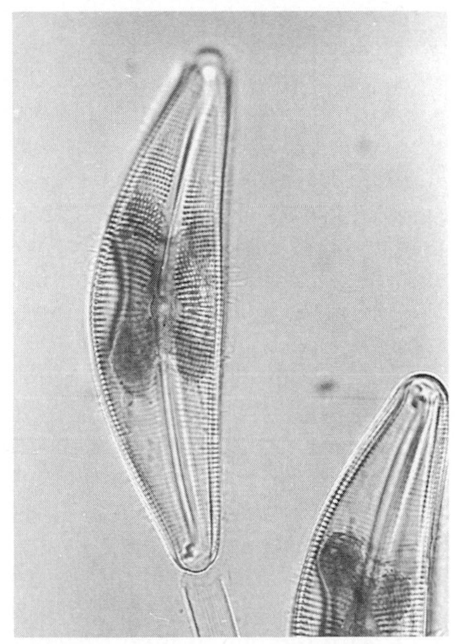

(c)

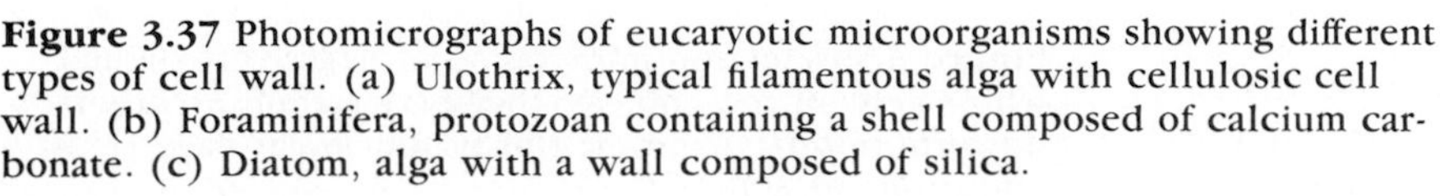

Figure 3.37 Photomicrographs of eucaryotic microorganisms showing different types of cell wall. (a) Ulothrix, typical filamentous alga with cellulosic cell wall. (b) Foraminifera, protozoan containing a shell composed of calcium carbonate. (c) Diatom, alga with a wall composed of silica.

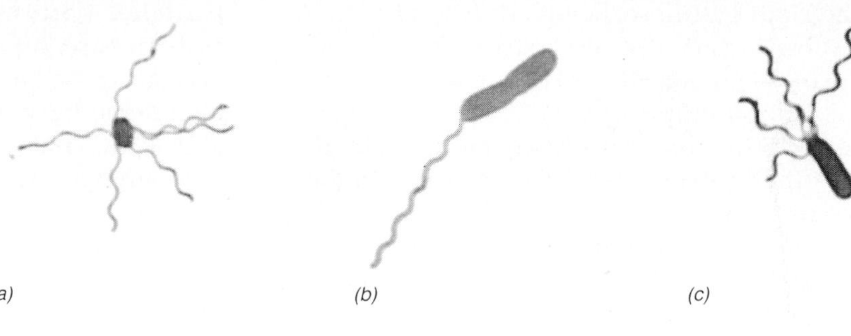

Figure 3.38 Light photomicrographs of bacteria stained with the Leifson flagella stain. (a) Peritrichous. (b) Polar. (c) Lophotrichous.

3.8 Flagella and Motility

Many bacteria are motile, and this ability to move independently is usually due to a special organelle of motility, the **flagellum** (plural, **flagella**). Certain bacterial cells can move along solid surfaces by *gliding* (see Section 19.13) and certain aquatic microorganisms can regulate their position in the water column by inflating or deflating small gas filled structures called gas vesicles (see Section 3.11). However, the majority of motile procaryotes move by means of flagella. Many eucaryotic microorganisms exhibit motility through the activity of **flagella** and **cilia**. Cilia resemble eucaryotic flagella in many ways, but are shorter and more numerous. In cell wall-less eucaryotes (for example amoebas and slime molds), movement occurs by a continual process of extending and retracting portions of the cells referred to as *pseudopodia*, into which cytoplasm streams.

In all cases of motility in microorganisms, the process results in the cell being able to reach different regions of its microenvironment. In the struggle for survival, movement to a new location may mean the difference between survival and death of the cell. But, as in any physical process, cell movement is closely tied to an energy expenditure. We begin now with a detailed consideration of flagellar motility in procaryotes, and proceed to compare the action of flagella in bacteria with their activity as locomotive organelles in eucaryotes.

Bacterial flagella Bacterial flagella are long, thin appendages free at one end and attached to the cell at the other end. They are so thin (about 20 nm) that a single flagellum can never be seen directly with the light microscope, but only after staining with special flagella stains (Figure 3.38). Flagella are also readily seen with the electron microscope by negative staining (Figure 3.39).

Flagella are arranged differently on different bacteria. In **polar flagellation** the flagella are attached at one or both ends of the cell. Occasionally a tuft of flagella may arise at one end of the cell, an arrangement called "lophotrichous" (lopho- means "tuft"; *trichous* means "hair"). Tufts of flagella of this type can often be seen in the living state by dark-field microscopy (Figure 3.40). In **peritrichous flagellation** the flagella are inserted at many places around the cell surface (*peri* means "around"). The type of flagellation is often used as a characteristic in the classification of bacteria (see Section 18.9).

Flagellar structure Flagella are not straight but helically shaped; when flattened, they show a constant length between two adjacent curves, called the *wavelength*, and this wavelength is constant for each species. Bacterial flagella are composed of protein subunits; the protein is called **flagellin**. The shape and wavelength of the flagellum are determined by the structure of the flagellin protein, and a change in the structure of the flagellin can lead to a change in the morphology of the flagellum.

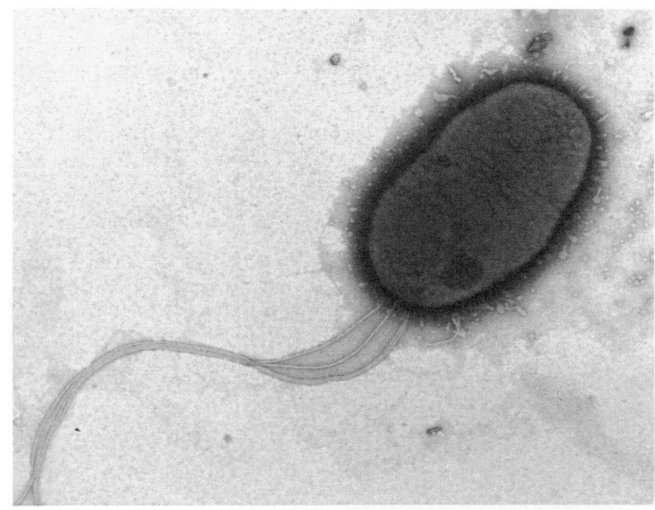

Figure 3.39 Electron micrograph of negatively stained whole cells of the photosynthetic bacterium *Ectothiorhodospira mobilis*. Note the tuft of four polar flagella. Magnification, 30,000×.

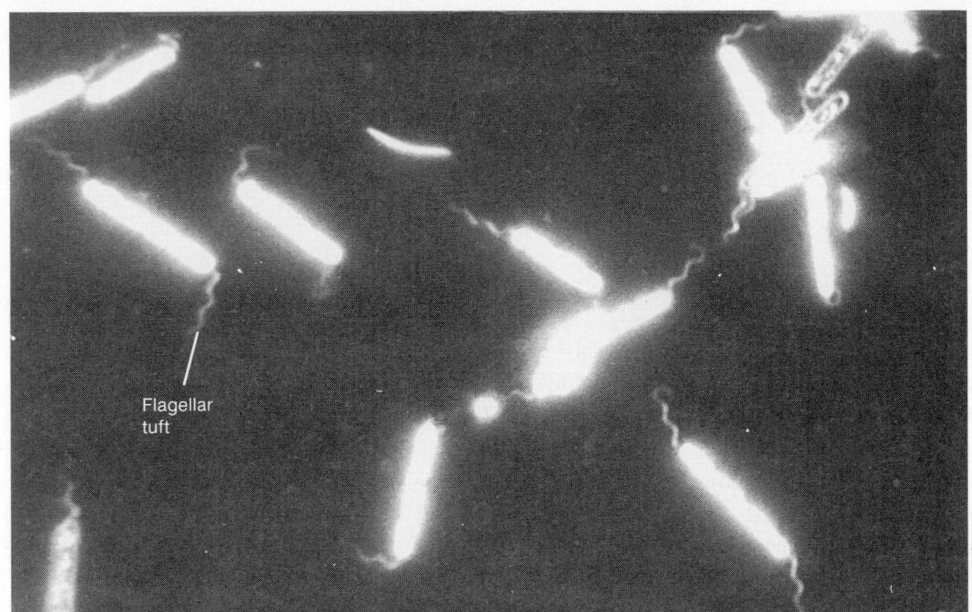

Figure 3.40 Dark-field photomicrograph of a group of large rod-shaped bacteria with flagellar tufts at each pole. Magnification, 1900×.

The basal region of the flagellum is different in structure from the rest of the flagellum. There is a wider region at the base of the flagellum called the *hook* (Figure 3.41). Attached to the hook is the **basal body**, a complex structure involved in the connection of the flagellar apparatus to the cell envelope. The hook and basal body are composed of proteins different from those of the flagellum itself. The basal body consists of a small central rod which passes through a system of rings. In Gram-negative bacteria, the outer pair of rings is associated with the lipopolysaccharide and peptidoglycan layers of the cell wall, and the inner pair of rings is located within or just above the plasma membrane (Figure 3.41). In Gram-positive bacteria, which lack the outer lipopolysaccharide layer, only the inner pair of rings is present.

Flagellar growth The individual flagellum grows not from the base, as does an animal hair, but from the tip. Flagellin molecules formed in the cell apparently pass up through the hollow core of the flagellum and add on at the terminal end. The synthesis of a flagellum from its flagellin protein molecules occurs by a process called ***self-assembly***: all of the information for the final structure of the flagellum resides in the protein subunits themselves. Growth of the flagellum occurs more or less continuously un-

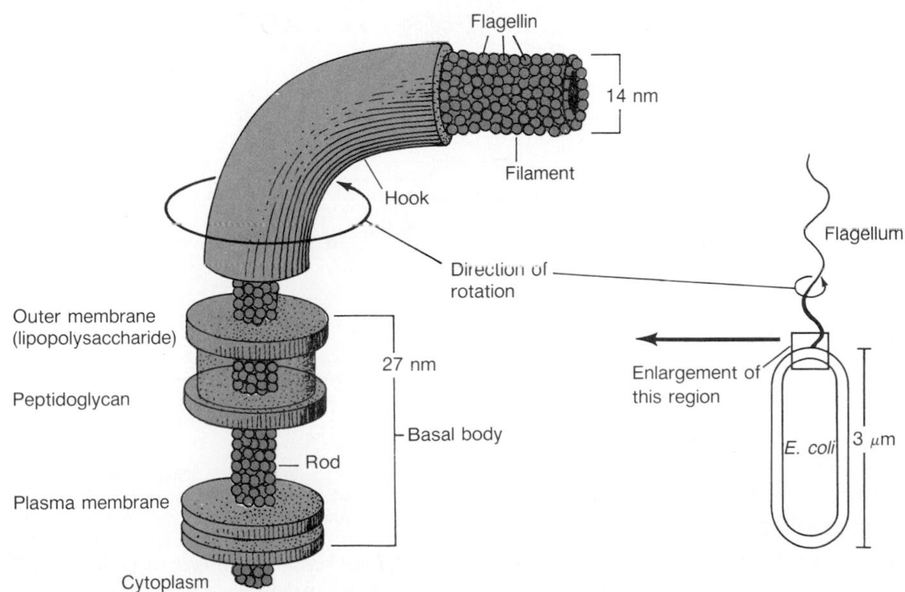

Figure 3.41 An interpretive drawing of the probable manner of attachment of the flagellum in a Gram-negative bacterium.

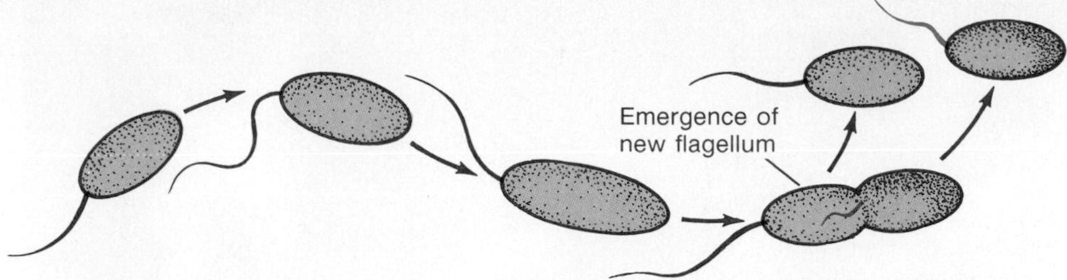

Figure 3.42 Synthesis of flagella during cell division of a polarly-flagellated bacterium.

til a maximum length is reached; however, if a portion of the tip is broken off, it is regenerated.

When a cell divides, the two daughter cells must acquire in some way a full complement of flagella. In polarly flagellated organisms, the process of cell division probably occurs as shown in Figure 3.42, the new flagellum forming at the location where cell division has just occurred. In a monopolarly flagellated cell, the two poles of the cell probably differ in some way so that the flagellum is formed at one pole and not at the other. In peritrichously flagellated organisms, the relation of cell division to flagella synthesis probably occurs by preexisting flagella being distributed equally between the two daughter cells, and new flagella being synthesized and filling in the gaps.

Flagellar movement How is motion imparted to the flagellum? Each individual flagellum is actually a rigid structure, which does not flex but moves by rotation, in the manner of a propeller. Some evidence for this conclusion was obtained by observing the behavior of cells that were tethered by their flagella to microscope slides. It was observed that such cells rotated around the point of attachment, at rates of revolution consistent with those inferred for flagellar movement in free-swimming cells. In another experiment, a flagellum was visualized by coating it with small latex beads. When the cells were attached to slides, the latex beads could be observed to rotate very rapidly.

It seems likely that the rotary motion of the flagellum is imparted from the basal body, which must act in some way like a motor. It is likely that the two inner rings located at the membrane rotate in relation to each other, so that (to extend the motor analogy) one of the rings could be considered to be the rotor, the other the stator.

The motions of polar and lophotrichous organisms are different from those of peritrichous organisms. Peritrichously flagellated organisms generally move and rotate in a straight line in a slow, stately fashion. Polar organisms, on the other hand, move more rapidly, spinning around and dashing from place to place. The different behavior of flagella on polar and peritrichous organisms is illustrated in Figure 3.43.

The average velocities of several bacteria are given in Table 3.3 and it can be seen that rates vary from about 20 μm to 80 μm/s. A speed of 50 μm/s is equivalent to 0.0001 mile/h, which seems slow. However, it is more reasonable to compare velocities in terms of number of cell lengths moved per second. The fastest animal, the cheetah, is about 4 ft long and moves at a maximum rate of 70 miles/h, or about 25 lengths per second. It can be seen from Table 3.3 that rates of bacterial movement, often around 10 or more lengths per second, are about as significant as those of higher organisms.

Eucaryotic flagella Eucaryotic flagella are long filamentous structures that are attached to one end of the cell and move in a whiplike manner to impart motion to the cell. Some organisms have only a single flagellum; others have more than one. The number and arrangement of flagella are important characteristics in classifying various groups of algae, fungi, and protozoa.

It should be emphasized that flagella of eucaryotes are quite different in structure from those of pro-

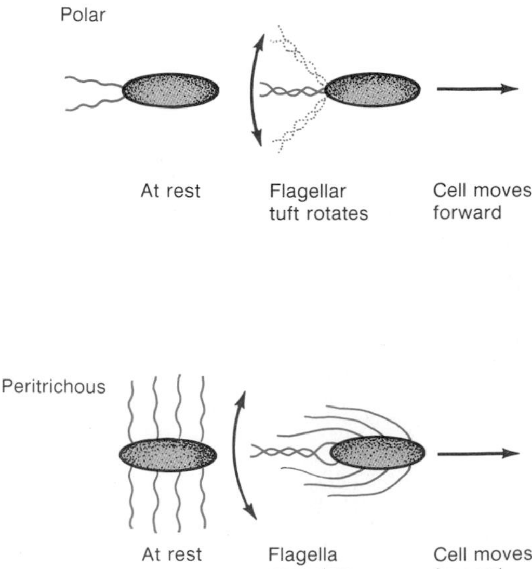

Figure 3.43 Manner of flagellar movement in polarly and peritrichously flagellated organisms.

Table 3.3 Velocities of several common bacteria

Organism	Flagellation	Cell length (μm)	Velocity (μm/s)	Lengths moved per second
Pseudomonas aeruginosa	Polar	1.5	55.8	37
Chromatium okenii	Lophotrichous	10	45.9	5
Thiospirillum jenense	Lophotrichous	35	86.5	2
Escherichia coli	Peritrichous	2	16.5	8
Bacillus licheniformis	Peritrichous	3	21.4	7
Sarcina ureae	Peritrichous	4	28.1	7

From Vaituzis, Z., and R.N. Doetsch. 1969. Appl. Microbiol. 17:584.

caryotes, even though the same word is used for both. Eucaryotic flagella are much more complex in structure. Flagella of eucaryotes contain two central fibers surrounded by nine peripheral fibers, each of which is doublet in nature (Figure 3.44). The whole unit is surrounded by a membrane and the fibers are embedded in an organic matrix. Each fiber is in reality a tube, called a **microtubule**, which is composed of a large number of protein molecules called **tubulin** (Figure 3.44*c*). Each microtubule is about the same diameter as a procaryotic flagellum, but they are composed of different proteins. Tubulin is a protein of 5000 molecular weight composed of two subunits. These subunits are arranged in an alternating, helical fashion up the microtubule axis (Figure 3.44*d*). When observed in cross section, each tubulin microtubule appears to consist of a helix composed of exactly 13 rows of tubulin molecules (Figure 3.44*c*).

Attached to each of the outer doublet fibers are molecules of another protein, **dynein** (Figure 3.44*c*). It is this protein that is involved in the conversion of chemical energy (from adenosine triphosphate, ATP) into the mechanical energy of flagellar movement. It is thought that movement is imparted to the flagellum by coordinated sliding of the fibers toward or away from the base of the cell, in a way analogous to that in which muscle filaments slide during muscular contraction. The manner in which the flagellum moves determines the direction of cellular movement, as is illustrated in Figure 3.45. The maximum rate of movement of flagellated eucaryotic microorganisms ranges from 30 to 250 μm/s.

A clear distinction can thus be made between the action of flagella in procaryotic and eucaryotic cells. The procaryotic flagellum is a relatively simple structure firmly embedded in the cell wall and membrane, and functions as a propeller by rotating either clockwise or counterclockwise. The eucaryotic flagellum, on the other hand, is structurally complex and drives the cell forward by a whiplike motion due to the sliding of microtubules. In addition, while procaryotic flagella rotate at the expense of the proton electrochemical gradient, in eucaryotes microtubule sliding (and ultimately flagellar movement) is driven by ATP. To sum, although procaryotic and eucaryotic flagella result in the same action, that is, cell movement, they are structurally and functionally distinct.

Cilia **Cilia** (singular, **cilium**) are similar to eucaryotic flagella in fine structure but differ in being shorter and more numerous (Figure 3.46). In microorganisms, cilia are found primarily in one group of protozoa, appropriately called the ciliates. (Cilia are widespread among the cells of higher animals, however.) A single protozoan cell of a *Paramecium* species has between 10,000 and 14,000 cilia. These organelles are fairly rigid structures and beat about 10 to 30 strokes per second. The rate of movement of ciliates varies from 300 to 2500 μm/s for different species. Thus ciliated organisms move much more rapidly than flagellated ones.

Cytoplasmic streaming and amoeboid movement **Cytoplasmic streaming** is readily detected with the light microscope in the larger eucaryotic cells by observing the behavior of various cytoplasmic particles; they are seen to move all together in a definite pattern, suggesting that they are being carried passively by the movement of the cytoplasm. Rates of cytoplasmic streaming vary greatly, depending on the organism and environmental conditions; values from 2 μm/s to greater than 1000 μm/s have been recorded.

In cells without walls (for example, amoebas, slime molds), cytoplasmic streaming can result in **amoeboid movement** (Figure 3.47), so called because it is the movement characteristic of the amoebas. During movement, a temporary projection of the protoplast called a **pseudopodium** develops, into which cytoplasm streams. Cytoplasm flows forward because the tip of the pseudopodium is less contracted and hence less viscous, while the rear is contracted and viscous; thus cytoplasm takes the path of least resistance. Amoeboid motion requires a solid surface along which the protoplasm can move.

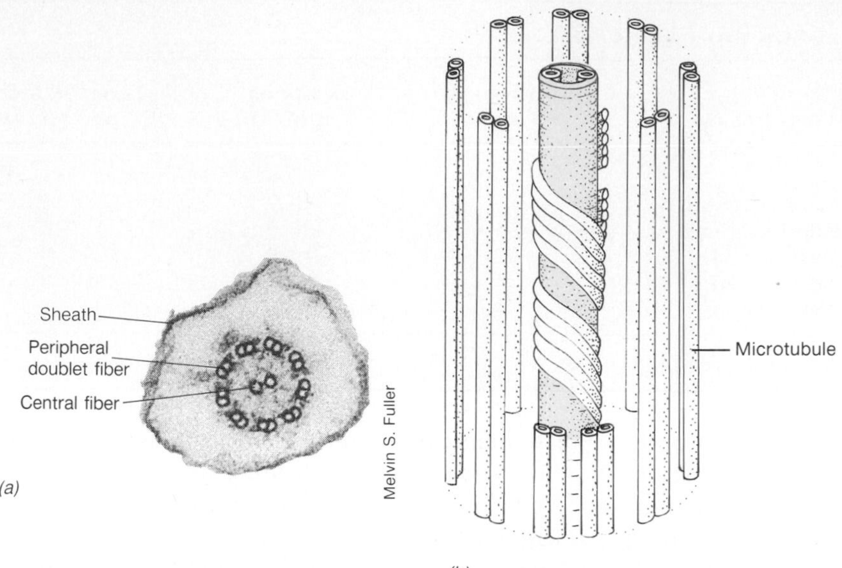

Figure 3.44 (a) Electron micrograph of a cross section of the flagellum of the zoospore of the fungus *Blastocladiella emersonii* showing the outer sheath, the outer nine fibers, and the central pair of single fibers. Magnification, 132,000×. (b) Interpretive diagram of the arrangement of microtubules in a eucaryotic flagellum. The outer sheath is not shown. (c) Cross section showing construction of the outer fiber. Each microtubule is about 20 nm in diameter. (d) The tubulin proteins of microtubules are actually made up of two subunits, and they are arranged in a helical fashion up the microtubule axis, as illustrated.

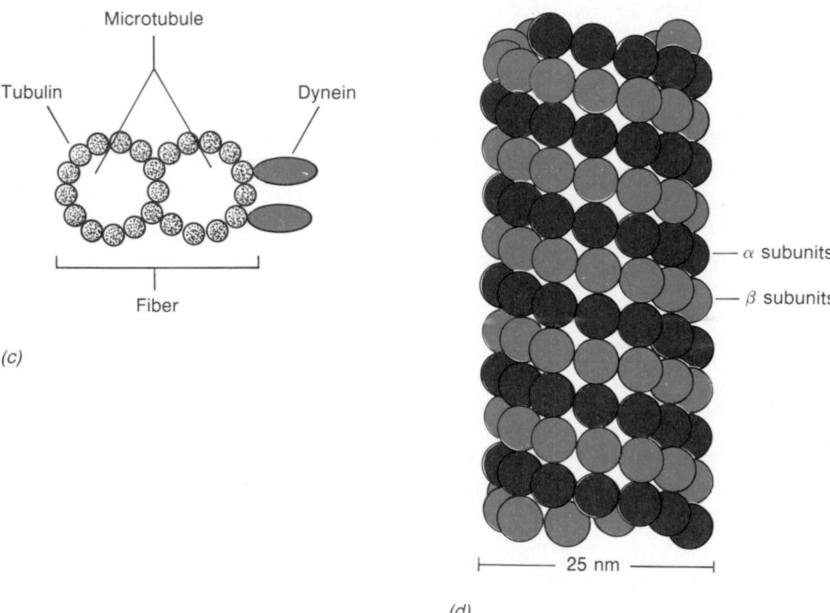

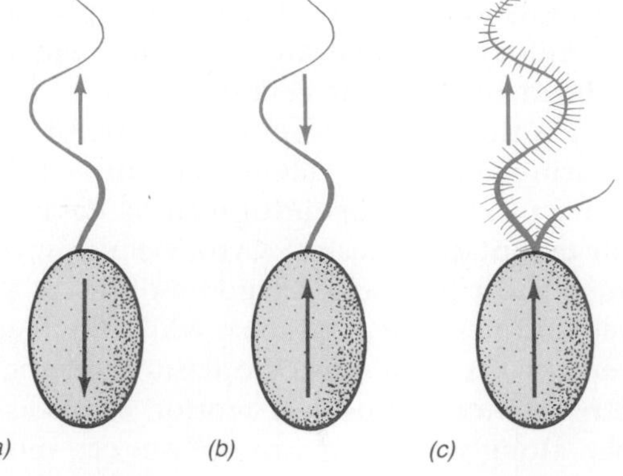

Figure 3.45 Manner in which flagella of different types impart motility to eucaryotic cells. (a) Wave directed away from cell pushes cell in opposite direction; type of movement found in dinoflagellates and animal spermatozoa. (b) Wave directed toward the cell pulls the cell in the direction opposite to the wave; type of movement found in trypanosomes and other flagellated protozoa. (c) Flagellum contains stiff lateral projections called mastigonemes. Wave directed away from cell pulls the cell in the same direction as the wave; type of movement found in chrysophytes.

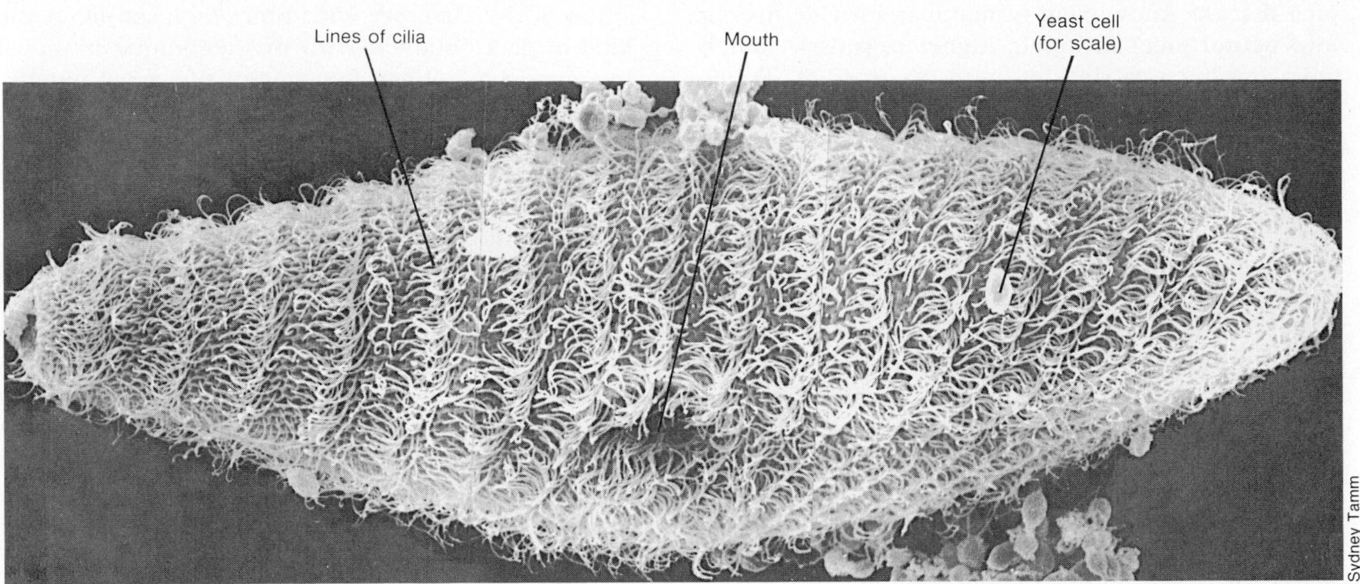

Figure 3.46 Scanning electron micrograph of the ciliate *Paramecium*. Note the large number and orderly arrangement of the cilia. The animal is shown swimming forward (to the right), and the ciliary wave travels from posterior (left) to anterior. Magnification, 860×.

The mechanism of cytoplasmic movement is different from that of flagellar movement. Microtubules are not involved in cytoplasmic movement; however, filaments of the protein actin have been shown to play a part in movement. **Actin** is found as a thin layer of filaments just beneath the plasma membrane of all eucaryotic cells. These actin filaments are arranged in tight bundles to yield an actin "gel." The hydrolysis of ATP in some way drives the movement of actin filaments against one another in the gel and in so doing drives amoeboid movement. Cytoplasmic streaming and amoeboid movement are quickly arrested when cells are treated with the actin binding drug phalloidin; this is strong evidence for the involvement of actin in these motility processes.

3.9 Chemotaxis in Bacteria

Chemotaxis is the movement of an organism toward or away from a chemical. Positive chemotaxis refers to movement *toward* a chemical and is usually exhibited when the chemical is of some benefit to the cell (for example, a nutrient). Negative chemotaxis is movement *away* from a chemical, usually one that is harmful. Chemicals that induce positive chemotaxis are called **attractants**, and chemicals that induce negative chemotaxis are called **repellents**. Although both procaryotes and eucaryotes exhibit chemotactic responses, the phenomenon has been studied most extensively in motile bacteria.

Chemotaxis is a behavioral phenomenon and suggests some kind of nervous response. There has been considerable interest in studying behavioral mechanisms in such simple organisms as bacteria, with the

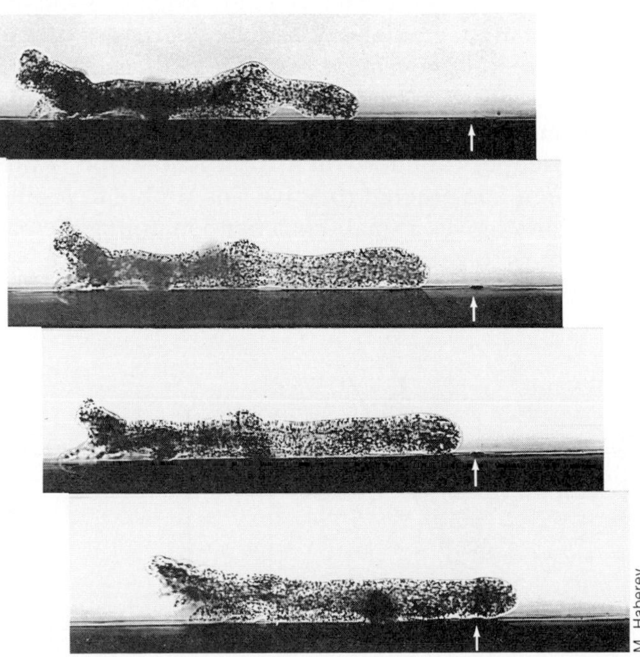

Figure 3.47 Side view of a moving amoeba, *Amoeba proteus*, taken from a film, the time interval between frames being 2 seconds. The arrows point to a fixed spot on the surface. Magnification, 76×.

idea that the knowledge gained may provide insight into neural mechanisms in higher organisms. Much work has been carried out with two species, *Escherichia coli* and *Salmonella typhimurium*, and the following discussion deals only with these two organisms.

Bacterial chemotaxis can be most easily demonstrated by immersing a small glass capillary containing an attractant into a suspension of motile bacteria which does not contain the attractant. From the tip of the capillary, a chemical gradient is set up into the surrounding medium, with the concentration of chemical gradually decreasing with distance from the tip. If the capillary contains an attractant, the bacteria will move toward the capillary, forming a swarm around the open tip (Figure 3.48); and, subsequently, many of the motile bacteria will move into the capillary. Of course, some bacteria will move into the capillary even if it contains a solution of the same composition as the medium, because of random movements; but if an attractant is within, the concentration of bacteria within the capillary can be many times higher than the external concentration. On the other hand, if the capillary contains a repellent, the concentration of bacteria within the capillary will be considerably less than the concentration outside. The problem is to determine how these tiny cells are able to "sense" the chemical gradient and move toward or away from it.

To explain chemotaxis, we must first consider the behavior of a single bacterial cell when it is moving in a chemical gradient. When viewed under the microscope, the movement of single cells is extremely hard to follow with the eye, because it is so rapid. Two methods are employed. The first uses a special microscope, called a **tracking microscope**, which has been constructed to solve this problem. It automatically moves a small chamber containing the bacteria so as to keep a particular cell fixed in space, and a record is generated in three dimensions of the position of the chamber with time. An example of the kind of data obtained with this technique is shown in Figure 3.49. When movements of a large number of cells are analyzed with the tracking microscope, it can be concluded that bacterial behavior can be divided into two actions, called **runs** and **twiddles**. When the organism runs, it swims steadily in a gently curved path. Then it twiddles, that is, stops and jiggles in place. Then it runs off again in a new direction. A twiddle is a random event, but occurs on the average of about once per second and lasts about a tenth of a second. Following the twiddle, the direction for the next run is almost random. Thus, by means of runs and twiddles, the organism moves randomly but does not go anywhere. Now, if a chemical gradient of an attractant is present, the random movements become biased. As the organism experiences higher concentrations of the attractant, the twiddles are less frequent, and the runs are longer (Figure 3.49*b*). The net result of this situation is that the organism moves up the concentration gradient by increasing the length of runs that take it in a direction favoring a higher concentration of the attractant.

In the second microscopic observation method, a microscopic field is photographed continuously for a short period of time (usually less than 1 second) while illumination is provided intermittently via a rapidly flashing (stroboscopic) light. If a cell is moving during the period of photographic exposure, it will be photographed in several different places, giving the appearance of a dashed line (Figure 3.50*a*), a so-called *motility track*. On the other hand, if a cell is twiddling during the exposure, it will be photographed several times in basically the same place, giving the appearance of a large dot (Figure 3.50*b*). It is these motions which combine to cause the net motion described in Figure 3.50.

It has been well established that bacteria do not, in the ordinary sense, respond to the *spatial* gradient itself, but instead respond to the *temporal* gradient

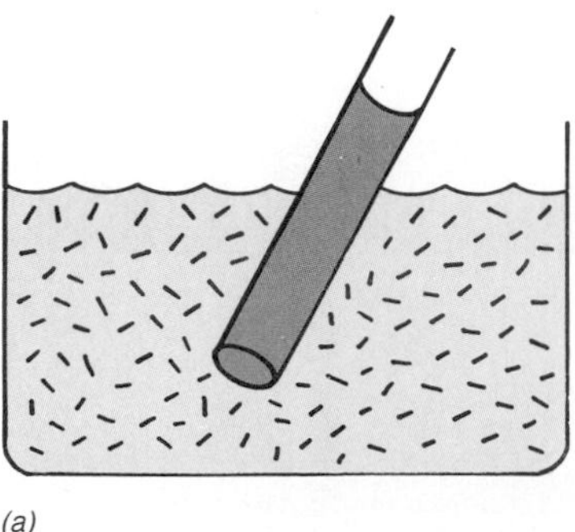
(a)

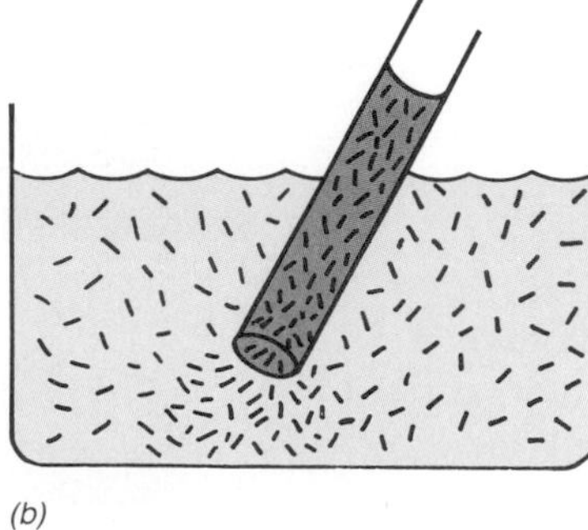
(b)

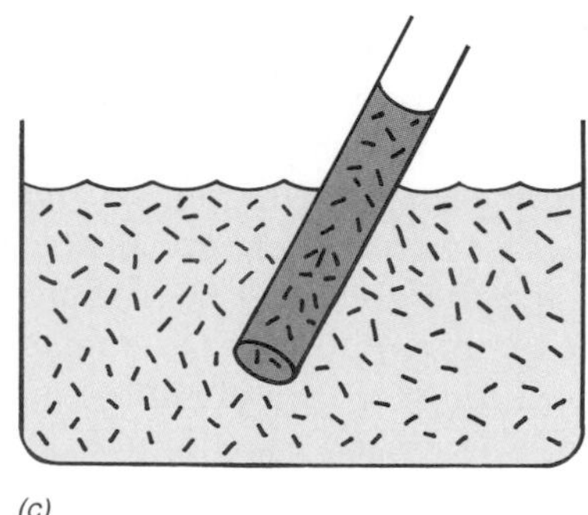
(c)

Figure 3.48 Capillary technique for studying chemotaxis in bacteria. (a) Insertion of capillary into a bacterial suspension. (b) Accumulation of bacteria in a capillary containing an attractant. (c) Control, capillary contains a salt solution which is neither an attractant nor a repellent.

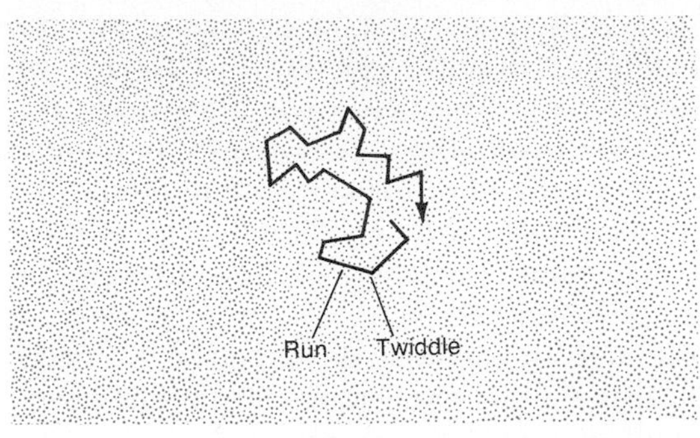

(a)

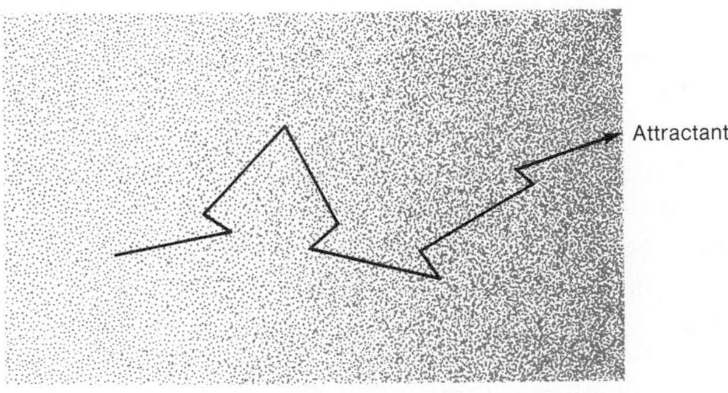

(b)

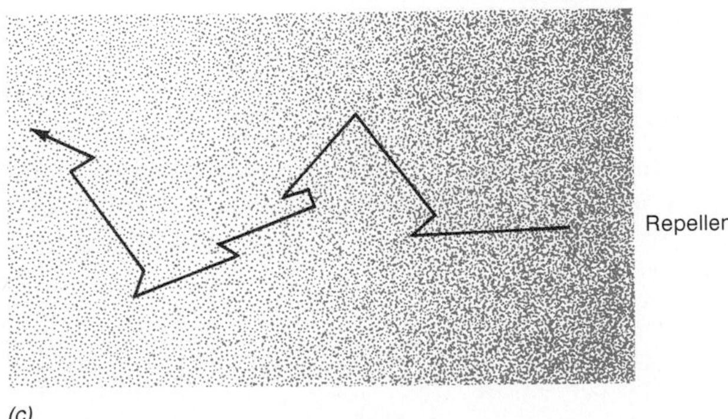

(c)

Figure 3.49 Diagrammatic representation of *Escherichia coli* movement, as analyzed with the tracking microscope. These drawings are two-dimensional projections of the three-dimensional movement. (a) Random movement of a cell in a uniform chemical field. Each run is followed by a twiddle, and the twiddles occur fairly frequently. (b) Directed movement toward a chemical attractant. The runs still go off in random directions, but when the run is up the chemical gradient, the twiddles occur less frequently, and the runs are longer. When runs occur away from the chemical, twiddles occur more frequently. The net result is movement toward the chemical. (c) Directed movement away from a chemical repellent.

that develops as they move through the medium. In other words, if the bacterium is, by chance, moving toward the capillary, it experiences with time a progressively higher concentration of the substance, and this temporal increase in chemical concentration brings into play the mechanism to decrease twiddles, thus increasing runs.

(a)

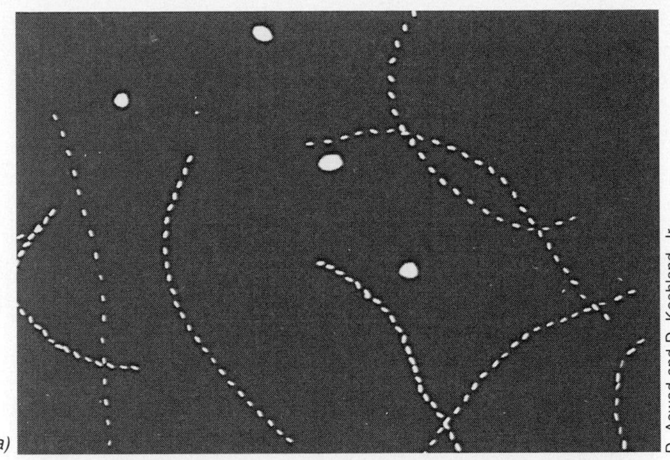

(b)

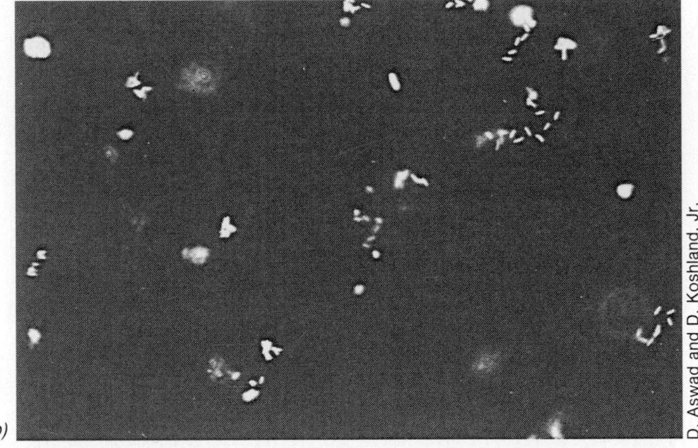

Figure 3.50 Chemotaxis in bacteria, as photographed with intermittent stroboscopic illumination. (a) Cells responding to a gradient with long runs and infrequent twiddles. (b) Cells in absence of gradient showing short runs and many twiddles.

We discussed in Section 3.8 the overall manner of movement of bacterial flagella. The specific movement of bacterial flagella in relation to the run-twiddle phenomenon also has been observed in experiments. In peritrichous organisms, such as *E. coli* and *S. typhimurium*, forward movement occurs when the flagellar bundle comes together. For forward movement, flagella within the bundle rotate counterclockwise, and the cell is propelled in a direction away from the flagella (Figure 3.43). This behavior results in a run. On the other hand, during a twiddle, the flagella rotate clockwise, and the flagellar bundle falls apart, resulting in cessation of forward motion. Bacterial flagella are rigid helices, and the direction of twist of the helix is left-handed, which results in the flagella coming together in a bundle when rotated counterclockwise and flying apart when rotated clockwise. Thus the chemical sensing mechanism in some manner controls which direction the flagella rotate.

How do bacteria use temporal changes in chemical concentrations to control flagellar rotation? Through a variety of genetic and biochemical studies, especially with the use of chemotaxis mutants, it has been shown that bacteria have specific proteins that "sense" the presence of the compound. Some of these proteins, called **chemoreceptors**, are located in the periplasm. Although a chemoreceptor is fairly specific for the compound which it combines with, this specificity is not absolute. For example, the galactose chemoreceptor also recognizes glucose and fucose, and the mannose chemoreceptor also recognizes glucose.

A second set of proteins, called *methyl-accepting chemotaxis proteins* (MCPs) are involved in translating chemotactic signals from the chemoreceptors to the flagellar motor. In the case of amino acid or dipeptide attractants, a special MCP itself (rather than a periplasmic chemoreceptor) binds the substance to initiate chemotactic events. MCPs are membrane proteins and are so named because they become methylated following the binding of substrate by the chemoreceptor (Figure 3.51). MCPs are transmembrane proteins that bind the periplasmic chemoreceptor following its binding of the chemoattractant. The latter event cause methylation of the MCP by cytoplasmic methylating proteins (which use S–adenosylmethionine as methyl donor); up to four methyl groups can be added to each MCP molecule. Methylation catalyzes the MCP to produce a chemical mediator that diffuses to the flagellar motor and causes it to continue rotating in the same direction (Figure 3.51). This action serves to lengthen runs and diminish twiddles. When the cell is in a position in which it can no longer sense a concentration gradient, the MCPs are methylated by cytoplasmic demethylating, and this event stops further production of the flagellar-stimulating chemical mediator (Figure 3.51). This serves to decrease the length of runs and increase the number of twiddles.

From the study of chemotactic mutants it has been determined that the movement away from chemorepellents involves the same mechanism as that involved in movement toward chemoattractants, however the sequence of events is reversed; binding of a repellent by a chemoreceptor inactivates rather than activates the MCP, effectively decreasing the concentration of the diffusable mediator. This leads to clockwise flagellar rotation and an increase in the number of twiddles. Although the chemical nature of the mediator molecule is unknown, it is presumed to be a small, easily diffusable molecule that can respond quickly to varying membrane events.

If a bacterium is presented with both an attractant and a repellent at the same time, it is forced to "choose" which compound to respond to. By carrying out experiments with pairs of such compounds at various concentrations, it has been possible to show that the nature of the response depends on the concentrations of the two agents in relation to the affinity of the chemoreceptors operating on each compound. Thus, if the concentration of *repellent* is relatively high, then the bacteria move away, whereas if the

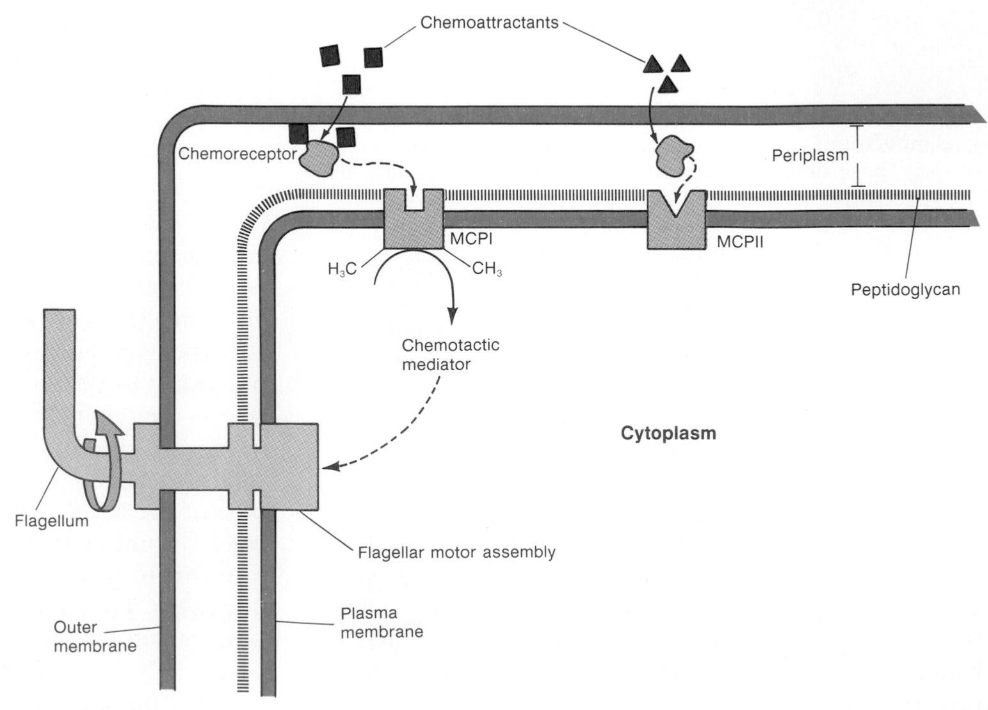

Figure 3.51 Interaction of chemoreceptors and methyl-accepting chemotaxis proteins (MCP) to initiate chemotaxis. Following the binding of the attractant to the chemoreceptor, interaction with the correct MCP initiates methylation; up to four methyl groups can be added to a single MCP molecule. Once methylated, the MCP produces a chemotactic mediator which diffuses to the flagellum and promotes continued flagellar rotation, thus diminishing twiddles. The net result is continued movement in the direction of increasing concentration of the attractant.

concentration of *attractant* is relatively high, they move toward the attractant even though the repellent may be harmful.

3.10 Procaryotic Cell Surface Structures and Cell Inclusions

Fimbriae and pili Fimbriae and pili are structures that are somewhat similar to flagella but are not involved in motility. **Fimbriae** are considerably shorter than flagella and are more numerous (Figure 3.52). Chemically, fimbriae consist of protein. Not all organisms have fimbriae, and the ability to produce them is an inherited trait. The functions of fimbriae are not known for certain in all cases, but there is some evidence that they enable organisms to stick to inert surfaces, or to form pellicles or scums on the surfaces of liquids.

Pili are similar structurally to fimbriae but are generally longer and only one or a few pili are present on the surface. Pili can be visualized under the electron microscope because they serve as specific receptors for certain types of virus particles, and when coated with virus can be easily seen (Figure 3.53). There is strong evidence that pili are involved in the mating process in bacteria, as will be discussed in Section 7.5. Pili are also significant in attachment to human tissues by some pathogenic bacteria.

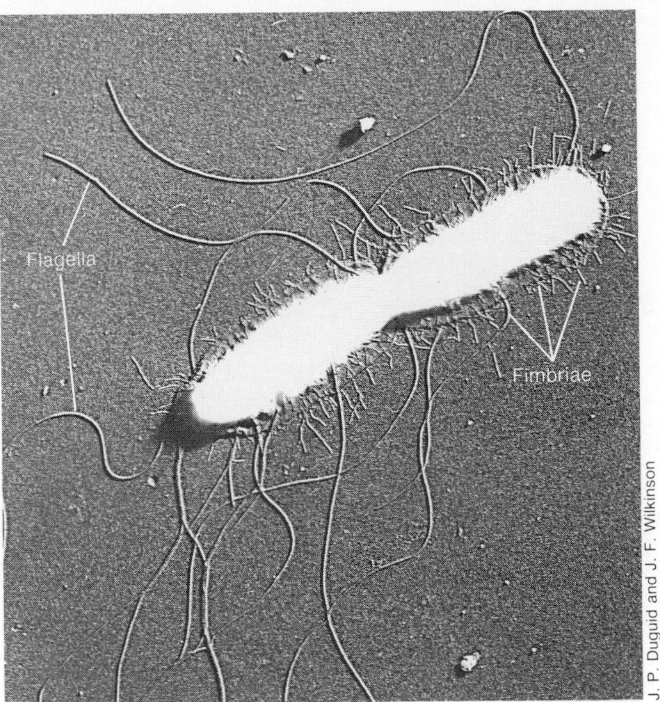

Figure 3.52 Electron micrograph of a metal-shadowed whole cell of *Salmonella typhosa*, showing flagella and fimbriae. Magnification, 15,100×.

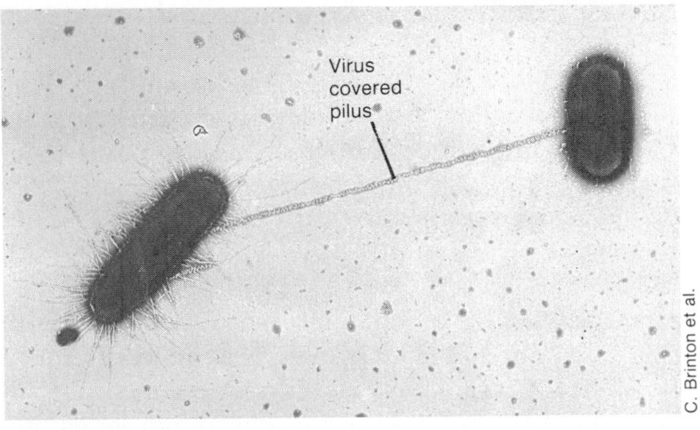

Figure 3.53 The presence of pili on an *E. coli* cell is revealed by the use of viruses which specifically adhere to the pilus.

Capsules and slime layers Most procaryotic organisms secrete on their surfaces slimy or gummy materials (Figure 3.54). A variety of structures consist of extracellular polysaccharide. The terms **capsule** and **slime layer** are sometimes used but the more general term **glycocalyx** is also applied. Glycocalyx is defined as the polysaccharide-containing material lying outside the cell. The glycocalyx varies in different organisms, but usually contains glycoproteins and a large number of different polysaccharides, including polyalcohols and amino sugars. The glycocalyx may be thick or thin, rigid or flexible, depending on its chemical nature in a specific organism. The rigid layers are organized in a tight matrix which excludes particles, such as India ink; this form is the traditional capsule. If the glycocalyx is more easily deformed, it will not exclude particles and is more difficult to see; this arrangement is that of the traditional slime layer.

Glycocalyx layers serve several functions in bacteria. Outer polysaccharide layers play an important role in the attachment of certain pathogenic microorganisms to their hosts. As we will see in Section 11.8, pathogenic microbes that enter the animal body by specific routes usually do so because of binding reactions which occur between outer cell surface components (such as the glycocalyx) and specific host tissues. The glycocalyx plays other roles as well. There is some evidence that encapsulated bacteria are more difficult for phagocytic cells of the immune system (see Section 11.14) to recognize and subsequently destroy. In addition, since outer polysaccharide layers probably bind a significant amount of water, there is reason to believe a glycocalyx layer plays some role in resistance to desiccation.

Inclusions and storage products Granules or other inclusions are often seen within cells. Their nature differs in different organisms, but they almost always function in the storage of energy or structural

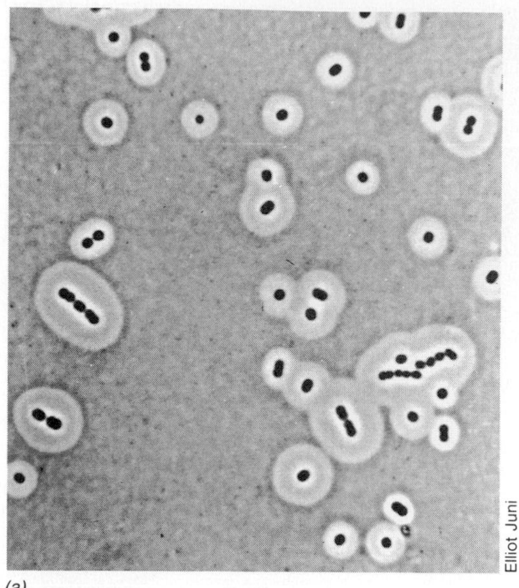

(a)

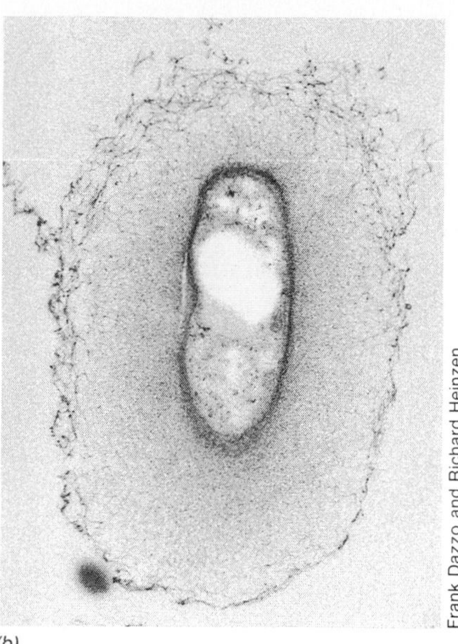

(b)

Figure 3.54 Bacterial capsules. (a) Demonstration of the presence of a capsule by negative staining with India ink observed by phase-contrast microscopy. The India ink does not penetrate the capsule so that it is revealed in outline as a light structure on a dark background. *Acinetobacter* sp. (b) Electron micrograph of a thin section of a *Rhizobium trifolii* cell stained with ruthenium red to reveal the capsule. Magnification, 16,500×.

building blocks. Inclusions can often be seen directly with the light microscope using special staining procedures, but their contrast can usually be increased by using dyes. Inclusions often show up very well with the electron microscope.

In procaryotic organisms, one of the most common inclusion bodies consists of **poly-β-hydroxybutyric acid** (PHB), a compound that is formed from β-hydroxybutyric acid units. The monomers of this acid are connected by ester linkages, forming the long PHB polymer, and these polymers aggregate into granules. With the electron microscope the positions of these granules can often be seen as light areas that do not scatter electrons, surrounded by a nonunit membrane. The granules have an affinity for fat-soluble dyes such as Sudan black and can be identified tentatively with the light microscope by staining with this compound. Poly-β-hydroxybutyric acid can be positively identified by extraction and chemical analysis. The PHB granules are a storage depot for carbon and energy.

Another storage product is **glycogen**, which is a starchlike polymer of glucose subunits (we discussed the chemistry of glycogen in Section 2.4). Glycogen granules are usually smaller than PHB granules, and can only be seen with the electron microscope, but the presence of glycogen in a cell can be detected in the light microscope because the cell appears a red-brown color when treated with dilute iodine, due to a glycogen-iodine reaction.

Many microorganisms accumulate large reserves of inorganic phosphate in the form of granules of **polyphosphate**. These granules are stained by many basic dyes; one of these dyes, toluidin blue, becomes reddish violet in color when combined with polyphosphate. This phenomenon is called *metachromasy* (color change), and granules that stain in this manner are often called **metachromatic granules**.

A variety of procaryotes are capable of oxidizing reduced sulfur compounds such as hydrogen sulfide, thiosulfate and the like. These oxidations are linked to either reactions of energy metabolism (Sections 16.7 and 16.9) or biosynthesis (Section 16.6), but in both instances **elemental sulfur** frequently accumulates inside the cell in large readily visible granules (see Sections 16.9 and 19.1). The granules of elemental sulfur remain as long as a source of reduced sulfur is still present. However, as the reduced sulfur source becomes limiting, the sulfur in the granules is oxidized, usually to sulfate, and the granules slowly disappear as this reaction proceeds.

3.11 Gas Vesicles

A number of procaryotic organisms that live a floating existence in lakes and the sea produce **gas vesicles**, which confer buoyancy upon the cells. The most dramatic instances of flotation due to gas vesicles are seen in cyanobacteria (blue-green algae) that form massive accumulations (blooms) in lakes (Figure 3.55*a*). Gas-vesiculate cells rise to the surface of the lake and are blown by winds into dense masses. When a sample of water containing these organisms is placed in a bottle, within minutes the organisms float to the surface, whereas if the gas vesicles are collapsed, the cells just as rapidly settle to the bottom (Figure 3.55*b,c*). Gas vesicles are also formed by certain purple and green photosynthetic bacteria (see Section 19.1) and by some nonphotosynthetic bacteria that live in lakes and ponds.

Gas vesicles are spindle-shaped structures, hollow but rigid, that are of variable lengths but constant

(a)

(b)

(c)

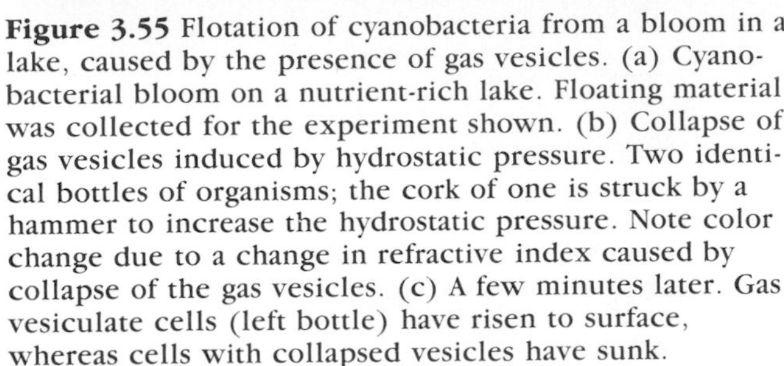

Figure 3.55 Flotation of cyanobacteria from a bloom in a lake, caused by the presence of gas vesicles. (a) Cyanobacterial bloom on a nutrient-rich lake. Floating material was collected for the experiment shown. (b) Collapse of gas vesicles induced by hydrostatic pressure. Two identical bottles of organisms; the cork of one is struck by a hammer to increase the hydrostatic pressure. Note color change due to a change in refractive index caused by collapse of the gas vesicles. (c) A few minutes later. Gas vesiculate cells (left bottle) have risen to surface, whereas cells with collapsed vesicles have sunk.

diameter. They are present in the cytoplasm and may number from a few to hundreds per cell. The gas vesicle membrane is an exception to the rule that membranes are composed of lipid bilayers. The gas vesicle membrane is composed only of protein, and consists of repeating protein subunits that are aligned to form a rigid structure. The gas vesicle membrane is impermeable to water and solutes, but permeable to gases, so that it exists as a gas-filled structure surrounded by the constituents of the cytoplasm (Figure 3.56). The rigidity of the gas vesicle membrane is essential for the structure to resist the pressures exerted on it from without; it is probably for this reason that it is composed of a protein able to form a rigid membrane rather than of lipid, which would form a fluid and a highly mobile membrane. However, even the gas vesicle membrane cannot resist high hydrostatic pressure, and can be collapsed, leading to a loss of buoyancy. The presence of gas vesicles can be determined by either bright-field or phase-contrast microscopy (Figure 3.57), but their identity is never certain unless they disappear when the cells are subjected to high hydrostatic pressure.

Gas vesicle proteins isolated from the vesicles of different procaryotes show a number of common features. Gas vesicle proteins are about 20,000 in molecular weight and are highly hydrophobic. Vesicle proteins also lack sulfur-containing amino acids, and are rather low in aromatic amino acids. Almost 50 percent of the amino acids from hydrolysates of gas vesicle proteins are nonpolar amino acids, with valine, alanine, and leucine predominating. Immunological experiments suggest that the amino acid sequences of various gas vesicle proteins are similar as well; antiserum raised against the vesicles from one organism generally cross-react with vesicles from another.

X-ray diffraction studies have shown that the majority of the nonpolar amino acids of the gas vesicle protein face the gas surface side of the vesicle, whereas the hydrophilic amino acids face the cytoplasm. This arrangement would facilitate the physical restrictions placed on gas vesicles: they must be watertight on their inner surface yet be able to exist in an aqueous solution.

3.12 Bacterial Endospores

The discovery of bacterial endospores (discussed in Section 1.3) was of immense importance to microbiology. Knowledge of such remarkably heat-resistant

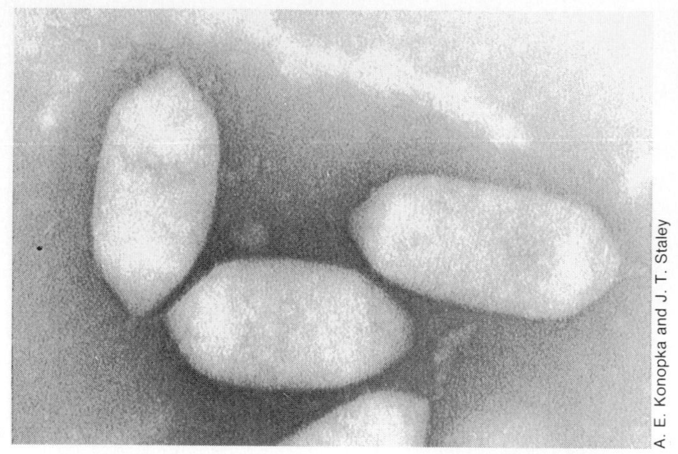

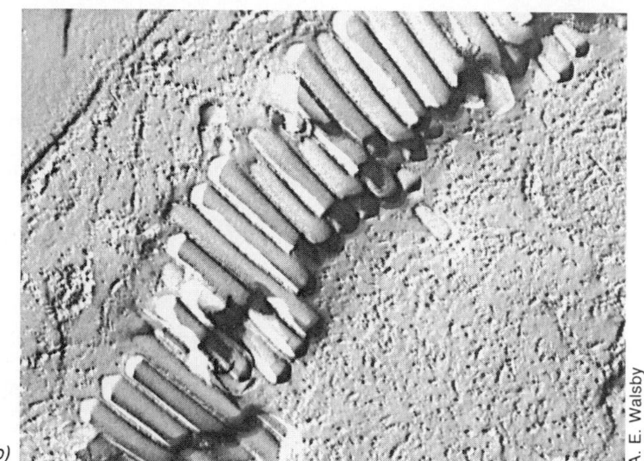

Figure 3.56 Electron micrographs of gas vesicles. (a) Vesicles purified from the bacterium *Microcyclus aquaticus* and examined in negatively stained preparations. Magnification 280,000×. (b) Vesicles of the cyanobacterium *Anabaena flos-aquae* examined in freeze fracture preparations of cells. Magnification, 48,000×.

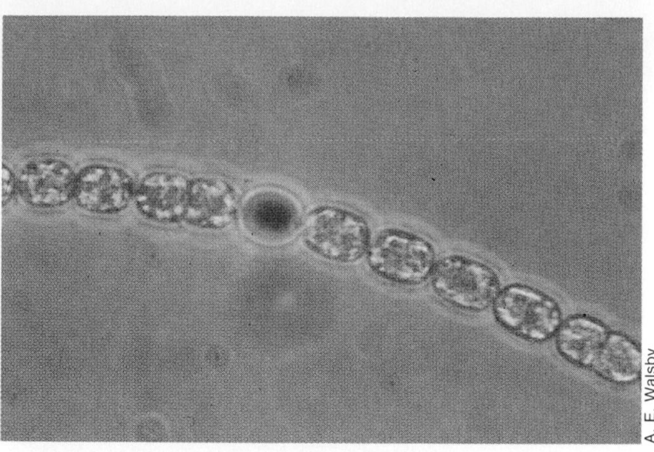

Figure 3.57 Gas vesicles of the cyanobacterium *Anabaena flos-aquae*. The cell in the center (a heterocyst) lacks gas vesicles. In the other cells, the vesicles are seen as phase-bright objects which scatter light.

forms was essential for the development of adequate methods of sterilization, not only of culture media but also of foods and other perishable products. Although many organisms other than bacteria form spores, the bacterial **endospore** is unique in its degree of heat resistance. Endospores are also resistant to other harmful agents such as drying, radiation, acids, and chemical disinfectants. The life cycle of a spore-forming organism is illustrated in Figure 3.58.

Endospores (so called because the spore is formed within the cell) are readily seen under the light microscope as strongly refractile bodies. Spores are very impermeable to dyes, so that occasionally they are seen as unstained regions within cells that have been stained with basic dyes such as methylene blue. To stain spores specifically, special spore-staining procedures must be used. The structure of the spore as seen with the electron microscope is vastly different

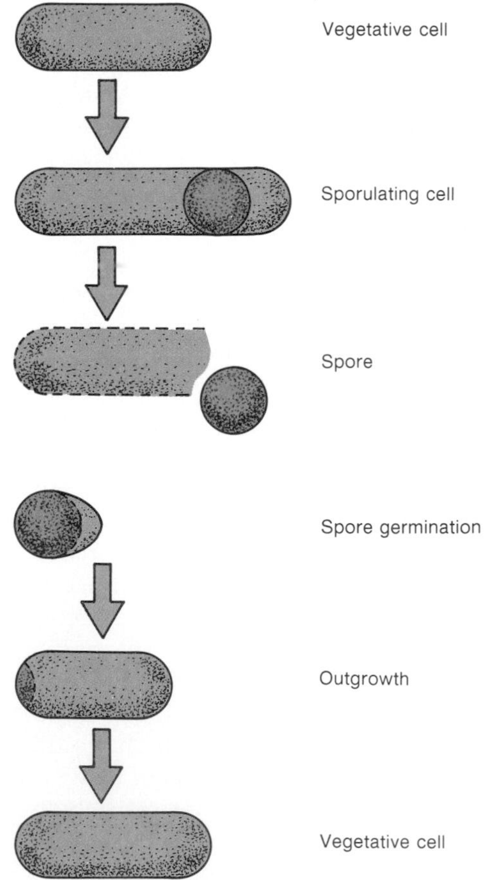

Figure 3.58 Life cycle of a spore-forming bacterium.

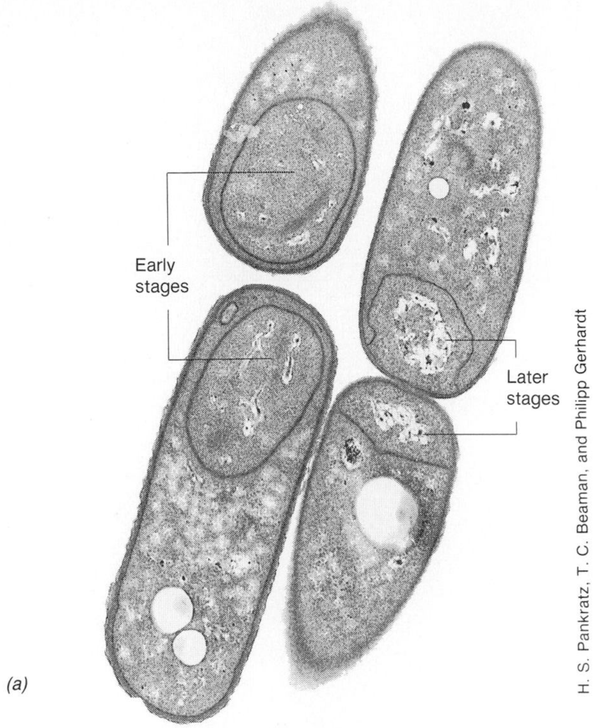

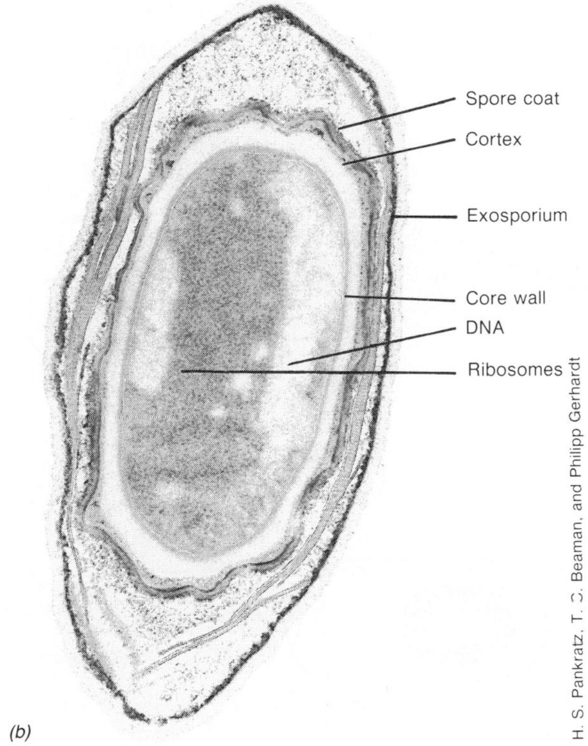

Figure 3.59 Electron microscopy of the bacterial spore. (a) Formation of spores within vegetative cells of *Bacillus megaterium*. Magnification, 17,500×. (b) Mature spore of *B. megaterium*. Magnification, 35,640×.

from that of the vegetative cell, as shown in Figure 3.59. The structure of the spore is much more complex than that of the vegetative cell in that it has many layers. The outermost layer is the **exosporium**, a thin, delicate covering. Within this is the **spore coat**, which is composed of a layer or layers of wall-like material. Below the spore coat is the **cortex**, and inside the cortex is the **core**, which contains the usual cell wall (core wall), cell membrane, nuclear region, and so on. Thus the spore differs structurally from the vegetative cell primarily in the kinds of structures found outside the core wall.

One chemical substance that is characteristic of endospores but is not in vegetative cells is **dipicolinic acid** (DPA) (Figure 3.60). This substance has been found in all endospores examined, and it is probably located primarily in the core. Spores are also high in calcium ions, most of which are also associated with the core, probably in combination with dipicolinic acid. There is good reason to believe that the association of calcium and dipicolinic acid has some role in conferring the unusual heat resistance on bacterial endospores.

The structural changes that occur during the conversion of a vegetative cell to a spore can be studied readily with the electron microscope. Under certain conditions (mainly as a result of nutrient exhaustion), instead of dividing, the cell undergoes the complex series of events leading to spore formation; these are illustrated in Figure 3.61. Aspects of the physiology of endospore formation are discussed in Section 9.9.

An endospore is able to remain dormant for many years, but it can convert back into a vegetative cell (**spore germination**) in a matter of minutes (see Figure 3.58 and Section 1.3). This process involves two steps: cessation of dormancy and outgrowth. The first is initiated by some environmental trigger such as heat. A few minutes of heat treatment at 60 to 70°C will often cause dormancy to cease. The first indications of spore germination are loss in refractility of the spore, increased stainability by dyes, and marked decrease in heat resistance. The spore visibly swells and its coat is broken. The new vegetative cell now pushes out of the spore coat, a process called **outgrowth**, and begins to divide. The spore wall and spore coat eventually disintegrate through the action of lytic enzymes.

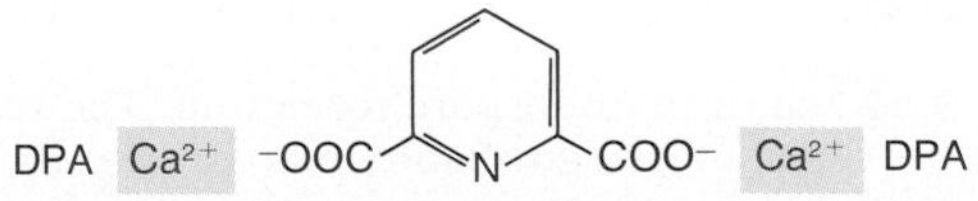

Figure 3.60 Dipicolinic acid (DPA). Ca^{2+} ions associate with the carboxyl groups to form a complex.

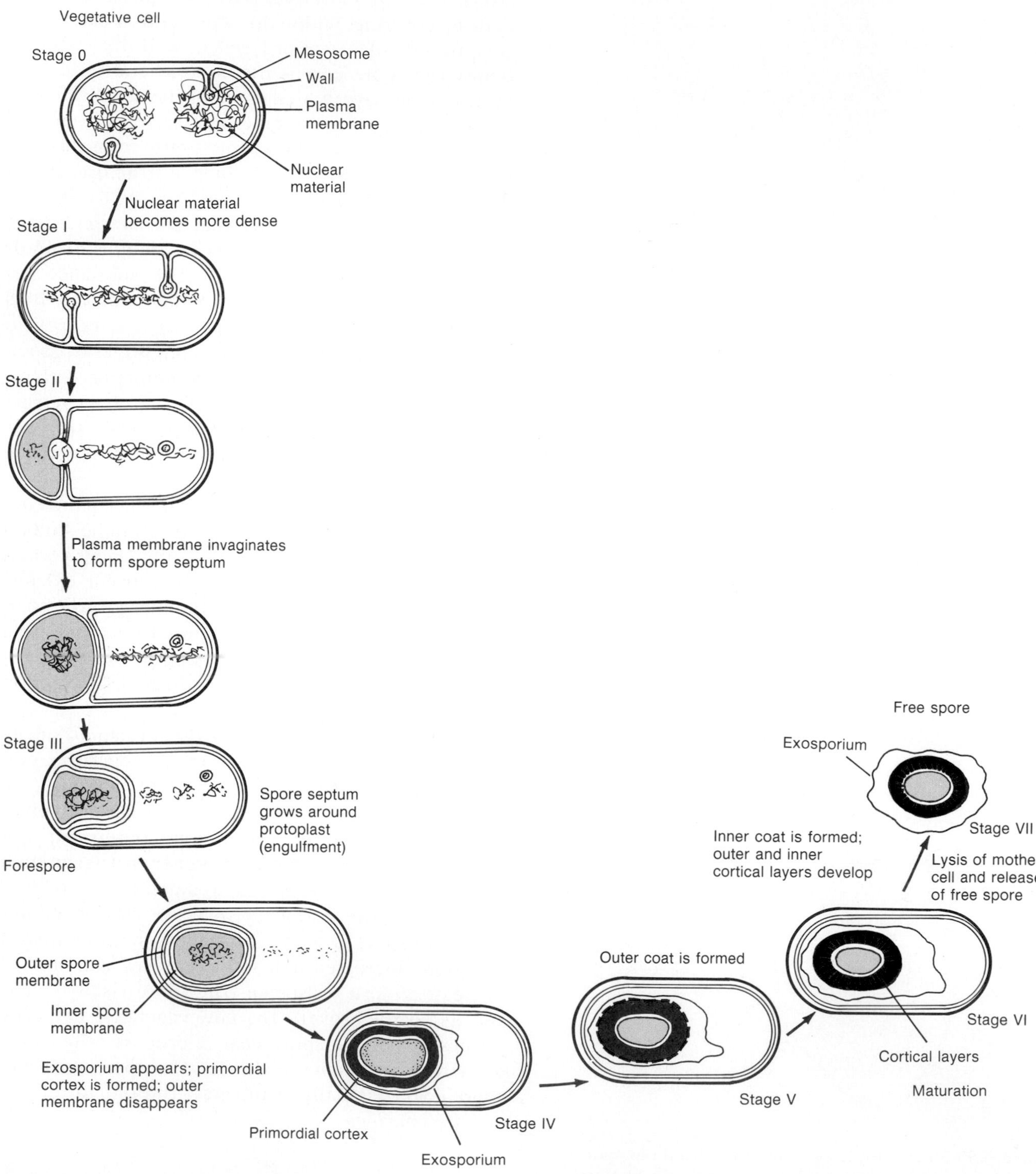

Figure 3.61 Stages in endospore formation. The stages listed (0 through VII) are those most clearly distinguishable microscopically and are used in studies on the kinetics of the sporulation process.

Cell Fractionation

Our understanding of the molecular structure of microbial cells has developed with improvements in cell fractionation. If one wishes to study the chemistry of bacterial cell walls, for example, it is necessary to prepare large amounts of cell walls free from other cellular structures, such as membranes and ribosomes. A number of methods have been developed for separating procaryotic and eucaryotic cells into their respective cellular fractions. These involve such techniques as centrifugation, chromatography, and electrophoresis. Since procaryotes are in general much smaller than eucaryotes, slightly different methods are used with any particular type of cell. For example, the following protocol would be useful for obtaining bacterial cell membranes:

1. **Cell extraction** A thick cell suspension is broken by mechanical means (sonication, grinding, or by placing the cells under high pressure [>20,000 psi] and releasing the pressure quickly, causing the cells to rupture). The cells will break apart violently to yield a **cell extract**.

2. **Removal of debris** The broken cell suspension is then subject to centrifugal force in a **centrifuge**. At first, a rather low centrifuge speed would be chosen so as to sediment only unbroken cells and large cell fragments. A typical protocol might be to centrifuge at 10,000×g for 10 minutes. The material at the bottom of the centrifuge tube is referred to as the **pellet**, while the material in the liquid above the pellet is referred to as the **supernatant** (see figure below).

3. **Sedimentation of cell components** The supernatant from the first centrifugation will contain membrane fragments, ribosomes, and all cytoplasmic (soluble) cell constituents. The supernatant is then recentrifuged, this time at much higher speed in an **ultracentrifuge**. An ultracentrifuge is a high speed version of a normal centrifuge except that the spinning rotor operates in an evacuated chamber to reduce noise and wind friction and to maintain careful temperature control at high speeds. A typical protocol might be to centrifuge at 100,000×g for 1–2 hours. The pellet would contain membranes and the supernatant would contain ribosomes and other cytoplasmic constituents.

For eucaryotic cells, following cell breakage, a centrifugation of 1000×g for 10 minutes would yield cell nuclei and any unbroken cells. A recentrifugation of the supernatant from the first step at 20,000×g for 20 minutes would yield mitochondria and chloroplasts. A centrifugation of the supernatant from the second step at 100,000×g for 1 hour would yield membranes, while a centrifugal force of 150,000×g for 3 hours would be required to pellet ribosomes.

4. **Fractionation of cell components** Several biochemical techniques have been used to separate cellular macromolecules. Proteins are the most abundant macromolecules in the cell and the separation of cellular proteins from one another is a common procedure in many microbial studies. One useful technique has been the ultracentrifugation of the protein mixture in a **sucrose gradient**—the lowest concentration of sucrose is on the top and the highest on the bottom. By centrifuging in such a gradient for several hours the proteins will migrate downward in the gradient until they reach a density equal to their own. They will remain stationary at this point even upon repeated centrifugation. Since proteins differ in their molecular weight and thus have different densities, fractions col-

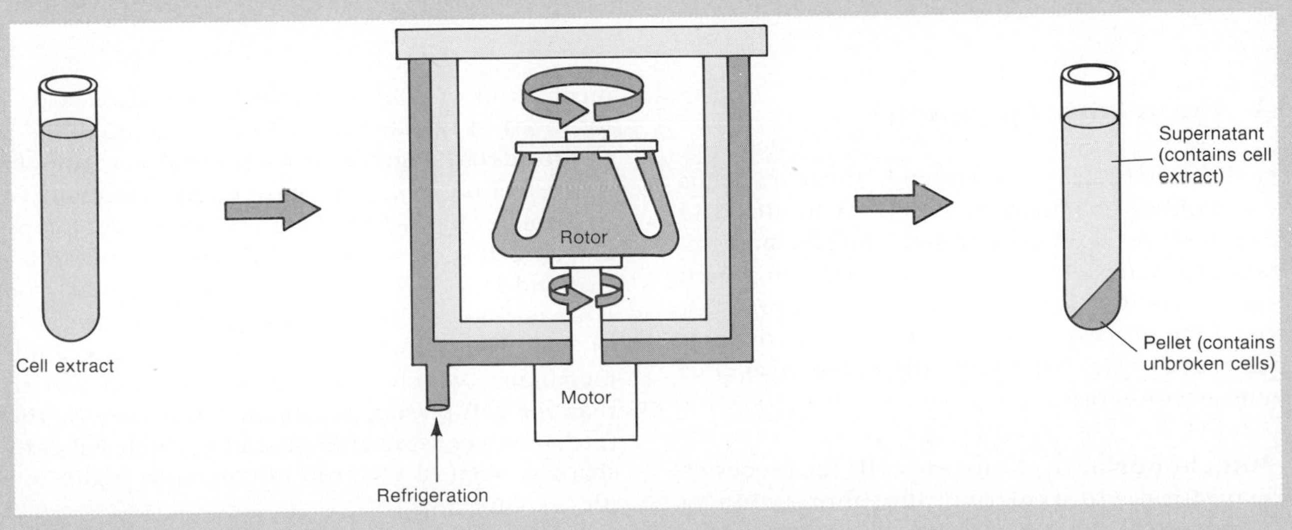

lected from the sucrose gradient will each contain a series of proteins of roughly one molecular size:

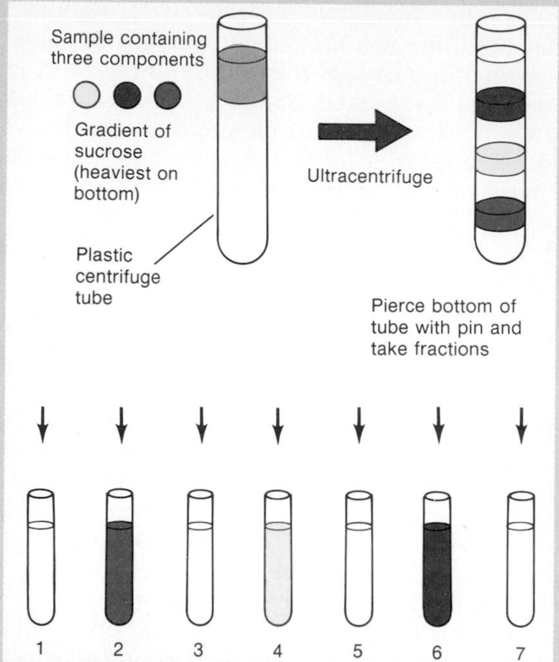

Chromatography is also a popular means of separating proteins from cell extracts. Many types of chromatography have been developed, but column chromatography is a popular method of protein purification. A glass tube is filled with a permeable solid matrix immersed in a solvent, usually a buffer solution of some type. Following application of the sample a continuous flow of solvent is passed through the column and the protein mixture begins to separate.

The principle of separation in chromatography varies. Some column packings separate macromolecules on the basis of differences in their molecular weight. Others separate by exploiting differences in charge, while still others contain substrates of particular enzymes bound to the column packing. In the latter process, called **affinity chromatography**, the enzyme of interest will stick to the bound substrate molecule on the column while other proteins pass through. The enzyme can then be removed (eluted) from the column by passing a solution of the specific molecule through the column.

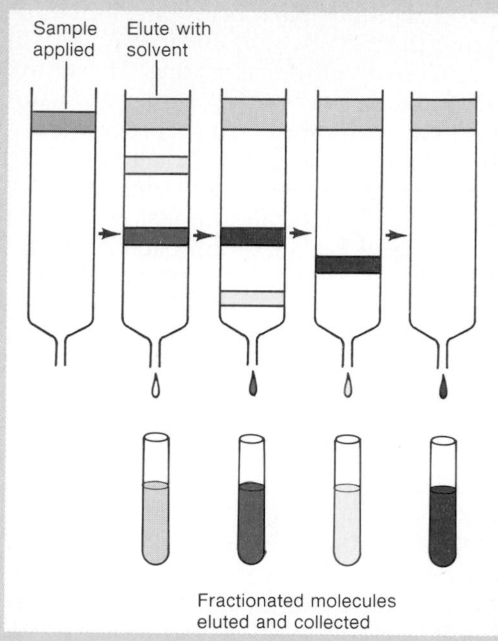

Electrophoresis is another method of separating cell fractions, especially proteins and nucleic acids. The principle of electrophoresis is quite simple: molecules will migrate in an electric field according to their net charge and size and shape. We discuss the electrophoresis of nucleic acids in Chapter 5. For determining protein size (molecular weight), the protein is first denatured in a detergent solution and then sub-

3.13 Eucaryotic Organelles

As we discussed briefly in Chapter 1, eucaryotic cells have a number of important functions localized in discrete bodies called **organelles**. The two most important organelles are the *mitochondria*, in which energy metabolism is carried out, and *chloroplasts*, in which the process of photosynthesis is carried out in plants and algae. We discuss these two organelles in more detail here.

Mitochondria In eucaryotic cells the processes of respiration and oxidative phosphorylation (a mechanism of ATP formation, see Section 4.9) are localized in membrane-enclosed structures, the **mitochondria** (singular, **mitochondrion**). Mitochondria are of bacterial size, most often rods about 1 μm in diameter by 2–3 μm long (Figure 3.62). A typical animal cell such as a liver cell can contain 1000 mitochondria, but the number per cell depends somewhat on the cell type; a yeast cell may have as few as two mitochondria per cell. The mitochondrial membrane, which lacks sterols, is much less rigid than the cell plasma membrane. Mitochondria therefore show a considerable plasticity which makes their shape as seen in electron micrographs highly variable. (Figure 3.62)

jected to electrophoresis; detergent-denatured proteins lose their unique charge characteristics and tend to migrate more readily in an electric field strictly according to size. (Recall that although a denatured protein loses its unique three-dimensional structure, it retains its primary structure. Thus the mass of a protein is unaffected by denaturation. See Section 2.7.) By comparing the extent of migration of a protein versus standard proteins of known molecular weight, it is possible to get an accurate estimate of the molecular weight of the sample protein.

In practice, the denatured proteins are applied to a thin (~2–3 mm) slab of a gel-like material called polyacrylamide. The **polyacrylamide gel** is fixed in a plastic container and is subject to an electric current for several hours.

The gel is removed and stained with a protein-specific dye to visualize the bands in the preparation. In this way, the investigator can analyze cells for the variety of proteins they contain.

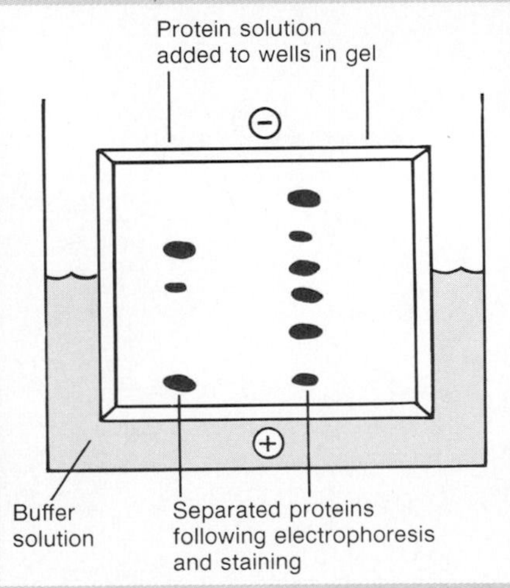

Cell fractionation techniques have become more refined with the passing years and new techniques are being introduced constantly. The idea of cell fractionation, of course, is to be able to study a particular component of the cell. By developing new and more refined techniques those interested in the molecular basis of life are able to generate increasingly pure cell fractions and are thus able to perform more critically defined experiments.

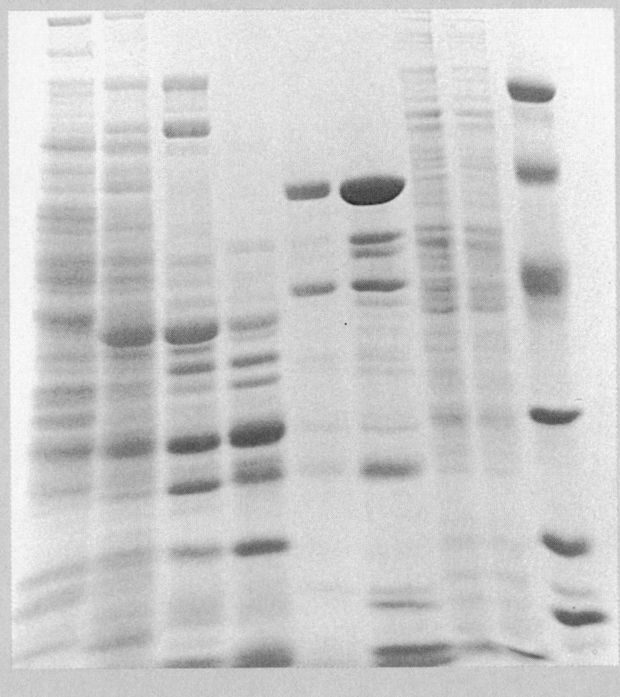

Example of proteins separated on a polyacrylamide gel

The mitochondrial membrane is constructed in a manner similar to other membranes; a bilayer of phospholipid with embedded proteins. However, unlike the plasma membrane, the mitochondrial membrane is rather permeable. Protein channels are present in the outer membrane that allow passage of any molecule of molecular weight less than approximately 10,000. It is for this reason that ATP, produced within the mitochondrion, can move to the cytoplasm where it is used in energy-requiring reactions. In addition to the outer membrane, mitochondria possess a system of folded inner membranes called *cristae*. These inner membranes are the site of enzymes involved in respiration and ATP production and of specific transport proteins that regulate the passage of metabolites into and out of the *matrix* of the mitochondrion (Figure 3.62). With high resolution electron microscopy, it is possible to see that the cristae are complex structures, consisting of small round particles attached to the membrane by short stalks. These particles are the membrane-bound ATPases, a complex series of proteins that couple the proton electrochemical gradient to the production of ATP (Section 4.12). The matrix of the mitochondrion is rich in protein and is gel-like in consistency. It contains proteins involved in the oxidation of pyruvate and fatty acids and also contains certain of the citric acid cycle enzymes (see Section 4.10). The space between the inner and outer

Figure 3.62 Structure of the mitochondrion. (a) The structure of the inner membrane complex is shown in the insert. (b) and (c) Transmission electron micrographs of mitochondria from rat tissue showing the variability in morphology of typical mitochondria.

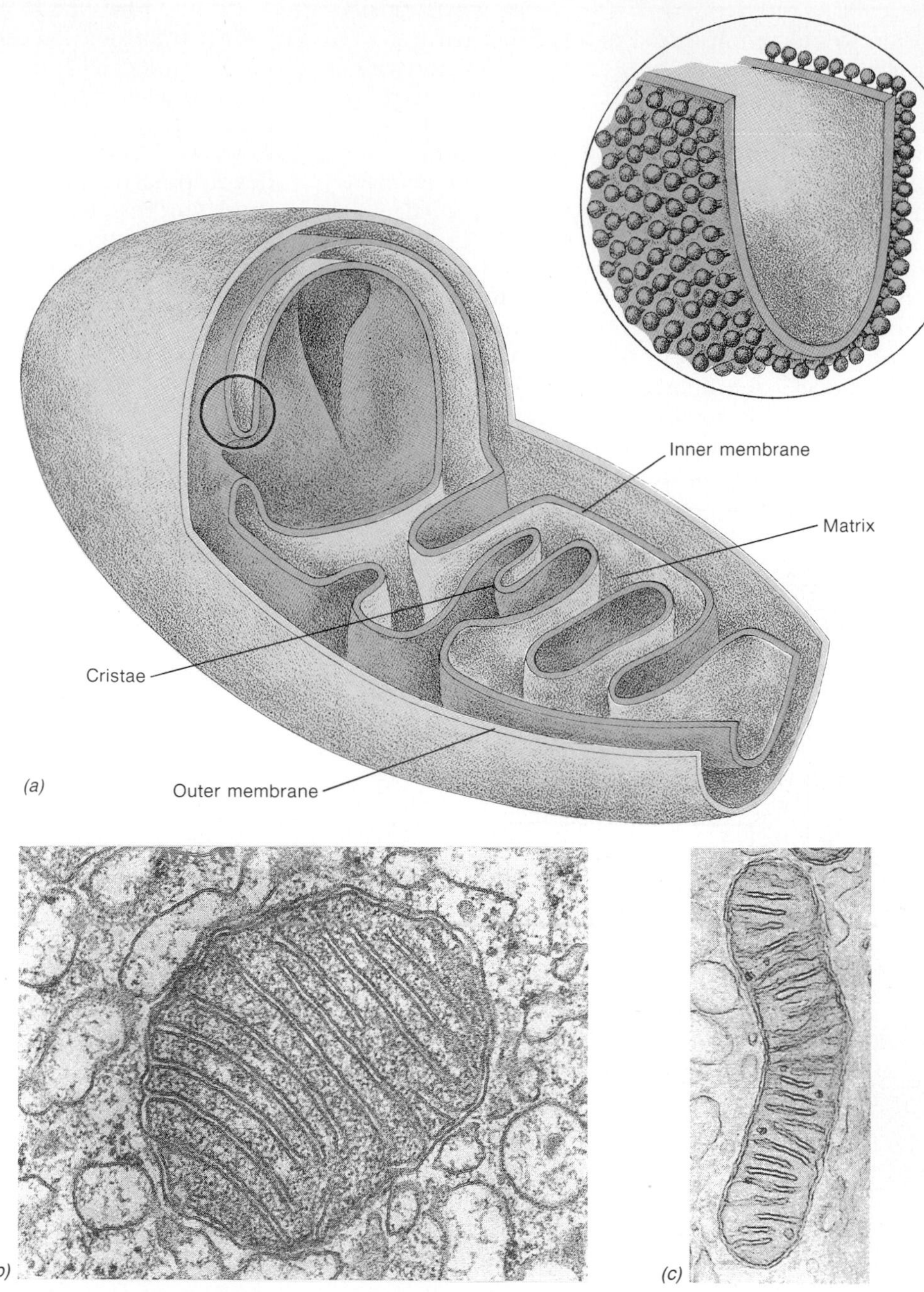

mitochondrial membranes contains a small number of proteins involved in converting ATP into certain other high energy-phosphate intermediates used either within the mitochondrial matrix or in the cell cytoplasm.

Chloroplasts Chloroplasts are green chlorophyll-containing organelles found in all eucaryotic organisms able to carry out photosynthesis. Chloroplasts of many algae are quite large and hence are readily visible with the light microscope (Figure 3.63). The size, shape, and number of chloroplasts vary markedly but, unlike mitochondria, they are generally much larger than bacteria.

Like mitochondria, chloroplasts have a very permeable outer membrane, a much less permeable inner membrane, and an intermembrane space. The inner membrane surrounds the lumen of the chloroplast, called the *stroma*, but is not folded into cristae like the inner membrane of the mitochondrion. Instead chlorophyll, photosynthetic-specific proteins, the photosynthetic electron transport chain, and all other components needed for photosynthesis are located in a series of flattened membrane discs called **thylakoids** (Figure 3.64). The thylakoid membrane is highly impermeable to ions and other metabolites, as befits a membrane whose function is to pump out protons to establish the proton gradient necessary for

ATP synthesis (see Section 4.12). In green algae and green plants, thylakoids are usually associated in stacks of discrete structural units called *grana*. The fact that among algae grana are found only in the green algae (*Chlorophyceae*) suggests that this algal group may have been the evolutionary forerunner of the higher plants.

The chloroplast stroma contains large amounts of the enzyme ribulose bisphosphate carboxylase. This enzyme is the key enzyme of the Calvin Cycle, the series of reactions by which most photosynthetic organisms convert CO_2 into organic form (see Section 16.6). Ribulose bisphosphate carboxylase makes up over 50 percent of the total chloroplast protein and serves to produce phosphoglyceric acid, a key compound in the biosynthesis of glucose (see Sections 4.17 and 16.6). The permeability of the outer chloroplast membrane allows glucose and ATP produced during photosynthesis to diffuse into the cytoplasm where they can be used to build new cell material.

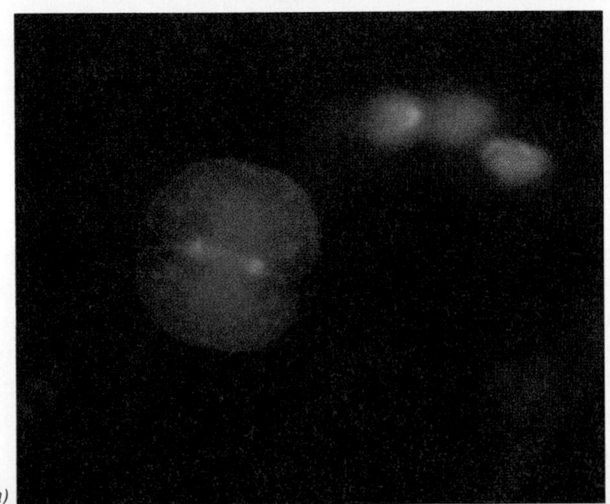

(a)

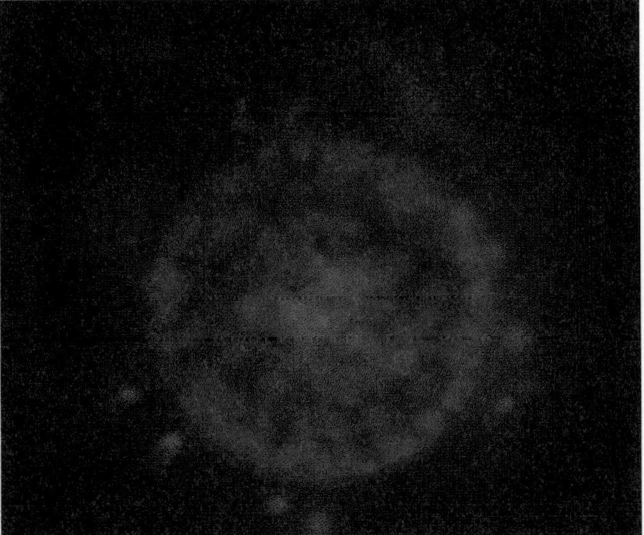

(b)

Figure 3.63 Photomicrographs by fluorescence microscopy of algal cells, showing the presence of chloroplasts The chlorophyll of the chloroplasts fluoresces red. (a) *Cosmarium*. (b) *Stephanodiscus*.

3.14 Relationships of Chloroplasts and Mitochondria to Bacteria

On the basis of their relative autonomy and morphological resemblance to bacteria, it was suggested long ago that mitochondria and chloroplasts were descendents of ancient procaryotic organisms. This proposal of *endosymbiosis* (endo means "within") says that eucaryotes arose from the invasion of one (larger) cell by a procaryotic cell. Several pieces of evidence strongly suggest that this scenario is correct:

1. Mitochondria and chloroplasts contain DNA. Although most organellar functions are coded for by nuclear DNA, a few organellar proteins are coded from within the organellar genome. Mitochondrial and chloroplast DNA exists in covalently closed circular form, as it does in bacteria (see Section 3.15).
2. Mitochondria and chloroplasts contain their own ribosomes. Ribosomes, the cells' protein synthesis "factories" (see Section 5.10) exist in either a large form (80S), typical of the *cytoplasm* of eucaryotic cells, or a smaller form (70S), unique to eubacteria and archaebacteria. Mitochondrial and chloroplast ribosomes are 70S in size, the same as that of bacteria.
3. Many of the antibiotics that kill or inhibit eubacteria by specifically interfering with 70S ribosome function also inhibit protein synthesis in mitochondria and chloroplasts (see Box).
4. Phylogenetic studies of bacteria and other organisms by comparative ribosomal RNA sequencing (see Section 18.3) have shown convincingly that the chloroplast and mitochondrion are of eubacterial lineage. These studies have clearly shown

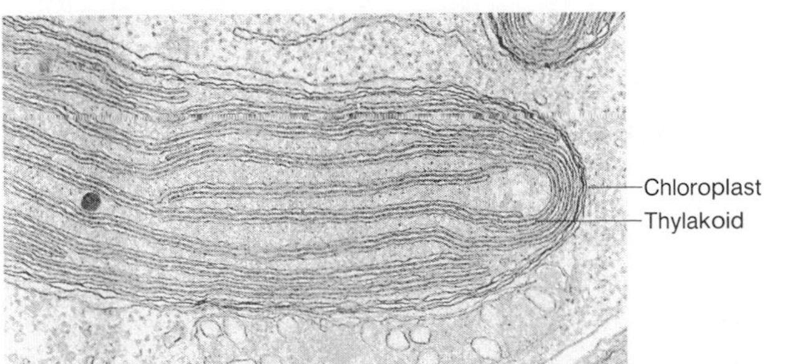

T. Slankis and S. Gibbs

Figure 3.64 Electron micrograph showing a chloroplast of the alga *Ochromonas danica*. Note that each thylakoid consists of three parallel membranes. Magnification, 29,000×.

Mitochondria and the Antibiotic Connection

Is it of any practical significance to know that mitochondria are procaryotic in character? Yes: A tragic use of an antibiotic in medicine is linked to the mitochondrion. This antibiotic, *chloramphenicol*, was widely used in the early days of antibiotic therapy because it was thought to be nontoxic. However, infants and certain other groups turned out to be unusually sensitive to chloramphenicol, and a number of deaths due to blood anemia occurred before the general use of this antibiotic in medicine ceased. What was going on here? We now know that chloramphenicol specifically affects the ribosomes of procaryotic cells, thus inhibiting protein synthesis. In eucaryotes, the ribosomes in the cytoplasm are unaffected by chloramphenicol, but those in the mitochondria, being procaryotic in character, are attacked. Once the connection between chloramphenicol and the procaryotic ribosome was discerned, it made sense that under certain conditions, chloramphenicol might inhibit eucaryotic cells. The cells inhibited in eucaryotes by chloramphenicol are those that are multiplying rapidly, such as blood-forming cells of the bone marrow, where *new* mitochondria are being synthesized at a rapid rate. With this understanding of the connection between eucaryotic mitochondria and procaryotes, the use of chloramphenicol in medicine ceased, except for special cases where it is the only antibiotic that works. Many other antibiotics in clinical use, for example streptomycin, tetracycline, and erythromycin, also interfere specifically with 70S ribosome function, but since these antibiotics are not taken up by eucaryotic cells, mitochondrial ribosomes are unaffected.

that the eucaryotic cell is an association of two organisms. The mitochondrion and chloroplast are descendents of different eubacterial groups because their ribosomal RNA sequences closely match those of certain eubacteria; the same techniques show that the cytoplasmic component of eucaryotes evolved totally independently.

Presumably organelles evolved (following endosymbiotic events) by the progressive loss of more and more of their genetic independence, eventually becoming functionally specialized and dependent on their cytoplasmic host cell. The result is the chloroplasts and mitochondria we see today. Although the endosymbiotic hypothesis can probably never be rigorously proven, substantial molecular evidence remains in organelles today to identify them as having once been procaryotic cells.

3.15 Arrangement of DNA in Procaryotes

We learned in Chapter 2 that DNA is a double-stranded molecule formed of complementary, antiparallel strands of polynucleotides. In eucaryotes (see Section 3.16) DNA exists as a linear structure, generally compacted by winding around basic proteins called histones. Although bacterial DNAs associate with basic proteins, true histones are not present, and procaryotic DNA is not arranged in linear fashion. DNA in bacteria is actually circular, present in covalently-closed circular structures which fold back upon themselves extensively. Such a compact structure can fit readily inside the cell. Bacterial DNA is not surrounded by a membrane typical of the eucaryotic nucleus, although it does tend to aggregate and is thus usually visible in a distinct region of the cell when observed in thin sections of bacteria with the electron microscope (Figure 3.65).

The DNA of the bacterium *Escherichia coli*, if opened and linearized, would be about 1 mm in length. This amount of DNA corresponds to about 4.2×10^6 base pairs and agrees nicely with the known molecular weight of the *E. coli* genome of about 2.7×10^9 daltons. To package this much DNA in an organism only 2–3 μm long obviously requires extensive folding of the molecule. Bacterial DNA undergoes extensive folding to yield a highly twisted form called a **supercoil** (Figure 3.66); supercoiled DNA takes on a considerably more compact shape than its freely circularized counterpart. Except for a recent discovery of positively supercoiled DNA in an archaebacterium, all eubacterial (and eucaryotic) DNA examined has been shown to exist in a negative supercoiled form. Negative or positive in this connection refers to the direction of supercoiling in relation to the long axis of the molecule.

The DNA of the mitochondrion and chloroplast is also circular and is probably arranged in some sort of supercoiled form. Although the genome size of the chloroplast is considerably larger than that of the mitochondrion (Figure 3.67), both the mitochondrial

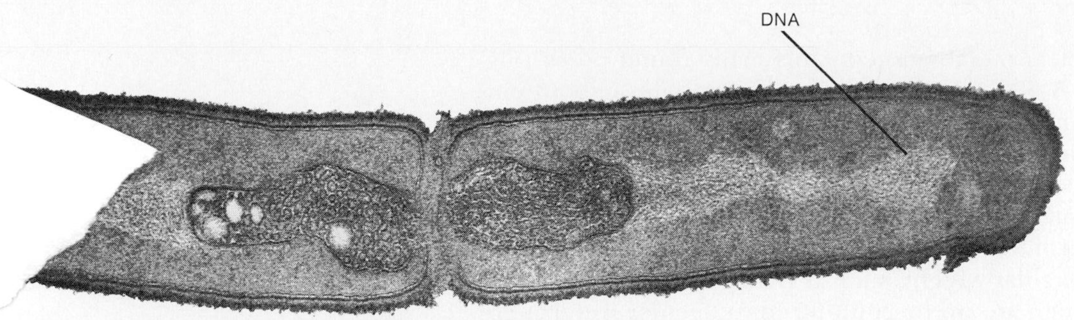

...ectron micrograph of a thin section of a typical bacterium, Gram- ...*llus subtilis*. The cell has just divided and two membrane-contain- ...res are attached to the cross wall. Note the light region in the middle ... DNA. Magnification, 31,200×.

...d chloroplast genomes are 10–100 times *smaller* than that of *E. coli*. Nevertheless, the amount of circular DNA present in organelles is probably sufficient to require some sort of folding and packaging.

3.16 The Eucaryotic Nucleus

One of the distinguishing characteristics of a eucaryote is that its genetic material (DNA) is organized in chromosomes contained in a membrane-enclosed structure, the nucleus. In many eucaryotic cells the nucleus is a large organelle many micrometers in diameter, easily visible with the light microscope, even without staining. In smaller eucaryotes, however, special staining procedures often are required to see the nucleus.

The key genetic processes of DNA replication and RNA synthesis (transcription) occur in the nucleus whereas the process of protein synthesis (translation) occurs in the cytoplasm. We will learn in Chapter 5 of the distinctly different arrangement of DNA in genes in eucaryotic and procaryotic cells. Eucaryotic cells contain much more DNA than bacteria (see Figure 3.67) and, following the transfer of the genetic blueprint into the intermediary carrier, RNA, in the process of transcription (see Section 5.8), eucaryotic RNAs are "processed" in the nucleus by cutting out any RNA that does not actually code for a protein product. The nucleus clearly serves as both a storehouse and processing factory of genetic information. In bacteria, which lack a membrane-bounded nucleus, the processes of transcription and translation are tightly coupled (see Section 5.11), and thus the opportunity for extensive RNA processing typical of eucaryotes does not arise.

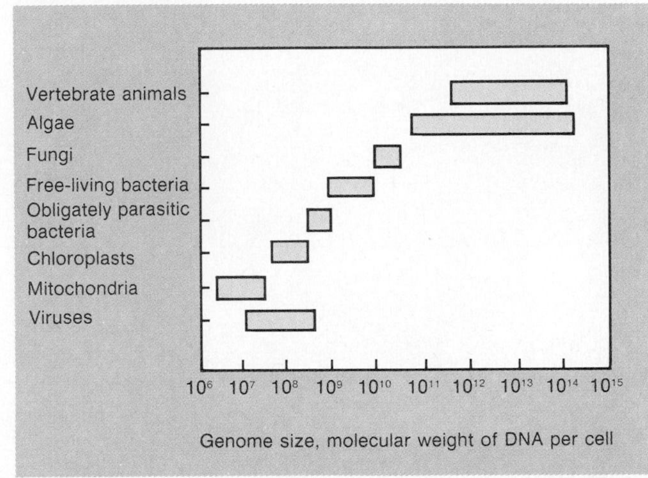

Figure 3.67 Range of genome sizes in various groups of organisms.

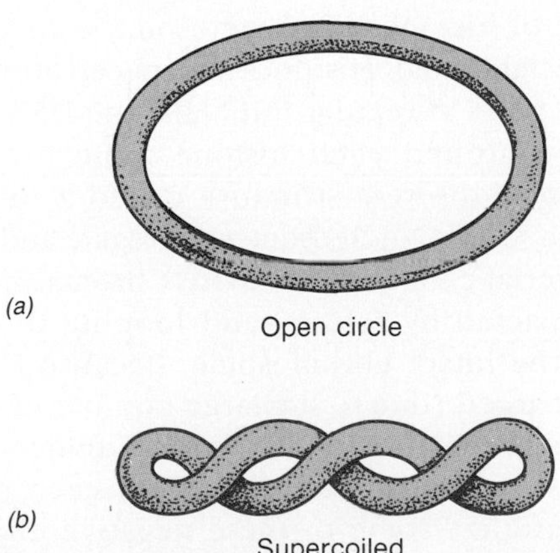

Figure 3.66 Bacterial DNA. (a) Open circular form. (b) Supercoiled form. Note that in either case the DNA is present in a covalently closed form.

Nuclear structure The nuclear membrane consists of a pair of parallel unit membranes separated by a space of variable thickness. The inner membrane is usually a simple sac, but the outer membrane is in many places continuous with cytoplasmic membranes. The dual membrane arrangement does, however, facilitate functional specificity, because the inner and outer membranes specialize in interactions with the nucleoplasm or cytoplasm, respectively. The nuclear membrane contains many round pores (Figure 3.68), which are formed from holes in both unit membranes at places where the inner and outer membranes are joined. The pores are about 9 nm wide and permit the facile passage in and out of the nucleus of macromolecules up to about 60,000 molecular weight. However, certain proteins larger than 60,000 molecular weight, such as DNA and RNA polymerases (which are up to 200,000 in molecular weight) are

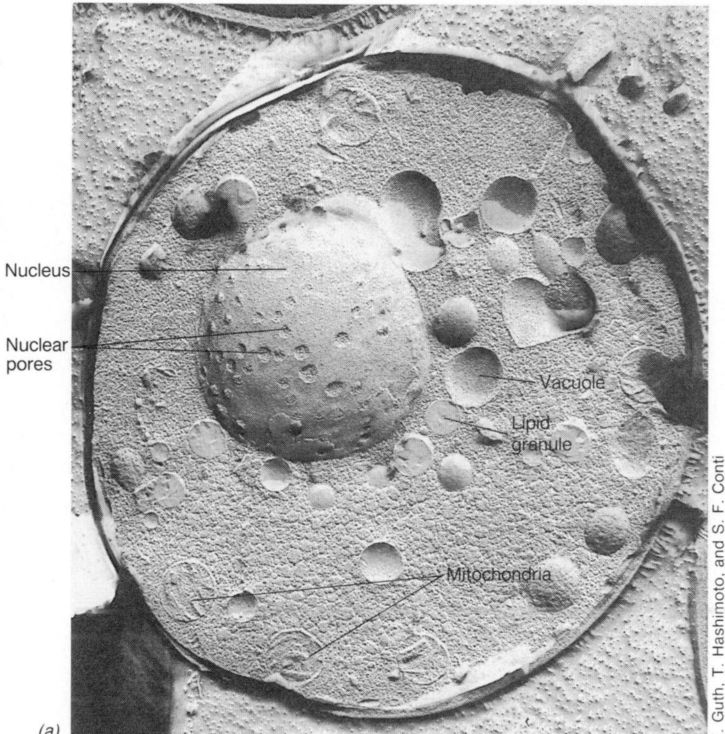

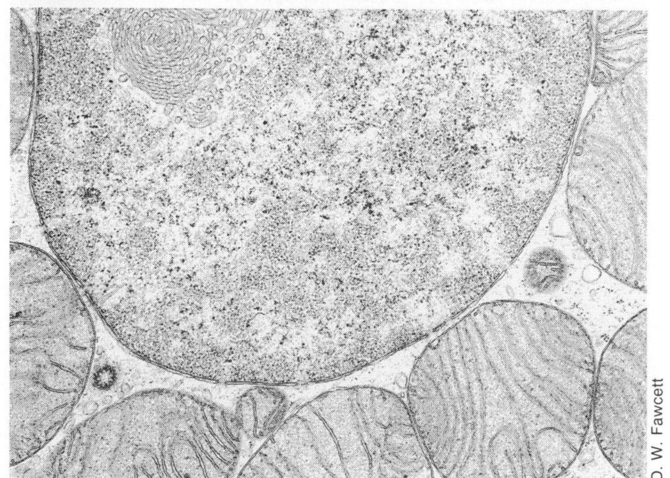

Figure 3.68 The nucleus and nuclear pores. (a) Electron micrograph of a yeast cell by the freeze-etch technique, showing a surface view of the nucleus. Magnification, 16,650×. (b) Thin section of mouse adipose tissue showing a portion of the nucleus and several mitochondria. Magnification, 35,000×. Note the pores in the nuclear membrane in both (a) and (b).

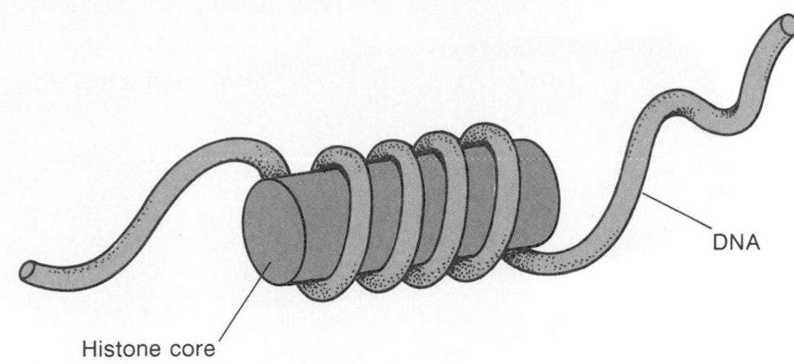

Figure 3.69 Packing of DNA around histones in the formation of a nucleosome.

also able to pass into the nucleus from their sites of synthesis in the cytoplasm. Whether these pass through nuclear pores or are threaded through the lipid bilayer by specific transport proteins is not known. A typical animal cell nucleus contains 3000–4000 nuclear pores.

A structure often seen within the nucleus is the **nucleolus**, an area rich in ribonucleic acid (RNA) and which is the site of ribosomal RNA synthesis. The small and large subunits of the eucaryotic ribosome are synthesized in the nucleolus and are exported to the cytoplasm where the complete (80S) ribosome is assembled and functions in protein synthesis.

Chromosomes and DNA As we discussed in the previous section, the DNA of procaryotes is contained primarily in a single molecule of free DNA, whereas in eucaryotes, DNA is present in more complex structures, the *chromosomes*. **Chromosome** means "colored body," for chromosomes were first seen as structures colored by certain stains. Many chromosome stains involve dyes that react strongly with basic (that is, cationic) proteins called **histones**, which in eucaryotes often are attached to the DNA. Chromosomes also usually contain small amounts of RNA.

The role of histones in chromosome contraction seems well established. Histones are spaced along the DNA double helix at regular intervals, the DNA itself being wound around each histone molecule. The packing forms a discrete structure called a **nucleosome** (Figure 3.69). Nucleosomes aggregate and form a fibrous material called **chromatin**. Chromatin itself can be compacted by folding and looping to eventually form the intact chromosome. Because DNA is negatively charged (due to the large number of phosphate groups present), there is a strong tendency for various parts of the molecule to repel each other. Histones neutralize some of these negative charges, permitting contraction of the chromosomes. The histone content of chromosomes is the same whether the chromosomes are contracted or highly extended; however, histone proteins are chemically modified by phosphorylation, acetylation, and methylation at the time of contraction. These chemical modifications change the charge on the histones, thus altering repulsive forces.

In each chromosome, the DNA is a single linear molecule to which histones (and other proteins) are attached. In yeast (and probably many other microorganisms), the length of the DNA in a single chromosome is actually shorter than a linearized procaryotic chromosome. For instance, the total amount of DNA per yeast cell is only three times that of *Escherichia coli*, but yeast has 17 chromosomes, so that the average yeast DNA molecule is much shorter than the *E. coli* DNA. Seen under the electron microscope, yeast DNA molecules all appear linear, showing no evidence of the circularity seen in bacteria. However, in higher organisms the length of the DNA molecule in a single chromosome is many times longer than procaryotic DNA if it were opened and linearized.

The DNA content per nucleus varies from species to species, in much the same way as does nuclear size. In addition, the chromosome number also varies greatly, from just a few to many hundreds. Another variable feature is genome size, which is the actual number of distinct genes per cell. Even with the same amount of DNA per cell, genome size can vary from organism to organism, because many genes are often present in multiple copies. The genome size of various eucaryotes was compared with that of viruses and procaryotes in Figure 3.67.

3.17 Cell Division and Sexual Reproduction

Cell division in eucaryotes is a highly ordered process and has been divided by cell biologists into four distinct phases (Figure 3.70). The G_1 (G stands for "gap") phase is a period of renewed biosynthetic activity following a previous cell division. The S phase is the beginning of DNA synthesis and continues until the complement of chromosomes in the nucleus has doubled (the process of mitosis, see also Figure 1.18). The period from the point of completion of DNA synthesis until the cell actually begins to divide is defined as the G_2 phase. Finally the M (mitosis) stage begins, which is the period of nuclear division. Following nuclear division the cytoplasm is partitioned into two cells and the division cycle is complete. Although the length of the total cell division cycle in different eucaryotes is highly variable, the major variability is observed in the length of the G_1 phase.

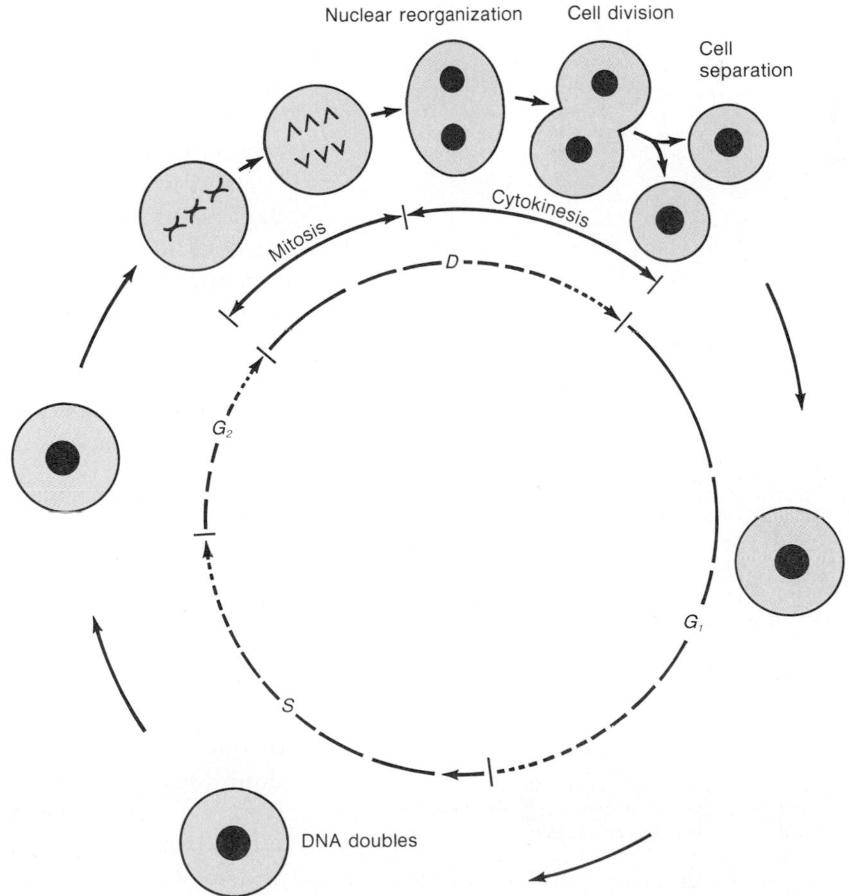

Figure 3.70 The cell-division cycle in a eucaryote. S, period during which DNA synthesis takes place; D, period of mitosis and cell division; G_1, growth period before S; G_2, growth period after S. The lengths of the various phases vary among organisms. In some organisms, G_2 is absent.

Cell division in most bacteria follows the process of **binary fission**. In a growing culture of a rod-shaped bacterium such as *Escherichia coli*, for example, cells are observed to elongate to approximately twice the length of an average cell and then form a partition which eventually separates the cell into two daughter cells (Figure 3.71). This partition is referred to as a *septum* and is the result of the inward growth of the plasma membrane and cell wall from opposing directions until the two daughter cells are pinched off (Figure 3.71). During the growth cycle all cellular constituents increase in number such that each daughter cell receives a complete chromosome and sufficient copies of all other macromolecules, monomers and inorganic ions to exist as an independent cell. The partitioning of the replicated DNA molecule between the two daughter cells depends on the DNA remaining attached to the cell membrane during division (Figure 3.71).

The time required for a complete growth cycle in bacteria is highly variable and is dependent upon a number of factors, both nutritional and genetic. Under the best nutritional conditions the bacterium *E. coli* can complete the cycle in about 20 minutes; a few bacteria can grow even faster than this but many grow much slower. The control of cell division is a complex process and appears intimately tied to chromosomal replication events. We discuss in Chapter 9 the replication of the bacterial chromosome in relation to cell division, and how the copies of the chromosome are partitioned to each daughter cell.

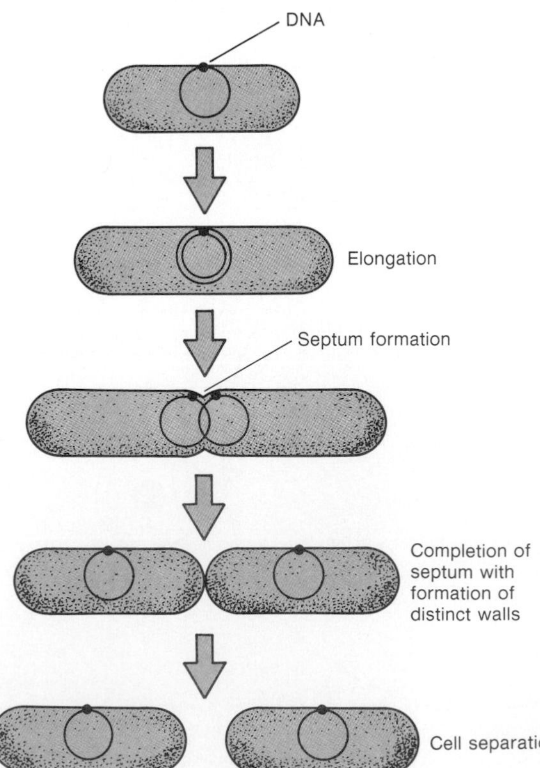

Figure 3.71 The process of binary fission in a rod-shaped bacterium.

Sexual reproduction Sexual reproduction in eucaryotic microorganisms involves conjugation, the coming together and fusing of two cells called **gametes**, which are analogous to the sperm and egg of multicellular eucaryotes. The conjugating gametes form a single cell called a **zygote**, and the nucleus of the zygote usually results from fusion of the nuclei of the two gametes. The zygote nucleus thus has twice the chromosome complement of the gametes. The chromosome number of the gametes is called the **haploid** number, and the zygote then has twice that many, the **diploid** number.

Meiosis is the process by which the change from the diploid to the haploid state is brought about. Whereas mitosis causes a *doubling* of the chromosome number and yields two progeny cells with the original chromosome number, the process of meiosis *reduces* the number of chromosomes by one-half and yields haploid precursors of gametes. The chromosomes of diploids always occur in pairs, called homologs, and during meiosis one chromosome of each pair goes to each of the two gametes formed. This regular assortment occurs because the homologs pair specifically at the time of meiosis, and one of each pair is then pulled to each pole by the spindle fibers, after which division of the cell into two cells occurs. A second division then occurs resulting in the formation of four haploid gametes.

Multicellular eucaryotes are usually diploid, the haploid phase being present only in the short-lived germ cells (sperm and eggs). In eucaryotic microorganisms the extent of development of the haploid and diploid phases varies. In many eucaryotic microorganisms the predominant growth phase is haploid, and the diploid phase is transitory; in others the diploid phase dominates, and in still others both haploid and diploid phases occur independently.

The advantages of sexual reproduction Sexually reproducing microorganisms undergo an alternation of haploid and diploid generations. The major benefit of this type of reproduction, as compared with a vegetative cycle, is that reduction division (meiosis) followed by reestablishment of the diploid chromosome number following fertilization, allows for tremendous reorganization of genetic material. Instead of one cell dividing to form two identical cells by doubling all of its cellular constituents, a cell derived from the fusion of gametes has received genetic material from two independent sources. This is obviously a major way of generating genetic diversity. Also, during meiosis, genetic recombination can occur by chromosomal crossing over. This phenomenon enhances genetic reassortment because each of the gametes produced will be genotypically unique.

Diploid organisms reproducing by sexual means can greatly benefit from the fact that one of the two copies of each gene can mutate and serve as the source of new genetic diversity. Beneficial mutations from different individuals of the same species can come together by sexual reproduction and thus be spread quickly among the population. In asexually reproducing procaryotes, on the other hand, independently derived beneficial mutations compete with one another in the two cell populations that spawned them and are not united until a sexual process occurs (if it occurs at all).

Processes similar to sexual reproduction also occur in procaryotes, but by mechanisms that are quite distinct from the process in eucaryotes. First, the process is quite fragmentary, almost never involving whole chromosome complements of the two cells. Second, the DNA is transferred in only one direction, from a donor to a recipient. Third, the mechanisms by which DNA transfer occurs are specialized. Three distinct types of mechanisms for DNA transfer have been recognized: 1) *Conjugation*, in which DNA transfer occurs as a result of cell-to-cell contact. Conjugation is the bacterial process that most closely resembles sex in eucaryotes. 2) *Transduction*, in which DNA transfer is mediated by viruses. 3) *Transformation*, in which free DNA is involved. In transformation, the donor cell generally lyses, releasing DNA into the medium, and some of this free (naked) DNA is taken up by recipient cells. We discuss the details of these various DNA transfer processes in Chapter 7.

Although biologists have struggled for years to understand why sexual reproduction is so successful, the general conclusion is that sexual processes and the reshuffling of genes that it entails, must increase the probability of survival and success of the organism in its environment. In the case of bacteria, it should be pointed out that our knowledge of bacterial sexuality is quite limited. Although laboratory evidence suggests that sexuality is not the norm in bacteria, in nature this may not be the case. Indeed evidence from studies of the bacterial transfer of antibiotic resistance genes located on small genetic elements called plasmids (see Section 5.5), suggest that sexual processes among bacteria in nature may be widespread. Thus, reshuffling of genes by sexual processes in bacteria may actually be an important mechanism for generating genetic diversity in these organisms as well.

3.18 Comparisons of the Procaryotic and Eucaryotic Cell

At this stage it might be useful to draw comparisons between the procaryotic and eucaryotic cell. It should be clear by now that there are profound differences in the structures of these two cell types. One important distinction is that eucaryotes have many types of cellular functions segregated into membrane-containing structures. We discussed mitochondria and chloroplasts earlier and Table 3.4 lists a number of other membranous structures.

Table 3.5 groups these differences into several categories of which the most important are: nuclear structure and function, cytoplasmic structure and organization, and forms of motility.

Table 3.4 Membrane-containing structures in eucaryotes

Structure	Characteristics	Function
Mitochondria	Bacteria-size, complex internal membrane arrays	Energy generation: respiration
Chloroplasts	Green, chlorophyll-containing, many shapes, often quite large	Photosynthesis
Endoplasmic reticulum	Not a distinct organelle, extensive array of internal membranes	Protein synthesis
Golgi bodies	Membrane aggregates of distinct structure	Secretion of enzymes and other macromolecules
Vacuoles	Round, membrane-enclosed bodies of low density	Food digestion: food vacuoles; waste product excretion: contractile vacuoles
Lysosomes	Submicroscopic membrane-enclosed particles	Contain and release digestive enzymes
Peroxisomes	Submicroscopic membrane-enclosed particles	Photorespiration in plants
Glyoxysomes	Submicroscopic membrane-enclosed particles	Enzymes of glyoxylate cycle
Nucleus	Large, generally centrally located	Contains genetic material

Table 3.5 Comparison of the procaryotic and eucaryotic cell

	Procaryotes	Eucaryotes
Groups	Eubacteria, archaebacteria	Algae, fungi, protozoa, plants, animals
Nuclear structure and function:		
Nuclear membrane	Absent	Present
Nucleolus	Absent	Present
DNA	Single molecule, not complexed with histones (other DNA in plasmids)	Present in several chromosomes, usually complexed with histones
Division	No mitosis	Mitosis; mitotic apparatus with microtubular spindle
Sexual reproduction	Fragmentary process; no meiosis; usually only portions of genetic complement reassorted	Regular process; meiosis; reassortment of whole chromosome complement
Cytoplasmic structure and organization:		
Plasma membrane	Usually lacks sterols	Sterols usually present
Internal membranes	Relatively simple; mesosomes	Complex; endoplasmic reticulum; Golgi apparatus
Ribosomes	70S in size	80S, except for ribosomes of mitochondria and chloroplasts, which are 70S
Simple membranous organelles	Absent	Present in vacuoles, lysosomes, microbodies (peroxisomes)
Respiratory system	Part of plasma membrane or mesosome; mitochondria absent	In mitochondria
Photosynthetic apparatus	In organized internal membranes or vesicles; chloroplasts absent	In chloroplasts
Cell walls	Present (in most) composed of peptidoglycan, other polysaccharide, protein, glycoprotein	Present in plants, algae, fungi; absent in animals, protozoa; usually polysaccharide
Endospores	Present (in some), very heat resistant	Absent
Gas vesicles	Present (in some)	Absent
Forms of motility:		
Flagellar movement	Flagella of submicroscopic size; each flagellum composed of one fiber of molecular dimensions; flagella rotate	Flagella or cilia; microscopic size; composed of microtubular elements arranged in a characteristic pattern of nine outer doublets and two central singlets; do not rotate
Nonflagellar movement	Gliding motility; gas vesicle-mediated	Cytoplasmic streaming and amoeboid movement; gliding motility
Microtubules	Probably absent	Widespread: present in flagella, cilia, basal bodies, mitotic spindle apparatus, centrioles
Size	Generally small, usually <2 μm in diameter	Usually larger, 2 to >100 μm in diameter

Several differences listed in Table 3.5 deserve further comment. Microtubules are widespread in eucaryotes, although they seem to be absent in procaryotes. Microtubules are associated with overt motile systems such as flagella and cilia and with mechanisms of chromosome movement in the mitotic apparatus. In these cases the microtubules arise from special structures, basal bodies and centrioles, which are of similar structure. Microtubules are probably not involved in amoeboid motion or cytoplasmic streaming, but in microorganisms without walls they may be involved in maintenance of nonspherical cell shapes. Because of the universal presence of microtubules in eucaryotes, and their apparent absence in procaryotes, it may be that the ability to form microtubules was an early acquisition of the evolving eucaryotic cell.

Another difference between procaryotes and eucaryotes deserving emphasis is the size of their respective ribosomes. Procaryotes have ribosomes with a sedimentation constant of 70S, which are composed of two subunits having constants of 50S and 30S. (The sum of 50 and 30 in this instance is 70 rather than 80 because the sedimentation constant is a function of the density of a particle which is mass per volume.) Cytoplasmic ribosomes of eucaryotes have a sedimentation constant of 80S, and these are composed of two subunits with sedimentation constants of 60S and 40S. The ribosomes of the mitochondria and chloroplasts of eucaryotes are 70S, however.

Although Table 3.5 emphasizes the great *structural* differences between procaryotes and eucaryotes, we should remember that these groups of organisms are chemically quite similar. Both contain proteins, nucleic acids, polysaccharides, and lipids of similar composition, and both use the same kinds of metabolic machinery. Thus it is not in their building blocks that procaryotes differ most strikingly from eucaryotes, but in how these building blocks are assembled to make the cell.

The student should also keep in mind the recent phylogenetic discoveries when thinking about procaryotes and eucaryotes: although eubacteria and archaebacteria are both procaryotic, their relationship to one another at the molecular level is probably not much greater than the relationship of either to eucaryotes. It is therefore likely that continued study of structure/function relationships, especially at the molecular level, of all types of procaryotic and eucaryotic cells, will greatly benefit our understanding of the procaryotic/eucaryotic dichotomy, and contribute to our further understanding of cell biology in general.

Study Questions

1. Calculate the surface/volume ratio of a spherical cell 10 μm in diameter, a cell 1 μm in diameter, and a rod-shaped cell 0.5 μm in diameter by 2 μm long. What are the consequences of these differences in surface/volume ratio for cell function?
2. From what you know about the nature of the bacterial cell wall, explain why a rod-shaped bacterial cell becomes a *spherical* structure when its wall is removed under conditions such that cell lysis cannot occur.
3. Describe in a single sentence the manner by which a unit membrane is formed from phospholipid molecules. How would the membrane formation process differ if the phospholipid molecules were dissolved in a nonaqueous solvent rather than water? How does a *membrane vesicle* differ from a unit membrane? How is such a vesicle *similar* to a unit membrane?
4. Explain in a single sentence why ionized molecules do not readily pass through the membrane barrier of a cell. How *do* such molecules get through the cell membrane?
5. Why is the bacterial cell wall rigid layer called "peptidoglycan"? What are the chemical reasons for the rigidity that is conferred on the cell wall by the peptidoglycan structure?
6. Since a single peptidoglycan molecule is very thin, explain in chemical terms how the very *thick* peptidoglycan-containing cell wall of the Gram-positive bacteria is formed.
7. Both lysozyme and penicillin bring about bacterial cell lysis, but by quite different mechanisms. Describe the mechanism by which each of these agents causes cell lysis. Be sure to include in your explanation an exact description of why *lysis* occurs.
8. Many bacteria are resistant to penicillin action. Can you think of several reasons why this might be?
9. Write a short explanation for the differences in Gram-stainability of Gram-positive and Gram-negative bacteria in terms of their cell wall structures.
10. List several functions for the outer wall layer in Gram-negative bacteria.
11. What is osmosis? Why does it occur? Of what significance is osmosis for the stability of the bacterial cell?

12. Write a clear explanation (two or three sentences) for why sucrose is able to stabilize bacterial cells from lysis by lysozyme.
13. In a few sentences, write an explanation for how a motile bacterium is able to "sense" the direction of an attractant and move toward it. Recall that bacteria are not able to distinguish *concentration* differences from one end of the cell to the other end.
14. In a few sentences, indicate how the bacterial endospore differs from the vegetative cell in structure, chemical composition, and ability to resist extreme environmental conditions.
15. The discovery of the bacterial endospore was of great practical importance. Why?
16. Describe one *chemical* difference between the eucaryotic and procaryotic membrane. Can you offer an explanation for why this chemical difference might exist?
17. Compare and contrast the chemical composition of the eubacterial and the archaebacterial membrane. Be sure to include a discussion of the chemical differences in *linkage* of these two kinds of membranes.
18. Water molecules penetrate cell membranes fairly readily but hydrogen ions (protons) do not, even though a proton is smaller than a water molecule. Why?
19. List three kinds of *membrane proteins* and give a short explanation for the function of each.
20. Compare and contrast the following processes: *passive diffusion*, *facilitated diffusion*, *group translocation*, and active *transport*. For each of these processes, include a discussion of specificity, energy requirement, and transport against a concentration gradient.
21. Compare and contrast the *flagellum* of procaryotic and eucaryotic cells. Include in your answer a discussion of both structure and function.
22. How does a *cilium* differ from a flagellum? In what way(s) are they similar?
23. Construct an experimental procedure that you could use for isolating and purifying the cell membrane of a bacterial cell. How would you determine that the preparation you obtained consisted solely of membranes?
24. List several properties by which *mitochondria* and *chloroplasts* are similar. List several ways in which they differ.
25. List two ways in which the DNA of the eucaryote differs from the DNA of the procaryote. List two ways in which the DNA in these two kinds of organisms is similar.
26. Set up a table following the format of Table 3.5 with the second and third columns blank, and then fill in the blanks. As you do so, think back to the figures in this chapter that illustrate the properties being considered.

Supplementary Readings

Alberts, B., D. Bray, J. Lewis, M. Raff, K. Roberts, and **J. D. Watson**. 1983. *The Molecular Biology of the Cell*. Garland Publishing, New York. A substantial textbook with extensive coverage of cell structure, with emphasis on the eucaryotic cell.

Annual Review of Biochemistry. Annual Reviews, Inc., Palo Alto. This annual volume always has several good reviews on cell biology, with emphasis on deep biochemical aspects.

Darnell, James E., H. Lodish, and **D. Baltimore**. 1986. *Molecular Cell Biology*. Scientific American Books, New York. The "other" cell biology book, almost a clone of Alberts et al., but with its own strengths (and weaknesses).

De Duve, C. 1984. *A Guided Tour of the Living Cell*. Volume One. Scientific American Books, New York. A cute, colorful, and frequently entertaining discussion of cell biology. Primary focus on the eucaryotic cell.

Harold, F. M. 1986. *The Vital Force: A Study of Bioenergetics*. W. H. Freeman, New York. An incisive and highly readable textbook that weaves discussion of cell structure into its coverage of membrane function. The best modern treatment of membrane structure and function.

Watson, J. D., N. H. Hopkins, J. W. Roberts, J. A. Steitz, and **A. M. Weiner**. 1987. *Molecular Biology of the Gene*, 4th edition. Benjamin/Cummings, Menlo Park. Volume I of this two-volume update is strong on weak bonds and has useful overview of the principles of cell biology.

4 Nutrition, Metabolism, and Biosynthesis

A key feature of a living organism is its ability to organize molecules and chemical reactions into specific structures and systematic sequences. The ultimate expression of this organization is the ability of a living organism to replicate itself. The term **metabolism** is used to refer to all the chemical processes taking place within a cell. The word *metabolism* is derived from the Greek word *metabole* which means *change*, and we can think of a cell as continually changing as it carries out its life processes. Although the cell appears under the microscope to be a fixed and stable structure, it is actually a dynamic entity, continually undergoing change, as a result of all the chemical reactions which are constantly taking place.

Microbial cells are built of chemical substances of a wide variety of types, and when a cell grows, all of these chemical constituents increase in amount. The basic chemical elements of a cell come from outside the cell, from the **environment**, but these chemical elements are transformed by the cell into the characteristic constituents of which the cell is composed.

The chemicals from the environment of which a cell is built are called **nutrients**. Nutrients are taken up into the cell and are changed into cell constituents. This process by which a cell is built up from the simple nutrients obtained from its environment is called **anabolism**. Because anabolism results in the biochemical synthesis of new material, it is often called **biosynthesis**.

Biosynthesis is an energy-requiring process, and each cell must thus have a means of obtaining energy. This energy is obtained from the environment, and three kinds of energy sources are used: light, inorganic chemicals, and organic chemicals. Although a number of organisms obtain their energy from light, most microorganisms obtain energy from chemical compounds. Chemicals used as energy sources are broken down into simpler constituents, and as this breakdown occurs, energy is released. The process by which chemicals are broken down and energy released is called **catabolism**. Cells also need energy for other cell functions, such as cell movement (motility).

We thus see that there are two basic kinds of chemical transformation processes occurring in cells, the building-up processes called *anabolism* and the breaking-down processes called *catabolism*. Metabolism is thus the collective result of anabolic and catabolic reactions.

A simplified overview of cell metabolism is shown in Figure 4.1, which depicts how catabolic reactions supply energy needed for cell functions, and how anabolic (biosynthetic) reactions bring about the synthesis of cell components from nutrients. Note that in anabolism, nutrients from the environment are converted into *cell components* whereas in catabolism, nutrients from the environment are converted into *waste products*. Catabolic

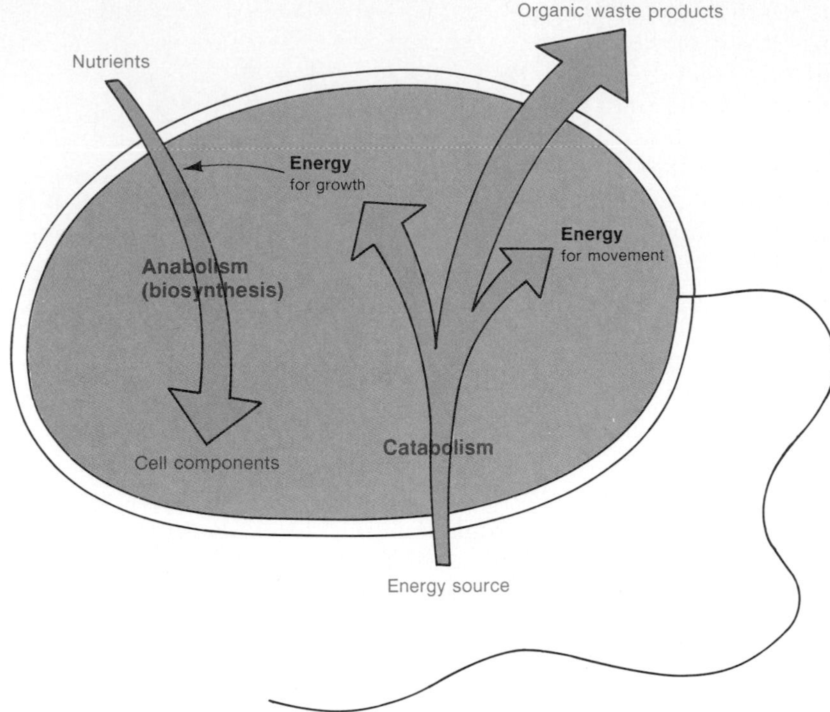

Figure 4.1 A simplified view of the major features of cell metabolism.

reactions result in the *release* of energy whereas anabolic reactions result in the *consumption* of energy. In this chapter, we will consider some of the anabolic and catabolic processes used by microorganisms.

As we have noted, three different kinds of energy sources are used. It is conventional to place microorganisms into classes, depending on the sources of energy which they use, and these classes are summarized in Table 4.1. All of the terms used to describe these classes employ the combining form *troph*, derived from a Greek word meaning *to feed*. Thus, organisms which use *light* as an energy source are called **phototrophs** (*photo* is from the Greek for *light*). Organisms which use *inorganic chemicals* as energy sources are called **lithotrophs** (*litho* is from the Greek for *rock*). Organisms which use *organic chemicals* as energy sources are called **heterotrophs** (which means, literally, feeding from sources other than oneself). Because they use organic compounds, heterotrophs are sometimes called *organotrophs*. Most of the organisms which we deal with in microbiology use organic compounds as energy sources, and are hence heterotrophs. The material in the present chapter will deal with heterotrophic metabolism and we reserve for Chapter 16 a discussion of the utilization of light and inorganic chemicals as energy sources.

A knowledge of cell metabolism is useful in understanding the biochemistry of microbial growth.

Table 4.1 Terms used to describe various nutritional types of microorganisms.

A. Three kinds of energy sources	
Energy source	**Term used**
Light	Phototroph
Inorganic chemicals	Lithotroph
Organic chemicals	Heterotroph
B. Two kinds of carbon sources	
Carbon source	**Term used**
Inorganic (carbon dioxide)	Autotroph
Organic	Heterotroph
C. Some mixed terms	
Photoautotroph: Use light and inorganic carbon	
Photoheterotroph: Use light and organic carbon	
Lithotrophic heterotroph: Use inorganic energy source and organic carbon	

Energy is needed for macromolecular synthesis and for the variety of chemical reactions needed for cell growth. Also, a knowledge of metabolism aids in developing useful laboratory procedures for culturing microorganisms, and in developing suitable procedures for preventing the growth of unwanted microbes. Because many of the important practical consequences of microbial growth, such as infectious disease, are linked to microbial metabolism, a knowledge of cellular metabolism is of great use in applied and medical microbiology. Even the formation of metabolic waste products is of interest. For instance, one important waste product produced by yeast during catabolism is *ethanol*, the key constituent of alcoholic beverages (wine, beer, whisky, etc.). We will have further to say about the formation of ethanol by yeast later in this chapter.

4.1 Nutrition and Biosynthesis

Chemical composition of a cell During our discussion of cellular structure in Chapter 3, some aspects of the chemical makeup of various cell constituents were presented. An overview of the chemical structures themselves was presented in Chapter 2. It may be useful at this point to summarize what is known about the overall chemical composition of a cell (Table 4.2). Cells contain large numbers of small molecules, such as water, inorganic ions, and organic substances, but consist primarily of large molecules, the polymers of the cell, the most important of which are the proteins and nucleic acids. The cell may obtain most of the small molecules from the environment in preformed condition, whereas the large molecules are synthesized in the cell. However, many small molecules are also synthesized within the cell, from substances obtained from the environment.

Although there are 92 naturally occurring elements, virtually the whole mass of a cell consists of only four types of atoms: carbon, oxygen, hydrogen, and nitrogen. In addition, a number of other atoms are quantitatively less important, but functionally very important. These include phosphorus, calcium, magnesium, sulfur, iron, zinc, manganese, copper, molybdenum, and cobalt, which are present in microbial cells, but in far lesser amounts than C, H, O, and N. With the exception of molybdenum, the elements of microbial life are all of atomic number 30 or less.

Water accounts for the bulk of the weight of a cell (around 70 percent) and in any chemical analysis this water must first be removed by drying. Macromolecules together make up the bulk of the remaining dry weight, proteins being the most abundant class of macromolecules (Table 4.2). Inorganic ions and monomers constitute the balance of the cell dry weight. Chemically, therefore, living organisms have evolved around a relatively few types of chemical elements. But the possibilities for polymerizing even a restricted number of different small molecules is enormous, leading to the various molecular combinations that are at the heart of the diversity of life on Earth.

Table 4.2 Chemical composition of a bacterial cell[a]

Molecule	Percent wet weight	Percent dry weight[b]	Different kinds
Water	70	—	1
Total macromolecules	26	96	~1500
Protein	15	60	~1100
Polysaccharide	3	5	2[c]
Lipid	2	9	4[d]
DNA	1	3	1
RNA	5	19	~500
Total monomers	3	3	~350
Amino acids and precursors	0.5	0.5	~100
Sugars and precursors	2	2	~50
Nucleotides and precursors	0.5	0.5	~200
Inorganic ions	1	1	18
Totals	100%	100%	

[a]Data taken from Ingraham, J.L., O. Maaløe, and F.C. Neidhart. 1983. Growth of the bacterial cell (Sunderland, Massachusetts: Sinauer Associates, Inc.) and from other sources.
[b]Dry weight of a cell of *E. coli* $\simeq 2.8 \times 10^{-13}$ g.
[c]Assuming peptidoglycan and glycogen to be the major polysaccharides present.
[d]There are several classes of phospholipids, each of which exists in many kinds because of variability in fatty acid composition between species and also due to different growth conditions.

Nutrition Substances in the environment used by organisms for catabolism and anabolism are called **nutrients**. Nutrients can be divided into two classes: (1) essential nutrients, without which a cell cannot grow; and (2) useful but dispensable nutrients, which are used if present but are not essential. Some nutrients are the building blocks from which the cell makes macromolecules and other structures, whereas other nutrients serve only for energy generation without being incorporated directly into the cellular material; sometimes a nutrient can play both roles. The required substances can be divided into two groups, macronutrients and micronutrients, depending on whether they are required in large or small amounts.

Macronutrients We learned in Chapter 2 that the major constituent of a microbial cell is **carbon**. Not surprisingly, therefore, a bacterium requires more carbon than any other nutrient. Carbon and a few other elements needed in major amounts are classed together as nutritional **macronutrients**. Most procaryotes require an organic compound of some sort as their source of carbon. Organic compounds are also commonly used as energy sources. Glucose, for example, is used by a large number of bacteria as both a carbon and energy source. Nutritional studies have shown that many bacteria can assimilate a variety of different organic carbon compounds and use them to make new cell material. Amino acids, fatty acids, organic acids, sugars, nitrogenous bases, aromatic compounds, and countless other organic compounds have been shown to be used by one bacterium or another.

After carbon, the next most abundant element in the cell is **nitrogen**. A typical bacterial cell is 12–15 percent nitrogen (by dry weight) and nitrogen is a major constituent of proteins and nucleic acids. Nitrogen is also present in the complex polysaccharide peptidoglycan, the rigid layer of the cell wall of most bacteria (see Section 3.5). Nitrogen can be found in nature in both organic and inorganic forms. As a major constituent of amino acids and nitrogenous bases, nitrogen is available to organisms in this form from the breakdown and mineralization of dead organisms. The bulk of available nitrogen in nature is in *in*organic form, either as ammonia (NH_3) or nitrate (NO_3^-) (see Section 17.14). Most bacteria are capable of using ammonia as the sole nitrogen source and many can also use nitrate. The atmosphere contains a vast store of nitrogen as nitrogen gas, N_2. However, nitrogen gas only serves as a nitrogen source for certain bacteria, the nitrogen-fixing bacteria (see Sections 16.24 and 17.14).

Phosphorus occurs in nature in the form of organic and inorganic phosphates and is used by the cell primarily in nucleic acids and phospholipids. Most, if not all, microorganisms utilize inorganic phosphate ($PO_4^{=}$) for growth. Organic phosphates occur very often in nature, however, and they can be utilized following the action of cell enzymes called *phosphatases*. The latter hydrolyze the organic phosphate ester, releasing free inorganic phosphate.

Sulfur is absolutely required by organisms because of its structural role in the amino acids cysteine and methionine (see Section 2.7) and because it is present in a number of vitamins (thiamine, biotin, and lipoic acid). Sulfur undergoes a number of chemical transformations in nature carried out exclusively by microbes (see Section 17.15), and is available to organisms in a variety of forms. Most cell sulfur originates from inorganic sources, either sulfate ($SO_4^{=}$) or sulfide (HS^-).

Potassium is universally required. A variety of enzymes, including some of those involved in protein synthesis, are specifically activated by potassium. **Magnesium** functions to stabilize ribosomes, cell membranes, and nucleic acids, and is also required for the activity of many enzymes, especially those involving phosphate transfer. Thus, relatively large amounts of magnesium are required for growth. **Calcium**, which is not essential for the growth of many microorganisms, occasionally plays a role in stabilizing the bacterial cell wall and plays a key role in the heat stability of bacterial endospores. Even though calcium and magnesium are closely related in the Periodic Table, calcium cannot replace magnesium in most of its roles in the cell.

Sodium is required by some but not all organisms, and its need may reflect the habitat of the organism. For example, seawater has a high sodium content, and marine microorganisms generally require sodium for growth, whereas closely related freshwater forms are usually able to grow in the absence of sodium. The extremely halophilic (salt-loving) halobacteria require large amounts of Na^+, approaching the limits of solubility of NaCl. For example, cells of *Halobacterium*, an extremely halophilic archaebacterium, actually disintegrate in solutions containing less than 2.5 molar NaCl; *Halobacterium* grows optimally in 4.3 molar NaCl (over five times saltier than the oceans). The biology of this interesting organism is discussed in Section 19.33.

Iron is required in fairly large amounts, although not at the level of other macronutrients. Iron is found in a number of enzymes in the cell, especially those involved in respiration (see later).

Common sources of these major elements needed for biosynthesis are summarized in Table 4.3.

Micronutrients (trace elements) Although required in just tiny amounts, micronutrients are never-

Table 4.3 Common forms of the major elements needed for biosynthesis of cell components

Element	Usual form in the environment
C	Carbon dioxide (CO_2) (autotrophs)
	Organic compounds (heterotrophs)
H	Water (H_2O)
	Organic compounds
O	Water (H_2O)
	Oxygen gas (O_2)
N	Ammonia (NH_3)
	Nitrate (NO_3^-)
	Organic compounds (e.g., amino acids)
P	Phosphate (PO_4^{3-})
S	Hydrogen sulfide (H_2S)
	Sulfate (SO_4^{2-})
	Organic compounds (e.g., cysteine)
K	K^+
Mg	Mg^{2+}
Ca	Ca^{2+}
Na	Na^+
Fe	Fe^{3+}
	Organic iron complexes

theless just as critical to the overall nutrition of a microorganism as are macronutrients.

Cobalt is needed only for the formation of vitamin B_{12}, and if this vitamin is added to the medium, cobalt is no longer needed. **Zinc** plays a structural role in many enzymes including carbonic anhydrase, alcohol dehydrogenase, and RNA and DNA polymerases. Zinc apparently excels at holding together protein subunits in the proper configuration for enzyme activity, and is thought to play this role in the enzymes noted. **Molybdenum** is present in certain enzymes called molybdoflavoproteins, which are involved in assimilatory nitrate reduction. It is also present in nitrogenase, the enzyme involved in N_2 reduction (N_2 fixation, see Section 16.24). **Copper** plays a role in certain enzymes involved in respiration. Like iron, the copper ion is a site for reaction with O_2. **Manganese** is an activator of many enzymes that act on phosphate-containing compounds. Manganese is also found in certain superoxide dismutases, enzymes of critical importance for detoxifying toxic forms of oxygen. Manganese also plays an important role in green plant photosynthesis. **Nickel** is present in enzymes called hydrogenases that function to either take up or evolve H_2. **Tungsten** and **selenium** are required by those bacteria capable of metabolizing formate. Selenium is present as part of the enzyme formate dehydrogenase.

Now that we have surveyed briefly the chemical constituents of a microbial cell, we turn to the biochemical reactions that are involved in the major cell functions. Because life is carbon-based, our discussion focuses primarily on the carbon (organic) compounds of the cell, and the reactions by which these are synthesized. The following discussion is in two parts, in keeping with the distinction we have been making between energy metabolism (catabolism) and biosynthesis (anabolism). Because biosynthesis requires energy, we begin with a discussion of energy metabolism.

4.2 Energy

Three forms of energy are useful biologically: chemical, electrical, and light energy. **Chemical energy** is the energy released when organic or inorganic compounds are oxidized, and is the primary source of energy for heterotrophs and lithotrophs (see Table 4.1). **Light energy** is the primary or sole energy source of the phototrophs. We discuss the utilization of light energy in Chapter 16. **Electrical energy** is produced when electrons move from one place to another, as when an electric current flows through a battery, and is expressed as a flow of current due to a difference in voltage (electrical potential) between two points. Although cells do not use electrical energy from the environment, they produce electrical energy as part of their energy-generating process. In keeping with the idea that life is basically a chemical process, we find that the large majority of energy reactions within a cell involve chemical energy. Even when light is used as an energy source, it is first converted into chemical energy before it is actually used by the cell.

The first law of thermodynamics tells us that energy can be converted from one form to another, but can neither be created nor destroyed. Because its many forms are interconvertible, energy is most conveniently expressed by a single energy unit. Although a variety of units exist, in biology the most commonly used energy units are the kilocalorie (kcal) and the kilojoule. A kilocalorie is defined as the quantity of heat energy necessary to raise the temperature of 1 kilogram of water by 1°C. One kilocalorie is equivalent to 4.184 kilojoules. We will use the convention kilocalorie throughout this book.

Chemical reactions are accompanied by *changes* in energy. The amount of energy involved in a chemical reaction is expressed in terms of the gain or loss of energy during the reaction. There are two expressions of the amount of energy released during a chemical reaction, abbreviated H and G. H, called enthalpy, expresses the total amount of energy released during a chemical reaction. However, some of the energy released is not available to do useful work, but instead is lost as heat energy. G, called **free energy**, is used to express the energy released that is available to do useful work. The change in free energy during a reaction is expressed as $\Delta G^{0\prime}$, where the symbol "Δ" should be read to mean "change in." Reactant and product concentrations and pH (when H^+ is a reactant or product) will affect the observed free energy changes. The superscripts "0" and "′" mean that a given free-energy value was obtained under "standard" conditions: pH 7, 25°C, all reactants and products initially at 1 molar concentration. If in the reaction:

$$A + B \rightarrow C + D$$

the $\Delta G^{0\prime}$ is *negative*, then free energy is released and the reaction as drawn will occur spontaneously; such reactions are called **exergonic**. If, on the other hand $\Delta G^{0\prime}$ is *positive*, the reaction will not occur spontaneously, but the reverse reaction (to the left) will occur spontaneously; such reactions are called **endergonic**.

In an exergonic reaction, the reaction proceeds until the concentration of products builds up, and then the reverse reaction, the conversion of products back to reactants, increases. An equilibrium is eventually reached in which the forward and reverse reactions are exactly balanced. This balanced condition does not mean that reactant and product occur in equal concentrations. The concentration of products and reactants at equilibrium is related to the free energy of the reaction. If the reaction proceeds with a *large* negative $\Delta G^{0\prime}$, then the equilibrium is far toward the products and very little of the reactants will remain. In contrast, if the reaction proceeds with a *small* negative $\Delta G^{0\prime}$, then at equilibrium there are nearly equal amounts of products and reactants. By determining the concentrations of products and reactants at equilibrium, it is possible to calculate the free-energy yield of any reaction (see Appendix 1 for further details).

In addition to speaking of the free energy yield of reactions, it is also necessary to talk about the free energy of individual substances. This is the *free energy of formation*, the energy yielded or energy required for the formation of a given molecule from the elements. By convention, the free energy of formation (G^0_f) of the elements (for instance, C, H_2, N_2) is zero. If the formation of a compound from the elements proceeds exergonically, then the free energy of formation of the compound is negative (energy is released), whereas if the reaction is endergonic (energy is required) then the free energy of formation of the compound is positive. A few examples of free energies of formation are: H_2O (water), −57 kcal/mole; CO_2 (carbon dioxide), −94.5 kcal/mole; H_2 (hydrogen), 0 kcal/mole (by definition); NH_4^+ (aqueous ammonia), −19 kcal/mole; N_2O (nitrous oxide), +24.7 kcal/mole; glucose −219 kcal/mole; CH_4 (methane), −12 kcal/mole. For most compounds G^0_f is negative, reflecting the fact that compounds tend to form spontaneously from the elements. Again, the relative probabilities of different reactions (formations in this case) can be derived from comparison of the respective energies of formation. Thus we see that glucose (G^0_f, −219 kcal/mole) is more likely to form from carbon, hydrogen, and oxygen than is methane (G^0_f, −12 kcal/mole) to form from carbon and hydrogen. The positive G^0_f for nitrous oxide (+24.7 kcal/mole) tells us that this molecule will not form spontaneously but rather will decompose to nitrogen and oxygen. The free energies of formation of a number of compounds of microbiological interest are given in Appendix 1.

Using free energies of formation, it is possible to calculate the *change* in free energy occurring in a given reaction. For a simple reaction, such as A + B → C + D, $\Delta G^{0\prime}$ is calculated by subtracting the *sum* of the free energies of formation of the reactants (in this case A and B) from the products (C and D). Thus:

$$\Delta G^{0\prime} \text{ of } A + B \rightarrow C + D$$
$$= G^0_f\,[C + D] - G^0_f\,[A + B]$$

The saying "products minus reactants" summarizes the necessary steps for calculating changes in free energy during chemical reactions. However, it is necessary to balance the reaction chemically before free energy calculations can be made. Appendix 1 details the steps involved in calculating free energies for any hypothetical reaction.

4.3 Activation Energy, Catalysis, and Enzymes

Free-energy calculations tell us only what conditions will prevail when the reaction or system is at equilibrium; they do not tell us how long it will take for equilibrium to be reached. The formation of water from gaseous oxygen and hydrogen is a good example. We have already seen that this reaction is quite favorable (energy of formation of −57 kcal/mole). However, if we were to simply mix O_2 and H_2 together, no measurable reaction would probably occur within our lifetime. The explanation is that the rearrangement of oxygen and hydrogen atoms to form water requires that the chemical bonds of the reactants be broken first. The breaking of bonds requires energy and this energy is referred to as **activation energy**. Activation energy is the amount of energy (in kcal) required to bring all molecules in a chemical reaction to the reactive state. For a reaction that proceeds with a net release of free energy (that is, an exergonic reaction), the situation is as diagramed in Figure 4.2.

The idea of activation energy leads us to the concept of catalysis. A **catalyst** is a substance which serves to *lower* the activation energy of a reaction. A catalyst serves to *increase* the rate of reaction even though it itself is not changed. It is important to note that catalysts do not affect the free energy change of a reaction; catalysts only affect the *speed* at which reactions proceed.

Most reactions in living organisms will not occur at appreciable rates without catalysis. The catalysts of biological reactions are proteins called **enzymes**. Enzymes are highly specific in the reactions which they catalyze. That is, each enzyme catalyzes only a single type of chemical reaction, or in the case of certain enzymes, a class of closely related reactions. This specificity is related to the precise three-dimensional structure of the enzyme molecule. In an enzyme-catalyzed reaction, the enzyme temporarily combines with the reactant, which is termed a **substrate** of the enzyme. The enzyme is generally much larger than the substrate(s) and the combination of

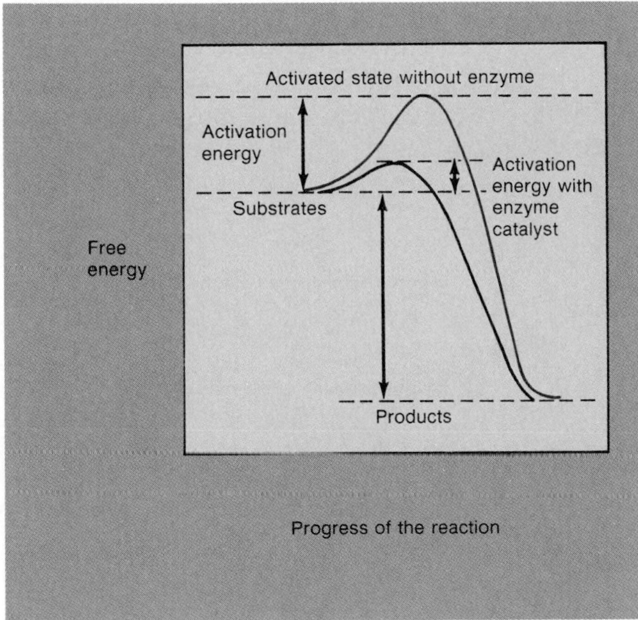

Figure 4.2 The concept of activation energy. Chemical reactions may not proceed spontaneously even though energy would be released, because the reactants must first be *activated*. Once activation has occurred, the reaction then proceeds spontaneously. Catalysts such as enzymes lower the required activation energy.

enzyme and substrate(s) depends on so-called weak bonds, such as hydrogen bonds, van der Waal's forces, and hydrophobic interactions (see Section 2.7) to join the enzyme to the substrate. The small portion of the enzyme to which substrates bind is referred to as the **active site** of the enzyme.

Binding of substrate to enzyme produces a so-called **enzyme-substrate complex**, and the stress or strain placed on the substrate(s) following enzyme binding reduces the activation energy required to make the reaction proceed from substrates to product (Figure 4.2). Note that the reaction depicted in Figure 4.2 is exergonic, since the free energy of formation of the substrate is greater than that of the product. Enzymes can also catalyze endergonic reactions, converting energy-poor substrates into energy-rich products. Here again, an activation energy barrier must be overcome. In addition, in the case of an endergonic reaction, sufficient free energy must be put *into* the system to raise the energy level of the substrates to that of the products. Although theoretically all enzymes are reversible in their action, in practice, enzymes catalyzing highly exergonic or highly endergonic reactions are essentially unidirectional. If a particularly exergonic reaction needs to be reversed during cellular metabolism, a distinctly different enzyme is frequently involved in the reaction.

As we have discussed, enzymes are proteins, polymers of subunits called amino acids (see Section 2.7). Each enzyme has a specific three-dimensional shape. The linear array of amino acids (primary structure) will fold and twist into a specific configuration recognized as the secondary and tertiary structure. The precise enzyme conformation may be seen more easily in a computer-generated space-filling model (Figure 4.3). In this example, the large cleft is the site where the substrate binds.

Many enzymes contain small nonprotein molecules which participate in the catalytic function. These small enzyme-associated molecules are divided into two categories on the basis of the nature of the association with the enzyme, *prosthetic groups* and *coenzymes*. **Prosthetic groups** are bound very tightly to their enzymes, usually permanently. If an enzyme contains a prosthetic group, the protein part alone is called an *apoenzyme* and the complete enzyme, formed when the prosthetic group is attached, is called a *holoenzyme*. The heme group present in cytochromes is an example of a prosthetic group; cytochromes will be described in detail later in this chapter. **Coenzymes** are bound rather loosely to enzymes and a single coenzyme molecule may associate with a number of different enzymes at different times during growth. Coenzymes serve as intermediate carriers

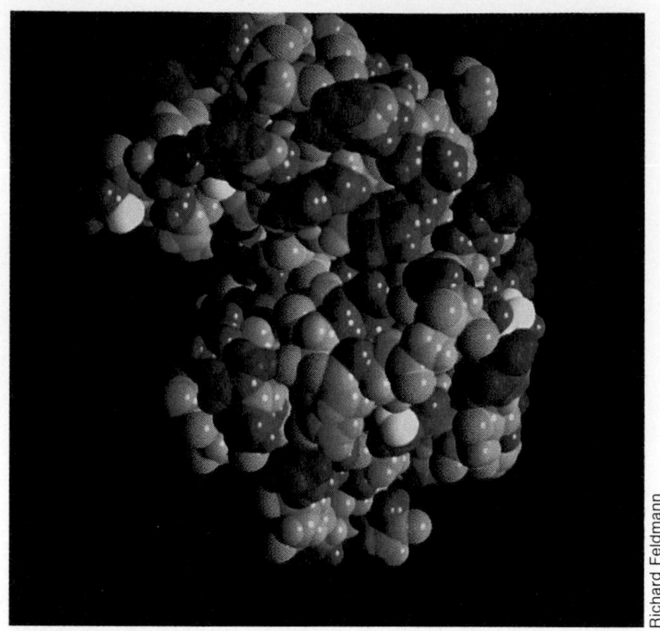

Figure 4.3 Computer-generated space-filling model of the enzyme lysozyme. The substrate-binding site is in the large cleft.

of small molecules from one enzyme to another. Most coenzymes are derivatives of vitamins.

Enzymes are generally named either for the substrate they bind or for the chemical reaction which they catalyze, by means of the combining form *-ase*. Thus cellul*ase* is an enzyme that attacks cellulose, glucose oxid*ase* is an enzyme that catalyzes the oxidation of glucose, and ribonucle*ase* is an enzyme that decomposes ribonucleic acid.

4.4 Oxidation-Reduction

The utilization of chemical energy in living organisms involves **oxidation-reduction reactions**. Chemically, an oxidation is defined as the *removal* of an electron or electrons from a substance. A reduction is defined as the *addition* of an electron (or electrons) to a substance. In biochemistry, oxidations and reductions frequently involve the transfer of not just electrons, but whole hydrogen atoms. A hydrogen atom consists of an electron plus a proton. We will on occasion need to distinguish between those oxidation-reduction reactions which involve electrons only or hydrogen atoms only, but the energetic concepts developed here will use the transfer of hydrogen atoms as primary examples. The student should also note the important fact that many oxidation-reduction reactions do *not* involve molecular oxygen (O_2). Instead, it is the transfer of hydrogen atoms (or electrons) that is important. For example, hydrogen gas, H_2, can release electrons and protons and become oxidized:

$$H_2 \rightarrow 2e^- + 2H^+$$

However, electrons cannot exist alone in solution; they must be part of atoms or molecules. The equation as drawn thus gives us chemical information but does not itself represent a real reaction. The above reaction is a *half reaction*, a term which implies the need for a second reaction. This is because for any oxidation to occur, a subsequent reduction must also occur. For example, the oxidation of H_2 could be coupled to the reduction of O_2, in a second reaction:

$$\tfrac{1}{2}O_2 + 2e^- + 2H^+ \rightarrow H_2O$$

This half reaction, which is a reduction, when coupled to the oxidation of H_2, yields the following overall balanced reaction:

$$H_2 + \tfrac{1}{2}O_2 \rightarrow H_2O$$

In reactions of this type, we will refer to the substance oxidized, in this case H_2, as the **electron donor**, and the substance reduced, O_2, as the **electron acceptor** (Figure 4.4). The key to understanding biological oxidations and reductions is to keep straight the proper half reactions—there must always be one reaction which involves an electron donor and another reaction involving an electron acceptor.

Reduction potentials Substances vary in their tendencies to give up electrons and become oxidized. This tendency is expressed as the **reduction potential** of the substance. This potential is measured electrically in reference to a standard substance, H_2. By

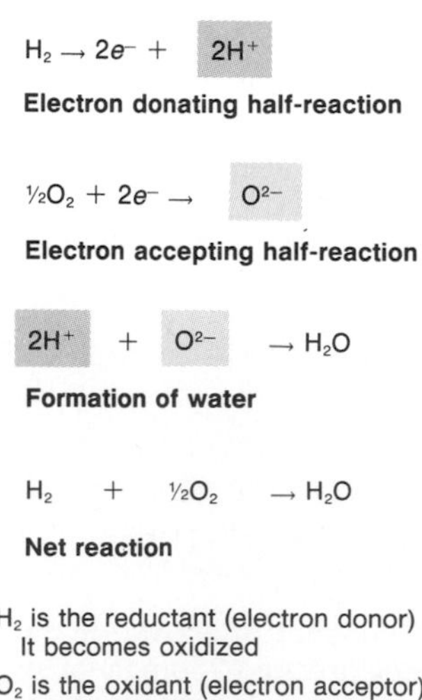

Figure 4.4 Example of a coupled oxidation-reduction reaction.

relating all potentials to a standard, it is possible to express potentials for various half reactions on a single scale, making possible ready comparisons between various reactions. By convention, reduction potentials are expressed for half reactions written with the oxidant on the left, that is, as *reductions*. Thus, oxidant + e^- ⟶ reduced product. If hydrogen ions are involved in the reaction, as is often the case, then the reduction potential will to some extent be influenced by the hydrogen ion concentration (pH). By convention in biology, reduction potentials are given for neutrality (pH 7), since the cytoplasm of the cell is neutral or nearly so. Using these conventions, the reduction potential of

$$\tfrac{1}{2}O_2 + 2H^+ + 2e^- \longrightarrow H_2O$$

is +0.816 V, and that of

$$2H^+ + 2e^- \longrightarrow H_2$$

is −0.421 V.

Oxidation-reduction pairs and coupled reactions Most molecules can serve as both electron donors and electron acceptors at different times, depending upon what other substances they react with. The same atom on each side of the arrow in the half reactions can be thought of as representing an oxidation-reduction (O–R) pair, such as $2H^+/H_2$, $\tfrac{1}{2}O_2/H_2O$. When writing an O–R pair, the *oxidized* form will always be placed on the left.

In constructing coupled oxidation-reduction reactions from their constituent half reactions, it is simplest to remember that the reduced substance of an O–R pair whose reduction potential is more negative *donates* electrons to the oxidized substance of an O–R pair whose potential is more positive. Thus, in the redox pair $2H^+/H_2$ which has a potential of −0.42 V, H_2 has a great tendency to *donate* electrons. On the other hand, in the redox pair $\tfrac{1}{2}O_2/H_2O$, which has a potential of +0.82 V, H_2O has a very slight tendency to donate electrons, but O_2 has a great tendency to *accept* electrons. It follows then that in the coupled reaction of H_2 and O_2, hydrogen will serve as the electron *donor*, and become oxidized, and oxygen will serve as the electron *acceptor* and become reduced (Figure 4.4). Even though by chemical convention both half reactions are written as reductions, in an O–R reaction one of the two half reactions must be written as an oxidation and therefore will proceed in the reverse direction. Thus note that in Figure 4.4, the oxidation of H_2 to $2H^+ + 2e^-$ is reversed from the formal half reaction, written as a reduction.

The electron tower A convenient way of viewing electron transfer in oxidation-reduction reactions is to imagine a vertical tower (Figure 4.5). The tower represents the range of reduction potentials for O–R pairs, from the most negative at the top to the most positive at the bottom. The reduced substance in the pair at the top of the tower has the ***greatest*** amount of potential energy (roughly, the energy that it took to lift the substance to the top), and the reduced

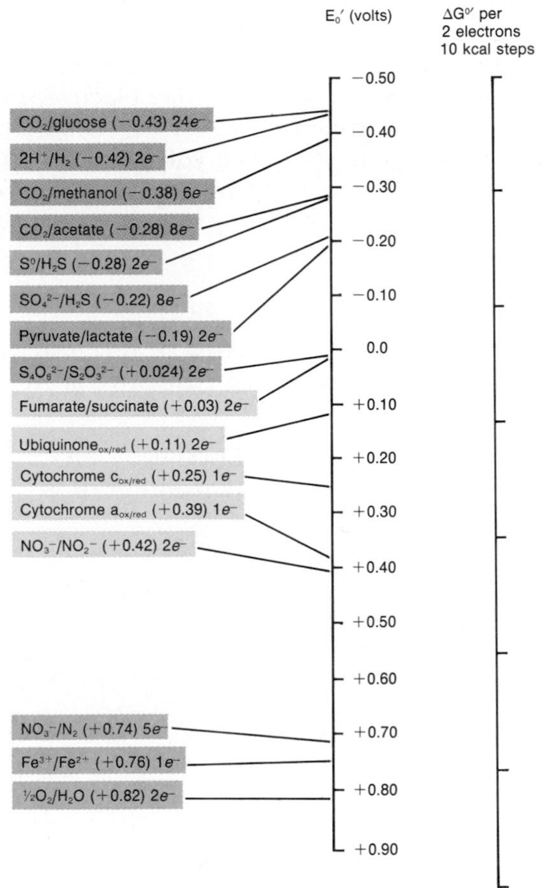

Figure 4.5 The electron tower. Redox pairs are arranged from the strongest reductants (negative reduction potentials) at the top to the strongest oxidants (positive reduction potentials) at the bottom. As electrons are removed from the top of the tower, they can be "caught" by acceptors at various levels. The farther the electrons fall before they are caught, the greater the difference in reduction potential between electron donor and electron acceptor, and the more energy is released. On the right, the energy released is given in 10 kcal/mol steps, assuming 2 electron transfers in each case. Some of the half-reactions indicated involve the transfer of several electrons, for example the CO_2/glucose couple. This reaction is included as a biologically important example of an overall process. The actual reduction of CO_2 to glucose involves several smaller redox reactions.

substance at the bottom of the tower has the *least* amount of potential energy. On the other hand, the oxidized substance in the O–R pair at the top of the tower has the *least* tendency to accept electrons, whereas the oxidized substance in the pair at the bottom of the tower has the *greatest* tendency to accept electrons.

As electrons from the electron donor at the top of the tower fall, they can be "caught" by acceptors at various levels. The difference in electrical potential between two substances is expressed as $\Delta E_o'$. The farther the electrons drop before they are caught, the greater the amount of energy released; that is, $\Delta E_o'$ is proportional to $\Delta G^{0\prime}$. O_2, at the bottom of the tower, is the final acceptor (or put in other terms, is the most powerful oxidizing agent). In the middle of the tower, the O–R pairs can act as either electron donors or acceptors. For instance, the $2H^+/H_2$ couple has a reduction potential of -0.42 volts. The fumarate/succinate couple has a potential of $+0.02$ volts. Hence, the oxidation of hydrogen can be coupled to the reduction of fumarate:

$$H_2 + \text{fumarate} \rightarrow \text{succinate}$$

On the other hand, the oxidation of succinate to fumarate can be coupled to the reduction of NO_3^- or $\frac{1}{2}O_2$:

$$\text{Succinate} + NO_3^- \rightarrow \text{fumarate} + NO_2^-$$
$$\text{Succinate} + \tfrac{1}{2}O_2 \rightarrow \text{fumarate} + H_2O$$

Hence under certain conditions (e.g., anaerobic in the presence of H_2) fumarate can act as an electron acceptor and under other conditions (e.g., anaerobic in the presence of NO_3^-, or aerobically) succinate can act as an electron donor. Indeed, all of the transformations involving fumarate and succinate described here are carried out by various microorganisms under certain nutritional and environmental conditions.

In catabolism the electron donor is often referred to as an **energy source**. It is necessary to remember, however, that it is the *coupled* oxidation-reduction process which actually releases energy. As discussed in the context of the electron tower, the amount of energy released in an O–R reaction depends on the nature of *both* the electron donor and the electron acceptor: the greater the difference between reduction potentials of the two half reactions, the more energy there will be released upon their coupling (see Appendix 1).

4.5 Electron Carriers

In the cell, the transfer of electrons in an oxidation-reduction reaction from donor to acceptor involves one or more intermediates, referred to as **carriers**. When such carriers are used, we refer to the starting compound as the **primary electron donor** and to the final acceptor as the **terminal electron accep-**

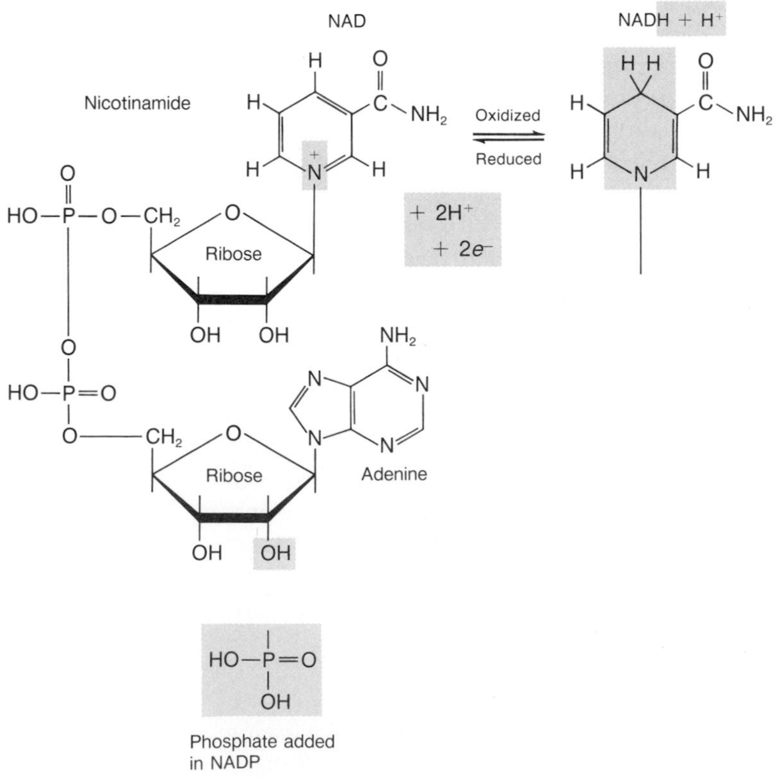

Figure 4.6 Structure of the oxidation-reduction coenzyme nicotinamide adenine dinucleotide (NAD). In NADP, a phosphate group is present, as indicated. Both NAD and NADP undergo oxidation-reduction as shown.

tor. The net energy change of the complete reaction sequence is determined by the *difference* in reduction potentials between the primary donor and the terminal acceptor. The transfer of electrons through the intermediates involves a series of oxidation-reduction reactions, but the energy change from these individual steps must add up to the value obtained by considering only the starting and ending compounds.

The intermediate electron carriers may be divided into two general classes: those freely diffusible, and those firmly attached to enzymes in the cell membrane. The fixed carriers function in membrane-associated electron transport reactions and are discussed in Section 4.11. Freely diffusible carriers include the coenzymes NAD^+ (nicotinamide-adenine dinucleotide) and $NADP^+$ (NAD-phosphate) (Figure 4.6). NAD^+ and $NADP^+$ are ***hydrogen atom*** carriers and always transfer two hydrogen atoms to the next carrier in the chain. Such hydrogen atom transfer is referred to as a dehydrogenation.*

The reduction potential of the NAD^+/NADH and the $NADP^+$/NADPH pairs is −0.32 V, which places them fairly high on the electron tower, that is, as good electron *donors*. Although the NAD^+ and $NADP^+$ couples have the same reduction potentials, they generally serve in different capacities in the cell. NAD^+ is directly involved in energy-generating (catabolic) reactions, whereas $NADP^+$ is involved primarily in biosynthetic (anabolic) reactions.

Coenzymes increase the diversity of possible O–R reactions by making it possible for chemically dissimilar molecules to be coupled as initial electron donor and ultimate electron acceptor, the coenzyme acting as intermediary. As we have discussed, most biological reactions are catalyzed by specific enzymes which can only react with a limited range of sub-

*Strictly speaking NAD^+ or $NADP^+$ carry two electrons and one proton, the second H^+ coming from solution. Therefore, $NAD^+ + 2e^- + 2H^+$ actually yields NADH + H^+. However, for simplicity, we will write "NADH + H^+" as NADH.

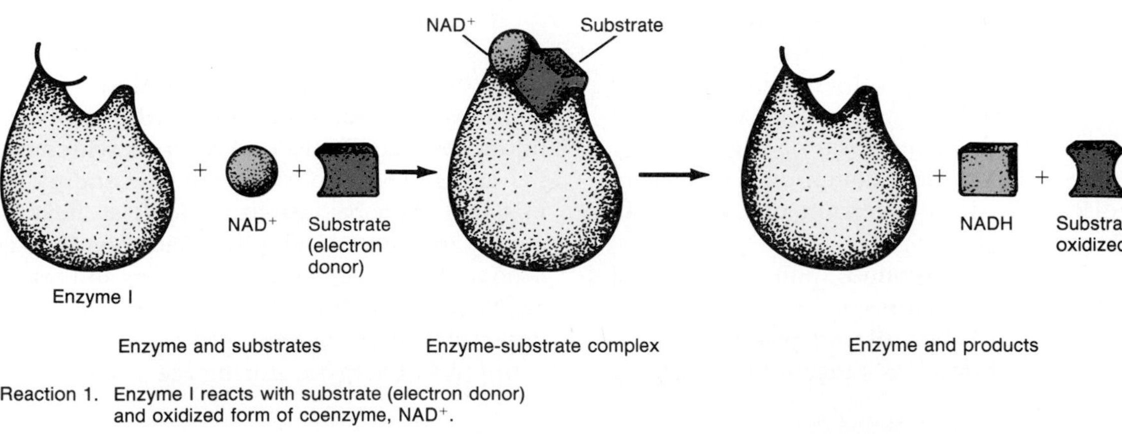

Reaction 1. Enzyme I reacts with substrate (electron donor) and oxidized form of coenzyme, NAD^+.

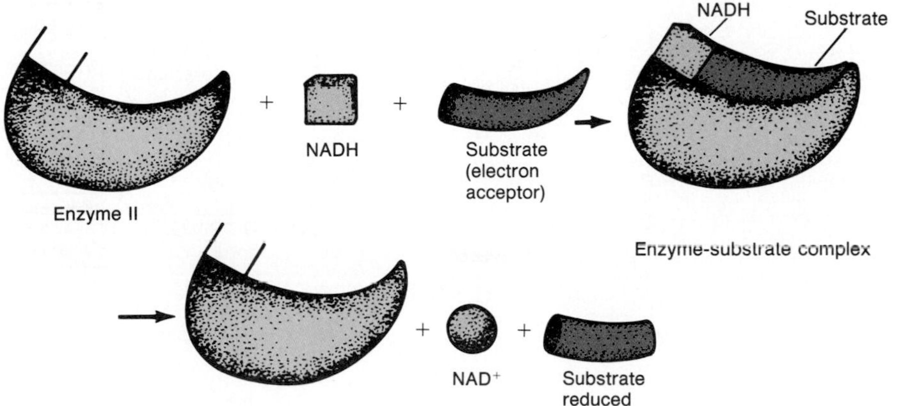

Reaction 2. Enzyme II reacts with substrate (electron acceptor) and reduced form of coenzyme, NADH.

Figure 4.7 Schematic example of an oxidation-reduction reaction involving the coenzyme, NAD^+/NADH.

strates. Oxidation-reduction reactions may be considered to proceed in three stages: removal of electrons from the primary donor, transfer of electrons through a series of electron carriers, and addition of electrons to the terminal acceptor. Each step in the reaction is catalyzed by a different enzyme, each of which binds to its substrate and to its specific coenzyme. Figure 4.7 is a schematic diagram showing the functioning of a coenzyme. Note that after a coenzyme has performed its chemical function in one reaction it can diffuse through the cytoplasm until it collides with another enzyme that requires the coenzyme in that form. Following conversion of the coenzyme back to its original form, the whole process can be repeated again.

4.6 High Energy Phosphate Compounds and Adenosine Triphosphate (ATP)

Energy released as a result of oxidation-reduction reactions must be conserved for cell functions. In living organisms, chemical energy released in O–R reactions is most commonly transferred to a variety of phosphate compounds in the form of **high-energy phosphate bonds**, and these compounds then serve as intermediaries in the conversion of energy into useful work.

In phosphorylated compounds, phosphate groups are attached via oxygen atoms in *ester* linkage, as illustrated in Figure 4.8. Not all phosphate ester linkages are high-energy bonds. As a means of expressing the energy of phosphate bonds, the free energy released when water is added and the bond is hydrolyzed can be given. As seen in Figure 4.8, the $\Delta G^{0\prime}$ of hydrolysis of the phosphate bond in glucose-6-phosphate is only -3.3 kcal/mole, whereas the $\Delta G^{0\prime}$ of hydrolysis of the phosphate bond in phosphoenolpyruvate is -14.8 kcal/mole, over four times that of glucose-6-phosphate. Thus phosphoenolpyruvate is considered a high-energy compound while glucose-6-phosphate is not. (Free energies of hydrolysis for a number of other high-energy phosphate compounds are given in Appendix 1.)

Adenosine triphosphate (ATP) The most important high energy phosphate compound in living organisms is adenosine triphosphate (ATP). ATP serves as the prime energy carrier in living organisms, being generated during exergonic reactions and being used to drive endergonic reactions. The structure of ATP is shown in Figure 4.8, where it can be seen that two of the phosphate bonds of ATP have high free energies of hydrolysis.

It should be emphasized that although we express the energy of the high-energy phosphate bonds in terms of the free energy of hydrolysis, in actuality it is undesirable for these bonds to hydrolyze in cells in the absence of a second reaction which can utilize the energy released, since the free energy of hydrolysis will then be lost to the cell as heat. The free energy of high energy phosphate bonds is generally used to drive biosynthetic reactions and other aspects of cell function through carefully regulated processes in which the energy released from ATP hydrolysis is coupled to energy-requiring reactions.

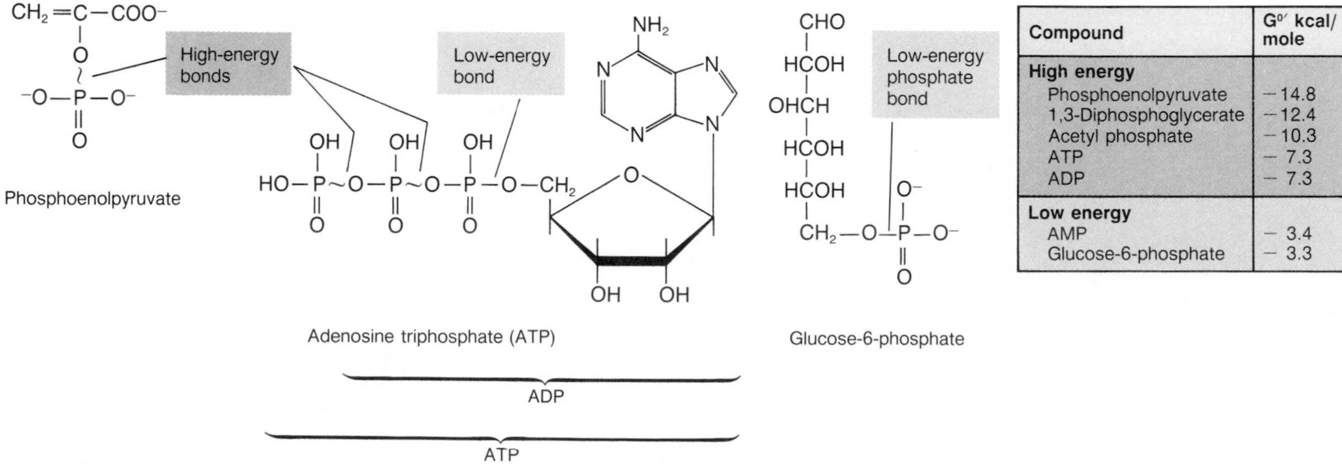

Compound	$G^{0\prime}$ kcal/mole
High energy	
Phosphoenolpyruvate	−14.8
1,3-Diphosphoglycerate	−12.4
Acetyl phosphate	−10.3
ATP	− 7.3
ADP	− 7.3
Low energy	
AMP	− 3.4
Glucose-6-phosphate	− 3.3

Figure 4.8 High-energy phosphate bonds. The table shows the free energy of hydrolysis of some of the key phosphate esters, indicating that some of the phosphate ester bonds are of higher energy than others. Structures of three of the compounds are given to indicate the position of low-energy and high-energy bonds.

4.7 Energy Release in Biological Systems

Organisms use a wide variety of energy sources in the synthesis of the high-energy phosphate bonds in ATP. Both chemical energy and light energy are used to synthesize ATP. Among chemical substances, both organic and inorganic compounds may serve as electron donors in heterotrophic and lithotrophic bacteria, respectively. Similarly, a number of different electron acceptors can be used in coupled redox reactions, including both organic and inorganic compounds. In the following pages we will be concerned with the mechanisms by which ATP is synthesized as a result of oxidation-reduction reactions involving *organic* compounds. Metabolism of organic compounds is the source of energy for all animals and for the vast majority of microorganisms and we begin our study of energy yielding reactions here. In Chapter 16 we will consider mechanisms of energy generation by phototrophic and lithotrophic organisms.

The series of reactions involving the oxidation of a compound is called a **biochemical pathway**. The pathways for the oxidation of organic compounds and conservation of energy in ATP can be divided into two major groups: (1) **fermentation**, in which the O–R process occurs in the *absence* of any added terminal electron acceptors; and (2) **respiration**, in which molecular oxygen serves as the terminal electron acceptor.

4.8 Fermentation

In the absence of externally supplied electron acceptors, many organisms perform internally balanced oxidation-reduction reactions of organic compounds with the release of energy, a process called **fermentation**. There are many different types of fermentations but under fermentative conditions only *partial* oxidation of the carbon atoms of the organic compounds occurs and therefore only a small amount of the potential energy available is released. The oxidation in a fermentation is coupled to the subsequent reduction of an organic compound generated from catabolism of the initial fermentable substrate; thus, no externally supplied electron acceptor is required (Figure 4.9).

ATP is produced in fermentations by a process called **substrate-level phosphorylation**. In substrate-level phosphorylation, ATP is synthesized during specific enzymatic steps in the catabolism of the organic compound. This is in contrast to **oxidative phosphorylation** (discussed later), where ATP is produced via membrane mediated events not con-

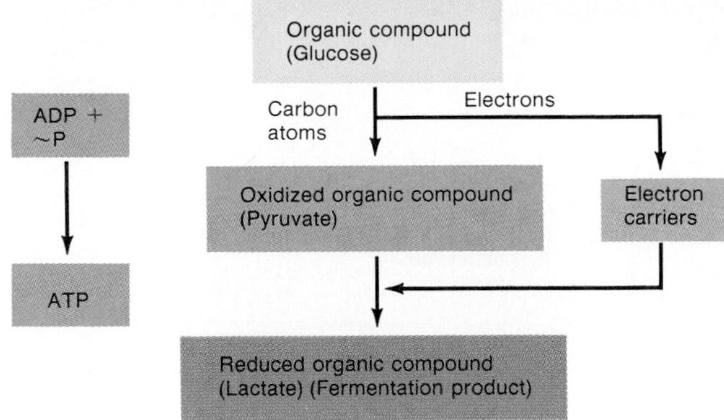

Figure 4.9 Carbon and electron flow in fermentation.

nected directly to the metabolism of specific substrates.

An example of fermentation is the catabolism of glucose by a lactic acid bacterium:

$$\text{glucose} \rightarrow 2 \text{ lactate}$$
$$C_6H_{12}O_6 \rightarrow 2C_3H_6O_3$$

Note that the product, lactic acid, has the same proportion of hydrogen and oxygen as glucose. Likewise, the catabolism of glucose by yeast in the absence of oxygen:

$$\text{glucose} \rightarrow 2 \text{ ethanol} + 2 \text{ carbon dioxide}$$
$$C_6H_{12}O_6 \rightarrow 2C_2H_6O + 2CO_2$$

Note that in the yeast reaction some of the carbon atoms end up in CO_2, a more *oxidized* form than the carbon atom in the starting molecule, glucose, while other carbon atoms end up in ethanol, which is more *reduced* (that is, it has more hydrogens and electrons per carbon atom) than glucose.

The energy released in the ethanol or lactic acid fermentation of glucose (-57 and -29 kcal/mole, respectively) is conserved by substrate-level phosphorylations in the form of high-energy phosphate bonds in ATP, with a net production of *two* such bonds in each case. We discuss now the biochemical steps involved in the fermentation of glucose to ethanol, and the manner in which some of the energy released is conserved in high-energy phosphate bonds.

Glucose fermentation: oxidation and ATP production The biochemical pathway for the breakdown of glucose can be divided into three major parts (Figure 4.10). Stage I is a series of preparatory rearrangements, reactions that do not involve oxidation-reduction and do not release energy, but which lead to the production of two molecules of the key intermediate, *glyceraldehyde-3-phosphate*. In Stage II, ox-

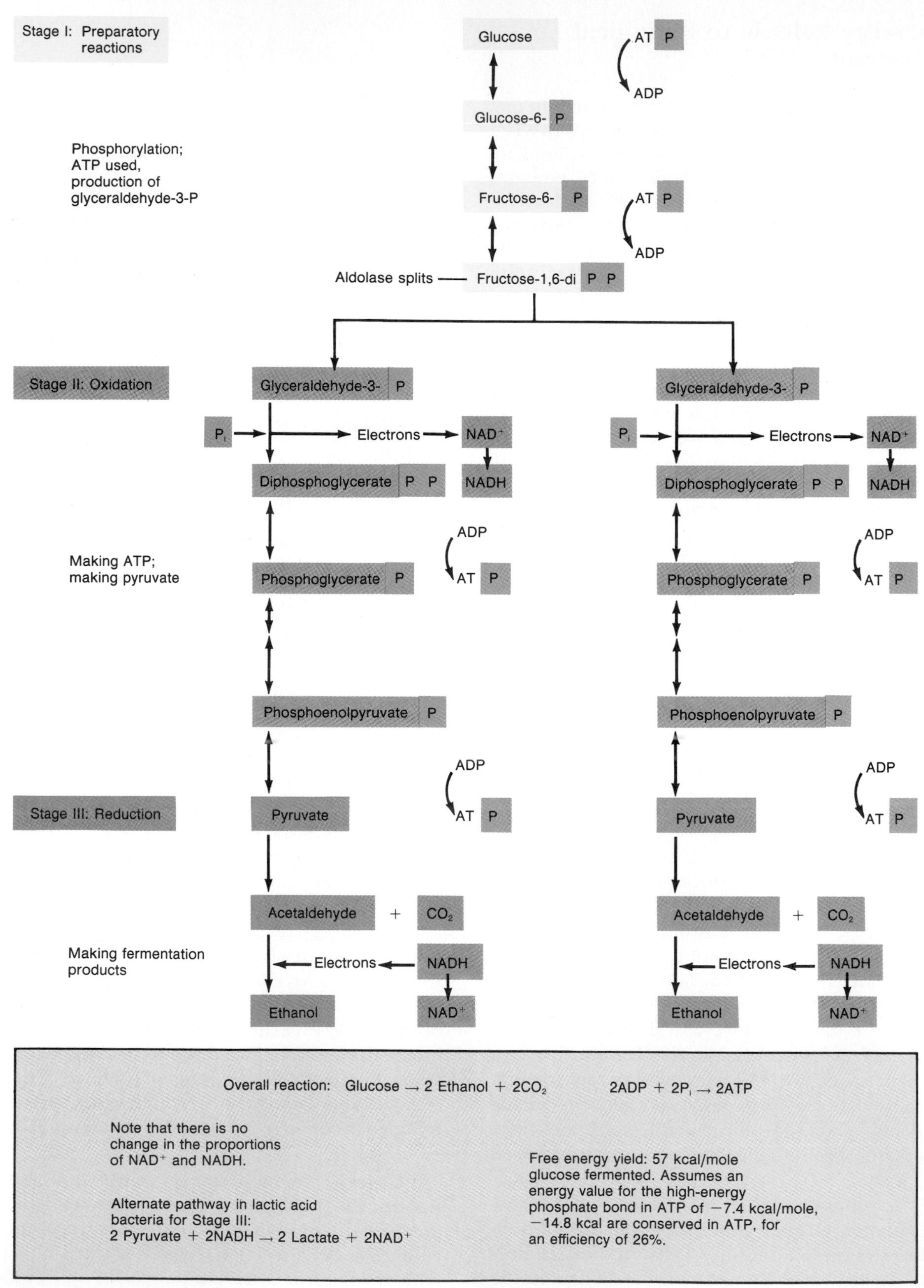

Figure 4.10 Embden Meyerhof Pathway, the sequence of enzymatic reactions in the conversion of glucose to pyruvate and then to fermentation products. The steps from glucose to pyruvate are sometimes collectively called glycolysis.

idation-reduction occurs, high-energy phosphate bond energy is produced in the form of ATP, and two molecules of pyruvate are formed. In Stage III, a second oxidation-reduction reaction occurs, and fermentation products (e.g., ethanol and CO_2, or lactic acid) are formed. The biochemical pathway from glucose to pyruvate is called **glycolysis** (literally, the splitting of glucose), and it is sometimes called the Embden-Meyerhof-Parnas pathway after its discoverers.

Initially, glucose is phosphorylated by ATP yielding glucose-6-phosphate. Phosphorylation reactions of this sort often occur preliminary to oxidation. The initial phosphorylation of glucose *activates* the molecule for the subsequent reactions. Glucose-6-phosphate is converted into its isomer, fructose-6-phosphate, and a second phosphorylation leads to the production of fructose-1,6-diphosphate, which is a key intermediate product. The enzyme *aldolase* catalyzes the splitting of fructose-1,6-diphosphate into two three-carbon molecules, glyceraldehyde-3-phosphate and its isomer, dihydroxyacetone phosphate.* Note that thus far, there have been no oxidation-reduction reactions, all the reactions including the consumption of ATP proceeding without electron transfers.

The *sole* oxidation reaction of glycolysis occurs in the conversion of glyceraldehyde-3-phosphate to 1,3 diphosphoglyceric acid. In this reaction (which occurs twice, once for each molecule of glyceraldehyde-3-phosphate), an enzyme containing the coenzyme NAD^+ accepts two hydrogen atoms and is converted into NADH. Simultaneously, each glyceraldehyde-3-P molecule is phosphorylated by addition of a molecule of inorganic phosphate. This energetically favorable reaction in which inorganic phosphate has been converted to organic form sets the stage for the next process, the step in which ATP is actually formed. High-energy bond formation is possible because each of the phosphates on a molecule of 1,3-diphosphoglyceric acid represents a high-energy phosphate bond. The synthesis of ATP occurs when each molecule of 1,3-diphosphoglyceric acid is converted to 3-phosphoglyceric acid, and later on in the pathway, when each molecule of phosphoenolpyruvate is converted to pyruvate.

In glycolysis, *two* ATP molecules were consumed in the two phosphorylations of glucose, and *four* ATP molecules were synthesized (two from each 1,3-diphosphoglyceric acid converted to pyruvate). Thus the *net gain* to the organism is two molecules of ATP per molecule of glucose fermented.

*There is an enzyme that catalyzes the interconversion of dihydroxyacetonephosphate and glyceraldehyde-3-phosphate. For simplicity we consider here only glyceraldehyde-3-phosphate since it is the compound which is further metabolized.

Glucose fermentation: reductive steps During the formation of two molecules of 1,3-diphosphoglyceric acid, two molecules of NAD^+ were reduced to NADH (see Figure 4.10). The cell has only a limited supply of NAD^+, and if all of it were converted to NADH, the oxidation of glucose would stop; the continued oxidation of glyceraldehyde-3-phosphate can proceed only if there is a molecule of NAD^+ present to accept released electrons. This "roadblock" is overcome in fermentation by the oxidation of NADH back to NAD^+ through reactions involving the reduction of pyruvate to any of a variety of **fermentation products**. In the case of yeast, pyruvate is reduced to ethanol with the release of CO_2. In lactic acid bacteria (or in muscle tissue rendered anaerobic by vigorous exercise) pyruvate is reduced to lactic acid (see lower part of Figure 4.10). Many routes of pyruvate reduction in various fermentative procaryotes are known (discussed in Chapter 16), but the net result is the same; NADH must be returned to the oxidized form, NAD^+, for the energy-yielding reactions of fermentation to continue. As a diffusible coenzyme, NADH can move away from the enzyme which oxidizes glyceraldehyde-3-phosphate, attach to an enzyme which reduces pyruvate to lactic acid, and diffuse away once again following conversion to NAD^+ to repeat the cycle all over again (see Figure 4.10).

In any energy-yielding process, oxidation must balance reduction, and there must be an electron acceptor for each electron removed. In this case, the *reduction* of NAD^+ at one enzymatic step is coupled with its *oxidation* at another. The final product(s) must also be in oxidation-reduction balance with the starting substrate, glucose. Hence, the products discussed here, ethanol plus CO_2 or lactic acid are in electrical balance with the starting glucose.

Glucose fermentation: net and practical results The ultimate result of glycolysis is the consumption of glucose, the net synthesis of two ATP, and the production of fermentation products. For the organism the crucial product is ATP, which is used in a wide variety of energy-requiring reactions, and fermentation products are merely waste products. However, the latter substances would hardly be considered waste products by the distiller, the brewer, or the cheesemaker. The anaerobic fermentation of glucose by yeast is a means of producing ethanol, the desired product in alcoholic beverages, and the production of lactic acid from glucose by lactic acid bacteria is the initial step in the production of fermented milk products including cheeses. For the baker, on the other hand, the production of CO_2 by yeast fermentation is the essential step in the leavening of bread. We discuss the large scale production of fermentation products for commercial benefit in Chapter 10.

4.9 Respiration

We have discussed above the metabolism of glucose as it occurs in the *absence* of external electron acceptors. A relatively small amount of energy is released in this process (and few ATP molecules synthesized). This small energy release may be understood in terms of the electron tower (see Figure 4.5) and the formal principles of oxidation-reduction reactions. Fermentation processes yield little energy for two reasons: (1) the carbon atoms in the starting compound are only partially oxidized, and (2) the difference in reduction potentials between the primary electron donor and terminal electron acceptor is small. However, if O_2 or some other external terminal acceptor is present, all the substrate molecules can be oxidized completely to CO_2, and a far *higher* yield of ATP is theoretically possible. The process by which a compound is oxidized using O_2 as external electron acceptor is called **respiration** (Figure 4.11).

The greater energy release during respiration occurs because respiring cells surmount the two limitations just listed for fermentation: (1) the carbon atoms in the starting compound can be completely oxidized to CO_2; and (2) the terminal electron acceptor has a relatively positive reduction potential, leading to a large net difference in potentials between primary donor and terminal acceptor and therefore the synthesis of much ATP.

Our discussion of respiration will deal with the biochemical mechanisms involved in both the carbon and electron transformations: (1) the biochemical pathways involved in the transformation of organic carbon to CO_2 and (2) the way electrons are transferred from the organic compound to the terminal electron acceptor driving ATP synthesis.

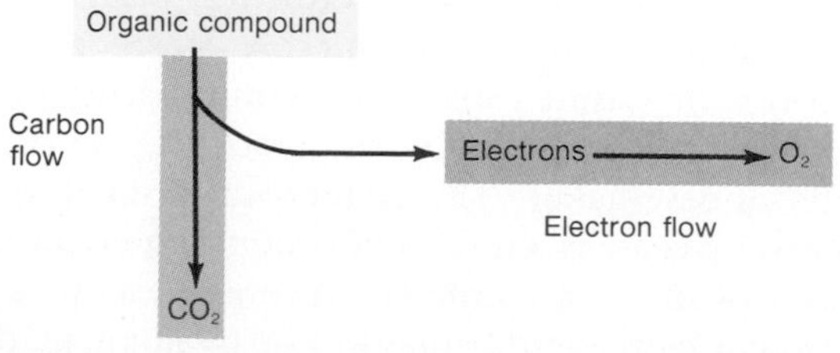

Figure 4.11 Carbon and electron flow in respiration.

4.10 Carbon Flow: The Tricarboxylic Acid Cycle

The early steps in the respiration of glucose involve the same biochemical steps as those of glycolysis (see Figure 4.10). As we noted, a key intermediate in glycolysis is pyruvate. Whereas in glycolysis, pyruvate is converted to fermentation products, in respiration pyruvate is oxidized fully to CO_2. One important pathway by which pyruvate is completely oxidized to CO_2 is called the **tricarboxylic acid cycle** (TCA cycle) as outlined in Figure 4.12.

Pyruvate is first decarboxylated, leading to the production of one molecule of NADH and an acetyl moiety coupled to coenzyme A (acetyl–CoA, Figure 4.13). The acetyl group of acetyl–CoA combines with the four-carbon compound oxalacetate, leading to the formation of citric acid, a six-carbon organic acid, the energy of the high-energy acetyl–CoA bond being used to drive this synthesis (Figure 4.12). Dehydration, decarboxylation, and oxidation reactions follow, and two additional CO_2 molecules are released. Ultimately, oxalacetate is regenerated and can serve again as an acetyl acceptor, thus completing the cycle.

For each pyruvate molecule oxidized through the cycle, three CO_2 molecules are released (Figure 4.12), one during the formation of acetyl–CoA, one by the decarboxylation of isocitrate, and one by the decarboxylation of α-ketoglutarate. As in fermentation, the electrons released during the oxidation of intermediates in the TCA cycle are usually transferred initially to enzymes containing the coenzyme NAD^+. However, respiration differs from fermentation in the manner in which NADH is oxidized. The electrons from NADH, instead of being transferred to an intermediate such as pyruvate, are transferred to oxygen or other terminal electron acceptors through the action of an electron transport system, forming NAD^+ and a reduced product. We discuss some of these electron transport components in the next section.

4.11 Electron Transport Systems

Electron transport systems are composed of membrane-associated electron carriers. These systems have two basic functions: (1) to accept electrons from an electron donor and transfer them to the electron acceptor and (2) to conserve some of the energy released during electron transfer by synthesis of ATP.

Several types of oxidation-reduction enzymes and electron transport proteins are involved in electron transport: (1) NADH dehydrogenases, which transfer hydrogen atoms from NADH; (2) riboflavin-contain-

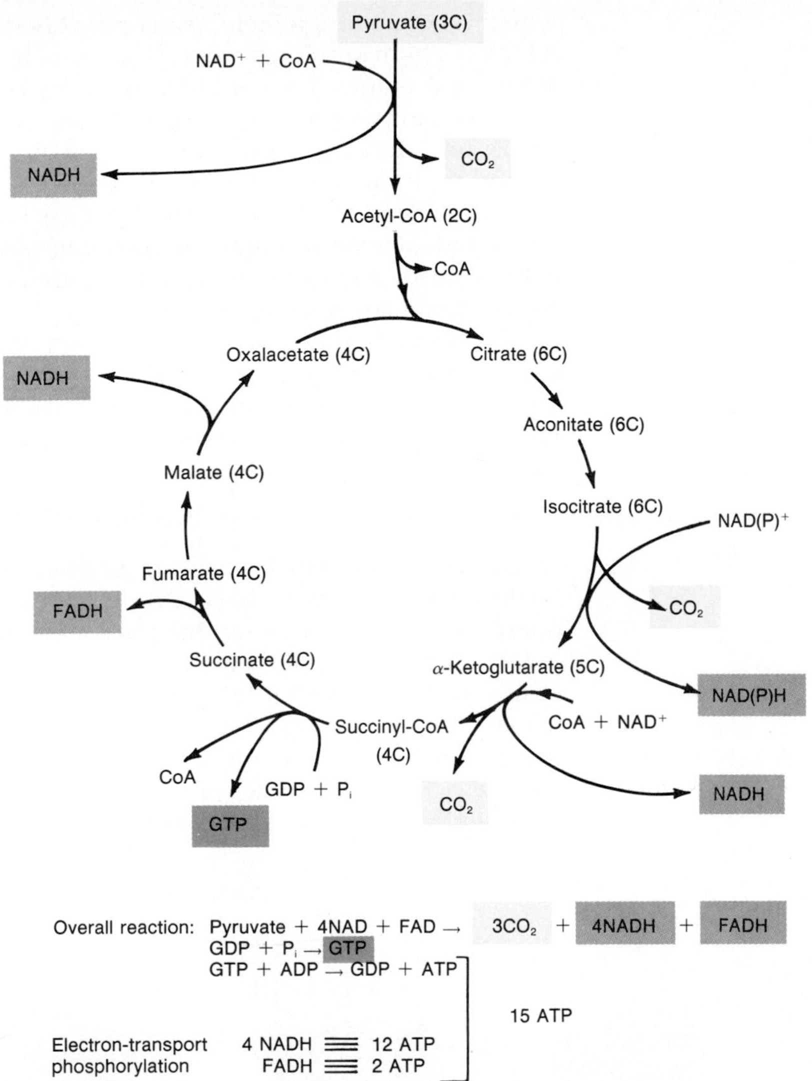

Figure 4.12 The tricarboxylic acid cycle. The three-carbon compound, pyruvate, is oxidized to CO_2, with the electrons used to make NADH and FADH. The reoxidation of NADH and FADH leads to the generation of a proton-motive force and thus to the synthesis of ATP. The cycle actually begins when the 2-carbon compound acetyl-CoA condenses with the 4-carbon compound oxalacetate, to form the 6-carbon compound citrate. Through a series of oxidations and transformations, this 6-carbon compound is ultimately converted back to the 4-carbon compound oxalacetate, which then passes through another cycle with the next molecule of acetyl-CoA. The overall balance sheet is shown at the bottom. The reducing equivalents formed (NADH and FADH) are shown on the left and right sides of the cycle.

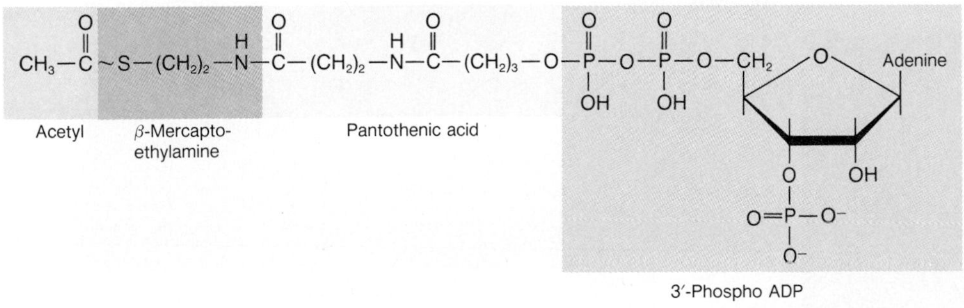

$CH_3-C(=O)\sim S-(CH_2)_2-NH-C(=O)-(CH_2)_2-NH-C(=O)-(CH_2)_3-O-P(=O)(OH)-O-P(=O)(OH)-O-CH_2$ — Adenine

Acetyl · β-Mercapto-ethylamine · Pantothenic acid · 3′-Phospho ADP

Figure 4.13 Structure of acetyl-CoA.

ing electron carriers, generally called flavoproteins (such as FAD); (3) iron-sulfur proteins, and (4) cytochromes, which are proteins containing an iron-porphyrin ring called heme. In addition, one class of nonprotein electron carriers is known, the lipid soluble quinones, sometimes called coenzymes Q. These electron transport components are either embedded in the membrane matrix or positioned on one side of the membrane or the other in an ordered arrangement which permits energy conservation. We discuss each of these classes of electron transport components below.

Flavoproteins are proteins containing a derivative of riboflavin (Figure 4.14); the flavin portion,

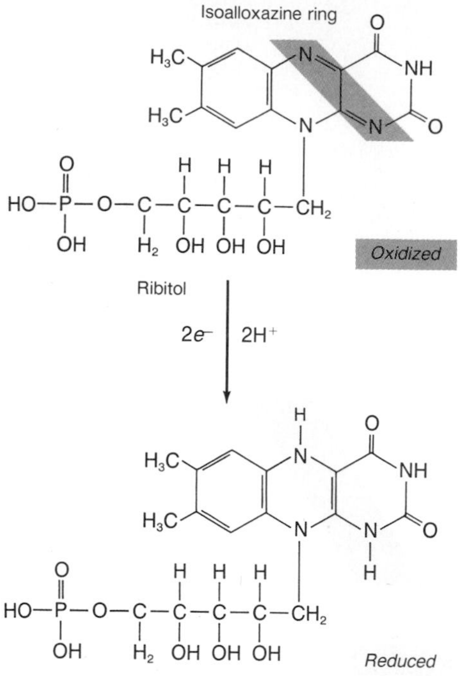

Figure 4.14 Flavin mononucleotide (FMN) (riboflavin phosphate). Note that $2H^+$ are taken up when the flavin becomes reduced, and $2H^+$ are given off when the flavin becomes oxidized.

which is bound to a protein, is the prosthetic group, which is alternately reduced as it accepts hydrogen atoms, and oxidized when electrons are passed on. Note that flavoproteins accept and donate hydrogen atoms and electrons, respectively. Two flavins are known, flavin mononucleotide (FMN) and flavin-adenine dinucleotide (FAD), in which FMN is linked to ribose and adenine through a second phosphate. Riboflavin, also called vitamin B_2, is a required organic growth factor for some organisms.

The **cytochromes** are proteins with iron-containing porphyrin rings attached to them (Figure 4.15). They undergo oxidation and reduction through loss or gain of a *single electron* by the iron atom at the center of the cytochrome:

$$\text{Cytochrome–Fe}^{2+} \rightleftharpoons \text{Cytochrome–Fe}^{3+} + e^-$$

Several classes of cytochromes are known, differing in their reduction potentials. One cytochrome can transfer electrons to another that has a more positive reduction potential and can itself accept electrons from cytochrome with a less positive reduction potential. The different cytochromes are designated by letters, such as cytochrome *a*, cytochrome *b*, cytochrome *c*. The cytochromes of one organism may differ slightly from those of another, so that there are

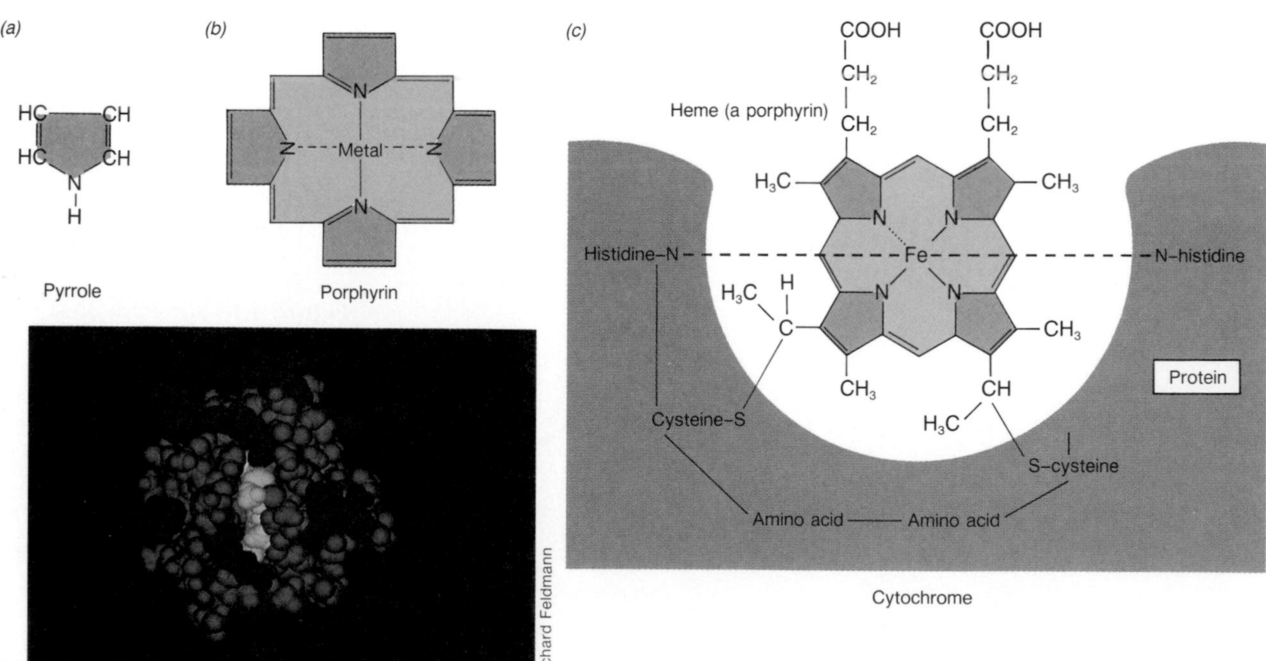

Figure 4.15 Cytochrome and its structure. (a) Structure of the pyrrole ring. (b) Four pyrrole rings are condensed, leading to the formation of the porphyrin ring. Various metals can be incorporated into the porphyrin ring system. (c) In the cytochromes, the porphyrin ring is covalently linked via disulfide bridges to cysteine molecules in the protein. Note the presence of iron in the center of the ring. (d) Computer-generated model of cytochrome c. The protein completely surrounds the porphyrin ring (light color) in the center.

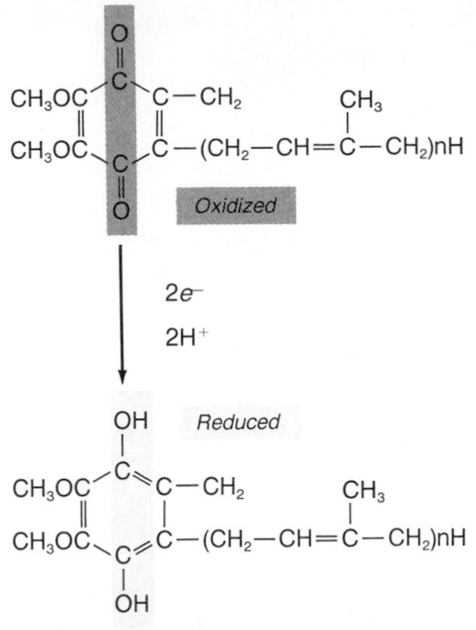

Oxidized

$2e^-$

$2H^+$

Reduced

Figure 4.16 Structure of oxidized and reduced forms of coenzyme Q, a quinone. The five-carbon unit in the side chain (an isoprenoid) occurs in a number of multiples. In bacteria, the most common number is n = 6; in higher organisms, n = 10. Note that $2H^+$ are taken up when the quinone becomes reduced, and $2H^+$ are given off when the quinone becomes oxidized.

designations such as cytochrome a_1, cytochrome a_2, and so on.

In addition to the cytochromes, where iron is bound to heme, several **nonheme iron-sulfur proteins** (abbreviated *nhFe*) are associated with the electron transport chain in several places. The reduction potentials of iron-sulfur proteins vary over a wide range depending on the number of iron atoms and sulfur atoms present and how the iron centers are attached to protein. Thus different iron-sulfur proteins can serve at different points in the electron transport process. Like cytochromes, iron-sulfur proteins carry *electrons* only, not hydrogen atoms.

The **quinones** (Figure 4.16) are lipid-soluble substances involved in electron transport. Some quinones found in bacteria are related to vitamin K, a growth factor for higher animals. Like flavoproteins, quinones serve as *hydrogen atom* acceptors and *electron* donors.

4.12 Energy Conservation in the Electron Transport System

The overall process of electron transport in the electron transport chain is shown in Figure 4.17. There are three places in the chain where large amounts of free energy are released: between NAD and flavin, between cytochrome *b* and cytochrome *c*, and be-

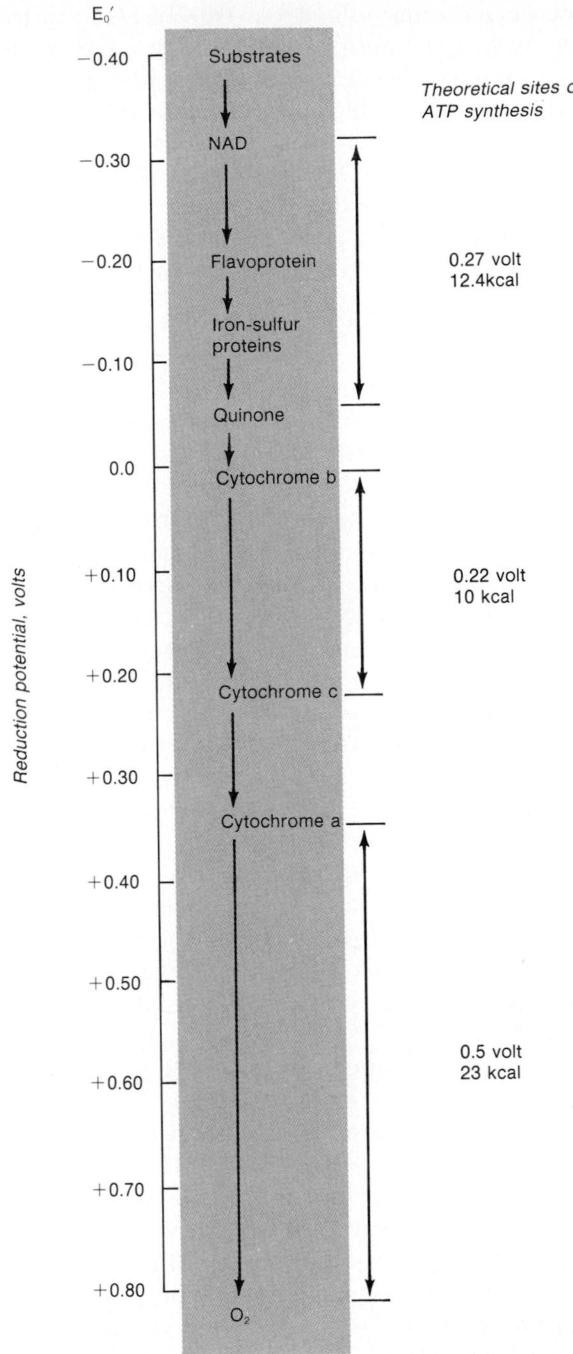

Figure 4.17 One example of an electron-transport system, leading to the transfer of electrons from substrate to O_2. This particular sequence is typical of the electron transport chain of the mitochondrion and some bacteria (for example, *Paracoccus denitrificans*). The chain in *Escherichia coli* lacks cytochromes *c* and *a*, and instead electrons go directly from cytochrome *b* to cytochrome *o* or *d* which act as terminal oxidases. By breaking up the complete oxidation into a series of discrete steps, energy conservation is possible, and ATP synthesis can occur. Experimental studies and calculations of differences in reduction potentials show that there are three places in the chain where sufficient drop in potential occurs to permit ATP synthesis.

tween cytochrome *a* and O_2. During electron transport, sufficient energy is released at each of these locations to synthesize a high-energy phosphate bond in ATP, a process referred to as **oxidative phosphorylation**. Remember that electrons are not transferred from NADH to O_2 directly, but rather go through a number of intermediate O–R reactions. This multiple-step process is essential in the mechanism by which ATP is synthesized (see below).

Generation of a proton motive force To understand the manner by which electron transport is linked to ATP synthesis, we must first discuss the manner in which the electron transport system is oriented in the cell membrane. The overall structure of the membrane was outlined in Section 3.3 (see Figure 3.8). It was shown there that proteins are embedded in the lipid bilayer and that the orientation of proteins in the membrane is asymmetric, some being accessible from the outside and others accessible from the inside. The electron transport carriers discussed above are oriented in the membrane in such a way that there is a *separation* of protons from electrons during the transport process. Hydrogen atoms, removed from hydrogen atom carriers such as NADH on the *inside* of the membrane, are separated into electrons and protons, the electrons returning to the cytoplasmic side of the membrane by specific carriers, and the protons being extruded into the external environment (Figure 4.18). At the termination of the electron transport system, the electrons are passed to the final electron acceptor (in the case of aerobic respiration, this is O_2) and reduce it.

When O_2 is reduced to H_2O it requires H^+ from the cytoplasm to complete the reaction, and these protons originate from the dissociation of water into H^+ and OH^-; $H_2O \rightarrow H^+ + OH^-$. The consumption of H^+ thus causes a net formation of OH^- on the inside of the membrane. Despite their small size, neither H^+ nor OH^- freely permeate the membrane so that equilibrium cannot be spontaneously restored. Thus, although electron transport to O_2 does produce water, the elements of water, H^+ and OH^-, are actually formed on opposite sides of the membrane. The net result is the generation of a *pH gradient* and an electrical potential across the membrane, with the *inside* of the cytoplasm electrically negative and alkaline, and the *outside* of the membrane electrically positive and acidic. This pH gradient and electrical potential represent an energized state of the membrane (something like a battery), and can be used by the cell to do useful work. In the same way that the energized state of a battery is expressed as its electromotive force (in volts), the energized state of a membrane can be expressed as a **proton motive force** (also in volts). The energized state of the membrane induced as a result of electron transport processes can be used directly to do useful work such as ion transport (see Section 3.4) or flagellar rotation (Figure 3.41), or it can be used to drive the formation of high-energy phosphate bonds in ATP, as will be described below. The key proteins involved in proton gradient formation across the membrane are the NAD and flavin enzymes and the quinones, for these sub-

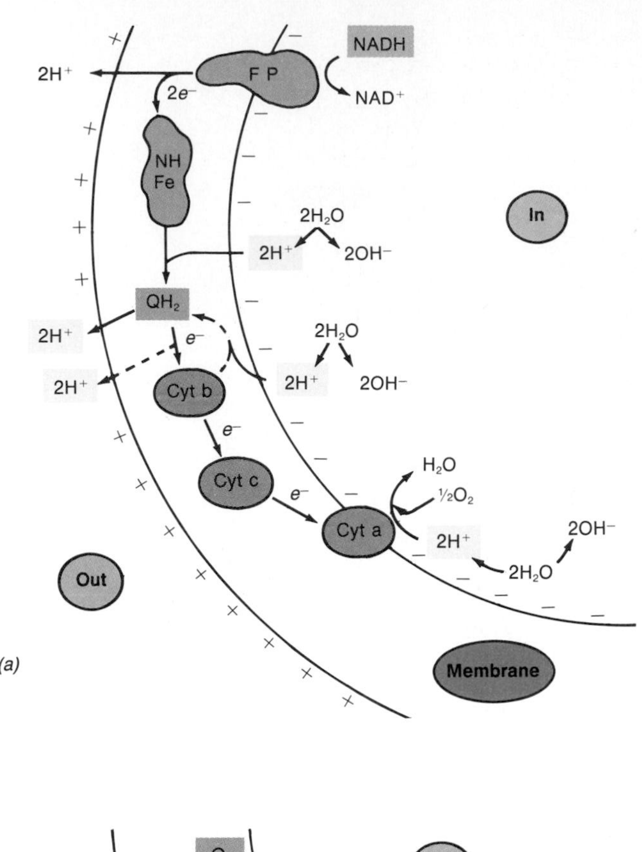

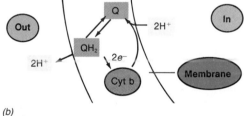

Figure 4.18 Generation of a proton-motive force. (a) Orientation of electron carriers in the bacterial membrane, showing the manner in which electron transport can lead to charge separation and generation of a proton gradient. The + and − charges originate from the protons (H^+) or hydroxyl ions (OH^-) accumulating on opposite sides of the membrane. (b) The Q-cycle—electrons shuttle between coenzyme Q and cytochrome *b* and increase the H^+ extrusion yield. FP, flavoprotein; nhFe, nonheme iron protein; Q, coenzyme Q; cyt, cytochromes. IN and OUT refer to the cytoplasm and the environment, respectively.

stances are both proton and electron carriers. Iron-sulfur proteins and cytochromes, on the other hand, carry electrons only, not hydrogen atoms. These electron carriers are oriented vectorially (directionally) in the membrane in such a way that protons are discharged to the environment when electrons are transferred to the acceptor, resulting in the accumulation of OH^- in the cytoplasm (Figure 4.18).

The series of oxidation-reduction reactions occurring during electron transport may be analyzed by examining each pair of carriers sequentially (Figure 4.18*a*). Following the donation of two hydrogen atoms from NADH to FAD, two H^+ are extruded when FADH donates two electrons (only) to an iron-sulfur protein. An additional pair of protons is extruded when coenzyme Q passes electrons (only) to cytochrome *b* (Figure 4.18). Also, because they have about the same reduction potentials, coenzyme Q and cytochrome *b* can pass electrons and hydrogen atoms back and forth across the membrane, resulting in a type of proton "shuttle" which yields at least two more H^+ to the outside of the membrane (the Q-cycle, Figure 4.18*b*). Finally, the reduction of 1/2 O_2 to H_2O consumes two H^+, but these of course are not extruded. All of the required H^+ ions are generated from the dissociation of water yielding OH^- on the inside surface of the membrane (Figure 4.18).

The electron transport scheme shown in Figure 4.18 is just one of many different carrier sequences observed in different organisms. However, the important feature of all of them is the generation of a *proton gradient*, acidic outside and alkaline inside. The gradient results in a proton motive force (see next Section) which actually drives the synthesis of ATP.

The proton-motive force and ATP synthesis How is the proton-motive force used to synthesize ATP? An important component of this process is a membrane-bound enzyme, an *ATPase*, which contains two parts, a multi-subunit headpiece present on the inside of the membrane, and a proton-conducting tailpiece that spans the membrane (Figure 4.19). This enzyme catalyzes a reversible reaction between ATP and ADP + P_i (inorganic phosphate) as shown in Figure 4.19. Operating in one direction, this enzyme catalyzes the formation of ATP by allowing the controlled reentry of protons across the energized membrane. Just as the formation of the proton gradient was energy driven, the carefully controlled dissipation of the proton-motive force releases energy, some of which is used to synthesize ATP in the process called oxidative phosphorylation.

ATPase can also catalyze the reverse reaction, that is, the hydrolysis of ATP and the extrusion of $2H^+$ to the outer portion of the membrane (Figure 4.19*b*).

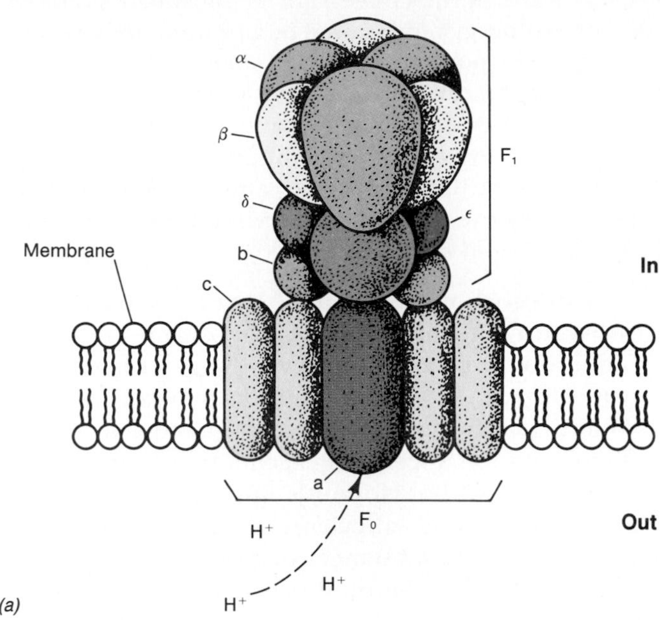

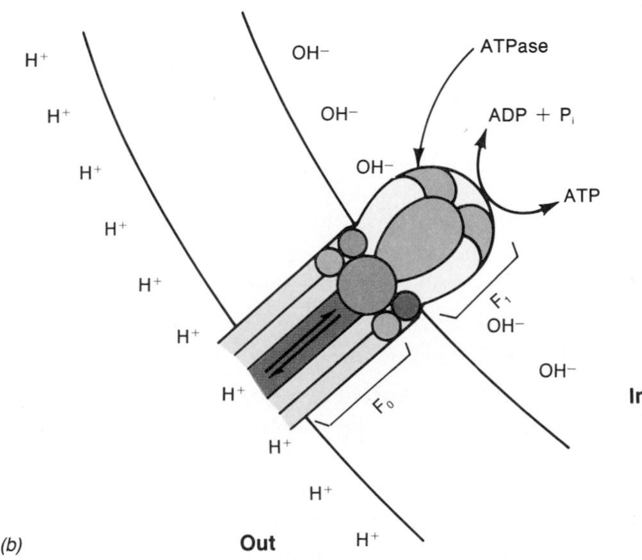

Figure 4.19 Structure and function of the membrane-bound ATPase, which acts as a proton channel between the cytoplasm and the cell exterior (environment). (a) Structure. F1 consists of five polypeptides, $\alpha_3\ \beta_3\ \gamma\ \delta\ \epsilon$. This is the catalytic protein, responsible for the interconversion of ATP and ADP + Pi. F_0 is integrated in the membrane and consists of three polypeptides, a, b_2, c_4. It is responsible for channeling protons across the membrane. (b) ATPase action. As protons flow in, the dissipation of the proton gradient drives ATP synthesis from ADP + Pi. The reverse reaction, ATP $\rightarrow$ ADP + Pi, drives the extrusion of protons to the cell exterior.

This results in the conversion of phosphate bond energy into energy represented by a proton-motive force. Thus, high-energy phosphate bonds and the proton-motive force can be looked at as different forms of cell energy. Because a membrane potential is used by the cell to drive a number of different reactions, transport and motility being chief among them, ATPases are present even in organisms which do not respire, e.g. lactic acid bacteria.

Uncouplers and inhibitors of oxidative phosphorylation A variety of chemical agents, called **uncouplers**, have been found to inhibit the synthesis of ATP during electron transport without inhibiting the electron-transport process itself. Examples of such uncoupling agents are dinitrophenol, dicumarol, carbonylcyanide-m-chlorophenylhydrazone (CCCP), and salicylanilide. All of these agents are lipid-soluble substances that can pick up protons and pass through the lipid matrix of the membrane. They thus promote leakage of protons across the membrane causing dissipation of the proton-motive force, and hence inhibition of ATP synthesis.

Various chemicals inhibit electron transport by interfering with the action of electron carriers. **Inhibitors** generally stop both electron flow and ATP synthesis. Carbon monoxide, for example, combines directly with the terminal cytochrome, cytochrome oxidase, and prevents the attachment of oxygen. Cyanide (CN^-) and azide (N_3^-) bind tightly to the iron of the porphyrin ring of the cytochromes and prevent its oxidation and reduction. The antibiotic antimycin A inhibits electron transport between cytochromes *b* and *c*. As one might expect, all of these inhibitors and uncouplers are powerful cell poisons.

Contrast of respiration between procaryotes and eucaryotes So far we have been discussing electron transport and the proton-motive force in abstract terms. However, in the cell these processes occur in definite structures. There are important differences between procaryotes and eucaryotes in where these processes occur.

In procaryotes, the electron transport components are embedded in the plasma membrane, and the development of a proton-motive force occurs across this membrane. Thus, in procaryotes, protons are actually excreted outside the cell into the environment, resulting in a mild acidification of the external milieu.

In eucaryotes, the processes of electron transport and ATP synthesis occur in the mitochondrion. As we saw in Section 3.13, the mitochondrion is a membrane-bounded structure which also has extensive internal membranes and it is across these internal membranes that the proton-motive force develops. The ATPase structure (Figure 4.19) is a part of this internal membrane and ATP synthesis occurs within the mitochondrial matrix. The ATP synthesized then diffuses out into the cytoplasm through the permeable external membrane, where it is used in various biosynthetic reactions.

4.13 The Balance Sheet of Aerobic Respiration

The net result of reactions of the tricarboxylic acid cycle is the complete oxidation of pyruvic acid to three molecules of CO_2 with the production of four molecules of NADH and one molecule of FADH. The NADH and FADH molecules can be reoxidized through the electron transport system, yielding up to three ATP molecules per molecule of NADH and two ATP per molecule of FADH. In addition, the oxidation of α-ketoglutarate to succinate involves a substrate-level phosphorylation, producing guanosine triphosphate (GTP), which is later converted to ATP. Thus a total of 15 ATP molecules can be synthesized for each turn of the cycle. Since the oxidation of glucose yields two molecules of pyruvic acid, a total of 30 molecules of ATP can be synthesized in the citric acid cycle. Also, when oxygen is available the two NADH molecules produced during glycolysis can be reoxidized by the electron transport system, yielding six more molecules of ATP. Finally, two molecules of ATP are produced by substrate-level phosphorylation during the conversion of glucose to pyruvic acid so that aerobes can form up to 38 ATP molecules from one glucose molecule in contrast to the two molecules of ATP produced fermentatively.

If we assume that the high-energy phosphate bond of ATP has an energy of about 7 kcal/mole, then 266 kcal of energy can be converted to high-energy phosphate bonds in ATP by the complete oxidation of glucose to CO_2 and H_2O. Since free energy calculations show that the total amount of energy available from the complete oxidation of glucose by oxygen is 688 kcal/mole, aerobic respiration is about 39 percent efficient, the rest of the energy being lost as heat.

If an organism were 100 percent efficient, all of the energy yield of a biochemical reaction would be conserved in the form of high-energy bonds or a membrane potential. However, organisms are not 100% efficient and a part of the energy yield is not conserved but is lost as heat. Thus we must distinguish between energy release—the total energy yield of a reaction—and energy conservation—that energy available to the organism to do work. Interestingly, although the *yield* of ATP from fermentation is very low, the *efficiency* of fermentations can be reasonably high. The lactic fermentation [glucose ⟶ 2 lactate], for example, releases 29 kcal/mole and drives the synthesis of two ATP's, for an efficiency of nearly 50%. Thus the energetic problems of fermentations revolve around the low *amount* of energy released, not necessarily in the thermodynamic efficiency of the conversion process.

In addition to its function as an energy-yielding mechanism, the tricarboxylic acid cycle provides key intermediates for biosynthesis. Oxalacetate and α-ketoglutarate lead to several amino acids (see Section 4.18), succinyl-CoA is the starting point for porphyrin biosynthesis, and acetyl–CoA provides the material for fatty acid synthesis.

4.14 Turnover of ATP and the Role of Energy-Storage Compounds

It can be calculated that for the synthesis of 1 gram of cell material (wet weight), about 20 millimoles (mmoles) of ATP would need to be consumed. Since the intracellular concentration of ATP is only about 2 millimolar, ATP obviously has only a *catalytic* role during growth; it is continually being broken down and resynthesized. It has been calculated that during the time that a cell doubles its ATP pool must turn over about 10,000 times.

Related to this short life of ATP is the fact that if ATP is not immediately used for energy growth and biosynthesis, it is hydrolyzed by reactions not yielding energy. For long-term storage of energy, most organisms produce insoluble organic polymers that can later be oxidized for the production of ATP. The glucose polymers starch and glycogen are produced by many microorganisms, both procaryotic and eucaryotic, and poly-β-hydroxybutyrate (PHB) is produced by many procaryotes. These polymers often are deposited within the cells in large granules that can be seen with the light or electron microscope (see Section 3.10). In the absence of an external energy source, the cell may then oxidize this energy-storage material and thus be able to maintain itself, even under starvation conditions.

Polymer formation has a twofold advantage to the cell. Not only is potential energy stored in a stable form, but also insoluble polymers have little effect on the internal osmotic pressure of cells. If the same number of units were present as monomers in the cell, the high solute concentration would increase cellular osmotic pressure, resulting in the inflow of water and possible swelling and lysis. A certain amount of energy is lost when a polymer is formed from monomers, but this disadvantage is more than offset by the benefits to the cell.

The Lowly Yeast Cell

The aerobic and anaerobic processes of energy generation may seem dull and prosaic, but they are at the basis of one of the most exciting discoveries of the human race, alcoholic fermentation. Although many bacteria form alcohol, it is a simple eucaryote which is most commonly exploited, the yeast *Saccharomyces cerevisiae*. Found in various sugar-rich environments such as fruit juices and nectar, yeasts have the ability to carry out the two opposing modes of metabolism discussed in this chapter, fermentation and respiration. When oxygen is present, yeasts grow efficiently on the sugar substrate, making lots of yeast cells and CO_2. However, when O_2 is absent, yeasts switch to an anaerobic metabolism, resulting in a reduced cell yield but satisfying amounts of alcohol. Every home wine maker or brewer is an amateur microbiologist, even without realizing it (see Box, Chapter 10). When grapes are squeezed to make juice, small numbers of yeast cells present on the grapes in the vineyard are transferred to the must. During the first several days of the wine-making process, these yeast cells grow by respiration, but consume O_2, making the juice anaerobic. As soon as the oxygen is depleted, fermentation can begin and the important process of alcohol formation takes over. This switch from aerobic to anaerobic metabolism is crucial, and special care must be taken to make sure air is kept out of the fermenting vessel. Lots of things can go wrong in wine making. The wrong yeast might get started, insufficient sugar in the juice will result in too low an alcohol concentration, and spoilage can occur by bad yeasts or bacteria. Wine is only one of many products made with yeast. Others include beer, whisky, vodka, and gin. Even alcohol for motor fuel is made with yeast in parts of the world where sugar is plentiful and petroleum is in short supply (such as Brazil). Yeast also serves as the leavening in bread, although here it is not the alcohol that is important but CO_2, the other product of the alcohol fermentation. We discuss yeast biotechnology in some detail in Chapter 10.

4.15 Alternate Modes of Energy Generation

This chapter is dealing only with energy generation by either fermentation or respiration of organic compounds, but we must note briefly here that microorganisms have many other possibilities for obtaining energy. These alternate modes of energy metabolism will be discussed in Chapter 16, but they will be summarized here to give the student an overview of catabolic processes.

One alternate mode of energy generation is a variation on respiration in which electron acceptors other than oxygen are used. Because of the analogy to respiration, these processes are classified under the heading of **anaerobic respiration**. Electron acceptors used in anaerobic respiration include nitrate (NO_3^-), sulfate (SO_4^{2-}), carbonate (CO_3^{2-}), and even other organic compounds. Because of their positions on the electron tower (see Figure 4.5), less energy is released when these electron acceptors are used instead of oxygen. However, the utilization of these alternate electron acceptors permits microorganisms to develop in environments where oxygen is absent. The contrasts between aerobic and anaerobic respiration are presented in Figure 4.20. We discuss a number of examples of anaerobic respiration in Chapter 16.

A second mode of energy generation involves the use of *inorganic* rather than organic chemicals. As we have noted, organisms able to use inorganic chemicals as energy sources are called **lithotrophs** (literally, *rock eating*). Examples of inorganic energy sources include hydrogen sulfide (H_2S), hydrogen gas (H_2), and ammonia (NH_3). Lithotrophic metabolism most commonly involves aerobic respiratory processes such as those described in this chapter, but using an inorganic energy source rather than an organic one (Figure 4.20). Lithotrophs have electron transport components like heterotrophs and form a proton-motive force. However, one important distinction between lithotrophs and heterotrophs is in their sources of carbon for biosynthesis. Heterotrophs can generally use compounds such as glucose as carbon sources as well as energy sources, but lithotrophs cannot, of course, use their inorganic energy compounds as sources of carbon. Most lithotrophs utilize carbon dioxide as carbon source and are, hence, **autotrophs** (see Table 4.1). We discuss lithotrophic metabolism in some detail in Chapter 16.

A large number of microorganisms, as well as higher plants, are *photosynthetic*, using light as an energy source. We call such organisms **phototrophs** (literally, *light eating*). The mechanisms by which light is used as an energy source are unique and complex, but the underlying result is the generation of a proton-motive force which can be used in the synthesis of ATP. Most phototrophs use energy conserved in ATP in the assimilation of carbon dioxide as the carbon source for biosynthesis. Such phototrophs are also, hence, autotrophs. However, as we will see in Chapter 16, photosynthesis in microorganisms has some special features and complications. There are, for instance, two types of photosynthesis in microorganisms, that similar to that of higher plants, and a unique type of photosynthesis found only in certain procaryotes.

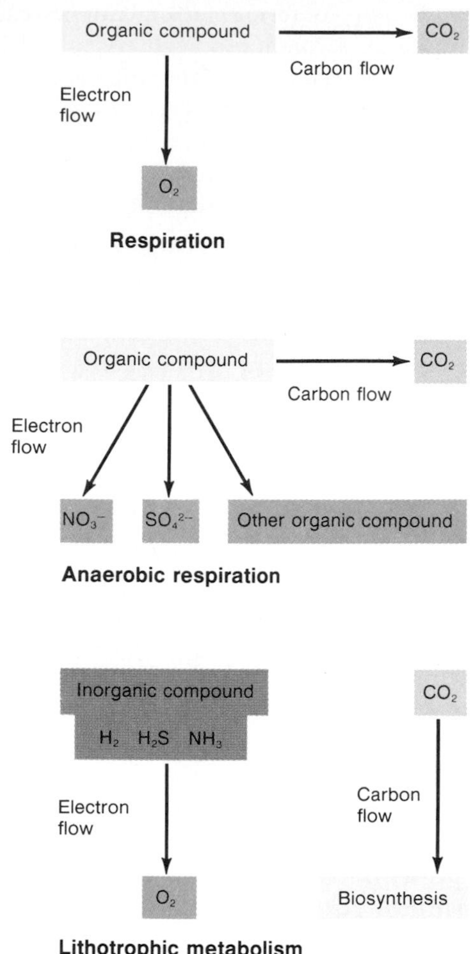

Figure 4.20 Electron and carbon flow in respiration, anaerobic respiration, and lithotrophic metabolism.

One overall conclusion that can be stated at this point is that in terms of energy metabolism, microorganisms show an amazing diversity. In the present chapter, we have presented the simplified picture that is obtained if one only examines the energy metabolism of common heterotrophs. The fascinating diversity in the microbial world will be considered in detail in Chapters 16 through 19.

4.16 Biosynthetic Pathways: Anabolism

We have now discussed in some detail **catabolism**, the biochemical processes by which heterotrophic microorganisms obtain energy. We now turn to the other side of the coin, **anabolism**, the biochemical processes by which microorganisms build up from nutrients the vast array of chemical substances of which they are composed (Figure 4.1). We outlined the chemical composition of the cell in Table 4.2 and noted that the dominant constituents (other than water) were macromolecules. Macromolecules, as we know (see Sections 2.4–2.7) are made by the polymerization of monomers. Polysaccharides are polymers built up of sugar monomers, proteins are built up from amino acids, and nucleic acids are polymers of nucleotides. We now summarize in the next sections the key biosynthetic reactions by which these small molecule building blocks are formed. Note that in this chapter we are discussing only the biosynthesis of *small molecules* (monomers). The biosynthesis of *macromolecules* is discussed in the next chapter.

Energy for anabolism is provided by ATP and the membrane potential. Energy conserved during catabolic reactions in ATP and the membrane potential is consumed during biosynthesis (Figure 4.21). However, although energy is required in many biosynthetic reactions, the focus of biosynthesis is not on energy but on carbon, and on the intermediates occurring during the buildup of cell constituents from simple starting materials. In Sections 4.8 and 4.10 we described the intermediates in the oxidation of glucose in both the glycolytic and the tricarboxylic acid

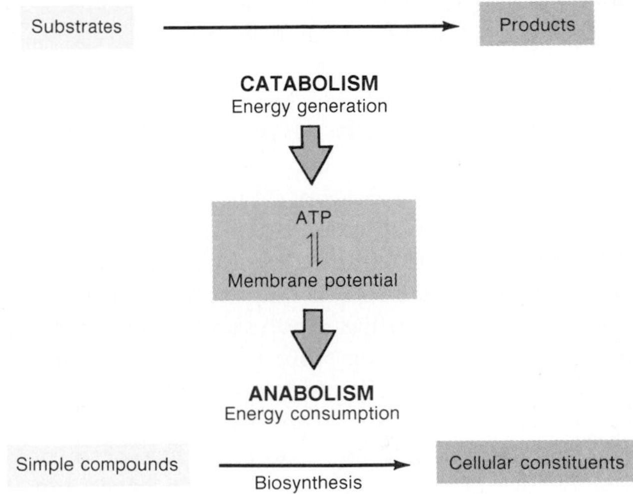

Figure 4.21 Scheme of anabolism and catabolism showing the key role of ATP and the membrane potential in integrating the processes.

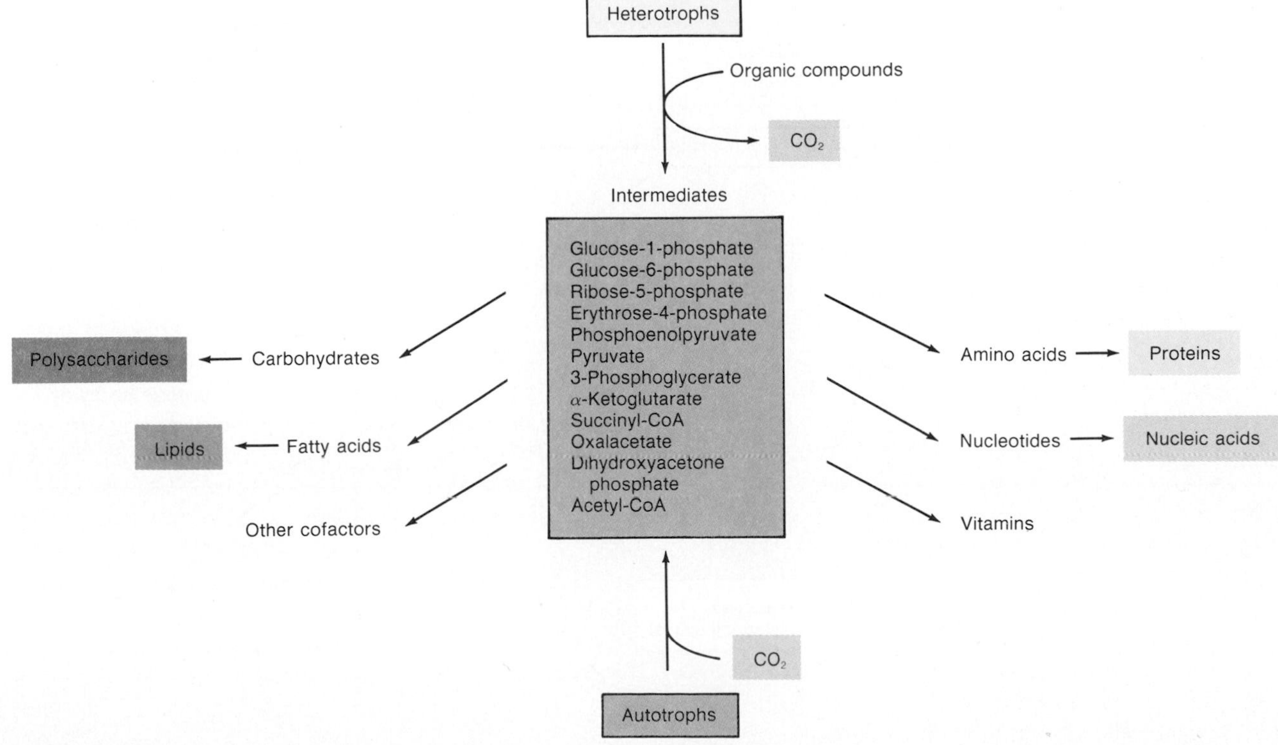

Figure 4.22 The key central metabolic intermediates produced in catabolism and used in anabolism.

cycles, but in those sections our main concern was not with the ultimate fate of these intermediates but with how their transformations led to the formation of ATP. Now we are concerned with how these intermediates are used in biosynthesis. Some of the same intermediates that are formed during catabolism are used during anabolism. These key intermediates, summarized in Figure 4.22, are actually very few in number.

4.17 Sugar Metabolism

As shown in Table 4.2, a small but significant amount of the cell dry weight is composed of polysaccharides. Polysaccharides are key constituents of the cell walls of many organisms, and in the bacteria, the peptidoglycan cell wall has a polysaccharide backbone. In addition, cells often store energy in the form of the polysaccharides **glycogen** and **starch**. The sugars in polysaccharides are primarily six carbon sugars called **hexoses**. The most common hexose in polysaccharides is **glucose**, which, as we have seen, is also a readily available energy source for microorganisms. In addition to the hexoses, two five-carbon sugars, **pentoses**, are key constituents of nucleic acids, **ribose** in RNA and **deoxyribose** in DNA. It is convenient to separate our discussion into two parts, dealing separately with hexoses and pentoses.

Hexoses Six-carbon sugars needed for biosynthesis can be obtained either from the environment or can be synthesized within the cell from nonsugar starting materials. A summary of the main pathways is given in Figure 4.23. It can be seen that two key intermediates of hexose metabolism are **glucose-6-phosphate** and **uridine diphosphoglucose (UDP-glucose)**. We already have discussed glucose-6-phosphate in some detail when discussing the process of glycolysis (see Figure 4.10). Glucose-6-phosphate serves as a key intermediate in the oxidation and fermentation of glucose as an energy source, in which case it is converted either to carbon dioxide or fermentation products. Glucose-6-phosphate can also feed into the pathways for polysaccharide synthesis, in which case it is converted to UDP-glucose (UDPG), a nucleoside diphosphate sugar. (The structure of UDPG is uracil-ribose-P-P-glucose. Note that there is an additional role for uracil; it is also one of the bases in RNA.) UDPG is an activated form of glucose which is synthesized from uridine triphosphate (UTP) and glucose-1-phosphate, as outlined in Figure 4.23. UDPG serves as the starting material not only for the synthesis of glucose-containing polysaccharides, but for the synthesis of other nucleoside diphosphate sug-

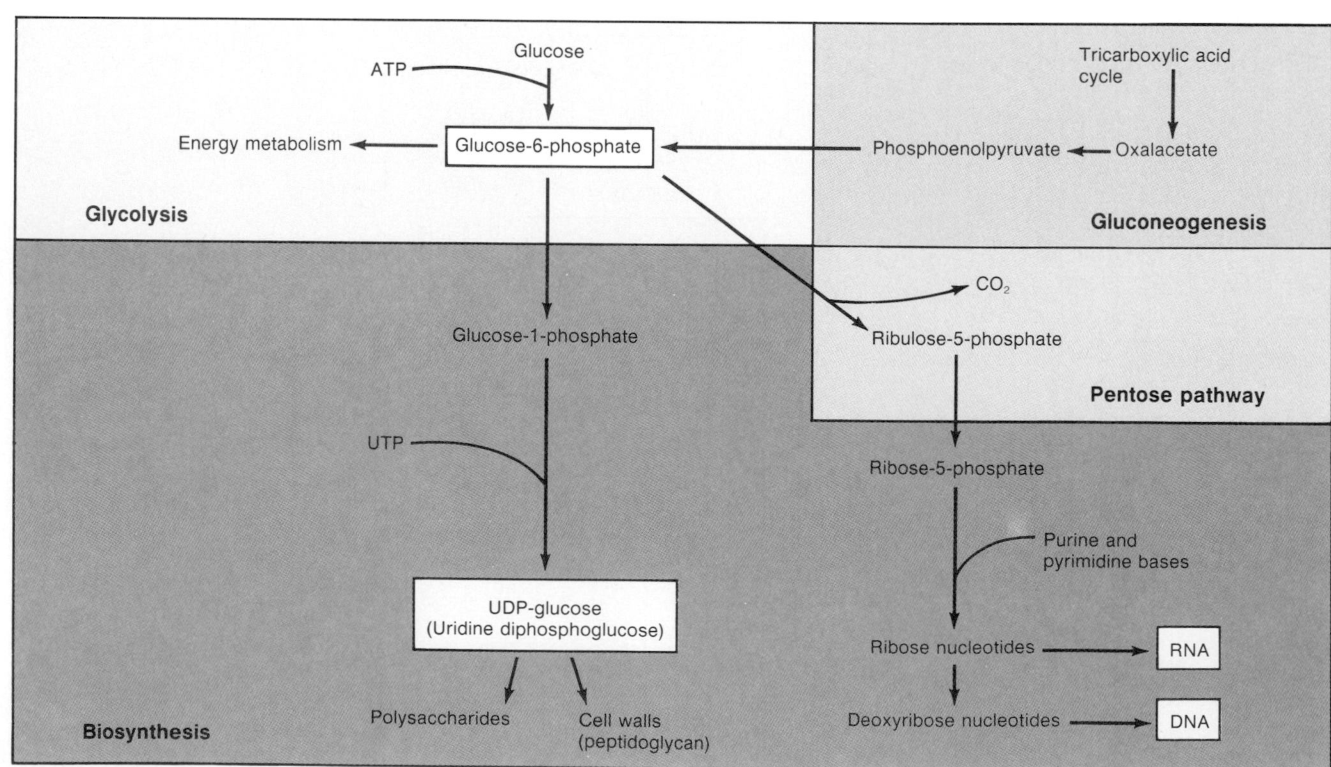

Figure 4.23 Hexose and pentose biosynthesis: summary of the main routes.

ars needed in biosynthesis. Thus, while glucose-6-phosphate is the central intermediate in glucose *catabolism*, UDPG is the central intermediate in glucose *anabolism*.

When a microorganism has available a hexose as an energy and carbon source, this hexose can also generally serve as the precursor of the hexoses needed for biosynthesis. However, many microorganisms are able to grow using nonhexose carbon sources. In such situations, the hexose needed for biosynthesis of cell walls and other hexose-containing polymers must be synthesized within the cell, a process called **gluconeogenesis**. Gluconeogenesis is the creation of new glucose molecules from noncarbohydrate precursors. The starting material for gluconeogenesis is phosphoenolpyruvate (PEP), one of the key intermediates in glycolysis. As we saw in Section 4.8 (see Figure 4.10), PEP is formed in the lower part of the glycolytic pathway, and by *reversal* of this pathway (but using a different enzyme than that used in glycolysis), glucose-6-phosphate can be formed. Where does the PEP needed for gluconeogenesis come from? There are a number of ways in which PEP can be formed, but a major one is by decarboxylation of oxalacetate, which itself is a key intermediate of the tricarboxylic acid cycle.

Pentoses In most instances, the five-carbon sugars are formed by the removal of one carbon atom from a hexose (Figure 4.23). Several different pathways are known. One of the common pathways is the oxidative decarboxylation of glucose-6-phosphate, yielding CO_2 and the five-carbon intermediate ribulose-5-phosphate. Ribulose-5-phosphate is converted to two other pentose sugars, one of which is ribose-5-phosphate, the pentose needed for nucleotide synthesis. Once a ribose nucleotide is formed, it can feed directly into RNA synthesis. For DNA synthesis, deoxyribose is needed and this is synthesized by enzymatic removal of an oxygen from the ribose of a ribose nucleotide. These various reactions are summarized in Figure 4.23.

4.18 Amino Acid Biosynthesis

As we have noted, there are 20 amino acids common in proteins. Those organisms that cannot obtain some or all amino acids preformed from the environment must synthesize them from other sources. The structures of the amino acids were discussed in Section 2.7 and illustrated in Figure 2.13. There are two aspects of amino acid biosynthesis: the synthesis of the **carbon skeleton** of each amino acid, and the manner by which the amino group is incorporated.

Amino acids can be grouped into *families* based on the precursor used for the synthesis of the carbon skeleton and these families are summarized in Figure 4.24. Note that the precursor molecules are ones

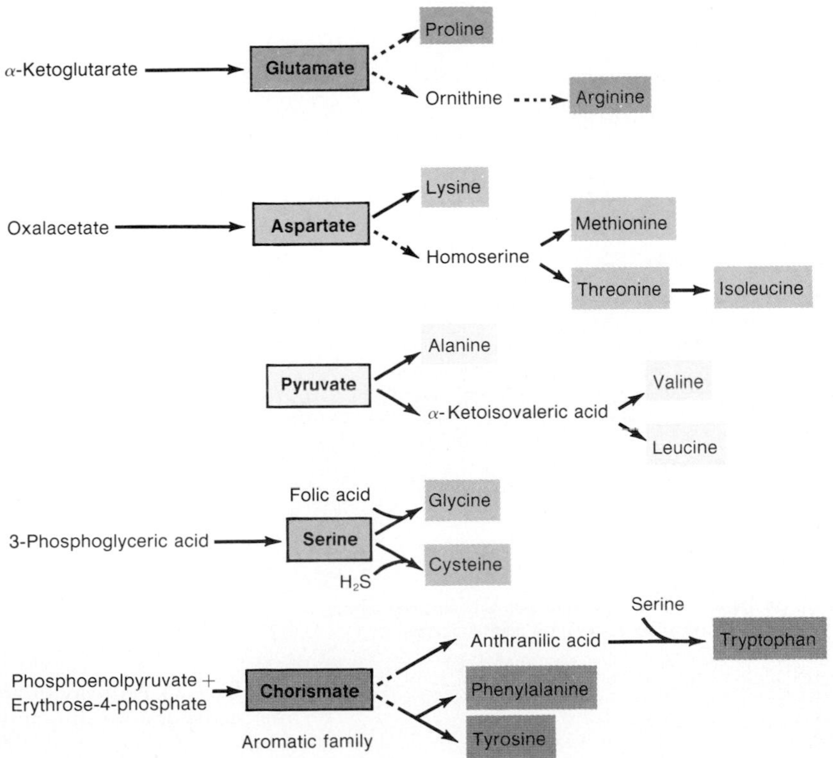

Figure 4.24 Amino acid families. The amino acids on the right are derived from the starting materials on the left. The compounds in boxes are the parent molecules of each amino acid family.

which we have encountered before in glycolysis and the TCA cycle, and were summarized in Figure 4.22.

The attachment of the amino group to the carbon skeleton can either be done at the very end, after the carbon skeleton is completely synthesized, or at some intermediate step. For instance, **glutamate**, the first amino acid of the glutamate family, is formed by direct addition of the amino group to the carbon skeleton, but since the other amino acids of the glutamate family are derived from glutamate, their amino group is, of course, added at an intermediate stage.

What is the source of the amino group for the amino acids? Many microorganisms utilize **ammonia** as the source of nitrogen, and there are several enzymes which are able to catalyze the addition of ammonia to a carbon skeleton. One of the most widespread enzymes involved in ammonia incorporation is **glutamate dehydrogenase**, an enzyme which uses NADH as a source of reducing power for the synthesis of glutamate from α-ketoglutaric acid and ammonia. The role of glutamate dehydrogenase in amino acid biosynthesis is illustrated in Figure 4.25*a*. Once ammonia has been incorporated into the amino group of glutamate, the amino group can be transferred to other carbon skeletons by an enzyme called **transaminase**. In the transamination process, the amino group of glutamate is exchanged for the keto group of an α-keto acid, leading to the formation of a new amino acid (from the α-keto acid) and regenerating α-ketoglutaric acid from glutamate (Figure 4.25*b*). By successive functioning of glutamate dehydrogenase and transaminase, ammonia can be assimilated into a variety of amino acids (Figure 4.25*c*). Not all organisms use the pathway illustrated in Figure 4.25. In some cases, other amino acid dehydrogenases replace glutamate dehydrogenase, and in still other cases a completely different enzyme, ***glutamine synthetase***, is involved in ammonia assimilation. These al-

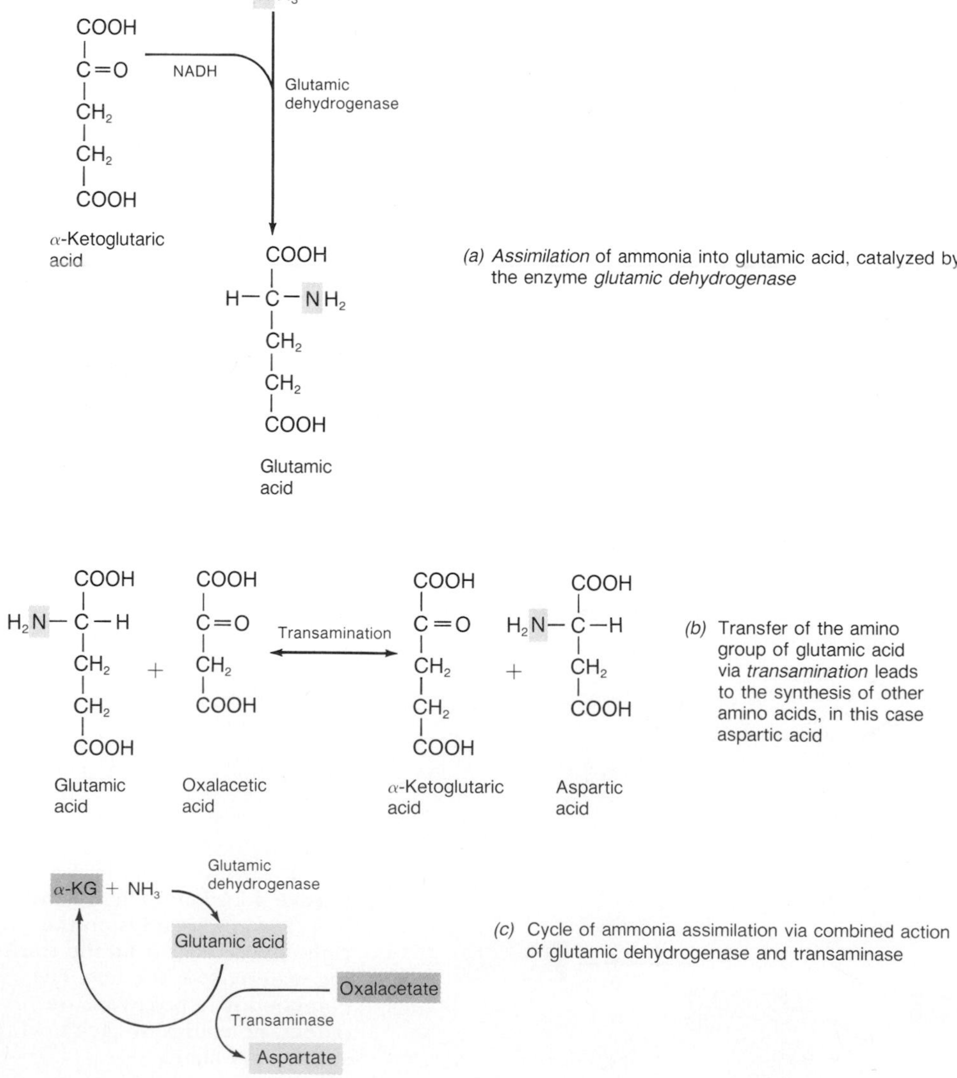

(a) *Assimilation* of ammonia into glutamic acid, catalyzed by the enzyme *glutamic dehydrogenase*

(b) Transfer of the amino group of glutamic acid via *transamination* leads to the synthesis of other amino acids, in this case aspartic acid

(c) Cycle of ammonia assimilation via combined action of glutamic dehydrogenase and transaminase

Figure 4.25 Pathway of ammonia incorporation into amino acids.

ternate pathways are discussed in Section 16.24 and the complete pathways of amino acid biosynthesis can be found in biochemistry textbooks.

4.19 Purine and Pyrimidine Biosynthesis

Purines and pyrimidines are components of nucleic acids, as well as of many vitamins and coenzymes (for example, ATP, NAD). The terminology of purines and pyrimidines was discussed in Section 2.6, where the structures of the key compounds were presented. The biosynthesis of purines and pyrimidines is surprisingly complex, and the details are beyond the scope of the present chapter and can be found in biochemistry textbooks. Here, we provide simply an overview.

Purine synthesis The purine ring is built up almost atom by atom using carbons and nitrogens derived from amino acids, CO_2, and formyl groups (themselves derived from the amino acid serine), as outlined in Figure 4.26*a*. The starting material for purine biosynthesis is a ribose-phosphate compound to which the various atoms are added. Thus, in the case of purine nucleosides, the sugar residue is present from the beginning.

The two carbon atoms in the purine ring derived from formyl groups are added in reactions involving a coenzyme derivative of the vitamin *folic acid*. This coenzyme is involved in a number of reactions involving addition of one-carbon units, and is of additional interest because a number of microorganisms cannot synthesize their own folic acid but must have it provided from the environment. Additionally, certain antimicrobial drugs called *sulfonamides* are specific inhibitors of folic acid synthesis. Sulfonamide inhibition of growth can be mostly overcome by providing the end products of pathways in which folic acid serves as a coenzyme, among which is a purine (see Section 9.20 for a detailed discussion of sulfonamide action).

The first purine ring formed is that of *inosinic acid*, which serves as an intermediate for the formation of the two key purine derivatives, adenylic acid and guanylic acid, as shown in Figure 4.26*c*.

Pyrimidine biosynthesis The origin of the atoms of the pyrimidine ring is shown in Figure 4.27*a*. In contrast to the purine ring, the complete pyrimi-

CO_2 · Glycine · Amino group of aspartate · Formyl group ($HCOO^-$) (from folic acid) · Formyl group (from folic acid) · Amide nitrogen of glutamine

(a) The basic precursors of the purine skeleton

Inosinic acid, the first purine · Ribose—Ⓟ

(b)

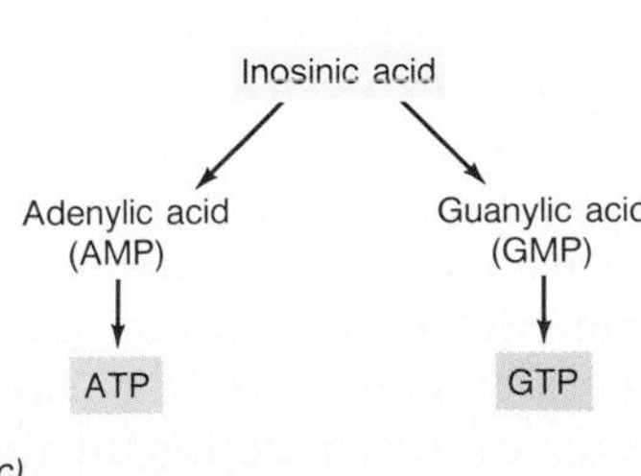

(c)

Figure 4.26 Purine biosynthesis.

NH_3 · Aspartic acid · CO_2 · Orotic acid

(a) Basic precursors of the pyrimidine skeleton

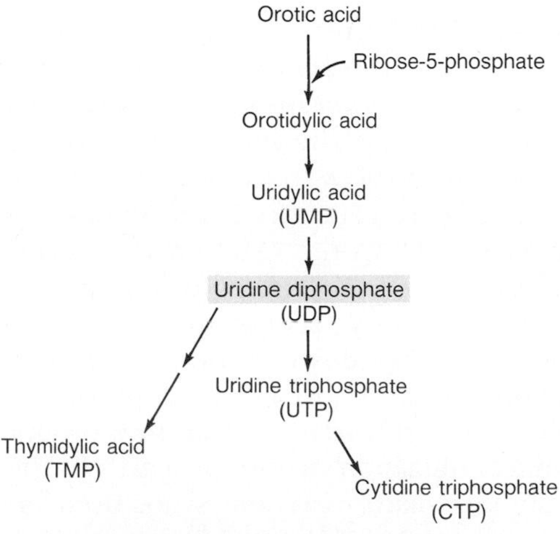

(b) Biosynthesis of the pyrimidine nucleotides

Figure 4.27 Pyrimidine biosynthesis.

dine ring (orotic acid) is built up before the sugar is added to form a nucleoside. After the addition of ribose and phosphate to orotic acid, the other pyrimidines are successively synthesized (Figure 4.27*b*). Complete details of pyrimidine biosynthesis are beyond this text. The biosynthesis of the deoxyribose sugar was discussed briefly in Section 4.17.

4.20 Regulation of Enzyme Activity

In the previous pages we have discussed a few of the key enzymatic reactions occurring during anabolism and catabolism. There are, of course, hundreds of different enzymatic reactions occurring simultaneously during a single cycle of cell growth. One thing that should be very clear is that not all these enzymatic reactions occur in the cell to the same extent. Some compounds are needed in large amounts and the reactions involved in their synthesis must therefore occur in large amounts. However, other compounds are needed only in small amounts and the reactions involved in their synthesis must occur only in small amounts. The need for regulation of biochemical reactions is clear. How does regulation occur?

Enzyme amount or enzyme activity? There are two major modes of regulation, one which controls the amount (or even the complete presence or absence) of an enzyme, and the other which controls the *activity* of preexisting enzyme. Regulation of enzyme amount, by the phenomena called **induction** and **repression**, will be discussed after we have discussed molecular biology and protein synthesis in Chapter 5. We discuss here briefly the processes involved in regulating the *activity* of enzymes.

Product inhibition A simple mechanism by which an enzymatic reaction may be regulated is by a process called *product inhibition*. As we have seen, an enzyme combines with its *substrate*, the reaction occurs, and the *product* is released. Because enzymatic reactions are generally *equilibrium reactions*, as product builds up the reaction can occur in the *reverse* direction, from product to substrate. This, in a sense, is a regulatory process, since the rate of product formation slows down as product builds up. At equilibrium, there is just as much product turning into substrate as substrate turning into product.

Product inhibition occurs primarily if the product of the enzymatic reaction is not used in subsequent reactions. But if the product is itself the substrate for another enzyme, then it will be continually removed and product inhibition will not occur. Note that in product inhibition it is the enzyme which *formed* the product which is also inhibited by the product.

Feedback inhibition A major mechanism for the control of enzymatic activity involves the phenomenon of **feedback inhibition**. Feedback inhibition is seen primarily in the regulation of whole biosynthetic pathways, such as, for instance, the pathway involved in the synthesis of an amino acid or purine. As we have seen, such pathways involve many enzymatic steps, and the final product, the amino acid or nucleotide, is many steps removed from the starting substrate. Yet, this final product is able to feed back to the first step in the pathway and regulate its own biosynthesis. How?

In feedback inhibition the amino acid or other end product of the biosynthetic pathway inhibits the activity of the *first* enzyme in this pathway. Thus, as the end product builds up in the cell, its further synthesis is inhibited. If the end product is used up, however, synthesis can resume (Figure 4.28).

How is it possible for the end product to inhibit the activity of an enzyme that acts on a compound quite unrelated to it? This occurs because of a property of the inhibited enzyme known as **allostery**. An allosteric enzyme has two important combining sites, the *active* site, where the substrate binds, and the *allosteric* site, where the inhibitor (sometimes called an "effector") binds reversibly. When an inhibitor binds at the allosteric site, the conformation of the enzyme molecule changes so that the substrate no longer binds efficiently at the active site (Figure 4.29). When the concentration of the inhibitor falls, equilibrium favors the dissociation of the inhibitor from the allosteric site, returning the active site to its catalytic shape. Allosteric enzymes are very common in both biosynthetic and degradative pathways, and are especially important in branched pathways. For example, the amino acids proline and arginine are both synthesized from glutamic acid. Figure 4.30 shows that these two amino acids can control the first enzyme unique to their own synthesis without affecting the other, so that a surplus of proline, for example, will not cause the organism to be starved for arginine.

In addition, some biosynthetic pathways are regulated by the use of **isozymes** (short for isofunctional enzymes: *iso* means same or constant). These enzymes catalyze the same reaction, but are subject to different regulatory control. An example is the synthesis of the aromatic amino acids as summarized in Figure 4.24 (see also Figure 10.3). Three different isozymes catalyze the first reaction in this pathway and each enzyme is regulated independently by each of the three different end product amino acids. Unlike

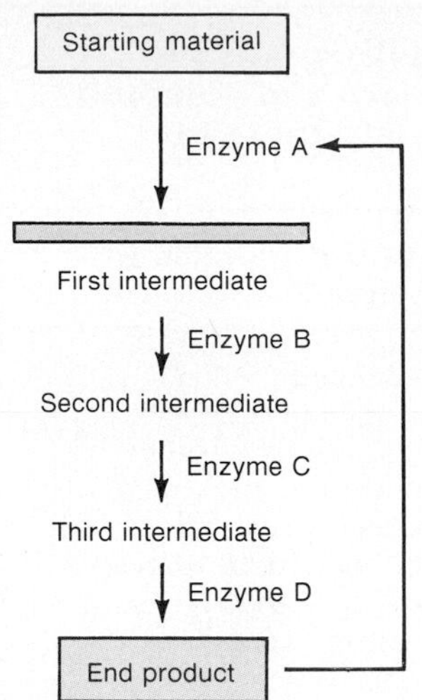

Figure 4.28 Feedback inhibition. The activity of the first enzyme of the pathway is inhibited by the end product, thus controlling production of end product.

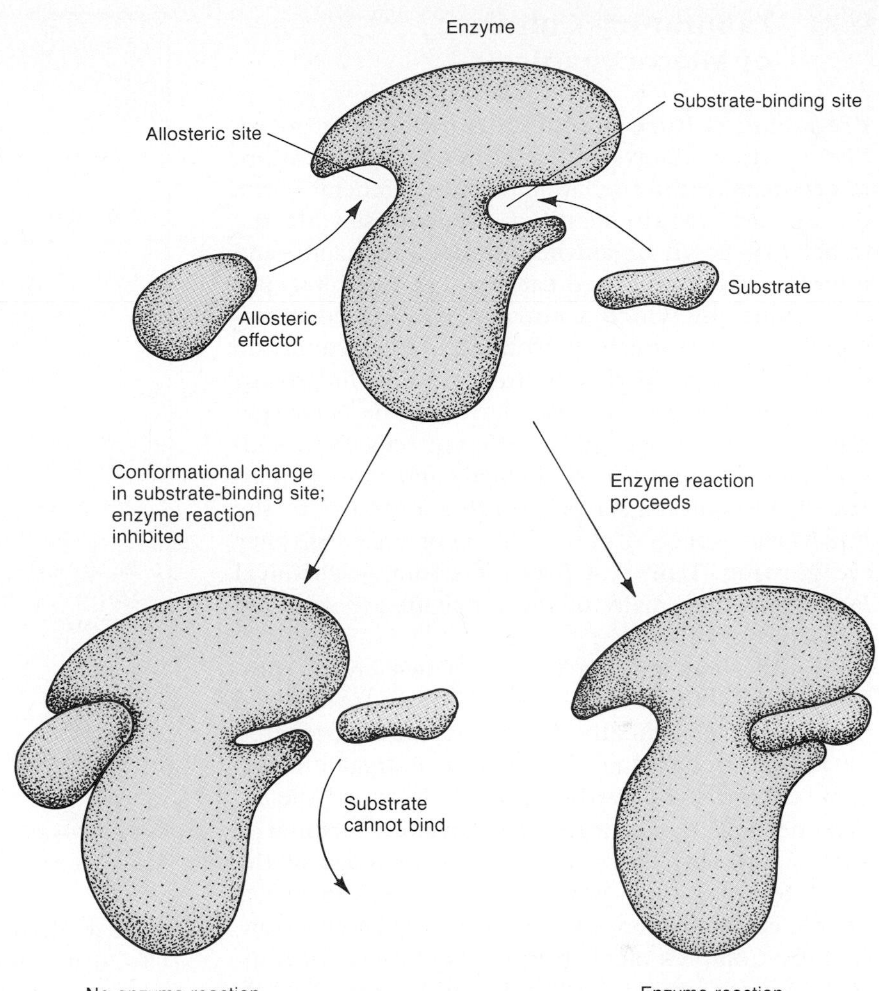

Figure 4.29 Mechanism of enzyme inhibition by allosteric effector. When the effector combines with the allosteric site, the conformation of the enzyme is altered so that the substrate can no longer bind.

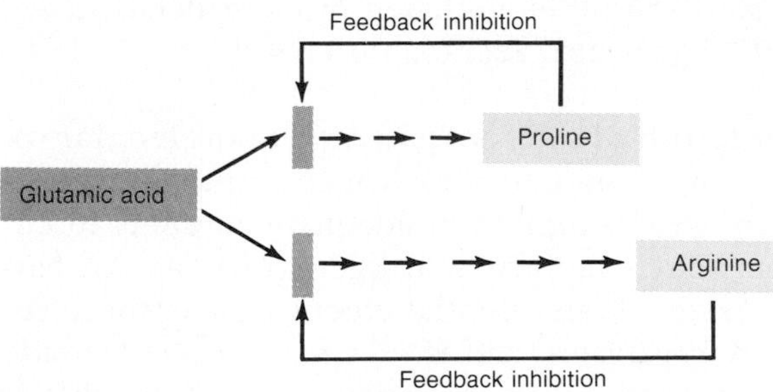

Figure 4.30 Feedback inhibition (dashed lines) in a branched biosynthetic pathway.

the earlier examples of feedback inhibition where inhibitors completely stopped an enzyme activity, in this case the total amount of the initial enzyme activity is diminished in a stepwise fashion and falls to zero only when all three products are present in excess.

Occasionally a branched biosynthetic pathway relies on a single enzyme (rather than several isozymes) to carry out the first step of the pathway. In this case regulation occurs by *concerted* feedback inhibition. In concerted feedback inhibition a buildup in each of the end products is required to affect an allosteric change in the first enzyme of the pathway. This is simply another way of insuring that a buildup of only one of the products does not affect synthesis of other products of the pathway.

These examples are representative of the major regulatory patterns observed; there are other, rather elegant examples of regulation now known. It is hypothesized that the phenomenon of feedback inhibition has evolved because efficient control of the rate of enzyme activity enables an organism to quickly adapt to changing environments. Examples of how the rates of synthesis of various end products are regulated can be found in some of the detailed pathways given in biochemistry textbooks.

4.21 Laboratory Culture of Microorganisms

Preparing culture media A knowledge of microbial nutrition allows the microbiologist to culture microorganisms in the laboratory. In a general sense, the nutrient requirements of all organisms are the same; that is, all organisms require the macro- and micronutrients discussed earlier in this chapter. But the manner by which a nutrient is obtained differs widely from one organism to another. The proportion of each nutrient added to a culture medium for growth of any microorganism must, therefore, be fairly specific. Literally thousands of different recipes for culture media have been published, and most experienced microbiologists pay careful attention to the nutritional aspects of the organism or organisms they are growing. Table 4.4 lists the common chemical forms by which many of the nutrients are supplied in culture media.

Culture media are generally of two broad types: **chemically defined (synthetic)** or **undefined (complex)**. Chemically defined media are prepared by adding precise amounts of pure inorganic or organic chemicals to distilled water. Therefore, the *exact* chemical composition of a defined medium is known. In many cases, however, knowledge of the exact composition is not critical. In these instances undefined media may suffice, or may even be advantageous. Complex media frequently employ crude digests of substances such as casein (milk protein), beef, soybeans, yeast cells, or a number of other highly nutritious, yet nutritionally undefined, substances. Such digests are available commercially in powdered form and can be weighed out rapidly and added to culture media. Hence, if the goal is simply to obtain growth of cells, it is frequently easier and quicker to use a complex medium. In most cases the use of complex ingredients in a culture medium negates the necessity for adding purified chemicals. Analyses of yeast extract, beef extract and various other constituents usually reveal sufficient levels of each required nutrient, including trace elements. However, an important disadvantage of using complex media is the loss of control over the precise nutrient specifications of the medium.

Culture media must be rendered **sterile** (free of all life forms, including spores and viruses) before use, and this is frequently done by heat treatment, generally in a device called an *autoclave*. An **autoclave** is an enlarged pressure cooker which allows one to heat objects under pressure to temperatures above the boiling point of water. Liquids are generally heated with steam at a pressure sufficient so that the temperature can reach 121°C. At such a temperature a small volume of liquid will be sterilized in 15–20 minutes. Autoclaving is necessary for sterilization because many bacterial endospores are resistant to the temperatures reached by boiling (100°C). Heat sterilization is discussed in more detail in Section 9.15 (see also Section 1.15).

Table 4.4 Nutrient requirements of microorganisms and common means to satisfy them in culture

Nutrient	Chemical form supplied in culture media
Carbon	*Organic*—defined media: glucose, acetate, pyruvate, malate, hundreds of other compounds; complex media: yeast extract, beef extract, peptone, many other digests *Inorganic*—CO_2, HCO_3^-
Nitrogen	*Organic*—Amino acids, nitrogenous bases *Inorganic*—NH_4Cl, $(NH_4)_2SO_4$, KNO_3, N_2
Phosphorus	KH_2PO_4, Na_2HPO_4
Sulfur	Na_2SO_4, H_2S
Potassium	KCl, K_2HPO_4
Magnesium	$MgCl_2$, $MgSO_4$
Sodium	NaCl
Iron	$FeCl_3$, $Fe(NH_4)(SO_4)_2$, iron chelates
Micronutrients	$CoCl_2$, $ZnCl_2$, Na_2MoO_4, $CuCl_2$, $MnSO_4$,$NiCl_2$, Na_2SeO_4, Na_2WO_4, Na_2VO_4

Relation of microorganisms to molecular oxygen One important chemical constituent that must be frequently supplied in adequate amounts to culture media is molecular oxygen (O_2). As we have seen, oxygen is an essential electron acceptor in respiratory organisms, and if it is not supplied in adequate amounts, such organisms will not be able to obtain energy.

Air contains 20 percent oxygen and is the common source of oxygen for microbial growth. Oxygen gas has, however, only a limited solubility in water. When microorganisms are cultivated on the surfaces of agar plates, aeration is generally of no problem, but when liquid cultures are used, the limited solubility of oxygen presents a serious problem which must be overcome by vigorous aeration of the culture medium.

One way in which microorganisms are classified is in their relationships with oxygen. Two major terms used are **aerobe**, for organisms which use oxygen, and **anaerobe**, for organisms that do not use oxygen. Organisms that *must* have oxygen supplied as a terminal electron acceptor are called **obligate aerobes**. These organisms are unable to obtain energy by fermentation or other anaerobic processes. **Obligate anaerobes**, in contrast, do not use O_2 and will not grow in its presence. Many obligate anaerobes are actually ***killed*** by O_2, even after brief exposure. We discuss oxygen toxicity in detail in Section 9.14. A third very important group of microorganisms are called **facultative**. These organisms are capable of growing *either* aerobically or anaerobically. Yeast discussed in this chapter is a good example of a facultative aerobe. Yeast can grow anaerobically by fermentation, producing alcohol and CO_2, and it can grow aerobically by respiration, producing only CO_2. Whether yeast grows aerobically or anaerobically depends, in the first instance, on the availability of O_2. If O_2 is present, yeast will grow aerobically, producing a large amount of ATP, and consequently a large amount of cell mass. If O_2 is absent and a fermentable energy source such as glucose is present, yeast can grow by fermentation, produce much less cell mass, but produce alcohol and CO_2. (Why does yeast produce so much less cell mass by fermentation than by respiration? Refer back to Section 4.13).

Aerotolerant anaerobes are organisms which cannot use O_2 but are not harmed by it either. The lactic acid bacteria, which produce lactic acid by the fermentation of glucose, are the best examples of aerotolerant anaerobes (see Section 19.25). **Microaerophilic** organisms are aerobes which only tolerate reduced levels of O_2. These organisms require molecular oxygen but if too much is present they are harmed. The selection of the proper aeration conditions is absolutely essential for successful cultivation of microorganisms. Much of the skill of the microbiologist is displayed in the selection of appropriate aeration conditions. When we discuss large-scale cultivation of microorganisms in an industrial context (see Chapter 10), we will see that proper aeration is not only vitally important for a successful culture, but even has economic implications.

Provision of trace elements The micronutrients (trace elements) are special cases in the preparation of culture media. As we have seen earlier in this chapter, most of these minerals are required in extremely small amounts. Very often, microorganisms appear to grow in culture media to which these trace elements have not even been added. This is because contaminating amounts of trace metals are already present, derived from the glassware, distilled water, and other chemicals (even those labeled "ultra-pure" may have significant amounts of trace elements). Even such an inert material as the cotton plug used to stopper a culture tube or flask may contribute trace elements to the culture medium. Often, only after the glassware has been scrupulously cleaned and the culture ingredients highly purified is it possible to demonstrate a trace-element requirement for a microbial culture. Nevertheless, the following trace elements have been shown to be required by one or more microorganisms: cobalt, zinc, copper, manganese, molybdenum, nickel, tungsten, and selenium. The biochemical role played by each of these trace elements was discussed in Section 4.1.

Iron is a special case, as it is considered a trace mineral but is required in larger amounts than other trace minerals. As we have seen, iron plays a major role in cellular respiration, being a key ingredient of the cytochromes and iron-sulfur proteins involved in electron transport. Because most inorganic iron salts are highly insoluble, providing an adequate supply of iron to a culture medium presents some difficulties. One way of supplying iron is to complex it with an organic chemical called a **chelating agent**. Two common chelating agents used to provide iron to culture media are ethylene diamine tetraacetic acid (EDTA) and nitrilotriacetic acid (NTA). If the medium contains complex organic materials such as yeast extract or peptone, natural iron chelators are already present, so that no special care need be taken to provide soluble iron. It is mainly when completely synthetic media are used that iron deficiency presents a problem.

Many organisms produce specific iron-binding organic compounds, called **ironophores**, which solubilize iron salts and transport iron into the cell. One major group of ironophores are derivatives of **hydroxamic acid**, which bind iron very strongly, as shown in Figure 4.31. Once the iron-hydroxamate complex has passed into the cell, the iron is released and the hydroxamate can pass out of the cell and be utilized again for iron transport. In some bacteria, the iron-binding compounds are not hydroxamates but phenolic acids. Enteric bacteria such as *Escherichia coli* and *Salmonella* spp. produce complex phenolic derivatives called **enterobactins** or **enterochelins**. As we will see in Chapter 11, availability of iron has important consequences in the ability of many harmful (pathogenic) bacteria to grow in the body.

Growth factors Growth factors are specific organic compounds that are required in very small amounts and that cannot be synthesized by some cells. Substances frequently serving as growth factors are vitamins, amino acids, purines, and pyrimidines. Although most organisms are able to synthesize all of these compounds, certain others require them pre-

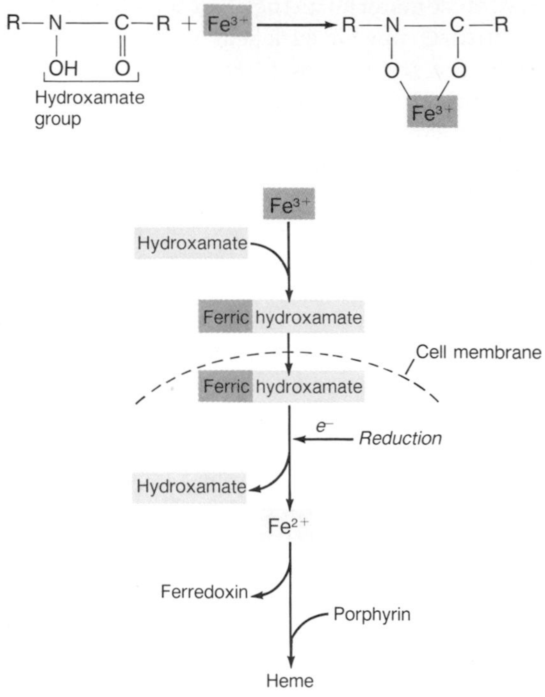

Figure 4.31 Function of iron-binding hydroxamate in the iron nutrition of an organism living in an environment low in iron.

Table 4.5 Vitamins and their functions

Vitamin	Function
p-Aminobenzoic acid	Precursor of folic acid
Folic acid	One-carbon metabolism; methyl group transfer
Biotin	Fatty acid biosynthesis; β-decarboxylations; CO_2 fixation
Cobalamin (B_{12})	Reduction of and transfer of single carbon fragments; synthesis of deoxyribose
Lipoic acid	Transfer of acyl groups in decarboxylation of pyruvate and α-ketoglutarate
Nicotinic acid (niacin)	Precursor of NAD; electron transfer in oxidation-reduction reactions
Pantothenic acid	Precursor of coenzyme A; activation of acetyl and other acyl derivatives
Riboflavin	Precursor of FMN, FAD in flavoproteins involved in electron transport
Thiamin (B_1)	α-Decarboxylations; transketolase
Vitamins B_6 (pyridoxal-pyridoxamine group)	Amino acid and keto acid transformations
Vitamin K group; quinones	Electron transport; synthesis of sphingolipids
Hydroxamates (terregens factor, coprogen, mycobactin, etc.)	Iron-binding compounds; solubilization of iron and transport into cell
Coenzyme M (Co-M)	Required by certain methanogens; plays a role in methanogenesis

formed from the environment. Hence, they must be added to the culture medium. If a complex organic material such as yeast extract or peptone is used in a culture medium, most or all of the potential growth factors will already be supplied, but in synthetic media, provision of proper growth factors is often of great importance. Because growth factor requirements are often introduced into organisms via genetic mutation, a detailed understanding of growth factor nutrition is essential in most genetic manipulations of microorganisms. Without an understanding of microbial growth factor nutrition, most of our sophisticated knowledge of genetics (Chapters 5–8) would not have come about.

Vitamins Vitamins are defined as organic compounds required in small amounts for growth and function that do not serve as either energy sources or building blocks of macromolecules. The first growth factors to be discovered and studied in any detail were the vitamins. Most vitamins function as parts of coenzymes (see, for instance, Figures 4.6, 4.13, and 4.14) and these are summarized in Table 4.5. Many microorganisms are able to synthesize all of the components of their coenzymes, but some microorganisms are unable to do so and must be provided with certain parts of these coenzymes in the form of vitamins. The lactic acid bacteria, which include the genera *Streptococcus* and *Lactobacillus* (see Section 19.25), are renowned for their complex vitamin requirements, which are even greater than those of humans. The vitamins most commonly required by microorganisms are thiamine (vitamin B_1), biotin, pyridoxine (vitamin B_6), and cobalamin (vitamin B_{12}).

Amino acids As we have seen, proteins are composed of **amino acids** and these amino acids must either be synthesized by an organism or obtained preformed from the environment. There are 20 different amino acids present in proteins (see Section 2.7), all of which must be available for cell growth. Many microorganisms require specific amino acids. Inability to synthesize an amino acid is related to lack of the enzymes needed for its synthesis. We summarized the pathways for amino acid biosynthesis in Figure 4.24. In some cases, an amino acid requirement exists because only a single enzyme in a pathway is missing, whereas in other cases enzymes of the whole pathway may be absent.

The required amino acids can usually be supplied in the culture medium as the chemically synthesized amino acids. If a complex material such as yeast extract or peptone is added to the medium, it will usually contain all of the 20 amino acids, either in the form of the free amino acids or in the form of small peptides or proteins.

Purines and pyrimidines As we have seen, purines and pyrimidines are key constituents of the building blocks of the nucleic acids. They are also present in certain coenzymes. Purines and pyrimidines can either be provided in the medium as the free base or in the form of a nucleoside (the free base with the ribose or deoxyribose sugar attached). The nucleotide forms of the purines and pyrimidines, containing the phosphate group, are generally not available to cells because the phosphate group makes the compound unable to pass the cell membrane.

Culture media We can now summarize the material of this section by presenting the detailed chemical compositions of two synthetic culture media, one simple, one complex (Table 4.6). The simple culture medium is one that will support the growth of *Escherichia coli* and a number of other enteric bacteria. The biosynthetic capabilities of *Escherichia coli* are quite impressive, as shown by the fact that only a *single* organic compound, glucose, need be added to the medium.

The other culture medium presented in Table 4.6 is one that satisfies the requirements of a bacterium that has complex nutritional requirements, the lactic acid bacterium, *Leuconostoc mesenteroides*. Which organism has more biosynthetic capacity, *Escherichia coli* or *Leuconostoc mesenteroides*? Obviously, *Escherichia coli*, since its ability to grow on a simple culture medium means that it has the ability to synthesize *all* of its organic cellular constituents from a single carbon compound.

As we have indicated, the ability to synthesize a particular monomeric building block such as an amino acid or nucleic acid base requires the presence in the cell of the suite of enzymes of that particular biosynthetic pathway. If any one enzyme of the pathway is missing, then the organism will not grow unless this

Table 4.6 Synthetic culture media for microorganisms with simple and complex nutritional requirements

Culture medium for *Escherichia coli*		Culture medium for *Leuconostoc mesenteroides*	
K_2HPO_4	7.0 g	K_2HPO_4	0.6 g
KH_2PO_4	2.0 g	KH_2PO_4	0.6 g
$(NH_4)_2SO_4$	1.0 g	NH_4Cl	3 g
$MgSO_4$	0.1 g	$MgSO_4$	0.1 g
$CaCl_2$	0.02 g	Glucose	25 g
Glucose	10 g	Sodium acetate	20 g
Trace elements (Fe, Co, Mn, Zn, Cu, Ni, Mo)	2–10 μg each		
Distilled water	1000 ml	Amino acids (alanine, arginine, asparagine, aspartate, cysteine, glutamate, glycine, histidine, isoleucine, leucine, lysine, methionine, phenylalanine, proline, serine, threonine, tryptophan, tyrosine, valine)	100–200 μg of each
		Purines and Pyrimidines (adenine, guanine, uracil, xanthine)	10 mg of each
		Vitamins (biotin, folate, nicotinic acid, pyridoxal, pyridoxamine, pyridoxine, riboflavin, thiamine, pantothenate, para-aminobenzoic acid)	0.01–1 mg of each
		Trace elements	2–10 μg each
		Distilled water	1000 ml

monomeric building block is provided, in some way, from the environment. Since the instructions for the synthesis of each enzyme of a biosynthetic pathway are found in the genetic code of the organism, our attention naturally focuses now on genes and how they work. The next group of chapters will thus deal with the extremely important topic of molecular genetics.

Study Questions

1. In the following list of substances indicate which ones could serve as energy sources: ferrous iron, O_2, CO_2, NH_4^+, SO_4^{2-}, NO_2^-, NO_3^-, H_2S, glucose, methane, ferric iron. In this same list indicate which compounds could serve as electron acceptors. Both donor and acceptor.
2. The following is a series of coupled electron donors and electron acceptors. Using the data given in Figure 4.5, order this series from most energy-yielding to least energy-yielding. H_2/Fe^{3+}, H_2S/O_2, methanol/NO_3^- (producing NO_2^-), H_2/O_2, Fe^{2+}/O_2, NO_2^-/Fe^{3+}, H_2S/NO_3^-.
3. What is an electron carrier? Give three examples of electron carriers and indicate their oxidized and reduced forms.
4. Which of the following reactions is exergonic and which endergonic: glucose + ATP yielding glucose-6-phosphate + ADP; acetyl phosphate yielding acetate + phosphate; ethanol + carbon dioxide yielding glucose; $H_2 + \frac{1}{2}O_2$ yielding H_2O. H_2O yielding $H_2 + \frac{1}{2}O_2$; $Fe^{2+} + SO_4^{2-}$ yielding Fe^{3+} + H_2S.
5. Give an example of an electron donor and electron acceptor which could function in each of the following processes: fermentation, respiration, anaerobic respiration.
6. An experimenter has isolated a mutant organism which is blocked in glycolysis at the step between acetaldehyde and ethanol. This organism is no longer able to grow anaerobically on glucose but is still able to grow when O_2 is present. Give a possible biochemical explanation for this observation.
7. Iron plays an important role within the cell in energy generating processes. Give three examples in which iron plays a role as an electron carrier. How is iron provided as a nutrient in culture media?
8. In the following list indicate which substance is a coenzyme and which is a prosthetic group: nicotinamide adenine dinucleotide (NAD), adenosine diphosphate (ADP), heme, riboflavin, iron (in ferredoxin).
9. Explain how the "sidedness" of the plasma membrane is critical for the generation of a proton-motive force.
10. The synthesis of ATP from ADP and inorganic phosphate is an endergonic reaction. How then is the membrane-bound ATPase able to bring about ATP synthesis?
11. The chemical inhibitors dinitrophenol and cyanide both affect the energy-generation process, but in quite different ways. Compare and contrast the modes of action of these two inhibitors.
12. Much more energy is available from glucose respiration than from glucose fermentation. However, the law of conservation of energy states that energy is neither created nor destroyed. Can you give an explanation for where all the energy in the glucose molecules that was not released in the fermenting organism might have gone to?
13. Work through the energy balance sheets for fermentation and respiration and account for all sites of ATP synthesis. Organisms can obtain about 15 times more ATP when growing aerobically on glucose than anaerobically. Write one sentence which accounts for this difference.
14. Cells frequently produce energy-storage polymers, either organic or inorganic. Since ATP is the primary energy currency of the cell, why is some of the energy from ATP converted into storage polymers? Why are polymers formed rather than equivalent monomers?
15. Explain why CO_2 could not serve as an energy source. Explain why it *could* serve as an electron acceptor. When CO_2 serves as an electron acceptor, to what substance is it often converted?
16. In aerobic respiration, O_2 serves as the ultimate electron acceptor. To what substance is O_2 always converted as it accepts electrons?
17. Knowing the function of the cytochrome system, could you imagine an organism that could live if it completely lacked a cytochrome system? If it lacked only part of the cytochrome system? How?
18. Give in simplified form the equation for the oxidation-reduction reaction which takes place in the cytochrome molecule. Can you suggest a role for the portions of the cytochrome molecule which are not involved in oxidation-reduction?
19. Indicate the electron donor (energy source) and the electron acceptor for each of the following cases: oxidation of glucose with air, reduction of nitrate in the presence of glucose, oxidation of glucose in the absence of air, oxidation of H_2S in air.
20. A substance such as glucose can serve in an aerobic organism both as an energy source and as a building block of cell substance. What are the names of the two types of metabolic processes involved in these two disparate functions? What is the fate of the carbon atoms of those glucose molecules that are used in energy generation? List three groups of carbon compounds that are building blocks of cell substance that are derived from glucose.

21. Pyruvate is an important intermediate in both catabolic and anabolic reactions. Describe one catabolic reaction in which pyruvate serves as a key intermediate. Describe an anabolic reaction in which pyruvate is converted to an amino acid.
22. Examine the data in Table 4.2. Why is it accurate to say that a bacterial cell consists primarily of macromolecules? What class of macromolecules constitutes the largest weight in the cell? What class of macromolecules is present in the cell in the largest number? From what you know about energy, where would you say that the largest amount of energy in the cell is stored?
23. There are a number of bacteria which are able to obtain energy and building blocks by attacking other bacteria. From what you know about bacterial cell structure, energy generation, and biosynthesis, describe in biochemical terms the processes by which a whole bacterial cell could be consumed by another bacterium for nutrition, biosynthesis, and energy generation. In this regard, the information given in Table 4.2 should be of value.
24. Can you suggest why it is necessary that polysaccharides such as starch and cellulose must be digested *outside* the cell whereas disaccharides such as lactose and sucrose can be digested *inside* the cell?
25. There are bacteria which *synthesize* glycogen and there are also bacteria which *utilize* glycogen. However the biochemical pathways for glycogen synthesis and utilization are distinct. Describe these two distinct types of pathways. Can you see any advantage in terms of cell function if separate pathways are used for these two processes?
26. The role of NAD in energy generation has been emphasized. NADP also plays important roles as a coenzyme in cell function. List several of these roles and give a rationale for why NADP might be used instead of NAD.
27. The cell wall peptidoglycan of *E. coli* contains two important hexose sugars. When *E. coli* is growing on glucose, some of the glucose molecules can be converted directly into these cell wall sugars. However, when *E. coli* is growing on alanine, these sugars must be synthesized *de novo*. Give an overall scheme (exact biochemical compounds are not needed) by which alanine can be converted into hexose. What is this process called?
28. Figure 4.22 has indicated that there are only a very few intermediate compounds which serve as the starting points for anabolism. For each of the amino acid families, list the intermediates from Figure 4.22 which are involved in the biosynthetic reactions.
29. Purine and pyrimidine metabolism can be approached in the same way that amino acid metabolism was just approached in the above question. List the comparable intermediates in purine and pyrimidine metabolism.

Supplementary Readings

Dawes, E. A. 1986. *Microbial Energetics.* Blackie and Sons, Ltd., Glasgow. An excellent short monograph dealing specifically with the energetics of bacteria and how energetic principles relate to growth.

Gottschalk, G. 1986. *Bacterial Metabolism*, 2nd edition. Springer-Verlag, New York. A college level text dealing exclusively with bacterial topics. Contains an indepth treatment of anaerobic metabolism.

Harold, F. M. 1986. *The Vital Force: A Study of Bioenergetics.* W. H. Freeman and Co., New York. An excellent treatment of the principles of bioenergetics, especially in terms of membrane-mediated events. Includes coverage of photosynthesis, bacteriorhodopsin-mediated ATP production, and membrane activities of eucaryotic cells as well. Highly recommended.

Hellingwerf, K. J. and **W. N. Konigs**. 1985. The energy flow in bacteria: The main free energy intermediates and their regulatory role. *Advances in Microbial Physiology* 26:125–154.

Nichols, D. G. 1982. *Bioenergetics: An Introduction to the Chemiosmotic Theory.* Academic Press, New York. An excellent summary of the chemiosmotic model.

Slayman, C. A. 1985. Proton chemistry and the ubiquity of proton pumps. *Bioscience* 35:16–17. This is the lead-off article in a series devoted to proton pumps and the evolution of proton pumping systems.

Stryer, L. 1981. *Biochemistry*, 2nd ed. W. H. Freeman and Co., San Francisco. Excellent coverage of basic energetics, enzymes and other aspects of biochemistry. Especially pertinent are Chapters 11–14.

Thauer, R. K. and **J. G. Morris**. 1984. Metabolism of chemotrophic anaerobes: Old views and new aspects. Pages 123–168 in *The Microbe.* 1984. Part II. Prokaryotes and Eukaryotes. *36th Symposium Society General Microbiology.* Cambridge University Press, Cambridge, England. Excellent overview of anaerobic metabolism, chemiosmosis, growth yields, CO_2 fixation by anaerobes, and other aspects of anaerobic metabolism.

5 Macromolecules and Molecular Genetics

We begin now a study of the genetics of microorganisms which will extend over the next four chapters. **Genetics** is the discipline which deals with the mechanisms by which traits are passed from one organism to another. As we noted in Chapter 1, two hallmarks of life are *energy transformation* and *information flow*. In the previous chapter we dealt with the problem of energy transformation: *metabolism*. In the present chapter and in those which follow we deal with the problem of information flow: *genetics*.

The study of genetics is central to an understanding of the variability of organisms and the evolution of species. Genetics is also a major research tool in attempts to understand the molecular mechanisms by which cells function. Genetics and biochemistry work together in the continuing quest to discern the ultimate basis of life.

Genetics also provides us with approaches to the construction of new organisms of potential use in human affairs. As such, genetics has provided us with some of the most important advances in agriculture, medicine, and industry. An understanding of genetic mechanisms makes it possible for researchers to manipulate species and construct new organisms. It also makes it possible for scientists and physicians to develop means for controlling the important infectious diseases of humankind. We will have much to say about the application of genetics to human affairs in subsequent chapters.

What is a gene? A **gene** may be defined as an entity which specifies the structure of a single protein polypeptide chain. We discussed the chemistry of proteins in Chapter 2 and noted that a polypeptide is composed of a series of amino acids connected in peptide linkage. There are 20 different amino acids present in proteins and a single polypeptide chain will usually have several hundred separate amino acid residues. The gene is the element of information which specifies the sequence of amino acids of the protein. The main purpose of this chapter is to explain and expand on this definition.

Genetic phenomena involve three types of macromolecules: *deoxyribonucleic acid (DNA)*, the genetic material of the cell, *ribonucleic acid (RNA)*, the intermediary or messenger, and *protein*, whose individual components are the functional entities of the living cell. During growth, all three types of macromolecules are synthesized. DNA is *replicated*, leading to the synthesis of exact copies (Figure 5.1). The information in DNA is also *transcribed* into complementary sequences of nucleotide bases in RNA (Figure 5.1); this RNA, containing the *information* for the amino acid sequence of the protein, is called **messenger RNA** (*mRNA*). Messenger RNA is then *translated*, using the specific protein-synthesizing machinery of the ribosomes, and the translation product is **protein**. Because the steps from DNA to RNA to protein involve the transfer of information, these macro-

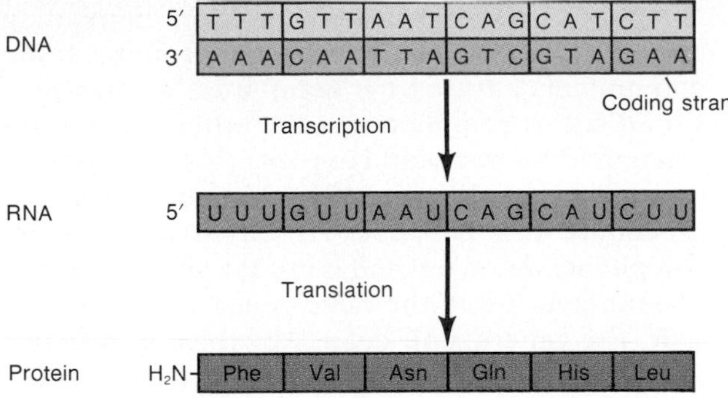

Figure 5.1 The three elements of macromolecular synthesis. Note that in any particular region only one of the two strands of the DNA double helix is usually transcribed.

molecules are often called **informational macromolecules** to distinguish them from macromolecules such as polysaccharides and lipids which are large but are not informational.

In what form is the information stored in the informational macromolecules? In DNA and RNA the information is encoded in the *base sequence* of the purine and pyrimidine residues of the polynucleotide chain. When we discuss the information content of a nucleic acid we thus speak of the *coding* properties of this material. The amino acid sequence of the polypeptide is *coded* by the sequence of purine and pyrimidine bases within the nucleic acid, with three bases coding for a single amino acid.

Importance Microbial genetics is important for a number of reasons:

1. Gene function is at the basis of cell function, and basic research in microbial genetics is necessary to understand how microbes function.
2. Microbes provide relatively simple systems for studying genetic phenomena, and are thus useful tools in attempts to decipher the mechanisms underlying the genetics of all organisms.
3. Microbes are used for the isolation and duplication of specific genes from other organisms, a technique called **molecular cloning**. In molecular cloning, genes are manipulated and placed in a microbe where they can be induced to increase in number.
4. Microbes produce many substances of value in industry, such as antibiotics, and genetic manipulations can be used to increase yields and improve manufacturing processes. Also, genes of higher organisms that specify the production of particular substances, such as human insulin, can be transferred by molecular cloning into microorganisms, and the latter used for the production of these useful substances.
5. Many diseases are caused by microorganisms, and genetic traits underlie these harmful activities. By understanding the genetics of disease-causing microbes, we can more readily control them and prevent their growth in the body. Viruses, although agents of disease, can be thought of as genetic elements, and understanding the genetics of viruses helps us to control virus disease.

Procaryotic and eucaryotic genetics Procaryotes have relatively simple genetic systems. Their chromosomes are single DNA molecules (see Section 3.15). Mechanisms for the transfer of genes from one procaryotic cell to another are also simple and easy to study. Eucaryotes, even the simplest eucaryotic microbes, are much more complex structurally, and their chromosomes are present in larger numbers (see Section 3.16). They have fairly complex mechanisms of sexual reproduction for bringing about genetic recombination between organisms. Thus in this chapter we will be making contrasts between procaryotes and eucaryotes. We will develop first procaryotic genetics and will then highlight those features that are different in eucaryotes.

5.1 Overview

What transformations do informational macromolecules undergo during cell growth and division? It is the purpose of the present section to outline briefly some of the processes that will be discussed in this chapter.

A cell is an integrated system containing a large number of specific macromolecules. When a cell divides and forms two cells, all of these macromolecules are duplicated. The fidelity of duplication is very high, although occasional errors do occur. The molecular processes underlying cell growth can be divided into a number of stages which are described briefly here.

1. ***Replication*** The DNA molecule is a **double helix** of two long chains (see Section 2.6). During replication, DNA, containing the master *genetic code*, duplicates. The products of DNA replication are two molecules, each a double helix, the two strands thus becoming four strands.
2. ***Transcription*** DNA does not function directly in protein synthesis, but through an RNA intermediate. The transfer of the information to RNA is called **transcription**, and the RNA molecule carrying the information is called **messenger RNA (mRNA)**. Messenger RNA molecules frequently contain the instructions for making more than one protein. In most cases, at any particular location on the chromosome, only one strand of the DNA is transcribed, and the information of this strand is then contained in the mRNA.

3. ***Genetic code*** The specific sequence of amino acids in each protein is directed by a specific sequence of bases in DNA. It takes *three* bases to code for a single amino acid, and each triplet of bases is called a **codon**. There is a one-to-one correspondence between the base sequence of a gene and the amino acid sequence of a protein (Figure 5.1). A change in a single base changes the codon and can lead to a change in the amino acid in the protein. Such changes, which are frequently detrimental, are called **mutations**. Microorganisms carrying mutations (mutants) are essential in genetic research, as they make possible the location of genes, through the construction of genetic crosses between related organisms.
4. ***Translation*** The genetic code is translated into protein by means of the protein-synthesizing system. This system consists of **ribosomes, transfer RNA**, and a number of enzymes. The ribosomes are the structures to which messenger RNA attaches. Transfer RNA (tRNA) is the key link between codon and amino acid. There is one or more separate tRNA molecules corresponding to each amino acid, and the tRNA has a triplet of three bases, the **anticodon**, which is *complementary* to the codon of the messenger RNA. An enzyme brings about the attachment of the correct amino acid to the correct tRNA.
5. ***Regulation*** Not all proteins are synthesized at equal rates. Complex systems of regulation exist which control the rates of synthesis of proteins. Some proteins, called *inducible*, are synthesized only when small molecules called **inducers** are present. Usually the inducer is a substrate of the enzyme, and induction thus ensures that the enzyme is only formed when it is needed. Another class of enzymes, called *repressible*, are synthesized only in the absence of specific small molecules, generally biosynthetic products. Enzyme repression and enzyme induction have as their basis the same underlying mechanisms, which is the regulation of mRNA synthesis.
6. ***Complementation*** When two mutant strains are genetically crossed (mated), if the two mutations are in the same gene the resulting hybrid should still be mutant, whereas if they are in different genes, the mutant phenotype should be eliminated. This type of experiment is called a **complementation test**, because one is determining whether the two mutations are able to complement each other. The two mutations can be on the same or on separate chromosomes. If the two mutations are each on separate chromosomes, they would be said to be in **trans** configuration. On the other hand, if the two mutations were on the *same* chromosome, then they would be said to be in **cis** configuration. Behavior is different for two mutants in *cis* or *trans* configuration, depending on whether they code for the same proteins. If the two mutants affected the same protein, then they would fail to complement each other when in *cis* configuration, but would in *trans* (because in trans there is a "good" gene for each mutation). The procedure just described is called a **cis/trans** complementation test and is used to define whether two mutants are in the same genetic (functional) unit. The genetic unit defined by the cis/trans test is called a **cistron**. In a sense, the cistron is equivalent to the gene, but the two terms are not exactly the same and the term cistron is generally used when it is necessary to speak precisely. As noted, two mutations in the *same* cistron *cannot* complement each other so that if complementation is found to exist, this implies that the two mutations lie in *different* cistrons (that is, different genes).
7. ***Eucaryotic gene structure and function*** Each eucaryotic chromosome consists of a single DNA molecule bound to proteins called **histones** (see Section 3.16). Eucaryotes generally have much more DNA per cell than procaryotes, and the DNA is present in a number of separate chromosomes. Transcription and translation are spatially separated in eucaryotes. Transcription occurs in the nucleus and the RNA molecules move to the cytoplasm for translation. The genes of eucaryotes are frequently split, with partial noncoding regions separating the coding regions. The coding sequences are called **exons**, and the intervening noncoding regions **introns**. Both intron and exon regions are transcribed into RNA, and the functional mRNA is subsequently formed by enzymatic removal of noncoding regions. A summary contrasting genetic phenomena in procaryotes and eucaryotes is given in Figure 5.2.

Although the genetic processes of eucaryotes differ markedly from those of procaryotes, the genetics of eucaryotic organelles such as mitochondria and chloroplasts much more closely resembles that of procaryotes. From an evolutionary viewpoint, genetic and macromolecular studies provide strong support for the idea that mitochondria and chloroplasts of eucaryotes have arisen from procaryotic cells by a process of endosymbiosis (see Sections 7.10 and 18.1).

We now begin a more detailed discussion of the macromolecular phenomena relating to genetics. This chapter can be thought of as divided into three parts: (1) Nucleic acid structure and synthesis, Sections 5.2–5.9; (2) Protein synthesis, Sections 5.10–5.14; (3) Mutation and the genetic code, Sections 5.15–end. We begin with a detailed discussion of nucleic acid structure.

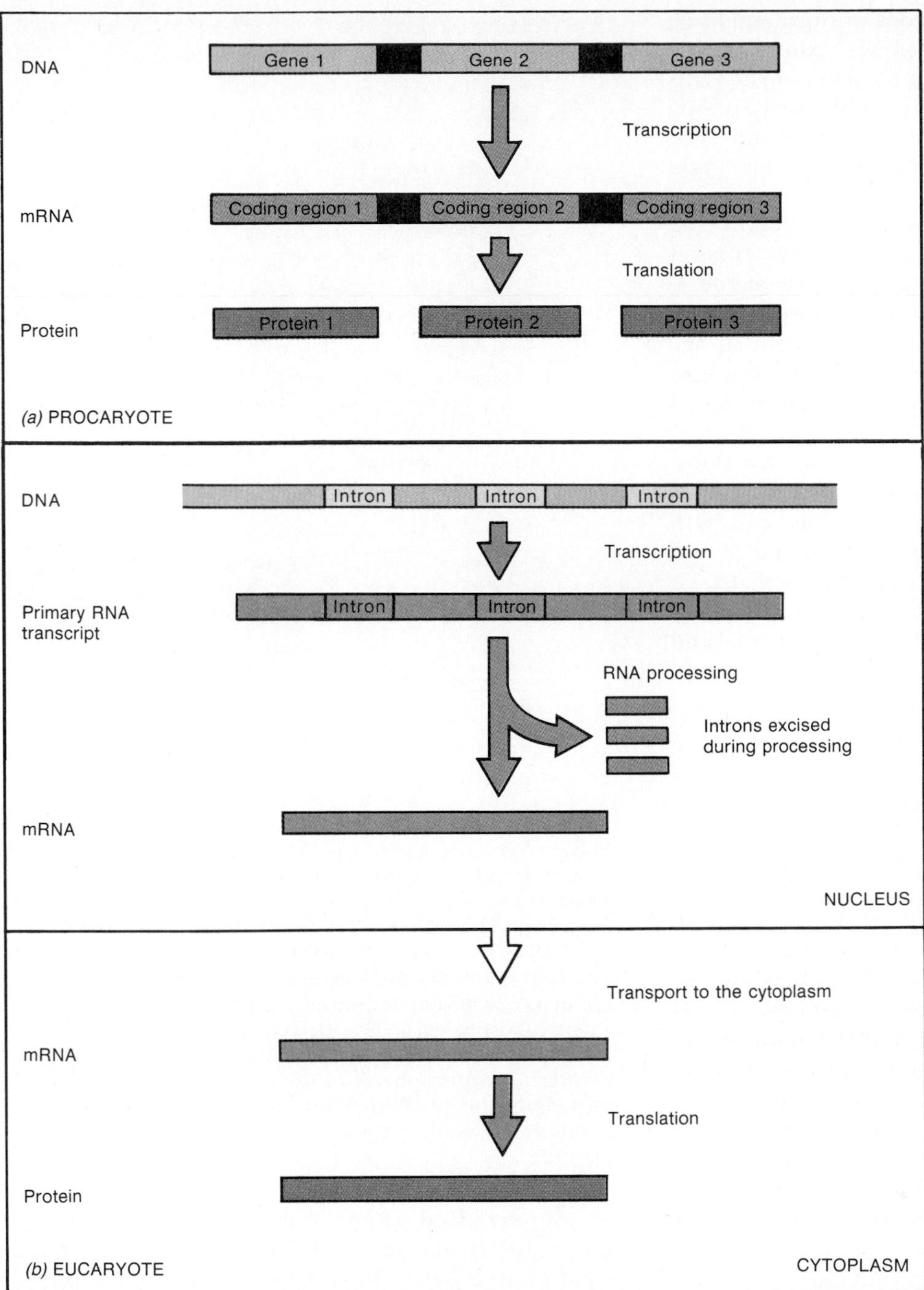

Figure 5.2 Contrast of information transfer in procaryotes and eucaryotes. (a) Procaryote. A single mRNA often contains more than one coding region (such mRNAs are called *polycistronic*). (b) Eucaryote. Noncoding regions (*introns*) are removed from the primary RNA transcript before translation.

5.2 DNA Structure

We discussed the structure of nucleic acids in Chapter 2. In the present chapter we deal with two main phenomena: 1) how DNA is replicated; 2) how DNA is studied experimentally. This chapter will describe details of DNA structure, the chemistry of DNA replication, some of the enzymes involved in DNA replication, and the procedures for determining DNA structure. With this information as a basis, we will then be able to turn to a discussion of how the information of DNA is copied into RNA.

As we have noted, only four different nucleic acid bases are found in DNA: adenine (A), guanine (G), cytosine (C), and thymine (T). The information is stored in DNA in the *sequence* of bases along the polynucleotide chain. The sequence of bases is expressed in terms of the **genetic code**, with a sequence of three bases coding for a single amino acid. As already shown in Figure 2.11, the backbone of the DNA chain consists of alternating units of phosphate and the sugar *deoxyribose*; connected to each sugar is one of the nucleic acid *bases*. Note especially the numbering system for the positions of sugar and base; the phosphate connecting two sugars spans from the 3′ of one sugar to the 5′ of the adjacent sugar. This numbering system is frequently used in discussing DNA replication and should be kept in mind. The phosphate linkage in DNA is a diester, since a single phosphate is connected by ester linkage to two separate sugars. At one end of the DNA molecule the sugar has a phosphate on the 5′ hydroxyl, whereas at the other end the sugar has a free hydroxyl at the 3′ position. The biochemistry of DNA replication is shown in Figure 5.3. As seen, the precursor of the new unit added is a deoxyribonucleoside *tri*phosphate. Replication of DNA proceeds by insertion of a new base-sugar-phosphate unit at the free 3′ (hydroxyl) end, with the subsequent loss of two phosphates (deoxyribonucleoside *mono*phosphate); thus DNA synthesis always occurs starting at the end with the 5′ phosphate and proceeding toward the 3′ end (5′ → 3′). As we will see, this requirement that DNA synthesis always proceeds 5′ → 3′ has important consequences in the replication of DNA in both cells and viruses.

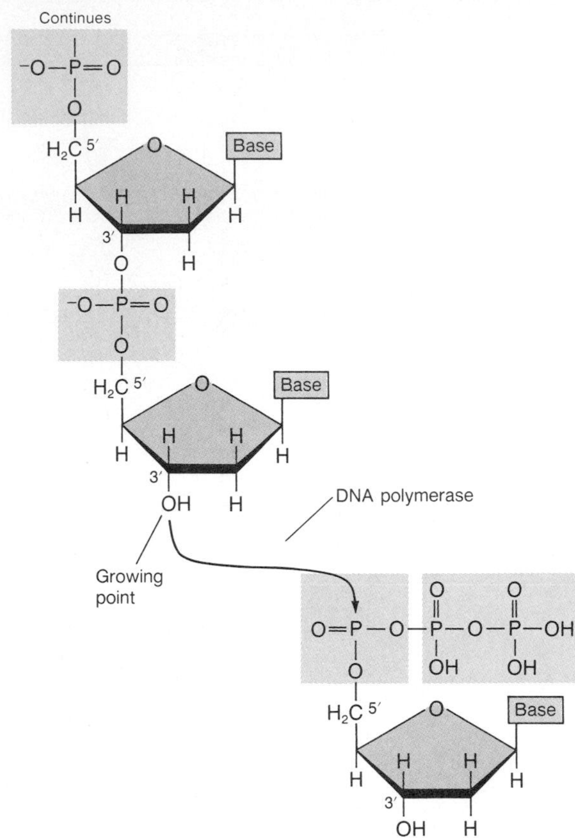

Figure 5.3 Structure of the DNA chain and mechanism of growth by addition from a deoxyribonucleotide triphosphate at the 3′ end of the chain. Growth always proceeds from the 5′ phosphate to the 3′ hydroxyl end. The enzyme DNA polymerase catalyzes the addition reaction. The four deoxyribonucleotides that serve as precursors are deoxythymidine triphosphate (dTTP), deoxyadenosine triphosphate (dATP), deoxyguanosine triphosphate (dGTP), and deoxycytidine triphosphate (dCTP). The two terminal phosphates of the triphosphate are split off as pyrophosphate (PP_i). Thus, two high-energy phosphate bonds are consumed upon the addition of a single nucleotide.

Size and molecular weight of DNA The size of a DNA molecule can be expressed in terms of its molecular weight, but since a single nucleotide has a molecular weight of around 400, and DNA polynucleotides are many nucleotides long, the molecular weight mounts up rapidly. (Small viruses, for instance, may have molecular weights in the millions; cells in the billions.) A more convenient way of expressing the sizes of DNA molecules is in terms of the number of thousands of nucleotide bases per molecule. Thus, a DNA molecule with 1000 bases would contain 1 *kilo*base of DNA. If the DNA were a double helix, then one would speak of ***kilobase pairs***. Thus, a double helix 5000 bases in length would have a length of 5 kilobase pairs. Kilobase is abbreviated ***kb*** and kilobase pairs is abbreviated ***kb pairs***. The bacterium *Escherichia coli* has about 4000 kilobase pairs of DNA.

In the chromosome, DNA does not exist as a single polynucleotide, but as two polynucleotide strands which are not identical but **complementary**. The complementarity of DNA molecules arises because of the specific pairing of the purine and pyrmidine bases: adenine always pairs with thymine and guanine always pairs with cytosine. This **double-stranded** molecule is arranged in a helix, the **double helix** (Figure 5.4). In this double helix, DNA has two distinct grooves, the *major groove* and the *minor groove*. There are many important proteins which interact specifically with DNA (as we shall see later in this chapter and in Chapter 6). In general, these proteins interact predominantly with the major groove, where there is a considerable amount of space.

Supercoiled DNA The conventional form in which the structure of DNA is illustrated is a rather rigid double helix. In this form, it is impossible for the DNA to be packed into a cell. For instance, the DNA of *Escherichia coli* is 1000 times as long as the *E. coli* cell! The problem with packing nucleic acid into viruses is even greater—see Chapter 6. How is it possible to pack so much DNA into such a little space? The solution: *supercoiling*. What is supercoiling? It is a state in which the DNA is folded upon itself so that it becomes highly twisted, as shown in Figure 5.5. Supercoiling puts the DNA molecule under torsion. (Take a rubber band and twist it about itself. This twisting generates a tightly coiled structure which is under considerable torsion. This torsion is only held, however, if the circular structure is maintained. Cut the twisted rubber band and see what happens!)

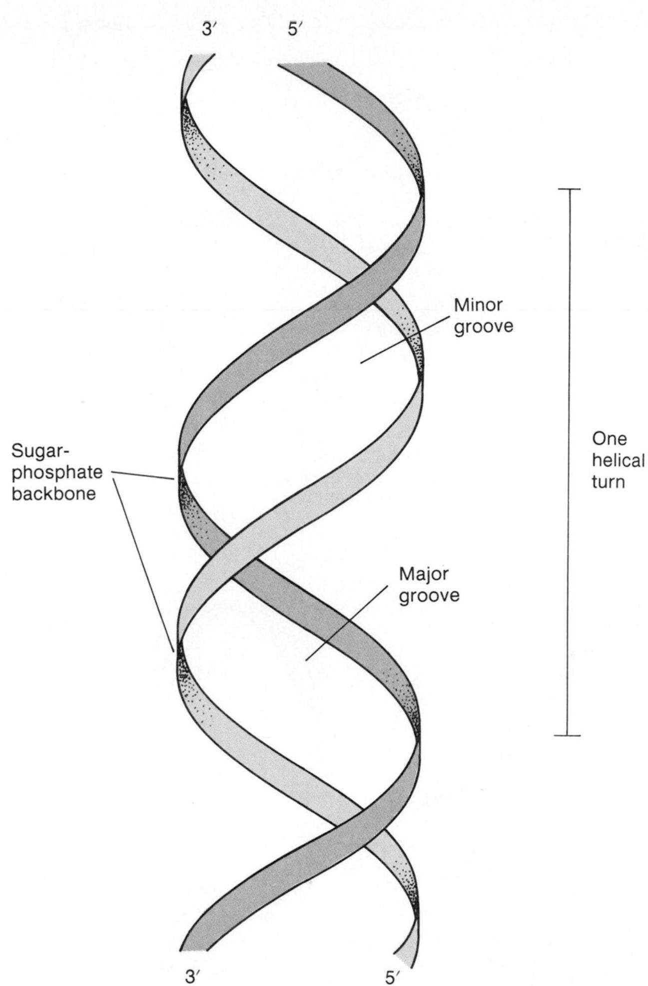

Figure 5.4 Diagram of DNA, showing the overall arrangement of the double helix. Note the locations of the major and minor grooves.

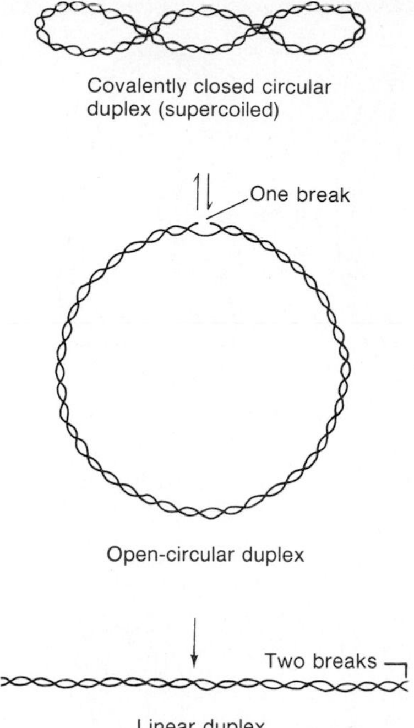

Figure 5.5 Supercoiled, open-circular, and linear duplex forms of DNA. Small plasmid DNA molecules isolated from cells and viruses are primarily in the supercoiled form.

DNA can be supercoiled in either a positive or negative direction. **Negative supercoiling** occurs when the DNA is twisted about its axis in the *opposite* direction from that of the right-handed double helix. It is in this form that supercoiled DNA is found in nature.

How is supercoiling brought about? There is a special enzyme called a **topoisomerase** (*topoisomerase II*, also called *DNA gyrase*) which introduces supercoiling. The process can be thought to occur in several stages. First, the circular DNA molecule is twisted, then a break occurs where the two chains come together, then the broken double helix is resealed on the opposite side of the break (Figure 5.6*a*). Note the derivation of the name *topoisomerase. Topology* is the branch of mathematics which deals with the properties of geometrical figures that are unaltered when the figures are twisted or contorted. We are dealing here with the topology of DNA and a topoisomerase is an enzyme which affects this topology. Of some interest is the fact that two antibiotics which act on bacteria, *nalidixic acid* and *novobiocin*, inhibit the action of DNA gyrase.

There is another enzyme that is able to *remove* supercoiling in DNA. This enzyme, called *topoisomerase I*, causes the passage of one single strand of the double helix through the other, as illustrated in Figure 5.6*b*. Through the action of these topoisomerases, the DNA molecule can be alternately coiled and relaxed. Because coiling is necessary for packing the DNA into the confines of a cell and relaxing is necessary so that DNA can be replicated, these two complementary processes clearly play an important role in the behavior of DNA in the cell.

Some important features of DNA structure As we have noted, complementary base pairing results in the association of the two single strands of DNA into the double helix. Although the base sequence along the DNA strand is primarily determined by the genetic code, there are frequently base sequences in DNA that are present not because of their coding properties but because they influence the secondary structure of DNA, or the way in which DNA interacts with proteins. A common structure in DNA is a region where the base sequence is symmetrical about an imaginary axis. Thus:

5′ . . . A T C | G A T . . . 3′
3′ . . . T A G | C T A . . . 5′

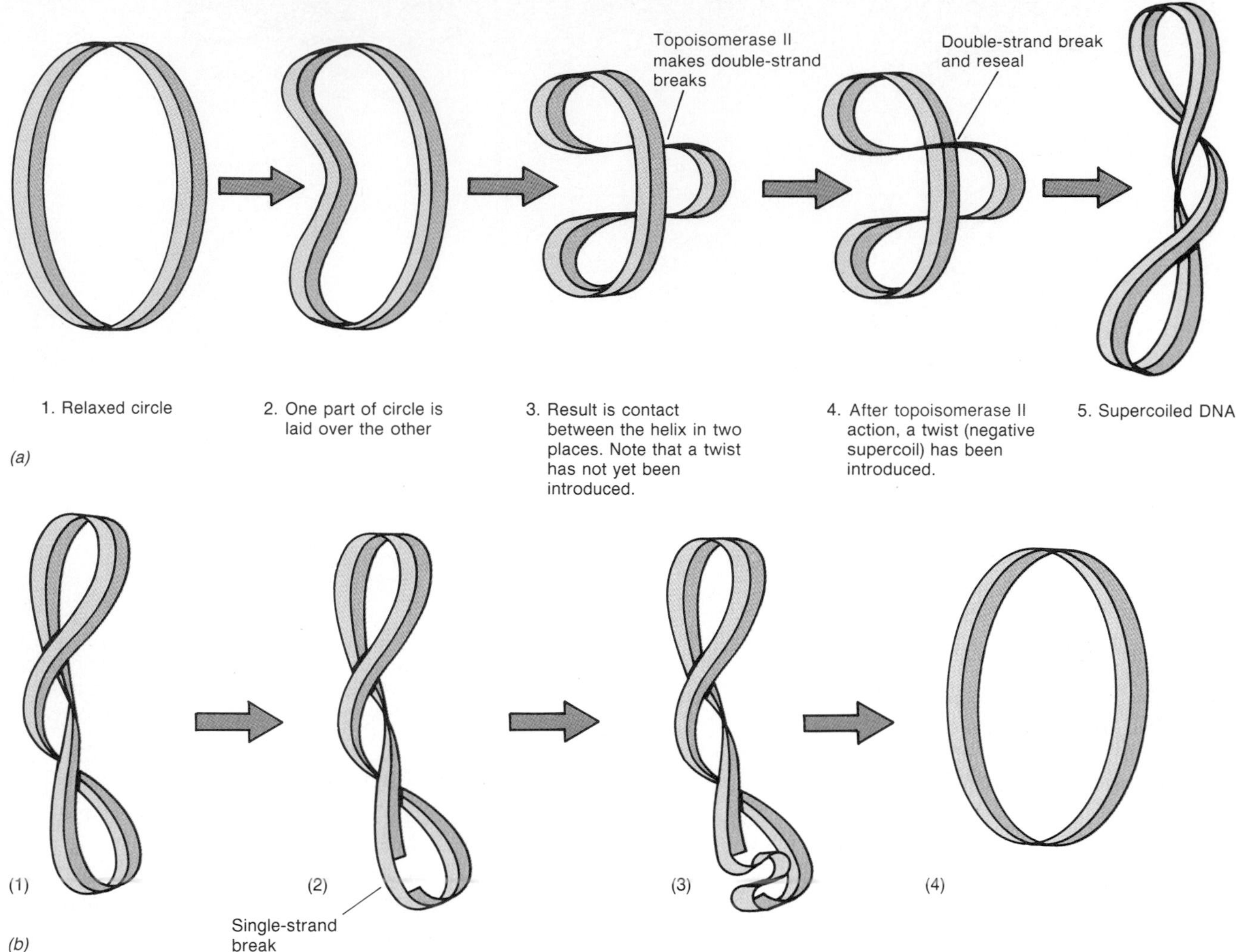
Topoisomerase II makes double-strand breaks
Double-strand break and reseal
1. Relaxed circle
2. One part of circle is laid over the other
3. Result is contact between the helix in two places. Note that a twist has not yet been introduced.
4. After topoisomerase II action, a twist (negative supercoil) has been introduced.
5. Supercoiled DNA
(a)
(1)
(2)
(3)
(4)
Single-strand break
(b)

Figure 5.6 Introduction of supercoiling in a circular DNA. (a) By action of topoisomerase II, which makes double-strand breaks. 1) Relaxed circle. 2) One part of circle is laid over the other. 3) Result is contact between the double helix at two places. Note that as yet a twist has not been introduced. 4) After topoisomerase II action, a twist (negative supercoil) has been introduced in the DNA. (b) Topoisomerase I passes one single strand through another, resulting in the removal of supercoils in DNA. 1) Supercoiled DNA. 2) Single strand break. 3) Other strand is passed through the break and the strand resealed. 4) DNA untwists into a relaxed configuration.

Note that the lower strand has the same sequence as the upper strand, if read in the opposite direction. The dash in the middle indicates the axis of symmetry. Such a structure is called a **palindrome**. (A palindrome is a sequence of characters which reads the same when read from either right or left. For instance: *Sex at noon taxes* or *Able was I ere I saw Elba*. The term *palindrome* is derived from the Greek, *to run back again*.) Because such sequences are repeated in inverse order they are commonly called **inverted repeats**. Such inverted repeats are common recognition sites for enzymes and other proteins which bind specifically with DNA. We will discuss a number of such proteins later.

The palindromic structure illustrated above is a *single* inverted repeat. Frequently, DNA may have two nearby inverted repeats, a situation which can lead to the formation of secondary structure in DNA. As shown in Figure 5.7, nearby inverted repeats can lead to the formation of **stem-loop** (cruciform) structures in a DNA double helix. Such inverted repeats are common sites at which proteins interact with DNA (see later).

Another common situation in DNA is when single-strands at the *ends* of DNA molecules are complementary. This leads to the possibility that the two ends can find each other and associate by complementary base pairing, as illustrated in Figure 5.8 for

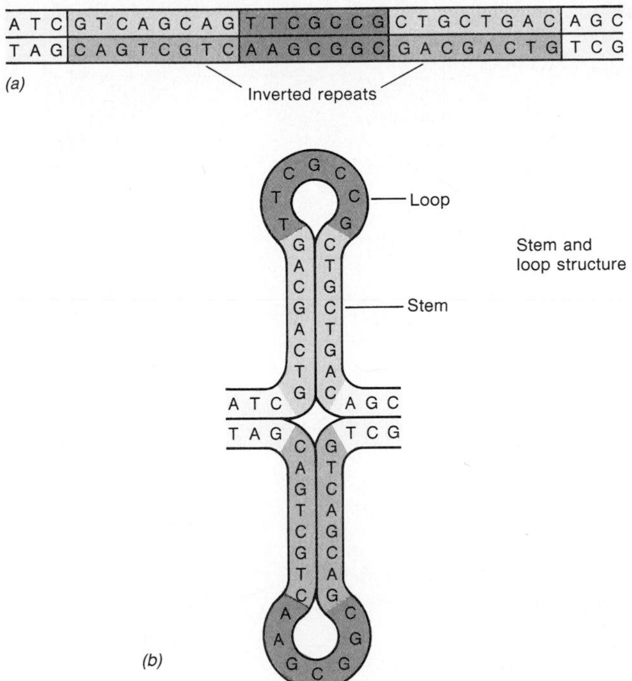

Figure 5.7 Inverted repeats and the formation of a stem and loop structure. (a) Nearby inverted repeats in DNA. (b) Formation of *stem-loop* structures (cruciform structures) by pairing of complementary bases on the *same* strand.

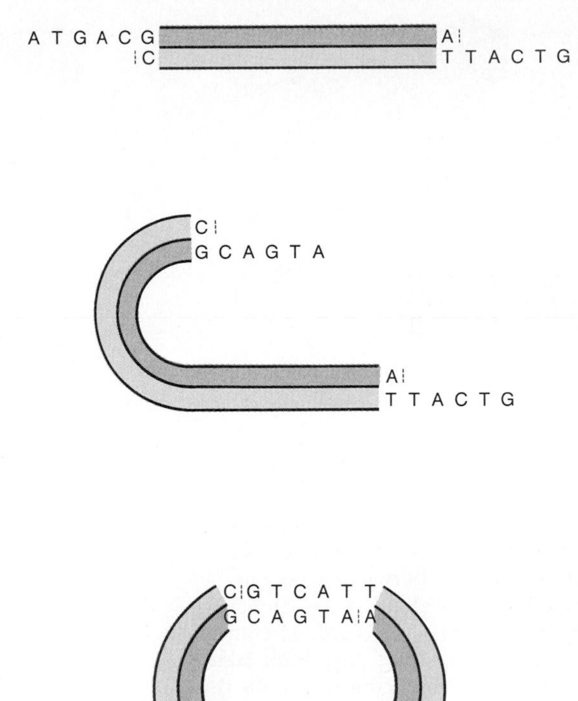

Figure 5.8 Linear DNA with complementary single-strand ends ("sticky ends") can cyclize by base pairing of the complementary ends.

the formation of a circle. DNA which has single-strand complementary sequences at the ends is said to have "sticky ends." Circular DNA molecules are very commonly found in bacteria and viruses (see Chapters 6 and 7).

Because of the enormous length of the DNA in a cell, it can almost never be handled experimentally as a complete unit. The mere manipulation of DNA in the test tube leads to its fragmentation into molecules of smaller size. The need to study the size and shape of DNA is evident. Some of the tools for studying DNA are present in the Box in this chapter. As described, a useful technique for studying the sizes of DNA molecules is electrophoresis. As noted, the nucleic acid molecules migrate through the pores of the gel at rates depending upon their molecular weight or molecular shape. Small molecules, or compact molecules, migrate more rapidly than large or loose molecules. In one figure in the Box, a number of DNA fragments have been spread out in the gel.

Interactions of chemicals with DNA A number of organic chemicals interact specifically with DNA, altering its structure and affecting its biological properties. Many chemicals associate with DNA by becoming inserted between adjacent base pairs along the chain, a phenomenon called *intercalation* (Figure 5.9). Chemicals which intercalate are generally planar molecules which can fit between adjacent bases without causing disruption of the Watson-Crick hydrogen bonding. However, the base pairs must become unstacked vertically to allow for intercalation, so that the sugar-phosphate backbone is distorted and the regular helical structure is destroyed. As we will see later (Section 5.17), one consequence of intercalation is that the reading frame of the code can be changed, ultimately resulting in formation of a faulty protein.

Examples of intercalating chemicals are the acridine dyes such as *acriflavine*, and *acridine orange* (Figure 5.9) and *ethidium* (see Box). Such compounds serve as useful tools in studying the structure and function of DNA and in visualizing DNA experimentally. Some of the planar molecules which intercalate into DNA are cancer-inducing agents, *carcinogens*, or mutation-inducing agents, *mutagens*. We will discuss carcinogens and mutagens later in this chapter (see Sections 5.18 and 5.19).

Some antibiotics combine strongly and specifically with DNA. We will discuss later the details of how DNA-binding antibiotics act, but note here one important group of antibiotics, the *actinomycins*. These antibiotics not only intercalate into the double helix, but have peptide side chains which attach to the major groove (Figure 5.9).

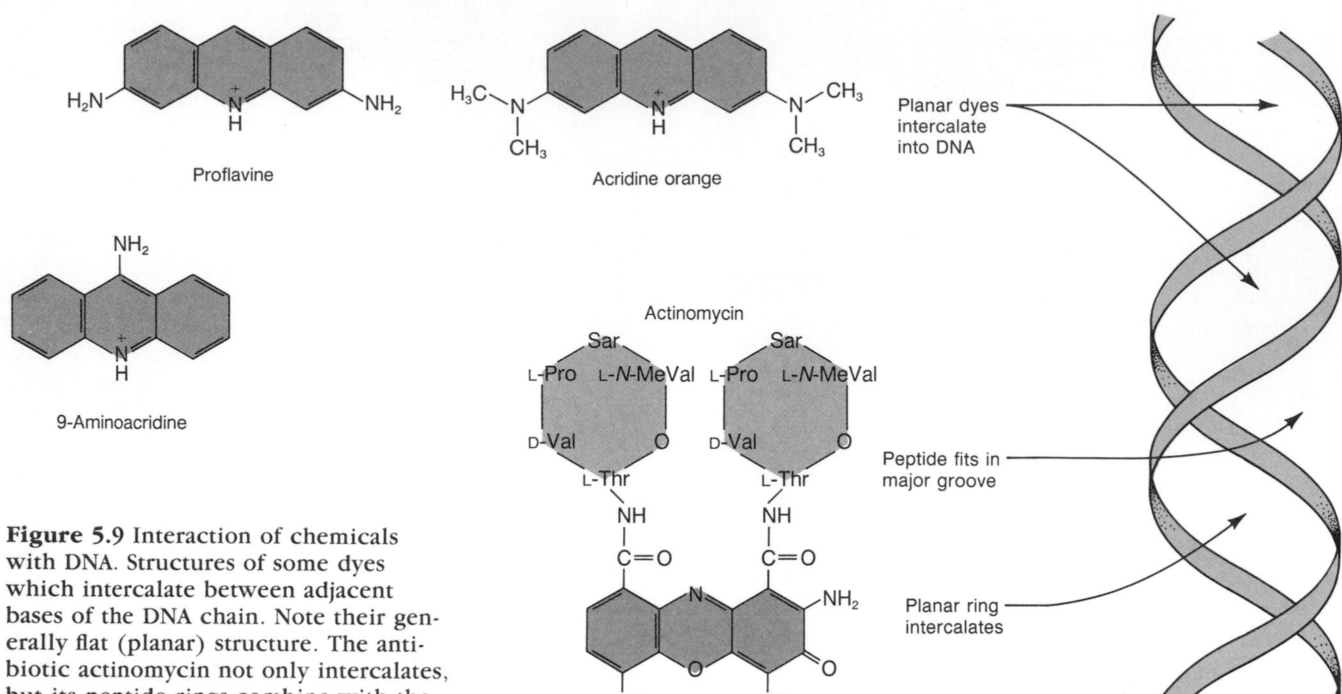

Figure 5.9 Interaction of chemicals with DNA. Structures of some dyes which intercalate between adjacent bases of the DNA chain. Note their generally flat (planar) structure. The antibiotic actinomycin not only intercalates, but its peptide rings combine with the phosphate-sugar backbone.

Hybridization of nucleic acids By **hybridization** is meant the artificial construction of a double-stranded nucleic acid by complementary base pairing of two single-stranded nucleic acids. The procedure for constructing nucleic acid hybrids is shown in the Box. Both DNA:DNA and DNA:RNA hybrids can be made. There must be a high degree of complementarity between two single-stranded nucleic acid molecules if they are to form a stable hybrid. In the most common use of hybridization in genetic engineering, one of the molecules is radioactive and formation of hybrids is detected by observing the formation of double-stranded molecules containing some of the radioactivity.

One of the most common uses of hybridization is to detect DNA sequences that are complementary to mRNA molecules. Detection of DNA:RNA hybridization is usually done with membrane filters constructed of cellulose nitrate. RNA does not stick to these filters, but single-stranded DNA and DNA:RNA hybrids do. The single-stranded *DNA probe* is first immobilized on the filter and the radioactive RNA added. After appropriate incubation, the unhybridized RNA is washed out and the radioactivity still bound to the filter is measured (see Box).

Interaction of nucleic acids with proteins Of great importance is the interaction of nucleic acids with proteins. Protein-nucleic acid interactions are central to replication, transcription, and translation, and to the regulation of these processes. Two general kinds of protein-nucleic acid interactions are noted: nonspecific and specific, depending upon whether the protein will attach *anywhere* along the nucleic acid, or whether the interaction is sequence-specific. As an example of proteins that do not interact in a sequence-specific fashion, we mention the **histones**, proteins which are extremely important in the structure of the eucaryotic chromosome (see Section 3.16), although less significant in procaryotes. Histones are relatively small proteins that have a high proportion of positively charged amino acids (arginine, lysine) (Figure 5.10). DNA, as we have noted, is a polynucleotide, and has a high proportion of negatively charged phosphate groups. Since the phosphate groups are on the outside of the DNA double helix, DNA is a negatively charged molecule. Histones, because of their positive charge, combine strongly and relatively nonspecifically with the negatively charged DNA. In the eucaryotic cell there is generally enough histone so that all of the phosphate groups of the DNA are covered. Association of histones with DNA leads to the formation of nucleosomes, the unit particles of the eucaryotic chromosome (see Section 3.16).

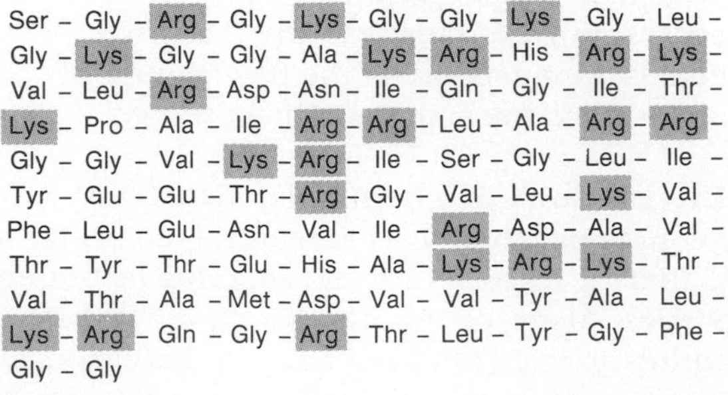

Figure 5.10 Structure of a histone protein. The positively charged amino acids are marked.

Working with DNA and RNA: The Tools

Our knowledge of molecular biology and genetics has depended on the development of adequate research tools. Advances in knowledge of how nucleic acids work have generally been tied to the development of new methods. We discuss here some of these methods.

1. **Extraction and purification of DNA** The first requirement is a sample of DNA free of other cellular chemicals. We discussed cell fractionation methods in Chapter 3. The steps in the purification of DNA are shown here. The aqueous solution in the final step is treated with RNAse to remove RNA. Proteins are then removed by use of denaturing solvents (usually phenol). By repeating the purification steps a number of times, a solution can be obtained that is virtually free of any components other than DNA.

Note that the solution of DNA obtained never consists of native DNA molecules of the length found in the cell. The purification process causes the DNA to break down into fragments of various (random) lengths. If the DNA has been handled gently during purification, the lengths of the fragments will be about one-hundredth of the length of the whole chromosome.

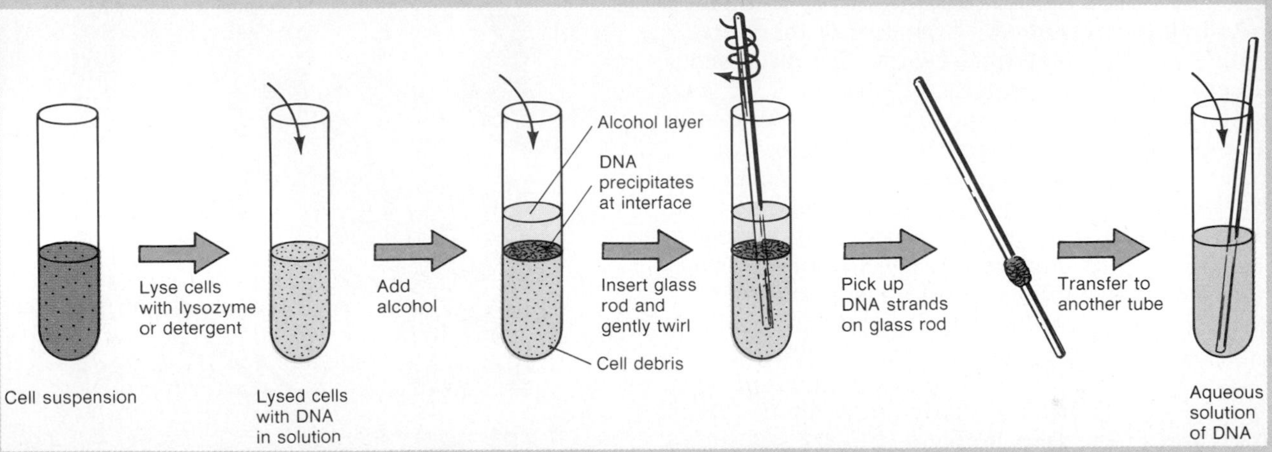

2. **Detecting the presence of DNA** There are several methods for detecting the presence of DNA in a solution. One of the most widely used is by its absorption of ultraviolet radiation. DNA absorbs strongly ultraviolet radiation at a wavelength of 260 nm. The absorption is due to the purine and pyrimidine bases. As seen in the figure, double-stranded DNA absorbs less strongly than single-stranded DNA. This is because the interaction between the bases on the opposite strands of the double-stranded DNA (hydrogen bonding) reduces the ultraviolet absorbance.

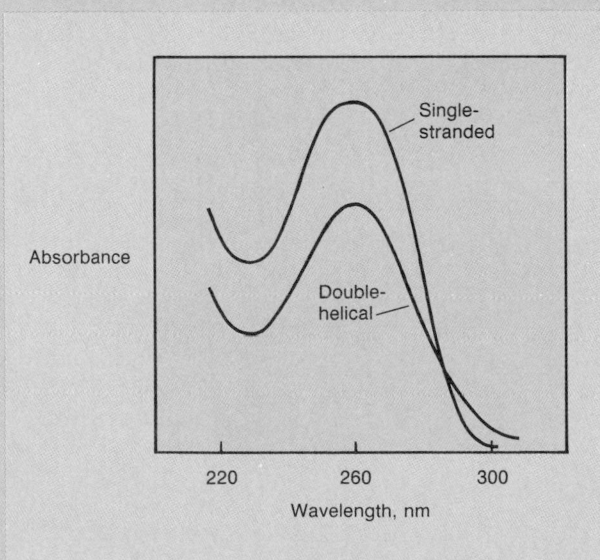

3. **Density-gradient centrifugation of DNA** DNA molecules vary in density, depending on their exact chemical composition. DNA molecules with higher content of guanine plus cytosine (GC) are denser than molecules with low GC. The density of DNA can be determined by centrifugation at very high speed in a *gradient* of cesium chloride (CsCl). The DNA solution is layered on top of a solution of CsCl and centrifuged at high speed for several days, until equi-

librium is reached. The CsCl forms a density gradient from top to bottom of the tube and DNA molecules form bands at appropriate densities. At equilibrium, the DNA molecules become positioned in the gradient at positions corresponding to their densities. Observation of the centrifuge tube with ultraviolet radiation after the centrifugation reveals bands of DNA. This method is called the *buoyant density method* and permits both determination of density and separation of molecules of differing density.

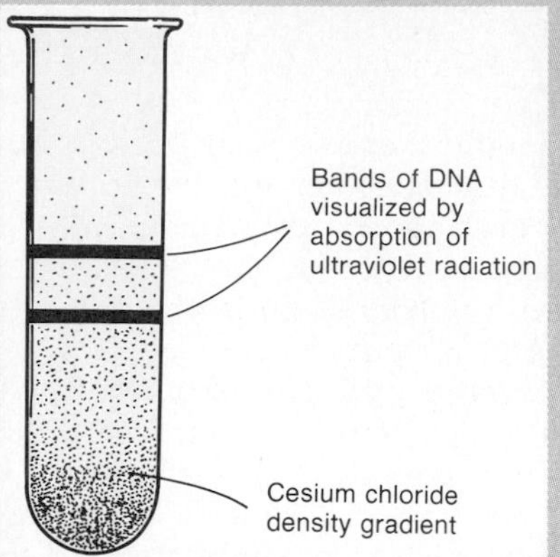

4. **Gel electrophoresis** One of the most widespread methods of studying nucleic acids is gel electrophoresis. Introduction of electrophoresis methods has revolutionized research on molecular genetics. *Electrophoresis* is the procedure by which charged molecules are allowed to migrate in an electrical field, the rate of migration being determined by the size of the molecules and their electrical charge. We discussed electrophoresis of proteins in Chapter 3. In gel electrophoresis, the nucleic acid is suspended in a gel, usually made of polyacrylamide or agarose. The gel is a complex network of fibrils and the *pore size* of the gel can be controlled by the way in which the gel is prepared. The nucleic acid molecules migrate through the pores of the gel at rates depending upon their molecular weight and molecular shape. Small molecules, or compact molecules, migrate more rapidly than large or loose molecules. After a defined period of time of migration (usually a few hours), the locations of the DNA molecules in the gel are assessed by making the DNA molecules fluorescent and observing the gel with ultraviolet radiation.

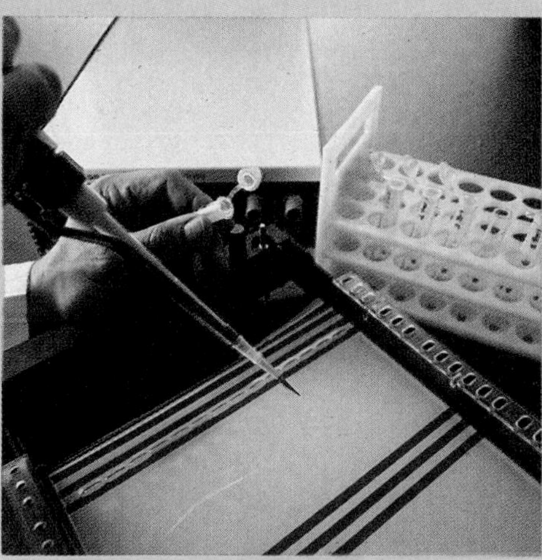

Applying a sample to a gel

Shown here is a photograph of an electrophoresis apparatus. The horizontal frame, made of lucite plastic, holds the gel. The ends of the gel are immersed in buffer which makes an electrical connection to the power supply (shown in rear). The gel is observed after electrophoresis by use of ultraviolet radiation. In each lane, a mixture of DNA fragments had been applied. A computerized scanner can be used to locate the positions of the DNA bands.

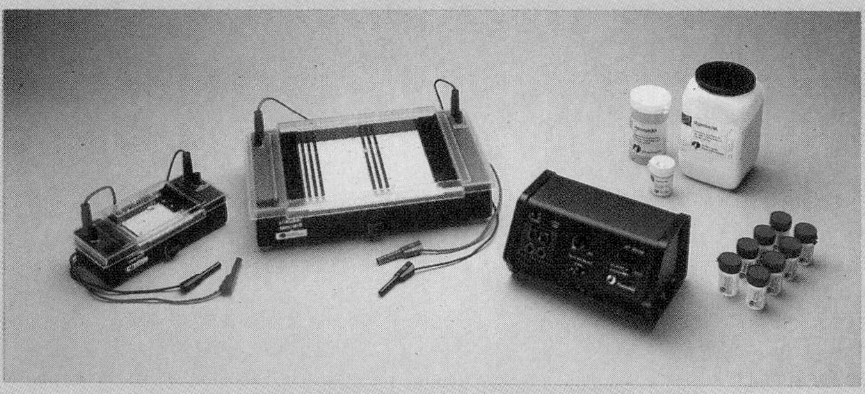

Apparatus for gel electrophoresis of nucleic acids

5. **Detecting DNA by fluorescence** When nucleic acids are treated with dyes which are fluorescent and which are able to combine firmly with the nucleic acid chain, the nucleic acid is rendered fluorescent. The dye *ethidium bromide* is widely used to render DNA fluorescent because it combines tightly within the DNA molecule. Ethidium bromide interacts with double-stranded DNA. If the DNA is now observed with an ultraviolet source, it will be seen to fluoresce.

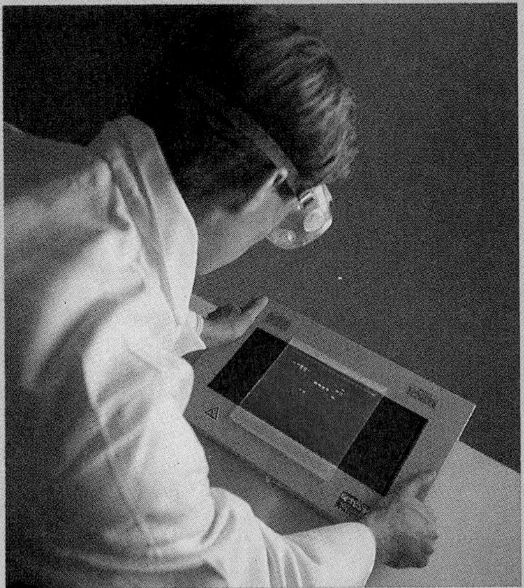

Using fluorescence to locate nucleic acid bands on a gel

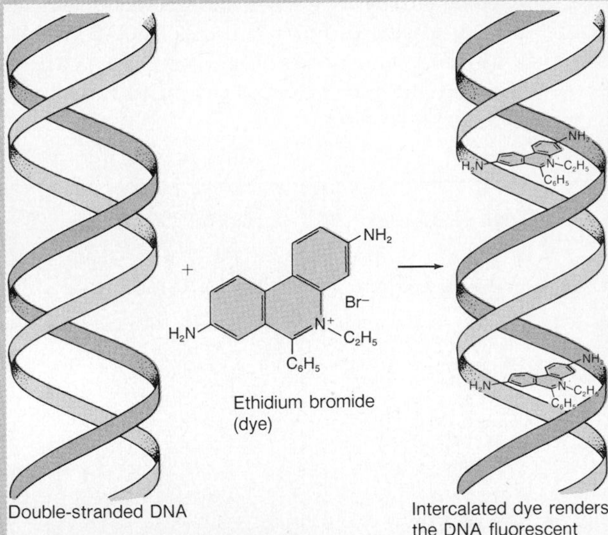

6. **Making nucleic acids radioactive** Radioactivity is widely used in nucleic acid research because radioactivity can be detected in extremely tiny amounts. Radioactive nucleic acids can either be detected directly with Geiger or scintillation counters, or indirectly via their effect on photographic film (autoradiography). Autoradiography of radioactive nucleic acids is one of the most widely used techniques in molecular genetics because it can be applied to the detection of nucleic acid molecules during gel electrophoresis.

a) A nucleic acid can be made radioactive by incorporation of radioactive phosphate during nucleic acid synthesis. If radioactive phosphate is added to a culture while nucleic acid synthesis is taking place, the newly synthesized nucleic acid becomes radioactively labeled.

$^{32}PO_4 \longrightarrow$ ^{32}P-labeled nucleotides $\longrightarrow$
^{32}P-labeled nucleic acid

b)End-labeling of DNA that contains a free hydroxyl group at the 5′ position can be done, using radioactive ATP labeled in the third phosphate. The enzyme polynucleotide kinase specifically attaches the third phosphate of ATP to the free hydroxyl group at the 5′ end of the molecule. End-labeling is an extremely useful technique as it permits labeling of preformed molecules. By tracing the radioactivity through subsequent chemical steps, the end of the molecule can be followed.

^{32}P-P-P-adenosine + HO-deoxyribose-DNA $\longrightarrow$
^{32}P-O-deoxyribose-DNA + ADP

7. **Effect of temperature on nucleic acids** As we have noted, double-stranded nucleic acid molecules are held together by large numbers of weak (hydrogen) bonds. These bonds break when the nucleic acid is heated, but the covalent bonds holding the polynucleotide chains together are unaffected. As shown in part 2 above, double-stranded molecules show lower ultraviolet absorbance than single-stranded molecules. Therefore, if the ultraviolet absorbance of a nucleic acid solution is measured while it is being heated, the increase in absorbance when the double-stranded molecules are converted into single-stranded molecules will show the temperature at which strand separation occurs. Strand separation brought about by heat is generally called *melting*. The stronger the double-strands are held together, the higher will be the temperature of melting. Because guanine-cytosine base pairs are stronger than adenine-thymine base pairs (three hydrogen bonds for GC pairs, only two for AT pairs), the higher the GC content,

the higher the melting temperature. The taxonomic significance of determining the G + C content of an organism's DNA is discussed in Chapter 18. The figure shows the change in absorbance at 260 nm when a solution containing double-stranded DNA is gradually heated. The mid-point of the transition, called T_m, is a function of the GC content of the DNA. If the heated DNA is allowed to cool slowly, the double-stranded native DNA may reform.

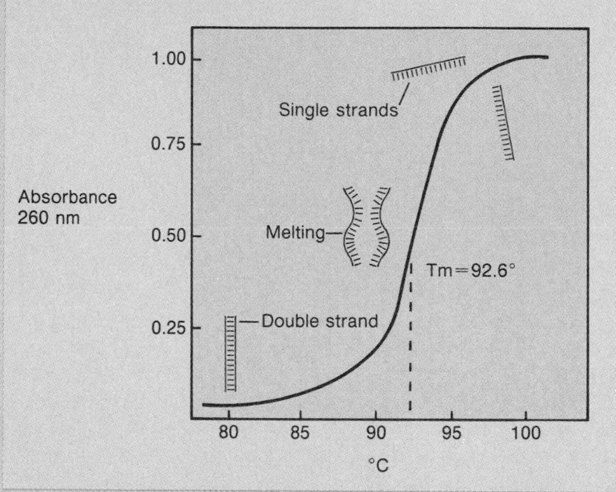

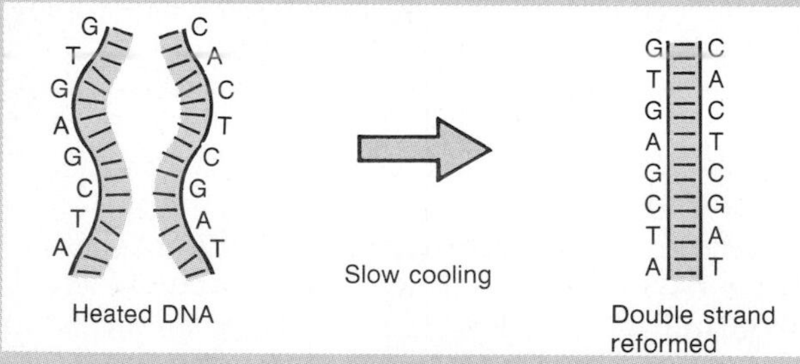

8. **Nucleic acid hybridization** By *hybridization* is meant the artificial construction of a double-stranded nucleic acid by complementary base pairing of two single-stranded nucleic acids. (These are side-to-side hybrids, not the end-to-end hybrids formed by genetic recombination or genetic engineering.) If a DNA solution that has been heated (see above) is allowed to cool slowly, many of the complementary strands will reassociate and the original double-stranded complex reforms, a process called *reannealing*. The reannealing only occurs if the base sequences of the two strands are complementary. Thus, nucleic acid hybridization permits the formation of artificial double-stranded hybrids of either DNA, RNA, or DNA-RNA. Nucleic acid hybridization provides a powerful tool for studying the genetic relatedness between nucleic acids. It also permits the detection of pieces of nucleic acid that are complementary to a single-stranded molecule of known sequence. Such a single-stranded molecule of known sequence is often called a **probe**. For instance, a radioactive nucleic acid probe can be used to *locate* in an unknown mixture a nucleic acid sequence complementary to the probe. Detection of nucleic acid hybridization is usually done with membrane filters constructed of cellulose nitrate. Double-stranded DNA and RNA do not adhere to these filters, but single-stranded DNA and DNA-RNA hybrids do adhere. The nonradioactive DNA (called *driver DNA*) is in great excess over the radioactive *probe nucleic acid*.

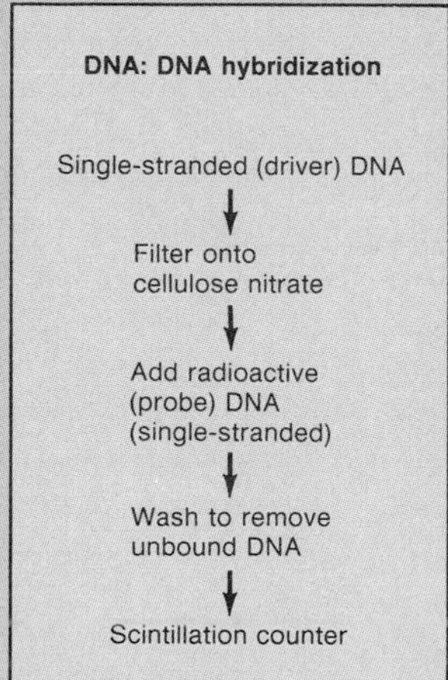

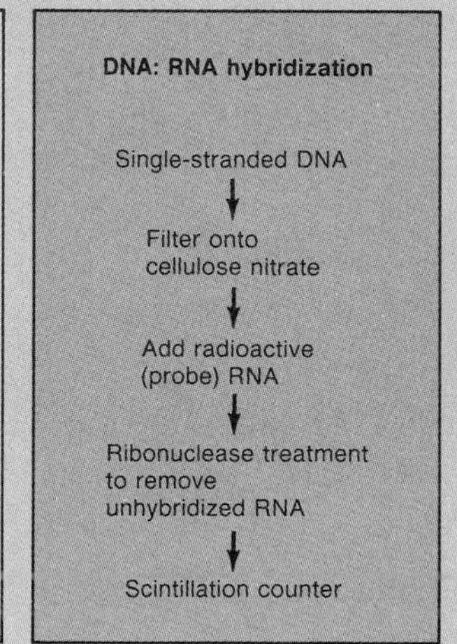

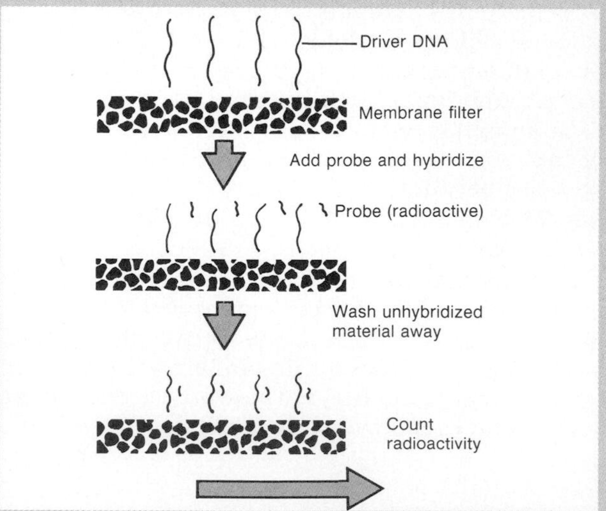

Hybridization can also be done after gel electrophoresis. The nucleic acid molecules are transferred by blotting from the gel to a sheet of membrane filter material, and the probe then added to the filter. (The procedure when DNA is in the gel and RNA is the probe is often called a *Southern blot procedure*, named for the scientist E.M. Southern, who first developed it.) When RNA is in the gel and DNA is the probe, the procedure is called a *Northern blot*. A Western blot (sometimes called an immunoblot) involves protein-antibody binding rather than nucleic acids; see Section 13.10. (There are no Eastern blots!) The figure also shows the use of a nucleic acid probe to search for complementary sequences in a mixture. The DNA fragments have been spread out by gel electrophoresis and then transferred to the membrane filter. The RNA probe, which is radioactively labeled, is allowed to reanneal to the DNA on the filter and its position is determined by autoradiography.

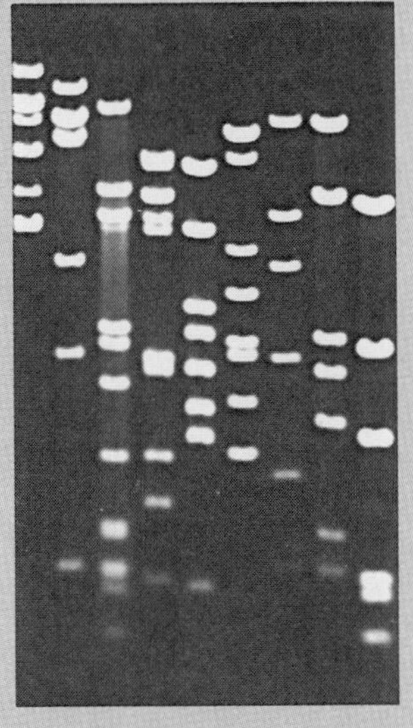

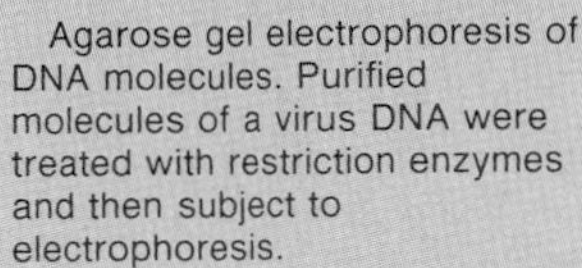

Agarose gel electrophoresis of DNA molecules. Purified molecules of a virus DNA were treated with restriction enzymes and then subject to electrophoresis.

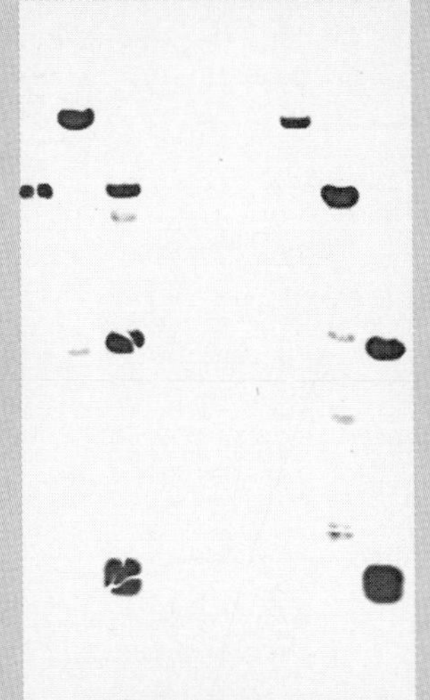

Southern blotting of the DNA gel shown to the left. After blotting, hybridization with radioactively labeled mRNA was carried out. The positions of the bands have been detected by X ray autoradiography. Note that only some of the DNA fragments have sequences complementary to the labeled mRNAs.

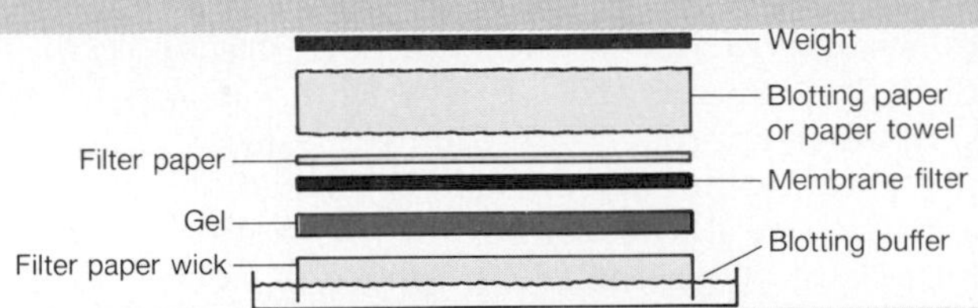

Laying the membrane filter on the gel

9. **Determining the sequence of DNA** Although the base sequences of both DNA and RNA can be determined, it turns out for chemical reasons that it is easier to sequence DNA. Even automatic machines are now available for determining the sequences of DNA molecules. Appropriate treatments are used to generate DNA fragments that end at the four bases, the ends of the fragments being radioactive. Then the fragments are subjected to electrophoresis so that molecules with one nucleotide difference in length are separated on the gel. This electrophoresis procedure involves *four* separate lanes, one for fragments broken at each of the four bases of the DNA, adenine, guanine, cytosine, and thymine. The positions of these fragments are located by autoradiography and from a knowledge of which base is represented by each lane, the sequence of the DNA can be read off.

Two different procedures have been developed to accomplish the above, called the *Maxam-Gilbert* and the *Sanger dideoxy* procedures. In the Maxam-Gilbert procedure for sequencing DNA the piece of DNA to be sequenced is made radioactive by labeling at the end with ^{32}P. The DNA is then treated with chemicals which break the DNA preferentially at each of the four nucleotide bases, under conditions in which only one break per chain is made. (Thus, there are four separate test tubes prepared.) After electrophoresis and autoradiography, the sequence can be read directly. The number of nucleotides in each fragment is shown on the figure.

In the Sanger dideoxy procedure for sequencing DNA the sequence is actually determined by making a *copy* of the single-stranded DNA, using the enzyme *DNA polymerase*. This enzyme uses deoxyribonucleoside triphosphates as substrates and adds them to a *primer*. In the incubation mixtures (four separate test tubes) are small

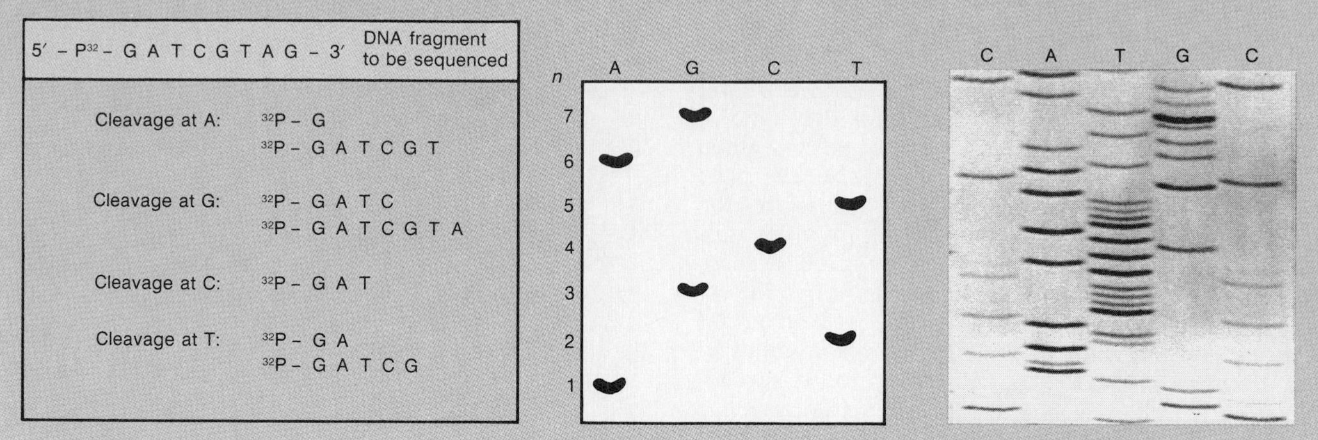

Maxam-Gilbert procedure

Photograph of an actual sequencing gel

amounts of each of the dideoxy analogs of the deoxyribonucleoside triphosphates in radioactive form. Because the dideoxy sugar lacks the 3′ hydroxyl, continued lengthening of the chain cannot occur. The dideoxy analog thus acts as a *specific chain termination reagent*. Radioactive fragments of variable length are obtained, depending on the incubation conditions. Electrophoresis of these fragments is then carried out and the positions of the radioactive bands determined by autoradiography. By aligning the four dideoxynucleotide lanes and noting the vertical position of each fragment relative to its neighbor, the sequence of the DNA copy can be read directly from the gel.

Another approach based on the Sanger principle is to use fluorescent labels instead of radioactivity, one fluorescent color for each of the four bases. Then the electrophoresis can be done in one lane instead of four, with the fragments allowed to run off the bottom of the gel, where their fluorescence color is measured with a spe-

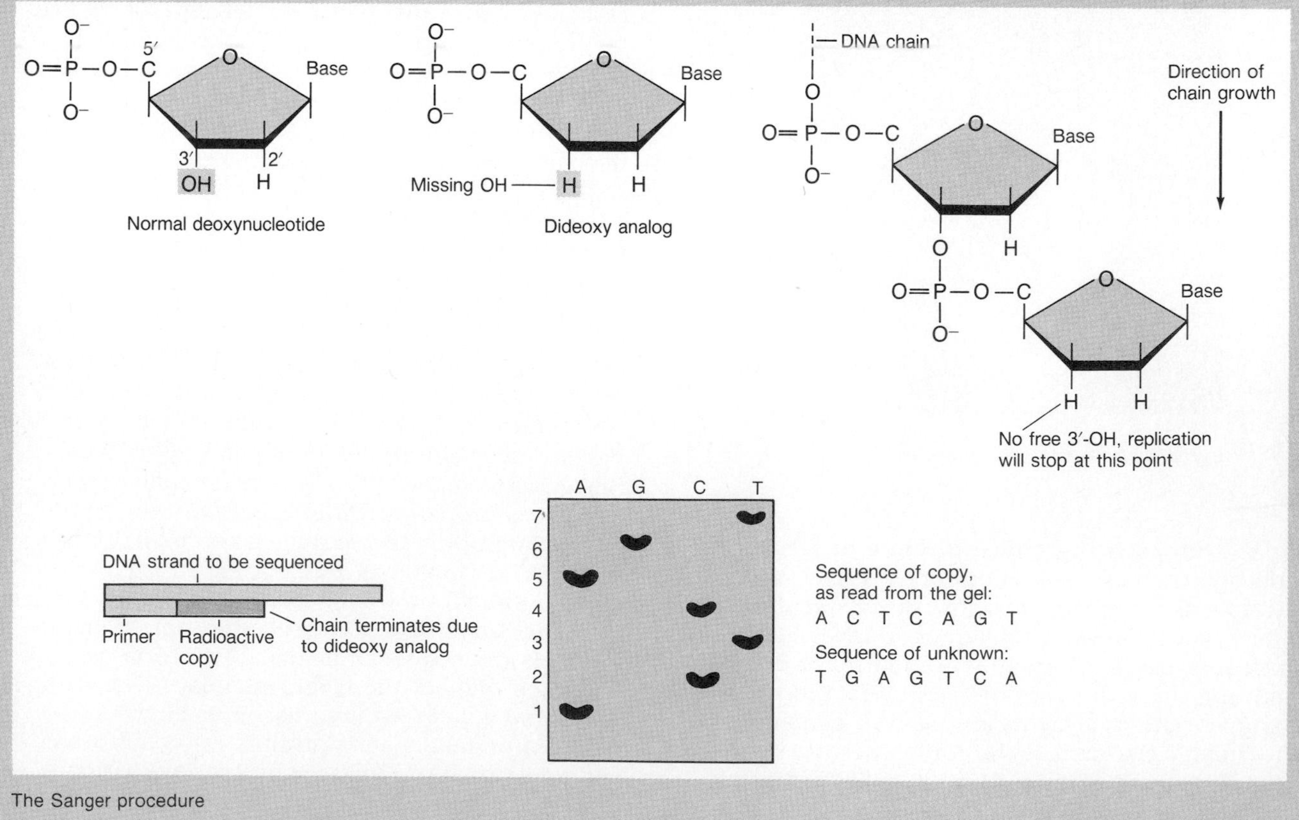

The Sanger procedure

cial laser fluormeter. This procedure makes it possible to *automate* the sequencing process.

A major advantage of the Sanger method is that it can be used to sequence RNA as well as DNA. To sequence RNA, a single-stranded DNA copy is made (using the RNA as the template) by the enzyme reverse transcriptase. By making the single-stranded DNA in the presence of dideoxynucleotides, various sized DNA fragments are generated suitable for Sanger-type sequencing. From the sequence of the DNA, the RNA sequence is deduced by base-pairing rules. The Sanger method has been instrumental in rapidly sequencing ribosomal RNAs for use in studies on microbial evolution (see Chapter 18).

For determining the DNA sequence of a long molecule, such as a whole gene, it is necessary to proceed in stages. First, the DNA is broken into small overlapping fragments and the sequence of each fragment determined. Using the overlaps as a guide, the sequence of the whole molecule can be deduced.

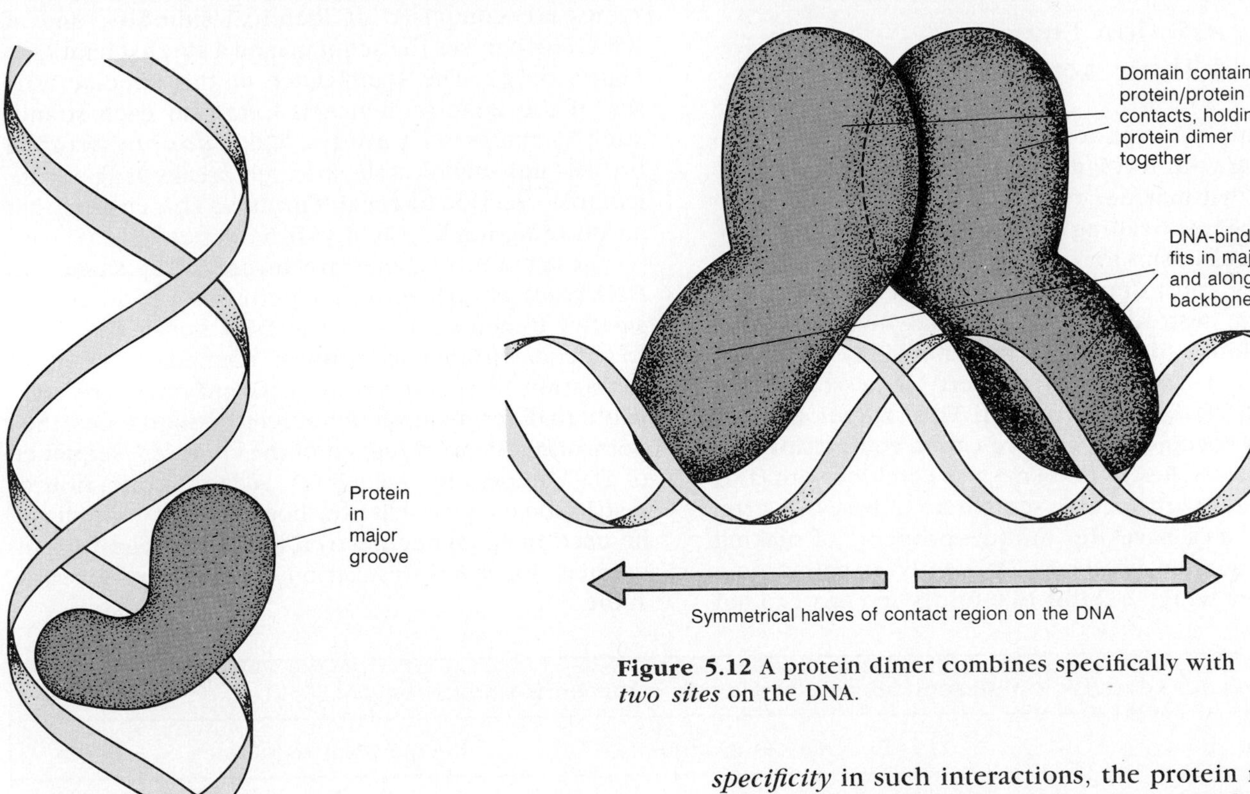

Figure 5.11 Manner in which part of a protein molecule can fit into the major groove of the DNA double helix.

Figure 5.12 A protein dimer combines specifically with *two sites* on the DNA.

There are also a number of proteins that interact with DNA in a sequence-specific manner. These interactions occur by association of the amino acid side chains of the proteins with the bases as well as with the phosphate and sugar molecules of the DNA. The major groove, because of its size, is an important site of protein binding (Figure 5.11). In order to achieve *specificity* in such interactions, the protein must interact with more than one nucleic acid base, frequently several. We have already described a structure in DNA called an *inverted repeat* (see Figure 5.7). Such inverted repeats are frequently the locations at which protein molecules combine specifically with DNA. Proteins which interact specifically with DNA are frequently *dimers*, composed of two identical polypeptide chains. On each polypeptide chain is a region, called a *domain*, which will interact specifically with a region of DNA in the major groove. A consideration of this type of interaction provides an explanation for the palindromic nature of the base sequence at the point in the DNA where the protein interacts: in this way, *each* of the polypeptides of the protein dimer combines with each of the DNA strands (Figure 5.12). Note, however, that the protein does not recognize the specific base sequence of the DNA. Rather, it recognizes contact points such as electrostatic charges and hydrophobic regions that are associated with specific base sequences.

Once a protein combines at a specific site on the DNA, a number of outcomes can occur. In some cases, all the protein does is *block* some other process, such as transcription (see Section 5.8). In other cases, the protein is an enzyme which carries out some specific action on the DNA. The most interesting of such proteins are the *restriction enzymes*, enzymes which specifically *cut* DNA at sites near where they combine. We discuss these interesting enzymes in the next section.

5.3 Restriction Enzymes and Their Action on DNA

Organisms are occasionally faced with the problem of coping with foreign DNA, generally derived from viruses, that may derange cellular metabolism or initiate processes leading to cell death. Although a number of mechanisms for coping with foreign DNA exist, one of the most dramatic is that which results in its enzymatic destruction. The enzymes involved in the destruction of foreign DNA are remarkably specific in their action, an essential property if destruction of cellular DNA is to be avoided. This marvelous class of highly specific enzymes are called **restriction endonucleases**. Restriction enzymes combine with DNA only at sites with specific sequences of bases. Restriction enzymes have the unique property of making double-stranded breaks in DNA only at palindrome sequences which exhibit two-fold symmetry around a given point (see Section 5.2). Thus one restriction endonuclease of *Escherichia coli* called EcoRI has the following recognition sequence:

```
          ↓
5′ . . . G–A–A–¦–T–T–C– . . . 3′
3′ . . . C–T–T–¦–A–A–G– . . . 5′
                     ↑
```

The cleavage sites are indicated by arrows, and the axis of symmetry by a dashed line.

Nucleotide sequences with inverted repeats, such as are recognized by restriction enzymes, have been found to be widespread in DNA. The palindromes recognized by restriction enzymes are relatively short, and probably are cleaved because the restriction enzymes are composed of identical subunits, one of which recognizes the sequence on a single chain (see Figure 5.12). The significance of this specificity is that if the same sequence is found on each strand, such enzymes will always make *double*-stranded breaks, and such double-stranded breaks are not subject to correction by repair enzymes. This ensures that an invading nucleic acid will be destroyed.

Restriction enzymes are of great importance in DNA research, because they permit the formation of smaller fragments from large DNA molecules. Such fragments with defined termini, created as a result of the action of specific restriction enzymes, are amenable to determination of nucleotide sequences, thus permitting the working out of the complete sequence of DNA molecules (see Box). A large collection of restriction enzymes has now been built up, which can be used in sequence determination. Recognition sequences for a few restriction enzymes are given in Table 5.1.

Table 5.1 Recognition sequences of a few restriction endonucleases

Organism	Enzyme designation	Recognition sequence*
Escherichia coli	EcoRI	G↓AA*TTC (methylated: second A)
Escherichia coli	EcoRII	↓CC*AGG (methylated: second C)
Haemophilus influenzae	HindII	GTPy↓PuAC* (methylated: C)
Haemophilus hemolyticus	HhaI	GC*G↓C (methylated: first C)
Bacillus subtilis	BsuRI	G↓C*C (methylated: first C)
Brevibacterium albidum	BalI	TGG↓C*CA (methylated: first C)
Thermus aquaticus	TaqI	T↓CGA* (methylated: A)

Arrows indicate the sites of enzymatic attack. Asterisks indicate the site of methylation (modification). G = guanine; C = cytosine; A = adenine; T = thymine; Pu = any purine; Py = any pyrimidine. Only the 5′ → 3′ sequence is shown.

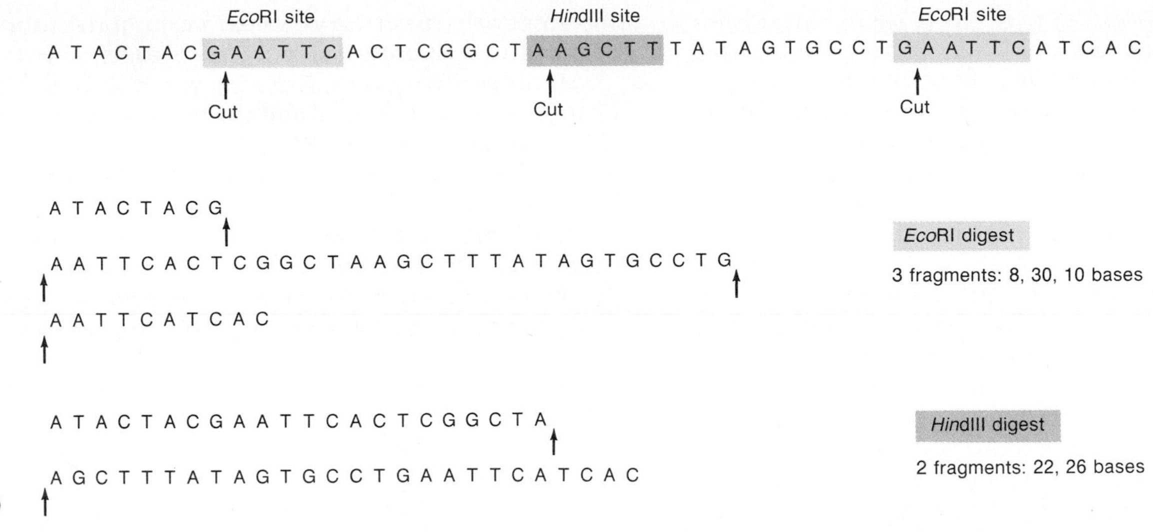

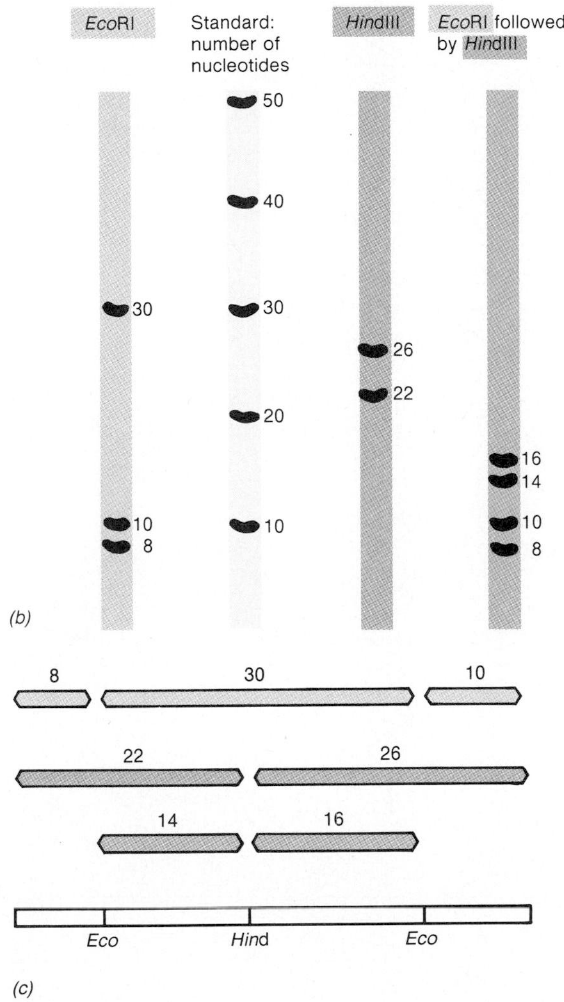

Figure 5.13 Restriction enzyme analysis of DNA. (a) A 48 base sequence, with 2 EcoRI and 1 HindIII restriction sites. Note that the enzyme cuts double-stranded DNA. For simplicity, only a single strand is shown. (b) Results of electrophoresis of digest with each of the enzymes. Note that when the EcoRI digest is then digested with HindIII, the 30 base fragment is affected, showing it has a HindIII site. (c) By orientation of the overlapping fragments, it is possible to deduce a restriction enzyme map of the DNA molecule.

Another use of certain restriction enzymes is that they permit the conversion of DNA molecules into fragments which can be joined by DNA ligase. This enables laboratory researchers to create artificial genes, as will be discussed in Chapter 8.

An integral part of the cell's restriction mechanism is the **modification** of the specific sequences on its *own* DNA so that they are not attacked by its own restriction enzymes. Such modification generally involves methylation of specific bases within the recognition sequence so that the restriction nuclease can no longer act. Thus, for each restriction enzyme there must also be a modification enzyme, the two enzymes being closely associated. For example, the sequence recognized by the EcoRII restriction enzyme (also see Table 5.1) is:

C–C–A–G–G
G–G–T–C–C

and modification of this sequence results in methylation of two cytosines:

  m
C–C–A–G–G
G–G–T–C–C
        m

Note that a given nucleotide sequence can be a substrate for either a restriction enzyme or a modification enzyme but not both. This is because modification makes the sequence unreactive with restriction enzyme, and action of restriction enzyme destroys the recognition site of the modification enzyme.

Restriction enzymes are such important tools in modern molecular genetic research that they have become widely available commercially. A number of companies purify and market restriction enzymes of a variety of specificities. (One company sells over 100 different restriction enzymes.) By referring to a catalog of base sequences, a research worker can generally obtain a restriction enzyme that will cut at or near a particular site. On the other hand, if it is known that a particular piece of DNA has been cut by a particular restriction enzyme, then, by reference to the base sequence which that restriction enzyme cuts, it is possible to deduce the base sequence around that site. This provides a powerful tool for studying DNA molecules, and restriction enzyme analyses will be mentioned in a number of places in this book.

Restriction enzyme analysis of DNA As noted, a DNA molecule can be cut at a specific location by a given restriction enzyme. Because the base sequences recognized by restriction enzymes are 4–6 nucleotides long, there will generally be only a restricted number of such sequences in a piece of DNA. After cleaving the DNA (Figure 5.13*a*), the fragments are separated by agarose gel electrophoresis, as shown in Figure 5.13*b*. The distance migrated by any band of DNA in such a gel can be determined by calibrating the electrophoresis system with DNA molecules of known size. By judicious use of several restriction enzymes of different specificities, and by use of overlapping fragments, it is possible to construct a **restriction enzyme map** in which the positions cut by each of the several restriction enzymes can be designated (Figure 5.13*c*).

Several procedures are now available for determining the base sequences of DNA molecules. In fact, automatic machines are available for sequencing DNA. Details are presented in the Box. By successively determining the sequences of small overlapping fragments of DNA, it is possible to determine the sequences of very large pieces of DNA. The sequences are now known for hundreds of genes, as well as for the complete genome of many viruses (see Chapter 6).

5.4 DNA Replication

The problem of DNA replication can be simply put: the nucleotide base sequence residing in each long DNA molecule must be precisely duplicated to form a copy of the original molecule. The cell has solved this seemingly complex problem in an elegant fashion: by means of *complementary base pairing*. As we have discussed (see Figure 2.12), the base adenine pairs specifically with thymine and guanine pairs with cytosine. If the DNA double strand opens up, a new strand can be begun as the complement of each of the parent strands. After replication, a partitioning of the strands must occur. As shown in Figure 5.14, partitioning could be either *conservative*, in which the newly formed double helix contains *both* of the new strands, or *semiconservative*, in which the two progeny double helices consist of one progeny and one parent molecule. Studies in which DNA molecules were labeled during the replication process have shown clearly that DNA replication, in both eucaryotes and procaryotes, is *semiconservative*.

An interesting point: For a short while after replication, the parent strand and the progeny strand differ chemically because the parent strand is methylated (see Section 5.3 for a discussion of methylation in the restriction/modification system). The progeny strand ultimately also becomes methylated, but for a short while after synthesis it is not. This chemical difference is used by so-called *proofreading* functions in the cell, to *check* the newly synthesized strand

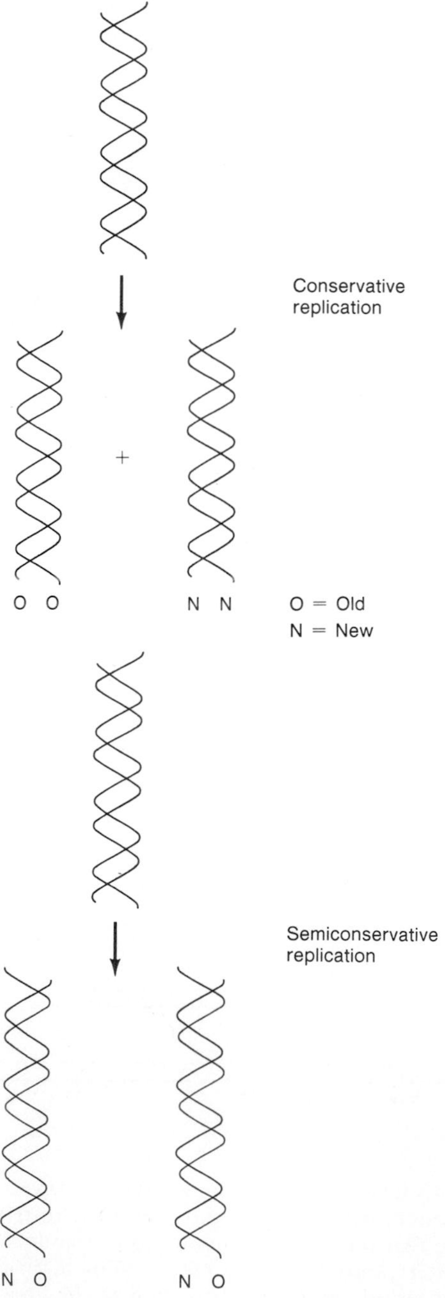

Figure 5.14 DNA replication contrasting conservative and semiconservative processes. In both procaryotes and eucaryotes the process is always semiconservative.

for possible mispairing. Any mispaired bases are cut out by the proofreading system, after which a DNA repair enzyme fills in the remaining gaps.

Primer or template? The DNA molecule which is copied into the complement is called a *template*. A template is a preformed pattern which is copied, but the *new* DNA molecule is not connected to the *old* DNA molecule. On the other hand the new strand as it grows serves as a **primer** or point of origin, a site at which the building blocks for further lengthening of the new strand are attached.

The chemistry of DNA places some important restrictions on the manner by which this priming function occurs. The precursor of each new nucleotide in the chain is a nucleoside *tri*phosphate, of which the two terminal phosphates are removed and the internal phosphate is attached covalently to the growing chain (see Figure 5.3). The addition of the nucleotide to the growing chain requires the presence of a free hydroxyl group, and such a free hydroxyl group is only available at the 3′ end of the molecule. This chemical restriction leads to an important law which is at the basis of many facets of DNA replication: *DNA replication always proceeds from the 5′ phosphate to the 3′ hydroxyl end.*

Most of the information on the mechanism of DNA replication has been obtained in the bacterium *Escherichia coli*, and the following discussion will deal with this organism. The enzyme which catalyzes the addition of the nucleotides is called DNA polymerase III. This enzyme is able to catalyze the addition of a nucleotide to the growing chain but cannot synthesize DNA *de novo* from a mixture of nucleotides. DNA polymerase III can only add on to preexisting DNA. How, then, does DNA polymerase III get started? A primer is needed, and the primer is actually RNA rather than DNA. When the double helix opens up at the beginning of replication, an RNA polymerizing enzyme first acts, resulting in the formation of this RNA primer. A specific RNA polymerizing enzyme, called *primase*, participates in primer synthesis. At the growing end of this RNA primer is a 3′–OH group to which DNA polymerase III can add the first deoxyribonucleotide. Once priming has begun, continued extension of the molecule occurs as DNA rather than RNA. Thus, the newly synthesized molecule has a structure such as that shown in Figure 5.15. The RNA primer is subsequently removed and replaced with DNA through the action of another enzyme, DNA polymerase I.

Initiation of DNA synthesis Initiation of DNA synthesis generally does not occur at the ends but internally. In most cases, the place where DNA synthesis first takes place is a specific location, the so-called **origin of replication**. The origin of replication consists of a specific base sequence of about 300 bases which is recognized by specific initiation proteins. At the origin of replication, the DNA double helix opens up and the initiation of DNA replication occurs on the two single strands. As replication proceeds, the site of replication, called the **replication fork**, moves down the DNA.

Replication is frequently bidirectional from the origin of replication, as shown in Figure 5.16*a*. In circular DNA, such as that of bacteria, replication leads to the formation of characteristic structures, visible under the electron microscope, called **theta structures** (Figure 5.16*b*). Under other conditions, linear DNA molecules are synthesized from the circular form first made, leading to the formation of a replicating structure called a **rolling circle** (Figure 5.16*c*). A rolling circle can arise because a nick in one of the two strands of the circle leads to the initiation of a single strand on only *one* of the two strands of the circle. Continued rotation of the circle leads to the synthesis of a linear single-stranded structure (which can be subsequently converted to a double strand by another mechanism). Note that rolling circle replication is *not* semiconservative. Rolling circle replication is commonly found in some bacterial virus systems (see Chapter 6) as well as in bacteria that are carrying out the conjugation process (see Section 7.5).

Leading and lagging strands We return now to the replication fork and the biochemistry of DNA replication. The details of events at the replication fork are illustrated in Figure 5.17. At the replication fork the DNA double helix unwinds and a small single-stranded region is formed. This single-stranded region is complexed with a special protein, the *single-strand binding protein*, which stabilizes the single-stranded DNA, preventing the formation of intrastrand hydrogen bonds.

Figure 5.17 reveals an important difference between replication of the two strands, which arises from the fact that DNA replication always proceeds from 5′ phosphate to 3′ hydroxyl (always adding a *new* nucleotide to the 3′–OH of the growing chain). On the strand growing from the 5′ phosphate → 3′ hydroxyl, called the **leading strand**, DNA synthesis can occur continuously, because there is always a free

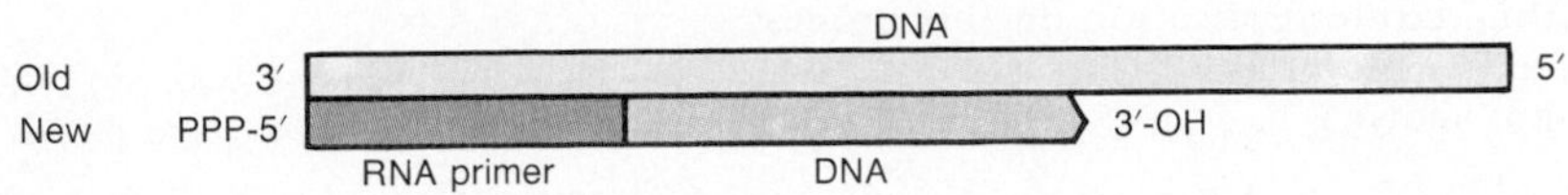

Figure 5.15 Structure of the RNA-DNA combination which results at the initiation of DNA synthesis.

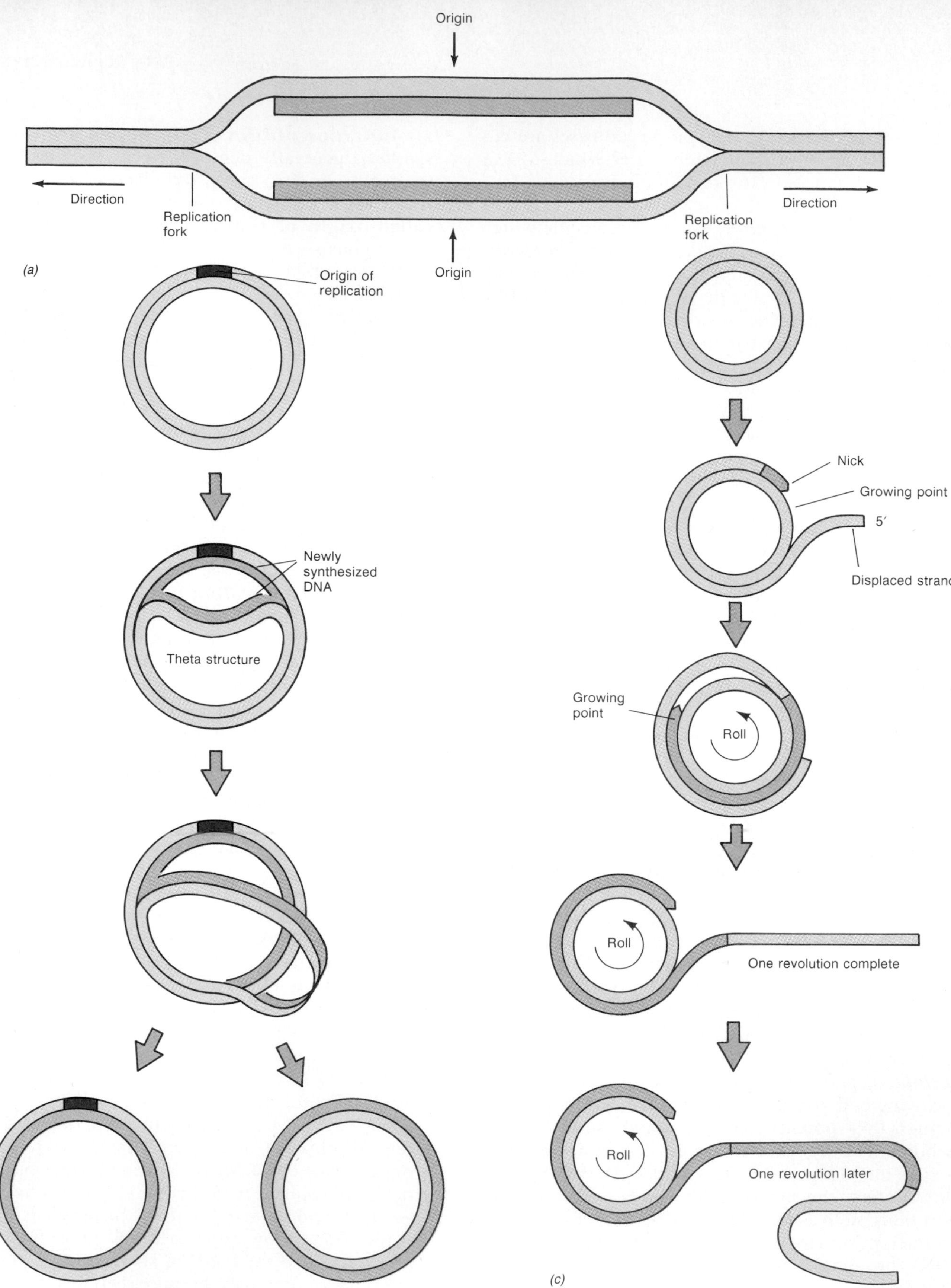

Figure 5.16 (a) Direction of replication of DNA is frequently bidirectional from the origin. (b) In circular DNA, replication leads to the formation of structures resembling the Greek letter *theta*. (c) Rolling circle replication. In this process, a multimeric single-stranded tail is generated (which can be converted to a duplex form by synthesis of a complement; not shown).

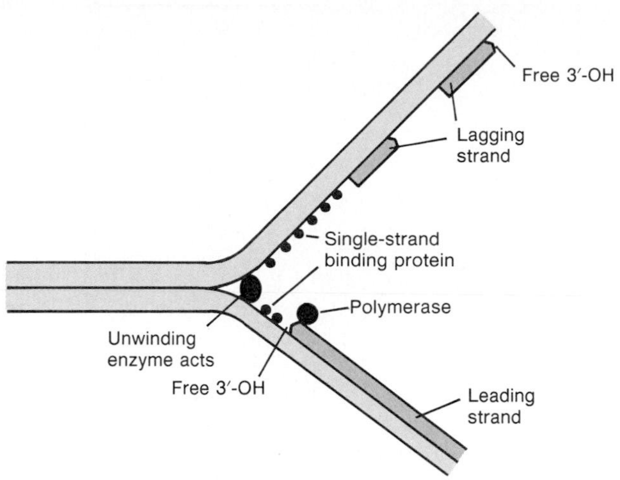

Figure 5.17 Events at the DNA replication fork.

3′–OH at the replication fork to add a new nucleotide. But on the strand growing from the 3′ hydroxyl → 5′ phosphate, called the **lagging strand**, DNA synthesis must occur discontinuously (because there is no 3′–OH at the replication fork for a new nucleotide to attach). Where is the 3′–OH on this strand? At the *opposite* end, *away* from the point of replication. Therefore, on the lagging strand, DNA synthesis must *occur discontinuously*, little fragments of RNA primer being added to provide free 3′–OH groups and then deoxyribonucleotides added until the DNA polymerase reaches the previously synthesized DNA. *Ligation* of the pieces of DNA then occurs and the RNA primer is removed. Finally, the gaps remaining where the RNA has been removed are filled in, resulting in a completed molecule on the lagging strand.

While DNA synthesis is continuing at the replication fork, changes in the coiling of the DNA will be occurring, modified by unwinding enzymes and topoisomerases (see Section 5.2). Unwinding is obviously an essential feature of DNA replication, and because supercoiled DNA is under strain, it unwinds more easily than DNA that is not supercoiled. Thus, by regulating the degree of supercoiling, topoisomerases regulate the process of replication (and also transcription, see later).

DNA synthesis and cell division As we have noted, the DNA of procaryotes is contained in a single long molecule, which is arranged in the form of a circle. Replication of the DNA molecule begins at one point and moves bidirectionally at a constant rate around the DNA circle (Figure 5.16*b*). The exact manner in which the DNA copies are partitioned into the two cells is not exactly understood, but one idea is that each DNA molecule is attached at one point to the cell membrane and after replication one attachment point moves toward each of the developing cells. Cross-wall formation then occurs, followed by cell separation (see Figure 3.71).

In eucaryotes, which have multiple chromosomes of considerably greater length, replication begins at a number of sites simultaneously and moves in both directions from each site. Clearly, the advantage of multiple initiation sites is that it permits replication of much larger amounts of DNA during reasonably short time spans.

Enzymes affecting DNA We have now mentioned a number of enzymes and other proteins which combine or act on DNA. A summary of some of these proteins is given in Table 5.2.

5.5 Genetic Elements

A genetic element is a particle or structure containing genetic material. A number of different kinds of genetic elements have been recognized. Although the main genetic element is the chromosome, other genetic elements are found and play important roles in

Table 5.2 Enzymes affecting DNA

Name	Action	Function in the cell
Restriction endonuclease	Cuts DNA at specific base sequences	Destroys foreign DNA
DNA ligase	Links DNA molecules	Completion of replication process
DNA polymerase I	Attaches nucleotides to the growing DNA molecule	Fills gaps in DNA, primarily for DNA repair
DNA polymerase III	Attaches nucleotides to the growing DNA molecule	Replication of DNA
DNA gyrase (topoisomerase)	Increases the twisting pattern of DNA, promoting supercoiling	Strand separation during replication
DNAse	Degrades DNA to nucleotides	General destruction of DNA
DNA methylase	Places methyl groups on DNA bases, thus inhibiting restriction endonuclease action	Modifies cellular DNA so that it is not affected by its own restriction endonuclease

gene function in both procaryotes and eucaryotes (Figure 5.18). Two key properties of genetic elements are (1) their ability for self-replication, and (2) their genetic coding properties.

Although single DNA molecules can be thought of as genetic elements, most genetic elements have a more complex structure than pure DNA. This structure is often associated with the condensation of DNA into compact bodies, a process necessary to fit the extremely long DNA molecules into the confines of a reasonable space. However, in all condensed genetic elements, the colinearity of the DNA molecule is preserved. When replication or transcription occurs, the condensed molecule must, at least temporarily, be unfolded.

Chromosome The single procaryotic chromosome, which is arranged in a circular fashion, contains most of the genetic information of the bacterial cell. In eucaryotes, more than one chromosome is present, the number constant within a species but varying widely between species. The overall structure of a eucaryotic chromosome can be visualized as a series of compact, highly folded units called **nucleosomes**, each containing about 200 DNA base pairs, separated by linkers of less extensively complexed DNA (see Section 3.16 and Figure 3.69). Within each nucleosome, the DNA molecule is wound around a cluster of histone molecules organized in a precise and repeatable pattern. One function of the nucleosome structure is to permit packing of the long DNA molecules within the cell. For instance, if all of the DNA of the chromosomes of a human cell were stretched end to end, the length would be 180 cm, yet this DNA is packed into 46 chromosomes in a total length of 200 μm.

In procaryotes, there is no membrane separating the chromosomes from the cytoplasm, and there is the possibility of close and perhaps even direct association between the DNA and the protein-synthesizing machinery (ribosomes, transfer RNA, and so on). In eucaryotes, chromosomes are located inside the nucleus, and only at the time of division, when the nuclear membrane breaks down, are the chromosomes free in the cell. Because of this partitioning of chromosomes within the nucleus, transcription and translation are spatially separated (see Figure 5.2). Transcription of DNA occurs within the nucleus and the messenger RNA molecules are transported out of the nucleus to the cytoplasm, where translation occurs. In procaryotes, transcription and translation can occur simultaneously, and it is even possible for translation to be initiated at one end of a messenger RNA molecule before transcription is complete at the other end. On the other hand, in eucaryotes, the initial RNA molecule, called the **primary transcript**, is extensively modified before it is finally translated, as we will discuss in Section 5.8.

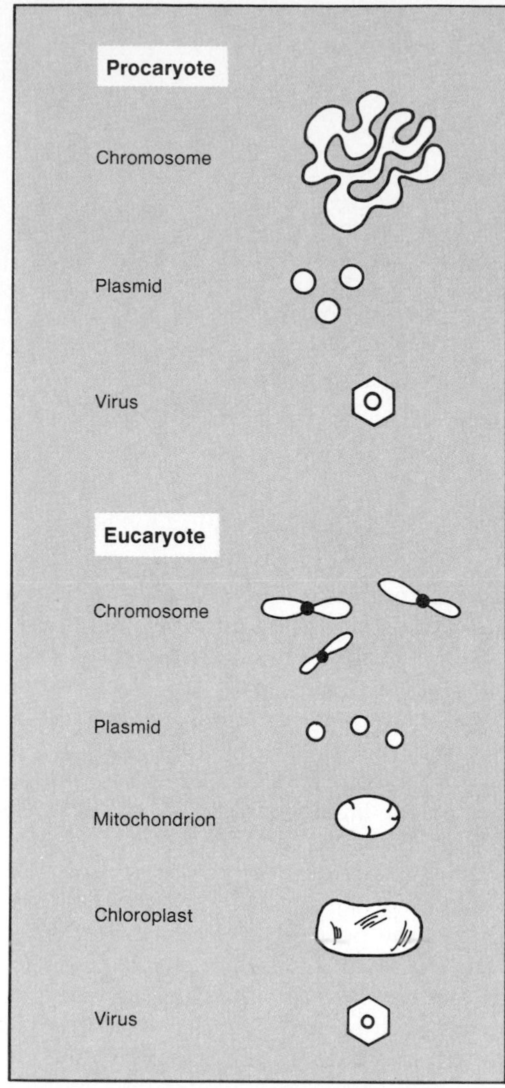

Figure 5.18 Kinds of genetic elements.

The chromosomal organization in eucaryotes involves two features not found in procaryotes:

1. **Split genes** Eucaryotic genes are discontinuous along the DNA, with noncoding sequences inserted between the sequences that actually code for protein. These noncoding intervening sequences are called **introns**, and the coding sequences are called **exons**. The number of introns per gene is variable, and ranges from none to over 50. During transcription, both introns and exons are copied, and the intron sequences are subsequently cut out and removed when the messenger RNA is processed into its final form in the cytoplasm.
2. **Repetitive sequences** Eucaryotes generally contain much more DNA per genome than is needed to code for all of the proteins required for cell function. Eucaryotic DNA can be divided into several classes. **Single copy DNA** contains the cod-

ing sequences for the main proteins of the cell. **Moderately repetitive DNA**, found in a few to relatively large numbers of copies, codes for some major macromolecules of the cell: histones, immunoglobulins (involved in immune mechanisms, as discussed in Chapter 12), ribosomal RNA, transfer RNA. **Highly repetitive (satellite) DNA**, is found in a very large number of copies. In humans, about 20 to 30 percent of the DNA is found in repetitive sequences, and almost all eucaryotic DNAs studied have some repeated sequences. The function of highly repetitive DNA is unknown.

Nonchromosomal genetic elements A number of genetic elements which are not part of the chromosome have been recognized. Some nonchromosomal genetic elements which we will discuss briefly here include viruses, plasmids, mitochondria, and chloroplasts.

Viruses are genetic elements, either DNA or RNA, which control their own replication and transfer from cell to cell. Viruses are of special interest because they are often (but not always) responsible for disease states. We discuss viruses in Chapter 6 and virus diseases in Chapter 15.

Plasmids are small circular DNA molecules that exist and replicate separately from the chromosome (see Section 7.4). Plasmids differ from viruses in that they do not cause cellular damage (generally they are beneficial). Although plasmids have been recognized in only a few eucaryotes, they have been found in most procaryotic genera. Some plasmids are excellent genetic vectors, and find wide use in gene manipulation and genetic engineering, as outlined in Chapter 8.

Mitochondria and **chloroplasts** are nonchromosomal genetic elements found in eucaryotes. As we discussed in Chapter 3, the mitochondrion is the site of respiratory enzymes, and plays a major role in energy generation in most eucaryotes. The chloroplast is a green, chlorophyll-containing structure which is the site of photosynthetic ATP formation. From a genetic viewpoint, mitochondria and chloroplasts can be viewed as independently replicating genetic elements. However, these organelles are much more complex than plasmids and viruses, since they contain not only DNA, but a complete machinery for protein synthesis, including their own ribosomes, transfer RNA, and all of the other components necessary for translation and formation of functional proteins. One intriguing feature of mitochondria and chloroplasts is that despite the fact that they contain many genes and a complete translation system, they are not independent from the chromosomes, since most proteins are coded not by the organelle DNA but by chromosomal DNA. We discuss the genetics of mitochondria in Section 7.11.

5.6 Rearrangement of Genes

Up to now we have discussed DNA as if the double helix remained linearly intact. Indeed, DNA is amazingly stable, as would be expected for a molecule which has the responsibility of maintaining the code of life. But it is well established that genetic exchanges between pieces of DNA do occur. Such genetic exchanges are at the basis of the whole process of genetic recombination, the process by which pieces of DNA coding for the same gene are exchanged. However, in the present case we are discussing not exchange between homologous genes, but a much more random, albeit powerful, type of genetic exchange. We are dealing here with the phenomenon of **movable genetic elements**, which involves a type of recombination called **transposition**.

We will deal with the genetic aspects of transposition in Chapters 6, 7, and 8; here we are discussing transposition primarily as a *molecular event*. Movable genetic elements contain special DNA sequences which are essential for transposition to occur. Two types of transposable elements have been recognized, **insertion sequences** (abbreviated IS), which encode only functions involved in transposition, and **transposons**, which are mobile genetic elements that contain not only IS but additional genes unrelated to transposition (for instance, genes bringing about antibiotic resistance). Insertion sequences and transposons are found in both procaryotes and eucaryotes, as well as in viruses which infect procaryotes and eucaryotes.

Transposons are thus segments of DNA which have the ability to move to other sites as *discrete units*. The characteristic structure of a transposon is that it has a unique gene or set of genes in the center (which usually codes for one or more detectable genetic characteristics) flanked by insertion sequences (Figure 5.19*a*). In addition to any detectable genes carried by the transposon, the transposable element also carries a gene for a *transposase enzyme*, coded by the insertion sequence. Within the insertion sequences at the ends of the transposon are short DNA repeats, either direct repeats or inverted repeats. The repeats, which play a role in the transposition process, may be short (as few as 38 base pairs) or long (upward of 1400 base pairs).

When a transposon becomes inserted into another DNA (the target DNA), the sequence in the target DNA at the site of integration is duplicated. This target DNA sequence was not present in the transposon, but the transposon has brought about a duplication of this DNA by the insertion process (Figure 5.19*a*). The duplication of the target sequence apparently arises because single-strand breaks are generated by the transposase (Figure 5.19*b*). The transposon is then attached to the single-strand ends that have been generated, and repair of the single-strand portions results

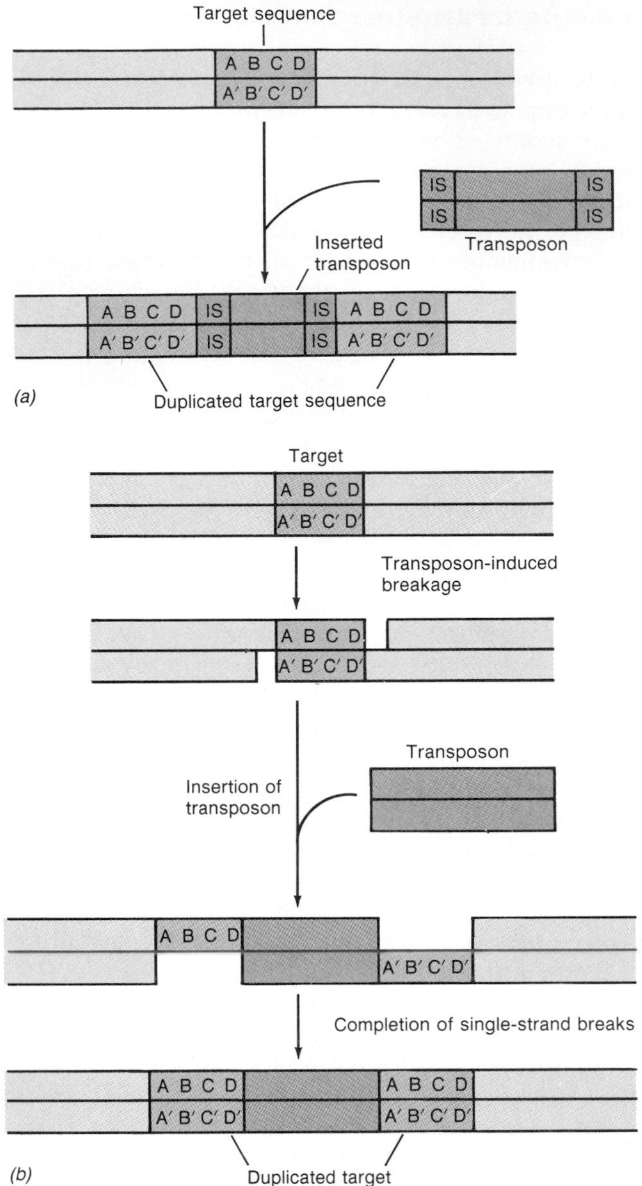

Figure 5.19 Transposons and the transposition process. (a) Insertion of a transposon generates a duplication of the target sequence. Note the presence of IS at the ends of the transposon. (b) A schematic diagram indicating how target sequences might be duplicated. For simplicity, the IS are not marked.

in the duplication. We will discuss genetic aspects of transposons in Section 7.7, and will present some viruses which behave as transposons in Sections 6.15 and 6.24. Several cancer viruses are known to act as transposons.

5.7 Synthetic DNA

Techniques are available for the synthesis of short fragments of DNA of specified base sequence. Synthetic DNA is widely used in molecular genetics, especially in genetic engineering (see Chapter 8), but also in basic research. The procedures for synthesis of DNA can be completely automated so that over a two day period, an oligonucleotide of 30–35 bases can be easily made, and oligonucleotides up to 70 bases in length can be made if necessary. For the synthesis of longer polynucleotides, the oligonucleotide fragments can be joined enzymatically, using DNA ligase.

DNA is synthesized in a *solid-phase procedure*, in which the first nucleotide in the chain is fastened to an insoluble porous support (such as a silica gel with particles around 50 μm in size). The overall procedure, the chemical details of which need not concern us here, is shown in Figure 5.20. Several chemical steps are needed for the addition of each nucleotide. After each step is completed, the solid phase containing the growing oligonucleotide chain is removed from the reaction mixture by filtration or centrifugation and the series of reactions repeated for the addition of the next nucleotide. Each step in the synthesis requires about 30 minutes, so that a pentadecamer (15 nucleotides in length) can easily be made in one day. Once the desired length is achieved, the oligonucleotide is removed from the solid phase support and purified to eliminate byproducts.

Synthetic DNA molecules are widely used as **probes** in genetic engineering, to detect, via nucleic acid hybridization, specific DNA sequences. Synthetic DNA is also used extensively in basic research on the molecular genetics of viruses and cells. We will de-

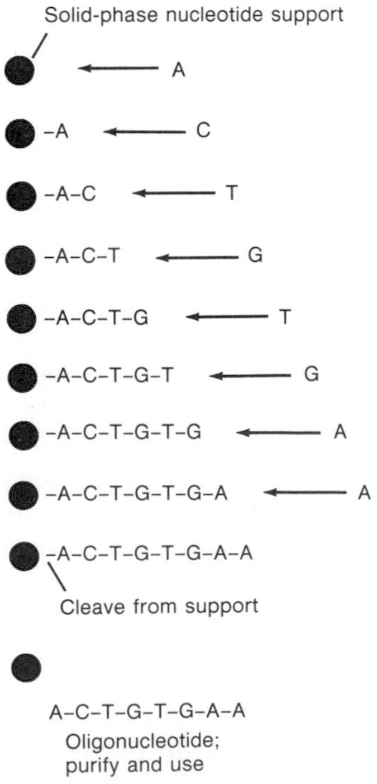

Figure 5.20 Solid-phase procedure for synthesis of a DNA fragment of defined sequence.

scribe later (Section 5.18 and Section 8.9) how synthetic DNA is used in a procedure called **site-directed mutagenesis** to create mutations at specific locations on the genome.

5.8 RNA Structure and Function

Ribonucleic acid (RNA) plays a number of important roles in the expression of genetic information in the cell. Three major types of RNA have been recognized: messenger RNA (mRNA), transfer RNA (tRNA), and ribosomal RNA (rRNA). There are three key differences between the chemistry of RNA and that of DNA: (1) RNA has the sugar *ribose* instead of *deoxyribose*; (2) RNA has the base *uracil* instead of the base *thymine*; and (3) except in certain viruses, RNA is not double stranded. A change from *deoxyribose* to *ribose* affects some of the chemical properties of a nucleic acid, and enzymes which affect DNA in general have no effect on RNA, and vice versa. The change from *thymine* to *uracil* does not affect base pairing, as the two nucleotide bases pair with adenine equally well.

It should be emphasized that RNA acts at two levels, genetic and functional. At the genetic level, RNA can carry the genetic information from DNA (messenger RNA, mRNA; or in the case of RNA viruses, play a direct genetic function). At the functional level, RNA acts as a macromolecule in its own right, serving a structural role in ribosomes (rRNA) or an amino acid transfer role in protein synthesis (tRNA). We discuss both roles of RNA in this section.

The formation of messenger RNA The transcription of the genetic information from DNA to messenger RNA (mRNA) is carried out through the action of the enzyme **RNA polymerase**, which catalyzes the formation of phosphodiester bonds between ribonucleotides. RNA polymerase requires the presence of DNA, which acts as a template. The precursors of mRNA are the ribonucleoside triphosphates, ATP, GTP, UTP, and CTP, which are polymerized with the release of the two high-energy phosphate bonds.

In most cases, the DNA template for RNA polymerase is a double-stranded DNA molecule, but only *one* of the two strands is transcribed. The enzyme RNA polymerase differs markedly between eubacteria, archaebacteria, and eucaryotes. The following discussion deals only with eubacterial RNA polymerase, which has the simplest structure (and for which the most is known). See Section 18.8 for a comparison of eubacterial, archaebacterial, and eucaryotic RNA polymerases.

The eubacterial RNA polymerase is a complex protein consisting of four subunits, each of which has a specific function in the enzyme action. All eubacterial RNA polymerases studied seem to be closely related in subunit structure. The enzyme from *Escherichia coli* has four subunits designated β, β', α, and σ (sigma), with α appearing in two copies. The subunits interact to form the active enzyme, but the sigma factor is not as tightly bound as the others, and easily dissociates, leading to the formation of what is called the *core enzyme* ($\alpha_2\beta\beta'$). The core enzyme alone can catalyze the formation of mRNA, and the role of sigma is in the recognition of the appropriate site on the DNA for the initiation of RNA synthesis. The process of mRNA synthesis involving RNA polymerase and sigma is illustrated in Figure 5.21.

RNA polymerase is a large protein, and forms contacts with the DNA over many bases simultaneously. We have discussed the general problem of protein-nucleic acid interactions in Section 5.2 and noted that the interactions are often sequence-specific. How does the RNA polymerase recognize the proper region to *start* the transcription process? The binding of the RNA polymerase to the DNA occurs at particular sites called **promoters**. Note that only *one* strand of the DNA double helix is transcribed at a time. Which strand is transcribed is determined by the orientation of the promoter sequence. RNA polymerase travels away from the promoter region, synthesizing mRNA as it moves.

Once the RNA polymerase has bound, the process of transcription can proceed. In this process, the DNA double helix *opens up* at the location where RNA polymerase is bound (Figure 5.21). As the polymerase moves, it causes the DNA to unwind in short segments, transcription of these segments occurs, and the DNA double helix closes up again. As a result of this transient unwinding, the bases of the coding region are *exposed* and then can be copied into the mRNA complement. Thus, the promoter *points* the RNA polymerase in one or the other direction. If a region of DNA has two nearby promoters on opposite strands, then transcription of one of the promoters will occur in one direction on one of the strands and transcription of the other will occur in the opposite direction on the other strand.

The first base in the mRNA is almost always a purine, adenine or guanine. The building blocks of mRNA are nucleoside triphosphates. The three phosphates of the initial nucleoside triphosphate remain intact, subsequent nucleotides being added to the 3′–OH of the ribose (Figure 5.22*a*). Thus, mRNA synthesis proceeds in the direction from 5′ phosphate to 3′ hydroxyl. Once a small portion of mRNA has been formed, the *sigma* factor dissociates; most of the elongation is therefore carried out by the core enzyme (Figure 5.21). Thus, *sigma* is involved only in the formation of the initial RNA polymerase-DNA complex. As the newly synthesized mRNA dissociates from the DNA, the opened DNA closes into the original double helix.

As important as initiation is *termination*. **Ter-**

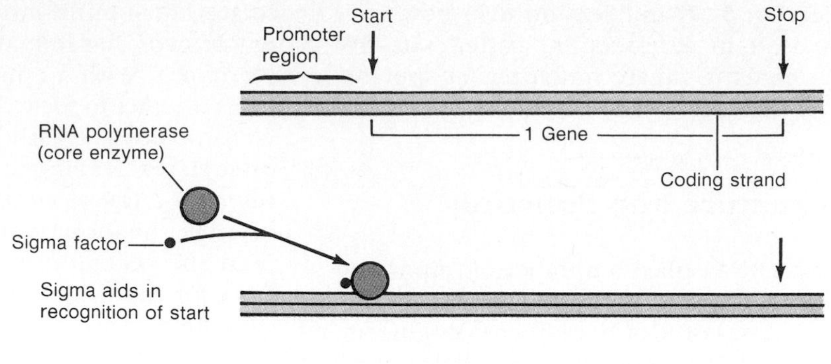

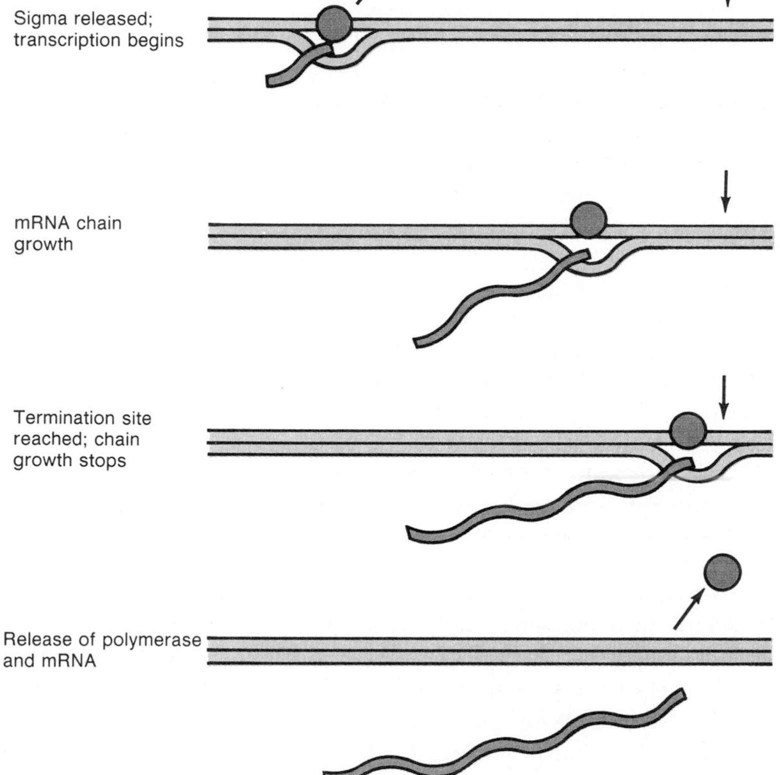

Figure 5.21 Steps in messenger RNA synthesis. The start and stop sites are specific nucleotide sequences on the DNA. RNA polymerase moves down the DNA chain, causing temporary opening of the double helix and transcription of one of the DNA strands. When a termination site is reached chain growth stops and the mRNA and polymerase are released.

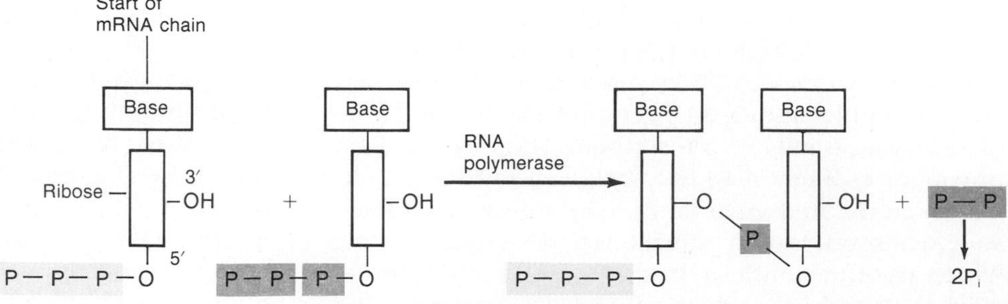

Figure 5.22 Chemistry of messenger RNA synthesis.

mination of mRNA synthesis occurs at specific sites on the DNA. Termination occurs at specific base sequences on the DNA. A common termination sequence on the DNA is one containing an inverted repeat with a central nonrepeating segment (see Section 5.2 and Figure 5.7 for an explanation of inverted repeats). Such a sequence is capable of intrastrand base pairing, leading to the formation of a stem and loop configuration. When such a sequence is copied, the mRNA can form a hairpin structure (Figure 5.23)

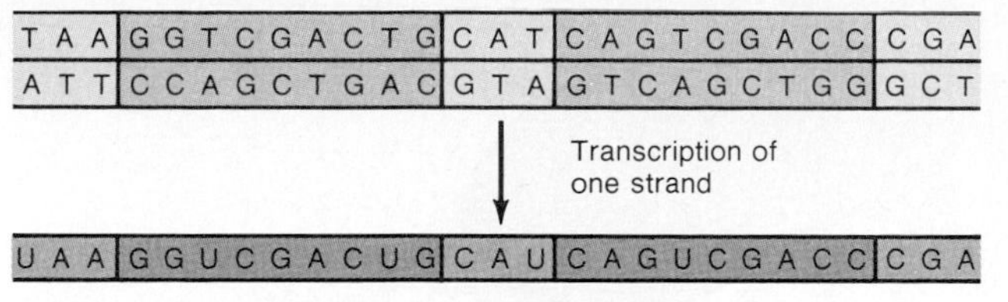

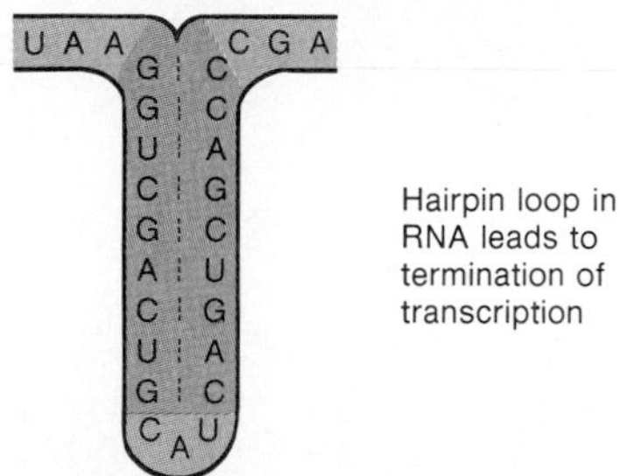

Figure 5.23 Inverted repeats in transcribed DNA leads to formation of hairpin loop in the mRNA, resulting in termination of transcription.

which in some way can lead to termination of transcription. Other termination sites are regions where a GC-rich sequence is followed by an AT-rich sequence. Such kinds of structures lead to termination without addition of any extra factors.

Another kind of mRNA termination which has been recognized involves a distinct protein called *rho*. Rho does not bind to polymerase or DNA but binds tightly to RNA, and after binding to the free end of the growing mRNA chain may move down the chain toward the DNA. Once polymerase has paused at a termination site, rho can then cause the mRNA and polymerase to leave the DNA, thus terminating transcription. Whether termination is rho-dependent or rho-independent, termination is ultimately determined by specific sequences on the DNA.

Promoters As we have noted, the promoter plays a key role in the initiation of mRNA synthesis. Promoters are specific DNA sequences at which RNA polymerase enzymes attach. The sequences of a number of promoters have been determined (Figure 5.24). In order for the RNA polymerase to recognize the promoter, the promoter must first have bound the proper *sigma* subunit. There are several different *sigma* factors, and different *sigmas* recognize preferentially different promoters.

Two sequences of the promoter have been recognized. Both are upstream of the start of transcription. One is a region 10 bases before the start of transcription, the −10 region (called the *Pribnow box*), that contains the sequence TATG or TTAA or TATA or another closely related sequence. A second part of the recognition site is −35 bases from the start of transcription. Different promoters in *E. coli* show slightly different sequences at the Pribnow box (Figure 5.24). Some promoter sequences are more effective than others in binding RNA polymerase. The more effective promoters are called *strong promoters* and are of considerable value in genetic engineering, as will be discussed in Chapter 8.

Specific inhibitors of RNA polymerase action A number of antibiotics and synthetic chemicals have been shown to specifically inhibit mRNA synthesis. A group of antibiotics called the *rifamycins* inhibits by attacking the *beta* subunit of the RNA polymerase enzyme. Rifamycin has marked specificity for procaryotes, but also inhibits RNA synthesis in chloroplasts and mitochondria of some eucaryotes. Rifamycin has been an especially useful tool in studying nucleic acid synthesis in virus-infected cells. A group of antibiotics called the *streptovaricins* is related to the rifamycins in structure and function. *Streptolydigin* is an antibiotic that also inhibits the beta subunit of RNA polymerase, but at a different site than rifamycin. Another chemical, *amanitin*, specifically inhibits mRNA synthesis in eucaryotes without affecting procaryotes.

Actinomycin (see Figure 5.9) inhibits RNA synthesis by combining with DNA and blocking elongation. Actinomycin binds most strongly to DNA at guanine-cytosine base pairs, fitting into the groove on the double strand where RNA is synthesized.

Messenger RNA may code for more than one protein In many cases a single mRNA molecule may contain the information for several proteins. In bacteria, it is frequently the case that the genes coding for related enzymes occur together in a cluster. A mRNA coding for such a group of genes or cistrons is called a **polycistronic mRNA**. In these situations of polycistronic mRNA, the RNA polymerase proceeds down the chain and transcribes the whole series of cistrons into a single long mRNA molecule. Subsequently, when this polycistronic mRNA participates in protein synthesis (see Section 5.10), the several proteins coded by the mRNA may be synthesized together.

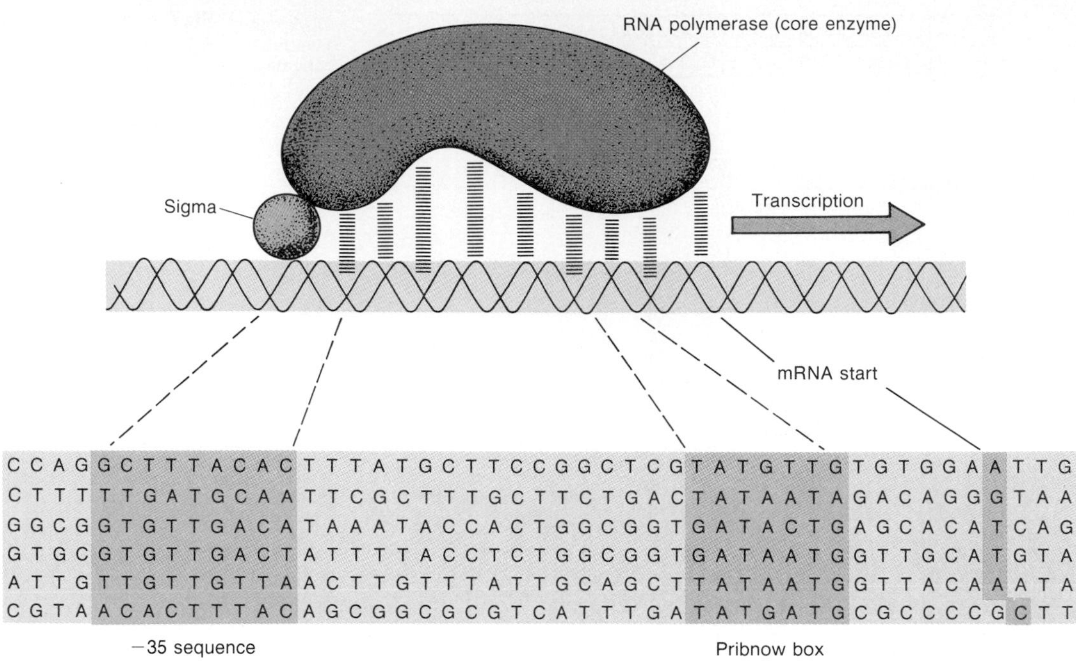

Figure 5.24 The interaction of RNA polymerase with the promoter. Shown below the diagram are 6 different promoter sequences identified in *E. coli*. The contacts of the RNA polymerase with the −35 sequence and the Pribnow box are shown. Transcription begins at a unique base just downstream from the Pribnow box.

Operons We will discuss regulation of mRNA synthesis later in this chapter, but introduce here the concept of the operon. An **operon** is a complete unit of gene expression, generally involving genes coding for several proteins on a polycistronic mRNA. The transcription of the mRNA for an operon is under the control of a specific region of the DNA, the **operator**, which is adjacent to the coding region of the first cistron in the operon. The operator is nearby and may overlap with the promoter (Figure 5.25). How does the operator participate in the regulation of the transcription process? Within the cell are specific proteins, known as **repressor proteins**, which are able to bind to specific operators. If the repressor protein has attached to the operator, then transcription of that operon cannot occur. We discuss the details of repressor action in Section 5.11.

RNA processing In many cases, the primary RNA transcript is not used directly, but is converted into the active RNA by means of special enzymes. The conversion of a *precursor* RNA into a *mature* RNA is called **RNA processing**, and is found in both procaryotes and eucaryotes. For instance, in procaryotes, tRNA and rRNA are made initially as long precursor molecules which are then cut at several places to make the final mature RNAs. In eucaryotes, and less commonly in procaryotes, mRNA also undergoes extensive processing. As discussed in Section 5.5, the genes of eucaryotes are split, with noncoding intervening sequences, *introns*, separating the coding regions called *exons*. The primary RNA transcript must be extensively processed to remove the noncoding regions before the translation process can be initiated (Figure 5.26). Although, as shown in Figure 5.26*a*,

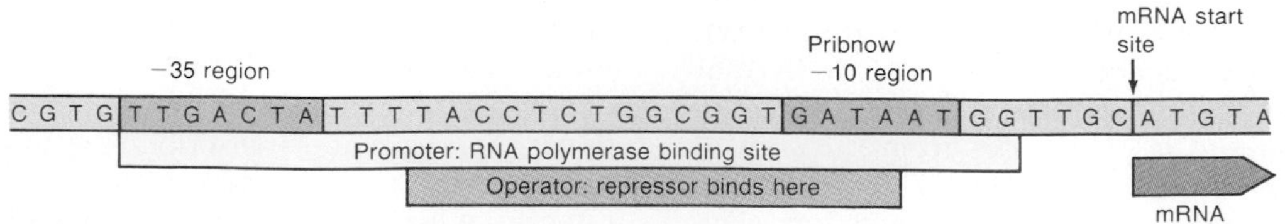

Figure 5.25 DNA base sequence of an operator/promoter region. Note how the operator region overlaps the location where RNA polymerase binds. (The sequence shown is one of the operators of bacteriophage lambda.)

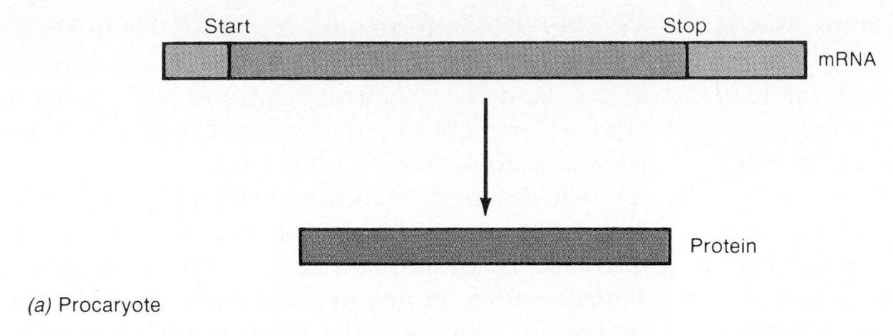

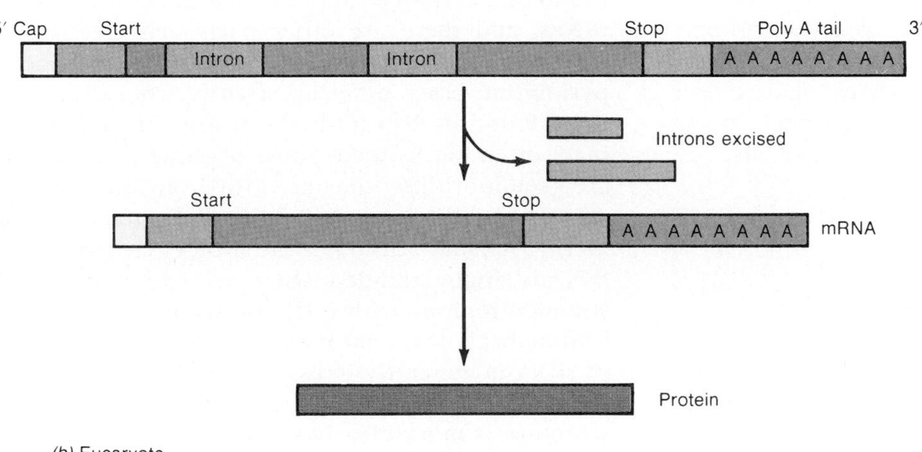

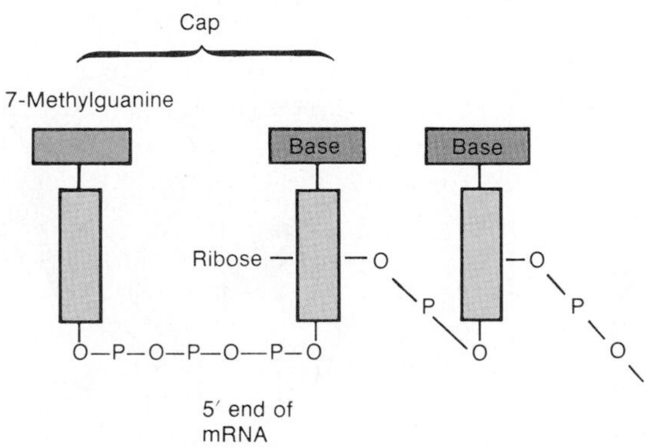

Figure 5.26 Contrast of mRNA in procaryotes and eucaryotes. (a) Procaryote: messenger RNA is synthesized in final form. (b) In eucaryotes, the mature mRNA is often derived from a primary transcript by excision of noncoding intervening sequences (intron). Eucaryotic mRNA also has a modified guanine cap at the 5′ phosphate end and a poly A tail at the 3′ hydroxyl end. (c) Structure of the cap.

procaryote mRNA is generally translated directly, a number of cases of introns are now also known in procaryotes.

Another major feature of eucaryotic mRNA is that after transcription but before transfer to the cytoplasm, the mRNA molecules have added to them a long stretch of adenine nucleotides at one end, the *poly A tail*, and a methylated guanine nucleotide at the 5′ phosphate end, called the *cap* (Figure 5.26*c*). Also, a number of the bases in mRNA are methylated after transcription and the 2′ hydroxyl group of the ribose is occasionally methylated.

We thus see that RNA synthesis is a complex and dynamic process that involves considerably more than the simple transcription of a DNA template.

RNA as a catalyst We have emphasized the role of proteins in biocatalysis but certain types of ribonucleic acid (RNA) can act as catalysts as well. In one case, an RNA intron serves a catalytic role in joining two adjacent exons into the final messenger RNA for translation. In the process the RNA intron becomes excised and is degraded. In several bacteria an RNA catalyst has been discovered and named RNAse P. RNAse P is a small (370–400 nucleotides) single-stranded RNA that functions in the cell as a tRNA processing enzyme. RNAse P modifies primary tRNA transcripts (which contain short oligonucleotide segments on both the 3′ and 5′ ends of the molecule that must be removed to generate the active tRNA). RNAse P is associated with a small basic protein which has no catalytic role in the RNA processing steps, but is required to stabilize the catalytic RNA.

Why is RNAse P an RNA and not a protein? It is hypothesized that the nature of the reactions catalyzed by RNAse P, the pruning of oligonucleotides from several structurally highly related molecules, could be performed by a molecule with a great deal of conformational flexibility, like RNA, and would not require the specificity of proteins. The secondary structures of RNAs are in general less highly folded than proteins and thus presumably less able to discriminate closely related molecules. Although the vast majority of enzymatically catalyzed reactions in bacteria are carried out by proteins, certain reactions are apparently catalyzed by RNA. Perhaps upon closer examination other reactions which do not require highly specific recognition processes will also be found to be RNA catalyzed.

5.9 Transfer RNA

The translation of the RNA message into the amino acid sequence of protein is brought about through the action of transfer RNA (tRNA) (Figure 5.27). Transfer RNA is an adaptor molecule having two specificities, one for a codon on mRNA, the other for an amino acid. The transfer RNA and its specific amino acid are brought together by means of an enzyme, the *amino acid activating enzyme*.

The detailed structure of tRNA is now well understood. There are about 60 different types of tRNA in bacterial cells and 100 to 110 in mammalian cells. Transfer RNA molecules are short, single-stranded molecules, with lengths (among different tRNAs) of 73 to 93 nucleotides. When compared, it has been found that certain structural parts are constant for all tRNAs, and there are other parts that are variable. Transfer RNA molecules also contain some purine and pyrimidine bases differing slightly from the normal bases found in RNA in that they are chemically modified, often methylated. Some of these unusual bases are pseudouridine, inosine, dihydrouridine, ribothymidine, methyl guanosine, dimethyl guanosine, and methyl inosine. Although the molecular structure of tRNA is single stranded, there are extensive double-stranded regions within the molecule, as a result of folding back of the molecule on itself. The structure of tRNA is generally drawn in cloverleaf fashion, as shown in Figure 5.27*a*, but the three-dimensional structure is more clearly shown in Figure 5.27*b*.

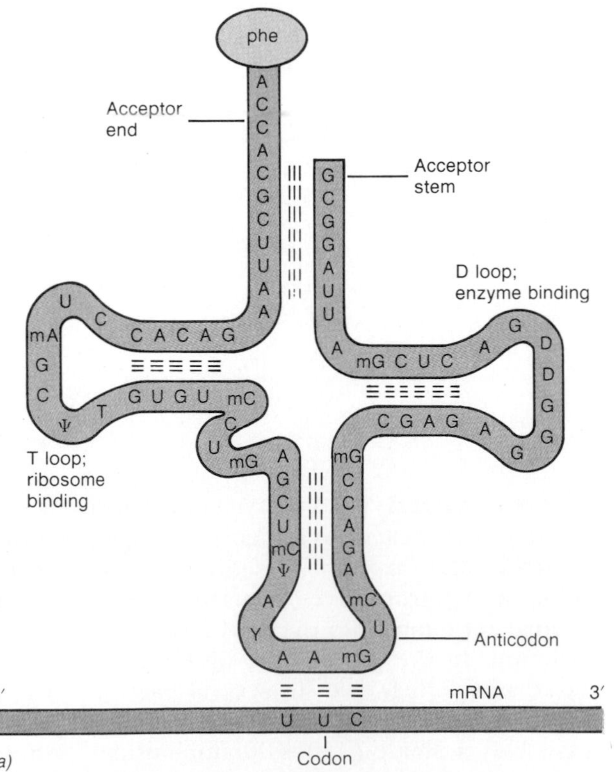

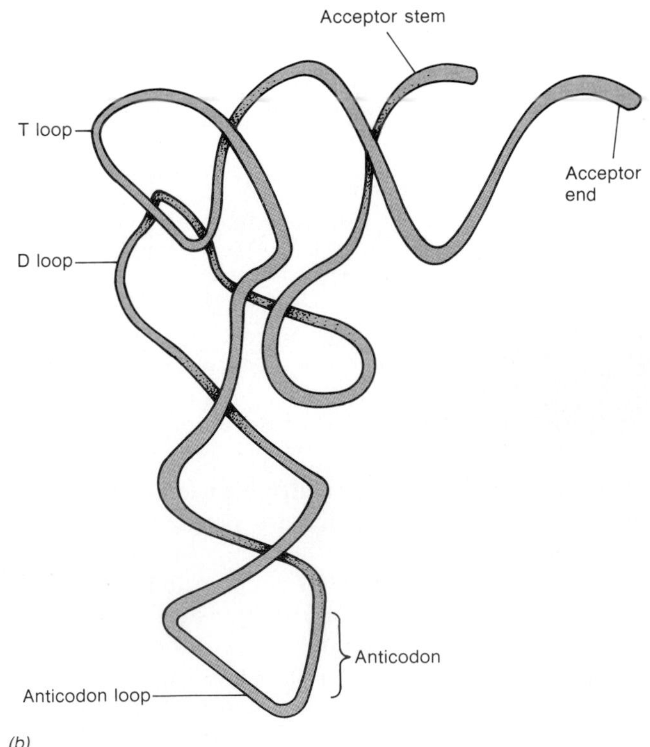

Figure 5.27 Structure of a transfer RNA, yeast phenylalanine tRNA. (a) The conventional cloverleaf structure. The amino acid is attached to the ribose of the terminal A at the acceptor end. Abbreviations: A, adenine; C, cytosine; U, uracil; G, guanine; ψ, pseudouracil; D, dihydrouracil; m, methyl; p, phosphate (terminal); Y, a modified purine. (b) In actuality the molecule folds so that the D loop and TψC loops are close together and associate by hydrophobic interactions.

Several parts of the tRNA molecule have specific functions in the translation process. One of the variable parts of the tRNA molecule contains the anticodon, the site recognizing the codon on the mRNA. The location of this **anticodon loop** is shown in Figure 5.27. There are just *three* nucleotides in the anticodon loop that are specifically involved in the recognition process. Another loop called the TψC loop (ψ is the symbol for pseudouridine) is believed to interact with the ribosome, whereas the D loop may be involved in binding to the activating enzyme. There is also one region at the end of the chain where a sequence of three nucleotides projects from the rest of the molecule. The sequence of these three nucleotides is always the same, cytosine-cytosine-adenine (CCA), and it is to the ribose sugar of the terminal A that the amino acid is covalently attached, via an ester linkage. From this acceptor portion of the tRNA, the amino acid is transferred to the growing polypeptide chain on the ribosome by a mechanism that will be described in the next section.

There is specific enzymatic machinery for activating the amino acid and placing it on the correct tRNA. The enzymes carrying out this process are called **aminoacyl-tRNA synthetases**, and they have the important function of recognizing *both* the amino acid *and* the specific tRNA for that amino acid. It should be emphasized at this point that the fidelity of this recognition process is crucial, for if the wrong amino acid is attached to the tRNA, it may be inserted in the improper place in the protein, leading to the synthesis of a faulty protein.

The specific chemical reaction between amino acid and tRNA catalyzed by the aminoacyl-tRNA synthetase involves *activation* of the amino acid by reaction with ATP:

$$\text{Amino acid} + \text{ATP} \rightleftharpoons \text{amino acid-AMP} + \text{P–P}$$

The amino acid-AMP intermediate formed normally remains bound to the enzyme until collision with the appropriate tRNA molecule, and the activated amino acid is then transferred to the tRNA:

$$\text{Amino acid-AMP} + \text{tRNA} \rightleftharpoons \text{amino acid-tRNA} + \text{AMP}$$

The pyrophosphate (P–P) formed in the first reaction is split by a pyrophosphatase, forming two molecules of inorganic phosphate. Since ATP is used and AMP is formed, *two* high-energy phosphate bonds are required for the activation of an amino acid. Once activation has occurred, the amino acid-tRNA leaves the synthetase and migrates through the cell to the ribosome-mRNA protein synthesizing machinery. The mechanism of protein synthesis is discussed in the next section.

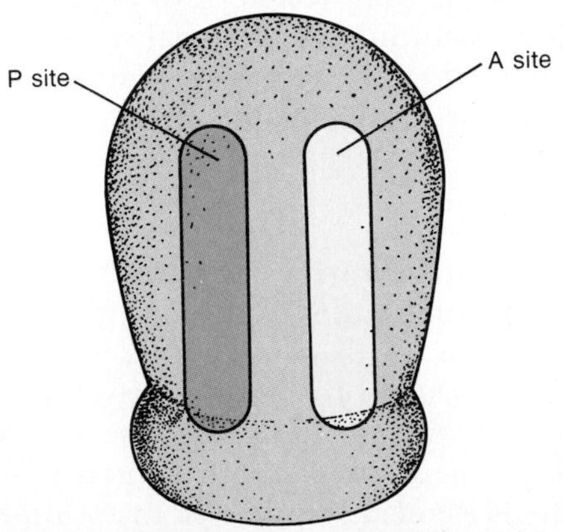

Ribosome structure

Property	Eubacterium	Eucaryote
Overall size	70S	80S
Small subunit	30S	40S
Number of proteins	~21	~30
RNA size	16S	18S
Nucleotides in RNA	1500	2000
Large subunit	50S	60S
Number of proteins	~34	~50
RNA size	23S	28S
Nucleotides in RNA	3000	5000

There is also a small RNA, the 5S RNA, which has 120 nucleotides, and is present in both procaryote and eucaryote. Ribosomes of mitochondria and chloroplasts of eucaryotes are similar to eubacterial ribosomes.

Figure 5.28 Structure of the ribosome, showing the position of the A or acceptor site (right) and the P or peptide site. Details of the ribosome subunits are shown below.

5.10 Translation: The Process of Protein Synthesis

It is the amino acid *sequence* that determines the structure of the final active protein. The key problem of protein synthesis is thus the placement of the proper amino acid at the proper place in the polypeptide chain. This is the role of the protein-synthesizing machinery of the cell.

Steps in protein synthesis Ribosomes are the site of protein synthesis. Each ribosome is constructed of two subunits. In eubacteria, the ribosome sedimentation constants are of 30S (Svedberg units) and 50S (Figure 5.28). Each large subunit consists of a number of individual proteins, 21 in the 30S ribosome and 34 in the 50S ribosome. The actual synthesis of a protein involves a complex cycle in which the various ribosomal components play specific roles.

Although a continuous process, protein synthesis can be thought of as occurring in a number of discrete steps: **initiation, elongation, termination-release**, and **polypeptide folding**. These steps are outlined in Figure 5.29. In addition to mRNA, tRNA, and ribosomes, the process involves a number of proteins designated initiation, elongation, and termination factors. Also guanosine triphosphate (GTP) provides energy for the process.

Initiation always begins with a free 30S ribosome subunit, and an **initiation complex** forms consisting of a 30S ribosome subunit, mRNA, formylmethionine tRNA, and initiation factors. To this initiation complex a 50S subunit is added to make the active 70S ribosome. At the end of the process, the released ribosome separates again into 30S and 50S subunits. Just preceding the initiation codon on the mRNA is a sequence of from three to nine nucleotides (the Shine-Dalgarno sequence) which is involved in binding of the message to the ribosome. This ribosome binding site is complementary to the 3′ end of the 16S RNA of the ribosome, and it is thought that base pairing ensures effective formation of the ribosome-mRNA complex.

In eubacteria, initiation always begins with a special amino acyl-tRNA, which is formylmethionine-tRNA. Subsequently, the formyl group at the N-terminal end of the polypeptide is removed. In eucaryotes and archaebacteria, initiation begins with methionine instead of formylmethionine.

The mRNA is attached to the 30S subunit and the tRNA holding the growing polypeptide chain is attached to the 50S subunit (Figure 5.29). There are two sites on the 50S subunit, designated P and A. Site A, the **acceptor** site, is the site where the new AA–tRNA first attaches. Site P, the **peptide** site, is the site where the growing peptide is held by a tRNA. During peptide bond formation, the peptide moves to the tRNA at the A site as a new peptide bond is formed. Then after the now empty tRNA at the P site is removed, the tRNA holding the peptide moves (is translocated) from the A site back to the P site, thus opening up the A site for another AA–tRNA.

The termination of protein synthesis occurs when a codon is reached which does not specify an AA–tRNA. There are three codons of this type and they are called **nonsense codons** (see Section 5.15); they serve as the stopping points for protein synthesis.

Secretory proteins Many proteins are used outside the cell and must somehow get from the site of synthesis on cytoplasmic ribosomes through the cell membrane. In procaryotes, periplasmic enzymes and extracellular enzymes are secretory proteins, and in eucaryotes various digestive enzymes and enzymes of the lysosome are in this category.

How is it possible for a cell selectively to transfer some proteins across a membrane, while leaving most proteins in place in the cytoplasm? This is explained by the **signal hypothesis**, which states that secretory proteins are synthesized with an extra N-terminal peptide sequence, some 15 to 20 amino acids in length, which is called the **signal sequence**. In this signal sequence, hydrophobic amino acids predominate and permit the enzyme to be threaded through the hydrophobic lipid membrane. In many cases, the ribosomes that synthesize secretory proteins are bound directly to the cell membrane, so that the protein is formed and passes through the membrane simultaneously. Once the protein has been secreted, the signal sequence is removed by a peptidase enzyme, an example of the process of **post-translational modification**.

The study of protein secretion has important practical implications for genetic engineering (see Chapter 8). If bacteria are genetically engineered to serve as agents for the production of foreign proteins, it is desirable to manipulate the signal sequence in order to arrange for the desired protein to be excreted so that it can be readily isolated and purified.

Effect of antibiotics on protein synthesis A large number of antibiotics inhibit protein synthesis. Some of these antibiotics are medically useful, whereas others have proved ineffective in treating diseases but are still of interest as research tools. One interesting aspect of antibiotic action is selectivity. Selectivity is especially apparent in antibiotics acting on the ribosome, and antibiotics affecting ribosome function in procaryotes have no effect on cytoplasmic ribosomes in eucaryotes. A brief summary of the modes of action of certain antibiotics is given in Table 5.3. Note that antibiotics not only act at different stages of protein synthesis, but they usually affect only one of the two ribosomal subunits. Also, most antibiotics act against either eubacteria or archaebacteria, a few against eucaryotes, and some act against both. Since mitochondria and chloroplasts have ribosomes of the procaryotic type, it is of interest that antibiotics inhibiting protein synthesis in eubacteria also generally inhibit protein synthesis in mitochondria and chloroplasts. This is one piece of evidence suggesting that mitochondria and chloroplasts might originally have been derived by intracellular infection of a eucaryotic cell by procaryotes (see Sections 3.14 and 18.1).

One interesting inhibitor of protein synthesis is the diphtheria toxin, an agent involved in the pathogenesis of the disease diphtheria. As discussed in Section 11.10, diphtheria toxin inactivates an elongation factor and hence is a powerful inhibitor of protein synthesis in eucaryotes and archaebacteria.

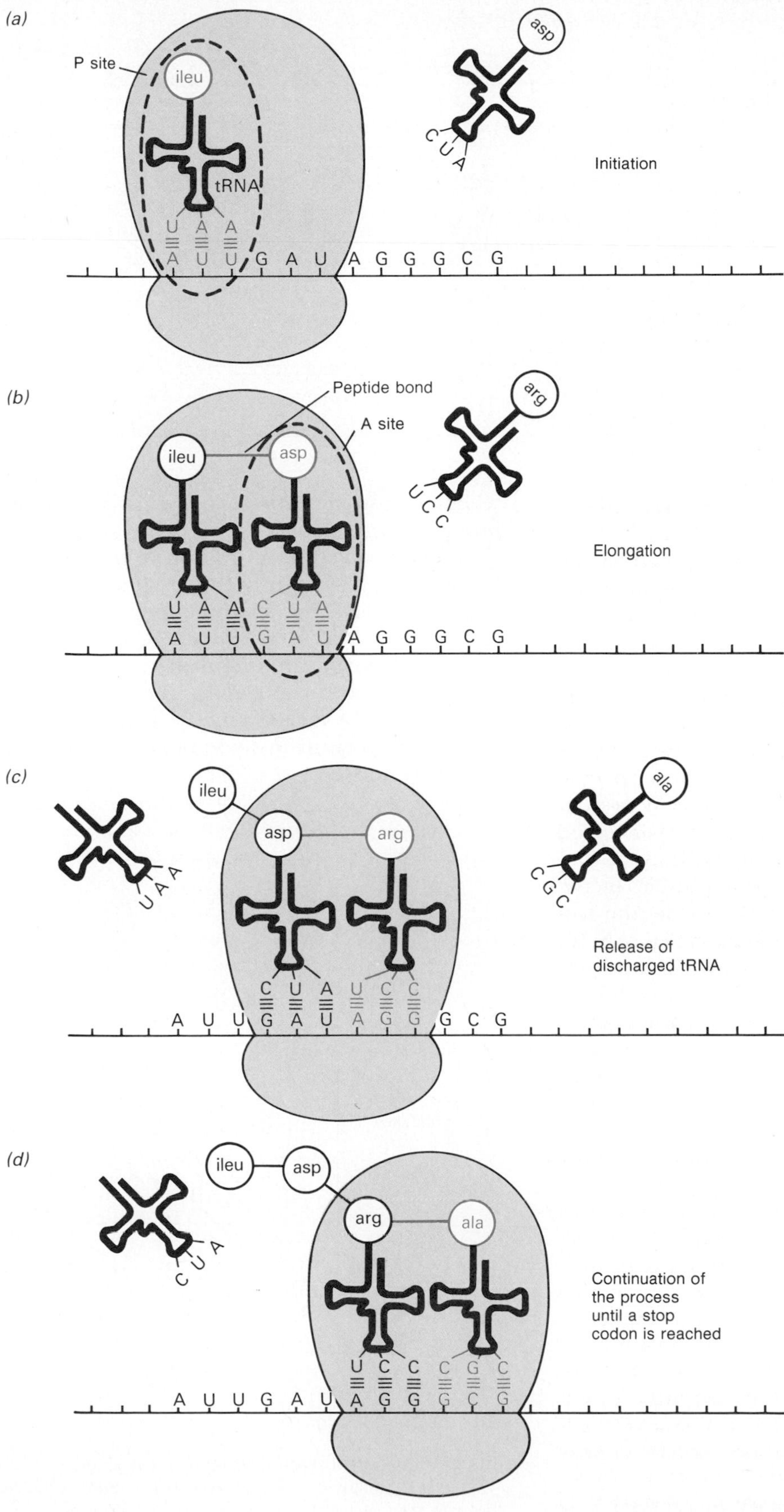

Figure 5.29 Translation of the information from messenger RNA (mRNA) into the amino acid sequence of protein. The whole process occurs on the surface of the ribosome. (a) Interaction between codon and anticodon brings into position the correct tRNA carrying the amino acid. (b) A peptide bond forms between amino acids on adjacent tRNA molecules. (c) and (d) As the ribosome moves to the next codon a new tRNA attaches and the old tRNA, now free of amino acid, leaves. The result is a growing chain of amino acids.

Table 5.3 Effects of antibiotics on protein synthesis

	Function				
Target	**Initiation**	**Codon recognition**	**Peptide formation**	**Elongation**	**Termination**
Large subunit	*Cycloheximide*		*Puromycin*	*Cycloheximide*	
			Chloramphenicol **Erythromycin**	*Diphtheria toxin* *Anisomycin*	
Small subunit	**Streptomycin**	**Streptomycin** **Neomycin** **Tetracycline**			**Tetracycline**

Those antibiotics printed in bold face act on eubacteria, those in italics act on eucaryotes. Anisomycin and diphtheria toxin act also on archaebacteria.

5.11 Regulation of Protein Synthesis

Not all enzymes are synthesized by the cell in the same amounts, some enzymes being present in far greater numbers of molecules than other enzymes. Clearly, the cell is able to regulate enzyme synthesis. Even more dramatic, the synthesis of many enzymes is greatly influenced by the environment in which the organism is growing, most particularly by the presence or absence of specific chemical substances. Often the enzymes catalyzing the synthesis of a specific product are not produced if this product is present in the medium. For example, the enzymes involved in the formation of the amino acid arginine are synthesized only when arginine is *not* present in the culture medium; external arginine *represses* the synthesis of these enzymes. As can be seen in Figure 5.30, if arginine is added to a culture growing exponentially in a medium devoid of arginine, growth continues at the previous rate, but the formation of the enzymes involved in arginine synthesis stops. Note that this is a *specific* effect, as the syntheses of all other enzymes in the cell are found to continue at the same rates as previously.

Enzyme repression is a very widespread phenomenon in bacteria—it occurs during synthesis of a wide variety of enzymes involved in the biosynthesis of amino acids, purines, and pyrimidines. In almost all cases it is the final product of a particular biosynthetic pathway that represses the enzymes of this pathway. In these cases repression is quite specific, and the process usually has no effect on the synthesis of enzymes other than those involved in a single biosynthetic pathway. The value to the organism of enzyme repression is obvious, since it effectively ensures that the organism does not waste energy synthesizing unneeded enzymes.

A phenomenon complementary to repression is **enzyme induction**, the synthesis of an enzyme only when a substrate is present. Figure 5.31 shows this process in the case of the enzyme β-galactosidase, which is involved in utilization of the sugar lactose. If lactose is absent from the medium the enzyme is not synthesized, but synthesis begins almost immediately after lactose is added. Enzymes involved in the breakdown of carbon and energy sources are often inducible. Again, one can see the value to the organism of such a mechanism, as it provides a means whereby the organism does not synthesize an enzyme until it is needed.

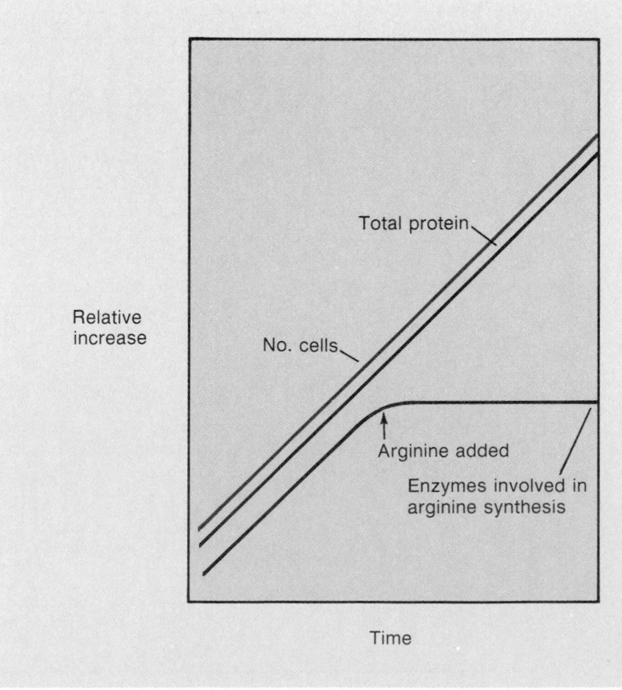

Figure 5.30 Repression of enzymes involved in arginine synthesis by addition of arginine to the medium. Note that the rate of total protein synthesis remains unchanged.

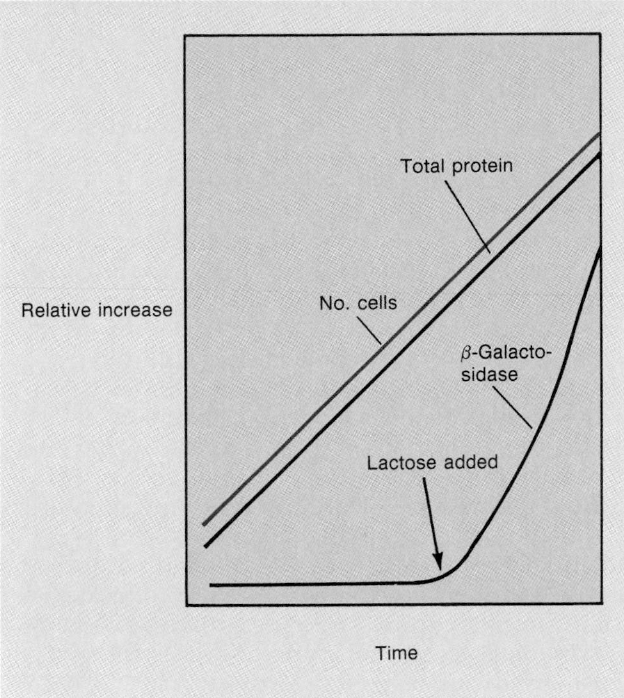

Figure 5.31 Induction of the enzyme β-galactosidase upon the addition of lactose to the medium. Note that the rate of total protein synthesis remains unchanged.

The substance that initiates enzyme induction is called an **inducer**, and a substance that represses enzyme production is called a **corepressor**; these substances, which are always small molecules, are often collectively called **effectors**. Not all inducers and corepressors are substrates or end products of the enzymes involved. For example, analogs of these substances may induce or repress even though they are not substrates of the enzyme. Thiomethylgalactoside (TMG), for instance, is an inducer of β-galactosidase even though it cannot be hydrolyzed by the enzyme. In nature, however, inducers and corepressors are probably normal cell metabolites.

Mechanism of induction and repression As we have just learned, the overall process of protein synthesis can be divided into two stages, transcription and translation. Although in principle enzyme induction and repression could affect either stage, in bacteria control is exerted mainly at the stage of *transcription*. Thus when an enzyme inducer is added, it initiates enzyme synthesis by causing the formation of the mRNA that codes for the particular enzyme. When a substance (corepressor) that causes enzyme repression is added, it causes an inhibition of mRNA formation.

How can inducers and corepressors affect transcription in such a specific manner? They do this indirectly by combining with specific proteins, the **repressors**, which then in turn affect mRNA synthesis. In the case of a repressible enzyme, it is thought that the corepressor (for example, arginine) combines with a specific **repressor protein** that is present in the cell (Figure 5.32). The repressor protein is an allosteric protein (see Section 4.20), its configuration being altered when the corepressor combines with it. This altered repressor protein can then combine with a specific region of the DNA at the initial end of the gene, the **operator region**, where synthesis of mRNA is initiated (see Section 5.7). If this occurs, the synthesis of mRNA is blocked, and the protein or proteins specified by this mRNA cannot be synthesized. If the mRNA is polycistronic, all of the proteins coded for by this mRNA will be repressed. A series of genes all regulated by one operator is called an **operon**.

For induction the situation described above is reversed. The specific repressor protein is active in the *absence* of the inducer, completely blocking the synthesis of mRNA, but when the inducer is added it combines with the repressor protein and inactivates it. Inhibition of mRNA synthesis being overcome, the enzyme can be made (Figure 5.33). Induction and repression thus both have the same underlying mechanism, the *inhibition* of the synthesis of mRNA by the action of specific repressor proteins that are them-

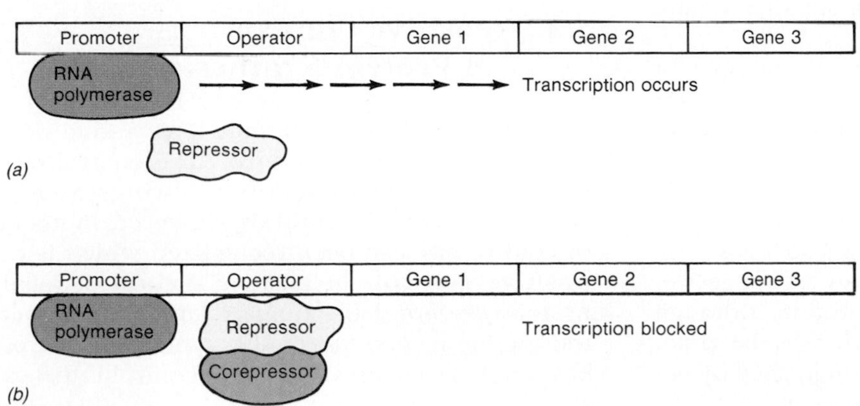

Figure 5.32 The process of enzyme repression. (a) Transcription of the operon occurs because the repressor is unable to bind to the operator. (b) After a corepressor (small molecule) binds to the repressor, the repressor now binds to the operator and blocks transcription; mRNA and the proteins it codes for are not made.

Inducers and Repressors

The phenomenon of enzyme induction has had a long history in microbial physiology, and detailed studies of this phenomenon have played a major role in understanding macromolecular synthesis and its regulation. The concepts that have been developed from studies on enzyme induction in bacteria have also been widely applied to eucaryotic cells and to an understanding of cancer. Emile Duclaux, an associate of Pasteur, first reported in 1899 that a fungus, *Aspergillus niger*, only produced the enzyme invertase (which hydrolyzes sucrose) when growing on a sucrose medium. Similar observations were made by other workers in the early 1900s. H. Karström of Finland first studied these phenomena systematically in the 1930s. He found that certain enzymes were always present, irrespective of the culture medium, whereas other enzymes were only formed when their substrates were present. Karström termed the enzymes formed all the time *constitutive*, and those formed only when substrate was present *adaptive*. In the 1940s, the French microbiologist Jacques Monod began a systematic study of bacterial growth and discovered the phenomenon he termed *diauxic growth*. From his study of diauxy, Monod turned to a more detailed study of the process of enzyme adaptation. For many years, the interpretation of enzyme adaptation had been complicated by the fact that the substrate of the enzyme seemed to be involved in the process. However, with the discovery of nonmetabolizable substances, such as thiomethylgalactoside (TMG), which brought about enzyme formation even though they were not substrates of the enzyme, it became clear that enzyme adaptation and enzyme function were two different things. Because the word *adaptation* has certain confusing connotations (among other things, adaptation does not distinguish between the selection of mutant strains from a culture and the synthesis of a new enzyme in a preexisting genotype), the term has been abandoned, and the word *induction* is used instead. Among the most critical experiments carried out by Monod and his colleagues was that which showed that enzyme induction re-

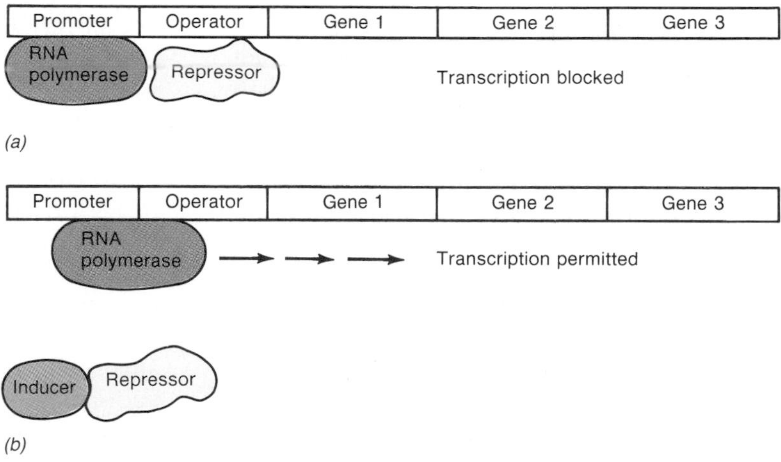

Figure 5.33 The process of enzyme induction. (a) A repressor protein binds to the operator region and blocks the action of RNA polymerase. (b) Inducer molecule binds to the repressor and inactivates it. Transcription by RNA polymerase occurs and a mRNA for that operon is formed.

selves under the control of specific small-molecule inducers and repressors.

It should be emphasized that not all enzymes of the cell are inducible or repressible. Those that are synthesized continuously are called **constitutive**. In fact, even inducible or repressible enzymes may become constitutive by genetic mutation. Two kinds of constitutive mutants are known: those that no longer produce the repressor protein, and those whose operator regions are altered so that they no longer respond to the action of the repressor protein. Although the net result of both of these mutations is the same, the two kinds of mutants can be distinguished by genetic tests.

5.12 Positive Control of Protein Synthesis

Induction and repression constitute a kind of regulation called *negative control*. The controlling element—the repressor protein—brings about the *repression* of mRNA synthesis. However, another type of control has also been recognized which is called **positive control**. In positive control, a regulator protein *promotes* the binding of RNA polymerase, thus acting to *increase* mRNA synthesis. One of the best studied examples of positive control is that called *catabolite repression*.

sulted in the *synthesis of a new protein* and not the activation of some preexisting protein. This suggested that the enzyme inducer was somehow causing differential gene action.

While the phenomenon of enzyme induction had been known for a long time, enzyme repression, the specific inhibition of enzyme synthesis, was not discovered until 1953, being reported simultaneously by the laboratories of Monod, Edward A. Adelberg and H. Edwin Umbarger in Berkeley, and Donald D. Woods at Oxford, England. It soon became clear that enzyme induction and repression, although having opposite effects, were manifestations of a similar mechanism. Biochemical studies on enzyme synthesis provided considerable insight into the phenomena of induction and repression, but by themselves they probably never would have led to a final understanding of the picture. The introduction of the techniques of bacterial genetics was essential, and some of the aspects of this will be discussed in Chapter 7. Among the major contributors to this genetic approach was Francois Jacob, a student and colleague of Monod at the Pasteur Institute. Jacob isolated a large number of mutants in the lactose pathway and analyzed these genetically. The results led to the conclusion that induction and repression were under the control of specific proteins, called *repressors*, which were coded by regulatory genes. These regulatory genes were associated with, but were distinct from, the structural genes coding for the specific enzyme proteins. Finally, Walter Gilbert and Benno Müller-Hill at Harvard University developed a cell-free system of assaying for repressor protein, and carried out a biochemical isolation and purification process for the repressor. The proof that the repressor was indeed a protein not only confirmed the theory of Jacob and Monod, but provided the approaches necessary for studying the manner in which proteins are able to specifically interact with defined sequences on DNA molecules. Biochemical and genetic studies on RNA polymerase then clarified the manner in which the information in DNA is transcribed into RNA. The most significant result of this important fundamental research is that it showed that regulation of enzyme synthesis in bacteria occurs at the level of transcription rather than at the level of translation.

Catabolite repression In catabolite repression the syntheses of a variety of unrelated enzymes are inhibited when cells are grown in a medium that contains an energy source such as glucose. Catabolite repression has been called the **glucose effect** because glucose was the first substance shown to initiate it, although in some organisms glucose does not cause enzyme repression. Catabolite repression occurs when the organism is offered a catabolizable energy source in the presence of a more readily catabolizable energy source, such as glucose. One consequence of catabolite repression is that it can lead to so-called **diauxic growth** if the two energy sources are present in the medium at the same time and if the enzyme needed for utilization of one of the energy sources is subject to catabolite repression. Growth first occurs on the one energy source, and there is then a temporary cessation before growth is resumed on the other energy source. This phenomenon is illustrated in Figure 5.34 for growth on a mixture of glucose and lactose. The enzyme β-galactosidase, which is responsible for utilization of lactose, is inducible, but its synthesis is also subject to catabolite repression. Thus, as long as glucose is present in the medium β-galactosidase is not synthesized; the organism grows only on the glucose and leaves the lactose untouched. When the glucose is exhausted, catabolite repression is abolished. After a lag β-galactosidase is synthesized and growth on lactose can occur.

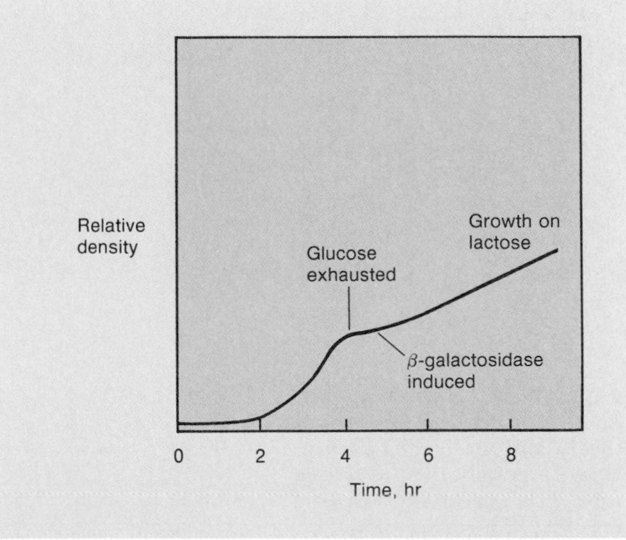

Figure 5.34 Diauxic growth on a mixture of glucose and lactose. Glucose represses the synthesis of β-galactosidase. After glucose is exhausted, a lag occurs until β-galactosidase is synthesized, and then growth can resume on lactose.

How does catabolite repression work? One mechanism of catabolite repression involves an additional type of control on transcription of the operon, at the level of the RNA polymerase. As discussed in Section

5.8, RNA polymerase begins transcription by binding to the promoter site on the DNA. However, in the case of catabolite-repressible enzymes, binding of polymerase only occurs if another protein, called **catabolite activator protein** (CAP), has bound first. An allosteric protein, CAP only binds if it has first bound a small-molecular-weight substance called *cyclic adenosine monophosphate* or **cyclic AMP**. Cyclic AMP (Figure 5.35) has been shown to be a key element in a variety of control systems, not only in bacteria but in higher organisms. Cyclic AMP is synthesized from ATP by an enzyme called *adenylate cyclase*, which is widespread in bacteria, and glucose either inhibits the synthesis of cyclic AMP or stimulates its breakdown. When glucose is present, the cyclic AMP level in the cell is low, and binding of RNA polymerase to the promoter does not occur. Thus catabolite repression is really a result of a deficiency of cyclic AMP, and can be overcome by adding this compound to the medium. The manner in which cyclic AMP is involved in catabolite repression is diagrammed in Figure 5.36.

Cyclic AMP has been shown to promote the synthesis of a number of enzymes in *E. coli*, mostly involved in catabolic processes. Cyclic AMP has a number of regulatory roles in eucaryotes that do not involve catabolite repression and is also an extracellular signal for the aggregation process in certain cellular slime molds (see Section 1.9).

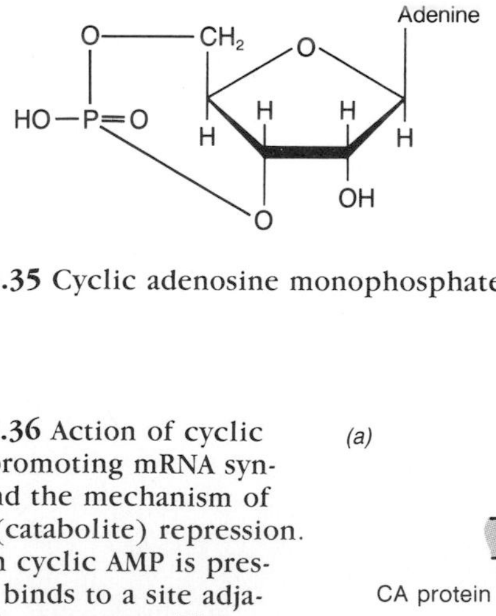

Figure 5.35 Cyclic adenosine monophosphate (cyclic AMP).

We have now described two quite different mechanisms for the regulation of protein synthesis. The mechanism involving the lactose repressor and operator is an example of **negative control**. Under normal conditions, the function of RNA polymerase is *blocked*, and this block can be overcome if an inducer is present. In the case of catabolite repression, a mechanism of **positive control** exists, since the catabolite activator protein normally promotes the binding of RNA polymerase, thus acting to *increase* mRNA synthesis. But this action can be overcome if cyclic AMP is limiting. There are a number of other positive control proteins which, by binding to specific promoters, bring about the synthesis of specific mRNAs. In negative control, if the regulatory protein is destroyed or inactivated, protein synthesis then occurs, whereas in positive control, if the regulatory protein is destroyed, protein synthesis does not occur.

5.13 Attenuation

Another element of control, called **attenuation**, has been recognized in some operons controlling the biosynthesis of amino acids. The best studied case is that involving biosynthesis of the amino acid *tryptophan*. The tryptophan operon contains the structural genes for five proteins of the tryptophan biosynthetic pathway, plus regulatory sequences at the beginning of the operon. In addition to the promoter and operator regions, there is an additional sequence, called the **leader sequence**, within which is a region called the **attenuator**, which codes for a tryptophan-rich peptide. If tryptophan is plentiful in the cell, the leader peptide will be synthesized. On the other hand, if tryptophan is in short supply, the tryptophan-rich leader peptide will not be synthesized. The striking fact is that synthesis of the leader peptide results in *termination* of transcription of the tryptophan struc-

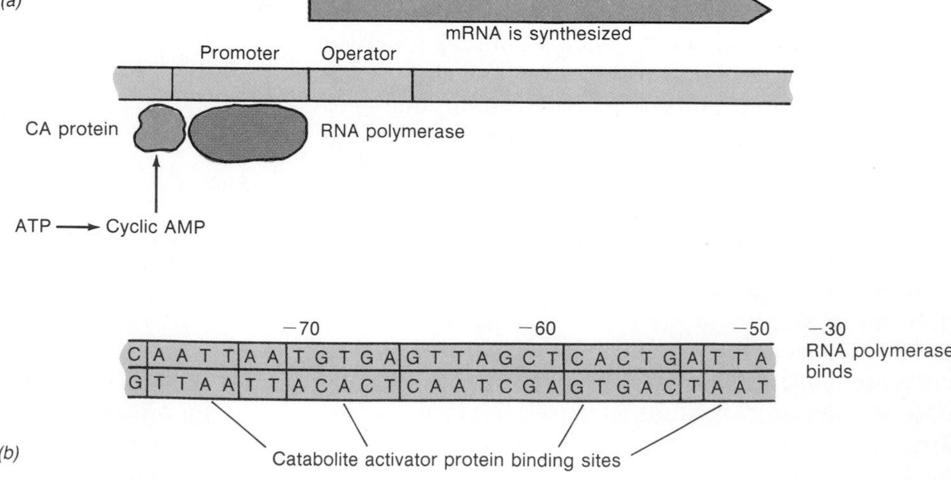

Figure 5.36 Action of cyclic AMP in promoting mRNA synthesis, and the mechanism of glucose (catabolite) repression. (a) When cyclic AMP is present, CAP binds to a site adjacent to the promoter. This stimulates the binding of RNA polymerase. When glucose is present, cyclic AMP is broken down and CAP cannot bind. mRNA synthesis then does not occur. (b) Base sequence of binding site of catabolite activator protein, upstream from the site where RNA polymerase binds. When CAP binds, the binding of RNA polymerase is stimulated.

tural genes, whereas if synthesis of the leader peptide is blocked by tryptophan deficiency, transcription of the tryptophan structural genes can occur.

How does *translation* of the leader peptide regulate *transcription* of the tryptophan genes downstream? Somehow or other, further action of RNA polymerase in transcription is blocked by translation of the leader peptide. One important hypothesis to explain this suggests that transcription and translation must be occurring virtually simultaneously (Figure 5.37). Thus, while transcription of downstream DNA sequences is still proceeding, translation of sequences already transcribed has begun. Apparently, as the mRNA is released from the DNA, the ribosome binds to it and translation begins. Attenuation occurs (RNA polymerase quits working on the mRNA) because a portion of the newly formed mRNA folds into a double-stranded loop which signals cessation of RNA polymerase action (see Figure 5.23). There is a transcription site in the leader called the **transcription pause site**, at which RNA polymerase temporarily slows down if tryptophan is plentiful. In this case, translation of the leader peptide can occur and this slows down transcription sufficiently so that the double-stranded loop can form and signal RNA polymerase to cease functioning. If tryptophan is in short supply, however, translation of the leader peptide cannot occur and the RNA polymerase moves past the termination site and begins transcription of the tryptophan structural genes. Thus we see that there is a highly integrated system in which transcription and translation interact, with the rate of transcription being influenced by the rate of translation. The double-loop structure formed by mRNA is brought about because two stretches of nucleotide bases near each other are complementary.

Thus, in the tryptophan biosynthetic pathway, two distinct mechanisms for the regulation of transcription exist, repression and attenuation. Repression is a mechanism that has large effects on the rate of enzyme synthesis, whereas attenuation brings about a finer control. Working together, these two mechanisms precisely regulate the synthesis of the biosynthetic enzymes, and hence the biosynthesis of tryptophan. Attenuation has also been shown to occur in the biosynthetic pathway for histidine, and for some other amino acids and essential metabolites as well.

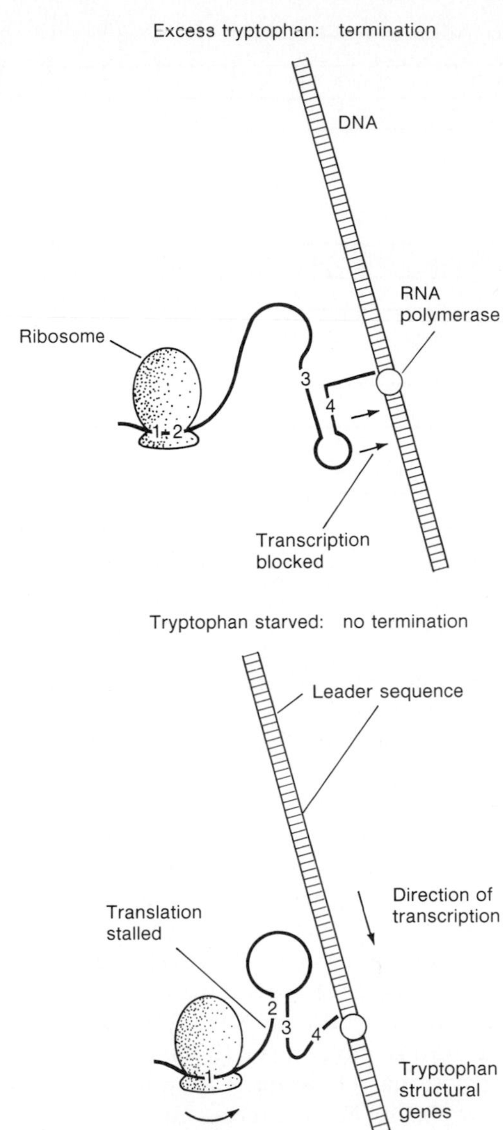

Figure 5.37 Model for attenuation in the *Escherichia coli* tryptophan operon. The leader peptide is coded by regions 1 and 2 of the mRNA. Two regions of the growing mRNA chain are able to form double-stranded loops, 2:3 and 3:4. Under conditions of excess tryptophan, the ribosome translates the complete leader peptide, so that region 2 cannot pair with region 3. Regions 3 and 4 then pair to form a loop which blocks RNA polymerase. If translation is stalled due to tryptophan starvation, loop formation via 2:3 pairing occurs, loop 3:4 does not form, and transcription proceeds past the leader sequence.

5.14 Contrasts Between Gene Expression in Procaryotes and Eucaryotes

The major genetic aspects of gene expression in procaryotes for a particular genetic region, that for the enzyme β-galactosidase (*lac* operon), are summarized in Figure 5.38. Because of the lack of compartmentation in procaryotes, the processes of transcription and translation are *integrated*. Also, the messenger RNA of procaryotes is frequently polycistronic, with more than one protein being translated from the same message.

In eucaryotes, on the other hand, transcription and translation take place in separate parts of the cell and the integration seen in procaryotes is lacking. How about induction and repression in eucaryotes?

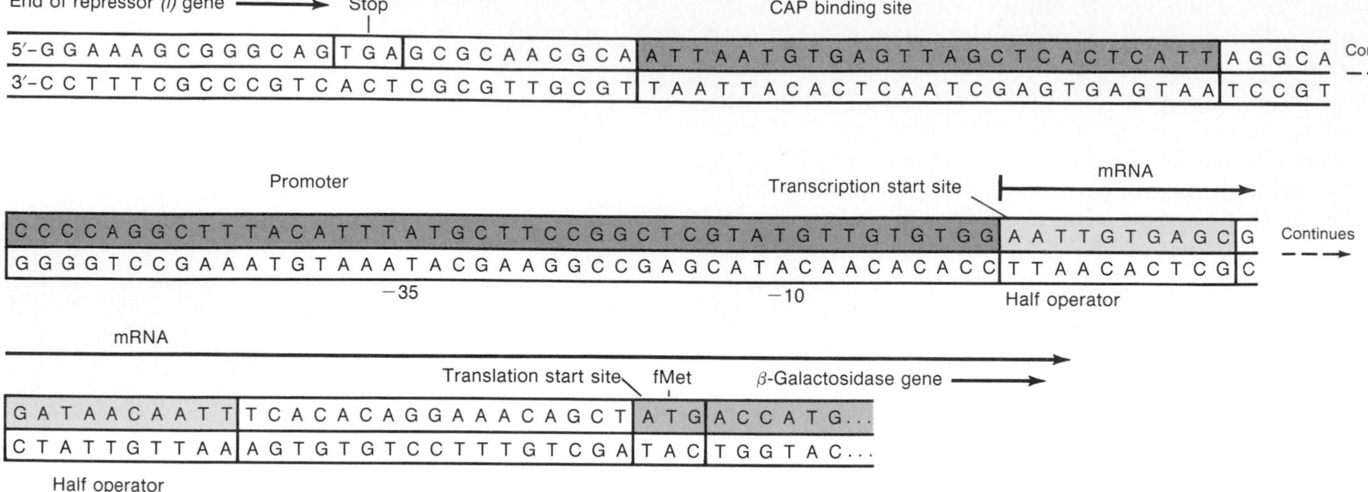

Figure 5.38 The genetic elements involved in regulation of a single enzyme, β-galactosidase.

Although many eucaryotes do respond to repression, there is no good evidence for the kind of negative control so commonly found in procaryotes. However, positive control, which is rare in procaryotes, is common in eucaryotes. If operons exist in eucaryotes, they involve the control of only single enzymes, rather than the multi-enzyme control systems so commonly seen in procaryotes, and there is no evidence for messenger RNA molecules that contain the coding sequences for several enzymes. However, eucaryotes frequently make single large proteins (polyproteins) from a single monocistronic mRNA and these proteins are then cleaved into several smaller active proteins. This posttranslational cleavage, common in eucaryotes, is rare or absent in procaryotes.

Contrasts between repression and feedback inhibition We have presented in this chapter some of the details by which the *synthesis* of enzymes is regulated. It should be clear that if a needed enzyme is not synthesized, then the process which this enzyme catalyzes cannot occur in the cell. Thus regulation of enzyme *synthesis* provides an important mechanism for regulating cell metabolism. However, regulation of enzyme synthesis is a relatively slow mechanism of control, since a period of time is needed after synthesis of an enzyme is brought about before that enzyme is present in the cell in sufficient amounts to affect metabolism.

We have described in Section 4.20 another mechanism of regulation of cell metabolism, called *feedback inhibition*, which affects not enzyme synthesis but enzyme *activity*. In feedback inhibition, one of the metabolites of a pathway inhibits a critical step (frequently an early step) in that pathway. In contrast to induction and repression, feedback inhibition provides an *immediate* mechanism of metabolic control, since it acts on preexisting enzyme. Thus, feedback inhibition can be viewed as a mechanism for finely regulating cell metabolism, whereas induction and repression are coarser mechanisms. Working together, these mechanisms result in an efficient regulation of cell metabolism, so that energy is not wasted by carrying out unnecessary reactions.

5.15 The Genetic Code

As noted, a triplet of three bases encodes a specific amino acid. It is conventional to present the genetic code in the mRNA rather than in the DNA, because it is with the mRNA that the translation process occurs. The 64 possible codons of mRNA are presented in Table 5.4. Note that in addition to the codons specifying the various amino acids, there are also special codons for the start (AUG) and stop (UAA, UAG, UGA) of the message.

Perhaps the most interesting feature of the genetic code is that a single amino acid may be encoded by several different but related base triplets. This means that in most cases there is no one-to-one correspondence between the amino acid and the codon—knowing the amino acid at a given location does not mean that the codon in the DNA at that location is automatically known.* The property of a code in which a one-to-one correspondence between word and code is lacking is called **degeneracy** (a term derived from the technical field of cryptography).

What is the significance of degeneracy? Degen-

*The reverse is true, however. Knowing the DNA codon, one can specify the amino acid in the protein. This permits the determination of amino acid sequences from DNA base sequences.

Table 5.4 The genetic code as expressed by triplet base sequences of mRNA*

Codon	Amino acid	Codon	Amino acid	Codon	Amino acid	Codon	Amino acid
UUU	Phenylalanine	CUU	Leucine	GUU	Valine	AUU	Isoleucine
UUC	Phenylalanine	CUC	Leucine	GUC	Valine	AUC	Isoleucine
UUG	Leucine	CUG	Leucine	GUG	Valine	AUG (start)†	Methionine
UUA	Leucine	CUA	Leucine	GUA	Valine	AUA	Isoleucine
UCU	Serine	CCU	Proline	GCU	Alanine	ACU	Threonine
UCC	Serine	CCC	Proline	GCC	Alanine	ACC	Threonine
UCG	Serine	CCG	Proline	GCG	Alanine	ACG	Threonine
UCA	Serine	CCA	Proline	GCA	Alanine	ACA	Threonine
UGU	Cysteine	CGU	Arginine	GGU	Glycine	AGU	Serine
UGC	Cysteine	CGC	Arginine	GGC	Glycine	AGC	Serine
UGG	Tryptophan	CGG	Arginine	GGG	Glycine	AGG	Arginine
UGA	None (stop signal)	CGA	Arginine	GGA	Glycine	AGA	Arginine
UAU	Tyrosine	CAU	Histidine	GAU	Aspartic	AAU	Asparagine
UAC	Tyrosine	CAC	Histidine	GAC	Aspartic	AAC	Asparagine
UAG	None (stop signal)	CAG	Glutamine	GAG	Glutamic	AAG	Lysine
UAA	None (stop signal)	CAA	Glutamine	GAA	Glutamic	AAA	Lysine

**The codons in DNA are complementary to those given here. Thus U here is complementary to the A in DNA, C is complementary to G, G to C, and A to T. The nucleotide on the left is toward the 5′ end of the mRNA.*

†AUG codes for N*-formylmethionine at the beginning of mRNA.*

eracy means that either (1) a single tRNA molecule can pair with more than one codon or (2) there is more than one tRNA for each amino acid. Actually, both situations are true. For some amino acids there is more than one tRNA molecule (there are, for instance, six tRNA molecules which carry the amino acid leucine). However, there is not a tRNA corresponding to every possible anticodon. In some cases, tRNA molecules form accurate base pairs at only the first two positions of the codon, tolerating some mismatch at the third position. This mismatch phenomenon, called **wobble**, is illustrated in Figure 5.39. (However, even with wobble, no single tRNA molecule can recognize more than three codons.)

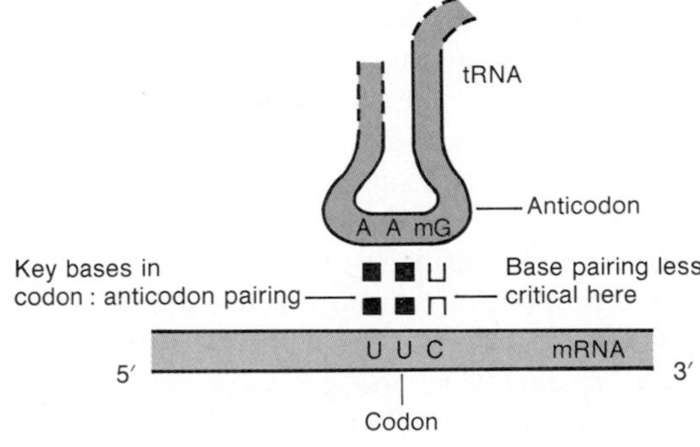

Figure 5.39 The wobble concept: base pairing is less critical for the third base of the codon.

Start and stop codons As seen in Table 5.4, a few triplets do not correspond to any amino acid. These triplets (UAA, UAG, UGA) are called **nonsense codons** or **stop codons** and they function as "punctuation," signalling the termination of translation of the gene coding for a specific protein. If no tRNA molecules correspond to these nonsense codons, no amino acid can be inserted, and the polypeptide is terminated and released from the ribosome.

What is the mechanism by which the proper starting point for translation is found? As we discussed in Section 5.10, there is a **start codon, AUG** which codes for the amino acid *N*-formylmethionine (or in eucaryotes and archaebacteria, methionine). The importance of having a well-defined starting point is readily understood if we consider that with a triplet code it is absolutely essential that translation begin at the correct location, since if it does not, the whole **reading frame** would be shifted and an entirely different protein (or no protein at all) would be formed (see later).

Open reading frames We have shown that a protein is coded by a specific segment of DNA, but that for the protein to be synthesized, the DNA segment must be transcribed into mRNA. The transcription of DNA into mRNA requires the presence, just upstream of the transcription start site, of a promoter. How does an experimenter know that the transcribed mRNA codes for a protein? One approach is to examine the base sequence of the mRNA to see whether there is present, near the beginning of the sequence, a *start codon* such as AUG. If such a start codon is followed by a long base sequence before a *stop codon* is present, then it is likely that this mRNA codes for a protein. Such a base sequence is called an **open reading frame** (because the start codon AUG sets the proper reading frame for translation), and a computer can be programed to scan such long base sequences in DNA data bases to look for open reading frames. The search for open reading frames is very useful in genetic engineering (see Chapter 8) when one has cloned an unknown piece of DNA and is not certain whether or not it codes for protein. The computer search for open reading frames permits the researcher to locate genes that were previously unsuspected to exist. For instance, several stretches of DNA of unknown function in adenovirus (see Section 6.23) were found upon computer analysis to have long open reading frames, and these regions were subsequently shown to be transcribed into mRNA and translated into protein.

In a sense, the term *open reading frame* is equivalent to the term *gene*. If a piece of DNA lacks a detectable open reading frame, it is unlikely that this piece of DNA codes for protein. Actually, in cells, especially in eucaryotic cells, a large amount of noncoding DNA is present.

Translation error Another problem in the translation of the genetic code is that errors sometimes occur. This means that a coding triplet of the mRNA may be "read" improperly and the wrong amino acid inserted. Amino acids whose codons differ by only a single base, for example, phenylalanine (UUU) and leucine (UUA), are most likely to be involved in errors because occasionally only two of the three bases are involved in codon-anticodon recognition because of "wobble" discussed above. In rare instances leucine may be added to the growing polypeptide instead of phenylalanine even when the codon is UUU. In the normal cell these rare errors probably occur in only a small number of all the protein molecules and hence have no detrimental effect. However, certain antibiotics which act on ribosomes, such as streptomycin, neomycin, and related substances (see Table 5.3), increase error to such an extent that many protein molecules in the cell are abnormal and the cell can no longer function properly. Other conditions that increase error are changes in cation concentration, pH, and temperatures that are not optimal for cell growth.

Overlapping genes Although the evidence is strong that the nucleotide sequence specifying one product is separate and distinct from the sequence specifying another product, studies on the small bacterial virus ϕX174 have shown that this virus has insufficient genetic information in nonoverlapping genes to code for all of the proteins necessary for its reproduction, and that genetic economy is introduced by using the same piece of DNA for the coding of more than one product. This phenomenon of overlapping genes is illustrated in Figure 6.21. It is a process made possible by reading of the same nucleotide sequence in two different phases, beginning at different sites.

5.16 Mutants and Their Isolation

A mutation is observed experimentally as a sudden inheritable change in the phenotype of an organism. It is common to refer to a strain isolated from nature as a **wild type**, and strains isolated from the wild type via mutation as **mutants**. We can distinguish between two kinds of mutations, selectable and unselectable. An example of an unselectable mutation is that of loss of color in a pigmented bacterium (Figure 5.40*a*). White colonies have neither an advantage nor a disadvantage over the pigmented parent colonies when grown on agar plates (there may be a selective advantage for pigmented organisms in nature, however). This means that the only way we can detect such mutations is to examine large numbers of colonies and look for the "different" ones.

A selectable mutation, on the other hand, confers upon the mutant an advantage under certain environmental conditions, so that the progeny of the mutant cell are able to outgrow and replace the parent. An example of a selectable mutation is drug resistance: an antibiotic-resistant mutant can grow in the presence of antibiotic concentrations that inhibit or kill the parent (Figure 5.40*b*). It is relatively easy to detect and isolate selectable mutants by choosing the appropriate environmental conditions.

Virtually any characteristic of a microorganism can be changed through mutation. Nutritional mutants can be detected by the technique of **replica plating** (Figure 5.41). Using sterile velveteen cloth, an imprint of colonies from a master plate is made onto an agar plate lacking the nutrient. The colonies of the parental type will grow normally, whereas those of the mutant will not. Thus, the inability of a colony to grow on the replica plate (Figure 5.40*d*) will be a signal that it is a mutant. The colony on the master plate corresponding to the vacant spot on the replica plate (Figure 5.40*e*) can then be picked, purified, and characterized. A nutritional mutant that has a requirement for a growth factor is often called an **auxotroph**, the wild-type parent from which the auxotroph was derived is called a **prototroph**.

An ingenious method which is widely used to isolate mutants that require amino acids or other growth factors is the **penicillin-selection method**. Ordinarily, mutants that require growth factors are at a disadvantage in competition with the parent cells, so that there is no direct way of isolating them. However, penicillin kills only growing cells, so that if

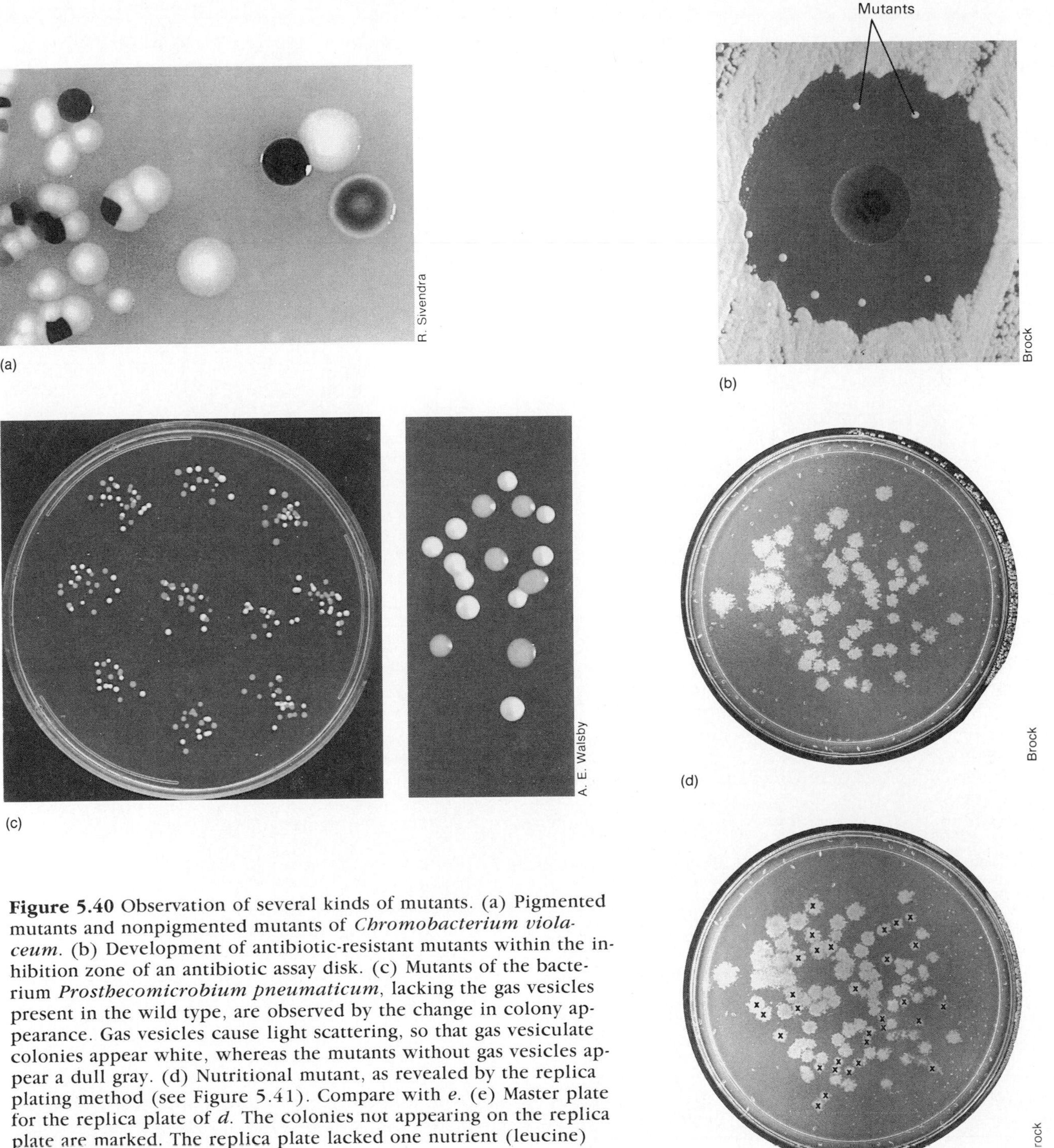

Figure 5.40 Observation of several kinds of mutants. (a) Pigmented mutants and nonpigmented mutants of *Chromobacterium violaceum*. (b) Development of antibiotic-resistant mutants within the inhibition zone of an antibiotic assay disk. (c) Mutants of the bacterium *Prosthecomicrobium pneumaticum*, lacking the gas vesicles present in the wild type, are observed by the change in colony appearance. Gas vesicles cause light scattering, so that gas vesiculate colonies appear white, whereas the mutants without gas vesicles appear a dull gray. (d) Nutritional mutant, as revealed by the replica plating method (see Figure 5.41). Compare with *e*. (e) Master plate for the replica plate of *d*. The colonies not appearing on the replica plate are marked. The replica plate lacked one nutrient (leucine) present in the master plate.

penicillin is added to a population growing in a medium lacking the growth factor required by the desired mutant, the parent cells will be killed, whereas the nongrowing mutant cells will be unaffected. Thus, after preliminary incubation in the absence of growth factor in penicillin-containing medium, the population is washed free of penicillin and transferred to plates containing the growth factor. Among the colonies that grow up (including some wild-type cells that have escaped penicillin killing) should be some representing growth factor mutants.

Some of the most common kinds of mutants and the means by which they are detected are listed in Table 5.5.

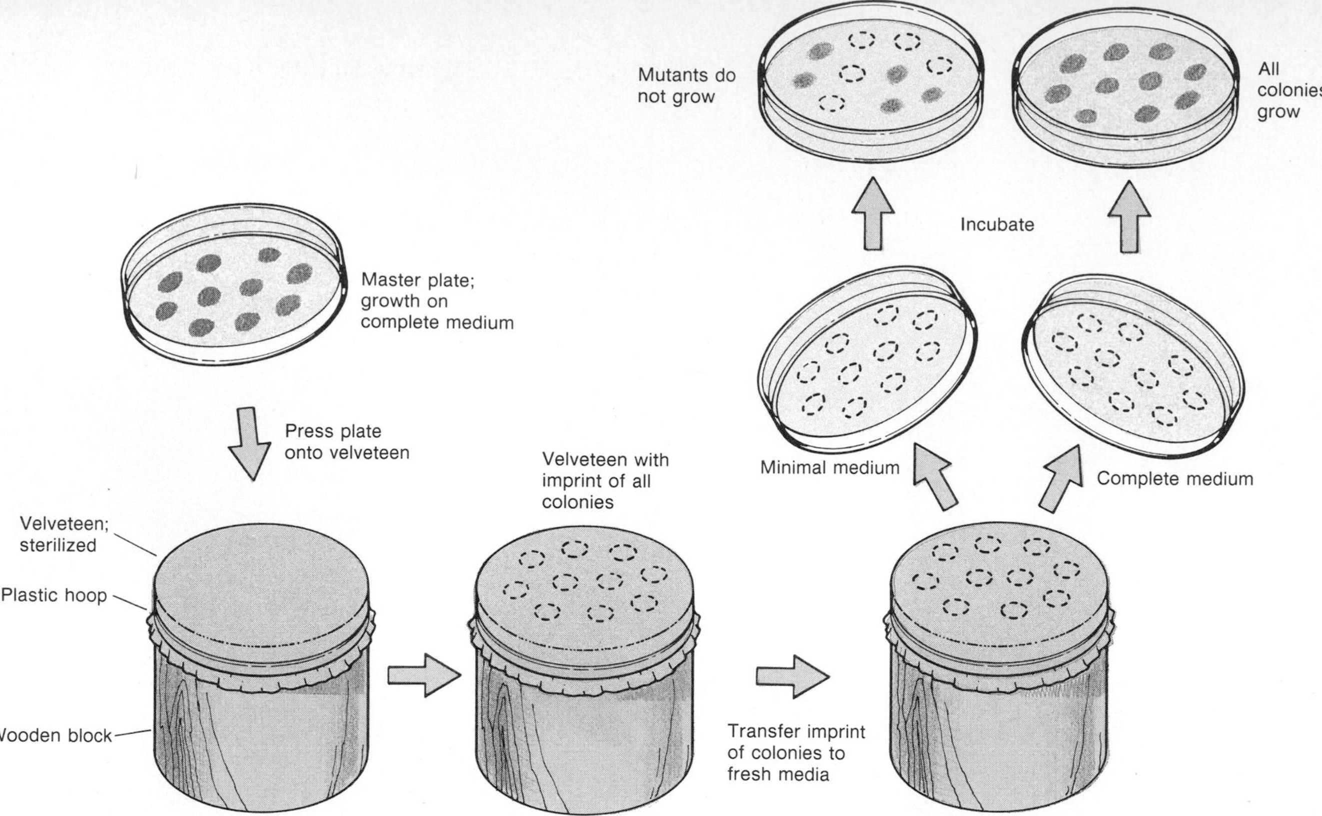

Figure 5.41 Replica-plating method for detecting nutritional mutants.

Table 5.5 Kinds of mutants

Description	Nature of change	Detection of mutant
Nonmotile	Loss of flagella; nonfunctional flagella	Compact colonies instead of flat, spreading colonies
Noncapsulated	Loss or modification of surface capsule	Small, rough colonies instead of larger, smooth colonies
Rough colony	Loss or change in lipopolysaccharide outer layer	Granular, irregular colonies instead of smooth, glistening colonies
Nutritional	Loss of enzyme in biosynthetic pathway	Inability to grow on medium lacking the nutrient
Sugar fermentation	Loss of enzyme in degradative pathway	Do not produce color change on agar containing sugar and a pH indicator
Drug resistant	Impermeability to drug or drug target is altered or drug is detoxified	Growth on medium containing a growth-inhibitory concentration of the drug
Virus resistant	Loss of virus receptor	Growth in presence of large amounts of virus
Temperature sensitive	Alteration of any essential protein so that it is more heat sensitive	Inability to grow at a temperature normally supporting growth (e.g., 37°C) but still growing at a lower temperature (e.g., 25°C)
Cold sensitive	Alteration in an essential protein so that it is inactivated at low temperature	Inability to grow at a low temperature (e.g., 20°C) that normally supports growth

5.17 The Molecular Basis of Mutation

Mutations arise because of changes in the *base sequence* in the DNA. In most cases, mutations that occur in the base sequence of the DNA lead to changes in the organism; these changes are mostly harmful, although beneficial changes do occur occasionally.

Mutation can be either spontaneous or induced. **Spontaneous mutations** can occur as a result of the action of natural radiation (cosmic rays, etc.) which alter the structure of bases in the DNA. Spontaneous mutations can also occur during replication, as a result of errors in the pairing of bases, leading to changes in the replicated DNA. In fact, spontaneous mutations in any specific gene occur about once in every 10^6 to 10^{10} replications. Thus in a normal, fully grown culture of organisms having approximately 10^8 cells/ml, there will probably be a number of different mutants in each milliliter of culture.

How Was the Genetic Code Cracked?

During the years 1953–1961 the Cold War between the United States and the Soviet Union intensified, the first Russian satellite was sent into orbit, and Egypt took over the Suez Canal, beginning a destabilization of the Middle East that has lasted until this day. Years of displeasure? Not in molecular biology! During the years 1953–1961, the Watson-Crick structure for DNA was announced, ribosomes and transfer RNA were discovered and their role in protein synthesis demonstrated, the colinearity of gene and protein was proved, the operon concept was developed, and the genetic code was cracked! Exciting and revolutionary years for molecular biology.

Although all of these discoveries were seminal for the advances in molecular biology which we know today, the one development which came as the biggest surprise was the cracking of the genetic code. Everyone working in the field of molecular genetics had thought that even if the mechanism of protein synthesis was known in some detail, it would take many years to crack the genetic code. Yet, the code was cracked quickly, and surprisingly easily, once a key breakthrough had been made. This key breakthrough was the discovery that synthetic polynucleotides containing only *single* code words would bring about the synthesis in cell-free ribosome-tRNA-containing systems of proteins with *single* amino acids. For instance, a polynucleotide containing only uracil (poly U) brought about the synthesis of polyphenylalanine. The conclusion was therefore made that one of the code words for phenylalanine was UUU. If adenine (poly A) was used, the protein that was synthesized was polylysine (AAA thus being a code word for lysine) (see Table 5.4 for the genetic code).

This simple and ingenious experiment using synthetic polynucleotides, done by Marshall Nirenberg and co-workers at the National Institutes of Health, was soon elaborated and extended (with modifications) to other polynucleotides. Gobind Khorana and co-workers developed chemical methods for synthesizing various polynucleotides. By 1965, the complete genetic code was known. What had been thought to be a job of generations was accomplished in just a few years. Although in retrospect it might seem simple and obvious to use synthetic polynucleotides to crack the genetic code, at the time that Nirenberg and colleagues did it, the experiment was anything but obvious. In fact, it was a stroke of genius, an accomplishment that was quickly recognized by the awarding of a Nobel Prize to Nirenberg and Khorana.

Why was it so important that the genetic code be cracked? At first, knowing the genetic code was not much more than a curiosity, but with the advent of DNA sequencing methods, recombinant DNA technology and synthetic DNA, knowledge of the code could soon be applied to a vast array of important problems in molecular genetics. Knowing the code makes possible the construction of *synthetic genes* that do completely new things. It is even possible to work backwards, deducing the amino acid sequence of a protein from a knowledge of the nucleotide sequence of a gene. Such is the power of modern DNA technology that it is actually easier to figure out the amino acid sequence of a protein in this indirect way than directly from a chemical analysis of the protein itself.

It is a common thing in science that great leaps forward are made in unexpected ways. Cracking the genetic code was one of those great leaps that has provided us with the marvelous sophisticated knowledge that we have today.

Point mutations Point mutations are changes of single bases; an adenine base may be replaced by a guanine or thymine, for instance. What is the result of a point mutation? First, the error in the DNA is transcribed into mRNA, and this erroneous mRNA in turn is used as a template and translated into protein. The triplet code that directs the insertion of an amino acid via a transfer RNA will thus be incorrect. What are the consequences?

In interpreting the results of mutation, we must first recall that the code is degenerate (see Section 5.15). Because of degeneracy, not all mutations result in changes in protein. This is illustrated in Figure 5.42, which shows several possible results when a single *tyrosine codon* undergoes mutation. As seen, a point mutation from UAC to UAU would have no apparent effect, since UAU is an additional tyrosine codon. Such mutations are called **silent mutations**. Note that silent mutations always occur in the third base of the codon. As seen in Table 5.4, changes in the first or second base of the triplet much more often lead to significant changes in the protein. For instance, a single base change from UAC to CAC (Figure 5.42) would result in a change in the protein from tyrosine to histidine. If the change occurred at a critical point in the polypeptide chain, the protein could be inactive, or of reduced activity. Another possible result of a point mutation would be the formation of

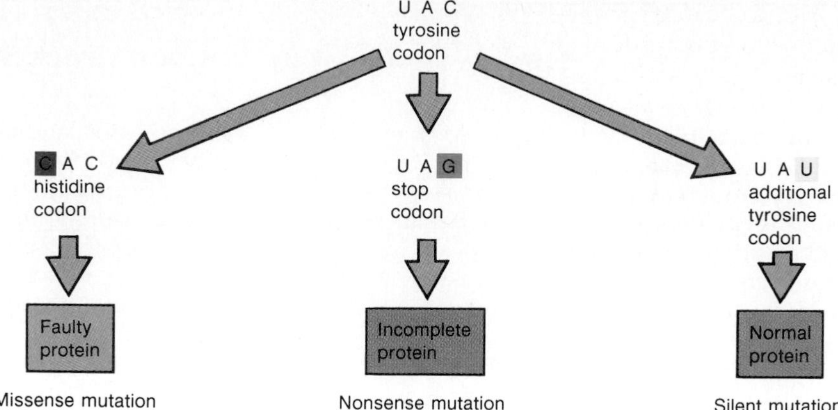

Figure 5.42 Possible effects of a point mutation: three different protein products from a single base change.

a *stop codon*, which would result in premature termination of translation, leading to an incomplete protein which would almost certainly not be functional (Figure 5.42). Mutations of this type are called **nonsense mutations** because the change is to a nonsense codon (see Section 5.15).

Thus, not all mutations that cause amino acid substitution necessarily lead to nonfunctional proteins. The outcome depends greatly on where in the polypeptide chain the substitution has occurred, and on how it affects the folding and the catalytic activity of the protein.

Deletions and insertions **Deletions** are due to elimination of portions of the DNA of a gene. A deletion may be as simple as the removal of a single base, or it may involve hundreds of bases. Deletion of a large segment of the DNA results in complete loss of the ability to produce the protein. Such deletions cannot be restored through further mutations, but only through genetic recombination. Indeed, one way in which large deletions are distinguished from point mutations is that the latter are usually reversible through further mutations, whereas the former usually are not.

Insertions occur when new bases are added to the DNA of the gene. Insertions can involve only a single base or many bases. Generally, insertions do not occur by simple copy errors as do deletions, but arise as a result of mistakes that occur during genetic recombination (discussed in Chapter 7). Many insertion mutations are due to the insertion of specific identifiable DNA sequences 700 to 1400 base pairs in length called **insertion sequences** or **insertion elements**. The behavior of such insertion sequences was described in Section 5.5 and is discussed in detail in Chapters 6, 7, and 8.

Frame-shift mutations Since the genetic code is read from one end in consecutive blocks of three bases, any deletion or insertion of a new base results in a **reading-frame shift**, and the translation of the gene is completely upset (Figure 5.43). Partial restoration of gene function can often be accomplished by insertion of another base near the one deleted. After correction, depending on the exact amino acids coded by the still faulty region and the region of the protein involved, the protein formed may have some biological activity or even be completely normal.

Polarity As we have discussed, a cluster of genes may be transcribed together in a single mRNA. One consequence of the fact that translation occurs from one end is that nonsense mutations near the beginning of translation (the 5′ end) will completely block the translation of all the successive genes, whereas nonsense mutations farther down will have fewer effects. This phenomenon is called **polarity**. Polarity gradients arise because when a ribosome reaches a chain-terminating nonsense codon, an incomplete polypeptide is released; the ribosome generally detaches from its mRNA template, so that the following genes are not translated. Because of polarity, nonsense mutations at the 3′ end will generally have no effect on genes transcribed at the 5′ end.

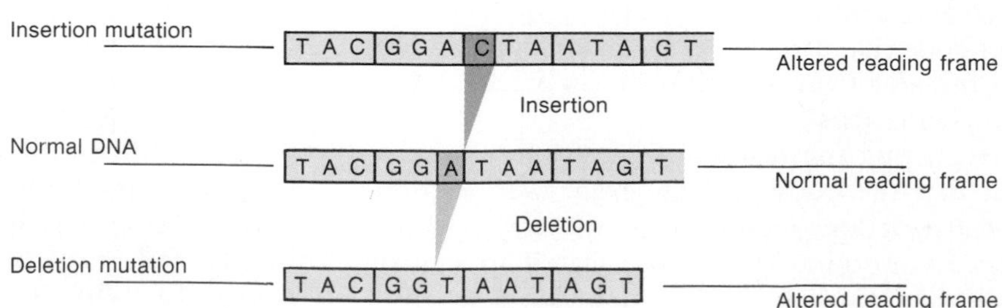

Figure 5.43 Shifts in reading frame caused by insertion or deletion mutations.

Back mutations or reversions Many but not all mutations are reversible. A "revertant" is operationally defined as a strain in which the wild-type phenotype that was lost in the mutant is restored. Revertants can be of two types. In first-site revertants, the mutation that restores activity occurs at the same site at which the original mutation occurred. In second-site revertants, the mutation occurs at some different site in the DNA. Several mechanisms for second-site mutations are known: (1) a mutation somewhere else in the same gene can restore enzyme function, such as in a reading-frame shift mutation; (2) a mutation in another gene may restore the wild-type phenotype (*suppressor mutation*, see below); and (3) the mutation may result in the production of another enzyme that can replace the mutant one by introducing a metabolic pathway different from that used by the mutant enzyme. In this last type no production of the original enzyme occurs although it does in the other types.

Suppressor mutations are new mutations that suppress the original one indirectly. One way in which a suppressor mutation can arise is through an alteration in a transfer RNA anticodon, so that it recognizes the wrong (but right!) mutant codon. In this way, the proper amino acid can be added to the polypeptide chain, at least some of the time. It is usually characteristic of suppressor mutations of this type that normal function is not completely restored; only some of the polypeptides formed on a single mRNA are normal. Therefore, revertants arising due to suppressor mutations often do not grow as rapidly as the wild type. They can also be distinguished from true revertants by genetic mapping, since the locus of the suppressor gene is in a different region of the genome from that of the original mutant gene.

Rates of mutation There are wide variations in the rates at which various kinds of mutations occur. Some types of mutations occur so very rarely that they are almost impossible to detect whereas others occur so frequently that they often present difficulties for the experimenter who is trying to maintain a genetically stable stock culture.

Spontaneous mutations occur at frequencies around 10^{-5} per generation. This means that there is one chance in 100,000 that a mutation will arise at some location of a given gene during one cell cycle. Transposition events may occur more frequently, around 10^{-4}. On the other hand, the occurrence of a nonsense mutation is much less frequent, 10^{-6}–10^{-8}, because only a few codons can mutate to nonsense codons. Missense codons also occur at about the same frequency as nonsense mutations. The experimental detection of events of such rarity is obviously difficult and much of the skill of the microbial geneticist is applied to increasing the efficiency of mutation detection. As we will see in the next section, it is possible to significantly increase the rate of mutation by the use of mutagenic treatments.

5.18 Mutagens

It is now well established that a wide variety of chemicals can induce mutations. An overview of some of the major chemical mutagens and their modes of action are given in Table 5.6. Several classes of chemical mutagens occur. A variety of chemical mutagens are **base analogs**, resembling DNA purine and pyrimidine bases in structure, yet showing faulty pairing properties. When one of these base analogs is incorporated into DNA, replication may occur normally most of the time, but occasional copying errors occur, resulting in the incorporation of the wrong base into the copied strand. During subsequent segregation of this strand, the mutation is revealed.

A variety of chemicals react directly on DNA, causing changes in one or another base, which results in faulty pairing or other changes (Table 5.6). Such chemicals differ in their action from the base analogs in that the chemicals reacting on DNA are able to introduce direct changes even in nonreplicating DNA, whereas the base analogs only act after incorporation during replication. One interesting group of chemicals, the acridines, are planar molecules which become inserted between two DNA base pairs, thereby pushing them apart. During replication, an extra base can then be inserted in acridine-treated DNA, lengthening the DNA by one base and thus shifting the reading frame.

Radiation Several forms of radiation are highly mutagenic. We can divide mutagenic radiation into two main categories, ionizing and nonionizing. Although both kinds of radiation are used in microbial genetics, nonionizing radiations find the widest use and will be discussed first.

The most important type of nonionizing radiation which is mutagenic is **ultraviolet radiation, UV**. The UV radiation of most interest genetically is short wavelength UV, around 260 nm, because the purine and pyrimidine bases of the nucleic acids absorb UV radiation of this wavelength strongly. One particular effect of UV radiation is the formation of thymine dimers, a state in which two adjacent thymine molecules are covalently joined, so that proper replication of the DNA cannot occur.

Many microorganisms have repair enzymes which correct damage such as thymine dimers induced in DNA by UV radiation. Damage in a given region of the DNA molecule is usually only on one of the two strands, and enzymes exist which will hydrolyze the phosphodiester bonds at each end of the damaged region and thus excise the altered material. Other enzymes then catalyze the insertion of new nucleo-

Table 5.6 Chemical mutagens and their modes of action

Agent	Action	Result
Base analogs:		
5-Bromouracil	Incorporated like T; occasional faulty pairing with G	A–T pair → G–C pair Occasionally G–C → A–T
2-Aminopurine	Incorporated like A; faulty pairing with C	A–T → G–C Occasionally G–C → A–T
Chemicals reacting with DNA:		
Nitrous acid (HNO_2)	Deaminates A,C	A–T → G–C G–C → A–T
Hydroxylamine (NH_2OH)	Reacts with C	G–C → A–T
Alkylating agents:		
Monofunctional (e.g., ethyl methane sulfonate)	Put methyl on G; faulty pairing with T	G–C → A–T
Bifunctional (e.g., nitrogen mustards, mitomycin, nitrosoguanidine)	Cross-link DNA strands; faulty region excised by DNase	Both point mutations and deletions
Intercalative dyes (e.g., acridines, ethidium bromide)	Insert between two base pairs	Reading-frame shift
Radiation:		
Ultraviolet	Pyrimidine dimer formation	Repair may lead to error or deletion
Ionizing radiation (e.g., X rays)	Free-radical attack on DNA, breaking chain	Repair may lead to error or deletion

tides that are complementary to the undamaged strand. One kind of repair enzyme works only when the damaged cell is activated by visible light; recovery of UV radiation damage through the action of this system is called *photoreactivation*. Other repair enzymes work in the absence of light (*dark reactivation*).

The type of UV radiation source most frequently used in mutation work is the germicidal lamp, which emits large amounts of UV radiation in the 260 nm region. One property of UV radiation is that it penetrates poorly through glass and other materials. It even penetrates poorly through microbial cultures, so that when UV radiation is used to induce mutation in microbial cultures, it is essential that the culture density be kept very low. A dose of UV radiation is used which brings about 90–95 percent killing of the cell population, and mutants are then looked for in the survivors. If much higher doses of radiation are used, the density of viable cells will be too low, whereas if lower doses are used, insufficient damage to the DNA will have been induced. UV radiation is a very useful tool in isolating mutants of microbial cultures.

Ionizing radiation includes short wavelength rays such as X rays, cosmic rays, and gamma rays. These radiations cause water and other substances to ionize, and mutagenic effects are brought about indirectly through this ionization. Among the potent chemical species formed by ionizing radiation are chemical free radicals, of which the most important is the hydroxyl radical, OH·. Free radicals react with and inactivate macromolecules in the cell, of which the most important is DNA. DNA is probably no more sensitive to ionizing radiation than other macromolecules, but since each DNA molecule contains only one copy of most genes, inactivation can lead to a permanent effect. At low doses of ionizing radiation, only a few hits on DNA occur, but at higher doses multiple hits occur, leading to the death of the cell. In contrast to UV radiation, ionizing radiation penetrates readily through glass and other materials. Because of this, ionizing radiation is used frequently to induce mutations in animals and plants (where its penetrating power makes it possible to reach the germ cells of these organisms readily), but because ionizing radiation is more dangerous to use and is less readily available, it finds less use with microorganisms (where penetration with UV is not a problem).

Mutations that arise from the SOS response Many kinds of mutations arise as a result of faulty repair of damage induced in DNA by various agents. Conditions that cause damage to DNA include ultraviolet radiation, ionizing radiation, thymine starvation, alkylating agents (such as ethyl methane sulfonate and nitrosoguanidine), and certain antibiotics (nalidixic acid, mitomycin C). A complex cellular mechanism, called the **SOS regulatory system**, is activated as a result of DNA damage, initiating a number of DNA repair processes. In the SOS system, DNA repair occurs in the absence of template instruction, resulting in the creation of many errors, hence many mutations.

In the SOS regulatory system, DNA damage serves as a distress signal to the cell, resulting in the coordinate derepression (activation) of a number of cellular functions involved in DNA repair. The SOS system is normally repressed by the LexA protein, but LexA is inactivated by RecA, a protease that is activated as a result of DNA damage (Figure 5.44). Since the DNA repair mechanisms of the SOS system are inherently error-prone, many mutations arise. Thus, through the SOS regulatory system, DNA damage by various agents such as chemicals and radiation leads to mutagenesis.

The SOS system senses the presence in the cell of DNA damage and the repair mechanisms are activated. But once the DNA damage has been repaired, the SOS system is switched off, and further mutagenesis ceases. In addition to its effect on cellular mutagenesis, the SOS regulatory system plays a central role in the regulation of temperate virus replication, as will be discussed in Section 6.14.

It should be emphasized that not all DNA repair occurs in the absence of template instruction. Cells generally have a constitutive DNA repair system which does require template instruction and does lead to proper DNA repair. This system apparently works most of the time, but is not sufficient to repair the large amounts of damage done by some of the agents mentioned above.

Site-directed mutagenesis So far, the mutations that we have been discussing have been randomly directed at the genome of the microbial cell. Recombinant DNA technology and the use of synthetic DNA make it possible to induce *specific* mutations in *specific* genes. The procedures for carrying out mutagenesis of specific sites in the genome are called site-directed mutagenesis, and will be discussed in detail in Chapter 8. Here we briefly indicate the overall principle of site-directed mutagenesis.

If the DNA coding for a specific gene has been isolated and its sequence determined, it is then possible to construct a modified form of this gene in which a specific base or series of bases has been changed. Imagine a single strand of this DNA covering the region of interest. A piece of synthetic DNA, an oligonucleotide 20–30 nucleotides long, which is complementary to this gene but which differs by the one base of interest, is allowed to hybridize with the single strand. Using DNA ligase and polymerase, the rest of this strand can now be completed, resulting in a double strand, one strand of which contains the sequence of interest. Such a double-stranded DNA can now be inserted into a recipient cell (using procedures described in Chapter 8), and *mutants* selected. Such mutants will then differ by the desired change, and will represent specific site-directed mutations. Further details of site-directed mutagenesis will be presented in Section 8.9.

5.19 Mutagenesis and Carcinogenesis

The sensitivity by which selectable mutations can be detected in bacterial populations is very high, because large populations of cells can be conveniently handled. Thus, even though mutations are rare events, they can be studied and quantified in bacteria. One striking observation that has been made is that many mutagenic chemicals are also carcinogenic, causing cancer in higher animals or humans. Although a discussion of cancer is beyond the scope of this section, in the present context the key point is that there seems to be a strong correlation between the mutagenicity of many chemicals and their carcinogenicity. This has led to the development of bacterial mutagenesis tests as screening procedures for potential carcinogenicity of various compounds. The variety of chemicals, both natural and artificial, which the human population comes into contact with through industrial exposure is enormous. There is considerable need for simple

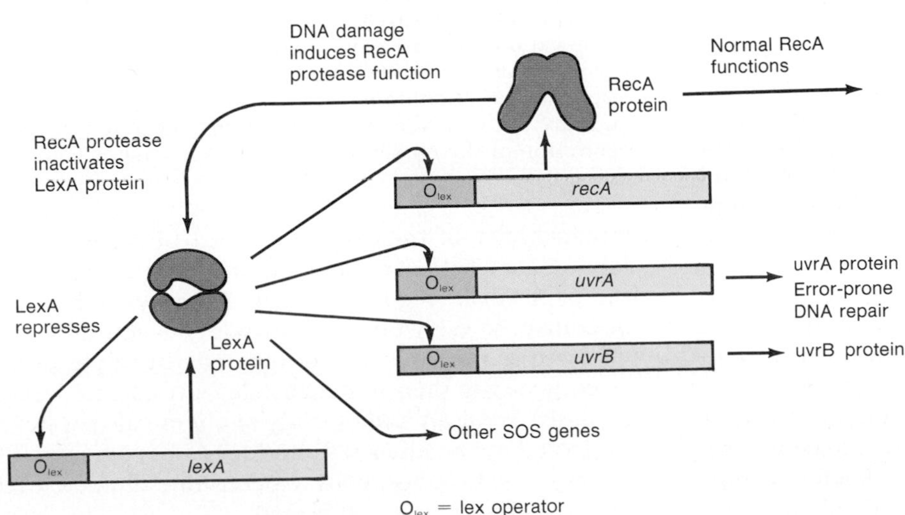

Figure 5.44 Mechanism of the SOS response. DNA damage results in conversion of RecA protein into a protease which cleaves LexA protein. LexA protein normally represses the activities of the recA gene and the DNA repair genes uvrA and uvrB. With LexA inactivated, these genes become active.

tests to ascertain the safety of such compounds. There is good evidence that a large proportion of human cancers have environmental causes, most likely through the agency of various chemicals, making the detection of chemical carcinogens urgent. It does not necessarily follow that because a compound is mutagenic, it will be carcinogenic. The correlation, however, is quite high, and the knowledge that a compound is mutagenic in a bacterial system serves as a warning of possible danger. Similarly, the fact that a compound is not mutagenic in a bacterial system does not mean that it is not carcinogenic, since the bacterial system cannot detect all compounds active in higher animals. The development of bacterial tests for carcinogenic screening has been carried out primarily by a group at the University of California in Berkeley, under the direction of Bruce Ames, and the mutagenicity test for carcinogens is sometimes called the *Ames test* (Figure 5.45).

The standard way to test chemicals for mutagenesis has been to measure the rate of *back* mutations in strains of bacteria that are auxotrophic for some nutrient. It is important, of course, that the original mutation be a point mutation, so that reversion can occur. When such an auxotrophic strain is spread on a medium lacking the required nutrient (for example, an amino acid or vitamin), no growth will occur, and even very large populations of cells can be spread on the plate without formation of visible colonies. However, if back mutants are present, those cells will be able to form colonies. Thus if 10^8 cells are spread on the surface of a single plate, even as few as 10 to 20 back mutants (revertants) can be detected by the 10 to 20 colonies they will form.

Although the simple testing of chemicals for mutagenesis in bacteria has been carried out for a long time, two elements have been introduced in the Ames test to make it a much more powerful test. The first of these is the use of strains of bacteria lacking DNA repair enzymes, so that any damage that might be induced in DNA is not corrected. The second important element in the Ames test is the use of liver enzyme preparations to convert the chemicals into their active mutagenic (and carcinogenic) forms. It has been well established that many potent carcinogens are not directly carcinogenic or mutagenic, but undergo chemical changes in the human body, which convert them into active substances. These changes take place primarily in the liver, where enzymes (mixed-function oxygenases) normally involved in detoxification cause formation of epoxides or other activated forms of the compounds, which are then highly reactive with DNA. In the Ames test, a preparation of enzymes from rat liver is first used to activate the compound. Next the activated complex is taken up on a filter-paper disk, which is placed in the center of a plate on which the proper bacterial strain has been overlayed. After overnight incubation, the mutagenicity of the compound can be detected by looking for a halo of back mutations in the area around the paper disk (Figure 5.45). It is always necessary, of course, to carry out this test with several different concentrations of the compound, because compounds vary in their mutagenic activity and are lethal at higher levels. A wide variety of chemicals has been subjected to the Ames test, and it has become one of the most useful prescreens to determine the potential carcinogenicity of a compound.

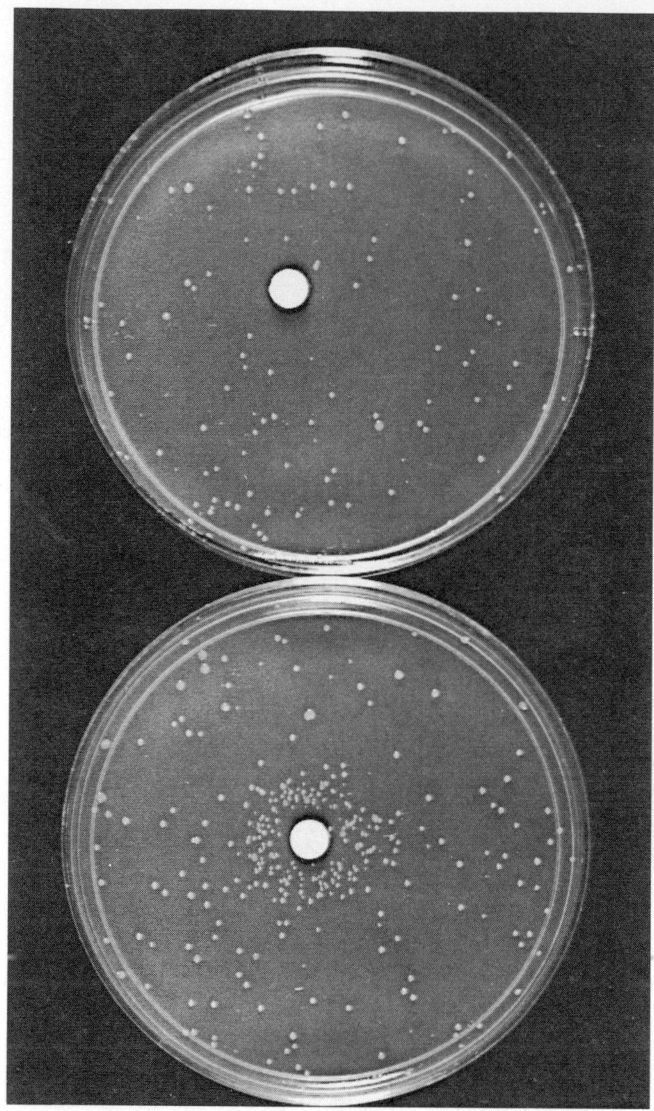

Figure 5.45 The Ames test is used to evaluate the mutagenicity of a chemical. Both plates were inoculated with a culture of a histidine-requiring mutant of *Salmonella typhimurium*. The medium does not contain histidine, so that only cells that revert back to wild type can grow. Spontaneous revertants appear on both plates, but the chemical on the filter paper disc in the test plate has caused an increase in the mutation rate, as shown by the large number of colonies surrounding the disk. Revertants are not seen very close to the disk because the concentration of the mutagen is so high there that it is lethal.

Study Questions

1. A gene can be defined in both functional (informational) and chemical terms. Write a one-sentence definition for each of these qualities. What problem do you have with your chemical definition of a gene when you consider the phenomenon of split genes?
2. DNA molecules which are A-T rich separate into two strands more easily when the temperature is raised than do DNA molecules which are G-C rich. Write an explanation for this observation based on the properties of A-T and G-C base pairing.
3. From a biochemical point of view, why is it essential that DNA replication occurs in the direction $5' \rightarrow 3'$? What problem does this pose for replication of the two separate strands of DNA?
4. Describe the mechanism by which the two enzymes, DNA polymerase and DNA ligase, function together to effect DNA synthesis.
5. The DNA molecule in *Escherichia coli* is 1200 μm long whereas the *E. coli* cell is only 2 μm long. Using your knowledge of DNA folding properties, explain how *partitioning* of the replicated DNA of the *E. coli* cell occurs during the cell division process.
6. What are restriction enzymes? What is the prime function of a restriction enzyme in the cell which produces it (that is, why do cells have restriction enzymes)? How does a restriction enzyme differ from a DNase? How is it that the restriction enzyme in a cell does not cause degradation of that cell's DNA?
7. Nucleic acid hybridization is at the basis of modern genetic techniques. Write a short explanation for each of the following statements:
   a. The strength of the DNA hybrid is greater if two long than if two short DNA molecules are involved.
   b. Even if the base sequences are not *exactly* complementary, hybridization can still occur between two relatively long DNA molecules, whereas this is less likely to occur if the molecules are short.
   c. Suppose you had available a pure mRNA and were given a pure double-stranded DNA containing the gene which coded for this mRNA. How could you use hybridization to isolate the DNA strand against which the mRNA had been made? (Assume you have some way of immobilizing the mRNA.)
8. We have indicated three ways in which eucaryotic mRNA differs from procaryotic mRNA. Write a short description of each of these three ways.
9. Describe one way in which mitochondria and chloroplasts are similar to bacterial plasmids and one way in which they differ, from a genetic point of view.
10. What would be the consequence (in terms of both mRNA and protein synthesis) if the *promoter* region for the gene coding for the enzyme β-galactosidase was deleted from the DNA? If the *base sequence* of the promoter were changed so that the binding of RNA polymerase was weaker?
11. From your understanding of how the antibiotic *actinomycin* acts, draw a graph to indicate what would happen to β-galactosidase synthesis if *actinomycin* were added just before the inducer lactose. If the *actinomycin* were added a short while after the inducer were added?
12. What would be the consequence (in terms of protein synthesis) if the structure of the *anticodon* region of a tRNA were altered so that it no longer recognized the mRNA codon? This question would have a different answer if the coding for that particular amino acid was degenerate. Why?
13. One type of *suppressor mutation* involves a change in tRNA. Draw a diagram with coding sequences and amino acid sequences which indicates how this occurs.
14. Explain why a *nonsense* codon functions as a stopping point for protein synthesis. Why doesn't a nonsense codon also function as a stopping point for mRNA synthesis?
15. Many bacterial mRNA molecules are polycistronic, each mRNA coding for more than one protein. Imagine an mRNA which codes for two proteins with an intervening noncoding region between the two coding regions. From your understanding of how the translation process works, explain why the end result would be two separate proteins rather than one mixed (hybrid) protein.
16. The start and stop sites for mRNA synthesis (on the DNA) are different than the start and stop sites for protein synthesis (on the mRNA). Explain.
17. Imagine you had isolated a new antibiotic which inhibited the translocation process in protein synthesis. Explain what would happen, in terms of mRNA synthesis, polypeptide synthesis, and cell growth, when such an antibiotic was added to a growing culture.
18. Streptomycin is known to affect the fidelity of the codon recognition process, so that wrong amino acids are occasionally inserted into the growing polypeptide chain. Would you expect this type of inhibition to result in bacteriostatic or bactericidal action? Why?
19. A constitutive mutant is a strain which makes continuously a protein which in the wild type is inducible. Describe two ways in which a change in a DNA molecule could lead to the development of a constitutive mutant. How could these two types of constitutive mutants be distinguished genetically?
20. What is diauxic growth? Why is there a lag in growth after glucose is exhausted before growth on lactose commences (what is happening during the lag)?
21. Overlapping genes make it possible for a small piece of DNA to code for more than one protein. What would be the result if a small deletion occurred in the middle of a DNA molecule that was involved in an overlapping gene? What kinds of results would occur if a point mutation occurred?
22. What would be the result (in terms of protein synthesis) if RNA polymerase initiated *transcription* 1 base upstream of its normal starting point? Why?

What would be the result (in terms of protein synthesis) if *translation* began 1 base downstream of its normal starting point? Why?

23. Although a large number of mutagenic agents (chemicals) are known, no chemical is known which induces mutation for a single gene only (gene-specific mutagenesis). From what you know about mutagens, explain why it is unlikely that a gene-specific mutagen could be found. Site-specific mutation is possible. Why isn't a site-specific mutagen also a gene-specific mutagen?
24. Explain how it is possible for a frame-shift mutation early in a gene to be corrected by another frame-shift mutation farther along the gene.
25. Two methods for determining the sequence of a DNA molecule are the Maxam-Gilbert and the Sanger methods. In what ways do these two methods differ? In what ways are they similar?
26. Suppose you were going to determine the DNA sequence of a gene of 1000 nucleotides. This is too long for direct determination by the Maxam-Gilbert or Sanger method. Describe a procedure you could use that would make it possible to determine the sequence of this gene in stages. In completing your answer, write a diagram that lists in sequence the various steps you would use.
27. What is a topoisomerase and what is its function? Contrast the actions of topoisomerase I and II.
28. Describe in chemical terms the manner by which a protein such as RNA polymerase or *lac* repressor combines in a *sequence-specific* manner with the DNA. Explain how symmetry in the DNA relates to the symmetry of the protein molecule.
29. Digestion of a short piece of DNA 15 nucleotides long with the restriction enzyme TaqI resulted in fragments of 4, 6, and 8 nucleotides. When the same DNA was digested with the restriction enzyme SalI, fragments of 3 and 12 were obtained. Digestion first with TaqI followed by SalI resulted in fragments of 3, 3, 4, and 5, whereas digestion of the SalI digest with TaqI resulted in no change (the fragments remained 3 and 12). From these data, write a restriction enzyme map of the whole 15 nucleotide fragment.
30. A classic experiment to show that DNA replication was semiconservative was that of Mathew Meselson and Franklin Stahl. These workers used the heavy isotope of nitrogen, N^{15}, to label DNA during the replication process. DNA labeled with N^{15} can be separated from regular DNA by ultracentrifugation. In the Meselson-Stahl experiment, cells whose DNA was fully labeled with N^{15} were transferred to a medium containing regular (light) N^{14}. As replication proceeded, DNA containing light nitrogen was obtained and could be separated in the ultracentrifuge. Three kinds of molecules can be anticipated: both strands heavy (initial parent), half heavy–half light, and both strands light. Describe the anticipated result of this experiment after the first round of replication, if the replication process were conservative or semiconservative. Label each of the two strands obtained as to whether it would be heavy or light. Describe the anticipated results after a further round of replication.
31. A structure commonly seen in circular DNA during replication is called a *theta structure*. Draw a diagram of the replication process and show how a theta structure could arise.

Supplementary Readings

Alberts, Bruce, Dennis Bray, Julian Lewis, Martin Raff, Keith Roberts, and **James D. Watson**. 1983. *Molecular Biology of the Cell*. This book has become a standard reference book for cellular molecular biology. Although the emphasis on eucaryotes is strong, procaryotes are not slighted.

Cech, Thomas R. 1987. The chemistry of self-splicing RNA and RNA enzymes. *Science* 236: 1532-1539. A review of the role of RNA as a catalyst.

Darnell, James, Harvey Lodish, and **David Baltimore**. 1986. *Molecular Cell Biology*. Scientific American Books, New York. One of the "big" cell biology textbooks, with strong emphasis on macromolecular processes in eucaryotes.

Freifelder, David. 1986. *Molecular Biology*, 2nd edition. Jones and Bartlett, Boston. This widely used text is a useful reference for most of the concepts presented in the present chapter.

Kornberg, Arthur. 1980. *DNA Replication*. W. H. Freeman, San Francisco. This advanced monograph, written by the discoverer of DNA polymerase, provides the best detailed treatment of the subject.

Lewin, Benjamin. 1986. *Genes*. John Wiley, New York. Although somewhat quirky in style and approach, this text provides a better treatment of macromolecular processes in procaryotes than do the other texts listed.

Neidhardt, Frederick C., John L. Ingraham, K. Brooks Low, Boris Magasanik, Moselio Schaechter, and **H. Edwin Umbarger**, Editors. 1987. *Escherichia coli and Salmonella typhimurium: Cellular and Molecular Biology*. American Society for Microbiology, Washington, D. C. An advanced treatise written by many experts with chapters on all aspects of the molecular biology of these important organisms.

Watson, James D., John Tooze, and **David T. Kurtz**. 1983. *Recombinant DNA. A Short Course*. Scientific American Books, New York. Although this short, accessible book deals primarily with recombinant DNA, its early chapters provide a useful background to the fundamentals of macromolecules.

Watson, James D., Nancy H. Hopkins, Jeffrey W. Roberts, Joan A. Steitz, and **Alan M. Weiner**. 1987. *Molecular Biology of the Gene*, 4th edition. Volume I. Benjamin-Cummings Publishing Co., Menlo Park. The Watson book has been the standard textbook on macromolecules and DNA for over 20 years. This edition provides an up-to-date treatment of the current status of the field.

6

Viruses

A **virus** is a genetic element containing either DNA or RNA that can alternate between two distinct states, intracellular and extracellular. In the extracellular state, a virus is a submicroscopic particle containing nucleic acid surrounded by protein and occasionally containing other components. In this extracellular state, the **virus particle**, also called the **virion**, is metabolically inert and does not carry out respiratory or biosynthetic functions. The virion is the structure by which the virus genome is carried from the cell in which the virion has been produced to another cell where the viral nucleic acid can be introduced and the **intracellular state** initiated. In the intracellular state, **virus reproduction** occurs: the virus genome is produced and the components which make up the virus coat are synthesized. When a virus genome is introduced into a cell and reproduces, the process is called **infection**. The cell that a virus can infect and in which it can replicate is called a **host**. The virus redirects preexisting host machinery and metabolic functions necessary for virus replication.

Viruses may thus be considered in two ways: as agents of *disease* and as agents of *heredity* (Figure 6.1). As agents of disease, viruses can enter cells and cause harmful changes in these cells, leading to disrupted function or death. We discuss virus diseases in Chapter 15. As agents of heredity, viruses can enter cells and initiate permanent, genetic changes that are usually not harmful and may even be beneficial. In many cases, whether a virus causes disease or hereditary change depends upon the host cell and on the environmental conditions.

Viruses are smaller than cells, ranging in size from 0.02 μm to 0.3 μm. A common unit of measure for viruses is the nanometer (abbreviated nm), which is 1000 times smaller than a μm and one million times smaller than a millimeter. The range of sizes of some common viruses is shown in Figure 6.2.

Viruses are classified initially on the basis of the hosts they infect. Thus we have animal viruses, plant viruses, and bacterial viruses. Bacterial viruses, sometimes called *bacteriophages* (or *phage* for short, from the Greek *phago* meaning *to eat*), have been studied primarily as convenient model systems for research on the molecular biology and genetics of virus reproduction. Many of the basic concepts of virology were first worked out with bacterial viruses and subsequently applied to viruses of higher organisms. Because of their frequent medical importance, *animal viruses* have been extensively studied. The two groups of animal viruses most studied are those infecting insects and those infecting warm-blooded animals. *Plant viruses* are often important in agriculture but have been less studied than animal viruses. Although viruses are known which infect eucaryotic microorganisms, they have been little studied. In the present chapter, we discuss the structure,

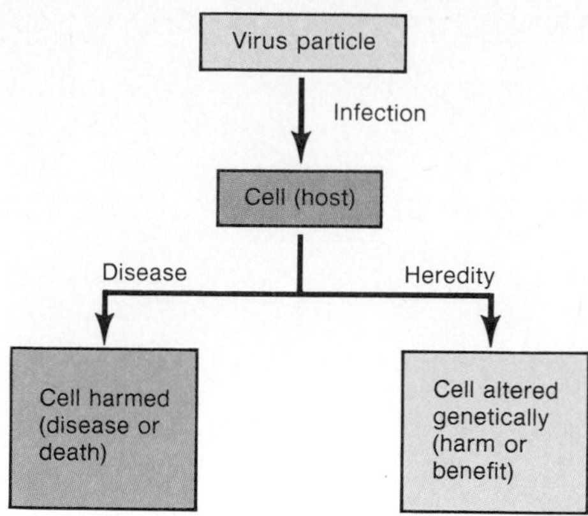

Figure 6.1 Virus infection: the two-fold path.

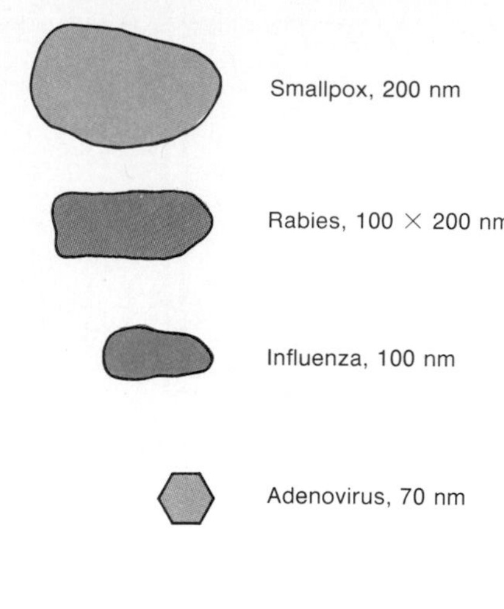

Figure 6.2 Relative sizes of some common viruses infecting humans. DNA viruses are green, RNA viruses are red.

replication, and genetics of viruses infecting bacteria and warm-blooded animals.

In a nutshell:

1. The **virus genome** consists of either RNA or DNA. The genome is surrounded by a *coat* of *protein* (and occasionally other material). When the virus genome is inside the coat it is called a *virus particle* or *virion*.
2. Viruses lack independent metabolism. They multiply only inside living cells, using the host cell metabolic machinery. Some virus particles do contain enzymes, however, that are under the genetic control of the virus genome. Such enzymes are only produced during the infection cycle.
3. When a virus multiplies, the genome becomes released from the coat. This process occurs during the infection process. The present chapter is divided into three parts. The first part deals with basic concepts of virus structure and function. The second part deals with the nature and manner of multiplication of the bacterial viruses (bacteriophages). In this part we introduce the basic molecular biology of virus multiplication. The third part deals with important groups of animal viruses, with emphasis on molecular aspects of animal virus multiplication. Important virus diseases are discussed later in this book, primarily in Chapter 15.

6.1 The Nature of the Virus Particle

Virus particles vary widely in size and shape. As we have stated, some viruses contain RNA, others DNA. We have discussed nucleic acids in previous chapters and have noted that the DNA of the cell genome is in the double-stranded form. Some viruses have double-stranded DNA whereas others have single-stranded DNA (Figure 6.3).

We have also noted in Section 5.8 that the RNA of the cell is generally in the single-stranded configuration. Interestingly, although single-stranded RNA viruses are more common, viruses are known in which the RNA is in the double-stranded form.

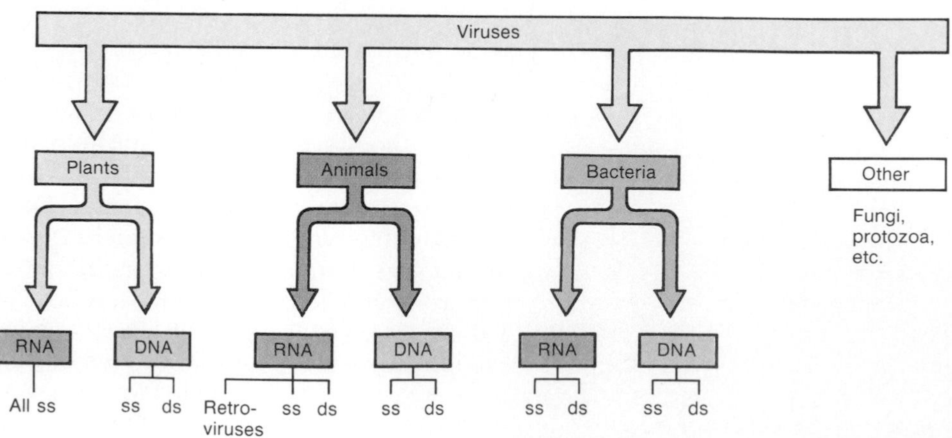

Figure 6.3 Diversity of viruses. ss: single stranded; ds: double stranded.

The Name "Virus"

The word *virus* originally referred to any poisonous emanation, such as the venom of a snake, and later came to be used more specifically for the causative agent of any infectious disease. Pasteur often referred to bacteria that caused infectious diseases as viruses. By the end of the nineteenth century, a large number of bacteria had been isolated and shown to be causal agents of specific infectious diseases, but there were some diseases for which a bacterial cause had not been shown. One of these was foot-and-mouth disease, a serious skin disease of animals. In 1898, Friedrich Loeffler and Paul Frosch presented the first evidence that the cause of foot-and-mouth disease was an agent so small that it could pass through filters that could hold back all known bacteria. That the agent was not an ordinary toxin could be shown by the fact that it was active at very low dilution and could be transmitted in filtered material from animal to animal. Loeffler and Frosch concluded "that the activity of the filtrate is not due to the presence in it of a soluble substance, but due to the presence of a causal agent capable of reproducing. This agent must then be obviously so small that the pores of a filter which will hold back the smallest bacterium will still allow it to pass. . . .If it is confirmed by further studies. . .that the action of the filtrate. . .is actually due to the presence of such a minute living being, this brings up the thought that the causal agents of a large number of other infectious diseases. . .which up to now have been sought in vain, may also belong to this smallest group of organisms."

A year later, the Dutch microbiologist Martinus Beijerinck published his work on tobacco mosaic disease, a crippling leaf disease of tobaccos and tomatoes. In 1892, D. Ivanowsky of Russia had first shown that the causal agent of tobacco mosaic disease was filterable, but Beijerinck went much further and provided strong evidence that although the causal agent was filterable, it had many of the properties of a living organism. He called the agent a *Contagium vivum fluidum*, a living germ that is soluble. He postulated that the agent must be incorporated into the living protoplasm of the cell in order to reproduce, and that its reproduction must be brought about with the reproduction of the cell. This postulate comes very close to our current understanding of how viruses reproduce. Beijerinck also noted that there were other plant diseases for which causal agents had not been isolated, and these might also be caused by filterable agents. Soon a number of other filterable agents were shown to be the causes of both plant and animal diseases. Such agents came to be called **filterable viruses**, but as further work on these agents was carried out the word "filterable" was graduable dropped. Today, the original meaning of "virus" has been forgotten, and the word is now used to refer to the kinds of agents discussed in this chapter. Bacterial viruses were first discovered by the British scientist F. W. Twort in 1915, and independently by the French scientist F. d'Herelle in 1917, who called them *bacteriophages* (from the combining form *phago* meaning "to eat"). Although bacteriophages are viruses, the name "phage" is still widely used to refer to this particular class of filterable infectious agents.

The structures of virions (virus particles) are quite diverse. Viruses vary widely in size, shape, and chemical composition. The nucleic acid of the virion is always located within the particle, surrounded by a protein coat called the *capsid*. The terms *coat, shell*, and *capsid* are often used interchangeably to refer to this outer layer. The protein coat is always formed of a number of individual protein molecules, called *protein subunits*, (sometimes called *capsomeres*) which are arranged in a precise and highly repetitive pattern around the nucleic acid (Figure 6.4). A few viruses have only a single kind of protein subunit, but most viruses have several chemically distinct kinds of protein subunits which are themselves associated in specific ways to form larger assemblies called *morphological units*. It is the morphological unit which is seen with the electron microscope. Genetic economy dictates that the variety of virus proteins be kept small, since virus genomes do not have sufficient genetic information to code for a large number of different kinds of proteins.

The information for proper aggregation of the protein subunits into the morphological units is contained within the structure of the subunits themselves, and the overall process of assembly is thus called **self-assembly**. A single virion generally has a large number of morphological units.

The complete complex of nucleic acid and protein, packaged in the virus particle, is called the virus **nucleocapsid**. Although the virus structure just described is frequently the total structure of a virus particle, a number of animal viruses (and a few bacterial viruses) have more complex structures. These viruses are *enveloped* viruses, in which the nucleo-

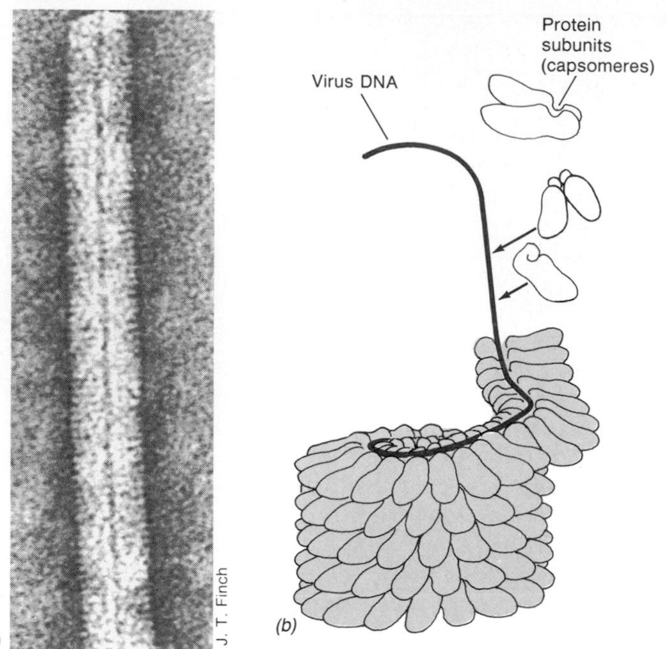

Figure 6.4 Structure and manner of assembly of a simple virus, tobacco mosaic virus. (a) Electron micrograph at high resolution of a portion of the virus particle. (b) Assembly of the tobacco mosaic virion. The RNA assumes a helical configuration surrounded by the protein capsomeres. The center of the particle is hollow.

capsid is enclosed in a membrane (Figure 6.5). **Virus membranes** are generally lipid bilayer membranes (see Section 3.3), but associated with these membranes are often virus-specific proteins. Inside the virion are often one or more **virus-specific enzymes**. Such enzymes usually play roles during the infection and replication process, as we will discuss later in this chapter.

Virus symmetry The nucleocapsids of viruses are constructed in highly symmetrical ways. Symmetry refers to the way in which the protein morphological units are arranged in the virus shell. When a symmetrical structure is rotated around an axis, the same form is seen again after a certain number of degrees of rotation. Two kinds of symmetry are recognized in viruses which correspond to the two primary shapes, rod and spherical. Rod-shaped viruses have helical symmetry and spherical viruses have icosahedral symmetry.

A typical virus with **helical symmetry** is the tobacco mosaic virus (TMV) that was illustrated in Figure 6.4. This is an RNA virus in which the 2130 identical protein subunits (each 158 amino acids in length) are arranged in a helix. In TMV, the helix has 16 1/2 subunits per turn and the overall dimensions of the virus particle are 18×300 nm. The lengths of helical viruses are determined by the length of the nucleic acid, but the width of the helical virus particle is determined by the size and packing of the protein subunits.

An **icosahedron** is a symmetrical structure roughly spherical in shape which has 20 faces. Icosahedral symmetry is the most efficient arrangement for subunits in a closed shell because it uses the smallest number of units to build a shell. The simplest arrangement of morphological units is 3 per face, for a total of 60 units per virus particle (Figure 6.6). The

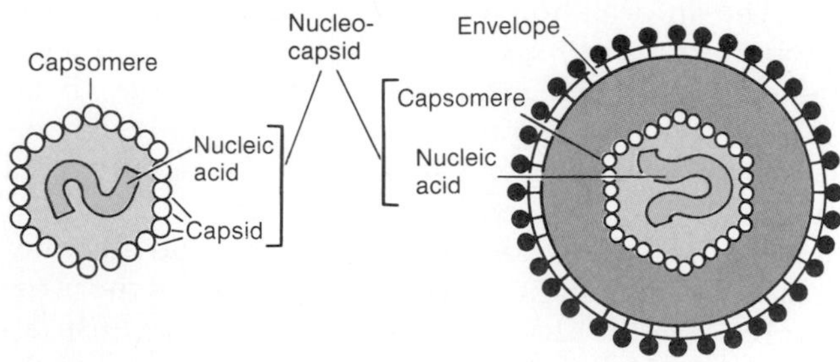

Figure 6.5 Comparison of naked and enveloped virus, two basic types of virus particles.

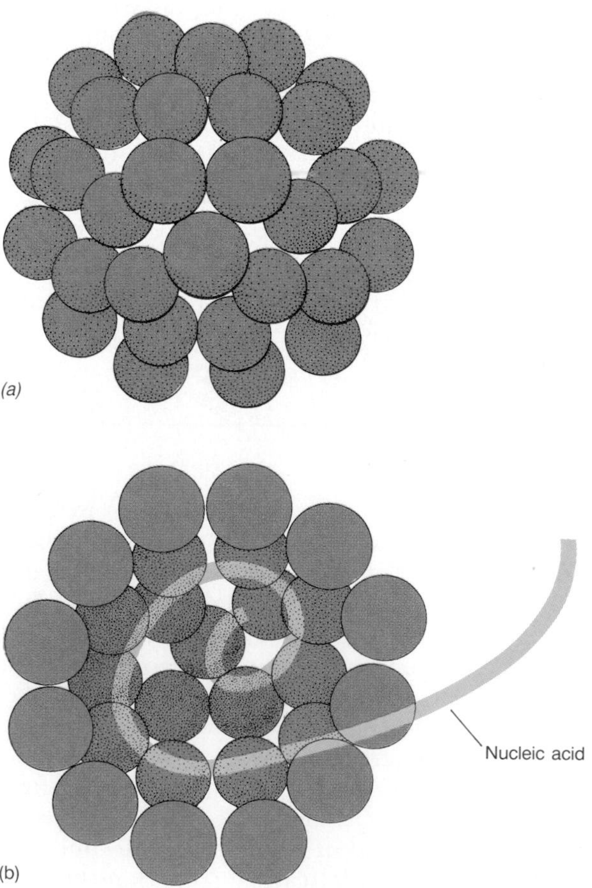

Figure 6.6 A simple icosahedral virus. Each face has three subunits. A single subunit consists of one or more proteins. (a) Whole virus particle. (b) Virus particle opened up; nucleic acid released.

three units at each face can be either identical or different. Most viruses have more nucleic acid than can be packed into a shell made of just 60 morphological units. The next possible structure which permits close packing contains 180 units and many viruses have shells with this configuration. Other known configurations involve 240 units and 420 units.

To help understand icosahedral symmetry, a model can be made following the instructions given in Figure 6.7. When discussing symmetry, one speaks of *axes of rotation*. A flat triangle shape, for instance, has one three-fold axis of symmetry, since there are three possible rotations that will lead to the exact configuration seen originally. Three dimensional objects such as viruses can have more than one axis of symmetry. An icosahedron, for instance, has three different axes of symmetry, two-fold, three-fold, and five-fold (see Figure 6.7). When a rod is placed through the two-fold axis of symmetry (one of the edges) in the model, the model can be turned once around this axis (1/2 way or 180°) to obtain the same configuration again. When the rod is placed through one of the three-fold axes of symmetry (one of the faces), the model can be turned three times, and if the rod is placed through one of the five-fold axes of symmetry (one of the vertices) the model can be turned five times.

In all cases, the characteristic structure of the virus is determined by the structure of the protein subunits of which it is constructed. Self-assembly leads to the final virus particle. An electron micrograph of a typical icosahedral virus is shown in Figure 6.8*a*.

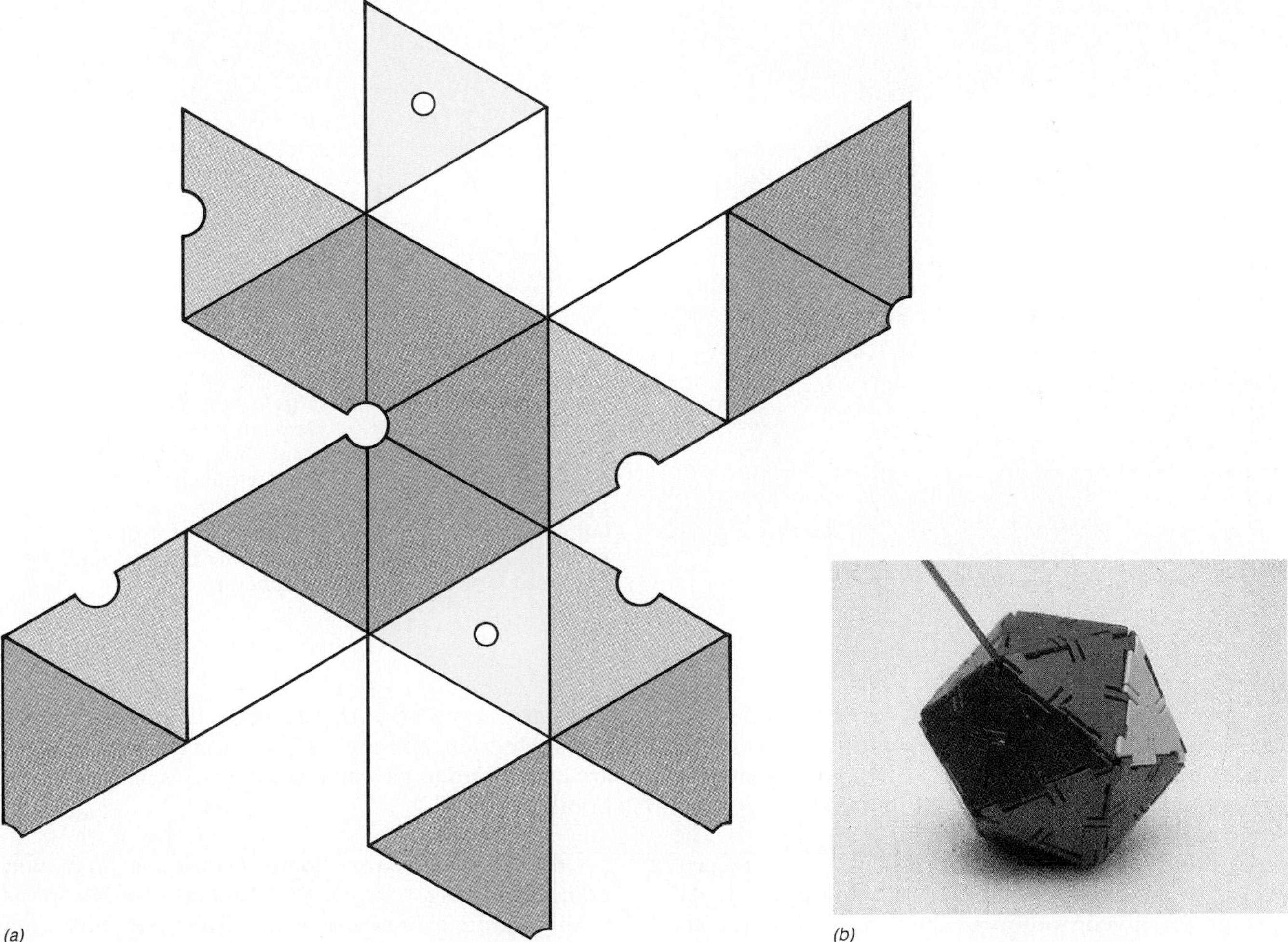

Figure 6.7 Demonstration of icosahedral symmetry. (a) A pattern that can be used to make a model of an icosahedral virus. Make a photocopy of this figure, cut it out, and fold at every line. Then tape the adjoining faces together to make a three-dimensional structure. (A copy machine which enlarges will provide a copy which can be more easily folded.) When the structure is folded, the axes of symmetry will be evident. Cut the paper at these axes, as marked, before folding. A metal rod or wire can then be inserted completely through the center of the model to study each axis. (b) Photograph of a model showing a rod through the axis of five-fold symmetry.

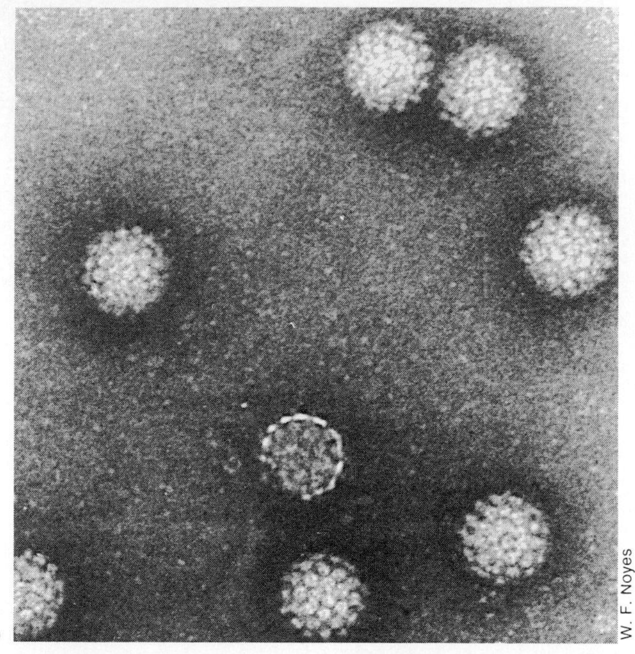
(a) W. F. Noyes

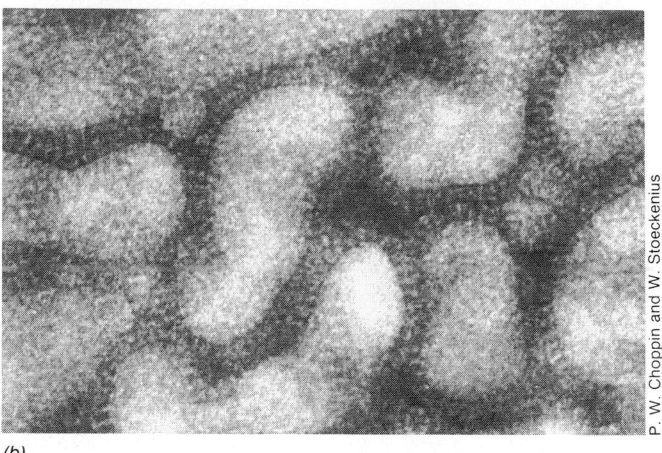
(b) P. W. Choppin and W. Stoeckenius

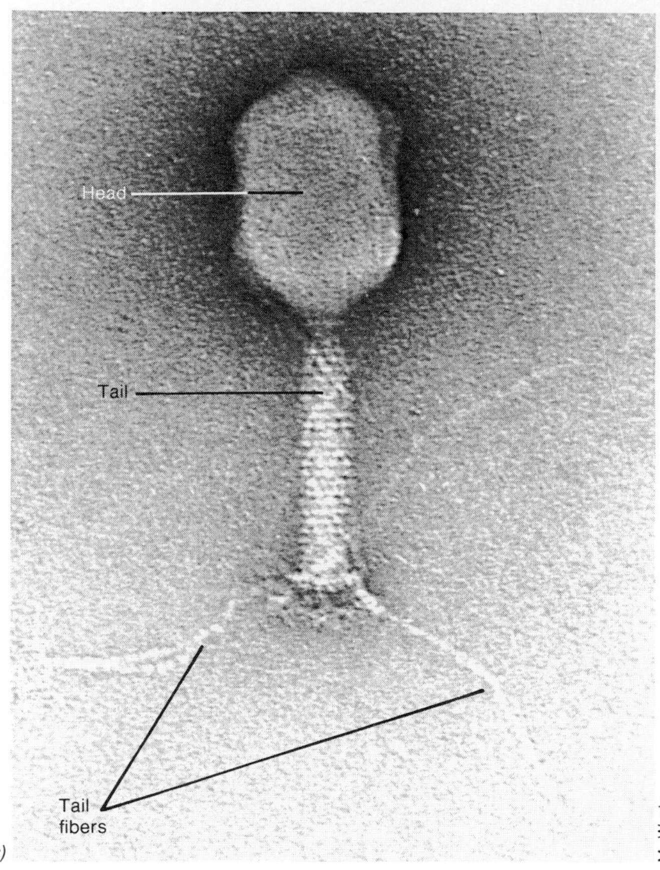

(c) M. Wurtz

Figure 6.8 Electron micrographs of various particles. (a) Human wart virus, a virus with icosahedral symmetry. The individual particles are about 55 nm in diameter. (b) Influenza virus, an enveloped virus. The individual particles are about 80 nm in diameter. (c) Bacterial virus (bacteriophage) T4 of *Escherichia coli*. Note the complex structure. The tail components are involved in attachment of the virion to the host and infection of the nucleic acid. The head is about 60 nm in diameter.

Enveloped viruses Many viruses have complex membranous structures surrounding the nucleocapsid (Figure 6.8*b*). Enveloped viruses are common in the animal world (for example, influenza virus), but some enveloped bacterial viruses are also known. The virus envelope consists of a lipid bilayer with proteins, usually glycoproteins, embedded in it. Although the glycoproteins of the virus membrane are encoded by the virus, the lipids are derived from the membranes of the host cell. The symmetry of enveloped viruses is expressed not in terms of the virion as a whole but in terms of the nucleocapsid present inside the virus membrane.

What is the function of the membrane in a virus particle? We will discuss this in detail later but note that because of its location in the virion, the membrane is the structural component of the virus particle that interacts first with the cell. The specificity of virus infection, and some aspects of virus penetration, are controlled in part by characteristics of virus membranes.

Complex viruses Some virions are even more complex, being composed of several separate parts, with separate shapes and symmetries. The most complicated viruses in terms of structure are some of the bacterial viruses, which possess not only icosahedral heads but helical tails (Figure 6.8*c*). In some bacterial viruses, such as the T4 virus of *Escherichia coli*, the tail itself is a complex structure. For instance, T4 has almost 20 separate proteins in the tail, and the T4 head has several more proteins. In such complex viruses, assembly is also complex. For instance, in T4 the complete tail is formed as a subassembly, and then the tail is added to the DNA-containing head. Finally, tail fibers formed from another protein are added to make the mature, infectious virus particle (see detailed discussion of T4 assembly in Section 6.13).

The virus genome We have stated that the virus genome consists of either DNA or RNA, never both. Viruses differ in size, amount, and character of their nucleic acid. Both single-stranded and double-stranded nucleic acid is found in viruses, and the amount of nucleic acid per virion may vary greatly from one virus type to another. In general, in enveloped viruses the nucleic acid constitutes only a small part of the mass of the virus particle (1–2 percent), whereas in nonenveloped viruses the percent of the particle which is nucleic acid is much larger, often 25–50 percent.

Interestingly, the nucleic acid in some viruses is not present in a single molecule, the genome being segmented into several or many molecules. For instance, *retroviruses*—causal agents of some cancers and AIDS, among other diseases—have two identical RNA molecules, influenza virus has 8 RNA molecules of sizes varying over about three-fold, and some other animal viruses have even more RNA molecules. The manner in which these various pieces of nucleic acid are replicated in the cell and then assembled into mature virions is of considerable interest—how do all these nucleic acid pieces end up together in one particle? We will consider this question in Section 6.16.

Enzymes in viruses We have stated that virus particles do not carry out metabolic processes. Outside of a host cell, a virus particle is metabolically inert. However, some viruses do contain enzymes which play roles in the infectious process. For instance, many viruses contain their own nucleic acid polymerases which transcribe the viral nucleic acid into messenger RNA once the infection process has begun. The retroviruses are RNA viruses which replicate inside the cell as DNA intermediates. These viruses possess an enzyme, an RNA-dependent DNA polymerase called *reverse transcriptase*, which transcribes the information in the incoming RNA into a DNA intermediate. It should be noted that reverse transcriptase is unique to the retroviruses and is not found in any other viruses or in cells.

A number of viruses contain enzymes which aid in release of the virus from the host cells in the final stages of the infectious process. One group of such enzymes, called *neuraminadases*, break down glycosidic bonds of glycoproteins and glycolipids of the connective tissue of animal cells, thus aiding in the liberation of the virus. Virions infecting some bacteria possess an enzyme, *lysozyme* (see Section 3.5), which hydrolyzes the cell wall, causing lysis of the host cell and release of the virus particles. We will discuss some of these enzymes in more detail later.

6.2 The Classification of Viruses

As we have noted, viruses can be classified into broad groups depending on their hosts. For instance, there are plant viruses, animal viruses, and bacterial viruses (Figure 6.3). A number of viruses infecting insects are also known and although viruses are known for fungi, protozoa, and algae, these viruses have been so little studied that no classification has been developed. In the present chapter, we discuss only the animal (primarily mammalian) and bacterial viruses, and we discuss here briefly how these two groups of viruses are classified.

Classification of bacterial viruses In the bacterial viruses, a formal classification scheme is rarely used. Rather, each bacterial virus is designated in terms of its principal bacterial host, followed by an arbitrary alphanumeric. Thus, we speak of T4 virus of *Escherichia coli* or P22 virus of *Salmonella typhimurium*. An overview of some of the major types of bacterial viruses is given later (see especially Figure 6.17). We should note, however, that although a bacterial virus may be designated in reference to its principal host, the actual host range of the virus may be broader. Thus, bacteriophage Mu, generally studied with *Escherichia coli*, also infects *Citrobacter* and *Salmonella*.

Classification of animal viruses We should note first that classification of animal viruses presents some major differences from the classification of organisms. The conventional approach to classification of organisms, involving hierarchical categories such as species, genera, families, etc., has been applied only to animal viruses. Even here, the higher levels of classification are not used. The highest level of animal virus classification is the virus family. Virus families are designated by terms ending in *-viridae*. Thus, the group of poxviruses is called the *Poxviridae* and the herpesviruses are called the *Herpesviridae*. A summary of all the families of human and animal viruses is given later in this chapter (see especially Figure 6.40). Note that the major criteria used in classifying animal viruses are the type of nucleic acid, the presence or absence of an envelope, and, for certain families, the manner of replication.

Virus genera are designated by terms ending in *-virus*. Thus, among the *Poxviridae* those poxviruses which infect fowl are called by the genus name *Avipoxvirus*. Note that frequently in the animal viruses, the genus is defined based on the host which the virus infects.

Except in a few cases, virus species have not been formally designated, but would refer to specific virus entities that have been recognized. At present, virus species are only designated by common names, such as mumps virus, poliovirus 1, and smallpox virus. For instance, in the virus genus *Orthopoxvirus* two virus species currently recognized are vaccinia and smallpox, but are not given Latin names. At present, it does not appear useful to use Latin names for virus species.

When contemplating the problem of virus classification, we can be truly impressed with the enormous diversity of viruses. Undoubtedly, many new viruses are awaiting discovery, although most undiscovered viruses will probably be considered members of existing virus families.

6.3 The Virus Host

Because viruses only replicate inside living cells, research on viruses requires use of appropriate hosts. For the study of bacterial viruses, pure cultures are used either in liquid or on semi-solid (agar) medium. Because bacteria are so easy to culture, it is quite easy to study bacterial viruses and this is why such detailed knowledge of bacterial virus reproduction is available.

With animal viruses, the initial host may be a whole animal which is susceptible to the virus, but for research purposes it is desirable to have a more convenient host. Many animal viruses can be cultivated in **tissue** or **cell cultures**, and the use of such cultures has enormously facilitated research on animal viruses.

Cell cultures A cell culture is obtained by enabling growth of cells taken from an organ of the experimental animal. Cell cultures are generally obtained by aseptically removing pieces of the tissue in question, dissociating the cells by treatment with an enzyme which breaks apart the intercellular cement, and spreading the resulting suspension out on the bottom of a flat surface, such as a bottle or petri dish. The cells generally produce glycoprotein-like materials that permit them to adhere to glass surfaces. The thin layer of cells adhering to the glass or plastic dish, called a *monolayer*,is then overlayed with a suitable culture medium and the culture incubated. The culture media used for cell cultures are generally quite complex, employing a number of amino acids and vitamins, salts, glucose, and a bicarbonate buffer system. To obtain best growth, addition of a small amount of blood serum is usually necessary, and several antibiotics are generally added to prevent bacterial contamination.

Some cell cultures prepared in this way will grow indefinitely, and can be established as *permanent cell lines*. Such cell cultures are most convenient for virus research because continuously available cell material can be available for research purposes. In other cases, indefinite growth does not occur but the culture may remain alive for a number of days. Such cultures, called *primary cell cultures*, may still be useful for virus research, although of course new cultures will have to be prepared from fresh sources from time to time.

Cancer Cancer is a cellular phenomenon of uncontrolled growth that is sometimes induced by virus infection. Most cells in a mature animal, although alive, do not divide extensively, apparently because of the presence of growth-inhibiting factors which prevent them from initiating cell division. Under a variety of pathological conditions, among which is included infection by certain viruses, growth inhibition is overcome and the cells begin to divide uncontrollably. Under some conditions, this extensive cellular growth is so excessive that the animal body is virtually consumed by cancer cells: the animal dies. Cancerous growth is thus due to a derangement in the control of cellular growth, and is of great medical as well as theoretical interest.

The tumorigenic or cancer-causing ability of viruses can often be detected by observing the induction in cell cultures of uncontrolled growth. In cell cultures, the general arrangement of the cells is as a *monolayer*, arising because growth generally ceases when the cells, as a result of growth, come in contact with each other. Cancer cells have altered growth requirements and continue to grow, piling up to form a small *focus of growth*. By observing for the induction of such foci of growth from virus infection, it is possible to observe the tumorigenic properties of viruses.

In some cases, cell culture monolayers can not be obtained but whole organs, or pieces of organs, can be cultured. Such **organ cultures** may still be useful in virus research, since they permit growth of viruses under more or less controlled laboratory conditions.

6.4 Quantification of Viruses

In order to obtain any significant understanding of the nature of viruses and virus replication, it is necessary to be able to *quantify* the number of virus particles. Virus particles are almost always too small to be seen under the light microscope. Although virus particles can be observed under the electron microscope, the use of this instrument is cumbersome for routine study. In general, viruses are quantified by measuring their effects on the host cells which they infect. It is common to speak of a *virus infectious unit*, which is the smallest unit that causes a detectable effect when placed with a susceptible host. By determining the number of infectious units per volume of fluid, a measure of virus quantity can be obtained. We discuss here several approaches to assessment of the virus infectious unit.

Large-scale Purification of Viruses

Modern virus research often requires the preparation of large amounts of pure virus particles for chemical studies or for preparation of antibodies. We describe here briefly the steps involved in the purification of polio virus, one of the most widely studied viruses affecting humans.

Before laboratory work is begun with a human virus, it is essential to immunize all laboratory workers against the virus. Vaccines for polio are widely available and most workers will probably already be immune, but reimmunization is an essential precaution.

The steps in virus purification can be briefly listed: 1) Growth of the virus in large-scale equipment; 2) removal of the virus-enriched fluid; 3) concentration of the virus material by precipitation; 4) final purification of the virus by density-gradient centrifugation procedures. We now present some details of each of these steps.

1. **Growth of the virus in large-scale equipment** In order to obtain sufficient virus for chemical studies, large volumes of virus must be prepared. Distinct procedures are available for the cultivation of each kind of virus. Polio virus is cultivated in a human or primate cell culture system, either on monolayers of cells growing in large bottles or in cell suspension in bottles that are gently agitated. Under favorable virus growth conditions, between 10 and 1000 infectious units of virus are produced per host cell, leading to titers of 10^7–10^9 infectious units per milliliter.

2. **Removal of virus-enriched culture fluid** Virus growth and release is generally accompanied by cell degeneration, so that at the time virus replication is complete, most of the cells in the culture have become converted into cell debris. To obtain the complete release of all virus particles, the cells are further disrupted by several cycles of freeze/thaw. The virus-rich fluid is then pipetted from the culture bottles and transferred to centrifuge tubes. A low-speed centrifugation removes large particles of cell debris.

3. **Concentration of the virus particles by precipitation** Because viruses are proteinaceous, they can be precipitated by methods that will precipitate proteins. Polio virus can be precipitated with the use of high concentrations of ammonium chloride (0.4 g per ml of culture fluid). The procedure is carried out at refrigeration temperature, the ammonium chloride being added to the culture fluid slowly with stirring. The solution will become cloudy with precipitated virus proteins. The precipitate is sedimented by low-speed centrifugation (2000 × gravity for 1–2 hours). The precipitate containing the virus particles is then resuspended in a small volume of phosphate-buffered saline. This procedure concentrates the virus about 10-fold and over 99 percent of the virus particles are recovered in the precipitate.

4. **Final purification by density-gradient centrifugation** Polio virus can be purified either by centrifugation through a density gradient of sucrose (described in the Box in Chapter 3) or by cesium chloride density gradient centrifugation (described in the Box in Chapter 5). With the latter procedure, the virus particles are banded in the centrifuge tube in a manner similar to that illustrated for DNA in the Box in Chapter 5. The centrifugation must be carried out at very high speed, 120,000 × gravity, and for many hours, preferably overnight. The liquid in the centrifuge tube is removed through the bottom of the tube as small drops, each of which is assayed for virus titer. Under appropriate centrifugation conditions, all of the virus particles will be contained in one or two fractions, and should be extremely pure.

5. **Crystallization of polio virus** If large amounts of highly purified virus are allowed to concentrate under appropriate conditions, crystals of virus can be obtained, as shown in the figure below. Such crystals provide favorable material for chemical analysis, and can be subjected to structural analysis by X ray diffraction. The computer-generated model of polio virus illustrated later in this chapter was obtained using highly purified polio virus crystals.

Poliovirus crystals arranged in a crystalline sheet

Plaque assay When a virus particle initiates an infection upon a layer or lawn of host cells which is growing spread out on a flat surface, a zone of *lysis* or *growth inhibition* may occur which results in a clearing of the cell growth. This clearing is called a **plaque**, and it is assumed that each plaque has originated from one virus particle.

Plaques are essentially "windows" in the lawn of confluent cell growth. With bacterial viruses, plaques may be obtained when virus particles are mixed into a thin layer of host bacteria which is spread out as an agar overlay on the surface of an agar medium (Figure 6.9*a*). During incubation of the culture, the bacteria grow and form a turbid layer which is visible to the naked eye. However, wherever a successful virus infection has been initiated, lysis of the cells occurs, resulting in the formation of a clear zone, called a *plaque* (Figure 6.9*b*).

The plaque procedure also permits the isolation of pure virus strains, since if a plaque has arisen from one virus particle, all the virus particles in this plaque are probably genetically identical. Some of the particles from this plaque can be picked and inoculated into a fresh bacterial culture to establish a pure virus line. The development of the plaque assay technique was as important for the advance of virology as was Koch's development of solid media (Section 1.18) for bacteriology.

Plaques may be obtained for animal viruses by using animal cell-culture systems as hosts. A monolayer of cultured animals cells is prepared on a plate or flat bottle and the virus suspension overlayed. Plaques are revealed by zones of destruction of the animal cells (Figure 6.10).

Phage dilution
Top agar
Bacterial cells
Pour mixture onto agar plate
Nutrient agar plate
Sandwich of top agar and nutrient agar
Incubate
Phage plaques
Lawn of host cells
(a)

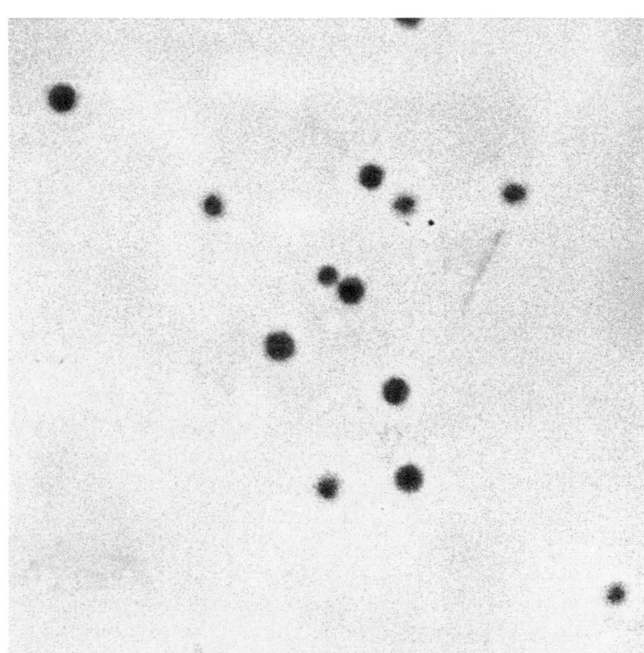
(b)

Figure 6.9 Quantification of bacterial virus by plaque assay using the agar overlay technique. A dilution of a suspension containing the virus material is mixed in a small amount of melted agar with the sensitive host bacteria, and the mixture poured on the surface of a nutrient agar plate. The host bacteria, which have been spread uniformly throughout the top agar layer, begin to grow, and after overnight incubation will form a *lawn* of confluent growth. Each virus particle that attaches to a cell and reproduces may cause cell lysis, and the virus particles released can spread to adjacent cells in the agar, infect them, be reproduced, and again lead to lysis and release. The size of the plaque formed depends on the virus, the host, and conditions of culture. (b) Photograph of plaques. The plaques shown are about 1–2 mm in diameter.

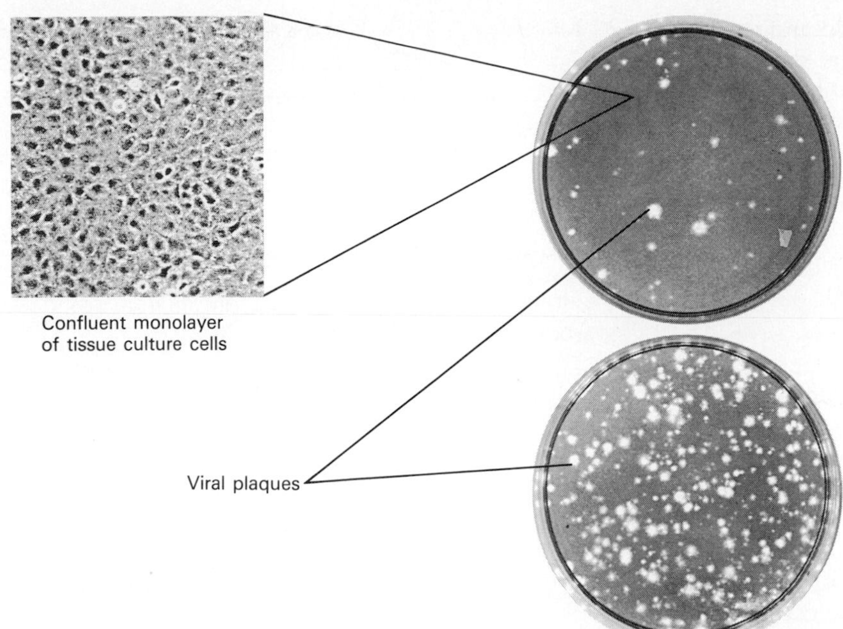

Figure 6.10 Cell cultures in monolayers within petri plates. Note the presence of plaques where virus lysis has occurred. Also shown is a photomicrograph of a cell culture.

In some cases, the virus may not actually destroy the cells, but cause changes in morphology or growth rate which can be recognized. For instance, tumor viruses may not destroy cells but cause the cells to grow faster than uninfected cells, a phenomenon called *transformation*. As we have noted, in a tissue culture monolayer, these transformed cells gradually develop into a recognizable cluster of cells called a *focus of infection* (Figure 6.11). By counting foci of infection, a quantitative measure of virus may be obtained. We discuss cancer viruses further later in this chapter.

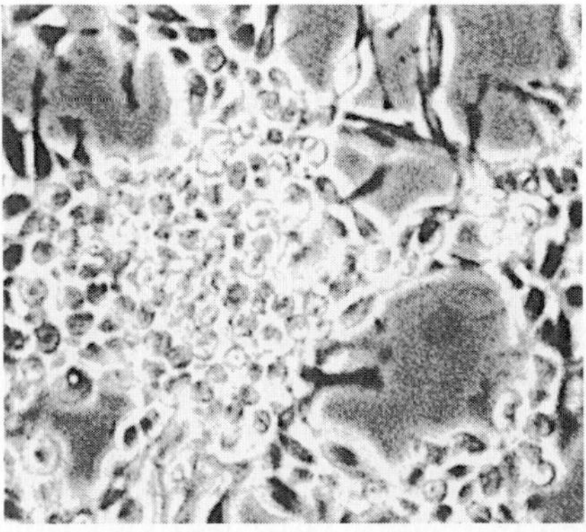

Figure 6.11 Microscopic appearance of a cell culture in which some of the cells have been transformed into tumor cells. The normal cells are elongated because they spread out on the glass culture dish. The tumor cells have lost contact inhibition and have piled up to make a small clump.

Efficiency of plating One important concept in quantitative virology involves the idea of *efficiency of plating*. Counts made by plaque assay are always lower than counts made with the electron microscope. The efficiency with which virus particles infect host cells is almost never 100 percent and may often be considerably less. This does not mean that the virus particles which have not caused infection are inactive. It may merely mean that under the conditions used, successful infection with these particles has not occurred. Although with bacterial viruses, efficiency of plating is often higher than 50 percent, with many animal viruses it may be very low, 0.1 or 1 percent. Why virus particles vary in infectivity is not well understood. It is possible that the conditions used for quantification are not optimal. Because it is technically difficult to count virus particles with the electron microscope, it is difficult to assess the actual efficiency of plating, but the concept is important in both research and medical practice. Because the efficiency of plating is rarely close to 100 percent, when the plaque method is used to quantify virus, it is accurate to express the titer of the virus suspension not as the number of virion units, but as the number of *plaque-forming units*

Animal infectivity methods Some viruses do not cause recognizable effects in cell cultures but cause death in the whole animal. In such cases, quantification can only be done by some sort of titration in infected animals. The general procedure is to carry out a serial dilution of the unknown sample, generally at ten-fold dilutions, and samples of each dilution are injected into numbers of sensitive animals. After a suitable incubation period, the fraction of dead and live animals at each dilution is tabulated and an *end point dilution* is calculated. This is the dilution at which, for example, *half* of the injected animals die. Although such serial dilution methods are much more cumbersome and much less accurate than cell culture methods, they may be essential for the study of certain types of viruses.

6.5 General Features of Virus Reproduction

The basic problem of virus replication can be simply put; the virus must somehow induce a living host cell to synthesize all of the essential components needed to make more virus particles. These components must then be assembled into the proper structure and the new virus particles must escape from the cell and infect other cells. The various phases of this replication process in a bacteriophage can be categorized in seven steps (Figure 6.12):

1. **Attachment** (adsorption) of the virion to a susceptible host cell;
2. **Penetration** (injection) into the cell by the virion or by its nucleic acid;
3. **Early steps in replication** of the virus nucleic acid, in which the host cell biosynthetic machinery is altered as a prelude to virus nucleic acid synthesis. Virus-specific enzymes may be made;
4. **Replication** of the virus nucleic acid;
5. **Synthesis of protein subunits** of the virus coat;
6. **Assembly** of nucleic acid and protein subunits (and membrane components in enveloped viruses) into new virus particles;
7. **Release** of mature virus particles from the cell (lysis).

These stages in virus replication are recognized when virus particles infect cells in culture and are illustrated in Figure 6.13, which exhibits what is called a **one-step growth curve.** In the first few minutes after infection, a period called the *eclipse* occurs, in which the virus nucleic acid has become separated from its protein coat so that the virus particle no longer exists as an infectious entity. Although virus nucleic acid may be infectious, the infectivity of virus nucleic acid is many times lower than that of whole virus particles because the machinery for bringing the virus genome into the cell is lacking. Also, outside the virion the nucleic acid is no longer protected from deleterious activities of the environment as it was when it was inside the protein coat.

The eclipse is the period during which the stages of virus multiplication occur. This is called the *latent period*, because no infectious virus particles are evident. Finally, maturation begins as the newly synthesized nucleic acid molecules become assembled inside protein coats. During the *maturation* phase, the titer of active virus particles inside the cell rises dramatically. At the end of maturation, *release* of mature virus particles occurs, either as a result of cell lysis or because of some budding or excretion process. The number of virus particles released, called the *burst size*, will vary with the particular virus and the particular host cell, and can range from a few to a few thousand. The timing of this overall virus replication cycle varies from 20–30 minutes in many

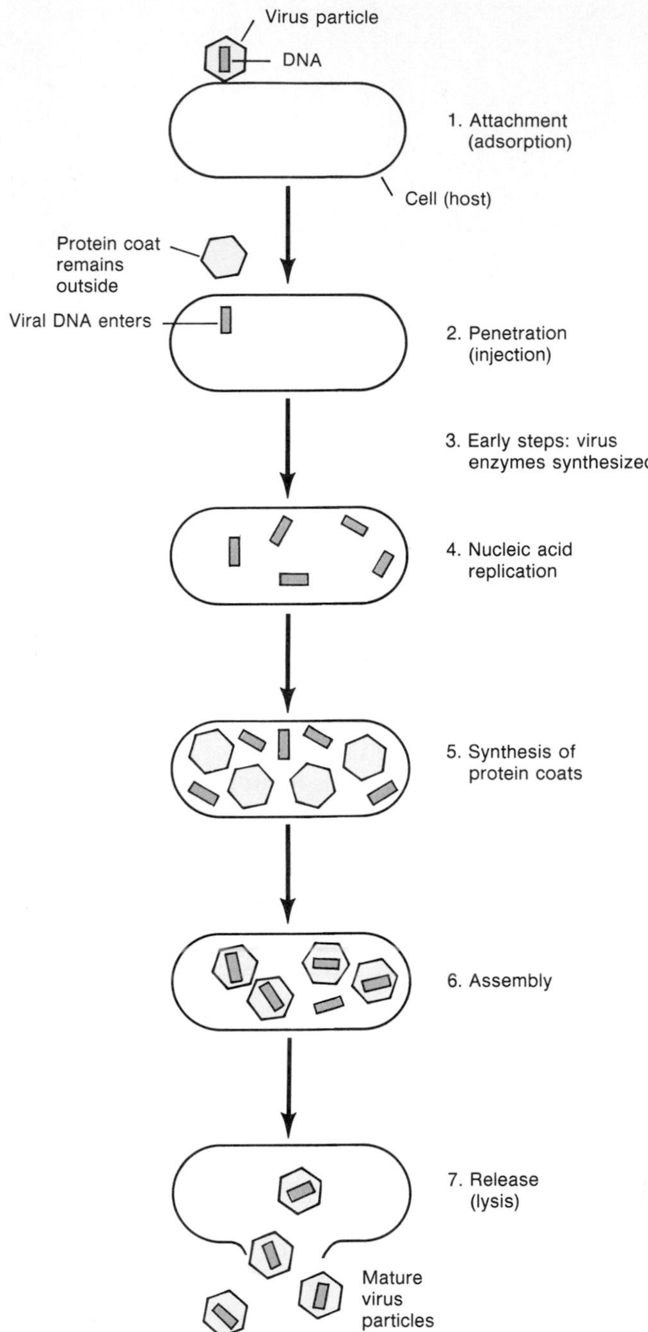

Figure 6.12 The replication cycle of a bacterial virus. The general stages of virus replication are indicated.

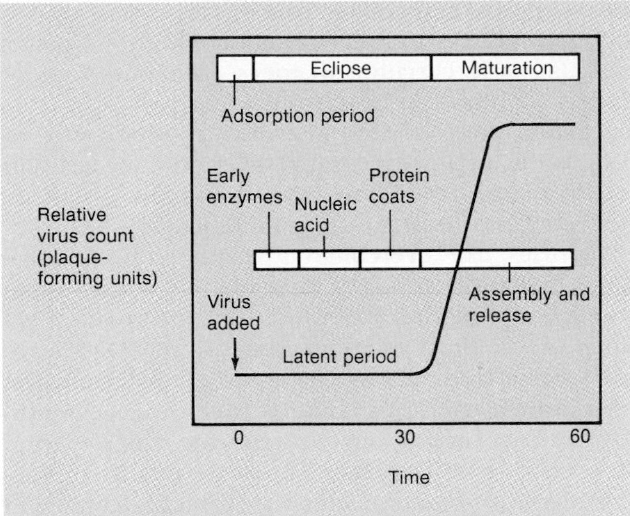

Figure 6.13 The one-step growth curve of virus replication. This graph displays the results of a single round of viral multiplication in a population of cells. Following absorption, the infectivity of the virus particles disappears, a phenomenon called *eclipse*. This is due to the uncoating of the virus particles. During the *latent period*, replication of viral nucleic acid and protein occurs. Then follows the *maturation period*, when virus nucleic acid and protein are assembled into mature virus particles. At this time, if the cells are broken up, active virus can be detected. Finally, *release* occurs, either with or without cell lysis. The timing of the one-step growth cycle varies with the virus and host. With many bacterial viruses, the whole cycle may be complete in 30–60 minutes, whereas with animal viruses 12–24 hours are usually required for a complete cycle.

bacterial viruses to 8–40 hours in most animal viruses. We now consider each of the steps of the virus multiplication cycle in more detail.

6.6 Early Events of Virus Multiplication

In order to discuss the stages of virus multiplication, we must return briefly to a consideration of the virus genome. As we have noted, virus genomes consist of either DNA or RNA, and both single-stranded and double-stranded forms of each of these nucleic acids is known to occur in different viruses. In the case of DNA viruses, the nucleic acid may be in either a linear or a circular form. The nucleic acid of RNA viruses is always in a linear form. Some virus nucleic acids also contain covalently linked polypeptides or amino acids which play roles in replication. In addition, in some RNA viruses the genome is not present in a single molecule, but may be divided over two or many nucleic acid molecules. Even more complicated, once inside the cell, the genetic information present in the virus genome may be transfered to another nucleic acid molecule. To avoid confusion, we restrict the term *virus genome* to the nucleic acid found in the virion (virus particle).

As we have noted, the outcome of a virus infection is the synthesis of viral nucleic acid and viral protein coats. In effect, the virus takes over the biosynthetic machinery of the host and uses it for its own synthesis. A few enzymes needed for virus replication may be present in the virus particle and may be introduced into the cell during the infection process, but the host supplies everything else: energy-generating system, ribosomes, amino-acid activating enzymes, transfer RNA (with a few exceptions), and all soluble factors. The virus genome codes for all new proteins. Such proteins would include the coat protein subunits (of which there are generally more than one kind) plus any new virus-specific enzymes.

Attachment There is a high specificity in the interaction between virus and host. The most common basis for host specificity involves the attachment process. The virus particle itself has one or more proteins on the outside which interact with specific cell surface components called *receptors*. The receptors on the cell surface are normal surface components of the host, such as proteins, polysaccharides, or lipoprotein-polysaccharide complexes, to which the virus particle attaches. In the absence of the receptor site, the virus cannot adsorb, and hence cannot infect. If the receptor site is altered, the host may become resistant to virus infection. However, mutants of the virus can also arise which are able to adsorb to resistant hosts.

In general, virus receptors carry out normal functions in the cell. For example, in bacteria some phage receptors are pili or flagella, others are cell-envelope components, and others are transport binding proteins. The receptor for influenza virus is a glycoprotein found on red blood cells and on cells of the mucous membrane of susceptible animals, whereas the receptor site of poliovirus is a lipoprotein. However, many animal and plant viruses do not have specific attachment sites at all and the virus enters passively as a result of phagocytosis or some other endocytotic process.

Penetration The means by which the virus penetrates into the cell depends on the nature of the host cell, especially on its surface structures. Cells with cell walls, such as bacteria, are infected in a different manner from animal cells, which lack a cell wall. The most complicated penetration mechanisms have been found in viruses that infect bacteria. The bacteriophage T4, which infects *E. coli*, can be used as an example.

The structure of the bacterial virus T4 was shown in Figure 6.8*c*. The particle has a **head**, within which the viral DNA is folded, and a long, fairly complex

tail, at the end of which is a series of tail fibers. During the attachment process, the virus particles first attach to the cells by means of the tail fibers (Figure 6.14). These tail fibers then contract, and the core of the tail makes contact with the cell envelope of the bacterium. The action of a lysozyme-like enzyme results in the formation of a small hole. The tail sheath contracts and the DNA of the virus passes into the cell through a hole in the tip of the tail, the majority of the coat protein remaining outside. The DNA of T4 has a total length of about 50 μm, whereas the dimensions of the head of the T4 particle are 0.095 μm by 0.065 μm. This means that the DNA must be highly folded and packed very tightly within the head.

With animal cells, the whole virus particle penetrates the cell, being carried inside by endocytosis (phagocytosis or pinocytosis), an active cellular process. We describe some of these processes in detail later in this chapter.

Virus restriction and modification by the host We have already seen that one form of host resistance to virus arises when there is no receptor site on the cell surface to which the virus can attach. Another and more specific kind of host resistance involves destruction of the viral nucleic acid after it has been injected. This destruction is brought about by host enzymes that cleave the viral DNA at one or several places, thus preventing its replication. This phenomenon is called *restriction*, and is part of a general host mechanism to prevent the invasion of foreign nucleic acid. We discussed *restriction enzymes* and their action in some detail in Section 5.3 and noted that their cellular role was in defense against foreign DNA. Restriction enzymes are highly specific, attacking only certain sequences (generally four or six base pairs). The host protects its own DNA from the action of restriction enzymes by *modifying* its DNA at the sites where the restriction enzymes will act. Modification of host DNA is brought about by methylation of purine or pyrimidine bases.

Viruses can overcome host restriction mechanisms by modifications of their nucleic acids so that they are no longer subject to enzymatic attack. Two kinds of chemical modifications of viral DNA have been recognized, glucosylation and methylation. The bacteriophages T2, T4, and T6 have their DNA glucosylated to varying degrees, and the glucosylation prevents or greatly reduces nuclease attack. In bacteriophage *lambda* the amino groups of adenine and cytosine bases are methylated by an enzyme that uses S-adenosylmethionine as methyl donor. Many other viral nucleic acids have been found to be modified by methylation but glucosylation has been found only in the T-even bacteriophages (bacteriophages T2, T4, and T6). It should be emphasized that modification of viral nucleic acid occurs after replication has occurred and the modified bases are not copied directly. The enzymes for methylation are actually present in the host before infection, and hence are not virus-induced functions. These host modification enzymes probably have as their main role the modification of host DNA so that it can be transferred without inactivation into other cells during genetic recombination (see Chapter 7).

The ability to modify nucleic acid is not found

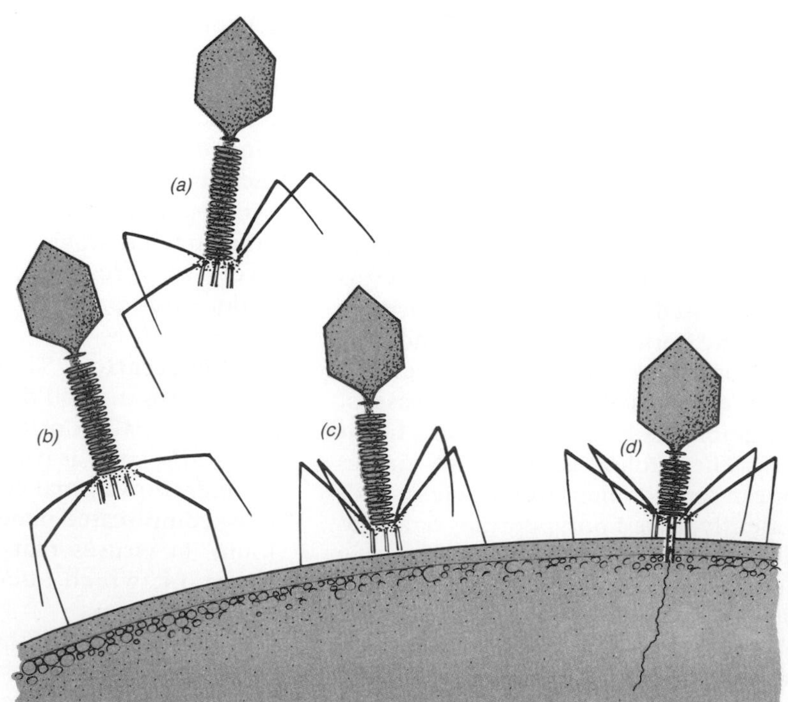

Figure 6.14 Attachment of T4 bacteriophage particle to the cell wall of *E. coli* and injection of DNA: (a) Unattached particle. (b) Attachment to the wall by the long tail fibers. (c) Contact of cell wall by the tail pin. (d) Contraction of the tail sheath and injection of the DNA.

in all strains that support the growth of a given virus. Thus, when bacteriophage *lambda* is grown on *E. coli* strain C it is not modified (*E. coli* strain C lacks both the *lambda* modification and restriction enzymes), and nucleic acid of virus grown on strain C is destroyed when it enters *E. coli* strain K-12, which does have the restriction enzyme. However, strain K-12 also has the modification enzyme, and, if *lambda* is grown on K-12, its nucleic acid is modified and it will infect both strains K-12 and C equally well. However, if *lambda* is grown on a K-12 strain made methionine deficient, methylation cannot occur and the phage particles released are unable to replicate in K-12. In the case of the T-even phages, glucosylation requires uridine diphosphoglucose (UDPG), and if a T-even phage is grown on a host deficient in UDPG its nucleic acid is not glucosylated and it is unable to replicate in susceptible cells.

As we discussed in Chapter 5, a knowledge of modification and restriction systems is of considerable practical utility in studying DNA chemistry. We discuss the use of restriction enzymes in genetic engineering in Chapter 8. So far, no evidence exists that either modification or restriction occurs in eucaryotic organisms.

Virus messenger RNA In order for the new virus-specific proteins to be made from the virus genome, it is necessary for new virus-specific RNA molecules to be made. Exactly how the virus brings about

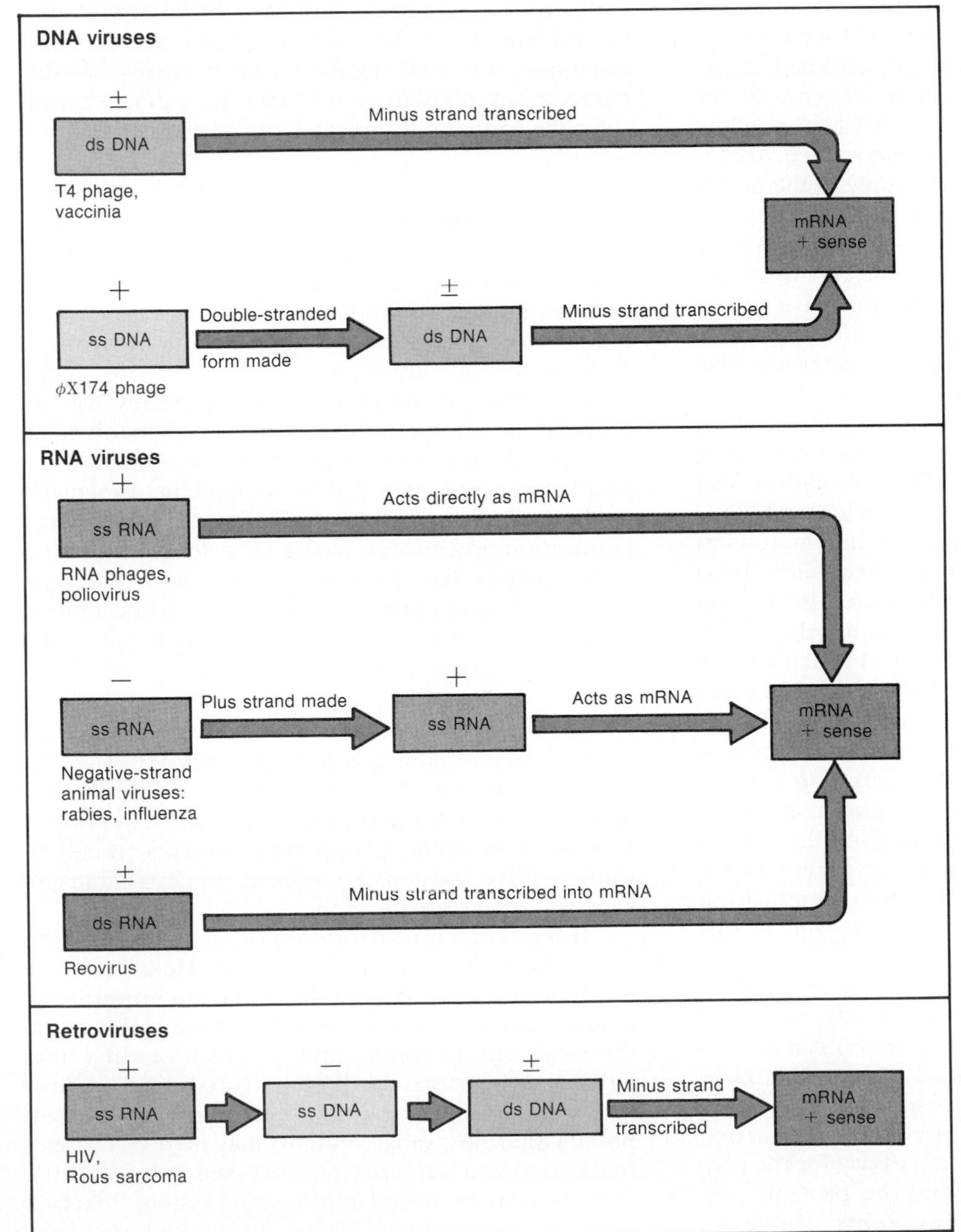

Figure 6.15 Formation of mRNA after infection of viruses of different types. The chemical sense of the mRNA is considered as plus (+). The senses of the various virus nucleic acids are indicated as + if the same as mRNA, as − if opposite, or as ± if double stranded. Examples are indicated next to the virus nucleic acids.

new mRNA synthesis depends upon the type of virus, and especially upon whether its genetic material is RNA or DNA, and whether it is single-stranded or double-stranded. The essential features of mRNA synthesis were discussed in Chapter 5, and it was shown that mRNA represents a complementary copy made by RNA polymerase of one of the two strands of the DNA double helix. Which copy is read into mRNA depends upon the location of the appropriate promoter, since the promoter points the direction that the RNA polymerase will follow. In cells (uninfected with virus) all mRNA is made on the DNA template, but with RNA viruses the situation is obviously different.

A virus-specific RNA → RNA polymerase is needed, since the cell RNA polymerase will generally not copy double-stranded RNA (and ribosomes are not able to translate double-stranded RNA either). A wide variety of modes of viral mRNA synthesis are outlined in Figure 6.15. By convention, the chemical sense of the mRNA is considered to be of the *plus* (+) configuration. The sense of the viral genome nucleic acid is then indicated by a *plus* if it is the same as the mRNA and a *minus* if it is of oppposite sense. As seen in Figure 6.15, if the virus has double-stranded DNA (ds DNA), then mRNA synthesis can proceed directly as in uninfected cells. However, if the virus has a single-stranded DNA (ss DNA), then it is first converted to ds DNA and the latter serves as the template for mRNA synthesis with the cell RNA polymerase.

If the virus has double-stranded RNA (ds RNA), this RNA serves as a template in a manner analogous to DNA. There are three classes of viruses with ss RNA and they differ in the mechanism by which mRNA is synthesized. In the simplest case, the incoming viral RNA is the *plus* sense and hence serves directly as mRNA, and copies of this viral RNA are also copied to make further mRNA molecules. In another class, the viral RNA has a *minus* (−) sense. In such viruses a copy is made (*plus* sense) and this copy becomes the mRNA. In the case of the retroviruses (causal agents of certain kinds of cancers and AIDS), a new phenomenon called **reverse transcription** is seen, in which virion ss RNA is copied to a double-stranded DNA (through a ss DNA intermediate) and the ds DNA then serves as the template for mRNA synthesis (thus: ss RNA → ss DNA → ds DNA). Retrovirus replication is of unusual interest and complexity, and is discussed in Section 6.24.

Viral proteins Once mRNA is made, viral proteins (for example, enzymes, capsomers) can be synthesized. The proteins synthesized as a result of virus infection can be grouped into two broad categories, the enzymes synthesized soon after infection, called the **early enzymes**, which are necessary for the replication of virus nucleic acid, and the proteins synthesized later, called the **late proteins**, which include the proteins of the virus coat. Generally, both the time of appearance and the amount of these two groups of virus proteins are regulated. The early proteins are enzymes which, because they act catalytically, are synthesized in smaller amounts and the late proteins, often structural, are made in much larger amounts.

Virus infection obviously upsets the regulatory mechanisms of the host, since there is a marked overproduction of nucleic acid and protein in the infected cell. In some cases, virus infection causes a complete shutdown of host macromolecular synthesis while in other cases host synthesis proceeds concurrently with virus synthesis. In either case, the regulation of virus synthesis is under the control of the virus rather than the host. There are several elements of this control which are similar to the host regulatory mechanisms discussed in Chapter 5, but there are also some uniquely viral regulatory mechanisms. We discuss various regulatory mechanisms when we consider the individual viruses later in this chapter.

6.7 Viral Genetics

Viruses exhibit genetic phenomena similar to those of cells. Studies of viral genetics have played a significant role in understanding many aspects of genetics at the molecular level. In addition, knowledge of the basic phenomena of viral genetics has increased our understanding of processes involved in virus replication. Understanding these processes has also led to some practical developments, especially in the isolation of viruses which are of use in immunization procedures. Most of the detailed work on viral genetics has been carried out with bacteriophages, because of the convenience of working with these viruses. We mention here briefly some of the types of genetic phenomena of viruses.

Mutations Much of our knowledge of viral reproduction and how it is regulated has depended on the isolation and characterization of virus mutants. Several kinds of mutants have been studied in viruses: host-range mutants, plaque-type mutants, temperature-sensitive mutants, nonsense mutants, transposons, and inversions.

Host-range mutations are those that change the range of hosts that the virus can infect. Host resistance to phage infection can be due to an alteration in receptor sites on the surface of the host cell, so that the virus can no longer attach, and host-range mutations of the virus can then be recognized as virus strains able to attach to and infect these virus-resistant hosts. Other host-range mutants may involve changes in the viral and host enzymes involved in replication, or in the restriction and modification systems (Section 5.3).

Plaque-morphology mutations are recognized as changes in the characteristics of the plaques formed when a phage infects cells in the conventional agar-plate technique (Figure 6.16). Characteristics of the plaque, such as whether it is clear or turbid, and its size, are under genetic control. The underlying basis of plaque morphology lies in processes taking place during the virus multiplication cycle, such as the rate of replication and the rate of lysis. Under appropriate experimental conditions plaque morphology can be a highly reproducible characteristic of the virus. The advantage of plaque mutants for genetic studies is that they can be easily recognized on the agar plate, but a disadvantage is that there is no convenient way of selecting for them among the large background of normal particles.

Temperature-sensitive mutations are those which allow a virus to replicate at one temperature and not at another, due to a mutational alteration in a virus protein that renders the protein unstable at moderately high temperatures. For instance, temperature-sensitive mutants are known in which the phage will not be replicated in the host at 43°C but will at 25°C, although the host functions at both temperatures. Such mutations are called *conditionally lethal*, since the virus is unable to reproduce at the higher temperature, but replicates at the lower temperature.

Nonsense mutations change normal codons into nonsense codons (see Section 5.15). In viruses, nonsense mutations are recognized because hosts are available that contain suppressors (see Section 5.17) able to read nonsense codons. The virus mutant will be able to grow in the host containing the suppressor, but not in the normal host.

Transpositions were discussed in Section 5.17; several viruses are known which act essentially as transposons and transposition events involving viruses can lead to their genetic change (see Section 6.15).

Genetic recombination in viruses The availability of virus mutants makes possible the investigation of genetic recombination. If two virus particles infect the same cell, there is a possibility for genetic exchange between the two virus genomes during the replication process. If recombination does occur, the progeny of such a mixed infection should include not only the parental types, but recombinant types as well. With appropriate mutants, it is possible to recognize both the parental types and the recombinants and to study the events involved in the recombination process. Genetic recombination in viruses is an extremely complex process to analyze because recombination does not occur as a single discrete event during mixed infection, but may occur over and over again during the replication cycle. It has been calculated that the T-even bacteriophages undergo, on the average, four or five rounds of recombination during a single infection cycle. By detailed and careful analysis of a wide variety of virus crosses, it has been possible to construct genetic maps of a number of bacterial viruses. Such maps have provided important information about the genetic structure of viruses. We present a few genetic maps when discussing specific viruses later in this chapter.

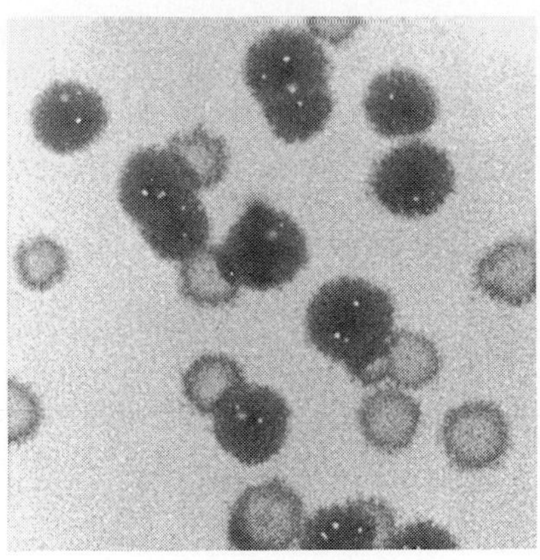

Figure 6.16 Plaque mutants of phage *lambda* in *Escherichia coli.* The turbid plaques are due to normal wild-type phage particles, and the clear plaques are due to mutant particles. A few host colonies from resistant mutants are seen within the clear plaques. The technique used here was similar to that described in Figure 6.9. The largest plaques are a few millimeters in diameter.

Genetic recombination arises by exchange of homologous segments of DNA between viral genomes, most often during the replication process. The enzymes involved in recombination are DNA polymerases, endonucleases, and ligases, which also play a role in DNA repair and synthesis processes (see Section 5.4).

Phenotypic mixing During studies on genetic recombination between viruses, another phenomenon was discovered which superficially resembles recombination but has a quite different basis. Phenotypic mixing occurs when the DNA of one virus is incorporated inside the protein coat of a different virus. For phenotypic mixing to occur, the two viruses must be closely related, so that the protein coat is of proper construction of the packaging of either viral DNA. As an example of phenotypic mixing, in phage T2 of *E. coli* there is a gene called the *h* gene which controls host specificity through modification of the tail fibers of the phage. If a mixed infection is set up with two T2 phages, mutant T2*h* and wild-type T2h^+, tail fibers of *h* specificity may be incorporated onto the particles containing DNA of h^+ specificity. Since it is the *h* function of the tail fibers that affects attachment, these mixed particles will show *h* specificity during the next round of infection, even though they contain h^+ DNA, but the particles resulting from this second round of infection will phenotypically become h^+, because the DNA has been unchanged.

6.8 General Overview of Bacterial Viruses

The range of bacterial viruses is illustrated in Figure 6.17. Most of the bacterial viruses which have been studied in any detail infect bacteria of the enteric group, such as *Escherichia coli* and *Salmonella typhimurium*. However, viruses are known that infect a variety of procaryotes, both eubacteria and archaebacteria. A few bacterial viruses have lipid envelopes but most do not. However, many bacterial viruses are structurally complex, with head and complex tail structures. As we illustrated in Figure 6.14, the tail is involved in the injection of the nucleic acid into the cell.

We now discuss some of the bacterial viruses for which molecular details of the multiplication process are known. Although these bacterial viruses were first studied as *model systems* for understanding general features of virus multiplication, as discussed in Chapter 8, many of them now serve as convenient tools for *genetic engineering*. Thus, the information on bacterial viruses is not only valuable as background for the discussion of animal viruses, but is essential for the material presented in the next two chapters on microbial genetics and genetic engineering.

It should already be clear from what has been stated that a great diversity of viruses exist. It should therefore not be surprising that there is also a great diversity in the manner by which virus multiplication occurs. Interestingly, many viruses have special features of their nucleic acid and protein synthesis processes that are not found in cells. In the present chapter, we are only able to present some of the major types of virus replication patterns, and must skip some of the interesting exceptional cases.

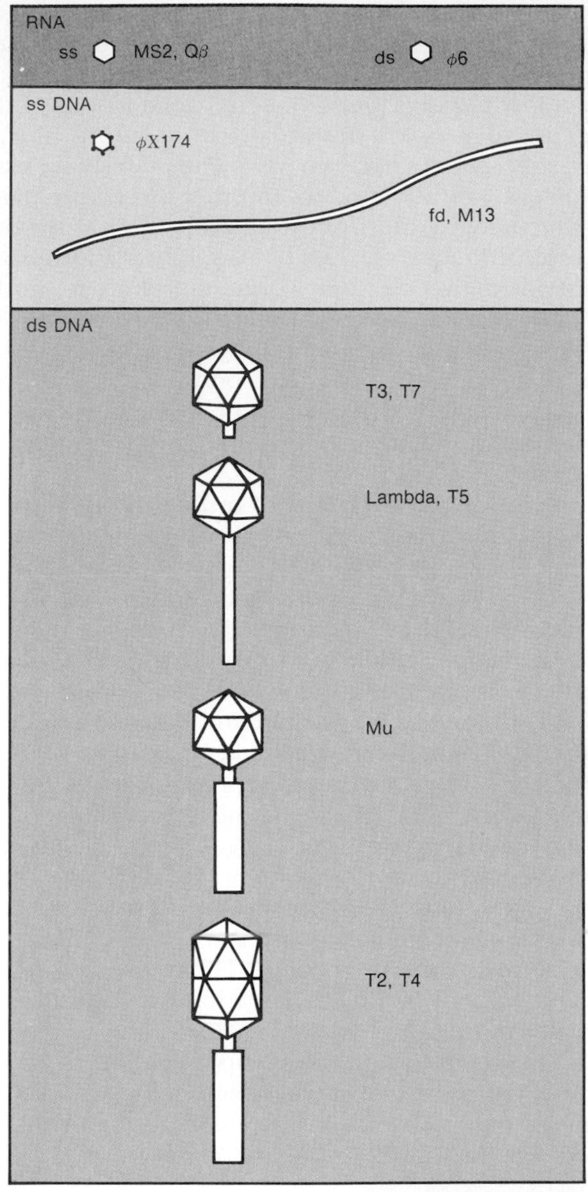

Figure 6.17 Schematic representations of the main types of bacterial viruses. Those discussed in detail are fd, M13, φX174, MS2, T4, lambda, T 7, and Mu. Sizes are to approximate scale.

6.9 RNA Bacteriophages

A number of bacterial viruses have RNA genomes. The best-known bacterial RNA viruses have single-stranded RNA. Interestingly, the bacterial RNA viruses known in the enteric bacteria group infect only bacterial cells which behave as gene donors (males) in genetic recombination (the interesting concept of male and female bacteria is discussed in Chapter 7). This restriction to male bacterial cells arises because these viruses infect bacteria by attaching to *male-specific pili* (Figure 6.18). Since such pili are absent on female cells, these RNA viruses are unable to attach to the females, and hence do not initiate infection in females.

The bacterial RNA viruses are all of quite small size, about 26 nm in size, and they are all icosahedral, with 180 copies of coat protein per virus particle. The complete nucleotide sequence of several RNA phages are known. In the RNA phage MS2, which infects *Escherichia coli*, the viral RNA is 3,569 nucleotides long. The virus RNA, although single stranded, has extensive regions of secondary and tertiary structure (Figure 6.19). The RNA strand in the virion has the plus (+) sense, acting directly as mRNA upon entry into the cell (see Figure 6.15).

The genetic map is shown in Figure 6.20*a* and the flow of events of MS2 multiplication is shown in Figure 6.20*b*. The infecting RNA goes to the host ribosome, where it is translated into four (or more) proteins. The four proteins that have been recognized are **maturation protein** (A-protein; present in the mature virus particle as a single copy), **coat protein, lysis protein** (involved in the lysis process which

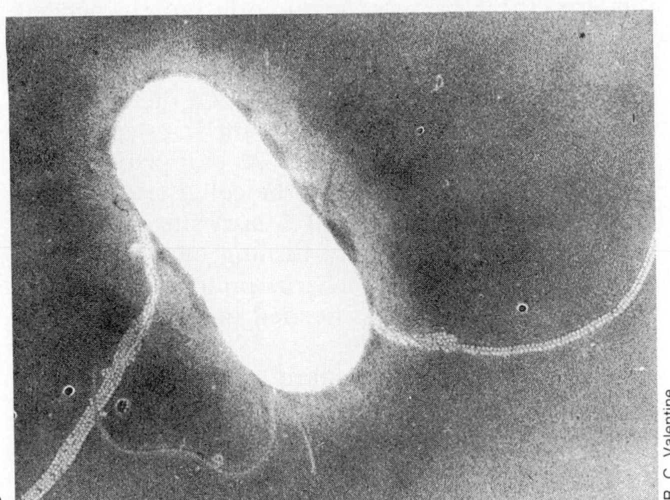

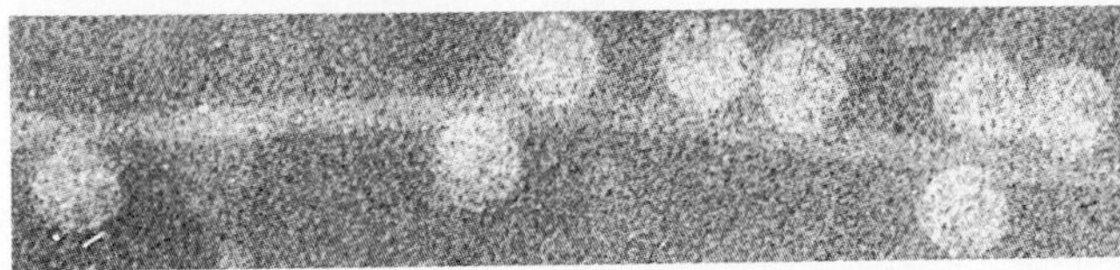

Figure 6.18 Electron micrographs of a male bacterial cell infected with a small RNA phage. (a) Note that the phage particles have attached to the pilus (male-specific pilus). (b) Close up of a pilus with virus particles.

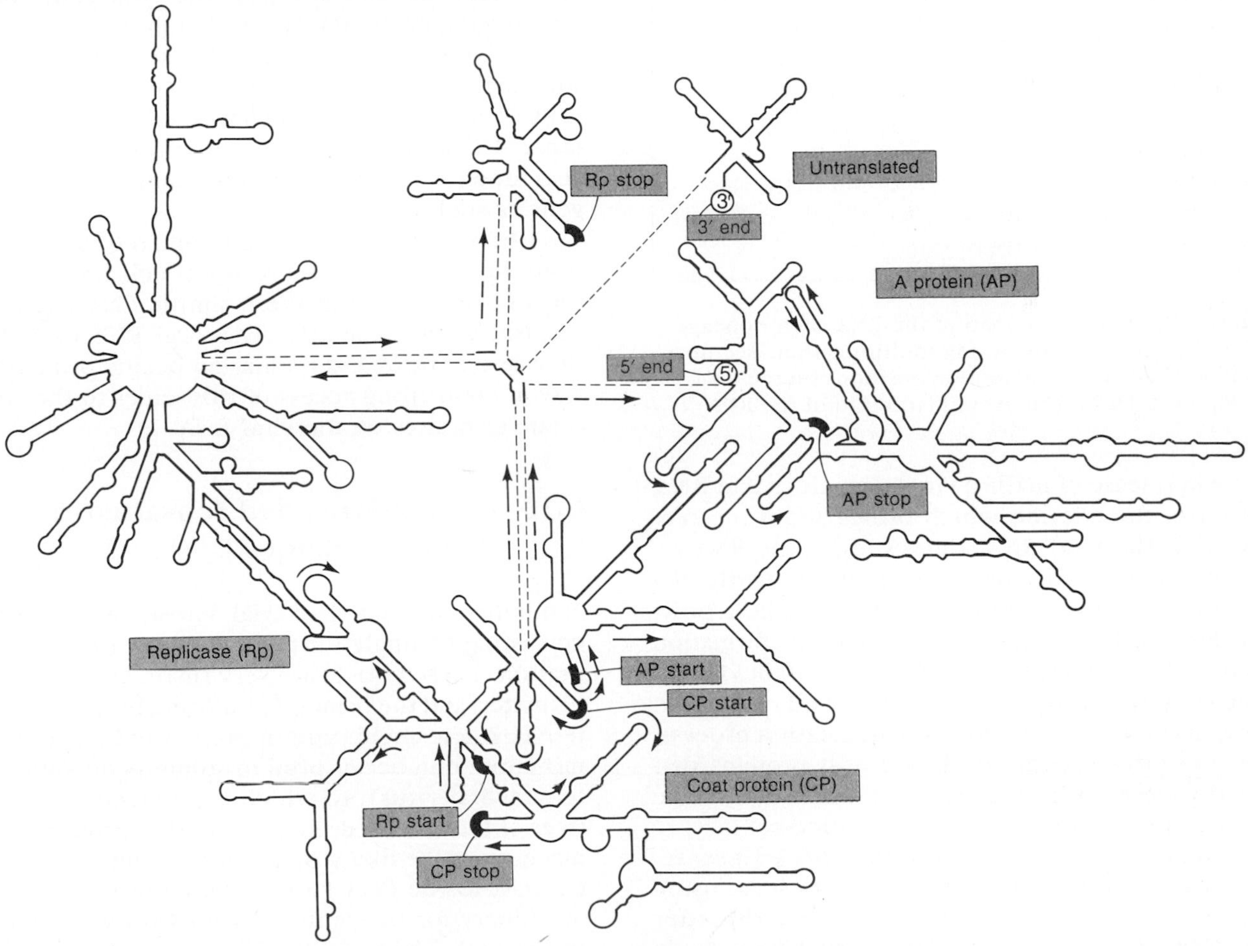

Figure 6.19 The RNA of bacteriophage MS2. The molecule is single stranded but there are extensive regions of complementary bases, so that pairing within the strand leads to the secondary structure shown. Note that the start sites for three coding regions are in the same part of the folded molecule. See Figure 6.20*a* for an overview of the genetic map.

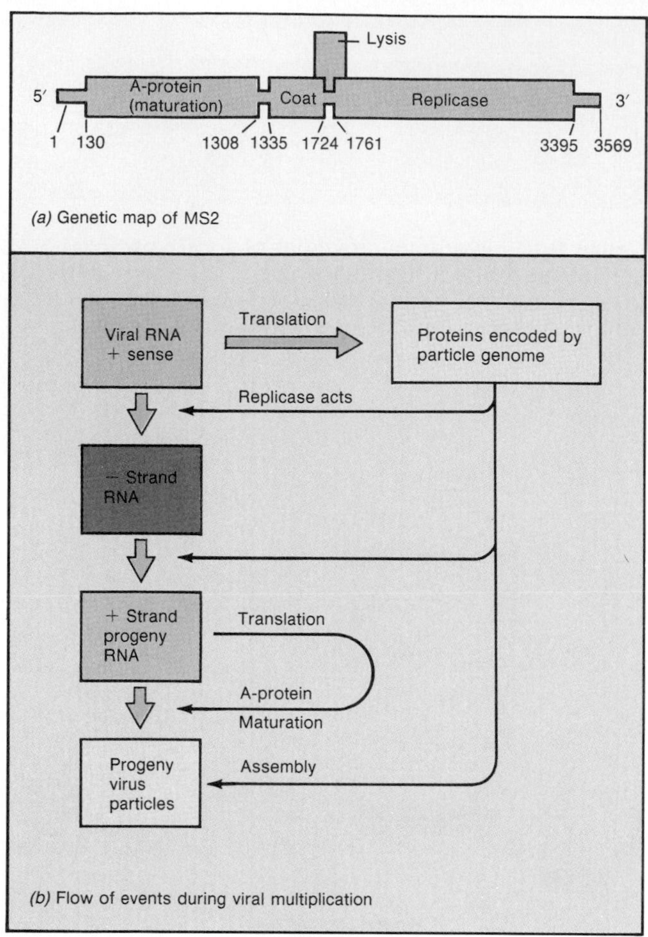

Figure 6.20 (a) Genetic map of the RNA bacteriophage MS2. (b) Flow of events during multiplication. The numbers in *a* refer to the nucleotide positions on the RNA (see Figure 6.19 for the overall structure of the RNA).

results in release of mature virus particles), and **RNA replicase**, the enzyme which brings about the replication of the viral RNA. Interestingly, the RNA replicase is a composite protein, composed partly of a virus-encoded polypeptide and partly of host polypeptides. The host proteins involved in the formation of active viral replicase are *ribosomal protein S1* (one of the subunits of the 30S ribosome), and elongation factors Tu and Ts, involved in the translation process. Thus, the virus appears to co-opt host proteins that normally have entirely distinct functions and make them become part of active viral replicase.

As noted, the viral RNA is of the *plus* (+) sense. Replicase synthesizes RNA of *minus* (−) sense using the infecting RNA as template (see Figure 6.15). After *minus* RNA has been synthesized, *plus* RNA is made from this *minus* RNA. The newly made *plus* RNA strands now serve as messengers for virus protein synthesis. The gene for the maturation protein is at the 5′ end of the RNA. Translation of the gene coding for the maturation protein (needed in only one copy per virus particle), occurs only from the newly formed *plus*-strand RNA as the replication process occurs. In this way, the amount of maturation protein needed is limited. As the virus RNA is made, it folds into a complex form with extensive secondary and tertiary structure, as shown in Figure 6.19. Of the four AUG start sites, the most accessible to the translation process is that for the coat protein. As coat protein molecules increase in number in the cell, they combine with the RNA around the AUG start site for the replicase protein, effectively turning off synthesis of replicase. Thus, the major virus protein synthesized is coat protein, which is needed in 180 copies per RNA molecule.

Another interesting feature of MS2 RNA virus is that the fourth virus protein, the *lysis* protein, is coded by a gene which overlaps with both the coat protein gene and the replicase gene (see genetic map in Figure 6.20*a*). The start of this lysis gene is not directly accessible to ribosomes. As the ribosome passes over the coat protein gene, a frame shift occasionally occurs, resulting in reading of the lysis gene. By restricting the efficiency of translation in this way, premature lysis of the cell is probably avoided. Only after sufficient coat protein is available for the assembly of mature virus particles, does lysis commence. (In another RNA phage, Qβ, the maturation protein itself also functions as a lysis protein, and a separate lysis gene as such is not present.)

Ultimately, assembly occurs and release of virions from the cell occurs as a result of cell lysis. The features of replication of these simple RNA viruses are themselves fairly simple. The viral RNA itself functions as an mRNA and regulation occurs primarily by way of controlling access of ribosomes to the appropriate start sites on the viral RNA.

6.10 Single-Stranded Icosahedral DNA Bacteriophages

A number of small bacterial viruses have genomes consisting of single-stranded DNA in circular configuration. These viruses are very small, about 25 nm in diameter, and the principle building block of the protein coat is a single protein present in 60 copies (the minimum number of protein subunits possible in an icosahedral virus), to which are attached at the vertices of the icosahedron several other proteins which make up spike-like structures (see Figure 6.17). In contrast to the RNA viruses, much of the enzymatic machinery for the replication of DNA already exists in the cell. These small DNA viruses possess only a limited amount of genetic information in their genomes, and the host cell DNA replication machinery is used in the replication of virus DNA.

The most extensively studied virus of this group is the phage designated ϕX174, which infects *Escherichia coli*. ϕX174 is of special interest because it was the first genetic element shown to have **overlapping genes**. As we have noted in Chapter 5, the genomes of cells are organized in linear fashion, with the gene coding for each protein separate from that for all other genes. In very small viruses such as ϕX174 there is insufficient DNA to code for all virus-specific proteins. ϕX174 has solved this problem by the use of overlapping genes. Thus, parts of certain nucleotide sequences are read twice, in different directions and in different reading frames (Figure 6.21). It should be noted that although the use of overlapping genes makes possible more efficient use of genetic information, it seriously complicates the evolution process, since a mutation in a region of gene overlap may affect two genes simultaneously.

As seen in the genetic map, the sequences of genes D and E overlap each other, gene E being contained completely *within* gene D. In addition, the termination codon of gene D overlaps the initiation codon of gene J by one nucleotide. The reading frame of gene E is therefore in a different phase (starting point) from that of gene D. Obviously, any mutation in gene E will also lead to an alteration in the sequence of gene D, but whether a given mutation affects one or both proteins will depend on the exact nature of the alteration (because the genetic code is degenerate; see Section 5.15). Other instances of gene overlap through use of overlapping reading frames in ϕX174 DNA are genes A/B, K/B, K/C, K/A, A/C, and D/E. Additionally, a small gene A protein, called A* protein, is formed by *reinitiation of translation* (not transcription) within gene A mRNA, with A* protein being read and terminated from the same mRNA reading frame as A protein.

The DNA of ϕX174 consists of a circular single-stranded molecule of 5386 nucleotide residues. The DNA of ϕX174 was the first DNA to be completely sequenced, a remarkable achievement when it was accomplished by Sanger and colleagues in 1977. Now, DNA sequencing is a routine procedure (see Chapter 5). The replication process of such a circular single-stranded DNA molecule is of considerable general interest, since cellular DNA replicates always in the double-stranded configuration (see Section 5.4). The DNA strand in the virion is referred to as the plus (+) strand and the complementary strand the *minus* (−) strand. Upon infection, the viral *plus* strand becomes separated from the protein coat; entrance into the cell is accompanied by the conversion of this single-stranded DNA into a double-stranded form called the *replicative form* (RF) DNA (Figure 6.21*b*). Cell-coded proteins involved in the conversion of viral DNA into RF consist of the enzyme *RNA primase* and *DNA polymerase, ligase*, and *gyrase*. No virus-coded proteins are involved in the conversion of single-stranded DNA to RF. The RF is a closed, double-stranded, circular DNA which has extensive supercoiling.

As we discussed in Section 5.4, DNA replication differs between the leading strand and the lagging strand of the DNA double helix. In cells, replication of the lagging strand involves the formation of short *RNA primers* by action of an enzyme called *RNA primase* (or *primase* for short). Such RNA primers are made at intervals on the lagging strand and are then removed and replaced with DNA by DNA polymerase.

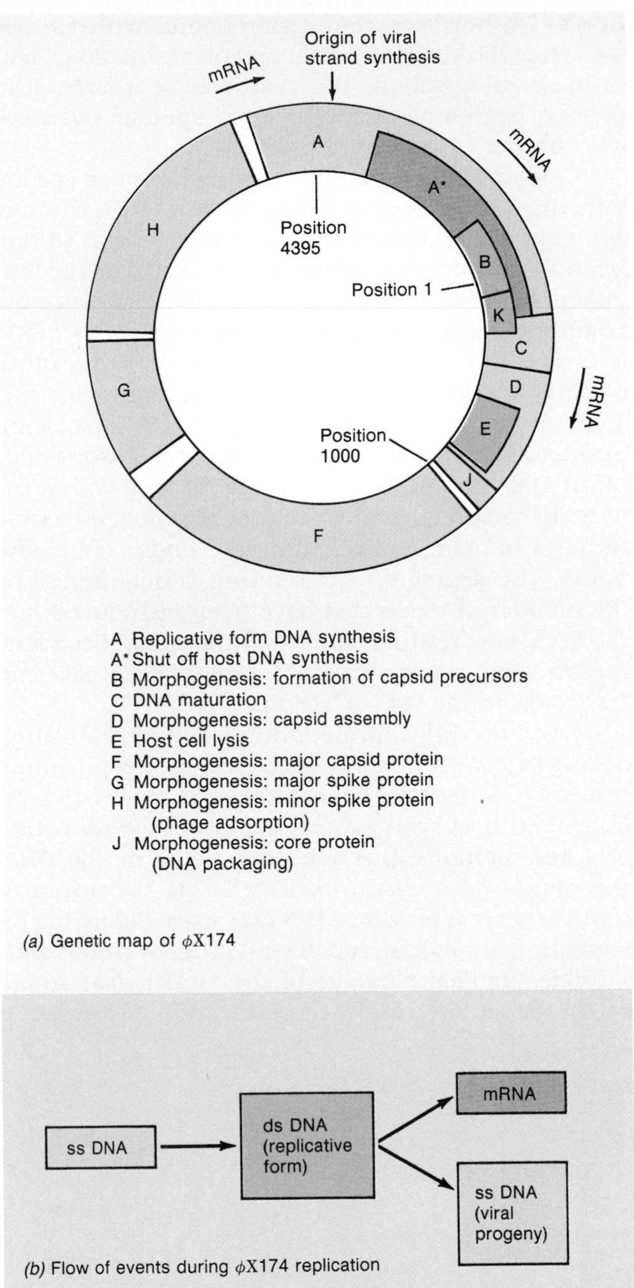

Figure 6.21 Bacteriophage ϕX174, a single stranded DNA phage. (a) Genetic map. Note the regions of gene overlap (A/B, K/B, K/C, K/A, A/C, and D/E). Intergenic regions are not colored. Protein A* is formed using only a part of the coding sequence of gene A by reinitiation of translation (see text). (b) Flow of events in ϕX174 multiplication.

In ϕX174, however, replication begins with a single stranded closed circle, a rather atypical situation. First, primase brings about the synthesis of a short RNA primer, beginning at one or more specific initiation sites on the DNA (Figure 6.22*a*).

Once priming of DNA synthesis has been carried out, the RNA primer is replaced with DNA through action of *DNA polymerase*. Continuation of DNA replication around the closed circle leads to the formation of the complete double-stranded RF. Once the complete second strand has been formed, its circle is closed with *DNA ligase* and a *DNA gyrase* introduces twists that result in supercoiling (Figure 6.22*b*). As we have discussed in Section 5.2, DNA gyrase introduces supercoils by cutting one of the two strands of the DNA double helix, holding the two ends apart without rotation, passing a distant region of the circle through the cut, and resealing the ends (see Figure 5.6*a*). The degree of supercoiling is determined by the number of twists that have been introduced into the DNA. One result of supercoiling is that it converts the DNA into a more compact form where it takes up less room in the cell or virion.

Once the RF is formed, nucleic acid replication occurs by conventional semiconservative replication, resulting in the formation of new RF molecules. As in general DNA synthesis, *initiation* of the formation of a new strand begins at a unique site on the DNA, the *origin of replication*. In ϕX174, the origin of replication is at residue 4395 (see map, Figure 6.21). Formation of single-stranded viral progeny begins with a single-stranded cleavage of the viral (*plus*) strand of the RF at the origin of replication. Cleavage is brought about by a protein called *gene A protein*; this protein also makes a covalent bond to the 5′ P of the viral strand. Asymmetric replication by the rolling circle mechanism (see Figure 5.16) results in the formation of single-stranded molecules that will become the virus progeny (Figure 6.22*b*). When the growing viral strand reaches unit length (5386 residues for ϕX174), gene A protein cleaves and then ligates the two ends of the newly synthesized single strand to give a viral single-stranded DNA. Early in infection, viral single-stranded progeny DNA are rapidly converted into RF by the mechanism already described above. Later in infection, when coat protein has accumulated, the single-stranded DNA is packaged into virions.

Viral mRNA synthesis is directed by the RF. Synthesis of mRNA begins at several major promoters on the RF, and terminates at a number of sites (see map, Figure 6.21). The polycistronic mRNA molecules are than translated into the various phage proteins. The strengths of the several promoters vary, so that some proteins are made in smaller amounts than others. Each promoter activates the transcription of a functionally related set of genes. As we have noted, protein A and protein A* are both made from the same gene, protein A* arising as a result of translation of a secondary initiation site internal to the A mRNA. Further, as we have noted, several proteins are made from mRNA transcripts formed from different reading frames from the same DNA sequences. One can truly be impressed by the efficiency with which such a small genome as that of ϕX174 can have multiple uses.

Ultimately, assembly of mature virus particles oc-

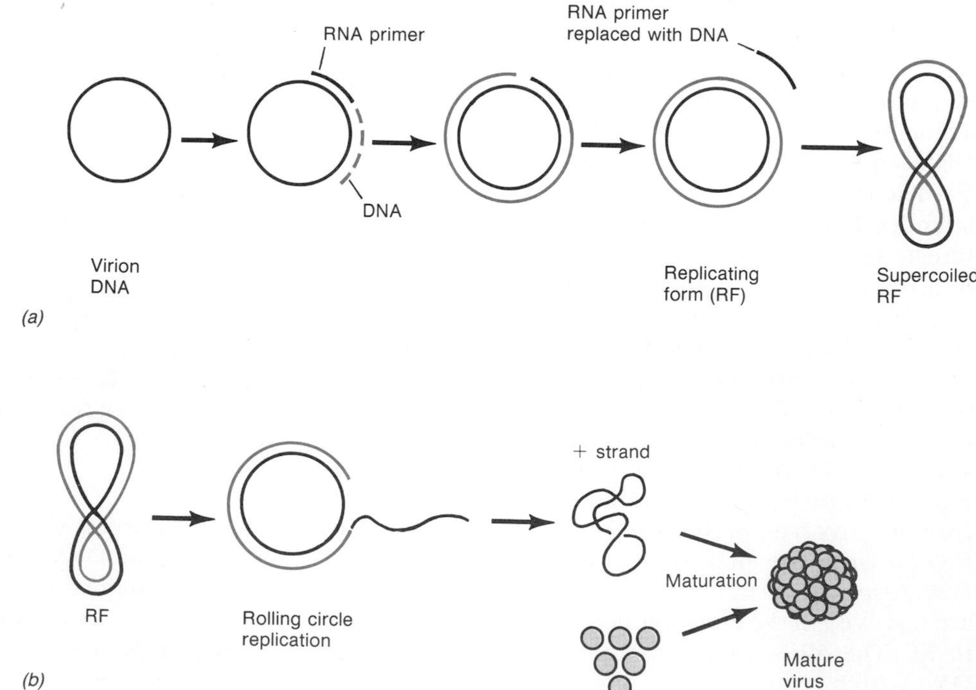

Figure 6.22 Events in the replication of ϕX174 DNA. (a) Formation of the replicating double-stranded DNA (RF). (b) Rolling circle replication leads to the formation of virus progeny.

curs. Release of virions from the cell occurs as a result of cell lysis, which involves the participation of gene E protein.

6.11 Single-Stranded Filamentous DNA Bacteriophages

Quite distinct from ϕX174 are the filamentous DNA phages, which have helical rather than icosahedral symmetry. The most studied member of this group is phage M13, which infects *Escherichia coli*, but related phages include f1 and fd. As with the small RNA viruses, these filamentous DNA phages only infect male cells, entering after attachment to the male-specific pilus. Interestingly, even though these phages are linear (filamentous) they possess *circular* DNA (Figure 6.23). The DNA is not self-complementary, however, so that the two adjacent halves of the molecule which run up and down the virus particle form loops at the ends but exhibit very little if any base pairing. Phage M13 has found extensive use as a cloning vector and DNA sequencing vehicle in genetic engineering (see Section 8.4). The virion of M13 is only 6 nm in diameter but is 860 nm long. These phages have the additional unique property of being released from the cell without killing the host cell. Thus, a cell infected with phage M13 or fd can continue to grow, all the while releasing virus particles. Virus infection causes a slowing of cell growth, but otherwise a cell is able to coexist with its virus. Plaques are thus seen only as areas of thinner cell growth in the bacterial lawn.

Many aspects of DNA replication in filamentous phages are similar to that of ϕX174 (described above in Section 6.10). The unique property, release without cell killing, can be briefly discussed. The release from the cell occurs by a budding process in which the virus particle is always released from the cell with the end containing the A protein first (Figure 6.24). Interestingly, the orientation of the virus particle across the cell membrane is the same for its entry and exit from the cell. There is no accumulation of intracellular virus particles; the assembly of mature virus particles occurs on the inner cell membrane and virus assembly is coupled with the budding process.

Several features of these phages make them useful as cloning and DNA sequencing vehicles. First, they have single-stranded DNA, which means that sequencing can be carried out by the Sanger dideoxynucleotide method (see Box, Chapter 5). Second, as long as infected cells are kept in the growing state they can be maintained indefinitely with cloned DNA, so that a continuous source of the cloned DNA is available. Third, there is an intergenic space which does not code for protein and can be replaced by variable amounts of foreign DNA.

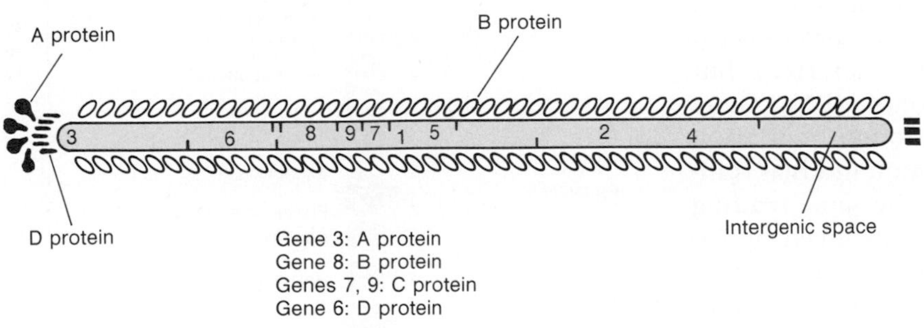

Figure 6.23 The filamentous single-stranded DNA bacteriophage fd. Orientation of the proteins and genes in the virion. Note the intergenic space which contains the origin of DNA synthesis. Gene cloning is done in this intergenic space.

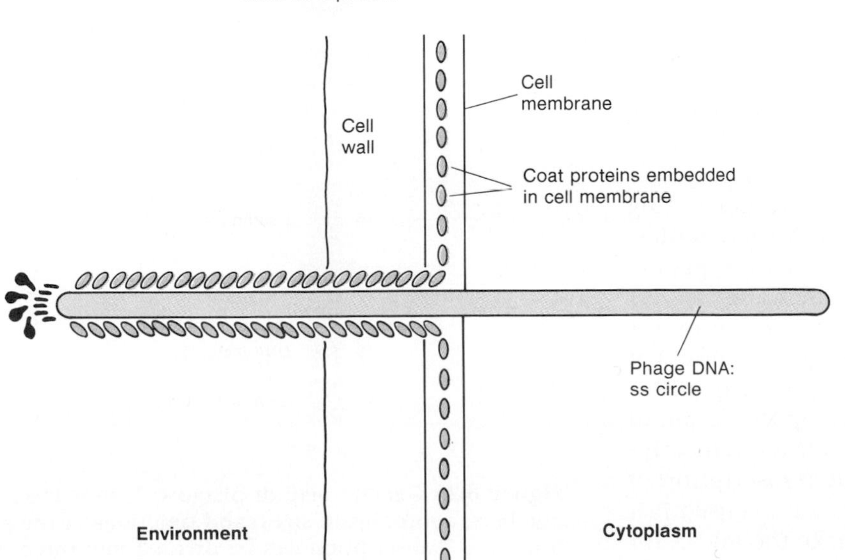

Figure 6.24 Illustration of the manner by which the virion of a filamentous single-stranded phage (such as M13 or fd) leaves an infected cell without lysis. The A protein passes first through the membrane at a site on the membrane where coat protein molecules have first become imbedded. The intracellular circular DNA is coated with dimers of another protein, gp5, which is displaced by coat protein as the DNA passes through the intact membrane.

6.12 Double-Stranded DNA Bacteriophages

Many bacterial viruses have genomes containing double-stranded DNA. Such viruses were the first bacterial viruses discovered, and have been the most extensively studied. With such a range of double-stranded DNA viruses, a wide variety of replication systems are present. In the present section, we discuss the best studied and most representative of the group, T4 and T7. The simpler, T7, will be discussed first.

Bacteriophage T7 Bacteriophage T7 and its close relative T3 are relatively small DNA viruses that infect *Escherichia coli.* (Some strains of *Shigella* and *Pasteurella* are also hosts for phage T7.) The virus particle has an icosahedral head and a very small tail (see Figure 6.17). The virus particle is fairly complex, with 5 different proteins in the head and 3–6 different proteins in the tail. One tail protein, the tail fiber protein, is the means by which the virus particle attaches to the bacterial cell surface. Only female cells of *Escherichia coli* can be infected with T7; male cells can be infected but the multiplication process is terminated during the latent period.

The nucleic acid of the T7 genome is a linear double-stranded molecule of 39,936 base pairs. The complete genome has been sequenced, and the sequence information has permitted discernment of gene structure and features of gene regulation. About 92 percent of the DNA of T7 codes for proteins. At least 25 separate genes have been characterized, but not all genes are separately coded on the DNA. Gene overlap occurs for several genes through translation in different reading frames and through internal reinitiation with one or more genes in the same reading frame. Further genetic economy is achieved by internal frame shifts within certain genes to yield longer proteins. The genetic map of T7 is shown in Figure 6.25.

When the phage particle attaches to the bacterial cell, the DNA is injected in a linear fashion, with the genes at the "left end" of the genetic map entering the cell first. Several genes at the left end of the DNA are transcribed immediately by a cell RNA polymerase, using three closely spaced promoters, generating a set of overlapping polycistronic mRNA molecules. These mRNA molecules are then cleaved by a specific RNase of the cell at 5 sites, thus generating smaller mRNA molecules which code for one to four proteins each. One of these proteins is an RNA polymerase that copies double-stranded DNA. Two other early mRNA molecules code for proteins which stop the action of host RNA polymerase, thus turning off the transcription of the early genes as well as the transcription of host genes. Thus, a host RNA polymerase is used just to copy the first few genes and to make the mRNA for the phage-specific RNA polymerase, and this phage-specific RNA polymerase is then involved in the major RNA transcription processes of the phage. This T7 RNA polymerase uses a new set of promoters that are distributed along the left-center and center portions of the genome (see Figure 6.25). It is thus seen that regulation of T7 has both negative and positive control: *negative*, by means of the formation of proteins that stop host RNA polymerase and thus shut off transcription of the early T7 genes that are recognized by this enzyme, and *positive*, by means of the formation of the new RNA polymerase which recognizes the rest of the T7 promoters. We also note that T7 is an example of a virus which strongly affects host transcription and translation processes, by producing proteins which turn off transcription of host genes. The virus also has genes coding for enzymes which degrade host cell DNA, and nucleotides from such degraded DNA

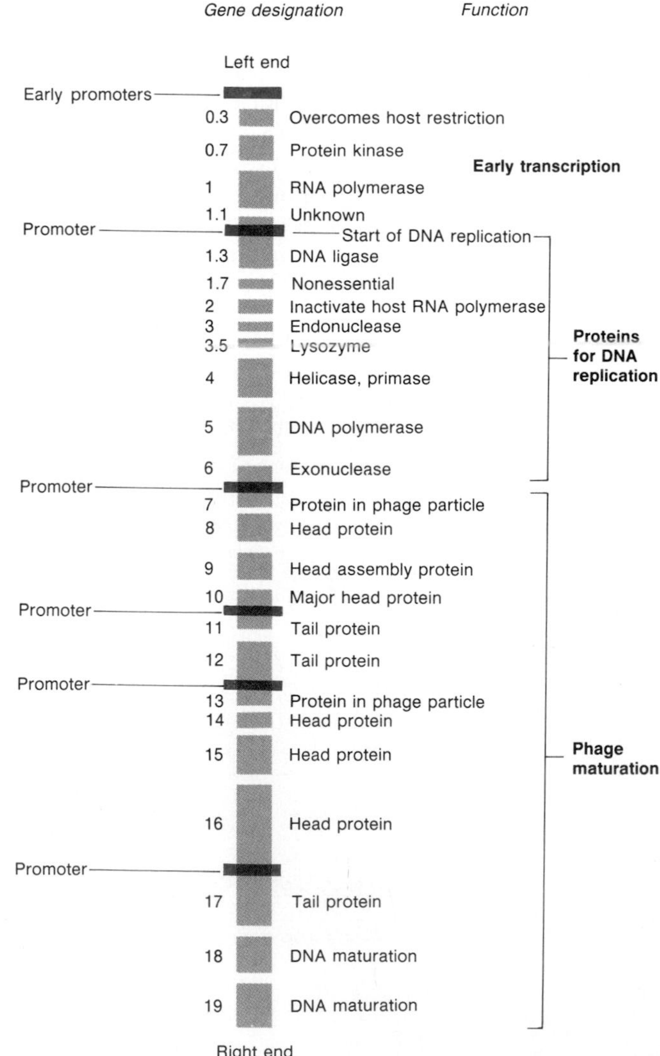

Figure 6.25 Genetic map of phage T 7, showing gene numbers, approximate sizes, and functions of the gene products. Transcription can be divided into three segments (see text). The genes are designated as numbers.

end up in virus progeny. Obviously, such a virus has profound pathological effects on its host cell.

As seen in the genetic map, the genes after gene 1.1, transcribed by the T7 RNA polymerase, code for proteins that are involved in T7 DNA synthesis, the formation of virus coat proteins, and assembly. Three classes of T7 proteins are formed: class I, made 4–8 minutes after infection, which use the cell RNA polymerase; class II, made 6–15 minutes after infection, which are made from T7 RNA polymerase and are involved in DNA metabolism; class III, made from 6 minutes to lysis, which are transcribed by T7 RNA polymerase and which code for phage assembly and coat protein. This sort of sequential pattern, commonly seen in many large double-stranded DNA phages, results in an efficient channeling of host resources, first toward DNA metabolism and replication, then on to formation of virus particles and release of virus by cell lysis.

DNA replication in T7 begins at an origin of replication (shown in Figure 6.25) at which DNA synthesis is initiated, and DNA synthesis proceeds *bidirectionally* from this origin (Figure 6.26). In both directions, an RNA primer (not shown in the figure)

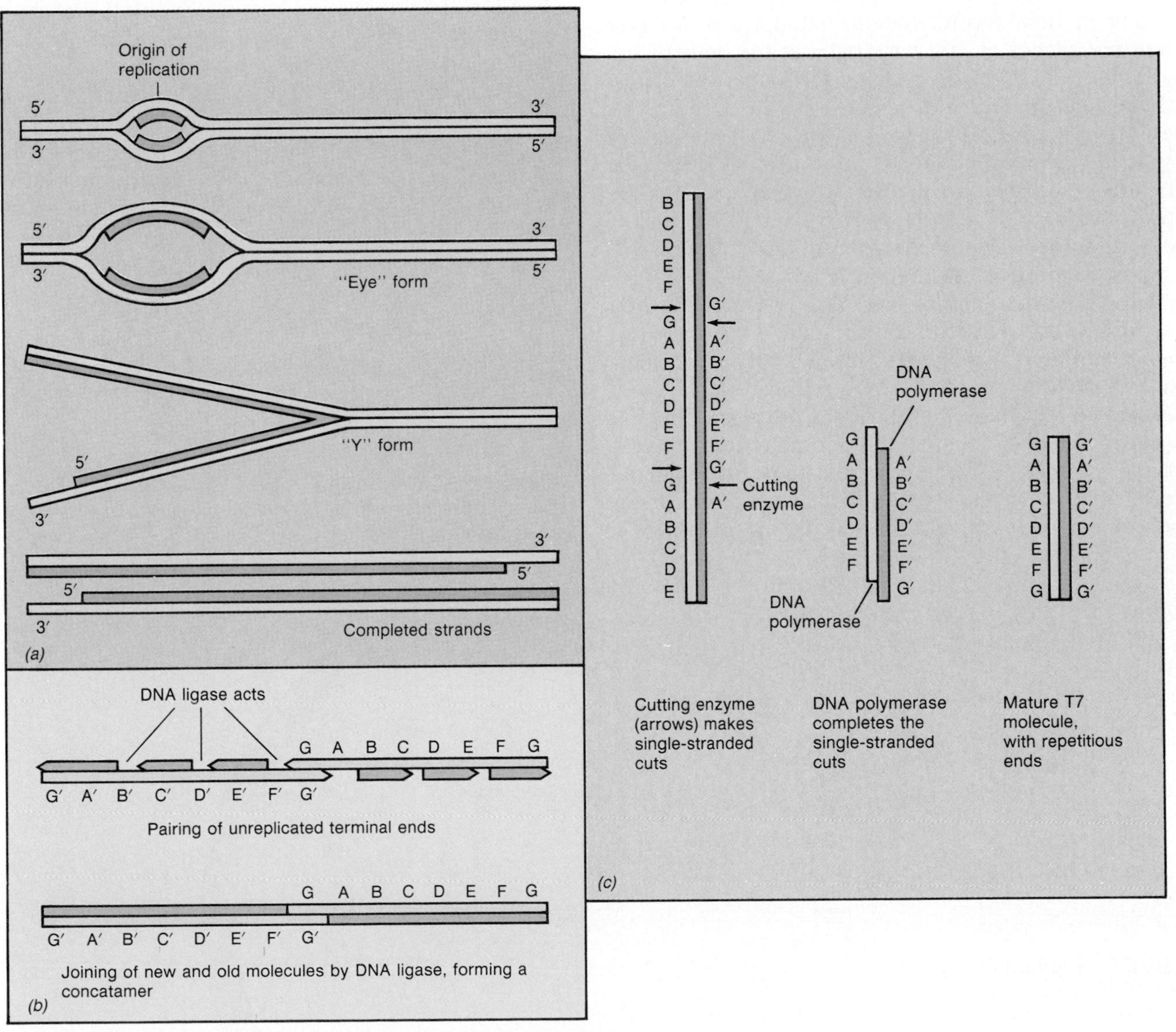

Figure 6.26 DNA replication in bacteriophage T 7, a linear, double-stranded DNA. (a) Bidirectional replication of linear T 7 DNA giving rise to intermediate "eye" and "Y" forms. (b) Formation of concatamers by joining DNA molecules at the unreplicated terminal ends. The designation of the gene is arbitrary. (c) Production of mature viral DNA molecules from long T 7 concatamers by action of cutting enzyme, an endonuclease. *Left*: the enzyme makes single-stranded cuts of specific sequences (arrows); *center*: DNA polymerase completes the single-stranded ends; *right*: the mature T 7 molecule, with repetitious ends.

is involved, but the enzyme involved in the synthesis of this primer is different for primer synthesis in the leftward and rightward direction. In the *rightward direction*, the RNA primer is synthesized by T7 RNA polymerase, whereas in the *leftward direction*, a virus-specific enzyme, T7 primase (gene 4 protein) is used. Both primers are then elongated by T7 polymerase. Replicating molecules of T7 DNA can be recognized under the electron microscope by their characteristic structures. Because the origin of replication is near the left end, Y-shaped molecules are frequently seen, and earlier in replication, bubble-shaped molecules appear.

A structural feature of the T7 DNA which is important in DNA replication is that there is a *direct terminal repeat* of 160 base pairs at the ends of the molecule. In order to replicate DNA near the 5′ terminus, RNA primer molecules have to be removed before replication is complete. There is thus an unreplicated portion of the T7 DNA at the 5′ terminus of each strand (see lower part of Figure 6.26*a*). The opposite single 3′ strands on two separate DNA molecules, being complementary, can pair with these 5′ strands, forming a DNA molecule twice as long as the original T7 DNA (Figure 6.26*b*) . The unreplicated portions of this end-to-end bimolecular structure are then completed through the action of DNA polymerase and DNA ligase, resulting in a linear bimolecule, called a *concatamer*. Continued replication can lead to concatamers of considerable length, but ultimately a cutting enzyme slices each concatamer at a specific site, resulting in the formation of virus-sized linear molecules with repetitious ends (Figure 6.26*c*).

We thus see that T7 has a much more complex replication scheme than that seen for the other bacterial viruses discussed earlier.

6.13 Large Double-Stranded DNA Bacteriophages

One of the most extensively studied groups of DNA viruses is the group sometimes called the T-even phages, which include the phages T2, T4, and T6. These phages are among the most complicated in terms of both structure and manner of multiplication replication. In the present section, we will discuss primarily bacteriophage T4, the phage of this group for which the most information is available.

The virus particle of phage T4 is structurally complex (Figure 6.8*c*). It consists of an icosahedral head which is elongated by the addition of one or two extra bands of protein hexamers, the overall dimensions of the head being 85 × 110 nm. To this head is attached

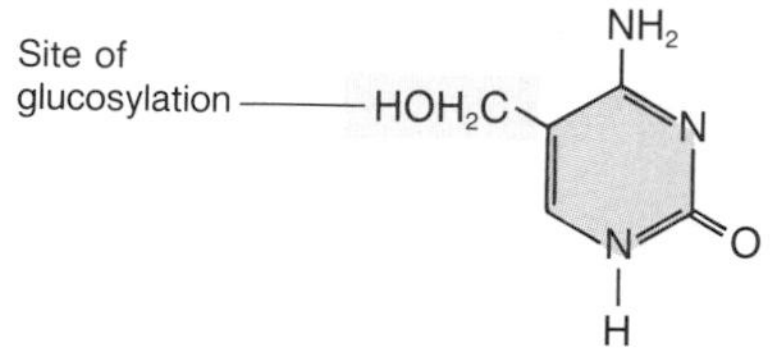

Figure 6.27 The unique base in the DNA of the T-even bacteriophages, 5-hydroxymethylcytosine. The site of glucosylation is shown.

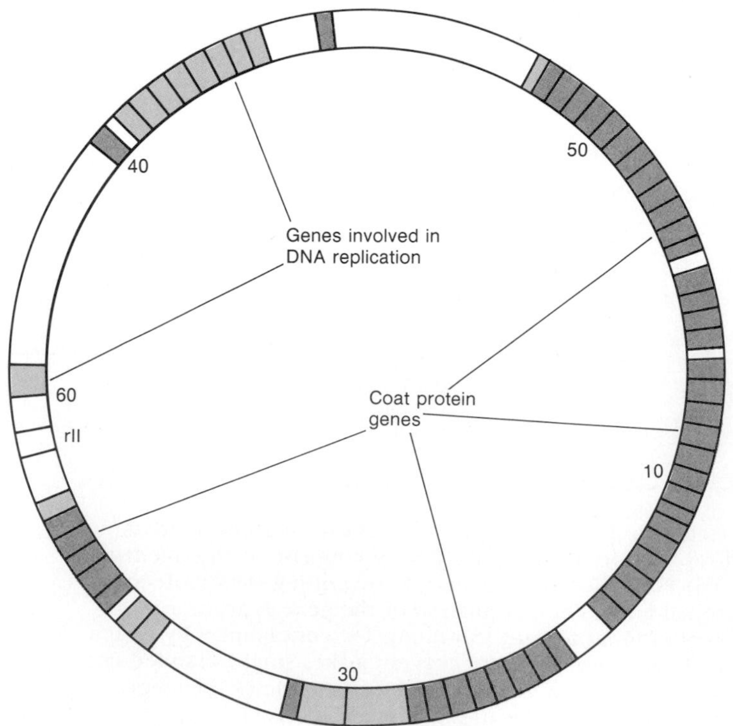

Figure 6.28 Simplified genetic map of T4. Late genes with morphogenetic functions (coat proteins and assembly), and genes with functions in DNA replication are identified. Note that although the genetic map is represented as a circle, the DNA itself is actually linear (see the discussion of the phenomenon of circular permutation in the text).

a complex tail consisting of a helical tube (25 × 110 nm) to which are connected a sheath, a connecting "neck" with "collar" and "whiskers," and a complex base plate with pins, to which are attached long jointed tail fibers. All together, the virus particle has over 25 distinct types of proteins.

The infection cycle of a susceptible cell with T4 was described briefly in Section 6.5 (see Figures 6.12–6.14). As we noted, the DNA of T4 has a total length about 650 times longer than the dimension of the head. This means that the DNA is highly folded and packed very tightly within the head.

The genome structure of T4 is quite complex. The DNA is large, with a molecular weight of about 120×10^6, and is chemically distinct from cell DNA, having a unique base, *5-hydroxymethylcytosine* instead of *cytosine* (Figure 6.27). The hydroxyl groups of the 5-hydroxymethylcytosine are modified by addition of glucosyl residues. This *glucosylated* DNA is resistant to virtually all restriction endonucleases of the host. Thus, this virus-specific DNA modification plays an important role in the ability of the virus to attack a host cell. Over 160 separate genes have been recognized in T4, of which the functions of 120 are known. These genes code not only for the complex array of coat proteins, but for a variety of enzymes and other proteins involved in the replication process itself.

The genetic map of T4 is generally represented as a circle, even though the DNA itself is linear (Figure 6.28). This "genetic circularity" arises because the DNA of the phage exhibits a phenomenon called *circular permutation*. This arises because in different T4 phage particles, the sequence of bases at each end differs (although for a given molecule the same base sequence occurs at both ends). This structure, a consequence of the way the T4 DNA replicates (see later), results in an appearance of genetic circularity even though the DNA itself is linear.

mRNA synthesis and regulation in bacteriophage T4 In bacteriophage T4, the details of regulation of replication are more complex than those of T7, but involve primarily *positive* control. T4 is a much larger phage than T7 and has many more genes and phage functions. In addition, the DNA of T4 contains the unusual base, 5-hydroxymethylcytosine (Figure 6.27) and some of the OH groups of this base are glucosylated. Thus enzymes for the synthesis of this unusual base and for its glucosylation must be formed after phage infection, as well as formation of an enzyme that breaks down the normal DNA precursor deoxycytidine triphosphate. In addition, T4 codes for a number of enzymes that have functions similar to those host enzymes in DNA replication, but are formed in larger amounts, thus permitting faster synthesis of T4-specific DNA. In all, T4 codes for over 20 new proteins that are synthesized early after infection. It also codes for the synthesis of several new tRNAs, whose function is presumably to read more efficiently T4 mRNA.

Overall, the T4 genes can be divided into three groups, for early, middle, and late proteins. The **early proteins** are the enzymes involved in DNA replication. The **middle proteins** are also involved in DNA replication. For instance, a DNA unwinding protein (DNA gyrase) is formed which destabilizes the DNA double helix, forming short single-stranded regions at which DNA synthesis can be initiated. The **late proteins** are the head and tail proteins and the enzymes involved in liberating the mature phage particles from the cell. In T4, there is no evidence for a new phage-specific RNA polymerase, as in T7. The control of T4 mRNA synthesis involves the production of proteins that modify the specificity of the host RNA polymerase so that it recognizes different phage promoters. The early promoter, present at the beginning of the T4 genome, is read directly by the host RNA

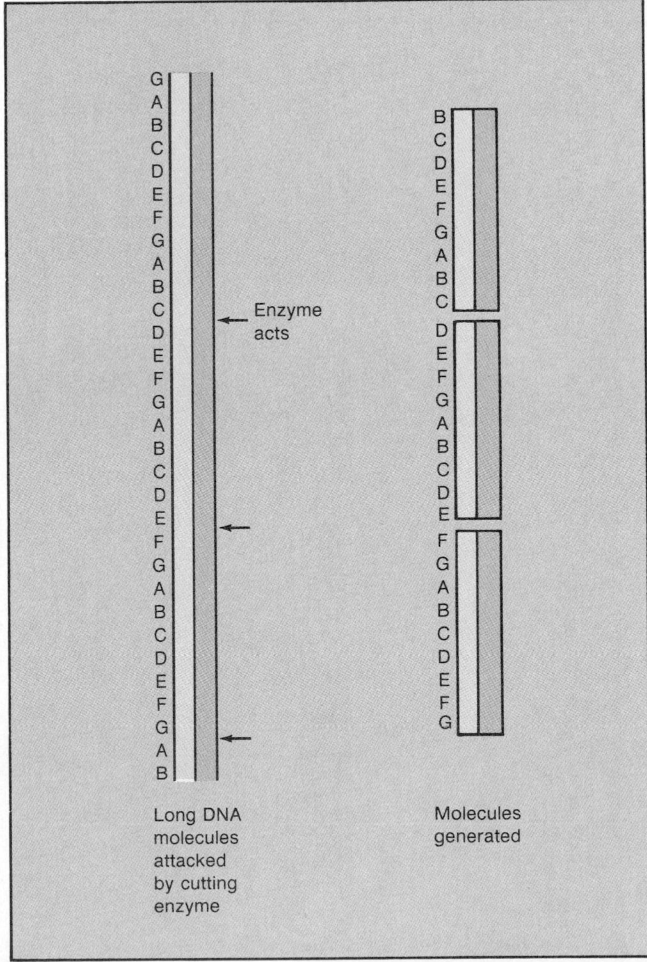

Figure 6.29 Generation in T4 of viral length molecules with permuted sequences by a cutting enzyme, which cuts off constant lengths of DNA irrespective of the sequence. Left, arrows, sites of enzyme attack. Right, molecules generated.

polymerase, and involves the function of host *sigma* factor. Host RNA polymerase moves down the chain until it reaches a stop signal. One of the early proteins blocks host sigma factor action. The early protein combines with the RNA polymerase core enzyme, and when this protein builds up, initiation of early phage genes is stopped. The RNA polymerase cores are now available to combine with new phage-specific activators, which control the transcription of the middle and late genes. The middle genes are generally transcribed along the same DNA strand as the early genes, but the late genes are transcribed along the opposite strand.

DNA replication The process of DNA replication in T4 is similar to that in T7, but in T4, the cutting enzyme which forms virus-sized fragments does not recognize specific locations on the long molecule, but rather cuts off head-full packages of DNA irrespective of the sequence. Thus each virus DNA molecule not only contains repetitious ends, but the nucleotide sequences at the ends of different molecules are different, although each molecule contains the complete sequence of viral genes (Figure 6.29). As shown,the cutting process results in the formation of DNA molecules with permuted sequences at the ends.

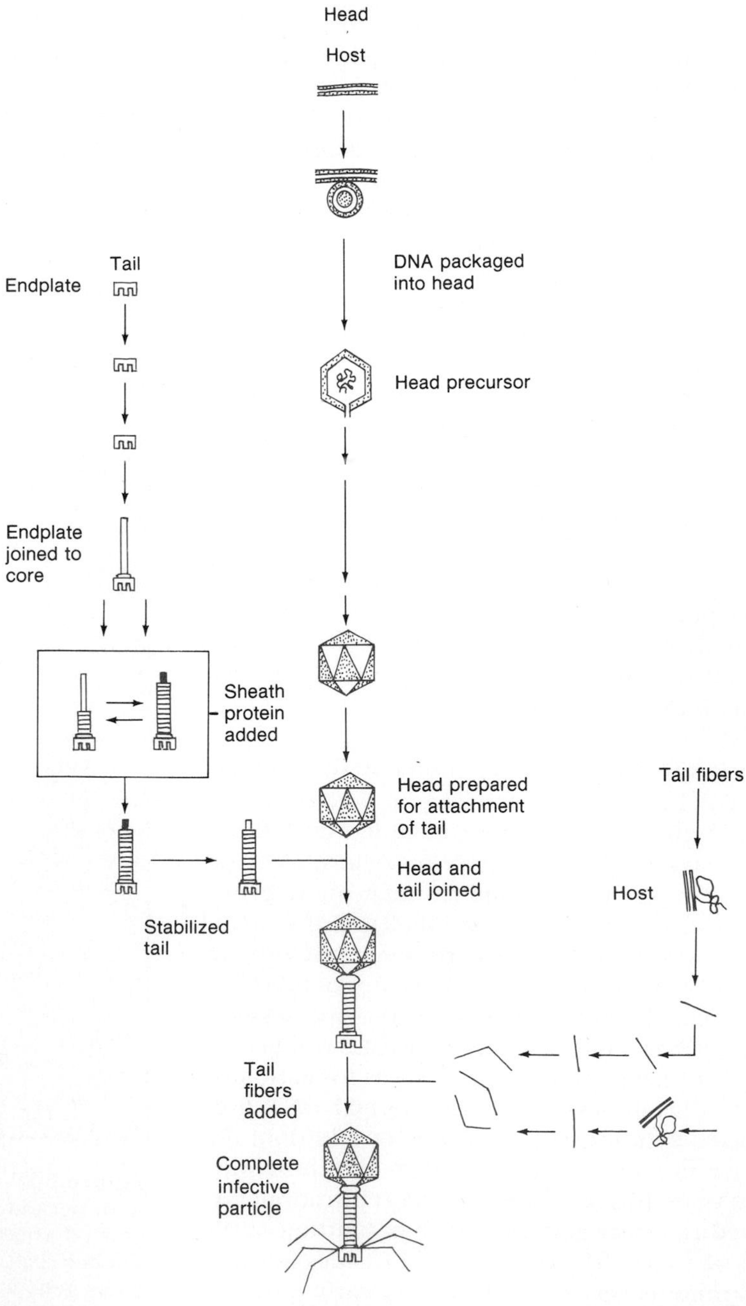

Figure 6.30 Steps in the assembly of a T4 bacterial virus.

The Phage Group

Historically, the T-even phages provided the study material for early research of the "Phage Group," a group of research workers from various universities and research institutions who spent their summers working together at the Cold Spring Harbor Laboratory on Long Island. The key members of the Phage Group, Max Delbrück, Salvador Luria, and A. D. Hershey, subsequently shared the Nobel Prize for their pioneering work. Among important concepts first uncovered from research on the T-even phages: only the nucleic acid of the virus entered the cell during infection (a discovery which provided key support for the hypothesis that DNA is the genetic material); the existence of genetic recombination in viruses; the phenomenon of restriction and modification (which led to the discovery of the restriction enzymes so important for genetic engineering); the presence in viruses of unique virus-encoded gene functions; the distinction between early and late viral functions; the phenomenon of phenotypic mixing. The first ideas of how viruses cause killing of host cells were also developed from research on the T-even phages. Selecting the T-even phages as a model system and concentrating work on them was greatly responsible for the remarkable success of the Phage Group.

Assembly and lysis In the case of T4, assembly of heads and tails occur on independent pathways. DNA is packaged into the assembled head and the tail and tail fibers are added subsequently (Figure 6.30). Somehow, the DNA is packed tightly and inserted into the empty phage head.

Exit of the virus from the cell occurs as a result of cell lysis. The phage codes for a lytic enzyme, the *T4 lysozyme*, which causes an attack on the peptidoglycan of the host cell. The burst size of the virus (the average number of phage particles per cell) depends upon how rapidly lysis occurs. If lysis occurs early, then a smaller burst size occurs, whereas slower lysis leads to a higher burst size. The wild type phage exhibits the phenomenon of *lysis inhibition*, and therefore has a large burst size, but *rapid lysis mutants*, in which lysis occurs early, show smaller burst sizes.

6.14 Temperate Bacterial Viruses: Lysogeny

Most of the bacterial viruses described above are called **virulent** viruses, since they usually kill the cells they infect. However, many other viruses, although also often able to kill cells, frequently have more subtle effects. Such viruses are called **temperate**. Their genetic material can become integrated into the host genome and is thus duplicated along with the host material at the time of cell division, being passed from one generation of bacteria to the next. Under certain conditions these bacteria can spontaneously produce virions of the temperate virus, which can be detected by their ability to infect a closely related strain of bacteria. Such bacteria, which appear uninfected but have the hereditary ability to produce phage, are called **lysogenic**.

With most temperate phages, if the host simply makes a copy of the viral DNA, lysis does not occur; but if complete virion particles are produced, then the host cell lyses. In a lysogenic bacterial culture at any one time, a small fraction of the cells, 0.1 to 0.0001 percent, produce virus and lyse, while the majority of the cells do not produce virus and do not lyse. Although only rarely do cells of a lysogenic strain actually produce virus, every cell has the potentiality. Lysogeny can thus be considered a genetic trait of a bacterial strain.

An overall view of the life cycle of a temperate bacteriophage is shown in Figure 6.31. The temperate virus does not exist in its mature, infectious state inside the cell, but rather in a latent form, called the **provirus** or **prophage** state. In considering virulent viruses we learned that the DNA of the virulent virus contains information for the synthesis of a number of enzymes and other proteins essential to virus reproduction. The prophage of the temperate virus carries similar information, but in the lysogenic cell this information remains dormant because the expression of the virus genes is blocked through the action of a specific repressor coded for by the virus. As a result of a *genetic switch* (see later), the repressor is inactivated, virus reproduction occurs, the cell lyses, and virus particles are released.

A lysogenic culture can be treated so that most or all of the cells produce virus and lyse. Such treatment, called **induction**, usually involves the use of agents such as ultraviolet radiation, nitrogen mustards, or X rays, known to damage DNA and activate the SOS system (see Section 5.18). However, not all prophages are inducible; in some temperate viruses, prophage expression occurs only by natural events.

Although a lysogenic bacterium may be susceptible to infection by other viruses, it cannot be in-

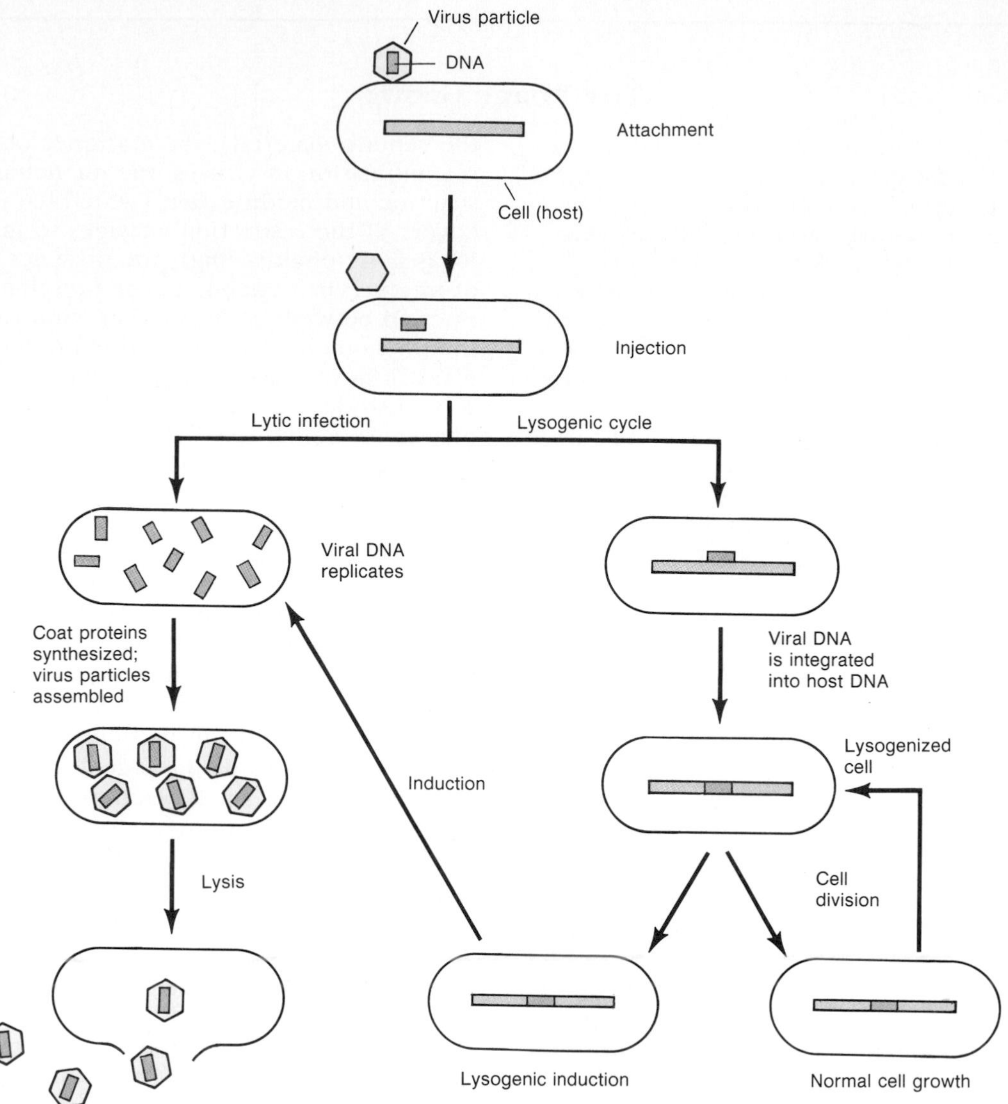

Figure 6.31 Consequences of infection by a temperate bacteriophage. The alternatives upon infection are integration of the virus DNA into the host DNA (lysogenization) or replication and release of mature virus (lysis). The lysogenic cell can also be induced to produce mature virus and lyse.

fected by virus particles of the type for which it is lysogenic. This **immunity**, which is characteristic of lysogenized cells, is conferred by the intracellular repression mechanism (see above) under the control of virus genes.

It is sometimes possible to eliminate the lysogenic virus (to "cure" the strain) by heavy irradiation or treatment with nitrogen mustards. Among the few survivors may be some cells that have been cured. Presumably the treatment causes the prophage to detach from the host chromosome and be lost during subsequent cell growth. Such a cured strain is no longer immune to the virus and can serve as a suitable host for study of virus replication.

How is it possible to determine whether a strain is lysogenic? A sensitive host is necessary—that is, a strain closely related to the presumed lysogenic strain but not infected with the prophage. In practice, a large number of related strains are obtained, either from nature or from culture collections. The presumed lysogenic strain is cultured in a suitable medium where it grows, infrequently releasing phage. The titer of phage particles in an induced lysogenic culture is typically 10^6–10^8/ml. Irradiation can be used to attempt to induce the prophage to replicate. After further incubation, the culture is filtered to remove live bacteria, and the filtrate is tested against the various test strains using the agar overlay technique described for use in assaying virus particles. If plaques are seen, it can be assumed that virus particles are present and that the strain is lysogenic. Sometimes a large number of strains must be tested to find a sensitive host. Most bacteria isolated from nature are lysogenic for one or more viruses.

Consequences of temperate virus infection What happens when a temperate virus infects a nonlysogenic organism? The virus may inject its DNA and initiate a reproductive cycle similar to that described for virulent viruses (left side, Figure 6.31, see also Figure 6.12), with the infected cell lysing and releasing more virus particles. Alternatively, when the virus injects its DNA, **lysogenization** may occur instead: the viral DNA becomes incorporated into the bacterial genetic material and the host bacterium is converted into a lysogenic bacterium (right side, Figure 6.31). In lysogenization the infected cell thus becomes genetically changed. Sensitive cells can undergo either lysis or lysogenization; which of these occurs is often determined by the action of a complex repression system, as will be described below. We thus see that the temperate virus can have a dual existence. Under one set of conditions, it is an independent entity able to control its own replication, but when its DNA is integrated into the host genetic material, replication is then under the control of the host.

Regulation of lambda reproduction One of the best studied temperate phages is lambda, which infects *E. coli*, and our knowledge of the molecular mechanisms involved in lysogenization and lytic processes in this phage is very advanced. Morphologically, lambda particles look like those of many other bacteriophages (Figure 6.32). The virus particle has an icosahedral head 64 nm in diameter, and a tail 150 nm which has helical symmetry. Attached to the tail is a single 23 nm long fiber. In addition to the major proteins of the coat, there are a number of minor coat proteins.

The nucleic acid of lambda consists of a linear double-stranded molecule of 31×10^6 molecular weight. At the 5′ terminus of each of the single-strands is a single-stranded tail of 12 nucleotides in length which are complementary (the ends of the DNA are said to be *cohesive*). Thus, when the two ends of the DNA are free in the host cell, they associate and form a double-stranded circle. In the circular form the DNA contains 48,502 base pairs, and its complete sequence is known.

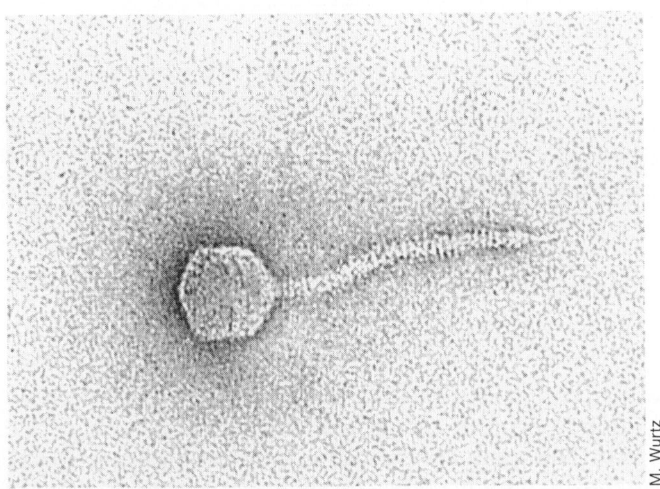

M. Wurtz

Figure 6.32 Electron micrograph by negative staining of a lambda bacteriophage particle.

Lysis or lysogenization? If lysogeny occurs, then the phage genes are maintained stably in the lysogenic state until a *switch* occurs and they are converted with high efficiency into a second state in which *lytic growth* occurs. This process is called *lysogenic induction* (center, Figure 6.31). How does the *genetic switch* from lysogeny to lysis occur?

The lambda genome has two sets of genes, one controlling lytic growth, the other lysogenic growth. Upon infection, genes promoting both lytic growth and lysogenic integration are expressed. Which pathway succeeds is determined by the competing action of these early gene products and by the influence of host factors. To understand how this works, we need to present the genetic map of lambda (Figure 6.33). The genetic map, although actually linear, can thus be oriented as a circle (because of the cyclization via the cohesive ends mentioned above). The lambda map consists of several operons, each of which controls a set of related functions. Some of the phage genes are transcribed by RNA polymerase from one strand of the double helix, and others are transcribed from the other strand. Upon injection, transcription of the phage genes which code for the lambda repressor occurs, and if repressor builds up before lytic functions are expressed, lytic reproduction is blocked. The repressor protein blocks the transcription of all later lambda genes, thus preventing expression of the genes involved in the lytic cycle.

In a lysogenic cell, a *single* phage gene is expressed continuously, the gene which codes for the lambda repressor protein. This repressor protein, which is coded for by a gene called cI, binds to two operators on the lambda DNA and thereby turns off the transcription of all the other genes of the phage genome. This is the *negative* control function of the lambda repressor. In addition, the lambda repressor turns on its *own* synthesis. This is the *positive* control function of the lambda repressor. Thus, by promoting its own synthesis, the lambda repressor ensures that no other genes except the gene coding for itself is made. In a lysogenic cell, there will usually be only one copy of the lambda genome, but about 100 active molecules of repressor protein. Therefore, there is almost always excess repressor to bind to lambda DNA and prevent the transcription of the genes necessary for lambda growth and reproduction.

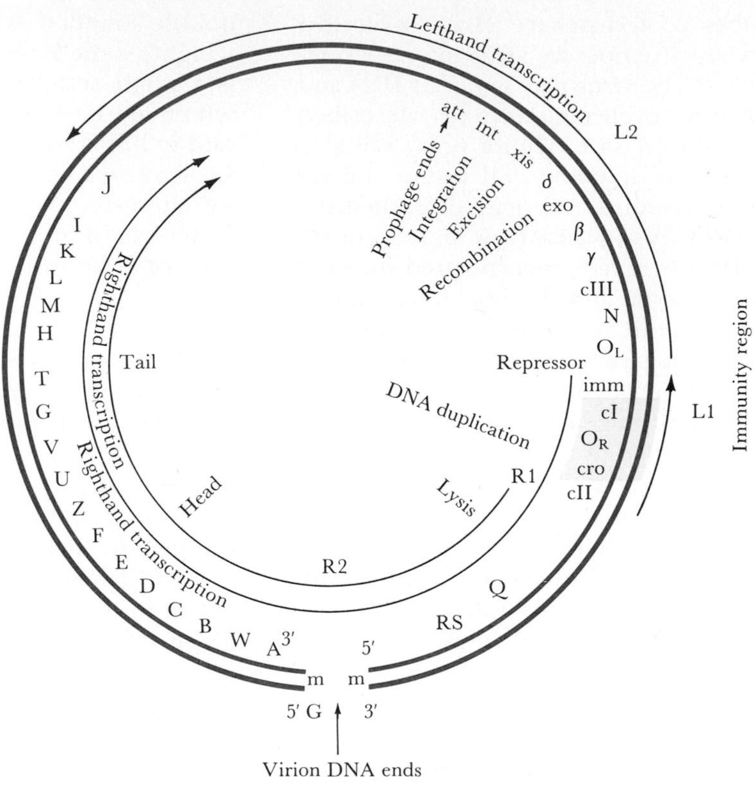

Figure 6.33 Genetic and molecular map of lambda. The two complementary DNA strands are indicated with 3′ and 5′ ends. The genes are designated by letters. att, attachment site for phage to host chromosome. DNA strand 1 is transcribed leftward (counterclockwise), with a 5′ G at its left cohesive end (m), and two operons are indicated as L1 and L2. DNA strand r is transcribed rightward (clockwise), with a 5′ A at its right cohesive end (m′); and two operons are indicated as R1 and R2. Genes of special interest: cI, repressor protein; O_R, operator for R1 transcription; O_L, operator for L2 transcription; cro, gene for second repressor, which represses transcription of L1, L2, and R1; immunity region, where cI, cro, O_R, and O_L are located; N, positive regulator counteracting rho dependent termination.

Lytic growth of lambda How, then, does lambda virus multiplication occur? In a lysogenic cell, multiplication of lambda occurs only after the repressor is inactivated. As we have noted, agents which induce lysogenic cells to produce phage are agents which damage DNA, such as ultraviolet irradiation, X rays, or DNA-damaging chemicals such as the nitrogen mustards. Upon DNA damage, a host defense mechanism called the SOS response (see Section 5.18) is brought into play (Figure 6.34). An array of 10–20 bacterial genes is turned on, some of which help the bacterium survive radiation. However, one result of DNA damage is that a bacterial protein called RecA (normally involved in genetic recombination) is turned into a special kind of protease which participates in the destruction of the lambda repressor. With lambda repressor destroyed, the inhibition of expression of lambda lytic genes is abolished. We should note that the protease activity of RecA, brought about by DNA damage, normally plays an important role in the cell's response to DNA damaging agents, by participating in the breakdown of a host protein, LexA, which represses a set of host genes involved in DNA repair (Section 5.18). Induction of bacteriophage lambda is thus an indirect consequence of the SOS response.

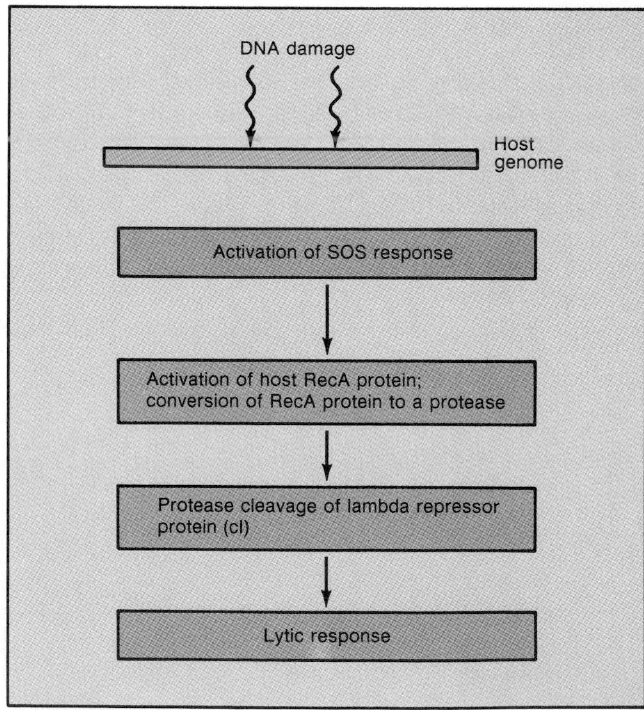

Figure 6.34 Activation of the host SOS response leads to *lysis* of a lysogenic cell.

Once the lambda repressor has been inactivated, the positive and negative control exerted by this repressor are abolished, and new transcriptional events can be initiated. The most important transcriptional event is that involved in the synthesis of another lambda protein called Cro, coded by a gene called *cro*. The gene *cro* is located almost adjacent to the gene cI which codes for lambda repressor (Figure 6.35).

The key to the genetic switch lies in the close

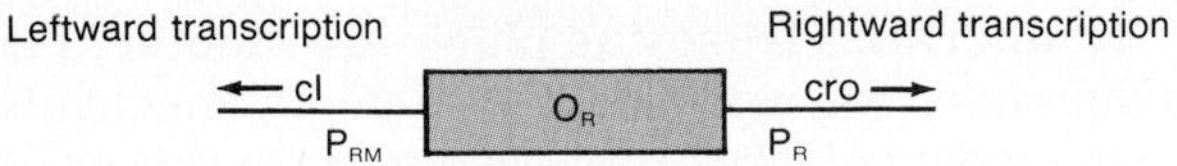

Figure 6.35 Two back-to-back promoters in the region of cI and cro control the genetic switch. When cI is present, it activates its own synthesis and ***blocks*** transcription of cro. When cI is inactivated, transcription of cro can occur, resulting in the lytic cycle. The cI (repressor) protein combines with the operator, O_R. See Figure 6.33 for the location of this region on the lambda genome.

proximity of the regulator genes for repressor and *cro* protein. These two genes are transcribed in opposite directions, beginning at different start points. In the region separating these two genes are two kinds of sites, promoters and operators, to which each of the proteins of the switch can bind. When lambda repressor is bound to its operator, it covers the *cro* promoter, whereas when Cro is bound, it covers one of the cI promoters. As we have noted (see Section 5.8), the direction in which transcription occurs on a DNA double-strand (and hence which of the two strands is read) depends upon the promoter. A promoter essentially points the RNA polymerase in the proper direction. In the case of lambda, the cI promoter points RNA polymerase "leftward," whereas *cro* promoter points the polymerase "rightward."

The lambda system provides one of the best studied examples of a genetic switch, in which one or the other of two competing genetic functions occurs. Which of the two genetic functions gets the upper hand will depend initially on chance events, but once one of the two functions has become established, it prevents the action of the other function. Only under unusual circumstances, such as when induction occurs, would the dominant genetic function be superseded.

Integration Integration of lambda DNA into the host chromosome occurs at a unique site on the *E. coli* genome. Integration occurs by insertion of the virus DNA into the host genome (thus effectively lengthening the host genome by the length of the virus DNA). As illustrated in Figure 6.36, upon in-

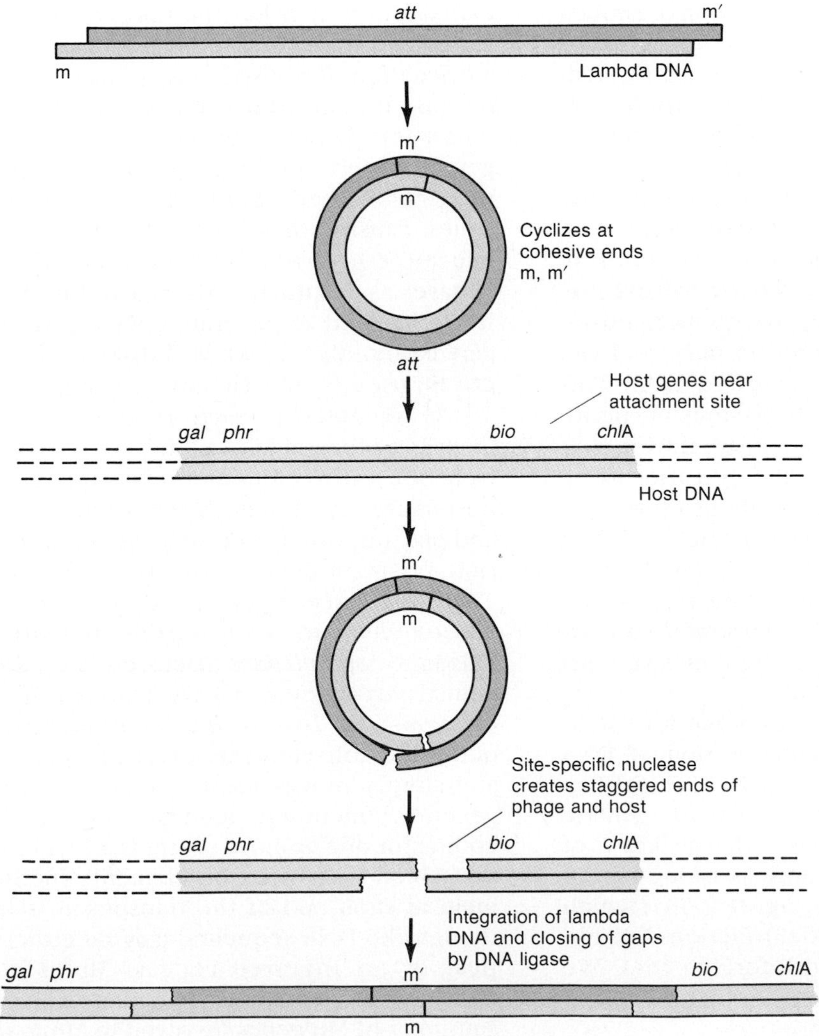

Figure 6.36 Integration of lambda DNA into the host. See the genetic map, Figure 6.33, for details of the gene order. Integration always occurs at a specific site on the host DNA, involving a specific attachment site (*att*) on the phage. Some of the host genes near the attachment site are given. A site-specific enzyme (integrase) is involved, and specific pairing of the complementary ends results in integration of phage DNA.

jection, the cohesive ends of the linear lambda molecule find each other and form a circle, and it is this circular DNA which becomes integrated into the host genome. To establish lysogeny, genes *cI* and *int* (Figure 6.33) must be expressed. As we have noted, the *cI* gene product is a protein which represses early transcription and thus shuts off transcription of all later genes. The integration process requires the product of the *int* gene, which is a site-specific topoisomerase catalyzing recombination of the phage and bacterial attachment sites (labeled *att* in Figures 6.33 and 6.36).

During cell growth, the lambda repression system prevents the expression of the integrated lambda genes except for the gene *cI*, which codes for the lambda repressor. During host DNA replication, the integrated lambda DNA is replicated along with the rest of the host genome, and transmitted to progeny cells. When release from repression occurs (see above), the lambda productive cycle occurs.

Replication Replication of lambda DNA occurs in two distinct fashions during different parts of the phage production cycle. Initially, liberation of lambda DNA from the host results in replication of a circular DNA, but subsequently linear concatamers are formed, which replicate in a different way. Replication is initiated at a site close to gene *O* (Figure 6.33) and from there proceeds in opposite directions (bidirectional symmetrical replication), terminating when the two replication forks meet. In the second stage, generation of long linear concatamers occurs, and replication occurs in an asymmetric way by **rolling circle replication** (see Figure 5.16). In this mechanism, replication proceeds in one direction only, and can result in very long chains of replicated DNA. This mechanism is efficient in permitting extensive, rapid, relatively uncontrolled DNA replication; thus it is of value in the later stages of the phage replication cycle when large amounts of DNA are needed to form mature virions. The long concatamers formed are then cut into virus-sized lengths by a DNA cutting enzyme. In the case of lambda, the cutting enzyme makes staggered breaks at specific sites on the two strands, twelve nucleotides apart, which provide the cohesive ends involved in the cyclization process.

Lambda is one of the agents of choice for use as a cloning vector for artificial construction of DNA hybrids with restriction enzymes. It has several features that make it an excellent system for genetic engineering. One feature of lambda that makes it of special use for cloning is that there is a long region of DNA, between genes *J* and *att* (Figure 6.33), which does not seem to have any essential functions for replication, and can be replaced with foreign DNA. We describe the use of lambda as a cloning vector in Section 8.3.

Temperate viruses as plasmids Another class of temperate viruses have quite a different mechanism for maintenance of the prophage state. In this group, viruses resemble plasmids. They do *not* actually become integrated into the host chromosome, but instead replicate in the cytoplasm as circular DNA molecules. Among such viruses is bacteriophage P1 of *Escherichia coli*. Although in broad features, such viruses resemble the temperate viruses such as lambda just discussed, at the level of virus replication they are, of course, quite different. Interestingly, although the plasmid prophage is not physically connected to the host DNA, phage DNA replication is closely coordinated with cell division, since only one copy of the prophage is present per host chromosome. The phage repressor is somehow involved in this regulation process.

6.15 A Transposable Phage: Bacteriophage Mu

One of the more interesting bacteriophages is that called Mu, which has the unusual property of replicating as a transposable element (see Section 5.5 for a discussion of transposons). This phage is called Mu because it is a *mutator* phage, inducing mutations in a host into which it becomes integrated. This mutagenic property of Mu arises because the genome of the virus can become inserted into the middle of host genes, causing these genes to become inactive (and hence the host which has become infected with Mu behaves as a mutant). Mu is a useful phage because it can be used to generate a wide variety of mutants very easily. Also, as we will discuss in Chapter 8, Mu can be used in genetic engineering.

As we noted in Section 5.6, a transposable element is a piece of DNA which has the ability to move from one site to another as a discrete element. Transposons are found in both procaryotes and eucaryotes, and play important roles in genetic variation (see Section 7.7 for a detailed discussion of transposition). There are three types of transposable elements: *insertion elements, transposons*, and viruses like Mu. (Also the *retroviruses* discussed later, are a group of animal viruses which have features of transposable elements). An *insertion element* is the simplest type of transposable element: it carries no detectable genes and simply moves itself around. A *transposon* is a genetic element with a unique piece of DNA, usually coding for one or more proteins, to which is attached, at each end, an insertion element. The insertion element at each end of the transposon is identical, although the DNA sequence may be either a direct repeat or an inverted repeat. Mu is a very large transposable element, carrying insertion elements and a number of Mu genes involved in Mu multiplication.

Structurally, bacteriophage Mu is a large double-stranded DNA virus, with an icosahedral head, a helical tail, and 6 tail fibers. The DNA of Mu is arranged inside the virus head as a linear double-stranded molecule. It has a molecular weight of 25 × 10^6 (38–39 kilobases). The genetic map of Mu is shown in Figure 6.37*a*. It can be seen that the bulk of the genetic information is involved in the synthesis of the head and tail proteins, but that important genes at each end are involved in replication and immunity. At the left end of the Mu DNA are 50–150 base pairs of host DNA and at the right end are 500–3000 base pairs of host DNA. These host DNA sequences are not unique and represent DNA adjacent to the location where Mu had become inserted into the host genome. When a Mu phage particle is formed, a length of DNA containing the Mu genome just large enough to fill the phage head is cut out of the host, beginning at the left end. The DNA is rolled in until the head is full but the place at the right end where the DNA is cut varies from one phage particle to another. For that reason, as shown on the genetic map, there is a variable sequence of host DNA at the right-hand end of the phage (right of the *attR* site) which represents the host DNA that has become packaged into the phage head. Each phage particle arising from a single infected cell will have a different amount of host DNA, and the host DNA base sequence of each particle from the same cell will be different. In some cases, completely empty Mu heads become filled with purely

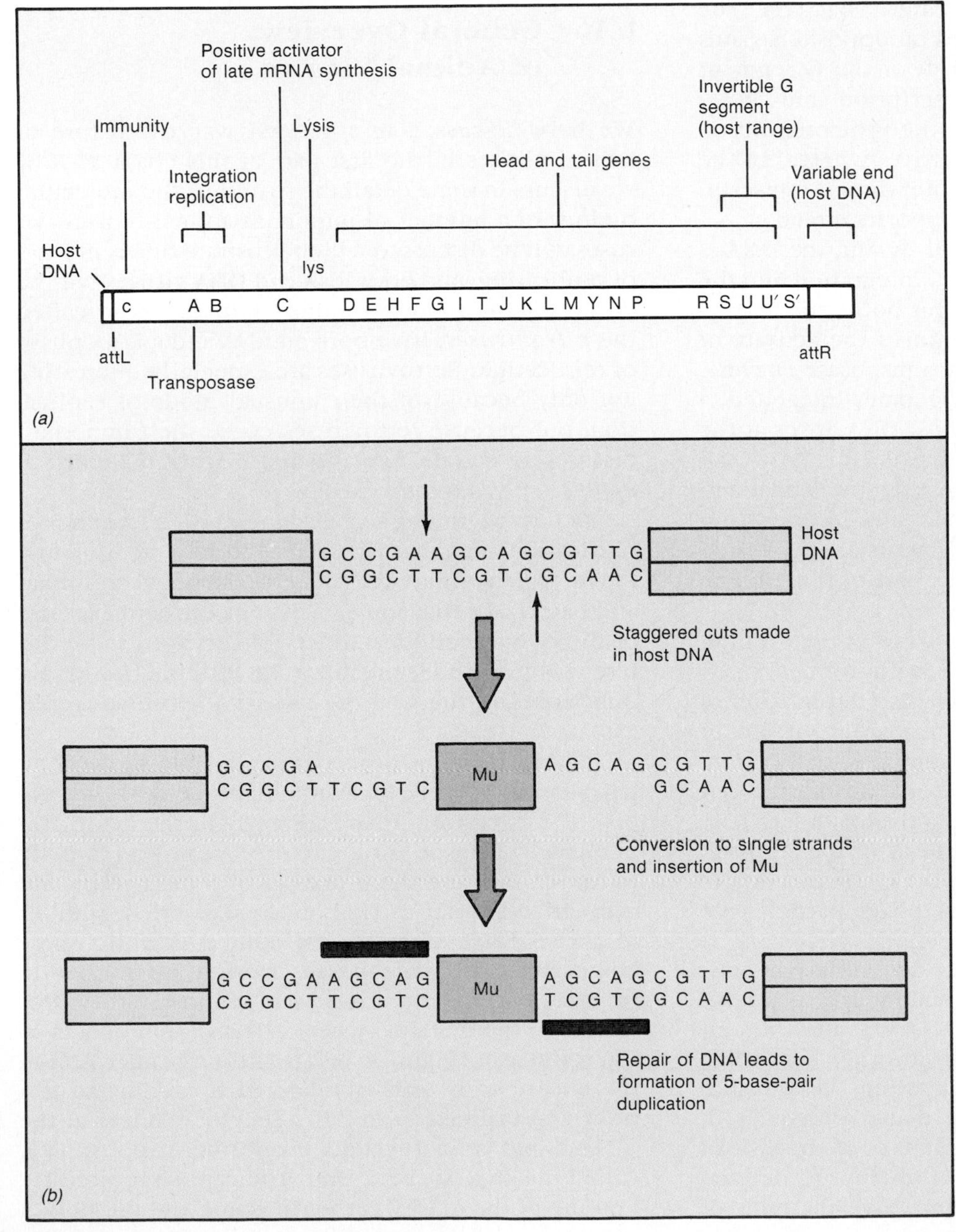

Figure 6.37 Bacteriophage Mu. (a) Genetic map of Mu. See text for details. (Confusingly, there are two G's, the G gene and the invertible G segment. These are different G's.) (b) Integration of Mu into the host DNA, showing the generation of a five-base-pair duplication of host DNA.

host DNA. Such particles can transfer host genes from one cell to another, a process called *transduction* (see Section 7.3).

As shown in the genetic map (Figure 6.37), a specific segment of the Mu genome called G (distinct from the G gene) is invertible, being present in either the orientation designated SU, or in the inverted orientation U′S′. The orientation of this segment determines the kind of tail fibers that are made for the phage. Since adsorption to the host cell is controlled by the specificity of the tail fibers, the host range of Mu is determined by which orientation of this invertible segment is present in the phage. If the G segment is in the orientation designated G^+, then the phage particle will infect *Escherichia coli* strain K-12. If the G segment is in the G^- orientation, then the phage particle will infect *Escherichia coli* strain C or several other species of enteric bacteria. The two tail fiber proteins are coded on opposite strands within this small G segment. Left of the G segment is a promoter that directs transcription into the G segment. In the orientation G^+, the promoter directing transcription of S and U is active, whereas in the orientation G^-, a different promoter directs transcription of genes S′ and U′ on the opposite strand.

Upon infection of a host cell by Mu, the DNA is injected. In contrast with lambda, integration into the host genome of Mu is essential for both lytic and lysogenic growth. Integration requires the activity of the A gene product, which is a transposase enzyme. At the site where the Mu DNA becomes integrated, a 5 base pair duplication of the host DNA arises at the target site. As shown in Figure 6.37*b*, this host DNA duplication arises because staggered cuts are made in the host DNA at the point Mu becomes inserted, and the resulting single-stranded segments are converted into the double-stranded form as part of the integration process.

Lytic growth of Mu can occur either upon initial infection, if the *c* gene repressor is not formed, or by induction of a lysogen. In either case, replication of Mu DNA involves repeated transposition of Mu to multiple sites on the host genome. Initially, transcription of only the early genes of Mu occurs, but after gene C protein, a positive activator of late RNA synthesis, is expressed, the synthesis of the Mu head and tail proteins occurs. Eventually, expression of the lytic function occurs and mature phage particles are released.

Because Mu integrates at a wide variety of host sites, it can be used to induce mutants at many locations. Also, Mu can be used to carry into the cell genes that have been derived from other host cells, a form of in vivo genetic engineering. In addition, modified Mu phage have been made artificially in which some of the harmful functions of Mu have been deleted. These phages, called Mini-Mu, are deleted for significant portions of Mu but have the ends of the phage in normal orientation. Mini-Mu phages are usually defective, unable to form plaques, and their presence must be ascertained by the presence of other genes which they carry. One set of Mini-Mu phages containing the β-galactosidase gene of the host (called Mud-*lac*, *d* for *defective*) can be detected in the integrated state if the lac gene is oriented properly in relation to a host promoter. Under these conditions, the host cell will form the enzyme β-galactosidase, which can be detected in colonies by a special color indicator. β-galactosidase-positive colonies from a β-galactosidase-negative host are thus an indication that Mud-*lac* infection has occurred. We thus see that phage Mu provides a useful tool for geneticists, as well as being an interesting bacteriophage in its own right.

6.16 General Overview of Animal Viruses

We have discussed in a general way the nature of animal viruses in the first part of this chapter. Now we discuss in some detail the structure and molecular biology of a number of important animal viruses. Viruses will be discussed which illustrate different ways of replicating, and both RNA and DNA viruses will be covered. One group of animal viruses, those called the *retroviruses*, have both an RNA and a DNA phase of replication. Retroviruses are especially interesting not only because of their unusual mode of replication, but because retroviruses cause such important diseases as certain *cancers* and *acquired immunodeficiency syndrome* (AIDS).

Before beginning our discussion of the manner of replication of animal viruses, we should mention first the important differences which exist between animal and bacterial cells. Since virus replication makes use of the biosynthetic machinery of the host, these differences in cellular organization and function imply differences in the way the viruses themselves replicate.

Bacteria, being procaryotic, do not show compartmentation of the biosynthetic processes. The genome of a bacterium relates directly to the cytoplasm of the cell. Transcription into mRNA can lead directly to translation, and the processes of transcription and translation are not carried out in separate organelles (Figure 6.38*a*). Animal cells, being eucaryotic, show compartmentation of the transcription and translation processes. Transcription of the genome into mRNA occurs in the nucleus, whereas translation occurs in the cytoplasm (Figure 6.38*b*). The messenger RNA in the eucaryote is usually modified by adding to it a **poly (A) tail** of 100 to 200 adenylic residues at the 3′ end and a methylated guanosine triphosphate, called the **cap**, at the 5′ end. The cap is required for binding of the mRNA to the ribosome and the poly *A*

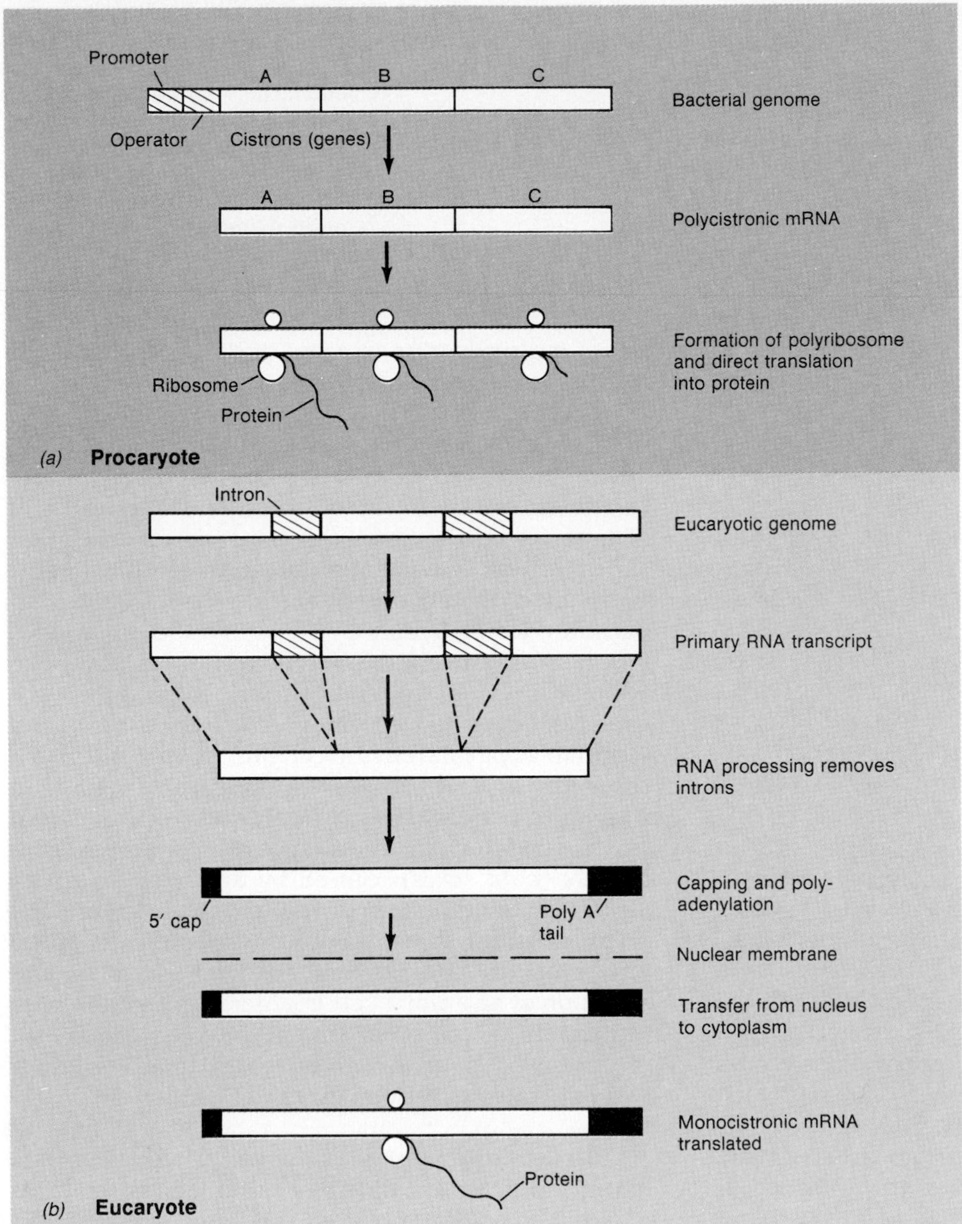

Figure 6.38 Comparison of protein synthesis processes in procaryote and eucaryote. (a) Procaryote. (b) Eucaryote.

tail may be involved in subsequent RNA processing and transfer of the mature mRNA from the nucleus to the cytoplasm. All of the protein-synthesizing machinery of the eucaryotic cell, the ribosomes, tRNA molecules, and accessory components, is in the cytoplasm, and the mature mRNA associates with the protein-synthesizing apparatus once it leaves the nucleus.

Also, as we have discussed in Section 5.8, the genes of eucaryotes are often *split*, with noncoding regions called *introns* separating coding regions (*exons*). Transcription of both the coding and noncoding regions of a gene occurs and an RNA, called the *primary RNA transcript*, is formed and is subsequently converted into the mature mRNA by a mechanism called *RNA processing*, in which the noncoding regions are excised (the cap and tail remain after RNA processing is complete). After processing, the mature mRNA is translated into protein (Figure 6.38*b*). One important distinction between eucaryotic and procaryotic mRNA is that procaryotic mRNA is generally *polycistronic*, with more than one coding region present in a single mRNA molecule, whereas eucaryotic mRNA is monocistronic. In procaryotes, during the translation process the ribosomal machinery moves down the mRNA past a stop site and initiates translation of another gene without ever leaving the mRNA. Although eucaryotic mRNA is usually monocistronic, this does not mean that only a single type of protein molecule results from the translation of a eucaryotic mRNA. Frequently, the eucaryotic mRNA codes for a single, large multifunctional protein, called a *poly-*

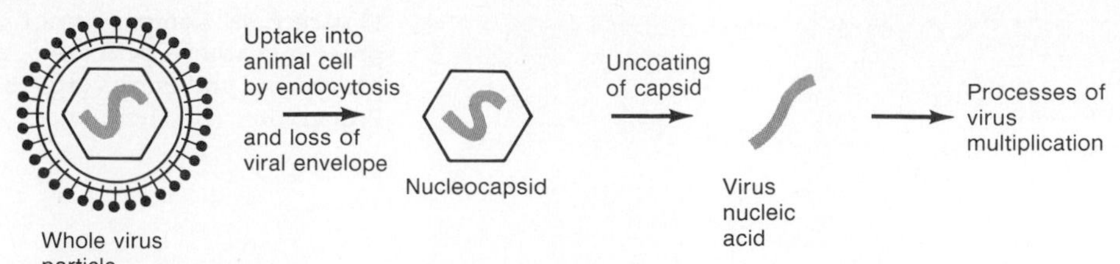

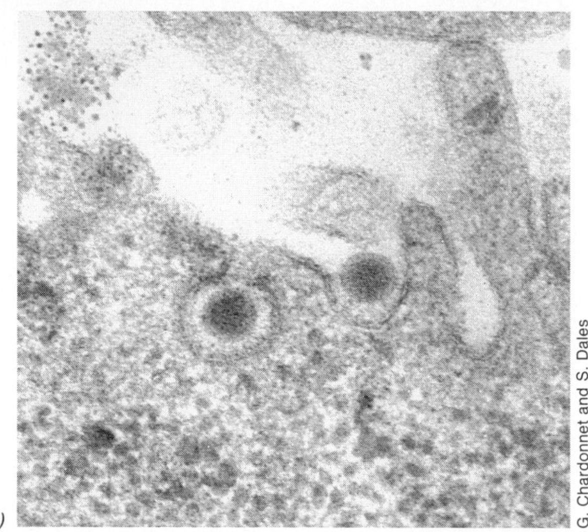

Figure 6.39 Uptake of an enveloped virus particle by an animal cell. (a) The process by which the viral nucleocapsid is separated from its envelope. (b) Electron micrograph of adenovirus particles entering a cell. Each particle is about 70 nm in diameter.

protein, which may subsequently be cleaved by a specific protease into several distinct enzymes. In other cases, the polyprotein may remain as a single multifunctional polypeptide.

We might also note another important difference between animal and bacterial cells. Bacterial cells have rigid cell walls containing peptidoglycan and associated substances (see Chapter 3). Animal cells, on the other hand, lack cell walls. This difference is important for the way by which the virus genome enters and exits the cell. In bacteria, the protein coat of the virus remains on the outside of the cell and only the nucleic acid enters. In animal viruses, on the other hand, uptake of the virus often occurs by endocytosis (pinocytosis or phagocytosis), processes which are characteristic of animal cells, so that the whole virus particle enters the cell (Figure 6.39). The separation of animal virus genomes from their protein coats then occurs inside the cell.

Classification of animal viruses Various types of animal viruses are illustrated in Figure 6.40. Most of the animal viruses which have been studied in any detail have been those which have been amenable to cultivation in cell cultures. As seen, animal viruses are known with either single-stranded or double-stranded DNA or RNA. Some animal viruses are enveloped, others are naked. Size varies greatly, from those large enough to be just visible in the light microscope, to those so tiny that they are hard to see well even in the electron microscope. In the following sections, we will discuss characteristics and manner of multiplication of some of the most important and best-studied animal viruses.

Consequences of virus infection in animal cells Viruses can have varied effects on cells. **Lytic infection** results in the destruction of the host cell (Figure 6.41). However, there are several other possible effects following viral infection of animal cells. In the case of enveloped viruses, release of the viral particles, which occurs by a kind of budding process, may be slow and the host cell may not be lysed. The cell may remain alive and continue to produce virus over a long period of time. Such infections are referred to as **persistent infections** (Figure 6.41). Viruses may also cause **latent infection** of a host. In a latent infection, there is a delay between infection by the virus and the appearance of symptoms. Fever blisters (cold sores), caused by the herpes simplex virus (see Section 6.21), result from a latent viral infection; the symptoms reappear sporadically as the virus emerges from latency. The latent stage in viral infection of an animal cell is generally not due to the integration of the viral genome into the genome of the animal cell, as is the case with latent infections by temperate bacteriophages.

Viruses and cancer A number of animal viruses have the potential to change a cell from a normal cell to a cancer cell (Figure 6.41 and Table 6.1). We discussed the essential nature of cancer briefly in Section 6.3. This process, called **transformation**, can be induced by infections of animal cells with certain kinds of viruses. One of the key differences between normal cells and cancer cells is that the latter have different requirements for growth factors. Rapidly growing cells pile up into accumulations that are visible in culture as **foci of infection**. Because cancerous cells in the animal body have fewer growth requirements, they grow profusely, leading to the formation of large masses of cells, called *tumors*. The term **neoplasm** is often used in the medical literature to describe malignant tumors.

Not all tumors are seriously harmful. The body is able to wall off some tumors so that they do not spread; such noninvasive tumors are said to be **benign**. Other tumors, called **malignant**, invade the body and destroy normal body tissues and organs. In advanced

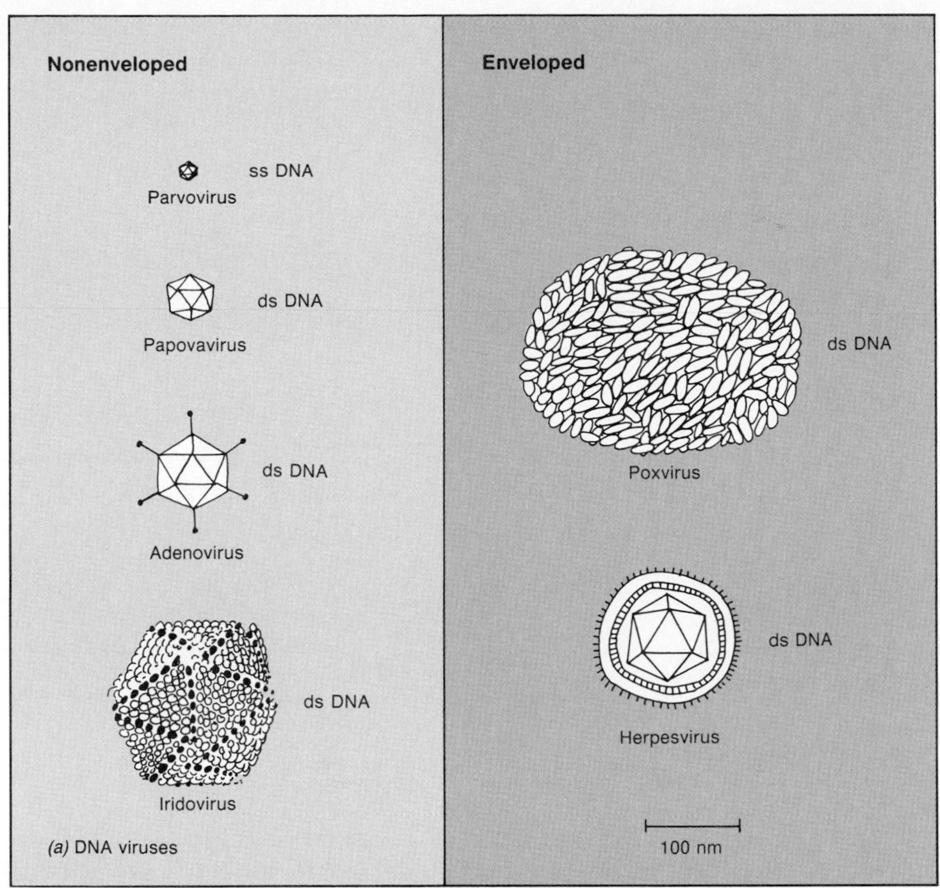

Nonenveloped

ss RNA
Picornavirus

RNA
Orbivirus

ds RNA
Reovirus

100 nm

(b) RNA viruses

Enveloped all ss RNA

Togavirus

Rhabdovirus

Orthomyxovirus

Bunyavirus

Coronavirus

Arenavirus

Retrovirus

Paramyxovirus

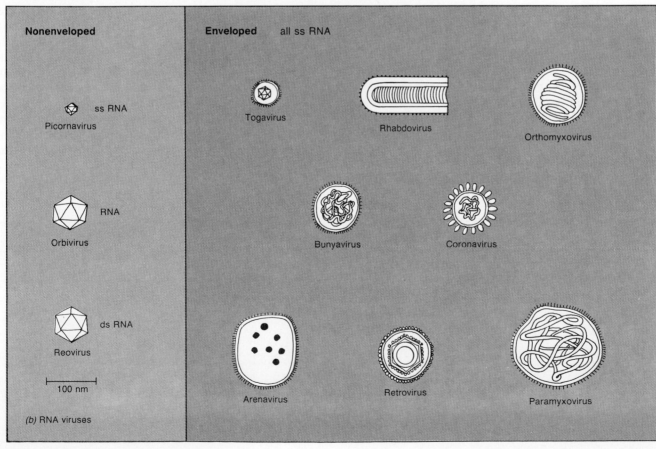

Figure 6.40 The shapes and relative sizes of vertebrate viruses of the major taxonomic groups. Bar = 100 nm.

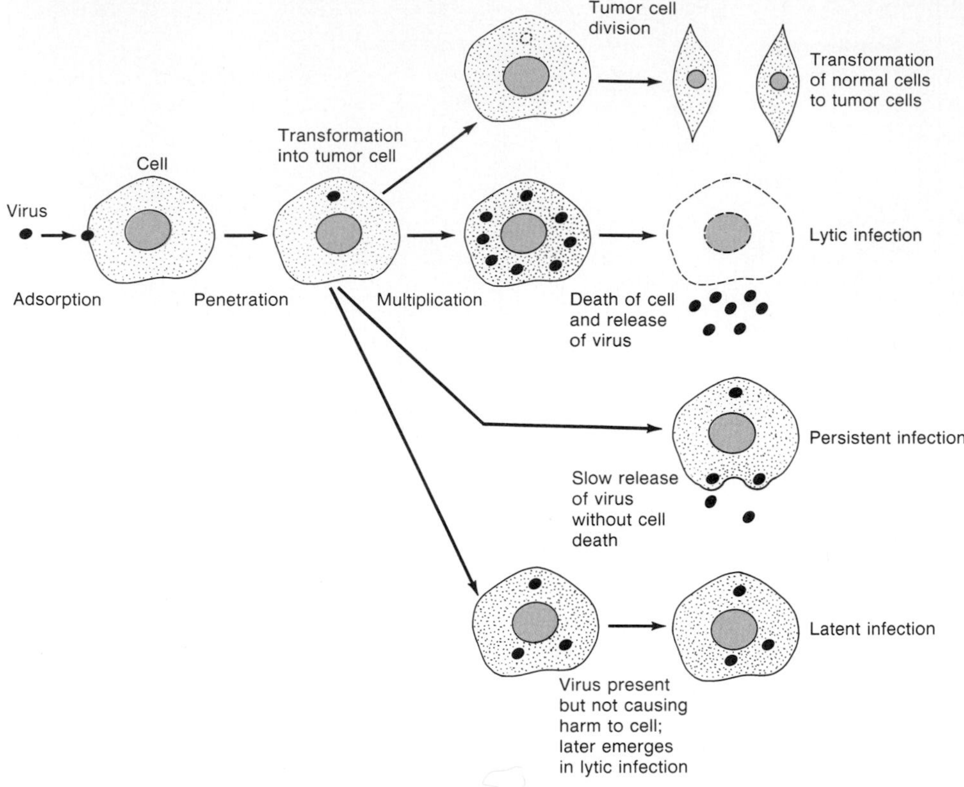

Figure 6.41 Possible effects that animal viruses may have on cells they infect.

stages of cancer, malignant tumors may develop the ability to spread to other parts of the body and initiate new tumors, a process called **metastasis**.

How does a normal cell become cancerous? The process can be broken down into several stages. In the first step, *initiation*, genetic changes in the cell occur. This step may be induced by certain chemicals, called *carcinogens*, or by physical stimuli, such as ultraviolet radiation or X rays. Certain viruses also bring about the genetic change that results in initiation of tumor formation. Once initiation has occurred, the potentially cancerous cell may remain dormant, but under certain conditions, generally involving some environmental alteration, the cell may become converted into a tumor cell, a process called **promotion**. Once a cell has been promoted to the cancerous condition, continued cell division can result in the formation of a tumor.

Although the ability of viruses to cause tumors in animals has been proved for many years, the relationship of viruses to cancer in humans has, in most cases, been uncertain. It is difficult to prove the viral origin of a human cancer because of the difficulties of carrying out the necessary experimentation. However, it is now well established that certain specific kinds of human tumors do have a viral origin. A summary of some of the human cancers with definite viral origins is given in Table 6.1.

Table 6.1 Some human cancers which may be caused by viruses

Cancer	Virus	Family	Genome
Adult T-cell leukemia	Human T-cell leukemia virus (type 1)	Retrovirus	RNA
Burkitt's lymphoma	Epstein-Barr virus	Herpes	DNA
Nasopharyngeal carcinoma	Epstein-Barr virus	Herpes	DNA
Hepatocellular carcinoma (liver cancer)	Hepatitis B virus	Not yet classified	DNA
Cervical cancers (?)	Herpes simplex type 2 virus (?)	Herpes	DNA
Skin and cervical cancers	Papilloma virus	Papova	DNA

Replication cycle of animal viruses In Section 6.5 we described the *one-step growth curve* of a bacterial virus. One-step growth curves are also obtained for animal viruses. Such curves exhibit the same overall features that are found in the bacterial viruses, but the rate of multiplication is much slower. A typical one-step growth curve of an animal virus is given in Figure 6.42. This figure also presents data on the length of the eclipse period and the total duration of the multiplication cycle for several well-studied animal viruses. Note that the multiplication cycles of animal viruses range from around eight hours to as long as 48 hours.

In the rest of this chapter, we discuss in detail the molecular events in the replication of some well-studied animal viruses.

Figure 6.42 One-step growth curve of animal viruses.

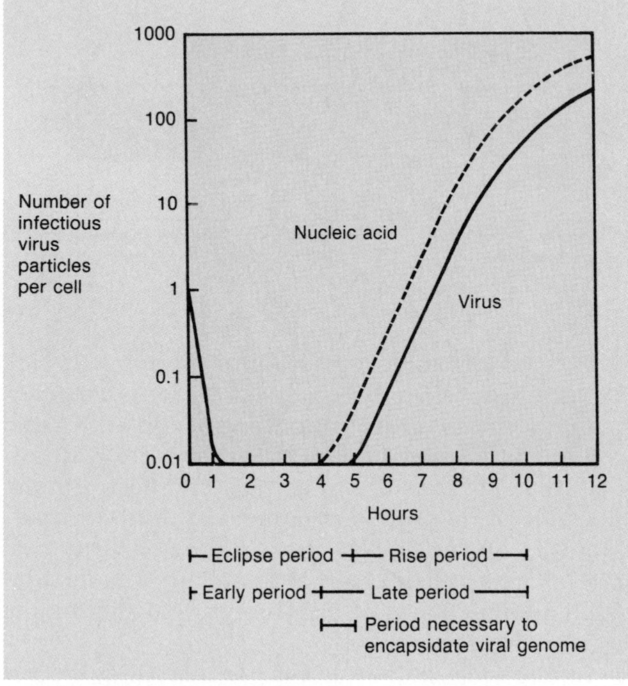

Approximate duration of eclipse period and of entire multiplication cycle		
Virus	**Eclipse period (hours)**	**Total multiplication cycle (hours)**
Poxvirus: vaccinia virus	4	24
Herpesvirus: herpes simplex virus	3–5	24–36
Adenovirus	8–10	48
Papovavirus: polyoma virus	12–14	48
Poliovirus	1–2	6–8
Togavirus: Sindbis virus	2	10
Reovirus	4	15
Orthomyxovirus: influenza virus	3–5	18–36
Rhabdovirus: vesicular stomatitis virus	2	8–10

6.17 Small RNA Animal Viruses

An important group of animal viruses is the *picornavirus family*, which contains such important viruses infecting humans as the *polioviruses*, the *cold viruses*, and the *hepatitis A virus*. These viruses are called *picornaviruses* because they are very small viruses (30 nm in diameter, *pico* means small) and contain single-stranded RNA. The virus particle has a simple icosahedral structure with 60 morphological units per virus particle, each unit consisting of four distinct proteins (Figure 6.43*a*).

The RNA is a linear molecule of about 7500 bases in length and has a molecular weight of 2.6×10^6 daltons. At the 5′ terminus of the viral RNA is a protein, called the *VPg protein*, which is attached covalently to the RNA. At the 3′ terminus of the RNA is a poly (A) tail. RNA molecules which lack the poly (A) tail are not infectious, but the VPg protein is not essential for infectivity.

An overview of the manner of multiplication of poliovirus is illustrated in Figure 6.43*b*. The RNA of the virus acts directly as a messenger RNA. It is therefore of plus sense, as we outlined in Figure 6.15. The virus RNA is monocistronic but codes for all of the proteins of the virus in a single polyprotein that is later cleaved into the individual proteins. The coat proteins are coded for by sequences at the 5′ end of the molecule and the proteins necessary for replication are coded for at the 3′ end. The whole replication process occurs in the cytoplasm.

At initial infection, the virus particle attaches to a specific receptor on the surface of a sensitive cell and enters the cell. Once inside the cell, the virus particle is uncoated, and the free RNA associates with ribosomes to form a polysome. The viral RNA is then translated from a single initiation point several hundred bases from the 5′ end into a large protein precursor, the polyprotein mentioned above, which has a molecular weight of about 240,000. This giant protein is then cleaved into about 20 smaller pro-

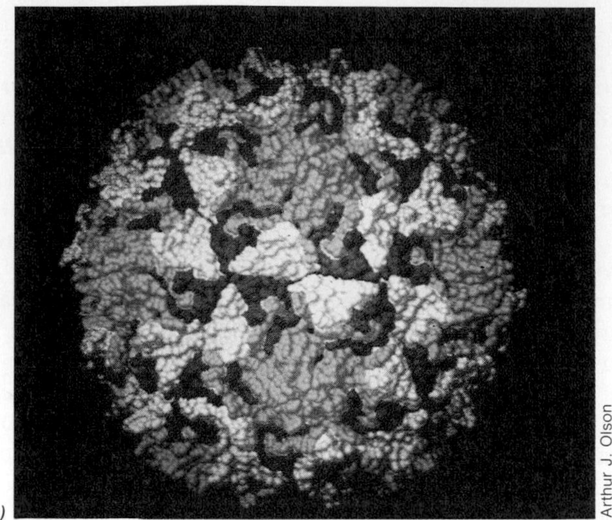

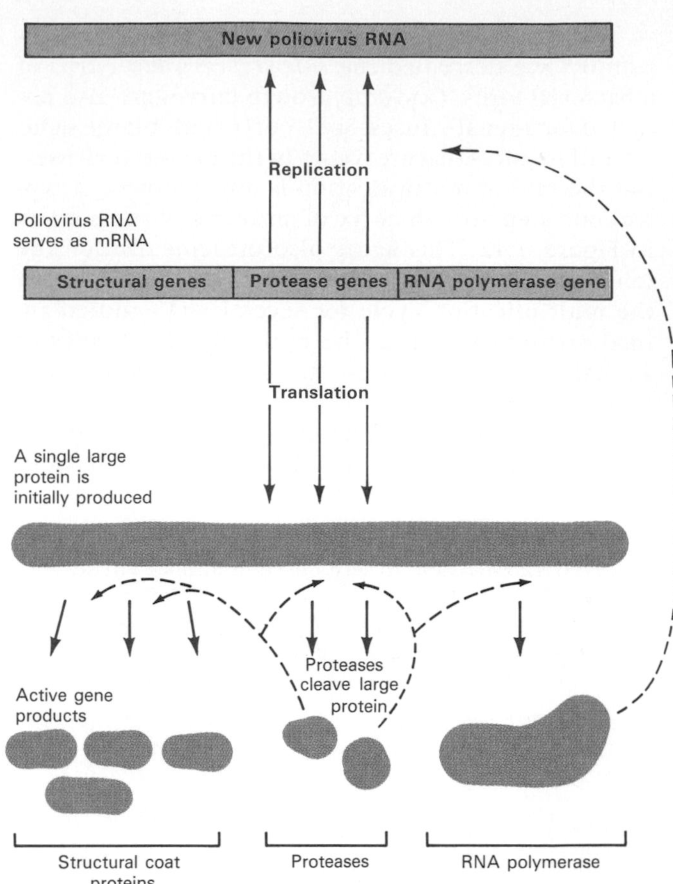

Figure 6.43 (a) The poliovirus particle. This is a computer model based on electron diffraction analysis of virus crystals. The various structural proteins are shown in distinct colors. (b) The reproduction of poliovirus. The single-stranded RNA of the virus is translated directly as a messenger RNA, with the production of one large protein molecule. This protein is cleaved, leading to the production of the active viral proteins, including the structural coat protein and the RNA polymerase which brings about the replication of the poliovirus RNA. The assembly of intact poliovirus from coat protein molecules and RNA then follows.

teins, among which are the four *structural proteins* of the virus particle, the RNA-linked *protein VPg*, an *RNA polymerase* which is responsible for synthesis of minus-strand RNA, and at least one *virus-coded protease* which carries out the cleavage process. This cleavage process, called *posttranslational cleavage*, occurs in a wide variety of animal viruses as well as in normal cell metabolism in animal cells.

Replication of poliovirus Replication of viral RNA begins within a short time after infection and is catalyzed by the RNA polymerase (replicase) made in the process described in the previous paragraph. This replicase transcribes the viral RNA, of *plus* complementarity, into an RNA molecule of *minus* complementarity. The poly (A) tail of the virus RNA becomes a poly (U) tail of the *minus* strand. This minus strand then serves as a template for repeated transcription of progeny plus strands. Some of the progeny *plus* strands may again be transcribed into *minus* strands, and as many as 1000 *minus* strands may subsequently be present in the cell. From these *minus* strands, as many as a million *plus* strands may ultimately be formed. Both the *plus* and *minus* strands become covalently linked to the tiny protein VPg (only 22 amino acids long), and it is thought that VPg serves as a primer for transcription, or is added after synthesis. Viral RNA molecules which serve as mRNA lack VPg and are not assembled into mature virus particles.

Once virus multiplication begins, host RNA and protein synthesis are inhibited. Host protein synthesis is inhibited as a result of cleavage of a host protein, the cap-binding protein complex required for translation of capped mRNAs, by a protease. Virus coat proteins are made by cleavage of precursor molecules which are much longer than the protein subunits of the virus particle.

At one time, polio was a major infectious disease of humans, but the development of an effective vaccine (see Chapter 12) has brought the disease completely under control.

6.18 Negative-Strand RNA Viruses

In a number of RNA viruses of animals, the RNA does not serve directly as a messenger, but is transcribed into a complement which functions as the mRNA. As discussed in Section 6.6, it is conventional to express the complementarity of the mRNA as *plus*, so that if the viral RNA is of opposite complementarity it is called *minus*. This group of viruses is then called *minus* strand or *negative-strand* RNA viruses.

Rhabdoviruses The most important human pathogen which is a negative-stranded RNA virus is the rabies virus, which causes the important disease rabies in animals and humans (see Section 15.8). Rabies virus is called a *rhabdovirus*, from *rhabdo* meaning *rod*, which refers to the shape of the virus particle. Another rhabdovirus which has been extensively studied is vesicular stomatitis virus (VSV), a virus which causes the disease *vesicular stomatitis* in cattle, pigs, horses, and sometimes in humans.

The rhabdoviruses are enveloped viruses, with an extensive and rather complex lipid envelope surrounding the nucleocapsid (Figure 6.44). The virus particle is bullet-shaped, about 70 nm in diameter and 175 nm long. The nucleocapsid is helically symmetrical and makes up only a small part of the virus particle weight (about 2–3 percent of the virus particle is RNA, in contrast to the 30 percent or more RNA in nonenveloped RNA viruses). Not only does the virus contain an extensive lipid envelope, but it contains several enzymes which are essential for the infection process: *RNA polymerase, RNA methylase*, and some capping enzymes. The presence of RNA polymerase is interesting because these negative-strand viruses cannot act as messengers directly, but must first be transcribed into the plus complement. Since host enzymes which transcribe RNA into RNA do not exist, it is understandable that the virus particle must possess its own RNA polymerase.

The RNA of the rhabdoviruses is transcribed inside the cytoplasm of the cell into two distinct kinds of RNA (Figure 6.45). The first type of RNA synthesis results in a series of messenger RNAs which are made from the various genes of the virus. The second is a *plus* strand RNA which is a *copy* of the complete viral genome. These long *plus* strand RNAs then serve as templates for the synthesis of the *negative-strand* RNA molecules of the new crop of virus particles (yet to come). Each mRNA is monocistronic, coding for a single protein. A mechanism exists to ensure that transcription stops at the end of each virus gene, and that a series of adenylic acid residues (a poly (A) tail) is put on the end of the RNA. Once the mRNA for the

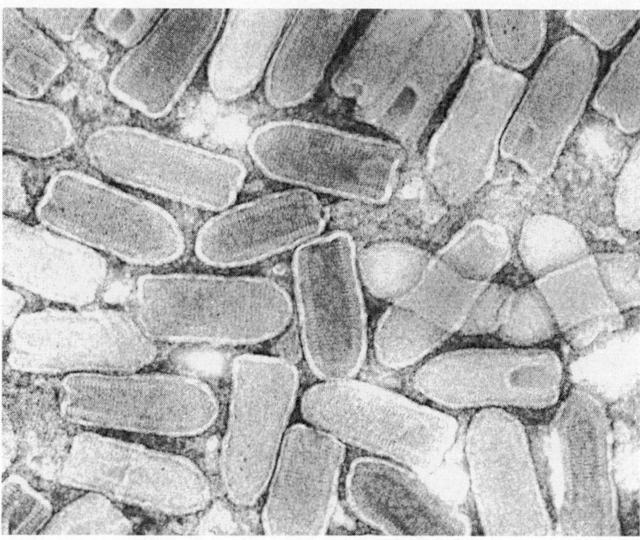

Figure 6.44 Electron micrograph of a rhabdovirus (vesicular stomatis virus). A particle is about 65 nm in diameter.

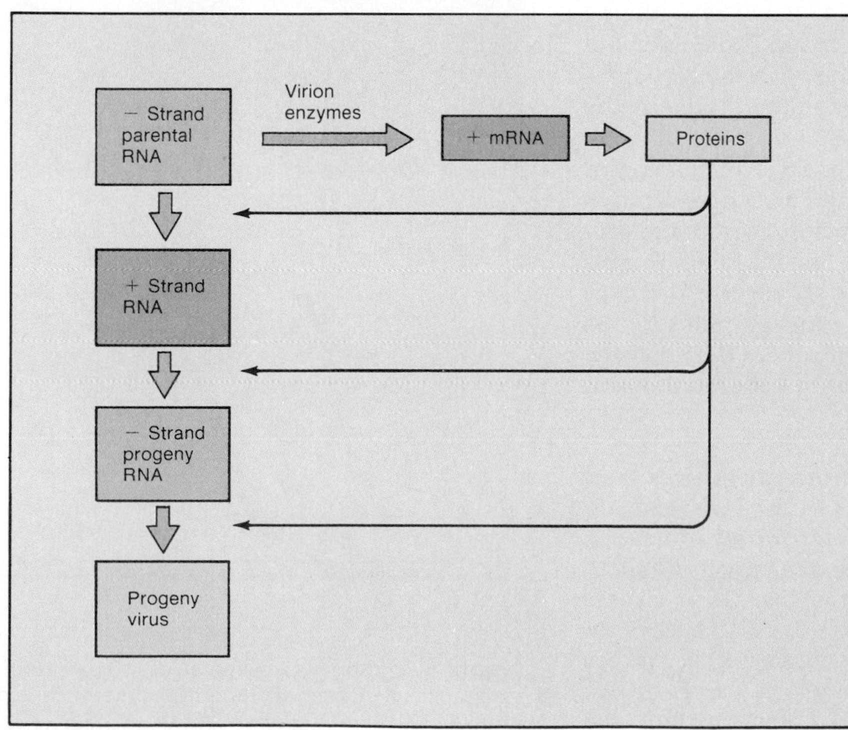

Figure 6.45 Flow of events during multiplication of a negative-strand RNA virus.

virus RNA polymerase is made in this primary transcription process, synthesis of the virus RNA polymerase can begin, leading to the formation of many *plus*-strand RNA molecules, both messengers and genomic (viral) RNA templates.

Translation of viral mRNAs leads to the synthesis of viral coat proteins, and copying of full-length *plus*-strand RNA leads to the formation of full-length *negative*-strand molecules. *Assembly* of an enveloped virus is considerably more complex than assembly of a simple virus particle. Two kinds of coat proteins are formed, *nucleocapsid proteins* and *envelope proteins*. The nucleocapsid is formed first, by association of the nucleocapsid protein molecules around the viral RNA. These nucleocapsid protein molecules are synthesized on ribosome complexes in the cytoplasm.

The *envelope proteins* of the virus are synthesized in quite a different way than the nucleocapsid proteins. First, we should note that the envelope proteins of the virus, like other proteins that associate with membranes, possess hydrophobic amino acid leader sequences 15- to 20-amino acids long at their amino terminal ends, which are cleaved off as soon as they have penetrated the lipid bilayer. Hydrophobic leader sequences ensure that viral envelope proteins will associate with membranes. The envelope proteins are synthesized on ribosome complexes that are themselves associated with membranes. As these proteins are synthesized, *sugar residues* are added, leading to the formation of *glycoproteins*. Such glycoproteins, characteristic of membrane-associated proteins, are transported to the cell membrane, where they replace host membrane proteins. Nucleocapsids then migrate to the areas on the cell membrane where these virus-specific glycoproteins exist, recognizing the virus glycoproteins with great specificity. The nucleocapsids then become aligned with the glycoproteins and bud through them, becoming coated by the glycoproteins in the process. The final result is an enveloped virus particle which has a nucleocapsid center and a surrounding membrane whose lipid is derived from the host cell but whose membrane proteins are encoded by the virus. The budding process itself does not cause detectable damage to the cell, which may continue to release virus particles in this way for a considerable period of time. (Host damage does occur, but is brought about by other, unknown, factors.)

Influenza and other orthomyxoviruses Another group of *negative strand* viruses of great importance is the group called the *orthomyxoviruses*, which contains the important human virus *influenza*. The term *myxo* refers to the fact that these viruses interact with the *mucus* or *slime* of cell surfaces. In the case of influenza virus, this mucus is the mucus membrane of the respiratory tract, as these viruses are transmitted primarily by the respiratory route (see Section 15.4). The term *ortho* has been added to the influenza virus group to distinguish this group from another group of negative-strand viruses, the *paramyxovirus group*. The paramyxoviruses, which include such important human viruses as those causing mumps and measles are actually similar in molecular biology to rhabdoviruses. The orthomyxoviruses have been extensively studied over many years, beginning with early work during and after the 1918 pandemic of influenza which caused the deaths of millions of people world-wide (see Section 15.4).

The orthomyxoviruses are enveloped viruses in which the viral RNA is present in the virus in a number of separate pieces. The genome of the orthomyxoviruses is thus said to be a **segmented genome**. In the case of influenza A virus, the genome is segmented into *eight* linear single-stranded molecules ranging in size from 0.35 to 1×10^6 daltons. The influenza virus nucleocapsid is of helical symmetry, about 6–9 nm in diameter and about 60 nm long. This nucleocapsid is embedded in an envelope which has a number of viral-specific proteins as well as lipid derived from the host (Figure 6.46).

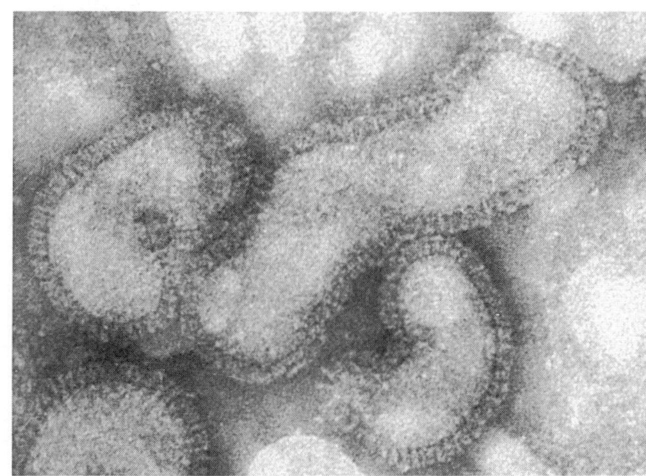

(a)

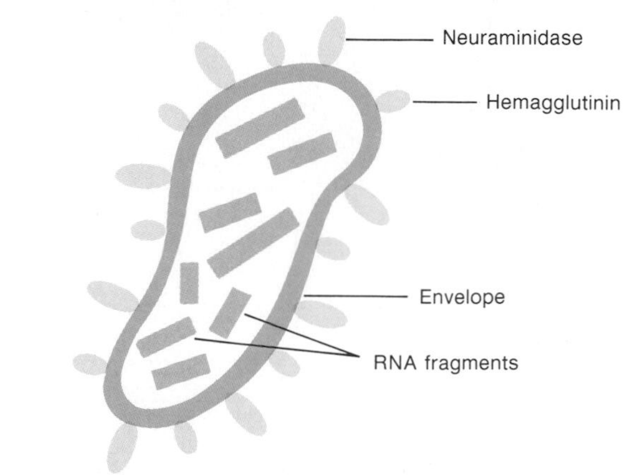

(b)

Figure 6.46 Influenza virion structure. (a) Electron micrograph. (b) Diagram, showing some of the components.

Because of the way influenza virus buds out, the virus has no defined shape and is said to be *polymorphic*. There are spikes on the outside of the envelope which interact with the host cell surface. One spike is called a *hemagglutinin* because it causes agglutination of red blood cells. (Agglutination is a process by which cells are caused to clump when they are mixed with an antibody or other protein or polysaccharide molecule which combines specifically with a substance on the cell surface, as described in Sections 12.8 and 13.9.) If the cells undergoing agglutination are red blood cells, then the process is called *hemagglutination*. (*Hema* is the combining form referring to *blood*.) The red blood cell is not the type of host cell which the virus normally infects, but contains on its surface the same type of membrane component, chemically characterized as *sialic acid*, which the mucous membrane cells of the respiratory tract contain. Thus, the red blood cell is merely a convenient cell type for measurement of agglutination activity. An important feature of the influenza virus hemagglutinin is that antibody directed against this hemagglutinin will *prevent* the virus from infecting a cell. Thus, antibody directed against the hemagglutinin *neutralizes* the virus, and this is the mechanism by which immunity to influenza is brought about during the immunization process (see Sections 12.13 and 15.4).

A second type of spike on the virus surface is an enzyme called *neuraminidase* (see Figure 6.46). Neuraminidase breaks down a mucopolysaccharide component of the plasma membrane called *sialic acid*, a derivative of neuraminic acid. Neuraminidase appears to function primarily in the virus assembly process, destroying host membrane sialic acid that would otherwise block assembly or become incorporated into the mature virus particle.

In addition to the neuraminidase, the virus possesses two other enzymes, *RNA polymerase*, which is involved in the conversion of negative to positive strand (as already discussed for the rhabdoviruses), and an *RNA endonuclease*, which cuts a primer from capped mRNA precursors.

The virus particle enters via the process of *endocytosis*. Once inside the cytoplasm, the nucleocapsid becomes separated from the envelope and migrates to the nucleus. *Replication* of the viral nucleic acid then occurs in the nucleus. *Uncoating* results in the activation of the virus RNA polymerase. The mRNA molecules formed from the virus RNA are *transcribed* in the nucleus, using as primers oligonucleotides which have been cut from the 5′ ends of

Table 6.2 The eight influenza virus genes and the proteins they encode

Gene	Size of gene: mol wt	Size of gene: No. of nucleotides	Protein	Size of protein mol wt	Approx. no. of protein molecules per virus particle	Approx. percent of virus particle protein	Function of protein
1	750,000	2341	P1	96,000	16	1	Initiation of transcription
2	730,000	2300	P2	87,000	16	1	Cap-binding protein
3	715,000	2233	P3	85,000	16	1	Elongation of transcription
4	565,000	1765	HA (HA1)	36,000	750	25	Hemagglutinin spike. HA is converted into HA1 and HA2. The size of the carbohydrate groups on HA1 and HA2 are 11,500 and 1400 daltons, respectively
			HA (HA2)	27,000	750		
5	500,000	1565	NP	56,000	1000	30	Structural protein of helical nucleocapsid
6	450,000	1413	NA	50,000	160	4	Neuraminidase
7	330,000	1027	M1	27,000	3000	38	Matrix protein
			M2	11,000	0		Nonstructural protein of unknown function
8	285,000	890	NS1	26,000	0		Nonstructural protein of unknown function
			NS2	12,000	0		Nonstructural protein of unknown function

newly synthesized capped cellular mRNAs. Thus, the viral mRNAs have 5′ caps. The poly (A) tails of the viral mRNAs are added and the virus mRNA molecules move to the cytoplasm.

Although influenza virus RNA replicates in the nucleus, influenza virus *proteins* are synthesized in the cytoplasm. There are *eight* virus proteins, each of which is coded for by a separate RNA of the split virus genome (Table 6.2) Some of these proteins are involved in virus RNA replication and others are structural proteins of the virus particle. The overall strategy of virus RNA synthesis resembles that of the rhabdoviruses, with primary transcription resulting in the formation of *plus*-stranded templates for the formation of progeny *minus*-stranded molecules. Details of assembly are still uncertain. One possibility is that the nucleocapsid proteins are transported from the cytoplasm to the cell nucleus, where *assembly* of the nucleocapsids occurs. The assembled nucleocapsids then migrate to the *plasma membrane* where the hemagglutinin and neuraminidase are present. The formation of the complete enveloped virus particle occurs by a budding out process, as was described for the rhabdoviruses. The neuraminidase plays a role in permitting the liberation of the virus particle from the cell, and virus liberation does not occur if neuraminidase activity is inhibited. It is thought that the neuraminidase removes neuraminic acid from oligosaccharide units of the hemagglutinin protein, thus eliminating the affinity which the virus has for the host cell surface.

The segmented genome of the influenza virus has some important practical consequences. Influenza virus, and other viruses of this family, exhibit a phenomenon called **antigenic shift**, in which pieces of the RNA genome from two separate virus particles that have infected the same cell become associated. This results in a change in the surface antigens (coat proteins) of the virus, making the virus resistant to antibody that has been formed as a result of an immunization process (see Section 15.4). This antigenic shift makes it possible for the newly formed virus to infect hosts that the parent could not have infected. Antigenic shift is thought to bring about major epidemics of influenza.

6.19 Double-Stranded RNA Viruses—Reoviruses

The reoviruses, an important family of animal viruses, have a genome consisting of double-stranded RNA. The name *reovirus* is an acronym, derived from the terms "*r*espiratory, *e*nteric, *o*rphan." The term "orphan" was applied because the first viruses of this group to be isolated from humans were not associated with any specific disease syndrome. However, viruses of the reovirus group are known to infect a variety of mammals as well as insects and plants. The mammalian reoviruses first isolated are now classified in the virus genus *Orthoreovirus*. To date, no specific human syndrome has been associated with the Orthoreovirus group, although these viruses are commonly isolated from persons with an array of inconsequential respiratory and/or gastrointestinal illnesses (as well as being isolated from apparently healthy people). A more important genus of reoviruses is the genus *Rotavirus*, a member of which has been implicated in an important diarrheal disease of infants. *Rotavirus* is probably the most common cause of diarrhea in infants from 6 to 24 months of age. Rotaviruses are also known to cause diarrhea in young animals.

As noted, the RNA of the reoviruses is double stranded. This is the only group of animal viruses with double-stranded RNA, all other RNA groups having single-stranded RNA (Figure 6.40). The reovirus particle consists of a nonenveloped nucleocapsid 60–80 nm in diameter, with a *double* shell of icosahedral symmetry (Figure 6.47). Three proteins make up the outer capsid and four proteins the inner. Present within the virus particle are several enzymes: *RNA polymerase, nucleotide phosphohydrolase*, and enzymes that participate in the *capping process* of messenger RNA, the *RNA methyltransferase* and *guanyl transferase*.

However, the unique thing about the reoviruses is their *genome*, which consists of 10–13 molecules of double-stranded RNA. We have already discussed the segmented genome of the single-stranded RNA influenza viruses, but when dealing with a double-stranded RNA, unique features occur. The *plus* strands of the reovirus double-stranded RNA carry methylated caps on their 5′ ends, and the *minus* strands carry ppG−. The 3′ termini do not have the poly (A) tails that we have mentioned as characteristic of mature mRNA molecules. However, inside the core of the

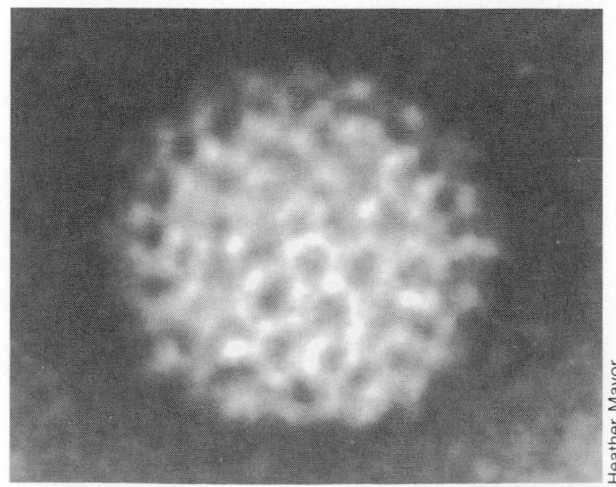

Figure 6.47 A reovirus particle. The diameter of the particle is about 70 nm.

virus particle there are approximately 3000 molecules of single-stranded oligonucleotides consisting mainly of AMP residues.

Each molecule of RNA in the genome codes for a single protein, although in a few cases the protein formed is cleaved to give the final product. *Replication* occurs exclusively in the *cytoplasm* of the host. The double-stranded RNA is inactive as mRNA and the first step in reovirus replication is the *transcription* of the *minus* strand into mRNA. This transcription occurs on just four of the RNA molecules of the genome and then, after some translation, all of the RNA molecules are transcribed. Transcription of the first four RNA molecules is brought about by the RNA polymerase present in the virus particle. The mRNAs do not contain poly (A) but they are capped early in infection. As we have noted, in eucaryotes poly (A) tails are added *before* the mRNA is transfered from the nucleus to the cytoplasm. Since reovirus mRNA is formed in the cytoplasm, the absence of poly (A) tails is explicable.

Replication of the reovirus seems to occur within an intracellular equivalent of the viral core, called the *subviral particle*, which remains intact in the cell. Each of the 10 capped single-stranded *plus* RNAs is assembled into this double-stranded RNA synthesizing body. The capped single-stranded *plus* RNAs act as templates for the synthesis of the progeny *minus* genomic RNAs, yielding progeny double-stranded viral RNAs. The progeny double-stranded RNAs are further encapsidated and when enough viral capsid proteins are present, mature virus particles are assembled. Details of the release process are not known, but since the viruses are not enveloped, no special mechanism need be present.

In the initial infection process, the virus particle binds to a cellular protein which itself plays a regulatory role. This receptor protein serves as the *beta-adrenergic receptor* (a hormone receptor of the cell). The specificity of virus infection derives from the receptor which the virus uses for attachment. This is a clever virus adaptation, since the host cannot alter the receptor without affecting a normal host function. Once attachment has occurred, the virus enters the cell and is transported into lysosomes. Within the lysosome the outer shell of the virus particle is modified by removal of two proteins and cleavage of another by lysosomal enzymes. This uncoating process activates the viral transcriptase and hence initiates the viral replication process.

6.20 Replication of DNA Viruses of Animals

Most animal viruses with DNA genomes contain double-stranded DNA (one group, the parvoviruses, contain single-stranded DNA). Among these DNA viruses of animals, the four major families are the poxviruses, the herpesviruses, the adenoviruses, and the papovaviruses. Of these, all replicate in the nucleus except for the poxviruses, which have the unique character (for DNA viruses) of replicating in the cytoplasm. In the present section, we discuss the replication of each of these families briefly.

Our knowledge of the manner of replication of the DNA viruses infecting animals has been markedly advanced by the use of recombinant DNA techniques. The use of restriction enzymes to make unique cuts in the viral genomes has made it possible to separate each genome into a number of fragments which can be ordered and thus used in the construction of a genetic map. We have discussed the general approach to restriction enzyme mapping in Section 5.3.

Polyoma-like viruses Some viruses of this group (also sometimes called *papovaviruses*) have the interesting ability of inducing *cancer* in animals. One of the DNA viruses of animals, the *polyoma-like virus* SV40, was one of the first genetic elements to be studied by genetic engineering techniques, and has been extensively used as a *vector* for moving genes into cells. When a virus of the polyoma-SV40 group infects a host cell, one of two modes of replication can occur, depending upon the type of host cell. In some types of host cells, known as *permissive* cells, virus infection results in a productive infection, with the formation of new virus particles and the lysis of the host cell. In other types of host cells, known as *nonpermissive*, efficient multiplication does not occur, but the virus DNA becomes integrated into some of the host cells, thereby creating new, genetically different cells. Such cells frequently show loss of growth inhibition and are called *transformed* cells. *Polyoma virus* causes tumors in mice or hamsters and its name comes from the fact that it causes tumors of a wide variety of tissues (hence the name: *poly* means *many* and *oma* means *tumor*). Virions of this group are simple nonenveloped particles with icosahedral heads (see Figure 6.8*a* for a virus of similar structure). One of these viruses, *SV40,* was first isolated from apparently normal monkey kidney cells. Virus SV40 received its name from the fact that it infects monkeys (simians); thus, *simian virus 40.* Since much recent work has been carried out with SV40, the discussion here will be restricted to this virus.

SV40 virus The SV40 virion is a simple nonenveloped particle of 45 nm diameter with an icosahedral head containing 72 protein subunits. The capsid has one major protein and two minor proteins, and there are no enzymes in the virus particle. Four host-derived *histone proteins* are found complexed with the viral DNA. We have discussed histone proteins during our discussion of chromosome structure

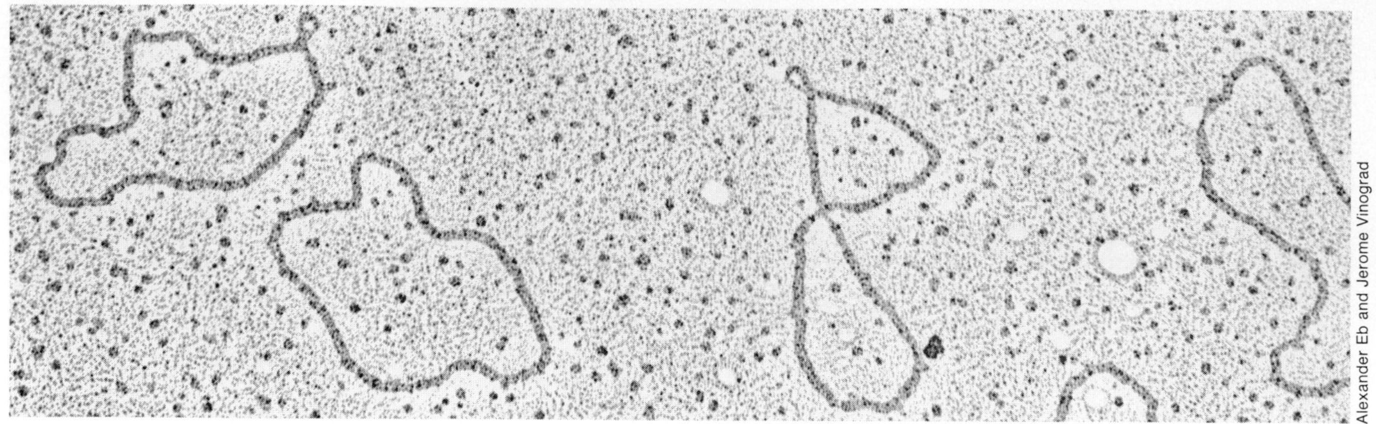

Figure 6.48 Electron micrograph of circular DNA from a tumor virus. The length of each circle is about 1.5 μm.

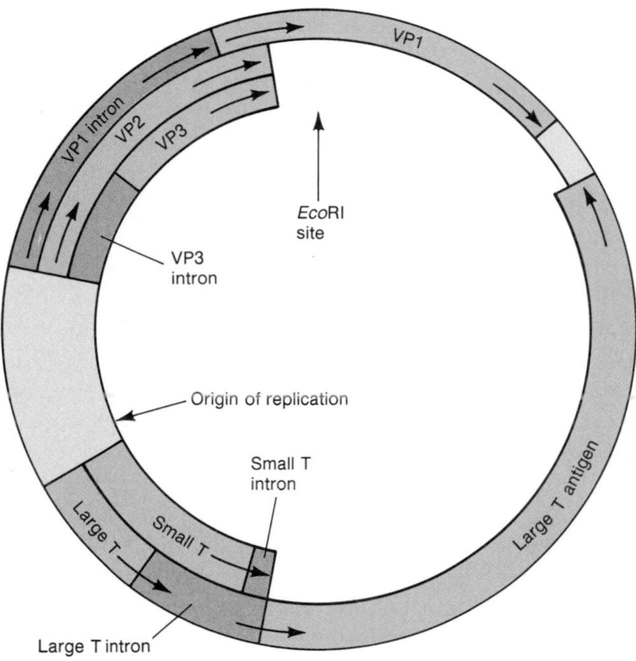

Figure 6.49 Genetic map of SV40.

(see Section 3.16) and have noted that histones play a role in neutralizing the negative charge originating from the phosphates of chromosomal DNA and aiding in the packing of the DNA into more compact configurations.

The genome of SV40 consists of one molecule of double-stranded DNA of 5243 base pairs (molecular weight, 3×10^6 daltons). The DNA is circular (Figure 6.48) and exists in a supercoiled configuration within the virus particle. The complete base sequence of SV40 has been determined and the genetic map is known in some detail (Figure 6.49).

The nucleic acid is replicated in the nucleus but the proteins are synthesized in the cytoplasm. Final assembly of the virus particle occurs in the nucleus. The replication of the DNA of these viruses closely resembles the replication of single-stranded bacteriophages such as ϕX174 discussed earlier (see Section 6.10). The replication phase can be divided into two distinct phases, *early* and *late*. During the *early phase*, about 50 percent of the viral DNA, in the region called the early genes, is transcribed. A single RNA molecule, the primary transcript, is made, but it is processed into *two species of mRNA*, a large one and a small one. A cell DNA-directed RNA polymerase carries out this transcription process. The DNA of SV40 has *introns* that are excised out of the primary RNA transcript. In the cytoplasm the viral mRNA is translated with the formation of two proteins. One of these proteins, the T-antigen, binds to the site on the parental DNA which is the *origin of replication*. The viral DNA is too small to code for its own DNA polymerases; host DNA polymerases are used. Replication occurs in a bidirectional fashion (so-called *theta* replication, see Figure 5.16) from a single replication origin. The process involves the same events that have already been described for host cell DNA replication (see Section 5.4): RNA primer synthesis, formation of discontinuous DNA fragments on the lagging strand, gap filling, ligase action, and supercoiling the DNA through the action of gyrase. Because of the circular nature of SV40 DNA, the formation of concatamers or larger ring forms does not occur.

Late mRNA molecules are synthesized using the strand complementary to that used for early mRNA synthesis. Transcription begins at several positions in a small region near the origin of replication, leading to the synthesis of very large primary RNA transcripts. These RNAs are then processed by splicing and polyadenylation to give multiple forms of mRNA corresponding to the three coat proteins. These genes overlap; part of the nucleotide sequence contains information for all three proteins (see Figure 6.49). These late mRNA molecules are transported to the cytoplasm and translated into the viral coat proteins.

These proteins are then transported back into the nucleus, where *virion assembly* takes place. The mechanism by which virus particles are released from the cell is not known.

In *nonpermissive hosts*, transformation to *tumor cells* can occur. In such hosts, the early phase of viral replication occurs but the late phase does not. Once the viral DNA has entered the nucleus and transcription of the early genes has taken place, there is a great stimulation in all of the host cell's biosynthetic activities involved in cellular DNA replication and mitosis. However, no replication of viral DNA occurs. The characteristic of nonpermissive cells is that viral replication cannot occur in them but infection can occur. It is generally found that if a cell type can support viral multiplication, then the cell will not be transformed; if the virus cannot multiply, the cell is often transformed. In the transformed cell, the viral DNA becomes stably integrated into the DNA of the host cell (Figure 6.50). In the integrated state, the viral DNA can only replicate as a *cellular gene*. Integration can occur at many sites in the cellular and viral genome. In this integrated form, two viral proteins are made which are essential for the maintenance of a stably integrated viral DNA, but none of the other viral proteins are synthesized. Some virus-transformed cells become converted into cells capable of producing virus, a process which probably involves excision of the viral genome from the host genome. It is also possible to detect host cells which have reverted to the noncancerous state, not by having lost the viral DNA but by changing the host regulatory processes in such a way that the cell no longer replicates like a transformed cell.

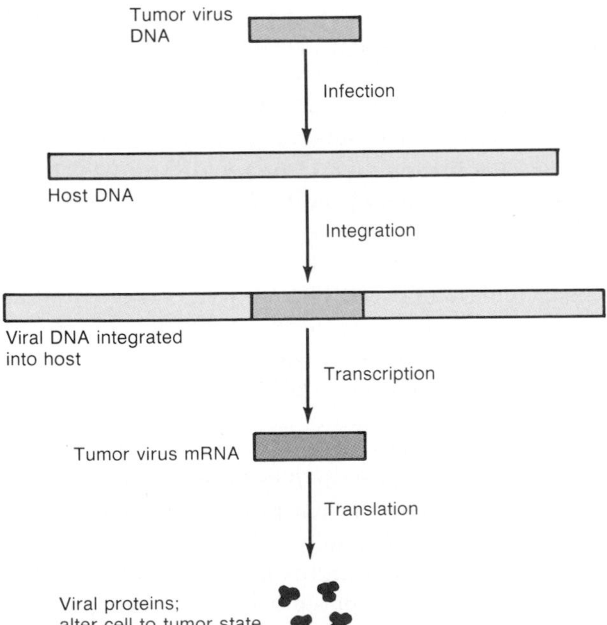

Figure 6.50 General scheme of molecular events involved in cell transformation by a DNA tumor virus. All or portions of viral DNA are incorporated into host cell DNA. The viral genes that encode transforming information are transcribed and processed to viral mRNA molecules that are transported to the cytoplasm. Here they are translated to form transforming proteins or T antigens that code for functions that can convert host cells into cancer cells.

A study of the manner of replication of polyoma and SV40 viruses has given some important insights into the manner by which viruses bring about the cancerous state in host cells. However, we note that DNA viruses of other groups can also cause the cancerous transformation. Also, the important family of viruses, the *retroviruses*, are also cancer viruses but have a completely different mode of replication (see Section 6.24).

6.21 Herpesviruses

The herpesviruses are a large group of viruses which cause a wide variety of diseases in humans and animals. Among the diseases caused by members of the herpesvirus group are fever blisters (cold sores), venereal herpes, chickenpox, shingles, and infectious mononucleosis; a number of these diseases are discussed in Chapter 15. Some herpesviruses also cause cancer. One of the interesting features of some herpesviruses is their ability to remain *latent* in the body for long periods of time, becoming active only under conditions of stress. Both *herpes simplex*, the virus which causes *fever blisters*, and *varicella-zoster virus*, the cause of *shingles*, are able to remain latent in the neurons of the sensory ganglia, from which they are able to emerge to cause infections of the skin.

An important group of herpesviruses are tumorigenic, causing clinical forms of cancer. One herpesvirus which is tumorigenic is the *Epstein-Barr virus*, which causes *Burkitt's lymphoma*, a common tumor among children in Central Africa and New Guinea. Burkitt's lymphoma was among the first human cancers to have been shown to be closely linked to virus infection.

The **herpesvirus particle** is structurally complex, consisting of four distinct morphologic units. The following information applies to herpes simplex virus type 1, the causal agent of fever blisters. Some other herpesviruses may lack certain of these structural elements. Herpes simplex type 1 is an enveloped virus of about 150 nm in diameter (Figure 6.51*a*). In the center of the virus is an electron dense *core* which consists of double-stranded DNA. Surrounding this core is the nucleocapsid, of icosahedral symmetry, which consists of 162 capsomeres, each of which is composed of a number of distinct proteins. Outside the nucleocapsid is an electron dense amorphous material which is called the *tegument*, a

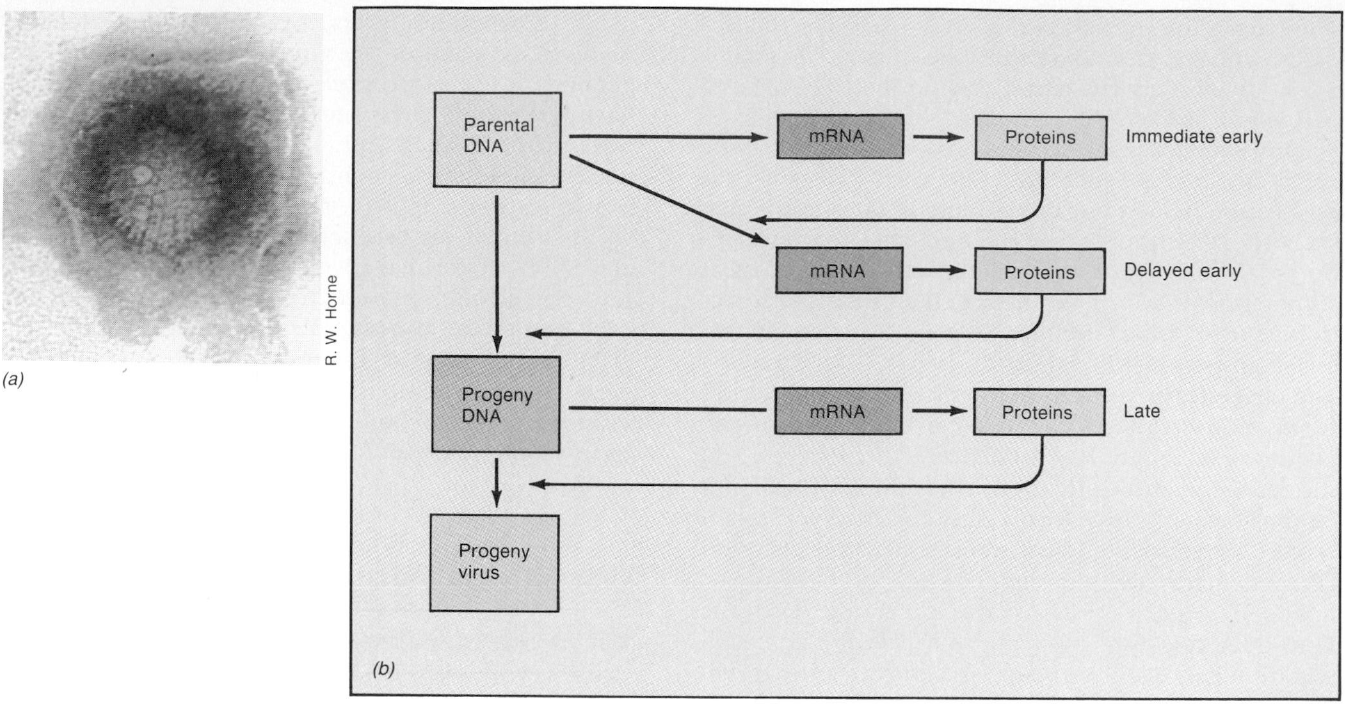

Figure 6.51 Herpesvirus. (a) Electron micrograph of a herpesvirus particle. The diameter of the particle is about 150 nm. (b) Flow of events in multiplication of herpes simplex virus.

fibrous structure which is unique to the herpesviruses. Surrounding the tegument is an *envelope* whose outer surface contains many small *spikes*. At least two enzymes have been detected in the virus particle, an *ATPase* and a *protein kinase* (this latter is an enzyme which phosphorylates proteins). It is not known how these enzymes function in the virus multiplication cycle. A large number of separate proteins are present within the virus particle, but not all of them have been characterized.

The *genome* of herpes simplex virus consists of one large linear double-stranded DNA molecule of 150 kilobases with a molecular weight of about 100×10^6 daltons. The DNA of herpes simplex virus is about 30 times larger than that of SV40. The DNA shows terminal repeat sequences, and there are also repeated sequences within the molecule. There are two regions, L and S (for *large* and *small*) which comprise 82 and 18 percent of the viral DNA, respectively. Each of these components consists of unique base sequences, bracketed by inverted repeat sequences. There are four forms of the viral DNA, differing in the orientations of the unique base sequences. These four configurations arise because of intramolecular recombination within a small region of the DNA. As a further indication of the complexity of herpes simplex, the DNA of this virus codes for over *100 separate polypeptides*.

Infection occurs by attachment of virus particles to specific cell receptors, and the fusion of the cell plasma membrane with the virus envelope, thus releasing the nucleocapsid into the cell. The nucleocapsids are transported to the nucleus, where viral DNA is uncoated. Components of the virus particle inhibit macromolecular synthesis by the host.

There are *three classes of messenger RNAs*, immediate early, which code for five regulatory proteins, delayed early, which code for DNA synthesis proteins, including thymidine kinase and DNA polymerase, and late, which code for the proteins of the virus particle (Figure 6.51*b*). During the *immediate early stage*, about one-third of the viral genome is transcribed by a host cell RNA polymerase. Initial RNA transcription is in the form of primary transcripts which must be processed, including splicing out some noncoding introns. The mRNAs are capped, methylated, and polyadenylated, after which they move to the cytoplasm for translation. As noted, this early mRNA codes for certain regulatory proteins which appear to stimulate the synthesis of the delayed early proteins. The second stage, *delayed early*, only occurs after the early proteins have been made. During this stage, about 40 percent of the viral genome is transcribed. It is thought that transcription of the delayed early genes is initiated at new promoters which are recognized by this new RNA polymerase. Among the 10 proteins so far characterized from the delayed early stage are a *DNA polymerase*, enzymes involved in synthesis of *deoxyribonucleotides*, and a *DNA-binding protein*. These enzymes are all involved in the process of viral DNA synthesis.

Viral DNA synthesis itself takes place in the

Virus Surprises

Viruses have provided scientists with lots of surprises. Every time scientists have thought they had a clear picture of the nature and diversity of viruses, some new discovery has complicated matters. Viruses were originally thought to be tiny living organisms, too small to be seen with the light microscope, but fundamentally like other organisms. Then, in the 1930s, Wendell Stanley *crystallized* the first virus, tobacco mosaic virus (TMV), dispelling for all time this idea of the nature of viruses. Stanley's TMV crystals made possible a study of the chemical composition of a virus. Stanley's chemical analysis showed that the TMV crystals were composed of only one thing, protein. This made a lot of sense because enzymes were proteins, and biochemists had come to believe that proteins contained the secret of life itself. Indeed, genes were thought to be simply very large proteins. Imagine everyone's surprise when a group of English scientists showed that Stanley was wrong and that there was something else in his TMV crystals, a small amount of RNA! In the 1930s, no one had thought nucleic acids were of much importance, but continued work eventually showed that it was not the protein but the RNA that was the vital component of TMV. Workers at Berkeley, California, in the late 1950s showed that highly purified TMV RNA alone could bring about infection in susceptible plants. The role of RNA in genetic phenomena thus became firmly established. In fact, the discovery that the RNA of TMV could be infectious on its own (without the protein coat) provided some of the strongest evidence of the importance of the nucleic acid component of a virus particle. Studies of a number of other plant viruses showed that they also contained RNA as their genetic material.

Bacteriophages were another surprise to scientists. Chemical analysis of a wide variety of phages showed that they contained DNA instead of RNA. Soon the idea became common that in plant viruses the genetic material was RNA whereas in bacteriophages it was DNA. Indeed, some scientists even developed evolutionary models to explain why bacteriophages *had* to have DNA as their genetic material. Imagine everyone's surprise when Norton Zinder and some of his students discovered an *RNA* bacteriophage! And then a plant virus was discovered that had *DNA* instead of RNA. Also among the animal viruses, representatives were discovered that had either DNA or RNA. The fanciful evolutionary ideas had to be greatly modified.

But still, there seemed to be no virus that involved *both* DNA *and* RNA. It is understandable, then, that scientists strongly resisted the idea first put forward by Howard Temin that in a virus which caused cancer in chickens, *rous sarcoma virus*, the virus particles contained RNA but replicated in the cell as DNA. This suggestion, which was soon proved to be true, turned out to be one of the biggest virus surprises of all. We now know that there is a whole group of these viruses, called *retro*viruses, which seem to operate in reverse, transferring information from RNA to DNA rather than the other way around as do cells. However, *retroviruses* seemed to be little more than curiosities until one was found which caused disease in humans. This is, of course, the human immunodeficiency virus (HIV), which causes the disease *acquired immunodeficiency syndrome*, better known as *AIDS*.

Two different disease entities, *viroids* and *prions*, have been recently discovered to be even simpler in structure than viruses, but like viruses each apparently depends upon host cells for reproduction. *Viroids* are small pieces of RNA which are not complexed with any protein. The best studied viroid causes a disease of potatoes. It is composed of 359 bases and thus has ten times less genetic material than the smallest known virus. There is just enough genetic information to code for one or two small proteins. *Prions* are small proteins which appear to be self-replicating but yet are completely devoid of nucleic acid. One prion studied contains only 250 amino acids and is about 100 times smaller than the smallest virus. Prions also appear to cause disease, and scrapie in sheep and Creutzfeldt-Jakob disease in humans have been linked to these agents. Prions could thus represent the first exception to the rule that all self-replicating entities contain nucleic acid as the genetic material. One explanation for how prions might reproduce is by activating a latent gene in the host which codes for the prion. It is interesting that the discovery of prions brings us back full circle to Stanley's first work when it was thought that TMV consisted only of protein.

Are there still *more* fundamental surprises awaiting discovery out there in the world of virology?

nucleus, probably by a rolling circle mechanism, occurring simultaneously from three origins. Long concatamers are formed which become processed to viral length DNA during the assembly process itself (in a manner similar to that described for DNA bacteriophages, see Sections 6.12 and 6.13).

Viral nucleocapsids are assembled in the nucleus and acquisition of the virus envelope occurs via a budding process through the *inner membrane of the nucleus*. The mature virus particles are subsequently released through the endoplasmic reticulum to the outside of the cell. Thus, the assembly of this enveloped DNA virus differs markedly from that of the enveloped RNA viruses, which were assembled on the plasma membrane instead of the nuclear membrane.

6.22 Pox Viruses

These are the most complex animal viruses known, and have some characteristics that approach those of primitive organisms. However, the pox viruses are not able to metabolize and do depend upon the host for the complete machinery of protein synthesis. These viruses are also unique in that they are DNA viruses which replicate in the *cytoplasm*. Thus, a host cell infected with a pox virus exhibits DNA synthesis outside of the nucleus, something that otherwise only occurs in intracellular organelles such as mitochondria.

Pox viruses have been important medically as well as historically. *Smallpox* was the first virus to be studied in any detail, and was the first virus for which a vaccine was developed (described by Edward Jenner in 1798). Of special interest in recent years, by diligent application of this vaccine on a worldwide basis, the disease smallpox has been *eradicated*, the first infectious disease to have been done so. Other pox viruses of importance are *cowpox* and *rabbit myxomatosis virus*, an important infectious agent of rabbits and one which has been responsible for a devastating destruction of the Australian rabbit population (see Section 14.6). Some pox viruses also cause tumors, but these tumors are generally benign.

The poxviruses are very large, so large that they can actually be seen under the light microscope. Most research has been done with *vaccinia virus*, the source of smallpox vaccine. The vaccinia particle is a brick-shaped structure about 400 × 240 × 200 nm in size. The virus particle is covered on its outer surface with tubules or filaments arranged in a membrane-like pattern, although the virus does not have a lipid membrane, the outer envelope consisting of protein. Within this there are two lateral bodies of unknown composition and a core, the *nucleocapsid*, which contains DNA bounded by a layer of protein subunits.

The poxvirus genome consists of double-stranded DNA. The pox DNA is interesting because the two strands of the double helix are crosslinked at the ends, due to the formation of phosphodiester bonds between adjacent strands.

Vaccinia virus particles are taken up into cells via a phagocytic process, from which the cores are liberated into the cytoplasm. Interestingly, *uncoating* of the virus genome requires the action of a new protein that is synthesized after infection. This protein is encoded by viral DNA and the gene specifying this protein is transcribed by an *RNA polymerase* that is present *within* the virus particle. In addition to this uncoating gene, a number of other viral genes are transcribed. The primary transcripts are turned into mRNAs by capping and polyadenylation while they are still inside the virus core.

Once the vaccinia DNA is fully uncoated, the formation of *inclusion bodies* within the cytoplasm begins. Within these inclusion bodies, transcription, replication, and encapsidation into progeny virus particles occurs. Each infecting particle initiates its own inclusion body, so that the number of inclusions depends on the multiplicity of infection. Progeny DNA molecules form a pool from which individual molecules are incorporated into virus particles. There are complex mechanisms in vaccinia for regulating transcription and translation of early and late genes, but space does not permit discussion here.

Mature virus particles accumulate in the cytoplasm. There seems to be no specific release mechanism and most virus particles are released only when the infected cell disintegrates.

6.23 Adenoviruses

The adenoviruses are a major family of DNA-containing viruses which have unique and interesting properties. The term *adeno* is derived from the Latin for *gland* and refers to the fact that these viruses were first isolated from the tonsils and adenoid glands of humans. Adenoviruses cause generally mild respiratory infections in humans, and a number of such viruses are isolated from apparently healthy individuals. Some adenoviruses also cause tumors in animals and cause transformation of animal cells in cell culture. However, the clinical significance of adenoviruses is uncertain; their main interest here is as models for studies of DNA replication and transformation.

The adenoviruses are structurally complex. A computer model of the structure of an adenovirus particle is shown in Figure 6.52. (See Figure 1.11*a* for an electron micrograph of adenovirus crystals.) The icosahedral virus has a diameter of 75 nm and its coat consists of 11 to 15 distinct proteins. The major coat proteins are called *hexons* and constitute the major proteins of the faces of the icosahedron.

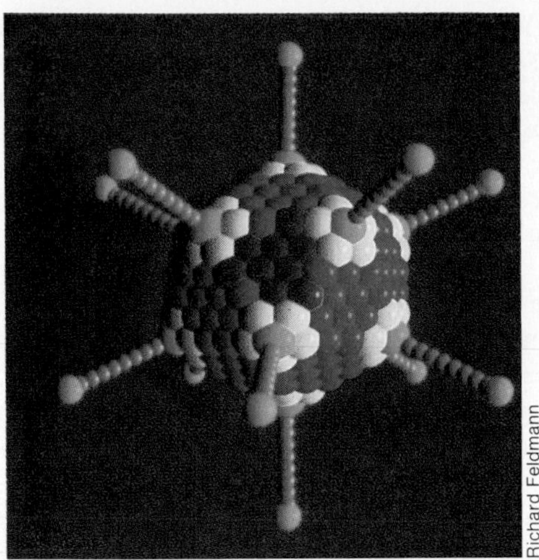

Figure 6.52 Structure of an adenovirus particle. Computer-generated model of the virus, showing the external structure and arrangement of coat proteins.

The hexons themselves are actually trimers composed of three identical polypeptides. At the five vertices of the icosahedron are 12 *pentons*, and attached to these pentons are characteristic fibers which are responsible for the attachment of the virus particle to the host cell. There are also *internal proteins* associated with the DNA which function in the neutralization of the negative charge of the DNA molecule. There is no envelope, and no enzymes have been found associated with the virus particle.

The *genomes* of the adenoviruses consists of linear double-stranded DNA. The molecular weight of the DNA is about 20×10^6 daltons. Attached in covalent linkage to the 5′ terminus of the DNA is a protein component which is essential for infectivity of the DNA. The DNA has inverted terminal repeats of 100–1800 base pairs (this varies with the virus strain). The DNA of the adenoviruses is six to seven times the size of the DNA of SV40.

The *genetic map* of adenovirus has been determined by restriction enzyme analysis coupled with analysis of messenger RNA transcription. At least 10 distinct transcriptional units have been recognized, occurring on both strands. In different places, there are also regions of overlap as well as noncoding regions (introns) which are removed from the primary transcripts in processing steps.

Replication of the virus DNA occurs in the *nucleus* (Figure 6.53). After the virus particle has been transported to the nucleus, the core is released and converted into a viral DNA-histone complex. *Early transcription* is carried out by an RNA polymerase of the host and a number of primary transcripts are made. The transcripts are spliced, capped, and polyadenylated giving several different mRNAs. In some cases, more than one separate mRNA arises from a single promoter, the differentiation arising at the processing stage. The mRNAs move to the cytoplasm where translation occurs. This system of transcription and processing provides a means by which any given region of the DNA can be used over and over again, thus leading to considerable economy in the amount of DNA needed to code for all the viral proteins.

The early proteins are involved in regulation of DNA replication; the later proteins are the virus coat proteins. *Viral DNA replication* uses a virus-encoded protein as a primer and another virus-encoded protein as DNA polymerase. However, host cell DNA polymerase may also function in adenovirus DNA replication. Also, two host proteins, a site-specific *DNA binding protein*, and a *topoisomerase I*, are involved in DNA replication. For the replication of a *linear* DNA molecule such as that of adenovirus, initiation of replication could begin at either end or at both ends simultaneously. In the case of the adenoviruses, replication begins at *either* end, the two strands being replicated asynchronously. Replication proceeds in the standard 5′ to 3′ direction by a *modified rolling circle* mechanism. The products of a round of replication are double-stranded and single-stranded mol-

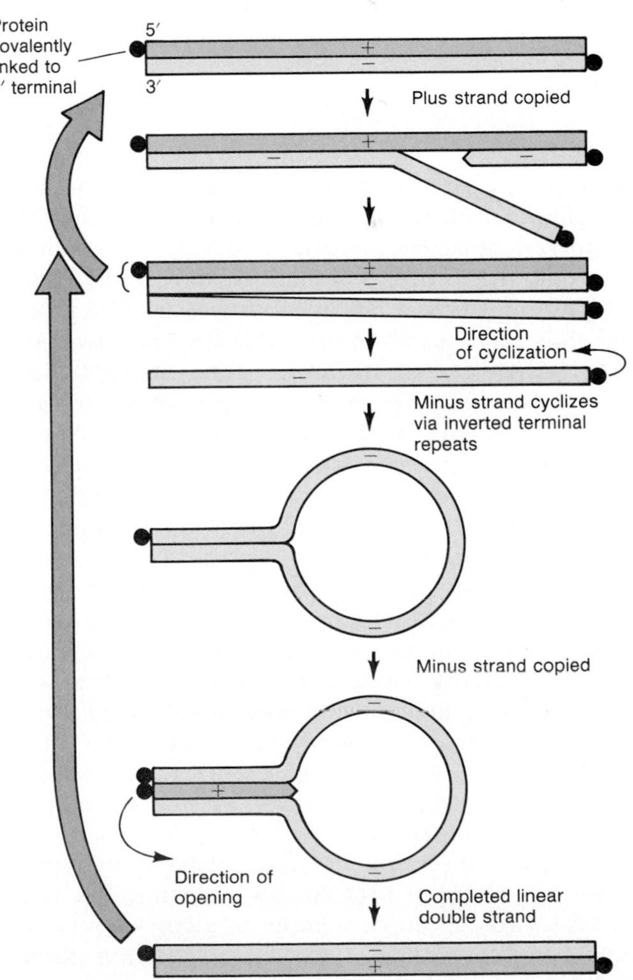

Figure 6.53 Replication of adenovirus DNA. See text for details.

ecules. The latter then cyclizes by means of the inverted terminal repeats, and a new complementary strand is synthesized beginning from the 5′ end, the products being another double-stranded molecule (Figure 6.53). This mechanism of replication is unique because it does not involve the formation of discontinuous fragments of DNA on the lagging strand, as in conventional DNA replication.

The adenovirus replication process just described is that which is found in permissive cells, in which a lytic infection cycle occurs. In nonpermissive cells, *transformation* to tumor cells sometimes occurs. There is little or no DNA replication in nonpermissive cells and no synthesis of viral coat proteins. Instead, the DNA becomes randomly *integrated* into the host genome and is replicated along with host DNA. This transformation process thus resembles that already described for SV40 (see Section 6.20).

Certain adenoviruses which infect humans have been shown to cause cancer in mice. Using a sensitive nucleic acid hybridization technique (see Box, Chapter 5), a search has been made for adenovirus DNA sequences in human tumors. So far, no evidence that human tumors contain adenovirus DNA sequences has been found.

6.24 Retroviruses

We now discuss one of the most interesting and complex families of animal viruses, the **retroviruses**. The term *retro* means *backwards*, and is derived from the fact that these viruses appear to have a backwards mode of replication. The retroviruses are RNA viruses but they *replicate by means of a DNA intermediate*. The retroviruses are of interest for a number of reasons. First, they were the first viruses shown to cause *cancer*, and have been studied most extensively for their carcinogenic characteristics. Second, one retrovirus, the one causing *acquired immunodeficiency syndrome (AIDS)*, has only been known since the early 1980s but has already become a major public health problem. Third, the retrovirus genome can become specifically integrated into the host genome, by way of the DNA intermediate, and this integration process is being studied as a means of introducing *foreign* genes into a host, a process called *gene therapy*. Finally, the retroviruses have an enzyme, called *reverse transcriptase*, which copies RNA sequences into DNA, and this enzyme has become a major tool in *genetic engineering*.

As we will see, the retroviruses have some properties like those of RNA viruses and some like those of DNA viruses. They resemble to a considerable extent movable genetic elements such as transposons, and can be considered to be *escaped cellular movable elements*. In this respect, the retroviruses resemble bacterial viruses such as *Mu* (see Section 6.15). We should note that the use of reverse transcription is not restricted to the retroviruses, because hepatitis B virus (a human virus) and cauliflower mosaic virus (a plant virus) also use reverse transcription in their replication processes. But in contrast to the retroviruses, these latter viruses encapsidate the DNA genome rather than the RNA genome as retroviruses do.

The retroviruses are enveloped viruses of uncertain symmetry (Figure 6.54*a*). There are a number of proteins in the virus coat, including at least two envelope proteins and seven internal proteins, four of which are structural and three enzymatic. The enzymatic activities found in the virus particle are *reverse transcriptase, DNA endonuclease (integrase)*, and a *protease*.

The *genome* of the retrovirus is unique. It consists of *two* identical single-stranded RNA molecules, of *plus* complementarity, which are hydrogen-bonded together by means of base pairing with specific cellular tRNA molecules (Figure 6.54*b*). Each of the two RNAs is 9300 nucleotides in length. The 5′ terminus of the RNA is capped and the 3′ terminus is polyadenylated, so that the RNA is capable of acting directly as mRNA. A *genetic map* of the most studied retrovirus, that of Rous sarcoma, is shown in Figure 6.54*c*. Key regions of this map are those called *gag*, coding for internal proteins, *pol*, coding for DNA polymerase and integrase, *env*, coding for envelope proteins, and *src*, the cancer transformation gene. The long terminal repeat (LTR) shown on the map plays an essential role in the replication process.

The overall process of replication of a retrovirus can be summarized in the following steps (Figure 6.55):

1. **Entrance** into the cell.
2. **Reverse transcription** of *one* of the two RNA genomes into a single-stranded DNA which is subsequently converted to a linear double-stranded DNA by host enzymes.
3. **Formation of a closed circular DNA** in the nucleus by ligation of the linear DNA.
4. **Integration** of the DNA copy into the host genome.
5. **Transcription** of the viral DNA, leading to the formation of viral mRNAs and progeny viral RNA.
6. **Encapsidation** of the viral RNA into nucleocapsids in the cytoplasm.
7. **Budding** of enveloped virus particles at the plasma membrane.
8. **Release** of virus particles from the surface of the cell.

We now discuss some aspects of the retrovirus multiplication process in detail. As we have noted, the first step after the entry of the RNA genome into the cell is the transcription of the RNA into a DNA copy,

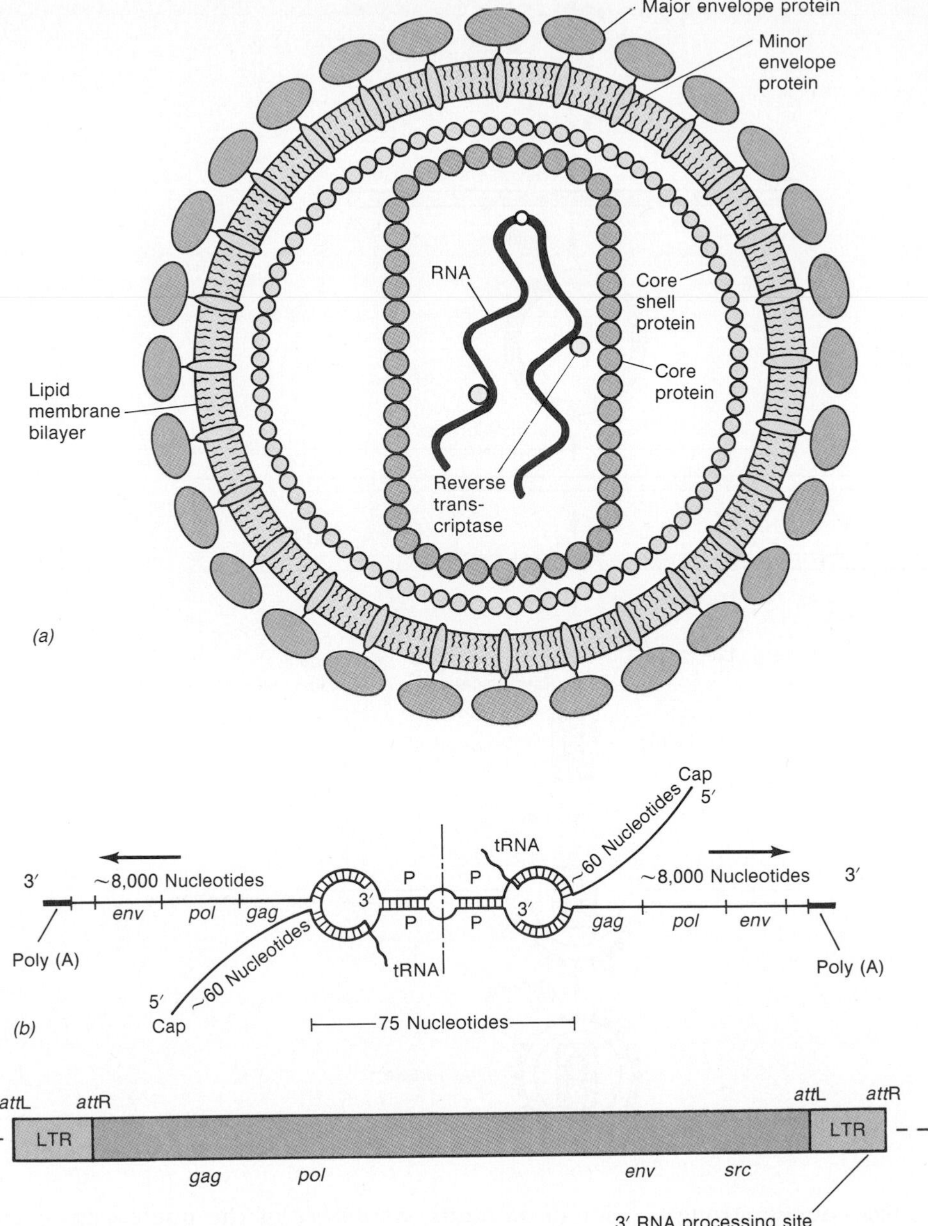

Figure 6.54 Retrovirus structure and function. (a) Structure of a retrovirus. (b) Structure of a retrovirus genome, showing the manner by which the two identical viral RNA molecules are held together. (c) Detailed genetic map, showing the location of the long terminal repeats (LTR). See text for details.

using the enzyme reverse transcriptase present in the virus particle. The DNA, a linear double-stranded molecule, is synthesized in the cytoplasm. An outline of the transcription of viral RNA into DNA is given in Figure 6.56.

The enzyme reverse transcriptase is essentially a DNA polymerase, and like all DNA polymerases, it needs a primer for DNA synthesis. The primer for retrovirus reverse transcription is a specific *cellular transfer RNA (tRNA)*. The type of tRNA used as primer depends on the virus. In the case of Rous sarcoma virus, the tRNA used is the *tryptophan* tRNA. Using the tRNA primer, the 100 or so nucleotides at the 5′ terminus of the RNA are transcribed into DNA. Once transcription reaches the 5′ end of the RNA, the transcription process stops. In order to copy the remaining RNA, which is the bulk of the RNA of the virus, a different mechanism comes into play. First, terminally redundant RNA sequences at the ends of the molecule are removed by the action of another enzymatic activity of reverse transcriptase, *ribonuclease H*. This leads to the formation of a small single-stranded DNA which is complementary to the RNA segment at the *other end* of the viral RNA. The small single-stranded piece of DNA then hybridizes with the other end of the viral RNA molecule, where copying of the viral RNA sequences continues. As summarized in Figure 6.56, continued action of reverse transcriptase leads to the formation of a double-stranded DNA molecule which has long terminal repeats (LTR) at

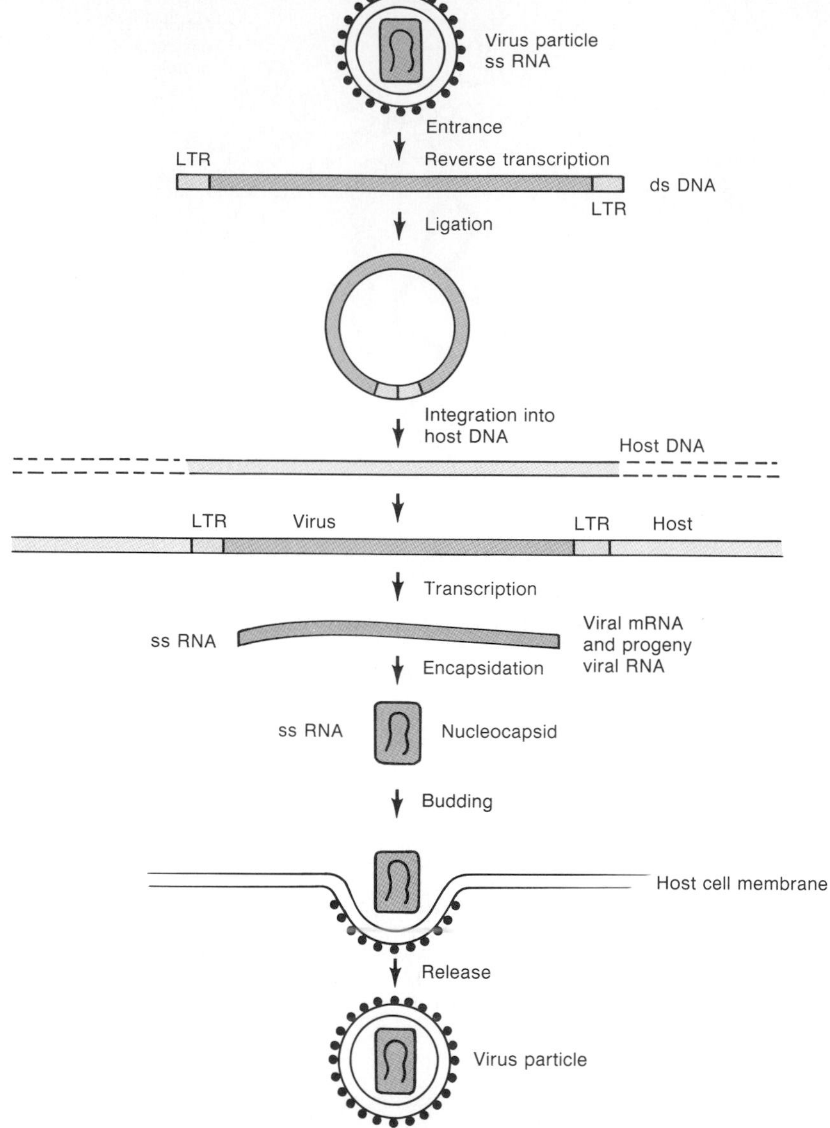

Figure 6.55 Replication process of a retrovirus.

each end (Figure 6.54*c*). These LTRs contain strong promoters of transcription and are involved in the integration process. The *integration* of the viral DNA into the host genome is analogous to the integration of Mu or a bacterial transposon into a bacterial genome. Integration can occur anywhere in the cellular DNA and once integrated, the element, now called a *provirus*, is a stable genetic element.

The integrated proviral DNA is transcribed by a cellular RNA polymerase into transcripts that are capped and polyadenylated. These RNA transcripts may either be encapsidated into virus particles or they may be processed and translated into virus proteins. Virus proteins are made initially as a large primary *gag* protein which is split by proteolytic action into the capsid proteins. Occasionally, read-through past *gag* leads to the transcription of *pol*, the reverse transcriptase gene.

When the virus proteins have accumulated in sufficient amounts, *assembly* of the nucleocapsid can occur. This encapsidation process leads to the formation of nucleocapsids which move to the plasma membrane for final assembly into the enveloped virus particles.

Not all retroviruses cause cancer, but tumorigenic retroviruses are quite common. Most tumorigenic retroviruses known cause *sarcomas and acute leukemia*; they possess high oncogenic potential. Infection with one of these viruses can cause cellular transformation, leading to the formation of a tumor. Why are these viruses tumorigenic? It appears that they possess a transforming gene, or *oncogene*, which codes for a protein that brings about cellular transformation. This gene, known in Rous sarcoma virus as the *src* gene (*src* for *sarcoma*, see Figure 6.54*c*) encodes for a phosphoprotein which possesses protein kinase activity. Protein kinases bring about the phosphorylation of proteins, and protein phospho-

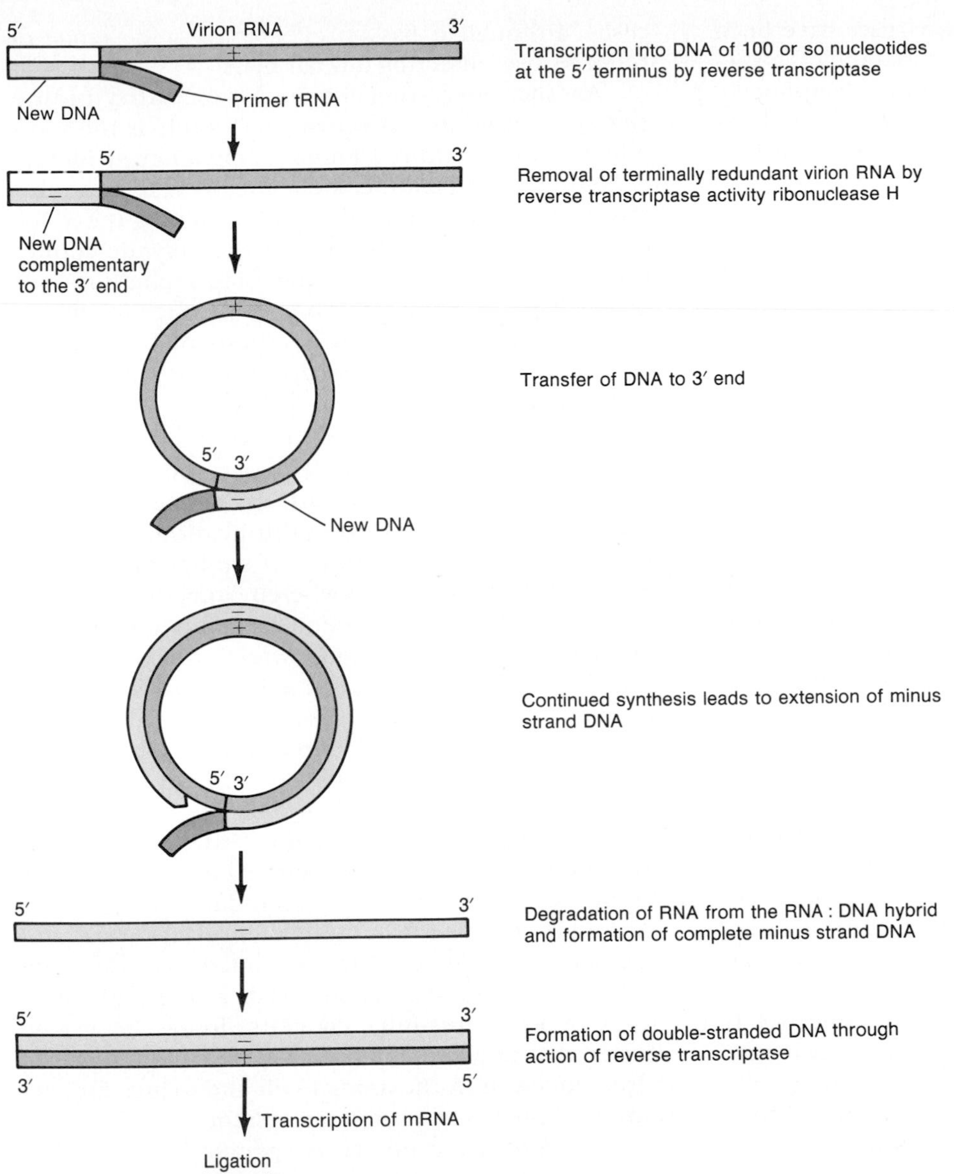
5′
Virion RNA
3′
New DNA
Primer tRNA
Transcription into DNA of 100 or so nucleotides at the 5′ terminus by reverse transcriptase
5′
3′
Removal of terminally redundant virion RNA by reverse transcriptase activity ribonuclease H
New DNA complementary to the 3′ end
Transfer of DNA to 3′ end
5′ 3′
New DNA
Continued synthesis leads to extension of minus strand DNA
5′ 3′
5′
3′
Degradation of RNA from the RNA : DNA hybrid and formation of complete minus strand DNA
5′
3′
Formation of double-stranded DNA through action of reverse transcriptase
3′
5′
Transcription of mRNA
Ligation
Integration

Figure 6.56 Overall steps in the formation of double-stranded DNA from retrovirus single-stranded RNA.

rylation is one mechanism for regulating the activity of proteins in eucaryotes. Depending on the protein, phosphorylation can either increase or decrease the activity of a protein. Transforming genes analogous to *src* have also been detected in human cancer cells. Interestingly, similar genes have also been detected in *normal* (that is, noncancerous) cells. These cell *src* sequences have been found not only in mammalian cells but also in the cells of insects and yeast, suggesting that these sequences are of fundamental importance in the regulation of cell growth. Presumably, retroviruses are able to incorporate such normal *src* sequences and can cause them to mutate. Retroviruses are thus the agents by which such genes are transferred from cell to cell. In normal cells, such *src* sequences may not be expressed, but the *src* sequences brought into the cell from another source may be expressed and lead to the formation of cancer.

Genetic engineering with retroviruses occurs naturally by the incorporation of oncogene sequences. It is also possible to use modified retroviruses to incorporate foreign genes into cells. This is possible because substitution of such foreign genes for essential virus functions can lead to the production of virus particles carrying the foreign gene. Such particles are capable of being integrated into the host genome, but are incapable of replicating or causing cancer. There is currently considerable interest in using retroviruses as *vectors* for human gene therapy, but there are considerable difficulties in making this work at the practical level. Further discussion of this topic would take us too far from the field of microbiology.

As we have noted, one of the newly discovered retroviruses is the virus causing AIDS. Although retroviruses had been known for a long time, it is only in the past decade or so that retroviruses have been found to infect humans. The AIDS virus infects a specific cell type in the human, a kind of T lymphocyte that is vital for the immune system. We discuss the medical and immunological aspects of AIDS in Sections 14.4 and 15.7.

6.25 Therapy for Virus Infections

We have presented in considerable detail the molecular events involved in the multiplication of viruses and we can now ask whether this sophisticated knowledge presents any clues as to how virus infections might be controlled. In the present section we discuss two approaches to virus therapy that have been studied in some detail, chemical inhibition and interferon action.

Chemical inhibition of animal viruses Since a virus depends on its host cell for many functions of virus replication, it is difficult to inhibit virus multiplication without at the same time affecting the host cell itself. Because of this, the spectacular medical successes following the discovery of antibacterial agents have not been followed by similar success in the search for specific antiviral agents. A few antiviral compounds are successful in controlling virus infections in laboratory situations (Table 6.3) and certain of these have been used in restricted clinical cases; but no substance has yet been found with more than limited practical use.

One interesting inhibitor listed in Table 6.3 is rifamycin, which is an inhibitor of RNA polymerase in procaryotes but not in eucaryotes (see Section 5.8). The RNA polymerase of vaccinia and other poxviruses is apparently inhibited by rifamycin, since this antibiotic specifically inhibits the replication of these viruses, although it has no effect on a wide range of other viruses affecting animal cells.

Another interesting chemical is **azidothymidine** (AZT), an inhibitor of retroviruses such as the virus which causes acquired immunodeficiency syndrome (AIDS). Azidothymidine is chemically related to thymidine but is a dideoxy derivative, lacking the 3′ hydroxyl (thus analogous to the dideoxynucleotides used in the Sanger DNA sequencing technique—see Box, Chapter 5). AZT inhibits multiplication of retroviruses by blocking the synthesis of the DNA intermediate (reverse transcription) and has shown promise in the treatment of AIDS.

Interferon Interferons are antiviral substances produced by many animal cells in response to infection by certain viruses. They are low-molecular-weight proteins that prevent viral multiplication. They were first discovered in the course of studies on virus interference, a phenomenon whereby infection with one virus interferes with subsequent infection with another virus, hence the name *interferon*. It has been found that interferons are formed in response not only to live virus but also to virus inactivated by radiation or to viral nucleic acid. Interferon is produced in larger amounts by cells infected with viruses of low virulence, whereas little is produced against highly virulent viruses. Apparently the virulent viruses inhibit cell protein synthesis before any interferon can be produced. Interferon is also induced by a variety of double-stranded RNA molecules, either natural or synthetic, and since double-stranded RNA does not exist in uninfected cells but exists as the replicative form in RNA-virus-infected cells, it has been suggested that double-stranded RNA serves as a signal of virus infection in the animal cell and brings into action the interferon-producing system.

Interferons are not virus specific but *host* specific; that is, an interferon produced by one type of

Table 6.3 Stages of virus replication at which chemical inhibition of virus action is known to occur

Stage of replication	Chemical	Virus
Free virus	Kethoxal	Influenza virus
Adsorption	None known	—
Entry of nucleic acid (uncoating)	Adamantanamine	Influenza virus
	Carbobenzoxypeptides	Measles
	3-Methylisoxazole compounds	Rhinoviruses (cold viruses)
Nucleic acid replication	Benzimidazole, guanidine	Poliovirus
	5-Fluorodeoxyuridine (FUDR)	Herpesvirus
	5-Iododeoxyuridine (IUDR)	Herpesvirus
	Acyclovir	Herpesvirus
	Rifamycin	Vaccinia virus
	Azidothymidine	Retrovirus (AIDS)
Maturation (or late protein synthesis)	Isatin-thiosemicarbazone	Smallpox virus
Release	None known	—

animal (for example, chicken) in response to influenza virus will also inhibit multiplication of other viruses in the same species but will have little or no effect on the multiplication of influenza virus in other animal species. Interferon has little or no effect on uninfected cells; thus it seems to inhibit viral synthesis specifically. It acts by preventing RNA synthesis directed by virus, thus inhibiting synthesis of virus-specific proteins.

Interferons have been of interest as possible antiviral agents, and possibly also as anticancer agents. Their use as therapeutic agents was long hindered by the difficulty and expense of producing large quantities, but genetic engineering techniques (see Chapter 8) have now made possible the production of interferon on a commercial scale, and many clinical trials are under way.

Study Questions

1. Define the term *host* as it relates to viruses.
2. Define *virus*. What are the minimal features needed to fit your definition? Viruses can be defined in either biological or chemical terms. Give both kinds of definitions.
3. Under some conditions, it is possible to obtain nucleic acid-free protein coats (*capsids*) of certain viruses. These capsids look very similar under the electron microscope to complete virus particles. What does this fact tell you about the role of the virus nucleic acid in the virus assembly process? Would you expect such virus particles to be infectious? Why?
4. Write a paragraph describing the events which occur on an agar plate containing a bacterial lawn which result when a single bacteriophage particle causes the formation of a *bacteriophage plaque*.
5. Suppose you had been the first person to discover bacterial viruses. You observe clear plaques on a bacterial lawn and hypothesize that these plaques arise because of the action of a virus (remember that viruses affecting plants and animals had already been characterized). Describe the experiments you would carry out: (1) to show that the clear zones (plaques) were really due to an infectious agent; (2) that the agent was subcellular. On one of your plates, two different types of plaque appear. Describe the experiments you would carry out to show that each plaque type is due to a separate virus type.
6. *Chemotherapeutic agents* are lacking for most virus diseases. From what you know about the stages of virus multiplication, write (in general terms) how you might try to assay (detect) a chemotherapeutic agent which acts at each of these stages.
7. Several stages in *virus multiplication* involve specific host biochemical machinery. Why is it less likely that successful (that is, nontoxic) chemotherapeutic agents could be found that would affect these stages?
8. How might an agent which affects the *virus release process* still be an effective chemotherapeutic agent, even though cells which have passed through the *assembly* stages are effectively dead?
9. Two major classes of *host mutants* resistant to virus infection are known, those not absorbing virus and those to which virus absorbs and injects but the nucleic acid is destroyed inside. Suppose you had a bacterial host mutant resistant to phage T4, how would you determine which class it belonged to?
10. Describe how a *restriction endonuclease* might play a role in resistance to bacteriophage infection? Why could a restriction endonuclease play such a role whereas a generalized DNase could not?
11. Put together a diagram describing mRNA synthesis and nucleic acid replication for each of the following virus types: RNA tumor virus, reovirus, poliovirus, T4 phage, ϕX174 phage. For each diagram, be sure to indicate the complementarity of both the virus nucleic acid and its mRNA.
12. Given a wild-type strain and two mutant strains of bacteriophage T2 infecting *E. coli*: one of the virus mutants is a host-range mutant and the other is a plaque-morphology mutant. Describe an experiment that would demonstrate the occurrence of genetic recombination. Be sure to indicate both how you would set up the experiment and how you would assess the results (that is, assay for recombinant types). How would you distinguish genetic recombination from phenotypic mixing?
13. Examine the graph which illustrates a one-step growth curve (Figure 6.13). Several stages are listed. Answer the following questions in reference to the stages indicated on the graph:
    a. In which stage does attachment of virus particles to cells occur?
    b. In which stage are early enzymes formed?
    c. In which stage does virus nucleic acid replication predominantly occur?
    d. In which stage does cell lysis occur?
    e. If it were to occur, in which stage would virus genetic recombination likely occur?
    f. In which stage would the self-assembly process predominantly occur?
14. Suppose you wanted to determine whether a bacterial culture (strain) you were studying was *lysogenic*. You have no information about the genetics of this strain, but you have access to a large number of other strains of the same bacterial species. How would you proceed? Suppose you had no other strains of the same species. Is there any way you might get an idea of lysogenicity?
15. One characteristic of *temperate bacteriophages* is that they cause turbid rather than clear plaques on bacterial lawns. Can you think why this might be? (Remember the process by which a bacterial plaque develops.)
16. *Virulent mutants* of temperate bacterial viruses are

known. How might you detect such virulent mutants based on observations of their plaque morphology? How might such mutants differ genetically from the wild type? (Examine the genetic map of *lambda*.)

17. Suppose a mutant of *lambda* temperate bacteriophage was isolated which lacked the *att* site. How might its phenotype be expressed? Suppose a host mutant were isolated (that is, a bacterial mutant) which lacked the site on its DNA where the *lambda* DNA attached. What result would you predict when an infection of this host with *lambda* virus was brought about?
18. Describe the role which the *lambda repressor protein* plays in the *lambda* infection process. How do repressor protein and the *cro* protein interact?
19. From what you know about cell function and virus function, would you say viruses are living or dead? Why?
20. Read the material on *prions* in the Box in this Chapter. How does the existence of the prion alter the validity of the definition of a virus which you gave in question 2?
21. Chemicals are known which inhibit the *attachment* of animal viruses to their host cells. How might a knowledge of *viral symmetry* (virion structure) help in either finding chemicals of this type, or determining the precise mode of action of such chemicals?
22. When considering the *inhibition of virus penetration* by chemical agents, *naked* and *enveloped* viruses could be expected to behave differently. Why?
23. A number of small viruses have developed unusual features that make the same genetic material serve more than one function. Several such approaches have been discussed in this chapter. List three.
24. Describe the process by which a single-stranded filamentous DNA bacteriophage is released from the cell. Why is it that even though this process does not result in lysis, plaques can be detected in conventional bacteriophage plaque assays?
25. Compare and contrast DNA replication in ϕX174, the M13-fd group, T7, Mu, and adenovirus.
26. The T-even bacteriophages have developed a different *modification-restriction system* than that found in other systems. What is this system and how does it work?
27. *Replication of DNA in bacteriophage Mu* differs significantly from replication of DNA in the other double-stranded DNA phages discussed. Discuss. Why is it stated that Mu has some similarities to a transposon?
28. Compare and contrast virus multiplication in *positive-strand* and *negative-strand* RNA viruses of animals.

Supplementary Readings

Bishop, J. M. 1987. The molecular genetics of cancer. *Science* 235:305–311. A concise review of how cancer arises, with emphasis on the role of retroviruses.

Fields, B. N. (editor). 1985. *Virology*. Raven Press, New York. A large and very detailed treatment of all aspects of virology with emphasis on animal and human viruses. Each chapter written by an expert.

Fraenkel-Conrat, Heinz, and **P. C. Kimball**. 1982. *Virology*, Prentice-Hall, Englewood Cliffs, N.J. A short textbook of virology.

Fraenkel-Conrat, Heinz, and **Robert R. Wagner** (editors). *Comprehensive Virology*, Plenum Press, New York. A multivolume work which presents detailed papers on all aspects of virology.

Fraenkel-Conrat, Heinz. 1985. *The Viruses. Catalogue, Characterization, and Classification*. Plenum Publishing Company, New York. This volume is a definitive catalog of virus taxonomy. Electron micrographs of most virus families are given.

Hendrix, R. W., J. W. Roberts, F. W. Stahl, and **R. A. Weisberg** (editors). 1983. *Lambda II*. Cold Spring Harbor Laboratory, New York. Collection of research papers devoted solely to *lambda*.

Hughes, S. S. 1977. *The Virus: a History of the Concept*. Heinemann, London. A brief and interesting history of the development of the virus concept.

Joklik, W. K. 1985. *Virology*, second edition. Appleton-Century-Crofts, New York. Covers similar material to Fields but shorter and more concise.

Kornberg, A. 1980. *DNA Replication* (also a 1982 supplement). W. H. Freeman, New York. The best treatment of the chemical basis of DNA replication.

Luria, S. E., Darnell, J. E., Baltimore, D., and **Campbell, A.** 1978. *General Virology*, third edition. John Wiley, New York. The *classic* virology textbook, but now seriously dated. Still of use for an overview of fundamental concepts.

Matthews, R. E. F. 1979. Classification and nomenclature of viruses. *Intervirology*, 12, no. 3-5. An overview of the classification of viruses.

Ptashne, M. 1986. *A Genetic Switch*, Blackwell Scientific Publications, Palo Alto. A short book which deals primarily with how *lambda* is regulated.

Temin, H. 1976. The DNA provirus hypothesis. *Science* 192: 1075-1080. An account of the discovery of reverse transcription in retroviruses.

Watson, J. D., N. H. Hopkins, J. W. Roberts, J. A. Steitz, and **A. M. Weiner**. 1987. *Molecular Biology of the Gene, Volume I, General Principles*, fourth edition. Benjamin-Cummings Publishing Company, Menlo Park. Has extensive coverage of how viruses replicate.

White, D. O., and **F. J. Fenner**. 1986. *Medical Virology*, third edition. Academic Press, Orlando. An excellent textbook and reference on medical aspects of virology.

Zinder, N. D. (editor). 1975. *RNA Phages*. Cold Spring Harbor Laboratory, New York. Collection of research papers on these interesting viruses.

Microbial Genetics

Now that we have introduced the main features of molecular genetics of cells and viruses, we can turn to a discussion of how genetic material is transferred from one organism to another. Gene transfer can occur in a number of different ways, and if it is accompanied by genetic recombination, it can lead to the formation of new organisms.

Genetic recombination is the process by which genetic elements contained in two separate genomes are brought together in one unit. Through this mechanism, new genotypes can arise even in the absence of mutation. Since the genetic elements brought together may enable the organism to carry out some new function, genetic recombination can result in adaptation to changing environments. While mutation usually brings about only a very small amount of genetic change in a cell, genetic recombination usually involves much larger changes. Entire genes, sets of genes, or even whole chromosomes, are transferred between organisms. As a result, offspring are not exactly like either parent; they are **hybrids** and contain a combination of the traits exhibited by each parent. In eucaryotes, genetic recombination is a regular process which occurs as a result of sexual reproduction (see Section 3.17). Bacteria do not carry out an exactly analogous process, but they do have important ways of undergoing genetic exchange. There are distinct differences between genetic recombination in eucaryotes and procaryotes.

Genetic recombination is an important research tool in dissecting the genetic structure of an organism. It is also of major importance in the construction of new organisms for practical applications, a major activity in the field of genetic engineering. We present the basic principles of bacterial genetics in this chapter and then show in the next chapter how these principles apply to research in genetic engineering.

7.1 Kinds of Genetic Recombination

At the molecular level, two kinds of genetic recombination have been recognized, which are called *general recombination* and *site-specific recombination*. General recombination, which occurs widely in both procaryotes and eucaryotes, results in genetic exchange between homologous DNA sequences on separate chromosomes. In this process, homologous base-pairing occurs over an extended length of the two DNA molecules. Thus, identical or nearly identical sequences on the two recombining molecules are required. The process of genetic exchange which occurs during sexual reproduction (meiosis) is an example of general recombination.

Site-specific recombination differs from general recombination in that

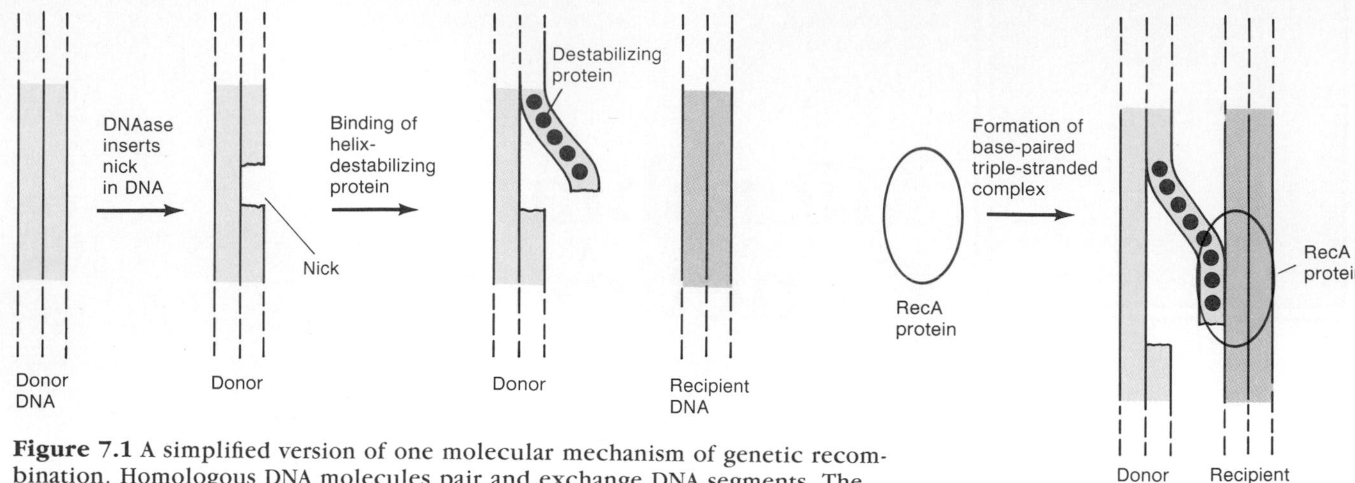

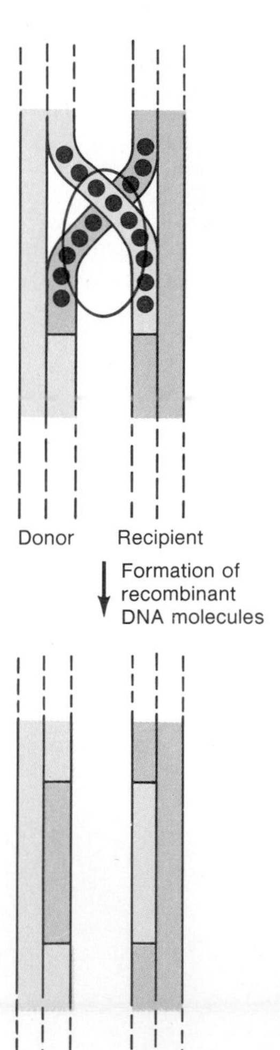

Figure 7.1 A simplified version of one molecular mechanism of genetic recombination. Homologous DNA molecules pair and exchange DNA segments. The mechanism involves breakage and reunion of paired segments. Two proteins, a helix-destabilizing protein and RecA protein are involved. The diagram is not to scale: pairing can occur over hundreds of bases.

DNA homology is not required. Instead, short, specific nucleotide sequences are recognized by a recombination enzyme, and base-pairing between adjacent molecules is not required. The process is called *site-specific recombination* because the recombination enzyme recognizes specific nucleotide sequences (sites) present on one or both of the recombining DNA molecules. It is at these sequences that recombination occurs. We discussed one type of site-specific recombination when we discussed the mechanism by which bacteriophage *lambda* is integrated into the host chromosome (see Section 6.14). Another type of site-specific recombination is that involving transposons (see Sections 5.6, 6.15, and 7.7). We discuss the site-specific recombination that occurs during transposition later in this chapter.

Molecular events in general recombination At the molecular level, recombination has been studied only in procaryotes and viruses. The process has been too complicated to analyze in eucaryotes. In bacteria, general recombination involves the participation of a specific protein called the RecA protein, which is specified by the *recA* gene. Bacteria which are mutant in *recA* show markedly decreased levels of general recombination.

As noted, general recombination involves the *pairing* of DNA molecules over long stretches. An overall molecular mechanism of general recombination is shown in Figure 7.1. The process begins with a *nick* in one of the DNA molecules, which leads to the formation of a short single-stranded segment. A special protein, the *helix-destabilizing protein*, combines at this site and aids in opening up of the DNA double helix. Next, in the presence of RecA protein, approach of a strand of DNA homologous in that region leads to *pairing* of the two double-stranded molecules *via* the single-stranded segment. This pairing is not shown to scale in Figure 7.1; it can occur over hundreds of bases. Following pairing, *exchange* of homologous DNA molecules can occur, leading to the formation of recombinant DNA structures.

Note that this mechanism for the formation of recombinant DNA structures is a completely natural mechanism which occurs extensively within the cell. Whether or not it leads to the formation of new genotypes will depend upon whether the two molecules undergoing recombination differ genetically in regions outside the region of recombination. Within limits, general recombination can be thought of as occurring at random sites throughout the genome. Thus, the probability of recombination occurring between two genes is proportional to their distance. This fact is useful for genetic mapping, that is, determining the position of genes on chromosomes, since the farther two genes are apart, the more likely they will show recombination.

In order for new genotypes to arise as a result of genetic recombination, it is essential that two genetically distinct molecules are present in the same cell at the time that genetic recombination occurs. In eucaryotes, the two distinct molecules are brought together as the result of the sexual reproduction process, a process which occurs as part of the regular life cycles of most eucaryotic organisms (see Section 7.9). In procaryotes, genetically distinct DNA molecules are brought together in less regular ways, but the process of genetic recombination is no less important. We discuss the various mechanisms by which DNA molecules are incorporated into bacterial cells in the first part of this chapter.

In procaryotes genetic recombination is observed because of the insertion into a recipient cell of a fragment of genetically different DNA derived from a donor cell; next the integration of this DNA fragment or its copy into the genome of the recipient cell must occur. We shall define briefly here the three means by which the DNA fragment is introduced into the recipient: (1) **transformation** is a process by which free DNA is inserted directly into a competent recipient cell; (2) **transduction** involves the transfer of bacterial DNA from one bacterium to another within a temperate or defective virus particle; and (3) **conjugation (mating)** involves DNA transfer via actual cell-to-cell contact between the recipient and the donor cell. The last process most closely resembles sexual recombination in eucaryotes, but differs from it in several fundamental respects. These processes are contrasted in Figure 7.2.

Detection of recombination In order to detect physical exchange of DNA segments, the products of recombination must be phenotypically different from the parents. Furthermore, genetic recombination in procaryotes usually is a rare event, occurring in only a small percentage of the population. Because of its rarity, special techniques are usually necessary to detect its occurrence. One must usually use as recipients strains that possess selectable characteristics such as the inability to grow on a medium on which the recombinants can grow. Various kinds of selectable and nonselectable markers (such as drug resistance, nutritional requirements, and so on) were discussed in Section 5.16. The exceedingly great sensitivity of the selection process is shown by the fact that 10^8 or more bacterial cells can be spread on a single plate

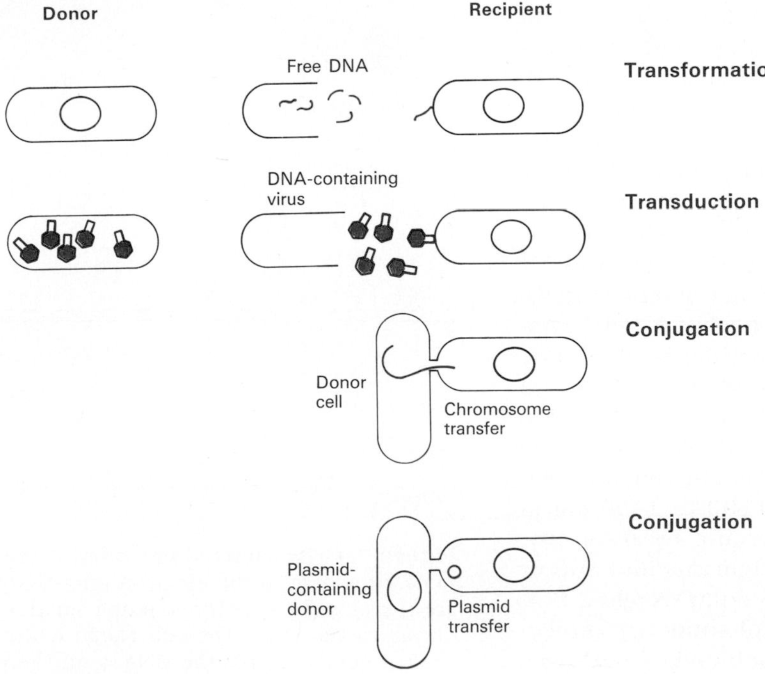

Figure 7.2 Processes by which DNA is transferred from donor to recipient bacterial cell. Just the initial steps in transfer are shown. For details of how the DNA is integrated into the recipient, see text.

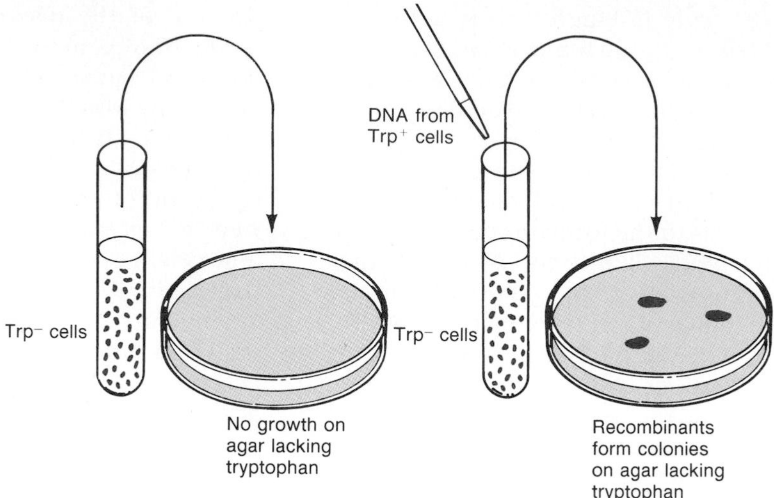

Figure 7.3 How a selective medium can be used to detect rare genetic recombinants among a large population of nonrecombinants. On the selective medium only the rare recombinants form colonies. Procedures such as this, which offer high resolution for genetic analyses, can ordinarily be used only with microorganisms.

and, if proper selective conditions are used, no parental colonies will appear, whereas even a few recombinants can form colonies (Figure 7.3). The only requirement is that the *reverse* mutation rate for the characteristic selected must be low, since revertants will also form colonies. This problem can often be overcome by using double mutants, since it will be very unlikely that two back mutations will occur in the same cell. Much of the skill of the bacterial geneticist is exhibited in the choice of proper mutants and selective media for efficient detection of genetic recombination.

7.2 Genetic Transformation

As we have noted, genetic transformation is a process by which free DNA is incorporated into a recipient cell and brings about genetic change. The discovery of genetic transformation in bacteria was one of the outstanding events in biology, as it led to experiments proving without a doubt that DNA was the genetic material. Chemical purification of the *transforming principle* (see Box, Chapter 5) showed that it consisted of pure DNA. This discovery became the keystone of molecular biology and modern genetics.

A number of bacteria have been found to be transformable, including both Gram-negative and Gram-positive species. However, even within transformable genera, only certain strains, called *competent*, are transformable. Since the DNA of procaryotes is present in the cell as a long single molecule, when the cell is gently lysed, the DNA pours out (Figure 7.4). Because of its extreme length (1100 to 1400 μm in *Escherichia coli*), the DNA molecule breaks easily; even after gentle extraction it fragments into 100 or more pieces (*E. coli* DNA of molecular weight 2.8×10^9 is converted into fragments of about 10^7 molecular weight). Since the DNA which corresponds to a single gene has a molecular weight of the order of 1×10^6 (corresponding to about 1000 nucleotides), each of the fragments of purified DNA will have about 10 genes. Any cell will usually incorporate only a few DNA fragments so that only a small proportion of the genes of one cell can be transferred to another by a single transformation event.

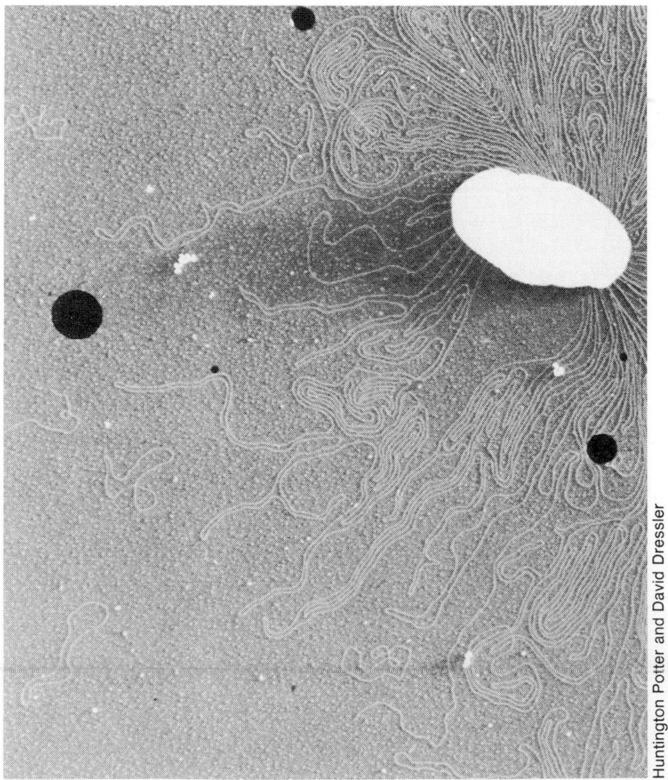

Figure 7.4 The bacterial chromosome and bacterial plasmids, as shown in the electron microscope. The plasmids are the circular structures, much smaller than the main chromosomal DNA. The cell (large white structure) was broken gently so that the DNA would remain intact.

Competence A cell that is able to take up a molecule of DNA and be transformed is said to be **competent**. Only certain strains are competent; the ability seems to be an inherited property of the organism. Further, competence is affected by the physiological state of the cells and the medium in which they are grown, and it varies with the stages of the growth cycle. There is a brief period, during middle exponential phase, when the competence of the population rises dramatically and then just as rapidly falls. Good evidence exists that during this brief period of competence the surfaces of the cells change so that DNA is now able to be bound to them. This surface change can also be brought about by an enzyme-like factor produced by the cells themselves, which appears in the culture medium at about the time competence appears; moreover, adding this factor to noncompetent cells of the same strain can induce them to convert into competent cells. At the same time that the peak of competence is reached, virtually all of the cells are able to take up DNA.

Increasing transformation efficiency Many organisms are transformed only poorly or not at all. Transformation has been found to be much more efficient in strains that are deficient in certain DNases, which would normally destroy incoming DNA. Determination of how to induce competence may involve considerable empirical study, with variation in culture medium, temperature, and other factors. The nature of the cell surface must be of importance in determining whether a cell can take up DNA. For genetic engineering (see Chapter 8), *Escherichia coli*, a Gram-negative bacterium, must be transformed. It has been found that if *E. coli* is treated with high concentrations of calcium ions and then stored overnight in the cold, it becomes transformable at low efficiency. With proper procedures it is possible to select *E. coli* transformants. Why the calcium treatment works is not known, but this procedure also works with some other Gram-negative bacteria. Presumably, with sufficient research most bacteria could be made transformable, at least at low efficiency. High efficiency transformation is found, however, only in a few bacteria.

Uptake of DNA During the transformation process, competent bacteria first bind DNA reversibly; soon, however, the binding becomes irreversible. Competent cells bind much more DNA than do noncompetent cells—as much as 1000 times more. As we noted earlier, the sizes of the transforming fragments are much smaller than that of the whole genome; in *Streptococcus pneumoniae* each cell can incorporate only about 10 molecules of molecular weight 1×10^7. The DNA fragments in the mixture compete with each other for uptake, and if excess DNA that does not contain the genetic marker is added, a decrease in the number of transformants occurs. In preparations of transforming DNA, only about 1 out of 100 to 200 DNA fragments contains the marker being studied. Thus at high concentrations of DNA, the competition between DNA molecules results in saturation of the system so that even under the best conditions it is impossible to transform all of the cells in a population for a given genetic marker. The maximum frequency of transformation that has so far been obtained is about 10 percent of the population; actually the values usually obtained are between 0.1 and 1.0 percent. The minimum concentration of DNA yielding detectable transformants is about 0.00001 μg/ml (1×10^{-5} μg/ml), which is so low that it is undetectable chemically.

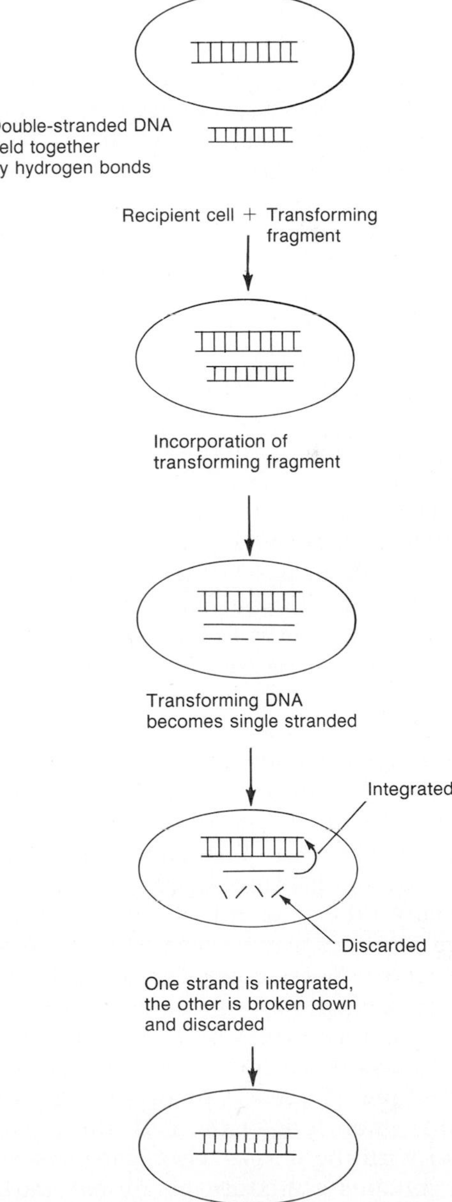

Figure 7.5 Integration of transforming DNA into a recipient cell.

Integration of incorporated DNA Soon after incorporation, one of the strands is degraded, while the other is integrated into the genome of the recipient (Figure 7.5). During replication of this hybrid DNA, one parental and one recombinant DNA molecule are formed. Upon segregation at cell division, the latter will be present in the transformed cell. The steps in the transformation process can be summarized briefly: (1) pairing of incoming DNA with the homologous sequence of the recipient chromosome; (2) single-stranded nicking by a specific DNA cutting enzyme which opens up the DNA double helix; (3) insertion of the incoming strand; and (4) connection of the incoming strand by action of DNA ligase.

7.3 Transduction

In transduction, DNA is transferred from cell to cell through the agency of viruses. Genetic transfer of host genes by viruses can occur in two ways. The first, called **specialized transduction**, occurs only in some temperate viruses; a specific group of host genes is integrated directly into the virus genome—usually replacing some of the virus genes—and is transferred to the recipient during lysogenization. In the second, called **generalized transduction**, host genes derived from virtually any portion of the host genome become a part of the DNA of the mature virus particle in place of, or in addition to, the virus genome. The transducing virus particle in both cases is **defective** and not able to cause lysis of the host, probably because of a lack of some virus genes.

Transduction has been found to occur in a variety of bacteria. Not all phages will transduce, and not all bacteria are transducible; but the phenomenon is sufficiently widespread for us to assume that it plays an important role in genetic transfer in nature.

Generalized transduction In generalized transduction, virtually any genetic marker can be transferred from donor to recipient. Generalized transduction was first discovered and extensively studied in the bacterium *Salmonella typhimurium* with phage P22, but is known to occur in *Escherichia coli* and many other bacteria. An example of how transducing particles may be formed is given in Figure 7.6. When the population of sensitive bacteria is infected with a phage, the events of the phage lytic cycle may be initiated. In a lytic infection, the host DNA often breaks down into virus-sized pieces, and some of these pieces become incorporated inside virus particles. Upon lysis of the cell, these particles are released with the normal virus particles, so that the lysate contains a mixture of normal and transducing virus particles. When this lysate is used to infect a population of recipient cells, most of the cells become infected with normal virus particles. However, a small proportion of the population receives transducing particles, whose DNA can now undergo genetic recombination with the host DNA. Since only a small proportion of the particles in the lysate are of the defective transducing type and since each particle contains only a small fragment of donor DNA, the probability of a defective phage particle containing a particular gene is quite low, and usually only about 1 cell in 10^6 to 10^8 is transduced for a given marker.

Phages that form transducing particles can be either temperate or virulent, the main requirement being that they do not cause complete degradation of the host DNA. The detection of transduction is most certain when the multiplicity of phage to host is low, so that a host cell is infected with only a single phage particle; with multiple infection, the cell may be killed by the normal particles.

Another example of generalized transduction is that which occurs in bacteriophage Mu, whose life cycle was described in Section 6.15.

Specialized transduction Generalized transduction is a rare genetic event, but in the specialized transduction to be discussed now a very efficient transfer by phage of a specific set of host genes can be arranged. The example we shall use, which was the first to be discovered and is the best understood today, involves transduction of the galactose genes by the temperate phage *lambda* of *E. coli*.

As we discussed in Section 6.14, when a cell is lysogenized by *lambda*, the phage genome becomes integrated into the host DNA at a specific site. The region in which *lambda* integrates is immediately adjacent to the cluster of host genes that control the enzymes involved in galactose utilization (Figure 6.36), and the DNA of *lambda* is inserted into the host DNA at that site. From then on, viral DNA replication is under host control. Upon induction, (for example, by ultraviolet radiation) the viral DNA separates from the host DNA by a process that is the reverse of integration (Figure 7.7). Ordinarily when the lysogenic cell is induced, the *lambda* DNA is excised as a unit. Under rare conditions, however, the phage genome is excised incorrectly. Some of the adjacent bacterial genes (the galactose cluster) are excised along with phage DNA. At the same time, some phage genes are left behind. One type of altered phage particle, called **lambda dg** (dg means "defective galactose"), is defective and does not make mature phage. However, if another phage called a **helper** is used together with *lambda dg* in a mixed infection, then the defective phage can be replicated and can transduce the galactose genes. The role of the helper phage is to provide those functions missing in the defective particles. Thus the culture lysate obtained contains a few *lambda dg* particles mixed in with a large number of normal *lambda* particles.

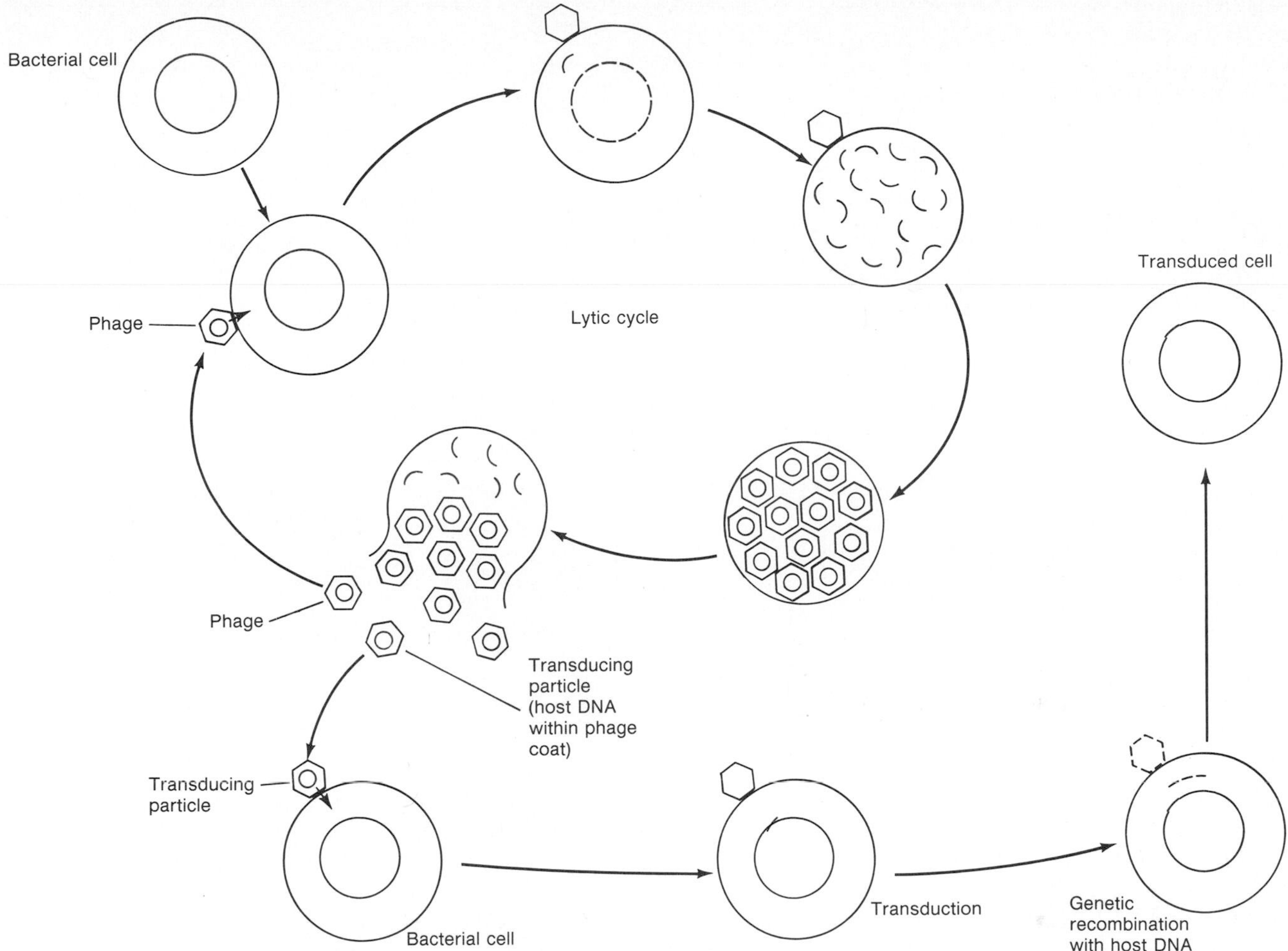

Figure 7.6 Generalized transduction: one possible mechanism by which virus (phage) particles containing host DNA can be formed.

If a galactose-negative culture is infected at high multiplicity with such a lysate, and gal^+ transductants are selected, many are double lysogens, carrying both *lambda* and *lambda dg*. When such a double lysogen is induced, a lysate is produced containing about equal numbers of *lambda* and *lambda dg*. Such a lysate can transduce at high efficiency, although only for a restricted group of *gal* genes.

If the phage is to be viable, there is a maximum limit to the amount of phage DNA that can be replaced with host DNA, since sufficient phage DNA must be retained in order to provide the information for production of the phage protein coat and for other phage proteins needed for lysis and lysogenization. However, if a helper phage is used together with the defective phage in a mixed infection, then even less information is needed in the defective phage for transduction. Only the *att* (attachment) region, the *cos* site (for packaging), and the replication origin are needed for production of a transducing particle, provided a helper is used (see the genetic map of *lambda* in Section 6.14).

One important distinction between specialized and generalized transduction is in how the transducing lysate can be formed. In specialized transduction this *must* occur by induction of a lysogenic cell whereas in generalized transduction it can occur either in this way or by infection of a nonlysogenic cell by the temperate phage, with subsequent phage replication and cell lysis.

Although we have discussed specialized transduction only in the *lambda-gal* system, phage *lambda* and its relatives (ϕ80, P22) have been widely used to form specialized transducing phages covering many specific regions of the *E. coli* genome.

Phage conversion This is a phenomenon analogous in some ways to specialized transduction. When a normal temperate phage (that is, a nondefective one) lysogenizes a cell and its DNA is converted into the prophage state, the lysogenic cell is immune to further infection by the same type of phage. In certain cases other phenotypic alterations can be detected in the lysogenized cell, which seem to be unrelated to

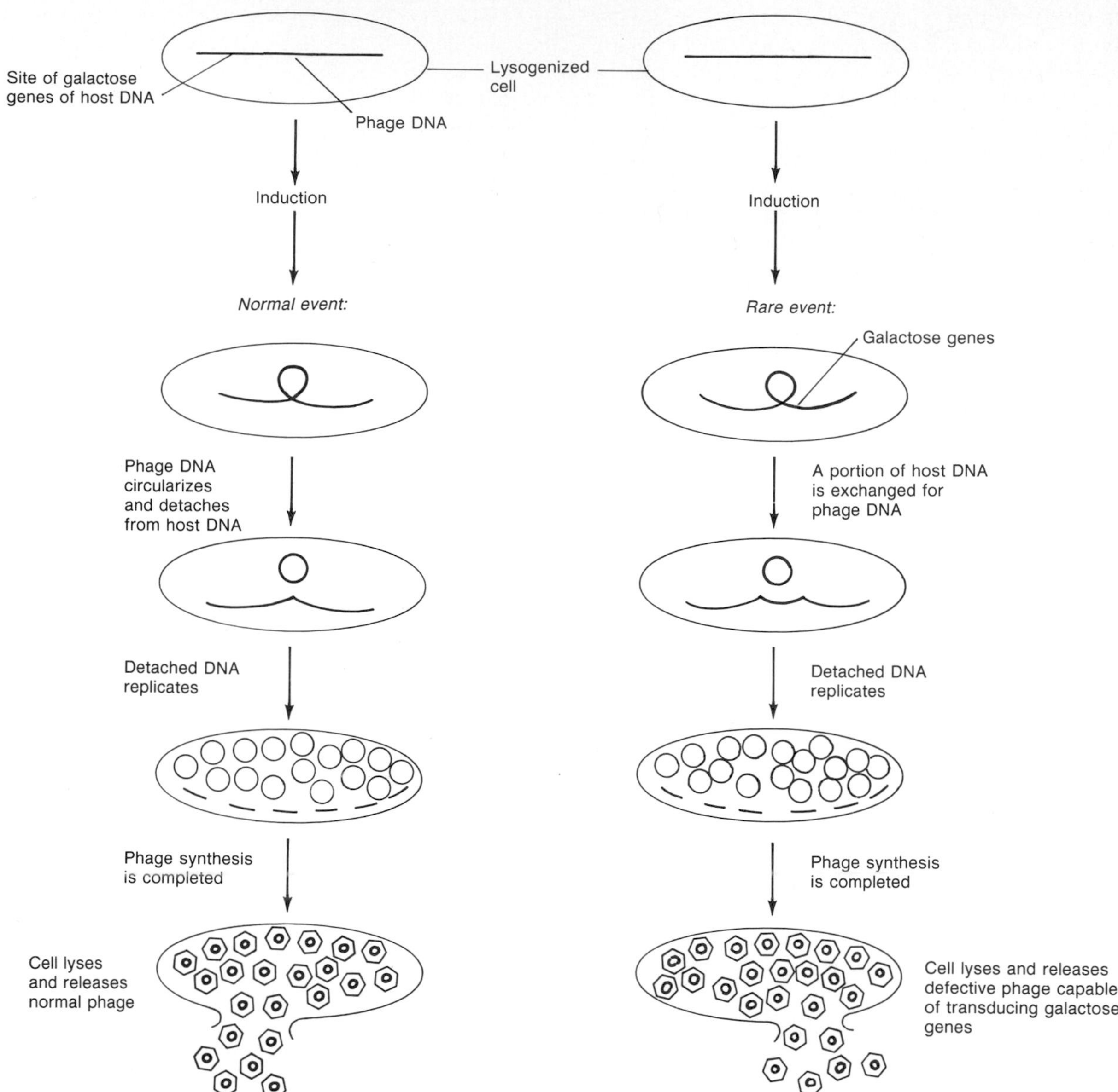

Figure 7.7 The production of particles transducing the galactose genes in an *E. coli* cell lysogenic for *lambda* virus.

the phage immunity system. Such a genetic change, which is brought about through lysogenization by a normal temperate phage, is called **phage conversion**. Two cases have been especially well studied. One involves a change in structure of a polysaccharide on the cell surface of *Salmonella anatum* upon lysogenization with phage ϵ^{15}. The second involves the conversion of nontoxin-producing strains of *Corynebacterium diphtheriae* to toxin-producing, pathogenic strains, upon lysogenization with phage β (we discuss the disease diphtheria in Section 15.2). In these situations the information for production of these new materials is apparently an integral part of the phage genome and hence is automatically and exclusively transferred upon infection by the phage and lysogenization.

Lysogeny probably carries a strong selective value for the host cell, since it confers resistance to infection by viruses of the same type. Phage conversion seems also to be of considerable evolutionary significance, since it results in efficient genetic alteration of host cells. Many bacteria isolated from nature are lysogenic. It seems reasonable to conclude, therefore, that lysogeny is the normal state of affairs and may often be essential for survival of the host in nature.

7.4 Plasmids

Before we discuss the third method of genetic transfer, **conjugation**, we must introduce a new kind of genetic element called the **plasmid**. Plasmids are circular genetic elements that reproduce autonomously and have an extrachromosomal existence (Figure 7.4). An outline of various aspects of plasmid biology is given in Figure 7.8. Most plasmids can be eliminated from the cell by various treatments (called *curing*) without lethal effect on the cell, and many plasmids can be transmitted from cell to cell by means of the conjugation process, which is discussed below. Some plasmids also have the ability to become integrated into the chromosome, and under such conditions their replication comes under control of the chromosome. Subsequently, an integrated plasmid may be released and then resume independent replication. The sequence of independent replication, chromosome integration, and escape from integration are phenomena equivalent to those undergone by the temperate bacteriophage (Section 6.14). This similarity is actually more than a mere analogy; by the definition given here many temperate phages are also plasmids.

Many plasmids carry genes that control the production of toxins and provide resistance to antibiotics and other drugs. Many plasmids also carry genes that control the process of conjugation, such as genes which alter the cell surface to permit cell-to-cell contact and genes which bring about the transfer of DNA from one cell to another. Plasmids which govern their own transfer by cell-to-cell contact are called **conjugative**, but not all plasmids are conjugative. Transmissability by conjugation is controlled by a set of genes within the plasmid called the **sex factor**. The presence of a sex factor in a plasmid can have another important consequence, if the plasmid becomes integrated into the chromosome. In that case, the plasmid can mobilize the transfer of chromosomal DNA

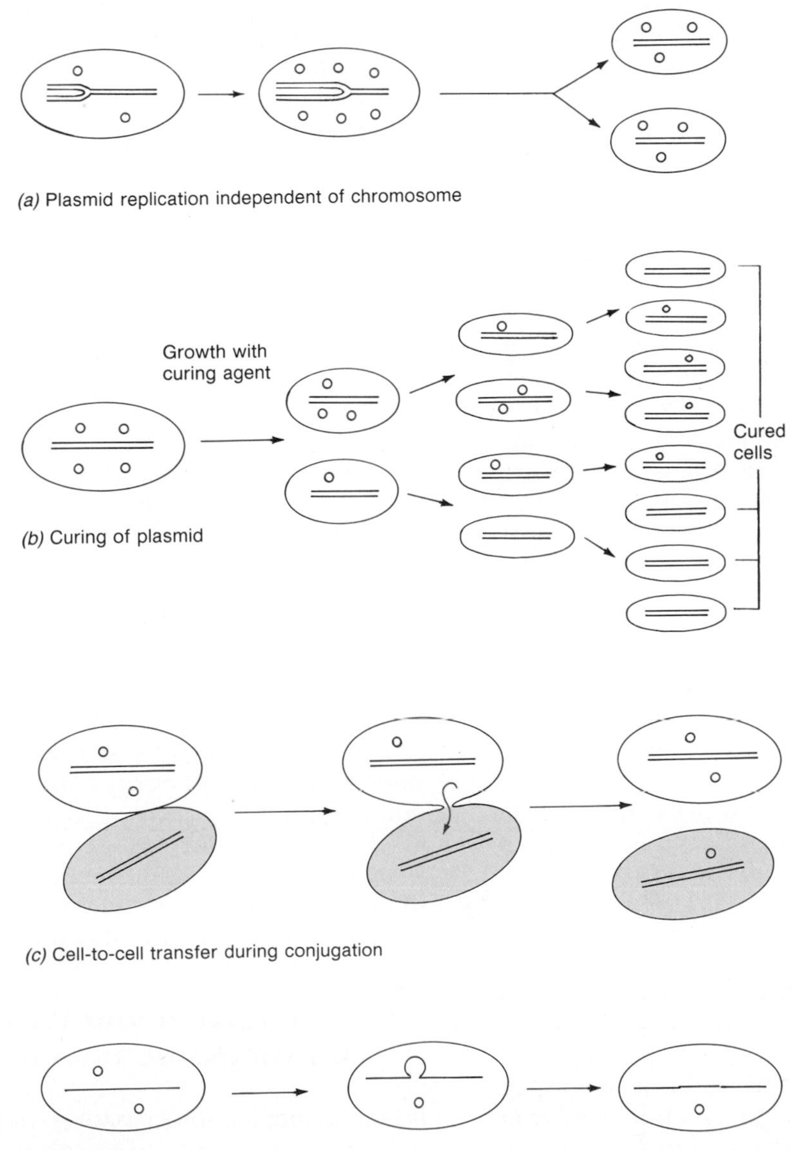

Figure 7.8 Plasmid biology.

from one cell to another. Bacteria that transfer large amounts of chromosomal DNA during conjugation are called *Hfr* (high frequency of recombination). Although considerably rarer than bacteria transferring only plasmid DNA, Hfr bacteria are of considerable interest and importance, since the study of Hfr bacteria permits an analysis of the genetic organization of the whole bacterial chromosome.

Physical evidence of plasmids Plasmids are generally less than one-twentieth the size of the chromosome. As usually isolated, the plasmid DNA is a double-stranded closed circle (Figure 7.4). When isolated, this circular double helix is twisted about itself in a form called a **supercoil** (see Figure 5.5). A single break (called a "nick") in one of the two strands causes the supercoil to convert to an open circular form, and if breaks occur in both strands at the same place, a linear duplex structure will be formed. Most of the plasmid DNA isolated from cells is in the supercoiled configuration, which would be the most compact form within the cell. Isolation of plasmid DNA can generally be readily accomplished by making use of certain physical properties of supercoiled DNA molecules. Such molecules are more compact than nicked or linear duplexes and hence sediment more rapidly in an ultracentrifuge. Further, supercoiled molecules show a high sedimentation rate in liquids such as alkaline sucrose, which ordinarily separate the strands of linear duplexes. They also show less affinity for certain dyes such as ethidium bromide, which can become inserted into the DNA molecule. Since binding of ethidium bromide to DNA lowers its density, supercoiled plasmid DNA assumes a different position in the gradient than chromosomal DNA when placed in a cesium chloride density gradient. In this way, the plasmid DNA can be easily separated from chromosomal DNA in cell extracts which have been treated with saturating amounts of ethidium bromide. Plasmid DNA molecules of different sizes can also be readily separated by electrophoresis on agarose gels (see Box, Chapter 5). This technique provides a good opportunity to analyze the action of endonucleases and other enzymes on plasmid DNA.

Plasmid DNA can be observed under the electron microscope, and one type of evidence for the existence and molecular size of a plasmid is observation of its structure and measurement of its length in electron micrographs (see Figure 7.4). Plasmid DNA molecules vary widely in molecular weight, from 3×10^6 to greater than 10^8. The larger plasmids are more often conjugative than the smaller ones.

Curing of plasmids One of the common features of plasmids is that they can be eliminated from host cells by various treatments. This process, termed **curing**, apparently results from inhibition of plasmid replication without parallel inhibition of chromosome replication, and as a result of cell division the plasmid is diluted out (Figure 7.8). Curing may occur spontaneously, but it is greatly increased by use of acridine dyes which become inserted into DNA, as well as by other treatments that affect DNA, such as ethidium bromide, thymine starvation (in thymine-deficient mutants), ultraviolet or ionizing radiation, and heavy metals. Growth at temperatures above the optimum may also result in elimination of the plasmid.

Replication of plasmids In order for a plasmid to be replicated in a cell, the plasmid must have an origin of replication (see Section 5.4). Although we have emphasized that plasmids are independently replicating genetic elements, replication of plasmids is under cellular control. The number of plasmid molecules per cell is determined to some extent by the cell and by environmental conditions. The manner of replication of plasmids is of interest not only in understanding the way in which plasmids maintain autonomy in the cell, but perhaps as a simplified model for how the much more complex chromosomal DNA is replicated. Some plasmids are present in the cell in only a few copies, perhaps one to three, and they probably replicate in a manner similar to that already described for the chromosome. This involves initiation of replication at a single point and bidirectional symmetrical replication around the circle. Because of the small size of the plasmid DNA, the whole replication process occurs very quickly, perhaps in one-tenth or less of the total time of the cell division cycle. The enzymes involved in plasmid replication are normal cell enzymes, so that the genetic elements within the plasmid itself, which control its replication, may be concerned primarily with the control of the timing of the initiation process and with the apportionment of the replicated plasmids between daughter cells. There is evidence for an association of the plasmid DNA with the cell membrane during replication, and this association probably plays some role in the partitioning of the replicated plasmid DNAs between daughter cells at division, but membrane association need not occur at all times to ensure plasmid retention.

Plasmids also occur in eucaryotes, although less is known about them. We discuss yeast plasmids later in this chapter. Plasmids find wide use in genetic engineering, as will be discussed in Chapter 8.

7.5 Conjugation and Plasmid: Chromosome Interactions

Bacterial conjugation or mating is a process of genetic transfer that involves cell-to-cell contact. The genetic material transferred may be a plasmid, or it may be a portion of the chromosome mobilized by a plasmid. In conjugation, one cell, the donor, transmits genetic information to another cell, the recipient.

Selection for recombinants formed by conjugation is accomplished by plating the mating mixture on culture media that allow growth of only the recombinant cells with the desired genotype. For instance, in the experiment shown in Figure 7.9, an Hfr donor which is sensitive to streptomycin (Str^s) and contains the genes coding for enzymes needed for synthesis of the amino acids threonine and leucine (T^+ and L^+) and for utilization of the energy source lactose (lac^+) is mated with a recipient cell which lacks these genes but is resistant to streptomycin (Str^r). Selective media contain streptomycin so that only recipient cells can grow. The composition of each selective medium is varied depending on which genotypic characteristics are desired in the recombinant, as shown in the figure. The frequency of the process is measured by counting the colonies which grow on the selective media.

In the conjugation process, specific pairing between donor and recipient cells must occur. The donor cell, by virtue of its possession of a conjugative plasmid, possesses a surface structure, the **sex pilus** (see Figure 7.10), which is involved in pair formation and in the transfer of DNA. It is thought that the sex pili make possible specific contact between donor and recipient cells and then retract, pulling the two cells together so that a conjugation bridge can form through or on which DNA passes from one cell to another. Although the recipient cells lack sex pili, they must have some sort of recognition mechanism on their surface, as pair formation for conjugation generally occurs only between strains of bacteria that are closely related.

Conjugative plasmids possess the genetic information to code for sex pili and for some proteins needed for DNA transfer. In *E. coli*, one class of conjugative plasmids, called *F factors* (F for fertility), possesses the additional property of being able, occasionally, to become integrated into the chromo-

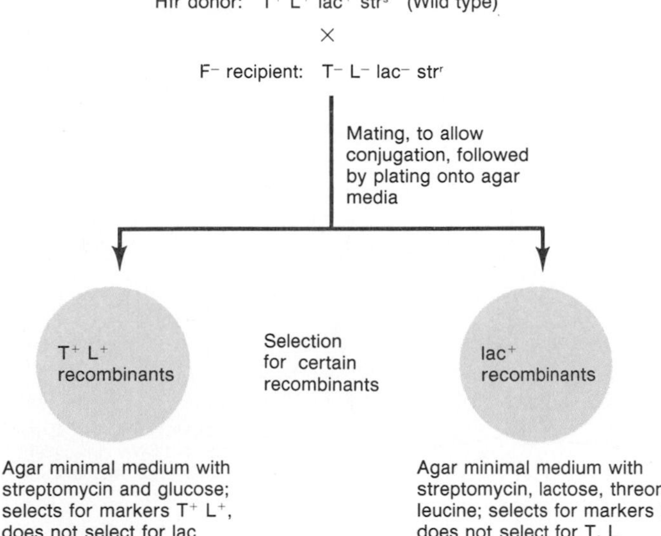
Hfr donor: T⁺ L⁺ lac⁺ strˢ (Wild type)
×
F⁻ recipient: T⁻ L⁻ lac⁻ strʳ
Mating, to allow conjugation, followed by plating onto agar media
T⁺ L⁺ recombinants
Selection for certain recombinants
lac⁺ recombinants
Agar minimal medium with streptomycin and glucose; selects for markers T⁺ L⁺, does not select for lac
Agar minimal medium with streptomycin, lactose, threonine, leucine; selects for markers lac⁺, does not select for T, L

Figure 7.9 Laboratory procedure for the detection of genetic conjugation. Symbols: T, threonine; L, leucine; *lac*, lactose; *str*, streptomycin. Note that each medium selects for specific classes of recombinants.

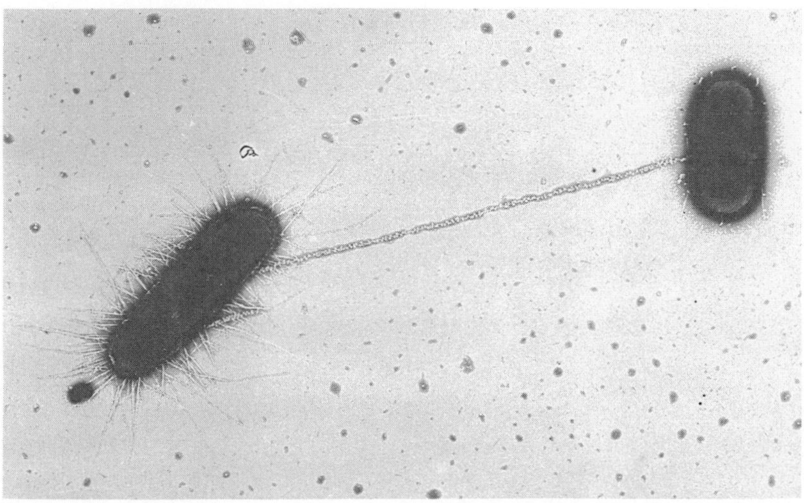

Figure 7.10 Direct contact between two conjugating bacteria is first made via a pilus. The cells are then drawn together for the actual transfer of DNA.

some and to mobilize the chromosome so that it can be transferred during cell-to-cell contact. When the F factor is integrated into the chromosome, large blocks of chromosomal genes can be transmitted, and genetic recombination between donor and recipient is then very extensive. As mentioned earlier, bacterial strains that possess a chromosome-integrated F factor and do show such extensive genetic recombination are called Hfr (for high frequency of recombination). When the F factor is not integrated into the chromosome, it behaves as a conjugative plasmid. Cells possessing an unintegrated F factor are called F^+, and strains which can act as recipients for F^+ or Hfr are called F^-. F^- cells lack the F-factor plasmid; in general, cells that contain a plasmid are very poor recipients for the same or closely related plasmids. We thus see that the presence of the F factor results in three distinct alterations in the properties of a cell: (1) ability to synthesize the F pilus; (2) mobilization of DNA for transfer to another cell; (3) alteration of surface receptors so that the cell is no longer able to behave as a recipient in conjugation.

Mechanism of DNA transfer during conjugation DNA synthesis is necessary for DNA transfer to occur, and the evidence suggests that one of the DNA strands is derived from the donor cell and the other is newly synthesized in the recipient during the transfer process. A mechanism of DNA synthesis in certain bacteriophages, called **rolling circle replication**, was presented in Figure 5.16. This model best explains DNA transfer during conjugation, and a possible mechanism for this process is outlined in Figure 7.11. The whole series of events is probably triggered by cell-to-cell contact, at which time the plasmid DNA circle opens, and one parental and one

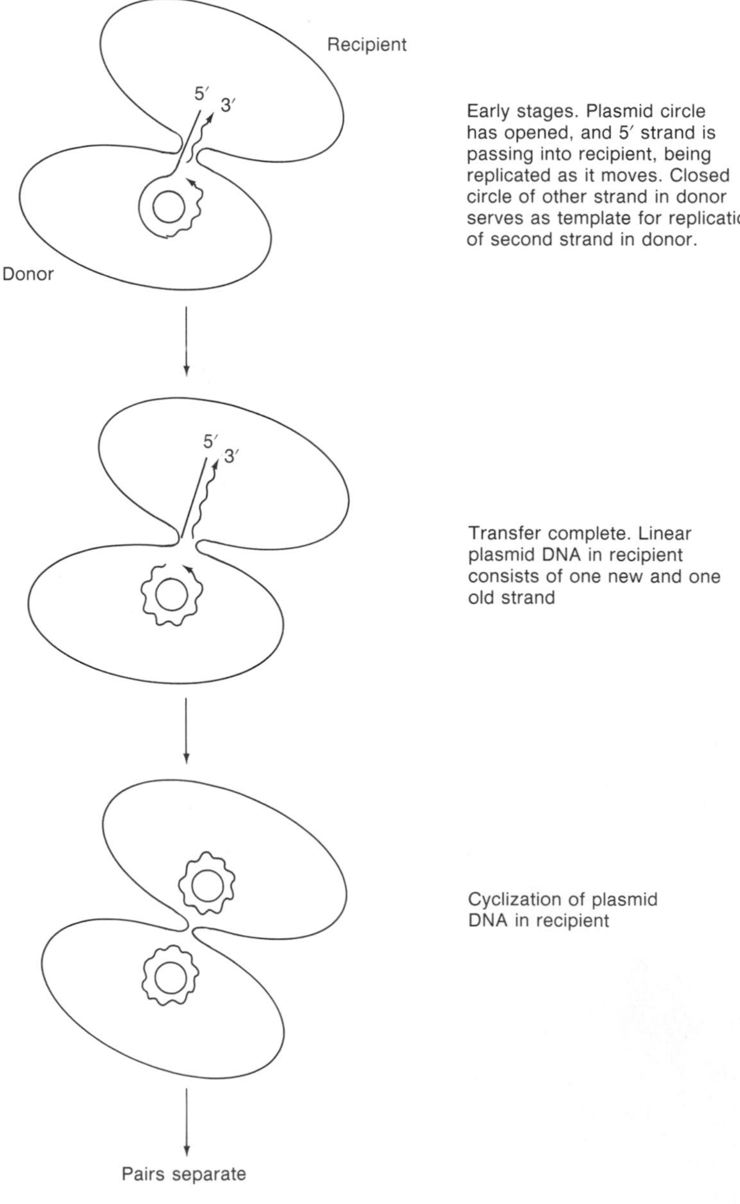

Recipient
5′
3′
Donor
Early stages. Plasmid circle has opened, and 5′ strand is passing into recipient, being replicated as it moves. Closed circle of other strand in donor serves as template for replication of second strand in donor.
5′
3′
Transfer complete. Linear plasmid DNA in recipient consists of one new and one old strand
Cyclization of plasmid DNA in recipient
Pairs separate

Figure 7.11 Rolling circle DNA replication during transfer of plasmid DNA from donor to recipient. Compare with Figure 5.16. Only the plasmid DNA is shown.

Escherichia coli, the Best-Known Procaryote

By far, the best-known procaryotic organism is the intestinal bacterium *Escherichia coli*. Indeed, there are those who say that we know more about the biology of *E. coli* than about any other living organism, even including the human *Homo sapiens*. The structure and function of *E. coli* is often considered the archetype of *all* living organisms. Why is *E. coli* so well known? There is nothing especially unusual about this organism. It is a run-of-the-mill procaryote, a eubacterium, a common inhabitant of the human intestine.

E. coli is so well known primarily because it is easy to work with in the laboratory. Even those who have difficulty with sterile technique and other bacteriological procedures can generally work with *E. coli* without difficulty. *E. coli* grows rapidly, has simple nutritional requirements, and does a fair bit of interesting biochemistry and physiology. But it is something else which has made it so useful in modern biology, sex. *E. coli* was the first bacterium in which the process of sexual reproduction was discovered, being first recognized in 1946 by Joshua Lederberg and Edward L. Tatum. With sexual reproduction, the possibility of doing real genetics is available. Genetic crosses could be carried out and genetic properties analyzed. Another valuable property of *E. coli* is its ability to support the growth of a whole range of bacterial viruses, which made it possible to study in detail the nature of viruses and virus multiplication. Thus, with its favorable laboratory properties, its suitability for studies of virology, and its sexual reproduction, biochemists and molecular biologists were able to probe deep into the nature of life, their work leading ultimately to the sophisticated understanding we now have of molecular biology and virology (see Chapters 5–8).

Escherichia coli was first isolated in 1885 by the German bacteriologist, Theodor Escherich, as a normal inhabitant of the intestinal tract. Escherich named the organism *Bacterium coli*, the name reflecting the rod-shape of the cell (*Bacterium* means rod-shaped) and its intestinal habitat (*coli* for colon). The genus name *Bacterium* subsequently was changed to *Escherichia* in honor of its discoverer. Although Escherich's strains, and most other strains of *E. coli*, are harmless, some *E. coli* strains are pathogenic, causing diarrhea and urinary tract infections. As we will see in Chapter 11, the pathogenic *E. coli* strains differ in significant ways from the harmless intestinal strains.

It is curious that although *E. coli* is the organism of choice for genetic engineering and biotechnological research (see Chapter 8), it has never been a major organism of use in industrial microbiology. The large-scale cultivation of living organisms for industrial purposes (see Chapter 10) has involved a quite different group of microorganisms. For complex and not especially revealing reasons, *E. coli* is not nearly as suitable for large-scale cultivation as yeast, *Streptomyces*, and *Bacillus*. Indeed, there are many who worry about the possible use of *E. coli* in industry, because of its potential pathogenicity. Industrialists now find themselves in a dilemma, trying to decide whether they should attempt to "tame" *E. coli* for industrial purposes, or to find an organism that would not only have *E. coli*'s favorable laboratory properties, but suitable industrial properties as well.

new strand are transferred. According to this model, transfer can only occur if DNA synthesis can also occur, and this has been shown experimentally by use of chemicals which specifically inhibit DNA synthesis. The model also accounts for the fact that if the DNA of the donor is labeled, some labeled DNA is transferred to the recipient, but only a *single* labeled strand is transferred. With the rolling circle mechanism, the donor cell also duplicates its plasmid at the time transfer occurs, so that at the end of the process both donor and recipient possess completely formed plasmids.

The high efficiency of the DNA transfer process is shown by the fact that under appropriate conditions, virtually every recipient cell which pairs can acquire a plasmid.

Infectiousness of conjugative plasmids If the plasmid genes can be expressed in the recipient, the recipient itself then soon becomes a donor and can then transfer the plasmid to other recipients. Conjugative plasmids can spread rapidly between populations, behaving in the manner of infectious agents (Figure 7.12). Thus plasmid-mediated drug resistance frequently has been called **infectious drug resistance**. The infectiousness of conjugative plasmids is of major ecological significance, since a few plasmid-positive cells introduced into an appropriate population of recipients, if they contain genes which confer a selective advantage, can convert the whole recipient population into a plasmid-bearing population in a short period of time. Under optimal conditions, the rate of spread of a conjugative plasmid

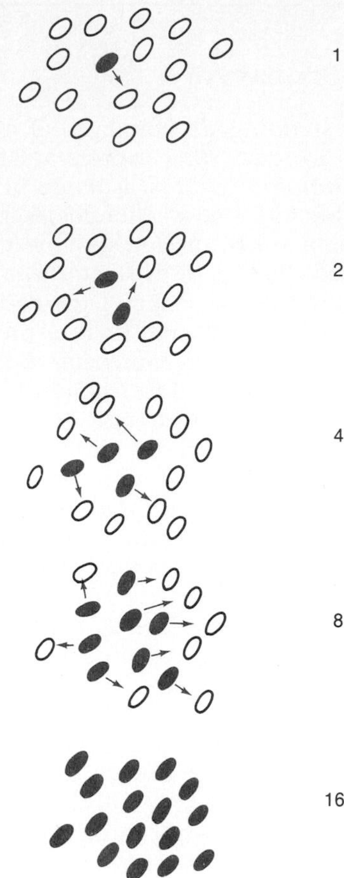

Figure 7.12 Infectiousness of a conjugative plasmid.

through a population can be exponential, resembling a bacterial growth curve (see Section 9.4). The widespread occurrence of infectious drug resistance in clinical medicine has led to some serious problems in the chemotherapy of infectious disease (see Section 11.17).

Formation and behavior of Hfr strains As noted above, an F factor can become integrated into the chromosome and mobilize the chromosome for conjugation. There are several specific sites in the chromosome at which F factors can be integrated, and these sites, called IS (for "insertion sequence") represents regions of homology between chromosome and F factor DNA (see Section 7.7 for a discussion of insertion sequences). As seen in Figure 7.13, integration of F factor involves insertion in the chromosome at the specific site. In the particular Hfr shown, the integration site is between the chromosomal genes *pro* and *lac*. The site on the F factor (*origin*) at which transfer will initiate during conjugation is indicated by an arrow. At the time of specific cell pairing, the chromosome opens at the origin, and the host genes are inserted into the recipient beginning with the gene downstream from the origin (Figure 7.14).

The origin of Hfr strains is thus seen as a result of the integration of F factor into the chromosome. Since a number of distinct insertion sites are present, a number of distinct Hfr strains are possible. A given Hfr strain always donates genes in the same order, beginning with the same position, but Hfr strains of independent origin transfer genes in different sequences. During normal cell division, the DNA of the Hfr replicates normally, but at the time of pairing with an F^- cell, a DNA strand from the Hfr is inserted into the F^- cell, and replication occurs by the rolling circle process. After transfer, the Hfr strain still remains Hfr, since it has retained a copy of the transferred genetic material.

Although Hfr strains transmit chromosomal genes at high frequency, they usually do not convert F^- cells to F^+ or Hfr, because the entire F factor is only rarely transferred. On the other hand, F^+ cells efficiently convert F^- to F^+ because of the infectious nature of the F plasmid.

At some insertion sites, the F factor is integrated with the origin in one direction, whereas at other sites the origin is in the opposite direction. The direction in which the F factor is inserted determines which of the chromosomal genes will be inserted first into the recipient. The manner in which a variety of Hfr strains can arise is illustrated in Figure 7.15. By use of various Hfr strains, it has been possible in *E. coli* to determine the arrangement and orientation of a large number of chromosomal genes, as will be described in Section 7.8.

Transfer of chromosomal genes to F Under rare conditions, integrated F factors may be excised from the chromosome, and the possibility exists for

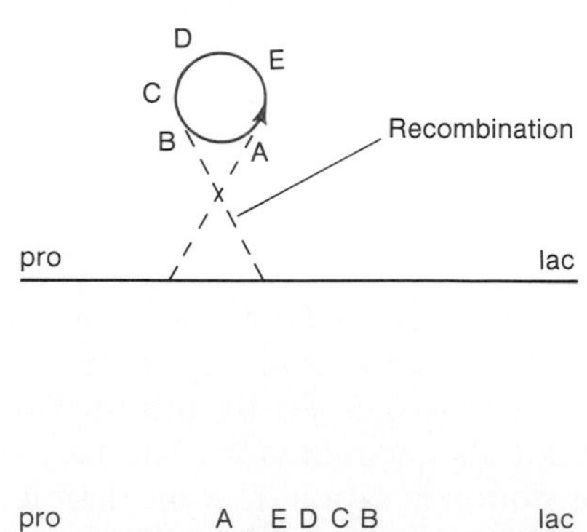

Figure 7.13 Integration of an F factor into the chromosome with the formation of an Hfr. The insertion of F factor occurs at a variety of specific sites, the one here being between chromosomal genes *pro* and *lac*. The letters on the F factor represent arbitrary genes. The arrow indicates the origin of transfer, with the arrow as the leading end. The site in F factor at which pairing with the chromosome occurs is between A and B. Thus, in this Hfr *pro* would be the first chromosomal gene to be transferred and *lac* would be among the last.

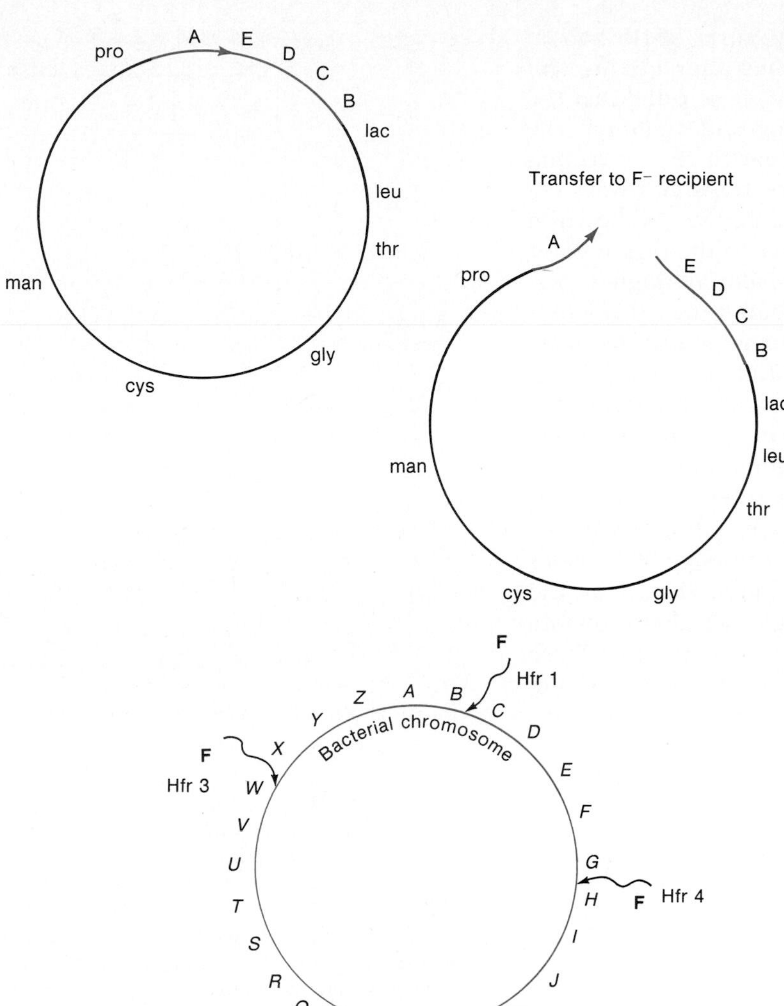

Figure 7.14 Breakage of Hfr chromosome at the origin and transfer of DNA to the recipient. Replication occurs during transfer, as illustrated in Figure 7.11.

Figure 7.15 Manner of formation of different Hfr strains, which donate genes in different orders and from different origins. The bacterial chromosome is a circle (*a*) that can open at various locations, at which F particles become attached. The gene orders are shown in (*b*).

the incorporation at that time of chromosomal genes into the liberated F factor. Such F factors containing chromosomal genes are called F′ (F prime). These F factors differ from normal F factors in that they contain identifiable chromosomal genes, and they transfer these genes at high frequency to recipients.

Oriented transfer of Hfr and the phenomenon of interrupted mating The measurement of bacterial conjugation is usually done by use of parental strains having properties that can be selected against on agar plates (see Figure 7.9). The recipient is usually resistant to an antibiotic and is auxotrophic for one or more nutritional characters. The donor is antibiotic sensitive but prototrophic for the nutritional characters. With proper agar media, only the recombinants can grow and the large background of nonrecombinants eliminated.

The oriented transfer of chromosomal genes from Hfr to F⁻ was most clearly shown by a procedure called **interrupted mating**. The mating pairs are not strongly joined and can be easily separated by agitation in a mixer or blender. If mixtures of Hfr and F⁻ cells are agitated at various times after mixing and the genetic recombinants scored, it is found that the longer the time between pairing and agitation, the more genes of the Hfr will appear in the F⁻ recombinant. It is also found that gene transfer always occurs in a specific order in a specific Hfr. As shown in Figure 7.16, genes present closer to the origin enter the F⁻ first and are always present in higher percentage of the recombinants than genes that enter late. In addition to showing that gene transfer from donor to recipient is a sequential process, experiments of this kind provide a method of determining the order of the genes on the bacterial DNA (genetic mapping). The arrangement of gene loci on the chromosome is called a **genetic map** (see Section 7.8).

Genetic recombination between Hfr genes and F⁻ genes in the F⁻ cell requires the presence of enzymes in the recipient cell. This has been shown by the isolation of mutants of F⁻ strains, which are unable to form recombinants when mated with Hfr. These mutants, called *rec⁻* (recombination-minus), are deficient in the RecA protein (see Figure 7.1). They are also unusually sensitive to ultraviolet radiation, and are deficient in enzymes involved in dark repair of DNA.

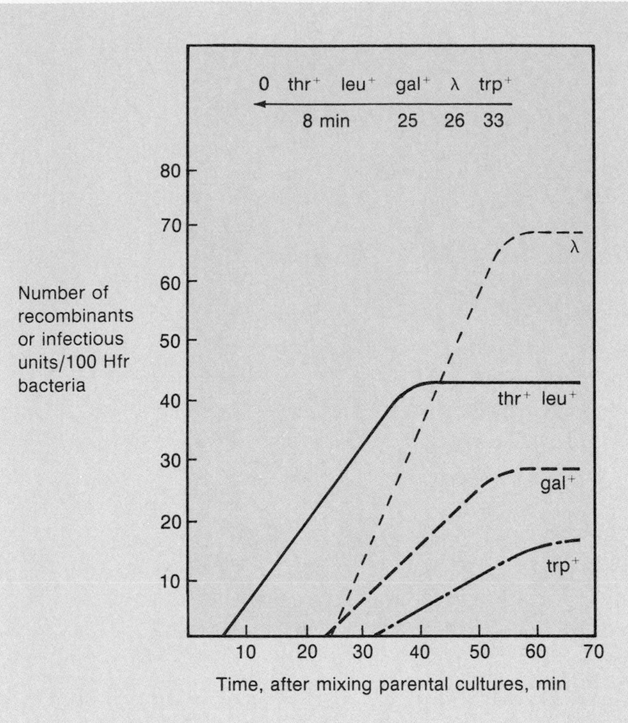

Figure 7.16 Rate of formation of recombinants containing different genes after mixing Hfr and F⁻ bacteria by the process known as interrupted mating. The location of the genes along the Hfr chromosome is shown in the small diagram. Note that the genes closest to the origin are the first ones detected in the recombinants. The experiment is done by mixing Hfr and F⁻ cells under conditions in which essentially all Hfr cells find mates. The F⁻ recipient was streptomycin resistant but auxotrophic for the markers being scored. The Hfr donor was streptomycin sensitive. At various intervals, samples of the mixture are shaken violently to separate the mating pairs and plated on a selective medium in which only the recombinants can grow and form colonies.

7.6 Kinds of Plasmids and Their Biological Significance

It should be clear from our discussion of plasmids that these genetic elements are of great importance as tools for understanding a wide variety of genetic phenomena in procaryotes. Plasmids also help explain a number of biological and ecological phenomena, and it is now evident that many ideas of bacterial taxonomy and ecology will have to be modified to account for the presence and activity of plasmids. We discuss a few of these aspects in the present section.

F and I pili At least two kinds of pili, called F and I pili, are known to be involved in cell-to-cell transfer of plasmids. Two classes of RNA phages are known to infect cells which carry transmissable plasmids. These phages can be used to demonstrate the presence on the cell of either F or I pili (see Figures 6.18 and 7.10). The two kinds of pili can also be distinguished immunologically. F pili are involved in the transfer of F factor and some antibiotic resistance plasmids. F pili are also present on Hfr cells. I pili are involved in the transfer of other antibiotic resistance plasmids, colicin-determining plasmids, and others.

Resistance-transfer factors Among the most widespread and well-studied groups of plasmids are the resistance-transfer factors (R factors), which confer resistance to antibiotics and various other inhibitors of growth. R factors were first discovered in Japan in strains of enteric bacteria which had gained resistance to a number of antibiotics (multiple resistance), and have since been found in other parts of the world. The emergence of bacteria resistant to several antibiotics is of considerable medical significance, and was correlated with the increasing use of antibiotics for the treatment of infectious diseases. Soon after these resistant strains were isolated it was shown that they could transfer resistance to sensitive strains via cell-to-cell contact. This is probably the reason for the rapid rise of multiply resistant strains, since it would be unlikely for resistance to a number of antibiotics to develop simultaneously by mutation and selection. The infectious nature of the R factors permits rapid spread of the characteristic through populations.

Two classes of R factors are known, those which use the F pilus as agent of transfer and those which use the I pilus. A variety of antibiotic-resistance genes can be carried by an R factor; those most commonly observed carry resistance to four antibiotics (tetracycline, chloramphenicol, streptomycin, sulfonamides), but some may have fewer and others more resistance genes. R factors with genes for resistance to kanamycin, penicillin, and neomycin are known, while others confer resistance to inhibition by the metals nickel, cobalt, or mercury. Many drug-resistant elements on R factors are transposable elements and can be used in transposon mutagenesis (see Section 7.7).

Genes for characteristics not related to antibiotic resistance may also be carried by R factors. Paramount among these is the production of F or I pili, but R factors also carry genes permitting their own replication and genes controlling production of proteins that prevent the introduction of other related plasmids. Thus the presence of one R factor inhibits the introduction of another of the same type, a phenomenon known as incompatibility (see below).

Because R factors are able to undergo genetic recombination, genes from two R factors can be integrated into one. Plasmid recombination is one means by which multiply drug-resistant organisms might have first arisen. The genes in an R factor are carried in a definite order, as are chromosomal genes, and this order can be mapped by genetic recombination. Many R factors and F factors use the same pilus for conjugation, and recombination can occur between F and R factors.

Streptomycin

Phosphorylation
Adenylylation

(In aminoglycoside antibiotics with free amino group, inactivation also by *N*-acetylation)

Chloramphenicol

Acetylation

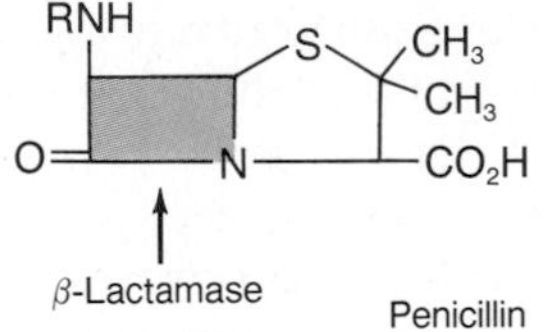

Figure 7.17 Sites at which antibiotics are attacked by enzymes coded for by R-factor genes.

Biochemical mechanism of resistance mediated by R factors The isolation in the laboratory of mutants resistant to antibiotics generally results in the selection of mutants in chromosomal genes. On the other hand, the majority of drug-resistant bacteria isolated from patients contain the drug-resistant genes on R factors. The biochemical mechanism of R-factor resistance is different from that of chromosomal resistance. In most cases, antibiotic resistance mediated by chromosomal genes arises because of a modification of the *target* of antibiotic action (e.g., a ribosome). On the other hand, R-factor resistance is in most cases due to the presence in the R factor of genes coding for new enzymes which *inactivate* the drug (Figure 7.17). For instance, a number of antibiotics are known which have similar chemical structures, containing aminoglycoside units. Among the aminoglycoside antibiotics are streptomycin, neomycin, kanamycin, and spectinomycin. Strains carrying R factors conferring resistance to these antibiotics contain enzymes that chemically modify these aminoglycoside antibiotics, either by phosphorylation, acetylation, or adenylylation, the modified drug then lacking antibiotic activity. In the case of the penicillins, R-factor resistance is due to the formation of penicillinase (β-lactamase), which splits the β-lactam ring, thus destroying the molecule. Chloramphenicol resistance mediated by R factor arises because of the presence of an enzyme that acetylates the antibiotic. Thus, the fact that R factors can confer multiple antibiotic resistance does not imply that the mode of action of the R-factor genes is similar. The presence of multiple antibiotic resistance is due to the fact that a single R factor contains a variety of genes coding for different antibiotic inactivating enzymes.

Toxins and other virulence characteristics We will discuss in Chapter 11 the physiological and genetic characteristics of microorganisms that enable them to colonize hosts and set up infections, which can lead to harm. In the present context, we merely note the two major characteristics involved in virulence: (1) the ability of microbes to attach to and colonize specific sites in the host; and (2) the formation of substances (toxins, enzymes, and other molecules), which cause damage to the host. It has now been well established that at least in some kinds of pathogenicity due to *E. coli*, each of these virulence characteristics is carried on plasmids. Enteropathogenic *E. coli* is characterized by an ability to colonize the small intestine and to produce a toxin that causes symptoms of diarrhea. Colonization requires the presence of a cell surface protein called the K antigen, which confers on the cells the ability to attach to epithelial cells of the intestine. At least two toxins in enteropathogenic *E. coli* are known to be carried by plasmid: the *hemolysin,* which lyses red blood cells, and the *enterotoxin,* which induces extensive secretion of water and salts into the bowel. It is the enterotoxin which is responsible for the induction of diarrhea, as will be discussed in Chapter 11. Since other virulence factors of *E. coli* are carried by chromosomal genes, it is not clear why these specific factors are carried by plasmids, but the existence of such plasmids raises the question of how widespread plasmid-related virulence is in microorganisms.

In *Staphylococcus aureus*, a number of virulence-conferring properties are either known or suspected to be plasmid-linked; *S. aureus* is noteworthy for the variety of enzymes and other extracellular proteins it produces that are involved in its virulence, and the production of *coagulase, hemolysin, fibrinolysin,* and *enterotoxin* is thought to be plasmid-linked. In addition, the yellow pigment in *S. aureus*, which is probably involved in its ability to resist the destructive action of singlet oxygen in the phagocyte (see Sections 11.14 and 11.15), may also be plasmid-linked.

Presently, extensive research is underway to determine how significant plasmid-mediated virulence is in the microbial world. The implications of this research are widespread in medicine and plant pathology.

Bacteriocins Most bacteria produce agents that inhibit or kill closely related species; these agents are called **bacteriocins** to distinguish them from the antibiotics, which have a wider spectrum of activity. Bacteriocins are a diverse group of substances, usually proteins and frequently of high molecular weight. The ability to produce some bacteriocins is inherited via plasmids. Those bacteriocins produced by *E. coli* called *colicins*, have been best studied; the elements controlling their production are called Col factors. Some Col factors are conjugative, and they are found to fall in two groups, those transmitted by the F pilus and those transmitted by the I pilus. In general, Col factors are not transmitted as rapidly through populations as F or R factors.

Incompatibility of plasmids A phenomenon of considerable importance both in research on plasmids and in the evolution and ecology of plasmids is compatibility. When a plasmid is inserted into a cell which already carries another plasmid, a common observation is that the second plasmid may not be maintained and is lost during subsequent cell replication. The two plasmids are said to be **incompatible**. Plasmid incompatibility is controlled by genes in the plasmid, and a number of incompatibility groups have been recognized, the plasmids of one incompatibility group excluding each other but being able to coexist with plasmids from other groups. Evidence suggests that the plasmids of one incompatibility group have been derived from a common ancestral source.

Biological and taxonomic significance of plasmids A summary of plasmids which have been identified and organisms for which evidence of plasmids exist is given in Table 7.1. It is likely that virtually all bacterial groups possess plasmids. Techniques are now available for detecting the presence in an organism of plasmid-like DNA, and these techniques have been applied to a wide variety of bacteria. In some cases, virtually every strain tested has been shown to have plasmid-like DNA. Almost certainly, such DNA molecules control some genetic functions for the cell, although it is frequently difficult to relate the presence of the physical structure with a genetic function or functions. The only certain way of showing this is to transfer the plasmid to another strain, where the DNA can be distinguished by its physical characteristics, and to show that at the same time some genetic characteristic of the donor has also been transferred. Some plasmids are much more promiscuous than others, being able to be transferred to a variety of bacteria, some even quite unrelated.

As we have noted, many plasmids are not transmissible by cell-to-cell contact. In *Staphylococcus aureus* plasmid transfer never occurs by cell-to-cell transfer, and the demonstration of the existence of plasmids is much more difficult. Evidence for plasmids comes from curing experiments, and from the transfer of plasmids via transduction. On the other hand, cell-to-cell transfer of plasmids is extremely common in Gram-negative bacteria, not only within the enteric group, which has been so widely studied, but in other groups, such as the pseudomonads. In fact, some pseudomonad plasmids are transferrable to a wide variety of other Gram-negative bacteria. Some pseudomonad plasmids have been shown to transfer the genetic information for biochemical pathways for the degradation of unusual organic com-

Table 7.1 Types of plasmids*

Type	Organisms
Conjugative plasmids	F-factor, *Escherichia coli*; pf*dm*, K, *Pseudomonas*; P, *Vibrio cholerae*
R plasmids:	
Wide variety of antibiotics	Enteric bacteria, *Staphylococcus*
Resistance to mercury, cadmium, nickel, cobalt, zinc, arsenic	
Bacteriocin and antibiotic production	Enteric bacteria; *Clostridium; Streptomyces*
Physiological functions:	
Lactose, sucrose, urea utilization, nitrogen fixation	Enteric bacteria
Degradation of octane, camphor, naphthalene, salicylate	*Pseudomonas*
Pigment production	*Erwinia, Staphylococcus*
Virulence plasmids:	
Enterotoxin, K antigen, endotoxin	*Escherichia coli*
Tumorigenic plasmid	*Agrobacterium tumefaciens*
Adherence to teeth (dextran)	*Streptococcus mutans*
Coagulase, hemolysin, fibrinolysin, enterotoxin	*Staphylococcus aureus*

**Plasmids have now been found in most bacterial genera.*

pounds, such as camphor, octane, and naphthalene. These substances are highly insoluble, and the plasmid-coded enzymes convert them into more soluble substrates that are on the main line of biochemical catabolism, such as acetate, pyruvate, and isobutyrate, the catabolism of which is coded by chromosomal genes.

Origin of plasmids Although specific evidence for the origin of multiple drug resistant R factors is not available, a number of lines of circumstantial evidence suggest that plasmids with R-factor type character existed before the antibiotic era, and that the widespread clinical use of antibiotics provided selective conditions for the evolution of R factors with one or more antibiotic-resistance genes. Indeed, a strain of *E. coli* which was freeze-dried in 1946 was found to contain a plasmid with genes conferring resistance to tetracycline and streptomycin, even though neither of these antibiotics were used clinically until several years later. Also, strains carrying R-factor genes for resistance to a number of semisynthetic penicillins were shown to exist before the semisynthetic penicillins had been synthesized. Of perhaps even more ecological significance, R factors conferring antibiotic resistance have been detected in some nonpathogenic Gram-negative soil bacteria. In the soil, such resistance may confer selective advantage, since the antibiotic-producing organisms (*Streptomyces, Penicillium*) are also normal soil organisms (however, there is no real evidence that the common antibiotics are produced or act in the soil). Thus, it seems reasonable to conclude that R factors are not a recent phenomenon, but existed to a small extent in the natural bacterial population before the antibiotic era, and that widespread use of antibiotics provided selective conditions for the rapid spread of these R factors. The evolution of R factors is thus completely understandable, albeit frightening in the implications for the continued success of human medicine to combat rampant infections with the ever increasing use of larger and larger doses of antibiotics.

Engineered plasmids The techniques of genetic engineering, discussed in the next chapter, have made possible the construction in the laboratory of a limitless number of new, artificial plasmids. Incorporation into artificial plasmids of genes from a wide variety of sources has made possible the transfer of genetic material across virtually any species barrier. It is even possible to synthesize completely new genes and introduce them into plasmids. Such artificial plasmids are useful tools in understanding plasmid structure and function, as well as for the more practical aims of genetic engineering that are discussed in the next chapter. In order to create an artificial plasmid, the main requirement is that the plasmid genes involved in replication and maintenance of the plasmid in the cell be connected to the new material of interest. Artificial plasmids are introduced into appropriate hosts via transformation and are selected for by means of antibiotic resistance characters. The ability to create artificial plasmids has greatly expanded the possibilities for plasmid research.

7.7 Transposons and Insertion Sequences

The genes of living organisms are not static, but are capable under certain conditions of moving around. The process by which a gene moves from one place to another on the genome is called **transposition**, and is an important process in evolution and in genetic analysis. We discussed briefly the molecular events in transposition in Sections 5.5 and 5.6.

First, we should emphasize that transposition is a *rare* event, occurring at frequencies of 10^{-5} to 10^{-7} per generation. Thus, the genes of living organisms are relatively stable. It is only because of the extreme sensitivity of genetic analysis in bacteria (see Section 7.8) that transposition is detectable with ease in these organisms. In higher eucaryotes, where population sizes are small and selectable markers less common, it is much more difficult to detect transposition, although transposable elements are definitely known in eucaryotes.

Several kinds of transposable elements have been recognized in procaryotes and eucaryotes. A brief summary is given in Table 7.2.

Not all genes are capable of transposition. Rather, transposition of genes is linked to the presence with these genes of special genetic elements. In the most common type, transposition is linked to the presence of special base sequences called *insertion sequences* (see Section 5.6). **Insertion sequences** are short, specific segments of DNA which have the ability to move to other sites on the genome as discrete units.

Transposons are composite movable elements containing paired insertion sequences which flank other genetic regions. In such structures, the paired insertion sequences can serve as the driving force for the transposition of the other genes present in the transposons (see Figure 5.19).

The simplest transposable elements are the insertion sequences themselves, which seem to lack genes other than those capable of bringing about their own transposition. Insertion sequences are short segments of DNA, around 1000 nucleotides long, that can become integrated at specific sites on the genome. Insertion sequences, abbreviated IS, are found in both host and plasmid DNA, as well as in certain bacteriophages (see the discussion of bacteriophage Mu in Section 6.15). Several distinct IS's have been characterized, and each is designated by a number which identifies its type: IS1, IS2, IS3, etc. IS's are scattered about the chromosome, and strains vary in the number and frequency of IS's. For instance, one strain of *Escherichia coli* has eight copies of IS1 and five copies of IS2.

Table 7.2 Transposable elements

Procaryote	Eucaryote
Insertion sequence: IS	Yeast: *sigma*
Transposon: Tn	Yeast: Ty
	Fruit fly: copia, P
	Maize: Ac
Virus: Mu	Retrovirus: Rous sarcoma, Human immunodeficiency virus (AIDS)

Required for integration of an IS is an enzyme, called *transposase*, which may be coded for by the IS element, but may also be coded by the host chromosome or by the plasmid or phage to which the IS is attached. Another requirement for an IS is the presence on the DNA of short *inverted terminal repeats*. These repeats range in length from around 40 in simple IS to greater than 1000 base pairs in composite transposons; each IS has a specific number of base pairs in its terminal repeats. Such inverted terminal repeats are involved in the transposition process (see below).

The mechanism of transposition Interestingly, the transposing element does not *move* from its present site. Rather, the transposon is *duplicated*, and the duplicated site becomes inserted at another location. Thus, after the transposition event is completed, *one copy* of the transposing element *remains* at the original site and *another copy* of the transposing element is found at the new site. During this whole transposition process, the source transposon *remains* at its original site; at no time does the source transposon become free in the cell.

Although many of the molecular details of transposition are uncertain, and different transposons appear to have different mechanisms, one model for this process is illustrated in Figure 7.18. As seen, single-strand cuts are made at the ends of the transposon (at the sites of the inverted repeats), and staggered single-strand cuts are made at the target site. The transposon now becomes joined to the target site via the single-stranded ends, leading to the formation of a composite structure called a *cointegrate*. Replication repair then brings about the filling in of the single-strand gaps in the target site. This process results in the formation of *direct repeats* in the target site at the ends of the transposon (in addition to the inverted repeats of the transposon). The final event is the *resolution* of the cointegrate structure, leading to the release of the original transposon and the presence of a new copy of the transposon at the target site. Now that the transposon is present at the new target site, this transposon can also serve as another source of transposition.

It should be emphasized that transposition is essentially a *recombination event*, but one which occurs outside of the regular genetic recombination system of the cell. It involves the special protein *transposase* rather than the RecA protein which is involved in general recombination.

Transposon mutagenesis If the insertion site for a transposable element is *within* a bacterial gene, insertion of the transposon will result in loss of linear continuity of the gene, leading to mutation (Figure

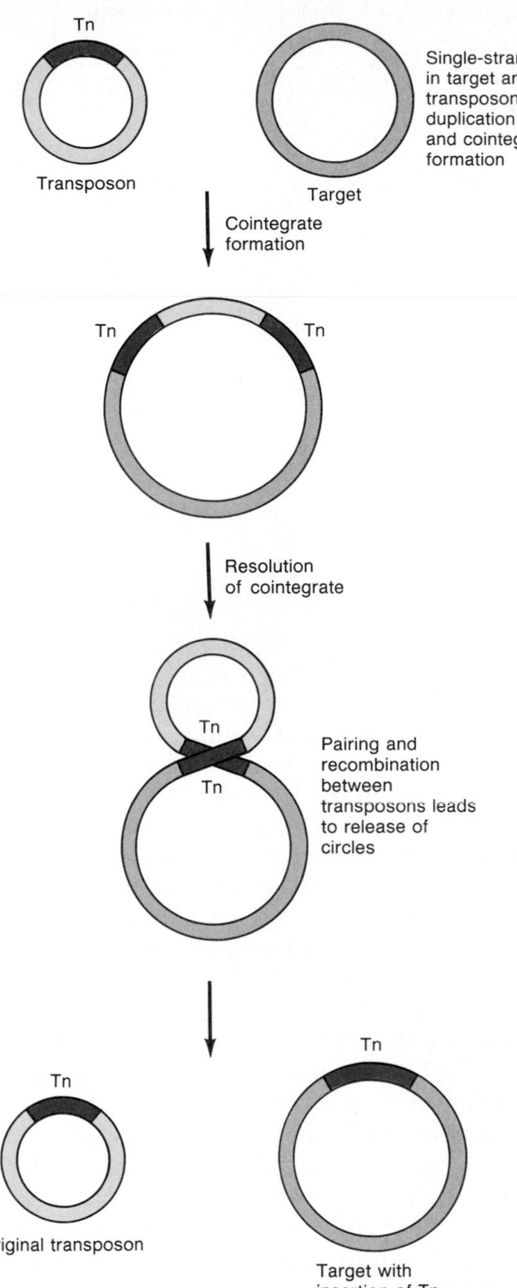

Figure 7.18 The mechanism of transposition. After the formation of single-strand cuts, a cointegrate structure arises by association of the two molecules. After recombination, resolution of the cointegrate structure leads to the release of the original transposon and duplication of the transposon in the target molecule. See Figure 5.19*b* for an explanation of how the duplication process occurs.

7.19). Transposons thus provide a facile means of creating mutants throughout the chromosome. The most convenient element for transposon mutagenesis is one containing an antibiotic resistance gene. Clones containing the transposon can then be selected by the isolation of antibiotic resistance colonies. If the antibiotic resistance clones are selected on rich medium where all auxotrophs can grow, they can be subsequently screened on minimal medium supplemented with various growth factors, to determine if a growth factor is required.

Transposons are also useful for incorporating an auxotrophic gene marker into a wild-type organism. Normally, auxotrophic recombinants cannot be isolated by positive selection, but if the auxotrophic marker to be introduced contains a transposable element with an antibiotic resistance marker, then one can select for antibiotic-resistant clones, a positive selection procedure, and automatically obtain clones which have incorporated the auxotrophic marker.

Two transposons widely used for mutagenesis are Tn5, which confers neomycin and kanamycin resistance, and Tn10, which contains a marker for tetracycline resistance.

Invertible DNA and the phenomenon of phase variation An interesting phenomenon of genetic change that has been recognized in some bacteria involves the **inversion** of a segment of DNA from one orientation to the other. When the segment is oriented in one direction, a particular gene is expressed, whereas when it is oriented in the opposite direction, a different gene is expressed. This "flip-flop" mechanism provides an interesting example of regulation of gene activity. The process by which gene inversion occurs is another example of *site-specific recombination*.

The best studied case of gene inversion is that called *phase variation*, which has been well studied in bacteria of the genus *Salmonella*. These enteric bacteria are motile by means of peritrichous flagella. As we have noted (see Section 3.8), bacterial flagella are composed of a single protein type. As a result of phase variation, the flagellar protein can be of one or two separate types. (The value to the bacterium of changing its flagellar type will be discussed in Chapter 12.) Each *Salmonella* cell has two genes, H1 and H2, coding for the two different flagellar proteins,

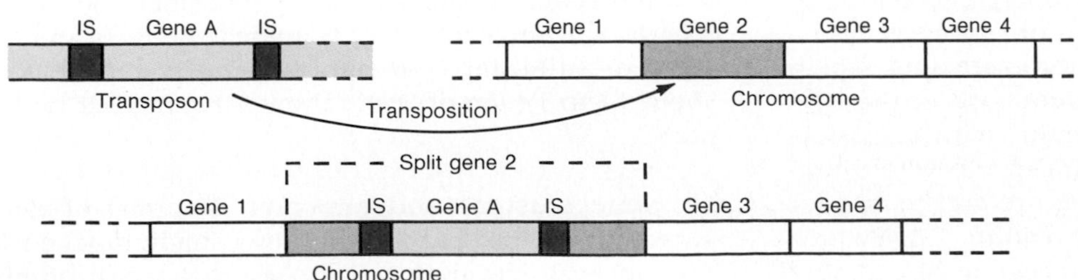

Figure 7.19 Transposon mutagenesis. The transposon moves into the middle of gene 2. Gene 2 is now split by the transposon and is inactivated. Gene A of the transposon may be expressed.

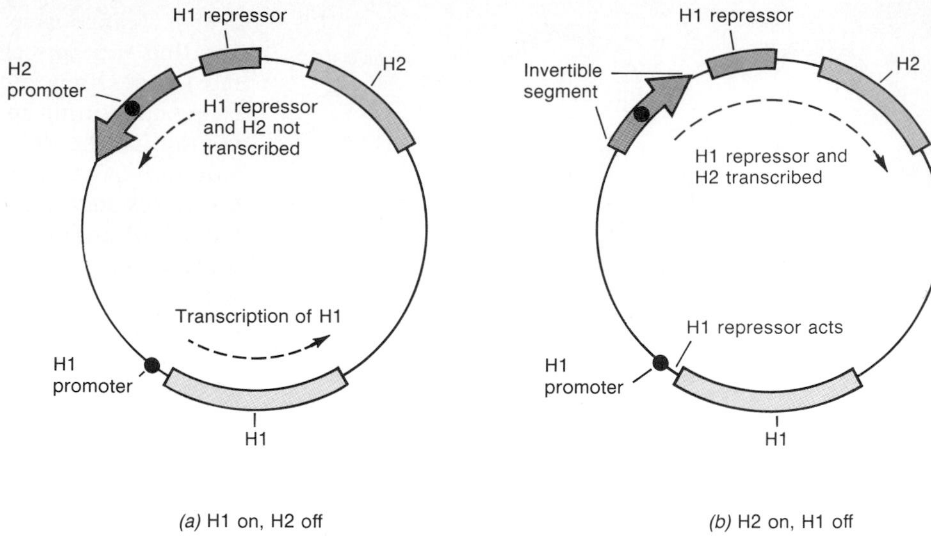

Figure 7.20 Site-specific inversion, the mechanism by which phase variation in *Salmonella* flagella is brought about. The dotted lines show the location and direction of transcription. (a) When the invertible segment is in one orientation, the H2 promoter points away from the H1 repressor and H2 gene; H1 is expressed. (b) In the opposite orientation, H1 repressor is made, turning off H1. At the same time, H2 is expressed. The result is a "flip-flop" between two alternate states.

but only one of the two genes is expressed at any one time. Thus, an individual bacterial cell will make either H1-type flagella or H2-type flagella.

The invertible element involved in expression of the flagellar proteins is a 970 base pair segment in the H2 gene (Figure 7.20). When the invertible segment is in one orientation, H2 gene is transcribed, but in addition, another gene is transcribed which codes for a protein which represses transcription of gene H1. Thus, when H2 is expressed, H1 is turned off. On the other hand, when the invertible segment is in the opposite orientation, the genes for H2 and the H1 repressor are no longer expressed, so that H1 can now be transcribed and expressed.

Invertible segments are also known in other genetic systems. We described an invertible segment which is involved in host range of bacteriophage Mu in Section 6.15. Regulation by rearrangement also occurs in eucaryotes: we discuss the regulation of mating type in yeast in Section 7.11 and we discuss the complex genetic rearrangements involved in the production of antibodies (defense against infection) in Section 12.6.

7.8 An Overview of the Bacterial Genetic Map

The procedure of interrupted mating (Figure 7.16) can be used to map the locations of various genes on the chromosome. Different Hfr strains are used, which initiate DNA transfer from different parts of the DNA (Figure 7.15). By using Hfr strains with origins in different sites, it is possible to map the whole bacterial gene complement. A circular reference map for *E. coli* strain K-12 is shown in Figure 7.21, which also indicates the leading transfer regions of a number of Hfr strains that have been used in genetic mapping. The map distances are given in minutes of transfer, with 100 minutes for the whole chromosome and with "zero time" arbitrarily set as that at which the first gene transfer (for the ***threonine*** gene) can be detected from strain Hfr-H, the first Hfr strain isolated.

Although interrupted mating experiments provide the best means of obtaining an overall picture of the bacterial genetic map, they are less convenient for mapping closely linked genes than genetic mapping by transduction. Bacteriophage P1 has been used extensively in *E. coli* to fill in the gaps in the genetic map, since it transduces fairly large segments of DNA, equivalent to about 2 minutes on the map. Transduction is especially useful for determining the order and location of genes that are closely linked, since interrupted mating experiments do not permit separation of genes that are very close together. By use of such transductional analysis, it has been determined that many genes that code for proteins involved in a single biochemical pathway are closely linked. Some of these gene clusters are indicated on the map (see Figure 7.21).

Electron microscopic studies on heteroduplex DNA molecules of known genetic composition make it possible to correlate genetic map distances with physical distances on the DNA. The total length of the *E. coli* genome, 100 minutes on the genetic map, is equivalent to 4.1×10^6 base pairs, corresponding to a molecular weight of 2.7×10^9. Since the total length of DNA of an *E. coli* nuclear body is about 1100 to 1400 μm, 1 minute of transfer represents about 11 to 14 μm of DNA (about 40 genes per minute).

Gene clusters and operons Mapping of the genes that control the enzymes of a single biochemical pathway has shown that these genes are often

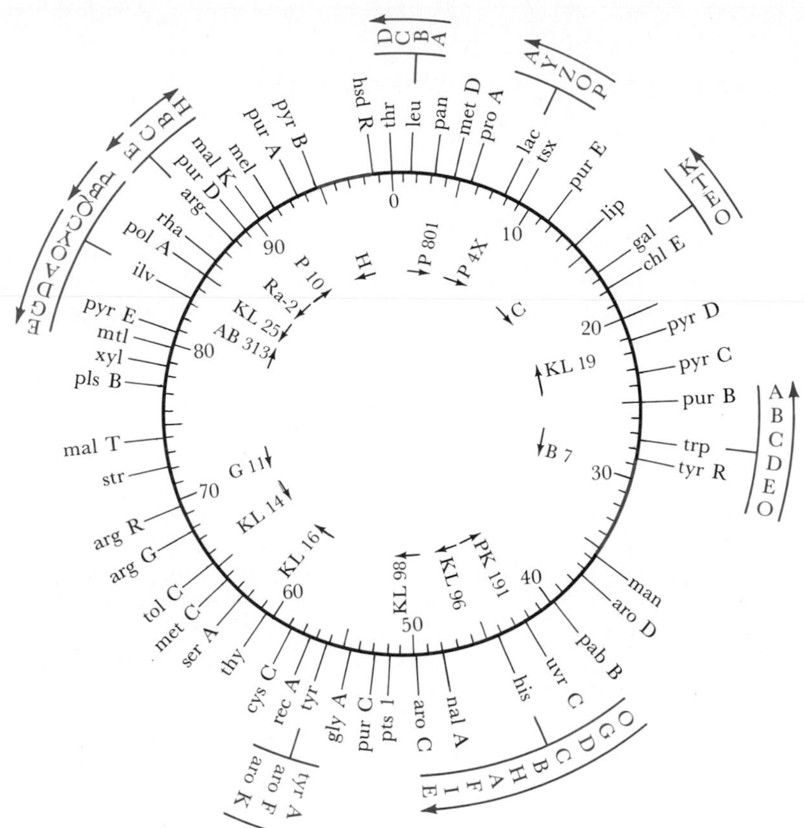

Figure 7.21 Circular reference map of *E. coli* K-12. The large numbers refer to map position in minutes, relative to the *thr* locus. From the complete linkage map, 52 loci were chosen on the basis of greatest accuracy of map location, utility in further mapping studies, and/or familiarity as long-standing landmarks of the *E. coli* K-12 genetic map. Inside the circle, the leading transfer regions of a number of Hfr strains are indicated. Arrows above operons indicate the direction of messenger RNA transcription for these genes.

clustered or closely linked. The gene clusters for several biochemical pathways on the *E. coli* chromosome are shown in Figure 7.21; letters are used to indicate the genes for the specific enzymes of the pathway (for instance, the galactose cluster at minute 17, with genes *K, T,* and *E*). It has also been found that all of the enzymes of a single gene cluster are often affected simultaneously by induction or repression. These related observations have been of considerable importance in the development of the *operon concept,* explaining not only how enzyme induction or repression can be brought about, but also how the synthesis of a series of related enzymes can be simultaneously controlled (Section 5.8).

The transcription of some operons proceeds in one direction along the chromosome, whereas with other operons transcription is in the opposite direction. The direction of mRNA transcription of the operons listed in Figure 7.21 is shown by arrows above the gene clusters. Since transcription always occurs in a 5′ to 3′ direction (see Section 5.8), this implies that transcription off of either strand of DNA can occur and thus that operons exist on each of the strands of the DNA double helix.

Although many enzymes controlling specific biochemical pathways are linked and their synthesis is regulated by operons, this is not always the case. For some pathways, individual enzymes are at different locations on the chromosome, with the same operator present at each site. In such a situation, coordinate regulation may occur even though the enzymes do not map together.

7.9 Genetics in Eucaryotic Microorganisms

We now turn from a discussion of genetic mechanisms in bacteria to a consideration of analogous processes in eucaryotes. Genetic recombination in eucaryotes differs in many ways from that in procaryotes. The complex nuclear organization of eucaryotes and the existence in each nucleus of a number of chromosomes lead to more regular mechanisms of gene assortment and segregation. Thus, genetic recombination in eucaryotes rarely results from fragmentary processes such as those that have been described earlier for procaryotes. Instead, the mechanism of recombination is more precise and predictable. Although the word "sex" can be applied to bacteria only in a very broad sense, it is used quite aptly to describe the genetic recombinational process of eucaryotic microorganisms. Unlike bacteria, whose genomes are usually single DNA molecules, eucaryotic genomes are usually segmented into a number of chromosomes. While bacterial chromosomes are usually circular, eucaryotic chromosomes are built upon linear DNA molecules.

The Origins of Bacterial Genetics

Although genetic recombination in eucaryotes had been known for a long time, the discovery of genetic recombination in bacteria by transformation, transduction, and conjugation has been a relatively recent event. Of the three recombination processes, the discovery of transformation was the most significant as it provided the first direct evidence that DNA is the genetic material. The first evidence of bacterial transformation was obtained by the British scientist Fred Griffith in the late 1920s. Griffith was working with *Streptococcus pneumoniae* (pneumococcus), a bacterium which owes its ability to invade the body in part due to the presence of a polysaccharide capsule. Mutants can be isolated which lack this capsule and are thus unable to cause infection; such mutants are called R strains, because their colonies appear rough on agar, in contrast to the smooth appearance of capsulated strains. A mouse infected with only a few cells of an S (smooth) strain will succumb in a day or two to pneumococcus infection, whereas even large numbers of R cells will not cause death when injected. Griffith showed that if heat-killed S cells were injected along with living R cells, a fatal infection ensued, and the bacteria isolated from the dead mouse were S types. A number of different polysaccharide capsules were known in different pneumococcus S strains, and it was possible to do this experiment with heat-killed S cells from a type different from that from which the R strain was derived. Since the isolated living S cells always had the capsule type of the heat-killed S cells, the R cells had been transformed into a new type, and the process had all the properties of a genetic event. The molecular explanation for the transformation of pneumococcus types was provided by Oswald T. Avery and his associates at Rockefeller Institute in New York, by a series of studies carried out during the 1930s, culminating in the now classic paper by Avery, McCarty, and MacLeod in 1944. Avery and his co-workers showed that under certain conditions the transformation process could be carried out in the test tube rather than the mouse, and that a cell-free extract of heat-killed cells could induce transformation. By a long series of painstaking biochemical experiments, the active fraction of cell-free extracts was purified and was shown to consist of DNA. The transforming activity of purified DNA preparations was very high, and only very small amounts of material were necessary. Subsequently, Rolin Hotchkiss, Harriet Ephrussi, and others at Rockefeller showed that transformation could occur in pneumococcus not only for capsular characteristics, but for other genetic characteristics of the organism, such as antibiotic resistance and sugar fermentation. In the 1950s, transformation was also shown to occur in *Haemophilus, Neisseria, Bacillus*, and a variety of other organisms. In 1953, James Watson and Francis Crick announced their model for the structure of DNA, providing a theoretical framework for how DNA could serve as the genetic material. Thus, two types of studies, the bacteriological and biochemical ones of Avery, and physical-chemical ones of Watson and Crick, solidified the concept of DNA as the genetic material. In the subsequent years, this work has led to the whole field of molecular genetics.

Although bacterial transformation resulted from an essentially accidental discovery, bacterial conjugation was initially shown to occur by Joshua Lederberg and E.L. Tatum in 1946, through experiments carefully designed to determine if a sexual process might occur in bacteria. Because it appeared that the process, if present, would be quite rare (no microscopical evidence for bacterial mating had ever been seen, although such evidence can easily be obtained in eucaryotes), Lederberg developed a method which involved the use of nutritional mutants of *Escherichia coli*. Fortunately, he isolated these mutants in strain K-12, one of the few strains now known to contain (in the wild-type state) the F factor. The principle was to mix two strains, one requiring biotin and methionine, the other requiring threonine and leucine, and plate the mixture on a minimal medium lacking all four growth factors. Neither parental type could grow on this medium, but any recombinants could, and when about 10^8 cells were plated, a small but significant number of colonies was obtained. Strains with two separate nutritional requirements were employed since it would be unlikely that spontaneous back mutation of both genes would occur in a single cell. Thus the only explanation for the phenomenon was some sort of genetic recombination. To show that the process required cell-to-cell contact, and hence could not be a type of transformation, it was shown that when culture filtrates or extracts were separated by a sintered glass disk, permeable to macromolecules but not to cells, recombination did not occur. Although initially conjugation appeared to be a very rare event, by the early 1950s a strain of *E. coli* had been isolated by the Italian scientist L.L. Cavalli-Sforza, while he was working in Lederberg's laboratory, which showed a high frequency of recombination. The British physician William Hayes, who independently isolated an Hfr strain, then showed that genetic transfer during recombination between Hfr and F^- was a one-way event, with the Hfr serving as donor. The interrupted mating experiment and the demonstration of the circular genetic map of *E. coli* were then carried out by Elie Wollman and Francois Jacob, working with Jacques Monod at the Pasteur Institute in Paris. The distinction between Hfr and F^+ was made by Lederberg, who also showed that F^+ behaved in an infectious manner. Lederberg coined the term *plasmid* in the 1950s, to describe such apparently extrachromosomal genetic elements, although the term did not find wide usage until the 1970s, when infectious drug resistance became a major medical problem.

Bacterial transduction was discovered by the American scientist Norton Zinder when he was working at the University of Wisconsin as a graduate student with Lederberg on genetic recombination in *Salmonella typhimurium*. The original motivation for this work was to show that conjugation occurred in an organism other than *E. coli*, and the techniques involved isolation of mutants and quantification of recombination by observing colony growth on minimal medium. However, although evidence of recombination was obtained, it could be shown that cell-to-cell contact was not required. Although this suggested a type of transformation, the process was not affected by DNase, and the gene transfer agent behaved like a bacteriophage. The gene transfer agent could be purified by the same procedures used to purify virus particles, and transduction occurred only with recipient cells that had receptor sites for the virus in question. Further, transducing activity could be eliminated by treatment of a lysate with substances able to adsorb the virus, such as sensitive cells or antibodies. Thus, in all cases, transducing activity and virus activity behaved in similar ways. Zinder and Lederberg coined the word "transduction" to refer to any genetic recombinational process that was only fragmentary and did not involve cell-to-cell contact, intending in this way to encompass processes involving either free DNA (transformation) or phage, but subsequently the word *transduction* has been applied only to virus-mediated genetic transfer.

The relationship between lysogeny and bacterial genetics was initially discerned by the French scientist Andre Lwoff, at the Pasteur Institute, who showed by careful microculture experiments that the ability to produce bacteriophage was a hereditary characteristic of bacteria. The discovery of transduction and conjugation soon resulted in the merging of many lines of investigation, since Ester Lederberg showed that *E. coli* K-12 carried a virus, called *lambda*, which was shown by Morse and Lederberg to transduce at high frequency the *gal* (galactose) genes, and Jacob and Wollman showed that, as a prophage, *lambda* behaved as a chromosomal element during conjugation. Jacob and Wollman soon recognized the analogy between *lambda* and the F factor, and with the physical evidence of the presence of extrachromosomal circular DNA molecules in cells, the concept of the plasmid became established. Work on plasmid biology, which only burgeoned in the 1970s, is too recent for historical treatment; however, bacterial genetics will probably be as exciting in the future as it has been in the past.*

*Key references to this section: Dubos, R. 1976. The professor, the institute, and DNA. Rockefeller University Press, New York. An account of Oswald T. Avery's research career, culminating in the discovery that DNA was the genetic material. Also provides brief historical background on other aspects of bacterial genetics. McCarty, M. 1985. *The Transforming Principle: Discovering that Genes are Made of DNA*. W. W. Norton, New York. A personal account of one of the key members of Avery's team.

Alternation of generations Eucaryotes exhibit an alternation of generations: in the *haploid* phase the number of chromosomes per cell is called *n*, and in the *diploid* phase, *2n* (Figure 7.22). Thus, in the fruit fly *Drosophila melanogaster* a haploid cell contains 4 chromosomes and the diploid 8. In humans, haploid and diploid cells contain 23 and 46 chromosomes, respectively. Multicellular plants and animals are usually diploid, with the haploid phase present only in the germ cells (sperm and eggs, also called *gametes*), whose life spans are transitory. In eucaryotic microorganisms, an alternation of generations also occurs, and structures equivalent to sperm and eggs are formed, but the extent of the development of the haploid and diploid phases varies. In many species, the predominant vegetative phase is haploid and the diploid phase is transitory; in other species the diploid phase dominates; and in still others both haploid and diploid phases have independent vegetative existences.

In diploid cells, two copies of each gene are present, one on each of the two homologous chromosomes. The term *allele* is used to refer to the two alternate forms of the *same gene* present on the two homologous chromosomes. If a mutation should occur in one allele, the other can continue to be ex-

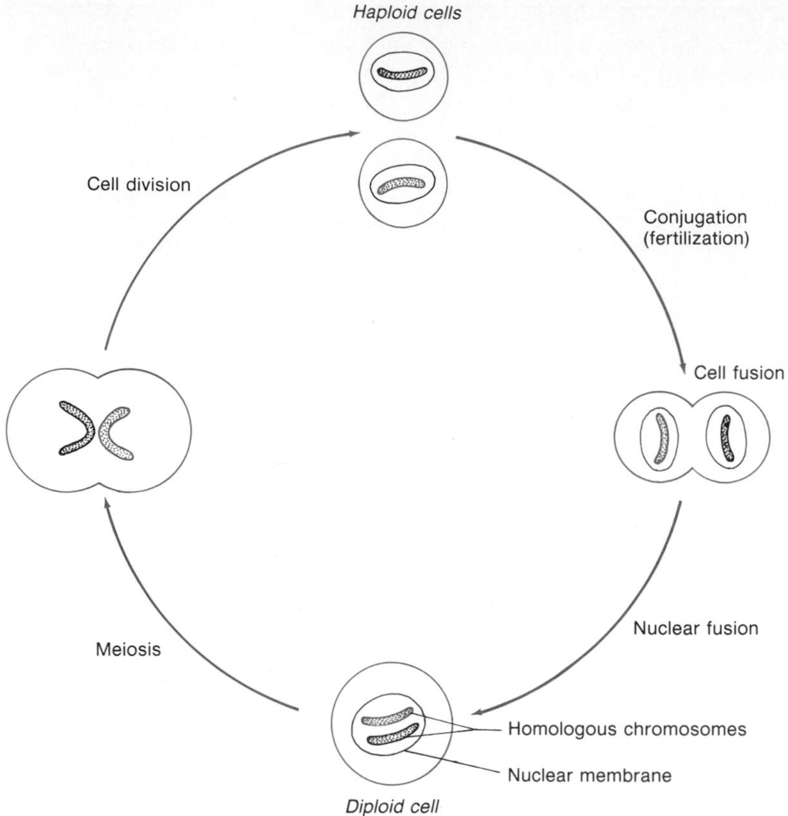

Figure 7.22 Alternation of haploid and diploid generations.

pressed, so that the effect of the mutation may not be evident. Thus, the expression of one allele may be masked by the other. The gene that is expressed is called *dominant* to the other allele, which is called *recessive*. Diploidy presents difficulties in genetic research since isolation of mutants is much easier in haploid than in diploid organisms, because only one form of the gene is present, whose effect can be directly expressed.

Meiosis We discussed *mitosis* in Section 3.17. **Meiosis** is the process by which the change from the diploid to the haploid state is brought about (see Section 3.17). When meiosis occurs, the two sets of homologous chromosomes are segregated into two separate cells, so that the number of chromosomes is reduced from $2n$ to n, yielding haploid cells that are the precursors of the germ cells. During the meiotic process, there is a *pairing* between homologous chromosomes, followed by a segregation of one of each of the pairs into a separate cell. The pairing of homologous chromosomes is a highly specific process. If genetic recombination between homologous chromosomes occurs, it occurs at the pairing stage. Each of the pairs becomes attached to special structures, called *spindle fibers*, which also are attached at the poles of the cell. One of each pair of chromosomes is then pulled toward the pole and a separation into two cells occurs. Once segregation of chromosomes is complete, each set of n chromosomes becomes segregated into a separate cell. After the first meiotic division, there is a second division of each of the cells, leading to the formation of four gametes, two of each type. By definition, eggs are formed by females and sperm are formed by males, but in many microorganisms there is no clear sexual distinction, but only different mating types (see Section 7.11).

Genetic consequences of meiosis If the diploid homologues are genetically identical for a particular allele, the cell is said to be *homozygous*. Under such conditions, the haploid cells formed by meiosis will also be identical. However, if the two homologues differ, the cell is said to be *heterozygous*. Under such conditions, the four haploid gametes will not be identical. The precise genetic constitution of the haploid cells will depend upon events occurring during the meiotic process. We need not discuss the details of segregation, but note that it is from studies on how characters are segregated that genetic mapping of eucaryotic organisms is carried out.

Once gametes are formed, **mating** of gametes of different type can occur, leading to the formation of diploids. The first diploid cell formed as a result of mating of haploid gametes is called a *zygote*. The mating process involves a specific association of gametes from related organisms, but the two gametes need not be of identical genetic constitution. The diploid cell formed can then become the forerunner of a new population of genetically identical diploid organisms.

Somatic genetics In multicellular eucaryotes, the cells that form gametes are called *germ cells*; the propagation of the species depends on them. Other cells of the body may die, but as long as germ cells remain alive, the species can be propagated. Those cells of the body that do not participate in the formation of gametes are called *somatic cells* (the term *soma* means *body*). Although somatic cells are not involved in genetic phenomenon related to the maintenance of the species, such cells do undergo mutation, and within the body of the plant or animal distinct populations of genetically distinct cell types can arise as a result of mutation followed by cell growth. In a sense, the body of the multicellular plant or animal can be considered to be analogous to a population of microbial cells; both can be composed of genetically distinct cell types that have arisen as a result of mutation. An analysis of somatic genetic phenomena in eucaryotes is of considerable interest, as it has important implications for cancer (see Section 6.16) and immunology (see Section 12.6).

7.10 Cytoplasmic Inheritance

Although most genetic characters are contributed by chromosomal genes, mitochondria and chloroplasts contain separate genetic systems that are required for their own replication. As we have noted, mitochondria are the organelles in which the respiratory functions occur in eucaryotes and chloroplasts are the photosynthetic organelles (see Section 3.13). Most of the genetic information for the biogenesis of mitochondria and chloroplasts resides in the nucleus, but these organelles also have their own genetic systems which code for some functions. In addition, mitochondria and chloroplasts have their own protein translation system which is involved in the synthesis of the few proteins which these organelles synthesize independently of the nuclear system.

As a eucaryotic cell grows, the number of mitochondria and chloroplasts increases by a process of organellar division—an existing organelle divides and forms two new organelles. Mitochondria and chloroplasts are never made *de novo* (that is, from scratch); they always arise from preexisting mitochondria and chloroplasts. Thus, if a cell lacks an organelle, its offspring will lack this same type of organelle. Because mitochondria and chloroplasts only arise from preexisting organelles, the genetic characteristics of the organelle in a new cell are determined by the genetic characteristics of the organelle of its parent. Thus, a whole genetic system exists in the cytoplasm of the eucaryotic cell, side by side with the nuclear genetic system. The inheritance of mitochondrion and chloroplast characteristics is therefore termed *cytoplasmic inheritance*.

The genomes of mitochondria and chloroplasts are present in relatively small circular DNA molecules. A brief summary of the molecular genetics of organelles is given in Table 7.3. One interesting feature of organellar DNA is that much of this DNA does *not* code for proteins. Only a few proteins are actually coded for by the organellar genetic system. However, many of the tRNA molecules, as well as the ribosomal RNA molecules, are coded by organellar DNA.

Many of the organelle proteins are transported into the organelle from the cytoplasm. We noted in Section 3.13 that organelles possess rather permeable outer membranes so that transport of macromolecules and other substances that would not normally traverse the cytoplasmic membrane occurs. However, organisms vary in how many proteins are coded for by organellar genes. In the human mitochondrion, for instance, 13 different polypeptides are coded by mitochondrial genes, whereas in yeast only 8 proteins have been linked to mitochondrial genes. We discuss some of the details of inheritance for yeast mitochondria in Section 7.11.

An important feature of organellar genetic systems is sensitivity to antibiotics which inhibit these same processes in procaryotes. Thus, protein synthesis in mitochondria and chloroplasts is inhibited by chloramphenicol, an antibiotic which acts on specific protein subunits of the 30S and 50S ribosomes (see Table 5.3 and the Box in Chapter 3). Because of this, chloramphenicol has some toxic effects on eucaryotic cells, even though its activity is generally restricted to bacteria. We discuss the selective toxicity of antibiotics in Chapter 9.

A hypothesis that has been widely discussed is that mitochondria and chloroplasts arose initially from

Table 7.3 Some key features of the genetic systems of eucaryotic organelles

Genome	Small, generally circular DNA Histones absent Structure of genome resembles that of eubacteria
Genetic code	Some differences from the universal code Much of the DNA is noncoding
Translation apparatus	Ribosomes 70S in size, like eubacteria; subunits are 30S and 50S Sensitive to antibiotics that affect eubacterial ribosomes Polypeptide starts with formylmethionine, like eubacteria

bacteria which somehow infected and then grew inside eucaryotic cells (see also Sections 3.14 and 18.1). According to this hypothesis, such bacteria would have initially performed the physiological roles that later became associated with the organelles, and these bacteria could therefore be considered **endosymbiotic bacteria**. Later, as a result of genetic changes, these endosymbiotic bacteria may have lost most of their genetic systems, and the functions coded by these systems were taken over by the nucleus. In some cases, there may have been actual transfer of genes from the endosymbiotic bacteria to the nucleus, whereas in other cases preexisting nuclear genes may have taken over bacterial functions. However the evolutionary process of mitochondria and chloroplasts might have occurred, the end result would have been organelles with restricted parts of their necessary genetic systems. Presumably, the genetic coding information that has remained in the organelle is that for proteins that cannot be readily transported from the cytoplasm into the organelle (perhaps because they are so large or so hydrophobic that they cannot reach their target organelle).

We should emphasize that many of the structures found in eucaryotes do not arise by division of preexisting structures and do not have their own genetic systems. Thus, structures such as lysosomes, Golgi apparatus, flagella, cilia, etc., all arise *de novo* and are completely under nuclear control. Only mitochondria and chloroplasts seem to have their own genetic systems, albeit only to a partial extent. We discuss further aspects of mitochondrial biogenesis in yeast in the next section.

7.11 Yeast Genetics

More is known about the molecular genetics of yeast than about almost any other eucaryote. This is because yeast is an extremely favorable organism for studies of genetic phenomena. The yeast that has been most studied is *Saccharomyces cerevisiae*, the common baker and brewer's yeast. This yeast is not only a useful model system for studies on eucaryotic genetics, but is an important organism of commerce, and therefore studies of its genetics can be expected to have practical significance (see Section 10.12).

A yeast grows as a single cell, and each yeast cell is capable of serving as a gamete. Because most yeasts are haploid and unicellular, isolation of mutants is quite straightforward, and a large variety of mutant types are known. By mating mutants, genetic analysis of yeast can be carried out.

The life cycle of a typical yeast is shown in Figure 7.23. Many yeasts have two separate *mating types*, which can be considered analogous to male and female. However, the two mating types of a yeast are alike in structure and can only be told apart by allowing them to mate. Mating type is itself a genetic characteristic in yeast (and actually under rather complex control, as will be described later in this section). Upon mating of opposite types, a diploid cell is formed. In many yeasts, this diploid cell is capable of growing vegetatively, leading to the formation of a population of genetically identical, albeit diploid, cells. Under certain conditions, diploid cells of such a population can undergo meiosis and form haploid gametes. Two distinct types of gametes are formed, of opposite mating type. From a single diploid cell, a structure containing four such gametes is formed, two of each mating type. The cell in which the gametes are formed is called an *ascus*, and the cells within the ascus are called *ascospores*. Yeast ascospores generally possess some resistance to environmental conditions, and exhibit a type of dormancy, but they are not as resistant to environmental conditions as the endospores of bacteria which we have discussed in Section 3.12. Ultimately, the ascus breaks down, releasing the ascospores. Each ascospore is capable of *germination*, and can become the forerunner of a new population of haploid cells.

One important advantage of yeast for genetic analysis is that hybridization of yeasts is fairly straightforward. After mating and ascospore formation has occurred, the experimenter can dissect the four ascospores from the ascus, use each ascospore as the forerunner of a separate culture, and analyze the cultures so obtained for phenotypic characters that were in the parent. In this way, it is possible to *map* genes in yeast. The yeast *Saccharomyces cerevisiae* has 17 chromosomes, and extensive genetic maps have been prepared for these chromosomes. However, despite the large number of chromosomes, the genome size of the yeast cell is small; the average chromosome contains only about 1×10^6 basepairs of DNA and the genome is thus only about four times the size of that of *Escherichia coli*. Another advantage of yeast is that it can be transformed using exogenous DNA, and plasmids are available for use in genetic engineering. Also, some very interesting analyses of mitochondrial inheritance (see Section 7.10) have been carried out.

Mating type genetics of yeast As we have noted, yeast has two mating types which are indistinguishable except through their behavior in mating. The two mating types of *Saccharomyces cerevisiae* are designated α and a. Cells of type α mate only with cells of type a and whether a cell is α or a is itself determined genetically. However, although a yeast cell line generally remains either α or a, haploid yeast cells are periodically able to *switch* their mating type from one to the other. (One consequence of this switching is that a pure culture of a single mating type can ultimately form diploids. Note that it is not appropriate to call a yeast cell *bisexual*. Under a given

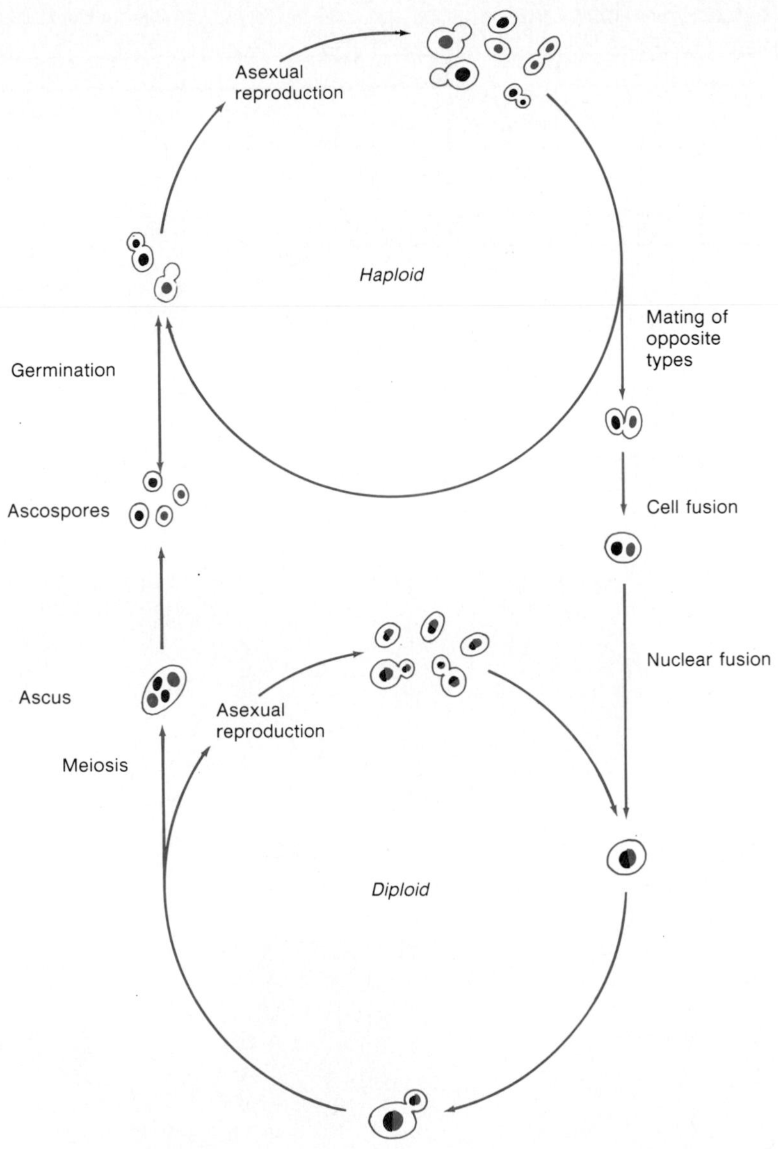

Figure 7.23 Life cycle of a typical yeast, *Saccharomyces cerevisiae.*

set of conditions, it is either one or the other mating type, never both.)

The switch in mating type from α to a and back to α has at its basis the behavior of a mobile genetic element, somewhat reminiscent of transposition or phase variation in *Salmonella* (see Section 7.7 for discussion of both topics). The phenomenon in yeast is illustrated in Figure 7.24. There is a single active genetic locus, called the MAT (for *mating type*) locus, at which either gene α or gene a can be inserted. At this active locus, the MAT promoter controls the transcription of whichever of the two mating type genes is present. Thus, if gene α is at that locus, then the cell is mating type α, whereas if gene a is at that locus, the cell is mating type a. *Somewhere else on the yeast genome* are copies of both genes, α and a, which are not expressed. These *silent copies* serve as the source of the gene that is inserted when the switch occurs. When the switch occurs, the appropriate gene, α or a, is copied from its silent site and then inserted into the MAT location, *replacing* the gene which is already present. Thus, the old gene is excised out of the locus and discarded and the new gene is inserted. This mechanism has been called the *cassette mechanism*, because each gene can be considered analogous to a tape cassette that is inserted into the "reading head," the place on the chromosome where active transcription takes place.

What do genes α and a do? At least one function has been identified. Yeast cells that are undergoing mating excrete *peptide hormones* called α factor and a factor. These hormones bind to cells of opposite mating type and bring about changes in the cell surfaces of these cells so that cells of opposite mating type will associate and fuse. It seems that α cells have receptors on their surfaces only for a factor, whereas a cells have receptors only for α factor. Once two cells of opposite mating type have associated, a com-

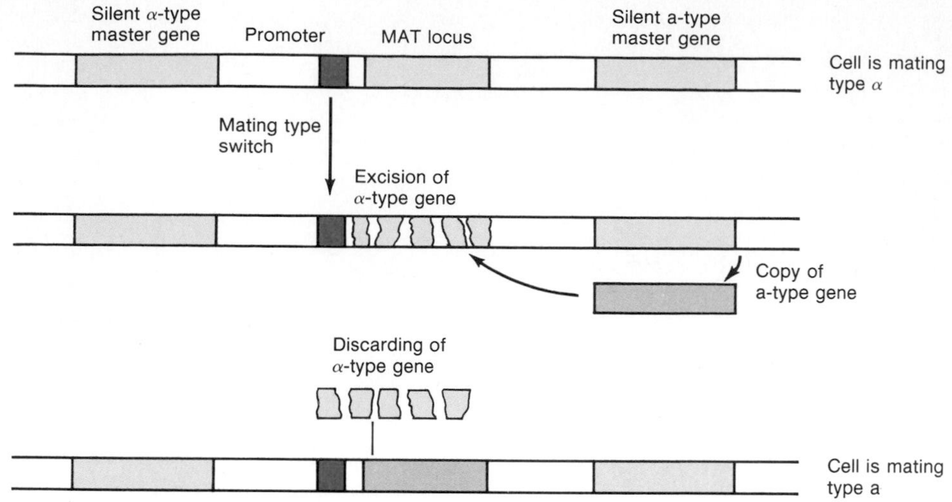

Figure 7.24 The cassette mechanism that is involved in the switch in yeast from mating type α to a and back again. Whichever "cassette" is inserted at the active locus (reading head) determines the mating type. The process shown is reversible, so that type a can revert to type α.

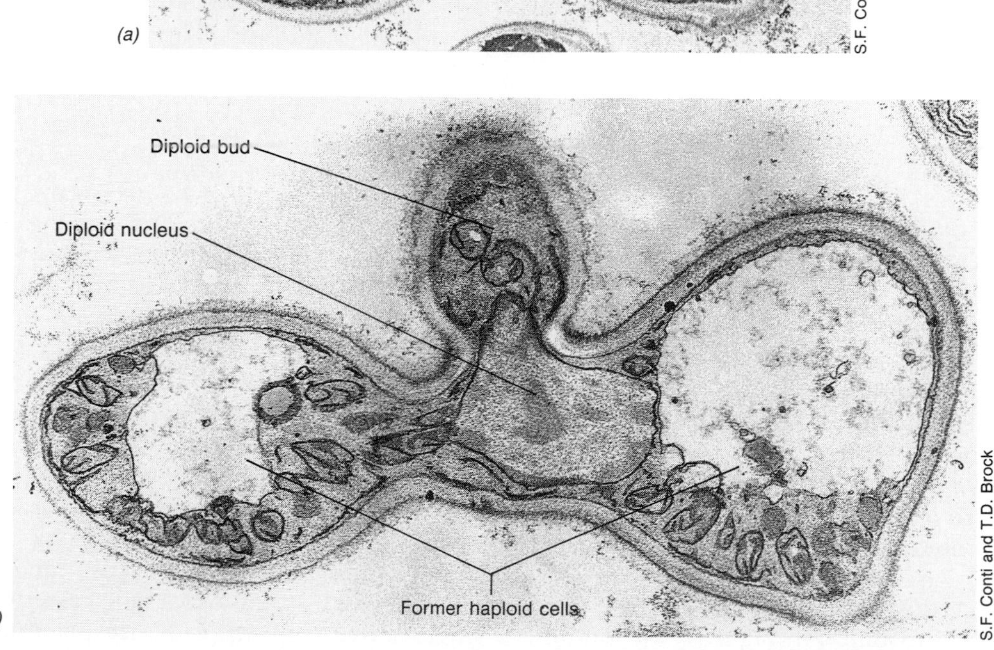

Figure 7.25 Electron micrographs of the conjugation process in a yeast, *Hansenula wingei*. Both magnifications, 12,000×. (a) Two conjugating cells, which have fused at the point of contact and have sent out protuberances toward each other. (b) Late stage of conjugation. The nuclei of the two cells have fused and the diploid bud has formed at a right angle to the conjugation tube. This bud will eventually separate from the conjugants and become the forerunner of a diploid cell line.

plex series of events is initiated which leads to fusion of these two cells as well as their nuclei, resulting in the formation of a diploid zygote (Figure 7.25).

Mitochondrial inheritance in yeast We have discussed the nature and structure of mitochondria in Chapter 3 and mitochondrial inheritance in Section 7.10. We noted that mitochondria carry out some of their own biosynthetic functions, including the synthesis of certain proteins. The translation apparatus of the mitochondrion shares a number of properties with the procaryotic translation system: ribosomal size, sensitivity to certain antibiotics, starting codon of *N*-formylmethionine instead of methionine as in eucaryotes. One strange property of the mitochondrial translation apparatus is that certain codons of the "universal genetic code" (see Table 5.4), are translated differently in the mitochondrion. For instance, the codon UGA, which is a *stop* codon in both bacteria and eucaryotes, is read as a *tryptophan* codon by yeast mitochondria, and the codon AUA, which is an *isoleucine* codon in cells, is translated as *methionine* by mitochondria. There are also some other differences in translation. These codon differences are related to differences in the tRNAs, which are encoded within the mitochondria from DNA present in the mitochondria.

One of the most interesting genetic analyses using yeast is that on *mitochondrial inheritance*. Although mitochondria receive some of their informational macromolecules preformed from the cytoplasm, others are synthesized within the mitochondrion, using genes present in the mitochondrion. A complete system for DNA replication is present within the mitochondrion, and mitochondria undergo a type of division process analogous to that undergone by bacteria. It has been possible to develop mutant yeast strains in which the mutations have occurred in the mitochondrion rather than in the nucleus, and by genetic analysis, the inheritance of the mitochondria themselves can be studied. Because inheritance of genetic characteristics via mitochondria occurs outside the nucleus, and outside the process of mitosis and meiosis, it can be considered a form of **cytoplasmic inheritance** (see Section 7.10). (It is sometimes called *non-Mendelian inheritance*, to indicate that it does not follow Mendel's laws.)

An example of the manner in which mitochondrial characteristics are inherited in yeast is shown in Figure 7.26. As seen, when two yeast cells of opposite mating type which also differ in mitochondrial characteristcs are mated, the outcome depends upon which of the two types of mitochondria replicates most rapidly during subsequent cell divisions. In one class of mitochondrial mutants, called *petite*, large deletions in the mitochondrial DNA has led to abolition of all mitochondrial protein synthesis. Such mitochondria are nonfunctional, and yeast cells containing such mitochondria are unable to carry out respiration (however, they are still able to grow anaerobically by fermentation processes). If a petite yeast cell is mated with a wild type yeast cell, the diploid zygote will contain both types of mitochondria, but since the mutant mitochondria do not replicate as effectively as the wild type mitochondria, they lose out in the competition during subsequent cell divisions. Ultimately, the cell lines derived from each of the four ascospores will all be wild type (Figure 7.26*a*). On the other hand, there are also petite mutants of yeast in which the mutation has occurred in a nuclear gene (since most of the proteins of the mitochondria are coded for by nuclear genes rather than mitochondrial genes). Crosses involving nuclear petites show conventional Mendelian inheritance (Figure 7.26*b*).

Genetic analysis and DNA analysis has shown that all of the yeast mitochondrial genes are localized in the mitochondrial DNA. Interestingly, most of the yeast

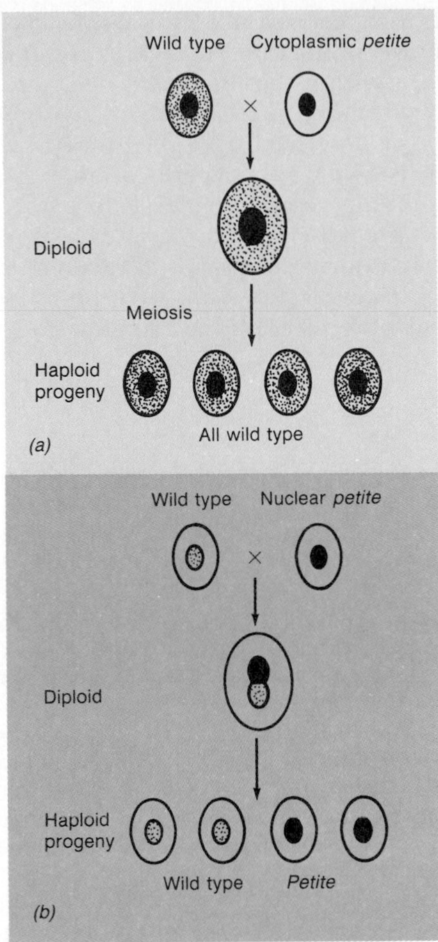

Figure 7.26 Mitochondrial inheritance. Outcome of crosses between **petite** and wild-type yeast: (a) cytoplasmic **petite**; (b) nuclear **petite**.

mitochondrial DNA does not code for proteins; its function is unknown. Proteins coded for by yeast mitochondrial DNA include cytochrome b, cytochrome c oxidase, ATPase, and one ribosome protein. In addition, a number of tRNA molecules are encoded by yeast mitochondrial DNA, as are the ribosomal RNA molecules (but not the ribosomal proteins). There is some reason to believe that during evolution, some genes have moved from mitochondrial DNA to nuclear DNA, and vice versa (see Section 7.10).

Yeast plasmids We have defined plasmids as DNA molecules which replicate independently from chromosomal DNA (see Section 7.4). The DNA of plasmids is generally circular. Do plasmids exist in eucaryotes? Circular DNA molecules have been observed in animal cells in culture, but whether these are plasmids or contaminating viruses is unclear. However, one eucaryotic plasmid that has been well studied is the so-called 2 μm circle in yeast. Although no genes have been mapped to the 2 μm circle, otherwise this piece of DNA has all the characteristics of a plasmid.

Most yeasts contain the 2 μm circle DNA. It is an autonomously replicating ring which is found in multiple copies within the nucleus. The 2 μm circle is packaged within the nucleus into nucleosomes, using histone protein derived from the nucleus. The 2 μm circle has its own origin of replication, but uses replication proteins derived, at least in part, from the nucleus. Segregation of the plasmid from mother to daughter cell during mitosis is at random, but if the number of 2 μm circles drops to a low value, an increased rate of replication can bring the copy number back up to 30 to 50 per cell. There is no evidence that this plasmid ever becomes integrated into the nuclear DNA.

Although the 2 μm plasmid is a useful model for studying DNA replication in yeast, its greatest value appears to be as a vector for cloning foreign genes into yeast. By use of appropriate treatment, it is possible to transform yeast cells using this plasmid, and hence to incorporate genes of interest. The value of vectors such as this for genetic engineering is discussed in Chapter 8.

Study Questions

1. Why is it difficult in a single experiment using transformation to transfer to a cell a large number of genes?
2. From what you know about cell wall structure of bacteria (Chapter 3), explain the problem a DNA molecule would encounter if the transformation process were to occur.
3. Suppose you had available only a virulent phage for a bacterium of interest. Why might such a phage not be suitable for carrying out a genetic transduction experiment? Can you think of any way this phage might be experimentally turned into a transducing phage?
4. Suppose you were given the task of developing a genetic transduction system for an industrial organism of interest. You have a large collection of bacterial strains but no phages. Describe the steps you would use to develop such a transduction system. (Hint: You must not only obtain transducing phage but also a collection of bacterial mutants to be able to follow genetic markers.)
5. The physical presence of a plasmid can be determined by use of electrophoresis on gels. How could you use such a procedure to show that a curing agent is able to eliminate the plasmid from the bacterial cell?
6. Two bacterial strains are available, one resistant to streptomycin, the other resistant to penicillin. Suppose the resistance to one of these two antibiotics was chromosomal, the other associated with a conjugative plasmid. Describe experiments that would determine which antibiotic resistance was plasmid controlled.
7. How could you determine whether a plasmid conferring antibiotic resistance had become integrated into the chromosome?
8. Describe an experiment that could be used to show that a conjugative plasmid was associated with the presence of a sex pilus.
9. Explain how it is possible to use the *interrupted mating procedure* to determine the relative order of genes on a bacterial chromosome.
10. Explain why the insertion of a transposon leads to mutation.
11. Mutants isolated by transposon mutagenesis are never point mutants. Why?
12. The most useful transposons for isolating a variety of bacterial mutants are transposons which contain antibiotic-resistance genes. Why are such transposons of such great use for this purpose?
13. Compare and contrast the genetic mechanisms in procaryotes and eucaryotes. Is there an analogy to the mitochondrial genetic element in procaryotes?
14. Compare and contrast the "flip-flop" process for *Salmonella* phase variation and the "cassette" mechanism for yeast mating type switching.

Supplementary Readings

Alberts, B., D. Bray, J. Lewis, M. Raff, K. Roberts, and **J. D. Watson**. 1983. *Molecular Biology of the Cell.* Garland Publishing Co., New York. Good comparative coverage of procaryotic and eucaryotic genetics, with extensive treatment of genetic mechanisms in yeast.

Freifelder, David. 1987. *Microbiol Genetics.* Jones and Bartlett Publishers, Boston. Primarily dealing with bacteria and bacteriophage, this is a thorough textbook of genetic phenomena.

Lin, E. C. C., R. Goldstein, and **M. Syvanen**. 1984. *Bacteria, Plasmids, and Phages.* Harvard University Press, Cambridge. A fairly brief paperback textbook.

Watson, J. D., N. H. Hopkins, J. W. Roberts, J. A. Steitz, and **A. M. Weiner**. 1987. *Molecular Biology of the Gene, Fourth Edition.* Benjamin-Cummings, Menlo Park, CA. An up-to-date treatment of the molecular genetics of procaryotes and eucaryotes.

Watson, J. D., J. Tooze, and **D. T. Kurtz**. 1983. *Recombinant DNA. A Short Course.* Scientific American Books, New York. The first few chapters provide useful background on procaryotic genetics.

Gene Manipulation and Genetic Engineering

The concepts of molecular genetics, described in the three preceding chapters, have made possible the development of sophisticated procedures for the isolation, manipulation, and expression of genetic material, a field called **genetic engineering**. Genetic engineering has applications in both basic and applied research. In basic research, genetic engineering techniques are used to study the mechanisms of gene replication and expression in procaryotes, eucaryotes, and their viruses. Some of the most important basic discoveries of molecular genetics were made using genetic engineering techniques. For applied research, genetic engineering permits the development of microbial cultures capable of producing valuable products, such as human insulin, human growth hormone, interferon, vaccines, and industrial enzymes. The potential of genetic engineering for commercial application seems limitless.

It is the purpose of this chapter to discuss the general principles of genetic engineering and to show how these principles can be applied. This chapter builds on the material discussed in the previous three chapters, and it may be desirable from time to time to refer back to these chapters to recall details of the processes of molecular genetics.

Underlying all other aspects of genetic engineering is the process of isolating and purifying specific genes, or **gene cloning**. Having large amounts of pure DNA allows characterization and manipulation of the gene and its product. With cloned genes, we can determine the *nucleotide sequence* of the gene, from which we can derive, through the genetic code, the *amino acid sequence* of its protein product. The cloned DNA itself can be used as a *probe* to determine the structure of more complex DNA molecules like the human genome. By moving genes from organisms that are difficult or dangerous to grow into well-characterized microorganisms, valuable biological substances can be produced cheaply and in quantities that were unthinkable until the advent of genetic engineering. By changing the sequence of a cloned gene in a predetermined way, genetic engineers can literally design new, useful biological products that are simply unavailable from natural sources.

In this chapter we will discuss the tools and processes involved in creating and purifying desired genes by techniques using *in vitro* recombination, or **recombinant DNA**. We will then describe how genetic engineering is used to produce large quantities of desired gene products, and how the products themselves can be altered by **site-directed mutagenesis**. We will present examples of practical results derived from genetic engineering, and will conclude this chapter with a review of the principles that underlie the new biotechnology.

Practical aspects of biotechnology will be discussed in Chapter 10.

8.1 Gene Cloning Systems

Gene cloning is at the base of most genetic engineering procedures. As we have mentioned above, the purpose of gene cloning is to isolate large quantities of specific genes in pure form. While it might be theoretically possible to physically isolate pure DNA fragments with single genes from a restriction digest of chromosomal DNA, a little reflection will demonstrate the impracticality of such an approach. Consider that even for a genetically simple organism like *Escherichia coli*, a specific gene represents 1–2 kilobases out of a genome of about 5,000 kb. An average gene in *Escherichia coli* coli is thus less than 0.05 percent of the total DNA in the cell. In humans the problem is even worse, since the average genes are not much bigger than in *Escherichia coli*, but the genome is 1000 times larger! In contrast, the DNA of bacteriophage *lambda* is only 50 kb, and the DNA of some plasmids is less than 5 kb. In these genetic elements, the average gene constitutes 2–40 percent of the DNA.

Thus, the basic strategy of gene cloning is to move the desired gene from a large complex genome to a small, simple one. Specialized transducing phages provide an example of how genes can be enriched (in the case of *lambda dg*, the *gal* operon, see Chapter 7) by moving them to simpler genetic elements. Purifying genes by constructing specialized transducing phages is limited, however, to genes adjacent to the sites where temperate phages integrate into bacterial chromosomes. This will hardly do for most bacterial genes, let alone genes from mammals or plants.

Fortunately, our knowledge of DNA chemistry and enzymology allows us to break and join DNA molecules *in vitro*. This process is known as ***in vitro* recombination**. Restriction enzymes, DNA ligase, and synthetic DNA are important tools used for *in vitro* recombination, and were discussed in Chapter 5.

Gene cloning can be divided into several steps:

1. Isolation and fragmentation of source DNA. This can be total genomic DNA from an organism of interest, DNA synthesized from an RNA template by reverse transcriptase (see Section 6.24), or even DNA synthesized from nucleotides *in vitro*. If genomic DNA is the source, it is generally cut with restriction enzymes to give a mixture of fragments.
2. Joining to a **cloning vector** with DNA ligase. The small, independently replicating genetic elements used to purify genes are known as **cloning vectors**. Cloning vectors are generally designed to allow integration of foreign DNA at a restriction site that cuts the vector in a way that does not affect its replication. If the source DNA and the vector are cut with the same restriction enzyme, then joining can be mediated by annealing of the single-stranded regions called "sticky ends." Blunt ends generated by different restriction enzymes can also be joined using special techniques, and different sticky ends or blunt ends can be joined by the use of synthetic DNA **linkers** or **adapters**. The properties of cloning vectors are discussed in Sections 8.2–8.4.
3. Incorporation into a **host**. The recombinant DNA can be introduced into a host organism by DNA transformation or by infection with phage particles made by *in vitro* packaging. Incorporation of the DNA into the host gives rise to a mixture of clones containing the desired clone along with any other clones that were generated by joining the source DNA to the vector. Such a mixture is known as a **clone library**, since many different clones can be purified from the mixture depending on the gene of interest of the source organism.
4. Detection and purification of the desired clone. Techniques for finding the right clone will be discussed in Section 8.6.
5. Production of large numbers of cells or bacteriophage containing the desired clone for isolation and study of the cloned DNA.

8.2 Plasmids as Cloning Vectors

We have discussed plasmids in some detail in Sections 7.4–7.6. Plasmids have very useful properties as cloning vectors. These properties include: (1) small size, which makes the DNA easy to isolate and manipulate intact; (2) circular DNA, which makes the DNA more stable during chemical isolation; (3) independent origin of replication, so that replication in the cell proceeds outside of cell control; (4) can be present in the cell in several or numerous copies, making amplification of the DNA possible; (5) frequently have

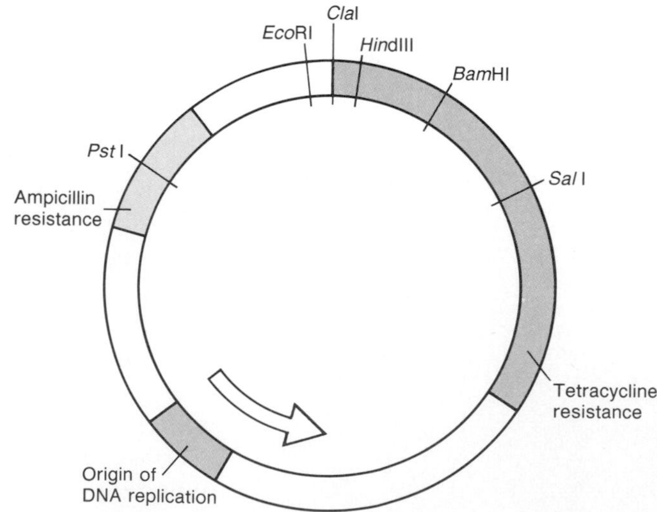

Figure 8.1 The structure of plasmid pBR322, a typical cloning vector, showing the essential features. The arrow shows the direction of DNA replication from the origin.

antibiotic resistance genes, making detection and selection of plasmid-containing clones easier.

Although in the natural environment plasmids are generally transferred by cell-to-cell contact, plasmid cloning vectors generally have been crippled in their ability to transfer conjugatively, for safety reasons. However, transfer in the laboratory can be brought about by transformation procedures. Depending on the host-plasmid system, replication of the plasmid may be under tight cellular control, in which case only a few copies are made, or under relaxed cellular control, in which case a large number of copies are made. Achievement of large copy number is often important in gene cloning, and by proper selection of the host-plasmid system and manipulation of cellular macromolecule synthesis, plasmid copy numbers of several thousand per cell can be obtained.

An example of a suitable cloning plasmid is pBR322, which replicates in *Escherichia coli* (Figure 8.1). Plasmid pBR322 has a number of characteristics that make it suitable as a cloning vehicle:

1. It is relatively small, with a molecular weight of 2.6×10^6.
2. It is stably maintained in its host (*Escherichia coli*) in relatively high copy number, 1–20 copies/cell.
3. It can be amplified to very high number (1000–3000 copies per cell, about 40 percent of the genome!) by inhibition of protein synthesis by addition of chloramphenicol.
4. It is easy to isolate in the supercoiled form by use of cesium chloride/ethidium bromide gradient (see Box, Chapter 5).
5. A large amount of foreign DNA can be inserted; up to 10 kilobases.
6. The complete base sequence of this 4362-nucleotide-long plasmid is known, making it possible to locate sites where restriction enzymes can act.
7. There are *single* cleavage sites for restriction enzymes PstI, SalI, EcoRI, HindIII, and BamHI. The EcoRI site is outside the genes of the plasmid. It is important that only a single recognition site for at least one restriction enzyme is available, so that treatment with that enzyme will open the plasmid but not chop it into pieces.
8. It has two antibiotic resistance markers, ampicillin and tetracycline, which permit ready selection of hosts containing the plasmid.
9. It can be used in transformation easily.

The use of plasmid pBR322 in gene cloning is shown in Figure 8.2. As seen, the BamHI site is within the

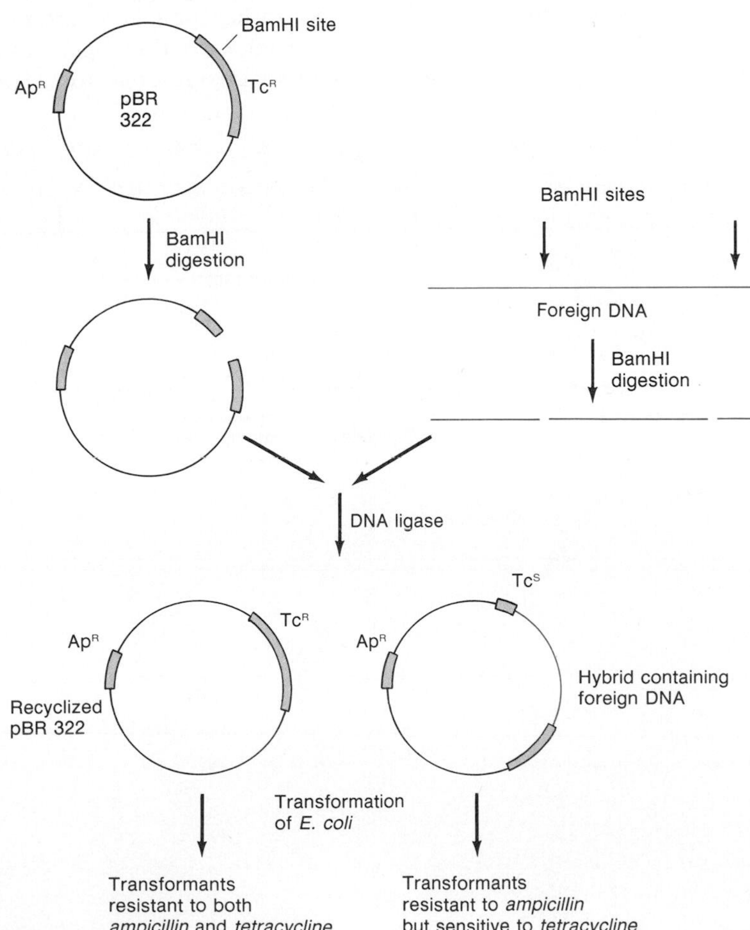

Figure 8.2 The use of plasmid pBR322 as a cloning vector, showing how insertion of foreign DNA causes inactivation of tetracycline resistance, permitting easy isolation of transformants containing the cloned DNA fragment.

gene for tetracycline resistance, and the PstI site is within the gene for ampicillin resistance. If a piece of foreign DNA is inserted into one of these sites, the antibiotic resistance conferred by this site is lost, a phenomenon called **insertional inactivation**. Insertional inactivation is used to detect the presence of foreign DNA within the plasmid. Thus, if pBR322 is digested with BamHI, linked with foreign DNA, and then transformed bacterial clones are isolated, those clones which are both ampicillin resistant and tetracycline resistant *lack* the foreign DNA (the plasmid incorporated into these cells represents DNA that had recyclized without picking up foreign DNA), whereas those cells still *resistant* to ampicillin but *sensitive* to tetracycline *contain* the plasmid with inserted foreign DNA. Since ampicillin resistance and tetracycline resistance can be determined independently on agar plates, isolation of bacteria containing the desired clones and elimination of cells not containing the plasmid can readily be accomplished.

8.3 Bacteriophages as Cloning Vectors

We have discussed temperate bacteriophages, and bacteriophage *lambda* in particular, in Section 6.14. Bacteriophage *lambda* is a useful cloning vector because its molecular genetics is well known, and because DNA can be efficiently packaged in phage particles, which can be used to infect suitable host cells. *Lambda* has a complex genetic map (see Figure 6.33) and a large number of gene functions. However, the central third of the *lambda* genome, between genes *J* and *N* is not essential for lytic growth and can be substituted with foreign DNA.

Modified *lambda* phages Wild-type *lambda* is not too suitable for a cloning vector as it has too many restriction enzyme sites, making it difficult to introduce unique single cuts. To avoid this difficulty, modified *lambda* phages have been constructed which can be used in cloning. In one set of modified *lambda* phages, the so-called Charon phages, unwanted restriction enzyme sites have been removed by point mutation, deletion, or substitution. In those variants that have only a single restriction site, a foreign piece of DNA can be *inserted*, whereas in variants with two sites, the foreign DNA can *replace* the *lambda* DNA. The latter variants, called **replacement vectors**, are especially useful for cloning large DNA fragments.

Further, foreign genes have been introduced into some of these modified *lambda* phages so that it is easier to select or detect clones containing the phage. As an example, the phage Charon 10 is compared with the wild-type *lambda* in Figure 8.3*a*. In this Charon phage, the gene for β-galactosidase has been inserted into the phage. When Charon 10 replicates on a lactose-negative strain of *Escherichia coli*, β-galactosidase is synthesized from the phage gene and the presence of lactose-positive plaques can be detected by use of a color indicator agar (Figure 8.3*b*). If a foreign gene is inserted *into* the β-galactosidase gene, the lactose-positive character is lost. Such lactose-nega-

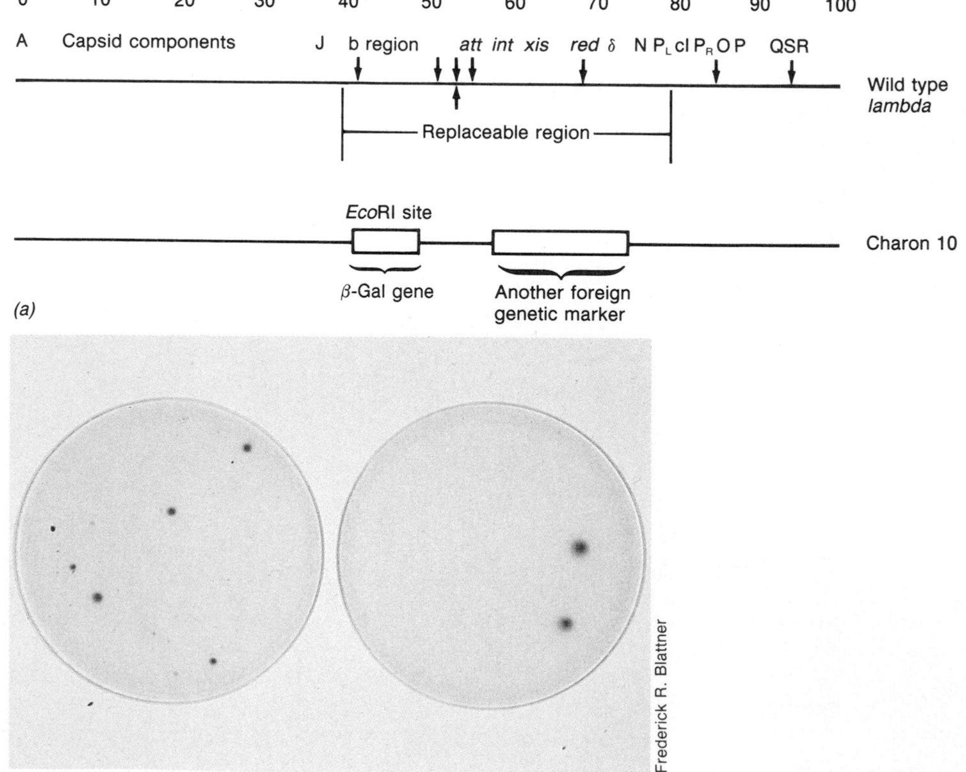

Figure 8.3 Gene cloning with *lambda*. (a) Abbreviated genetic map of bacteriophage *lambda*. (See also Figure 6.33.) Charon 10 is a derivative of *lambda* in which two foreign genes have replaced nonessential DNA of *lambda*. One of these foreign genes codes for the enzyme β-galactosidase, which permits detection of clones containing this phage. (b) Color indicator agar and its use in gene cloning in *lambda*. Plaques formed on a lawn of *Escherichia coli* growing on a plate containing a chemical derivative of galactose (5-chloro-4-homo-3-indolyl-β-D-galactoside, called *Xgal*). The colored plaques arise because cells infected with *lambda* containing the *lac* gene produce β-galactosidase, which hydrolyzes *Xgal* and liberates the colored compound 5-bromo-4-chloro-3-indole.

tive plaques can be readily detected as colorless plaques among a background of colored plaques.

Steps in cloning with *lambda* Cloning with *lambda* replacement vectors involves the following steps (Figure 8.4):

1. Isolation of the vector DNA from phage particles and digestion with the appropriate restriction enzyme.
2. Connection of the two *lambda* fragments to the foreign DNA using DNA ligase. Conditions are chosen so that molecules are formed of a length suitable for packaging into phage particles.
3. Packaging of the DNA by adding cell extracts containing the head and tail proteins and allowing the formation of viable phage particles.
4. Infection of *Escherichia coli* and isolation of phage clones by picking plaques on a host strain.
5. Checking recombinant phage for the presence of the desired foreign DNA sequence, by nucleic acid hybridization procedures or observation of genetic properties.

Selection of recombinants is less of a problem with *lambda* than with plasmids because (1) the efficiency of transfer of recombinant DNA into the cell by *lambda* is very high, and (2) *lambda* fragments which have not received new DNA are too small to be incorporated into phage particles.

Although *lambda* is a useful cloning vector, there are limits on how much DNA can be inserted. Viability of phage particles is low if the DNA is longer than 105 percent of normal *lambda* DNA, so that really large fragments (greater than 20,000 bases) cannot be efficiently cloned. Also, *lambda* finds little use as an expression vector, since it replicates to high copy number only during the lytic cycle.

8.4 Other Vectors

A number of other vectors have been developed which are useful for various purposes in recombinant DNA technology. It is understandable that in the early phases of the development of recombinant DNA, the vectors used were those that had been in most widespread use in research. However, it does not necessarily follow that a genetic system that is useful in basic research automatically provides the best system for the development of a cloning vector. So far, the systems described above, and systems modified from these, have found the greatest use, but this will not necessarily be true in the future. In this section, we discuss briefly several other kinds of vectors, and describe some useful modifications.

Bacteriophage M13 We have discussed the nature and life cycle of bacteriophage M13 in Section 6.11. This filamentous phage contains single-stranded DNA, and replicates without killing its host. Mature particles of M13 are released from host cells by a budding process, and it is possible to obtain infected cultures which can provide continuous sources of phage DNA. An important feature of M13 is its single-stranded DNA. In the Sanger procedure for DNA sequencing (see Box, Chapter 5), single-stranded DNA is needed, and DNA cloned into M13 thus provides a ready source of this single-stranded DNA. Also, single-

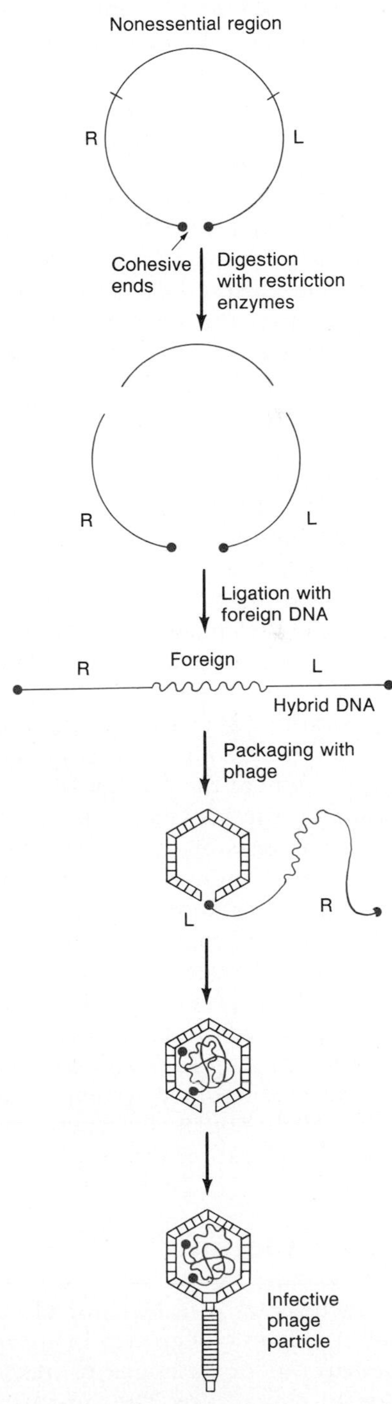

Figure 8.4 The use of bacteriophage *lambda* as a cloning vector. See text for details.

stranded DNA is very useful as a probe for detecting other nucleic acid sequences in transfer procedures such as the Southern blot, and M13 permits ready production of such single-strand DNA probes.

However, in order to use M13 for cloning, a double-stranded form must be available, since restriction enzymes only work on double-stranded DNA. Double-stranded M13 DNA can be obtained by isolation from infected cells, since M13 replicates in the host as a double-stranded *replicative form* (see Section 6.11). However, the wild type M13 has no good regions for inserting foreign DNA so that in order to convert M13 into a cloning vector, it must be modified by introduction of suitable promoters and linkers.

We now describe the manner in which M13 is used in cloning. The replicative double-stranded DNA is isolated from the infected host and treated with a restriction enzyme. The foreign DNA, also double stranded, is treated with the same restriction enzyme. Upon ligation, double-stranded M13 molecules are obtained which contain the foreign DNA. When these molecules are packaged in the bacteriophage and introduced into the cell, they replicate and in time mature bacteriophage particles containing single-stranded DNA molecules are produced. Although two kinds of single-stranded DNA molecules are obtained, only one kind of DNA is packaged into mature phage. Which of the two foreign strands the mature phage will contain will depend upon the *orientation* with which the strand was inserted. Since foreign DNA will be inserted (in separate phage molecules) in either orientation, *both* strands of the foreign DNA can be cloned.

The single-stranded M13 DNA containing the foreign DNA can then be used in DNA sequencing. Since the base sequence where the foreign DNA is inserted is known (based on the specificity of the restriction enzyme used), it is possible to use a synthetic oligonucleotide complementary to this region as a primer, and hence determine the sequence of the whole DNA downstream from this point. In this way, M13 derivatives have proved extremely useful in sequencing foreign DNA, even rather long molecules. (The sequence of the whole bacteriophage *lambda* DNA, 48,513 bases in length, was determined rather quickly in this way.) A modified system for producing single-stranded DNA has been constructed by placing an M13 origin of replication into a plasmid vector.

8.5 Hosts for Cloning Vectors

The ideal characteristics of a host for cloned genes are: rapid growth, capable of growth in cheap culture medium, not harmful or pathogenic, transformable by DNA, and stable in culture. The most useful hosts for cloning are microorganisms which grow well and for which a lot of genetic information is available, such as the bacteria *Escherichia coli* and *Bacillus subtilis* and the yeast *Saccharomyces cerevisiae.*

Although most molecular cloning has been done in *Escherichia coli*, there are perceived disadvantages in using this host. *Escherichia coli* presents dangers for large-scale production of products derived from cloned DNA, since it is found in the human intestinal tract and is potentially pathogenic (see Box). Also, even nonpathogenic strains produce endotoxins which could contaminate products, an especially bad situation with pharmaceutical injectibles. Finally, *Escherichia coli* retains extracellular enzymes in the periplasmic space, making isolation and purification potentially difficult. However, modified *Escherichia coli* strains have been developed in which most of these problems have been eliminated. Because of the extensive knowledge of its genetics and biochemistry, *Escherichia coli* remains the organism of choice for most cloning studies.

The Gram-positive organism *Bacillus subtilis* can be used as a cloning vehicle. *Bacillus subtilis* is not potentially pathogenic, does not produce endotoxin, and excretes proteins in the medium. Although the technology for cloning in *B. subtilis* is not nearly as well developed as for *Escherichia coli*, plasmids suitable for cloning have been constructed. Transformation is a well-developed procedure in *B. subtilis*, and a number of bacteriophages that can be used as DNA vectors are known. Some distinct disadvantages to *B. subtilis* as a cloning host exist, however. Plasmid instability is a real problem, and it is hard to maintain plasmid replication over many culture transfers. Also, foreign DNA is not well maintained in *B. subtilis* cells, so that the cloned DNA is often unexpectedly lost. Finally, few good regulation systems are known in *B. subtilis*, making its use as an expression vector (see later) difficult.

Cloning in *eucaryotic microorganisms* has some important uses, especially for understanding details of gene regulation in eucaryotic systems. The yeast *Saccharomyces cerevisiae* is the best known genetically (see Section 7.11), and is being extensively studied as a cloning host. Plasmids are known in yeast, and transformation using genetically engineered DNA can be accomplished. However, gene expression in eucaryotes is not well understood, making the development of high-yielding systems difficult. The ability to clone appropriate genetic material in yeast will advance our understanding of the complex transcription and translation systems of eucaryotes, and so should provide a better foundation for basic research.

For many purposes, gene cloning in *mammalian cells* would be desirable. Mammalian cell culture systems can be handled in some ways like microbial cultures, and find wide use in research on human genetics, cancer, infectious disease, and physiology. Vaccine production (see Section 10.15) is also done

Hazards of Genetic Engineering

The development of the technology for moving DNA from one organism into another organism, called *recombinant DNA technology*, has had significant impact on science, business, and the lives of humans in general. The ability to study specific fragments of foreign DNA in a well-understood microorganism could, for example, aid scientists in understanding the molecular basis of many of the diseases that afflict humans. Industry was interested in such technology, because of its tremendous potential for improving production. For the first time, a fast-growing microorganism could be used to produce the proteins encoded by genes of a higher organism. For example, insulin, normally made by humans and other animals, can be produced in large amounts by a bacterium into which the human insulin gene has been inserted. The biotechnology industry has attracted extensive capital investment because of its potential to create new products of predicted high demand. The same recombinant DNA technology, however, could also be used either accidentally or intentionally to create new organisms harmful to humans. For instance, it is now possible to create a potentially harmful organism by combining characteristics of a particularly harmful microbe (e.g., a pathogen) and a normal inhabitant of the human body.

However, the scientists who developed this new technology recognized the potential dangers and took the first steps toward the regulation of recombinant DNA technology. Perhaps they saw a parallel with the development of destructive nuclear weapons which followed the discovery of nuclear fission. Following the scientists' early lead, official guidelines for recombinant DNA research were adopted by an agency of the United States government, the National Institutes of Health. With these guidelines in place, research on recombinant DNA technology has developed at an impressive rate, and the tools of recombinant DNA research now find wide use in almost all phases of biomedical investigation. However, guidelines for the *release* of genetically engineered microorganisms into the environment are still under development.

using mammalian cell cultures. The DNA virus SV40 (see Section 6.20), a virus causing tumors in primates, has been developed as a cloning vector into human tissue culture lines. SV40 virus has double-stranded *circular* DNA and the entire nucleotide sequence is known. Derivatives of SV40 which do not induce cancer have been developed which permit cloning of mammalian genes, and expression of these genes has been obtained. SV40 or similar mammalian cloning vectors should prove very useful in understanding the events involved in gene expression in these complex organisms.

The *retroviruses* (see Section 6.24) can also be used to introduce genes into mammalian cells, since these viruses replicate through a DNA form which becomes integrated into the host chromosome. It is also possible to obtain the incorporation of free DNA into mammalian cells under certain conditions, a process called *transfection*.

One important advantage of mammalian cells as hosts for cloning vectors is that they already possess the complex RNA and protein processing systems that are involved in the production of gene products in higher organisms, so that these systems do not have to be engineered into the vector, as they need to be if production of the desired product is to be carried out in a procaryote. A disadvantage of mammalian cells as hosts is that they are expensive and difficult to produce under the large-scale conditions needed for a practical system.

8.6 Finding the Right Clone

A crucial step in recombinant DNA technology is finding the right clone among the mixture of clones that have been created by the recombinant DNA procedure. The foreign DNA used in the cloning procedure will contain a large number of genes, only one of which will be the gene of interest. Although it is relatively easy to select for colonies which are maintaining the cloning vector (using antibiotic resistance markers of the vector), it is much more difficult to select the clone which has the gene of interest. Note that we are dealing here with procedures for examining colonies of bacteria growing on agar plates and detecting in these colonies small amounts of the protein or gene of interest. It is the purpose of the present section to discuss possible approaches to finding the right clone. We consider first the situation in which the protein is *expressed* in the cloning host. Then we discuss the situation, rather common, in which the protein is not expressed and we must look for the presence of the DNA itself.

If the foreign gene is expressed in the cloning host If the foreign gene is expressed (that is, synthesized) in the cloning host, then procedures can be used which look for the presence of this protein in recombinant colonies. The cloning host must *not* produce the protein being studied. If we are looking for clones which express the gene, then we are looking for the rare colonies in which this protein is present. If the protein is one that the cloning host normally produces, then the host used must be defective, that is, mutant, for the gene of interest. Then, when the foreign gene is incorporated, the expression of this foreign gene can be detected by complementation (see Section 5.1). Obviously, if the host already expressed a protein with the same activity, there would be a large background of this activity against which the protein produced via the foreign gene could not be detected. If the protein is a mammalian protein that is not normally produced in host bacteria, then the host may be naturally defective. A striking example is the cloning of luminescence genes from a marine bacterium into *Escherichia coli*. The desired clones glowed in the dark.

If the protein is an enzyme that carries out a measurable reaction, then it may be possible to screen colonies for this enzymatic activity. For instance, colonies producing the enzyme β-galactosidase can be detected by treating them with a β-galactoside which produces a colored compound when hydrolyzed. If the cloning host is lactose negative, nonrecombinants will be colorless and the rare recombinants will be visible as colored colonies or sectors (Figure 8.5).

Antibody as a method of detecting the protein If the protein is not an enzyme with a detectable function, then a different approach is needed. In the present section we discuss an approach which can potentially work with any protein of interest. It involves the use as a reagent of an antibody which is specific for the protein of interest. We will discuss antibodies and immunology in Chapter 12. For our present purposes, we note that an antibody is a serum protein produced by a mammalian system which combines in a highly specific way with another protein, the ***antigen***. In the present case, the protein of interest is the antigen, and this protein is used to produce in an experimental animal an antibody. Since the antibody combines specifically with the antigen, if the antigen is present in one or more colonies on the plate, then the locations of these colonies can be determined by observing the binding of the antibody. Because only a small amount of the protein (antigen) will be present in the colonies, only a small amount of antibody will be bound, so that a highly sensitive procedure for detecting bound antibody must be available. In practice, a system involving radioactive labelling is used, and the presence of bound antibody is detected by means of autoradiography using X ray film. Other extremely sensitive techniques for measuring antigen:antibody interactions are discussed in Section 12.9.

The whole procedure is outlined in Figure 8.6. As seen, the replica plating procedure (see Figure 5.41) is used to make a duplicate of the master plate, and all of the manipulations are done on this master plate. After the duplicate colonies have grown up, they are partially lysed to release the protein (antigen) of interest. The radioactive mixture containing the antibody is then added and the antibody-antigen reaction allowed to proceed. Unbound radioactivity is washed off and a piece of X ray film is placed over the plate and allowed to become exposed. If a radioactive colony is present, a spot on the X ray film will be observed after the X ray film was developed. The location of this spot on the film then corresponds to a location on the master plate where a colony is present which produces the protein. This colony can now be picked from the master plate and cultured.

One limitation of this procedure is that an antibody must be available which is *specific* for the protein in question. As we will see in Chapter 12, production of antibody can be readily done by injecting the protein (antigen) into an animal, but the protein injected must be pure, otherwise more than one antibody will be formed. Thus, one must have previously purified the protein. (The use of monoclonal antibodies simplifies this requirement a little; see Chapter 12.)

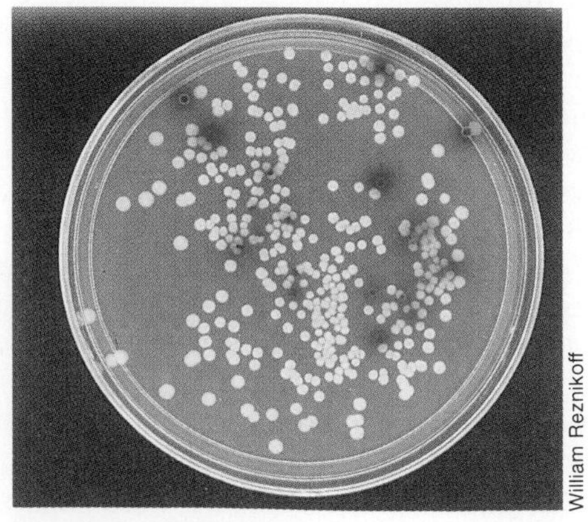

(a)

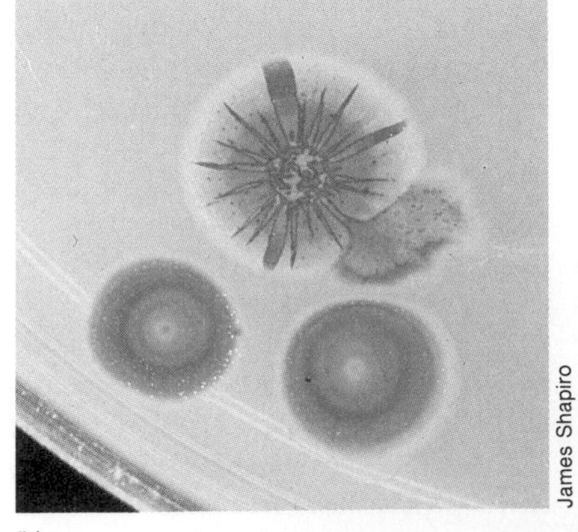

(b)

Figure 8.5 Detection of *lac*$^+$ colonies on an agar medium containing the compound *Xgal* (see Figure 8.3*b*). (a) A few *lac*$^+$ colonies or *lac*$^+$ sectors among a background of *lac*$^-$ colonies. (b) β-galactosidase production by *Escherichia coli* containing a *lac*$^+$ transposable element (Mud *lac*, derived from bacteriophage Mu; see Section 6.15). In the present case, the Mud *lac* transposon inserts randomly into the *E. coli* chromosome, and when it is in proper orientation in relation to a promoter, β-galactosidase is expressed. This explains the sectored appearance of the colonies.

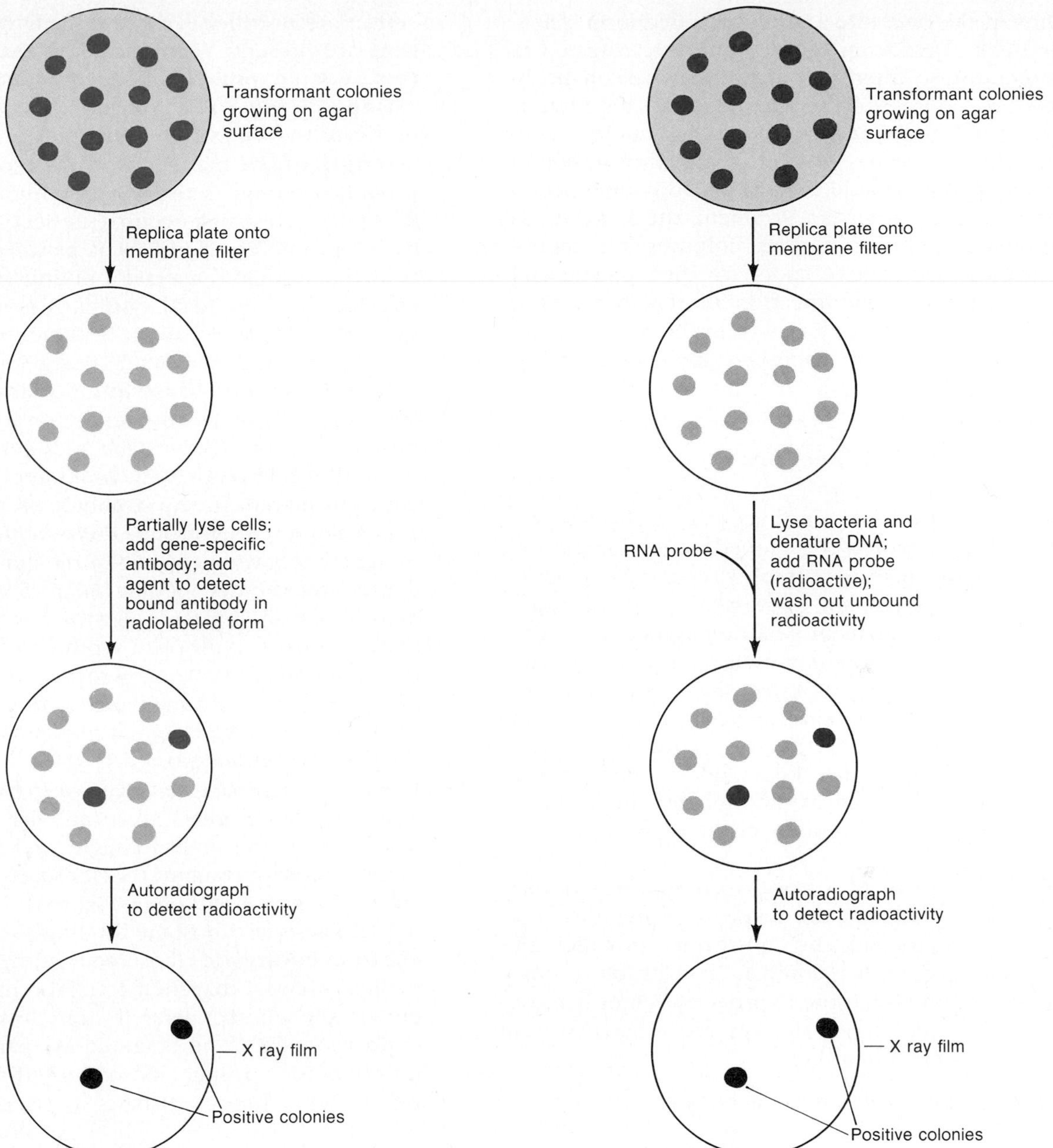

Figure 8.6 Method for detecting production of protein by use of specific antibody.

Figure 8.7 Method for detection of recombinant clones by colony hybridization with a radioactive RNA probe.

Nucleic acid probes: searching for the gene itself Suppose the gene is not expressed in the cloning host. How does one detect its presence in colonies? The most general way is to use a nucleic acid **probe** which contains a key part of the base sequence of the gene of interest. As we have discussed (see Section 5.2), nucleic acid hybridization can be used as a specific means of detecting polynucleotides with specific sequences. Either DNA or RNA can be used as probe. The general procedure is to have the probe nucleic acid in a labelled form, generally with radioactive phosphate, and allow single-stranded probe to hybridize with single-stranded nucleic acid derived from the gene. Because of specific complementary base pairing, two single-stranded polynucleotides will hybridize only if they are fairly complementary. By using appropriate hybridization conditions, it is possible to obtain binding of the radioactive probe only to the nucleic acid of interest.

The way in which a nucleic acid probe can be used to detect the presence of recombinant DNA in colonies is shown in Figure 8.7. The procedure, **colony hybridization**, again makes use of replica plating to produce a duplicate of the master plate. The colonies which have grown up on the duplicate plate are lysed in place to release their nucleic acid and to convert the DNA into a single-stranded form. Then, some of the DNA from each colony is transferred to a nitrocellulose filter on which the actual hybridization procedure will be carried out. This filter is then treated with radioactive probe nucleic acid (either RNA or DNA) and after removal of unbound radioactive nucleic acid the filter is now subjected to autoradiography. After development, the X ray film is examined for radioactive spots, colonies corresponding to these radioactive spots are then picked and studied further. A modification of this procedure, avoiding the use of a radioactive probe, has been developed for clinical microbiology and is discussed in Section 13.11.

8.7 Expression Vectors

For practical developments it is essential that systems be available in which the cloned genes can be *expressed*. Organisms have complex regulatory systems, and many genes are not expressed all of the time. One of the major goals of genetic engineering is the development of vectors in which high levels of gene expression can occur. An **expression vector** is a vector which not only contains the desired gene but also contains the necessary regulatory sequences so that expression of the gene is kept under control of the genetic engineer. Some of the elements involved in gene expression are summarized in Figure 8.8.

Requirements of a good expression system Many factors influence the level of expression of a gene, and a vector must be constructed in which all these factors are under control. In addition, a host must be used in which the expression vector is most effective. We summarize the key requirements of a good expression system here.

1. **Number of copies** of the genes per cell. In general, more product is made if many copies of the gene are present. Vectors such as small plasmids (for example, pBR322) and temperate phage (for example, *lambda*) are valuable because they can replicate to large copy number.
2. **Strength of the transcriptional promoter**. The promoter region is the site at which binding of RNA polymerase first occurs (see Section 5.8), and native promoters in different genes vary considerably in RNA polymerase binding strength. For engineering a practical system, it is important to include a *strong* promoter in the expression vector. The DNA region around 10 and 35 nucleotides before the start of transcription (called the −10 and −35 *regions*) is especially important in the promoter. Many *Escherichia coli* genes are controlled by relatively weak promoters, and eucaryotic promoters function poorly or not at all in *Escherichia coli*. Strong *Escherichia coli* promoters that have been used in the construction of expression vectors include *lac*uv 5 (which normally controls β-galactosidase), *trp* (which normally controls tryptophan synthetase), *tac*, a synthetic hybrid of the *trp* and *lac* promoters, *lambda* P_L (which normally regulates *lambda* virus production), and *ompF* (which regulates production of outer membrane protein).
3. Presence of the bacterial **ribosome binding site**. The transcribed mRNA must bind firmly to the ribosome if translation is to begin, and an early part of the transcript contains the ribosome-binding site (Shine-Dalgarno sequence, Figure 8.8 and Section 5.10). Bacterial ribosome binding sites are different from eucaryotic ribosome binding sites and it is thus essential that the bacterial region be present in the cloned gene if high levels of gene expression are to be obtained. As part of the requirement for proper ribosome binding is the necessity for a proper distance between the ribo-

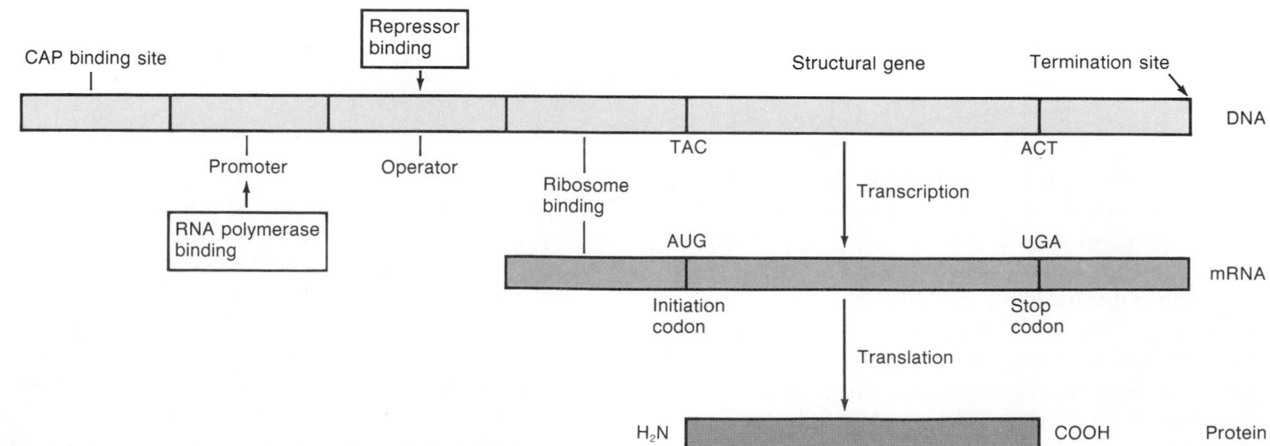

Figure 8.8 Factors affecting the expression of cloned genes in bacteria. Sequences and signals that must be appropriate for high levels of gene expression are indicated.

some binding site and the translation initiation codon. If these sites are too close or too far apart, the gene will be translated at low efficiency.

4. Avoidance of **catabolite repression** (see Section 5.12).
5. Proper **reading frame**. Because of the way the source DNA is being fused into the vector, three possible reading frames (see Section 5.17) could be obtained, only one of which is satisfactory. One approach to obtaining the correct reading frame is the use of three vectors, each having the restriction site into which new DNA will be inserted positioned such that the insert will be in a different reading frame. The gene is inserted into all three vectors and the one which gives proper expression is selected by testing.
6. **Codon usage**. There is more than one codon for most of the 20 amino acids (see Table 5.4), and some codons are used more frequently than others. Codon usage is partly a function of the concentration of the appropriate tRNA in the cell. A codon frequently used in a mammalian cell may be used less frequently in the organism in which the gene is being cloned. Insertion of the appropriate codon would be difficult because it would have to be changed in all locations in the gene. One approach would be to engineer the host so that it used the required codon more frequently.
7. **Fate of the protein** after it is produced. Some proteins are susceptible to degradation by intracellular proteases and may be destroyed before they can be isolated. Excreted proteins must have the signal sequence attached (see Section 5.10) if they are to move through the cell membrane. Some eucaryotic proteins are toxic to the procaryotic host, and the host for the cloning vector may be killed before sufficient amount of the product is synthesized. Further engineering of either the host or the vector may be necessary to eliminate these problems.

The skill of the genetic engineer is thus essential in the construction of an appropriate vector which can be (1) efficiently incorporated into the proper host, (2) replicated to high copy number, (3) efficiently transcribed, and (4) efficiently translated. Many mammalian proteins are completely unexpressed when their genes are first cloned in *Escherichia coli*, but with appropriate manipulation of the vector expression can sometimes be achieved. The best example is the production on a commercial scale of human insulin in *Escherichia coli*, as described in Section 8.10.

The role of regulatory switches in expression vectors Unfortunately, many proteins that are of commercial interest are toxic to the bacterial hosts. Therefore, for large-scale production of a protein, it is very desirable that the synthesis of the protein be under the direct control of the experimenter. The ideal situation is to be able to grow up the culture containing the expression vector until a large population of cells is obtained, each containing a large copy number of the vector, and then turn on expression in all copies simultaneously by manipulation of a regulatory switch.

We have discussed regulatory controls of gene expression in Sections 5.11 and 5.12. Recall the major importance of the repressor/operator system in regulating gene transcription. A strong repressor can completely block the synthesis of the proteins under its control by binding to the operator region. Repressor function can be turned off at the chosen time by adding an inducer, allowing the transcription of the proteins controlled by the operator.

For the repressor-operator system to work as a regulatory switch for the production of a foreign protein, it is desirable to retain in the expression vector a fragment of the structural gene and the operator controlled by the repressor, to which the source gene is fused. This permits proper arrangement of the sequence of genetic elements: promoter-operator-ribosome binding site-structural gene so that efficient transcription and translation can occur. The presence of a fragment of the normal protein can help render the foreign protein stable and capable of being excreted.

The construction of plasmid expression vectors containing the regulatory components of the *lac* operon provides one means of providing a suitable regulatory switch. As we have discussed in Section 5.11, the *lac* operon is switched on by inducers such as lactose or related β-galactosides. Phasing of cell growth and protein synthesis can thus be achieved by allowing growth to proceed in the absence of inducer until a suitable cell density is achieved, then adding inducer to bring about protein synthesis. Plasmids have been constructed containing the *lac* promoter, ribosome binding site, and operator. When the desired gene is inserted into such a system, expression can then be achieved by adding *lac* inducer. Production of high levels of mammalian proteins have been achieved in *Escherichia coli* using such expression vectors (for instance: 10,000 to 15,000 molecules of beta globin per cell; 5000 to 10,000 molecules of human interferon per cell).

Using *lambda bacteriophage*, it is possible to integrate the engineered *lambda* genome into the *Escherichia coli* chromosome, then induce replication of the *lambda* cloning vector by inactivating the *lambda* repressor. Under suitable conditions, replication of *lambda* will be followed by expression of the gene for the foreign protein. For large-scale production, a more suitable system for *lambda* induction is the use of a *lambda* repressor which is temperature sensitive. By raising the temperature of the culture to the proper value (usually 8 to 10°C higher than the growth temperature), the *lambda* repressor is inactivated and *lambda* replication begins. Once released from repression, other genes that have been inserted into the *lambda* genome will also be expressed.

8.8 The Cloning and Expression of Mammalian Genes in Bacteria

We have now laid out the principles behind the development of systems for obtaining the production of foreign genes in an organism such as *Escherichia coli*. How are these principles put together in the engineering of a desired expression system? In the human or some other complex organism, the sought-for gene will be buried within a large mass of DNA containing unwanted genes, as well as repetitive DNA sequences that have no coding function at all (see Section 5.5). Also, most eucaryotic genes are split, with noncoding introns interspersed among the coding exons (see Section 5.1). How can a gene be selected from this complex mixture? Although a number of approaches are available, this is the least well developed stage of genetic engineering. Not only skill, but also intuition and good luck play big roles in a successful outcome. A summary of approaches and procedures is given in Figure 8.9.

We start first with a consideration of the desired product and where it is produced in the human body. A human being is a highly differentiated organism, and many genes are expressed only in certain organs or tissues. The hormone *insulin*, for instance, is produced only in the pancreas. Of course, the insulin gene is found in all tissues and organs, but it is only *expressed* in this one organ.

Reaching the gene via messenger RNA One approach is to get to the gene through its mRNA. A major advantage of using mRNA is that the noncoding information present in the DNA (introns) has been removed (see Section 5.8). The isolated mRNA is used to make complementary DNA (cDNA), by means of reverse transcription (see Section 6.24). It is likely that a tissue expressing the gene will contain large amounts of the desired mRNA, although except in rare cases this will certainly not be the only mRNA produced. In a fortunate situation, where a single mRNA dominates a tissue type, extraction of mRNA from that tissue provides a useful starting point for gene cloning.

In a typical mammalian cell, about 80 to 85 percent of the RNA is ribosomal, 10 to 15 percent is transfer RNA and other low-molecular-weight RNAs, and 1 to 5 percent is messenger RNA. Although low in abundance, the mRNA in a eucaryote is identifiable because of the poly A tails found at the 3′ end (see Section 5.8). In maturing red blood cells, for instance, where virtually the only protein made is the globin portion of hemoglobin, from 50 to 90 percent of the poly A-containing cytoplasmic RNA consists of globin mRNA. By passing a poly A-rich RNA extract over a chromatographic column containing poly T fragments (linked to a cellulose support), most of the mRNA of the cell can be separated from the other cellular RNA by the specific pairing of A and T bases. Elution of the RNA from the column then gives a preparation greatly enriched in mRNA.

Reaching the gene via the protein In most cases, the sought-for messenger RNA is present in only a small amount, mixed with many other mRNAs. Under these conditions it cannot be isolated directly. Two basic approaches, reverse translation and polyribosome precipitation, will be discussed here.

The most widely used approach to cloning low-abundance mRNA is to make a **synthetic DNA** which is complementary to part of the mRNA, and then use this DNA as a *probe* in a Northern blot procedure (see Box, Chapter 5) to pull out by hybridization the mRNA of interest. This requires that a partial or complete sequence of the protein be known. Then, from a consideration of the genetic code, the nucleotide sequence of a section of the DNA is deduced, and this piece of DNA is synthesized.

The procedure by which the nucleotide sequence is deduced from the protein sequence is called **reverse translation**. (Note that *reverse translation* is not a cellular process but a mental exercise of the genetic engineer.) The procedure of reverse translation is illustrated in Figure 8.10. From the genetic code, the nucleotide sequence of a section of the DNA is deduced, and this piece of DNA is synthesized. Unfortunately, degeneracy of the genetic code (see Section 5.15) somewhat complicates the problem. Most amino acids are coded for by more than one codon, and codon usage varies from organism to organism. The best section of DNA to synthesize is one which corresponds to a part of the protein rich in amino acids specified by only a single codon (methionine, AUG; tryptophan, UGG) or by two codons (phenylalanine; UUU, UUC; tyrosine; UAU, UAC; histidine; CAU, CAC) since this will increase the chances that the synthesized DNA will be complementary or nearly complementary to the mRNA of interest. If the complete amino acid sequence of the protein is not known, then the sequence used is generally one at the amino terminus of the protein, since it is at the amino terminus that sequencing of the protein begins.

Polyribosome precipitation involves the separation from the tissue of *ribosome complexes* that are in the process of synthesizing the desired protein, using an antibody specific for this protein. As we describe in Chapter 12, antibodies are proteins made in response to foreign proteins which are able to combine with and specifically precipitate them. How is an antibody directed against a *protein* used to detect

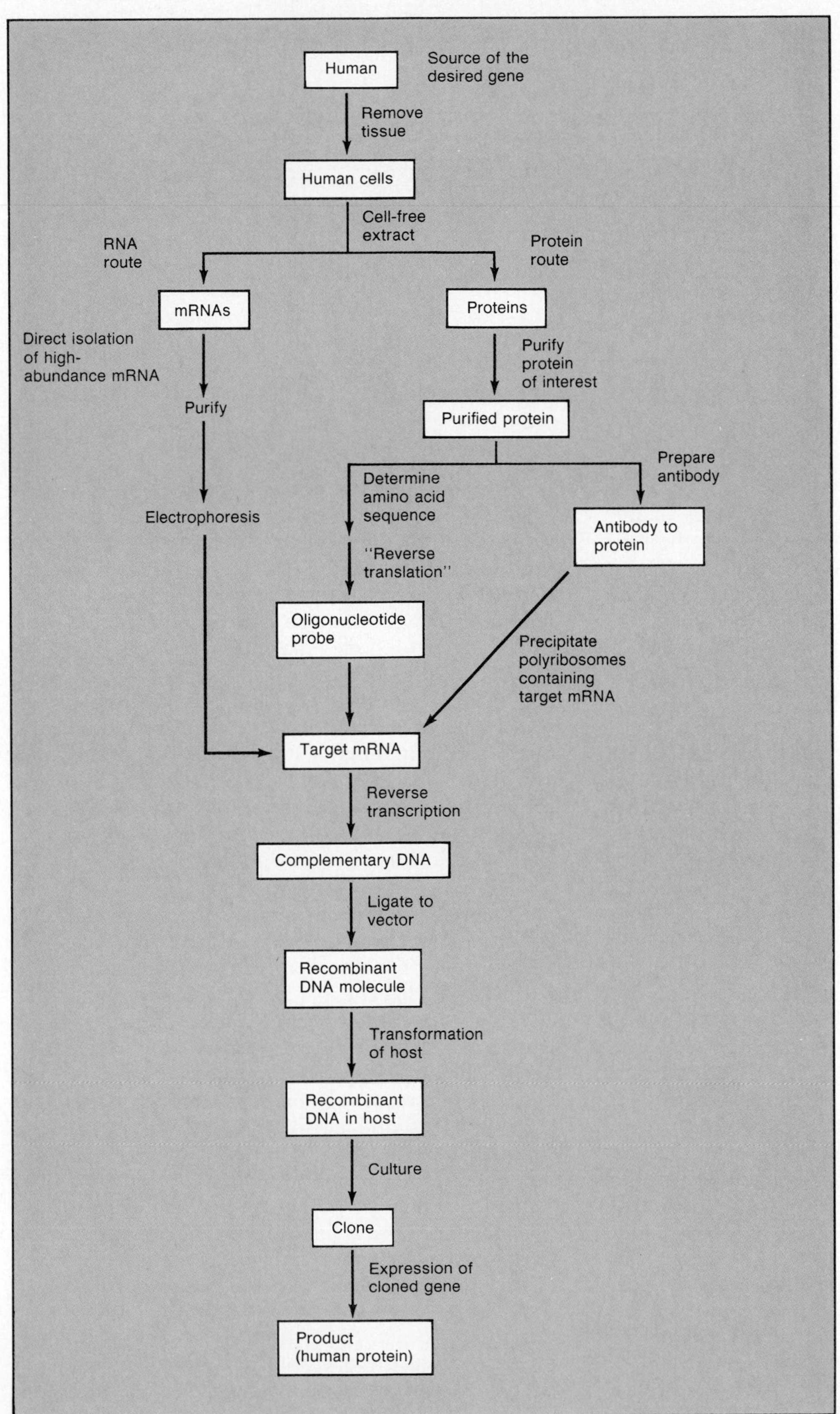

Figure 8.9 Several routes to the isolation of mammalian genes.

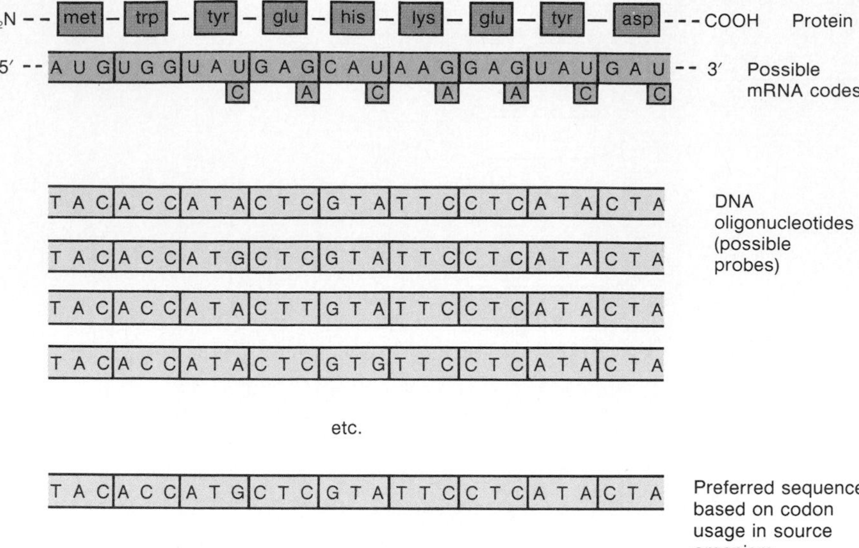

Figure 8.10 Reverse translation: deducing the best sequence of an oligonucleotide probe from the amino acid sequence of the protein. Because of degeneracy, many probes are possible. If codon usage by the same organism is known, then a preferred sequence can be selected. It is not essential that complete accuracy is achieved, since a small amount of mismatch can be tolerated.

an *mRNA*?

In the ribosome complex, polypeptides of the protein of interest are still complexed with the protein-synthesizing machinery (which contains, among other things, the sought-for mRNA). When the antibody precipitates the protein in the ribosome complex, the mRNA also is precipitated. After isolation of the mRNA, it can be used to prepare complementary DNA as described in this section.

Synthesis of complementary (cDNA) from mRNA Once the RNA message has been isolated, it is necessary to convert the information into DNA. This is accomplished by use of the enzyme **reverse transcriptase**, which we have discussed in Section 6.24. This remarkable enzyme, an essential component of retrovirus replication, copies information from RNA into DNA (Figure 8.11). As we noted, this enzyme requires a primer in order for it to begin working (in retrovirus infection the primer is a tRNA). In the present procedure, an oligo-dT primer is used which is complementary to the poly A tail of the mRNA. The oligo-dT primer is hybridized with the mRNA and then reverse transcriptase is allowed to act (Figure 8.11). As seen, the newly synthesized DNA copy has a hairpin loop at its end, arising because when the enzyme completes copying the mRNA it starts to copy the newly synthesized DNA. This hairpin loop, which is probably an artifact of the test tube reaction, provides a convenient primer for the synthesis of the second DNA strand. The resultant double-stranded DNA, with the hairpin loop intact, is then cleaved by a single-strand-specific nuclease to produce the sought-for double-stranded DNA complementary to the mRNA. This double-stranded DNA can then be inserted into a plasmid or other vector for cloning. The detection of specific clones makes use of the procedures discussed in Section 8.6.

Synthesis of the complete gene If the protein is small enough, or is of sufficient economic interest to justify a major effort, the complete gene can be synthesized chemically. Chemical synthesis not only permits the acquisition of genes that cannot be obtained otherwise, but also permits synthesis of *modified genes* which may make new proteins of utility. The techniques for the synthesis of DNA molecules are now well developed, and it is possible to synthesize genes coding for proteins 100 to 200 amino acid residues in length (300 to 600 nucleotides). The synthetic approach was used for the production of the human hormone insulin in bacteria, as discussed in Section 8.10. We will discuss the use of synthetic DNA in mutagenesis work in a later section.

Expression of mammalian genes in bacteria For the cloned mammalian gene to be expressed in a bacterium, it is essential that the mammalian gene be inserted adjacent to a strong *promoter* and that a bacterial *ribosome-binding site* be present. Also, it is essential that the *reading frame* be correct.

One method of providing a ribosome-binding site in the proper reading frame is to arrange for the mammalian DNA sequence to be expressed as part of a **fusion protein** which contains a short procaryotic sequence at the amino end and the desired eucaryotic sequence at the carboxyl end. Although less desirable for some purposes, fusion proteins are often more stable in bacteria than unmodified eucaryotic proteins, and the bacterial peptide portion can often be removed by chemical treatment after purification of the fused protein. One advantage of making a fusion protein is that the bacterial portion can contain the bacterial sequence coding for the signal peptide that enables transport of the protein across the cell membrane (see Section 5.10), making possible the development of a bacterial system which not only syn-

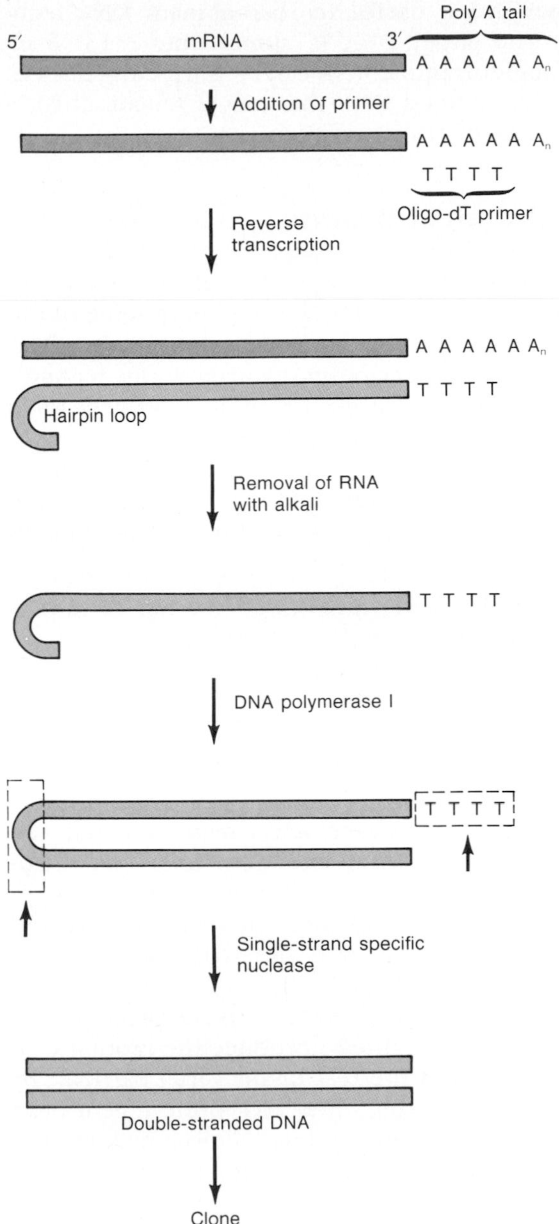

Figure 8.11 Steps in the synthesis of complementary DNA (cDNA) from an isolated mRNA, using the retroviral enzyme reverse transcriptase.

thesizes the mammalian protein but actually excretes it. However, the main advantage of making a fusion protein is that the reading frame is ensured to be correct.

To obtain an expression vector that will bring about the synthesis of an *unfused* protein, it is essential that the eucaryotic gene be fused exactly to the initiation codon that is just downstream from the ribosome-binding site. In many mammalian proteins, the mature and active form lacks some of the amino acids present at the initiation site because they are cleaved after synthesis by a post-translational modification. Thus, to obtain synthesis of the active form of the protein directly, it is necessary to fuse to the initiation codon just the sequence coding for the final protein, and place this in proper position downstream of the ribosome-binding sequence. If high efficiency of translation is to be obtained, an intervening region of noncoding DNA of the proper length must be placed between the ribosome-binding sequence and the initiation codon. Several plasmids have been constructed which contain built-in promoters and suitable restriction enzyme sites, so that the proper tailoring of coding sequence to initiation site and ribosome-binding site can be achieved. Human proteins that have been expressed at high yield under the control of bacterial regulatory systems include human growth hormone, insulin, virus antigens, interferon, and somatostatin (see Section 8.10).

8.9 In Vitro and Site-Directed Mutagenesis

Recombinant DNA technology has opened up a whole new field of mutagenesis. Whereas conventional mutagens (see Section 5.18) act at random, by use of synthetic DNA and recombinant DNA techniques it is possible to construct mutants at precisely determined sites on genes. Proteins made from such mutants can be expected to have different properties than the wild-type proteins, properties that may be predicted from a knowledge of protein structure. Two approaches will be discussed here: total synthesis of mutants, and site-directed mutagenesis using synthetic oligonucleotides.

Total synthesis of mutants Synthetic DNA technology can be used to synthesize a complete gene with a genetic change inserted at the desired location. Although a complete gene of 3000 or so nucleotides would be difficult to synthesize in one piece, it is possible to synthesize smaller portions of the gene and then to link these together to produce the final gene. By addition of appropriate sites at the ends of the gene, this gene can then be linked into a vector and cloned. The possibilities of this approach seem virtually limitless, although because of the expense involved, it is essential that one have a carefully considered rationale for the particular sequence.

Site-directed mutagenesis This approach, although more indirect than that just described, is much less expensive and provides for considerable flexibility. The basic procedure is to synthesize a short oligodeoxyribonucleotide containing the desired base change and to allow this to pair with a single-stranded DNA containing the gene of interest. Pairing will be complete except for the short region of mismatch. Then the short single-stranded fragment of the synthetic oligonucleotide is now extended using DNA polymerase, copying the rest of the gene. The double-stranded molecule then obtained is inserted into a cloning host by transformation and mutants selected by a procedure already described (see Section 8.6). The mutant obtained is then used in the production of the modified (mutant) protein.

The whole procedure of site-directed mutagenesis is illustrated in Figure 8.12. As seen, one must begin with the gene of interest cloned into a single-stranded DNA. A widely used vector for site-directed mutagenesis is bacteriophage M13, which, as we have seen (Section 6.11 and Section 8.4) has some properties which are useful in recombinant DNA technology. The target DNA is cloned into M13, from which single-stranded DNA can be purified with ease. Since cells infected with this phage remain alive, a ready source of DNA is available.

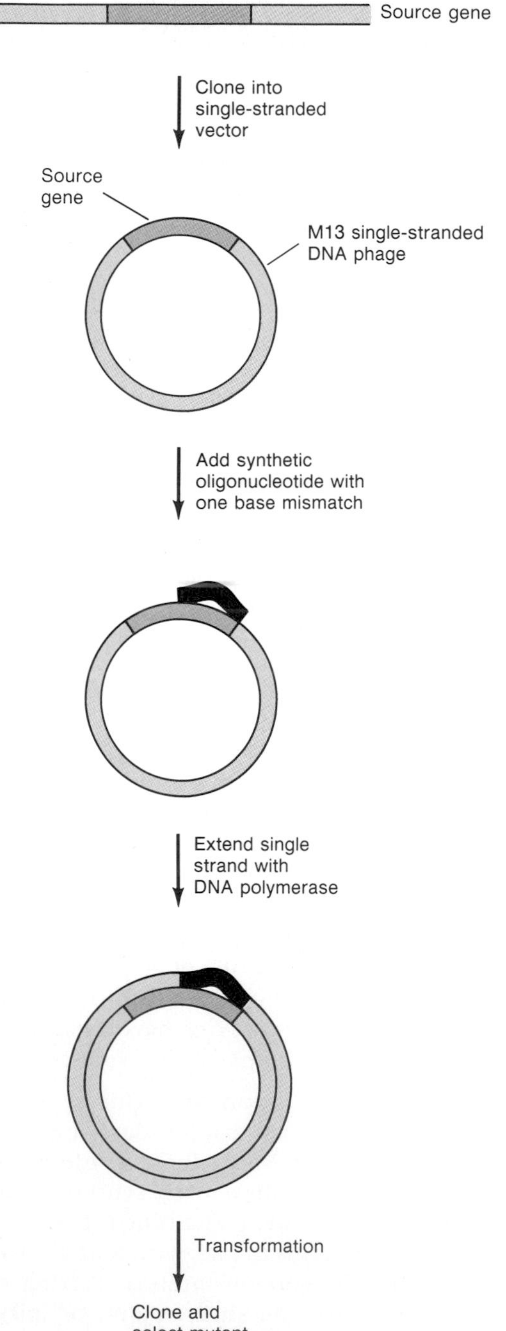

Figure 8.12 Site-directed mutagenesis, using short synthetic oligodeoxyribonucleotide fragments.

8.10 Practical Results

We discuss many of the specifics of microbial biotechnology in Chapter 10. We note here some of the main practical implications of recombinant DNA technology, and describe in detail how this technology was used to develop a practical process for a particular protein, human insulin.

A number of commercial products have been developed using genetic engineering techniques. Three main areas of interest for commercial development are:

1. **Microbial fermentations.** A number of important products are made industrially using microorganisms, of which the antibiotics are the most significant (see Section 10.6). Genetic engineering procedures can be used to manipulate the antibiotic-producing organism in order to obtain increased yields.
2. **Virus vaccines.** A vaccine is a material which can induce immunity to an infectious agent (see Chapter 12). Frequently, killed virus preparations are used as vaccines and there is always a potential danger to the patient if the virus has not been completely inactivated. Since the active ingredient in the killed virus vaccine is the protein coat, it would be desirable to produce the protein coat separately from the rest of the virus particle. By genetic engineering, viral coat protein genes can be cloned and expressed in bacteria, making possible the development of safe, convenient vaccines.
3. **Mammalian proteins.** A number of mammalian proteins are of great medical and commercial interest. Some of these are discussed in Section 10.15. In the case of human proteins, commercial production by direct isolation from tissues or fluids is complicated and expensive, or even impossible. By cloning the gene for a human protein in bacteria, its commercial production is possible.

In the rest of this section we discuss several applications of genetic engineering, including the production of viral vaccines, the production of human insulin, and the potential for genetic engineering advances in veterinary medicine, agriculture, and human genetics.

Virus vaccine As we discussed in Chapter 6, a virus particle consists of a nucleic acid core surrounded by a coat containing one or more specific proteins. For vaccine purposes, it is the protein coat which is of interest and the gene for the virus protein is cloned.

Virus particles are generally produced by growing the virus in tissue culture. From a purified virus preparation, the nucleic acid of the virus is isolated. If a DNA virus is under study, then direct cloning of the virus DNA can be undertaken. With an RNA virus, a complementary DNA (cDNA) must first be made. The isolated nucleic acid is then fragmented with restriction enzymes and the fragments inserted into a suitable cloning vector, generally a plasmid, using DNA ligase. It is essential to find a restriction enzyme which does not cut within the gene of interest. Provision for proper reading frame and ribosome-binding site must be arranged, as outlined in Figure 8.10. The hybrid plasmids are then inserted into bacteria by transformation and an antibiotic resistance character is used to select for plasmid-containing colonies. A large number of colonies are isolated and each one is tested for the production of the virus antigen. Once a clone of the virus protein gene has been obtained, further manipulation can be used to increase yield of the protein, and make it easier to purify. The ultimate goal is the development of a system which produces the antigenic virus protein in high yield. The purified protein is then used as a vaccine in humans without the dangers attendant on using killed virus particles.

Human insulin One of the most dramatic examples of the value of genetic engineering is the production of human insulin. Insulin is a protein produced in the pancreas that is vital for the regulation of carbohydrate metabolism in the body. Diabetes, a disease characterized by insulin deficiency, afflicts millions of people. The standard treatment for diabetes is periodic injections of insulin, and because insulins of most mammals are similar in structure, it is possible to treat human diabetes by use of insulin isolated commercially from beef or pork pancreas. However, nonhuman insulin is not as effective as human insulin, and the isolation process is expensive and complex. Cloning of the human insulin gene in bacteria has hence been carried out.

Insulin in its active form consists of two polypeptides (A and B) connected by disulfide bridges (Figure 8.13*a*). These two polypeptides are coded by separate parts of a single insulin gene. The insulin gene codes for *preproinsulin*, a longer polypeptide containing a signal sequence (involved in excretion of the protein), the A and B polypeptides of the active insulin molecule, and a connecting polypeptide that is absent from mature insulin. *Proinsulin* is formed from preproinsulin and the conversion of proinsulin to insulin involves the enzymatic cleavage of the connecting polypeptide from the A and B chains.

Two approaches have been used to obtain production of human insulin in bacteria: (1) production of proinsulin and conversion to insulin by chemical cleavage, and (2) production in two separate bacterial cultures of the A and B chains, and joining of the two chains chemically to produce insulin. Because the insulin protein is fairly small, it was more convenient with either approach to synthesize the proper DNA sequence chemically rather than to attempt to isolate the insulin gene from human tissue. There are 63 bases coding for the A chain and 90 bases coding for the B chain (Figure 8.13*b*). In proinsulin there are an additional 105 bases for the peptide which connects the A and B chains. When the polynucleotides were synthesized, suitable restriction enzyme sites were placed at each end so that the polynucleotides could be ligated to plasmid pBR322. To obtain effective expression, the synthesized genes were fused to suitable *Escherichia coli* promoters, either *lac* or *trp*, in a manner such that a portion of the β-galactosidase or tryptophan synthetase protein was synthesized. An important advantage of making the fused protein is that the fusion product is much more stable in *Escherichia coli* than insulin itself. The *trp* fusion in particular results in the formation of an insoluble protein that precipitates inside *Escherichia coli*, thus preventing *Escherichia coli* proteases from breaking it down. Finally, a nucleotide triplet coding for methionine was placed at the region joining the *trp* or *lac* gene to the insulin gene. The reason for this is that the chemical reagent *cyanogen bromide* specifically cleaves polypeptide chains at methionine residues, permitting recovery of the insulin product once the fused protein has been isolated from the bacteria. Insulin itself does not contain methionine and hence is unaffected by cyanogen bromide treatment.

If the proinsulin route is used, then the proinsulin isolated from the bacteria via cyanogen bromide treatment is converted to insulin by disulfide bond formation, followed by enzymatic removal of the connecting peptide of proinsulin. Proinsulin naturally folds so that the cysteine residues are opposite each other (Figure 8.13*a*), and chemical treatment then causes the formation of disulfide cross links. Once this has been accomplished, the connecting peptide can be removed by treatment with the proteases trypsin and carboxypeptidase B, which have no effect on insulin itself.

If insulin is produced by way of the separate A and B peptides, then each of the fusion proteins is isolated from a separate bacterial culture and the chains then released by cyanogen bromide cleavage. The cleaved chains are then connected by use of chemical treatment that results in disulfide bond formation. Under appropriate conditions, a yield of at least 60 percent of the theoretical yield can be obtained.

The final product, biosynthetic human insulin, is identical in all respects to insulin purified from the human pancreas, and can be marketed commercially. The principles discovered in the development of the insulin process should find wide use in other applications of genetic engineering.

Plant agriculture Recombinant DNA technology has many uses in agriculture. Genetic manipulation of plants has conventionally been a slow and difficult task, but recombinant DNA technology promises revolutionary changes. It is possible to use plant cell culture procedures to select clones of plant cells that are genetically changed and then, with proper treatments, induce these cell cultures to make whole plants which can be propagated vegetatively or by seeds. Recombinant DNA technology enters into this approach because one can transform plant cells with free DNA, or insert foreign DNA into plant cells via the bacterium *Agrobacterium tumefaciens* (discussed in Chapter 17). This organism causes crown gall (Section 17.23), a disease resulting in the formation of tumors. The bacterium contains a large plasmid called the Ti plasmid that is responsible for virulence. Part of this plasmid becomes integrated into the plant DNA where its replication then comes under the control of the plant. The Ti plasmid can thus be used as a vehicle for introducing foreign genes into plant cells, and since it can be grown both in the bacterium and in the plant it appears to have considerable promise as a cloning vector. This permits transferring plant genes across plant species barriers, making it possible to get around some of the

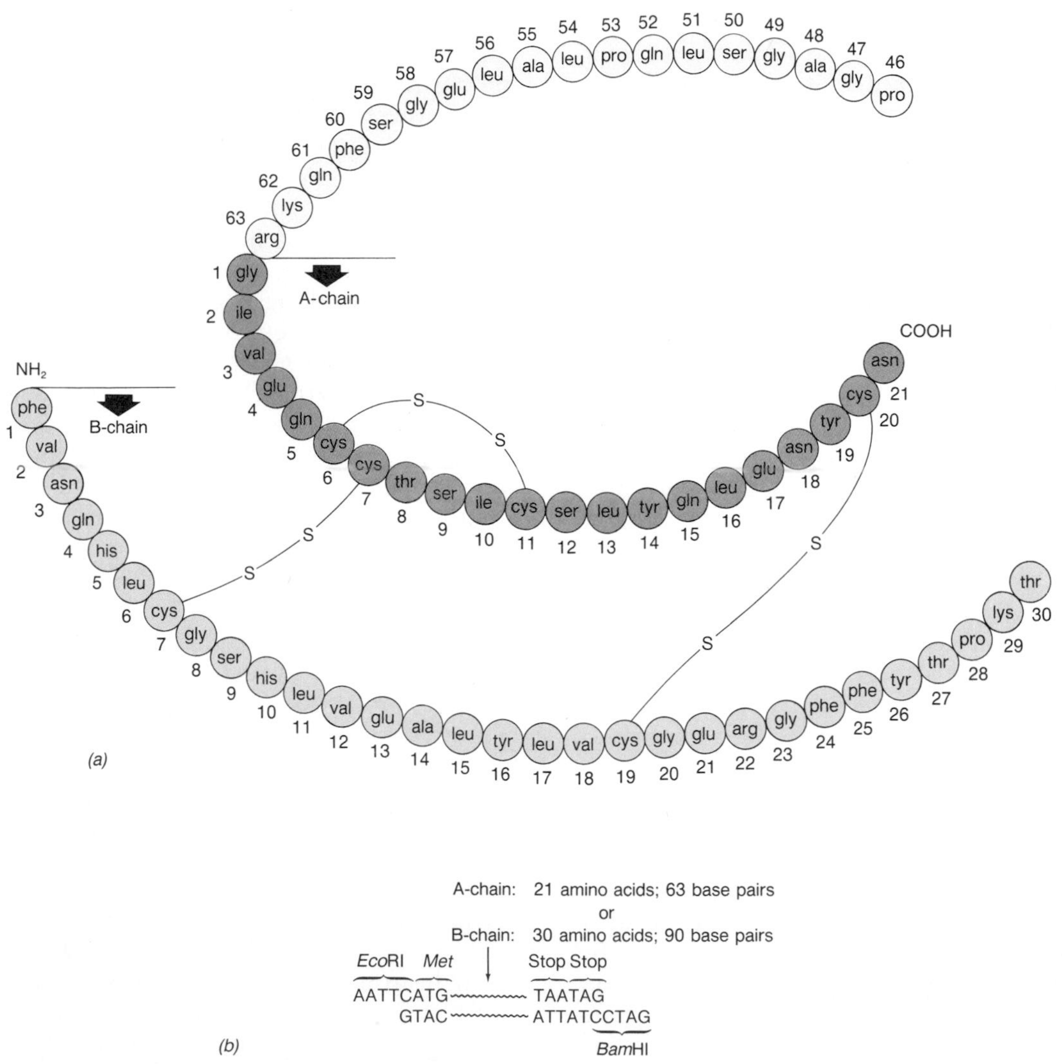

Figure 8.13 Genetic engineering for the production of human insulin in bacteria. (a) Structure of human proinsulin. (b) Chemical synthesis of the insulin gene and suitable linkers, permitting cloning and expression. The synthesized fragments were linked via restriction sites EcoRI and BamHI to plasmid pBR322. The methionine coding sequence was inserted to permit chemical cleavage of the A and B chains from the fused protein made in the bacteria, since the reagent cyanogen bromide specifically cleaves at methionine residues and insulin does not contain methionine. Two stop signals were incorporated at the downstream (carboxyl) end of the coding sequence.

complications that were present in conventional plant breeding.

Human genetics Recombinant DNA technology is finding wide use in research on human genetics. A detailed discussion of human genetics is beyond the scope of this book, but a few general remarks on the utility of recombinant DNA technology can be made.

There are several areas in human genetics in which recombinant DNA technology finds utility. We have already discussed cancer and the role of viruses in carcinogenesis in Chapter 6. Another important area is in the study of genetic diseases of humans. A vast number of such diseases are known, but except in rare cases little is known about the molecular bases. By use of recombinant DNA technology, coupled with conventional genetic studies (following family inheritance, etc.), it is possible to localize particular defects to particular chromosomes and to particular locations on chromosomes. With the use of recombinant DNA technology it is possible to clone the region containing the genetic defect and then to make comparisons between the base sequence in wild type and in genetically altered chromosomes. From such studies, even in the absence of knowledge of the enzyme defect, it may be possible to obtain information about the genetic change. In a sense, this is looking for a gene without knowing what the gene does, an approach which is backwards from that normally used in molecular genetics, but an approach which is essential in an organism such as *Homo sapiens* where conventional genetics (breeding, mutagenesis) cannot be done. Already, the genes for Huntington's disease and Duchennes muscular dystrophy have been localized using these techniques.

8.11 Genetic Engineering as a Microbial Research Tool

Up until now we have presented the principles of genetic engineering in the light of practical goals and processes. However, genetic engineering technology finds many uses in basic research on microorganisms. Gene cloning and the engineering of new microbial strains provides some of the best ways in which to understand basic microbial processes such as structure-function relationships, cell growth, enzyme regulation, and microbial ecology. A novel microbial strain can be created which differs from the wild type in a defined manner, and the underlying process can then be studied. In this way, one can observe the importance of a particular gene product for a basic microbial process under precisely controlled conditions.

Genetic engineering also permits the researcher to "tag" an organism in such a way that it can be readily traced through the environment. For instance, *Escherichia coli* has been engineered in such a way that it produces the luciferase enzyme of *Photobacterium*. Because of the presence of this enzyme, the engineered *Escherichia coli* becomes luminescent, and its colonies therefore glow in the dark. One can, then, detect on agar plates colonies of the engineered *Escherichia coli* by their luminescence among a large background of other colonies. This glow-in-the-dark system can be used in a wide variety of genetically engineered microorganisms.

Another approach to tracking organisms through the environment has been to introduce the β-galactosidase gene and use indicator agar plates (see Section 8.6) to search for colonies producing this enzyme (blue colonies).

8.12 Principles at the Basis of Genetic Engineering

We have presented the fundamentals of genetic engineering and have shown how the approaches used have been derived from an understanding of basic concepts of molecular genetics. We now summarize the principles of genetic engineering by relating current knowledge back to the basic information presented in Chapters 5 through 7.

The following developments were essential for the perfection of genetic engineering:

1. **DNA chemistry**: development of procedures for isolation, sequencing, and synthesis of DNA.
2. **DNA enzymology**: discovery of restriction endonucleases, DNA ligases, and DNA polymerases.
3. **DNA replication**: understanding how DNA replication occurs, and the importance of DNA vectors capable of independent replication.
4. **Plasmids**: discovery of plasmids, and determination of the mechanisms by which plasmids replicate.
5. **Temperate bacteriophage**: understanding how replication and/or integration is controlled in the DNA of temperate bacteriophages.
6. **Transformation**: discovery of methods for getting free DNA into cells.
7. **RNA chemistry and enzymology**: understanding how to work with messenger RNA, how eucaryotic mRNA is constructed, and the importance of RNA processing in the formation of mature eucaryotic mRNA.
8. **Reverse transcription**: the discovery of the enzyme *reverse transcriptase* in retroviruses and its development as a means for transcribing information from mRNA back into DNA.
9. **Regulation**: understanding the factors involved in the regulation of transcription, including the discovery of promoter sites and operon control.

10. **Translation**: understanding the steps involved in translation, the importance of ribosome-binding sites on the mRNA, the role of the initiation codon, and the importance of a proper reading frame.
11. **Protein chemistry**: development of methods for isolation, purification, assay, and sequencing of proteins.
12. **Protein excretion and post-translational modification**: understanding how proteins are built with signal sequences that are removed during or after excretion. Discovery of other kinds of post-translational modification of proteins, such as the removal of polypeptides at the initiation end of the protein.
13. **The genetic code**: the discovery of the genetic code and the determination that it was the same in all organisms. The understanding of the importance of proper reading frame, and that certain codons were less frequently used in some organisms than in others.

Study Questions

1. What are the characteristics of plasmids that make them especially useful for gene cloning? Why aren't all plasmids equally useful for cloning?
2. What are the characteristics of temperate bacteriophages that make them especially useful for gene cloning? Why aren't virulent bacteriophages suitable for gene cloning?
3. What are the essential features of a cloning vector? Expression vector? Why aren't all cloning vectors also expression vectors?
4. If *insertional inactivation* is used to detect the presence of an introduced plasmid in a bacterial cell, why is it desirable to have *two* antibiotic-resistance markers in the plasmid?
5. If a plasmid is used as a cloning vector, why is it essential that only a few sites be present that are recognized by the restriction enzyme used for the cloning process? What would be the result if the plasmid contained a large number of recognition sites for this restriction enzyme?
6. Suppose you were given the task of constructing a plasmid suitable for gene cloning in an organism of industrial interest. List the characteristics which such a plasmid should have. List the steps you would use to develop such a plasmid.
7. Why is a knowledge of the genetic map of *lambda* desirable in order to be able to use *lambda* properly as a cloning vector?
8. Suppose you were working with an organism that was intractable to genetic transformation with free DNA. How might you still be able to use this organism as a cloning host?
9. Suppose you have just determined the DNA base sequence for an especially strong promoter in *E. coli* and you are interested in incorporating this sequence into an expression vector. Describe the steps you would use. What precautions would be necessary to be sure that this promoter actually worked as expected in its new location?
10. Explain why the use of a regulatory switch is desirable for the large-scale production of a protein.
11. Suppose you are interested in using Southern blot analysis to detect the presence of the nitrogenase gene (*nif*) in a newly isolated organism. You have cloned the *nif* gene from a known nitrogen-fixing organism and can obtain large amounts of the *nif* DNA in radioactive form. How would you use Southern blot analysis to detect *nif* in your new organism? Could you absolutely conclude that if *nif* were present in your new organism, that this organism was capable of nitrogen fixation? Why?
12. Compare and contrast reverse transcription and reverse translation. Which of these processes would you not expect to occur in a mammalian organism?
13. You have just discovered a fantastic protein in human blood which is an effective cure for cancer, but it is only present in extremely small amounts. Describe the steps you would use to obtain production of this protein in *E. coli*.

Supplementary Readings

Harris, T. J. R. 1983. Expression of eukaryotic genes in *E. coli*. R. Williamson, ed. *Genetic Engineering, vol. 4*, pages 128–155. Academic Press, New York. Detailed review of the procedures that have been used for obtaining expression of eucaryotic genes in bacteria.

Maniatis, T., E. F. Fritsch, and **J. Sambrook**. 1982. *Molecular Cloning: A Laboratory Manual*. Cold Spring Harbor Laboratory, Cold Spring Harbor, N. Y. 545 pp. A detailed manual describing many procedures for cloning and obtaining expression of cloned genes. Brief discussions of the principles behind each procedure.

Old, R. W., and **S. P. Primrose**. 1987. *Principles of Gene Manipulation: An Introduction to Genetic Engineering*, 4th edition. Blackwell Scientific Publications, Oxford. Brief textbook on genetic engineering procedures.

Sherman, F., G. R. Fink, and **J. B. Hicks**. 1987. *Methods in Yeast Genetics. A Laboratory Course Manual*. Cold Spring Harbor Laboratory, Cold Spring Harbor, New York.

Watson, J. D., J. Tooze, and **D. T. Kurtz**. 1983. *Recombinant DNA. A Short Course*. Scientific American Books, New York.

Growth and Its Control

We have discussed in previous chapters the molecular biology and biochemistry of the microbe, and now we place the concepts advanced in these chapters in the broader context of how a cell grows. **Growth** is defined as an increase in the number of cells or an increase in cellular mass. Growth is an essential component of microbial function, as any given cell has only a finite life span in nature, and the species is maintained only as a result of continued growth. In most practical situations involving microorganisms, we are also concerned with growth. Control of microbial action requires a knowledge of the control of growth.

In the present chapter, we present first some general ideas about the molecular basis of microbial growth, then discuss the growth of microbial populations, then consider some of the factors influencing growth, and finally consider some of the practical aspects. Most of our discussion concerns the bacterial cell, since most information is available about bacteria and most practical aspects deal with control of bacterial growth.

9.1 Cell Growth

The bacterial cell is essentially a synthetic machine which is able to duplicate itself. The synthetic processes of bacterial cell growth involve as many as 1000–2000 chemical reactions of a wide variety of types. Some of these reactions involve energy transformations. Other reactions involve biosynthesis of small molecules, the building blocks of macromolecules, as well as the various cofactors and coenzymes needed for enzymatic reactions. These various metabolic reactions were discussed in Chapter 4.

The main reactions of the cell synthesis process itself are *polymerization reactions*, the processes by which polymers (macromolecules) are made from monomers. The major reactions of macromolecular synthesis have been discussed in Chapter 5: DNA synthesis, RNA synthesis, protein synthesis. Once polymers are made, the stage is set for the final events of cell growth: assembly of macromolecules to form cellular structures such as the cell envelope, flagella, ribosomes, inclusion bodies, enzyme complexes, etc. Such assembly occurs both by spontaneous processes, which we have termed *self-assembly* (see Section 6.6), as well as by specific mechanisms. What specific mechanisms? Proteins which are to be assembled in components outside the plasma membrane, for instance, must be excreted, and protein excretion occurs by specific mechanisms involving the participation of signal peptide sequences which must be cleaved off as the proteins pass through the membrane (see Section 5.10). Polyribosomes are formed as the result of the association of ribosome subunits with polycistronic messenger RNA, again involving specific mechanisms of assembly.

In most microorganisms, growth continues until the cell divides into two new cells, a process sometimes called *binary fission* (*binary* to express the fact that *two* cells have arisen from one cell). Microbial growth therefore usually results in an increase in cell number, the new cells formed eventually attaining the same size as the original (parent) cell.

It is important to distinguish between the growth of individual cells and the growth of populations of cells. The growth of a cell is an increase in its size and weight and is usually a prelude to cell division. Population growth, on the other hand, is an increase in the *number* of cells as a consequence of cell growth and division. It is extremely difficult to study the growth of individual cells quantitatively, especially in microorganisms, because analytical techniques are not sensitive enough for use with these small structures. Therefore, almost all growth studies of microorganisms are population studies involving very large numbers of organisms. Although such studies have proven very valuable, it must always be remembered that the answers obtained reflect the average condition of all cells in the population; individual variations between cells cannot be detected in this way.

9.2 Population Growth

Growth is defined as an increase in the *number* of microbial cells or an increase in microbial mass. **Growth rate** is the change in cell number or mass *per unit time*. In unicellular microorganisms, growth usually involves an increase in cell number. A single cell continually increases in size until it is approximately double its original size; then cell division occurs, resulting in the formation of two cells the size of the original cell. During this cell-division cycle, all the structural components of the cell double. The interval for the formation of two cells from one is called a **generation**, and the time required for this to occur is called the **generation time**. The generation time is thus the time required for the cell number to double. Because of this, the generation time is also sometimes called the *doubling time*. Note that during a single generation, both the cell number and cell mass have doubled. Generation times vary widely among organisms. Most bacteria have generation times of 1 to 3 hours, but a few are known that divide in as little as 10 minutes. At the other extreme, some slow-growing protozoa and algae have generation times of 24 hours or more.

Because of the rapid exponential growth of many microorganisms, large populations of cells develop quickly, and one is often forced to deal with very large numbers. For instance, cell numbers in the millions, hundred millions, and even billions occur quite often. Since it is difficult to handle such large numbers, the microbiologist makes use of scientific notation which employs exponents of 10. Thus we express 1,000,000 (one million) as 10^6; 10,000,000 as 10^7; 100,000,000 as 10^8, and 1,000,000,000 (one billion) as 10^9. To express a number such as 5,000,000 the figure is written as the unit integer multiplied by the proper power of 10: thus 5,000,000 = 5.0×10^6; 25,000,000 = 2.5×10^7; and 700,000,000 = 7.0×10^8.

A growth experiment beginning with a single cell having a doubling time of 30 minutes is presented in Figure 9.1. This pattern of population increase, where the number of cells increases by a constant factor during each unit time period, is referred to as **exponential growth** (see Equation 1 in Appendix 2). When the cell number is graphed on arithmetic coordinates as a function of elapsed time, one obtains

Time hours	Number of cells
0	1
0.5	2
1	4
1.5	8
2	16
2.5	32
3	64
3.5	128
4	256
4.5	512
5	1,024
.	.
.	.
.	.
10	1,048,576

(a)

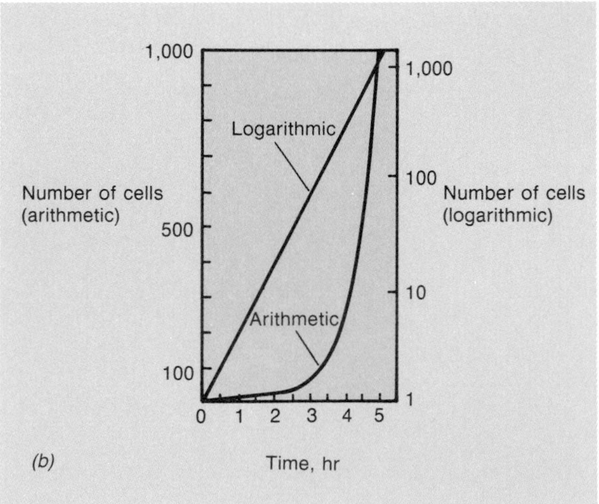

Figure 9.1 The rate of growth of a microbial culture. (a) Data for a population which doubles every 30 minutes. (b) Data plotted on an arithmetic and a logarithmic scale.

a curve with a constantly increasing slope. Examination of curved lines is not convenient and population results are usually *transformed* by taking the logarithm of each data point. The $\log_{10}$ values are presented in Figure 9.1 in a graph in which cell number is plotted logarithmically and time is plotted arithmetically (a semilogarithmic graph), resulting in a straight line. This semilogarithmic graph is convenient and simple to use for calculating generation time from a set of results. The doubling time may be read directly from the graph (Figure 9.2). For some purposes it is useful to express growth mathematically. This is discussed in Appendix 2.

One of the characteristics of exponential growth is that the rate of increase in cell number is slow initially but increases at an ever faster rate. This results, in the later stages, in an explosive increase in cell numbers. A practical implication of exponential growth is that when a nonsterile product such as milk is allowed to stand under conditions such that microbial growth can occur, a few hours during the early stages of exponential growth are not detrimental, whereas standing for the same length of time during the later stages is disastrous.

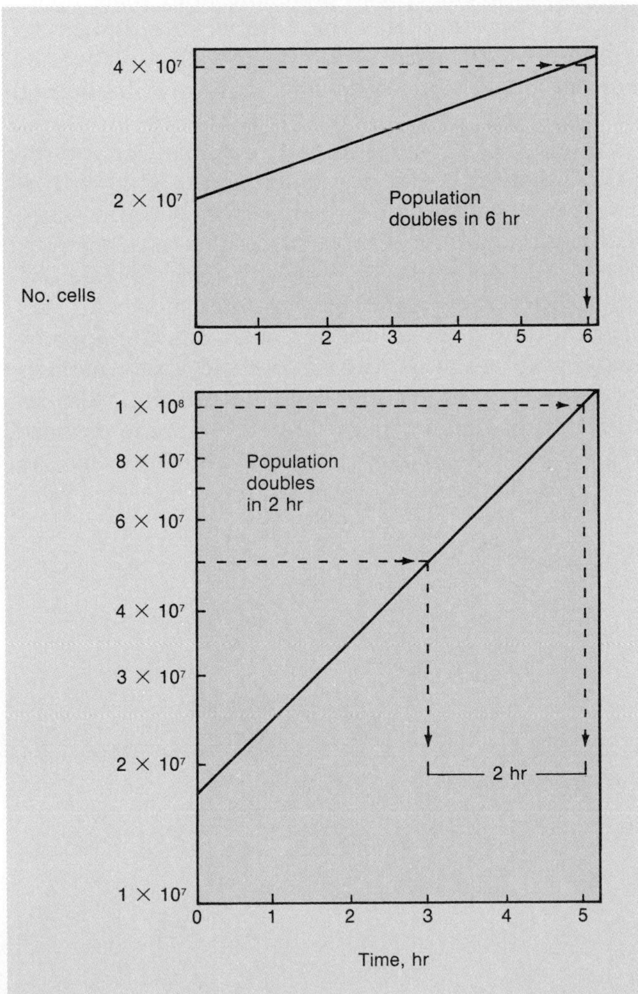
No. cells
4 × 10⁷
2 × 10⁷
Population doubles in 6 hr
0 1 2 3 4 5 6
1 × 10⁸
8 × 10⁷
6 × 10⁷
4 × 10⁷
3 × 10⁷
2 × 10⁷
1 × 10⁷
Population doubles in 2 hr
2 hr
0 1 2 3 4 5
Time, hr

Figure 9.2 Method of estimating the generation times of exponentially growing populations with generation times of 2 hours and 6 hours, respectively.

9.3 Measurement of Growth

Growth is measured by following changes in number of cells or weight of cell mass. There are several methods for counting cell number or estimating cell mass, suited to different organisms or different problems.

Total cell count The number of cells in a population can be measured by counting under the microscope, a method called the **direct microscopic**

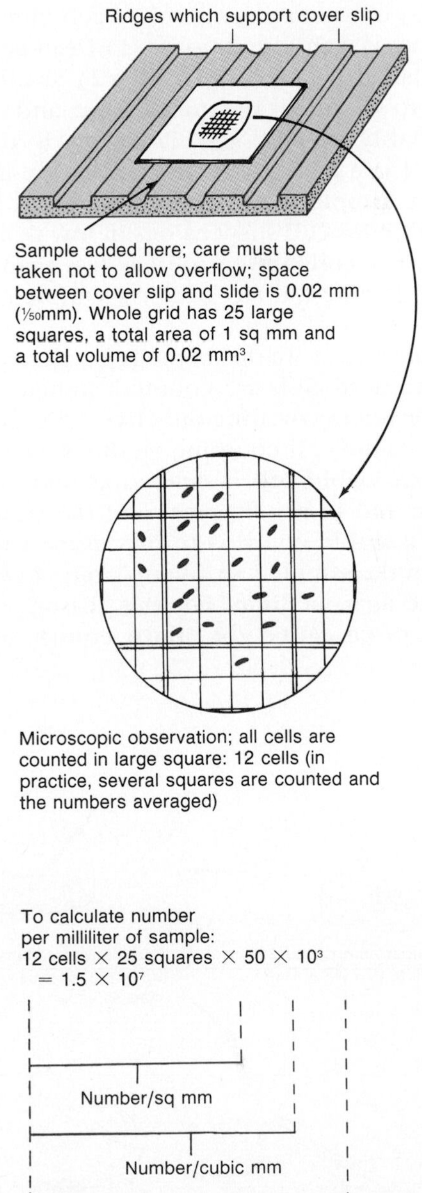
Ridges which support cover slip
Sample added here; care must be taken not to allow overflow; space between cover slip and slide is 0.02 mm (1/50mm). Whole grid has 25 large squares, a total area of 1 sq mm and a total volume of 0.02 mm³.
Microscopic observation; all cells are counted in large square: 12 cells (in practice, several squares are counted and the numbers averaged)
To calculate number per milliliter of sample:
12 cells × 25 squares × 50 × 10³ = 1.5 × 10⁷
Number/sq mm
Number/cubic mm
Number/cubic cm (ml)

Figure 9.3 Direct microscopic counting procedure using the Petroff-Hausser counting chamber.

count. Two kinds of direct microscopic counts are done, either on samples dried on slides or on samples in liquid. With liquid samples, special *counting chambers* must be used. In such a counting chamber, a special grid is marked on the surface of the glass slide, with squares of known small area (Figure 9.3). Over each square on the grid is a volume of known size, very small but precisely measured. The number of cells per unit area of grid can be counted under the microscope, giving a measure of the number of cells per small chamber volume. Converting this value to the number of cells per milliliter of suspension is easily done by multiplying by a conversion factor based on the volume of the chamber sample.

Direct microscopic counting is tedious but is a quick way of estimating microbial cell number. However, it has certain limitations: (1) Dead cells are not distinguished from living cells. (2) Small cells are difficult to see under the microscope, and some cells are probably missed. (3) Precision is difficult to achieve. (4) A phase-contrast microscope is required when the sample is not stained. (5) The method is not suitable for cell suspensions of low density. With bacteria, if a cell suspension has less than 10^6 cells per milliliter, few if any bacteria will be seen.

Viable count In the methods just described, both living and dead cells are counted. In many cases we are interested in counting only live cells, and for this purpose viable cell counting methods have been developed. A viable cell is defined as one that is able to divide and form offspring, and the usual way to perform a viable count is to determine the number of cells in the sample capable of forming *colonies* on a suitable agar medium. For this reason, the viable count is often called the **plate count**, or **colony count**.

There are two ways of performing a plate count: the spread plate method and the pour plate method (Figure 9.4). With the **spread plate method**, a volume of culture no larger than 0.1 ml is spread over the surface of an agar plate, using a sterile glass spreader. The plate is then incubated until the colonies appear, and the number of colonies is counted. It is important that the surface of the plate be dry so that the liquid that is spread soaks in. Volumes greater than 0.1 ml should never be used, since the excess liquid will not soak in and may cause the colonies to coalesce as they form, making them difficult to count. In the **pour plate method** (Figure 9.4), a known volume of 0.1 to 1.0 ml of culture is pipetted into a sterile petri plate; melted agar medium is then added and mixed well by gently swirling the plate on the table top. Because the sample is mixed with the molten agar medium, a larger volume can be used than with the spread plate; however, with the pour plate the organism to be counted must be able to briefly withstand the temperature of melted agar, 45°C.

With both the spread plate and pour plate methods, it is important that the number of colonies developing on the plates not be too large, since on crowded plates some cells may not form colonies and the count will be erroneous. It is also essential that the number of colonies not be too small, or the statistical significance of the calculated count will be low. The usual practice, which is most valid statistically, is to count only those plates that have between 30 and 300 colonies. To obtain the appropriate colony number, the sample to be counted must thus usually be diluted. Since one rarely knows the approximate viable count ahead of time, it is usually necessary to make more than one dilution. Several ten-fold dilutions of the sample are commonly used (Figure 9.5). To make a ten-fold dilution, one can

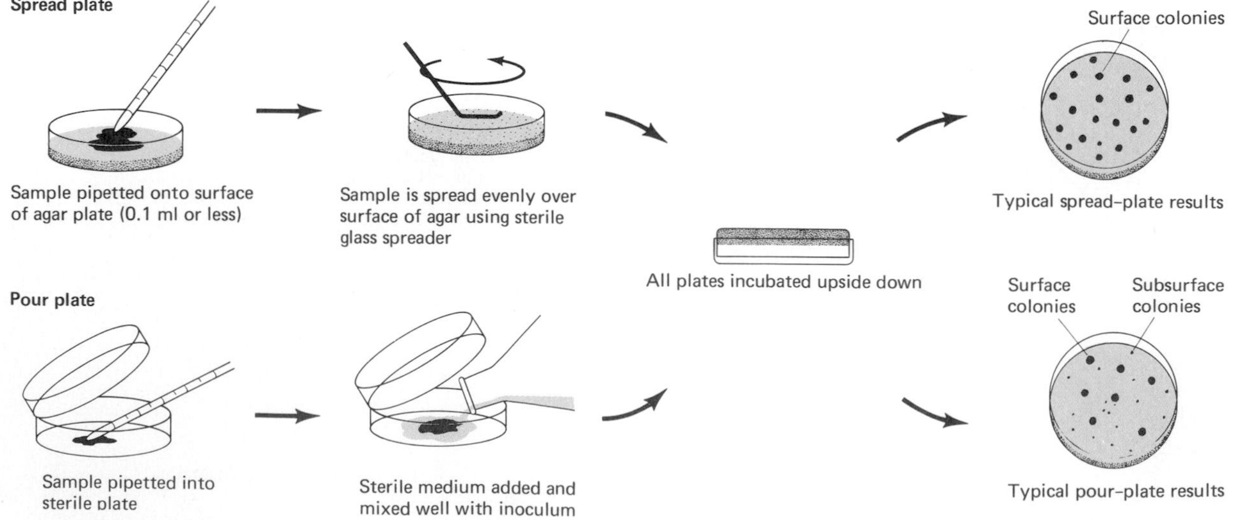

Figure 9.4 Two methods of performing a viable count (plate count).

mix 0.5 ml of sample with 4.5 ml of diluent, or 1.0 ml with 9.0 ml diluent. If a hundred-fold dilution is needed, 0.05 ml can be mixed with 4.95 ml diluent, or 0.1 ml with 9.9 ml diluent, or of course two successive ten-fold dilutions may be made. In most cases, such *serial dilutions* are needed to reach the final dilution desired. Thus, if a $1/10^6$ (1 to 1,000,000) dilution is needed, this can be achieved by making three successive $1/10^2$ (1 to 100) dilutions or six successive ten-fold dilutions. The liquid used for making the dilutions is important. It is best if it is identical with the liquid used in making the solidified medium, although for economy, it is often possible to use a sterile solution of inorganic salts or a phosphate buffer. It is important when making dilutions that a *separate* sterile pipette be used for each dilution, even of the same sample. This is because in the initial sample, which contains the largest number of organisms, not all the organisms will be washed out of the pipette when its contents are expelled. Organisms sticking to the pipette will be washed out in later dilutions and can cause serious error in the final count obtained.

The number of colonies obtained on a plate will depend not only on the inoculum size but also on the suitability of the culture medium and the incubation conditions used; also, it will depend on the length of incubation. The cells deposited on the plate will not all develop into colonies at the same rate, and if a short incubation time is used, less than the maximum number of colonies will be obtained. Furthermore, the size of colonies often varies. If some tiny colonies develop, they may be missed during the counting. It is usual to determine incubation conditions (medium, temperature, time) that will give the maximum number of colonies and then to use these conditions throughout. Viable counts are usually subject to large error, and if accurate counts are desired great care must be taken and many replicate plates must be prepared. Note that two or more cells in a clump will form only a single colony, so that a viable count may be erroneously low. To more clearly state the result, viable counts are often expressed as the number of *colony-forming units* rather than as the number of *viable cells* (since a colony-forming unit may contain one or more cells).

Despite the difficulties involved with viable counting, the procedure gives the best information on the number of viable cells, so it is widely used. In food, dairy, medical, and aquatic microbiology, viable counts are used routinely. The method has the virtue of high sensitivity: samples containing only a very few cells can be counted, thus permitting sensitive detection of contamination of products or materials.

Cell mass For many studies it is desirable to estimate the weight of cells rather than the number. Net weight can be measured by centrifuging the cells, and weighing the pellet of cells obtained. Dry weight is measured by drying the centrifuged cell mass before weighing, usually by placing it overnight in an oven at 100 to 105°C. Dry weight is usually about 10 to 15 percent of the wet weight.

The average weight of a single cell can be calculated by measuring the weight of a large number of cells and dividing by the total number of cells present, as measured by direct microscopic count. Procaryotic cells range in dry weight from less than 10^{-15} g to greater than 10^{-11} g, and the cells of eucaryotic microorganisms range from about 10^{-11} g to about 10^{-7} g.

A simpler and very useful method for obtaining a relative estimate of cell mass is by use of *turbidity*

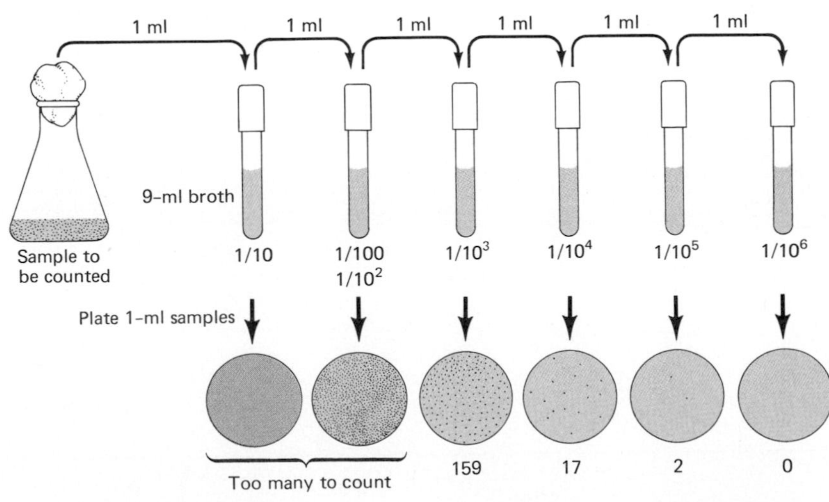

Figure 9.5 Procedure for viable count using serial dilutions of the sample.

measurements. A cell suspension looks turbid because each cell scatters light. The more cell material present, the more the suspension scatters light and the more turbid it will be. Turbidity can be measured with an electrically operated device called a *colorimeter*, or *spectrophotometer*. With such a device, the turbidity is expressed in units of *absorbance*. For unicellular organisms, absorbance is proportional to cell number as well as cell weight, and turbidity readings can thus be used as a substitute for counting. To perform cell counts in this way, a standard curve must be prepared for each organism studied, relating cell number to cell mass or absorbance. Turbidity is a much less sensitive way of measuring cell density than is viable counting but has the virtues that it is quick, easy, and does not destroy the sample. Such measurements are used widely to follow the rate of growth of cultures, since the same sample can be checked repeatedly.

9.4 The Growth Cycle of Populations

The hypothetical culture examined in Figure 9.1 reflects only part of the growth cycle of a microbial population. A typical *growth curve* for a population of cells is illustrated in Figure 9.6. This growth curve can be divided into several distinct phases, called the **lag phase, exponential phase, stationary phase**, and **death phase**.

Lag phase When a microbial population is inoculated into a fresh medium, growth usually does not begin immediately, but only after a period of time called the *lag phase*, which may be brief or extended, depending on conditions. If an exponentially growing culture is inoculated into the same medium under the same conditions of growth, a lag is not seen, and exponential growth continues at the same rate. However, if the inoculum is taken from an old (stationary phase) culture and inoculated into the same medium, a lag usually occurs even if all of the cells in the inoculum are alive. This is because the cells are usually depleted of various essential coenzymes or other cell constituents, and time is required for resynthesis. A lag also ensues when the inoculum consists of cells that have been damaged (but not killed) by treatment with heat, radiation, or toxic chemicals, due to the time required for the cells to repair the damage.

A lag is also observed when a population is transferred from a rich culture medium to a poorer one. This happens because for growth to occur in a particular culture medium the cells must have a complete complement of enzymes for the synthesis of the essential metabolites not present in that medium. On transfer to a new medium, time is required for synthesis of the new enzymes.

Exponential phase The *exponential phase* of growth has already been discussed. As noted, it is a consequence of the fact that each cell divides to form two cells, each of which also divides to form two more cells, and so on. Most unicellular microorganisms grow exponentially, but rates of exponential growth vary greatly. For instance, the organism causing typhoid fever, *Salmonella typhi*, grows very rapidly in culture, with a generation time of 20 to 30 minutes, whereas the tubercle bacterium, *Mycobac-*

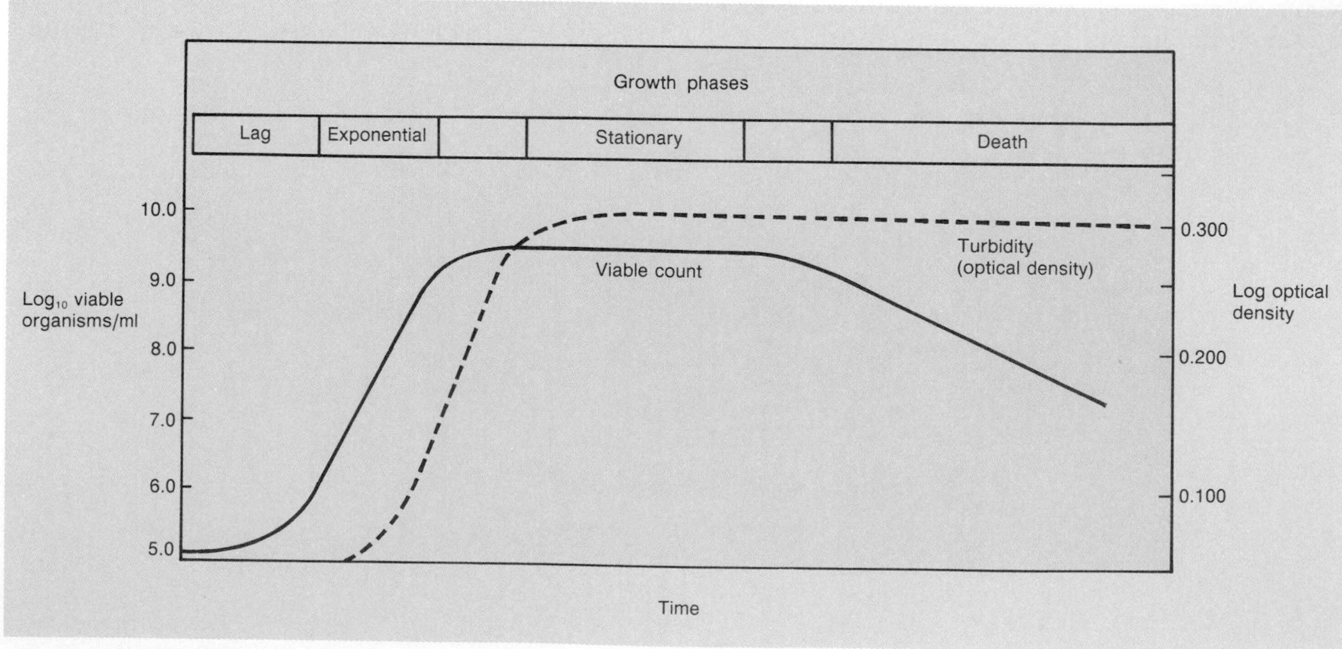

Figure 9.6 Typical growth curve for a bacterial population.

terium tuberculosis, grows slowly, with only one or two doublings per day. The rate of exponential growth is influenced by environmental conditions (temperature, composition of the culture medium) as well as by characteristics of the organism itself. In general, bacteria grow faster than eucaryotic microorganisms, and small eucaryotes grow faster than large ones.

Stationary phase In a closed system, exponential growth cannot occur indefinitely. One can calculate that a single bacterium with a generation time of 20 minutes would, if it continued to grow exponentially for 48 hours, produce a population that weighed about 4,000 times the weight of the earth! This is particularly impressive since a single bacterial cell weighs only about one-trillionth of a gram. Obviously, something must happen to limit growth long before this time. What generally happens is that either an essential nutrient of the culture medium is used up or some waste product of the organism builds up in the medium to an inhibitory level and exponential growth ceases. The population has reached the **stationary phase**.

In the stationary phase there is no net increase or decrease in cell number. However, although no growth occurs in the stationary phase, many cell functions may continue, including energy metabolism and some biosynthetic processes. Certain cell metabolites, called *secondary metabolites*, are produced primarily in the stationary phase (see Section 10.3).

Death phase If incubation continues after a population reaches the stationary phase, the cells may remain alive and continue to metabolize, but often they die. If the latter occurs, the population is said to be in the *death phase*. During this death phase the total count (as measured by a direct microscopic count) may remain constant but the viable count

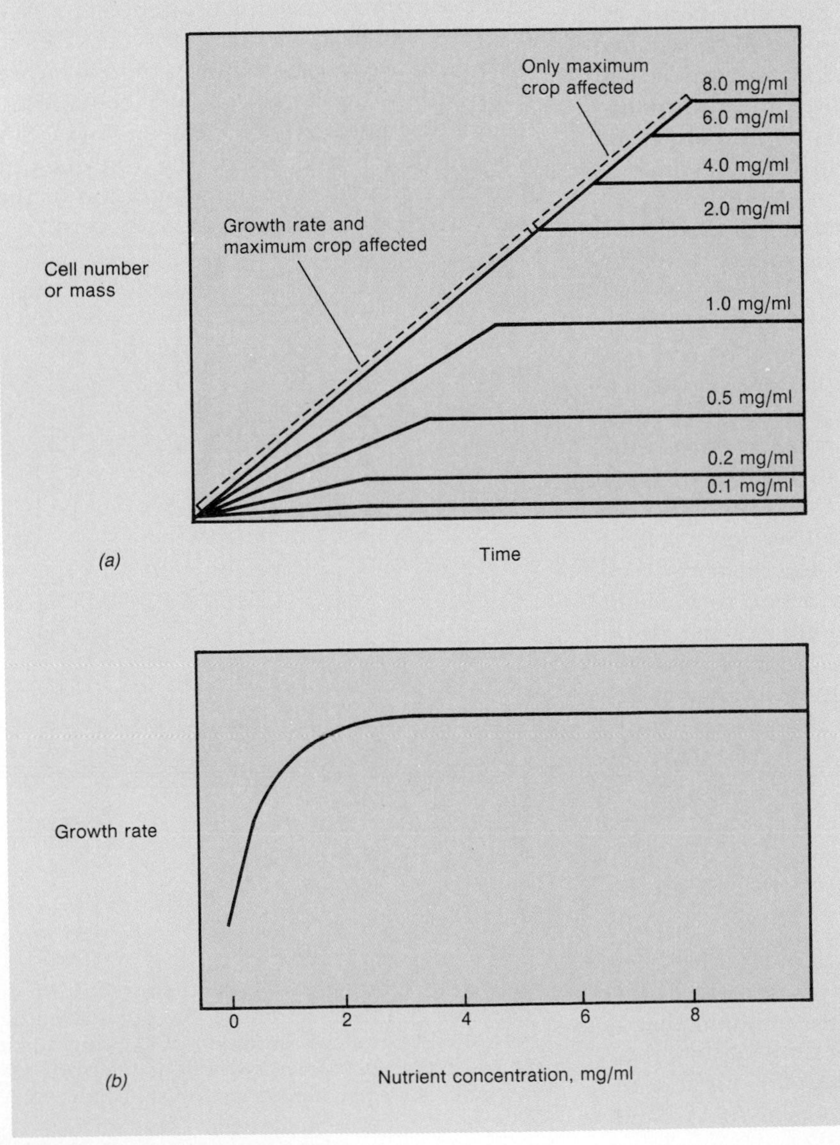

Figure 9.7 Relationship between nutrient concentration, growth rate, and maximum crop. (a) Growth curves at different nutrient concentrations. At low nutrient concentrations both growth rate and maximum crop are affected. (b) Effect of nutrient concentration on growth rate. The data from part *a* are replotted.

slowly decreases. In some cases death is accompanied by cell **lysis**, leading to a decrease in the direct microscopic count concurrent with the drop in viable count. A discussion of the effects of germicidal chemicals on viability will be found later in this chapter.

It must be emphasized that these phases are reflections of the events in a population, not of individual cells. The terms lag phase, exponential phase, stationary phase, or death phase do not apply to individual cells, but only to populations of cells.

9.5 Effect of Nutrient Concentration on Growth

Nutrient concentration can affect either growth rate or total growth, as shown in Figure 9.7. At very low concentrations of the nutrient, the rate of growth is reduced, whereas at moderate and higher levels of nutrient, growth rates are identical but the total growth (sometimes called the **total** or **maximum crop)** is limited. As the nutrient concentration is increased still further, a concentration will be reached that is no longer limiting, and further increases then no longer lead to increases in total crop. At this point, some nutrient or environmental factor other than the one being varied is limiting to further increases in growth.

The effect of nutrient concentration on *total growth* is easy to understand, since much of the nutrient is converted into cell material, and if the amount of nutrient is limited, the amount of cell material will also be limited. The reason for the effect of very low nutrient concentrations on *growth rate* is less certain. One idea is that at these low nutrient concentrations the nutrient cannot be transported into the cell at sufficiently rapid rates to satisfy all of the metabolic demands for the nutrient. Conceivably not all carrier sites are fully occupied (see Section 3.4 for a discussion of the role of carriers in transport). The shape of the curve relating growth rate to nutrient concentration (Figure 9.7*b*) resembles a saturation process of the kind also seen for active transport. This ability to alter the growth rate by nutrient concentration is used in the operation of the type of continuous-culture apparatus called a **chemostat**, described in the next section.

9.6 Continuous Culture

Our discussion of population growth thus far has been confined to closed or *batch cultures*, growth occurring in a fixed volume of a culture medium that is continually being altered by the actions of the growing organisms until it is no longer suitable for growth. In the early stages of exponential growth in batch cultures, conditions may remain relatively constant but in later stages drastic changes usually occur. For many studies, it is desirable to keep cultures in constant environments for long periods, and this is done by employing *continuous cultures*. A continuous culture is essentially a flow system of constant volume to which medium is added continuously and from which a device allows continuous removal of any overflow. Once such a system is in equilibrium, cell number and nutrient status remain constant, and the system is said to be in **steady state**.

The most common type of continuous-culture device used is called a **chemostat** (Figure 9.8), which permits control of both the population density and the growth rate of the culture. Two elements are used in the control of a chemostat—the flow rate and the concentration of a limiting nutrient, such as carbon, energy or nitrogen source, or growth factor. As described in Figure 9.7*b*, at low concentrations of an essential nutrient, growth rate is proportional to the nutrient concentration. However, at these low nutrient concentrations, the nutrient is quickly used up by growth. In batch cultures growth ceases at the time the nutrient is used up, but in the chemostat continuous addition of fresh medium containing the limiting nutrient permits continued growth. Since at the low nutrient levels used, the limiting nutrient is quickly assimilated, its concentration in the chemostat vessel itself is always virtually zero.

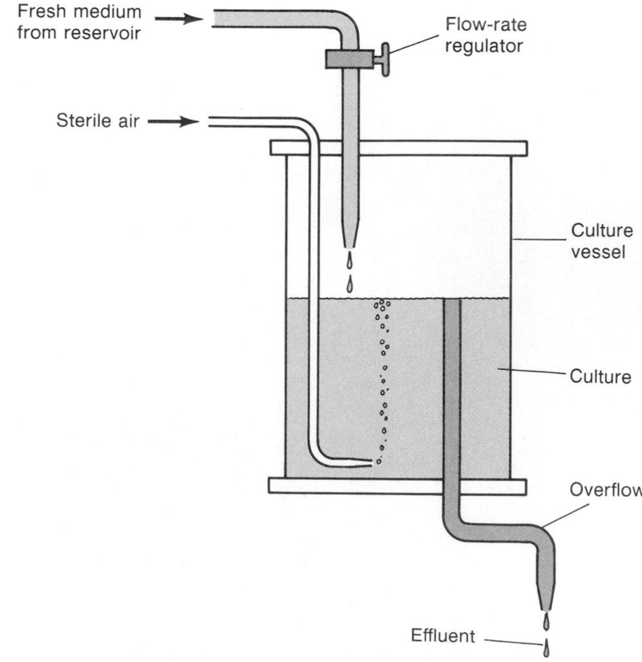

Figure 9.8 Schematic for a continuous-culture device (chemostat). In such a device, the population density is controlled by the concentration of limiting nutrient in the reservoir, and the growth rate is controlled by the flow rate. Both parameters can be set by the experimenter.

Effects of varying dilution rate and concentration of the inflowing growth-limiting substrate are given in Figure 9.9. As seen, there are rather wide limits over which dilution rate will control growth rate, although at both very low and very high dilution rates the steady state breaks down. At high dilution rates, the organism cannot grow fast enough to keep up with its dilution, and the culture is washed out of the chemostat. At the other extreme, at very low dilution rates, a large fraction of the cells may die from starvation, since the limiting nutrient is not being added fast enough to permit maintenance of cell metabolism. There is probably a minimum amount of energy necessary to maintain cell structure and integrity, called **maintenance energy**, and this nutrient used for maintenance energy is not available for biosynthesis and cell growth. Thus at very low dilution rates steady-state conditions will probably not be maintained, and the population will slowly wash out (not shown in Figure 9.9).

The cell density in the chemostat is controlled by the level of the limiting nutrient. If the concentration of this nutrient in the incoming medium is raised, with dilution rate remaining constant, cell density will increase although growth rate will remain the same and the steady-state concentration of the nutrient in the culture vessel will still be virtually zero. Thus, by adjusting dilution rate and nutrient level, the experimenter can obtain at will a variety of population densities growing at a variety of growth rates. The actual shape of the curve for bacterial concentration given in Figure 9.9 will depend on the organism, the environmental conditions, and the limiting nutrient used.

9.7 Growth and Macromolecular Synthesis

Balanced and unbalanced growth During exponential growth, all biochemical constituents are being synthesized at the same relative rates (Figure 9.10*a*), a condition called *balanced growth*. Figure 9.10*b* shows what happens to the growth rates and the rates of synthesis of RNA, DNA, and protein if the culture is transferred to a different medium, in which growth can occur at a faster rate than previously. (These are called "step-up" conditions.) Immediately upon transfer to the richer medium the rate of *RNA synthesis increases*, and somewhat later the rates of DNA

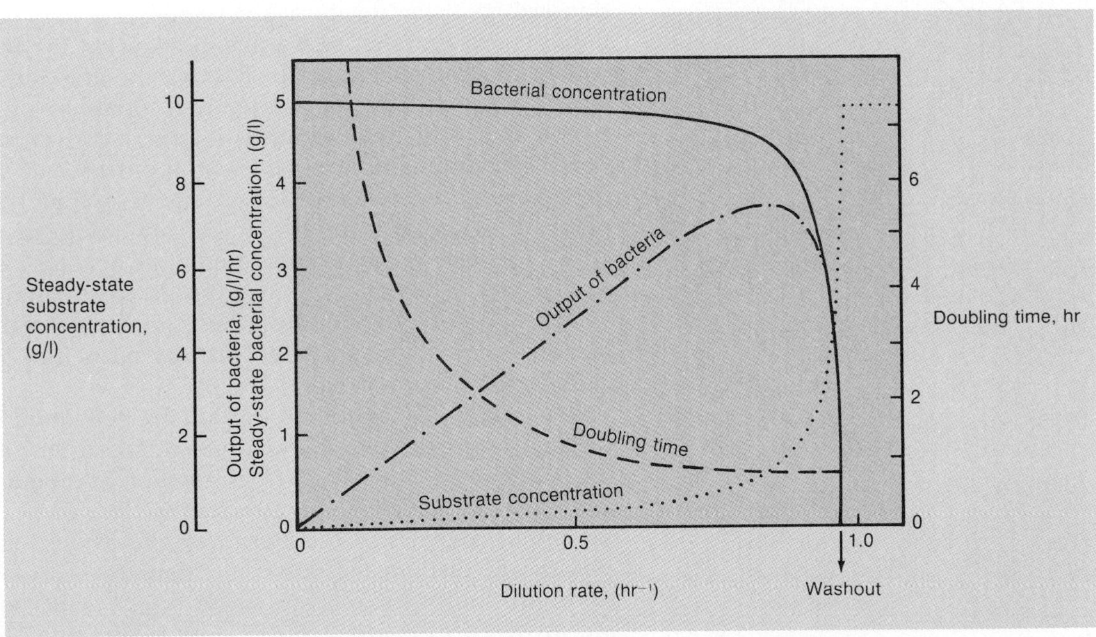

Figure 9.9 Steady-state relationships in the chemostat. The dilution rate is determined from the flow rate and the volume of the culture vessel. Thus, with a vessel of 1000 ml and a flow rate through the vessel of 500 ml/hr, the dilution rate would be 0.5 hr^{-1}. Note that at high dilution rates, growth cannot balance dilution, the population washes out, and the substrate concentration rises to a maximum (since there are no bacteria to use the inflowing substrate). However, throughout most of the range of dilution rates shown, the population density remains constant and the substrate concentration remains at a very low value. Note that although the population density remains constant, the growth rate (doubling time) varies over a wide range. Thus, the experimenter can obtain populations with widely varying growth rates, without affecting population density.

and protein synthesis also increase. The rate of cell division also steps up after a longer lapse of time, and eventually the rates of synthesis of all components are in balance again. In the initial period after the transfer to the new medium we have conditions of *unbalanced growth* since not all cell constituents are being synthesized at the same rate. If the reverse experiment (a "step down") is performed, the rate of RNA synthesis decreases immediately, while the rate of DNA and protein synthesis and the rate of cell division continue at the previous (more rapid) rate, later decreasing. These results suggest that, since the rate of RNA synthesis is the first to change during step-up or step-down conditions, it may be the key factor controlling growth rate.

Although there are several kinds of RNA in the cell (see below), the largest fraction consists of RNA in ribosomes. During the step-up conditions, there is an increased rate of synthesis of ribosomes, leading to a greater number of ribosomes per cell. Since protein synthesis occurs on ribosomes, it is thus understandable that the rate of protein synthesis should start to change after the rate of ribosome synthesis increases. The efficiency with which a ribosome acts in protein synthesis appears to remain constant at different growth rates. Thus, growth is controlled at the most basic level by the *number* of ribosomes present per cell.

A specialized type of unbalanced growth, that involving enzyme induction and repression, was discussed in detail in Section 5.11. In induction/repression processes, synthesis of only a single or small group of proteins is regulated. In the next section, we discuss more dramatic growth regulatory processes.

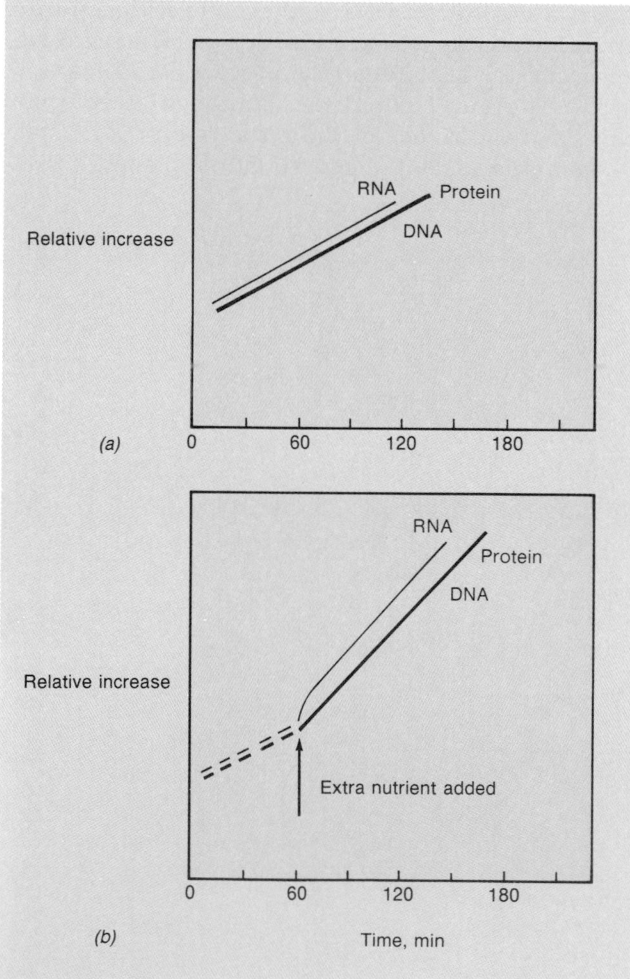

Figure 9.10 Changes in RNA, DNA, and protein content of a population during growth: (a) balanced growth, (b) step-up conditions. The arrow indicates the time at which extra nutrient was added to speed up growth. Note that, during the step-up, the rate of RNA synthesis increases first, followed by increases in rates of protein and DNA syntheses.

Global control of macromolecular synthesis Shifts between one and another environmental conditions occur frequently for microorganisms, and it is not surprising that special regulatory processes exist which come into play under such conditions. For instance, when the bacterium *Escherichia coli* is starved for phosphate, over 80 different genes are transcribed, bringing about the synthesis of new proteins. These proteins play roles in adapting the bacterium to a phosphate-deficient environment. Such a set of genes that becomes active in response to a particular environmental stimulus is called a *stimulon*. Several stimulons have been identified in *E. coli*: nitrogen starvation (over 40 separate genes), high temperature stress (around 16 genes), low temperature stress (6 genes), oxygen deficiency or anaerobiosis (18 genes). The SOS response discussed in Section 5.18 also involves a stimulon. Because these control mechanisms occur on a wide cellular basis they are called *global control mechanisms*.

Not all the genes of a stimulon are regulated by the same control system. Each stimulon consists of a set of genes that share a specific regulatory element. Such a set of genes, all controlled by the same regulatory element, is called a *regulon*. We should emphasize the distinction between a regulon and an operon. An operon involves a single set of genes that are involved in a single physiological function (for instance, lactose utilization, synthesis of a specific amino acid) and which respond to the same operator. A regulon involves *unlinked* genes which can be distributed widely throughout the genome and which are not necessarily related functionally.

What is the molecular mechanism by which many unrelated genes are regulated? In the case of the response to high temperature, the so-called *heat-shock response*, it is known that a new *sigma* factor, *sigma*^{32}, increases dramatically in amount after the temperature shift. Recall that *sigma* is a component of the

RNA polymerase which brings about the synthesis of messenger RNA (see Section 5.8). For most *E. coli* genes, *sigma*^{70} is the component which activates transcription, but RNA polymerase containing *sigma*^{70} does not transcribe effectively the heat shock genes. Instead, *sigma*^{32} plays this role. *sigma*^{32} increases dramatically in amount when *E. coli* is placed at temperatures slightly above its normal growth temperature. Heat shock gene promoters which are in front of the groups of heat shock genes are transcribed when this new *sigma* factor is present. The promoters recognized by RNA polymerase containing *sigma*^{32} have a particular base sequence in the -35 and -10 regions. Thus, in this particular stimulon, a change in environment leads to the synthesis of a new *sigma* factor which modifies RNA polymerase in such a way that it transcribes a series of genes that is not normally transcribed. The factor *sigma*^{32} is very unstable in the cell, so that as soon as the conditions which brought about its synthesis are removed, its activity dramatically decreases and the heat shock genes are no longer transcribed. Note that a regulatory mechanism of this type, in which a new protein (*sigma*^{32}) promotes transcription is a *positive control* mechanism, in contrast to the *negative control* of induction and repression discussed in Section 5.11. Not all stimulons may have at their base a special *sigma* factor, but a mechanism involving positive control is most likely involved.

DNA synthesis and the cell cycle As noted in Section 5.4, DNA synthesis is initiated at a single point of the chromosome, and proceeds bidirectionally around the circle. In cells growing at normal or slow rates, synthesis of DNA takes about two-thirds of a generation time. Thus in *Escherichia coli* growing with a generation time of 60 minutes, 40 minutes are required for DNA synthesis. Since the *E. coli* DNA is about 1100 μm long, this gives a rate of replication of about 27 μm/minute. In slower growing cultures, DNA synthesis takes longer, but it still requires only two-thirds of a generation time. In fast-growing cultures, the situation is more complicated, since new rounds of replication begin before the old round is completed. Thus in rapidly growing cells, several replicating points are present in the cell at one time, the exact number depending on how fast the organism is growing. As seen in Figure 9.11, fast-growing cells will have more DNA copies than cells with normal growth rate, but there are always twice as many DNA copies at the end of each cycle as at the beginning. When the DNA copies are partitioned into the daughter cells they are usually already partly replicated before the new round of replication begins.

In eucaryotes, DNA synthesis occurs simultaneously in the various chromosomes. Further, on each chromosome there are a number of separate sites where DNA synthesis can proceed simultaneously. The histone proteins present on the chromosomes probably affect the accessibility of the DNA to polymerase enzymes, and hence influence initiation of DNA replication.

RNA synthesis About 80 percent of the total RNA in a growing cell is ribosomal RNA, and most of the remainder is transfer RNA. Despite its importance, messenger RNA is only a small fraction of the total, and in bacteria it is unstable, being rapidly synthesized, used for several rounds of protein synthesis, and then broken down. Thus in studies of RNA synthesis during growth, it is primarily ribosomal RNA that is examined. As opposed to DNA, ribosomal RNA synthesis is essentially continuous throughout the growth cycle. In eucaryotes, RNA synthesis stops during mitosis, probably because at this stage the chromosomes are condensed and transcription cannot occur.

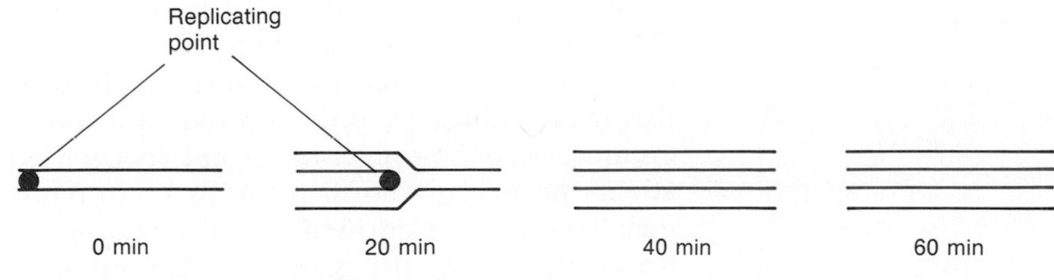

(a) Doubling time greater than 40 minutes; one replication point

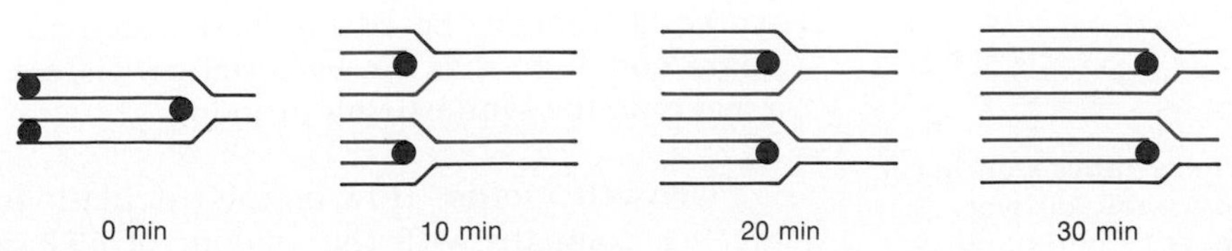

(b) Fast growth rate (for example, less than 40 minutes); multiple replication points

Figure 9.11 Replication of DNA during the cell cycle in the bacterium *Escherichia coli* as a function of the growth rate. The times given are from initiation of a new cycle. For simplicity, the DNA is shown as a linear structure, but in reality it is a closed circle.

Protein synthesis The synthesis of specific enzymes during the cell cycle is of considerable interest, since the time of increased enzyme amount may be related to when the gene that codes for it has been duplicated. Two broad groups of enzymes have been recognized, those whose synthesis is continuous throughout the cell cycle and those that are synthesized periodically. Most enymes are synthesized periodically at one or more stages during the cycle. If an enzyme with a periodic pattern is stable, its increase in activity during the cell cycle shows a step pattern (Figure 9.12*a*), whereas if the enzyme is unstable, it shows a peak pattern (Figure 9.12*b*). Enzymes synthesized continuously show either an exponential (Figure 9.12*c*) or a linear increase, probably depending on the manner of formation. In some cases, it has been shown that periodic synthesis of an enzyme is related to doubling of the gene coding for it, and since genes are linearly arranged along the chromosome, different enzymes show periodic increases at different times of the cell cycle. However, in a number of other cases, periodic increase in enzyme is not related to increased number of genes (the gene dosage), but to oscillatory events related to repression control systems. In the case of inducible enzymes, the degree of inducibility often increases just after the gene coding for the enzyme has doubled.

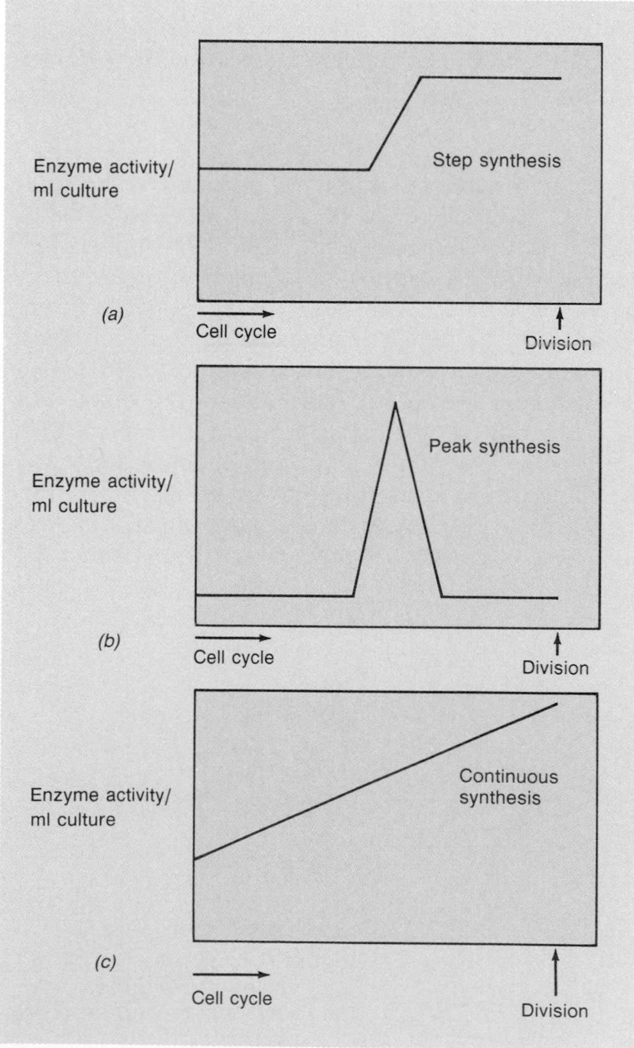

Figure 9.12 Three patterns of enzyme synthesis during the cell cycle. (a) Periodic synthesis, stable enzyme. (b) Periodic synthesis, unstable enzyme. (c) Continuous synthesis.

9.8 The Energetics of Growth

Growth is an energy-using process. We discussed energy-yielding and biosynthetic processes in Chapter 4, and macromolecular synthesis in Chapter 5. We can now attempt to present an integrated picture of the relations between energy-yielding and energy-consuming processes. The analysis is based on the demand for and synthesis of ATP, the energy currency of living organisms.

We can divide energy-consuming processes into several distinct types: (1) formation of building blocks (monomers) from CO_2 or other carbon sources; (2) polymerization of monomers, leading to formation of RNA, DNA, proteins, polysaccharides, and lipids; (3) active transport, movement, and other mechanical events. At present we are unable to estimate the energy requirements for the third type, although they are probably small compared to those of the first two types.

The energy required for synthesis of cell monomers can be calculated from the biosynthetic pathways, but if a complex culture medium is used, all or virtually all the cell monomers will be provided preformed. Under these conditions, the energy source is used primarily for polymerization reactions, and from knowledge of the macromolecular composition of the cell and the amount of energy needed for polymerization, the energy expenditure for polymer synthesis can be calculated. This is given in Table 9.1, where the basis upon which the various calculations are made is also given. It should be understood that these values are only approximate since the macromolecular composition of cells varies. Also, not all of the energy-requiring steps in polymerization may have been identified yet. The results show that for every 100 mg dry mass of cells synthesized, over 3,000 μmoles of ATP are required for polymerization. As seen, the bulk of the energy is required for protein synthesis. Indeed, based on these calculations, we might conclude that the bacterial cell is essentially a machine for synthesizing protein.

Growth yields How do the calculations in Table 9.1 compare with the amount of ATP actually required for growth? The ATP required for growth

Table 9.1 Energy expenditure for polymer synthesis

		Monomer units in polymer		ATP required	
Substance	Approximate dry weight, percent	Average molecular weight	Micromoles/ 100 mg cells	Per monomer*	Micromoles ATP/100 mg cells
Protein	60	110	545	5	2,725
Nucleic acid	20	300	67	5	335
Lipid	5	262	19	1	19
Polysaccharide	5	166	30	2	60
Peptidoglycan	10	1,000	10	10	100
Total					3,239 or 31 g cell material per mole ATP

**Calculation of ATP required per monomer polymerized:*
Protein: amino acid activation: 2ATP; ribosome function: 2GTP = 2ATP; mRNA turnover: 1ATP.
Nucleic acid: nucleotide formation from free base: 3ATP; polymerization: 2ATP.
Lipid: formation of glycerol ester from fatty acyl-CoA and glycerophosphate: 1ATP.
Polysaccharide: formation of uridine diphosphosugar: 2ATP.
Peptidoglycan: activation of five amino acids: 5ATP; UDP-muramic acid: 2ATP; UDP-N-acetylglucosamine: 2ATP; lipid carrier: 1ATP.

can be estimated if the growth yield (Y) per unit energy source used is measured, provided we can calculate how much ATP is generated from the energy source. With fermentative organisms, the ATP yield from the energy source can usually be calculated directly. For instance, when glucose is fermented via the glycolytic pathway, two moles of ATP are generated per mole of glucose used (see Section 4.8).

Growth may be examined in more detail by measurement of cell yield (that is, biomass produced) in comparison to amount of ATP generated during catabolism. The parameter calculated is called Y_{ATP}, for cell yield per ATP generated. Experimental determination of these values is done most easily with fermentative organisms in which the amount of ATP produced per molecule of substrate metabolized can be deduced from knowledge of the catabolic pathway used by the organism under study. For example, the fermentation of glucose to lactic acid via the glycolytic pathway leads to the formation of two ATP from each glucose (Section 4.8). The actual experiment consists of culturing the organism anaerobically in a rich medium which contains all necessary monomers, and with a single fermentable electron donor such as glucose. At the end of the experiment, the amount of glucose consumed is determined and the dry weight of new cell material is measured. It should be noted that this procedure mainly measures the energetic cost of assembling macromolecules, which is only part (but undoubtedly the major part) of the overall growth process. Table 9.2 presents yield data from some fermenting organisms.

Table 9.2 Molar growth yields for anaerobic growth of fermentative organisms using glucose as electron donor

Organism	$Y_{substrate}$ (grams dry weight per mole substrate used)	ATP yield (moles ATP per mole of substrate)*	Y_{ATP} (grams dry weight per mole ATP)
Streptococcus faecalis	20	2	10
Streptococcus lactis	19.5	2	9.8
Lactobacillus plantarum	18.8	2	9.4
Saccharomyces cerevisiae	18.8	2	9.4
Zymomonas mobilis	9	1	9
Aerobacter aerogenes	29	3	9.6
Escherichia coli	26	3	8.6

**ATP yield is based on a knowledge of the pathway by which glucose is fermented in the different organisms.*

Note that the Y_{ATP} values in Table 9.2 are quite similar, between 9 and 10, even though the ATP yield per mole of substrate varies from 1 to 3. Since 1 mole of ATP has the potential for forming 32 g of polymers in a typical bacterial cell (Table 9.1), fermenting bacterial cells appear to synthesize only one-third of their potential. This difference probably reflects (1) our lack of knowledge of some energy-requiring steps; *and* (2) wastage of energy by the cell. For example, active transport and motility are energy-requiring activities for which precise energy costs cannot be accurately estimated. One must also consider the concept of **maintenance energy**, an expenditure necessary to repair damage to vital structures such as the cell membrane and cell wall.

9.9 Cellular Differentiation and Morphogenesis

Induction and repression are simple examples of *cellular differentiation*; the induced or repressed cell differs from its forerunner. Many microorganisms have evolved much more complicated differentiation mechanisms that include not only changes in enzyme content but changes in cellular structure as well. Sometimes the structural changes are relatively simple, such as the production of a flagellum or stalk, but in other cases they are more complex, as in the formation of spores and cysts. Sometimes these structural changes are related to sexual reproduction, but in many cases they occur in the absence of sex. The study of the manner by which organisms effect changes in cell structure is often called *morphogenesis*. In Chapter 19 we shall see a number of examples of morphogenesis. Here we simply indicate some of the principles involved and analyze one well-studied case, the formation of the heat-resistant bacterial endospore.

Changes in cell structure can involve simple changes in shape, but they are often much more complicated and involve the formation of new layers of wall, capsule, or sheath. In all cases, changes in structure must be preceded by changes in enzyme activity. If a new outer-cell-wall polymer is made, enzymes will be involved in both the synthesis of the monomers and their polymerization. Since the forerunner did not have this polymer, this means the enzymes were either not present or were inactive. As we have seen, enzyme activity can be controlled by allosteric inhibition or stimulation brought about by end products or other small molecules (see Section 4.20), and such feedback control provides a direct and immediate means of altering enzyme activity. On the other hand, more permanent and more extensive control of enzyme activity occurs through induction and repression mechanisms, which result in changes in the amount of enzyme protein itself. Although both kinds of mechanisms are probably involved in morphogenetic changes, induction and repression are probably of considerably greater importance than allosteric control.

The trigger for morphogenetic change can be an alteration in an environmental factor, such as a seasonal variation in light or temperature. In a number of cases, the trigger is starvation; when the population begins to be depleted of an essential nutrient, sporulation or cyst formation may occur. In some cases, the trigger may be buildup in the culture of some metabolite that, upon reaching a threshhold level, induces morphogenesis. Once the trigger has brought about the primary induction of morphogenesis, the later stages may occur in the complete absence of further environmental change. New enzymes may appear or disappear in defined sequences, and through their activity changes in structure occur. In many cases, the sequence of enzyme synthesis is so regular that it has been said to be *programmed*.

Underlying this programming of enzyme synthesis is *differential gene action*. It is clear from our discussion of global control mechanisms (see Section 9.7), as well as from our discussion of the more refined control mechanisms of individual operons (see Section 5.11 and 5.12), that mechanisms exist which can bring about the expression of a whole series of genes. Although molecular details are lacking in most cases of cellular differentiation, we can see in principle how programming of gene expression might occur.

Endospore formation One example of cellular differentiation is that of endospore formation in bacteria, during which the vegetative cell is converted into a nongrowing, heat-resistant structure, the endospore. The differences between the endospore and the vegetative cell are profound (Table 9.3), and sporulation involves a very complex series of events. Bacterial sporulation does not occur when cells are dividing exponentially, but only when growth ceases owing to the exhaustion of an essential nutrient. For instance, if a culture growing on glucose as an energy source exhausts the glucose in the medium, vegetative growth ceases and several hours later spores begin to appear. If more glucose is added to the culture just at the end of the growth period, sporulation is inhibited. Glucose probably prevents sporulation through catabolite repression (see Section 5.12), inhibiting the synthesis of the specific enzymes involved in forming the spore structures. Glucose is not the only substance repressing spore formation; many other energy sources can also do this. We thus see that growth and sporulation are opposing processes. It seems reasonable to assume that the elaborate control mechanisms in spore-forming bacteria ensure that in nature sporulation will occur when conditions are no longer favorable for growth.

The structural changes during sporulation were described in Section 3.12, and are summarized in Fig-

Table 9.3 Differences between bacterial endospores and vegetative cells

	Vegetative cell	Spore
Structure	Typical Gram-positive cell	Thick spore cortex Spore coat Exosporium (some species)
Microscopic appearance	Nonrefractile	Refractile
Chemical composition:		
Calcium	Low	High
Dipicolinic acid	Absent	Present
PHB	Present	Absent
Polysaccharide	High	Low
Protein	Lower	Higher
Parasporal crystalline protein (some species)	Absent	Present
Sulfur amino acids	Low	High
Enzymatic activity	High	Low
Metabolism (O_2 uptake)	High	Low or absent
Macromolecular synthesis	Present	Absent
mRNA	Present	Low or absent
Heat resistance	Low	High
Radiation resistance	Low	High
Resistance to chemicals and acids	Low	High
Stainability by dyes	Stainable	Stainable only with special methods
Action of lysozyme	Sensitive	Resistant

ure 9.13. As seen, certain morphological changes begin to occur within three hours after growth ceases, and shortly after that refractile spores begin to appear.

During exponential growth, glucose is fermented and organic acids accumulate, leading to a drop in pH. When glucose is exhausted, the population converts from fermentative to oxidative growth owing to the synthesis of enzymes of the citric acid cycle (see Section 4.10), and the organic acids are now used as energy sources. At this point there is a marked increase in oxygen uptake, and an increase in pH due to the removal of the organic acids from the medium. At about this stage, the spore septum begins to appear. There is now a considerable breakdown of proteins within the cytoplasm in the region outside the developing spore, and the amino acids released are used in synthesis of spore proteins and as additional energy sources. From now on, all the energy for sporulation must come from endogenous sources since those in the medium are exhausted. This energy comes from both protein and poly-β-hydroxybutyrate (PHB). New mRNA molecules, which are probably responsible for synthesis of the new spore enzymes, are formed throughout the stages of the sporulation process until the mature spore is completed. The whole process of spore formation, from the time exponential growth ceases until the fully mature spore is released, takes many hours.

Most of the research on the molecular basis of

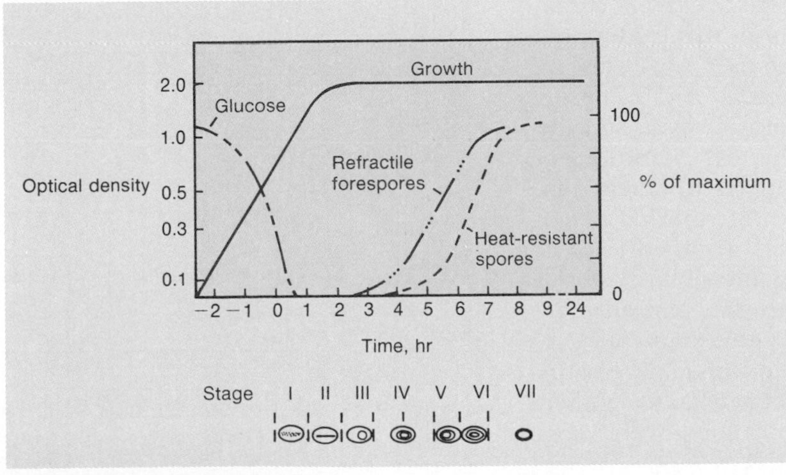

Figure 9.13 Relation of endospore formation to the growth curve. The events are initiated (time 0) by the exhaustion of glucose from the medium. The drawings indicate various morphological states, and relate to the structural sequence outlined in Figure 3.63.

endospore formation has been done with the organism *Bacillus subtilis*, for which good genetic systems are available. A large number of mutants of this organism have been isolated which are unable to sporulate, and physiological and genetic study of these mutants has shown that there are at least 45 separate sporulation loci, distributed among the various stages of sporulation that are diagrammed in Figure 9.13. Some of these loci consist of only single genes, whereas others consist of contiguous clusters of genes. The latter can be considered *sporulation operons.*

Regulation of sporulation occurs at the levels of both transcription and translation, but transcriptional control is at the basis of the overall process. Hybridization experiments have shown that new species of mRNA appear and disappear during various stages of the sporulation process. The RNA polymerase of vegetative cells does not transcribe sporulation genes effectively and a new RNA polymerase activity appears when sporulation begins. This change in activity is not associated with a change in the core enzyme (see Section 5.8) but with changes in the *sigma* factor. These changes involve two overall processes: reduction in activity of vegetative *sigma*; appearance of new *sigmas*. The inhibition of vegetative *sigma*, which turns off the synthesis of vegetative genes, is due to the synthesis early in the sporulation process of an *antisigma* factor which interferes with the binding of vegetative *sigma* to RNA polymerase. In addition, a whole series of other *sigma* factors are known which are successively synthesized and destroyed during various phases of the sporulation process. The details of how these various *sigma* factors themselves are regulated are still to be worked out, but the evidence is clear that *differential gene action* through the transcription process is at the basis of cellular differentiation.

9.10 Effect of Environmental Factors on Growth

Up to now we have described growth of microorganisms under hypothetical, essentially ideal, conditions. That analysis provides a useful introduction, but experimental work with actual organisms requires consideration of many factors. It is to be expected that the activities of microorganisms are greatly affected by the chemical and physical conditions of their environments. Understanding environmental influences helps us to explain the distribution of microorganisms in nature and makes it possible for us to devise methods for controlling microbial activities and destroying undesirable organisms. Not all organisms respond equally to a given environmental factor. In fact, an environmental condition may be harmful to one organism, and actually beneficial to another. However, organisms can tolerate some adverse conditions under which they cannot grow, and hence we must distinguish between the effects of environmental conditions on the viability of an organism and effects on growth, differentiation, and reproduction.

9.11 Temperature

Temperature is one of the most important environmental factors influencing the growth and survival of organisms. It can affect living organisms in either of two opposing ways. As temperature rises, chemical and enzymatic reactions in the cell proceed at more rapid rates and growth becomes faster. However, above a certain temperature, proteins, nucleic acids, and other cellular components are sensitive to high temperatures and may be irreversibly denatured. Usually, therefore, as the temperature is increased within a given range, growth and metabolic function increase up to a point where inactivation reactions set in. Above this point cell functions fall sharply to zero. Thus we find that for every organism there is a **minimum temperature** below which growth no longer occurs, an **optimum temperature** at which growth is most rapid, and a **maximum temperature** above which growth is not possible (Figure 9.14). The optimum temperature is always nearer the *maximum* than the minimum. These three temperatures, often called the **cardinal temperatures**, are generally characteristic for each type of organism, but are not completely fixed, as they can be modified by other factors of the environment. The maximum temperature most likely reflects the inactivation discussed above. However, the factors controlling an organism's minimum temperature are not as clear. As mentioned earlier (Section 3.4), the cell membrane must be in a fluid state for proper functioning. Perhaps the minimum temperature of an organism results from "freez-

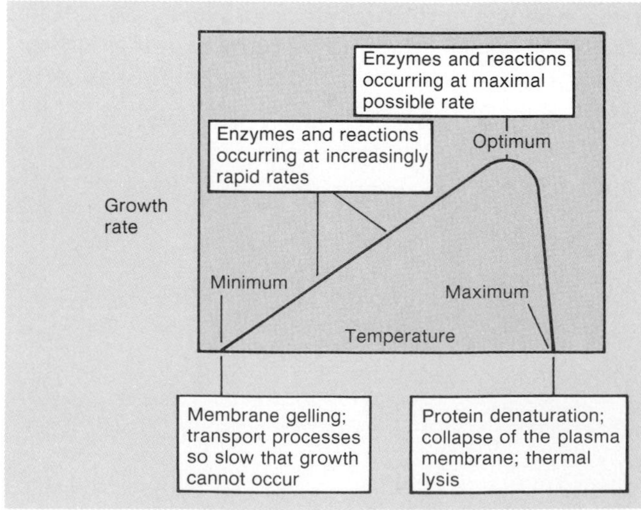

Figure 9.14 Effect of temperature on growth rate and the molecular consequences.

ing'' of the cell membrane so that it no longer functions properly in nutrient transport or proton gradient formation. This explanation is supported by experiments in which the minimum temperature for an organism can be altered to some extent by causing it to incorporate different fatty acids into its phospholipids, thereby changing the fluidity. It is also observed that the cardinal temperatures of different microorganisms differ widely; some microbes have temperature optima as low as 5 to 10°C and some as high as 90 to 100°C. The temperature range throughout which growth occurs is even wider than this, from below freezing (−12°C) to greater than boiling (the organism *Pyrodictium brockii* has a temperature *optimum* of 105°C!). No single organism will grow over this whole temperature range. However, the usual range for a given organism is about 30 to 40 degrees, although some have a much broader temperature range than others.

Although there is a continuum of organisms, from those with very low temperature optima to those with high temperature optima, it is possible to broadly distinguish four groups of organisms: **psychrophiles**, with low-temperature optima, **mesophiles**, with mid-range-temperature optima, **thermophiles**, with high-temperature optima, and **extreme thermophiles**, with very high-temperature optima (Figure 9.15). These temperature distinctions are made for convenience and the precise numbers should not be taken as absolutes. Mesophiles are found in warm-blooded animals and in terrestrial and aquatic environments in temperate and tropical latitudes. Psychrophiles and thermophiles are found in unusually cold or unusually hot environments, respectively. Extreme thermophiles are found in hot springs, geysers, or deep-sea vents (see below).

Cold environments Much of the world has fairly low temperatures. The oceans, which make up over half of the earth's surface, have an average temperature of 5°C, and the depths of the open oceans have temperatures of around 1 to 2°C. Vast land areas of the Arctic and Antarctic are permanently frozen, or are unfrozen only for a few weeks in summer. These cold environments are rarely sterile, and some microorganisms can be found alive and growing at any low temperature at which liquid water still exists. Even in many frozen materials there are usually microscopic pockets of liquid water present where microbes can grow. It is important to distinguish between environments that are cold throughout the year and those that are cold only in winter. The latter, characteristic of continental temperate climates, may have summer temperatures as high as 30°C, and winter temperatures far below 0°C. Such highly variable environments are much less favorable for cold-adapted organisms than are the constantly cold environments found in polar regions, at high altitudes, and in the depths of the oceans.

As noted earlier, organisms that are able to grow at low temperatures are called **psychrophiles** or **cryophiles**. A number of definitions of psychrophile exist, the most common being that a psychrophile is any organism able to grow at 0°C. This is a rather imprecise definition because many organisms with temperature optima in the 20s or 30s grow, albeit slowly, at 0°C. A current definition is that a true or strict psychrophile is an organism with an optimal temperature for growth of 15°C or lower, and a minimal temperature for growth at 0°C or lower. Organisms which grow at 0°C but have optima of 25 to 30°C have often been called facultative psychrophiles, but a more compact term is *psychrotroph*.

Psychrophilic algae are often seen on the surfaces of snowfields and glaciers in such large numbers that they impart a distinctive red or green coloration to the surface (Figure 9.16*a*). The most common snow alga is the eucaryote *Chlamydomonas nivalis*; its

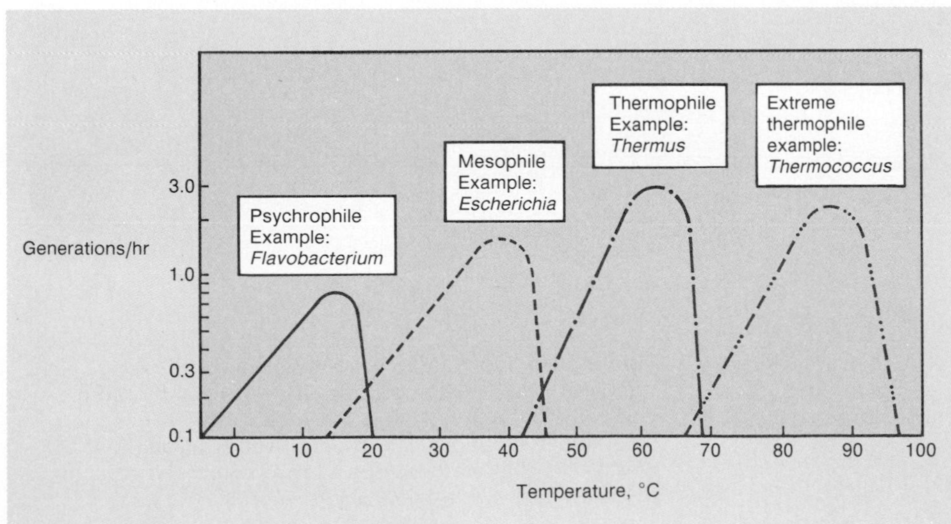

Figure 9.15 Relation of temperature to growth rates of a psychrophile, a mesophile, a thermophile, and an extreme thermophile.

brilliant red spores are responsible for the red color (Figure 9.16*b*). The alga probably grows within the snow as a green-pigmented vegetative cell, and then sporulates; as the snow dissipates by melting, erosion, and vaporization, the spores become concentrated on the surface. Snow algae are most commonly seen on permanent snowfields in mid- to late summer, and are especially common in sunny dry areas, probably because in more rainy areas they are washed away from the snowfields.

Psychrotrophs (facultative psychrophiles) are much more widely distributed than strict psychrophiles and can be isolated from soils and water in temperate climates. As we noted, they grow best at a temperature between 25 and 30°C, with a maximum of about 35°C. Since temperate environments do warm in summer it is understandable that they cannot support the very sensitive obligate psychrophiles, the warming essentially providing a selective force favoring facultative psychrophiles and excluding the obligate forms. It should be emphasized that although psychrotrophs do grow at 0°C, they do not grow very well, and one must often wait several weeks before visible growth is seen in culture media. Various genera of bacteria, fungi, algae, and protozoa have members that are psychrotrophic.

Meat, milk and other dairy products, cider, vegetables, and fruits, when stored in refrigerated areas, provide excellent habitats for the growth of psychrotrophic organisms. Growth of bacteria and fungi in foods at low temperatures can lead to changes in the quality of the food and eventually to spoilage. The lower the temperature, the less rapidly does spoilage occur, but only when food is frozen solid is microbial growth impossible. The frozen food industry owes its great development in recent years to the much greater keeping qualities of frozen foods over merely refrigerated foods.

High-temperature environments As noted earlier, organisms that grow at temperatures above 45 to 50°C are called **thermophiles**. Temperatures as high as these are found in nature only in certain restricted areas. For example, soils subject to full sunlight are often heated to temperatures above 50°C at midday, and darker soils may become warmed even to 70°C, although a few inches under the surface the temperature is much lower. Fermenting materials such as compost piles and silage usually reach temperatures of 60 to 65°C. However, the most extensive and extreme high-temperature environments are found in nature in association with volcanic phenomena.

Many hot springs have temperatures around boiling, and steam vents (fumaroles) may reach 150 to 500°C. Geothermal vents in the bottom of the ocean have temperatures of 350°C or greater (see Section 17.9 for a discussion of the deep sea vents). Hot

(a)

Katherine M. Brock

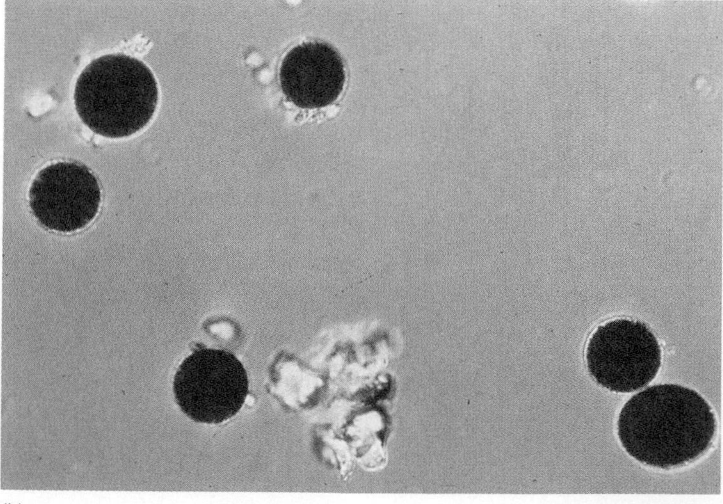

(b)

Figure 9.16 Snow algae. (a) Snowbank in the Sierra Nevada, California, with red coloration caused by the presence of snow algae. Pink snow such as this is common on summer snow banks at high altitudes throughout the world. (b) Photomicrograph of red-pigmented cells of the snow alga *Chlamydomonas nivalis*.

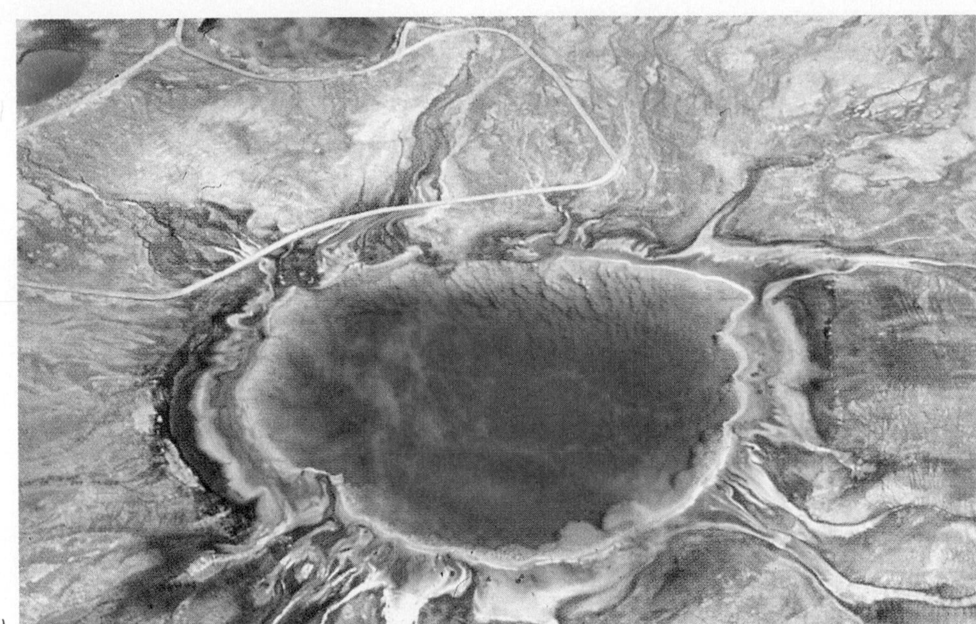

(a)

(b)

Figure 9.17 Cyanobacterial growth in hot springs in Yellowstone National Park. (a) Aerial photograph of a very large boiling spring, Grand Prismatic Spring. The orange color in the outflow channel is due to the rich carotenoid pigments of bacteria and cyanobacteria. (b) Characteristic V-shaped pattern formed by cyanobacteria at the upper temperature for photosynthetic life, 70–73°C. The pattern develops because the water cools more rapidly at the edges than in the center of the channel. The spring flows from the back of the picture toward the foreground.

Table 9.4 Upper temperature limits for growth of living organisms

Group	Approximate upper temperature (°C)
Animals	
Fish and other aquatic vertebrates	38
Insects	45–50
Ostracods (crustaceans)	49–50
Plants	
Vascular plants	45
Mosses	50
Eucaryotic microorganisms	
Protozoa	56
Algae	55–60
Fungi	60–62
Procaryotic microorganisms	
Cyanobacteria (blue-green algae)	70–73
Photosynthetic bacteria	70–73
Chemolithotrophic bacteria	>100
Heterotrophic bacteria	>100

springs occur throughout the world but are especially concentrated in western United States, New Zealand, Iceland, Japan, the Mediterranean region, Indonesia, Central America, and central Africa. The area with the largest single concentration of hot springs in the world is Yellowstone National Park, Wyoming. Although some springs vary in temperature, others are very constant, not varying more than 1 to 2°C over many years.

Many of these springs are at the boiling point for the altitude (92 to 93°C at Yellowstone, 99 to 100°C at locations where the springs are close to sea level). As the water overflows the edges of the spring and flows away from the source, it gradually cools, setting up a thermal gradient. Along this gradient, microorganisms develop (Figure 9.17), with different species growing in the different temperature ranges. By studying the species distribution along such thermal gradients and by examining hot springs and other ther-

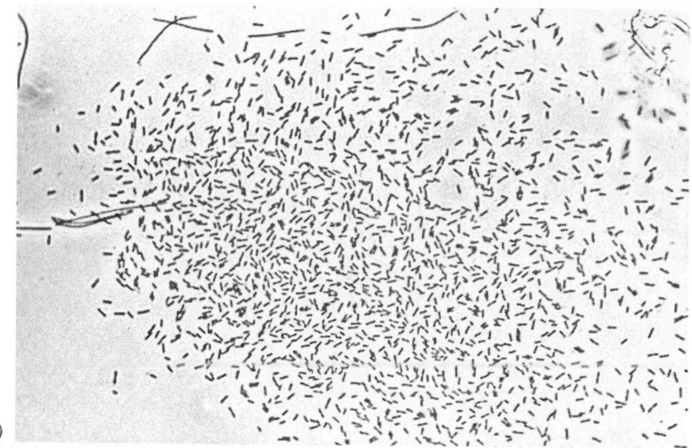

Figure 9.18 Bacterial growth in boiling water. (a) A typical small boiling spring in Yellowstone National Park. This spring is superheated, having a temperature 1 to 2 degrees above the boiling point. The mineral deposits around the spring consist mainly of silica. (b) Photomicrograph of a bacterial microcolony that developed on a microscope slide immersed in a boiling spring such as that shown above. Magnification, 380×.

mal habitats at different temperatures around the world it is possible to determine the upper temperature limits for each kind of microorganism (Table 9.4). From this information we conclude that (1) procaryotic organisms in general are able to grow at temperatures higher than those at which eucaryotes can grow, (2) nonphotosynthetic organisms are able to grow at higher temperatures than can photosynthetic forms, and (3) structurally less complex organisms can grow at higher temperatures than can more complex organisms. However, it should be emphasized that not all organisms from a group are able to grow near the upper limits for that group. Usually only a relatively few species or genera are able to function successfully near the upper temperature limit.

Bacteria can grow over the complete range of temperatures in which life is possible, but no one organism can grow over this whole range. As mentioned above, each organism is limited to a restricted range of perhaps 30°C, and it can grow well only within a still narrower range. Even within the thermophilic group of bacteria there are differences. Some thermophilic bacteria have temperature optima for growth at 55°C, others at 70°C, and still others at 100°C; organisms with a temperature optimum of 80°C or greater are referred to as *extreme* thermophiles. In most boiling springs (Figure 9.18) it is found that a variety of bacteria grow, often surprisingly rapidly. The growth of such bacteria can be easily studied by immersing microscope slides into the spring and retrieving them after a few days. Microscopic examination of the slides generally reveals small or large colonies of bacteria (Figure 9.18*b*) that have developed from single bacterial cells which attached to and grew on the glass surface.

Ecological studies of organisms living in boiling springs have shown that growth rates are surprisingly rapid, doubling times of 2 to 7 hours having been found. By use of radioactively labeled substrates it has been possible to show that bacteria living at these high temperatures have optimum temperatures near those of their environments. Both aerobic and anaerobic bacteria have been found living at high temperatures, and many morphological types exist (Table 9.5). Those extreme thermophiles that have been best studied are members of the archaebacteria and are discussed in detail in Sections 18.6, 18.8, and the latter part of Chapter 19. One anaerobe, isolated from an undersea thermal vent, has an optimum of 105°C.

Thermophilic bacteria related to those living in hot springs have also been found in artificial thermal environments. The hot-water heater, domestic or industrial, usually has a temperature of 55 to 80°C and is a favorable habitat for the growth of thermophilic bacteria. Organisms resembling *Thermus aquaticus*, a common hot spring organism, have been isolated from the hot water of many installations. Electric power plants, hot industrial process water, and other artificial thermal sources probably also provide sites where thermophiles can grow.

Table 9.5 Genera of thermophilic bacteria

Genus	Number of species	Temperature range (°C)
Phototrophic bacteria		
Cyanobacteria	16	55–70 (One strain, 74)
Purple bacteria	1	55–60
Green bacteria	1	70–73
Gram-positive bacteria		
Bacillus	15	50–70
Clostridium	11	50–75
Lactic acid bacteria	5	50–65
Actinomycetes	23	55–75
Other eubacteria		
Thiobacillus	3	50–60
Spirochete	1	54
Desulfotomaculum	7	37–55
Gram-negative aerobes	7	50–75
Gram-negative anaerobes	4	50–75
Archaebacteria		
Methanogens	4	55–95
Sulfur-dependent	10	55–110
Thermoplasma	1	37–55

9.12 Acidity and Alkalinity (pH)

Acidity or alkalinity of a solution is expressed by its **pH value** on a scale in which neutrality is pH 7 (Figure 9.19). Those pH values that are less than 7 are *acidic* and those greater than 7 are *alkaline* (or *basic*). It is important to remember that pH is a *logarithmic function*; a change of one pH unit represents a *tenfold* change in hydrogen ion concentration. Thus vinegar (pH near 2) and household ammonia (near pH 11) differ in hydrogen ion concentration by one billion times.

The pH of a culture medium is adjusted by adding an alkaline compound, such as sodium hydroxide, if the pH is too acid, or an acidic compound, such as hydrochloric acid, if the pH is too alkaline. Because microorganisms usually cause changes in the pH of their environments as they grow, it is often desirable to add to the culture medium a pH *buffer*, which acts to keep the pH relatively constant. Such pH buffers work only over a narrow pH range; hence, different buffers must be selected for different pH regions. For near neutral pH ranges (pH 6 to 7.5), phosphate makes an excellent buffer. Indicator dyes are often added directly to culture media and can then indicate not only the initial pH of the medium but also changes in pH that result from the growth or activity of microorganisms.

Each organism has a pH range within which growth is possible, and usually has a well-defined pH optimum. Most natural environments have pH values between 5 and 9, and organisms with optima in this range are most common. Only a few species can grow at pH values of less than 2 or greater than 10. Organisms that live at low pH are called *acidophiles*. Most bacteria grow best at neutral or alkaline pH, although a few acid-tolerant bacteria exist. Fungi as a group tend to be more acid-tolerant than bacteria. Many fungi grow optimally at pH 5 or below, and a few grow quite well at pH values as low as 2, although their interior pH is close to neutrality.

Although most bacteria grow best at neutral pH,

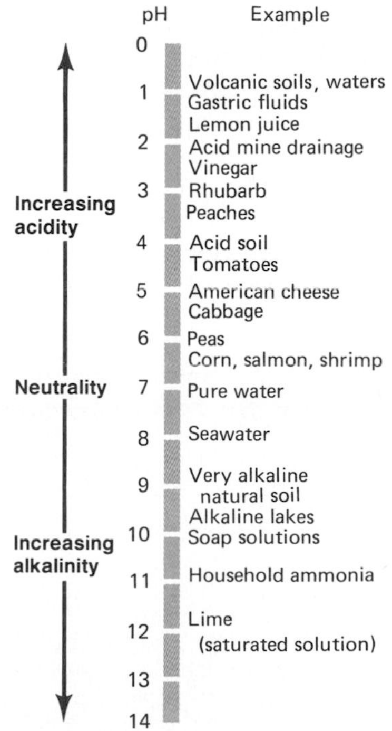

Figure 9.19 The pH scale.

a few acidophilic bacteria exist. In fact, some of these bacteria are *obligate* acidophiles, unable to grow at all at neutral pH. Obligately acidophilic bacteria include several species of the eubacterial genus *Thiobacillus*, and several genera of archaebacteria, including *Sulfolobus* and *Thermoplasma*. *Thiobacillus* and *Sulfolobus* exhibit an interesting property: they oxidize sulfide minerals and produce sulfuric acid. We discuss the role of these organisms in mining processes in Sections 17.16 and 17.17. It is strange to consider that for obligate acidophiles a neutral pH is actually toxic! Probably the most critical factor for obligate acidophily is the plasma membrane. When the pH is raised to neutrality, the plasma membrane of obligate acidophilic bacteria actually dissolves, and the cells lyse, suggesting that high concentrations of hydrogen ions are required for membrane stability.

9.13 Water Availability

All organisms require water for life, and water availability is one of the most important factors affecting the growth of microorganisms in natural environments. Water availability does not depend only on the water content of the environment because various solid substances and surfaces are able to absorb water molecules more or less tightly and hence render them unavailable. Also, solutes such as salts and sugars that are dissolved in water have an affinity for water, and the water associated with such solutes becomes unavailable to organisms.

Water availability is generally expressed in physical terms such as **water activity** or **water potential**. In food microbiology where water availability is frequently of major concern (see Section 9.21), the concept of water activity is used. Water activity, abbreviated a_w, is expressed as a ratio of the vapor pressure of the air over a substance or solution divided by the vapor pressure at the same temperature of pure water. Thus values of a_w vary between 0 and 1. Some representative values are given in Table 9.6. Water activities in agricultural soils generally range between 0.900 and 1.000.

Water diffuses from a region of high water concentration to a region of lower water concentration. Thus, if pure water and a salt solution are separated by a semi-permeable membrane, water will diffuse *from* the pure water *into* the salt solution. The process by which water diffuses across a semi-permeable membrane is called *osmosis*. In most cases, the cytoplasm of a cell has a higher solute concentration than the environment, so that water will diffuse into the cell. However, if a cell is present in an environment of low water activity, there will be a tendency for water to flow out of the cell. Thus, when a cell is immersed in a solution of low water activity, such as salt or sugar solution, it *loses* water to the environment. The salt or sugar solution can, in effect, be considered analogous to a *dry* environment.

From an ecological viewpoint, osmotic effects are of interest mainly in habitats with high concentrations of salts. Seawater contains about 3.5 percent sodium chloride plus small amounts of many other minerals and elements. Microorganisms found in the sea usually have a specific requirement for the sodium ion, in addition to living optimally at the water activity of seawater. Such organisms are often called **moderate halophiles** (*moderate* to distinguish them from the extreme halophiles discussed below). Marine microorganisms are generally inhibited at both higher and lower concentrations of salt (Figure 9.20). Marine microorganisms require sodium ions for the stability of the cell membrane and in addition many of their enzymes require sodium ions for activity.

Most organisms are unable to cope with environments of very low water activity and either die or remain dormant. A few specialized organisms are able to live under conditions of low water activity and these organisms are fairly important in applied microbiology. Organisms capable of living in very salty environments are called **extreme halophiles** (*extreme* to distinguish them from the moderate halophiles living in the sea); organisms able to live in environments high in sugar are called **osmophiles**, and organisms able to live in very dry environments are called **xerophiles**. Examples of these various organisms are given in Table 9.6. How do these orga-

Table 9.6 Water activity for several materials

Water activity, a_w	Material	Some organisms growing at stated water activity
1.000	Pure water	*Caulobacter, Spirillum*
0.995	Human blood	*Streptococcus, Escherichia*
0.980	Seawater	*Pseudomonas, Vibrio*
0.950	Bread	Most Gram-positive rods
0.900	Maple syrup, ham	Gram-positive cocci
0.850	Salami	*Saccharomyces rouxii* (yeast)
0.800	Fruit cake, jams	*Saccharomyces bailii, Penicillium* (fungus)
0.750	Salt lake, salt fish	*Halobacterium, Halococcus*
0.700	Cereals, candy, dried fruit	Xerophilic fungi

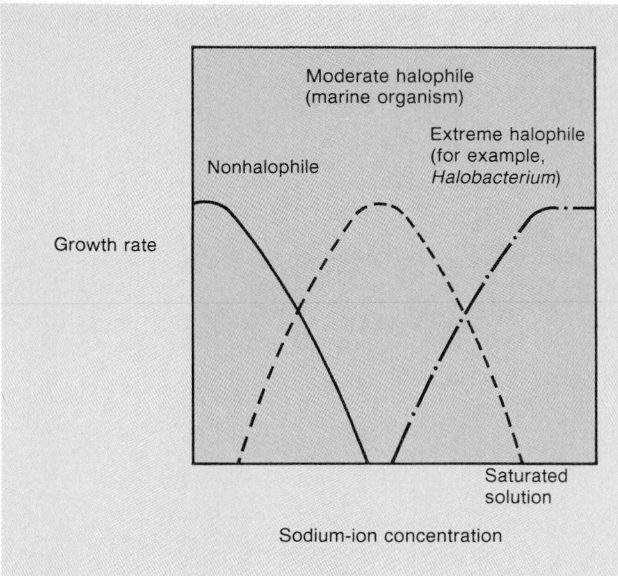

Figure 9.20 Effect of sodium ion concentration on growth of microorganisms of different salt tolerances.

nisms grow under conditions of low water activity? When an organism grows in a medium with a low water activity, it must perform work to extract water from the medium. However, organisms are not able to transport water molecules themselves; water movement is strictly a physical process. Thus, a cell can only obtain water by increasing its internal solute concentration. An increase in internal solute concentration can be done either by pumping ions into the cell from the environment, or by producing inside the cell via metabolism an organic solute. Organisms are known which have each of these mechanisms. Several examples are given below.

The solute which is used inside the cell for adjustment of cytoplasmic water activity must be nontoxic; such compounds are called **compatible solutes**. Several examples of compatible solutes can be mentioned here. (1) Gram-positive cocci such as *Staphylococcus* are able to live in environments of moderately low water activity (see Table 9.6). These organisms use the amino acid *proline* as compatible solute. (A common enrichment procedure for *Staphylococcus* is a medium containing 6 percent sodium chloride.) Many Gram-negative bacteria able to tolerate similar salt concentrations use the amino acid *glutamate* as a compatible solute. (2) Halobacteria (for example, *Halobacterium halobium*) are archaebacteria which live in very saline habitats such as salt lakes or salt evaporating ponds. These organisms use *potassium* ions as the compatible solute, concentrating the latter from the environment until the interior of the cell is nearly saturated with potassium. (3) Some osmotolerant yeasts can live in the presence of high sucrose or salt concentrations. Such organisms present problems in food spoilage of very sugary materials such as syrups or very salty solutions such as soy sauce. These osmotolerant yeasts use polyalcohols such as *sorbitol* or *ribitol* as compatible solutes. (4) The halophilic alga *Dunaliella salina* lives in extremely salty lakes, such as Great Salt Lake or salt-evaporating ponds. As its compatible solute, this alga synthesizes *glycerol*. (5) Lichens are composite organisms composed of a fungus and alga living together (see Section 1.7). Lichens are able to live directly in the atmosphere, where water activities are often quite low (see Figure 1.25). They are able to extract water from the atmosphere by production of a polyalcohol such as *mannitol* or *sorbitol* as compatible solute.

9.14 Oxygen

As discussed in Chapter 4, microorganisms vary in their need for, or tolerance of, oxygen. Microorganisms can be divided into several groups depending on the effect of oxygen, as outlined in Table 9.7. Organisms which lack a respiratory system will not be able to use oxygen as terminal electron acceptor. Such organisms are called **anaerobes**, but there are two kinds of anaerobes, those which can tolerate oxygen even though they cannot use it, and those which are killed by oxygen. Organisms killed by oxygen are called *obligate* (or strict) *anaerobes*. Although the reason why obligate anaerobes are killed by oxygen is not clear, one idea is that obligate anaerobes are unable to detoxify some of the products of oxygen metabolism. When oxygen is reduced, several toxic products, hydrogen peroxide (H_2O_2), superoxide (O_2^-), and hydroxyl radical ($OH\cdot$) are formed. Aerobes have enzymes which decompose these products, whereas anaerobes seem to lack all or some of these enzymes.

For the growth of many aerobes, it is necessary to provide extensive aeration. This is because O_2 is

Table 9.7 Terms used to describe O_2 relations of microorganisms

Group	O_2 effect
Aerobes	
Obligate	Required
Facultative	Not required, but growth better with O_2
Microaerophilic	Required, but at levels lower than atmospheric
Anaerobes	
Aerotolerant	Not required, and growth no better when O_2 present
Obligate (strict) anaerobes	Harmful or lethal

only poorly soluble in water, and the O_2 used up by the organisms during growth is not replaced fast enough by diffusion from the air. Forced aeration of cultures is therefore frequently desirable and can be achieved either by vigorously shaking the flask or tube on a shaker or by bubbling sterilized air into the medium through a fine glass tube or porous glass disc. Usually aerobes grow much better with forced aeration than when O_2 is provided by simple diffusion.

For anaerobic culture, the problem is to *exclude* oxygen. One of the more difficult techniques in microbiology is the maintenance of anaerobic conditions: oxygen is ubiquitous in the air. Obligate anaerobes vary in their sensitivity to oxygen, and a number of procedures are available for reducing the O_2 content of cultures—some simple and suitable mainly for less sensitive organisms, others more complex but necessary for the most fastidious obligate anaerobes.

Bottles or tubes filled completely to the top with culture medium and provided with tightly fitting stoppers will provide anaerobic conditions for organisms not too sensitive to small amounts of oxygen. It is also possible to add a chemical which reacts with oxygen and excludes it from the culture medium. Such a substance is called a *reducing agent* because it reduces oxygen to water. A good example is *thioglycollate*, which is added to a medium commonly used to test whether an organism is aerobic, facultative, or anaerobic (Figure 9.21). After thioglycollate reacts with oxygen throughout the tube, oxygen can only penetrate near the top of the tube where the medium contacts air. Aerobes grow only at the top of the tube. Facultative organisms grow throughout the tube, but best near the top. Anaerobes grow only near the bottom of the tube, where oxygen cannot penetrate. An indicator dye, such as *resazurin*, is usually added to the medium because the dye will change color in the presence of oxygen and thereby indicate the degree of penetration of oxygen into the medium.

To remove all traces of O_2 for the culture of very fastidious anaerobes, it is possible to place an O_2-consuming gas in a jar holding the tubes or plates. One of the simplest devices for this is the *anaerobic jar*, a heavy-walled jar with a gas-tight seal, within which tubes, plates, or other containers to be incubated are placed. The air in the jar is replaced with hydrogen gas (H_2), and in the presence of a chemical catalyst the traces of O_2 left in the vessel or culture medium are consumed, thus leading to anaerobic conditions. (See Section 13.1 for a discussion of the use of the anaerobic jar.)

For the most fastidious anaerobes, such as the methanogens, it is necessary not only to carefully remove all traces of O_2 but also to carry out all manipulations of cultures in an anaerobic atmosphere, as these organisms are frequently killed by even brief exposure to O_2. In one technique, all manipulations

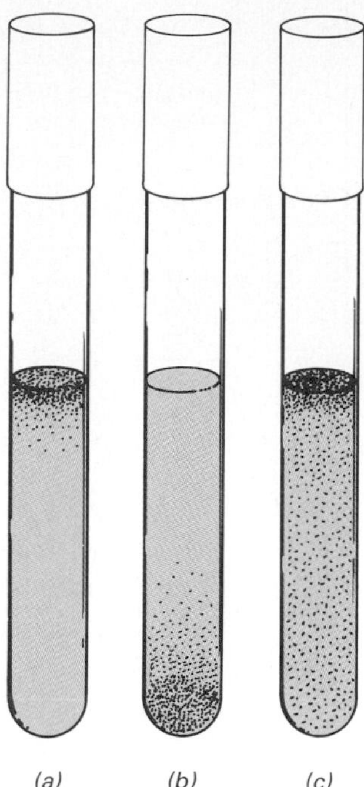

Figure 9.21 Aerobic, anaerobic, and facultative growth, as revealed by the position of microbial colonies within tubes of a culture medium. A small amount of agar has been added to keep the liquid from becoming disturbed. (a) Oxygen only penetrates a short distance into the tube, so that aerobes grow only at the surface. (b) Anaerobes, being sensitive to oxygen, only grow away from the surface. (c) Facultative anaerobes are able to grow either in the presence or absence of oxygen and grow throughout the tube.

are carried out under a tiny jet of O_2-free hydrogen or nitrogen gas which is directed into the culture vessel when it is open, thus driving out any O_2 that might enter. For extensive research on anaerobes, special boxes fitted with gloves, called *anaerobic glove boxes*, permit work with open cultures in completely anaerobic atmospheres (Figure 9.22).

Toxic forms of oxygen To understand oxygen toxicity, it is necessary to consider the chemistry of oxygen. Molecular oxygen, O_2, is unique among diatomic elements in that two of its outer orbital electrons are unpaired. It is partly because it has two unpaired electrons that oxygen has such a high reduction potential and is such a powerful oxidizing agent. However, the two electrons are in separate outer orbitals and have parallel spins, whereas most other molecules have antiparallel spins. Since reactions between molecules occur much more readily if the spins of the outer electrons are the same, molecular oxygen is not immediately reactive in many situations, but needs activation.* The unactivated, ground state of

*The common occurrence of iron in respiratory systems may be due to the fact that ferrous iron does readily undergo redox reaction with O_2.

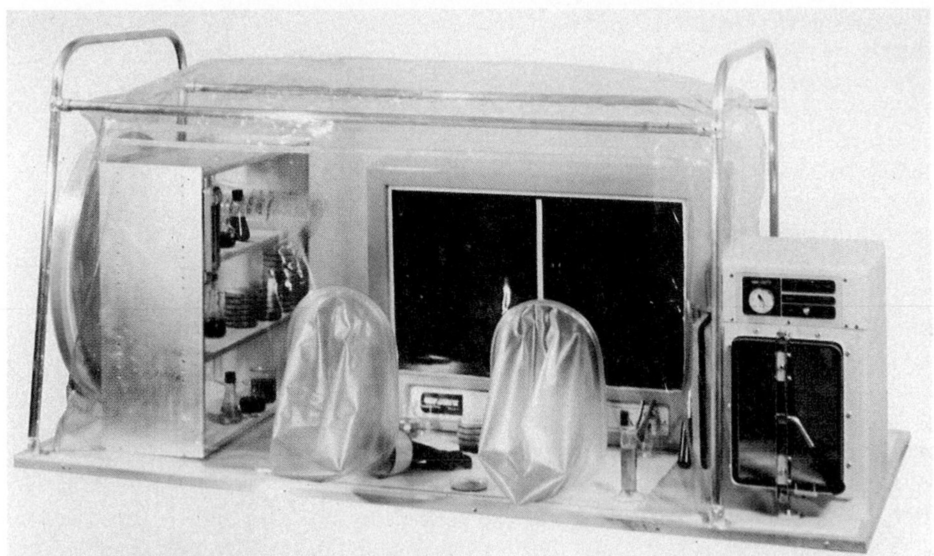

Figure 9.22 Special anaerobic hood for manipulating and incubating cultures under anaerobic conditions.

oxygen, which is the most common, is called **triplet oxygen**.

Singlet oxygen is a higher energy form of oxygen. It is formed when the two outer electrons achieve antiparallel spins, either in the same orbital or in separate orbitals. Singlet oxygen is extremely reactive and is one of the main forms of oxygen that is toxic to living organisms. Note that no additional electrons have been added to convert triplet to singlet oxygen, but energy has been added to change the spin of one of the electrons. Singlet oxygen is produced chemically in a variety of ways and is an atmospheric pollutant, being one of the components present in smog. Singlet oxygen may also be produced biochemically, either spontaneously or via specific enzyme systems. The most common biochemical means of singlet-oxygen production involves a reaction of triplet oxygen with visible light. This involves a dye molecule which acts as a mediator by absorbing light. Another important means of generating singlet oxygen is through the lactoperoxidase and myeloperoxidase enzyme systems. These systems are present in milk, saliva, and body cells (phagocytes) that eat and digest invading microorganisms. When a phagocyte ingests a microbial cell (see Figure 11.25), the peroxidase system is activated and singlet oxygen is generated. Due to the high toxicity of singlet oxygen, the phagocytes thus are able to kill invading organisms. The high reactivity of singlet oxygen means that if it is present in a biological system, a wide variety of uncontrolled and undesirable oxidation reactions can occur, leading to oxidative destruction of vital cell components such as phospholipids in the plasma membrane.

The reduction of O_2 to $2H_2O$, which occurs during respiration (see Section 4.11), requires addition of four electrons (Figure 9.23). This reduction usually occurs by single electron steps, and the first product formed in the reduction of O_2 is **superoxide** anion, O_2^-. Superoxide is probably formed transiently in small amounts during normal respiratory processes, and it is also produced by light through mediation of a dye and by one-electron transfers to oxygen. Flavins, flavoproteins, quinones, thiols, and iron-sulfur proteins (see Section 4.11) all carry out one-electron reductions of oxygen to superoxide. Superoxide is highly reactive and can cause oxidative destruction of lipids and other biochemical components. It has the longest life of the various oxygen intermediates and may even pass from one cell to another. As we will see below, superoxide is probably important in explaining the oxygen sensitivity of obligate anaerobes.

The next product in the stepwise reduction of oxygen is peroxide, O_2^{2-}. Peroxide anion is most familiar in the form of hydrogen peroxide, H_2O_2, which is sufficiently stable that it can be used as an item of chemical commerce. Peroxide is commonly formed biochemically during respiratory processes by a two-electron reduction of O_2, generally mediated by flavoproteins. Hydrogen peroxide is probably produced in small amounts by almost all organisms growing aerobically.

$O_2 + e^- \rightarrow O_2^-$ Superoxide

$O_2^- + e^- + 2H^+ \rightarrow H_2O_2$ Hydrogen peroxide

$H_2O_2 + e^- + H^+ \rightarrow H_2O + OH\cdot$ Hydroxyl radical

$OH\cdot + e^- + H^+ \rightarrow H_2O$ Water

Overall: $O_2 + 4e^- + 4H^+ \rightarrow 2H_2O$

Figure 9.23 Four-electron reduction of O_2 to water by stepwise addition of electrons. All of the intermediates formed are reactive and toxic to cells.

Hydroxyl free radical, OH·, is the most reactive of the various oxygen intermediates. It is the most potent oxidizing agent known and is capable of attacking any of the organic substances present in cells. Hydroxyl radical is formed as a result of the action of ionizing radiation, and is probably one of the main agents in the killing of cells by X rays and gamma rays. Hydroxyl radical can also be produced as shown in Figure 9.23.

Enzymes that act on oxygen derivatives With such an array of toxic oxygen derivatives, it is perhaps not surprising that organisms have developed enzymes that destroy certain oxygen products (Figure 9.24). The most common enzyme in this category is **catalase**, the activity of which is illustrated in Figure 9.25. Another enzyme that acts on hydrogen peroxide is **peroxidase** (Figure 9.24*b*), which requires the presence of a reducing substance, usually NADH. Superoxide is destroyed by the enzyme **superoxide dismutase** (Figure 9.24*c*), which combines two molecules of superoxide to form one molecule of hydrogen peroxide and one molecule of oxygen. Superoxide dismutase and catalase working together can thus bring about the conversion of superoxide back to oxygen.

Anaerobic microorganisms Environments of low reduction potential (that is, anaerobic environments) include muds and other sediments of lakes, rivers, and oceans; bogs and marshes; waterlogged soils; canned foods; intestinal tracts of animals; the oral cavity of animals, especially around the teeth; certain sewage-treatment systems; deep underground areas such as oil pockets; and some underground waters (see Chapter 17 for description of some of these habitats). In most of these habitats, the low reduction potential is due to the activities of organisms, mainly bacteria, that consume oxygen during respiration. If no replacement oxygen is available, the habitat becomes anaerobic.

So far as is known, obligate anaerobiosis occurs in only two groups of microorganisms, the bacteria and the protozoa. The best known obligately anaerobic bacteria belong to the genus *Clostridium*, a group of Gram-positive spore-forming rods. Clostridia are widespread in soil, lake sediments, and intestinal tracts, and are often responsible for spoilage of canned foods. Other obligately anaerobic bacteria are found among the methanogenic bacteria, the sulfate-reducing bacteria (see Chapter 19), and many of the bacteria which inhabit the animal gut (see Section 17.1). Among obligate anaerobes, however, the sensitivity to oxygen varies greatly; some organisms are able to tolerate traces of oxygen whereas others are not.

(*a*) Catalase:

$$H_2O_2 + H_2O_2 \rightarrow 2H_2O + O_2$$

Hydrogen peroxide

(*b*) Peroxidase:

$$H_2O_2 + NADH + H^+ \rightarrow 2H_2O + NAD^+$$

(*c*) Superoxide dismutase:

$$O_2^- + O_2^- + 2H^+ \rightarrow H_2O_2 + O_2$$

Superoxide

Figure 9.24 Enzymes acting on toxic oxygen species. Catalases (a) and peroxidases (b) are generally porphyrin-containing proteins, although some flavoproteins may act in this manner. Superoxide dismutases (c) are metal-containing proteins, the metal being either copper, zinc, manganese, or iron.

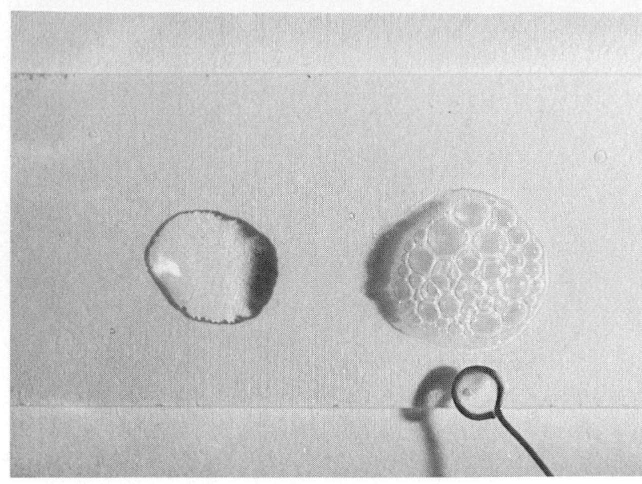

Figure 9.25 Method for testing a microbial culture for the presence of catalase. A heavy loopful of cells from an agar culture was mixed on a slide with a drop of 30 percent hydrogen peroxide. The immediate appearance of bubbles is indicative of the presence of catalase. The bubbles are O_2, produced by the reaction: $H_2O_2 + H_2O_2 \rightarrow 2H_2O + O_2$.

9.15 Control of Microbial Growth: Heat Sterilization

Up until now, we have been discussing environmental factors that influence growth, with emphasis on *promoting* the growth process. The control of microbial growth is necessary in many practical situations. Control can be effected either by ***killing*** organisms or by ***inhibiting*** their growth. The complete killing of all organisms is generally called **sterilization** and is brought about by use of heat, radiation, or chemicals. Sterilization means the complete destruction of all microbial cells present; once a product is sterilized, it will remain sterile indefinitely if it is properly sealed (see the illustration of Pasteur's swan-necked flask in Figure 1.47).

Heat sterilization One of the most important and widely used agents for sterilization is heat. Let us consider the effects of temperature on viability. As temperature rises past the maximum temperature for growth, lethal effects become apparent. As shown in Figure 9.26, death from heating is an exponential or first-order function and occurs more rapidly as temperature is raised. These facts have important practical consequences. If we wish to kill every cell, that is to ***sterilize*** a population, it will take longer at lower temperatures than at higher temperatures. It is thus necessary to select the time and temperature that will sterilize under stated conditions.

The first-order relationship shown in Figure 9.26 means that the rate of death is proportional at any instant only to the concentration of organisms at that instant and that the time taken for a definite fraction (for example, 90 percent) of the cells to be killed is independent of the initial concentration. This means, to a first approximation, that heat treatment has no cumulative effect on a cell, and that the length of time a cell has been exposed to heat prior to death has no perceivable effect on it.

The rate of inactivation may deviate from a simple first-order reaction for a number of reasons (Figure 9.27). If the cell suspension is clumped, every cell in the clump must be inactivated before the colony-forming ability of the clump will be inactivated, so that the rate of death during the early stages of heating will be slower than later (Figure 9.27*a*). Second, if the suspension is composed of two populations with different heat resistances, the more heat-sensitive cells will be inactivated first, and then the inactivation rate will slow to a new rate during the time when the more resistant organism is being inactivated (Figure 9.27*b*). If inactivation of spores is being studied, germination of the spore is usually activated by heat, which sometimes results in an initial rise in the viable count before a subsequent first-order inactivation (Figure 9.27*c*).

The most useful way of characterizing heat inactivation is to determine the time at which a defined fraction of the population is killed. For a variety of reasons the time required for a ten-fold reduction in

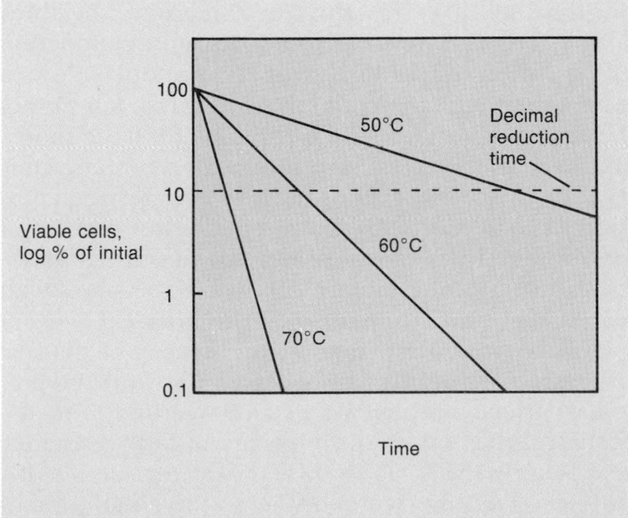

Figure 9.26 Effect of temperature on viability of a mesophilic bacterium.

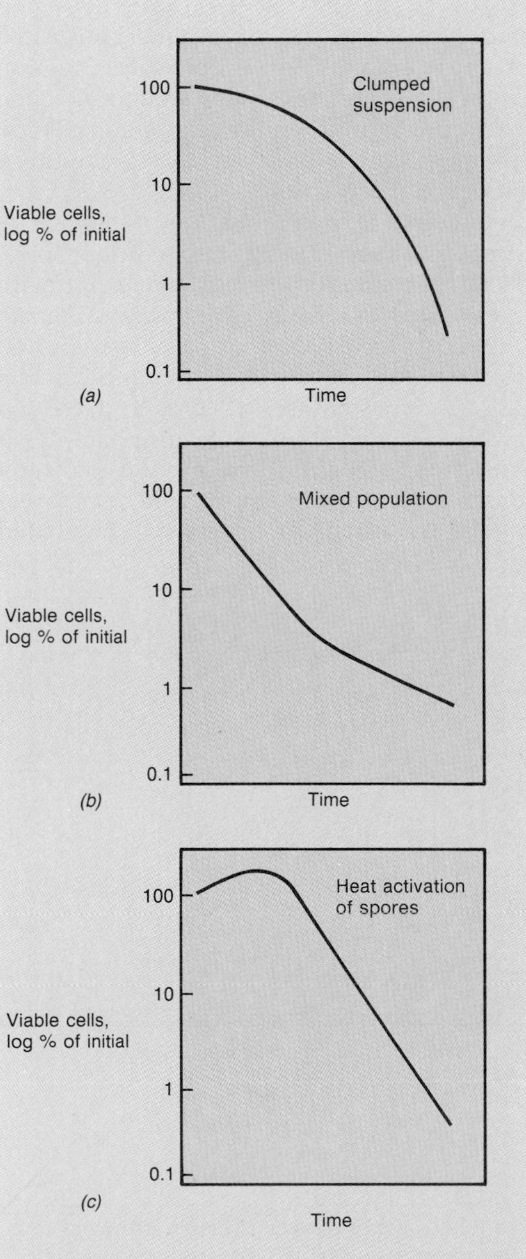

Figure 9.27 Deviations from first-order heat inactivation. Curves shown represent responses at single temperatures. Compare with Figure 9.26.

the population density at a given temperature, called the **decimal reduction time** or D, is the most useful parameter. Over the range of temperatures usually used in food sterilization, the relationship between D and temperature is essentially exponential, so that when the logarithm of D is plotted against temperature a straight line is obtained (Figure 9.28). The slope of the line provides a quantitative measure of the sensitivity of the organism to heat under the conditions used, and the graph can be used in calculating process times for sterilization, such as in canning operations. Organisms and spores vary considerably in heat resistance. For instance, in the autoclave, a temperature of 121°C may be reached. Under these conditions, spores from a very heat-resistant organism may require 4 to 5 minutes for a decimal reduction, whereas spores of other types show a decimal reduction in 0.1 to 0.2 minute at this temperature. Vegetative cells may require only 0.1 to 0.5 minute at 65°C for a decimal reduction.

Although accurate, determination of decimal reduction times is a fairly lengthy procedure, since it requires making a number of viable count measurements. A less satisfactory but easier way of characterizing the heat sensitivity of an organism is to determine the **thermal death time**, the time at which all cells in a suspension are killed at a given temperature. This is done simply by heating samples of this suspension for different times, mixing the heated suspensions with culture medium, and incubating. When all cells are killed, no growth will be evident in the incubated samples. Of course, the thermal death time determined in this way will depend on the size of the population tested, since a longer time will be required to kill all cells in a large population than in a small one. If the number of cells is standardized, then it is possible to compare the heat sensitivities of different organisms by comparing their thermal death times. When the logarithm of the thermal death time is graphed versus temperature, a straight line similar to that shown in Figure 9.28 is obtained.

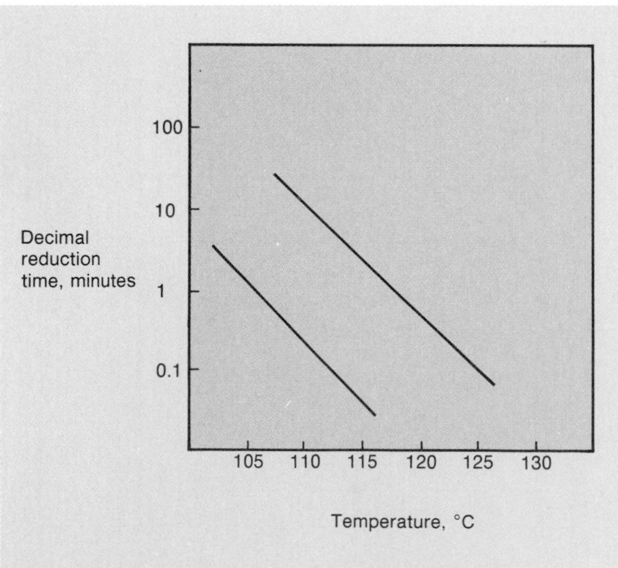

Figure 9.28 Relationship between temperature and rate of killing as indicated by the decimal reduction time. Data for such a graph are obtained from curves such as those given in Figure 9.26. The upper line in the figure represents data for a very heat-resistant organism.

The nature of the medium in which heating takes place influences the killing of vegetative cells or spores. Microbial death is more rapid at acidic pH values, and for this reason acid foods such as tomatoes, fruits, and pickles are much easier to sterilize than more neutral foods such as corn and beans. High concentrations of sugars, proteins, and fats usually increase the resistance of organisms to heat, while high salt concentrations may either increase or decrease heat resistance, depending on the organism. Dry cells (and spores) are more heat resistant than moist ones; for this reason, heat sterilization of dry objects always requires much higher temperatures and longer times than does sterilization of moist objects. (Even some multicellular desert animals can survive brief heating to 100°C if they are dry.)

Bacterial endospores (see Section 3.12) are the most heat-resistant structures known: they are able to survive conditions of heating that would rapidly kill vegetative cells of the same species. Although it is not completely certain, it seems likely that a major factor in heat resistance is the amount and state of *water* within the endospore, since dry cells are much more resistant to heat than wet cells. During endospore formation, the protoplasm is reduced to a minimum volume as a result of the accumulation of metal ions, such as Ca^{2+} (high in spores), and dipicolinic acid (present only in spores), which lead to formation of a gel-like structure. At this stage, the thick cortex forms around the protoplast and contraction of the cortex results in a shrunken protoplast low in water. The water content of the protoplast determines the degree of heat resistance of the spore. If the spores of a bacterial species have high water content, they have low heat resistance, and the heat resistance of spores can be varied by altering the water content of the spores. Water moves freely in and out of spores, so that it is not the impermeability of the spore wall that affects water content, but the physical state of the spore protoplast, that is, the degree of gel-like structure. The tightly fitting cortex prevents the protoplast from imbibing water and swelling. The importance of an intact spore cortex in heat resistance is shown by the fact that, when germination occurs, loss of refractility (which is due to the dissolution of the cortex) and decrease in heat resistance parallel each other.

Pasteurization and sterilization Pasteurization is a process using mild heat to reduce the microbial populations in milk and other foods that are exceptionally heat sensitive. It is named for Louis Pasteur, who first used heat for controlling the spoilage of wine (see Box, Chapter 1). It is not synonymous with sterilization since not all organisms are killed. Originally, pasteurization of milk was used to kill pathogenic bacteria, especially the organisms causing tuberculosis, brucellosis, and typhoid, but it was discovered that the keeping qualities of the milk were also improved. Today, since milk rarely comes from cows infected with the pathogens mentioned above, pasteurization is used primarily because it improves the keeping qualities.

Pasteurization of milk is usually achieved by passing the milk continuously through a heat exchanger where its temperature is raised quickly to 71°C and held there for 15 seconds, and then is quickly cooled, a process called **flash pasteurization**. Occasionally pasteurization is done by heating milk in bulk at 63 to 66°C for 30 minutes and then quickly cooling. However, flash pasteurization is more satisfactory in that it alters the flavor less, kills heat-resistant organisms better, and can be carried out on a continuous-flow basis, thus making it adaptable to large dairy operations.

Sterilization by heat involves treatment that results in the complete destruction of all organisms, and since bacterial endospores are ubiquitous, sterilization procedures are designed to eliminate them. This requires heating at temperatures above boiling and the use of steam under pressure (Figure 9.29*a*). The usual procedure is to heat at 1.1 kg/cm^2 (15 lb/in^2) steam pressure, which yields a temperature of 121°C. Heating is usually done in an autoclave, although if only small batches of material need be sterilized, a home pressure cooker is quite satisfactory and much less expensive. At 121°C, the time of autoclaving to achieve sterilization is generally considered to be 10 to 15 minutes (Figure 9.29*b*). If bulky objects are being sterilized, heat transfer to the interior will be slow, and the heating time must be sufficiently long so that the material is at 121°C for 10 to 15 minutes.

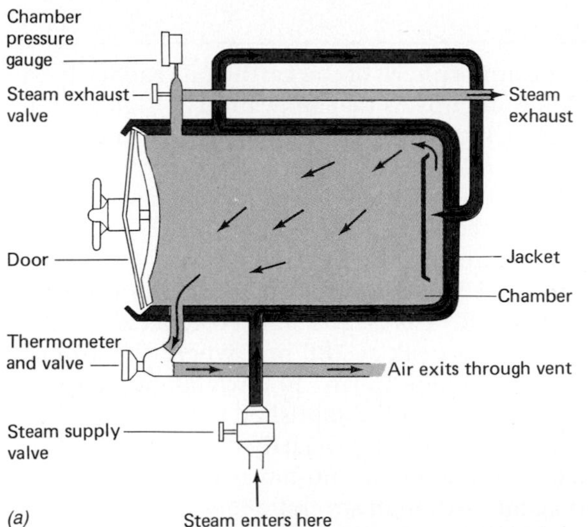

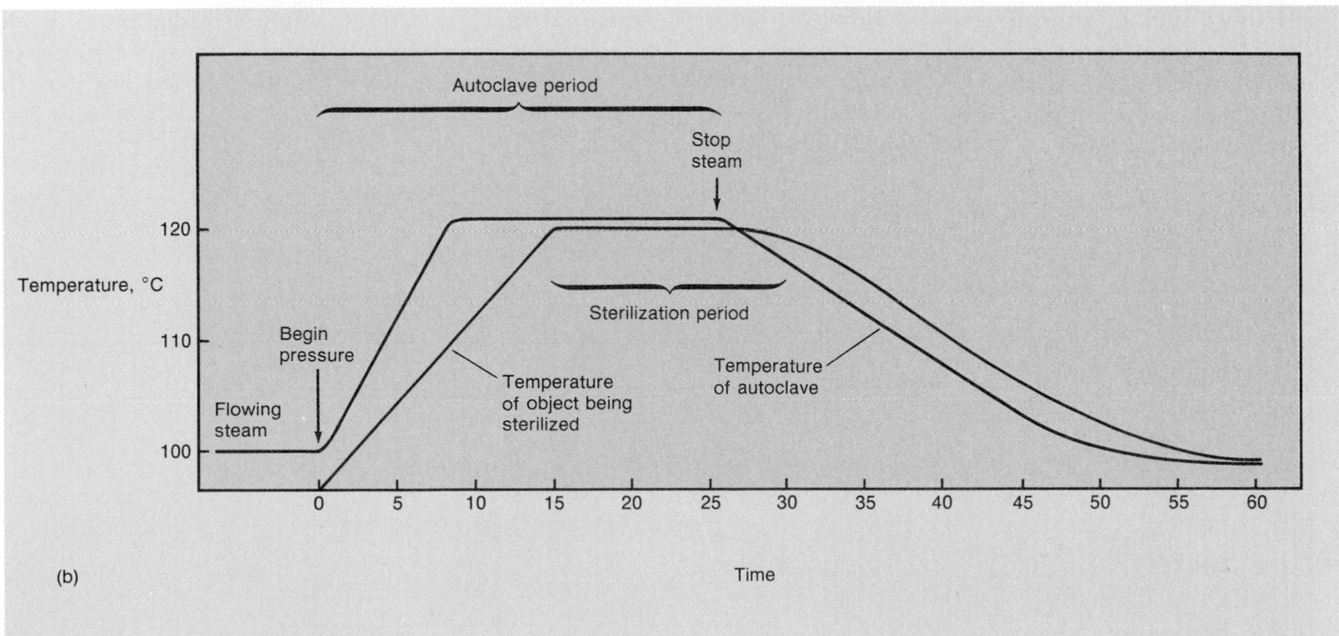

Figure 9.29 Use of the autoclave for sterilization. (a) The flow of steam through an autoclave. (b) A typical autoclave cycle. Shown is the sterilization of a fairly bulky object. Note that the temperature of the object rises more slowly than the temperature of the autoclave.

When the Pasteurizer Fails

Within a two-week period in the spring of 1985, about 18,000 persons in northern Illinois and surrounding states experienced severe intestinal disease due to infection with a bacterium called *Salmonella typhimurium*. The outbreak was traced to milk provided by a single dairy plant operated by a large grocery store chain. The bacterium was easily cultivated from patients and from the milk, using a culture medium selective for *Salmonella*. All of the cultures obtained were of a single strain which was highly resistant to the antibiotics ampicillin and tetracycline. The outbreak was the largest in the United States associated with a single *Salmonella* strain and illustrates how widely a consumed product, once contaminated, can spread its evil. *Salmonella* is found in dairy cattle and in raw milk, but is killed by the pasteurization process. Careful engineering study at the suspected dairy plant showed that a defective valve had allowed unpasteurized milk to contaminate pasteurized milk. The economic impact of this outbreak on the company was enormous: millions of dollars were forfeited in lost revenue, legal fees were huge, and the milk processing plant closed its doors permanently. This story illustrates two points: the pasteurization process in a large dairy is complex; and antibiotic-resistant bacteria are pervasive in modern society.

9.16 Control of Microbial Growth: Radiation

Some types of radiation cause death of living organisms and are of value in the sterilization and control of microorganisms in various materials. We distinguish between two kinds of radiations: ionizing and nonionizing (Figure 9.30). *Electromagnetic* radiations that have effects on living organisms include ultraviolet, visible, and infrared radiation. The latter two are usually beneficial and are used as sources of energy by photosynthetic organisms, whereas *ultraviolet radiation* is usually lethal. **Ionizing radiations** include X rays, cosmic rays, and emanations from radioactive materials, such as gamma rays.

Ultraviolet (UV) radiation Ultraviolet radiation is of considerable microbiological interest because of its frequent lethal action on microorganisms. Although the sun emits intense UV radiation of all wavelengths, only that of longer wavelengths penetrates the atmosphere of the earth and reaches its surface. Microorganisms that are wafted to high altitudes or dragged into space on the outside of rockets are quickly killed by the UV there. Shorter-wavelength UV radiation is produced in germicidal lamps for killing microorganisms.

The purine and pyrimidine bases of the nucleic acids absorb UV radiation strongly, and the absorption for DNA and RNA is at 260 nm. Proteins also absorb UV, but have a peak at 280 nm, where the aromatic amino acids (tryptophan, phenylalanine, tyrosine) absorb. It is now well established that killing of cells by UV radiation is due primarily to its action on DNA so that UV radiation at 260 nm is most effective as a lethal agent. Although several effects are known, one well-established effect is the induction in DNA of **thymine dimers,** a state in which two adjacent thymine molecules are covalently joined, so that replication of the DNA cannot occur (Figure 9.31*a*). The genetic effect of UV radiation on DNA was discussed in Section 5.18.

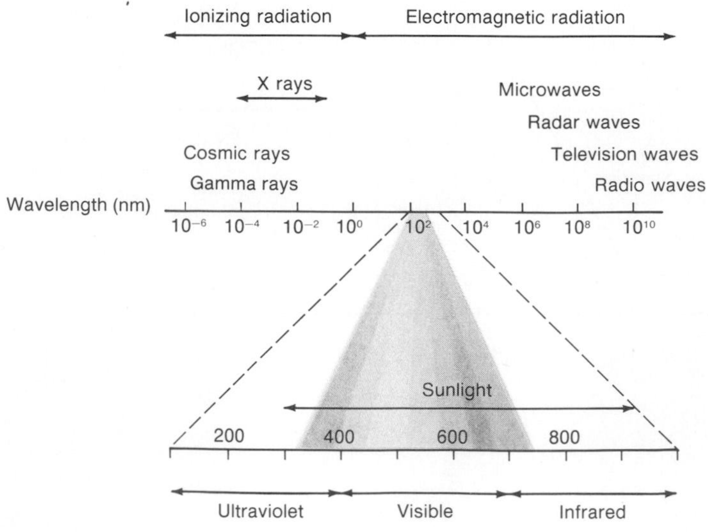

Figure 9.30 Wavelengths of radiation.

Many microorganisms have repair enzymes which correct damage induced in DNA by UV radiation. Damage in a given region of the DNA molecule is usually only on one of the two strands, and enzymes exist that will hydrolyze the phosphodiester bonds at each end of the damaged region and thus excise the altered material. Other enzymes then catalyze the insertion of new nucleotides that are complementary to the undamaged strand. One kind of repair enzyme works only when the damaged cell is activated by visible light in the blue region of the spectrum; recovery of UV radiation damage through action of these enzymes is called **photoreactivation** (Figure 9.31*b*). Other repair enzymes work in the absence of light **(dark reactivation)**. Because of the existence of these repair enzymes, a given dose of ultraviolet radiation will cause death only in the event that the damage it induces is greater than that which can be repaired.

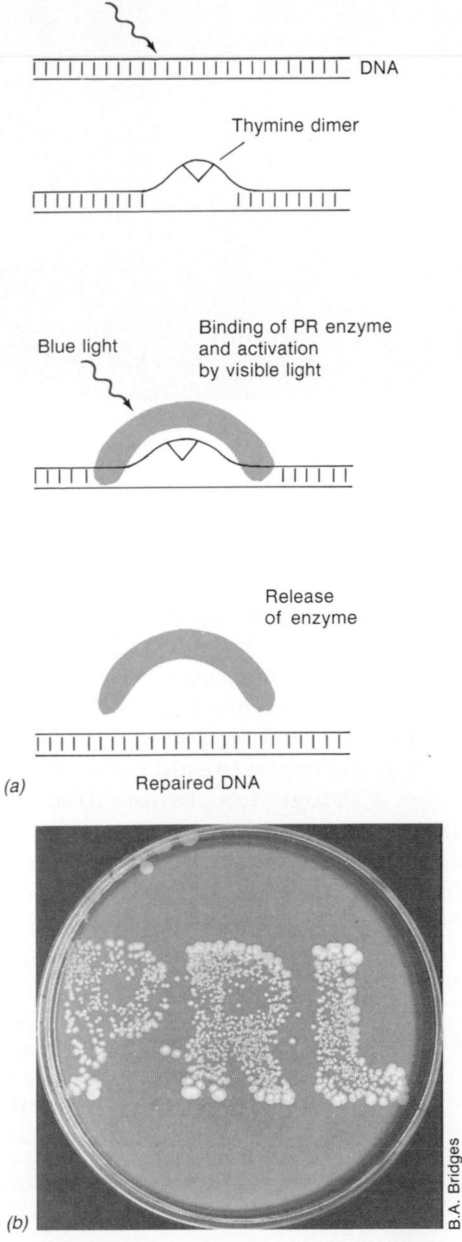

Figure 9.31 The phenomenon of photoreactivation. (a) Mechanisms of ultraviolet damage and repair. Thymine dimers formed as a result of UV damage are repaired by a photoreactivation (PR) enzyme, which is activated by blue (visible) light. (b) Demonstration of photoreactivation of *Escherichia coli*. About 10^7 bacteria were spread over the surface of an agar plate and irradiated with a dose of UV sufficient to reduce survival to less than 1 in 10^7. The areas within the letters PRL were exposed to visible light for 20 minutes before the plate was incubated, resulting in recovery of viability.

Ionizing radiation Ionizing radiation does not kill by directly affecting the cell constituents, but indirectly by inducing in the medium reactive chemical radicals (free radicals), of which the most important is the hydroxyl radical, OH· (see Section 9.14). Free radicals can react with and inactivate sensitive macromolecules in the cell. Ionizing radiation can act on all cellular constituents, but death usually results from effects on DNA. Dark reactivation of damage due to ionizing radiation occurs, but photoreactivation is absent. DNA is probably no more sensitive to ionizing radiation than other structures, but since each DNA molecule contains only one copy of most genes, inactivation of any critical gene can lead to death, whereas inactivation of single protein molecules, which are almost always present in multiple copies, will not cause death.

Bacteria vary markedly in sensitivity to ionizing radiation. At one end are sensitive organisms such as *Pseudomonas*, and at the other are resistant organisms such as *Micrococcus* and *Streptococcus*. One organism, *Deinococcus radiodurans*, is unusually resistant to ionizing radiation and is isolated frequently from irradiated foods and other products treated with ionizing radiation (see Section 19.24).

Ionizing radiation may be used to sterilize materials such as drugs, foods, and other items that are heat sensitive. In contrast to ultraviolet radiation, ionizing radiation is penetrating and hence can be used on products even after they have been packaged. Sources of ionizing radiation are X-ray machines, radioisotopes such as cobalt-60, and neutron piles in atomic energy installations. The most convenient commercial sources for radiation sterilization are cobalt-60 radioisotopes, which require no maintenance or input of energy once installed.

Strict precautions must be used in handling of ionizing radiation, since it is highly lethal for humans as well as microorganisms. Protective lead shielding is essential, and the doses received by operators must be carefully monitored. Ionizing radiation has been shown to be effective in the sterilization of foods, and its use may become widespread in the future (Figure 9.32).

Figure 9.32 Food being sterilized by ionizing radiation. The large pallets of food, wrapped in plastic film, are moved into the room on rails. The radiation source is in the middle between the two pallets. The room is about 5 by 7 meters in size.

9.17 Filter Sterilization

Although heat is the most common and effective way of sterilizing liquids, it cannot be used for the sterilization of heat-sensitive fluids. An especially valuable technique for sterilizing such materials is filtration. The filter is a device with pores too small for the passage of microorganisms but large enough to allow the passage of the liquid. The range of particles involved in sterilization is rather large. Some of the largest microbial cells are greater than 10 μm in diameter, but at the lower end of the size scale certain bacteria are less than 0.3 μm in diameter. Filters are also frequently used in the field of virology, and here we are dealing with even smaller particles—as small as 10 nm (see Figure 6.2). (Historically, filtration was used to demonstrate the existence of virus-sized infectious particles.)

There are three major types of filters, which are illustrated in Figure 9.33. One of the oldest types of filters used is the **depth filter**. A depth filter is a fibrous sheet or mat made from paper, asbestos, or glass fibers, which is constructed of a random array of overlapping fibers (Figure 9.33*a*). The depth filter functions primarily because particles get trapped in the tortuous paths created throughout the depth of the structure. Although it is possible to manufacture depth filters that will remove particles as small as viruses and bacteria, such filters have considerable limitations. In practice, depth filters are often used as *prefilters* to remove larger particles from a solution so that clogging does not occur in the final filter ster-

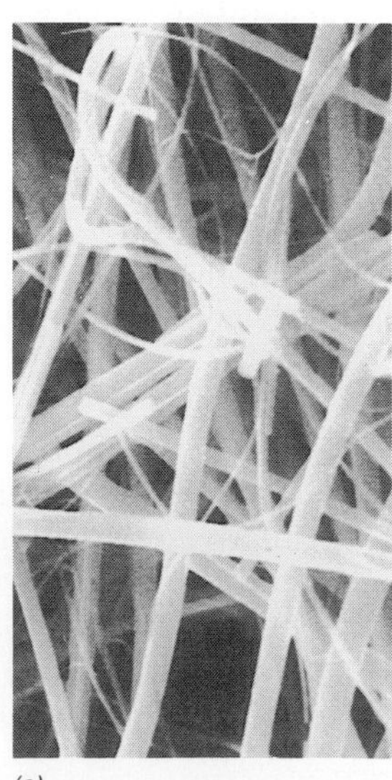
(a)

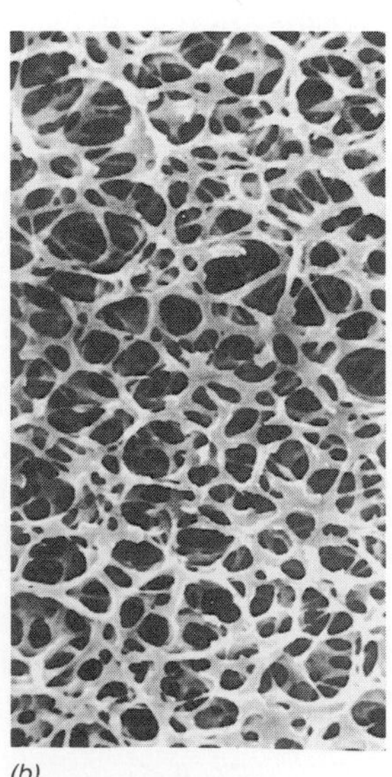
(b)

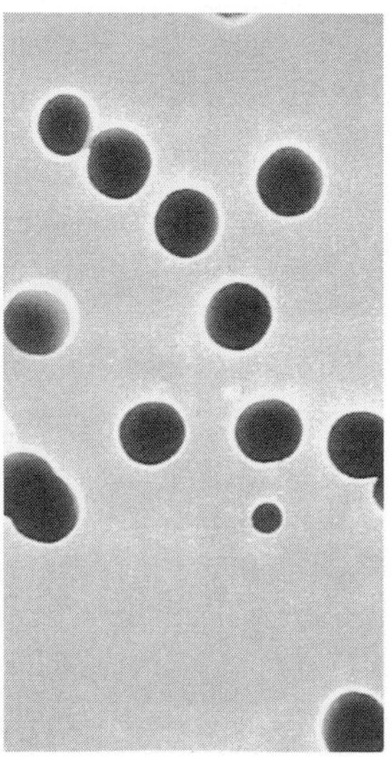
(c)

Figure 9.33 Comparison of the structure of (a) a depth filter, (b) a conventional membrane filter, and (c) a Nuclepore filter.

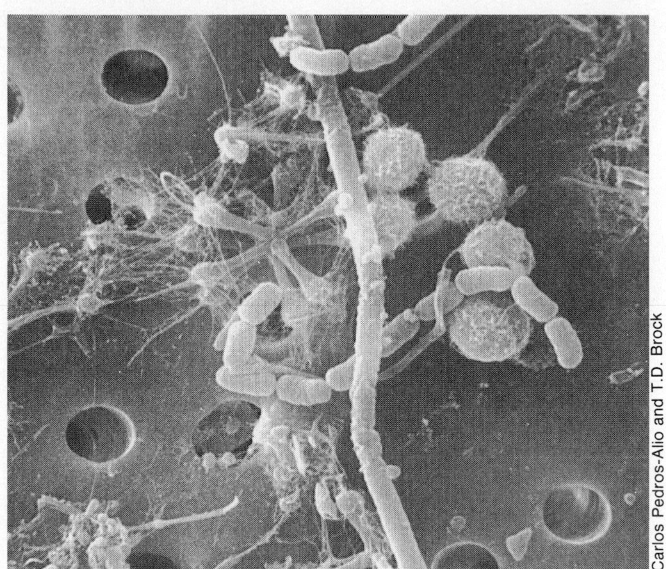

Figure 9.34 Scanning electron micrograph of aquatic bacteria and algae trapped on a Nuclepore filter. The pore size is 5 μm. Magnification, 2100×.

ilization process. They are also used in the filter sterilization of air in industrial processes (see Section 10.5).

The most common type of filter for sterilization in the field of microbiology is the **membrane filter** (Figure 9.33*b*). The membrane filter is a tough disc, generally composed of cellulose acetate or cellulose nitrate, which is manufactured in such a way that it contains a large number of tiny holes. The membrane filter differs from the depth filter in that the membrane filter functions more like a sieve, trapping many of the particles on the surface of the filter. Membrane filters are open structures with about 80–85 percent of the filter occupied by space. This openness provides for a relatively high fluid flow rate. However, the conventional membrane filter does not function completely as a sieve, since many particles penetrate into the filter matrix before being removed.

The third type of filter in common use is the **nucleation track (Nuclepore) filter**. These filters are created by treating very thin polycarbonate films (10 μm in thickness) with nuclear radiation and then etching the film with a chemical. The radiation causes localized damage in the film and the etching chemical then acts to enlarge these damaged locations into holes. The sizes of the holes can be precisely controlled by the strength of the etching solution and the time allowed for etching to proceed. A typical Nuclepore filter has very uniform holes that are arranged almost vertically through the thin film (Figure 9.33*c*). Nuclepore filters function as true sieves, removing all particles greater than the size of the holes. However, because of their lower porosity, the flow rate through Nuclepore filters is lower than through membrane filters, and clogging occurs more rapidly. Nuclepore filters are very commonly used in scanning electron microscopy of microorganisms, since an organism of interest can be easily removed from a liquid by filtration, and the particles are all held in a uniform plane on the top of the filter where they can be readily studied in the microscope (Figure 9.34).

The use of a membrane filter for the sterilization of a liquid is illustrated in Figure 9.35. The filter apparatus is generally sterilized separately from the filter, and the apparatus assembled aseptically at the time of filtration. The arrangement shown in Figure 9.35 is suitable for small volumes of liquid. For large-volume sterile filtration, the membrane filter material is arranged in a cartridge and placed in a stainless steel housing. Large-volume filtration of heat-sensitive fluids is very commonly done in the pharmaceutical industry.

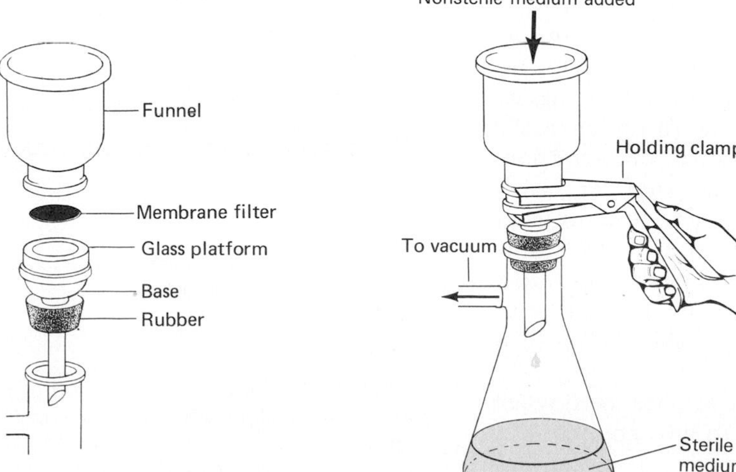

Figure 9.35 Use of a membrane filter for sterilization of a liquid.

9.18 Chemical Control of Microbial Growth

An **antimicrobial agent** is a chemical that kills or inhibits the growth of microorganisms. Such a substance may be either a synthetic chemical or a natural product. Agents that kill organisms are often called "cidal" agents, with a prefix indicating the kind of organism killed. Thus, we have **bactericidal, fungicidal**, and **algicidal** agents. A bactericidal agent, or bactericide, kills bacteria. It may or may not kill other kinds of microorganisms. Agents that do not kill but only inhibit growth are called "static" agents, and we can speak of **bacteriostatic, fungistatic**, and **algistatic** agents.

The distinction between a static and cidal agent is often arbitrary, since an agent that is cidal at high concentrations may be static at lower concentrations. To be effective, a static agent must be continuously present, since if it is removed or its activity neutralized, the organisms present in the product may initiate growth if conditions are favorable.

Antimicrobial agents can vary in their **selective toxicity**. Some act in a rather nonselective manner and have similar effects on all types of cells. Others are far more selective and are more toxic to microorganisms than to animal tissues. Antimicrobial agents with selective toxicity are especially useful as *chemotherapeutic agents* in treating infectious diseases, as they can be used to kill disease-causing microbes without harming the host. They will be described later in this chapter.

Effect of antimicrobial agents on growth Antimicrobial agents affect growth in a variety of ways, and a study of the action of these agents in relation to the growth curve is of considerable aid in understanding their modes of action. The most accurate way of observing the effect of a chemical is to add it at an inhibitory concentration to an exponentially growing culture and continue to sample the culture for growth and viability. For measurement of growth, turbidity is most convenient, although direct microscopic counting may also be useful in some cases. For measurement of viability, some procedure is necessary by which the inhibitory activity of the chemical is nullified, since if growth-inhibitory concentrations are carried over into the medium used for viable counting, even viable cells will not be able to grow and form colonies. In many cases, dilution for viable counting is sufficient to eliminate growth-inhibitory properties of the agent, but in some cases the organisms may have to be removed from the chemical by filtration or centrifugation.

Three distinct kinds of effects are observed when an antimicrobial agent is added to an exponentially growing bacterial culture, bacteriostatic, bactericidal, and bacteriolytic. As described in the preceding section, a **bacteriostatic** effect is observed when growth is inhibited, but no killing occurs (Figure 9.36*a*). Bacteriostatic agents are frequently inhibitors of protein synthesis and act by binding to ribosomes. The binding, however, is not tight, and when the drug concentration is lowered the agent becomes free from the ribosome and growth is resumed. The mode of action of protein synthesis inhibitors is discussed in Section 5.10. **Bactericidal** agents prevent growth and induce killing, but lysis or cell rupture does not occur (Figure 9.36*b*). Bactericidal agents generally bind tightly to their cellular targets and are not removed by dilution. **Bacteriolytic** agents induce killing by cell lysis, which is observed as a decrease in cell numbers or in turbidity after the agent is added (Figure 9.36*c*). Bacteriolytic agents include antibiotics that inhibit cell-wall synthesis, such as penicillin (see

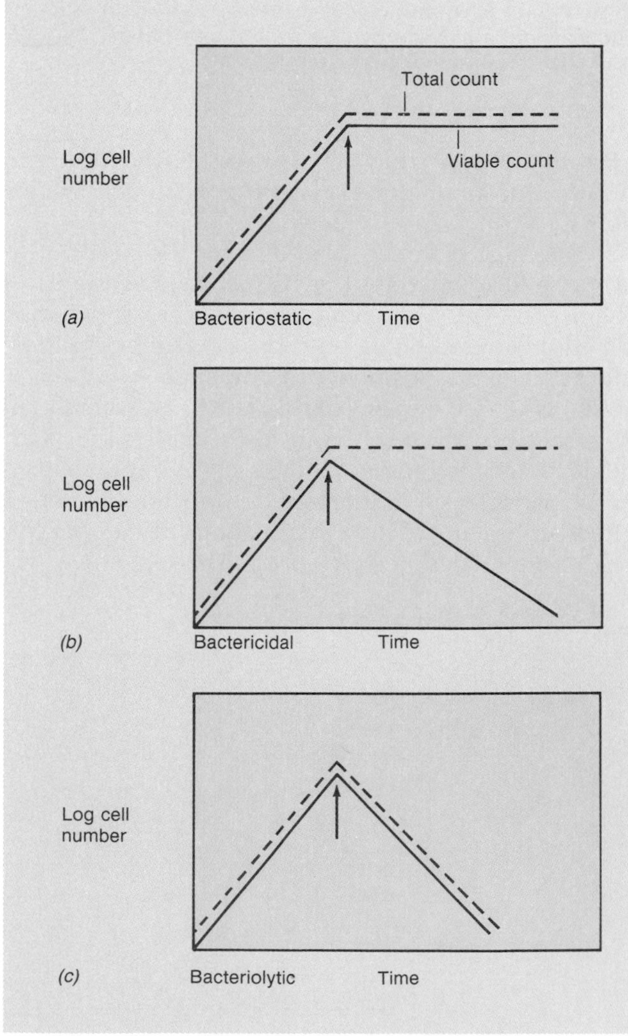

Figure 9.36 Three types of action of antimicrobial agents. At the time indicated by the arrow, a growth-inhibitory concentration was added to the exponentially growing culture. Note the relationships between viable and total cell counts.

Sections 3.5 and 9.20), as well as agents that act on the cell membrane.

Measuring antimicrobial activity Antimicrobial activity is measured by determining the smallest amount of agent needed to inhibit the growth of a test organism, a value called the **minimum inhibitory concentration** (MIC). In one method of determining the MIC, a series of culture tubes is prepared, each tube containing medium with a different concentration of the agent, and all tubes of the series are inoculated. After incubation, the tubes in which growth does not occur (indicated by absence of visible turbidity) are noted, and the MIC is thus determined (Figure 9.37). This simple and effective procedure is often called the *tube dilution technique*. The MIC is not an absolute constant for a given agent, since it is affected by the kind of test organism used, the inoculum size, the composition of the culture medium, the incubation time, and the conditions of incubation, such as temperature, pH, and aeration. If all conditions are rigorously standardized, it is possible to compare different antimicrobials and determine which is most effective against a given organism or to assess the activity of a single agent against a variety of organisms. Note that the tube dilution method does not distinguish between a cidal and a static agent, since the agent is present in the culture medium throughout the entire incubation period.

Another commonly used procedure for studying antimicrobial action is the **agar diffusion method** (Figure 9.38). A petri plate containing an agar medium evenly inoculated with the test organism is prepared. The test plate may be inoculated by pouring an overlay of agar containing the test organism, or by swabbing the medium surface with a broth culture of the test organism. Known amounts of the antimicrobial agent are added to filter-paper discs which are then placed on the surface of the agar. During incubation, the agent diffuses from the filter paper into the agar; the further it gets from the filter paper, the smaller the concentration of the agent. At some distance from the disc the MIC is reached. Past this point growth occurs, but closer to the disc growth is absent. A **zone of inhibition** is thus created, and its size can be measured with a ruler; the diameter will be proportional to the amount of antimicrobial agent added to the disc.

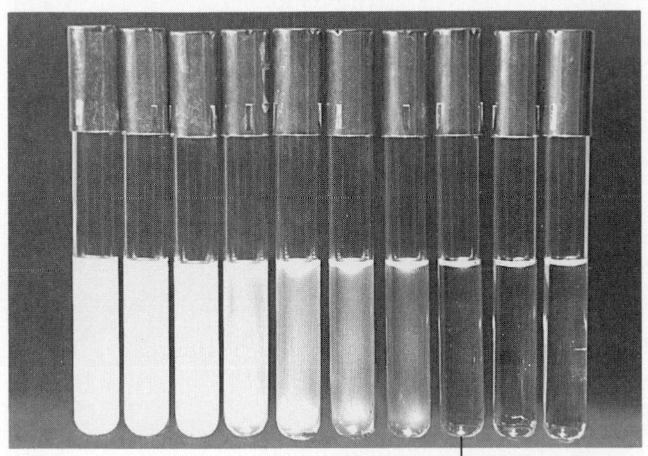

Figure 9.37 Antibiotic assay by tube dilution, permitting detection of the *minimum inhibitory concentration, MIC*. A series of increasing concentrations of antibiotic is prepared in the culture medium. Each tube is inoculated and incubation allowed to proceed. Growth occurs in those tubes with antibiotic concentration below the MIC.

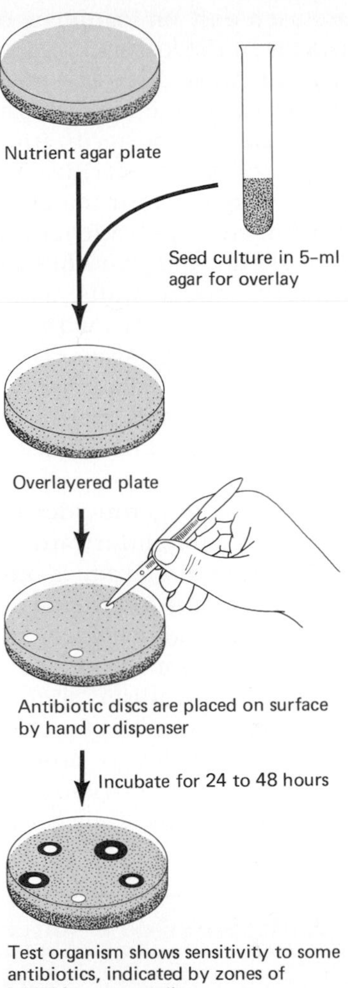

Figure 9.38 Agar diffusion method for assaying antibiotic activity.

9.19 Germicides, Disinfectants, and Antiseptics

We have defined germicides as chemical agents that kill microorganisms. We have differentiated antiseptics, which can be used on the skin, from disinfectants, which are used only on inanimate objects. The quantitative aspects of the killing of microorganisms with chemical agents were discussed in the previous section. Germicides have wide use in situations where it is impractical to use heat for sterilization. Hospitals find it frequently necessary to sterilize heat-sensitive materials, such as surgical instruments, thermometers, lensed instruments, polyethylene tubing and catheters, and inhalation and anaesthesia equipment. In the food industry, floors, walls, and surfaces of equipment must often be treated with germicides to reduce the load of microorganisms. Drinking water is commonly disinfected to reduce or eliminate any potentially harmful organisms, and treated wastewater is generally disinfected before it is discharged into the environment.

Although the testing of germicides in laboratory situations is relatively straightforward, in practical cases, the determination of efficacy is often very difficult. This is because many germicides are neutralized by organic materials, so that germicidal concentrations are not maintained for sufficient time. Further, bacteria and other microorganisms are often encased in particles, and the penetration of a chemical agent to the viable cells may be slow or absent. Also, bacterial spores are much more resistant to any germicides than are vegetative cells. Thus germicide effectiveness must ultimately be determined under the intended conditions of use. It should be emphasized that germicidal treatments do not necessarily sterilize. **Sterility** is defined as the complete absence of living organisms, and sterilization with chemicals often requires long contact periods under special conditions. In most cases, the use of germicides ensures only that the microbial load is reduced significantly, although perhaps with the hope that pathogenic organisms are completely eliminated. However, bacterial endospores as well as vegetative cells such as those of *Mycobacterium tuberculosis*, the causal agent of tuberculosis, are very resistant to the action of many germicides, so that even the complete elimination of pathogens by germicidal treatment may not occur.

A summary of the most widely used antiseptics and disinfectants, and their modes of action, is given in Table 9.8.

In addition to their wide use in the medical field, antiseptic and disinfectant chemicals find many important uses in industry. In industry, the concern is generally to prevent microbial deterioration of materials. Any organic material is potentially subject to microbial attack. Table 9.9 summarizes some of the industries in which chemicals are used to control microbial growth.

Table 9.8 Antiseptics and disinfectants

Agent	Use in health-related fields	Mode of action
Antiseptics		
Organic mercurials	Skin	Combines with SH groups of proteins
Silver nitrate	Eyes of newborn to prevent gonorrhea	Protein precipitant
Iodine solution	Skin	Iodinates tyrosine residues of proteins
Alcohol (70% ethanol in water)	Skin	Lipid solvent and protein denaturant
Bis-phenols (hexachlorophene)	Soaps, lotions, body deodorants	Disrupts cell membrane
Cationic detergents (quaternary ammonium compounds)	Soaps, lotions	Interact with phospholipids of membrane
Hydrogen peroxide (3% solution)	Skin	Oxidizing agent
Disinfectants		
Mercuric dichloride	Tables, bench tops, floors	Combines with SH groups
Copper sulfate	Algicide in swimming pools, water supplies	Protein precipitant
Iodine solution	Medical instruments	Iodinates tyrosine residues
Chlorine gas	Purification of water supplies	Oxidizing agent
Chlorine compounds	Dairy, food industry equipment	Oxidizing agent
Phenolic compounds	Surfaces	Protein denaturant
Cationic detergents (quaternary ammonium compounds)	Medical instruments; food, dairy equipment	Interact with phospholipids
Ethylene oxide (gas)	Temperature-sensitive laboratory materials such as plastics	Alkylating agent
Ozone	Purifying drinking water	Strong oxidizing agent

Table 9.9 Industries which use chemicals to control microbial growth

Industry	Chemicals	How used
Paper	Organic mercurials, phenolics	To prevent microbial growth during manufacture
Leather	Heavy metals, phenolics	Antimicrobial agents are present in the final product
Plastic	Cationic detergents	To prevent growth of bacteria on aqueous dispersions of plastics
Textile	Heavy metals, phenolics	To prevent microbial deterioration for fabrics exposed in the environment such as awnings, tents
Wood	Phenolics	To prevent deterioration of wooden structures
Metal working	Cationic detergents	To prevent growth of bacteria in aqueous cutting emulsions
Petroleum	Mercurics, phenolics, cationic detergents	To prevent growth of bacteria during recovery and storage of petroleum and petroleum products
Air conditioning	Chlorine, phenolics	Prevent growth of bacteria (e.g. *Legionella*) in cooling towers
Electrical power	Chlorine	Prevent growth of bacteria in condensors and cooling towers
Nuclear	Chlorine	Prevent growth of radiation resistant bacteria in nuclear reactors

9.20 Chemotherapeutic Agents

The discussion above has dealt primarily with chemical agents that are used to inhibit microbial growth outside the human body. Most of the chemicals mentioned were too toxic to be used in the body, although antiseptics can be used on the skin. For control of infectious disease, agents which can be used internally are essential. Such agents are called **chemotherapeutic agents** and play major roles in modern medicine. We discuss the uses of chemotherapeutic agents for treatment of infectious disease in Chapter 11. Here we present simply some general concepts. The key requirement of a successful chemotherapeutic agent is *selective toxicity*, the ability to inhibit bacteria or other microorganisms without affecting the body.

Growth-factor analogs In Section 4.21 we discussed growth factors and defined them as specific chemical substances required in the medium because the organism cannot synthesize them. Substances exist that are related to growth factors and that act to block the utilization of the growth factor. These "growth-factor analogs" usually are structurally similar to the growth factors in question but are sufficiently different so that they cannot duplicate the work of the growth factor in the cell. The first of these to be discovered were the *sulfa drugs*, the first modern chemotherapeutic agents to specifically inhibit the growth of bacteria; they have since proved highly successful in the treatment of certain diseases. The simplest sulfa drug is sulfanilamide (Figure 9.39*a*). Sulfanilamide acts as an analog of *p*-aminobenzoic acid (Figure 9.39*b*), which is itself a part of the vitamin folic acid (Figure 9.39*c*). In organisms that synthesize their own folic acid, sulfanilamide acts by blocking the synthesis of folic acid. Sulfanilamide is active against bacteria but not against higher animals because bacteria synthesize their own folic acid, whereas higher animals obtain folic acid preformed in their diet.

The concept that a chemical substance can act as a competitive inhibitor of an essential growth factor has had far-reaching effects on chemotherapeutic research, and today analogs are known for various vitamins, amino acids, purines, pyrimidines, and other compounds. A few examples are given in Figure 9.40.

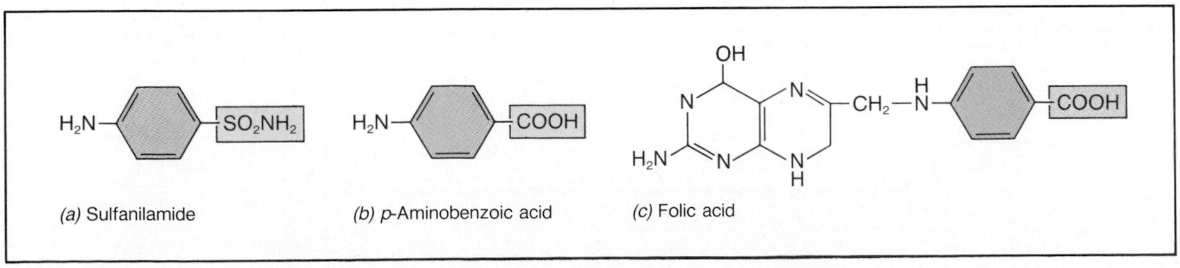

(a) Sulfanilamide (b) *p*-Aminobenzoic acid (c) Folic acid

Figure 9.39 (a) The simplest sulfa drug, sulfanilamide. It is an analog of *p*-aminobenzoic acid (b), which itself is part of the growth factor folic acid (c).

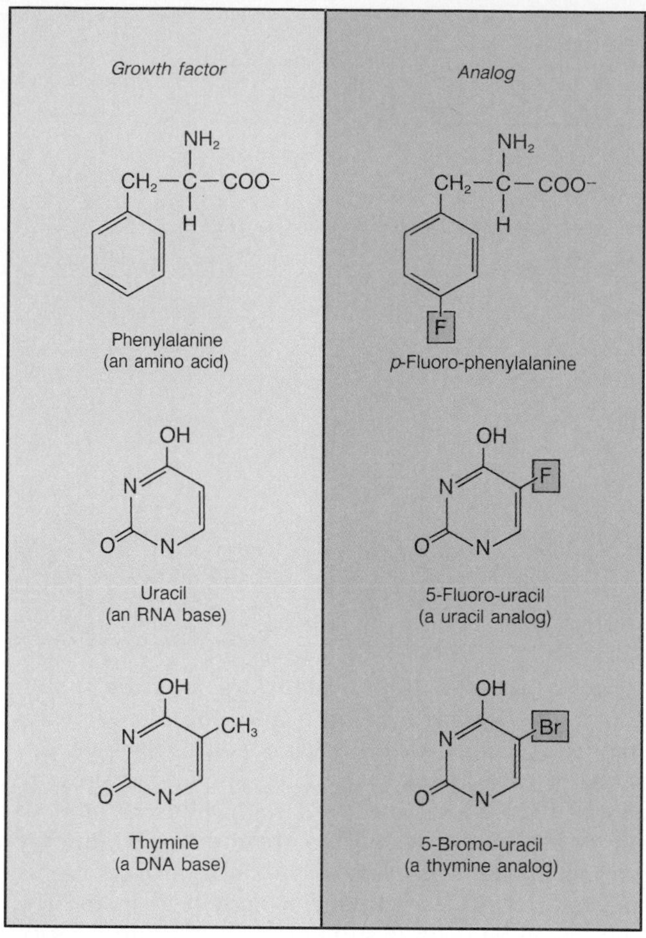

Figure 9.40 Examples of growth-factor analogs.

In these examples, the analog has been formed by addition of fluorine or bromine. Fluorine is a relatively small atom and does not alter the overall shape of the molecule, but it changes the chemical properties sufficiently so that the compound does not act normally in cell metabolism. Fluorouracil resembles the nucleic-acid base uracil; bromouracil resembles another base, thymine. This is because the bromine atom is considerably larger than the fluorine atom and thus resembles the methyl group of thymine.

Analogs of purines and pyrimidines also act as inhibitors of virus replication. We discussed in Section 6.25 a few antiviral compounds which function in this way. It is of interest that the original concept of competitive inhibition was developed from studies on the mode of action of sulfa drugs.

Antibiotics Antibiotics are chemical substances produced by certain microorganisms which are active against other microorganisms. Antibiotics constitute a special class of chemotherapeutic agents, distinguished most importantly by the fact that they are natural products (products of microbial activity) rather than synthetic chemicals (products of human activity). Antibiotics constitute one of the most important classes of substances produced by large-scale microbial processes. The industrial aspects of antibiotics, how antibiotics are discovered, and some details of specific antibiotics, are discussed in Chapter 10. The uses of antibiotics in treatment of disease are discussed in Chapter 11. Here we present a broad overview of the antibiotics and describe some of the methods of testing antibiotic activity.

A very large number of antibiotics have been discovered, but probably less than 1 percent of them have been of great practical value in medicine. Those few which have been useful have had a dramatic impact on the treatment of many infectious diseases. Further, some antibiotics can be made more effective by chemical modification; these are said to be *semisynthetic*.

The sensitivity of microorganisms to antibiotics varies (Figure 9.41). Gram-positive bacteria are usually more sensitive to antibiotics than are Gram-negative bacteria, although conversely, some antibiotics act only on Gram-negative bacteria. An antibiotic that acts upon both Gram-positive and Gram-negative bacteria is called a **broad-spectrum antibiotic**. In general, a broad-spectrum antibiotic will find wider medical usage than a *narrow-spectrum antibiotic*, which acts on only a single group of organisms. A narrow-

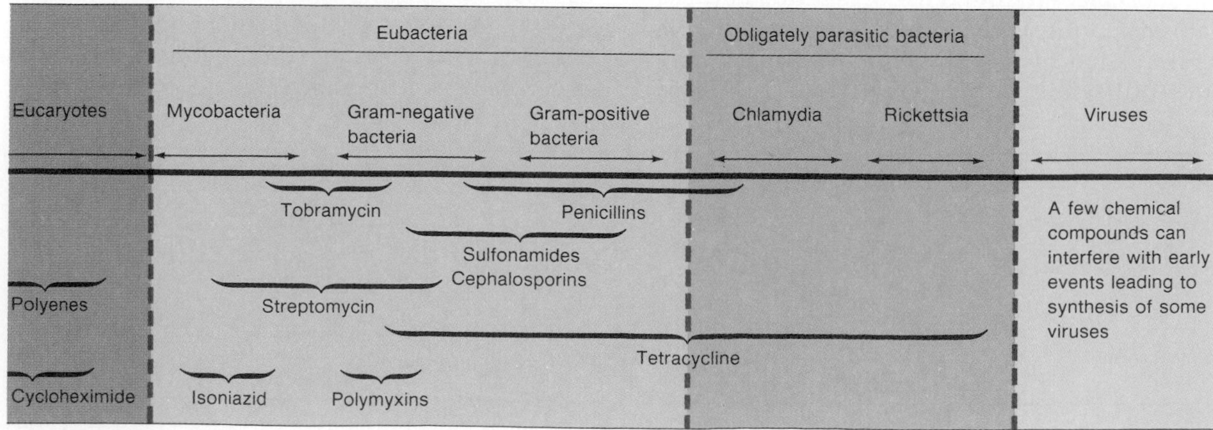

Figure 9.41 Ranges of action of antibiotics and other agents.

Table 9.10 Chemotherapeutic agents and their modes of action

Target in cell	Type of agent	Examples
Bacterial cell wall	Penicillins	Penicillin G, ampicillin, methicillin
	Cephalosporins	Cephalothin, cephamycin
Cell membrane		
Procaryotes	Polymyxins	Polymyxin B
Eucaryotes	Polyenes	Nystatin, amphotericin
Protein synthesis		
Procaryotes	Nitroaromatic	Chloramphenicol
	Aminoglycosides	Streptomycin, tobramycin
	Tetracyclines	Tetracycline, chlortetracycline
	Macrolides	Erythromycin
	Lincomycins	Lincomycin, clindamycin
Eucaryotes	Glutarimide	Cycloheximide
RNA synthesis	Rifamycins	Rifampin
Growth factor analog	Sulfonamides	Sulfanilamide
Unknown	Pyridine	Isoniazid
	Benzofuran	Griseofulvin

spectrum antibiotic may, however, be quite valuable for the control of certain kinds of diseases. Some antibiotics have an extremely limited spectrum of action, being toxic to only one or a few bacterial species.

Antibiotics and other chemotherapeutic agents can be grouped based on chemical structure or on mode of action. In bacteria, the important targets of chemotherapeutic action are the cell wall, the cell membrane, and the biosynthetic processes of protein and nucleic acid synthesis. Some chemotherapeutic agents work because they mimic important growth factors needed in cell metabolism. Examples of each type are given in Table 9.10.

Antibiotic resistance It is obvious from the discussion above that not all antibiotics act against all microorganisms. Some microorganisms are resistant to some antibiotics. Antibiotic resistance can be an inherent property of a microorganism, or it can be acquired. There are several reasons why microorganisms may have inherent resistance to an antibiotic: (1) The organism may lack the structure which an antibiotic inhibits. For instance, some bacteria, such as mycoplasmas, lack a typical bacterial cell wall and are resistant to penicillins. (2) The organism may be impermeable to the antibiotic. (3) The organism may be able to change the antibiotic to an inactive form. This is very frequently the case in penicillin resistance. A group of enzymes called *penicillinases* or *β-lactamases* destroy the β-lactam ring of penicillin, thus rendering it inactive. Penicillinases are produced by a wide variety of bacteria but some penicillins are less susceptible to penicillinase action. A major objective in the creation of semisynthetic penicillins is to decrease the sensitivity to penicillinase action (see Section 10.6).

Because of the development of antibiotic resistance, testing of bacteria isolated from clinical material for antibiotic sensitivity must be carried out. Details of the sensitivity testing of clinical isolates is described in Sections 11.17 and 13.3.

9.21 Some Aspects of Food Microbiology

We complete this chapter with a short discussion of how some of the principles presented here can be applied to an important practical problem, the prevention of microbial growth in foods. Microorganisms are responsible for some of the most serious kinds of food poisonings and toxicities (see Section 15.11), and also cause spoilage of a wide variety of food and dairy products. Because of this, a knowledge of microbial growth in foods is of great value.

Food spoilage Food spoilage is probably the most serious economic problem in the food processing industry. Foods are subjct to attack by microorganisms in a variety of ways, and such attack is generally harmful to the quality of the food. Foods are organic, and hence provide adequate nutrients for the growth of a wide variety of heterotrophic bacteria. What determines whether a food will support the growth of a microorganism? The physical and chemical characteristics of the food and how it is stored determine its degree of susceptibility to microbial attack, the kinds of microorganisms that will affect it, and the kinds of spoilage that will result. Foods can be classified into three major categories: (1) *Highly perishable foods* such as meats, fish, poultry, eggs, milk, and most fruits and vegetables. (2) *Sem-*

iperishable foods such as potatoes, some apples, and nuts. (3) *Stable* or *nonperishable foods* such as sugar, flour, rice, and dry beans. How do these three categories differ? To a great extent, perishability is a function of *moisture content*, which is related, as we have seen, to water activity, a_w. The stable foods are the ones with *low* water activity, and these can generally be stored for considerable lengths of time without any undue effects. The perishable and semiperishable foods are those with high water activity. In order for these latter foods to be kept, they must be stored under conditions such that microbial growth is slow or does not occur at all.

It is useful, in discussing microbial growth in foods, to think of the standard growth curve, as has been illustrated in Figure 9.6. The lag phase may be of variable duration in a food, depending upon the contaminating organism and its previous growth history. The rate of growth during the exponential phase will depend upon the temperature, the available nutrients, and other conditions of growth. The time required for the population density to reach a significant level in a given food product will depend on both the initial inoculum and how rapidly growth occurs during the exponential phase. However, it is only when the microbial population density reaches a substantial level that harmful effects are usually observed. Indeed, throughout much of the exponential phase of growth, population densities may be low enough that no perceptible effect can be observed, but because of the nature of exponential growth (see Figure 9.1), it is only the *last* doubling or so that leads to problems. Thus, for much of the period of microbial growth in a food, the observer is unaware of impending problems.

Food preservation One of the most crucial factors affecting microbial growth in food is **temperature**. In general, the *lower* the temperature of storage, the *less* rapid the spoilage rate, although, as we have seen, psychrophilic microorganisms are able to grow well at refrigerator temperatures. Therefore, storage of perishable food products for long periods of time is only possible at temperatures below freezing. Freezing greatly alters the physical structure of many food products and therefore cannot be universally used, but it is widely and successfully used for the preservation of meats and many vegetables and fruits. Freezers providing temperatures of $-20°C$ are most commonly used for storing frozen products. At such temperatures, storage for weeks or months is possible, but even at these temperatures some microbial growth may occur, usually in pockets of liquid water trapped within the frozen mass. Also, nonmicrobial chemical changes in the food may still occur. For very long-term storage, lower temperatures than $-20°C$ are necessary, such as $-80°C$ (dry ice temperature), but maintenance at such low temperatures is expensive and consequently is not used for routine food storage.

It should be emphasized that many microorganisms are not killed upon freezing, so that foods to be frozen should be of the best quality and free of spoilage. If a contaminated product is frozen, much of its microbial content will remain alive, and when it is thawed many of the organisms will still be present and may cause problems.

High-temperature treatments are widely used to reduce microbial loads or sterilize foods. We discussed the pasteurization process in Section 9.15 and discuss the canning process later in this Section.

Another major factor affecting microbial growth in food is **pH** or **acidity**. Foods vary widely in pH, although most are neutral or acidic. As we have seen, microorganisms differ in their ability to grow under acidic conditions. It is an important practical matter that most food spoilage bacteria do not grow at pH values below 5. It is for this reason that acid is often used in food preservation, in the process called *pickling*. Foods commonly pickled include cucumbers (sweet, sour, and dill pickles), cabbage (sauerkraut), and some meats and fruits. The food can be made acid either by addition of vinegar or by allowing acidity to develop directly in the food through microbial action, in which case the product is called a **fermented food**. The microorganisms involved in food fermentations are acid-tolerant bacteria, the lactic acid bacteria, the acetic acid bacteria, and the propionic acid bacteria. But even these bacteria are not able to continue developing once the pH drops below 4–5, so that the food fermentation is self-limiting. Vinegar, frequently added to food to lower the pH, is essentially dilute acetic acid. Vinegar itself is a product of the action of acetic acid bacteria; its industrial production is discussed in Section 10.10.

Since microorganisms do not grow at low water activities, microbial growth can be controlled by lowering water activity of the product by drying or by adding salt. Sun drying is the least expensive way of drying foods and is appropriate if the climate is right. Some foods can be successfully dried with heat, although generally with significant changes in flavor or quality. The least damaging way of drying foods is freeze-drying (lyophilization), but this is quite expensive and can only be justified if the food has a high economic value. Milk, meats, fish, vegetables, fruits, and eggs are all preserved by drying.

A number of foods are preserved by addition of salt or sugar to lower water activity. Foods preserved by addition of sugar are mainly fruits (jams, jellies, and preserves). Salted products are primarily meats and fish. Sausage and ham are preserved by salt, although these products vary in water activity depending on how much salt is added and how much the meat has been dried. Several famous sausages such as *landjaeger* can be stored indefinitely in the absence of refrigeration, but are fairly salty to eat.

(a)

(b)

(c)

(d)

Figure 9.42 Changes in cans as a result of microbial spoilage. (a) Normal can; note that the top of the can is indented due to negative pressure (vacuum) inside. (b) Slight swell resulting from minimal gas production. Note that the lid is slightly raised. (c) Severe swell due to extensive gas production. Note the great deformation of the can. This can is potentially dangerous, and could explode if dropped or hit! (d) The can shown in (c) was dropped and the gas pressure resulted in a violent explosion. Note that the lid has been torn apart.

Canning Canning is a process in which a food is sealed and heated so as to kill all living organisms, or at least to assure that there will be no growth of residual organisms in the can. Canning is hence a type of heat sterilization, and the principles already presented in Section 9.15 apply. When the can is properly sealed and heated, the food should remain stable and unspoiled indefinitely, even when stored in the absence of refrigeration. Home-canned foods are usually prepared in glass containers whereas commercial products are most often in tin-coated steel cans. In any canning process, the seal itself is a critical part. The heating process itself is done by submersion in water, generally under pressure.

The temperature/time relationships for canning depend upon the type of food, especially its pH, the size of the container, and the consistency or bulkiness of the food. Because heat must penetrate completely to the center of the food within the can, heating times must be longer for large cans or very viscous foods. Acid foods can often be canned effectively by heating just to boiling, 100°C, whereas nonacid foods must be heated to autoclave temperatures. It is not desirable to heat foods much longer than necessary, since prolonged heating affects the nutritional and eating qualities of the food. It should also be noted that the term "sterilize" is not completely accurate when used in relation to canning. The numbers of organisms are reduced greatly during the heating process, and in most cases the product is probably sterile, but in fact if the initial load of organisms in the food is high, then not every cell may be killed. Heating times long enough to guarantee absolute sterility of every can would alter the food so greatly that it would likely be unpalatable. Thus, a fine balance is generally kept between reduction in microbial numbers, eradication of all pathogens, and palatability.

The environment inside a can is anaerobic and microbial growth in a canned food frequently is because of the activity of fermentative organisms which produce extensive amounts of gas. This results in a change in the physical appearance of the can, which is often visible by observation of the outside of the can (Figure 9.42). If microbial growth is extensive enough, the gas pressure build-up can actually result in the destruction of the can. Because many of the anaerobic bacteria which grow in canned foods are powerful toxin producers (see Section 11.10 for a discussion of bacterial exotoxins), food from a visibly altered can should never be eaten.

Chemical food preservation Although chemicals should never be used in place of careful food sanitation, there are a number of antimicrobial chemicals that find wide use in the control of microbial growth of foods. These are summarized in Table 9.11. The chemicals listed in this table are all in a category called by the U.S. Food and Drug Administration "Generally Regarded as Safe" (GRAS). These chemicals have been used for many years with no evidence of human toxicity. However, the introduction of one of these chemicals into food today would be impossible without completion of a massive safety testing program.

Table 9.11 Chemical food preservatives

Chemical	Foods
Sodium or calcium propionate	Bread
Sodium benzoate	Carbonated beverages, fruit juices, pickles, margarine, preserves
Sorbic acid	Citrus products, cheese, pickles, salads
Sulfur dioxide	Dried fruits and vegetables
Formaldehyde (from food-smoking process)	Meat, fish
Sodium nitrite	Smoked ham, bacon

Study Questions

1. Examine the graph describing the relationship between growth rate and temperature. Give an explanation in biochemical terms why the optimum temperature for an organism is closer to its maximum than its minimum.
2. Would you expect to find a psychrophilic microorganism alive in a hot spring? Why? It is frequently possible to isolate thermophilic microorganisms from cold-water environments. Give an explanation how this can be.
3. Even though extreme thermophilic bacteria are resistant to killing by the temperatures used in the food canning process, they never present a problem in food spoilage. Why?
4. Give an explanation why a culture of *Streptococcus* might not exhibit a simple first-order thermal death curve.
5. Even though heat sterilization is quite effective with many foods, give an explanation why it is still important to begin with a food that has been prepared under relatively clean conditions so that its microbial load is low.
6. What is *thermal death time*? How would the presence of bacterial endospores affect the thermal death time?
7. Write an explanation in molecular terms for how an obligate halophile is able to arrange so that water molecules flow *into* the cell.
8. Contrast an aerotolerant and an aerophobic anaerobe in terms of sensitivity and ability to grow in the presence of oxygen O_2. How does an aerotolerant anaerobe differ from a microaerophile?
9. In terms of oxygen relations (obligate anaerobe, obligate aerobe, facultative aerobe, microaerophile) list the kinds of organisms that you would expect to find in each of the following environments of the human body: skin, heart muscle, teeth, stomach, lungs, large intestines.
10. Compare and contrast the enzymes *catalase* and *superoxide dismutase* from the following points of view: substrates, oxygen products, organisms containing, role in oxygen tolerance.
11. Define growth and describe one direct and one indirect method by which growth can be measured. Make sure that the methods you choose agree with your definition.
12. Describe the growth pattern of a population of bacterial cells from the time this population is first inoculated into fresh medium. How may the growth pattern differ when it is measured by total count or by viable count?
13. Describe briefly the process by which a single cell develops into a visible colony on an agar plate. With this explanation as a background, describe the principle behind the viable count method.
14. Starting with 4 bacterial cells in a rich nutrient medium, with a one-hour lag phase and a 20-minute generation time, how many cells would there be after one hour. After 2 hours? After 2 hours if one of the initial 4 cells was dead?
15. Describe the growth pattern of a population of bacterial cells from the time they are first inoculated into fresh medium. How may the growth pattern differ when it is measured by total count or by viable count?
16. How would you distinguish bacteriostatic from bactericidal actions of an antibiotic? Bactericidal from bacteriolytic?
17. Define *antibiotic*. How do antibiotics differ from other antimicrobial agents?
18. Briefly describe two different ways to determine the *minimum inhibitory concentration* of an antibiotic. Explain why this is not necessarily the same concentration needed to inhibit microorganisms within the body.

Supplementary Readings

American Public Health Association. 1985. *Standard Methods for the Examination of Water and Wastewater, 16th edition*. American Public Health Association, Washington, D. C. A useful reference providing methods commonly used in measuring the numbers of microorganisms.

Block, S. S., ed. 1977. *Disinfection, Sterilization, and Preservation, 2nd edition*. Lea and Febiger, Philadelphia. A standard reference on chemical disinfection, with separate chapters on each of the different groups of agents.

Brock, T. D., ed. 1986. *Thermophiles: General, Molecular, and Applied Microbiology*. John Wiley, N. Y. The best treatment available of thermophilic bacteria, with emphasis on extreme thermophiles.

Dworkin, M. 1985. *Developmental Biology of the Bacteria*. Benjamin/Cummings Publishing Company, Menlo Park, CA. A relatively elementary textbook on special features of bacterial growth. Chapter 3 deals with endospore formation.

Gerhardt, P. ed. 1981. *Manual of Methods for General Bacteriology*. American Society for Microbiology, Washington, D. C. Detailed, up-to-date presentation of bacteriological procedures with extensive references.

Ingraham, J. L., O. Maaløe, and **F. C. Neidhardt.** 1983. *Growth of the Bacterial Cell*. Sinauer Associates, Sunderland, Mass. A detailed textbook on bacterial growth with emphasis on molecular aspects.

Lennette, E. H., A. Balows, W. J. Hausler, Jr., and **H. J. Shadomy.** 1985. *Manual of Clinical Microbiology, 4th edition*. American Society for Microbiology, Washington, D. C. A good reference for information on measurement of antibiotic activity and sensitivity.

10 Microbial Biotechnology

Biotechnology is the discipline which deals with the use of living organisms or their products in large-scale industrial processes. *Microbial biotechnology* is that aspect of biotechnology which deals with processes involving microorganisms. Microbial biotechnology is an old field which has been rejuvenated in recent years because of the development of genetic engineering techniques. Microbial biotechnology originally began with alcoholic fermentation processes, such as those for making beer and wine. Subsequently, microbial processes were developed for the production of pharmaceutical agents such as antibiotics, the production of food additives such as amino acids, the production of enzymes, and the production of industrial chemicals such as butanol and citric acid.

All the industrial processes just described were enhancements of ones the microbes were already capable of carrying out. But now we are in a new era of microbial biotechnology due to the advent of *gene* technology. Gene technology permits a completely new approach to microbial biotechnology in which the microorganism is *engineered* to produce a substance which it normally would not be capable of producing. Thus, for instance, a process for manufacturing the hormone *insulin* has been developed by inserting the human insulin gene into a bacterium (see Section 8.10).

We can thus divide microbial biotechnology into two distinct phases:

1. *Traditional microbial technology*, which involves the large-scale manufacture by microorganisms of products that they are normally capable of producing. In traditional biotechnology, the industrial microbiologist's task is primarily to modify the organism or the process so that the highest yield of the desired product is obtained.
2. *Microbial technology with engineered organisms*, which involves the use of microorganisms into which *foreign* genes have been inserted. In this *new* biotechnology, the industrial microbiologist works closely with the genetic engineer in the development of a suitable organism which not only produces a novel product of interest, but is capable of being cultivated on the *large scale* needed for commercial exploitation.

In the present chapter, we focus on both traditional and contemporary aspects of microbial biotechnology, emphasizing the unique problems and requirements of large-scale industrial cultivation of microorganisms. We will show that there are many difficulties in moving from the laboratory to the large-scale industrial tank. We will also discuss in some detail a number of the key large-scale processes that are currently in use, with emphasis on those aspects of these processes that are of unusual interest to the industrial microbiologist.

Fermentor or fermenter? The term *fermentation* was defined in Section 4.8 as an energy-yielding process in which an organic compound is used as both electron donor and electron acceptor. Fermentations discussed in Chapter 4 included the alcohol fermentation and the lactic acid fermentation. An *organism* that carries out such a fermentation is generally called a *fermenter*.

In the present chapter, the term *fermentation* is used in a much broader sense and we introduce a new term, *fermentor*. In industrial microbiology, the term *fermentation* refers to *any* large-scale microbial process, whether or not it is biochemically a fermentation. In fact, most industrial fermentations are aerobic. The *tank* in which the industrial fermentation is carried out is called a **fermentor** (note the *e* in *fermenter* and the *o* in *fermentor*).

10.1 Industrial Microorganisms

Industrially useful microbes are a unique and quite specific subset of all of the microorganisms that are available on this planet. Whereas microbes isolated from nature exhibit cell growth as their main physiological property, industrial microorganisms are organisms which have been selected carefully so that they manufacture one or more specific *products*. Even if the industrial microorganism is one which has been isolated by traditional techniques, it becomes a highly "modified" organism before it enters large-scale industries. To a great extent, industrial microorganisms are metabolic specialists, capable of producing specifically and to high yield particular metabolites. In order to achieve this high metabolic specialization, industrial strains are genetically altered by mutation or recombination. Minor metabolic pathways are usually repressed or eliminated. Metabolic imbalance frequently is present. Industrial microorganisms may show poor growth properties, loss of ability to sporulate, and altered cellular and biochemical properties. Although industrial strains may grow satisfactorily under the highly specialized conditions of the industrial fermentor, they may show poor growth properties in competitive environments in nature.

Where do industrial strains come from? The ultimate source of all strains of industrial microorganisms is, of course, the natural environment. But through the years, as large-scale microbial processes have become perfected, a number of industrial strains have been deposited in culture collections. When a new industrial process is patented, the applicant for the patent is required to deposit a strain capable of carrying out the process in a recognized culture collection. There are a number of culture collections which serve as the repositories of microbial cultures (Table 10.1). Although these culture collections can serve as ready sources of cultures, it should be understood that most industrial companies will be reluctant to deposit their *best* cultures in culture collections.

Strain improvement As we have noted, the initial source of an industrial microorganism is the natural environment, but the original isolate is greatly modified in the laboratory. As a result of this modification, progressive improvement in the yield of a product can be anticipated. The most dramatic example of such progressive improvement is that of penicillin, the antibiotic produced by the fungus *Penicillium chrysogenum*. When penicillin was first produced on a large scale, yields of 1–10 μg/ml were obtained. Over the years, as a result of strain improvement coupled with changes in the medium and growth conditions, the yield of penicillin has been increased to about 85,000 μg/ml! It is of interest that all of this 85,000-fold increase in yield was obtained by mutation, selection, mating; no genetic engineering manipulations were involved.

Table 10.1 Culture collections which supply cultures of industrial microorganisms

Abbreviation	Name	Location
ATTC	American Type Culture Collection	Rockville, MD USA
CBS	Centraalbureau voor Schimmelcultur	Baarn, Netherlands
CCM	Czechoslovak Collection of Microorganisms	J.E. Purkyne University, Brno, Czechoslovakia
CDDA	Canadian Department of Agriculture	Ottawa, Canada
CIP	Collection of the Institut Pasteur	Paris, France
CMI	Commonwealth Mycological Institute	Kew, UK
DSM	Deutsche Sammlung von Mikroorganismen	Göttingen, Federal Republic of Germany
FAT	Faculty of Agriculture, Tokyko University	Tokyo, Japan
IAM	Institute of Applied Microbiology	University of Tokyo, Japan
NCIB	National Collection of Industrial Bacteria	Aberdeen, Scotland
NCTC	National Collection of Type Cultures	London, UK
NRRL	Northern Regional Research Laboratory	Peoria, IL USA

Listed here are just a few of the general culture collections. Many universities and research laboratories maintain collections of specific microbial groups.

What are the requisites of an industrial microorganism? A microorganism suitable for industrial use must, of course, produce the substance of interest, but there is much more than that. The organism must be available in pure culture, must be genetically stable, and must grow in large-scale culture. It must also be possible to maintain cultures of the organism for a long period of time in the laboratory and in the industrial plant. The culture should preferably produce spores or some other reproductive cell form so that the organism can be easily inoculated into large fermentors.

An important characteristic is that the industrial organism should grow rapidly and produce the desired product in a relatively short period of time. Rapid growth and product formation are desirable for several reasons: (1) Expensive large-scale equipment will not be tied up so long if the product is produced quickly. (2) If the organism grows rapidly, contamination of the fermentor will be less likely to occur. (3) If the organism grows rapidly, it will be much easier to control the various environmental factors of the fermentor.

An important requisite of an industrial microorganism is that it should not be harmful to humans or economically important animals or plants. Because of the large population size in the industrial fermentor, and the virtual impossibility of avoiding contamination of the environment outside the fermentor, a pathogen would present potentially disastrous problems. In addition to being nonpathogenic, the industrial organism should be free of toxins, or if toxins are produced, they should be inactivated readily.

Another important requisite of an industrial microorganism is that it should be possible to remove the cells of the organism from the culture medium relatively easily. In the laboratory, cells are removed from culture media primarily by centrifugation, but centrifugation may be difficult or expensive on a large scale. The most favorable industrial organisms are those of large cell size, since larger cells settle rapidly from a culture or can be easily filtered out with relatively inexpensive filter materials. Fungi, yeasts, and filamentous bacteria, are preferred. Unicellular bacteria, because of their small size, are difficult to separate from a culture fluid.

Finally, an industrial microorganism should be amenable to genetic manipulation. In traditional microbial biotechnology increased yields have been obtained genetically primarily by means of mutation and selection. Mutation is most effective for microorganisms that live vegetatively in the haploid state, and for microorganisms which form uninucleate cells. With diploid or multinucleate organisms, mutation of one genome will not lead to a mutant which can be readily isolated. With filamentous fungi, genetic treatment must be preferably of the spores, because the filaments themselves are not capable of facile genetic manipulation. It is also desirable for the industrial organism to be capable of genetic recombination, either by a sexual or by some sort of parasexual process. Genetic recombination permits the incorporation in a single genome of genetic traits from more than one organism. However, many industrial strains have been greatly improved genetically without any use of genetic recombination.

10.2 Kinds of Industrial Products

Microbial products of industrial interest can be of several major types:

1. The *microbial cells* themselves may be the desired product. This is the case for *yeast* cultivated for food and baking, and for *mushrooms* cultivated for food. Processes have also been developed for raising *bacteria* and *algae* as food sources, but these processes have not become commercially significant. The term *single-cell protein* is sometimes used to refer to microbial cells as industrial products, because the protein content of the microbes is high and is often the characteristic of most commercial interest. Another group of industrial products in which the microbial cells themselves are of interest are those that are used for *inoculation*. For instance, bacteria of the genus *Rhizobium* are frequently inoculated onto leguminous seeds in order to encourage the formation of nitrogen-fixing root nodules (see Section 17.24), and cultures of *lactic acid bacteria* are sold as inoculants for fermented dairy and sausage products.
2. *Enzymes* produced by microorganisms may be the desired product. A number of important enzymes used commercially are produced in large scale by microbial processes, including starch-digesting enzymes (amylases), protein-digesting enzymes (proteases, rennin), and fat-digesting enzymes (lipases). An important industrial enzyme is *glucose isomerase*, used in large amounts industrially for the production from glucose of fructose sweetener. Another important microbial enzyme is *penicillin acylase*, used industrially in the manufacture of semi-synthetic penicillins (see Section 10.6).
3. *Microbial metabolites* constitute some of the most important industrial products. Such metabolites may be relatively major fermentation products such as ethanol, acetic acid, or lactic acid; key growth factors such as amino acids or vitamins; or pharmacologically active agents such as antibiotics, steroids, or alkaloids. The pharmacologically active agents are generally in the category called *secondary metabolites*. Secondary metabolites are compounds that are not produced during the primary growth phase but produced when the cul-

ture is entering stationary phase. Secondary metabolites are some of the most important and interesting industrial products, and an understanding of the nature of secondary metabolism is important in developing new processes (see Section 10.3).

Classification of microbial products by use In addition to classifying microbial products in terms of source, we can classify them in terms of their particular use. Major industries which make wide use of industrial microbiology include the *pharmaceutical, agricultural, food, chemical*, and *environmental protection* industries. We outline briefly major products of interest to most of these industrial types and will discuss them in more detail later in this chapter.

Pharmaceutical products of microbial origin From an economic viewpoint, the pharmaceutical industry is probably the most important industry using microorganisms. Virtually all of the medically important *antibiotics* are produced by microorganisms, and antibiotics are one of the major economic forces in the pharmaceutical industry. Another major class of pharmaceutical agents produced by microorganisms are the *steroid hormones*.

It is in the pharmaceutical industry that some of the most interesting applications of the new gene technology have been made (see Section 10.6 for details). Major products that either are currently manufactured or are anticipated to be manufactured via genetically engineered microbes are human hormones such as *insulin* and *human growth hormone*, antiviral and antitumor agents such as *interferons*, pharmacologically active proteins such as *lymphokines* and *neuroactive peptides*; blood products such as the *blood-clotting factors*; a variety of *vaccines*, and *monoclonal antibodies* of use either for diagnostic medicine or therapy. Because of the potential high economic value of pharmaceutical agents, it can be anticipated that many of the new advances in biotechnology will occur in the pharmaceutical field.

Agricultural uses of microbial biotechnology Many of the pharmaceutical uses mentioned above will also apply to veterinary medicine. In addition, industrial microbiology has had some impact on plant agriculture and more can be expected in the future. We have already mentioned the use of *Rhizobium* inoculants to encourage the formation of root nodules. The new biotechnology is also being used to *engineer* the plant itself for desirable properties, either disease resistance, more rapid growth, or more efficient production of a desired plant product (see Sections 8.10 and 17.23). In plant genetic engineering, microorganisms are being used as *gene vectors*, to transfer genes from one plant to an unrelated plant. The microorganism most widely used for plant genetic engineering is *Agrobacterium tumefaciens*, a plant pathogen which possesses a plasmid, the Ti plasmid, which can become incorporated into the genome of many kinds of plants (see Section 17.23).

Specialty chemicals and food additives Another major area of industrial microbiology is that in which microorganisms are employed to manufacture speciality chemicals and food additives. The most important products in this category are the *amino acids*, a number of which are produced economically in large amounts via microbial processes. The amino acids most commonly produced microbially are: *glutamic acid, lysine, tryptophan, aspartic acid*, and *phenylalanine*. Several vitamins are also produced microbially, including riboflavin, vitamin B_{12}, and ascorbic acid (vitamin C).

Commodity chemicals and energy production Commodity chemicals are defined as those chemicals that have low monetary value and hence are sold primarily in bulk. Commodity chemicals differ from specialty chemicals, discussed above, in that the latter sell at significantly higher prices. Commodity chemicals are used primarily as starting materials for chemical synthesis of more complex molecules. Because commodity chemicals are sold at relatively low prices, it is essential that processes for production of these chemicals operate as economically as possible.

A few commodity chemicals are made commercially using microorganisms, but the majority of commodity chemicals are made industrially by strictly chemical processes, generally from petroleum. Commodity chemicals are simple molecules structurally, which is the reason why chemical synthesis on an industrial scale is possible. Generally, microbial processes can be competitive with chemical processes for commodity chemical synthesis if the target compound is produced in quite high yields, or if the starting material for the microbial process is inexpensive. The commodity chemicals that have been made or are made microbiologically are ethanol, acetic acid, lactic acid, and glycerol. Of these, *ethanol* is the most important.

10.3 Growth and Product Formation in Industrial Processes

In Section 9.4 we discussed the microbial growth process and described the various stages: lag, log, and stationary phase. In the present section, we discuss the microbial growth process as it occurs in an *industrial* process. We are concerned here primarily with those processes in which a microbial metabolite is the desired product. We have noted that there are two basic types of microbial metabolite, those called *primary* and those called *secondary*. A *primary me-*

tabolite is one that is formed during the primary growth phase of the microorganism whereas a *secondary metabolite* is one that is formed near the end of the primary growth phase, frequently at or near the *stationary phase* of growth. The contrast between a primary metabolite and a secondary metabolite is illustrated in Figure 10.1.

Primary microbial metabolites A typical microbial process in which the product is formed during the primary growth phase is the *alcohol (ethanol) fermentation*. Ethanol is a product of anaerobic metabolism of yeast and certain bacteria (see Section 4.8), and is formed as part of energy metabolism. Since growth can only occur if energy production can occur, ethanol formation occurs in parallel with growth. A typical alcohol fermentation, showing the formation of microbial cells, ethanol, and sugar utilization, is illustrated in Figure 10.2*a*.

Secondary microbial metabolism A more interesting, albeit more complex, type of microbial industrial process is one in which the desired product is not produced during the primary growth phase, but at or near the onset of the stationary phase. Me-

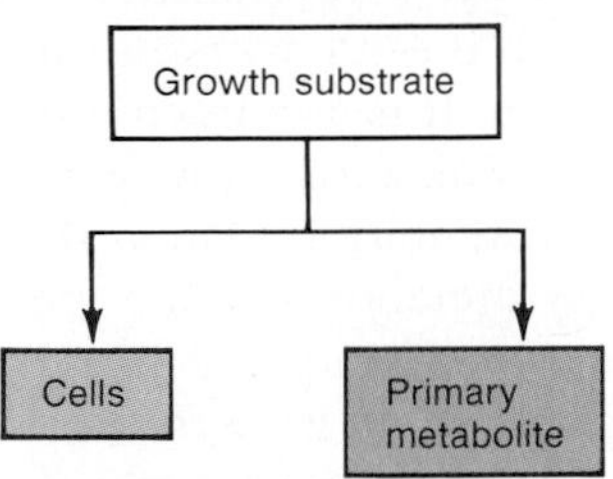

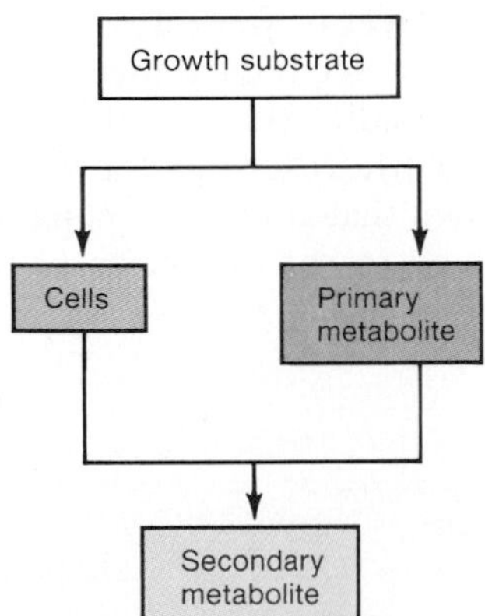

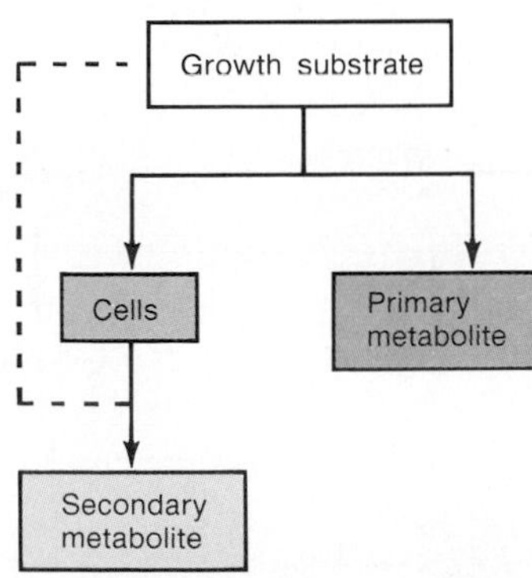

Figure 10.1 Contrast between primary and secondary metabolites

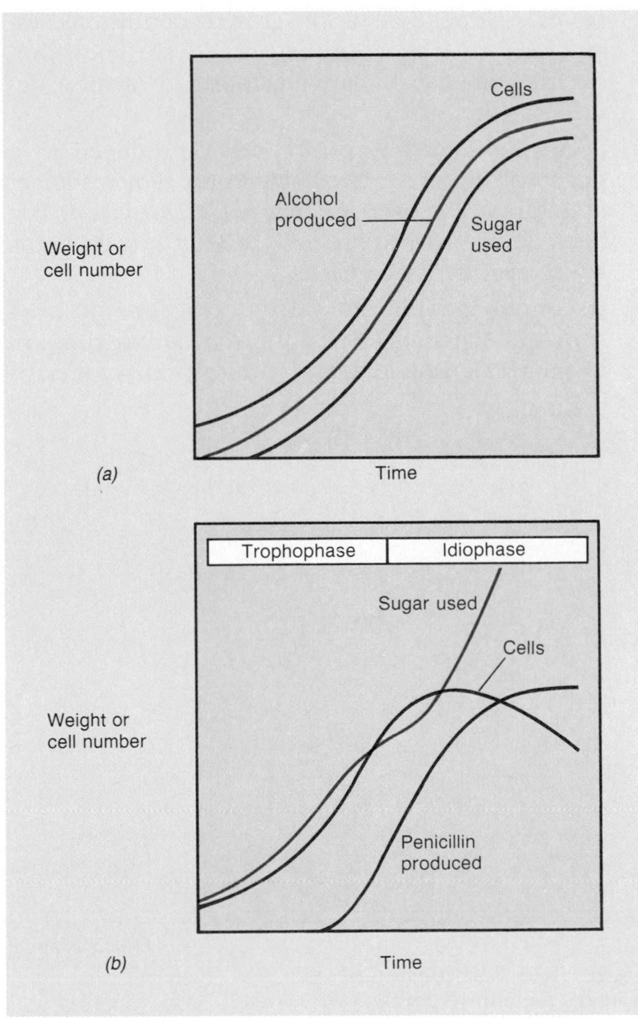

Figure 10.2 Contrasts between primary and secondary metabolism. (a) Primary metabolism: the formation of alcohol by yeast. (b) Secondary metabolism: the formation of penicillin by the fungus *Penicillium chrysogenum*, showing the separation of the growth phase (trophophase) and the production phase (idiophase). Note that in (b), most of the product is produced *after* growth has entered the stationary phase. Note that in both (a) and (b) the y axes are arithmetic rather than log.

tabolites produced during the stationary phase are frequently called **secondary metabolites,** and are some of the most common and important metabolites of industrial interest. The kinetics of a typical process of secondary metabolism, the penicillin process, is shown in Figure 10.2*b*.

Whereas primary metabolism is generally similar in all cells, secondary metabolism shows distinct differences from one organism to another. The following characteristics of secondary metabolites have been recognized:

1. Each secondary metabolite is only formed by a relatively few organisms.
2. Secondary metabolites are seemingly not essential for growth and reproduction.
3. The formation of secondary metabolites is extremely dependent upon growth conditions, especially on the composition of the medium. Repression of secondary metabolite formation frequently occurs.
4. Secondary metabolites are often produced as a group of closely related structures. For instance, a single strain of a species of *Streptomyces* has been found to produce 32 related but different anthracycline antibiotics.
5. It is often possible to get dramatic *overproduction* of secondary metabolites, whereas primary metabolites, linked as they are to primary metabolism, can usually not be overproduced in such a dramatic manner.

Trophophase and idiophase In secondary metabolism, the two distinct phases of metabolism are called the *trophophase* and the *idiophase* (Figure 10.2*b*). The *trophophase* is the *growth* phase (*tropho-* is a combining form meaning "growth"), whereas the metabolite production phase is the *idiophase* (*idio-* is a combining form from the Greek *idios* meaning "one's own"—related to the familiar word "idiot"). Although it is an oversimplification to think of only *two* phases, it is a convenient simplification, since it helps us think about how we should *manage* the industrial fermentation. Thus, if we are dealing with a secondary metabolite, then we should ensure that appropriate conditions are provided during the trophophase for excellent growth, and then we should ensure that conditions are properly altered at the appropriate time in order to assure excellent product formation.

The best known and most extensively studied secondary metabolites are the antibiotics.

Relation between primary and secondary metabolism Most secondary metabolites are complex organic molecules that require a large number of specific enzymatic reactions for synthesis. For instance, it is known that at least 72 separate enzymatic steps are involved in the synthesis of the antibiotic

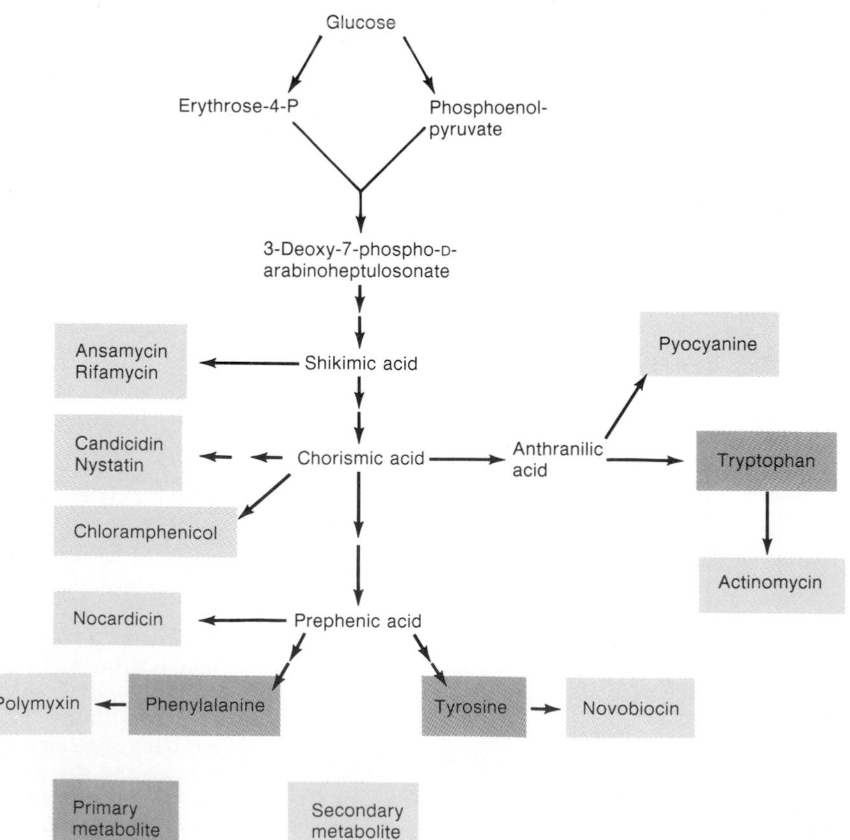

Figure 10.3 Relationship of the primary metabolic pathway for the synthesis of the aromatic amino acids (see Section 4.18) and formation of a variety of secondary metabolites containing aromatic rings. Note that this is a composite scheme of processes occurring in a variety of microorganisms: no one organism produces all these secondary metabolites.

tetracycline and over 25 steps are involved in the synthesis of *erythromycin*, none of which are reactions occurring during primary metabolism. The metabolic pathways of these secondary metabolites do arise out of primary metabolism, however, since the starting materials for secondary metabolism come from the major biosynthetic pathways. This is summarized in Figure 10.3 which shows the interrelationship of the main primary metabolic pathway for aromatic amino acid synthesis with the secondary metabolic pathways for a variety of antibiotics.

10.4 The Characteristics of Large-Scale Fermentations

The vessel in which the industrial process is carried out is called a *fermentor*. Fermentors can vary in size from the small 5–10 l laboratory scale (Figure 10.4*a*) to the enormous 400,000 l industrial scale. The size of the fermentor used depends on the process and how it is operated. Processes operated in batch mode require larger fermentors than processes operated continuously or semi-continuously. A summary of fermentor sizes for some common microbial fermentations is given in Table 10.2.

Industrial fermentors can be divided into two major classes, those for *anaerobic* processes and those for *aerobic* processes. Anaerobic fermentors require little special equipment except for removal of heat generated during the fermentation, whereas aerobic fermentors require much more elaborate equipment to ensure that mixing and adequate aeration are achieved. The present discussion will be confined to aerobic fermentors.

The construction of an aerobic fermentor Large-scale industrial fermentors are almost always constructed of stainless steel. Such a fermentor is es-

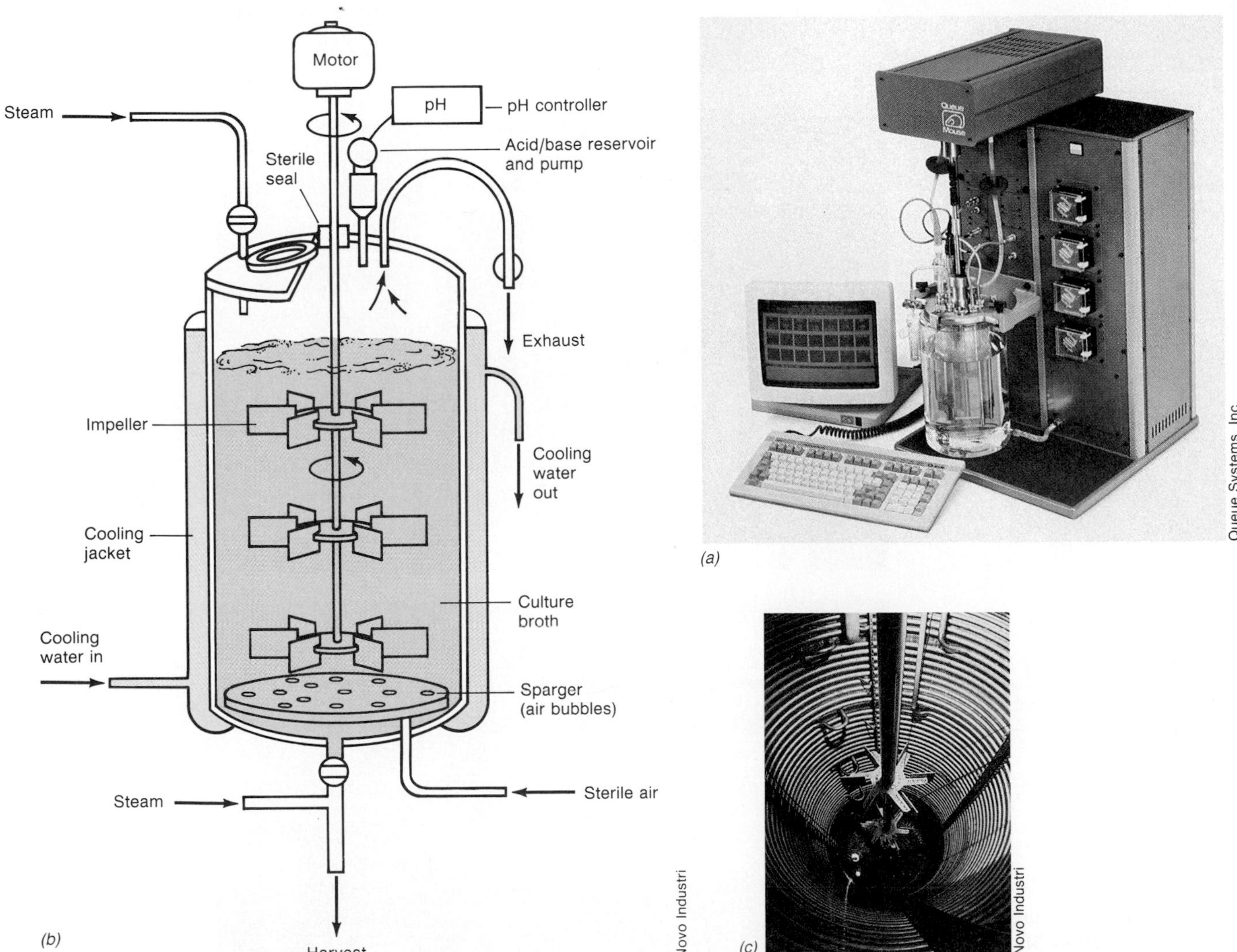

Figure 10.4 An industrial fermentor. (a) Photograph of a small research model. (b) Diagram of a fermentor, illustrating construction and facilities for aeration and process control. (c) Photograph of the inside of a large fermentor, showing the impeller and internal coils.

sentially a large cylinder, closed at the top and bottom, into which various pipes and valves have been fitted (Figure 10.4*b*). Because sterilization of the culture medium and removal of heat are vital for successful operation, the fermentor generally has an external *cooling jacket* through which steam or cooling water can be run. For very large fermentors, insufficient heat transfer occurs through the jacket so that *internal coils* must be provided, through which either steam or cooling water can be piped.

Table 10.2 Fermentor sizes for various processes

Size of fermentor, liters	Product
1– 20,000	Diagnostic enzymes, substances for molecular biology
40– 80,000	Some enzymes, antibiotics
100–150,000	Penicillin, aminoglycoside antibiotics, proteases, amylases, steroid transformations, amino acids
–450,000	Amino acids (glutamic acid)

An important part of the fermentor is the aeration system. With large-scale equipment, transfer of oxygen from the gas to the liquid is a very difficult process and elaborate precautions must be taken to ensure proper aeration. Oxygen is poorly soluble in water, and in a fermentor with a high microbial population density, there is tremendous oxygen demand by the culture. Two separate installations are used to ensure adequate aeration: an aeration device, called a *sparger*, and a stirring device called an *impeller* (Figure 10.4*b*). The sparger is a device, often a series of holes in a metal ring or a nozzle, through which air can be passed into the fermentor under high pressure. The air enters the fermentor as a series of tiny bubbles, from which the oxygen passes by diffusion into the liquid. A good sparging system should ensure that the bubbles are of very small size, so that diffusion of oxygen from the bubble into the liquid can occur readily.

(a) Novo Industri

(b) Novo Industri

Figure 10.5 (a) A large industrial fermentation plant. Only the tops of the fermentors are visible. (b) Computer control room for a large fermentation plant.

In small fermentors use of a sparger alone may be sufficient to ensure adequate aeration, but in industrial-sized fermentors, *stirring* of the fermentor with an impeller is essential (Figure 10.4*c*). Stirring accomplishes two things: it mixes the gas bubbles through the liquid; and it mixes the organism through the liquid, thus ensuring uniform access of the microbial cells to the nutrients. One of the most common stirrers is a flat blade or flat disk, fastened to a center shaft, which is rotated at high rate by a shaft attached to a motor. In order to ensure most effective mixing by the impeller, *baffles* are generally installed vertically along the inside diameter of the fermentor. As it is stirred by the impeller, the liquid passes the baffles and is broken up into smaller patches. Fluid dynamics in fermentors is extremely complex but is important to understand, because effective design and operation of a fermentor depend on adequate mixing.

The shaft which drives the impeller is attached to the motor by way of a drive shaft which must penetrate into the fermentor from outside. Because of the need to maintain sterility, it is vital that the seal which connects the drive shaft to the motor be arranged so that contaminants cannot pass through (see Figure 10.4*b*). A typical large-scale fermentor installation is shown in Figure 10.5.

Process control and monitoring Any microbial process must be monitored to ensure that it is proceeding properly, but it is especially important that industrial fermentors be monitored carefully, because there is such a major expense involved. In most cases, it is not only necessary to measure growth and product formation, but to *control* the process by altering environmental parameters as the process proceeds. Environmental factors that are frequently monitored include oxygen concentration, pH, cell mass, and product concentration. It is also essential, in many cases, to control the fermentor for foaming, and to adjust the temperature (either cooling or heating, depending upon the process).

Computer control of the fermentation process Computers now play important roles in fermentor process control (Figure 10.5*b*). They are used in two major ways: (1) In *acquisition of data* that reflect the changes taking place during the fermentation process. (2) In *control* of various environmental factors that must be adjusted or altered during the fermentation.

During the growth and product formation process in a large-scale fermentation, it is essential, if the fermentation is to be operated properly, that data be obtained as the process is actually taking place. Data acquisition occurring in this way is spoken of as **online acquisition**, and is of special value when such data can be processed by a computer. In some cases, it may be necessary to convert raw values exhibited by a fermentor probe into values that can be interpreted in microbiological or biochemical terms, and in such cases the computer is admirably suited. In addition, the computer can be used to *graph* the data that are being acquired, providing to the fermentor operator a visual indication of the progress of the fermentation. Finally, the computer can be used to *store* the data in computer-readable form, so that the data can be transferred to another system, or analyzed in more sophisticated form off-line.

An even more sophisticated use of computers is in **on-line control** of the fermentation process. For instance, it may be desirable to change one of the environmental parameters as the fermentation progresses, or to feed a nutrient at a rate which exactly balances growth. The computer can be used to process the data used to measure growth, and then to *decide* (from instructions provided) when and how much nutrient to add. In this way, nutrient is added when needed, and not before, thus avoiding potential diversion of nutrient from the desired product into unwanted products.

10.5 Scale-Up of the Fermentation Process

One of the most important and complicated aspects of industrial microbiology is the transfer of a process from small-scale laboratory equipment to large-scale commercial equipment, a procedure called **scale-up**. An understanding of the problems of scale-up are extremely important, because rarely does a microbial process behave the same way in large-scale fermentors as in small-scale laboratory equipment (Figure 10.6).

Why does a microbial process differ between large-scale and small-scale equipment? Why is a knowledge of scale-up essential? Mixing and aeration are much easier to accomplish in the small laboratory flask than in the large industrial fermentor. As the *size* of the equipment is increased, the *surface/volume* ratio changes (the principle of surface to volume ratio was explained in Figure 3.1); the large fermentor has much more volume for a given surface area. Since gas transfer and mixing depend more on surface exposed than on fermentor volume, it is obviously more difficult to mix the big tank than the small flask. Oxygen transfer especially is much more difficult to obtain in a large fermentor, again due to the different surface/volume ratio. Because most commercial fermentations are aerobic, effective oxygen transfer is essential. With the rich culture media used in industrial processes, high biomass is obtained, leading to high oxygen demand. If aeration is reduced, even for a short period, the culture may experience partial anaerobiosis, with serious consequences in terms of product yield. There are pockets within the large fer-

Figure 10.6 (a) A bank of small research fermentors, used in process development. (b) A large bank of outdoor industrial scale fermentors (240 m^3) used in commercial production of amino acids in Japan.

mentor where mixing is less efficient, and if microbial cells travel into such pockets, they experience different environmental conditions than they would in the bulk fluid. In the laboratory flask, such pockets do not exist.

Scale-up of an industrial process is the task of the ***biochemical engineer***, who is familiar with gas transfer, fluid dynamics, mixing, and thermodynamics. The role of the industrial microbiologist in the scale-up process is to work closely with the biochemical engineer to ensure that all parameters needed for a successful fermentation are understood, and that microbial strains appropriate for a large-scale fermentation are available. Very often it is found that a strain or culture medium which works well in a laboratory flask does not work well in the big fermentor. For instance, a strain may produce a polymer which increases markedly the viscosity of the culture, thus indirectly affecting aeration efficiency. The industrial microbiologist would then be required to develop a strain which did not cause this increase in viscosity.

In transferring an industrial process from the laboratory to the commercial fermentor, several stages can be envisioned: (1) Experiments in the *laboratory flask* which are generally the first indication that a process of commercial interest is possible. (2) The *laboratory fermentor*, a small-scale fermentor, generally of glass and generally of 5–10 liter size, in which the first efforts at scale-up are made (Figures 10.5*a* and 10.6*a*). In the laboratory fermentor, it is possible to test variations in medium, temperature, pH, etc. inexpensively, since little cost is involved for either equipment or culture medium. (3) The *pilot plant stage*, usually carried out in equipment of 300–3000 liter size. Here, the conditions more closely approach the commercial scale, however cost is not yet a major factor. In the pilot plant fermentor, careful instrumentation and computer control is desirable, so that the conditions most similar to those in the laboratory fermentor can be obtained. (4) The *commercial fermentor* itself, generally of 10,000 to 400,000 liters (Figure 10.6*b*).

It is generally found in scale-up studies on aerobic fermentations that the stirring and aeration rates of the fermentor are best kept *constant* as the size of the fermentor is increased. Thus, if an oxygen transfer rate into the fermentor of 200 mM O_2/l/hr is required to obtain optimal yield in a small fermentor, then stirring and aeration in the large fermentor should be adjusted to assure this same rate. To do this, it will probably require much more rapid stirring as well as a higher pressure of the inlet air. Because stirring is a mechanical process that can be monitored in terms of power, one approach to scale-up is to maintain constant power to the fermentor when going from small- to large-scale equipment.

10.6 Antibiotics

Of the microbial products manufactured commercially, probably the most important are the antibiotics. As we have discussed in Section 9.20, antibiotics are chemical substances that are produced by microorganisms which kill or inhibit the growth of other microorganisms. The development of antibiotics as agents for treatment of infectious disease

Table 10.3 Some antibiotics produced commercially

Antibiotic	Producing microorganism	Type
Bacitracin	*Bacillus subtilis*	Endospore-forming bacterium
Cephalosporin	*Cephalosporium* sp.	Fungus
Chloramphenicol	Chemical synthesis (formerly *Streptomyces venezuelae*)	
Cycloheximide	*Streptomyces griseus*	Actinomycete
Cycloserine	*S. orchidaceus*	Actinomycete
Erythromycin	*S. erythreus*	Actinomycete
Griseofulvin	*Penicillium griseofulvin*	Fungus
Kanamycin	*S. kanamyceticus*	Actinomycete
Lincomycin	*S. lincolnensis*	Actinomycete
Neomycin	*S. fradiae*	Actinomycete
Nystatin	*S. noursei*	Actinomycete
Penicillin	*Penicillium chrysogenum*	Fungus
Polymyxin B	*Bacillus polymyxa*	Endospore-forming bacterium
Streptomycin	*S. griseus*	Actinomycete
Tetracycline	*S. rimosus*	Actinomycete

has probably had more impact on the practice of medicine than any other single development.

Antibiotics are products of secondary metabolism. Although they are produced in relatively small concentrations in most industrial fermentations, because of their high therapeutic activity and consequently high economic value, they can be produced commercially by microbial fermentation. Many antibiotics can be synthesized chemically, but because of the chemical complexity of the antibiotics and the great expense attendant on chemical synthesis, rarely is it possible for chemical synthesis to compete with microbial fermentation.

Commercially useful antibiotics are produced primarily by filamentous fungi and by bacteria of the actinomycete group. A listing of the most important antibiotics produced by large-scale industrial fermentation is given in Table 10.3. Frequently, a number of chemically related antibiotics exist, so that *families* of antibiotics are known. Antibiotics can hence be classified according to their chemical structure, as outlined in Table 10.4. Most antibiotics employed medically are used to treat bacterial diseases, although a few antibiotics are known which are effective against fungal diseases. The economic significance of the antibiotics is shown by the fact that over 100,000 *tons* of antibiotics are produced per year, with gross sales substantially over 4 billion dollars.

Table 10.4 Classification of antibiotics according to their chemical structure. An example of each is given in parentheses

1. Carbohydrate-containing antibiotics	
Pure sugars	(Nojirimycin)
Aminoglycosides	(Streptomycin)
Orthosomycins	(Everninomicin)
N-Glycosides	(Streptothricin)
C-Glycosides	(Vancomycin)
Glycolipids	(Moenomycin)
2. Macrocyclic lactones	
Macrolide antibiotics	(Erythromycin)
Polyene antibiotics	(Candicidin)
Ansamycins	(Rifamycin)
Macrotetrolides	(Tetranactin)
3. Quinones and related antibiotics	
Tetracyclines	(Tetracycline)
Anthracyclines	(Adriamycin)
Naphthoquinones	(Actinorhodin)
Benzoquinones	(Mitomycin)
4. Amino acid and peptide antibiotics	
Amino acid derivatives	(Cycloserine)
β-Lactam antibiotics	(Penicillin)
Peptide antibiotics	(Bacitracin)
Chromopeptides	(Actinomycins)
Depsipeptides	(Valinomycin)
Chelate-forming peptides	(Bleomycins)
5. Heterocyclic antibiotics containing nitrogen	
Nucleoside antibiotics	(Polyoxins)
6. Heterocyclic antibiotics containing oxygen	
Polyether antibiotics	(Monensin)
7. Alicyclic derivatives	
Cycloalkane derivatives	(Cycloheximide)
Steroid antibiotics	(Fusidic acid)
8. Aromatic antibiotics	
Benzene derivatives	(Chloramphenicol)
Condensed aromatic antibiotics	(Griseofulvin)
Aromatic ether	(Novobiocin)
9. Aliphatic antibiotics	
Compounds containing phosphorous	(Fosfomycins)

The search for new antibiotics Over 5000 antibiotic substances are known, and around 300 new antibiotics are discovered yearly. Are there more antibiotics waiting to be discovered? Almost certainly yes, because most of the microorganisms which have been examined for ability to produce antibiotics are members of a few genera, such as *Streptomyces, Penicillium*, and *Bacillus*. Many antibiotic researchers believe that a vast number of new antibiotics will be discovered if other groups of microorganisms are examined. It seems also likely that genetic engineering techniques will permit the artificial construction of new antibiotics, once the details of the gene structure of antibiotic-producing microorganisms have been discerned.

However, the main way in which new antibiotics have been discovered has been by *screening*. In the screening approach, a large number of isolates of possible antibiotic-producing microorganisms are obtained from nature in pure culture, and these isolates are then tested for antibiotic production by seeing whether they produce any diffusible materials that are inhibitory to the growth of test bacteria. The test bacteria used are selected from a variety of bacterial types, but are chosen to be representative of or related to bacterial pathogens. The classical procedure for testing new microbial isolates for antibiotic production is the cross-streak method, first used by Fleming in his pioneering studies on penicillin (Figure 10.7). Those isolates which show evidence of antibiotic production are then studied further to determine if the antibiotics they produce are new. In most current screening programs, most of the isolates obtained produce *known* antibiotics, so that the microbiologist must quickly identify producers of known antibiotics and discard them. Once an organism producing a new antibiotic is discovered, the antibiotic is produced in large amounts, purified, and tested for toxicity and therapeutic activity in infected animals. Most new antibiotics *fail* these animal tests, but a few pass them successfully. Ultimately, a very few of these new antibiotics prove useful medically and are produced commercially. The discovery of a new antibiotic of commercial significance is not unlike looking for the proverbial "needle in a haystack."

Steps toward commercial production An antibiotic that is to be produced commercially must first be produced successfully in large-scale industrial fermentors. We have discussed in general the problems of scale-up earlier in this chapter. One of the most important tasks is the development of efficient purification methods. Because of the small amounts of antibiotic present in the fermentation liquid, elaborate methods for the extraction and purification of the antibiotic are necessary (Figure 10.8). If the antibiotic is soluble in an organic solvent which is im-

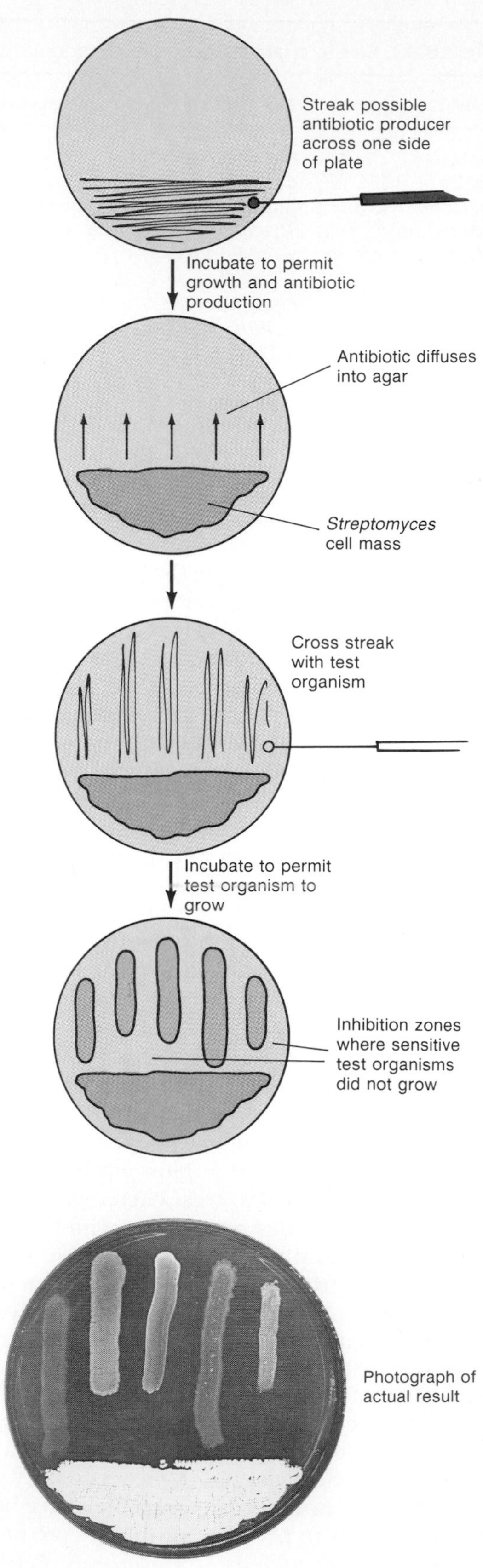
Streak possible antibiotic producer across one side of plate
Incubate to permit growth and antibiotic production
Antibiotic diffuses into agar
Streptomyces cell mass
Cross streak with test organism
Incubate to permit test organism to grow
Inhibition zones where sensitive test organisms did not grow
Photograph of actual result

Figure 10.7 Method of testing an organism for the production of antibiotics. The producer (*Streptomyces*) is streaked across one-third of the plate and the plate incubated. After good growth is obtained, the test bacteria are streaked perpendicular to the *Streptomyces*, and the plate was further incubated. The failure of several organisms to grow near the *Streptomyces* indicates that the *Streptomyces* produced an antibiotic active against these bacteria. Test organisms left to right: *Escherichia coli*; *Bacillus subtilis*; *Staphylococcus aureus*; *Klebsiella pneumoniae*; *Mycobacterium smegmatis*.

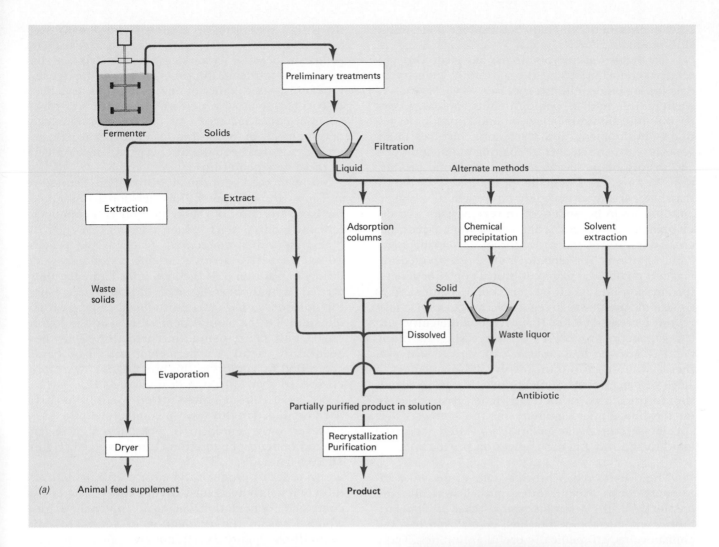

Fermenter
Preliminary treatments
Solids
Filtration
Liquid
Alternate methods
Extraction
Extract
Adsorption columns
Chemical precipitation
Solvent extraction
Waste solids
Solid
Dissolved
Waste liquor
Evaporation
Antibiotic
Partially purified product in solution
Dryer
Recrystallization Purification
(a) Animal feed supplement
Product

miscible in water, it may be relatively simple to purify the antibiotic because it is possible to extract the antibiotic into a small volume of the solvent and thus concentrate the antibiotic easily. If the antibiotic is not solvent-soluble, then it must be removed from the fermentation liquid by adsorption, ion exchange, or chemical precipitation. In all cases, the goal is to obtain a crystalline product of high purity, although some antibiotics do not crystallize readily and are difficult to purify. A related problem is that cultures often produce other end products, including other antibiotics, and it is essential to end up with a product consisting of only a single antibiotic. The purification chemist may be required to develop methods for eliminating undesirable byproducts, but in some cases it may be necessary for the microbiologist to find strains that do not produce such undesirable chemicals.

Rarely do antibiotic-producing strains just isolated from nature produce the desired antibiotic at sufficiently high concentration so that commercial production can begin immediately. One of the main tasks of the industrial microbiologist is to isolate new high-yielding strains. Another goal of strain selection

(b)

Figure 10.8 Purification of an antibiotic. (a) Overall process of extraction and purification. (b) Installation for the solvent extraction of an antibiotic from fermentation broth.

is to obtain strains which do not produce undesirable side products.

The industrial microbiologist has made very significant contributions to the antibiotic industry by developing high-yielding processes. As we have noted earlier, the yield of penicillin was increased over 85,000-fold by strain selection and appropriate medium development. Strain selection involves mutagenesis of the initial culture, plating of mutant types, and testing of these mutants for antibiotic production. In most cases, mutants produce *less* antibiotic than the parent, so that only rarely would a higher-yielding strain be obtained. In recent years, the development of genetic engineering techniques has greatly improved the procedures for seeking high-yielding strains. The technique of *gene amplification* makes it possible to place additional copies of genetic information into a cell by means of a vector such as a plasmid. Alterations in regulatory processes also may permit increased yields. However, one difficulty with using genetic procedures for increasing antibiotic yield is that the biosynthetic pathways for the synthesis of most antibiotics involve large numbers of steps with many genes, and it is not clear which genes should be altered to increase yields. Basic research on the biochemical pathway for the synthesis of a specific antibiotic may go farther to lead to increased yield than blind mutation/selection programs.

The β lactam antibiotics One of the most important groups of antibiotics, both historically and medically, is the β lactam group. The β lactam antibiotics include the penicillins, cephalosporins, and cephamycins, all medically useful antibiotics. These antibiotics are called β lactams because they contain the β lactam ring system (Figure 10.9), a complex two-membered heterocylic ring system. As we have discussed in Section 3.5, β lactam antibiotics act by inhibiting peptidoglycan synthesis in bacterial cell walls. The target of these antibiotics is the transpeptidation reaction involved in the cross-linking step of peptidoglycan biosynthesis. Because this reaction is unique to bacteria, the β lactam antibiotics have high specificity and relatively low toxicity.

The first β lactam antibiotic discovered, **penicillin G**, is active primarily against Gram-positive bacteria. Its action is restricted to Gram-positive bacteria primarily because Gram-negative bacteria are impermeable to the antibiotic. As a result of extensive research, a vast number of new penicillins have been discovered, some of which are quite effective against Gram-negative bacteria. One of the most significant developments in the antibiotic field over the past several decades has been the discovery and development of these new penicillins.

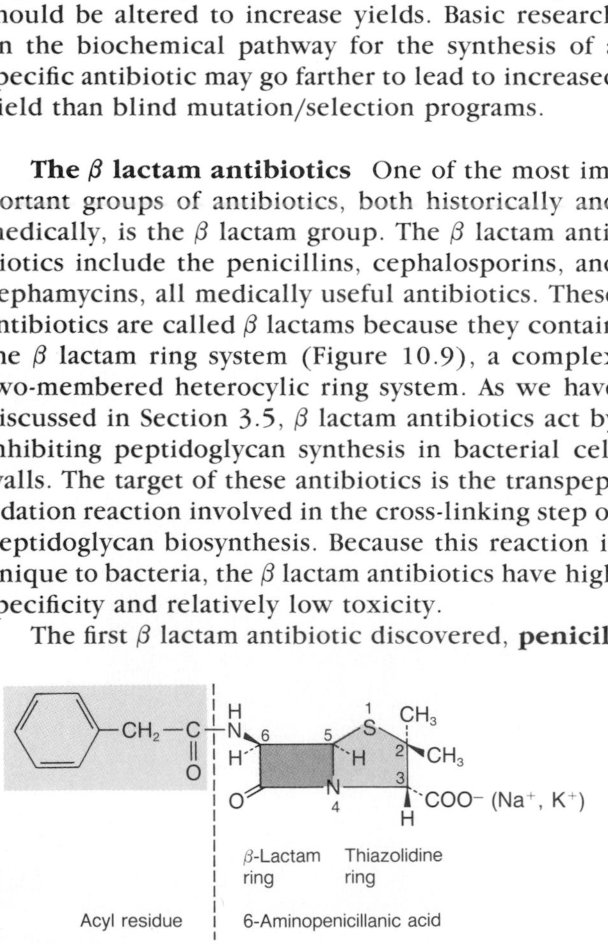

Figure 10.9 The structure of penicillin G, the most important naturally occurring penicillin. Note the β-lactam ring system. The acyl residue is variable.

How do the various penicillins differ in structure and how are new penicillins developed? As noted, the basic structure of the penicillins is 6-aminopenicillanic acid (6-APA), which consists of a thiazolidine ring with a condensed β-lactam ring (Figure 10.9). The 6-APA carries a variable acyl moiety (side chain) in position 6. If the penicillin fermentation is carried out without addition of side-chain precursors, the *natural penicillins* are produced (Figure 10.10, left side). Of the natural penicillins listed in Figure 10.10, only benzylpenicillin (penicillin G) is therapeutically useful. The fermentation can be better controlled by adding to the broth a *side-chain precursor*, so that only *one* desired penicillin is produced. Over 100 such **biosynthetic penicillins** have been produced in this way. In commercial processes, however, only penicillin G, penicillin V, and very limited amounts of penicillin O are produced (Figure 10.10, right side).

In order to produce the most useful penicillins, those with activity against Gram-negative bacteria, a combined fermentation/chemical approach is used which leads to the production of **semisynthetic penicillins** (Figure 10.10, bottom right). The starting material for the production of such semisynthetic penicillins is penicillin G, which serves as the source of the 6-APA nucleus. Pencillin G is split either chemically or enzymatically (using penicillin acylase) and the 6-APA obtained is then coupled chemically to another side chain. As seen in Figure 10.10, the semisynthetic penicillins have significantly improved medical properties.

Production methods for β-lactam antibiotics Penicillin G is produced using a submerged fermentation process in 40,000–200,000 liter fermentors. Pencillin production is a highly aerobic process and efficient aeration is necessary. Pencillin is a typical secondary metabolite, as illustrated in Figures 10.2*b* and 10.11. During the growth phase, very little penicillin is produced but once the carbon source has been exhausted, the penicillin production phase begins. By feeding with various culture medium components, the production phase can be extended for several days.

Certain culture media permit high penicillin yields whereas other culture media are quite effective for growth but lead to little product formation. A

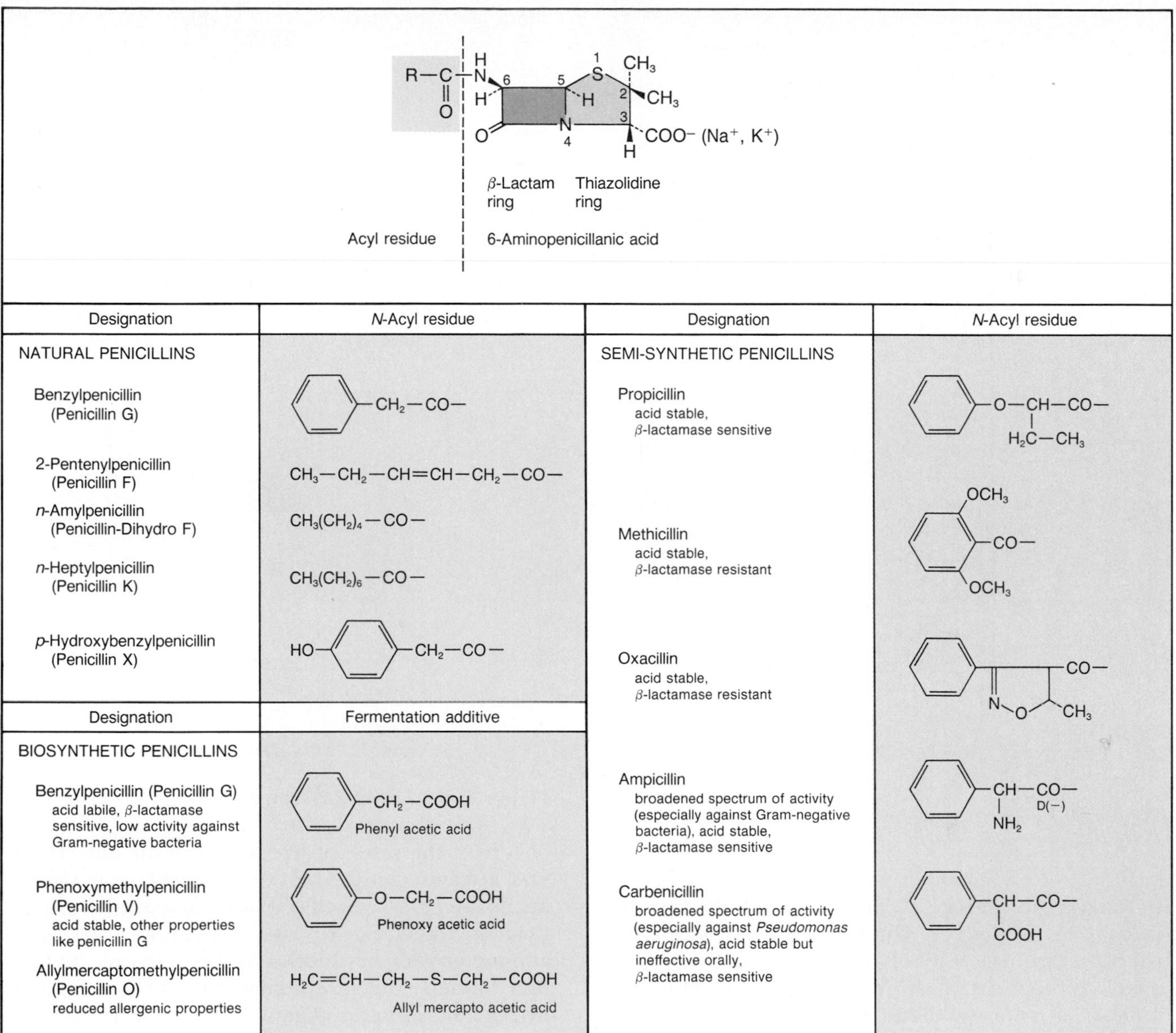

Designation	N-Acyl residue
NATURAL PENICILLINS	
Benzylpenicillin (Penicillin G)	—CH_2—CO—
2-Pentenylpenicillin (Penicillin F)	CH_3—CH_2—CH=CH—CH_2—CO—
n-Amylpenicillin (Penicillin-Dihydro F)	$CH_3(CH_2)_4$—CO—
n-Heptylpenicillin (Penicillin K)	$CH_3(CH_2)_6$—CO—
p-Hydroxybenzylpenicillin (Penicillin X)	HO— —CH_2—CO—

Designation	Fermentation additive
BIOSYNTHETIC PENICILLINS	
Benzylpenicillin (Penicillin G) acid labile, β-lactamase sensitive, low activity against Gram-negative bacteria	—CH_2—COOH Phenyl acetic acid
Phenoxymethylpenicillin (Penicillin V) acid stable, other properties like penicillin G	—O—CH_2—COOH Phenoxy acetic acid
Allylmercaptomethylpenicillin (Penicillin O) reduced allergenic properties	H_2C=CH—CH_2—S—CH_2—COOH Allyl mercapto acetic acid

Designation	N-Acyl residue
SEMI-SYNTHETIC PENICILLINS	
Propicillin acid stable, β-lactamase sensitive	—O—CH—CO— ; H_2C—CH_3
Methicillin acid stable, β-lactamase resistant	OCH_3 ; —CO— ; OCH_3
Oxacillin acid stable, β-lactamase resistant	—CO— ; N O ; CH_3
Ampicillin broadened spectrum of activity (especially against Gram-negative bacteria), acid stable, β-lactamase sensitive	—CH—CO— D(−) ; NH_2
Carbenicillin broadened spectrum of activity (especially against *Pseudomonas aeruginosa*), acid stable but ineffective orally, β-lactamase sensitive	—CH—CO— ; COOH

Figure 10.10 Structure of some of the natural penicillins, the most important biosynthetic penicillins, and several semisynthetic penicillins.

major ingredient of most penicillin-production media is *corn steep liquor*, a product of the corn wet milling industry. This component serves as the nitrogen source, as well as a source of other growth factors. The carbon source is generally *lactose*, which is not used as rapidly as glucose and does not cause catabolite repression. Glucose can be used as a carbon source, but when glucose is used, it must be fed *slowly*, to avoid catabolite repression. The acyl side chain of penicillin G is the phenylacyl moiety, and yields of penicillin G can be markedly increased if phenylacetic acid or phenoxyacetic acid is fed continuously as a precursor. If a pencillin with another acyl side is desired, then its appropriate precursor should be fed.

Penicillin is excreted into the medium and less than 1 percent remains with the fungal filaments. After the fungal mycelium is removed by filtration, the pH of the medium is lowered and the antibiotic extracted from the filtered broth with amyl or butyl acetate. After concentration into the solvent, the antibiotic is back-extracted into an alkaline aqueous medium, concentrated further, and crystallized. Highly purified penicillin can be readily obtained in this way. When penicillin first became available commercially, its price was extremely high, but it is frequently said today that the package that the penicillin is in costs more than the ingredients within!

Cephalosporins are β-lactam antibiotics containing a dihydrothiazine instead of a thiazolidine ring

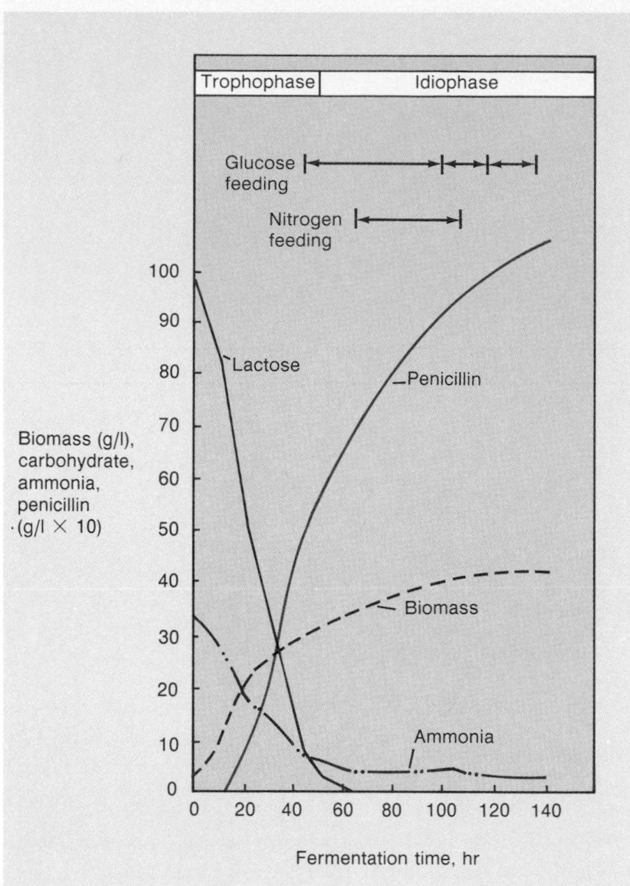

Figure 10.11 Kinetics of the penicillin fermentation with *Penicillium chrysogenum*.

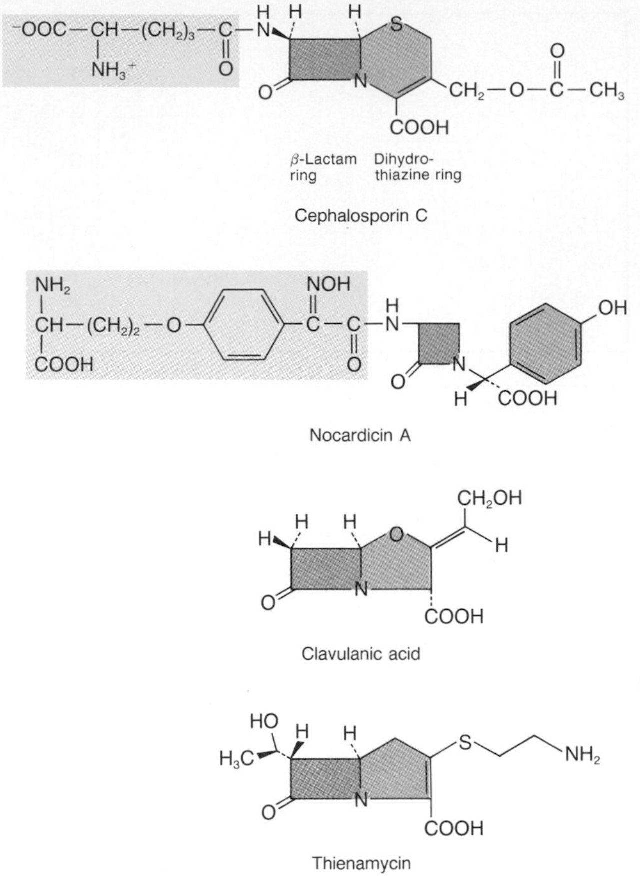

β-Lactam ring Dihydrothiazine ring

Cephalosporin C

Nocardicin A

Clavulanic acid

Thienamycin

Figure 10.12 Newer β-lactam antibiotics.

system (Figure 10.12). Cephalosporins were first discovered as products of the fungus *Cephalosporium acremonium*, but a number of other fungi also produce antibiotics with this ring system. In addition, a number of semi-synthetic cephalosporins are produced. Cephalosporins are valued not only because of their low toxicity but also because they are broad-spectrum antibiotics.

Intensive screening for new β-lactam antibiotics has led to the development of compounds whose structures are different from both the penicillins and the cephalosporins. Included in this category are nocardicin, clavulanic acid, and thienamycin (Figure 10.12).

Aminoglycoside antibiotics Aminoglycosides are antibiotics which contain amino sugars bonded by glycosidic linkage (see Section 2.4) to other amino sugars. A number of clinically useful antibiotics are aminoglycosides, including *streptomycin* (see Section 7.6) and its relatives, *kanamycin, gentamicin,* and *neomycin*. The aminoglycoside antibiotics are used clinically primarily against Gram-negative bacteria. Streptomycin has also been used extensively in the treatment of tuberculosis. Historically, the discovery of the value of streptomycin for tuberculosis was a major medical advance, as this was the first antibiotic discovered capable of controlling this dreaded infectious disease. However, none of the aminoglycoside antibiotics find as wide use today as they formerly did. Streptomycin has been supplanted by several synthetic chemicals for tuberculosis, due to the fact that streptomycin causes several serious side effects and because bacterial resistance readily develops. The use of aminoglycosides for Gram-negative infections has been less significant since the development of the semisynthetic penicillins (see above) and the tetracyclines (see below). The aminoglycoside antibiotics are now considered reserve antibiotics, used primarily when other antibiotics fail. We discussed the mode of action of streptomycin in Section 9.20.

One of the interesting features of the aminoglycoside antibiotics is the *regulation of their biosynthesis*. As seen in Figure 10.13, the three moieties of streptomycin are synthesized by separate pathways, and the subunits brought together at the end. Streptomycin is synthesized as a typical secondary metabolite. At least several of the enzymes involved are synthesized only at the end of the trophophase.

One aspect of the regulation involves the production of an inducer called *factor A* (Figure 10.14).

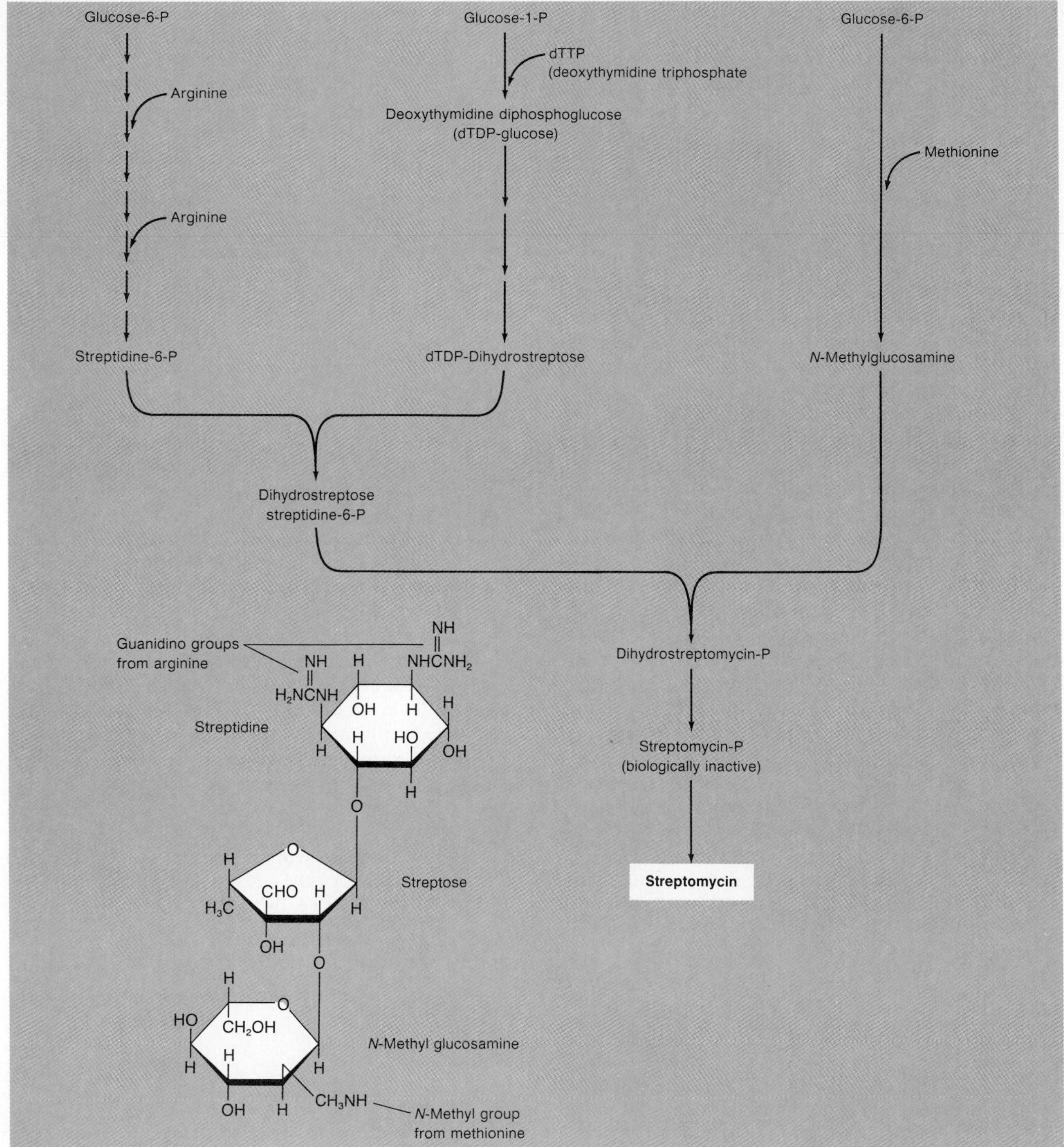

Figure 10.13 Streptomycin biosynthesis in *Streptomyces griseus.*

Factor A is not chemically related to streptomycin, but is involved somehow in carbohydrate metabolism. Key enzymes in streptomycin biosynthesis are not synthesized until factor A concentration builds up, thus explaining how this factor might play a role in secondary metabolism. During the growth phase, factor A that is synthesized is excreted and gradually builds up in the medium. Only when the concentration reaches a critical level does factor A begin to act and induce the synthesis of the key streptomycin biosynthesis enzymes. The importance of A-factor for streptomycin biosynthesis is shown by the fact that A-factor-negative mutants lose their capacity for streptomycin biosynthesis, but if synthetic A-factor is

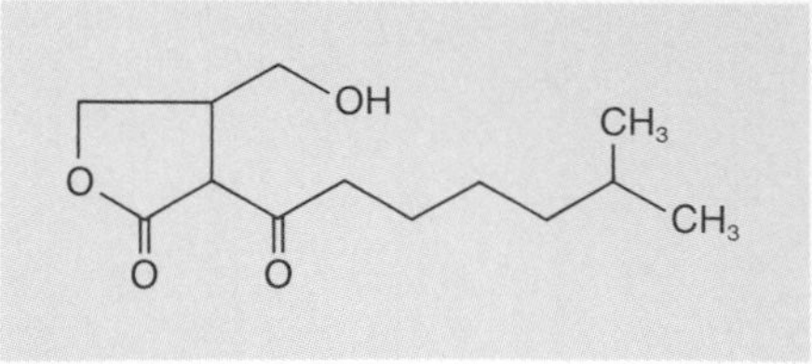

Figure 10.14 Structure of Factor A from *Streptomyces griseus*.

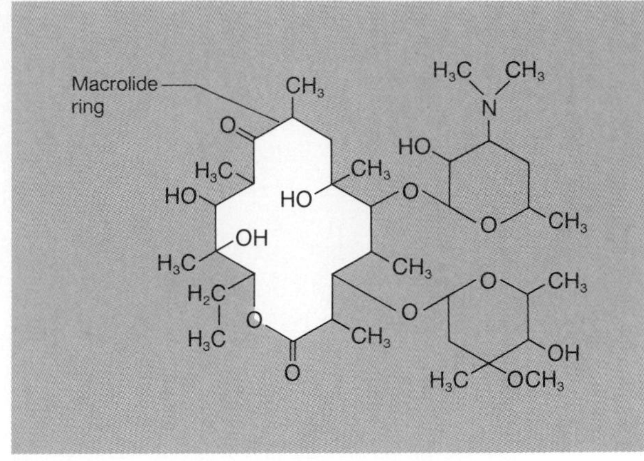

Figure 10.15 Structure of erythromycin, a typical macrolide antibiotic.

added to such a mutant, streptomycin biosynthesis is restored. However, A-factor is not itself a precursor of streptomycin, as shown by the fact that the addition of a tiny amount of pure A-factor, 1 μg, can lead to the production of 1 g of streptomycin. It should be noted, however, that streptomycin biosynthesis is not regulated *solely* by A-factor, since other regulatory processes also affect streptomycin biosynthesis.

Macrolide antibiotics Macrolide antibiotics contain large lactone rings connected to sugar moieties (Figure 10.15). Variation in both the macrolide ring and the sugar moieties are known, so that a large variety of macrolide antibiotics exist. The best known macrolide antibiotic is *erythromycin*, but other macrolides include *oleandomycin, spiramycin*, and *tylosin*. We discussed the mode of action of erythromycin in Section 9.20. Erythromycin is commonly used clinically in place of penicillin in those patients allergic to penicillin or other β-lactam antibiotics. Erythromycin has been particularly valuable in treating cases of legionellosis (see Section 15.2), because of the exquisite sensitivity of the causative agent, the bacterium *Legionella pneumophila*, to this antibiotic.

The complexity of the macrolide antibiotics is evident from their structure. Over 25 unique enzymatic steps are known to be involved in erythromycin biosynthesis. Regulation of biosynthesis occurs in a number of ways. Glucose and phosphate inhibition are known to occur, and erythronolide B, one of the intermediates in erythromycin biosynthesis, inhibits its own production. End product inhibition by erythromycin itself of certain key enzymes is also known to occur.

Tetracyclines The tetracyclines are an important group of antibiotics which find widespread medical use. They were some of the first so-called *broad-spectrum* antibiotics, inhibiting almost all Gram-positive and Gram-negative bacteria. The basic structure of the tetracyclines consists of a naphthacene ring system (Figure 10.16). To this ring is added any of several constituents. *Chlortetracycline*, for instance, has a chlorine atom whereas *oxytetracycline* has an additional hydroxy (OH) group and no chlorine (Figure 10.16). All three of these antibiotics are pro-

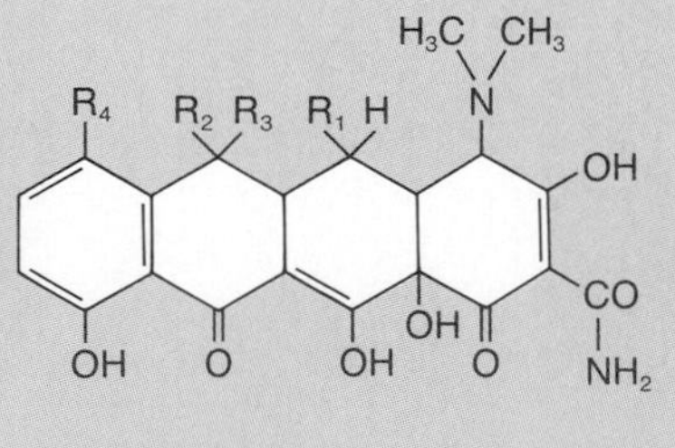

Tetracyclines	R_1	R_2	R_3	R_4	Production strain
Tetracycline	H	OH	CH_3	H	*S. aureofaciens* (in chloride-free medium) or through chemical modification of chlortetracycline
7-Chlortetracycline (Aureomycin)	H	OH	CH_3	Cl	*S. aureofaciens*
5-Oxytetracycline (Terramycin)	OH	OH	CH_3	H	*S. rimosus*

Figure 10.16 Structure of tetracycline and important derivatives.

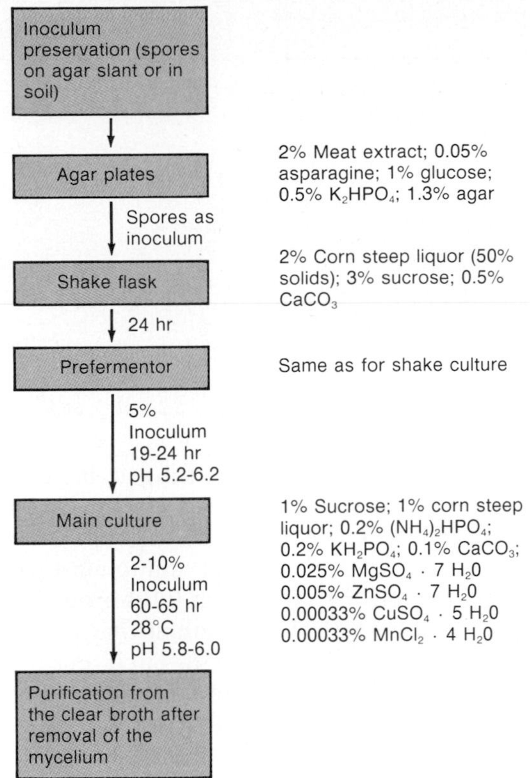

Figure 10.17 Production scheme for chlortetracycline with *Streptomyces aureofaciens*.

duced microbiologically, but there are also semisynthetic tetracyclines on the market, in which new constituents have been inserted chemically into the napthacene ring system. The mode of action of the tetracyclines was discussed in Section 9.20.

The tetracyclines and the β lactam antibiotics are the two most important groups of antibiotics in the medical field. The tetracyclines also find use in veterinary medicine, and in some countries are also used as nutritional supplements for poultry and swine. At one time, chlortetracycline was used to preserve fish, being added to the ice with which fish were refrigerated when they were caught at sea, but such nonmedical uses of medically important antibiotics are now discouraged because of the potential danger of development of antibiotic resistance (see Section 11.17).

The biosynthesis of a tetracycline involves a large number of enzymatic steps. In the case of chlortetracycline, as many as 72 intermediate products may be involved, most of which are only known in a very general way. Studies on the genetics of *Streptomyces aureofaciens*, the producer of chlortetracycline, has shown that over 300 genes are involved! With such a large number of genes, regulation of antibiotic biosynthesis is obviously quite complex. Rational approaches to the development of high-yielding strains are probably also far in the future.

Repression of chlortetracycline synthesis by both glucose and phosphate is known to occur. Phosphate repression is especially significant, so that the culture medium used in commercial production must be run with restricted phosphate concentrations. A production scheme for chlortetracycline is shown in Figure 10.17.

10.7 Vitamins and Amino Acids

Vitamins and amino acids are growth factors that are often used pharmaceutically or are added to foods. Several important vitamins and amino acids are produced commercially by microbial processes.

Vitamins Vitamins are used as supplements for human food and animal feeds. Production of vitamins is second only to antibiotics in terms of total sales of pharmaceuticals—more than $700 million per year. Most vitamins are made commercially by chemical synthesis. However, a few are too complicated to be synthesized inexpensively but fortunately they can be made by microbial fermentation. Vitamin B_{12} and riboflavin are the most important of this class of vitamins.

Vitamin B_{12} is synthesized in nature exclusively by microorganisms. The requirements of animals for this vitamin are satisfied by food intake or by absorption of the vitamin produced in the gut of the animal by intestinal microorganisms. Humans, however, must obtain vitamin B_{12} from food or as a vitamin supplement, since even if it is synthesized by microorganisms in the large intestine, it does not pass from the large intestine into the blood stream. Microbial strains are used that have been specifically selected for their high yields of the vitamin. Members of the bacterial genus *Propionibacterium* give yields of the vitamin ranging from 19 to 23 mg/liter in a two-stage process, while another bacterium, *Pseudomonas denitrificans*, produces 60 mg/liter in a one-stage process which uses sugar-beet molasses as the carbon source. Vitamin B_{12} contains cobalt as an essential part of its structure, and yields of the vitamin are greatly increased by addition of cobalt to the culture medium.

Riboflavin is synthesized by many microorganisms, including bacteria, yeasts, and fungi. The fungus *Ashbya gossypii* produces a huge amount of this vitamin (up to 7 grams per liter) and is therefore used for most of the microbial production processes. In spite of this good yield, there is great economic competition between this microbiological process and chemical synthesis.

Amino acids Amino acids have extensive uses in the food industry, as feed additives, in medicine, and as starting materials in the chemical industry (Table 10.5). The most important commercial amino acid is **glutamic acid**, which is used as a flavor enhancer.

Table 10.5 Amino acids used in the food industry

Amino acid	Foods	Purpose
Glutamate (MSG)	Various foods	Flavor enhancer
Aspartate and alanine	Fruit juices	"Round off" taste
Glycine	Sweetened foods	Improve flavor
Cysteine	Bread	Improves quality
	Fruit juices	Antioxidant
Tryptophan + histidine	Various foods, dried milk	Antioxidant, prevents rancidity
Aspartame (made from phenylalanine + aspartic acid)	Soft drinks, etc.	Low-calorie sweetener
Lysine	Bread (Japan)	Nutritive additive
Methionine	Soy products	Nutritive additive

Two other important amino acids, **aspartic acid** and **phenylalanine**, are the ingredients of the artificial sweetner, **aspartame**, an important constituent of diet soft drinks and other foods which are sold as sugar-free products. **Lysine**, an essential amino acid for humans, is produced by the bacterium *Brevibacterium flavum* for use as a food additive.

Although most of the amino acids can be made chemically, chemical synthesis results in the formation of the optically inactive DL forms. If the biochemically important L form is needed, then an enzymatic or microbiological method of manufacturing is needed (see Figure 10.6*b*). Microbiological production of amino acids can be either by *direct fermentation*, in which the microorganism produces the amino acid in a standard fermentation process, or by *enzymatic synthesis*, in which the microorganism is the source of an enzyme, and the enzyme is then used in the production process. The present discussion will be restricted to direct fermentation processes.

Regulation of amino acid biosynthesis We have discussed amino acid biosynthesis and regulation by feedback inhibition and repression in Chapters 4 and 5. Because the amino acids are used by microorganisms as building blocks of proteins, strict regulation of amino acid production occurs. Overproduction of an amino acid in a normal growth process would be wasteful of energy and nutrients. The role of the industrial microbiologist is to find ways to bypass the microbial regulatory mechanisms.

As we have discussed (see Section 4.18), the biosynthesis of each amino acid involves a number of steps, generally beginning with a central component of cell metabolism such as a citric acid cycle intermediate or a glycolysis intermediate. It has been determined from studies on the biosynthesis of the various amino acids that the *first* enzymatic step unique to a particular biosynthetic pathway is usually subject to feedback inhibition by the end product of this pathway (see Section 4.20 and Figures 4.28 and 4.30). Thus, as the concentration of an amino acid increases, the activity of the enzyme leading to the synthesis of this amino acid is inhibited, resulting in the reduction in synthesis. One way to achieve overproduction of a desired amino acid is to obtain a mutant in which the first unique enzyme in that amino acid pathway is no longer subject to feedback inhibition. This can often be done by isolating a mutant that is *resistant* to an analog of that amino acid. Such an approach works because the amino acid analog is recognized by the allosteric site of the biosynthetic enzyme, so that the amino acid analog inhibits growth of the organism. A mutant able to grow in the presence of the analog is often one in which the first biosynthetic enzyme is resistant to feedback inhibition. Thus, if one wishes to obtain a mutant capable of overproducing tryptophan, a mutant resistant to a tryptophan analog such as 5-methyl tryptophan could be isolated. In some cases, it may be desirable to isolate mutants resistant to a variety of amino acid analogs.

In addition to feedback-resistant mutants, mutants no longer sensitive to repression can also be isolated. As we have noted, repression results from the interaction of the amino acid (or other small molecule metabolite) with a repressor protein, activating the repressor and causing it to inhibit transcription of the mRNA for that gene. A mutant which either lacks the repressor altogether or makes a repressor unable to bind its corepressor, will make the mRNA all the time and should hence *overproduce* the enzyme involved in the amino acid biosynthesis.

A third factor that is very important, if a commercial production process is to be obtained, is to obtain *excretion* of the amino acid into the culture medium. In general, organisms do not excrete essential metabolites such as amino acids. By arranging for excretion, unusually high concentrations, which might cause feedback inhibition or repression even in resistant mutants, will not occur inside the cells. An interesting procedure for obtaining high excretion rate of an amino acid is that used for the production of *glutamic acid*. Production and excretion of excess glutamic acid is dependent upon cell permeability. The organism producing glutamic acid, *Corynebacterium glutamicum*, requires the vitamin *biotin*, an essential cofactor in fatty acid biosynthesis. Deficiency in biotin leads to cell membrane damage (as a result of poor phospholipid production), and under these conditions, intracellular glutamic acid is excreted. The medium used for commercial glutamic acid production contains sufficient biotin to obtain good growth, after which biotin deficiency ensues and glutamic acid is excreted. The discovery of the involvement of biotin deficiency in glutamic acid production permitted the development of rational methods for medium formulation for production of this amino acid.

10.8 Microbial Bioconversion

One of the most far-reaching discoveries in industrial microbiology is the understanding that microorganisms can be used to carry out specific chemical reactions that are beyond the capabilities of organic chemistry. The use of microorganisms for this purpose is called **bioconversion** and involves the growth of the organism in large fermentors, followed by the addition at an appropriate time of the chemical to be converted. Following a further incubation period during which the chemical is acted upon by the organism, the fermentation broth is extracted, and the desired product purified. Although in principle bioconversion may be used for a wide variety of processes, its major use has been in the production of certain steroid hormones (Figure 10.18).

We discussed the role of steroids in eucaryotic cell membranes in Section 3.3. Steroids are also important hormones in animals which regulate various metabolic processes. Some steroids are also used as drugs in human medicine. One group, the adrenal cortical steroids, reduce inflammation and hence are effective in controlling the symptoms of arthritis and allergy. Another group, the estrogens and androgenic steroids, are involved in human fertility, and some of these can be used in the control of fertility. Steroids can be obtained by complete chemical synthesis, but this is a complicated and expensive process. Certain key steps in chemical synthesis can be carried out more efficiently by microorganisms, and commercial production of steroids usually has at least one microbial step.

10.9 Enzyme Production by Microorganisms

Each organism produces a large variety of enzymes, most of which are produced only in small amounts and are involved in cellular processes. However, certain enzymes are produced in much larger amounts by some organisms, and instead of being held within the cell, they are excreted into the medium. Extracellular enzymes are usually capable of digesting insoluble nutrient materials such as cellulose, protein, and starch, the products of digestion then being transported into the cell, where they are used as nutrients for growth. Some of these extracellular enzymes are used in the food, dairy, pharmaceutical, and textile industries and are produced in large amounts by microbial synthesis (Table 10.6). They are especially useful because of their specificity and efficiency when catalyzing reactions of interest at moderate temperature and pH. Similar reactions achieved by chemical means would generally require extreme conditions of temperature or pH and be less efficient and less specific.

Enzymes are produced commercially from both fungi and bacteria. The production process is usually aerobic, and culture media similar to those used in antibiotic fermentations are employed. The enzyme itself is generally formed in only small amounts during the active growth phase but accumulates in large amounts during the stationary phase of growth. As we have seen (Section 5.11), induced enzymes are produced only when an appropriate inducer is present in the medium. The potential for the production of useful enzymes has improved markedly in recent years because of the increased ease with which genes can be manipulated.

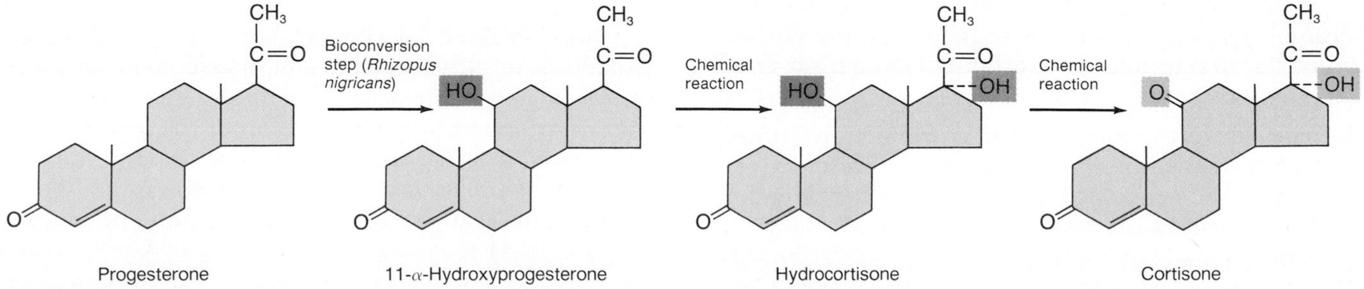

Figure 10.18 Cortisone production using a microorganism. The first reaction is a typical microbial bioconversion, the formation of 11 α-hydroxy progesterone from progesterone. This highly specific oxidation, carried out by the fungus *Rhizopus nigricans*, bypasses a difficult chemical synthesis. All of the other steps from progesterone to the steroid hormone cortisone are performed chemically.

Table 10.6 Microbial enzymes and their application

Enzyme	Source	Application	Industry
Amylase (starch digesting)	Fungi	Bread	Baking
	Bacteria	Starch coatings	Paper
	Fungi	Syrup and glucose manufacture	Food
	Bacteria	Cold-swelling laundry starch	Starch
	Fungi	Digestive aid	Pharmaceutical
	Bacteria	Removal of coatings (desizing)	Textile
Protease (protease-digesting)	Fungi	Bread	Baking
	Bacteria	Spot removal	Dry cleaning
	Bacteria	Meat tenderizing	Meat
	Bacteria	Wound cleansing	Medicine
	Bacteria	Desizing	Textile
	Bacteria	Household detergent	Laundry
Invertase (sucrose-digesting)	Yeast	Soft-center candies	Candy
Glucose oxidase	Fungi	Glucose removal, oxygen removal	Food
		Test paper for diabetes	Pharmaceutical
Pectinase	Fungi	Pressing, clarification	Wine, fruit juice
Rennin	Fungi	Coagulation of milk	Cheese

The microbial enzymes produced in the largest amounts on an industrial basis are the bacterial proteases, used as additives in laundry detergents. By 1969, 80 percent of all laundry detergents contained enzymes, chiefly proteases, but also amylases, lipases, reductases, and other enzymes. However, because of the recognition of allergies to these proteins in production personnel and consumers, the use of proteases was drastically reduced in 1971, with annual world sales falling from \$150 million to \$50 million. This market is gradually returning, as special processing techniques such as microencapsulation have been developed to ensure dust-free preparations.

Other important enzymes manufactured commercially are amylases and glucoamylases, which are used in the production of glucose from starch. The glucose so produced can then be acted upon by glucose isomerase to produce fructose (which is sweeter than either glucose or sucrose) resulting in the final production of a high-fructose sweetener from corn, wheat, or potato starch. The use of this process in the food industry has been increasing, especially in the production of soft drinks.

Three reactions, each catalyzed by a separate microbial enzyme, operate in sequence in the conversion of corn starch into the product called **high-fructose corn syrup**.

1. The enzyme **α-amylase** brings about the initial attack on the starch polysaccharide, shortening the chain, and reducing the viscosity of the polymer. This is called the *thinning reaction*.
2. The enzyme **glucoamylase** produces glucose monomers from the shortened polysaccharides, a process called *saccharification*.
3. The enzyme **glucose isomerase** brings about the final conversion of glucose to fructose, a process called *isomerization*.

All three enzymes are produced industrially by microbial fermentation. The end product of this series of reactions is a syrup containing about equal amounts of glucose and fructose which can be added directly to soft drinks and other food products. The savings in the United States from using domestic corn instead of imported sucrose has been over \$1 billion a year.

The demand for high-fructose corn syrup is likely to increase. To date, the enzyme processes in use have been developed without the use of genetic engineering, but it is likely that recombinant-DNA technology will not only permit increased production of the present enzymes, but also will permit the development of completely new enzyme processes. The soft drink may continue to appear in its familiar container, but its contents will represent a refined product of the biotechnologist's art.

Another enzyme of commercial significance is **microbial rennin**. It has been used in place of calf's rennin for cheese production since 1965. It is much simpler and less expensive to produce than calf's rennin, and seems to be equally effective in cheese production.

Immobilized enzymes For use in industrial processes, it is frequently desirable to convert soluble enzymes into some sort of immobilized state. Immobilization not only makes it easier to carry out the enzymatic reaction under large-scale conditions, it generally stabilizes the enzyme to denaturation. There are three basic approaches to enzyme immobilization (Figure 10.19):

1. **Cross-linkage** or polymerization of enzyme molecules. Linkage of enzyme molecules with each other is usually done by chemical reaction with a bifunctional cross-linking agent such as glutaraldehyde. Cross-linking of enzymes involves the

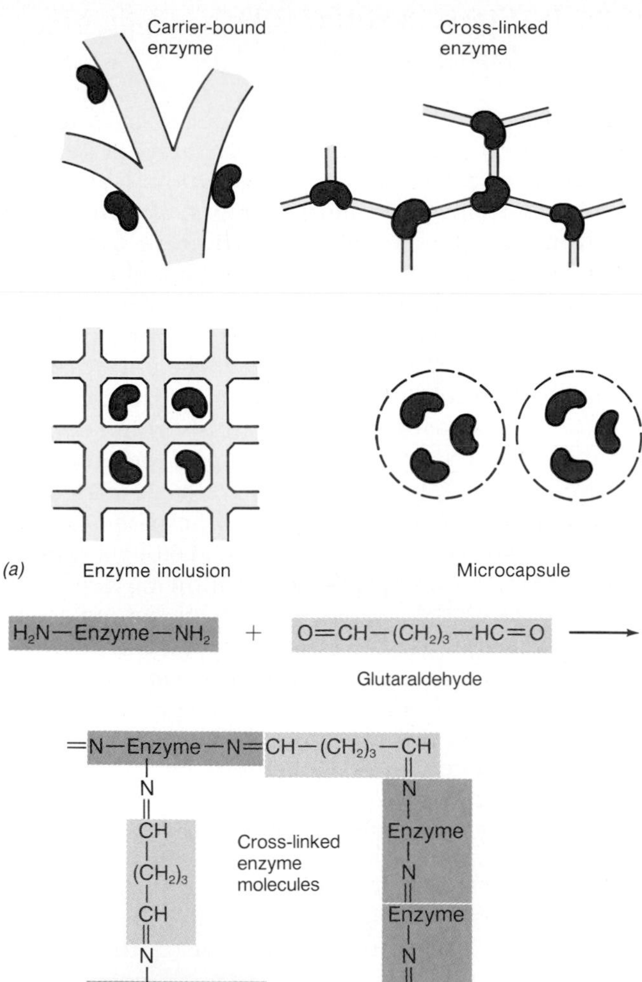

Figure 10.19 Immobilized enzymes. (a) Procedures for the immobilization of enzymes. (b) Procedure for cross linking with glutaraldehyde.

chemical reaction of amino groups of the enzyme protein with a bifunctional reagent such as glutaraldehyde (Figure 10.19*b*). If the reaction is carried out properly, the enzyme molecules can be linked in such a way that most enzymatic activity is maintained.

2. **Bonding** of the enzyme to a carrier. The bonding can be through adsorption, ionic bonding, or covalent bonding. Carriers used include modified celluloses, activated carbon, clay minerals, aluminum oxide, and glass beads.
3. **Enzyme inclusion**, which involves incorporation of the enzyme into a *semi-permeable membrane*. Enzymes can be enclosed inside microcapsules, gels, semi-permeable polymer membranes, or fibrous polymers such as cellulose acetate.

Each of these procedures has advantages and disadvantages and the procedure used would depend on the enzyme and on the industrial process.

10.10 Vinegar

Vinegar is the product resulting from the conversion of ethyl alcohol to acetic acid by acetic acid bacteria, members of the genera *Acetobacter* and *Gluconobacter*. Vinegar can be produced from any alcoholic substance, although the usual starting materials are wine or alcoholic apple juice (cider). Vinegar can also be produced from a mixture of pure alcohol in water, in which case it is called *distilled* vinegar, the term "distilled" referring to the alcohol from which the product is made rather than the vinegar itself. Vinegar is used as a flavoring ingredient in salads and other foods, and because of its acidity, it is also used in the pickling of foods. Meats and vegetables properly pickled in vinegar can be stored unrefrigerated for years.

The *acetic acid bacteria* are an interesting group of eubacteria (see Section 19.17); do not confuse these aerobic acetic acid bacteria with the anaerobic *acetogenic* bacteria (Section 19.9). The aerobic acetic acid bacteria differ from most other aerobes in that they do not oxidize their energy sources completely to CO_2 and water (Figure 10.20). Thus, when provided with ethyl alcohol, they oxidize it only to acetic acid, which accumulates in the medium. Acetic acid bacteria are fairly acid tolerant and are not killed by the acidity that they produce. There is a high oxygen demand during growth, and the main problem in the production of vinegar is to ensure sufficient aeration of the medium.

There are three different processes for the production of vinegar. The **open-vat** or **Orleans method** was the original process and is still used in France where it was developed. Wine is placed in shallow vats with considerable exposure to the air, and the acetic acid bacteria develop as a slimy layer on the top of the liquid. This process is not very efficient, since the only place that the bacteria come in contact with both the air and substrate is at the surface. The second process is the **trickle method** in which the contact between the bacteria, air, and substrate is increased by trickling the alcoholic liquid over beechwood twigs or wood shavings that are packed loosely in a vat or column while a stream of air enters at the

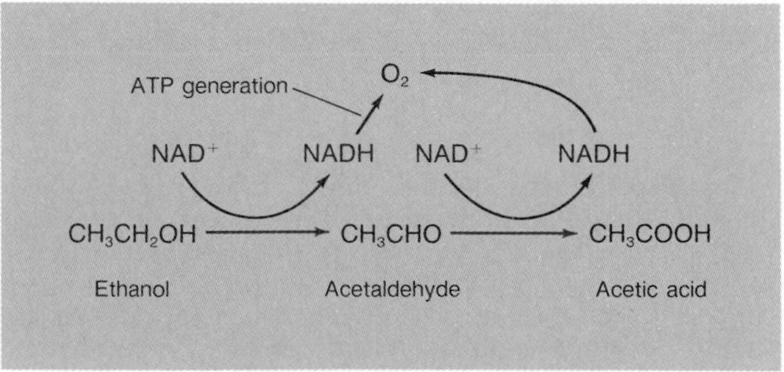

Figure 10.20 Oxidation of ethanol to acetic acid.

bottom and passes upward. The bacteria grow upon the surface of the wood shavings and thus are maximally exposed both to air and liquid. The vat is called a vinegar generator (Figure 10.21), and the whole process is operated in a continuous fashion. The life of the wood shavings in a vinegar generator is long, from 5 to 30 years, depending on the kind of alcoholic liquid used in the process. The third process is the **bubble method**. This is basically a submerged fermentation process, such as already described for antibiotic production. Efficient aeration is even more important with vinegar than with antibiotics, and special highly efficient aeration systems have been devised. The process is operated in a continuous fashion: alcoholic liquid is added at a rate just sufficient to balance removal of vinegar, while most of the alcohol is converted to acetic acid. The efficiency of the process is high, and 90 to 98 percent of the alcohol is converted to acid. One disadvantage of the bubble method is that the product must undergo more filtering to remove the bacteria, whereas in the open-vat and trickle methods the product is virtually clear of bacteria since the cells are bound in the slimy layer in the former and adhere to the wood chips in the latter.

Although acetic acid can be easily made chemically from alcohol, the microbial product, vinegar, is a distinctive material, the flavor being due in part to other substances present in the starting material. For this reason, the fermentation process has not been supplanted by a chemical process.

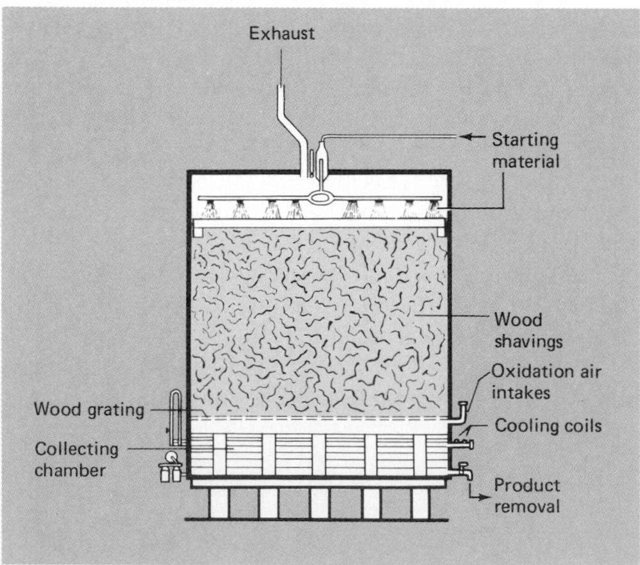

Figure 10.21 Diagram of one kind of vinegar generator. The alcoholic juice is allowed to trickle through the wood shavings and air is passed up through the shavings from the bottom. Acetic acid bacteria develop on the wood shavings and convert alcohol to acetic acid. The acetic acid solution accumulates in the collecting chamber and is removed periodically. The process can be run semicontinuously.

10.13 Citric Acid and Other Organic Compounds

Many organic chemicals are produced by microorganisms in sufficient yields so that they can be manufactured commercially by fermentation. *Citric acid*, used widely in foods and beverages, *itaconic acid*, used in the manufacture of acrylic resins, and *gluconic acid*, used in the form of calcium gluconate to treat calcium deficiencies in humans and industrially as a washing and softening agent, are produced by fungi. *Sorbose*, which is produced when *Acetobacter* oxidizes sorbitol, is used in the manufacture of *ascorbic acid*, vitamin C. (In fact, this sorbitol-sorbose reaction is the only biological step in the otherwise entirely nonbiological chemical synthesis of ascorbic acid.) *Gibberellin*, a plant growth hormone used to stimulate growth of plants, is produced by a fungus. *Dihydroxyacetone*, produced by allowing *Acetobacter* to oxidize glycerol, is used as a suntanning agent. *Dextran*, a gum used as a blood-plasma extender and as a biochemical reagent, and *lactic acid*, used in the food industry to acidify foods and beverages, are produced by lactic acid bacteria. *Acetone* and *butanol* can be produced in fermentations by *Clostridium acetobutylicum* but are now produced mainly from petroleum by strictly chemical synthesis.

Citric acid Of the foregoing, citric acid is perhaps the most interesting product to consider here since it was one of the earliest successful aerobic fermentation products. Citric acid was formerly made commercially in Italy and Sicily by chemical purification from citrus fruits, and for many years, Italy held a world monopoly on citric acid, which resulted in relatively high prices. This monopoly was broken when the microbiological process using the fungus *Aspergillus niger* was developed, and the price of citric acid fell drastically. Today, virtually all citric acid is produced by fermentation. Citric acid is a *primary* metabolic product formed in the tricarboxylic acid cycle (Figure 4.12), but with certain organisms excretion of large amounts of citric acid can be obtained. The process is carried out in large aerated fermentors, using a molasses-ammonium salt medium. One of the key requirements for high citric acid yields is that the medium must be low in iron since the citric acid is produced by the fungus specifically to scavenge iron from an iron-poor environment; therefore, most of the iron is removed from the medium before it is used. There has been considerable fundamental research on the citric-acid fermentation process, and some of this work has led to great improvements in the efficiency of the industrial process.

In the trophophase, part of the added glucose is used for the production of mycelium and is converted through respiration into CO_2. In the idiophase, the

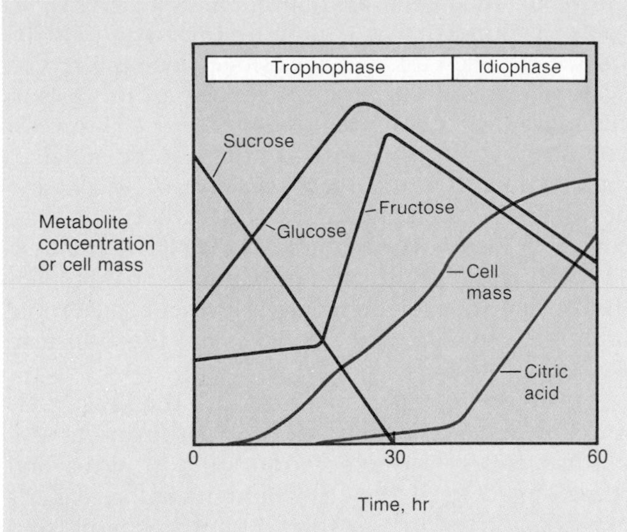

Figure 10.22 Kinetics of the citric acid fermentation.

rest of the glucose is converted into the organic acids and during this phase there is a minimal loss through respiration (Figure 10.22). The media used for citric acid production have been highly perfected over the many years that the commercial process has been underway. A 15–25 percent sugar solution is converted during fermentation. A variety of starting materials can be used as carbohydrate sources: starch from potatoes, starch hydrolysates, glucose syrup from saccharified starch, sucrose of different levels of purity, sugar cane syrup with two-thirds of the sucrose converted into invert sugar, sugar cane molasses, sugar beet molasses. If starch is used, amylases formed by the producing fungus or added to the fermentation broth hydrolyze the starch to sugars.

Historically, the development of a submerged process for citric acid was of great importance because it was the first *aerobic* industrial fermentation. The technology for manufacturing aerobic fermentors was perfected with the citric acid process. This technology was then applied to penicillin and the other economically much more important antibiotic fermentations. Thus, we owe some of our current success with large-scale production of antibiotics to the pioneering work that was done on the citric acid fermentation.

10.12 Yeasts in Industry

Yeasts are the most important and the most extensively used microorganisms in industry (Table 10.7). They are cultured for the cells themselves, for cell components, and for the end products that they produce during the alcoholic fermentation. Yeast cells are used in the manufacture of bread, and also as sources of food, vitamins, and other growth factors. Large-scale fermentation by yeast is responsible for the production of alcohol for industrial purposes but yeast is better known for its role in the manufacture of alcoholic beverages: beer, wine, and liquors. Production of yeast cells and production of alcohol by yeast are two quite different processes industrially, in that the first process requires the presence of oxygen for maximum production of cell material and hence is an *aerobic* process, whereas the alcoholic fermentation process is *anaerobic* and takes place only in the absence of oxygen. However, the same or similar species of yeasts are used in virtually all industrial processes. The yeast *Saccharomyces cerevisiae* was derived from wild yeast used in ancient times for the manufacture of wine and beer. The yeasts currently used are descendants of the early *S. cerevisiae.* Since they have been cultivated in laboratories for such a long time, there has been ample opportunity for selection of strains according to particular desirable properties. It is possible to genetically alter yeasts in the laboratory, using genetic hybridization and cloning methods to produce new strains that contain desirable qualities from two separate parent strains. By the techniques of genetic engineering, it is now also possible to transfer a desired gene from one organism into another, thus improving strains by direct intervention.

Baker's yeast The baker uses yeast as a leavening agent in the rising of the dough prior to baking. A secondary contribution of yeast to bread is its flavor. In the leavening process, the yeast is mixed with the moist dough in the presence of a small amount of sugar. The yeast converts the sugar to alcohol and CO_2, and the gaseous CO_2 expands, causing the dough

Table 10.7 Industrial uses of yeast and yeast products

Production of yeast cells
Baker's yeast, for bread making
Dried food yeast, for food supplements
Dried feed yeast, for animal feeds
Yeast products
Yeast extract, for culture media
B vitamins, Vitamin D
Enzymes for food industry; invertase, galactosidase
Biochemicals for research; ATP, NAD, RNA
Fermentation products from yeast
Ethanol, for industrial alcohol
Glycerol
Beverage alcohol
Beer
Wine
Whiskey
Brandy
Vodka
Rum

to rise. When the bread is baked, the heat drives off the CO_2 (and incidentally, the alcohol) and holes are left within the bread mass, thus giving bread its characteristic light texture. That yeast contributes more to bread than CO_2 is shown by the fact that dough raised with baking powder, a chemical source of CO_2, produces a quite different product than dough raised by yeast. Only the latter bears the name *bread*.

In early times, the bread maker obtained yeast from a nearby brewery, since yeast is a by-product of the brewing of beer. Today, however, baker's yeast is specifically produced for bread making. The yeast is cultured in large aerated fermentors in a medium containing molasses as a major ingredient. Molasses, which is a by-product of the refining of sugar from sugar beets or sugar cane, still contains large amounts of sugar that serves as the source of carbon and energy. Molasses also contains minerals, vitamins, and amino acids used by the yeast. To make a complete medium for yeast growth, phosphoric acid (a phosphorus source) and ammonium sulfate (a source of nitrogen and sulfur) are added.

Fermentation vessels for baker's yeast production range from 40,000 to 200,000 liters. Beginning with the pure stock culture, several intermediate stages are needed to build up the inoculum to a size sufficient to inoculate the final stage (Figure 10.23). Fermentors and accessory equipment are made of stainless steel and are sterilized by high-pressure steam. The actual operation of the fermentor requires special control to obtain the maximum amount of yeast. It has been found that it is undesirable to add all the molasses to the tank at once, since this results in a sugar excess, and the yeast converts some of this surplus sugar to alcohol rather than turning it into yeast cells. Therefore, only a small amount of the molasses is added initially, and then as the yeast grows and consumes this sugar, more is added. Because yeast grows exponentially, the molasses is added at an exponential rate. Careful attention must be paid to aeration, since if insufficient air is present, anaerobic conditions develop, and alcohol is produced instead of yeast.

At the end of the growth period, the yeast cells are recovered from the broth by centrifugation. The cells are usually washed by dilution with water and recentrifuged until they are light in color. Baker's yeast is marketed in two ways, either as compressed cakes or as a dry powder. *Compressed yeast* cakes are made by mixing the centrifuged yeast with emulsifying agents, starch, and other additives that give the yeast a suitable consistency and reasonable shelf life and then forming the product into cubes or blocks of various sizes for domestic or commercial use. A yeast cake will contain about 70 percent moisture and about 2×10^{10} cells per gram. Compressed yeast must be stored in the refrigerator so that its activity is maintained. Yeast marketed in the dry state for baking is usually called *active dry yeast*. The washed yeast is mixed with additives and dried under vacuum

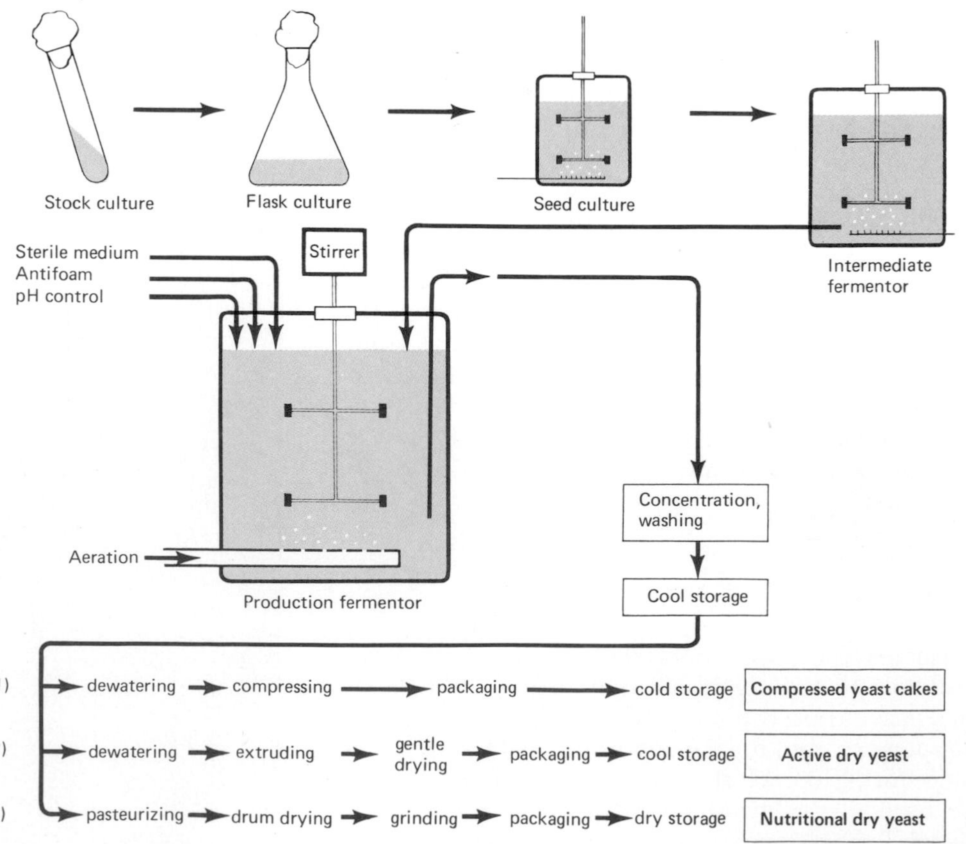

Figure 10.23 Stages in commercial yeast production.

at 25 to 45°C for a 6-hour period, until its moisture is reduced to about 8 percent. It is then packed in airtight containers, such as fiber drums, cartons, or multiwall bags, sometimes under a nitrogen atmosphere to promote long shelf life. Active dry yeast does not exhibit as great a leavening action as compressed fresh yeast but has a much longer shelf life.

10.13 Alcohol and Alcoholic Beverages

The use of yeast for the production of alcoholic beverages is an ancient process. Most fruit juices undergo a natural fermentation caused by the wild yeasts which are present on the fruit. From these natural fermentations, yeasts have been selected for more controlled production, and today, alcoholic-beverage production is a large industry. The most important alcoholic beverages are wine, produced by the fermentation of fruit juice; beer, produced by the fermentation of malted grains; and distilled beverages, produced by concentrating alcohol from a fermentation by distillation. The biochemistry of alcohol fermentation by yeast was discussed in Section 4.8.

Wine Wine is a product of the alcoholic fermentation by yeast of fruit juices or other materials that are high in sugar. Most wine is made from grapes, and unless otherwise specified, the word *wine* refers to the product resulting from the fermentation of grape juice. Wine manufacture occurs in parts of the world where grapes can be most economically grown. The greatest wine-producing countries, in order of decreasing volume of production, are Italy, France, Spain, Algeria, Argentina, Portugal, and the United States. Wine manufacture originated in Egypt and Mesopotamia well before 2000 B.C. and spread from there throughout the Mediterranean region, which is still the largest wine-producing area in the world. Other parts of the world where wine is extensively produced often have a climate similar to that of the Mediterranean, for example, California (Figure 10.24), Chile, South Africa, and Australia. There are a great number of different wines, and their quality and character vary considerably. Dry wines are wines in which the sugars of the juice are practically all fermented, whereas in sweet wines, some of the sugar is left or additional sugar is added after the fermentation. A fortified wine is one to which brandy or some other alcoholic spirit is added after the fermentation; sherry and port are the best-known fortified wines. A sparkling wine is one in which considerable carbon dioxide is present, arising from a final fermentation by the yeast directly in the bottle.

The grapes are crushed by machine, and the juice, called *must*, is squeezed out. Depending on the grapes used and on how the must is prepared, either white or red wine may be produced (Figure 10.25). A white wine is made either from white grapes or from the juice of red grapes from which the skins, containing the red coloring matter, have been removed. In the making of red wine, the *pomace* (skins, seeds and pieces of stem) is left in during the fermentation. In addition to the color difference, red wine has a stronger flavor than white because of the presence of larger amounts of chemicals called *tannins*, which are extracted into the juice from the grape skins during the fermentation.

(a)

The Christian Brothers Winery

(b)

(c)

Figure 10.24 Commercial wine making. (a) Equipment for transporting grapes into the winery for crushing. (b) Large tanks where the main wine fermentation takes place. (c) Barrels where the aging process takes place.

The yeasts involved in wine fermentation are of two types: the so-called wild yeasts, which are present on the grapes as they are taken from the field and are transferred to the juice, and the cultivated wine yeast, *Saccharomyces ellipsoideus*, which is added to the juice to begin the fermentation. One important distinction between wild yeasts and the cultivated wine yeast is their alcohol tolerance. Most wild yeasts can tolerate only about 4 percent alcohol and when the alcohol concentration reaches this point, the fermentation stops. The wine yeast can tolerate up to 12 to 14 percent alcohol before it stops growing. In unfortified wine, the final alcoholic content reached is determined partly by the alcohol tolerance of the yeast and partly by the amount of sugar present in the juice. The alcohol content of most unfortified wines ranges from 10 to 12 percent. Fortified wines such as sherry have an alcohol content as high as 20 percent, but this is achieved by adding distilled spirits such as brandy. In addition to the lower alcohol content produced, wild yeasts do not produce some of the flavor components considered desirable in the final product, and hence the presence and growth of wild yeasts during fermentation is unwanted.

It is the practice in many wineries to kill the wild yeasts present in the must by adding sulfur dioxide at a level of about 100 ppm. The cultivated wine yeast is resistant to this concentration of sulfur dioxide and is added as a starter culture from a pure culture grown

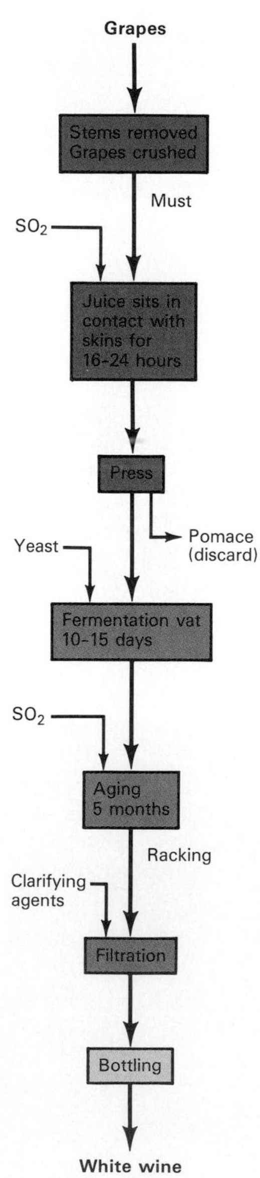

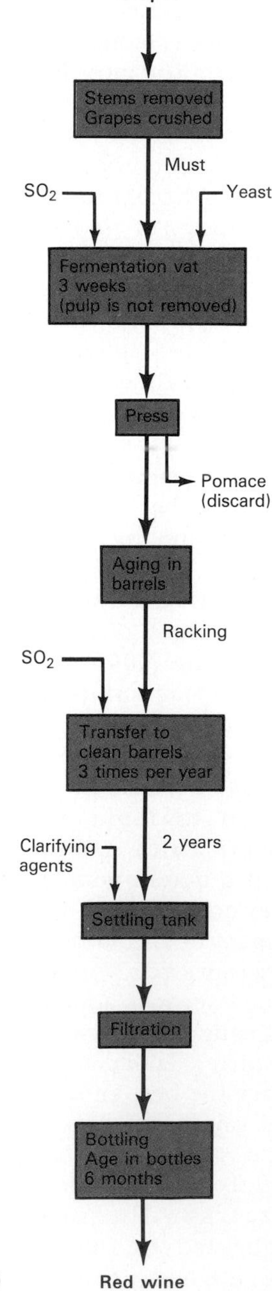

Figure 10.25 Procedure in making wine. (a) White wine. (b) Red wine.

on sterilized or pasteurized grape juice. During the initial stages, air is present in the liquid and rapid aerobic growth of the yeast occurs; then, as the air is used up, anaerobic conditions develop and alcohol production begins. The fermentation may be carried out in vats of various sizes, from 50-gallon casks to 55,000-gallon tanks, made of oak, cement, stone, or glass-lined metal. Temperature control during the fermentation is important, since heat produced during metabolism potentially raises the temperature above the point where yeast can function. Temperatures must be kept below 29°C, and the finest wines are produced at lower temperatures, from 21 to 24°C. Temperature control is best achieved by using jacketed tanks through which cold water is circulated. The fermentor must be constructed so that the large amount of carbon dioxide produced during the fermentation can escape but air cannot enter; this is often accomplished by fitting the tank with a special one-way valve.

With a red wine, after 3 to 5 days of fermentation, sufficient tannin and color have been extracted from the pomace, and the wine is drawn off for further fermentation in a new tank, usually for another week or two. The next step is called *racking*; the wine is separated from the sediment (called *lees*), which contains the yeast and organic precipitate, and is then stored at lower temperature for aging, flavor development, and further clarification. The final clarification may be hastened by addition of materials called fining agents, such as casein, tannin, or bentonite clay, or the wine may be filtered through diatomaceous earth, asbestos, or membrane filters. The wine is then bottled and either stored for further aging or sold. Red wine is usually aged for several years or more after bottling, but white wine is sold without much aging. During the aging process, complex chemical changes occur, resulting in improvement in flavor and odor, or *bouquet*.

Brewing The manufacture of alcoholic beverages made from malted grains is called *brewing*. Typical malt beverages include beer, ale, porter, and stout. *Malt* is prepared from germinated barley seeds, and it contains natural enzymes that digest the starch of grains and convert it into sugar. Since brewing yeasts are unable to digest starch, the malting process is essential for the preparation of a fermentable material from cereal grains. Malted beverages are made in many parts of the world but are most common in areas with cooler climates where cereal grains grow well and where wine grapes grow poorly.

The fermentable liquid from which beer and ale are made is prepared by a process called *mashing*. The grain of the mash may consist only of malt, or other grains such as corn, rice, or wheat may be added. The mixture of ingredients in the mash is cooked and allowed to steep in a large mash tub at warm temperatures. There are a number of different methods of mashing, involving heating at different temperatures for various lengths of time; the particular combination of temperature and time used will considerably influence the character of the final product. During the heating period, enzymes from the malt cause digestion of the starches and liberate sugars and dextrins, which are fermented by the yeast. Proteins and amino acids are also liberated into the liquid, as are other nutrient ingredients necessary for the growth of yeast.

After cooking, the aqueous extract, which is called *wort*, is separated by filtration from the husks and other grain residues of the mash. *Hops*, an herb that is derived from the female flowers of the hops plant, is added to the wort at this stage. Hops is a flavoring ingredient, but it also has antimicrobial properties, which probably help to prevent contamination in the subsequent fermentation. The wort is then boiled for several hours (Figure 10.26), during which time desired ingredients are extracted from the hops, proteins present in the wort that are undesirable from the point of view of beer stability are coagulated and removed, and the wort is sterilized. Heating is accomplished either by passing steam through a jacketed kettle or by direct heating of the kettle from below by fire. Then the wort is filtered again and cooled and then transferred to the fermentation vessel.

Brewery yeast strains are of two major types: the top-fermenting and the bottom-fermenting yeasts. The main distinction between the two is that **top-fermenting yeasts** remain uniformly distributed in the fermenting wort and are carried to the top by the CO_2 gas generated during the fermentation, whereas **bottom yeasts** settle to the bottom. Top yeasts are used in the brewing of ales, and bottom yeasts are used to make the lager beers. The bottom yeasts are usually given the species designation *Saccharomyces carlsbergensis*, and the top yeasts are called *S. cerevisiae*. Fermentation by top yeasts usually occurs at higher temperatures (14–23°C) than does that by bottom yeasts (6–12°C) and is accomplished in a shorter period of time (5 to 7 days for top fermentation versus 8 to 14 days for bottom fermentation). After completion of lager beer fermentation by bottom yeast (Figure 10.26*b*), the beer is pumped off into large tanks where it is stored at a cold temperature (about −1°C) for several weeks (in German, *lager* means "to store"). Lager beer is the most widely manufactured type of beer and is made by large breweries in the United States, Germany, Scandinavia, the Netherlands, and Czechoslovakia. Top-fermented ale is almost exclusively a product of England and certain former British colonies. After its fermentation, the clarified ale is stored at a higher temperature (4 to 8°C), which assists in the development of the characteristic ale flavor.

(a) (b)

Figure 10.26 Brewing beer. (a) The brew kettle is being filled with wort. (b) The aging process is carried out in these large tanks.

Home Brew

The skills of the brewer can be applied by anyone who is willing to learn sterile technique and the principles of microbiology. The amateur brewer can make many kinds of beer, from English bitters and India pale ale to German bock and Russian Imperial stout. The necessary equipment and supplies can be purchased from a local beer and winemakers shop (the Home Wine and Beer Trade Association, 604 N. Miller Road, Valrico, FL 33594, can supply the address of a nearby shop).

The brewing process can be divided into three basic stages: making the wort, carrying out the fermentation, bottling and aging. The character of the brew depends upon many factors: the proportion of malt, sugar, hops, and grain; the kind of yeast; the temperature and duration of the fermentation; and how the aging process is carried out. The instructions provided here are for a simple and relatively fool-proof beer (so-called single-stage fermentation).

The fermentor itself consists of a 5 gallon glass jar or carboy which can be fitted with a tightly-fitting closure. In order to have a good quality beer, it is essential that *everything* be sterilized that comes into contact with the wort. This includes the fermentor, tubing, stirring spoon, and bottles. The best procedure is to use a sterilizing rinse consisting of 1–2 ounces of liquid bleach in 5 gallons of water. Soak the items for 15 minutes, then rinse lightly with hot water, or air dry.

1. **Making the wort**. In commercial brewing, the wort is made by producing fermentable sugars and yeast nutrients from malt, sugar, and hops. The process is complex and relatively difficult to carry out satisfactorily. Many home brewers do make their own wort from malt, but a reasonably satisfactory beer can be made with hop-flavored malt extract purchased ready-made. Malt extracts come in a variety of flavors and colors, and the kind of beer will depend upon the type of malt extract used. A simple recipe for making the wort uses 5-6 pounds of hop-flavored malt extract and 5 gallons of water. The malt extract and 1.5 gallons of water are brought to a boil for 15 minutes in an enamel or stainless steel container (aluminum heating kettles must be avoided because of the inhibiting action of metals leached from aluminum containers). The hot wort is then poured into 3.5 gallons of clean, cold water which has already been added to the fermentor. After the temperature has dropped below 30°C the yeast can be added to initiate the fermentation.

2. **Carrying out the fermentation**. The process by which yeast is added to the wort is called *pitching*. Brewer's yeast can be purchased

Distilled alcoholic beverages Distilled alcoholic beverages are made by heating a fermented liquid at a high temperature which volatilizes most of the alcohol. The alcohol is then condensed and collected, a process called *distilling*. A product much higher in alcohol content can be obtained than is possible by direct fermentation. Virtually any alcoholic liquid can be distilled, and each yields a characteristic distilled beverage. The distillation of malt brews yields *whiskey*, distilled wine yields *brandy*, and distillation of fermented molasses yields *rum*. The distillate contains not only alcohol but also other volatile products arising either from the yeast fermentation or from the mash itself. Some of these other products are desirable flavor ingredients, whereas others are undesirable substances called *fusel oils*. To eliminate the latter, the distilled product is almost always aged, usually in wood barrels. During the aging process, fusel oils are removed, and desirable new flavor ingredients develop. The fresh distillate is usually colorless, whereas the aged product is brown or yellow. The character of the final product is partly determined by the manner and length of aging, and the whole process of manufacturing distilled alcoholic beverages is highly complex. To a great extent, the process is carried out by traditional methods that have been found to yield an adequate product, rather than by scientifically proven methods. Whiskey was originally almost exclusively an Anglo-Saxon (or Gaelic) product. A number of distinct whiskeys exist, usually associated with a country or region. Each of these has a characteristic flavor, owing to the local practices of fermenting, distilling, and aging. Even the word has local spellings, "whisky" being the Scottish, English, and Canadian spelling, and "whiskey" the Irish and United States spelling.

10.14 Food From Microorganisms

Microorganisms can be grown to produce food for humans, and we have just discussed the production of food and feed yeast. In recent years, there has been considerable interest in the expanded production of microorganisms as food, especially in parts of the world where conventional sources of food are in short

as active dry yeast from the home brew supplier. Different yeasts are available for producing different kinds of beer. If the brewing is carried out in the summer time when the temperature is higher, a yeast suitable for a high-temperature fermentation should be used. Add two packs of fresh beer yeast to the cooled wort and cover the fermentor with a rubber stopper into which a plastic hose has been inserted. The hose is directed into a bucket containing water. During the initial 2-3 days of the fermentation, large amounts of CO_2 will be given off which will exit through the hose. The water trap is to prevent wild yeasts or bacteria from the air from getting back into the fermentor. After about 3 days, the activity will diminish as the fermentable sugars are used up. At this time, the rubber stopper and hose are replaced with an inexpensive fermentation lock. The fermentation lock, which can be purchased at the home brew store, prevents contamination while permiting the small amount of gas still being produced to escape. Allow the beer to ferment for 7–10 days at 10–15°C or higher. The fermentation should begin within 24 hours after pitching the yeast. If it does not, the yeast used may not have been active, or the temperature too high when pitching was carried out.

3. **Bottling and aging**. The fermentation should be allowed to proceed for the full 7–10 days, even if the vigorous fermentation action has ceased earlier. Most of the yeast should have settled to the bottom of the fermentor. Carefully siphon the beer off the yeast layer, allowing it to run into glass beer bottles. The bottles themselves should have been sanitized first. Take care that the yeast at the bottom of the fermentor does not get stirred up, and leave the yeast-rich liquid at the bottom. The bottles used should accept standard crown caps, and new, clean caps should be used. Before capping, add 3/4 teaspoon of corn sugar syrup to each bottle. Be certain not to add more than 3/4 teaspoon of syrup, because if excess sugar is added, the build up of carbon dioxide in the bottles may cause the bottles to burst. Once the bottles are capped, turn each one upside down once to mix the sugar syrup, then allow the beer to age upright at room temperature for at least 7–10 days. If another large container is available, a better way of adding the sugar is to siphon the beer into this second container, add the proper amount of sugar for the whole brew, dissolve, and then siphon into the bottles. After this aging period, the beer may be stored at cooler temperature.

All homemade beer has a natural yeast sediment in the bottom of the bottles. When drinking the beer, pour off the supernatant, leaving the heavy residue in the bottom of the bottle for the beer fairy. The beer will improve if it is allowed to age for several weeks. Aging tends to make beer smoother.

Prosit!

supply. Perhaps the most important potential use of microorganisms is not as a complete diet for humans but as a protein supplement. It is usually protein that is in shortest supply in food, and it is in the production of protein that microorganisms are perhaps the most successful. In many cases, microbial cells contain greater than 50 percent protein, and in at least some species, this is complete protein, that is, it contains all of the amino acids essential to humans. The protein produced by microbes as food has been called *single-cell protein* to distinguish it from the protein produced by multicellular animals and plants.

The only organism presently used as a source of single-cell protein is yeast, as already mentioned, but algae, bacteria, and fungi have also been considered. The following are desirable properties that an organism should possess to be most useful as a source of single-cell protein: (1) rapid growth; (2) simple and inexpensive medium; (3) efficient utilization of energy source; (4) simple culture system; (5) simple processing and separation of cells; (6) nonpathogenic; (7) harmless when eaten; (8) good flavor; (9) high digestibility; (10) high nutrient content. Unfortunately, no organism currently known meets all these criteria.

There has been considerable research in Europe, Japan, and the United States on the use of microbial systems as sources of single-cell protein. Although not yet developed on a commercial scale, methylotrophic bacteria (see Section 19.7) have been used as cattle feed supplements and may be used in the future to increase the protein content of human food products. Since they grow on methane as sole energy source, methylotrophic bacteria can be cultured relatively cheaply on waste methane derived from anaerobic digestion processes. Chemical analyses of cells of methylotrophs show them to have high protein content, and especially high levels of sulfur-containing amino acids.

Mushrooms Several kinds of *fungi* are sources of human food, of which the most important are the mushrooms. Mushrooms have been used as food by humans for thousands of years. Both wild mushrooms and those grown commercially in special mushroom beds are now used, although only the latter are produced and eaten extensively. The manner of formation of the mushroom fruiting body was illustrated in Figure 1.24.

(a) American Mushroom Institute

(b) American Mushroom Institute

(c) Bob Harris

(d) Bob Harris

Figure 10.27 Commercial mushroom production. (a) An installation for *Agaricus bisporus*, the common commercial mushroom of the western world. (b) Close up of a mushroom flush. (c) Shiitake, *Lentinus edulus*, the most common commercial mushroom of the orient, but finding increasing production in the west (see also Figure 1.23). A large Japanese installation where the mushroom is cultivated on hardwood logs. (d) Close-up of the shiitake mushroom.

The mushroom commercially available in most parts of the world is *Agaricus bisporus*, and it is generally cultivated in mushroom farms. The organism is grown in special beds, usually in buildings where temperature and humidity are carefully controlled (Figure 10.27*a*). Since light is not necessary, mushrooms may even be grown in basements of homes. Other favored spots for mushroom culture are caves. Beds are prepared by mixing soil with a material very rich in organic matter, such as horse manure, and these beds are then inoculated with mushroom "spawn." The spawn is actually a pure culture of the mushroom fungus that has been grown in large bottles on an organic-rich medium. In the bed, the mycelium grows and spreads through the substrate, and after several weeks it is ready for the next step, the induction of mushroom formation. This is accomplished by adding to the surface of the bed a layer of soil called "casing soil." The appearance of mushrooms on the surface of the bed is called a "flush" (Figure 10.27*b*), and when flushing occurs the mushrooms must be collected immediately while still fresh. After collection they are packaged and kept cool until brought to market. Several flushes will take place on a single bed, and after the last flush the bed must be cleaned out and the process begun again.

Another widely cultured mushroom is **shiitake**, *Lentinus edulus*. The most widely cultivated mushroom in the Orient, shiitake is now finding expanding interest in North America. Shiitake is a cellulose-digesting fungus that grows well on hardwood trees and is cultivated on small logs (Figure 10.27*c*). The logs are soaked in water to hydrate them, then inoculated by inserting plugs of spawn into small holes drilled in the logs. The fungus grows through the log and after about a year forms a flush of fruiting bodies (see Figure 1.23). Shiitake has the advantage that it can be cultivated on waste or scrap wood. It is also said to be much tastier than *Agaricus bisporus*.

There have been some attempts to raise the mushroom fungus in aerated fermentation vats. In this case, the characteristic mushroom structure does not develop, but the mushroom flavor does, and the goal would be to use such material as a flavor ingredient in soups and sauces. However, no commercial process for deep-vat mushroom culture is yet available. Filamentous fungi other than mushrooms have also been tested for commercial deep-vat culture but have not been found satisfactory.

Although mushrooms make flavorful food, their digestibility and nutritional value are not very high. They are low in protein content and deficient in certain essential amino acids; they are also not exceptionally rich in vitamins. The mushrooms and the filamentous fungi are definitely inferior to yeast as food sources, although they serve as valuable flavoring ingredients.

10.15 Production of Mammalian Products by Engineered Microorganisms

As we have noted in the introduction to this chapter, microbial biotechnology deals not only with the large-scale production of conventional microbial products, but with the production of substances not normally produced by means of genetically engineered microorganisms. Although to date most industrial processes involve conventional microbial products, a few products are made using engineered organisms, and it is likely that this will be a major future development. The greatest interest is in the production of mammalian proteins and peptides by means of microorganisms, since many such mammalian materials have potentially high value as pharmaceutical agents, and they are expensive and difficult to produce by presently available means. If the gene which controls the production of a mammalian protein can be cloned into a microorganism, and good expression of this gene obtained, then an industrial process for making this protein can likely be developed.

A list of proteins with possible pharmaceutical applications which are being developed using gene engineering technology is given in Table 10.8. As seen, most of the proteins of use have hormonal or immunological activity. In addition to these proteins, a number of other proteins with potential pharmaceutical activity are of industrial interest.

Table 10.8 Some mammalian and virus genes expressed in bacteria

Protein	Function
Interferon	Antiviral agent, anticancer (?)
Insulin	Treatment of diabetes
Serum albumin	Transfusion applications
Growth hormone	Growth defects
Urokinase	Blood-clotting disorders
Parathyroid hormone	Calcium regulation
Human viruses (Hepatitis B, cytomegalovirus, influenza, AIDS)	Vaccines
Animal viruses (foot-and-mouth disease)	Vaccines
Human growth hormone	Corrects genetic deficiency in dwarfs
Interleukin-2	Promotes T-cell growth; immunotherapy
β-Endorphin	Pain relief

Study Questions

1. In what ways do industrial microorganisms differ from conventional microorganisms? In what ways are they similar?
2. Describe some of the techniques that can be used to improve strains of industrial microorganisms.
3. List three major types of industrial products that can be obtained with microorganisms and give two examples of each type.
4. Give an example of a *commodity chemical* produced by a microorganism and describe briefly the process by which this chemical is manufactured.
5. Compare and contrast *primary* and *secondary metabolites* and give an example of each. List several molecular explanations for why some metabolites are secondary rather than primary.
6. Define *trophophase* and *idiophase*.
7. How does an industrial fermentor differ from a laboratory culture vessel?
8. Discuss the problems of *scale up* from the viewpoints of *aeration*, *sterilization*, and *process control*. Why is sterility so much more important in an industrial fermentor than in a laboratory fermentor?
9. Describe the vinegar process and give two ways by which aeration is achieved.
10. List five examples of *antibiotics* that are important industrially. For each of these antibiotics, list the producing organism, the general chemical structure, and the mode of action.
11. List the various stages that occur during the industrial production of an antibiotic.
12. Why are the β lactam antibiotics so important medically? Compare and contrast *natural, biosynthetic*, and *semi-synthetic* β lactam antibiotics.
13. Describe briefly the unique aspect of the regulation of *streptomycin biosynthesis*.
14. List the unique characteristics of the *tetracycline antibiotics*.
15. Explain why the yield of *vitamin* B_{12} can be markedly improved by addition of cobalt to the fermentation medium.
16. What unusual characteristics must an organism have if it is to overproduce and excrete an amino acid such as *glutamic acid*.
17. Define *microbial bioconversion* and give an example. Explain why the chemical reactions involved in microbial bioconversions are preferably carried out microbially rather than chemically.
18. List four different kinds of *enzymes* that are produced commercially. For each enzyme, list the organism that is used in commerical production, the action of the enzyme, and how the enzyme is used in commerce.
19. Describe the stages involved in the production of *high-fructose syrup* and explain the role of an enzyme in each step.
20. Why is it desirable to *immobilize* enzymes? Give examples of three different immobilization procedures and describe how each is carried out.
21. In this chapter we have discussed at least six *foods or food-related products* that are manufactured at least in part with microorganisms. List as many of these six as you can and for each explain precisely the microbial role.
22. Why are yeasts of such great industrial importance?
23. In this chapter we have discussed at least four *beverages* that are manufactured at least in part with microorganisms. List as many of these as you can and for each explain precisely the microbial role.
24. In what way is the manufacture of *beer* similar to the manufacture of *wine*? In what ways do these two processes differ?
25. Explain how *genetically engineered microorganisms* can play unique roles in industrial microbiology, roles not played by conventional microorganisms.

Supplementary Readings

Aiba, S., A. E. Humphrey, and **N. F. Millis.** 1973. *Biochemical Engineering, Second edition*. Academic Press, New York. This textbook is the best all around treatment of the physical processes that occur during industrial fermentation. Heavily mathematical, but with a strong dose of practical sense.

Atkinson, B. and **F. Mavituna.** 1983. *Biochemical Engineering and Biotechnology Handbook*. Macmillan, New York. A table-packed reference book which serves as a useful source for looking up specific industrial processes and concepts.

Bailey, J. E. and **D. F. Ollis.** 1977. *Biochemical Engineering Fundamentals*. McGraw-Hill Book Co., New York. A solid textbook of biochemical engineering with emphasis on quantitative and mathematical aspects.

Crueger, W. and **A. Crueger.** 1984. *Biotechnology. A Textbook of Industrial Microbiology*. English edition edited by Thomas D. Brock. Sinauer Associates, Sunderland, Mass. A concise textbook of industrial microbiology which emphasizes the economically important microbial processes. Has excellent chapters on the various large-scale processes and a good chapter on fermentor design and scale-up.

Demain, A. L. and **N. A. Solomon** (editors). 1986. *Manual of Industrial Microbiology and Biotechnology*. American Society for Microbiology, Washington. One of the respected manuals of the American Society for Microbiology, this detailed work represents a collection of chapters written by more than 30 authors. The emphasis is on principles that apply to a wide variety of processes. An advanced work of considerable value to the specialist in industrial microbiology.

11 Host-Parasite Relationships

This chapter begins a major section of the book which deals with the roles of microorganisms in infectious disease. The material presented builds on basic concepts of cell chemistry (Chapter 2), microbial structure (Chapter 3), metabolism (Chapter 4), genetics (Chapters 5 and 7), and growth (Chapter 9). Infectious disease is the most important applied aspect of microbiology, and is the one about which the most is known. In the present chapter we discuss the general principles regarding how microorganisms grow in the animal body. In the next chapter, we discuss immunology, which concerns the specific processes by which the animal body counteracts infectious disease. Then in Chapter 13 we discuss the use of microbiological tools and techniques for the diagnosis of infectious disease and for the selection of proper chemotherapeutic procedures. This is followed in Chapter 14 by a general discussion of how infectious disease spreads through populations and how a knowledge of the spread of disease can aid in disease control. Finally, Chapter 15 presents discussion of each of the major microbial diseases.

A word about the organisms we will be dealing with. Infectious diseases are caused by viruses, bacteria, fungi, and protozoa. We have already discussed viruses and how they interact with their hosts in Chapter 6. The present chapter outlines our knowledge of how nonviral microorganisms affect the body, with emphasis on the bacteria, the most important microbial pathogens. Although viruses will not be discussed in the present chapter, in the following chapters viruses will be discussed along with the bacteria, fungi, and protozoa.

The animal body is in continual contact with microorganisms. Literally billions of bacterial cells are present in and on the human body and most play beneficial, indeed sometimes even essential, roles in the overall health of the person. These organisms are collectively referred to as the "normal" flora, and represent species which have developed an intimate relationship with certain tissues of the animal body.

A **parasite** is an organism that lives on or in a second organism, called the **host**. In some cases, the parasite has little or no harmful effect on the host and its presence may be inapparent. Such situations are called *commensal*. In some cases, however, the parasite brings about damage or harm to the host: such organisms are called **pathogens**. The relationships between host and parasite is dynamic, since each modifies the activities and functions of the other. The outcome of the host-parasite relationship depends on the **pathogenicity** of the parasite, that is, on its ability to inflict damage, and on the resistance or susceptibility of the host. The term **virulence** is a quantitative term used to indicate the *degree* of pathogenicity of the parasite and is usually expressed as the dose or cell number that will damage or kill a certain fraction of the inoculated animals within a given time period. Neither the virulence of the parasite nor the resistance of the host are constant factors, however, each varying under the influence of external factors or as a result of the host-parasite relationship itself.

Infection refers to the growth of microorganisms in the host. Infection is not synonymous with disease because infection does not always lead to injury of the host, even if the pathogen is potentially virulent. In a diseased state the host is harmed in some way, whereas infection refers to any situation in which a microorganism is established and growing in a host, whether or not the host is harmed.

The ability to cause infectious disease is one of the most dramatic properties of microorganisms. The understanding of the physiological and biochemical basis of infectious disease has led to therapeutic and preventive measures that have had far-reaching influence on medicine and human affairs. We begin this chapter by considering the normal flora of the healthy human adult. By understanding the microbial ecology of the human body we will be in a better position to appreciate the competitive forces which govern the success or failure of a potential pathogen in initiating disease.

11.1 Microbial Interactions with Higher Organisms

Animal bodies provide favorable environments for the growth of many microorganisms. They are rich in organic nutrients and growth factors required by heterotrophs, they provide relatively constant conditions of pH and osmotic pressure, and warm-blooded animals have highly constant temperatures. However, an animal body should not be considered as one uniform microbial environment throughout. Each region or organ differs chemically and physically from other regions and thus provides a selective environment where certain kinds of microbes are favored over others. In the higher animal, for instance, the skin, respiratory tract, gastrointestinal tract, and so on, each provide a wide variety of microenvironments in which different microorganisms can grow selectively. Further, higher animals possess a variety of defense mechanisms that act in concert to prevent or inhibit microbial invasion and growth. The microorganisms that ultimately colonize successfully are those that have developed ways of circumventing these defense mechanisms. Interestingly, some animals have developed mechanisms to encourage the growth of beneficial microorganisms.

Actually, it is often difficult to determine whether a relationship between a microorganism and a higher organism is beneficial, harmful, or neutral. This is because the outcome of an interaction may be influenced by external factors, so that under certain conditions the relationship may be beneficial, whereas under other conditions it may be neutral or harmful.

Our discussion here will emphasize warm-blooded animals, especially mammals, as it is for this group that we have the most information. Microorganisms are almost always found in those regions of the body exposed to the outside world, such as the skin, oral cavity, respiratory tract, intestinal tract, and genitourinary tract. They are not found normally in the organs and blood and lymph systems of the body; if microbes are found in any of these latter areas in significant quantities, it is usually indicative of a disease state.

11.2 Normal Flora of the Skin

An average human adult has about two square meters of skin surface containing a variety of different microenvironments: dry, moist, clean, dirty, and so on, depending on personal hygienic habits and the frequency with which the skin surface is washed. Figure 11.1 indicates diagrammatically the anatomy of the skin and suggests some regions in which bacteria may live. The skin surface itself is not a favorable place for microbial growth, as it is subject to periodic drying. Only in certain areas of the body, such as the scalp, face, ear, underarm regions, genitourinary and anal regions, and palms and interdigital spaces of the toes, are moisture conditions on the skin surface sufficiently high to support resident microbial populations; in these regions characteristic surface microbial floras do exist.

Most skin microorganisms are associated directly or indirectly with the sweat glands, of which there

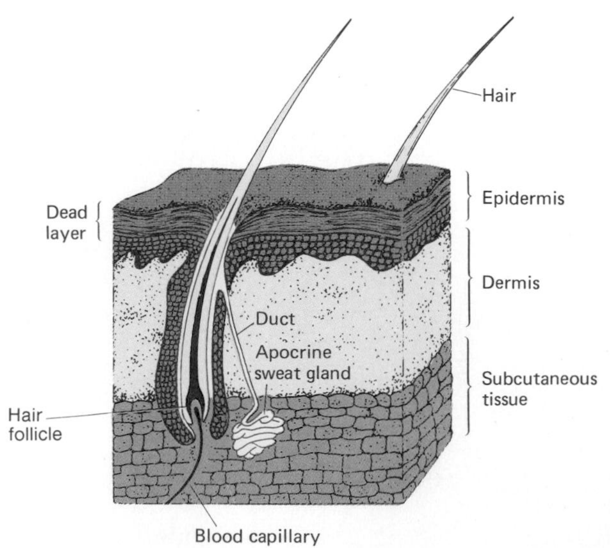

Figure 11.1 Anatomy of the human skin. Microbes are associated primarily with the sweat ducts and the hair follicles.

are several kinds. The **eccrine glands** are not associated with hair follicles and are rather unevenly distributed over the body, with denser concentrations on the palms, finger pads, and soles of the feet. They are the main glands responsible for the perspiration associated with body cooling. Eccrine glands seem to be relatively devoid of microorganisms, perhaps because of the extensive flow of fluid, since when the flow of an eccrine gland is blocked, bacterial invasion and multiplication do occur. The **apocrine glands** are more restricted in their distribution, being confined mainly to the underarm and genital regions, the nipples, and the umbilicus. They are inactive in childhood and become fully functional only at puberty. Bacterial populations on the surface of the skin in these warm, humid places are relatively high, in contrast to the situation on the smooth surface skin. Underarm odor develops as a result of bacterial activity on the secretions of the apocrines; aseptically collected apocrine secretion is odorless but develops odor upon inoculation with certain bacteria isolated from the skin. Each hair follicle is associated with a **sebaceous gland** which secretes a lubricant fluid. Hair follicles provide an attractive habitat for microorganisms; a variety of aerobic and anaerobic bacteria, yeasts, and filamentous fungi inhabit these regions, mostly within the area just below the surface of the skin. The secretions of the skin glands are rich in microbial nutrients. Urea, amino acids, salts, lactic acid, and lipids are present in considerable amounts. The pH of human secretions is almost always acidic, the usual range being between pH 4 and 6.

The microorganisms of the normal flora of the skin can be characterized as either **residents** or **transients**. Residents are organisms that are able to multiply, not merely survive, on the skin. The skin as an external organ is continually being inoculated with transients, virtually all of which are unable to multiply and usually die. The normal flora of the skin consists primarily of Gram-positive bacteria restricted to a few groups. These include several species of *Staphylococcus* and a variety of both aerobic and anaerobic corynebacteria. Of the latter, *Propionibacterium acnes* is ordinarily a harmless resident but can incite or contribute to the condition known as acne. Gram-negative bacteria are almost always minor constituents of the normal flora, even though such intestinal organisms as *Escherichia coli* are being continually inoculated onto the surface of the skin by fecal contamination. It is thought that the lack of success of Gram-negative bacteria is due to their inability to compete with Gram-positive organisms that are better adapted to the skin; if the latter are eliminated by antibiotic treatment, the Gram-negative bacteria can flourish. Yeasts are uncommon on the skin surface, but the lipophilic yeast *Pityrosporum ovalis* is occasionally found on the scalp. The death of microorganisms inoculated onto the surface of the skin is thought to result primarily from two factors: the skin's low moisture content and low pH due to its organic acid content. Those organisms that survive and grow are able to resist these adverse conditions.

11.3 Normal Flora of the Oral Cavity

The oral cavity, despite its apparent simplicity, is one of the more complex and heterogeneous microbial habitats in the body. This cavity includes the teeth and tongue and the central space that they fill. The teeth can be viewed as a surface upon which saliva and materials derived from the food adsorb, rather than as a direct source of microbial nutrients. Although saliva is the most pervasive source of microbial nutrients in the oral cavity, in point of fact it is not an especially good microbial culture medium. Saliva contains about 0.5 percent dissolved solids, about half of which are inorganic (mostly chloride, bicarbonate, phosphate, sodium, calcium, potassium, and trace elements); the predominant organic constituents of saliva are proteins, such as salivary enzymes, mucoproteins, and some serum proteins. Small amounts of carbohydrates, urea, ammonia, amino acids, and vitamins are also present. A number of antibacterial substances have been identified in saliva, of which the most important are the enzymes lysozyme and lactoperoxidase. Lactoperoxidase, an enzyme present in both milk and saliva, kills bacteria by a reaction involving chloride ions and H_2O_2, in which singlet oxygen is probably generated (see Sections 9.14 and 11.14). The pH of saliva is controlled primarily by a bicarbonate buffering system and varies between 5.7 and 7.0, with a mean near 6.7. The composition of saliva varies from individual to individual, and even within the same individual variations due to physiological and emotional factors are seen. Despite the activity of antibacterial substances, the presence of food particles and epithelial debris makes the oral cavity a very favorable microbial habitat.

The teeth and dental plaque The tooth consists of a mineral matrix of calcium phosphate crystals (enamel), within which the living tissue of the tooth (dentin and pulp) is present (Figure 11.2). In the infant, the lack of teeth is probably of considerable importance in determining the nature of the microbial flora. Bacteria found in the mouth during the first year of life are predominantly aerotolerant anaerobes such as streptococci and lactobacilli, but a variety of other bacteria, including some aerobes, can occur in small numbers. When the teeth appear, there is a pronounced shift in the balance of the microflora toward anaerobes, and a variety of bacteria specifically adapted for growth on surfaces and in crevices of the teeth develop (see Figure 11.2). A film forms on the surface of the teeth, called **dental plaque**, consisting mainly of bacterial cells surrounded by a polysac-

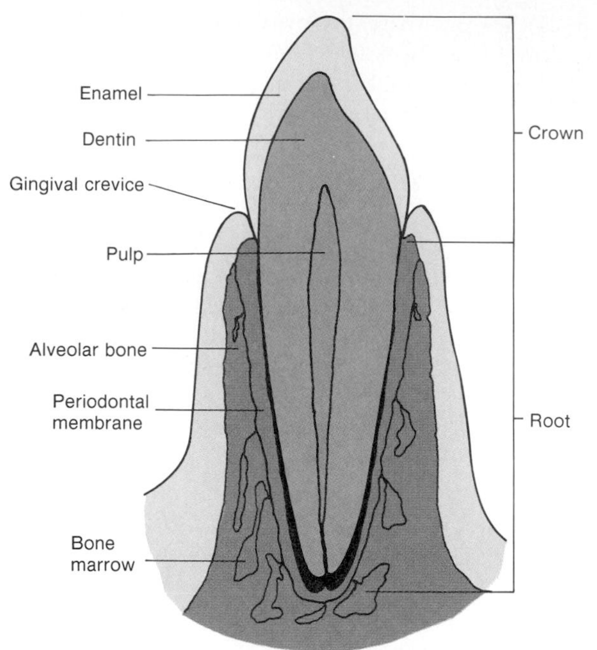

Figure 11.2 Section through a tooth showing the surrounding tissues which anchor the tooth in the gum.

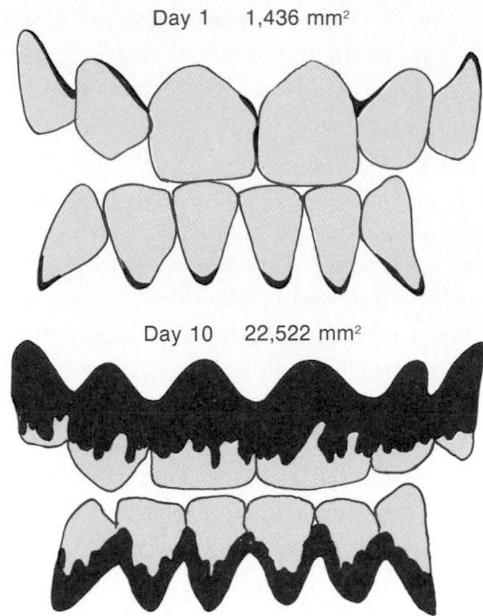

Figure 11.3 Distribution of dental plaque, as revealed by use of a disclosing agent, on brushed and unbrushed teeth. The numbers give the total area of dental plaque.

charide matrix. Dental plaque can be readily observed by staining with dyes such as basic fuchsin or erythrosin; substances that stain the plaque are called *disclosing agents*. If effective tooth brushing is practiced, dental plaque is present only in crevices which are protected from the brush, but plaque rapidly accumulates if brushing is stopped (Figure 11.3).

Dental plaque consists mainly of filamentous bacteria closely packed and extending out perpendicular to the surface of the tooth, embedded in an amorphous matrix. These filamentous organisms are usually classified as *Fusobacterium*. They are obligate anaerobes on initial isolation but after subculturing become microaerophilic; they ferment carbohydrates to lactic acid. Associated with these predominant filamentous organisms are streptococci, spirochetes, diphtheroids, Gram-negative cocci, and others. The anaerobic nature of the flora may seem surprising, considering that the mouth has good accessibility to oxygen. It is likely that anaerobiosis develops through the action of facultative bacteria growing aerobically upon organic materials on the tooth, since the dense matrix of the plaque greatly decreases the diffusion of oxygen onto the tooth surface. The microbial populations of the dental plaque are thus seen to exist in a microenvironment partly of their own making, and are probably able to maintain themselves in the face of wide variations in the macroenvironment of the oral cavity.

The bacterial colonization of smooth tooth surfaces occurs as a result of firm attachment of single bacterial cells, followed by growth in the form of microcolonies. Beginning with a freshly cleaned tooth surface, the first event is the formation of a thin organic film as a result of the attachment of acidic glycoproteins from the saliva. This film then provides a firmer attachment site for the colonization and growth of bacterial microcolonies (Figure 11.4). The colo-

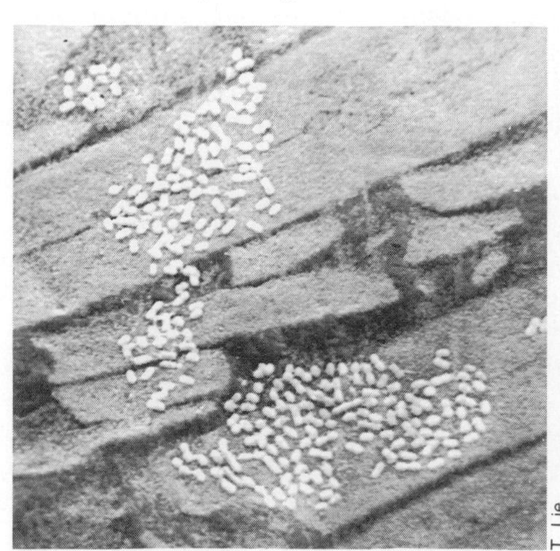

(a)

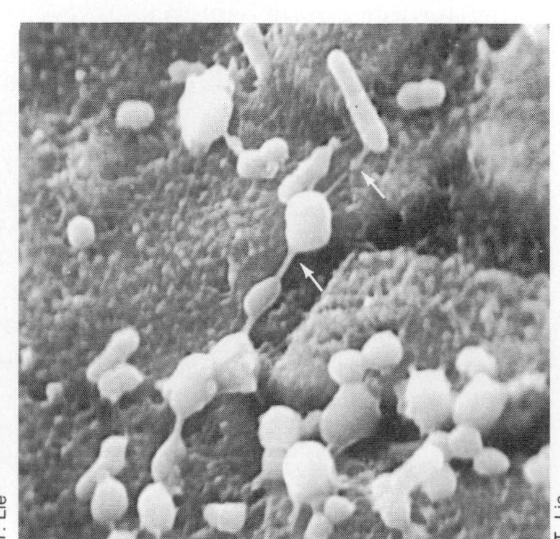

(b)

Figure 11.4 (a) Bacterial microcolonies growing on a model tooth surface inserted into the mouth for 6 hours. (b) Higher magnification of the preparation in *a*. Note the diverse morphology and the slime material holding the organisms together.

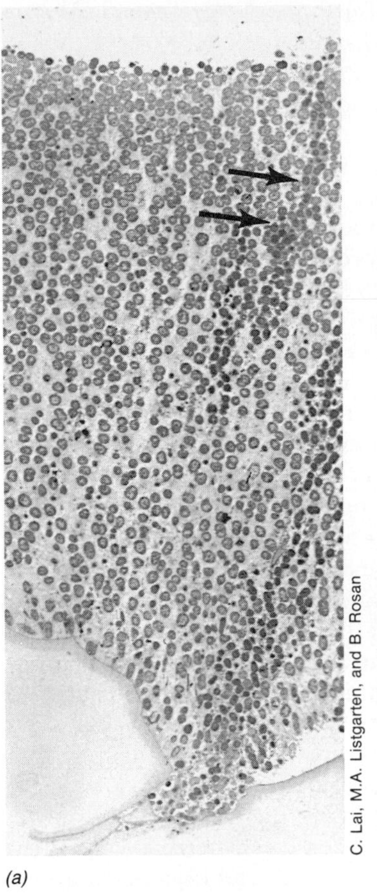

(a)

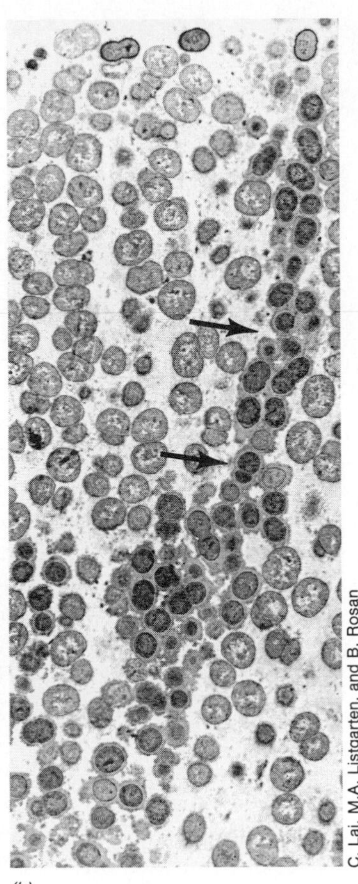

(b)

Figure 11.5 Electron micrographs of thin sections of dental plaque. Bottom is the base of the plaque; top is the portion exposed to oral cavity. (a) Low-power electron micrograph. Organisms are predominantly streptococci. The species *S. sobrinus* has been labeled by an antibody-microchemical technique, and these cells appear darker than the rest. They are seen as two distinct lines (arrows). Magnification, 2200×. (b) Higher-power electron micrograph showing the region with *S. sobrinus* cells (dark; arrow). Note the extensive glycocalyx (see Section 11.8) surrounding the *S. sobrinus* cells. Magnification, 9000×.

nization of this glycoprotein film is highly specific, and only a few species of *Streptococcus* (primarily *S. sanguis, S. sobrinus*, and *S. mitis*) are involved. As a result of extensive growth of these organisms, a thick bacterial zone is formed (Figure 11.5*a* and *b*). Subsequently, the plaque is colonized by other organisms, and in mature plaque filamentous organisms such as *Actinomyces* may actually predominate.

Dental caries The role of the oral flora in tooth decay (dental caries) has now been well established through studies on germ-free animals, although the precise mechanisms of the process are still under study. The smooth surfaces of the teeth that are exposed to frequent cleaning by the tongue, cheek, saliva, or toothbrush or to the abrasive action of food mastication are relatively resistant to dental caries. The tooth surfaces in crevices, where food particles can be retained, are the sites where tooth decay predominates. The shape of the teeth is an important factor in the degree to which such crevices develop; dogs are highly resistant to tooth decay because the shape of their teeth does not favor retention of food. Diets high in sugars are especially cariogenic because lactic acid bacteria ferment the sugars to lactic acid, which causes decalcification of the enamel (see Figure 11.2) of the tooth. Once breakdown of the hard tissue has begun, proteolysis of the matrix of the tooth enamel occurs, through the action of proteolytic enzymes released by bacteria. Microorganisms penetrate further into the decomposing matrix, but the later stages of the process may be exceedingly slow and are often highly complex.

Two organisms that have been implicated in dental caries are *S. sobrinus* and *S. mutans*, both lactic acid-producing bacteria. *S. sobrinus* is able to colonize smooth tooth surfaces because of its specific affinity for salivary glycoproteins, and this organism is probably the primary organism involved in tooth decay of smooth surfaces. *S. mutans* is found predominantly in crevices and small fissures, and its ability to attach seems to be related to its ability to produce a dextran polysaccharide, which is strongly adhesive (Figure 11.6). *S. mutans* dextran is only produced when sucrose is present, by means of the enzyme dextransucrase (see Figure 11.6). Sucrose is a common sugar in the human diet, and its ability to act as a substrate for dextransucrase may be one reason that sucrose is highly cariogenic.

Susceptibility to tooth decay varies greatly among individuals and is affected by inherent traits in the individual as well as by diet and other extraneous factors. Studies of the distribution of oral streptococci have shown a direct correlation between the presence of *S. mutans*, and to a lesser degree *S. sobrinus*, in humans and the extent of dental caries. In the United States and Western Europe, for example, 80–90 percent of all people have their teeth colonized by *S.*

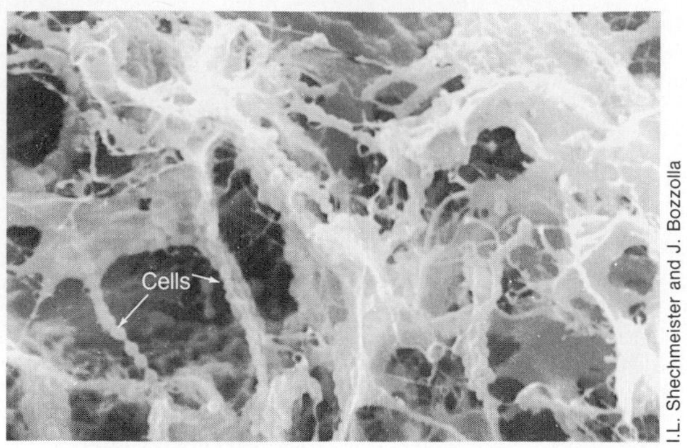

Figure 11.6 Scanning electron micrograph of the cariogenic bacterium *Streptococcus mutans*. The sticky dextran material can be seen as masses of filamentous particles. Magnification, 13,920×.

mutans and dental caries is a nearly universal phenomenon. By contrast, dental caries does not occur in Tanzanian children, presumably because of dietary factors, and *S. mutans* is absent from the plaque of these individuals.

The structure of the calcified tissue also plays an important role in the extent of dental caries. Incorporation of fluoride into the calcium phosphate crystal matrix makes the latter more resistant to decalcification by acid, hence the use of fluorides in drinking water or dentrifices to aid in controlling tooth decay. Although tooth decay is an infectious disease, we tend to place it in a different category from other infectious diseases. However, microorganisms in the mouth can also cause infections that are more typically disease states, such as periodontal disease, gingivitis, infections of the tooth pulp, and so on.

11.4 Normal Flora of the Gastrointestinal Tract

The anatomy of the gastrointestinal tract is shown in Figure 11.7. The human gastrointestinal tract, the site of food digestion, consists of the stomach, small intestine, and large intestine. The pH of stomach fluids is low, about pH 2. The stomach can thus be viewed as a microbiological barrier against entry of foreign bacteria into the intestinal tract. Although the bacterial count of the stomach contents is generally low, the walls of the stomach are often heavily colonized with bacteria. These are primarily acid-tolerant lactobacilli and streptococci and can be seen in large numbers in histological sections or by scanning electron microscopy of the stomach epithelium (Figure 11.8). These bacteria appear very early after birth of an animal, being well established by the first week. In humans, under abnormal conditions such as cancer of the stomach that produce higher pH values, a characteristic microbial flora consisting of yeasts and bacteria (genera *Sarcina* and *Lactobacillus*) may develop.

The **small intestine** is separated into two parts, the duodenum and the ileum. The former, being adjacent to the stomach, is fairly acidic and resembles the stomach in its microbial flora, although it may lack heavy populations on the epithelium. From the duodenum to the ileum the pH gradually becomes more alkaline and bacterial numbers increase. In the lower ileum, bacteria are found in the intestinal cavity (the lumen), mixed with digestive material. Cell numbers of 10^5–10^7 per gram are common.

In the **large intestine**, bacteria are present in enormous numbers, so much so that this region can be viewed as a specialized fermentation vessel. Many bacteria live within the lumen itself, probably using as nutrients some products of the digestion of food. Facultative aerobes, for example *Escherichia*, are present but not abundantly so with respect to other bacteria; total counts of facultative aerobes are generally less than 10^7 per gram of intestinal contents. The activities of facultative aerobes consume any oxygen present, making the environment of the large intestine strictly anaerobic and favorable for the profuse growth of obligate anaerobes. The proportion of anaerobes to aerobes is often grossly underestimated, owing to the difficulties of cultivating the anaerobes. In recent years better anaerobic techniques have been developed, and organisms previously uncultured are now being isolated. Many of these bacteria are long,

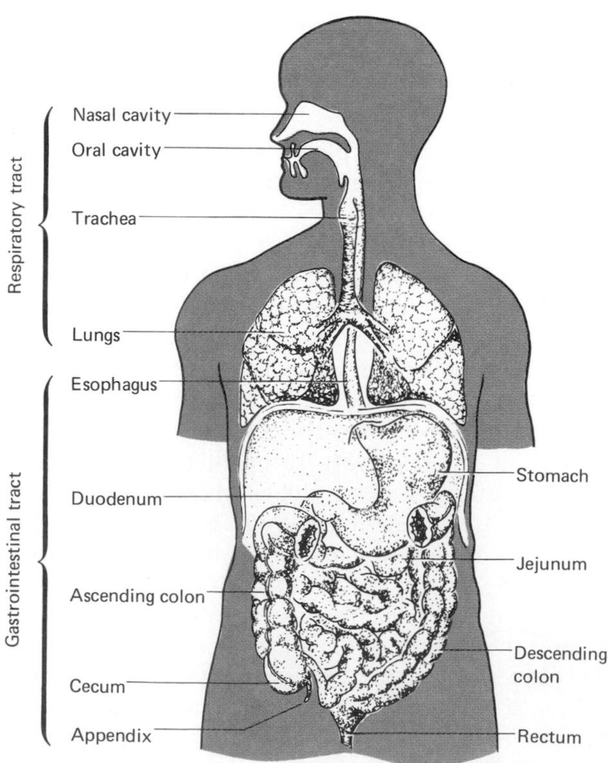

Figure 11.7 General anatomy of the gastrointestinal and respiratory tracts.

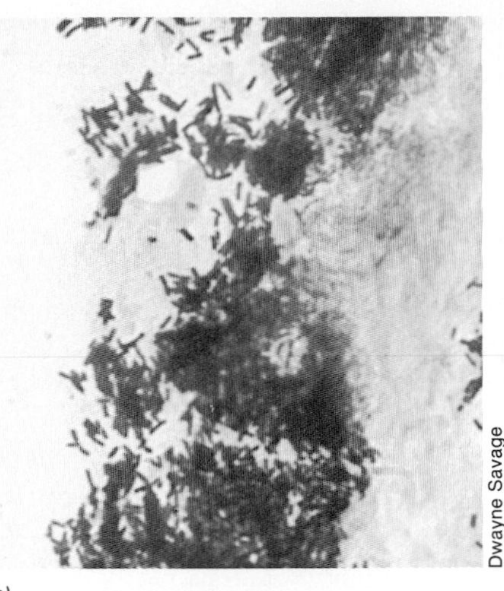

(a)

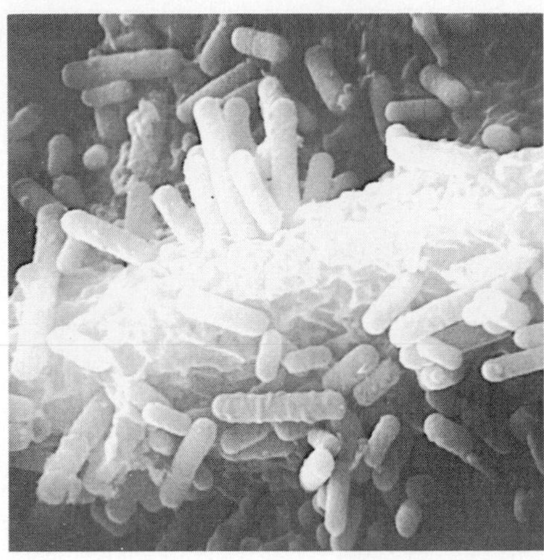

(b)

Figure 11.8 Light microscopy and scanning electron microscopy of the stomach microflora of a mouse. (a) Section through a Gram-stained preparation of the stomach wall of a 14 day-old mouse, showing extensive development of lactic acid bacteria (*Lactobacillus* sp.) in association with the epithelial layer. Magnification 670×. (b) Bacterial community on the surface of the keratinized epithelium of the stomach of an adult mouse as observed by scanning electron microscopy. Magnification 4760×.

thin, Gram-negative rods, with tapering ends (called *fusiform*) and can be seen with the scanning electron microscope attached end-on to small indentations in the intestinal wall (Figure 11.9). Other obligate anaerobes include species of *Clostridium* and *Bacteroides*. The total number of obligate anaerobes is enormous. Counts of 10^{10} to 10^{11} cells per gram of intestinal contents are not uncommon, with various species of *Bacteroides* accounting for the majority of intestinal obligate anaerobes. In addition, *Streptococcus faecalis* is almost always present in significant numbers.

The intestinal flora of the newborn becomes established early. In breast-fed human infants the flora is often fairly simple, consisting largely of *Bifidobacterium* spp. (formerly called *Lactobacillus bifidus*). In bottle-fed infants the flora is usually more complex. The flora is conditioned partly by the fact that the infant's main source of food is milk, which is high in the sugar lactose. The reason that the flora of breast-fed infants differs from that of bottle-fed ones is not completely understood, but it is known that human milk contains a dissacharide amino sugar that is required as a growth factor by *Bifidobacterium*. As the infant ages and its diet changes, the composition of the intestinal flora also changes, ultimately approaching that of the adult.

The normal flora of the gastrointestinal tract varies among species. For example, in guinea pigs lactobacilli make up 80 percent of the intestinal flora, whereas the same organisms are only minor components of human gastrointestinal flora. The gut flora in humans can also vary qualitatively depending on the diet. Persons who consume a considerable amount of animal protein (meat) show higher numbers of *Bacteroides* and lower numbers of coliforms and lactic acid bacteria than those on a vegetable diet. An overview of the microorganisms of the gastrointestinal tract is given in Figure 11.10.

The intestinal flora has a profound influence on the animal, carrying out a wide variety of metabolic reactions (Table 11.1). Not all microorganisms carry out these reactions, and changes in the intestinal flora due to diet or disease may thus affect the animal. Of special note in Table 11.1 are the roles of the intestinal flora in modifying compounds of the bile, the bile acids. Bile acids are steroids produced in the liver and excreted in the intestine via the gall bladder.

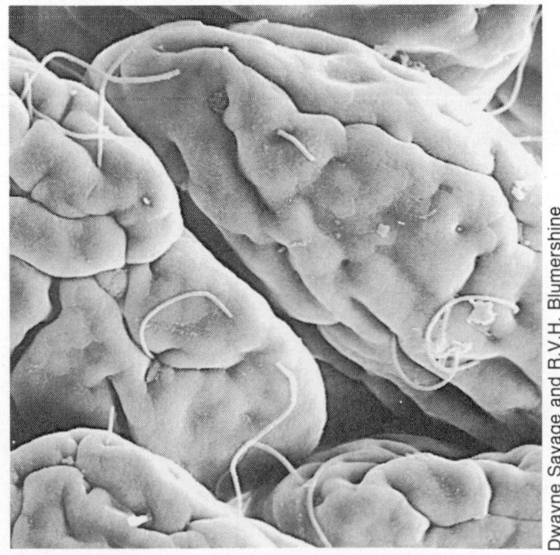

(a)

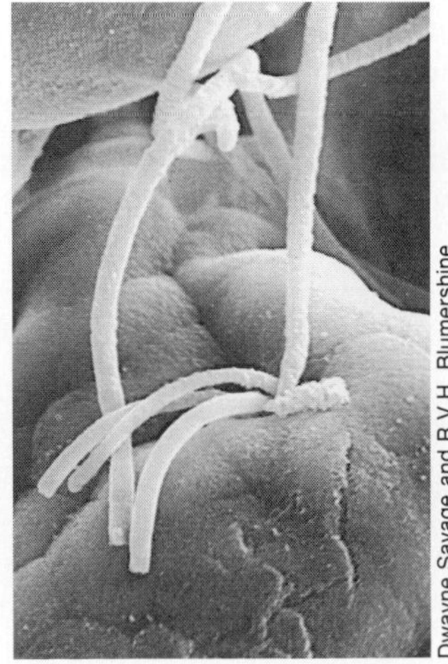

(b)

Figure 11.9 Scanning electron micrographs of the microbial community on the surface of the columnar epithelium in the mouse ileum. (a) An overview at low magnification. Note the long, filamentous organisms lying on the surface. Magnification, 600×. (b) Higher magnification, showing several filaments attached at a single depression. Note that the attachment is at the end of the filaments only. Magnification, 2800×.

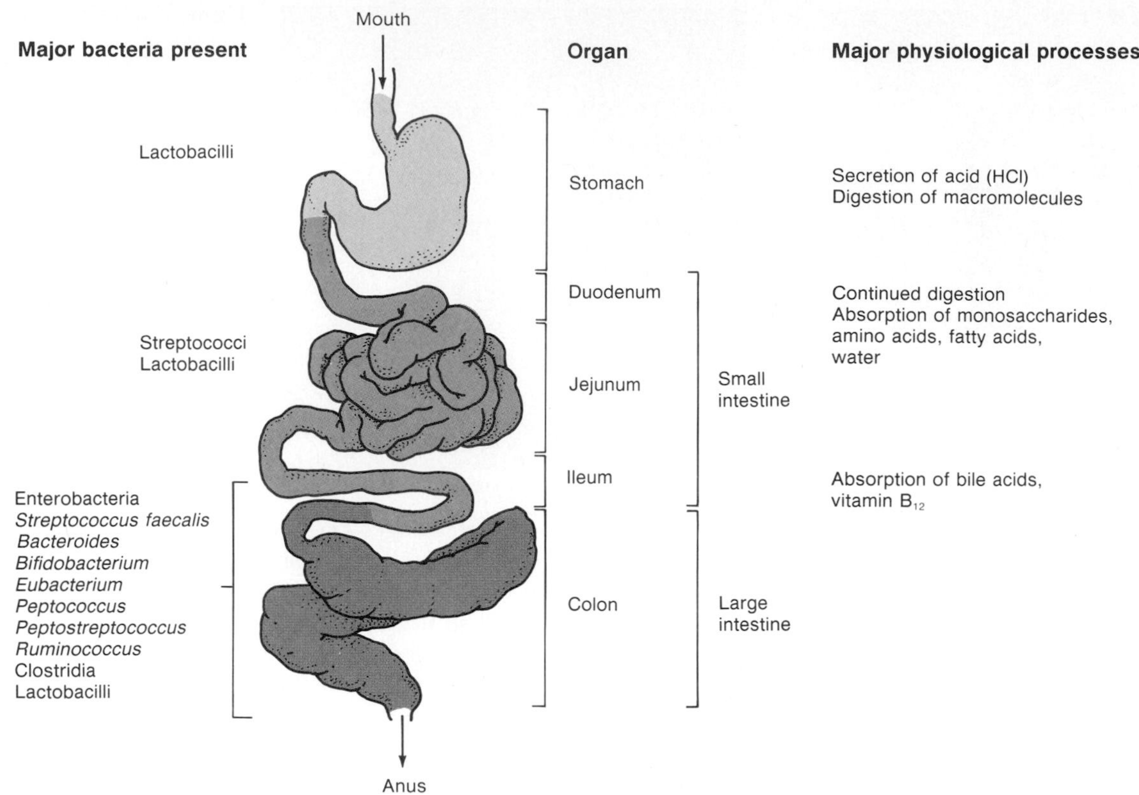

Figure 11.10 A detailed view of the human gastrointestinal tract showing functions and the distribution of microorganisms.

Their role is to promote emulsification of fats in the diet so that these can be effectively digested. Intestinal microbes cause a variety of transformations of these bile acids so that the materials excreted in the feces are quite different from bile acids. Other products of microbial fermentation are the odor-producing substances listed in Table 11.1. Composition of the microbial flora as well as diet influences the amount of gas and the amount of odoriferous materials present. Further discussion of the role of the gastrointestinal flora in the animal will be given in Section 11.6 in the discussion of germ-free animals.

One rather dramatic instance of a beneficial role of a component of the intestinal flora has been the discovery of significant numbers of nitrogen-fixing bacteria of the species *Klebsiella pneumoniae* in the intestines of New Guineans subsisting on a diet in which 80 to 90 percent of their calories are derived from sweet potatoes, a food virtually devoid of protein nitrogen. This bacterium fixes N_2 anaerobically and is a constant member of the intestinal flora of these people. Apparently sufficient nitrogen is fixed and passes through the intestinal wall into the bloodstream so that these individuals can subsist on a diet that would be unsuitable otherwise.

During the passage of food through the gastrointestinal tract, water is withdrawn from the digested material, and it gradually becomes more concentrated and is converted into feces. Bacteria, chiefly dead ones, make up about one-third of the weight of fecal matter. Organisms living in the lumen of the large intestine are continuously being displaced downward by the flow of material, and if bacterial numbers are to be maintained, those bacteria that are lost must be replaced by new growth. Thus, the large intestine resembles in some ways a chemostat (see Section 9.6).

Table 11.1 Biochemical and metabolic contributions of the intestinal microflora

Vitamin synthesis	Product: thiamin, riboflavin, pyridoxine, B_{12}, K
Gas production	Product: CO_2, CH_4, H_2 (and N_2 from air)
Odor production	Product: H_2S, NH_3, amines, indole, skatole, butyric acid
Organic acid production	Product: acetic, propionic, butyric acids
Nitrogen fixation	Agent: *Klebsiella pneumoniae* (in humans on high-carbohydrate low-protein diet)
Glycosidase reactions	Enzyme: β-glucuronidase, β-galactosidase, β-glucosidase, α-glucosidase, α-galactosidase
Sterol metabolism	Process: esterification, dehydroxylation, oxidation, reduction, inversion

The time needed for passage of material through the complete gastrointestinal tract is about 24 hours in humans; the growth rate of bacteria in the lumen is one to two doublings per day.

When an antibiotic is given orally it may inhibit the growth of the normal flora as well as pathogens; continued movement of the intestinal contents then leads to loss of the preexisting bacteria and the virtual sterilization of the intestinal tract. In the absence of the normal flora the environmental conditions of the large intestine change, and there may become established opportunistic microorganisms such as antibiotic-resistant *Staphylococcus, Proteus*, or the the yeast *Candida albicans* which usually do not grow in the intestinal tract because they cannot compete with the normal flora. Occasionally, establishment of these opportunistic pathogens can lead to a harmful alteration in digestive function or even to disease. In mice, for example, a single injection of penicillin or streptomycin makes the animal ten thousand times more susceptible to experimental *Salmonella* infections. Following antibiotic therapy the normal flora eventually becomes reestablished, but often only after a considerable period.

Intestinal gas The gas produced within the intestines, called *flatus*, is the result of the action of fermentative and methanogenic microorganisms. Some foods which are poorly absorbed in the stomach and intestines can be metabolized by fermentative bacteria resulting in the production of hydrogen (H_2) and carbon dioxide (CO_2). In many individuals, methanogenic bacteria then convert some of the H_2 and CO_2 to methane gas (CH_4). Methane is produced by a number of different bacteria called *methanogens* which are all members of the archaebacteria (see Section 19.34). Although all humans have some methanogens in their intestinal tract, about one-third of the American population has an active methane-producing microbial flora in the intestines. Curiously, there is no pattern, such as inheritance, age, or diet, which can be associated with the presence in the intestinal tract of methanogenic bacteria.

Normal adult humans expel a few hundred milliliters of gas per day via the rectum and in their breath. More than half of this gas is nitrogen (N_2) which enters the body in swallowed air and passes unchanged into the intestines, but the rest is microbially produced. Diet can have a dramatic effect on the amount and type of gas produced. If large quantities of beans are consumed, total gas production increases about tenfold. It is thought that a bean polysaccharide that is not digested by humans passes into the intestines, where fermentative bacteria convert it to H_2 and CO_2. Although a high-bean diet can cause a gassy condition in most humans, some people inherently have a large amount of flatulence, even on a low-bean diet, probably due to their inability to digest or absorb certain fermentable sugars. A common reason for gassiness is poor absorption of the milk sugar lactose. If individuals with this abnormality do not eliminate lactose from their diets, their intestinal bacteria are stimulated to produce gas.

11.5 Normal Flora of Other Body Regions

Respiratory tract The gross anatomy of the respiratory tract was shown in Figure 11.7. In the **upper respiratory tract** (throat, nasal passages, and nasopharynx) microorganisms live primarily in areas bathed with the secretions of the mucous membranes. Bacteria enter the upper respiratory tract from the air in large numbers during breathing, but most of these are trapped in the nasal passages and expelled again with the nasal secretions. The resident organisms most commonly found are staphylococci, streptococci, diphtheroid bacilli, and Gram-negative cocci. Each person generally has a characteristic flora, which may remain constant over extended periods of time. Potentially harmful bacteria, such as *Staphylococcus aureus, Streptococcus pneumoniae, Streptococcus pyogenes*, and *Corynebacterium diptheriae* are often part of the normal flora of the nasopharynx of healthy individuals.

The **lower respiratory tract** (trachea, bronchi, and lungs) is essentially sterile, in spite of the large numbers of organisms potentially able to reach this region during breathing. Dust particles, which are fairly large, are filtered out in the upper respiratory tract. As the air passes into the lower respiratory tract, its rate of movement decreases markedly, and organisms settle onto the walls of the passages. These walls are lined with ciliated epithelium, and the cilia, beating upwards, push bacteria and other particulate matter toward the upper respiratory tract where they are then expelled in the saliva and nasal secretions. Only droplet nuclei smaller than 10 μm in diameter are able to reach the lungs.

Genitourinary tract The main anatomical features of the male and female genitourinary tracts are shown in Figure 11.11. In both male and female, the bladder itself is usually sterile, but the epithelium of the urethra is colonized by anaerobic Gram-negative rods and cocci. The vagina of the adult female generally is weakly acidic and contains significant amounts of the polysaccharide glycogen. A *Lactobacillus*, sometimes called *Döderlein's bacillus*, which ferments glycogen and produces acid, usually is present in the vagina and may be responsible for the acidity. Other organisms, such as yeasts (*Torulopsis* and *Candida* species), streptococci, and *E. coli*, may also be present. Before puberty, the female vagina is alkaline and does not produce glycogen, Död-

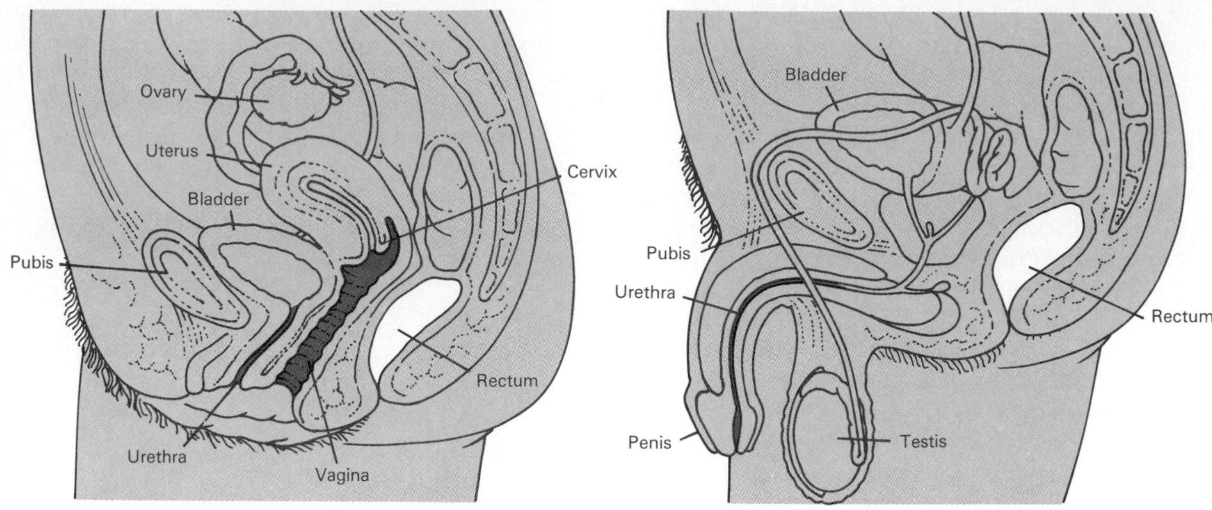

Figure 11.11 Anatomy of the genitourinary tracts of the human female and male, showing regions (color) where microorganisms often grow.

erlein's bacillus is absent, and the flora consists predominantly of staphylococci, streptococci, diphtheroids, and *E. coli*. After menopause, glycogen disappears, the pH rises, and the flora again resembles that found before puberty.

11.6 Germ-Free Animals

For a number of kinds of studies on the role of the normal flora, **germ-free animals** are useful. Colonies of germ-free animals are now widely established and are finding uses in many research studies. Establishing a germ-free animal colony is sometimes simplified by the fact that mammals frequently are microbially sterile until birth, so that if the fetus is removed aseptically just before the time of expected birth, germ-free offspring can often be obtained. These newborns must then be placed in germ-free isolators (Figure 11.12), and all air, water, food or other objects entering the isolators must be sterile. Within the isolators the infants must then be hand-fed until they have developed to a stage where they can feed themselves. Once established, a germ-free colony can be maintained by continued mating between germ-free males and females. With birds, establishment of germ-free colonies is easier, as the inside of the egg is usually sterile and the newly hatched chick is able to feed itself immediately. Germ-free colonies of mice, rats, guinea pigs, rabbits, hamsters, monkeys, and chickens have been established and similarly, germ-free individual lambs and pigs have been kept for considerable periods of time. Raising large animals in the germ-free condition is obviously much more difficult than the raising of small animals, since large isolators are expensive and difficult to keep sterile. Germ-free mice are now available commercially from laboratory-animal supply houses.

Germ-free individuals differ from normal animals in several important respects. Structures and systems which participate in defense against bacterial invaders, are poorly developed in germ-free animals. Also some organs that would normally have natural populations of bacteria are often reduced in size or capacity in the germ-free animal. However, in the germ-free guinea pig, rat, and rabbit the cecum is greatly enlarged (Figure 11.13). In the normal rodent the cecum is the part of the intestinal tract that has the largest bacterial population; in the germ-free state the cecum might lack the stimulus to evacuate caused by

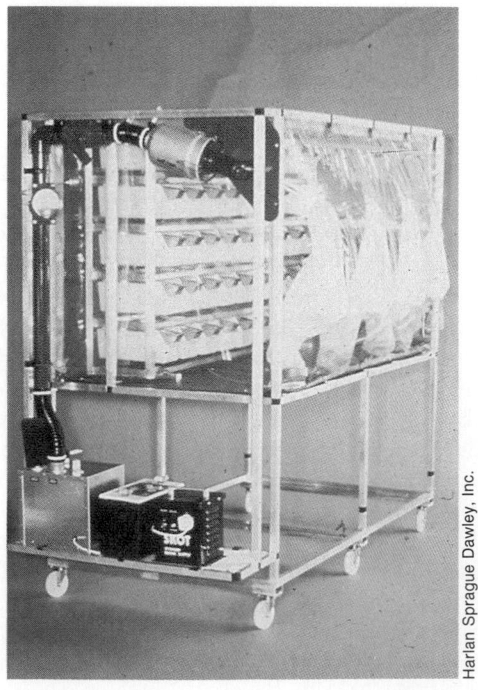

Figure 11.12 A germ-free isolator.

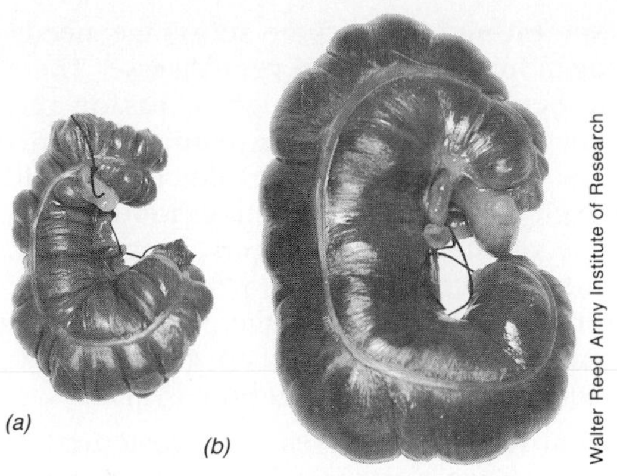

Figure 11.13 Comparison of the cecum size of a normal (a) and a germ-free (b) rodent.

the bacteria and hence continuously fill up. Furthermore, the whole intestinal wall of the germ-free animal is thin and unresponsive to mechanical stimuli. The conclusion seems inescapable that bacteria are necessary for the normal development of the intestine.

Germ-free animals also differ in nutritional needs from conventional ones. For instance, vitamin K, which usually is not required by conventional animals, is required by germ-free animals. *Escherichia coli* synthesizes vitamin K; when this organism is established in the intestinal tract of germ-free animals, the vitamin K deficiency disappears, thus showing that *E. coli* probably is responsible for synthesizing this vitamin in the conventional animal.

Germ-free animals are much more susceptible to bacterial infections than are conventional animals. Organisms like *Bacillus subtilis* and *Micrococcus luteus*, which are harmless to conventional animals, are harmful to germ-free forms. It is difficult to infect conventional animals with *Vibrio cholerae* (the causal agent of cholera) and *Shigella dysenteriae* (the cause of bacterial dysentery); yet these organisms can readily be established in germ-free animals. It is likely that the normal flora of the conventional animal has a competitive advantage in the intestinal tract, preventing establishment of opportunistic pathogens; with the normal flora gone, the foreign organisms can become established easily. Also, germ-free animals do not show tooth decay (dental caries) even if they are fed a diet high in sucrose, which is very conducive to tooth decay in conventional animals. However, if germ-free animals are inoculated with a pure culture of certain *Streptococcus* strains, especially *S. mutans*, isolated from the teeth, tooth decay does occur on a high sucrose diet. Hence, we can conclude that the conventional animal does derive considerable benefit from its normal flora, even though the normal flora causes some harmful effects (dental caries, for example).

11.7 Nonspecific Host Defenses

Pathogens are generally not part of the normal flora, and the host is generally able to prevent the growth of pathogens without affecting the normal flora. Many of the mechanisms responsible for the suppression of pathogens are not clearly understood. However, it is clear that animals have a number of innate "resistance factors" which serve to suppress the growth of pathogenic microbes. These resistance factors can be divided into two categories: *specific* host defenses, which are directed against individual species or strains of pathogens, and *nonspecific* host defenses, directed against a variety of pathogens. In this chapter we consider the major *nonspecific* host defenses that have been identified as important in preserving the healthy state of the host. In the following chapter we consider *specific* host defenses—the immune response.

Host immunity The ability of a particular pathogen to cause disease in different animal species is highly variable. In *rabies*, for instance, death usually occurs in all species of mammals once symptoms of the disease develop. Nevertheless, certain animal species are much more susceptible to rabies than others. Raccoons and skunks, for example, are extremely susceptible to rabies infection as compared to opossums, which are rarely linked to wild cases of rabies (see data of Table 14.4). *Anthrax*, which can infect a variety of animals but in particular herbivores, causes disease symptoms varying from mild pustules in humans to a fatal blood poisoning in cattle. Birds, on the other hand, are naturally immune to anthrax. Also, diseases of warm-blooded animals are rarely transmitted to cold-blooded species and vice-versa. Why should this be so?

Immunity to certain diseases and susceptibility to others is an innate property of a given species and is probably governed by a number of different complex and interdependent factors. Differences in physiology and nutrition from species to species as well as anatomical differences are important here, as is variation in tissue surface receptors as discussed below. The net result, however, is the interesting and sometimes perplexing observation that different animal species, occasionally even closely related species, may show completely different susceptibilities to the same disease agent.

Age, stress, and diet **Age** is an important factor in susceptibility to infectious disease. Infectious diseases are frequently more common in the very young and in the aged. In the infant, for example, development of an intestinal microflora occurs quite quickly, but the normal flora of a young infant is not the same as that of the adult. Before the development of an adult flora, and especially in the days immediately following birth, pathogens have a greater opportunity to become established and produce disease. Thus, neonatal diarrhea caused by enteropathogenic strains of *E. coli* (see Section 15.11) or *Pseudomonas aeruginosa*, is frequently encountered in infants under the age of one year. These organisms can be transmitted from the mother, where they may be causing no ill effects, because they have established a stable residency as part of the mother's flora. The undeveloped state of the infant's microflora provides poor competition for pathogenic species. In the elderly certain diseases may develop because of a malfunctioning immune system or because of anatomical changes associated with old age. For example, enlargement of the prostrate gland, a common condition in elderly males, is often associated with increased incidence of urinary tract infections. Many of the infections characteristic of young children or the aged are thus uncommon in healthy adults. The latter have a well-developed, characteristic normal flora (see Section 11.4), and have already developed immunity to a number of infectious agents. The stability of the gut flora in healthy adults is therefore a major barrier to colonization of pathogens.

Stress can predispose a normally healthy adult to disease. Although not well understood, the effects of stress are known to affect health. In studies with rats and mice it has been shown that fatigue, exertion, poor diet, dehydration, or drastic climatic changes increase the incidence and severity of infectious diseases. For example, rats subjected to intense physical activity for long periods of time show a higher mortality rate from experimental *Salmonella* infections than do well-rested animals. Also, subclinical, undiagnosed infections can become full-blown infections when the animal is placed under stressful conditions. It is hypothesized that the interaction of hormones with the immune system may play a role in stress-mediated disease. It is well known that hormonal balances change dramatically when an animal is placed under stressful conditions. The hormone *cortisone*, for example, is produced at much higher levels in times of stress than during calm periods, and this hormone is an effective anti-inflammatory agent. Suppression of inflammation removes one of the defenses an animal has against disease (see Section 11.16). Other chemical messengers also fluctuate in response to stress and these may play a role here as well.

Diet plays a role in host resistance. The correlation between famine and infectious disease has been known for centuries. Protein shortages may alter the composition of the normal flora, thus allowing opportunistic pathogens a better chance to multiply. At the other extreme, overeating may be harmful as well. Studies of clostridial diseases of sheep, in particular bloats caused by excessive gas accumulation, indicate that constant overeating can affect the composition of the normal flora, leading to massive development of bacterial species normally present in low numbers.

Not eating a particular substance needed by a pathogen may serve to prevent disease. The best example here is the effect sucrose has on the development of dental caries. As explained in Section 11.3, absence of sucrose from the diet (along with good oral hygiene) virtually eliminates tooth decay. In the absence of sucrose, the highly cariogenic bacteria *Streptococcus mutans* and *Streptococcus sobrinus* are unable to synthesize their gummy outer surface polysaccharide needed to keep the cells attached to the teeth and gums (see Section 11.3).

Anatomical defenses The structural integrity of tissue surfaces poses a barrier to penetration by microorganisms. When tissues remain healthy and intact, potential pathogens must not only bind to tissue surfaces but also grow at these sites before traveling elsewhere in the body. Intact sufaces form an effective barrier to colonization, while microbial access to damaged surfaces is more easily obtained. Resistance to colonization is due to the production of harmful substances by the host and to various mechanical actions which serve to disrupt colonization. A summary of the major anatomical defenses is shown in Figure 11.14.

The **skin** is an effective barrier to the penetration of microorganisms. *Sebaceous glands* in the skin (see Section 11.2) secrete fatty acids and lactic acid which lower skin pH and retard microbial growth. Microorganisms inhaled through the *nose* or *mouth* are removed by the action of ciliated epithelial cells in the mucous surfaces of the nasopharynx and tracheal regions. Cilia beat and push bacterial cells upward until they are caught in oral secretions and are either expectorated or are swallowed and killed in the stomach. Potential pathogens entering the host via the oral route must first survive the *acidity* of the stomach

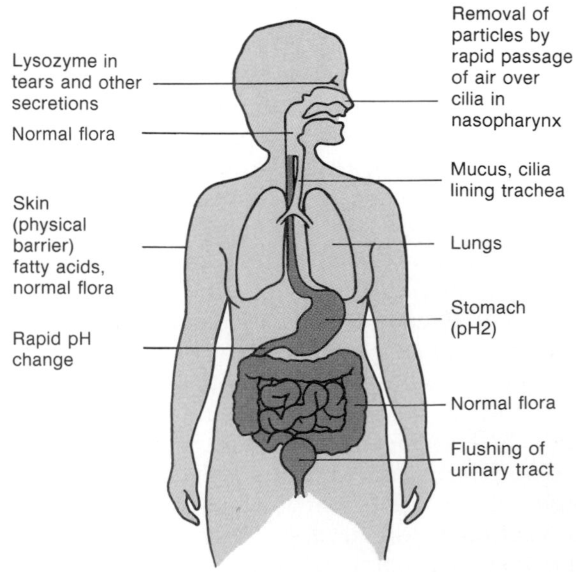

Figure 11.14 Summary of nonspecific barriers to infection.

(which is around pH 2) and then successfully compete with the increasingly abundant resident microflora present in the small intestine (which is around pH 5) and finally into the large intestine (pH 6–7). The latter organ contains bacterial numbers around 10^{10} per gram of intestinal contents in the healthy adult (see Section 11.4).

In a healthy adult, the kidney and the surface of the eye are constantly bathed with secretions containing lysozyme which markedly reduce microbial populations. Other tissues such as the spleen, thymus, and brain secrete basic proteins which have antibacterial activity. Extracellular tissues such as blood plasma also contain bactericidal substances. For example, serum proteins called β-lysins bind and destroy microbial cells. β-lysins are basic proteins which act by disrupting the bacterial cytoplasmic membrane leading to leakage of cytoplasmic constituents and death of the cell.

The chemical and physical barriers to microbial colonization present in the normal host combine to prevent routine colonization of host tissues by pathogenic microorganisms. However effective these defenses may be, certain pathogens are able to overcome them, especially in a weakened host. Damage to physical barriers and deleterious changes in other nonspecific defenses can quickly lead to growth of the pathogen and initiation of the diseased state.

Tissue specificity Unless introduced into the circulatory system where spread to other tissues is readily possible, most pathogens must first establish themselves at the site of infection. If the site is not compatible with their nutritional and environmental needs the organisms will not multiply. Thus, if cells of *Clostridium tetani* were ingested they would not bring about the disease tetanus since they are killed by the acidity of the stomach. If, on the other hand, *C. tetani* were introduced into a deep flesh wound, the organism might grow in the anaerobic zones created by localized tissue destruction and produce the potent toxin characteristic of tetanus (see Section 11.10). By contrast, enteric bacteria such as *Salmonella* and *Shigella* do not cause skin infections but successfully colonize the intestinal tract.

If a pathogen gains access to tissues, it must overcome the next barrier to infection, that of colonization. Colonization requires that the pathogen bind to specific tissue surface receptors and overcome any chemical defenses produced by the tissues. Studies of kidney disease in man and other animals are particularly instructive in this connection. Although many species of bacteria can infect kidney tissue, *Corynebacterium renale* in cattle and *Escherichia* and *Proteus* in humans are responsible for most serious kidney infections, often leading to irreversible damage to kidney tissue. One of the reasons these species predominate has to do with their strong *urease* production. Animal kidneys produce *urea* as the main form of excretable nitrogen, and urea is toxic to many microorganisms at relatively low concentrations. Urease, the enzyme which breaks down urea into NH_3 and CO_2, detoxifies the urea and releases a readily available source of nitrogen for growth of the pathogen. Why *C. renale* mainly infects cattle and *Proteus* infects humans probably reflects differences in tissue receptors (see Section 11.8) in the two kidney tissue types. Table 11.2 summarizes a number of examples of tissue specificity.

The compromised host We see that there is a complex array of properties that are brought into play in the body at the onset of microbial infection. In addition to the nonspecific factors just discussed, there is the complex and highly specific immune system which will be discussed in the next chapter. Due to various circumstances, one or more of these resistance mechanisms may be lost, thus increasing the probability of infection. The term *compromised host* is used to refer to hosts in which one or more resistance mechanisms is malfunctioning and in which the probability of infection is therefore increased.

Hospital patients are often compromised hosts and many patients have weakened resistance to infectious disease because of their illness. Many hospital procedures such as catheterization, hypodermic injection, spinal puncture, and biopsy carry with them the risk of introducing pathogens into the patient. Surgical procedures are a major hazard, since not only are highly susceptible parts of the body exposed to

Table 11.2 Tissue specificity as a factor in bacterial disease

Disease	Tissue infected	Bacterium responsible
Diphtheria	Throat epithelium	*Corynebacterium diphtheriae*
Gonorrhea	Urogenital epithelium	*Neisseria gonorrhoeae*
Cholera	Small intestine epithelium	*Vibrio cholerae*
Pyelonephritis	Kidney medulla	*Proteus* spp.
Dental Caries	Oral epithelium	*Streptococcus mutans, Streptococcus sobrinus*
Spontaneous Abortion (cattle)	Placenta	*Brucella abortus*

sources of contamination but the stress of surgery often diminishes the resistance of the patient to infection. Finally, in organ transplant procedures, drugs are used which suppress the immune system (to prevent rejection of the transplant, as discussed in Sections 12.4 and 12.10), and these immunosuppressive drugs will greatly increase susceptibility to infection. Thus, many hospital patients with noninfectious primary ailments (cancer, heart disease) die of microbial infection because they are compromised hosts.

But compromised hosts exist even outside the hospital. Smoking, excess consumption of alcohol, intravenous drug usage, lack of sleep, poor nutrition, are all conditions that can, in certain situations, produce a compromised host. Infection itself may lead to the compromise of the host. For instance, the virus causing acquired immunodeficiency syndrome (AIDS) destroys one type of cell involved in the immune response (helper T cells, see Section 12.3). AIDS patients are hence unable to mount effective resistance to infection; death is generally due to some infectious agent.

Finally, there are certain genetic conditions that may compromise the host. Several genetic diseases are known which eliminate important parts of the immune system. Individuals with such conditions frequently die at an early age, not from the genetic condition itself, but from microbial infection.

Medical practice, including antimicrobial therapy (see Section 11.7), has greatly reduced the risk of infection in compromised hosts, but has not eliminated it.

11.8 Entry of the Pathogen into the Host

A pathogen must first gain access to host tissues and multiply before damage can be done. In most cases this requires that the organism penetrate the skin, mucous membranes, or intestinal epithelium, surfaces which normally act as microbial barriers. Passage through the skin into the subcutaneous layers almost always occurs through wounds; only in rare instances is there any evidence that pathogens can penetrate through the unbroken skin.

Specific adherence Most microbial infections begin on the mucous membranes of the respiratory, alimentary, or genitourinary tracts. There is considerable evidence that bacteria or viruses able to initiate infection are able to adhere specifically to epithelial cells (Figure 11.15). The evidence for specificity is of several types. First, an infecting microbe does not adhere to all epithelial cells equally, but shows selectivity by adhering to cells in the particular region of the body where it normally gains entrance. For example, *Neisseria gonorrhoeae*, the causative agent of the venereal disease gonorrhea, adheres much more strongly to urogenital epithelia than to other tissues. Second, adherence is strain specific: a bacterial strain that normally infects humans will adhere more strongly to the appropriate human epithelial cells than to similar cells in another animal (for example, the rat), whereas a strain which specifically colonizes the rat will adhere more firmly to rat cells than to human cells. Thus there is a *host-strain* specificity for adherence. Many bacteria possess specific surface macromolecules that bind to complementary receptor molecules on the surfaces of certain animal cells, thus promoting specific and firm adherence. Certain of these macromolecules are polysaccharide in nature and form a sticky meshwork of fibers called the bacterial **glycocalyx** (Figure 11.15*b*). The glycocalyx is important not only in attaching bacterial cells to eucaryotic cell surfaces, but also in adherence between bacterial cells (Figure 11.15*b*). In addition, fibriae (see Section 3.10) may be important in the attachment process, since it now

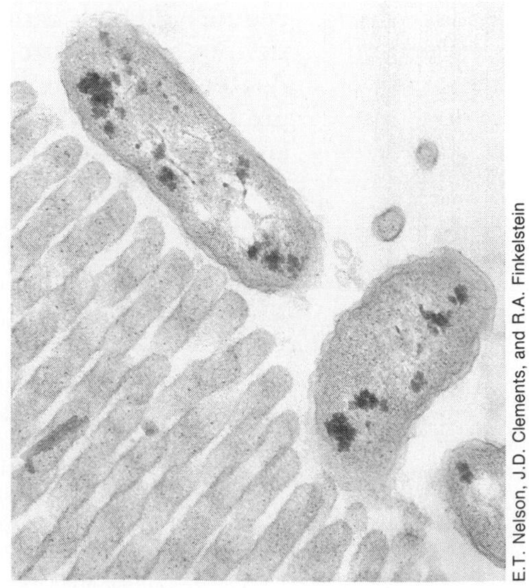

(a)

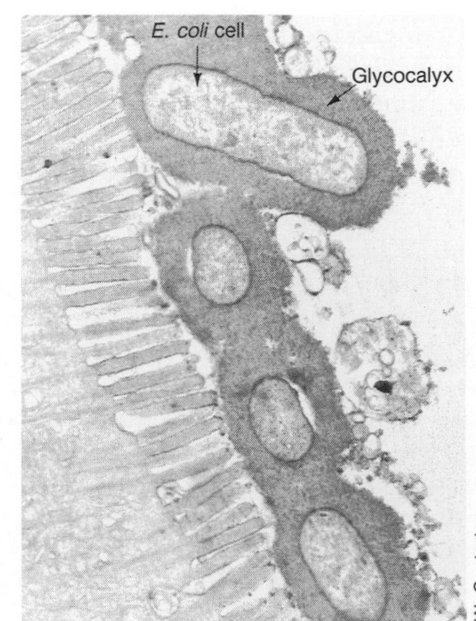

(b)

Figure 11.15 Adherence of pathogens to animal tissues. (a) Transmission electron micrograph of a thin section of *Vibrio cholerae* adhering to the brush border of the rabbit villus. Note the absence of the outer layer (glycocalyx). (b) Enteropathogenic *E. coli* in fatal model infection in the newborn calf. The bacterial cells are attached to the brush border of the calf villus via an extensive glycocalyx. Magnification 11,700×.

appears that fimbriae of *N. gonorrhoeae* play a key role in the attachment of this organism to urogenital epithelium.

Evidence of the specific interaction between mucosal epithelium and pathogen comes from studies on diarrhea caused by *Escherichia coli.* Most strains of *E. coli* are nonpathogenic and are part of the normal flora of the *large* intestine. A few strains (only a handful of the 160 different *E. coli* serotypes known) are enteropathogenic, possessing the ability to colonize the *small* intestine and initiate diarrhea. Such strains possess a specific surface structure, the K antigen, an acidic polysaccharide capsule, which is involved in specific attachment to intestinal mucosa (see Figure 11.15*b*). As we discussed in Section 7.6, the ability to produce the K antigen is controlled by a conjugative plasmid. We shall see in Section 11.11 that colonization of the small intestine alone is not sufficient to initiate symptoms of diarrhea; another factor, the enterotoxin, must also be formed. Thus two kinds of *E. coli* can be recognized: pathogenic strains, which are able to adhere to the mucosal surface of the small intestine and cause disease symptoms; and "normal" *E. coli*, strains which are unable to adhere to the small intestine or produce enterotoxin, which grow in the large intestine (cecum and colon), and which enter into a long-lasting symbiotic relationship with the mammalian host.

A summary of major factors in microbial adherence is given in Table 11.3.

Penetration In some diseases, the pathogen can remain localized at the mucosal surface and initiate damage by liberating toxins, as described later. Examples include whooping cough, caused by *Bordetella pertussis*, diphtheria, caused by *Corynebacterium diphtheriae*, and cholera, caused by *Vibrio cholerae*. However, in most cases the pathogen penetrates the epithelium and either grows in the submucosa, or spreads to other parts of the body where growth is initiated. Access to the interior of the body may occur in those areas where lymph glands are near the surface, such as the nasopharyngeal region, the tonsils, and the lymphoid follicles of the intestine. Many times small breaks or lesions in the mucous membrane in one of these regions will permit an initial entry. Motility may be of some value to the invader, although many pathogens are nonmotile. Among factors inhibiting establishment of the pathogen is the normal microbial flora itself, with which the pathogen must compete for nutrients and living space. If the normal flora is altered or eliminated by antimicrobial therapy, successful colonization by a pathogen may be easier to accomplish.

11.9 Colonization and Growth

The initial inoculum is rarely sufficient to cause damage; a pathogen must *grow* within host tissues in order to produce a disease condition. If the pathogen is to grow it must find in the host appropriate nutrients and environmental conditions. Temperature, pH, and reduction potential are environmental factors that affect pathogen growth, but of most importance is the availability of microbial nutrients in host tissues. Although at first thought it might seem that a vertebrate animal would be a nutritional paradise for microbes, not all nutrients are in plentiful supply, and there is probably considerable selectivity in determining what kinds of organisms can grow. Soluble nutrients such as sugars, amino acids, and organic acids may often be in short supply and organisms able to utilize high-molecular-weight components may be favored. Not all vitamins and other growth factors are necessarily in adequate supply in all tissues at all times. *Brucella abortus*, for example, can grow slowly in most tissues of infected cattle, but grows rapidly only in the placenta, where it causes bovine abortion. This specificity is due to the elevated concentration of erythritol found in the placenta, which greatly

Table 11.3 Major adherence factors used to facilitate attachment of microbial pathogens to host tissues*

Factor	Example
Sticky outer capsule (glycocalyx)	Dental caries—binding to tooth surface by *Streptococcus mutans*
Other surface polysaccharides	Enteropathogenic *Escherichia coli*—K antigen binds receptor on small intestinal mucosa
Adherence proteins	M-protein on surface of *Streptococcus pyogenes* binds receptor on respiratory mucosa
Lipoteichoic acid	Along with M-protein of *Streptococcus pyogenes*—facilitates binding to respiratory mucosal receptor
Fimbriae (pili)	Gonorrhea—pili on *Neisseria gonorrhoeae* facilitates binding to urogenital epithelium

**For the most part, receptor sites on host tissues are glycoproteins or complex lipids such as gangliosides or globosides.*

stimulates growth of *B. abortus*.

Trace elements may also be in short supply and influence establishment of the pathogen. In the latter category, the most evidence exists for the influence of **iron** on microbial growth. A specific protein called transferrin, present in animals, binds iron tightly and transfers it through the body. Such is the affinity of this protein for iron that microbial iron deficiency may be common; indeed, administration of a soluble iron salt to an infected animal may greatly increase the virulence of some pathogens. As we noted in Section 4.21, many bacteria produce iron-chelating compounds which help them to obtain iron from the environment.

Localization in the body After initial entry, the organism often remains localized and multiplies producing a small focus of infection such as the boil, carbuncle, or pimple that commonly arises from *Staphylococcus* infections of the skin. Alternatively, the organisms may pass through the lymphatic vessels and be deposited in lymph nodes. If an organism reaches the blood it will be distributed to distant parts of the body, usually concentrating in the liver or spleen. Spread of the pathogen through the blood and lymph systems can result in a generalized (systemic) infection of the body, with the organism growing in a variety of tissues. If extensive bacterial growth in tissues occurs, some of the organisms are usually shed into the bloodstream in large numbers, a condition called **bacteremia**. However, generalized infection of this type is rare; a more common situation is for the pathogen to localize in a specific organ.

Enzymes involved in invasion Streptococci, staphylococci, pneumococci, and certain clostridia produce **hyaluronidase**, an enzyme that promotes spreading of organisms in tissues by breaking down hyaluronic acid, a polysaccharide that functions in the body as a tissue cement. Production of this enzyme may therefore enable these organisms to spread from an initial focus. Streptococci and staphylococci also produce a vast array of proteases, nucleases, and lipases which serve to depolymerize host proteins, nucleic acids, and fats, respectively. Clostridia that cause gas gangrene produce **collagenase**, which breaks down the collagen network supporting the tissues; the resulting dissolution of tissue is a factor in enabling these organisms to spread through the body. Fibrin clots are often formed by the host in a region of microbial invasion and serve to wall off the organism and prevent its spread through the body. Some organisms are able to produce fibrinolytic enzymes to dissolve these clots, and make further invasion possible. One such fibrinolytic substance, produced by streptococci, is known as **streptokinase**. On the other hand, some organisms produce enzymes that actually promote fibrin clotting, which causes localization of the organism rather than its spread. The best studied fibrin-clotting enzyme is **coagulase**, produced by pathogenic staphylococci, which causes the fibrin material to be deposited on the cocci and may offer them protection from attack by host cells. The fibrin matrix produced as a result of coagulase activity probably accounts for localization of many staphylococcal infections as boils and pimples.

Hemolysins Various pathogens produce proteins that are able to act on the animal cell membrane, including cell lysis and hence cell death. The action of these toxins is most easily detected with red blood cells, hence they are often called **hemolysins**; in probably all cases, however, they also work on cells other than erythrocytes. The production of such toxins is most readily demonstrated by streaking the organism on a blood agar plate. During growth of the colonies, some of the hemolysin is released and lyses the surrounding red blood cells, typically clearing a zone (Figure 11.16*a*). Some hemolysins have been shown to be enzymes that attack the phospholipid of the host cell membrane. Because the phospholipid lecithin (phosphatidylcholine) is often used as a substrate, these enzymes are called **lecithinases** or **phospholipases** (Figure 11.16*b*). Since the cell membranes of all organisms, both procaryotes and eucaryotes, contain phospholipids, hemolysins that are phospholipases sometimes destroy bacterial as well as animal cell membranes. Some hemolysins are not phospholipases, however. Streptolysin O, a hemolysin produced by streptococci, affects the sterols

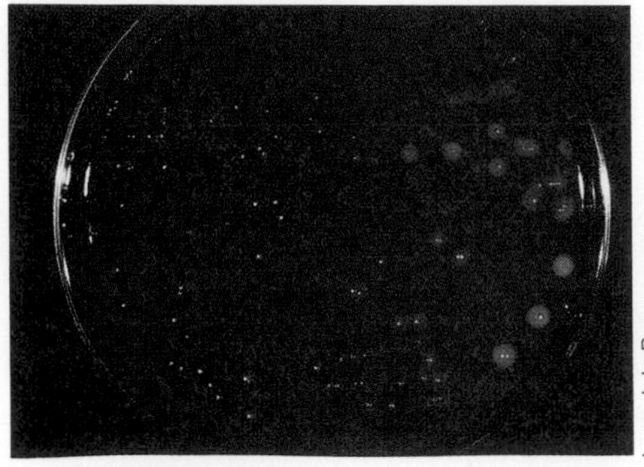

(a)

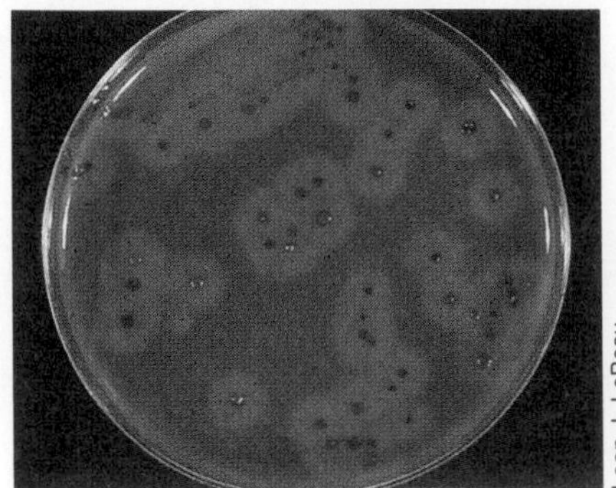

(b)

Figure 11.16 (a) Zones of hemolysis around colonies of *Streptococcus pyogenes* growing on a blood agar plate. (b) Action of phospholipase around colonies of *Clostridium perfringens*, growing on an agar medium containing egg yolk.

Fulfilling Koch's Postulates

To show that a particular microbial species is the causal agent of a particular infectious disease requires the use of Koch's postulates, discussed in Section 1.18. Briefly, the postulates are: 1) Constant association of the pathogen with the disease condition; 2) cultivation of the organism away from the body; 3) inoculation of the pure culture into experimental animals and initiation of the disease again with its characteristic symptoms; 4) reisolation of the pathogen from the experimental animal. A key postulate is thus the induction of the infectious disease in an experimental animal using a pure culture. But how can Koch's postulates be fulfilled if the disease is restricted to humans? A number of important pathogens are so specialized that they only infect humans. In such cases, an animal model is not possible and the complete steps prescribed by Koch's postulates cannot be carried out. Infectious diseases for which an animal model has been difficult or impossible to develop include cholera, dysentery, syphilis, and acquired immunodeficiency syndrome (AIDS). How, then, can a researcher provide that final proof needed?

Laboratory accidents have often provided the evidence needed to make the final link between pathogen and disease. Courageous scientists working in the laboratory are always extremely careful when working with pathogens, especially those for which no treatment is available. But accidents do happen, and if the laboratory worker develops the specific symptoms attributed to the pathogen, this is strong proof indeed that the pathogen is the causal agent. For relatively mild infectious diseases, the scientist may actually perform a self-inoculation with the suspected organism and then eagerly wait for symptoms to develop. (It is, of course, completely unethical to experiment on other human beings, and government regulations and standard laboratory practice require that any human experimentation be approved first by a recognized human experimentation committee.)

Robert Koch himself had trouble demonstrating his postulates when he worked on the disease *cholera*. When he began his cholera work, he had already had dramatic success isolating the causal agent of tuberculosis, *Mycobacterium tuberculosis*, a pathogen which can be readily transferred to the guinea pig. However, the causal agent of cholera, *Vibrio cholerae*, only induces its characteristic symptoms in humans. Koch was able to fulfill his first two postulates through careful bacteriology and pathology studies in India, but without an animal model he was thwarted in his attempts to make a final proof. Back in Germany, Koch met great resistance to his "bacterial theory for cholera" from Max von Pettenkofer, a distinguished physician at Munich who believed that cholera was associated with particular kinds of soil conditions. Pettenkofer, a pioneer in the field of environmental medicine, was so convinced that he was correct that he and one of his co-workers actually swallowed cultures of Koch's *V. cholerae*. The co-worker suffered a mild case of cholera, but Pettenkofer himself only experienced a slight diarrhea and continued to believe for the rest of his life that "Koch's bacillus" was only a contributory factor to cholera rather than the ultimate cause. Even today, a satisfactory animal model for cholera does not exist, and although there is no doubt that *V. cholerae* is *the* cause of this disease, experiments in such areas as vaccine development and chemotherapy require the use of human volunteers.

of the host cell membrane, and its action is neutralized by addition of cholesterol or other sterols. **Leukocidins** are lytic agents capable of lysing white blood cells (see later) and hence serve to decrease host resistance (see Section 11.14).

11.10 Exotoxins

The ways in which pathogens bring about damage to the host are diverse. Only rarely are symptoms of a disease due to the presence of large numbers of microorganisms, per se. Although a large mass of cells can block vessels or heart valves or clog the air passages of the lungs, in most cases more specific factors are involved. Many pathogens produce *toxins* that are responsible for all or much of the host damage.

Toxins released extracellularly as the organism grows are called **exotoxins**. These toxins may travel from a focus of infection to distant parts of the body and hence cause damage in regions far removed from the site of microbial growth. Table 11.4 provides a summary of the properties and actions of some of the best known exotoxins.

Diphtheria toxin The toxin produced by *Corynebacterium diphtheriae*, the causal agent of diphtheria, was the first exotoxin to be discovered. It is a protein of molecular weight 58,000 which differs markedly in its action on different animal species; rats and mice are relatively resistant, whereas humans, rabbits, guinea pigs, and birds are susceptible. Diphtheria toxin is very potent in its action, with only a single molecule being required to cause the death of a single cell. It binds irreversibly to the cell, and within a few hours the cell loses its ability to synthesize protein because the toxin interferes with the protein synthesis process by blocking transfer of an amino acid from a tRNA to the growing peptide chain. The toxin specifically inactivates one of the elongation factors (elongation factor 2) involved in growth of the polypeptide chain (see Figure 11.17) by catalyzing the attachment of the adenosine diphosphate ribose moiety of NAD to the elongation protein (Figure 11.17). The elongation protein is ADP-ribosylated at a single amino acid residue, a modified histidine molecule called *diphthamide*; following ADP-ribosylation the activity of elongation factor 2 drops dramatically.

Diphtheria toxin is formed only by strains of *C. diphtheriae* that are lysogenized by phage β, and its production is coded for by genetic information present in the phage genome, as shown by the fact that mutants of phage β can be isolated which cause the production of toxin molecules with altered proteins. Nontoxigenic and hence nonpathogenic strains of *C. diphtheriae* can be converted to pathogenic strains by infection with the phage (the process of phage conversion, see Section 7.3).

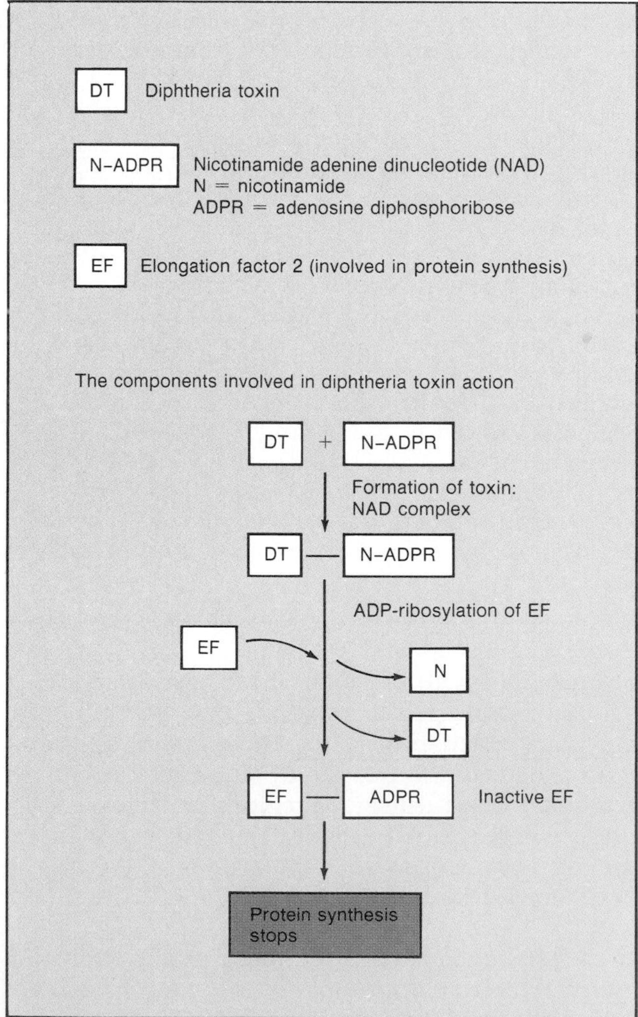

Figure 11.17 Catalysis by diphtheria toxin of attachment of the adenyldiphosphoribosyl (ADPR) portion of NAD to elongation factor 2, leading to inhibition of protein synthesis.

The gene coding for diphtheria toxin has been cloned and sequenced. The toxin as excreted by *C. diphtheriae* cells is a single polypeptide containing 535 amino acids. Following binding to the host cell, the polypeptide is cleaved by a protease into two fragments. Fragment A (193 amino acids) enters the cell and acts to disrupt protein synthesis, while the remaining piece, Fragment B (342 amino acids), is discarded. Before cleavage, Fragment B is believed to promote specific binding of the toxin to the host cell, and, following cleavage, it assists in the entry of Fragment A into the host cytoplasm.

A nongenetic factor of significance for toxin production is the concentration of iron present in the environment in which the bacteria are growing. In media containing sufficient iron for optimal growth no toxin is produced, whereas when the iron concentration is reduced to suboptimal levels, toxin production occurs. Some evidence exists that the role of iron is to bind to and activate a regulatory protein in *C. diphtheriae* (that is, act as a positive control element, see Section 5.12), and that the iron-binding protein combines with a control region of the DNA of β phage and prevents expression of the diphtheria toxin gene. When iron is absent, the regulatory protein does not act, and toxin synthesis can occur. Mutants in the regulatory region are known which synthesize toxin even in the presence of high iron concentrations. The disease diphtheria is discussed in Section 15.2.

Exotoxin A of *Pseudomonas aeruginosa* has been shown to have an action quite similar to diphtheria toxin, also transferring the ADP-ribosyl portion of NAD to elongation factor 2.

Tetanus and botulinum toxins These toxins are produced by two species of obligately anaerobic bacteria, *Clostridium tetani* and *C. botulinum*, which are normal soil organisms that occasionally become involved in disease situations in animals. *Clostridium tetani* grows in the body in deep wound punctures that become anaerobic, and although it does not invade the body from the initial site of infection, the toxin it produces can spread and cause a severe neurological disturbance, which can result in death. *Clostridium botulinum* rarely if ever grows directly in the body, but it does grow and produce toxin in improperly canned foods. Ingestion of the food without proper heat treatment results in neurological disease and death.

Table 11.4 Exotoxins produced by certain bacteria pathogenic for humans

Organism	Disease	Toxin	Action
Clostridium botulinum	Botulism	Neurotoxin	Flaccid paralysis
C. tetani	Tetanus	Neurotoxin	Spastic paralysis
C. perfringens	Gas gangrene, food poisoning	α-Toxin	Hemolysis (lecithinase)
		β-Toxin	Hemolysis
		γ-Toxin	Hemolysis
		δ-Toxin	Hemolysis
		θ-Toxin	Hemolysis (cardiotoxin)
		κ-Toxin	Collagenase
		λ-Toxin	Protease
		Enterotoxin	Alters permeability of intestinal epithelium
Corynebacterium diphtheriae	Diphtheria	Diphtheria toxin	Inhibits protein synthesis (see Figure 11.17)
Staphylococcus aureus	Pyogenic infections (boils, etc.), respiratory infections, food poisoning	α-Toxin	Hemolysis
		Leukocidin	Destroys leukocytes
		β-Toxin	Hemolysis
		γ-Toxin	Kills cells
		δ-Toxin	Hemolysis, leukolysis
		Enterotoxin	Induces vomiting, diarrhea
Streptococcus pyogenes	Pyogenic infections, tonsillitis, scarlet fever	Streptolysin O	Hemolysis
		Streptolysin S	Hemolysis
		Erythrogenic toxin	Causes scarlet fever rash
Vibrio cholerae	Cholera	Enterotoxin	Induces fluid loss from intestinal cells (see Figure 11.20)
Escherichia coli (enteropathogenic strains only)	Gastroenteritis	Enterotoxin	Induces fluid loss from intestinal cells
Shigella dysenteriae	Bacterial dysentery	Neurotoxin	Paralysis, hemorrhage
Yersinia pestis	Plague	Plague toxin	Kills cells
Bordetella pertussis	Whooping cough	Whooping cough toxin	Kills cells
Pseudomonas aeruginosa	Various *P. aeruginosa* infections	Dermonecrotic toxin	Kills cells

The tetanus toxin is a protein of molecular weight 67,000. Upon entry into the central nervous system this toxin becomes fixed to nerve synapses, binding specifically to one of the lipids. To understand how it induces its characteristic spastic paralysis, it is essential to discuss briefly how normal nerve function in muscle contraction operates. As seen in Figure 11.18, in normal muscle contraction, nerve impulses from the brain travel through the spinal cord and initiate a sequence of events at the motor end plate (the structure forming a junction between a muscle fiber and its motor nerve), which results in muscle contraction. When one set of muscles contracts, the opposing set of muscles becomes stretched. A stretch-sensitive receptor in the opposing muscles would cause neurons enervating these muscles to fire and oppose the stretching, but the firing is inhibited by impulses from an inhibitory nerve. Thus the opposing set of muscles relaxes. Tetanus toxin binds specifically to these inhibitory motoneurons and blocks the inhibitory action resulting in contraction of both sets of muscles. This leads to a spastic paralysis, as the two types of muscles attempt to oppose each other. Since the flexor muscles are usually dominant over the extensor muscles, the symptoms in advanced tetanus are generalized flexor muscle contractions. If the muscles of the mouth are involved, the prolonged spasm restricts the mouth's movement, resulting in the condition known as **lockjaw**. If the respiratory muscles are involved, death may be due to asphyxiation.

Botulinum toxin is an even more potent neurotoxin, perhaps the most poisonous substance known. One milligram of pure botulinum toxin is enough to kill more than 1 million guinea pigs. At least six distinct botulinum toxins have been described and the

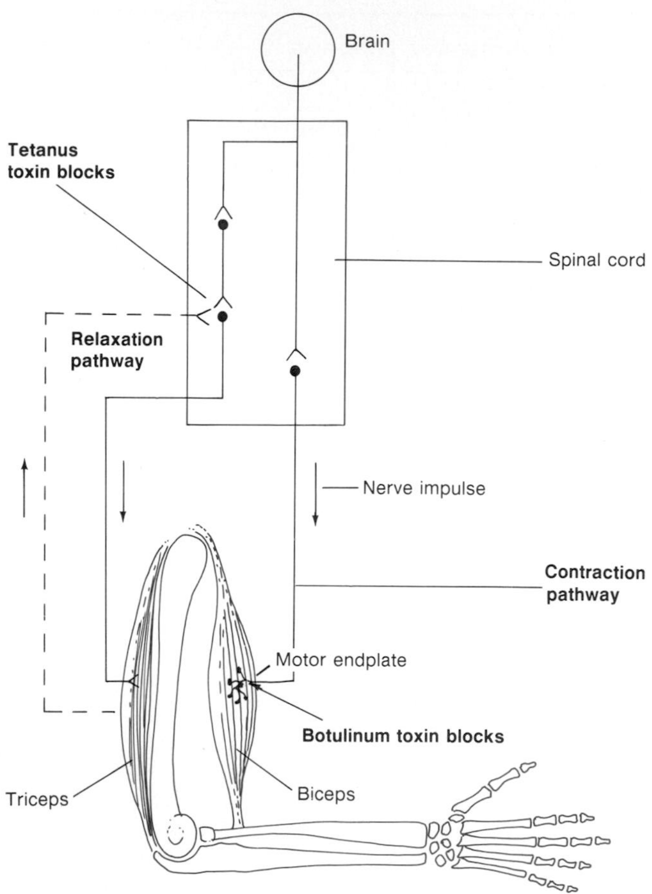

Figure 11.18 Contrasting actions of tetanus and botulinum toxins. See text for details.

syntheses of at least two of these are coded for by genes residing on lysogenic bacteriophages specific for *C. botulinum*. The major toxin is a protein of about 150,000 molecular weight which readily forms complexes with nontoxic botulinum proteins to give an active species of almost 1,000,000 molecular weight. Toxicity occurs due to binding of the toxin to presynaptic terminal membranes at the nerve-muscle junction, blocking the release of acetylcholine. Since transmission of the nerve impulse to the muscle is by means of acetylcholine action, muscle contraction is inhibited, causing a flaccid paralysis. The fatality rate from botulism poisoning can approach 100 percent, but can be significantly reduced by quick administration of antibody against the toxin (antitoxin, see Section 12.13) and by use of the artificial respirator to prevent breathing failure. Death in cases of botulism is usually due to respiratory or cardiac failure.

11.11 Enterotoxins

Enterotoxins are exotoxins that act on the *small* intestine, generally causing massive secretion of fluid into the intestinal lumen, leading to the symptoms of diarrhea. The action of enterotoxins has been most commonly studied using animal models, of which the most precise is the ligated ileal loop. For most work, rabbits have been used, but pigs have also been studied. The procedure is done under anaesthesia, an incision being made and segments of the ileum being tied off with sutures. Experimental materials, such as cultures, culture filtrates, or purified toxins, are injected into one or more of the segments (called loops), and control segments are either uninjected or are injected with sterile saline. After 18 to 24 hours, the animal is sacrificed and the ileal loops removed. Visual observation readily indicates the segments in which positive enterotoxin action has occurred (Figure 11.19), but for more precise evaluation, the fluid from each segment is removed and its volume determined. Because of the expense and time involved in the ileal loop assay, a number of tissue culture assays have been developed, but it is always essential to check these simpler models with the ileal loop model, since the latter duplicates the clinical action of the enterotoxin. Certain enterotoxins, such as the heat-stable enterotoxin of enterotoxic strains of *E. coli* (see Section 15.11), can be assayed by measuring accumulated fluid in the stomachs of mice.

Enterotoxins are produced by a variety of bacteria, including the food-poisoning organisms *Staphylococcus aureus, Clostridium perfringens*, and *Bacillus cereus*, and the intestinal pathogens *Vibrio cholerae, Escherichia coli*, and *Salmonella enteritidis*. The *S. aureus* and *E. coli* enterotoxins are plasmid encoded. It is likely that the plasmid which codes for the enterotoxin of *E. coli* also codes for synthesis

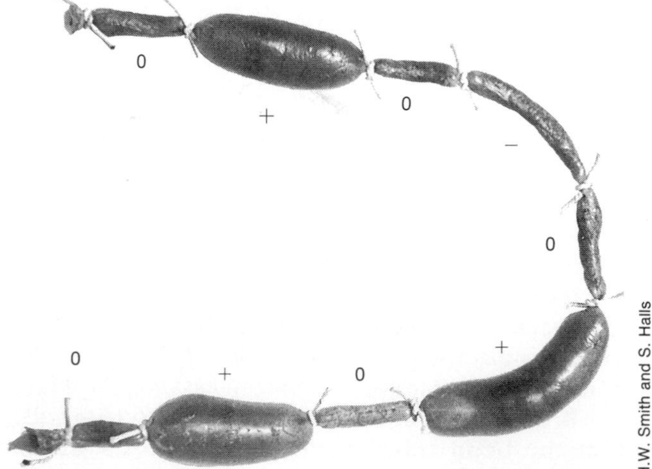

Figure 11.19 Action of *Escherichia coli* enterotoxin in the ligated ileum of the pig. The various segments were isolated by sutures, and inoculations made. Inoculated with enterotoxin-producing culture: + segments. Inoculated with culture not producing enterotoxin: − segments. Uninoculated segment: 0 segments. Enterotoxin action can be seen visually, but for precise quantification, the amount of fluid accumulating in each segment can be determined by removal with a syringe.

of the specific surface antigens that are essential for attachment of enteropathogenic *E. coli* to intestinal epithelial cells (see Section 11.8).

Cholera toxin The enterotoxin produced by *V. cholerae*, the causal agent of cholera, is the best understood. Cholera toxin is a protein consisting of three polypeptides, the A_1, A_2, and B polypeptide chains. Chains A_1 and A_2 are connected together covalently by a disulfide bridge to make a dimer called subunit A, and this is loosely associated with a variable number of B chains. The B subunit contains the binding site by which the cholera toxin combines specifically with a ganglioside (complex glycolipid) in the epithelial cell membrane, but the B subunit itself does not cause alteration in membrane permeability. Rather, the toxic action of cholera toxin is found in the A_1 chain, which activates the cellular enzyme *adenyl cyclase*, causing the conversion of ATP to cyclic AMP.

As we have discussed in Section 5.12, cyclic AMP is a specific mediator of a variety of regulatory systems in cells. In mammals, cyclic AMP is involved in the action of a variety of hormones, as well as in synaptic transmission in the nervous system, and in inflammatory and immune reactions of tissues, including allergies. Although the A_1 subunit of cholera toxin is responsible for activation of adenyl cyclase, A_1 itself must first be activated by an enzymatic activity of the cell, which requires NAD and ATP. In the action of cholera enterotoxin, the increased cyclic AMP levels bring about the active secretion of chloride and bicarbonate ions from the mucosal cells into the intestinal lumen. This change in ionic balance leads to the secretion into the lumen of large amounts of water (Figure 11.20). In the acute phase of cholera, the rate of secretion of water into the small intestine is greater than reabsorption of water by the large intestine, so that massive fluid loss occurs. Cholera victims generally die from extreme dehydration, and the best treatment for the disease is the oral administration of electrolyte solutions containing solutes (Figure 11.20) to make up for the loss of fluid and ions. It is of some interest that at the molecular level cholera enterotoxin has a mode of action (formation of cyclic AMP) identical to some normal mammalian hormones, and it has been suggested that cholera toxin may represent an ancestral hormone. Since cholera enterotoxin activates adenyl cyclase in a variety of cells and tissues, pathological manifestations of cholera toxin thus seem to be related more to the specific site at which it binds, the epithelial cells of the small intestine, than to its activation of adenyl cyclase. Indeed, it has been shown that purified B subunits devoid of adenyl cyclase activity can actually prevent the action of cholera enterotoxin, if they are administered first, because they bind to the specific cholera receptors on the mucosal cells and block the binding of the complete toxin.

There is good evidence that the enterotoxins produced by enteropathogenic *Escherichia* and *Salmonella* have similar modes of action to the cholera toxin, and it is of interest that antibody against cholera enterotoxin also inactivates these other enterotoxins. Molecular cloning and sequencing of the cholera toxin

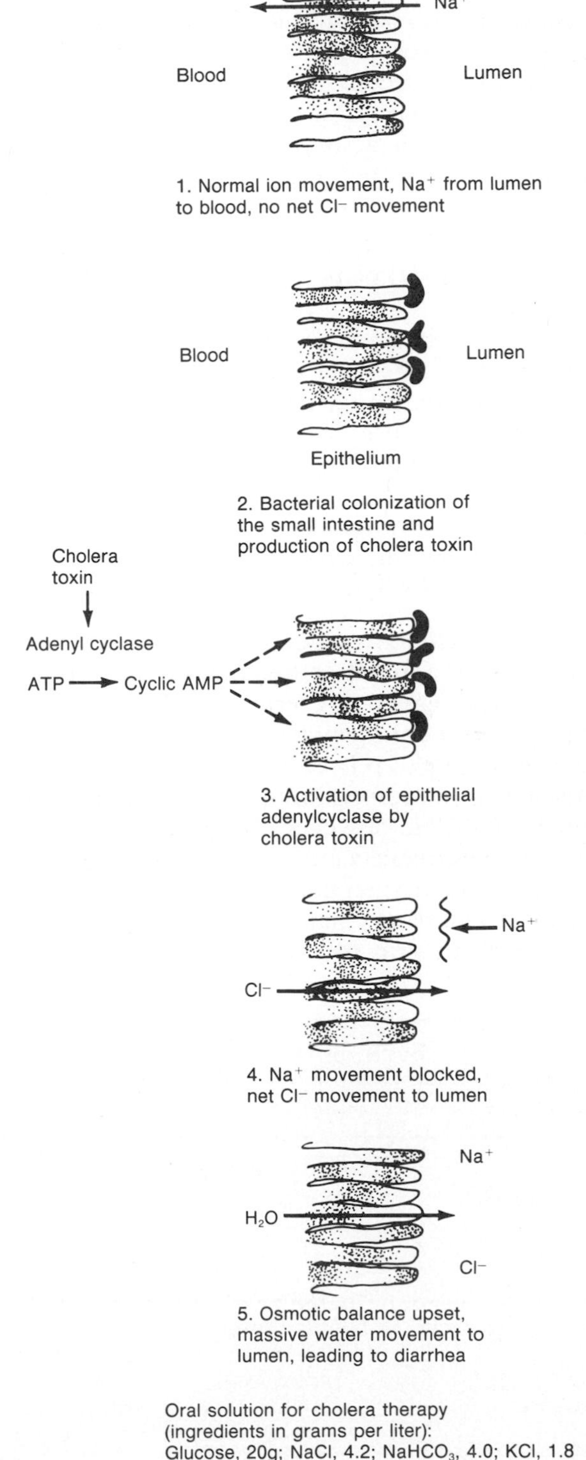

Figure 11.20 Action of cholera enterotoxin.

gene further supports this relationship: the cholera toxin gene shows greater than 75 percent sequence homology with the gene coding for the heat-labile enterotoxin produced by enteropathogenic *E. coli.* Also, the active component of *Escherichia* enterotoxin is also activated by a cellular enzyme system requiring ATP and NAD. As discussed in Section 7.6, *Escherichia* enterotoxin is controlled by a conjugative plasmid, but the enterotoxin gene of *Vibrio cholerae* is chromosomal, although transmissible by conjugation. However, the enterotoxins produced by the food-poisoning bacteria (*S. aureus, C. perfringens, B. cereus*) may be quite different in their modes of action, since their action is at least partly systemic and is not limited to alterations in intestinal permeability alone.

11.12 Endotoxins

Gram-negative bacteria produce lipopolysaccharides as part of the outer layer of their cell walls (Figures 3.27 and 3.28), which under many conditions are toxic. These are called **endotoxins** because they are generally cell bound and released in large amounts only when cells lyse. In most cases, *endotoxin* can be equated with lipopolysaccharide toxin. The major differences between exotoxins and endotoxins are listed in Table 11.5.

Endotoxin structure and function When injected into an animal, endotoxins cause a variety of physiological effects. Fever is an almost universal symptom of endotoxemia, because endotoxin stimulates host cells to release proteins called *pyrogens*, which affect the temperature-controlling center of the brain. In addition, however, the animal may develop diarrhea, experience a rapid decrease in lymphocyte, leukocyte, and platelet numbers, and enter into a generalized inflammatory state. Large doses of endotoxin can cause death, primarily through hemorrhagic shock and tissue necrosis. In general, the toxicity of endotoxins is much *lower* than that of exotoxins. For instance, in the mouse the LD_{50} is 200 to 400 μg per mouse for an average endotoxin preparation, whereas the LD_{50} for botulinum toxin is about 0.000025 μg per mouse.

The overall structure of lipopolysaccharide is diagrammed in Figure 11.21. Lipopolysaccharide consists of lipid A, a core polysaccharide, which in *Salmonella* is the same for many species, consisting of ketodeoxyoctonate, seven-carbon sugars (heptoses), glucose, galactose, and *N*-acetylglucosamine, and the *O-polysaccharide*, a highly variable molecule which usually contains galactose, glucose, rhamnose, and mannose, and generally contains one or more unusual dideoxy sugars such as abequose, colitose, paratose, or tyvelose (Figure 11.21). The sugars of the O-polysaccharide are connected in four- to five-sugar sequences (often branched) which then repeat to form the complete molecule (Figure 11.21). Lipid A is not a normal glycerol lipid, but instead the fatty acids are connected by ester linkage to *N*-acetylglucosamine. Fatty acids frequently found in the lipid include β-hydroxymyristic, lauric, myristic, and palmitic acids.

Most of the evidence suggests that it is the lipid portion of the lipopolysaccharide that is responsible for toxicity, and that the polysaccharide acts mainly to render the lipid water soluble. However, experimental studies have shown quite clearly that the entire endotoxin complex, which contains both polysaccharide and lipid (Figure 11.21), is required to obtain a response in experimental animals. Thus, the involvement of polysaccharide in endotoxin function cannot be ruled out. Endotoxins have been studied primarily in the genera *Escherichia, Shigella*, and especially *Salmonella*.

Limulus assay for endotoxin Endotoxins are pyrogens (fever inducers, see Section 11.16) and antibiotics and glucose solutions to be administered

Table 11.5 Basic properties of exotoxins and endotoxins

Property	Exotoxins	Endotoxins
Chemical properties	Proteins, excreted by certain Gram-positive or Gram-negative bacteria; generally heat labile	Lipopolysaccharide/lipoprotein complexes (see Figure 11.21); released upon cell lysis as part of the outer membrane of Gram-negative bacteria; extremely heat stable
Toxicity	Highly toxic, often fatal	Weakly toxic, rarely fatal
Immunogenicity	Highly immunogenic; stimulate the production of neutralizing antibody (antitoxin)	Relatively poor immunogen; immune response not sufficient to neutralize toxin
Toxoid potential	Treatment of toxin with formaldehyde will destroy toxicity, but treated toxin (toxoid) remains immunogenic	None
Fever potential	Do not produce fever in host	Pyrogenic, often produce fever in host

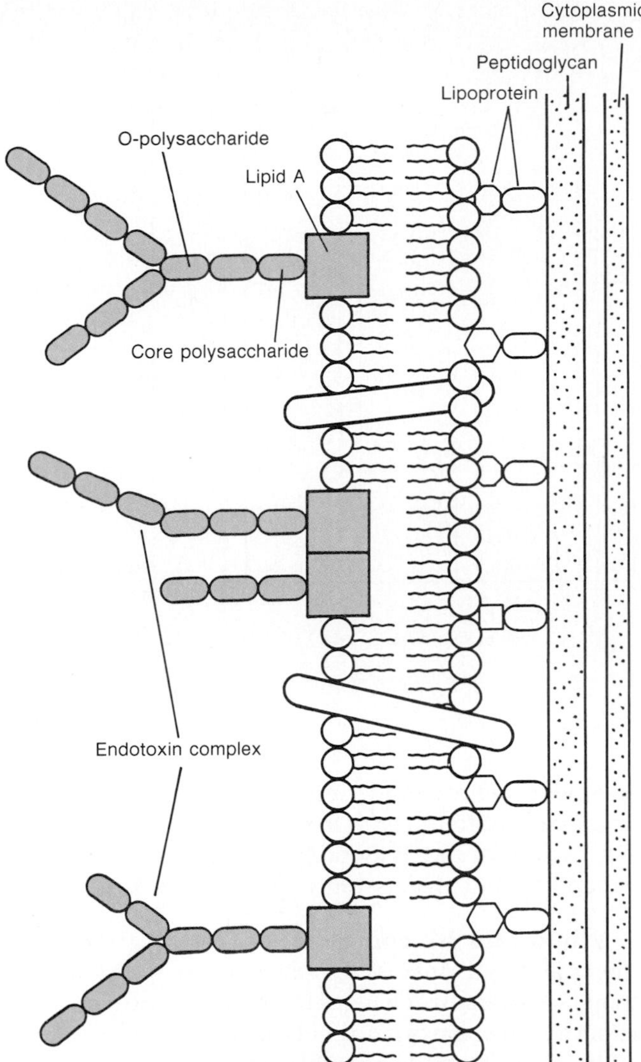

Figure 11.21 Diagrammatic structure of lipopolysaccharide showing the endotoxin complex.

intravenously must be pyrogen-free. An endotoxin assay of unusual sensitivity has been developed using lysates of amoebocytes from the horseshoe crab, *Limulus polyphemus*. Although the mechanism of this assay is not known, it has been shown that endotoxin specifically causes the gelation or precipitation of the lysate, and the degree of reaction can be measured by a spectrophotometer, an instrument that compares turbidities of various samples. A measurable reaction can be obtained with as little as 10 to 20 picograms/ml of lipopolysaccharide (a picogram is 10^{-12} g or 10^{-6} μg). Apparently the active component of the *Limulus* extract reacts with the lipid component of lipopolysaccharide. The *Limulus* assay has been used to detect the presence of minute quantities of endotoxin in serum, cerebrospinal fluid, drinking water, and fluids used for injection.

The limulus test is so exquisitely sensitive that considerable care must be taken to avoid contamination of the equipment, solutions, and reagents used, since Gram-negative bacteria abound in the laboratory and clinical environment, for example, as contaminants in the distilled water. In clinical work, detection of endotoxin by the *Limulus* assay in serum or cerebrospinal fluid can be taken as evidence of Gram-negative infection of these body fluids.

11.13 Virulence and Attenuation

Virulence is a quantitative term that refers to the relative ability of a parasite to cause disease, that is, its *degree* of pathogenicity. Virulence is determined by the *invasiveness* of the organism and by its *toxigenicity*. Both are quantitative properties and may vary over a wide range from very high to very low. An organism that is only weakly invasive may still be virulent if it is highly toxigenic. A good example of this is the organism *Clostridium tetani*. The cells of this organism rarely leave the wound where they were first deposited; yet they are able to bring about death of the host because they produce the potent tetanus exotoxin, which can move to distant parts of the body and initiate paralysis. On the other hand, a weakly toxigenic organism may still be able to produce disease if it is highly invasive. *Streptococcus pneumoniae* is not known to produce any potent toxin, but is able to cause extensive damage and even death because it is highly invasive, being able to grow in the lung tissues in enormous numbers and initiate responses in the host that lead to disturbance of lung function. These two organisms exemplify the extremes of invasiveness and toxigenicity; most pathogens fall somewhere between these two extremes.

The virulence of a pathogen can be estimated from experimental studies on the number of cells needed to kill a certain fraction of experimental animals. Highly virulent pathogens frequently show little difference in the number of cells required to kill 100 percent of the population as compared to the number required to kill a smaller fraction of the population. This is illustrated in Figure 11.22 for experimental *Streptococcus* and *Salmonella* infections in mice. Only a few cells of *Streptococcus pneumoniae* are required to establish a fatal infection in mice. Once established, however, the organism is so virulent that the number of cells required to kill 100 percent of the population is virtually the same as that required to kill only 50 percent of the population (Figure 11.22). By contrast, in order to kill 100 percent of an experimental mouse population using *Salmonella typhimurium*, also a mouse pathogen but a much less virulent one, over 1000 times as many cells must be injected per mouse compared with that necessary to kill 50 percent of the population.

It is often observed that, when pathogens are kept in laboratory culture and not passed through animals

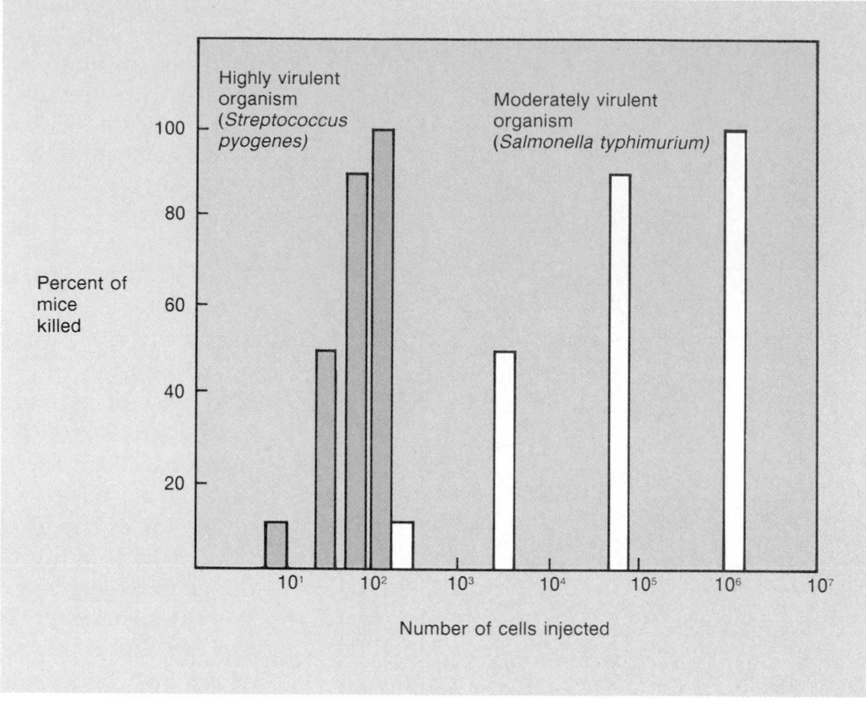

Figure 11.22 Comparison of the number of cells of *Streptococcus pyogenes* or *Salmonella typhimurium* required to kill various fractions of a mouse population.

for long periods, their virulence is decreased or even completely lost. Such organisms are said to be **attenuated.** Attenuation probably occurs because nonvirulent mutants grow faster and, through successive transfers to fresh media, such mutants are selectively favored. Attenuation often occurs more readily when culture conditions are not optimal for the species. If an attenuated culture is reinoculated into an animal, virulent organisms are sometimes reisolated, but in many cases loss of virulence is permanent. Attenuated strains find frequent use in the production of vaccines (see Section 12.13).

11.14 Interaction of Pathogens with Phagocytic Cells

The host is anything but passive to attack by invading microorganisms. Two mechanisms of host defense against infection exist, which generally can be classified as **cellular** and **humoral** (see Chapter 12). Humoral defenses involve agents that are soluble (or at least noncellular). Such agents include antibodies, a large class of specific protein molecules which will be discussed in the next chapter, and several enzymes and chemical substances, which act as nonspecific antimicrobial agents. The best known enzyme involved in defense is *lysozyme*, which is found in tears, nasal secretions, saliva, mucus, and tissue fluids. As discussed in Section 3.5, lysozyme hydrolyzes the peptidoglycan layer of bacteria, causing lysis and death. Cellular defense mechanisms involve the activity of *phagocytes* (literally, "cells that eat"), which are able to ingest and destroy invading microbes. Phagocytes are found both in tissues and in body fluids, such as blood and lymph. In this section we shall concentrate our discussion on cellular defenses.

Blood and its components Many of the substances and cells involved in defense are found in the blood. Additionally, changes in blood constituents and properties are sensitive reflections of disease states, and because blood is a readily available material for clinical analysis, many analytical procedures involve sampling of blood. Blood consists of cellular and noncellular components. The most numerous cells in the blood are *red blood cells (erythrocytes)*, which are non-nucleated cells that function to carry oxygen from the lungs to the tissues. The **white blood cells**, or **leukocytes**, include a variety of phagocytic cells, as well as cells called **lymphocytes** which are involved in antibody production and cell-mediated immunity. Red blood cells outnumber leukocytes by roughly a factor of one thousand. **Platelets** are small cell-like constituents that lack a nucleus and play an important role in preventing leakage of blood from a damaged blood vessel. Platelets clump together to form a temporary plug in a damaged vessel until a permanent clot forms through the action of various clotting agents, some of which are released from the platelets themselves. When the cells and platelets are removed from blood, the remaining fluid is called **plasma**. An important component of plasma is fibrinogen, a clotting agent, which undergoes a complex set of reactions during the formation of the fibrin clot. Clotting can be prevented by addition of anticoagulants, such

The Discoverers of the Main Bacterial Pathogens

The history of the discovery of the microbial role in infectious disease has been described briefly in Chapter 1. Once the concept of specific microbial disease agents had been clarified, and the procedures for culture of microorganisms had been developed, it was a relatively simple procedure to isolate a large number of microbial pathogens. The two decades after the enunciation of Koch's postulates were indeed fruitful for medical microbiology. The rapid development of this field is indicated by the table below, which lists the main pathogens isolated during the "golden age of bacteriology."

Year	Disease	Organism	Discoverer
1877	Anthrax	*Bacillus anthracis*	Koch, R.
1878	Suppuration	*Staphylococcus*	Koch, R.
1879	Gonorrhea	*Neisseria gonorrhoeae*	Neisser, A. L. S.
1880	Typhoid fever	*Salmonella typhi*	Eberth, C. J.
1881	Suppuration	*Streptococcus*	Ogston, A.
1882	Tuberculosis	*Mycobacterium tuberculosis*	Koch, R.
1883	Cholera	*Vibrio cholerae*	Koch, R.
1883	Diphtheria	*Corynebacterium diphtheriae*	Klebs, T. A. E.
1884	Tetanus	*Clostridium tetani*	Nicolaier, A.
1885	Diarrhea	*Escherichia coli*	Escherich, T.
1886	Pneumonia	*Streptococcus pneumoniae*	Fraenkel, A.
1887	Meningitis	*Neisseria meningitidis*	Weichselbaum, A.
1888	Food poisoning	*Salmonella enteritidis*	Gaertner, A. A. H.
1892	Gas gangrene	*Clostridium perfringens*	Welch, W. H.
1894	Plague	*Yersinia pestis*	Kitasato, S., Yersin, A. J. E. (independently)
1896	Botulism	*Clostridium botulinum*	van Ermengem, E. M. P.
1898	Dysentery	*Shigella dysenteriae*	Shiga, K.
1900	Paratyphoid	*Salmonella paratyphi*	Schottmüller, H.
1903	Syphilis	*Treponema pallidum*	Schaudinn, F. R., and Hoffman, E.
1906	Whooping cough	*Bordetella pertussis*	Bordet, J., and Gengou, O.

as potassium oxalate, potassium citrate, heparin, or sodium polyanetholsulfonate. Plasma is stable only when such an anticoagulant is added. When plasma is allowed to clot, the fluid components left behind, called **serum**, consist of all of the proteins and other dissolved materials of the plasma except fibrin. Since serum contains antibodies, it is widely used in immunological investigations (see Chapter 12).

Blood is pumped by the heart through a network of arteries and capillaries to various parts of the body and is returned through the veins (Figure 11.23). The circulatory system carries not only nutrients (including O_2) but also the components of the blood which are involved in host resistance to infection. At the same time, the circulatory system facilitates the spread of pathogens to various parts of the body.

Lymph is a fluid similar to blood but which lacks red cells. There is a separate circulatory system for lymph, called the **lymphatic system** within which lymph flows (Figure 11.23). Fluids in tissues drain into lymphatic capillaries, then into **lymph nodes**, found at various locations throughout the system, which filter out microorganisms and other particulate materials. Specialized white blood cells found in abundance in the lymphatic system, called **macrophages**, actually carry out the filtering action, as will be described below. In addition to filtering foreign particles, lymph nodes may be sites of infection, since organisms that are collected there by the filtering mechanisms may then proliferate. Lymphatic fluid eventually flows into the circulatory system via the thoracic duct.

Lymphocytes, another special type of cell found within the lymphatic system, are involved in the immune response. Their role in antibody formation will be discussed in Chapter 12.

Phagocytes Some of the leukocytes found in whole blood are phagocytes, and phagocytes are also found in various tissues and fluids of the body. Phag-

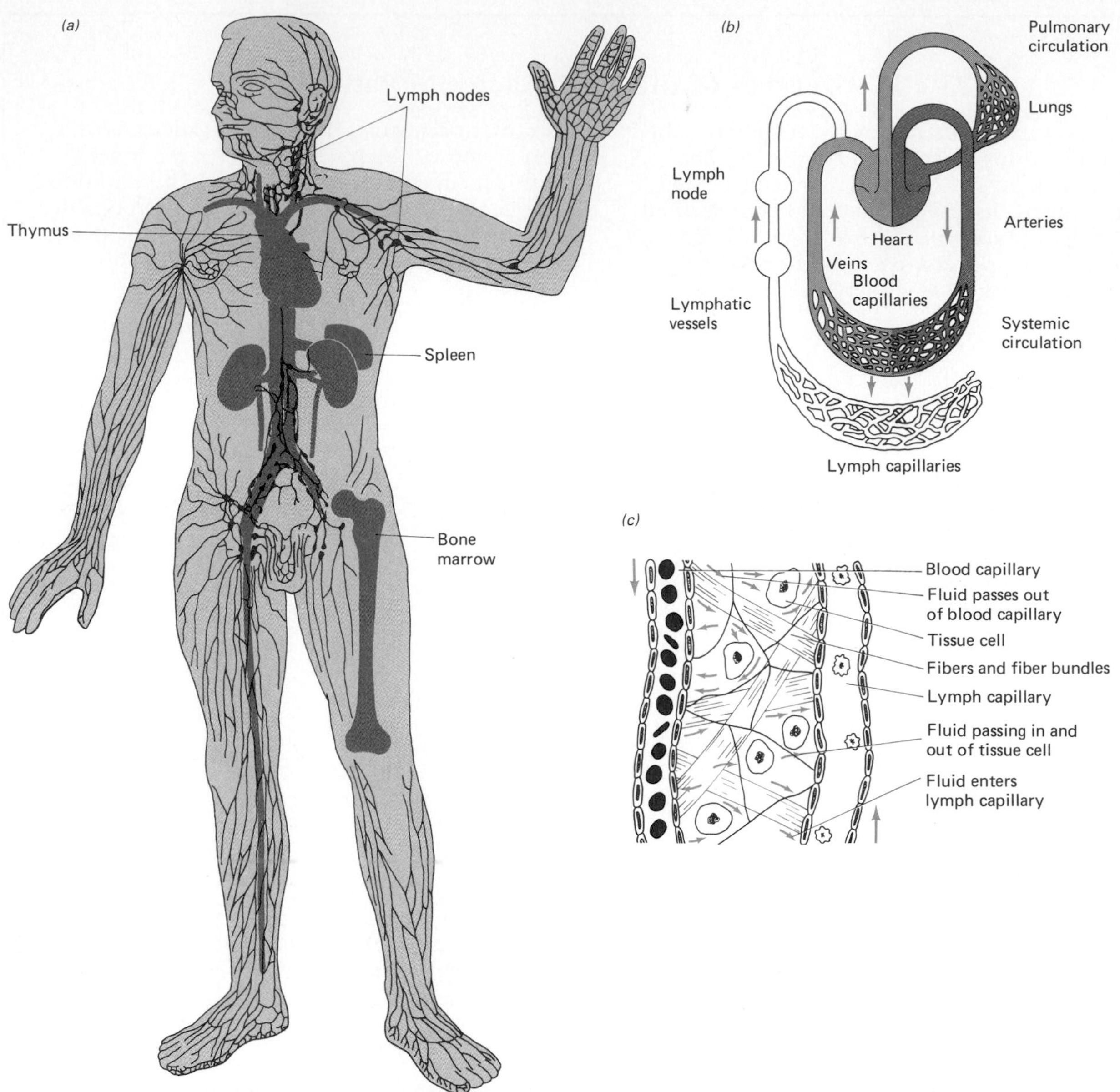

Figure 11.23 The blood and lymph systems. (a) Overall view of the major lymph systems, showing the locations of major organs. (b) Diagramatic relationship between the lymph and blood systems. Blood flows from the veins to the heart, then to the lungs where it becomes oxygenated, then through the arteries to the tissues. (c) Connection between the blood and lymph systems is shown microscopically. Both blood and lymph capillaries are closed vessels but are permeable to water and salts. Changes in capillary permeability, as a result of infection or other source of inflammation, can increase greatly the movement of fluid into the tissue spaces.

ocytes are usually actively motile by amoeboid action. Attracted to microbes by chemotactic phenomena, the phagocytes engulf the microbes and kill and digest them. One group of phagocyte, the **polymorphonuclear leukocytes** (sometimes abbreviated PMN), are small, actively motile cells containing many distinctly staining membranous granules called *lysosomes* (Figure 11.24*a*). These granules contain several bactericidal substances and enzymes, such as hydrogen peroxide, lysozyme, proteases, phosphatases, nucleases, and lipases. PMN leukocytes are short-lived cells that are found predominantly in the bloodstream and bone marrow and appear in large numbers during the acute phase of an infection. They can move rapidly, up to 40 μm/min, and are attracted chemotactically to bacteria and cellular components by im-

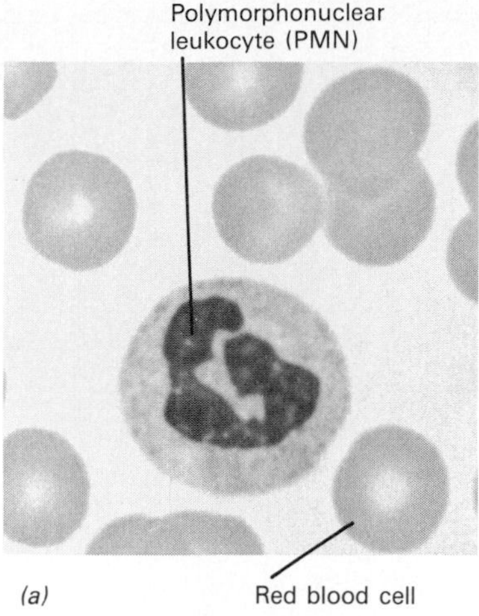

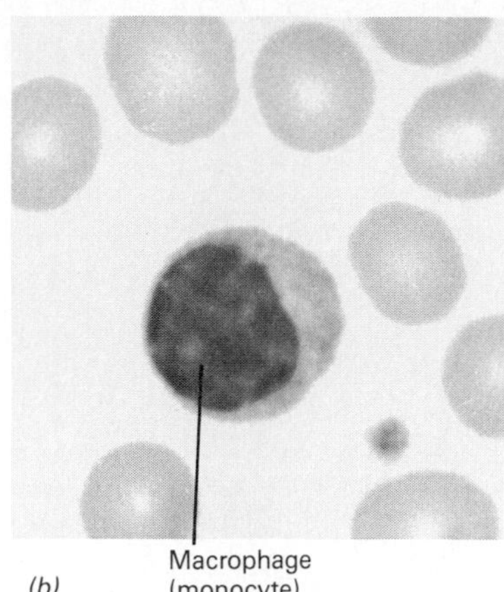

Figure 11.24 Two major phagocytic cell types. (a) Polymorphonuclear (PMN) leukocyte. (b) Macrophage (monocyte). Magnifications, 1500×.

mune mechanisms described in Chapter 12. Because they appear in the blood in large numbers during acute infection, they can serve as indicators of infection. Formation of granulocytes is severely retarded by ionizing radiation, leading to severe bacteremia from bacteria of the normal intestinal flora, such as *E. coli.*

The other group of phagocytes, the **macrophages** (also called **monocytes**), are phagocytic cells that can readily be distinguished from PMN leukocytes by their nuclear morphology (Figure 11.24*b*). They contain few lysosomes and hence do not appear granular when stained. Macrophages are of two types: *migrating cells*, which are found free in the blood and lymph, and *fixed cells*, which are found embedded in various tissues of the body and have only limited mobility. Macrophages are long-lived cells that play roles in both the acute and chronic phases of infection and in antibody formation (see Section 12.5).

The total system of phagocytic cells and organs is often called the **reticuloendothelial system** (RE system); it consists of the circulating PMN leukocytes and monocytes and phagocytes fixed to connective tissue or the endothelial layer of blood capillaries.

Phagocytosis As noted, phagocytes are attracted chemotactically to invading microbes. Some aspects of the chemotactic process involve the action of the complement system and lymphocytes, and will be discussed in Chapter 12. As a result of chemotactic attraction, large numbers of phagocytes are often seen around foci of infection, and such phagocytic accumulations are often the first indication of the presence of infection. Phagocytes work best when they can trap a microbial cell upon a surface, such as a vessel wall, a fibrin clot, or even particulate macromolecules. After adhering to the cell, an invagination of the phagocyte's plasma membrane envelopes the foreign cell, and the entire complex is pinched off and enters the cytoplasm in a phagocytic vacuole called a *phagosome*. The microbial cell is then released in a region of the phagocyte containing granules, and the latter are disrupted and release enzymes, which digest and destroy the invader (Figure 11.25).

During the process of phagocytosis, the metabolism of the granulocyte converts from aerobic pathways to anaerobic glycolysis. Glycolysis results in the formation of lactic acid and a consequent drop in pH; this lowered pH is partly responsible for the death of the microbe, and it also provides an environment in which the hydrolytic enzymes, all of which have acid pH optima, can act more effectively. The initial act of phagocytosis conditions a cell so that it is more efficient in subsequent phagocytic action—a cell that has recently phagocytized can take up bacteria about 10 times better than a cell that has not.

Oxygen-dependent phagocytic killing As we discussed in Section 9.14, various biochemical reactions can lead to the formation of toxic forms of oxygen: hydrogen peroxide (H_2O_2), superoxide anion (O_2^-), and singlet oxygen ($^1O_2$). Phagocytic cells make use of toxic forms of oxygen in killing ingested bacterial cells. Superoxide, formed by the reduction of O_2 by NADPH oxidase, reacts at the acid pH of the phagocyte to yield singlet oxygen and hydrogen peroxide (H_2O_2). The phagocytic enzyme myeloperoxidase forms hypochlorous acid (HOCl) from chloride ions and H_2O_2, and the HOCl reacts with a second molecule of H_2O_2 to yield additional singlet oxygen. The combined action of these oxygen-dependent phagocyte enzymes can form sufficient levels of toxic oxygen species to actually kill ingested bacterial cells. These reactions occur within the phagosome; this insures that constituents of the phagocytic cell itself are not damaged by the toxic oxygen produced. The action of phagocytic cells in oxygen-mediated killing is summarized in Figure 11.26.

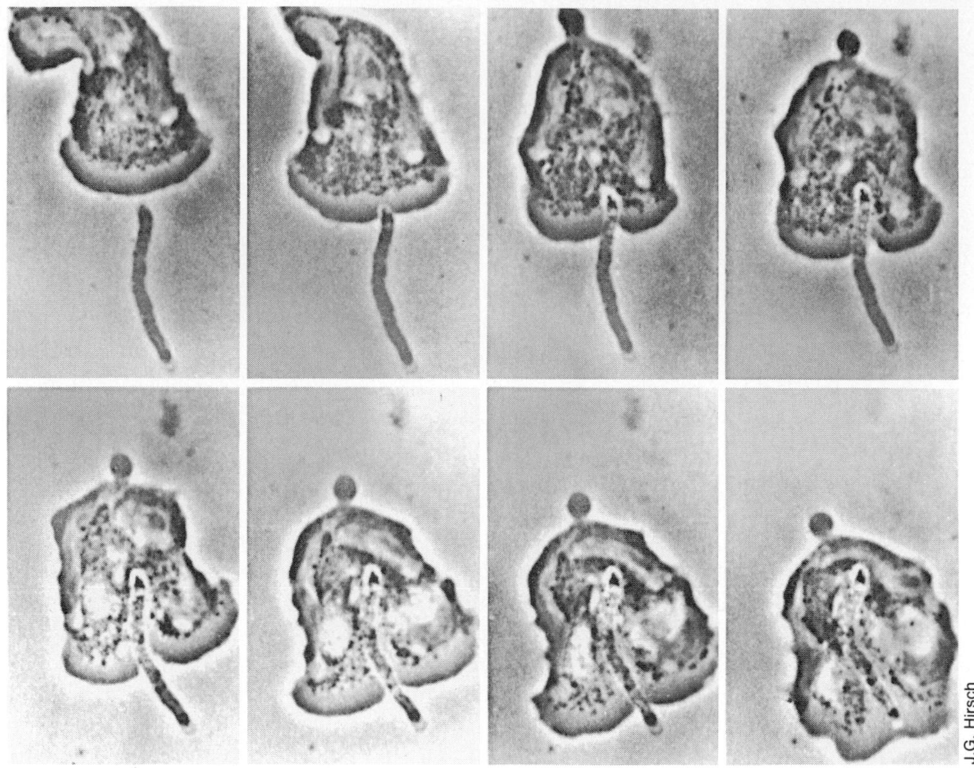

Figure 11.25 Phagocytosis: engulfment and digestion of a *Bacillus megaterium* cell by a human phagocyte, observed by phase-contrast microscopy. Magnification, 1250×.

As we discussed in Section 9.14, carotenoids quench singlet oxygen, and bacteria containing carotenoids are much more resistant to killing within phagocytes than those that do not contain carotenoids. It is of interest that *Staphylococcus aureus*, a pathogen which commonly causes infections where large numbers of phagocytes consequently develop, contains yellow carotenoids (hence the name *aureus*), and this bacterium is quite resistant to phagocytic killing.

Leukocidins Some pathogens produce proteins called **leukocidins** which destroy phagocytes. In such cases the pathogen is not killed when ingested but instead kills the phagocyte and is released alive. Pathogenic streptococci and staphylococci are the major leukocidin producers. Destroyed phagocytes make up much of the material of pus, and organisms that produce leukocidins are therefore usually **pyogenic** (pus-forming) and bring about characteristic abscesses. One leukocidin produced by certain strains

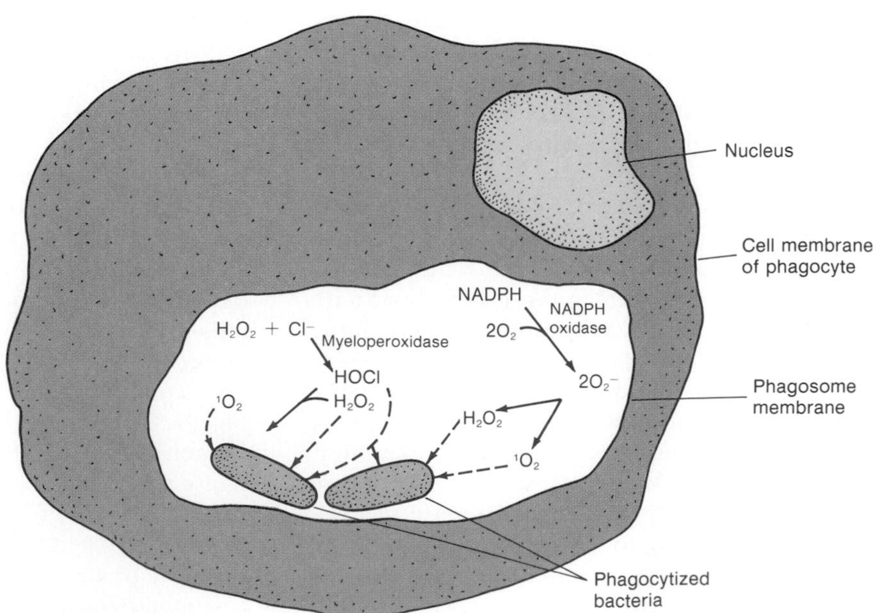

Figure 11.26 Action of phagocyte enzymes in generating toxic oxygen species, hydrogen peroxide (H_2O_2), hypochlorous acid (HOCl), and singlet oxygen ($^1O_2$).

of *Staphylococcus aureus* binds to the cytoplasmic membrane of phagocytic cells and increases permeability. This protein also stimulates membrane fusions between the cytoplasmic membrane and the membranes which surround lytic granules; such fusions effectively transport the contents of the granules outside the phagocytic cell and away from the invading pathogen.

Capsules and phagocytosis An important mechanism for defense against phagocytosis is the bacterial capsule (see Section 3.10). Capsulated bacteria are frequently highly resistant to phagocytosis, apparently because the capsule prevents in some way the adherence of the phagocyte to the bacterial cell. The clearest case of the importance of a capsule in permitting invasion is that of *Streptococcus pneumoniae.* If only a few cells of a capsulated strain of this species are injected into a mouse, an infection is initiated that leads to death within a few days. On the other hand, noncapsulated mutants of capsulated strains are completely avirulent, and injection of even large numbers of bacteria usually causes no disease. If a noncapsulated strain is transformed genetically with DNA from a capsulated strain (see Section 7.2), both capsulation and virulence are restored at the same time. Furthermore, enzymatic removal of the capsule renders the organism noninvasive. Surface components other than capsules can also inhibit phagocytosis. Pathogenic streptococci produce on both the cell surface and fimbriae a specific protein called the *M protein*, which apparently alters the surface properties of the cell in such a way that the phagocyte cannot act.

11.15 Growth of the Pathogen Intracellularly

One group of organisms, the **intracellular parasites**, are readily phagocytized but are not killed, nor do they kill the phagocyte. Instead, they can remain alive for long periods of time and even reproduce within the phagocyte. A number of major diseases are caused by intracellular pathogens (Table 11.6). In most situations the pathogen enters the host cell in the phagocytic vacuole (phagosome), and the pathogen grows within this structure until the phagosome bursts. Intracellular parasites are a diverse group. Some, such as *Mycobacterium tuberculosis, Salmonella typhi* (cause of typhoid fever), and *Brucella* species are facultative parasites, and can live either intracellularly or extracellularly. In acute infections they multiply in the extracellular body fluids, but in chronic conditions they may live only intracellularly. Intracellular seclusion of certain pathogens may be one of the means by which chronic disease "carriers" remain infectious despite being clinically free of symptoms. When growing intracellularly the organism is protected from immune mechanisms of the host and is less susceptible to drug therapy. Some important intracellular parasites are unable to grow outside of living cells and are called obligate intracellular parasites. Included in this category are the viruses, the chlamydias, the rickettsias, and some protozoa, such as the one that causes malaria (see Sections 15.10, 19.22, and 19.23).

The intracellular environment differs markedly from the extracellular environment in physical and chemical characteristics (Table 11.7), being much

Table 11.6 Major diseases caused by pathogens capable of intracellular growth

Disease	Organism
Rocky Mountain Spotted Fever	*Rickettsia rickettsii*
Typhus	*Rickettsia prowazekii*
Chlamydia venereal syndromes	*Chlamydia trachomitis*
Trachoma	*Chlamydia trachomitis*
Brucellosis	*Brucella abortus*
Tuberculosis	*Mycobacterium tuberculosis*
Leprosy	*Mycobacterium leprae*
Listeriosis	*Listeria monocytogenes*
Malaria	*Plasmodium vivax*
Tularemia	*Francisella tularensis*

Table 11.7 Major chemical differences between the intracellular and extracellular mammalian environment

Character	Intracellular	Extracellular (Plasma)
Ca^{2+}	Very low	0.0025 M
Mg^{2+}	0.02 M	0.0015 M
Na^+	0.01 M	0.15 M
K^+	0.15 M	0.005 M
Cl^-	0.003 M	0.1 M
HPO_4^{2-}	0.05 M	0.002 M
HCO_3^-	0.01 M	0.03 M
Overall ionic strength	High	About 75% of intracellular ionic strength
Organic nutrients: cofactors, acids, and high-energy compounds	High levels	Relatively low levels

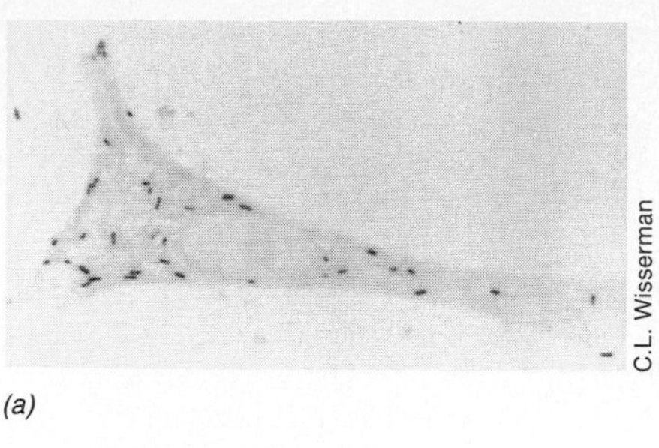

(a)

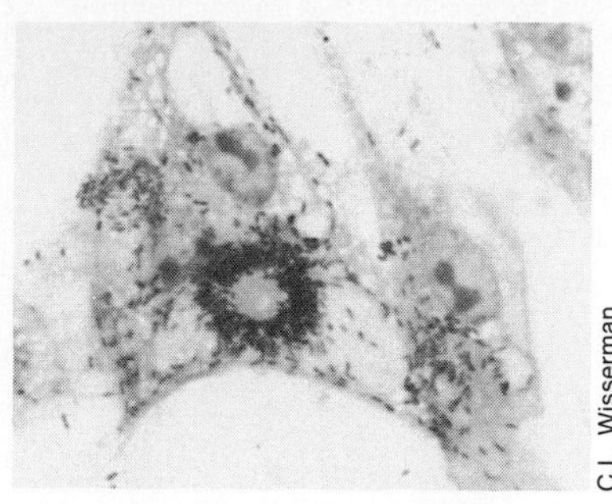

(b)

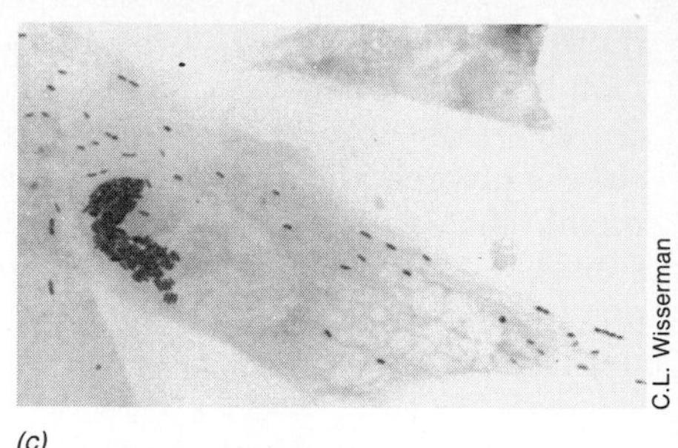

(c)

Figure 11.27 Intracellular growth of *Rickettsia rickettsii* in chicken embryo cells. The preparations were stained and examined by the light microscope. (a) Early stage of growth, 28 hours after infection. Note the generally dispersed cytoplasmic distribution. (b) Extensive intracytoplasmic colonies after 120 hours incubation. The dark mass represents extensive development of rickettsiae in a ring or doughnut radially arranged around a cytoplasmic vacuole. (c) Intranuclear growth. Note the compact intranuclear mass and the dispersed cytoplasmic distribution. 45 hours after incubation. Not all cells showed intranuclear growth.

richer in organic nutrients and cofactors; the advantages to a parasite of living intracellularly are thus partly nutritional. In addition to the normal microbial nutrients, the intracellular environment also provides ATP, which is completely absent extracellularly; some of the obligate intracellular parasites are unable to generate their own ATP from organic compounds and require preformed ATP provided by the host (see Section 19.23).

Some intracellular pathogens (for example, *Listeria monocytogenes, Rickettsia rickettsii*) grow within the cell nucleus as well as the cytoplasm. The intracellular growth of *Rickettsia rickettsii* has been studied in some detail. The organism is able to penetrate the plasma membrane readily and initiate growth (Figure 11.27*a* and *b*), but progeny are not restricted to the intracellular environment, and readily pass out of the plasma membrane without causing damage to the host cell. Thus, despite sustained growth of the intracellular parasite, massive accumulation within the host cell does not always occur. The cells released from the cytoplasm readily infect adjacent cells, thus resulting in a rapidly spreading infection. However, if growth occurs in the nucleus, the progeny do not leave, so that sustained growth results in an extensive bacterial colonization of the nucleus (Figure 11.27*c*). Thus the rickettsial traffic is bidirectional across the plasma membrane and dominantly monodirectional across the nuclear membrane. One significant aspect of these observations is that release of rickettsial cells from the host can occur without apparent damage to the host cell.

11.16 Inflammation

The tissues of the animal react to infection and to mechanical injury by an **inflammatory response**, the characteristic symptoms of which are redness, swelling, heat, and pain. The initial effect of the foreign stimulus is to cause local dilation of blood vessels and an increase in capillary permeability. This results in an increase flow of blood and passage of fluid out of the circulatory system into the tissues, causing swelling (edema) (Figure 11.28). Phagocytes also pass through the capillary walls to the inflamed area. Initially granulocytes appear, followed later by macrophages. Within the inflamed area, a fibrin clot is usually formed, which in many instances will localize the invading microbe. Pathogens that produce fibrinolytic enzymes may be able to escape and continue to invade the body.

Complement is a series of serum enzymes which participate in several different immunological reactions. Complement protein C-5 (see Section 12.12) has been identified as a major trigger of the inflammatory process. Protein C-5 induces contraction of smooth muscle, increases vascular permeability, and catalyzes the release of inflammatory substances. Protein C-5 also induces the release of tissue degrading enzymes which play a major role in the tissue destruction that usually accompanies inflammation.

Inflammation is one of the most important and ubiquitous aspects of host defense against invading microorganisms and is present in a small way virtually continuously. However, inflammation is also an important aspect of microbial pathogenesis since the inflammatory response elicited by an invading microorganism can result in considerable host damage.

Fever The healthy human body maintains a surprisingly constant temperature. Over an average 24 hour period, body temperature varies only over the narrow range of 1–1.5°C. However, individuals vary in their "normal" temperatures, and although 37°C is assumed, normal temperature in some individuals may be as low as 36°C or as high as 38°C. Also, body

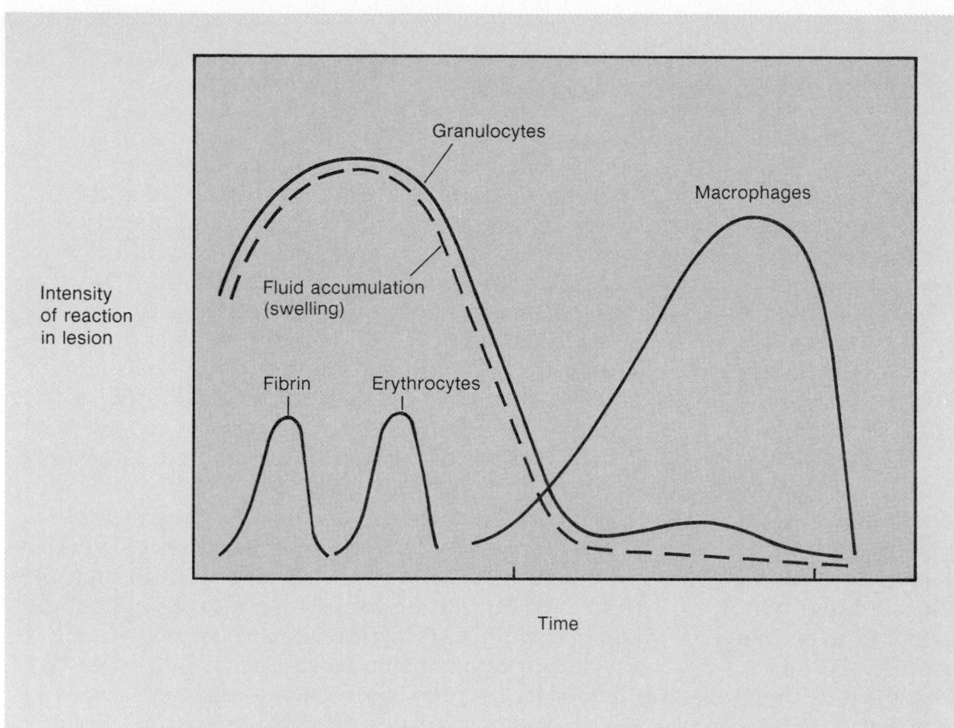

Figure 11.28 Sequence of events in a typical inflammatory response.

temperature varies with amount of physical activity, and can be as much as 2°C below normal in sleep and as much as 4°C above normal during strenuous exercise.

Fever is defined as an abnormal increase in body temperature. Although fever can be caused by noninfectious disease, most fevers are caused by infection. Diseases in which fever is elicited are called "febrile" and a substance or condition which induces fever is called "pyrogenic" (pyro is a combining form meaning "fire"). (Note that the word "pyrogenic" should not be confused with the word "pyogenic," meaning pus-inducing.)

At least one reason why fever occurs during many infectious diseases is that certain products of pathogenic organisms are pyrogenic. The most well studied pyrogenic agents are the endotoxins of Gram-negative bacteria, which are often exceedingly potent in eliciting fever (see Section 11.12). However, many organisms which do not produce pyrogens are able upon infection to cause fever, and in these cases it is thought that the fever-inducing principle is produced by the body itself as a result of its reaction to the infection. These pyrogens, called *endogenous pyrogens*, are released by phagocytic cells during the phagocytosis process. Other body cells which produce pyrogens include neutrophils, eosinophils, and monocytes. One endogenous pyrogen is a protein which affects the thermoregulatory control centers in the hypothalamus and results in an increase of the normal 37°C (98.6°F) human body temperature. Slight temperature increases benefit the host by accelerating phagocytic and antibody responses, while strong fevers of 40°C (104°F) or greater may benefit the pathogen if host tissues are further damaged. Pathogens vary in the degree to which they induce inflammation; in some cases a major reaction by the host to only a minor microbial invasion may occur, resulting in serious damage to the host.

Three kinds of characteristic fever patterns have been recognized in infectious disease. (1) *Continuous fever* is that condition in which the body temperature remains elevated over a whole 24-hour period and the total range of variation in temperature is less than 1°C. Continuous fever is seen in *typhoid fever* and *typhus fever*. (2) An *intermittent fever* is one in which the body temperature is abnormal over the whole of a 24-hour period, but the daily range shows swings greater than 1°C. This occurs in some *pyogenic infections* and *tuberculosis*. (3) A *remittent fever* is one in which the temperature is normal for part of the day, and then rises above normal. Most infectious diseases elicit remittent fever, although the condition is typical of *malaria*. *Relapsing fever* is an intermittent fever in which the temperature remains normal for a long period of time, followed by a new attack of fever. This is characteristic of an extensive but incomplete recovery from an infectious disease, the remittent fever arising when the disease flairs up again.

Pathogenic Fungi

The pathogens that we have discussed in this chapter have been exclusively bacterial. However, some human diseases are also caused by fungi. Most fungi are harmless to humans, growing in nature on nonliving and dead organic materials and only a few fungi are pathogenic to animals. It has been estimated that only about 50 species cause human disease, and the incidence of fungal infections is relatively low. Infectious diseases caused by fungi are called *mycoses*, and three kinds of mycoses have been recognized: *superficial*, involving only the skin; *subcutaneous*, when the protective barrier of the skin is broken and fungal growth occurs beneath; and *systemic*, which involves fungal growth in internal organs of the body. Many mycoses are due to infections caused by opportunistic fungi which are ordinarily harmless and are able to initiate serious infection only in compromised hosts. In the United States, the most widespread fungal infections are *histoplasmosis*, caused by *Histoplasma capsulatum*, and *coccidioidomycosis* ("San Joaquin Valley fever"), caused by *Coccidioides immitis*. These are both respiratory diseases in which the opportunisitic fungus, normally living in the soil, becomes inoculated and grows in the lungs. Histoplasmosis is primarily a disease of rural areas in midwestern United States, especially in the Ohio and Mississippi River valleys. Most cases are mild and may be mistaken for more common respiratory infections. San Joaquin Valley fever is generally restricted to desert regions of southwestern United States. The fungus lives in desert soils and the spores are disseminated on dry windblown particles which are inhaled. In some areas in the southwestern United States as many as 80 percent of the inhabitants may be infected, although suffering no apparent ill effects.

Effective chemotherapy against fungal infections is generally lacking. Most antibiotics which inhibit fungi also inhibit other eucaryotic organisms, including the human host. One antibiotic, amphotericin B, is widely used to treat systemic fungal infections of humans, but harmful side effects of its use may occur. For fungal infections of the skin, the antibiotic *griseofulvin* has been used. This antibiotic is adminstered orally and after entering the blood stream passes to the skin where it can bring about an inhibition of fungal growth.

Some pathogenic fungi and the diseases they cause

Disease	Causal organism	Main disease foci
Superficial mycoses (dermatomycoses)		
Ringworm	*Microsporum*	Scalp of children
Favus	*Trichophyton*	Scalp
Athlete's foot	*Epidermophyton*	Between toes, skin
Jock itch		Genital region
Subcutaneous mycoses		
Sporotrichosis	*Sporothrix schenckii*	Arms, hands
Chromoblastomycosis	Several genera	Legs, feet
Systemic mycoses		
Cryptococcosis	*Cryptococcus neoformans*	Lungs, meninges
Coccidioidomycosis	*Coccidioides immitis*	Lungs
Histoplasmosis	*Histoplasma capsulatum*	Lungs
Blastomycosis	*Blastomyces dermatitidis*	Lungs, skin
Candidiasis (opportunistic)	*Candida albicans*	Oral cavity, intestinal tract

11.17 Chemotherapy

Chemical agents able to cure infectious diseases are generally called **chemotherapeutic agents**. The hallmark of a chemotherapeutic agent is *selective toxicity*, that is, toxicity to the pathogen but not to the host. We have discussed antimicrobial action in some detail in Chapters 5 and 9, and industrial aspects in Chapter 10. Here we discuss the principles involved in the use of antimicrobial agents in the therapy of infectious disease. The action of antimicrobial agents in the test tube (*in vitro*) may be quite different from the action of such agents in the animal (*in vivo*). An agent highly effective in vitro may be

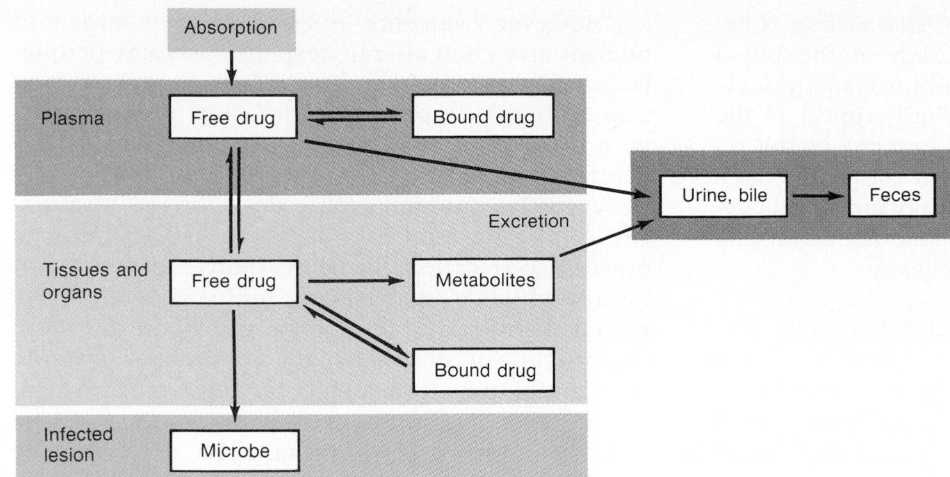

Figure 11.29 Movement of drug through various compartments of the body.

completely ineffective in vivo, for the animal body is not a neutral environment for chemical agents, and many agents are modified so that they are no longer biologically active.

Drug distribution and metabolism in the body When a drug is administered it becomes distributed through various compartments of the body, as outlined in Figure 11.29. After initial absorption into the blood, a major portion of the drug may be found bound to plasma proteins. For some agents, as much as 90 percent of the drug may be bound to blood proteins. This binding is reversible, but bound drug is not microbially active. Because of protein binding, the minimum inhibitory concentration of an antimicrobial agent when tested in plasma or serum may be three- to four-fold higher than when tested in culture medium. However, the bound drug is not inactivated and can be looked upon as a reservoir, to be released when the concentration of the free drug is lowered. In the tissues, the drug may be metabolized, and the metabolites are generally less active than the administered drug. The most common organ in which drug metabolism occurs is the liver. From the viewpoint of the body, the drug is a foreign agent, and metabolic systems for natural detoxification function to detoxify many drugs. Detoxification enzymes are generally less well developed in infants, so that drugs may be much more toxic (per unit body weight) in infants than in older individuals.

Excretion of the drug and its metabolites generally occurs rapidly. Two main routes of excretion exist: renal (kidney) excretion to the urine, and hepatic (liver) excretion to the bile, which is excreted with the feces. Minor routes of excretion are through sweat, saliva, and milk (lactating animals).

The characteristic pattern of drug absorption and excretion is illustrated by data for penicillin in Figure 11.30. As seen, after injection there is a rapid increase in concentration of antibiotic in the blood, followed

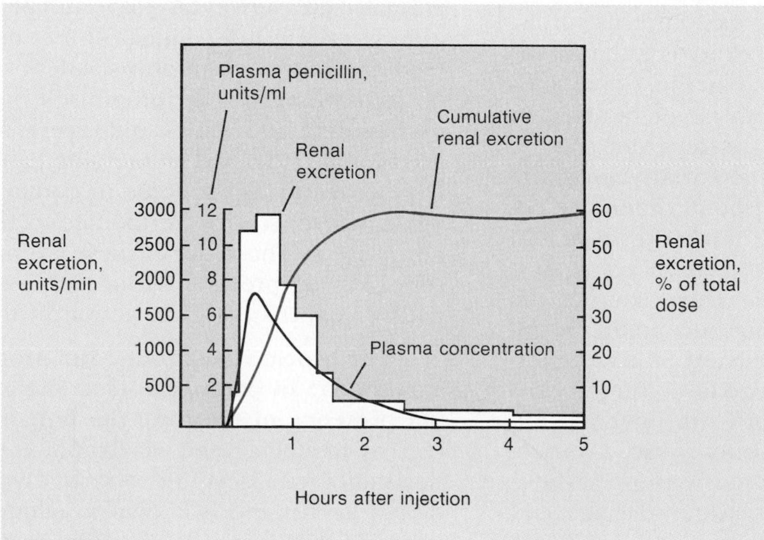

Figure 11.30 Kinetics of penicillin absorption and excretion in a normal adult human. One intramuscular injection of 300,000 units of aqueous penicillin G was given at zero time. Samples of blood and urine were assayed periodically for antibiotic concentration. The results given are average values found in humans with good renal function. Note that the peak plasma level is reached within 15 to 30 minutes, and that about 60 percent of the dose was excreted in the urine within 5 hours.

by a gradual fall in concentration as the drug is excreted. The antibiotic concentration in the blood reaches a peak within 15 to 30 minutes, and quickly thereafter large amounts of antibiotic appear in the urine. Within 5 hours about 60 percent of the injected dose has been eliminated in the urine. Such data indicate that if bacteriostatic or bactericidal concentrations of an antibiotic are to be maintained in the body, *periodic doses* must be given.

Drug toxicity Virtually all drugs have some toxicity (that is, cause some harm) to the host, and a knowledge of host toxicity is vital for the intelligent use of chemotherapeutic agents. Two broad classes of toxicity are recognized: acute and chronic. Acute toxicity is expressed by pathological manifestations observed within a few hours after administration of a single dose. Generally, acute toxicity occurs as a result of drug overdosage.

Chronic toxicity is expressed by gradual changes, which take place during continuous administration of a drug over a long period of time. A variety of toxic manifestations require considerable time to develop, and if the course of treatment with the agent must be extended, chronic toxicity must be taken into consideration. For example, extended treatment with the antibiotic streptomycin in humans is thought to cause inner ear problems which can result in deafness. However, for treatment of infectious disease, very long courses of treatment are the exception rather than the rule, although for tuberculosis and certain other chronic infectious diseases, therapy may continue over many months or years. Also, some antimicrobial agents are used continuously as prophylactic agents (to prevent future attacks of the disease) in certain high-risk patients, and under these conditions, chronic toxicity becomes an important consideration. However, some manifestations of chronic toxicity may be exhibited in even as short a time as 1 to 2 weeks.

Although antibiotics are usually selective in their action, their toxicity for animals and humans varies, and a knowledge of antibiotic toxicity is essential for the wise use of antibiotics in medicine. Some antibiotics are so toxic that they can never be used in humans, others have a limited toxicity and may be used with care, while others are essentially nontoxic. This is in part related to the degree of difference between the cells of the invading microorganism and the cells of the host. Antibiotics which act against specific components of bacteria (for example the unique cell wall), are usually not very toxic for eucaryotic host cells. On the other hand, it is more difficult to selectively injure protozoan or fungal cells in the human body because both the infecting microbial cells and host cells are eucaryotic. A variety of toxic reactions can be observed within the host, including intestinal disturbances, kidney damage, and deafness.

However, even antibiotics which are nontoxic to humans may elicit allergic responses in some people. For example, 5 to 10 percent of the human population is allergic to penicillin and must not be treated with this antibiotic. Thus, a wide variety of factors determine the usefulness of an antibiotic. The ratio between the minimum toxic dose for the host and the minimum effective dose (to kill the microorganism) is called the **therapeutic index**. Antibiotics which are particularly useful generally have a high therapeutic index; that is, they are far more toxic to the microbe than to the host. A good example of an antibiotic with a high therapeutic index is penicillin. Some antibiotics are less useful because the doses required to harm the microbe also harm the host; they have a low therapeutic index. Such antibiotics may still have a useful role, however, if they are the only agents which act against a particular pathogen.

Testing of antimicrobial agents in vivo Once the acute and chronic toxicity of a new antimicrobial agent has been determined, in vivo tests in experimental animals are initiated. Animal models can be developed for many infectious diseases, and such models are then used to evaluate chemotherapeutic agents. For economy and ease of operation, the mouse is the preferred experimental animal. Groups of mice can be infected with a pathogen, and the effect of drug administration observed. The in vivo testing of antibiotics in experimental animals is an important procedure, and much can be learned about drug action from careful observation of such studies. The effect of the drug can be measured either by simply counting the number of infected animals that do not die after various periods of time, or by actually measuring the growth of the pathogen in the animal, and determining if the drug inhibits growth. The precise experimental design will depend upon the pathogen and upon the aim of the experiment. It should be emphasized, however, that animal models do not closely mimic most human infections, since animal models make use of highly virulent strains, and large doses of pathogens are often used, so that the course of the infection is quick and severe. Also, the animals used are of undefined immunological background, but have probably never come in contact with the pathogen before, so that antibodies to the pathogen are not present. The value of these animal models is that they provide precise, reproducible systems for evaluating quantitatively the action of the antibiotic.

Antibiotic resistance Mutations to drug resistance occur in pathogens (see Section 5.16) and in the presence of the drug the mutant form has a selective advantage and may replace the parent type. Resistance can develop to virtually all chemotherapeutic agents and is known to occur in vivo as well as in vitro (Figure 11.31). The resistant strain may be just as virulent as the parent, may not be controllable by other chemotherapeutic agents, and may be passed on to other individuals.

It is now well established that the uncontrolled use of antibiotics is leading to the rapid development of antibiotic resistance in disease-causing microorganisms. Parallel to the history of discovery and clinical use of the many known antibiotics has been the emergence of bacteria which resist their action. This is, in fact, a major reason why we continually seek new antibiotics and attempt to modify existing ones through chemical alterations (see Section 10.6). There are numerous examples of the association between the use of antibiotics and the development of resistance. The example in Figure 11.32*a* shows a correlation between the number of tons of antibiotics used and the percentage of bacteria resistant to each antibiotic which were isolated from patients with diarrheal disease. In general, the more an antibiotic was used, the more bacteria were resistant to it.

There are many examples of diseases in which the drug recommended for treatment has changed due to the increased resistance of the microbe causing the disease. A classic example is the development of resistance to penicillin in *Neisseria gonorrhoeae*, the bacterium which causes gonorrhea (Figure 11.32*b*). The explanation for these observations is probably an ecological one. To grow in the presence of an antibiotic an organism must develop resistance to it. Thus, resistant microorganisms are *selected for* by the presence of the antibiotic in the environment. Surveys have shown that antibiotics are used far more often than is necessary. For instance, during the 1960s antibiotic production in the United States more than tripled although the population grew only 11 percent. The increased use of antibiotics was in part a result of poor prescription practices by physicians, who prescribed antibiotics when unnecessary. Antibiotics are also used extensively in agricultural practices both as growth-promoting substances in animal feeds and as prophylactics (to prevent the occurrence of disease rather than to treat an existing one). Several recent food poisoning outbreaks have been blamed on the use of antibiotics in animal feeds (see Box, Chapter 17). By overloading the various environments with antibiotics, we create more stress for microorganisms in general. Rapid development of drug resistance may be the result. Resistance can be minimized if drugs are used only for serious diseases and are given in sufficiently high doses so that the population level is reduced before mutants have a chance to appear. Resistance can also be minimized by combining two unrelated chemotherapeutic agents, since it is likely that a mutant resistant to one will still be sensitive to the other. With the increasing prevalence of resistance transfer factors in pathogenic bacteria (see Section 7.6), however, multiple antibiotic therapy is proving less attractive as a clinically useful strategem.

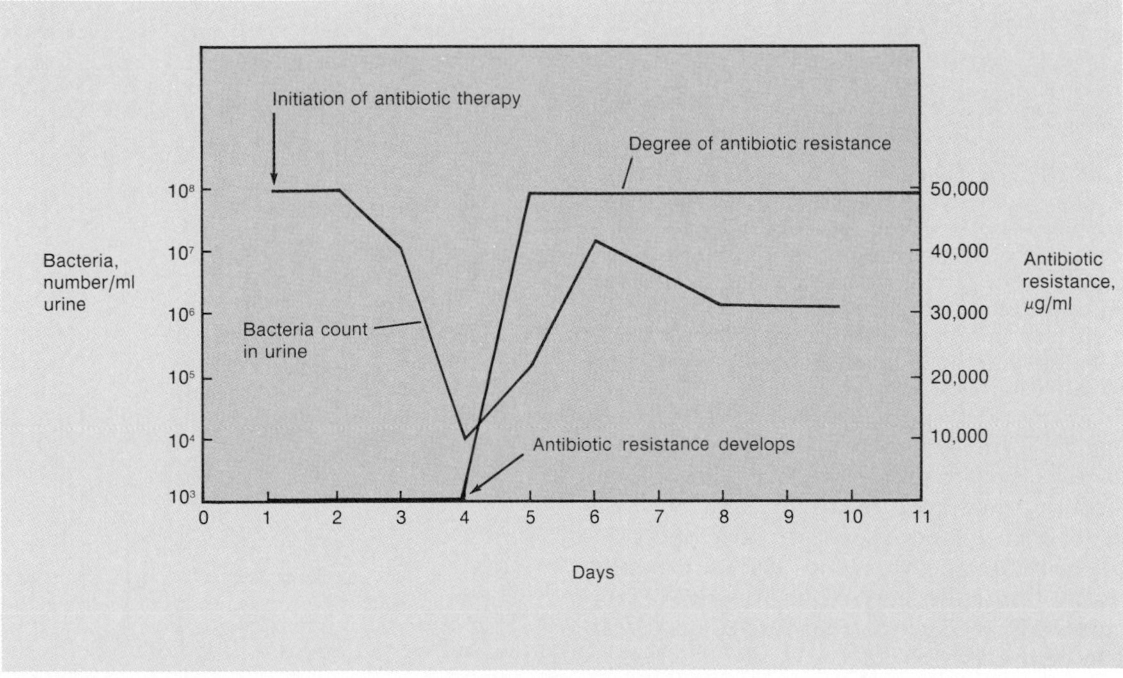

Figure 11.31 Development of an antibiotic-resistant mutant in a patient undergoing antibiotic therapy. The patient, suffering from chronic pyelonephritis caused by a Gram-negative bacterium, was treated with streptomycin by intramuscular injection. Antibiotic sensitivity is expressed as the minimum inhibitory concentration of the antibiotic in micrograms per milliliter.

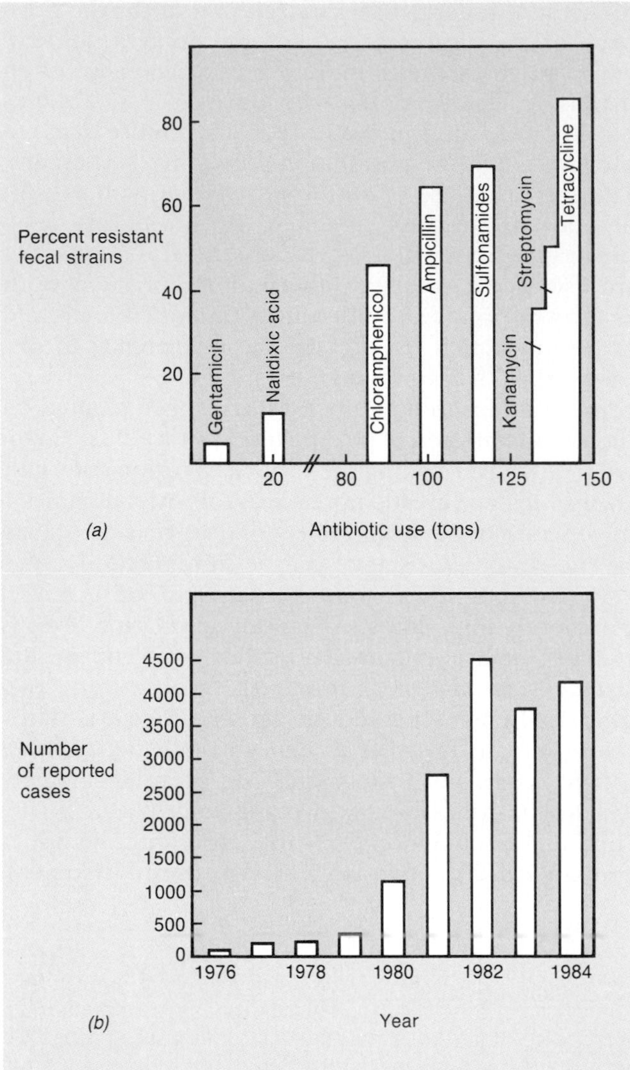

Figure 11.32 The emergence of antibiotic-resistant bacteria. (a) Relationship between antibiotic use and the percentage of bacteria isolated from diarrheal patients resistant to the antibiotic. Note that those antibiotics which have been used in the largest amounts, as indicated by the amount of antibiotic produced commercially, are those for which antibiotic-resistant strains are most frequent. (b) Increase in the incidence of bacteria causing gonorrhea which are resistant to penicillin.

Some organisms, such as the streptococci, do not seem to develop drug resistance readily in vivo, whereas the staphylococci are notorious for developing resistance to chemotherapeutic agents. The rise of strains resistant to antibiotics because of the presence of conjugative plasmids (resistance transfer factors) is illustrated in Figure 11.33.

Clinical uses of antimicrobial agents Chemotherapeutic agents vary widely in the range of organisms they attack. A summary of the actions of the most important antimicrobial agents was given in Figure 9.41. This figure included information for both antibiotics and synthetic chemicals. Note that only a few antimicrobial agents are effective against intracellular parasites such as the chlamydia and rickettsiae. An agent which is to be effective against these intracellular parasites must penetrate host cells, and host cells are impermeable to many of the common antimicrobial agents.

Although there are presently available a wide variety of relatively nontoxic antimicrobial agents, only a restricted number actually are in use in chemotherapy. Many agents that are effective against organisms in vitro have no effect in an infected host. There are several possible explanations for this: (1) the drug might be destroyed, inactivated, bound to body proteins, or too rapidly excreted; (2) the drug might remain at the injection site or might not penetrate as far as the site of infection; or (3) the parasite in vivo might be different from the parasite in vitro, perhaps showing different physiological properties or possibly growing intracellularly, where it is protected from the action of the drug.

In addition to helping to cure diseases actively in progress, drugs may be given to prevent future infections in individuals who are unusually susceptible, a procedure called **chemoprophylaxis**. For example, penicillin is used to prevent streptococcal sore throats in rheumatic fever patients, since these streptococcal infections often lead to a recurrence of rheumatic fever symptoms.

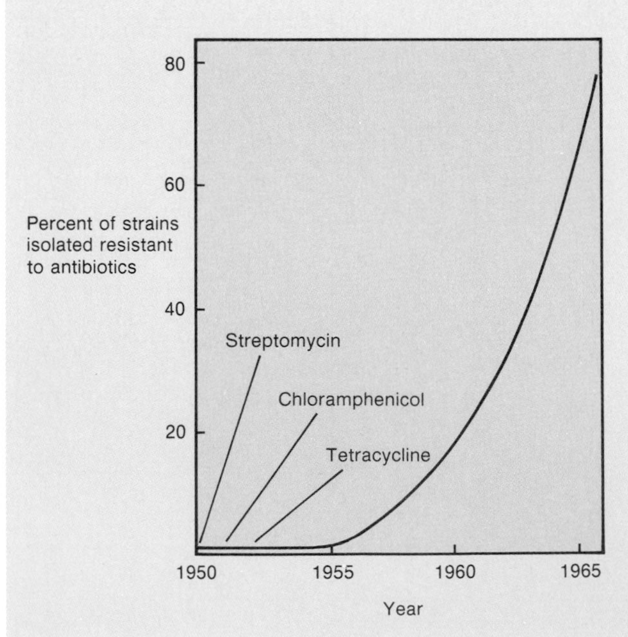

Figure 11.33 Rise in antibiotic-resistant *Shigella* in Japan following introduction of antibiotic therapy. Arrows indicate the years when the three antibiotics were introduced. SM, streptomycin; CM, chloramphenicol; TC, tetracycline.

Study Questions

1. Distinguish between a parasite and a pathogen. Between infection and disease.
2. Which parts of the human body are normally heavily colonized with microorganisms? What parts are normally devoid of microorganisms?
3. Distinguish between the resident and transient microflora of a body habitat. How could you distinguish resident and transient microorganisms experimentally?
4. Why is a hair follicle a likely site of microbial entry into the body through the skin? What other ways can you imagine that bacteria might enter the body through the skin?
5. Write a brief discussion of the steps involved in the development of dental plaque.
6. Obligately anaerobic bacteria are very common in the large intestine, yet they are only able to grow there if facultatively aerobic bacteria are also present. Explain.
7. Certain antibiotics, even antibiotics whose mode of action is bacteriostatic rather than bacteriocidal, sterilize the intestinal tract. Explain how a bacteriostatic antibiotic could bring about this result.
8. Design an experiment to show the likely route of infection of a urinary tract pathogen.
9. What microbial structures have been implicated as important adherence factors for pathogenic microorganisms?
10. Describe the ways in which enteropathogenic strains of *E. coli* differ from normal strains of *E. coli*; include a discussion of structural and ecological variables.
11. What do the substances hyaluronidase, collagenase, streptokinase, and coagulase have in common?
12. Define and contrast: exotoxin, enterotoxin, endotoxin. Give two examples of each and the name of the organism producing each.
13. Although mutants incapable of producing exotoxins are relatively easy to isolate, those incapable of producing endotoxin are much more difficult to isolate. From what you know of the structure and function of these two classes of toxins, explain why this difference might be.
14. Describe the manner by which fibrin formed by the host in response to a bacterial infection can reduce the severity of that infection. How do bacteria counteract host resistance due to fibrin clotting?
15. Compare and contrast the cell composition and protein content of whole blood, serum, and plasma. How are plasma and serum prepared from whole blood?
16. Compare and contrast macrophages and granulocytes. What is the principal roles of each in the body?
17. Although most bacteria ingested into phagocytic cells are killed, some types actually replicate. Explain. Give three examples of bacteria capable of phagocytic growth.
18. A number of new antibiotics which show promise as clinically useful agents based on *in vitro* tests turn out to be worthless in treating the same bacterium *in vivo*. List at least three reasons why this might be.
19. Explain why the following statement is erroneous: Since the discovery of antibiotics it is no longer necessary to immunize persons for any bacterial disease because antibiotics can control any outbreaks of the disease.
20. What significance for our understanding of infectious diseases do you see in the fact that large numbers of bacteria and viruses may occur in a healthy human being?
21. For each of the following exotoxins, describe (a) the organism producing; (b) its mode of action on the host; (c) its role in pathogenicity; (d) how its effects can be counteracted: 1) diphtheria toxin; 2) tetanus toxin; 3) botulinus toxin; 4) cholera toxin.

Supplementary Readings

Beachey, E. H. (ed.) 1980. *Bacterial Adherence.* Chapman and Hall, Ltd. London. A collection of chapters written by experts on adherence of bacteria to various animal surfaces.

Davis, B. D., R. Dulbecco, H. N. Eisen, and **H. S. Ginsberg.** 1988. *Microbiology, 4th ed.* Harper & Row, Hagerstown, Md. A brief treatment of host-parasite relationships can be found in part 4 of this standard medical school textbook. A good reference source for information on diseases caused by specific groups of organisms.

Drasar, B. S., and **P. A. Barrow.** 1985. *Intestinal Microbiology.* American Society for Microbiology, Washington, D.C. A short monograph on the intestinal microflora of different regions of the human gastrointestinal tract.

Lennette, E. H., A. Balows, W. J. Hausley, Jr., and **H. J. Sodomy.** 1985. *Manual of Clinical Microbiology, 4th ed.* American Society for Microbiology, Washington, D.C. The standard reference work of clinical microbiological procedures.

Linton, A. H. (ed.) 1982. *Microbes, Man and Animals: The natural history of microbial interactions.* John Wiley & Sons, New York. An excellent collection of essays on interactions between host and parasite, illustrated with a number of specific examples.

12 Immunology and Immunity

Higher animals possess a highly sophisticated mechanism, the **immune response**, for developing resistance to specific microorganisms or viruses. The immunological response occurs because the body has a general system for the neutralization of foreign macromolecules and microbial cells. Foreign substances which elicit the immune response are called **antigens**. As a result of antigen stimulation, the immune system produces specific proteins called **antibodies** or specific cells called **activated T cells**. Since an invading microorganism contains a variety of macromolecules foreign to the host, antibodies and activated T cells are generated against it and are able to recognize the foreign material and bring about its destruction.

Three major features characterize the immune system: specificity, memory, and tolerance. (1) The **specificity** of the antigen-antibody or antigen-T cell interaction is unlike any of the host-resistance mechanisms described in Chapter 11. Phagocytosis, inflammation, and the other host resistance mechanisms discussed in Chapter 11 develop against virtually any invading microorganism, even those the host has never had any encounter with before, whereas in the immune response, a specific interaction must occur for *each* new invader. (2) Once the immune system produces a specific type of antibody or activated T cell, it is capable of producing more of the same antibody or T cell more rapidly and in larger amounts. This capacity for **memory** is of major importance in resistance of the host to subsequent reinfection or in the protection to the host provided by vaccination. (3) **Tolerance** exists because macromolecules on the surfaces of body cells are also potentially antigenic and would be materially altered if antibodies and activated T cells to these body cells were produced during an immune response. The development of tolerance ensures that an immune response to body cells does not normally occur. The immune system is thus able to distinguish "self" from "nonself" (foreign) materials. Some problems related to the ability of the immune system to distinguish self from nonself include blood transfusion complications, tissue transplant rejection, allergy, and autoimmune disease (all discussed later in this chapter).

Immunity based on antibodies is called **humoral immunity** and immunity based on activated T cells is called **cell-mediated immunity**. We shall discuss first the nature and formation of antibodies, and then show how they confer specific resistance to infection. Later, we will discuss the concepts of cell-mediated immunity.

Several features of the humoral response will be listed here and then discussed in some detail below.

1. Antibodies are formed against a variety of foreign macromolecules, but ordinarily not against macromolecules of the animal's own tissues; thus the animal is able to distinguish between its own ("self") and foreign ("nonself") macromolecules.

2. In virtually every case, antibody against a foreign macromolecule is formed only if the animal is challenged with the foreign substance.
3. Many but not all foreign macromolecules elicit the immunological response; those that do are called **immunogens**.
4. There is a high specificity in the interaction between antibody and antigen; antibodies made against one antigen may react against structurally related antigens but usually with reduced efficiency.
5. Antigen-antibody reactions are manifested in a wide variety of ways, depending on the nature of the antigen, the antibody, and the environment in which the reaction takes place.
6. Not all antigen-antibody reactions are beneficial; some, such as those involved in allergy and autoimmune reactions are harmful.
7. The high specificity of antigen-antibody reactions makes them useful in many research and diagnostic procedures.
8. Microorganisms and viruses that invade the host contain large numbers of different macromolecules that can act as antigens and the immunological response can be made the basis of specific immunization procedures for the prevention and control of specific diseases.

12.1 Immunogens and Antigens

Immunogens are substances that, when administered to an animal in the appropriate manner, induce an immune response. The immune response may involve either antibody production, the activation of specific immunologically-competent cells (called *activated T cells*), or both. **Antigens** are substances that react with either antibodies or activated T cells and most antigens are also immunogens. However, some substances are antigenic while not being true immunogens. For example, **haptens** are low-molecular weight substances that combine with specific antibody molecules but do not themselves induce antibody formation. Haptens include such molecules as sugars, amino acids, and small polymers. Thus the terms antigen and immunogen, although often used interchangeably, do not mean exactly the same thing.

An enormous variety of macromolecules can act as immunogens under appropriate conditions. These include virtually all proteins and lipoproteins, many polysaccharides, some nucleic acids, and certain of the teichoic acids. One important requirement is that the molecules must be of fairly *high* molecular weight, usually greater than 10,000. However, the antibody is directed not against the antigenic macromolecule as a whole, but only against restricted portions of the molecule that are called its **antigenic determinants** (Figure 12.1). Chemically, antigenic determinants include sugars, amino acid side chains, organic acids and bases, hydrocarbons, and aromatic groups. Antibodies are formed most readily to determinants that project from the foreign molecule or to terminal residues of a polymer chain. In proteins, for example, the majority of antibodies are made to surface determinants, because the surface contains a continuum of antigenic sites. A region of as few as 4–5 amino acids can define an antigenic determinant on a protein. Also, the surface of a protein can and frequently will have many overlapping antigenic determinants. A cell or virus is a mosaic of proteins, polysaccharides, and other macromolecules, each of which is a potential antigen. Each antigen of the cell is also a mosaic of side chains and residues, each of which is a potential antigenic determinant. The immunological response to an invading microbe or virus is thus a complex phenomenon.

In general, the specificity of antibodies is comparable to that of enzymes, which are able to distinguish between closely related substances. For instance, antibodies can distinguish between the sugars glucose and galactose, which differ only in the position of the hydroxyl group on carbon 4. However, specificity is not absolute, and an antibody will react at least to some extent with determinants related to the one that induced its formation. The antigen which induced the antibody is called the **homologous antigen** and others, if any, that react with the antibody are called **heterologous antigens**.

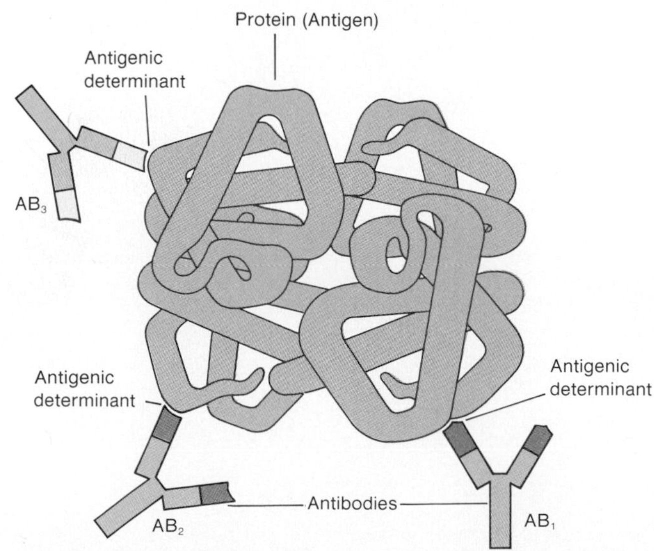

Figure 12.1 Antigens and antigenic determinants. Antigens may contain several different antigenic determinants, each capable of reacting with a specific antibody.

12.2 Immunoglobulins (Antibodies)

Antibodies are found predominantly in the serum fraction of the blood, although they may also be found in other body fluids, as well as in milk. Serum is the fluid portion of the blood that is left when the blood cells and the materials responsible for clotting (fibrin, platelets, and various cofactors, see Section 11.14) are removed. Serum containing antibody is often called **antiserum**. When serum proteins are separated by movement in an electric field (electrophoresis), four predominant fractions are seen: serum albumin and alpha, beta, and gamma globulins. Antibody activity occurs predominantly in the gamma globulin fraction, which is composed of many distinct proteins. Those gamma globulin molecules with antibody properties are called **immunoglobulins** (abbreviated Ig) and can be separated into five major classes on the basis of their physical, chemical, and immunological properties: **IgG, IgA, IgM, IgD**, and **IgE** (Table 12.1). Immunoglobulin class IgG has been further resolved into four immunologically distinct subclasses called IgG_1, IgG_2, IgG_3, and IgG_4. The basis for this separation will be discussed below. Antibody molecules specific for a given antigenic determinant are found in each of the several classes, even in a single immunized individual. Upon initial immunization, it is IgM, a pentameric immunoglobulin with a molecular weight of about 970,000, that first appears; IgG appears later. In most individuals about 80 percent of the immunoglobulins are IgG proteins, and these have therefore been studied most extensively.

Immunoglobulin structure Immunoglobulin G is the most common circulating antibody and thus we will discuss its structure in some detail. Immunoglobulin G has a molecular weight of about 160,000 and is composed of four polypeptide chains (Figure 12.2). Both intrachain and interchain disulfide (S—S) bridges are present (Figure 12.2). The two light (short) chains are identical in amino acid sequence, as are the two heavy (longer) chains. The molecule as a whole is thus symmetrical (Figure 12.2). Each light chain consists of about 212 amino acids and each heavy chain consists of about 450 amino acids.

Table 12.1 Properties of the immunoglobulins

Class designation	Molecular weight	Proportion of total antibody (percent)	Concentration in serum (mg/ml)	Antigen binding sites	Properties	Distribution
IgA	160,000	15–20	3	2	Major secretory antibody	Secretions (saliva, colostrum, serum), cellular and blood fluids; exists as a monomer in serum and as a dimer in secretions
	385,000 (secretory form)		0.05	4		
IgG	146,000	75	13	2	Major circulating antibody; four subclasses exist: IgG_1, IgG_2, IgG_3, IgG_4; binds complement; antitoxin	Extracellular fluid; blood and lymph; crosses placenta
IgM	970,000 (pentamer)	10	1.5	10	First antibody to appear after immunization; binds complement	Blood and lymph only
IgD	184,000	1	0.03	2	Minor circulating antibody; heat labile; high carbohydrate content	Blood and lymph; lymphocyte surfaces
IgE	188,000	0.002–0.05	0.00005	2	Involved in allergic reactions: contains mast cell binding fragment	Blood and lymph only

When an IgG molecule is treated with a thiol-containing *reducing agent* and the proteolytic enzyme *papain* under carefully controlled conditions, it breaks into several fragments. The two fragments containing the complete light chain plus the amino terminal half of the heavy chain are the portions that combine with antigen and are called *Fab* fragments (*f*ragment of *a*ntigen *b*inding), whereas the fragment containing the carboxyl terminal half of both heavy chains, called Fc (fragment *c*rystallizable), does *not*

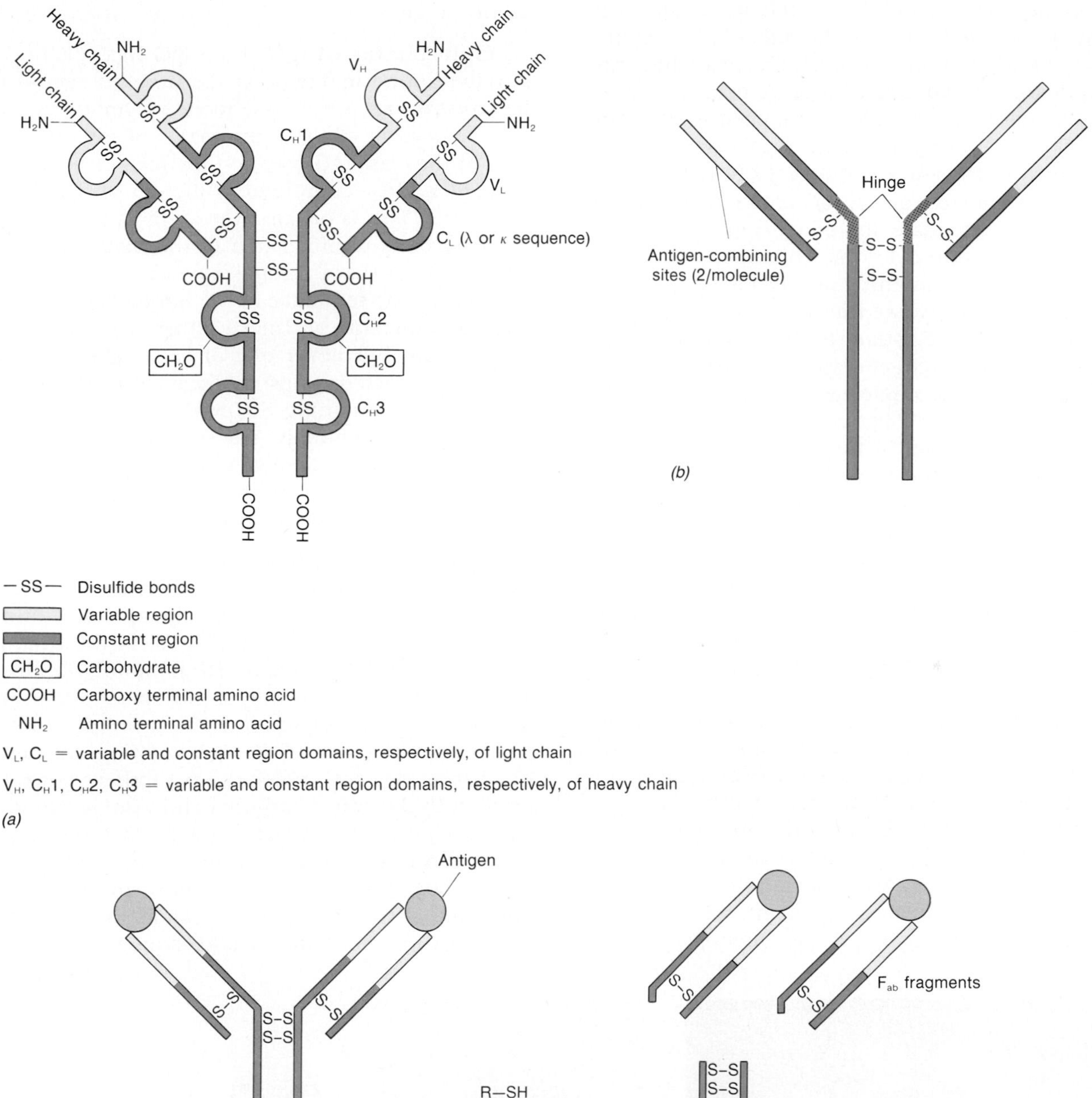

Figure 12.2 Structure of immunoglobulin G (IgG). (a) Structure showing disulfide linkages within and between chains. (b) Alternative structural diagram which deletes the intrachain disulfide bonds to simplify the diagram. (c) Effect of papain treatment on immunoglobulin structure. R—SH represents one of several organic thiols which will react with selected S—S bonds of the immunoglobulin.

combine with antigen (Figure 12.2*c*). Therefore, each antibody molecule of the IgG class contains *two antigen combining sites* (called *bivalent*). This bivalency is of considerable importance in understanding the manner in which some antigen-antibody reactions occur (see below). Within *Fab*, the antigen binding site is found in a small region of the amino terminal portion of both the heavy and the light chains (Figure 12.2). Immunoglobulins also contain small amounts of complex carbohydrates consisting mainly of hexose and hexosamine, which are attached to portions of the heavy chain (Figure 12.2); the carbohydrate is not involved in the antigen binding site.

Although the view of the IgG molecule shown in Figure 12.2 is adequate for conveying the general structure of this molecule, the student should recall that since immunoglobulins are proteins, they are likely to be highly twisted and folded in the final antibody molecule. This fact is illustrated in Figure 12.3 where a computer generated model and an electron micrograph of a single IgG molecule is shown. Note that although the basic Y-shaped structure is apparent in both views of the IgG molecule shown in Figure 12.3, the twisting and folding characteristic of the secondary and tertiary structure of proteins is readily apparent in Figure 12.3*a*. We will see later how the higher order structure of the polypeptide chains that combine to form an intact immunoglobulin expose a unique binding site on each antibody molecule and how this binding site is ultimately responsible for the specificity of antigen-antibody reactions.

Present in serum in larger amounts than immunoglobulins of other classes, IgG found in normal serum represents a complex mixture of antibody molecules directed against a number of different antigenic determinants. To effectively study the chemistry of immunoglobulins, it is necessary to have large amounts of a single kind. Certain antibody-producing cell types that have become malignant, referred to as **myeloma tumors**, usually produce homogenous antibody, and humans suffering from myelomas produce huge amounts of antibody of a single antigenic specificity (**monoclonal antibody**, see Section 12.7). Amino acid sequence studies of these monospecific immunoglobulins have greatly advanced the understanding of the structure of IgG and humoral immunity in general.

Light chains of IgG Each IgG light chain contains two amino acid regions, the *variable region* and the *constant region.* The sequence of amino acids in a major portion of the light chains of immunoglobulins of the class IgG may be identical, even in IgG's directed against completely different antigenic determinants. This is because the amino acid sequence in the carboxy terminal half of the light chain constitutes one of two specific sequences, referred to as the *lambda* (λ) sequence or the *kappa* (κ) sequence. One IgG molecule will have either two λ chains or two κ chains but never one of each. Thus two IgG molecules directed against different antigens may have light chains containing 50 percent or more amino acid sequence homology. The sequence in the amino terminal region (antigen binding site) of their light chains will, however, be different, since they react with different antigens (Figure 12.2).

Heavy chains of IgG Each IgG heavy chain contains four amino acid regions, one variable and three constant regions (referred to as variable and constant "domains," respectively). Analogous to the situation that exists in the light chain, all immunoglobulins of the class IgG have a portion of their heavy chain (the carboxy terminal region) in which the amino acid sequence is identical (C_H1, C_H2, and C_H3, see Figure 12.2) from one IgG molecule to another, as well as a region in the amino terminal end (antigen binding site V_H, Figure 12.2) where considerable amino acid sequence variation occurs from one IgG to the next. An antibody molecule is thus a multi-subunit protein consisting of four polypeptide chains each of which contains a region of amino acid sequence variability

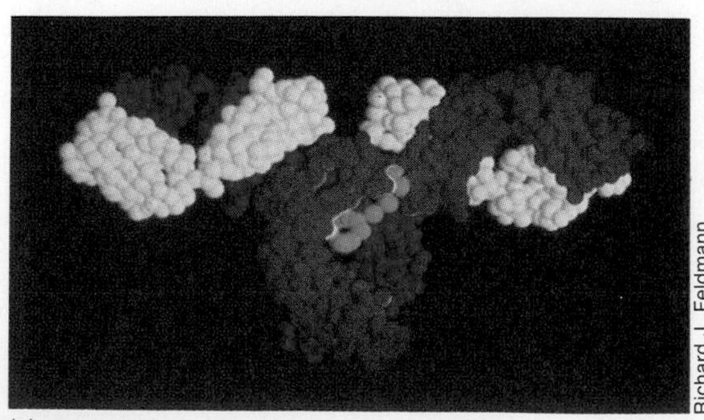
Richard J. Feldmann
(a)

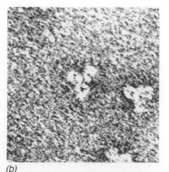

(b)

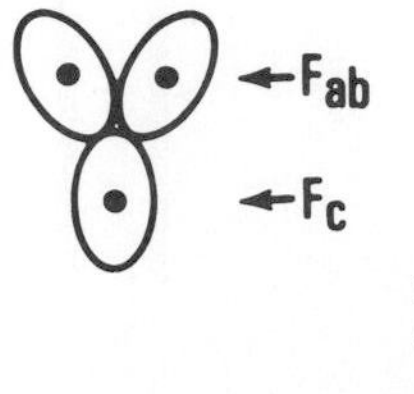

Craig Smith

Figure 12.3 Three dimensional views of immunoglobulin IgG. (a) Computer-generated model of IgG, the heavy chains are shown in red and dark blue, the light chains in green and light blue. The light blue between the F_c portions of the two heavy chains represents attached carbohydrate. (b) Electron micrograph of a single molecule of IgG.

(**variable region**) and a region of constant sequence (**constant region**) for *all* immunoglobulin molecules of a given *class*. The great specificity of a given antibody molecule for a particular antigen lies in the unique three-dimensional structure of the antigen binding site dictated by the amino acid sequence in the variable regions of the heavy and light chains (see Figure 12.3 and Figure 12.4).

Other classes of immunoglobulins How do immunoglobulins of the other classes differ from IgG? κ or λ sequences are found in the light chains of immunoglobulins of all classes, but distinct sequence differences exist in the heavy chain constant region domains. The heavy-chain *constant* regions of a given immunoglobulin molecule can have five distinctly different amino acid sequences called gamma, alpha, Mu, delta, or sigma, and these sequences constitute the carboxy terminal three-fourths of the heavy chains of immunoglobulins of the class IgG, IgA, IgM, IgD, or IgE, respectively. Each antibody of the class IgM, for example, will contain a stretch of amino acids in its heavy chain which constitutes the Mu sequence. If two immunoglobulins of *different classes* react with the same antigenic determinant, their entire light chains and the variable regions of their heavy chains would be identical, but their class-determining sequences, specific to their *heavy* chains, would be different. It is not unusual in a typical immune response to observe the production of antibodies of two different classes to the same antigenic determinant.

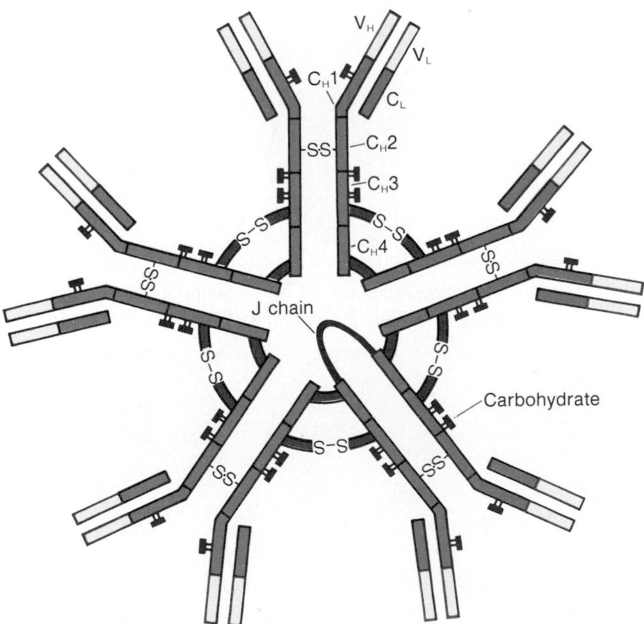

Figure 12.4 Structure of IgM, a large immunoglobulin with five molecules (a pentamer). Note that each heavy chain has four rather than three constant regions and that the five molecules are themselves held together by disulphide bonds. Also note that 10 antigen binding sites are available.

The structure of **immunoglobulin M** (IgM) is shown in Figure 12.4. It is usually found as an aggregate of five immunoglobulin molecules attached as shown in Figure 12.4 by short peptides and accounts for 10 percent of the total serum immunoglobulins. Each heavy chain of IgM contains an extra constant region domain (C_H4), and IgM in general is very carbohydrate rich. IgM is the first class of immunoglobulin made in a typical immune response to a bacterial infection (see Figure 12.11), but immunoglobulins of this class are generally of low affinity. The latter problem is compensated to some degree, however, by the high *valency* of the pentameric IgM molecule; 10 binding sites are available for interaction with antigen (Table 12.1). The term *avidity* is used to describe the *strength* of multivalent antigen-binding molecules; thus, IgM is said to be of low affinity but high avidity.

Immunoglobulin A (IgA) is of interest because it is present in body secretions. IgA is the dominant antibody in all fluids bathing organs and systems in contact with the outside world: saliva, tears, breast milk and colostrum, gastrointestinal secretions, and mucus secretions of the respiratory and genitourinary tracts. IgA is also present in serum, but the IgA of secretions has an altered molecular structure, consisting of a dimeric immunoglobulin attached to a protein high in carbohydrate, called the secretory piece, and a second protein called the J chain (Figure 12.5). These proteins help hold the dimeric immunoglobulin molecule together and possibly aid in the passage of IgA into secretions. Because of its location, IgA probably provides a first line of immunologic attack against bacterial invaders, which as we have seen (see Section 11.8), first become established on tissue surfaces.

Immunoglobulin E (IgE) is found in serum in extremely small amounts (in an average human about 1 of every 50,000 serum immunoglobulin molecules is IgE). Despite its low concentration it is important, since immediate-type hypersensitivities (allergies; see

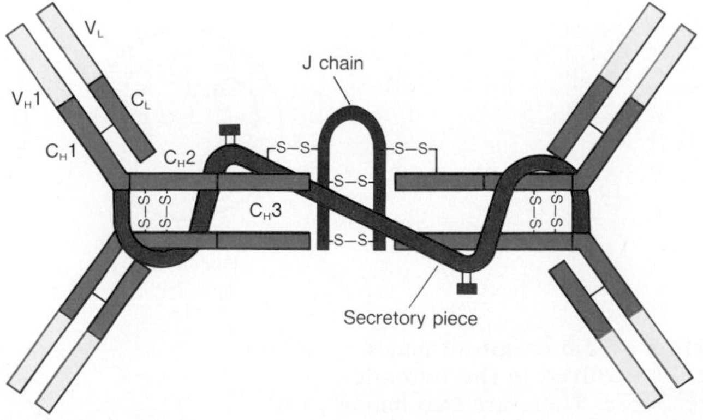

Figure 12.5 Structure of human secretory immunoglobulin A (IgA). Although most IgA is monomeric, the secretory form of IgA is a dimeric molecule.

Section 12.11) are mediated by IgE, following the binding of these antibodies to certain cell types which respond by releasing chemicals that elicit allergic reactions. The molecular weight of an IgE molecule is significantly higher than that of other immunoglobulins (Table 12.1) because, like IgM, it contains an additional constant region. This region is thought to function in binding IgE to mast cell surfaces (see Section 12.11), an important prerequisite for certain allergic reactions.

Immunoglobulin D (IgD) is also present in low concentrations (but 10 times higher than that of IgE), and its function in the overall immune response is unclear. Recent experiments have shown that IgD is abundant on the surfaces of antibody-producing cells (lymphocytes, see next section) and IgD may play a role in binding antigen as a signal to the lymphocyte to begin antibody production.

12.3 Cells of the Immune System

The immune system enlists the activity of a number of organs and cell types which interact in various ways to elicit the observed immune response. All cells of the immune system arise from undifferentiated stem cells in the bone marrow and differentiate into functionally distinct cell types during a maturation process in association with specific host tissues (Figure 12.6). Although almost a dozen different cell types have been recognized, we limit our discussion here to lymphocytes and macrophages, because it is these cells which are primarily involved in specific immune responses. Because they are differentiated lymphocytic cells, we will discuss the nature of **plasma cells** and **memory cells** when we present the details of antibody production. Properties of nonspecific phagocytic cells were discussed in Section 11.14.

Lymphocytes The cells responsible for immunoglobulin production are a class of white blood cells referred to as **lymphocytes**. These cells are dispersed throughout the body and are one of the most prevalent mammalian cell types (the average adult human has about 10^{12} lymphocytes). Two types of lymphocytes, B lymphocytes (B cells) and T lymphocytes (T cells), are involved in immune responses.

Immunoglobulins are made only by lymphocytes of the B type, while T cells play a variety of alternative

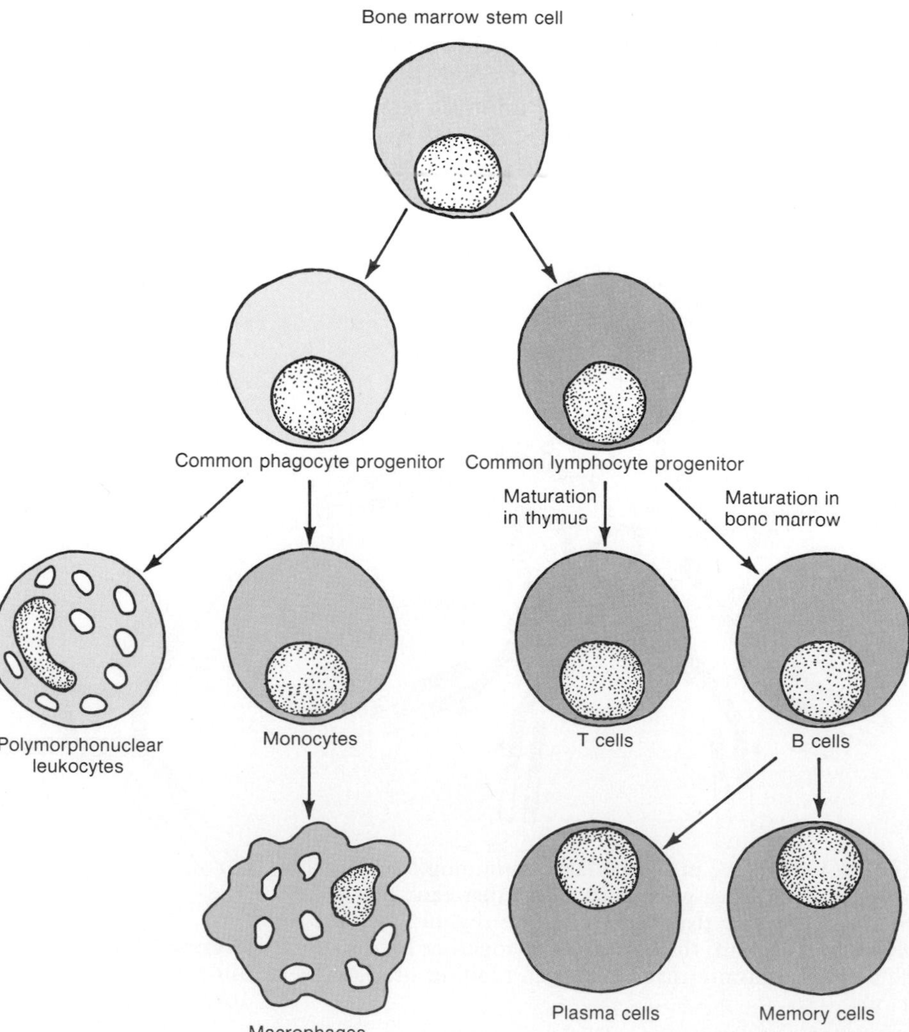

Figure 12.6 Origin of major cells involved in the immune response. There are two major lines, one generating phagocytic cells, the other generating lymphocytic cells that participate directly in immune responses.

roles in the overall immune response. Both B and T cells are derived from stem cells in the bone marrow, and their subsequent differentiation is determined by the organ within which they become established (Figure 12.6). B cells probably mature in the bone marrow and T cells in the thymus (hence the designation "T"). Several subsets of T lymphocytes are known. T-helper (T_H) cells stimulate B cells to produce high levels of immunoglobulin; in most cases little if any antibody is made without T_H cell interaction (see Section 12.5). T-suppressor (T_S) cells perform a regulatory function in their ability to suppress immunoglobulin production by B cells once the antigen has been removed, and also function to control immune responses against self antigens (see Section 12.11). Cytotoxic (T_C) cells secrete toxic substances that can kill foreign cells, and together with T-delayed-typed hypersensitivity (T_{DTH}) cells, play primary roles in the cellular immune responses (Table 12.2 and Section 12.10). In birds, B cells differentiate in an organ called the *bursa of Fabricius* (hence the designation "B"). A similar organ has not been found in mammals, and it is thought that B cells differentiate in lymph nodes, in lymphoid structures such as the appendix or tonsils, or more likely in the bone marrow itself.

Macrophages Macrophages are large phagocytic cells which are capable of ingesting antigens and destroying them as well as cooperating with lymphocytes in the production of specific antibody. Phagocytic cells are of two types, **monocytes** and **polymorphonuclear granulocytes**. The latter are usually referred to as "polymorphs," and have been given a variety of trivial names depending on the histological staining properties of their internal lytic granules (see Section 11.14). Monocytes can differentiate to become **macrophages**. The term macrophage is generally used to describe phagocytes of monocytic lineage that are *fixed* to tissue surfaces, while the term *monocyte* is used to describe freely circulating phagocytes. Macrophages are abundant in lymphoid tissue and in the spleen, while monocytes are abundant in the blood and lymph.

If an antigen penetrates an epithelial surface it will eventually come in contact with phagocytic cells. As discussed in Chapter 11, phagocytic cells such as macrophages can engulf particles as large as bacterial cells and kill them by releasing lytic substances (proteases, nucleases, lipases, and lysozyme) from cytoplasmic vesicles within the cell. Lysis of a bacterial cell releases a variety of distinct bacterial antigens within the macrophage, and these, as well as antigens ingested directly, can be processed by the macrophage and used to initiate the early steps in antibody synthesis. In brief, we will see that macrophages imbed foreign antigens on their cell surface where they can come in contact with specific T cells and B cells; antigen recognition by T cells and B cells is the first step in antibody production (see Section 12.5).

Macrophages and monocytes are *nonspecific* cells. Unlike T cells and B cells, phagocytic cells cannot distinguish between antigens; any foreign substance is ingested, whether antigenic or not. However, since many macromolecules are antigens and foreign cells contain numerous antigens, the majority of particles phagocytized by macrophages or monocytes will in-

Table 12.2 Comparison of B and T cells

T cells	B cells
Origin: bone marrow	Origin: bone marrow
Maturation: thymus	Maturation: lymphoid tissue or bone marrow; bursa in birds
Long-lived: months to years	Short-lived: days to weeks
Mobile	Relatively immobile (stationary)
No complement receptors	Have complement receptors
No immunoglobulins on surface	Immunoglobulins on surface
Restricted antigenic specificity	Restricted antigenic specificity
Proliferate upon antigenic stimulation	Proliferate upon antigenic stimulation into plasma cells and memory cells
Produce cell-mediated immunity (T_{DTH} and T_c cells)	Synthesize immunoglobulin (antibody)
Show delayed hypersensitivity (T_{DTH} cells)	
Produce lymphokines (T_{DTH} cells)	
Participate in transplantation immunity (T_c cells)	
Help in antibody production by B cells (T_H cells)	
React against intracellular bacteria, viruses, and parasites (cellular immunity) (T_c cells)	
Perform as killer T-lymphocytes in cell-mediated immunity (T_c cells)	
Modulate degree of humoral response (T_s cells)	
Suppress differentiation of "self" B-cell clones (T_s cells)	

deed be antigenic. The action of macrophages in antigen processing and antigen presentation is a key step in the overall process of antibody production, because the vast majority of antigens can only stimulate lymphocytes through macrophages acting as intermediaries.

12.4 Histocompatability Antigens and T-Cell Receptors

Among the many molecules on the surfaces of animal cells there is a group of proteins that are highly specific for each animal species, or even each strain within a species. These proteins are referred to as **major histocompatability complex** (MHC) **antigens**. These molecules can be thought of as species or strain "markers," molecular reference points by which cells from one species can be differentiated from those of another. MHC antigens are the major target molecules for transplantation rejection; if tissues from one animal are immunologically rejected when transplanted to a second, then their major histocompatability antigens are likely to differ. The MHC of mice and humans have been the best characterized, and we focus here on the MHC of humans.

Structure of the human major histocompatability complex The major histocompatability complex consists of three groups of proteins, known as Class I, Class II, and Class III MHC antigens. **Class I antigens** consist of two polypeptides (Figure 12.7), one of which is coded for by genes in the MHC region of human chromosome number 6. The other Class I polypeptide, called *β-2 microglobulin,* is coded for by non-MHC linked genes. The MHC-coded polypeptide is a glycoprotein firmly anchored in the cell membrane; β-2 microglobulin is a small polypeptide noncovalently linked to the MHC Class I protein (Figure 12.7).

Class II MHC antigens consist of two noncovalently-linked glycosylated polypeptides, called α and β. Like Class I molecules, these polypeptides are imbedded in the plasma membrane and project outward as surface markers (Figure 12.7). **Class III MHC proteins** consist of the C4 and C2 complement proteins (see Section 12.12). A comparison of Class I and Class II MHC antigens is shown in Figure 12.7.

MHC antigens are not structurally identical in all representatives of a given species. Instead, different individuals frequently show subtle differences in amino acid sequence of their MHC antigens. These minor sequence variations, referred to as *polymorphisms*, occur in all classes of histocompatability antigens. By contrast, *major* structural changes in a given MHC antigen exist between different animal species and is the major reason why tissues from one animal species are generally not accepted by another when tissue grafts or organ transplants are made between species.

Class I MHC antigens are found on the surfaces of *all* nucleated cells and on the surfaces of platelets. In mice, but not in humans, Class I proteins are also found on the surfaces of erythrocytes (red blood cells). Class II antigens are found only on the surface of B lymphocytes, certain types of T lymphocytes, and macrophages. Class III MHC proteins are not normally found on cell surfaces, but instead are soluble proteins capable of binding to cell surfaces during complement activation (see Section 12.12).

Functions of the major histocompatability complex In the normal animal, T cells, especially cytotoxic T cells (T_c), are constantly interacting with foreign proteins or other foreign antigens. In the discrimination of "self" from "nonself" it is thought that MHC proteins function as *reference points* for the body's immune system, readily identifiable molecules that help T_c cells discriminate between cells that belong to the animal and those that are foreign and should be destroyed. How then does a T_c cell recognize a foreign antigen? When foreign antigens are taken up by a host cell, the cell proceeds to "process"

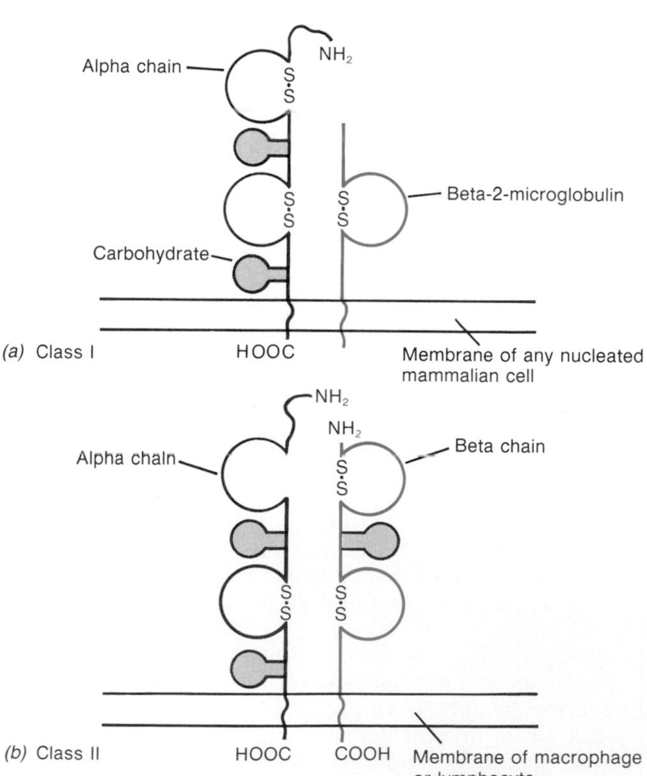

Figure 12.7 Structure of major histocompatibility (MHC) antigens. (a) Class I type. (b) Class II type. Note that Class I molecules are present on the surfaces of all nucleated cells and that Class II molecules are only present on certain cells.

the antigen, and ends up inserting the antigen, or a portion of the antigen, on its cell surface. Viral infection of an animal cell, for example, leads to the imbedding of viral antigens on the infected cell's surface (Figure 12.8). T_c cells, surveying the cell population for foreign antigens, will eventually encounter the virus-infected cell and recognize the viral antigen as foreign. In distinguishing between "self" and "nonself," it appears that MHC Class I proteins on host cells are recognized by complementary Class I *receptors* on the T_c cell surface. While this recognition event is occurring, the presence of the foreign antigen is detected when the antigen binds to a receptor molecule on the T_c cell surface. When contact between a T_c cell receptor and a foreign antigen has been made, a third polypeptide on the T_c cell surface is activated, and this induces the cell killing functions of the T_c cell (see Section 12.10).

By contrast, T-helper cells recognize MHC Class II molecules present on the surface of phagocytic cells and certain lymphocytes as molecular markers. We will see that T_H cells need to recognize these cell types in order to initiate a humoral immune response. T-helper cells contain a receptor which recognizes both MHC Class II antigens and a specific foreign antigen (see below). Once a T-helper cell is activated by contact with foreign antigen, it secretes molecules that stimulate antibody production of specific B-cell clones (see Section 12.5). In summary, then, MHC molecules serve as *specific identifying markers* for host cells such that cytotoxic and helper T cells can "see" a foreign antigen in relation to a reference molecule (the MHC antigen). Figure 12.9 summarizes how MHC class I and II molecules interact with antigen on cell surfaces to initiate T cell-mediated functions.

T-cell receptors: types and distribution T cells play a variety of complex roles in the overall immune response. Although T cells do not produce antibody, they do recognize antigen, and this recognition process is due to specific antigen receptor molecules located on T-cell surfaces. Two major subpopulations of T cells have been identified: T4 lymphocytes and T8 lymphocytes. The two groups can be distinguished by whether they contain one or the other of two major identifying antigens on their surfaces. The T4 and T8 antigens are present in addition to the MHC/foreign antigen receptor on the T-cell surface (Figure 12.9).

Lymphocytes of the T4 type are classified as T-helper (T_H) cells. Their major function is to recognize and bind foreign antigen and secrete substances that stimulate a complementary clone of B lymphocytes to divide and begin producing specific antibody (see Section 12.5). T8 lymphocytes are either cytotoxic T cells (T_c), whose function is to identify foreign cells or host cells displaying foreign antigens and kill them, or suppressor T cells (T_s), T cells which function to modulate the intensity of the immune response (see Section 12.11). However, besides specific identifying molecules and MHC receptors, T cells can recognize specific antigens. How is this accomplished at the molecular level?

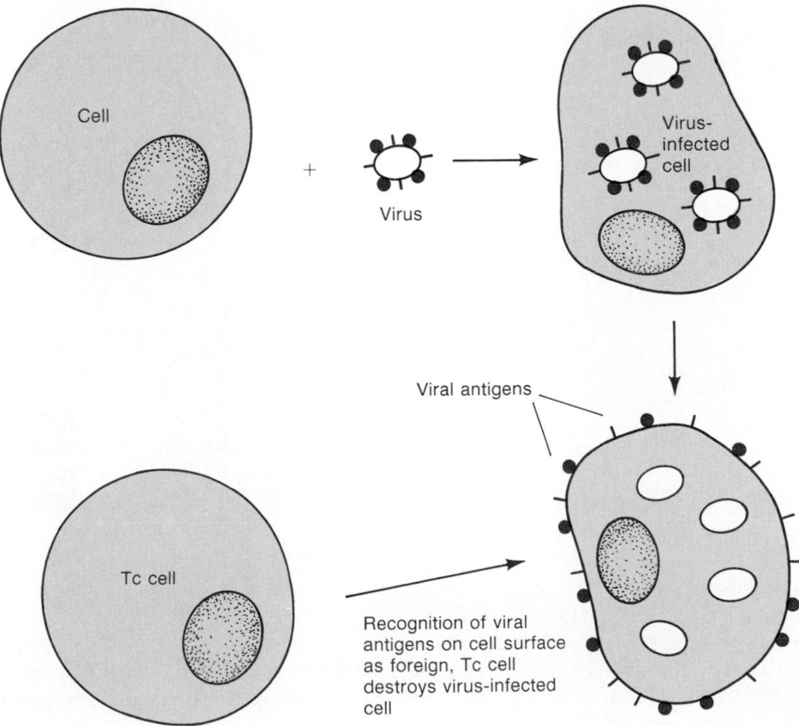

Figure 12.8 Manner in which a T cell recognizes a foreign antigen. In this case, a virus-infected cell is used as an example.

Structure of T-cell antigen receptor Antigen receptors have to be highly specific—so specific in fact that their specificity must rival that of antibodies themselves. Although T-cell receptors are not antibody molecules, they resemble antibody molecules in many ways. Indeed, T-cell receptors and antibodies have much in common and are clearly evolutionarily-related molecules (see Section 12.6).

The T-cell antigen receptor consists of two disulfide-linked polypeptides, called α and β. In analogy to the structure of antibodies, T-cell antigen receptors contain regions of constant and regions of highly variable amino acid sequence. Also as in antibody molecules, it is the *variable* region of the T-cell receptor which actually binds antigen. In contrast to antibodies, however, each polypeptide of the T-cell receptor contains only one constant region; antibody molecules contain one constant region on each light chain and three to four constant regions on each heavy chain (see Section 12.2). A comparison of the T-cell antigen receptor with a corresponding antibody molecule is shown in Figure 12.10.

T-cell receptors are integral cell membrane proteins, as are the MHC antigens and those antibody molecules which serve as B-cell receptors (see next section). Studies of polypeptide folding have shown that the secondary structure of T-cell receptor constant regions consists of relatively rigid pleated sheets (β-sheet, see Section 2.7), while the variable regions (antigen binding sites) are α-helices, with a much greater capacity to flex and fold. The latter form of secondary structure is ideal for generating highly specific binding sites. The constant and variable regions of antibody molecules also show β-sheet and α-helical folding, respectively.

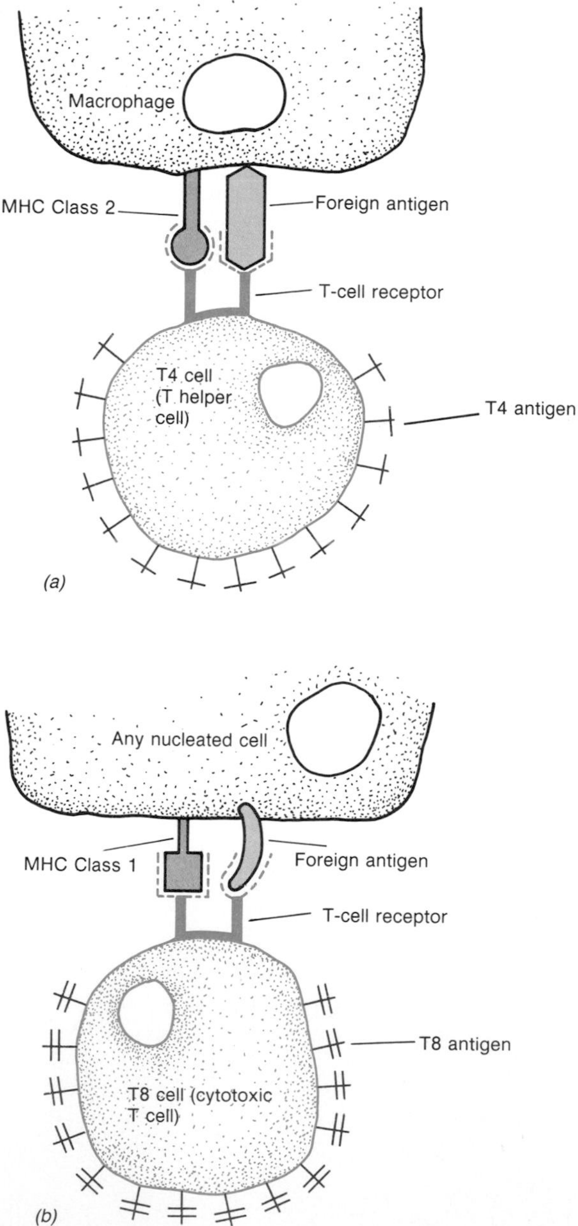

Figure 12.9 (a) Interaction of T4 lymphocytes with MHC Class II cells. (b) Interaction of T8 lymphocytes with MHC Class I cells. Note that the two T cells themselves are antigenically unique due to distinct T4 or T8 surface antigens.

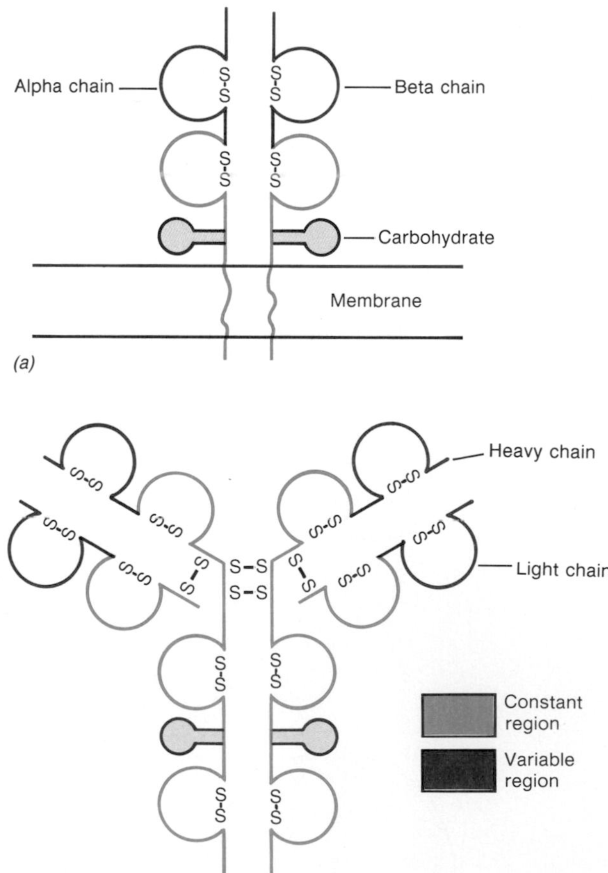

Figure 12.10 Structural comparison of the T-cell receptor with an immunoglobulin. Note the presence of variable regions.

12.5 The Mechanism of Antibody Formation

How is it possible for animals to produce such a large variety of specific proteins, the antibodies, in response to invasion by foreign macromolecules, viruses, and cells? The problem of antibody formation can be divided into two separate questions: (1) How do cells of the immune system interact to produce antibodies when stimulated by an antigen? (2) How is antibody diversity generated?

The production of antibodies in response to stimulation by antigen is a complex process involving T cells, B cells, and macrophages, and involves the intimate interaction of the various cell surface molecules discussed in the previous section. Macrophages act nonspecifically to ingest foreign antigens and present them to T cell:B cell pairs. Following recognition, the T cell then releases substances that stimulate B cells to begin antibody synthesis. Although the initial interaction between antigen and macrophage is nonspecific, all further steps in the process of antibody production are *highly* specific; receptor molecules on the surfaces of both T cells (T-cell receptor, see Section 12.4) and B cells (antibody molecules, see later) help ensure that antibodies made in response to an antigen will indeed be specific for that antigen.

The genetic control of antibody production is also a highly complex process. An animal is capable of making over 10 million structurally distinct antibody molecules. Although at first sight this seems like an enormous demand on an organism's genetic coding potential, we will see that only a relatively small amount of DNA is required to code for this immense antibody diversity because of a phenomenon known as **gene rearrangement**. During development of lymphocytic cells in the bone marrow, random gene rearrangements and deletions occur in each B cell to eventually yield two transcriptional units, one of which codes for the synthesis of a specific heavy chain and one for a light chain of the antibody molecule. The number of possible gene rearrangements, even of a relatively small number of genes, is sufficient to account for the diversity of antibody molecules (see Section 12.6). Similar rearrangements presumably occur during T cell development to yield the diversity of T-cell receptors observed. We now detail the steps in antibody production, beginning with the injection of an antigen and ending with the production of a specific antibody that will react with that antigen.

Exposure to antigen In considering the mechanism of antibody formation, we must first explore what happens to the antigen in the body. Antigens are carried to all parts of the body in the blood and lymph systems, which were illustrated in Figure 11.23. The main site of antigen localization in the body are the lymph nodes, the spleen, and the liver. It has been well established that antibodies are formed in both the spleen and lymph nodes; the liver is not involved. If the antigen is injected intravenously, the spleen is the site of greatest antibody formation, whereas subcutaneous, intradermal, and intraperitoneal injections lead to antibody formation in lymph nodes. Fragments of lymph node or spleen from immunized animals can continue to produce antibody when placed in tissue culture or when injected into other, nonimmunized, animals.

Following the first introduction of an antigen, there is a lapse of time (latent period) before any antibody appears in the blood, followed by a gradual increase in *titer* (that is, concentration) and then a slow fall. This reaction to a single injection is called the **primary antibody response** (Figure 12.11).

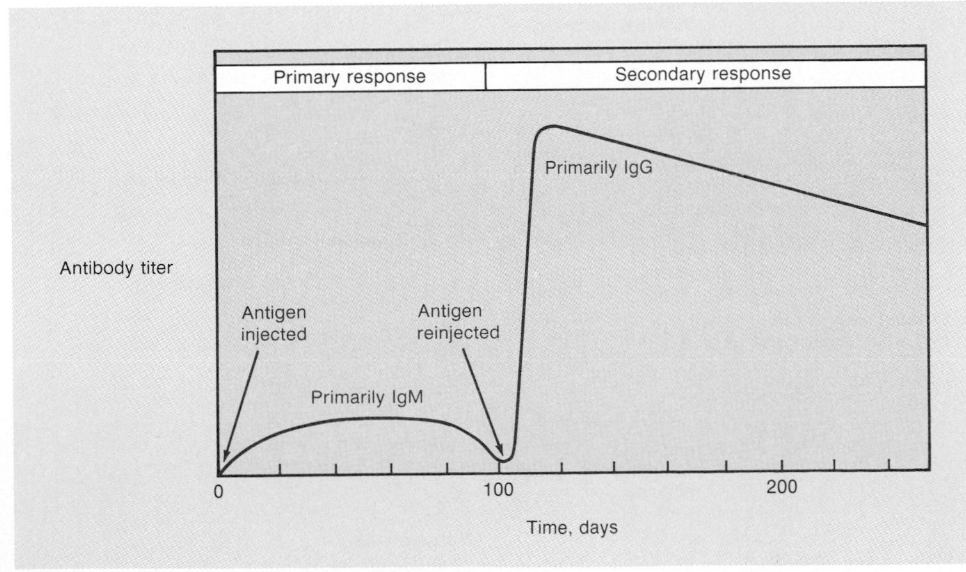

Figure 12.11 Primary and secondary antibody responses.

When a second exposure to antigen is made some days or weeks later, the titer rises rapidly to a maximum 10 to 100 times above the level achieved following the first exposure. This large rise in antibody titer is referred to as the **secondary antibody response** (see Figure 12.11). With time the titer slowly drops again, but later injections can bring it back up. The secondary response is the basis for the vaccination procedure known as a "booster shot," (for example, the yearly rabies shot given to domestic animals) to maintain high levels of circulating antibody specific for a certain antigen.

Antibody production How do B cells, T cells, and macrophages act together to produce a humoral response? Macrophages act nonspecifically, phago-

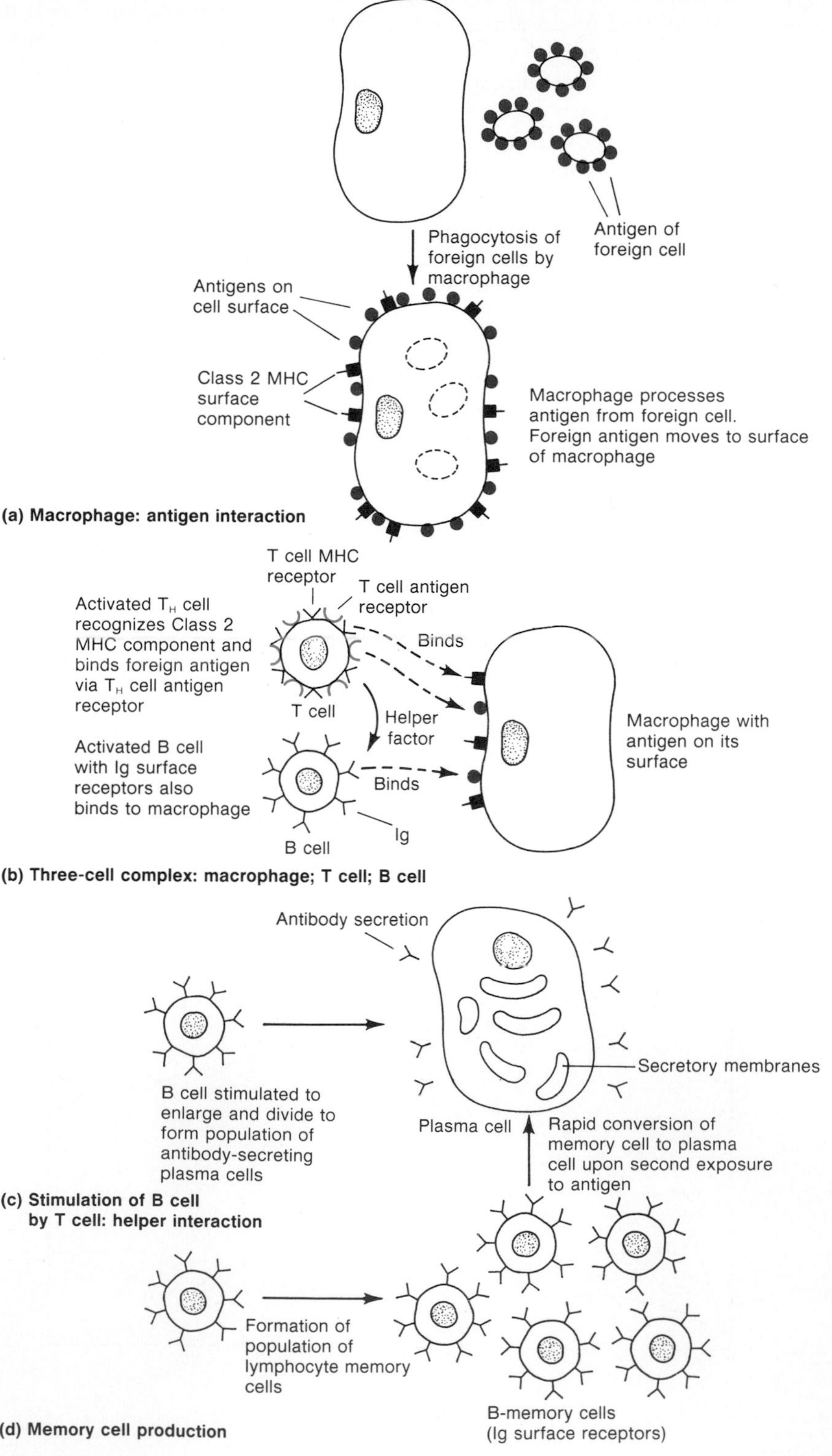

Figure 12.12 Interaction between B and T cells and macrophages to produce a line of antibody-secreting plasma cells and a population of B-memory cells.

cytizing antigens so that antigenic materials become attached to macrophage cell surfaces where they can interact with other cell types (Figure 12.12). Macrophages are very sticky and attach well to surfaces; their stickiness probably also promotes the attachment of B and T cells. In this fashion, macrophages present the antigen to B and T lymphocytes, thus initiating the process of antibody production. Both B and T cells contain specific antigen receptors on their cell surfaces. The receptors on B cell surfaces are the antibody molecules that the cells are genetically programmed to produce.* T-cell receptors (see Section 12.4) recognize both MHC Class II antigens and a specific foreign antigen. Following binding of the T cell to the foreign antigen expressed on macrophage surfaces, the T cell releases a soluble "helper factor" which stimulates B cell differentiation. The nature of antigen-specific T-cell helper factors is not completely understood. However, it is possible that T-cell helper factors are simply released T-cell receptors, possibly also containing portions of the original antigen. If this is true, it would account for how a specific T cell activates only a certain B cell and not others when it releases its helper factor (the activated B cells presumably recognize the antigen bound to the released T-cell receptor). Antigen-specific T-cell helper factors are distinct from other T-cell factors, such as lymphokines. Lymphokines act as *nonspecific* activators of cells in the immune system, including B cells (see Section 12.10).

The T-cell helper factor stimulates growth and division of the B cell to form a clone of B cells, each capable of producing identical antibody molecules. Further differentiation of the activated B cell population then occurs, resulting in the formation of large antibody-secreting cells called **plasma cells** and a special form of B cell called a **memory cell** (see Figure 12.12). Plasma cells are relatively short-lived (less than 1 week) but excrete large amounts of antibody during this period. Memory cells, on the other hand, are very long-lived cells, and upon second exposure to antigen, these cells are quickly transformed into plasma cells and begin secreting antibody. This accounts for the rapid and more abundant antibody response observed upon antigenic stimulation the second time (secondary response; see Figure 12.11).

Certain antigens can stimulate low-level antibody production in the *absence* of previous T cell interactions (these are the so-called T-independent antigens). Most T-independent antigens are large polymeric molecules with repeating antigenic determinants (for example, polysaccharides). The immunoglobulins produced to T-independent antigens are usually of the class IgM and are of low affinity. In addition, B cells which respond to T-independent antigens do not have immunogenic memory.

The discussion above describes the general principles of antibody production. Each antigen (strictly speaking, each antigenic determinant) will catalyze, generally through the action of T-helper cells, the expansion of a *different* B-cell line, the B-cell type in each case being genetically capable of producing antibodies that react specifically with that antigen. In this fashion it is thought that the normal animal can respond to at least several million distinct antigens by expanding a specific B-cell clone in response to stimulation by antigen.

Thus we see that the production of specific antibodies involves the interaction of a number of distinct cell types. Although antibodies are clearly made by B cells, very few antibodies would ever be made without the cooperation of T cells. Likewise, in the absence of macrophages to serve as antigen scavenging and presenting cells, T cells would not function. The production of antibody is therefore a series of cascading events where each cell type is intimately dependent upon the activity of others. We now turn our attention to the genetic orchestration of these immune events and investigate the molecular diversity of antibodies and T-cell receptors at the DNA level.

12.6 Genetics and Evolution of Immunoglobulins and T-Cell Receptors

Gene organization If one B lymphocyte produces one type of immunoglobulin, it should theoretically require only one gene to code for the two identical light chains and one gene for the two identical heavy chains; the difference in immunoglobulin production of one B cell from another resides in differential gene expression. Based on all available biological precedents, the above two statements were thought to be correct as they applied to Ig production. However, they are not. A single light or heavy polypeptide chain is coded for by several genes and, although the genetic information present in *immature* B lymphocytes as they are formed in the stem cells of the bone marrow is identical, *mature* B cells are not genetically identical. How can this be accounted for in genetic terms?

Recent research has shown that the variable regions of immunoglobulin heavy and light chains are more complex than that shown in Figure 12.2. Although variation in amino acid sequences is readily apparent in the variable region of different immunoglobulins, amino acid variability is especially extreme in several so-called **hypervariable regions** (Figure 12.13). It is at these hypervariable sites that

*One B lymphocyte may produce immunoglobulins of more than one class, but each will react with only one antigenic determinant (that is, their variable regions will be identical).

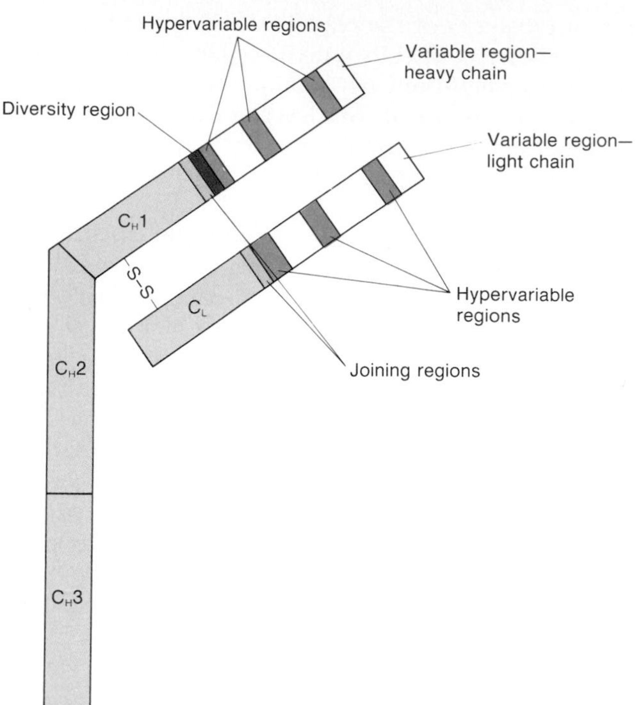

Figure 12.13 Detailed structure of variable regions of immunoglobulin light and heavy chains. Only one-half of a typical immunoglobulin molecule is shown. C_H, C_L, constant regions of heavy and light chains, respectively.

combination with antigen actually occurs. Each variable region in the light and heavy chains is thought to have three hypervariable regions with a portion of the third region on the heavy chain being coded for by a distinct gene called the *D* (for "diversity") *gene*, with the first two hypervariable sites being coded for by the variable region gene itself (Figure 12.14). In addition, at the site of joining between the diversity region and the constant region DNA, there is a stretch of nucleotides, about 40 bases in length, called the *J* (for "joiner") region that is coded for by a distinct gene (*J gene*). Finally, the class-defining constant region of the immunoglobulin molecule is coded for by its own gene, the *C* (for "constant") *gene*. Light chains are coded for by their own variable region genes, joining region genes, and constant region genes, but do not contain diversity regions as in the heavy chains (Figure 12.14).

Gene processing The origin of cells of the immune system was presented in Figure 12.6. All the genetic information required to make antibodies against a virtually unlimited pool of antigens is thought to exist in each lymphocyte as it is formed in the stem cells of the bone marrow. Each *immature* B cell is thought to contain about 150 light-chain variable region genes and five distinct joining sequences, while 100–200 variable region genes, 4 joining sequences, and about 50 diversity region genes are thought to exist for the heavy chains (Figure 12.14). These genes are not located adjacent to one another, but are separated by noncoding sequences (introns) typical of gene arrangements in eucaryotes (see Section 5.1). During maturation of B lymphocytes, genetic recombination occurs, resulting in the construction of an **active heavy gene** and an **active light gene** that are transcribed and translated to make the heavy and light chains of the immunoglobulin molecule, respectively (Figure 12.14). *Randomly selected* V, J, and D segments are fused by enzymes that delete all intervening DNA. Each "set" of genes, *variable* (V) region genes, *constant* (C) region genes, and *diversity* (D) and *joining* (J) genes, reside on separate chromosomes. The active gene (containing an intervening sequence between the VDJ genes and the constant region gene) is transcribed, and the resulting primary RNA transcript spliced to yield the final mRNA (Figure 12.14).

The final light and heavy gene structure of any given B cell is therefore a matter of *chance rearrangement,* and all possible gene combinations appear equally probable. The number of possible gene combinations is such that over 10^7 different antibody molecules theoretically can be made. By the time each B cell has matured, however, the gene rearrangements will have produced a B cell capable of making antibodies with only a *single* specificity. Put another way, each B cell makes only antibody molecules with identical variable regions.

Additional antibody diversity occurs as the result of mutations arising during formation of the active heavy chain gene. The DNA splicing mechanism appears to be rather imprecise and frequently varies the site of VDJ fusions by a few nucleotides. This, of course, is sufficient to change an amino acid or two, and this apparently sloppy splicing mechanism leads to even greater antibody diversity!

One of the truly fascinating discoveries in biology in the last few years has been the realization that such an enormous number of different antibody molecules can be genetically coded for by a relatively small amount of DNA, as a result of genetic recombination. As alluded to above, the discovery of gene rearrangements in lymphocytic cells has had a marked impact on our understanding of eucaryotic genetics. Biologists have always assumed that every cell of a multicellular organism is genetically identical (barring any mutation, of course) and that the differences between cell types were a function of differential gene expression. This clearly is not the case with B lymphocytes, as it is likely that several million (if not billion) genetically distinct lymphocytes exist in the mammalian body. In addition, the molecular biological dogma that states that one gene codes for the production of one polypeptide has also had to be abandoned in the case of antibody molecules. For example, although an immunoglobulin heavy chain consists of a single polypeptide, no fewer than four distinct genes are required to code for this molecule. For light chains at least three genes are required. It is not known whether lymphocytes are the only eucaryotic cell line in which gene rearrangements are commonplace, but it is hoped that an understanding of the mechanisms which direct these genetic shifts may help our understanding of the complex changes involved in cellular differentiation in animals.

Clonal selection The selection of certain B cells for expansion and subsequent antibody production is to a major degree a function of the antigenic history

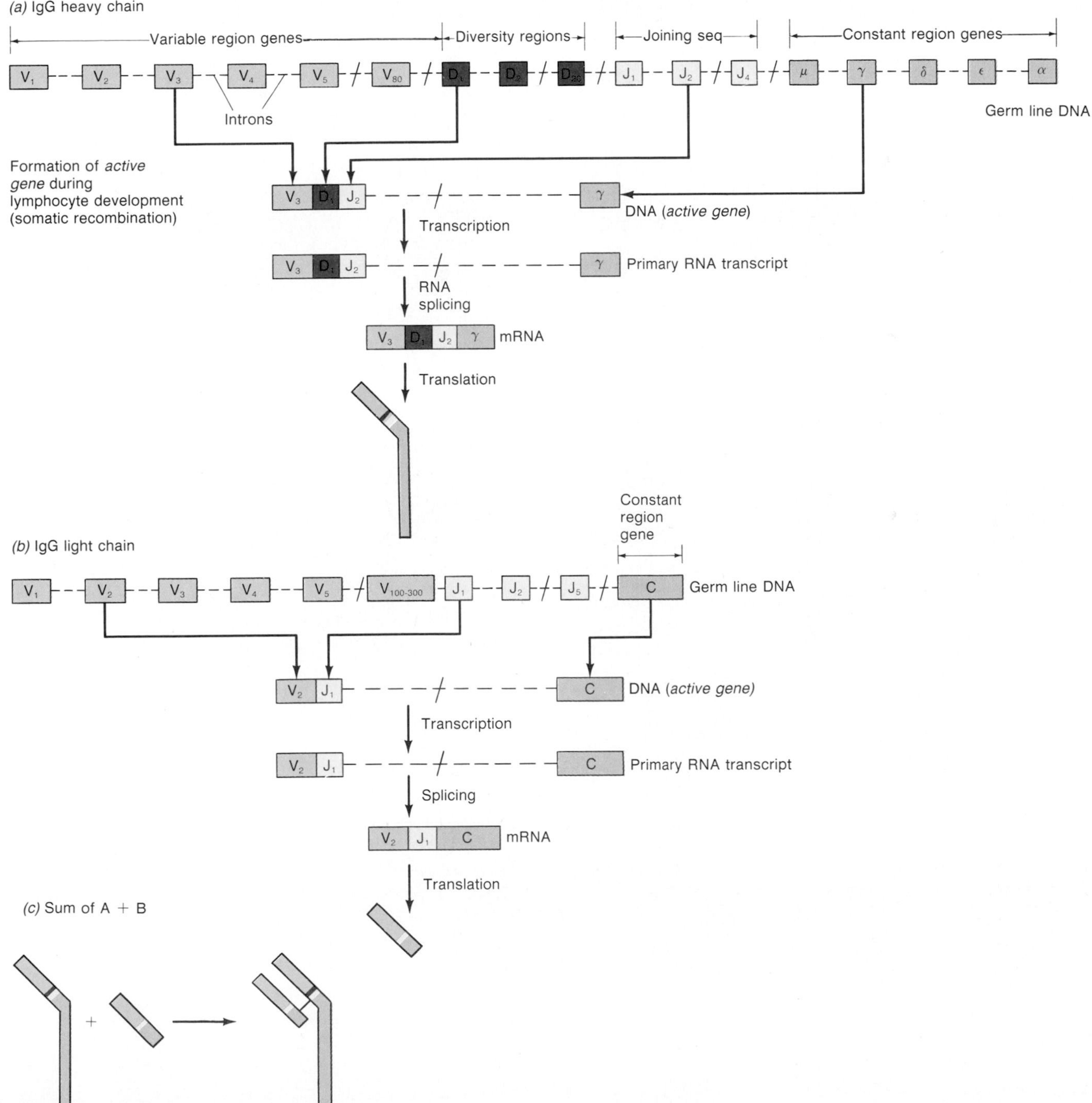

Figure 12.14 Immunoglobulin gene arrangement in immature lymphocyte and the mechanism of active gene formation. (a) Heavy chain. (b) Light chain. (c) Formation of one-half an antibody molecule.

of the animal. From the diverse pool of mature B lymphocytes discussed above, specific B cells are stimulated by antigen to expand and form a clone of B cells, all of identical genetic makeup (Figure 12.15). This idea, known as **clonal selection**, predicts that many B-cell types will remain "silent" within the animal's body due to lack of exposure to the correct antigen. Presumably, however, this large pool of antibody diversity remains available for future selective expansion when reactive antigens are encountered.

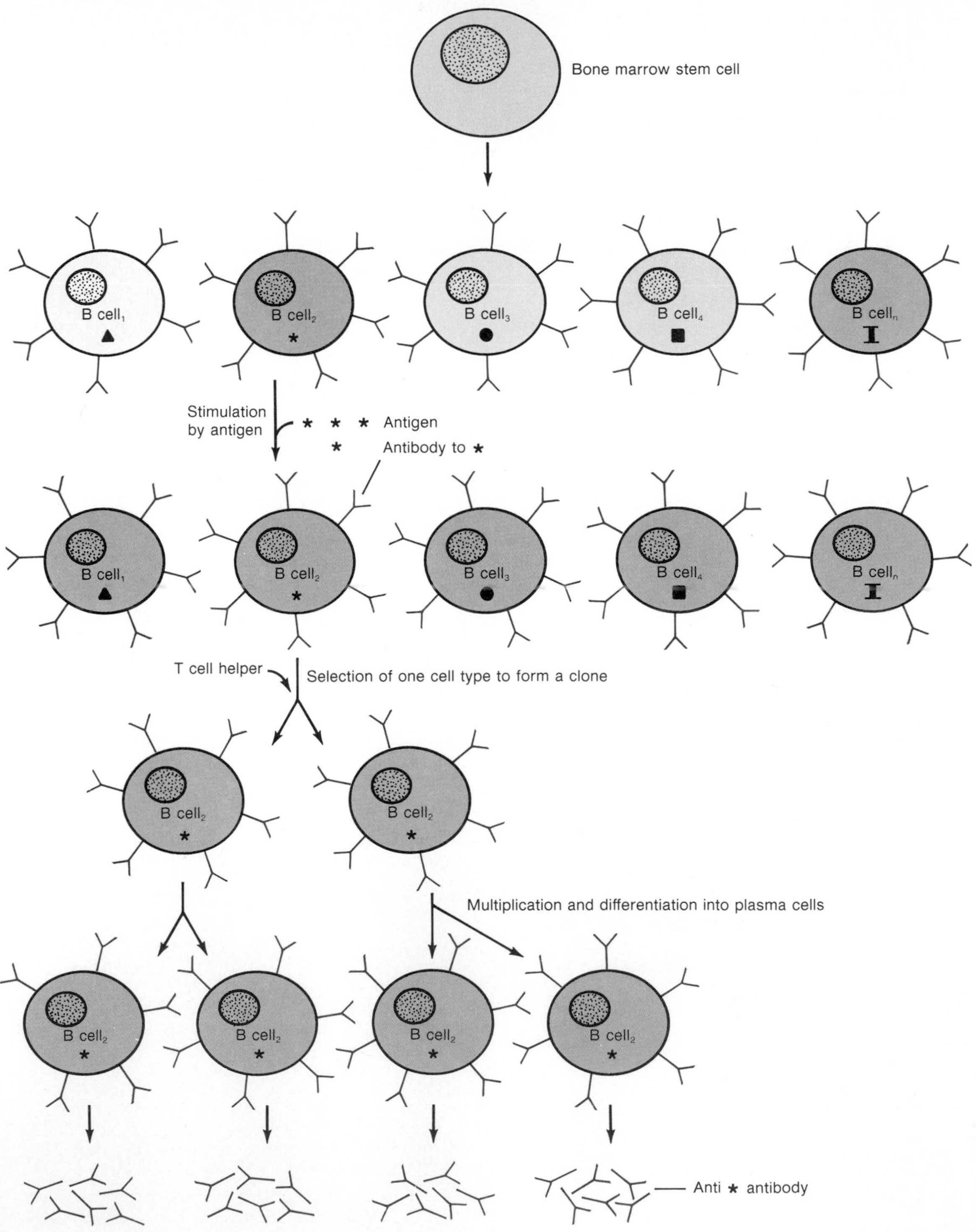

Figure 12.15 Clonal selection. The expansion of a particular antibody-producing cell line after stimulation with antigen.

Genetics of T-cell receptors Like B cells, T cells must recognize antigens in a very specific way. Indeed, the T-cell receptor rivals antibody in terms of its antigenic specificity. How is this accounted for in genetic terms?

T-cell receptors contain regions of "constant" and regions of "variable" amino acid sequences. Therefore, in analogy to the genetics of antibody production, it is likely that a number of genes are involved in coding for the variable regions of T-cell receptors. From studies of T-cell receptors in mice, it has been shown that about 20 variable region genes exist, and that "joining" and "diversity" genes exist, at least for the β-chain of the receptor. Preliminary estimates from studies of mouse T-cell receptor α chains suggest that hundreds of variable and joining genes exist to code for the different amino acid sequences observed in the variable region of the α chain; no diversity genes are present in the DNA coding for the mouse α chain.

The general picture of variable, joining, and diversity genes found in the mouse seems to hold for the human system as well. From what is known in the mouse system thus far, the mixing and matching of all possible genetic combinations for the α and β chains of the mouse T-cell receptor, should allow coding for the synthesis of about 10^7 different T-cell receptors. Not surprisingly, this is exactly the order of magnitude predicted for the number of possible *antibody* types produced by different B cells. It thus appears that a specific T cell receptor exists for nearly every distinct immunoglobulin type. Could T-cell receptors and antibody molecules have common evolutionary roots?

Evolution of molecules of the immune system From an evolutionary standpoint, Class I and II MHC proteins, T-cell antigen receptors, and immunoglobulin molecules share a number of common structural features, and may represent increasingly complex stages in the evolution of the immune system. Molecular cloning and sequencing of genes coding for immunoglobulins and T-cell receptors strongly suggests that a family of genes has evolved from a primordial gene or set of genes that originally coded for very simple, and presumably much less specific, immune molecules. This can be visualized by comparing the structures of the molecules discussed thus far in this chapter (Figure 12.16). By gene duplication, mutation, and insertion of new sequences, the genes coding for Class I and Class II major histocompatability antigens could have evolved to yield the genetic organization necessary to code for T-cell receptors, with their increased specificity due to the presence of variable regions (Figure 12.16). Evolutionary modifications of T-cell receptors through the acquisition of genes coding for additional constant regions would then have led to the immunoglobulins. It is likely that during evolution of the immune system, selection was based on the need for molecules of greater and greater specificity, culminating in the highly specific immunoglobulins and T-cell receptors. Simpler molecules (such as MHC antigens and Thy-1) were presumably retained, because they could play other roles in the immune response that did not require the great specificity of T-cell receptors or antibodies (the molecule Thy-1 has not been discussed in this chapter, but is simply one of many T-cell surface antigens; B cells do not contain the Thy-1 antigen).

Because the molecules of the immune system are obviously related in an evolutionary sense, the term gene "superfamily" has been proposed to describe the genes which code for these molecules. The gene superfamily presumably arose from evolutionary selection on genes coding for primitive immune molecules in ancient organisms, and this genetic progression is apparent today when the immune system of various animals is compared. For example, a com-

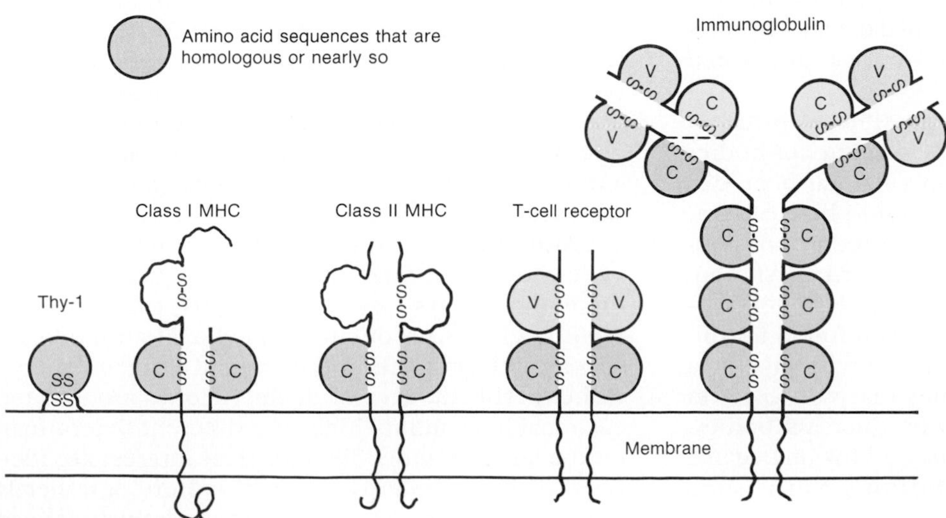

Figure 12.16 Comparative structure of several molecules of the immune system. Note those regions of homologous or nearly homologous amino acid sequences that strongly suggest evolutionary relatedness. C = constant regions; V = variable regions.

parison of immune reactions in vertebrates and invertebrates suggests that only portions of the gene superfamily exist in invertebrates. Invertebrates are capable of eliciting "primitive" immune responses, such as secreting antibacterial substances and carrying out phagocytosis. Echinoderms, for example, secrete lymphokines and proteins with antibacterial activity similar to antibodies. However, invertebrates do not show humoral or cellular immunity based on T and B lymphocyte activity, as do the vertebrates. In addition, some higher invertebrates can actually reject grafted foreign tissues, suggesting that MHC-like molecules may be present on their cell surfaces. But it is only in the vertebrates that we see the highly developed MHC-dependent immune functions, and the complex genetic system required to code for the many molecules involved. Although studies of the genetics of the immune response in lower animals have barely begun, the known relationships between the immune response of vertebrates and invertebrates lends support to the gene superfamily concept of how the immunological system of higher animals has evolved.

12.7 Monoclonal Antibodies

A typical immune response to a given antigen results in the production of a broad spectrum of antibodies of various classes, affinities, and specificities for the determinants present on the antigen. The antibodies directed toward a particular determinant will represent only a small portion of the total antibody pool. An antiserum containing this mixture of antibodies to a given antigen is referred to as a **polyclonal antiserum**. However, for certain research and medical applications, it would be desirable to be able to obtain antibodies of a *single* specificity. Methods for accomplishing this are now available and a variety of monospecific or **monoclonal antibodies** have been generated for use in research and clinical medicine. Table 12.3 compares the properties of antibodies prepared against an antigen in the usual way—by preparing a polyclonal antiserum—with monoclonal antibodies.

Theoretically, it should be possible to isolate a single B lymphocyte capable of forming antibodies and grow it in cell culture. From what has been discussed thus far in this chapter it should be clear that such a B cell would produce monospecific antibodies. However, for unknown reasons normal lymphocytes are not easy to grow and maintain in cell culture, hence this approach has not been found useful. However, B lymphocytes can be fused with myeloma (tumor) cells to form B cell lines that will grow in culture, yet retain the ability to produce antibodies. This technique of B cell/myeloma cell fusion became known as the **hybridoma technique**, and is summarized in Figure 12.17.

Table 12.3 Characteristics of monoclonal and polyclonal systems

Polyclonal	Monoclonal
Contains antibodies recognizing many determinants on an antigen	Contains an antibody recognizing only a single determinant
Various classes of antibodies are present (IgG, IgM, etc.)	Single antibody class (IgG only)
To make a specific system, highly purified antigen is necessary	Can make a specific system using an impure antigen
Reproducibility and standardization difficult	Highly reproducible

How are monoclonal antibodies made? First, a mouse is immunized with the antigen of interest and a period of weeks allowed for B-cell clones to be selected and begin producing antibody by the normal sequence of events. Then, spleen tissue, rich in B lymphocytes, is removed from the mouse and the B cells fused with malignant B cells that have been previously inactivated so as not to produce their own antibodies (Figure 12.18). Although true fusions represent only a small fraction of the total cell population remaining in the mixture, the addition of the compounds hypoxanthine and thymidine to the medium (the so-called *HAT* medium) was found to strongly select for *fused* hybrids. This occurs because unfused myeloma cells are unable to use the metabolites hypoxanthine and thymidine to bypass a metabolic block caused by the cell poison, aminopterin. By contrast, fused hybrid cells are able to use hypoxanthine and thymidine and thus grow normally in HAT medium. Unfused B cells can also use these metabolites, but die off in a week or two anyway, because they are unable to grow in culture for more than a few cell divisions. Following fusion, the most important job is to identify the clones of interest; that is, if individual fused cells are placed in the wells of microtiter plates and allowed to grow and produce antibody (Figure 12.18), which ones are producing the antibodies of interest?

Although not a simple matter, it turns out that a variety of techniques, most notably ELISAs and radioimmunoassays (see Section 12.9), can be used to identify clones producing monoclonal antibodies to specific determinants of the original antigen. In a typical experiment, several distinct clones are isolated, each making a monoclonal to a different determinant on the antigen. Once the clones of interest are identified they can be grown in cell culture, a rather laborious procedure that yields only relatively low an-

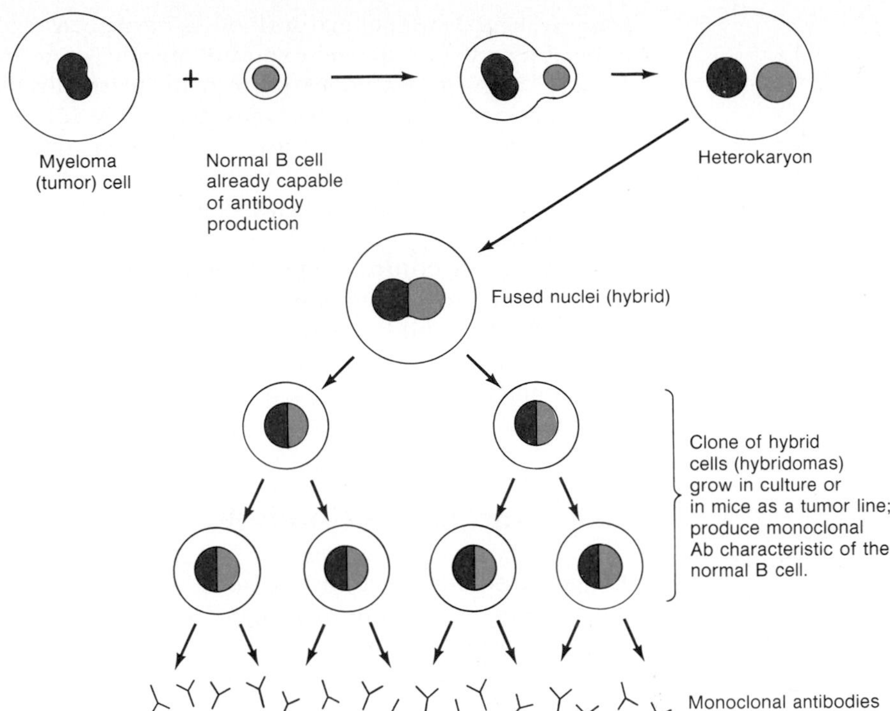

Figure 12.17 Basic theory behind monoclonal antibody formation.

tibody titers, or they can be injected into mice to perpetuate the fused cell lines as a mouse myeloma. As a mouse tumor line, hybridomas secrete large amounts of monoclonal antibody (in mice over 10 milligrams of pure monoclonal antibody can be obtained per milliliter of mouse peritoneal fluid). The specific hybridomas of interest can also be stored in the frozen state and later thawed and injected back into mice when more of a specific monoclonal is needed.

Monoclonal antibodies are extremely useful reagents for research and medical science. For research purposes, monoclonal antibodies directed against specific cell markers, for example the T4 and T8 lymphocyte markers (see Section 12.4) can be used to separate mixtures of these cells from one another. By attaching a fluorescent dye to the monoclonal antibody, those cells to which the monoclonal antibody binds will now fluoresce. Fluorescing cells can be separated from their nonfluorescing counterparts with a fluorescence activated cell sorter (FACS), an instrument employing a laser beam to place an electrical charge on fluorescing molecules (in this case fluorescing cells). Following laser treatment, the application of an electrical field to the cell mixture will orient fluorescing and nonfluorescing cells to opposite ends of the electrical field where they can be collected in separate receptacles. FACS machines have given the immunologist the ability to separate complex mixtures of immune cells into highly enriched populations of one type or the other for use in immunological research. Other research applications of monoclonal antibodies include studies of the active site of enzymes or the combining site of antibodies themselves by observing the effect on an enzyme's activity or an antibody's binding ability when treated with monoclonal antibodies to specific determinants on the molecule. Monoclonal antibodies show great promise in the immunological typing of bacteria and in the identification of cells containing foreign surface antigens (for example, a virus infected cell). Monoclonal antibodies have also been used in genetic engineering for measuring levels of gene products not detectable by other methods.

Of considerable importance are the applications of monoclonal antibodies to clinical diagnostics and medical therapeutics. We discuss the use of monoclonal antibodies in clinical diagnostics in Section 13.7. In therapeutics, perhaps the boldest application of monoclonal antibodies will be in the detection and treatment of human malignancies. Malignant cells contain a variety of surface antigens not expressed on the surfaces of normal cells. These include differentiation antigens expressed during fetal development but not in the normal adult, and unique glycolipid cell surface antigens (the latter were originally detected by monoclonal antibodies prepared against random tumor cell surface antigens). Because monoclonal antibodies prepared against cancer-related antigenic determinants should be able to distinguish between normal and malignant cells, monoclonal antibodies may serve as vehicles for directing toxins or radioisotopes to malignant cells. Such specificity would greatly improve cancer chemotherapy, be-

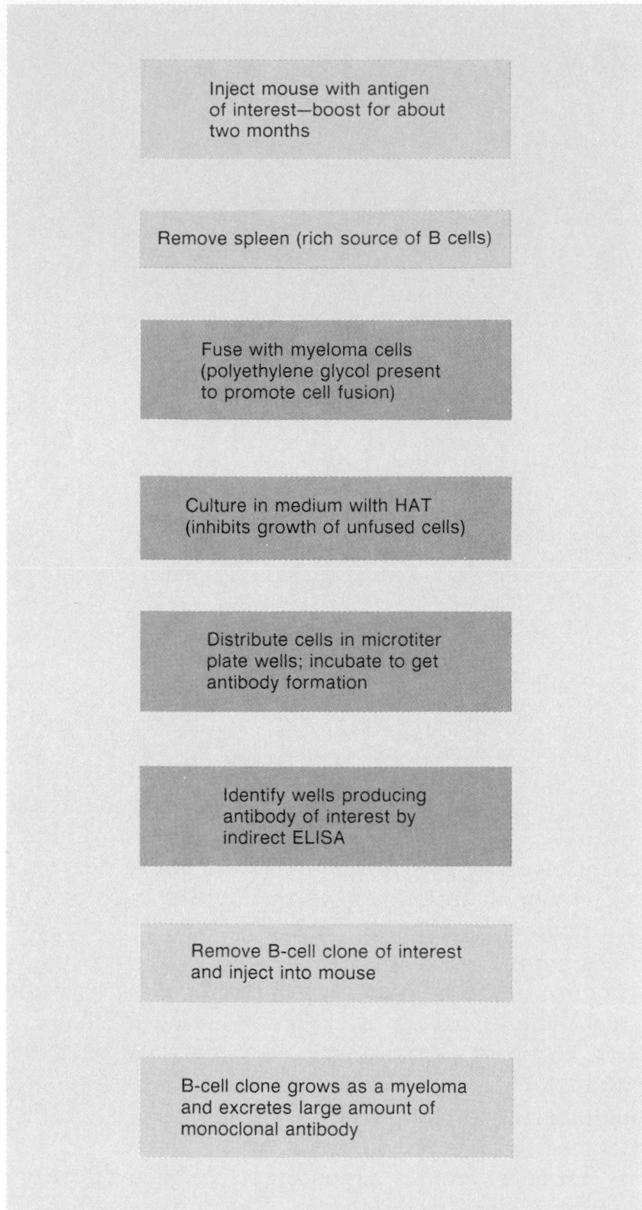

Inject mouse with antigen of interest—boost for about two months
Remove spleen (rich source of B cells)
Fuse with myeloma cells (polyethylene glycol present to promote cell fusion)
Culture in medium with HAT (inhibits growth of unfused cells)
Distribute cells in microtiter plate wells; incubate to get antibody formation
Identify wells producing antibody of interest by indirect ELISA
Remove B-cell clone of interest and inject into mouse
B-cell clone grows as a myeloma and excretes large amount of monoclonal antibody

Figure 12.18 Production of monoclonal antibodies.

cause damage to normal cells by the toxic agents necessary to kill cancerous cells is one of the major problems with conventional cancer chemotherapy.

Monoclonal antibodies have also been prepared that can distinguish human transplantation antigens, thus improving the specificity of the tissue matching process critical for successful transplantations. Monoclonal antibodies also show great promise for increasing the specificity of a variety of conventional clinical tests including blood typing, rheumatoid factor determination, and even pregnancy determination; the latter test employs monoclonal antibodies prepared against specific hormones associated with the pregnant state. Other monoclonals are being developed for *in vivo* diagnostics. For example, radioactively tagged monoclonal antibodies have been used experimentally to detect exposed myosin in heart muscle; myosin is a major protein in heart muscle cells that normally only becomes detectable following damage to the heart muscle. A myosin-specific monoclonal antibody would therefore be useful for diagnosing the extent of heart damage following a heart attack.

As with any clinical reagent, the greater the specificity the more useful it is in clinical diagnostics. Monoclonal antibodies now offer the physician and clinical microbiologist the ultimate in immunological specificity. (See Box for a discussion of the use of monoclonal antibodies for urine testing for drugs.)

12.8 Antigen-Antibody Reactions

Antigen-antibody reactions are most easily studied in vitro using preparations of antigens and antisera. The study of antigen-antibody reactions in vitro is called **serology**, and is especially important in clinical diagnostic microbiology. A variety of different kinds of serological reactions can be observed (Table 12.4), depending on the natures of the antigen and antibody and on the conditions chosen for reaction. Serology has many ramifications, only a few of which will be discussed here.

Regardless of the final visible or otherwise measurable result, the principal of all antigen-antibody reactions lies in the *specific* combination of determinants on the antigen with the *variable* region of the antibody molecule (Figure 12.19). Recall that the recognition of antigen by antibody is ultimately governed by the higher order structure of the polypeptide chains in the antibody molecule. The folding itself is dictated by the primary structure—the amino acid sequence—which in turn is controlled by the genetic constitution of the B cell. The end result of this complex process is a specific antibody molecule which combines with a specific antigenic determinant as shown in Figure 12.19. This combination serves to distort the antigen sufficiently to either reduce or even totally eliminate the biological activity of the antigen.

Neutralization Neutralization of microbial toxins by specific antibody can occur when toxin molecules and antibody molecules directed against the toxin combine in such a way that the active portion of the toxin that is responsible for cell damage is blocked (Figure 12.20). Neutralization reactions of this type are known for a wide variety of exotoxins, including most of those listed in Table 11.4. Reactions of a similar nature also occur between viruses and their specific antibodies. For instance, antibodies directed against the protein coats of viruses may prevent the adsorption of the viruses to host cells. Neu-

Table 12.4 Types of antigen-antibody reactions

Location of antigen	Accessory factors required	Reaction observed
Soluble	None	Precipitation
On cell or inert particle	None	Agglutination
Flagellum	None	Immobilization or agglutination
On bacterial cell	Complement	Lysis
On bacterial cell	Complement	Killing
On erythrocyte	Complement	Hemolysis
Toxin	None	Neutralization
Virus	None	Neutralization
On bacterial cell	Phagocyte, complement	Phagocytosis (opsonization)

tralizing antibodies are also known for a variety of enzymes; these act by combining with the enzyme at or near the active site and prevent formation of the enzyme-substrate complex. Neutralizing antibody requires only a single antigen-combining site for action and thus two antigen molecules may be bound by one antibody molecule. An antiserum containing neutralizing antibody against a toxin is sometimes referred to as an **antitoxin** (see Section 12.13).

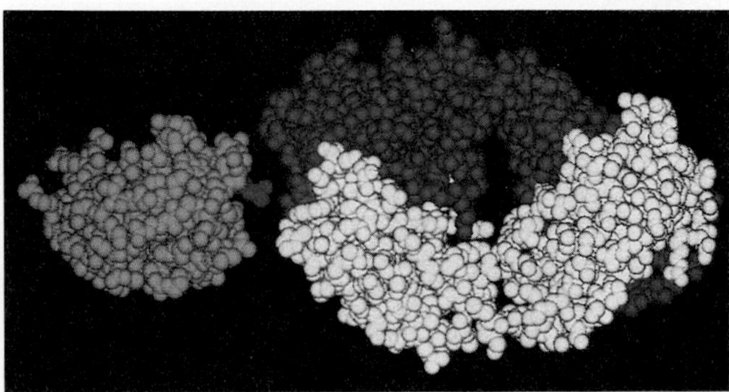

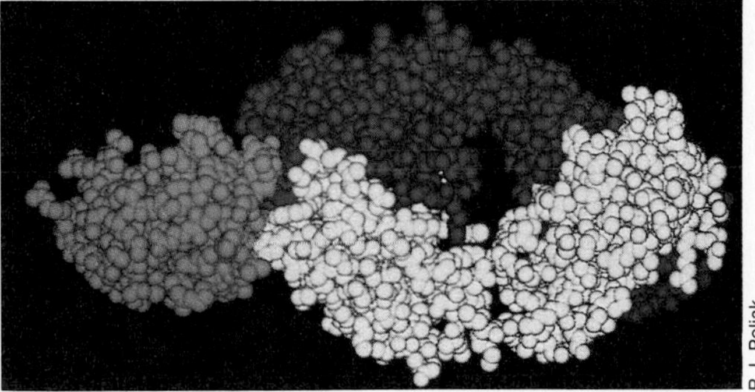

Figure 12.19 Structure of the combining site of an antigen and antibody. The antigen (lysozyme) is in green. The variable region of the antibody heavy chain is shown in blue, the light chain in yellow. The amino acid shown in red is a glutamine residue of lysozyme. The glutamine residue fits into a pocket on the antibody molecule, but the basis of antigen/antibody recognition involves contacts made between several other amino acids on both the antibody and antigen as well.

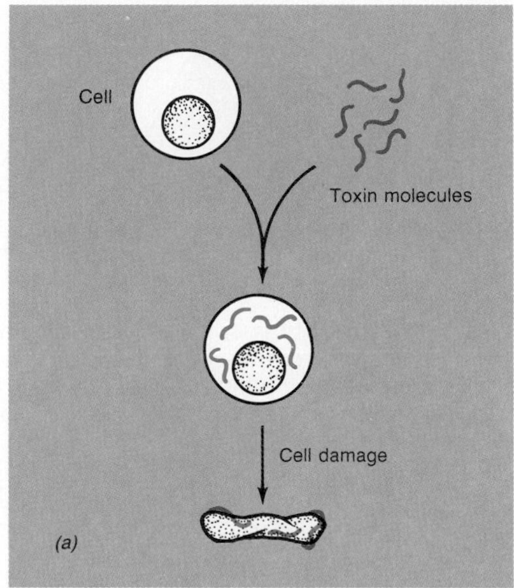

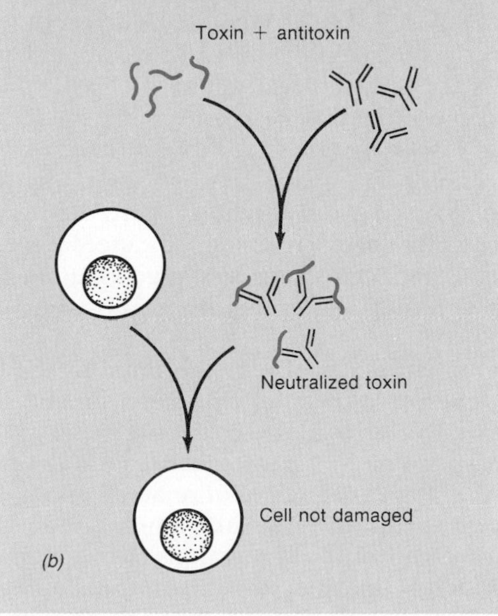

Figure 12.20 The antibody as antitoxin. (a) Toxin action. (b) Mechanism of neutralization of a toxin by antitoxin.

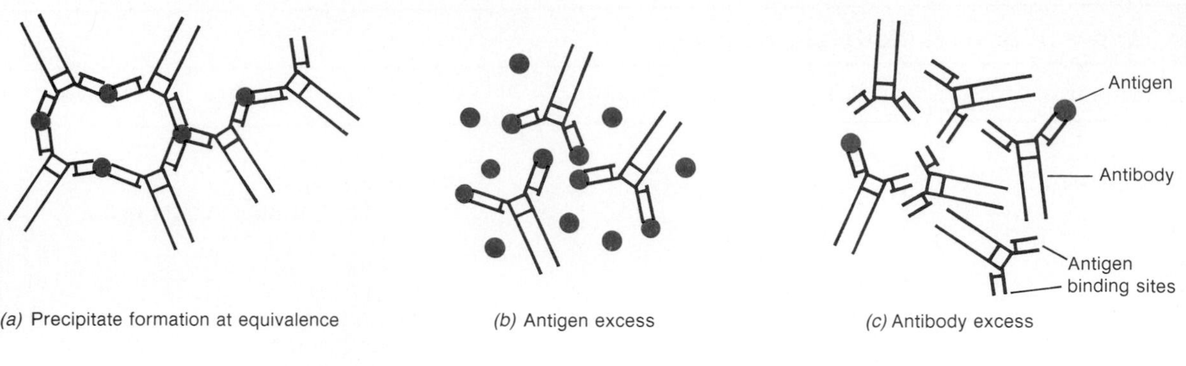

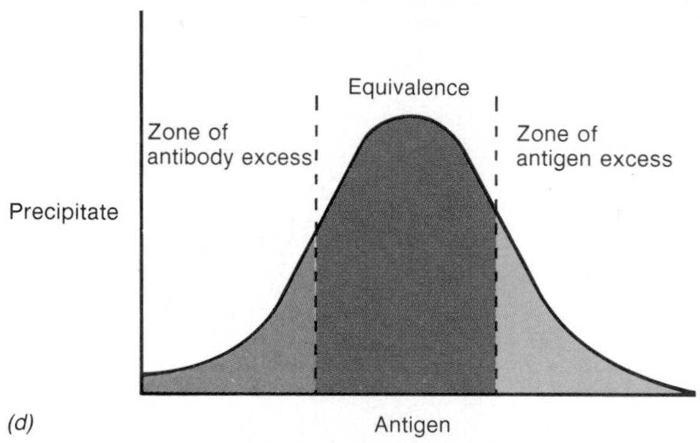

Figure 12.21 The antibody as precipitin. (a) Formation of precipitate between a soluble antigen and antibody at equivalence. No precipitate is observed when either (b) antigen or (c) antibody is in excess. (d) Results portrayed graphically with antibody constant and varying amounts of antigen added.

Precipitation Since antibody molecules generally have two combining sites (that is, are bivalent), it is possible for each site to combine with a separate antigen molecule, and if the antigen also has more than one combining site, a **precipitate** may develop consisting of aggregates of antibody and antigen molecules (Figure 12.21). Precipitation reactions for a wide variety of soluble polysaccharide and protein antigens are known. Because they are easily observed in vitro, precipitation reactions are very useful serological tests, especially in the quantitative measurement of antibody concentrations. Precipitation occurs maximally only when there are optimal proportions of the two reacting substances since, when either reactant occurs in excess, the formation of large antigen-antibody aggregates is not possible (Figure 12.21).

Precipitation can also be inhibited by the hapten corresponding to the antigenic determinant. In fact, hapten inhibition is an extremely useful way of studying the specificity of serological reactions since the inhibiting power of a series of related haptens can be compared quantitatively. Such studies show that although antigen-antibody reactions have a high degree of specificity, this specificity is not absolute, and a given antibody will almost always react with a few structurally-related antigens. Also, polyclonal antisera contain mixtures of antibodies with varying specificities, and such heterogeneity must be taken into consideration in any serological study. Cross reactions can often be eliminated or minimized by allowing the antiserum to react first with heterologous antigens and then removing the precipitate that forms (a process referred to as *adsorption*). This will remove antibodies specific for the heterologous antigen and will leave behind in the supernatant those antibodies which do not react with heterologous antigens but do react with the homologous antigen.

Precipitation reactions carried out in agar gels have proved to be of considerable utility in the study

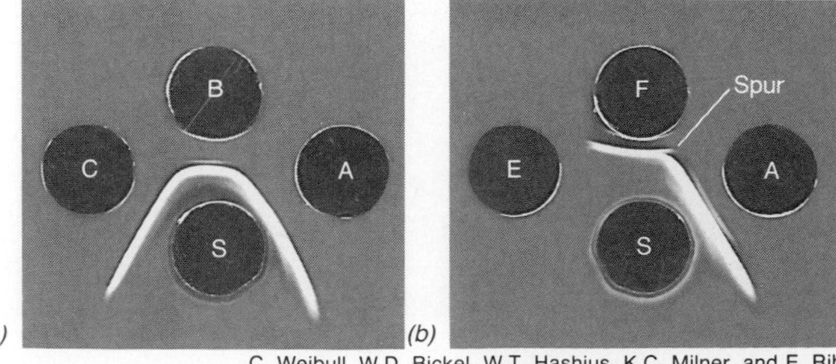

Figure 12.22 Formation of precipitation bands in agar gels due to reactions between antibody and soluble antigens. Wells labeled S contain antiserum to *Proteus mirabilis.* Wells labeled A, B, C contain soluble extracts of *P. mirabilis* cells. In (a), the major antigen present is identical in all three wells and a line of identity is observed. In (b), antigen E does not react, and antigen F shows a band of partial identity with antigen A.

of the specificity of antigen-antibody reactions. Both antigen and antibody diffuse out from separate wells cut in the agar gel and precipitation bands form in the region where antibody and antigen meet in equivalent proportions (Figure 12.22). The shapes of the precipitation bands and the distance from the wells are characteristic for the reacting substances, and it is possible to determine whether two antigens reacting against antibodies in an antiserum are identical or not by observing the bands formed when the two antigens are placed in adjacent wells near the antiserum well (Figure 12.22). For example, if two antigens in adjacent wells are identical they will form a fused precipitin band. This is referred to as a *line of identity*. If, on the other hand, adjacent wells contain one antigen in common but one well contains a second antigen, a *line of partial identity* will form (Figure 12.22). The small protruding precipitin line (representing a reaction between the antiserum and the second antigen) is referred to as a *spur* (Figure 12.22).

Agglutination If an antigen is not in solution but is present on the *surface* of a cell or other particle, an antigen-antibody reaction can lead to clumping of the particles, called **agglutination**. When foreign cells are injected into an animal, antibodies are formed against a wide variety of their macromolecular constituents, including those in the cytoplasm as well as on the surface of the cells. Only the surface antigens are involved in agglutination, however, since only these are exposed in the intact cell. Considerable attention has been focused on the chemical nature of the surface macromolecules that might be involved in agglutination of bacterial cells. Antibodies against both the flagella (so-called H antigens) and the lipopolysaccharide layer (somatic or O antigens) will agglutinate bacterial cells, but the nature of the clump differs; that of flagellar agglutination is much looser and more flocculent (Figure 12.23).

Soluble antigens can be adsorbed to or coupled chemically to cells or other particulate structures such as latex beads or colloidal clay, and they can then be detected by agglutination reactions, the cell or particle serving only as an inert carrier. This greatly increases the ability to detect the presence of antibodies against soluble antigens, since agglutination is much more sensitive than precipitation.

Agglutination of red blood cells is referred to as *hemagglutination*. Red blood cells contain a variety of antigens, and people differ in the antigens present on their red blood cells. The classification of blood antigens is called *blood typing* and is a good example of an agglutination reaction. The principle of blood typing is that antibodies, specific for particular erythrocyte surface antigens, cause red blood cells to visibly clump. The human red cell antigens called A, B, and D are most commonly tested for. In the United States 42 percent of all humans contain major group A antigen (Type A blood), 8 percent contain antigen B (Type B), 3 percent contain both A and B (Type AB), and 47 percent contain neither antigen group A *or* B on their erythrocyte surfaces and are called Type O. Humans also contain antibodies to certain blood group antigens. Type A humans contain antibodies to group B antigens, while Type B individuals have antibodies to group A antigens. Type AB individuals contain neither A nor B antibodies while Type O's have both. Antibodies against A and B antigens are called *natural antibodies*, since they appear to be produced even in the absence of antigenic stimulation. The explanation for this resides in the fact that the A and B antigens contain polysaccharide determinants which, in addition to being present on red blood cells, are also found in certain common foods (the antigens in food and those on the red blood cell are said to *cross-react*). Since such foods are frequently eaten, antigenic stimulation against A and B substances occurs throughout life. Self-recognition insures that anti-

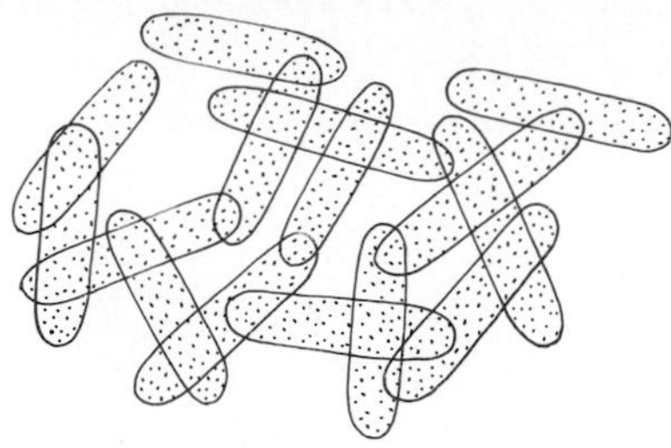

(a) Antiserum against somatic antigen

(b) Antiserum against flagellar antigen

Figure 12.23 The antibody as agglutinin. (a) Appearance of agglutinated bacterial cells when antibody is directed against somatic, or O, antigen. (b) Appearance of agglutinated bacterial cells when antibody is directed against flagellar, or H, antigen.

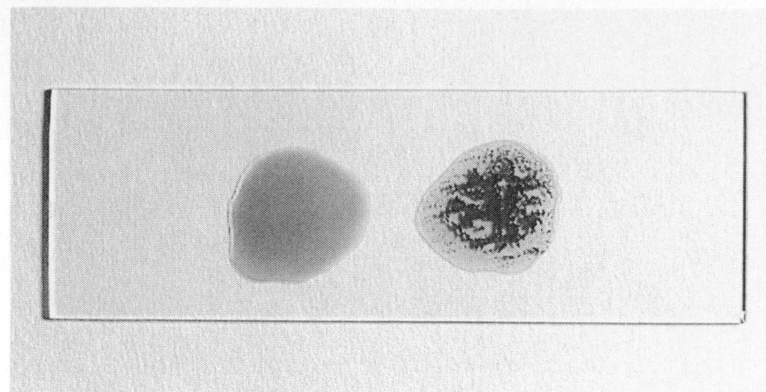

Blood	Serum	
	Anti A	Anti B
Type O	No aggl.	No aggl.
Type A	Aggl.	No aggl.
Type B	No aggl.	Aggl.
Type AB	Aggl.	Aggl.

Figure 12.24 Typing of human blood as to ABO group by an agglutination assay of whole blood—"blood typing." Also given are expected results with different kinds of blood types.

body is not formed against A or B if that antigen is also on the red blood cells, but antibody will be formed against either of these antigens if they are not on the red cells.

The reason for blood typing before blood transfusion is to prevent the blood-cell agglutination that would occur if blood containing a particular antigen were transfused into a person containing an antibody against this antigen. If agglutination occurred, the clumps of cells could lodge in blood vessels or arteries and block the flow of blood, causing serious illness or death.

The presence or absence of the A and B antigens or antibodies is the basis for the ABO blood grouping assay (Figure 12.24). A small drop of blood is placed on a microscope slide containing antiserum prepared against either the A or B antigens of human erythrocytes (Figure 12.24). These commercially available high titer antisera are prepared by injecting an animal such as a goat with human red blood cells of the A or B type. The agglutination reaction observed in the ABO blood grouping test is rapid and easy to interpret (Figure 12.24), hence this simple assay is used routinely to group human blood. At least 15 different blood group antigens have been described in man, the ABO system being one of the most important.

Fluorescent antibodies Antibody molecules can be made fluorescent by attaching to them fluorescent organic compounds such as rhodamine B or fluorescein isothiocyanate (Figure 12.25). This does not alter the specificity of the antibody significantly, but makes it possible to detect the antibody bound

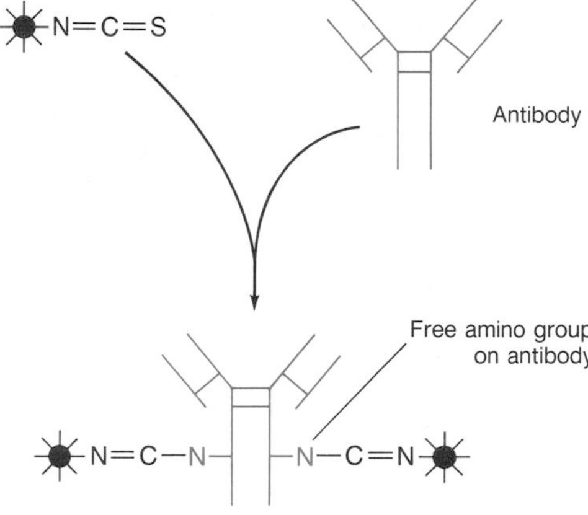

Figure 12.25 Fluorescent antibody. Preparation of fluorescent antibody by coupling the fluorescent dye fluorescein isothiocyanate to the antibody protein.

to cell or tissue surface antigens by use of the fluorescence microscope (Figure 12.26). Cells to which fluorescent antibodies have bound emit a bright fluorescent color, usually green, red, or yellow, depending on the dye used. Fluorescent antibodies have been of considerable utility in diagnostic microbiology since they permit the study of immunological reactions on single cells. The fluorescent-antibody technique is very useful in microbial ecology as one of

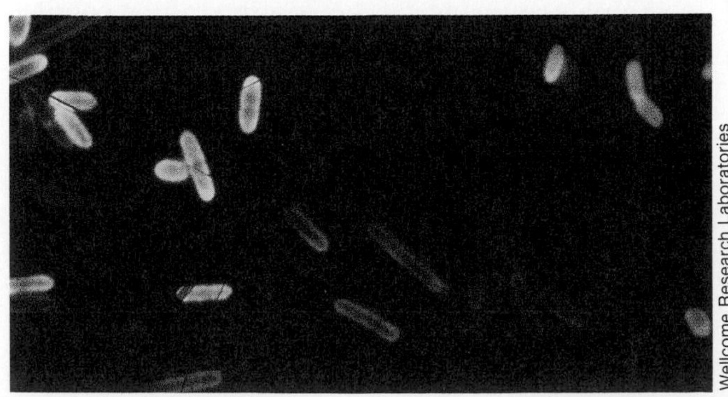
Wellcome Research Laboratories

Figure 12.26 Fluorescent antibody reactions. Cells of *Clostridium septicum* were treated with antibody conjugated with fluorescein isothiocyanate, which fluoresces green. Cells of *Clostridium chauvei* were stained with antibody conjugated with rhodamine B, which fluoresces red.

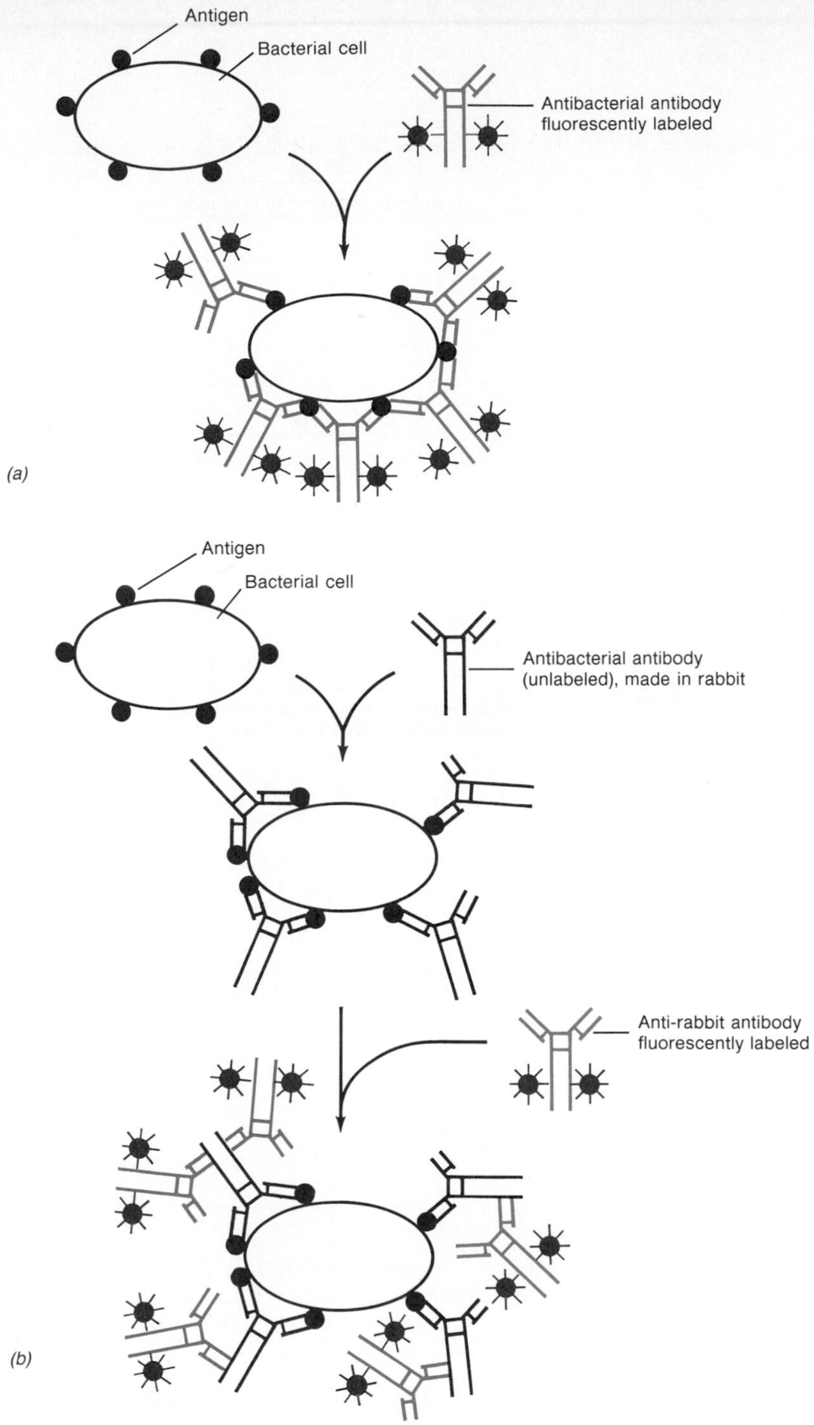

Figure 12.27 Methods of using fluorescent antibodies to detect bacterial surface antigens. (a) Direct staining method. (b) Indirect staining method.

the few methods for identifying microbial cells directly in the natural environments in which they live.

Two distinct fluorescent antibody procedures, the **direct** and the **indirect** staining methods, are used (Figure 12.27). In the direct method, the antibody against the organism is itself fluorescent. In the indirect method, the presence of a nonfluorescent antibody on the surface of the cell is detected by the use of a fluorescent antibody directed against the nonfluorescent antibody (Figure 12.27). This is possible because immunoglobulins, as all proteins, are antigenic, and the immunoglobulins of one animal species can induce antibody formation in those of another. For this reason, fluorescent goat anti-rabbit immunoglobulin can be used to detect the presence of rabbit immunoglobulin previously adsorbed to cells. One of the advantages of the indirect staining method is that it eliminates the need to make the fluorescent antibody for each organism of interest.

12.9 ELISA and Radioimmunoassay

The specificity of antibodies are such that the limiting factor in most of the immunological reactions discussed thus far is not specificity, but *sensitivity*. That is, a reaction between an antigen and an antibody, no matter how specific, is essentially useless for clinical or research purposes if it is not possible to measure its occurrence. Because of their exquisite sensitivity,

Antibody
1. Antibody to antigen is bound to microtiter well

Antigen
2. Blood sample added as source of antigen

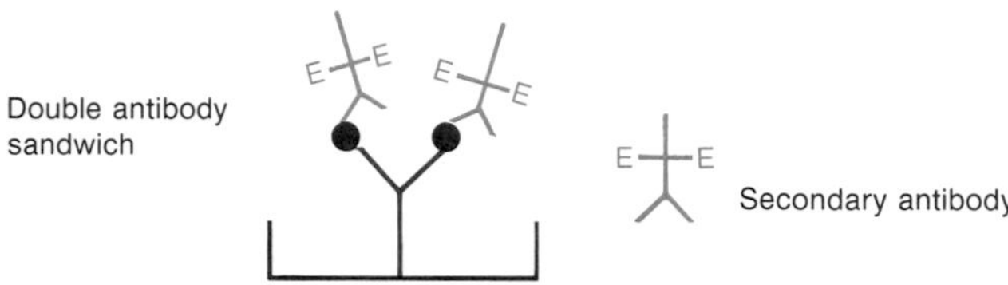
Double antibody sandwich
E
Secondary antibody
3. Antibody labeled with enzyme (E) added. Enzyme-labeled antibody binds to antigen

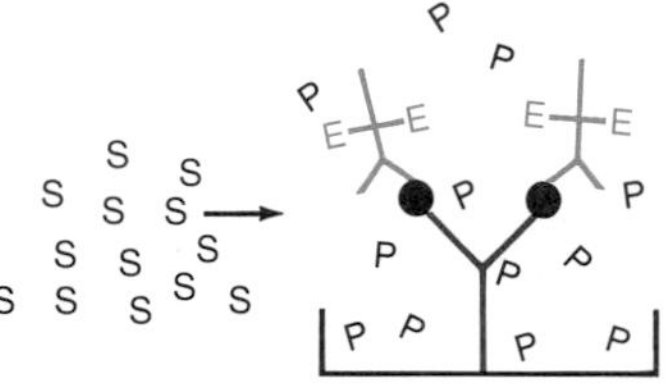
S
P
E
4. Substrate (S) for enzyme is added. Colored product of enzyme formed (P); color is proportional to amount of antigen

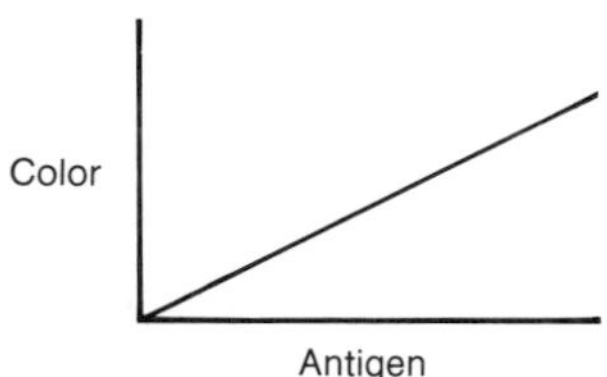
Color
Antigen

Figure 12.28 Enzyme-linked immunosorbent assay (ELISA). Direct ELISA for detecting antigen (double antibody sandwich method). The antigen is the "meat," the two antibodies the "bread."

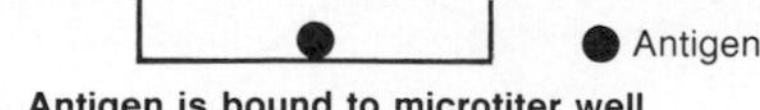
Antigen
1. Antigen is bound to microtiter well

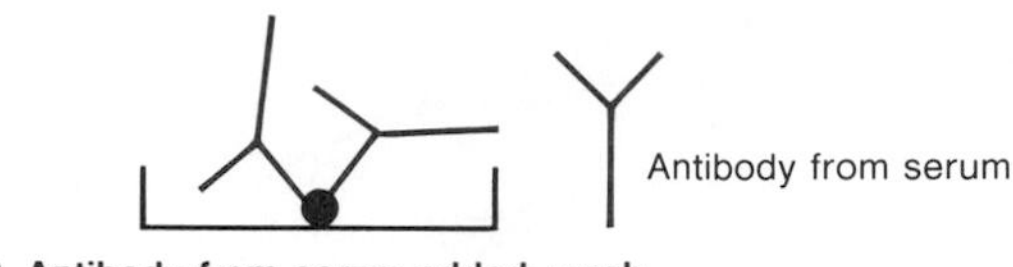
Antibody from serum
2. Antibody from serum added; wash

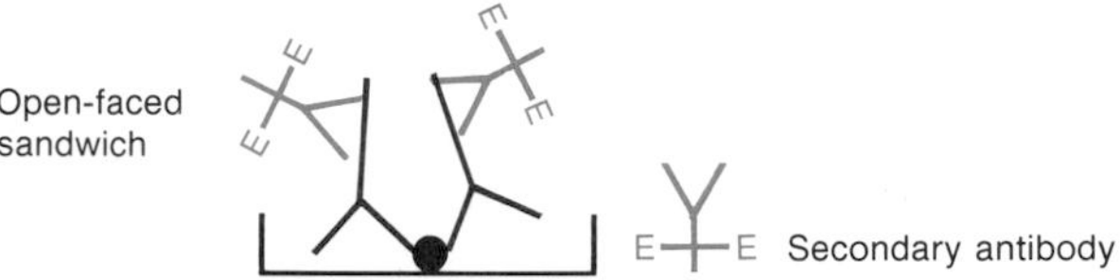
Open-faced sandwich
E
Secondary antibody
3. Anti-human immunoglobulin antibody labeled with enzyme (E) added; wash

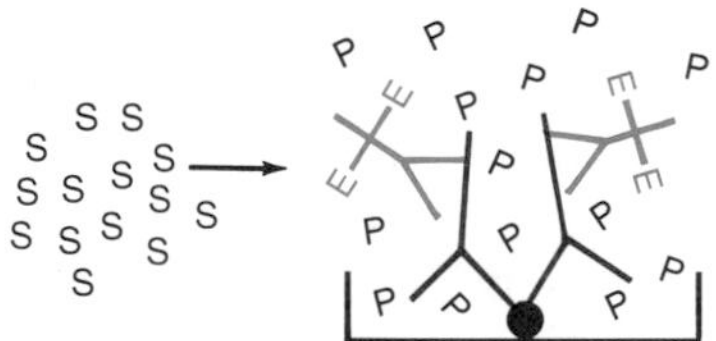
S
P
E
4. Add enzyme substrate (S). Color formed (P) is proportional to amount of antibody in serum

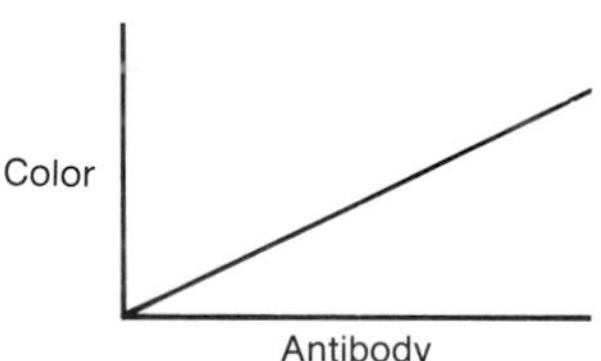
Color
Antibody

Figure 12.29 Indirect ELISA for detecting specific antibodies (open-faced sandwich method). The antigen is the "bread," the first antibody is the "mayonnaise," the second antibody is the "meat."

radioimmunoassay (RIA) and enzyme-linked immunosorbent assay (ELISA) are the two most widely used immunological techniques. These methods employ radioisotopes and enzymes, respectively, as ligands to antibody molecules. Because radioactivity and the products of certain enzymatic reactions can be measured in very small amounts, the attachment of radioactive or enzyme ligands to antibody molecules serves to *decrease* the amount of antigen-antibody complex required in order to detect that a reaction has occurred. This increased sensitivity has been extremely helpful in clinical diagnostics and research, and has opened the door to the development of a variety of new immunological tests, previously impossible because the methods available were not sufficiently sensitive.

ELISA The covalent attachment of enzymes to antibody molecules creates an immunological tool possessing both high specificity and high sensitivity. The technique, called ELISA (*e*nzyme-*l*inked *i*mmunosorbent *a*ssay), makes use of antibodies to which enzymes have been covalently bound such that the enzyme's catalytic properties and the antibody's recognition properties are retained. Typical linked enzymes include peroxidase, alkaline phosphatase, and β-galactosidase, all of which catalyze reactions whose products are colored and can be measured in very low amounts.

Two basic ELISA methodologies have been developed, one for detecting antigen (*direct* ELISA) and the other for detecting antigens or antibodies (*indi-*

The Microtiter Plate and Immunology

A number of modern immunoassays have been developed that require that either the antigen or the antibody be bound to a solid support. A variety of solid phase carriers such as latex beads or plastic tubes have been used, but the plastic, disposable *microtiter plate* has proved the most successful solid support in modern immunoassays. Techniques such as the ELISA test and radioimmunoassay were designed to be used with microtiter plates.

Microtiter plates are small plastic trays containing a number of wells.

Perkin-Elmer Cetus

The plates are made of polystyrene or polypropylene, both of which are transparent plastics. Proteins, either antigens or antibodies, adsorb to the plastic surface of the microtiter plate as a result of interactions between hydrophobic regions of the protein and the nonpolar plastic surface. Once bound, the proteins are not easily removed by washing or other manipulations and thus bound proteins can be treated with various reagents and washed repeatedly without being removed from the plate surface. The standard microtiter plate contains 96 wells, which allows for several replicates of each sample to be run. For quantitation, several dilutions are prepared. Although only small amounts of protein can be adsorbed to each well, the sensitivity of ELISA or radioimmunoassays are such that only small amounts of antigen or antibody are needed anyway.

Automatic machines are available for adding reagents and preparing dilutions. Depending upon whether the microtiter plate assay is an ELISA or a radioimmunoassay, special machines are available to read the adsorbance of a colored product (ELISA) or amount of radioactivity (radioimmunoassay) in each of the wells of the plate. Such automation allows these immunoassays to be performed both rapidly and routinely at relatively low cost. The microtiter plate has therefore helped to revolutionize research and diagnostic immunoassays. ELISAs in particular have become part of the everyday activities of the clinical laboratory (see Section 13.8).

rect ELISA). For detecting antigen by a **direct ELISA**, such as detecting virus particles from a blood or feces sample, the so-called *double-antibody sandwich technique* is used. In this procedure the antigen is "trapped" (sandwiched) between two layers of antibodies (Figure 12.28). The specimen is added to the wells of a microtiter plate previously coated with antibodies specific for the antigen to be detected. If the antigen (virus particles) is present in the sample it will be trapped by the antigen binding sites on the antibodies. After washing unbound material away, a second antibody containing a conjugated enzyme is added. The second antibody is also specific for the antigen so it will bind to any remaining exposed determinants. Following a wash, the enzyme activity of the bound material in each microtiter well is determined by adding the substrate of the enzyme. The color formed is proportional to the amount of antigen originally present (Figure 12.28).

To detect *antibodies* in human serum, an **indirect ELISA** is used (Figure 12.29). The microtiter wells in this case are coated with antigen and a serum sample added. If antibodies to the antigen are present in the serum, they will bind to the antigen in the wells. Following a wash, a second antibody is added. The second antibody is a rabbit or goat anti-human IgG antibody preparation containing a conjugated enzyme (this antiserum is prepared by injecting a goat or a rabbit with human IgGs; these will be recognized as foreign and anti-human IgG antibodies made). Following the addition of enzyme substrate a color is formed, and the amount of circulating human antibody to the specific antigen is quantitated from the intensity of the color reaction (Figure 12.29).

Since ELISAs require very little in the way of expensive equipment and are highly sensitive, they are widely used in clinical laboratories. Direct ELISAs have been developed for detecting hormones, drugs (see Box), viruses, and a variety of other substances or disease agents in human blood and other tissues. In-

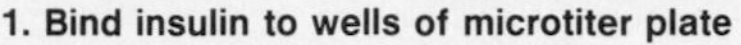
1. Bind insulin to wells of microtiter plate

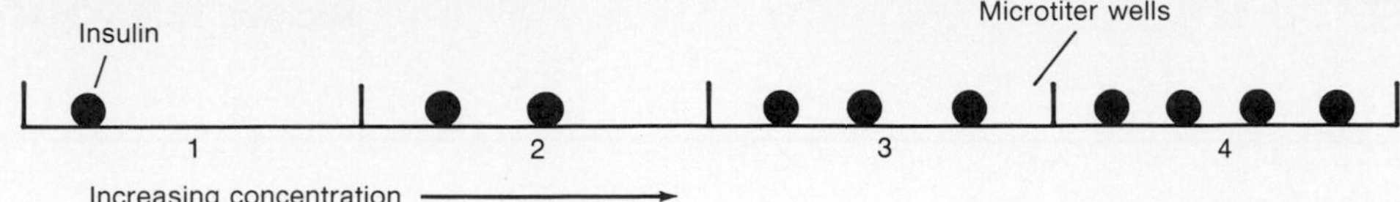
Insulin
Microtiter wells
1
2
3
4
Increasing concentration

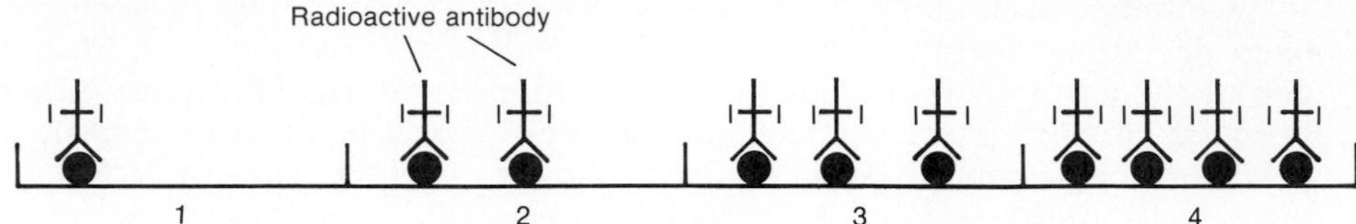
2. Add excess anti-insulin antibodies that are labeled with 125Iodine; wash to remove unbound antibody
Radioactive antibody
1
2
3
4
3. Count radioactivity in gamma radiation counter. This establishes a standard curve with known amounts of antigen (insulin).

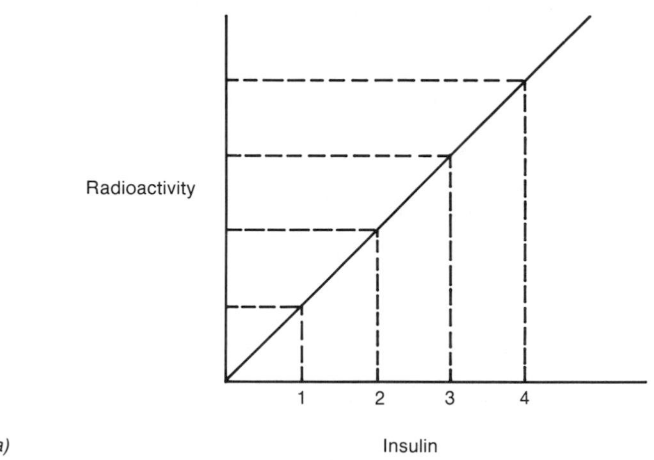
Radioactivity
1
2
3
4
(a)
Insulin

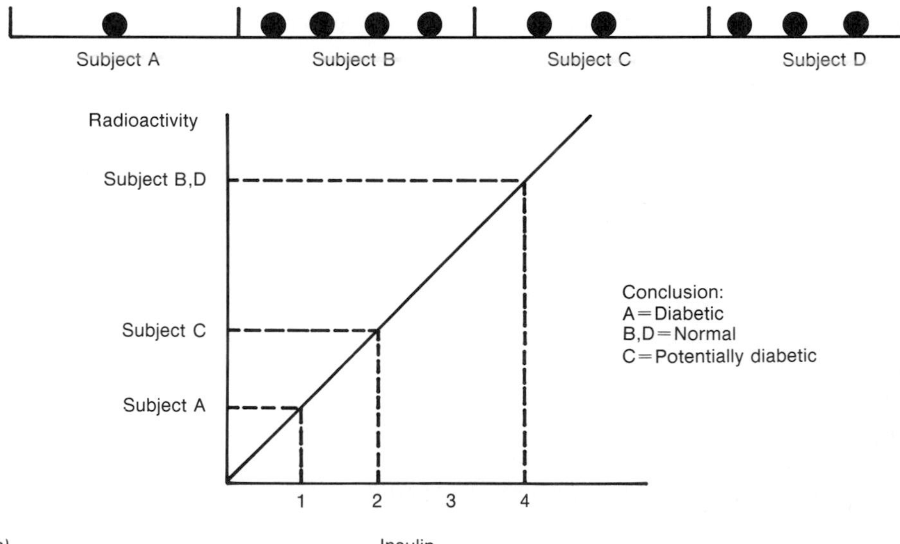
Subject A
Subject B
Subject C
Subject D
Radioactivity
Subject B,D
Subject C
Subject A
Conclusion:
A=Diabetic
B,D=Normal
C=Potentially diabetic
1
2
3
4
(b)
Insulin
Test serum samples for insulin content. Bind diluted serum to wells; add radioactive anti-insulin antibody; wash and count. Compare with standard curve to determine insulin levels.

Figure 12.30 Radioimmunoassay (RIA). Using RIA to detect insulin levels in human serum. (a) Establishment of a standard curve. (b) Determination of the insulin concentration in a serum sample.

direct ELISAs have been developed to detect serum antibodies to *Treponema pallidum* (the causative agent of syphilis), *Vibrio cholerae* (cholera), *Salmonella* species, German measles virus, and HIV (the virus which causes acquired immunodeficiency syndrome, AIDS). New ELISAs are being developed and marketed constantly. The specificity, rapidity, and inexpensive nature of ELISAs have insured their prominence in clinical and veterinary medicine. Several additional clinical examples of ELISA technology are given in Section 13.8.

Radioimmunoassay Various proteins, many of which serve as hormones or other specific mediators, can be assayed directly from patient serum by a procedure known as radioimmunoassay (RIA). The principle of RIA is based on a comparison of the binding of specific antibodies to a pure antigen as compared with the amount of antibody binding to an unknown sample of antigen (from patient serum). The detection of serum insulin levels is a good example of how a radioimmunoassay works.

To detect insulin, an insulin "standard curve" is prepared by binding various amounts of pure human insulin to the wells of a microtiter plate. A constant amount of anti-human insulin antibody labeled with ^{125}I is then added to the wells. With increasing antigen (insulin, in this case), there will be an increasing amount of anti-insulin antibody bound until a plateau is eventually reached (Figure 12.30). As more anti-insulin antibody is bound, more radioactivity will be present in the precipitates in the well, and the amount of radioactivity in each well can be measured with a device called a gamma counter, since ^{125}I is a gamma emitter.

To determine the amount of insulin in serum, a carefully measured serum sample is bound to microtiter wells (in place of the insulin used in the standard curve) and treated with the radioactive anti-insulin antibody. The amount of radioactivity in the precipitates of each well will be proportional to the amount of insulin present in the patient's serum, and the latter can be calculated by comparing the amount of radioactive antibody bound to that bound in the standard (see Figure 12.30). Low or undetectable insulin levels are diagnostic of the disease *diabetes*. Other antigens of medical importance detected by RIA include human growth hormone, glucagon, vasopressin, testosterone, and many other serum proteins difficult to quantitate by other means because of their low serum concentrations.

Radioimmunoassay also can be used to detect rare serum antibodies. For example, although not usually a medical problem because they are present in serum in such small amounts, circulating antibodies of the class IgE are occasionally found in substantial amounts in hyperallergic patients. To detect these antibodies, various amounts of human IgE are bound to a microtiter plate and titrated with ^{125}I-labeled anti-human IgE. A standard curve is generated that can be used to quantitate the level of serum IgE by repeating the assay using the serum sample as the source of IgE. Note in this case that the antibodies to be detected are actually serving as antigens in the assay system.

12.10 Cell-Mediated Immunity

The term "cell-mediated immunity" was originally used to describe immune phenomena other than antibody-mediated events. However, cellular immunity now must be considered in a broader context, because as we have seen, even the production of antibodies involves the activities of many different cell types. The term "cellular immunity" is therefore now used to describe any immune response in which cells of the immune system are directly involved, but in which antibody production or activity is of minor importance.

Cellular immunity differs from antibody-mediated immunity in that the immune response cannot be transferred from animal to animal by antibodies or serum containing antibodies, but can be transferred by lymphocytes removed from the blood. The lymphocytes that function in transfer of cellular immunity are T cells, which have first been activated by previous antigenic stimulation. Most of the lymphocytes circulating in the blood are T cells; B cells are not generally circulating but instead are localized in lymphoid tissues.

The overall series of events which occur when an animal is exposed to an antigen is diagrammed in Figure 12.31. Thus far we have focused only on antibody-mediated events, and have only discussed the activities of T cells, primarily T-helper cells, in connection with their roles in promoting antibody synthesis. We now turn our attention to the activities of other subsets of T cells, in particular cytotoxic and delayed type hypersensitivity T cells and natural killer cells, and discuss the role each plays in cellular immunity.

Cytotoxic T lymphocytes In addition to macrophages, a subset of T lymphocytes, called cytotoxic T cells (T_c cells), also are involved in the destruction of foreign cells. As previously mentioned in connection with the major histocompatibility complex (MHC, see Section 12.4), T_c cells recognize foreign antigens within the context of MHC Class I molecules. Any cell carrying a foreign antigen, such as those introduced by incompatible tissue grafts, or cells harboring viruses, can be lysed by T_c cells (recognition of viral antigens by T_c cells was discussed in Section 12.4 in connection with the T-cell receptor). If tissue grafts are made between two different species, or even between two strains of the same species with insufficiently similar histocompatibility antigens, cytotoxic T cells respond by killing the foreign tissue, resulting in rejection of the graft. In humans, tissue cross-matching to ensure that the major histocompatibility antigens are identical (or nearly so) in donor and recipient is now a routine clinical procedure and is required for successful skin grafting or organ transplants.

Contact between a T_c cell and the target cell is required for lysis, although the exact mechanism of lysis is not known. Unlike the situation in which antibody and complement work together to form holes in the cell membrane of the target cells (see Section

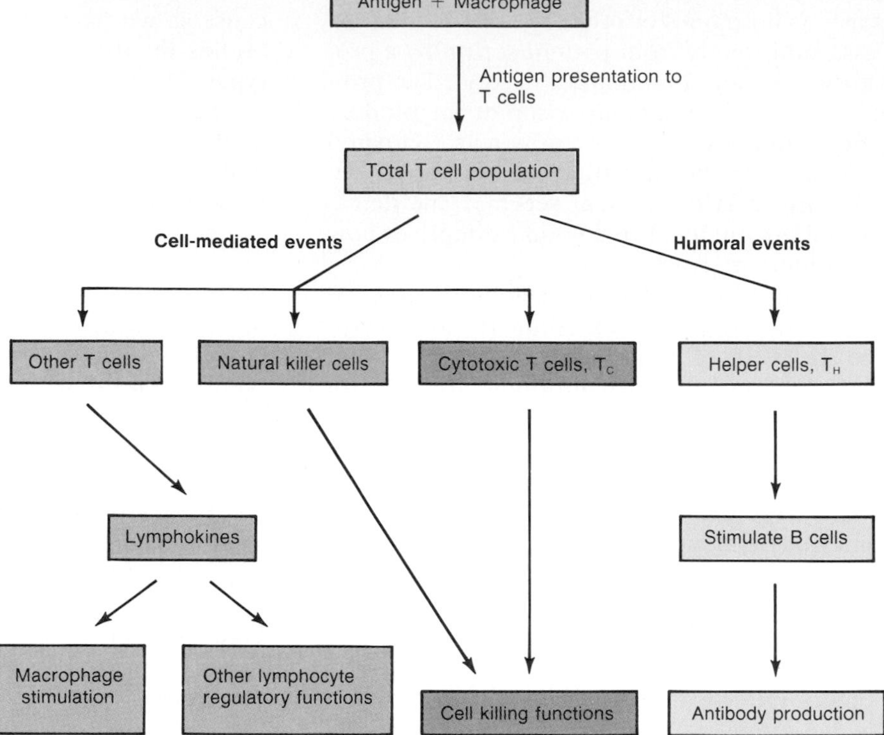

Figure 12.31 Overall view of cell-mediated immunity and its relationship to humoral immunity.

12.12) cells attacked by cytotoxic T cells literally disintegrate shortly after coming in contact with the T_c cells. Interestingly, in contrast to the nonspecific nature of activated macrophage killing, T_c cells are only effective at lysing the specific target cells containing the foreign antigen; killing of other foreign cells occurs at much lower frequencies.

Natural killer cells An additional class of lymphocytes distinct from cytotoxic T cells, play a role in destroying foreign cells. Natural killer cells contain neither T-specific (T4 or T8) nor B-specific antigens as surface markers, and are currently classified as "null" lymphocytes (that is, neither of the T nor B class). Nevertheless, natural killer cells resemble cytotoxic T cells in their ability to kill foreign cells, but differ from T_c cells in that they are able to kill in the *absence* of stimulation by specific antigen. Natural killer cells are capable of destroying malignant and virus infected cells in vitro without previous exposure or contact with the foreign antigen. However, like cytotoxic T cells, natural killer cells must first bind to the cell containing a foreign antigen before releasing a cytotoxic factor that actually kills the cells.

Lymphokines T lymphocytes, macrophages, other leukocytes, and natural killer cells can also act to eliminate invading microorganisms and other foreign material by releasing molecules called **lymphokines**. These can have several functions. For example, the lymphokines *interleukin-2, interleukin-3*, and α and γ *interferons* are involved in the regulation of other lymphocytic cells, but are not produced in response to specific antigens. Other lymphokines, such as *chemotactic factor, macrophage activation factor*, and *migration inhibition factor* act to modulate the function of phagocytic cells. Chemotactic factor attracts macrophages to the site of antigen concentration, macrophage activation factor stimulates macrophages to become more effective killers, and migration inhibition factor prevents the migration of macrophages away from the antigen. Other lymphokines serve to induce inflammatory responses or serve as toxic proteins for tumor cells (Figure 12.32).

Skin reactions in cellular immunity The best example of a cellular immune response is the development of immunity to the causal agent of tuberculosis, *Mycobacterium tuberculosis*. This cellular immune response was first discovered by Robert Koch during his classical work on tuberculosis and has been widely studied. Antigens derived from the bacterium, when injected subcutaneously into an animal previously immunized with the same antigen, elicit a characteristic skin reaction which develops only after a period of 24 to 48 hours. (In contrast, skin reactions to antibody-mediated responses, as seen in conventional allergic reactions, develop almost immediately after antigen injection.) In the region of the injected antigen, T cells become stimulated by the antigen and release lymphokines, which attract large numbers of macrophages (see Figure 12.32). The macrophages are responsible for the ingestion and digestion of the invading antigen. The characteristic skin reaction seen

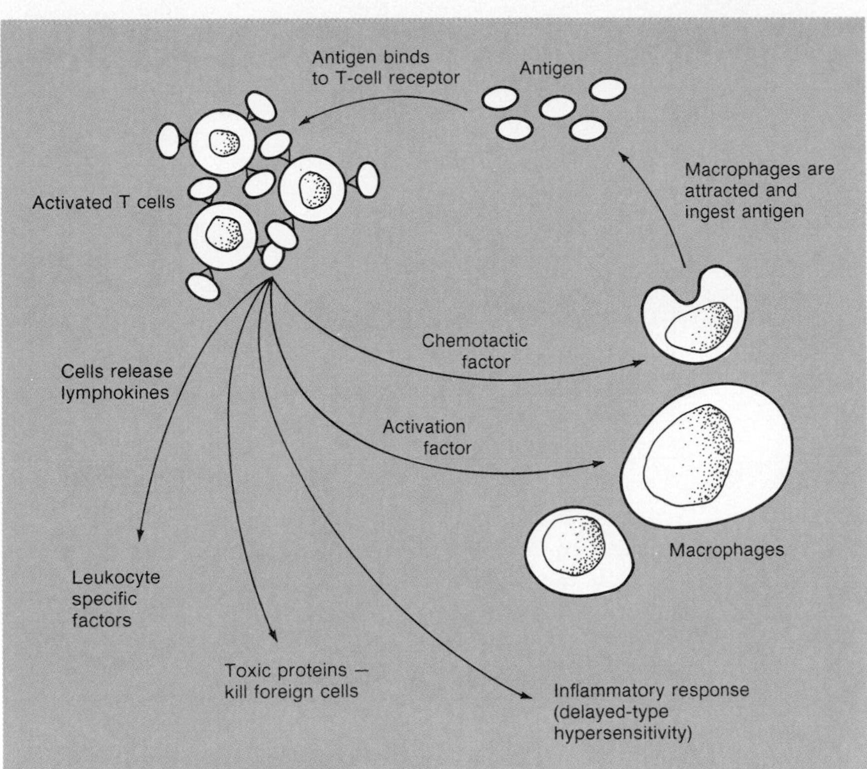

Figure 12.32 An overview of lymphokine activity in cellular immunity.

at the site of injection is a result of an inflammatory response arising as a result of the release of lymphokines by the activated T cells. This skin response serves as the basis for the **tuberculin test** for determining prior exposure to *M. tuberculosis*.

A number of microbial infections elicit cellular immune reactions. In addition to tuberculosis, these include leprosy, brucellosis, psittacosis (all caused by bacteria), mumps (caused by a virus), and coccidioidomycosis, histoplasmosis, and blastomycosis (caused by fungi). In all of these cellular immune reactions, characteristic skin reactions are elicited upon injection of antigens derived from the pathogens, and these skin reactions can be used in diagnosis of prior exposure to the pathogen. Immunity to tumors, rejection of transplants, and some drug allergies also involve cell-mediated immune responses.

Macrophage activation Macrophages play a central role in both antibody-mediated and cell-mediated immunity. In antibody production, macrophages bind and "present" antigen to T-helper cells, which leads to T_H activation of B cells (see Section 12.5). As phagocytic cells, however, macrophages also take up and kill certain foreign cells, and these properties can be stimulated by T cells. A key property of activated macrophage-stimulating T cells is that they can cause changes in macrophages so that the macrophages are able to kill intracellular bacteria that would normally multiply. As we have noted (Section 11.15), some bacteria are able to survive and multiply within macrophages, whereas most bacteria taken into macrophages are killed and digested. Bacteria multiplying within macrophages include *M. tuberculosis, M. leprae* (causal agents of tuberculosis and leprosy, respectively), *Listeria monocytogenes* (causal agent of listeriosis) and various *Brucella* species (causal agents of undulant fever and infectious abortion). Animals given a moderate dose of *M. tuberculosis* are able to overcome the infection and become immune, because of the development of a T cell-mediated immune system. Surprisingly, such immunized animals are also immune to infection by an unrelated organism such as *Listeria*, and it can be shown that macrophages in the immunized animal have been somehow changed so that they more readily kill the secondary invader. The macrophages have become activated in some way. An example of the difference in activity of normal and activated macrophages is illustrated in Figure 12.33. This nonspecific immunity to intracellular bacterial infection can be induced by various delayed-type hypersensitivity reactions (see below). When activated T cells come into contact with antigen, the T cells release lymphokines that modify some macrophages so that they are more able to kill ingested microorganisms. The macrophages that are altered by interaction with activated T cells are probably cells that are in an early stage of development. These unusual "killer" macrophages can be recognized because they contain large numbers of hydrolytic granules, suggesting that they have a more highly developed system for killing foreign cells. Certain of these granules contain hydrogen peroxide (H_2O_2), and when this compound is re-

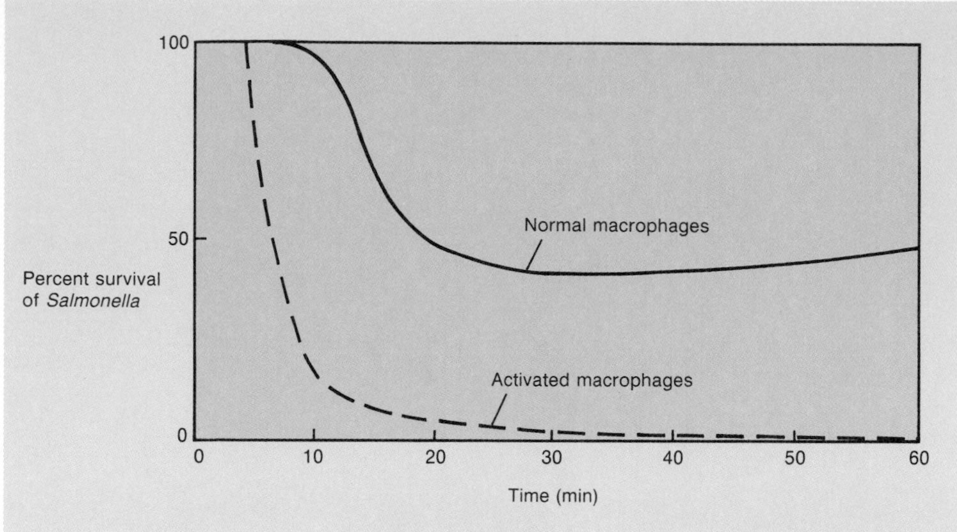

Figured 12.33 Experiment illustrating the activation of macrophages in cell-mediated immunity. The experiment measures the rate at which opsonized *Salmonella typhimurium* is killed by normal and activated macrophages.

leased it causes oxidative killing of foreign cells (see Section 11.14).

It is of some interest that macrophages not only kill foreign pathogens, but are also involved in the destruction of foreign mammalian cells. This shows up in the development of transplantation immunity, and is a major problem in the transplantation of tissues from one person to another. Tumor cells contain some specific antigens not seen on normal cells, and tumors can function in a manner like self-inflicted transplants. There is considerable evidence that tumor cells are recognized as foreign and are destroyed by macrophages and cytotoxic T lymphocytes.

Delayed-type hypersensitivity Hypersensitivity can be defined as a reaction or series of reactions in which tissue damage results due to a previously immunized state. The term hypersensitivity implies a heightened reactivity to an antigen, and may be either antibody mediated (see Section 12.11) or cell mediated. **Delayed-type hypersensitivity** is cell mediated and involves the activities of a special subset of T cells called T-delayed type hypersensitivity (T_{DTH}) cells, with symptoms appearing several hours to a few days after exposure to the allergen. Delayed-type hypersensitivity reactions occur against invasion by certain microorganisms and as a result of skin contact with certain sensitizing chemicals. The latter phenomenon, known as **contact dermatitis**, is responsible for the majority of common allergic skin disorders in humans, including those to poison ivy, cosmetics, and certain drugs or chemicals. Delayed-type hypersensitivities are mediated by T cells of the T_{DTH} and cytotoxic (T_c) subclasses. Shortly after exposure to the allergen the skin feels itchy at the site of contact, and within several hours reddening and swelling appear, indicative of a general inflammatory response. Macrophages, white blood cells, and cytotoxic T cells are attracted to the site of contact by lymphokines secreted by T_{DTH} cells, and localized tissue destruction occurs due to the lytic and phagocytic activities of these cells.

12.11 Allergy, Immune Tolerance, and Autoimmunity

Another form of hypersensitivity is known in which an immune response occurs more quickly than in the delayed-type response. This form of hypersensitivity is referred to as **immediate-type hypersensitivity** and is mediated by *antibodies* instead of activated T cells. Immediate-type hypersensitivities are frequently referred to as **allergies** and, depending on the individual and the antigen, may cause mild or extremely severe, even life-threatening reactions.

Allergy (anaphylaxis) From 10 to 20 percent of the human population suffer from immediate-type hypersensitivities, involving allergic (anaphylactic) reactions to specific allergens such as pollens, animal dander, and a variety of other agents (Table 12.5). In a typical anaphylactic reaction an allergen will elicit (upon first exposure) the production of immunoglobulins of the class IgE. Instead of circulating like immunoglobulins of the class IgG or IgM, IgE molecules tend to become attached via their cell binding fragment to the surfaces of mast cells and basophils (Figure 12.34). **Mast cells** are nonmotile connective tissue cells found adjacent to capillaries throughout the body, and **basophils** are motile white blood cells (leukocytes) that make up about 1 percent of the total leukocyte population. Upon subsequent exposure to antigen, the IgE molecules attached to these cells bind the antigen which triggers the release from mast cells and basophils of several allergic mediators. There is a requirement that an antigen bridge at least two IgE molecules on the cell

Table 12.5 Common immediate-type hypersensitivities (allergies)

Pollen and fungal spores (hay fever)
Asthma
Insect bites
Penicillin and other drug allergies
Certain food allergies
Animal dander
Mites in house dust

to initiate release of these active substances. The primary chemical mediators are **histamine** and **serotonin** (both are modified amino acids), but other mediators have been characterized and most are small peptides. The release of histamine and serotonin causes dilation of blood vessels and contraction of smooth muscle which initiate the typical symptoms of the immediate-type allergic response. These symptoms include, among others, difficulty in breathing, flushed skin, copious mucus production, and itchy, watery eyes. In general, the symptoms are relatively short-lived, but once initially sensitized by an allergen an individual can respond repeatedly upon subsequent exposure to the antigen. Depending on the individual (a genetic propensity to allergy exists) the magnitude of the anaphylactic reactions may vary from mild symptoms (or none), to such severe symptoms that the individual goes into a state of **anaphylactic shock**. In humans, the latter is characterized by severe respiratory distress, capillary dilation (causing a sharp drop in blood pressure), and flushing and itching. If severe cases of anaphylactic shock are not treated immediately with large doses of adrenalin to counter smooth muscle contraction and promote breathing, death can occur.

Immune tolerance The clonal selection hypothesis (Section 12.6) assumes that clones of B and T cells destined to make antibody or elicit cell-mediated responses against self-antigens are somehow suppressed in the mature animal. How might this occur?

The development of tolerance in B cells is thought to be due to the *suppression* of clones of B cells destined to produce self-reacting (auto) antibodies. This is thought to occur when immature B cells are exposed to self antigens in the developing fetus (Figure 12.35). In some way this constant exposure makes B cells destined to produce autoantibodies tolerant of these antigens, and prevents normal maturation of the B cell into a competent antibody-producing cell. By arresting maturation of self-reacting B cell lines, these B cells remain suppressed in later life, and self antigens are therefore unable to elicit an antibody response in the mature animal (Figure 12.35*a*).

The major means of T cell tolerance involves the *selective deletion* of self-reacting T cell clones (Figure 12.35*b*). For any given antigen, several antigen-specific T cells exist capable of responding to that antigen; T-helper, T-suppressor and cytotoxic T cells

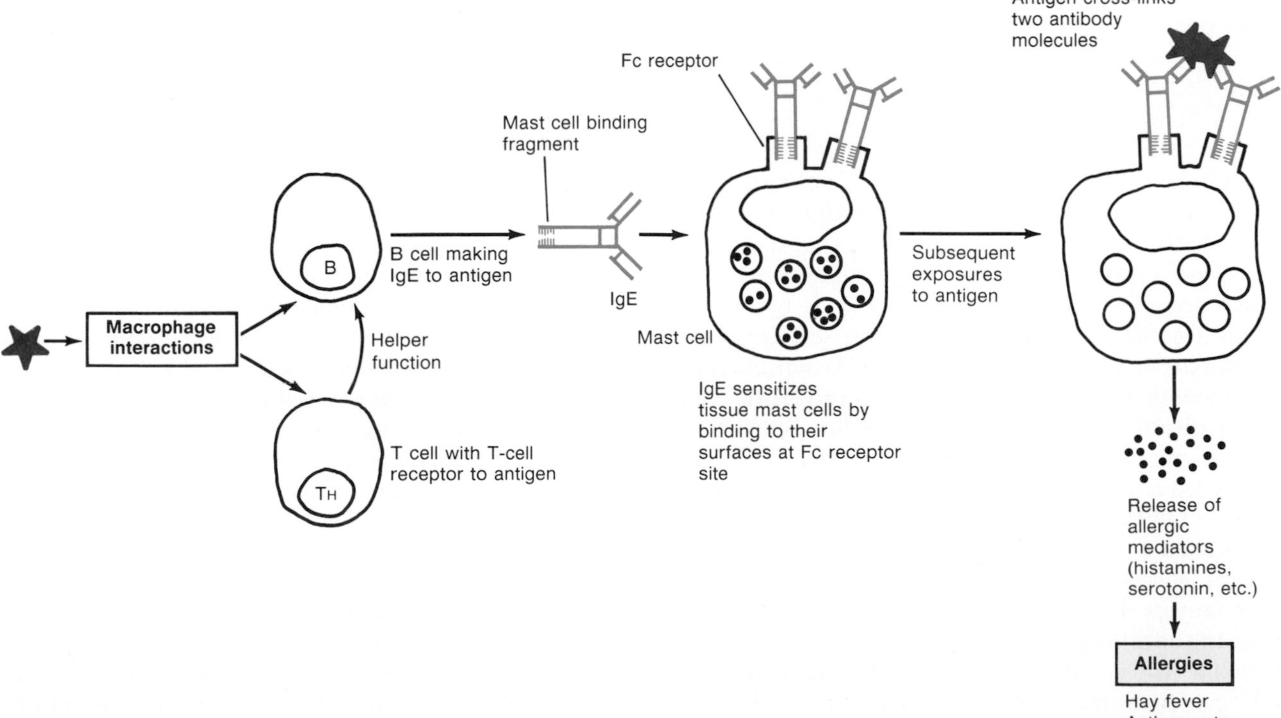

Figure 12.34 Immediate (antibody-mediated) hypersensitivity.

Figure 12.35 Immune tolerance. (a) Mechanisms of immune tolerance of B cells. (b) T cell tolerance. See text for discussion.

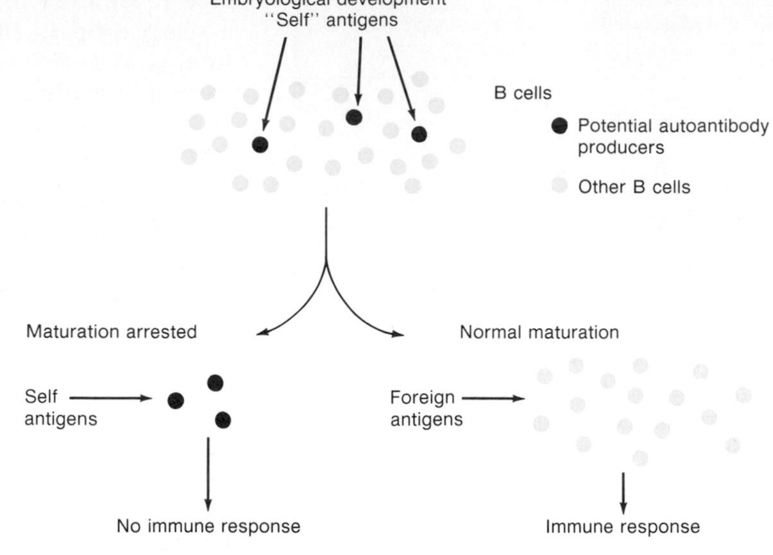

T cells
Potential self-reacting T cells
Other T cells
T-suppressor cells
Selective deletion
Control by T-suppressor cells
(b)

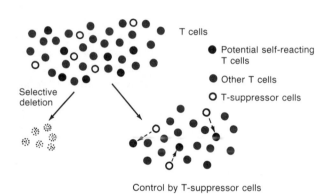

can all be involved. However, T_H or T_C cells that respond to self antigens are generally undetectable in the blood. By contrast, T_S cells that respond to self antigens are easily detectable. This suggests that T cells can somehow be selectively deleted; T_H and T_C (but not T_S) cells capable of attacking self antigens are eliminated, while subsets of T cells which recognize foreign antigens remain unaffected. How this selective elimination occurs is not yet known.

T-cell tolerance to self antigens can also be induced by the suppressor activities of T_S cells (Figure 12.35*b*). One of the roles of T_S cells is to suppress the activities of T-helper cells. This presumably occurs by release from T_S cells of antigen-specific suppressor factors which specifically affect T_H cells, in analogy to the antigen-specific helper factors released by T_H cells which stimulate specific B cells (see Section 12.5). By suppressing the activity of T_H clones destined to stimulate autoantibody-producing B cell clones, T-suppressor cells control both the activities of specific T cells and their B cell complements in a single suppressive event.

Thus, several routes to immunological tolerance apparently exist. B cell tolerance results from suppression (but not deletion) of autoantibody-producing B cell clones, while T cell tolerance involves deletion or suppression of self-reacting T cell clones. If an animal is to have a highly specific, but not self-reacting, immune system, careful checks and balances must be in place, and thus it is not surprising that tolerance to self antigens is immunologically complex.

Autoimmune diseases In some individuals the activities of T and B lymphocytes destined to react with self antigens remain unchecked. This can lead to one or more of a series of immunological disorders referred to as **autoimmune diseases**. A number of diseases have now been recognized as manifestations of the autoimmune response (Table 12.6). Depend-

Table 12.6 Some autoimmune diseases of humans

Disease	Organ or area affected	Mechanism
Juvenile diabetes	Pancreas	Autoantibodies against surface and cytoplasmic antigens of islets of Langerhans
Myasthenia gravis	Skeletal muscle	Autoantibodies against acetylcholine receptors on skeletal muscle
Goodpastures' syndrome	Kidney	Autoantibodies against basement membrane of kidney glomeruli
Rheumatoid arthritis	Cartilage	Autoantibodies against self IgG antibodies, leading to cartilage breakdown
Hashimoto's disease (hypothyroidism)	Thyroid	Autoantibodies to thyroid surface antigens
Male infertility (some cases)	Sperm cells	Autoantibodies agglutinate host sperm cells
Pernicious anemia	Intrinsic factor	Autoantibodies prevent absorption of vitamin B_{12}
Systemic lupus erythematosis	DNA, cardiolipin, nucleoprotein, blood clotting factors	Massive autoantibody response to various cellular constituents
Addison's disease	Adrenal glands	Autoantibodies to adrenal cell antigens
Allergic encephalomyelitis	Brain	Cell-mediated response against brain tissue
Multiple sclerosis	Brain	Cell-mediated and autoantibody response against central nervous system

ing on the specific disorder, the problem may be due to the production of harmful antibodies or, more rarely, to a cellular immune response to self constituents. Certain autoimmune diseases are highly organ specific. For example, in *Hashimoto's disease* autoantibodies are made against thyroglobulin, the major iodine-containing protein in the thyroid. In *juvenile diabetes,* autoantibodies against the insulin-producing cells, the islets of Langerhans, are observed. Other disorders, such as *systemic lupus erythematosis,* involve a large scale production of autoantibodies against virtually all self constituents (including DNA!). Specific organ autoimmune diseases are more easily controlled clinically because the substance produced, for example thyroxin in hypothyroidism or insulin in diabetes, can be supplied in pure form. More generalized syndromes, such as systemic lupus, can be controlled by immunosuppressive therapy, but this approach is not without risk, because of the increased chance of opportunistic infections.

Certain autoimmune diseases clearly involve the activity of T lymphocytes. *Allergic encephalomyelitis* and *multiple sclerosis* appear to be due to cellular autoimmune responses involving activated T lymphocytes which attack brain tissue. In the case of multiple sclerosis, autoantibodies to basic myelin proteins (the substances which surround nerve fibers) also can be demonstrated. As more becomes known about control of the immune response it is likely that certain other disorders will be recognized as manifestations of autoimmunity. Hopefully, however, a better understanding of the immune response also will yield methods of controlling or eliminating autoimmune diseases.

12.12 Complement and Complement Fixation

In addition to the antigen-antibody and cell-mediated immune reactions described above, other types of reactions exist that are of great importance in immunity. Before these reactions are discussed, however, it will be necessary to introduce another group of serum substances, called **complement**. Complement acts in concert with specific antibody to bring about several kinds of antigen-antibody reactions that would not otherwise occur. These reactions are described in Table 12.4.

Complement is not an antibody but is a series of enzymes found in blood serum that attack bacterial cells or other foreign cells, causing lysis or leakage of cellular constituents as a result of damage to the cell membrane. These enzymes, found even in unimmunized individuals, are normally inactive, but become active when an antibody-antigen reaction occurs. In fact, a major function of antibody is to recognize invading cells and activate the complement system for attack. There is considerable economy in an arrangement such as this, since a wide variety of antibodies, each specific for a single antigen, can call into action the complement enzymatic machinery; thus the body does not need separate enzymes to attack each kind of invading agent.

Certain components of complement are very heat labile, being destroyed by heating at 55°C for 30 minutes; antibody, on the other hand, is heat stable. The usual procedure for studying complement-requiring antigen-antibody reactions is to destroy the comple-

Urine Testing for Drug Abuse

Many government agencies and private employers carry out drug testing programs in an effort to control drug abuse in the workplace. If a person uses a drug, metabolites of the drug are excreted into the urine. Testing urine for the presence of such metabolites will thus permit detection of drug use. Such drug testing is an example of how science can be applied to legal matters in the discipline called forensic science. Drugs that are commonly tested include cannabinoids (found in marijuana), cocaine, phencyclidine, amphetamines, opiates, steroids, and barbiturates.

Because of the minute quantities of drugs or drug metabolites that occur in the urine, extremely sensitive detection methods are needed, but these methods must also be extremely specific. Immunological procedures are among the most sensitive and specific methods known for testing urine. For a drug immunoassay, an antibody must first be prepared against the drug or drug metabolite to be assayed. The antibody thus specifically recognizes an antigenic determinant of the drug. The assay is based on the principle of competition between a labeled antigen present in the drug-testing system and an unlabeled antigen (the drug or drug metabolite in the urine) for binding sites on the specific antibody. Two types of immunoassays are usually employed for drug testing, radioimmunoassay (RIA) and enzyme-linked immunoassay (ELISA). A widely used RIA system for drug testing is the Abuscreen RIA manufactured by Roche Diagnostics (Nutley, NJ). The ELISA most widely used is the EMIT system manufactured by the Syva Company of Palo Alto, CA.

In RIA drug testing, known amounts of radiolabeled drug are added to a urine sample together with known amounts of an antibody specific for the drug being tested. Following standard RIA procedures, the presence of the drug is measured by determining the radioactivity of the antibody bound to a solid substrate. As the concentration of drug in the urine goes up, the radioactivity bound goes down. Concentrations of drug as low as 1–5 μg/ml can be detected and the time to complete an assay is from 1 to 5 hours. Although the RIA test is extremely sensitive, it has the disadvantage that a radioactive material must be used, and the reagents and instrumentation are costly. However, the RIA test is very suitable for large-scale screening programs, since automatic pipetting and counting procedures can be used.

In the ELISA drug testing procedure, the antigen (drug) is covalently linked to an enzyme, commonly glucose-6-phosphate dehydrogenase, whose activity can be detected by a simple and sensitive colorimetric assay. If the enzyme, via its linked drug ligand, becomes associated with the antibody, it loses its enzymatic activity, but if it remains free it can react with its substrate. Urine is mixed with a reagent consisting of antibody to the drug, the glucose-6-phosphate dehydrogenase–drug derivative, and glucose-6-phosphate. If drug is present in the urine, the

ment activity of the antiserum by heating, and then add some fresh serum from an unimmunized animal as a source of complement. In this way one can set up reaction mixtures with defined amounts of complement.

Some reactions in which complement participates include (1) bacterial lysis, especially in Gram-negative bacteria, when specific antibody combines with antigen on bacterial cells in the presence of complement; (2) microbial killing, even in the absence of lysis; and (3) phagocytosis, which may not occur during infection if the invading microorganism possesses a capsule or other surface structure that prevents the phagocyte from acting. When specific antibody combines with the cell in the presence of complement, the cell is changed in such a way that phagocytosis can occur. (This process in which antibody plus complement renders a cell susceptible to phagocytosis is sometimes called **opsonization**).

Activation of the complement system Complement is a system of 11 proteins, designated C1, C2, C3, and so on. Activation of complement occurs only with antibodies of the IgG and IgM classes; when such antibodies combine with their respective antigens they are altered in such a way that the first component of complement, C1 (which is really a complex of three subunits called C1q, C1r, and C1s), combines with the F_c portion of the antibody (Figure 12.36*a*1). Protein C4 then combines with the C1 complex and is converted by the enzymatic activity of C1 to an active fragment. This activated C4 binds C2, and the C2 molecule is then converted into an activated complex, C4-C2, which possesses enzymatic activity. The C4-C2 complex binds directly to the cell membrane adjacent to the C1-antibody binding site (Figure 12.36*a*2). The complement protein C3 is the substrate for the C4-C2 complex, and when C3 becomes activated, C5 can combine with it (Figure 12.36*a*3). Following C5 binding, C6 and C7 bind and the C5-C6-C7 complex binds to the cell membrane and this event leads to the formation of a small hole in the membrane (Figure 12.36*a*4). The binding of additional C9 (a maximum of 6 molecules even-

attachment of the enzyme–drug derivative to the antibody is inhibited, but if drug from the urine has prevented binding of enzyme-linked drug to the antibody, then the enzyme is free to react with its substrate. Thus, the more drug present in the subject's urine, the more intense the color. The ELISA procedure has the advantage that the analysis time is short, and the color change can be detected simply. Also, potentially hazardous radioactive materials are not required. However, the ELISA procedure is generally less sensitive than the RIA procedure. The ELISA procedure is especially valuable where a small number of samples are to be analyzed at the site where the urine is collected (for instance, away from a laboratory). In some cases, the reagents have been incorporated into paper strips, providing means for urine testing that can be used even by those lacking laboratory experience (see photo).

Either polyclonal or monoclonal antibodies can be used for urine testing. Monoclonal antibodies have the advantage of better specificity, but may not necessarily have as high an affinity for the drug as a selected polyclonal antibody. The selection of an antibody will depend on the specificity and sensitivity needed, the cost of the product, and the ease of use.

One problem with the use of any of these immunological methods for urine testing is cross reaction to other chemicals that might be innocently present in the urine. For instance, the pain killer ibuprofen will cross react with mari-

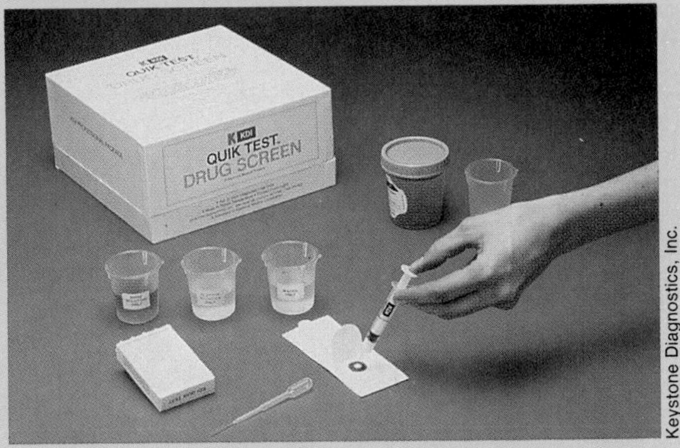

Keystone Diagnostics, Inc.

juana metabolites in some tests, and the antihistamine drug diphenhydramine can cross react in the ELISA test for methadone. Because of such false-positive reactions, it is recommended that positive tests be confirmed by independent tests, such as the use of gas chromatography or high-performance liquid chromatography.

Immunological tests for drugs in the urine have provided a new and potentially widespread technique for controlling the use of illegal drugs in the workplace. Urine drug testing is another example of the power and utility of immunology in modern society.*

*Reference: Hawks, R.L. and C.N. Chiang (editors). 1986. *Urine Testing for Drugs of Abuse.* National Institute of Drug Abuse Research Monograph 73, Department of Health and Human Services, 5600 Fishers Lane, Rockville, MD.

tually bind) enlarges the hole causing the loss of internal ions and a large influx of water. The sequence and some of the resulting reactions are summarized in Figure 12.36. As seen, reactions at the C3 level result in chemotactic attraction of phagocytes to invading agents, and to phagocytosis following opsonization. Reaction at C5 also leads to leukocyte attraction. The terminal series of reactions from C5 through C9 results in cell lysis and death. Lysis itself results from a destruction of the integrity of the cell membrane, leading to the formation of holes through which cytoplasm can leak (Figure 12.37).

The complement system is thus seen to act in cascade fashion, the activation of one component resulting in the activation of the next. In summary, the steps are (1) binding of antibody to antigen (initiation); (2) recognition of antigen-antibody complex by C1; (3) C4-C2 binding to an adjacent membrane site; (4) activation of C3; (5) formation of the C5-C6-C7 complex, causing attraction of leukocytes; and (6) formation of the C8-C9 complex causing cell lysis. This pathway is called the *classical* pathway, to distinguish it from the *properdin* pathway discussed below.

A variety of complement reactions The complement system is involved in other immunological reactions besides those leading to the destruction of invading organisms. The reactions of importance here include certain aspects of the inflammatory response and the production of proteins called **anaphylatoxins** which produce symptoms resembling those of allergic responses (see Section 11.16). Anaphylatoxins consist of proteolytic fragments of C3 and C5, called C3a and C5a, respectively, and these substances can act directly on smooth muscle, causing contraction of the uterus, trachea, arteries, atrium, and intestine (C3a and C5a also serve as leukocyte chemotactic factors). Anaphylatoxins also cause release of histamines following binding to tissue mast cells. Histamine induces inflammation and increases the permeability of capillaries (Section 11.16), enabling leukocytes and fluid to escape into the tissue cells. One serious type of allergic reaction, **anaphy-**

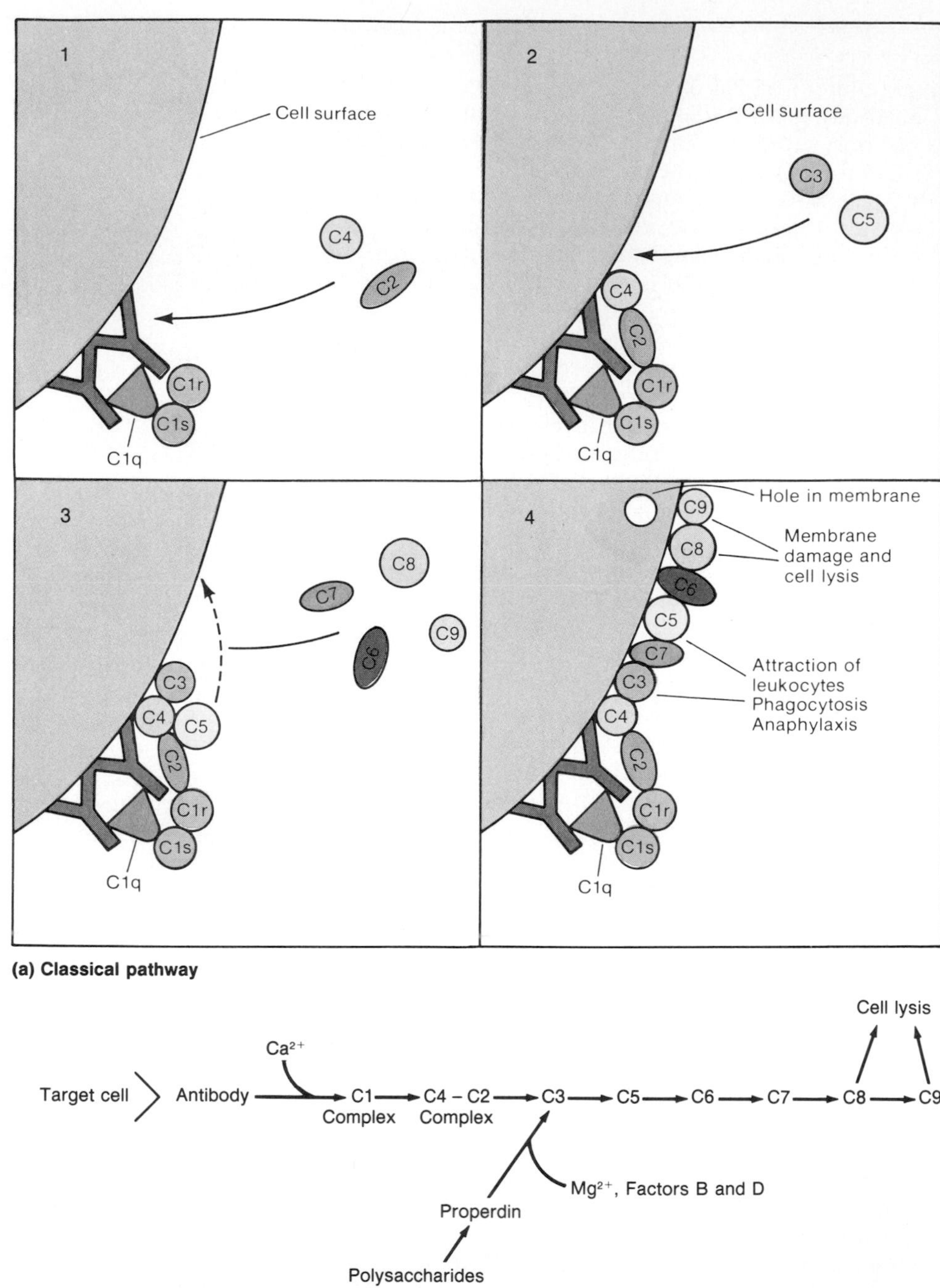

Figure 12.36 The complement system: (a) The sequence, orientation, and activity of the various components as they interact to lyse a cell. Panel 1, binding of the antibody recognition unit C1q, and other C1 proteins; panel 2, the C4-C2 complex; panel 3, the C4-C2-C3-C5 complex, after activation the C-5 unit travels to an adjacent membrane site; panel 4, binding of C6, C7, C8, and C9, resulting in membrane damage. (b) The relation of the classical to the properdin pathways.

lactic shock (see Section 12.11), may also involve components of the complement system, although most such allergic responses involve the activity of immunoglobulins of the class IgE.

In addition to the activation of the complement system beginning with C1, which was described above, there is another means of activating C3 which bypasses the C1 complex and C2 and C4. This is the "alternate pathway" of complement activation, commonly referred to as the **properdin** system (Figure 12.36*b*). Properdin is a protein complex in normal blood serum (it may also be formed as a result of antigenic stimulation) that combines with bacterial polysaccharides or certain aggregated immunoglobulins to activate C3 and C5 directly, leading to opsonization and the production of inflammatory reactions. The importance of the properdin system is that it leads to defense against invading organisms without the production of specific antibodies; thus it plays a role in innate (nonspecific) immunity.

Another important reaction in which complement participates is hemolysis, the lysing of red blood cells. If an antiserum against erythrocytes is mixed with a suspension of erythrocytes and some normal serum added as a source of complement, lysis occurs within 30 minutes on incubation at 37°C. Erythrocyte

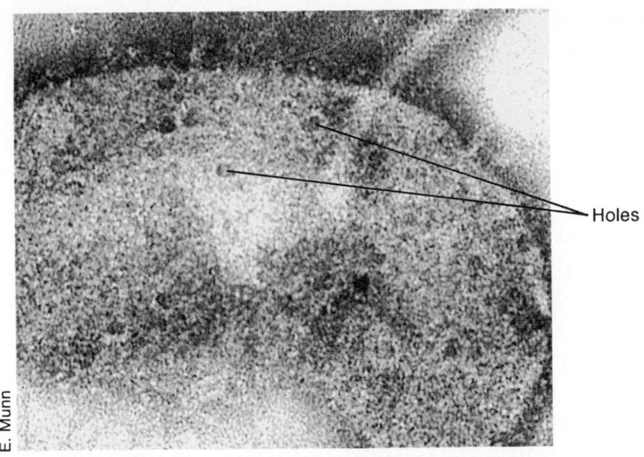
Holes
E. Munn

Figure 12.37 Electron micrograph of a negatively stained preparation of *Salmonella paratyphi*, showing holes created in the cell envelope as a result of reaction between cell-envelope antigens, specific antibody, and complement.

hemolysis is often used to test for the presence of complement in unknown sera or to measure complement fixation (see below).

Complement is necessary in the bactericidal and lytic actions of antibodies against many Gram-negative bacteria. (Interestingly, Gram-positive bacteria are not killed by specific antibody, in either the presence or absence of complement, although Gram-positive bacteria are opsonized.) Probably death (including that from lysis) involves antibodies against antigens on the surface of the cell; complement perhaps brings about an actual change in the cell surface, possibly by an enzymelike reaction, after antibodies have prepared the way. No cytocidal or lytic effect is seen when cells and complement are mixed alone, whereas if antibody has bound to cells first, death or lysis occurs rapidly after complement is added.

As noted, complement is necessary for opsonization, the promotion of phagocytosis by antibodies against bacterial capsules. This does not require the whole complement pathway, but only the reactions through C3. Opsonization occurs because the binding of C3 causes the cells to adhere to phagocytes. Phagocytes have C3 receptors on their surfaces and it is presumably these receptor molecules which are involved in adherence of the C3-coated bacterial cell to the phagocyte.

Complement fixation An important property of the complement system is that the components are enzymatically altered during reaction, so that they will no longer react in a new sequence of reactions. Complement thus appears to be used up during antibody-antigen reactions. This is called **complement fixation** and occurs whenever an IgG or IgM antibody reacts with antigen in the presence of complement, even if complement is not required in the reaction. Complement fixation is measured by assaying the concentration of complement remaining after an antibody-antigen reaction (Figure 12.38). After the initial reaction to permit complement fixation, an indicator system is added consisting of sheep red blood cells and antibody to the red blood cells. In the absence of complement, lysis of the red cells will not take place, but if complement is present, the normal series of reactions leading to cell lysis occurs. Thus, if complement has *not* been fixed, lysis occurs, but if complement *has* been fixed, lysis *will not* occur. Sheep cells are used since their lysis is readily observed by eye. Appropriate controls must be set up to be sure that nothing in the system is inactivating complement nonspecifically. By measuring complement fixation, one has a means of determining whether an antigen-antibody reaction has taken place, even if less sensitive immunological assays such as precipitation or agglutination have not produced a visible reaction.

Antigen bound

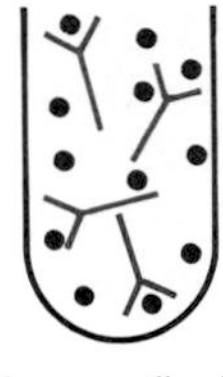
Wrong antibody; antigen not bound
1. Mix antibody ( ) and antigen (●)

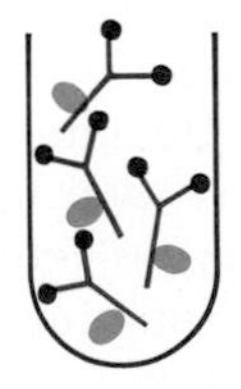
Complement fixed

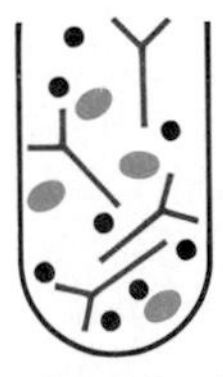
Complement not fixed
2. Add complement ( )

Complement previously fixed. Sheep red blood cells remain intact

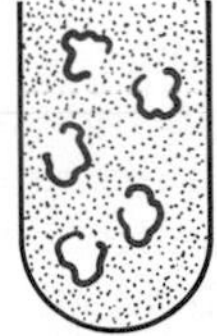
Free complement available. Sheep red blood cells lysed
3. Add indicator system (sheep red blood cells and anti-sheep antibody). Observe for lysis of red blood cells.

Figure 12.38 Principle of the complement fixation test. An antigen-antibody reaction must occur before complement will be fixed.

12.13 Applied Immunology: Immunization against Infectious Diseases

Although many aspects of immunological reactions do not concern immunity to infectious disease, the major role of antibodies is in protecting the animal from the consequences of infection. The importance of antibodies in disease resistance is shown dramatically in individuals with the genetic disorder **agammaglobulinemia**, in whom antibodies are not produced because their B cells are defective. Such individuals are unusually sensitive to bacterial diseases, and in the days before antibiotic therapy, few of them survived infancy. The general lack of an antibody response is also observed in those suffering from acquired immunodeficiency syndrome (AIDS), however, in this case the problem is not due to defective B cells. Instead, AIDS patients suffer from a virtually total cessation of T-helper cell activities (see Section 15.7). Although T cells themselves do not make antibodies, the crucial importance of T cells, in particular T-helper cells, in the production of antibodies is clearly evident in AIDS patients: the inability to mount an antibody response leads to the eventual death of AIDS patients from unusual infections rarely observed in the population at large (see Section 15.7).

The induction of specific immunity to infectious diseases provided one of the first real triumphs of the scientific method in medicine, and was one of the outstanding contributions of microbiology to the treatment and prevention of infectious diseases. An animal or human being may be brought into a state of immunity to a disease in either of two distinct ways. (1) The individual may be given injections of an antigen that is known to induce formation of antibodies, which will confer a type of immunity known as **active immunity** since the individual in question produced the antibodies itself. (2) The individual may receive injections of an antiserum that was derived from another individual, who had previously formed antibodies against the antigen in question. The second type is called **passive immunity** since the individual receiving the antibodies played no active part in the antibody-producing process.

An important distinction between active and passive immunity is that in *active* immunity the immunized individual is fundamentally changed, since it is able to continue to make the antibody in question, and it will exhibit a secondary or booster response if it later receives another injection of the antigen in question. Active immunity often may remain throughout life. A *passively* immunized individual will never have more antibodies than it received in the initial injection, and these antibodies will gradually disappear from the body; moreover, a later inoculation with the antigen will not elicit a booster response. Active immunity is usually used as a *prophylactic* measure, to protect a person against future attack by a pathogen. Passive immunity is usually *therapeutic*, designed to cure a person who is presently suffering from the disease. For example, tetanus toxoid (see the following section) actively immunizes an individual against future encounters with *Clostridium tetani* exotoxin, while tetanus antiserum (antitoxin; see below) is administered to passively immunize an individual suspected of coming in contact with *C. tetani* exotoxin via growth of the organism in a penetrating wound.

Vaccination The material used in inducing active immunity, the antigen or mixture of antigens, is known as a **vaccine**. (The word derives from Jenner's vaccination process using cowpox virus; *vacca* is Latin for "cow.") However, to induce active immunity to toxin-caused diseases, it is clearly not desirable to inject the toxin itself; to overcome this problem, many exotoxins can be modified chemically so that they retain their antigenicity but are no longer toxic. Such a modified exotoxin is called a **toxoid**. One of the common ways of converting toxin to toxoid is by treating it with formaldehyde, which blocks some of the free amino groups of the toxin. Toxoids are usually not such efficient antigens as the original exotoxin but have the advantage that they can be given safely and in high doses. When immunization against whole microorganisms is necessary, such as for endotoxin-producing organisms, the microorganism in question may first be killed by agents such as formaldehyde, phenol, or heat, and the dead cells then injected. Endotoxin-caused diseases for which vaccines are made routinely are whooping cough and typhoid. Formaldehyde treatment is also used to inactive viruses in preparing some vaccines, such as the Salk polio vaccine.

Immunization with live cells or virus is usually more effective than with dead or inactivated material. Often it is possible to isolate a mutant strain of a pathogen which has lost its virulence but which still retains the immunizing antigens; strains of this type are called **attenuated strains** (see Section 11.13).

A summary of vaccines available for use in humans is given in Table 12.7

Vaccination practices Infants possess antibodies derived from their mothers and hence are relatively immune to infectious disease during the first six months of life. However, it is desirable to immunize infants for key infectious diseases as soon as possible, so that their own active immunity can replace the passive immunity received from the mother. However, infants have a rather poorly developed ability to form antibodies, so that immunization is not begun until a few months after birth. As discussed in Section 12.5 a single injection of antigen does not

Table 12.7 Available vaccines for infectious diseases in humans

Disease	Type of vaccine used
Bacterial diseases:	
Diphtheria	Toxoid
Tetanus	Toxoid
Pertussis	Killed bacteria (*Bordetella pertussis*)
Typhoid fever	Killed bacteria *Salmonella typhi*)
Paratyphoid fever	Killed bacteria (*Salmonella paratyphi*)
Cholera	Killed cells or cell extract (*Vibrio cholerae*)
Plague	Killed cells or cell extract (*Yersinia pestis*)
Tuberculosis	Attenuated strain of *Mycobacterium tuberculosis* (BCG)
Meningitis	Purified polysaccharide from *Neisseria meningitidis*
Bacterial pneumonia	Purified polysaccharide from *Streptococcus pneumoniae*
Typhus fever	Killed bacteria (*Rickettsia prowazekii*)
Viral diseases:	
Smallpox	Attenuated strain (Vaccinia)
Yellow fever	Attenuated strain
Measles	Attenuated strain
Mumps	Attenuated strain
Rubella	Attenuated strain
Polio	Attenuated strain (Sabin) or inactivated virus (Salk)
Influenza	Inactivated virus
Rabies	Inactivated virus (human) or attenuated virus (dogs and other animals)

Table 12.8 Some recommended immunization procedures for infants and children*

Age	Immunization	Comments
During the first year of life	Diphtheria, pertussis, tetanus (DPT)	Given at 2, 4, and 6 months
	Oral polio (OP)	Given at 2, 4, and sometimes 6 months
During the second year	Mumps, measles, rubella (MMR)	Given at 15 months
	DPT booster	Given at 18 months
	OP booster	Given at 18 months
At 4–6 years	DPT booster, OP booster	May be given at 3 years on entering nursery school
At 14–16 years	Adult-type tetanus and diphtheria booster	Pertussis omitted

**These are recommended practices in some areas. Actual practice may vary in the details.*

lead to a high antibody titer; it is desirable therefore to use a series of injections, so that a *high titer* of antibody is developed. A typical vaccination schedule for children from 2 months of age to the mid-teens is given in Table 12.8.

The importance of immunization procedures in controlling infectious diseases is well established. Upon introduction of a specific immunization procedure into a population, the incidence of the disease often drops markedly (Figure 12.39). The degree of immunity obtained by vaccination varies greatly, depending on the individual and on the quality and quantity of the vaccine. However, rarely is life-long immunity achieved by means of a single injection, or even a series of injections, and the population of antibody-producing cells induced by immunization will gradually disappear from the body. One way in which antigenic stimulation occurs even in the absence of immunization is by periodic subclinical infection. A natural infection will result in a rapid booster response, leading to both a further increase in activated antibody-producing cells and to production of antibody, which will attack the invading pathogen. It is not known how long immunity will last in the complete absence of antigenic stimulation, but probably the immune period varies considerably from person to person.

Immunization procedures are not only beneficial to the individual, but are effective public health procedures, since disease spreads poorly through a pop-

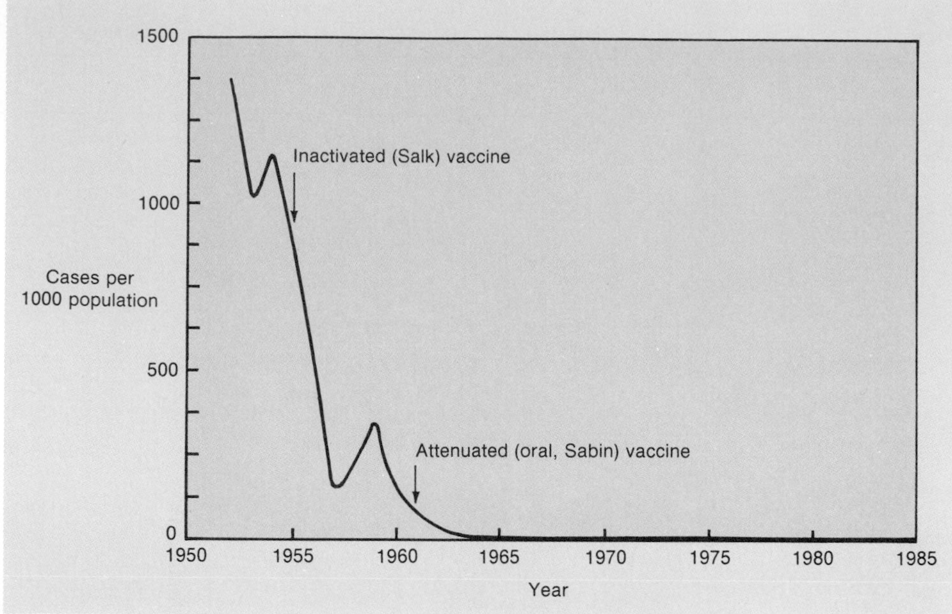

Figure 12.39 Cases of polio in the United States 1950–1983, showing the consequences of the introduction of polio vaccine.

ulation in which a large proportion of the individuals are immune (See Section 14.6).

Passive immunity The material used in inducing passive immunity—the serum containing antibodies—is known as a *serum*, an *antiserum*, or an *antitoxin* (the last applies to a serum containing antibodies directed against a toxin). Antisera are obtained either from large immunized animals, such as the horse, or from human beings who have high antibody titers (that is, who are **hyperimmune**). The antiserum or antitoxin is standardized to contain a known amount of antibody, using some internationally agreed upon arbitrary unit of antibody titer; a sufficient number of units of antiserum must be inoculated to neutralize any antigen that might be present in the body. Sometimes the gamma-globulin fraction of pooled human serum is used as a source of antibodies. This contains a wide variety of antibodies that normal people have formed through the years by artificial or natural exposure to various antigens; it is used when hyperimmune antisera are not available, but in recent years its routine use is declining because of the threat of AIDS contamination.

The most common use of passive immunization is in the prevention of infectious hepatitis (due to hepatitis A virus, see Section 15.11). Pooled human gamma globulin contains fairly high titers of antibody against hepatitis A virus, due to the fact that infection with hepatitis A virus is widespread in the population. Travelers to areas where incidence of infectious hepatitis is high, such as North and tropical Africa, the Middle East, Asia, and parts of South America, may be given prophylactic doses of pooled human gamma globulin. A single dose of 0.02 ml/kg body weight should protect for up to 2 months, but for more prolonged exposures doses at repeated intervals should be given. Pooled human gamma globulin may also be of value in the therapy of infectious hepatitis, if given early in the incubation period.

Antibody titers and the diagnosis of infectious disease In the diagnosis of an infectious disease, isolation of the pathogen is not always possible, and one alternative is to measure antibody *titer* to a suspected pathogen. The principle here is that if an individual is infected with the suspected pathogen, the antibody titer to that pathogen should be elevated. Antibody titer can be measured by agglutination, precipitation, ELISA, or immobilization methods, or by complement fixation, depending upon the situation. The general procedure is to set up a series of dilutions of the serum (usually twofold dilutions: 1:2, 1:4, 1:8, 1:16, 1:32, and so on) and to determine the *highest* dilution at which the serological reaction occurs.

It should be emphasized that a single measure of antibody titer does not indicate active infection. Many antibodies remain at high titer for long times after infection, so that in order to establish that an acute illness is due to a particular agent, it is essential to show a *rise* in antibody titer in successive samples of serum from the same patient. Frequently, the antibody titer will be low during the acute stage of the infection and rise during convalescence (Figure 12.40). Such a rise in antibody titer is the best indication that the illness was due to the suspected agent. Measurement of such a rise in antibody titer is also useful in diagnosis of infectious diseases of a rather chronic nature, such as typhoid fever and brucellosis. In some cases, however, the mere presence of antibody may be sufficient to indicate infection. This is the case for a pathogen which is quite rare in a population, so that the presence of antibody is suf-

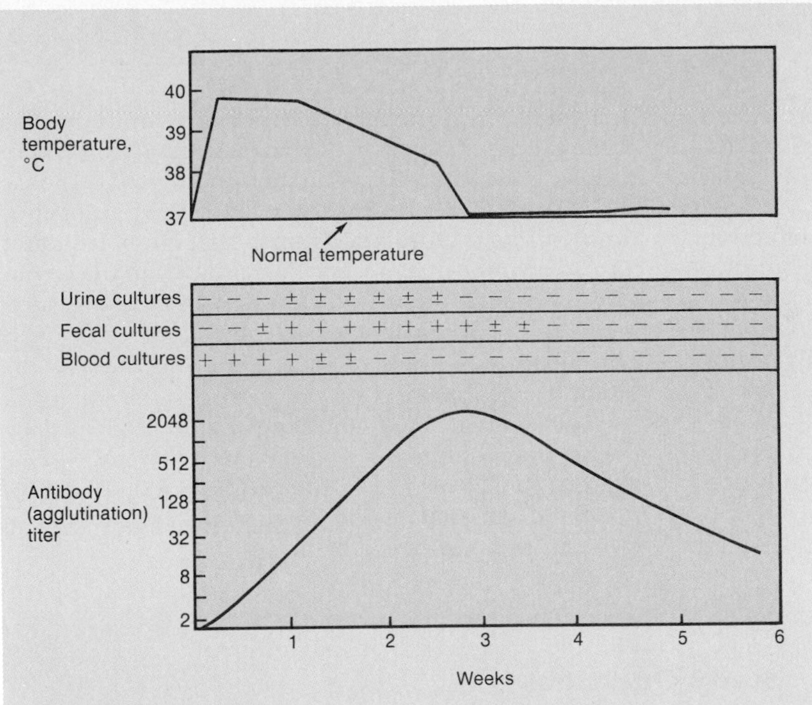

Figure 12.40 The course of infection in a typical untreated typhoid fever patient. Measurement of body temperature provides a measure of the course of clinical symptoms. The antibody titer was measured by determining the highest dilution (two-fold series) causing agglutination of a test strain of *Salmonella typhi*. Presence of viable bacteria in blood, feces, and urine was determined from periodic cultures. Note that the pathogen clears from the blood as the antibody titer rises, and clearance from feces and urine requires longer time. Body temperature gradually drops to normal as the antibody titer rises. The data given do not represent a single patient, but are a composite of the picture seen in large numbers of patients.

Table 12.9 Some clinical immunological procedures for diagnosis of infectious disease

Pathogen or disease	Antigen	Serological procedure*
Streptococcus (group A)	Streptolysin O (exotoxin)	Neutralization of hemolysis
	DNase (extracellular protein)	Neutralization of enzyme
Neisseria meningitidis	Capsular polysaccharide	Passive hemagglutination (*N. meningitidis* polysaccharide adsorbed to red cells)
	N. meningitidis cells	Indirect fluorescent antibody
Salmonella	O or H antigen	Agglutination (Widal test)
		ELISA
Vibrio cholerae	O antigen	Agglutination
		Bactericidal (in presence of complement)
		ELISA
Brucella	Cell wall antigen	Agglutination
		ELISA
Corynebacterium diphtheriae	Toxin	Skin test (Schick test)
Mycobacterium tuberculosis	Tuberculin (partially purified bacterial proteins, PPD)	Skin test (Tuberculin test)
		ELISA
Syphilis (*Treponema pallidum*)	Cardiolipin-lecithin-cholesterol	Flocculation [Venereal Disease Research Laboratory (VDRL) test]
	T. pallidum antigens	ELISA
	T. pallidum cells	Indirect fluorescent antibody (FTA)
Rickettsial diseases (Q fever, typhus, Rocky Mountain Spotted fever)	Killed rickettsial cells	Complement fixation or cell agglutination tests
		ELISA
Influenza virus	Influenza virus suspensions	Complement fixation test
	Nasopharynx cells containing influenza virus	Immunofluorescence
AIDS	Human immunodeficiency virus (HIV)	ELISA

**Except for the skin tests, the serum of the patient is assayed for antibody against the specific antigen by the methods shown.*

ficient to indicate that the individual has experienced an infection. A relevant example here is AIDS. As discussed in Section 13.8, an extremely sensitive and highly reliable ELISA test is now available for detecting antibodies to the AIDS virus. Upon infection with HIV, an antibody response is mounted, but, unlike most diseases where antibody titers *increase* in the later stages of the disease, in AIDS the loss of T-helper cell function (see Section 15.7) actually causes titers to *decrease* as the disease progresses. Nevertheless, the exquisite sensitivity of ELISA allows detection of even very low antibody titers, and thus this immunoassay can be used to routinely screen blood samples for signs of AIDS.

Another point to emphasize is that not all infections result in formation of systemic antibody. If a pathogen is extremely localized in its action, there may be little induction of an immunological response and no rise in antibody titer, even if the pathogen is proliferating profusely at its site of infection. It should be also obvious that presence of antibody in the serum may have been due to vaccination. In fact, measurement of rise in antibody titer during vaccination is one of the best ways of indicating that the vaccine being used is effective.

Some of the most common immunological diagnostic procedures are outlined in Table 12.9. The student should also refer to Sections 13.6–13.10 for further discussion of the theory and application of clinical immunological tests.

Study Questions

1. What are the major differences between a host resistance mechanism such as phagocytosis, and humoral and cell-mediated immunity?
2. To what substances are antibodies made and to what substances are they not made? What are the necessary prerequisites for antibody formation from the standpoint of the antigen?
3. Define antigen. Immunogen. Immunoglobulin. Antibody. Can you define antibody without reference to antigen and vice versa?
4. Distinguish between an immunogen, an antigen, and a hapten. How could you experimentally determine that something was a hapten but not an immunogen?
5. Draw the structure of an IgG molecule labeling the heavy and light chains. Which portion of the molecule binds antigen? Which portion defines the class of the immunoglobulin? Which portions of the molecule have identical amino acid sequences? Be specific and be sure to consider all possible sequence identities and nonidentities within the molecule.
6. What is the major immunological role of each class of immunoglobulins?
7. Compare and contrast T and B lymphocytes with macrophages in terms of function, location, and immunological specificity. What is the origin of all of these cell types and how do they end up taking on such different functions?
8. What is the basis of immune recognition that allows the lymphocytes to distinguish between "self" and "nonself?"
9. Compare and contrast MHC class I and class II antigens as to bodily location, composition, and overall dimensions of the molecules. Do MHC antigens consist solely of protein?
10. Discuss the major steps in the production of antibody starting with the introduction of antigen into the animal and resulting in a humoral immune response. What is the importance of MHC antigens in antibody production?
11. Compare and contrast an undifferentiated B cell, a memory cell, and a plasma cell? Which cell type(s) actually excretes significant amounts of antibody?
12. Describe on the gene level the mechanism for generating antibody diversity.
13. Discuss the structural similarities and differences between MHC class I and II antigens, T-cell receptors, and immunoglobulins. Which molecules are most specific and why? What evidence exists to suggest these molecules form an evolutionary progression?
14. How does the genetics of T-cell receptors resemble those of immunoglobulins? How would you predict the genes for MHC antigens are arranged and what similarity would they have to those of T-cell receptors and immunoglobulins?
15. Contrast the production of monoclonal antibodies with the techniques involved in preparing a polyclonal serum. What are some of the advantages of using monoclonal antibodies for research or diagnostic work?
16. Describe four types of antigen-antibody reaction and explain how each might be useful in immunity to infectious disease. Give examples in each case.
17. How does agglutination differ from precipitation? How could a soluble protein be agglutinated?
18. Why is it possible in an emergency to use human type O blood in transfusions to any blood type (type O = universal donor) while people with type AB blood can receive blood from any blood type (type B = universal recipient)?
19. Describe the differences between a direct and indirect fluorescent antibody procedure. What are the advantages of the indirect test?
20. What is the major advantage of an ELISA or radioimmunoassay over a precipitation reaction? Which assay(s) requires more antigen and antibody?
21. What major advantages does the ELISA test have over radioimmunoassay?

22. Compare and contrast the various types of T cells and other cells involved in cell-mediated immunity.

23. What are lymphokines and what do they do?

24. What are the major differences between delayed-type and immediate-type hypersensitivities?

25. List three autoimmune diseases. In each case what is the immunological basis of the disease and how, if possible, can the disease be treated?

26. Write a one sentence definition which clearly describes the nature and function of *complement*.

27. Compare and contrast active and passive immunity from the following points of view: manner of induction (substance used), function in infectious disease resistance, long-term retention, value in vaccination.

28. Why are infants not vaccinated immediately after birth? Discuss at least three reasons.

29. List five common vaccination procedures and indicate the nature of the antigen, how it is prepared, and the way in which antibody against this antigen confers immunity.

30. Explain why a *rise* in antibody titer is a better indication of active infectious disease than the mere *presence* of antibody in the serum. Exactly how is a rise in antibody titer measured in a patient?

31. Why is it incorrect to say that the mere presence of an antibody indicates disease and that the mere presence of a pathogen indicates disease?

Supplementary Readings

Hood, L. E., I. L. Weissman, W. B. Wood, and **J. H. Wilson.** 1984. *Immunology, 2nd ed.* The Benjamin Cummings Publishing Company, Menlo Park, California. A textbook of immunology stressing molecular concepts.

Kimball, J. W. 1986. *Introduction to Immunology, 2nd ed.* Macmillan Publishing Co., New York. An excellent immunology text suitable for undergraduates, covering all major topics in immunology.

Leder, P. 1982. *The genetics of antibody diversity.* Sci. Am. 246:102–115. A concise review of the fascinating events involved in the rearrangement of genes in B-lymphocytes.

Marrack, P., and **J. Kappler.** 1986. *The T cell and its receptor.* Sci. Am. 254:36–45. An excellent and well illustrated account of the T-cell receptor and the evolution of the major lymphocytic recognition molecules.

Milstein, C. 1986. *From antibody structure to immunological diversification of immune response.* Science 231:1261–1268. A review of antibody diversity, monoclonal antibodies and messenger RNA processing by the Nobel Laureate who developed the techniques for producing monoclonal antibodies. Highly recommended.

Pollock, R. R., J.-L. Teilland, and **M. D. Scharff.** 1984. *Monoclonal antibodies. A powerful tool for selecting and analyzing mutations in antigens and antibodies.* Ann. Rev. Microbiol. 28:389–417. An overview of monoclonal antibodies and their research applications.

Roitt, I. M. 1984. *Essential Immunology, 5th ed.* Blackwell Scientific, Oxford. Probably the best short treatment of immunology, with emphasis on the general processes of immune functions rather than molecular details.

Roitt, I. M., J. Brostoff, and **D. K. Male.** 1985. *Immunology.* C. V. Mosby Co., St. Louis. A well-illustrated textbook of immunology for advanced undergraduates.

Sites, D. P., J. D. Stobo, H. H. Fudenberg, and **J. V. Wells** (eds.) 1984. *Basic and Clinical Immunology, 5th ed.* Lange Medical Publications, Los Altos, California. An advanced text/laboratory manual of immunological techniques useful in the clinical laboratory.

Smith, H. R., and **A. D. Steinberg.** 1983. *Autoimmunity. A perspective.* Ann. Rev. Immunol. 1:175–210. An overview of autoimmune diseases.

13 Clinical and Diagnostic Microbiology

The most important activity of the microbiologist in medicine is to isolate and identify the causal agents of infectious disease. This major area of microbiology is called **clinical microbiology**. In recent years, this field has greatly expanded because of the increasing awareness of the importance of precise identification of the pathogen for proper treatment of infectious disease, and because of the availability of newer sophisticated methods. Clinical laboratories are generally able to isolate, identify, and determine the antibiotic sensitivity of most routinely-encountered pathogenic bacteria within 72 hours of sampling. However, recent advances in rapid diagnostic methods have made it possible to identify some pathogens in less than one day, and in only a few hours in certain cases. Diagnostic methods based on immunology and molecular biology have reduced this time even more, and in many cases a pathogen can be positively identified without culturing the organism at all. Molecular methods have also greatly improved the diagnosis of viral and parasitic infections, diseases that are typically difficult to pinpoint because of the difficulty of culturing the agent. The clinical microbiologist works with and advises the physician in matters relating to the diagnosis and treatment of infectious diseases.

13.1 Isolation of Pathogens from Clinical Specimens

The physician, on the basis of careful examination of the patient, may decide that an infectious disease is present. Samples of infected tissues or fluids are then collected for microbiological, immunological, and molecular biological analyses (Figure 13.1). Depending upon the kind of infection, materials collected may incude blood, urine, feces, sputum, cerebrospinal fluid, or pus. A sterile swab may be passed across a suspected infected area (Figure 13.2). The swab is then streaked over the surface of an agar plate or placed directly in a liquid culture medium. Small pieces of living tissue may be aseptically removed (biopsy). Table 13.1 summarizes current recommendations. The sample must be carefully taken under aseptic conditions so that contamination is avoided. Once taken, the sample is analyzed as soon as possible. If it cannot be analyzed immediately, it is usually refrigerated to slow down deterioration. In the rest of this section, we describe some of the most common microbiological procedures used in the clinical laboratory.

Blood cultures **Bacteremia** means the presence of bacteria in the blood (see Section 11.14). Because of the extreme danger of bacteremia, the rapid and accurate identification of the causative agent(s) is one of the

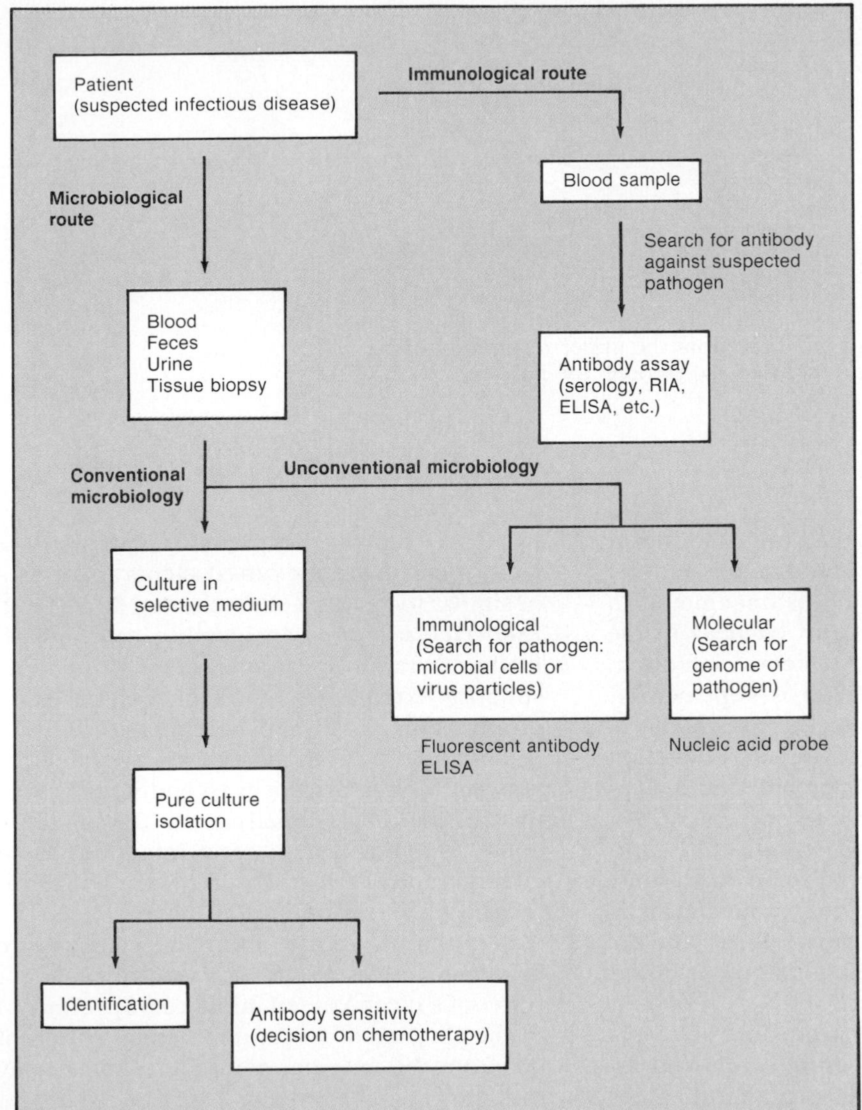

Figure 13.1 Clinical and diagnostic methods used for infectious disease.

Table 13.1 Recommended agar media for primary isolation purposes in a clinical microbiology laboratory[a]

	Media[b]				
Specimen	**Blood agar**	**Enteric agar**	**CA**	**TM**	**Anaerobic**
Fluids: chest, abdomen, pericardium	+	+	+	−	+
Feces: rectal swabs	+	+	+	−	−
Surgical tissue biopsies: lung, lymph nodes	+	+	−	−	+
Throat: swabs, sputum, tonsil, nasopharynx	+	+	+	−	−
Genitourinary swabs: urethra, vagina, cervix	+	+	+	+	−
Urine	+	+	−	−	−
Blood	+	−	−	−	+
Swabs: wounds, abscesses, exudates	+	+	+	−	+

[a]*Data from Lennette, E.H., A. Balows, W.J. Hausler, Jr., and H.J. Shadomy. 1985. Manual of Clinical Microbiology, 4th edition, ASM, Washington, D.C.*
[b]*Descriptions: Blood agar, 5 percent whole sheep blood added to trypticase soy agar; Enteric agar, either eosin-methylene blue agar, or MacConkey agar; CA, chocolate (heated blood) agar; TM, Thayer-Martin agar; Anaerobic, thioglycolate-containing blood agar or supplemented thioglycolate agar incubated anaerobically.*

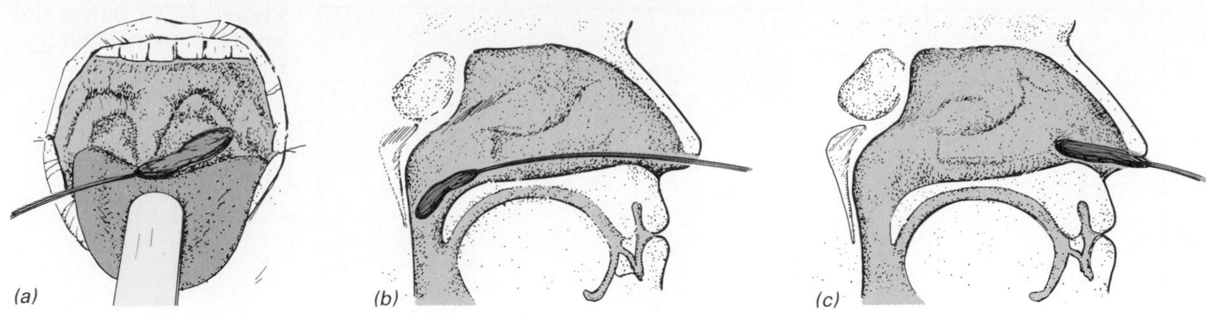

Figure 13.2 Methods for obtaining specimens from the upper respiratory tract. (a) Throat swab. (b) Nasopharyngeal swab passed through the nose. (c) Swabbing the inside of the nose.

most important duties of the clinical microbiologist. Bacteria are cleared from the bloodstream rapidly (Section 11.14) so that bacteremia is uncommon in healthy individuals, and presence of bacteria in the blood is generally indicative of systemic infection. The classic type of blood infection is **septicemia**, resulting from a virulent organism entering the blood from a focus of infection, multiplying, and traveling to various body tissues to initiate new infections. Septicemia is indicated by the presence of severe systemic symptoms, usually with fever and chills, followed by prostration. In many disease situations, culture of the blood provides the only immediate way of isolating and identifying the causal agent, and diagnosis may thus depend on careful and proper blood culture.

The standard blood culture procedure is to remove 10 ml of blood aseptically from a vein and inject it into a blood culture bottle containing an anticoagulant and an all-purpose culture medium. Two cultures are set up with one bottle being incubated aerobically and one anaerobically. Media used are all relatively rich containing protein digests and other complex ingredients. Blood culture bottles are incubated at 35°C and examined daily for up to 7 days. Clinically significant bacteria are generally recovered within this period.

Because a certain amount of skin contamination is probably unavoidable during initial drawing of the blood, even under optimal conditions a contamination rate of 2 to 3 percent can be expected. Because of this, identification of positive blood cultures is essential, and contamination may be indicated if certain organisms commonly found on the skin are isolated, such as *Staphylococcus epidermidis*, coryneform bacteria, or propionibacteria, although even these organisms can occasionally cause infection of the wall of the heart (subacute bacterial endocarditis). Thus considerable microbiological and clinical experience is necessary in interpreting blood cultures.

Urine cultures Urinary infections are very common, and because the causal agents are often identical or similar to bacteria of the normal flora (for example, *Escherichia coli*), considerable care must be taken in the bacteriological analysis of urine. Since urine supports extensive bacterial growth under many conditions, fairly high cell numbers are often found in urinary infection. In most cases, the infection occurs as a result of an organism ascending the urethra from the outside. Occasionally, even the bladder may become infected. Urinary tract infections are the most common form of *nosocomial* (hospital-acquired) infection (see Section 14.7).

Significant urinary infection generally results in bacterial counts of 10^5 or more organisms per milliliter of a clean-voided midstream specimen, whereas in the absence of infection, contamination of the urine from the external genitalia (almost unavoidable to some extent) results in less than 10^3 organisms per milliliter. The most common organisms are members of the enteric bacteria, with *E. coli* accounting for about 90 percent of the cases. Other organisms include *Klebsiella, Enterobacter, Proteus, Pseudomonas*, and *Streptococcus faecalis*. *Neisseria gonorrhoeae*, the causal agent of gonorrhea, does not grow in the urine itself, but in the urethral epithelium, and must be diagnosed by different methods (see below). Because of the relatively high number of organisms found in a significant urinary infection, direct microscopic examination of the urine is of considerable value and is recommended as part of the basic procedure. A small drop of urine is allowed to dry on a microscope slide and a Gram stain performed. If more than 10^5 organisms per milliliter are present, there are usually one or more bacteria per microscope field (100 × objective). Cultural examination of the specimen must be by quantitative means, because contamination of the specimen by small numbers of bacteria from the external parts of the body is almost unavoidable. Ideally two media should be used in the initial culture of urine, blood

Table 13.2 Colony characteristics of frequently isolated Gram-negative rods cultured on various clinically useful agars[a]

Organism	Agar media[b]			
	EMB	MC	SS	BS
Escherichia coli	Dark center with greenish metallic sheen	Red or pink	Red to pink	Mostly inhibited
Enterobacter	Similar to *E. coli*, but colonies are larger	Red or pink	White or beige	Mucoid colonies with silver sheen
Klebsiella	Large mucoid, brownish	Pink	Red to pink	Mostly inhibited
Proteus	Translucent, colorless	Transparent, colorless	Black center, clear periphery	Green
Pseudomonas	Translucent, colorless to gold	Transparent, colorless	Mostly inhibited	No growth
Salmonella	Translucent, colorless to gold	Translucent, colorless	Opaque	Black to dark green
Shigella	Translucent, colorless to gold	Transparent, colorless	Opaque	Brown or inhibited

[a]*Modified from Lennette, E.H., A. Balows, W.J. Hausler, Jr., and H.J. Shadomy, eds. 1985. Manual of Clinical Microbiology, 4th ed. American Society for Microbiology, Washington, D.C.*
[b]*BS, Bismuth Sulfite agar, EMB, eosin-methylene blue agar; MC, MacConkey agar; SS,* Salmonella-Shigella *agar.*

agar as a nonselective general medium, and a medium selective for enteric bacteria, such as MacConkey or EMB agar. Either of these latter media permit the initial differentiation of lactose fermenters from nonfermenters, and the growth of Gram-positive organisms such as *Staphylococcus* (a common skin contaminant) is inhibited. Organisms isolated can be identified, and antibiotic susceptibility tests performed. Experienced clinical microbiologists may make a tentative identification of an isolate by observing the color and morphology of colonies of the suspected pathogen grown on various selective media as described in Table 13.2. Such an identification must be followed up with more detailed analyses, of course, but clinical microbiologists will use this information in conjunction with more detailed test results in order to make a positive identification. The nucleic acid/immunological probes described in Section 13.8 and 13.9 allow for identification of the pathogen on molecular rather than physiological/morphological grounds.

Fecal cultures Proper collection and preservation of feces is important in the isolation of intestinal pathogens. During storage, the pH of feces drops and thus an extended delay between sampling and sample processing must be avoided. This is especially critical for the isolation of *Shigella* and *Salmonella* species, both of which are rather sensitive to acid pH. Stool samples, collected from feces freshly voided onto paper toweling attached to the toilet bowl, are placed in a vial containing phosphate buffer for transport to the lab. If a patient has a bloody or pus-containing stool, the attending technician should be sure to sample this material; such discharges generally contain a large number of the organisms of interest. In the case of suspected food-or-water-borne infections, stool samples should be inoculated into a variety of selective media (see Section 13.2) for the isolation of specific bacteria or characterization of intestinal parasites. Positive identifications would be made by the techniques described in Sections 13.6–13.11.

Laboratory diagnosis of gonorrhea Although rarely fatal, gonorrhea is one of the most common infectious diseases, and laboratory procedures are central to its diagnosis. The causal agent, *Neisseria gonorrhoeae* (referred to clinically as the *gonococcus*) colonizes mucosal surfaces of the urethra, uterine cervix, anal canal, throat, and conjunctiva. The organism is quite sensitive to drying, and because of this it is transmitted almost exclusively by direct person-to-person contact, usually by sexual intercourse. The major goal of public health measures to control gonorrhea involves identification of asymptomatic carriers, and this requires microbiological analysis. Because the organism is a Gram-negative coccus, and similar organisms are not very common in the normal flora of the urogenital tract, direct microscopy of

Gram-stained material is of value. For example, observation of Gram-negative diplococci in a urethral discharge or in a vaginal or cervical smear is considered presumptive evidence for gonorrhea.

For direct microscopy, smears are prepared from urethral swabs and conventional Gram stains made. During visualization under the oil-immersion lens, attention should be paid to polymorphonuclear leukocytes (granulocytes), since in acute gonorrhea such cells will frequently contain groups of Gram-negative, kidney-shaped diplococci (Figure 13.3), and there will be few if any other types of organisms present.

Cultural procedures have a higher degree of sensitivity than microscopical analyses. Most media for the culture of *N. gonorrhoeae* contain heated blood or hemoglobin (referred to as *chocolate agar* because of its deep brown appearance), the heating causing the formation of a precipitated material, which is quite effective in absorbing toxic products present in the medium. After streaking, the plates must be incubated in a humid environment in an atmosphere containing 3 to 7 percent CO_2 (CO_2 is required for growth of gonococci). The plates are examined after 24 and 48 hours, and portions of colonies should be immediately tested by the oxidase test, since all *Neisseria* are oxidase-positive (see Section 13.2). Oxidase-positive Gram-negative diplococci growing on chocolate agar can be presumed to be gonococci if the inoculum was derived from genitourinary sources, but definite identification requires determination of carbohydrate utilization patterns or by immunological means (see Section 13.7). *Neisseria gonorrhoeae* degrades glucose with acid production, but does not produce acid from maltose, sucrose, or lactose, and this pattern is considered a confirmation of identification. A second primary isolation medium, called *Thayer-Martin agar*, also is used for isolation of *N. gonorrhoeae* (Figure 13.3*b*). This medium incorporates the antibiotics vancomycin, nystatin, and colistin, to which most clinical isolates of *N. gonorrhoeae* are naturally resistant.

Because cells of *N. gonorrhoeae* are so delicate, drying and contact with antiseptics during sample collection must be avoided and isolation procedures must be initiated as soon as possible after specimen collection. On the other hand, cells of *N. gonorrhoeae* can remain viable for long periods in moist samples, hence laboratory personnel must exercise great care in not inadvertently contaminating themselves.

Culture of anaerobes Obligately anaerobic bacteria are common causes of infection and will be completely missed in clinical diagnosis unless special precautions are taken for their isolation and culture. We have discussed anaerobes in general in Section 9.14, and we noted that many anaerobes are extremely susceptible to oxygen. Because of this, specimen collection, handling, and processing require special attention if it is thought that an obligate anaerobe is involved. There are several habitats in the body (for example, the intestinal tract and the oral cavity, see Sections 11.3 and 11.4) that are generally anaerobic, and in which obligately anaerobic bacteria can be found as part of the normal flora. However, other parts of the body can become anaerobic as a result of tissue injury or trauma, which results in reduction of blood supply to the injured site, and such anaerobic sites can then become available for colonization by obligate anaerobes. In general, the pathogenic anaerobic bacteria are part of the normal flora and are only opportunistic pathogens, although two important pathogenic anaerobes, *Clostridium tetani* (causal agent of tetanus) and *C. perfringens* (causal agent of gas gangrene), both sporeformers, are predominantly soil organisms.

With anaerobic culture, the microbiologist is presented not only with the usual problems of obtaining and maintaining an uncontaminated specimen, but also with ensuring that the specimen not come in contact with air. If only exudate can be collected, the best procedure is to remove the specimen directly from the infected tissue by aspiration with a syringe

(a)
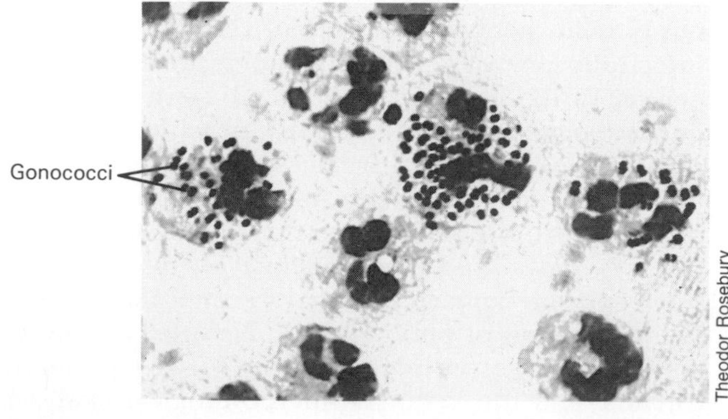

(b)

Figure 13.3 (a) Photomicrograph of *Neisseria gonorrhoeae* within human granulocytes. Note how many cells are in pairs as diplococci (arrows). Magnification, 1200×. (b) Colonies of *N. gonorrhoeae* growing on Thayer-Martin agar.

and needle, care being taken to be certain that no air is drawn into the syringe. It is preferable to obtain a sample of infected tissue rather than just exudate, since viability will be maintained much longer in tissue. Although swabs can be used to collect material, this is not generally recommended, because it is difficult to avoid aeration; also the number of organisms obtained on the swab is much lower than those obtained in tissue or exudate. The collected sample must be immediately placed in a tube containing oxygen-free gas, preferably containing a small amount of a dilute salts solution containing the redox indicator resazurin. This dye is colorless when reduced, and becomes pink when oxidized, thus any oxygen contamination of the specimen will be quickly indicated. If a proper anaerobic transport tube is not available, the syringe itself can be used to transport the specimen, the needle being inserted into a sterile rubber stopper so that no air is drawn into the syringe.

For anaerobic incubation, agar plates are placed in a sealed jar which is made anaerobic by either replacing the atmosphere in the jar with an oxygen-free gas mixture (a mixture of N_2 and CO_2 is frequently employed), or by adding some compound to the enclosed vessel which removes O_2 from the atmosphere (Figure 13.4). Alternative means of isolating anaerobes include the use of prereduced culture media or anaerobic hoods. The latter are large gas impermeable bags filled with an oxygen-free gas such as nitrogen or hydrogen and which are fitted with an airlock for inserting and removing cultures (see Figure 9.22). The advantage of an anaerobic glove box is that manipulations can be done as one would normally perform them on a laboratory bench. However, because of their expense, anaerobic glove boxes are not used extensively in clinical laboratories, but are in widespread use in research laboratories which specialize in anaerobic microorganisms (see Figure 9.22).

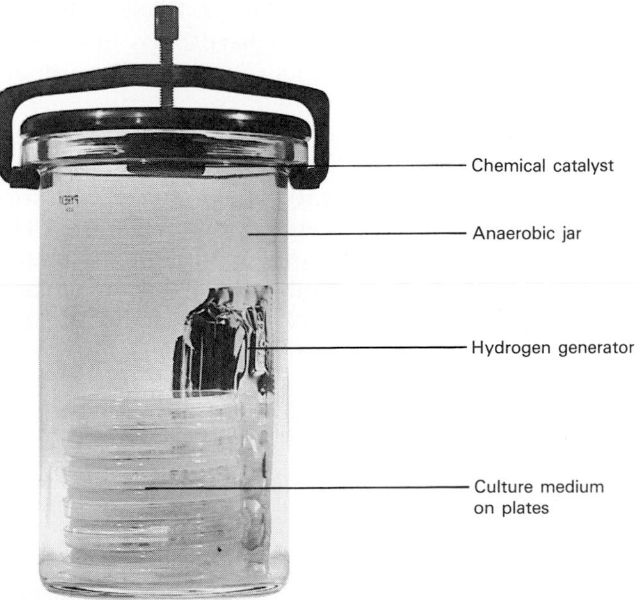

Figure 13.4 Special jar for incubating cultures under anaerobic conditions.

In general, media for anaerobes do not differ greatly from those used for aerobes, except that they are generally richer in organic constituents, and contain reducing agents (usually cysteine or thioglycolate) and a redox indicator such as resazurin. Once positive cultures have been obtained, it is essential that they be characterized and identified, to be certain that the isolate is not a member of the normal flora.

13.2 Growth-Dependent Identification Methods

If the inoculation of a primary medium results in bacterial growth, the clinical microbiologist goes to work immediately trying to identify the organism or organisms present. If the primary enrichment medium yields a culture containing a single morphological type, the organism is probably in pure culture. However, if a liquid enrichment was used, agar plates are generally streaked to insure culture purity. From here on, the methodology used for positively identifying a pathogenic agent will depend not only on the properties of the organism (e.g., Gram reaction, suspected identity), but also on the resources and facilities available to the clinical microbiology lab. We discuss here these growth-dependent identification methods.

Growth on selective/differential media Based on growth in primary isolation media, it is frequently possible to restrict the unknown pathogen to a relatively few choices. Clinical microbiologists have at their disposable dozens of diagnostically useful culture media to which their unknown can be inoculated. In many large hospitals and clinics, media are purchased from commercial sources, which ensures quality control and reliable testing in different clinical settings (Figure 13.5). Many of these media are available in miniaturized kits containing a number of different media in separate wells, all of which can be inoculated at one time (Figure 13.5*d* and *e*).

The battery of media employed are either selective, differential or both. A *selective medium* is one to which compounds have been added to selectively inhibit the growth of certain microorganisms but not others. A *differential medium* is one to which some sort of indicator, usually a dye, has been added, which allows the clinician to differentiate between various chemical reactions carried out during growth. Eosin-methylene blue (EMB) agar, for example, is a widely used selective and differential medium. EMB agar is used for the isolation of Gram-negative enteric bacteria. The methylene blue is present to inhibit Gram-positive bacteria; although the mechanism is unclear,

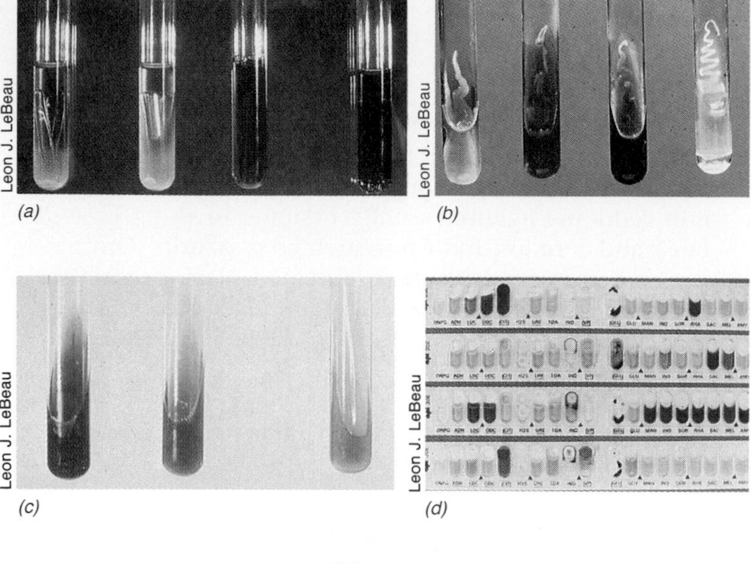

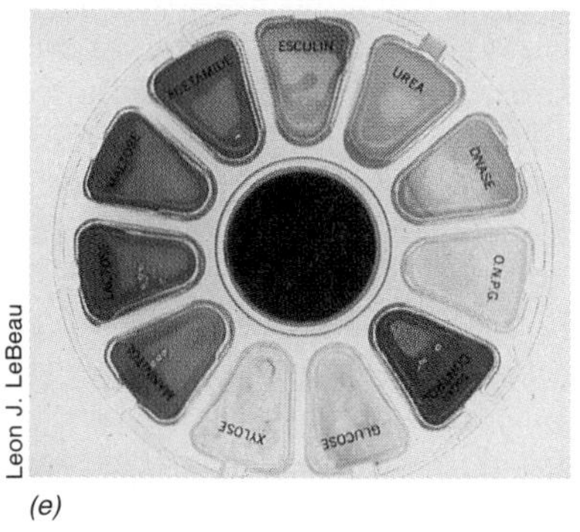

Figure 13.5 Growth-dependent diagnostic methods used for the identification of clinical isolates by color changes in various diagnostic media. (a) Use of a differential medium to assess sugar fermentation. Acid production is indicated by color change of the pH-indicating dye added to the medium. If gas production occurs a bubble appears in the inverted vial. From left to right: acid, acid and gas, negative, uninoculated. (b) A conventional diagnostic test for enteric bacteria in a medium called *triple sugar iron agar*. The medium is inoculated both on the surface of the slant and by stabbing into the butt. The medium contains a small amount of glucose and a large amount of lactose and sucrose. Organisms able to ferment only the glucose will cause acid formation only in the butt, whereas lactose or sucrose-fermenting organisms will also cause acid formation in the top. Gas formation is indicated by the breaking up of the agar in the butt. Hydrogen sulfide formation (from either protein degradation or from reduction of thiosulfate in the medium) is indicated by a blackening due to reaction of the H_2S with ferrous iron in the medium. From left to right: fermentation of glucose and another sugar; hydrogen sulfide formation; no reaction; fermentation of glucose only. (c) Measurement of citrate utilization by *Salmonella* on Simmons citrate agar. The change in pH causes a change in the color of the indicating dye. From left to right: positive, negative, uninoculated. (d) Miniaturized kits used for the rapid identification of clinical isolates. The principle is the same as in *a* but the whole arrangement has been miniaturized so that a number of tests can be run at the same time. Four separate strips, each with a separate culture, are shown. (e) Another arrangement of a miniaturized test kit. This one permits diagnosis of sugar utilization in nonfermentative organisms.

small amounts of this dye effectively inhibit the growth of most Gram-positive bacteria. Eosin is a dye which responds to changes in pH, going from colorless to black under acidic conditions. EMB contains lactose and sucrose, but not glucose, as energy sources, and colonies of lactose-fermenting (generally enteric) bacteria, such as *E. coli, Klebsiella*, and *Enterobacter*, acidify the medium and have black centers with a greenish sheen. Colonies of lactose nonfermenters, such as *Salmonella, Shigella*, or *Pseudomonas* are translucent or pink.

In the battery of tests performed to help identify an organism many different biochemical reactions can be measured. The most important tests are summarized in Table 13.3. In effect, these tests measure the presence or absence of *enzymes* involved in catabolism of the substrate or substrates added to the differential medium. Fermentation of sugars is measured by incorporating pH dyes which change color upon acidification (Figure 13.5*a*). Production of hydrogen gas and/or carbon dioxide during sugar fermentation is assayed by observing for gas bubbles trapped in the agar (Figure 13.5*a*). Hydrogen sulfide production is indicated following growth in a medium containing ferric iron. If sulfide is produced, ferric iron complexes with H_2S to form a black precipitate of iron sulfide (Figure 13.5*b*). Utilization of citric acid, a six carbon acid containing three carboxylic acid groups, is accompanied by a pH rise, and a specific dye incorporated into this test medium changes color as conditions become alkaline (Figure 13.5*c*). Literally hundreds of differential tests have been developed for clinical use, but only about 20 are used routinely (Figure 13.5*d*).

The typical reaction patterns for large numbers of strains of various pathogens have been published, and in the modern clinical microbiology laboratory all of this information is stored in a computer. The results of the differential tests on an unknown pathogen are entered and the computer will make the best match by comparing the characteristics of the unknown with the species in the databank. For many organisms, as few as three or four key tests are all that are required to make an unambiquous identification. In cases of a dubious match, however, more sophisticated identification procedures may be called for, especially if the course of chemotherapy is a critical one.

Making a diagnosis Many companies market their own versions of growth-dependent rapid identification systems (Figure 13.5*d* and *e*). Such systems

Table 13.3 Important clinical diagnostic tests for bacteria

Test	Principle	Procedure	Most common use
Carbohydrate fermentation	Acid and/or gas during fermentative growth with sugars or sugar alcohols	Broth medium with carbohydrate and phenol red as pH indicator; inverted tube for gas	Enteric bacteria differentiation (also several other genera or species separations with some individual sugars)
Catalase	Enzyme decomposes hydrogen peroxide, H_2O_2	Add drop of H_2O_2 to dense culture and look for bubbles (O_2) (Figure 9.25)	*Bacillus* (+) from *Clostridium* (−); *Streptococcus* (−) from *Micrococcus/ Staphylococcus* (+)
Citrate utilization	Utilization of citrate as sole carbon source, results in alkalinization of medium	Citrate medium with bromthymol blue as pH indicator, look for intense blue color (alkaline pH)	*Klebsiella-Enterobacter* (+) from *Escherichia* (−), *Edwardsiella* (−) from *Salmonella* (+)
Coagulase	Enzyme causes clotting of blood plasma	Mix dense liquid suspension of bacteria with plasma, incubate, and look for sign of clot in few minutes	*Staphylococcus aureus* (+) from *S. epidermidis* (−)
Decarboxylases (lysine, ornithine, arginine)	Decarboxylation of amino acid releases CO_2 and amine	Medium enriched with amino acid. Bromcresol purple pH indicator. Alkaline pH if enzyme action, indicator becomes purple	Aid in determining bacterial group among the enteric bacteria
β-Galactosidase (ONPG) test	Orthonitrophenyl-β-galactoside (ONPG) is an artificial substrate for the enzyme. When hydrolyzed, nitrophenol (yellow) is formed	Incubate heavy suspension of lysed culture with ONPG, look for yellow color	*Citrobacter* and *Arizona* (+) from *Salmonella* (−). Identifying some *Shigella* and *Pseudomonas* species
Gelatin liquefaction	Many proteases hydrolyze gelatin and destroy the gel	Incubate in broth with 12% gelatin. Cool to check for gel formation. If gelatin hydrolyzed, tube remains liquid upon cooling	To aid in identification of *Serratia, Pseudomonas, Flavobacterium, Clostridium*
Hydrogen sulfide (H_2S) production	H_2S produced by breakdown of sulfur amino acids or reduction of thiosulfate	H_2S detected in iron-rich medium from formation of black ferrous sulfide (many variants: Kliger's iron agar, triple sugar iron agar, also detect carbohydrate fermentation)	In enteric bacteria, to aid in identifying *Salmonella, Arizona, Edwardsiella,* and *Proteus*
Indole test	Tryptophan from proteins converted to indole	Detect indole in culture medium with dimethylaminobenzaldehyde (red color)	To distinguish *Escherichia* (+) from *Klebsiella-Enterobacter* (−); *Edwardsiella* (+) from *Salmonella* (−)
Methyl red test	Mixed-acid fermenters produce sufficient acid to lower pH below 4.3	Glucose-broth medium. Add methyl red indicator to a sample after incubation	To differentiate *Escherichia* (+, culture red) from *Enterobacter* and *Klebsiella* (usually −, culture yellow)

(continued)

Table 13.3 Important clinical diagnostic tests for bacteria (continued)

Test	Principle	Procedure	Most common use
Nitrate reduction	Nitrate as alternate electron acceptor, reduced to NO_2^- or N_2	Broth with nitrate. After incubation, detect nitrite with α-naphthylamine-sulfanilic acid (red color). If negative, confirm that NO_3^- still present by adding zinc dust to reduce NO_3^- to NO_2^-. If no color after zinc, then $NO_3^- \rightarrow N_2$	To aid in identification of enteric bacteria (usually +)
Oxidase test	Cytochrome *c* oxidizes artificial electron acceptor: tetramethyl (or dimethyl)-*p*-phenylenediamine	Broth or agar. Oxidase-positive colonies on agar can be detected by flooding plate with reagent and looking for blue or brown colonies	To separate *Neisseria* and *Moraxella* (+) from *Acinetobacter* (−). To separate enteric bacteria (all −) from pseudomonads (+). To aid in identification *Aeromonas* (+)
Oxidation-fermentation (O/F) test	Some organisms produce acid only when growing aerobically	Acid production in top part of sugar-containing culture tube; soft agar used to restrict mixing during incubation	To differentiate *Micrococcus* (aerobic acid production only) from *Staphylococcus* (acid anaerobically). To characterize *Pseudomonas* (aerobic acid production) from enteric bacteria (acid anaerobically)
Phenylalanine deaminase test	Deamination produces phenylpyruvic acid, which is detected in a colorimetric test	Medium enriched in phenylalanine. After growth, add ferric chloride reagent, look for green color	To characterize the genus *Proteus* and the *Providencia* group
Starch hydrolysis	Iodine-iodide gives blue color with starch	Grow organism on plate containing starch. Flood plate with Gram's iodine, look for clear zones around colonies	To identify typical starch hydrolyzers such as *Bacillus* spp.
Urease test	Urea ($H_2N{-}CO{-}NH_2$) split to $2NH_3 + CO_2$	Medium with 2% urea and phenol red indicator. Ammonia release raises pH, intense pink-red color	To distinguish *Klebsiella* (+) from *Escherichia* (−). To distinguish *Proteus* (+) from *Providencia* (−)
Voges-Proskauer test	Acetoin produced from sugar fermentation	Chemical test for acetoin using α-napthol	To separate *Klebsiella* and *Enterobacter* (+) from *Escherichia* (−). To characterize members of genus *Bacillus*

are frequently designed for use in identifying enteric bacteria, because enterics are frequently implicated in routine urinary tract and intestinal infections. Infections of the urinary tract caused by *E. coli* or other enteric bacteria are the leading cause of hospital infections (see Section 14.7). If a urinary tract infection is suspected but no organism(s) can be cultured, a highly sensitive urine screening test is available that detects very low numbers of bacterial cells in urine by measuring the amount of ATP present; ATP concentration is proportional to cell biomass. Using firefly luciferin/luciferase (see Section 17.4), the concentration of ATP in a urine sample can be accurately determined in less than one hour, and the number of cells per ml calculated. The test is very useful for diagnosing low-grade urinary tract infections where microscopy or attempts to culture an organism may not reveal the presence of bacteria.

Other growth-dependent rapid identification kits are available for other bacterial groups or even for single bacterial species. For example, commercial kits containing a battery of tests have been developed for specifically identifying the following bacteria: *Staphylococcus aureus, Streptococcus pyogenes, Neisseria gonorrhoeae, Haemophilus influenzae*, and *Mycobacterium tuberculosis*. Other kits are available for identification of the pathogenic yeasts *Candida albicans* and *Cryptococcus neoformans*.

The decision to use a specific diagnostic test is usually made by the clinical microbiologist following consultation with the physician. The clinical microbiologist will take into consideration the nature of the tissue sampled, basic characteristics (especially the Gram stain) of pure cultures obtained, and previous experience with similar cases. For example, an enteric identification kit would be useless in identifying a Gram-positive coccus isolated from an abscess. Instead, an *S. aureus* or *S. pyogenes* kit would be used to make a positive identification.

13.3 Testing Cultures for Antibiotic Sensitivity

In medical practice, microbial cultures are isolated from diseased patients to confirm diagnosis and to aid in decision on therapy. Determination of the sensitivity of microbial isolates to antimicrobial agents is one of the most important tasks of the clinical microbiologist.

The sensitivity of a culture can be most easily determined by an agar diffusion method. Federal regulations of the Food and Drug Administration (FDA) now control the procedures used for sensitivity testing in the United States and similar regulations exist in other countries. The recommended procedure is called the *Kirby-Bauer method*, after the workers who developed it (Figure 13.6). A plate of suitable culture medium is inoculated by spreading an aliquot of culture evenly across the agar surface. Filter-paper discs containing known concentrations of different antimicrobial agents are then placed on the plate. The concentration of each agent on the disc is specified so that zone diameters of appropriate size will develop to indicate sensitivity or resistance. After incubation, the presence of inhibition zones around the discs of the different agents is noted. Table 13.4 presents typical zone sizes for several antibiotics. Zones observed on the plate are measured and compared to standard data, to determine if the isolate is truly sensitive to a given antibiotic.

Another procedure for antibiotic sensitivity testing involves an *antibiotic dilution assay,* either in culture tubes or in the wells of a microtiter plate (Figure 13.6*e*). (We discussed the use of microtiter plates in the Box in Chapter 12.) A series of two-fold dilutions of each antibiotic are made in the wells and then all wells are inoculated with the same test or-

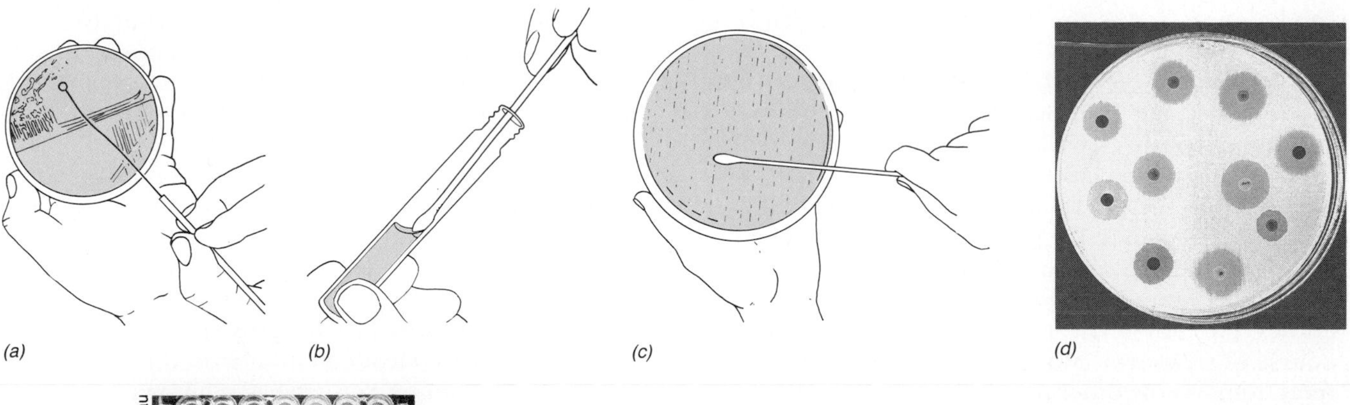

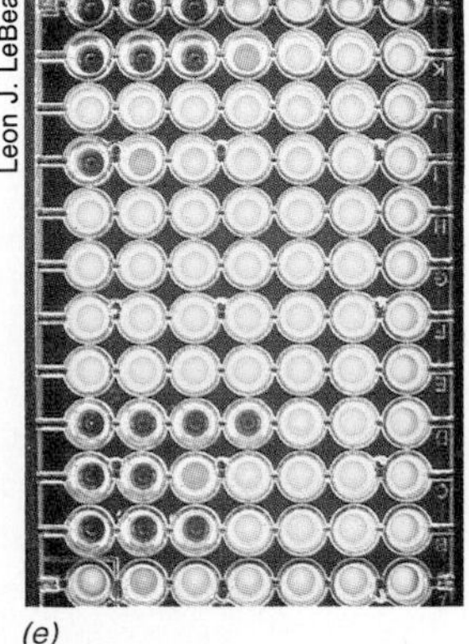

Figure 13.6 The Kirby-Bauer procedure for determining the sensitivity of an organism to antibiotics. (a) A colony is picked from an agar plate. It is inoculated into a tube of liquid culture medium and allowed to grow to a specified density. (b) A swab is dipped in the liquid culture. (c) The swab is streaked evenly over a plate of sterile agar medium. (d) Discs containing known amounts of different antibiotics are placed on the plate. After incubation, inhibition zones are observed. The susceptibility of the organism is determined by reference to a chart of zone-sizes (see Table 13.4). (e) Antibiotic sensitivity determined by the dilution method on a microtiter plate. The organism is *Pseudomonas aeruginosa*. Each row has a different antibiotic.

Table 13.4 Typical zone sizes for some common antibiotics

Antibiotic	Amount on disc	Inhibition zone diameter (mm)		
		Resistant	Intermediate	Sensitive
Ampicillin[a]	10 μg	11 or less	12–13	14 or more
Ampicillin[b]	10 μg	28 or less	—	29 or more
Cephoxitin	30 μg	14 or less	15–17	18 or more
Cephalothin	30 μg	14 or less	15–17	18 or more
Chloramphenicol	30 μg	12 or less	13–17	18 or more
Clindamycin	2 μg	14 or less	15–16	17 or more
Erythromycin	15 μg	13 or less	14–17	18 or more
Gentamicin	10 μg	12 or less	13–14	15 or more
Kanamycin	30 μg	13 or less	14–17	18 or more
Methicillin[c]	5 μg	9 or less	10–13	14 or more
Neomycin	30 μg	12 or less	13–16	17 or more
Nitrofurantoin	300 μg	14 or less	15–16	17 or less
Penicillin G[c]	10 Units	28 or less	—	29 or more
Penicillin G[d]	10 Units	11 or less	12–21	22 or more
Polymyxin B	300 Units	8 or less	9–11	12 or more
Streptomycin	10 μg	11 or less	12–14	15 or more
Tetracycline	30 μg	14 or less	15–18	19 or more
Trimethoprim-sulfamethoxazole	1.25/23.75 μg	10 or less	11–15	16 or more
Tobramycin	10 μg	12 or less	13–14	15 or more

[a]*For Gram-negative organisms and enterococci.*
[b]*For staphylococci and highly penicillin-sensitive organisms.*
[c]*For staphylococci.*
[d]*For organisms other than staphylococci. Includes some organisms, such as enterococci and some Gram-negative bacilli, that may cause systemic infections treatable by high doses of Penicillin G.*

ganism. After incubation, the inhibition of growth by the various antibiotics can be observed. Sensitivity can be expressed as the highest dilution which completely inhibits growth.

Because of the widespread occurrence of antibiotic resistance (see Section 11.17), an antibiotic sensitivity test is essential for an appropriate isolate from each patient. Data such as those in Table 13.4 are useful to the physician in choosing the best antibiotic for a specific bacterial infection. Fortunately, many potentially serious pathogens are highly susceptible to a number of different antibiotics and this allows the physician considerable latitude in the course of treatment. Most *Pseudomonas aeruginosa* infections, on the other hand, are very difficult to treat with the majority of common antibiotics, and the physician's choice is limited to a rather restricted group of drugs.

13.4 Diagnostic Virology

The procedures that we have discussed so far in this chapter are suitable only for microbial infections, where it is possible to cultivate a suspected pathogen in a lifeless culture medium. What about viruses? We discussed in Chapter 6, the wide variety of viruses which infect animals (including humans) and we discuss some of the important viral diseases of humans in Chapter 15. Here we present briefly laboratory procedures that can be used in the diagnosis of viral diseases by direct culture of the virus. Later in this chapter we present immunological and molecular genetic procedures that are also available.

First, we should note that laboratory cultivation of viruses from clinical materials is more difficult, time-consuming, and specialized than the cultivation of microbial pathogens. This is because viruses only grow in living cells. We discussed in Section 6.3 the use of cell cultures for the growth of viruses, and such cultures are most commonly used in diagnostic virology. A common cell line is a human diploid fibroblast culture called WI-38, which grows rapidly and reproducibly in cell culture medium. Another cell line sometimes used is HeLa, a culture derived initially from a human cancer but now so long in culture that it has greatly changed its character (it is no longer diploid, for instance). In addition to these two cell cultures, which can be maintained in the laboratory indefinitely, cultures are also made from Rhesus monkey kidneys. Monkey kidney cell lines are called *primary*, because they are not maintained by successive transfer in the laboratory. Primary monkey kidney cells support the rapid growth of a number

Table 13.5 Some laboratory procedures used in the diagnoses of viral diseases

Condition	Possible viral cause	Samples to obtain	Inoculation procedure
Upper respiratory infection	Rhinovirus	Nasopharyngeal or tracheal fluid (aspirate)	Human fibroblast culture
	Coronavirus		
	Adenovirus		
Pneumonia	Influenza	Nasopharyngeal fluid or swab	Human fibroblast cultures or embryonated eggs
Measles	Measles virus	Nasopharyngeal fluid or swab	Monkey kidney cells
Vesicular rash	Herpes simplex	Vesicular fluid by aspiration	Human fibroblast culture
Diarrhea	Rotavirus (infants)	Feces or rectal swab	Look for characteristic virus particles with the electron microscope
	Norwalk agent (adults)		
Nonbacterial meningitis	Enterovirus	Spinal fluid	Human fibroblast or monkey kidney cultures
	Mumps		
	Herpes simplex		

Immunological methods are also widely used in the diagnosis of viral infections (see Sections 13.6–13.10).

of viruses infecting humans and are therefore of value in initial isolation of unknown viruses, but are less easy for routine use in the clinical laboratory. It should be emphasized that diagnostic virology is much more technical than diagnostic microbiology and is therefore only carried out in specialized laboratories (such as those associated with large medical centers).

A summary of a few laboratory procedures used in diagnostic virology is given in Table 13.5. It should be noted that most of these procedures are used only under special circumstances. For routine virus infections, diagnosis will be made by symptoms or by serological procedures (see Section 13.6).

13.5 Safety in the Clinical Laboratory

By their very nature, clinical laboratories are areas in which potentially dangerous biological specimens must be handled on a routine basis. Hence, a defined protocol for handling clinical samples must be established to avoid laboratory accidents.

Studies of laboratory-associated infections have indicated that most such infections do not result from known exposures or accidents, but instead from routine handling of patient specimens. Infectious aerosols, generated during processing of the specimen, are the most likely cause of laboratory infections. In attempts to minimize the exposure of clinicians to infectious agents and to thereby reduce the number of nonaccident associated laboratory infections, well-run clinical laboratories stress the following biological safety rules:

1. Laboratories handling hazardous materials restrict access to laboratory or support personnel only. These individuals should have knowledge of the biological risks involved in the clinical laboratory and act accordingly.
2. Effective procedures for decontaminating infectious materials or wastes, including specimens, inoculated media, bacterial cultures, tissue cultures, experimental animals, glassware, instruments, and surfaces must be in place and be practiced without compromise.
3. Personnel working with hazardous infectious agents or vaccines (for example, rabies, polio, or diphtheria-pertussis-tetanus vaccines) must be properly vaccinated against the agent.
4. All clinical specimens should be considered potentially infectious and handled in the appropriate manner. This is especially important for preventing laboratory acquired hepatitis, because of the relative frequency with which hepatitis viruses are present in clinical specimens.
5. All pipetting must be done with automatic pipetting devices (not by mouth), and devices such as syringes, needles, and clinical centrifuges must always be used with proper biological containment equipment.
6. Animals should only be handled by trained laboratory personnel and anesthetics or tranquilizers

should be used to avoid injury to both personnel and animals.

7. Laboratory personnel should always wear safety clothing including laboratory coats or uniforms, and wear additional protective clothing (such as rubber gloves and breathing masks) when appropriate. All laboratory personnel should practice good personal hygiene in respect to hand washing. Eating and drinking are never permitted in the clinical laboratory.

The safety rules outlined above should be the norm for all clinical laboratories. Specialized clinical laboratories may have additional rules to ensure a safe work environment. For example, if laboratory personnel handle extremely hazardous material (such as the causative agent of tuberculosis, *Mycobacterium tuberculosis*) on a routine basis, the laboratory would be fitted with special features, such as negatively pressurized rooms and air filters, to prevent the accidental release of the pathogen from the laboratory. In the final analysis, however, it is the attitude of the personnel that make the laboratory a safe or an unsafe place to work. Any clinical laboratory is a potentially hazardous place for untrained personnel or those unwilling to take the necessary steps to prevent laboratory-acquired infection.

13.6 Fluorescent Antibodies

As we have seen in Chapter 12, antibodies are highly specific in their interactions with antigens. This specificity can be used in many ways in the diagnosis of infectious disease and a wide variety of immunological methods are employed by the clinical microbiologist. Some immunological methods can be applied directly to a clinical specimen, thus avoiding the need for a culture of the organism.

We described the theory of fluorescent antibodies in Section 12.8. Recall that fluorescent dyes can be attached to antibody molecules. By injecting whole bacterial cells or cell fractions (for example, cell walls, surface proteins, flagella, pili), of known pathogens into experimental animals, banks of reference antisera can be prepared which can be made fluorescent for diagnostic use. Commercially available antisera prepared against several different antigens from known pathogenic microorganisms are readily available for clinical use. In the fluorescent antibody test, a smear of the suspected pathogen is allowed to react with a specific fluorescent antibody and observed under a fluorescent microscope. If the pathogen contains surface antigens against which the fluorescent antiserum was prepared (i.e., the suspected pathogen is identical to or immunologically closely related to the cells used to generate the antibodies), the cells will fluoresce (Figure 13.7). Organisms immunologically unrelated to the control organism generally do not react or react only weakly.

Fluorescent antibody methodology can also be applied directly to infected host tissues to permit diagnosis long before primary isolation techniques can yield a suspected organism. For example, in diagnosing legionellosis (see Section 15.2) a positive diagnosis can be made by staining biopsied lung tissue with fluorescent antibodies prepared against cell walls of *Legionella pneumophila*, the causative agent of legionellosis. Likewise, a fluorescent antibody against the capsule of *Bacillus anthracis* can be used in the microscopic diagnosis for anthrax. Fluorescent antibody reactions can also be used in diagnosis of viral infection (Figure 13.7*b* and *c*). Fluorescent antibodies therefore provide useful, rapid, and highly specific clinical tests that can often be of great benefit in guiding the physician to the proper diagnosis.

Immunodiagnoses using fluorescent antibody techniques are not without their pitfalls, however.

Figure 13.7 Examples of the use of fluorescent antibodies in clinical microbiology. (a) Immunofluorescent stained cells of *Legionella pneumophila*, the cause of legionellosis. Magnification, 1000×. (b) Detection of virus-infected cells by immunofluorescence. Human B-lymphotrophic virus (HBLV) infected spleen cells were incubated with serum which contained antibodies to HBLV from a patient with a lymphoproliferative disorder. Cells were then treated with fluorescein isothiocyanate conjugated anti-human IgG antibodies. HBLV-infected cells fluoresce bright yellow. Cells in the background did not react with patient's serum. (c) Detection of viral membrane antigen. HBLV-infected as well as uninfected cells were treated with patient serum and detected by immunofluorescence as described in (b). An HBLV-infected cell with patchy surface fluorescence is shown.

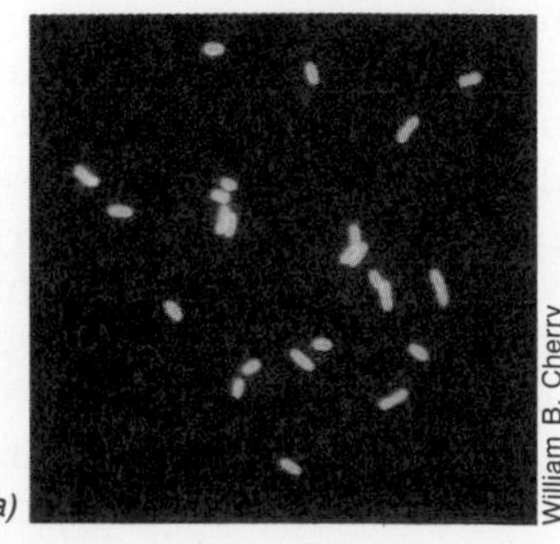

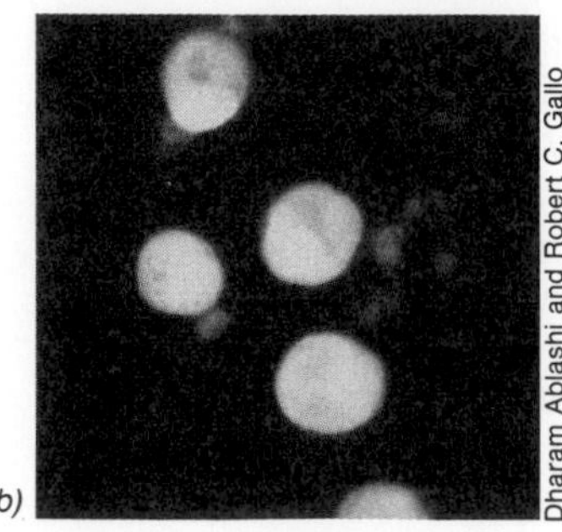

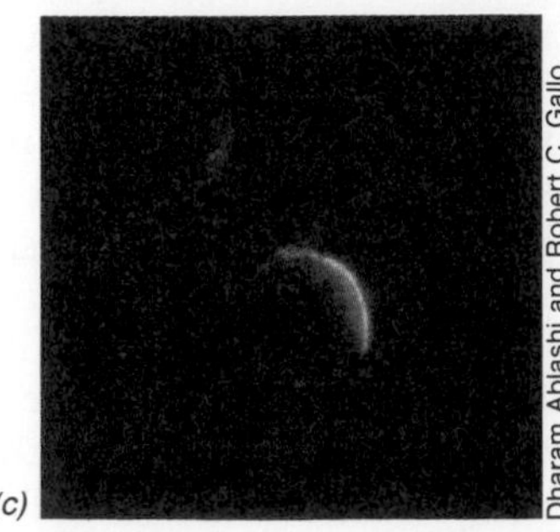

Nonspecific staining can be a problem because of surface antigens that may cross-react between various bacterial species, some of which may be members of the normal flora. This is particularly a problem among enteric bacteria, where antigens derived from lipopolysaccharide, although sometimes highly variable, are frequently sufficiently similar among species to cause binding or partial binding of the fluorescent probe. The clinical microbiologist must therefore be careful to perform all the necessary controls using nonspecific (premmune) sera, and confirm all positive immunofluorescent findings by other immunological or microbiological tests.

13.7 Monoclonal Antibodies

Monoclonal antibody technology has greatly reduced the problems of cross-reactivity in immunodiagnoses. We discussed monoclonal antibodies in Section 12.7. As noted, a monoclonal antibody is generally highly specific for a *single* antigenic determinant and hence is very useful in immunodiagnosis. Fluorescein-labeled monoclonal antibodies for use in fluorescent microscopy are already available commercially against *Chlamydia trachomitis* membrane proteins. These monoclonals react with *C. trachomitis* but are so specific they fail to react with even the closely related species, *C. psittaci. C. trachomitis*, the causative agent of a variety of venereal syndromes (see Section 15.6), is an obligate intracellular parasite and not easily cultured for a variety of technical reasons. Use of the fluorescent anti-*C. trachomitis* monoclonal on cervical *scrapings* or urethral or vaginal *exudates* now makes positive identification of chlamydial infections almost routine. Other monoclonals have been developed against outer membrane proteins of *Neisseria gonorrhoeae*. These probes are not only monospecific, reacting only with *N. gonorrhoeae*, but are also capable of differentiating *strains* of *N. gonorrhoeae*. The use of fluorescent monoclonal antibodies therefore eliminates much of the cross-reactivity problem observed when polyclonal sera are used.

The power of monoclonal antibodies in microbiological diagnostics should be emphasized. With presently available technology, it is possible to generate monoclonal antibodies which react with only a certain bacterial species, or even with only a certain *strain* of a species. In addition, viral antigens can be detected with the appropriate monoclonals. For example, both fluorescent and enzyme-conjugated monoclonal antibodies (subsequently assayed by ELISA, see next section) have been developed for the diagnosis of Herpes infections and the typing of Herpes virus obtained from clinical specimens. Hence, monoclonal antibodies are useful for broad screening purposes as well as highly detailed analyses. However, the generation of monoclonal antibodies is not a trivial exercise, taking anywhere from 4 to 6 months. The majority of monoclonal antibodies used in medical diagnostics are therefore commercially prepared, and in a sense can be considered biologically-produced assay reagents. The great promise for both medical diagnostics and the treatment of diseases using monoclonal antibodies (see Section 12.7) has encouraged the formation of biotechnology companies specializing in production of monoclonal antibodies.

13.8 Diagnostic Procedures Using ELISA Tests

The enzyme-linked immunosorbent assay (ELISA), described in Section 12.9, has been frequently applied to problems in clinical diagnostics. Both direct and indirect ELISAs have been developed. Recall that *direct* ELISAs employ enzyme-conjugated antibodies, frequently monoclonal antibodies, to detect *antigen* in a patient specimen, while *indirect* ELISAs employ enzyme-conjugated antibodies to recognize *antibodies* from a patient's serum which bind a specific antigen. In either case, the high specificity, extreme sensitivity, and low cost of ELISA tests make them excellent tools for clinical diagnostics. An example of a direct ELISA procedure for detection of virus antigens is given in Figure 13.8.

An indirect ELISA assay for the immunodiagnosis of AIDS Since the discovery that the causative agent of AIDS, a virus named HIV (human immunodeficiency virus) (see Section 15.7), could be transmitted by bodily fluids, including blood, a screening test was needed to test blood samples to ensure that HIV was not being inadvertently transmitted during blood transfusions or through the transfer of blood products. An ELISA test is used for the routine screening of blood for signs of exposure to HIV (and hence possible AIDS).

The AIDS ELISA test measures *antibodies* to the HIV virus, rather than detecting the HIV itself or HIV proteins, as does the immunoblot procedure (see Section 13.10). Even though AIDS is a disease that severely cripples the body's immune system, initial infection with HIV leads to the production of antibodies to several HIV antigens, in particular those of the HIV envelope. These antibodies can be detected by the AIDS ELISA test (Figure 13.9).

To carry out an AIDS ELISA test, microtiter plates are first coated with a disrupted preparation of HIV particles; about 200 ng of disrupted HIV is all that is required. Following a brief incubation period to ensure binding of the antigens to the surface of the microtiter wells, a diluted serum sample is added and the mixture incubated to allow any potential HIV-specific antibodies to bind to HIV antigens. To detect the presence of antigen-antibody complexes, a second antibody is then added. This second antibody is

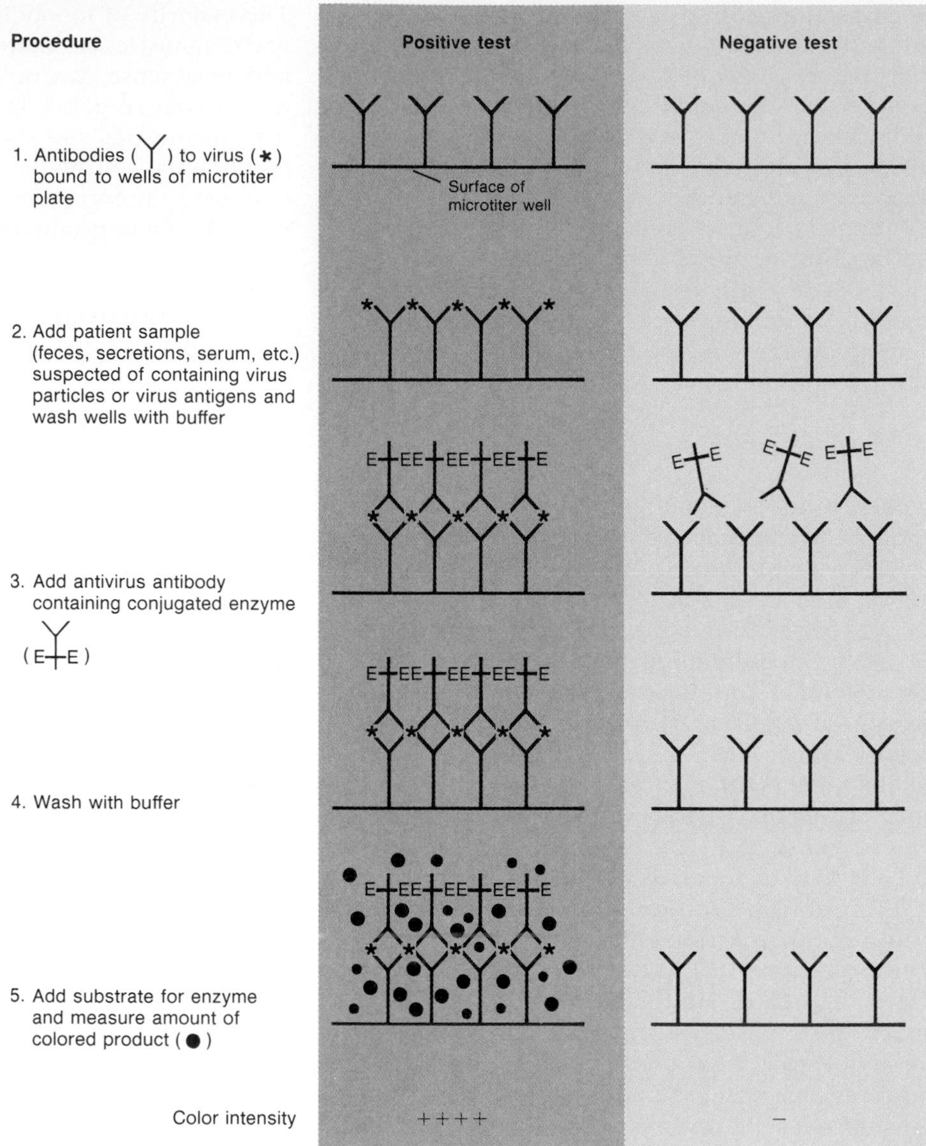
Procedure
Positive test
Negative test
1. Antibodies ( ) to virus (*) bound to wells of microtiter plate
Surface of microtiter well
2. Add patient sample (feces, secretions, serum, etc.) suspected of containing virus particles or virus antigens and wash wells with buffer
3. Add antivirus antibody containing conjugated enzyme
(E E)
4. Wash with buffer
5. Add substrate for enzyme and measure amount of colored product (●)
Color intensity
++++
−

Figure 13.8 Detection of viruses by a direct ELISA test.

an enzyme conjugated antihuman IgG preparation. Following a brief incubation period with the second antibody and a washing step to remove any unbound second antibody, the enzyme activity is assayed (the antihuman IgG antibodies will bind to any HIV-specific IgG antibodies previously bound to the HIV antigen preparation). A color is obtained in the enzyme assay in proportion to the amount of antihuman IgG antibody bound. The binding of the second antibody is an indication that antibodies from the patient's serum recognized the HIV antigens, and thus that the patient has antibodies to HIV (and hence, possibly exposed to HIV). Control sera (known to be HIV-negative) are run in parallel with any samples to measure the extent of background absorbance in the assay.

The AIDS ELISA represents a rapid, highly sensitive and specific method for detecting possible cases of AIDS and a number of companies have developed their own versions of the test. Since ELISAs in general are highly adaptable to mass screening and automation and require very little in the way of equipment, the AIDS ELISA can be used as a standard screening method for blood. The only drawbacks to using the AIDS ELISA for AIDS screening is that the test occasionally gives false positives and that advanced cases of AIDS may actually be missed by the technique. The latter occurs because in advanced AIDS cases the patient's serum contains few if any antibodies to the AIDS virus (or to any antigen, for that matter). However, a different technique, called Western blotting (also called immunoblotting, see Section 13.10), has been developed to circumvent both of these problems, and is used clinically if there is any doubt as to the accuracy of the AIDS ELISA.

Other ELISA tests of clinical importance Besides the AIDS ELISA described above for detecting

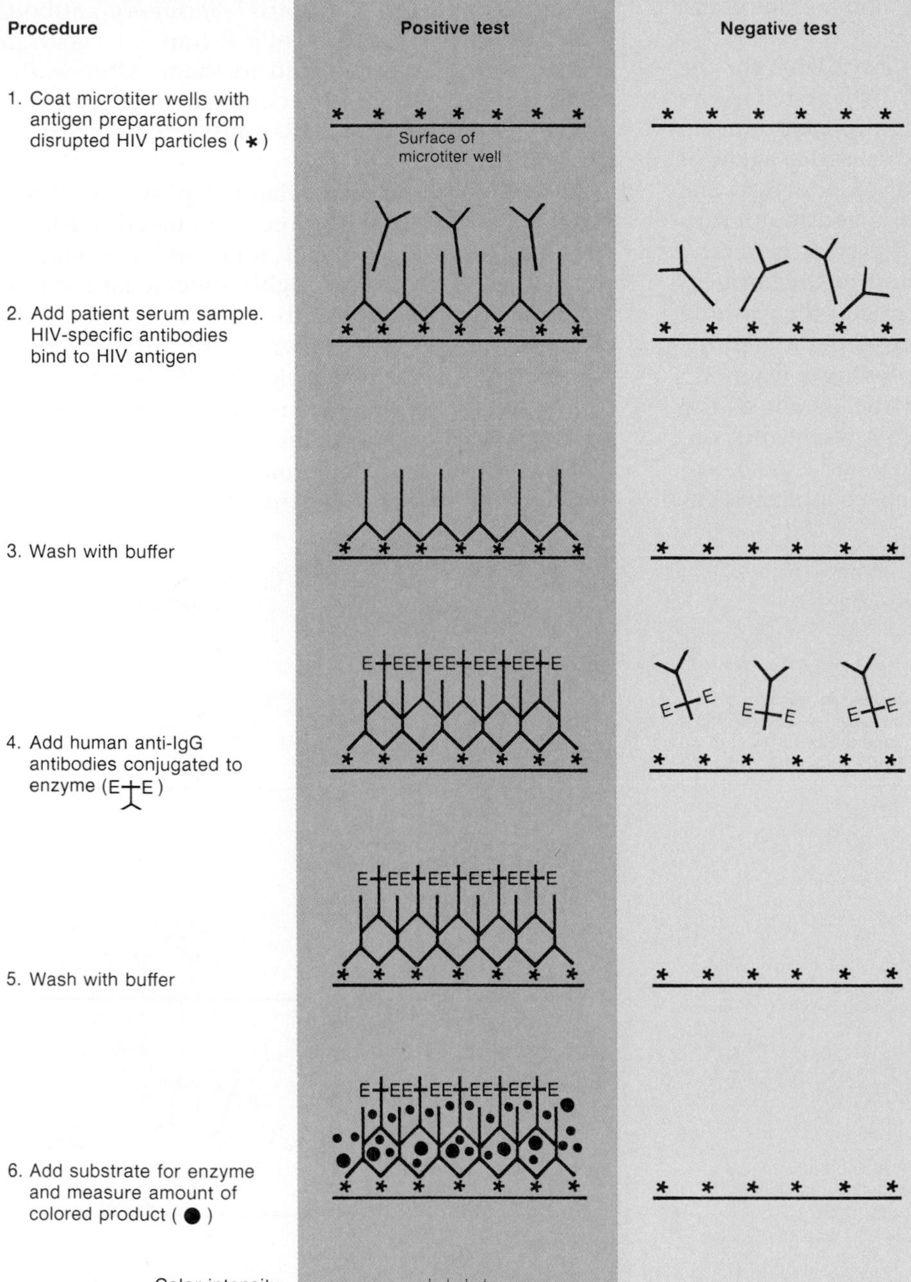
Procedure
Positive test
Negative test
1. Coat microtiter wells with antigen preparation from disrupted HIV particles (*)
Surface of microtiter well
2. Add patient serum sample. HIV-specific antibodies bind to HIV antigen
3. Wash with buffer
4. Add human anti-IgG antibodies conjugated to enzyme (E┼E)
5. Wash with buffer
6. Add substrate for enzyme and measure amount of colored product (●)
Color intensity
++++
−

Figure 13.9 Indirect ELISA test for detecting antibodies to human immunodeficiency virus (HIV), the causal agent of acquired immunodeficiency syndrome (AIDS).

antibodies to a particular antigen in a patient's serum, a variety of direct ELISAs have been developed for detecting *antigens* in a patient's serum. In this type of ELISA test, antibody to a particular antigen is first bound to the well of a microtiter plate. For example, in the detection of human viruses (for example rotavirus, a common cause of gastroenteritis in infants and young children), antibodies prepared against virus surface antigens are first bound to the microtiter plate (using the procedure illustrated in Figure 13.8). Following a wash, a diluted sample from the patient is added, the preparation washed again, and an antivirus antiserum containing a conjugated enzyme added. A positive sample (one containing virus) will show high enzyme activity because the virus particles from the patient are trapped in the well by the antivirus antibodies and then serve as sites for attachment of enzyme-conjugated antivirus antibodies. A negative sample will show no enzyme activity because no virions are present. Viral detection by ELISA techniques is especially useful clinically because many clinical laboratories are unable to bear the expense of maintaining the tissue cultures and cell lines required to grow viruses (see Section 13.4). Viruses currently detected using ELISA techniques include rotavirus, HIV, hepatitis viruses, rubella virus, bunya virus, measles and mumps viruses, and parainfluenza virus.

An extremely rapid ELISA test is available for the diagnosis of syphilis (Figure 13.10). The test employs metal beads coated with an antigen preparation from cells of *Treponema pallidum*, the causative agent of syphilis (see Section 15.6). Serum is added to the antigen-coated beads which are arranged in microtiter wells. Serum from a syphilitic patient generally contains antibodies which will bind to the antigen-coated beads. Since in all ELISA protocols the washing of wells is the most time-consuming step, the syphilis ELISA assay is accelerated by employing a magnetic transfer device that lifts the metal beads out of the wells, moves them rapidly through a wash solution, and incubates them briefly with goat anti-human immunoglobulin antibodies containing a conjugated enzyme (Figure 13.10). If anti-*Treponema* antibodies were present in the patient's serum, the goat anti-human antibodies will bind to them. After washing again, the activity of the conjugated enzyme is assayed; a positive ELISA test is considered confirmatory for syphilis.

The commercially available syphilis ELISA kit is provided complete with precoated metal beads, positive and negative control sera, and the magnetic transferring device. This highly specific and sensitive means of diagnosing syphilis takes just 4 hours from the time the blood is drawn to the time the results are known. Similar magnetic bead ELISAs have been developed for detecting antibodies to HIV, the causative agent of acquired immune deficiency disease (AIDS, see Section 15.7) and as an alternate means of detecting human rotavirus.

Procedure

1. Add serum sample to microtiter well containing magnetic beads coated with *Treponema pallidum* antigens (★)
2. Wash beads using magnetic transfer/washing device
3. Add anti-human IgG containing conjugated enzyme (Y E+E)
4. Repeat washing step (2, above)
5. Add substrate of enzyme and measure colored product (●)

Color intensity

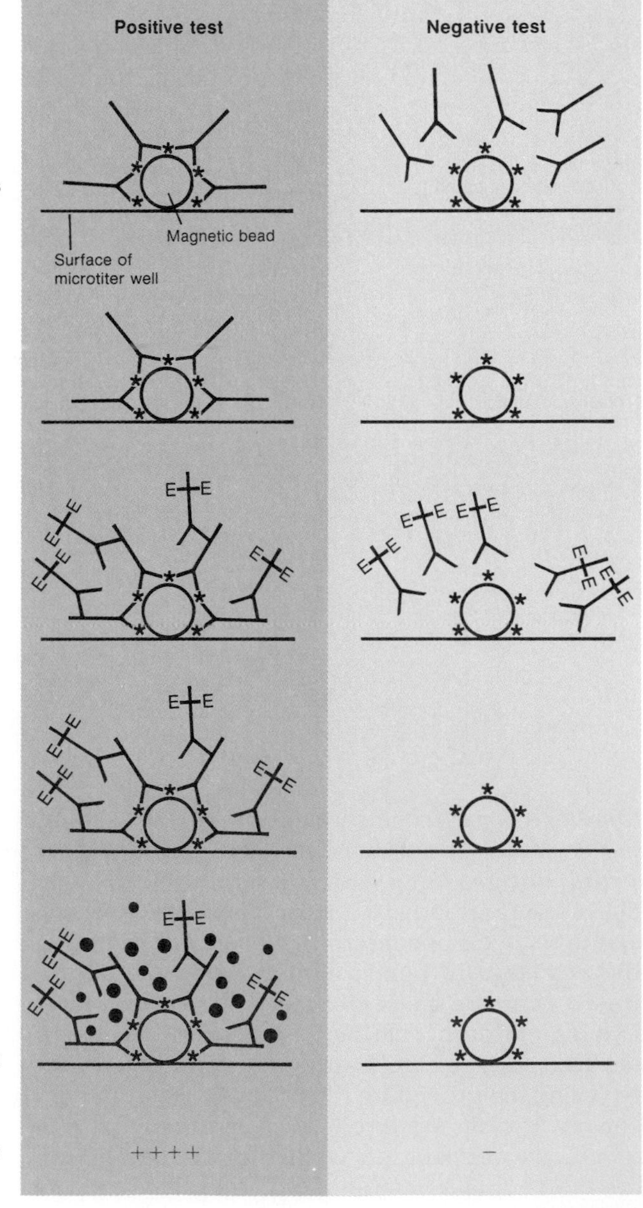

Figure 13.10 A rapid indirect ELISA test for syphilis using antigen-coated metal beads.

ELISAs have also been developed for detecting antibodies to *Candida* (yeast), and antibodies to a variety of parasites, including those causing amebiasis, chagas disease, schistosomiasis, toxoplasmosis, and malaria. The rapidity and low costs of running ELISA tests is of great appeal since time and money are usually at a premium. But it is the extreme *sensitivity* of ELISAs that really make them important immunodiagnostic tools. New ELISA tests are marketed each year and many of them are rapidly replacing older methods.

Besides the application to AIDS antibody detection, *indirect* ELISAs have been developed for detecting antibodies to a variety of clinically important bacteria. Although not meant to be a complete list, ELISAs for detecting serum antibodies to *Salmonella* (gastrointestinal problems), *Yersinia* (plague), *Brucella* (brucellosis), a variety of rickettsias (Rocky Mountain Spotted Fever, typhus, Q-fever), *Vibrio cholerae* (cholera), *Mycobacterium tuberculosis* (tuberculosis) and *Mycobacterium leprae* (leprosy), and *Legionella pneumophila* (legionellosis) have been developed. In addition, ELISAs designed to detect bacterial toxins such as cholera toxin, enteropathogenic *E. coli* toxin, and *Staphylococcus* enterotoxin, have been developed. In the case of bacterial toxins, the toxins serve as antigens and are detected by a *direct* ELISA method.

13.9 Agglutination Tests Used in the Clinical Laboratory

Agglutination is a visible antigen: antibody reaction due to the binding of a particulate antigen by antibody. Agglutination reactions were discussed in Section 12.8, the well known ABO blood grouping reaction serving as a prime example. However, many other agglutination reactions have been adapted for clinical use for the detection of antigens or antibodies associated with certain disease. Although not as sensitive a test as most ELISAs, agglutinations remain useful in clinical diagnostics as inexpensive and highly specific and rapid immunoassays.

Latex bead agglutination The agglutination of antigen- or antibody-coated latex beads by complementary antigen or antibody from patient serum is an alternate method for an inexpensive and rapid diagnosis. Small (0.8 μm) latex beads coated with a specific antigen or antibody are mixed with patient serum on a microscope slide and incubated for a short period. If the antibody complementary to the molecule bound to the bead surface is present in the patient's serum, the milky-white latex suspension will be visibly agglutinated. Latex agglutination is also used to detect bacterial surface antigens by mixing a small amount of a bacterial colony with antibody-coated latex beads. For example, a commercially available suspension of latex beads containing antibodies to Protein A and Clumping Factor, two molecules found exclusively on the surface of *Staphylococcus aureus*, is virtually 100 percent accurate in identifying clinical isolates of *S. aureus*. Unlike traditional tests for *S. aureus*, many of which are growth-dependent assays, identification of *S. aureus* by the latex bead assay takes only 30 seconds. Other latex bead agglutination assays have been developed to identify pathogenic streptococci, *Neisseria gonorrhoeae, Haemophilus influenzae*, and the yeasts *Cryptococcus neoformans* and *Candida albicans*.

A very widely used latex agglutination assay is that for detecting specific serum antibodies for *rheumatoid factor*, a protein associated with the autoimmune disease, *rheumatoid arthritis* (see Section 12.11). Coated latex beads are mixed with whole blood and agglutination scored versus positive and negative control sera run in parallel. Latex bead assays are simple and specific. In addition, the inexpensive nature of the assays make them suitable for large scale screening purposes as well; the widespread use of the rheumatoid test is a good example of this. Because they require no expensive equipment, they should be especially useful in developing countries.

13.10 Immunoblot Procedures

Antibodies can be used in clinical diagnostics to identify proteins associated with specific pathogens. The procedure employs two important biochemical techniques discussed previously: (1) the separation of proteins on polyacrylamide gels (see Box, Chapter 3), and (2) the transfer (blotting) of proteins from gels to nitrocellulose paper (see Box, Chapter 5). Since the latter technique, originally devised to transfer DNA from gels to nitrocellulose paper was termed a "Southern" blot, protein blotting and the subsequent identification of the proteins by specific antibodies is sometimes called the "Western" blot technique.

The immunoblot is a very sensitive method for detecting specific proteins in complex mixtures. In the first step of an immunoblot, a protein mixture is subjected to electrophoresis on a polyacrylamide gel. This separates the proteins into several distinct bands, each of which represents a single protein type of specific molecular weight (Figure 13.11). The proteins are then transferred to nitrocellulose paper by an electrophoretic transfer process that forces the proteins out of the gel and onto the paper. At this point the powerful specificity of antibodies is employed to identify a protein of diagnostic importance. Antibodies raised against a protein or group of proteins from a pathogen are added to the nitrocellulose blot. Following a short incubation period to allow the anti-

1. Denature proteins by boiling in detergent

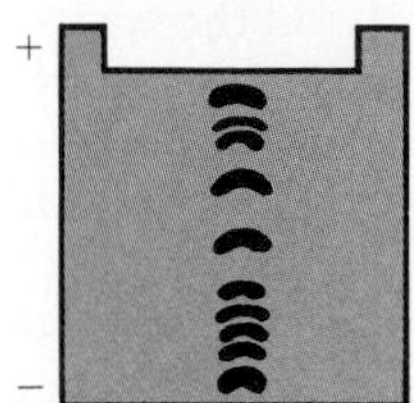

2. Subject mixture to electrophoresis; proteins separate by molecular weight

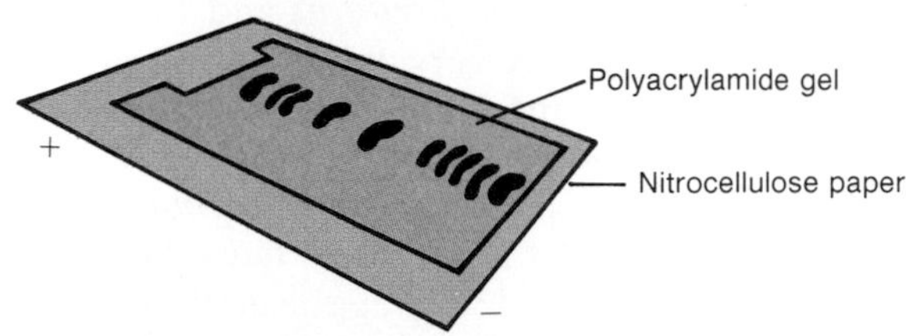

3. Blot the separated proteins from the gel to nitrocellulose paper

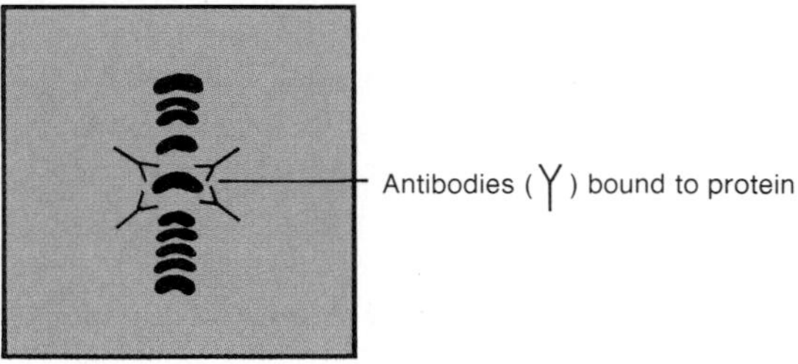

4. Treat nitrocellulose paper containing blotted proteins with antibodies; each antibody recognizes and binds to a specific protein

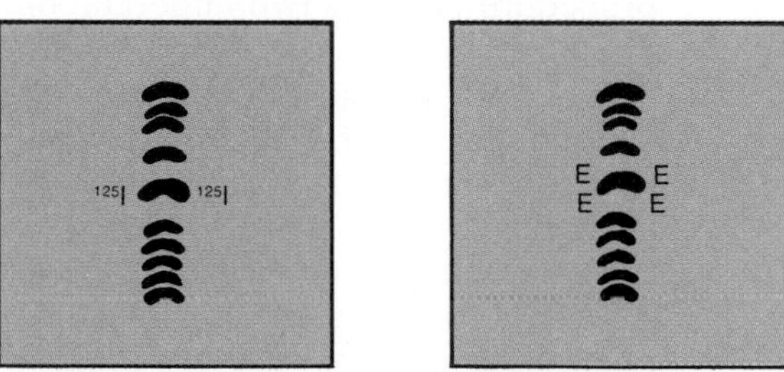

5. Add marker to bind to antigen : antibody complexes, either (left) radioactive *Staphylococcus* protein A—^{125}I, or (right) antibody containing conjugated enzyme

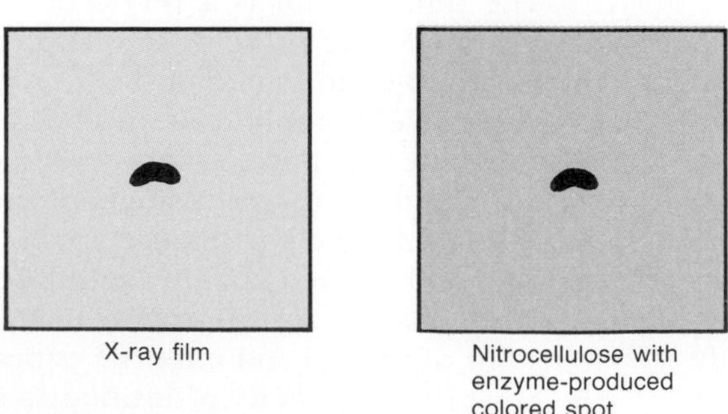

6. Develop blot by either exposing blot to X-ray film (left), or adding substrate of enzyme (right). Visualize dark spot where antibody bound to antigen on blot.

(a)

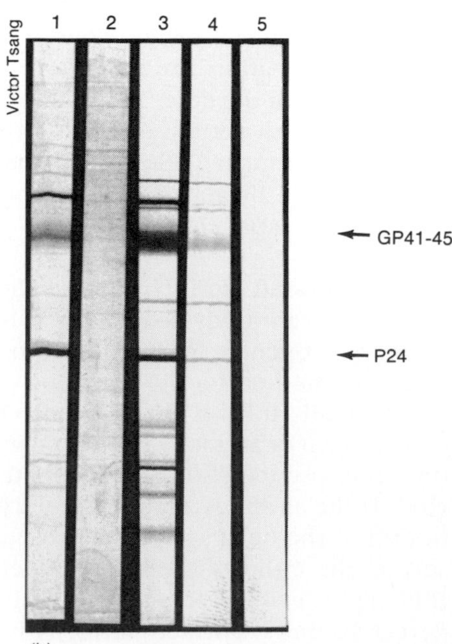

(b)

Figure 13.11 The Western Blot (Immunoblot) and its use in the diagnosis of AIDS. (a) Protocol for an immunoblot. (b) Developed AIDS immunoblot. The proteins P24 and GP41-45 are coat proteins of the virus and are diagnostic for HIV. Lane 1, positive control serum (from known AIDS patients); lane 2, negative control serum (from healthy volunteer); lane 3, strong positive from patient sample; lane 4, weak positive from patient sample; lane 5, reagent blank to check for background binding.

bodies to bind, a radioactive marker which binds antigen:antibody complexes is added. The most common radioactive marker used is *Staphylococcus* protein A iodinated with radioactive iodine, ^{125}I. Protein A has a strong affinity for antigen:antibody complexes and binds firmly to them. Once the radioactive marker has bound, its vertical position on the blot can be detected by exposing the nitrocellulose blot to X-ray film; the gamma rays emitted by the ^{125}I expose the film only in the region where the radioactive antibody has bound to antigen:antibody complexes (Figure 13.11).

For clinical applications, where the use of radioactivity is not advisable, the immunoblot technique has been modified to employ ELISA technology (see Section 13.8) for detection of bound antigen:antibody complexes. Following treatment of the blotted proteins with specific antibody, the paper is washed and then treated with a second antibody which binds to the first. For example, if antibodies raised in a rabbit were used in the first step, then the second antibody could be a goat anti-rabbit antibody. Covalently attached to this second antibody is the enzyme peroxidase. The original antigen:antibody complexes are visualized when the enzyme is assayed, because the product of the reaction leaves an insoluble brown precipitate on the nitrocellulose filter at any spot where the goat antibodies bound to the rabbit antibodies. By comparing the location of the brown spots on the nitrocellulose paper with the position of spots run from controls, a protein associated with a given pathogen can be positively identified from a clinical sample; presence of the specific protein is evidence of exposure to the disease.

Immunoblots have had a significant clinical impact on the diagnosis and confirmation of cases of AIDS. Since an immunoblot is more laborious and costly than the ELISA test, AIDS-specific ELISAs have been widely used for screening blood supplies for the presence of antibodies to the AIDS virus (see Section 13.8). However, although highly accurate, all of the commercially available AIDS ELISAs occasionally yield false positive results, that is, they occasionally indicate the presence of AIDS antibodies in a patient's serum even though the patient has never been exposed to the virus. ELISA positive blood samples are generally tested a second time by ELISA, and if the sample tests positive again, an immunoblot is used to confirm the ELISA results; the immunoblot AIDS test is generally not prone to false positives.

The AIDS immunoblot is designed, like the AIDS ELISA, to detect the presence of *antibodies* to HIV in a serum sample. However, unlike the AIDS ELISA, where false positives occasionally occur due to problems with cross-reactions, the AIDS immunoblot assays for antibodies to *specific* HIV antigens. To perform an AIDS immunoblot, a purified preparation of HIV is treated with the detergent sodium dodecyl sulfate (SDS), which solubilizes HIV proteins and also renders the virus inactive. HIV proteins are then resolved by polyacrylamide gel electrophoresis, a technique described in the Box in Chapter 3. The HIV proteins are then blotted from the gel onto sheets of nitrocellulose paper (Figure 13.11*b*). At least seven major HIV proteins are resolved by electrophoresis and two of them, designated P24 and GP41-45, are used as diagnostic proteins in the AIDS immunoblot. Protein P24 is the HIV core protein and proteins GP41-45 are thought to be HIV coat proteins (the structure of HIV is discussed in Section 6.24).

Following blotting of the proteins, the nitrocellulose strips are incubated with a serum sample previously identified as AIDS positive by the AIDS ELISA. If the sample is truly AIDS positive, antibodies against HIV proteins will be present and will bind to the HIV proteins separated on the nitrocellulose paper (Figure 13.11*b*). To detect whether antibodies from the serum sample have bound to HIV antigens, a *detecting antibody*, anti-human IgG conjugated to the enzyme peroxidase, is added to the strips. After allowing a suitable period for binding of the detecting antibody, the strip is washed to remove any unbound antibodies and the substrate for peroxidase is added (Figure 13.11*a*).

If detecting antibody has bound, the activity of the conjugated enzyme will form a brown precipitate on the strip at the site of antibody binding (Figure 13.11*b*). A confirmation of AIDS is made if the position of the bands obtained in strips using a serum sample correspond to the position of bands formed from a serum sample of a known AIDS patient; negative control sera are also run but show no reaction (Figure 13.11*b*).

A positive AIDS immunoblot is considered clinical confirmation of exposure to HIV because antibodies to *specific* HIV antigens are measured directly. Although the intensity of the bands obtained in the AIDS immunoblot varies somewhat from sample to sample (see Figure 13.11*b*), the interpretation of an AIDS immunoblot is generally unequivocal, and thus the test is valuable in distinguishing false positives from real positives initially identified by ELISA. To make the AIDS immunoblot clinically accessible, nitrocellulose strips containing inactivated HIV antigens (previously separated by electrophoresis) are available commercially. These strips can be incubated directly with a serum sample and subsequently treated with the detecting antibody. This alleviates the need for clinical microbiology laboratories to grow their own HIV as a source of antigen for the AIDS immunoblot.

Although the AIDS immunoblot represents the first large scale clinical use of immunoblot technology, because of the powerful specificity afforded antibodies in combination with protein separation techniques, it is likely that the immunoblot will find a number of additional applications in the clinical microbiology laboratory.

13.11 Diagnostic Techniques Using Nucleic Acids

With recent developments in nucleic acid technology it is now possible to use DNA in the clinical laboratory as a diagnostic tool for identifying suspected pathogenic bacteria. Thus far three approaches have been used in the clinical setting: plasmid "fingerprinting," DNA:DNA hybridization, and DNA:RNA hybridization.

Plasmid fingerprinting Plasmid fingerprinting involves the characterization of plasmids obtained from cells of a suspected pathogen. As discussed in Chapter 7, certain plasmids have been identified as coding for antibiotic resistances (R factors, see Section 7.6), toxin production (e.g., the enterotoxins of *E. coli*, see Section 11.11), and pathogenic determinants, such as tissue-specific binding factors and other determinants required for invasion and colonization by the pathogen (see Sections 11.8 and 11.9). But plasmids are also of use in the *identification* of bacterial strains and hence have value in clinical diagnosis.

Most bacteria contain one or more plasmids and enteric bacteria typically contain several. Each type of plasmid is a covalently closed circle of DNA and has a characteristic molecular weight. Plasmids of different sizes can be separated by electrophoresis on agarose gels (see Box, Chapter 5). To do a plasmid fingerprint, cells are lysed and the plasmid DNA separated from chromosomal DNA by a specific extraction procedure. The DNA is subject to electrophoresis on the agarose gel and then the gel is stained with ethidium bromide to visualize the DNA (Figure 13.12). The rate of migration of the plasmid DNA in the gel is *inversely* proportional to its molecular weight; that is, small plasmids migrate farther than large plasmids. The positions and number of plasmid bands in the gel (the plasmid "fingerprint") are characteristic of a given species, and the greater the number of plasmids present, the easier species differentiation becomes. The plasmid fingerprint of a given organism is diagnostically useful; major differences in plasmid size and/or number between the isolated pathogen and reference strains are generally indicative of little taxonomic affiliation. Identical plasmid fingerprints of reference and sample organisms are highly suggestive of relatedness, but cannot be interpreted as definitive. Problems can be encountered if the particular strain of pathogen tested has picked up plasmids from other species or has lost plasmids characteristic of the reference strain. Plasmid fingerprinting is therefore used as an adjunct method, complementing other methods of identification, especially if the latter produce ambiguous results. Of importance to the clinical laboratory, however, is the fact that the sensitivity of plasmid fingerprinting is very high. For a typical analysis all that is needed is around one ml of a liquid culture or one or two colonies from a plate.

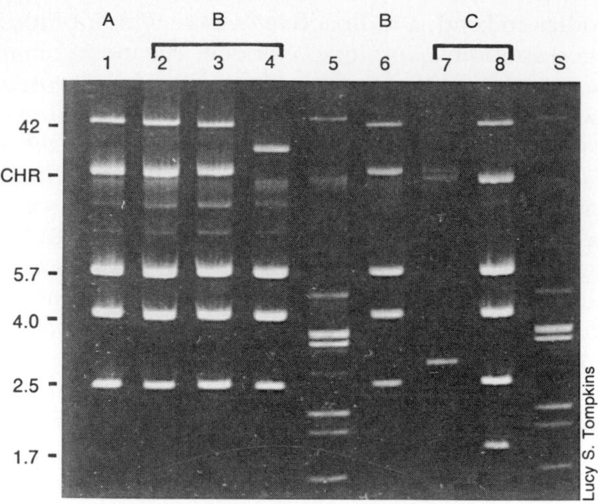

Figure 13.12 Plasmid fingerprints of *Staphylococcus epidermidis* isolates recovered from three patients suffering from prostetic valve endocarditis. Patient A had the same strain of *S. epidermidis* as Patient B, but Patient B had, in addition, two unique strains (lanes 4 and 6). Patient C showed two unique strains. The numbers on the left refer to the major plasmid sizes (in megadaltons [10^6 daltons]). The CHR band is fragmented chromosomal DNA, S, molecular weight standards.

Nucleic acid probes One of the most powerful approaches to the detection and identification of microorganisms or viruses is nucleic acid hybridization. We discussed the mechanics of nucleic acid hybridization in Section 5.2. The basic principle behind nucleic acid hybridization is that the closer the similarity in nucleotide *sequence* between two nucleic acid molecules, the stronger they will bind together. Central to the use of the nucleic acid hybridization technique in clinical microbiology is the *nucleic acid probe*. A **probe** is a strand of nucleic acid which has a sequence characteristic of a given organism. The probe nucleic acid, usually DNA, is generally obtained by a cloning procedure (see Chapter 8). The probe must consist of a sequence unique to the particular organism being sought. For example, DNA probes have been developed which can detect the enterotoxin gene sequence of *Escherichia coli*, thus permitting the differentiation of enterotoxigenic strains from *E. coli* strains that are part of the normal flora. Other DNA probes have been developed to detect specific viral sequences and sequences from parasitic protozoans.

The probe technique has two major advantages: (1) It is highly *specific*, permitting distinction of a particular strain or group of strains from all other organisms. (2) It is very *sensitive*, permitting detection of very tiny amounts of nucleic acid even when this nucleic acid is surrounded by large amounts of extraneous nucleic acid. This great sensitivity makes it possible in many cases to detect the presence of an organism or virus even *without culture*. This not only decreases significantly the time necessary for a diagnosis, but makes it possible to detect organisms that, for technical reasons, cannot be readily cultured (such as protozoan parasites and a number of viruses).

How does the nucleic acid probe technique work? The probe must be labeled in some way so that its hybridization to the unknown DNA can be measured. For research purposes, labeling the probe with radioactive phosphorous ($^{32}PO_4^{2-}$) is suitable, but for clinical applications, the use of radioactivity is undesirable. Radioactive materials require special handling and complex safety procedures and are subject to stringent government regulation. One approach which eliminates the need for radioactivity involves the use of **biotin-labeled probes**. Biotin is a vitamin which acts as a carrier of carbon dioxide in fatty acid biosynthesis, but the present use of biotin has nothing to do with its vitamin function. Rather, biotin is used because it has a free carboxyl group which can be used to chemically couple the vitamin to free amino groups of cytosine bases of the nucleic acid and because there is a protein, *avidin* (present in egg whites), which combines specifically and to high efficiency with biotin. Since enzymes can be easily attached to avidin, the amount of DNA:DNA hybridization occurring can be measured using a variation of ELISA technology; enzyme linked avidin substitutes for enzyme linked antibodies.

For diagnostic use, the biotin-labeled DNA probe is mixed with single-stranded DNA prepared from a

1. Mix biotin-labeled (B) probe DNA with excess DNA from unknown pathogenic organism. Allow hybridization to occur

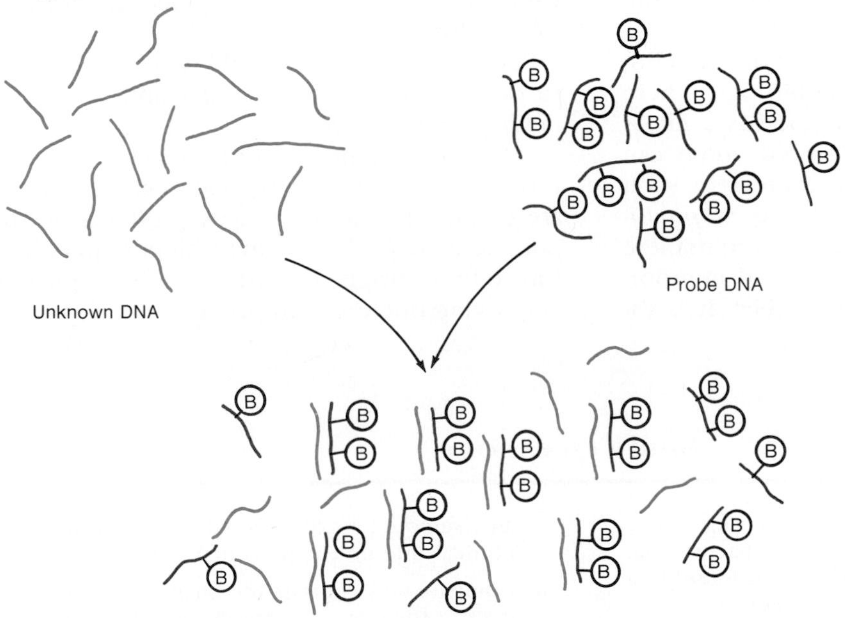

2. Filter to collect double-stranded hybrid DNA. Add enzyme-conjugated avidin)–A–

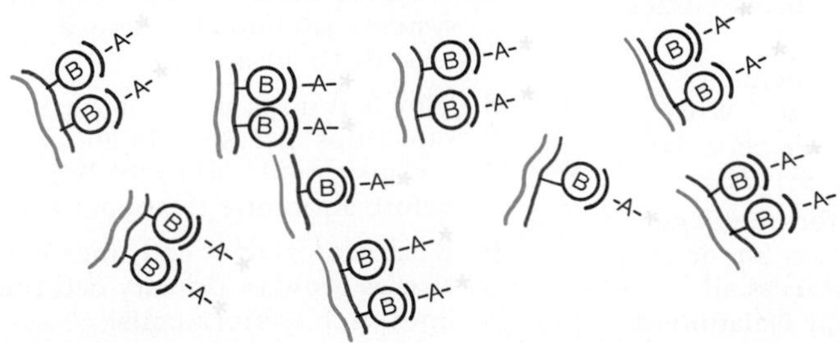

3. Add substrate of conjugated enzyme and measure colored product. Color intensity is proportional to amount of avidin bound, which is proportional to the percent hybridization of unknown DNA with the biotinylated probe.

Figure 13.13 Use of biotinylated probes to measure DNA:DNA hybridization analyses

clinical sample suspected to contain the target sequence. After hybridization, the double-stranded hybrids are separated from single-stranded molecules by filtering. The filter containing the double strands is then reacted with the enzyme-linked avidin (the enzyme used is usually alkaline phosphatase or peroxidase, enzymes whose activity can be readily measured by color release from a suitable substrate). The degree of sequence homology is proportional to the enzyme activity (color intensity) observed (Figure 13.13). Homologous controls are run to define the enzyme activity equivalent to 100 percent binding, and in a matter of a few minutes the percent sequence homology in the genomic DNAs of several organisms can be determined. From the data obtained it is often possible to identify the suspected pathogen.

With current technology, it is possible to detect about 10^{-18} moles of nucleic acid in a single sample.This means that it is possible to detect around 1000 cells or virus particles in a single sample. Although not as sensitive as direct culture (which can under favorable conditions detect less than 10 cells per sample), the sensitivity is quite sufficient for many clinical procedures, especially when used in conjunction with subsequent culture.

Nucleic acid probes can be used not only in clinical microbiology, but in food microbiology, where control of the microbial quality of food is a major activity. Because culture is not required, foods can be quickly analyzed for microbial content of such important pathogens as *Salmonella* and *Staphylococcus*. The limiting factor in the use of the nucleic acid probe technique in clinical and food microbiology is the availability of suitable probes. It is essential that the probe be *specific* for the organism of interest. We discuss some uses of the probe technique in the study of microbial evolution in Chapter 18.

Hybridization with RNA also promises to become a useful clinical tool. As discussed in detail in Chapter 18, analysis of ribosomal RNA (rRNA) sequences have been used to discern evolutionary relationships of bacteria. Although a highly conserved molecule, rRNA sequences highly diagnostic of particular organisms or groups of related organisms have been identified. In the case of *Mycobacterium tuberculosis*, the causative agent of tuberculosis (see Section 15.3), a complementary DNA (cDNA) has been prepared that hybridizes strongly to rRNA from *M. tuberculosis* and related pathogenic mycobacteria. The probe can be used to detect *M. tuberculosis* by extracting RNA from a sputum or saliva sample and testing it for hybridization to the cDNA probe. The cDNA probe is labeled with ^{125}I, a radioisotope, in this case, because biotin hybridization methods are interfered with by substances in sputum or saliva. The DNA:RNA hybridization test for tuberculosis totally eliminates the need for culturing the organism, a long and tedious process, and can be used in conjunction with the ELISA tuberculosis test (see Section 13.8) to effect a highly reliable diagnosis. Other rRNA probes are available to detect certain mycoplasmas and streptococci of medical importance.

The development of nucleic acid probes is a rapidly expanding enterprise and commercial concerns are constantly marketing new products; in this case nucleic acid probes of ever higher specificity. The future looks bright indeed for rapid diagnostic tools employing nucleic acid probes.

Study Questions

1. Describe the standard procedure for obtaining and culturing a blood sample for evidence of bacteremia. What would you conclude if the organism found in a blood culture was a *Staphylococcus* species?
2. Why is the *number* of bacterial cells in urine rather than simply the *presence* of cells in urine of significance to the microbiological analysis of urine? What organism is responsible for most urinary tract infections?
3. Describe the microscopic and cultural evidence required to make a diagnosis of gonorrhea? What is the advantage of Thayer-Martin agar over chocolate agar for the isolation of *Neisseria gonorroheae*?
4. Describe the major procedures used for culturing anaerobic microorganisms in liquid medium or on petri plates. Why is it important to process all clinical samples quickly, especially for isolation of anaerobes?
5. What is a selective medium? How does a selective medium differ from a differential medium? Is EMB agar selective or differential? How is it used in the clinical microbiology laboratory?
6. Compare and contrast the changes undergone by pH dyes in the tests for carbohydrate fermentation and in the test for citrate utilization. Could the same dye be used in both tests? Why or why not?
7. What are the advantages of rapid identification systems as clinical diagnostic procedures? Discuss at least three advantages.
8. Why is it not a common medical practice to treat an infectious disease with antibiotics before isolating the suspected pathogen? What steps are usually taken before antibiotic treatment begins?
9. What is a primary cell line? Why do some animal viruses grow in primary cell lines but not in cell lines such as HeLa cells?
10. How are most laboratory-associated infections contracted? What actions can be taken to prevent them?

11. Design a fluorescent antibody assay for confirming an initial diagnosis of "strep throat" (*Streptococcus pyogenes* is the causative agent of "strep throat"). Discuss all aspects of the assay including preparation of antisera, necessary controls, and clinical interpretation.
12. What are the advantages of using fluorescent monoclonal antibodies instead of polyclonal antibodies in clinical diagnostics?
13. Design an ELISA test for detecting Hepatitis A virus in feces. Would such a test require anti-human IgG antibodies? Why or why not?
14. List the advantages of an ELISA test for diagnosing syphilis over older methods such as the cardiolipin VDRL test (the VDRL test is a precipitin test where cardiolipin antigens from *Treponema pallidum* react with antibodies from a serum sample).
15. What is the function of latex beads in the agglutination tests developed for several pathogens? What advantages do latex agglutination assays have over ELISA? What disadvantages?
16. What advantages does the AIDS Western Blot have over the AIDS ELISA for the diagnosis of the disease AIDS? Why is the AIDS Western Blot only used for confirmation of AIDS and not for screening?
17. What is a plasmid fingerprint and how is it obtained? What does a plasmid fingerprint say about sequence homology between plasmids?
18. What are the major advantages of using DNA probes in diagnostic microbiology? Discuss at least four aspects of probe technology that benefit clinical medicine.
19. How is the vitamin biotin involved in DNA probe technology?
20. Although antibodies are not used with DNA probes, how have some of the principles of ELISA been applied to DNA:DNA hybridization using biotinylated probes?

Supplementary Readings

Caskey, C. T. 1987. Disease diagnosis by recombinant DNA methods. *Science* 236:1223–1229. A concise review of recombinant methods for diagnosis of both infectious and noninfectious diseases.

Ewing, W. H. 1986. *Edward's and Ewing's Identification of Enterobacteriaceae, 4th ed.* Elsevier Science Publishing Co., Inc., New York. The definitive description of identification methods for enteric bacteria.

Finegold, S. M., and **E. J. Baron.** 1986. *Bailey and Scott's Diagnostic Microbiology, 7th ed.* The C. V. Mosby Company, St. Louis, MO. A good source of clinical microbiological methods and diagnostic and identification procedures.

Jawetz, E., J. L. Melnick, and **E. A. Adelberg.** 1986. Review of Medical Microbiology, 16th ed. Lange Medical, Los Altos, CA. A brief but very up-to-date treatment of infectious diseases and clinical procedures.

Lennette, E. H., A. Balows, W. J. Hausler, Jr., and **H. J. Shadomy**, 1985. *Manual of Clinical Microbiology, 4th ed.* American Society for Microbiology, Washington, D. C. The standard reference work of clinical microbiology.

Macario, A. J., and **E. C. de Macario** (eds.) 1985. *Monoclonal Antibodies Against Bacteria.* Academic Press, New York. A comprehensive overview of the production of monoclonal antibodies and their use in antibacterial therapy as well as research.

Epidemiology and Public Health Microbiology

In Chapters 11 and 12 we considered the general principles of how microorganisms cause infectious disease and how the host responds to microbial onslaught. In Chapter 13, we discussed the methods and procedures used in the diagnosis of microbial diseases. In the present chapter, we consider how a pathogen spreads from an infected individual to others and the whole problem of infectious disease in populations. An understanding of the principles put forward in this chapter is vital in attempts to control the spread of infectious disease. Thus, we are dealing here with *public health*. One measure of the success of microbiology in the control of infectious disease is shown by the data presented in Figure 1.42, which compared the present causes of death with those at the beginning of the century. As seen, microbial diseases are no longer the major threats to public health that they once were.

14.1 The Science of Epidemiology

The most visible aspect of microbial disease is the actual diseased individual. However, animals and humans do not live isolated, but in populations, and when we consider infectious diseases in populations, some new factors arise. The study of infectious disease in populations is part of the field of **epidemiology**.

To continue existing in nature the pathogen must be able to grow and reproduce. For this reason, an important aspect of the epidemiology of any disease is a consideration of how the pathogen maintains itself in nature. In most cases the pathogen cannot grow outside the host, and if the host dies, the pathogen will also die. Pathogens that kill the host before they are transmitted to a new host would thus become extinct. This raises the question of why pathogens occasionally kill their hosts. Actually, a well-adapted parasite lives in harmony with its host, taking only what it needs for existence, and causing only a minimum of harm. Serious host damage most often occurs when new races of pathogens arise for which the host has not developed resistance, or when the resistance of the host changes because of the factors discussed in Chapter 11. Pathogens are selective forces in the evolution of the host, just as hosts are selective forces in the evolution of pathogens. When equilibrium between host and pathogen exists, both are able to live more or less harmoniously together.

The epidemiologist is concerned with methods for tracing the spread of pathogens. By use of these methods, the epidemiologist can assess the microbial safety of various agents and trace the path of a disease to identify its origin and mode of transmission. The epidemiologist relies heavily on statistical analyses of data obtained from clinical studies, disease reporting, insurance questionaires, and interviews with patients. The science of epidemiology has frequently been referred to as "medical ecology," because the study of a disease in populations is really a study of a disease in its natural environment. This is in contrast to the clinical or laboratory study of disease, where the focus is on treating the individual patient. However, knowledge of both the clinical aspects and ecological aspects of a given disease are important if public health measures to control diseases are to be effective.

14.2 Terminology

A number of terms having specific meanings are used by the epidemiologist to describe patterns of disease. The **prevalence** of a disease in a population is defined as the proportion (or percentage) of diseased individuals in a population at any one time. The **incidence** of a disease is the *number* of diseased individuals in a population at risk. A disease is said to be **epidemic** when it occurs in an unusually high number of individuals in a community at the same time; a **pandemic** is a worldwide epidemic. By contrast, an **endemic** disease is one which is constantly present, usually at low incidence, in a population. In an endemic disease, the pathogen may not be highly virulent, or the majority of the individuals may be immune, so that the incidence of disease is low. However, as long as an endemic situation lasts, there will remain a few individuals who may serve as reservoirs of infection.

Sporadic cases of a disease occur when individual cases are recorded in geographically remote areas, implying that the incidences are not related. A disease **outbreak**, on the other hand, occurs when several cases are observed, usually in a relatively short period of time, in an area previously experiencing only sporadic cases of the disease. Finally, *subclinical infection* is used to describe diseased individuals who show no or only mild symptoms. Subclinically infected individuals are frequently identified as **carriers** of a particular disease, because even though they themselves show few or perhaps no symptoms, they may still be actively carrying and shedding the pathogenic agent (see Section 14.3).

Mortality and morbidity In practice, the incidence and prevalence of disease is determined by obtaining statistics of illness and death. From these data a picture of the public health in a population can be obtained. The population under consideration could range in size from the total global population of humans down to the population of a localized region of a country or district. Public health varies from region to region, as well as with time; thus a picture of public health at a given moment provides only an instantaneous picture of the situation. By continuing to examine health statistics over many years, it is possible to assess the value of various public health protective measures influencing the incidence of disease.

Mortality expresses the incidence of *death* in the population. Infectious diseases were the major causes of death in 1900, whereas currently they are of much less significance, and diseases such as heart disease and cancer are of greater importance. However, the current situation could rapidly change if a breakdown in public health measures were to occur.

Morbidity refers to the incidence of *disease* in populations and includes both fatal and nonfatal diseases. Clearly, morbidity statistics will more precisely define the health of the population than mortality statistics, since many diseases that affect health in important ways have only a low mortality. The major causes of illness are quite different from the major causes of death. Major illnesses are acute respiratory diseases (the common cold, for instance) or acute digestive system conditions, which are generally due to infectious causes.

14.3 Disease Reservoirs

Reservoirs are sites in which viable infectious agents remain alive and from which infection of individuals may occur. Reservoirs may be either animate or inanimate. Table 14.1 lists some common human diseases and their reservoirs. Some pathogens are primarily saprophytic (living on dead matter) and only incidentally infect and cause disease. A good example is *Clostridium tetani* (the causal agent of tetanus) whose normal habitat is the soil. Infection by this organism is not essential for its continued existence and can be considered only an accidental event; if all susceptible hosts died, *C. tetani* would still be able to survive in nature.

More critical, from the viewpoint of epidemiology, are pathogens whose only reservoirs are living organisms. In these cases, the reservoir is an essential component of the cycle by which the infectious agent maintains itself in nature. Some infections occur only in humans, so that maintenance of the cycle involves person-to-person transmittal. This type of pathogen cycle is the most common for the following diseases: most viral and bacterial respiratory diseases, sexually transmitted diseases, staphylococcal and streptococcal infections, diphtheria, typhoid fever, and mumps.

A large number of infectious diseases that occur in humans also occur in animals. Diseases which occur primarily in animals but are occasionally transmitted to people are called **zoonoses**. Because public health measures for animal populations are much less highly developed, the infection rate for these diseases will be much higher in animals, and animal-

Table 14.1 Major reservoirs of infection for a number of human diseases

Disease	Major reservoir of infection
Anthrax, brucellosis	Cattle, swine, goats, horses, sheep
Salmonellosis	Domestic and wild animals, especially poultry, whole eggs and egg products; water polluted with sewage
Botulism	Soil, contaminated food
Giardiasis	Beaver, muskrat, marmot
Malaria	*Anopheles* mosquito
Plague	Wild rodents
Psittacosis	Parakeets, parrots, pigeons, other birds
Rabies	Wild and domestic carnivores, especially skunks and raccoons
Respiratory diseases (viral, bacterial)	Infected humans, human carriers
Rocky Mt. Spotted Fever	Ticks, rabbits, mice
Syphilis, gonorrhea, AIDS, other sexually transmitted diseases	Infected humans
Tetanus	Soil, intestine
Tuberculosis	Humans, dairy cattle
Tularemia	Wild animals, especially rabbits
Typhoid fever	Human carriers, infected humans

to-animal transmission will be the rule. However, occasionally transmission will be from animal to human. It would be very unlikely, in such diseases, for transmission to also occur from person-to-person. Thus the maintenance of the pathogen in nature depends on animal-to-animal transfer. It should be obvious that control of a zoonosis in the human population in no way eliminates it as a public health problem. Indeed, more effective human control can generally be achieved through elimination of the disease in the animal reservoir. Marked success has been achieved in the control of two of the diseases that occur primarily in domestic animals, bovine tuberculosis and brucellosis. Pasteurization of milk was of considerable importance in the prevention of the spread of bovine tuberculosis to humans, since milk was the main vehicle of transmission.

Certain infectious diseases have more complex cycles, involving an obligate transfer from animal to human to animal. These are organisms with a complex life cycle and are either metazoans (for example, tapeworms) or protozoa (for example, malaria, see Section 15.10). In such cases, control of the disease in the population can be either through control in humans or in the alternate animal host.

Carriers A carrier is an infected individual not showing obvious signs of clinical disease. Carriers are potential sources of infection for others, and are thus of considerable significance in understanding the spread of disease. Carriers may be individuals in the incubation period of the disease, in which case the carrier state precedes the development of actual symptoms. Carriers of this sort are prime sources of infectious agents for respiratory infections, since they are not yet aware of their infection and so will not be taking any precautions against infecting others. Such persons can be considered **acute carriers**, because the carrier state will only last for a short while. More significant from the public health standpoint are chronic carriers, who may remain infected for long periods of time. **Chronic carriers** may be either individuals who had a clinical disease and recovered, or they may have had an infection that remained inapparent throughout (see Box).

Carriers can be identified by routine surveys of populations, using cultural, radiological (chest X ray), or immunological techniques. In general, carriers are only sought among groups of individuals who may be sources of infection for the public at large, such as food handlers in restaurants, groceries, or processing plants. Two diseases in which carriers have been of most significance are typhoid fever and tuberculosis, and routine surveys of food handlers for inapparent cases of these diseases are sometimes made. (Foodborne diseases will be considered in Section 15.11).

14.4 Epidemiology of AIDS: An Example of How Epidemiological Research is Done

Cases of Acquired Immunodeficiency Syndrome (AIDS), the viral-mediated infectious disease that severely cripples the body's immune system (see Section 15.7), were first reported in the United States in 1981. As seen in Figure 14.1, since that time the number of new AIDS cases diagnosed *each year* has approximately doubled. Initial clinical observations suggested an unusually high AIDS prevalence among homosexual men and intravenous drug abusers. This is turn strongly implicated a transmissible agent, presumably transferred during sexual activity or by con-

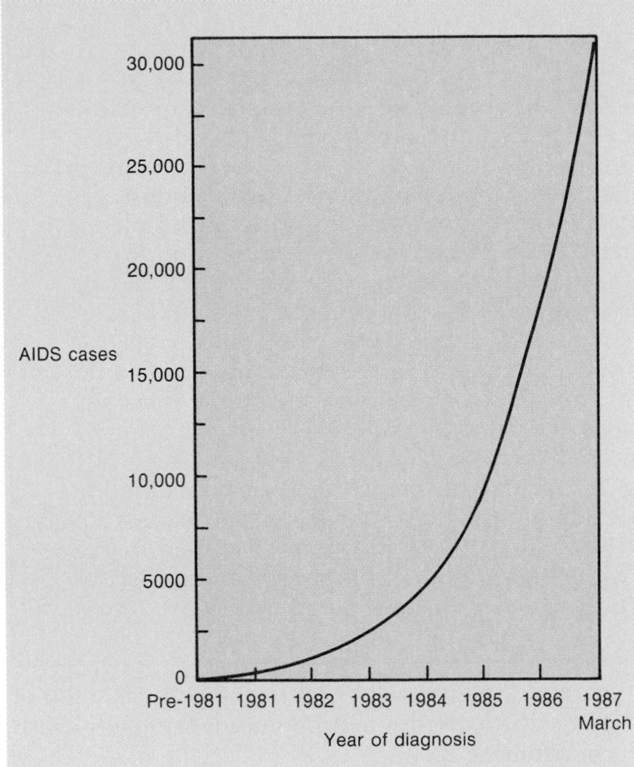

Figure 14.1 Incidence of acquired immunodeficiency syndrome (AIDS) in the United States. Each point represents the number of new cases diagnosed within that calendar year.

taminated needles. The finding that individuals requiring blood or blood products were also at high risk strengthened the case for a transmissable agent.

Within one year the accumulated epidemiological evidence pointed strongly to a virus as the causative agent of AIDS. Shortly thereafter two research teams, one French and the other American, isolated viruses from AIDS patients that were subsequently shown to be the cause of AIDS (human immunodeficiency virus, HIV, see Section 15.7). Statistics such as those shown in Table 14.2 identified several "high-risk" groups: homosexual men, intravenous drug abusers, and individuals who regularly required blood or blood products, such as hemophiliacs. The association of AIDS cases with *specific groups* was an important epidemiological finding, because it implied that the disease was not transmitted from person-to-person by alternative modes of transmission, such as casual contact, respiratory means, or contaminated food or water. Instead, incidence statistics suggested that the transmission of body fluids from one individual to another was required to transmit AIDS. Subsequent studies of AIDS patients have shown HIV to be present in blood and semen.

Thus far, AIDS has affected mainly homosexual men and the geographic distribution of the disease—highest incidence in New York City or San Francisco where there are large homosexual communities—correlates well with the observed trend. Further epidemiological surveys have shown that homosexual men with multiple sexual partners are more likely to contact AIDS than monogamous homosexual males. This undoubtedly reflects the increased probability of contacting an infected individual when engaging in sexual activity with multiple partners.

The prevalence of AIDS in hemophiliacs increases with the severity of the hemophilia, especially if the patient requires blood factor concentrates obtained from pooled blood plasma. Most type A hemophiliacs belong to this group and have been at extremely high risk for contracting AIDS (see Table 14.2). However, sensitive and highly specific immunoassays are now available for screening donated blood for antibodies to HIV (see Sections 13.8 and 13.10); the presence of AIDS antibodies correlates strongly with either the presence of HIV in body fluids or a condition referred to as AIDS-related-complex (ARC). Persons testing positive for ARC constitute generally subclinical cases that have become serologically positive for HIV but have not yet contracted clinical AIDS; that is, the immune system of the ARC patient is still functioning normally. Individuals diagnosed as having ARC can still transmit the virus and are therefore discouraged from donating blood. A small number of AIDS cases have been diagnosed in infants not fitting any of the high risk groups, but whose mothers were subsequently diagnosed to have AIDS. It is now known that HIV can also be transmitted in the uterus or from mother's milk. Incidence of AIDS in heterosexuals is correlated primarily with degree of sexual promiscuity.

Table 14.2 Incidence rates of acquired immunodeficiency disease (AIDS) in United States*

Group	AIDS rate/ 100,000 population
Single men	
United States (overall)	14.3
Residents of New York, New York	263
Residents of San Francisco, California	340
Intravenous drug users	
United States (overall)	167
New York, New York	261
Hemophiliacs	
Type A (hereditary factor VIII deficiency)	346
Type B (hereditary factor IX deficiency)	35

**Data given are for the year 1984, typical of the early phase of the AIDS epidemic.*

The Tragic Case of Typhoid Mary

The classic example of a chronic carrier was the woman known as "Typhoid Mary," a cook in New York City and Long Island in the early part of this century. Typhoid Mary (her real name was Mary Mallon) was employed in a number of households and institutions, and as a cook she was in a central position to infect large numbers of people. Eventually she was tracked down by Dr. George Soper after an extensive epidemiological investigation of a number of typhoid outbreaks revealed that she was the likely source of contamination. When her feces was examined bacteriologically, it was found that she had very high numbers of the typhoid bacterium, *Salmonella typhi.* She remained a carrier for many years, probably because her gallbladder was infected, and organisms were continuously being excreted from there into her intestine. Public health authorities offered to remove her gallbladder but she refused the operation, and to prevent her from continuing to serve as a source of infection she was imprisoned. After almost 3 years in prison, she was released on the pledge that she would not cook or handle food for others and that she was to report to the health department every three months. She promptly disappeared, changed her name, and cooked in hotels, restaurants, and sanitariums, leaving behind a wake of typhoid fever. After 5 years she was captured as a result of the investigation of an epidemic at a New York hospital. She was again arrested and imprisoned and remained in custody on North Brother Island in the East River of New York City for 23 years. She died in 1938, 32 years after epidemiologists had first discovered she was a chronic typhoid carrier.

Epidemiological studies of AIDS have led to research on vaccine development and other forms of specific antiviral therapy. The identification of high risk groups has promoted public education campaigns designed to inform members of these groups on how to prevent AIDS transmission (see Box, Chapter 15). In addition, the public has been informed that AIDS is not transmitted by routes other than the transfer of body fluids. A major job of the epidemiologist and those in public health is therefore educating the public on how to prevent disease transmission and dispelling the myths that frequently surround various diseases. In AIDS, as in other sexually transmitted diseases, public health measures offer the most cost-effective approach to control of the disease.

14.5 Modes of Infectious Disease Transmission

Epidemiologists follow the incidence of a disease by correlating geographical, seasonal, and age-group distribution of a disease with possible modes of transmission. A disease limited to a restricted geographical location, for example, may suggest a particular vector. A marked seasonality to a disease is often indicative of certain modes of transmission, such as in the case of chickenpox or measles, where the number of cases jumps sharply when children enter school and come in close contact (see Figures 14.4 and 15.16). The age-group distribution of a disease can be an important epidemiological statistic, frequently suggesting or eliminating particular routes of transmission. Sometimes the data accumulated by the epidemiologist are for noninfectious diseases such as cancer, heart, and lung disease because the general study of disease statistics will sometimes point to a likely cause or contributing factor.

Different pathogens have different modes of transmission, which are usually related to the habitats of the organisms in the body. For instance, respiratory pathogens are generally airborne, whereas intestinal pathogens are spread by food or water. If the pathogen is to survive in nature it must undergo transmission from one host to another. Thus pathogens generally have evolved features or mechanisms that permit or ensure transmittal. Transmission involves three stages: (1) escape from the host, (2) travel, and (3) entry into a new host. We give here a brief overview of transmission mechanisms and several of these will be discussed in detail for certain diseases in the next chapter.

Direct host-to-host transmission Host-to-host transmission occurs whenever an infected host transmits the disease to a susceptible host by any of a number of means. These include transmission by the respiratory route and by direct contact. Transmission by infectious droplets is the most frequent means by which upper respiratory infections such as the common cold or influenza are propagated. However, some pathogens are so sensitive to environmental influences that they are unable to survive for significant periods of time away from the host. Such pathogens are transmitted from host-to-host by direct contact. The best examples of pathogens transmitted in this way are those responsible for sexually transmitted diseases, such as *Treponema pallidum* (syphilis) and *Neisseria gonorrhoeae* (gonorrhea). These agents are extremely sensitive to drying and do not survive away from the body, even for a few moments. Intimate person-to-person contact, such as kissing or sexual intercourse, provides a direct means for the transmission of such pathogens. However, it should be obvious that such intimate transfer can only occur if the viable pathogen is present on the transmitting person at the body site that comes in direct contact with that of the recipient. Thus the pathogens causing sexually transmitted diseases live in genitalia, the mouth, or the anus, since these are the only sites involved in intimate person-to-person contact.

Direct contact is also involved in the transmittal of skin pathogens, such as staphylococci (boils and pimples) or fungi (ringworm). Because these pathogens are relatively resistant to environmental influences such as drying, intimate person-to-person contact is not the only means of transmission, as it is with the sexually transmitted diseases. It could also be considered that many respiratory pathogens are transmitted by direct means, since they are spread by droplets emitted as a result of sneezing or coughing, and many of these droplets do not remain airborne for long. Transmission, therefore, almost always requires close, although not necessarily intimate, person-to-person contact.

Indirect host-to-host transmission Indirect transmission can occur by either living or inanimate means. Living agents transmitting pathogens are called **vectors**; they are generally arthropods (for example, insects, mites, or fleas) or vertebrates (for example, dogs, rodents). Arthropod vectors may not actually be hosts for the disease, but merely carriers of the agent from one host to another. Large numbers of arthropods obtain nourishment by biting, and if the pathogen is present in the blood, the arthropod vector will receive some of the pathogen and may transmit it when biting another individual. In some cases, the pathogen actually replicates in the arthropod, which is then considered an alternate host, and such replication leads to a buildup of the inoculum, increasing the probability that a subsequent bite will lead to infection. Inanimate agents such as bedding, toys, books, or surgical instruments, which come in contact with people only casually, can also transmit disease. These inanimate objects are collectively referred to as **fomites**. Inanimate objects such as food and water are referred to as disease **vehicles**. Fomites can be disease vehicles, but major epidemics originating from a single source are usually traced to food and water, since these are actively consumed in large amounts.

The nature of epidemics Two major types of epidemics can be distinguished: common source and propagated (person-to-person). These two types are contrasted in Figure 14.2. A **common-source epidemic** arises as the result of infection (or intoxication) of a large number of people from a contaminated source, such as food or water. Usually such

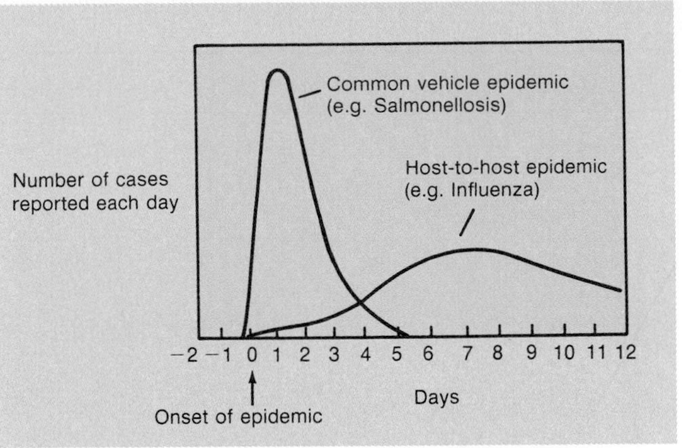

Figure 14.2 The shape of the epidemic curve helps to distinguish the likely origin. In a common vehicle outbreak, such as from contaminated food or water, the curve is characterized by a sharp rise to a peak, with a rapid decline, which is less abrupt than the rise. Cases continue to be reported for a period approximately equal to the duration of one incubation period of the disease. In a host-to-host epidemic, the curve is characterized by a relatively slow, progressive rise, and the cases will continue to be reported over a period equivalent to several incubation periods of the disease.

contamination occurs because of a malfunction in some aspect of the distribution system providing food or water to the population. Food- and waterborne diseases are primarily *intestinal* diseases, the pathogen leaving the body in fecal material, contaminating food or water via improper sanitary procedures, and then entering the intestinal tract of the recipient during ingestion. Because food- and waterborne diseases are some of those which are most amenable to control by public health measures, we shall discuss them in some detail in Chapter 15. The shape of the incidence curve for a common-source outbreak is characterized by a sharp rise to a peak, since a large number of individuals succumb within a relatively brief period of time. The common-source outbreak also declines rapidly, although the decline is less abrupt than the rise. Cases continue to be reported for a period of time approximately equal to the duration of one incubation period of the disease.

In a **propagated** or **person-to-person epidemic**, the curve of disease incidence shows a relatively slow, progressive rise (Figure 14.2), and a gradual decline. Cases continue to be reported over a period of time equivalent to several incubation periods of the disease. The epidemic may have been initiated by the introduction of a single infected individual into a susceptible population, and this individual has infected one or a few people in the population. Then the pathogen replicated in these individuals, reached a communicable stage, and was transferred to others, where it replicated and again became communicable. Table 14.3 summarizes the major epidemic diseases observed today.

Table 14.3 Major epidemic diseases and their control

Disease	Infective organism	Infection sources	Control measures
Common-vehicle epidemics*			
Typhoid Fever	*Salmonella typhi*	Contaminated food and water	Decontamination of water sources; pasteurization of milk; vaccination
Paratyphoid	*Salmonella paratyphi*	Contaminated food and water	Control of public water sources, food vendors and handlers; vaccination
Bacillary Dysentery	*Shigella dysenteriae*	Contaminated food and water	Detection and control of carriers; inspection of food handlers; decontamination of water supplies
Brucellosis	*Brucella melitensis*	Milk or meat from infected animals	Pasteurization of milk; control of infection in animals
Host-to-host epidemics			
Respiratory			
Diphtheria	*Corynebacterium diphtheriae*	Human cases and carriers; infected food and fomites	Immunization with diphtheria toxoid; quarantine of infected individuals
Meningococcal Meningitis	*Neisseria meningitidis*	Human cases and carriers	Treatment of all exposed with sulfadiazine (if isolated strain is susceptible)
Pneumococcal Pneumonia	*Streptococcus pneumoniae*	Human carriers	Treatment and rest for all upper respiratory infections; isolation of cases for period of communicability
Rocky Mountain Spotted Fever	*Rickettsia rickettsii*	Infected wild rodents; wood and dog ticks	Wearing protective clothing to avoid tick attachment; routine body surveilance for ticks; use of insect repellant
Tuberculosis	*Mycobacterium tuberculosis*	Sputum from human cases; contaminated milk	Early and extensive treatment of all cases with isoniazid; pasteurization of milk
Whooping Cough	*Bordetella pertussis*	Human cases	Immunization with *B. pertussis* vaccine; case isolation
Measles	Measles virus	Human cases	Measles vaccine
German Measles	Rubella virus	Human cases	Rubella vaccine; avoidance of contact between infected individual and pregnant women
Influenza	Flu virus	Human cases	Vaccine (recommended in certain cases only)
Venereal			
Gonorrhea	*Neisseria gonorrhoeae*	Urethral and vaginal secretions	Chemotherapy of carriers and potential contacts; case tracing and treatment
Syphilis	*Treponema pallidum*	Infected exudate or blood	Case identification by serological tests; treatment of seropositive individuals
Chlamydia	*Chlamydia tracomitis*	Urethral, vaginal and anal secretions	Testing for *C. trachomitis* during routine pelvic examinations; treatment of infected individuals and all potential contacts

**Some Common-Vehicle Diseases can also be spread from host-to-host.*

14.6 Nature of the Host Community

The colonization of a susceptible, unimmunized host by a parasite may first lead to an explosive infection and an epidemic. As the host population develops resistance, however, the spread of the parasite is checked, and eventually a balance is reached in which host and parasite are in equilibrium. A subsequent genetic change in the parasite could lead to the formation of a more virulent form, which would then initiate another explosive epidemic until the host again responded, and another balance was reached. In effect, the host and parasite are affecting each other's evolution; that is, the host and parasite are *coevolving*.

Introduction of a parasite into a virgin population An excellent experimental example of the consequences of the introduction of virulent pathogens into a susceptible population is one in which a virus was introduced for purposes of population control in the wild rabbits of Australia.

The wild rabbit was introduced into Australia from Europe in 1859 and quickly spread until it was overrunning large parts of the continent. Myxoma virus was discovered in South American rabbits, which are of a different species. In South America the virus and its hosts are apparently in equilibrium, as the virus causes only minor symptoms. However, this same virus was found to be extremely virulent to the European rabbit and almost always caused a fatal infection. The virus is spread from rabbit to rabbit by mosquitoes and other biting insects, and hence is capable of rapid spread in areas where appropriate insect vectors are present. The virus was introduced into Australian rabbits in 1950, with the aim of controlling the rabbit population. Within several months, the virus was well established in the population and spread over an area in Australia as large as all of Western Europe. The disease showed a marked seasonal pattern, rising to a peak in the summer when the mosquito vectors were present and declining in the winter. Research on the epidemiology of myxoma virus was initiated as a model of a virus-induced epidemic by Australian workers, under the direction of Professor Frank Fenner. Isolations of the virus from wild rabbits were carried out, and the virus strains isolated were characterized for virulence with laboratory rabbits. At the same time, baby rabbits were removed from their dens before infection could occur and reared in the laboratory. Then these rabbits were challenged with standard virulent strains of myxoma virus to determine their susceptibility. The results of this large scale model study are shown in Figure 14.3.

During the early years of the epidemic as high as 99 percent of the infected rabbits were dying. However, within seven years both the virus and the rabbit population had changed. By 1957 rabbit mortality had dropped to about 90 percent, and the virus isolated was of decreased virulence. In addition, changes were noted in the resistance of the rabbit. In parts of Australia where the virus had been first introduced, the remaining rabbit population had been subjected to selective pressure by the virus for seven years. Because of the rapidity with which rabbits multiply, a number of generations of rabbits developed, and the surviving population had withstood at least five years of intensive selective pressure, surviving epidemics in which over 90 percent of their cohorts had been killed. As seen in Figure 14.3, within five years the resistance of the rabbits had increased dramatically. It should be emphasized that this resistance is innate and is not due to immunological responses, for the rabbits tested had been removed from their mothers at birth and had never been in contact with the virus. Their resistance was due to some genetic change in the animal, which made it less susceptible to myxoma virus.

As a result of introduction of myxoma virus, the Australian rabbit population was greatly controlled, but the genetic changes in virus and host served to prevent a complete eradication of the rabbit from Australia. A steady state rabbit population of about 20 percent of that present before the introduction of myxoma virus was observed (Figure 14.3). The virus was thus a major factor in population control, but was unable to totally eliminate rabbits because of coevolutionary events on behalf of both host and parasite. From the present point of view, the significance of the Australian experiment is that it reveals how quickly an equilibrium is reached between host and parasite. The manner in which the malaria parasite has affected biochemical evolution in humans is another case in point and will be discussed in Section 15.10.

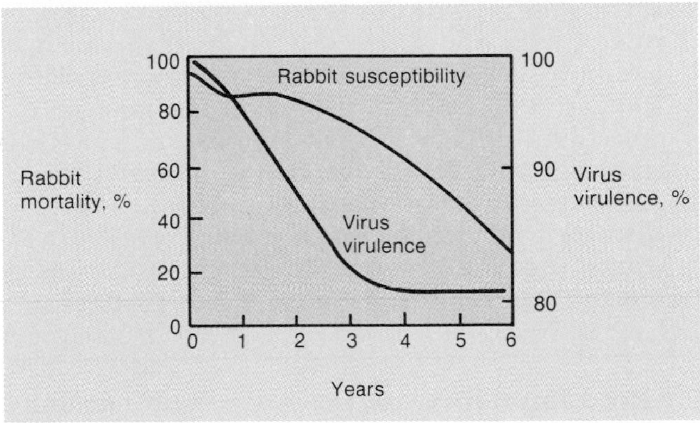

Figure 14.3 Changes in virulence of myxoma virus and in susceptibility of the Australian rabbit during the years after the virus was first introduced into Australia in 1950. Virus virulence is given as the average mortality in a standard laboratory breed of rabbits, for virus recovered from the field during each year. Rabbit susceptibility was determined by removing young rabbits from their dens and challenging with a virus strain of moderately high virulence, which killed 90 to 95 percent of normal laboratory rabbits.

Snow on Cholera

The importance of drinking water as a vehicle for the spread of cholera was first shown in 1855 by British physician John Snow, who at that time was even without any knowledge of the bacterial causation of the disease. Snow's study is one of the great classics of epidemiology and serves as a model for how a careful study can lead to clear and meaningful conclusions.

In London, the water supplies to different parts of the city were from different sources and were transmitted in different ways. In a large area south of the Thames River, across the river from Westminster Abbey and Parliament Building, the water was supplied to houses by two competing private water companies, the Southwark and the Vauxhall Company and the Lambeth Company. It was the water of the former company that was the major vehicle for the transmission of cholera. When Snow began to suspect the water supply of the Southwark and Vauxhall Company, he made a careful survey of the residence of every death in this district and determined which company supplied the water to that residence. In some parts of the area served by these two companies, each had a monopoly, but in a fairly large area the two companies competed directly, each having run independent water pipes along the various streets. Houses had the option of connecting with either supply, and the distribution of houses between the two companies was random. The clear-cut results of Snow's survey were completely convincing, even to those skeptical about the importance of polluted water in the transmission of cholera: in the first 7 weeks of the epidemic, there were 315 deaths per 10,000 houses supplied by the Southwark and Vauxhall Company, and only 37 per 10,000 houses supplied by the Lambeth Company. In the rest of London, there were 59 deaths per 10,000 houses, showing that those supplied by the Lambeth Company had fewer deaths than the general population. In the districts where each company had exclusive rights, it could of course be argued that it was not the water, but some other factor (soil, air, general layout of houses and so on), which might have been responsible for the differences in disease incidence, but in the districts where the two companies competed, all of these other factors were the same, yet the incidence was high for those supplied with Southwark and Vauxhall water and low for those supplied with Lambeth water. Snow attempted to relate these differences in disease incidence to the sources of the waters used by the two companies. As he suspected that the excrement and evacuations from cholera patients were highly infectious, he considered that sewage contamination of the water supply might exist. In those days, sewage treatment did not exist and raw sewage was dumped directly into the Thames River. The Southwark and Vauxhall Company obtained its water supply from the Thames right in the heart of London, where much opportunity for sewage contamination could occur, while the Lambeth Company obtained its water from a point on the river considerably above the city, and hence was relatively free of pollution. It is almost certain that it was this difference in source which accounted for the difference in disease incidence. In Snow's words: "As there is no difference whatever, either in the houses or the people receiving the supply of the two Water Companies, or in any of the physical conditions with which they are surrounded, it is obvious that no experiment could have been devised which would more thoroughly test the effect of water supply on the progress of cholera than this. . . . The experiment, too, was on the grandest scale. No fewer than three hundred thousand people of both sexes, of every age and occupation, and of every rank and station, from gentlefolk down to the very poor, were divided into two groups without their choice, and, in most cases, without their knowledge; one group being supplied with water containing the sewage of London, and, amongst it, whatever might have come from cholera patients, the other group having water quite free from such impurity."

Herd immunity An analysis of herd immunity is of great importance in understanding the role of immunity in the development of epidemics. **Herd immunity** is a concept used to explain the resistance of a group to invasion and spread of an infectious agent, due to immunity of a high proportion of the members of the group. If the proportion of immune individuals is sufficiently high then the whole population will be protected. The fraction of resistant individuals necessary to prevent an epidemic is higher for a highly virulent agent or one with a long period of infectivity, and lower for a mildly virulent agent or one with a brief period of infectivity.

The proportion of the population that must be immune to prevent infection of the rest of the population can be estimated from data on poliovirus immunization in the United States. From epidemiological studies of the incidence of polio in large populations, it appears that if a population is 70 percent immunized, the disease will be essentially absent in the population. Clearly, these immunized individuals are protecting the rest of the population. (One of the consequences of herd immunity is that the decreased incidence of the disease leads to a reduction in motivation of many individuals to become immunized; the proportion of immunized individuals decreases with time, leading again to an increase in the likelihood of disease.) For a highly infectious disease such as smallpox, the proportion of immunes necessary to confer herd immunity has been estimated to be 90 to 95 percent. A value of about 70 percent had been estimated for diphtheria, but further study of several small diphtheria outbreaks has shown that in densely settled areas a much higher proportion may have to be immunized to prevent development of an epidemic. Apparently in dense populations, person-to-person transmission can occur even if the agent is not highly infectious. In the case of diphtheria an additional complication arises because immunized persons can harbor the pathogen (inapparent infection) and thus act as chronic carriers, serving as potential sources of infection.

Cycles of disease The concepts of propagated epidemics and herd immunity can also explain why certain diseases occur in *cycles.* A good example of a cyclical disease is measles, which, in the days before the measles vaccine, occurred in a high proportion of school children. Since the measles virus is transmitted by the respiratory route, its infectivity is high in crowded situations such as schools. Upon entry into school at age five, most children would be susceptible, so that upon the introduction of measles virus into the school, an explosive propagated epidemic would result. Virtually every individual would become infected and develop immunity, and as the immune population built up, the epidemic died down. Measles showed an annual cycle (Figure 14.4) probably because a new group of nonimmune children arrived each year; the phasing of the epidemic would be related to the time of the year at which school began after the summer vacation. The consequences of vaccination for the development of measles epidemics are also illustrated in Figure 14.4.

14.7 Hospital-Acquired (Nosocomial) Infections

The hospital may not only be a place where sick people get well, it may also be a place where sick people get sicker. The fact is that cross-infection from patient to patient or from hospital personnel to patients presents a constant hazard. Hospitals are especially hazardous for the following reasons: (1) Many patients have weakened resistance to infectious disease because of their illness (compromised hosts; see Section 11.7). (2) Hospitals must of necessity treat patients suffering from infectious disease, and these patients may be reservoirs of highly virulent pathogens. (3) The crowding of patients in rooms and wards increases the chance of cross-infection. (4) There is much movement of hospital personnel from patient to patient, increasing the probability of transfer of pathogens. (5) Many hospital procedures, such as catheterization, hypodermic injection, spinal puncture, and removal of samples of tissues or fluids for diagnosis (biopsy), carry with them the risk of introducing pathogens to the patient. (6) In maternity wards of hospitals, newborn infants are unusually susceptible to certain kinds of infection. (7) Surgical procedures are a major hazard, since not only are highly susceptible parts of the body exposed to sources of contamination but the stress of surgery often diminishes the resistance of the patient to infection. (8) Many drugs used for immunosuppression (for instance, in organ transplant procedures) increase susceptibility to infection. (9) Use of antibiotics to control infection carries with it the risk of selecting antibiotic-resistant organisms, which then cannot be controlled if they cause further infection.

Hospital pathogens Hospital infections are often called *nosocomial infections (nosocomium* is the Latin word for hospital). Hospital infections are partly due to the prevalence of diseased patients, but are largely due to the presence of pathogenic microorganisms which are selected for by the hospital environment. Hospital infections most commonly occur during surgery; pediatric services have the fewest

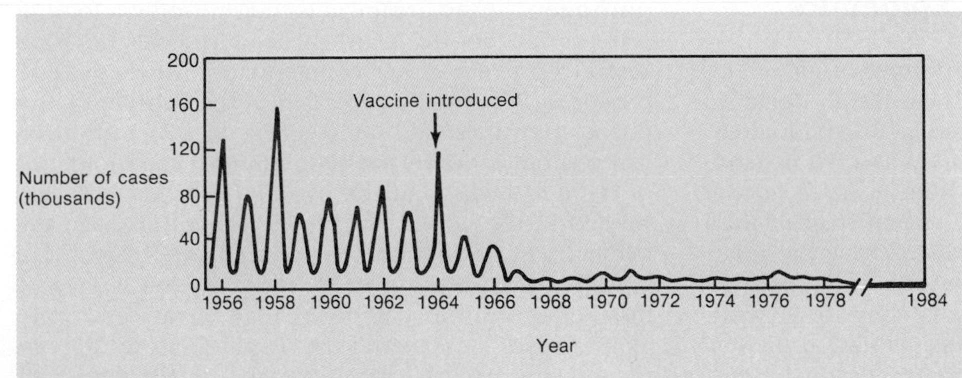

Figure 14.4 Cycles of disease, as illustrated by measles. Reported cases of measles by 4-week periods in the United States. The sharp decline in incidence after introduction of immunization is also seen.

problems. All parts of the body are subject to these infections, but urinary tract infections are most common. *Escherichia coli* is the most common cause of urinary tract infections in hospitals, but other enteric bacteria are commonly implicated as well.

One of the most important and widespread hospital pathogens is *Staphylococcus aureus*. It is most commonly associated with skin, surgical, and lower respiratory tract infections, and is a particular problem in infections acquired by newborns in the hospital. Certain strains of unusual virulence have been so widely associated with hospital infections that they are sometimes designated "hospital staphylococci." The habitat of these staphylococci is the upper respiratory tract, usually the nasal passages, and they often become established as "normal flora" in hospital personnel. In such healthy personnel the organism may cause no disease, but these symptomless carriers may be a source of infection of susceptible patients. Since staphylococci are resistant to drying, they survive for long periods on dust particles and other inanimate objects and can subsequently infect patients. Because of the potential seriousness of infection with hospital staphylococci, careful application of principles of hospital sanitation is necessary.

Pseudomonas aeruginosa is important in causing infections of the lower respiratory and urinary tracts. It is also an important cause of infections in burn patients who have lost their primary barrier to skin infection. *P. aeruginosa* exhibits one of the most significant features complicating the treatment of nosocomial infections, *drug resistance*. Isolates of this bacterium from patients with hospital infections are commonly resistant to many antibiotics. A somewhat lower degree of resistance has been noted among *S. aureus* isolates (with occasional highly-resistant strains not uncommon), whereas *E. coli* isolates generally remain sensitive to antibiotics. Antibiotic-resistant pathogenic bacteria in hospitals generally contain plasmids coding for *multiple drug resistance*, as described in Section 7.6.

14.8 Public Health Measures for the Control of Epidemics

An understanding of the epidemiology of an infectious disease generally makes it possible to develop methods for the control of the disease. **Public health** refers to the health of the population as a whole, and to the activities of public health authorities in the control of epidemics. It should be emphasized that the incidence of many infectious diseases has dropped dramatically over the past 100 years not because of any specific control methods, but because of general increases in the well-being of the population. Better nutrition, less crowded living quarters, and lighter work loads have probably done as much as public health measures to control diseases such as tuberculosis. However, diseases such as typhoid fever, diphtheria, brucellosis, and poliomyelitis owe their low incidence to active and specific public health measures. Finally, one infectious disease, gonorrhea, is much more prevalent today than it was 50 years ago. This increased incidence can be linked directly to changes in sexual behavior of the population. From an understanding of sources of infection and mechanisms of transmission, specific measures can be devised for control of diseases such as gonorrhea.

Controls directed against the reservoir If the disease occurs primarily in domestic animals, then infection of humans can be prevented if the disease is eliminated from the infected animal population. Immunization procedures or slaughter of infected animals may be used to wipe out the disease in animals. These procedures have been quite effective in eliminating brucellosis and bovine tuberculosis from humans. Not incidentally, the health of the domestic animal population is also increased, with likely economic benefits to the farmer. However, a program for eradication of a disease in a domestic animal is expensive, and the government must compensate farmers for lost animals. In the long run, the money spent will be returned many times over in reduced medical costs. When the reservoir is a wild animal (for example, tularemia, plague, see Section 15.10), then eradication is much more difficult. Rabies is a disease that occurs both in wild and domestic animals, but is transmitted to domestic animals primarily by wild animals. Thus control of rabies can be achieved by immunization of domestic animals, although this will never lead to eradication of the disease completely. Indeed, statistics show that the majority of rabies cases are in wild rather than domestic animals (Table 14.4). In England, where wild animal populations are fairly low, rabies has been virtually eradicated by strict laws and quarantine requirements for newly introduced animals. When humans are the reservoir (for example, gonorrhea), then control and eradication are much more difficult, especially if there are asymptomatic carriers.

Controls directed against transmission of the pathogen If the organism is transmitted via food or water, then public health procedures can be instituted to prevent either contamination of these vehicles or to destroy the pathogen in the vehicle. Water purification methods (see Section 15.12) have been responsible for dramatic reductions in the incidence of typhoid fever, and the pasteurization of milk has helped in the control of bovine tuberculosis in humans. Food protection laws have been devised that greatly decrease the probability of transmission of a number of enteric pathogens to humans. Transmission of respiratory pathogens is much more difficult to prevent. Attempts at chemical disinfection of air

Table 14.4 Reported cases of rabies in animals, United States

Domestic animals	Cases	Wild animals	Cases
Cattle	152	Skunks	2081
Cats	135	Bats	1038
Dogs	85	Foxes	139
Other domestic	59	Raccoons	1820
		Other wild	58
Total:	431	Total:	5136
Total domestic and wild: 5567			
Percent domestic: 7.7%			
Percent wild: 92.3%			

From Morbidity and Mortality Weekly Report, Annual Summary, 1984, issued March, 1986. Centers for Disease Control, Atlanta, Ga.

have been unsuccessful. In Japan, many individuals wear face masks when they have upper respiratory infections to prevent transmission to others, but such methods, although effective, are voluntary, and would be difficult to institute as public health measures.

Immunization procedures Immunization has been the prime means by which smallpox, diphtheria, tetanus, pertussis (whooping cough), and poliomyelitis have been eliminated. As we have discussed in Section 14.6, 100 percent immunization is not necessary in order to control the disease in a population, although the percentage needed to ensure control varies with the disease and with the condition of the population (for example, crowding).

The application of public health measures can lead to a classic confrontation between the rights of the private individual and the welfare of the population. Many diseases could be readily controlled if known control procedures were universally applied, but marked resistance exists to universal application of public health measures, such as immunization, which directly affect the individual. Over the past several decades, the proportion of children vaccinated for diphtheria, tetanus, pertussis, and polio has been slowly decreasing, apparently because the public has become less fearful of contracting these diseases due to their very low incidence in the population. However, since none of these diseases has been eradicated from the United States (indeed, the reservoir of tetanus is the soil, so that it will never be eradicated), and with a decrease in proportion of individuals immunized, the protection afforded by herd immunity (see Section 14.6) may be overcome, and diseases such as diphtheria, pertussis, and polio could reappear in epidemic form.

Also serious is the fact that many adults are inadequately immunized, either because they received low-titer vaccines when they were children or because their immunity has gradually disappeared. Surveys in the United States have indicated that up to 80 percent of adults may lack solid immunity to important childhood diseases. When these so-called childhood diseases occur in adults, they can have severe effects. If a woman contracts rubella (a virus disease; see Section 15.4) during pregnancy, the unborn child can be seriously impaired. Measles and polio are much more serious diseases in adults than in children.

All adults are advised to review their immunization status, checking their medical records (if available) to ascertain dates of vaccinations. *Tetanus* vaccinations should be given routinely at least every 10 years. Surveys of adult populations have shown that more than 10 percent of adults under the age of 40 and over 50 percent of those over 60 are not protected. *Measles* immunity also needs to be reviewed. People born before 1957 probably had measles as children and are immune. Those born after 1956 may have been vaccinated but the effectiveness of early vaccines was variable and solid immunity may not be present, especially if the vaccine was given before 1 year in age. Revaccination for polio is not recommended for adults unless they are traveling to countries in Africa and Asia where polio is still prevalent.

Vaccination practices and procedures have been discussed in Chapter 12 and those for particular infections in Chapter 15.

Quarantine Quarantine involves the limitation of the freedom of movement of individuals with active infections to prevent spread of disease to other members of the population. The time limit of quarantine is the longest period of communicability of the given disease. Quarantine must be done in such a manner that effective contact of the infected individual with those not exposed is prevented. Quarantine is not as severe a measure as strict isolation, which is used for unusually infectious diseases in hospital situations.

At one time, quarantine was required for a number of infectious diseases of childhood, such as measles, chickenpox, and mumps, and residences in which quarantined children were housed had placards affixed to the outside. Such measures were found to have little public health significance in the control of the spread of these diseases, and quarantine is no longer required, although it is still advisable as much as possible to prevent contact of infected children with other, possibly susceptible, children.

By international agreement six diseases are considered quarantinable: smallpox, cholera, plague, yellow fever, typhoid fever, and relapsing fever. Although smallpox has been eliminated from the world (see Box), quarantine for the other five diseases is still mandated. Each of these is considered a highly serious disease and particularly communicable. Thus it is essential to quarantine an infected individual for the period of communicability.

The Eradication of Smallpox

Smallpox, one of the scourges of humankind, is the first major infectious disease of humans to be completely eradicated. The basis for worldwide eradication was the availability of a highly effective live virus vaccine, and the use of this vaccine throughout the world. The eradication of smallpox can be considered to be the most dramatic example of the power of immunization, and constitutes a stunning display of worldwide cooperation. The campaign began first in the developed countries, and the disease was completely eliminated from North America by the end of the 1960s. In 1966 the World Health Assembly decided to begin an intensified effort of global eradication of smallpox, coordinated by the World Health Organization. Areas such as South America, Africa, India, and Indonesia were selected as major regions in wich efforts were to be concentrated. Initial attempts to *mass-vaccinate* populations were much less successful than later efforts involving a strategy called *surveillance-and-containment*. This required vaccination only of those people in contact with a diseased individual, to prevent the immediate spread of the virus. Despite interruption by cultural barriers, climate, and warfare, the areas where smallpox was endemic were reduced during the 1970s. By 1977, the countries in Africa which were the last to have smallpox cases were declared free of smallpox, and in 1980 the World Health Assembly certified that smallpox had been eradicated worldwide. Vaccinations for smallpox are now no longer required anywhere in the world.

14.9 Global Health Considerations

The United States can be considered typical of those countries where public-health protection is highly developed. Other countries with similar characteristics include Japan, Australia, New Zealand, Israel, and the European countries. In quite another category as far as infectious disease is concerned are the developing countries, a category that includes most of the countries in Africa, Central and South America, and Asia. In these countries, infectious diseases are still major causes of death.

Infectious disease in developing countries As shown in Figure 14.5 and Table 14.5 there is a sharp contrast in the degree of importance of infectious diseases as causes of death in developing versus developed nations. In developing regions of the world, infectious diseases account for between 30 and 50 percent of deaths whereas infectious diseases only account for about 4–8 percent of deaths in developed regions (see Table 14.5). Diseases which were leading causes of death in the United States nearly a century ago, such as tuberculosis and gastroenteritis, are still leading causes of death in developing countries (Figure 14.5). Furthermore, the majority of deaths due to infectious disease in developing regions occur among infants and children. Thus, the age bracket of deaths due to infectious diseases in developing versus developed countries is also dramatically different.

The distinct differences in the health status of people in different regions of the world is due in part

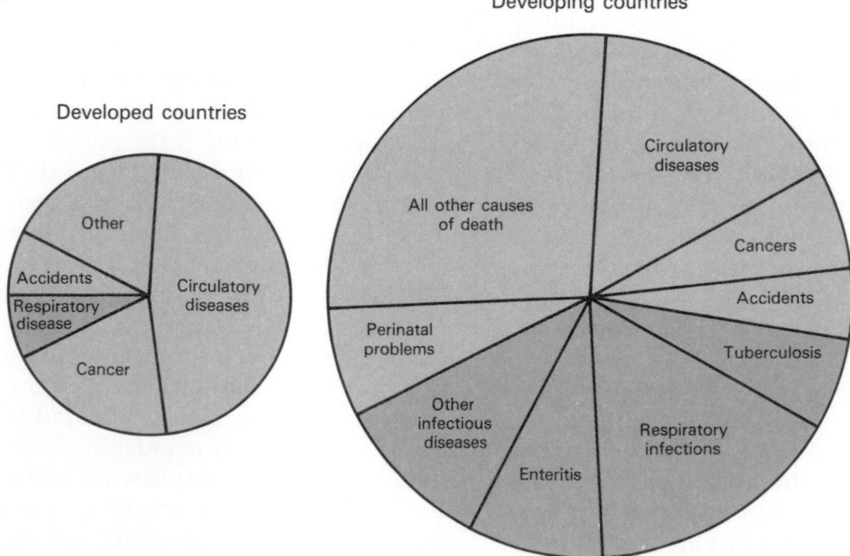

Figure 14.5 Leading causes of death in developed and developing countries. Infectious diseases are shown in color. The sizes of the circles are proportional to the relative number of deaths which occurred. The data are typical of recent years.

Table 14.5 Causes of death in developed and developing countries

	Developed countries		Developing countries			
Cause of death	Americas	Europe	Americas	Southeast Asia	Africa	Eastern Mediterranean
Infectious disease	3.6*	8.6	31.1	43.9	49.8	44.5
Cancer	21.5	18.1	9.0	4.4	2.9	4.2
Circulatory diseases	54.5	53.8	24.5	15.6	11.7	14.1
Accidents	8.4	5.6	6.3	4.3	3.8	4.1

**Percentage of deaths due to various causes in developed and developing countries in 1980. Data from World Health Statistics Annual, 1984, World Health Organization.*

to general nutritional deficiency in individuals in undeveloped countries and to a lower overall standard of living. As discussed in Section 11.7, factors such as physical stress and diet play important roles in the ability of the host to ward off infection. Thus, it is not surprising that in undeveloped countries death from infection is more likely. In addition, the generally lower levels of public health protection in undeveloped countries makes infection more likely in the first place. Statistics on disease in developed countries show that control of many diseases is possible. However, statistics on the worldwide incidence of disease show that infectious disease remains an important problem about which we must all be concerned.

Travel to endemic areas The high incidence of disease in many parts of the world is also a concern for people traveling to such areas. It is possible to be immunized against many of the diseases which are endemic in foreign countries. Recommendations for immunization of U.S. citizens traveling abroad are shown in Table 14.6. Many foreign countries currently require immunization certificates for cholera and yellow fever, but many other immunizations are only recommended for people who are expected to be at high risk. There is also risk in many parts of the world of exposure to diseases for which there is no vaccine available. These include amebiasis, dengue fever, encephalitis, giardiasis, malaria, and typhus. Travelers are advised to take reasonable precautions such as avoiding insect or animal bites, only drinking water which has been properly treated, and undergoing chemotherapeutic programs after exposure is suspected.

Table 14.6 Immunizations required or recommended for U.S. travelers abroad

Disease	Destination	Recommendation
Cholera	Many central African nations, India, Pakistan, S. Korea, Albania, Malta	*Vaccination required* if entering from or continuing to endemic areas
Yellow fever	Many Central and South American, and African countries	*Vaccination often required* for entry; or if entering from or continuing to endemic areas
Plague	Mostly rural mountainous or upland areas of Africa, Asia and South America	*Vaccination recommended* if direct contact with rodents is anticipated
Infectious hepatitis (A)	Specific tropical areas and many developing countries	*Passive immunization recommended* for long stays
Serum hepatitis (B)	Africa, Indochina, eastern and southern Europe, USSR, Central and South America	*Vaccination recommended*
Typhoid fever	Many African, Asian, Central and South American countries	*Vaccination recommended*

From Health Information for International Travelers, 1984, U.S. Department of Health and Human Services.
Vaccinations are not required, but are recommended, for diphtheria, pertussis, tetanus, polio, measles, mumps, and rubella. Most U.S. citizens are already immunized through normal immunization practices.

Study Questions

1. How would a physician and an epidemiologist likely differ in their perspective on infectious diseases?
2. How could a disease have relatively low incidence but at the same time be highly prevalent?
3. Distinguish between *morbidity* and *mortality*. How would you get statistics on morbidity? Mortality?
4. What is an *epidemic*? How is it distinguished from an *endemic* disease? *Pandemic*?
5. What is the disease status of someone who is identified as a disease *carrier*?
6. What are the possibilities for control of an infectious disease in which the primary reservoir is a domestic animal? A wild animal such as a rat? A wild animal such as a fox?
7. In the United States, even a small number of cases of the disease *polio* would be considered to be an epidemic whereas if even a large number of cases of *chickenpox* occurred it would not likely be considered an epidemic. Explain.
8. What are the consequences for public health control of infectious disease for those diseases in which most cases occur as inapparent infections? How would you determine the incidence of a disease which caused many inapparent infections (refer back to Table 12.9 for a clue).
9. What lines of evidence led epidemiologists to conclude that the disease *AIDS* was not transmitted by casual contact between humans?
10. Compare and contrast the epidemiological picture of a disease like *measles* with a disease like *botulism*.
11. What is a disease *vector*? What is a disease *vehicle*? What is a *fomite*? Briefly discuss three different diseases in which you correctly use these terms.
12. Based on the consideration of four kinds of individuals: infected individuals, immune individuals, uninfected individuals, and carriers; describe how a disease cycle might occur for a respiratory pathogen.
13. Although *myxoma* virus is highly fatal for rabbits, explain why the introduction of this virus into the Australian rabbit population did not kill off all the rabbits.
14. If most of the population is immune for a particular infectious disease, the unimmunized remainder of the population rarely becomes ill even though they are not immunized. Explain.
15. How does the concept of *herd immunity* depend to some extent on the virulence of the pathogen?
16. *Rabies* is almost nonexistent in dogs (see Table 14.4), yet it is still necessary to ensure that all dogs are vaccinated. Explain.
17. Why does a disease like *chickenpox* show a periodic cycle of disease whereas other diseases, such as *venereal* diseases, do not?
18. What are *nosocomial* infections? Why are they so common? How has the advent of antibiotics actually increased the number of nosocomial infections?
19. Why is *Staphylococcus aureus* a common nosocomial pathogen? *Pseudomonas aeruginosa*?
20. What are the major public health measures taken to prevent disease epidemics? Which measures are most effective? Which methods do you think would be the most expensive and why?
21. Why would it be necessary to *quarantine* an individual suffering from *smallpox* (assuming an isolated case was discovered)?
22. What are the major reasons why public health measures are more effective and the overall level of public health higher in developed countries than in undeveloped countries?

Supplementary Readings

Anderson, R. M., and **R. M. May** (eds.) 1982. *Population Biology of Infectious Diseases*. Dahlem Workshop on Population Biology of Infectious Disease Agents, 1982, Berlin. Springer-Verlag, Berlin.

Bacon, P. J. (ed.) 1985. *Population Dynamics of Rabies in Wildlife*. Academic Press, New York. A recent updating of the incidence of rabies in wild animal populations worldwide.

Benenson, A. L. (ed.) 1985. *Control of Communicable Diseases in Man, 14th ed.* American Public Health Association, Washington, D.C. A concise and useful reference to the complete spectrum of infectious diseases of humans, with emphasis on diagnostic and public health measures.

Bitton, G. 1980. *Introduction to Environmental Virology*. John Wiley and Sons, New York. An excellent source on transmission and survival of pathogenic viruses.

Centers for Disease Control. Morbidity and Mortality Weekly Report. Atlanta, GA. Issued weekly. This publication, available in most large libraries, gives an instantaneous view of epidemiological problems in the United States. Lists incidence of all major infectious diseases by geographic region and gives updates on particular disease problems. An Annual Summary is published yearly providing an overview of the incidence of past and present infectious diseases. A good source of graphs, tables, and statistical analyses of disease trends.

Fenner, F. 1983. *Biological control as exemplified by smallpox eradication and myxomatosis*. Proc. R. Soc. London B 218:259–285. An excellent historical account of the control of the Australian rabbit population and the global eradication of smallpox.

Snow, John. 1936. *Snow on cholera*. A reprint of two papers by John Snow, M.D. The Commonwealth Fund New York. A reprinting of Snow's 1855 study of cholera, the first epidemiological investigation, and a fascinating detective story. Highly recommended.

15

Major Microbial Diseases

Within the microbial world only a relatively few species are pathogens. The majority of microorganisms carry out important transformations of elements or are closely associated with plants or animals in stable, beneficial relations (see Chapter 17). However, as we have seen, pathogenic organisms can have profound effects on animals and humans. In this chapter we consider the major human diseases and will group together those microbial diseases that affect a specific human organ system. We will continually make reference to modes of transmission in our discussion of diseases, because the pathology of a disease is best considered in light of its ecology. For example, streptococcal sore throat and influenza are diseases whose etiological agents are completely different, one a bacterium and the other a virus. Yet, both diseases are transmitted from person-to-person primarily by an airborne route. Hence, if we group diseases by their *modes of transmission*, what at first looks like a long list of unrelated diseases quickly shrinks to a short list of ecologically related diseases. We will also refer back to material in Chapters 11–14 from time to time, because the biology of specific pathogenic determinants and the nature of the immune responses are considered in detail in these chapters.

15.1 Airborne Transmission of Pathogens

Air usually is not a suitable medium for the growth of microorganisms; those organisms found in air are derived from soil, water, plants, animals, people, or other sources. In outdoor air, soil organisms predominate. Microbial numbers indoors are considerably higher than those outdoors, and the organisms are mostly those commonly found in the human respiratory tract.

Outdoors the source of virtually all airborne bacteria is the soil. Windblown dust carries with it significant microbial populations, which can be carried long distances. Indoors, the main source of airborne microbes is the human respiratory tract. Most of these organisms survive only poorly in air, so that effective transmittal to a suitable habitat (another human) occurs only over short distances. However, certain human pathogens (*Staphylococcus, Streptococcus*) are able to survive under dry conditions fairly well and so may remain alive in dust for long periods of time. Gram-positive bacteria are in general more resistant to drying than Gram-negative bacteria, and it may be for this reason that Gram-positive bacteria are more likely to be involved in air dispersal. Spore-forming bacteria are also resistant to drying, but are generally not present in humans in the spore form. Other airborne microbes found to be derived from soil are also Gram-positive (for example, *Micrococcus*). The reason Gram-positive bacteria are more re-

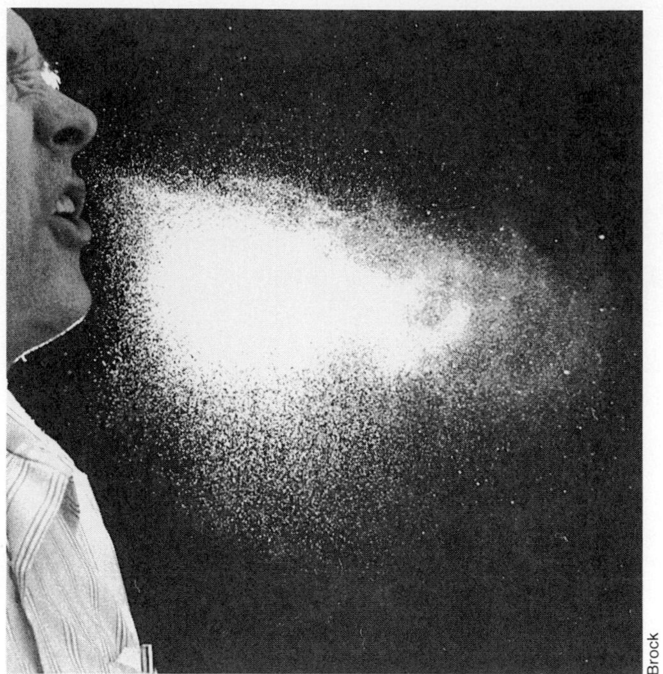

Figure 15.1 High-speed photograph of an unstifled sneeze.

sistant to drying than Gram-negative ones probably relates to the stability to drought effected by the thicker and more rigid Gram-positive cell wall.

An enormous number of droplets of moisture are expelled during sneezing (Figure 15.1), and a considerable number are expelled during coughing or even merely talking. Each infectious droplet has a size of about 10 μm and contains one or two bacteria. The speed of the droplet movement is about 100 m/s (over 200 miles/h) in a sneeze and about 16 to 48 m/s during coughing or loud talking. The number of bacteria in a single sneeze varies from 10,000 to 100,000. Because of the small size of the droplets, the moisture evaporates quickly in the air, leaving behind a nucleus of organic matter and mucus, to which bacterial cells are attached.

Respiratory infection The average human breathes several million cubic feet of air in a lifetime, much of it containing microbe-laden dust, which is a potential source of inoculum for upper-respiratory infections caused by streptococci and staphylococci resistant to drying. The speed at which air moves through the respiratory tract varies, and in the lower respiratory tract the rate is quite slow. As the air slows down, particles in it stop moving and settle, the larger particles first and the smaller ones later, and as seen in Figure 15.2, in the tiny bronchioles of the lung only particles below 3 μm will be present. As also shown in Figure 15.2, different organisms reach different levels in the tract, thus accounting for the differences in the kinds of infections that occur in the upper and lower respiratory tracts.

15.2 Respiratory Infections: Bacterial

A variety of bacterial pathogens affect the respiratory tract and, as discussed above, because their mode of transmission is air, they are predominantly Gram-positive bacteria. Because secondary problems associated with an initial bacterial respiratory infection can often be quite serious, it is important to swiftly diagnose and treat these infections to prevent damage to host tissues. Fortunately, most respiratory bacterial pathogens respond readily to antibiotic therapy and many can also be controlled by immunization programs. Nevertheless, bacterial respiratory infections are still

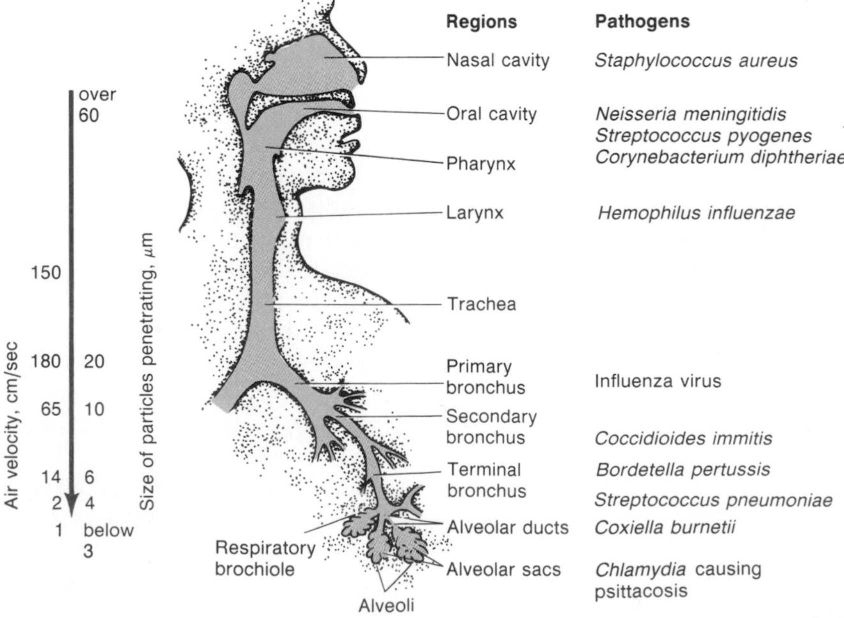

Figure 15.2 Characteristics of the respiratory system of humans and locations at which various organisms generally initiate infections.

Table 15.1 Medically-important streptococci

Streptococcus species	Habitat	Disease involvement
S. pyogenes	Human upper respiratory tract	Acute pharyngitis; bacteremia
S. pneumoniae	Humans	Lobar pneumonia
S. mutans	Oral cavity of warm-blooded animals	Dental caries
S. faecalis	Intestinal tract of warm-blooded animals; dairy products	Urinary tract infection; bacteremia

rather common and we begin here with a consideration of one of the most common bacterial respiratory pathogens, *Streptococcus* species.

Streptococcal diseases Streptococci are Gram-positive cocci which typically grow in elongated chains (see Section 19.25). As lactic acid bacteria, some species of streptococci play economically important roles in the production of fermented milk products. Other species, however, are potent human pathogens. Table 15.1 lists the major medically significant streptococci. Of particular relevance to our discussion here are *S. pyogenes* and *S. pneumoniae*.

S. pyogenes and *S. pneumoniae* are frequently isolated from the nasopharyngeal microflora of healthy adults. Although numbers of *S. pyogenes* and *S. pneumoniae* are usually low here, if the host's defenses are weakened, or a new, highly virulent strain is introduced, acute streptococcal bacterial infections are possible. *S. pyogenes* is the leading cause of streptococcal pharyngitis, so-called "strep throat" (the pharynx is the tube which connects the oral cavity to the larynx and the esophagus). Most isolates from clinical cases of strep throat produce a toxin which lyses red blood cells, a condition called *β-hemolysis* (see Figure 11.16*a*). Streptococcal pharyngitis is characterized by intense inflammation of the mucous membranes of the throat, a mild fever, and a general feeling of malaise. *S. pyogenes* can also cause related infections of the inner ear (otitis media), the tonsils (tonsillitis), the mammary glands (mastoditis), and infections of the superficial layers of the skin, a condition referred to as impetigo (however, most cases of impetigo are caused by *Staphylococcus aureus*, see Figure 15.3).

About half of the clinical cases of severe sore throat turn out to be due to *S. pyogenes*, the remainder being of viral origin. An accurate diagnosis is important since if the sore throat is due to a virus, the application of antibacterial antibiotics would be useless, whereas if the sore throat is due to *S. pyogenes*, immediate antibiotic therapy is indicated. This is because occasional cases of streptococcal sore throat can lead to two more serious streptococcal syndromes, *scarlet fever* and *rheumatic fever*.

Certain strains of *S. pyogenes* carry a lysogenic bacteriophage which codes for production of a potent exotoxin, responsible for most of the symptoms of scarlet fever. The erythrogenic toxin causes a pink-red rash to develop (Figure 15.4) and also acts to damage small blood vessels and initiate fever. The condition is acute and easily treated with antibiotics. However, a prolonged case of scarlet fever may predispose the individual to rheumatic fever, and this condition is of great medical concern. Rheumatic fever is caused by unlysogenized strains of *S. pyogenes* which contain cell surface antigens sufficiently similar to certain human cell-surface antigens that when an immune response to the invading pathogen is made, the antibodies produced also damage host tissues, in particular those of the heart and joints. In addition, kidney damage from rheumatic fever occurs from the deposition of antigen:antibody complexes on the basement membrane of kidney glomeruli. In essence, then, rheumatic fever is a type of autoimmune disease—antibodies reacting with self constituents. Rheumatic fever patients must keep a close watch in later life for symptoms of heart or kidney damage.

The other major pathogenic streptococcal species, *S. pneumoniae*, causes lung infections which often develop as secondary infections to other respiratory disorders. A characteristic of *S. pneumoniae* is that cells are typically present in pairs (or short chains) and are surrounded by a large capsule (see Figure 15.5). The capsule enables the cells to resist phagocytosis, and capsulated strains of *S. pneumoniae* are hence very invasive. Cells invade alveolar tissues (lower respiratory tract) of the lung and elicit a strong host inflammatory response. Reduced lung function can result from accumulation of phagocytic cells and fluid, and the *S. pneumoniae* cells can spread from the focus of infection as a bacteremia, resulting in bone and inner ear infections and occasionally endocarditis. Streptococcal pneumonia is a serious infection, untreated cases having a mortality rate of about 30 percent. However, most strains of *S. pneumoniae* and *S. pyogenes* respond dramatically to penicillin therapy, and penicillin remains the clinical drug of choice. An increasing incidence of penicillinase-producing streptococci has been reported lately, and erythromycin is the next best therapeutic agent.

The potential for serious host damage following a streptococcal sore throat has encouraged the development of clinical methods for rapid identification of *S. pyogenes*. At least two immunological techniques have been developed thus far that can detect surface proteins unique to *S. pyogenes* by latex bead

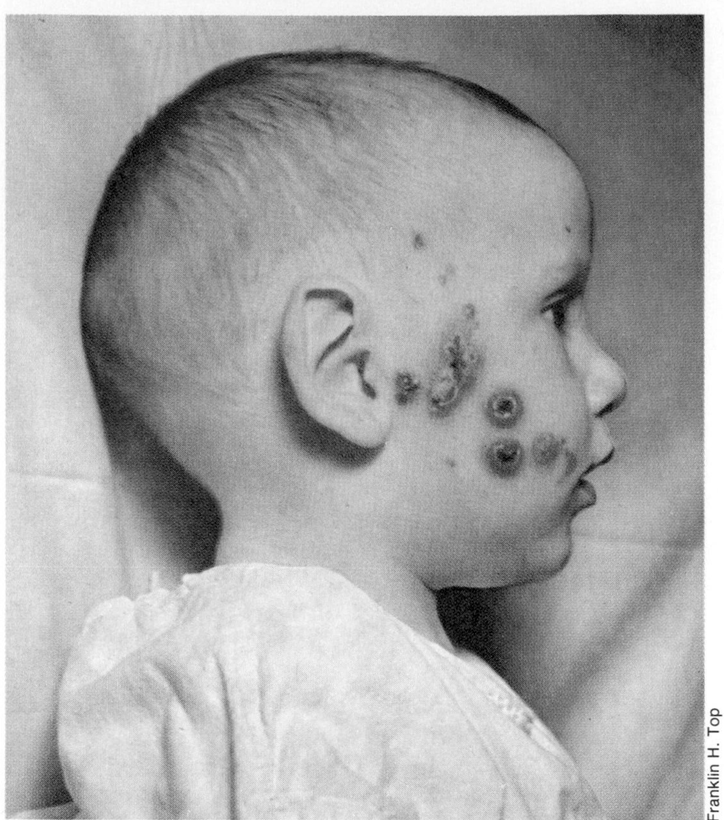

Figure 15.3 Typical lesions of impetigo, commonly caused by *Streptococcus pyogenes* or *Staphylococcus aureus*.

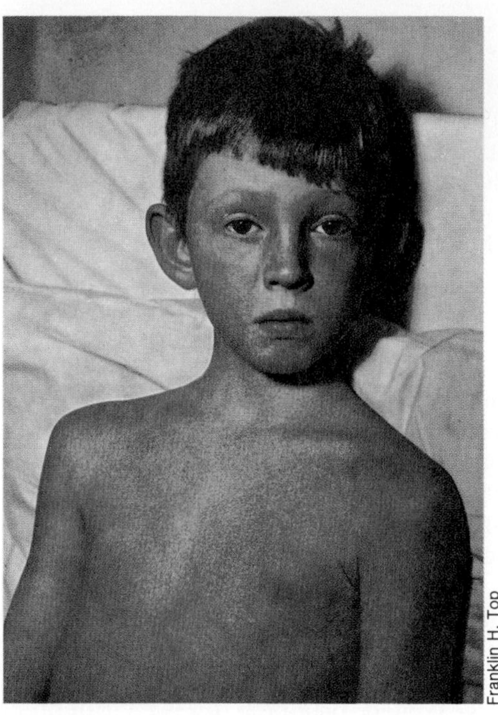

Figure 15.4 The typical rash of scarlet fever, due to the action of an erythrogenic toxin produced by *Streptococcus pyogenes*.

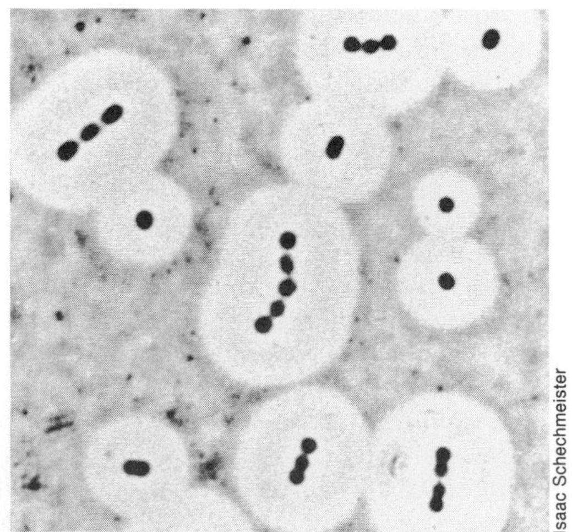

Figure 15.5 India ink negatively stained preparation of cells of *Streptococcus pneumoniae*. Note the extensive capsule surrounding the cells.

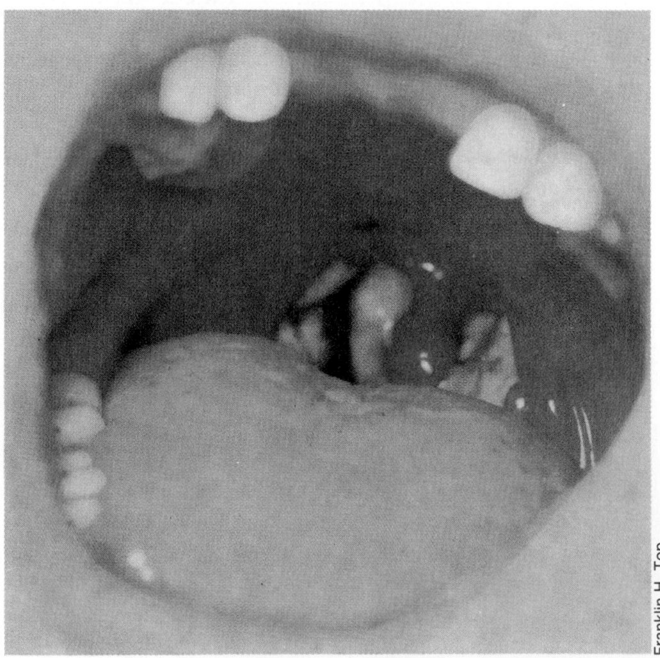

Figure 15.6 Pseudomembrane in an active case of diphtheria caused by the bacterium *Corynebacterium diphtheriae*.

agglutination or by fluorescent antibody staining (see Sections 13.6 and 13.9) directly from a patient throat swab. Such procedures allow the confident initiation of immediate antibiotic therapy in order to avoid complications such as rheumatic fever.

Diphtheria *Corynebacterium diphtheriae*, the causative agent of diphtheria, is a Gram-positive irregular rod (see Section 19.29 for detailed description of coryneform bacteria). *C. diphtheriae* enters the body via the respiratory route with cells lodging in the throat and tonsils. Although limited information is available concerning the mechanism of adherence of *C. diphtheriae* to these tissues, the organism produces a neuraminidase capable of splitting *N*-acetylneuraminic acid (a component of glycoproteins found on animal cell surfaces), and this may enhance the invasion process. The inflammatory response of throat tissues to *C. diphtheriae* infection results in formation of a characteristic lesion called a *pseudomembrane* (Figure 15.6), which consists of damaged host cells and cells of *C. diphtheriae*. As described in Section 11.10, certain strains of *C. diphtheriae* are lysogenized by bacteriophage β, and these strains produce a powerful exotoxin, the diphtheria toxin. Diphtheria toxin inhibits eucaryotic protein synthesis (see Sections 11.10 and 18.8).

The pseudomembrane that forms in diphtheria may block the passage of air, and death from diphtheria is usually due to a combination of the effects of partial suffocation and tissue destruction by exotoxin. Although at one time a major cause of childhood mortality, diphtheria is now rarely encountered because an effective vaccine is available. This vaccine is made by treating the diphtheria exotoxin with formalin to yield an immunogenic, yet nontoxic, toxoid (see Section 12.13 for general discussion of toxoids).

Diphtheria toxoid is part of the so-called "DPT" (diphtheria, pertussis, tetanus) vaccine, administered several times in the first year of life (see Section 12.13). To determine whether an individual is immune to diphtheria the so-called *Schick Test* has been used. This test involves injecting a diluted diphtheria toxin suspension under the skin, usually on the forearm, and observing for formation of a reddening inflammatory response due to neutralization of the toxin by circulating antibodies. A patient diagnosed as having diphtheria by culture of *C. diphtheriae* from a pseudomembrane in the throat is usually treated simultaneously with antibiotics and diphtheria antitoxin (an antitoxin contains neutralizing antibodies formed in another animal, see Section 12.13 for discussion of antitoxins). Penicillin, erythromycin, or gentamycin are generally effective in diphtheria therapy. Studies of recent outbreaks of diphtheria have shown that early administration of both antibiotics and antitoxin appears necessary for effective control of the disease.

Legionellosis (Legionnaire's disease) This disease, caused by the organism *Legionella pneumophila*, derives its name from the fact that it was first recognized as a disease entity from an outbreak of pneumonia occurring during a convention of the American Legion in the summer of 1976. *Legionella* is a thin Gram-negative rod (Figure 15.7) with complex nutritional requirements, including an unusually high iron requirement, and is immunologically distinct from any known pathogen associated with respiratory infections. *Legionella* can be detected by immunofluorescence techniques (see Section 13.6 and Figure 13.7*a*) and can be isolated from many terrestrial and aquatic habitats as well as from patients suffering from legionellosis. Epidemiological studies have showed that *Legionella* is a common inhabitant of cooling towers of air conditioning units and that infectious *Legionella* aerosols from such sources can spread the organisms to humans. Curiously, however, although apparently spread via an airborne route, no evidence of direct person-to-person transmission of *Legionella* has been obtained. Consistent with these findings is the fact that cases of legionellosis peak in the late summer months (Figure 15.8) when air conditioners are most extensively used. This is in contrast to an airborne disease such as chickenpox, which is highly contagious and peaks in the winter months (see Figure 15.17) when people are more frequently indoors and in close contact. More recently, a trend to late fall/early winter peaks of Legionaire's disease has been observed (Figure 15.8) and if this continues it may suggest alternative vehicles for transmission of the disease.

Legionella infections may be totally asymptomatic and occasionally result in only mild symptoms such as headache and fever. In the majority of cases where pneumonia develops patients are frequently elderly individuals whose resistance has been pre-

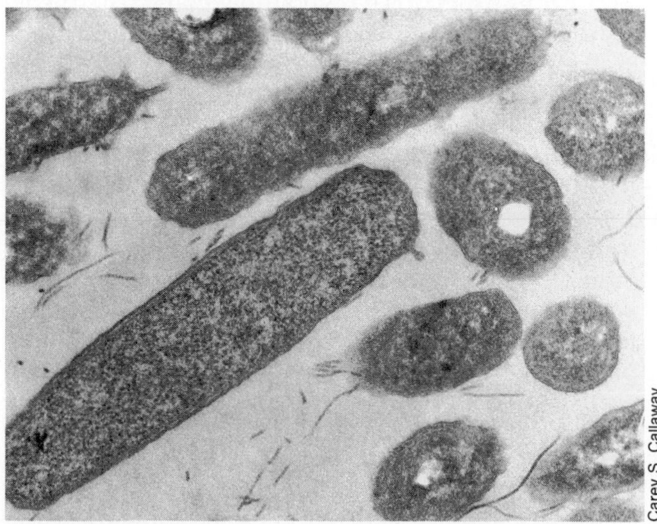
Carey S. Callaway

Figure 15.7 Transmission electron micrograph of *Legionella pneumophila*.

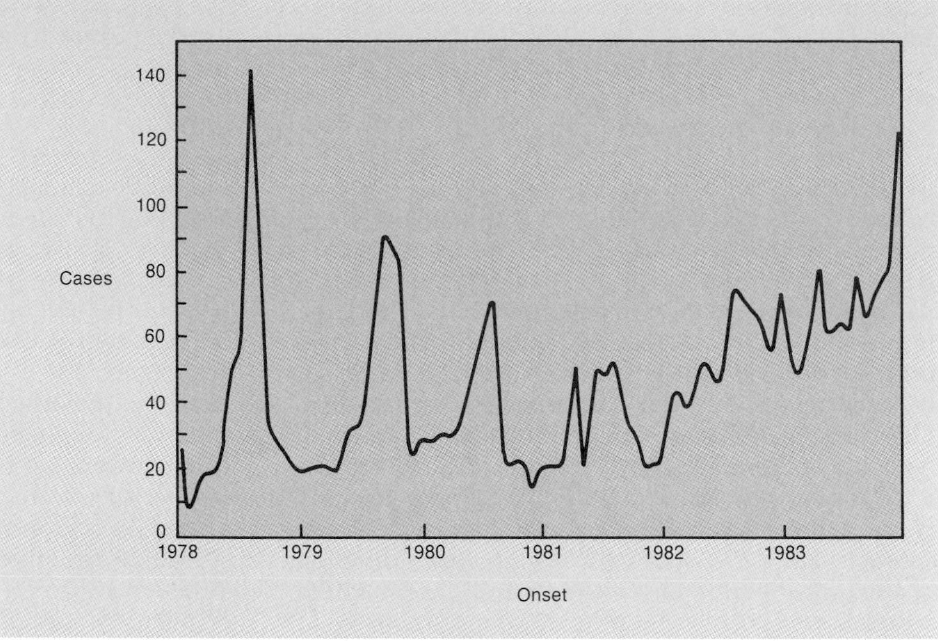

Figure 15.8 Seasonal distribution of cases of legionellosis, 1978–1983. Note the tendency for cases to peak in mid-summer.

viously compromised. In addition, certain serotypes of *Legionella* (8 are known) are more strongly associated with the pneumonic form of the illness than others. Prior to the onset of pneumonia, intestinal disorders are common, followed by high fever, chills, and muscle aches. These symptoms precede the dry cough and chest and abdominal pains typical of legionellosis. Death, if it occurs, is usually due to respiratory failure. Clinical detection of *L. pneumophila* is now straightforward, because of fluorescent antibody and other highly specific immunological probes (see Figure 13.7*a*). *Legionella pneumophila* is sensitive to the antibiotics rifampicin and erythromycin, and intravenous administration of erythromycin is the treatment of choice in most cases.

Whooping cough Whooping cough (pertussis) is an acute, highly infectious respiratory disease generally observed in children under one year of age (Figure 15.9). Whooping cough is caused by a small, Gram-negative, strictly aerobic coccobacillus, *Bordetella pertussis*. The organism attaches to cells of the upper respiratory tract by producing a specific adherence factor called *filamentous hemagglutinin antigen* which recognizes a complementary molecule on the surface of host cells. Once attached, *B. pertussis* grows and produces pertussis exotoxin that induces synthesis of cyclic adenosine monophosphate (cyclic AMP, see Figure 5.35), which is at least partially responsible for events that lead to host tissue damage. *B. pertussis* also produces an endotoxin, which also may induce some of the symptoms of whooping cough. Clinically, whooping cough is characterized by a recurrent, violent cough that usually lasts up to six weeks. The spasmodic coughing gives the disease its name, for a whooping sound results from the patient inhaling in deep breaths to obtain sufficient air.

A vaccine consisting of killed whole cells of *B. pertussis* is part of the routinely administered DPT vaccine (see Section 12.13) and thus fewer than 50 cases of whooping cough/100,000 population are reported each year in the United States (Figure 15.9). However, because of undesirable side effects including local swelling and redness, fever, and occasional more serious problems such as encephalitus and convulsions in certain recipients of this vaccine, a "second generation" pertussis vaccine, containing purified cell fractions of *B. pertussis*, is now being tested

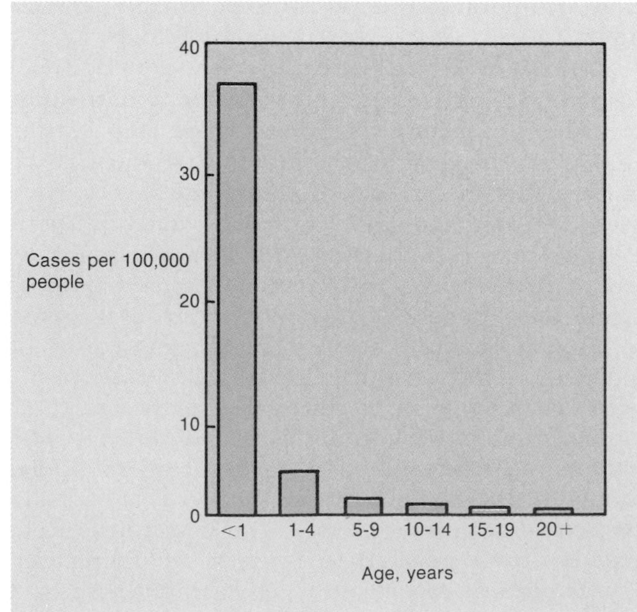

Figure 15.9 Reported cases of whooping cough (pertussis) by age group, United States, 1983.

in the United States and is already in wide use in Japan.

Diagnosis of whooping cough can be made by fluorescent antibody staining of throat smears or by actually culturing the organism. For culture of *B. pertussis*, the "cough-plate" method is used. The patient is asked to cough directly into a blood-glycerol-potato extract agar plate (although not selective, this medium supports good recovery of *B. pertussis*). Alternatively, throat and nose swabs (if streaked immediately after sampling) can be used. β-hemolytic colonies containing small Gram-negative coccobacilli are tested for *B. pertussis* by a latex bead agglutination test, or are stained with an anti-*B. pertussis* fluorescent antibody for positive identification. Cultures of *B. pertussis* are killed by the antibiotics ampicillin, tetracycline, and erythromycin; the efficacy of antibiotics *in vivo*, however, has been questioned. Since a patient with whooping cough remains infectious for up to two weeks following commencement of antibiotic therapy, it is generally considered that the immune response is as important, if not more so, than antibiotics, in the elimination of *B. pertussis* from the body.

15.3 *Mycobacterium* and Tuberculosis

The famous German microbiologist Robert Koch isolated and described the causative agent of tuberculosis, *Mycobacterium tuberculosis*, in 1882. At one time, tuberculosis was the single most important infectious disease of humans and accounted for one-seventh of all deaths. Even at present, over 20,000 new cases of tuberculosis are diagnosed each year.

The microbiology of *M. tuberculosis* is discussed in Section 19.31. The interaction of the human host and *M. tuberculosis* is an extremely complex phenomenon, being determined in part by the virulence of the strain but probably more importantly by the specific and nonspecific resistance of the host. Cell-mediated immunity plays an important role in the development of disease symptoms. It is convenient to distinguish between two kinds of human tuberculosis infection: *primary* and *postprimary* (or reinfection). Primary infection is the first infection that an individual receives and often results from inhalation of droplets containing viable bacteria derived from an individual with an active pulmonary infection. Dust particles that have become contaminated from sputum of tubercular individuals are another source of primary infection. The bacteria settle in the lungs and grow. A delayed-type hypersensitivity reaction (see Section 12.10) results in the formation of aggregates of activated macrophages, called *tubercles*, characteristic of tuberculosis. In a few individuals with low resistance the bacteria are not effectively controlled, and an acute pulmonary infection occurs, which leads to the extensive destruction of lung tissue, the spread of the bacteria to other parts of the body, and death.

In most cases, however, acute infection does not occur, and the infection remains localized and is usually inapparent; later it subsides. But this initial infection hypersensitizes the individual to the bacteria or their products and consequently alters the response of the individual to subsequent infections. A diagnostic test, called the **tuberculin test**, can be used to measure this hypersensitivity. If *tuberculin*, a protein fraction extracted from *M. tuberculosis*, is injected intradermally into a hypersensitive individual, it elicits at the site of injection a localized immune reaction characterized by hardening and reddening of the site 1 to 2 days after injection. An individual exhibiting this reaction is said to be *tuberculin positive*, and many healthy adults give positive reactions as a result of previous inapparent infections. A positive tuberculin test does not indicate active disease but only that the individual has been exposed to the organism at some time.

It is in tuberculin-positive individuals that the postprimary type of tuberculosis infection can occur. When renewed pulmonary infections occur in tuberculin-positive individuals, they are usually chronic infections that involve destruction of lung tissue, followed by partial healing and a slow spread of the lesions within the lungs. Spots of destroyed tissue may be revealed by X-ray examination (Figure 15.10), but viable bacteria are found in the sputum only in individuals with extensive tissue destruction. In many cases, symptoms in tuberculin-positive individuals are a result of reactivation and growth of bacteria that have remained alive and dormant in the lungs for long periods of time. Malnutrition, overcrowding, stress, and hormonal imbalance often are factors predisposing an individual to secondary infection.

Chemotherapy of tuberculosis has been a major factor in control of the disease and has resulted in

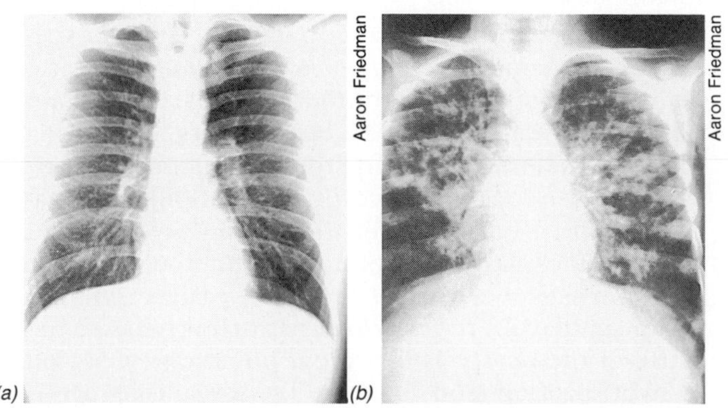

Figure 15.10 X-ray photographs. (a) Normal chest X ray. The faint white lines are arteries and blood vessels. The heart is visible as a white bulge in the lower right quadrant. (b) An advanced case of pulmonary tuberculosis; white patches indicate areas of disease.

Isoniazid

Nicotinamide

Figure 15.11 Structure of Isoniazide (isonicotinic acid hydrazide), an effective chemotherapeutic agent for tuberculosis. Note the structural relationship to nicotinamide.

the virtual elimination of tuberculosis sanatoriums, which once were common in the countryside. The initial success in chemotherapy occurred with the introduction of streptomycin, but the real revolution in tuberculosis treatment came with the discovery of isonicotinic acid hydrazide (isoniazid, INH, Figure 15.11), a nicotinamide derivative virtually specific for mycobacteria. This agent is not only effective and free from toxicity, but inexpensive and readily absorbed when given orally. Although the mode of action of INH is not completely understood, it is known that it affects in some way the synthesis of mycolic acid by *Mycobacterium* (mycolic acid is a complex lipid which complexes to the peptidoglycan of the mycobacterial cell wall, see Section 19.31). INH may act by mimicking the activity of a structurally related molecule, nicotinamide (Figure 15.11), becoming incorporated in place of nicotinamide, thus inactivating enzymes requiring this compound for activity. Treatment of mycobacteria with very small amounts of INH (as little as 5 picomoles per 10^9 cells) results in complete inhibition of mycolic acid synthesis, and continued incubation results in a complete loss of outer membrane areas of the cell, a loss of cellular integrity, and death. Following treatment with INH, mycobacteria lose their acid-alcohol fastness, in keeping with the role of mycolic acid in this staining property (see Section 19.31). Resistance to INH is uncommon.

Other pathogenic mycobacteria Another species of the genus *Mycobacterium, M. bovis*, is pathogenic for both man and other animals. A common pathogen of dairy cattle, *M. bovis* enters humans via the intestinal tract typically from the ingestion of raw milk. After a localized intestinal infection, the organism eventually spreads to the respiratory tract and initiates the classic symptoms of tuberculosis. The question as to whether *M. bovis* is really a different organism than *M. tuberculosis* has recently arisen because of the nearly 100 percent homology observed in hybridization studies of the two organisms' DNA. Pasteurization of milk and elimination of diseased cattle have essentially eradicated bovine-to-human transmission of tuberculosis.

M. leprae is the causative agent of the ancient and dreaded disease, *leprosy* (Hansens disease). Unfortunately, *M. leprae* has never been grown on artificial media. It can be grown in mice but the typical human symptoms of leprosy are not observed. The only experimental animal which has been successfully used is the armadillo. The symptoms of leprosy are the characteristic folded, bulb-like lesions on the body, especially on the face and extremities (Figure 15.12) due to growth of *M. leprae* cells in the skin. In severe cases the disfiguring lesions lead to destruction of peripheral nerves and loss of motor function. Leprosy can be treated with the drug dapsone (disulfone 4,4′-sulfonylbistenzamine), although, as for tuberculosis, extended drug therapy is required to effect a cure.

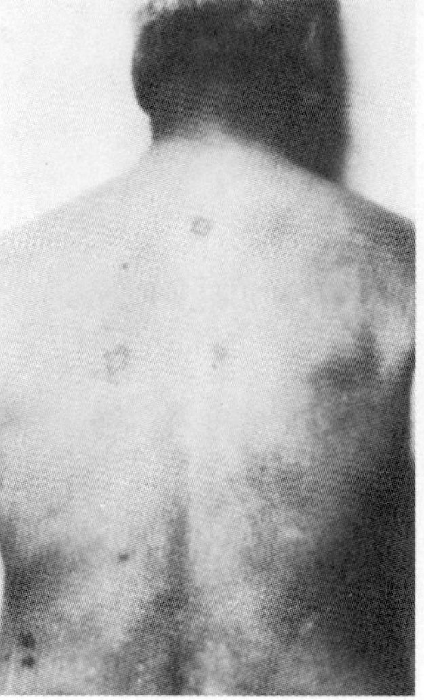

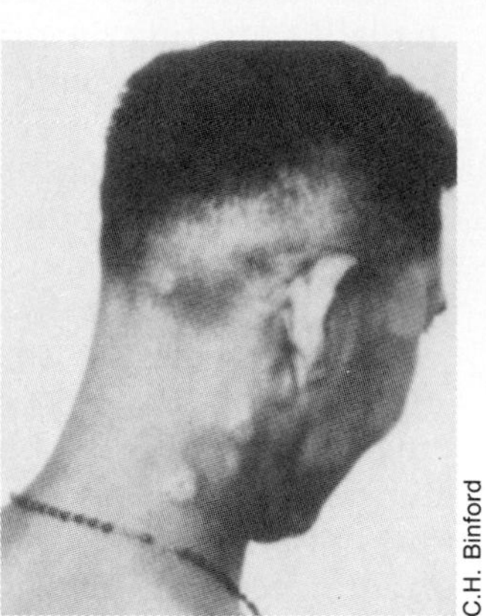

Figure 15.12 Leprosy lesions on the skin due to infection with *Mycobacterium leprae*.

Leprosy is not a highly contagious disease. Little is known of the pathogenicity of *M. leprae* or even of its mode of transmission, though it would seem likely that it is transmitted by direct contact. The bacterium grows within macrophages, causing an intracellular infection which can result in an enormous population of bacteria within the skin. In many areas of the world the incidence of leprosy is very low, although in ancient times it was apparently much more common, perhaps due to crowding and poor sanitation. In other areas, such as tropical areas, the incidence is higher; leprosy remains a medical problem for some 15 million people worldwide.

15.4 Respiratory Infections: Viral

As was pointed out in Chapter 6, viruses, by their very nature, are less easily controlled by chemotherapeutic means than bacteria or other living microorganisms. Since the growth of viruses is intimately tied to host cell functions, it is difficult to specifically attack viruses by medical therapy without causing at least some harm to host cells as well. Not surprisingly, therefore, we see that the most prevalent infectious

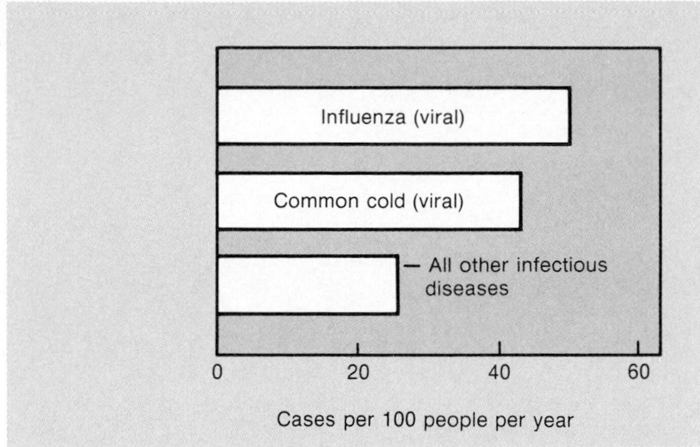

Figure 15.13 Viruses are the leading causes of acute illness in the United States. The data are typical of recent years.

diseases today are of viral etiology (Figure 15.13). On the other hand, many viral diseases are acute, self-limiting infections that are rarely fatal in the normal healthy adult. In addition, serious viral diseases such as smallpox and rabies have been effectively controlled by immunization. We begin here by describing two widely experienced viral infections, the common cold and influenza, and proceed to discuss measles, mumps, and chickenpox; these viral diseases all share in common the transmission of infectious droplets by an airborne route.

The common cold The common cold is one of the most prevalent diseases of children and adults. The symptoms include rhinitis (inflammation of the nasal region, especially the mucous membranes), nasal obstruction, watery nasal discharges, and a general feeling of malaise. Rhinoviruses (single-stranded RNA viruses of the picornavirus group, see Section 6.17), are the dominant etiologic agents of the common cold. Over 100 different serotypes of rhinoviruses are known and hence long-term immunity to the common cold via vaccination or previous exposure is not to be expected. Another group of single-stranded RNA viruses, the coronaviruses, are responsible for about 20 percent of all colds in adults.

Although airborne transmission of the common cold virus is suspected of being a major means of spreading the infection, transmission experiments with human volunteers suggest that direct contact and/or fomite transmission (see Section 14.5) is also an important means of transmission. Little can be done to effectively treat the common cold. However, rest and the ingestion of fluids is thought to promote a quicker recovery. Symptoms of the common cold, especially nasal discharges, can be moderated by a variety of antihistamine drugs.

Influenza Influenza is caused by an RNA virus of the orthomyxovirus group (see Section 6.18). Influenza virus is an enveloped structure, the RNA genome being surrounded by an envelope made up of protein, a lipid bilayer, and external glycoproteins (Figure 15.14; see also Figure 6.46). Human influenza virus exists in nature only in humans. It is transmitted from person-to-person through the air, primarily in droplets expelled during coughing and sneezing. The virus infects the mucous membranes of the upper respiratory tract and occasionally invades the lungs. Symptoms include a lowgrade fever for 3 to 7 days, chills, fatigue, headache, and general aching. Recovery is usually spontaneous and rapid. Most of the serious consequences of influenza infection do not occur because of the viral infection, but because bacterial invaders may be able to set up secondary infections in persons whose resistance has been lowered. Especially in infants and elderly people, influenza is often followed by bacterial pneumonia; death, if it occurs, is usually due to the bacterial infection.

Influenza often occurs in pandemics. Early pandemics, of which the one in 1918 is the most famous, occurred before knowledge was sufficiently advanced to make careful analysis possible, but the 1957 pandemic of the so-called Asiatic flu provided an opportunity for careful study of how a worldwide epidemic develops (Figure 15.15). The epidemic probably arose when a mutant virus strain of marked virulence and differing from all previous strains in antigenicity, appeared in the population. Since immunity to this strain was not present, the virus was able to advance rapidly throughout the world. It first appeared in the interior of China in late February 1957 and by early April had been brought to Hong Kong by refugees. It spread from Hong Kong along air and naval routes and was apparently transferred to San Diego, California, by naval ships. An outbreak occurred in Newport, Rhode Island, on a naval vessel in May. Other out-

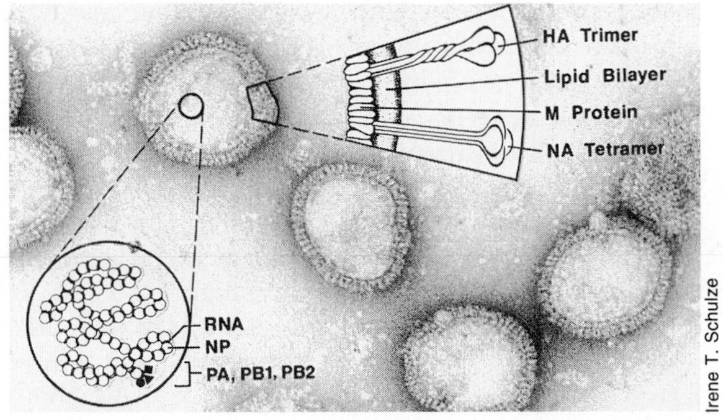

Figure 15.14 Electron micrograph of the influenza virus, showing the location of the major viral coat proteins and the nucleic acid. Magnification, 300,000×. HA, hemagglutinin, three copies make up the HA coat spike; NA, neuraminidase, four copies make up the NA coat spike; M, coat protein; NP, nucleoprotein; PA, PB1, PB2, other internal proteins some of which may have enzymatic functions.

Figure 15.15 Route of spread of a major influenza epidemic, the Asian flu pandemic of 1957.

breaks occurred in various parts of the United States. Peak incidence occurred in the last two weeks of October, during which time 22 million new cases developed. From this period on there was a progressive decline.

The genetic material of influenza, RNA, is arranged in a highly unusual manner. As discussed in Section 6.18, the influenza virus genome is *segmented,* with genes found on each of eight distinct fragments of its single-stranded RNA (Figure 6.46). Such an arrangement facilitates the reassortment of genes with genes from a different strain of influenza virus, since more than one strain of influenza virus can infect a cell at one time. This allows recombinant viruses to crop up at frequent intervals and is probably responsible for the generally unsuccessful attempts to control influenza by vaccination. This reassortment of genes in different strains of influenza virus usually manifests itself in the phenomenon called *antigenic shift*. Antigenic shift refers to modifications in the protein coat of the virions, especially to two proteins important in the attachment and eventual release of virus from host cells, hemagglutinin and neuraminidase, respectively (see Figure 15.14). Immunity to influenza in humans is largely dependent on the production of secretory antibody (IgA, see Section 12.2), especially to antigenic determinants of the hemagglutinin and neuraminidase proteins.

Once a strain of influenza virus has passed through the population, a majority of the people will be immune to that strain and it will be impossible for a strain of similar antigenic type to cause an epidemic for about three years. There is some suggestion that in 1918 (the year of an unusually serious epidemic), the strain may have originated from a related virus which infects swine (swine flu), and the 1957 strain may have arisen from a similar animal reservoir (perhaps a wild animal) somewhere in Asia. Although a vaccine can be prepared to any strain, the large number of strains and the phenomenon of antigenic shift make it difficult to prevent influenza epidemics.

Measles Measles (rubeola) virus causes an acute highly infectious childhood disease characterized by nasal discharges, redness of the eyes, cough, and fever. The measles virus is a paramyxovirus (see Section 6.18), and enters the body in the nose and throat by airborne transmission and quickly leads to systemic viremia. As the disease progresses the fever and cough intensify and a rash appears; in most cases measles lasts a total of 7–10 days. Circulating antibodies to measles virus are measurable about 5 days after initiation of infection, and the activities of both serum antibodies and cytotoxic T-lymphocytes (see Section 12.10) combine to eliminate the virus from the system. As a consequence of measles, a variety of complications may occur, including inner ear infection, pneumonia, and, in rare cases, measles encephalomyelitis. The latter can cause neurological disorders and a form of epilepsy. Measles encephalomyelitis is a serious complication of measles infection because it has a mortality rate of nearly 20 percent.

Although once a common childhood illness, measles generally occurs nowadays in rather isolated outbreaks because of widespread vaccination programs begun in the mid 1960s (Figure 15.16*a*). In the United States most school systems require proof of measles vaccination before allowing children to enroll. The potential complications and highly infectious nature of measles mandates that this disease be tightly con-

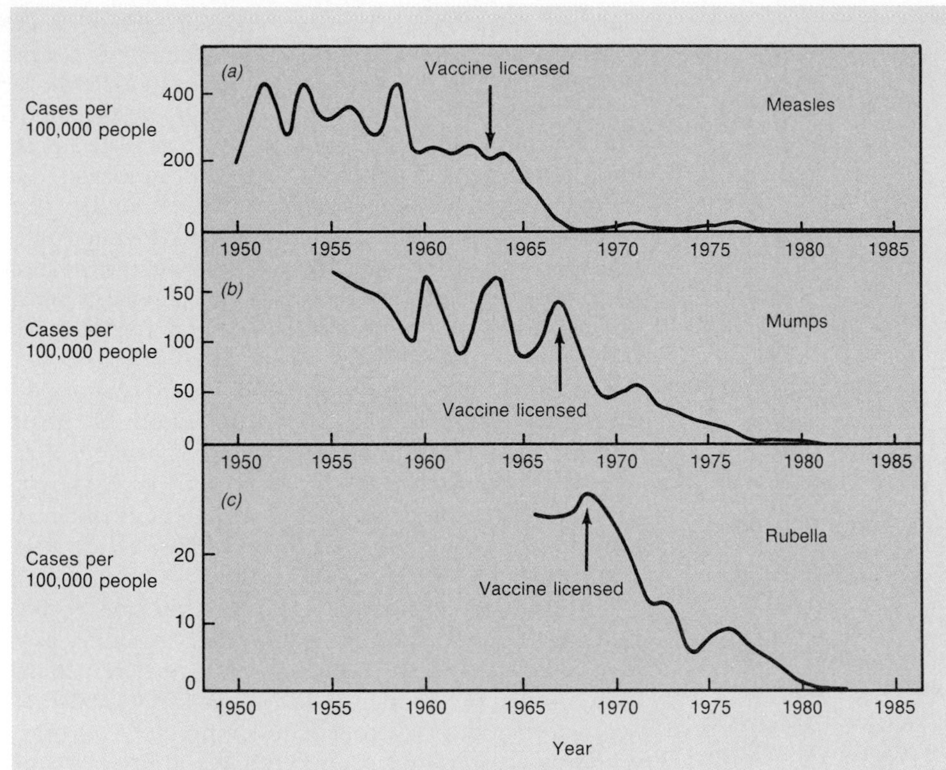

Figure 15.16 The effect of vaccines on the incidence in the United States of the major childhood virus diseases now controlled by the MMR (measles, mumps, rubella) vaccine. (a) Measles. (b) Mumps. (c) Rubella.

trolled. Active immunization is affected with the MMR (measles, mumps, and rubella) vaccine (see Tables 12.7 and 12.8). A childhood case of measles generally confers life-long immunity.

Mumps Mumps is caused by a different paromyxovirus than that causing measles, but shares with the disease measles a highly infectious character. Mumps is spread by airborne droplets, and the disease is characterized by inflammation of the salivary glands leading to swelling of the jaws and neck. The virus spreads through the blood stream and may infect other organs including the brain, testes, and pancreas. The host immune response produces antibodies to mumps virus surface proteins and this generally leads to a quick recovery. A live, attenuated mumps vaccine is highly effective in preventing the disease (see Figure 15.16*b*). Hence, like measles, the incidence of mumps has also been greatly reduced in the last two decades, with mumps epidemics usually restricted to those individuals who for one reason or another did not receive the MMR vaccine during childhood.

Rubella Rubella (*German measles*) is caused by an RNA virus of the togavirus group (see Section 6.17). The symptoms of the disease resemble those of measles, but are generally milder. German measles is less contagious than true measles and thus a good proportion of the population has never been infected. However, during the first 3 months of pregnancy rubella virus can infect the fetus by placental transmission and cause a host of serious fetal abnormalities. Thus, it is important that pregnant women not be exposed to, vaccinated against, or contract German measles during this period. Rubella can cause stillbirth, or deafness, heart and eye defects, and brain damage in live births. For this reason routine childhood vaccination against German measles should be practiced. An attenuated, live virus vaccine is administered with attenuated measles and mumps viruses in the MMR vaccine mentioned above (see Figure 15.16*c*).

Chickenpox and shingles Chickenpox (varicella) is a common childhood disease caused by a Herpes virus. Chickenpox is highly contagious and is transmitted by infectious droplets, especially when susceptible individuals are in close contact. The incidence of chickenpox shows a disease cycle typical of a respiratory infection (Figure 15.17). In school children, for example, close confinement during the winter months leads to the spread of chickenpox by airborne secretions from infected classmates and through contact with contaminated fomites. The virus enters the respiratory tract, multiplies, and is quickly disseminated via the bloodstream resulting in a systemic papular rash that quickly heals, rarely leaving disfiguring marks. For unknown reasons, most chickenpox vaccines have been poorly immmunogenic, but a successful and highly protective vaccine has just been marketed in Japan.

Varicella virus can apparently relocate to nerve cells and exist in a latent (inapparent) infection for years. The virus occasionally migrates from these res-

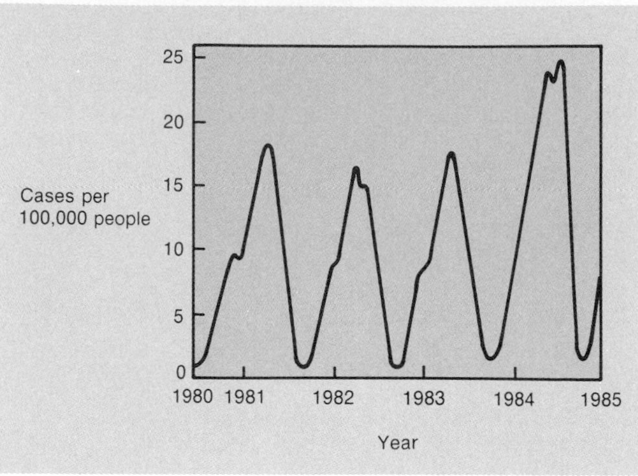

Figure 15.17 Reported cases of chickenpox by month in the United States 1981–1984. Note the marked early spring seasonality of chickenpox typical of a respiratory disease.

ervoirs to the skin surface causing a painful skin infection referred to as **shingles** (zoster). Shingles most commonly strikes immune suppressed individuals or the elderly. Studies with human volunteers suggest that T-cells are important in destroying the virus. The prophylactic use of human hyperimmune globulin prepared against the virus is useful for preventing the onset of symptoms of shingles. Such therapy is only advised for patients where secondary infections occasionally associated with shingles, such as pneumonia or encephalitis, may be life threatening.

15.5 Why Are Respiratory Infections So Common?

Although we have described a number of different respiratory infections, some bacterial and some viral, what do they have in common? Respiratory infections are generally transmitted by airborne droplets. The infectious agents occur in finely dispersed form in air, dust, or on the surfaces of fomites, and the diseases typically show marked seasonality. In addition, because they are easily transmitted by normal human activities, respiratory infections are the most common infectious diseases observed today (at least in developed countries).

Respiratory infections are frequently difficult to control from a public health standpoint. There are many reasons for this. First, many respiratory infections are caused by viruses; antibiotics, and indeed most other therapeutic agents, are completely ineffective in controlling viral diseases. Two, most respiratory diseases are acute, showing a rapid onset of symptoms and generally a rapid recovery. Although this may seem desirable, in many ways it isn't. Patients suffering from the common cold, for example, are frequently communicable before symptoms suggest they are actually ill; chickenpox can spread in explosive fashion among school children for the same reason. Third, effective vaccines are not available for a number of respiratory diseases. This can be due to unusual molecular biological properties of the pathogen, such as the antigenic drift and genomic rearrangements of the influenza virus that essentially generate new strains (from an immunological standpoint) every few years, or because vaccines against certain respiratory pathogens, for example against *Streptococcus pyogenes*, would actually harm the host due to cross-reacting antigens on bacterial and host cell surfaces. Fourth, many respiratory infections are transmitted from healthy carriers to susceptible hosts. This is very common in the case of streptococcal infections, because about 20 percent of the adult population carry, apparently without ill effect, β-hemolytic strains of *Streptococcus pyogenes* in their nose and throat.

Respiratory diseases therefore represent a difficult challenge for public health officials. Since many respiratory diseases cause discomfort but are not life threatening, the tendency for many individuals suffering a respiratory illness is to simply "wait it out," while carrying on normal everyday activities. Unfortunately, such individuals are frequently unaware that they are actively infecting others. Because of this and the other problems discussed above, no practical means of control exists for many respiratory infections. It indeed appears that pathogens transmitted by the respiratory route have evolved a highly successful strategy for maintaining themselves in nature.

15.6 Sexually Transmitted Diseases

Many sexually transmitted (venereal) diseases are known and the causative agents can be either bacterial, viral, or protozoan. Table 15.2 summarizes the major sexually-transmitted diseases encountered in medical practice today. Control of sexually-transmitted diseases presents an unusually difficult public health problem. Effective control of venereal infections requires the frank exchange of information between public health officials and those infected. It is frequently necessary to deal with emotionally sensitive issues, such as contacting sexual partners of a diseased individual or attempting to alter firmly ingrained sexual practices. We begin our discussion here with gonorrhea, because, despite the value of antibiotics in treating this disease, the prevalence of inapparent infections and the advent of birth-control pills have made gonorrhea of nearly epidemic proportions (Figure 15.18). Syphilis, on the other hand, is much less of a public health problem, from the standpoint of the number of cases encountered (Figure 15.18). To understand why gonorrhea, but not syphilis, has become epidemic, it is necessary to

Table 15.2 Summary of some sexually transmitted diseases and treatment guidelines

Disease	Causative organisms*	Recommended treatment†
Gonorrhea	*Neisseria gonorrhoeae* (B)	Penicillin (tetracycline or spectinomycin for penicillin-resistant strains)
Syphilis	*Treponema pallidum* (B)	Penicillin
Chlamydia trachomatis infections	*Chlamydia trachomatis* (B)	Tetracycline
Nongonococcal urethritis	*Chlamydia trachomatis* (B) or *Ureaplasma urealyticum* (B)	Tetracycline
Lymphogranuloma venereum	*C. trachomatis* (B)	Tetracycline
Chancroid	*Haemophilus ducreyi* (B)	Erythromycin
Genital herpes	Herpes Simplex Type 2 (V)	No known cure, symptoms can be controlled with topical application of acyclovir (see Figure 15.23).
Anogenital warts (venereal warts)	Papilloma virus (certain strains) (V)	10–25% podophyllin in tincture of benzoin (alternatively, warts can be removed surgically)
Trichomoniasis	*Trichomonas vaginalis* (P)	Metronidazole
Acquired immune deficiency syndrome (AIDS)	Human immunodeficiency virus (V)	None currently available; nucleotide base analog azidothymidine clinically successful in treatment (but not cure)

**B, bacterium; V, virus; P, protozoan.*
†U.S. Department of Health and Human Services, Public Health Service Recommendations. From: Sexually transmitted diseases treatment guidelines. Morbidity and Mortality Weekly Report, Suppl. 31(25), August 20, 1982.

briefly consider the host-parasite relationships of these two infections.

Gonorrhea Gonorrhea is one of the most widespread human diseases, and in spite of the availability of excellent drugs it is still a common disease even in countries where the cost of drugs is no economic problem. As opposed to syphilis, gonococcal infections rarely result in serious complications or death. The disease symptoms of gonorrhea are quite different in the male and female. In the female the symptoms are usually a mild vaginitis that is difficult to distinguish from vaginal infections caused by other organisms, and the infection may easily go unnoticed; in the male, however, the organism causes a painful infection of the urethral canal (refer to Figure 11.11 for a description of the genitourinary tract). The organism is killed quite rapidly by drying, sunlight, and ultraviolet light; this extreme sensitivity probably explains in part the venereal nature of the disease, the organism being transmitted from person-to-person only by intimate direct contact. In addition to gonorrhea, the organism also causes eye infections in the newborn and adult. Infants may become infected in the eyes during birth to diseased mothers. Prophylactic treatment of the eyes of all newborns with silver nitrate or an ointment containing penicillin is generally mandatory and has helped to control the disease in infants.

We discussed the clinical microbiology of the

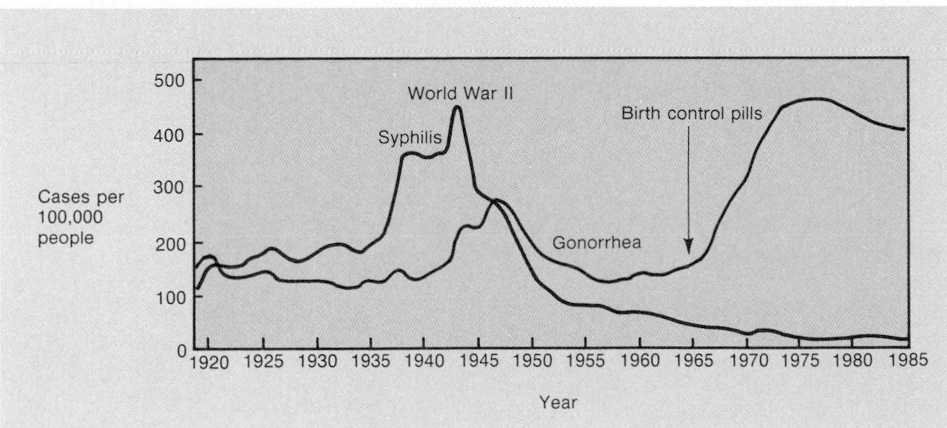

Figure 15.18 Sexually transmitted diseases: reported cases of gonorrhea and syphilis per 100,000 population, United States. After 1940, military cases are excluded.

causal agent, *Neisseria gonorrhoeae*, in Section 13.1, and the general bacteriology of the genus *Neisseria* is discussed in Section 19.21.

N. gonorrhoeae enters the body by way of the mucous membranes of the genitourinary tract (Figure 15.19*a*), customarily being transmitted from a member of the opposite sex. Treatment of the infection with penicillin has been successful in the past, with a single injection usually resulting in elimination of the organism and complete cure. For many years, all isolates of *N. gonorrhoeae* were found to be sensitive to penicillin or ampicillin (a semisynthetic penicillin), and sensitivity testing was not necessary. However, strains of *N. gonorrhoeae* resistant to penicillin are now widespread, and this resistance is due to a plasmid-encoded penicillinase. The incidence of **penicillinase-producing** *N. gonorrhoeae* has increased dramatically since the discovery of these strains in the mid-1970s (Figure 15.19*b*). Fortunately, however, the majority of penicillinase-producing strains respond to alternative antibiotic therapy, with spectinomycin being the drug of choice in most cases. Where penicillin therapy is effective, it generally results in a more rapid cure in males than in females.

Despite the ease with which gonorrhea can be cured, the incidence of gonococcus infection remains relatively high. The reasons for this are threefold: (1) acquired immunity does not exist, hence repeated reinfection is possible (whether this is due to lack of local immunity or to the fact that at least 16 distinct serotypes of *N. gonorrhoeae* have been isolated is not known); (2) the widespread use of oral contraceptives. The latter cause a mimicking of the pregnant state, which results, among other things, in a lack of glycogen production in the vagina and a raising of the vaginal pH. Lactic acid bacteria, normally found in the adult vagina (see Section 11.5) fail to develop under such circumstances, and this allows *N. gonorrhoeae* transmitted from an infected partner to colonize more easily than in an acidic vagina containing lactobacilli; and (3) symptoms in the female are frequently such that the disease may go totally unrecognized, and a promiscous infected female can serve as a reservoir for the infection of many males. The disease could be controlled if the sexual contacts of infected persons were quickly identified and treated, but it is often difficult to obtain the necessary information and even more difficult to arrange treatment. The incidence of gonorrhea correlates closely with the promiscuity of the society; elimination of this disease is extremely difficult because, despite notable advances in drug therapy, gonorrhea remains a social rather than a medical problem.

Syphilis The venereal disease syphilis is potentially much more serious than gonorrhea, but because of differences in pathobiology, it has been better controlled (Figure 15.18). Syphilis is caused by a spirochete, *Treponema pallidum*, an obligate anaerobe that has been very difficulty to cultivate (Figure 15.20). We mentioned the clinical immunology and diagnostic methods for syphilis in Table 12.9, and

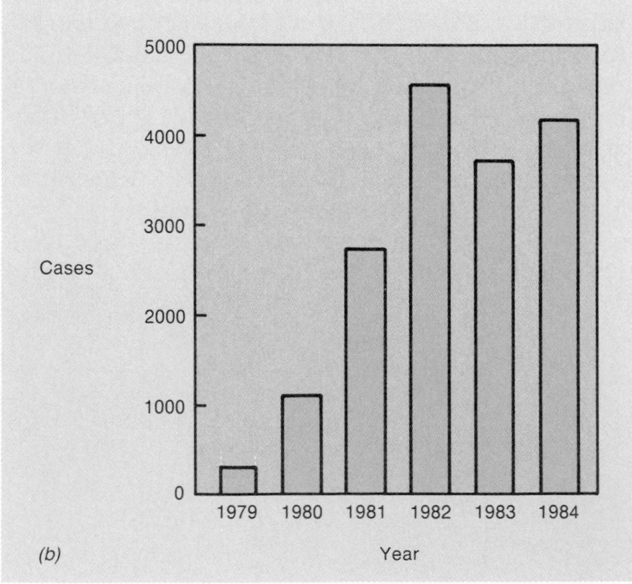

Figure 15.19 The causative agent of gonorrhea, *Neisseria gonorrhoeae,* and the incidence of penicillinase production in this organism. (a) Scanning electron micrograph of the microvilli of human fallopian tube mucosa showing how cells of *N. gonorrhoeae* attach to the surfaces of epithelial cells. Note the distinct diplococcus morphology. Magnification, 10,000 ×. (b) Reported penicillinase-producing *Neisseria gonorrhoeae* (PPNG), 1979–1984.

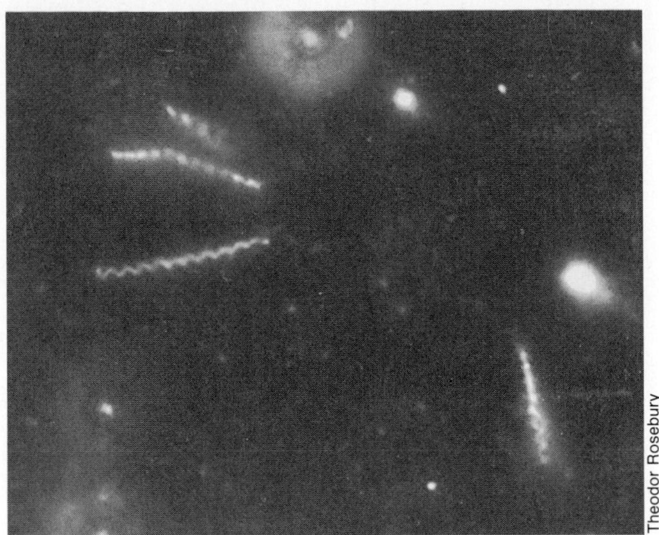

Figure 15.20 The spirochete of syphilis, *Treponema pallidum* observed by dark field microscopy of an exudate. *T. pallidum* cells measure 0.15 μm × 10–15 μm.

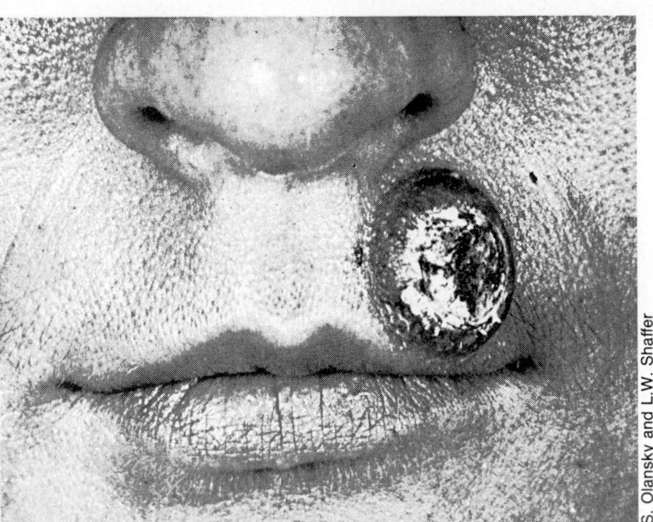

Figure 15.21 Primary syphilis lesion.

the biology of the spirochetes is discussed in Section 19.12.

The disease syphilis exhibits variable symptoms. The organism does not pass through unbroken skin, and initial infection most probably takes place through tiny breaks in the epidermal layer. In the male, initial infection is usually on the penis, whereas in the female it is most often in the vagina, cervix, or perineal region. In about 10 percent of the cases infection is extragenital, usually in the oral region. During pregnancy, the organism can be transmitted from an infected woman to the fetus; the disease acquired in this way by an infant is called **congenital syphilis**. *T. pallidum* multiplies at the initial site of entry and a characteristic *primary* lesion known as a **chancre** (Figure 15.21) is formed within 2 weeks to 2 months. Dark field microscopy of the exudate from syphilitic chancres often reveals the actively motile spirochetes. In most cases the chancre heals spontaneously and the organisms disappear from the site. Some, however, spread from the initial site to various parts of the body, such as the mucous membranes, the eyes, joints, bones, or central nervous system, and extensive multiplication occurs. A hypersensitive reaction to the treponeme takes place, which is revealed by the development of a generalized skin rash; this rash is the key symptom of the *secondary* stage of the disease. At this stage the patient's condition may be highly infectious, but eventually the organism disappears from secondary lesions and infectiousness ceases. The subsequent course of the disease in the absence of treatment is highly variable. About one-fourth of the patients undergo a spontaneous cure and another one-fourth do not exhibit any further symptoms, although the infection may persist. In about half of the patients the disease enters the *tertiary* stage, with symptoms ranging from relatively mild infections of the skin and bone to serious or fatal infections of the cardiovascular system or central nervous system. Involvement of the nervous system is the most serious phase of the illness, since generalized paralysis or other severe neurological damage may result. In the tertiary stage only very few organisms are present, and most of the symptoms probably result from hypersensitivity reactions (see Sections 12.10 and 12.11) to the spirochetes.

Penicillin is highly effective in syphilis therapy and the early stages of the disease can usually be controlled by a series of injections over a period of 1 to 2 weeks. In the secondary and tertiary stages treatment must extend for longer periods of time. It is interesting that the incidence of syphilis has remained low during the current epidemic of gonorrhea. Almost certainly, this is due to the fact that syphilis shows similar symptoms in male and female, so that both sexes tend to obtain treatment. The inapparent infections in females, so common in gonorrhea, do not occur with syphilis.

Chlamydial infections Although not officially *reportable* diseases as is the case for gonorrhea or syphilis, it now appears that a host of venereal diseases can be ascribed to the obligate intracellular parasite *Chlamydia trachomatis*, with the total incidence of such diseases probably greatly outnumbering that of gonorrhea or syphilis. *C. trachomatis* also causes a serious disease of the eye called trachoma (see Section 19.23), but the strains responsible for venereal infections consist of a group of *C. trachomatis* serotypes distinct from those causing trachoma. Chlamydial infections may also be transmitted to the newborn from contamination in the birth canal, causing newborn conjunctivitis and respiratory distress.

Chlamydial nongonococcal urethritis (NGU) is

one of the most frequently observed venereal diseases today. *C. trachomatis* causes urethritis in males and urethritis, cervicitis, and pelvic inflammation in females. In both the male and female inapparent chlamydial infections are common, and this undoubtedly accounts for the prevalence of chlamydial infections in sexually active adults. In a small percentage of cases, uncontrolled chlamydial NGU can lead to serious complications, including testicular swelling and prostate inflammation in men and pelvic inflammatory diseases and fallopian tube blockage in women. The latter can cause sterility and nonuteral pregnancies.

Chlamydial NGU is relatively difficult to diagnose by traditional isolation and identification methods. To expedite diagnoses a variety of immunological tests have been developed for identifying *C. tracomatis* from a vaginal/pelvic swab or from discharges. These clinical tests include fluorescent monoclonal antibodies and various ELISA tests for detecting specific *C. trachomatis* antigens. If a chlamydial infection is suspected, treatment is begun with tetracycline or erythromycin; penicillin is ineffective against most strains of *C. trachomatis* because the organisms lack peptidoglycan, the target of penicillin (see Section 19.23).

Chlamydial NGU is frequently observed as a secondary event following gonorrhea infection. If both *N. gonorrhoeae* and *C. trachomatis* are transmitted to a new host in a single event, treatment of gonorrhea with penicillin is usually successful, but does not eliminate the chlamydia. Although cured of gonorrhea, such patients are still communicable for chlamydia and eventually experience an apparent recurrence of gonorrhea which is instead a case of NGU. It is therefore advisable for patients undergoing gonorrhea therapy to be checked for complete recovery from symptoms, because a latent chlamydial infection will require additional chemotherapy.

Lymphogranuloma venereum is a venereal syndrome also caused by *C. tracomatis*. The disease, which occurs most frequently in males, consists of a swelling of the lymph nodes in and about the groin. From the infected lymph nodes chlamydial cells may travel to the rectum and cause a painful inflammation of rectal tissues called *proctitus*. Because of the potential for regional lymph node damage and the complications of proctitis, lymphogranuloma verereum is considered to be one of the most serious sexually transmitted chlamydial syndromes.

Herpes We discussed the molecular biology of herpes viruses in Section 6.21. Herpes simplex virus infections are responsible for a number of blisterlike conditions in humans. **Herpesvirus type 1** (HV1) is generally associated with cold sores and fever blisters in and around the mouth and lips (Figure 15.22). The incubation period of HV1 infections is short (3 to 5 days) and the lesions heal without treatment in 2 to 3 weeks. Relapses of HV1 infections are relatively common and it is thought that the virus is spread primarily via the respiratory route. Latent herpes infections are apparently quite common with the virus persisting in low numbers in nerve tissue. Recurrent acute herpes infections may thus be due to a periodic triggering of virus activity.

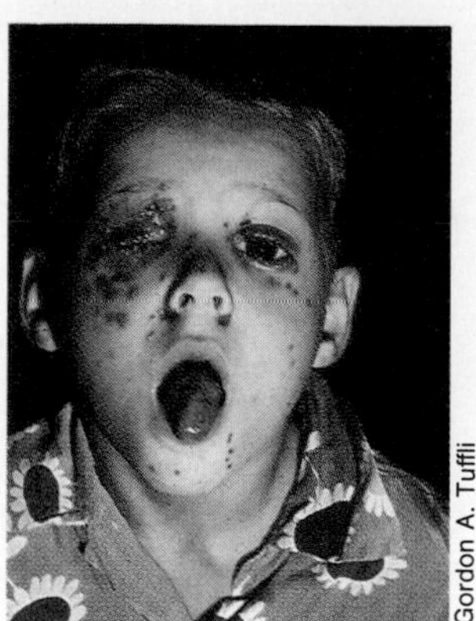

Figure 15.22 A severe case of Herpes blisters due to infection with Herpesvirus type I.

Herpesvirus type 2 (HV2) infections are associated primarily with the anogenital region, where the virus causes painful blisters on the penis of males or the cervix, vulva, or vagina of females. HV2 infections are transmitted by direct sexual contact, and the disease is most easily transmitted during the active blister stage rather than during periods of inapparent (presumably latent) infection.

Genital herpes infections are incurable at the present time, although a limited number of drugs have been found successful in controlling the infectious blister stages. The guanine analog **acyclovir** (Figure 15.23) is particularly effective in limiting the shed of active virus from blisters and promoting the healing of blisterous lesions. Acyclovir acts by specifically interfering with herpesvirus DNA polymerase, hence inhibiting viral DNA replication.

The overall health implications of genital herpes infections are not yet understood, but epidemiological studies have shown a significant correlation between genital herpes infections and cervical cancer in females. In addition, herpesvirus type 2 can be

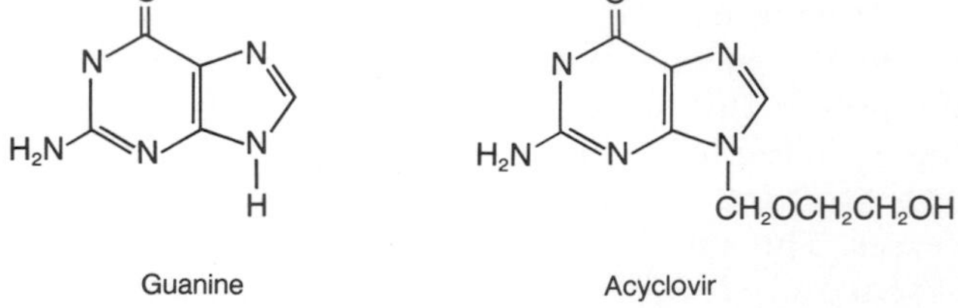

Guanine

Acyclovir ($CH_2OCH_2CH_2OH$ side chain)

Figure 15.23 Structure of guanine and the guanine analog acyclovir. Acyclovir has been used therapeutically to control genital herpes (HV2) blisters. See text for mode of action.

Herpes Simplex Virus in Shakespeare

Romeo and Juliet: Act I, Scene IV
Mercutio O! then, I see, Queen Mab hath been with you.
Benvolio Queen Mab! What's she?
Mercutio She is the fairies' midwife . . .
. . . she gallops night by night
Through lovers' brains, and then they dream of love . . .
O'er ladies lips, who straight on kisses dream;
Which oft the angry Mab with blisters plagues.

transmitted to the newborn at birth by contact with herpetic lesions in the birth canal. The disease in the newborn varies from latent infections with no apparent damage to systemic disease which can result in death. Severely affected infants who survive may suffer permanent brain damage. To avoid infection in the newborn, delivery by caesarean section is advised for pregnant women showing genital herpes lesions.

Trichomoniasis Nongonococcal urethritis may also be caused by infections with the protozoan *Trichomonas vaginalis*. Although many protozoa produce resting cells called *cysts, T. vaginalis* does not produce cysts and transmission must be person-to-person, generally by sexual intercourse. However, cells of *T. vaginalis* can survive a few hours outside the host, provided they do not dry out. Thus transmission of *T. vaginalis* by contaminated toilet seats, sauna benches, and paper towels occasionally occurs. *T. vaginalis* infects the vagina in women, the prostate and seminal vesicles of men, and the urethra of both males and females.

Many cases of trichomoniasis are totally asymptomatic. In fact, asymptomatic cases are the rule in males. In women trichomiasis is characterized by a vaginal discharge, vaginitis, and painful urination. The organism is commonly found in females, surveys indicating from 25–50 percent of sexually active women being infected; only about 5 percent of men are infected. Trichomoniasis is diagnosed by preparing a wet mount of fluid discharged from the patient and examining microscopically for motile protozoa. The antiprotozoal drug *metronidazole* is particularly effective in treating trichomoniasis.

It is important in the control of trichomoniasis that the male partner of an infected female be examined for *T. vaginalis* and treated if necessary since promiscuous asymptomatic males can serve as reservoirs, transmitting the infection to several females. Transmission of *T. vaginalis* between either infected partner can be prevented by the use of condoms.

15.7 Acquired Immune Deficiency Syndrome (AIDS)

Because of its fatal consequences, the disease AIDS has received considerable attention since it was first recognized as a clinical syndrome in 1981. We discussed the epidemiology of AIDS in Section 14.4 and the clinical diagnosis of AIDS in Sections 13.8 and 13.10. At present, over 50,000 cases of AIDS have been reported in the United States alone and the outlook is for this number to rise dramatically unless an effective treatment becomes available soon.

AIDS was first suspected of being a disease of the immune system when a startling increase in the number of so-called "opportunistic" infections was observed. Opportunistic infections are defined as infections rarely observed in humans with normal immune responses (see Section 11.7 for a discussion of the *compromised host*). Opportunistic pathogens in effect take advantage of the "opportunity" afforded by reduced host resistance to initiate an infection that would otherwise be checked by the host's immune system. The occurrence of opportunistic infections in otherwise healthy individuals was one of the first epidemiological clues that a new immune disorder was appearing in the population, and indeed opportunistic infections are important in the overall clinical picture of AIDS because they are the leading cause of death in AIDS patients. The most common AIDS-associated opportunistic infections include pneumonia caused by the protozoan *Pneumocystis carinii* (Figure 15.24*a*) and other protozoal infections such as cryptosporidiosis, caused by *Cryptosporidium* species (Figure 15.24*b*), and toxoplasmosis, caused by *Toxoplasma gondii* (Figure 15.14*c*), systemic yeast infections due to *Cryptococcus neoformans* (Figure 15.24*d*), *Candida albicans* (Figure 15.24*e*), and *Histoplasma capsulatum* (Figure 15.24*f*), viral infections due to herpes simplex (Figure 6.51*a*) or cytomegalovirus, tuberculosis and other mycobacterial infections (Figure 15.24*g*), and enteric helminthic infections due to *Strongyloides stercoralis* (Figure 15.24*h*). *Pneumocystis* pneumonia is by far the most common opportunistic pathogen encountered, being observed in nearly two-thirds of all cases of clinical AIDS.

Besides the opportunistic infections associated with many AIDS cases, a rare form of cancer called *Kaposi's sarcoma* (Figure 15.25) is also observed in many AIDS patients. Kaposi's sarcoma is a cancer of the cells lining blood vessel walls and is diagnosed from the purplish patches it leaves on the surface of the skin (Figure 15.25). The incidence of Kaposi's sarcoma is much higher in homosexual men suffering from AIDS than in the other high risk groups identified (see Table 14.2), but the reasons for this are not known.

The disease AIDS is caused by human immuno-

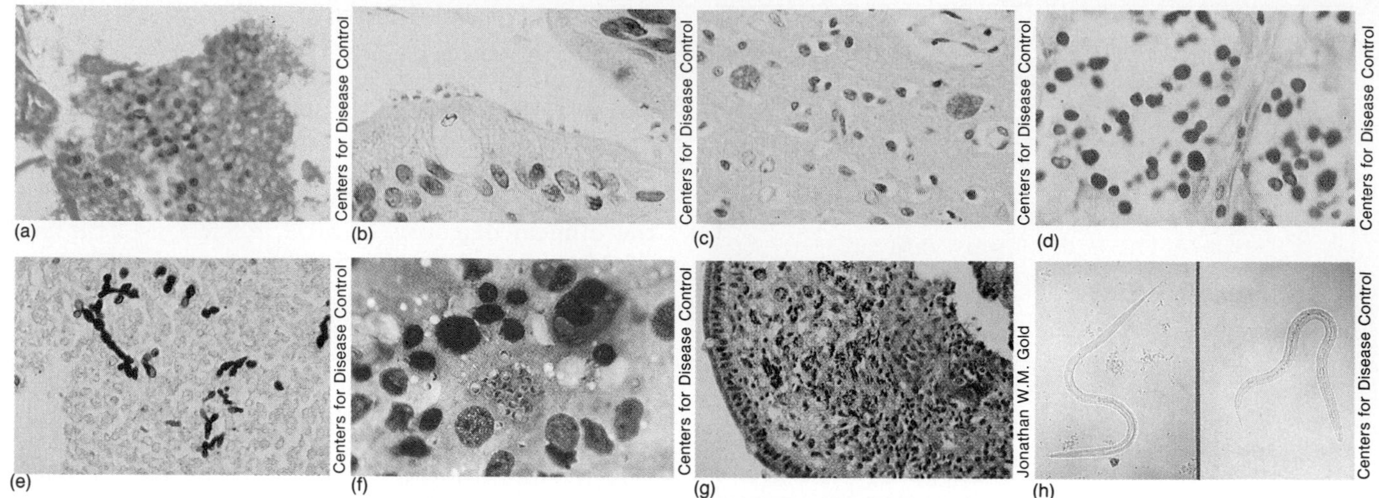

Figure 15.24 Opportunistic pathogens associated with cases of acquired immunodeficiency disease (AIDS). (a) *Pneumocystis carinii*, from patient with pulmonary pneumocystosis. (b) *Cryptosporidium* sp., from biopsy of small intestine. (c) *Toxoplasma gondii*, from brain tissue of patient with toxoplasmosis. (d) *Cryptococcus neoformans*, from liver tissue of patient with cryptococcosis. (e) *Candida albicans*, from heart tissue of patient with systemic *Candida* infection. (f) *Histoplasma capsulatum*, from liver tissue of patient with histoplasmosis. (g) Mycobacterial infection of small bowel, acid-fast stain. (h) *Strongyloides stercoralis*, filariform larvae.

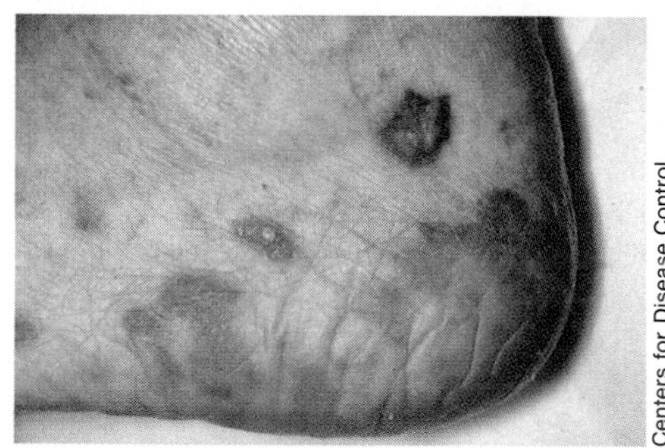

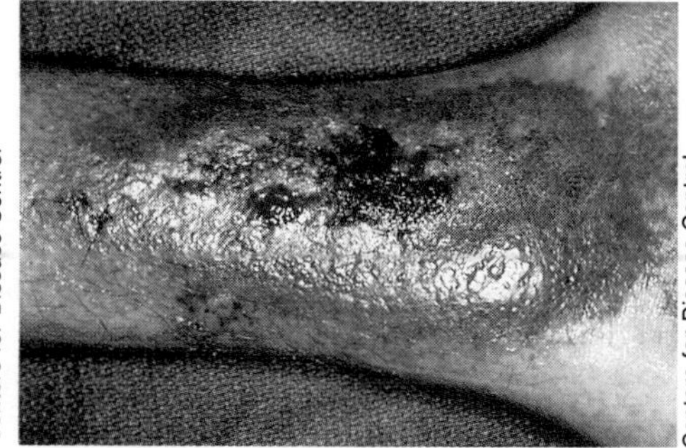

Figure 15.25 Kaposi's sarcoma lesions as they appear on (a) the heel and lateral foot, and (b) the distal leg and ankle.

deficiency virus (HIV), a retrovirus (see Section 6.24 and Figure 6.54*a*). HIV has been variously called human T-lymphotrophic virus type III (HTLV-III) isolated by an American group, and lymphoadenopathy associated virus (LAV), isolated by a French group, but the accepted name is now **human immunodeficiency virus (HIV)**. Although retroviral infections in humans are quite rare, in the United States it is estimated that HIV has infected over two million people. In central Africa, especially in Zaire where HIV is thought to have originated, 10–20 percent of the general population is infected with HIV. However, *infection* with HIV is *not* synonomous with the *disease* AIDS. Of those infected with HIV only 10–30 percent appear to succumb to full-blown clinical AIDS, which suggests that a large number of healthy HIV carriers exist in which the virus is kept in check by the activities of the immune system. Shortly after infection with HIV, circulating antibodies specific for HIV antigens appear. We discussed in Section 13.8 the means by which HIV exposure is detected clinically using ELISA and Western Blot techniques. But in certain individuals exposed to HIV the immune system becomes progressively crippled due to a suppression by HIV of a specific group of T-lymphocytes, and this state of immune incompetence allows the establishment of opportunistic infections and cancers that invariably kill the individual.

With this introduction let us now turn our attention to the disease process itself and see at the molecular and cellular levels how HIV effectively dismantles the human immune system.

HIV T-lymphocyte interactions Human immunodeficiency virus (HIV) has the specific capacity to infect the T4-class of T-lymphocytes; infected T4 lymphocytes produce high numbers of spherical particles, about 90–120 nm in diameter (Figure 15.26). Blood monocytes can also be infected by HIV without any apparent ill effect and this may serve as an HIV reservoir for infecting other cells. Recall that T4 lymphocytes include the T-helper and T-delayed type hypersensitivity subsets (Section 12.3). T4 cells respond to foreign antigens bound in association with

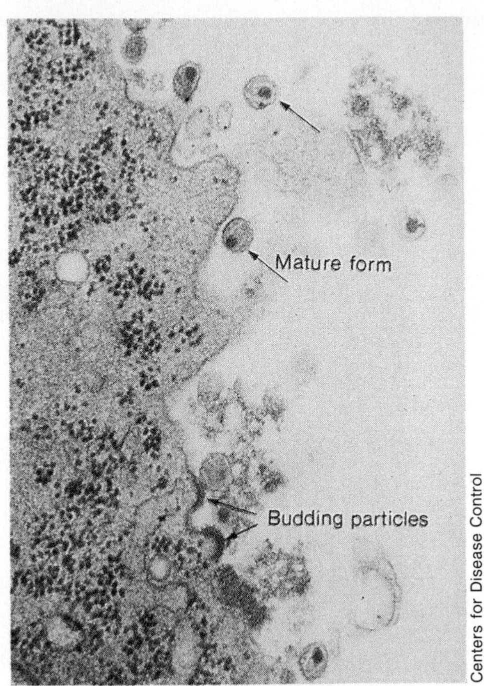

Figure 15.26 Transmission electron micrograph of a thin section of a T4 lymphocyte releasing AIDS virus particles. Cells were from a hemophiliac patient who developed AIDS. Particles are 90–120 nm in diameter.

class II major histocompatability (MHC) antigens on the surface of macrophages (see Section 12.4). Once activated by a properly presented antigen, normal T-helper cells secrete lymphokines, helper factors, and other protein substances that stimulate particular clones of B-lymphocytes to multiply and secrete antibody specific for the given antigen (see Section 12.5). HIV specifically infects T cells of the T4 class; it has been shown that the T4 antigen (the identifying marker on the surface of T4 cells) serves as the cell surface receptor molecule for HIV. In T-cell cultures, HIV infection initiates events leading to the depletion of most T4 cells in the mixture. But why is the immune system unable to eliminate HIV?

Although highly immunogenic, HIV may escape elimination by the host's immune system because of its propensity to undergo antigenic drift (see Section 15.4), the process of high-frequency mutational events that sufficiently modifies the surface structure of the virus to make previously formed antibodies immunologically impotent. The initial immune response of an HIV infected individual is thus insufficient to prevent continual infection by the virus. In up to 30 percent of those infected, however, the immune system becomes deranged and leads to the clinical symptoms of AIDS. What occurs in these individuals to effectively destroy their immune systems?

A clue to answering this question emerges from a consideration of the opportunistic infections observed in clinical cases of AIDS. T4 lymphocytes play an important role in both antibody-mediated and delayed-type hypersensitivity immune phenomena, and both of these processes are severely restricted in AIDS patients. Systemic yeast and mycobacterial infections (see Figure 15.24), for example, clearly point to a loss of T_{DTH} function, because this T-cell subset plays an important role in controlling such infections in healthy individuals (see Section 12.10). Many of the other opportunistic pathogens associated with AIDS are effectively eliminated in a healthy host by antibody-mediated reactions. Therefore, loss of T_H and T_{DTH} cell function must be the result of either direct cell death following HIV infection or the cessation of normal growth and division processes. Indeed, both mechanisms may occur.

Unlike many viruses which actually kill and lyse their host cells, HIV-infected T4 lymphocytes remain viable but stop growing, effectively becoming diluted out by normal growth and division of other T-cell subsets. In a healthy adult human, T4 cells constitute about 70 percent of the total T-lymphocyte population: in AIDS victims T4 cells are virtually undetectable. An additional means of T4 cell elimination is thought to occur because HIV antigens are expressed on the surface of infected T4 cells. These foreign antigens are recognized as "non-self" by other T cells, including cytotoxic T cells, and the latter may actually destroy HIV infected T4 cells, further depleting the already shrinking population.

As T4 numbers decline, there is also a decline in the level of interleukin-2, a protein produced by T4 cells that stimulates the clonal expansion of T-cells in general (see Section 12.10). This decline leads to widespread losses in overall T-cell function. In effect, infection of T4 cells by the AIDS virus effectively shuts down the whole immune system.

Further study of the molecular biology of HIV and T4 lymphocytes have shown that virus-infected T4 cells secrete a protein called soluble suppressor factor (SSF), a protein showing a remarkable resemblance to one of the coat proteins of HIV itself. It is hypothesized that a DNA sequence in the T4 lymphocyte genome, rather than AIDS virus RNA itself, actually codes for SSF. This T-cell SSF DNA sequence presumably originated from an ancient retroviral infection. Infection with HIV in some way triggers host transcription, translation, and secretion of SSF. Soluble suppressor factor serves to block certain T-cell-dependent immune responses and could be a contributing factor in the overall loss of immune function.

Treatment and diagnosis of AIDS In the final analysis, opportunistic pathogens (see Figure 15.24) or malignant transformations (see Figure 15.25) create life-threatening situations which eventually kill AIDS patients. Unfortunately, there is currently no known cure for AIDS, although a number of experi-

Sexual Activity and AIDS

Sexual promiscuity has always been associated with venereal disease, but the rise of the AIDS epidemic, discussed in this chapter and elsewhere in this book, has focussed new attention on the dangers of multiple sex partners and on the high risk associated with certain sex practices. AIDS, caused by the human immunodeficiency virus (HIV), is only one type of sexually transmitted disease. Others include gonorrhea, syphilis, herpes simplex, nonspecific urethritis (caused by *Chlamydia*), protozoal vaginitis (caused by *Trichomonas vaginalis*), fungal vaginitis (caused by *Candida albicans*), and venereal warts (caused by another virus). Some of these sexually transmitted diseases have been associated with human society for all of recorded history. The unique thing about AIDS is that it is new and that it is almost uniformly fatal. There are neither drugs to cure AIDS nor vaccines to prevent it, and it is unlikely that either drugs or vaccines will be available in the near future. We do not at this time know how extensive the AIDS epidemic is, since the long latent period means that many people now infected (and perhaps infectious) have not yet exhibited symptoms. Some public health officials believe that the AIDS cases so far seen are only a small part of the AIDS cases that will eventually appear, and that over the next 10 years there will be a massive increase in the incidence of the disease, as latent cases develop into full-blown AIDS.

Because AIDS is linked to certain sex practices, prevention means avoidance of these sex practices. The United States Surgeon General has issued a report which makes specific recommendations that individuals can follow if they wish to reduce the likelihood of AIDS infection. Among the recommendations are:

1. Avoid mouth contact with penis, vagina, or rectum.
2. Avoid all sexual activities which could cause cuts or tears in the linings of the rectum, vagina, or penis.
3. Avoid sexual activities with individuals from high-risk groups. These include prostitutes (both male and female), homosexual or bisexual individuals, and intravenous drug users.
4. If a person has had sex with a member of one of the high-risk groups, a blood test should be made to determine if infection with HIV has occurred. If the test is positive, then it is essential that sexual partners of an AIDS-positive individual be protected by use of a condom during sexual intercourse.

mental drugs are being tested to control the condition and reduce mortality. The most promising drugs discovered thus far are chemicals which inhibit the activity of *reverse transcriptase*, the enzyme which converts the genetic information (which resides in single-stranded RNA) of retroviruses, such as HIV, into a complementary DNA copy (see Section 6.24). For example, the nucleotide base analog **azidothymidine** (AZT, Figure 15.27) is an effective inhibitor of HIV replication because it closely resembles thymidine but lacks the correct attachment point for the next nucleotide in the chain and thus serves as a *DNA chain terminator*. The drug *suramin* is also a reverse transcriptase inhibitor. A major problem has been observed with all experimental AIDS drugs in that the compounds are invariably toxic when used on a continuing basis. Nevertheless, drugs such as AZT have yielded dramatic results in clinical trials and may serve as temporary AIDS treatments until a true cure for this disease is discovered.

The genetic variability of HIV has thus far hampered the development of an AIDS vaccine. HIV apparently can undergo a rather large number of mutational events and still retain the ability to disrupt lymphocyte function. Indeed, studies of the RNA isolated from strains of HIV obtained from different AIDS patients have shown that only about 70 percent RNA sequence homology is necessary for the strains to maintain the HIV phenotype. However, a DNA fragment (cDNA) has been produced complementary to the RNA sequence coding for an HIV envelope protein called gp120. This protein appears to be highly conserved across strains of HIV, and is the best candidate identified thus far as a target for an AIDS vaccine. The cDNA fragment coding for HIV protein gp120 has been cloned into *E. coli* and the resulting protein is currently being tested for its immunogenicity in experimental animals. If successful, such a protein produced in *E. coli* would provide a completely safe

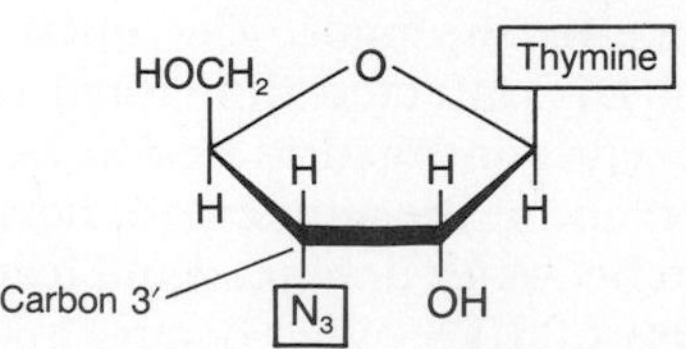

Figure 15.27 Structure of azidothymidine (AZT), a drug used for AIDS therapy. Because AZT lacks a hydroxyl group on carbon 3′, when it is incorporated into the growing DNA chain, replication ceases.

It is important to emphasize that AIDS is *not* just a disease of male homosexuals. In certain cultures, AIDS is almost as common in women as in men. The disease seems to be linked primarily to promiscuous sexual activities, which include not only male homosexuality but also female prostitution.

Is it possible, then, to have sex without incurring the risk of AIDS? Certain sex practices do seem to be inherently much safer than others. Safe sex practices include dry kissing (mouths closed), mutual masturbation (in the absence of breaks in the skin), and intercourse protected by a condom. Dangerous sex practices include wet kissing (mouths open), masturbation where breaks in the skin occur, oral sex (either male or female), and unprotected sexual intercourse (either vaginal or anal). The U.S. Surgeon General has strongly recommended that if the health status of the partner is unknown that a condom be used for all sex practices in which exchange of bodily fluids occurs.

The AIDS epidemic has focussed anew on the condom, a device that has been used for many years. Condoms have always played two roles in sexual activity: disease protection, and prevention of pregnancy. Although it should be firmly stated that the best way of avoiding AIDS is avoidance of dangerous sex practices, if sexual intercourse is to be carried out with an individual whose infectiousness is unknown, then a condom should be used. The U.S. Surgeon General strongly recommends the use of condoms for all extramarital sexual activity. In certain countries, advertising campaigns to promote the use of condoms are already widespread.

It is very important to emphasize that moralistic statements alone (prescriptions for monogamy, abstinence, avoidance of sexual activity outside of matrimony), will *not* control the AIDS epidemic. Epidemiological studies on all previously known sexually transmitted diseases have shown that fear of disease is not, by itself, sufficient to prevent sexual activities that put an individual at risk for a sexually transmitted disease. The sex urge in some individuals is so strong that it will suppress the fear of disease. Every individual must therefore take the responsibility for protecting himself or herself from this widespread and extremely dangerous infectious disease.

For more information on prevention of AIDS, see the *Surgeon General's Report on Acquired Immune Deficiency Syndrome*, U.S. Department of Health and Human Services. For more information on protection against AIDS, the Public Health Service has established a toll-free telephone number, called the PHS AIDS Hotline. The number to call is 800-342-2437.

AIDS vaccine, eliciting in the host circulating antibodies and T-cell subsets that could remove HIV from the body before it could do any damage. Other potential vaccines include strains of HIV containing genomic deletions that leave the virus able to infect but unable to kill T4 cells. Strains totally unable to infect are also being explored, but the liability risks associated with use of any type of "live" virus AIDS vaccine has made their clinical usage unlikely. Additionally, it should be remembered that a vaccine strategy is *not* useful in actually *treating* clinical AIDS, because the immune system in such patients is already malfunctioning.

Clinical AIDS, the pre-AIDS condition called AIDS-Related-Complex (ARC), or any previous exposure to HIV can be sensitively diagnosed by immunological means. Both radioimmunoassay (RIA, see Section 12.9), and ELISA tests (see Section 12.9 and 13.8) have been developed for screening blood samples to detect anti-HIV antibodies. The ELISA test has proven particularly valuable for large-scale screening of donated blood to prevent transfusion-associated HIV. Statistics have shown that about 0.25 percent of all blood donated by volunteer donors in the United States tests HIV-positive in the ELISA assay. A positive AIDS ELISA test must be confirmed by a second procedure called immunoblotting (Western blotting), a technique which combines the analytical tools of protein purification and immunology (see Section 13.10). Although not foolproof because both methods can miss HIV carriers who have been very recently infected but who have not yet produced detectable levels of circulating antibodies to HIV, these immunological tools have served to greatly improve the safety of the blood supply. Indeed statistics suggest that the current risk of contracting AIDS through contaminated blood or blood products is now very low.

The ecology of sexually-transmitted diseases

As we have seen, a variety of microorganisms have been implicated as causative agents of sexually-transmitted diseases. What properties do these organisms share in common that limit their distribution to the human genitourinary tract and their mode of transmission to sexual activity? For one, unlike respiratory infections where large numbers of infectious particles may be expelled by an individual, sexually transmitted pathogens are generally *not* shed in large numbers other than during sexual activity. Consequently, transmission will be limited to physical contact, generally during sexual intercourse. In addition, many sexually-transmitted pathogens are very sensitive to drying. Their habitat, the human genitourinary tract, is generally a moist environment. Thus, these organisms colonize moist niches, and have apparently lost the ability to survive outside the animal host. The effect of drying is most pronounced in *Neisseria gonorrhoeae* and *Treponema pallidum*—both of these organisms are killed easily when dried. The possibility of contracting gonorrhea or syphilis from activities other than intercourse is therefore rather slim.

Many sexually-transmitted diseases show inapparent infections in individuals of one sex or the other, and this problem complicates attempts by public health officials to identify sources of infection. In addition, the widespread use of oral contraceptives has tended to reduce the use of condoms, which have traditionally served to prevent disease transmission. And finally, the perceived moral stigma of contracting a sexually-transmitted disease discourages some infected individuals from seeking prompt medical care. The irony of this situation lies in the fact that most sexually-transmitted diseases can be treated with antibiotics or other therapeutic agents.

15.8 Rabies

Animals can contract a number of infectious diseases, some of which can be passed to humans. Through the use of effective vaccination practices and good veterinary care, domestic animal populations are generally maintained in good overall health standing. But when a new pathogen, or a new strain of a pathogen currently under control, is introduced into the domestic animal population, infections can spread quickly, leading to substantial economic losses. This situation is different in the wild animal population. Wild animals cannot be routinely vaccinated, and do not receive veterinary care. Thus animal diseases (zoonoses) cycle through wild animal populations on a periodic basis.

Most zoonoses are of little serious consequence to human health. Rabies is an exception. Rabies is one of the best examples of a disease which occurs primarily in animals but under rare accidental conditions occurs in humans. Rabies is caused by a single-stranded virus of the rhabdovirus family (see Section 6.17), which attacks the central nervous system of most warm-blooded animals, almost invariably leading to death if not treated. Rabies is a rhabdovirus (see Figure 6.44). The virus enters the body through a bite wound from a rabid animal. The virus multiplies at the site of inoculation and then travels to the central nervous system. The incubation period for the onset of symptoms is highly variable, depending on the size, location, and depth of the wound, and the actual number of viral particles transmitted in the bite. In dogs, the incubation period averages 10–14 days. In humans, up to 9 months may elapse before the onset of rabies symptoms. The virus proliferates in the brain (especially in the thalamus and hypothalamus), leading to fever, excitation, dilation of the pupils, excessive salivation, and anxiety. A fear of swallowing (hydrophobia) develops from uncontrollable spasms of the throat muscles; death eventually results from respiratory paralysis.

A rabies vaccine was developed by Louis Pasteur in his classic work on immunology, and today such highly effective rabies vaccines are available that less than 5 human rabies death are usually reported in the United States each year. Because of the long incubation period, treatment of an infected human before the onset of symptoms is almost always successful in preventing rabies. An animal which makes an unprovoked bite of a human is generally held 10 days for observation and if no signs of rabies are apparent, treatment of the human with rabies vaccine or rabies immune serum (anti-rabies virus antibodies obtained from a hyperimmune individual) is generally not undertaken. However, if the animal is rabid, the patient will be passively immunized with rabies immune

Table 15.3 Guidelines for treating possible human exposure to rabies virus

I. Unprovoked bite by a domestic animal	
Animal suspected of rabies	*Animal not suspected of rabies*
1. Sacrifice animal and test for rabies	1. Hold for 10 days—if no symptoms, do not treat human
2. Begin treatment of human immediately*	2. If symptoms develop—treat human immediately*
II. Bite by wild carnivore (e.g., skunk, bat, fox, raccoon, coyote, etc.)	
1. Carefully attempt to restrain animal for sacrifice and testing.	
2. If no. 1 is unsuccessful, assume animal is rabid—treat human immediately.	
III. Bite by wild rodent, squirrel, livestock, rabbit	
Consult local or state public health officials about possible recent cases of rabies transmitted by these animals (these animals rarely transmit rabies). If no reports, do not treat human.	

**All bites should be thoroughly cleansed with soap and water. Treatment is generally effected by combination of rabies immune globulin and human diploid cell rabies vaccine (3–5 injections intramuscularly).*

globulin (injected at both the site of the bite and intramuscularly), and also will be vaccinated with attenuated rabies virus, preferably virus grown in human cell lines. Because the use of a viral vaccine always carries with it some risk of actually transmitting the disease, a "second generation" rabies vaccine is now under development. The most promising lead thus far is a vaccine prepared by recombinant DNA techniques involving the cloning and expression in *E. coli* of highly immunogenic rabies virus surface glycoproteins. Such a vaccine would be a completely safe means of controlling rabies in both animals and humans. A summary of guidelines for treating possible human exposure to rabies is shown in Table 15.3.

Rabies has been effectively controlled in domestic animal populations by vaccination, but it is still a problem in the United States because of the apparently large wild animal reservoir of rabies virus (see Table 14.4). Because of the highly fatal nature of the disease and the effective treatment available, any unprovoked animal bites must be taken quite seriously. A wild animal suspected of being rabid should be captured, sacrificed, and examined immediately for evidence of rabies in order to expedite treatment of those exposed. Characteristic virus inclusion bodies in the cytoplasm of nerve cells, called *Negri bodies*, are taken as confirmation of rabies. Fluorescent antibody preparations (see Section 12.8) which recognize rabies-infected brain tissue are also very useful for confirming a diagnosis of rabies.

15.9 Insect-Transmitted Diseases: Rickettsias

The rickettsias are small bacteria that have a strictly intracellular existence in vertebrates, usually in mammals, and are also associated at some point in their natural cycle with blood-sucking arthropods such as fleas, lice or ticks. We discuss the biology of rickettsias in Section 19.22. Rickettsias cause a variety of diseases in humans and animals, of which the most important are typhus fever, Rocky Mountain Spotted Fever, scrub typhus (tsutsu-gamushi disease), and Q fever. Rickettsias take their name from Howard Ricketts, a scientist at the University of Chicago who first provided evidence for their existence, but who unfortunately died from infection with the rickettsia that causes typhus fever, *Rickettsia prowazekii*. Most rickettsias have not been cultured in nonliving media and hence must be considered obligate intracellular parasites, although there is nothing so unusual about their physiology as to suggest that they cannot eventually be grown in vitro. (The causal agent of trench fever, *R. quintana*, has been cultured in a cell-free medium.) Rickettsias have been cultivated in laboratory animals, lice, mammalian tissue cultures, and the yolk sac of chick embryos. In animals, growth takes place primarily in phagocytic cells (see Section 11.15).

Typhus fever (epidemic typhus) Typhus fever is caused by *Rickettsia prowazekii*. Epidemic typhus is transmitted from host to host by the common body or head louse. Typhus can be a serious disease. During World War I an epidemic of typhus spread throughout eastern Europe and claimed almost three million lives. Typhus has been frequently a problem in military troops during wartime. Cells of *R. prowazekii* are introduced through the skin when the puncture caused by the louse bite becomes contaminated with louse feces, the major source of rickettsial cells. During an incubation period of from one to three weeks, the organism multiples inside cells lining the small blood vessels, following which symptoms of typhus (fever, headache, and general body weakness) begin to appear. Five to nine days later a characteristic *rash* is observed which begins in the armpits and generally spreads throughout the body *except* for the face, palms of the hands and soles of the feet. If untreated, complications from typhus may develop involving damage to the central nervous system, lungs, kidneys, and heart. Several drugs are effective against rickettsias, tetracycline and chloramphenicol being the most commonly used to control *R. prowazekii*.

Murine typhus, caused by *R. typhi*, is much less common than epidemic typhus. The disease occurs primarily in rodents and is transmitted to humans by feces from the rat flea. Symptoms are similar to those of epidemic typhus, including the formation of a rash. Unlike epidemic typhus, however, murine typhus is rarely fatal, although kidney problems are observed in untreated cases. Prevention of murine typhus depends on effective rat control and thus the disease is more common in areas of poor sanitation.

Rocky Mountain Spotted Fever Rocky Mountain Spotted Fever was first recognized in the western United States around the turn of the century, but is actually more common today in the southeastern United States. Rocky Mountain Spotted Fever is caused by *Rickettsia rickettsii* and is transmitted to humans by various species of ticks, most commonly the dog and wood ticks (Figure 15.28). Humans acquire the pathogen from tick fecal matter which gets injected into the body during a bite, or by rubbing infectious material into the skin by scratching. Cells of *R. rickettsii*, unlike other rickettsias, grow within the nucleus of the host cell as well as in host cell cytoplasm (Figure 15.28). Following an incubation period of from 3–12 days an abrupt onset of symptoms occurs, including fever and a severe headache. Three to five days later, a rash is observed that is present on the palms of the hands and soles of the feet (the location of rickettsial rashes are therefore of some diagnostic

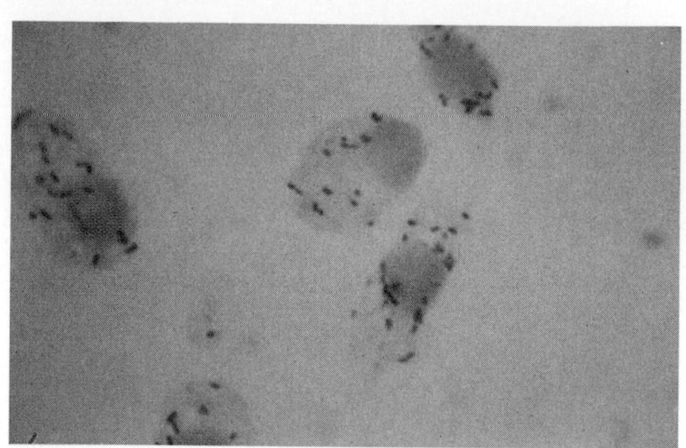

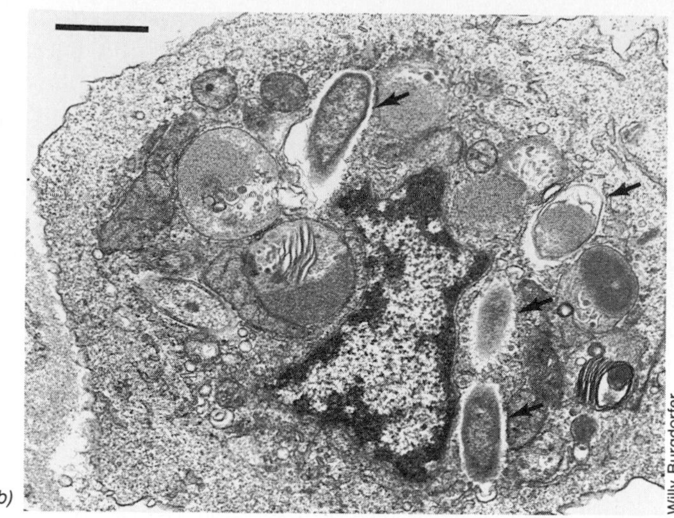

(a) (b)

Willy Burgdorfer

Figure 15.28 *Rickettsia rickettsii*, the causative agent of Rocky Mt. Spotted Fever. (a) Cells of *R. rickettsii*, growing in the cytoplasm and nucleus of tick hemocytes. Magnification, 1000×. (b) Cells of *R. rickettsii* in a granular hemocyte of an infected wood tick, *Dermacentor andersoni*. Transmission electron micrograph, bar = 1 μm.

value). Gastrointestinal problems such as diarrhea or vomiting are usually observed as well, and the clinical symptoms of Rocky Mountain Spotted Fever may exist for over two weeks if the disease is left untreated. Tetracycline or chloramphenicol generally promotes a prompt recovery from Rocky Mountain Spotted Fever if administered early in the course of the infection.

Q fever Q fever is a pneumonia-like infection caused by an obligate intracellular parasite, *Coxiella burnetii*, related to the rickettsias (see Section 19.23). Although not transmitted to humans directly by an insect bite, the agent of Q fever is transmitted to animals by insect bites, and various arthropod species serve as a reservoir of infection (the "Q" in Q fever stands for *query*, because when cases of the disease were first observed, no pathogenic agent could be clearly implicated). Domestic animals generally have inapparent infections, but may shed large quantities of *Coxiella burnetii* in their urine, feces, milk, and other secretions.

Cattle ticks are frequently implicated in transmitting the infection to dairy herds, where the disease Q fever can be a considerable occupational hazard to farmers and others involved in animal husbandry. Outbreaks of Q fever in humans have usually been traced to improperly pasteurized milk. With the recognition of Q fever as a potential milk pathogen, dairy processing facilities have modified their pasteurization protocol to ensure that pasteurized milk is *Coxiella*-free. Because it is a somewhat thermotolerant organism, the "holding method" of milk pasteurization, originally set by federal law at 60°C for 20 minutes, was raised to 62.8°C for 30 minutes to ensure elimination of *C. burnetii*.

Symptoms of Q fever are quite variable. They include an influenza-like illness, prolonged fever, headache and chills, chest pains, and pneumonia. Although Q fever is rarely fatal, complications arising from a *C. burnetii* infection may be severe. In particular, endocarditis (inflammation of the lining of heart and heart valves) is associated with many cases of Q fever. Q fever endocarditis frequently occurs many months or even years after the primary infection, the organisms remaining dormant in liver cells during the interim. Damage to heart valves as a result of Q fever endocarditis weakens the heart and may necessitate heart valve replacement surgery in later life.

An obligate intracellular parasite, *C. burnetii* is difficult to isolate and culture. However, a diagnosis of Q fever can readily be made by immunological tests designed to measure host antibodies to the organism (an indirect ELISA test has made the diagnosis of Q fever an almost routine procedure). *C. burnetii* infections respond quite dramatically to the antibiotic tetracycline, and therapy is usually begun quickly in any suspected human case of Q fever in order to prevent heart damage by latent *C. burnetii* infections. Q fever is one of the infectious diseases that has been studied as a possible agent for biological warfare (see Box).

General comments In the past, ricketsial infections have been difficult to diagnose because the characteristic rash associated with many rickettsial diseases may be mistaken for measles, scarlet fever, or adverse drug reactions. Clinical confirmation of rickettsial diseases has now been greatly aided by the introduction of specific immunological reagents. These include polyclonal or monoclonal antibodies

Principal Agents of Biological Warfare

During World War II, extensive research was carried out by the United States and British governments on the use of infectious agents for biological warfare (sometimes euphemistically abbreviated BW). In the United States, BW research was concentrated at Fort Detrick, a military base in Maryland, and at a military installation near Dugway, Utah. In Great Britain, BW research was carried out at laboratories of the Microbiological Research Establishment, Porton, Salisbury. Some BW research in the United States continued through the so-called Cold War period, but has now been completely terminated by international agreement. However, from time to time, suspicions that one or another country has been engaged in research on biological warfare have been aroused, and the introduction of techniques for gene cloning and genetic engineering have raised the spectre of a new and more sophisticated type of biological warfare. What is biological warfare, and what are the requirements of an effective agent? How can biological warfare be defended against?

The chief requirement of an effective biological weapon is that the organism be highly infectious by the respiratory route, thus permitting effective airborne dispersal. The organism should form spores or in some other way be resistant to drying. The organism should be highly infectious, so that each viable unit will be able to initiate an infection. And it should be possible to produce the organism easily in very large quantities. It is also desirable that an effective vaccine against the agent be available, so that the military forces employing the agent can avoid getting infected. These requirements and limitations mean that the principal agents of biological warfare will be bacteria or fungi rather than viruses, since bacteria can be readily cultured in large fermentation vats.

One of the most widely studied agents for biological warfare has been the bacterium *Bacillus anthracis*, which causes the disease anthrax. Because this organism is an endospore-forming bacterium, it can survive long periods of time in the dormant state. It is able to initiate a respiratory infection which, if untreated, is normally fatal. It is also able to penetrate through the skin. One "trouble" with *B. anthracis* is that it is quite sensitive to antibiotics, so that an enemy population could protect itself, but antibiotic-resistant mutants can be readily isolated. During the peak of research on biological warfare, the U.S. Army did extensive studies on the genetics of antibiotic resistance in this organism, presumably in order to develop strains of *B. anthracis* that could not be defended against. The British military actually did field tests with *B. anthracis* during World War II on a small isolated island, Gruinard Island, off the coast of Scotland, and subsequent studies showed that the agent remained viable in the soil for many years. Indeed, this island must still remain uninhabited, although sheep were reintroduced finally in 1987 to observe whether the island might be safe.

Another agent that has been studied for possible use in BW is *Yersinia pestis*, the causal agent of plague. This organism is highly infectious and causes a serious incapacitating disease that is often fatal (see Section 15.10). The organism can infect by either the respiratory or oral route, and can be readily cultivated in the laboratory. A related organism, *Francisella tularensis*, causes the disease tularemia, which although not highly fatal is seriously incapacitating.

One rickettsial disease that has been extensively studied as a BW agent is Q fever, caused by the organism *Coxiella burnetii*. This organism is also transmitted by the respiratory route, and produces a stable resting cell. It is so highly infectious that inhalation of a single viable organism may be enough to initiate infection. Although mortality is low, the fever and debilitation caused by this agent could be sufficient to immobilize a military force.

The fungus *Coccidioides immitis* has also been studied as a BW agent. It causes the respiratory disease coccidioidomycosis, also known as San Joaquin Valley fever. The spores of the fungus will remain alive for long periods of time in dry soil and are disseminated on windblown particles. One disadvantage of this organism is that effective antibiotics are not available, and a suitable vaccine has yet to be developed.

In some forms of BW, it is not necessary that the agent actually multiply in the population. A sufficiently toxic substance could be used like a chemical poison, being dispersed by bombs into an enemy population. For this purpose, the botulinum toxin, produced by *Clostridium botulinum*, has been considered. This toxin is the most poisonous substance known, capable of causing death at a vanishingly small dose. The mortality rate from botulinum poisoning is high, from 60–70%, meaning that the agent would be highly effective if it could be delivered to the target population. One disadvantage of the botulinum toxin is that it decomposes relatively rapidly in air, so that its effects would not remain in the environment for very long. However, for certain types of military operation this short-lived character might be desirable, since it would mean that the attacking force could quickly move into the territory once the enemy force had been terminated.

A number of other organisms that have been studied as potential BW agents are *Brucella melitensis* (brucellosis), *Vibrio cholera* (cholera), *Pseudomonas mallei* (glanders), *Pseudomonas pseudomallei* (melioidosis), and *Rickettsia prowazeki* (epidemic typhus).

One encouraging note: the Fort Detrick laboratory where BW research was carried out was converted in the early 1970's into a cancer research laboratory.

that can detect rickettsial surface antigens by immunofluorescence, latex bead agglutination assays, complement fixation, and ELISA analyses (see Sections 13.6–13.9 for description of the principles of these assays). Control of most rickettsial diseases requires control of the vectors: lice, fleas, and ticks. For humans traveling in wooded or grassy areas, the use of insect repellants on the exposed extremities usually prevents tick attachment. Firmly attached ticks should be removed gently with forceps, care being taken to remove all of the mouth parts. Solvents, such as gasoline or ethanol applied to a tick from a saturated swab will usually expedite removal of the tick. Although a vaccine is available for the prevention of typhus, the few cases reported do not warrant its general administration. No vaccines are currently available for the prevention of Rocky Mountain Spotted Fever or Q fever, however an experimental Q fever vaccine, using a mixed *C. burnetii* antigen preparation, is now undergoing clinical trials in human volunteers and thus far looks promising.

15.10 Insect-Transmitted Diseases: Malaria and Plague

Malaria is a disease caused by a protozoan, a member of the Sporozoa group. The malaria parasite is one of the most important human pathogens and has played an extremely significant role in the development and spread of human culture. Indeed, as we will see, malaria has even affected the human evolutionary process. Over 300,000 people worldwide are affected with malaria.

Four species of sporozoans infect humans, of which the most widespread is *Plasmodium vivax*. This parasite carries out part of its life cycle in humans, and part in the mosquito; the mosquito is the vector by which the parasite spreads from person to person. Only mosquitoes of the genus *Anopheles* are involved, and since these primarily inhabit warmer parts of the world, malaria occurs predominantly in the tropics and subtropics. Malaria did not exist in the northern regions of North America prior to settlement by Europeans, but was of great incidence in the South, where appropriate breeding grounds for the mosquito existed. The disease has always been associated with swampy low-lying areas, and the name *malaria* is derived from the Italian words for "bad air."

The life cycle of the malaria parasite is complex (Figure 15.29). One stage of the cycle occurs in humans, with the liberation into the bloodstream of a stage infective for mosquitoes. If a mosquito bites an infected person, the life cycle can be completed in the mosquito, with the formation of a stage infective for humans.

As illustrated in Figure 15.29, the human host is infected by plasmodial *sporozoites*, small elongated cells produced in the mosquito, which localize in the salivary gland of the insect. The sporozoites replicate in the liver, where they become transformed into a stage referred to as a *schizont*, which subsequently enlarges and segments into a number of small cells called *merozoites*; these cells are liberated from the liver into the blood stream. Some of the merozoites infect red blood cells (erythrocytes). The cycle in erythrocytes proceeds as in the liver and usually

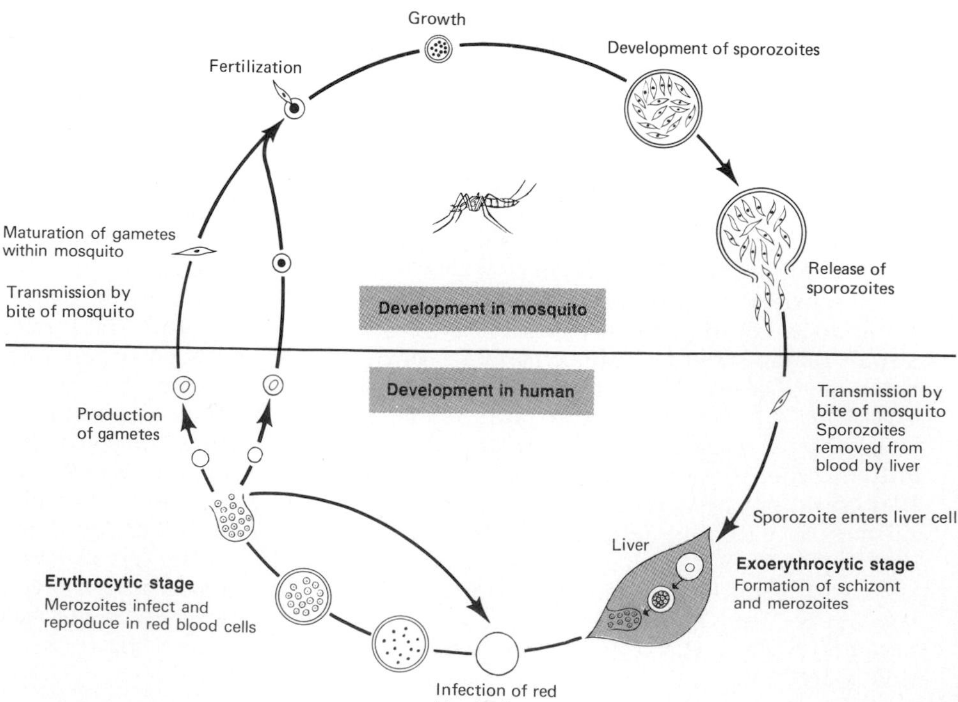

Figure 15.29 Life cycle of the malaria parasite, *Plasmodium vivax*.

repeats at regular intervals, 48 hours in the case of *P. vivax*. It is during this period that the characteristic symptoms of malaria, fever (up to 40°C), followed by chills, occur, the chills occurring when a new brood of cells is liberated from erythrocytes. Vomiting and severe headache may accompany the fever-chill cycles, but asymptomatic periods generally alternate with periods in which the characteristic symptoms are present. Because of the loss of red blood cells, malaria generally causes anemia and some enlargement of the spleen as well.

Not all the cells liberated from red cells are able to infect other erythrocytes; those which cannot, called *gametocytes*, are infective only for the mosquito. If these gametocytes happen to be ingested when another insect of the proper species of *Anopheles* bites the infected person, they mature within the mosquito into *gametes*. Two gametes fuse, and a zygote is formed; the zygote migrates by amoeboid motility to beneath the outer wall of the insect's intestine, where it enlarges and forms a number of sporozoites. These are liberated, some of them reaching the salivary gland of the mosquito, from where they can be inoculated into another person; and the cycle begins again.

Conclusive evidence for the diagnosis of malaria in humans is obtained by examining blood smears for the presence of infected cells. *Chloroquine* is the drug of choice for treating parasites *within* red blood cells, but this quinine derivative does not kill stages of the malarial parasite which reside *outside* erythrocytes. The related drug *primaquine* effectively eliminates the latter and thus together treatment of malarial patients with chloroquine and primaquine can affect a complete cure. However, because recurrences of malaria many years after a primary infection are not uncommon, it is likely that small numbers of sporozoites can reside in the liver protected from the effects of quinine drugs and release merozoites months or years later to reinitiate the disease.

Plasmodium vivax has been very difficult to grow in culture, and because of its specificity for humans, experimental infection has also been difficult. For these reasons, most of the experimental work on malarial parasites has been done with species of *Plasmodium* that infect birds or rats, and it is with these forms that most of the studies on the development of new drugs have been carried out. Recently, scientists have been able to grow *P. falciparum* in laboratory culture and this organism has therefore served as a model system for experimental malarial research.

Eradication of malaria Although quinine derivatives are effective in treating human cases of malaria, because of the obligatory alternation of hosts, *control* of malaria can be best effected by elimination of the *Anopheles* mosquito. Two approaches to mosquito control are possible: elimination of the habitat by drainage of swamps and similar areas, and elimination of the mosquito by insecticides. Both approaches have been extensively used. The marked drop in incidence of malaria as a result of mosquito eradication programs in the United States is shown in Figure 15.30. During the 1930s, about 33,000 miles of ditches were dug in 16 southern states, removing 544,000 acres of mosquito breeding area. Millions of gallons of oil were also used to spread on swamps, with the purpose of cutting off the oxygen supply to mosquito larvae. With the discovery of the insecticide dichlorodiphenyltrichloroethane (DDT, see Figure 17.40), chemical control of both larvae and adult mosquitoes was possible. During World War II, the Public Health Service organized an Office of Mosquito Control in War Areas, and because many U.S. military

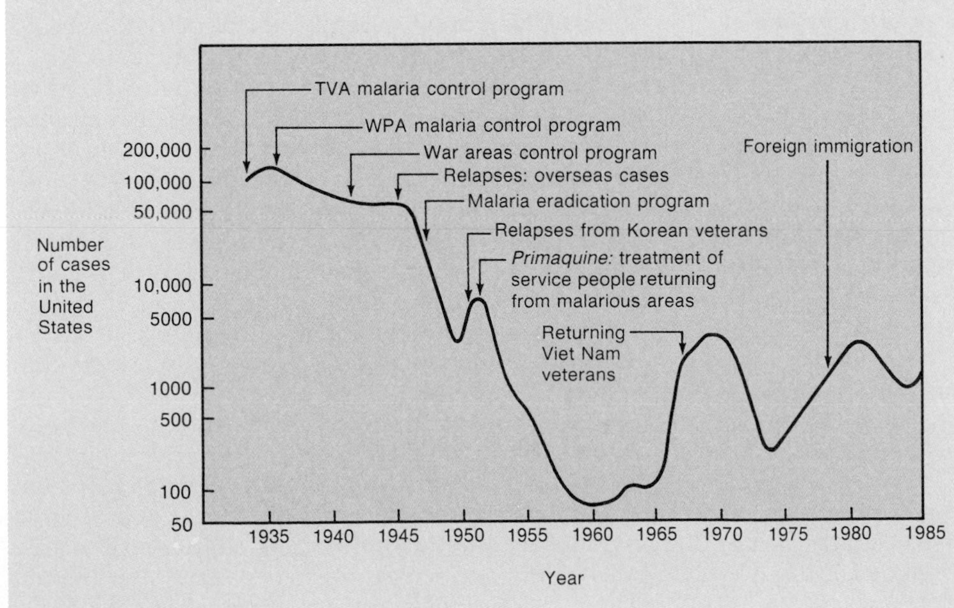

Figure 15.30 Dramatic decrease in incidence of malaria in the United States as a result of mosquito eradication programs. Note that the scale is logarithmic, so that the rise during the Vietnam War era is really quite minor.

bases were in the southern states, this organization carried out an extensive eradication program in the United States as well as overseas. In 1946, Congress established a 5-year malaria eradication program, which involved drug treatment of humans and spraying of mosquito areas with DDT to a point where malaria transmission could not occur. As a further indication of the success of this program, in 1935, there were about 4000 deaths from malaria in the United States, and by 1952, the number of deaths had been reduced to 25.

In other parts of the world, eradication has been much more difficult, but the same control measures are used. Despite some of the environmental problems associated with DDT (see Section 17.20), it is still an effective agent for mosquito control and has been most responsible for the control of malaria in large areas of Africa and South America.

Malaria and biochemical evolution of humans One of the most interesting discoveries about malaria concerns the mechanism by which human beings in regions of the world where it is endemic acquire resistance to *Plasmodium* infections. Malaria has undoubtedly been endemic in Africa for thousands of years. In West Africans, resistance to malaria caused by *P. falciparum* is associated with the presence in their red cells of hemoglobin S, which differs from normal hemoglobin A only in a single amino acid in each of the two identical halves of the molecule. In hemoglobin S a neutral amino acid, valine, is substituted for an acidic amino acid, glutamic acid. Red cells containing hemoglobin S have reduced affinity for oxygen, and the malaria parasite, having a highly aerobic metabolism, cannot grow as well in these red cells as it can in normal ones. An additional consequence is that individuals with hemoglobin S are less able to survive at high altitudes, where oxygen pressures are lower, but in tropical lowland Africa this disadvantage is not manifested. In West Africans, resistance to another malarial parasite, *P. vivax*, is associated with the presence of another abnormal hemoglobin, hemoglobin E. In certain Mediterranean regions where malaria is endemic, resistance to *P. falciparum* is associated with a deficiency in the red cells of the enzyme glucose-6-phosphate dehydrogenase (GPD). The faulty GPD leads to raised levels of oxidants which damage parasite membranes. The malaria parasite has thus been a factor in the biochemical evolution of human beings. Other microbial parasites have also probably promoted evolutionary changes in their hosts, but in no case do we have such clear evidence as in the case of malaria.

Plague Pandemic occurrences of **plague** in the past have been responsible for more human deaths than any other infectious disease. Plague is caused by a Gram-negative, facultatively anaerobic rod called *Yersinia pestis* (see Section 19.20). Plague is a natural disease of domestic and wild rodents, but rats appear to be the primary disease reservoir. Most infected rats die soon after symptoms begin, but a low proportion develop a chronic infection and can serve as a ready source of virulent *Y. pestis*. The majority of cases of human plague in the United States originate in the southwestern states, where the disease is endemic among wild rodents.

Plague is transmitted by the rat flea (*Xenopsylla cheopsis*), which ingests *Y. pestis* cells by sucking blood from an infected animal. Cells multiply in the flea's intestine and can be transmitted to a healthy animal in the next bite. As the disease spreads, rat mortality becomes so great that infected fleas seek new hosts, including humans. Once in humans, cells of *Y. pestis* usually travel to the lymph nodes, where they cause the formation of swollen areas referred to as *buboes*, and for this reason the disease is frequently referred to as **bubonic plague**. The buboes become filled with *Y. pestis*, but antiphagocytic surface antigens and the distinct capsule on cells of *Y. pestis* prevent them from being phagocytized. Secondary buboes form in peripheral lymph nodes and cells eventually enter the bloodstream, causing a generalized septicemia. Multiple hemorrhages produce dark splotches on the skin. If not treated prior to the bacteremic stage, the symptoms of plague, including extreme lymph node pain, prostration, shock, and delirium, usually cause death within 3 to 5 days.

The pathogenesis of plague is not clearly understood, but it is known that cells of *Y. pestis* produce a number of antigenically-distinct molecules, including toxins, that undoubtedly contribute to the disease process. The "V" and "W" antigens of *Y. pestis* cell walls are protein/lipoprotein complexes which serve to prevent phagocytosis. Other envelope proteins are also present. An exotoxin called *murine toxin*, because of its extreme toxicity for mice, is produced by all virulent strains of *Y. pestis*. Murine toxin serves as a respiratory inhibitor by blocking mitochondrial electron transport reactions at the point of coenzyme Q. Although it is not clear that murine toxin is involved in the pathogenesis of human plague (because murine toxin is highly toxic for certain animal species but not for others), the symptoms it produces in mice, systemic shock, liver damage, and respiratory distress, are similar to those in human cases of plague. *Y. pestis* also produces a highly immunogenic *endotoxin* which may play a role in the disease process as well.

Pneumonic plague occurs when cells of *Y. pestis* are either inhaled directly or reach the lungs during bubonic plague. Symptoms are usually absent until the last day or two of the disease when large amounts of bloody sputum are emitted. Untreated cases rarely survive more than 2 days. Pneumonic plague, as one might expect, is a highly contagious disease, and can spread rapidly via the respiratory route if infected individuals are not immediately quarantined. **Septicemic plague** involves the rapid spread of *Y. pestis* throughout the body without the formation of buboes, and usually causes death before a diagnosis can be made.

Plague can be successfully treated if swiftly diagnosed. Although *Y. pestis* is naturally resistant to penicillin, most strains are sensitive to streptomycin, chloramphenicol, or the tetracyclines. If treatment is begun soon enough, mortality from bubonic plague can be reduced to as few as 1 to 5 percent of those infected. Pneumonic and septicemic plague can also be treated, but these forms progress so rapidly that antibiotic therapy in the latter stages of the disease is usually too late. Although potentially a devastating disease, an average of fewer than 10 cases of human plague are reported each year in the United States; this is undoubtedly due to improved sanitary practices and the overall control of rat populations.

15.11 Foodborne Diseases

There are two categories of foodborne disease, *food poisoning,* caused by toxins produced by microorganisms before eating, and *food infection*, caused by growth of the microorganisms in the human body after the contaminated food has been eaten. Table 15.4 gives a breakdown of foodborne illness in the United States. As seen, the majority of the cases are bacterial.

In addition to passive transfer of pathogens in food, active growth of a pathogen may also occur (for example, due to improper storage) leading to marked increases in the microbial load. We have discussed some aspects of heat sterilization and pasteurization, and food microbiology in Sections 9.15 and 9.20. A summary of the major food-poisoning bacteria is given in Table 15.5. We begin our discussion here with the common food poisonings, *staphylococcal* and *perfringens,* and proceed to a discussion of the most serious of all poisonings, *botulism*. We will then turn our attention to the common food infection, *salmonellosis.*

Table 15.4 Foodborne disease outbreaks due to microorganisms[a]

Responsible microorganism	Outbreaks, percent[b]	Cases, percent[c]
Bacteria		
Salmonella sp.	39	31
Staphylococcus aureus	25	38
Clostridium perfringens	13.5	14
Clostridium botulinum	10.5	0.3
Shigella sp.	5	4.5
Campylobacter jejuni	2.5	7
Vibrio parahaemolyticus	2	0.2
Yersinia enterocolitica	0.5	4
Virus		
Hepatitis A	2	1

[a]*Compiled from data provided by the Centers for Disease Control.*
[b]*Of 600 outbreaks reported in recent years.*
[c]*Of 8000 cases in a single year.*

Staphylococcal food poisoning The most common food poisoning is caused by the Gram-positive coccus, *Staphylococcus aureus*. This organism produces an enterotoxin (see Section 11.11) that is released into the surrounding medium or food; if food containing the toxin is ingested, severe reactions are observed within a few hours, including nausea with vomiting and diarrhea. The kinds of foods most commonly involved in this type of food poisoning are custard- and cream-filled baked goods, poultry, meat and meat products, gravies, egg and meat salads, puddings, and creamy salad dressings. If such foods are kept refrigerated after preparation, they remain relatively safe, as *Staphylococcus* is unable to grow at low temperatures. In many cases, however, foods of this type are kept warm for a period of hours after preparation, such as in warm kitchens or outdoors at summer picnics. Under these conditions, *Staphylococcus,* which might have entered the food from a food handler during preparation, grows and produces enterotoxin. Many of the foods involved in staphylococcal food poisoning are not cooked again before eating, but even if they are, this toxin is relatively stable to heat and may remain active. Staphylococcal food poisoning can be prevented by careful sanitation methods so that the food does not become inoculated, by storage of the food at low temperatures to prevent staphylococcal growth, and by the discarding of foods stored for any period of hours at warm temperatures.

Perfringens food poisoning *Clostridium perfringens* is a major cause of food poisoning in the United States. *C. perfringens* produces an enterotoxin (see Section 11.11) which catalyzes diarrhea and intestinal cramps, but not vomiting, in those infected. The symptoms last for about 24 hours and fatalities are rare. The disease results from the ingestion of a large dose ($>10^8$ cells) of *C. perfringens* and is most frequently associated with the consumption of tainted meat or meat products. The bacteria form spores in the intestine and during the sporulation process the enterotoxin is produced and released. The toxin alters the permeability of the intestinal epithelium which leads to the symptoms described above. The onset of symptoms begins some 8–22 hours after consumption of the contaminated food. *C. perfringens*

Table 15.5 The major food-poisoning/food infection bacteria

Organism	Products involved
Salmonella spp.	Chicken, other meats; milk and cream, eggs
Staphylococcus aureus	Meat dishes, desserts
Clostridium perfringens	Cooked and reheated meats and meat products
Vibrio parahaemolyticus	Sea foods
Bacillus cereus	Rice and other starchy foods
Clostridium botulinum	Home-canned vegetables, smoked fish
Campylobacter jejuni	Chicken, milk
Yersinia enterocolitica	Pork, milk

is quite common in a variety of cooked and uncooked foods, especially meat, poultry and fish, and in soil and sewage. The organism is present naturally in low numbers in the human gut, but apparently is not a successful enough competitor to reach large numbers. Large doses of *C. perfringens* are most likely to be obtained from meat dishes where the meat has been cooked in bulk lots (heat penetration in these situations is often slow and insufficient) and then left at 20–40°C for short periods. Spores of *C. perfringens* germinate and grow quickly in the meat and can easily reach the required dosage in a brief period.

Botulism Botulism is the most severe type of food poisoning; it is usually fatal, and occurs following the consumption of food containing the exotoxin produced by the anaerobic bacterium ***Clostridium botulinum***. This bacterium normally lives in soil or water, but its spores may contaminate raw foods before harvest or slaughter. If the foods are properly processed so that the *C. botulinum* spores are killed, no problem arises; but if viable spores are present, they may initiate growth and even a small amount of the resultant neurotoxin can render the food poisonous. We discussed the nature and action of botulinum toxin in Section 11.10 (see also Figure 11.18). The toxin itself is destroyed by heat (80°C for 10 minutes) so that a properly cooked food should be harmless, even if it did originally contain the toxin. Most cases of botulism occur as a result of eating foods that are not cooked after processing. Canned vegetables and beans are often used without cooking in making cold salads. Similarly, smoked fish and meat and most of the vacuum-packed sliced meats are often eaten directly, without heating. If these products contain the botulinum toxin, then ingestion of even a small amount will result in this severe and highly dangerous type of food poisoning.

Infant botulism occurs occasionally when spores of *C. botulinum* are ingested. If the infant's normal flora is not well developed or if the infant is undergoing antibiotic therapy, the spores may germinate and *C. botulinum* cells grow and release toxin. Honey as a vehicle of infant botulism has been suspected in a number of cases.

Salmonellosis Although sometimes called a food poisoning, gastrointestinal disease due to foodborne *Salmonella* is more aptly called ***Salmonella* food infection**, because symptoms arise only after the pathogen grows in the intestine (hence symptoms can be experienced as late as several *days* after eating a contaminated food). The symptoms of salmonellosis include the sudden onset of headache, chills, vomiting, and diarrhea, followed by a fever that lasts a few days. Diagnosis is from symptoms and by culture of the organism from feces. Virtually all species of *Salmonella* are pathogenic for humans: one, *S. typhi*, causes the serious human disease typhoid fever, while a small number of other species cause foodborne gastroenteritis. The ultimate sources of the foodborne salmonellas are humans and warm-blooded animals. The organism reaches food by contamination from food handlers; or in the case of foods such as eggs or meat, the animal which produced the food may be the source of contamination. The foods most commonly implicated are meats and meat products (such as meat pies, sausage, cured meats), poultry, eggs, and milk and milk products. If the food is properly cooked, the organism will be killed and no problem will arise, but many of these products are eaten uncooked or partially cooked. *Salmonella* strains causing foodborne gastroenteritis are often traced to products made with uncooked eggs such as custards, cream cakes, meringues, pies, and eggnog. Previously cooked foods that have been warmed and held without refrigeration or canned foods held for a while after opening often support the growth of *Salmonella* if they have become contaminated by an infected food handler. *Salmonella* infection is more common in summer than in winter, probably because environmental conditions are more favorable for growth of salmonellae in foods. The seasonal distribution of *Salmonella* infections in humans is shown by the frequency of isolation in the *Salmonella* Surveillance Program, illustrated in Figure 15.31.

Campylobacter *Campylobacter* is a Gram-negative, curved rod which grows at reduced oxygen tensions, that is, as a microaerophile. Two major species are recognized, *C. jejuni* and *C. fetus*, and together

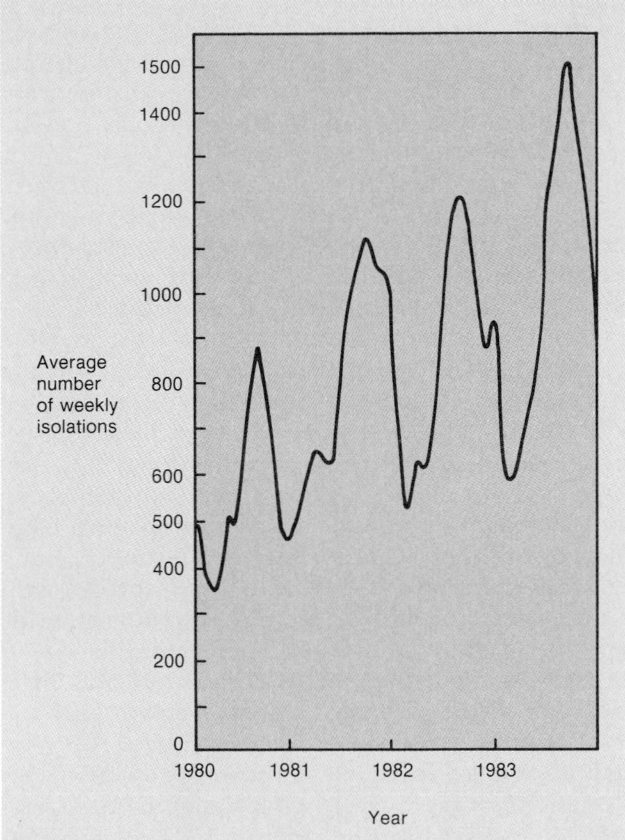

Figure 15.31 Seasonal distribution of *Salmonella* infection as revealed by isolations from humans in the *Salmonella* Surveillance Program of the Centers for Disease Control. Note the marked increase in isolations in mid to late summer.

these species are thought to account for the majority of cases of bacterial diarrhea in children. *C. fetus* is also of economic importance because it is a major cause of sterility and spontaneous abortion in cattle and sheep. The symptoms of *Campylobacter* infection include a strong fever (usually greater than 40°C), nausea, abdominal cramps, and a watery, frequently bloody, stool. Diagnosis requires isolation of the organism from stool samples and identification by growth dependent tests or immunological assays. Because of the frequency with which *C. jejuni* infections are observed in infants, a variety of selective media and highly specific immunological methods have been developed (see Chapter 13) for positive identification of this organism.

Campylobacter is transmitted to man via contaminated food, most frequently in poultry, pork, raw clams and other shellfish, or by a water route in surface waters not subjected to chlorination. Poultry is a major reservoir of *C. jejuni* and virtually all chicken and turkey carcasses contain this organism (Table 15.6). Beef carcasses, on the other hand, rarely harbor *Campylobacter*. Proper washing of uncooked poultry (and any utensils coming in contact with uncooked

Table 15.6 Prevalence of *Campylobacter jejuni* in processed poultry and red meats

Source	Positive for *C. jejuni* (%)
Chicken	
Whole carcass	85
Liver	85
Gizzard	89
Turkey	
Whole carcass	92
Heart, liver, and gizzards	33
Beef-whole carcass	2
Sheep-whole carcass	24
Swine-whole carcass	56

Data from: Doyle, M.P. 1984. Campylobacter *in Foods, p. 164–180 in* Campylobacter *infection in man and animals, J-P. Butzlen (ed.), CRC Press, Boca Raton, FL.*

poultry) and thorough cooking of the meat eliminates the problem of *Campylobacter* infection. *Campylobacter* species also infect domestic animals such as dogs, causing a milder form of diarrhea than that observed in humans. Not surprisingly, therefore, a number of *Campylobacter* infections in infants have been traced to contact between the infant and puppies; diarrhea in puppies is rather common and is frequently caused by *Campylobacter*. A recent study has shown that about half of all puppies carry *Campylobacter*.

Hepatitis Infectious hepatitis (hepatitis A) is a viral-mediated inflammation of the liver caused by a picornavirus (see Section 6.17). The virus is transmitted primarily through fecal contamination of water, food, or milk. Hepatitis can be subclinical in mild cases, or it can lead to severe liver damage in chronic infections. The virus spreads from the intestine via the bloodstream to the liver and usually results in jaundice, a yellowing of the skin and eyes, and a browning of the urine due to stimulation of bile pigment production by infected liver cells. An immune response is initiated against the hepatitis A virus and this eventually brings the condition under control. However, in severe cases, permanent loss of a portion of liver function can occur.

The most significant food vehicles for infectious hepatitis are shellfish (oysters and clams) harvested from waters polluted with human feces. As filter feeders, shellfish living in such environments tend to concentrate the hepatitis virus. Only raw shellfish are a problem because hepatitis A virus is destroyed by heating. Therefore the best means of controlling hepatitis transmission include sound sanitary practices, especially sewage treatment, the prevention of fecal contamination of food products by infected food handlers, and avoiding the consumption of uncooked shellfish.

Serum hepatitis (hepatitis B) is caused by a *DNA-containing* hepatitis virus transmitted primarily by infected blood or blood products. Serum hepatitis frequently results in more severe liver damage than infectious hepatitis, leading to death in up to 10 percent of those infected; hepatitis A rarely causes death. Those at high risk for serum hepatitis include individuals that require frequent blood transfusions, dialysis patients, and intravenous drug abusers. As for infectious hepatitis, no specific therapy for serum hepatitis is currently available. However, due to the seriousness of the infection, a vaccine against serum hepatitis is now available. Use of the serum hepatitis vaccine is recommended for those in the high risk groups already mentioned, as well as for health personnel who come in frequent contact with blood or blood products. A variety of immunological tests, especially ELISAs (see Section 12.9), have been devised for the diagnosis of infectious and serum hepatitis by either antibody detection or direct viral detection methods.

Traveler's diarrhea Traveler's diarrhea is an extremely common enteric infection in North Americans and Europeans traveling to developing countries. The primary causal agent is enteropathogenic *Escherichia coli* (EEC), although *Salmonella* and *Shigella* are sometimes implicated. The *E. coli* is an enterotoxin-producing strain, and as noted in Section 11.11, the gene for enterotoxin is often plasmid-linked. The K antigen, necessary for successful colonization of the small intestine, is also generally plasmid-linked (Section 11.8).

Several studies have been done on groups of U.S. citizens traveling in Mexico. Such studies have shown that the infection rate is often quite high, greater than 50 percent, and that the prime vehicles are foods, such as uncooked vegetables (for example, lettuce in salads) and water. The impressively high infection rate in travelers has been shown to be due to the fact that the local population has a marked immunity to the infecting strains, due undoubtedly to the fact that they have lived with the agent for a long period of time. Secretory antibodies present in the bowel may prevent successful colonization of the pathogen in local residents, but when the organism colonizes the intestine of a non-immune person, it finds a hospitable environment. Also, stomach acidity, so often a barrier to intestinal infection, may not be able to act if only small amounts of liquid are consumed (as, for instance, the melting ice of a cocktail), since small amounts of liquid induce rapid emptying of the stomach and hence pass through so quickly that stomach acidity may have no effect on an enteric pathogen present in the liquid.

Assessing microbial content of foods All fresh foods will have some viable microbes present. The purpose of assay methods is to detect evidence of abnormal microbial growth in foods or to detect the presence of specific organisms of public health concern, such as *Salmonella, Staphylococcus*, or *Clostridium botulinum*. With nonliquid foods, preliminary treatment is usually required to suspend in a fluid medium those microorganisms embedded or entrapped within the food. The most suitable method for treatment is high-speed blending. Examination of the food should be done as soon after sampling as possible, and if examination cannot begin within 1 hour of sampling the food should be refrigerated. A frozen food should be thawed in its original container in a refrigerator and examined as soon as possible after thawing is complete. For *Salmonella* several selective media are available (see Section 13.2), and tests for its presence are most commonly done on animal food products, such as raw meat, poultry, eggs, and powdered milk, since *Salmonella* from the animal may contaminate the food. For staphylococcal counts, a medium high in salt (either sodium chloride or lithium chloride at a final concentration of 7.5 percent) is used, since of the organisms present in foods, staphylococci are the only common ones tolerant of such levels of salt. Since *Staphylococcus aureus* is responsible for one of the most common types of food poisoning, staphylococcal counts are of considerable importance. See Section 9.21 for a discussion of food preservation methods.

15.12 Waterborne Diseases

Pathogenic organisms transmitted by water Organisms pathogenic to humans that are transmitted by water include bacteria, viruses, and protozoa (Table 15.7). Organisms transmitted by water usually grow in the intestinal tract and leave the body in the feces. Fecal pollution of water supplies may then occur, and if the water is not properly treated, the pathogens enter a new host when the water is consumed. Because water is consumed in large quantities, it may be infectious even if it contains only a small number of pathogenic organisms.

Probably the most important pathogenic *bacteria* transmitted by the water route are *Salmonella typhi*, the organism causing typhoid fever, and *Vibrio cholerae*, the organism causing cholera. Although the causal agent of typhoid fever may also be transmitted by contaminated food (see Section 15.11) and by direct contact from infected people, the most common and serious means of transmission is the water route. Typhoid fever has been virtually eliminated in many parts of the world, primarily as a result of the development of effective water treatment methods. However, typhoid fever does still occur occasionally, usually in the summer months, when swimmers are active in polluted water supplies. A breakdown in water pu-

Table 15.7 Water-borne disease outbreaks due to microorganisms[a]

Disease	Causal agent	Outbreaks, percent[b]	Cases, percent[c]
Bacteria			
Typhoid fever	*Salmonella typhi*	10	0.5
Shigellosis	*Shigella* species	9	9
Salmonellosis	*Salmonella paratyphi*, etc.	3	12
Gastroenteritis	*Escherichia coli*	0.3	2.5
	Campylobacter species	0.3	2.5
Viruses			
Infectious hepatitis	Hepatitis A virus	11	1.6
Poliomyelitis	Polio virus	0.2	0.01
Diarrhea	Norwalk virus	1.5	2
Protozoa			
Dysentery	*Entamoeba histolytica*	0.1	0.05
Giardiasis	*Giardia lamblia*	7	13
Unknown etiology			
Gastroenteritis		57	58

[a]*Compiled from data provided by the Centers for Disease Control.*
[b]*Of 650 outbreaks in recent decades.*
[c]*Of 150,000 cases over the same period.*

rification methods, contamination of water during floods, earthquakes, and other disasters, or cross contamination of water pipes from leaking sewer lines occasionally results in epidemics of typhoid fever. Water that has been contaminated can be rendered safe for drinking by boiling for 5 to 10 minutes or by adding chlorine as discussed later. Usually, contaminated water is safe to use for laundry or other domestic cleaning purposes, provided care is taken that it is not swallowed.

The causal agent of cholera is transmitted only by the water route. At one time cholera was common in Europe and North America, but the disease has virtually been eliminated from these areas by effective water purification. The disease is still common in Asia, however, and travelers from the West are advised to be vaccinated for cholera. Both *V. cholerae* and *S. typhi* are eliminated from sewage during proper sewage treatment and hence do not enter water courses receiving treated sewage effluent. More frequent than typhoid, but less serious a disease, is salmonellosis caused by species of *Salmonella* other than *S. typhi* (see Section 15.11). As seen in Table 15.7, the largest number of cases of waterborne bacterial disease in the United States have been due to salmonellosis.

Enteric bacteria are effectively eliminated from water during the water purification process (see below), so that they should never be present in properly treated drinking water. Most outbreaks of waterborne disease in the United States are due to breakdowns in treatment systems, or as a result of post-contamination in the pipelines. This latter problem can be controlled by maintaining a detectable level of free chlorine in the pipelines.

Viruses transmitted by the water route include poliovirus and other viruses of the enterovirus group, as well as the virus causing infectious hepatitis. Poliovirus has several modes of transmittal, and transmission by water may be of serious concern in some areas. This is one of the pathogens more commonly encountered during the summer months in polluted swimming areas.

Infectious hepatitis is caused by a virus (hepatitis A virus), which resembles the enteroviruses, but hepatitis A virus has not been successfully cultured so that its characteristics are poorly known. Although one of the most serious waterborne viral disease agents at present, hepatitis is also transmitted in foods (see Table 15.4), and probably most of the infectious hepatitis cases arise from foodborne rather than waterborne means (see Section 15.11)

Because viruses are acellular, they are more stable in the environment and are not as easily killed as bacteria. However, both poliovirus and infectious hepatitis virus are eliminated from water by proper treatment practices, and the maintenance of 0.6 ppm free chlorine in a water supply will generally ensure its safety.

Cholera We discussed the action of cholera enterotoxin in Section 11.11. The disease cholera is caused by *Vibrio cholerae*, a Gram-negative, curved rod transmitted almost exclusively via contaminated water. Cholera enterotoxin catalyzes a life-threatening diarrhea which can result in dehydration and death unless the patient is given fluid and electrolyte therapy. The disease has swept the world in seven major pandemics, the most recent in 1961. Today the disease is virtually absent from the western world, although it is still common in areas where sewage treatment is either not practiced or poorly performed.

Two major biotypes of *V. cholerae* have been recognized, the *classic* and the *El Tor* types. Each biotype has two major serotypes. The classic strain, such as the *V. cholerae* strain first isolated by Robert Koch in 1883 was more prevalent in cholera outbreaks before 1960, while the El Tor strain has been more frequently observed since that time. Following ingestion of a substantial inoculum, the *V. cholerae* cells take up residence in the *small* intestine. Studies with human volunteers have shown that the acidity of the stomach is responsible for the large inoculum needed to initiate cholera. Although the ingestion of 10^{10} cholerae vibrios is usually sufficient to cause cholera, human volunteers given bicarbonate to neutralize gastric acidity developed the disease when only 10^4 cells were administered. Far lower cell numbers are required to initiate infection if administered with food, as well. Cholera vibrios attach firmly to small intestinal epithelium and grow and release enterotoxin. The enterotoxin causes large loss of fluids, 20 liters per day not being uncommon in a fulminant case of cholera. If untreated, the mortality rate can be as high as 60 percent. Intravenous or oral liquid replacement therapy is the major means of treatment (see Section 11.11 and Figure 11.20). Streptomycin or tetracycline may shorten the course of cholera but antibiotics are of little benefit without simultaneous fluid replacement.

Control of cholera depends primarily on satisfactory sanitation measures, particularly in the treatment of sewage and the purification of drinking water. In the last ten years sporadic outbreaks of cholera (10–20 cases total) have been reported in the United States, and some evidence exists that shellfish may be an alternate vehicle. Cholera is heavily endemic in India, Pakistan, and Bangladesh, with occasional epidemics fluorishing from time to time. Although most of these cases are due to water contaminated with human feces, cholera is also spread within households in these regions by fecally contaminated food.

Giardiasis Giardiasis is an acute form of gastroenteritis caused by the protozoan parasite, *Giardia lamblia* (Figure 15.32). *G. lamblia* is a flagellated protozoan that is transmitted to humans primarily by contaminated water, although foodborne and even venereal transmission of giardiasis has been documented. The protozoal cells, called trophozoites (Figure 15.32*a*) produce a resting stage called a cyst (Figure 15.32*b*), and this is the primary form transmitted by water. Cysts germinate in the gastrointestinal tract and catalyze the symptoms of giardiasis: an explosive, foul-smelling, watery diarrhea, and intestinal cramps, flatulence, nausea, and malaise. The foul-smelling nature of the diarrhea and the absence of blood or mucous in the stool are diagnostically helpful in distinguishing giardiasis from diarrhea of bacterial or viral origin. The drugs quinacrine and metronidazole are useful in treating the disease.

Between 1965–1981, 53 waterborne outbreaks of giardiasis affecting over 20,000 people were reported in the United States. Most outbreaks have occurred in undeveloped or mountainous regions where surface water sources have been used for drinking purposes. *Giardia* cysts are fairly resistant to chlorine and many outbreaks have been associated with water systems using only chlorination as a means of water purification. Water subjected to proper sedimentation/filtration (see below) in addition to being chlorinated is generally free of *Giardia* cysts. Many cases of giardiasis have been associated with drinking untreated water in wilderness areas. Although water may look "crystal clear," it may contain fecal material from animals harboring *Giardia*. Studies of wild animals have indicated that beavers and muskrats are major carriers of *Giardia* and may transmit cells or cysts to water supplies. As a safety precaution, all water consumed from natural rivers and streams, for example during a camping or hiking trip, should either be filtered or boiled. Boiling is the preferred method of rendering water microbially safe.

D.E. Feely, S.L. Erlandsen, and D.G. Case S.L. Erlandsen

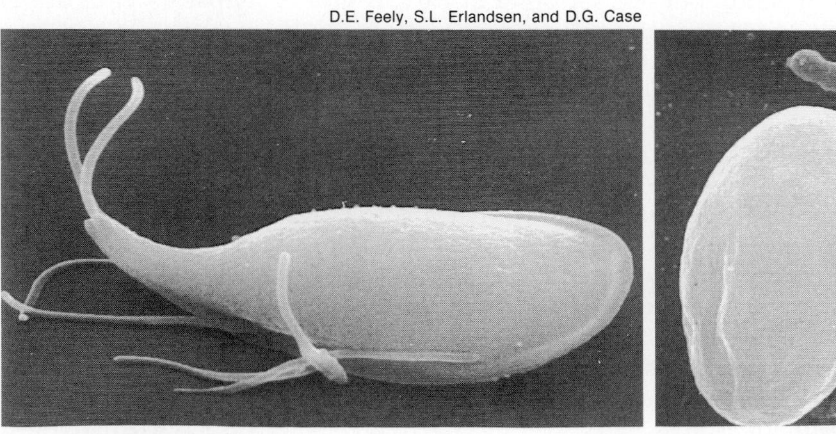

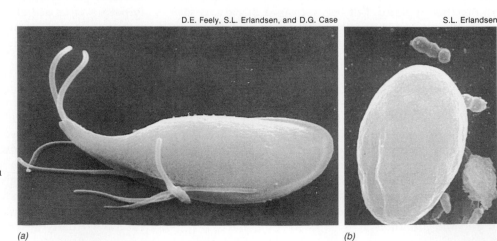

Figure 15.32 Scanning electron micrographs of the parasite *Giardia*. (a) Motile trophozoite. (b) Cyst. Magnifications: (a) 9200×. (b) 7000×.

Amebic dysentery A number of different amoebas inhabit the tissues of humans and other vertebrates, usually in the oral cavity or intestinal tract. Certain of these are pathogenic. *Entamoeba histolytica*, the causative agent of amebic dysentery, is a common pathogenic protozoan transmitted to man primarily by contaminated water and occasionally by the foodborne route. *E. histolytica* is an *anaerobic* amoeba, the trophozoites lacking mitochondria. Like *Giardia*, the trophozoites of *Entamoeba* produce cysts. Cyst germination occurs in the intestine and cells grow both on and in intestinal mucosal cells. Continued growth leads to ulceration of intestinal mucosa causing diarrhea and severe intestinal cramps. Diarrhea is replaced by a condition referred to as *dysentery*, characterized by the passage of intestinal exudates, blood, and mucous. If not treated, trophozoites of *E. histolytica* can migrate to the liver, lung, and brain. Growth in these tissues can cause abcesses and other tissue damage.

Amebic dysentery can be treated with the drugs metronidazole and chloroquine, but experience has shown that amebicidal drugs are not always effective. Spontaneous cures do occur, implying that the host immune system plays some role in ending the infection. However, protective immunity is not afforded primary infection because reinfection is not uncommon. Amebic dysentery is rather easily diagnosed by examining stool samples for the morphologically distinct cysts of *E. histolytica*. Amebic dysentery occurs at very low incidence in those regions which practice sanitary sewage treatment. Ineffective sewage treatment or use of untreated surface waters for drinking purposes are the usual scenarios for cases of amebic dysentery.

Indicator organisms and the coliform test As we have just discussed, a number of important diseases are waterborne. Even water which looks clear and pure may be sufficiently contaminated with pathogenic microorganisms to be a health hazard. Some means are necessary to ensure that drinking water is safe. One of the main tasks of water microbiology is the development of laboratory methods which can be used to detect the microbiological contaminants that may be present in drinking water. It is not usually practical to examine water directly for the various pathogenic organisms that may be present. As stated earlier, a wide variety of organisms may be present, including bacteria, viruses, and protozoa. To check each drinking water supply for each of these agents would be a difficult and time-consuming job. In practice, *indicator* organisms are used instead. These are organisms associated with the intestinal tract, whose presence in water indicate that the water has received contamination of an intestinal origin. The most widely used indicator is the **coliform group** of organisms. This group is defined in water bacteriology as all the aerobic and facultatively anaerobic, Gram-negative, nonspore-forming, rod-shaped bacteria that ferment lactose with gas formation within 48 hours at 35°C. This is an operational rather than a taxonomic definition, and the coliform group includes a variety of organisms, mostly of intestinal origin. In practice, the coliform organisms are almost always members of the enteric bacterial group (see Section 19.20). The coliform group includes the organism *Escherichia coli* a common intestinal organism, plus the organism *Klebsiella pneumoniae*, a less common intestinal organism. The definition also currently includes organisms of the species *Enterobacter aerogenes*, not generally associated with the intestine.

The coliform group of organisms are suitable as indicators because they are common inhabitants of the intestinal tract, both of humans and warm-blooded animals, and are generally present in the intestinal tract in large numbers. When excreted into the water environment, the coliform organisms eventually die, but they do not die at any faster rate than the pathogenic bacteria *Salmonella* and *Shigella*, and both the coliforms and the pathogens behave similarly during water purification processes. Thus it is likely that if coliforms are found in a water sample, the water has received fecal contamination and may be unsafe for drinking purposes. There are very few organisms in nature that meet the definition of the coliform group that are not associated with the intestinal tract. It should be emphasized that the coliform group includes organisms derived not only from humans but from other warm-blooded animals. Since many of the pathogens (for example, *Salmonella, Leptospira*)

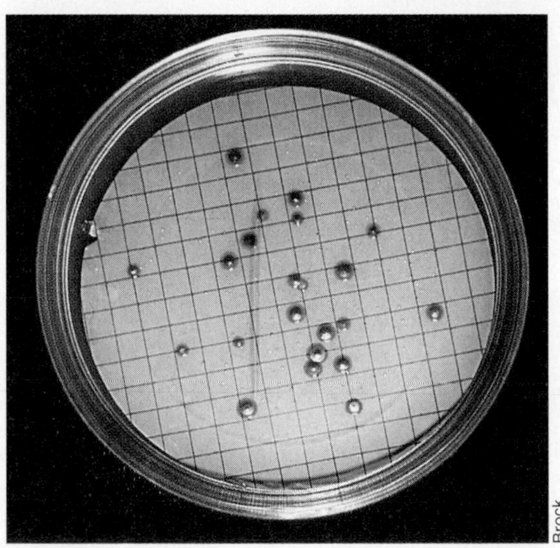

Figure 15.33 Coliform colonies growing on a membrane filter. A drinking water sample has been passed through the filter and the filter placed on a culture medium which is both selective and differential for lactose-fermenting bacteria (coliforms). The dark color of the colonies is characteristic of coliforms. A count of the number of colonies gives a measure of the coliform count of the original water sample.

found in warm-blooded animals will also infect humans, an indicator of both human and animal pollution is desirable.

There are two types of procedures that are used for the coliform test. These are the **most-probable-number** (MPN) procedure and the **membrane filter** (MF) procedure. The MPN procedure employs liquid culture medium in test tubes, the samples of drinking water being added to the tubes of media. In the more common MF procedure, the sample of drinking water is passed through a sterile membrane filter, which removes the bacteria (see Figure 9.35), and the filter is then placed on a culture medium for incubation. When using the membrane filter method with drinking water, at least 100 ml of water should be filtered, although in clean water systems, even larger volumes could be filtered. After filtration of a known volume of water, the filter is placed on the surface of a plate of a culture medium highly selective for coliform organisms (see Section 13.2). The coliform colonies (Figure 15.33) are counted and from this value the number of coliforms in the original water sample can be determined. In well-regulated water systems, coliforms will always be negative.

Louisville Water Company

Ohio River

River pumping station

Coagulation basin

Sedimentation basins

Chlorination

Filter buildings

Underground clear-water reservoir

(a)

Metropolitan Water District of Southern California

(b)

Metropolitan Water District of Southern California

(c)

Figure 15.34 Water purification plants. (a) Aerial view of water treatment plant of Louisville, Ky. The arrows indicate direction of flow of water through the plant. (b) Los Angeles, California obtains most of its water from distant sources. This photograph shows a large open viaduct bringing Colorado River water across the Mojave Desert. (c) One of several water filtration plants for the City of Los Angeles.

Water purification It is a rare instance when available water is of such a clarity and purity that no treatment is necessary before use. Water treatment is carried out both to make the water safe microbiologically and to improve its utility for domestic and industrial purposes. Treatments are performed to remove pathogenic and potentially pathogenic microorganisms and also to decrease turbidity, eliminate taste and odor, reduce or eliminate nuisance chemicals such as iron or manganese, and soften the water to make it more useful for the laundry.

The kind of treatment that water is given before use must depend on the quality of water supply. A typical treatment installation for a large city using a source water of poor quality is shown in Figure 15.34*a*. Water is first pumped to **sedimentation basins**, where sand, gravel, and other large particles settle out. A sedimentation basin should be used only if the water supply is highly turbid, since it has the disadvantages that algal growth may occur in the basin, adding odors and flavors, and that pollution of the water by surface runoff may occur. Bacteria may grow in the bottom mud and add further problems.

Most water supplies are subjected to **coagulation**. Chemicals containing aluminum and iron are added, which under proper control of pH form a flocculent, insoluble precipitate that traps organisms, absorbs organic matter and sediment, and carries them out of the water. After the chemicals are added in a mixing basin, the water containing the coagulated material is transferred to a settling basin where it remains for about 6 hours, during which time the coagulum separates out. Around 80 percent of the turbid material, color, and bacteria are removed by this treatment.

After coagulation, the clarified water is usually filtered to remove the remaining suspended particles and microbes. Filters can be of the slow or rapid sand type. **Slow sand filters** are suitable for small installations such as resorts or rural places. The water is simply allowed to pass through a layer of sand 2 to 4 ft deep. Eventually the top of the sand filter will become clogged and the top layer must be removed and replaced with fresh sand. **Rapid sand filters** are used in large installations. The rate of water flow is kept high by maintaining a controlled height of water over the filter. When the filter becomes clogged, it is clarified by backwashing, which involves pumping water up through the filter from the bottom. From 98 to 99.5 percent of the total bacteria in raw water can be removed by proper settling and filtration.

Chlorination is the most common method of ensuring microbiological safety in a water supply. In sufficient doses it causes the death of most microorganisms within 30 minutes. In addition, since most taste- and odor-producing compounds are organic, chlorine treatment reduces or eliminates them. It also oxidizes soluble iron and manganese compounds, forming precipitates which can be removed. Chlorine can be added to water either from a concentrated solution of sodium or calcium hypochlorite or as a gas from pressure tanks. The latter method is used most commonly in large water treatment plants, as it is most amenable to automatic control.

When chlorine reacts with organic materials it is used up. Therefore, if a water supply is high in organic materials, sufficient chlorine must be added so that there is a residual amount left to react with the microorganisms after all reactions with organic materials have occurred. The water plant operator must perform chlorine analyses on the treated water to determine the residual level of chlorine. A chlorine residual of about 0.2 to 0.6 μg/ml is an average level suitable for most water supplies.

After final treatment the water is usually pumped to storage tanks, from which it flows by gravity to the consumer. Covered storage tanks are essential to ensure that algal growth does not take place (no light is available) and that dirt, insects, or birds do not enter. Because of low rainfall, some cities have to transport water long distances (Figure 15.34*b*). In such cases, water *quantity* may appear more important than water *quality*, but water purification procedures must still be followed carefully (Figure 15.34*c*).

Drinking water standards Drinking water standards in the United States are specified under the Safe Drinking Water Act, which provides a framework for the development by the Environmental Protection Agency (EPA) of drinking water standards. Current standards prescribe that when the MF technique is used, 100-ml samples must be filtered, and the number of coliform bacteria shall not exceed any of the following: (1) 1 per 100 ml as the arithmetic mean of all samples examined per month; (2) 4 per 100 ml in more than one sample when less than 20 are examined per month; or (3) 4 per 100 ml in more than 5 percent of the samples when 20 or more are examined per month. Water utilities report their results to the EPA and if they are not meeting the prescribed standards they must notify the public and take steps to correct the problem. Many smaller communities and even large cities sometimes fail to meet the standards.

Public health significance of drinking water purification Today the incidence of waterborne disease is so low that it is difficult to appreciate the significance of treatment practices and drinking water standards. Most intestinal infection today is not due to transmission by the water route, but via food (see Section 15.11). It was not always so. At the beginning of the twentieth century, effective water treatment practices did not exist, and there were no bacteriological methods for evaluating the health significance

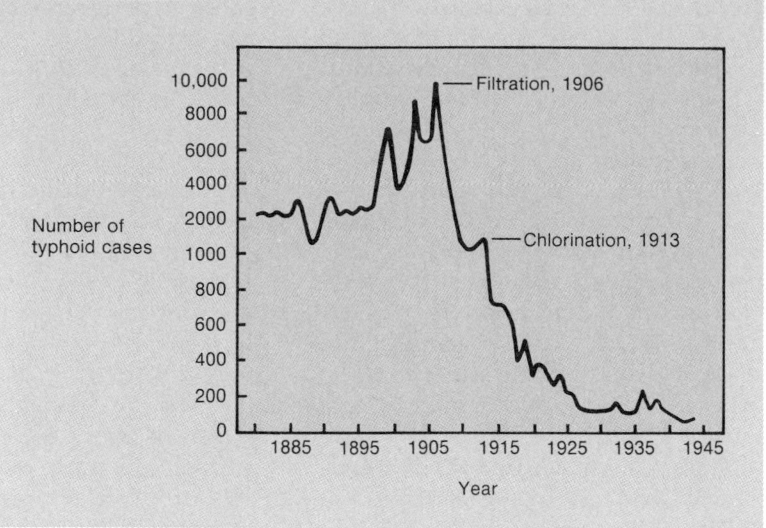

Figure 15.35 The dramatic effect of water purification on incidence of water-borne disease. The graph shows the incidence of typhoid fever in the City of Philadelphia during the early part of the twentieth century. Note the marked reduction in incidence of the disease after the introduction of filtration and chlorination.

of polluted drinking water. The first coliform counting procedures were introduced about 1905. Up until then, water purification, if practiced at all, was primarily for aesthetic purposes, to remove turbidity. Actually, turbidity removal by filtration provides for a significant decrease in the microbial load of a water, so that filtration did play a part in providing safer drinking water. But filtration alone was of only partial value, since many organisms passed through the filters. It was the discovery of the efficacy of chlorine as a water disinfectant in about 1910, which was to have major impact. Chlorine is so effective and so inexpensive that its use spread widely, and it is almost certain that the practice of chlorination was of major significance in reducing the incidence of waterborne disease. However, the effectiveness of chlorination would not have been realized, and the necessary doses could not have been determined, if the standard methods for assessing the coliform content of drinking water had not been developed. Thus engineering and microbiology moved forward together.

The significance of filtration and chlorination for ensuring the safety of drinking water cannot be overemphasized. Figure 15.35 illustrates the dramatic drop in incidence of typhoid fever in a major American city after these two water purification procedures were introduced. Similar results were obtained in other major cities. The dramatic improvement in the health of the American people in the early decades of the twentieth century was due to a very large extent to the establishment of satisfactory water purification procedures.

Study Questions

1. Give an explanation why many *respiratory bacterial pathogens* are Gram-positive.
2. Most sneezes generate an *aerosol* of mucous droplets containing microorganisms. Why are infections due to sneeze transmission more likely to be *upper* rather than *lower* respiratory tract infections?
3. Explain how *rheumatic fever* can be considered a type of *autoimmune* response.
4. The *DPT* and *MMR vaccines* are routinely administered to children at several times from infancy through early childhood (see Table 12.8). What is the composition of each of these so-called *trivalent vaccines* and to what diseases do they offer protection?
5. From what has been described about the reservoir of *Legionella pneumophila*, explain why this disease shows an atypical disease cycle as compared with other respiratory diseases.
6. The *Schick test* and the *tuberculin test* are two immunological procedures in wide use for the diagnosis of two important respiratory diseases. Compare and contrast these two tests in terms of (1) the disease being diagnosed, (2) the nature of the antigen used in the test, (3) the nature of the immunological response leading to a positive or negative test result.
7. Why does the disease *tuberculosis* usually lead to a permanent reduction in the respiratory capacity of an individual while most other respiratory pathogens generally cause only temporary respiratory problems?
8. Why are *colds* and *influenza* such common respiratory diseases? Discuss at least five reasons why this might be so.
9. Discuss the molecular biology and immunological basis of *antigenic shift* and the reasons why this phenomenon effectively thwarts control measures for influenza. Why do cycles of influenza occur about every 3–5 years?
10. Compare and contrast the diseases *measles, mumps*, and *rubella*. Include in your discussion a description

of the pathogen, major symptoms encountered, and any potentially serious consequences of these infections. Why is it essential that females be vaccinated against rubella at an early age?

11. Discuss at least three reasons why the incidence of *gonorrhea* rose dramatically in the mid 1960s while that of *syphilis* did not.
12. Despite the ease with which gonorrhea can be diagnosed and cured, its incidence remains high. Why?
13. What *sexually-transmitted disease* is most prevalent in the United States today? Why is the causative agent of this disease difficult to identify by traditional clinical microbiological methods and what modern techniques are used to confirm a diagnosis?
14. For the sexually-transmitted diseases listed below describe the methods for treating the infection. In each case is the treatment an effective cure? Why or why not? Why is trichomoniasis not treatable with antibiotics such as penicillin and tetracycline?
    1. Gonorrhea
    2. Syphilis
    3. Chlamydial infections
    4. Genital herpes
    5. Trichomoniasis
15. Describe how HIV effectively shuts down most aspects of both humoral immunity and cell-mediated immunity. Are there any aspects of the immune system that remain functional in cases of AIDS?
16. What is the basis for treatment of AIDS? Why does it work? Is this treatment a cure? Why or why not?
17. Describe the sequence of events you would take if your child received an unprovoked bite from a stray dog.
18. Discuss at least three common properties of the disease agents and/or the disease process for the following diseases: *Rocky Mountain Spotted Fever, Typhus Fever, Q Fever.*
19. The classic symptoms of *malaria*, fever followed by chills, are related to activities of the pathogen *Plasmodium vivax*. Describe the growth stages of this pathogen in the human host and relate these to the classic symptoms. Why is a person of Western European descent more susceptible to malaria than a person of West African descent?
20. Contrast *food infection* and *food poisoning*. Give an example of each and indicate how each of these could be effectively controlled.
21. Write a scenario for a typical food poisoning outbreak due to *Staphylococcus*. (That is, what kind of conditions might lead to the development of the toxic food and how might the outbreak arise?)
22. Compare and contrast *perfringens* and *botulinum* food poisoning. Discuss similarities and differences between the organisms involved, the toxins produced, and the symptoms observed.
23. Why is the incidence of *Salmonella* infection higher in the human population in the summer than in the winter?
24. What common modes of transmission are involved in salmonellosis and *Campylobacter* food infections?
25. Why was the development of the *coliform test* necessary in order to perfect effective methods for drinking water purification?
26. Describe a typical *water purification plant* and indicate the role of the various stages in the process.
27. Describe the steps used to carry out a coliform test on a sample of drinking water. What are the acceptable limits for coliforms in drinking quality water?

Supplementary Readings

Baron, S. 1986. *Medical Microbiology*. Addison-Wesley, Menlo Park, CA. An advanced level medical microbiology text; very complete and up-to-date.

Cukor, G., and **N. R. Blackow.** 1984. *Human viral gastroenteritis*. Microbiological Reviews 48:157–179.

Crowlery, L. V. 1983. *Introduction to Human Disease*. Wadsworth Health Science Division, Wadsworth, Inc., Belmont, CA. A textbook intended for students in the health professions. Has extensive coverage of noninfectious diseases such as cancers, organ diseases, etc., as well as covering basic infectious diseases.

Farthing, C. F., S. E. Brown, R. C. D. Staughton, J. J. Cream, and **M. Mühlemann**. 1986. *A Colour Atlas of AIDS: Acquired Immunodeficiency Syndrome*. Wolfe Medical Publications, Ltd., London. An excellent colored atlas containing descriptions of clinical manifestations of AIDS and AIDS related diseases. Includes material on clinical methods for diagnosis of AIDS.

Felman, Y. M. 1986. *Sexually-transmitted Diseases*. Churchill Livingston, New York. A collection of well illustrated chapters, written by experts on the complete spectrum of sexually-transmitted diseases.

Gallo, R. C. 1987. *The AIDS virus*. Scientific American 256:46–56. An overview of HIV and the acquired immune deficiency syndrome.

Institute of Medicine, National Academy of Sciences. 1986. *Confronting AIDS—Directions for Public Health, Health Care, and Research*. National Academy Press, Washington, D.C. The report of a National Academy of Sciences (USA) committee on all aspects of AIDS including epidemiology, pathogenesis, social aspects, research needs, and medical/legal aspects of the acquired immune deficiency syndrome.

Jolik, W. K., H. P. Willett, and **D. B. Amos.** 1984. *Zinsser Microbiology, 10th ed.* Appleton-Century-Crofts, New York. A standard medical school textbook of microbiology. Quite detailed and of considerable reference value.

16

Metabolic Diversity among the Microorganisms

Up until now, we have been discussing primarily microorganisms which can be said to have a "conventional" metabolism. These have been microbes that use organic compounds as energy sources, primarily via aerobic (respiratory) metabolism. In Chapter 4, we discussed briefly alternate modes of energy metabolism and indicated that there were interesting types of metabolism which we would return to later in this book. We discuss in this chapter some of these alternate microbial "life styles."

The organisms we are about to discuss are some of the most widespread and significant organisms on the planet earth. They play major roles in the functioning of the whole biosphere, and they are of great importance in agriculture and in other aspects of human affairs, as we will discuss in the next chapter.

Microorganisms show impressive diversity of metabolic processes. The discussion in this chapter will build on earlier discussions of microbial metabolism from Chapter 4. Figure 16.1 summarizes the basic types of energy metabolism and lists some of the terms that have been used. As seen, energy metabolism can be divided into two broad categories, that in which the energy source is *chemical*, and that in which *light energy* is involved. Many microorganisms resemble animals and humans, using *organic* chemicals as energy sources. We call these organisms **heterotrophs** or **organotrophs**. An interesting collection of microorganisms, exclusively procaryotic, use *inorganic* chemicals as energy sources. These organisms are called **lithotrophs** (from the Latin *litho-* meaning *rock*). These lithotrophic microorganisms will be discussed in some detail later in this chapter.

A major class of living organisms, called **phototrophs**, use *light* as an energy source. Phototrophs include, of course, the green plants, but also many microorganisms, both procaryotic and eucaryotic. Because of the great importance of the process of photosynthesis in the functioning of the biosphere, and the inherent interest in the mechanics of the utilization of light energy, we discuss photosynthesis in some detail in this chapter. We will see that there are two distinct types of photosynthesis, that typified by the green plants but also found in both eucaryotic and procaryotic microorganisms, and that found exclusively in special groups of bacteria, the purple and green bacteria.

Organisms which use inorganic chemicals or light as energy sources are frequently able to develop in the complete absence of organic materials, using *carbon dioxide* as their sole source of carbon. The term **autotroph** (meaning literally, *self-feeding*) is sometimes applied to organisms able to obtain all the carbon they need from inorganic sources. Note that autotrophy does not refer to the *energy* source used, but to the *carbon* source. Autotrophs are of great importance in the functioning of the biosphere because

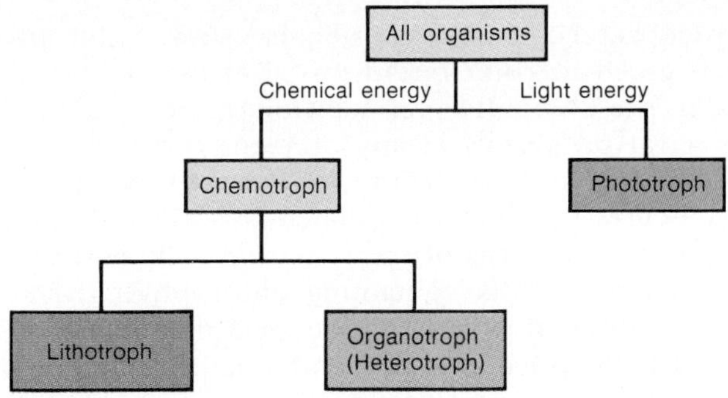

Figure 16.1 Classification of organisms in terms of energy.

they are able to bring about the synthesis of organic matter from inorganic (nonliving) sources. Because humans and other animals *require* organic carbon, the life of the biosphere itself is dependent upon the activities of autotrophic organisms. The process by which carbon dioxide is used as a sole carbon source is sometimes called *CO_2 fixation*. The details of this important and intricate biochemical process, which is at the basis of biosphere function, will also be discussed in this chapter.

It might be useful at this point to review briefly the concept of oxidation/reduction discussed in Section 4.4, especially the terms *oxidation, reduction, electron acceptor, electron donor,* and *electron transport*. We will find a number of occasions in the present chapter to use these terms, and will assume that the student is familiar with them from Section 4.4. A summary of reduction potentials of some redox pairs important in the present chapter is given in Table 16.1 (full details can be found in Appendix 1).

Table 16.1 Reduction potentials of some redox pairs of biochemical and microbiological importance

Redox pair	$E_o'(V)$
SO_4^{2-}/HSO_3^-	−0.52
$2H^+/H_2$	−0.41
Ferredoxin ox/red	−0.38
$S_2O_3^{2-}/HS^- + HSO_3^-$	−0.36
NAD/NADH (or NADP/NADPH)	−0.32
Cytochrome c_3 ox/red	−0.29
CO_2/acetate$^-$	−0.29
S^0/HS^-	−0.27
SO_4^{2-}/HS^-	−0.25
CO_2/CH_4	−0.24
Pyruvate$^-$/lactate$^-$	−0.19
HSO_3^-/HS^-	−0.11
Fumarate/succinate	+0.030
Cytochrome *b* ox/red	+0.030
Ubiquinone ox/red	+0.11
Cytochrome c_1 ox/red	+0.23
NO_2^-/NO	+0.36
Cytochrome a_3 ox/red	+0.385
NO_3^-/NO_2^-	+0.43
Fe^{3+}/Fe^{2+}	+0.77
O_2/H_2O	+0.82
N_2O/N_2	+1.36

See Appendix 1 for details.

In addition to alternate styles of energy metabolism, we also discuss in this chapter a number of specialized aspects of heterotrophic metabolism that we have passed over lightly earlier. For instance, our discussion of respiratory metabolism has centered primarily on molecular oxygen, O_2, as electron acceptor, yet there are a number of other electron acceptors that can participate in the electron transport process in particular groups of microorganisms. The process in which an electron acceptor other than molecular oxygen is used is called **anaerobic respiration** and will be one of the major features of metabolic diversity discussed in this chapter.

Our discussion of organic energy metabolism has dealt primarily with the utilization of *glucose* and related sugars, but there are many other organic compounds which serve as energy sources for heterotrophs. The utilization of these "unusual" organic compounds is not only of biochemical interest, but of applied interest, as we will see when we discuss petroleum microbiology, sewage treatment, and biodegradation in the next chapter. We discuss the biochemical pathways of the utilization of a number of these organic compounds in the present chapter.

Finally, although the major thrust of the present chapter is carbon, our discussion would not be complete without a consideration of one of the important "alternate life styles" of microbes which involves *nitrogen*. This is the utilization of nitrogen gas, N_2, as a source of nitrogen for biosynthesis, a process called **nitrogen fixation**. This important process of the biosphere participates in the recycling of nitrogen into living matter and is important not only agriculturally but in the total function of the biosphere.

16.1 Photosynthesis

One of the most important biological processes on earth is photosynthesis, the conversion of light energy into chemical energy. Most photosynthetic organisms are autotrophs, capable of growing on CO_2 as sole carbon source. Energy from light is thus used in the *reduction* of CO_2 to organic compounds. The ability to photosynthesize is dependent on the presence of special light-sensitive pigments, the *chlorophylls*, which are found in plants, algae, and some bacteria.

The growth of a phototrophic autotroph can be characterized by two distinct sets of reactions: the **light reactions**, in which light energy is converted into chemical energy, and the **dark reactions**, in which this chemical energy is used to reduce CO_2 to organic compounds. For autotrophic growth, energy is supplied in the form of ATP, while electrons for the reduction of CO_2 come from NADPH. The latter is produced by the reduction of $NADP^+$ by electrons originating from various electron donors to be discussed below.

The light reactions bring about the conversion of ***light*** energy into ***chemical*** energy in the form of ATP. Purple and green bacteria use light primarily to form ATP; they produce NADPH from reducing materials present in their environment, such as H_2S or organic compounds. Green plants, algae, and cyanobacteria, however, do not generally use H_2S or organic compounds to obtain reducing power. Instead, they obtain electrons for $NADP^+$ reduction from the oxidation of water molecules, producing O_2 as a byproduct (Figure 16.2). The reduction of $NADP^+$ to NADPH by these organisms is therefore a *light-mediated* event. Because molecular oxygen, O_2, is produced, the process of photosynthesis in these organisms is called **oxygenic** photosynthesis. In contrast, the purple and green bacteria do not produce oxygen; their process is called **anoxygenic** photosynthesis.

16.2 The Role of Chlorophyll and Bacteriochlorophyll in Photosynthesis

Photosynthesis occurs only in organisms that possess some type of ***chlorophyll***. Chlorophyll is a porphyrin, as are the cytochromes, but chlorophyll contains a *magnesium* atom instead of an iron atom at the center of the porphyrin ring. Chlorophyll also contains specific substituents bonded to the porphyrin ring and a long hydrophobic alcohol molecule. Because of this alcohol side chain, chlorophyll associates with lipid and hydrophobic proteins of photosynthetic membranes.

The structure of chlorophyll *a*, the principal chlorophyll of higher plants, most algae and of the cyanobacteria, is shown in Figure 16.3. Chlorophyll *a* is green in color because it absorbs red and blue light preferentially and transmits green light. The spectral properties of any pigment can best be expressed by its *absorption spectrum*, which indicates the degree to which the pigment absorbs light of different wavelengths of the spectrum. The absorption spectrum of cells containing chlorophyll *a* shows strong absorption of red light (maximum absorption at a wavelength of 680 nm) and blue light (maximum at 430 nm) (Figure 16.4*a*). We discussed the electromagnetic spectrum in Section 9.16.

There are a number of chemically different chlorophylls that are distinguished by their different absorption spectra. Chlorophyll *b*, for instance, absorbs maximally at 660 nm rather than at 680 nm. Many plants have more than one chlorophyll, but the most common are chlorophylls *a* and *b*. Among procaryotes, the cyanobacteria have chlorophyll *a*, but the

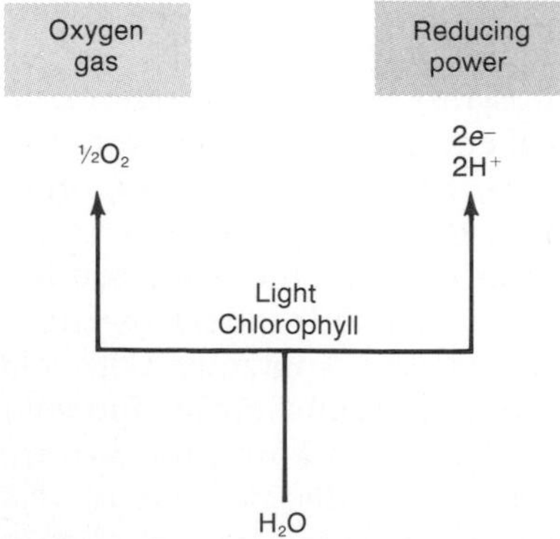

Figure 16.2 Production of O_2 and reducing power from H_2O.

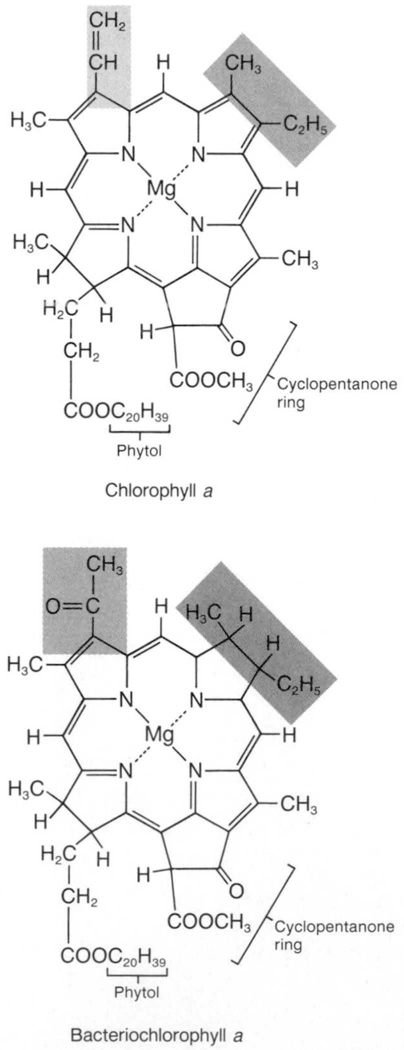

Figure 16.3 Structures of chlorophyll *a* and bacteriochlorophyll *a*. The two molecules are almost identical except for those portions shown in different colors.

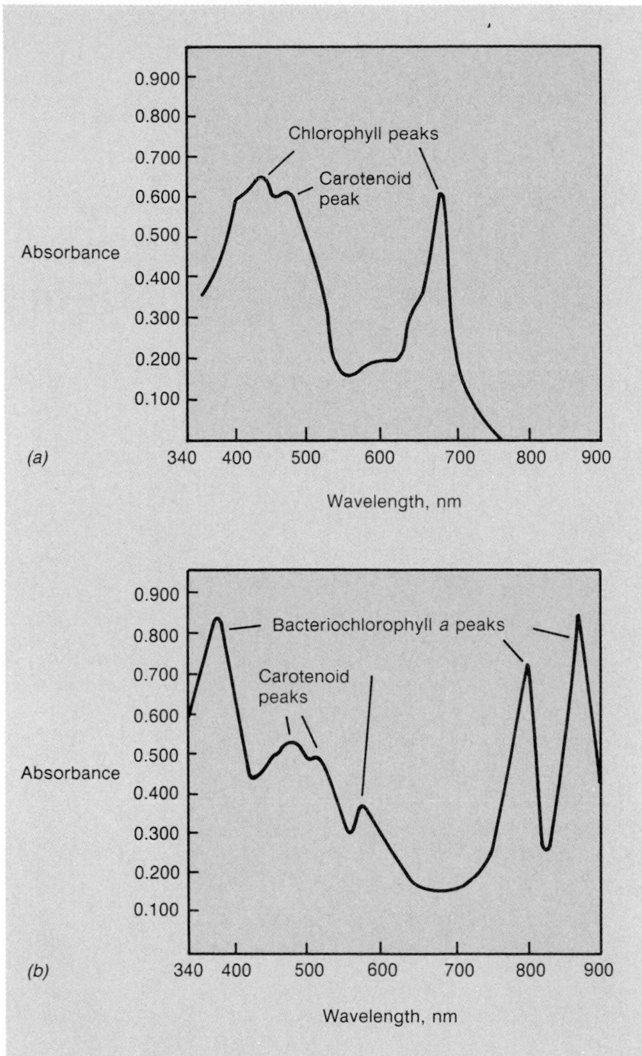

Figure 16.4 (a) Absorption spectrum of cells of the green alga *Chlamydomonas*. The peaks at 680 and 440 nm are due to chlorophyll *a*; the peak at 480 is due to carotenoids. (b) Absorption spectrum of cells of the photosynthetic bacterium *Rhodopseudomonas palustris*. Peaks at 870, 805, 600, and 360 nm are due to bacteriochlorophyll *a*; peaks at 525 and 450 are due to carotenoids.

purple and green bacteria have chlorophylls of slightly different structure, called the *bacteriochlorophylls*. Bacteriochlorophyll *a* (Figure 16.3*b*) from purple photosynthetic bacteria absorbs maximally around 850 nm (Figure 16.4*b*); other bacteriochlorophylls absorb maximally at 720–780, and one type of bacteriochlorophyll absorbs at 1020 nm (see also Figure 19.1).

Why do organisms have several kinds of chlorophylls absorbing light at different wavelengths? One reason appears to be to make it possible to use more of the energy of the electromagnetic spectrum. Only light energy which is absorbed will be used biologically, so that by having more than one chlorophyll, more of the incident light energy becomes available to the organism. By having different pigments, two unrelated organisms can coexist in a habitat, each using wavelengths of light that the other is not using. Thus, pigment diversity has ecological significance.

Just where are the chlorophyll pigments located inside the cell? These pigments, and all the other components of the light-gathering apparatus, are associated with special membrane systems, the **photosynthetic membranes**. The location of the photosynthetic membranes within the cell differs between procaryotic and eucaryotic microorganisms. In eucaryotes, photosynthesis is associated with special intracellular organelles, the **chloroplasts** (see Section 3.13 and Figures 3.63 and 3.64). The chlorophyll pigments are attached to sheetlike (lamellar) membrane structures of the chloroplast (Figure 16.5*a*). These photosynthetic membrane systems are called **thylakoids**. The thylakoids are so arranged that the chloroplast is divided into two regions, the matrix space, which surrounds the thylakoids, and the inner space within the thylakoid array. This arrangement makes possible the development of a light-driven pH gradient which can be used to synthesize ATP, as will be described in the next section.

Within the thylakoid membrane, the chlorophyll molecules are associated in groups consisting of about 200 to 300 molecules. Only some of these chlorophyll molecules participate *directly* in the conversion of light energy to ATP. These special chlorophyll molecules are referred to as **reaction center chlorophyll** and receive light energy by transfer from the more numerous *light-harvesting* (or *antenna*) chlorophyll molecules (Figure 16.6). The chlorophyll molecules are bound to proteins which control precisely their orientation in the membrane so that energy absorbed by one chlorophyll molecule can be efficiently transferred to another.

In procaryotes, chloroplasts are not present and the photosynthetic pigments are integrated into extensive internal membrane systems, as illustrated in Figure 16.7. In the purple and green bacteria, a single reaction center probably contains about 30 light-harvesting bacteriochlorophyll molecules.

Antenna chlorophyll molecules make possible a dramatic increase in the rate at which photosynthesis can be carried out. At the low light intensities prevailing in nature, reaction centers can only be excited about once per second, which would not be sufficient to carry out a significant photosynthetic process. The additional antenna chlorophyll molecules permit collection of light energy at a much more rapid rate. Since reaction-center chlorophyll absorbs light energy only over a very narrow range of the spectrum, antenna pigments also perform the additional function of spreading out the spectral range available for use.

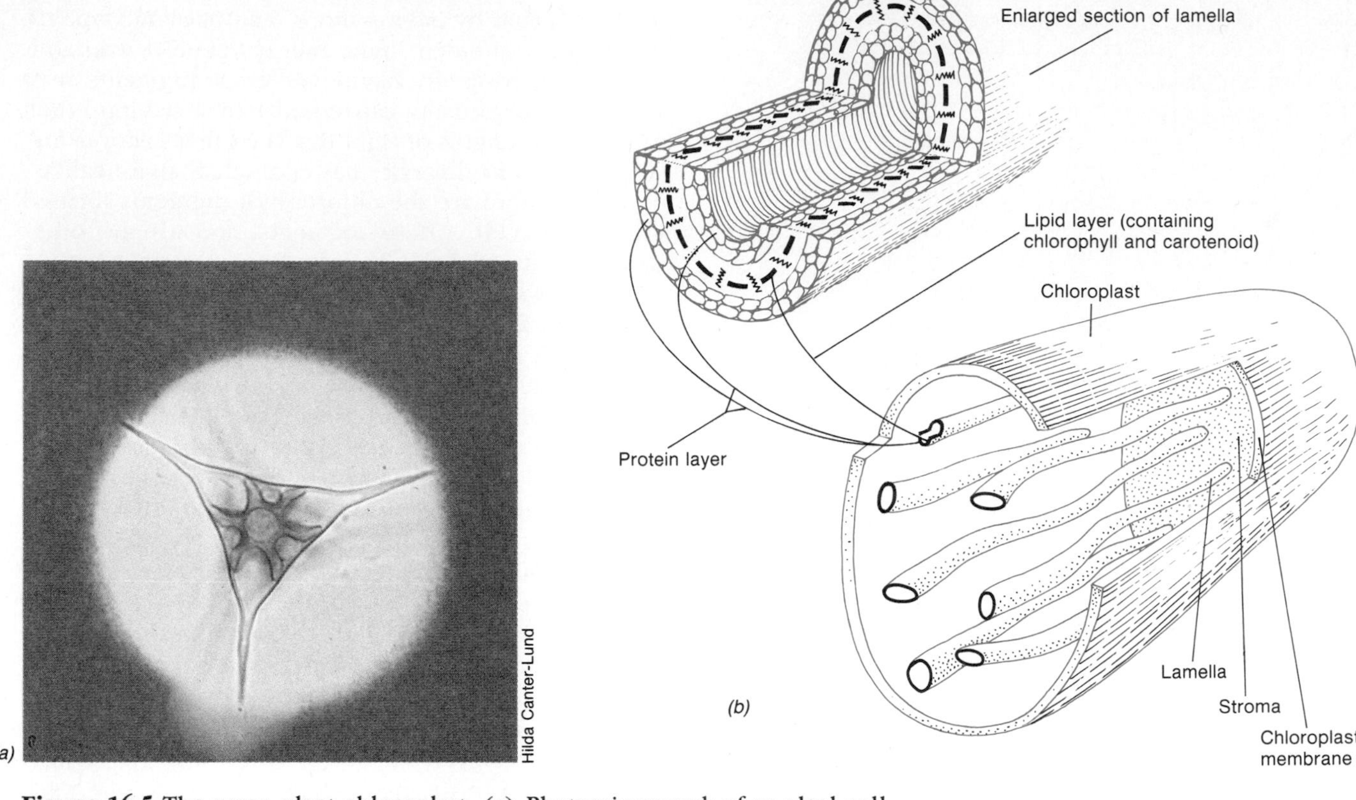

Figure 16.5 The green plant chloroplast. (a) Photomicrograph of an algal cell showing chloroplasts. (b) Details of chloroplast structure, showing how the convolutions of the thylakoid membranes define an inner space and a matrix space within the overall chloroplast envelope.

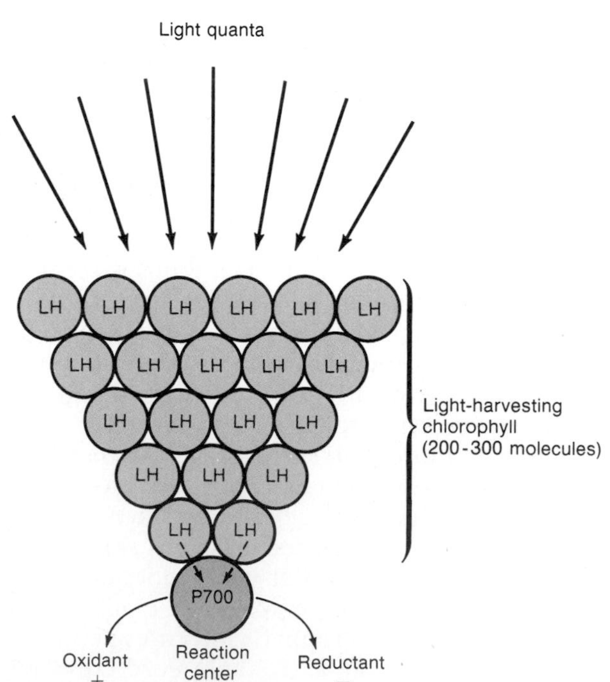

Figure 16.6 The photosynthetic unit and its associated reaction center. Light energy, absorbed by light-harvesting chlorophyll molecules, travels to the reaction center where the actual ejection of an electron occurs, generating a charge separation.

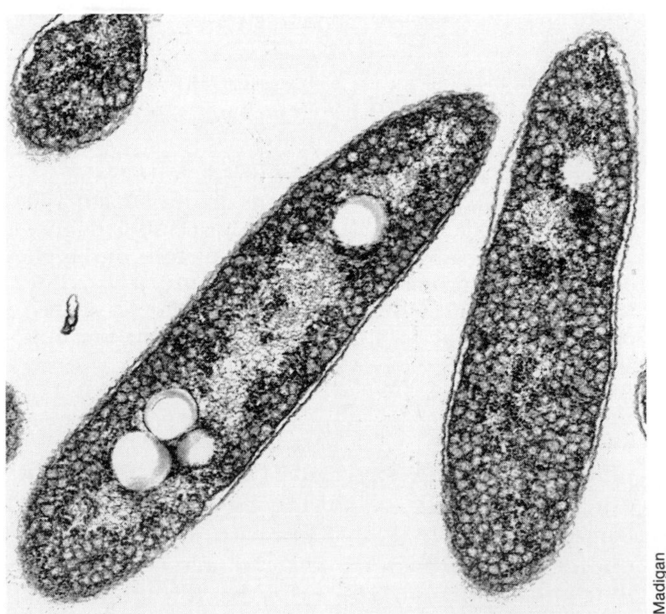

Figure 16.7 Chromatophores. Section through a cell of the phototrophic bacterium, *Rhodobacter capsulatus*, containing an abundance of vesicular photosynthetic membranes. The circular membranes arise by invagination of the cytoplasmic membrane. When the cell is broken, these invaginations pinch off, yielding sealed vesicles whose polarity is inverted with respect to the parent cell.

16.3 Anoxygenic Photosynthesis

The process of light-mediated ATP synthesis in all phototrophic organisms involves electron transport through a sequence of electron carriers. These electron carriers are arranged in the photosynthetic membrane in series from those with negative to those with more positive potentials. Conceptually, the process of photosynthetic electron flow resembles that of respiratory electron flow. As a matter of fact, in photosynthetic bacteria capable of aerobic (dark) growth, many of the same electron transport components are present in the membranes of photosynthetically and dark-aerobically grown cells. How, therefore, do photosynthetic and respiratory electron flow differ?

The key to photosynthesis lies in the ability of the light-sensitive pigments, chlorophyll or bacteriochlorophyll, to use the energy of light to push an electron from the pigment molecule *against* the thermodynamic gradient to reduce an acceptor molecule which has a very negative reduction potential. However, the way this is done differs in oxygenic and anoxygenic photosynthesis. As we have noted, purple and green bacteria do *not* produce O_2 during photosynthesis. Their photosynthesis is therefore said to be *anoxygenic*. Instead of water, they use a reduced compound from their environment as an electron donor for CO_2 reduction. Examples of electron donors used in anoxygenic photosynthesis are reduced sulfur compounds (for example, H_2S, elemental sulfur, thiosulfate), H_2 gas, or organic compounds (for example, succinate, malate, butyrate). During autotrophic growth, these bacteria oxidize the external electron donor and reduce $NADP^+$ to NADPH. The electron donor then becomes oxidized. For instance, during growth on H_2S elemental sulfur is produced which is deposited either outside the cells in the green bacteria (Figure 16.8*a*) or inside the cells in some of the purple bacteria (Figure 16.8*b*). If succinate is used as an electron donor, fumarate (a more oxidized compound) is produced.

An overall scheme of electron flow in bacterial photosynthesis is shown in Figure 16.9. The absorption of light excites the bacteriochlorophyll molecule and converts it into a *strong reductant*, sufficient to donate electrons to an acceptor molecule of very low potential (highly reducing). This acceptor, referred to as the intermediary carrier (I) is a bacteriopheophytin *a* molecule (a bacteriochlorophyll *a* molecule minus its magnesium atom); the E_0' of I is about -0.7 volts.

The effect of light thus far is to generate a strong reductant from a relatively weak reductant. Remember from the concept of the electron tower in Figure 4.5 that electrons spontaneously proceed from compounds with *negative* reduction potentials to those with *positive* reduction potentials, a transfer accompanied by the release of energy. Transfer of an electron from P870 to the bacteriopheophytin, on the other hand, is a transfer from the *bottom* of the electron tower to the *top*, but is made possible by the light-mediated conversion of P870 into a strong reductant. Reduction of the intermediary carrier by P870 thus represents work done on the system. From the

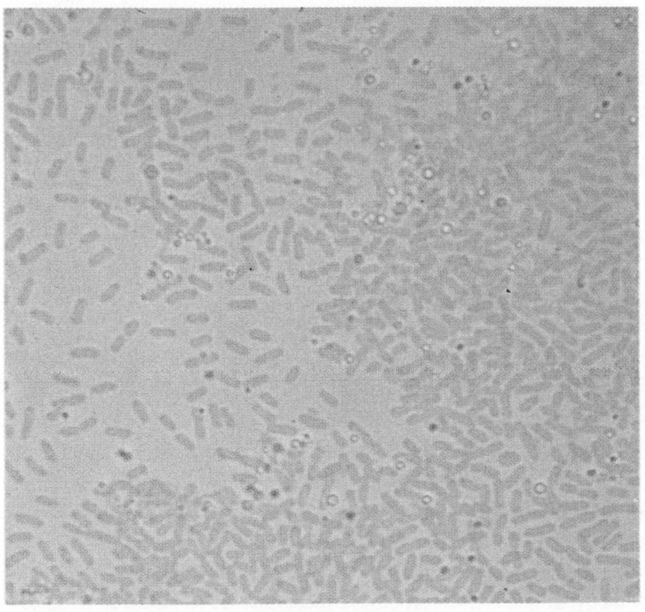

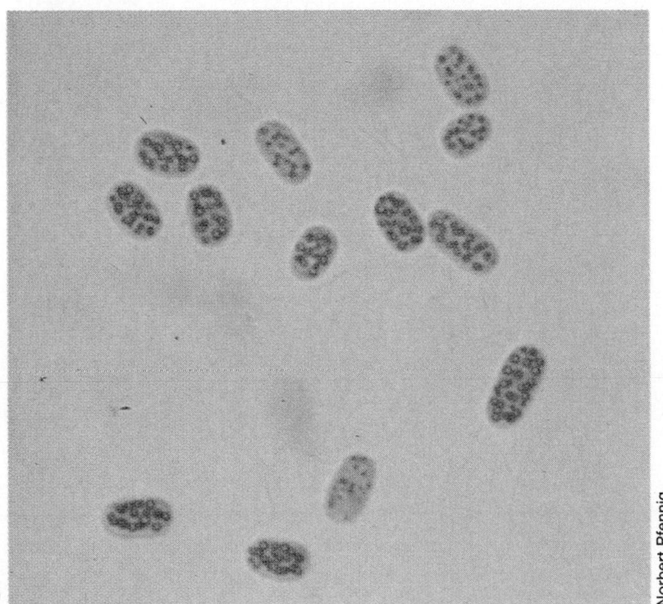

Figure 16.8 Photomicrographs of photosynthetic bacteria taken by bright field microscopy. (a) Green bacterium: *Chlorobium limicola.* The refractile bodies are sulfur granules deposited *outside* the cell. Magnification, 2000×. (b) Purple bacterium: *Chromatium okenii.* Notice the sulfur granules deposited *inside* the cell. Magnification, 1200×.

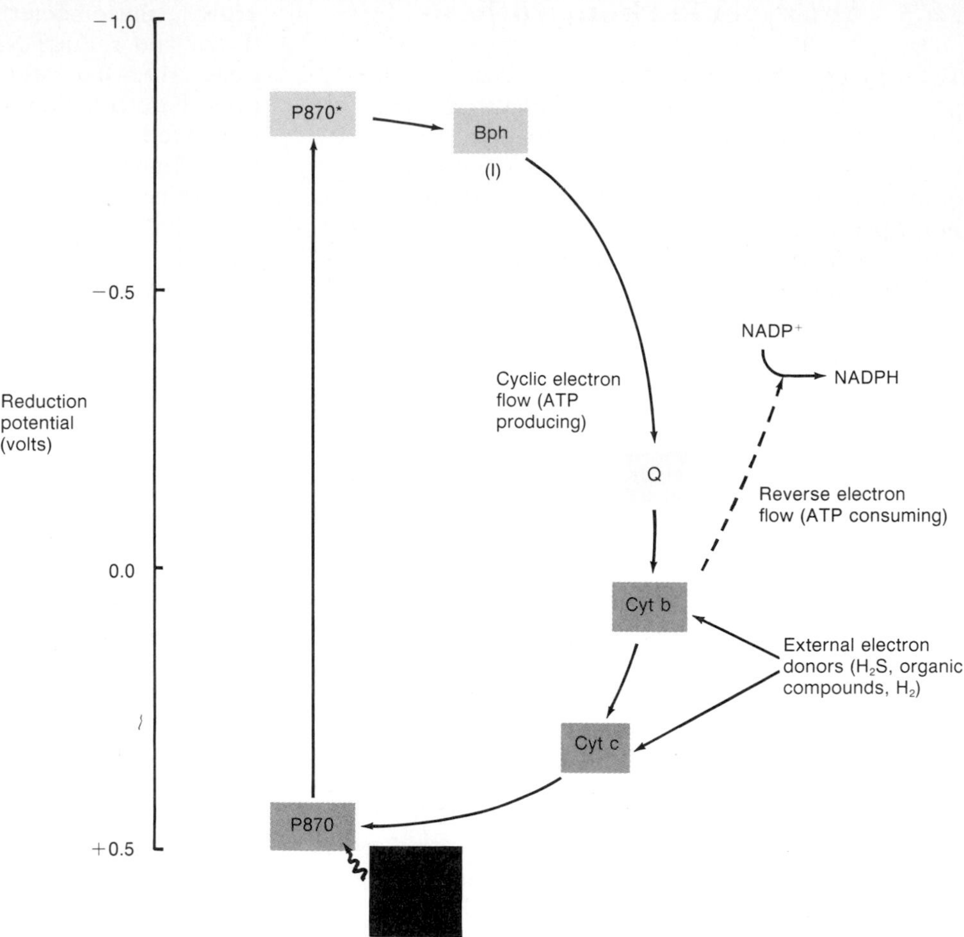

Figure 16.9 Electron flow in anoxygenic (bacterial) photosynthesis. Only a single light reaction occurs. P870, reaction center bacteriochlorophyll; Bph, bacteriopheophytin (intermediary carrier, I); Q, quinone; Cyt, cytochrome.

intermediary carrier electrons are donated to a quinone molecule, referred to as the *primary electron acceptor*, which has an E_o' of about -0.2 volts. From here electrons are transported through a chain containing iron-sulfur proteins and cytochromes (Figure 16.9). Synthesis of ATP occurs as a result of the formation of a proton gradient generated by proton extrusion during electron transport and the activity of ATPases in coupling the dissipation of the proton gradient to ATP formation (see Section 4.12 for a discussion of proton-motive force). The reaction series is completed when the electron returns to the oxidized P870 molecule, reducing it to its original ground state form ($E_o' + 0.5$ volts). It is then capable of absorbing the energy from another photon and repeating the process. This method of making ATP is called **cyclic photophosphorylation**, since electrons are repeatedly moved around a closed circle; *there is no net input or consumption of electrons* as in respiration.

It is extremely important that the light-driven redox reaction just described takes place *across* the photosynthetic membrane. The spatial relationships of the electron transport components in the bacterial chromatophore are illustrated in Figure 16.10. Note that protons are pumped to the *center* of the chromatophore, thus setting up the proton gradient which is used in ATP synthesis. The net result of the light reaction is the translocation of *two* protons across the membrane for each photochemically excited electron. It should be emphasized that the *only* thing light does in this process is create a strong reductant; the remaining reactions are *not* light dependent.

The importance of the chromatophore arrangement should be made clear. If the process of electron flow illustrated in Figure 16.10 would occur in a flat membrane, the protons pumped across the membrane would become part of the general external environment and a proton gradient would not develop. In the chromatophore arrangement, however, the proton gradient is maintained and can be employed to do useful work, such as the synthesis of ATP shown.

However, if an organism carrying out anoxygenic photosynthesis is going to grow with CO_2 as its sole or major carbon source, formation of ATP is not enough: reducing power (NADPH) must also be made. The purple and green bacteria use reduced compounds from their environment as sources of reducing power, but at least two ways are known for using these compounds to reduce $NADP^+$ to NADPH. The first involves direct transfer from a highly reduced compound to $NADP^+$. The best example is H_2. The

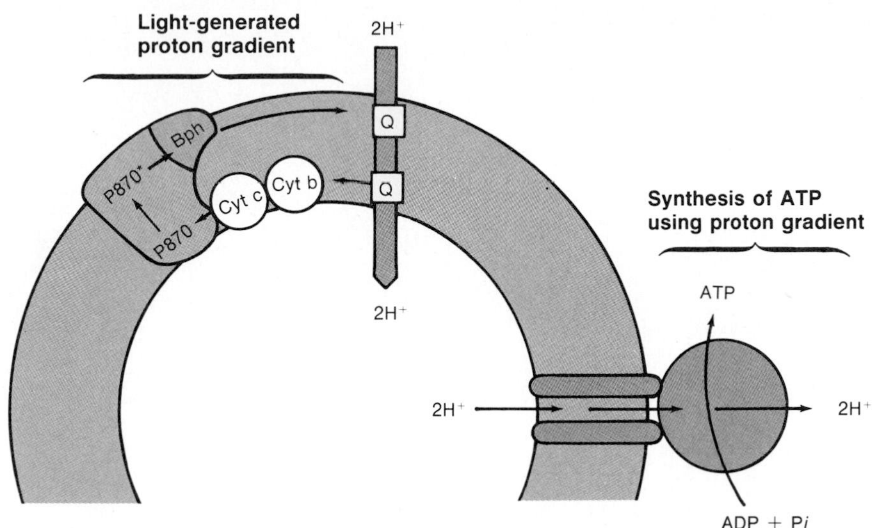

Figure 16.10 Electron flow and proton translocation in the photosynthetic membrane of a purple bacterium. The spatial relationships shown are as they would be found in the chromatophore (see Figure 16.7). The electron transport components are shown as in Figure 16.9. The light-generated proton gradient is used in the synthesis of ATP via the membrane-bound ATPase.

reduction potential of H_2 is sufficiently low (−0.42 volts) to reduce $NADP^+$ (−0.32 volts) directly, provided the organism has the enzyme *hydrogenase* to incorporate H_2 into the cell. However, many purple and green bacteria grow autotrophically using electron donors such as sulfide or thiosulfate, whose reduction potential is *more positive* than that of the $NADP^+$/NADPH couple. Under these conditions, electrons enter the photosynthetic electron transport chain usually at the level of cytochromes, and are forced backwards (against the electropotential gradient) by the hydrolysis of ATP, eventually reducing $NADP^+$ to NADPH. This process, called **reversed electron transport**, is necessitated by the need for a low potential reductant for CO_2 fixation, coupled with the reality of having to use electron donors of higher potential. Reversed electron transport is also a key feature of the synthesis of reducing power in lithotrophic procaryotes (see Section 16.12).

A brief overview of the characteristics of the purple and green bacteria is given in Table 16.2. Many of the purple and green bacteria are strict anaerobes, but one subgroup each of purple and green bacteria are able to grow aerobically in the dark as heterotrophs. However, *phototrophic* growth in purple and green bacteria is always strictly anaerobic, even in those organisms capable of dark heterotrophic growth. Purple bacteria are nutritionally diverse and use a variety of organic compounds as carbon sources as well as remaining capable of autotrophic growth when necessary. The term *photoheterotroph* is frequently used to describe the nutritional state of the organism when photosynthetic growth on organic compounds is occurring. The combining form "photo" indicates that *light* is serving as the energy source, while the word *heterotroph* implies that *organic carbon* is serving as the carbon source. Similarly, the term *photoautotroph* is used to describe growth photosynthetically on CO_2 as sole carbon source. As mentioned previously, some purple and green bacteria can also grow in the dark heterotrophically, using either fermentative or respiratory energy metabolism.

Table 16.2 Characteristics of the purple and green bacteria

Purple bacteria	Green bacteria
Bacteriochlorophyll *a* or *b*	Bacteriochlorophyll *c*, *d*, or *e*
Purple- or red-colored carotenoids	Yellow or brown-colored carotenoids
Photosynthetic membrane system in a series of lamellae or tubes, continuous with the plasma membrane	Majority of bacteriochlorophyll in chlorosomes; only bacteriochlorophyll *a* in plasma membrane

16.4 Oxygenic Photosynthesis

Electron flow in oxygenic phototrophs involves two distinct, but interconnected, photochemical reactions. Oxygenic phototrophs use light to generate both ATP *and* NADPH, the electrons for the latter arising from the splitting of water into oxygen and electrons (see Figure 16.2). The two systems of light reactions are called *photosystem I* and *photosystem II*, each photosystem having a spectrally distinct form of chlorophyll *a*. Photosystem I chlorophyll, called P700, absorbs light best at long wavelengths (far red light), whereas photosystem II chlorophyll, called P680, absorbs best at shorter wavelengths (near red light). The two forms of chlorophyll *a* are attached to specific proteins in the membrane and interact as shown in Figure 16.11.

The path of electron flow in oxygenic phototrophs roughly resembles the letter *Z* turned on its side, and scientists studying oxygenic photosynthesis have come to refer to the electron flow of oxygenic

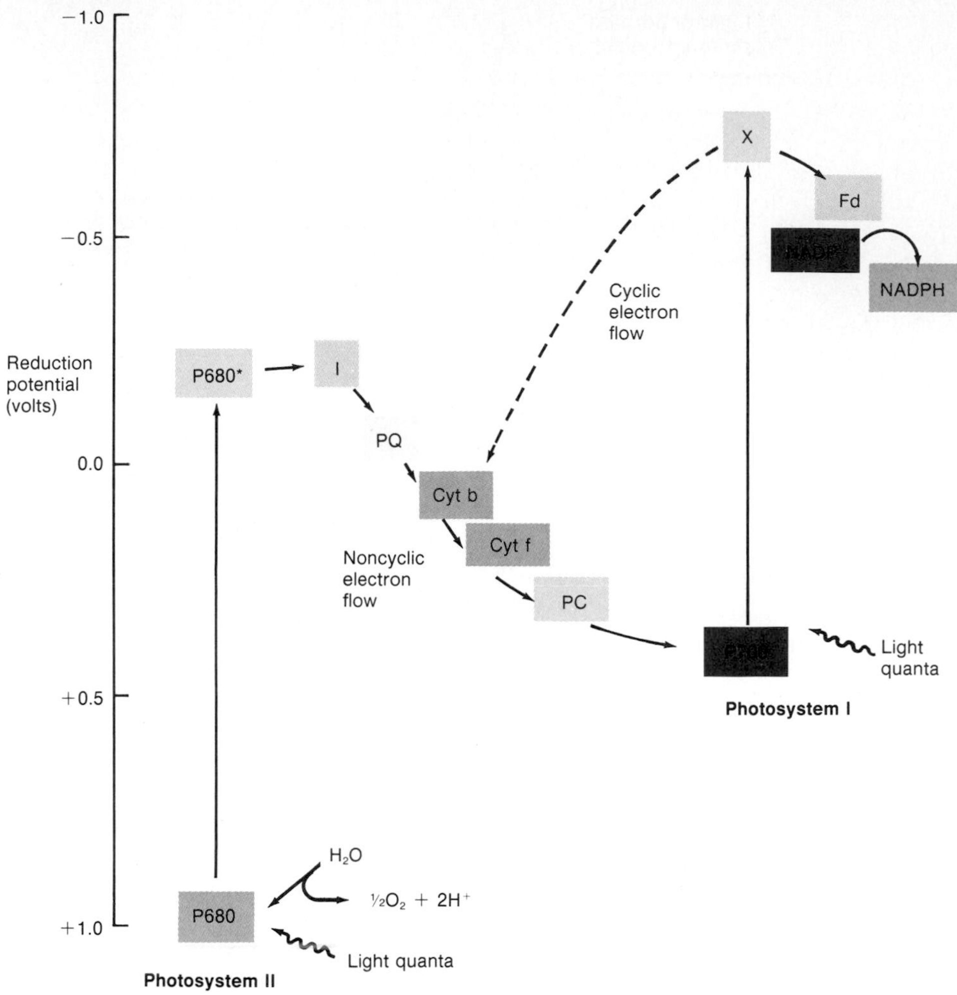

Figure 16.11 Electron flow in oxygenic (green plant) photosynthesis, the "Z" scheme. Two photosystems are involved, PSI and PSII. I, unidentified intermediary acceptor of PSII; PQ, plastoquinone; Cyt, cytochrome; PC, plastocyanin; X, unidentified acceptor of electrons in PSI; Fd, ferredoxin. P680 and P700 are the reaction center chlorophylls of PSII and PSI, respectively.

phototrophs as the "Z" scheme. We should first note that the reduction potential of the P680 chlorophyll *a* molecule in photosystem II is very high, about that of the O_2/H_2O couple (see Table 16.1). This is because the first step in oxygenic electron flow is the splitting of water into oxidizing and reducing equivalents (Figure 16.11), a thermodynamically unfavorable reaction. An electron from water is donated to the P680 molecule following the absorption of a quantum of light near 680 nm. Light energy converts P680 into a moderately strong reductant, capable of reducing an intermediary molecule about −0.2 volts. The nature of this molecule is unknown, but it may be a pheophytin *a* molecule (chlorophyll *a* without the magnesium atom). From here the electron travels through several membrane carriers including quinones, cytochromes, and a copper-containing protein called **plastocyanin**; the latter donates electrons to photosystem I. The electron is accepted by the reaction center chlorophyll of photosystem I, P700, which has previously absorbed light quanta and donated electrons to the primary acceptor of photosystem I; this acceptor has a very negative potential, about −0.75 volts. As in photosystem II, the primary acceptor of electrons from photosystem I has not been positively identified, but is thought to be a free radical form of chlorophyll *a*. At any rate, the acceptor in photosystem I, once reduced, is at a reduction potential sufficiently negative to reduce the iron-sulfur protein **ferredoxin**, which then reduces $NADP^+$ to NADPH (Figure 16.11).

ATP synthesis in oxygenic photosynthesis Besides the net synthesis of reducing power (i.e., NADPH), other important events have occurred while electrons flowed from one photosystem through another. During transfer of an electron from the acceptor in photosystem II to the chlorophyll molecule in photosystem I, electron transport occurred in a thermodynamically favorable (negative to positive) direction. This generates a membrane potential (a proton gradient) from which ATP can be produced. This type of ATP generation has been designated *noncyclic* photophosphorylation because the electron travels a noncyclic route from water to $NADP^+$. (Contrast cyclic photophosphorylation shown in Figure 16.9 with noncyclic photophosphorylation shown in Figure 16.11.) When sufficient reducing power is present,

ATP can also be produced in oxygenic phototrophs by *cyclic* photophosphorylation involving only photosystem I. This occurs when the primary acceptor of photosystem I, instead of reducing ferredoxin (and hence $NADP^+$), returns the electron to the P700 molecule via membrane bound cytochromes *b* and *f*. This flow creates a membrane potential and synthesis of additional ATP (see dashed scheme in Figure 16.11).

Photosystems I and II normally function together in the noncyclic, oxygenic process. However, under certain conditions many algae and some cyanobacteria are able to carry out cyclic photophosphorylation using only photosystem I, obtaining reducing power from sources other than water, in effect photosynthesizing anoxygenically as do purple and green bacteria. This alteration requires the presence of anaerobic conditions as well as a reducing substance, such as H_2 or H_2S. Under these conditions the electrons for CO_2 reduction come not from water but from the reducing substance. In algae H_2 is generally the reductant, and following a period of adaptation to anaerobic conditions, the enzyme hydrogenase is made and is used to assimilate H_2 which reduces $NADP^+$ to NADPH directly.

A number of cyanobacteria can use H_2S as an electron donor for anoxygenic photosynthesis. When H_2S is used, it is oxidized to elemental sulfur (S^0), and sulfur granules are deposited outside the cells similar to those produced by green sulfur bacteria. The filamentous cyanobacterium *Oscillatoria limnetica* has been found in sulfide-rich saline ponds where it carries out anoxygenic photosynthesis along with photosynthetic green and purple bacteria and produces sulfur as an oxidation product of sulfide (Figure 16.12). In cultures of *O. limnetica* electron flow from photosystem II is strongly inhibited by H_2S, thus necessitating anoxygenic photosynthesis if the organism is to survive in its sulfide-rich environment.

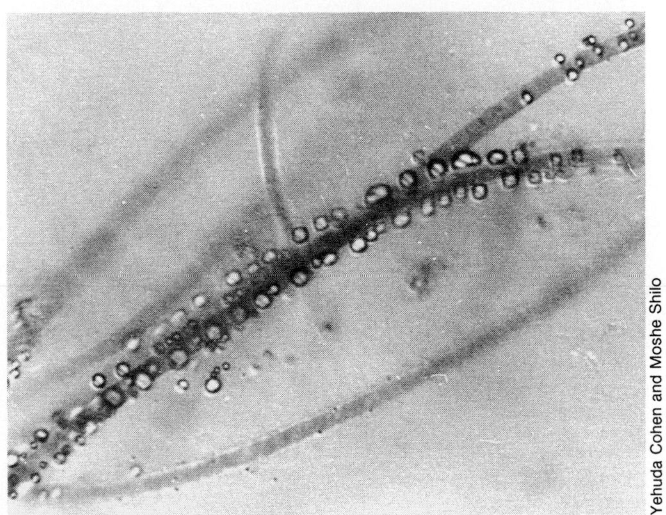

Figure 16.12 Cells of the cyanobacterium *Oscillatoria limnetica* grown anaerobically on sulfide as photosynthetic electron donor. Note the globules of elemental sulfur, the oxidation product of sulfide, formed *outside* the cells.

From an evolutionary point of view, the existence of cyclic photophosphorylation indicates a close relationship between green-plant and bacterial photosynthesis. Although organisms carrying out oxygenic photosynthesis have acquired photosystem II, and hence the ability to split H_2O, they still retain the ability under certain conditions to use photosystem I alone.

16.5 Additional Aspects of Photosynthetic Light Reactions

Accessory pigments Although chlorophyll or bacteriochlorophyll is obligatory for photosynthesis, photosynthetic organisms have other pigments that are involved, at least indirectly, in the capture of light energy. The most widespread accessory pigments are the **carotenoids**, which are almost always found in photosynthetic organisms. Carotenoids are water insoluble pigments firmly embedded in the membrane; the structure of a typical carotenoid is shown in Figure 16.13. Carotenoids have long hydrocarbon chains with alternating C–C and C=C bonds, an arrangement called a *conjugated* double bond system. As a rule, carotenoids are yellow, red, or green in color (see Figure 19.2*b*) and absorb light in the blue region of the spectrum (see Figure 16.4). Carotenoids are usually closely associated with chlorophyll in the photosynthetic apparatus, and there are approximately the same number of carotenoid as there are chlorophyll molecules. Carotenoids do not act directly in photophosphorylation reactions, but transfer, by way of fluorescence, some of the light energy they capture to chlorophyll. This transferred energy may thus be used in photophosphorylation in the same way as light energy captured directly by chlorophyll.

Cyanobacteria and red algae contain **phycobiliproteins**, which are accessory pigments that are red or blue in color. The red pigment, called phycoerythrin, absorbs light most strongly at wavelengths around 550 nm, whereas the blue pigment, phycocyanin, absorbs most strongly at 620 to 640 nm. Phy-

Figure 16.13 Structure of β-carotene, a typical carotenoid. The conjugated double-bond system is shown in color.

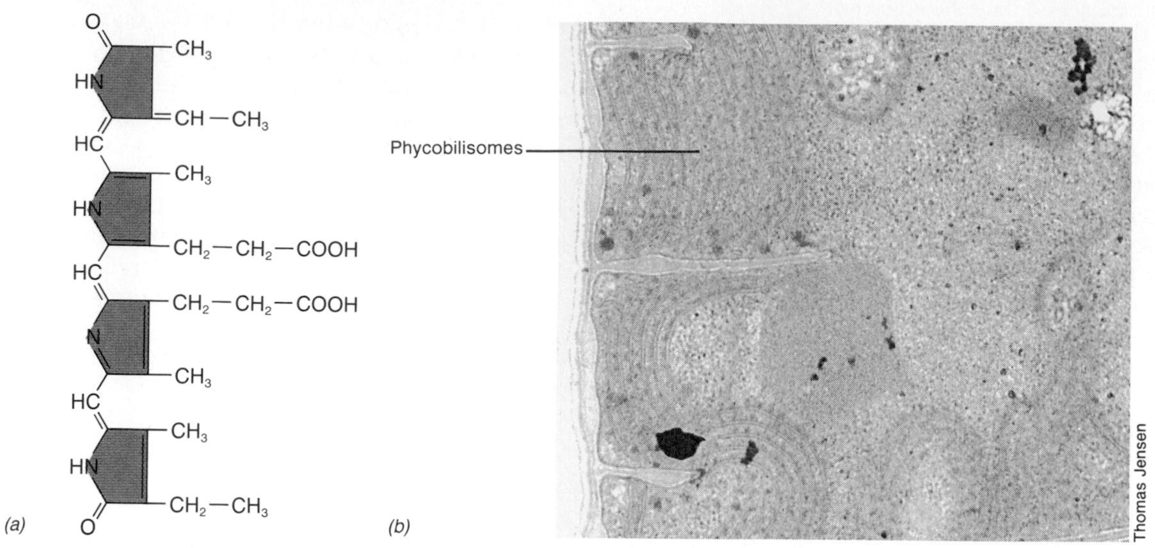

Figure 16.14 (a) A typical phycobilin. This compound is an open-chain tetrapyrrole, derived biosynthetically from a closed porphyrin ring by loss of one carbon atom as carbon monoxide. The structure shown is the prosthetic group of phycocyanin, a proteinaceous pigment found in cyanobacteria and red algae. (b) Electron micrograph of a thin section of the cyanobacterium *Microcoleus vaginatis*. Note the darkly staining ball-like phycobilisomes attached to the lamellar membranes (arrows).

cobiliproteins contain open-chain tetrapyrroles called *phycobilins* (Figure 16.14*a*), which are coupled to proteins. Phycobiliproteins occur as high molecular weight aggregates called *phycobilisomes*, attached to the photosynthetic membranes (Figure 16.14*b*). They are closely linked to the chlorophyll-containing system, which makes for very efficient energy transfer, approaching 100 percent, from biliprotein to chlorophyll.

The light-gathering function of accessory pigments seems to be of obvious advantage to the organism. Light from the sun is distributed over the whole visible range, yet chlorophylls absorb well in only a part of this spectrum. By having accessory pigments, the organism is able to capture more of the available light (Figure 16.15). Another function of accessory pigments, especially of the carotenoids, is as photoprotective agents. Bright light can often be harmful to cells, in that it causes various photooxidation reactions that can actually lead to the destruction of chlorophyll and of the photosynthetic apparatus itself. The accessory pigments absorb much of this harmful light and thus provide a shield for the light-sensitive chlorophyll. Since phototrophic organisms must by their very nature live in the light, the photoprotective role of the accessory pigments is of obvious advantage.

Energetics of photosynthesis and the quantum yield Energy from light does not come in continuous fluxes but in small discrete packets, each of which is called a **quantum**. A fundamental law of physics relates the energy in a quantum to the wavelength of light. The shorter the wavelength, the more energy available per quantum. If we take light of 680 nm, which is the wavelength absorbed by reaction-center chlorophyll in photosystem II, it can be calculated from the equation relating energy to wavelength that a mole of quanta has 41 kcal (a mole of quanta is Avogadro's number of quanta).

From free energy data (Section 4.4 and Appendix 1), the **quantum yield** (number of quanta needed to fix one molecule of CO_2) can be calculated. The

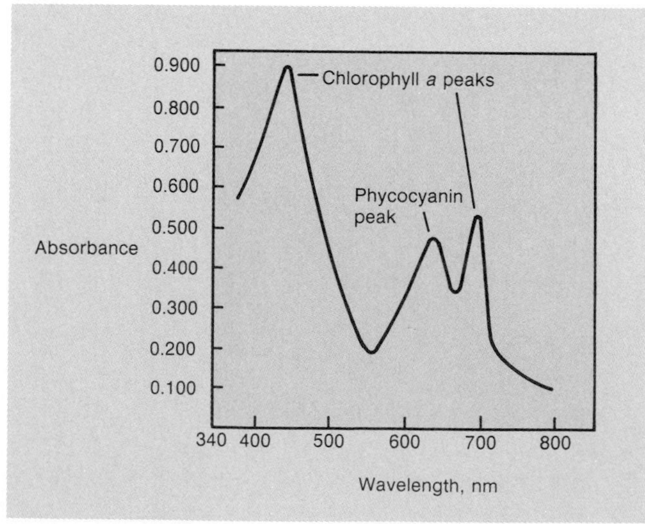

Figure 16.15 The absorption spectra of a cyanobacterium which has a phycobiliprotein as an accessory pigment. Note how the presence of these accessory pigments broadens the wavelengths of usable light energy. Compare with Figure 16.4.

synthesis of a mole of glucose from CO_2 and H_2O has a $\Delta G^{0\prime}$ of +686 kcal. Taking one-sixth of this, we can calculate that the fixation of 1 mole of CO_2 requires +114 kcal, and since a mole of quanta of red light has 41 kcal, it would take at least 3 quanta (114/41) to fix one molecule of CO_2. This value of 3 quanta per one molecule of CO_2 fixed is the maximum possible quantum yield of photosynthesis. Experimental studies using uniformly illuminated suspensions of algae and careful measurement of light absorbed and CO_2 fixed indicate that in practice it takes about *eight* quanta to fix one molecule of CO_2, which is about 38 percent of the theoretical value of 3 quanta.

Carrying this analysis further, we know from the balance sheet for photosynthetic CO_2 fixation (Section 16.6) that for each molecule of CO_2 fixed, two NADPH and three ATP are required. In noncyclic photophosphorylation (Figure 16.11), 2 quanta are required to eject each electron (one for each photosystem), but two electrons are required to reduce one NADP. Thus, to synthesize the two NADPH necessary for CO_2 fixation, four electrons must be donated, which is equivalent to 8 quanta. It also requires two electrons to synthesize one ATP, so that when 8 quanta are absorbed, two molecules of ATP can be synthesized. Thus 8 quanta must be absorbed to synthesize two NADPH and two ATP, and a quantum yield of 8 has been measured experimentally. This does not quite add up, since as we have just noted, three ATP are required to reduce one molecule of CO_2, and we have only made two ATP. Since the quantum yield of 8 has been measured experimentally, it seems reasonable to conclude that extra ATP is made somewhere during the process; perhaps in some manner by cyclic photophosphorylation.

If we consider the energetics of *anoxygenic* photosynthesis carried out by the photosynthetic bacteria, we find that much less energy is required to fix CO_2 into organic compounds than is required by oxygenic photosynthesis. This is because reducing power does not come from the splitting of water but from stronger reductants such as H_2 or H_2S. If we write the simplest equation for CO_2 fixation by photosynthetic bacteria using cyclic photophosphorylation:

$$6CO_2 + 12H_2S \rightarrow C_6H_{12}O_6 + 6H_2O + 12S$$

and calculate the free energy from the values given in Appendix 1, we find that the $\Delta G^{0\prime}$ is +97 kcal/mole, or +16 kcal per mole of CO_2 fixed. This is considerably *less* than the +114 kcal per mole of CO_2 required in green-plant photosynthesis. It is perhaps because of the small energy demand of anoxygenic photosynthesis that purple and green bacteria can exist in environments where light intensities are very low (see Section 19.1). It may also explain how these bacteria are able to function using only cyclic photophosphorylation to generate ATP.

Phototaxis Many photosynthetic microorganisms move toward light, a process called **phototaxis**. The advantage of phototaxis is that it allows the organism to orient itself for most efficient photosynthesis; indeed, if a light spectrum is spread across a microscope slide on which there are motile photosynthetic bacteria, the bacteria accumulate at those wavelengths at which their pigments absorb (Figure 16.16).

We discussed in Section 3.9 the details of chemotaxis, the movement of motile bacteria toward and away from chemicals. The mechanism of phototaxis is somewhat different. The bacteria exhibit a kind of response called a *shock movement* (from the German word "Schreckbewegung"). When a bacterial cell moves out of the light into the dark it stops suddenly (as if shocked) and reverses direction. This shock movement is the only response the bacterial cell makes to changing light conditions, but it is sufficient to keep the bacterial cell within light. Imagine a photosynthetic bacterium moving in a spot of light. As it swims around through the liquid, it may by accident wander out of the light field. Immediately, the cell stops, reverses its flagellar motor (and hence its direction), and thus reenters the light. Suppose now that we examine a cell which is completely removed from the light spot. It moves back and forth through the dark liquid, showing no special attraction to the light. (After all, it has no way of knowing the light spot is even there.) But if, in the course of its movement, it should wander into the light spot, it would remain in the light, since whenever it leaves the light spot, it reverses direction. Thus, even though bacteria do not swim *toward* the light, most of the cells end up in the light spot, and only a few remain in the

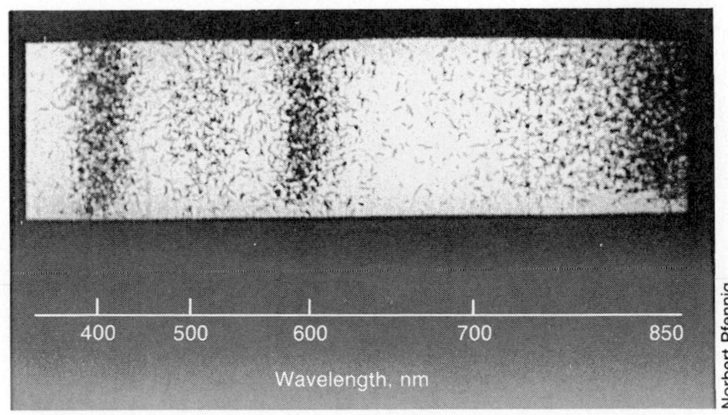

Figure 16.16 Phototactic accumulation of the photosynthetic bacterium *Thiospirillum jenense* at light wavelengths at which its pigments absorb. A light spectrum was displayed on a microscope slide containing a dense suspension of the bacteria; after a period of time, the bacteria had accumulated selectively and the photomicrograph was taken. The wavelengths at which accumulations occur are those at which bacteriochlorophyll *a* absorbs (compare with Figure 16.4*b*).

dark. This accumulation of photosynthetic cells in the light is therefore a biased random walk, as is the movement toward or away from chemicals (chemotaxis) discussed in Section 3.9. In addition to being able to distinguish light from dark, these bacteria can distinguish bright light from dimmer light. How? It is not the absolute amount of light to which they respond, but to differences in light intensity; phototactic bacteria can distinguish between two light sources that differ in intensity by only 5 percent. The accumulation of bacteria at different wavelengths of light, as illustrated in Figure 16.16, results because the pigments of the bacteria absorb light at certain wavelengths better than at others, so that the bacteria "perceive" these regions of the spectrum as being brighter. It is of historical interest that T. W. Engelmann, the first person to suggest that certain bacteria were photosynthetic, was led to this idea by observing that the purple bacteria he was studying accumulated preferentially at specific regions of the light spectrum.

16.6 Autotrophic CO_2 Fixation

In the first half of this chapter, we have discussed the light reactions of phototrophic autotrophs. We now discuss the so-called *dark* reactions, which concerns the biochemistry of the process by which autotrophs, both phototrophic and lithotrophic, convert CO_2 into organic matter.

Although all organisms require some of their carbon for cellular biosynthesis to come from CO_2, autotrophs can obtain *all* of their carbon from CO_2. The overall process is called *CO_2 fixation,* and all of the reactions of CO_2 fixation will occur in complete darkness, using ATP and reducing power (NADPH) generated either during the light reactions of photosynthesis or during oxidation of inorganic compounds (see Section 16.7). Most autotrophs so far examined have a special pathway for CO_2 reduction, the **Calvin cycle**, also referred to as the *reductive pentose* cycle. This cycle is summarized later, but first let us consider some of the key individual reactions.

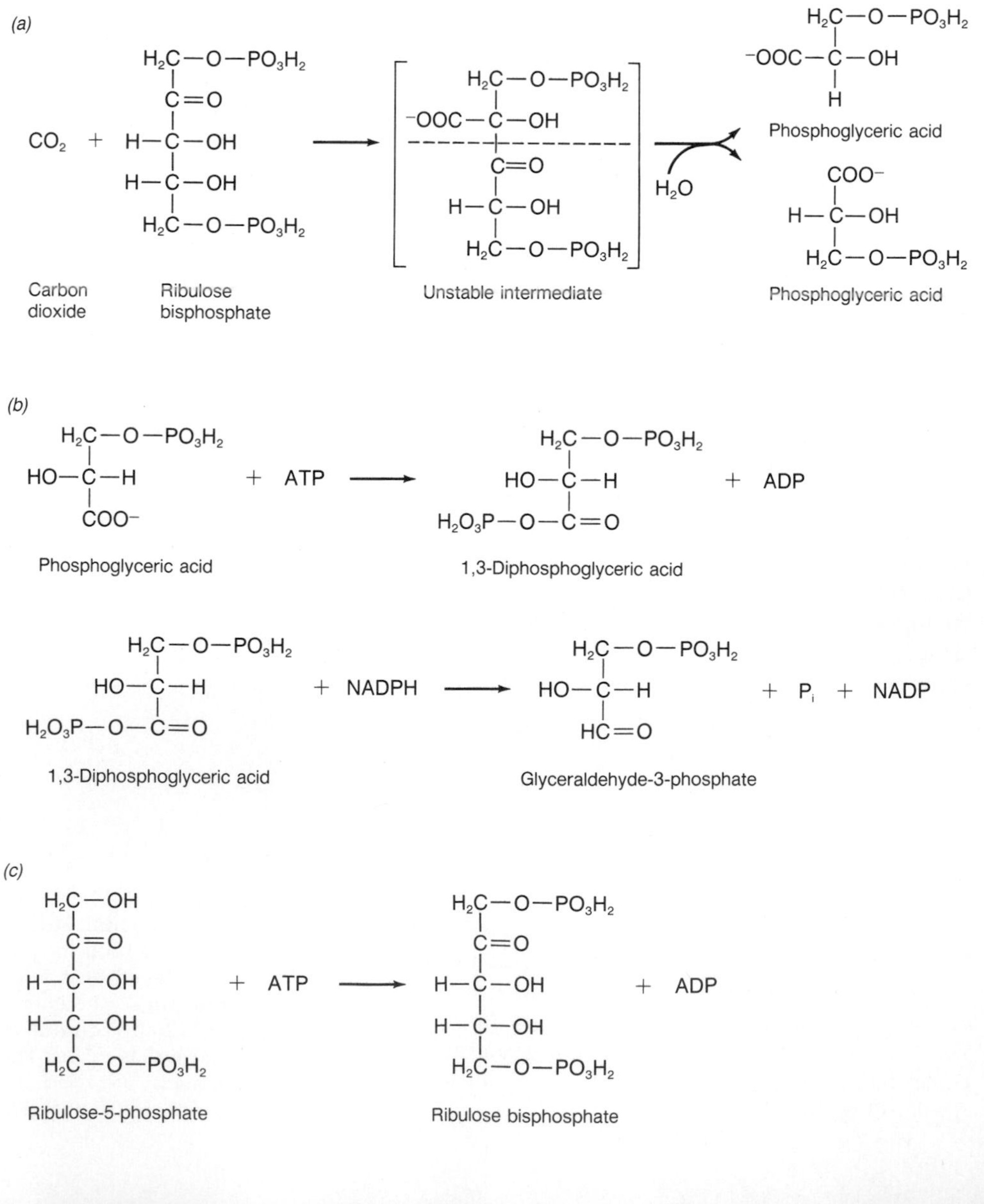

(a)

Carbon dioxide (CO_2) + Ribulose bisphosphate ($H_2C{-}O{-}PO_3H_2$ / $C{=}O$ / $H{-}C{-}OH$ / $H{-}C{-}OH$ / $H_2C{-}O{-}PO_3H_2$) → Unstable intermediate ($H_2C{-}O{-}PO_3H_2$ / $^-OOC{-}C{-}OH$ / $C{=}O$ / $H{-}C{-}OH$ / $H_2C{-}O{-}PO_3H_2$) + H_2O → Phosphoglyceric acid ($H_2C{-}O{-}PO_3H_2$ / $^-OOC{-}C{-}OH$ / H) + Phosphoglyceric acid (COO^- / $H{-}C{-}OH$ / $H_2C{-}O{-}PO_3H_2$)

(b)

Phosphoglyceric acid ($H_2C{-}O{-}PO_3H_2$ / $HO{-}C{-}H$ / COO^-) + ATP → 1,3-Diphosphoglyceric acid ($H_2C{-}O{-}PO_3H_2$ / $HO{-}C{-}H$ / $H_2O_3P{-}O{-}C{=}O$) + ADP

1,3-Diphosphoglyceric acid ($H_2C{-}O{-}PO_3H_2$ / $HO{-}C{-}H$ / $H_2O_3P{-}O{-}C{=}O$) + NADPH → Glyceraldehyde-3-phosphate ($H_2C{-}O{-}PO_3H_2$ / $HO{-}C{-}H$ / $HC{=}O$) + P_i + NADP

(c)

Ribulose-5-phosphate ($H_2C{-}OH$ / $C{=}O$ / $H{-}C{-}OH$ / $H{-}C{-}OH$ / $H_2C{-}O{-}PO_3H_2$) + ATP → Ribulose bisphosphate ($H_2C{-}O{-}PO_3H_2$ / $C{=}O$ / $H{-}C{-}OH$ / $H{-}C{-}OH$ / $H_2C{-}O{-}PO_3H_2$) + ADP

Figure 16.17 Key enzyme reactions of the Calvin cycle. (a) Reaction of the enzyme ribulose bisphosphate carboxylase. (b) Steps in the conversion of 3-phosphoglyceric acid (PGA) to glyceraldehyde-3-phosphate. Note that both ATP and NADPH are required. (c) Conversion of ribulose 5-phosphate to ribulose bisphosphate by the enzyme phosphoribulokinase.

Key enzymes of the Calvin cycle The first step in CO_2 reduction is the reaction catalyzed by the enzyme *ribulose bisphosphate carboxylase*, which involves a reaction between CO_2 and ribulose bisphosphate (Figure 16.17*a*) leading to formation of two molecules of *3-phosphoglyceric acid (PGA)*, one of which contains the carbon atom from CO_2. PGA constitutes the first identifiable intermediate in the CO_2 reductive process. The carbon atom in PGA is still at the same oxidation level as it was in CO_2; the next two steps involve reduction of PGA to the oxidation level of carbohydrate (Figure 16.17*b*). In these steps, *both* ATP and NADPH are required: the former is involved in the phosphorylation reaction that activates the carboxyl group, the latter in the reduction itself. The carbon atom from CO_2 is now at the reduction level of carbohydrate (CH_2O), but only one of the carbon atoms of glyceraldehyde phosphate has been derived from CO_2, the other two having arisen from the ribulose bisphosphate. Since autotrophs can grow on CO_2 as a sole carbon source, there must be a way by which carbon from CO_2 can become incorporated into the other positions in glyceraldehyde phosphate. Further, the ribulose bisphosphate that was used up in the ribulose bisphosphate carboxylase step must be regenerated. The remainder of the reactions of the Calvin cycle are concerned with these matters.

The Calvin cycle The overall scheme of the Calvin cycle is shown in Figure 16.18. The series of reactions leading to the synthesis of ribulose bisphosphate involve a number of sugar rearrangements. Through the action of enzymes which act on pentose phosphate compounds and enzymes of the glycolytic pathways, glyceraldehyde-3 phosphate is converted into ribulose-5 phosphate and subsequently into ribulose bisphosphate. However, because *six* CO_2 are needed to make *one* net C-6 compound for biosynthesis (see Figure 16.18) *six* ribulose bisphosphate molecules are required. These originate from a complex series of reactions in which 10 molecules of glyceraldehyde phosphate (total of 30 carbons) are rearranged by action of the aforementioned enzymes to yield six molecules of ribulose monophosphate (total of 30 carbons, see Figure 16.18). These reactions involve C-3, C-5, C-6, and C-7 intermediates, but the net result is the regeneration of six C-5 molecules. The final step in the regeneration of ribulose bisphosphate is the phosphorylation of ribulose-5-phosphate with ATP by the enzyme *phosphoribulokinase* (Figure 16.17*c*). This enzyme is another enzyme unique to the Calvin cycle.

To summarize the Calvin cycle, six molecules of CO_2 and six molecules of ribulose bisphosphate react to form twelve C-3 compounds. Two of the C-3 units go to biosynthesis as a 6-carbon sugar while the remaining ten are recycled to form six C-5 acceptors. Since we begin and end the reaction sequence with the same number of acceptor molecules, the cyclic nature of the Calvin cycle is thus apparent.

Biosynthetic reactions Most photosynthetic organisms live in nature in alternating light and dark regimes, but the primary products of the light reactions, ATP and NADPH, are short-lived and must be quickly used up since they cannot be stored. Photosynthetic organisms circumvent this difficulty by converting CO_2 into energy-rich storage products during the light cycle, then using these during the dark cycle. In algae and cyanobacteria, storage products are usually carbohydrates such as sucrose or starch, whereas in purple and green bacteria the main

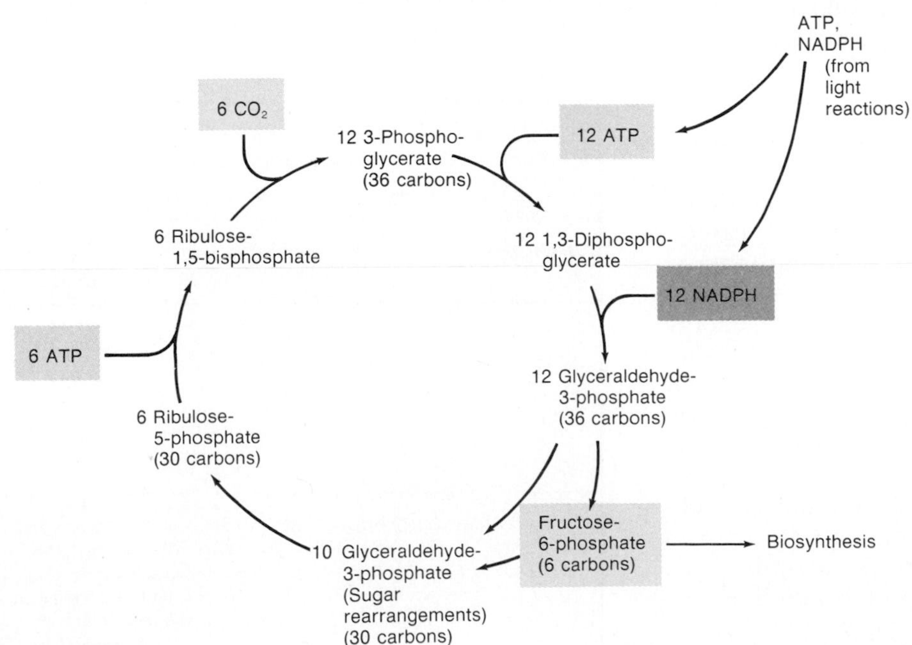

Figure 16.18 The Calvin cycle. For each *six* molecules of CO_2 incorporated, *one* fructose 6-phosphate is produced.

storage products are glycogen and poly-β-hydroxybutyric acid.

The balance sheet of the Calvin cycle Let us consider now the complete balance sheet for conversion of six molecules of CO_2 into one molecule of fructose-6 phosphate (Figure 16.18). Twelve molecules each of ATP and NADPH are required for the reduction of twelve molecules of PGA to glyceraldehyde phosphate, and six ATP molecules are required for conversion of ribulose phosphate to ribulose bisphosphate. Thus twelve NADPH and eighteen ATP are required to manufacture one fructose (glucose) molecule from CO_2.

Alternatives to the Calvin cycle The key enzymes of the Calvin cycle, *ribulose bisphosphate carboxylase* and *phosphoribulokinase*, are unique to autotrophs which fix CO_2 via the Calvin cycle. These enzymes have been found in virtually all photosynthetic organisms examined—plants, algae, and bacteria. They are also found in the lithotrophic autotrophic bacteria, such as the sulfur, iron, and nitrifying bacteria (see below). The enzyme *ribulose bisphosphate carboxylase* is sometimes stored inside the cell in large aggregates called *carboxysomes* (see Section 19.5)

However, three groups of autotrophs, the green bacteria, the methanogenic bacteria, and the acetogenic bacteria, do not use the Calvin cycle for CO_2 fixation. In these groups, Calvin cycle enzymes are not present. We discuss the pathways by which these groups of bacteria fix CO_2 in Sections 19.1, 19.9, and 19.34.

16.7 Lithotrophy: Energy from the Oxidation of Inorganic Electron Donors

Organisms that can obtain their energy from the oxidation of inorganic compounds are called **lithotrophs**. With minor exception (see Section 17.9), all known lithotrophs are procaryotes. Most lithotrophic bacteria are also able to obtain all of their carbon from CO_2 and they are therefore also autotrophs. As we have noted, two components are needed when an organism grows on CO_2 as sole carbon source: energy in the form of ATP, and reducing power. In lithotrophs, ATP generation is coupled to the *oxidation* of the electron donor and is hence in principle little different from that of conventional electron transport metabolism (see the discussion of oxidative phosphorylation in Section 4.12). Reducing power is obtained from the inorganic compounds (also used as energy sources) by *reverse electron transport reactions*, as discussed for phototrophic bacteria in Section 16.3

Energetics of lithotrophy A review of redox potentials illustrated in Table 16.1 will indicate that the oxidation of a number of inorganic compounds, when O_2 is used as electron acceptor, will provide sufficient energy for ATP synthesis. The further apart two half reactions are, the more energy can be released. The manner by which energy release can be calculated from reduction potentials is presented in Appendix 1. Here we simply give some examples for illustrative purposes. For instance, the difference in reduction potential between the H^+/H_2 couple and the $\frac{1}{2}\ O_2/H_2$ couple is −1.23 volts, which is equivalent to a free energy yield of −57 kcal/mole. On the other hand, the potential difference between the H^+/H_2 couple and the NO_3^-/NO_2^- couple is less, 0.84 volts, equivalent to a free energy yield of −39 kcal/mole. This is still quite sufficient for the production of ATP (the high-energy phosphate bond of ATP has a free energy of around −7 kcal/mole). However, a similar calculation will show that there is insufficient energy available from the oxidation of H_2S using CO_2 as electron acceptor.

From such energy calculations, it is possible to predict the kinds of lithotrophic reactions that might be found in nature. Since organisms obey the laws of thermodynamics, only reactions which are thermodynamically favorable will be potential energy-yielding reactions. However, just because a reaction is theoretically possible does not mean that it actually occurs. There may be evolutionary or ecological reasons why a particular lithotrophic reaction does not take place. Table 16.3 summarizes energy yields for some reactions which are known to be carried out by lithotrophic microorganisms. We discuss some of these lithotrophic processes briefly in the rest of this section, and the organisms involved are discussed in more detail in Chapter 19. We also discuss ecological aspects of lithotrophy in Chapter 17.

Table 16.3 Energy yields from the oxidation of various inorganic electron donors

Reaction	$\Delta G^{0\prime}$ (kcal/mole)
$HS^- + H^+ + \frac{1}{2}O_2 \rightarrow S^0 + H_2O$	−48.5
$S^0 + 1\frac{1}{2}O_2 + H_2O \rightarrow SO_4^{2-} + 2H^+$	−140.6
$NH_4^+ + 1\frac{1}{2}O_2 \rightarrow NO_2^- + 2H^+ + H_2O$	−62.1
$NO_2^- + \frac{1}{2}O_2 \rightarrow NO_3^-$	−18.1
$H_2 + \frac{1}{2}O_2 \rightarrow H_2O$	−54.6
$Fe^{2+} + H^+ + \frac{1}{4}O_2 \rightarrow Fe^{3+} + \frac{1}{2}H_2O$	−17.0

Data calculated from values in Appendix 1. Values for Fe^{2+} are for pH 2–3, others, for pH 7. The energy yield depends on both pH and concentrations of reactants, and the values given should be viewed only as average values, suitable primarily for comparative purposes.

16.8 Hydrogen-Oxidizing Bacteria

Hydrogen (H_2) gas is a common product of microbial metabolism and a number of lithotrophic bacteria are able to use H_2 as an energy source. A wide variety of H_2-oxidizing bacteria are known, differing in the electron *acceptor* they use. Those H_2-oxidizing bacteria which use electron acceptors other than oxygen carry out a type of anaerobic respiration and are discussed later in this chapter (Section 16.13). We discuss here the *aerobic* H_2-oxidizing species, which are the types most commonly referred to when the term *hydrogen bacteria* is used.

The hydrogen bacteria are facultative lithotrophs, capable of growing either as conventional heterotrophs on organic compounds, or by H_2 oxidation; taxonomically, the hydrogen bacteria belong to various groups of eubacteria (see Section 19.6). When growing autotrophically hydrogen bacteria show the following stoichiometry:

$$6H_2 + 2O_2 + CO_2 \rightarrow (CH_2O) + 5H_2O$$

where the formulation (CH_2O) signifies cell material at the same oxidation state as carbohydrate.

We discussed the free energy change occurring in the above listed reaction earlier in this chapter. The enzyme **hydrogenase** catalyzes the initial oxidation of H_2, the electrons being transferred to either NAD^+ or a quinone acceptor. Two kinds of hydrogenases have been recognized: soluble and membrane-bound. Only a few bacteria have both soluble and membrane-bound hydrogenases; most bacteria have either one or the other. Electrons introduced into the electron transport chain at the level of NADH or reduced quinone undergo electron transport reactions leading to the formation of a proton-motive force, as discussed earlier, and ATP formation occurs via the membrane-bound ATPase.

Many nonlithotrophic bacteria can oxidize H_2 without growing on it as a sole energy source. For instance, *Escherichia coli* can use H_2 as an energy source but cannot grow on CO_2 as sole carbon source. To obtain growth on H_2 therefore, *E. coli* must be given an organic compound as carbon source. Organisms which use an inorganic compound as energy source and an organic compound as carbon source are sometimes called *mixotrophs*. Many sulfate-reducing bacteria and most methanogenic bacteria also use H_2 as electron donor, as discussed later in this chapter.

When growing autotrophically, hydrogen bacteria fix CO_2 by the Calvin cycle. If organic compounds are present, the Calvin cycle enzymes are repressed and enzymes involved in the utilization of the organic compound induced. Regulation of lithotrophic metabolism in hydrogen bacteria is very precise and specific, but details of regulation in these hydrogen bacteria are not known.

16.9 Sulfur Bacteria

Many reduced sulfur compounds can be used as electron donors by a variety of colorless sulfur bacteria (called "colorless" to distinguish them from the bacteriochlorophyll-containing [pigmented] green and purple sulfur bacteria discussed earlier in this chapter). The most common sulfur compounds used as energy sources are hydrogen sulfide (H_2S), elemental sulfur (S^0), and thiosulfate ($S_2O_3^{2-}$). The final prod- of sulfur oxidation in most cases is sulfate (SO_4^{2-}) and the total number of electrons involved between H_2S (oxidation state, -2) and sulfate (oxidation state, $+6$) is 8 (see Table 16.6 for a summary of sulfur oxidation states). Less energy is available when one of the intermediate sulfur oxidation states is used:

$$H_2S + 2O_2 \rightarrow SO_4^{2-} + 2H^+ \quad \text{sulfide } -188.7 \text{ kcal/mole}$$

$$S^0 + H_2O + 1\ 1/2O_2 \rightarrow SO_4^{2-} + 2H^+ \quad \text{sulfur } -140.6 \text{ kcal/mole}$$

$$S_2O_3^{2-} + H_2O + 2O_2 \rightarrow 2SO_4^{2-} + 2H^+ \quad \text{thiosulfate } -97.7 \text{ kcal/mole}$$

The oxidation of the most reduced sulfur compound, H_2S, occurs in stages, and the first oxidation step results in the formation of elemental sulfur, S^0, a highly insoluble material. Some H_2S-oxidizing bacteria deposit the elemental sulfur formed inside the cell (Figure 16.19*a*). The sulfur deposited as a result of the initial oxidation is an energy reserve, and when the supply of H_2S has been depleted, additional energy can be obtained from the oxidation of the sulfur deposit.

When elemental sulfur is provided externally as an electron donor, the organism must grow attached to the sulfur particle because of the extreme insolubility of elemental sulfur (Figure 16.19*b*). By adhering to the particle, the organism can efficiently obtain the atoms of sulfur needed, which slowly go into solution. As these atoms are oxidized, more dissolve; gradually the sulfur particle is consumed.

Note that in the sulfur-oxidation reactions shown above, one of the products is H^+. Production of H^+ ions results in a lowering of the pH, and one result of the oxidation of reduced sulfur species is the acidification of the medium. The acid formed by the sulfur bacteria is sulfuric acid, H_2SO_4, and sulfur bacteria are often able to bring about a marked reduction in pH of the medium. Some sulfur bacteria have been found to lower the pH to values less than 2!

The electron transport system of a typical sulfur-oxidizing bacterium is illustrated in Figure 16.20*a*. As shown, electrons from reduced sulfur compounds

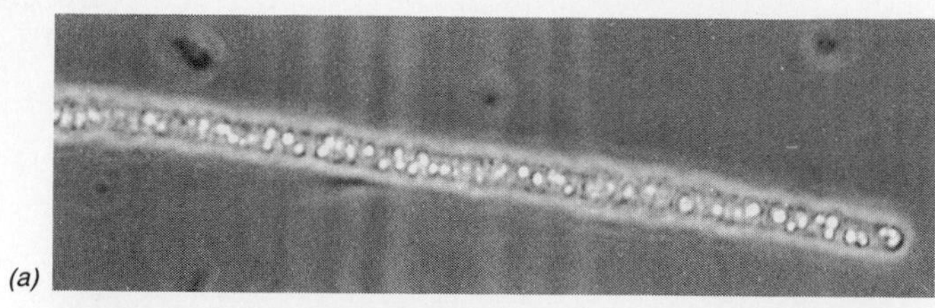

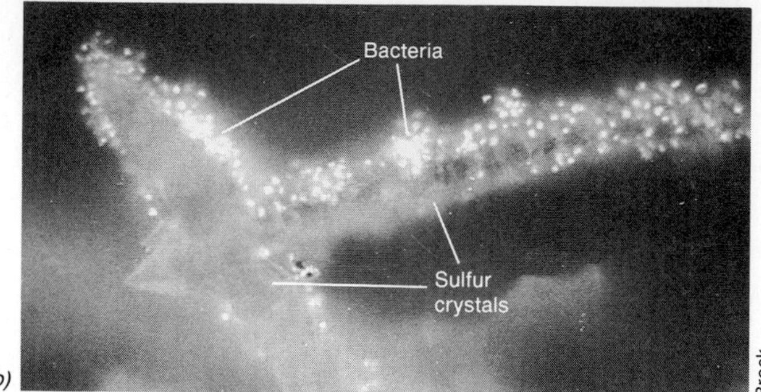

Figure 16.19 Sulfur bacteria. (a) Deposition of internal sulfur granules by *Beggiatoa*. Magnification, 2300×. (b) Attachment of the sulfur-oxidizing bacterium *Sulfolobus acidocaldarius* to a crystal of elemental sulfur. Visualized by fluorescence microscopy after staining the cells with the dye acridine orange. The sulfur crystal does not fluoresce. Magnification, 630×.

enter the chain at various points (depending upon their reduction potentials), and are transported to molecular oxygen, the whole process generating a ***membrane potential*** which leads to ATP production by oxidative phosphorylation. Electrons for CO_2 reduction come from reverse electron transport reactions eventually yielding NADPH. Note in Figure 16.20 how the relatively high reduction potentials of compounds like $S_2O_3^{2-}$ or S^0 result in a rather abbreviated electron transport chain for energy purposes but a rather long route to the reduction of NADPH; the latter is, of course, an ATP-dependent series of reactions. These biochemical limitations are in part responsible for the relatively poor growth yields of most sulfur lithotrophs.

When growing autotrophically, the sulfur bacteria use the Calvin cycle to fix CO_2. However, some sulfur bacteria are also able to grow using organic compounds. A few sulfur bacteria are also able to grow anaerobically on reduced sulfur compounds, utilizing nitrate (NO_3^-) as electron acceptor. Under these latter conditions, a type of anaerobic respiration is being carried out (see Section 16.13). Several sulfur-oxidizing bacteria appear to grow only mixotrophically, using H_2S as energy source and an organic compound as carbon source. In this category is *Beggiatoa*, a widespread organism in marine and freshwater sediments (see Box).

16.10 Iron-Oxidizing Bacteria

The aerobic oxidation of iron from the ferrous (Fe^{2+}) to the ferric (Fe^{3+}) state is an energy-yielding reaction for a few bacteria. Only a small amount of energy is available from this oxidation (see Table 16.3), and for this reason the iron bacteria must oxidize large amounts of iron in order to grow. Ferric iron forms very insoluble ferric hydroxide [$Fe(OH_3)$] precipitates in water (Figure 16.21*a*). Most iron-oxidizing bacteria also oxidize sulfur and are obligate acidophiles.

That most iron-oxidizing bacteria are acidophiles is explained in part by the chemistry of iron. At neutral pH, ferrous iron is not only insoluble, it rapidly oxidizes spontaneously to the ferric state. Thus, at neutral pH, ferrous iron is only stable under *anaerobic* conditions. At low pH, however, ferrous iron is both stable to spontaneous oxidation, and is soluble in water. Since spontaneous oxidation of ferrous iron at neutral pH is so rapid, significant amounts of ferrous iron do not accumulate under these conditions. This explains why most of the iron-oxidizing bacteria are obligately acidophilic.

The best known iron-oxidizing bacterium is *Thiobacillus ferrooxidans*, which is able to grow au-

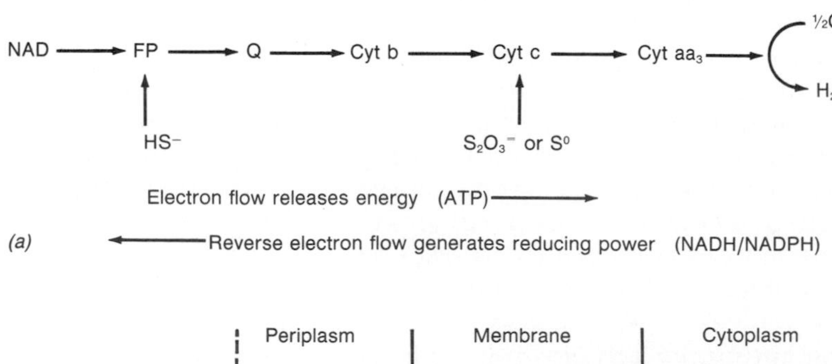

Figure 16.20 Electron flow in a *Thiobacillus* grown on (a) sulfur, thiosulfate, sulfide, or (b) ferrous iron.

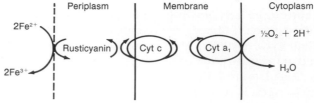

Winogradsky's Legacy

The discovery of autotrophy in lithotrophic bacteria was of major significance in the advance of our understanding of cell physiology, since it showed that CO_2 could be converted into organic carbon without the intervention of chlorophyll. Previously, it had been thought that only green plants converted CO_2 into organic form. The idea of lithotrophic autotrophy was first developed by the great Russian microbiologist Sergei Winogradsky, who did most of this work while he was still rather a junior scientist. Winogradsky studied sulfur bacteria because certain of the colorless sulfur bacteria (*Beggiatoa, Thiothrix*) are very large and hence are easy to study even in the absence of pure cultures. Springs with waters rich in H_2S are fairly common around the world, and Winogradsky studied several such springs in the Bernese Oberland district of Switzerland. In the outflow channels of sulfur springs, vast populations of *Beggiatoa* and *Thiothrix* develop, and suitable material for microscopic and physiological studies could be obtained by merely lifting up the white filamentous masses. Pure cultures were not needed for many studies. As Winogradsky noted, "This study of demipurity would be poor in an ordinary culture but is sufficient for a culture under the microscope, since it is possible to observe the development from day to day, almost from hour to hour, and see easily the presence of contaminants." Winogradsky first showed that the colorless sulfur bacteria were present only in water containing H_2S. As the water flowed away from the source, the H_2S gradually dissipated, and sulfur bacteria were no longer present. This suggested to him that their development was dependent on the presence of H_2S. Winogradsky then showed that by starving *Beggiatoa* filaments for a while, they lost their sulfur granules; he found, however, that the granules were rapidly restored if a small amount of H_2S was added. He thus concluded that H_2S was being oxidized to elemental sulfur. But what happened to the sulfur granules when the filaments were starved of H_2S? Winogradsky showed by some clever microchemical tests that when the sulfur granules disappeared, sulfate appeared in the medium. Thus he formulated the idea that *Beggiatoa* (and by inference other colorless sulfur bacteria) oxidize H_2S to elemental sulfur and subsequently to sulfate. Because they seemed to require H_2S for development in the springs, he postulated that this oxidation was the principal source of energy for these organisms.

Studies on *Beggiatoa* provided the first evidence that an organism could oxidize an inorganic substance as a possible energy source, and this was the origin of the concept of lithotrophy. However, the sulfur bacteria proved difficult to work with, primarily because there are a number of spontaneous chemical changes in sulfur compounds, which can also occur and confuse the study. Winogradsky thus turned to a study of the nitrifying bacteria, and it was with this group that he clearly was able to show that autotrophic fixation of CO_2 was coupled to the oxidation of an inorganic compound in the complete absence of light and chlorophyll. The process of nitrification had been known before Winogradsky's work from studies on the fate of sewage when added to soil. Two Frenchmen, T. Schloesing and A. Mutz, had shown that the process was due to living organisms. Winogradsky proceeded to isolate some of the bacteria, using completely mineral media in which CO_2 was the sole carbon source and ammonia was the sole electron donor. Because ammonia is chemically stable, it was easy to show that the oxidation of ammonia to nitrite, and subsequently to nitrate, is a strictly bacterial process. As no organic materials were present in the medium, it was also possible to show that organic matter (the bacterial cell material) was formed only from CO_2. If the ammonia was left out of the medium, no growth occurred. Careful chemical analyses showed that the amount of organic matter formed by the bacteria was proportional to the amount of ammonia or nitrite which they oxidized. Winogradsky concluded, "This [process] is contradictory to that fundamental doctrine of physiology which states that a complete synthesis of organic matter cannot take place in nature except through chlorophyll-containing plants by the action of light." His basic conclusion has been confirmed by a large number of subsequent studies. At least in one way, however, autotrophy in lithotrophs and phototrophs is similar, in that in both processes, the pathway of CO_2 fixation follows the same biochemical steps (the Calvin cycle) involving the enzyme ribulose bisphosphate carboxylase.

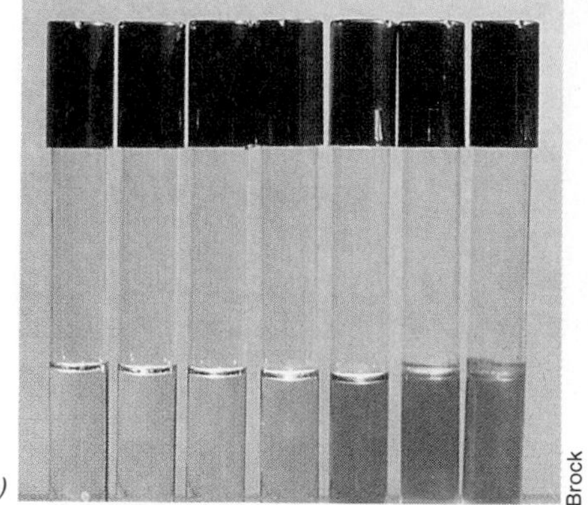

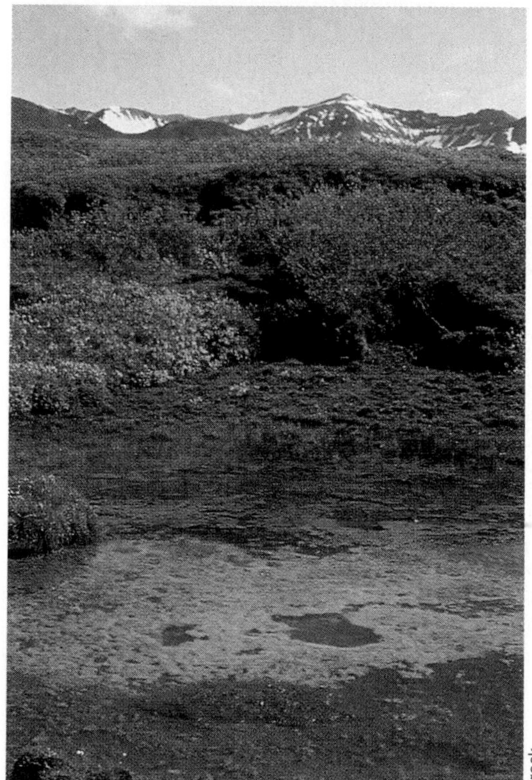

Figure 16.21 Iron-oxidizing bacteria. (a) Acid mine drainage, showing the confluence of a normal river and a creek draining a coal-mining area. The acidic creek is very high in ferrous iron. At low pH values, ferrous iron does not oxidize spontaneously in air, but *Thiobacillus ferrooxidans* carries out the oxidation. Insoluble ferric hydroxide and complex ferric salts precipitate, forming the precipitate called "yellow boy" by coal miners. (b) Cultures of *Thiobacillus ferrooxidans*. Shown is a dilution series, no growth in the tube on the left and increasing amounts of growth from left to right. Growth is evident from the production of Fe^{3+} which readily forms $Fe(OH)_3$ leading to the yellow orange color. (c)Extensive development of insoluble ferric hydroxide in a small pool draining a bog in Iceland. Iron deposits such as this are widespread in cooler parts of the world and are modern counterparts of the extensive bog-iron deposits of earlier geological eras. These ancient deposits are now the sources of much commercially mined iron ore. In the water-saturated bog soil, facultatively anaerobic bacteria reduce ferric iron to the more soluble ferrous state. The ferrous iron leaches into the drainage area surrounding the bog, where oxidation occurs, either spontaneously or through the agency of iron-oxidizing bacteria (*Gallionella* and *Leptothrix*), and the insoluble ferric hydroxide deposit is formed.

totrophically on either ferrous iron or reduced sulfur compounds. The electron transport system was illustrated in Figure 16.20*b*. This organism is very common in acid-polluted environments such as coal-mining dumps (Figure 16.21*a*), and we discuss its role in acid-mine pollution and mineral oxidation in Sections 17.16 and 17.17. Another iron-oxidizing bacterium is the archaebacterium *Sulfolobus*, which lives in hot, acid springs at temperatures up to the boiling point of water (see Figure 16.19*b*). This organism is nutritionally versatile, as it not only oxidizes ferrous iron, but also H_2S, elemental sulfur, and a variety of organic compounds. *Sulfolobus* can therefore grow either autotrophically or heterotrophically.

The bioenergetics of iron oxidation by *T. ferrooxidans* is of biochemical interest because of the high reduction potential of the Fe^{3+}/Fe^{2+} couple (+0.77 volts). The respiratory chain of *T. ferrooxidans* contains cytochromes of the *c* and a_1 types and a copper-containing protein called *rusticyanin*. Since the reduction potential of the Fe^{3+}/Fe^{2+} couple is so high, the route of electron transport to oxygen ($\frac{1}{2}$ O_2/H_2O, $E_0' = +0.88$ volts) will obviously be very short. Electrons from the oxidation of Fe^{2+} cannot reduce NAD^+, FAD^+ or many of the other components of the electron transport chain. How, then, do acidophilic iron-oxidizing bacteria make a living?

Biochemical studies of electron flow in *Thiobacillus ferrooxidans* suggest that acidophilic iron bacteria take advantage of the preexisting proton gradients of their environment for energy-generating purposes. Remember that the cytoplasm of any organism, including that of acidophiles, must remain near neutrality; in *T. ferrooxidans* the internal pH is about 6.0. However, the pH of the environment of *T. ferrooxidans* is much lower, near pH 2.0. This pH difference across the cytoplasmic membrane represents a membrane electrochemical gradient that can play a role in ATP synthesis. However, to retain a neutral environment, protons entering the cell through the proton translocating ATPase (driving the phosphorylation of ADP in the process) must be consumed. It is here that the oxidation of Fe^{2+} plays an important role.

The oxidation of Fe^{2+} to Fe^{3+} ($2Fe^{2+} + \frac{1}{2} O_2 + 2H^+ \rightarrow 2Fe^{3+} + H_2O$) is a *proton-consuming* reaction. Experimental evidence suggests that the reaction $\frac{1}{2} O_2 + 2H^+ \rightarrow H_2O$ occurs on the *inner* face of the cytoplasmic membrane while the reaction $Fe^{2+} \rightarrow Fe^{3+}$ occurs near the *outer* face of the membrane. Electrons from Fe^{2+} are accepted in the periplasmic space by rusticyanin (Figure 16.20*b*); as befits its periplasmic location, rusticyanin is completely stable and functions optimally at pH 2. Electrons are donated from rusticyanin to a high potential membrane-bound cytochrome *c* which subsequently transfers electrons to cytochrome a_1, the terminal oxidase (Figure 16.20*b*). Cytochrome a_1 donates electrons to $\frac{1}{2} O_2$ with the two protons required to form water coming from the cytoplasm. An influx of protons via the ATPase replenishes the proton supply, and as long as Fe^{2+} remains available the natural proton gradient across the *T. ferrooxidans* membrane can continue to drive ATP synthesis. Thus, although energy conservation in *T. ferrooxidans* results from a classical chemiosmotic ATPase reaction which couples the entry of protons to the synthesis of ATP, the proton gradient in *T. ferrooxidans* is *not* established as a result of electron transport, but instead is a simple consequence of the habitat of the organism!

As we discussed earlier, autotrophic CO_2 fixation requires the presence of both ATP and reducing power. Reducing power in the form of NADH or NADPH is formed by the process of *reversed electron transport* discussed earlier in this chapter, using energy from ATP to reverse electron flow and form NADH or NADPH. Because of the high potential of the electron donor this process takes a lot of iron oxidation, and it is therefore understandable that the cell yields of iron-oxidizing bacteria are quite low. In an environment where these organisms are living, their presence is signalled not by the formation of much cell material, but by the presence of massive amounts of ferric iron precipitation (Figure 16.21*a* and *b*). We discuss the important ecological roles of the iron-oxidizing bacteria in Chapter 17, and their taxonomy is discussed in Chapter 19.

In addition to the true iron-oxidizing bacteria just discussed, there are some other bacteria which have been traditionally considered to be iron bacteria but which are probably unable to grow autotrophically by ferrous iron oxidation. These bacteria, named *Gallionella ferruginea, Sphaerotilus natans,* and *Leptothrix ochracea*, live at near *neutral* pH values and are commonly found in environments where ferrous iron is moving from anaerobic to aerobic conditions (see Figure 16.21*c* for a typical site for these bacteria). As we have discussed, at neutral pH, ferrous iron is not stable in the presence of oxygen, and is rapidly oxidized to the ferric (insoluble) state. Because of this, the only neutral pH environments where ferrous iron is present are *interfaces* between anaerobic and aerobic conditions. *Gallionella ferruginea, Sphaerotilus natans,* and *Leptothrix ochracea* live at such interfaces, and are generally seen mixed in with the characteristic deposits which they form (Figure 16.22). If these organisms live autotrophically from ferrous iron oxidation is unclear but they are, nonetheless, interesting organisms of the iron-rich biotope. We discuss the taxonomy of these organisms in Chapter 19.

In addition to ferrous iron, the related metal *manganese* is also oxidized by a few bacteria at neutral pH. Manganese autotrophy has been much discussed but never proved. The most common manganese-oxidizing bacterium is *Leptothrix discophorus*, which is discussed in Section 19.14. We discuss the process of manganese oxidation itself in Section 17.16.

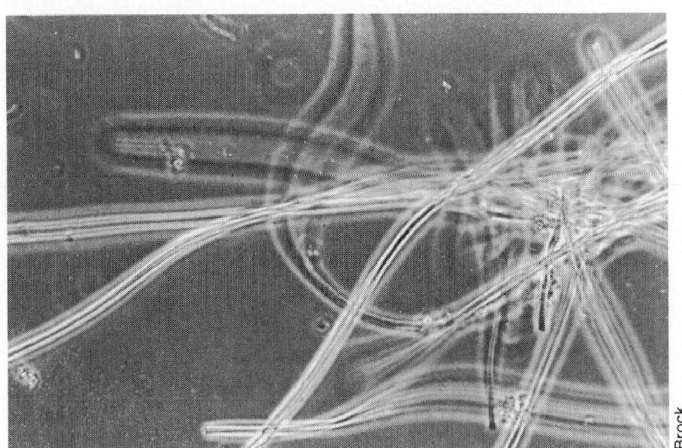

Figure 16.22 Phase-contrast photomicrograph of empty iron-encrusted sheaths of *Sphaerotilus* collected from seepage at the edge of a small swamp. Magnification, 1875×.

16.11 Ammonium and Nitrite Oxidizing Bacteria

The most common *inorganic nitrogen compounds* used as electron donors are ammonia (NH_3) and nitrite (NO_2^-), which are *aerobically* oxidized by the nitrifying bacteria. The nitrifying bacteria are eubacteria which are widely distributed in soil and water. One group of organisms (*Nitrosomonas* is one genus) oxidizes ammonia to nitrite, and another group (*Nitrobacter*) oxidizes nitrite to nitrate; the complete oxidation of ammonia to nitrate is thus carried out by members of these two groups of organisms acting in sequence. It is puzzling that the complete oxidation of ammonia to nitrate requires cooperation of two separate organisms, whereas the oxidation of H_2S to sulfate can be done by a single organism. No explanation for this difference exists. Nitrifying bacteria are widespread in the soil, and their significance in soil fertility and in the nitrogen cycle is discussed in Section 17.14; their morphology, taxonomy, physiology, and biochemistry is discussed in Section 19.4.

16.12 Autotrophy and ATP Production among Lithotrophs

As we have noted, when growing autotrophically lithotrophs must also produce reducing power in addition to ATP, if they are to grow with CO_2 as sole carbon source. Although NAD^+ can be reduced to NADH directly by H_2, all of the other inorganic electron donors have reduction potentials *higher* than that of NADH (Table 16.1). If the reduction potential of an electron donor is higher than that of NADH, there is no way in which its oxidation can be directly coupled to the reduction of NAD^+ to NADH. There is good evidence from studies with nitrifying, sulfur-oxidizing, and iron-oxidizing bacteria that NAD^+ is reduced by reversed electron transport as already discussed in Sections 16.3 and 16.10. In lithotrophic bacteria, reversed electron transport leads to the reduction of NAD^+ (or $NADP^+$) using energy from ATP generated from the oxidation of the inorganic electron donor. Most lithotrophs can grow with CO_2 as sole source of carbon and many appear to prefer autotrophic growth. Most lithotrophs will assimilate organic compounds if available but many are restricted in terms of the compounds they can catabolize. Other lithotrophs grow much better as heterotrophs than as lithotrophs and probably only grow lithotrophically in nature in the absence of sufficient organic nutrients.

ATP production by lithotrophs As noted in the discussions of the individual groups of lithotrophs, lithotrophic bacteria all possess electron transport chains containing cytochromes, quinones, and many of the major components described in aerobic heterotrophic organisms (see Section 4.11). However, with the exception of H_2, none of the lithotrophic electron donors are sufficiently negative in reduction potential to reduce NAD^+ directly and initiate electron flow to O_2 from NADH. Hence, electrons must feed into the electron transport chain at a point "downstream" from that of the NAD^+/NADH couple, at a point where the reduction potential of the inorganic electron donor O–R couple is more *negative* than that of the acceptor molecule. In most cases this turns out to be a cytochrome of some type and the situation for a sulfur-oxidizing bacterium like *Thiobacillus* was shown in Figure 16.20. As we noted, the situation is an extreme one for the iron-oxidizing bacteria because the reduction potential of their electron donor, Fe^{2+}, is already very high. Nitrite oxidizing bacteria face a similar problem. Nevertheless, lithotrophs manage to transport electrons to O_2 through a sufficient number of carriers to generate a proton gradient which is dissipated through the action of ATPase to yield ATP.

Table 16.3 summarizes energy yields from the aerobic oxidation of various lithotrophic electron donors. It can be seen that the energetics of hydrogen and sulfur oxidation are far more favorable than those of the nitrifying or iron-oxidizing bacteria, since ATP yield is directly proportional to the amount of energy released in a given oxidation and the amount of energy released is a direct function of the *difference* in reduction potential between the electron donor and electron acceptor couples, respectively. Because of the added burden of autotrophy most lithotrophs synthesize only small amounts of cell material while oxidizing huge amounts of substrate (Figure 16.23). The growth of *Nitrobacter* is a case in point. Cultures of *Nitrobacter* oxidize large amounts of NO_2^- before barely visible turbidity is achieved in a flask. The energetic demands of autotrophic growth puts strong demands on the small amount of ATP being produced. From an ecological standpoint, however, it should be remembered that lithotrophs are utilizing electron donors not catabolized by heterotrophic organisms. Hence, lithotrophs survive in nature without severe nutritional competition from the bulk of the microbial world.

Comparison of autotrophs and heterotrophs The energy metabolism and the terminology used in describing the energy relationships of autotrophs and heterotrophs are indicated in Table 16.4, which emphasizes the similarities and differences among various groups. Note that Table 16.4 deals only with energy sources and reductants, not with carbon sources. The ability of lithotrophs to assimilate organic compounds as carbon sources while using inorganic oxidations for energy makes possible the growth of these organisms as lithotrophic hetero-

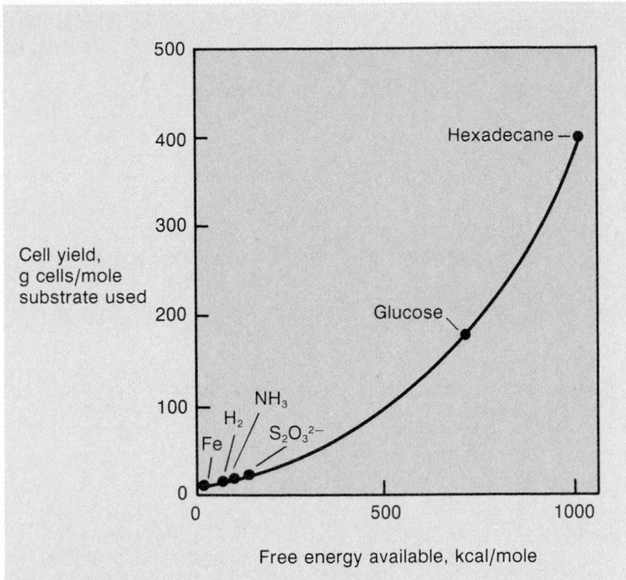

Figure 16.23 Comparison of cell yield (various microorganisms) on energy sources with different free energies of the reaction. In all cases, the electron acceptor is O_2.

trophs. This type of growth has also been referred to as **mixotrophic**. Some lithotrophs grow best under mixotrophic conditions, because the electron donor need be used only for generation of ATP, reducing power either not being needed or coming from the organic compound. At least one sulfur bacterium, *Beggiatoa*, seems to require mixotrophic conditions to use reduced sulfur compounds as electron donors. Most strains of *Beggiatoa* are unable to grow on a completely inorganic medium with reduced sulfur compounds and CO_2, but will use reduced sulfur compounds as electron donors when acetate is present. *Beggiatoa* will also grow on acetate as sole source of carbon and energy in the absence of reduced sulfur compounds. A number of flagellated algae (eucaryotes) also require organic compounds such as acetate for carbon, but will use light as an energy source. One bacterium, *Thiobacillus perometabolis*, is apparently deficient in CO_2 fixation ability while still being capable of lithotrophic oxidation of reduced sulfur compounds. This organism is not able to make ribulose bisphosphate carboxylase (see Section 16.6) and is appropriately described as an obligate mixotroph.

One group of photosynthetic bacteria, the nonsulfur purple bacteria, shows either *photoheterotrophic* or conventional *heterotrophic* metabolism, depending on environmental conditions. They are photosynthetic under anaerobic conditions and heterotrophic under aerobic conditions. In these organisms, O_2 inhibits formation of the photosynthetic pigments, so that under aerobic conditions the organisms become colorless and carry on a normal heterotrophic respiratory metabolism. Under anaerobic conditions in the light, O_2-mediated respiration cannot occur, of course, and bacteriochlorophyll is made and cyclic photophosphorylation can take place, although organic compounds can be used as sources of carbon. Under anaerobic conditions, some of these organisms can also grow slowly in the dark and then energy generation is by fermentation. The nonsulfur purple bacteria are thus one of the most versatile groups of organisms from the viewpoint of energy-generating mechanisms.

16.13 Anaerobic Respiration

We discussed in some detail in Chapter 4 the process of *aerobic* respiration. As we noted, molecular oxygen (O_2) serves as an external electron acceptor, accepting electrons from electron carriers such as NADH by way of an electron transport chain. We noted in Chapter 4 that a variety of other electron acceptors could be used instead of O_2, in which case the process was called *anaerobic respiration*. We now discuss some of the anaerobic respiration processes that have been recognized. In most cases, the energy source(s) used by organisms carrying out anaerobic respiration are *organic* compounds, but several lithotrophic organisms are also able to use electron acceptors other than O_2. Also, although most of the electron acceptors to be discussed are *inorganic*, several organic com-

Table 16.4 The basic types of energy metabolism

	Energy source	
Reductant	**Light: phototrophs**	**Chemical: chemotrophs**
Inorganic	Photosynthetic eucaryotes; cyanobacteria (reductant water) Purple and green bacteria (reductant mainly H_2S)	Sulfur bacteria; iron bacteria; nitrifying bacteria (lithotrophs)
Organic	Purple and green bacteria using organic reductants (photoheterotrophs)	Many procaryotes and eucaryotes (conventional heterotrophs)

pounds can also serve as electron acceptors. Most organisms carrying out anaerobic respiration are procaryotes and the chemical transformations which they carry out during their energy generation process are frequently of great importance ecologically or industrially, as we will discuss in Chapter 17.

The bacteria carrying out anaerobic respiration generally possess electron transport systems containing cytochromes. Their respiratory systems are thus analogous to those of conventional aerobes. In some cases, the anaerobic respiration process competes in the same organism with an aerobic one. In such cases, if O_2 is present, aerobic respiration is usually favored, and only when O_2 is depleted from the environment would the alternate electron acceptor be reduced. Other organisms carrying out anaerobic respiration are *obligate* anaerobes, and are unable to use O_2.

The energy released from the oxidation of an electron donor using O_2 as electron acceptor is higher than if the same compound is oxidized with an alternate electron acceptor. These energy differences can be perceived readily if the reduction potentials listed in Table 16.1 are examined. As discussed at the beginning of this chapter, and in detail in Appendix 1, theoretical energy yields can be calculated using differences in reduction potential between the electron donor (which is often at the reduction potential of the $NAD^+/NADH$ couple) and the electron acceptor. Because the O_2/H_2O couple is the most oxidizing, more energy is available when O_2 is used than when another electron acceptor is used. As noted in Table 16.1, other electron acceptors which are near O_2 are Fe^{3+}, NO_3^-, and NO_2^-. Farther up the scale are CO_2 and SO_4^{2-}. A summary of the most common types of anaerobic respiration is given in Figure 16.24.

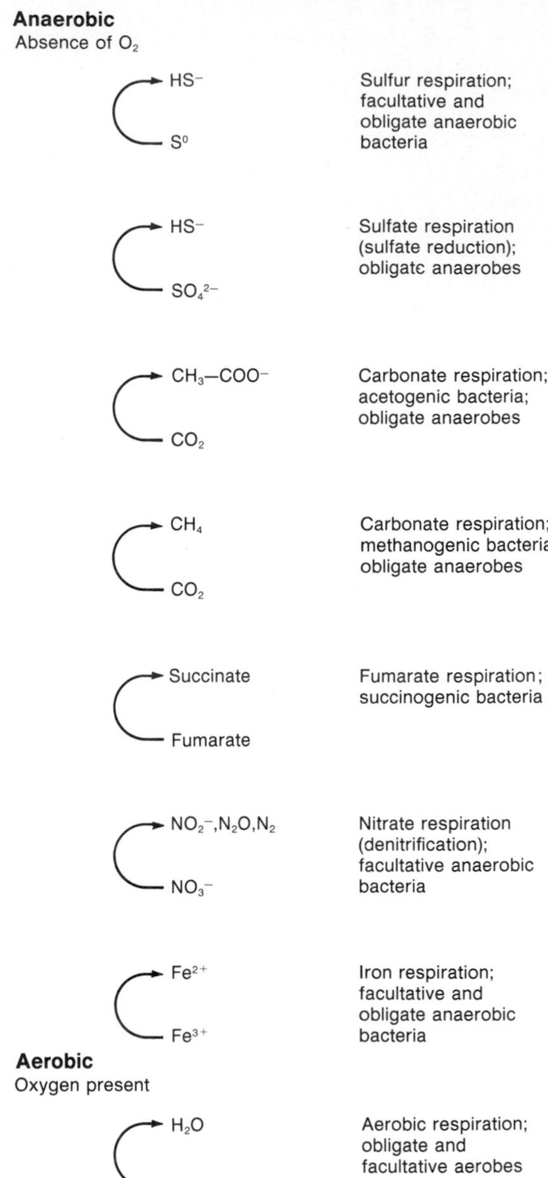

Figure 16.24 Various kinds of anaerobic respiration processes. The compounds are arranged in approximate order from most reducing (top) to most oxidizing.

Assimilative and dissimilative metabolism Inorganic compounds such as NO_3^-, SO_4^{2-}, and CO_2 are reduced by many organisms as sources of nitrogen, sulfur, and carbon, respectively. The end products of such reductions are amino groups (—NH_2), sulfhydryl groups (—SH), and organic carbon compounds (at the oxidation level of carbohydrate or lower), respectively. We discussed briefly the *nutrition* of microorganisms in Section 4.1 and noted that all organisms need sources of N, S, and C for growth. When an inorganic compound such as NO_3^-, SO_4^{2-}, and CO_2 is reduced for use as a nutrient source, it is said to be *assimilated*, and the reduction process is called *assimilative metabolism*. We emphasize here that assimilative metabolism is quite different from the use of compounds such as NO_3^-, SO_4^{2-}, and CO_2 as electron acceptors for energy metabolism. To distinguish these two kinds of reduction processes, the use of these compounds as electron acceptors in energy metabolism is called *dissimilative metabolism*. Assimilative and dissimilative metabolism differ markedly. In assimilative metabolism, only enough of the compound (NO_3^-, SO_4^{2-}, or CO_2) is reduced to satisfy the needs of the nutrient for growth. The atoms reduced end up within the cell material in the form of macromolecules. In dissimilative metabolism, a comparatively large amount of the electron acceptor is reduced, and the reduced product is *excreted* into the environment. Many organisms carry out assimilative metabolism of compounds such as NO_3^-, SO_4^{2-}, and CO_2 (for example, many bacteria, fungi, algae, and higher plants) whereas only a restricted variety of organisms carry out dissimilative metabolism (almost all procaryotes). We now discuss the main types of dissimilative metabolism.

16.14 Nitrate Reduction and the Denitrification Process

Inorganic nitrogen compounds are some of the most common electron acceptors in anaerobic respiration. A summary of the various inorganic nitrogen species with their oxidation states is given in Table 16.5. The most widespread inorganic nitrogen species in nature are ammonia and nitrate, both of which are formed in the atmosphere by inorganic chemical processes, and nitrogen gas, N_2, also an atmospheric gas, which is the most stable form of nitrogen in nature. We discuss *nitrogen fixation*, the utilization of N_2 as a nitrogen source, later in this chapter.

One of the most common alternative electron acceptors is nitrate, NO_3^-, which is converted into more reduced forms of nitrogen, N_2O, NO, and N_2. Because these three products of nitrate reduction are all gaseous, they can be lost from the environment, and because of this the process is called *denitrification*.

Assimilative nitrate reduction, in which nitrate is reduced to the oxidation level of ammonia for use as a nitrogen source for growth, and *dissimilative nitrate reduction*, in which nitrate is used as an alternative electron acceptor in energy generation, are contrasted in Figure 16.25. Under most conditions, the end product of dissimilatory nitrate reduction is N_2 or N_2O. The process is the main means by which gaseous N_2 is formed biologically, and since N_2 is much less readily available to organisms than nitrate as a source of nitrogen, denitrification is a detrimental process.

Table 16.5 Oxidation states of key nitrogen compounds

Compound	Oxidation state
Organic N ($R—NH_2$)	−3
Ammonia (NH_3)	−3
Nitrogen gas (N_2)	0
Nitrous oxide (N_2O)	+1 (average per N)
Nitrogen oxide (NO)	+2
Nitrite (NO_2^-)	+3
Nitrogen dioxide (NO_2)	+4
Nitrate (NO_3^-)	+5

The enzyme involved in the first step of nitrate reduction, *nitrate reductase*, is a molybdenum-containing enzyme. In general, assimilative nitrate reductases are soluble proteins which are ammonia repressed, whereas dissimilative nitrate reductases are membrane-bound proteins which are repressed by O_2 and synthesized under anaerobic (anoxic) conditions. Because O_2 inhibits the synthesis of dissimilative nitrate reductase, the process of denitrification is strictly an *anaerobic* process, whereas assimilative nitrate reduction can occur quite well under aerobic conditions. Assimilative nitrate reduction occurs in all plants and most fungi, as well as in many bacteria, whereas dissimilative nitrate reduction is restricted

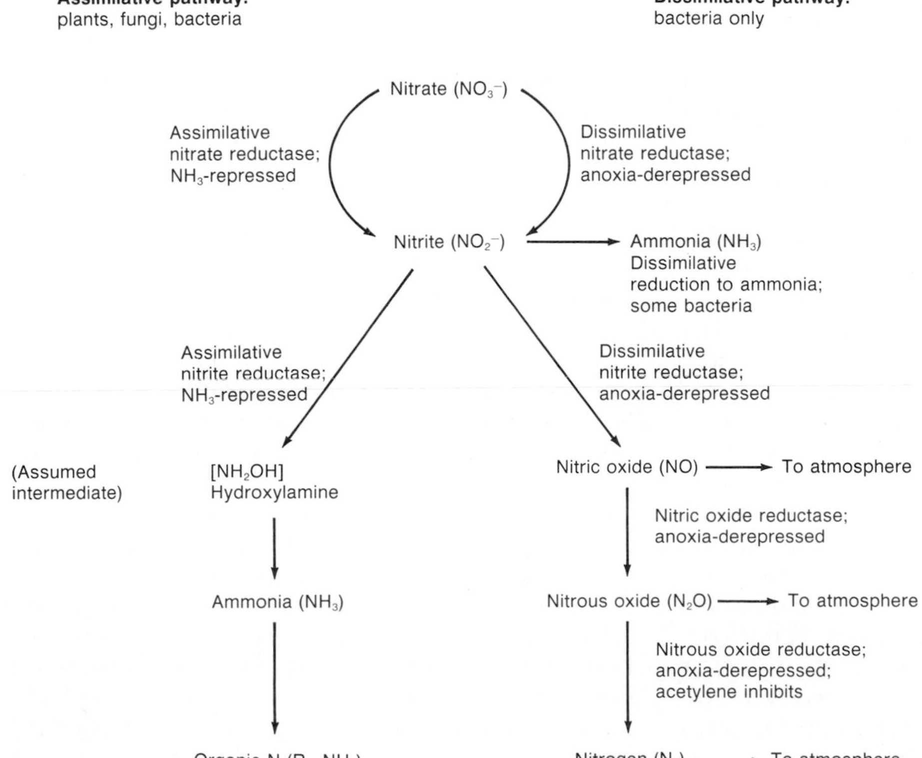

Figure 16.25 Comparison of assimilative and dissimilative processes for the reduction of nitrate.

to bacteria, although a wide diversity of bacteria can carry out this process. In all cases, the first product of nitrate reduction is nitrite, NO_2^-, and another enzyme, ***nitrite reductase***, is responsible for the next step. In the dissimilative process, two routes are possible, one to ammonia and the other to N_2. The route to ammonia is carried out by a fairly large number of bacteria, but is of less practical significance because it does not result in the formation of a gaseous product. There are also some bacteria which do not reduce nitrate but do reduce nitrite to ammonia. This may be a detoxification mechanism, since nitrite can be fairly toxic under mildly acidic conditions (nitrous acid is an effective mutagen, Table 5.6). The pathway to nitrogen gas proceeds via two intermediate gaseous forms of nitrogen, nitric oxide (NO) and nitrous oxide (N_2O). Several organisms are known which produce only N_2O during the denitrification process, while other organisms produce N_2 as the gaseous product. Because of their global significance, the formation of gaseous nitrogen compounds by denitrifying bacteria has been under considerable study (see Section 17.14).

The biochemistry of denitrification has been studied in most detail in *Escherichia coli*, an organism which is able to carry out only the first step of the process, the reduction of NO_3^- to NO_2^-. The enzyme nitrate reductase is membrane-bound and accepts electrons from a cytochrome *b* when O_2 is *absent* from the environment. A comparison of the electron transport chains in aerobic metabolism and nitrate respiration in *E. coli* is given in Figure 16.26. As seen, because of the reduction potential of the NO_3^-/NO_2^- couple, only one proton translocating process occurs instead of the two which occur in aerobic respiration. This is consistent with the lower reduction potential of the NO_3^-/NO_2^- couple as compared to the O_2/H_2O couple (see Table 16.1).

Enrichment culture of denitrifying bacteria is quite straightforward. A synthetic culture medium is used in which potassium nitrate is added as an electron acceptor and anaerobic conditions maintained. An energy source (electron acceptor) must be added which is not readily fermentable (see Section 16.18) so that fermentative organisms will not be selected. Suitable energy sources include ethanol, acetate, succinate, or benzoate. Because all denitrifying organisms are facultative aerobes, great care to maintain anaerobic conditions is not necessary. A simple glass-stoppered incubation vessel is sufficient. The residual oxygen present in the culture medium will be quickly used up and then the population will switch to anaerobic respiration. Once good growth is obtained, the culture should become quite turbid. If nitrogen gas is produced, bubbles may appear under the glass stopper. The most common type of bacterium enriched in this way is a *Pseudomonas* species, such as *Pseudomonas fluorescens*.

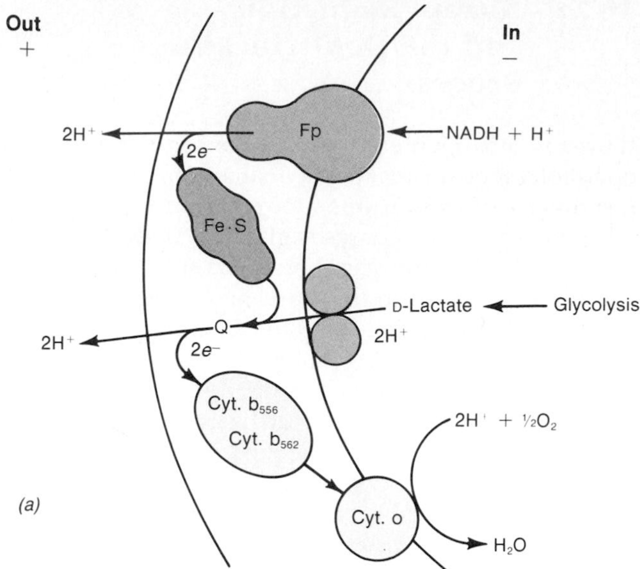

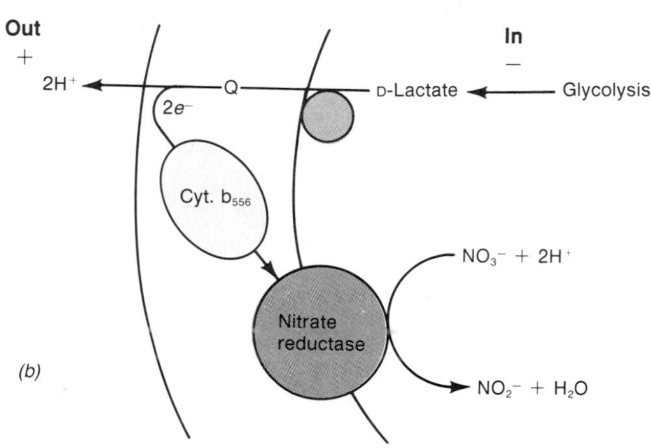

Figure 16.26 Electron transport processes in *Escherichia coli* when (a) O_2 or (b) NO_3^- are used as electron acceptors. Fp, flavoprotein; Q, coenzyme Q.

16.15 Sulfate Reduction

Several inorganic sulfur compounds are important electron acceptors in anaerobic respiration. A summary of the oxidation states of the key sulfur compounds is given in Table 16.6. Sulfate, the most oxidized form of sulfur, is one of the major anions in seawater and is used by the *sulfate-reducing bacteria*, a group which is widely distributed in nature. The end product of sulfate reduction is H_2S, an important natural product which participates in many biogeochemical processes (see Section 17.15). Again, it is important to distinguish between assimilative and dissimilative sulfate reduction. Many organisms, including higher plants, algae, fungi, and most bacteria use sulfate as a sulfur source for biosynthesis. But the ability to utilize sulfate as an *electron acceptor* for energy-generating processes is restricted to a very

Table 16.6 Sulfur compounds and sulfate reduction donors

Compound	Oxidation state
A. Oxidation states of key sulfur compounds	
Organic S (R—SH)	−2
Sulfide (H_2S)	−2
Elemental sulfur (S^0)	0
Thiosulfate ($S_2O_3^{2-}$)	+2 (average per S)
Tetrathionate ($S_4O_6^{2-}$)	+2 (average per S)
Sulfur dioxide (SO_2)	+4
Sulfite (SO_3^{2-})	+4
Sulfur trioxide (SO_3)	+6
Sulfate (SO_4^{2-})	+6
B. Some electron donors used for sulfate reduction	
H_2	Propionate
Lactate	Acetate
Pyruvate	Butyrate
Ethanol	Fatty acids
Fumarate	Benzoate
Malate	Indole
Choline	

special group of obligately anaerobic bacteria, the sulfate-reducing bacteria.

As the reduction potential in Table 16.1 shows, sulfate is a much less favorable electron acceptor than either O_2 or NO_3^-. However, sufficient energy to make ATP is available when an electron donor that yields NADH or FADH is used. Because of the less favorable energetics, growth yields are much lower for an organism growing on SO_4^{2-} than one growing on O_2 or NO_3^-. A list of some of the electron donors used by sulfate-reducing bacteria is given in Table 16.6. The first three compounds listed, H_2, lactate, and pyruvate, are used by a wide variety of sulfate-reducing bacteria; the others have more restricted use. However, a large variety of morphological and physiological types of sulfate-reducing bacteria are known; their characteristics and taxonomy are discussed in Section 19.8.

Biochemistry of sulfate reduction The reduction of SO_4^{2-} to hydrogen sulfide, an 8-electron reduction, proceeds biochemically through a number of intermediate stages. The sulfate ion is fairly stable and cannot be used without first being activated. Sulfate is activated by means of ATP. The enzyme *ATP sulfurylase* catalyzes the attachment of the sulfate ion to a phosphate of ATP, leading to the formation of **adenosine phosphosulfate, APS**, as shown in Figure 16.27. In dissimilative sulfate reduction, the sulfate ion of APS is reduced directly to sulfite (SO_3^{2-}), but in assimilative reduction, another P is added to APS to form **phosphoadenosine phosphosulfate (PAPS**, Figure 16.27*b*), and only then is the sulfate ion reduced. In both cases, the first product of sulfate

ATP

APS (Adenosine-5′-phosphosulfate) — Used in dissimilative metabolism

(a) PAPS (Phosphoadenosine-5′-phosphosulfate) — Used in assimilative metabolism

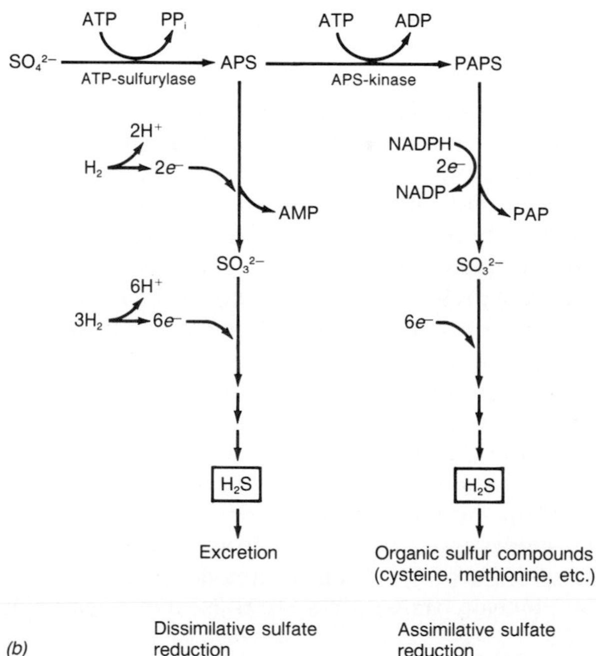

Figure 16.27 Biochemistry of sulfate reduction. (a) Two forms of *active sulfate*, adenosine-5′-phosphosulfate (APS) and phosphoadenosine-5′-phosphosulfate (PAPS). (b) Schemes of assimilative and dissimilative sulfate reduction.

reduction is *sulfite*, SO_3^{2-}. Once SO_3^{2-} is formed, the subsequent reductions proceed readily. Many organisms are able to reduce SO_3^{2-}, for use either as an electron acceptor or in the detoxification of sulfite (which is fairly toxic). In *assimilative* sulfate reduction, the H_2S formed is converted into organic sulfur in the form of amino acids, etc., but in dissimilative sulfate reduction, the H_2S is excreted.

As discussed in Section 17.15, H_2S is an important biogeochemical product. It can also be highly toxic, due to the fact that it combines with the iron of cytochromes and other essential iron-containing compounds in the cell.* Cyanide is more dangerous to humans than sulfide because it is odorless and hence toxic levels can build up unforeseen. The odor of H_2S is so strong that it is readily perceived in the environment before a toxic level is reached. However, human deaths due to sulfide toxicity are not uncommon, especially for sewage workers and others working in enclosed systems where H_2S might build up. One common detoxification mechanism for sulfide in the environment is by combination with iron, leading to the formation of the insoluble FeS. Organisms such as the sulfate-reducing bacteria are also susceptible to H_2S toxicity, but there is generally sufficient iron in the environments in which they live so that all H_2S formed is removed as FeS. The black color of many sediments where sulfate reduction is taking place is due to the accumulation of FeS. We discuss the enrichment culture of the sulfate-reducing bacteria in Section 19.8, and the ecology and biogeochemistry of sulfate reduction in Section 17.15.

Energy conservation in sulfate-reducing bacteria The sulfate-reducing bacteria carry out a cytochrome-based electron transport process, the electrons from the energy source being transferred to the sulfate ion in APS. The cytochrome of the sulfate-reducing bacteria is a unique type called cytochrome c_3. This cytochrome is not found in organisms using other electron acceptors. Other electron carriers in the electron transport chain of the sulfate-reducing bacteria include ferredoxin and flavodoxin. Type II sulfate-reducers (species capable of degrading acetate and other fatty acids, see Section 19.8) contain a cytochrome of the *b* type which is presumably involved in the electron transport chain of these species; cytochrome *b* is absent from species of sulfate-reducing bacteria which do not degrade fatty acids. The electron transport system is shown in Figure 16.28.

In electron transport in sulfate-reducing bacteria, hydrogen, H_2, either directly from the environment or via organic electron donors, transfers electrons to the enzyme *hydrogenase*, which is situated in the membrane in close association with cytochrome c_3. Due to the spatial arrangement of the electron transport components in the membrane, when the H atoms of H_2 are oxidized, the protons (H^+) remain *outside* the membrane, whereas the electrons are transferred *across* the membrane. In this way, a proton-motive force is set up which can be used for the synthesis of ATP. In the cytoplasm, the electrons are used in the reduction of APS.

*H_2S is comparable in toxicity to cyanide, HCN, which has a similar action—see Section 4.12.

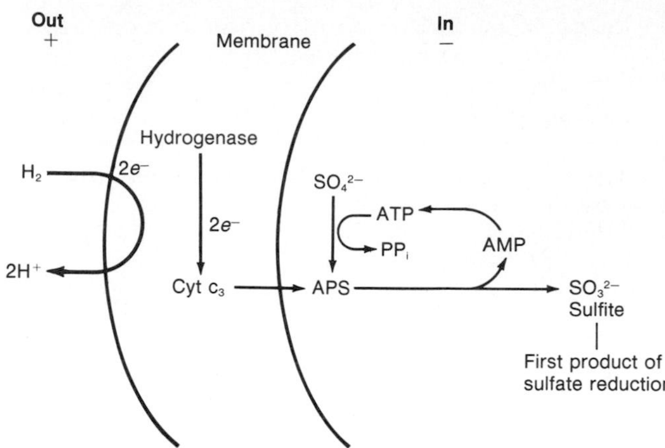

Figure 16.28 Electron transport and generation of a proton-motive force in a sulfate-reducing bacterium. In addition to external hydrogen (H_2) arising from the metabolism of fermentative anaerobes, this compound can also originate from the catabolism of organic compounds, such as lactate.

When sulfate-reducing bacteria grow on H_2/SO_4^{2-}, they are growing lithotrophically as H_2 bacteria (see Section 16.8). Some species may also grow autotrophically under these conditions, using CO_2 as sole source of carbon. However, most sulfate-reducing bacteria are heterotrophs, using various organic compounds as electron donors (some of which are illustrated in Table 16.6).

16.16 Carbonate (CO_2) as an Electron Acceptor

Carbonate (CO_2) is one of the most common inorganic anions in nature and is, of course, a major product of the energy metabolism of heterotrophs. (Seawater is essentially a bicarbonate buffer containing a number of mineral salts.) Several groups of bacteria are able to use CO_2 as an electron acceptor in anaerobic respiration. The most important CO_2-reducing bacteria are the **methanogens**, a major group of archaebacteria. Some of these organisms utilize H_2 as electron donor (energy source); the overall reaction is shown in Figure 16.29*a*. When growing on H_2/CO_2 as shown in Figure 16.29*a*, the methanogens are lithotrophic autotrophs. However, the process by which CO_2 is fixed is not the Calvin cycle of conventional autotrophs. We discuss the overall process of methanogenesis in Section 17.12 and the methanogens themselves in Section 19.34.

Another group of CO_2-reducing bacteria are the **acetogens**, which produce *acetate* rather than CH_4

$$4H_2 + H^+ + HCO_3^- \rightarrow CH_4 + 3H_2O \qquad -34 \text{ kcal/mole}$$

(a) Methanogenesis

$$4H_2 + H^+ + 2HCO_3^- \rightarrow \underset{\text{Acetate}}{CH_3\text{—}COO^-} + 4H_2O \qquad -27 \text{ kcal/mole}$$

(b) Acetogenesis

Figure 16.29 Two reactions in which CO_2 (bicarbonate, HCO_3^-) is used as an electron acceptor in anaerobic respiration.

from CO_2 and H_2. The overall reaction of acetogenesis is shown in Figure 16.29*b*. The acetogenic bacteria are not a defined taxonomic group, since they include the spore-forming bacterium *Clostridium aceticum* and the nonspore-forming *Acetobacterium woodii*. The acetogenic bacteria are discussed in detail in Section 19.9.

As seen in Figure 16.29, somewhat more energy is released when methane is formed from CO_2 reduction than when acetate is formed. This is consistent with the fact that methanogens are much more common in nature than acetogens. When growing on $H_2 + CO_2$, both methanogens and acetogens are growing autotrophically (they are also lithotrophs). Both groups are also obligate anaerobes. Note, however, that methanogens are not restricted to $H_2 + CO_2$; many methanogens grow and form methane from methanol, CH_3OH, formic acid, HCOOH, and acetate, CH_3COOH, as discussed later in this chapter and in Sections 17.12 and 19.34. As discussed in Section 19.9, most acetogenic bacteria can also grow heterotrophically by the fermentation of sugar.

16.17 Other Electron Acceptors for Anaerobic Respiration

In addition to nitrate, sulfate, and carbonate, a variety of other compounds both organic and inorganic, are used by one or another group of bacteria in anaerobic respiration. We discuss the most important in this section.

Ferric and manganese reduction Iron and manganese are related metal ions which are reduced by various bacteria under anaerobic conditions. It is not known definitely that these metals serve as functional electron acceptors for energy generation, but the reduction processes are so widespread that they are of interest even if usable energy is not obtained. The **ferric ion** (Fe^{3+}) is reduced to the ferrous ion (Fe^{2+}). This process is carried out by many organisms which reduce nitrate, and at least in some cases the same enzyme, *nitrate reductase*, functions in the reduction of both nitrate and Fe^{3+}. As noted in Section 16.10, ferric iron itself forms highly insoluble oxides at neutral pH. In addition, the ferrous ion oxidizes spontaneously at neutral pH, so that under aerobic conditions at conventional pH values, most of the iron in nature is present in the ferric form. Ferric ion is one of the most common ions present in soils and rocks. Under anaerobic conditions, if an appropriate electron donor (energy source) is present, such as an organic compound or H_2, reduction to the ferrous state can occur. The ferrous ion is more soluble than the ferric ion, so that iron reduction by microorganisms can lead to the solubilization of iron, an important geochemical process. As illustrated in Figure 16.21*c*, one type of iron deposit, called *bog iron*, is formed as a result of the activities of iron-reducing bacteria.

The metal *manganese* has a number of oxidation states, of which Mn^{4+} and Mn^{2+} are the most stable. Mn^{4+} forms highly insoluble compounds whereas Mn^{2+} is more soluble. Reduction of Mn^{4+} to Mn^{2+} is carried out by a variety of bacteria, many of which are conventional heterotrophs. We discuss the biogeochemistry of Mn^{4+} reduction in Section 17.16.

Organic electron acceptors Several organic compounds have been found to participate in anaerobic respiration. Note that many organic compounds also serve as electron acceptors in conventional heterotrophic metabolism. However, in the case of conventional metabolism, for example, during fermentation, the organic compound is produced *internally* as a part of the metabolic process and then reduced internally. In the present case, we are discussing organic compounds that serve as *major* electron acceptors when added *externally*. Several organic compounds that have been recognized as common electron acceptors for anaerobic respiration are listed in Table 16.7. Of those listed, the compound which has been most extensively studied is **fumarate**, which is reduced to **succinate**. An examination of the *tricarboxylic acid cycle* in Figure 4.12 will indicate that fumarate and succinate are important intermediates. Fumarate's role as an electron acceptor for anaerobic respiration derives from the fact that the fumarate/succinate couple has a favorable reduction potential (near 0 volts, see Table 16.1), which allows coupling of fumarate reduction with NADH oxidation. The energy yield is sufficient for the synthesis of 1 ATP. Bacteria able to use fumarate as an electron acceptor include *Vibrio (Wolinella) succinogenes* (which can

Table 16.7 Some organic compounds used as electron acceptors in anaerobic respiration

Acceptor	Reduced product
Fumarate	Succinate
Glycine	Acetate
Dimethyl sulfoxide	Dimethyl sulfide
Trimethylamine oxide	Trimethylamine

grow on H_2 as sole energy source using fumarate as electron acceptor), *Desulfovibrio gigas* (a sulfate-reducing bacterium which can also grow under non-sulfate-reducing conditions), some clostridia, *Escherichia coli*, and *Proteus rettgeri*. Another bacterium, *Streptococcus faecalis*, can use fumarate as an electron acceptor but does not couple this to oxidative phosphorylation. In the latter case, fumarate merely serves in the reoxidation of NADH that had been formed during glycolysis.

The compound **trimethylamine oxide** listed in Table 16.7 is an interesting electron acceptor. Trimethylamine oxide (TMAO) is an important organic compound in marine fish, where it serves in these animals as a means of excreting excess nitrogen, but a variety of bacteria are able to reduce TMAO to trimethylamine (TMA). TMA has a strong odor and flavor, and some of the spoiled odor that frequently occurs in marine fish is due to TMA produced by bacterial action. A variety of facultative anaerobic bacteria are able to utilize TMAO as an alternate electron acceptor. In addition, several nonsulfur purple bacteria are able to use TMAO as an electron acceptor for a kind of dark anaerobic metabolism. A compound analogous to TMAO is **dimethyl sulfoxide** (DMSO), which is reduced by a variety of bacteria to dimethyl sulfide (DMS). DMSO is a common natural product and is found in both marine and freshwater environments. DMS has a strong odor, and bacterial reduction of DMSO to DMS is signalled by the presence of the characteristic odor of DMS. A variety of bacteria, including *Campylobacter*, *Escherichia*, and many purple bacteria, are able to use DMSO as an electron acceptor in energy generation.

16.18 Metabolic Diversity among the Heterotrophs

We have defined heterotrophs as organisms that use *organic* compounds as energy sources. We discussed general aspects of heterotrophic energy metabolism in Chapter 4. In that discussion, we dealt primarily with the utilization of glucose and related hexose sugars as energy sources. Although hexoses are common energy sources, many other organic compounds are present in nature, as components of living organisms. The term *biomass* is frequently used to refer to the vast bulk of organic materials present in higher animals and plants. The breakdown of biomass and its recycling into simple organic compounds is carried out in nature primarily by microorganisms, via activities generally called *decomposition processes*. We discuss the ecological and biogeochemical aspects of decomposition, and the global carbon cycle, in Chapter 17. Here we are concerned with the microbiology and biochemistry of these processes.

Table 16.8 Kinds of organic compounds produced by living organisms

Natural products	
Sugars	Organic acids
Amino acids	Fatty acids
Purines	Alcohols
Pyrimidines	Phycobilins
Steroids	Porphyrins
Flavin (flavinoids)	Carotenoids
Quinones	
Aromatic rings	
Antibiotics (various structures)	
Aliphatic amines—choline	

A summary of the most common biochemicals that are components of biomass is given in Table 16.8. Most of these compounds are used by one or another microorganism as energy and/or carbon sources.

There are two problems an organism faces if it is to use a particular organic compound as an electron donor (that is, as an energy source): (1) Conserving some of the energy released as ATP; (2) Disposing of the electrons removed from the electron donor. With *respiratory organisms*, molecular oxygen, O_2, is used as the electron acceptor in electron transport. The electrons released from the energy source are transferred through an electron transport system to O_2, with the generation of a proton-motive force (PMF) and concomitant ATP synthesis (see Section 4.12). In addition to aerobic respiration, we have the various processes of *anaerobic respiration* which were discussed earlier in this chapter. In anaerobic respiration, electron acceptors other than O_2 can be utilized in the formation of a PMF. In all these cases, both aerobic and anaerobic, disposal of electrons uses the external electron acceptor, and ATP synthesis occurs primarily by means of an energized membrane and a PMF.

But in many anaerobic situations, alternate electron acceptors are not available. How does energy generation occur under these conditions? Here we are dealing with the various processes of *fermentation*, in which the electron donor is also the electron acceptor. We now discuss the diverse array of fermentative microorganisms.

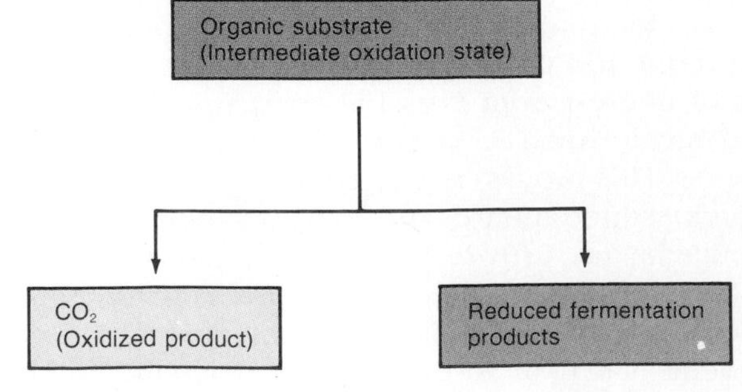

Figure 16.30 Overall process of fermentation.

Diversity of fermentative microorganisms We defined *fermentation* in Section 4.8 as the utilization of an energy source in the *absence* of an external electron acceptor. In fermentation, some of the molecules of the fermentable substance are *reduced*, whereas others are *oxidized*, usually to CO_2 (Figure 16.30). Overall, an oxidation-reduction balance must be maintained. We discussed the alcohol and lactic acid fermentations in Section 4.8 and showed how ATP synthesis occurred by way of *substrate-level phosphorylation*. ATP synthesis can also occur in fermentative organisms via generation of a proton-motive force, although many fermentative microorganisms conserve energy almost exclusively via substrate-level phosphorylation.

High-energy compounds and substrate-level phosphorylation Many fermentative microorganisms obtain energy only by way of substrate-level phosphorylation. Under these conditions, a biochemical mechanism must exist to couple the fermentation of the energy source to the synthesis of ATP. Coupling requires the intermediate formation of one or another high energy compound. A list of the major high-energy intermediates is given in Table 16.9. Although this list is not complete, it includes most recognized high-energy intermediates which are known to be formed during biochemical processes. Since the compounds listed in Table 16.9 can couple to ATP synthesis, if an organism can form one or another of these compounds during fermentative metabolism, it should be able to make ATP.

Pathways for the anaerobic breakdown of various fermentable substances to high-energy intermediates are summarized in Figure 16.31. It should be noted that this figure is organized by the high-energy compounds listed in Table 16.9. Thus, Figure 16.31 and Table 16.9 can be examined together.

Table 16.9 Energy-rich compounds involved in substrate-level phosphorylation

Name	Free energy of hydrolysis, $-\Delta G^{0\prime}$ (kcal/mole)
Acetyl-CoA	8.5
Propionyl-CoA	8.5
Butyryl-CoA	8.5
Succinyl-CoA	8.4
Acetylphosphate	10.7
Butyrylphosphate	10.7
1,3-Diphosphoglycerate	12.4
Carbamyl phosphate	9.6
Phosphoenolpyruvate	12.3
Adenosine-phosphosulfate	21
N^{10}-formyltetrahydrofolate	5.6

Data from Thauer, R.K., K. Jungermann, and K. Decker. 1977. Bacteriol. Rev. 41:100–180.

Energy yields of fermentative organisms We discussed in Section 9.8 the general energy requirements for growth. We defined the term Y_{ATP}, which expresses the cell yield per ATP generated. As seen in Table 9.2, Y_{ATP} values were around 10 grams dry weight of cells produced per mole of ATP synthesized. How much ATP can be produced by a fermentative organism? As we have seen, glucose fermenters produce 2–3 ATP per mole of glucose fermented (see Figure 4.10). This is about the maximum amount of ATP produced by fermentation; many other substrates provide less energy. The potential energy released from a particular fermentation can be calculated from the balanced reaction and from the free energy values that are given in Appendix 1. For instance, the fermentation of glucose to ethanol and CO_2 has a theoretical energy yield of -57 kcal/mole, enough to produce about 6 ATP. However, only 2 ATPs are produced, which implies that the organism operates at considerably less than 100 percent efficiency (see Section 4.13 for a discussion of energy efficiency).

Even with substrates where there is insufficient energy release to synthesize 1 ATP by substrate-level phosphorylation, it is still possible for a fermentative organism to develop. This is possible because some organisms are able to generate a proton-motive force during the fermentation process, and even though an energy yield of less than 1 ATP occurs, by repetitive translocation of protons across the membrane, a membrane can be energized sufficiently so that net ATP synthesis can occur. In some anaerobic bacteria, sodium ions, Na^+ are translocated across the membrane instead of protons (see Section 19.30), but the net result is the same, an energized membrane across which ATP synthesis can be carried out.

Oxidation-reduction balance In any fermentation reaction, there must be a *balance* between oxidation and reduction. The total number of electrons in the products on the right side of the equation must balance the number in the substrates on the left side of the equation. When fermentations are studied experimentally in the laboratory, it is conventional to calculate a fermentation balance to make certain that no products are missed. The fermentation balance can also be calculated theoretically from the oxidation states of the substrates and products (see Appendix 1 for the procedure for calculating oxidation states).

In a number of fermentations, electron balance is maintained by the *production of molecular hydrogen*, H_2. In H_2 production, protons (H^+) of the medium, derived from water, serve as electron acceptor. Production of H_2 is generally associated with the presence in the organism of *ferredoxin*, an electron carrier of low redox potential. The transfer of

Arginine Agmatine Allantoin (Pyrimidines) → Carbamyl-P → CAK → CO_2

Purines (Methanol) → Formyl-FH_4 → FTS → Formate

Crotonate Lysine (Glutamate) γ-Amino-butyrate → 3-Keto-acyl-CoA → Butyryl-CoA → Butyryl-phosphate → BK → Butyrate

Alcohols: Ethanol Ethylene glycol

Purines, pyrimidines; Organic acids: Lactate Acrylate Malate Fumarate (Succinate) Citrate; Amino acids: Alanine Glutamate Histidine Aspartate Glycine Serine Cysteine Tryptophan

Sugars: Hexoses → Pentoses

Hexose diphosphate pathway → Fructose-1,6-PP → Glycerate-1,3-PP → PGK → Glycerate-3-P → Phosphoenol-pyruvate → (PyrK) → Pyruvate

Hexose monophosphate pathway → Gluconate-6-P → Xylulose-5-P

Pyruvate → Acetyl-CoA → Acetyl-phosphate → AK → Acetate

Amino acids: Threonine Homoserine Homo-cysteine Methionine → 2-Keto-butyrate → Propionyl-CoA → Propionyl-phosphate → PK → Propionate

Amino acids: Leucine Isoleucine Valine → 3-Alkyl-pyruvate → 2-Alkyl-acetyl-CoA → 2-Alkyl-acetyl-phosphate → AKK → 2-Alkyl-acetate

Amino acids: Tyrosine Phenyl-alanine Trypto-phan → 3-Aryl-pyruvate → 2-Aryl-acetyl-CoA → 2-Aryl-acetyl-phosphate → 2-Aryl-acetate

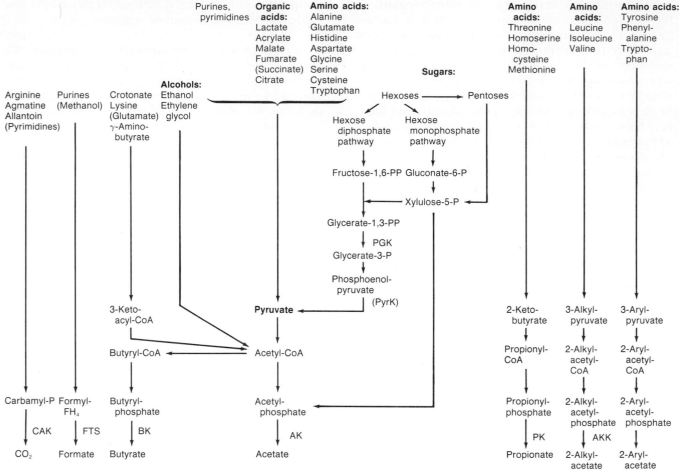

Figure 16.31 Pathways for the anaerobic breakdown of various fermentable substances. The sites of substrate-level phosphorylation are shown by the abbreviations: CAK, carbamyl phosphate kinase; FTS, formyltetrahydrofolate synthetase; AK, acetate kinase; PK, propionate kinase; BK, butyrate kinase; AKK, alkyl (aryl) acetate kinase; PGK, phosphoglycerate kinase; PyrK, pyruvate kinase.

electrons from ferredoxin to H^+ is catalyzed by the enzyme **hydrogenase**, as illustrated in Figure 16.32. We have already discussed earlier in this chapter the enzyme hydrogenase in reference to the *utilization* of hydrogen by sulfate-reducing bacteria. Here, hydrogenase is involved in the *production* of hydrogen.

The energetics of hydrogen production are actually somewhat unfavorable, so that most fermentative organisms only produce a relatively small amount of hydrogen along with other fermentation products. Hydrogen production thus primarily serves to maintain redox balance. If hydrogen production is prevented, for instance, then the oxidation-reduction balance of the other fermentation products will be directed more toward *reduced* products. Thus, many fermentative organisms that make H_2 produce both ethanol and acetate. Since ethanol is more reduced than acetate, its formation is favored when hydrogen production is inhibited.

Numerous anaerobic bacteria produce *acetate* as one of the products of fermentation. The production of acetate is advantageous, since it allows the organism to produce additional ATP by substrate-level phosphorylation. The key intermediate generated in acetate production is acetyl-CoA (see Table 16.9), a high-energy intermediate. Acetyl-CoA can be con-

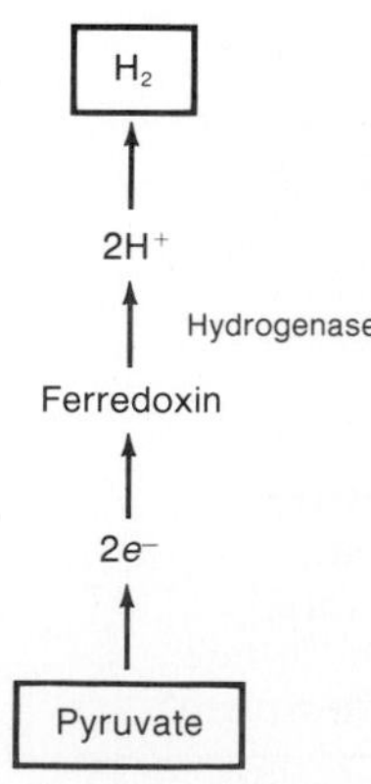

Figure 16.32 Production of molecular hydrogen from pyruvate.

verted to acetyl phosphate (also listed in Table 16.9), and the high-energy phosphate group of acetyl phosphate is subsequently transferred to ADP, leading to the formation of ATP. One of the main substrates that is converted to acetyl-CoA is pyruvate, a major product of glycolysis. However, the conversion of pyruvate to acetyl-CoA is an oxidation process, since pyruvate is more reduced than acetyl-CoA. The excess electrons generated must either be used to make a more reduced end product, or can be used in the production of H_2 as discussed above.

Classification of anaerobic fermentations Fermentations can be classified either in terms of the substrate fermented or in terms of the fermentation products formed. Many of the specific fermentation reactions of bacteria will be discussed when the individual groups are discussed in Chapter 19. Here we present an overview of anaerobic fermentations. Table 16.10 summarizes some of the main types of fermentations, as classified on the basis of *products* formed. Note some of the broad categories, such as alcohol, lactic acid, propionic acid, mixed acid, butyric acid, homoacetic acid, etc.

A number of fermentations are classified on the basis of the *substrate fermented* rather than the *fermentation product*. For instance, many of the spore-forming anaerobic bacteria (genus *Clostridium*) ferment *amino acids* with the production of acetate, lactate, ammonia, and H_2. Another *Clostridium* species, *C. acidi-urici*, ferments *purines* such as xanthine or adenine with the formation of acetate, formate, CO_2, and ammonia. Likewise, *C. proticum* ferments the *pyrimidine* protic acid to acetate, succinate, CO_2, and ammonia. Still other anaerobes ferment *aromatic* compounds. As an example, the bacterium *Pelobacter acidigallici* ferments the aromatic compound *phloroglucinol* (1,3,5-benzenetriol, $C_6H_6O_3$) via the following overall pathway:

$$\text{phloroglucinol } (C_6H_6O_3) + 3H_2O \rightarrow 3 \text{ acetate}^- + 3H^+ \qquad -34 \text{ kcal/mole}$$

Methane as the final product of anaerobic decomposition processes Many of the products of microbial metabolism listed in Table 16.10 are themselves energy sources for other fermentative organisms. It is reasonable that organisms might exist

Table 16.10 Examples of various microbial fermentations and some of the organisms carrying them out

Type	Overall reaction	Organisms
Alcohol fermentation	Hexoses $\rightarrow$ Ethanol + CO_2	Yeast *Zymomonas*
Homolactic fermentation	Hexose $\rightarrow$ lactic acid	*Streptococcus* Some *Lactobacillus*
Heterolactic fermentation	Hexose $\rightarrow$ Lactic acid Ethanol CO_2	*Leuconostoc* Some *Lactobacillus*
Propionic acid	Lactate $\rightarrow$ Propionate Acetate CO_2	*Propionibacterium* *Clostridium propionicum*
Mixed acid	Hexoses $\rightarrow$ Ethanol 2,3-Butanediol Succinate Lactate Acetate Formate H_2 + CO_2	Enteric bacteria *Escherichia* *Salmonella* *Shigella* *Klebsiella*
Butyric acid	Hexoses $\rightarrow$ Butyrate Acetate H_2 + CO_2	*Clostridium butyricum*
Butanol	Hexoses $\rightarrow$ Butanol Acetate Acetone Ethanol H_2 + CO_2	*C. butyricum*
Caproate	Ethanol + Acetate + CO_2 $\rightarrow$ Caproate + Butyrate + H_2	*C. kluyveri*
Homoacetic	H_2 + CO_2 $\rightarrow$ Acetate	*C. aceticum* *Acetobacterium*
Methanogenic	Acetate $\rightarrow$ CH_4 + CO_2	*Methanothrix* *Methanosarcina*

which are able to use the fermentation products of other organisms, since all these anaerobes would likely be found together in environments where fermentation was taking place. However, two fermentation products listed in Table 16.10 cannot be further fermented: CO_2 and CH_4, the most oxidized and the most reduced forms of carbon. But many of the compounds listed in Table 16.10 are fermented by other organisms (for example, succinate, lactate, ethanol). Fermentation of these "fermentation products" leads ultimately to the formation of acetate, H_2, and CO_2, which are themselves utilized by the *methanogenic bacteria* in the formation of methane. Thus, the terminal products of anaerobic decomposition are CH_4 and CO_2. It is to these two carbon compounds, one the most reduced, the other the most oxidized, to which all anaerobic decomposition processes ultimately lead (Figure 16.33).

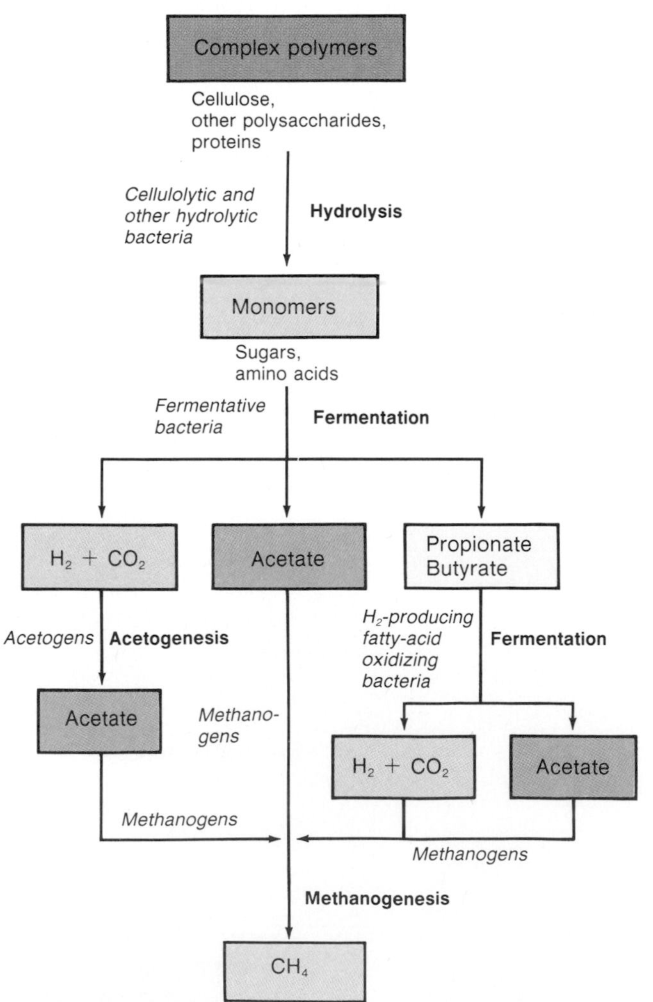

Figure 16.33 Overall process of anaerobic decomposition, showing the manner in which various groups of fermentative anaerobes act together in the conversion of complex organic materials ultimately to methane and CO_2.

Coupled fermentation reactions Even if a single fermentation reaction does not provide favorable energetics, this reaction may still be carried out by an organism if the *product* of the reaction can be used by another organism in an energetically favorable reaction. We are dealing here with the phenomenon of **syntrophy** (mutual feeding), where two organisms do something together that neither can do separately. The best studied syntrophic system in fermentative metabolism is the fermentation of ethanol to acetate and methane by two bacteria, an ethanol fermenter and a methanogen. This coupled scheme is illustrated in Figure 16.34. As seen, the ethanol fermenter produces hydrogen, H_2 and acetate, but this reaction has an unfavorable (positive) energy balance. However, the H_2 produced by the ethanol fermenter is consumed by the methanogen in an energetically favorable reaction. When the energies of these two reactions are summed, the overall reaction is favorable energetically. The reaction summarized here is an example of a phenomenon called **interspecies hydrogen transfer**, which is discussed in more detail in Section 17.12. Note that the methanogen is unable by itself to utilize ethanol so that both the ethanol fermenter and the methanogen both benefit from this relationship.

Recalcitrant compounds It has been a watchword of general microbiology that the power of microorganisms to utilize organic compounds is impressive. But are there limits? Can *any* organic compound be fermented by a microorganism? Are there natural products that are stable to anaerobic metabolism?

Two groups of natural organic materials appear to be refractory to anaerobic breakdown: lignin, and hydrocarbons. **Lignin** is a complex aromatic polymer of phenylpropane building blocks, held together by C—C and C—O—C (ether) linkages. Lignin is a major component of wood and is responsible for conferring

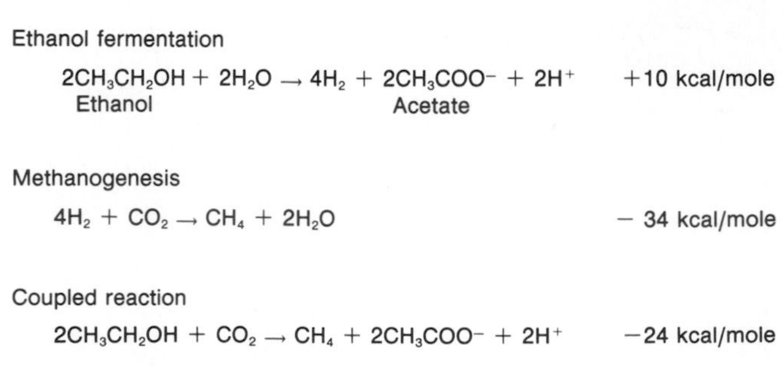

Figure 16.34 Fermentation of ethanol to methane and acetate by the coupling of an energetically unfavorable reaction (ethanol fermentation, +10 kcal/mole) with an energetically favorable reaction (methanogenesis, −34 kcal/mole). The overall energy yield is thus −24 kcal/mole. This is an example of *syntrophy;* it is also an example of *interspecies hydrogen transfer.*

rigidity on the cellulosic walls of woody plants. Lignin appears to be completely stable to anaerobic degradation and hence does not decompose in anaerobic habitats. *Coal* is an organic material that is ultimately derived from woody plants. Coal would not exist today if lignin did not stabilize woody material to anaerobic decay.

The other major natural materials that are refractory to anaerobic fermentation are the long-chain **aliphatic hydrocarbons**, such as hexadecane ($C_{16}H_{34}$) and octadecane ($C_{17}H_{38}$). These hydrocarbons, which are major constituents of petroleum, are either produced directly by plants and animals, or are derived chemically by reduction from long-chain fatty acids. Protected from decay by anaerobic conditions, these hydrocarbons are stable in nature and become the basis of petroleum reserves that are a major foundation of modern society.

Once lignin or hydrocarbons are brought into oxygen-containing environments, these compounds are readily broken down by microorganisms. Thus, it is only under anaerobic conditions that these materials are stable in the biosphere. We discuss aerobic hydrocarbon oxidation in Section 16.23, the global carbon cycle in Section 17.11, and petroleum microbiology in Section 17.19.

In addition to the vast array of natural organic compounds, many organic compounds exist which have been produced artificially by chemical synthesis, for industrial or agricultural purposes. Although some of these synthetic compounds are similar to natural ones, others are quite different chemically from anything produced by living organisms. These unusual synthetic compounds are of special microbiological interest because, since these compounds have never existed in natural environments, the evolution of microorganisms capable of degrading these compounds may not have occurred. These unusual synthetic compounds which are unrelated to anything found naturally are sometimes called **xenobiotics** (*xeno* is a combining form meaning *foreign*). Some of these compounds are stable under both aerobic and anaerobic conditions. A discussion of the metabolism of xenobiotics is given in Section 17.20.

We now discuss some special aspects of heterotrophic metabolism related to the biochemistry of particular groups or classes of organic compounds.

16.19 Sugar Metabolism

Hexose and polysaccharide utilization Sugars with six carbon atoms, called **hexoses**, are the most important electron donors for many heterotrophs and are also important structural components of microbial cell walls, capsules, slimes, and storage products. The most common hexose sources in nature are listed in Table 16.11 from which it can be seen that most are polysaccharides, although a few are disaccharides.

An overview of the main features of hexose metabolism was presented in Section 4.17 (see Figure 4.23). *Polysaccharides*, usually insoluble and too large to pass through cell membranes, are hydrolyzed by enzymes that are excreted outside the cell (*exoenzymes*). Two general kinds of enzymes are involved, those hydrolyzing hexose units from the ends of polysaccharide chains, called *exohydrolases*, and those attacking internal units, called *endohydro-*

Table 16.11 Sources of hexose sugars in nature*

Substance	Composition	Sources	Enzymes breaking down
Cellulose	Glucose polymer (β-1,4-)	Plants (leaves, stems)	Cellulases (β,1-4 glucanases)
Starch	Glucose polymer (α-1,4-)	Plants (leaves, seeds)	Amylase
Glycogen	Glucose polymer (α-1,4- and α-1,6)	Animals (muscle)	Amylase, phosphorylase
Laminarin	Glucose polymer (β-1,3-)	Marine algae (Phaeophyta)	β-1,3-Glucanase (laminarinase)
Paramylon	Glucose polymer (β-1,3-)	Algae (Euglenophyta and Xanthophyta)	β-1,3-Glucanase
Agar	Galactose and galacturonic acid polymer	Marine algae (Rhodophyta)	Agarase
Chitin	*N*-acetyl glucosamine polymer (β-1,4-)	Fungi (cell walls) Insects (exoskeletons)	Chitanase
Pectin	Galacturonic acid polymer (from galactose)	Plants (leaves, seeds)	Pectinase (polygalacturonase)
Sucrose	Glucose-fructose disaccharide	Plants (fruits, vegetables)	Invertase
Lactose	Glucose-galactose disaccharide	Milk	β-Galactosidase

**Each of these is subject to degradation by microorganisms.*

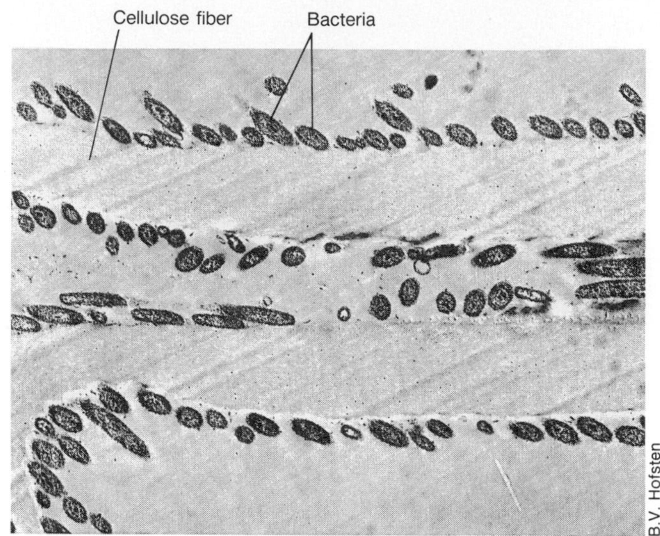

Figure 16.35 Attachment of cellulose-digesting bacteria, *Sporocytophaqa myxococcoides,* to cellulose fibers.

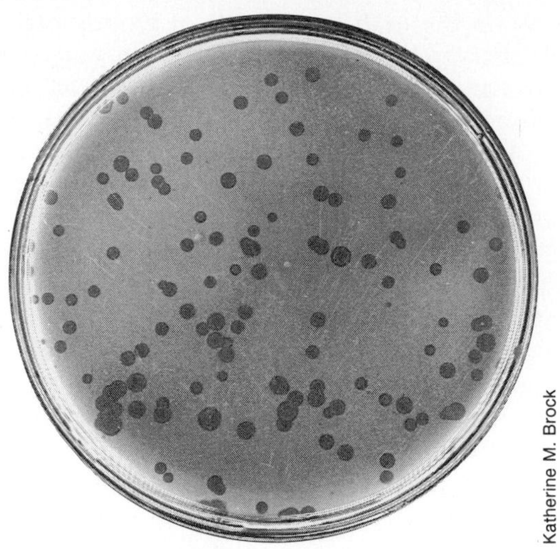

Figure 16.36 *Cytophaga hutchinsoni* colonies on cellulose-agar plate. Clear areas are where cellulose has been digested.

lases. Although both kinds of enzymes act independently, the endohydrolases, by creating more chain ends, greatly increase the rate at which the exohydrolases can act.

Although both starch and cellulose are composed of glucose units, they are connected differently (Table 16.11), and this profoundly affects their properties. Cellulose is much more insoluble than starch and is usually less rapidly digested. Cellulose forms long fibrils, and organisms that digest cellulose are often found closely associated with them (Figure 16.35). Many fungi are able to digest cellulose and these are mainly responsible for decomposition of plant materials on the forest floor. Among bacteria, however, cellulose digestion is restricted to only a few groups, of which the gliding bacteria such as *Cytophaga* (Figure 16.36), clostridia, and actinomycetes are the most common. Anaerobic digestion of cellulose is carried out by a few *Clostridium* species, which are common in lake sediments, animal intestinal tracts, and systems for anaerobic sewage digestion. Starch is digestible by many fungi and bacteria; this is illustrated for a laboratory culture in Figure 16.37. Starch-digesting enzymes, called *amylases*, are of considerable practical utility in many industrial situations where starch must be digested, such as the textile, laundry, paper, and food industries, and fungi and bacteria are the commercial sources of these enzymes.

All of the polysaccharides occurring extracellularly and utilized as substrates are broken down to monomeric units by hydrolysis. In contrast, the polysaccharides formed within cells as storage products are broken down not by hydrolysis but by **phosphorolysis**. This process, involving the addition of *inorganic* phosphate, results in the formation of hexose phosphate rather than the free hexose and may be

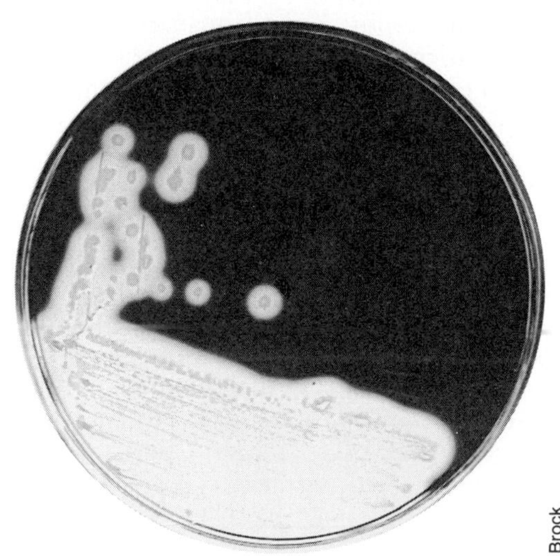

Figure 16.37 Demonstration of hydrolysis of starch by colonies of *Bacillus subtilis*. After incubation, the plate was flooded with Lugol's iodine solution. When starch hydrolysis occurred, the characteristic purple color of the starch-iodine complex is absent. Hydrolysis of starch occurs at some distance from the bacterial colonies because of the production of extracellular amylase, which diffuses into the surrounding medium.

summarized as follows for the degradation of starch, an α-1,4 polymer of glucose:

$$(C_6H_{12}O_6)_n + P_i \rightarrow (C_6H_{12}O_6)_{n-1} + \text{glucose-1-phosphate}$$

There are two advantages to this phosphorylation. First, the cell membrane is rather impermeable to phosphorylated sugars and they are therefore much less likely to be lost from the cell by diffusion. Second, there is a net energy savings from this phosphorolysis. As we saw in Section 4.8, an early step in glycolysis is the phosphorylation of glucose at the expense of a molecule of ATP. When glucose-1-phosphate is produced directly by phosphorolytic cleavage, the expenditure of one ATP is saved. This phosphorolytic cleavage represents the trapping of energy released from a very favorable reaction (starch phosphorolysis) in the form of hexose phosphate. Glucose-1-phosphate formed in this way may also be converted to glucose-6-phosphate by the enzyme *phosphoglucomutase* and then metabolized via the pentose-phosphate pathway or via glycolysis.

Many microorganisms can use *disaccharides* for growth (Table 16.11). *Lactose* utilization by microorganisms is of considerable economic importance because milk-souring organisms produce lactic acid from lactose. The utilization of lactose commonly involves the enzyme **β-galactosidase**, an enzyme of considerable interest in molecular biology because its genetics and mechanism of synthesis have been widely studied in *Escherichia coli* (see Chapters 5 and 7). Many bacteria living in the mammalian intestine form β-galactosidase, which enables them to metabolize some of the lactose that reaches the intestinal tract. *Sucrose*, the common disaccharide of higher plants, is usually first hydrolyzed to its component monosaccharides (glucose and fructose) by the enzyme *invertase*, and the monomers are then metabolized by normal pathways.

The microbial polysaccharide *dextran* is synthesized by some bacteria using the enzyme *dextransucrase*. The starting material is the disaccharide sucrose: an α-1,6-glucose polymer is formed from the glucose moiety of sucrose while fructose is liberated in the free state:

$$n \text{ Sucrose} \rightarrow \underset{\textit{Dextran}}{(\text{glucose})_n} + n \text{ fructose}$$

Hydrolysis of the glycosidic bond connecting the two sugars of sucrose proceeds exergonically and the energy released is used to drive dextran synthesis. In this way the requirement for a nucleoside sugar with its high-energy phosphate bond is eliminated. Dextran is formed in this way by the bacterium *Leuconostoc mesenteroides* and a few others, and the polymer formed accumulates around the cells as a massive slime or capsule (Figure 16.38). Since sucrose is required for dextran formation, no dextran is formed when the bacterium is cultured on a medium with glucose or fructose. The dextran formed by *L. mesenteroides* has been used medically as a blood plasma substitute, and can be produced commercially by allowing the purified enzyme to react with sucrose, the length of the polymer chains being determined by how long the reaction is allowed to proceed. Dextran is also one of many polysaccharides produced by bacteria that are involved in tooth decay (see Section 11.3). *Streptococcus mutans* for example, produces a strongly adhesive dextran polysaccharide which promotes attachment of the organism to the teeth and gums. Cells of *S. mutans*, imbedded in dextran (dental plaque) produce lactic acid by fermentation, which initiates the tooth decay process. Since dextran is formed only when sucrose is present, bacterial adherence should occur only when sucrose is present in the diet, and at least one way of controlling tooth decay is to reduce the sucrose intake by eliminating candies and other sweets.

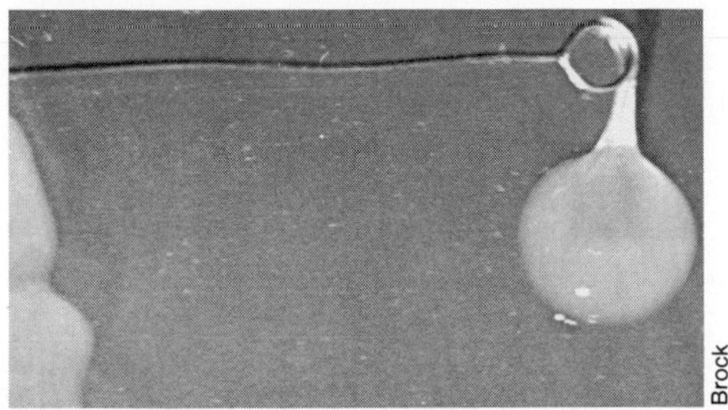

Figure 16.38 Slimy colony formed by dextran-producing bacterium, *Leuconostoc mesenteroides*, growing on a sucrose-containing medium. When the same organism is grown on glucose, the colonies are small and not slimy.

16.20 Organic Acid Metabolism

A variety of organic acids can be utilized by microorganisms as carbon sources and electron donors. The acids of the tricarboxylic acid cycle such as *citrate, malate,* and *succinate* are common natural products formed by plants and are also fermentation products of microorganisms. Because the citric acid cycle has major *biosynthetic* (see Section 4.18) as well as *energetic* (see Section 4.13) functions, the complete cycle or major portions of it are nearly universal in microbes. Thus, it is not surprising that many microorganisms are able to utilize these acids as electron donors and carbon sources. Aerobic utilization of four-, five-, and six-carbon acids can be accomplished by means of enzymes of the tricarboxylic acid cycle, with ATP formation by oxidative phosphorylation.

Anaerobic utilization of organic acids usually involves conversion to pyruvate and ATP formation via the phosphoroclastic reaction (see Section 16.18).

The glyoxylate cycle Utilization of two- or three-carbon acids as carbon sources cannot occur by means of the tricarboxylic acid cycle alone. This cycle can continue to operate only if the acceptor molecule, the four-carbon acid *oxalacetate* is regenerated at each turn of the cycle; any removal of carbon compounds for biosynthetic reactions would prevent completion of the cycle. When acetate is utilized, the oxalacetate needed to continue the cycle is produced through the **glyoxylate cycle** (Figure 16.39), so-called because glyoxylate is a key intermediate. This cycle is composed of most of the TCA cycle reactions plus two additional enzymes: *isocitrate lyase*, which splits isocitrate to succinate and glyoxalate, and *malate synthase*, which converts glyoxalate and acetyl-CoA to malate.

Biosynthesis through the glyoxylate cycle occurs as follows. The splitting of isocitrate into succinate and glyoxalate allows the succinate molecule to be drawn off for biosynthesis, since glyoxylate combines with acetyl-CoA to yield malate. Malate can be converted to oxalacetate to maintain the cyclic nature of the TCA cycle despite the fact that a C-4 intermediate (succinate) has been drawn off. The succinate molecule can be used directly in the production of porphyrins, be oxidized to oxalacetate and serve as a carbon skeleton for C-4 amino acids, or be converted (via oxalacetate and phosphoenolpyruvate) into glucose.

Pyruvate and C-3 utilization Three-carbon compounds such as pyruvate or compounds converted to pyruvate (for example, lactate or carbohydrates) also cannot be utilized as carbon sources through the tricarboxylic acid cycle alone. The oxalacetate needed to keep the cycle going is synthesized from pyruvate by the addition of a carbon atom from CO_2. In some organisms this step is catalyzed by the enzyme *pyruvate carboxylase*:

$$\text{Pyruvate} + \text{ATP} + CO_2 \longrightarrow \text{oxalacetate} + \text{ADP} + P_i$$

whereas in others it is catalyzed by *phosphoenolpyruvate carboxylase*:

$$\text{Phosphoenolpyruvate} + CO_2 \longrightarrow \text{oxalacetate} + P_i$$

These reactions replace oxalacetate that is lost when compounds of the tricarboxylic acid cycle are removed for use in biosynthesis, and the cycle can continue to function.

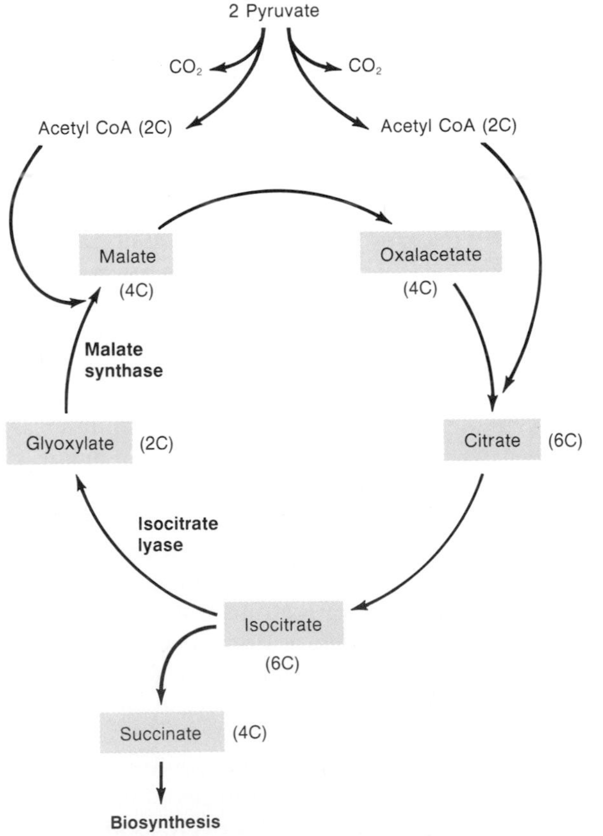

Figure 16.39 The glyoxylate cycle, leading to the synthesis of oxalacetate from acetate. Two unique reactions, isocitrate lyase and malate synthase, operate with a majority of the TCA cycle reactions.

16.21 Lipids As Microbial Nutrients

Fat and phospholipid hydrolysis Fats are esters of glycerol and fatty acids (see Section 2.5). Microbes utilize fats only after hydrolysis of the ester bond, and extracellular enzymes called **lipases** are responsible for the reaction. The end result is formation of glycerol and free fatty acids (Figure 16.40). Lipases are not highly specific, and attack fats containing fatty acids of various chain lengths. Phospholipids are hydrolyzed by specific enzymes called *phospholipases*, given different designations depending on which ester bond they cleave (Figure 16.40). Phospholipases

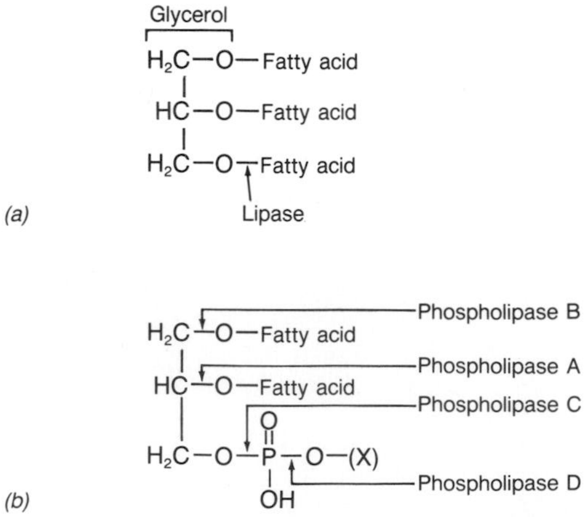

(a)
Glycerol
H_2C—O—Fatty acid
HC—O—Fatty acid
H_2C—O—Fatty acid
Lipase

(b)
H_2C—O—Fatty acid — Phospholipase B
HC—O—Fatty acid — Phospholipase A
Phospholipase C
H_2C—O—P(=O)(OH)—O—(X)
Phospholipase D

Figure 16.40 (a) Action of lipase on a fat. (b) Phospholipase action on phospholipid. The sites of action of the four distinct phospholipases *A, B, C,* and *D* are shown. X refers to a number of small organic molecules which may be at this position in different phospholipids. Compare this diagram to the more complete figure of a phospholipid in Figure 2.7.

A and B cleave fatty acid esters and thus resemble the lipases described above, but phospholipases C and D cleave *phosphate* ester linkages and hence are quite different types of enzymes. Attack by phospholipase C of *Clostridium perfringens* on the phospholipid of the red blood cell membrane results in lysis (**hemolysis**) and this enzyme is partly responsible for the symptoms of gas gangrene induced by this bacterium (see Section 11.9). As illustrated in Figure 11.16*b*, enzyme action is recognized by the zone of opalescence that develops in the medium surrounding the colony of a phospholipase-producing organism.

Fatty acid oxidation The fatty acids released by action of lipases and phospholipases are oxidized by a process called *beta oxidation*, in which two carbons of the fatty acid are split off at a time (Figure 16.41). In eucaryotes the enzymes are in the mitochondria, whereas in procaryotes, they have not been localized. The fatty acid is first activated with coenzyme A; oxidation results in the release of *acetyl-CoA* and the formation of a fatty acid shorter by two carbons. The process of beta oxidation is then repeated and another acetyl-CoA molecule is released. Two separate dehydrogenation reactions occur. In the first, electrons are transferred to flavin-adenine dinucleotide (FAD), whereas in the second they are transferred to NAD. Both of these coenzymes are reoxidized through electron transport phosphorylation with the synthesis of ATP. Most fatty acids have an even number of carbon atoms, and complete oxidation yields only acetyl-CoA. The acetyl-CoA formed is then oxidized by way of the TCA cycle or is converted into hexose and other cell constituents via the glyoxylate cycle. Fatty acids are good electron donors in respiration. For example, the complete oxidation of the 16-carbon fatty acid palmitic acid results in the net synthesis of 129 ATP molecules from electron-transport phosphorylation of reducing equivalents generated during the formation of acetyl-CoA from beta oxidations and from the oxidation of the acetyl-CoA units themselves through the TCA cycle.

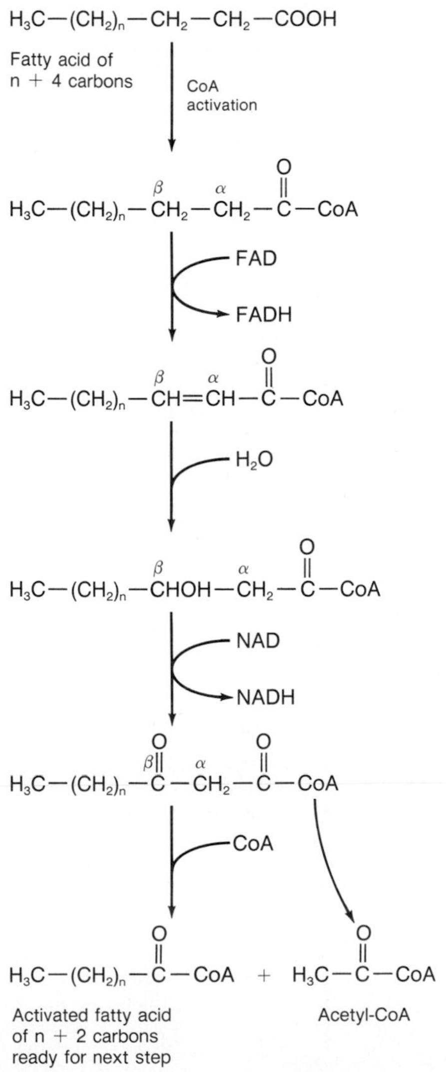

$H_3C{-}(CH_2)_n{-}CH_2{-}CH_2{-}COOH$

Fatty acid of n + 4 carbons

↓ CoA activation

$H_3C{-}(CH_2)_n{-}\overset{\beta}{C}H_2{-}\overset{\alpha}{C}H_2{-}\overset{O}{\overset{\|}{C}}{-}CoA$

↓ FAD → FADH

$H_3C{-}(CH_2)_n{-}\overset{\beta}{C}H{=}\overset{\alpha}{C}H{-}\overset{O}{\overset{\|}{C}}{-}CoA$

↓ H_2O

$H_3C{-}(CH_2)_n{-}\overset{\beta}{C}HOH{-}\overset{\alpha}{C}H_2{-}\overset{O}{\overset{\|}{C}}{-}CoA$

↓ NAD → NADH

$H_3C{-}(CH_2)_n{-}\overset{O}{\overset{\|}{C}}{-}CH_2{-}\overset{O}{\overset{\|}{C}}{-}CoA$ (β, α)

↓ CoA

$H_3C{-}(CH_2)_n{-}\overset{O}{\overset{\|}{C}}{-}CoA \; + \; H_3C{-}\overset{O}{\overset{\|}{C}}{-}CoA$

Activated fatty acid of n + 2 carbons ready for next step; Acetyl-CoA

Figure 16.41 Mechanism of beta oxidation of a fatty acid, which leads to successive formation of two-carbon fragments of acetyl-CoA.

16.22 Molecular Oxygen (O_2) as a Reactant in Biochemical Processes

We have discussed in some detail the role of O_2 as an electron acceptor in energy-generating reactions. Although this is by far the most important role of O_2 in cellular metabolism, O_2 also plays an interesting and important role in certain types of anabolic and catabolic reactions. Even though O_2 has a very positive reduction potential, it is not a very reactive molecule because of its peculiar chemical properties. In particular, O_2 exists in a triplet ground state with two unpaired electron spins (see Section 9.14). Carbon, on the other hand, exists in organic compounds in the singlet state, so that concerted reactions between singlet carbon and triplet oxygen are not possible. In order for oxygen to react with carbon, it must first be activated and converted into the singlet state. In biological systems this is generally done through interaction of O_2 with an enzyme containing a metal such as iron, Fe, or copper, Cu. Such metals have one or more *unpaired* electrons and thus direct reaction between these metals and O_2 is possible.

Oxygenases are enzymes that catalyze the incorporation of oxygen from O_2 into organic compounds. There are two kinds of oxygenases: *dioxygenases*, that catalyze the incorportion of *both* atoms of O_2 into the molecule; and *monooxygenases*, that catalyze the transfer of *only one* of the two O_2 atoms to an organic compound as a hydroxyl (OH) group, with the second atom of O_2 ending up as water, H_2O. Because monooxygenases catalyze the formation of hydroxyl groups (OH) in organic compounds, they are sometimes called *hydroxylases*. Because monooxygenases require a second electron donor to reduce the second oxygen atom of O_2 to water, they are also sometimes called *mixed-function oxygenases*. In most monooxygenases, the second electron donor is NADH or NADPH, although the direct coupling to O_2 is through a flavin which becomes reduced by the NADH or NADPH donor.

There are several types of reactions in living organisms that require O_2 as a reactant. One of the best examples is the involvement of O_2 in steroid biosynthesis. The formation of the fused steroid ring system (see Figure 3.9) requires the participation of molecular oxygen. Such a reaction can obviously not take place under anaerobic conditions, so that organisms which grow anaerobically must either dispense with this reaction or obtain the required substance (steroid) preformed from their environment. The requirement of O_2 as a reactant in biosynthesis is of considerable evolutionary significance, as molecular O_2 was originally absent from the atmosphere of the earth when life evolved and only became available after the evolution of cyanobacteria, the first photosynthetic organisms to produce O_2 (see Chapter 18). The role of O_2 in hydrocarbon utilization is discussed below.

16.23 Hydrocarbon Transformations

Hydrocarbons are organic carbon compounds containing only carbon and hydrogen and are highly insoluble in water. Low-molecular-weight hydrocarbons are gases, whereas those of higher molecular weight are liquids or solids at room temperature. Some hydrocarbons are aliphatic compounds, a class of carbon compounds in which the carbon atoms are joined in open chains. There is a tremendous variation among aliphatic hydrocarbons in chain length, degree of branching, and number of double bonds. Another important group of hydrocarbons contains the aromatic ring and can be viewed as derivatives of benzene.

Aliphatic hydrocarbons Only relatively few kinds of microorganisms (for example, *Nocardia, Pseudomonas, Mycobacterium*, and certain yeasts and molds) can utilize hydrocarbons for growth. Utilization of saturated aliphatic hydrocarbons is strictly an *aerobic* process: in the absence of O_2, saturated hydrocarbons are completely unaffected by microbes (see Section 16.18). This accounts for the fact that hydrocarbons in petroleum deposits have remained unchanged for millions of years. As soon as petroleum resources are brought to the surface and exposed to air, microbial oxidation of the hydrocarbons begins and the materials are eventually decomposed.

The initial oxidation step of saturated aliphatic hydrocarbons involves molecular oxygen (O_2) as a reactant, and one of the atoms of the oxygen molecule is incorporated into the oxidized hydrocarbon. This reaction is carried out by a **monooxygenase** (see Section 16.22), and a typical reaction sequence is that shown in Figure 16.42. The end product of the reaction sequence is acetyl-CoA. However, the initial oxidation is not at the terminal carbon in all cases. Oxidation may sometimes occur at the second carbon and then quite different subsequent reactions occur.

Aromatic hydrocarbons Many aromatic hydrocarbons can be used as electron donors aerobically by microorganisms, of which bacteria of the genus *Pseudomonas* have been the best studied. It has been demonstrated that the metabolism of these compounds, some of which are quite complex, frequently has as its initial stage the formation of either of two molecules, *protocatechuate* or *catechol* as shown in Figure 16.43*a*.

These single ring compounds are referred to as *starting substrates*, since oxidative catabolism proceeds only after the complex aromatic molecules have been converted to these more simple forms. Protocatechuate and catechol may then be further degraded to compounds which can enter the TCA cycle:

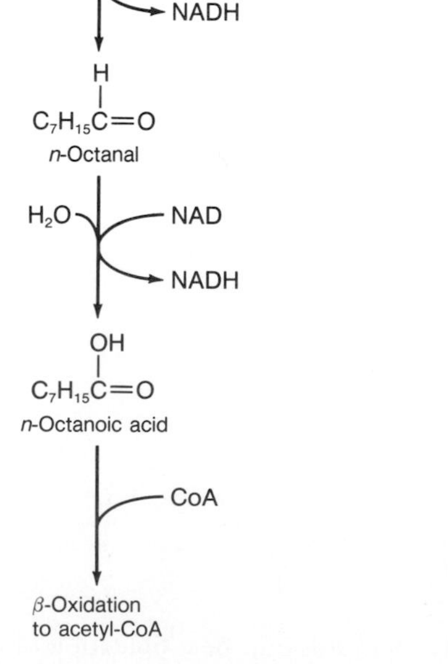

$C_7H_{15}{-}CH_3$ + NADH + O:O
n-Octane (O_2)

↓

$C_7H_{15}CH_2OH$ + NAD + H_2O
n-Octanol

↓ (NAD → NADH)

$C_7H_{15}C(H){=}O$
n-Octanal

↓ (H_2O; NAD → NADH)

$C_7H_{15}C(OH){=}O$
n-Octanoic acid

↓ (CoA)

β-Oxidation to acetyl-CoA

Figure 16.42 Steps in oxidation of an aliphatic hydrocarbon, catalyzed by a monooxygenase.

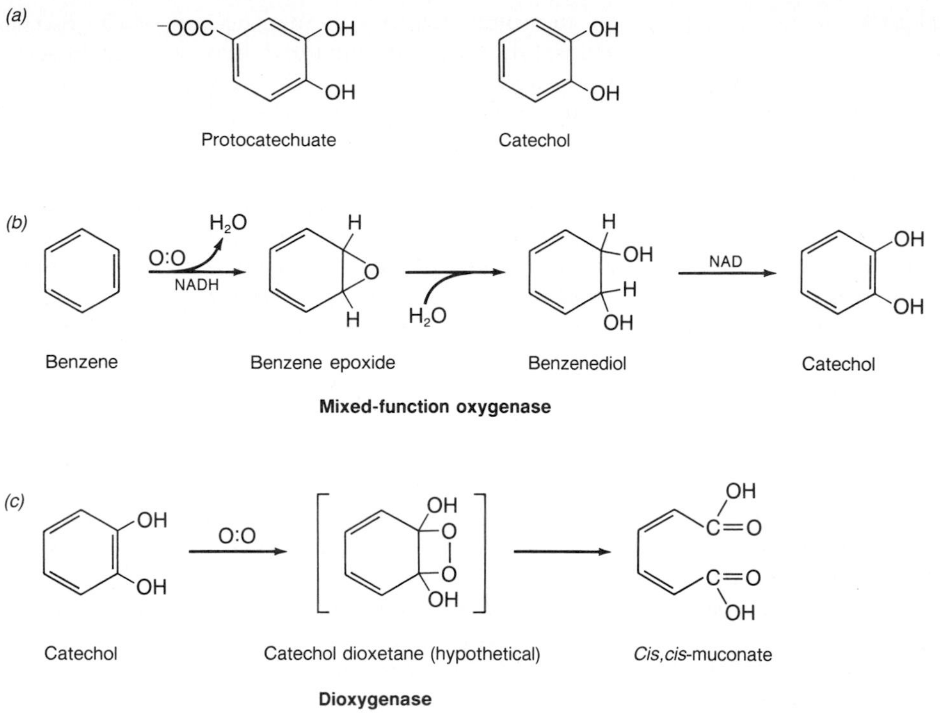

Figure 16.43 Roles of oxygenases in catabolism of aromatic compounds. (a) Protocatechuate and catechol, two common oxidation products of the benzene ring. (b) Hydroxylation of benzene to catechol by a mixed-function oxygenase in which NADH is the second electron donor. (c) Cleavage of catechol to *cis, cis*-muconate by a dioxygenase. Reactant oxygen atoms are shown in color in both reactions to demonstrate the different mechanisms.

succinate, acetyl-CoA, pyruvate. Several steps in the catabolism of aromatic hydrocarbons require oxygenases. Figures 16.43*b* and *c* show two different oxygenase-catalyzed reactions, one using a monooxygenase and the other a dioxygenase.

Aromatic compounds can also be degraded anaerobically, especially if they contain an atom of oxygen. Mixed cultures capable of degrading aromatic compounds have been shown to degrade benzoate and other substituted phenolic compounds yielding CH_4 and CO_2 as final products. Substituted phenolic compounds are also degraded by certain denitryifying, photosynthetic, and sulfate-reducing bacteria. The anaerobic catabolism of aromatic compounds proceeds by ***reductive*** rather than oxidative ring cleavage (Figure 16.44). This is reasonable, since anaerobically oxygen cannot be involved in the oxidation of aromatic compounds. Anaerobic catabolism involves ***ring reduction*** followed by ***ring cleavage*** to yield a straight chain fatty acid or dicarboxylic acid. These intermediates can be converted to acetyl-CoA and used for both biosynthetic and energy yielding purposes. In most cases of anaerobic aromatic catabolism observed thus far, at least one oxygen atom must already be present on the aromatic molecule before it will be catabolized rapidly. Benzene, for example, is degraded very slowly anaerobically, whereas benzoate is rapidly metabolized.

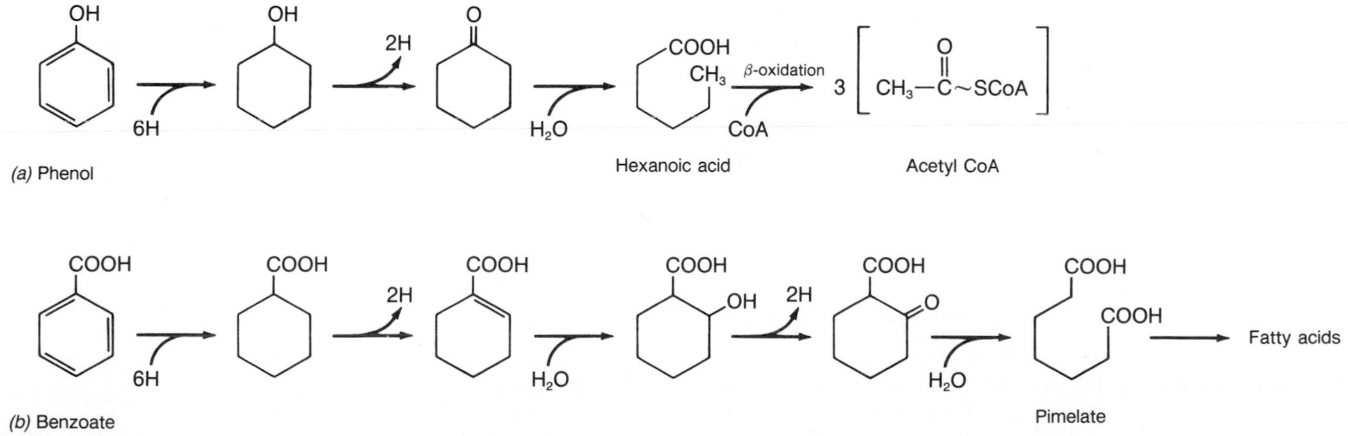

Figure 16.44 Anaerobic degradation of phenol and benzoate by reductive-ring cleavage.

16.24 Nitrogen Metabolism

We have discussed nitrogen nutrition in general terms in Sections 4.1 and 4.18. Here we discuss some of the special biochemistry and genetics of nitrogen utilization, with emphasis on the process of **nitrogen fixation**. Nitrogen can be obtained by microorganisms either from inorganic or organic forms. The most common inorganic nitrogen sources are nitrate and ammonia, but other inorganic sources used by certain microbes include cyanide (CN^-), cyanate (OCN^-), thiocyanate (SCN^-), cyanamide (NCN^{2-}), nitrite (NO_2^-), and hydroxylamine (NH_2OH). As discussed later, free nitrogen gas (N_2) is also used by a variety of bacteria.

We discussed the general biochemistry of the utilization of organic nitrogen in Sections 4.1 and 4.18. In that place we discussed enzymes such as *transaminases* and *amino-acid dehydrogenases* which catalyze the reductive assimilation of ammonia from the environment. Another very important means of incorporating ammonia is via the enzyme **glutamine synthetase**. As shown in Figure 16.45, this enzyme catalyzes the addition of ammonia to glutamate to yield *glutamine*. Then in a second reaction, the enzyme **glutamate synthase** catalyzes the transfer of the amide group of glutamine to α-ketoglutarate, forming two molecules of *glutamate*. When these two reactions are summed, the net reaction is the conversion of one molecule of α-ketoglutarate and one molecule of ammonia to glutamate, with consumption of one high-energy phosphate bond from ATP. In addition to its role in ammonia assimilation, glutamine also functions as an amino donor for a variety of reactions, most importantly in the synthesis of the purine ring (see Section 4.19).

A rather interesting control is exerted on the activity of the enzyme glutamine synthetase. Glutamine synthetase exists in two forms, the active form and an inactive form; the latter form occurs when adenyl residues from ATP are attached to the molecule, a process called *adenylylation*. Adenylylation of glutamine synthetase is promoted by high concentrations of ammonia, leading to inactivation of the enzyme. However, since glutamate dehydrogenase functions when ammonia concentrations are high, ammonia assimilation is shifted from the glutamine pathway to the glutamate pathway. When ammonia concentrations fall, glutamine synthetase becomes deadenylylated and becomes active again. Control via adenylylation is different from allosteric control described in Section 4.20, as adenylylation involves a *covalent modification* of the enzyme rather than merely a change in its conformation. The presumed function of adenylylation is to prevent the wasteful consumption of ATP by glutamine synthetase when ammonia levels are sufficiently high for the operation of amino acid dehydrogenases; the latter do not require ATP for their function. We discussed the utilization of nitrate and the whole process of assimilative nitrate reduction early in this chapter (see Section 16.14).

Nitrogen fixation The utilization of nitrogen gas (N_2) as a source of nitrogen is called *nitrogen fixation* and is a property of only certain bacteria and cyanobacteria. An abbreviated list of nitrogen-fixing organisms is given in Table 16.12 from which it can be seen that a variety of bacteria, both anaerobic and aerobic, fix nitrogen. In addition, there are some bacteria, called *symbiotic*, that fix nitrogen only when present in nodules or on roots of specific host plants. As far as is currently known, no eucaryotic organisms fix nitrogen. Symbiotic nitrogen fixation will be discussed in Section 17.24.

In the fixation process, N_2 is *reduced* to ammonium and the ammonium converted into organic form. The reduction process is catalyzed by the enzyme complex **dinitrogenase,** which consists of two separate protein components called Component I (**nitrogenase**) and Component II (**nitrogenase reductase**). Both components contain iron, and nitrogenase contains molybdenum as well. The molybdenum and iron in Component I are contributed by a co-factor known as FeMo-co, and the actual reduction of N_2 involves participation of this iron-molybdenum center. Owing to the stability of the N≡N triple bond,

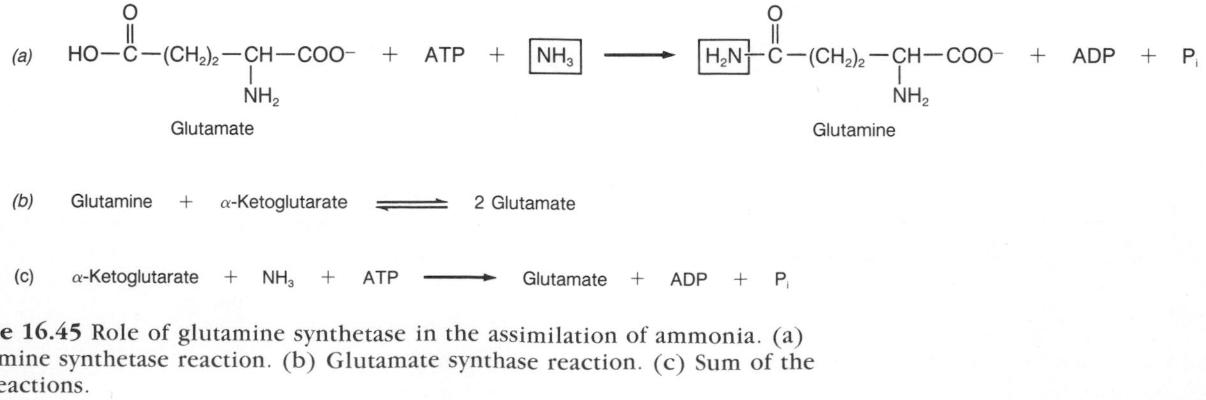

(a) $\text{HO—C(=O)—(CH}_2)_2\text{—CH(NH}_2)\text{—COO}^- + \text{ATP} + \boxed{\text{NH}_3} \longrightarrow \boxed{\text{H}_2\text{N}}\text{—C(=O)—(CH}_2)_2\text{—CH(NH}_2)\text{—COO}^- + \text{ADP} + \text{P}_i$

Glutamate → Glutamine

(b) Glutamine + α-Ketoglutarate $\rightleftharpoons$ 2 Glutamate

(c) α-Ketoglutarate + NH_3 + ATP ⟶ Glutamate + ADP + P_i

Figure 16.45 Role of glutamine synthetase in the assimilation of ammonia. (a) Glutamine synthetase reaction. (b) Glutamate synthase reaction. (c) Sum of the two reactions.

N_2 is extremely inert and its activation is a very energy-demanding process. Six electrons must be transferred to reduce N_2 to $2NH_3$, and several intermediate steps might be visualized; but since no intermediates have ever been isolated, it is now assumed that the three successive reduction steps occur with the intermediates firmly bound to the enzyme (Figure 16.46*a*). Nitrogen fixation is highly reductive in nature and the process is inhibited by oxygen since nitrogenase reductase is rapidly and irreversibly inactivated by O_2 (even when isolated from *aerobic* nitrogen-fixers). In aerobic bacteria, N_2 fixation occurs in the presence of O_2 in whole cells, but not in purified enzyme preparations, and it is thought that the dinitrogenase within the cell is in an O_2-protected microenvironment. Some bacteria and cyanobacteria able to grow both aerobically and anaerobically fix N_2 only under anaerobic conditions.

The electrons for nitrogen reduction are transferred to nitrogenase reductase from ferredoxin, the low-redox-potential electron carrier already described. In *Clostridium pasteurianum*, ferredoxin is reduced by phosphoroclastic splitting of pyruvate and ATP is synthesized at the same time. In addition to reduced ferredoxin, ATP is required for N_2 fixation in all organisms studied. The ATP requirement for nitrogen fixation is very high, about 4 to 5 ATP being split to ADP + P_i for each $2e^-$ transferred. ATP is apparently required in order to *lower* the reduction potential of the system sufficiently so that N_2 may be reduced. The reduction potential of nitrogenase reductase is -0.30 V and this is lowered to -0.40 V when the enzyme combines with ATP. Electrons are first transferred from ferredoxin to nitrogenase reductase (Figure 16.46), after which ATP is hydrolyzed and lowers the reduction potential. This complex then combines with nitrogenase and nitrogenase becomes reduced. Reduced nitrogenase can now convert N_2 to NH_3, but the biochemical details of the reduction are not known. Although only *six* electrons are necessary to reduce N_2 to $2NH_3$, careful measurements have shown that *eight* electrons are actually consumed, *two* electrons being lost as *hydrogen gas* (H_2) for each mole of N_2 reduced. The reason for this apparent wastage is not known, but good evidence exists to show that H_2 evolution is an intimate part of the reaction mechanism of dinitrogenase.

The genes for nitrogenase and nitrogenase reductase are part of a complex regulon (a large operon) called the *nif* regulon (Figure 16.46*b*). In addition the structural genes and the genes for FeMo-co, genes controlling the electron transport proteins and a number of regulatory genes are also present. Component I (nitrogenase) is a complex protein made up of two subunits, α (product of the *D* gene) and β (product of the *K* gene), each of which is present in two copies. Component II (nitrogenase reductase) is a protein dimer consisting of two identical subunits, the product of the H gene. FeMo-co, whose complete structure is not yet known, is synthesized through the participation of four separate genes, N,E, and B, as well as gene Q which controls a product involved in molybdenum uptake.

Dinitrogenase is subject to strict regulatory controls. Nitrogen fixation is inhibited by O_2 and by fixed nitrogen, including NH_3, NO_3^-, and amino acids. A major part of this regulation is at the level of transcription. The various transcriptional units of the *nif* regulon are shown in Figure 16.46*b*. Transcription of the *nif* structural genes is *activated* by the *nif*A

Table 16.12 Some nitrogen-fixing organisms

Free-living				Symbiotic	
Aerobes		Anaerobes			
Heterotrophs	Phototrophs	Heterotrophs	Phototrophs	Leguminous plants	Nonleguminous plants
Bacteria: *Azotobacter* spp. *Klebsiella* *Beijerinckia* *Bacillus polymyxa* *Mycobacterium flavum* *Azospirillum lipoferum* *Citrobacter freundii* Methylotrophs (various, but not all)	Cyanobacteria (various, but not all)	Bacteria: *Clostridium* spp. *Desulfovibrio* *Desulfotomaculum*	Bacteria: *Chromatium* *Chlorobium* *Rhodospirillum* *Rhodopseudomonas* *Rhodomicrobium* ***Rhodobacter*** *Heliobacterium*	Soybeans, peas, clover, locust, etc., in association with a bacterium of the genus *Rhizobium* or *Bradyrhizobium*	*Alnus, Myrica, Ceanothus, Comptonia, Casuarina*; in association with actinomycetes of the genus *Frankia*

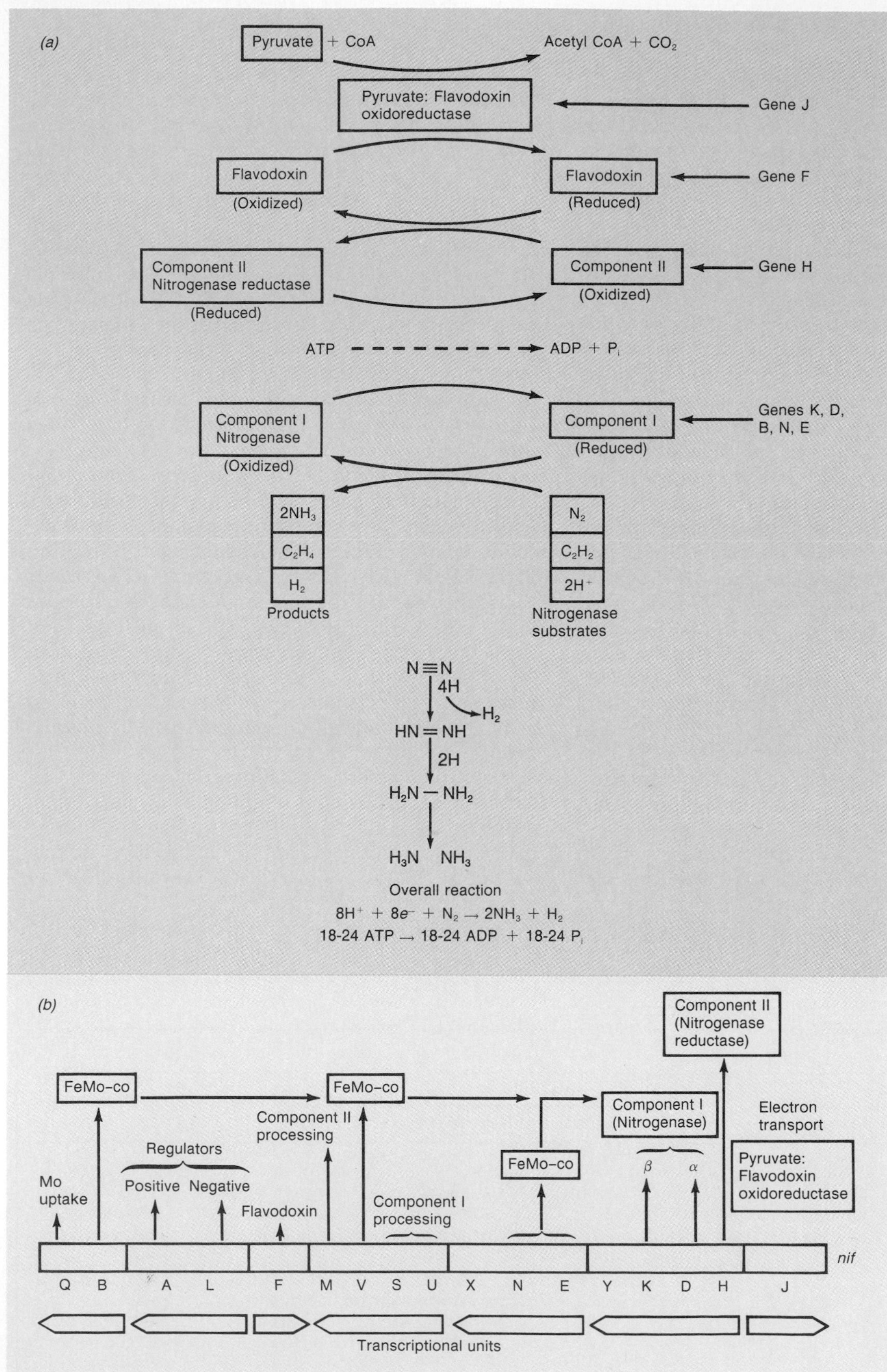

Figure 16.46 The *dinitrogenase* system. (a) Steps in nitrogen fixation: reduction of N_2 to $2NH_3$. (b) The genetic structure of the *nif* regulon (operon) in *Klebsiella pneumoniae*, the best studied nitrogen-fixing organism. The functions of some of the genes are uncertain. The mRNA transcripts (transcriptional units) are shown below the genes; arrows indicate the direction of transcription.

The Acetylene Reduction Assay for Dinitrogenase

The enzyme **dinitrogenase** has been one of the most difficult enzymes to work on, partly because it is so sensitive to oxygen inactivation, and partly because it has been difficult to assay. The natural substrate of nitrogenase, N_2, is in such high concentrations in the atmosphere that it is almost impossible to measure nitrogen fixation by measuring the change in the concentration of N_2 itself. Biochemists have often been able to circumvent such difficulties by using a radioactive tracer, but there is no convenient radioisotope of nitrogen. The only available isotope, $^{15}N_2$, is a *stable* isotope whose presence can only be detected by means of an expensive mass spectrometer. But until the acetylene reduction assay was developed, the only means of detecting nitrogenase activity was by means of ^{15}N.

The discovery that the enzyme nitrogenase also reduced acetylene was a classic example of serendipity (finding one thing while looking for something else). Painstaking research using the ^{15}N technique had permitted partial purification and characterization of the enzyme. Once partially purified material was available, it was natural to examine the specificity of nitrogenase by testing other substrates. The goal here was to determine how specific the nitrogenase system was. Because N_2 is a triple-bonded molecule, the triple-bonded substrate acetylene (C_2H_2) was also tested. Acetylene proved not only to be an excellent substrate for the enzyme, but it was reduced only to its double-bonded relative ethylene. Thus, when an organism containing active nitrogenase or a purified nitrogenase preparation was incubated with acetylene, ethylene accumulated in the gas phase, and the amount of ethylene produced was proportional to the amount of enzyme present. The ethylene concentration can be easily measured with a gas chromatograph, so that a simple and efficient assay for nitrogenase could be perfected. Thus, by studying the biochemistry of the enzyme, a very useful enzyme assay was by chance discovered.

Once the acetylene reduction assay was developed, it not only permitted extensive biochemical studies on the enzyme, it made it possible to study the nitrogen fixation process in various organisms as well as in natural environments. By use of this assay, the range of organisms capable of fixing nitrogen was rapidly extended. From an ecological viewpoint, the acetylene reduction assay has permitted extensive studies on the importance of the process of biological nitrogen fixation in natural and agricultural environments, as discussed in Section 17.14. Probably no discovery has benefitted research on biological nitrogen fixation more than the discovery of the acetylene reduction assay.

protein (positive regulation) and *repressed* by the *nifL* protein (negative regulation), although the details of how these proteins regulate transcription are not known.

The ammonia produced by dinitrogenase does not repress enzyme synthesis, of course, because as soon as it is made it is incorporated into organic form and used in biosynthesis. But when ammonia is in excess (as in environments high in ammonia), dinitrogenase synthesis is quickly repressed. This prevents the wastage of ATP by not making a product already present in ample amounts. In certain nitrogen-fixing bacteria, especially phototrophic bacteria, dinitrogenase *activity* is also regulated by ammonia, a phenomenon called the ammonia "switch-off" effect. In this case, excess ammonia catalyzes a covalent modification of dinitrogenase, resulting in a totally *inactive* enzyme. When ammonia again becomes limiting, this modified dinitrogenase is converted back to the active form and resumes N_2 fixation. Ammonia "switch-off" is thus a rapid and reversible method of controlling ATP consumption by dinitrogenase.

Dinitrogenase has been purified from a large number of nitrogen-fixing organisms and in all cases has been shown to be a two protein complex. It is of considerable evolutionary interest that nitrogenase from one organism will function with nitrogenase reductase from another organism. Even when the components are taken from organisms thought to be distantly related (for example, a photosynthetic and a nonphotosynthetic organism) the hybrid complex retains some activity. This can be interpreted to mean that the structures of the dinitrogenase components have not changed markedly during evolution, suggesting that the molecular requirements for N_2 reduction are fairly specific.

Dinitrogenase is not specific for N_2 but will also reduce cyanide (CN^-), acetylene ($HC{\equiv}CH$), and several other compounds. The reduction of acetylene by dinitrogenase is only a two-electron process and *ethylene* ($H_2C{=}CH_2$) is produced. The reduction of acetylene probably serves no useful purpose to the cell but provides the experimenter with a simple way of measuring the activity of nitrogen-fixing systems, since it is fairly easy to measure the reduction of acetylene to ethylene via gas chromatography (see Box).

This technique is now widely used to detect nitrogen fixation in unknown systems. Previously, it was not easy to prove that an organism fixed N_2; indeed, many claims for nitrogen fixation in microorganisms have been erroneous. The growth of an organism in a medium to which no nitrogen compounds have been added does not mean that the organism is fixing nitrogen from the air since traces of nitrogen compounds often occur as contaminants in the ingredients of culture media or drift into the media in gaseous form or as dust particles. Even distilled water may be contaminated with ammonia. One method of proving nitrogen fixation is to show a net increase in the total nitrogenous content of the medium plus its organisms after incubation; the assumption would be that the increased nitrogen could come only from N_2 from the air. A more accurate procedure is to use an isotope of nitrogen, ^{15}N, as a tracer.* The gas phase of a culture is enriched with ^{15}N, and after incubation the cells and medium are digested, the ammonia produced being distilled off and assayed for its ^{15}N content. If there has been a significant production of ^{15}N-labeled NH_3, it is proof of nitrogen fixation. However, the acetylene-reduction method is a more sensitive way of measuring nitrogen fixation and has mostly replaced the more difficult ^{15}N method. The sample, which may be soil, water, a culture, or a cell extract is incubated with acetylene, and the reaction mixture is later analyzed by gas chromatography for production of the gaseous substance ethylene. This method is far simpler and faster than other methods.

*^{15}N is not a radioactive isotope but a stable isotope, and its presence must be detected with the mass spectrometer, an expensive and rather cumbersome instrument.

Study Questions

1. In what nutritional class would you place an organism that used *glucose* as sole carbon and energy source? *Elemental sulfur* as energy source? How would we refer to the latter organism if it grew with CO_2 as sole carbon source? What is the energy source for *phototrophic* organisms?
2. What is the role of *light* in the photosynthetic process of *green and purple bacteria*? Of *cyanobacteria*? Compare and contrast the photosynthesis process in these two groups of procaryotes.
3. Compare and contrast the absorption spectrum of *chlorophyll a* and *bacteriochlorophyll a*. What wavelengths are preferentially absorbed by each pigment and how do the absorption properties of these molecules compare with the regions of the spectrum visible to our eye? Why are most plants green in color?
4. What are the functions of *light harvesting* and *reaction center* chlorophyll? Why would a mutant incapable of making light-harvesting chlorophylls (such mutants can be readily isolated) probably not be a successful competitor in nature?
5. How does light result in ATP production in an *anoxygenic phototroph*? In what ways are photosynthetic and respiratory electron flow similar? In what ways do they differ?
6. Where are the *photosynthetic pigments* located in a purple bacterium? A cyanobacterium? A green alga? Considering the function of chlorophyll pigments why couldn't they be located elsewhere in the cell, for example in the cytoplasm or in the cell wall?
7. How is *reducing power* made for *autotrophic growth* in a purple bacterium? In a cyanobacterium?
8. The growth rate of the phototrophic purple bacterium *Rhodobacter* is about twice as fast when the organism is grown in a medium containing malate as carbon source than when the organism is grown with CO_2 as carbon source (with H_2 as electron donor). Discuss the reasons why this occurs and list the nutritional class we would place *Rhodobacter* in when growing under each of the two different conditions.
9. How does the *reduction potential* of chlorophyll *a* in photosystem I and photosystem II differ? Why must the reduction potential of photosystem II chlorophyll be so high?
10. Discuss the nature of the evidence obtained from studies of the photosynthetic process of certain cyanobacteria that supports the hypothesis that these organisms evolved from anoxygenic phototrophs.
11. What is the major function of *carotenoids* and *phycobilins* in photosynthetic microorganisms?
12. Although physiologically distinct, lithotrophs and heterotrophs share a number of features in common with respect to the production of ATP. Discuss these common features along with reasons for why the growth yield of a heterotroph respiring glucose is so much higher than for a lithotroph respiring sulfur.
13. Based on a consideration of energetics, discuss why photosynthetic bacteria are able to thrive in nature even though their photosynthetic pigments absorb in regions of the spectrum such as the infrared, which contains relatively little energy.
14. What two enzymes are unique to organisms which carry out the Calvin cycle? What reactions do these enzymes carry out? What would be the consequence if a mutant arose that lacked either of these enzymes?
15. For conversion of 6 molecules of CO_2 into one fructose molecule, 18 molecules of ATP are required. Where in the Calvin cycle reactions are these ATPs consumed?
16. Compare and contrast the utilization of H_2S by a purple phototrophic bacterium and by a colorless sulfur bacterium. What role does H_2S play in the metabolism of each organism?

17. Discuss why the growth yield (grams of cells/mole of substrate) of *Thiobacillus ferrooxidans* is considerably *greater* when the organism is growing on elemental sulfur than on ferrous iron as energy source (assume the organism is growing autotrophically in both cases).
18. Growth yields of *Nitrosomonas* are generally much higher than for *Nitrobacter*. Why might this be so?
19. What is a mixotroph? Why will an organism capable of mixotrophy frequently grow better mixotrophically than autotrophically or even heterotrophically?
20. In *Escherichia coli* synthesis of the enzyme *nitrate reductase* is strongly repressed by oxygen. Based on bioenergetic arguments, why do you think this repression phenomenon might have evolved?
21. Discuss at least three major differences between assimilative and dissimilative nitrate reduction.
22. Why is the following statement, if interpreted literally, incorrect: "Anaerobic respiration is simply a process where an alternative electron acceptor is substituted for O_2 in the respiratory process." Although technically incorrect, why do you think many microbiologists think of anaerobic respiration in this way?
23. Define the term *substrate-level phosphorylation*. How does it differ from oxidative phosphorylation? Assuming an organism is facultative, what basic nutritional conditions will dictate whether the organism obtains energy from substrate-level rather than oxidative phosphorylation?
24. Although many different compounds are theoretically fermentable, in order to support a fermentative process an organic compound must be eventually converted into one of a relatively small group of molecules. What are these molecules and why must they be produced?
25. To a culture of *Escherichia coli* growing fermentatively you add 1 g/l of $NaNO_3$. Would you expect the growth *yield* of the culture to increase or decrease and why?
26. Why have hydrocarbons accumulated in large reservoirs on earth despite the fact that they are readily degradable microbiologically under certain conditions?
27. How are xenobiotic compounds defined? Give an example of a compound you think would qualify as a xenobiotic.
28. Compare and contrast the conversion of cellulose and starch to glucose units. What enzymes are involved and which process is the more energy efficient?
29. Although dextran is a glucose polymer, glucose cannot be used to make dextran. Explain. How is dextran synthesis important in oral hygiene?
30. How do monooxygenases differ from dioxygenases in the reactions they catalyze?
31. *Pseudomonas fluorescens* can grow on benzoate aerobically while the phototrophic bacterium *Rhodopseudomonas palustris* can grow on benzoate anaerobically. Compare and contrast the metabolism of benzoate by these two species focusing on the following considerations: requirement for oxygenases, initial reactions leading to the opening of the ring, and product(s) formed that can feed into central metabolic pathways.
32. How do ammonia levels in a cell control the activities of the enzyme glutamine synthetase? How does such control benefit the cell? Although ammonia controls the *activity* of glutamine synthetase, this is not a typical "feedback" inhibition phenomenon. Explain.
33. Write out the reaction catalyzed by the enzyme *dinitrogenase*. How many electrons are required in this reaction? How many are actually used? Explain.
34. What is the function of ATP in the nitrogen fixation process? How might ATP affect nitrogenase reductase?
35. What is the evidence that dinitrogenase is a highly conserved protein in an evolutionary sense?

Supplementary Readings

Gottschalk, G. 1986. *Bacterial Metabolism, 2nd ed.* Springer-Verlag, New York. This standard textbook of bacterial physiology covers many of the physiological processes discussed in this chapter. Includes especially good discussion of fermentation.

Ormerod, J. G. (ed.) 1983. *The Phototrophic Bacteria: Anaerobic Life in the Light.* University of California Press, Berkeley. A collection of review chapters devoted to various aspects of anoxygenic photosynthesis.

Postgate, J. R. 1982. *The Fundamentals of Nitrogen Fixation.* Cambridge University Press, New York. An excellent monograph which considers the process of nitrogen fixation, the properties of dinitrogenase, and the major groups of N_2-fixing microorganisms.

Schlegel, H. G. 1986. *General Microbiology, 6th ed.* Cambridge University Press, Cambridge, England. This English translation of a popular German textbook of microbiology devotes a considerable amount of space to fermentations, anaerobic respiration, lithotrophy, and phototrophy.

Starr, M. P., H. Stolp, H. G. Trüper, A. Balows, and **H. G. Schlegel** (eds.) 1981. *The Prokaryotes—A Handbook on Habitats, Isolation, and Identification of Bacteria.* A complete reference source on all groups of bacteria. Excellent chapters on phototrophs, lithotrophs, methylotrophs, nitrogen fixers, and various fermentative anaerobes. In many cases the chapters on the different groups of organisms are prefaced by an overview chapter on the particular metabolic process characteristic of the group.

Zehnder, A. J. B. (ed.) 1988. *Biology of Anaerobic Bacteria.* John Wiley and Sons, New York. A collection of review chapters covering all major groups of anaerobes and anaerobic processes.

17

Microbial Ecology

In this chapter we consider some of the activities of microorganisms in the world at large. Microorganisms play far more important roles in nature than their small sizes would suggest. To see properly the place of microorganisms in nature, we must first consider the concept of an **ecosystem**. Ecologists define an ecosystem as a total community of organisms, together with the physical and chemical environment in which they live. Each organism interacts with its physical and chemical environment and with the other organisms in the system, so that the ecosystem can be viewed as a kind of superorganism with the ability to respond to and modify its environment. A good example of an ecosystem is a *lake*. The sides of the lake define the boundaries of the ecosystem. Within these boundaries organisms live and carry out their activities, greatly modifying the characteristics of the lake as well as each other.

Energy enters an ecosystem mainly in the form of sunlight and is used by photosynthetic organisms in the synthesis of organic matter. Some of the energy contained in this organic matter is dissipated by the photosynthesizers themselves during respiration, and the rest is available to *herbivores*, which are animals that consume the photosynthesizers. Of the energy entering the herbivores, one portion is dissipated by them during respiration and the rest is used in synthesizing the organic matter of the herbivore bodies. Herbivores are themselves consumed by *carnivorous* animals and these carnivores are eaten by other carnivores, and so on. At each step in this chain of events, a portion of the energy is dissipated as heat. Any plants or animals that die, whether from natural causes, injury, or disease, are attacked by microorganisms and small animals, collectively called **decomposers**. These decomposers also utilize energy released by plants or animals in the form of excretory products. These reactions constitute a **food chain** or food web.

Although most of the energy fixed by photosynthesizers is ultimately dissipated as heat, the chemical elements that serve as nutrients usually are not lost from the ecosystem. For instance, carbon from CO_2 fixed by plants in photosynthesis is released during respiration by various organisms of the food chain and becomes available for reutilization by the plants. Nitrogen, sulfur, phosphorus, iron, and other elements taken up by plants are also released through the activity of the decomposers, and hence are made available for reassimilation by other plants. Thus, although energy flows *through* the ecosystem, chemical elements are cycled *within* the ecosystem. In some parts of the cycle the element is oxidized, whereas in other parts it is reduced; for many elements a **biogeochemical cycle** can thus be defined, in which the element undergoes changes in oxidation state as it is acted upon by one organism after another. In addition to this redox (oxidation-reduction) cycle, it is also possible to define a **transport cycle**, which describes the movement of an element from one place to another on earth, as for instance from land to air or from air to water. Such a transport cycle may or may not also involve a redox cycle. For instance, when oxidation or reduction leads to conversion of a nonvolatile substance to one that is volatile, the latter can then be transported to the air, so that the transport cycle is coupled to the redox cycle. In other cases, oxidation or reduction does not lead to a change in state, and has no influence on transport.

In order to evaluate the roles of microorganisms in ecosystems, it is essential to understand their precise natural habitats, and how their activities in these habitats are measured. We thus open this chapter with a discussion of these questions.

17.1 Microorganisms in Nature

The natural habitats of microorganisms are exceedingly diverse. Any habitat suitable for the growth of higher organisms will also permit microbial growth, but in addition, there are many habitats unfavorable to higher organisms where microorganisms exist and even flourish. Because microorganisms are usually invisible, their physical existence in an environment is often unsuspected. In many cases it is not the microorganisms themselves that we observe in a natural environment, but instead the *chemical* evidence of their existence; microbial action is usually of considerable importance for the function of the ecosystem.

Microenvironments Because microbes are small, microbial environments are also small, and within a single soil crumb or upon a single root surface there may be a variety of distinct **microenvironments**, each suitable for the growth of certain kinds of organisms but not of others. Thus when we think of the existence of microbes in nature we must learn to "think small." Any study of the microenvironment requires the use of the microscope, and by carefully examining samples of the habitat under the microscope it is possible to observe where microbes are flourishing. A simple procedure that permits use of ordinary microscopes is to immerse microscope slides in the environment (for example, soil or water).

The microscope slide serves as a surface upon which organisms can attach and grow, and after a period of time the slide is removed and examined, preferably with a phase-contrast microscope. The development of microbial colonies on glass slides (Figure 17.1) presumably is similar to the way the same organisms develop upon natural surfaces.

Surfaces Surfaces in general are of considerable importance as microbial habitats because nutrients from the environment adsorb to them; thus in the microenvironment of a surface the nutrient levels may be much higher than they are away from the surface. A chemical assay for a nutrient in a body of water (for example, lake, river, ocean), therefore, will not reveal the concentrations present on surfaces such as sand, silt, or rock within the body of water. On such surfaces microbial numbers and activity will usually be much higher than in the free water, due to these adsorption effects.

Nutrients Despite adsorption to surfaces, nutrient concentrations are usually much lower in nature than they are in the usual laboratory culture media, frequently 100 to 1000 times lower. Thus microorganisms in natural environments are often subjected to nutrient-poor conditions. Even where richer sources of nutrients are available, as for instance the leaf litter on the forest floor or an algal bloom in a lake, the nutrients are available only briefly, for they are rapidly colonized and consumed by microbes, which then are returned to a semistarvation diet. In the case of photosynthetic organisms, their energy source, light, is highly variable in amount from day to night and throughout the seasons of the year. Thus organisms in nature must be able to adapt to a feast-or-famine existence, probably one of the main reasons why the highly sophisticated regulatory mechanisms described in Chapter 5 have evolved.

Growth rates in nature Extended periods of exponential growth in nature are rare. Growth more often occurs in spurts, when substrate becomes available, followed by an extended stationary phase, after the substrate is used up. Even where extended growth periods do occur, they are rarely exponential, but bear more resemblance to growth in the chemostat (see Section 9.6). For instance, the generation time of *Escherichia coli* in the intestinal tract is about 12 hours (two doublings per day) whereas in culture it grows much faster, with a generation time of 20 to

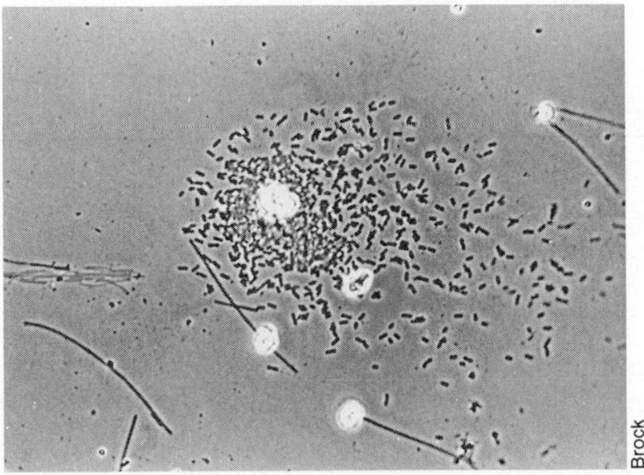

Figure 17.1 Bacterial microcolonies developing on a microscope slide immersed in a small river. The bright particles are mineral matter.

30 minutes (48 doublings per day). The marine bacterium *Leucothrix mucor*, an epiphyte of seaweeds, has a generation time of about 11 hours in nature and about 2 hours in culture. Some soil bacteria grow very much slower in nature than in the laboratory, and generation times as long as 1200 hours have been estimated in grassland soil. If these estimates are accurate it would mean that the bacteria are dividing only a few times per year.

However, all of these estimates for natural growth rates are averages, and it is possible (even likely) that growth rates are much faster for shorter periods of time, followed then by long periods when the organisms are not growing and are essentially dormant.

Plants and animals as microbial habitats Microorganisms inhabit the surfaces of higher organisms and in some cases actually live within plants and animals. Microorganisms frequently reach large numbers in such habitats and may benefit the plant or animal in a nutritionally significant way. The leaves and roots of plants may be heavily colonized by bacteria due to the excretion by plants of substances such as carbohydrates. Microbial populations in the root zone of the plant (rhizosphere, see Section 17.22) may be of benefit to the plant by producing vitamins, amino acids, and plant growth hormones, as well as removing toxic substances, such as H_2S.

Microorganisms inhabit both the surface of the skin and the intestinal tract of animals (see Sections 11.2 and 11.14). Large bacterial populations exist in the intestine and their metabolic processes may be absolutely essential to the livelihood of the animal. For example, without a rumen cellulolytic microflora, ruminant animals would not be able to digest cellulose (see Section 17.13 for a discussion of rumen microbiology). Intestinal microbial populations remain relatively constant over time in terms of both their numbers and species diversity, and this is a good indication that they are highly adapted to their habitat.

Many microbial associations with plants or animals clearly benefit both partners. Such is the case in the formation of root nodules in leguminous plants (see Section 17.24) and in the microbial activities of the rumen of ruminant animals (see Section 17.13); the highly beneficial processes of N_2 fixation and cellulose digestion, respectively, occur here. Leguminous plants and ruminants clearly benefit from their microbial associations, but the microorganisms benefit as well because they obtain a continuous supply of nutrients from the host. Such mutually beneficial associations, where microorganisms coexist with higher organisms to the benefit of both partners, are referred to as **symbioses** (singular, symbiosis). We will see many examples of microbial symbioses in this chapter, from casual, nonessential relationships, to extremely tight, highly interdependent associations. In essence, symbioses serve to maximize the talents of microorganisms and macroorganisms in order to improve the competitive fitness of each partner.

Microbial competition Living organisms respond not only to their environments but also to the presence and competitive influence of other microorganisms. Perhaps nowhere in the living world is competition more fierce than among the microorganisms. Population densities are high and generation times are short. Because of this, the microbe found in a given environment is probably the best adapted at that time to that environment.

Competition usually arises because both microorganisms are after some common resource which is present in limited amounts. In the simplest case, the outcome of a competitive interaction will depend on the relative growth rates of the two organisms; the organism which grows the faster will eventually replace the other. However, in many cases of competition, the activities of one organism *inhibit* the growth of the other organism. When one microorganism attacks another it is called *antagonism.* In the most common type of antagonism, one organism produces a substance, such as an *antibiotic*, which is active against the other organism. We discussed the nature and action of antibiotics in Sections 9.20 and 10.6. Antibiotic-producing microorganisms occur very frequently in natural habitats, especially in soil. Bacteria produce antibiotics active against other bacteria, as well as against fungi and other eucaryotic microorganisms. Further fungi produce antibiotics active against other fungi, as well as against bacteria. In some surveys, over 30 percent of all microbial cultures isolated from soils produced antibiotics.

Because of microbial competition, the diversity of microorganisms found in a given habitat is limited. However, because microbes live in very tiny microenvironments, two antagonist microorganisms can coexist in the same habitat.

There has been considerable interest in using the competitive activity of microorganisms in practical ways, such as inoculating a habitat with a harmless microorganism to prevent the development of a related harmful one. However, because microbes are so well adapted to their environments, it is very difficult to set up conditions so that an inoculated microbe will actually replace a native microbe. We will discuss later (see Section 17.22) an attempt to use microbial competition in agriculture.

17.2 Methods in Microbial Ecology

Microbial ecologists generally approach the study of microbial ecology in one of two ways. Either the study focuses on isolation, identification, and perhaps enumeration of the microbial populations present, or, alternatively, measures the *activity* of the microbial population directly in the natural habitat. The latter technique has several advantages. First and foremost it gives information on what the organisms are actually *doing* in their environment. Second, it allows the microbial ecologist to make perturbations in the environment and examine how these affect microbial activities. And third, by combining isolation and *in situ* experimentation techniques one can draw conclusions not only about what chemical reactions are occurring in the microbial habitat, but also about which organisms are responsible for the activities measured.

Microbial ecology cannot be carried out in the laboratory alone, although laboratory studies are frequently done in conjunction with field research. To do real microbial ecology one must go to the microbial habitat and measure the microbe's *activity* in its environment; *field studies* are the basis of microbial ecology. We thus begin here with an overview of enrichment and isolation approaches and proceed to discuss a number of methods currently in use for measuring microbial activities directly in natural environments.

17.3 Enrichment, Isolation and Counting Methods

Rarely does a natural environment contain only a single type of microorganism. In most cases, a variety of organisms are present and it is particularly challenging for the microbiologist to devise methods and procedures which will permit the isolation and culture of organisms of interest. The most common approach to this goal is the **enrichment culture technique**. In this method, a medium and set of incubation conditions are used that are *selective* for the desired organism and that are counterselective, or even inhibitory, for the undesired organisms. An overview of some successful enrichment culture procedures is given in Table 17.1.

Successful enrichment requires that an appropriate *inoculum* source be used. The dictum "Everything is everywhere, the environment selects" has some basis in fact, but it should be understood that a given handful of soil or mud will not automatically contain the organism sought. Thus, we begin an enrichment culture protocol by going to the appropriate habitat (see Table 17.1) and sampling to obtain an enrichment inoculum.

Enrichment cultures are frequently established by placing the inoculum directly into a highly selective medium; many common procaryotes can be isolated this way. For example, by pasteurizing the inoculum to kill vegetative cells, spore-forming bacilli can be isolated on starch agar plates. An anaerobic methylamine/nitrate enrichment is a nearly foolproof method for enriching for *Hyphomicrobium*. The nitrogen-fixing bacterium *Azotobacter* can be easily isolated aerobically in a glucose mineral salts medium lacking fixed forms of nitrogen (see Box), and so on. Literally hundreds of different enrichment approaches have been tried in the past and some of the more dependable ones are listed in Table 17.1. Both the culture medium *and* the general incubation conditions are listed because in devising enrichment culture protocols one should strive to optimize both parameters in order to have the best chance at obtaining the organism of interest.

The Winogradsky Column For isolation of purple and green phototrophic bacteria, the **Winogradsky Column** has traditionally been used. Named for the famous Russian enrichment culture microbiologist, Sergei Winogradsky, the column was devised by him in the 1880's to study soil microorganisms. The column is a miniature anaerobic ecosystem, and if properly established, can be an excellent long term source of all types of bacteria involved in nutrient cycling.

A Winogradsky Column can be prepared by filling a large glass cylinder about one-third full with organic-rich, preferably sulfide-containing, mud (Figure 17.2). Carbon substrates are first mixed with the mud, with the exact composition of the final product being left somewhat to the imagination. Organic ad-

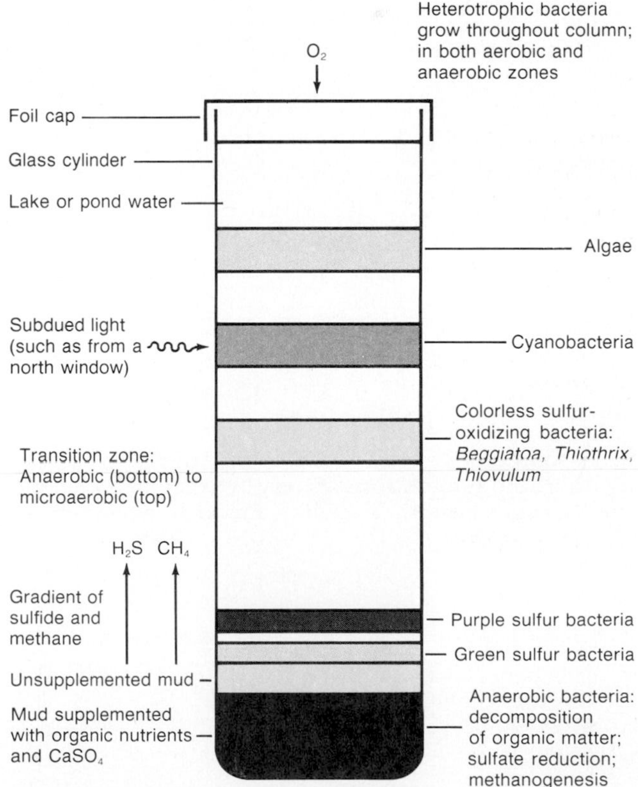

Figure 17.2 Assembly and chemistry of a Winogradsky column (idealized representation).

Table 17.1 Enrichment culture methods for bacteria

Light *Phototrophic bacteria: main C source, CO_2*

Aerobic incubation:		**Organisms enriched**	**Inoculum**
N_2 as N source		Cyanobacteria	Pond or lake water, sulfide-rich muds, stagnant water, moist decomposing leaf litter, moist soil exposed to light
Anaerobic incubation			
H_2 or organic acids; N_2 as sole nitrogen source		Nonsulfur purple bacteria	
H_2S as electron donor		Purple sulfur bacteria	
H_2S as electron donor		Green sulfur bacteria	

Dark *Lithotrophic bacteria: main C source, CO_2 (medium must lack organic C)*

Aerobic incubation **Electron donor**	**Electron acceptor**	**Organisms enriched**	**Inoculum**
NH_4^+	O_2	Nitrosofying bacteria (*Nitrosomonas*)	Soil, mud, sewage effluent
NO_2^-	O_2	Nitrifying bacteria (*Nitrobacter*)	
H_2	O_2	Hydrogen bacteria (various genera)	
H_2S, S^0, $S_2O_3^{2-}$	O_2	*Thiobacillus* spp.	
Fe^{2+}, low pH	O_2	*Thiobacillus ferrooxidans*	
Anaerobic incubation			
S^0, $S_2O_3^{2-}$	NO_3^-	*Thiobacillus denitrificans*	
H_2	NO_3^- + yeast extract	*Paracoccus denitrificans*	

Dark *Heterotrophic and methanogenic bacteria: main C source, organic compounds*

Aerobic incubation: respiration **Main ingredients**	**Electron acceptor**	**Organisms enriched**	**Inoculum**
Lactate + NH_4^+	O_2	*Pseudomonas fluorescens*	Soil, mud, decaying vegetation, lake sediments; pasteurize inoculum (80°C) for all *Bacillus* enrichments
Benzoate + NH_4^+	O_2	*Pseudomonas fluorescens*	
Starch + NH_4^+	O_2	*Bacillus polymyxa*, other *Bacillus* spp.	
Ethanol (4%) + 1% yeast extract pH 6.0	O_2	*Acetobacter, Gluconobacter*	
Urea (5%) + 1% yeast extract	O_2	*Sporosarcina ureae*	
Hydrocarbons (e.g., mineral oil) + NH_4^+	O_2	*Mycobacterium, Nocardia*	
Cellulose + NH_4^+	O_2	*Cytophaga, Sporocytophaga*	
Mannitol or benzoate, N_2 as N source	O_2	*Azotobacter*	
Anaerobic incubation: anaerobic respiration			
Main ingredients	**Electron acceptor**	**Organisms enriched**	**Inoculum**
Organic acids	KNO_3 (2%)	*Pseudomonas* (Denitrifying species)	Soil, mud
Yeast extract	KNO_3 (10%)	*Bacillus* (Denitrifying species)	
Organic acids	Na_2SO_4	*Desulfovibrio, Desulfotomaculum*	
Acetate, propionate, butyrate	Na_2SO_4	Type II sulfate reducers	
H_2	Na_2CO_3	Methanogenic bacteria (lithotrophic species only)	As above, or sewage digestor sludge, rumen contents
CH_3OH	Na_2CO_3	*Methanosarcina barkeri*	
CH_3NH_2	KNO_3	*Hyphomicrobium*	

Table 17.1 continued

Anaerobic incubation: fermentation Main ingredients		Organisms enriched	Inoculum
Glutamate or histidine	No exogenous electron acceptors added	*Clostridium tetanomorphum*	Mud, lake sediments, rotting plant material, dairy products (lactic and propionic acid bacteria), rumen or intestinal contents
Starch + NH_4^+		*Clostridium* spp.	
Starch, N_2 as N source		*Clostridium pasteurianum*	
Lactate + yeast extract		*Veillonella* spp.	
Glucose + NH_4^+		*Enterobacter*, other fermentative organisms	
Glucose + yeast extract (pH 5)		Lactic acid bacteria (*Lactobacillus*)	
Lactate + yeast extract		Propionic acid bacteria	

ditions that have been successfully used in the past include hay, shredded newsprint, sawdust, shredded leaves or roots, ground meat, hard boiled eggs, and even a dead mouse! The mud is also spiked with $CaCO_3$ and $CaSO_4$ to serve as a buffer and as a source of sulfate, respectively. The mud is packed tightly in the container, care being taken to avoid entrapping air. A small layer of unadulterated mud is then packed on top (Figure 17.2). The mud is then covered with lake, pond or ditch water and the top of the cylinder covered with aluminum foil. The cylinder is placed in a north window so as to receive adequate (but not excessive) sunlight and left to develop for a period of weeks (those in the Southern Hemisphere should use a south window).

In a typical Winogradsky Column, algae and cyanobacteria appear quickly in the upper portions of the water column and by producing O_2 help to keep this zone aerobic. Fermentative decomposition processes in the mud quickly lead to the production of organic acids, alcohols, and H_2, suitable substrates for sulfate-reducing bacteria. As a result of the production of sulfide, purple and green patches appear on the outer layers of the mud exposed to light, the purple patches, consisting of purple sulfur bacteria, frequently developing in the upper layers and the green patches, consisting of green sulfur bacteria, in the lower layers (this occurs because of differences in sulfide tolerance between green and purple bacteria, see Section 19.1). At the mud-water interface the water is frequently quite turbid and may be colored due to the growth of purple sulfur and purple nonsulfur bacteria (Figure 17.2). Sampling of the column is performed by inserting a long thin pipette (called a *Pasteur pipette*) into the column and removing a patch of colored material or a small amount of mud or water. These can be used to inoculate enrichment media.

Winogradsky columns have been used to enrich for a variety of procaryotes, both aerobes and anaerobes. The great advantage of a column, besides the ready availability of inocula for enrichment cultures, is that the column can be spiked with a particular compound whose degradation one wishes to study, and then allowed to select for an organism or organisms that can degrade it. In addition, since the Winogradsky column more nearly resembles the natural environment than does culture media, a variety of each physiological type of microorganism is more frequently observed in Winogradsky columns than in enrichment where a liquid culture medium is inoculated directly with a natural sample. In the latter approach, rapidly growing species frequently rise to the forefront leaving the slower growing species behind. Enrichment in the Winogradsky column followed by isolation in agar (see below) gives the enrichment culture microbiologist a fighting chance of obtaining the slower growing, but perhaps ecologically important species.

Chemostat enrichment cultures Enrichment cultures can also be carried out in the chemostat. It will be recalled (see Section 9.6) that the chemostat is a continuous culture device that allows for manipulation of both the population density and growth rate of the culture by controlling certain culture parameters. For use in enrichment culture, the chemostat is inoculated with a suitable inoculum source and run at a defined dilution rate. In time, the organism best adapted to the conditions used is favored. A major advantage of the chemostat as an enrichment device is that it allows the study of microbial competition at *low* substrate concentrations, similar to what organisms often encounter in nature. Competition for nutrients in the chemostat will select from the original mixed population the organism capable of growing *fastest* at the limiting substrate concentration and dilution rate chosen; other organisms are washed out because they are unable to compete.

The chemostat is thus a useful enrichment tool, especially for selecting for rapidly growing species. An excellent example of the use of the chemostat in enrichment and isolation is the recent isolation of a strain of *Pseudomonas cepacia* capable of degrading the herbicide 2,4,5-T. Although this compound is known to be extremely resistant to microbial attack (see Section 17.20), competition in the chemostat, where very low levels of 2,4,5-T could be supplied, eventually selected for an isolate capable of using the herbicide as a sole source of carbon and energy (see Figure 17.42). The chemostat is thus quite useful for isolating bacteria capable of degrading unusual substrates, especially substrates that are too toxic to be degraded at the concentrations normally supplied in batch cultures.

However, the organism enriched in a chemostat may not necessarily be the most ecologically important species. For example, slow growing species may be entirely missed in the chemostat even though they may play a key role in the overall chemical transformation(s) as they actually occur in nature. Also, in many natural habitats nutrients are not supplied at a constant rate; microbes in nature probably live a "feast or famine" existence. Specialist organisms which grow actively only during periods of large nutrient influxes may therefore be missed in chemostat enrichments as well, even though they may be the most ecologically relevant members of the microbial community.

From enrichments to pure cultures Frequently the objective of an enrichment culture study is to obtain a pure culture. *Pure* means free of foreign elements or of other living organisms. Hence a pure culture is a culture containing a *single kind* of microorganism. Any culture that contains more than one kind of microorganism is not a pure culture, and is, by definition, a *mixed* (or contaminated) *culture*. In examining a culture microscopically, it is almost impossible to detect a stray contaminant because the sensitivity of the light microscope is low. With the average 100× oil immersion lens, the field size is such that if the bacterial count is 10^6 cells/ml, there will be, on the average, only 1 cell per field. This means that if a contaminant numbered 10^4 cells per milliliter, one would have to examine 100 fields in order to find just one organism. Much more sensitive purity checks involve inoculation of a putative pure culture into a medium which favors growth of contaminants but not that of the culture of interest; growth in such media quickly indicates that the culture is not pure. On the other hand, absence of growth does not necessarily mean that contaminants are absent, since the culture medium used or incubation conditions chosen may not have been optimal for their growth. Thus critical determination that a culture is pure is not easy.

Pure cultures can be obtained in many ways, but the most frequently employed means are the streak plate, the agar shake, and liquid dilution methods. For organisms that grow well on agar plates, the **streak plate** is the method of choice (see Section 1.16 and Figure 1.43). By repeated picking and restreaking of a well-isolated colony, one usually obtains a pure culture which can then be transferred to a liquid medium. With proper incubation it is possible to purify both aerobes and anaerobes on agar plates via the streak plate method.

The *agar shake-tube* method involves the dilution of a mixed culture in tubes of molten agar, resulting in colonies embedded *in* the agar rather than on the surface of a plate. The shake tube method has been found useful for purifying particular types of microorganisms (for example, phototrophic sulfur bacteria and sulfate-reducing bacteria). For organisms that do not grow well in or on agar, purification can be obtained by successively diluting a cell suspension in tubes of liquid medium until a dilution is reached beyond which no growth is obtained. By repeating this *liquid dilution* procedure using the highest dilution showing growth as inoculum for a new set of dilutions, it is possible to obtain pure cultures, although the process frequently takes much longer than agar-based purification methods.

Enumerating microorganisms in nature A number of methods have been devised for counting microorganisms in pure culture, and these can easily be adapted for use on natural samples. For example, dilution and plate counts (see Section 9.3) can be used to enumerate bacteria from water or soil samples. Microorganisms can also be identified by direct microscopic examination of the habitat. However, special procedures are needed to make microorganisms visible in opaque habitats. In soil, for instance, organisms can be stained with any of a variety of different fluorescent dyes specific for living cells. *Acridine orange* is such a dye; this dye stains any organism containing DNA (that is, any cell that is alive). Although not foolproof, cells staining green with acridine orange are generally viable whereas cells staining orange are dead. Direct staining techniques are widely used for estimating total microbial numbers in soil and water.

The scanning electron microscope can also be used to visualize microbial cells on opaque soil surfaces. The scanning microscope offers much higher resolution and depth of field than any type of light microscope and is ideal for observing the various morphological forms present in a heterogenous sample such as soil (see Figure 17.14).

Fluorescent antibodies Fluorescent antibody staining is a highly specific method for identifying a single species (or even a single strain of a given species) in soil samples. We discussed the theory of fluorescent antibodies in Section 12.8. The great specificity of antibodies prepared against cell wall constituents of a particular organism can be exploited to identify that organism in a complex habitat such as soil (Figure 17.3). Fluorescent antibodies were used successfully to identify the causative agent of Legionnaires disease, *Legionella pneumophila* (see Section 15.2 and Figure 13.7*a*) in a variety of natural samples. Fluorescent antibodies are therefore most useful for studies involving the identification and tracking of a single species amidst a myriad of other species.

Use of nucleic acid probes A potentially very powerful approach to the assessment of the quanti-

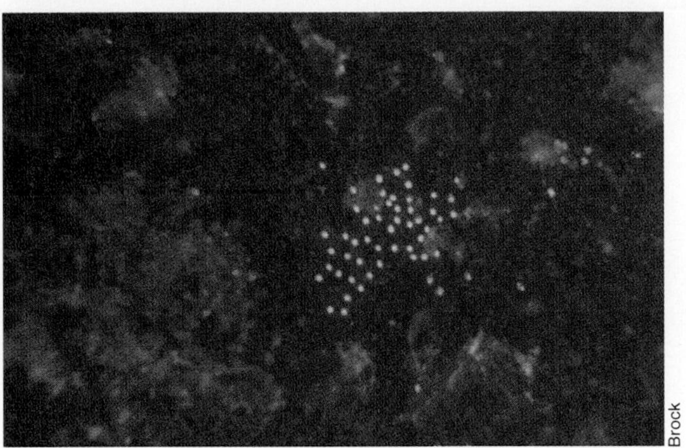

Figure 17.3 Visualization of bacterial microcolonies on the surface of soil particles by use of the fluorescent antibody technique. Notice some areas with bacteria localized on soil particles.

tative significance of specific microbes in nature is the use of nucleic acid probes. We have already discussed nucleic acid probe technology in some detail in Section 13.11 and the principles described there apply in the wider field of general microbiology ecology. (Clinical microbiology, as the student should readily recognize, is merely a specialized aspect of microbial ecology.)

The main problem with employing probe technology is developing a probe that discriminates the organism of interest from all other organisms in the environment. If the organism being studied is one which has already been inoculated into the environment, then it is likely that the researcher will know enough about the genetic structure of the organism to construct a suitable probe. Thus, probe technology is finding most use now in tracing the spread through the environment of organisms that are being released for agricultural or other practical reasons.

Problems with counting and identification methods In terms of their benefit to microbial ecology, all viable counting and identification methods are beset with two major problems. First, depending on the culture medium and incubation conditions chosen, only a *fraction* of the total microbial population will ever be counted. For example, media designed to count spore-forming bacteria could not possibly yield methanogenic bacteria, even though significant populations of both types of organisms may originally have been present in the sample. Also, any medium using organic materials would probably miss algal and cyanobacterial populations, while an autotrophic medium would completely miss heterotrophs. The point here is that no one culture medium can be expected to support growth of all physiological types of microorganisms that may be present in the habitat. Thus, "total count" data obtained from plate counts or other dilution methods must be interpreted in light of the cultural conditions chosen, for these methods yield quantitative information on only a fraction of the total microbial community.

Second, counting methods do not indicate the *activity* of the organisms in the natural habitat. Although the *numbers* of bacteria in a given habitat may be relatively high, they may be present in a dormant state and thus not be of significance in chemical transformations in the environment. Endospore-producing bacteria are good examples here. The isolation of large numbers of *Bacillus* cells by dilution and plating techniques may only indicate that the soil sample contained a large number of biochemically inert endospores. Dormancy among nonsporeformers is also a problem, although counts and ecological inpact are probably more closely correlated here than with spore-forming bacteria.

17.4 Measurements of Microbial Activity in Nature

It is relatively easy to perform counts of the numbers of microorganisms in natural environments, but as previously pointed out, counts do not necessarily reflect microbial activity. Several procedures are available which do indicate microbial activity. One of the most widely used is measurement of respiration, as either O_2 uptake or CO_2 production. A sample of soil or water is incubated in a closed container under simulated natural conditions, and changes in one of these gases is measured. Although adequate, this method is not very sensitive.

Measurement of ATP Where low microbial numbers or low activity exist, measurement of ATP levels is now widely used. All organisms produce ATP as a result of energy metabolism. As we saw in Section 4.14, ATP turns over rapidly in metabolizing cells and under starvation conditions cellular ATP levels dip to a low value. In ecological studies a sample of water or soil is treated to extract the ATP from the microbial cells, and the ATP level of the extract is measured. Sensitive methods are available for measuring ATP, the limit of detection being about 10^{-5} μg of ATP per liter of sample. The most popular method involves the measurement of light produced in the **luciferin/luciferase reaction**. Luciferin and luciferase are obtained from firefly lanterns and the amount of light produced when the enzyme, *luciferase*, acts on the substrate, *luciferin*, is proportional to the amount of ATP present. One μg of ATP is equivalent to about 250 μg of carbon in living organisms. Since ATP is lost very rapidly from dead or dormant organisms, ATP measurements essentially provide a measure of living biomass. ATP measurements have been made most frequently in the oceans, where microbial num-

The Rise of General Microbiology

Students of bacterial diversity owe a great debt to the Delft School of Microbiology, which was initiated by Martinus Beijerinck and continued by A. J. Kluyver and C. B. van Niel. Beijerinck, one of the world's greatest microbiologists (among other things, he was the first to characterize viruses), first devised the enrichment culture technique and used this technique in the isolation and characterization of a wide variety of bacteria. Subsequently, Kluyver and van Niel used the enrichment culture technique to isolate photosynthetic and lithotrophic bacteria, and to show the fascinating physiological diversity among the bacteria. Another important figure from the Dutch school was L. G. M. Baas-Becking, who carried out the first calculations of the energetics of photosynthetic and lithotrophic bacteria, and emphasized the importance of these organisms in geochemical processes. van Niel subsequently came to the United States, and was responsible for the training of a number of general bacteriologists who have carried on the Delft tradition, including the late R. Y. Stanier and M. Doudoroff, and Robert E. Hungate, who is still active at the University of California at Davis. In addition, a number of scientists visited van Niel's laboratory at Pacific Grove, California, in the 1950's and 1960's and learned his methods and enrichment approaches. In recent years, the Delft tradition has been carried on in its home country, The Netherlands, by Hans Veldkamp (at Groningen), and in Germany by Norbert Pfennig (at Konstanz), both of whom spent time in van Niel's laboratory.

An excellent example of the application of the enrichment culture technique can be found in the isolation of *Azotobacter*, a nitrogen-fixing bacterium discovered by Beijerinck in 1901 (see Section 19.16). Beijerinck was interested in knowing whether any aerobic bacteria capable of fixing nitrogen existed in the soil. The only other nitrogen-fixing organism known at that time, *Clostridium pasteurianum*, was an anaerobe discovered several years earlier by Sergei Winogradsky. To look for an aerobic nitrogen fixer, Beijerinck added a small amount of soil to an Erlenmeyer flask containing a thin layer of mineral salts medium devoid of ammonia, nitrate, or any other form of fixed nitrogen, and which contained mannitol as carbon source. Within 3 days a thin film developed on the surface of the liquid and the liquid became quite turbid. Beijerinck observed large, rod-shaped cells which appeared quite distinct from the spore-containing rods of *C. pasteurianum*. Beijerinck streaked agar plates containing phosphate and mannitol with the turbid liquid, incubated aerobically, and within 48 hours obtained large slimy colonies typical of *Azotobacter*. Pure cultures were obtained by picking and restreaking well-isolated colonies a number of times. Beijerinck assumed his new organism was using N_2 from air as its source of cell nitrogen and later proved this by showing total nitrogen increases in pure cultures of *Azotobacter* grown in the absence of fixed nitrogen.

The selective pressure of the enrichment approach used here should be evident. By omitting from his medium any source of combined nitrogen and incubating aerobically, Beijerinck placed severe constraints on the microbial population. Any organism that developed had to be able to both fix its own nitrogen and tolerate the presence of molecular oxygen. Beijerinck noted that if he placed too much liquid in his flasks, or employed more readily fermentable organic substrates, such as glucose or sucrose, in place of mannitol, his enrichment would frequently turn anaerobic and favor the growth of *Clostridium* rather than *Azotobacter*. The addition of ammonia or nitrate to the original enrichment never resulted in the isolation of *Azotobacter*, but only to a variety of non-nitrogen-fixing bacteria instead. Hence Beijerinck showed that *Azotobacter* has a strong selective advantage over other soil bacteria under a specific set of nutritional conditions, and in the process defined the key physiological properties of *Azotobacter*: aerobiosis and nitrogen fixation. Beijerinck also demonstrated that the composition of the growth medium as well as the incubation conditions employed were of paramount importance in the proper development of an enrichment culture.

bers are low and where very sensitive methods to detect microbial activity are needed; some representative data for ocean waters are given in Figure 17.4. It is obvious that the upper layers of ocean water where photosynthesis is occurring (the photic zone) is the most important from the standpoint of microbial ecology. This point will be emphasized in Section 17.8 where we consider microbial processes in the deep sea.

Use of radioisotopic tracers For the measurement of a specific microbial process in natural environments, **radioisotopes** are very useful. They provide extremely sensitive and highly specific means of measuring chemical processes (Figure 17.5). For instance, if photosynthesis is to be measured, the light-dependent uptake of $^{14}CO_2$ into microbial cells can be measured; if sulfate reduction is of interest, the rate of conversion of $^{35}SO_4^{=}$ to $H_2{}^{35}S$ can be assessed.

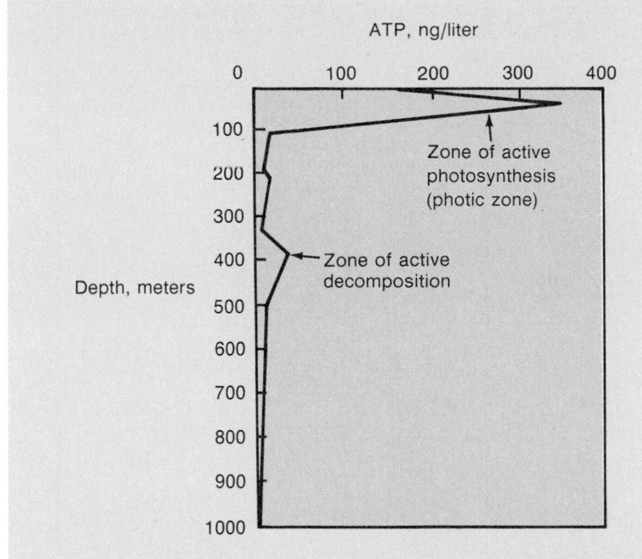

Figure 17.4 Deep sea distribution of ATP with depth in the ocean. Water samples were taken from the Pacific Ocean off the coast of California. Note the high ATP level in the surface layers where light permits growth of photosynthetic organisms, and another small peak of ATP caused primarily by bacteria growing on organic matter that accumulated at this depth.

Methanogenesis in natural environments can be studied by measuring the conversion of $^{14}CO_2$ to $^{14}CH_4$ in the presence of a suitable reductant like H_2, or by the conversion of methyl compounds like $^{14}CH_3OH$ or $^{14}CH_3NH_2$ to $^{14}CH_4$. Heterotrophic activity can be measured by following incorporation of ^{14}C-organic compounds. For example, the uptake of ^{14}C-glucose or ^{14}C-amino acids by heterotrophs in lake water can be studied by filtering the bacterial cells from suspension following an incubation period with the isotope. Alternatively, in soil one might measure the extent of $^{14}CO_2$ production from the added radioisotope.

Isotope methods are extremely valuable to the microbial ecologist and are probably the most widely used means of evaluating the activity of microbes in nature. However, because there is always the possibility that some transformation of a labeled compound might be due to a strictly chemical, rather than microbial process, it is essential when using isotopes to employ proper controls. The key control necessary, of course, is the ***killed cell control***. It is absolutely essential to show that the transformation being measured in nature is prevented by microbicidal agents or heat treatments that are known to block microbial action or kill the organisms. Formalin at a final concentration of 4 percent is frequently used as a chemical sterilant in microbial ecological studies.

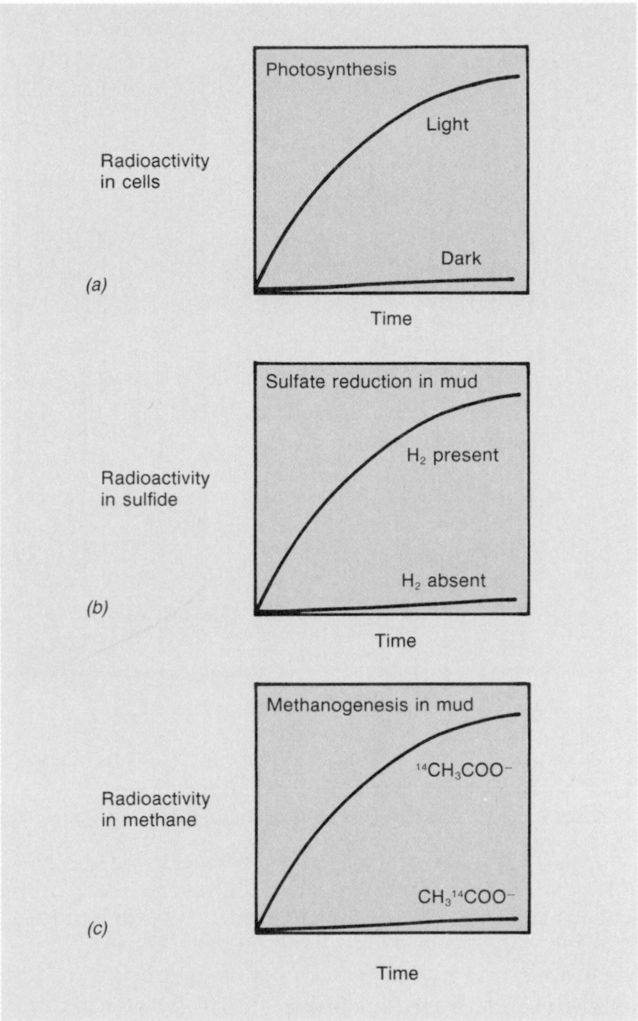

Figure 17.5 Use of radioisotopes to measure microbial activity in nature. (a) Photosynthesis measured in natural seawater with $^{14}CO_2$. (b) Sulfate reduction in mud measured with $^{35}SO_4{}^{2-}$. (c) Methanogenesis measured in mud with acetate labeled in either methyl ($^{14}CH_3COO^-$) or carboxyl ($CH_3{}^{14}COO^-$) carbon.

Microelectrodes We discussed earlier the fact that the environments of microorganisms are very tiny, that is, they are *microenvironments*. A number of microbial ecologists are making use of small glass electrodes, referred to as **microelectrodes**, to measure the activity of microorganisms in nature. Although limited to measurements of those chemical species that can be detected with an electrode, several types of electrodes have found use in field studies. Currently, the most popular microelectrodes are those that measure pH, oxygen, and sulfide.

As the name implies, microelectrodes are very small, the tips of the electrode ranging in diameter from 5–100 μm (Figure 17.6). The electrodes are carefully inserted into the material to be studied using a micromanipulator, a device that allows for precise movement of the electrode through distances of a millimeter or less. Microelectrodes have been used extensively in the study of chemical transformations and photosynthesis in microbial mats. The latter are layered microbial communities usually containing

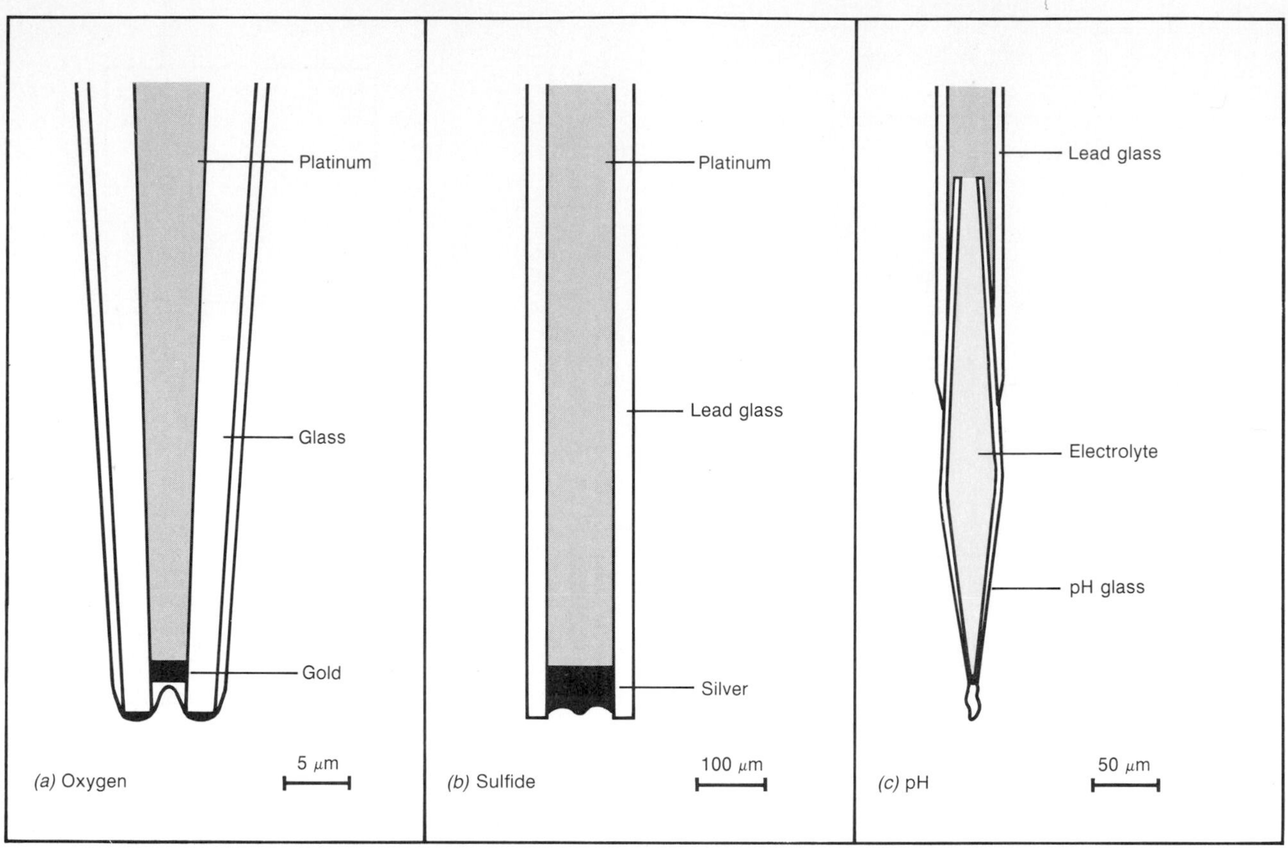

Figure 17.6 Schematic drawings of oxygen, sulfide, and pH microelectrodes. The principle of measurement of each electrode is the same as larger models even though the size of the electrode is greatly reduced.

cyanobacteria in the uppermost layer, photosynthetic bacteria in subsequent layers (until the mat becomes light-limited), and heterotrophic, especially sulfate-reducing, bacteria in the lower layers (Figure 17.7*a*). Microbial mats are dynamic systems where primary productivity occurring in the upper layers of the mat is balanced by decomposition from below. The metabolic activities of a microbial mat are analogous to those occurring in the Winogradsky column (see Section 17.3), except that in the mat the various autotrophic and heterotrophic processes are compressed into a much smaller area. The dense accumulations of biomass makes the microbial mat an ideal experimental subject. Large, actively metabolizing microbial populations allow for short term analyses of chemical transformations; in microbial ecology short term studies are frequently desirable because the ecosystem is disturbed for only a brief period.

Microbial mats are found in a variety of environments, especially in hot springs (Figure 17.7*a*) and shallow marine basins. As shown in a study of a microbial mat in Figure 17.7*b*, microelectrodes allow the microbial ecologist to make sequential measurements of a particular chemical parameter at ex-

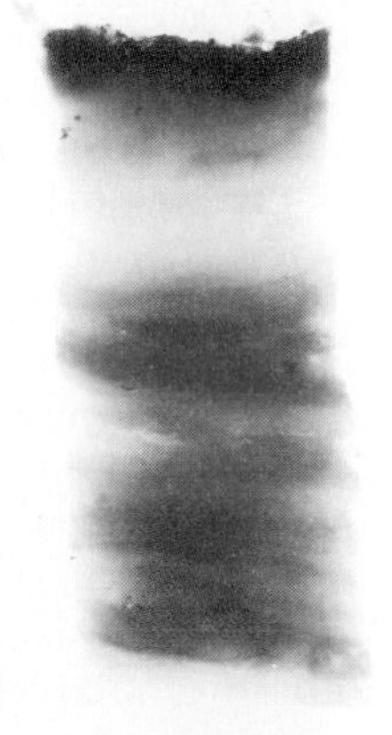

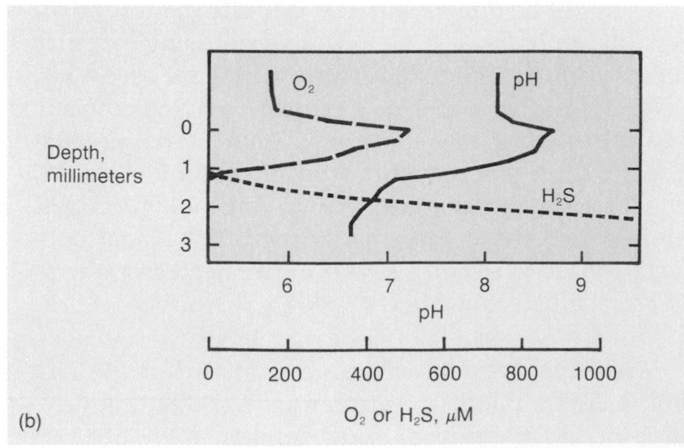

Figure 17.7 Microbial mats and the use of microelectrodes to study them. (a) Photograph of a core taken through the kind of hot spring microbial mat used in the experiment shown in *b*. Upper layer (dark green) contains cyanobacteria beneath which are several layers of photosynthetic bacteria (orange and yellow). The whole thickness of the mat is about 2 cm. (b) Oxygen, sulfide, and pH microprofiles in a hot spring microbial mat. Note the millimeter scale on the ordinate.

tremely fine intervals while passing through the mat. Microelectrodes can be immersed through a microbial mat in 200–300 μm increments, and have proven to be the only method of measuring the extremely sharp gradients of oxygen and sulfide that occur in these interesting microbial ecosystems.

Microelectrodes do have limitations, of course, a key one being their fragility; microelectrodes are not practical tools for the study of solid habitats such as dry soil. Another limitation is electrode technology; microelectrodes for several chemical species are not sufficiently sensitive to measure the concentration of the compound as it exists in nature. Hydrogen gas (H_2) is a good example of this. Although H_2 measurements in anaerobic ecosystems would greatly benefit ecological studies (see Section 17.12), currently available H_2 electrodes can only detect H_2 at levels far higher than those present in anaerobic ecosystems. Ammonia microelectrodes are available, but are only useful if the habitat to be studied is an alkaline one. Despite these limitations, technology is improving in the microelectrode area and the future looks bright for the eventual use of microelectrodes for measuring a variety of chemical species including dissolved organic compounds (such as acetate). Use of a bank of such microelectrodes could potentially define the major chemical transformations in a microbial habitat and thereby give a good picture of the dynamics of microbial interrelationships therein.

17.5 Stable Isotopes and Their Use in Microbial Biogeochemistry

We learned in Chapter 2 that different isotopes of most elements exist and that while certain isotopes are unstable and break down (due to radioactive decay, see above), others are stable and simply contain a different number of neutrons than the major isotopic form. In microbial ecology geochemists have teamed up with microbiologists to study biogeochemical transformations using ***stable isotopes***.

The two chemical elements which have proven most useful for stable isotope studies are carbon and sulfur. Carbon exists in nature primarily as ^{12}C. However, a small amount of carbon exists as ^{13}C. Likewise, sulfur exists primarily as ^{32}S, although some sulfur exists as ^{34}S. All of these are stable isotopes. The reason stable isotope measurements are useful in microbial ecology is that most biochemical processes favor the ***lighter*** isotope of either carbon or sulfur. That is, the heavier isotope is discriminated against when the elements are initially acted upon biochemically. This is particularly true of enzymes that bind CO_2 or $SO_4^{=}$. Thus, when CO_2 of known isotopic composition is fed to a photosynthetic organism, the cellular carbon becomes enriched in ^{12}C, while the CO_2 remaining in the medium becomes enriched in ^{13}C. Likewise, sulfide produced from the bacterial reduction of sulfate is much "lighter" than sulfide of strictly geochemical origin.

The isotope composition of any material is meaningless unless it is stated in relation to a standard. By international agreement the universal standard for carbon is limestone obtained from the fossil skeleton of a Cretaceous-age fish. The sulfur standard is a meteoritic FeS sample. Because isotopic fractionation occurs at the time of CO_2 fixation or $SO_4^{=}$ reduction, comparison of the stable carbon or sulfur isotope composition of a substance with the standard will indicate the likelihood of its original biological (as opposed to geochemical) origin. The ratio of $^{13}C/^{12}C$ or $^{34}S/^{32}S$ in a sample is determined in a mass spectrometer, an instrument capable of measuring the

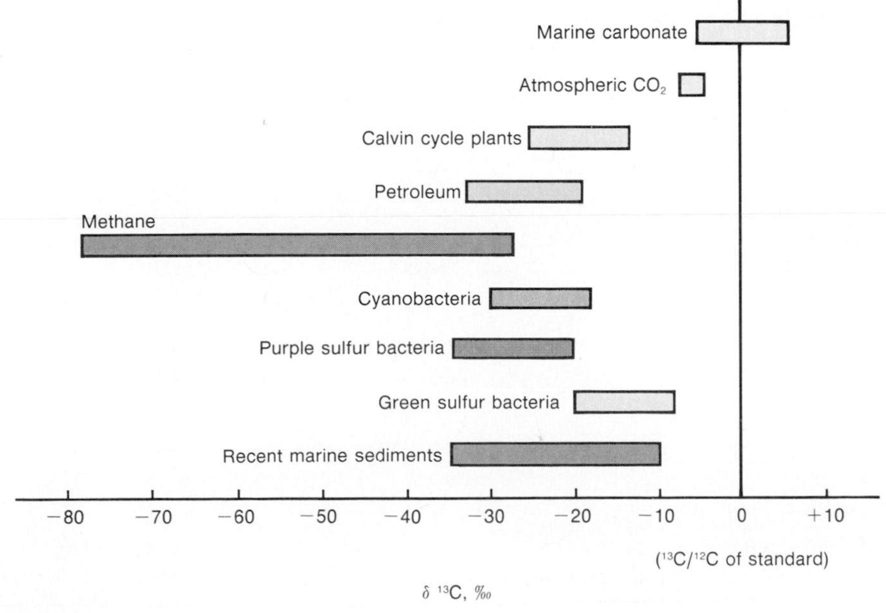

Figure 17.8 Carbon isotopic compositions of various substances. The values were calculated using the formula:

$$\frac{^{13}C/^{12}C \text{ sample} - ^{13}C/^{12}C \text{ standard}}{^{13}C/^{12}C \text{ standard}} \times 1000$$

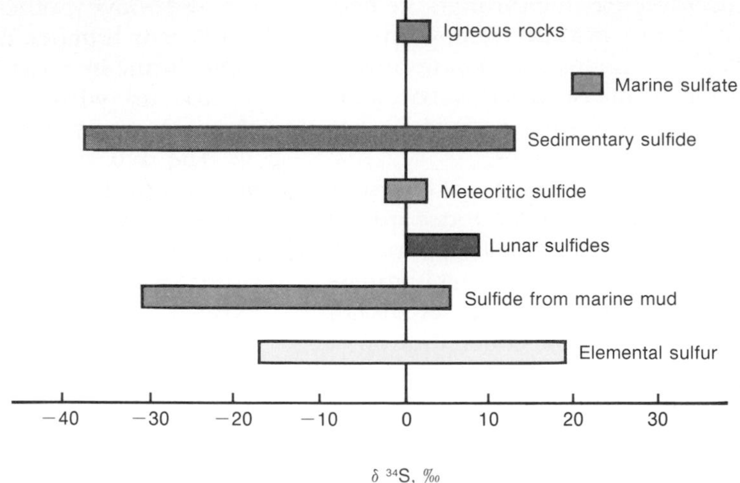

Figure 17.9 Summary of the isotope geochemistry of sulfur, indicating the range of values for ^{34}S and ^{32}S in various sulfur-containing substances. For the definition of $\delta^{34}S$ see the legend to Table 17.2. Note that sulfide and sulfur of biogenic origin tend to be depleted in ^{34}S.

proportion of each isotope in a sample.

The typical isotopic composition of different carbon samples relative to the limestone standard is shown in Figure 17.8. It is easy to see from these data that plant material and petroleum (which is derived from plant material), have similar isotopic compositions; carbon from both sources is isotopically lighter than the standard. Methane of biological origin can be extremely light, indicating that CO_2-reducing methanogens must fractionate drastically at the time of CO_2 binding (see Sections 17.12 and 19.34 for discussion of the biochemistry of methanogenesis). Marine carbonates, on the other hand, are clearly of geological origin, because they are much "heavier" than biogenic carbon (Figure 17.8). This large carbon reservoir has obviously not yet passed through photosynthetic organisms. Because of the vast differences in the proportion of ^{12}C and ^{13}C in carbon of biological and geological origins, the $^{13}C/^{12}C$ isotopic ratio of various geological strata can be used to detect the onset of living (in this case, autotrophic) processes.

The key role of sulfate-reducing bacteria in the formation of sulfur deposits is known from studies of the fractionation of $^{34}S/^{32}S$ (Figure 17.9). As compared to the meteoritic sulfur standard, sedimentary sulfide is highly enriched in ^{32}S. Nonbiogenic sulfide (as, for instance, from volcanic deposits) does not show this bias toward the lighter isotope (Figure 17.9). Also, biological oxidation of sulfide to sulfur or sulfate, either aerobically or anaerobically, shows a preference for the lighter isotope, although the fractionation is not nearly as great as that occurring during sulfate reduction (Table 17.2).

Stable isotope analyses have several uses in microbial ecology. Sulfur isotope analyses have been used to distinguish between biogenic and abiogenic

Table 17.2 Selective utilization of ^{32}S over ^{34}S (isotope fractionation) in certain parts of the microbiological sulfur cycle

Process	Starting substance	End product	Isotope fractionation*
Dissimilatory sulfate reduction (*Desulfovibrio*)	SO_4^{2-}	HS^-	−46.0
Dissimilatory sulfite reduction (*Desulfovibrio*)	SO_3^{2-}	HS^-	−14.3
Assimilatory sulfate reduction (*E. coli*)	SO_4^{2-}	Organic S	−2.8
Putrefaction and desulfurylation (*Proteus*)	Organic S	HS^-	−5.1
Lithotrophic sulfide oxidation (*Thiobacillus*)	HS^-	S^0	−2.5
	HS^-	SO_4^{2-}	−18.0
Phototrophic sulfide oxidation (*Chromatium*)	HS^-	S^0	−10.0
	HS^-	SO_4^{2-}	0

**Isotope fractionation is expressed as:*

$$\delta S^{34}\ {}^0/_{00} = \frac{S^{34}/S^{32}\ \text{sample} - S^{34}/S^{32}\ \text{standard}}{S^{34}/S^{32}\ \text{standard}} \times 1000$$

(This standard is an iron sulfide mineral from the Canyon Diablo Meteorite.)

ore (iron sulfides) and elemental sulfur deposits. Such information is extremely useful in interpreting the evolutionary record. Carbon isotopic analyses have been used to distinguish biogenic from nonbiogenic sediments, and oxygen analyses (using $^{18}O/^{16}O$ measurements) have been used to trace the earth's transition from an anaerobic to an aerobic environment. Stable isotopic analyses have also been used as evidence for the lack of living processes on the moon. For example, the data of Figure 17.9 show that the sulfide isotopic composition of lunar rocks closely approximates that of meteoritic sulfide and not that typical of biogenic sulfide. Recently, animal ecologists have used stable isotope analyses to study the food chains of aquatic insects. Reasoning that the isotopic composition of a given animal should approximate the isotopic composition of its major food source, it has been possible to trace the flow of carbon through an ecosystem in ways previously not possible.

17.6 Aquatic Habitats

Typical aquatic environments are the oceans, estuaries, salt marshes, lakes, ponds, rivers, and springs. Aquatic environments differ considerably in chemical and physical properties, and it is not surprising that their microbial species compositions also differ. The predominant photosynthetic organisms in most aquatic environments are microorganisms; in aerobic areas cyanobacteria and eucaryotic algae prevail, and in anaerobic areas photosynthetic bacteria are preponderant. Algae floating or suspended freely in the water are called **phytoplankton**; those attached to the bottom or sides are called **benthic algae**. Because these photosynthetic organisms utilize energy from light in the initial production of organic matter, they are called **primary producers**. In the final analysis, the biological activity of an aquatic ecosystem is dependent on the rate of primary production by the photosynthetic organisms. The activities of these organisms are in turn affected by the physical environment (for example, temperature, pH, and light) and by the kinds and concentrations of nutrients available. Open oceans are very low in primary productivity, whereas inshore ocean areas are high, with some lakes and springs being highest of all. The open ocean is infertile because the inorganic nutrients needed for algal growth are present only in low concentrations. The more fertile inshore ocean areas, on the other hand, receive extensive nutrient enrichment from rivers. There are, however, some open ocean areas that are rather fertile; these are places where winds or currents cause an extensive upwelling of deep ocean water, bringing to the surface nutrients from the bottom of the sea. It is because of such upwellings that areas off the coasts of California and Peru are so productive. The amount of economically important crops such as fish or shellfish is determined ultimately by the rate of primary production; lakes and inshore ocean areas are high in primary production and thus are the richest sources of fish and shellfish.

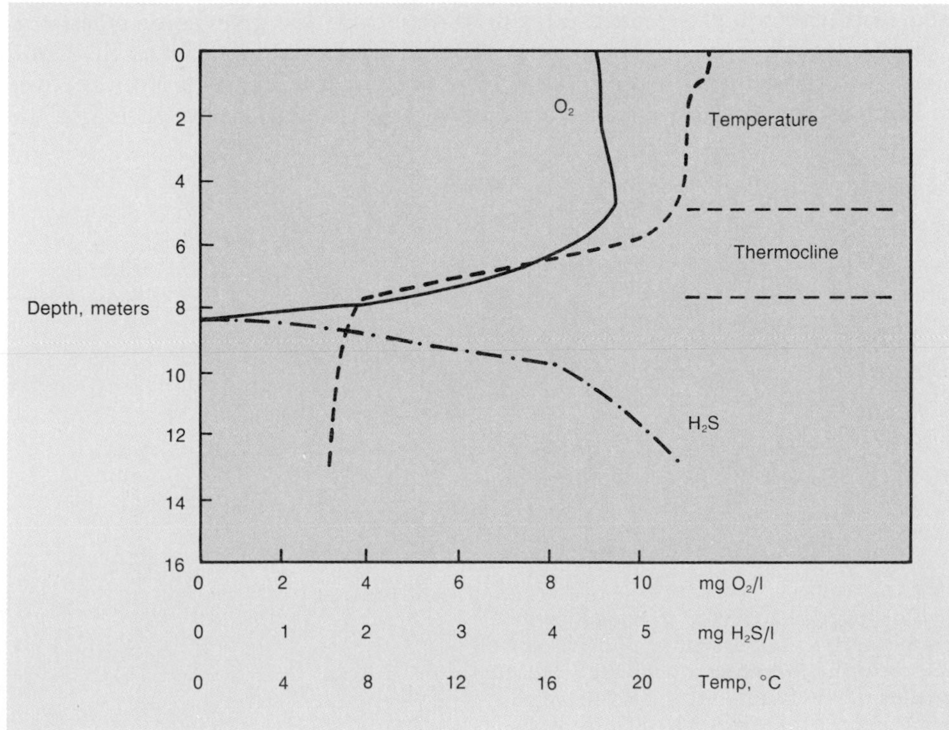

Figure 17.10 Development of anaerobic conditions in the depths of a temperate-climate lake as a result of summer stratification. The colder bottom waters are more dense and contain H_2S from bacterial sulfate reduction. The zone of rapid temperature change is referred to as the thermocline.

Oxygen relationships in lakes and rivers We discussed oxygen requirements and anaerobiosis in Section 9.14 and the production of oxygen via photosynthesis in Section 16.4. Although oxygen is one of the most plentiful gases in the atmosphere, it has limited solubility in water, and in a large water mass its exchange with the atmosphere is slow. Significant photosynthetic production of oxygen occurs only in the *surface layers* of a lake or ocean, where light is available. Organic matter that is not consumed in these surface layers sinks to the depths and is decomposed by facultative microorganisms, using oxygen dissolved in the water. Once the oxygen is consumed, the deep layers become anaerobic; here strictly aerobic organisms such as higher plants and animals cannot grow, and the bottom layers have a species composition restricted to anaerobic bacteria and a few kinds of microaerophilic animals. In addition, there is a conversion from a respiratory to a fermentative metabolism, with important consequences for the carbon cycle (see Sections 16.13–16.18 for a consideration of this transition).

Whether or not a body of water becomes depleted of oxygen depends on several factors. If organic matter is sparse, as it is in unproductive lakes or in the open ocean, there may be insufficient substrate available for heterotrophs to consume all the oxygen. Also important is how rapidly the water from the depths exchanges with surface water. Where strong currents or turbulence occurs, the water mass may be well mixed, and consequently oxygen may be transferred to the deeper layers. In many bodies of water in temperate climates, however, the water mass becomes stratified during the summer, with the warmer and less dense surface layers separated from the colder and denser bottom layers. After stratification sets in, usually in early summer, the bottom layers become anaerobic (Figure 17.10). In the late fall and early winter, the surface waters become colder and heavier than the bottom layers, and the water "turns over," leading to a reaeration of the bottom. Most lakes in temperate climates thus show an annual cycle in which the bottom layers of water pass from aerobic to anaerobic and back to aerobic.

Rivers The oxygen relations in a river are of particular interest, especially in regions where the river receives much organic matter in the form of sewage and industrial pollution. Even though the river may be well mixed because of rapid water flow and turbulence, the large amounts of added organic matter can lead to a marked oxygen deficit. This is illustrated in Figure 17.11. As the water moves away from the sewage outfall, organic matter is gradually consumed, and the oxygen content returns to normal. Oxygen depletion in a body of water is undesirable because most higher animals require O_2 and die under even very temporary anaerobic conditions. Further, conversion to anaerobic conditions results in the production by anaerobic bacteria of odoriferous compounds (for example, amines, H_2S, mercaptans, fatty acids), some of which are also toxic to higher organisms.

Biochemical oxygen demand Sanitary engineers term the oxygen-consuming property of a body of water its **biochemical oxygen demand** (BOD). The BOD is determined by taking a sample of water, aerating it well, placing it in a sealed bottle, incubating for a standard period of time (usually 5 days at 20°C), and determining the residual oxygen in the water at the end of incubation. Although it is a crude method, a BOD determination gives some measure of the amount of organic material in the water that could be oxidized by microorganisms. As a river recovers from contamination with an organic pollutant, the

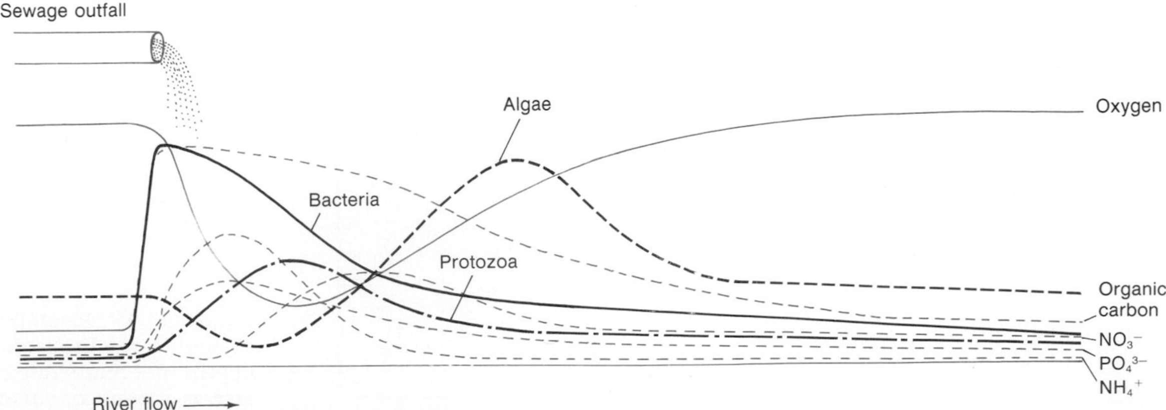

Figure 17.11 Chemical and biological changes in a river at various locations downstream from the source of a sewage outfall. Increased organic matter below the outfall leads to an increase in heterotrophic bacteria, a decrease in oxygen content, and an increase in NH_4^+. Farther downstream, NH_4^+ is oxidized to NO_3^- by nitrifying bacteria, and organic matter is oxidized by heterotrophs.

drop in BOD is accompanied by a corresponding increase in dissolved oxygen (Figure 17.11). Government regulatory agencies specify permissible BOD levels for effluents to be released in rivers or lakes.

We thus see that the oxygen and carbon cycles in a water body are greatly intertwined, and that heterotrophic microorganisms, mainly bacteria, play important roles in determining the biological nature and productivity of the body of water.

17.7 Terrestrial Environments

In the consideration of terrestrial environments, our attention inevitably turns to *soil* and *plants*, since it is within the soil and on or near plants that many of the key processes occur that influence the functioning of the ecosystem. The process of soil development involves complex interactions between the parent material (rock, sand, glacial drift, and so on), topography, climate, and living organisms. Soils can be divided into two broad groups—**mineral soils** and **organic soils**—depending on whether they derive initially from the weathering of rock and other inorganic material or from sedimentation in bogs and marshes. Our discussion will concentrate on mineral soils, the predominant soil in most areas.

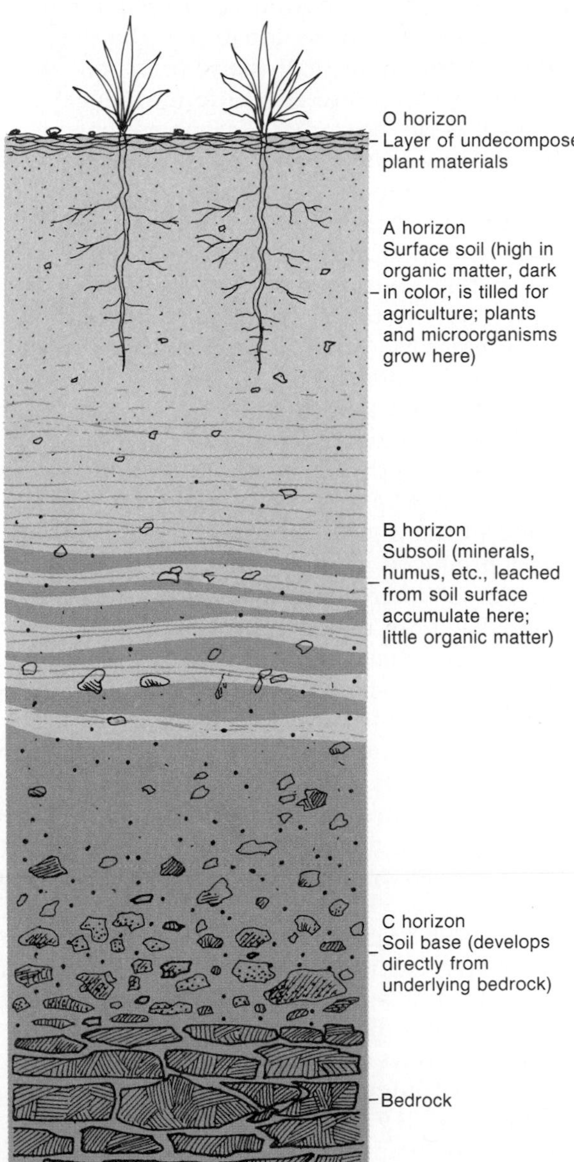

Figure 17.12 Profile of a mature soil. The soil "horizons" are soil zones as defined by soil scientists.

Soil formation Soils form as a result of combined physical, chemical, and biological processes. An examination of almost any exposed rock will reveal the presence of algae, lichens, or mosses. These organisms are able to remain dormant on the dry rock and then grow when moisture is present. They are photosynthetic and produce organic matter, which supports the growth of heterotrophic bacteria and fungi. The numbers of heterotrophs increase directly with the degree of plant cover of the rocks. Carbon dioxide produced during respiration by heterotrophs is converted into carbonic acid, which is an important agent in the dissolution of rocks, especially those composed of limestone. Many heterotrophs also excrete organic acids, which further promote the dissolution of rock into smaller particles. Freezing and thawing and other physical processes lead to the development of cracks in the rocks. In these crevices a raw soil forms, in which pioneering higher plants can develop. The plant roots penetrate farther into crevices and increase the fragmentation of the rock, and their excretions promote the development of a rhizosphere (the soil that surrounds plant roots) flora. When the plants die, their remains are added to the soil and serve as nutrients for an even more extensive

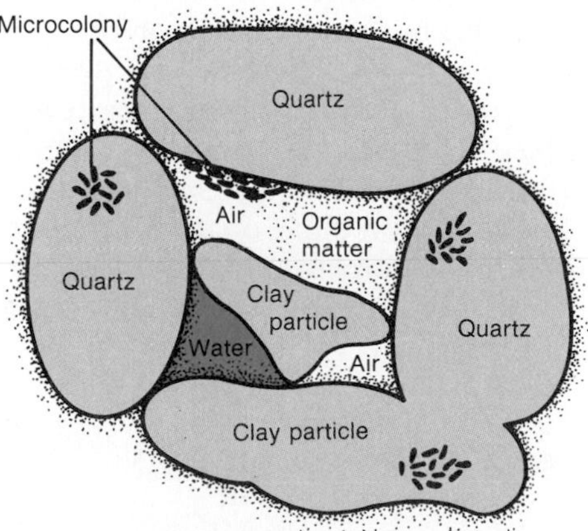

Figure 17.13 A soil aggregate composed of mineral and organic components, showing the localization of soil microorganisms. Very few microorganisms are found free in the soil solution; most of them occur as microcolonies attached to the soil particles.

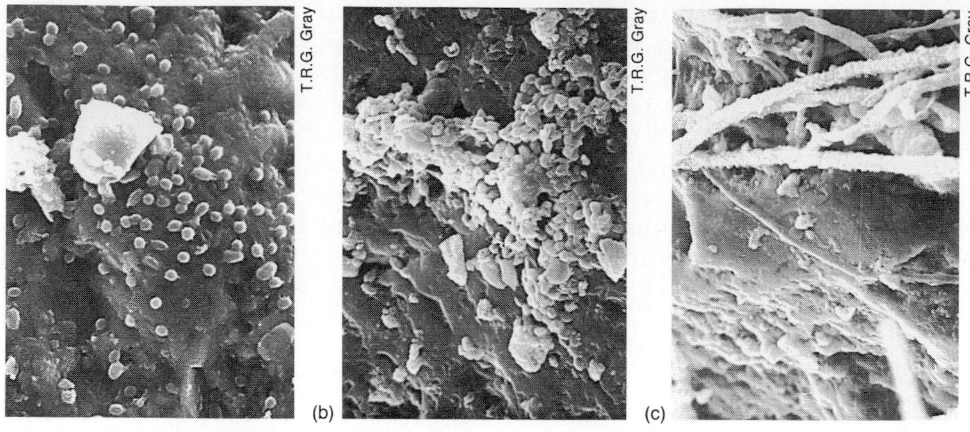

Figure 17.14 Visualization of microorganisms on the surface of soil particles by use of the scanning electron microscope. (a) Bacteria (magnification, 1950×). (b) Actinomycete spores (magnification, 1710×). (c) Fungus hyphae (magnification, 990×).

microbial development. Minerals are further rendered soluble, and as water percolates it carries some of these chemical substances deeper. As weathering proceeds, the soil increases in depth, thus permitting the development of larger plants and trees. Soil animals become established and play an important role in keeping the upper layers of the soil mixed and aerated. Eventually the movement of materials downward results in the formation of layers, and a typical soil profile becomes outlined (Figure 17.12). The rate of development of a typical soil profile depends on climatic and other factors, but it is usually very slow, taking hundreds of years. The picture given here is a general one, and marked variation among climates and geography exists.

Soil as a microbial habitat The most extensive microbial growth takes place on the surfaces of soil particles (Figure 17.13). To examine soil particles directly for microorganisms, fluorescence microscopes are often used, the organisms in the soil being stained with a dye that fluoresces (see Section 17.3). In effect each microbial cell is its own light source, and its shape and position on the surface of the particle can easily be seen. To observe a *specific* microorganism in a soil particle, **fluorescent-antibody staining** (see Sections 12.8 and 17.3, and Figure 17.3) can be used. Microorganisms can also be visualized excellently on such opaque surfaces as soil by means of the **scanning electron microscope** (Figure 17.14); the scanning electron microscope was discussed in Section 1.13.

17.8 Deep-Sea Microbiology

Since the oceans cover over three-quarters of the earth's surface, there is a natural interest among microbiologists in marine microbiology. However, because open ocean waters are relatively unproductive, microbial activity in these areas is not extensive. Nevertheless, marine microorganisms are of interest for several reasons, including their ability to grow at low-nutrient concentrations and cold temperatures and, in microorganisms inhabiting the deep sea, to withstand enormous hydrostatic pressures. We focus in this section on microbial life in the deep sea.

What is the deep sea? Oceanographers generally agree that visible light penetrates no further than about 300 meters in open ocean waters; this upper region is referred to as the **photic zone** (see Figure 17.4). Beneath the photic zone, down to a depth of about 1000 meters, considerable biological activity still occurs due to the action of animals and heterotrophic microorganisms. Water at depths greater than 1000 meters is, by comparison, biologically inactive, and has come to be known as the "deep sea." Greater than 75 percent of all ocean water is in the deep sea, primarily at depths between 1000–6000 meters.

Oceanic primary (photosynthetic) production is limited to the photic zone, and most of the organic matter produced is recycled within this zone by the activities of micro- and macroorganisms. It is estimated that only about 1 percent of the photosynthetically-produced organic matter reaches the ocean floor. This suggests that the deep sea is relatively nutrient poor and that it would not be expected to support abundant microbial life. With the exception of the thermal vents (see below) and regions in which significant organic matter is transported to the ocean floor by unusual mixing processes, this notion is correct; the majority of deep sea water is a biological wasteland.

For those organisms which inhabit the deep sea, three major environmental extremes must be reckoned with: low temperature, high pressure, and low nutrient levels. Below depths of about 100 meters ocean water stays a constant 2–3°C. We discussed the responses of microorganisms to changes in temperature in Section 9.11. As would be expected, bacteria isolated from depths below 100 meters are *psychrophilic*. Some are *extreme* psychrophiles, growing *only* in a narrow range near the *in situ* temperature (see below). Deep-sea microorganisms must also be able

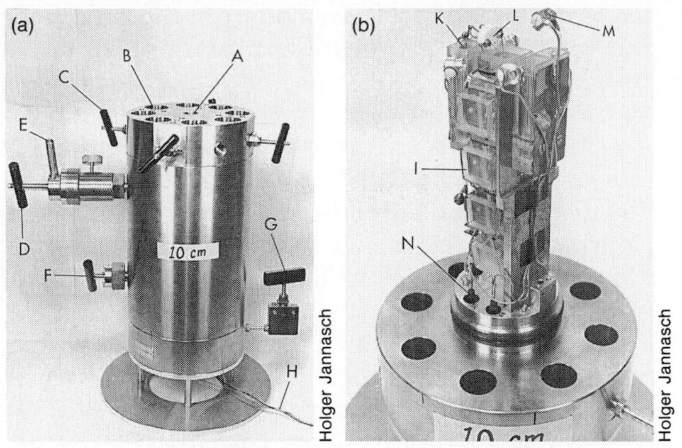

Figure 17.15 Photograph of pressurized sampling/culture device developed by microbiologists at Woods Hole Oceanographic Institution. (a) Isolation chamber assembled: A, viewing window; B, bolts; C, valves for sample transfer; D and E, handles to control streaking loop; F, handle for moving agar plates; G, pressurization valve; H, wires for sterilizing loop. (b) Chamber disassembled: I, storage bin for agar plates; K, storage bin for liquid vials; L, streaking loop; M, lamps for illumination; N, seals for wires.

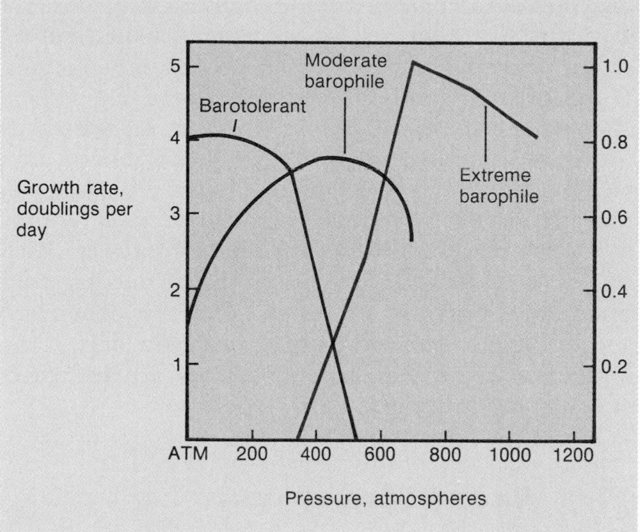

Figure 17.16 Growth of barotolerant, moderately barophilic and extremely barophilic bacteria. The extreme barophile was isolated from the Mariana Trench (10,500 meters). Note the much slower growth rate (at any pressure) of the extreme barophile (right ordinate) as compared to the barotolerant and moderately barophilic bacteria (left ordinate). Note also the inability of the extreme barophile to grow at low pressures.

to withstand the enormous hydrostatic pressures associated with great depths. As a rule of thumb, pressure increases by *one* atmosphere (14.7 psi) for every *10 meters* depth. Thus, an organism growing at a depth of 5000 meters must be able to withstand pressures of 500 atmosphere.

Barotolerant and barophilic bacteria Do deep-sea bacteria simply tolerate high pressure by virtue of compensatory physiological changes (that is, are they *barotolerant*), or, are they highly adapted to life at high hydrostatic pressures (*barophilic*)? To answer this question, it is necessary to sample, maintain, and culture organisms under the pressures they would experience in their natural habitat. Studies of bacterial growth in special pressurized culture vessels (Figure 17.15) showed that pure cultures of different heterotrophic bacteria derived from deep sea samples (up to about 4000 meters) and maintained at high pressure during all isolation and culture procedures were clearly only *barotolerant;* higher metabolic rates were observed at 1 atmosphere than at 400 atmospheres, although growth rates at the two pressures were about the same (Figure 17.16). However, none of the barotolerant isolates grew at pressures above 500 atmospheres. On the other hand, cultures derived from samples taken at greater depths, about 5000 meters, were found to be mildly *barophilic,* growing optimally at pressures of about 400 atmospheres, but still able to grow at 1 atmosphere (Figure 17.16).

Samples from even deeper oceanic areas (10,000 meters) have yielded cultures of **extreme (obligate) barophiles**. One strain studied in detail was found to grow fastest at a pressure of 700–800 atmospheres, and grew nearly as well at 1035 atmospheres, the pressure it was experiencing in its natural habitat (see Figure 17.16). The unique aspect of this extremely barophilic isolate was that it not only *tolerated* pressure, it actually *required* pressure for growth; the isolate would not grow at pressures of less than 500 atmospheres. Interestingly, however, this extreme barophile was not killed by decompression, because it could tolerate moderate periods of decompression; however, viability was lost if the culture was left for several hours in a decompressed state.

From microbiological studies of the deep sea conducted thus far it can be concluded that barotolerant bacteria inhabit ocean waters at depths of 2000–4000 meters. Below 4000 meters barophiles are observed. However, it is only at very great depths, below 7000 meters or so, that extreme barophiles are observed.

Barotolerant and barophilic bacteria are also cold-loving, that is, psychrophilic. This property appears to be more prevalent among extremely barophilic isolates. The extreme barophile described in Figure 17.16 was found to be quite sensitive to temperature; the optimal growth temperature was determined to be the environmental temperature of 2°C, and temperatures above 10°C significantly reduced viability.

Study of extreme barophiles may shed light on how pressure affects cellular physiology and biochemistry. In this connection it has been established that increased pressure decreases the binding capacity of enzymes for their substrates; presumably the enzymes of extreme barophiles must be folded in such a way as to minimize these pressure-related effects. Other potential pressure-sensitive cellular targets include protein synthesis and membrane phenomena such as transport. When shallow water bacteria are grown at increased pressure, many of the cells grow without dividing and form long filaments, suggesting that cell division processes are also affected by pressure. The rather slow growth rates of extreme barophiles (see Figure 17.16) are probably due to a combination of pressure effects on cellular biochemistry and to the fact that these organisms grow only at low temperatures, where reactions rates are decreased considerably to begin with.

17.9 Hydrothermal Vents

The general conception of the deep sea as a remote, low temperature, high pressure environment capable of supporting only the slow growth of barotolerant and barophilic bacteria is generally correct, but there are some amazing exceptions. A number of dense, thriving *invertebrate* communities supported by the activities of microorganisms exist clustered about thermal springs in deep waters throughout the world. Geophysical measurements have identified thermal springs in several locations on both the Atlantic and Pacific Ocean floors. Geologically, these springs are associated with so-called *ocean floor spreading centers*, regions where hot basalt and magma very near the sea floor cause the floor to slowly drift apart. Seawater seeping into these cracked regions mixes with hot minerals and is emitted from the springs (see Figure 17.17); because of their unique geophysical properties these springs have come to be known as **hydrothermal vents**.

From the standpoint of temperature, two major types of vents have been found. *Warm vents* emit hydrothermal fluid at temperatures of 6–23°C (into seawater at 2°C). *Hot vents* (usually referred to as "black smokers" because the mineral-rich hot water forms a dark cloud of precipitated material upon mixing with seawater; see below) emit hydrothermal fluid at 270–380°C (Figure 17.17). The flow rates of these two vent types are also characteristic; warm vents emit fluid at 0.5–2 cm/sec while hot vents have higher flow rates, about 1–2 meters/sec.

Using small pressurized submarines, researchers studied the organisms associated with the thermal vents. In close proximity to the thermal vents, thriving invertebrate communities have been found, with tube worms over two meters in length, and large numbers of giant clams and mussels (Figure 17.18). Ob-

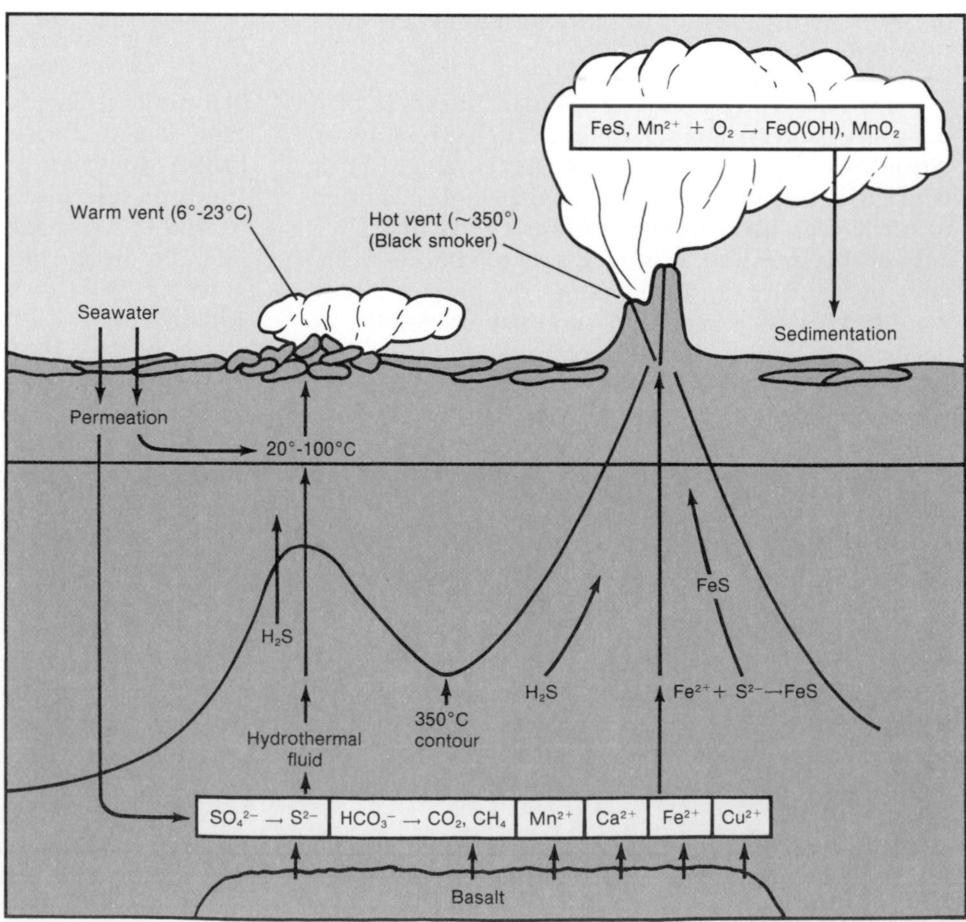

Figure 17.17 Schematic diagram showing the morphology and major chemical species occurring at warm vents and black smokers. At warm vents the hot hydrothermal fluid is cooled by cold seawater permeating the sediments. In hot vents, hot hydrothermal fluid reaches the sea floor directly.

Dudley Foster, Woods Hole Oceanographic Institution

James Agviar, Woods Hole Oceanographic Institution

Carl Wirsen, Woods Hole Oceanographic Institution

Figure 17.18 Invertebrates from habitats near deep-sea thermal vents. (a) Tube worms (family Pogonophora), showing the sheath (white) and plume (red) of the worm bodies. (b) Close-up photograph showing worm plume. (c) Mussel bed in vicinity of warm vent. Note yellow deposition of elemental sulfur in and around mussels.

serving this luxuriant animal diversity an immediate biological question comes to mind: How can such dense animal communities exist in the absence of photosynthetic primary producers? At a depth of 2600 meters hydrothermal vents remain in permanent darkness, thus the animals that live there cannot be grazing on indigenous algae or other photosynthetic organisms. In addition, the amount of detritus fallout from the photic zone is nowhere near that necessary to support such an extensive animal community. The energy source for vent ecosystems must thus be coming from the vents themselves.

What is the nature of the energy source(s)? Chemical analyses of hydrothermal fluid from the Galapagos vents show large amounts of H_2S, Mn^{2+}, H_2, CO, CH_4 and a variety of inorganic nutrients. Vents in the Guayamas Basin (Gulf of California) contain little H_2S, but do contain high levels of NH_4^+. Organic matter is not characteristic of the fluid emitted from any of the hydrothermal vents. From studies of vent chemistry it soon became clear that the animal communities were in some way dependent upon the activities of **lithotrophic bacteria** (see Section 16.7) growing at the expense of inorganic energy sources (H_2S, NH_4^+, H_2) emitted from the vents. Carbon dioxide, abundant in seawater as $CO_3^=$ and HCO_3^- was presumably being fixed into organic carbon by the lithotrophs, and the latter then formed the base of an extremely short food chain for hydrothermal vent animals. In summary, a unique, *light-independent* ecosystem had been found, driven by lithotrophic, rather than phototrophic, primary production.

Microorganisms in the hydrothermal vents Large numbers of sulfur-oxidizing lithotrophs such as *Thiobacillus, Thiomicrospira, Thiothrix*, and *Beggiatoa* (see Section 16.9) have been found in and around the vents. Samples collected from near the vents readily yielded cultures of these organisms, and *in situ* experiments showed CO_2 fixation and H_2S and $S_2O_3^=$ oxidation by natural populations of these bacteria. Other vents yielded nitrifying bacteria, hydrogen-oxidizing bacteria, iron and manganese-oxidizing bacteria, and methylotrophic bacteria, the latter presumably growing on the methane and carbon monoxide (CO) emitted from the vents (see Chapter 16 and Section 19.7 for discussion of some of these physiological groups). Table 17.3 summarizes the lithotrophic electron donors (and required electron acceptors) for lithotrophs suspected of playing a role in vent ecology. However, there is no direct evidence that the animals of the vents *eat* these lithotrophic bacteria. A different role of lithotrophs in animal nutrition is suggested in the next section.

Table 17.3 Lithotrophic bacteria of potential significance to hydrothermal vent primary productivity

Lithotroph	Electron donor	Electron acceptor
Sulfur-oxidizing bacteria	HS^-, S^0, $S_2O_3^=$	O_2, NO_3^-
Nitrifying bacteria	NH_4^+, NO_2^-, NO_2^-	O_2
Sulfur and sulfate-reducing bacteria	H_2	S^0, $SO_4^=$
Methanogenic bacteria	H_2	CO_2
Hydrogen-oxidizing bacteria	H_2	O_2, NO_3^-
Iron and manganese-oxidizing bacteria	Fe^{2+}, Mn^{2+}	O_2
Methylotrophic bacteria	CH_4, CO	O_2

See Sections 16.7–16.12 for a discussion of lithotrophs.

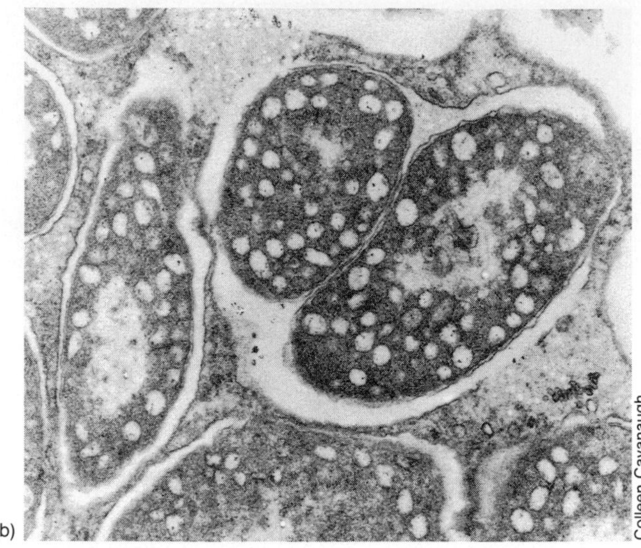

Figure 17.19 Lithotrophic sulfur-oxidizing bacteria associated with the trophosome tissue of tube worms from the hydrothermal vents. (a) Scanning electron microscopy of a trophosome tissue showing spherical lithotrophic sulfur-oxidizing bacteria. Magnification, 2000×. (b) Transmission electron micrograph of bacteria in trophosome tissue. The cells are frequently enclosed in pairs by an outer membrane of unknown origin.

Nutrition of animals living near hydrothermal vents In addition to the free-living lithotrophic bacteria, certain lithotrophs also live directly in association with animals of the thermal vents. Some hydrothermal vent animals are highly anatomically unusual. For instance, the two-meter long tube worms (see Figure 17.18) lack a mouth, gut, or anus, and contain a modified gastrointestinal tract consisting primarily of spongy tissue called the **trophosome** (Figure 17.19). The tube worm, *Riftia pachyptila*, exists anchored inside a protective outer tube called the *vestimentum* (see Figure 17.18*a*); the tube worms have thus been frequently referred to as "vestimentiferan" worms. The bright red plume (Figure 17.18*b*) is rich in blood vessels and serves as a trap for O_2 and H_2S (see below) for transport to the trophosome. Making up about 50 percent of the weight of the worm, trophosome tissue is loaded with sulfur granules, and microscopy of sectioned trophosome tissue shows large numbers of procaryotic cells (Figure 17.19), an average of 3.7×10^9 cells per gram of trophosome tissue. The large spherical cells observed in the trophosome resemble morphologically the marine sulfur-oxidizing bacterium *Thiovulum*. Trophosome tissue (containing these symbiotic sulfur bacteria) contains the enzyme *rhodanese*, a protein catalyst capable of oxidizing $S_2O_3^=$ to S^0 and $SO_3^=$. Also present are enzymes of the *Calvin cycle*, the pathway by which most autotrophic organisms fix CO_2 into cellular material (see Section 16.6).

Although the giant tube worm appears at first glance to be to an "autotrophic animal" it is in reality a heterotroph, living off the excretory products and dead cells of its lithotrophic (presumably autotrophic) symbionts. Similar conclusions can be reached concerning the nutrition of the giant clams and mussels (see Figure 17.18*c*), because rich sulfur oxidizing bacterial communities are found in the gill tissues of these animals as well. Use of nucleic acid sequencing techniques (see Box, Chapter 5) have shown that each vent animal harbors only one major species of bacterial symbiont but the species of symbiont varies among the different animal types.

Further study of the tube worms have shown that the animal contains unusual soluble hemoglobins that bind H_2S as well as O_2 and transports both substrates to the trophosome where they are released to the bacterial symbiont; this is apparently necessary to prevent H_2S from poisoning the animal. Furthermore, stable isotope analyses (see Section 17.5) of the elemental sulfur found within the bacterial symbionts have shown the $^{34}S/^{32}S$ isotope composition to be the same as the sulfide emitted from the vent. This ratio is distinctly different from that of seawater sulfate and serves as proof that geothermal sulfide is entering the worm.

The link between animal nutrition and other physiological groups of lithotrophs is possible, but thus far other associations have not been as well studied as the now well documented association between hydrothermal vent animals and sulfur-oxidizing lithotrophs. H_2-oxidizing and nitrifying bacteria have been isolated from tube worm trophosome tissue, but their nutritional significance to the animal is unknown. In certain vents, for example the ammonia-rich vents in the Guayamas Basin, vent animals may be supported

Robert D. Ballard

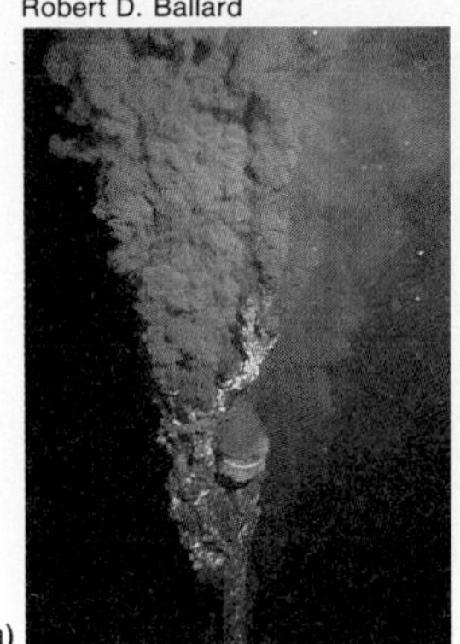

(a)

Dudley Foster, Woods Hole Oceanographic

(b)

Figure 17.20 Black smokers emitting sulfide- and mineral-rich water at temperatures of 350°C. In (a) the chimney is quite large, about one meter in length. In (b) the chimney is much smaller. Note the scientific equipment near the smoker in (b), giving a feeling for the relatively small size of the chimney.

by the activities of nitrifying bacteria. Methylotrophic symbionts may also play a nutritional role in hydrothermal vent animals, since methane-oxidizing bacteria (see Section 19.7) have been found which live in association with certain marine animals near natural gas seeps at relatively shallow depths in the Gulf of Mexico. Although not autotrophs, these symbionts clearly support growth of the animal, in this case by the oxidation of CH_4 as an energy source.

Other procaryotes, iron and manganese oxidizers for example, are probably not animal symbionts, but instead exist as free-living lithotrophic bacteria growing at the expense of reduced substances emitted from the vents. Nevertheless, these lithotrophs probably contribute something to overall primary productivity in the vent ecosystem, even if they are free-living forms. It is possible that free-living lithotrophs serve as a food source for fish frequently observed swimming near the vents.

Black smokers The great depths of the deep sea create huge hydrostatic pressures that affect the physical properties of water. At a depth of 2600 meters, water does not boil until it reaches a temperature of about 450°C. At certain vent sites superheated (but not boiling) hydrothermal fluid is emitted at a temperature of 270–380°C (see Figure 17.17) and could theoretically be a habitat for extremely thermophilic bacteria (see Section 9.11). The hydrothermal fluid emitted from black smokers contains abundant metal sulfides, especially iron sulfides, and cools quickly as it emerges into cold seawater. The precipitated metal sulfides form a tower referred to as a "chimney" about the source (Figure 17.20). Although it is very doubtful that bacteria actually live in the hot hydrothermal fluid for reasons discussed below, good evidence exists for the presence of thermophilic bacteria in the seawater/hydrothermal fluid *gradient* that forms as the hot water blends with cold ocean water. Reports of thermophilic lithotrophs and heterotrophs isolated from water estimated to be 120–150°C have been made, and sulfate- or sulfur-reducing bacteria capable of growth at 120°C (in culture) exist in the waters surrounding the chimneys. However, the prospects of finding "super thermophiles," organisms growing above 200°C, seem very unlikely as measurements of the stability of the monomeric constituents of informational macromolecules, such as amino acids and nucleotides, indicate that they are destroyed at temperatures above about 150°C (see Section 19.35). The best guess for the upper temperature of living organisms, therefore, seems to be 130–150°C; above this temperature life as we know it probably does not exist because of the lability of the basic building blocks of life.

Evolution of hydrothermal vent communities Hydrothermal vents have probably existed for over one billion years. Vent animal communities are also thought to be quite old; zoologists conclude that invertebrates have inhabited vent regions for the past 200 million years or so. The relatively constant, remote environment of the deep sea has allowed these animal communities to evolve in the absence of significant outside influences. Thus, although the animals resemble other marine invertebrates, the giant tube worms, clams, and mussels are sufficiently unique that zoologists have created new genera to accommodate them. Probably the most exciting aspect of vent ecology, however, is the fact that the entire ecosystem is supported by *nonphotosynthetic* primary production. Without lithotrophs the vent animal communities would not exist, just as we would not exist if it were not for sunlight driving the primary production we exploit as a food source. The abundant animal life surrounding hydrothermal vents is ample proof that primary production, regardless of how it is driven, is the foundation of all living systems.

The discovery of symbiotic lithotrophic bacteria in hydrothermal vent animal tissues has also awakened interest in the possibility that certain *surface* animal communities may use such a system as a food supplement or even as a primary food source. Small invertebrates including clams are frequently found on the surface of sulfide-rich intertidal muds. In the case of the gutless clam *Solemya*, bacterial symbionts, thought to be sulfur-oxidizing lithotrophs, were recently discovered within the animal's gill tissue. Naturally it was presumed that the bacteria played a role similar to that in the hydrothermal vent tube worms. Surprisingly, however, analyses of animal tissue have shown that sulfide can actually be oxidized by mitochondria in cells of clam gill tissue and that sulfide oxidation is coupled to the production of ATP. In essence, *Solymea* is a lithotrophic animal! The sulfur-oxidizing bacterial symbionts of *Solymea* probably serve to oxidize sulfide only when levels get too high. The sulfide-oxidizing mitochondrial system is rather sensitive to sulfide, but nevertheless, *Solymea* represents the first documented incidence of lithotrophy in animals. Perhaps lithotrophic reactions by other animals and by procaryotic animal symbionts are a more important source of energy than previously suspected. Indeed lithotrophy apparently is not strictly a procaryotic feat. It will be interesting to examine other animals known to tolerate sulfide-rich environments to see whether their tissues have also evolved lithotrophic mechanisms.

17.10 Global Biogeochemical Cycles

In the previous section we discussed the processes carried out by microorganisms at the bottom of the sea. In the present section, we consider the role of microorganisms in chemical processes on a global scale. The redox cycles for various elements, catalyzed to a greater or lesser extent by microorganisms, are intimately related to the transfer of elements from place to place on earth. Despite the enormous size of the earth, a considerable amount of information is available about cycling of elements on a global scale, and an understanding of these events is important in predicting long-term consequences of human perturbations of these cycles. To a first approximation, the earth can be divided into a series of compartments: ***atmosphere, land surface, oceans, and earth's crust***. Global cycling can then be expressed as the rate of movement of an element from one compartment to another (Figure 17.21). Over the short run, the major movement of an element from one place to another on earth occurs either as a gas in the atmosphere or as suspended or dissolved forms in water. Once a gas reaches the atmosphere, it can be carried with the general atmospheric circulation (winds, air currents, vertical circulation) around the earth.

This circulation within the atmosphere is relatively rapid, and microbial production of gases provides an important means of speeding up global cycling of the elements carbon, nitrogen, and sulfur. Movement in water is slower but still relatively quick. Elements are carried by streams and rivers to the sea, and there is a well-defined system of horizontal ocean currents, downwellings, and upwellings, which carry materials throughout the earth. Once an element sediments to the bottom and becomes incorporated into ocean sediments, its movement is much slower. Eventually it may become converted into rock (sedimentary rocks) and may then only reach the surface of the earth again when tectonic forces (that is, the forces of the earth's crustal structure) cause uplift and mountain building. As a result of rock weathering and soil erosion, elements are now being transferred from land to the oceans that had been deposited in oceanic sediments many millions of years ago.

Each box in Figure 17.21 can be considered to be a reservoir or compartment within which an element is stored for a definite period of time. The rate of movement from one compartment to another, called the ***flux***, provides a quantitative expression of the intensity of the biogeochemical cycle. The length of time an element remains in a compartment is expressed as its **residence time** which is defined as the amount of an element in a compartment at any given time divided by the rate of addition (or subtraction) of the element. Some components show short residence times in the atmosphere, whereas others have long residence times.

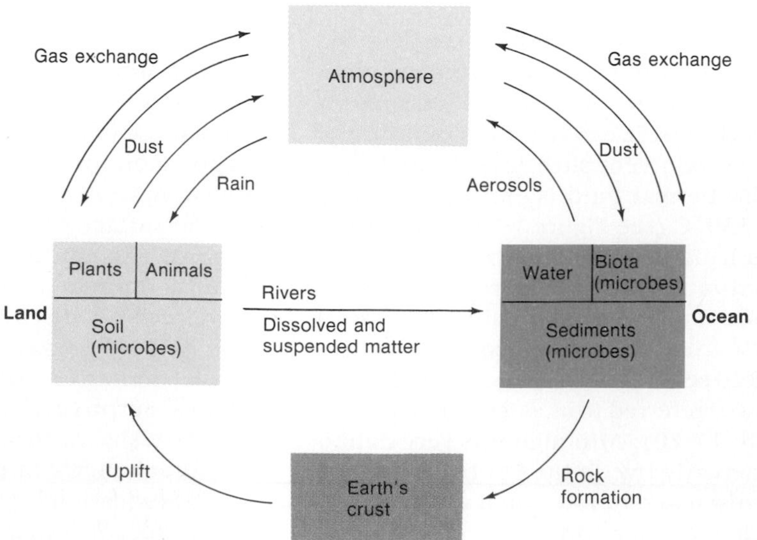

Figure 17.21 Global cycling of elements. Some of the major compartments are indicated.

17.11 Biogeochemical Cycles: Carbon

We discussed carbon metabolism and carbon transformations in some detail in Chapter 16. The *carbon* cycle is central to a discussion of all other biogeochemical cycles, because other cycles are driven by energy derived from photosynthesis or the breakdown of organic materials. A brief overview of the redox cycle for carbon is given in Figure 17.22. This shows that three oxidation states of carbon are of main significance, CH_4 (methane), the most reduced; $(CH_2O)_n$, a general formula for protoplasm, which is approximately at the oxidation level of carbohydrate; and CO_2 (carbon dioxide), the most oxidized form of carbon. Another substance, carbon monoxide, CO, oxidation state +2, is a minor component of the carbon cycle.

To appreciate the significance of the various steps in the carbon cycle, it is necessary to consider the compartment sizes and residence times for the various components on a global scale. These are summarized in Table 17.4.

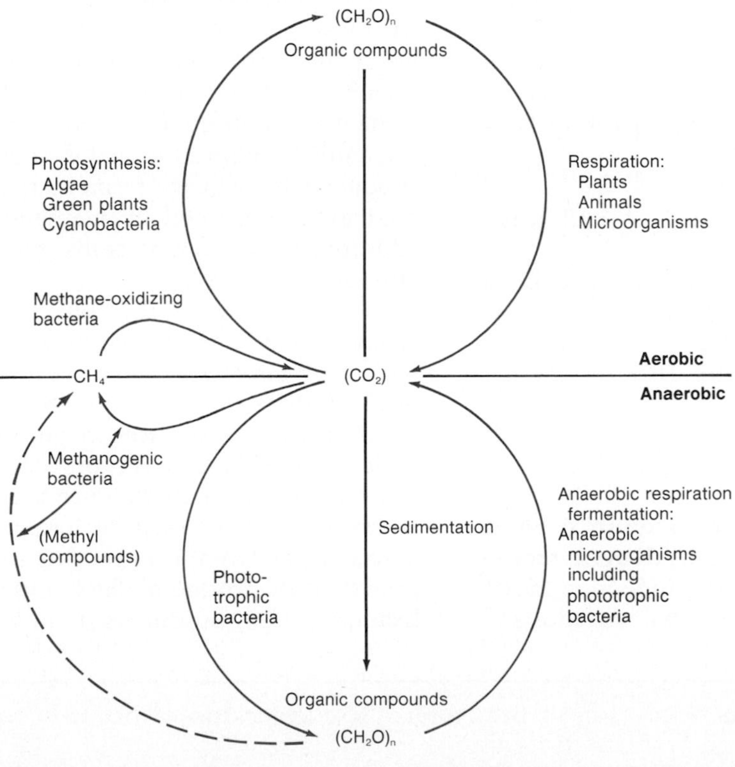

Figure 17.22 Redox cycle for carbon.

Table 17.4 Carbon reservoirs and residence times

Component	Reservoir (grams)	Residence time (years)	Main removal process
Atmosphere:			
CH_4 (1.6 ppm*)	$3\text{–}6 \times 10^{15}$	3.6	Photochemical oxidation in atmosphere
CO (0.1 ppm)	0.3×10^{15}	0.1	Photochemical oxidation in atmosphere
CO_2 (320 ppm)	670×10^{15}	4	Plant photosynthesis
Land:			
Living organic (mainly plants)	$500\text{–}800 \times 10^{15}$	16 (plants)	Death; grazing; predation
Dead organic	$700\text{–}1200 \times 10^{15}$	40	Microbial decomposition to CO_2
Oceans:			
Living organic	6.9×10^{15}	0.14	Death; grazing; predation
Dead organic	760×10^{15}	19	Microbial decomposition
Inorganic	$40{,}000 \times 10^{15}$	100,000	Formation of carbonate rocks; CO_2 exchange with the atmosphere
Sediments and rocks	$72{,}500{,}000 \times 10^{15}$	300×10^6	Weathering; fossil fuel burning

**Parts per million (μg gas/cm³ of air).*

17.12 Methane and Methanogenesis

Although methane (CH_4) is a relatively minor component of the global carbon cycle (Table 17.4), it is of great importance in many localized situations. In addition, it is of considerable microbiological interest because it is a product of anaerobic microbial metabolism. Methane production is carried out by a highly specialized group of organisms, the *methanogenic bacteria*, which are obligate anaerobes. We discussed methane formation as a type of anaerobic respiration in Section 16.16. Most methanogenic bacteria (or **methanogens**, for short) use CO_2 as their terminal electron acceptor in anaerobic respiration, converting it to methane; the electron donor used in this process is generally hydrogen, H_2. The overall reaction of **methanogenesis** in this pathway is as follows:

$$4H_2 + CO_2 \rightarrow CH_4 + 2H_2O \quad \Delta G^{0\prime} = -32 \text{ kcal}$$

which shows that CO_2 reduction to methane is an eight-electron process (four H_2 molecules). A few other substrates can be converted to methane, including methanol, CH_3OH; formate, $HCOO^-$; methyl mercaptan, CH_3SH; acetate, CH_3COO^-; and methylamines (see Section 19.34).

Anaerobic decomposition of organic carbon and the concept of interspecies hydrogen transfer We discussed this concept briefly in Section 16.18 during our discussion of fermentative metabolism. Here we present the significance for the whole problem of the anaerobic carbon cycle. Large-molecular-weight substances such as polysaccharides, proteins, and fats, are converted to CH_4 by the cooperative interaction of several physiological groups of bacteria. In many anaerobic environments the immediate precursors of CH_4 are H_2 and CO_2, these substrates being generated by the activities of fermentative anaerobes. For the conversion of a typical polysaccharide such as cellulose to methane as many as five major physiological groups of bacteria may be involved in the overall process (Figure 16.33 and Table 17.5). *Cellulolytic bacteria* cleave the high-molecular-weight cellulose molecule into cellobiose (glucose → glucose) and into free glucose. Glucose is then fermented by *fermentative anaerobes* to a variety of fermentation products, acetate, propionate, butyrate, H_2, and CO_2 being the major ones observed. Any H_2 produced in primary fermentative processes is immediately consumed by methanogenic bacteria. In addition, acetate can be converted to methane by certain methanogens.

Key organisms in the conversion of complex materials to methane are the *H_2-producing fatty acid-oxidizing bacteria*. These organisms use fatty acids or alcohols as energy sources but grow poorly or not at all on these substrates in pure culture. However, in association with a H_2-consuming organism (such as a methanogen or a sulfate-reducing bacterium), the H_2-producing bacteria grow luxuriantly. As explained below, H_2 consumption is critically important to the growth of H_2-producing fatty acid-oxidizing bacteria. Examples of H_2 producers include *Syntrophomonas*

Table 17.5 Major reactions occurring in the anaerobic conversion of organic compounds to methane

Reaction type	Reaction	Free energy change (Kcal/reaction) ΔG^{0a}	ΔG^{b}
Fermentation of glucose to acetate, H_2 and CO_2	Glucose + $4H_2O \rightarrow 2CH_3COO^- + 2HCO_3^- + 4H^+ + 4H_2$	−49.4	−76.2
Fermentation of glucose to butyrate, CO_2, and H_2	Glucose + $2H_2O \rightarrow C_4H_7O_2^- + 2HCO_3^- + 2H_2 + 3H^+$	−32.2	−67.8
Fermentation of butyrate to acetate and H_2	Butyrate + $2H_2O \rightarrow 2CH_3COO^- + H^+ + H_2$	+11.5	−4.2
Fermentation of propionate to acetate, CO_2 and H_2	Propionate + $3H_2O \rightarrow CH_3COO^- + HCO_3^- + H^+ + H_2$	+18.2	−1.3
Methanogenesis from $H_2 + CO_2$	$4H_2 + HCO_3^- + H^+ \rightarrow CH_4 + 3H_2O$	−32.5	−7.6
Methanogenesis from acetate	Acetate + $H_2O \rightarrow CH_4 + HCO_3^- + H^+$	−7.4	−5.9
Acetogenesis from $H_2 + CO_2$	$4H_2 + 2HCO_3^- + H^+ \rightarrow CH_3COO^- + 2H_2O$	−25.0	−1.7

Data taken from Zinder S. 1984. American Society for Microbiology News 50:294–298.

[a] *Standard conditions: solutes, 1 molar; gases, 1 atmosphere.*

[b] *Concentrations of reactants in typical anaerobic ecosystem: fatty acids, 1 mM; HCO_3^-, 20 mM; glucose, 10 μM; CH_4, 0.6 atm; H_2, 10^{-4} atm.*

and *Syntrophobacter**. *Syntrophomonas* oxidizes C_4 to C_7 fatty acids yielding acetate, CO_2 (if the fatty acid contains an odd number of carbon atoms), and H_2 (from the reduction of protons, see Table 17.5). *Syntrophobacter* specializes in propionate oxidation and generates acetate, CO_2, and H_2 (see Table 17.5). Although at first not appearing to be unusual reactions, these fatty acid conversions when written with all reactants at standard conditions (solutes, 1 M; gases, 1 atm) yield free energy changes that are *positive* in sign (see Table 17.5). That is, the ΔG^0 associated with these conversions are such that the reactions do *not* occur with the release of free energy.

Without energy release, how can these reactions support growth of the fatty acid-oxidizing bacteria? If growth of the H_2-producers occurs when H_2 is removed, the anaerobic oxidation of fatty acids to H_2 must somehow be coupled to ATP production. This implies that the presence or absence of H_2 affects the energetics of the reaction. A brief review of the principles of free energy given in Appendix 1 indicates that the *actual* concentration of reactants and products in a given reaction can drastically change the available free energy of a reaction. In a natural situation, reactants and products of fatty acid oxidation, acetate and H_2, are consumed by methanogenic (or sulfate-reducing) bacteria; measurements of H_2 in actively methanogenic ecosystems are generally below 10^{-3} atmospheres. As shown in Table 17.5, if the concentration of H_2 is kept very low, the free energy change associated with the oxidation of fatty acids to acetate plus H_2 by H_2 producing fatty acid-oxidizers now becomes *negative* in sign and sufficient in magnitude to drive the synthesis of ATP.

Hydrogen-consuming *acetogenic bacteria* such as *Acetobacterium* (see Section 16.16 and 19.9) assist in anaerobic digestion by consuming H_2 and supplying additional acetate for methanogenesis. However, the terminal reaction in anaerobic ecosystems is the production of methane (Figure 17.23). Nearly all known methanogens are capable of reducing CO_2 to CH_4 and the reaction is associated with a reasonable free energy release (see Table 17.5). Acetate is also a substrate for methanogenesis, but its role as an immediate precursor of methane depends somewhat on conditions in the ecosystem. In the rumen (see Section 17.13), for example, H_2-mediated methanogenesis is thought to predominate, while in anaerobic sewage bioreactors acetate is an important precursor of methane (see Section 17.21).

The term **interspecies hydrogen transfer** has been coined to describe the interdependent series of reactions involved in the anaerobic conversion of complex polymers to methane (see also Section 16.18). With the exception of the cellulolytic microflora, all of the remaining microbial groups are in some way dependent upon each other, with the interspecies transfer of reducing equivalents (that is, hydrogen) eventually terminating at the methane "sink." The interdependence of organisms in the degradation of cellulose can thus be summarized as follows:

1. Noncellulolytic fermentative anaerobes are dependent upon cellulolytic bacteria for their energy source, glucose.
2. Fatty acid-oxidizing bacteria are dependent on the primary fermenters for their energy sources, and on methanogens and acetogens (or sulfate reducers in sulfate-rich ecosystems) for consumption of H_2.
3. The methanogens, be they CO_2-reducers or acetate degraders, are dependent upon all the major groups to ultimately supply them with the raw materials for methanogenesis, primarily $H_2 + CO_2$, and acetate. Acetogens are more independent than methanogens, because they can also ferment sugars to acetate (see Section 19.9).

In most anaerobic ecosystems, the rate-limiting step in methanogenesis from organic compounds is not the terminal step of methane formation, but the steps involved in the production of acetate and H_2. Growth rates of the fatty acid-oxidizers are generally very slow. As soon as any H_2 is formed during fermentation it is quickly consumed by a methanogen, an acetogen, or a sulfate-reducer. The only situations in which H_2 accumulates in nature are when methanogenesis and sulfate reduction (see below) are inhibited in some way.

Methanogenic habitats Despite the obligate anaerobiosis and specialized metabolism of methanogens, they are quite widespread on earth (Figure 17.23). Although high levels of methanogenesis only occur in anaerobic environments, such as swamps and marshes, or the rumen (see later, Section 17.13), the process also occurs in habitats that normally might be considered aerobic, such as forest and grassland soils. In such habitats, it is likely that methanogenesis is occurring in anaerobic microenvironments, for example, in the midst of soil crumbs. An overview of the rates of methanogenesis in different kinds of habitats is given in Table 17.6. It should be noted that biogenic production of methane by the methanogenic bacteria exceeds considerably the production rate from gas wells and other abiogenic sources. Eructation by ruminants (see Section 17.13) is the largest single source of biogenic methane. In fact, it is likely that the 1–2 percent rise in atmospheric methane

*The genus names of these organisms reflect their dependence on *syntrophic* relationships with other bacteria (in this case H_2-consuming bacteria) for growth (see Figure 16.34). The term *syntrophic* means, literally, "eating together."

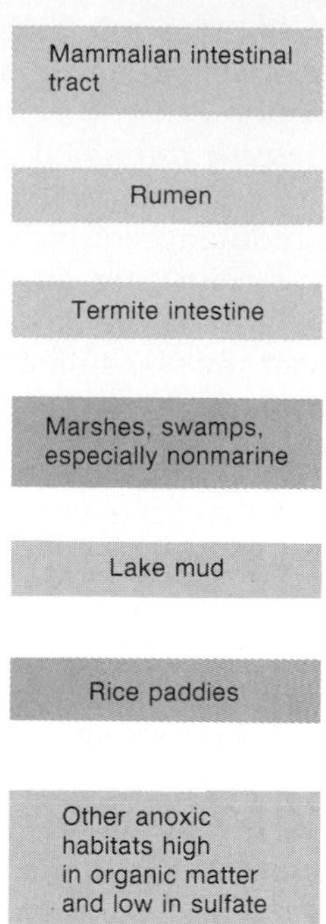

Figure 17.23 Habitats of methanogens.

levels observed over the past few years is due to an increase in the number of domestic ruminants on a worldwide basis.

Table 17.6 Sources of atmospheric methane[1]

Source	Mean annual production (Tg CH_4/year)[2]	
Biogenic		
Ruminants	85	
Paddy fields	45	
Swamps	35	
Termites	3.5	
Ocean/Lakes	4	
Other	7	
Abiogenic		
Biomass burning	75	
Pipeline losses	24	
Coal mining	30	
Automobile	1	
Volcanoes	0.5	
Totals:		% of total
Biogenic	180	58
Abiogenic	130	42

[1]*Data adapted from Seiler, W. 1984. Contribution of Biological Processes to the Global Budget of CH_4 in the Atmosphere. Pages 468–477* in *M. J. Klug and C. A. Reddy (eds.). Current Perspectives in Microbial Ecology, American Society for Microbiology, Washington, D.C.*
[2]*One Tg = 10^{12} g.*

Sulfate inhibition of methanogenesis Methanogenesis is much more common in freshwater and terrestrial environments than in the sea. The reason for this appears to be that marine waters and sediments contain rather high levels of sulfate, and sulfate-reducing bacteria (see Sections 16.15 and 19.8) effectively compete with the methanogenic population for available acetate and H_2, two major electron donors for sulfate-reducing bacteria:

$$4H_2 + SO_4^{2-} \rightarrow H_2S + 2H_2O + 2OH^- \quad \Delta G^{0\prime} = -39.3 \text{ kcal}$$

$$CH_3COO^- + SO_4^{2-} \rightarrow 2HCO_3^- + HS^- \quad \Delta G^{0\prime} = -11.3 \text{ kcal}$$

The biochemical basis for the success of sulfate-reducing bacteria in scavenging H_2 appears to lie in the increased *affinity* sulfate-reducing bacteria have for H_2 as compared to typical methanogens. When H_2 levels get below 5 to 10 μM, as they often do in sulfate-rich environments, methanogens are no longer able to grow since their H_2 uptake systems are unable to function at such low H_2 concentrations. Sulfate reducers, on the other hand, can grow at these low partial pressures of H_2, effectively preventing H_2-mediated methanogenesis. Sulfate reduction can also be a significant process in fresh water, but because the sulfate concentration of fresh water is so low, sulfate is rapidly depleted at the surface of anaerobic sediments; thus, throughout the bulk of the sediment, methanogenesis is the major process consuming H_2.

A kinetic mechanism is also responsible for the ability of sulfate-reducers to effectively compete with methanogens for acetate. The affinity for acetate of some sulfate reducers is over ten times that of methanogens, and this differential expands as acetate levels become limiting. In sulfate-rich environments, acetate levels are generally low; hence the majority of acetate consumed in these environments will be by sulfate-reducing, rather than methanogenic, bacteria.

In marine environments it appears that because of H_2 consumption by sulfate-reducers the major precursors of methane are methylated substrates, such as methylamines and methanol. Trimethylamine, a major excretory product of marine animals, is readily converted to CH_4 by the methanogens *Methanosarcina* and *Methanococcus*. Thus, methanogenesis in marine sediments is not supported by H_2 and acetate, major methanogenic substrates in other methanogenic ecosystems.

As seen in Table 17.4, the residence time of methane in the atmosphere is short, 3.6 years. The main process consuming methane in the atmosphere is photochemical oxidation to CO_2. Some methane formed in soils and anaerobic sediments never reaches the atmosphere because it is oxidized by another group of bacteria, the methanotrophic (methylotrophic) bacteria (see Section 19.7). We discuss the role of methanogenesis in sewage treatment and waste disposal in Section 17.21.

17.13 The Rumen Microbial Ecosystem

Ruminants are herbivorous mammals that possess a special organ, the **rumen**, within which the digestion of cellulose and other plant polysaccharides occurs through the activity of special microbial populations. Some of the most important domestic animals, the cow, sheep, and goat, are ruminants. Since the human food economy depends to a great extent on these animals, rumen microbiology is of considerable economic significance.

Rumen fermentation The bulk of the organic matter in terrestrial plants is present in insoluble polysaccharides, of which *cellulose* is the most important. Mammals, and indeed almost all animals, lack the enzymes necessary to digest cellulose, but all mammals that subsist primarily on grasses and leafy plants can metabolize cellulose by making use of microorganisms as digestive agents. Unique features of the rumen as a site of cellulose digestion are its relatively large size (100 to 150 liters in a cow, 6 liters in a sheep) and its position in the alimentary tract as the organ where ingested food goes first. The high constant temperature (39°C), constant pH (6.5), and anaerobic nature ($E_o = -0.4$ volts) of the rumen are also important factors in overall rumen function. The rumen operates in a more or less continuous fashion, and in some ways can be considered analogous to a microbial chemostat (see Section 9.6).

The relationship of the rumen to other parts of the ruminant digestive system is shown in Figure 17.24. Food enters the rumen mixed with saliva containing bicarbonate and is churned in a rotary motion during which the microbial fermentation occurs. This peristaltic action grinds the cellulose into a fine suspension which assists in microbial attachment. The food mass then passes gradually into the reticulum, where it is formed into small clumps called cuds, which are regurgitated into the mouth where they are chewed again. The now finely divided solids, well mixed with saliva, are swallowed again, but this time the material passes down a different route, ending in the abomasum, an organ more like a true (acidic) stomach. Here chemical digestive processes begin which continue in the small and large intestine.

Food remains in the rumen about 9 hours. During this period cellulolytic bacteria and cellulolytic protozoa hydrolyze cellulose to the disaccharide cellobiose and to free glucose units. The released glucose then undergoes a bacterial fermentation with the production of **volatile fatty acids** (VFAs), primarily *acetic, propionic*, and *butyric*, and the gases *carbon dioxide* and *methane* (Figure 17.25). The fatty acids pass through the rumen wall into the bloodstream and are oxidized by the animal as its main source of energy. In addition to their digestive functions, rumen microorganisms synthesize amino acids and vitamins that are the main source for the animal of these essential nutrients. The rumen contents after fermentation consist of enormous numbers of microbial cells (10^{10} to 10^{11} bacteria per milliliter of rumen fluid by direct microscopic counts) plus partially digested plant materials; these proceed through the gastrointestinal tract of the animal where they undergo further digestive processes similar to those of nonruminants. Many microbial cells formed in the rumen are digested in the gastrointestinal tract and serve as a major source of proteins and vitamins for the animal. Since many of the microbes of the rumen are able to grow on urea as a sole nitrogen source, it is often supplied in cattle feed in order to promote microbial protein synthesis. The bulk of this protein will end up in the animal itself. A ruminant is thus nutri-

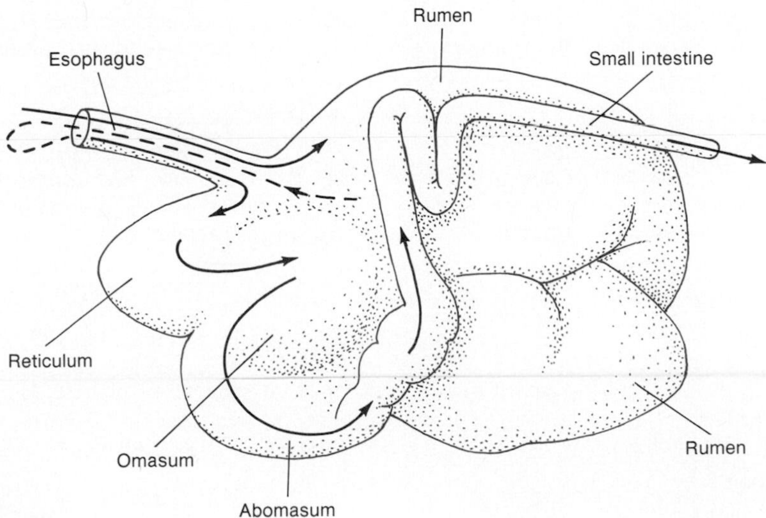

Figure 17.24 Schematic diagram of the rumen and gastrointestinal system of a cow. See text for details.

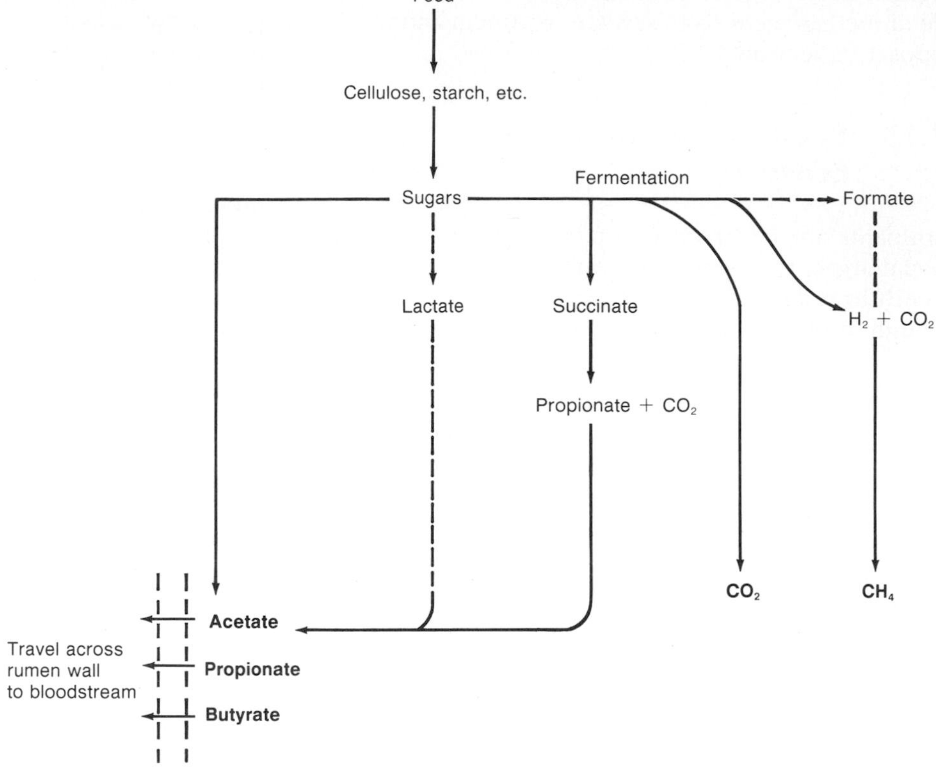

Figure 17.25 Biochemical reactions in the rumen. The end products are shown in boldface type; dashed lines indicate minor pathways. Steady state rumen levels of VFAs are: acetate, 60 mM; propionate, 20 mM; butyrate, 10 mM.

Table 17.7 Characteristics of some rumen bacteria

Organism	Gram stain	Shape	Motility	Fermentation products
Cellulose decomposers:				
Bacteroides succinogenes	Neg.	Rod	−	Succinate, acetate, formate
Butyrivibrio fibrisolvens	Neg.	Curved rod	+	Acetate, formate, lactate, butyrate, H_2, CO_2
Ruminococcus albus	Pos.	Coccus	−	Acetate, formate, H_2, CO_2
Clostridium lochheadii	Pos.	Rod (spores)	+	Acetate, formate, butyrate, H_2, CO_2
Starch decomposers:				
Bacteroides amylophilus	Neg.	Rod	−	Formate, acetate, succinate
Bacteroides ruminicola	Neg.	Rod	−	Formate, acetate, succinate
Selenomonas ruminantium	Neg.	Curved rod	+	Acetate, propionate, lactate
Succinomonas amylolytica	Neg.	Oval	+	Acetate, propionate, succinate
Streptococcus bovis	Pos.	Coccus	−	Lactate
Lactate decomposers:				
Selenomonas lactilytica	Neg.	Curved rod	+	Acetate, succinate
Peptostreptococcus elsdenii	Pos.	Coccus	−	Acetate, propionate, butyrate, valerate, H_2, CO_2
Pectin decomposer:				
Lachnospira multiparus	Pos.	Curved rod	+	Acetate, formate, lactate, H_2, CO_2
Methane producer:				
Methanobrevibacter ruminantium	Pos	Rod	−	CH_4 (from H_2 + CO_2 or formate)
Methanomicrobium mobile	Neg.	Rod	+	CH_4 (from H_2 + CO_2 or formate)

Nonmedical Uses of Antibiotics

A major nonmedical use of antibiotics in the United States is their addition to animal feed. The observation has consistently been made that low levels of antibiotics added to feeds stimulates animal growth, shortening the period required to get the animal to market. For example, addition of 25 mg of penicillin per pound of chicken feed saves 2×10^9 pounds of feed a year, due to more rapid weight gains and feeding efficiency. The antibiotics probably act by inhibiting organisms responsible for low-grade infections and by reducing intestinal epithelial inflammation. Studies with germ-free animals have borne this out. The growth of germ-free animals is not accelerated by antibiotic-supplemented feed. In addition, the intestinal wall of the normal animal is much thicker than that of the germ-free animal, probably because of low level inflammation caused by the bacterial flora. The lessening of inflammation in the gut wall of animals fed low levels of antibiotics probably promotes nutrient uptake and could account for the more efficient utilization of feed observed.

The problem with low level application of antibiotics to animal feeds is that an antibiotic resistant microflora is selected for by the constant exposure to antibiotics. The use of antibiotics in animal feed therefore serves to expand the gene pool of antibiotic resistance in nature. Since some of the members of the gut flora of animals also inhabit the human gut, the transmission of resistant flora from animals to humans is thus a real possibility. Indeed, studies of antibiotic resistance in human gut flora have shown that many strains of human enteric bacteria are multiply resistant.

Molecular biological studies of resistant strains of *Salmonella* isolated from poultry have shown that the resistance determinants reside on conjugative plasmids or transposons (see Sections 7.6 and 7.7). Resistance sequences are rapidly transferred between different species and even between different genera. Resistant organisms can then be transmitted to humans in contaminated meat or by contact with live animals.

Unfortunately, long-term studies of animals previously fed antibiotics and then put on antibiotic-free rations have shown that antibiotic resistant bacteria are not quickly lost from the gut. It is postulated that resistance genes have become part of stable plasmids in the gut flora, and, in the absence of counterselective forces, these resistance sequences have been maintained and will probably remain a part of the gut flora of animals for some time, even if supplementation of feeds with antibiotics was to stop immediately. Nonmedical use of antibiotics has therefore reinforced, albeit painfully, a simple lesson in microbial ecology: the environment selects the best adapted species.

Although the continued use of clinically useful antibiotics in animal feeds will undoubtedly serve to widen the dissemination of resistant gene sequences, it is not clear whether halting this practice will effectively solve the problem. Continued veterinary use of antibiotics may by itself maintain resistant animal microflora. However, in the hopes of reducing the spread of antibiotic resistance in Europe, most European countries have banned the use of antibiotics in animal feeds. In the United States antibiotics continue to be used in the cattle, poultry, and swine industries.

tionally superior to the nonruminant when subsisting on foods that are deficient in protein, such as grasses.

Rumen microorganisms The biochemical reactions occurring in the rumen are complex and involve the combined activities of a variety of microorganisms. Since the reduction potential of the rumen is −0.4 volts (the O_2 concentration at this highly reducing potential is 10^{-22} M!), anaerobic bacteria naturally dominate. Furthermore, since the conversion of cellulose to CO_2 and CH_4 involves a multistep microbial food chain, a variety of anaerobes can be expected (Table 17.7).

Several different rumen bacteria hydrolyze polymers such as cellulose to sugars and ferment the sugars to volatile fatty acids. *Bacteroides succinogenes* and *Ruminococcus albus* are the two most abundant cellulolytic rumen anaerobes. Although both organisms produce cellulases, *Bacteroides*, a Gram-negative bacterium, employs a periplasmic cellulase (see Section 3.5 for a discussion of the periplasm) to break down cellulose; thus the organism must remain attached to the cellulose fibril while digesting it. *Ruminococcus*, on the other hand, produces an extremely large (>2,000,000 molecular weight) cellulase that is excreted into the rumen contents where it degrades cellulose outside the bacterial cell proper. However, the end result is the same in both cases; free glucose is made available for fermentative anaerobes. If a ruminant is gradually switched from cellulose to a diet high in starch (grain, for instance), then starch-digesting bacteria such as *Bacteroides amylophilus* and *Succinomonas amylolytica* develop; on a low starch diet these organisms are in a minority. If an animal is fed on legume hay, which is high in pectin, then the pectin-digesting bacterium *Lachnospira multiparus* is a common member of the rumen flora.

17.14 Biogeochemical Cycles: Nitrogen

The element nitrogen, N, a key constituent of protoplasm, exists in a number of oxidation states (see Table 16.5). We discussed two major processes of microbial nitrogen transformation in Chapter 16: denitrification in Section 16.14 and nitrification in Section 16.11. These and several other nitrogen transformations are summarized in the redox cycle shown in Figure 17.26. Several of the key redox reactions of nitrogen are carried out in nature almost exclusively by microorganisms, so that microbial involvement in the nitrogen cycle is of great importance. Thermodynamically, nitrogen gas, N_2, is the most stable form of nitrogen, and it is to this form that nitrogen will revert under equilibrium conditions. This explains the fact that a major reservoir for nitrogen on earth is the atmosphere (Table 17.8). This is in contrast to carbon, in which the atmosphere is a relatively minor reservoir. The high energy necessary to break the N≡N bond of molecular nitrogen (Section 16.24) means that the utilization of N_2 is an energy-demanding process. Only a relatively small number of organisms are able to utilize N_2, in the process called **nitrogen fixation**; thus the recycling of nitrogen on earth involves to a great extent the more easily available forms, ammonia and nitrate. Globally, only about 3 percent of the net primary production of organic matter involves nitrogen derived from N_2. The remaining primary production uses nitrogen from "fixed" forms of nitrogen (nitrogen in combination with other elements). However, because N_2 constitutes by far the greatest reservoir of nitrogen available to living organisms, the ability to utilize N_2 is of great ecological importance. In many environments, productivity is limited by the short supply of combined nitrogen compounds, putting a premium on biological nitrogen fixation.

The global nitrogen cycle is given in Figure 17.27. Note that the chemical form in which transfer between compartments occurs is not given. Transfer in and out of the atmosphere is to a great extent as N_2, with a smaller amount of transfer as nitrous oxide, N_2O, and as gaseous ammonia, NH_3. Transfer between terrestrial and aquatic compartments is primarily as organic nitrogen, ammonium ion, and nitrate ion.

Nitrogen fixation We have discussed the biochemistry and microbiology of nitrogen fixation ($N_2 + 6H^+ + 6e^- \rightarrow 2NH_3$) in Section 16.24; and we discuss symbiotic nitrogen fixation by legumes in Section 17.24. Nitrogen fixation can also occur chemically in the atmosphere, to a small extent, via lightning discharges, and a certain amount of nitrogen fixation occurs in the industrial production of nitrogen fertilizers (labeled as industrial fixation in Figure 17.27). Some nitrogen fixation also occurs during artificial combustion processes, since air contains 78 percent N_2 by weight, and burning in air inevitably involves high-temperature combustion of some N_2 (to nitrogen oxides and ultimately to nitrate). However, as can be calculated from the fluxes given in Figure 17.27, about 85 percent of nitrogen fixation on earth is of *biological* origin. As can also be calculated from Figure 17.27, about 60 percent of biological nitrogen fixation occurs on land, and the other 40 percent in the oceans.

Denitrification We have discussed the role of nitrate as an alternative electron acceptor in Section 16.14. Assimilatory nitrate reduction, in which nitrate is reduced to the oxidation level of ammonia, for use as a nitrogen source for growth, and dissimilatory nitrate reduction, in which nitrate is used as an alternative electron acceptor in energy generation, were contrasted in Figure 16.25. Under most conditions, the end product of dissimilatory nitrate reduction is N_2 or N_2O, and the conversion of nitrate

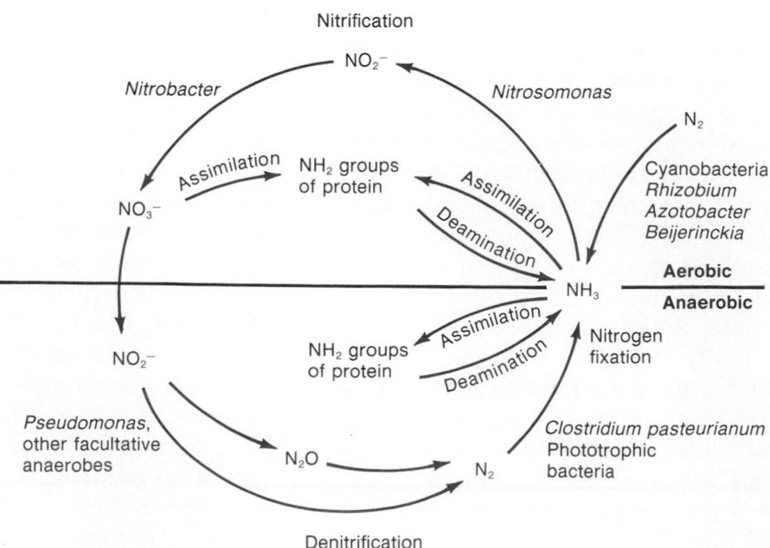

Figure 17.26 Redox cycle for nitrogen.

Table 17.8 Nitrogen reservoirs and residence times

Component	Reservoir (g)	Residence time
Atmosphere:		
$NH_3 + NH_4^+$	0.003×10^{15}	Few days to few months
N_2	$3{,}800{,}000 \times 10^{15}$	44×10^6 years
N_2O	13×10^{15}	12–13 years
NO_3^-	0.0005×10^{15}	2–3 weeks
Organic nitrogen	0.001×10^{15}	Approx. 10 days
NO	0.03×10^{15}	Approx. 1 month
Land:		
Plant biomass	$11\text{–}14 \times 10^{15}$	16 years
Animal biomass	0.2×10^{15}	
Soil organic nitrogen	300×10^{15}	1–40 years
Soil inorganic nitrogen	16×10^{15}	Less than 1 year
Oceans:		
Plant biomass	0.3×10^{15}	0.14 year
Animal biomass	0.17×10^{15}	
Dead organic matter		
Dissolved	5.30×10^{15}	
Particulate	$3\text{–}24 \times 10^{15}$	
N_2 (dissolved)	$22{,}000 \times 10^{15}$	220,000 years
N_2O	0.2×10^{15}	2.5 years
NO_3^-	570×10^{15}	
NO_2^-	0.5×10^{15}	
NH_4^+	7×10^{15}	
Sediments and rocks:		
Rocks	$190{,}000{,}000 \times 10^{15}$	
Sediments	$400{,}000 \times 10^{15}$	400×10^6 years (organic nitrogen of sediments, fossil fuels)
Coal deposits	120×10^{15}	

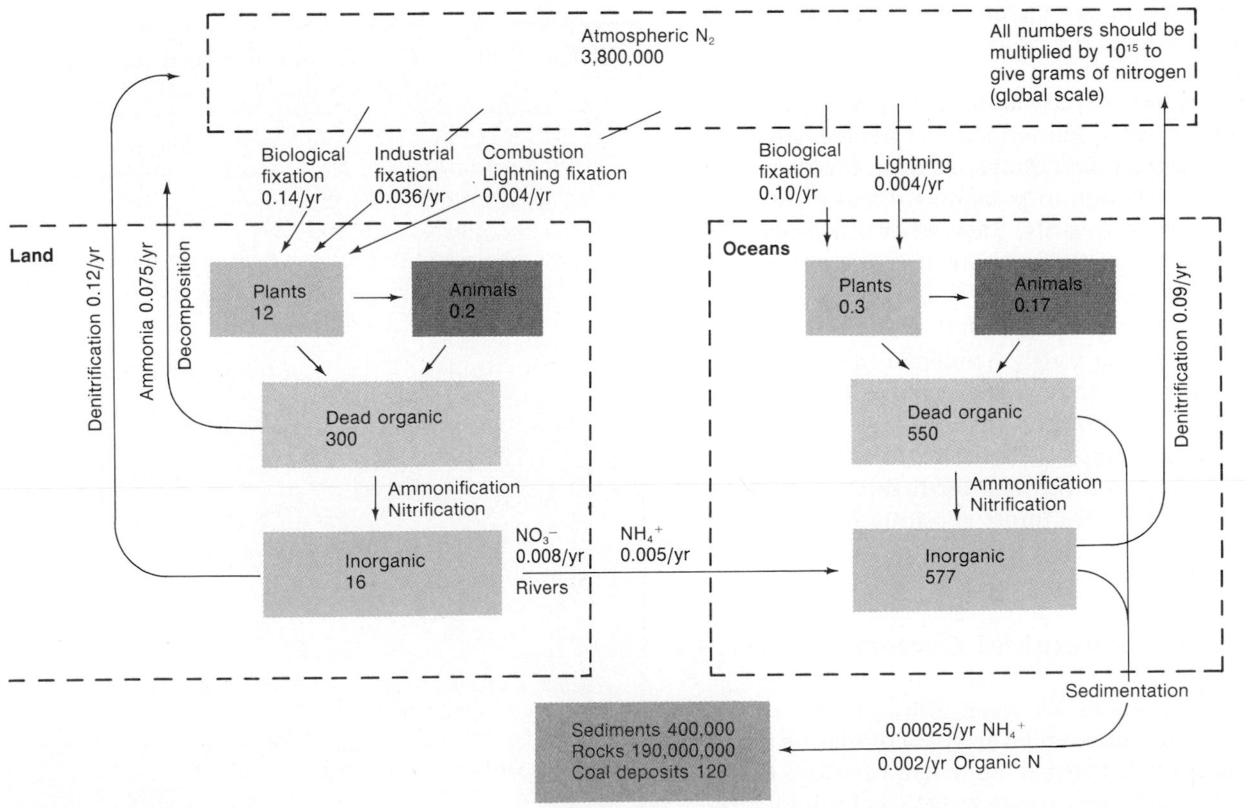

Figure 17.27 The global nitrogen cycle. Minor compartments and fluxes are not given. See also Table 17.8.

to gaseous nitrogen compounds is called **denitrification**. This process is the main means by which gaseous N_2 is formed biologically, and since N_2 is much less readily available to organisms than nitrate as a source of nitrogen, denitrification is a detrimental process.

Ammonia fluxes and nitrification Ammonia is produced during the decomposition of organic nitrogen compounds (**ammonification**) and exists at neutral pH as ammonium ion (NH_4^+). Under anaerobic conditions, ammonia is stable, and it is in this form that nitrogen predominates in anaerobic sediments. In soils, much of the ammonia released by aerobic decomposition is rapidly recycled and converted into amino acids in plants. Because ammonia is volatile, some loss can occur from soils (especially highly alkaline soils) by vaporization, and major losses of ammonia to the atmosphere occur in areas of dense animal populations (for example, cattle feedlots). On a global basis, ammonia constitutes only about 15 percent of the nitrogen released to the atmosphere, the majority of the rest being in the form of N_2 or N_2O (from denitrification).

In aerobic environments ammonia can be oxidized to nitrogen oxides and nitrate, but ammonia is a rather stable compound and strong oxidizing agents or catalysts are usually needed for the chemical reaction. However, a specialized group of bacteria, the *nitrifying bacteria,* are excellent catalysts, oxidizing ammonia to nitrate in a process called **nitrification** (see Section 16.11).

Nitrification is strictly an *aerobic* process and occurs readily in well-drained soils at neutral pH; it is inhibited by anaerobic conditions or in highly acidic soils. If materials high in protein, such as manure or sewage, are added to soils, the rate of nitrification is increased. Although nitrate is readily assimilated by plants, it is very water soluble and is rapidly leached from soils receiving high rainfall. Consequently, nitrification is not necessarily beneficial in agricultural practice. Ammonia, on the other hand, is cationic and consequently is strongly adsorbed to negatively-charged clay minerals. Anhydrous ammonia is now used extensively as a nitrogen fertilizer, but some of the added ammonia is almost certainly converted to nitrate by nitrifying organisms.

17.15 Biogeochemical Cycles: Sulfur

Sulfur transformations are even more complex than those of nitrogen due to the variety of oxidation states of sulfur and the fact that some transformations occur at significant rates *chemically* as well as biologically. We discussed some of the processes of sulfate reduction and lithotrophic sulfur oxidation in Chapter 16. The redox cycle for sulfur and the involvement of some microorganisms in sulfur transformations are given in Figure 17.28. Although a number of oxidation states are possible, only three form significant amounts of sulfur in nature,—2(sulfhydryl, R—SH and sulfide, HS^-), 0(elemental sulfur, S^0), and +6(sulfate). Sulfur reservoirs on earth are given in Table 17.9 and it can be seen that in contrast to nitrogen, the atmosphere is quite insignificant as a sulfur reservoir. The bulk of the sulfur of the earth is found in sediments and rocks, in the form of sulfate minerals (primarily gypsum, $CaSO_4$) and sulfide minerals (primarily pyrite, FeS_2), although the oceans constitute the most significant reservoir of sulfur for the biosphere (in the form of inorganic sulfate). The global transport cycle for sulfur is given in Figure 17.29, and some of the components of this cycle are discussed below.

Hydrogen sulfide and sulfate reduction A major volatile sulfur gas is *hydrogen sulfide.* As we saw in Section 16.15, this substance is formed primarily by the bacterial reduction of sulfate:

$$SO_4^{2-} + 8e^- + 8H^+ \rightleftharpoons H_2S + 2H_2O + 2OH^-$$

Table 17.9 Sulfur reservoirs and gaseous sulfur emmissions on earth

Component	Reservoir (g)
I. Reservoirs:	
Atmosphere:	
Hydrogen sulfide	9.6×10^{11}
Sulfur dioxide	6.4×10^{11}
Sulfate	16×10^{11}
Land	
Living organic (mainly plants)	$25{,}000–40{,}000 \times 10^{11}$
Dead organic	$35{,}000–60{,}000 \times 10^{11}$
Inorganic	Unknown
Oceans:	
Living organic	0.0000035×10^{11}
Dead organic	0.00038×10^{11}
Inorganic (sulfate)	$13{,}760{,}000{,}000 \times 10^{11}$
Sediments and rocks:	
Calcium sulfate (gypsum)	$63{,}000{,}000{,}000 \times 10^{11}$
Metal sulfides (predominantly pyrite FeS_2)	$47{,}000{,}000{,}000 \times 10^{11}$
II. Emissions:	
Oceans	390×10^{11}
Land	170×10^{11}
Salt Marshes	20×10^{11}

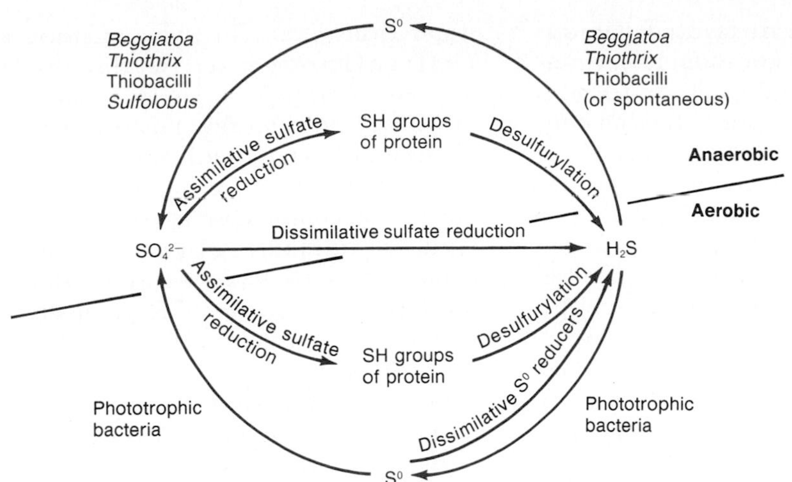

Figure 17.28 Redox cycle for sulfur.

The form in which sulfide is present in an environment depends on pH due to the following equilibria:

$$H_2S \text{ Low pH} \rightleftharpoons HS^- \text{ Neutral pH} \rightleftharpoons S^{2-} \text{ High pH}$$

At high pH, the dominant form is sulfide, S^{2-}. At neutral pH, HS^- predominates, and at acid pH, H_2S is the major species. For simplicity, the reactions will be written with HS^-, the major form at neutral pH, unless a gaseous product is involved, in which case H_2S is written. HS^- and S^{2-} are very water soluble, but H_2S is not and readily volatilizes. Even at neutral pH, some volatilization of H_2S from HS^- can occur, because there is an equilibrium between HS^- and H_2S, and as volatilization occurs, the reaction is pulled toward H_2S.

A wide variety of organisms can use sulfate as a sulfur source and carry out *assimilatory* sulfate reduction, converting the HS^- formed to organic sulfur, R—SH (see Figure 16.27*b*). HS^- is ultimately formed from the decomposition of this organic sulfur by putrefaction and desulfurylation (Figure 17.28), and this is a significant source of HS^- in fresh water. In the marine environment, because of the vast amount of sulfate present (Table 17.9), dissimilatory sulfate reduction is the main source of HS^-. Dissimilatory sulfate reduction, in which sulfate serves as an electron acceptor, is carried out by a variety of bacteria, collectively referred to as the ***sulfate-reducing bacteria*** (see Sections 16.15 and 19.8), which are all obligate anaerobes. It should be emphasized that the sulfate anion is very stable chemically, and its reduction does not occur spontaneously in nature under normal environmental conditions. Most sulfate reducers can also use S^0 as an electron acceptor, reducing it to H_2S.

As discussed in Section 17.12, there is a competition in nature between methanogenic and sulfate-reducing bacteria for available electron donors, especially H_2 and acetate, and as long as sulfate is pres-

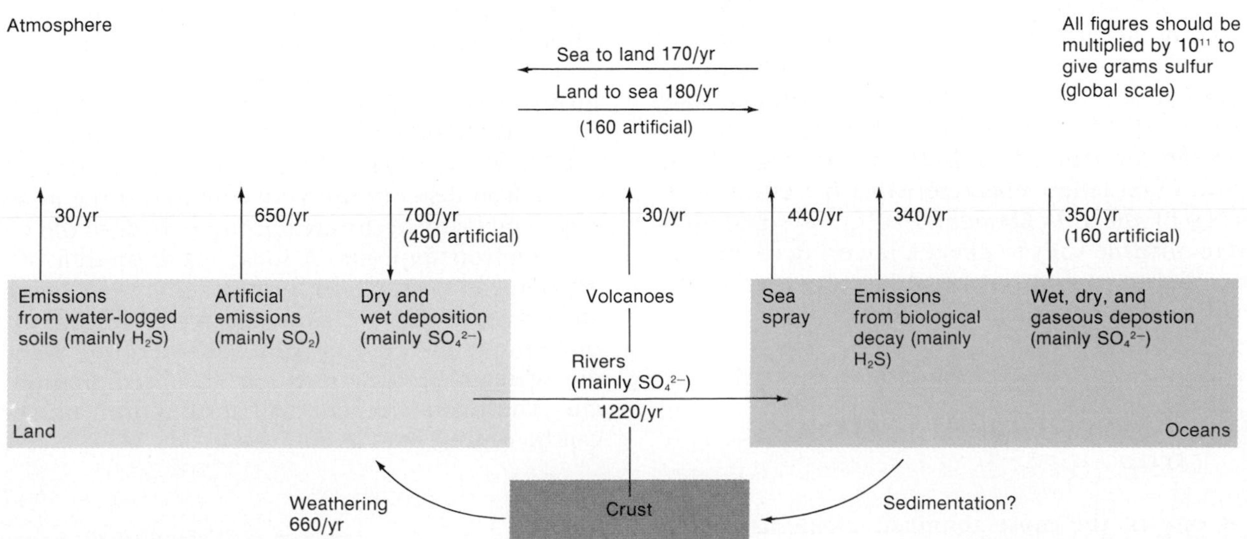

Figure 17.29 The global sulfur cycle.

ent the sulfate-reducing bacteria are favored. Because of the necessity of organic electron donors (or molecular H_2, which is itself derived from the fermentation of organic compounds), sulfate reduction only occurs extensively where significant amounts of organic matter are present. In many marine sediments, the rate of sulfate reduction is carbon limited, and the rate can be greatly increased by addition of organic matter. This is of considerable importance for marine pollution, since disposal of sewage, sewage sludge and garbage in the sea can lead to marked increases in organic matter in the sediments. Since HS^- is a toxic substance to many organisms, formation of HS^- by sulfate reduction is potentially detrimental.

Sulfide and elemental sulfur oxidation Sulfide (HS^-) rapidly oxidizes spontaneously at neutral pH, if O_2 is present (see Section 16.9). When HS^- reacts with O_2, oxidation generally does not proceed all the way to sulfate spontaneously, but stops either at elemental sulfur or thiosulfate. The precise product formed depends on the ratio of sulfide to oxygen, the initial concentration of sulfide, and the presence or absence of metals as catalysts. The sulfur-oxidizing bacteria are also able to catalyze the oxidation of sulfide, but because of the rapid spontaneous reaction, bacterial oxidation of sulfide only occurs in narrow zones or regions where H_2S rising from anaerobic areas meets O_2 descending from aerobic areas. If light is available, anaerobic oxidation of HS^- can also occur, catalyzed by the photosynthetic bacteria (see Sections 16.3 and 19.1), but this only occurs in restricted areas, usually in lakes, where sufficient light can penetrate to anaerobic zones.

Elemental sulfur, S^0, is chemically stable in most environments in the presence of oxygen, but is readily oxidized by sulfur-oxidizing bacteria. Although a number of sulfur-oxidizing bacteria are known, members of the genus *Thiobacillus* (see Section 16.9) are most commonly involved in elemental sulfur oxidation. Elemental sulfur is very insoluble, and the bacteria that oxidize it attach firmly to the sulfur crystals (see Figure 16.19). Oxidation of elemental sulfur results in the formation of sulfate and hydrogen ions, and sulfur oxidation characteristically results in a *lowering* of the pH. Elemental sulfur is sometimes added to alkaline soils to effect a lowering of the pH, reliance being placed on the ubiquitous thiobacilli to carry out the acidification process.

17.16 Biogeochemical Cycles: Iron

Iron is one of the most abundant elements in the earth's crust, but is a relatively minor component in aquatic systems because of its relative insolubility in water. Iron exists in two oxidation states, ferrous (+II) and ferric (+III).* We discussed iron oxidation and reduction in Sections 16.10 and 16.17. The form in which iron is found in nature is greatly influenced by pH and oxygen. Because of the high electrode potential of the Fe^{3+}/Fe^{2+} couple, 0.76 mV, the only electron acceptor able to oxidize ferrous iron is oxygen, O_2. At neutral pH, ferrous iron oxidizes spontaneously in air to ferric iron, which forms highly insoluble precipitates of ferric hydroxide and ferric oxides. Thus, at neutral pH, the only way the iron is maintained in solution is by chelation with organic materials (see below). Although ferrous iron is also fairly insoluble in water, it is considerably more soluble than ferric iron, so that reduction of ferric to ferrous iron results in some solubilization of iron.

Bacterial iron reduction and oxidation The bacterial reduction of ferric iron to the ferrous state is a major means by which iron is solubilized in nature. As we noted in Section 16.17, a number of organisms can use ferric iron as an electron acceptor. In addition to the bacterially catalyzed reduction, if hydrogen sulfide is present, as it is in many anaerobic environments, ferric iron is also reduced chemically to FeS (ferrous sulfide). Thus, there are complex interactions in many environments between the iron and sulfur cycles.

Ferric iron reduction is very common in waterlogged soils, bogs, and anaerobic lake sediments. In many waterlogged soils, the ferrous iron so formed is leached out of the soil, leaving the soil gray and mottled, a condition called **gleying**. (The faintly brownish or reddish color of most normal surface soils is due to the presence of oxidized iron.) Movement of iron-rich groundwater from anaerobic bogs or waterlogged soils can result in the transport of considerable amounts of iron. Once this iron-laden water reaches aerobic regions, the ferrous iron is quickly oxidized spontaneously and ferric compounds precipitate, leading to the formation of a brown deposit (see Figure 16.21*c*). Such brown deposits frequently are a serious problem in the pipes used to carry industrial or drinking water from deep wells. Iron deposits are very common at the edges of bogs, and many of the great iron-ore beds of the world are bog-iron deposits. A local manifestation of this phenomenon is the iron spring or iron seep, in which an extensive flow of iron-rich water results in the movement of ferrous iron considerable distance downstream before oxidation and precipitation occur. The overall reaction of ferrous iron oxidation can be represented as follows:

*Elemental iron, Fe^0, does not exist significantly in nature but is a widespread artificial product. It is subject to corrosion processes.

$$Fe^{2+} + 1/4O_2 + H^+ \rightarrow Fe^{3+} + 1/2H_2O$$
$$Fe^{3+} + 3OH^- \rightarrow Fe(OH)_3 \text{ precipitates}$$

Summation:

$$Fe^{2+} + 1/4O_2 + 2OH^- + 1/2H_2O \rightarrow Fe(OH)_3$$

Note that although the initial oxidation of ferrous iron consumes hydrogen ions and thus leads to a rise in pH, the hydrolysis of Fe^{3+} and formation of $Fe(OH)_3$ consumes hydroxyl ions and leads to an acidification of the medium. This is one way in which iron oxidation leads to the formation of acidic conditions in the environment.

Although ferric iron forms very insoluble hydroxides, some ferric iron can be kept in solution in natural waters by forming complexes with organic materials. If an organism is present that can oxidize the organic compound, then the iron present will precipitate. This is probably a major mechanism of iron precipitation in many neutral pH environments. In addition, at neutral pH, organisms such as *Gallionella* (see Figure 19.42) and *Leptothrix* (see Figure 19.64) contribute to the oxidation of ferrous iron, but is not clear whether the ferrous to ferric iron conversion serves as an energy source for these bacteria as it does for acidophilic thiobacilli (see below).

Ferrous iron oxidation at acid pH At acid pH values, spontaneous oxidation of ferrous iron to the ferric state is slow (see Section 16.10). However, the acidophilic lithotrophic organism *Thiobacillus ferrooxidans* is able to catalyze the oxidation (Figure 17.30); *T. ferrooxidans* oxidizes ferrous iron as its primary energy-generating process. Because very little energy is generated in the oxidation of ferrous to ferric iron (see Section 16.10), these bacteria must oxidize large amounts of iron in order to grow, and consequently even a small number of cells can be responsible for precipitating a large amount of iron.

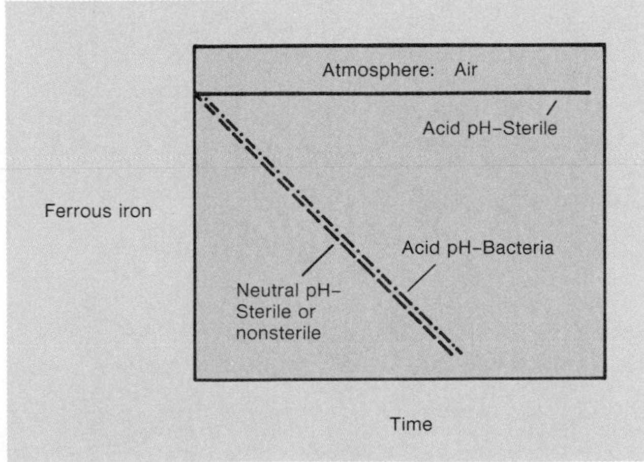

Figure 17.30 Oxidation of ferrous iron as a function of pH and the presence of the bacterium *Thiobacillus ferrooxidans*.

This iron-oxidizing bacterium, which is a strict acidophile, is very common in acid mine drainages and in acid springs, and is probably responsible for most of the iron precipitated at acid pH values.

Thiobacillus ferrooxidans lives in environments in which sulfuric acid is the dominant acid and large amounts of sulfate are present. Under these conditions, ferric iron does not precipitate as the hydroxide, but as a complex sulfate mineral called *jarosite*, which has the approximate formula of: $MFe_3(SO_4)_2(OH)_6$ where $M = K^+$, NH_4^+, or H^+. In culture media, M is usually K^+ or NH_4^+, but in nature, where these cations are low, M is generally H^+. Jarosite is a yellowish or brownish precipitate and is responsible for one of the manifestations of acid mine drainage, an unsightly yellow stain, called "yellow boy" by U. S. miners (see Figure 16.21*a*).

Pyrite oxidation One of the most common forms of iron and sulfur in nature is *pyrite*, which has the overall formula FeS_2. Pyrite is formed from the reaction of sulfur with ferrous sulfide (FeS) to form a highly insoluble crystalline structure, and pyrite is very common in bituminous coals and in many ore bodies (Figure 17.31). The bacterial oxidation of pyrite is of great significance in the development of acidic conditions in mines and mine drainages. Additionally, oxidation of pyrite by bacteria is of considerable importance in the process called microbial leaching of ores (see below). The oxidation of pyrite is a combination of spontaneous and bacterially catalyzed reactions. Two electron acceptors for this process can function: molecular oxygen (O_2) and ferric ions (Fe^{3+}). However, ferric ions are only present when the solution is acidic, at pH below about 2.5. At pH values above 2.5, ferric ion reacts with water to form the insoluble ferric hydroxide. When pyrite is first exposed, as in a mining operation, a slow spontaneous reaction with molecular oxygen occurs as shown in the following reaction:

$$FeS_2 + 3\ 1/2O_2 + H_2O \rightarrow Fe^{2+} + 2SO_4^{2-} + 2H^+$$

This reaction, called the *initiator reaction*, leads to the development of acidic conditions; once acidic conditions develop, the ferrous iron which is formed by the reaction is relatively stable in the presence of oxygen. However, *T. ferrooxidans* catalyzes under acid conditions the oxidation of ferrous to ferric ions. The ferric ions formed under these conditions, being soluble, can readily react with more pyrite to oxidize the pyrite to ferrous ions plus sulfate ions:

$$FeS_2 + 14Fe^{3+} + 8H_2O \rightarrow 15Fe^{2+} + 2SO_4^{2-} + 16H^+$$

The ferrous ions formed are again oxidized to ferric ions by the bacteria, and these ferric ions again

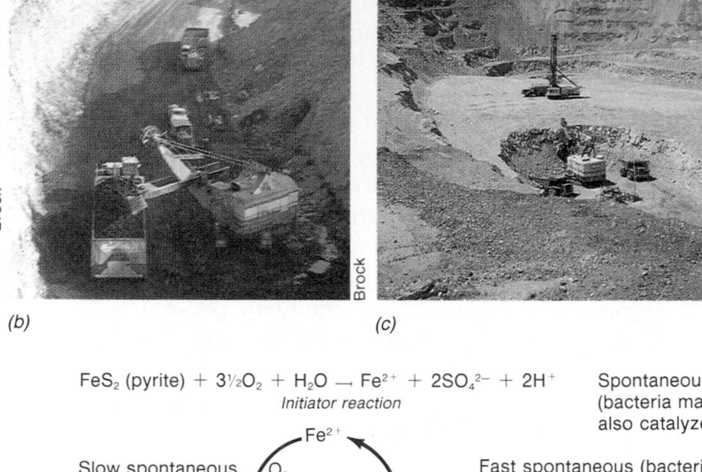

Figure 17.31 Pyrite-rich microbial habitats in bituminous coal and copper mining environments. (a) Bituminous coal mining operation in a strip mine. The shovel is removing the overburden to reach the coal seam. (b) The coal seam. Removal of the bituminous coal exposes the environment to air. The pyrite associated with the coal is colonized with iron-oxidizing bacteria. (c) Copper ore deposit rich in pyrite. Mining of the copper ore results in exposure of the formation to air.

FeS_2 (pyrite) + $3\frac{1}{2}O_2$ + H_2O → Fe^{2+} + $2SO_4^{2-}$ + $2H^+$ Spontaneous (bacteria may also catalyze)

Initiator reaction

Fe^{2+} → (O_2) Slow spontaneous, bacteria catalyze → Fe^{3+}

Fe^{3+} + FeS_2 → Fe^{2+} Fast spontaneous (bacteria may also catalyze)

Propagation cycle

Figure 17.32 Role of iron-oxidizing bacteria in the oxidation of the mineral pyrite.

react with more pyrite. Thus there is a progressive, rapidly increasing rate at which pyrite is oxidized, called the *propagation cycle*, as illustrated in Figure 17.32. Under natural conditions some of the ferrous iron generated by the bacteria leaches away, being carried by groundwater into surrounding streams. Since the pH of many of these acid mine drainages is quite low, often below pH 3, the ferrous iron is stable in the absence of bacteria; but because oxygen is present in the aerated drainage, bacterial oxidation of the ferrous iron takes place. Since this is occurring in the absence of pyrite, the ferrous iron is oxidized to the ferric state, and an insoluble ferric precipitate is formed, as described in the preceding section.

17.17 Microbial Mining

Bacterial oxidation of sulfide minerals is the major factor in the formation of **acid mine drainage**, a common environmental problem in coal mining regions (for example, see Figure 16.21*a*). The same bacterium, *Thiobacillus ferrooxidans*, also carries out a beneficial oxidation of sulfide minerals in the process called **microbial leaching**, which plays a major role in the concentration of copper from low-grade copper ores. In both cases, bacterial attack on pyrite is also involved, following the steps outlined in Section 17.16.

Acid mine drainage Not all coal seams contain iron sulfide; thus acid mine drainage does not occur in all coal mining regions. Where acid mine drainage

Figure 17.33 Acid mine drainage from a bituminous coal region. Note the bright-red color due to precipitated iron oxides. See also Figure 16.21a.

does occur (see Figure 17.31*a*), however, it is often a very serious problem. Mixing of acidic mine waters with natural waters in rivers and lakes causes a serious degradation in the quality of the natural water, since both the acid and the dissolved metals are toxic to aquatic life (see Figure 16.21*a*). In addition, such polluted waters are unsuitable for human consumption and industrial use.

We have outlined in Section 17.16 the steps in the bacterial oxidation of pyrite, and we noted that certain of the steps occurred chemically, but that the rate-limiting step, the oxidation of ferrous to ferric iron, occurred at acid pH only in the presence of the bacterium. The breakdown of pyrite leads ultimately to the formation of sulfuric acid and ferrous iron, and pH values can be as low as pH 2. The acid formed attacks other minerals in the rock associated with the coal and pyrite, causing breakdown of the whole rock fabric. A major rock-forming element, aluminum, is only soluble at low pH, and often large amounts of aluminum are brought into solution. The typical composition of an acid mine water is as follows: pH, 2 to 4.5; Fe^{2+}, 500 to 10,000 mg/liter; Al^{3+}, 100 to 2000 mg/liter; SO_4^{2-}, 1000 to 20,000 mg/liter; with small amounts of N, P, and trace elements.

The requirement for O_2 in the oxidation of ferrous to ferric iron helps to explain how acid mine drainage develops. As long as the coal is unmined, oxidation of pyrite cannot occur, since neither air nor the bacteria can reach it. When the coal seam is exposed, it quickly becomes contaminated with *T. ferrooxidans*, and O_2 is introduced, making oxidation of pyrite possible. The acid formed can then leach into the surrounding streams (Figure 17.33).

Since acid mine drainage would not develop in the absence of bacterial activity, it might seem logical to prevent its occurrence by using chemicals or other agents that kill or inhibit the growth of bacteria. A number of such agents are available, including antibiotics, antiseptics, and organic acids. None of these agents have yet proved of practical value in controlling acid mine drainage. They may be too expensive to use in the enormous quantities necessary to have significant impact, or it may be difficult or impossible to deliver them at appropriate concentrations to the active sites of acid production. To date, the only effective way of controlling acid mine drainage is to seal or cover acid-bearing material, and to use mining practices that keep air and bacteria from the significant sites. The acid produced from coal-mining operations can be neutralized by use of lime. However, after a time the lime particles become coated with a layer of ferric hydroxide, and their neutralizing power is thus reduced.

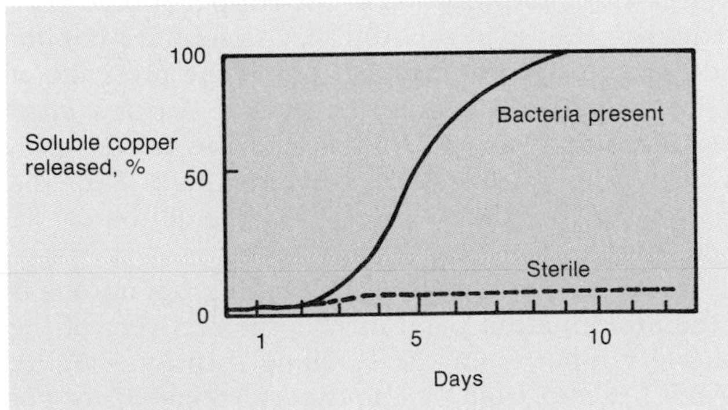

Figure 17.34 Effect of the bacterium *Thiobacillus ferrooxidans* on the leaching of copper from the mineral covellite. The leaching was done in a laboratory column, and the acid leach solution contained inorganic nutrients necessary for the development of the bacterium. The leaching activity was monitored by assaying for soluble copper in the leach solution at the bottom of the column. The leach solution was continuously recirculated, maintaining an essentially closed system.

Microbial leaching Sulfide forms highly insoluble minerals with many metals, and many ores used as sources of these metals are sulfides. If the concentration of metal in the ore is low, it may not be economically feasible to concentrate the mineral by conventional chemical means. Under these conditions, **microbial leaching** is frequently practiced. Microbial leaching is especially useful for *copper* ores, because copper sulfate, formed during the oxidation of the copper sulfide ores, is very water soluble. We have noted that sulfide itself, HS^-, oxidizes spontaneously in air. Most metal sulfides will also oxidize spontaneously, but the rate is very much slower than that of free sulfide. Bacteria such as *T. ferrooxidans* are able to catalyze a much faster rate of oxidation of the sulfide minerals, thus aiding in solubilization of the metal. The relative rate of oxidation of a copper mineral in the presence and absence of bacteria is illustrated in Figure 17.34. The susceptibility to oxidation varies among minerals, as seen in Table 17.10, and those minerals which are most readily oxidized are most amenable to microbial leaching.

Table 17.10 Sulfide minerals listed in order of increasing resistance to oxidation

Mineral	Formula
Pyrrhotite	FeS
Chalcocite	Cu_2S
Covellite	CuS
Tetrahedrite	$3Cu_2S \cdot Sb_2S_3$
Bornite	Cu_5FeS_4
Galena	PbS
Arsenopyrite	$FeAsS$
Sphalerite	ZnS
Pyrite	FeS_2
Enargite	$3Cu_2S \cdot As_2S_5$
Marcasite	FeS_2
Chalcopyrite	$CuFeS_2$
Molybdenite	MoS_2

(a) Brock

(b) Brock

(c) Brock

(d) Brock

Figure 17.35 The leaching of low-grade copper ores using bacteria. (a) A typical leaching dump. The low-grade ore has been dumped in a large pile. Pipes distribute the acidic leach water over the surface of the pile. The acidic water slowly percolates through the pile and exits at the bottom. (b) Effluent from a copper leaching dump. The acidic water is very rich in dissolved copper. (c) Recovery of dissolved copper by passage of the copper-rich water over metallic iron in a long flume. (d) A small pile of recovered copper metal removed from the flume, ready for further purification.

In the general microbial leaching process, low-grade ore is dumped in a large pile (the leach dump), and a dilute sulfuric acid solution (pH around 2) is percolated down through the pile (Figure 17.35*a*). The liquid coming out of the bottom of the pile (Figure 17.35*b*), rich in the mineral, is collected and transported to a precipitation plant (Figure 17.35*c*) where the metal is reprecipitated and purified. The liquid is then pumped back to the top of the pile and the cycle is repeated. As needed, more acid is added to maintain the low pH.

There are several mechanisms by which the bacteria can catalyze oxidation of the sulfide minerals. To illustrate these, examples will be used of the oxidation of two copper minerals, chalcocite, Cu_2S, in which copper has a valence of $+1$, and covellite, CuS, in which copper has a valence of $+2$. As illustrated in Figure 17.36, *T. ferrooxidans* is able to oxidize monovalent copper in chalcocite (CuS) to divalent copper, thus removing some of the copper in the soluble form, Cu^{2+}, and forming the mineral covellite. Note that in this reaction, there is no change in the valence of sulfide, the bacteria apparently utilizing the reaction Cu^+ to Cu^{2+} as a source of energy. This is analogous to the oxidation by the same bacterium of Fe^{2+} to Fe^{3+}.

A second mechanism, and probably the most important in most mining operations, involves an indirect oxidation of the copper ore with *ferric* ions that were formed by the bacterial oxidation of ferrous ions (Figure 17.36). In almost any ore, pyrite is present, and the oxidation of this pyrite (see Section 17.16) leads to the formation of ferric iron. Ferric iron is a very good oxidant for sulfide minerals, as has already been illustrated in the process of pyrite oxidation itself, and reaction of the copper sulfide with ferric iron results in the solubilization of the copper and the formation of ferrous iron. In the presence of O_2, at the acid pH values involved, *T. ferrooxidans* reoxidizes the ferrous iron back to the ferric form, so that it can oxidize more copper sulfide. Thus the process is kept going indirectly by the action of the bacterium on iron.

Another source of iron in leaching operations is at the precipitation plant used in the recovery of the soluble copper from the leaching solution (Figure 17.35*c*). Scrap iron, Fe^0, is used to recover copper from the leach liquid by the reaction shown in the lower part of Figure 17.36, and this results in the formation of considerable Fe^{2+}. In most leaching operations, the Fe^{2+}-rich liquid remaining after the copper is removed is conducted to an oxidation pond, where *T. ferrooxidans* proliferates and forms Fe^{3+}. Acid is added at the pond to keep the pH low, thus keeping the Fe^{3+} in solution, and this ferric-rich liquid is then pumped to the top of the pile and the

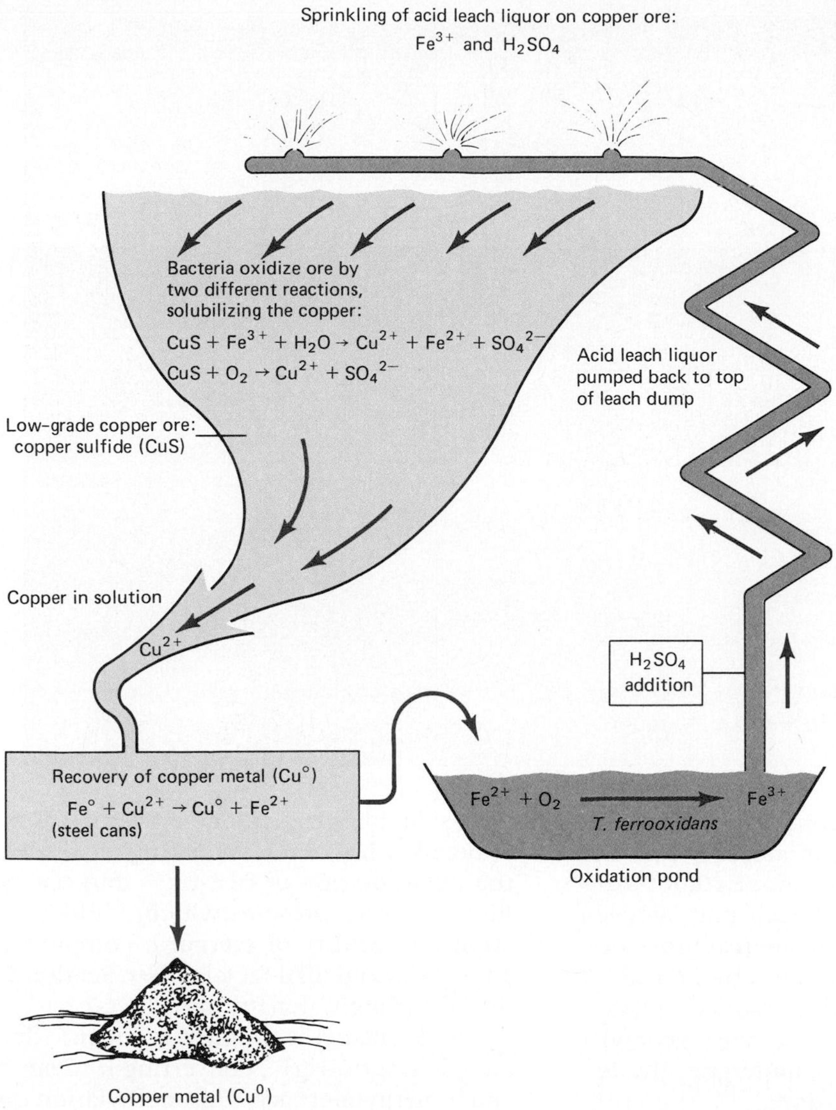

Figure 17.36 Arrangement of a leaching pile and reactions involved in the microbial leaching of copper sulfide minerals.

Fe^{3+} is available to oxidize more sulfide mineral.

Because of the huge dimensions of copper leach dumps, penetration of oxygen from air is poor, and the interior of these piles usually becomes anaerobic. Although most of the reactions written in Figure 17.36 require molecular O_2, it is also known that *T. ferrooxidans* can use Fe^{3+} as an electron acceptor in the absence of O_2, and thus catalyze the oxidation reactions anaerobically. Because of the large amounts of Fe^{3+} added to the leach solution from scrap iron, the process can continue to occur under anaerobic conditions.

17.18 Biogeochemical Cycles: Trace Metals and Mercury

Trace elements are those elements that are present in low concentrations in rocks, waters, and the atmosphere. Some trace elements (for example, cobalt, copper, zinc, nickel, molybdenum) are nutrients (see Section 4.1), but a number of trace elements in high concentrations are actually toxic to organisms. Of these toxic elements, several are sufficiently volatile so that they exhibit significant atmospheric transport, and hence are of some environmental concern. These include mercury, lead, arsenic, cadmium, and selenium. Many of these trace elements undergo redox reactions catalyzed by microorganisms, and several are also converted into organic form (alkylated) via microbial action. Because of environmental concern and significant microbial involvement, we discuss the biogeochemistry of the element mercury.

Mercury Although mercury is present in quite low concentrations in most natural environments, it is a widely used industrial product, and is the active component of many pesticides which have been introduced into the environment. Because of its unusual ability to be concentrated in living tissues and

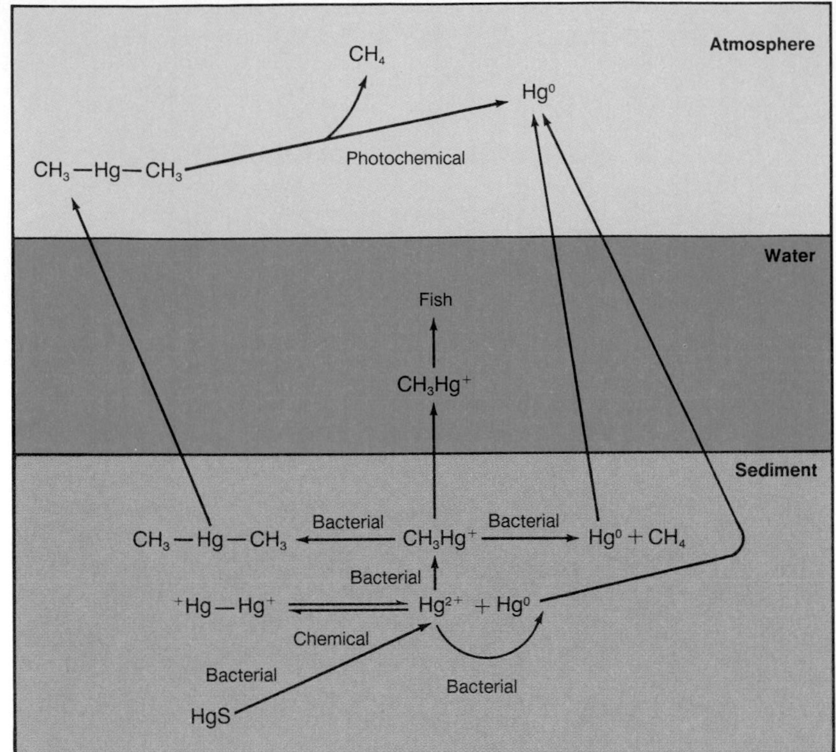

Figure 17.37 Bacterial activities influencing the global cycle of mercury.

its high toxicity, mercury is of considerable environmental importance. The mining of mercury ores and the burning of fossil fuels release about 40,000 *tons* of mercury into the environment each year; an even greater amount is released by geochemical processes. How is this mercury acted upon microbiologically?

In nature, mercury exists in three oxidation states, Hg^0, Hg^+, and Hg^{2+}. Monovalent mercury generally exists as the dimer, $^+Hg\text{–}Hg^+$. It undergoes the following chemical disproportionation:

$$^+Hg\text{–}Hg^+ \rightleftharpoons Hg^{2+} + Hg^0$$

The main mercury ore is the sulfide, HgS, called *cinnabar*. The solubility of HgS is extremely low, so that in anaerobic environments mercury will be present primarily in this form. But upon aeration, oxidation of HgS can occur, primarily by thiobacilli, leading to the formation of mercuric ion, Hg^{2+}. The soluble Hg^{2+} is quite toxic, but many bacteria have a detoxification mechanism for converting Hg^{2+} into elemental mercury, Hg^0. An NADP-linked Hg^{2+} reductase (that is frequently coded for by a plasmid) catalyzes the following reaction:

$$Hg^{2+} + NADPH + H^+ \rightleftharpoons Hg^0 + 2H^+ + NADP^+$$

A periplasmic Hg^{2+} "trapping" protein has been identified in a species of *Pseudomonas* that has a very high affinity for the mercuric ion. Hg^{2+} is trapped between two cysteine residues on the protein forming R–S–Hg–S–R; the mercuric ion is then transported across the plasma membrane where it is immediately reduced to Hg^0. Such a trapping mechanism prevents the incorporation of free Hg^{2+} into the cysteine residues of other proteins which could cause denaturation (the ability of mercuric compounds to act as antiseptics and disinfectants (see Section 9.19) is due to their protein denaturing effects).

Some bacteria employ a second detoxification mechanism for Hg^{2+}, converting it to methylmercury and dimethylmercury. This methylation involves a vitamin B_{12} coenzyme and can be represented as follows:

$$Hg^0 + CH_3 - B_{12} \rightarrow CH_3 - Hg^0 \quad \textit{methylmercury}$$

$$CH_3 - Hg^+ + CH_3 - B_{12} \rightarrow CH_3 - Hg - CH_3 \quad \textit{dimethylmercury}$$

Both methylmercury and dimethylmercury are lipophilic and tend to be concentrated in cellular lipids. Methylmercury is about 100 times more toxic than Hg^+ or Hg^{2+}, and is concentrated to a considerable extent in fish, where it acts as a potent neurotoxin, eventually causing death. In addition, dimethylmercury is volatile and can be transported to the atmosphere. The rate of formation of dimethylmercury is considerably slower than that of methylmercury, and the steady-state concentrations of the two methylated forms will depend on the rate of synthesis and rate of removal. Some microorganisms possess a detoxification mechanism for methylmercury, reducing it to Hg^0 plus methane. A summary of these transformations is given in Figure 17.37.

A variety of plasmids isolated from both Gram-positive and Gram-negative bacteria have been found to code for resistance to the effects of heavy metals. Certain antibiotic resistance plasmids also have genes for resistance to mercury and arsenic. Other plasmids code only for heavy-metal resistances. A large plasmid isolated from *Staphylococcus aureus* has been found to code for resistance to mercury, cadmium, arsenate, and arsenite. The mechanism of resistance to any specific metal varies. For example, arsenate and cadmium resistances are due to the action of enzymes that facilitate the immediate pumping out of any arsenate or cadmium ions incorporated, while resistance to mercury involves mechanisms of detoxifying the heavy metal.

17.19 Petroleum and Natural Gas (Methane) Biodegradation

Petroleum Microbial decomposition of petroleum and petroleum products is of considerable economic and environmental importance. Since petroleum is a rich source of organic matter and the hydrocarbons within it are readily attacked aerobically by a variety of microorganisms, it is not surprising that when petroleum is brought into contact with air and moisture, it is subject to microbial attack.

Hydrocarbon-oxidizing bacteria and fungi are the main agents responsible for decomposition of oil and oil products. A wide variety of bacteria, several molds and yeasts, and certain cyanobacteria and green algae have been shown to be able to oxidize hydrocarbons. Bacteria and yeasts, however, appear to be the prevalent hydrocarbon degraders in aquatic ecosystems. Small scale oil pollution of aquatic and terrestrial ecosystems from human as well as natural activities is very common, and hence it is not surprising that a diverse microbial population exists capable of using hydrocarbons as an electron donor.

Hydrocarbon-oxidizing microorganisms develop rapidly within oil films and slicks. However, as we saw in Section 16.23, aliphatic hydrocarbons are not fermentable. Thus, significant aliphatic hydrocarbon oxidation occurs only in the presence of O_2; if the oil gets carried into anaerobic bottom sediments, it will not be decomposed and may remain in place for many years (natural oil deposits in anaerobic environments are millions of years old). Even in aerobic environments, hydrocarbon-oxidizing microbes can act only if other environmental conditions, such as temperature, pH, and inorganic nutrients, are adequate. Because oil is insoluble in water and is less dense, it floats to the surface and forms slicks. Hydrocarbon-oxidizing bacteria are able to attach to insoluble oil droplets, and can often be seen there in large numbers (Figure 17.38). The action of these bacteria eventually leads to decomposition of the oil and dispersal of the slick.

Interfaces where oil and water meet often occur on a massive scale. It is virtually impossible to keep moisture from bulk storage tanks; it accumulates as a layer of water beneath the petroleum. Gasoline-storage tanks (Figure 17.39) are thus potential habitats for hydrocarbon-oxidizing microbes, which can accumulate and grow at the oil-water interface. Gasoline generally has additional chemicals (added to aid combustion and inhibit corrosion in engines) which inhibit the growth of microbes, but not all fuels have such additives.

Methane oxidation Methane is the simplest hydrocarbon. We discussed the biological formation of methane in Section 17.12. Methane is an important microbial substrate for a specific group of bacteria, the **methanotrophic bacteria**. We discuss the taxonomy and diversity of methanotrophic bacteria in Section 19.7; here we note some of their important ecological activities. Although a few bacteria are known which can oxidize methane anaerobically, anaerobic methane oxidation does not seem to be an important biogeochemical process. Thus, methane is to a great extent stable in anaerobic environments and

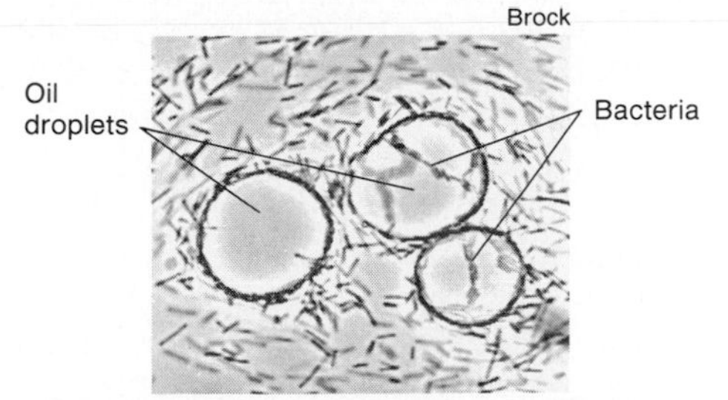

Figure 17.38 Hydrocarbon-oxidizing bacteria in association with oil droplets. The bacteria are concentrated in large numbers at the oil-water interface but are not within the droplet. Magnification, 610×.

Figure 17.39 Bulk fuel storage tanks, where massive microbial growth may occur at oil-water interfaces.

the methane formed through the action of the methanogenic bacteria often accumulates in deep subsurface deposits called *natural gas*. It is only when methane is brought back to the surface, where it is exposed to oxygen, O_2, that it is subject to attack by methanotrophs. Methanotrophs are responsible for the recycling of carbon from CH_4 back into CO_2 (see Figure 17.22). Methane is also of interest because it can be used as an energy source for the large-scale growth of microorganisms for use as possible food sources (single-cell protein, see Section 10.14).

17.20 Biodegradation of Xenobiotics

We defined the term *xenobiotic* in Section 16.18, noting that xenobiotics were chemically synthesized compounds that have never existed naturally. Thus, organisms capable of using them may not exist in nature. Some of the most widely distributed xenobiotics are the **herbicides** and **pesticides**. The use of herbicides and pesticides has increased dramatically in the past generation. These compounds are of a wide variety of chemical types, such as phenoxyalkyl carboxylic acids, substituted ureas, nitrophenols, chlorinated organic acids, phenylcarbamates, and others (Figure 17.40). Some of these substances are suitable as carbon sources and electron donors for certain soil microorganisms, whereas others are not. If a substance can be attacked by microorganisms, it will eventually disappear from the soil. Such degradation in the soil is usually desirable, since toxic accumulations of the compound are avoided. However, even closely related compounds may differ remarkably in their degradability, as is shown for 2,4-dichlorophenoxyacetic acid (2,4-D) and 2,4,5-trichlorophenoxyacetic acid (2,4,5-T) in Figure 17.41. The relative persistence rates of a number of herbicides is shown in Table 17.11. However, these figures are only approximate since a variety of environmental factors, such as temperature, pH, aeration, and organic-matter content of the soil, influence decomposition. Some of the chlorinated insecticides are so indestructible that they have persisted for over 10 years. Disappearance of a pesticide from an ecosystem does not necessarily mean that it was degraded by microorganisms, since pesticide loss can also occur by volatilization, leaching, or spontaneous chemical breakdown.

The organisms that are able to metabolize pesticides and herbicides are fairly diverse, including genera of both bacteria and fungi. Some pesticides serve well as carbon and energy sources and are oxidized completely to CO_2. However, other compounds are much more recalcitrant, and are attacked only slightly or not at all, although they may often be degraded either partially or totally provided some other organic material is present as primary energy source, a phenomenon called **co-metabolism**. Where the breakdown is only partial, the microbial degradation product of a pesticide may sometimes be even more toxic than the original compound.

The existence of organisms able to metabolize xenobiotics is of considerable evolutionary interest, since these compounds are completely new to the earth in the past 50 years or so. Observations on the rapidity with which organisms metabolizing new compounds arise can give us some idea of the rates

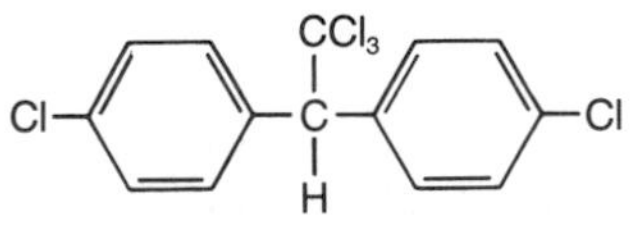

DDT; dichlorodiphenyltrichloroethane (a chlorinated insecticide)

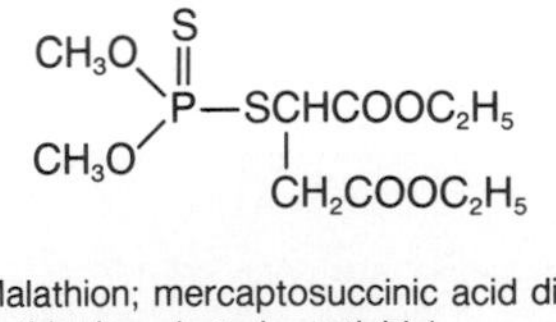

Malathion; mercaptosuccinic acid diethyl ester (a thiophosphate insecticide)

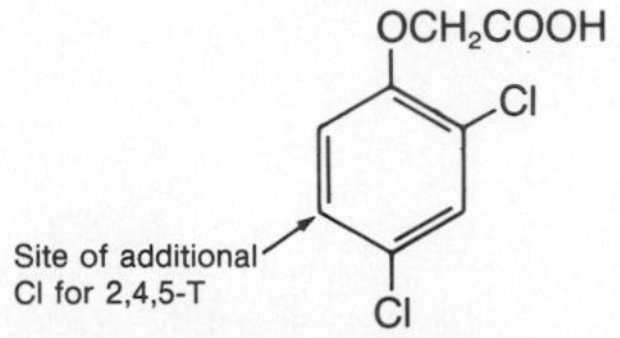

Site of additional Cl for 2,4,5-T

2,4-D; 2,4-dichlorophenoxy acetic acid (a chlorinated herbicide)

Figure 17.40 Some xenobiotic compounds. Although none of these compounds exist naturally, various microorganisms exist (or can be developed experimentally) which will break them down.

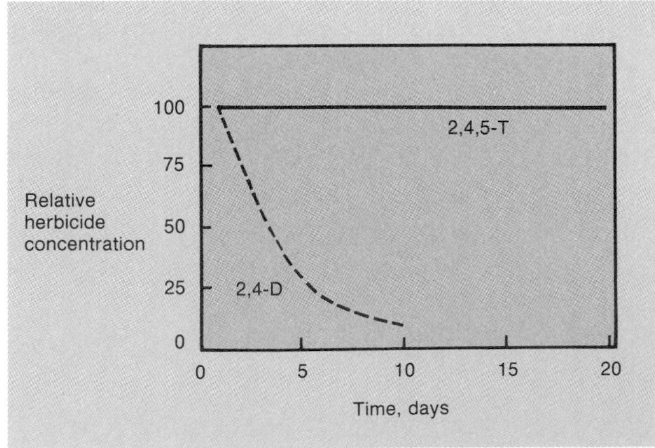

Figure 17.41 Rate of microbial decomposition in a soil suspension of two herbicides, 2,4-dichlorophenoxyacetic acid (2,4-D) and 2,4,5-trichlorophenoxyacetic acid (2,4,5-T). See Figure 17.40 for the structures of these two compounds.

Table 17.11 Persistence of herbicides and insecticides in soils

Substance	Time for 75–100% disappearance
Chlorinated insecticides:	
DDT[1,1,1-trichloro-2,2-bis(p-chlorophenyl)ethane]	4 years
Aldrin	3 years
Chlordane	5 years
Heptachlor	2 years
Lindane (hexachloro-cyclohexane)	3 years
Organophosphate insecticides:	
Diazinon	12 weeks
Malathion	1 week
Parathion	1 week
Herbicides:	
2,4-D(2,4-dichloro-phenoxyacetic acid)	4 weeks
2,4,5-T(2,4,5-trichloro-phenoxyacetic acid)	20 weeks
Dalapin	8 weeks
Atrazine	40 weeks
Simazine	48 weeks
Propazine	1.5 years

of microbial evolution in general. In Section 7.6 we discussed the evolution in past decades of plasmids conferring resistance to antibiotics. The evolution of pesticide-degrading bacteria seems to be a similar case. Chemostat enrichment cultures (see Section 17.3) have recently yielded bacteria capable of degrading 2,4,5-T. These organisms appear to be common *Pseudomonas* species which are now capable of growing on this pesticide as sole source of carbon and energy (Figure 17.42). Plasmids carrying the genetic information necessary to degrade the related, but much less recalcitrant herbicide, 2,4-D (Figure 17.41), have been known for some time. Portions of these plasmids appear to be conserved in a strain of *Pseudomonas cepacia* capable of growth on 2,4,5-T. There is good evidence that plasmids are important in the ability of this strain to degrade 2,4,5-T, but the complete genetics of the process has not yet been worked out. However, the point is that plasmids, or portions of plasmids that recombine with other genetic elements, perhaps as transposons, can catalyze the rapid evolution of new metabolic capabilities in bacteria previously unable to degrade the compound.

In many cases the genes for pesticide degradation are clustered on the plasmid and could thus be easily manipulated by genetic engineering techniques for dissemination to a large number of Gram-negative bacteria. In the specific case of 2,4,5-T, this could lead to a significant decrease in the persistance of this compound in soil, thus decreasing the toxic threat to wildlife, domestic animals and humans. Certain problems have been encountered in attempts to transfer such plasmids to other species, however, including the instability of the plasmid in certain organisms and the frequent loss of the plasmid if the organism is grown on substrates other than the pesticide. Assuming these problems could be overcome and new plasmids constructed which coded for enzymes degrading a variety of pesticides, a major xenobiotic problem might be overcome.

17.21 Sewage and Wastewater Microbiology

Wastewater treatments are processes in which microorganisms play crucial roles, and illustrate well some of the principles of biogeochemistry discussed above. Wastewaters are materials derived from domestic sewage or industrial processes, which for reasons of public health and for recreational, economic, and aes-

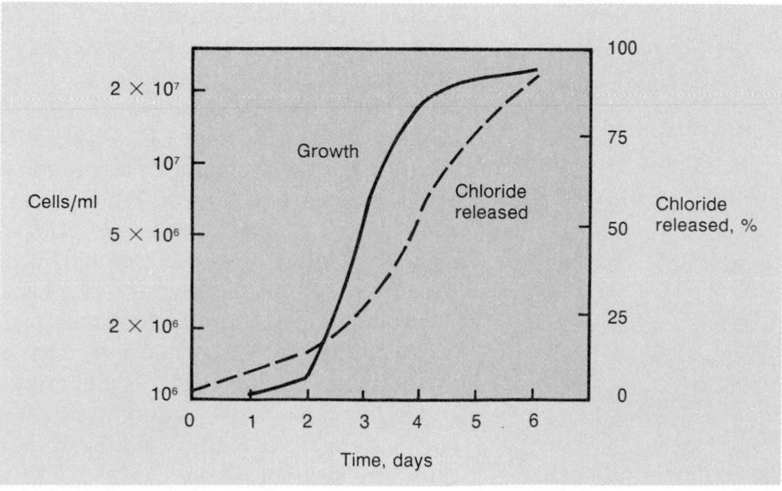

Figure 17.42 Growth of *Pseudomonas cepacia* on the herbicide 2,4,5-T as sole source of carbon and energy. The strain was enriched from nature using a chemostat to keep the concentration of herbicide low. Growth here is on 1.5 g/l of 2,4,5-T. The release of chloride from the molecule is indicative of degradation.

thetic considerations cannot be disposed of merely by discarding them untreated into convenient lakes or streams. Rather, the undesirable materials in the water must first be either removed or rendered harmless. Inorganic materials such as clay, silt, and other debris are removed by mechanical and chemical methods, and microbes participate only casually or not at all. If the material to be removed is *organic* in nature, however, treatment usually involves the activities of microorganisms, which oxidize and convert the organic matter to CO_2. Wastewater treatment usually also results in the destruction of pathogenic microorganisms, thus preventing these organisms from getting into rivers or other supply sources.

Levels of sewage treatment Sewage treatment is generally a multistep process employing both physical and biological treatment steps (Figure 17.43). **Primary treatment** of sewage consists only of physical separations. Sewage entering the treatment plant is passed through a series of grates and screens that remove large objects and then the effluent is left to settle for a number of hours to allow suspended solids to sediment.

Because of the high nutrient loads that remain in sewage effluent following primary treatment, those municipalities that treat sewage no further than the primary stage usually suffer from extremely polluted water when the sewage is dumped into adjacent waterways. This is why the majority of modern sewage processing plants employ **secondary treatment** processes to reduce the biochemical oxygen demand (BOD, see Section 17.6) of the sewage to reasonable levels before releasing it to natural waterways. Secondary treatment is intimately tied to microbiological processes as described in the following sections.

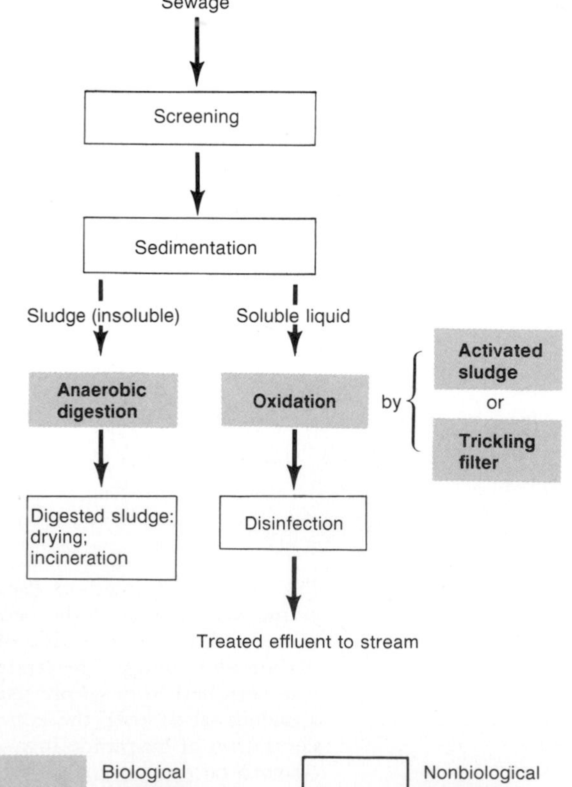

Figure 17.43 An overview of sewage treatment processes.

Tertiary treatment is the most complete method of treating sewage, but has not been widely adopted because it is so expensive. Tertiary treatment is a physiochemical process employing precipitation, filtration, and chlorination to sharply reduce the levels of inorganic nutrients, especially phosphate and nitrate, from the final effluent. Wastewater receiving proper tertiary treatment is unable to support extensive microbial growth, and in many cases is of such high quality that it has been pumped directly into the municipal water supply! We focus our discussion here on secondary treatment processes, for it is these that have the most microbial components and are the most widely practiced in sewage treatment facilities today.

Anaerobic secondary treatment processes Anaerobic sewage treatment involves a complex series of digestive and fermentative reactions carried out by a host of different bacterial species. The net result is the conversion of organic materials into CO_2 and methane gas (CH_4; see Sections 16.18 and 17.12), the latter of which can be removed and burned as a source of energy. Since both end products, CO_2 and CH_4, are volatile, the liquid effluent is greatly decreased in organic substances. The efficiency of a treatment process is expressed in terms of the percent decrease of the initial biochemical oxygen demand (BOD); the efficiency of a well-operated plant can be 90 percent or greater, depending on the nature of the organic waste.

Anaerobic decomposition is usually employed for the treatment of materials that have much *insoluble* organic matter, such as fiber and cellulose, or for concentrated industrial wastes. The process depends on interspecies hydrogen transfer reactions (see Section 17.12) and can be summarized as: (1) initial digestion of the macromolecular materials by extracellular polysaccharidases, proteases, and lipases to soluble materials; (2) fermentation of the soluble materials to fatty acids, H_2 and CO_2, by acid-producing fermentative organisms; (3) fermentation of the fatty acids to acetate, CO_2, and H_2; and (4) the conversion of H_2 plus CO_2 and acetate to CH_4 by methanogenic bacteria. Unlike in the rumen, where H_2 is the main precursor for methanogenesis, methanogens in bioreactors such as sludge digestors use *acetate* as the main methanogenic precursor. This is probably because the ruminant serves as an active sink for acetate and other fatty acids, whereas in the bioreactor acetate accumulates and is available for utilization by the methanogens.

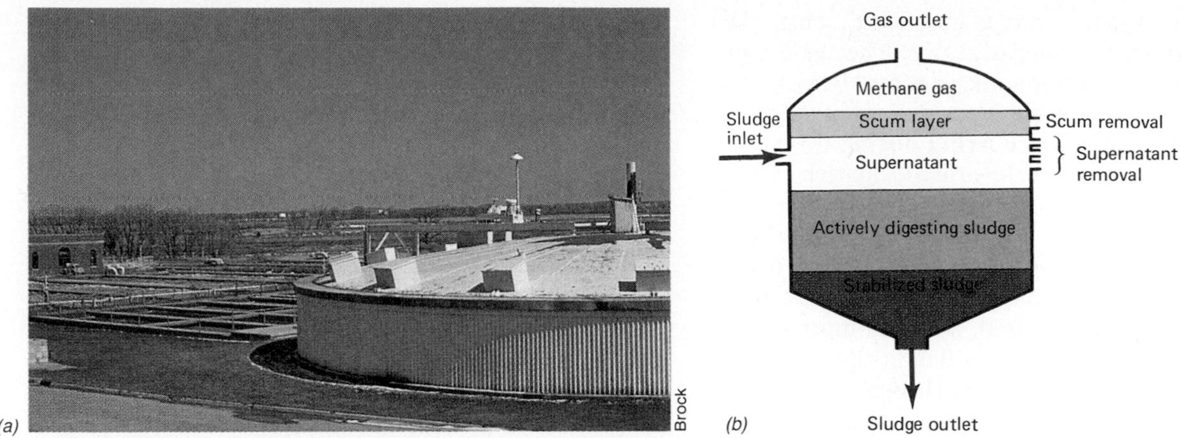

Figure 17.44 (a) Anaerobic sludge digestor. Only the top of the tank is shown; the remainder is underground. (b) Inner workings of an anaerobic sludge digestor.

An anaerobic decomposition process is operated semicontinuously in large enclosed tanks called **sludge digestors**, into which the untreated material is introduced and from which the treated material is removed at intervals (Figure 17.44). The retention time in the tank is on the order of 2 weeks to a month. The solid residue consisting of indigestible material and bacterial cells is allowed to settle and is removed periodically and dried for subsequent burning or burial.

Aerobic secondary treatment process There are several kinds of aerobic decomposition processes used in sewage treatment. A **trickling filter** is basically a bed of crushed rocks, about 6 ft thick, on top of which the liquid containing organic matter is sprayed. The liquid slowly trickles through the bed, the organic matter adsorbs to the rocks, and microbial growth takes place. The complete mineralization of organic matter to carbonate, ammonia, nitrate, sulfate, and phosphate occurs.

The most common aerobic treatment system is the **activated sludge** process. Here, the wastewater to be treated is mixed and aerated in a large tank (Figure 17.45). Some municipalities force pure oxygen through their activated sludge tanks to increase the levels of dissolved oxygen thereby increasing the rate of biodegradation. Slime-forming bacteria (primarily a species called *Zoogloea ramigera*) grow and form flocs (so-called **zoogloeas**), and these flocs form the substratum to which protozoa and other animals attach. Occasionally, filamentous bacteria and fungi also are present. The basic process of oxidation is similar to that in a trickling filter. The effluent containing the flocs is pumped into a holding tank or clarifier, where the flocs settle. Some of the floc material is then re-

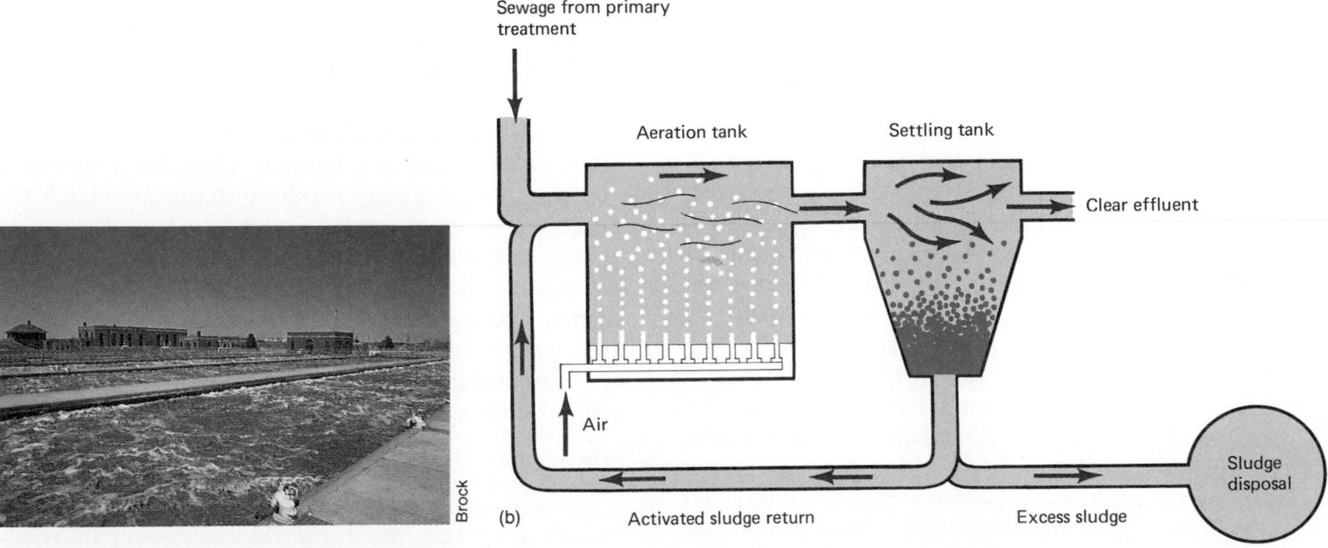

Figure 17.45 Activated sludge process. (a) Aeration tank of an activated sludge installation. (b) Inner workings of an activated sludge installation.

turned to the aerator to serve as inoculum, while the rest is sent to the anaerobic digestor. The residence time in an activated sludge tank is generally 5 to 10 hours, too short for complete oxidation of organic matter. The main process occurring during this short time is *adsorption* of soluble organic matter to the floc, and incorporation of some of the soluble materials into microbial cell material. The BOD of the liquid is thus considerably reduced by this process (75 to 90 percent), but the overall BOD (liquid plus solids) is only slightly reduced, because most of the adsorbed organic matter still resides in the floc. The main process of BOD destruction thus occurs in the anaerobic digestor to which the floc is transferred.

17.22 Plant-Microbe Interactions

As microbial habitats, plants are clearly vastly different from animals (see Chapter 11). Compared with warm-blooded animals, plants vary greatly in temperature, both diurnally and throughout the year and, compared with the complex circulatory system of animals, the communication system of the plant is only poorly developed, so that transfer of microorganisms from one part to another is relatively inefficient. The aboveground parts of the plant, especially the leaves and stems, are subjected to frequent drying, and for this reason many plants have developed waxy coatings that retain moisture and serve to keep out microorganisms. The roots, on the other hand, exist in an environment whose moisture is less variable and whose nutrient concentrations are higher. For this reason, the roots of plants are a main area of microbial action.

The **rhizosphere** is the region immediately outside the root; it is a zone where microbial activity is usually high. The bacterial count is almost always higher in the rhizosphere than it is in regions of the soil devoid of roots, often many times higher. This is because roots excrete significant amounts of sugars, amino acids, and vitamins, which promote such an extensive growth of bacteria and fungi that these organisms often form microcolonies on the root surface. There is some evidence that rhizosphere microorganisms benefit the plant by promoting the absorption of nutrients, but this is probably only of minor importance. The numbers and kinds of microorganisms on or near root surfaces is highly variable depending on the plant type and soil nutrient/water status.

The **phyllosphere** (also called the **phylloplane**) is the surface of the plant leaf, and under conditions of high humidity, such as develop in wet forests in tropical and temperate zones, the microbial flora of leaves may be quite high. In tropical rain forests, the phyllosphere bacteria often produce gums or slimes, which presumably help bacterial cells stick to leaves under the onslaught of heavy tropical rains. Many of the bacteria on leaves fix nitrogen (see Sections 16.24 and 17.24), and nitrogen fixation presumably aids these organisms in growing with the predominantly carbohydrate nutrients provided by leaves. Some of the nitrogen fixed by these bacteria may be absorbed by the plant, although there is little evidence that this occurs on a large scale, except for the stem-nodulating rhizobia discussed in Section 17.24.

One interesting effect of a leaf bacterium on plant leaves is the *ice-nucleation phenomenon*. The leaf-inhabiting bacterium *Pseudomonas syringae* brings about the formation of ice crystals on the leaves of plants at temperatures higher than would otherwise be the case, and hence plays a role in *frost damage*. When the environmental temperature drops below freezing, the water on the surfaces of plant leaves normally becomes supercooled, and freezing only occurs at temperatures well below the regular freezing point. If *P. syringae* cells are present, they serve as nuclei for the formation of ice crystals, so freezing occurs at a higher temperature. This ice nucleation process is brought about by a specific chemical component of the outer membrane layer of *P. syringae*. Thus, in this indirect way *P. syringae* brings about frost damage.

Because of the agricultural significance of frost damage, it would be desirable to prevent this frost-inducing action of *P. syringae*. One approach that has been considered is to develop a strain of *P. syringae* from which the gene for the ice nucleation structure has been deleted (*ice-minus* strain) and then spray plant leaves with this strain early in the season, before the leaves have become colonized with normal (*ice-plus*) *P. syringae*. The thought is that the ice-minus bacteria will become established first and prevent, by competitive exclusion, the establishment of ice-plus bacteria. Testing of such an approach is being carried out in frost-sensitive agricultural areas of northern California.*

Plant disease resistance At least three major microbial defense mechanisms have been identified in plants. Many plants produce toxic chemicals that either prevent microbial growth or actually kill invading microorganisms; examples include the phenols produced by onions and many cereal and root plants, hydrogen cyanide produced by flax, and the allyl sulfides produced by onions.

Phytoalexins are an important class of antimicrobial compounds produced by a variety of plants.

*The ice nucleation power of *Pseudomonas* species has also been put to positive use to enhance ice nucleation in the snow-making process in ski resorts. Killed bacteria (which still retain their ice-nucleating power) are mixed with the water before it is sprayed into the atmosphere by the snow-making equipment.

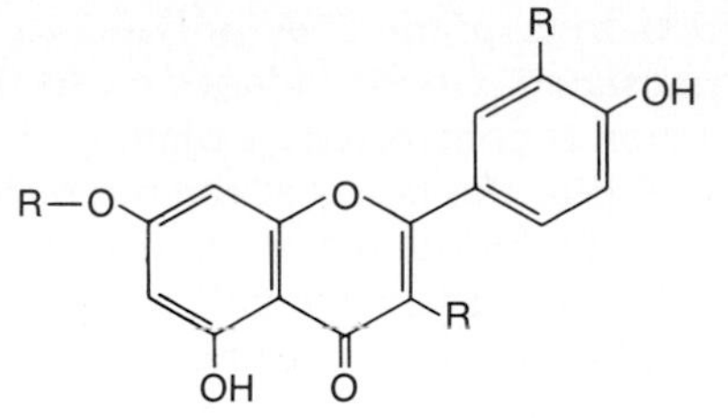

Figure 17.46 Parent structure of the isoflavanoid-derived phytoalexins—plant antimicrobial agents. The R groups represent substituents that can vary from one compound to another.

Phytoalexins are a chemically heterogenous group of low-molecular-weight aromatic compounds, a number of which are derived from isoflavanoids (Figure 17.46). Unlike the toxic compounds just mentioned which are produced by the plant nonspecifically, phytoalexins are not present in healthy plants but are produced in response to microbial invasion. Phytoalexin synthesis is induced by a variety of microbial components, including fungal β-glucans, glycoproteins, and acidic cell wall components. Phytoalexin synthesis is also elicited by pectin components released from the plant's own cell wall. The pectin elicitors are all low-molecular-weight polysaccharides, frequently of unusual chemical structure. Presumably these pectin fragments serve as a chemical signal to the plant that the plant cell wall has been damaged by the action of microbial pectinases.

Although poorly understood, plants also protect themselves from microbial attack by catalyzing the rapid killing of host cells in the vicinity of an infection site. This mechanism has been referred to as the **plant hypersensitive response**, and has been observed in a number of plant species. The hypersensitive response is characterized by water loss from the tissue surrounding the infection site leading to a collapse and flattening of the cells. The invading microorganisms are then trapped within the collapsed area, thus protecting the surrounding healthy tissues.

Thus, although plants lack the sophisticated microbial defense mechanisms of warm-blooded animals (see Chapters 11 and 12), they do have a variety of mechanisms for countering microbial invasion.

Bacterial-plant interactions With few exceptions, the relationships between *specific* bacteria and their plant hosts are poorly understood. Whether most heterotrophic soil bacteria develop only casual relationships with plants, or, alternatively, interact with plants in complex ways, is really not known. On the one hand, it could be argued that the beneficial aspects of plant-bacterial interactions are mainly slanted toward the bacterium: plants excrete relatively large amounts of various organic compounds and bacteria in the rhizosphere or phyllosphere can assimilate these compounds and use them as carbon and energy sources. On the other hand, one could envision complex, chemical interactions occurring between plants and bacteria, where the exchange of nutrients, growth factors, and even DNA, combine to play an intimate role in the success and survival of both partners. Our present relatively poor understanding of plant microbiology is such that neither scenario can be firmly supported or ruled out for *most* plant/bacterial relationships. However, in two instances, the interrelationship between plants and specific bacteria have been well characterized, even to the point of understanding the relationship in sophisticated molecular terms. These systems are the well-known *Rhizobium*-legume symbiosis, and the *Agrobacterium*-mediated crown gall disease of dicotyledenous plants. In the next two sections we discuss these interrelationships in some detail to give the student an appreciation for the potential of plant/bacterial interactions.

17.23 Agrobacterium/Plant Interactions

Agrobacterium and *Rhizobium* are closely related genera of Gram-negative flagellated rods capable of establishing tumorlike growths on certain plants. We discuss *Agrobacterium* here and *Rhizobium* in the next section. The genus *Agrobacterium* comprises organisms that cause the formation of tumorous growths on a wide variety of plants. The two species most widely studied are *A. tumefaciens* which causes *crown gall*, and *A. rhizogenes* which causes *hairy root*. Although plants often form benign accumulations of tissue, called a **callus**, when wounded, the growth induced by *A. tumefaciens* (Figure 17.47) is different in that the callus shows uncontrolled growth. It thus resembles tumor growth in animals, and considerable research on crown gall has been carried out with the idea that it may provide a model for how malignant growths occur in humans. *A. rhizogenes* causes a malignant mass of roots to emanate from the infected wound site, creating the condition known as hairy root.

In general, plant growth in tissue culture requires the presence of specific plant hormones, but crown gall or hairy root tissue is self-proliferating. Interestingly, once induced these tumors continue to grow in the absence of *Agrobacterium* cells. Thus, once *Agrobacterium* has brought about the induction of the tumorous condition, its presence is no longer necessary.

It is now well established that a large plasmid called the *Ti* (*t*umor *i*nduction) *plasmid* must be present in the *Agrobacterium* cells if they are to induce tumor formation. Following infection, a part of the Ti plasmid, called the *transfer DNA* (T-DNA), is integrated into the plant's genome. T-DNA carries the genetic information for tumor formation and also for the production of a number of modified amino acids

Figure 17.47 Photograph of tumor on tobacco plant caused by crown-gall bacteria of the genus *Agrobacterium*.

called **opines**. *Octopine* [N^2-(1,3-dicarboxyethyl)-L-arginine] and *nopaline* [N^2-(1,3-dicarboxypropyl)-L-arginine] are the two most common opines. Opines are produced by plant cells transformed by T-DNA and serve as a source of carbon and nitrogen for *Agrobacterium* cells. Studies with mutant strains of *A. tumefaciens* has shown that opine production or catabolism is not a prerequisite for tumor formation. The Ti plasmid also encodes genes for opine catabolism, several virulence factors, and for two phytohormones, **auxin** and **cytokinin**. As stimulators of cell growth auxin and cytokinin stimulate tumor development, and the ratio of one to the other during tumor development affects the final morphology of the tumor.

Plasmid transfer The mechanism of plasmid transfer from *Agrobacterium* to the plant is not yet known. However, it is known that *Agrobacterium cells* do not enter plant cells. Either the entire Ti plasmid or perhaps just the T-DNA portion of the Ti plasmid moves from bacterium to plant. There is some evidence that the T-DNA portion of the Ti plasmid is excised and transferred to the plant as a miniplasmid, that is, in covalently closed circular form (see Section 5.2 for a discussion of this form of DNA). In fact, it is hypothesized that one of the functions of the *vir* (*vir*ulence) genes known to reside on the Ti plasmid is to code for the proteins necessary for excision, recircularization, and transfer of the T-DNA to the plant.

The T-DNA integrates into the plant genome and is transcribed by a plant RNA polymerase and subsequently translated. Good evidence exists that T-DNA gene expression is controlled by plant regulatory elements; this despite the fact that the net result is damage to the plant! Obviously the relationship between plant and bacterium in crown gall disease is a highly developed and cooperative arrangement.

Recognition of *Agrobacterium* by the plant To initiate the tumorous state, cells of *Agrobacterium* must first attach to a wound site on the plant. Molecular evidence suggests that the recognition of *Agrobacterium* by plant tissue involves complementary receptor molecules on the surfaces of the bacterial and plant cells. It is likely that the plant receptor molecule is *pectin* and that the bacterial receptor is *lipopolysaccharide* (LPS). However, since most Gram-negative bacteria contain LPS, the structure of agrobacterial LPS obviously must be of unique character (there is some evidence that the Ti plasmid encodes genes affecting the composition of *A. tumefaciens* LPS).

Studies with mutants of *A. tumefaciens* have shown clearly that most functions necessary for attachment of the bacterium to plant are borne on the bacterial chromosome. Although LPS is thought to play a major role in attachment, low-molecular-weight carbohydrates called *β-glucans* (mentioned previously in connection with fungal stimulation of phytoalexins) are also thought to play some role in assisting agrobacteria in attachment to the plant wound site. Presumably β-glucans modify the bacterial cell wall in a way that facilitates binding to the plant cell. In addition to β-glucan and LPS, good evidence exists to show that attachment of *A. tumefaciens* to certain plant types, notably to carrot or tobacco tissue, is mediated by bacterial production of *cellulose* microfibrils. Following recognition and initial binding (mediated by LPS and β-glucan), the rapid synthesis of cellulose microfibrils anchors the inoculum to the wound site and literally entraps the bacterial cells, forming large bacterial aggregates on the plant cell surface. Although helpful in mediating colonization and rapid growth of agrobacteria at the site of infection, studies with mutants of *A. tumefaciens* incapable of synthesizing cellulose indicates that cellulose production is not necessary for induction of the tumorous state.

Genetic engineering with the Ti plasmid Because of its natural ability to transfer DNA to plant cells, *A. tumefaciens* has been used as a vector for introducing foreign DNA into plants by means of recombinant DNA techniques (see Section 8.10). The general protocol involves opening the Ti plasmid with restriction enzymes, inserting the foreign gene(s) of interest, ligating the plasmid, and transforming agrobacterial cells (Figure 17.48). By using Ti plasmids

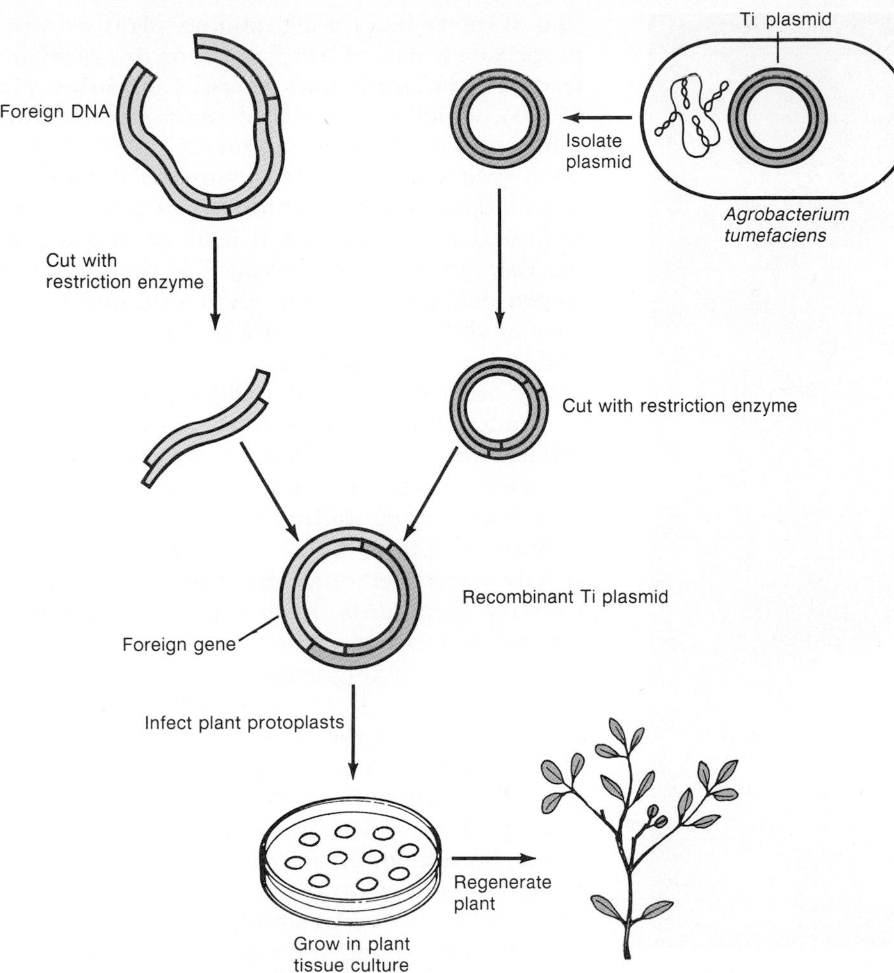

Figure 17.48 The use of *Agrobacterium* in genetic engineering.

containing mutations in key tumor function genes, it is possible to transform plant tissue cultures without initiating tumor formation.

Use of the Ti plasmid as a natural plant genetic engineer has been very successful with dicotyledenous plants, especially potatoes, tobacco, petunias, and tomatoes (Figure 17.49). A major thrust of research in this area has been attempts to increase the herbicide resistance of plants. In many cases herbicides act by inhibiting a key plant enzyme or protein necessary for photosynthesis. For example, the herbicide *glyphosate* kills plants by suppressing the activity of a key enzyme necessary for making aromatic amino acids. Such a herbicide kills both weeds and crop plants and thus must be used as a "preemergence herbicide," that is, before the crop plants emerge from the ground. However, glyphosate-resistant strains of the Gram-negative bacterium *Salmonella* (glyphosate inhibits the activity of the bacterial enzyme as well) have been obtained using standard genetic techniques, and the glyphosate resistance transferred to plants using the Ti plasmid. Tobacco plants containing the bacterial gene are capable of growing when sprayed with glyphosate.

Use of the Ti plasmid and other DNA transfer techniques to introduce foreign genes into plants is a major focus of plant biology today. A number of research groups at both universities and in the industrial sector are busy attempting to alter plants genetically in a variety of ways. Besides herbicide resistance studies, research projects are in progress aimed at improving the nutritional composition of plants, increasing resistance to plant pathogens, improving drought and salinity resistance, and understanding gene expression in plants. In the latter connection the introduction by an *A. tumefaciens* vector of the firefly luciferase gene into tobacco plants producing a plant that is now luminescent (Figure 17.49), offers plant biologists a visible marker for use as an indicator of gene expression, as a marker in genetic crosses, and as a general tool for studying plant molecular biology.

Initial excitement in plant molecular biology was created by proposals to genetically transfer major physiological processes, such as nitrogen fixation, from bacteria into cereal plants like corn and wheat. Although this idea is now thought to be unfeasible for a variety of reasons, such proposals clearly indicated the potential scope of human benefit to be derived from a better knowledge of plant microbiology and plant molecular biology. The crown gall-*A. tu-*

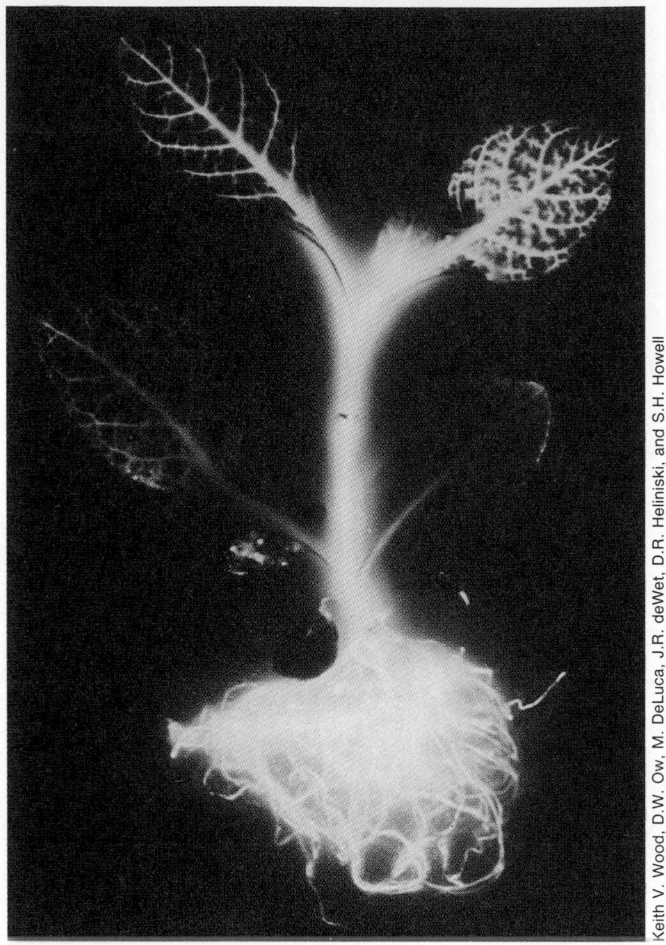

Keith V. Wood, D.W. Ow, M. DeLuca, J.R. deWet, D.R. Helinski, and S.H. Howell

Figure 17.49 Leaf from a tobacco plant, *Nicotiana tabacum*, that had been genetically engineered by insertion of the luciferase gene from the firefly, *Photinus pyralis*. Note that the leaf is now luminescent due to expression of the luciferase gene.

mefaciens system has served as an excellent model for understanding at the molecular level the process of genetic transfer between bacteria and plant. Further research in plant/bacterial interactions will likely expose new and perhaps even more useful associations which can be tapped to introduce foreign genes into plants and thereby further advance the field of plant molecular biology.

17.24 Rhizobium: Legume Symbiosis

One of the most interesting and important plant bacterial interactions is that between leguminous plants and bacteria of the genera *Rhizobium* and *Bradyrhizobium*. Legumes are a large group that includes such economically important plants as soybeans, clover, alfalfa, string beans, and peas, and are defined as plants which bear seeds in pods. *Rhizobium* and *Bradyrhizobium* are Gram-negative motile rods. Infection of the roots of a leguminous plant with the appropriate species of *Rhizobium* or *Bradyrhizobium* leads to the formation of **root nodules** (Figure 17.50), which are able to convert gaseous nitrogen into combined nitrogen, a process called nitrogen fixation (see Section 16.24). Nitrogen fixation by the legume-*Rhizobium* symbiosis is of considerable agricultural importance, as it leads to very significant increases in combined nitrogen in the soil. Since nitrogen deficiencies often occur in unfertilized bare soils, nodulated legumes are at a selective advantage under such conditions and can grow well in areas where other plants cannot (Figure 17.51).

Under normal conditions, neither legume nor *Rhizobium* alone is able to fix nitrogen; yet the interaction between the two leads to the development of nitrogen-fixing ability. In pure culture, the *Rhizobium* is able to fix N_2 alone when grown under strictly controlled *microaerophilic* conditions. Apparently *Rhizobium* needs some O_2 to generate energy for N_2 fixation, yet its nitrogenase (like those of other nitrogen-fixing organisms, see Section 16.24) is inactivated by O_2. In the nodule, precise O_2 levels are controlled by the O_2-binding protein **leghemoglobin**. This is a red, iron-containing protein which is always found in healthy, N_2-fixing nodules (Figure 17.52). Neither plant nor *Rhizobium* alone synthesizes leghemoglobin, but formation is induced through the interaction of these two organisms. Leghemoglobin serves as an "oxygen buffer" cycling between the oxidized (Fe^{3+}) and reduced (Fe^{2+}) forms, to keep free O_2 levels within the nodule at a low but constant level. The ratio of leghemoglobin-bound to free O_2 in the root nodule is on the order of 10,000:1.

Joe Burton

Figure 17.50 Soybean root nodules. The nodules develop by infection with *Bradyrhizobium japonicum*.

Figure 17.51 A field of unnodulated (left) and nodulated soybean plants growing in nitrogen-poor soil.

About 90 percent of all leguminous species are capable of becoming nodulated. There is a marked specificity between species of legume and strains of *Rhizobium*. A single *Rhizobium* strain is generally able to infect certain species of legumes and not others. A group of *Rhizobium* strains able to infect a group of related legumes is called a *cross-inoculation group*. The major rhizobial cross-inoculation groups are listed in Table 17.12. Even if a *Rhizobium* strain is able to infect a certain legume, it is not always able to bring about the production of nitrogen-fixing nodules. If the strain is *ineffective* the nodules formed will be small, greenish-white, and incapable of fixing nitrogen; if the strain is *effective*, on the other hand, the nodule will be large, reddish (Figure 17.52), and nitrogen-fixing. Effectiveness is determined by genes in the bacterium that can be lost by mutation or gained by genetic transformation (see "Genetics of nodule formation," below).

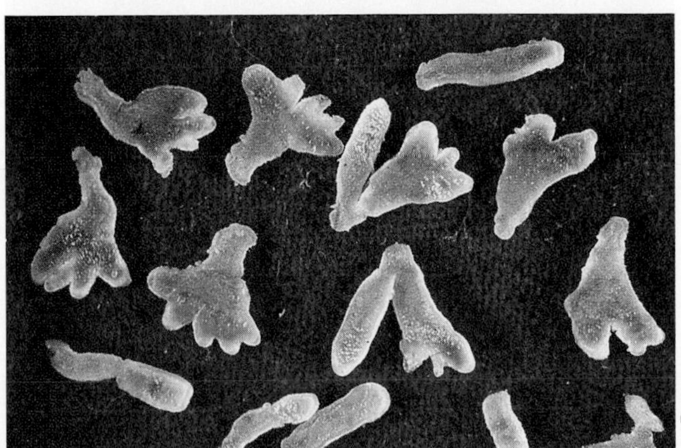

Figure 17.52 Sections of root nodules from the legume *Coronilla varia*, showing the reddish pigment leghemoglobin.

Stages in nodule formation The stages in the infection and development of root nodules are now fairly well understood (Figure 17.53). They include:

1. **Recognition** of the correct partner on the part of both plant and bacterium and **attachment** of the bacterium to root hairs.
2. **Invasion** of the root hair by the bacterial formation of an infection thread.
3. **Travel** to the main root via the infection thread.
4. Formation of deformed bacterial cells, **bacteroids**, within the plant cells and the development of the nitrogen-fixing state.
5. Continued plant and bacterial division and formation of the mature **root nodule**.

We now discuss some of these stages in nodule formation in more detail.

The roots of leguminous plants secrete a variety

Table 17.12 Major cross-inoculation groups of leguminous plants

Plant	Nodulated by	Growth rate*
Pea	*Rhizobium leguminosarum*	Fast
Bean	*Rhizobium phaseoli*	Fast
Clover	*Rhizobium trifolii*	Fast
Alfalfa	*Rhizobium meliloti*	Fast
Soybean	*Bradyrhizobium japonicum*	Slow

**In yeast extract-mannitol agar. Fast species generally have a generation time of less than 4 hours yielding gummy raised colonies 2–4 mm in diameter within 3–5 days. Slow species generally have a generation time of greater than 8 hours and yield granular colonies not exceeding 1 mm in diameter within 5–7 days.*

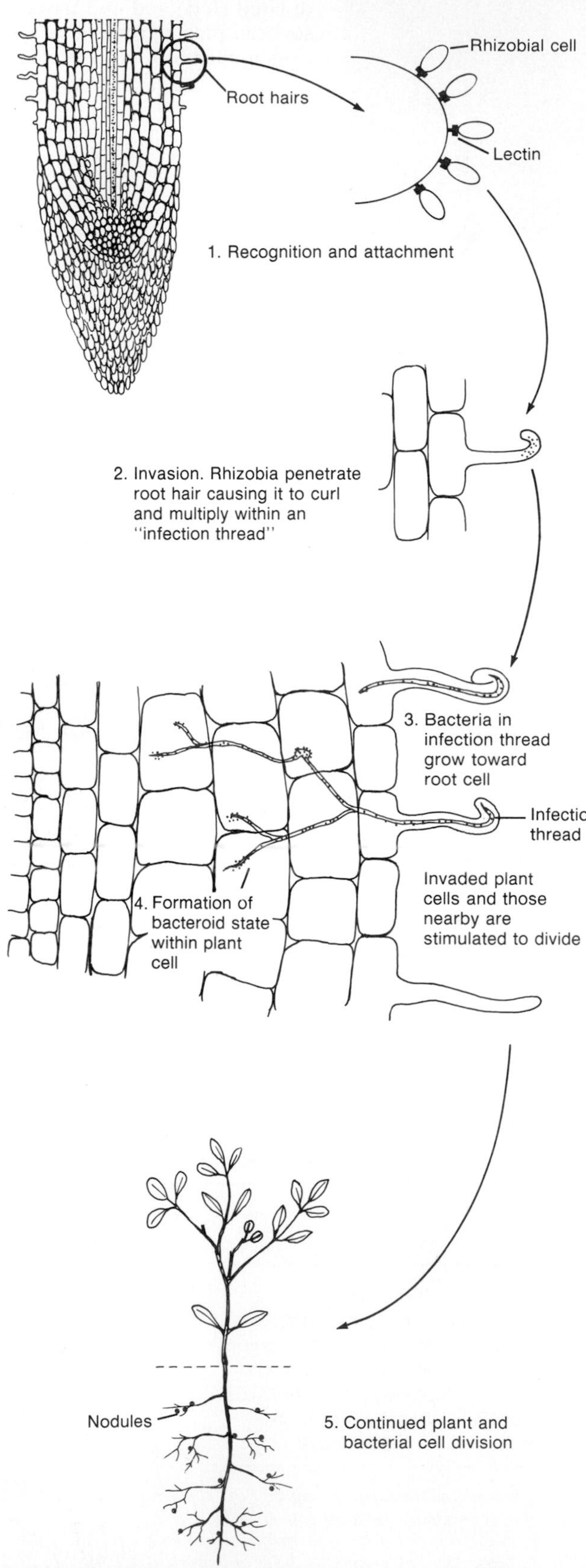

Rhizobial cell
Root hairs
Lectin
1. Recognition and attachment
2. Invasion. Rhizobia penetrate root hair causing it to curl and multiply within an "infection thread"
3. Bacteria in infection thread grow toward root cell
Infection thread
Invaded plant cells and those nearby are stimulated to divide
4. Formation of bacteroid state within plant cell
Nodules
5. Continued plant and bacterial cell division

Figure 17.53 Steps in the formation of a root nodule in a legume infected by *Rhizobium*.

of organic materials which stimulate the growth of a rhizosphere microflora. This stimulation is not restricted to the rhizobia but occurs with a variety of rhizosphere bacteria. If there are rhizobia in the soil, they grow in the rhizosphere and build up to high population densities. Attachment in the legume-*Rhizobium* symbiosis depends upon the surface macromolecule of the root hair surface which interacts with cell surface polysaccharides of the *Rhizobium* cell. This process has been studied in most detail in the white clover-*R. trifolii* symbiosis. The cells of *R. trifolii* have an extensive capsule and it has been shown that there is a specific binding of the bacterial cells to the root hairs of white clover. *R. trifolii* does not bind strongly to root hairs of other legumes, and other species of *Rhizobium*, not infective for white clover, do not bind strongly to white clover root hairs. A substance can be extracted and purified from white clover roots that specifically agglutinates *R. trifolii* cells, and this substance, a **lectin*** called *trifolin* is the agent that binds *R. trifolii* cells to the root hair. Root lectins which interact specifically with their respective *Rhizobium* species have also been reported for soybean, sweet clover, and pea.

Binding in the *R. trifolii*-clover symbiosis involves a polysaccharide portion of the *R. trifolii* outer layer, which contains the sugar 2-deoxyglucose. If 2-deoxyglucose is added to roots, the binding of *R. trifolii* is specifically blocked. The interaction of the polysaccharide capsular material from *R. trifolii* with the root hair surface has been studied by converting it into a fluorescent derivative and then observing fluorescence of root hairs when treated with this material. As seen in Figure 17.54, the root hairs bind

*Lectins are plant proteins with a high affinity for specific sugar residues.

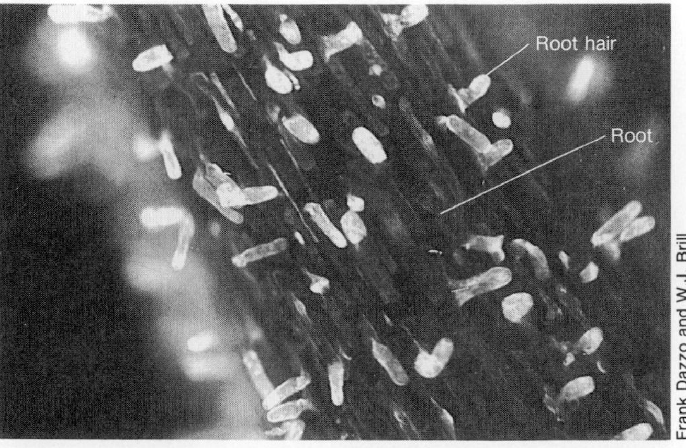

Root hair
Root
Frank Dazzo and W.J. Brill

Figure 17.54 Specific attachment to clover root hairs of the capsular lipopolysaccharide from *Rhizobium trifolii*. The LPS extracted from the bacteria was rendered fluorescent by conjugation with fluorescein isothiocyanate, and after treating roots with the preparation, they were viewed under a fluorescence microscope. The root hair tips are especially heavily stained.

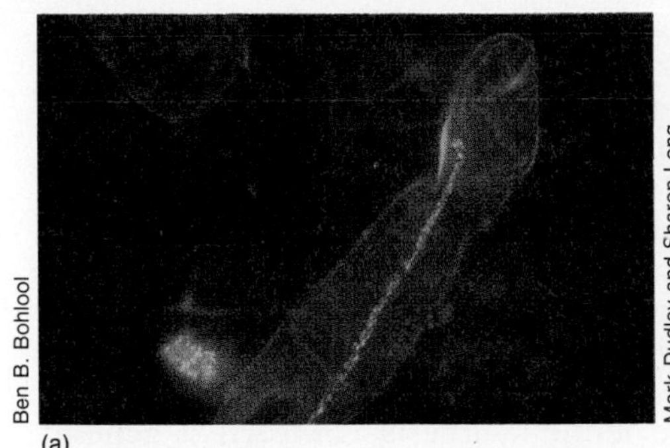

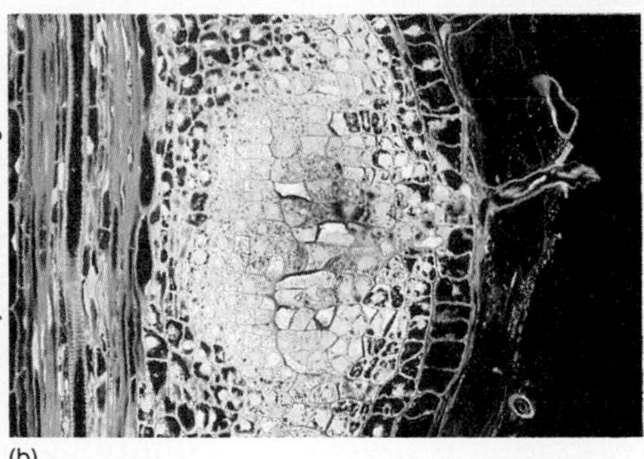

Figure 17.55 (a) An infection thread formed by ***Rhizobium trifolii*** on white clover (*Trifolium repens*). A number of bacteria can still be seen attached to the root hair. The infection thread consists of a cellulose deposit down which the bacteria move to the cortical cells of the root. Magnification, 880×. (b) Emerging alfalfa root nodule, resulting from ***Rhizobium meliloti*** infection. The infection thread can be seen as a blue staining cell in the upper right of the photograph.

this material more strongly than do the cells beneath the root hairs, and the binding seems to be especially strong near the root hair tips. It is known from other work that the initial penetration of *Rhizobium* cells into the root hair is via the root hair tip. Following binding, the root hair curls and bacteria enter the root hair and induce formation by the plant of an **infection thread**, which spreads down the root hair (Figure 17.55). Although cell wall polysaccharides appear to be critical for *R. trifolii*/white clover attachment, some evidence exists that pili (see Section 3.10) may also facilitate attachment here and in the *R. japonicum*/soybean attachment process as well.

Root cells adjacent to the root hairs subsequently become infected by rhizobia. If these cells are normal diploid cells they are usually destroyed by the infection, undergoing necrosis and degeneration; if they are *tetraploid* cells, however, they can become the forerunner of a nodule. There are always in the root a small number of tetraploid cells of spontaneous origin, and if one of these cells becomes infected, it is stimulated to divide. Progressive divisions of such infected cells leads to the production of the tumor-like nodule (Figure 17.56*a*). In culture, rhizobia produce substances called *cytokinins* which cause tetraploid cells to divide, and it is likely that production of cytokinins also occurs in the infected cells.

The bacteria multiply rapidly within the tetraploid cells and are transformed into swollen, misshapen, and branched forms called *bacteroids*. Bacteroids become surrounded singly or in small groups by portions of the plant cell membrane, called the *peribacteroid membrane* (Figure 17.56*b*). Only after the formation of bacteroids does nitrogen fixation begin. Eventually the nodule deteriorates, releasing bacteria into the soil. The bacteroid forms are incapable of division, but there are always a small number of dormant rod-shaped cells present. These now proliferate, using as nutrients some of the products of the deteriorating nodule, and can initiate the infection in other roots or maintain a free-living existence in the soil.

Biochemistry of nitrogen fixation in nodules As discussed in Section 16.24, nitrogen fixation involves the activity of the enzyme **dinitrogenase**, a large two-component protein containing iron and molybdenum. Dinitrogenase in nodules has characteristics similar to the enzyme of free-living N_2-fixing bacteria, including O_2 sensitivity and ability to reduce acetylene as well as N_2. Dinitrogenase is localized within the bacteroids themselves and is not released into the plant cytosol.

Bacteroids are totally dependent on the plant for supplying them with energy sources for N_2 fixation. The major organic compounds transported across the peribacteroid membrane and into the bacteroid proper are citric acid cycle intermediates, in particular the

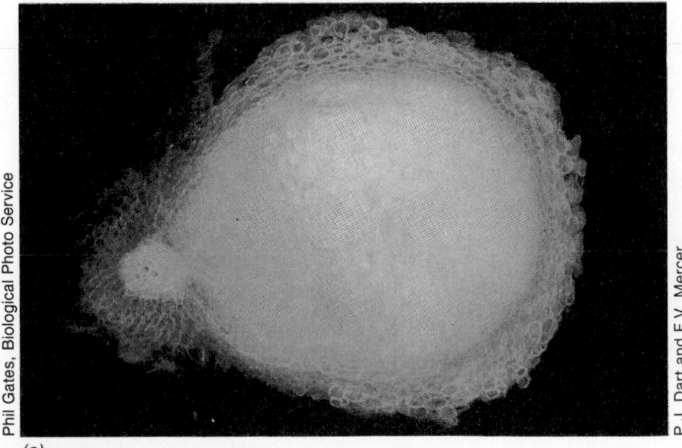

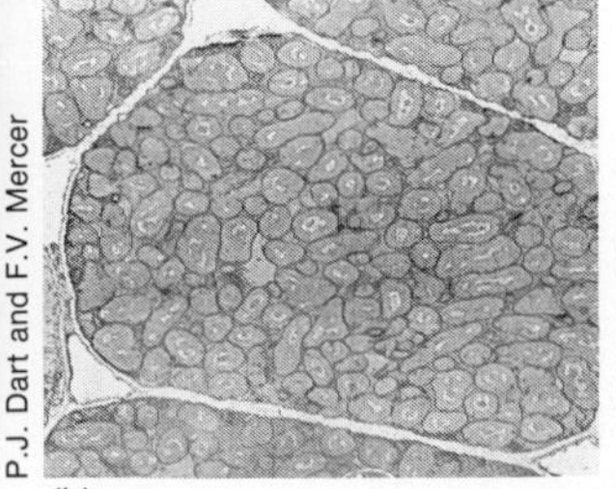

Figure 17.56 (a) Cross section through a legume root nodule, as seen by fluorescence microscopy. The darkly stained region contains plant cells filled with bacteria. Magnification, 110×. (b) Electron micrograph of a thin section through a single bacteria-filled cell of a subterranean clover nodule. Magnification, 2680×.

C-4 acids *succinate, malate,* and *fumarate* (Figure 17.57). These serve as electron donors for ATP production and as the ultimate source of electrons for the reduction of N_2. Although the precise reductant for N_2 fixation in the bacteroid is not known, the biochemical steps in the conversion of N_2 to NH_3 are probably similar to those in free-living N_2-fixing systems, suggesting that *pyruvate* (oxidized via flavodoxin) is the direct electron donor to bacteroid dinitrogenase (see Figure 17.57 and Section 16.24).

The first stable product of N_2 fixation is *ammonia*, and several lines of evidence suggest that assimilation of ammonia into organic nitrogen compounds in the root nodule is primarily carried out by the plant. Although bacteroids can assimilate some ammonia into organic form, the levels of ammonia assimilatory enzymes in bacteroids are quite low. By contrast, the ammonia assimilating enzyme *glutamine synthetase* (see Figure 16.45) is present in high levels in the plant cell cytoplasm. Hence, ammonia transported from the bacteroid to the plant cell can be assimilated by the plant as the amino acid *glutamine*. Besides glutamine, other nitrogenous compounds, in particular other amino acid amides, such as asparagine and 4-methyleneglutamine, and *ureides*, such as *allantoin* and *allantoic acid*, are synthesized by the plant and subsequently transported to plant tissues (see Figure 17.57).

Genetics of nodule formation The important O_2-binding protein in the root nodule, *leghemoglobin*, is genetically coded for in part by both the plant and the bacterium. The globin (protein) portion of leghemoglobin is coded for by plant DNA while heme synthesis is genetically directed by the bacterium. In addition to leghemoglobin, however, there are thought to be about 20 additional polypeptides that play crucial roles in the overall legume-*Rhizobium* symbiosis. Nitrogen fixation itself is clearly a process genetically directed by the bacterium (since several species of *Rhizobium* have been shown to fix N_2 in culture) while lectin synthesis is clearly a genetic property of the plant. Between the initial step of correct plant-bacterium recognition and the formation of an effective nodule, however, several genetically directed steps undoubtedly occur. There is good evidence that a number of nodulation-specific proteins are coded for by large plasmids present in most species of *Rhizobium* and *Bradyrhizobium*. Such a plasmid was first found in *R. leguminosarum*, but similar-size plasmids have now been identified in a variety of *Rhizobium* species. Proof that these plasmids dictate host specificity is shown by experiments in which the plasmid has been conjugatively transferred from one species of *Rhizobium* to another. When the plasmid is transferred from *R. leguminosarum* (whose host is pea) to *R. trifolii* (whose host is clover) for example, cells of the latter species become capable of effectively nodulating pea.

A number of nodule-specific proteins are coded for by the host plant. These proteins have been referred to as *nodulins*. Two classes of nodulins are

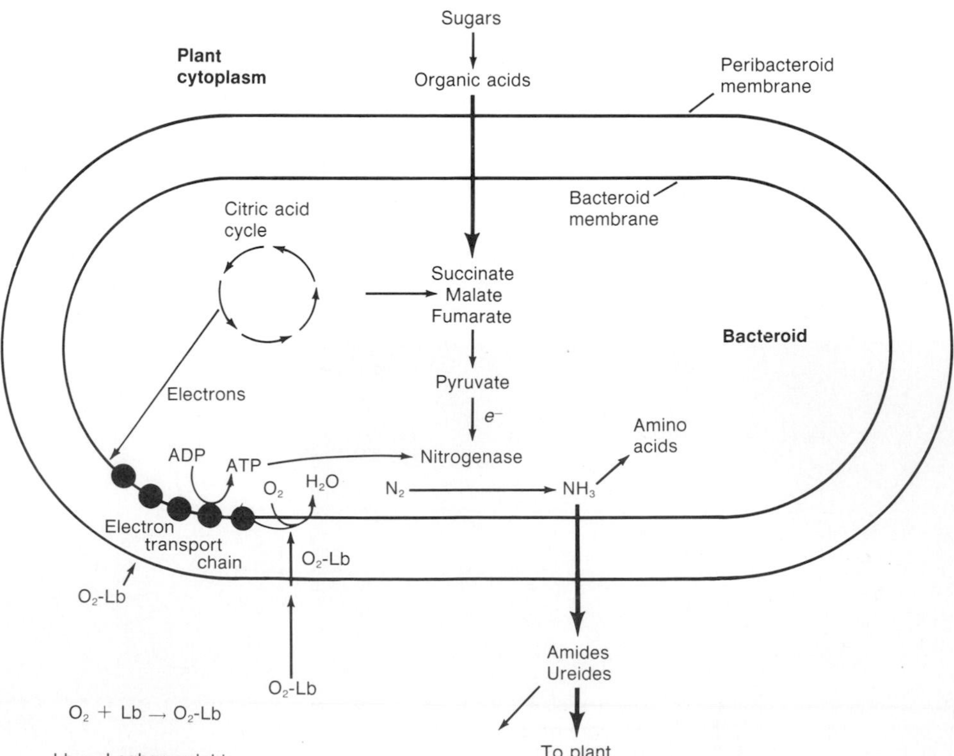

Figure 17.57 Schematic diagram of major metabolic reactions and nutrient exchanges in the bacteroid.

known. One class consists of proteins common to the root nodules of all leguminous plants and are apparently involved in supporting the process of N_2 fixation. The other class consists of proteins specific for biochemical processes unique to one type of nodule or another. These include specific enzymes involved in carbon or nitrogen metabolism in the nodule. Nodulins are not synthesized by the plant until infection by *Rhizobium*, but how the infection process triggers nodulin synthesis is not known.

The gene for **hydrogenase**, an enzyme whose role in *Rhizobium* is to take up H_2 for use as a reductant in nitrogen fixation, is also a plasmid coded protein.* Hydrogenase genes constitute part of the major nodulation plasmid. However, hydrogenases are not present in all rhizobia. Screening of wild-type strains has shown hydrogenase to be present in about 25 percent of all *R. japonicum* strains, 15 percent of *R. leguminosarum* strains, and to be absent from all strains of *R. meliloti* and *R. trifolii*. Field studies have shown that strains of rhizobia possessing hydrogenase increase total plant nitrogen levels about 10 percent more than strains unable to use H_2, because some of the otherwise wasted H_2 (from the activity of dinitrogenase, see Section 16.24) is recycled and used to reduce N_2. This fact has stimulated genetic engineering approaches to making all field strains of *Rhizobium* capable of H_2 uptake.

Finally, although there is little evidence that dinitrogenase genes are plasmid encoded in other nitrogen-fixing bacteria, certain dinitrogenase structural genes do reside on plasmids in *R. leguminosarum* and *R. meliloti*. It is thus clear that plasmid DNA is an important source of genetic information for symbiotic nitrogen fixation.

Stem-nodulating rhizobia Although most leguminous plants form nitrogen-fixing nodules on their *roots*, a few legume species bear nodules on their *stems*. Most stem-nodulated legumes are tropical plants capable of growth in waterlogged soils. Although thought to be members of the genus *Rhizobium*, stem-nodulating bacteria do not form root nodules on any normally root-nodulated legumes. Stem nodules usually form in the submerged portion of the stems or on portions of the stem just above the water level (Figure 17.58). From studies carried out thus far, the general sequence of events in formation of stem nodules strongly resembles that of root nodules. An infection thread is formed and bacteroid formation predates the N_2-fixing state. However, stem-nodulating rhizobia differ from root nodulating rhizobia in two important ways: (1) dinitrogenase from stem nodulating rhizobia is considerably *less* O_2 sensitive than the dinitrogenases of root-nodulating species, and (2)

*Certain species of *Rhizobium* are also capable of using H_2 as an electron donor for lithotrophic growth with O_2 as electron acceptor (see Section 16.8).

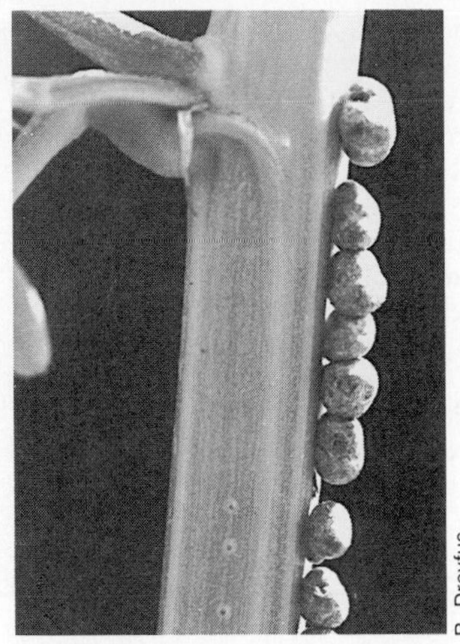

Figure 17.58 Stem nodules caused by stem-nodulating *Rhizobium*. The photograph shows the stem of the tropical legume *Sesbania rostrata*. On the left side of the stem are uninoculated sites, on the right identical sites inoculated with stem-nodulating rhizobia.

stem-nodulating species grow on N_2 as sole nitrogen source in pure culture. (In root nodulating strains of *Rhizobium*, N_2 fixation in pure culture requires that the medium be supplemented with some fixed nitrogen, usually an amino acid.) It is hypothesized that the increased tolerance to molecular oxygen of the dinitrogenase of stem-nodulating rhizobia may be a reflection of the fact that this enzyme in these species must function in plant cells containing a relatively high concentration of O_2 (due to photosynthesis in nearby plant cortical cells); this is obviously not a problem in root cells where photosynthesis does not occur. Also, it has been found that the leghemoglobin of stem nodules differs from that found in root nodules, which again probably reflects the increased O_2 levels found in stem nodules.

Stem-nodulated leguminous plants are quite widespread in tropical regions where soils are often nitrogen deficient due to leaching and intense biological activity. Further study of stem-nodulating rhizobia may yield important information regarding O_2 regulation of nitrogenase activity in the nodule environment.

Nonlegume nitrogen-fixing symbioses In addition to the legume-*Rhizobium* relationship, nitrogen-fixing symbioses occur in a variety of nonleguminous plants, involving microorganisms other than rhizobia. Nitrogen-fixing cyanobacteria form symbioses with a variety of plants. The water fern *Azolla* contains a species of heterocystous N_2-fixing cyanobacteria called *Anabaena azollae* within small pores of its fronds (Figure 17.59). *Azolla* has been used for centuries to enrich rice paddies with fixed nitrogen. Before planting rice, the farmer allows the surface of the rice paddy to become densely covered with the fern. As the rice plants grow they eventually

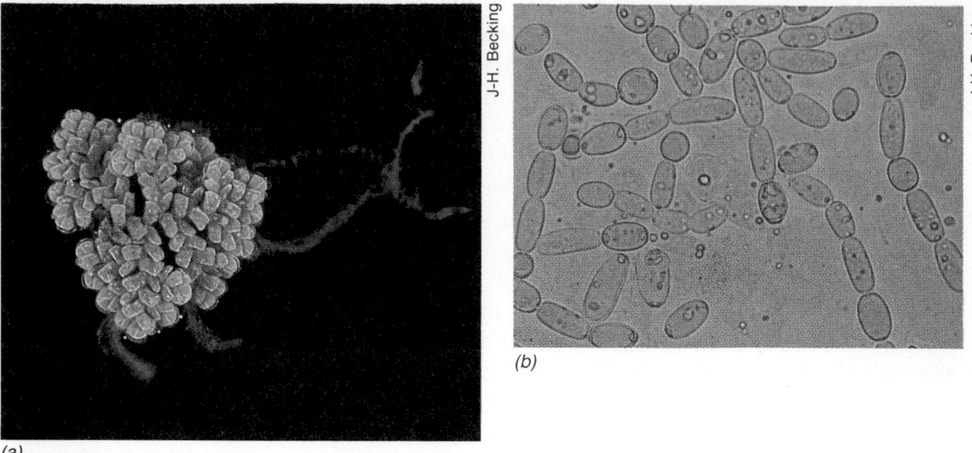

Figure 17.59 *Azolla/Anabaena* symbiosis. (a) Intact association showing a single plant of *Azolla pinnata*. The diameter of the plant is approximately 1 centimeter. (b) Cyanobacterial symbiont *Anabaena azollae* as observed in crushed leaves of *A. pinnata*. Magnification, 900×. Note the oval-shaped *heterocysts* (lighter color), the site of nitrogen fixation in the cyanobacterium.

crowd out the *Azolla-Anabaena* mixture, leading to death of the fern and release of fixed nitrogen which is assimilated by the rice plants. By repeating this process each growing season, the farmer can obtain high yields of rice without the need for nitrogenous fertilizers.

The alder tree (genus *Alnus*) has nitrogen-fixing root nodules (Figure 17.60) which harbor a filamentous, streptomycete-like organism which has now been cultured and named *Frankia*. *Frankia* grows very slowly in the laboratory but has been shown to fix nitrogen in culture. Although when assayed in cell extracts the dinitrogenase of *Frankia* is sensitive to molecular oxygen, like *Azotobacter*, intact cells of *Frankia* fix N_2 at full oxygen tensions. This implies that *Frankia* protects its dinitrogenase in some way. The development of dinitrogenase activity in *Frankia* nodules coincides with the differentiation of terminal swellings on the vegetative cells, called *vesicles* (Figure 17.60*b*), and studies with purified vesicle fractions indicate that most, if not all, *Frankia* dinitrogenase is localized in these vesicular structures (Figure 17.60*b*). The vesicles contain thick walls that apparently act as a barrier to O_2 diffusion, thus pois-

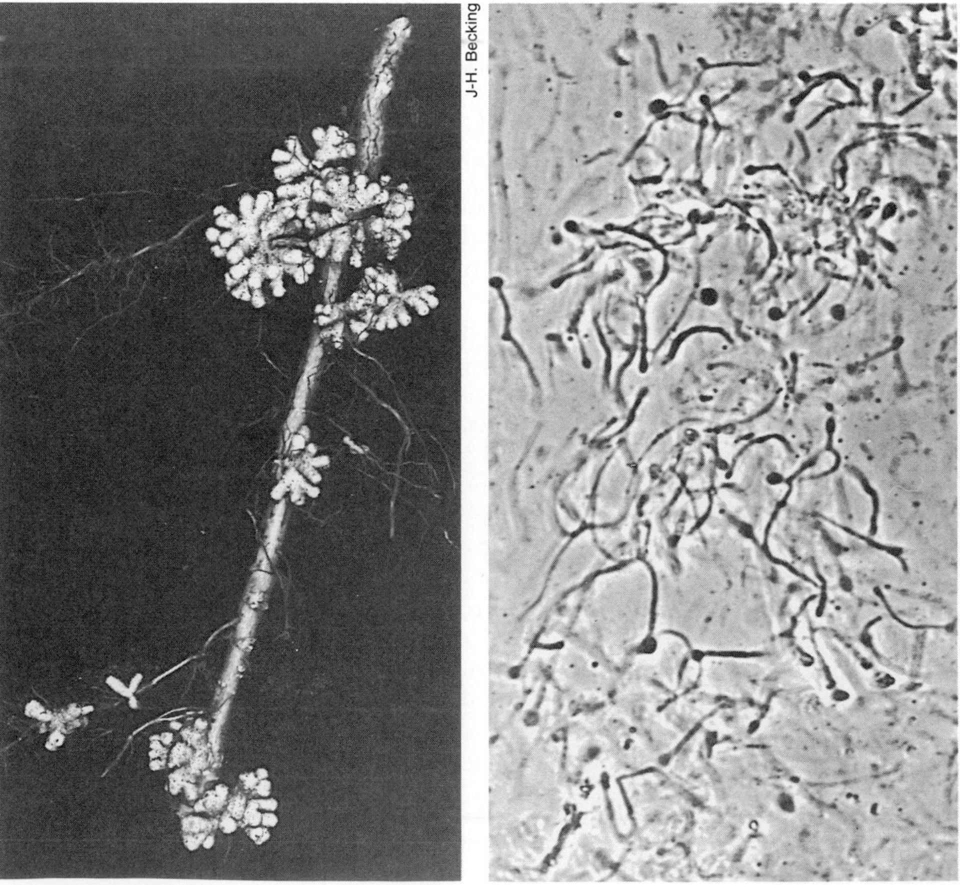

Figure 17.60 *Frankia* nodules and *Frankia* cells. (a) Root nodules of the common alder *Alnus glutinosa*. (b) *Frankia* culture purified from nodules of *Comptonia peregrina*. Note vesicles (spherical structures) on the tips of hyphal filaments.

ing the O_2 tension within vesicles at levels compatible with dinitrogenase activity.

Alder is a characteristic pioneer tree, able to colonize bare soils in nutrient-poor sites, and this is likely due to its ability to enter into a symbiotic nitrogen-fixing relationship with *Frankia*. A host of other small woody plants are nodulated by *Frankia*. This root nodule symbiosis has been reported in eight families of plants, many of which show no evolutionary relationships to one another. This suggests that the nodulation process in the *Frankia* symbiosis may be more of a generalized phenomenon than the highly specific process known in the *Rhizobium*/legume symbiosis.

Azospirillum lipoferum is a N_2-fixing bacterium that lives in a rather casual association with roots of tropical grasses. It is found in the rhizosphere, where it grows on products excreted from the roots. It also has the ability to grow around the roots of cultivated grasses, such as corn (*Zea mays*), and inoculation of corn with *A. lipoferum* may lead to small increases (~10 percent) in growth yield of the plant. Recent evidence suggests that *Azospirillum* can actually penetrate roots and grow *inter*cellularly between root cells. In such an environment it is likely that considerable fixed nitrogen and organic carbon flow from bacterium to plant and plant to bacterium, respectively. It should be recalled that the *Rhizobium*/legume symbiosis involves *intra*cellular (rather than *inter*cellular) growth of the bacterium. It is likely that many casual relationships of the *Azospirillum* type exist between bacteria and plants, although only in the case of the *Rhizobium* or *Frankia* nodule is easy recognition of a relationship possible.

The agronomic potential of legume and nonlegume nitrogen-fixing symbioses may be far greater than the agricultural applications made to date. Intensive research efforts are now under way to refine and improve known symbioses such as that of *Rhizobium* and the legume, and to better understand the basis of plant bacterium recognition, especially in the more generalized nodule-producing *Frankia* symbiosis. A major goal of research on symbiotic nitrogen fixers is to be able to stimulate plant growth in the absence of chemical fertilizers and to ultimately increase protein and fiber production on a global scale.

Study Questions

1. Compare and contrast a lake ecosystem with a hydrothermal vent ecosystem. How does *energy* enter each ecosystem? What is the basis of *primary production* in each ecosystem? What nutritional classes of organisms exist in each ecosystem and how do they feed themselves?
2. What is the basis of the *enrichment culture technique*? Why is an enrichment medium usually only suitable for the enrichment of a certain group or groups of bacteria?
3. What is the principle of the *Winogradsky column* and what types of organisms does it serve to enrich? How might a Winogradsky column be used to study the breakdown of a xenobiotic compound?
4. Compare and contrast the use of *fluorescent dyes* and *fluorescent antibodies* for use in enumerating bacteria in natural environments. What advantages and limitations do the use of each have?
5. What are the major advantages of *radioisotopic methods* for the study of microbial ecology? What type of *controls* (discuss at least two) would you include in a radioisotopic experiment to show $^{14}CO_2$ incorporation by photosynthetic bacteria or to show $^{35}SO_4^{2-}$ reduction by sulfate-reducing bacteria?
6. What information is obtained from knowledge of the *biochemical oxygen demand* (BOD) of a water source? Which should have a higher BOD, drinking water or raw sewage? Why?
7. How is the *deep sea* defined and why is an understanding of its biology of importance?
8. What is the difference between *barotolerant* and moderately *barophilic* bacteria? Between these two groups and *extreme barophiles*? What properties do barotolerant, moderately barophilic and extreme barophiles have in common?
9. ^{14}C-labeled cellulose is added to a vial containing a small amount of sewage sludge and sealed under anaerobic conditions. A few hours later $^{14}CH_4$ appears in the vial. Discuss what has happened to yield such a result.
10. How can organisms like *Syntrophobacter* grow when their metabolism is based on thermodynamically unfavorable reactions? How is *Syntrophobacter* usually grown in laboratory culture?
11. Why is *sulfate reduction* the main form of anaerobic respiration in marine environments whereas *methanogenesis* dominates in fresh waters? Does any methanogenesis occur in the marine environment? If so, how?
12. What is the *rumen* and how do the digestive processes operate in the *ruminant digestive tract*? What are the major benefits of a rumen system? What are the disadvantages? List some ruminant animals.
13. Compare and contrast the microbiological steps involved in the conversion of cellulose to methane in the rumen as compared to cellulose conversion in a sewage digestor or in lake sediments. What organisms are involved in each ecosystem and why?
14. Why can urea or ammonia serve as a *nitrogen source* for *ruminants* but not for humans?
15. Compare and contrast the processes of *nitrification* and *denitrification* in terms of the organisms involved, the environmental conditions which favor each process, and the changes in nutrient availability which accompany each process.

16. What organisms are involved in cycling *sulfur compounds* anaerobically? If sulfur lithotrophs had never evolved, would there be a problem in the microbial cycling of sulfur compounds?
17. Why are all *iron-oxidizing lithotrophs* obligate aerobes and why are most iron oxidizers acidophilic?
18. Explain how spontaneous chemical reactions can acidify a coal seam and how both chemical reactions and *Thiobacillus ferrooxidans* continues the production of acid thereafter.
19. How is *Thiobacillus ferrooxidans* useful in the mining of copper ores? What crucial step in the *indirect* oxidation of copper ores is carried out by *Thiobacillus ferrooxidans*? How is copper recovered from copper solutions produced by leaching?
20. What physical and chemical conditions are necessary for the rapid *microbial degradation of oil* in aquatic environments? Design an experiment that would allow you to test what conditions optimized the oil oxidation process.
21. What are *xenobiotic compounds* and why might microorganisms have difficulty catabolizing them? Why is the ability to degrade a xenobiotic compound sometimes lost in laboratory cultures?
22. Why is *sewage* that has received only primary treatment both a serious health hazard and a source of environmental pollution?
23. Compare and contrast aerobic and anaerobic means of *secondary sewage treatment*. What types of compounds are best degraded by each method? What are the major products of the treatment process in each case?
24. What methods are used by *plants* to resist attack by pathogens? How do these differ from the substances produced by animals for the same purpose?
25. Compare and contrast the production of a *plant tumor* by *Agrobacterium tumefaciens* and a *root nodule* by a *Rhizobium* species. In what ways are these structures similar? In what ways are they different? Of what importance are plasmids to the development of both structures?
26. What genetic information resides on the *Ti plasmid*? Which gene(s) could be deleted from the Ti plasmid without affecting tumorigenesis?
27. Describe the steps involved in tumorigenesis by *Agrobacterium tumefaciens*.
28. What ecological advantages do *leguminous plants* have over nonlegumes?
29. What substances of plant origin are found in *root nodules* of legumes that are required for the rhizobial symbiont to fix nitrogen?
30. Describe the steps in the development of root nodules on a leguminous plant. What is the nature of the recognition between plant and bacterium? How does this compare with recognition in the *Agrobacterium*/plant system?
31. How does nitrogen fixed by *Rhizobium* become part of the plant's proteins?

Supplementary Readings

Childress, J. J., H. Felbeck, and **G. N. Somero.** 1987. *Symbioses in the deep sea.* Sci. Amer. 256:114–120. An overview of the biology of deep-sea thermal vent animals with emphasis on how lithotrophy can support large animals.

Fletcher, M., and **T. R. G. Gray** (eds.) 1987. *Ecology of Microbial Communities.* Forty-first symposium of the Society for General Microbiology. Cambridge University Press, Cambridge, England. This volume contains chapters written by experts on modern research in microbial ecology. Particularly recommended is the first chapter, which discusses the present state and future prospects for microbial ecology.

Gibson, D. T. (ed.) 1984. *Microbial Degradation of Organic Compounds.* Marcel Dekker, Inc., New York. An excellent collection of review articles on various microbial processes in nature including the degradation of hydrocarbons, complex-ringed molecules, halogenated compounds, lignin, aromatic compounds, and many others.

Halverson, L. J., and **G. Stacey**. 1986. *Signal-exchanges in plant microbe interactions.* Microbiol. Rev. 50:193–225. An excellent review of *Agrobacterium*-crown gall and *Rhizobium*-legume interactions with a definite molecular slant. Highly recommended as a starting point for an in-depth study of either association.

Jannasch, H. W., and **M. J. Mottl**. 1985. *Geomicrobiology of deep-sea hydrothermal vents.* Science 229:717–725. An overview of the biogeochemistry of warm vents and black smokers.

Jannasch, H. W., and **C. D. Taylor**. 1984. *Deep-sea microbiology.* Ann. Rev. Microbiol. 38:487–514. A treatment of deep sea microbial life, principles of barophily and aspects of hydrothermal vent biology.

Pfennig, N., and **F. Widdel**. 1982. *The bacteria of the sulfur cycle.* Phil. Trans. R. Soc. Lond. B. 298:433–441. An excellent overview of all the major groups of bacteria which participate in sulfur cycling.

Prakask, P. K., and **A. G. Atherly**. 1986. *Plasmids of* Rhizobium *and their role in symbiotic nitrogen fixation.* Int. Rev. Cytol. 104:1–24. A short review focusing on the role of plasmid DNA in the *Rhizobium*/legume symbiosis.

Revsbech, N. P., and **B. B. Jorgensen**. 1986. *Microelectrodes: Their use in microbial ecology.* Adv. Microb. Ecol. 9:293–352. A comprehensive review of microelectrode technology and the use of microelectrodes in field research studies.

Staley, J. T., and **A. E. Konopka**. 1985. *Measurement of* in situ *activities of nonphotosynthetic microorganisms in aquatic and terrestrial habitats.* Ann. Rev. Microbiol. 39:321–346. An excellent treatment of the methods available for measuring bacterial activities in nature.

Tate, R. L. III (ed.) 1986. *Microbial Autecology: A Method For Environmental Studies.* John Wiley, New York. A description of methods useful for studying individual bacterial species in nature.

18 Molecular Systematics and Microbial Evolution

A recurrent theme throughout this book has been the enormous diversity of microorganisms. How did such diversity arise? It should be clear by now that the broad morphological, physiological, and ecological characteristics of the various groups of microorganisms are ultimately controlled by the *genetic* constitution of the organisms interacting with the *environments* of which they are a part. Thus microbial diversity arises as a result of mutation and genetic recombination processes, working in an environment which constantly selects. In other words, microbial diversity reflects the diversity of habitats on Earth suitable for life.

18.1 Evolution of the Earth and Earliest Life Forms

Origin of the earth The earth is about 4.5 billion years old as determined by radiodating measurements. The most useful techniques to measure dates of this magnitude are the potassium/argon, uranium/thorium, and rubidium/strontium dating methods. For various reasons, the *potassium/argon dating method* is considered most reliable. To date rocks by the potassium/argon method, the proportion of K^{40} and Ar^{40} in a rock sample is determined in a mass spectrometer, and using the figure of 1.26 billion years as the half-life of the $K^{40} \rightarrow Ar^{40}$ transition, estimates of the age of the rock are calculated. Using the K/Ar method, reliable estimates of the ages of rocks up to 5 billion years old can be made.

Our solar system is thought to have been formed about 4.5 billion years ago when a large, very hot star exploded, generating a new star (our sun), and the other components of our galaxy. Although no rocks dating to this period have yet been discovered on Earth, rocks dating back to over 3.5 *billion* years ago have been found in several locations on Earth. The oldest rocks discovered thus far are those of the Isua formation in Greenland, which date by the K/Ar method to 3.8 billion years ago. The Isua rocks are of three types: sedimentary, volcanic, and carbonate. The sedimentary composition of the Isua rocks is of particular evolutionary interest, because from our understanding of how modern sedimentary rocks are formed, the presence of sedimentary rocks 3.5 billion years old strongly suggests that liquid water was present at that time. The presence of liquid water in turn implies that conditions on Earth at that time were likely to be compatible with life as we know it. Other rocks of ancient origin include the Warawoona series in western Australia and the Fig Tree series in southern Africa; both rock formations are about 3.5 billion years old.

Evidence for microbial life on the early earth Good fossil evidence exists for microbial life in rocks 3.5 billion years old and younger. Most of

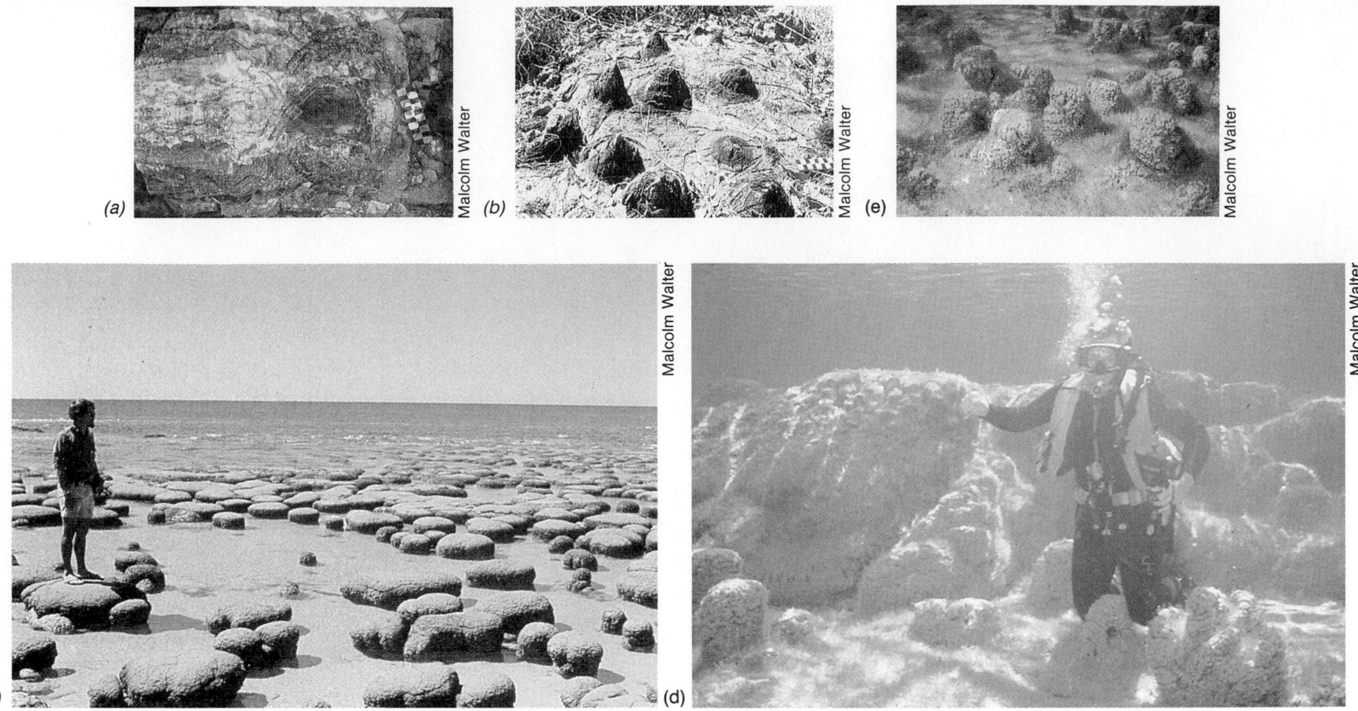

Figure 18.1 Ancient and modern stromatolites. (a) The oldest known stromatolite, found in a rock around 3.5 billion years old, from the Warrawoona Group in Western Australia. Shown is a vertical section through a laminated, hemispheroidal structure, which has been preserved in the rock. Scale, 10 cm. (b) Stromatolites of conical shape from 1.6 billion year old dolomite rock of the McArthur Basin of the Northern Territory of Australia. (c) Modern stromatolites in a warm marine bay, Shark Bay, Western Australia. (d) Underwater photograph of modern stromatolites growing in Shark Bay. The diver indicates the scale. Shown are large columns formed by a complex community of diatoms, cyanobacteria, and green algae, to which are attached various macroscopic algae. (e) Another view of large modern stromatolites from Shark Bay. Note the resemblance to the ancient stromatolites shown in (b).

the fossils in the oldest rocks resemble simple rod-shaped bacteria. But in certain rocks of 3.5 billion years old or younger, stromatolitic microfossils are present in abundance. *Stromatolites* are fossilized microbial mats consisting of layers of filamentous procaryotes containing trapped sediment (Figure 18.1*a* and *b*). We discussed some characteristics of microbial mats in Section 17.4. By comparison with modern stromatolites growing in shallow marine basins (Figure 18.1*c,d* and *e*) and in hot springs in various locations in the world, it is assumed that ancient stromatolites consisted of filamentous photosynthetic bacteria. Although modern stromatolites are frequently composed of filamentous *cyano*bacteria, this would have probably not been the case in the oldest stromatolites. Because the earth was still anoxic at this time (see below), stromatolites dating from 2.5 billion years or older were most certainly made by *anoxygenic* phototrophs (purple and green bacteria) rather than O_2-evolving cyanobacteria.

In rocks younger than about 2 billion years old, the variety and morphological diversity of fossil microorganisms is considerable. Figure 18.2 shows some photomicrographs of extremely thin sections of rocks containing structures remarkably similar to certain modern filamentous bacteria. Structures such as those shown in Figure 18.2*a* are surprisingly similar to certain filamentous green photosynthetic bacteria or cyanobacteria and provide strong evidence that procaryotes as a group had evolved an impressive morphological diversity earlier than two billion years ago.

The primitive earth The original atmosphere of the earth was devoid of significant amounts of O_2 and hence constituted a *reducing* environment. Besides H_2O, a variety of gases were present, the most abundant being CH_4, CO_2, N_2, and NH_3. In addition, trace amounts of CO and H_2 existed, as well as some H_2S. It is also likely that a considerable amount of hydrogen cyanide, HCN, was produced on the early earth when NH_3 and CH_4 reacted chemically to yield HCN. The most reliable geochemical estimates of the temperature of the early earth suggest that it was a

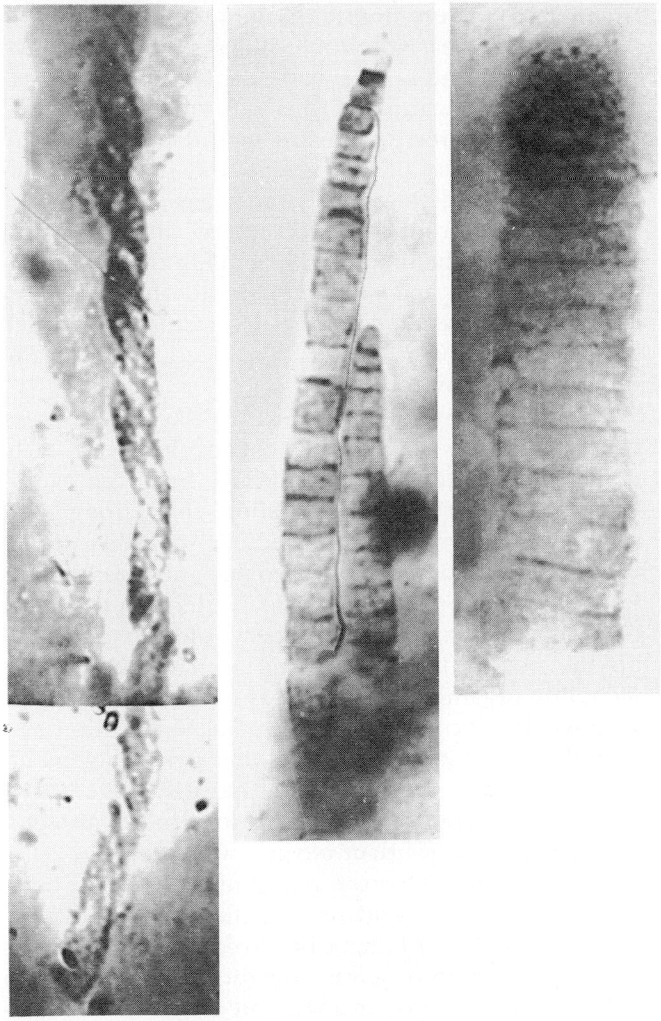

(a)

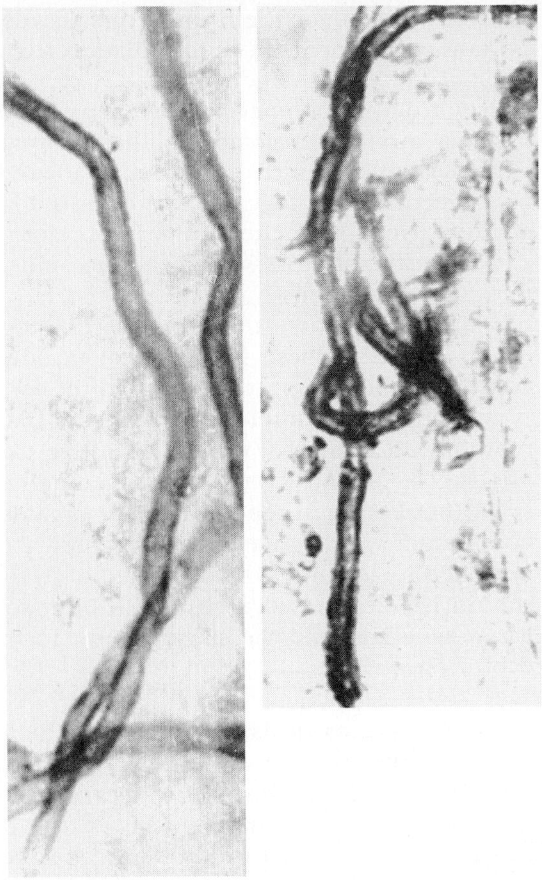

(b)

(c)

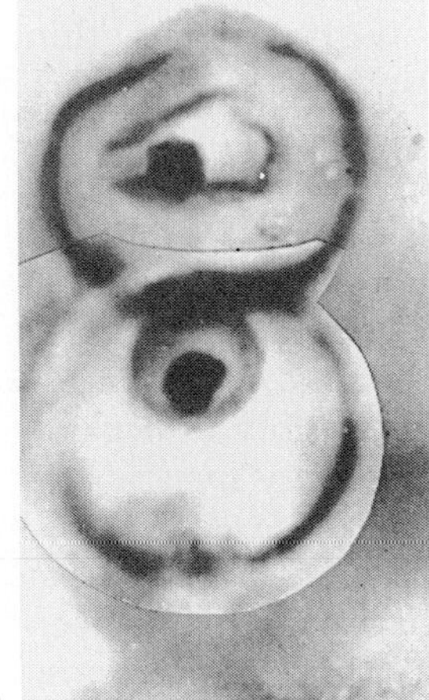

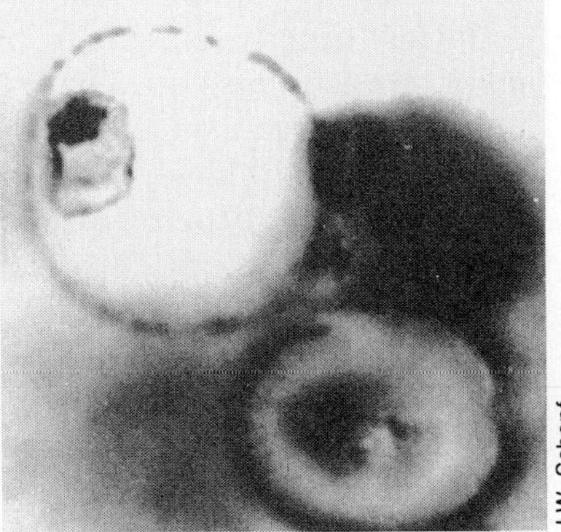

J.W. Schopf

Figure 18.2 The five photographs in (a) (magnification, 2000×) and (b) (magnification, 920×) show fossil procaryotic microorganisms found in the Bitter Springs formation, a rock formation in central Australia about a billion years old. These forms bear a striking resemblance to modern cyanobacteria or their colorless counterparts. The two photographs in (c) (magnification, 2000×) show fossils possibly of a eucaryotic alga. The cellular structure is remarkably similar to that of certain modern chlorophyta, such as *Chlorella* sp. These are from the same rock formation as are the procaryotic organisms.

much hotter planet than it is today. For the first half billion years of the earth's existence it is likely that the surface of the earth was greater than 100°C; thus free water probably did not exist on the early earth, but accumulated later as the earth cooled. It is thus likely that life originated on an earth much hotter than it is today, and therefore that the earliest life forms were thermophilic, or at least thermotolerant (this has implications for microbial evolution as discussed in Sections 18.6 and 19.35).

It is now well established that the synthesis of biologically important molecules can occur if reducing atmospheres containing the aforementioned gases are subjected to intense energy sources. Of the energy sources available on the primitive earth, the most important was ultraviolet (UV) radiation from the sun, but lightning discharges, radioactivity, and thermal energy from volcanic activity were also available. If gaseous mixtures resembling those thought to be present on the primitive earth are irradiated with UV or subjected to electric discharges in the laboratory, a wide variety of biochemically important molecules can be made, such as sugars, amino acids, purines, pyrimidines, various nucleotides, and fatty acids. It has also been shown that under prebiological conditions some of these biochemical building blocks could have polymerized, leading to the formation of polypeptides, polynucleotides, and the like. We can therefore imagine that on the primitive earth a rich mixture of organic compounds eventually accumulated, but in the absence of living organisms these compounds would have been stable and should have persisted for countless years. Thus, with time, there should have been an extensive accumulation of organic materials, and the stage was set for biological evolution.

Primitive organisms What was the first living organism like? One can envisage several paths leading to the first living organism. However, even very primitive organisms must have possessed the following: (1) **metabolism**, that is, the ability to accumulate, convert, and transform nutrients and energy; and (2) a **hereditary mechanism**, that is, the ability to replicate and transfer its properties to its offspring. Both of these features really require the development of a *cellular* structure. Such structures probably arose through the spontaneous aggregation of lipid and protein molecules to form membranous structures, within which were trapped polynucleotides, polypeptides, and other substances. This step may have occurred countless times to no effect, but just once the proper set of constituents could have become associated, and a primitive organism arose. This primordial organism would certainly have been structurally simple (that is, resembling *procaryotic* cells) and would have found itself surrounded by a rich supply of organic materials usable as nutrients for energy metabolism and growth.

Because of the reducing conditions prevailing at the time, the first primitive organisms must have been able to metabolize *anaerobically*. From what we know of metabolic diversity (see Chapter 16), therefore, the earliest organisms may have relied on either fermentative, photosynthetic, or anaerobic respiratory processes. However, since photosynthesis and anaerobic respiration as currently understood are processes intimately tied to complex membrane activities, and early cell membranes were probably structurally quite simple, it is most logical to assume that simple fermentative organisms were the first forms of life on earth. Such organisms were undoubtedly biochemically very simple, in the sense of possessing very few enzymes. No cytochrome system would have been present, and no cell wall, flagella, or other special morphological properties would have been necessary. The earliest organisms would probably have required a variety of growth factors and other complex organic nutrients. As time went on, however, growth and reproduction of these organisms would have depleted the organic nutrients in their environment. Mutation and selection could then have resulted in the appearance of new organisms with greater biosynthetic capacities, better adapted to the changing chemical environment.

To make survival in osmotically varying environments possible, a cell wall probably evolved fairly early. Another key step would have been the evolution of the first porphyrins, since this, along with refinements in membrane structure, could have led to the construction of cytochromes and an electron-transport chain able to carry on electron-transport phosphorylation. These events would have opened up the possibility of utilizing a variety of nonfermentable organic compounds as electron donors. In an environment in which fermentable organic compounds were being depleted, this development no doubt provided a distinct selective advantage. However, since conditions were still anaerobic, an electron acceptor other than oxygen was needed.

One possible electron acceptor would have been CO_2, present in abundance in the early earth's atmosphere. CO_2 could have been reduced to CH_4, permitting evolution of methanogenic bacteria, a group still widespread today in anaerobic environments. Another possible anaerobic electron acceptor is sulfate, which could have permitted the evolution of the sulfate-reducing bacteria, which reduce sulfate to sulfide. Although the methane bacteria lack cytochromes and use other unique electron carriers, the sulfate-reducing bacteria do possess a primitive type of cytochrome called cytochrome c_3 (see Section 16.15).

The eventual evolution of porphyrins in organisms capable of anaerobic respiration would have allowed development of light-sensitive magnesium tetrapyroles, the chlorophylls, and thus photosynthesis. Following this major event a great explosion of life could have occurred because of the availability of the enormous amount of energy from the visible light of the sun. The first photosynthetic organism was no doubt anaerobic, probably using light only for ATP synthesis, and using reduced compounds from its environment—such as H_2S—as sources of reducing power. Such an organism may have resembled one of the modern day purple or green sulfur bacteria.

The next major step in microbial evolution probably involved development of the second light reaction of green-plant photosynthesis, making it possible to use the plentiful supply of H_2O as an electron donor. Since both ATP and reduced pyridine nucleo-

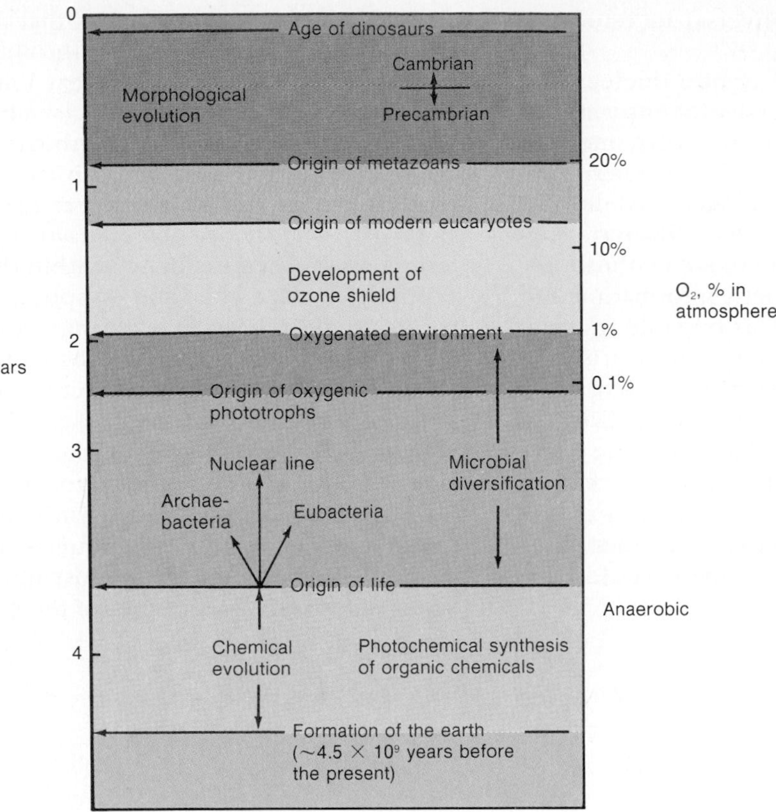

Figure 18.3 Major landmarks in biological evolution. The positions of the stages on the time scale are only approximate.

tides (NADH, NADPH) could now be made photosynthetically, light energy could be used more efficiently. Clearly such an organism would have considerable competitive advantage over other photosynthetic organisms. The first of this type was probably similar to one of the present-day cyanobacteria, some of which can grow completely anaerobically, although they produce O_2 photosynthetically. The evolution of an oxygen-producing photosynthetic apparatus had enormous consequences for the environment of the earth since, as O_2 gradually accumulated, the atmosphere changed from a reducing to an oxidizing type. With O_2 available as an electron acceptor, aerobic organisms evolved; these were able to obtain much more energy from the oxidation of organic compounds than could anaerobes (see Chapter 4). More energy was made available, and higher population densities could develop, increasing the chances for the appearance of new types of organisms. There is good evidence from the fossil record that, at about the time that the earth's atmosphere became highly oxidizing, there was an enormous burst in the rate of evolution, leading to the appearance of eucaryotic microorganisms and from them to metazoans and eventually to higher animals and plants (see Section 18.7).

Another major consequence of the appearance of O_2 was that it is the major source of **ozone** (O_3), a substance that provides a barrier preventing the intense ultraviolet radiation of the sun from reaching the earth. When O_2 is subject to short-wavelength ultraviolet radiation, it is converted to O_3, which strongly absorbs the wavelengths up to 300 nm. Until this ozone shield developed, evolution could only have continued in environments protected from direct radiation from the sun, such as under rocks or in the deeper parts of the oceans. After the photosynthetic production of O_2 and development of an ozone shield, organisms could have ranged generally over the surface of the earth, permitting evolution of a greater diversity of living organisms. A summary of the steps that could have occurred in biological evolution is shown in Figure 18.3.

Origin of modern eucaryotes From comparative nucleic acid sequencing studies we now know that the three main lines of descent, *eubacterial, archaebacterial*, and *nuclear*, were established relatively early in cellular evolution (see Section 18.4). However, the eucaryotic cell we know today probably does not resemble primitive eucaryotic cells. The nuclear line of descent (primitive eucaryotic cells) originally consisted of structurally quite simple cells, resembling modern day procaryotic cells in lacking mitochondria, chloroplasts, and a membrane-bound nucleus (see Section 18.7). Modern eucaryotic cells, with their distinctive cellular organelles, were the result of endosymbiotic events (see below) that took place billions of years after the divergence of the nuclear line of descent from the universal ancestor (see Figure 18.3 and Section 18.7).

It seems likely that the eucaryotic nucleus and mitotic apparatus arose as a necessity for ensuring the replication and orderly partitioning of DNA once the genome size had increased to the point where replication as one molecule was no longer feasible. The widespread occurrence of plasmids in bacteria suggests that even in procaryotes there are evolutionary advantages for segregation of genetic information into more than one DNA molecule. It is possible to imagine how separate chromosomes might have arisen in a primitive eucaryotic cell from plasmid-like structures, and become segregated within the cell into a membrane-enclosed nucleus. Probably spindle fibers and the mitotic apparatus would also have had to evolve at the same time. There is no obvious reason why this primitive eucaryote would have needed other typical eucaryotic organelles, and these could have arisen later.

Strong evidence now exists that the modern eucaryotic cell evolved in steps through the incorporation into cells from the nuclear line of descent of heterotrophic and phototrophic symbionts. This theory, referred to as the **endosymbiotic theory** of eucaryotic evolution has, through the years, gathered increasing experimental support (see Section 3.14 and 7.10). The theory postulates that an aerobic bacterium established residency within the cytoplasm of a primitive eucaryote and supplied the eucaryotic partner with energy in exchange for a stable, protected environment and a ready supply of nutrients (Figure 18.4). This aerobic bacterium would represent the forerunner of the present eucaryotic mitochondrion. In similar fashion, the endosymbiotic uptake of a photosynthetic procaryote would have made the primitive eucaryote photosynthetic and no longer dependent on organic compounds for energy production. The phototrophic endosymbiont would then be considered the forerunner of the chloroplast (Fig-

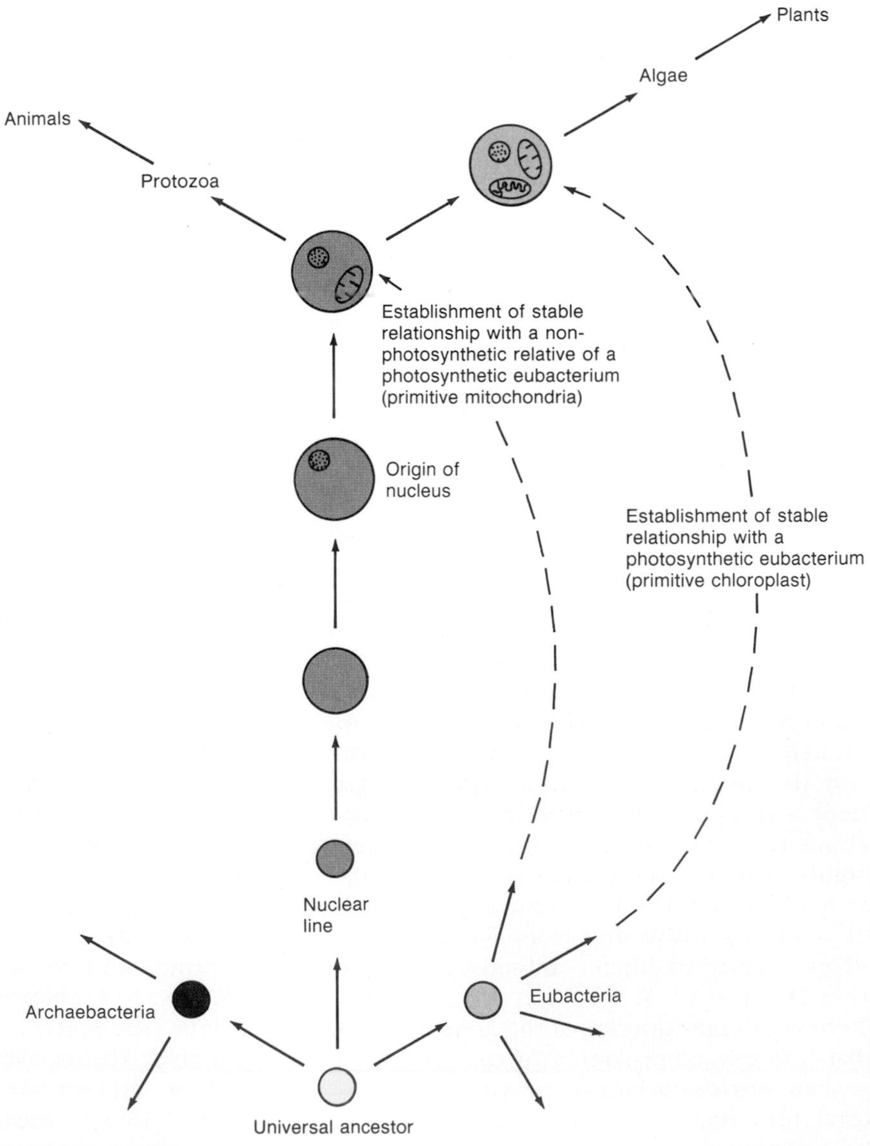

Figure 18.4 Origin of modern eucaryotes by endosymbiotic events.

ure 18.4). Following the acquisition of procaryotic endosymbionts, eucaryotic cells underwent an explosion in biological diversity. The period from about 1.5 billion years ago to the present saw the rise of multicellular eucaryotes, the metazoans, and diversification of this highly successful group culminating in the structurally complex higher plants and animals (see Section 18.7). Studies of the nucleic acids and ribosomes of eucaryotic cells and their organelles (see Section 5.1 and 18.4) have built an impressive case for the theory of endosymbiosis.

However, in order to evolve to the modern eucaryotic stage, primitive eucaryotic cells necessarily had to sacrifice certain procaryotic features such as genetic plasticity, structural simplicity, and the ability to adapt rapidly to new environments. Also, the greater complexity of the eucaryotic cell meant that it would face difficulty in adapting to life in extreme environments, such as thermal areas, where the procaryote is today preeminent. For these reasons the evolution of the eucaryote did not toll the death bell for eubacteria and archaebacteria. All three cell lines continued to evolve, serving as the forerunners of the various species we know today.

A point that deserves emphasis here is that *none* of the organisms living today are *primitive*. All extant life forms are *modern* organisms, well adapted to, and successful in, their ecological niches. Certain of these organisms may indeed be phenotypically similar to primitive organisms and may represent stems of the evolutionary tree that have not changed for millions of years (see Section 19.37); in this respect they are related to primitive organisms, but they are not themselves primitive. Furthermore, we must distinguish between primitive organisms and *simple* organisms, that is, those whose cell structures and biochemical potentialities are uncomplicated. The latter may represent organisms that have evolved little from their primitive ancestors or they may be organisms that evolved late and became "simple" through adaptive processes which selected for the reduction and loss of properties. For instance, consider the variety of bacteria pathogenic on warm-blooded animals. Many of these pathogens are so specialized that they are incapable of growing outside the body. Obviously, such organisms could not evolve until warm-blooded animals evolved, a relatively recent evolutionary event (see Figure 18.3). Once they were established, certain functions would no longer have been necessary and conceivably could have been lost through mutation, without affecting the success of the organisms.

Biological evolution and geological time scales In terms of geological time, the period from the origin of the first metazoans to the present represents only 1/6 of the total time that life has existed on Earth. Put another way, 5/6 of the earth's history was restricted to *microbial* life, the bulk of this period to *procaryotes only* (see Figure 18.3). However, because metazoans have left a considerable fossil record, our understanding of evolutionary events since the rise of the metazoans is much greater than our knowledge of evolutionary relationships among procaryotes. Fortunately, this is changing now with the advent of molecular methods for discerning evolutionary relationships. The main purpose of the remainder of this chapter is to present in some detail the modern approaches to an understanding of microbial evolution which are based on molecular biology and genetics. The emphasis will be on the molecular evolution of procaryotes, because procaryotes are the organisms with the greatest molecular diversity and for which the greatest amount of molecular biological information is known.

18.2 Molecular Approaches to Bacterial Phylogeny: Nucleic Acid Sequencing

In 1965, Emile Zuckerkandl and Linus Pauling published a seminal paper entitled "Molecules as Documents of Evolutionary History" that was to serve as a major catalyst for studies of molecular evolution. The paper hypothesized that informational macromolecules such as nucleic acids and proteins could serve as evolutionary chronometers. The authors argued that evolutionary distances between species can be measured by differences in the nucleic acid or amino acid *sequence* of homologous molecules. Put another way, the extent of sequence heterogeneity should be proportional to the number of stable mutational changes fixed in the DNA of the two organisms as they diverged from their common ancestor.

Zuckerkandl and Pauling's paper focused biologists' attention on the fact that comparative macromolecular sequencing could be used to determine the full range of phylogenetic relationships, including those of bacteria (provided the proper molecule was chosen). Selection of the proper molecule is important for several reasons. First, the molecule should be *universally distributed* across the group chosen for study. Second, it must be *functionally homologous* in each organism. Just as the old saying "you can't compare apples and oranges" has some philosophical validity, phylogenetic comparisons must start with molecules of *identical* function. One cannot compare the amino acid sequences of enzymes that carry out different reactions, or the nucleotide sequences of nucleic acids of different function, because functionally unrelated molecules would not be expected to show structural similarities. Third, it is crucial in sequence comparisons to be able to properly *align* the two molecules in order to identify regions of sequence homology and sequence heterogeneity. Finally, the sequence of the molecule chosen should change at a rate commensurate with the evolutionary distance measured. The broader the phylogenetic distance being measured, the slower must be the *rate* at which the sequence changes. A molecule which has undergone numerous sequence changes would be randomized over the evolutionary distances thought to exist in the bacterial world, and hence would be of no use as a molecular tool for determining evolutionary relationships.

18.3 Ribosomal RNAs as Evolutionary Chronometers

Because of the antiquity of the protein-synthesizing process, ribosomal RNAs appear to be excellent molecules for discerning evolutionary relationships among bacteria. Ribosomal RNAs (rRNAs) are ancient molecules, functionally constant, universally distributed, and moderately well conserved across broad phylogenetic distances. They are also readily purified from organisms, even without the use of cloning procedures. Recall the structure of the ribosome (see Figure 5.28). There are three ribosomal RNA molecules, which in procaryotes have sizes of 5S, 16S and 23S. The large bacterial rRNAs, 16S, and 23S rRNA (approximately 1500 and 3000 nucleotides, respectively) contain several regions of highly conserved sequence useful for obtaining proper sequence alignment, yet they contain sufficient sequence variability in other regions of the molecule to serve as excellent phylogenetic chronometers (Figure 18.5).

5S rRNA (Figure 18.6) has also been used for phylogenetic measurements, but its small size (~120 nucleotides), limits the information obtainable from this molecule. The secondary structure of ribosomal RNAs has remained relatively constant across the liv-

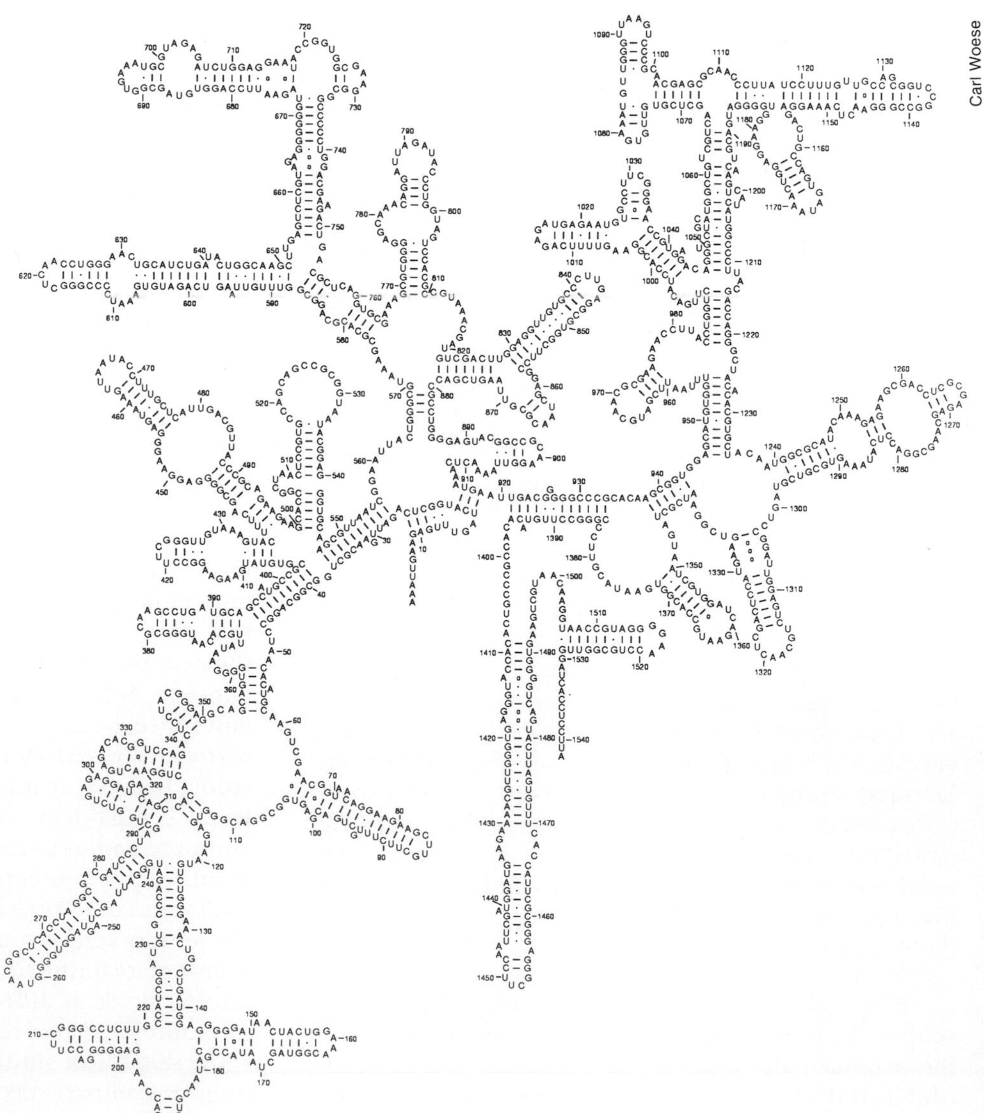

Figure 18.5 Primary and secondary structure of 16S ribosomal RNA (rRNA). This is the eubacterial form of *Escherichia coli*. Archaebacterial 16S rRNA has general similarities in secondary structure (folding) but numerous differences in primary structure (sequence).

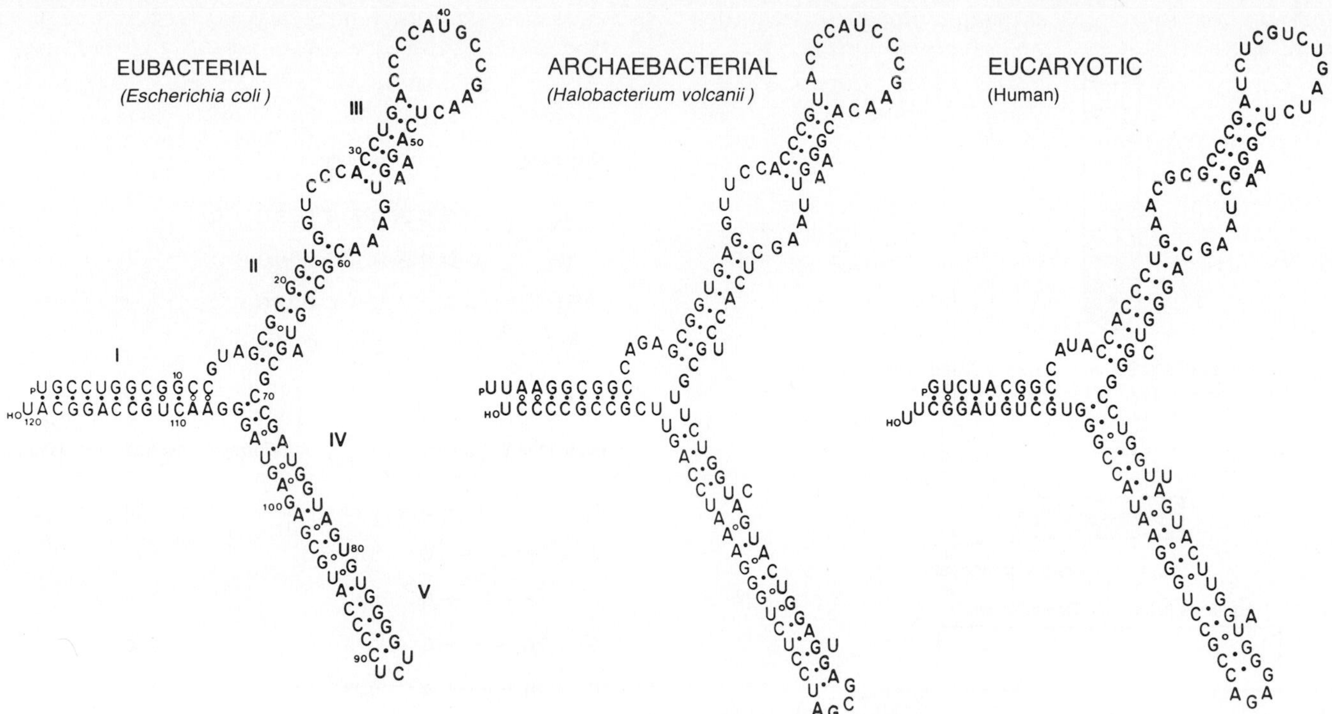

Figure 18.6 Primary and secondary structure of eubacterial, archaebacterial, and eucaryotic 5S rRNA. Notice the general similarities in secondary structure, but numerous differences in primary structure, between the two molecules.

ing kingdoms, and this supports the argument that ribosomal RNAs are ancient and functionally homologous.

Because 16S RNA is more experimentally manageable than 23S RNA, it has been used extensively to develop the phylogeny of procaryotes and to some extent, the eucaryotes. The accumulating data base of partial and complete 16S sequences now comprises over 500 different organisms. Use of 16S rRNA as a phylogenetic tool was pioneered in the early 1970s by Carl Woese at the University of Illinois (see Box), and it is now used by a number of groups throughout the world.

Complete sequence analyses Ribosomal RNAs are now relatively easy to sequence because the molecule can be directly sequenced from crude cell extracts using reverse transcriptase and the dideoxy sequencing method (see Box, Chapter 5). A typical rRNA sequencing experiment involves phenol extraction of a relatively small (0.2–1 g) amount of cells to remove protein and other non-nucleic acid material and release the RNA. Total RNA is then precipitated with alcohol and salt, and then a small DNA oligonucleotide primer of about 15–20 nucleotides in length *complementary* in base sequence to some highly conserved section in the 16S rRNA molecule, is added (Figure 18.7). The enzyme reverse transcriptase (see Box, Chapter 5, and Section 6.24) is then added along with a ^{32}P-labeled deoxyadenosine triphosphate and the other unlabeled deoxyribonucleotides, and then the mixture is divided into four identical portions. To each portion a small amount of a different 2′,3′ *di*deoxynucleotide is added. Reverse transcriptase will read the 16S rRNA template and begin making a DNA copy interrupted at various spots by the incorporation of *di*deoxynucleotides. The fragments are then sequenced by the dideoxy sequencing method (see Box, Chapter 5). From knowledge of the cDNA sequence obtained by dideoxy sequencing, the sequence of the original 16S rRNA can be deduced.

Phylogenetic trees from complete sequences The preferred method of generating phylogenetic trees from complete sequences is the so-called *distance-matrix method* (Figure 18.8). Two rRNA sequences are aligned and an **evolutionary distance (E_D)** calculated by recording the number of positions in the sequence at which the two *differ*. A statistical correction factor is then used to account for the possibility that multiple changes might have occurred which would lead to the same sequence. A matrix of evolutionary distances generated from sequence comparisons is then run through a computer program designed to produce phylogenetic trees from E_D measurements. In effect, this amounts to examining all

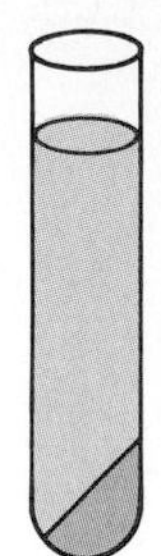

1. Centrifuge cells (300-500 mg wet weight) and break open cells in the presence of DNAase. All DNA destroyed.

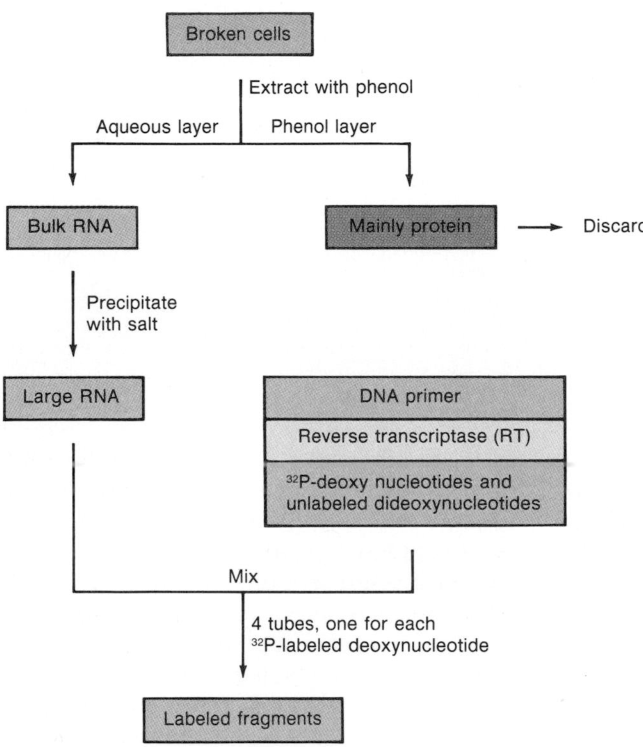

2. Preparation of labeled cDNA fragments for sequence determination

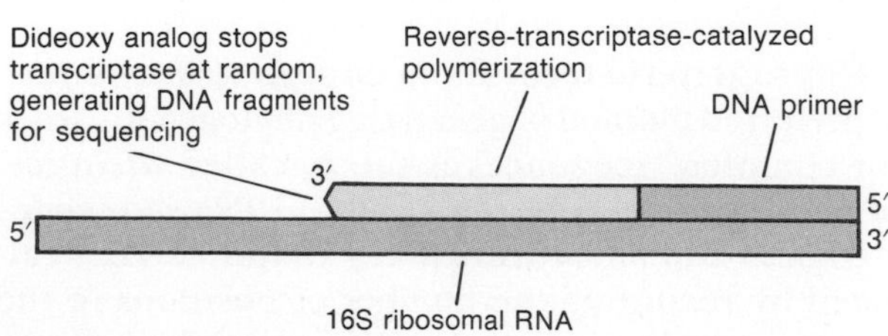

3. Determine DNA sequences by the Sanger method (see Box, page 163) and deduce the ribosomal RNA sequence from the cDNA sequence.

Figure 18.7 Making a cDNA copy of 16S rRNA using the retroviral enzyme, reverse transcriptase.

Organism	Sequence
A	C G U A G A C C U G A C
B	C C U A G A G C U G G C
C	C C A A G A C G U G G C
D	G C U A G A U G U G C C

(a)

	Evolutionary distance		Corrected evolutionary distance
E_D	A → B	0.25	0.30
E_D	A → C	0.33	0.44
E_D	A → D	0.42	0.61
E_D	B → C	0.25	0.30
E_D	B → D	0.33	0.44
E_D	C → D	0.33	0.44

(b)

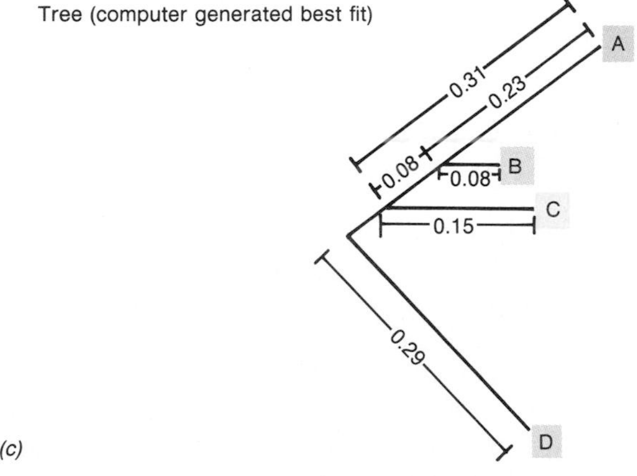

(c)

Figure 18.8 Preparing a phylogenetic distance-matrix tree from continuous (complete) 16S rRNA sequences. Just short sequences are used, for illustrative purposes, and it should be assumed that the sequences shown in (a) are representative of the entire 16S rRNA. The evolutionary distance (E_D) shown in (b) is the percent of *nonhomologous* sequence between the RNAs of two organisms. The corrected E_D is a statistical correction necessary to account for either back mutations to the original genotype or additional forward mutations at the same site. The tree (c) is ultimately generated by computer analysis of the data to give the best fit. The sum of the length of the branches separating any two organisms is proportional to the evolutionary distance between them.

Table 18.1 Signature sequences defining the three primary kingdoms of living organisms

Oligonucleotide sequence*	Approximate position†	Occurrence among		
		Archaebacteria	Eubacteria	Eucaryotes
CACYYG	315	0	>95	0
CYAAYUNYG	510	0	>95	0
AAACUCAAA	910	3	100	0
AAACUUAAAG	910	97	0	>90
NUUAAUUCG	960	0	>95	0
YUYAAUUG	960	100	<1	100
CAACCYYCR	1110	0	>95	0
UUCCCG	1380	0	>95	0
UCCCUG	1380	100	0	100
CUCCUUG	1390	100	0	0
UACACACCG	1400	0	>99	100
CACACACCG	1400	100	0	0

**Y = any pyrimidine base; R = any purine base; N = any purine* or *pyrimidine base.*
†Refer to Figure 18.5 for numbering scheme of 16S rRNA.

possible branching arrangements for the set of distances compared, arranging the branch lengths for each branching arrangement to optimally fit the data, and then selecting the best arrangement as the "correct" phylogenetic tree. The E_D separating any two organisms is directly proportional to the *total length* of the branches separating them. The principles of phylogenetic tree formation are shown in Figure 18.8 and a sample tree constructed from sequence data of four hypothetical organisms. As the number of sequences compared increases, the computational analyses quickly become unwieldy, so that virtually all matrix-distance trees developed are computer generated. Even the simple tree shown in Figure 18.8 required computer analysis to generate.

Signature sequences Over 80 percent of the 500 or so 16S rRNAs characterized, have been done by the cataloging method (see Box). Computer analysis of the catalogs has revealed **signature sequences**, oligonucleotide stretches unique to certain groups of organisms. Oligonucleotide signatures defining each of the three primary kingdoms (Table 18.1) have been found. Other signatures defining the major taxa within each kingdom have also been detected (see Sections 18.5 and 18.6). The "signatures" are generally found in defined regions of the 16S rRNA molecule, but are only revealed when computer scans of two aligned sequences are performed. The exclusitivity of signature sequences make them particularly useful for placing unknown organisms in the correct major phylogenetic group.

18.4 Microbial Phylogeny as Revealed by Ribosomal RNA Sequencing

Prior to the advent of molecular sequencing, all bacteria were classified in one large group, the procaryotes, which was considered to be quite distinct from the eucaryotes. However, comparative sequencing studies of 16S rRNA and, to a lesser extent, 5S RNA, has led to the astonishing and totally unsuspected finding that life on Earth is not fundamentally of *two* types, procaryotes and eucaryotes, but instead comprises at least *three* primary lineages, two of them bacterial (procaryotic), the other eucaryotic. Comparative sequencing further suggested that the two procaryotic lineages were no more related to one another (in an evolutionary sense) than either was to the eucaryotes, despite the fact that both "look like bacteria.". In recognition of this, two classes of procaryotes were defined which were called **eubacteria** and **archaebacteria**. A wealth of additional molecular data has subsequently accumulated to support the eubacterial/archaebacterial dichotomy (see Sections 18.5 through 18.7). Today, most microbiologists agree that archaebacteria and eubacteria differ profoundly in many of their fundamental biochemical properties.

Another major conclusion drawn from sequencing studies is that the eucaryotic line of descent is *not* of relatively recent origin, as previously assumed. Although the archaebacterial and eubacterial kingdoms were well established and highly diversified be-

The Oligonucleotide Catalog: A Shortcut to Molecular Phylogeny

Although the microbial phylogeny presented in this chapter is based on complete sequences of 16S ribosomal RNA molecules, sequence determination of a molecule as long as 16S RNA is a fairly lengthy procedure and shortcut methods are desirable. A simplifying approach is the determination of the base sequences of short oligonucleotides and the preparation for each organism of a *catalog* of oligonucleotide sequences. In this approach, radioactively labeled purified 16S rRNA is digested with a ribonuclease which cuts RNA at every guanine residue, and the resulting oligonucleotides separated by electrophoresis. Only oligonucleotides with five or more bases contain sufficient sequence information to permit reliable comparisons, and the base sequences of all such oligonucleotide fragments are determined. For each organism, a catalog (list) of oligonucleotide sequences is prepared, and comparisons made with all other organisms. The oligonucleotide catalogs from two organisms, organism A and organism B, can be related by calculating a similarity coefficient, the so-called S_{AB} value, defined as twice the number of nucleotides in identical oligonucleotides (of length five or greater) common to the two catalogs, divided by the total number of nucleotides in all oligonucleotides in the catalog (also length five or greater). S_{AB} values run from 1.0 for identical organisms to about 0.05 for randomly related sequences. Organisms with 90 percent sequence homology have S_{AB} values of about 0.70.

The S_{AB} values can be used to generate phylogenetic trees similar to those presented in this chapter for complete base sequences, and the trees obtained are quite similar to those obtained from complete sequence comparisons. In addition, preparation of an oligonucleotide catalog permits the detection of the characteristic *signature sequences*, those short oligonucleotide stretches unique to certain groups of organisms (see Table 18.1).

The use of 16S ribosomal RNA as a phylogenetic tool was pioneered in the early 1970s by Carl Woese, a microbial geneticist at the University of Illinois. The oligonucleotide catalog approach was the first method used by Woese to discern molecular relationships between organisms. Because of the huge molecular differences between archaebacteria and eubacteria, members of these two great groups of microorganisms were quickly recognized as different on the basis of their oligonucleotide catalogs. When two archaebacteria were compared, S_{AB} values of around 0.35 to 0.40 were obtained, whereas if an archaebacterium was compared with a eubacterium or a eucaryote, S_{AB} values of 0.10 or less were obtained. Although the archaebacteria were first recognized based on their distinct oligonucleotide catalogs, other major differences between archaebacteria and eubacteria were soon discovered (lipids, cell wall composition, RNA polymerase, etc.), as are discussed elsewhere in this chapter. The discovery of the archaebacteria by oligonucleotide cataloging stimulated research to determine complete base sequences for ribosomal RNA molecules, and this sequence approach has now virtually supplanted the cataloging approach. Carl Woese's pioneering contribution to the discovery of phylogenetic relationships between organisms has had far-reaching implications for bacterial taxonomy, implications extensively discussed in the present chapter. (See Woese reference at the end of this chapter for a detailed historical background.)

fore the rise of eucaryotic cells *with which we are familiar* (that is, as endosymbiotic arrangements of procaryotic cells living inside a larger cell, see Section 18.1), the *nuclear line of descent* (the pre-endosymbiotic eucaryote) appears to be as old as either of the bacterial lines (see Section 18.7 and Figure 18.4). Thus, the cell line that eventually gave rise to eucaryotes as we know them today, presumably is a descendent from the universal ancestor of all cells, just as the two bacterial cell lines are.

Overview of bacterial phylogeny While the earliest 16S rRNA sequence analyses were done by oligonucleotide cataloging (see Box), the more recent analyses are of either complete or nearly complete sequences. However, full sequences do not change the major conclusion resulting from the earlier approach: life is based on three major lines of descent. This living tripartite arrangement is illustrated in Figure 18.9 using a distance matrix tree to group the major representatives of each kingdom. The evolutionary distance along the scale gives a quantitative estimate of how much the three lines of descent have diverged from a common ancestor. The point that joins the three primary lines of descent does not necessarily represent the root of the tree, the position that corresponds to the universal ancestor. Where that point lies on this (unrooted) tree depends upon the relative *rates* of evolution in the three lineages—and those are not known with certainty. What is evident from the (unrooted) tree, however, is that the various lineages have evolved at different rates. No matter

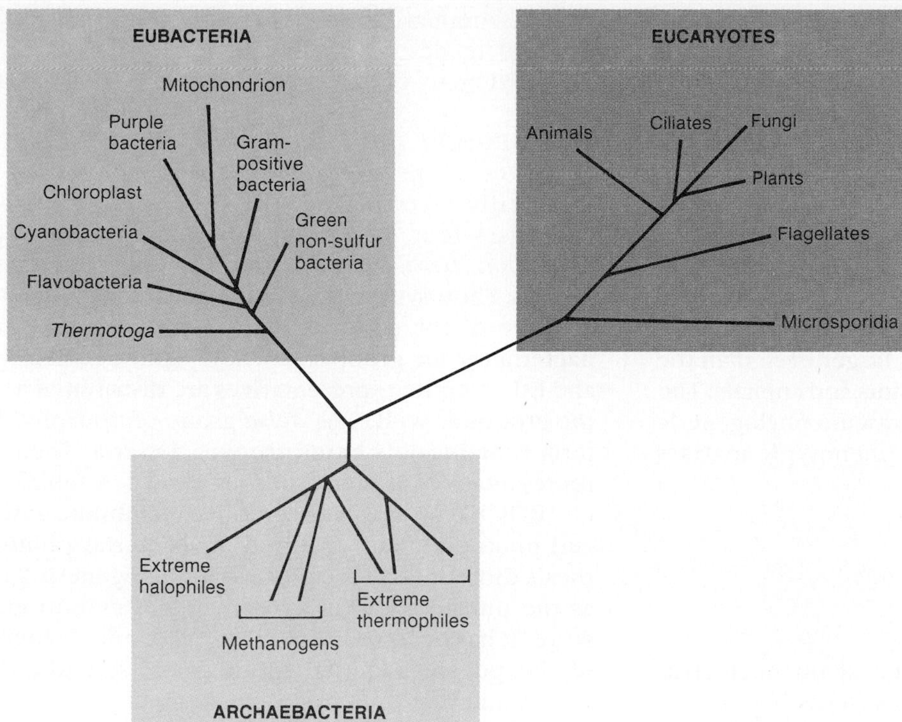

Figure 18.9 Universal phylogenetic tree determined from ribosomal RNA sequence comparison.

where the real root lies, the archaebacteria appear to be closer to the root than the organisms in either of the other groups. Assuming the root of the tree is somewhat centrally located as shown in Figure 18.9 (and assuming, of course, that the tree derived from rRNAs is representative of the evolution of the organism as a whole) then the archaebacteria would appear to be the most primitives (least evolved) of the three group phenotypes and the eucaryotes would be the least primitive, that is, the most derived. A roughly central root placement would then imply that the archaebacteria are a more ancient group than either of the others.

The unrooted tree in Figure 18.9 suggests that the eucaryotic lineage evolved some important characteristic or set of characteristics that allowed rapid evolutionary diversity. Perhaps this evolutionary event was something as simple as a larger cell size; this would have facilitated the larger amounts of DNA needed for the development of metazoan organisms, and produced an organism sufficiently large to accept procaryotic endosymbionts. Placement of the archaebacteria nearest the universal ancestor (implying that this group has evolved *less* than either of the others) is supported by the fact that many archaebacteria inhabit extreme environments, high temperature, low pH, high salinity, and so on (see Sections 19.33–19.37); these environmental extremes, especially that of high temperature, reflects the environmental conditions under which life originated (see Section 18.1). Thus, archaebacteria may well be evolutionary relics of the earth's earliest life forms.

The overall phylogenetic picture presented in Figure 18.9 also tells us about other evolutionary events. For example, it is clear that mitochondria and chloroplasts arose from endosymbiotic *eubacteria* that established stable relationships, perhaps several different times, with cells from the nuclear line of descent. Mitochondria appear to have arisen from a small group of organisms that includes the modern procaryotes *Agrobacterium, Rhizobium*, and the rickettsias. The latter organisms belong to a larger group of eubacteria called the purple bacteria (organisms such as *Rhodopseudomonas* and *Rhodobacter*). It is interesting to note that the mitochondrion, itself an intracellular symbiont, is specifically related to organisms (*Agrobacterium, Rhizobium*, and the rickettsias) that are capable of an intracellular existence (see Sections 17.23 and 17.24).

The other eucaryotic endosymbiont, the chloroplast, has originated from eubacteria as well. The evolutionary tree in Figure 18.9 shows that the chloroplast and cyanobacteria *shared* a common ancestor. Although sequence analysis of a number of cyanobacteria have made it clear that cyanobacteria share affinities with the green plant chloroplast, the green plant chloroplast apparently was *not* derived from within the cyanobacterial group. Instead, the chloroplast and cyanobacteria evolved from a common (photosynthetic) relative. The red algal chloroplast (which, unlike the green plant chloroplast contains phycobiliproteins, the cyanobacterial light harvesting pigments, see Section 19.2) seems to be a closer direct relative of cyanobacteria than the green plant chloroplast. Thus, although the precise roots of the green plant chloroplast are still unclear, they are clearly *eubacterial* in nature.

A further examination of the evolutionary tree shown in Figure 18.9 shows that evolutionary distances between phenotypically similar organisms can often be quite large, while the evolutionary distance between certain phenotypically quite different organisms is sometimes rather small. For example, the evolutionary distance between purple bacteria and green nonsulfur bacteria, both Gram-negative phototrophic bacteria, is quite large, larger even than the evolutionary distance between plants and animals! The lesson to be learned here is that drawing phylogenetic (evolutionary) conclusions from phenotypic analyses is often misleading.

18.5 Eubacteria

The phylogeny of eubacteria, based on nucleotide catalog analyses, suggests that eleven distinct groups exist. The members of some groups are phylogenetically highly related, similarity coefficients (S_{AB}, see Box) within the group being at 0.5 or higher. However, most of the groups are phylogenetically diverse, although coherent when compared with other groups. Some groups previously established by phenotypic criteria such as physiology or morphology, for example green nonsulfur bacteria or spirochetes, form distinct phylogenetic groups as well. Most groups, however, are a collage of physiological and morphological types of bacteria, emphasizing the fact that phenotypic markers are often poor phylogenetic criteria.

We summarize here the eubacterial groups and proceed to describe them in detail in Chapter 19. The phylogeny of eubacteria is summarized in Figure 18.10.

1. Purple phototrophic bacteria and their relatives This group is the largest and most physiologically diverse of all the eubacteria. The group comprises four distinct subdivisions, referred to as the *alpha, beta, gamma*, and *delta* purple bacterial groups. Photosynthetic purple bacteria form the heart of three of the subdivisions (thus the name purple bacteria for the group as a whole), but heterotrophic and lithotrophic representatives are distributed across the groups as well. The *delta* group of "purple" bacteria contains only heterotrophic bacteria. The major representatives of each group are shown in Table 18.2.

It is likely that the complex membrane systems and photopigments required for bacterial photosynthesis did not arise in so broad a phylogenetic group as the purple bacteria by convergent evolution. Instead, it has been proposed that the common ancestor of the purple bacterial group was itself photosynthetic, and that the physiological diversity characteristic of the purple bacterial group arose as the result of the exchange of photosynthetic capacity for other physiological processes (such as sulfate reduction and lithotrophy) better adapted to new and expanding ecological niches (Figure 18.11). Such a hypothesis nicely accounts for the specific phylogenetic relatedness of organisms like *Rhodopseudomonas* (a budding, *phototrophic* bacterium containing polar stacks of lamellar membranes) and *Nitrobacter* (a budding, *lithotrophic* bacterium containing polar stacks of lamellar membranes), or the specific phylogenetic relatedness of phototrophic and certain heterotrophic spirilla.

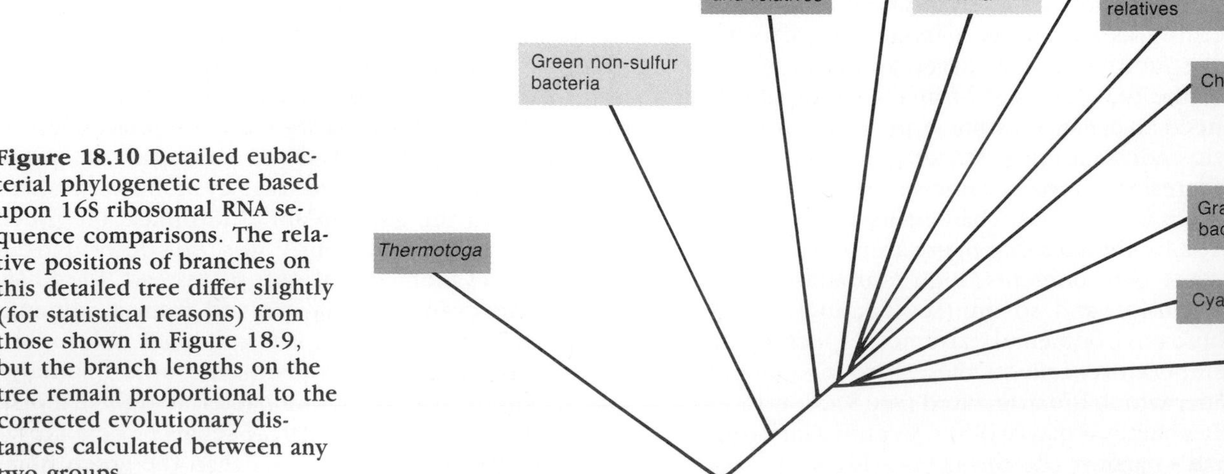

Figure 18.10 Detailed eubacterial phylogenetic tree based upon 16S ribosomal RNA sequence comparisons. The relative positions of branches on this detailed tree differ slightly (for statistical reasons) from those shown in Figure 18.9, but the branch lengths on the tree remain proportional to the corrected evolutionary distances calculated between any two groups.

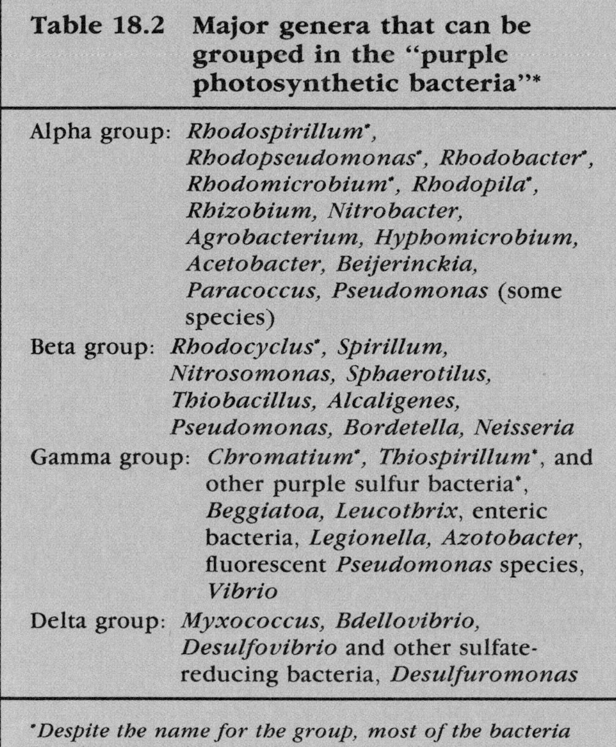

Table 18.2 Major genera that can be grouped in the "purple photosynthetic bacteria"*

Alpha group:	*Rhodospirillum**, *Rhodopseudomonas**, *Rhodobacter**, *Rhodomicrobium**, *Rhodopila**, *Rhizobium, Nitrobacter, Agrobacterium, Hyphomicrobium, Acetobacter, Beijerinckia, Paracoccus, Pseudomonas* (some species)
Beta group:	*Rhodocyclus**, *Spirillum, Nitrosomonas, Sphaerotilus, Thiobacillus, Alcaligenes, Pseudomonas, Bordetella, Neisseria*
Gamma group:	*Chromatium**, *Thiospirillum**, and other purple sulfur bacteria*, *Beggiatoa, Leucothrix*, enteric bacteria, *Legionella, Azotobacter*, fluorescent *Pseudomonas* species, *Vibrio*
Delta group:	*Myxococcus, Bdellovibrio, Desulfovibrio* and other sulfate-reducing bacteria, *Desulfuromonas*

**Despite the name for the group, most of the bacteria listed are nonphototrophic. Phototrophic representatives are shown by an asterisk after the name.*

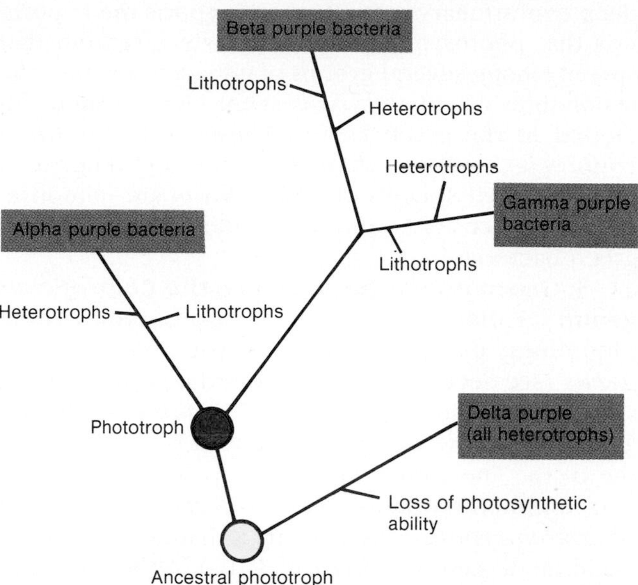

Figure 18.11 Divergence of the "purple" bacterial lineage to give rise to nonphotosynthetic relatives.

Nonsulfur purple bacteria are found only in the *alpha* and *beta* subdivisions of the purple bacterial group, while purple sulfur bacteria are found in the *gamma* subdivision (Table 18.2). Certain lithotrophic sulfide oxidizing bacteria which store sulfur intracellularly, for example members of the genus *Beggiatoa* (see Sections 16.9 and 19.5), are found within the *gamma* purple bacteria group. Sulfur lithotrophs carry out the oxidation of sulfide to sulfur and sulfate just as purple sulfur bacteria do, but do so as a means of obtaining *energy* rather than simply as a source of reducing power for CO_2 reduction (see Section 16.9). If the storage of sulfur intracellularly is a unique biochemical property, then one could imagine how the process of phototrophic sulfur storage could have been recruited as a mechanism to store sulfur for use as an *energy* source lithotrophically by *Beggiatoa* and its relatives. Indeed, recent experiments have shown that a variety of phototrophic sulfur bacteria are capable of lithotrophic (dark, microaerobic) growth at the expense of reduced sulfur compounds (see Section 19.1). Hence, the phylogenetic relationships between phototrophic and lithotrophic sulfide oxidizers are not surprising.

The *delta* group of purple bacteria is an extremely broad one that consists of physiologically diverse organisms that appear to have no recognizable phenotypic thread among them. The major subgroups include (a) the sulfate- and sulfur-reducing eubacteria, (b) the fruiting myxobacteria, and (c) the procaryotic predator, *Bdellovibrio*. The sulfate/sulfur reducers are strictly anaerobic heterotrophs that oxidize a variety of organic electron donors and reduce sulfate or sulfur to H_2S (see Sections 16.15 and 19.8). The fruiting myxobacteria are obligate aerobes that move by gliding motility and which aggregate under certain nutritional conditions to form complex arrangements of cells called fruiting bodies (see Section 19.13). *Bdellovibrio* is a small, spiral-shaped Gram-negative bacterium which acts as a predator of other Gram-negative bacteria, attaching and eventually moving into the periplasmic space of its prey and destroying it (see Section 19.11).

Finally, all of the enteric bacteric, most *Pseudomonas* species, the free-living and symbiotic N_2-fixers, and most lithotrophs are also phylogenetically related to the purple bacteria. Hence the purple bacterial lineage appears to have given rise to a wide diversity of ecologically and physiologically significant bacteria. The oligonucleotide sequence AAAUUCG is diagnostic of *alpha* purple bacteria, CYUUACACAUG (where Y is any base), is characteristic of the *beta* group, and ACUAAAACUCAAAG is present in the 16S RNA of most *delta* purple bacteria; no specific signature has been identified for the *gamma* group.

2. Green sulfur bacteria The phototrophic green sulfur bacterium *Chlorobium* is related to other eubacteria (including purple photosynthetic bacteria) at only the deepest phylogenetic levels (Figure 18.10). This lack of close relationship is not too surprising in light of the unique pigments, light-harvesting apparatus (chlorosomes), and autotrophic mechanisms used by these organisms (see Section 19.1). The finding that photosynthesis once again has deep evolutionary roots further supports the hypothesis that photosynthesis was an early invention that spread among several groups of eubacteria before the major lines of eubacterial descent were defined. Included in the green bacterial group besides *Chlorobium* is the genus *Chloroherpeton*, a gliding green bacterium; all species of *Chlorobium* are immotile. The sequence AUACAAUG provides a signature for green bacteria.

3. Green nonsulfur bacteria: the *Chloroflexus* group Following its initial isolation, studies of the physiology, pigments, and fine structure of *Chloroflexus* (see Section 19.1) suggested a close relationship to green sulfur bacteria (*Chlorobium*). Phylogenetically, however, nothing could be further from the truth. The *Chloroflexus* group, which includes two nonphotosynthetic genera, *Herpetosiphon*, and *Thermomicrobium*, represents a distinct line of eubacterial descent (see Figure 18.10). Although members of the *Chloroflexus* group are related to one another at S_{AB} (see Box) levels of 0.3 or greater, representatives of this group are only related to other groups at an S_{AB} level of 0.2 or less. *Chloroflexus* is clearly an evolutionary puzzle because of its lack of phylogenetic relationship to green sulfur bacteria with which it shares a number of phenotypic properties (see Section 19.1). Because most of the members of the *Chloroflexus* group are thermophiles, the thermophilic property may be of more significance in defining the phylogeny of *Chloroflexus* and its relatives than is their relationship to green bacteria. The sequence CCUAAUG provides a signature for *Chloroflexus*.

4. Cyanobacteria The cyanobacteria were discussed previously in connection with the origin of the chloroplast (see Section 18.4). Although the taxonomy of cyanobacteria is presently weighted toward morphological criteria (see Section 19.2), only a few morphologically defined cyanobacterial groups have firm phylogenetic standing. Both the branching and the heterocystous cyanobacteria appear to form rather tight phylogenetic groups; unicellular cyanobacteria by contrast are phylogenetically diverse. The eubacterial branch leading to the cyanobacteria is also the branch within which the precursor of the green plant chloroplast arose (see Figure 18.9). The nucleotide sequence AAUUUUYCG is a signature sequence for cyanobacteria.

5. *Planctomyces/Pirella* group The next eubacterial group to be described contains budding organisms which lack peptidoglycan, contain a proteinaceous cell wall, are obligately aerobic, and require very dilute culture media for growth. The *Planctomyces/Pirella* line forms a major branch of eubacteria (see Figure 18.10). Although lacking peptidoglycan, the presence of ester-based lipids and nonarchaebacterial rRNA sequences are consistent with the idea that the *Planctomyces* group is really eubacterial, albeit a branch quite distinct from the other groups (see also Section 19.10). The oligonucleotide CUUAAUUCG provides a signature for members of the *Planctomyces/Pirella* group.

6. Spirochetes The main spirochete cluster contains the genera *Spirochaeta, Borrelia*, and *Treponema*. These organisms may be free-living or host associated and many are well-known pathogens. The genus *Leptospira* forms a distinct branch within the spirochete group. Spirochetes have traditionally been grouped on the basis of their unique morphology and mode of motility (see Section 19.12), and these traits turn out to be valid phylogenetic indicators as well. The signature sequence AAUCUUG is highly diagnostic for one subgroup of spirochetes; the sequence UCACACYAYCYG (where Y is any base) is diagnostic for most, but not all spirochetes.

7. *Bacteroides/Flavobacterium* group This group consists of a number of genera forming a major phylogenetic line of Gram-negative eubacteria. The members of the genus *Bacteroides* form one subline within the group, while *Cytophaga* and *Flavobacterium* define a second. A mixture of physiological types shows up here, obligate anaerobes (*Bacteroides*) along with obligate aerobes (*Sporocytophaga*). However, the two subgroups are themselves relatively homogeneous. *Bacteroides* and *Fusobacterium* are the major genera in one subgroup; both are obligately anaerobic, primarily fermentative rods. The *Flavobacterium* subgroup includes *Cytophaga, Flexibacter, Sporocytophaga*, and *Saprospira*, all rod-shaped or filamentous organisms, primarily respiratory in physiology, but which move by gliding motility. The sequence UUACAAUG is the signature for members of the *Bacteroides* group; the sequence CCCCCACACUG is the signature for the gliding group.

8. Chlamydia This group consists thus far only of species of the obligate intracellular parasite, *Chlamydia*. Homology at the 95 percent level exists between the 16S rRNA sequences of the psittacosis organism, *C. psittaci*, and *C. trachomitis*, the causative agent of a variety of venereal infections and a form of acquired blindness called *trachoma* (see Section 15.6). On the basis of signature sequence comparisons, the chlamydial group shows a distant relationship to members of the *Planctomyces* group. It is interesting to note that members of both of these groups have cell walls that lack peptidoglycan. The biology of the chlamydia is discussed in Section 19.23.

9. *Deinococcus-Thermus* group This group really consists of only two well characterized genera, the Gram-positive, highly radiation-resistant *Deinococcus* (see Section 19.24) and the Gram-negative heterotrophic thermophile *Thermus*. Both *Deinococcus* and *Thermus* have an atypical cell wall that contains ornithine in place of diaminopimelic acid in its peptidoglycan (*Chloroflexus* also contains ornithine in its peptidoglycan). Interspecies S_{AB} (see

Box) values within the group are greater than 0.5, indicating that the group is tightly clustered and probably of relatively recent origin. The oligonucleotide CUUAAG is a signature for the *Deinococcus-Thermus* group.

10. Gram-positive eubacteria With the exception of *Deinococcus*, the Gram-positive eubacteria, both rods and cocci, form a relatively coherent phylogenetic cluster. Included here are the endospore formers, lactic acid bacteria, most anaerobic and aerobic cocci, the coryneform bacteria, the actinomycetes, and even the majority of mycoplasmas, which phylogenetically appear to be clostridia that lack cell walls. Surprisingly, certain Gram-negative phototrophic bacteria are related to these Gram-positive bacteria. Heliobacteria, a group consisting of *Heliobacterium, Heliospirillum*, and *Heliobacillus* (see Section 19.1), appear to be distant relatives of the clostridia. This finding is significant because we have seen that phototrophic bacteria and their relatives are distributed among a number of other eubacterial groups as well. The distribution of a major physiological property (such as photosynthesis) across broad evolutionary distances thus suggests that photosynthesis arose very early in procaryotic evolution.

All S_{AB} (see Box) values of the Gram-positive eubacteria are 0.25 or above and thus the property of Gram-positiveness is a phylogenetically valid property. Two main classes of Gram-positive eubacteria are recognized, those containing relatively low GC base ratios (clostridia, bacilli, lactic acid bacteria, see Sections 19.24–19.26), and those with high GC ratios (actinomycete group, see Section 19.28). The oligonucleotide signature CUAAAACUCAAAG is highly diagnostic for the high GC group. No low GC group oligonucleotide signature has been found. The major genera placed in the Gram-positive eubacterial group are shown in Table 18.3.

Table 18.3 Major genera of phylogenetically-related Gram-positive eubacteria

Group	Genera
Low GC group: *Clostridium* and relatives	*Clostridium, Bacillus, Thermoactinomyces, Sporosarcina, Acetobacterium,Streptococcus, Peptococcus, Lactobacillus, Staphylococcus, Ruminococcus, Mycoplasma, Acholeplasma, Spiroplasma*
High GC group: Actinomycetes	*Actinomyces, Bifidobacterium, Propionibacterium, Streptomyces, Nocardia, Actinoplanes, Arthrobacter, Corynebacterium, Mycobacterium, Micromonospora, Frankia, Cellulomonas, Brevibacterium*
Photosynthetic subdivision: *Heliobacterium*	

11. *Thermotoga* The final group of eubacteria consists of only a single genus, *Thermotoga*. Isolated thus far only from geothermally heated marine sediments, *Thermotoga* is the most thermophilic of all known *eu*bacteria. *Thermotoga* is a strictly anaerobic fermentative organism, capable of growth at temperatures from 55–90°C, with an optimum of about 80°C. *Thermotoga* produces unique lipids containing extremely long-chain fatty acids and its cell wall contains peptidoglycan. Although clearly a eubacterium, *Thermotoga* is phylogenetically positioned far from other eubacteria (Figure 18.10) and in the three-kingdom tree (Figure 18.9), seems to occupy an intermediate phylogenetic position between the remaining eubacteria and the archaebacteria. Because of its extremely thermophilic nature, the discovery of *Thermotoga* lends support to the hypothesis that the earliest bacterial phenotype, whether archaebacterial or eubacterial, may have been thermophilic (see Sections 19.35 and 19.37).

18.6 Archaebacteria

Archaebacteria can be divided into four relatively broad groups, one consisting of some methanogens and the extremely halophilic bacteria, one consisting of methanogens only, one containing the sulfur-dependent bacteria (also referred to as thermoacidophilic bacteria), and a final "group" consisting of the genus *Thermoplasma*. The methanogenic/halophilic group is a physiological enigma. The methanogens are a large group of obligately anaerobic bacteria unified by their ability to make methane (see Section 19.34). By contrast, the extreme halophiles are generally obligate aerobes and grow only at NaCl concentrations near saturation (see Section 19.33). The sulfur-dependent archaebacteria are mainly anaerobic heterotrophs that are also extreme thermophiles (see Section 19.35). The overall phylogeny of archaebacteria, as defined by a distance-matrix tree, is shown in Figure 18.12.

From the tree shown in Figure 18.12, it is obvious that several subgroups of methanogens exist; this is not surprising considering the phenotypic diversity of methanogenic bacteria (see Section 19.34). Although a highly specific and conserved biochemical process, methanogenesis is of relatively broad phylogenetic distribution. The extremely halophilic procaryotes (such as *Halobacterium*) and their relatives, are placed near the methanogens, more specifically near the genus *Methanobacterium. Thermoplasma acidophilum*, an acidophilic, thermophilic mycoplasma (see Section 19.36) is also found on the methanogen branch of the archaebacteria. The sequence AYUAAG is a signature for the methanogens while the

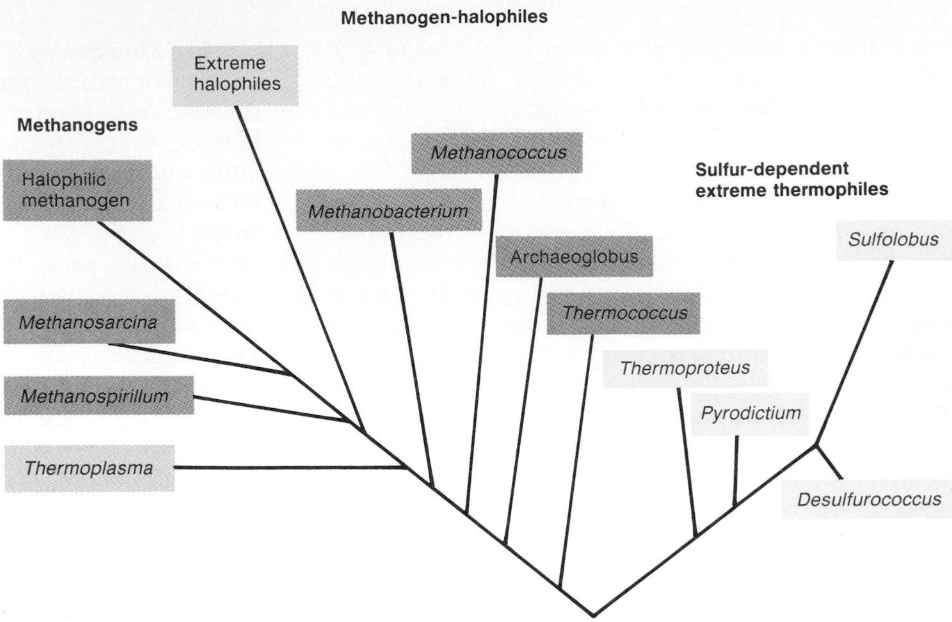

Figure 18.12 Detailed phylogenetic tree of the archaebacteria based upon 16S ribosomal RNA sequence comparisons. See note on Figure 18.10 for comparisons of detailed tree with the three-kingdom tree.

sequence AAUUAG provides a signature for extreme halophiles. The sequences AAAACUG and ACCCCA are signatures for *Thermoplasma*.

The other branch of the archaebacteria is the sulfur-dependent archaebacteria. Phenotypically, the group is united by a variety of striking properties. First, all sulfur-dependent archaebacteria are *extreme thermophiles*, most having growth temperature optima above 80°C including the extremely thermophilic organism, *Pyrodictium*. The latter organism has a temperature optimum of 105°C (see Section 19.35). Second, most sulfur-dependent archaebacteria require *elemental sulfur* for optimal growth. This is because these organisms are for the most part anaerobes that use sulfur as an electron *acceptor*, oxidizing various organic compounds as energy sources. One sulfur-dependent archaebacterium, *Sulfolobus*, uses elemental sulfur not as an electron acceptor but as an electron *donor* with oxygen or ferric iron serving as an electron acceptor (see Sections 16.9 and 16.17). *Archaeoglobus,* an extremely thermophilic *sulfate*-reducing archaebacterium, may represent a phenotypic bridge between the sulfur-dependent group and the methanogens (see Section 19.35). Sulfur is thus seen to play an intimate role in the biochemistry of sulfur-dependent archaebacteria. This fact, coupled with the thermophilic character of all members of this group, has led to the suggestion that sulfur-dependent archaebacteria represent an ancient metabolic phenotype. Since these extreme thermophiles appear to have evolved *more slowly* than other archaebacteria (see Figure 18.9), their phenotype would seem more like the ancient phenotype than the others (see also Section 19.37). Such conjecture is consistent with what we know of early Earth geochemistry, and the probable environmental conditions within which life arose (see Section 18.1). The sequence UAACACCAG and CACCACAAG are oligonucleotide signatures for sulfur dependent archaebacteria.

18.7 Eucaryotes

Molecular sequencing methods can also be applied to eucaryotic organisms. Evolutionary relationships between eucaryotic *macro*organisms have traditionally been deduced from an examination of the fossil record and comparisons with extant plants and animals. Because an abundance of morphological markers are present in *macro*organisms, their phylogeny, as deduced from this comparative approach, has turned out to be a fairly accurate reflection of their true phylogeny as determined from molecular sequencing studies. However, for eucaryotic *micro*organisms, many of the same problems which delayed discovery of the natural phylogeny of bacteria (for example, lack of structural diversity) apply here as well. Thus, molecular sequencing has greatly improved the construction of evolutionary trees for the lower eucaryotes.

A eucaryotic phylogenetic tree based on ribosomal RNA sequence comparisons is shown in Figure 18.13. Because eucaryotes contain *18S* rRNA instead of *16S* rRNA in their small ribosomal subunits, the tree shown in Figure 18.13 is constructed from sequence comparisons of this slightly larger, but functionally equivalent, RNA species. The eucaryotic tree suggests that evolution along the nuclear line of descent was not a continuous process, but instead occurred in major epochs. By comparing Figures 18.9

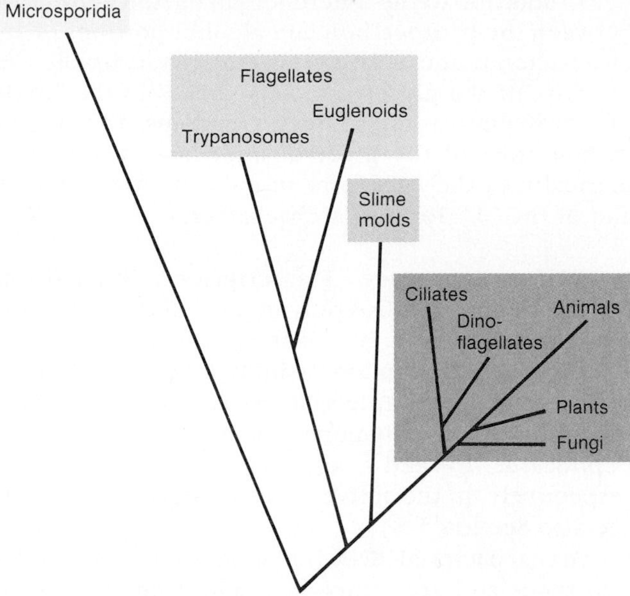

Figure 18.13 Eucaryotic phylogenetic tree based upon ribosomal RNA sequence comparisons. See note on Figure 18.10 for comparisons of detailed tree with the three-kingdom tree.

and 18.13, it is clear that *microsporidia*, a little-known group of highly unusual parasitic eucaryotes, are modern relatives of one of the first major eucaryotic cell lines, and that following the development of microsporidium-like organisms, several groups of eucaryotes radiated off the nuclear line of descent, culminating in the major eucaryotic lineages of plants and animals (Figure 18.13).

Microsporidia are of evolutionary importance because molecular sequencing suggests they are an ancient group of eucaryotes. Microsporidia are obligate parasites which live in association with representatives of virtually all groups of eucaryotes, from microorganisms to humans. Although they contain a membrane-bound nucleus, microsporidia lack mitochondria; in this connection microsporidia resemble the type of cell that first accepted stable endosymbionts (see Figure 18.4). Besides lacking organelles, microsporidia differ from modern eucaryotes in many other ways. Microsporidia lack 5.8S rRNA, the small RNA species found in the large subunit of the ribosome of all other eucaryotic organisms examined. In addition, the small ribosomal subunit of microsporidia contains an RNA of about 16S (that is, of bacterial dimensions) rather than the 18S rRNA of all other eucaryotes. In sum, it appears that microsporidia represent an evolutionary relic of the nuclear line of descent, the latter of which is now dominated by organisms containing mitochondria, chloroplasts, and other internal structures (see Figure 18.4).

The radial branches of the eucaryotic tree (Figure 18.13) represent later-evolving organisms like the plants and animals. The rapid radiation observed in this region of the nuclear lineage was likely triggered by some major geochemical or geophysical event. When the fossil record is compared with the eucaryotic phylogenetic tree derived from molecular sequencing, the rapid evolutionary radiation can be dated to about 1.5 billion years ago (see Figure 18.3). Geochemical evidence suggests that this is about the period in the Earth's history that significant *oxygen* levels were accumulating in the oceans, and perhaps the rapid evolutionary radiation of eucaryotes was catalyzed by the generalized onset of oxic conditions on the planet. The accumulation of significant O_2 levels allowed the development of an ozone shield (see Figure 18.3) which would have greatly expanded the number of habitats available for colonization (see Section 18.1). The explosive development of eucaryotes could then simply have been a response to the availability of new habitats for evolutionary development.

Although not nearly as many eucaryotes as procaryotes have been studied by ribosomal RNA sequencing, it is clear that nucleic acid sequencing holds promise for a better understanding of eucaryotic phylogeny, especially at the *microbial* level, and of planetary evolution in general. It will be exciting to see whether further examples of organelle-less eucaryotes, like the microsporidia, can be discovered, and if so, how they fit in the phylogenetic tree (other examples of mitochondria-less eucaryotes are known, such as a few obligately anaerobic protozoa, but their phylogenetic relationships are unknown). A major unanswered question in biology is what evolutionary pressures selected for the modern eucaryotic state when all other organisms (eubacteria and archaebacteria) remained procaryotic. Continued molecular sequencing could hold the answer to such important evolutionary questions.

18.8 Characteristics of the Primary Kingdoms

Although the primary kingdoms—eubacteria, archaebacteria, and the eucaryotes (nuclear line of descent)—were originally defined on the basis of comparative ribosomal RNA sequencing, subsequent studies, especially of the archaebacterial group, have shown that each kingdom can also be defined on the basis of a number of *phenotypic* properties. Some of these characteristics are unique to one kingdom, while others are found in two kingdoms to the exclusion of the third. Some characteristics bring the archaebacteria and eucaryotes together, others the archaebacteria and eubacteria, while still others are shared only by eubacteria and eucaryotes. Although no phenotypic property *shared* between kingdoms is of value for placing an organism in a particular kingdom, certain traits are unique to one or another kingdom and can thus be used as evidence for placing unknown organisms into the correct primary kingdom. We present here an overview of major phenotypic traits of phylogenetic value.

Cell walls Virtually all eubacteria have cell walls containing peptidoglycan (see Section 3.5). The only known exceptions are members of the *Planctomyces/Pirella* group (see Section 18.5), whose cell walls are made of protein, and the *Mycoplasma* group, which lack cell walls altogether. Eucaryotes never have peptidoglycan; their cell walls are composed of either cellulose or chitin (or are absent in protozoa and animal cells, see Section 3.7).

None of the archaebacteria have peptidoglycan cell walls. Instead, a collage of cell wall types is known (see Section 3.6). Most extreme halophiles, as well as the sulfur-dependent archaebacteria and a few methanogens, contain a glycoprotein wall. Certain extreme halophiles produce a sulfated heteropolysaccharide wall whose precise structure is unknown. A few methanogens make an extremely acidic (negatively charged) polysaccharide cell wall, but the majority of methanogens have a wall consisting only of protein. One group of methanogens produces a modified peptidoglycan. This material, referred to as *pseudopeptidoglycan*, resembles the peptidoglycan of eubacteria, but differs in the chemical nature of the cross-linking amino acids and the fact that it lacks muramic acid, the key amino sugar in peptidoglycan (see Figure 3.34 for the structure of pseudopeptidoglycan). Thus cell wall structure does have phylogenetic significance: peptidoglycan can be considered a characteristic molecule for eubacteria; glycoprotein for most archaebacteria. Cell walls of eubacteria, archaebacteria, and eucaryotes were considered in detail in Chapter 3.

Lipids The chemical nature of membrane lipids is perhaps the most useful of all nongenic criteria for differentiating archaebacteria from eubacteria. Eubacteria and eucaryotes produce membrane lipids with a backbone consisting of straight chain fatty acids hooked in *ester* linkage to a molecule of glycerol (see Figure 2.7). Although the nature of the fatty acid can can be highly variable, the key point is that the chemical linkage to glycerol is an **ester link**. By contrast, lipid hydrolysates of archaebacteria contain no free fatty acids, because archaebacterial lipids consist of **ether-linked** molecules (see Figure 3.10). In ester-linked lipids the fatty acids are straight chain (linear) molecules, whereas in ether-linked lipids branched chained hydrocarbons are present. In archaebacteria long chain branched hydrocarbons, either of the phytanyl or biphytanyl type are hooked by ether-linkage to glycerol molecules (see Figure 3.10).

In addition to the differences in chemical *linkage* between the hydrocarbon and alcohol portions of archaebacterial and eubacterial/eucaryotic lipids, the chirality of the glycerol moiety differs in the lipids of representatives of the three kingdoms. The central carbon atom of the glycerol molecule is stereoisomerically of the "R" form in eubacteria/eucaryotes and of the "L" form in archaebacteria.

RNA polymerase Transcription is carried out by DNA-dependent RNA polymerases in all organisms; DNA is the template, RNA is the product (see Section 5.8). Eubacteria contain a single type of RNA polymerase of rather simple quaternary structure. This is the classic RNA polymerase, containing *four* polypeptides, α, β, β', and σ, combined in a ratio of 2:1:1:1, respectively, in the active polymerase (Figure 18.14; see also Section 5.8).

Archaebacterial RNA polymerases are of at least two types and are structurally more complex than

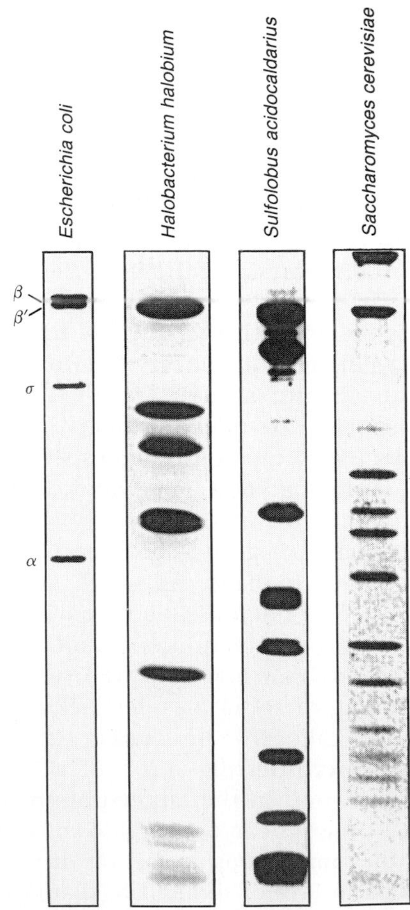

Figure 18.14 RNA polymerases from representatives of the three kingdoms, *E. coli* (eubacteria), *Halobacterium halobium* (archaebacteria), *Sulfolobus acidocaldarius* (archaebacteria), and *Saccharomyces cerevisiae* (eucaryote). The RNA polymerases have been denatured and separated by electrophoresis on a polyacrylamide gel. Largest subunits are on the top, smallest subunits are on the bottom.

those of eubacteria. The RNA polymerases of methanogens and halophiles contain *eight* polypeptides, five large ones and three smaller ones (Figure 18.14). Sulfur-dependent archaebacteria contain an even more complex RNA polymerase, consisting of at least *ten* distinct polypeptides (Figure 18.14). The major RNA polymerase of eucaryotes contains 10–12 polypeptides, and the relative sizes of the peptides coincide most closely with those from the sulfur-dependent archaebacterial group (Figure 18.14). At least two other RNA polymerases are known in eucaryotes, and each specializes in transcribing certain regions of the genome, one for rRNA, one for messenger RNA, and one for tRNA. In terms of phylogenetic signatures, the $\alpha_2\beta\beta'\sigma$ polymerase of eubacteria is highly diagnostic of this group; the remaining polymerases are of too great a complexity and variability to be phylogenetically definitive.

Features of protein synthesis Because of differences in ribosomal RNA sequences and several protein synthesis factors, it is not surprising that certain aspects of the protein synthesizing machinery differ in representatives of the three kingdoms. Although eubacterial and archaebacterial ribosomes are the same size (70S, as compared with the 80S ribosomes of eucaryotes), several steps in archaebacterial protein synthesis more strongly resemble those of eucaryotes than eubacteria. Recall that translation always begins at a unique codon, the so-called *start codon*. In eubacteria this start codon (AUG) calls for the incorporation of an initiator tRNA containing a modified methionine residue, *formyl*methionine (see Section 5.10). By contrast, in eucaryotes and in archaebacteria the initiator tRNA carries an *unmodified* methionine.

The exotoxin produced by *Corynebacterium diphtheriae* is a potent inhibitor of eucaryotic protein synthesis, because using ADP it ribosylates an elongation factor required to translocate the ribosome along the mRNA; the modified elongation factor is inactive (see Section 11.10). Diphtheria toxin also inhibits protein synthesis in archaebacteria, but does not affect this process in eubacteria.

Most antibiotics that specifically affect eubacterial protein synthesis do not affect archaebacterial (or eucaryotic) protein synthesis. The sensitivity of representatives of the three kingdoms to various protein synthesis inhibitors is shown in Table 18.4 (see also Section 5.10). *Fusidic acid*, an inhibitor of protein synthesis elongation factor II, affects both 70S and 80S ribosomes, and stops protein synthesis in all organisms *except* sulfur-dependent archaebacteria (Table 18.4). *Anisomycin* and its chemical derivatives interfere with 80S ribosome function; these agents also inhibit methanogen and extreme halophile (but not thermoacidophile) protein synthesis. *Streptomycin, chloramphenicol, erythromycin*, and the like, inhibit only eubacterial protein synthesis. Antibiotics of the *pulvomycin* class affect only procaryotic protein synthesis. Pulvomycin inhibits elongation factor Tu, which is apparently present in all 70S ribosomes. Antibiotics of the *neomycin/puromycin* group inhibit protein synthesis in organisms of all three kingdoms. Puromycin causes premature termination of protein synthesis, and this suggests that the interference is not particularly specific, because both 70S and 80S ribosomes are affected.

In agreement with the variability in RNA polymerase structure of representatives of the three kingdoms noted above (see above), the RNA polymerase inhibitor *rifampicin* affects only eubacterial protein synthesis. Rifampicin binds to a subunit of the eubacterial RNA polymerase ($\alpha_2\beta\beta'$ σ-type) and prevents initiation of transcription (see Section 5.10). And, although the genomic arrangement of archaebacteria is circular, like that of eubacteria, the DNA replication inhibitor *aphidicolin* inhibits DNA synthesis in both archaebacteria and eucaryotes, even though the

Table 18.4 Sensitivity of representatives of the three kingdoms to various protein synthesis inhibitors.*

	Archaebacteria		*Eubacteria*	*Eucaryote*
	Methanobacterium	*Sulfolobus*	*E. coli*	*Saccharomyces cerevisiae*
Fusidic acid, Sparsomycin	+	–	+	+
Anisomycin, Narciclasine	+	–	–	+
Cycloheximide	–	–	–	+
Erythromycin, Streptomycin, Chloramphenicol, Tetracycline	–	–	+	–
Virginiamycin Pulvomycin	+	–	+	–
Neomycin, Puromycin	+	+	+	+

A "+" indicates that protein synthesis (and growth) is inhibited. Data adapted from Bock, A., J. H. Hummel, and G. Schmid. 1984. Evolution of translation, p. 73–90. In: K. H. Schleifer and E. Stackebrandt (eds.), Evolution of Prokaryotes, Academic Press, London.

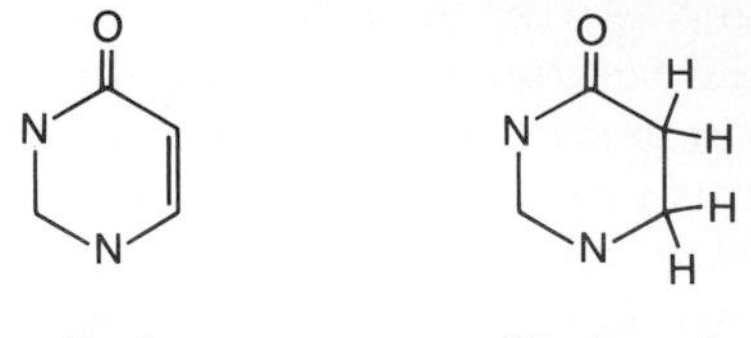

Uracil Dihydrouracil

Figure 18.15 Comparison of the structures of uracil and dihydrouracil. Dihydrouracil is found in eubacterial and eucaryotic RNAs, but not in archaebacterial RNAs (with one exception).

latter possess linear DNA arranged in chromosomes. Aphidicolin does not affect eubacterial DNA synthesis.

Although many of the antibiotics discussed in this section are not of clinical interest for a variety of reasons, they are clearly useful for studying evolution of the translational apparatus. The antibiotic studies have exposed similarities and differences between the ribosomes of organisms from each of the primary kingdoms. However, the sulfur-dependent archaebacteria seem to be unique in that typical representatives are resistant to virtually all antibiotics.

Other features defining the kingdoms A number of other more subtle differences can be listed which assist in delineating organisms at the kingdom level. For example, the number of modified nucleotides present in eucaryotic 18S rRNA (the functional equivalent of procaryotic 16S rRNA) is considerable. Depending on the eucaryote, 25–35 or more nucleotides are chemically modified. Interestingly, a similar number of modified nucleotides are found in representatives of the sulfur-dependent archaebacteria. Eubacteria and the methanogen/halophile/*Thermoplasma* archaebacteria, on the other hand, contain only 4–8 modified residues in their 16S rRNA. The modified base dihydrouracil (Figure 18.15) is present in the tRNAs of all eubacteria and eucaryotes but is absent from archaebacteria, except for one group of methanogens.

Table 18.5 summarizes the major points that define the three kingdoms. In most cases the differences are in fundamental biological molecules of structural and functional importance. This suggests that the evolution of the three cellular lines of descent from the universal ancestor (which can be thought of as the first attempt at speciation) was a seminal event, and produced major differences in the fundamental biochemistry of organisms that traveled the three different evolutionary paths.

18.9 Taxonomy, Nomenclature, and Bergey's Manual

Taxonomy is the science of classification and consists of two major subdisciplines, *identification* and *nomenclature*. Just as the chemist examines the properties of the chemical elements and is able to see relationships that made possible construction of the periodic table, so the taxonomist attempts to construct a rational scheme for classifying and naming organisms. It is important to distinguish between bacterial *taxonomy* and bacterial *phylogeny,* for the terms really mean different things. Bacterial taxonomy has traditionally relied on *phenotypic* analyses as the basis of classification. By contrast, because bacteria are so small and contain relatively few structural clues to their evolutionary roots, phylogenetic relationships between bacteria have only emerged from the *genotypic* analyses discussed in the previous sections.

Because practicing microbiologists need to identify and classify bacteria for a variety of practical reasons, bacterial taxonomy is still an important disci-

Table 18.5 Summary of major differentiating features between eubacteria, archaebacteria, and eucaryotes

Characteristic	Eubacteria	Archaebacteria	Eucaryotes
Membrane-bound nucleus	Absent	Absent	Present
Cell wall	Muramic acid present	Muramic acid absent	Muramic acid absent
Membrane lipids	Ester linked	Ether linked	Ester linked
Ribosomes	70S	70S	80S
Initiator tRNA	Formylmethionine	Methionine	Methionine
Introns in tRNA genes	No	Yes	Yes
Ribosome sensitivity to diphtheria toxin	No	Yes	Yes
RNA polymerases (see Figure 18.14)	One (4 subunits)	Several (8–12 subunits) each	Three (12–14 subunits each)
Sensitivity to chloramphenicol, streptomycin and kanamycin	Yes	No	No

pline. However, before we begin our study of bacterial classification in a *taxonomic* sense (that is, based on phenotypic criteria), we must first decide on the basic unit of classification to use. The taxonomic unit of microbiology is the *species*, which is usually defined from the characterization of several strains or clones. The use of the word *clone* in this sense can be taken to mean a population of *genetically identical* cells derived from a single cell. In a taxonomic study, one usually selects some subgroup for detailed study. It is most desirable to assemble a collection of strains of the subgroup under study, in order that the variability within the group can be recognized.

The species concept In microbiology, the basic taxonomic unit is the **species**. A species can be operationally defined as a collection of similar strains that differ sufficiently from other groups to warrant recognition as a basic taxonomic unit. Unfortunately, depending upon their philosophical outlook, taxonomists tend to interpret strain differences either very strictly or very casually. Taxonomists themselves can be classified into two broad groups called "splitters" or "lumpers" depending upon whether they create greater or fewer species designations in their taxonomic schemes. Regardless of the number of species finally recognized, the species concept is important because it gives the collected strains *formal* taxonomic identity. Groups of species are collected into **genera** (singular, **genus**). By analogy to the species, a genus can be defined as a collection of different species, each sharing some major property or properties that define the genus, but differing from one another at the species level by the presence or absence of other (usually less significant) characteristics.

Nomenclature and formal taxonomic standing Following the **binomial system** of nomenclature, all bacteria are given genus and species names. The genus name of an organism is frequently abbreviated to a single (capital) letter; the species name is never abbreviated. Thus, *Escherichia coli* is usually written *E. coli*. The genus and species names are either Latin or Greek derivatives of some descriptive property appropriate for the species, and are set in print in *italics*. For example, several species of the genus *Bacillus* have been described, including *Bacillus (B.) subtilis, B. cereus, B. stearothermophilus*, and *B. acidocaldarius*. The species names mean "slender," "waxen," "heat-loving," and "acid-thermal," respectively, and in each case speak to morphological or physiological/ecological traits characteristic of each organism.

When a new organism is isolated and thought to be unique, a decision must be made as to whether it is sufficiently different from other species to be described as a new species, or perhaps even sufficiently different from all described genera to warrant description as a new genus (in which a new species is automatically created). In order to achieve formal taxonomic standing as a new genus or species, a description of the isolate and the proposed name is published and a pure culture of the organism is deposited in an approved culture collection (usually the American Type Culture Collection [ATCC] or the Deutsche Sammlung von Mikroorganismen [DSM, German Collection of Microorganisms]). The deposited strain serves as the *type* strain of the new species or genus/species, and remains as the standard by which other strains thought to be the same can be compared.

Culture collections preserve the deposited culture, generally by freezing or freeze drying. This practice is very different from the botanical or zoological approach. These disciplines employ preserved (dead) specimens (either dried herbarium material or chemically fixed animal specimens) as the basis for comparison with proposed new species. Microbiologists rely on a *living type strain*, and this approach allows for more detailed and reproducible comparisons, especially at the molecular level.

Conventional bacterial taxonomy There are many ways in which to group bacteria. In conventional bacterial taxonomy a variety of characteristics of different strains or species are measured, and these traits are then used to group the organisms. Characteristics of taxonomic value that are widely used include: morphology, Gram reaction, nutritional classification (phototroph, heterotroph, lithotroph), cell-wall chemistry, presence of cell inclusions and storage products, capsule chemistry, pigments, nutritional requirements, ability to use various carbon, nitrogen, and sulfur sources, fermentation products, gaseous needs, temperature and pH requirements (and tolerances), antibiotic sensitivity, pathogenicity, symbiotic relationships, immunological characteristics, and habitat. Some of these characteristics may not be applicable or useful for the taxonomy of a particular group, and one must judge how extensive a compilation of data is needed to properly classify any given organism.

The *Gram stain* reaction is an especially useful differentiating characteristic. In traditional taxonomic arrangements bacteria have been grouped together as Gram-positive rods, Gram-negative cocci, Gram-negative rods, and so on. Interestingly, the Gram reaction turns out to be a property of fundamental importance for classifying bacteria phylogenetically as well as taxonomically, and it is usually one of the first tests performed in classifying a new bacterial isolate.

Cell morphology and special cell structures such as spores are widely used for taxonomic purposes. The bacterial endospore, for example, is a structure of such unique characteristics and practical importance, that all endospore-forming bacteria are classified together (both taxonomically and phylogenetically). A very large group of bacteria possess either a filamentous or coryneform (indeterminate) morphology, and are classified together as the Actinomycetes and related organisms. Several other groups that are separated on the basis of morphology are: the gliding bacteria, the sheated bacteria, budding or appendaged bacteria, and the spirochetes (see Chapter 19 for detailed descriptions of all of these groups).

The emphasis of early taxonomists on cell morphology and basic physiological characteristics, although useful from the standpoint of *identifying* bacteria, has ingrained in modern microbiologists the prejudice that cell morphology and physiology are characteristics of *phylogenetic* value as well. However, molecular sequencing studies have shown that morphology and physiology are often poor indicators of phylogenetic relatedness (for example, see Section 18.5).

Numerical taxonomy Numerical taxonomy is simply a more comprehensive version of conventional bacterial taxonomy and was designed to avoid some of the latter's pitfalls. Numerical taxonomy involves the systematic analysis of as many phenotypic properties as possible. An organism is scored as either positive or negative for a given phenotypic trait, and the number of positive matches between any two organisms is taken as an indication of their relationships. Numerical taxonomists will frequently test over 100 different phenotypic properties in comparing different organisms. As the number of characteristics scored positive or negative becomes larger, the taxonomic information emerging from a numerical taxonomy study becomes more reliable. Because numerical taxonomists like to compare as many characteristics as possible, numerical taxonomy is probably the most dependable form of conventional taxonomy. The numerical taxonomic approach also serves to build a data base upon which future isolates can be objectively compared. However, like conventional taxonomy, numerical taxonomy measures phenotypic properties only and thus is useful for classification and identification, but not as a phylogenetic tool.

Bergey's Manual The most extensive effort to classify bacteria is that of *Bergey's Manual of Systematic Bacteriology. Bergey's Manual* is a compendium of classical and molecular information on all recognized species of bacteria, and contains a number of dichotomous keys that are useful for identifying bacteria. Although still a taxonomic (rather than phylogenetic) treatise, the latest edition of *Bergey's Manual* has incorporated some molecular sequencing information in the descriptions of various bacterial groups. The latest edition of *Bergey's Manual* consists of four volumes. Volume I covers Gram-negative bacteria of medical or industrial significance; Volume II the Gram-positive bacteria of medical or industrial significance; Volume III the remaining Gram-negative bacteria including archaebacteria and cyanobacteria; and Volume IV the Gram-positive, filamentous spore-forming bacteria (Actinomycetes). The genera described in Bergey's Manual are given in an Appendix to this book. Although the latest edition of *Bergey's Manual* recognizes the contribution of molecular phylogeny to our understanding of bacterial evolution, *Bergey's Manual* retains many classical taxonomic groupings because it is not yet clear how molecular sequencing studies will affect the nomenclatural aspects of taxonomy.

18.10 Molecular Taxonomic Approaches: DNA Base Composition and Nucleic Acid Hybridization

Nucleic acid analyses have become part of conventional taxonomy, but the techniques involved differ from the nucleic acid *sequencing* methods discussed in Section 18.3. Determination of the base composition of the DNA of various organisms or the extent to which mixtures of DNA from two organisms hybridize, have been used to support or refute taxonomic conclusions based primarily on phenotypic criteria. Although useful in certain taxonomic circumstances, DNA base compositions and nucleic acid hybridization analyses are generally *not* suitable for determining evolutionary relationships, because neither method yields a data base of nucleic acid *sequences* suitable for phylogenetic analysis. Nevertheless, both techniques have some value for *taxonomic* classification and are thus considered briefly here.

DNA base composition In procaryotes, DNA base ratios, expressed as mole percent GC

$$\frac{G + C}{A + T + G + C} \times 100\%$$

vary over a wide range (Figure 18.16). GC ratios as low as 20 percent and as high as 78 percent are known among procaryotes. In fact, the range of GC ratios for procaryotes is greater than for any other kingdom of organisms (Figure 18.16).

Base compositions of DNA have been determined for a wide variety of bacteria and several correlations can be observed. (1) Organisms with highly similar phenotypes often possess similar DNA base ratios (although numerous exceptions have been noted). (2) If two organisms thought to be closely related by phenotypic criteria are found to have widely different base ratios, closer examination usually indicates that these organisms are not so closely related as was supposed. (3) Two organisms can have identical base

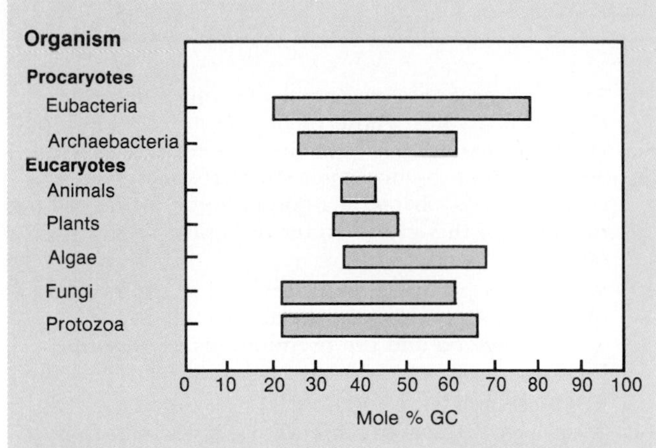

Figure 18.16 Ranges of DNA base composition of various organisms.

ratios and yet be quite unrelated (both taxonomically and phylogenetically), since a variety of *sequences* is possible with DNA of identical base composition.

The DNA base ratio is required for formal description of any organism proposed as either a new genus or new species. In reality, the GC ratio is an exclusionary determinant rather than a supportive one, because two organisms with similar GC ratios need not have similar DNA sequences. Indeed, DNA:DNA hybridization studies (see below) have often shown this to be the case. However, it has been calculated that two strains whose GC base ratio differs by more than 5 mole percent, have sufficiently different base sequences that they could not possibly belong to the same species. Thus, GC ratios are useful for determining *unrelatedness*, but cannot be used to support the claim that two organisms are related.

Phylogenetic studies using 16S ribosomal RNA sequence analyses (see Sections 18.3–18.7) have shown that although organisms with strong evolutionary ties often have similar GC ratios, this is frequently not the case. For example, the genus *Lactobacillus* is a rather phylogenetically coherent genus despite the fact that representative species have GC ratios that vary from 34 to 52 mole percent. Heterotrophic cocci, on the other hand, cluster into two major phylogenetic groups, the staphylococci and the micrococci. In this case, GC ratios have good phylogenetic predictive value as the staphylococci are a *low* GC group (30–40 mole percent GC), while the micrococci are a *high* GC group, (66–75 mole percent GC).

Nucleic acid hybridization A variety of nucleic acid hybridization analyses are possible. Either DNA or RNA from one organism can be hybridized to DNA from another, or RNA from one organism can be hybridized to RNA from another. We discussed the theory and methodology behind nucleic acid hybridization in Section 5.2. DNA homology experiments are generally used to detect similarities between closely related organisms (species level or below). Homology values above 60–70 percent are generally obtained for different strains within a species, while a minimum value of 20 percent sequence homology is required between the DNA of different species in order to support their inclusion in the same genus. DNA from unrelated genera, for example *Bacillus subtilis* and *Escherichia coli*, hybridize at very low levels, generally less than 10 percent (since the GC base ratios of these two species are nearly identical, about 50 mole percent GC, the value of hybridization studies over GC determinations should be apparent). The observation that different genera rarely show significant DNA sequence homology has been useful for detecting seriously misclassified organisms.

Hybridization using DNA from one organism and ribosomal RNA (rRNA) from a second has also been used in molecular taxonomic studies. Several studies have shown that organisms that show only a very low degree of DNA homology occasionally show a much higher degree of homology by DNA:rRNA hybridization. This is because the genes coding for rRNA have remained more conserved during evolution than bulk DNA. Indeed, DNA:rRNA analyses were one of the early indications that rRNA sequences were sufficiently highly conserved across broad evolutionary distances to act as phylogenetic tools for discerning deep evolutionary relationships. The nature of rRNA thus makes DNA:rRNA comparisons an appropriate tool for defining similarities between more distantly related organisms. Comparisons of DNA:rRNA hybridization with ribosomal RNA sequencing have shown the two techniques to yield similar phylogenetic information at the species, genus, and even the family levels. However, DNA:rRNA hybridization is not sufficiently sensitive, as is rRNA sequencing, for differentiating organisms at the kingdom level.

Taxonomy versus phylogeny We learned earlier in this chapter that molecular sequencing of universally distributed and relatively conserved molecules (like the ribosomal RNAs) are required to discern phylogenetic relationships. This is not the case in traditional taxonomy. Phenotypic analyses have dominated the field of bacterial taxonomy and still play an important role in identifying bacteria in a practical sense. Thus, two approaches to bacterial classification exist—phylogenetic and taxonomic. It is the goal of the editors of *Bergey's Manual* to eventually construct a classification of bacteria that allows for rapid and facile identification, but that is founded on evolutionary principles. Such a unified classification may never be possible. However, as we learn more about both the phenotypic and genotypic properties of bacteria we are bound to better understand bacterial evolution, and perhaps some day be able to answer the critical question of how life originated.

Study Questions

1. From the following terms, select a combination that could best describe the earliest types of organisms on Earth and discuss why or why not each term would be appropriate to describe early life forms: phototroph, thermophile, eucaryote, lithotroph, acidophile, heterotroph, psychrophile, autotroph, halophile, mesophile, aerobe, facultative, alkalinophile, procaryote, obligate intracellular parasite, free-living, anaerobe, microaerophile.
2. Why was the evolution of cyanobacteria of such importance to the further evolution of life on Earth?
3. Describe the evolutionary events leading to the rise of modern eucaryotes. What molecular evidence supports this hypothesis?
4. Why can macromolecules like nucleic acids or proteins serve as phylogenetic markers whereas polysaccharides and lipids cannot?
5. Why are ribosomal RNAs better molecules for phylogenetic studies than proteins like ferredoxins, cytochromes, or specific enzymes?
6. What are signature sequences and of what phylogenetic value are they? How are signature sequences discerned?
7. Describe the methods involved in obtaining complete 16S rRNA sequences.
8. What major evolutionary finding emerged from the study of ribosomal RNA sequences? How did this modify the classical view of evolution? How has this discovery served to change our thinking on the origin of eucaryotic organisms?
9. The term *archaebacteria* was chosen to describe those procaryotes whose phylogenetic relationships (by ribosomal RNA sequencing) to eubacteria were as low as that to eucaryotes. In retrospect, why was the term *archae*bacteria a good choice for describing members of this group (refer to Figure 18.9 for a clue)?
10. What major group of eubacteria led to the evolution of familiar Gram-negative bacteria such as the enteric bacteria and the pseudomonads? Describe how such groups could have been derived from this major eubacterial group.
11. What major lesson has microbiology learned from RNA sequencing concerning the use of *phenotypic* criteria in establishing evolutionary relationships?
12. Describe the two basic groups of Gram-positive bacteria and discuss what phenotypic and genotypic properties each group displays.
13. What major physiological/biochemical properties do archaebacteria share with eucaryotes? With eubacteria?
14. What major phenotypic properties are used to group organisms in classical bacterial taxonomy? Which, if any, of these properties have phylogenetic predictive value?
15. Why is nucleic acid hybridization of more value than GC base ratio determinations for genetic taxonomic studies? In what situations are GC base ratios of use in taxonomic studies?

Supplementary Readings

Gottschalk, G. (ed.) 1985. *Methods in Microbiology, Volume 18.* Academic Press, Inc., Orlando, FL. Detailed but very readable descriptions of the methodology involved in genetic phylogeny and taxonomy studies.

Olsen, G. J., **D. L. Lane**, **S. J. Giovannoni**, and **N. R. Pace**. 1986. *Microbial ecology and evolution: a ribosomal RNA approach.* Ann. Rev. Microbiol. 40:337–365. An overview of ribosomal RNA sequencing and its importance to phylogeny and a demonstration of how sequencing technology can be used to analyze natural populations of bacteria.

Schleifer, K. H., and **E. Stackebrandt.** 1983. *Molecular systematics of prokaryotes.* Ann. Rev. Microbiol. 37:143–187. An overview of the molecular techniques useful for the study of bacterial evolution and taxonomy.

Schliefer, K. H., and **E. Stackebrandt** (eds.) 1985. *Evolution of Prokaryotes.* Academic Press, Inc., Orlando, FL. The proceedings of a symposium on various aspects of molecular evolution.

Schopf, J. W. (ed.) 1983. *Earth's Earliest Biosphere—Its Origin and Evolution.* Princeton University Press, Princeton, NJ. A large collection of papers on early earth chemistry and the origin of life.

Stackebrandt, E., and **C. R. Woese**. 1981. *The evolution of prokaryotes.* Pages 1–31 *in* J. Charlile, J. F. Collins, and B. E. B. Mosely (eds.), *Molecular and Cellular Aspects of Microbial Evolution.* Thirty-second Symp. of the Soc. Gen. Microbiol., Cambridge University Press, Cambridge, England. A detailed review of bacterial phylogeny based on S_{AB} analyses.

Woese, C. R. 1987. Bacterial evolution. *Microbiological Reviews.* 51:221–271. An excellent summary of what ribosomal RNA sequencing has taught us about the evolution of both procaryotes and eucaryotes. Highly recommended as a starting point for a detailed study of molecular evolution.

Woese, C. R., and **R. S. Wolfe** (eds.) 1985. *Archaebacteria. Volume 8 of The Bacteria—A treatise on structure and function.* (J. R. Sokatch, L. N. Ornston, editors.) Academic Press, Inc., Orlando, FL. A series of chapters, each written by an expert, on the biology of archaebacteria. Highly recommended.

Zuckerkandl, E., and **L. Pauling**. 1965. *Molecules as documents of evolutionary history.* J. Theoretical Biology 8:357–366. The seminal paper in the field of molecular sequencing.

The Bacteria

In the preceding chapter we stressed the evolutionary relationships between bacteria and pointed out how phylogenetic groupings are frequently at odds with taxonomic classifications established on phenotypic grounds. In this chapter we will describe the major morphological and physiological groups of procaryotes in some detail.

The problem of how to classify bacteria within a framework that accounts for both genotypic and phenotypic information remains unsettled in the latest version of *Bergey's Manual*. The editors of *Bergey's Manual* have, for the most part, taken a *traditional* taxonomic approach to bacterial classification, and have grouped bacteria in the *Manual* into sections based on morphological, physiological/biochemical, and ecological criteria. Because the recent developments in our understanding of bacterial phylogeny have not yet affected bacterial taxonomy and nomenclature, we will follow the *Bergey's Manual* system here, and have divided this chapter into several sections, each dealing with a group of taxonomically related bacteria. The groups discussed in each section should not necessarily be considered phylogenetically related, although in many cases they are. However, to remain consistent with the recognized authority on bacterial classification, *Bergey's Manual*, the present discussion will focus on the morphological, physiological and ecological diversity of each of the groups considered. The complete grouping of bacteria in the four volumes of *Bergey's Manual* is given in Appendix 3.

19.1 Purple and Green (Phototrophic) Bacteria

Bacteria able to use light as an energy source comprise a large and heterogeneous group of organisms, grouped together primarily because they possess one or more pigments called *chlorophylls* and are able to carry out light-mediated generation of ATP, a process called **photophosphorylation**. Two major groups are recognized, the **purple and green bacteria** as one group, and the **cyanobacteria** as the other group. The basic distinction between the purple and green bacteria and the cyanobacteria is based on the photopigments and overall photosynthetic process. Cyanobacteria are *oxygenic* phototrophs, employing chlorophyll *a* and two photosystems in their photosynthetic process. Purple and green bacteria are *anoxygenic* phototrophs, employing bacteriochlorophyll (of several different types), but only *one* photosystem in their photosynthetic process.

The biochemistry of photosynthesis has been discussed in some detail in Chapter 16. Green-plant photosynthesis, which is also exhibited by the cyanobacteria, involves two light reactions, and H_2O serves as the electron

Pigment	R_1	R_2	R_3	R_4	R_5	R_6**	R_7	Long wave-length absorption maxima (nm)	
								In vivo	Extract (methanol)
Bacterio-chlorophyll *a*	$-C(=O)-CH_3$	$-CH_3$*	$-CH_2-CH_3$*	$-CH_3$	$-C(=O)-O-CH_3$	P/Gg	$-H$	805 830–890	771
Bacterio-chlorophyll *b*	$-C(=O)-CH_3$	$-CH_3$†	$=C(H)-CH_3$†	$-CH_3$	$-C(=O)-O-CH_3$	P	$-H$	835–850 1020–1040	794
Bacterio-chlorophyll *c*	$-C(H)(OH)-CH_3$	$-CH_3$	$-C_2H_5$ $-C_3H_7$†† $-C_4H_9$	$-C_2H_5$ $-CH_3$(?)	$-H$	F	$-CH_3$	745–755	660–669
Bacterio-chlorophyll c_s	$-C(H)(OH)-CH_3$	$-CH_3$	$-C_2H_5$	$-CH_3$	$-H$	S	$-CH_3$	740	667
Bacterio-chlorophyll *d*	$-C(H)(OH)-CH_3$	$-CH_3$	$-C_2H_5$ $-C_3H_7$ $-C_4H_9$	$-C_2H_5$ $-CH_3$(?)	$-H$	F	$-H$	705–740	654
Bacterio-chlorophyll *e*	$-C(H)(OH)-CH_3$	$-C(=O)-H$	$-C_2H_5$ $-C_3H_7$ $-C_4H_9$	$-C_2H_5$	$-H$	F	$-CH_3$	719–726	646
Bacterio-chlorophyll *g*	$-C(H)=CH_2$	$-CH_3$	$-C_2H_4$	$-CH_3$	$-C(=O)-O-CH_3$	F	$-H$	788, 670	765

**No double bond between C-3 and C-4; additional H-atoms are in position C-3 and C-4.*
†No double bond between C-3 and C-4; an additional H-atom is in position C-3.
***P = phytyl ester ($C_{20}H_{39}O-$); F = farnesyl ester ($C_{15}H_{25}O-$); Gg = geranylgeraniol ester ($C_{10}H_{17}O-$); S = stearyl alcohol ($C_{18}H_{37}O-$).*
‡Bacteriochlorophylls c, d, and e consist of isomeric mixtures with the different substituents on R_3 as shown.
From Gloe, A., N. Pfennig, H. Brockmann, and W. Trowitsch. 1975. Archives of Microbiology 102:103–109, and Gloe, A., and N. Risch. 1978. Archives of Microbiology 118:153–156.

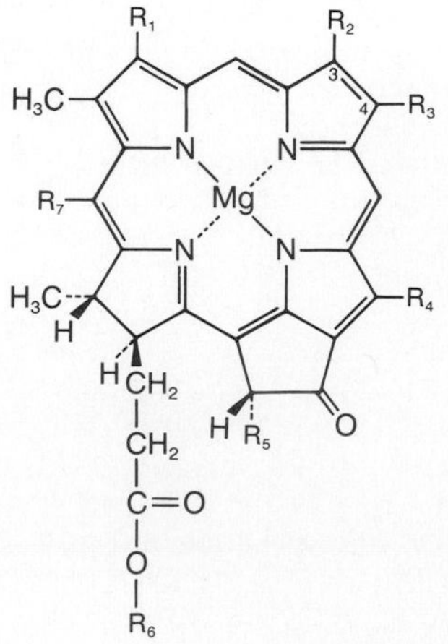

Figure 19.1 Structure of all known bacteriochlorophylls. The different substituents present in the positions R_2 to R_7 are given in the accompanying table.

donor. As a result of the photolysis of water, O_2 is produced. During photosynthesis by the purple and green bacteria, on the other hand, water is not photolysed, and O_2 is not produced. Because the purple and green bacteria are unable to photolyse water, they must obtain their reducing power for CO_2 fixation from a reduced substance in their environment. This can be either an organic compound, a reduced sulfur compound, or H_2. Purple and green bacteria can only grow photosynthetically under *anaerobic* conditions. Cyanobacteria, on the other hand, can develop readily under aerobic conditions with no reduced compounds present. Many purple and green bacteria can grow autotrophically with CO_2 as carbon source and a reduced sulfur compound or H_2 as reductant under anaerobic conditions in the light. Under microaerobic conditions, these same reductants serve as electron donors to support lithotrophic growth of many purple bacteria. Phototrophic growth is also possible, however, with light serving as energy source and an organic compound as a carbon source; under such conditions, CO_2 is only a minor source of cell carbon.

Some purple and green bacteria can also grow in the dark using an organic compound as electron donor, but under these conditions a suitable electron acceptor is necessary; O_2 is the most commonly used electron acceptor, but NO_3^-, elemental sulfur, thiosulfate, dimethylsulfoxide, and trimethylamine oxide can be used by certain organisms. A few purple bacteria can also grow in the dark (albeit slowly) by strictly fermentative metabolism. One interesting organism, a *Rhodocyclus* species, can even grow anaerobically via the fermentation of carbon monoxide: $CO + H_2O \rightarrow CO_2 + H_2$.

Bacteriochlorophylls We compared a typical bacteriochlorophyll with algal chlorophyll in Figure 16.3. A number of bacteriochlorophylls exist, differing in substituents on various parts of the porphyrin ring, and these are outlined in Figure 19.1. The various modifications lead to changes in the characteristic absorption spectra of the bacteriochlorophylls, so that an organism containing a certain bacteriochlorophyll is best able to utilize light of particular wavelengths. The ecological significance of ability to utilize different wavelengths of light was discussed in Section 16.5; it is likely that this selective absorption provides an evolutionary pressure for the development of organisms with various chlorophylls. The long wavelength absorption maxima of the *bacteriochlorophylls* are the most characteristic, and these are given in Figure 19.1, as measured in the living cell (in vivo) and in solvent extract. Although the in vivo absorption maximum is of most significance ecologically, from the viewpoint of characterizing the various purple and green bacteria taxonomically, the absorption spectrum in solvent extract is most convenient, because its measurement is easier. Thus to characterize the bacteriochlorophyll of a new isolate of a purple or green bacterium, a simple extract in methanol or other organic solvent is made and the absorption spectrum determined. From the long wavelength maximum listed in Figure 19.1, the likely identification of the bacteriochlorophyll can be made. For confirmation, it is desirable to carry out chromatographic studies on the isolated bacteriochlorophyll.

Classification The purple and green bacteria are a diverse group morphologically and phylogenetically (see Section 18.5). Morphologically, cocci, rods, vibrios, spirals, budding, and gliding types are known. Both polarly and peritrichously flagellated organisms are known in this group. It thus seems likely that the ability to grow phototrophically has developed in a wide variety of bacterial types, and that the only unifying thread among the whole group is the ability to carry out photophosphorylation. Traditionally, these bacteria were classified into three major groups: green sulfur bacteria, purple sulfur bacteria, and purple nonsulfur bacteria. However, research over the past several decades has shown that this classification is oversimplified. Most important, the distinction between the purple sulfur and the purple nonsulfur bacteria in relation to sulfur metabolism is no longer valid (see below). However, because the ecological roles played by purple sulfur and nonsulfur bacteria appear quite different, the distinction between the two groups is maintained in the current taxonomy. The major properties of purple and green bacteria, along with those of the heliobacteria, are as follows:

Group	Bacterio-chlorophylls	Photosynthetic membrane systems
Purple bacteria	Bchl *a*, Bchl *b*	Lamellae, tubes, or vesicles continuous with plasma membrane (see Figure 19.3*a* and *b*)
Green bacteria	Bchl *c*, Bchl *d*, or Bchl *e*, plus small amounts of Bchl *a*	Chlorosomes, attached to but not continuous with plasma membrane (see Figure 19.3*c*)
Helio-bacteria	Bchl *g*	Cytoplasmic membrane only (see Figure 19.3*d*)

Photosynthetic bacteria generally also produce **carotenoid pigments** (Figure 19.2), and the carotenoids of the purple bacteria differ from those of the green bacteria. Carotenoid pigments are responsible for the purple color of the purple bacteria, and mutants lacking carotenoids are blue-green in color (see Figure 19.2*b*). In fact, purple photosynthetic bacteria are frequently not purple, but brown, pink, brown-red, purple-violet, or orange-brown, depending on their carotenoid pigments (see Figure 19.2*b*). Also, many of the "green" bacteria are actually brown in color, due to their complement of carotenoids (Figure 19.2). Thus color is not a good criterion for use in identifying isolates as either green bacteria or purple bacteria.

Photosynthetic membrane systems A major difference between the green and purple bacteria is in the nature of the photosynthetic membrane system. In the purple bacteria, the photosynthetic pigments are part of an elaborate internal membrane system, connected to and produced from the plasma membrane; the membrane often occupies much of the cell interior. In some cases the membrane system is an array of flat sheets called **lamellae** (Figure 19.3*a*), whereas in others it consists of round tubes referred to as **vesicles** (Figure 19.3*b*). The membrane content of the cell varies with pigment content, which is itself affected by light intensity and presence of O_2. When cells are grown aerobically, synthesis of bacteriochlorophyll is repressed, and the organisms may be virtually devoid of photopigments as well as the internal membrane systems. (This experiment can, of course, only be done in those purple and green bacteria able to grow aerobically, such as members of the nonsulfur purple bacteria, and members of the genera *Chloroflexus, Thiocapsa, Chromatium*, and *Ectothiorhodospira*.) Consequently, phototrophic growth is only possible

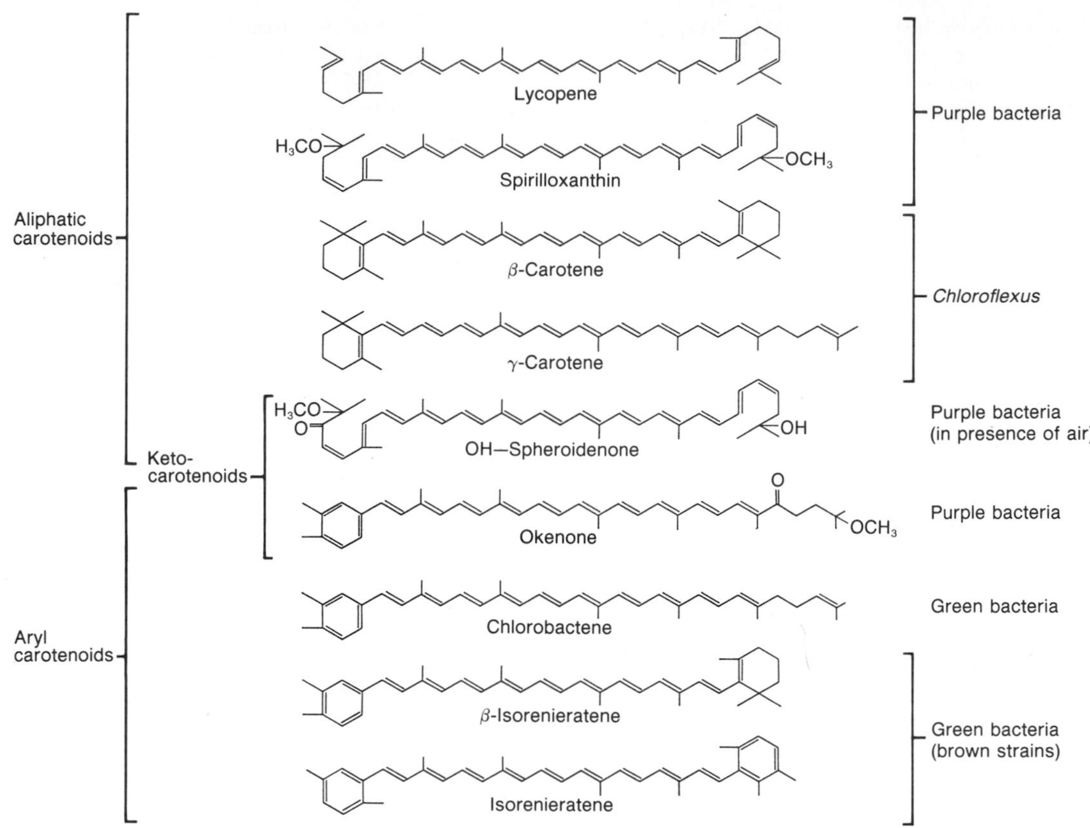

(a)

Group	Names	Color of organisms
1	Lycopene, rhodopin, spirilloxanthin	Orange-brown, brownish-red, pink, purple-red
2	Spheroidene, hydroxyspheroidene, spheroidenone, hydroxyspheroidenone, spirilloxanthin	Red (aerobic) Brownish red to purple (anaerobic)
3	Okenone, methoxylated keto carotenoids (*Rhodopila globiformis* only)	Purple-red
4	Lycopenal, lycopenol, rhodopin, rhodopinal, rhodopinol	Purple-violet
5	Chlorobactene, hydroxychlorobactene, β-isorenieratene, isorenieratene	Green (chlorobactene) Brown (isorenieratene)
6	β-carotene, γ-carotene	Orange-green

Based on Schmidt, K. 1978. Pages 729–750, and S. Liaaen-Jensen, 1978. Pages 233–247 in *R. K. Clayton and W. R. Sistrom (eds.),* The Photosynthetic Bacteria, *Plenum Press, New York. N.Y.*

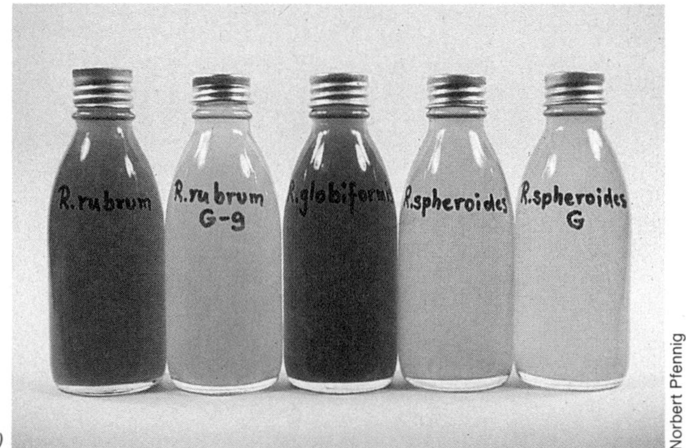

(b)

Norbert Pfennig

Figure 19.2 Major carotenoids of photosynthetic bacteria. (a) This figure gives a few representative structures of carotenoids of the various groups. A number of variants on the above structures also occur and are listed in the accompanying table. (b) Photographs of mass cultures of phototrophic bacteria showing the color of strains with various carotenoid pigments. The blue culture is a carotenoid—less mutant derivative of the organism *Rhodospirillum rubrum* showing how bacteriochlorophyll *a* is actually *blue* in color.

under *anaerobic* conditions, where bacteriochlorophyll synthesis can occur. Superimposed upon this O_2 effect is an effect of light intensity. Even under anaerobic conditions, when synthesis of the photosynthetic apparatus is not repressed, the level of the photopigments and internal membranes is affected by light intensity. At *high* light intensity, the synthesis of the photosynthetic apparatus is inhibited, whereas when cells are grown at *low* light intensity the bacteriochlorophyll content is high, and the cells are packed with membranes. (This increase in cell pigment content at low light intensities allows the organism to better utilize the available light.) Usually carotenoid pigment synthesis is coordinately regulated with the bacteriochlorophyll content.

An interesting group of bacteria which produce bacteriochlorophyll only under *aerobic* conditions have been described from marine environments. The aerobic phototrophs resemble nonsulfur purple bacteria, but they will not grow or produce bacteriochlorophyll anaerobically. Significant pigment levels are only observed when cells are grown aerobically in either the light or the dark. The function of bacteriochlorophyll in the aerobic phototrophs is not clear (presumably the pigment functions in photosynthesis but this has not been demonstrated), but it does appear clear that O_2 regulates pigment synthesis in these organisms in a manner opposite from that of typical purple bacteria. The majority of aerobic phototrophic bacteria have been placed in the newly created genus *Erythrobacter*, but a few isolates have been classified within the genus *Pseudomonas*.

In the green bacteria, the photosynthetic apparatus is structurally quite different, consisting of a series of cylindrically shaped structures called **chlorosomes** underlying and attached to the cell membrane (Figure 19.3*c*). These vesicles are enclosed within a thin membrane that does not have the usual bilayered appearance (it is referred to as a *nonunit* membrane). Bacteriochlorophylls *c*, c_s, *d*, or *e* (depending on the species) are present inside the chlorosomes, while Bchl *a* and components of the photosynthetic electron transport chain are located in the cytoplasmic membrane.

In heliobacteria, a group of strictly anaerobic nonsulfur bacteria that are green in color, neither internal membranes, typical of purple, nor chlorosomes, typical of green bacteria are observed. Bacteriochlorophyll in heliobacteria is presumably associated with the cytoplasmic membrane (Figure 19.3*d*). High resolution electron microscopy of thin sections of heliobacteria have shown the membrane to be studded with particles that may be the site of bacteriochlorophyll:protein complexes in these organisms.

Enrichment culture The purple and green sulfur bacteria are generally found in *anaerobic* zones of aquatic habitats, often where H_2S accumulates. These organisms can be enriched by duplicating the habitat in the laboratory (Table 19.1). A basal mineral salts medium is used, to which bicarbonate is added as a source of CO_2. Since many photosynthetic bacteria require vitamin B_{12}, this vitamin is usually added. The optimum pH range is 6 to 8, and the optimum temperature for mesophilic species is 25 to 30°C. Incubation is anaerobic, and a small amount of sodium sulfide (0.05 to 0.1 percent $Na_2S \cdot 9H_2O$) is added as photosynthetic electron donor. Selection of appropriate light conditions is important. Light intensities should not be too high, since these bacteria usually live in deep areas of lakes or other anaerobic habitats where light is low. Intensities between 20 and 100 footcandles are adequate. The quality

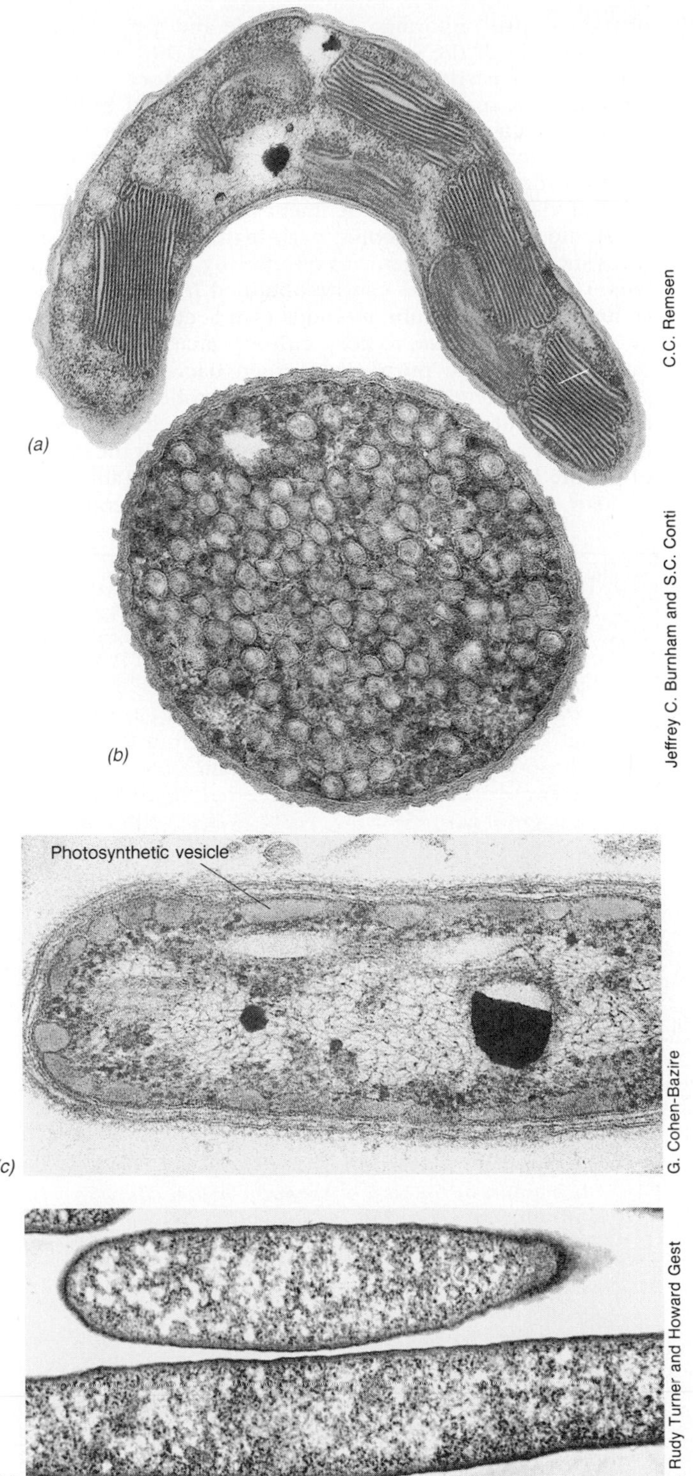

Figure 19.3 Membrane systems of photosynthetic bacteria as revealed by the electron microscope. (*a*) Purple photosynthetic bacterium, *Ectothiorhodospira mobilis*, showing the photosynthetic lamellae in flat sheets. (*b*) *Chromatium* sp., strain D, another purple photosynthetic bacterium, showing the photosynthetic membranes as individual vesicles. (*c*) Green photosynthetic bacterium, *Pelodictyon* sp., showing location of photosynthetic vesicles (chlorosomes). (*d*) *Heliobacterium chlorum*, showing the lack of internal membranes.

of light is also important. Most purple and green bacteria use radiation of the near infrared, 700 to 900 nm, which can be most easily obtained in the laboratory by use of conventional tungsten light bulbs. Fluorescent bulbs are usually deficient in infrared radiation and are not as satisfactory. After inoculation, culture tubes or bottles are incubated for several weeks and examined periodically for signs of visible growth. Enrichment cultures should appear pigmented, and microscopic examination of positive cultures should reveal organisms resembling purple and green bacteria. Pure cultures can be obtained from positive enrichments by shake tube methods (see Section 17.3), care being taken of course, to keep cultures anaerobic throughout. However, the purple and green bacteria are not as sensitive to oxygen as some other anaerobes, so that extreme precautions to maintain anaerobic conditions are usually not necessary. A widely used medium for the culture of purple and green sulfur bacteria is given in Table 19.1.

For *nonsulfur* purple bacteria, the sulfide concentration of the medium should be reduced to a much lower level, 0.01 to 0.02 percent $Na_2S \cdot 9H_2O$ (or sulfide can be eliminated altogether), and an organic substance added to provide a carbon source and/or electron donor. Because many nonsulfur purple bacteria have multiple growth-factor requirements, usually one or more B vitamins, addition of 0.01 percent yeast extract as a source of growth factors is recommended. The organic substrate used should be nonfermentable (to avoid enrichment of fermentative organisms such as clostridia); acetate, ethanol, benzoate, isopropanol, butyrate or dicarboxylic acids such as succinate are ideal.

Table 19.1 Pfennig's medium for the culture of phototrophic sulfur bacteria

Solution 1: 0.83 g $CaCl_2 \cdot 2H_2O$ in 2.5 liters H_2O. For marine organisms, add NaCl, 130 g.

Solution 2: H_2O 67 ml; KH_2PO_4, 1 g; NH_4Cl, 1 g; $MgCl_2 \cdot 2H_2O$, 1 g; KCl, 1 g; 3 ml vitamin B_{12} solution (2 mg/100 ml H_2O); 30 ml trace element solution (1 liter H_2O, ethylene diamine tetraacetic acid, 500 mg; $FeSO_4 \cdot 7H_2O$, 200 mg; $ZnSO_4 \cdot 7H_2O$, 10 mg; $MnCl_2 \cdot 4H_2O$, 3 mg; H_3BO_3, 30 mg; $CoCl_2 \cdot 6H_2O$, 20 mg; $CuCl_2 \cdot 2H_2O$, 1 mg; $NiCl_2 \cdot 6H_2O$, 2 mg; $Na_2MoO_4 \cdot 2H_2O$, 3 mg, pH 3).

Solution 3: Na_2CO_3, 3 g; H_2O, 900 ml. Autoclave in a container suitable for gassing aseptically with CO_2.

Solution 4: $Na_2S \cdot 9H_2O$, 3 g; H_2O, 200 ml. Autoclave in flask containing a Teflon-covered magnetic stirring rod.

Dispense the bulk of solution 1 in 67-ml aliquots in 30 screw-capped bottles of 100 ml capacity, and autoclave with screw caps loosely on. Autoclave the other solutions in bulk. After autoclaving, cool all solutions rapidly by placing the bottles in a cold water bath (to prevent lengthy exposure to air). Solution 3 is then gassed with CO_2 gas until it is saturated (about 30 minutes; pH drops to 6.2). Add this to cooled solution 2 and aseptically place 33 ml of this mixture in each bottle containing solution 1. Solution 4 is partially neutralized by adding dropwise (while stirring on a magnetic stirrer) 1.5 ml of sterile 2 M H_2SO_4. Add 5-ml portions of solution 4 to each bottle, fill the bottles completely with remaining solution 1, and tightly close. The final pH should be between 6.7 and 7.2. Store overnight to consume residual oxygen before using. (For some organisms, the sulfide concentration should be reduced. Use 2.5 ml instead of 5 ml of solution 4 per bottle.) It may be necessary to "feed" the cultures with sulfide (solution 4) from time to time, as the sulfide is used up during growth. To do this, remove 2.5 or 5 ml of liquid from a bottle and refill with an equal amount of sterile neutralized solution 4.

Originally described by Pfennig, N. 1965. Zentrabl. Bakteriol. Parasitenkd. Infektionskr. Hyg. Abt. 1, Supplementh. 1; 179. For an English version, see van Niel, C. B. 1971. Methods Enzymol. 23:3–28.

Heliobacteria can be isolated by taking advantage of a major property apparently shared by all representatives. All known heliobacteria grow well at temperatures up to 42°C. This temperature is restrictive for most other phototrophic bacteria. Hence, to enrich heliobacteria, anaerobic enrichments containing pyruvate or lactate and lacking sulfide are inoculated with soil (heliobacteria are very resistant to drying and thus dry soil will often yield strains of these organisms) and incubated in the light at 40–42°C.

It should be noted that many purple and green bacteria are incapable of assimilatory sulfate reduction, so that they must be given a source of reduced sulfur. The sulfide added to the medium thus not only uses up the remaining O_2 and serves as an electron donor, but provides this source of reduced sulfur. If an electron donor such as H_2 or an organic compound is to be used, and sulfide must for some reason be avoided, then it is necessary to add another source of reduced sulfur, such as methionine or cysteine (if yeast extract is added, it often serves as a source of reduced sulfur as well as a growth-factor source).

Nonsulfur purple bacteria These bacteria have been called the *nonsulfur* purple bacteria because it has been thought that they were unable to use sulfide as an electron donor for the reduction of CO_2 to cell material. But it is now known that sulfide can be used by most species provided the concentration is maintained at a low level. It appears that levels of sulfide utilized well by green or purple *sulfur* bacteria are toxic to most nonsulfur purple bacteria; hence the previous inability to demonstrate sulfide metabolism. However, nonsulfur purple bacteria differ from purple sulfur bacteria in three important ways: (1) they are unable to oxidize *elemental sulfur* to sulfate (although they can oxidize *sulfide* to sulfate without the intermediate accumulation of elemental sulfur); (2) if elemental sulfur is produced from the oxidation of sulfide, it is *not* stored inside the cells; and (3) they are facultative phototrophs, able to grow aerobically in the dark or anaerobically in the light. (It should be noted that often species of purple sulfur bacteria have the capacity to grow aerobically in darkness as well.) Some nonsulfur purple bacteria can also grow anaerobically in the dark, using fermentative metabolism. Because of the difficulty of providing nontoxic levels of sulfide, these organisms are generally cultured under phototrophic conditions with an organic compound as major carbon source. Most members of this group also require vitamins, so that yeast extract or some other source of vitamins is usually provided.

The morphological diversity of this group is typical of that of other purple and green bacteria (Table 19.2 and Figure 19.4), and it is clearly a heterogeneous group as it contains both polarly and peritrichously flagellated genera, the latter growing by budding.

Some nonsulfur purple bacteria have the ability to utilize methanol or formate as sole carbon source for phototrophic growth. When growing anaerobically with methanol, some CO_2 fixation is necessary, and the following stoichiometry is observed:

$$2CH_3OH + CO_2 \rightarrow 3(CH_2O) + H_2O$$

Cell material

Apparently, CO_2 is required because methanol is at a more reduced oxidation state than cell material, and the CO_2 serves as an electron sink (electron acceptor).

Enrichments for nonsulfur purple bacteria can be made highly selective by omitting fixed nitrogen sources such as ammonia or nitrate from the medium, and substituting an ample supply of gaseous nitrogen, N_2. Most nonsulfur purple bacteria are active N_2 fixers and grow well in a medium in which N_2 is the sole nitrogen source.

Purple sulfur bacteria Purple bacteria that deposit sulfur and oxidize it to sulfate are morphologically diverse (Table 19.3). The cell is usually larger than that of green bacteria, and in sulfide-rich environments may be packed with sulfur granules (Figure 19.5*a* and *b*; see also Figure 19.13), although in the smaller-celled genera the sulfur granules may not be so obvious (Figure 19.5*c*). Purple sulfur bacteria are commonly found in anaerobic zones of lakes as well as in sulfur springs; because of their conspicuous purple color they are often easily seen in large blooms or masses (see Figures 19.13 and 19.14), and in fact, blooms of purple sulfur bacteria were described in ancient literature. In general, purple sulfur bacteria grow at somewhat higher pH values than do the green bacteria. The genus

Table 19.2 Genera and characteristics of nonsulfur purple bacteria

Characteristics	Genus	DNA (mole % GC)
Spirals, polarly flagellated	*Rhodospirillum*	62–66
Rods, polarly flagellated; divide by budding	*Rhodopseudomonas*	62–72
Rods, divide by binary fission	*Rhodobacter*	65–71
Ovals, peritrichously flagellated; growth by budding and hypha formation	*Rhodomicrobium*	61–63
Ring-shaped, rod-shaped or spirilla	*Rhodocyclus*	64–72
Large spheres, pH optimum 5	*Rhodopila*	66

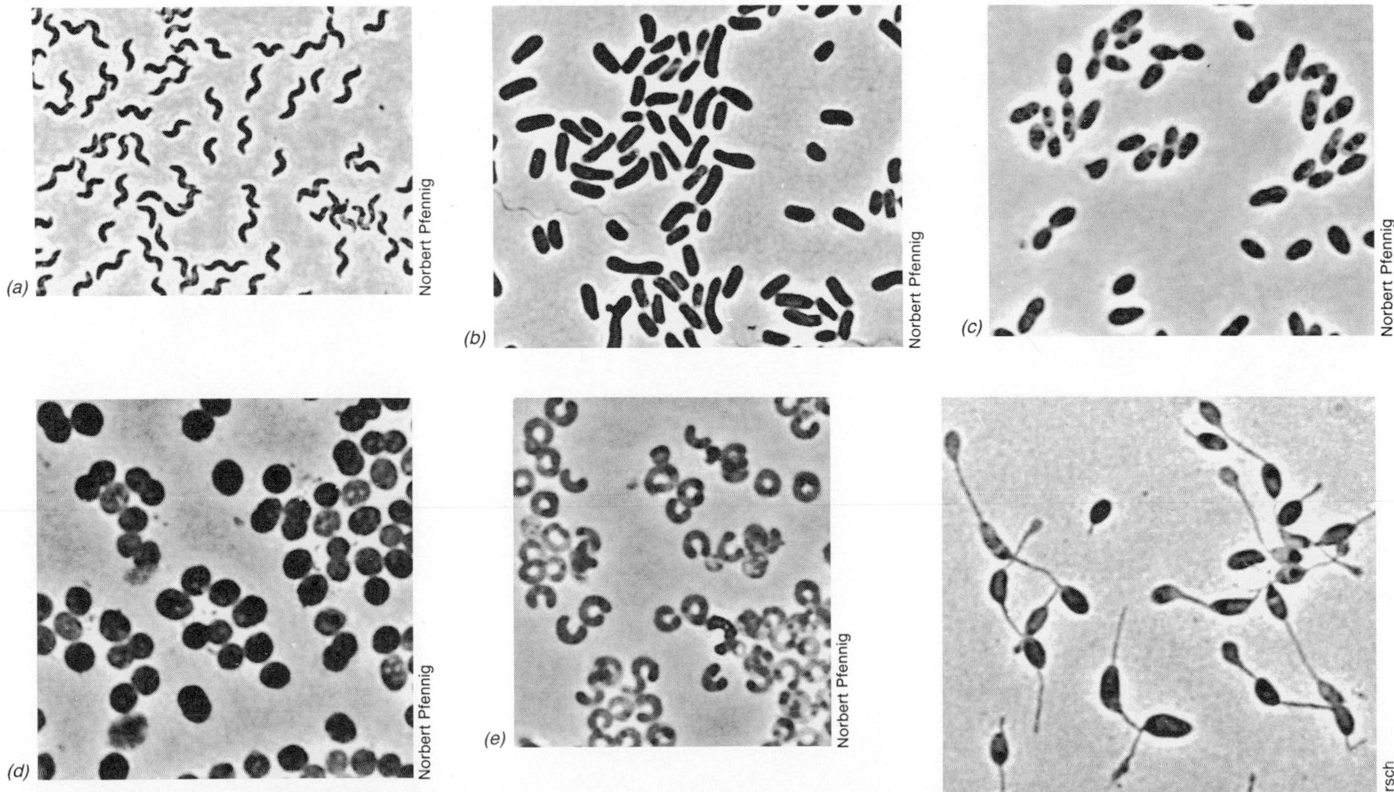

Figure 19.4 Representatives of the six genera of nonsulfur purple bacteria (see also Table 19.2). (a) *Rhodospirillum fulvum*. (b) *Rhodopseudomonas acidophila*. (c) *Rhodobacter sphaeroides*. (d) *Rhodopila globiformis*. (e) *Rhodocyclus purpureus*. (f) *Rhodomicrobium vannielii*.

Ectothiorhodospira is of interest because it deposits sulfur externally and is also halophilic, growing at sodium chloride concentrations approaching saturation, often at very high pH. It is found in saline lakes, soda lakes, salterns, and other bodies of water high in salt (see Section 19.33). Purple bacteria which contain bacteriochlorophyll *b* can be enriched by using infrared radiation instead of visible light, since bacteriochlorophyll *b* has an absorption maximum at 1025 nm, radiation which is invisible to the human eye.

Purple sulfur bacteria have a limited ability to utilize organic compounds as carbon sources for phototrophic growth. Acetate and pyruvate are utilized by all species, some species will use other organic acids, sugars, or ethanol. A few purple sulfur bacteria will grow lithotrophically in darkness with thiosulfate as electron donor, and *Thiocapsa* will grow heterotrophically on acetate.

Green bacteria At one time considered a relatively small group, it is now known that the green bacteria are quite diverse, including nonmotile rods, spirals, and spheres (green sulfur bacteria), and motile filamentous gliding forms (Figure 19.6 and Table 19.4). We discuss the gliding bacteria in some detail in Section 19.13. The gliding green bacteria (Figure 19.6*c–e*) could easily be classified with the rest of the gliding bacteria, but it is generally considered that the ability to grow phototrophically is a more fundamental characteristic; thus these gliding organisms are classified with the rest of the phototrophic bacteria. Several other green bacteria have complex appendages called **prosthecae**, and in this connection resemble the budding and/or appendaged bacteria (Section 19.10).

The green bacteria that live planktonically in lakes generally possess gas vesicles, whereas the species that live in the outflow of sulfur and hot springs, or in other benthic habitats, are not gas vesiculate. Members of one genus, *Pelodictyon*, consist of rods that undergo branching, and since the rods remain attached, a three-dimensional network is formed (Figure 19.6*b*).

Green sulfur bacteria are strictly anaerobic and obligately phototrophic, being unable to carry out respiratory

Table 19.3 Genera and characteristics of purple sulfur bacteria

Characteristics	Genus	DNA (mole % GC)
Sulfur deposited externally:		
Spirals, polar flagella	*Ectothiorhodospira*	62–70
Sulfur deposited internally:		
Do not contain gas vesicles		
Ovals or rods, polar flagella	*Chromatium*	48–70
Spheres, diplococci, tetrads, nonmotile	*Thiocapsa*	63–70
Spheres or ovals, polar flagella	*Thiocystis*	62–68
Large spirals, polar flagella	*Thiospirillum*	45
Contain gas vesicles		
Irregular spheres, ovals, nonmotile	*Amoebobacter*	65
Spheres, ovals, polar flagella	*Lamprocystis*	64
Rods; nonmotile; forming irregular network	*Thiodictyon*	65–66
Spheres; nonmotile; forming flat sheets of tetrads; not available in pure culture	*Thiopedia*	—

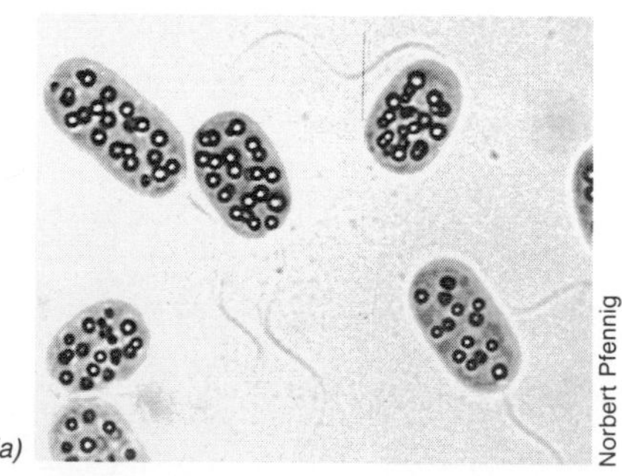
(a) Norbert Pfennig

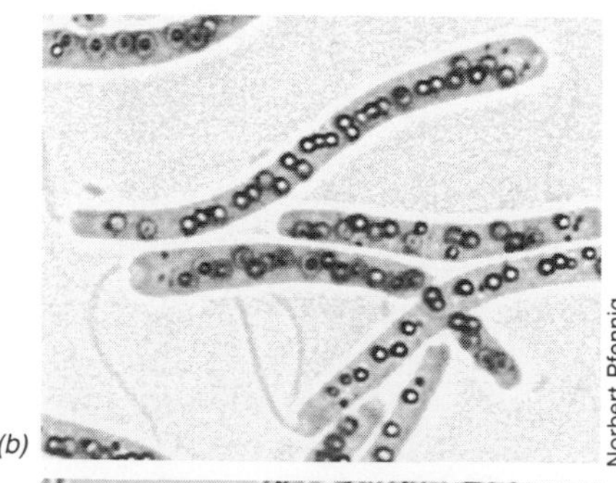
(b) Norbert Pfennig

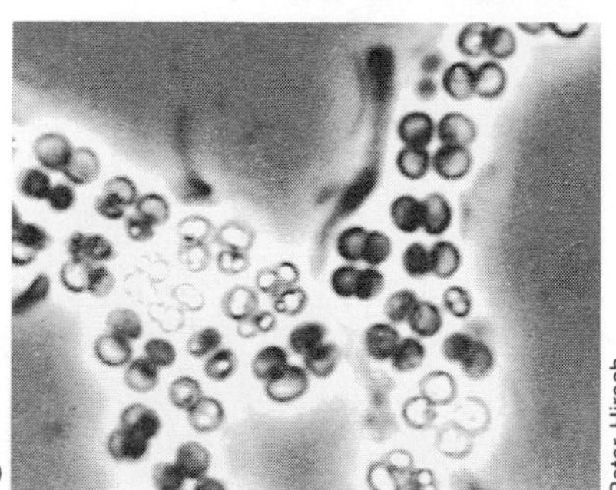
(c) Peter Hirsch

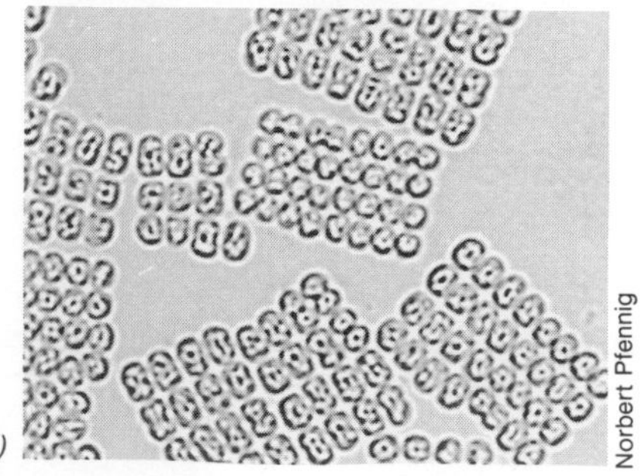
(d) Norbert Pfennig

(e) R. Scherer

Figure 19.5 Bright field photomicrographs of purple sulfur bacteria. (a) *Chromatium okenii*. Note the globules of elemental sulfur inside the cells. (b) *Thiospirillum jenense*, a very large, polarly flagellated spiral. Note the sulfur globules. (c) *Thiocapsa*. (d) *Thiopedia rosea*. (e) Scanning electron micrograph of a sheet of 16 cells of *Thiopedia rosea* showing the major division planes.

metabolism in the dark. Most green sulfur bacteria can assimilate simple organic substances for phototrophic growth, provided that a reduced sulfur compound is present as a sulfur source (since they are incapable of assimilatory sulfate reduction). Organic compounds used by these species include acetate, propionate, pyruvate, and lactate. *Chloroflexus* is much more versatile than most green sulfur bacteria, being able to grow heterotrophically in the dark under aerobic conditions, as well as phototrophically on a wide variety of sugars, amino acids, and organic acids, or photoautotrophically with H_2S or H_2 and CO_2.

Green sulfur bacteria share some photochemical characteristics with heliobacteria (see below for discussion of the latter group). The chain of electron carriers between the primary electron acceptor and the bacteriochlorophyll of green bacteria and heliobacteria is quite similar. Of particular interest is the fact that the primary electron acceptor in green bacteria and heliobacteria is poised at a reduction

Table 19.4 Genera and characteristics of green photosynthetic bacteria

Characteristics	Genus	DNA (mole % GC)
No gas vesicles:		
Straight or curved rods, nonmotile	*Chlorobium*	49–58
Spheres and ovals, nonmotile, forming prosthecae (appendages)	*Prosthecochloris*	50–56
	Ancalochloris	—
Filamentous, gliding	*Chloroflexus*	53–55
	Heliothrix	—
Filamentous, gliding, large diameter (5 μm)	*Oscillochloris*	—
Contain gas vesicles:		
Branching nonmotile rods, in loose irregular network	*Pelodictyon*	48–58
Spheres and ovals, nonmotile, forming chains	*Clathrochloris*	—
Rods, gliding	*Chloroherpeton*	45–48
Filamentous, gliding, large diameter (2–2.5 μm)	*Chloronema*	—

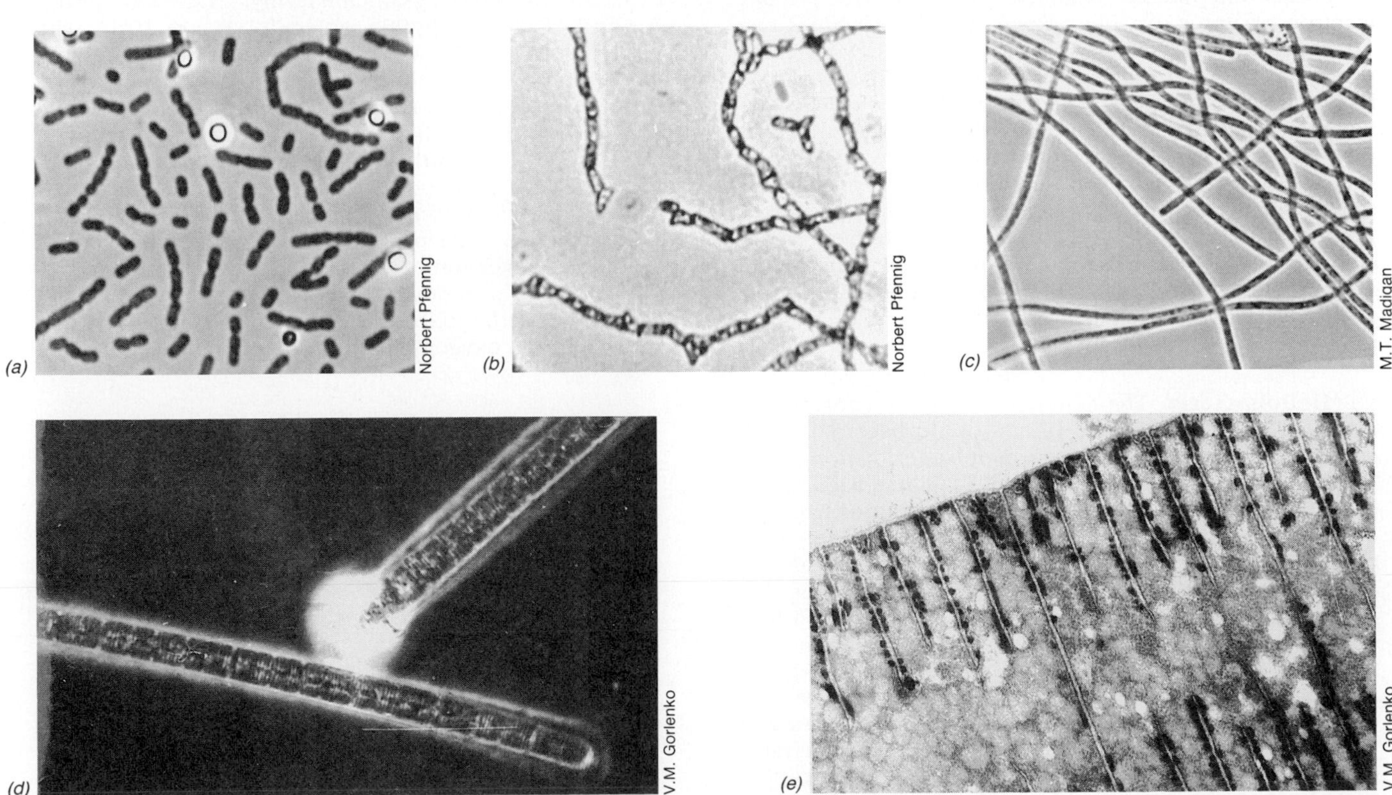

Figure 19.6 Green photosynthetic bacteria. (a) *Chlorobium limicola*. Note the sulfur granules deposited *extra*cellularly. (b) *Pelodictyon clathratiforme*, a bacterium forming a three-dimensional network. (c) *Chloroflexus aurantiacus*, a filamentous gliding bacterium. (d) *Oscillochloris*, a large filamentous, gliding green bacterium. Phase contrast. The brightly contrasting material is the holdfast. (e) Electron micrograph of *Oscillochloris*. The chlorosomes in this preparation are darkly stained.

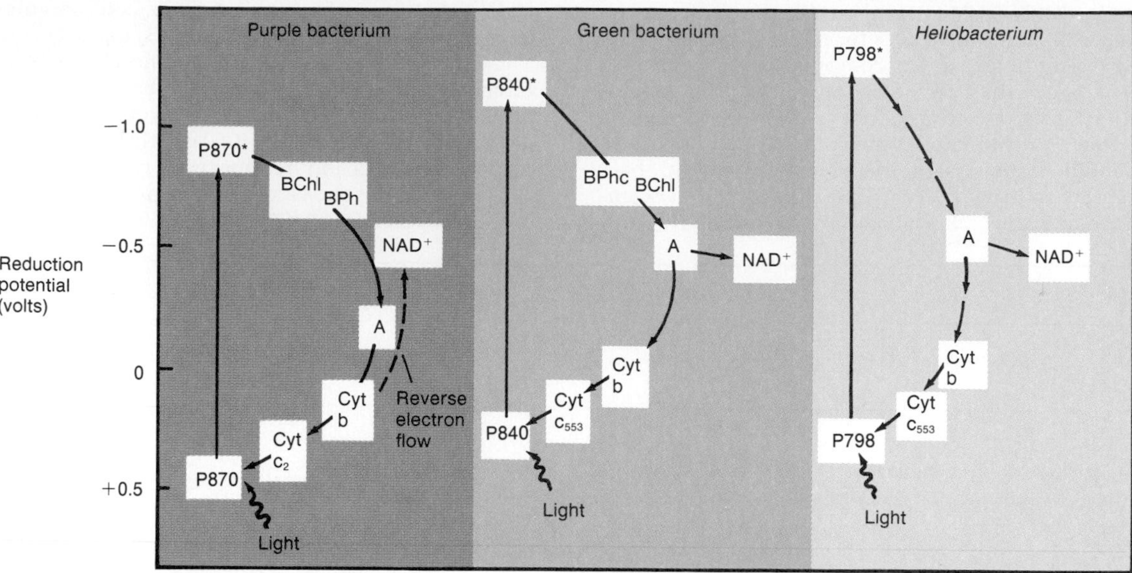

Figure 19.7 A comparison of electron flow in green sulfur bacteria, heliobacteria, and purple bacteria. Note how reverse electron flow in purple bacteria is necessitated by the fact that the primary acceptor (A) is more positive in potential than the NAD^+/NADH couple. BChl, bacteriochlorophyll; BPh, bacteriopheophytin.

potential of about −0.5 volts. This is much more reducing than the primary electron acceptor of purple bacteria, which has a reduction potential of about −0.15 volts (the primary electron acceptor in *Chloroflexus* closely resembles that of purple bacteria). The significance of the low potential primary acceptor in green sulfur bacteria and heliobacteria lies in the fact that such an acceptor, once reduced, is sufficiently negative to reduce NAD^+ directly; this alleviates the need for reverse electron flow as a means of reducing NAD^+ (Figure 19.7). Reverse electron flow is presumably the mechanism by which purple bacteria and *Chloroflexus* produce NADH (Figure 19.7). Thus, although they employ completely different bacteriochlorophylls in their photosynthetic reactions (see Table 19.1), green sulfur bacteria and heliobacteria are remarkably similar in terms of primary photochemical events.

Heliobacteria The heliobacteria are an entirely separate group of anoxygenic phototrophic bacteria which contain a structurally distinct form of bacteriochlorophyll, bacteriochlorophyll *g* (see Table 19.1). These bacteria are both physiologically and phylogenetically unique from the other phototrophic bacteria. The group consists of three genera, ***Heliobacterium, Heliospirillum***, and ***Heliobacillus***, with only a single species known for each genus. ***Heliobacterium*** (Figure 19.3*d*) is a gliding rod-shaped bacterium, ***Heliospirillum*** is a large motile spirillum, and ***Heliobacillus*** (Figure 19.8) is an actively motile rod. Nutritionally, heliobacteria are remarkably restricted. The only carbon sources utilized by these organisms are pyruvate and lactate (*Heliobacillus* will also use acetate and butyrate) and none of the heliobacteria appear capable of autotrophic growth. Although physiologically related to nonsulfur purple bacteria, heliobacteria are strictly anaerobic phototrophs and are unable to grow by respiratory means as is typical of nonsulfur purple bacteria.

Because of the unique structure of bacteriochlorophyll *g*, the isolation of heliobacteria has renewed interest in the evolution of photosynthesis. Figure 19.9 shows the structure of bacteriochlorophyll *g* and that of chlorophyll *a*. Unlike all other bacteriochlorophylls, but like chlorophyll *a*, bacteriochlorophyll *g* contains a vinyl ($H_2C{=}CH_2$) group on ring I of the tetrapyrole molecule (Figure 19.1). The only structural differences between bacteriochlorophyll *g* and chlorophyll *a* therefore lie in ring II of the tetrapyrole. Like other bacteriochlorophylls, ring II of bacteriochlorophyll *g* is ***reduced,*** unlike that of chlorophyll *a*. However, if cells of *Heliobacterium* (which are originally a brown-green in color) are exposed to air, the double bond across ring II of bacteriochlorophyll *g* becomes oxidized, and cultures of the organism slowly turn an emerald green in color, due to the conversion of bacteriochlorophyll *g* to chlorophyll *a*. This remarkable series of events has led to the suggestion that heliobacteria represent a bridge between anoxygenic and oxygenic phototrophs, a "missing link" so

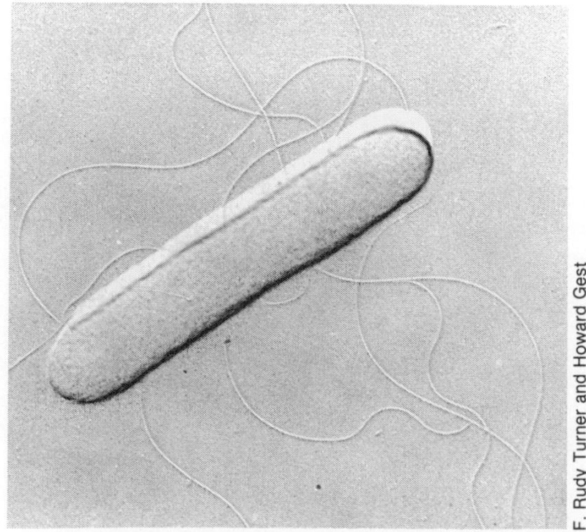

Figure 19.8 Electron micrograph of a representative species of heliobacteria, *Heliobacillus mobilis*. Notice the peritrichous flagellation.

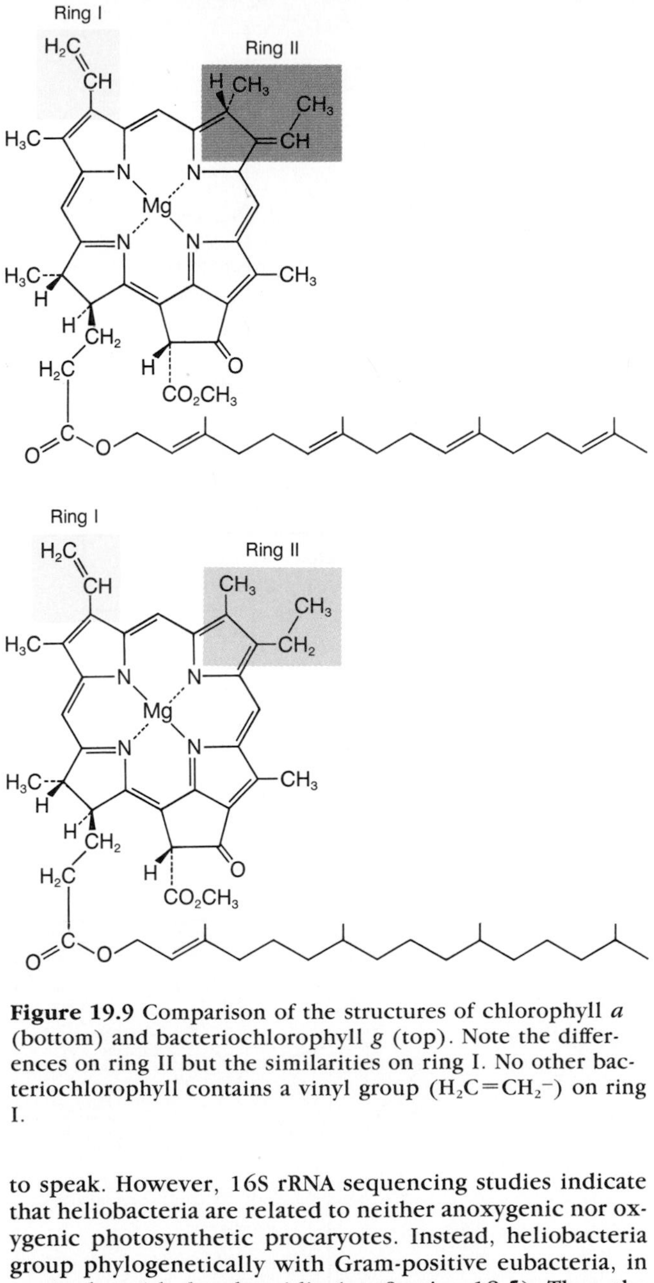

Figure 19.9 Comparison of the structures of chlorophyll *a* (bottom) and bacteriochlorophyll *g* (top). Note the differences on ring II but the similarities on ring I. No other bacteriochlorophyll contains a vinyl group ($H_2C{=}CH_2{-}$) on ring I.

to speak. However, 16S rRNA sequencing studies indicate that heliobacteria are related to neither anoxygenic nor oxygenic photosynthetic procaryotes. Instead, heliobacteria group phylogenetically with Gram-positive eubacteria, in particular with the clostridia (see Section 18.5). Thus, the structural relationship between bacteriochlorophyll *g* and chlorophyll *a* may simply be a fortuitous one. However, the discovery of heliobacteria has reinforced the idea that bacterial photosynthesis is an ancient and highly diverse process that has evolutionary roots in Gram-positive as well as Gram-negative bacteria.

Physiology of phototrophic growth The overall picture that emerges from a study of the comparative physiology of purple and green bacteria is that light is used primarily in the generation of ATP, and is not generally involved in the generation of reducing power, as it is in organisms exhibiting oxygenic photosynthesis. Possible exceptions to this rule exist with green sulfur bacteria and heliobacteria, as discussed above (see Figure 19.7). Whether or not a source of reducing power is actually needed depends upon the carbon source supplied. If CO_2 is the sole carbon source, then reducing power is needed, and this can come from a reduced sulfur compound or H_2. Reducing power for CO_2 fixation may also come from an organic compound, but if an organic compound is supplied the situation is more complex. This is because the organic compound will serve as a carbon source itself, so that CO_2 need not necessarily be reduced. Whether or not CO_2 is reduced when an organic compound is added will depend at least in part on the oxidation state of the organic compound. Compounds such as acetate, glucose, and pyruvate are at about the oxidation level of cell material and can thus be assimilated directly as carbon sources with no requirement for either oxidation or reduction. Fatty acids longer than acetate (for example, propionate, butyrate, caprylate) are more reduced than cell material, and some means of disposing of excess electrons is necessary, such as the reduction of CO_2. Thus the amount of CO_2 fixed by a purple or green bacterium growing with an organic compound will depend upon the oxidation state of the compound, and whether or not inorganic electron donors such as sulfide are also present.

Many purple and green bacteria can grow phototrophically using H_2 as sole electron donor, with CO_2 as carbon source. These organisms have a *hydrogenase* for activating H_2 for CO_2 reduction. However, since many phototrophic bacteria are incapable of carrying out assimilatory sulfate reduction, it is essential when testing for growth on H_2 to add a small amount of sulfide as a source of reduced sulfur. It may also be necessary to add vitamins for some strains.

In the absence of ammonia as nitrogen source, many phototrophic bacteria produce H_2 in the light. This has been shown to be due to the presence in these organisms of *nitrogenase*, the enzyme involved in N_2 fixation (see Section 16.24). All nitrogenase enzymes are able to reduce H^+ to H_2, and this reaction runs in competition with N_2 reduction. Since ammonia represses nitrogenase synthesis, production of H_2 by phototrophic bacteria only occurs when ammonia is absent.

Although the presence of the Calvin cycle (see Section 16.6) has been conclusively shown in many phototrophic purple bacteria, it is not the mechanism by which CO_2 is fixed in green sulfur bacteria or in *Chloroflexus*. In *Chlorobium* CO_2 fixation occurs by a reversal of steps in the citric acid cycle. *Chlorobium* contains two ferredoxin-linked enzymes, which will cause a reductive fixation of CO_2 into intermediates of the tricarboxylic acid cycle (Figure 19.10). The two ferredoxin-linked reactions involve the carboxylation of succinyl-CoA to α-ketoglutarate and the carboxylation of acetyl-CoA to pyruvate. Most of the other reactions in the reverse citric acid cycle in *Chlorobium* are normal enzymes of the cycle working in reverse of the normal oxidative direction of the cycle. One exception is *citrate lyase*, an ATP-dependent enzyme which cleaves citrate into acetyl-CoA and oxalacetate in green sulfur bacteria. In the oxidative direction citrate is produced from these same components by the enzyme *citrate synthase*.

Chloroflexus will grow autotrophically with either H_2 or H_2S as electron donor, however neither the Calvin cycle nor reverse citric acid cycle operate in this organism. Instead, two molecules of CO_2 are converted into acetyl-CoA by an unknown reaction mechanism in *Chloroflexus*, after which acetyl-CoA is converted to pyruvate, presumably by a ferredoxin-linked reaction similar to that in green sulfur bacteria (see Figure 19.10). This *total synthesis of acetate* also occurs in acetogenic bacteria, methanogenic bacteria, and in certain sulfate reducing bacteria. In these organisms CO_2 is fixed by the *acetyl-CoA pathway*, a complex mech-

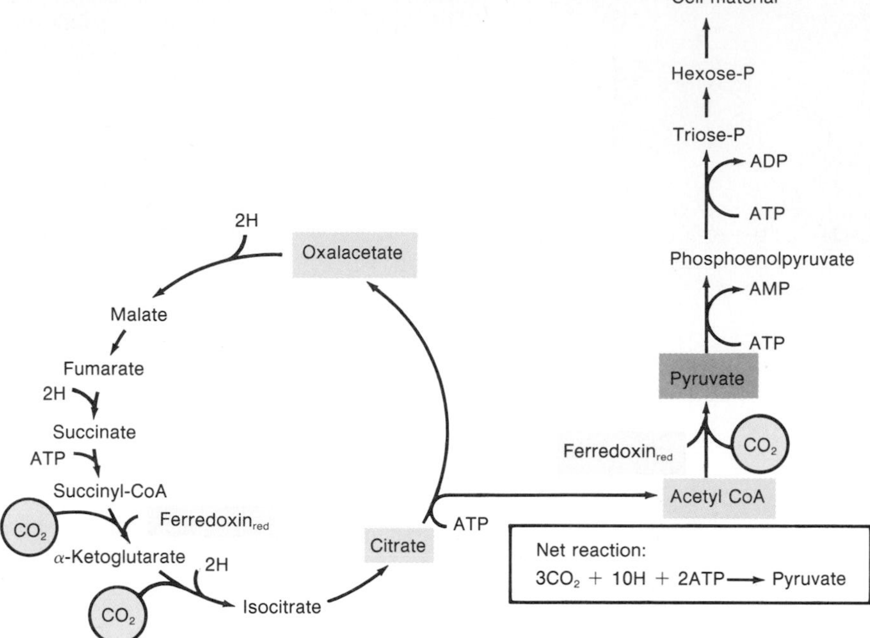

Figure 19.10 The reversed citric acid cycle is the mechanism of CO_2 fixation in *Chlorobium*. Ferredoxin$_{red}$ indicates carboxylation reactions requiring reduced ferredoxin. Starting from oxalacetate, each turn of the cycle results in three molecules of CO_2 being incorporated and pyruvate as the product. The cleavage of citrate regenerates the C-4 acceptor oxalacetate, and produces acetyl-CoA for biosynthesis.

anism involving tetrahydrofolate and corrinoid coenzymes (see Section 19.9 for discussion of this pathway). However, whether the acetyl-CoA pathway operates in *Chloroflexus* is unknown.

Sulfur metabolism in the purple and green bacteria Most of the purple and green bacteria are able to oxidize reduced sulfur compounds under anaerobic conditions, with the formation of sulfate. The most common reduced sulfur compounds used are *sulfide* and *thiosulfate*. Tetrathionate, which is an oxidation product of thiosulfate, is also used by some organisms, probably after it is reduced to thiosulfate. Elemental sulfur is frequently formed during the oxidation of sulfide or thiosulfate, and is either deposited inside or outside the cells. Elemental sulfur deposited inside the cells (see Figure 19.5) is readily available as a further source of reduced sulfur compounds, and externally deposited elemental sulfur (see Figure 19.6) may also be utilized, although probably less efficiently (because of its high insolubility and relative unavailability when present in the medium instead of inside the cells). The overall pathway of oxidation of reduced sulfur compounds in purple and green bacteria is shown in Figure 19.11. As seen either sulfide or thiosulfate is oxidized first to sulfite, SO_3^{2-}, and the enzymes adenylphosphosulfate reductase (APS reductase) and ADP-sulfurylase catalyze the oxidation of sulfite to sulfate. Note that a substrate-level phosphorylation occurs at this step, providing for the synthesis of a high-energy phosphate bond (in ADP).

It seems likely that elemental sulfur is not an obligatory intermediate between sulfide and sulfate, but merely a side product. Elemental sulfur is probably a *storage* product, formed when sulfide concentrations in the environment are high. Indeed, under limiting sulfide levels sulfide is oxidized directly to sulfate without formation of elemental sulfur. The formation of elemental sulfur as a storage product has been most clearly shown in *Chromatium*, where it is deposited inside the cells (Figure 19.5*a*; see also Figure 16.8*b*). Intracellular elemental sulfur in *Chromatium* can serve as an electron donor for phototrophic growth when sulfide is absent. Additionally, when *Chromatium* is placed in the dark, S^0 can also serve as an electron *acceptor*, being reduced to sulfide. Under these conditions electrons for reduction of S^0 comes from a glycogen storage product; thus *Chromatium* stores both its electron donor (glycogen) and electron acceptor (S^0) as intracellular polymers. If an external electron donor is present, such as acetate, S^0 can also serve as an electron acceptor for dark anaerobic metabolism

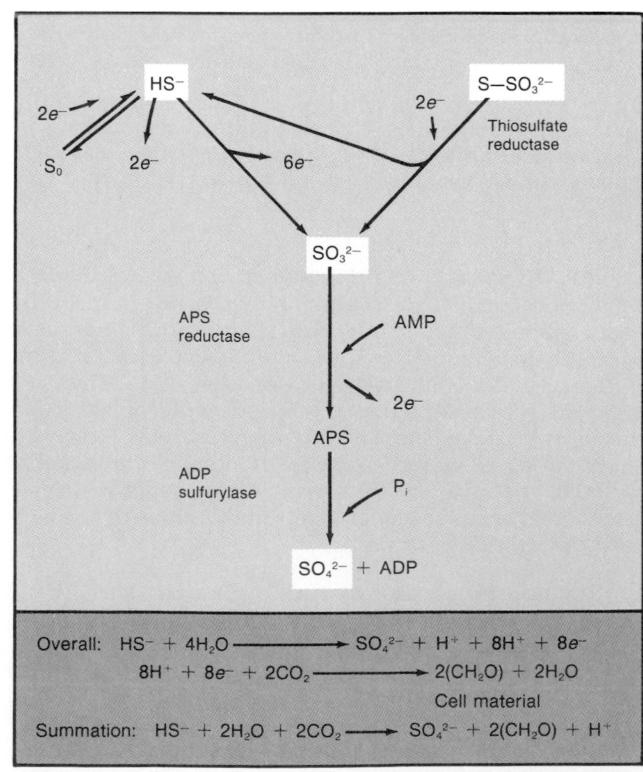

Figure 19.11 Pathways of oxidation of reduced sulfur compounds in phototrophic bacteria. Note that a substrate-level phosphorylation occurs via the enzyme ADP sulfurylase.

(but not growth) in the absence of intracellular storage polysaccharide.

Symbiotic associations and mixed culture interactions A two-membered system in which each organism does something for the benefit of the other has been called a **consortium**. An association of two organisms consisting of a large, colorless central bacterium, which is polarly flagellated, surrounded by smaller, ovoid- to rod-shaped green bacteria arranged in four to six rows has been called *Chlorochromatium aggregatum* (Figure 19.12*a*). Electron micrographs of thin sections of this association clearly show the chlorosomes of the green bacterial partner and the intimate relationship which exists between the two components of *C. aggregatum* (Figure 19.12*b*). The genus and species name are invalid in formal taxonomy because they refer not to a single organism, but to an association of two or more organisms, yet the association is seen quite commonly in lakes and in muds and is probably a rather specific one. The green organism in this association has been cultured and classified as *Chlorobium chlorochromatii*, but the large colorless central organism has not been cultured. A similar association in which the colored organism is a brown-pigmented *Chlorobium* has been called *Pelochromatium roseum*. A clue to the possible role of the colorless central organism comes from the common observation that the sulfide needed by purple and green bacteria can be derived from sulfate- or sulfur-reducing bacteria associated with them. Sulfate- and sulfur-reducing bacteria use organic compounds such as ethanol, lactate, formate, or fatty acids as electron donors and reduce sulfate or sulfur to sulfide. The sulfide produced can then be used by associated phototrophic bacteria, which in the light oxidize the sulfide back to sulfate. If some of the organic matter produced by the phototrophic bacterium is used by the sulfate reducer, a self-feeding system can develop, driven by light energy.

Ecology The greatest number of purple and green bacteria are found in the depths of certain kinds of lakes where stable conditions for growth occur. We discussed the development of thermal stratification in lakes in Chapter 17 (see Figure 17.10), after stratification occurs, stable anaerobic conditions may continue in the deep waters throughout the summer season. If there is a sufficient supply of H_2S and the lake water is sufficiently clear so that light penetrates to the anaerobic zone, a massive layer of purple or green bacteria can develop (Figure 19.13). This layer,

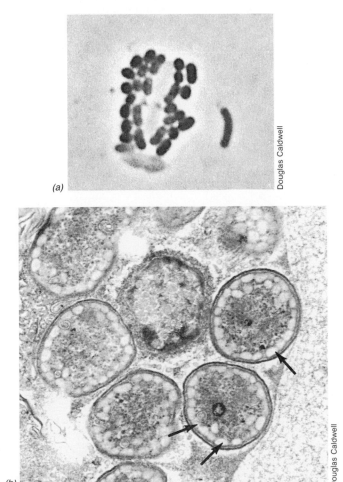

Figure 19.12 Phase contrast micrograph (a) and transmission electron micrograph (b) of the green bacterial consortium, *Chlorochromatium aggregatum*. In (a) the nonphotosynthetic central organism is much lighter in color than the phototrophic bacteria. Note the chlorosomes (arrows) in (b).

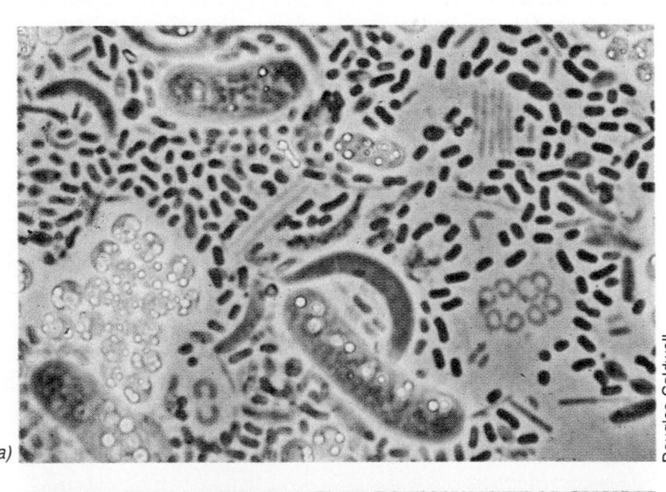

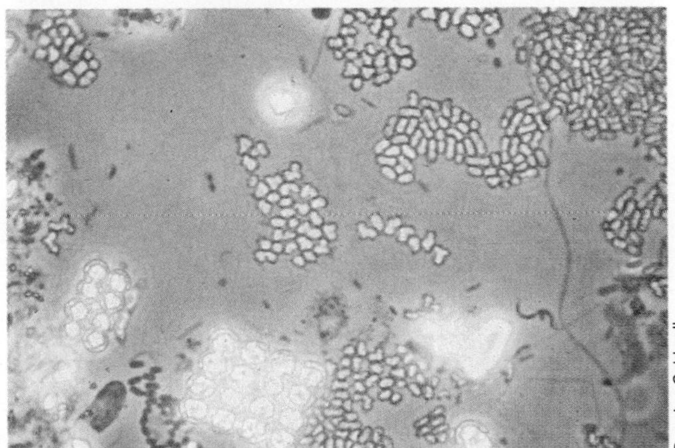

Figure 19.13 Phase contrast photomicrographs of layers containing purple sulfur bacteria (a), and green sulfur bacteria (b), from the water column of a small, stratified lake in Michigan. The purple bacteria include large-celled *Chromatium* species and *Thiocystis*. The green bacteria are predominately *Clathrochloris* filled with gas vesicles.

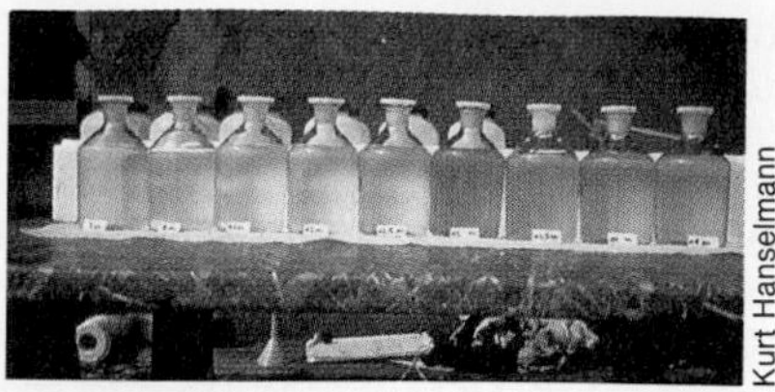

Figure 19.14 Water samples from various depths of a small meromictic alpine lake in Switzerland. During the summer months a layer of phototrophic purple sulfur bacteria forms at a depth of 13–13.5 meters.

hidden from view of the observer on the surface, can be studied by sampling water at various depths (Figure 19.14). The bacteria often form a distinct layer just at the depth where H_2S is first present (Figure 19.15). The most favorable lakes for development of these bacteria are those called **meromictic**, which are essentially permanently stratified due to the presence of denser (usually saline) water in the bottom. In such permanently stratified lakes, the bloom of purple or green bacteria may be present throughout the year. Often the photosynthetic activity of these bacteria is sufficiently great so that it is an important source of organic matter to the lake ecosystem. In some cases more photosynthesis may occur in the bacterial layer than in the surface algal layer. In general, blooms of purple and green bacteria are more common in small lakes and ponds than in large bodies of water, mainly because the stable stratification of large bodies of water is more affected by wind. The ideal lake for observing blooms of purple and green bacteria is a fairly deep small one nestled in a valley protected from strong winds with a fairly large amount of hydrogen sulfide in the bottom waters.

One of the easiest places to observe purple and green bacteria in nature is in sulfur springs, where massive blooms are often present a few inches below the surface of the water (Figure 19.16). Along the seacoast, blooms also occur in warm shallow pools of seawater not connected to the open ocean, where the activities of sulfate-reducing bacteria lead to production of large amounts of H_2S.

Hot springs serve as habitats for a variety of purple and green bacteria. *Chloroflexus* forms thick mats (frequently in association with cyanobacteria) in hot springs at temperatures from 40–70°C (see Figure 17.7*a* for photograph of a microbial mat). *Chlorobium* mats are occasionally found in high sulfide hot springs, generally at temperatures below 50°C. One species of the purple sulfur bacterium *Chromatium* grows in high sulfide springs at temperatures up to 57°C. A filamentous, photosynthetic gliding bacterium that resembles *Chloroflexus* but that lacks chlorosomes and bacteriochlorophyll c_s grows in certain hot springs in association with *Chloroflexus*. This organism, named *Heliothrix*, is unusual among phototrophic bacteria because it requires extremely high light intensities for growth. Although *Heliothrix* contains only bacteriochlorophyll *a* and thus should technically be considered a purple bacterium, the morphological similarity between *Heliothrix* and *Chloroflexus*, the ability to glide, and the fact that the two organisms have nearly identical carotenoids have been used

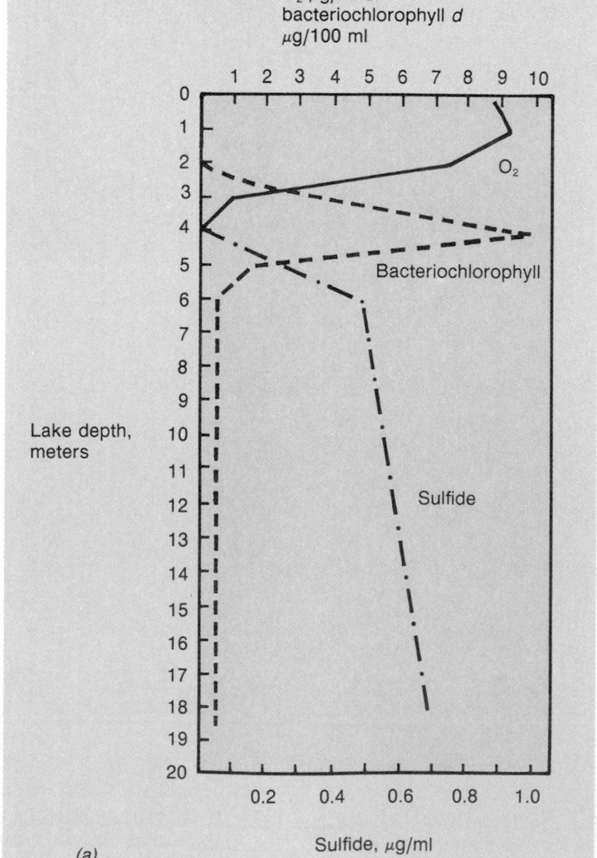

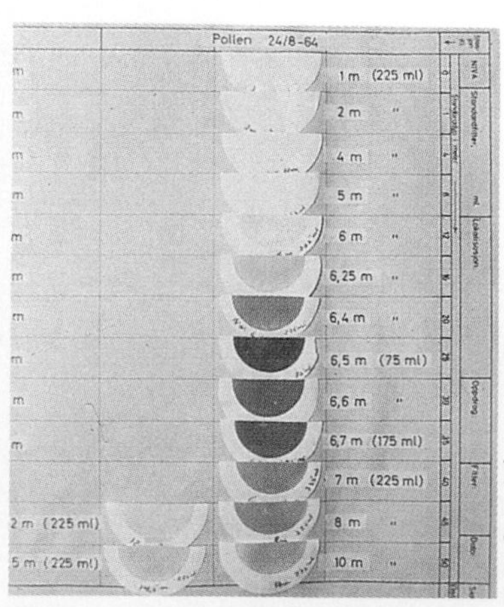

Figure 19.15 Vertical distribution of *Chlorobium* in stratified lakes. (a) Lake Mary, Wisconsin. The phototrophic bacterium forms a layer just at the top of the anaerobic zone. Bacterial population size is quantified by measuring bacteriochlorophyll *d* concentration. (b) Stratified Norwegian lake. Series of membrane filters through which were passed water samples taken at varying depths. At 6.4–6.6 meters a heavy cyanobacterial bloom is present, and at 6.7–8 meters the green sulfur bacterium *Pelodictyon luteolum* reaches its highest population density.

Figure 19.16 Massive accumulation of purple sulfur bacteria, *Thiopedia roseopersicinia*, in a spring in Madison, Wisconsin. The bacteria grow near the bottom of the spring pool and float to the top (by virtue of their gas vesicles) when disturbed. The green alga is the eucaryotic organism *Spirogyra*.

Figure 19.17 Cross section through a bacterial mat in the Sippewisset salt marsh near Woods Hole, Massachusetts. The top layer contains cyanobacteria, the pink layer phototrophic purple sulfur bacteria attached to sand grains, and the black layers sulfate-reducing bacteria.

to provisionally classify *Heliothrix* in with the other gliding green bacteria. Nucleic acid sequencing of ribosomal RNA (see Chapter 18) from *Heliothrix* and *Chloroflexus* supports this classification.

Mats of phototrophic bacteria also form in nonthermal environments. Cyanobacterial-photosynthetic bacterial mats are common in salt marsh and marine intertidal regions (Figure 19.17). The top layer of the mat usually contains filamentous cyanobacteria, and phototrophic purple (generally purple sulfur) bacteria lie underneath (Figure 19.17). The lower layers of these communities are typically black, due to active sulfate reduction which generates the sulfide necessary for development of the purple bacteria.

The purple and green bacteria are one of the most diverse groups of bacteria known and are of interest for a wide variety of reasons. They have provided extremely useful systems for studying fundamental aspects of the photosynthetic process (see Chapter 16), and are of great evolutionary significance (see Chapter 18). Their ecological roles may also be important, and their extreme diversity challenges the bacterial taxonomist. Much more work on this interesting group of bacteria would be desirable.

19.2 Cyanobacteria

The **cyanobacteria** comprise a large and heterogeneous group of phototrophic organisms that used to be referred to as *blue-green algae*. Cyanobacteria differ in fundamental ways from purple and green bacteria, most notably in the fact they are *oxygenic* phototrophs. Classification of cyanobacteria has improved in the last decade, due to the increased availability of pure cultures. Because of the procaryotic nature of these organisms, they are called the *cyanobacteria*, instead of blue-green algae, to indicate clearly that they are not eucaryotic phototrophs.

Structure The morphological diversity of the cyanobacteria is considerable (Figure 19.18). Both unicellular and filamentous forms are known, and considerable variation within these morphological types occurs. Recent taxonomic studies have divided the cyanobacteria into five major groups: unicellular dividing by binary fission (see Figure 19.18*a*); unicellular dividing by multiple fission (colonial, see Figure 19.18*b*); filamentous containing differentiated cells called heterocysts which function in nitrogen fixation (see Figures 19.18*d* and 19.20); filamentous nonheterocystous forms (see Figure 19.18*c*); and branching filamentous types (see Figure 19.18*e*). Cyanobacterial cells range in size from those of typical bacteria (0.5 to 1 μm in diameter) to cells as large as 60 μm in diameter (in the species *Oscillatoria princeps*). The latter has the largest cells known among the procaryotes.

The cyanobacteria differ in fatty acid composition from all other procaryotes. Other bacteria contain almost exclusively saturated and monounsaturated fatty acids (one double bond), but the cyanobacteria frequently contain unsaturated fatty acids with two or more double bonds.

The fine structure of the cell wall of some cyanobacteria is similar to that of Gram-negative bacteria and peptidoglycan can be detected in the walls. Many cyanobacteria produce extensive mucilaginous envelopes, or sheaths, that bind groups of cells or filaments together (see for example Figure 19.18*a*). The photosynthetic lamellar membrane system is often complex and multilayered (see Figure 16.14*b*), although in some of the simpler cyanobacteria the lamellae are regularly arranged in concentric circles around the periphery of the cytoplasm (Figure 19.19). Cyanobacteria have only one form of chlorophyll, chlorophyll *a*, and all of them also have characteristic biliprotein pigments, **phycobilins** (see Figure 16.14), which function as accessory pigments in photosynthesis. One class of phycobilins, the *phycocyanins*, are blue, absorbing light maximally at 625 to 630 nm, and together with the green chlorophyll *a*, are responsible for the blue-green color of the bacteria. However, some cyanobacteria produce *phycoerythrin*, a red phycobilin absorbing light maximally at 570 to 580 nm, and bacteria possessing this pigment are red or brown in color. Even more confusing, the eucaryotic red algae (Rhodophyta) are red because of phycoerythrin, but some species have phycocyanin instead and are blue-green.

Structural variation Among the cytoplasmic structures seen in many cyanobacteria are **gas vesicles** (see Section 3.11), which are especially common in species that live in open waters (planktonic species). Their function is probably to provide the organism with flotation (see Figure 3.55) so that it may remain where there is most light. Some

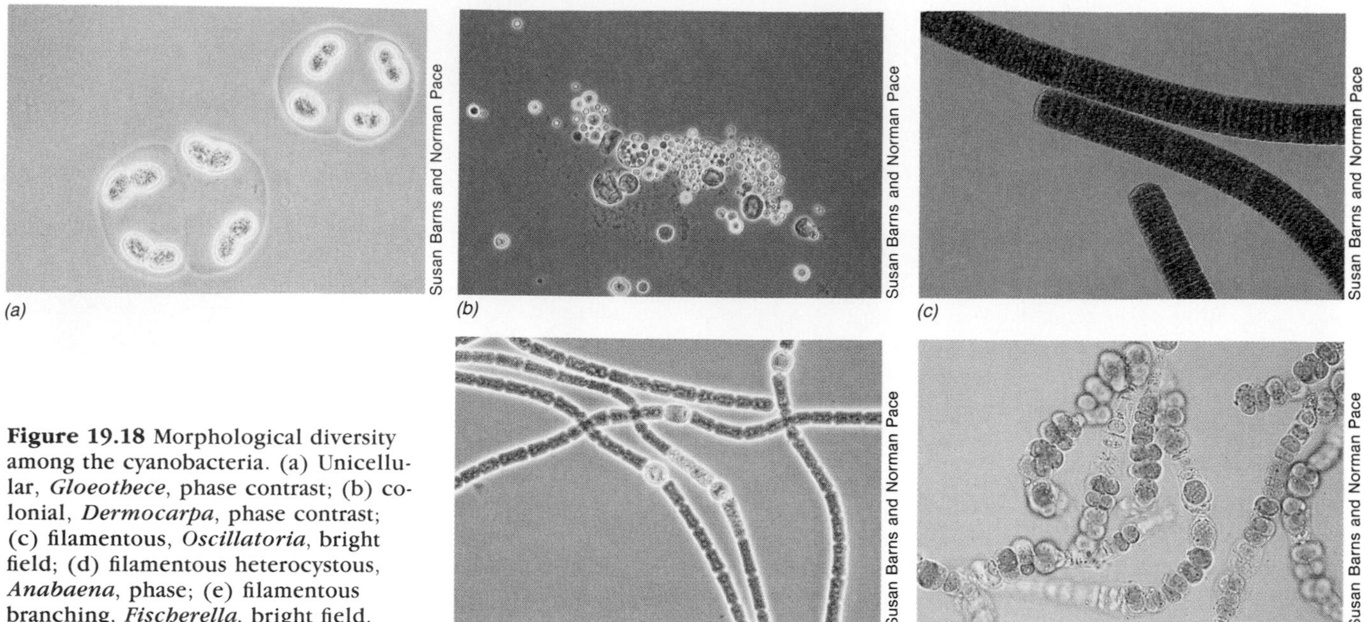

Figure 19.18 Morphological diversity among the cyanobacteria. (a) Unicellular, *Gloeothece*, phase contrast; (b) colonial, *Dermocarpa*, phase contrast; (c) filamentous, *Oscillatoria*, bright field; (d) filamentous heterocystous, *Anabaena*, phase; (e) filamentous branching, *Fischerella*, bright field.

cyanobacteria form **heterocysts**, which are rounded, seemingly more or less empty cells, usually distributed individually along a filament or at one end of a filament (Figure 19.20). Heterocysts arise from vegetative cells and are the sole sites of *nitrogen fixation* in heterocystous cyanobacteria. The heterocysts have intercellular connections with adjacent vegetative cells, and there is mutual exchange of materials between these cells, with products of photosynthesis moving from vegetative cells to heterocysts and products of nitrogen fixation moving from heterocysts to vegetative cells. Heterocysts are low in phycobilin pigments and *lack* photosystem II, the oxygen-evolving photosystem. They are also surrounded by a thickened cell wall containing large amounts of glycolipid, which serves to slow the diffusion of O_2 into the cell. Because of the reductive nature of nitrogen fixation and the oxygen lability of the enzyme dinitrogenase (see Section 16.24), it seems likely that the heterocyst, by maintaining an anaerobic environment, makes possible stabilization of the nitrogen-fixing system in organisms that are not only aerobic but also oxygen-producing. Indeed, some nonheterocystous filamentous cyanobacteria produce nitrogenase and fix nitrogen in normal vegetative cells if they are grown anaerobically. However, a few unicellular cyanobacteria of the sheath-forming *Gloeothece* type (Figure 19.18*a*) do not produce heterocysts but nevertheless fix nitrogen aerobically. Marine *Oscillatoria (Trichodesmium)* also fix nitrogen without heterocysts, and seem to produce a series of cells in the center of the filament that lack photosystem II activity; nitrogen fixation apparently occurs in this O_2-nonproducing region.

A structure called the **cyanophycin granule** can be seen in electron micrographs of many cyanobacteria. This structure is a simple polymer of aspartic acid, with each aspartate residue containing an arginine molecule:

asp—asp—asp—asp—asp—
arg arg arg arg arg

and can constitute up to 10 percent of the cell mass. It appears that this co-polymer serves as a nitrogen storage product in many cyanobacteria, and when nitrogen in the environment becomes deficient, this polymer is broken down and used. The phycobilin pigments can also constitute a major portion of the cell mass, up to 10 percent, and likewise serve as a nitrogen storage material, being broken

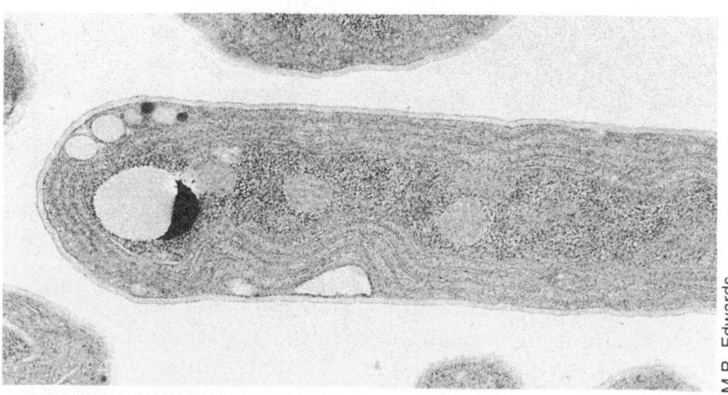

Figure 19.19 Electron micrograph of a thin section of the cyanobacterium *Synechococcus lividus*.

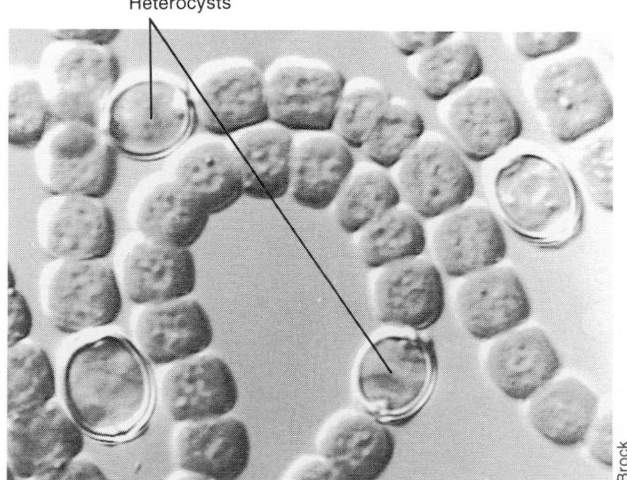

Figure 19.20 Heterocysts of the cyanobacterium *Anabaena* sp. by Nomarski interference contrast. Heterocysts are the sole site of nitrogen fixation in heterocystous cyanobacteria.

down under nitrogen starvation. Because of this, nitrogen-starved cyanobacteria often appear green instead of blue-green in color. Cyanophycin can apparently also serve as an *energy* reserve in cyanobacteria. Arginine, derived from cyanophycin, can be hydrolyzed to yield ornithine, with the production of ATP through the action of the enzyme *arginine dihydrolase*:

$$\text{arginine} + \text{ADP} + \text{P}_i + \text{H}_2\text{O} \rightarrow$$
$$\text{ornithine} + 2\text{NH}_3 + \text{CO}_2 + \text{ATP}$$

Arginine dihydrolase is present in many cyanobacteria and may function there as a source of ATP for maintenance purposes during dark periods.

Many, but by no means all, cyanobacteria exhibit *gliding motility*; flagella have never been found. The rate of gliding varies from so slow that it is not directly observable in the microscope to over 10 μm/s in *Oscillatoria princeps*. Gliding occurs only when the cell or filament is in contact with a solid surface or with another cell or filament. In some cyanobacteria gliding is not a simple translational movement but is accompanied by rotations, reversals, and flexings of filaments. Most gliding forms exhibit directed movement in response to light (phototaxis); it is usually positive although negative movement from bright light may also occur. Chemotaxis (see Section 3.9) may occur as well.

Among the filamentous cyanobacteria, fragmentation of the filaments often occurs by formation of **hormogonia** (Figure 19.21*a* and *b*), which break away from the filaments and glide off. In some species resting spores or **akinetes** (Figure 19.21*c*) are formed, which protect the organism during periods of darkness, drying, or freezing. These are cells with thickened outer walls; they germinate through the breakdown of the outer wall and outgrowth of a new vegetative filament. However, even the vegetative cells of many cyanobacteria are relatively resistant to drying or low temperatures.

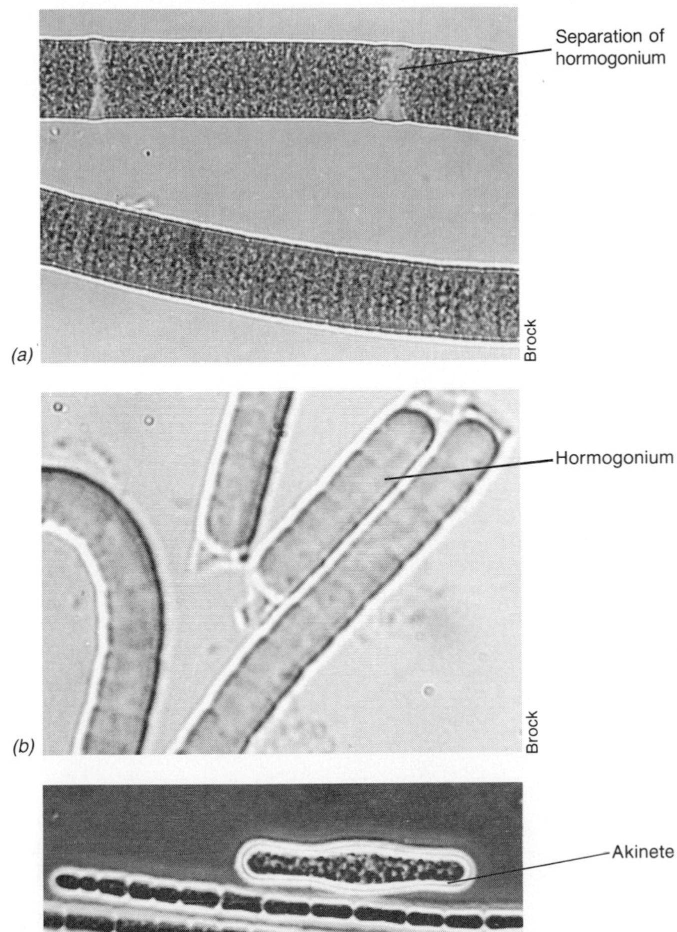

Figure 19.21 Structural differentiation in filamentous cyanobacteria. (a) Initial stage of hormogonium formation in *Oscillatoria*. Notice the empty spaces where the hormogonium is separating from the filament. (b) Hormogonium of a smaller *Oscillatoria* species. Notice that the cells at both ends are rounded. Nomarski interference contrast microscopy. (c) Akinete (resting spore) of *Anabaena* sp. by phase contrast.

Physiology The nutrition of cyanobacteria is simple. Vitamins are not required, and nitrate or ammonia is used as nitrogen source. Nitrogen-fixing species are also common. Most species tested are obligate phototrophs, being unable to grow in the dark on organic compounds. Some cyanobacteria can assimilate simple organic compounds such as glucose and acetate if light is present. Apparently they are unable to make ATP by oxidation of organic compounds, but if ATP is provided by means of photophosphorylation, organic compounds can be utilized as carbon sources. However, some species, mainly filamentous forms, can grow in the dark on glucose or other sugars, using the organic material as both carbon and energy source. As we discussed in Section 16.4, a number of cyanobacteria can carry out anoxygenic photosynthesis using only photosystem I, when sulfide is present in the environment.

Ecology and evolution Cyanobacteria are widely distributed in nature in terrestrial, freshwater, and marine habitats. In general they are more tolerant to environmental extremes than are eucaryotic algae and are often the dominant or sole photosynthetic organisms in hot springs (Table 9.4), saline lakes, and other extreme environments. Many members are found on the surfaces of rocks or soil. In desert soils subject to intense sunlight, cyanobacteria often form extensive crusts over the surface, remaining dormant during most of the year and growing during the brief winter and spring rains. Other cyanobacteria are common inhabitants of the soils of greenhouses. In shallow marine bays, where relatively warm seawater temperatures exist, cyanobacterial mats of considerable thickness may form. Freshwater lakes, especially those that are fairly rich in nutrients, may develop blooms of cyanobacteria. A few cyanobacteria are symbionts of liverworts, ferns, and cycads; a number are found as the photosynthetic component of lichens. In the case of the water fern *Azolla* (see Section 17.24), it has been shown that the cyanobacterial endophyte (a species of *Anabaena*) fixes nitrogen that becomes available to the plant. Some of the photosynthetic symbionts of corals and other invertebrates also are cyanobacteria.

Base compositions of DNA of a variety of cyanobacteria have been determined. Those of the unicellular forms vary from 35 to 71 percent GC, a range so wide as to suggest that this group contains many members with little relationship to each other. On the other hand, the values for the heterocyst formers vary much less, from 39 to 47 percent GC. Phylogenetically, cyanobacteria group along morphological lines in most cases (see Section 18.5). Filamentous heterocystous and nonheterocystous species form distinct groups, as do the branching forms. However, unicellular cyanobacteria are phylogenetically highly diverse, as representatives show evolutionary relationships to several different groups and show little phylogenetic relationships to one another.

The evolutionary significance of cyanobacteria is discussed in Section 18.1, and it is nearly certain that these organisms were the first oxygen-evolving photosynthetic organisms and were responsible for the initial conversion of the atmosphere of the earth from anaerobic to aerobic. Fossil evidence of cyanobacteria in the Precambrian is good, and there is evidence that cyanobacteria occupied vast areas of the earth in those ancient times. Although quantitatively much less significant today, the cyanobacteria still exist in considerable numbers and can be the dominant microbial member of certain communities, such as microbial mats (see Section 17.4).

19.3 Prochlorophytes

Prochlorophytes are procaryotic phototrophs that contain chlorophyll *a* and *b* but do *not* contain phycobilins. Prochlorophytes therefore resemble both cyanobacteria (because they are procaryotic and produce chlorophyll *a*) and the green plant chloroplast (because they contain chlorophyll *b* instead of phycobilins). *Prochloron*, the first prochlorophyte discovered, has served to stimulate renewed attempts to trace the evolutionary origin of the chloroplast using molecular approaches to phylogeny (see Section 18.2). Phenotypically, *Prochloron* is the best candidate yet discovered for the type of organism that led, following endosymbiotic events (see Section 18.1), to the green plant chloroplast.

Unfortunately, progress in the study of *Prochloron* has been rather slow because the organism has not yet been grown in laboratory culture. *Prochloron* is found in nature as a symbiont of marine invertebrates (didemnid ascidians) and all studies of the organism to date have relied on material collected from natural samples. Cells of *Prochloron* expressed from the cavities of didemnid tissue are roughly spherical in morphology (Figure 19.22), 8–10 μm in diameter. Electron micrographs of thin sections (Figure 19.22) show that *Prochloron* has an extensive thylakoid membrane system similar to that observed in the chloroplast (see Figure 3.64). Evidence that *Prochloron* is really a bacterium (eubacterium) is the presence of muramic acid in the cell walls, indicating that peptidoglycan is present.

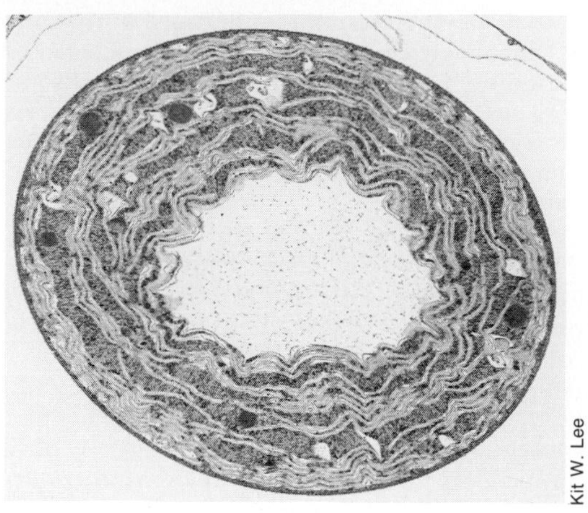

Figure 19.22 Electron micrograph of the prochlorophyte, *Prochloron*. Note the extensive intracytoplasmic membranes (thylakoids).

The ratio of chlorophyll *a*:chlorophyll *b* in *Prochloron* is about 4:1, slightly higher than the value of 2 or 3:1 typical of green algae. The carotenoids of *Prochloron* are similar to those of cyanobacteria, predominantly β-carotene and zeaxanthin. The GC base ratio of different samples of *Prochloron* isolated from different ascidians vary from 31–41 percent, indicating a fair bit of genetic heterogeneity. Different species of *Prochloron* probably exist, but confirmation of this must await laboratory culture and study of pure strains.

Another prochlorophyte has been discovered that can be easily grown in pure culture. This prochlorophyte, a filamentous organism (Figure 19.23), grows as one of the dominant phototrophic bacteria in several shallow Dutch lakes. Because of its filamentous morphology, this prochlorophyte has been given the genus name *Prochlorothrix*. Like *Prochloron, Prochlorothrix* is procaryotic and contains chlorophylls *a* and *b* and lacks phycobilins. The ratio of chlorophyll *a* to chlorophyll *b* is somewhat higher in *Prochlorothrix* than in *Prochloron* and the thylakoid mem-

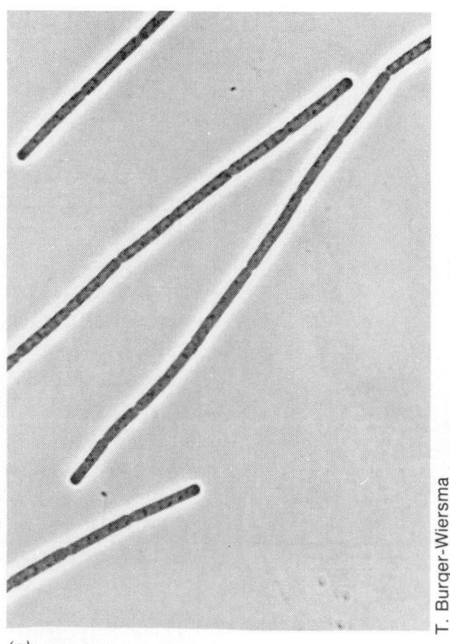

(a)

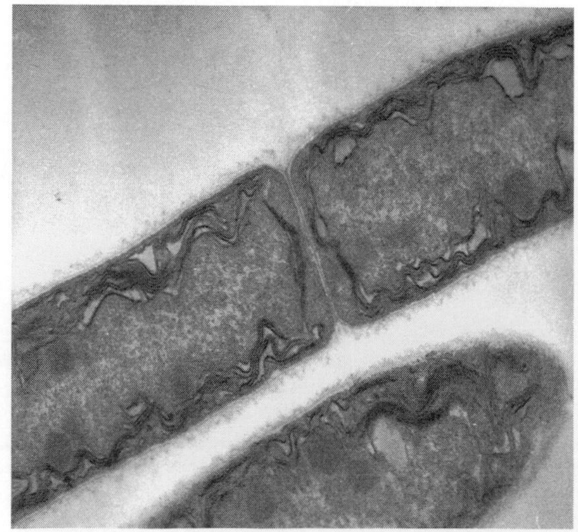

(b)

Figure 19.23 Phase and electron micrographs of the filamentous prochlorophyte, *Prochlorothrix*. (a) Phase contrast. (b) Electron micrograph of thin section showing arrangement of membranes.

branes are much less developed than in *Prochloron* (compare Figures 19.22 and 19.23).

Recalling our discussion of endosymbiosis (see Section 18.1) the evolutionary significance of prochlorophytes should be apparent. Until the discovery of prochlorophytes it was always assumed that the chloroplast originated from endosymbiotic association of cyanobacteria with a primitive eucaryotic cell. However, this hypothesis has never been scientifically satisfying for one major reason: how did the green plant chloroplast evolve the pigment composition it has today if it originated from a *cyanobacterial* endosymbiont that contained phycobilins instead of chlorophyll *b*? The hypothesis that *prochlorophytes* instead of cyanobacteria were the ancestors of the green plant chloroplast eliminates this major point of contention. However, comparative sequencing of 16S rRNA (see Section 18.4) does not show *Prochloron* or *Prochlorothrix* to be the immediate ancestors of the green plant chloroplast. Instead, prochlorophytes, cyanobacteria and the green plant chloroplast all share a common ancestor. Thus, although prochlorophytes have arisen from the same evolutionary roots as the chloroplast and are phenotypically highly related to chloroplasts, it appears that the closest free-living ancestor of chloroplasts, if such an organism is still extant, has defied isolation and culture.

19.4 Lithotrophs: The Nitrifying Bacteria

We discussed in Chapter 16 the conceptual basis of lithotrophy. Various lithotrophic bacteria are known, but they are all physiologically united by their ability to utilize *inorganic* electron donors as energy sources. Most lithotrophs are also capable of autotrophic growth and in this way share a major physiological trait with phototrophic bacteria and cyanobacteria. The best studied lithotrophs are those capable of oxidizing reduced sulfur and nitrogen compounds, and the hydrogen-oxidizing bacteria, and we focus here and in the next two sections on these groups.

Bacteria able to grow lithotrophically at the expense of reduced inorganic nitrogen compounds are called **nitrifying bacteria**. Several genera are recognized on the basis of morphology and the particular steps in the oxidation sequences that they carry out (Table 19.5). No lithotrophic organism is known that will carry out the complete oxidation of ammonia to nitrate; thus **nitrification** of ammonia in nature results from the sequential action of two separate groups of organisms, the **ammonia-oxidizing bacteria**, the **nitrosifyers** (Figure 19.24), and the **nitrite-oxidizing bacteria**, the true nitrifying (nitrate-producing) bacteria (Figure 19.25). Some heterotrophic bacteria and fungi will oxidize ammonia completely to nitrate, but the rate of the process is much less than that accomplished by the lithotrophic nitrifying bacteria, and may not be significant ecologically. Historically, the nitrifying bacteria were the first organisms to be shown to grow lithotrophically; Winogradsky showed that they were able to produce organic matter and cell mass when provided with CO_2 as sole carbon source.

Many of the nitrifying bacteria have remarkably complex internal membrane systems (see Figures 19.24 and 19.25), although not all genera have such membranes.

Biochemistry Ammonia oxidation occurs best at *high* pH values, and it is inferred that this is because the enzyme

Table 19.5 Characteristics of the nitrifying bacteria

Characteristics	Genus	DNA (mole % GC)	Habitats
Oxidize ammonia:			
Gram-negative rods, motile (polar flagella) or nonmotile; peripheral membrane systems:			
Short rods, pointed ends	*Nitrosomonas* (group 1)	51	Soil, sewage
Long rods, rounded ends	*Nitrosomonas* (group 2)	47–49	Fresh water
Long rods, rounded ends, chains	*Nitrosomonas* (group 3)	48–49	Marine
Large cocci, motile (peritrichous flagella) or nonmotile; membrane	*Nitrosococcus*	50–51	Soil, marine
Spirals, motile (peritrichous flagella); no obvious membrane system	*Nitrosospira*	54	Soil
Pleomorphic, lobular, compartmented cells; motile (petitrichous flagella)	*Nitrosolobus*	53–56	Soil
Oxidize nitrite:			
Short rods, reproduce by budding, occasionally motile (single subterminal flagellum); membrane system arranged as a polar cap	*Nitrobacter*	60–61	Soil, fresh water, marine
Long, slender rods, nonmotile; no obvious membrane system	*Nitrospina*	57	Marine
Large cocci, motile (one or two subterminal flagella); membrane system randomly arranged in tubes	*Nitrococcus*	61	Marine

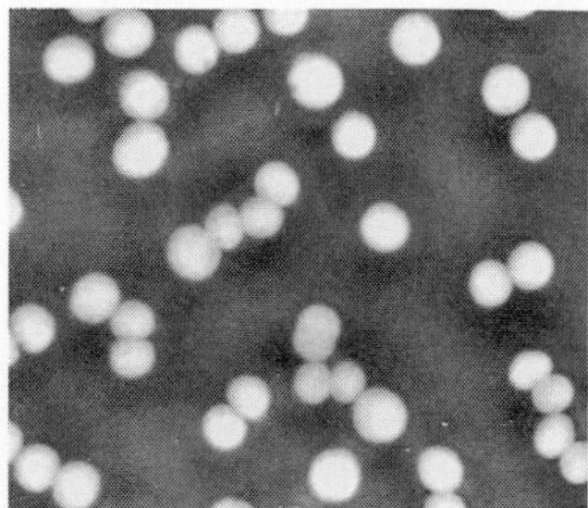

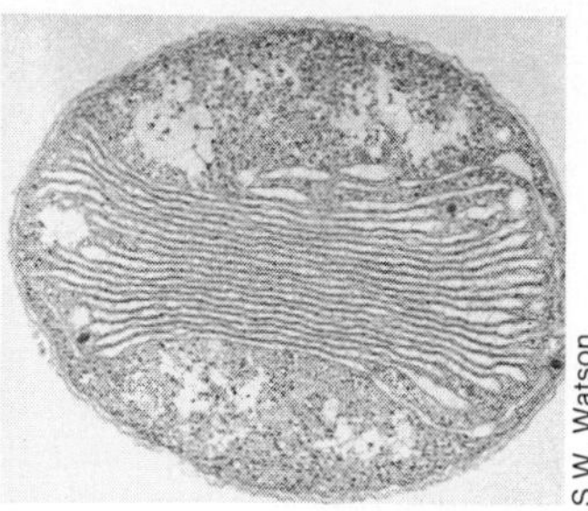

Figure 19.24 Phase-contrast photomicrograph and electron micrograph of the nitrosofying bacterium *Nitrosococcus oceanus*.

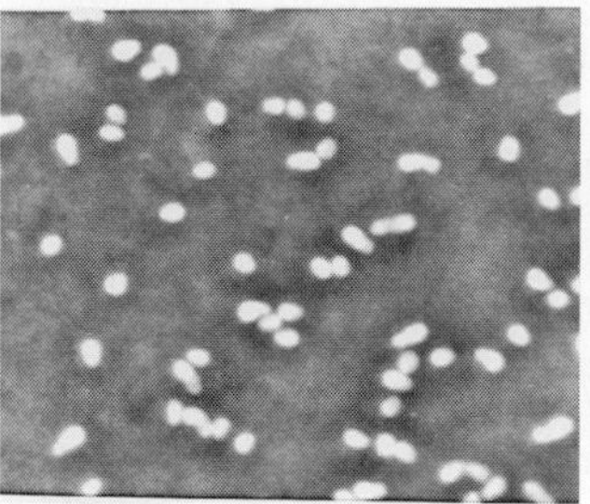

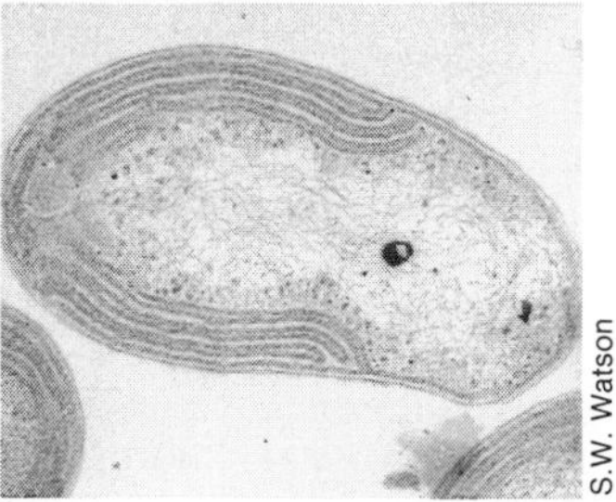

Figure 19.25 Phase-contrast photomicrograph and electron micrograph of the nitrifying bacterium *Nitrobacter winogradskyi*.

involved in the initial step uses the nonionized NH_3 rather than the NH_4^+ ion. Molecular oxygen is required for ammonia oxidation, the initial step involving a *mixed-function oxygenase*, which uses NADH as electron donor. The first product of ammonia oxidation is *hydroxylamine*, NH_2OH, and no energy is generated in this step (energy is actually used up via the oxidation of NADH) (Figure 19.26). Hydroxylamine is then oxidized to nitrite and ATP formation occurs at this step via electron-transport phosphorylation through a cytochrome system (Figure 19.26).

Only a single step is involved in the oxidation of nitrite to nitrate by the nitrite-oxidizing bacteria (Figure 19.26). This reaction is carried out by a *nitrite oxidase* system, the electrons being transported to O_2 via cytochromes, with ATP being generated by electron-transport phosphorylation. The energy available from the oxidation of nitrite to nitrate is only 18.1 kcal/mole, which is sufficient for the formation of two ATP. However, careful measurements of molar growth yields suggest that only *one* ATP is produced per each NO_2^- oxidized to NO_3^-. As we discussed in Section 16.7, the generation of reducing power (NADPH) for the reduction of CO_2 to organic compounds comes from ATP-driven reversed electron transport reactions, since the NO_3^-/NO_2^- reduction potential is too high to reduce $NADP^+$ directly.

Ecology Although rarely present in large numbers, the nitrifying bacteria are widespread in soil and water. They can be expected to be present in highest numbers in habitats where considerable amounts of ammonia are present, such as sites where extensive protein decomposition occurs (ammonification). Nitrifying bacteria develop especially well in lakes and streams which receive inputs of treated (or even untreated) sewage, because sewage effluents are generally high in ammonia. The classic picture downstream from a sewage outfall, as was shown in Figure 17.11, indicates high ammonia concentrations close to the outfall, and falling concentrations downstream as nitrification occurs and nitrate builds up. Because O_2 is required for ammonia oxidation, ammonia tends to accumulate in anaerobic habitats, and in stratified lakes nitrifying bacteria may develop especially well at the thermocline, where both ammonia and O_2 are present. Nitrification results in acidification of the habitat, due to the buildup of nitric acid (the situation is analogous to the buildup of sulfuric acid through the activities of sulfur-oxidizing bacteria). Since nitrous acid can form at acid pH values as a result of ammonia oxidation, the accumulation of this toxic (and mutagenic; see Table 5.6) agent in acidic environments can result in inhibition of further nitrification. In general, nitrification is much more extensive in neutral and alkaline than in acidic habitats, due partly to avoidance of nitrous acid accumulation and partly to the requirement of the nonionized NH_3 by the ammonia-oxidizing bacteria.

Nitrosofying bacteria

1. $NH_3 + O_2 + NADH + H^+ \rightarrow NH_2OH + H_2O + NAD^+$
2. $NH_2OH + O_2 \rightarrow NO_2^- + H_2O + H^+$

Sum: $NH_3 + 2O_2 + NADH \rightarrow NO_2^- + 2H_2O + NAD^+$
$\Delta G^{0\prime} = -65$ kcal/mole

Nitrifying bacteria

$NO_2^- + \frac{1}{2}O_2 \rightarrow NO_3^-$
$\Delta G^{0\prime} = -18$ kcal/mole

Figure 19.26 Reactions involved in the oxidation of inorganic nitrogen compounds.

Culture Despite the fact that the nitrifying bacteria have been known since the nineteenth century, work on these organisms has not been extensive, at least in part because of the difficulty of obtaining pure cultures and the rather slow growth of nitrifying bacteria in general. Enrichment cultures of nitrifying bacteria are readily obtained by using selective media containing ammonia or nitrite as electron donor, and bicarbonate as sole carbon source. Because of the inefficiency of growth of these organisms, visible turbidity may not develop even after extensive nitrification has occurred, so that the best means of monitoring growth is to assay for production of nitrite (with ammonia as electron donor) or disappearance of nitrite (with nitrite as electron donor). After one or two weeks incubation, chemical assays will reveal whether a successful enrichment has been obtained, and attempts can be made to get pure cultures by streaking on agar plates. The same problem arises when purifying nitrifying bacteria by colony selection on agar that was discussed with the methane-oxidizing bacteria: many common heterotrophs present in the enrichment will grow rapidly on the traces of organic matter present in the medium. Thus purification must be done by repeated picking and streaking, followed by testing to be certain that heterotrophic contaminants have not been taken. In addition, many nitrifying bacteria, especially the ammonia oxidizers, appear to be inhibited by the traces of organic material present in most agar preparations. Culture of these organisms on solid media sometimes requires the use of extensively washed, high purity agar, or the completely *inorganic* solidifying agent, **silica gel**. The latter compound was first used by Winogradsky to successfully grow the nitrosifyers on solid media. Because most nitrifying bacteria do not grow on organic compounds, the simplest way of checking for contamination is to inoculate a presumed nitrifying isolate into media containing organic matter. If growth occurs, it can be assumed that the isolate was not a nitrifier.

Most of the nitrifying bacteria appear to be obligate lithotrophs. *Nitrobacter* is an exception, however, and is able to grow, although slowly, on acetate as sole carbon and energy source. None of these bacteria require growth factors. Although the group is somewhat heterogeneous morphologically, it seems to be more homogeneous than the sulfur-oxidizing or photosynthetic bacteria, as shown by the fairly narrow range of DNA base composition (Table 19.5) and the similar biochemical properties.

19.5 Lithotrophs: Sulfur- and Iron-Oxidizing Bacteria

The ability to grow lithotrophically on reduced sulfur compounds is a property of a diverse group of microorganisms. Many of these organisms have not, however, been cultured, and precise information on some of them is minimal. Only four genera, *Thiobacillus, Thiomicrospira, Thermothrix*, and *Sulfolobus* have been consistently cultured so that our discussion here is restricted to these genera. With the exception of *Sulfolobus,* which is an archaebacterium and is discussed in Section 19.35, the remaining sulfur lithotrophs are eubacteria and phylogenetically fall into the purple bacteria group (see Sections 18.4 and 18.5). Two broad ecological classes of sulfur-oxidizing bacteria can be discerned, those living at neutral pH and those living at acid pH. Many of the forms living at acid pH also have the ability to grow lithotrophically using ferrous iron as electron donor. We discussed the biogeochemistry of these acidophilic sulfur- and iron-oxidizing bacteria in Sections 16.9 and 16.10.

Thiobacillus The genus *Thiobacillus* contains those Gram-negative, polarly flagellated rods that are able to derive their energy from the oxidation of elemental sulfur, sulfides, and thiosulfate (Table 19.6). Some pseudomonads have been confused with members of the genus *Thiobacillus*. Morphologically, *Thiobacillus* (Figure 19.27) is similar to *Pseudomonas*, but the two differ in that *Thiobacillus* can grow lithotrophically using reduced sulfur compounds. A few thiobacilli can also grow heterotrophically with organic electron donors, and under such conditions resemble pseudomonads. Further, a few pseudomonads (both marine and freshwater) can oxidize thiosulfate to tetrathionate:

$$2S\text{—}SO_3^{2-} \rightarrow {}^{2-}O_3S\text{—}S\text{—}S\text{—}SO_3^{2-} + 2e^-$$

and although they cannot grow solely from the energy obtained in this reaction, they do obtain some slight growth

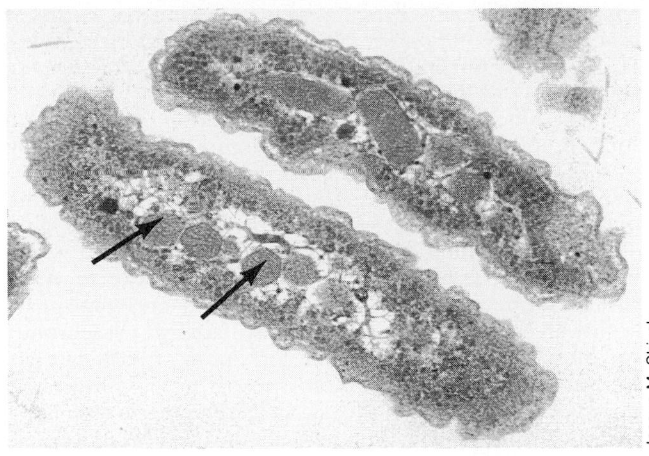

Figure 19.27 Transmission electron micrograph of the lithotrophic sulfur oxidizer *Thiobacillus neapolitanus*. Note the polyhedral bodies (carboxysomes) distributed throughout the cell (arrows). See text for details.

Table 19.6 Physiological characteristics of sulfur-oxidizing bacteria

	Lithotrophic electron donor	**Range of pH for growth**	**DNA (mole % GC)**
***Thiobacillus* species growing poorly in organic media**:			
1. *T. thioparus*	H_2S, sulfides, S^0. $S_2O_3^{2-}$	6–8	62–66
2. *T. denitrificans**	H_2S, S^0, $S_2O_3^{2-}$	6–8	63–67
3. *T. neapolitanus*	S^0, $S_2O_3^{2-}$	5–8	55–57
4. *T. thiooxidans*	S^0	2–5	52–57
5. *T. ferrooxidans*	S^0, sulfides, Fe^{2+}	1.5–4	53–60
***Thiobacillus* species growing well in organic media**:			
1. *T. novellus*	$S_2O_3^{2-}$	6–8	66–68
2. *T. intermedius*	$S_2O_3^{2-}$	3–7	64
Filamentous sulfur lithotrophs			
Beggiatoa	H_2S, $S_2O_3^{2-}$	6–8	37–43
Thiothrix	H_2S	6–8	—
Other genera			
Thiomicrospira†	$S_2O_3^{2-}$, H_2S	6–8	36–44
*Thermothrix**	H_2S, $S_2O_3^{2-}$, SO_3^-	6.5–7.5	—
*Sulfolobus***	H_2S, S^0	1–4	41

**Facultative aerobes; use NO_3^- as electron acceptor anaerobically.*
†One of its species is capable of using NO_3^- anaerobically
**Archaebacterium.

advantage from this process when growing on sugars or other organic compounds. The mechanism by which growth is stimulated during the oxidation of thiosulfate to tetrathionate is not known.

The biochemical steps in the oxidation of various sulfur compounds are summarized in Figure 19.28. Oxidation of sulfide and sulfur involves first the reaction of these substances with sulfhydryl groups of the cell such as glutathione with formation of a sulfide-sulfhydryl complex (Figure 19.28). The sulfide ion is then oxidized to sulfite (SO_3^{2-}) by the enzyme *sulfide oxidase*. There are two ways in which sulfite can be oxidized to produce high-energy phosphate bonds. In one, sulfite is oxidized to sulfate by a cytochrome-linked sulfite oxidase, with the formation of ATP via electron-transport phosphorylation. This pathway is universally present in thiobacilli. In the second, sulfite reacts with AMP; two electrons are removed, and adenosine phosphosulfate (APS) is formed. The electrons removed in either case are transferred to O_2 via the cytochrome system, leading to the formation of high-energy phosphate bonds through electron-transport phosphorylation. In addition, a substrate-level phosphorylation occurs, APS reacting with P_i and being converted to ADP and sulfate. With the enzyme adenylate kinase, two ADP can be converted to one ATP and one AMP. Thus oxidation of two sulfite ions via this system produces three ATP, two via electron-transport phosphorylation and one via substrate-level phosphorylation. The significance of the APS pathway in thiobacilli in general is unclear, however, since it has only been found in a few *Thiobacillus* species.

Thiosulfate ($S_2O_3^{2-}$), which can be viewed as a sulfide of sulfite (SSO_3^{2-}), is split into sulfite and sulfur (Figure 19.28). The sulfite is oxidized to sulfate with production of ATP, and the other sulfur atom is converted into insoluble elemental sulfur. Thus, when they oxidize thiosulfate, the thiobacilli produce elemental sulfur but when they oxidize sulfides they do not. The elemental sulfur produced can itself be oxidized later when the thiosulfate supply is exhausted. If thiosulfate is low, elemental sulfur does not accumulate, probably being oxidized as soon as it is formed.

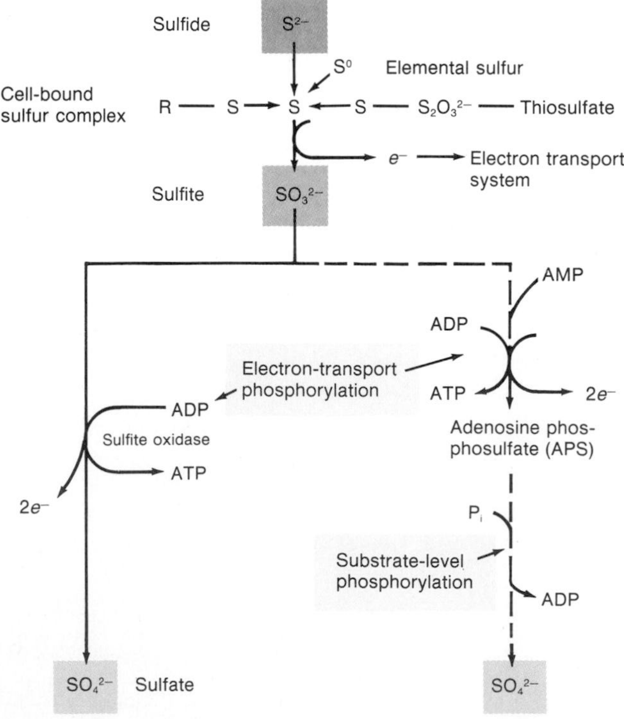

Figure 19.28 Steps in the oxidation of different compounds by thiobacilli. The sulfite oxidase pathway is thought to account for the majority of sulfite oxidized.

Enrichment cultures of thiobacilli are quite easy to prepare. Sulfur or thiosulfate is added to a basal salts medium with NH_4^+ as a nitrogen source and bicarbonate as a carbon source, and the medium is then inoculated with a sample of soil or mud. After aerobic incubation at room temperature for a few days, the liquid should appear turbid owing to the growth of thiobacilli. If thiosulfate has been used, droplets of amorphous sulfur will also be present. If elemental sulfur is used, many of the bacteria may be attached to the insoluble sulfur crystals, and the edges of such crystals should be examined for the presence of bacteria. The bacteria also attach to the crystals of metal sulfides such as PbS, HgS, or CuS if these are used as energy sources. From thiosulfate enrichment cultures, pure cultures might be obtained by streaking onto a solidified medium of the same composition. However, pure cultures are fairly difficult to obtain, since growth is usually slow and heterotrophic contaminants grow on small amounts of organic matter released by the thiobacilli. Another problem in obtaining pure cultures of thiobacilli that grow best at neutral pH, such as *T. thioparus*, is that sulfur oxidation leads to sulfuric acid production and a drop in pH, resulting in the death of the culture. Thus highly buffered media (2 to 10 g/liter of a mixture of K_2HPO_4 and KH_2PO_4) and frequent transfers of the culture are necessary. On the other hand, isolation of the acidophilic *T. thiooxidans* is relatively easy, since this organism is resistant to the acid it produces and most heterotrophic contaminants cannot grow at low pH values (2 to 5) where *T. thiooxidans* thrives.

Many isolates of acidophilic thiobacilli can also oxidize ferrous iron. The geochemical aspects of iron oxidation were discussed in Sections 17.16 and 17.17. At acid pH, ferrous iron is not readily oxidized spontaneously, and the acid-tolerant thiobacilli are the main agents in nature for the oxidation of ferrous iron in acidic environments. Those isolates which oxidize both iron and sulfur compounds are currently classified as the species *T. ferrooxidans*, the species *T. thiooxidans* being restricted to those acid-tolerant thiobacilli which cannot oxidize iron.

Thiomicrospira and Thermothrix Members of the genera *Thiomicrospira* and *Thermothrix* are thiosulfate-oxidizing bacteria which grow at neutral pH. *Thiomicrospira*, as its name implies, is a tiny, spiral-shaped bacterium. In fact, this organism is so small that it can be selectively enriched by inoculating a thiosulfate- or hydrogen sulfide-containing medium with the filtrate remaining from filtering a mud slurry through a membrane filter with a pore size of just 0.22 μm. Such a filter would retain organisms the size of *Thiobacillus*. The genus *Thiomicrospira* consists of two species; both are obligate lithotrophs and one species can grow anaerobically with nitrate as electron acceptor. *Thermothrix* is a filamentous bacterium which inhabits hot sulfur springs having a neutral or slightly acidic pH. Cultures of *Thermothrix* grow aerobically or anaerobically (with nitrate as electron acceptor) at temperatures between 55 and 85°C (optimum at 70°C). Like *Sulfolobus, Thermothrix* is capable of heterotrophic growth, and utilizes a variety of organic compounds, including glucose, acetate, and amino acids as electron donors.

Carbon metabolism and carboxysomes Several carbon nutritional classes of lithotrophs are known, from

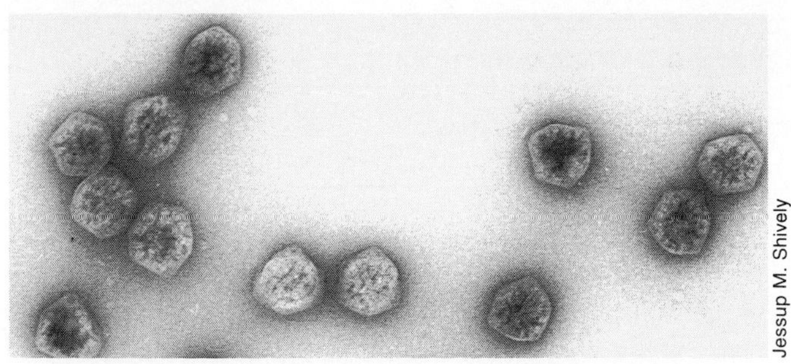

Figure 19.29 Polyhedral bodies (carboxysomes) purified from the lithotrophic sulfur oxidizer *Thiobacillus neapolitanus.*

pure autotrophs to pure heterotrophs. The sulfur lithotrophs discussed thus far, such as *Thiobacillus* are generally capable of *autotrophic* growth with CO_2 as sole carbon source. However, certain sulfur lithotrophs are incapable of autotrophic growth. Organisms such as *Beggiatoa*, certain pseudomonads, and one species of *Thiobacillus, T. perometabolis*, fall into this category. Most of the remaining lithotrophs are true autotrophs and presumably grow autotrophically in nature.

Studies of CO_2 fixation in *Thiobacillus* has shown that reactions of the Calvin cycle (see Section 16.6) are responsible for CO_2 fixation in this lithotroph. Thin sections of cells of various thiobacilli have revealed polyhedral cell inclusions scattered throughout the cell (see Figure 19.27). The inclusions are about 100 nm in diameter (Figure 19.29) and are surrounded by a thin, nonunit membrane. The polyhedral bodies have been referred to as *carboxysomes* and have been shown to consist of the enzyme *ribulose-1,5-bisphosphate carboxylase* (the key enzyme of the Calvin cycle) in crystalline form. Originally thought to be an inactive storage form of the enzyme, carboxysomes are now known to contain active ribulose bisphosphate carboxylase. Presumably the enzyme functions in CO_2 fixation both in the carboxysome as well as in the cytoplasm.

Although the reason for carboxysome formation is not clear, they may simply serve as a means of increasing the amount of ribulose bisphosphate carboxylase in the cell. Since the enzyme is in an insoluble form, the carboxysome is an effective mechanism for concentrating enzyme in the cell without affecting osmolarity. However, whatever their exact function, carboxysomes are common inclusions in a wide variety of autotrophic procaryotes. Carboxysomes have been observed in nitrifying bacteria, various sulfur lithotrophs, and in the cyanobacteria, but not in lithotrophic bacteria that require organic compounds as a carbon source or in phototrophic purple bacteria, many of which can grow heterotrophically. This suggests that carboxysomes are unique to organisms that have specialized in the autotrophic way of life.

19.6 Lithotrophs: Hydrogen-Oxidizing Bacteria

A wide variety of bacteria are capable of growing with H_2 as sole electron donor and O_2 as electron acceptor using the "knallgas" reaction, the reduction of O_2 with H_2:

$$2H_2 + O_2 \rightarrow 2H_2O$$

Many, but not all, of these organisms can also grow autotrophically (using reactions of the Calvin cycle to incorporate CO_2) and are grouped together here as the lithotrophic *hydrogen-oxidizing bacteria.* Both Gram-positive and Gram-negative hydrogen bacteria are known, with the best studied representatives classified in the genus *Pseudomonas* and *Alcaligenes* (Table 19.7). All hydrogen-oxidizing bacteria contain one or more *hydrogenase* enzymes that function to bind H_2 and use it to either produce ATP or serve as reducing power for autotrophic growth.

All hydrogen bacteria are *facultative lithotrophs,* meaning that they can also grow organotrophically with organic compounds as energy sources. This is a major distinction between hydrogen lithotrophs and many sulfur lithotrophs or nitrifying bacteria; most representatives of the latter two groups are *obligate lithotrophs*—growth does not occur in the absence of the inorganic energy source. By contrast, hydrogen lithotrophs can switch between lithotrophic and organotrophic modes of metabolism and presumably do so in nature as nutritional conditions warrant.

Energetics of hydrogen oxidation We briefly described the energetics of hydrogen-oxidizing bacteria in Section 16.8. The majority of hydrogen-oxidizing bacteria are obligate aerobes; a few can grow anaerobically with nitrate as electron acceptor. H_2 is taken up by hydrogenase and electrons shunted through an electron transport chain leading to the establishment of a membrane potential and a proton gradient; ATP is then produced by chemiosmotic mechanisms (see Section 4.12 for discussion of chemiosmosis and related topics). In *Alcaligenes eutrophus* (Figure 19.30), one of the best studied hydrogen bacteria, two distinct hydrogenases are present, one membrane bound and the other cytoplasmic (soluble). The membrane-bound enzyme is involved in energetics. Following binding of H_2 to the enzyme, hydrogen atoms are transferred from the hydrogenase to a quinone and from there through a series of cytochromes to O_2. Cytochromes of the *c* and a/a_3 types are present in *A. eutrophus.* In other organisms, such as *Paracoccus denitrificans*, a similar sequence is observed but *b*-type cytochromes are also present.

The second hydrogenase of *A. eutrophus* is involved in the production of *reducing power* for autotrophic growth. This soluble hydrogenase takes up H_2 and reduces NAD^+ to NADH directly. NADH can be converted into NADPH (by enzymes called *transhydrogenases*) for use as a reductant in the Calvin cycle, but many hydrogen lithotrophs use NADH itself, instead of NAD*P*H, as the reductant in Calvin cycle reactions. Because the reduction potential of H_2 is so low (−0.42 volts) it can be used to reduce NAD^+ directly in hydrogen bacteria like *A. eutrophus* that contain two hydrogenases. Thus in hydrogen bacteria that contain an NAD^+-reducing hydrogenase, ATP-dependent reverse electron flow reactions (see Section 16.7) are *not* required. Because of this, growth rates and cell yields of organisms like *A. eutrophus* are typically much higher than in lithotrophic bacteria that must use ATP to make NAD(P)H. Many hydrogen bacteria contain only a membrane-bound hydrogenase (see Table 19.7), however, and this enzyme does not couple to the reduction of NAD^+. In these organisms reverse electron flow mechanisms presumably operate as a means of generating NADH.

Physiology and isolation of hydrogen bacteria Most hydrogen bacteria prefer microaerophilic conditions when growing lithotrophically on H_2 because hydrogenases are oxygen-sensitive enzymes. Typically, oxygen levels of about 10 percent (half the level in air) support best growth. *Nickel* must be present in the medium for growth of hydrogen bacteria, because virtually all hydrogenases have Ni^{2+} in the active enzyme. A few hydrogen bacteria also fix

Table 19.7 Differential characteristics of species of hydrogen-oxidizing bacteria

Genus/species	Denitrification	Growth on fructose	Motility	DNA (mole % GC)	Other characteristics
Gram-negative					
Alcaligenes eutrophus	+	+	+	66	Membrane-bound and cytoplasmic hydrogenases
Alcaligenes ruhlandii	−	−	+	−	Membrane-bound and cytoplasmic hydrogenases
Aquaspirillum autotrophicum	−	−	+	61	Only membrane-bound hydrogenase present
Pseudomonas carboxydovorans	−	−	+	60	Only membrane-bound hydrogenase present; also oxidizes CO
Pseudomonas flava	−	+	+	67	Colonies are bright yellow
Seliberia carboxydohydrogena	−	?	+	58	Also oxidizes CO
Paracoccus denitrificans	+	+	−	66	Only membrane-bound hydrogenase present. Strong denitrifyer
Gram-positive					
Bacillus schlegelii	−	−	+	66	Produces endospores
Arthrobacter sp.	−	+	−	70	Only membrane-bound hydrogenase present
Mycobacterium gordonae	−	?	−	−	Acid fast, colonies yellow to orange

molecular N_2, and when growing on N_2 the organisms are quite oxygen sensitive because the enzyme nitrogenase needed for the reduction of molecular nitrogen (see Section 16.24) is an oxygen-sensitive enzyme. Unlike nitrifying bacteria or methanotrophs, hydrogen-oxidizing bacteria generally lack internal membranes; apparently the cytoplasmic membrane of hydrogen bacteria can integrate sufficient levels of hydrogenase to maintain high rates of H_2 oxidation.

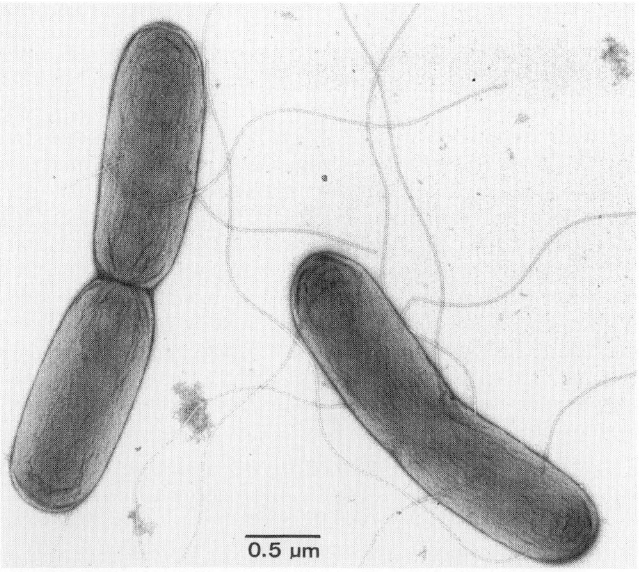

0.5 μm

Frank Mayer

Figure 19.30 Transmission electron micrograph of negatively stained cells of the hydrogen-oxidizing lithotroph *Alcaligenes eutrophus*.

Hydrogen bacteria are generally grown autotrophically, however all species will incorporate organic compounds if added to the medium. In some hydrogen bacteria organic compounds such as glucose or acetate strongly repress the synthesis of both Calvin cycle enzymes and hydrogenase. Some hydrogen bacteria will also grow on carbon monoxide, CO, as energy source, with electrons from the oxidation of CO to CO_2 entering the electron transport chain to drive ATP synthesis. CO-oxidizing bacteria grow autotrophically using Calvin cycle reactions to fix the CO_2 generated from the oxidation of CO.

Ecology of hydrogen bacteria The ecological importance of aerobic hydrogen-oxidizing bacteria is difficult to assess. This is because in natural environments there is great competition for H_2 among various anaerobic bacteria. Methanogens, sulfate-reducing bacteria, phototrophic bacteria, and certain denitrifying bacteria all compete for H_2 produced from fermentation. And, since hydrogen levels in actively methanogenic or sulfide-rich ecosystems are frequently very low, it is doubtful whether large amounts of H_2 ever reach aerobic habitats from sources in anoxic environments. It is perhaps for this reason that all aerobic hydrogen bacteria also grow organotrophically; a "backup" system is required if the primary energy source is not always available.

Little is known of the ecology of CO-oxidizing hydrogen bacteria although CO consumption by these organisms is probably ecologically significant. CO is oxidized both aerobically and anaerobically by various bacteria. Under anaerobic conditions CO is oxidized by certain photosynthetic bacteria, methanogens and sulfate-reducers. However, the most significant releases of CO (primarily from automobile exhaust, incomplete combustion of fossil fuels, and the catabolism of lignin) occur in oxic environments.

Thus aerobic CO-oxidizing bacteria in upper layers of soil probably represent the most significant sink for CO. Carbon monoxide-oxidizers are Gram-negative rods of uncertain taxonomic or phylogenetic affiliation and are capable of growth on organic compounds as well as on H_2 or CO. Unfortunately, because of the technical problems associated with growing organisms on a highly poisonous odorless gas such as CO, these organisms have not been intensively studied.

19.7 Methylotrophs and Methanotrophs

Methane, CH_4, is found extensively in nature. It is produced in anaerobic environments by methanogenic bacteria (see Section 17.12) and is a major gas of anaerobic muds, marshes, anaerobic zones of lakes, the rumen, and the mammalian intestinal tract. Methane is the major constituent of natural gas and is also present in many coal formations. It is a relatively stable molecule; but a variety of bacteria, the **methanotrophs**, oxidize it readily, utilizing methane and a few other one-carbon compounds as electron donors for energy generation and as sole sources of carbon. These bacteria are all aerobes, and are widespread in nature in soil and water. They are also of a diversity of morphological types, seemingly related only in their ability to oxidize methane. There has been considerable interest in methane-utilizing bacteria in recent years because of the possibility of using this simple and widely available energy source to produce bacterial protein as a food or feed supplement (single-cell protein).

In addition to methane, a number of other one-carbon compounds are known to be utilized by microorganisms. A list of these compounds is given in Table 19.8. From a biochemical viewpoint, these compounds share a key characteristic: they contain no carbon-carbon bonds, so that all carbon-carbon bonds of the cell must be synthesized de novo. Organisms that can grow using only one-carbon compounds are generally called **methylotrophs**. Many, but not all methylotrophs are also methanotrophs. From the viewpoint of carbon assimilation, methanotrophs and methylotrophs have something in common with autotrophs (Chapter 16), which also use a carbon compound lacking a carbon-carbon bond, CO_2. The two groups differ, however, in that the methylotrophs require a carbon compound *more reduced* than CO_2.

It is important to distinguish between methylotrophs and methanotrophs. A wide variety of bacteria are known which will grow on methanol, methylamine or formate, but not methane, and these bacteria are members of various genera of heterotrophs: *Hyphomicrobium, Pseudomonas, Bacillus, Vibrio*. In contrast, methanotrophs are unique in that they can grow not only on some of the more oxidized one-carbon compounds, but also on methane. The methane-oxidizing bacteria possess a specific enzyme system, *methane monooxygenase* for the introduction of an oxygen atom into the methane molecule, leading to the formation of methanol. It should be noted that although the methane-oxidizing bacteria can also oxidize more oxidized one-carbon compounds such as methanol and formate, initial isolation from nature requires the use of methane as a sole electron donor, since if one of these other one-carbon compounds is used in initial enrichment, a nonmethanotrophic methylotroph will almost certainly be isolated. All methane-oxidizing bacteria appear to be obligate C-1 utilizers, unable to utilize compounds with carbon-carbon bonds. Many nonmethanotrophic methylotrophs are facultative methylotrophs, also being able to utilize organic acids, ethanol, and sugars.

Methane-oxidizing bacteria are also unique among procaryotes in possessing relatively large amounts of **sterols**. As we noted in Chapter 3, sterols are found in eucaryotes as a functional part of the membrane system, but seem to be absent from most procaryotes. In methanotrophs, sterols may be an essential part of the complex internal membrane system (see below) that is involved in methane oxidation.

Classification An overview of the classification of methanotrophs and methylotrophs is given in Table 19.9. These bacteria were initially distinguished on the basis of morphology and formation of resting stages, but it was then found that they could be divided into two major groups depending on their internal cell structure and carbon assimilation pathway. Type I organisms assimilate one-carbon compounds via a unique pathway, the **ribulose monophosphate cycle**, whereas Type II organisms assimilate C-1 intermediates via the **serine pathway**. The requirement

Table 19.8 Substrates used by methylotrophic bacteria*

Substrates used for growth	Substrates oxidized, but not used for growth (cometabolism)
Methane, CH_4	Ammonium, NH_4^+
Methanol, CH_3OH	Ethylene, $H_2C{=}CH_2$
Methylamine, CH_3NH_2	Chloromethane, CH_3Cl
Dimethylamine, $(CH_3)_2NH$	Bromomethane, CH_3Br
Trimethylamine, $(CH_3)_3N$	Higher hydrocarbons (ethane, propane)
Tetramethylammonium, $(CH_3)_4N^+$	
Trimethylamine N-oxide, $(CH_3)_3NO$	
Trimethylsulfonium, $(CH_3)_3S^+$	
Formate, $HCOO^-$	
Formamide, $HCONH_2$	
Carbon monoxide, CO	
Dimethyl ether, $(CH_3)_2O$	
Dimethyl carbonate, $CH_3OCOOCH_3$	

A single isolate does not use all of the above, but at least one methylotrophic bacterium has been reported to oxidize each of the listed compounds.

Table 19.9 Some characteristics of methane-oxidizing bacteria

Organism	Morphology	Flagellation	Resting stage	Internal membranes*	Carbon assimilation pathway†	DNA (mole % GC)
Methylomonas	Rod	Polar	Cystlike body	I	Ribulose monophosphate	50–54
Methylobacter	Rod	Polar	Thick-walled cyst	I	Ribulose monophosphate	50–54
Methylococcus	Coccus	None	Cystlike body	I	Ribulose monophosphate	62
Methylosinus	Rod or vibrioid	Polar tuft	Exospore	II	Serine	63–66
Methylocystis	Vibrioid	None	Poly-β-hydroxy-butyrate-rich cyst	II	Serine	—
Methylobacterium	Rod	Nonmotile	None	II	Serine	58–66

**Internal membranes: Type I, bundles of disk-shaped vesicles distributed throughout the organism; Type II, paired membranes running along the periphery of the cell. See Figure 19.31.*
†See Figures 19.32 and 19.33.

for O_2 as a reactant in the initial oxidation of methane explains why all methanotrophs are obligate aerobes, whereas some organisms using methanol as electron donor can grow anaerobically (with nitrate or sulfate as electron acceptor). Both groups of methanotrophs contain extensive internal membrane systems, which appear to be related to their methane-oxidizing ability. Type I bacteria are characterized by internal membranes arranged as bundles of disk-shaped vesicles distributed throughout the organism (Figure 19.31*b*) whereas Type II bacteria possess paired membranes running along the periphery of the cell (Figure 19.31*a*). Type I methanotrophs are also characterized by a lack of a complete tricarboxylic acid cycle (the enzyme *α-ketoglutarate dehydrogenase* is absent), whereas Type II organisms possess a complete cycle. In addition, most Type II methanotrophs are capable of fixing molecular nitrogen whereas Type I organisms are not.

Biochemistry of methane oxidation Two aspects of the biochemistry of methanotrophs are of interest, the manner of oxidation of methane to CO_2 (and how this is coupled to ATP synthesis), and the manner by which one-carbon compounds are assimilated into cell material. The overall pathway of methane oxidation involves stepwise, two-electron oxidations:

$$\underset{\textit{Methane}}{CH_4} \xrightarrow{-30\text{ kcal}} \underset{\textit{Methanol}}{CH_3OH} \xrightarrow{-46\text{ kcal}} \underset{\textit{Formaldehyde}}{HCHO} \xrightarrow{-51\text{ kcal}}$$

$$\underset{\textit{Formate}}{HCOOH} \xrightarrow{-57\text{ kcal}} \underset{\textit{Bicarbonate}}{HCO_3^-}$$

The initial step in the oxidation of methane involves an enzyme called **methane monooxygenase** which is a mixed-function oxygenase. As we discussed in Section 16.22, oxygenase enzymes catalyze the incorporation of oxygen from O_2 into carbon compounds and seem to be widely involved in the metabolism of hydrocarbons. Oxygenases require a source of reducing power, usually NADH, but in *Methylosinus*, where the process has been most thoroughly studied, electrons come not from NADH but from cytochrome *c*. This cytochrome *c* is involved in recycling of electrons from methanol dehydrogenase (and possibly from

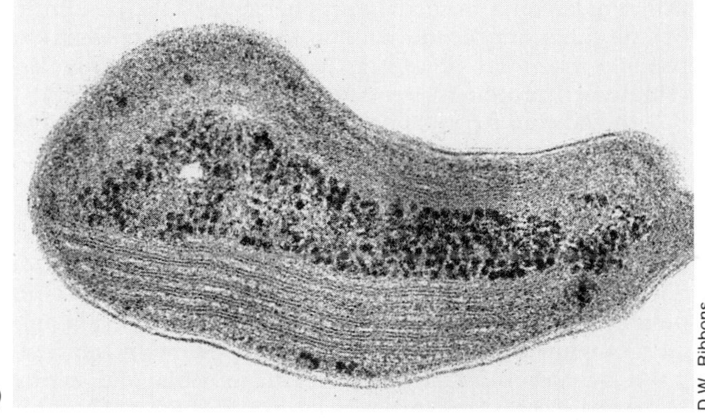

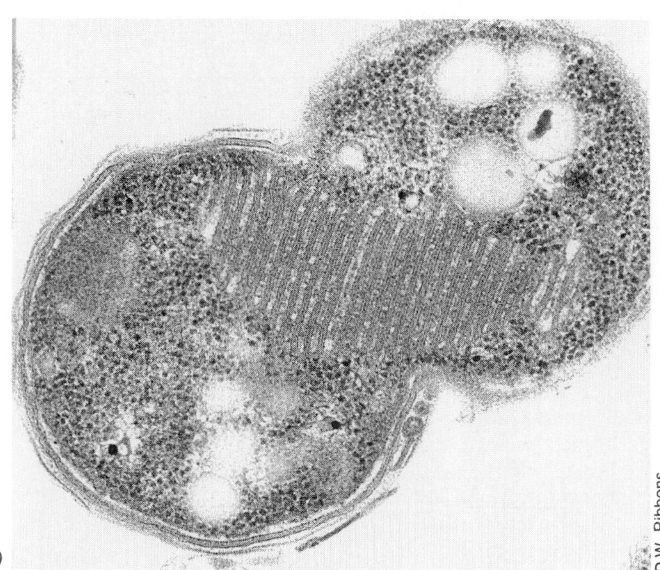

Figure 19.31 Electron micrographs of methane-oxidizing bacteria. (a) *Methylomonas methanica*, illustrating type II membrane system. (b) *Methylococcus capsulatus*, illustrating type I membrane system.

formaldehyde dehydrogenase). In this scheme, ATP synthesis could occur by electron-transport phosphorylation during the reduction of cytochrome *c* by either the methanol or formaldehyde dehydrogenases. This is possible because the reduction potentials of the $HCHO/CH_3OH$ and $HCOOH/HCHO$ couples are −182 and −450 mV, respectively, whereas the reduction potential of cytochrome *c* is +310 mV. No ATP synthesis occurs during the first step, the oxidation of methane to methanol, and this is consistent with the fact that growth yields of methane-oxidizing bacteria are the same whether methane or methanol is used as substrate. Thus, although considerable energy is potentially available in the oxidation of CH_4 to CH_3OH (about 30 kcal/mole), this energy is not available to organisms, apparently because no biochemical mechanism is available for conserving the energy of methane oxidation.

Biochemistry of one-carbon assimilation As noted above, all Type I methanotrophs possess the ribulose monophosphate pathway for carbon assimilation, whereas Type II methanotrophs have the serine pathway.

The serine pathway is outlined in Figure 19.32. It is not only present in Type II methanotrophs, but is also the pathway for C-1 incorporation in facultative methylotrophs (*Hyphomicrobium* and *Pseudomonas*, for example). In this pathway, a two-carbon unit, acetyl-CoA, is synthesized from one molecule of formaldehyde and one molecule of CO_2. The pathway requires the introduction of reducing power and energy in the form of *two* molecules each of NADH and ATP for each acetyl-CoA synthesized.

The ribulose monophosphate pathway, present in Type I methanotrophs, is outlined in Figure 19.33. It is more efficient than the serine pathway in that *all* of the carbon atoms for cell material are derived from formaldehyde, and since formaldehyde is at the same oxidation level as cell material no reducing power is needed. The ribulose monophosphate pathway requires the introduction of energy in the form of *one* molecule of ATP for each molecule of 3-phosphoglyceric acid synthesized (Figure 19.33).

Consistent with the lower energy requirements of the ribulose monophosphate pathway, the cell yield of Type I organisms from a given amount of methane is *higher* than the cell yield of Type II organisms. It seems clear that if a methane-oxidizing bacterium were to be used as a source of single-cell protein, that an organism with a ribulose monophosphate pathway would be preferable, since cell yields on methane are higher. However, since no energy is gained from the methane to methanol step, it is not certain that there is any advantage of producing single-cell protein from methane as opposed to methanol. Methanol, not being a gas, is much more convenient to use industrially than methane, so that methanol may be the preferred energy source for single-cell protein.

Ecology and isolation Methane-oxidizing bacteria are widespread in aquatic and terrestrial environments, being found wherever stable sources of methane are present. Methane produced in the anaerobic portions of lakes rises

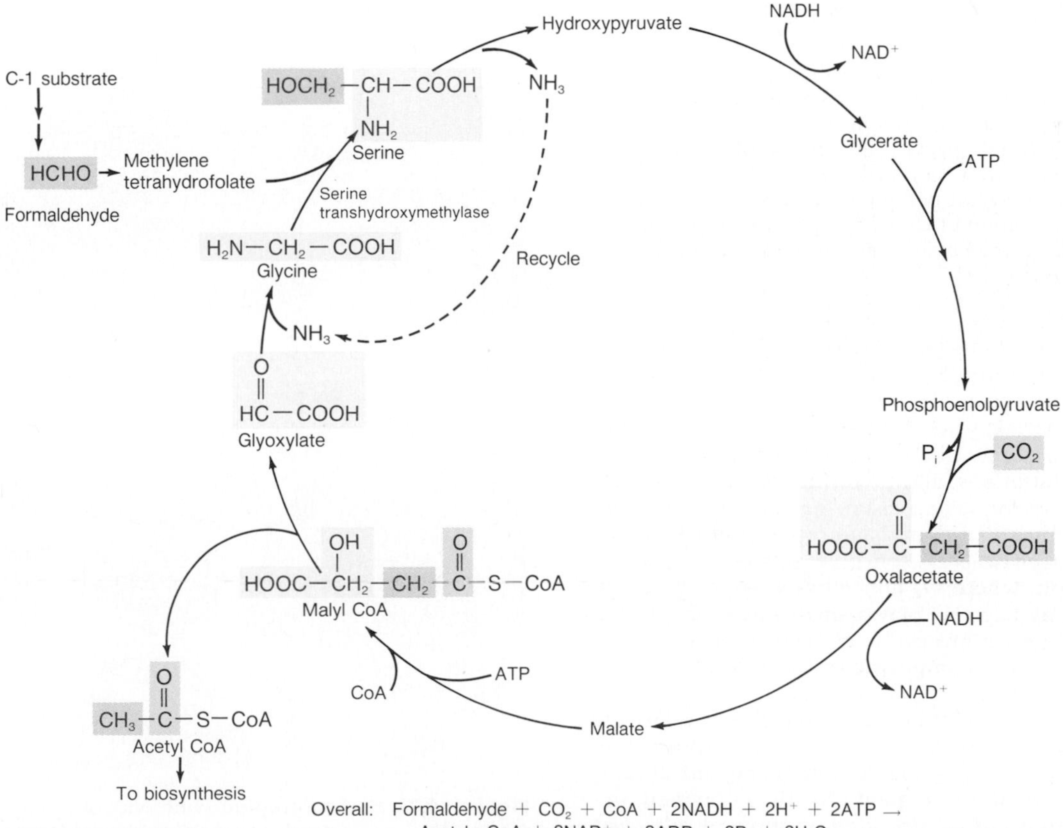

Figure 19.32 The serine pathway for the assimilation of C-1 units into cell material by Type II methylotrophic bacteria. The product of the pathway, acetyl-CoA, is used as the starting point for making new cell material. The key enzyme of the pathway is *serine transhydroxymethylase*.

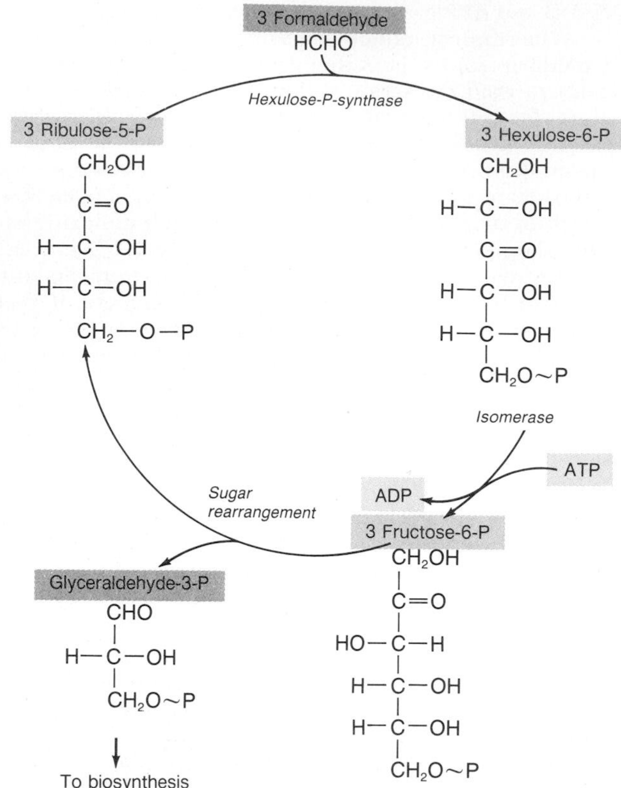

Figure 19.33 The ribulose monophosphate pathway for assimilation of one-carbon compounds, as found in Type I methylotrophic bacteria. The complete name of the hexulose sugar is D-erythro-L-glycero-3-hexulose-6-phosphate. Three formaldehydes are needed to carry the cycle to completion, with the net result being one molecule of glyceraldehyde-3-P. Regeneration of ribulose-5-phosphate and formation of phosphoglyceraldehyde occur via a series of pentose phosphate reactions.

through the water column, and methane oxidizers are often concentrated in a narrow band at the thermocline, where methane from the anaerobic zone meets oxygen from the aerobic zone. There is some evidence that although methane oxidizers are obligate aerobes, they are sensitive to O_2 at normal concentrations, and prefer microaerophilic habitats for development. One reason for this O_2 sensitivity may be that many aquatic methanotrophs simultaneously fix N_2, and nitrogenase is O_2 sensitive, so that optimal development occurs where O_2 concentrations are reduced. Methane-oxidizing bacteria play a small but probably important role in the carbon cycle, converting methane derived from anaerobic decomposition back into cell material (and CO_2).

The initial enrichment of methanotrophs is relatively easy, and all that is needed is a mineral salts medium over which an atmosphere of 80 percent methane and 20 percent air is maintained. Once good growth is obtained, purification is carried out by streaking on mineral salts agar plates, which are incubated in a jar with the methane-air mixture. Colonies appearing on the plates will be of two types, common heterotrophs growing on traces of organic matter in the medium, which appear in 1 to 2 days, and methane oxidizers, which appear after about a week. The colonies of many methane oxidizers are pink in color. Colonies of methane oxidizers should be picked when small, and purification by continued picking and restreaking is essential. It should be ephasized that cultures of methane oxidizers are often contaminated with common heterotrophs, or with methanol oxidizers, so that careful attention should be paid to the purity of an isolate before any detailed studies are undertaken.

Methanotrophs are able to oxidize ammonia, although they apparently cannot grow lithotrophically using ammonia as sole electron donor. The methane oxygenase also functions as an ammonia oxygenase, and a competitive interaction between the two substrates exists. For this reason, ammonia is generally toxic to methane oxidizers, and the preferred nitrogen source is nitrate. It has been speculated that perhaps the methane-oxidizing bacteria could have arisen from the nitrifying bacteria via a mutation causing the conversion of ammonia oxidase to methane oxidase. The fact that both groups of bacteria have elaborate internal membrane systems (see Section 19.4) supports such a theory.

Methanotrophic symbionts of animals A symbiotic association between methanotrophic bacteria and marine mussels has recently been discovered. The mussels live in the vicinity of hydrocarbon seeps in the Gulf of Mexico where methane is released in substantial amounts. Intact mussels as well as isolated mussel gill tissue consume methane at high rates in the presence of O_2. In the gill tissue of the mussel coccoid-shaped bacteria are present in high numbers (Figure 19.34). The bacterial symbionts con-

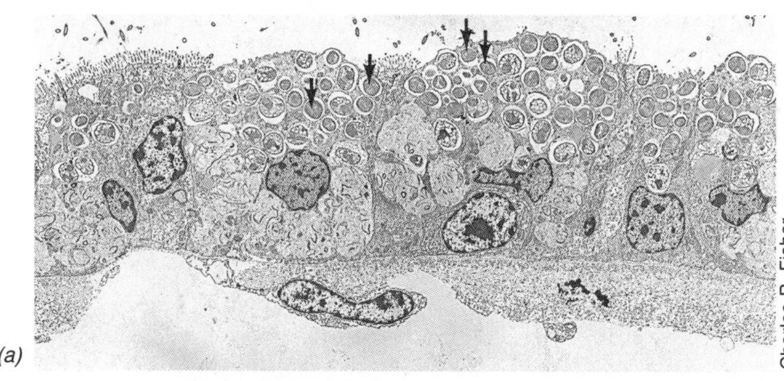

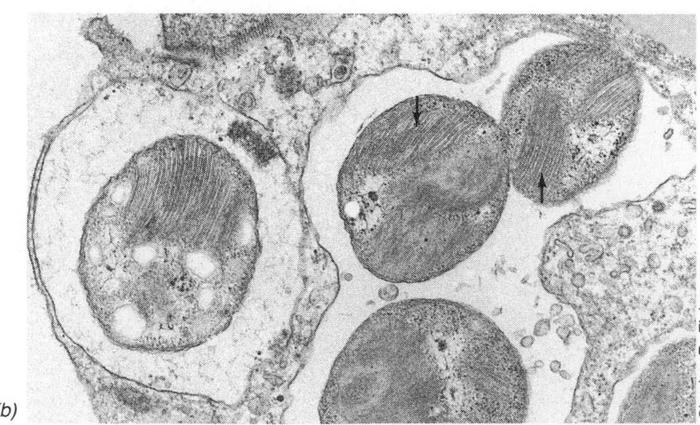

Figure 19.34 Methanotrophic symbionts of marine mussels. (a) Electron micrograph of a thin section at low magnification of gill tissue of a marine mussel which lives near hydrocarbon seeps in the Gulf of Mexico. Note the symbiotic methanotrophs (arrows) in the tissues. (b) High magnification view of gill tissue showing Type I methanotrophs. Note membrane bundles (arrows).

tain stacks of intracytoplasmic membranes (Figure 19.34*b*), typical of Type I methanotrophs. The symbionts are found in vacuoles within animal cells near the gill surface, which probably ensures an effective gaseous exchange with seawater. The oxidation of methane by mussel gill tissue is strictly O_2 dependent and totally inhibited by acetylene, a known poison of biological methane oxidation. This is consistent with the hypothesis that methane oxidation is actually carried out by the methanotrophic symbiont.

Stable isotope studies of mussel tissue suggest that methane serves as the major food source for the animal. Using $^{13}C/^{12}C$ analyses (see Section 17.5), it was shown that both gill tissues and deep tissues of the mussel had $^{13}C/^{12}C$ isotopic compositions nearly identical to that of the methane from geochemical sources being consumed from its environment. Presumably methane assimilated by the methanotrophs is distributed throughout the animal by the excretion of carbon compounds by the methanotrophs. The methanotrophic symbiosis is therefore conceptually similar to the symbiosis established between sulfide-oxidizing lithotrophs and hydrothermal vent tube worms and giant clams discussed in Section 17.9.

Animal/bacterial symbioses, such as the methanotrophic mussel symbiosis and the sulfide-oxidizing vent animal symbioses show that procaryotic cells can constitute the basis of a one-step food chain. Furthermore, discovery of these symbioses hint that arrangements between animals and bacteria may be more common than previously suspected. For example, methanotrophic symbioses are also likely to exist in certain hydrothermal vent animal communities where hydrothermal fluid rich in methane is emitted, and in fresh water mudflats and similar environments where biologically produced methane merges with aerobic zones.

19.8 Sulfate- and Sulfur-Reducing Bacteria

Sulfate is used as a terminal electron acceptor under anaerobic conditions by a heterogenous assemblage of bacteria which utilize organic acids, fatty acids, and alcohols as electron donors. Many of these organisms possess the enzyme hydrogenase and are thus able to use H_2 as an electron donor as well. Although morphologically diverse, the **sulfate-reducing bacteria** can be considered a physiologically unified group, in the same manner as the phototrophic or methanogenic bacteria. Ten genera of dissimilatory sulfate-reducing bacteria are currently recognized and they are placed in two broad physiological subgroups as outlined in Table 19.10. The genera in group I, *Desulfovibrio* (Figure 19.35*a*), *Desulfomonas, Desulfotomaculum,* and *Desulfobulbus* (Figure 19.35*c*), utilize lactate, pyruvate, ethanol, or certain fatty acids as carbon and energy sources, reducing sulfate to hydrogen sulfide. The genera in group II, *Desulfobacter* (Figure 19.35*d*), *Desulfococcus, Desulfosarcina* (Figure 19.35*e*), and *Desulfonema* (Figure 19.35*b*), specialize in the oxidation of fatty acids, particularly *acetate*, reducing sulfate to sulfide. The sulfate-reducing bacteria are all obligate anaerobes, and strict anaerobic techniques must be used in their cultivation, although they are not as fastidious with respect to oxygen as the methanogenic bacteria (see Section 19.34).

Sulfate-reducing bacteria are widespread in aquatic and terrestrial environments that become anaerobic due to active microbial decomposition processes. The best known genus is *Desulfovibrio* (Figure 19.35*a*), which is common in aquatic habitats or waterlogged soils containing abundant organic material and sufficient levels of sulfate. *Desulfotomaculum* consists of endospore-forming rods primarily found in soil, and one species is thermophilic. Growth and reduction of sulfate by *Desulfotomaculum* in certain canned foods leads to a type of spoilage called *sulfide stinker*. The remaining genera of sulfate reducers are indigenous to freshwater or marine anaerobic environments; *Desulfomonas* can also be isolated from the intestine.

The sulfate- and sulfur-reducing bacteria represent a rather broad phylogenetic group. By 16S rRNA sequence analysis the endospore-forming sulfate reducer *Desulfotomaculum* groups, as expected, with the *Clostridium* branch of the Gram-positive eubacteria (see Section 18.5 and 19.26). The remaining sulfate reducers group most closely with other Gram-negative bacteria that are phenotypically much different. The sulfate reducers *Desulfovibrio, Desulfobacter*, etc. are phylogenetically related to the bacterial predator *Bdellovibrio* (see Section 19.11) and the gliding myxobacteria (see Section 19.13). By contrast, species of the sulfur-reducing organism *Desulfuromonas* (see below) are very closely related to one another, but are not closely related (in an evolutionary sense) to sulfate-reducing bacteria. In general, then, it appears that the abil-

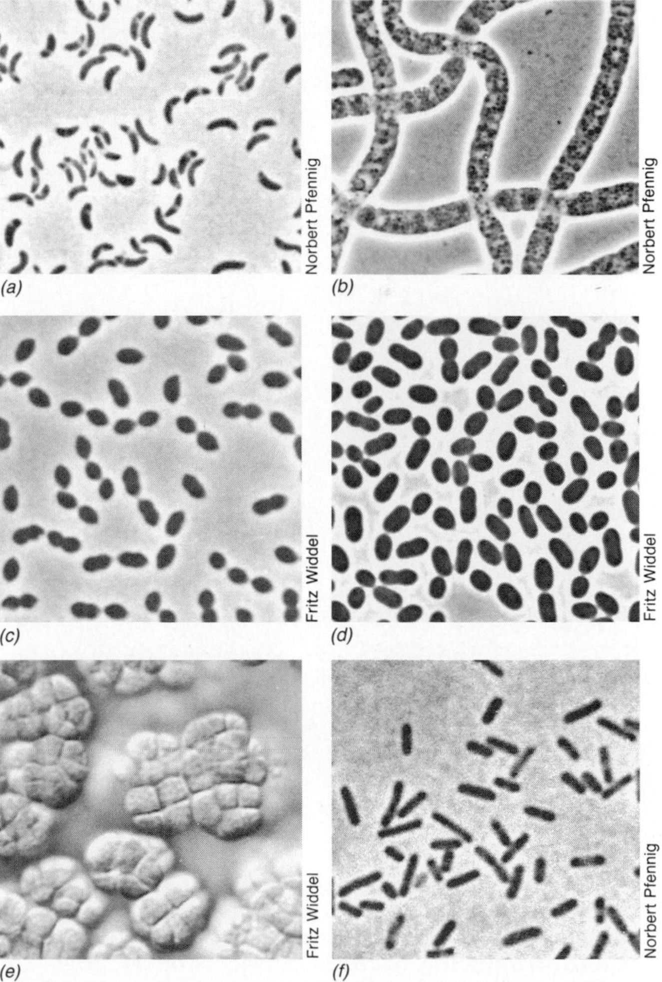

Figure 19.35 Photomicrographs by phase contrast of representative sulfate-reducing (a-e) and sulfur-reducing (f) bacteria. (a) *Desulfovibrio desulfuricans.* (b) *Desulfonema limicola.* (c) *Desulfobulbus propionicus.* (d) *Desulfobacter postgatei.* (e) *Desulfosarcina variabilis* (interference contrast microscopy). (f) *Desulfuromonas acetoxidans.*

Table 19.10 Characteristics of sulfate- and sulfur-reducing bacteria

Genus	Characteristics	DNA (mole % GC)
Group I sulfate reducers: Non-acetate oxidizers		
Desulfovibrio	Polarly flagellated, curved rods, no spores; Gram negative; contain desulfoviridin; seven species recognized, one thermophilic	46–61
Desulfomonas	Long, fat rods; nonmotile; no spores; Gram negative; contain desulfoviridin; habitat, intestinal tract; one species	66–67
Desulfotomaculum	Straight or curved rods; motile by peritrichous or polar flagellation; Gram negative; desulfoviridin absent; produce endospores; four species, one thermophilic; one species capable of utilizing acetate as energy source	37–46
Archaeoglobus	Archaebacterium; extreme thermophile, temperature optimum, 83°C; contains some unique coenzymes of methanogenic bacteria, makes small amount of methane during growth; H_2, formate, glucose, lactate, and pyruvate serve as electron donors, SO_4^{2-}, $S_2O_3^{2-}$, or SO_3^{2-} as electron acceptor; one species (see Section 19.35).	46
Desulfobulbus	Ovoid or lemon-shaped cells; no spores; Gram negative; desulfoviridin absent; if motile, by single polar flagellum; utilizes only propionate as electron donor with acetate + CO_2 as products; one species	59
Group II sulfate reducers: Acetate-oxidizers		
Desulfobacter	Rods; no spores, Gram negative; desulfoviridin absent; if motile, by single polar flagellum; utilizes only acetate as electron donor and oxidizes it to CO_2 via the citric acid cycle; one species	45
Desulfobacterium	Rods, some with gas vesicles, marine; capable of autotrophic growth via the acetyl-CoA pathway, three species	—
Desulfococcus	Spherical cells; nonmotile; Gram negative; desulfoviridin present, no spores; utilizes C_1–C_{14} fatty acids as electron donor with complete oxidation to CO_2; capable of autotrophic growth via the acetyl-CoA pathway; one species	57
Desulfonema	Large filamentous gliding bacteria; Gram positive; no spores; desulfoviridin present or absent; utilizes C_2–C_{12} fatty acids as electron donor with complete oxidation to CO_2; capable of autotrophic growth via the acetyl-CoA pathway; one species	34–41
Desulfosarcina	Cells in packets (sarcina arrangement); Gram negative; no spores; desulfoviridin absent; utilizes C_2–C_{14} fatty acids as electron donor with complete oxidation to CO_2; capable of autotrophic growth via the acetyl-CoA pathway; one species	—
Dissimilatory sulfur reducers:		
Desulfuromonas	Straight rods, single lateral flagellum; no spores; Gram negative; does not reduce sulfate; acetate, succinate, ethanol or propanol used as electron donor; obligate anaerobe; three species	50–63
Campylobacter	Curved, vibrio-shaped rods; polar flagella; Gram negative; no spores; unable to reduce sulfate but can reduce sulfur, sulfite, thiosulfate, nitrate, or fumarate anaerobically with acetate or a variety of other carbon/electron donor sources; facultative aerobe	40–42
Sulfur-dependent archaebacteria (see Section 19.35)		

ity to reduce sulfate or sulfur are phylogenetically distinct physiological markers.

Sulfur reduction A variety of bacteria are known that can reduce elemental sulfur to sulfide but are unable to reduce sulfate to sulfide. These organisms are referred to as **sulfur-reducing bacteria**. Members of the genus *Desulfuromonas* (Figure 19.35*f*) can grow anaerobically by coupling the *oxidation* of substrates such as acetate or ethanol to the *reduction* of elemental sulfur to hydrogen sulfide. However, the ability to reduce elemental sulfur, as well as other sulfur compounds such as thiosulfate, sulfite, or dimethyl sulfoxide (DMSO), is a widespread property of a variety of heterotrophic, generally facultatively aerobic bacteria (for example *Proteus, Campylobacter, Pseudomonas*, and *Salmonella*). *Desulfuromonas* differs from the latter in that it is an obligate anaerobe and utilizes only sulfur as an electron acceptor (see Table 19.10). In addition, certain sulfate-reducing bacteria are capable of substituting sulfur for sulfate as an electron acceptor for anaerobic growth.

Studies of sulfur-rich thermal environments have led to the discovery of a host of extremely thermophilic dissimilatory sulfur reducing archaebacteria (see Section 19.35). Although morphologically distinct and separated into several genera, these organisms are metabolically similar, all carrying out a heterotrophic metabolism with small peptides or glucose as electron donors and elemental sulfur as electron acceptor (a few species grow lithotrophically with H_2 as electron donor). Many of these organisms have growth temperature optima near 90°C or above, and from studies carried out thus far, it is clear that these extremely thermophilic anaerobes represent a distinct branch of dissimilatory sulfur reducers, probably sharing little else with organisms like *Desulfuromonas* except the ability to use sulfur as an electron acceptor. One extremely thermophilic archaebacterium, *Archaeoglobus,* has been shown to reduce sulfate but is incapable of reducing elemental sulfur (see Section 19.35). Clearly, sulfur compounds play a key role in anaerobic respiratory processes in thermal environments.

Physiology The range of electron donors used by sulfate-reducing bacteria is broad. Lactate and pyruvate are almost universally used, and many group I species will utilize malate, formate, and certain primary alcohols (for example, methanol, ethanol, propanol, and butanol). Some strains of *Desulfotomaculum* will utilize glucose but this is rather rare among sulfate reducers in general. Group I sulfate reducers oxidize their energy source to the level of acetate and excrete this fatty acid as an end product. Group II organisms differ from those in group I by their ability to oxidize fatty acids, lactate, succinate, and even benzoate in some cases all the way to CO_2. *Desulfosarcina, Desulfonema, Desulfococcus, Desulfobacterium, Desulfotomaculum*, and certain species of *Desulfovibrio* (see Figure 19.35), are unique in their ability to grow lithotrophically with H_2 as electron donor, sulfate as electron acceptor, and CO_2 as sole carbon source.

In addition to growth using sulfate as electron acceptor, many sulfate-reducing bacteria grow using *nitrate* (NO_3^-) as electron acceptor, reducing NO_3^- to ammonia, NH_3, or can use certain organic compounds for energy generation by fermentative pathways in the complete absence of sulfate or other terminal electron acceptors. The most common fermentable organic compound is *pyruvate*, which is converted via the phosphoroclastic reaction to acetate, CO_2, and H_2. With lactate or ethanol, insufficient energy is available from fermentation, so that sulfate is required, although if a H_2-utilizing organism (for instance, a methanogen) is present in a mixed culture, fermentation of these substrates by the sulfate reducer to acetate and H_2 can occur (the H_2 is immediately consumed by the methanogenic component of such a mixed culture to form CH_4). The value of sulfate as an electron acceptor can be readily demonstrated by comparing growth yields on pyruvate with and without sulfate. In the presence of sulfate, the growth yield is higher, because of the larger amount of energy available when pyruvate utilization is coupled with sulfate reduction.

Biochemistry The enzymology of sulfate reduction was discussed in Section 16.15. Recall that the first step in dissimilatory sulfate reduction is the formation of *adenosine phosphosulfate* (APS) from ATP and sulfate; the enzyme APS reductase then catalyzes the reduction of the sulfate moiety to sulfite (see Figure 16.27). Sulfite is then reduced to sulfide in a series of steps of which little is known biochemically. A sulfite reductase has been identified, but it is not clear whether this enzyme serves to reduce all or only some of the potential intermediates between sulfite and sulfide.

Sulfate-reducing bacteria contain a number of electron transfer proteins including cytochromes of the c_3 and *b*-types (see Section 16.15), ferredoxins, flavodoxins, and hydrogenases, although the actual function of many of the components (other than hydrogenase) remain obscure. Because many sulfate-reducers can use H_2 as an energy source, however, it is clear that the reduction of sulfate is linked to ATP synthesis via electron transport and chemiosmotic mechanisms (see Section 16.15 and Figure 16.28). The utilization of organic compounds as electron donors presumably involves the transfer of electrons through the same electron transport chain, thus generating a membrane potential.

Growth yield studies with sulfate-reducing bacteria have been used to determine the stoichiometry between ATP production and sulfate reduction. Studies with *Desulfovibrio* suggest that one (net) ATP is produced per sulfate reduced to sulfide and that three (net) ATPs are produced per sulfite reduced to sulfide. These values are in line with what is known concerning the conversion of sulfate to sulfite. During this conversion two high energy phosphate bond equivalents are consumed, since ATP and sulfate are the initial reactants and AMP and sulfite the final products (see Section 16.15). Growth yields when H_2 is the electron donor are substantially higher, provided cell carbon is supplied in organic form. Yields of autotrophically grown sulfate reducers are very low.

Certain sulfate reducers will grow completely autotrophically with CO_2 as sole carbon source, H_2 as electron donor, and sulfate as electron acceptor. As discussed in Section 19.9 in connection with the acetogenic bacteria, autotrophic sulfate reducers use the *acetyl-CoA pathway* for net incorporation of CO_2 into cell material (see Table 19.10). In addition, most group II sulfate reducers (those capable of oxidizing acetate to CO_2) *reverse* the steps of the acetyl-CoA pathway to yield CO_2 from acetate, and do not employ the more common citric acid cycle for this purpose.

Isolation and ecology The enrichment of *Desulfovibrio* is relatively easy on a lactate-sulfate medium to which ferrous iron is added. A reducing agent such as thioglycolate or ascorbate is also added. The sulfide formed from sulfate reduction combines with the ferrous iron to form black insoluble ferrous sulfide. This blackening not only serves as an indicator of the presence of sulfate reduction, but the iron ties up and detoxifies the sulfide, making possible growth to higher cell yields. The conventional procedure is to set up liquid enrichments, and after some growth has occurred as evidenced by blackening of the medium, purification is accomplished by streaking onto roll tubes or Petri plates in an anaerobic environment. Although the sulfate reducers are obligate anaerobes, they are not as rapidly inactivated by oxygen as the methane producers, so that streaking on plates can be done in air, provided a reducing agent is present in the medium and plates are quickly incubated in an anaerobic environment. Alternatively, agar "shake tubes" can be used for purification purposes. In the **shake tube method** a small amount of liquid from the original enrichment is added to a tube of molten agar growth medium, mixed thoroughly, and sequentially diluted through a series of molten agar tubes (see Section 17.3). Upon solidification, individual cells distributed throughout the agar form colonies, which can be removed aseptically and the whole process repeated until pure cultures are obtained. Colonies of sulfate-reducing bacteria are recognized by the black deposit of ferrous sulfide, and purified by further streaking.

19.9 Acetogenic Bacteria

Acetogenic bacteria are obligate anaerobes that utilize CO_2 as a terminal electron acceptor, producing acetate as the major product of anaerobic respiration; we discussed the overall formation of acetate by these organisms in Section 16.16. Electrons for the reduction of CO_2 to acetate can come from H_2, a variety of C-1 compounds, sugars, organic acids, alcohols, amino acids and certain nitrogenous bases. However the major unifying thread among acetogens is the pathway of CO_2 reduction. It appears that most, if not all, acetogens convert CO_2 to acetate by the **acetyl-CoA pathway** (see below), and in many acetogens autotrophic growth via the acetyl-CoA pathway also occurs. A list of the major organisms that produce acetate or oxidize acetate via the acetyl-CoA pathway are listed in Table 19.11. Organisms such as *Acetobacterium woodii* and *Clostridium aceticum* can grow both organotrophically or lithotrophically by carrying out a "homoacetic acid" fermentation of sugars or through the reduction of CO_2 to acetate with H_2 as electron donor, respectively. The stoichiometries observed are as follows:

1. $C_6H_{12}O_6 \rightarrow 3CH_3COOH$
2. $2CO_2 + 4H_2 \rightarrow CH_3\,COOH + 2H_2O$

Acetogens ferment glucose via the glycolytic pathway converting glucose into two molecules of pyruvate and 2 molecules of NADH (the equivalent of 4H). From this point, two molecules of acetate are produced as follows:

3. $2\text{ pyruvate} \rightarrow 2\text{ acetate} + 2CO_2 + 4H$

The third acetate of the homoacetate fermentation comes from the reduction of the two molecules of CO_2 generated in reaction 3 above, using the 4 electrons generated from glycolysis *plus* the 4 electrons produced during the oxidation of two pyruvates to two acetates (reaction 3). Starting from pyruvate, then, the overall production of acetate can be written as:

$$2\text{ pyruvate} + 4H \rightarrow 3\text{ acetate}$$

All acetogenic bacteria that produce and excrete acetate in *energy metabolism* are Gram-positive and many are classified in the genus *Clostridium*. A few other Gram-positive and many different Gram-negative bacteria utilize the acetyl-CoA pathway for *autotrophic* purposes, reducing CO_2 to acetate, which then serves as a source of cell carbon. The acetyl-CoA pathway functions in autotrophic growth for those sulfate-reducers capable of $H_2 + CO_2 + SO_4^{=}$—mediated autotrophy (see Section 19.8), and is also used by the methanogenic bacteria, most of which can grow autotrophically on $H_2 + CO_2$ (see Section 19.34 and Table 19.11). By contrast, certain bacteria employ the reactions of the acetyl-CoA pathway primarily in the *reverse* direction, as a means of *oxidizing* acetate. These include acetoclastic methanogens (see Section 19.34) and most type II (acetate-oxidizing) sulfate reducers (see Section 19.8).

Reactions of the acetyl-CoA pathway We will consider the acetyl CoA pathway here from the standpoint of organisms using H_2 as electron donor and CO_2 as electron acceptor for autotrophic growth. Unlike other autotrophic pathways, such as the Calvin cycle (see Section 16.6) and the reverse citric acid cycle (see Section 19.1), the acetyl-CoA pathway of CO_2 fixation is *not* a cycle. Instead it involves the direct reduction of CO_2 to acetyl-CoA; one molecule of CO_2 is reduced to the methyl group and the other molecule of CO_2 is reduced to the carbonyl group of acetate. A key enzyme of the autotrophic pathway is *carbon monoxide (CO) dehydrogenase*. CO dehydrogenase is found in a variety of anaerobic bacteria and also in some aerobic bacteria and catalyzes the following reaction:

$$CO + H_2O \rightleftharpoons CO_2 + 2H^+ + 2e^-$$

CO dehydrogenase from anaerobes contains nickel as a metal cofactor whereas that from aerobes contains molybdenum. The importance of CO dehydrogenase in the acetyl-CoA pathway is that it acts to reduce CO_2 to the level of CO, the CO ending up in the *carbonyl* position of acetate. The reactions of the acetyl-CoA pathway are shown in Figure 19.36. The methyl group of acetate originates from the reduction of CO_2 to a methyl group by a series of enzymatic reactions involving the coenzyme *tetrahydrofolate*. The latter is a common C-1 carrier in all organisms, but plays a special role in acetogenic organisms. CO_2 is first reduced to formate by the enzyme formate dehydrogenase, and then the formate is converted to formyl tetrahydrofolate (Figure

Table 19.11 Organisms employing the acetyl-CoA pathway of CO_2 fixation

I. Acetate synthesis the result of energy metabolism
- *Acetoanaerobium noterae*
- *Acetobacterium woodii*
- *Acetobacterium wieringae*
- *Acetogenium kivui*
- *Clostridium thermoaceticum*
- *Clostridium formicoaceticum*
- *Desulfotomaculum orientis*

II. Acetate synthesis in autotrophic metabolism
- Autotrophic acetogenic bacteria
- Autotrophic methanogenic bacteria (see Section 19.34)
- Autotrophic sulfate-reducing bacteria (see Section 19.8)

III. Acetate oxidation in energy metabolism
- *Reaction*: Acetate $+ 2H_2O \rightarrow 2CO_2 + 8H$
  - Group II sulfate reducers (other than *Desulfobacter*)
- *Reaction*: Acetate $\rightarrow CO_2 + CH_4$
  - Acetoclastic methanogens (*Methanosarcina, Methanothrix*)

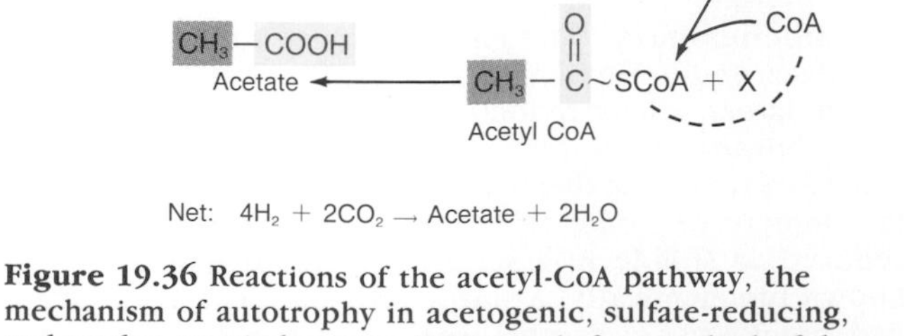

Figure 19.36 Reactions of the acetyl-CoA pathway, the mechanism of autotrophy in acetogenic, sulfate-reducing, and methanogenic bacteria. THF stands for tetrahydrofolate and B_{12} is vitamin B_{12} in an enzyme-bound intermediate. X also represents an enzyme-bound form of the compound. Note how the pathway is direct rather than cyclical.

19.36). The addition of two more pairs of electrons yields *methyl tetrahydrofolate*. The methyl group is then transferred from methyl tetrahydrofolate to an enzyme containing vitamin B_{12} as cofactor, and it is the latter compound that serves as the methyl donor in the final assembly of the acetyl-CoA molecule (Figure 19.36). The methyl corrinoid combines with CO to form an acetyl group bound to an enzyme (coenzyme A is then added to yield the final product, acetyl-CoA, Figure 19.36).

19.9 Budding and/or Appendaged (Prosthecate) Bacteria

This large and rather heterogeneous group contains bacteria which form various kinds of cytoplasmic extrusions: *stalks, hyphae,* or *appendages* (Table 19.12). Extrusions of these kinds, which are smaller in diameter than the mature cell and which contain cytoplasm and are bounded by the cell wall, are called **prosthecae** (singular, **prostheca**) (Figure 19.37). Of considerable interest in this group of bacteria is that cell division often occurs as a result of unequal cell growth. In contrast to cell division in the typical bacterium, which occurs by *binary fission* and results in the formation of two equivalent cells (Figure 19.38), cell division in the stalked and budding bacteria involves the formation of a new daughter cell with the mother cell retaining its identity after the cell division process is completed (Figure 19.38). The genera from Table 19.12 which show this unequal cell-division process, are indicated in Figure 19.38.

The critical difference between these bacteria and conventional bacteria is not the formation of buds or stalks, but the formation of a new cell wall from a single point (polar growth) rather than throughout the whole cell (intercalary growth). Several genera not normally considered to be budding bacteria show polar growth without differentiation of cell size (Figure 19.38). An important consequence of **polar growth** is that internal structures, such as membrane complexes, are not involved in the cell-division process, thus permitting the formation of more complex internal structures than in cells undergoing intercalary growth. Several additional consequences of polar growth include: *aging* of the mother cell occurs; cells are mortal (rather than immortal as in intercalary growth); cell division may be asymmetrical; the daughter cell at division is immature and must form internal or budding structures before it can divide; organisms showing polar growth have a potential for morphogenetic evolution not possible in cells with intercalary growth. Thus some of the most complex morphogenetic processes in the procaryotes are found in the budding and stalked bacteria.

Although the majority of budding and/or appendaged bacteria phylogenetically group with the purple bacteria group, specifically the alpha purple bacteria (see Section 18.5), the genera *Planctomyces* and *Pirella* do not. These two organisms are related to each other, but are phylogenetically unrelated to any other eubacterial group. *Planctomyces* and *Pirella* are also unusual because their cell walls consist primarily of protein. The walls of *Planctomyces* and *Pirella* contain large amounts of cysteine (cross-linked as cystine), and large amounts of proline as well. As would be expected of organisms lacking peptidoglycan, *Planctomyces* and *Pirella* are totally resistant to the antibiotics penicillin, cephalosporin, and cycloserine, all drugs which affect peptidoglycan synthesis.

Most budding and/or appendaged bacteria are aquatic; in nature many live attached to surfaces, their stalks or appendages serving as attachment sites. Many of the prosthecate forms are free-floating, and it is thought that their appendages serve as absorptive organs, making possible more

Table 19.12 Characteristics of stalked, appendaged (prosthecate), and budding bacteria

Characteristics	Genus	DNA (mole % GC)
Stalked bacteria, Stalk an extension of the cytoplasm and involved in cell division	*Caulobacter*	62–67
Stalked fusiform-shaped cells	*Prosthecobacter*	55–60
Stalked, but stalk an excretory product, not containing cytoplasm:		
Stalk depositing iron, cell vibrioid	*Gallionella*	–
Laterally excreted gelatinous stalk, not depositing iron	*Nevskia*	–
Pear-shaped or globular cells with long stalk	*Planctomyces*	50
Appendaged (prosthecate) bacteria:		
Single or double prosthecae	*Asticcacaulis*	55–61
Multiple prosthecae		
Short prosthecae, multiply by fission	*Prosthecomicrobium*	66–70
Flat, st -shaped cells	*Stella*	60
Long prosthecae, multiply by budding	*Ancalomicrobium*	70–71
Phototrophic	*Prosthecochloris*	50–56
With gas vesicles	*Ancalochloris*	–
Budding bacteria:		
Phototrophic, produce hyphae	*Rhodomicrobium*	61–64
Phototrophic, budding without hyphae	*Rhodopseudomonas*	61–64
Heterotrophic, pear-shaped, stalks lacking	*Pirella*	57
Heterotrophic, rod-shaped cells	*Blastobacter*	59–64
Heterotrophic, buds on tips of slender hyphae:		
Single hyphae from parent cell	*Hyphomicrobium*	59–67
Multiple hyphae from parent cell	*Pedomicrobium*	–

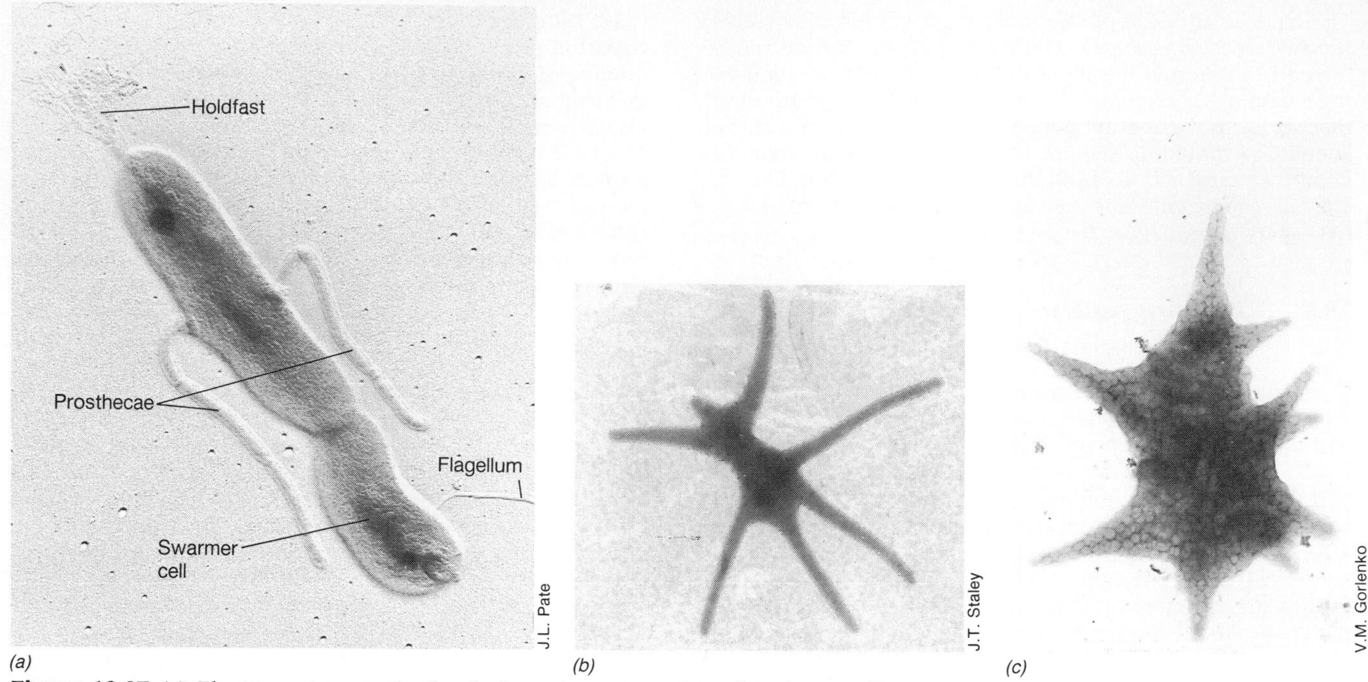

Figure 19.37 (a) Electron micrograph of a shadow-cast preparation of *Asticcacaulis biprosthecum*, illustrating the location and arrangement of the prosthecae. Note also the holdfast material and the swarmer cell in process of differentiation. (b) Electron micrograph of a negative-stained preparation of a cell of the prosthecate bacterium, *Ancalomicrobium adetum*. The appendages are cellular (i.e., prosthecae) because they are bounded by the cell wall and contain cytoplasm. (c) Electron micrograph of a whole cell of a prosthecate phototrophic green bacterium *Ancalochloris perfilievii*. The structures seen within the cell are gas vesicles.

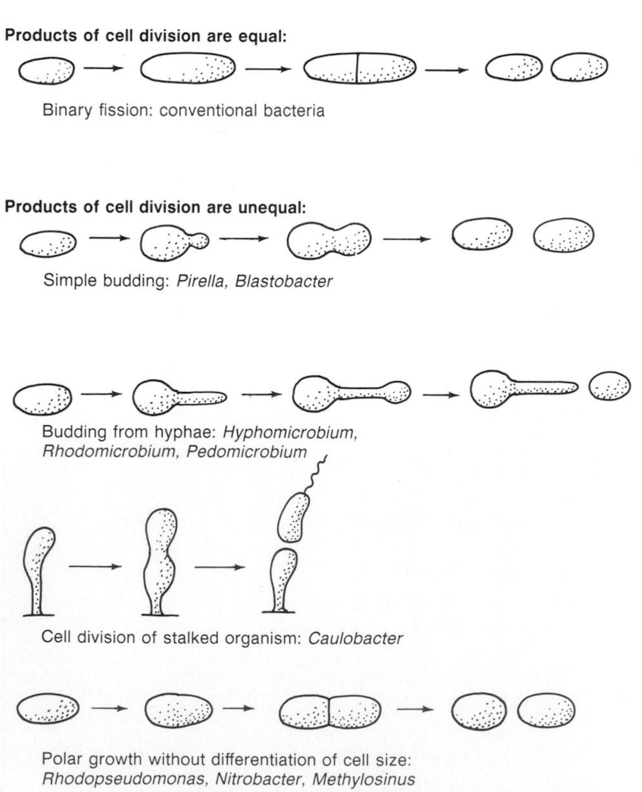

Figure 19.38 Contrast between cell division in conventional bacteria and in budding and stalked bacteria.

efficient growth in the nutritionally dilute aquatic environment. Calculations of surface:volume ratios (see Section 3.1) of appendaged bacteria suggest that a major function of bacterial appendages may be to *increase* the surface:volume ratio; such structures would serve to increase the cell's surface area and thus be of competitive advantage for survival in dilute environments. Many of the free-floating forms have gas vesicles, presumably an adaptation to the planktonic existence.

Stalked bacteria The stalked bacteria comprise a group of Gram-negative, polarly flagellated rods that possess a **stalk**, a structure by which they attach to solid substrates. Most members of this group are classified in the genus *Caulobacter*. Stalked bacteria are frequently seen in aquatic environments attached to particulate matter, plant materials, or other microorganisms; generally they are found attached to microscope slides that have been immersed in lake or pond water for a few days. When many *Caulobacter* cells are present in the suspension, groups of stalked cells are seen attached, exhibiting the formation of *rosettes* (Figure 19.39*a*). Electron-microscopic studies reveal that the stalk is not an excretion product but is an outgrowth of the cell, since it contains cytoplasm surrounded by cell wall and plasma membrane (Figure 19.39*b* and *c*). The *holdfast* by which the stalk attaches the cell to a solid substrate is at the tip of the stalk, and, once attached, the cell usually remains permanently fixed. Since the stalk is cytoplasmic, it is also a prostheca. A stalk which is *not* a prostheca can be seen in an electron micrograph of the budding bacterium *Planctomyces* (Figure 19.40). In this organism the stalk contains no cytoplasm or cell wall and is probably pro-

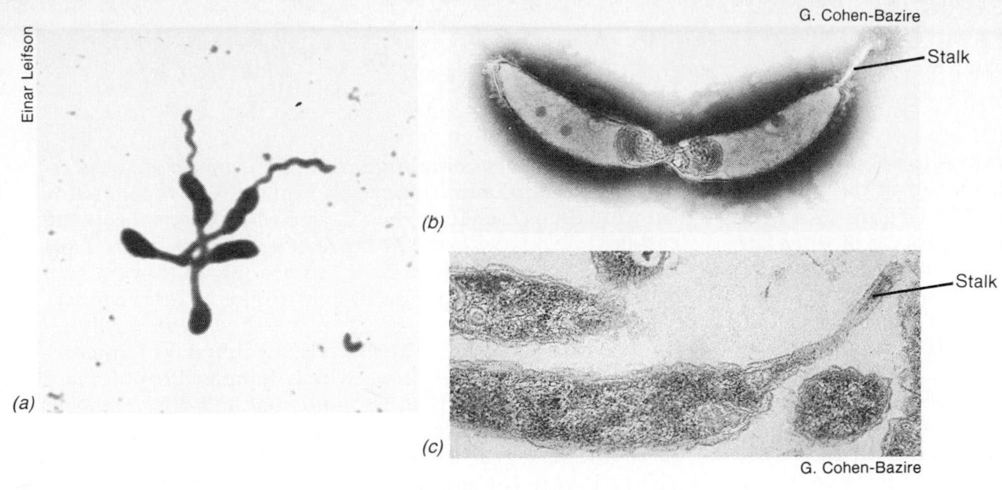

Figure 19.39 (a) A *Caulobacter* rosette. The five cells are attached by their stalks (prosthecae). Two of the cells have divided and the daughter cells have formed flagella. (b, c) Electron micrographs of *Caulobacter* cells. (b) Negatively stained preparation of a cell in division. (c) A thin section. Notice that the cytoplasmic constituents are present in the stalk region.

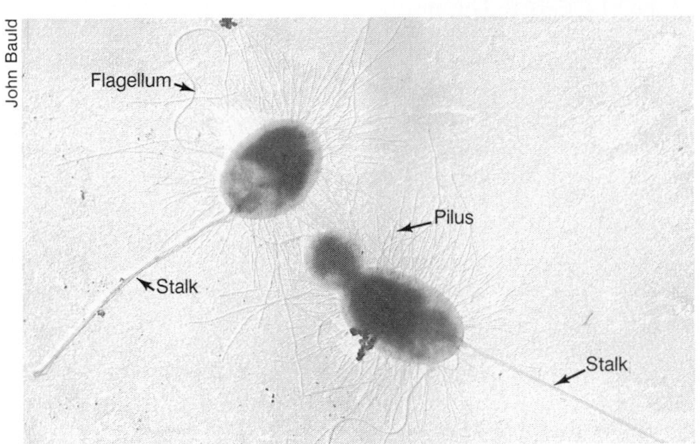

Figure 19.40 An electron micrograph of a metal-shadowed preparation of *Planctomyces maris*. Note the fibrillar nature of the stalk. Pili are also abundant. Note also the flagella (curly appendages) on each cell and the bud that is developing from the non-stalked pole of one cell.

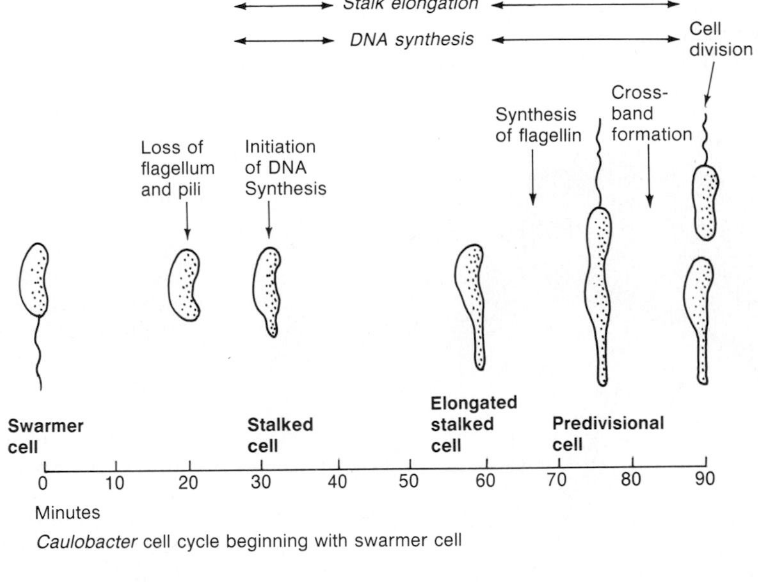

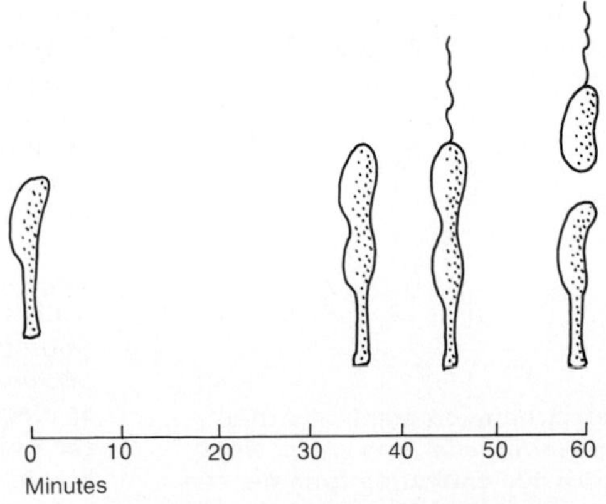

Figure 19.41 Stages in the *Caulobacter* cell cycle.

teinaceous in nature. The stalk functions to anchor the cell to surfaces via a holdfast located at the tip of the stalk.

The *Caulobacter* cell-division cycle (Figure 19.41) is of special interest because it involves a process of *unequal binary fission*. Cell division occurs by elongation of the cell followed by fission, a single flagellum forming at the pole opposite the stalk. The flagellated cell so formed, called a *swarmer*, separates from the nonflagellated mother cell, swims around, and settles down on a new surface, forming a new stalk at the flagellate pole; the flagellum then disappears (Figure 19.41). Stalk formation is a necessary precursor to cell division and is coordinated with DNA synthesis (Figure 19.41). The time span for the division of the stalked cell is thus shorter than the time span for division of the swarmer cell, owing to the requirement that a swarmer (flagellated) cell must synthesize a stalk before it divides. A crossband is produced in the stalk during or shortly after each cell division (Figure 19.41), so that the "age" of a stalked cell can be read from the number of stalk crossbands. The cell-division cycle in *Caulobacter* is thus more complex than simple binary fission since the stalked and swarmer cells have polar differentiation, and the cells themselves are structurally different.

A stalked organism sometimes classified with the caulobacters is *Gallionella*, which forms a twisted stalk containing ferric hydroxide (Figure 19.42). However, the stalk of *Gallionella* is not an integral part of the cell but is excreted from the cell surface. It contains an organic matrix on which the ferric hydroxide accumulates. *Gallionella* is frequently found in the waters draining bogs, iron springs, and other habitats where ferrous iron is present, usually in association with sheathed bacteria such as *Sphaerotilus*. In very acidic waters containing iron, *Gallionella* is not present, and acid-tolerant thiobacilli replace it. A recent study showing the presence of the enzyme ribulose bisphosphate carboxylase in *Gallionella* indicates that this neutral pH iron oxidizer may indeed be a lithotrophic autotroph (see Section 16.6 for discussion of autotrophy in lithotrophs).

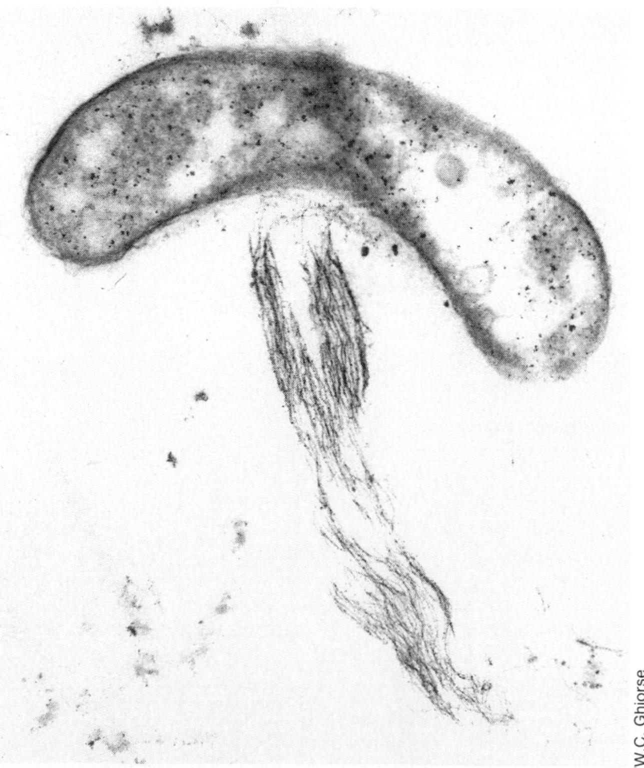

Figure 19.42 Transmission electron micrograph of a thin section of the iron bacterium *Gallionella ferruginea*. Note the twisted stalk of ferric hydroxide emanating from the center of the cell.

Budding bacteria The two best studied budding bacteria are *Hyphomicrobium*, which is heterotrophic, and *Rhodomicrobium*, which is phototrophic; both organisms release buds from the ends of long thin hyphae. The process of reproduction in a budding bacterium is illustrated in Figure 19.43. The mother cell, which is often attached by its base to a solid substrate, forms a thin outgrowth that lengthens to become a hypha, and at the end of the hypha a bud forms. This bud enlarges, forms a flagellum, breaks loose from the mother cell, and swims away. Later, the daughter cell loses its flagellum and after a period of maturation forms a hypha and buds. Further buds can also form at the hyphal tip of the mother cell. Many variations on this cycle are possible. In some cases the daughter cell does not break away from the mother cell but forms a hypha from its other pole. Complex arrays of cells connected by hyphae are frequently seen (Figure 19.44). In some cases a bud begins to form directly from the mother cell without the intervening formation of hypha, whereas in other cases a single cell forms hyphae from each end (Figure 19.44). The hypha is a direct cellular extension of the mother cell (Figure 19.45), containing cell wall, cytoplasmic membrane, ribosomes, and occasionally DNA.

Chromosomal replication events during the budding cycle are of interest (Figure 19.43). The DNA located in the mother cell replicates and then once the bud has formed, a copy of the circular chromosome is moved down the length of the hypha and into the bud. A cross-septum then forms, separating the still developing bud from the hypha and mother cell.

Hyphomicrobium has unique nutritional characteristics. Preferred carbon sources are *one-carbon* compounds such as methanol, methylamine, formaldehyde, formate, and even cyanide (CN^-). Growth on acetate, ethanol, or higher aliphatic compounds is usually slow, and growth does not occur at all on sugars or most amino acids. Urea, amides, ammonia, nitrite, and nitrate can be utilized as nitrogen sources; no vitamins are required. *Hyphomicrobium* not only is able to grow on very low concentrations of carbon, but will also grow in a liquid medium in the complete absence of any added carbon source, apparently obtaining its energy and carbon from volatile compounds present in the atmosphere. This has occasionally led to the erroneous conclusion that *Hyphomicrobium* is an autotroph, but there is no evidence that the enzymes for autotrophic CO_2 fixation (see Section 16.6) are present. *Hyphomicrobium* is widespread in freshwater, marine, and terrestrial habitats. Initial enrichment cultures can be prepared using a mineral-salts medium lacking organic carbon and nitrogen, to which a sample of natural material is added. After several weeks incubation, the surface film that develops is streaked out on agar medium containing methylamine or methanol as a sole carbon source. Colonies are then checked microscopically for the characteristic *Hyphomicrobium* cellular morphology. A fairly specific enrichment procedure for *Hyphomicrobium* uses methanol as electron donor with nitrate as electron acceptor under anaerobic conditions. Virtually the only denitrifying organism using methanol is *Hyphomicrobium*, so that this procedure selects this organism out of a wide variety of environments.

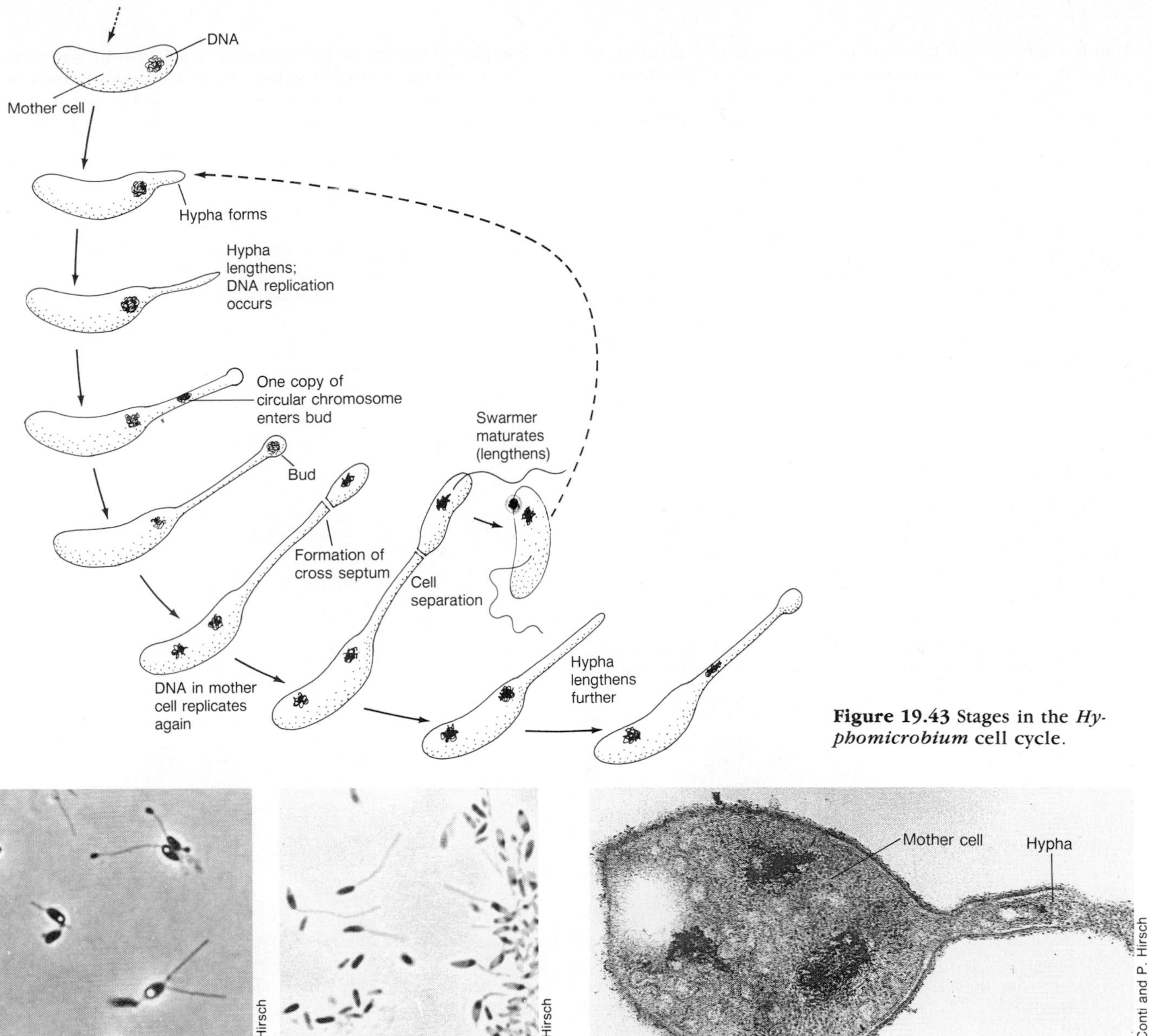

Figure 19.43 Stages in the *Hyphomicrobium* cell cycle.

Figure 19.44 Photomicrographs of *Hyphomicrobium*. (a, b) By phase contrast, showing typical fields. Notice the long hyphae and the occasional budding cell.

Figure 19.45 Electron micrograph of a thin section of a single *Hyphomicrobium* cell.

19.11 Spirilla

The division of genera of Gram-negative, motile, rod-shaped, and curved bacteria has presented some difficulties to the organizers of *Bergey's Manual*. A large number of organisms have been studied, and characteristics overlap among groups. Some of the key taxonomic criteria used are cell shape, size, kind of polar flagellation (single or multiple), relation to oxygen (obligately aerobic, microaerophilic, facultative), relationship to plants (as symbionts or plant pathogens) or animals (as pathogens), fermentative ability, and certain other physiological characteristics (nitrogen-fixing ability, ability to utilize methane as electron donor, halophilic nature, thermophilic nature, luminescence). The genera to be covered in the present discussion are given in Table 19.13.

Spirillum, Aquaspirillum, Oceanospirillum, and Azospirillum These are helically curved rods, which are motile by means of polar flagella (usually tufts at both poles, see Figure 19.46). The number of turns in the helix may vary from less than one complete turn (in which case the organism looks like a vibrio; see Section 19.19) to many turns. Spirilla with many turns can superficially resemble spirochetes, but differ in that they do not have an outer sheath and axial filaments, but instead contain typical bacterial flagella. Some spirilla are very large bacteria, and were seen by early microscopists. It is likely that van Leeuwenhoek first described *Spirillum* in the 1670s, and the genus was first created by the protozoologist Ehrenberg in 1832. The organism seen by these workers is now called *Spirillum volutans*, and is one of the largest bacteria known (Figure 19.46*a* and *b*). A phototrophic organism resembling *S. volutans* is *Thiospirillum* (see Figure 19.5*b*). Despite the fact that *S. volutans* has been known for a long time, it resisted cultivation in pure culture until fairly recently. The difficulty turned out to be that the organism is *microaerophilic*, requiring O_2, but being inhibited by O_2 at normal levels. In mixed culture, contaminants used up much of the oxygen and kept the level low, but once the organism was separated from its contaminants it succumbed to oxygen toxicity (see Section 9.14 for a discussion of O_2 toxicity). The simplest way of achieving microaerophilic growth of *S. volutans* is to use a medium such as nutrient broth in a closed vessel with a large head (gas) space, with an atmosphere of N_2, and to then inject sufficient O_2 from a hypodermic syringe to occupy 1 to 5 percent of the head space volume. Because *S. volutans* is dependent on O_2 for growth, high cell yields cannot be expected in this way, because the O_2 is quickly used up. For cultivation for phys-

Table 19.13 Characteristics of the genera of spiral-shaped bacteria*

Genus	Characteristics	DNA (mole % GC)
Spirillum	Cell diameter 1.7 μm; microaerophilic; fresh water	38
Aquaspirillum	Cell diameter 0.2–1.5 μm; aerobic; fresh water	50–65
Oceanospirillum	Cell diameter 0.3–1.2 μm; aerobic; marine (require 3% NaCl)	42–48
Azospirillum	Cell diameter 1 μm; microaerophilic; fixes N_2	69–71
Campylobacter	Cell diameter 0.2–0.8 μm; microaerophilic to anaerobic; pathogenic or commensal in humans and animals; single polar flagellum	29–35
Bdellovibrio	Cell diameter 0.25--0.4 μm; aerobic; predatory on other bacteria; single polar sheathed flagellum	42–50
Microcyclus	Cell diameter 0.5 μm; curved rods forming rings; nonmotile; aerobic; sometimes gas vesicuolate	34–68

**All are Gram-negative and respiratory but never fermentative.*

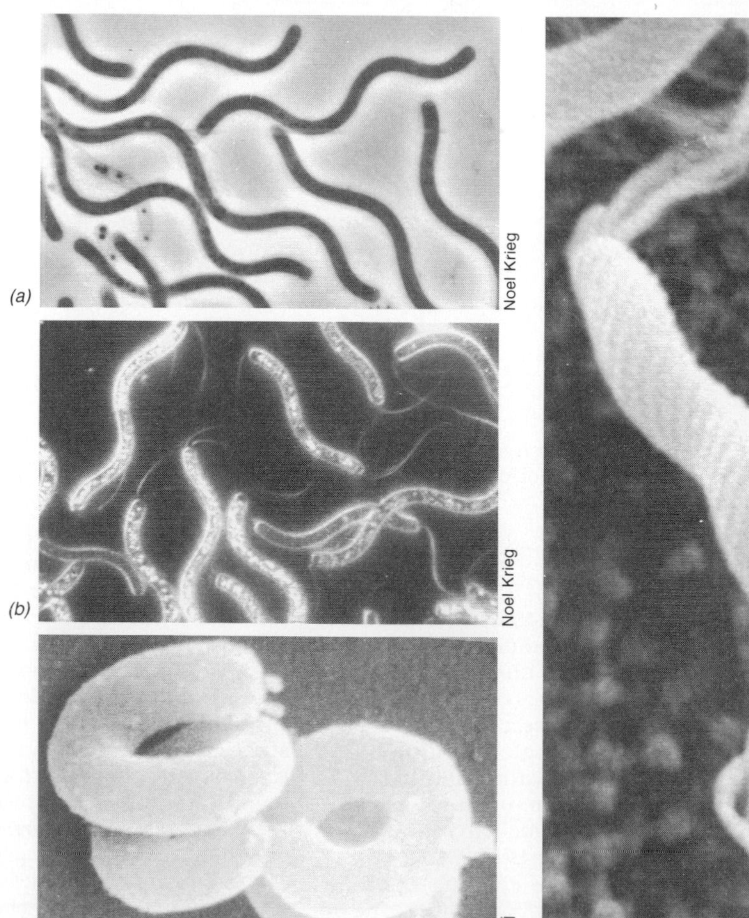

Figure 19.46 (a) Photomicrograph by phase contrast of *Spirillum volutans*, the largest spirillum. (b) *S. volutans*, by dark-field microscopy, showing flagellar bundles and volutin granules. (c) Scanning electron micrograph of an intestinal spirillum. Note the polar flagellar tufts and the spiral structure of the cell surface. (d) Scanning electron-micrograph of cells of *Microcyclus flavus*.

iological studies, a continuous stream of 1 percent O_2 in N_2 can be passed through the culture vessel. Another characteristic of *S. volutans* is the formation of prominent granules (volutin granules) consisting of polyphosphate (see Section 3.10).

Azospirillum lipoferum is a nitrogen-fixing organism, which was originally described and named *Sprillum lipoferum* by Beijerinck in 1922. It has become of considerable interest in recent years because this bacterium has been found to enter into a loose symbiotic relationship with tropical grasses and grain crops (see Section 17.24).

In the latest edition of *Bergey's Manual* the genus *Spirillum* includes only a single species, *S. volutans*, characterized by its microaerophilic character, large size, and formation of volutin granules. The small-diameter spirilla (which are not microaerophilic) have been separated into two genera, *Aquaspirillum* and *Oceanospirillum*, the former for freshwater forms and the latter for those living in seawater and requiring NaCl for growth (Table 19.13). At least 16 species of *Aquaspirillum* have been described and 9 species of *Oceanospirillum*, the various species being separated on physiological grounds. The student isolating heterotrophic bacteria from freshwater and marine environments will almost certainly obtain isolates of these genera, as they are among the most common organisms appearing in heterotrophic enrichments, and they are easily purified by streaking on agar. These organisms undoubtedly play an important role in the recycling of organic matter in aquatic environments. Evidence from enrichment culture studies using the chemostat suggests that the aquatic spirilla are adapted to the utilization of organic substrates at very low substrate concentrations, probably because they have very low affinity constants (K_m's) for uptake of organic nutrients.

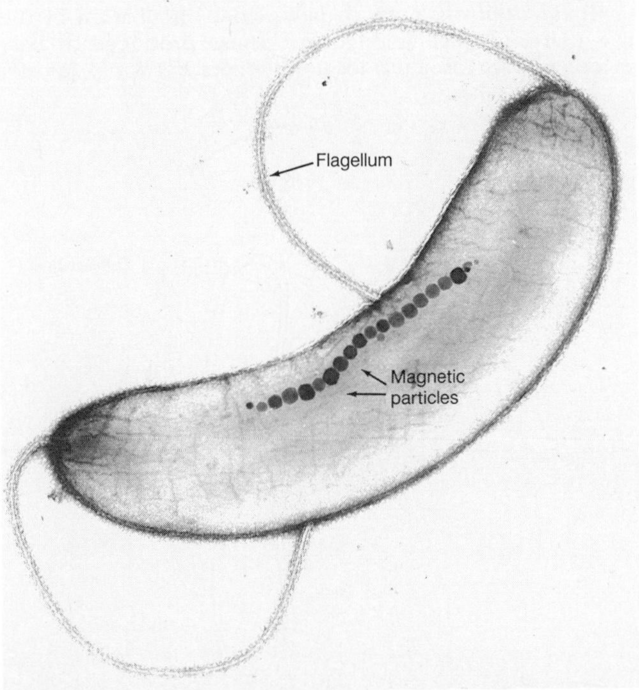

Figure 19.47 Negatively stained electron micrograph of a magnetotactic spirillum, *Aquaspirillum magnetotacticum*. This bacterium contains particles of Fe_3O_4 (magnetite) arranged in a chain; the particles serve to align the cell along geomagnetic lines.

Highly motile microaerophilic spirilla have been isolated from freshwater habitats that demonstrate a dramatic directed movement in a magnetic field referred to as **magnetotaxis**. In an artificial magnetic field magnetotactic spirilla quickly orient their long axis along the north-south magnetic moment of the field. Within the cells chains of 5–40 magnetic particles consisting of Fe_3O_4 are present (Figure 19.47), and it is thought that these serve as internal magnets that orient the cells along a specific magnetic field. Although the ecological role of bacterial magnets are unclear, it has been suggested that the ability to orient in a magnetic field would be of selective advantage in directing these microaerophilic organisms downwards toward microaerobic zones near the sediments. Magnetotactic bacteria have stimulated great interest among scientists interested in the field of magnetism, because like chemotaxis, magnetotaxis can be interpreted as a primitive behavioral response. Since magnetic phenomena have been implicated in a number of biological responses including bird migration, insect behavior, and other forms of animal navigation, magnetotactic bacteria may turn out to be ideal model systems for the study of biological magnetism.

Bdellovibrio These small vibroid organisms have the unique property of preying on other bacteria, using as nutrients the cytoplasmic constituents of their hosts. These **bacterial predators** are small, highly motile cells, which stick to the surfaces of their prey cells. Because of the latter property, they have been given the name *Bdellovibrio* (*bdello-* is a combining form meaning "leech"). A number of strains of *Bdellovibrio* have been found, each shows prey specificity, attacking some species of bacteria but not others; strains that attack enteric bacteria, photosynthetic bacteria, gliding bacteria, *Azotobacter* and *Rhizobium*, as well as many other Gram-negative bacteria have been isolated. In general, Gram-positive bacteria are not attacked. After attachment of a *Bdellovibrio* cell to its prey, the predator penetrates through the prey wall and replicates in the space between the prey wall and membrane (the periplasmic space), eventually forming a spherical structure called a **bdelloplast**. The stages of attachment and penetration are shown in electron micrographs in Figure 19.48. A schematic representation of the *Bdellovibrio* life cycle is shown in Figure 19.49.

As originally isolated, *Bdellovibrio* cells grow only upon the living prey, but it is possible to isolate mutants that are prey independent and are able to grow on complex organic media such as yeast extract-peptone. These strains are unable to utilize sugars as electron donors, but are proteolytic and can oxidize the amino acids liberated by protein digestion. Prey-dependent revertants can be reisolated from the prey-independent mutants by reinfecting a host strain.

Bdellovibrio is an obligate aerobe, obtaining its energy from the oxidation of amino acids and acetate (via the citric acid cycle). It apparently is unable to utilize sugars as electron donors. Studies have been carried out to determine the growth efficiency of *Bdellovibrio* as indicated by the Y_{ATP} value. As we noted in Section 9.8, most organisms exhibit a Y_{ATP} value of around 10. *Bdellovibrio*, on the other hand, exhibits a Y_{ATP} value of 18 to 26, considerably higher than that of free-living bacteria. This considerably higher efficiency of conversion of substrates into ATP could possibly be due to the more efficient growth process in the periplasmic environment of its host, and to the fact that *Bdellovibrio* assimilates nucleoside phosphates, fatty acids, and even whole proteins in some cases directly from its host without first breaking them down. However, these energy-

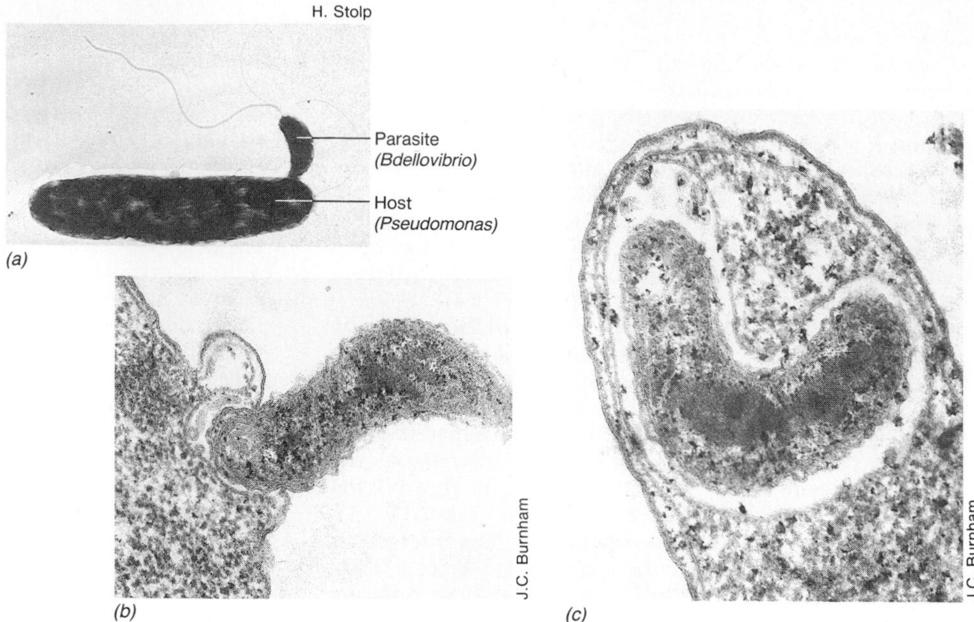

Figure 19.48 Stages of attachment and penetration of a prey cell by *Bdellovibrio*. (a) Electron micrograph of a shadowed whole-cell preparation showing *B. bacteriovorus* attacking *Pseudomonas*. (b, c) Electron micrographs of thin sections of *Bdellovibrio* attacking *Escherichia coli*: (b) early penetration; (c) complete penetration. The *Bdellovibrio* cell is enclosed in a membranous infolding of the prey cell (the bdelloplast), and replicates in the periplasmic space between wall and membrane.

sparing effects cannot account for all of the growth efficiency of *Bdellovibrio*, and other reactions must be sought. One idea is that *Bdellovibrio* is much more efficient in coupling energy-generating reactions to biosynthesis, so that energy is not wasted. It is clear that the predatory mode of existence has involved the development in *Bdellovibrio* of interesting and unusual biochemical processes.

Phylogenetically, bdellovibrios fall into the delta group of purple bacteria (see Section 18.5). Taxonomically, three species of *Bdellovibrio* are recognized as shown in Table 19.14. Nucleic acid studies in bdellovibrios have shown them to be a rather heterogeneous group. Both GC base ratio and DNA:DNA hybridization analyses suggest that bdellovibrios are genetically diverse (Table 19.14), even though they share a unique life style. In addition to being predators themselves, bdellovibrios, as for most bacteria, are subject to attack by bacteriophages. Nicknamed ***bdellophages***, most phages that plaque on lawns of *Bdellovibrio* cells are strictly lytic single-stranded DNA phages.

Members of the genus *Bdellovibrio* are widespread in soil and water, including the marine environment. Their detection and isolation require methods somewhat reminiscent of those used in the study of bacterial viruses. Prey bacteria are spread on the surface of an agar plate to form a lawn, and the surface is inoculated with a small amount of soil suspension that has been filtered through a membrane filter; the latter retains most bacteria but allows the small *Bdellovibrio* cells to pass. Upon incubation of the agar plate, plaques analogous to those produced by bacteriophages are formed at locations where *Bdellovibrio* cells

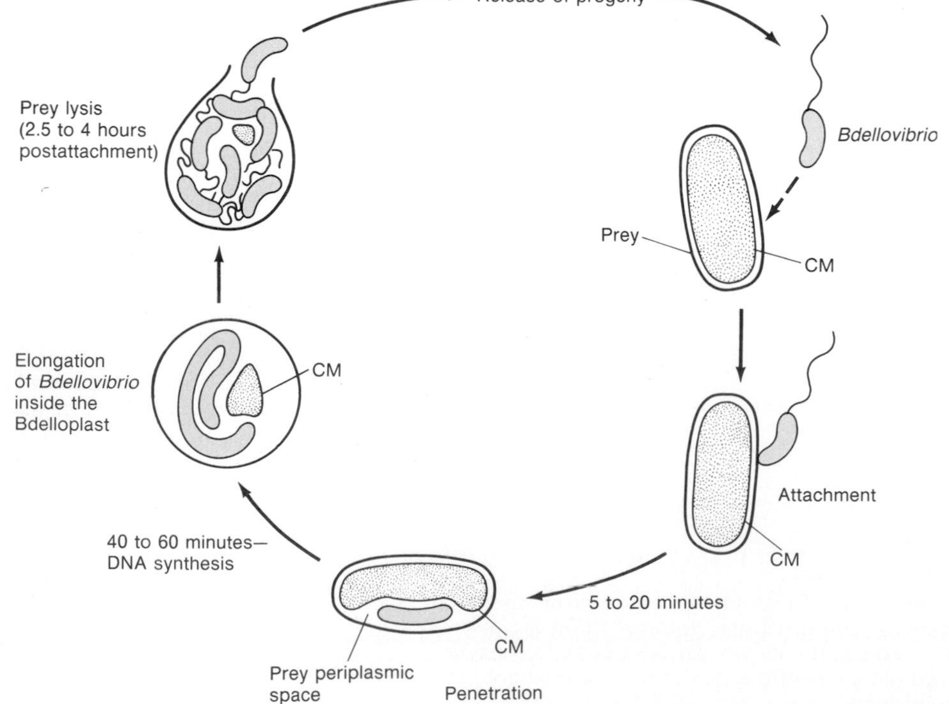

Figure 19.49 Developmental cycle of the bacterial predator, *Bdellovibrio bacteriovorus*. Following primary contact between a highly motile *Bdellovibrio* cell and a Gram-negative bacterium, attachment and penetration into the prey periplasmic space occurs. Once inside, *Bdellovibrio* elongates and within 4 hours progeny cells are released. The number of progeny cells released varies with the size of the prey bacterium, for example, 5 to 6 bdellovibrios are released from each infected *E. coli* cell, 20 to 30 for *Spirillum serpens*. CM = prey cytoplasmic membrane.

Table 19.14 Genetic relationships among species of *Bdellovibrio**

Species	DNA GC (mole %)	Genome size (× 10^9 daltons)	DNA hybridization to *B. bacteriovorus*	*B. stolpii*	*B. starrii*
B. bacteriovorus	50	1.3	100	28	26
B. stolpii	42	1.5	28	100	16
B. starrii	44	1.7	23	16	100
Escherichia coli	51	2.2	—	—	—

**Data from Torrella, F., R. Guerrero, and R. J. Seidler. 1978. Can. J. Microbiol. 24:1387–1394, and Stolp, H. 1981. The genus* Bdellovibrio. In: *M. P. Starr, et al. The Prokaryotes,* Springer-Verlag, New York.

are growing. Unlike phage plaques, which continue to enlarge only as long as the bacterial host is growing, *Bdellovibrio* plaques continue to enlarge even after the prey has stopped growing, resulting in large plaques on the agar surface. Pure cultures of *Bdellovibrio* can then be isolated from these plaques. *Bdellovibrio* cultures have been obtained from a wide variety of soils and are thus common members of the soil population. As yet, the ecological role of bdellovibrios is not known, but it seems likely that they play some role in regulating the population densities of their prey.

Microcyclus Members of the genus *Microcyclus* are ring-shaped and nonmotile (Figure 19.46*d*). They resemble very tightly curved vibrios and are widely distributed in aquatic environments. A photosynthetic counterpart to *Microcyclus* has been discovered and placed in the genus *Rhodocyclus* (see Section 19.1).

19.12 Spirochetes

The **spirochetes** are bacteria with a unique morphology and mechanism of motility. They are widespread in aquatic environments and in the bodies of animals. Some of them cause diseases of animals and humans, of which the most important is *syphilis*, caused by *Treponema pallidum*. The spirochete cell is typically slender, flexuous, helical (coiled) in shape, and often rather long (Figure 19.50). The "protoplasmic cylinder," consisting of the regions enclosed by the cytoplasmic membrane and the cell wall, constitutes the major portion of the spirochetal cell. Fibrils, referred to as **axial fibrils** or **axial filaments**, are attached to the cell poles and wrapped around the coiled protoplasmic cylinder (Figure 19.51). Both the axial fibrils and the protoplasmic cylinder are surrounded by a three-layered membrane called the "outer sheath" or "outer cell envelope" (Figure 19.51). The outer sheath and axial fibrils are usually not visible by light microscopy, but are observable in negatively stained preparations or thin sections examined by electron microscopy (see, for example, Figure 19.52).

From 2 to more than 100 axial fibrils are present per cell, depending on the type of spirochete. The ultrastructure and the chemical composition of axial fibrils are similar to those of bacterial flagella (see Section 3.8). As in typical bacterial flagella (Figure 3.41), basal hooks and paired disks are present at the insertion end. The shaft of each fibril is composed of a core surrounded by an "axial fibril sheath," so that the spirochete axial fibrils are in a sense analogous to sheathed flagella.

The axial fibrils play a significant role in spirochete motility. Each fibril is anchored at one end and extends for approximately two-thirds of the length of the cell. It is thought that the axial fibrils rotate rigidly, as do bacterial flagella (see Section 3.8). Since the protoplasmic cylinder is also rigid, whereas the outer sheath is flexible, if both axial fibrils rotate in the same direction, the protoplasmic cylinder will rotate in the *opposite* direction, as illustrated in Figure 19.51*b*. If the sheath is not in contact with a surface, it also rotates (see arrow in Figure 19.51*b*). This simple mechanism is all that is needed to generate the wide variety of motions exhibited by spirochetes. If the protoplasmic cylinder is helical (as in most spirochetes), then forward motion will be generated when the sheath is moving through a liquid or semisolid medium by the circumferential slip of the helix through the medium. If the sheath is in contact along its length with a solid surface, the protoplasmic cylinder may not be able to rotate so that the roll of the sheath will cause the cell to slide in a direction nearly parallel to the axis of the helix, generating a "creeping" motility. In free liquid, many narrow diameter spirochetes show flexing or lashing motions due to torque exerted at

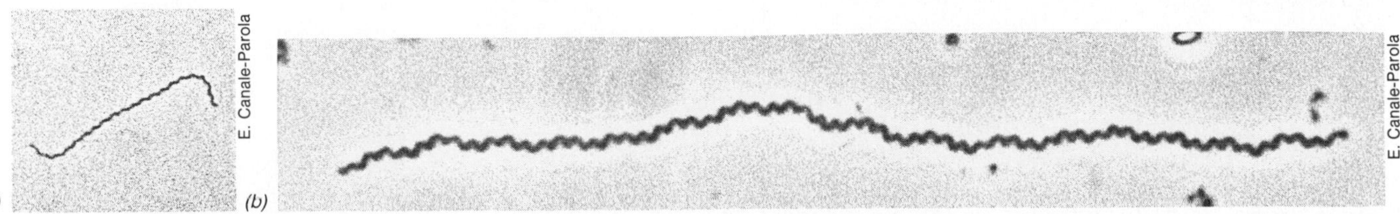

Figure 19.50 Two spirochetes at the same magnification, showing the wide size range in the group. (a) *Spirochaeta stenostrepta*, by phase-contrast microscopy. (b) *Spirochaeta plicatilis.*

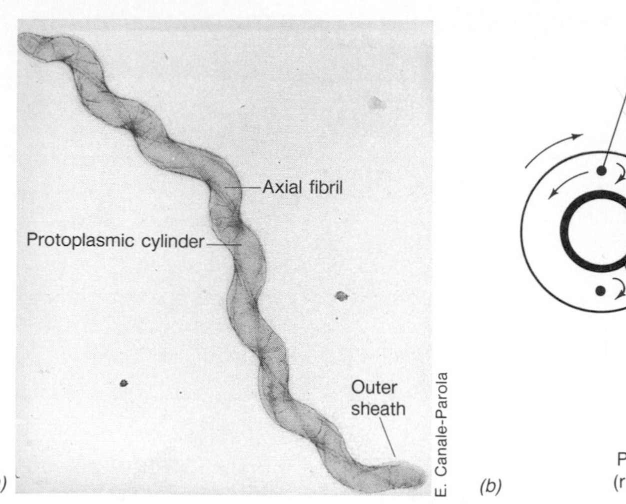

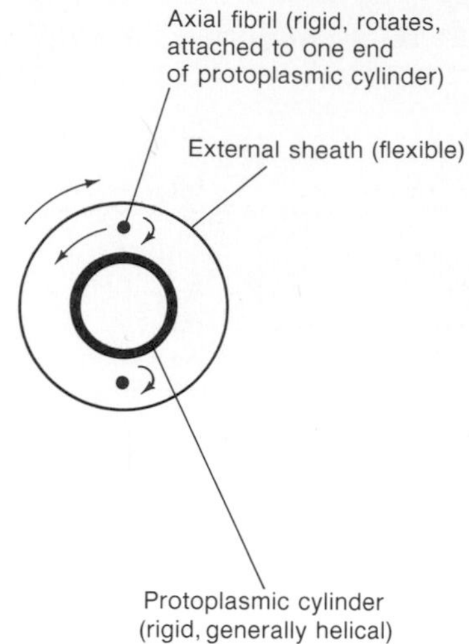

Figure 19.51 (a) Electron micrograph of a negatively stained preparation of *Spirochaeta zuelzerae*, showing the position of the axial filament. (b) Cross-section of a spirochete cell, showing the arrangement of the protoplasmic cylinder, axial fibrils, and external sheath, and the manner by which the rotation of the rigid axial fibril can generate rotation of the protoplasmic cylinder and (in opposite direction) rotation of the external sheath. If the sheath is free, the cell will rotate about its longitudinal axis and move along it. If the sheath is in contact with a solid surface, the cell will creep forward. See text for details.

the ends of the protoplasmic cylinder by the twisting axial fibrils.

In this model of motility, the external sheath plays a central role, as do the opposing axial fibrils. Similarly, the rigidity of the protoplasmic cylinder is essential in this mechanism. Interestingly, antibodies directed against the axial fibrils do *not* cause immobilization of whole cells, whereas antibodies against some component of the external sheath do cause immobilization, perhaps by stiffening the external sheath. It thus appears that despite superficial differences, spirochetes have fundamentally the same motility mechanism as other bacteria, namely the rotation of rigid flagellar fibrils attached in the cell membrane via a basal hook.

Spirochetes are classified into five genera primarily on the basis of habitat, pathogenicity, and morphological and physiological characteristics. Table 19.15 lists the major genera and their characteristics. Molecular sequencing studies of 16S rRNA from various spirochetes show that spirochetes form a cluster, albeit a loose one, phylogenetically. In addition, it has been found that most spirochetes are naturally resistant to the antibiotic rifampicin, an indication that their RNA polymerases may differ from those of other bacteria (see Section 5.8).

Spirochaeta* and *Cristispira The genus *Spirochaeta* includes free-living, anaerobic, and facultatively aerobic spirochetes. These organisms are common in aquatic en-

Table 19.15 Genera of spirochetes and their characteristics

Genus	Dimensions (μm)	General characteristics	Number of axial fibrils	DNA (mole % GC)	Habitat	Diseases
Cristispira	30–150 × 0.5–3.0	3–10 complete coils; bundle of axial fibrils visible by phase-contrast microscopy	>100	—	Digestive tract of molluscs; has not been cultured	None known
Spirochaeta	5–500 × 0.2–0.75	Anaerobic or facultatively aerobic; tightly or loosely coiled	2–40	50–66	Aquatic, free-living	None known
Treponema	5–15 × 0.1–1.5	Anaerobic, coil amplitude up to 0.5 μm	2–15	38–53	Commensal or parasitic in humans, other animals	Syphilis, yaws
Borrelia	3–15 × 0.2–0.5	Anaerobic; 5–7 coils of approx. 1 μm amplitude	Unknown	46	Humans and other mammals, arthropods	Relapsing fever
Leptospira	6–20 × 0.1	Aerobic; tightly coiled, with bent or hooked ends	2	35–53	Free-living or parasitic of humans, other mammals	Leptospirosis

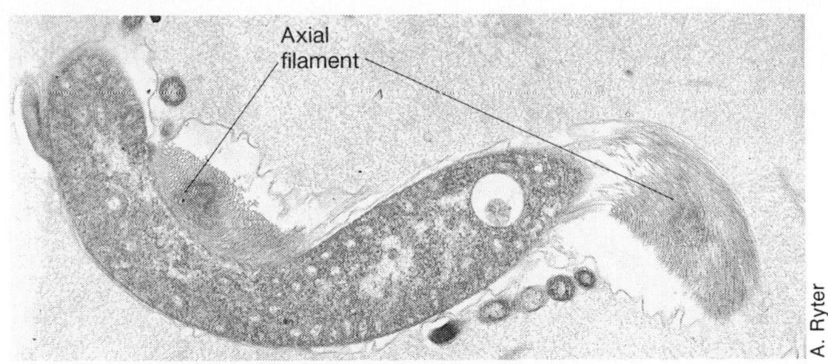

Figure 19.52 Electron micrograph of a thin section of *Cristispira*, a very large spirochete. Notice the numerous fibrils in the axial filament.

vironments, such as the water and mud of rivers, ponds, lakes, and oceans. One species of the genus *Spirochaeta* is *S. plicatilis* (Figure 19.50*b*), a fairly large organism that was the first spirochete to be discovered; it was reported by C. G. Ehrenberg in the 1830s. It is found in freshwater and marine H_2S-containing habitats, and is probably anaerobic. The axial fibrils of *S. plicatilis* are arranged in a bundle that winds around the coiled protoplasmic cylinder. From 18 to 20 axial fibrils are inserted at each pole of this spirochete. Another species, *S. stenostrepta*, has been cultured, and is shown in Figure 19.50*a*. It is an obligate anaerobe commonly found in H_2S-rich, black muds. It ferments sugars via the glycolytic pathway to ethanol, acetate, lactate, CO_2, and H_2. The species *S. aurantia* is an orange-pigmented facultative aerobe, fermenting sugars via the glycolytic pathway under anaerobic conditions, and oxidizing sugars aerobically mainly to CO_2 and acetate.

The genus *Cristispira* (Figure 19.52) contains organisms with a unique distribution, being found in nature primarily in the *crystalline style* of certain molluscs, such as clams and oysters. The crystalline style is a flexible, semisolid rod seated in a sac and rotated against a hard surface of the digestive tract, thereby mixing with and grinding the small particles of food. Being large spirochetes, the cristispiras can readily be seen microscopically within the style as they rapidly rotate forward and backward in corkscrew fashion. *Cristispira* may occur in both freshwater and marine molluscs, but not all species of molluscs possess them. Unfortunately, *Cristispira* has not been cultured, so that the physiological reason for its restriction to this unique habitat is not known. There is no evidence that *Cristispira* is harmful to its host; in fact, the organism may be more common in healthy than in diseased molluscs.

Treponema Anaerobic, host-associated spirochetes that are commensals or parasites of humans and animals are placed in the genus *Treponema*. Another genus, *Borrelia*, includes those spirochetes that cause relapsing fever, an acute febrile illness of people and animals (see below). *Treponema pallidum*, the causal agent of syphilis (see Section 15.6), is the best known species of *Treponema*. It differs in morphology from other spirochetes; the cell is not helical, but has a flat wave form. Furthermore, electron microscopy does not show the presence of an outer sheath surrounding both the axial fibrils and the protoplasmic cylinder. Apparently the axial fibrils of *T. pallidum* lie on the outside of the organism. The *T. pallidum* cell is remarkably thin, measuring approximately 0.2 μm in diameter. Living cells are clearly visible in the dark-field microscope or after staining with fluorescent antibody; dark-field microscopy has long been used to examine exudates from suspected syphilitic lesions (see Figure 15.20). In nature *T. pallidum* is restricted to humans, although artificial infections have been established in rabbits and monkeys. Although never grown in laboratory culture, it has now been established from animal studies that virulent *T. pallidum* cells (purified from infected rabbits) contain a cytochrome system, and may in fact be microaerophiles. This property further separates *T. pallidum* from the other species of *Treponema*, which, as noted, are obligate anaerobes. Meaningful taxonomic studies on this species will have to await its cultivation away from its host.

Those treponemes that have been cultured are non-pathogenic obligate anaerobes requiring complex media, and all grow slowly, with generation times of 12 to 18 hours. *Treponema pallidum* is quite sensitive to increased temperature, being rapidly killed by exposure to 41.5 to 42.0°C. This was once the basis of fever therapy for syphilis, that is, increasing the body temperature of the patient in order to kill the spirochete, but the procedure is now supplanted by antibiotic therapy. The heat sensitivity of *T. pallidum* is also reflected in the fact that the organism becomes most easily established in cooler sites of the body, such as the male genital organs, although once established in other areas of the body it will multiply there. Infection of rabbits is also most extensive in cooler sites such as the testicles or skin; indeed, artificial cooling of rabbit skin results in a dramatic increase in the number of lesions. The organism is rapidly killed by drying, and this at least partially explains why transmission of *T. pallidum* between persons is only by direct contact, usually sexual intercourse.

Other species of the genus *Treponema* are common commensal organisms in the oral cavity of humans and can generally be seen in material scraped from between the teeth and from the narrow space between the gums and the teeth. Various oral spirochetes can be cultivated anaerobically in complex media containing serum. Three species, *T. denticola, T. macrodentium*, and *T. oralis*, have been described, differing in morphology and physiological characteristics. *Treponema denticola* ferments amino acids such as cysteine and serine, forming acetate as the major fermentation acid and CO_2, NH_3, and H_2S. This spirochete can also ferment glucose, but in media containing both glucose and amino acids, the amino acids are used preferentially. The true relationship between *T. pallidum* and the remaining members of the genus *Treponema* may be a distant one, since the GC base ratio of *T. pallidum* is about 53 percent while other species of this genus cluster tightly between 38 to 40 percent.

Spirochetes are also found in the rumen. *Treponema saccharophilum* (Figure 19.53) is a large, *pectinolytic* spirochete found in the bovine rumen. *T. saccharophilum* is an obligately anaerobic bacterium that ferments pectin, starch, inulin, and other plant polysaccharides. This and other spirochetes may play an important role in the conversion of plant polysaccharides to volatile fatty acids, usable as energy sources by the ruminant (see Section 17.13).

Leptospira The genus *Leptospira* contains strictly *aerobic* spirochetes that use long-chain fatty acids (for example, oleic acid) as electron donor and carbon sources. With few exceptions these are the only substrates utilized by leptospiras for growth. The leptospira cell is thin, finely coiled, and usually bent at each end into a semicircular hook. At present only two species, *L. interrogans* and *L. biflexa*, are recognized in this genus. Strains of *L. inter-*

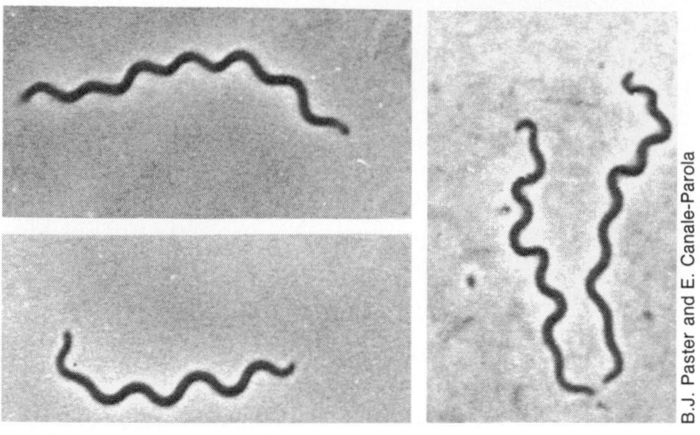

Figure 19.53 Phase contrast photomicrographs of *Treponema saccharophilum*, a large pectinolytic spirochete from the bovine rumen. Left, regularly coiled cells. Right, irregularly coiled cells.

rogans are parasitic for humans and animals, and *L. biflexa* includes the free-living aquatic leptospiras. Leptospiras of the parasitic complex are catalase positive, whereas aquatic leptospiras are catalase negative. Different strains are distinguished serologically by agglutination tests, and a large number of serotypes have been recognized. Rodents are the natural hosts of most leptospiras although dogs and pigs are also important carriers of certain strains. In human beings the most common leptospiral syndrome is called "Weil's disease"; in this disorder the organism usually localizes in the kidney. Leptospiras ordinarily enter the body through the mucous membranes or through breaks in the skin. After a transient multiplication in various parts of the body the organism localizes in the kidney and liver, causing nephritis and jaundice. The organism passes out of the body in the urine and infection of another individual is most commonly by contact with infected urine. Therapy with penicillin, streptomycin, or the tetracyclines is possible but may require extended courses to eliminate the organism from the kidney; this is probably because of the slow growth and protected location of the leptospiras. Domestic animals are vaccinated against **leptospirosis** with a killed virulent strain; dogs are usually immunized routinely with a combined distemper-leptospira-hepatitis vaccine. In humans, prevention is effected primarily by elimination of the disease from animals. The serotype that infects dogs, *L. interrogans* serotype *canicola*, does not ordinarily infect humans, but the strain attacking rodents, *L. interrogans* serotype *icterohaemorrhagiae*, does; hence elimination of rats from human habitation is of considerable aid in preventing the organism from reaching human beings.

Borrelia The majority of species in the genus *Borrelia* are animal or human pathogens. *B. recurrentis* is the causative agent of **relapsing fever** in humans and is transmitted via an insect vector, usually by the human body louse. Relapsing fever is characterized by a high fever and generalized muscular pain which lasts for 3 to 7 days followed by a recovery period of 7 to 9 days. Left untreated, the fever returns in two to three more cycles (hence the name relapsing fever) and causes death in up to 40 percent of those infected. Fortunately, the organism is quite sensitive to tetracycline, and if the disease is correctly diagnosed, treatment is straightforward. Other borrelia are of veterinary importance causing diseases in cattle, sheep, horses, and birds. In most of these diseases the organism is transmitted by ticks.

A *Borrelia* species, *B. burgdorferi,* is responsible for **Lyme disease.** *B. burgdorferi* is transmitted from animals to humans by the deer tick, *Ixodes ricinus.* Lyme disease is the most frequently diagnosed tick-borne disease in the United States and was named after a small town in Connecticut where the disease was first reported. Symptoms of Lyme disease include an acute headache, backache, chills, and fatigue, followed by a large rash at the site of the tick bite. Without treatment, the disease can progress to produce a type of rheumatoid arthritis, as well as cardiac difficulties and central nervous system damage. Lyme disease responds quickly to a variety of antibiotics including penicillin, erythromycin, and the tetracyclines, and its diagnosis has now been aided by immunological methods such as fluorescent antibodies (see Section 12.8). Although a relatively "new" disease (the first cases were reported in 1975), Lyme disease appears to be spreading quickly across the United States in areas where tick infestations are common.

19.13 Gliding Bacteria

A variety of bacteria exhibit gliding motility. These organisms have no flagella, but are able in some manner to move when in contact with surfaces. A few of the **gliding bacteria** are morphologically very similar to cyanobacteria and have been considered to be their nonphotosynthetic counterparts (phylogenetically, however, cyanobacteria and gliding bacteria are very different, see Section 18.5). All gliding bacteria are Gram-negative. One group of gliding bacteria, the fruiting myxobacteria, possesses the interesting property of forming multicellular structures of complex morphology called **fruiting bodies**. Table 19.16 gives a brief outline of some of the genera of gliding bacteria.

The mechanism of gliding by gliding bacteria appears complex, and it is likely that more than one mechanism is responsible. Gliding is most apparent when cells are on a solid surface. Some gliding bacteria rotate along their long axis while moving while others seem to keep only one side of the cell in contact with the surface while moving. Gliding motility is generally much slower than flagellar motility, but absolute rates of gliding are somewhat dependent on cell length; filamentous gliding bacteria usually move much faster than unicellular gliding bacteria. Although most hypotheses put forward to account for gliding have not been rigorously proven, two mechanisms of gliding have some experimental support. In *Cytophaga*, there is evidence that small rotating particles, presumably made of protein, lie between the cytoplasmic membrane and the Gram-negative outer membrane. Acting like miniature ball bearings, these particles rotate, presumably at the expense of a membrane potential or perhaps ATP directly, and this rotary motion slides the cell along the solid surface. By contrast, in *Myxococcus*, gliding seems to occur as a result of the secretion of a chemical surfactant which affects surface tension forces between the cell and the solid surface, and is apparently sufficient to slide the cell along the surface.

***Cytophaga* and related genera** Organisms of the genus *Cytophaga* are long slender rods, often with pointed ends, which move by gliding. Many digest cellulose, agar, or chitin. They are widespread in the soil and water, often being present in great abundance. The cellulose decomposers can be easily isolated by placing small crumbs of soil on pieces of cellulose filter paper laid on the surface of mineral agar. The bacteria attach to and digest the cellulose fibers, forming transparent spreading colonies that

Table 19.16 Characteristics of some genera of heterotrophic gliding bacteria

Characteristics	Genus	DNA (mole % GC)
Rods and nonseptate filaments:		
Unicellular, rod shaped, many digest cellulose, chitin, or agar	*Cytophaga* (no microcysts)	28–39
	Sporocytophaga (microcysts formed)	36
Helical or spiral shaped	*Saprospira*	35–48
Filamentous	*Microscilla, Flexibacter*	46–48
Septate filaments:		
Filamentous, heterotrophic or lithotrophic, producing S^0 granules from H_2S	*Beggiatoa*	37–43
	Thiofilum	—
Filamentous, heterotrophic; life cycle involving gonidia and rosette formation	*Leucothrix*	46–50
Filamentous, lithotrophic sulfur oxidizer; life cycle like *Leucothrix*	*Thiothrix*	—
Cells in chains or short filaments; heterotrophic; occurs in oral cavity or digestive tract of man and other animals	*Simonsiella, Alysiella*	41–55
Filamentous	*Vitreoscilla*	44–45
Rods, forming fruiting bodies:		
Unicellular, rod shaped; life cycle involving aggregation, fruiting-body formation, and myxospore formation	Fruiting myxobacteria; *Archangium, Chondromyces, Myxococcus, Polyangium*, etc. (see Table 19.17)	67–71

are usually yellow or orange in color. Microscopic examination reveals the bacteria aligned upon the surface of the cellulose fibrils. The cytophagas do not produce soluble, extracellular, cellulose-digesting enzymes (cellulases); the enzymes probably remain attached to the cell envelope, accounting for the fact that the cells must adhere to cellulose fibrils in order to digest them. Organisms of the genus *Sporocytophaga* are similar to *Cytophaga* in morphology and physiology, but form resting spherical structures called *microcysts*, similar to those produced by some fruiting myxobacteria (see below), although they are produced without formation of fruiting bodies. Despite its ability to form microcysts, *Sporocytophaga* is not related to the fruiting myxobacteria; its DNA base composition is similar to that of *Cytophaga*, but far removed from those of the fruiting myxobacteria (Table 19.16). In pure culture, *Cytophaga* can be cultured on agar containing embedded cellulose fibers, the presence of the organism being indicated by the clearing that occurs as the cellulose is digested (Figure 16.36).

Beggiatoa Organisms of this genus are morphologically similar to filamentous cyanobacteria. The filaments of *Beggiatoa* are usually quite large in diameter and long, consisting of many short cells attached end-to-end (Figure 19.54). In addition to moving by gliding, they can flex and twist so that many filaments may become intertwined to form a complex tuft. *Beggiatoa* is found in nature primarily in habitats rich in H_2S, such as sulfur springs, decaying seaweed beds, mud layers of lakes, and waters polluted with sewage, and in these habitats the filaments of *Beggiatoa* are usually filled with sulfur granules (Figures 19.54 and 19.55; see also Figure 19.56). *Beggiatoa* are also common inhabitants of hydrothermal vents (see Section 17.9). It was with *Beggiatoa* that Winogradsky first demonstrated that a living organism could oxidize H_2S to S^0 and then to SO_4^{2-}, leading him to formulate the concept of lithotrophy. However, most pure cultures of *Beggiatoa* so far isolated grow best *heterotrophically* on organic compounds such as acetate, succinate, and glucose, and when H_2S is provided as an electron donor, they still require organic substances for growth. On On the other hand, some marine strains of *Beggiatoa* have been isolated and shown to be true autotrophs, but this does not seem to be common. Organisms that can use inorganic compounds or light as energy sources but cannot use CO_2 as sole carbon source have been called **mixotrophs** (see Section 16.12).

An interesting habitat of *Beggiatoa* is the rhizosphere of plants (rice, cattails, and other swamp plants) living in flooded, and hence anaerobic, soils. Such plants pump oxygen down into their roots, so that a sharply defined bound-

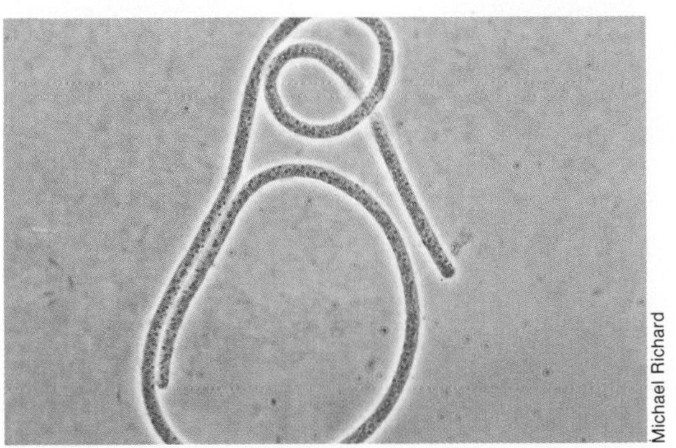

Figure 19.54 Phase-contrast photomicrograph of a *Beggiatoa* species.

Figure 19.55 Filamentous sulfur-oxidizing bacteria in a small stream. The filamentous cells twist together to form thick streamers and the white color is due to the abundant elemental sulfur content of the cells.

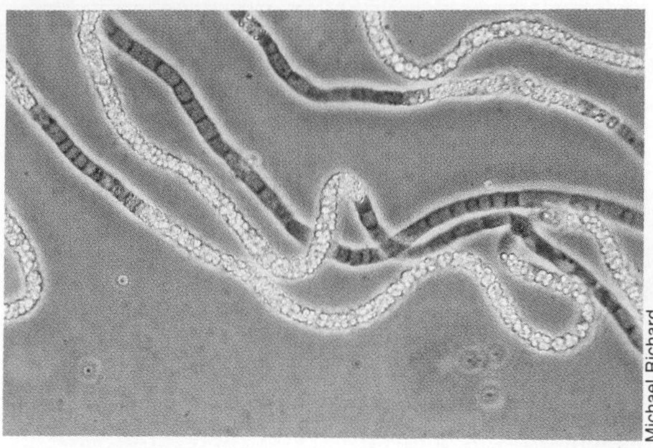

Figure 19.56 Phase-contrast photomicrograph of the filamentous sulfur-oxidizing bacterium *Thiofilum,* isolated from a sewage treatment plant. Note the abundant elemental sulfur granules in some of the cells.

ary develops at the root surface between O_2 on the root and H_2S in the soil. *Beggiatoa* (and probably other sulfur bacteria) develops at this boundary, and it has been suggested that *Beggiatoa* plays a beneficial role for the plant by oxidizing and thus detoxifying hydrogen sulfide. The growth of *Beggiatoa* is greatly stimulated by the addition to culture media of the enzyme *catalase* (which converts hydrogen peroxide into water and oxygen), and since plant roots contain catalase, it has been suggested that the plant promotes the growth of *Beggiatoa* in its rhizosphere via catalase production, thus leading to the development of a loose mutualistic relationship between the plant and the bacterium.

Beggiatoa and a related filamentous sulfide-oxidizer called *Thiofilum* (Figure 19.56) can cause major settling problems in sewage treatment facilities and in industrial waste lagoons such as from canning, paper pulping, brewing, and milling. These problems are generally referred to as ***bulking*** and occur when filamentous bacteria overgrow the normal flora of the waste system, producing a loose detrital floc instead of the normal and more easily settling tight floc. If bulking occurs the wastewater remains improperly treated because the effluent discharged is still high in BOD; in sewage treatment, for example, bulking occurs when *Beggiatoa* and *Thiofilum* replace *Zoogloea* in the activated sludge process (see Section 17.21).

Leucothrix* and *Thiothrix These two genera are related in cell structure and life cycle. *Thiothrix,* a lithotroph that oxidizes H_2S, is probably an obligate lithotroph, although this has not been proved with pure cultures. *Leucothrix* is the heterotrophic counterpart of *Thiothrix*, and since its members have been amenable to cultivation, the details of its life cycle and physiology are fairly well established. *Leucothrix* is a filamentous organism that has been found in nature only in marine environments, where it grows most commonly as an epiphyte on marine algae.

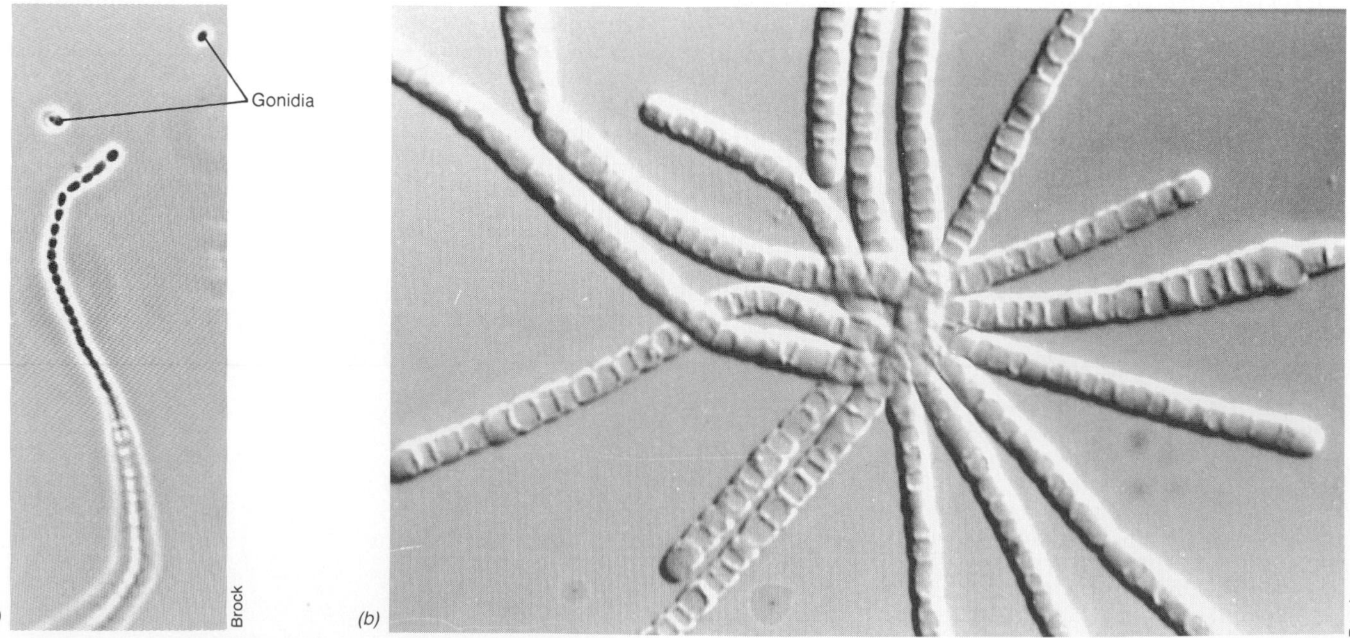

Figure 19.57 *Leucothrix mucor.* (a) Filaments showing multicellular nature and release of gonidia. (b) Rosette composed of several multicellular filaments. Nomarski interference contrast.

Leucothrix filaments are 2 to 5 μm in diameter and may reach lengths of 0.1 to 0.5 cm. The filaments have clearly visible cross walls, and cell division is not restricted to either end but occurs throughout the length of the filament. The free filaments never glide (thus distinguishing them from *Beggiatoa*), although they occasionally wave back and forth in a jerky fashion. Under environmental conditions unfavorable to rapid growth, individual cells of the filaments become round and form ovoid structures called *gonidia*, which are released individually, often from the tips of the filaments (Figure 19.57*a*). The gonidia are able to glide in a jerky manner when they come into contact with a solid surface. They settle down on solid surfaces, synthesize a holdfast, and through growth and successive cell divisions form new filaments. Presumably in nature the gonidia are elements of dispersal, enabling the organism to spread to other areas. If there are high concentrations of gonidia, individual cells may aggregate, probably because of mutual attraction; they then synthesize a holdfast that causes their ends to adhere in a rosette, and new filaments grow out (Figure 19.57*b*). Rosette formation is found in both *Leucothrix* and *Thiothrix* and is an important means of distinguishing these organisms from many other filamentous bacteria.

Fruiting myxobacteria The fruiting myxobacteria exhibit the most complex behavioral patterns and life cycles of all known procaryotic organisms. The vegetative cells of the fruiting myxobacteria are simple, nonflagellated, Gram-negative rods that glide across surfaces and obtain their nutrients primarily by causing the lysis of other bacteria. Under appropriate conditions a swarm of vegetative cells aggregate and construct "fruiting bodies," within which some of the cells become converted into resting structures called **myxospores**. (A myxospore is defined as a resting cell contained in a fruiting body. A myxospore enclosed in a hard slime capsule is called a **microcyst**.) It is the ability to form complex fruiting bodies that distinguishes the myxobacteria from all other procaryotes. Since the vegetative cells of fruiting myxobacteria look like those of nonfruiting gliding bacteria, it is only through observation of the fruiting bodies that these organisms can be identified (see Table 19.17).

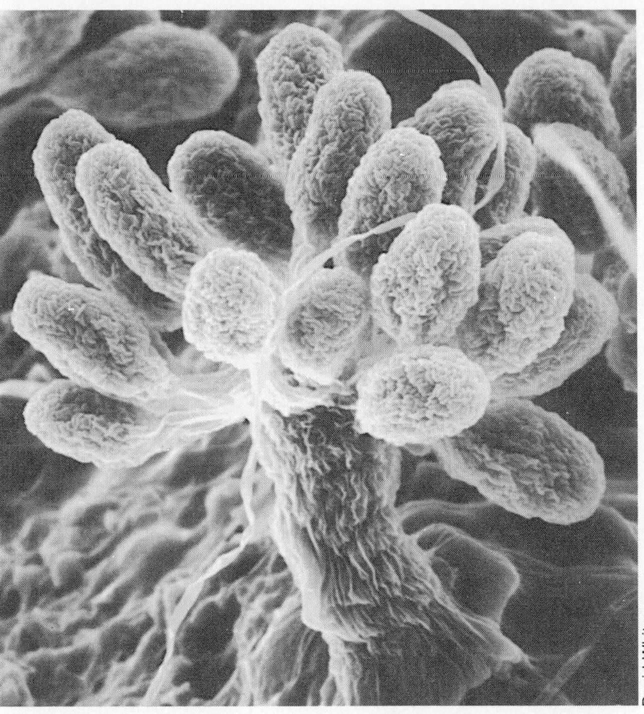

Figure 19.58 Scanning electron micrograph of a fruiting body of the gliding myxobacterium *Stigmatella aurantiaca* growing on a piece of wood. Note the individual cells visible in each fruiting structure.

The fruiting bodies of the myxobacteria vary from simple globular masses of myxospores in loose slime to complex forms with a fruiting-body wall and a stalk (Figure 19.58). The fruiting bodies are often strikingly colored (Table 19.17). Occasionally they can be seen with a hand lens or dissecting microscope on pieces of decaying wood or plant material. Fruiting bodies of myxobacteria often develop on dung pellets (for example, those of the sheep or rabbit) after they have been incubated for a few days in a moist chamber. Although the vegetative cells are common

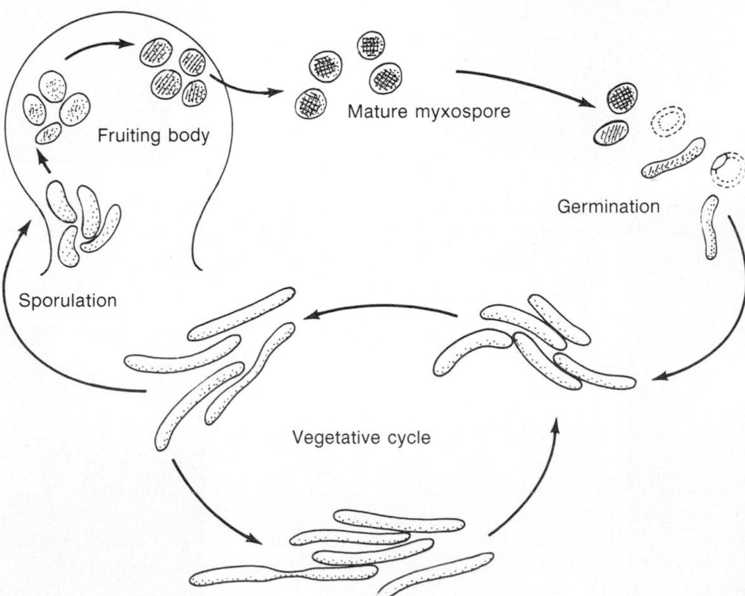

Figure 19.59 Life cycle of *Myxococcus xanthus*.

Table 19.17 Classification of the fruiting myxobacteria

Characteristics	Genus	DNA (mole % GC)
Vegetative cells tapered, microcysts produced		
Spherical or oval microcysts, fruiting bodies usually without well-defined sporangia or stalks	*Myxococcus*	68–71
Rod-shaped microcysts:		
Microsysts not contained in sporangia, fruiting bodies without stalks	*Archangium*	67–68
Microcysts contained in cysts with well-defined walls:		
Fruiting bodies without stalks	*Cystobacter*	68
Stalked fruiting bodies, single sporangia	*Melittangium*	—
Stalked fruiting bodies, multiple sporangia	*Stigmatella*	67–68
Vegetative cells not tapered (blunt rounded ends); microcysts not produced, myxospores resemble vegetative cells; sporangia always produced:		
Fruiting bodies without stalks, myxospores rod shaped	*Sorangium*	—
	Polyangium	—
Fruiting bodies without stalks, myxospores coccoid	*Nannocystis*	70–71
Stalked fruiting bodies	*Chondromyces*	69–71

Based on Reichenbach, H., and M. Dworkin. 1981. Pages 328–355 **in** *M. P. Starr, H. Stolp, H. G. Trüper, A. Balows, and H. G. Schlegel (eds.)* The Prokaryotes: A Handbook on Habitats, Isolation, and Identification of Bacteria, Vol. I. *Springer-Verlag, New York.*

Typical fruiting bodies of selected myxobacteria

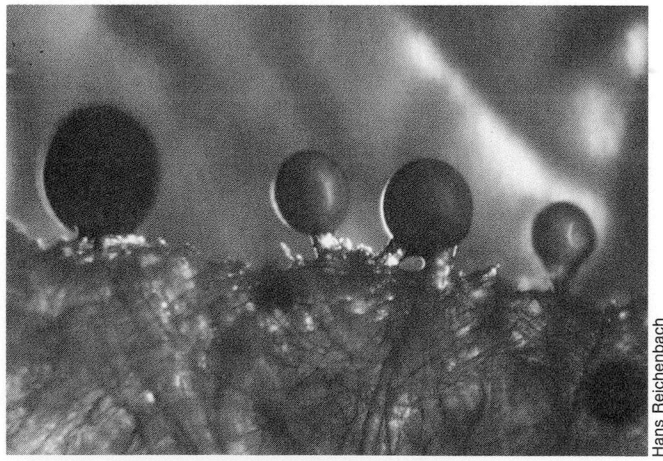

Myxococcus fulvus, about 125μm high

Myxococcus stipitatus, about 170μm high

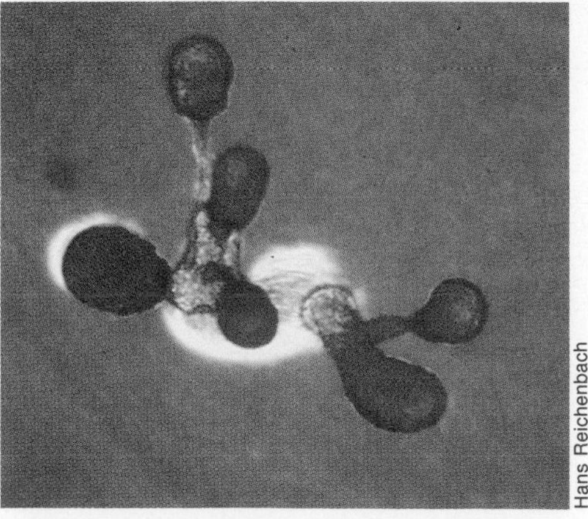

Mellitangium erectum, about 50μm high

Stigmatella aurantiaca about 150μm high

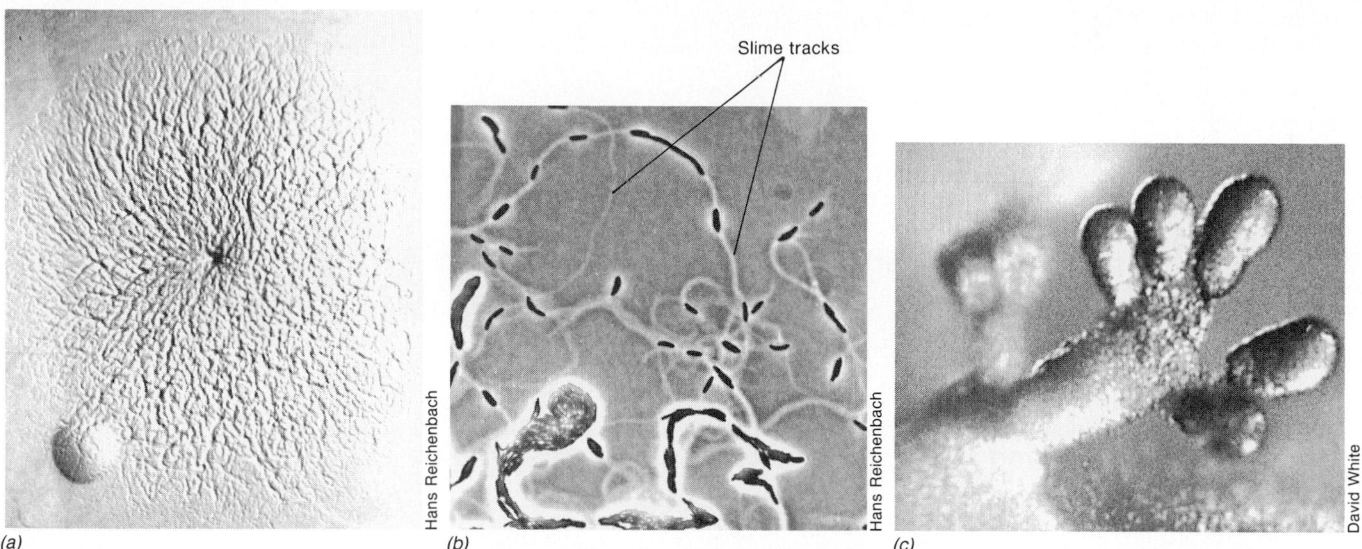

Figure 19.60 (a) Photomicrograph of a swarming colony (9-mm diameter) of *Myxococcus xanthus* on agar. (b) Single cells of *M. fulvus* from an actively gliding culture, showing the characteristic slime tracks on the agar. (c) Scanning electron micrograph of a fruiting body of *Stigmatella aurantiaca*.

in soils, the fruiting bodies themselves are less common. An effective means of isolating fruiting myxobacteria is to prepare Petri plates of water agar (1.5 percent agar in distilled water with no added nutrients) on which is spread a heavy suspension of any of several bacteria that the myxobacteria can lyse and use as a source of nutrients (for example, *Micrococcus luteus* or *E. coli*). In the center of the plate a small amount of soil, decaying bark, or other natural material is placed. Myxobacteria in the inoculum lyse the bacterial cells and use their liberated products as nutrients; as they grow, they swarm out across the plate from the inoculum site. After several days to a week, the plates are examined under a dissecting microscope for myxobacterial swarms or fruiting bodies, and pure cultures are obtained by transfer to organic media of cells from the fruiting bodies or from the edge of the swarm.

The life cycle of a typical fruiting myxobacterium is shown in Figure 19.59. The vegetative cells are typical Gram-negative rods and do not reveal in their fine structure any clue as to their gliding motility or to their ability to aggregate and form fruiting structures. The vegetative cells of many strains grow poorly or not at all when first dispersed in liquid medium but can often be adapted to growth in liquid by making several passages through shaken liquid growth medium. A vegetative cell usually excretes a slime, and as it moves across a solid surface it leaves a slime trail behind (Figure 19.60*a*). This trail is preferentially used by other cells in the swarm so that often a characteristic radiating pattern is soon created, with cells migrating along slime trails (Figure 19.60*b*). The fruiting body ultimately formed (Figure 19.60*c*) is a complex structure formed by the differentiation of cells in the stalk region and in the myxospore-bearing head. A wide variety of Gram-positive and Gram-negative bacteria, as well as fungi, yeasts, and algae, can be used as food sources. A few fruiting myxobacteria can also use cellulose. Many myxobacteria can be grown in the laboratory on media containing peptone or casein hydrolysate, which provides organic nutrients in the form of amino acids or small peptides; carbohydrates do not ordinarily promote or stimulate growth, and vitamins are usually not required. Inability to utilize carbohydrates reflects the lack of certain enzymes of the Embden-Meyerhof pathway. The organisms are typical aerobes with a well-developed citric acid cycle and cytochrome system.

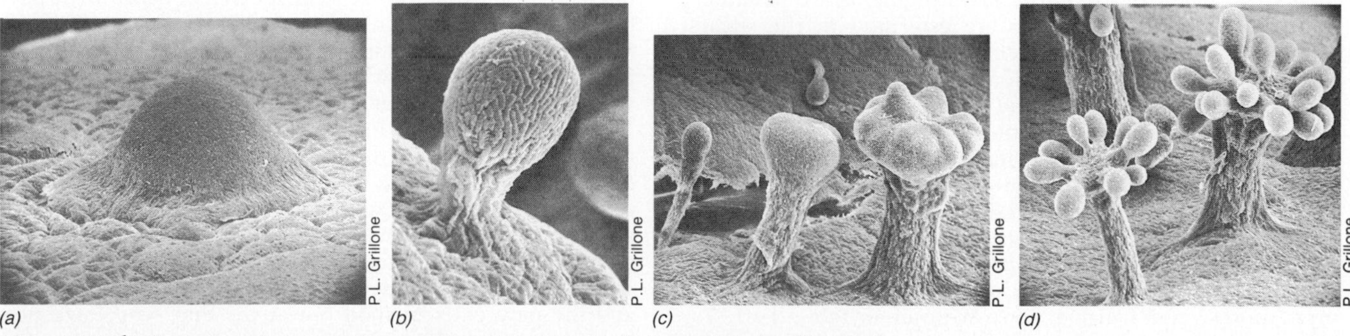

Figure 19.61 Scanning electron micrographs of fruiting body formation in *Chondromyces crocatus*. (a) Early stage, showing aggregation and mound formation. (b) Initial stage of stalk formation. Slime formation in the head has not yet begun so that the cells of which the head is composed are still visible. (c) Three stages in head formation. Note that the diameter of the stalk also increases. (d) Mature fruiting bodies.

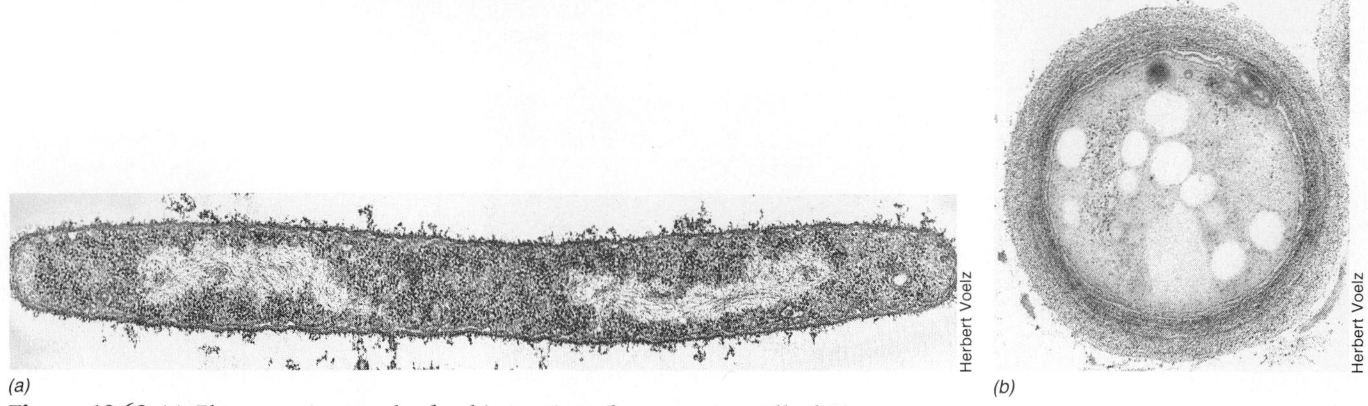

(a) (b)

Figure 19.62 (a) Electron micrograph of a thin section of a vegetative cell of *M. xanthus*. (b) Myxospore (microcyst) of *M. xanthus*, showing the multilayered outer wall.

Fruiting-body formation does not occur so long as adequate nutrients for vegetative growth are present, but upon exhaustion of amino acids the vegetative swarms begin to fruit. Cells aggregate, possibly through a chemotactic response, with the cells migrating toward each other and forming mounds or heaps (Figure 19.61*a*). A single fruiting body may have 10^9 or more cells. As the cell mounds become higher, the differentiation of the fruiting body into stalk and head begins (Figure 19.61*b* and *c*). Figure 19.61*d* clearly illustrates the differentiation of the fruiting body into stalk and head. The stalk is composed of nonliving slime, within which a few cells may be trapped. The majority of the cells accumulate in the fruiting-body head and undergo differentiation into *myxospores*. Some genera form encapsulated myxospores sometimes called *microcysts* (Figure 19.62 and Table 19.17). And, in some genera, the myxospores are enclosed in larger walled structures called **cysts**. Compared to the vegetative cell, the microcyst is more resistant to drying, sonic vibration, UV radiation, and heat, but the degree of heat resistance is much less than that of the bacterial endospore. Typically, the microcyst can withstand 58 to 60°C for 10 to 30 minutes, temperatures that would kill the vegetative cell. It seems likely that the main function of the microcyst is to enable the organism to survive desiccation during dispersal or during drying of the habitat. The microcyst germinates by a localized rupture of the capsule, with the growth and emergence of a typical vegetative rod.

Myxobacteria are usually colored by carotenoid pigments (see Table 19.17). The main pigments are peculiar carotenoid glycosides, which are esterified in the sugar moiety with fatty acids. Pigment formation is promoted by light, and at least one function of the pigment is photoprotection. Since in nature, the myxobacteria usually form fruiting bodies in the light, the presence of these photoprotective pigments is understandable. In the genus *Stigmatella*, light greatly stimulates fruiting body formation, and it is thought that light catalyzes production of a pheromone that initiates the aggregation step. The fruiting myxobacteria are classified primarily on morphological grounds using characteristics of the vegetative cells, the myxospores, and fruiting body structure (Table 19.17). However, many of the genera outlined in Table 19.17 have not been studied in much detail, and further study may alter ideas on how they should be classified. Phylogenetically, gliding myxobacteria belong to the purple bacterial group (see Section 18.5).

The fruiting myxobacteria provide experimental material for the study of a number of interesting problems in developmental microbiology, microbial ecology, and microbial evolution.

19.14 Sheathed Bacteria

Sheathed bacteria are filamentous organisms with a unique life cycle involving formation of flagellated swarmer cells within a long tube or sheath. Under certain (generally unfavorable) conditions, the swarmer cells move out and become dispersed to new environments, leaving behind the empty sheath. Under favorable conditions, vegetative growth occurs within the filament, leading to the formation of long, cell-packed sheaths. Sheathed bacteria are common in freshwater habitats that are rich in organic matter, such as polluted streams, trickling filters, and activated sludge plants, being found primarily in flowing waters. In habitats where reduced iron or manganese compounds are present the sheaths may become coated with a precipitate of ferric hydroxide or manganese oxide (see, for example, Figure 16.22). Iron precipitation is probably due to chemical reactions, but some sheathed bacteria have the ability to oxidize manganous ions to manganese oxide. Two genera are currently recognized: *Sphaerotilus*, in which manganese oxidation does not occur, and *Leptothrix*, whose members do oxidize Mn^{2+}. A single species of *Sphaerotilus* is recognized, *S. natans*, but several species of *Leptothrix* have been discerned, distinguished primarily on size, flagellation of swarmers, and some other morphological characteristics. Most of our discussion will concern *S. natans*, the organism which has been most extensively studied.

The *Sphaerotilus* filament is composed of a chain of rod-shaped cells with rounded ends enclosed in a closely fitting sheath. This thin and transparent sheath is difficult to see when it is filled with cells, but when the filament is partially empty the sheath can easily be seen by phase-contrast microscopy (Figure 19.63*a*) or by staining. The cells within the sheath divide by binary fission (Figure 19.63*b*), and the new cells pushed out at the end synthesize new sheath material. Thus the sheath is always formed at the *tips* of the filaments. Individual cells are 1 to 2 μm wide by 3 to 8 μm long and stain Gram-negatively. Eventually cells are liberated from the sheaths, probably when the nutrient supply is low. These free cells are actively motile, the flagella being arranged lophotrichously (in a bundle at one pole) (Figure 19.63*c*). Probably the flagella are synthesized before the cells leave the sheath and, if so, may even aid in their liberation. It is thought that the swarmer cells then migrate, settle down, and begin to grow, each swarmer being the forerunner of a new filament. The sheath, which is devoid of muramic acid or other components of the peptidoglycan cell wall, is a protein-polysaccharide-lipid complex, possibly analogous to the capsules formed by many Gram-negative bacteria but differing in that it forms a linear structure.

Sphaerotilus cultures are nutritionally versatile, able to use a wide variety of simple organic compounds as carbon and energy sources, with inorganic nitrogen sources. Many strains require vitamin B_{12}, a substance frequently needed by aquatic microorganisms. Befitting its habitat in flowing waters, *Sphaerotilus* is an obligate aerobe.

As we noted, *Sphaerotilus* is widespread in nature in aquatic environments receiving rich organic matter. *Sphaerotilus* blooms often occur in the fall of the year in streams and brooks when leaf fall causes a temporary increase in the organic content of the water. Its filaments are the main component of a microbial complex that sanitary engineers call "sewage fungus," which is the fungus-like filamentous slime found on the rocks in streams receiving sewage pollution. In activated sludge plants (see Section 17.21), *Sphaerotilus* growth, like that of *Beggiatoa* and *Thiofilum* (see Section 19.13), is often responsible for the detrimental condition called "bulking." The tangled masses of *Sphaerotilus* filaments so increase the bulk of the sludge that it does not settle properly, thus presenting difficulties in sludge clarification.

The oxidation of Fe^{2+} and Mn^{2+} by members of the *Sphaerotilus-Leptothrix* group has been the topic of considerable research and controversy. Because members of this group live at neutral pH in aerobic habitats, where iron and manganese often oxidize spontaneously, it has been difficult to show a specific effect of these organisms on the oxidation process. Although at one time, the *Sphaerotilus-Leptothrix* group was considered to be a specific group of iron bacteria, there is currently no basis for this classification. However, the ability of *Sphaerotilus* and *Leptothrix* to cause precipitation of iron oxides on their sheaths is well established. Such iron-encrusted sheaths are frequently seen in iron-rich waters (Figure 19.64). The process whereby iron deposition occurs is as follows: in iron-rich waters, ferrous iron is often held in solution as a chelate with organic materials such as humic and tannic acids. The sheathed bacteria can take up these soluble chelates, oxidize the organic compound, liberating the ferrous ions, which then oxidize spontaneously and become precipitated, generally in the region of the sheath (see Figure 19.64).

In the case of Mn^{2+}, specific oxidation by *Leptothrix*, but not by *Sphaerotilus*, is known to occur. (Oxidation of Mn^{2+} is not unique to *Leptothrix*, as a number of other bacteria, as well as fungi and yeasts, are able to carry out

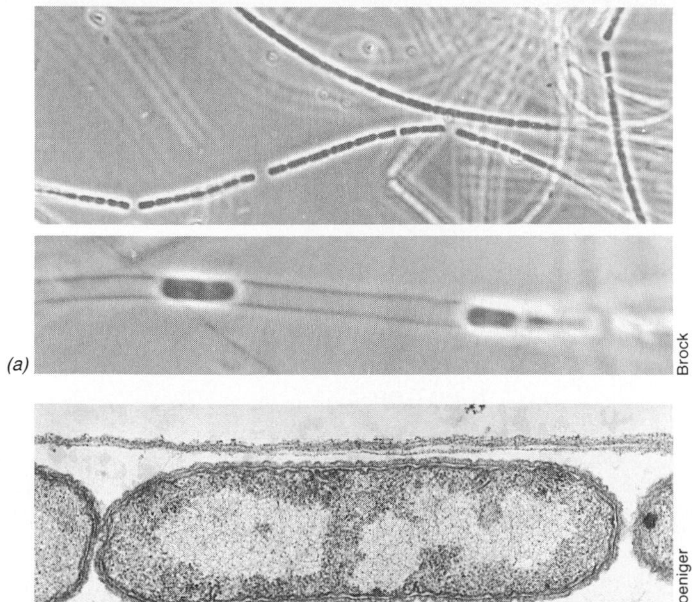

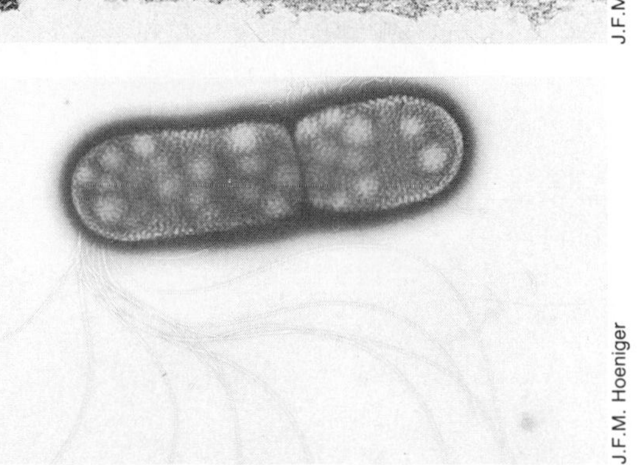

Figure 19.63 *Sphaerotilus natans.* (a) Phase-contrast photomicrographs of material collected from a polluted stream. Active growth stage (above) and swarmer cells leaving the sheath. (b) Electron micrograph of a thin section through a filament. (c) Electron micrograph of a negatively stained swarmer cell. Notice the polar flagellar tuft.

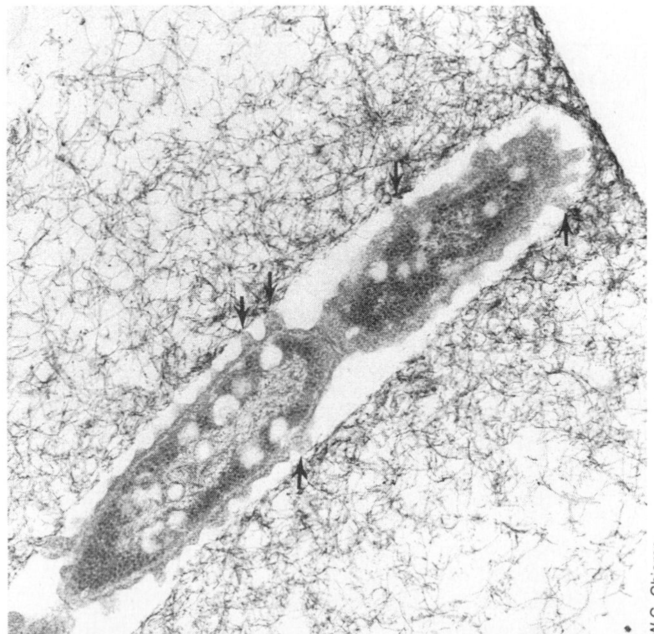

Figure 19.64 Transmission electron micrograph of a thin section of *Leptothrix* sp. in a sample from a ferromanganese film in a swamp in Ithaca, New York. Note the protuberances of the cell envelope which contact the sheath (arrows).

this process.) At pH values below 8, Mn^{2+} does not oxidize spontaneously, so that a significant biological oxidation is possible. However, even with Mn^{2+}, there is no evidence that *Leptothrix* obtains *energy* from the process, either lithotrophically or mixotrophically. *Leptothrix* is unable to grow on completely inorganic media containing Mn^{2+} as sole electron donor for energy generation and when organic compounds are present, careful analysis of growth yields shows no increase in yield due to the added Mn^{2+}. Although *Leptothrix* produces a protein which catalyzes Mn^{2+} oxidation, it is unclear as yet what benefit the organism derives from this oxidation process.

19.15 The Pseudomonads

All of the genera in this group are straight or slightly curved rods with *polar* flagella (Figure 19.65). They are heterotrophic aerobes, and never show a fermentative metabolism. (Fermentative organisms with polar flagella are generally classified in the genera *Aeromonas* or *Vibrio*, as indicated in Section 19.19). The most important genus is *Pseudomonas*, discussed in some detail here. Other genera include *Xanthomonas*, primarily a plant pathogen responsible for a number of necrotic plant lesions and which is characterized by its yellow-colored pigments; *Zoogloea*, characterized by its formation of an extracellular fibrillar polymer, which causes the cells to aggregate into distinctive flocs (this organism is a dominant component of activated sludge, Section 17.21), and *Gluconobacter*, characterized by its incomplete oxidation of sugars or alcohols to acids, such as the oxidation of glucose to gluconic acid or ethanol to acetic acid (this organism is discussed briefly with the other acetic acid bacteria in Section 19.17). Phylogenetically, the various genera of pseudomonads scatter within the purple bacterial group (see Section 18.5). Presumably the pseudomonads are derived from ancestral phototrophic bacteria that dispensed with photosynthesis in evolving to colonize habitats in which the ability to carry out anoxygenic photosynthesis was not a significant advantage (for example, in soil and on the surfaces of plants and animals).

Table 19.18 The genus *Pseudomonas*

General characteristics:
Straight or curved rods but not vibrioid; size 0.5–1.0 μm by 1.5–4.0 μm; no spores; Gram-negative; polar flagella: single or multiple; no sheaths, appendages or buds; respiratory metabolism, never fermentative, although may produce small amounts of acid from glucose aerobically; use low-molecular-weight organic compounds, not polymers; some are lithotrophic, using H_2 or CO as sole electron donor; some can use nitrate as electron acceptor anaerobically; some can use arginine as energy source anaerobically.

Minimal characteristics for identification:
Gram-negative, straight or slightly curved; no spores; motile (always); polar flagella (flagellar stain); oxidative-fermentative medium with glucose: open, acid produced; closed, acid not produced; gas not produced from glucose (distinguishes them easily from enteric bacteria and *Aeromonas*); oxidase, almost always positive (enterics are oxidase negative); catalase always positive; photosynthetic pigments absent (distinguishes them from nonsulfur purple bacteria); indole, negative; methyl red negative; Voges-Proskauer, negative

The genus *Pseudomonas* The distinguishing characteristics of members of the genus *Pseudomonas* are given in Table 19.18. Also given in this table are the minimal characteristics needed to identify an organism as a pseudomonad. Key identifying characteristics are the absence

(a)

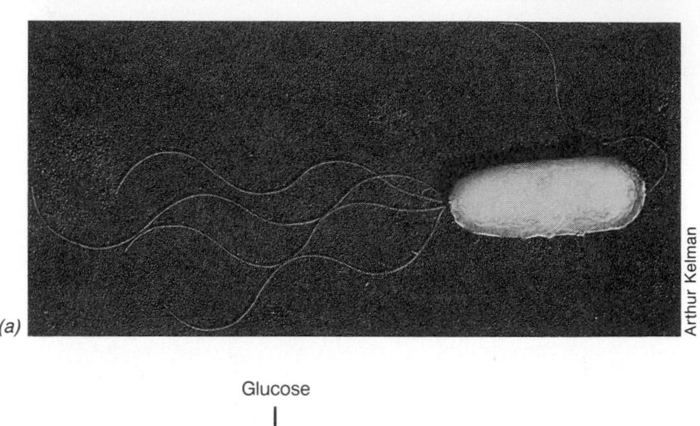

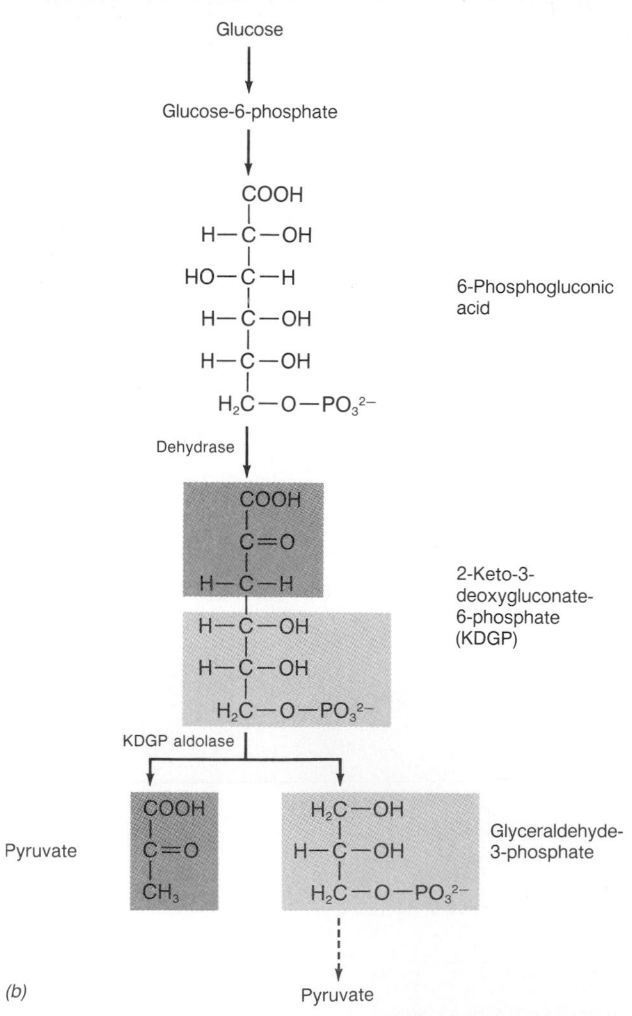

(b)

Figure 19.65 Typical pseudomonad morphology and a biochemical pathway common to pseudomonads. (a) Shadow cast preparation of *Pseudomonas coronafaciens*. (b) The Entner-Doudoroff pathway, the major means of glucose catabolism in pseudomonads.

of gas formation from glucose, and the positive oxidase test, both of which help to distinguish pseudomonads from enteric bacteria (Section 19.20). Also, the genus *Rhodopseudomonas* (see Section 19.1) is distinguished from *Pseudomonas* by the presence of photosynthetic pigments when the former is grown anaerobically in the light. (Since *Rhodopseudomonas* will grow aerobically in the dark without producing photosynthetic pigments, it could be mistaken for a pseudomonad.)

The species of the genus *Pseudomonas* are defined on the basis of various physiological characteristics, as outlined in Tables 19.19 and 19.20. Some species (for example, *P. aeruginosa*) are quite homogeneous, so that all isolates fit into a very narrow range of distribution of characteristics, whereas other species are much more heterogeneous. The taxonomy of the genus *Pseudomonas* has been greatly clarified by DNA hybridization studies, and the subgroups given in Table 19.19 are based on DNA homologies.

Representatives of the genus *Pseudomonas* have in most cases very simple nutritional requirements and grow organotrophically at neutral pH and at temperatures in the mesophilic range. One of the striking properties of the pseudomonads is the wide variety of organic compounds used as carbon sources and as electron donors for energy generation. Some strains utilize over *100* different compounds, and only a few strains utilize fewer than 20. As an example of this versatility, a single strain of *P. aeruginosa* can make use of many different sugars, fatty acids, dicarboxylic acids, tricarboxylic acids, alcohols, polyalcohols, glycols, aromatic compounds, amino acids, and amines, plus miscellaneous organic compounds not fitting into any of the above categories. On the other hand, pseudomonads are generally unable to break down polymers into their component monomers. The pseudomonads are ecologically important organisms in soil and water and are probably responsible for the aerobic degradation of many soluble compounds derived from the breakdown of plant and animal materials.

A few pseudomonads are pathogenic (Table 19.20). Among the fluorescent pseudomonads, the species *P. aeruginosa* is frequently associated with infections of the urinary and respiratory tracts in humans. *Pseudomonas aeruginosa* infections are also common in patients receiving treatment for severe burns. *Pseudomonas aeruginosa* is able to grow at temperatures up to 43°C, whereas the nonpathogenic species of the fluorescent group can grow only at lower temperatures. *Pseudomonas aeruginosa* is not an obligate parasite, however, since it can be readily isolated from soil, and as a denitrifier (see Section 17.14) it plays

Table 19.19 Characteristics of subgroups and species of the genus *Pseudomonas*

Group	Characteristics	DNA (mole % GC)
Fluorescent subgroup:	Most produce water-soluble, yellow-green fluorescent pigments; do not form poly-β-hydroxybutyrate; single DNA homology group	
P. aeruginosa	Pyocyanin production, growth at up to 43°C, single polar flagellum, denitrification	67
P. fluorescens	Do not produce pyocyanin or grow at 43°C; tuft of polar flagella	59–61
P. putida	Similar to *P. fluorescens* but do not liquefy gelatin and do grow on benzylamine	62
P. syringae	Lack arginine dihydrolase, oxidase negative, pathogenic to plants	58–60
Acidovorans subgroup:	Nonpigmented, form poly-β-hydroxybutyrate, tuft of polar flagella, do not use carbohydrates; single DNA homology group	
P. acidovorans	Uses muconic acid as sole carbon source and electron donor	67
P. testosteroni	Use testosterone as sole carbon source	62
Pseudomallei-cepacia subgroup:	No fluorescent pigments, tuft of polar flagella, form poly-β-hydroxybutyrate; single DNA homology group	62
P. cepacia	Extreme nutritional versatility; some strains pathogenic to plants	67–68
P. pseudomallei	Cause melioidosis in animals; nutritionally versatile	69
P. mallei	Cause glanders in animals; nonmotile; nutritionally restricted	69
Diminuta-vesicularis subgroup:	Single flagellum of very short wavelength, require vitamins (pantothenate, biotin, B_{12})	
P. diminuta	Nonpigmented, do not use sugars	66–67
P. vesicularis	Carotenoid pigment, use sugars	66
Miscellaneous species:		
P. solanacearum	Plant pathogens	66–68
P. saccharophila	Grow lithotrophically with H_2, digest starch	69
P. maltophilia	Require methionine, do not use NO_3^- as N source, oxidase negative	67

Table 19.20 Pathogenic pseudomonads

Species	Relationship to disease
Animal pathogens	
P. aeruginosa:	Opportunistic pathogen, especially in hospitals; in patients with metabolic, hematologic, and malignant diseases; hospital-acquired infections from catheterizations, tracheostomies, lumbar punctures, and intravenous infusions; in patients given prolonged treatment with immunosuppressive agents, corticosteroids, antibiotics and radiation; may contaminate surgical wounds, abscesses, burns, ear infections, lungs of patients treated with antibiotics; primarily a soil organism
P. fluorescens:	Rarely pathogenic, as does not grow well at 37°C; may grow in and contaminate blood and blood products under refrigeration
P. maltophilia:	An ubiquitous, free-living organism that is an occasional opportunist in humans
P. cepacia:	Onion bulb-rot; has also been isolated from humans and from environmental sources of medical importance
P. pseudomallei:	Causes melioidosis, a disease endemic in animals and humans in southeast Asia
P. mallei:	Causes glanders, a disease of horses that is occasionally transmitted to humans
P. stutzeri:	Often isolated from humans and environmental sources; may live saprophytically in the body
Plant pathogens	
P. solanacearum:	Plant pathogenic, causing wilts of many cultivated plants (e.g., potato, tomato, tobacco, peanut)
P. syringae:	Plant pathogen, attacks foliage causing chlorosis and necrotic lesions on leaves; rarely found free in soil
P. marginalis:	Causes soft rot of various plants; active pectinolytic species
Xanthomonas:	Causes necrotic lesions on foliage, stems, fruits; also causes wilts and tissue rots; rarely found free in soil

an important role in the nitrogen cycle in nature. As a pathogen it appears to be primarily an opportunist, initiating infections in individuals whose resistance is low. In addition to urinary infections it can also cause systemic infections, usually in individuals who have had extensive skin damage by burns. The organism is naturally resistant to many of the widely used antibiotics, so that chemotherapy is often difficult. Resistance is frequently due to a *resistance transfer factor* (*R factor;* see Section 7.6), which is a plasmid carrying genes coding for detoxification of various antibiotics. *P. aeruginosa* is commonly found in the hospital environment and can easily infect patients receiving treatment for other illnesses (see Section 14.7 for a discussion of hospital-acquired infections). Polymyxin, an antibiotic not ordinarily used in human therapy because of its toxicity, is effective against *P. aeruginosa* and can be used with caution.

Many pseudomonads, as well as a variety of other Gram-negative bacteria, metabolize glucose via the Entner-Doudoroff pathway (Figure 19.65*b*). Two key enzymes of the Entner-Doudoroff pathway are *6-phosphogluconate dehydrase* and *ketodeoxyglucosephosphate aldolase* (Figure 19.65*b*). A survey for the presence of these enzymes in a wide variety of bacteria has shown that they are absent from all Gram-positive bacteria (except a few *Nocardia* isolates) and are generally present in bacteria of the genera *Pseudomonas, Rhizobium*, and *Agrobacterium*, as well as in some isolates of several other genera of Gram-negative bacteria.

Phytopathogens Certain species of *Pseudomonas* and the genus *Xanthomonas* are well-known plant pathogens (phytopathogens). In many cases these organisms are so highly adapted to the plant environment that they can rarely be isolated from other habitats, including soil. Phytopathogens frequently inhabit nonhost plants (where disease symptoms are not apparent) and from here become transmitted to host plants and initiate infection. Disease symptoms vary considerably depending on the particular phytopathogen and host plant, and are generally due to the release by the bacterium of plant toxins, lytic enzymes, plant growth factors, and other substances that destroy or distort plant tissue. In many cases the disease symptoms are highly diagnostic of the type of phytopathogen, and are actually used in the taxonomy of phytopathogenic pseudomonads. Thus, *P. syringae* is frequently isolated from leaves showing chlorotic lesions, whereas *P. marginalis* is a typical "soft-rot" pathogen, infecting stems and shoots, but rarely leaves.

19.16 Free-Living Aerobic Nitrogen-Fixing Bacteria

A variety of organisms which primarily inhabit the soil are capable of fixing N_2 *aerobically* (Table 19.21). The genus *Azotobacter* comprises large, Gram-negative, obligately aerobic rods capable of fixing N_2 nonsymbiotically (see Figure 19.66). The first member of this genus was discovered by the Dutch microbiologist M. W. Beijerinck early in the twentieth century, using an enrichment culture technique with a medium devoid of a combined nitrogen source (see Box, Chapter 17). Although capable of growth on N_2, *Azotobacter* grows more rapidly on NH_3; indeed, adding NH_3 actually *represses* nitrogen fixation (see Section 16.24). Much work has been done in seeking to evaluate the role of *Azotobacter* in nitrogen fixation in nature, especially in comparison to the anaerobic organism *Clostridium pasteurianum* and the symbiotic organisms of the genus *Rhizobium. Azotobacter* is also of interest because it has the

Table 19.21 Genera of free-living aerobic nitrogen-fixing bacteria

Genus	Characteristics	DNA (mole % GC)
Azotobacter	Large rod; produces cysts; primarily found in neutral to alkaline soils	63–66
Azomonas	Large rod; no cysts; primarily aquatic	53–59
Azospirillum	Microaerophilic rod; associates with plants	69–71
Beijerinckia	Pear-shaped rods with large lipid bodies at each end; produces extensive slime; inhabits acidic soils	54–60
Derxia	Rods; form coarse, wrinkled colonies	70

highest respiratory rate (measured as the rate of O_2 uptake) of any living organism. In addition to its ecological and physiological importance, *Azotobacter* is of interest because of its ability to form an unusual resting structure called a *cyst* (Figure 19.66*b*).

Azotobacter cells are rather large for bacteria, many isolates being almost the size of yeasts, with diameters of 2–4 μm or more. Pleomorphism is common, and a variety of cell shapes and sizes have been described. Some strains are motile by peritrichous flagella. On carbohydrate-containing media, extensive capsules or slime layers are produced by free-living N_2-fixing bacteria (Figure 19.67). *Azotobacter* is able to grow on a wide variety of carbohydrates, alcohols, and organic acids. The metabolism of carbon compounds is strictly oxidative, and acids or other fermentation products are rarely produced. All members fix nitrogen, but growth also occurs on simple forms of combined nitrogen: ammonia, urea, and nitrate.

Despite the fact that *Azotobacter* is an obligate aerobe, its nitrogenase is as O_2 sensitive as all other nitrogenases (Section 16.24). It is thought that the high respiratory rate of *Azotobacter* (mentioned above) has something to do with protection of nitrogenase from O_2. The intracellular O_2 concentration is kept low enough by metabolism so that inactivation of nitrogenase does not occur.

The remaining genera of free-living N_2 fixers include *Azomonas*, a genus of large rod-shaped bacteria that resemble *Azotobacter* except that they do not produce cysts and are primarily aquatic, and *Beijerinckia* and *Derxia* (Figure 19.68), two genera that grow well in acidic soils.

Like bacterial endospores, *Azotobacter* cysts (Figure 19.66*b*) show negligible endogenous respiration and are resistant to desiccation, mechanical disintegration, and ultraviolet and ionizing radiation. In contrast to endospores, however, they are *not* especially heat resistant, and they are not completely dormant since they rapidly oxidize exogenous energy sources. Treatment of cysts with metal-binding agents such as citrate results in solubilization of the outer layers, presumably because the structure of the outer layer is stabilized by metals. The central body is then liberated in viable form. The central body does not possess the resistance characteristics of the cyst, thus suggesting that it is the cyst coat that confers resistance. The carbon source of the medium greatly influences the extent of cyst formation, butanol being especially favorable; compounds related to butanol, such as β-hydroxybutyrate, also promote cyst formation.

The species *Azotobacter chroococcum* was the first N_2-fixing bacterium shown capable of growth on N_2 in the complete *absence* of molybdenum. When cells of *A. chroo-*

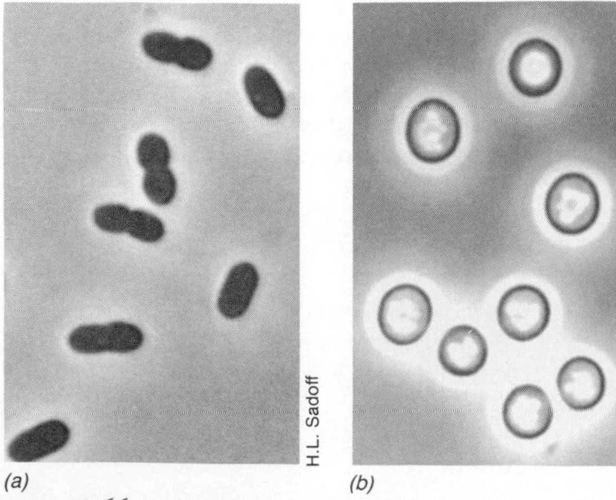

Figure 19.66 *Azotobacter vinelandii*: vegetative cells (a) and cysts (b) by phase-contrast microscopy.

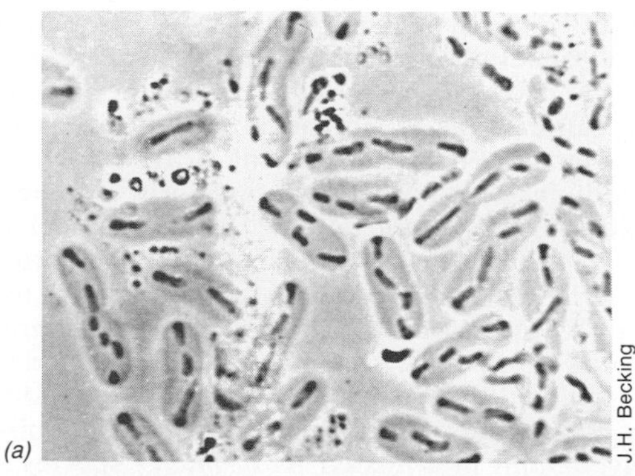

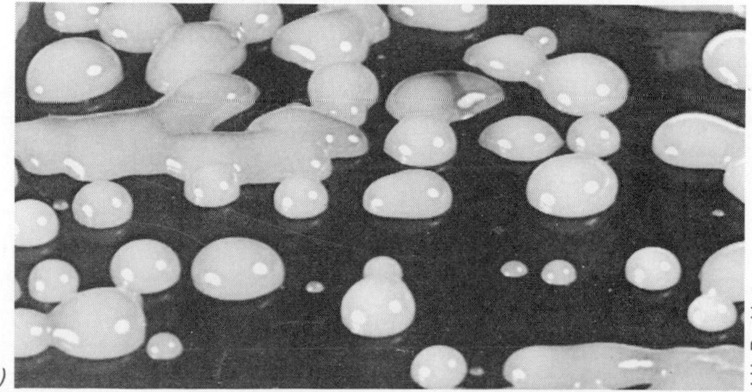

Figure 19.67 Examples of slime production by free-living N_2-fixing bacteria. (a) Cells of *Derxia gummosa* encased in slime. (b) Colonies of *Beijerinckia* species growing on a carbohydrate-containing medium. Note the raised glistening appearance of the colonies due to abundant capsular slime.

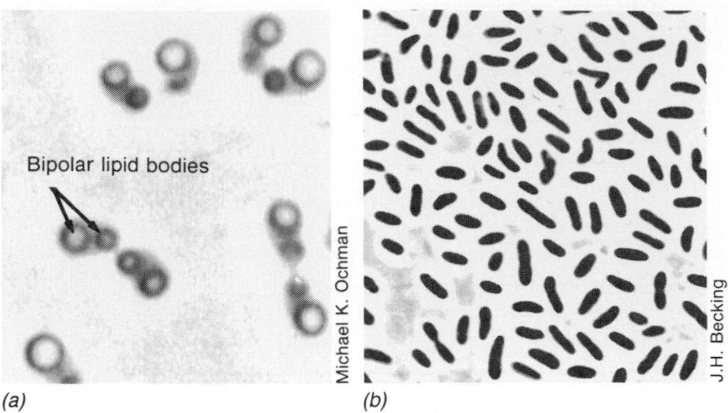

Figure 19.68 Phase-contrast photomicrographs of two genera of acid-tolerant, free-living N_2-fixing bacteria. (a) *Beijerinckia indica.* The cells are roughly pear-shaped and contain a large globule of poly-β-hydroxybutyrate at each end. (b) *Derxia gummosa.*

coccum are placed in a mineral medium lacking ammonia and molybdenum but containing the metal vanadium, a second form of dinitrogenase containing *vanadium* in place of *molybdenum* is produced and functions to support growth of the organism on N_2. Like the molybdenum enzyme (see Section 16.24), vanadium dinitrogenase consists of *two* proteins, one of which contains iron, the second iron and vanadium, and reduces N_2 to NH_3, H^+ to H_2, and C_2H_2 to C_2H_4. However, the rates of substrate reduction are considerably *lower* by the vanadium dinitrogenase than in the molybdenum enzyme, indicating that the vanadium enzyme is a less effective catalyst. Interestingly, the vanadium dinitrogenase also reduces C_2H_2 to ethane, C_2H_6. Although only about 3 percent of the acetylene reduced goes to ethane, the ability of the vanadium dinitrogenase to catalyze the complete reduction of C_2H_2 to C_2H_6 can serve as a test for detecting N_2-fixing bacteria capable of synthesizing this alternative dinitrogenase. Molybdenum dinitrogenases produce ethylene, C_2H_4, but not ethane, C_2H_6.

Vanadium dinitrogenase is apparently not the major dinitrogenase of *A. chroococcum*, because addition of molybdenum to the culture medium represses synthesis of the vanadium dinitrogenase, and instead the conventional dinitrogenase is formed. It is not known whether vanadium dinitrogenases are universal among N_2-fixing bacteria, however it is known that vanadium stimulates N_2 fixation in a number of organisms and that vanadium can also substitute for molybdenum in a related *Azotobacter* species, *A. vinelandii*, and in the heterocystous cyanobacterium, *Nostoc muscorum*. Using the production of ethane as an assay, vanadium dinitrogenase has also been discovered in *Clostridium pasteurianum*. Hence, vanadium-based dinitrogenases may be a widespread "backup" system for fixing N_2 in molybdenum-deficient habitats.

19.17 Acetic Acid Bacteria

As originally defined, the **acetic acid bacteria** comprised a group of Gram-negative, aerobic, motile rods that carried out an *incomplete* oxidation of alcohols, leading to the accumulation of organic acids as end products. With *ethanol* as a substrate, *acetic acid* is produced; hence the derivation of the common name for these bacteria. Another property is the relatively high tolerance to acidic conditions, most strains being able to grow at pH values lower than 5. This acid tolerance should of course be essential for an organism producing large amounts of acid. The genus name *Acetobacter* was originally used to encompass the whole group of acetic acid bacteria, but it is now clear that the acetic acid bacteria as so defined are a heterogeneous assemblage, comprising both peritrichously and polarly flagellated organisms. The *polarly* flagellated organisms are related to the pseudomonads, differing mainly in their acid tolerance and their inability to carry out a complete oxidation of alcohols. These organisms are now classified in the genus *Gluconobacter* (Table 19.22).

The genus *Acetobacter* now comprises the *peritrichously* flagellated organisms; these have no definite relationship to other Gram-negative rods. In addition to flagellation, *Acetobacter* differs from *Gluconobacter* in being able to further oxidize the acetic acid it forms to CO_2. This difference in ability to oxidize acetic acid is related to the presence of a complete citric acid cycle. *Gluconobacter*, which *lacks* a complete citric acid cycle, is unable to oxidize acetic acid, whereas *Acetobacter*, which has all enzymes of the cycle, can oxidize it. Because of these differences in oxidative potential, gluconobacters are sometimes called *underoxidizers*; the acetobacters, *overoxidizers*.

The acetic acid bacteria are frequently found in association with alcoholic juices, and probably originally evolved from pseudomonads inhabiting sugar-rich flowers and fruits where a yeast-mediated alcoholic fermentation is common. Acetic acid bacteria can often be isolated from alcoholic fruit juices such as cider or wine. Colonies of acetic acid bacteria can be recognized on $CaCO_3$-agar plates containing ethanol, the acetic acid produced causing a dissolution and clearing of the insoluble $CaCO_3$ (Figure 19.69). Cultures of acetic acid bacteria are used in commercial production of vinegar (see Section 10.10).

In addition to ethanol, these organisms carry out an incomplete oxidation of such organic compounds as higher alcohols and sugars. For instance, glucose is oxidized only to gluconic acid, galactose to galactonic acid, arabinose to arabonic acid, and so on. This property of underoxidation is exploited in the manufacture of ascorbic acid (vitamin C). Ascorbic acid can be formed from sorbose, but sorbose is difficult to synthesize chemically. It is, however, conveniently obtainable microbiologically from acetic acid bacteria, which oxidize sorbitol (a readily available sugar

Table 19.22 **Differentiation of *Acetobacter*, *Gluconobacter*, and *Pseudomonas***

Character	*Acetobacter*	*Gluconobacter*	*Pseudomonas*
Flagellation	Peritrichous	Polar	Polar
Growth at pH 4.5	+	+	−
Oxidation of ethanol to acetic acid at pH 4.5	+	+	−
Complete citric acid cycle	Present	Absent	Present
DNA (mole % GC)	53–64	56–64	58–70
Number of species	4	1	27

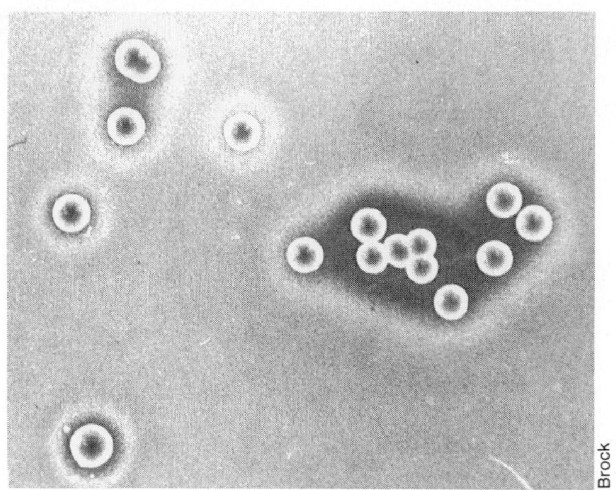

Figure 19.69 Photograph of colonies of *Acetobacter aceti* on calcium carbonate agar containing ethanol as energy source. Note the clearing around the colonies due to the dissolution of calcium carbonate by the acetic acid produced by the bacteria.

alcohol) only to sorbose, a process called *bioconversion,* see Section 10.8. The use of acetic acid bacteria makes the manufacture of ascorbic acid economically feasible.

19.18 Zymomonas and Chromobacterium

The genera *Zymomonas* and *Chromobacterium* are grouped together in *Bergey's Manual* as facultatively aerobic Gram-negative rods of uncertain taxonomic affiliation. Phylogenetically, both organisms are members of the purple bacterial group (see Section 18.5). The best known *Chromobacterium* species is *C. violaceum*, a bright purple pigmented organism (Figure 19.70) found in soil and water and occasionally in pyogenic infections of man and other animals. *C. violaceum* and a few other chromobacteria produce the purple pigment *violacein* (Figure 19.71), a water insoluble pigment that has antibiotic properties and is only produced in media containing the amino acid tryptophan. *Chromobacterium* is a facultative aerobe, growing fermentatively on sugars and aerobically on a variety of carbon sources.

The genus *Zymomonas* consists of large Gram-negative rods (2 to 6 μm long by 1 to 1.4 μm wide), which carry

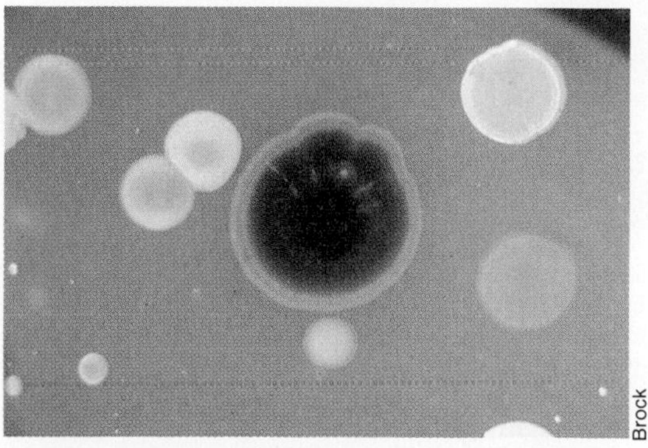

Figure 19.70 A large colony of *Chromobacterium violaceum* growing among other colonies on an agar plate.

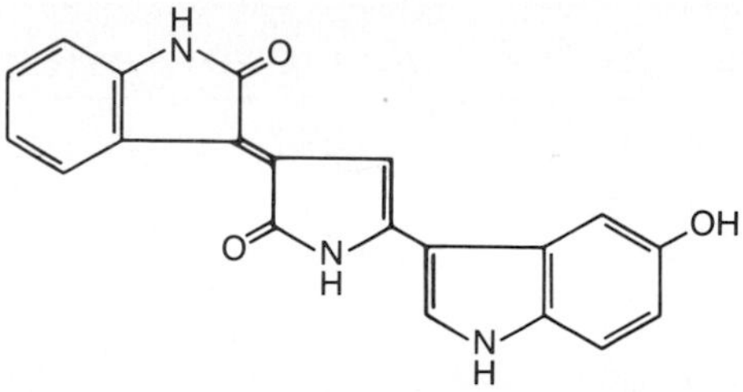

Figure 19.71 Structure of the pigment *violacein,* produced by various species of the genus *Chromobacterium.*

out a vigorous fermentation of sugars to ethanol. Although not all strains are motile, if motility occurs it is by lophotrichous flagella. *Zymomonas* is a common organism involved in alcoholic fermentation of various plant saps, and in many tropical areas of South and Central America, Africa, and Asia, it occupies a position similar to that of *Saccharomyces cerevisiae* (yeast) in North America and Europe. *Zymomonas* has been shown to be involved in the alcoholic fermentation of agave in Mexico, and palm sap in many tropical areas. It also carries out an alcoholic fermentation of sugar cane juice and honey. Although *Zymomonas* is rarely the sole organism involved in these alcoholic fermentations, it is often the dominant organism, and is probably responsible for the production of most of the ethanol (the desired product) in these beverages. *Zymomonas* is also responsible for spoilage of fruit juices such as apple cider and perry. It also may be a constituent of the bacterial flora of spoiled beer, and may be responsible for the production in beer of an unpleasant odor of rotten apples (due to traces of acetaldehyde and H_2S).

Zymomonas is distinguished from *Pseudomonas* by its fermentative metabolism, its microaerophilic to anaerobic nature, oxidase negativity, and by other molecular taxonomic characteristics. It resembles most closely the acetic-acid bacteria, specifically *Gluconobacter* because of its polar flagellation, and it is often found in nature associated with the acetic-acid bacteria. This is of interest because *Zymomonas* ferments glucose to ethanol whereas the acetic acid bacteria ferment ethanol to acetic acid. Thus the acetic-acid bacteria may depend upon the activity of *Zymomonas* for the production of their growth substrate, ethanol. Like the acetic-acid bacteria, *Zymomonas* is quite tolerant of low pH. Unlike yeast, which ferments glucose to ethanol via the Embden-Meyerhof (glycolytic) pathway, *Zymomonas* employs the Entner-Doudoroff pathway. This pathway is active in many pseudomonads as a means of catabolizing glucose (see Figure 19.65).

Zymomonas is of interest to the ethanol industry because it shows higher rates of glucose uptake and ethanol production and gives a higher yield of ethanol than many yeasts. *Zymomonas* is also rather tolerant of high ethanol concentrations (up to 10 percent), but is not quite as tolerant as some of the best yeast strains which can grow up to 12–15 percent ethanol. However, the fact that *Zymomonas* is a *Gram-negative bacterium*, and can thus be readily manipulated genetically, makes it an attractive candidate for use by ethanol production industries.

19.19 Vibrio and Related Genera

The *Vibrio* group is placed adjacent to the enteric bacteria (see Section 19.20) in *Bergey's Manual* because it contains Gram-negative, facultatively aerobic rods and curved rods

Table 19.23 Distinguishing characteristics of genera of the family Vibrionaceae and related genera*

Characteristic	*Vibrio*	*Aeromonas*	*Photobacterium*	*Plesiomonas*
Morphology	Straight or curved rods, single polar flagellum	Straight rods, single polar flagellum	Straight rods, 1–3 polar flagella	Straight rods with round ends, 2–5 polar flagella
Sodium required for growth	+	–	+	–
Gas production	–	+	+	–
Sensitivity to vibriostat (2,4-diamino-6,7-diisopropyl pteridine)	+	–	+	+
Luminescence	+ or –	–	+	–
DNA (mole % GC)	38–51	57–63	40–44	51
Pigment	None	None or brown	None	None

**All are Gram-negative rods, straight or curved, facultative aerobes with a fermentative metabolism (O/F medium), but oxidase positive. They are predominantly aquatic organisms.*

which possess a fermentative metabolism. Most of the members of the *Vibrio* group are polarly flagellated, although some members are peritrichously flagellated. One key difference between the *Vibrio* group and enteric bacteria is that members of the former are oxidase *positive* (see Table 13.3), whereas members of the latter are oxidase *negative*. Although *Pseudomonas* is also polarly flagellated and oxidase positive, it is *not* fermentative, and hence can be separated from the vibrios by simple sugar fermentation tests.

Most vibrios and related bacteria are aquatic, found either in freshwater or marine habitats, although one important organism, *Vibrio cholerae*, is pathogenic for humans. The group contains four genera, *Vibrio, Aeromonas, Photobacterium*, and *Plesiomonas*, and the major characteristics of each are given in Table 19.23.

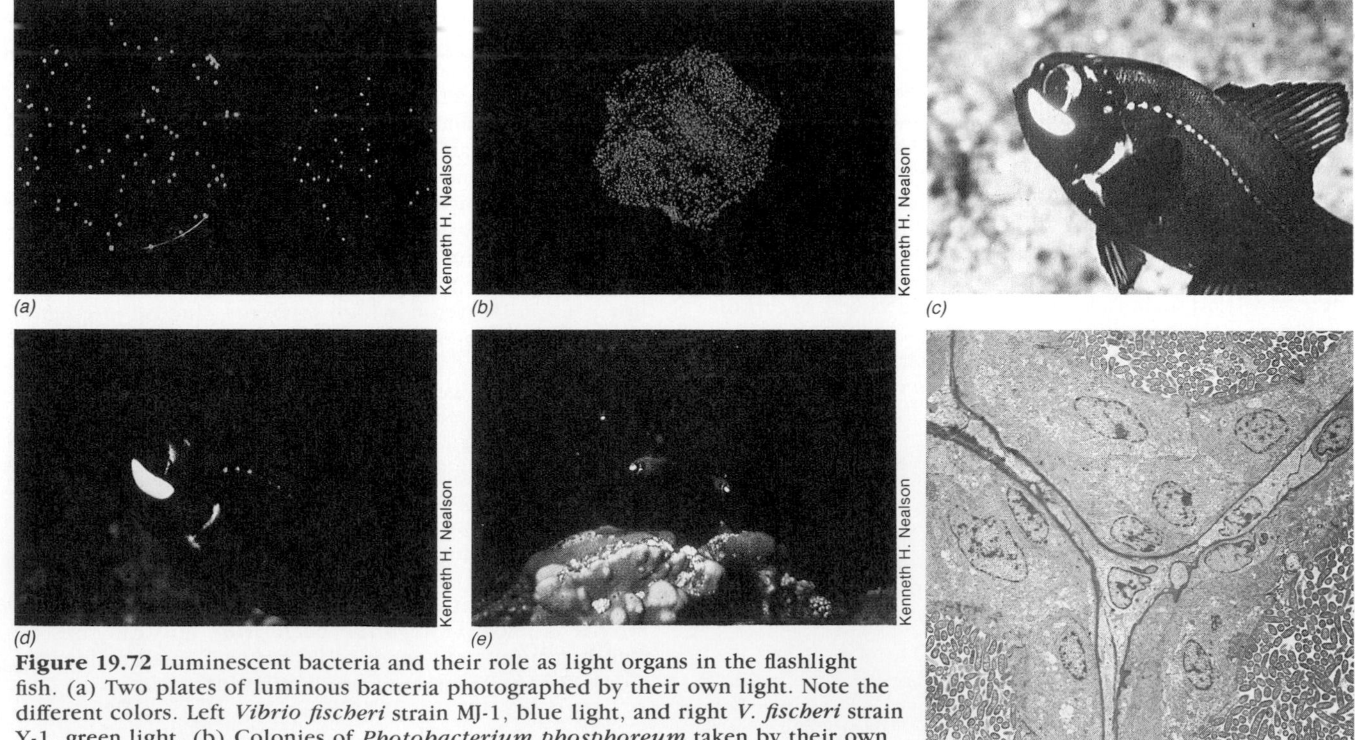

Figure 19.72 Luminescent bacteria and their role as light organs in the flashlight fish. (a) Two plates of luminous bacteria photographed by their own light. Note the different colors. Left *Vibrio fischeri* strain MJ-1, blue light, and right *V. fischeri* strain Y-1, green light. (b) Colonies of *Photobacterium phosphoreum* taken by their own light. (c) The flashlight fish *Photoblepharon palpabratus*; the bright area is the light organ containing luminescent bacteria. (d) Same fish photographed by its own light. (e) Underwater photograph taken at night of *P. palpabratus* in coral reefs in the Gulf of Eilat. (f) Electron micrograph of a thin section through the light-emitting organ of *P. palpabratus*, showing the dense array of luminescent bacteria.

Vibrio cholerae is the specific cause of the disease *cholera* in humans (see Section 15.12); the organism does not normally infect other hosts. Cholera is one of the most common infectious human diseases in undeveloped countries and one that has had a long history. The organism is transmitted almost exclusively via water, and studies on its distribution in the nineteenth century played a major role in demonstrating the importance of water purification in urban areas (see Box, Chapter 14). We discussed the pathogenesis of *V. cholerae* in Section 11.11. *Vibrio cholerae* is capable of good growth at a pH of over 9, and this characteristic is frequently employed in the selective isolation and identification of this organism.

Vibrio parahemolyticus is a marine organism. It is a major cause of gastroenteritis in Japan (where raw fish is widely consumed) and has also been implicated in outbreaks of gastroenteritis in other parts of the world, including the United States. The organism can be frequently isolated from seawater or from shellfish and crustaceans, and its primary habitat is probably marine animals, with human infection being a secondary development.

Luminescent bacteria A number of Gram-negative, polarly flagellated rods possess the interesting property of emitting light (luminescence). Most of these bacteria have been classified in the genus *Photobacterium* (Table 19.23, see Figure 19.72*b*), but a few *Vibrio* isolates are also luminescent (Figure 19.72*a*). Most **luminescent bacteria** are marine forms, usually found associated with fish. Some fish possess a special organ in which luminescent bacteria grow (Figure 19.72*c–f*). Other luminescent marine bacteria live saprophytically on dead fish. A good way of isolating luminescent bacteria is to incubate a dead marine fish for 1 or 2 days at 10 to 20°C; the luminescent bacterial colonies that usually appear on the surface of the fish can be easily seen and isolated (Figure 19.72*a* and *b*). (To see luminescence readily, one should observe the material in a completely dark room after the eyes have become adapted to the dark.)

Although *Photobacterium* isolates are facultative aerobes, they are luminescent only when O_2 is present. The amount of O_2 required is quite low, however, and in fact bacterial luminescence can be used as a relatively sensitive way of detecting small amounts of O_2 in a solution. If bacteria are incubated anaerobically for a while and air is then introduced into the culture, a bright flash of light results, following which the light intensity decreases to a low steady-state value. It is thought that under anaerobic conditions a component required for luminescence (probably NADH) accumulates in amounts higher than normal and that its rapid utilization when O_2 is introduced produces the flash.

Two specific components are needed for bacterial luminescence, the enzyme **luciferase**, and a long-chain aliphatic aldehyde (for example, *dodecanal*); flavin mononucleotide (FMN) and O_2 are also involved. The primary electron donor is NADH, and the electrons pass through FMN to the luciferase. The reaction can be expressed as:

$$FMNH_2 + O_2 + RCHO \xrightarrow[\text{luciferase}]{} FMN + RCOOH + \textit{light}$$

The light generating system constitutes a bypass route for shunting electrons from $FMNH_2$ to O_2, without involving other electron carriers such as quinones and cytochromes.

The enzyme luciferase shows a unique kind of regulatory synthesis called **autoinduction**. The luminous bacteria produce a specific substance, the *autoinducer*, which accumulates in the culture medium during growth, and when the amount of this substance has reached a critical level, induction of the enzyme occurs. Thus cultures of luminous bacteria at low cell density are not luminous, but become luminous when growth reaches a sufficiently high density so that the autoinducer can accumulate and function. Because of the autoinduction phenomenon, it is obvious that free-living luminescent bacteria in seawater will not be luminous because the autoinducer could not accumulate, and luminescence only develops when conditions are favorable for the development of high population densities. Although it is not clear why luminescence is density-dependent in free-living bacteria, in symbiotic strains of luminescent bacteria (see Figure 19.72), the rationale for density-dependent luminescence is clear: luminescence only develops when sufficiently high population densities are reached in the light organ of the fish to allow a visible flash of light.

19.20 Enteric Bacteria

The **enteric bacteria** comprise a relatively homogeneous phylogenetic group within the gamma purple bacteria (see Section 18.5), and are characterized phenotypically as follows: Gram-negative, nonsporulating rods, nonmotile or motile by *peritrichous* flagella (Figure 19.73), facultative aerobes, oxidase *negative* with relatively simple nutritional requirements, fermenting sugars to a variety of end products. The phenotypic characteristics used to separate the enteric bacteria from other bacteria of similar morphology and physiology are given in Table 19.24.

Among the enteric bacteria are many strains pathogenic to humans, animals, or plants as well as other strains of industrial importance. Probably more is known about *Escherichia coli* than about any other bacterial species (see Box, Chapter 7).

Because of the medical importance of the enteric bacteria, an extremely large number of isolates have been studied and characterized, and a fair number of distinct genera have been defined. Despite the fact that there is marked genetic relatedness between many of the enteric bacteria, as shown by DNA homologies and genetic recombination, separate genera are maintained, largely for practical rea-

Table 19.24 Defining characteristics of the enteric bacteria

General characteristics: Gram-negative rods; motile by peritrichous flagella, or nonmotile; nonsporulating; facultative aerobes, producing acid from glucose; catalase positive; oxidase negative; usually reduce nitrate to nitrite (not to N_2)
Key tests to distinguish enteric bacteria from other bacteria of similar morphology: oxidase test, enterics always negative—separates enterics from oxidase-positive bacteria of genera *Pseudomonas, Aeromonas, Vibrio, Alcaligenes, Achromobacter, Flavobacterium, Cardiobacterium*, which may have similar morphology; nitrate reduced only to nitrite, (assay for nitrite after growth)—distinguishes enteric bacteria from bacteria that reduce nitrate to N_2 (gas formation detected), such as *Pseudomonas* and many other oxidase-positive bacteria; ability to ferment glucose—distinguishes enterics from obligately aerobic bacteria

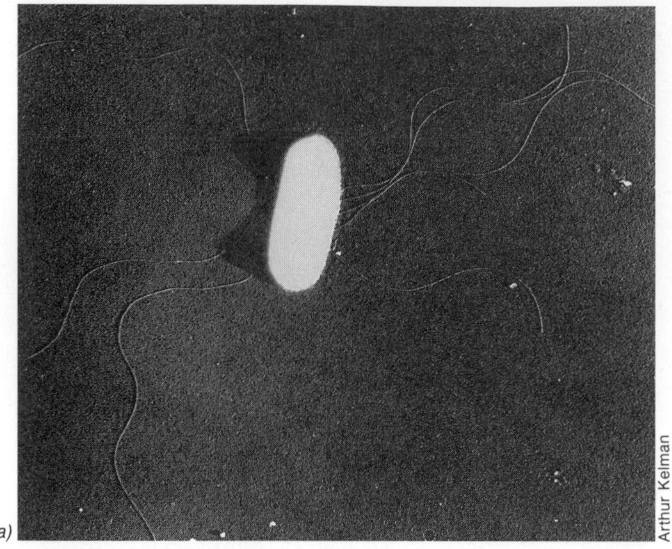

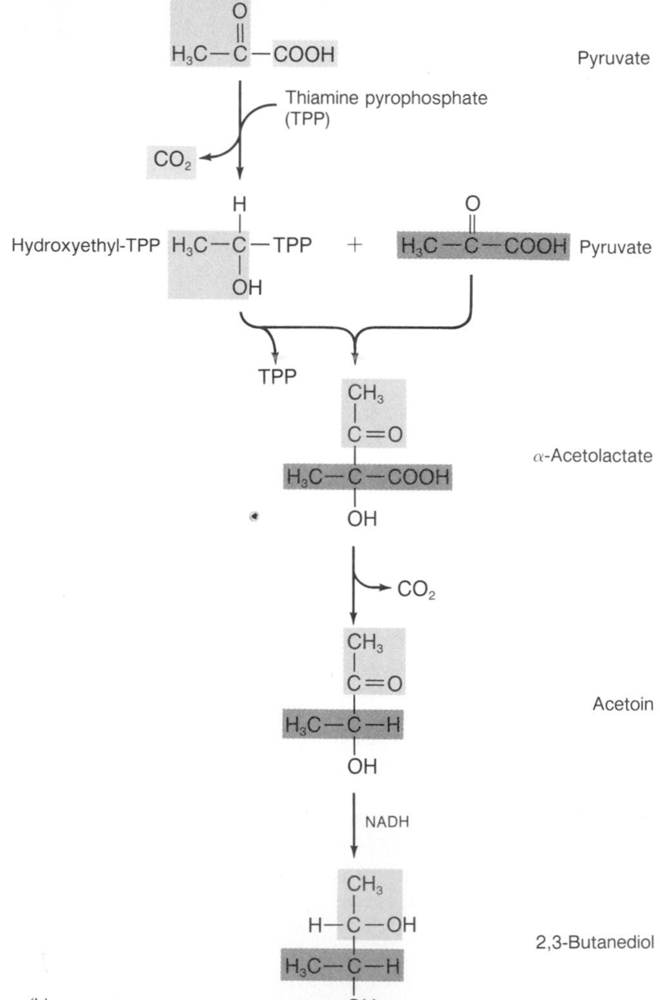

Figure 19.73 (a) Electron micrograph of a shadow-cast preparation of cells of the butanediol-producing enteric bacterium, *Erwinia carotovora*. Note the peritrichously arranged flagella. (b) Biochemical pathway for formation of butanediol from two molecules of pyruvate by butanediol fermenters.

sons. Since these organisms are frequently cultured from diseased states, some means of recognizing and designating them is necessary.

One of the key taxonomic characteristics separating the various genera of enteric bacteria is the type and proportion of fermentation products produced by anaerobic fermentation of glucose. Two broad patterns are recognized, the **mixed-acid fermentation** and the **2,3-butanediol fermentation** (Figure 19.74). In mixed-acid fermentation, *three* acids are formed in significant amounts—acetic, lactic, and succinic—and ethanol, CO_2, and H_2 are also formed, but *not* butanediol. In butanediol fermentation, *smaller* amounts of acids are formed, and butanediol, ethanol, CO_2, and H_2 are the main products. As a result of a mixed-acid fermentation *equal* amounts of CO_2 and H_2 are produced, whereas with a butanediol fermentation *more* CO_2 than H_2 is produced. This is because mixed-acid fermenters produce CO_2 only from formic acid by means of the enzyme system *formic hydrogenlyase*:

$$HCOOH \rightarrow H_2 + CO_2$$

and this reaction results in equal amounts of CO_2 and H_2. The butanediol fermenters also produce CO_2 and H_2 from formic acid, but they produce additional CO_2 during the reactions that lead to butanediol.

A variety of diagnostic tests and differential media are used to separate the various genera in the two broad groups of enteric bacteria, and these are listed in Table 19.25. On the basis of these and other tests, the genera can be defined, as outlined in Tables 19.26 and 19.27.

Table 19.25 Key differential media and tests for classifying enteric bacteria*

Triple sugar iron (TSI) agar
- Butt reactions: acid (yellow), gas
- Slant reactions: acid (yellow), alkaline (red)
- H_2S reaction: blackening of butt (in the absence of acid butt reaction)

Urea medium: urea agar with phenol red indicator; look for alkaline reaction due to urease action on urea, liberating ammonia

Citrate medium: growth on citrate as sole energy source

Indole test:peptone broth with high tryptophan content; assay for indole

Voges-Proskauer (VP) test: glucose-containing broth; assay for acetoin

Methyl red (MR) test: buffered glucose-peptone broth; measure pH drop with methyl red indicator

KCN medium: broth with 75 μg/ml potassium cyanide; look for growth

Phenylalanine agar: contains 0.1 percent phenylalanine; after growth, look for phenylpyruvic acid production (indicative of phenylalanine deaminase) by adding ferric chloride solution and looking for green color formation

Utilization of carbon sources: mannitol, tartrate, mucate, acetate, dulcitol, sorbitol, adonitol, inositol

**Several of these tests are depicted in color in Figure 13.5*

(a) **Mixed acid fermentation,** *e.g., E. coli*

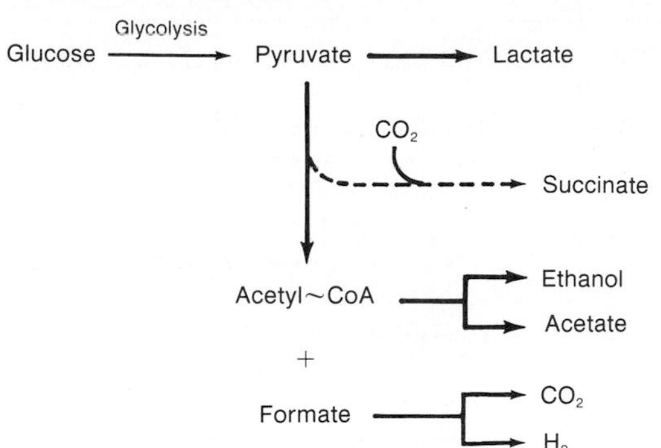

Typical products (molar amounts)

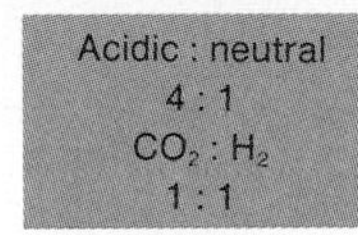

(b) **Butanediol fermentation,** *e.g., Enterobacter*

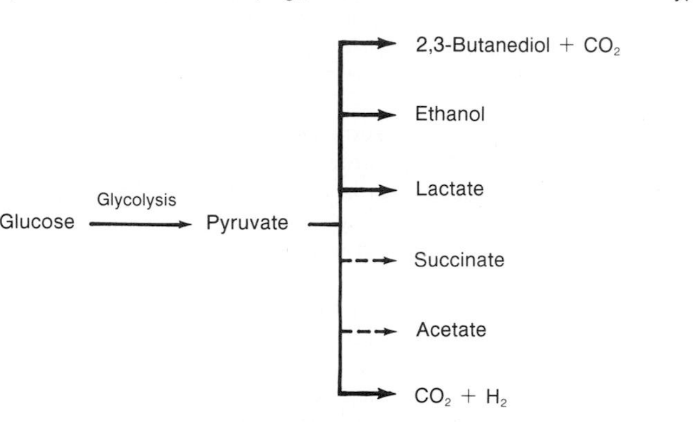

Typical products (molar amounts)

Acidic : neutral
1 : 6
$CO_2 : H_2$
5 : 1

Figure 19.74 Distinction between mixed acid and butanediol fermentation in enteric bacteria. The bold arrows indicate reactions leading to major products.

Table 19.26 Key diagnostic reactions* used to separate the various genera of mixed-acid fermenters

Genus	H_2S(TSI)	Urease	Indole	Motility	Gas from glucose	β-Galactosidase
Escherichia	−	−	+	+ or −	+	+
Shigella	−	−	+ or −	−	−	+ or −
Edwardsiella	+	−	+	+	+	−
Salmonella	+	−	−	+	+	+ or −
Arizona	+	−	−	+	+	+
Citrobacter	+ or −	−	−	+	+	+
Proteus	+ or −	+	+ or −	+	+ or −	−
Providencia	−	−	+	+	−	−
Yersinia	−	+	−	†	−	+

Genus	KCN	Citrate	Mucate	Tartrate	Phenylalanine deaminase	DNA (mole % GC)
Escherichia	−	−	+	+	−	50–51
Shigella	−	−	−	−	−	50
Edwardsiella	−	−	−	−	−	50
Salmonella	−	+ or −	+ or −	+ or −	−	50–53
Arizona	−	+		−	−	50
Citrobacter	+ or −	+	+	+	−	50
Proteus	+	+ or −			+	39–50
Providencia	+	+			+	41
Yersinia		−			−	45–47

**See Tables 19.24 and 19.25 for the procedures for these diagnostic reactions. Tartrate: utilization as carbon source; mucate: fermentation.*
†*Motile when grown at room temperature; nonmotile at 37°C.*

Table 19.27 Key diagnostic reactions used to separate the various genera of 2,3-butanediol producers

Genus	Ornithine decarboxylase	Gelatin hydrolysis	Temperature optimum (°C)	Pigmentation	Motility	Lactose	DNase	Sorbitol	DNA (mole % GC)
Klebsiella	−	−	37–40	None	−	+	−	+	54–59
Enterobacter	+	slow	37–40	Yellow (or none)	+	+	−	+	52–59
Serratia	+	+	37–40	Red (or none)	+	−	+	−	53–59
Erwinia	−	+ or −	27–30	Yellow (or none)	+	+ or −	−	+	50–58
Hafnia	+	−	35	None	+	−	−	−	52–57

Positive identification of enteric bacteria often presents considerable difficulty because of the large number of strains that have been characterized. Almost any small set of diagnostic characteristics will fail to provide clear-cut distinctions between genera if a large number of strains are tested, because exceptional strains will be isolated. In advanced work in clinical laboratories, identification is now based on computer analysis of a large number of diagnostic tests, carried out using miniaturized rapid diagnostic media kits (see Section 13.2), with consideration being given for variable reactions of exceptional strains. Thus the separation of genera given in Tables 19.26 and 19.27 must be considered approximate, for it will always be possible to isolate a strain that does not possess one or another characteristic normally considered positive for the genus as a whole. With these limitations in mind, an even more simplified separation of the key genera is found in Figure 19.75. This key will permit quick decision on the likely *genus* in which to place a new isolate.

Escherichia Members of the genus *Escherichia* are almost universal inhabitants of the intestinal tract of humans and warm-blooded animals, although they are by no means the dominant organisms in these habitats. *Escherichia* may play a nutritional role in the intestinal tract by synthesizing vitamins, particularly vitamin K. As a facultative aerobe, this organism probably also helps consume oxygen, thus rendering the large intestine anaerobic. Wild-type *Escherichia* strains rarely show any growth-factor requirements and are able to grow on a wide variety of carbon and energy sources such as sugars, amino acids, organic acids, and so on. Some strains of *Escherichia* are pathogenic. The latter strains of *Escherichia* have been implicated in diarrhea in infants, occasionally occurring in epidemic proportions in children's nurseries or obstetric wards, and *Escherichia* may also cause urinary tract infections in older persons or in those whose resistance has been lowered by surgical treatment or by exposure to ionizing radiation. Enteropathogenic strains of *E. coli* are becoming more frequently implicated in dysenterylike infections and generalized fevers (see Sections 11.11 and 15.11). As noted, these strains form *K antigen*, permitting attachment and colonization of the *small* intestine, and *enterotoxin*, responsible for the symptoms of diarrhea.

1	MR +; VP − (mixed-acid fermenters)	2
	MR −; VP + (butanediol producers)	7
2	Urease +	*Proteus*
	Urease −	3
3	H_2S (TSI) +	4
	H_2S (TSI) −	6
4	KCN +	*Citrobacter*
	KCN −	5
5	Indole +; Citrate −	*Edwardsiella*
	Indole −; Citrate +	*Salmonella*
6	Gas from glucose	*Escherichia*
	No gas from glucose	*Shigella*
7	Nonmotile; Ornithine −	*Klebsiella*
	Motile; Ornithine +	8
8	Gelatin +; DNAse +	*Serratia* (red pigment)
	Gelatin slow; DNAse −	*Enterobacter*

Figure 19.75 A very simplified key to the main genera of enteric bacteria. Note that only the most common genera are given. See text for precautions in use of this key. Diagnostic tests for use with this figure are given in Table 19.25. Other characteristics of the genera are given in Tables 19.26 and 19.27.

Shigella The shigellas are very closely related to *Escherichia*; in fact they are so similar that they are able to undergo genetic recombination with each other and are susceptible to some of the same bacteriophages. In contrast to *Escherichia*, however, *Shigella* is commonly pathogenic to humans, causing a rather severe gastroenteritis usually called *bacillary dysentery*. *S. dysenteriae* is transmitted by food and waterborne routes and is capable of invading intestinal epithelial cells. Once established, it produces both an endotoxin and a neurotoxin that exhibits enterotoxic effects.

Salmonella *Salmonella* and *Escherichia* are quite closely related; the two genera have about 45 to 50 percent of their DNA sequences in common. However, in contrast to *Escherichia*, members of the genus *Salmonella* are usually pathogenic, either to man or to other warm-blooded animals. In humans the most common diseases caused by salmonellas are *typhoid fever* and *gastroenteritis* (see Sections 15.11 and 15.12). The salmonellas are characterized immunologically on the basis of three cell-surface antigens, the O, or cell-wall (somatic) antigen; the H, or flagellar, antigen; and the Vi (outer polysaccharide layer) antigen, found primarily in strains of *Salmonella* causing typhoid fever. The O antigens are complex lipopolysaccharides that are part of the endotoxin structure of these organisms (see Section 11.12). We discussed the chemical structure of these lipopolysaccharides in Section 3.5 (see Figure 3.27). The genus *Salmonella* contains over 1000 distinct types having different antigenic specificities in their O antigens. Additional antigenic subdivisions are based on the antigenic specificities of the flagellar (H) antigens. There is little or no correlation between the antigenic type of a *Salmonella* and the disease symptoms elicited, but antigenic typing permits tracing a single strain involved in an epidemic.

Proteus The genus *Proteus* is characterized by rapid motility and by production of the enzyme *urease*. It is a frequent cause of urinary-tract infections in people and rarely may cause enteritis. Because of its active urea-splitting ability, *Proteus* has been implicated in several kidney infections, see Section 11.7. The species of *Proteus* probably do not form a homogeneous group, as is indicated by the fact that DNA base compositions vary over a fairly wide range (39 to 50 percent GC). Because of the rapid motility of *Proteus* cells, colonies growing on agar plates often exhibit a characteristic **swarming** phenomenon. Cells at the edge of the growing colony are more rapidly motile than are those in the center of the colony; the former move a short distance away from the colony in a mass, then undergo a reduction in motility, settle down, and divide, forming a new crop of motile cells that again swarm. As a result, the mature colony appears as a series of concentric rings, with higher concentrations of cells alternating with lower concentrations.

Although all enteric bacteria can use nitrate as alternate electron acceptor anaerobically, *Proteus* has the additional ability to use several *sulfur* compounds as electron acceptors for anaerobic growth: thiosulfate, tetrathionate, and dimethylsulfoxide.

Yersinia The genus *Yersinia* has been created to accommodate former members of the genus *Pasteurella* that are obviously members of the enteric group. The genus *Yersinia* consists of three species, *Y. pestis*, the causal agent of the ancient and dread disease bubonic plague (see Section 15.10); *Y. pseudotuberculosis*, the causal agent of a tuberculosis-like disease of the lymph nodes in animals (and rarely in humans); and *Y. enterocolitica*, the causal agent of an intestinal infection (and also occasionally systemic infections) in humans and animals. Although the latter two species are rarely involved in fatal infections, *Yersinia pestis* was responsible for the so-called "Black Death" that ravaged Europe during the fourteenth century killing over *one-fourth* of the European population. Other pandemics of bubonic plague occurred in subsequent centuries causing great suffering and countless thousands of deaths. The differentiation of *Yersinia* from other mixed-acid fermenters is given in Table 19.26. In recent years, the prevalence of *Y. enterocolitica* as a waterborne and foodborne pathogen has been realized, and now that methods are available for its rapid isolation and recognition, it is being detected with increasing frequency.

Butanediol fermenters The butanediol fermenters are genetically more closely related to each other than to the mixed-acid fermenters, a finding that is in agreement with the observed physiological differences. Their DNA base composition is higher, 50 to 59 percent GC, and genetic recombination does not occur between the two groups. However, it is possible to transfer plasmids from a mixed-acid fermenter to a butanediol fermenter; the plasmid replicates in the latter but does not become integrated into the chromosome. The genus designation *Aerobacter*, at one time widely used for one group of the butanediol fermenters, has been abandoned, and now the genus name *Enterobacter* is used. A current classification of this group is outlined in Table 19.27.

One species of *Klebsiella, K. pneumoniae*, occasionally causes pneumonia in humans, but klebsiellas are most commonly found in soil and water. Most *Klebsiella* strains fix N_2 when growing anaerobically, a property not found among other enteric bacteria. The genus *Serratia* is also a butanediol producer, and forms a series of red pyrrole-containing pigments called **prodigiosins**. This pigment, a linear tripyrrole, is of interest because it contains the pyrrole ring also found in the pigments involved in energy transfer: porphyrins, chlorophylls, and phycobilins. There is no evidence that prodigiosin plays any role in energy transfer, however, and its exact function is unknown. Species of *Serratia* can be isolated from water and soil as well as from the gut of various insects and vertebrates and occasionally from the intestines of man.

19.21 Neisseria and Other Gram-Negative Cocci

This group comprises a diverse collection of organisms, related by Gram stain, morphology, lack of motility, nonfermentative aerobic metabolism, and similar DNA base composition. The five genera *Neisseria, Moraxella, Branhamella, Kingella*, and *Acinetobacter* are distinguished as outlined in Table 19.28. At one time all of these organisms were classified in the genus *Neisseria*, but detailed taxonomic work including genetic and DNA homology studies has revealed that several genera are warranted. A major distinction is made on the basis of oxidase reaction. The oxidase-positive organisms (such as *Neisseria gonorrhoeae*, see Figure 13.3*b*) are unusually sensitive to penicillin, being

Table 19.28 Characteristics of the genera of Gram-negative cocci

Characteristics	Genus	DNA (mole % GC)
I. Oxidase-positive, penicillin sensitive:		
Cocci; complex nutrition, utilize carbohydrates, obligate aerobes	*Neisseria*	46–53
Grow in simple media; do not utilize carbohydrates	*Branhamella*	40–47
Rods, forming cocci in stationary phase; generally no growth factor requirements, generally do not utilize carbohydrates; exhibit "twitching" motility	*Moraxella*	40–47
II. Oxidize-negative, penicillin resistant: some strains can utilize a restricted range of sugars, and some exhibit "twitching" motility	*Kingella*	47–55
	Acinetobacter	38–47

inhibited by 1 μg/ml, differing in this way from most other Gram-negative bacteria, including *Acinetobacter*.

In the genus *Neisseria* the cells are always cocci (see Figure 15.19*a*), whereas the cells of the other genera are rod shaped, becoming coccoid in the stationary phase of growth. This has led to designation of these organisms as **coccobacilli**. Organisms of the genera *Neisseria, Kingella, Branhamella*, and *Moraxella* are commonly isolated from animals, and some of them are pathogenic, whereas organisms of the genus *Acinetobacter* are common soil and water organisms, although they are occasionally found as parasites of certain animals. Some strains of *Moraxella* and *Acinetobacter* possess the interesting property of **twitching motility**, exhibited as brief translocative movements or "jumps," covering distances of about 1 to 5 μm. We discussed the clinical microbiology of *Neisseria gonorrhoeae* in Section 13.1 and the disease gonorrhea in Section 15.6.

19.22 Rickettsias

The rickettsias are Gram-negative, coccoid or rod-shaped cells in the size range of 0.3 to 0.7 μm wide by 1 to 2 μm long. They are *obligate intracellular parasites* (Figure 19.76*a*) and the causative agents of such diseases as typhus fever, Rocky Mountain Spotted Fever, and Q-fever (see Section 15.9). Electron micrographs of thin sections of rickettsias show profiles with a normal bacterial morphology (Figure 19.76*b*); both cell wall and cell membrane are visible. The cell wall contains muramic acid and diaminopimelic acid. Both RNA and DNA are present, and the DNA is in the normal double-stranded form, with a GC content varying from 29 to 33 percent in various species of the genus *Rickettsia* and 43 percent for *Coxiella burnetii* (the causal agent of Q fever). The rickettsias divide by normal binary fission, with doubling times of about 8 hours. The penetration of a host cell by a rickettsial cell is an active process, requiring both host and parasite to be alive and metabolically active. Once inside the phagocytic cell, the bacteria multiply primarily in the cytoplasm and continue replicating until the host cell is loaded with parasites (see Figures 11.27 and 19.76), at which time the host cell bursts and liberates the bacteria into the surrounding fluid.

(a)
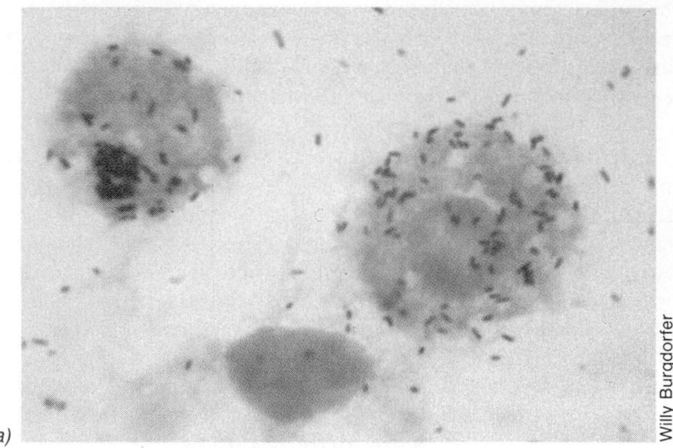
Willy Burgdorfer

(b)
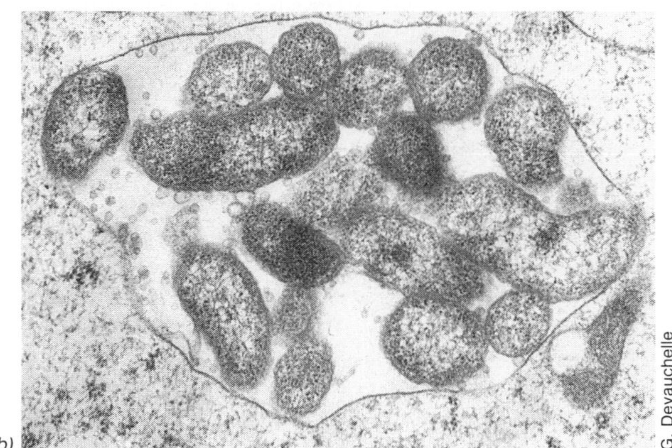
G. Devauchelle

Figure 19.76 Rickettsias growing within host cells. (a) *Rickettsia rickettsii* in tunica vaginalis cells of the vole, *Microtus pennsylvanicus*. (b) Electron micrograph of cells of *Rickettsiella popilliae* within a blood cell of its host, the beetle *Melolontha melolontha*. Notice that the bacteria are growing within a vacuole within the host cell.

Table 19.29 Major properties of rickettsias and comparison of rickettsias with chlamydias and viruses

Property	Rickettsias	Chlamydias	Viruses
Structural:			
Nucleic acid	RNA and DNA	RNA and DNA	Either RNA or DNA, never both
Ribosomes	Present	Present	Absent
Cell wall	Muramic acid, DAP	Muramic acid may or may not be present, D-alanine	No wall
Structural integrity during multiplication	Maintained	Maintained	Lost
Biosynthetic:			
Macromolecular synthesis	Carried out	Carried out	Only with use of host machinery
ATP-generating system	Present	Absent	Absent
Sensitivity to antibacterial antibiotics	Sensitive	Sensitive	Resistant

Table 19.30 Characteristics of representative *Rickettsia* and *Coxiella* species

Species	Rickettsial group	Arthropod host	Intracellular location	DNA Mole % GC	DNA percent hybridization to *R. rickettsii* DNA
R. rickettsii	Spotted Fever	Tick	Cytoplasm and nucleus	32–33	100
R. prowazekii	Typhus	Louse	Cytoplasm	29–30	53
R. typhi	Typhus	Flea	Cytoplasm	29–30	36
C. burnetii	Q-Fever	Tick	Vacuoles	43	—

Much attention has been directed to the metabolic activities and biochemical pathways of rickettsias in an attempt to explain why they are obligate intracellular parasites. Since biochemical studies must be done with large populations of cells, and since these populations can be obtained only by growing the parasites in animal cells, much effort has been expended on devising methods for purifying rickettsias and for separating them from any contaminating host tissues that might confuse metabolic or biochemical studies. Many rickettsias possess a highly distinctive energy metabolism, being able to oxidize only one amino acid, *glutamate*, and being unable to oxidize glucose, glucose-6-phosphate, or organic acids. However, *Coxiella burnetii* is able to utilize both glucose and pyruvate as electron donors. Rickettsias possess a complete cytochrome system and are able to carry out electron-transport phosphorylation, using NADH as electron donor. They are also able to synthesize at least some of the small molecules needed for macromolecular synthesis and growth, while they obtain the rest of their nutrients from the host cell. There is some suggestion that the host also provides some key coenzymes, such as NAD^+ and coenzyme A. Such large coenzymes do not usually penetrate readily into bacteria that live independently, and there is evidence that the rickettsial membrane is looser and more "leaky" than those of other bacteria. For a fastidious organism capable of infecting other cells rich in nutrients and cofactors, such a permeable membrane would be advantageous (recall that the mitochondrial membrane is also "leaky," for probably the same reasons, see Section 3.13). A summary of the biochemical properties of rickettsias is given in Table 19.29.

If the membranes of the rickettsias are indeed unusually leaky, the organisms may die quickly when out of their hosts, and this may explain why they must be transmitted from animal to animal by arthropod vectors. When the arthropod obtains a blood meal from an infected vertebrate, rickettsias present in the blood are inoculated directly into the arthropod, where they penetrate to the epithelial cells of the gastrointestinal tract, multiply, and appear later in the feces. When the arthropod feeds upon an uninfected individual, it then transmits the rickettsias either directly with its mouthparts or by contaminating the bite with its feces. However, the causal agent of Q fever, *C. burnetii* (see Section 15.9), can also be transmitted to the respiratory system by aerosols. *Coxiella burnetii* is the most resistant of the rickettsias to physical damage, which probably explains its ability to survive in air. A summary of the major properties of species of *Rickettsia* and *Coxiella* are given in Table 19.30.

The phylogenetic relationships of the rickettsias has recently been determined. By 16S rRNA sequencing criteria, the rickettsias clearly group within the purple bacteria (see Section 18.5), more specifically with the plant pathogen *Agrobacterium*. Although at first this may appear puzzling, it should be remembered that *Agrobacterium*, like the rickettsias, has evolved close associations with eucaryotic cells, being responsible for the plant disease crown gall (see Section 17.23). The 16S rRNA sequences of rickettsias and *Agrobacterium tumefaciens* are about 95 percent homologous; those of *E. coli* or *B. subtilis* and the rickettsias are less than 80 percent homologous. The evolutionary relationships between rickettsias and intracellular symbionts such as *A. tumefaciens* therefore suggests that rickettsias evolved from plant-associated bacteria. Perhaps the rickettsias were originally *plant* pathogens that evolved to be associated with animals following transfer from plants to animals by insect vehicles.

19.23 Chlamydias

Organisms of the genus *Chlamydia* probably represent a further stage in degenerate evolution from that discussed above for the rickettsias, since the chlamydias are obligate parasites in which there has been an even greater loss of metabolic function. In fact, for many years the chlamydias were considered to be large viruses rather than bacteria, and only since the nature of virus replication has been well understood has the bacterial nature of the chlamydias been firmly established. Many chlamydias are smaller than some of the true viruses, such as the smallpox virus, but the chlamydias divide by binary fission, and do not replicate in the manner of viruses. Two species of *Chlamydia* are recognized (Table 19.31), *C. psittaci*, the causative agent of the disease *psittacosis*, and *C. trachomatis*, the causative agent of trachoma and a variety of other human diseases (Table 19.32). **Psittacosis** is an epidemic disease of birds that is occasionally transmitted to humans and causes pneumonialike symptoms. **Trachoma** is a debilitating disease of the eye characterized by vascularization and scarring of the cornea. Trachoma is the leading cause of blindness in humans. Other strains of *C. trachomatis* infect the genitourinary tract, and chlamydial infections are one of the leading sexually transmitted diseases in society today (see Section 15.6). A comparison of the properties of *C. psittaci* and *C. trachomatis* is shown in Table 19.31.

Besides being disease entities, the chlamydias are intriguing because of the biological and evolutionary problems they pose. The bacterial nature of the chlamydias was first suspected when it was discovered that, unlike viruses, they were susceptible to antibiotics whose action is restricted to the bacteria. Biochemical studies showed that the chlamydias have Gram-negative bacterial cell walls (although peptidoglycan may not be present) and they have both DNA and RNA. Electron microscopy of thin sections

Table 19.31 Differential characteristics of species of the genus *Chlamydia*

Characteristic	*C. trachomatis*	*C. psittaci*
Hosts	Humans	Birds, lower mammals, occasionally humans
Usual site of infection	Mucous membrane	Multiple sites
Human-to-human transmission	Common	Rare
Mole % GC	41.9–44%	41.3
Percent homology to *C. trachomatis* DNA by DNA: DNA hybridization	100	10
DNA, molecular weight (*E. coli* = 2×10^9)	6.6×10^8	3.6×10^8

of infected cells shows forms that clearly are undergoing binary fission (Figure 19.77). The biosynthetic capacities of the chlamydias are very restricted, however, even more so than that of the rickettsias. This raised the interesting question of the limits to which evolutionary loss of function can be pushed while independence of macromolecular function is still retained. Although no convincing evidence exists that all the chlamydias are closely related, they are currently classified in a single genus; but the name *chlamydia* should be viewed more as a convenience than as a taxonomic entity.

The life cycle of a typical member of the genus *Chlamydia* is shown in Figure 19.78. Two cellular types are seen in a typical life cycle: a small, dense cell, called an **elementary body**, which is relatively resistant to drying and is the means of *dispersal* of the agent, and a larger, less dense cell, called a **reticulate body**, which divides by binary fission and is the *vegetative* form. Elementary bodies are nonmultiplying cells specialized for transmission, whereas reticulate bodies are noninfectious forms that specialize in intracellular multiplication. Unlike the rickettsias just discussed above, the chlamydias are not transmitted by arthropods but are primarily *airborne* invaders of the respiratory system–hence the significance of resistance to drying of elementary bodies. When a virus infects a cell, it loses its structural integrity and liberates nucleic acid. When an elementary body enters a cell, however, although it changes form, it remains a structural unit and enlarges and begins to undergo binary fission. A reticulate body is seen in Figure 19.77. After a number of divisions, the vegetative cells are converted into elementary bodies that are released when the host cell disintegrates and can then infect other cells. Generation times of 2 to 3 hours have been reported, which are considerably faster than those found for the rickettsias.

As we noted, chlamydia cells have a chemical composition similar to that of other bacteria. Both RNA and DNA are present. The DNA content of a chlamydia corresponds to about twice that of vaccinia virus and one-eighth to one-tenth that of *Escherichia coli*. At least some of the RNA is in the form of ribosomes, and, like those of other procaryotes, the ribosomes are 70S particles composed of one 50S and one 30S unit.

The metabolic properties of chlamydias purified from infected cells have been studied by methods similar to those used with the rickettsias. The biosynthetic capacities of the chlamydias are much more limited than are those of the rickettsias (see Table 19.29). Although macromolecular syntheses occur in the chlamydias, no energy-generating system is present and the cells obtain their ATP from the host. They are able to oxidize glucose to pentose by means of enzymes of the pentose phosphate pathway, but they do not obtain energy from this oxidation. Chlamydias synthesize their own folic acid, D-alanine, lysine, and probably a number of other small molecules. One interesting feature of chlamydias is that their proteins are deficient in the amino

Table 19.32 Chlamydial diseases

Human diseases caused by *C. trachomatis*
Trachoma
Inclusion conjunctivitis
Otitis media
Infant pneumonia
Nongonococcal urethritis (males)
Urethral inflammation (females)
Lymphogranuloma venereum
Cervicitis
Diseases caused by *C. psittaci* in nonhuman hosts
Avian chlamydiosis (Parrots, parakeets, pigeons, turkeys, geese, other birds)
Seminal vesiculitis (Sheep, cattle)
Pneumonia (Kittens, lambs, calves, piglets, foals)
Conjunctivitis (Lambs, calves, piglets, cats)
Synovial tissue arthritis (Lambs, calves, foals, piglets)

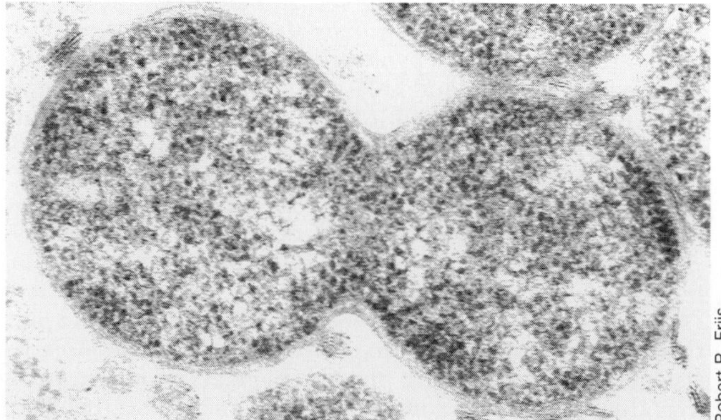

Figure 19.77 Electron micrograph of a thin section of a dividing cell (reticulate body) of *Chlamydia psittaci*, a member of the psittacosis group, within a mouse tissue culture cell.

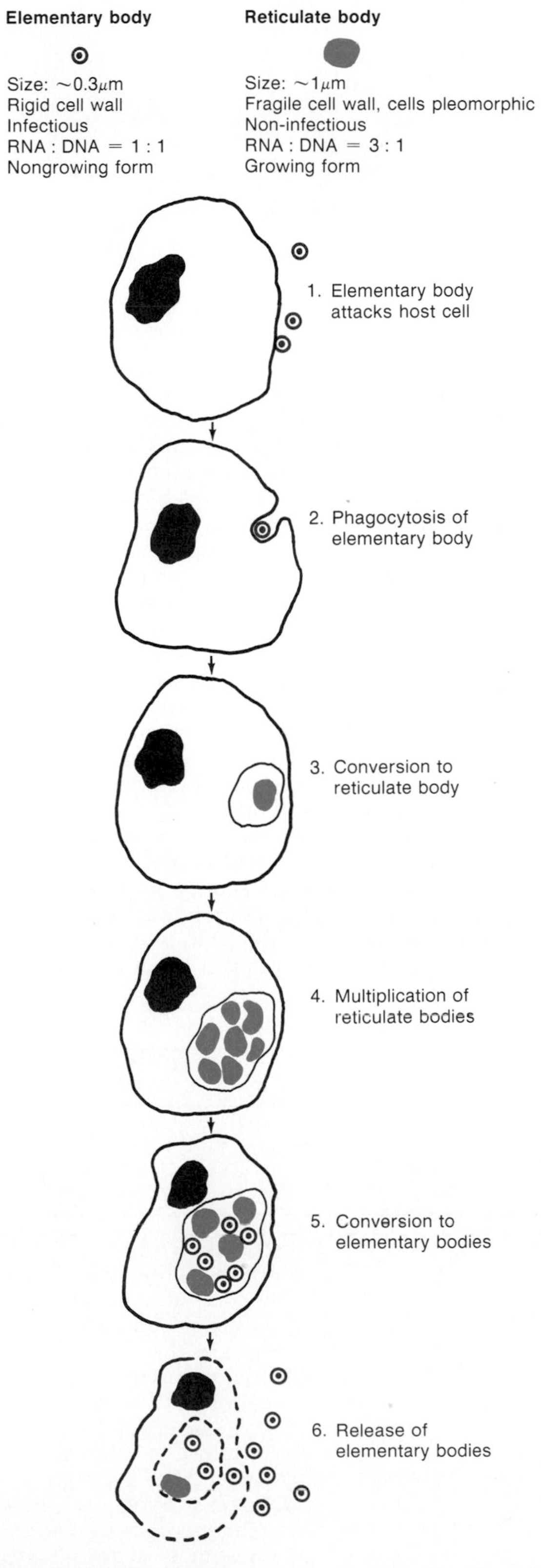

acids arginine and histidine, which are found in most other organisms as well as in viruses. In keeping with this lack, neither amino acid is required for multiplication. From the limited biosynthetic capacities of chlamydias it is easy to see why they are obligate parasites. Their inability to manufacture ATP means that at the least they are energy parasites of their hosts; probably they also obtain from these hosts a variety of other coenzymes, as well as low-molecular-weight building blocks. The chlamydias have the simplest biochemical abilities known among cellular organisms.

19.24 Gram-Positive Bacteria: Cocci

With the exception of *Deinococcus*, all Gram-positive eubacteria form a phylogenetically coherent group. Two major subgroups of Gram-positive bacteria are apparent from ribosomal RNA sequencing studies: the "Clostridium" subgroup, consisting of the endosporeformers, the lactic acid bacteria, and most Gram-positive cocci; these organisms generally have a fairly *low* mole percent GC in their DNA, and the "Actinomycetes" subgroup, which essentially consists of genera of actinomycetes, most of which have a rather ***high*** GC content, and the genus *Propionibacterium*. We begin here with the Gram-positive cocci, common organisms to the beginning student in microbiology.

The Gram-positive cocci (Figure 19.79) include bacteria with widely differing physiological characteristics. The major genera of Gram-positive cocci and some differentiating characteristics are given in Table 19.33. The genus *Streptococcus* is not considered here but instead is discussed in Section 19.25 along with the other lactic acid bacteria.

Staphylococcus* and *Micrococcus *Staphylococcus* and *Micrococcus* (Figure 19.80) are both aerobic organisms with a typical respiratory metabolism. They are catalase positive, and this test permits their distinction from *Streptococcus* and some other genera of Gram-positive cocci.

As we discussed in Section 9.13, the Gram-positive cocci are relatively resistant to reduced water potential, and tolerate drying and high salt fairly well. Their ability to grow in media with high salt provides a simple means for isolation. If an inoculum is spread on an agar plate containing a fairly rich medium containing about 6 to 7.5 percent NaCl, and the plate incubated aerobically, Gram-positive cocci will often form the predominant colonies. Often, these organisms are pigmented, and this provides an additional aid in selecting Gram-positive cocci.

The two genera *Micrococcus* and *Staphylococcus* can easily be separated based on the oxidation/fermentation test. *Micrococcus* is an obligate aerobe, and produces acid from glucose only aerobically, whereas *Staphylococcus* is a facultative aerobe, and produces acid from glucose both aerobically and anaerobically. Their DNA base compositions are also widely different: *Micrococcus*, 66 to 73 percent GC; *Staphylococcus*, 30 to 38 percent GC.

Staphylococci are common parasites of humans and animals, and occasionally cause serious infections. In humans, two major forms are recognized, *S. epidermidis*, a nonpigmented, nonpathogenic form that is usually found on the skin or mucous membranes, and *S. aureus*, a yellow pigmented form that is most commonly associated with pathological conditions, including boils, pimples, pneumonia, osteomyelitis, meningitis, and arthritis. We listed the exotoxins of *S. aureus* in Table 11.4. One of the significant

Figure 19.78 The infectious cycle of chamydia. The whole cycle takes about 48 hours.

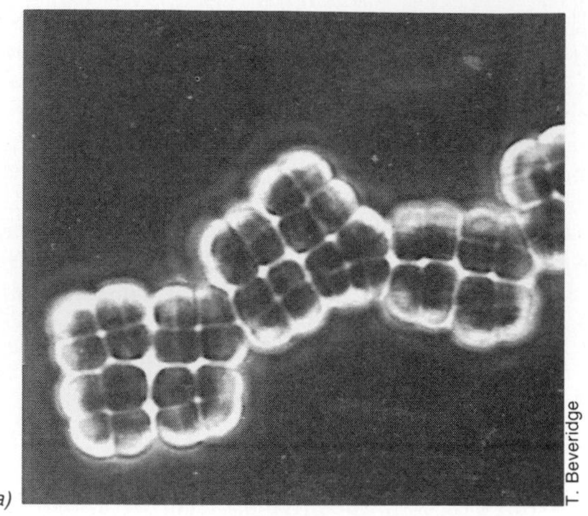

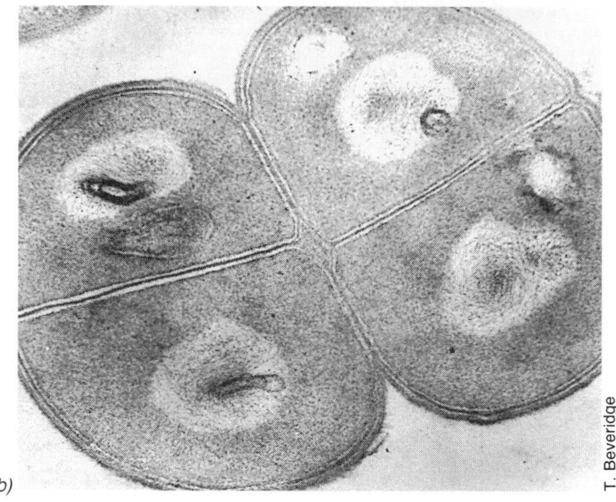

Figure 19.79 (a) Phase-contrast photomicrograph of cells of a typical Gram-positive coccus *Sarcina* sp. (b) Electron micrograph of a thin section.

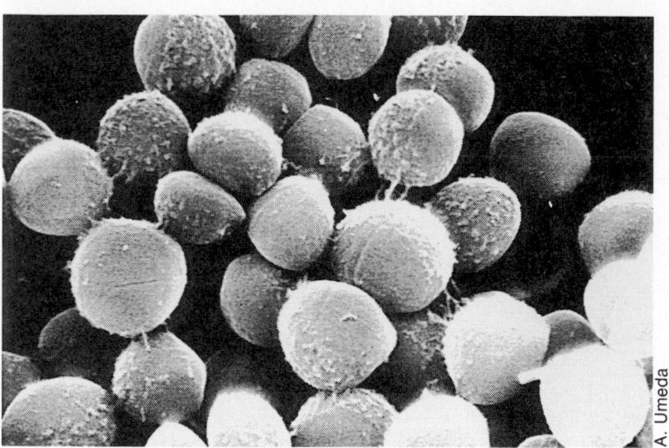

Figure 19.80 Scanning electron micrograph of typical *Staphylococcus*, showing the irregular arrangement of the cell clusters.

exotoxins is **coagulase**, an enzymelike factor that causes fibrin to coagulate and form a clot. Strains of *S. aureus* are generally coagulase *positive,* whereas *S. epidermidis* is coagulase *negative*. We discussed the possible role of the yellow carotenoid pigment of *S. aureus* in resistance to phagocytosis in Section 11.14, and staphylococcal food poisoning was discussed in Section 15.11.

Certain strains of *S. aureus* have been implicated as the agents responsible for the so-called **toxic shock syndrome** (TSS), a severe infection characterized by high fever, vomiting, and diarrhea. Toxic shock occurs most frequently in a small percentage of menstruating women who use tampons. Blood and mucus in the vagina become colonized by hemolytic *S. aureus* from the skin, and the presence of a tampon may concentrate this material, creating ideal microbial growth conditions.

The symptoms of TSS are thought to be due to an *exotoxin* released by *S. aureus*, which causes a considerable blood pressure drop followed shortly by the aforementioned symptoms. Toxin production in pure cultures of *S. aureus* has been shown to be linked to *magnesium* levels; low magnesium levels seem to favor toxin production. It

Table 19.33 Distinguishing features of Gram-positive cocci

Genus	Motility	Arrangement of cells	Growth by fermentation	DNA (mole % GC)	Other characteristics
Micrococcus	−	Clusters, tetrads	−	64–75	Strict aerobes
Staphylococcus	−	Clusters, pairs	+	30–39	Only genus to contain teichoic acid in cell wall
Stomatococcus	−	Clusters, pairs	+	56–61	Only genus containing a capsule
Planococcus	+	Pairs, tetrads	−	39–52	Motile, primarily marine
Sarcina	−	Cuboidal packets of 8 or more cells	+	28–31	Extremely pH tolerant, grows from pH 2.9. Cellulose in cell wall
Deinococcus	−	Pairs, tetrads	−	62–70	Phylogenetically unique
Ruminococcus	+	Pairs, chains	+	39–46	Obligate anaerobe; inhabits rumen, cecum, and large intestine of many animals

is hypothesized that tampons serve to chelate (bind) Mg^{2+} in the vagina, thus triggering toxin production by *S. aureus*. A small number of cases of TSS have been fatal. Women can markedly reduce their risk of TSS by either not using tampons (especially so-called "maximally absorbent" tampons) or by alternating tampons with other products such as sanitary napkins or shields.

Deinococcus The genus *Deinococcus* was created to accommodate several species of Gram-positive cocci previously classified in the genus *Micrococcus*. Besides their unique phylogenetic stature (see Section 18.5) species of *Deinococcus* differ from other Gram-positive cocci in a number of interesting chemical and physiological properties. The peptidoglycan of deinococci contains *ornithine* instead of *lysine* as the diamino acid in the cross-linking peptide. The cell walls of deinococci are structurally complex and consist of several layers, including an *outer membrane* layer (Figure 19.81) normally only present in Gram-negative bacteria (see Section 3.5). The outer membrane of *Deinococcus* is chemically unique and does not contain heptoses and lipid A typical of that of Gram-negative bacteria.

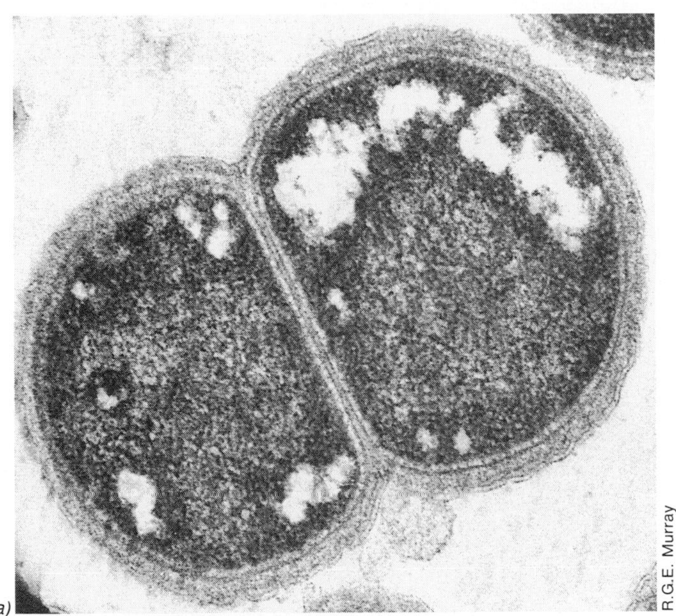

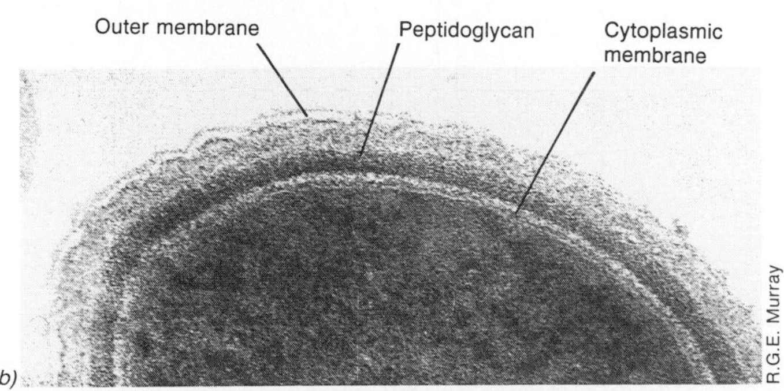

Figure 19.81 The radiation resistant coccus, *Deinococcus radiodurans*. (a) Transmission electron micrograph of *D. radiodurans*. Note the outer membrane layer. (b) High magnification micrograph of wall layer.

Most deinococci are bright red or pink in color because of the variety of carotenoids found in these organisms, and many strains are highly resistant to ultraviolet radiation and to desiccation. Resistance to radiation can be used to advantage for isolating deinococci. These remarkable organisms can be isolated from soil, ground meat, dust, and filtered air following exposure of the sample to intense ultraviolet (or even gamma) radiation and plating on a rich medium containing tryptone and yeast extract. Since many strains of *Deinococcus radiodurans* are even more resistant to radiation than bacterial endospores, treatment of a sample with strong doses (1–2 megarads) of radiation effectively sterilizes the sample of organisms other than *D. radiodurans*, making isolation of deinococci relatively straightforward.

In addition to impressive radiation resistance, *D. radiodurans* is resistant to the mutagenic effects of many highly mutagenic chemicals. Studies of the mutability of *D. radiodurans* have shown it to be highly efficient in repairing damaged DNA. Several different DNA repair enzymes exist in *D. radiodurans* to repair breaks in single- or double-stranded DNA and to excise and repair thymine dimers formed by the action of ultraviolet light. These efficient repair mechanisms have actually hindered studies of the molecular genetics of *D. radiodurans*. Because it is difficult to induce stable mutations in the DNA of *D. radiodurans*, it has been nearly impossible to obtain *mutants* of this organism for genetic study! The only chemical mutagens that seem to work on *D. radiodurans* are agents like nitrosoguanidine, which induce *deletions* in DNA; deletions are apparently not repaired as efficiently as point mutations in this organism. Consistent with the fact that *D. radiodurans* is an extremely radiation resistant bacterium, strains of this organism have been isolated from near atomic reactors and other potentially lethal radiation sources.

19.25 Lactic Acid Bacteria

The lactic acid bacteria are characterized as Gram-positive, usually nonmotile, nonsporulating bacteria that produce lactic acid as a major or sole product of fermentative metabolism. Members of this group lack porphyrins and cytochromes, do not carry out electron-transport phosphorylation, and hence obtain energy only by ***substrate-level phosphorylation***. All of the lactic acid bacteria grow anaerobically. Unlike many anaerobes, however, most lactics are not sensitive to O_2 and can grow in its presence as well as in its absence; thus they are **aerotolerant anaerobes**. Some strains are able to take up O_2 through the mediation of flavoprotein oxidase systems, producing H_2O_2, although most strains lack catalase and most dispose of H_2O_2 via alternative enzymes referred to as *peroxidases* (see Section 9.14). No ATP is formed in the flavoprotein oxidase reaction, but the oxidase system can be used for reoxidation of NADH generated during fermentation. Most lactic acid bacteria can obtain energy only from the metabolism of sugars and related fermentable compounds, and hence are usually restricted to habitats in which sugars are present. They usually have only limited biosynthetic ability, and their complex nutritional requirements include needs for amino acids, vitamins, purines, and pyrimidines.

Traditionally, the lactic acid bacteria were considered a single group containing both cocci and rods. Some workers prefer to separate the group; however, we continue to keep both rods and cocci together in the present text, recognizing the obvious physiological similarities of the group as a whole.

Homo- and heterofermentation One important difference between subgroups of the lactic acid bacteria lies in the nature of the products formed during the fermentation of sugars. One group, called **homofermentative**, produces virtually a single fermentation product, *lactic acid*, whereas the other group, called **heterofermentative**, produces other products, mainly *ethanol* and CO_2 as well as lactate. Abbreviated pathways for the fermentation of glucose by a homo- and a heterofermentative organism are shown in Figure 19.82. The differences observed in the fermentation products are determined by the presence or absence of the enzyme **aldolase**, one of the key enzymes in *glycolysis* (see Figure 4.10). The heterofermenters, lacking aldolase, cannot break down fructose diphosphate to triose phosphate. Instead, they oxidize glucose-6-phosphate to 6-phosphogluconate and then decarboxylate this to pentose phosphate, which is broken down to triose phosphate and acetylphosphate by means of the enzyme **phos-**

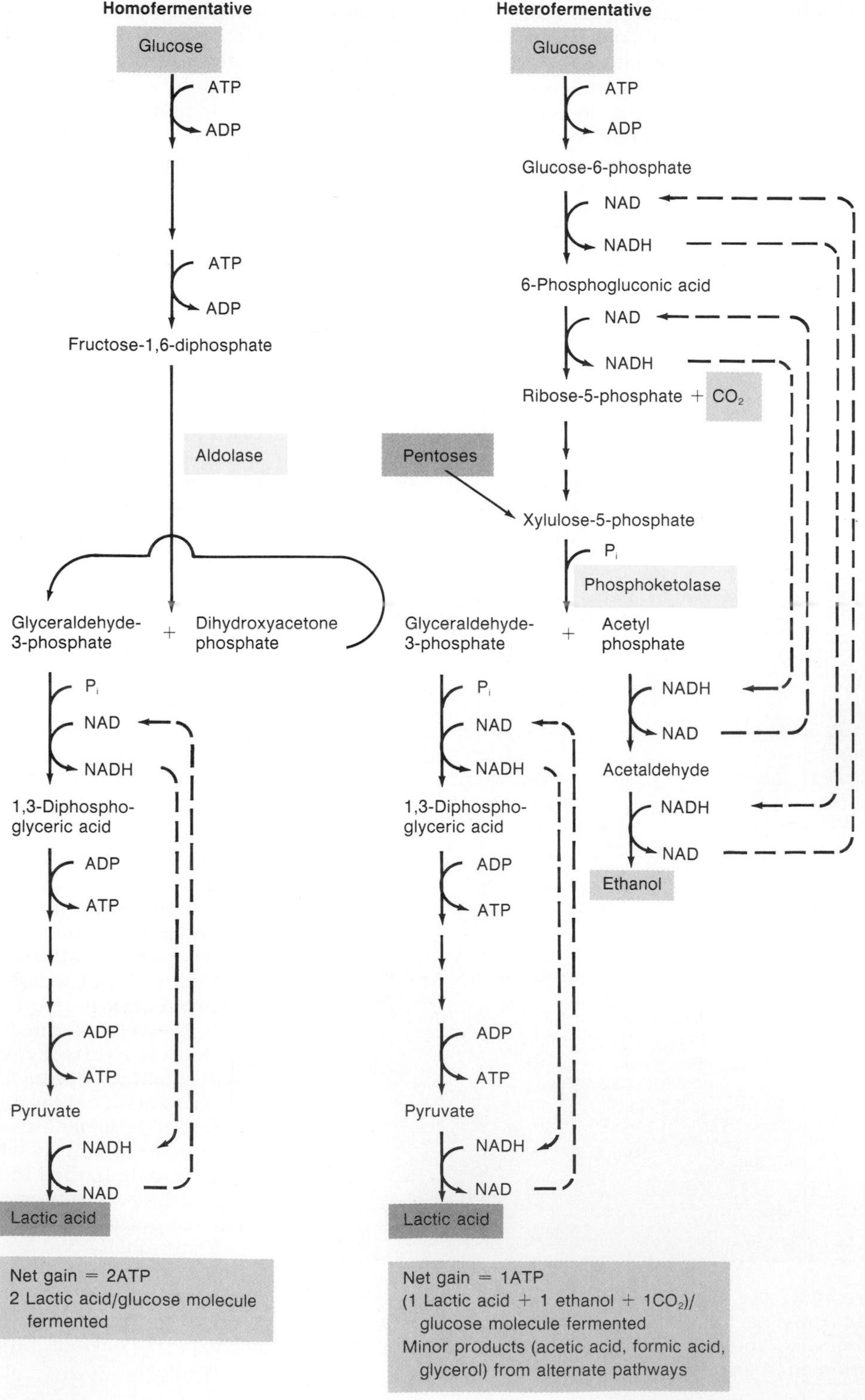

Figure 19.82 The fermentation of glucose in homofermentative and heterofermentative lactic acid bacteria.

phoketolase. Triose phosphate is converted ultimately to lactic acid with the production of 1 mole of ATP, while the acetylphosphate accepts electrons from the NADH generated during the production of pentose phosphate and is thereby converted to ethanol *without* yielding ATP. Because of this, heterofermenters produce only *1 mole* of ATP from glucose instead of the 2 moles produced by homofermenters. This difference in ATP yield from glucose is reflected in the fact that homofermenters produce twice as much cell mass as heterofermenters from the same amount of glucose. Because the heterofermenters decarboxylate 6-phosphogluconate, they produce CO_2 as a fermentation product, whereas the homofermenters produce little or no CO_2; therefore one simple way of detecting a heterofermenter is to observe for production of CO_2 in laboratory cultures. At the enzyme level, heterofermenters are characterized by the lack of aldolase and the presence of phosphoketolase. Many strains of heterofermenters can use O_2 as an electron acceptor with a reduced flavoprotein serving as electron donor. In this reaction half of the NADH generated from the oxidation of glucose to ribose is transferred to a flavin and on to O_2. Acetylphosphate can then be converted to acetate instead of being reduced to ethanol, and an additional ATP is synthesized.

The various genera of lactic acid bacteria have been defined on the basis of cell morphology, DNA base composition, and type of fermentative metabolism, as is shown in Table 19.34. Members of the genera *Streptococcus, Leuconostoc*, and *Pediococcus* have fairly similar DNA base ratio compositions; in addition, there is very little variation from strain to strain. The genus *Lactobacillus* on the other hand, has members with widely diverse DNA compositions and hence does not constitute a homogeneous group.

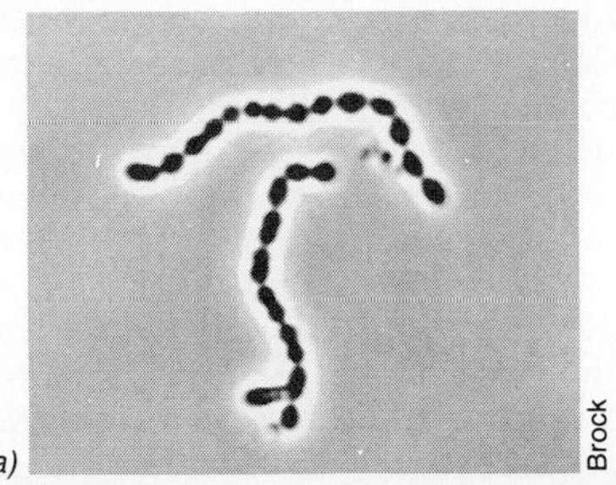

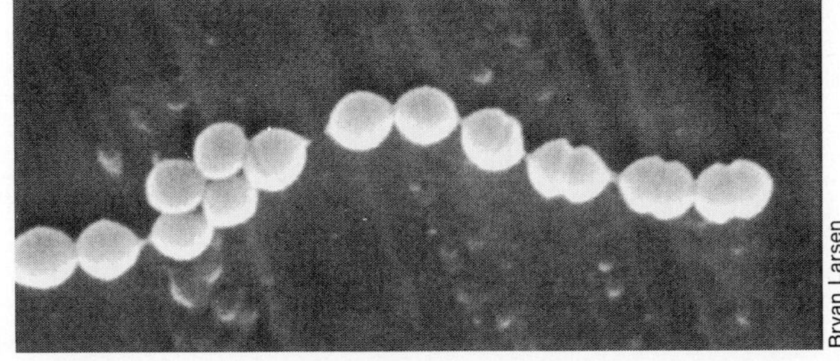

Figure 19.83 Phase contrast and scanning electron micrographs of *Streptococcus* sp. (a) *Streptococcus lactis.* (b) *Streptococcus* sp.

Streptococcus and other cocci The genus *Streptococcus* (Figure 19.83) contains a wide variety of species with quite distinct habitats, whose activities are of considerable practical importance to humans. Some members are pathogenic to people and animals (see Section 15.2). As

Table 19.34 Differentiation of the principal genera of lactic acid bacteria

Genus	Cell form and arrangement	Fermentation	DNA (mole % GC)
Streptococcus	Cocci in chains	Homofermentative	34–46
Leuconostoc	Cocci in chains	Heterofermentative	38–41
Pediococcus	Cocci in tetrads	Homofermentative	34–42
Lactobacillus	(1) Rods, usually in chains	Homofermentative	32–53
	(2) Rods, usually in chains	Heterofermentative	34–53

Table 19.35 Differential characteristics of streptococci

Group	Antigenic (Lancefield) groups	Representative species	Type of hemolysis on blood agar	Good growth at 10°C	Good growth at 45°C	Survive 60°C for 30 min	Growth in: Milk with 0.1% methylene blue	Growth in: Broth with 40% bile	Habitat
Pyogenes	A,B,C,F,G	*S. pyogenes*	Lysis (β)	−	−	−	−	−	Respiratory tract, systemic
Viridans	Not grouped	*S. mutans*	Greening (α)	−	+	−	−	−	Mouth, intestine
Fecal (enterococci)	D	*S. faecalis*	Lysis (β), greening (α), or none	+	+	+	+	+	Intestine, vagina, plants
Lactic	N	*S. cremoris*	None	+	−	+	+	+	Plants, dairy products

producers of lactic acid, certain streptococci play important roles in the production of buttermilk, silage, and other fermented products (see Section 9.21).

The genus *Streptococcus* is subdivided into a number of groups of related species on the basis of characteristics enumerated in Table 19.35. *Hemolysis* on blood agar is of considerable importance in the subdivision of the genus. Colonies of those strains producing streptolysin O or S are surrounded by a large zone of complete red blood cell hemolysis, a condition called **β hemolysis**. On the other hand, many streptococci that do not produce hemolysins cause the formation of a greenish or brownish zone around their colonies, which is due not to true hemolysis but to discoloration and loss of potassium from the red cells. This type of reaction has classically been referred to as **α hemolysis**. The streptococci are also divided into *immunological* groups based on the presence of specific carbohydrate antigens. These antigenic groups (or **Lancefield groups** as they are commonly known, named for Rebecca Lancefield, a pioneer in *Streptococcus* taxonomy), are designated by letters; A through O are currently recognized. Those β-hemolytic streptococci found in human beings usually contain the group A antigen, which is a cell-wall polymer containing *N*-acetylglucosamine and rhamnose. The fecal streptococci contain the group D antigen, a glycerol teichoic acid containing glucose side chains. Group B streptococci are usually found in association with animals and are a cause of mastitis in cows. Streptococci that are found in milk, the so-called lactic streptococci, are of antigen group N.

Placed in the genus *Leuconostoc* are cocci that are morphologically similar to streptococci but are *heterofermentative*. Strains of *Leuconostoc* also produce the flavoring ingredients diacetyl and acetoin by breakdown of citrate and have been used as starter cultures in dairy fermentations, but their place has now been taken by *S. diacetilactis*. Some strains of *Leuconostoc* produce large amounts of dextran polysaccharides (α-1,6-glucan) when cultured on sucrose (see Figure 16.38). Dextrans produced by *Leuconostoc* have found some medical use as plasma extenders in blood transfusions. Other strains of *Leuconostoc* produce fructose polymers called *levans*.

Lactobacillus Lactobacilli are typically rod-shaped, varying from long and slender to short bent rods (Figure 19.84). Most species are homofermentative, but some are heterofermentative. The genus has been divided into three major subgroups (Table 19.36).

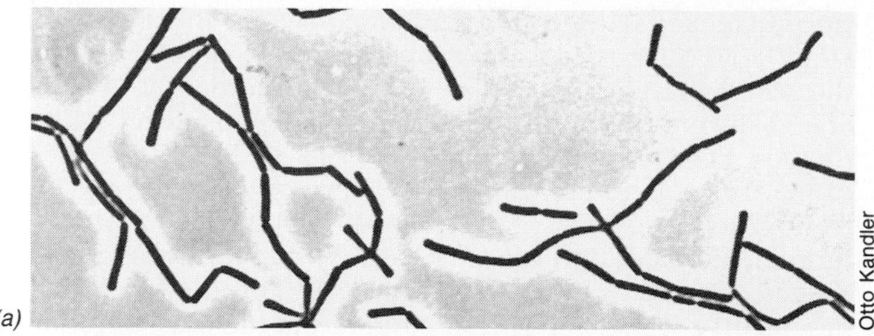
(a) Otto Kandler

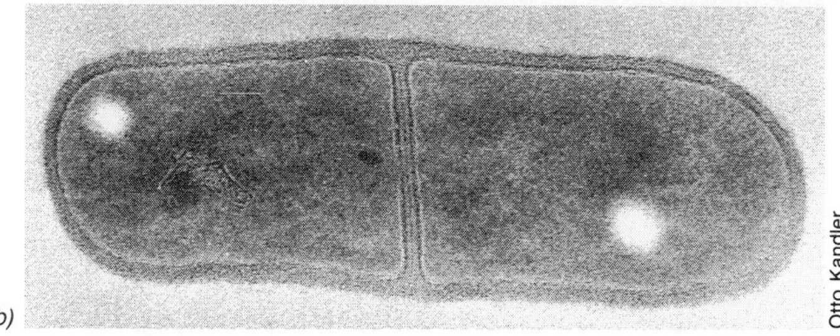
(b) Otto Kandler

Figure 19.84 Phase contrast and transmission electron micrographs of *Lactobacillus* species. (a) *L. acidophilus*. (b) *L. brevis*.

Lactobacilli are often found in dairy products, and some strains are used in the preparation of fermented products. For instance, *L. delbrueckii* is used in the preparation of yogurt, *L. acidophilus* (Figure 19.84) in the production of acidophilus milk, and other species are involved in the production of sauerkraut, silage, and pickles (see Section 9.21). The lactobacilli are usually more resistant to acidic conditions than are the other lactic acid bacteria, being able to grow well at pH values around 4–5. Because of this, they can be selectively isolated from natural materials by use of carbohydrate-containing media of acid pH, such as tomato juice-peptone agar. The acid resistance of the lactobacilli enables them to continue growing during natural lactic fermentations when the pH value has dropped too low for other lactic acid bacteria to grow, and the lactobacilli are therefore responsible for the final stages of most lactic acid fermentations. The lactobacilli are rarely if ever pathogenic.

Table 19.36 Characteristics of subgroups in the genus *Lactobacillus*

Characteristics	Species	DNA (mole % GC)
Homofermentative:		
Lactic acid the major product (>85% from glucose)		
No gas from glucose; aldolase present		
(1) Grow at 45°C but not at 15°C, long rods; glycerol teichoic acid	*L. delbrueckii*	50
	L. acidophilus	32–37
(2) Grow at 15°C, variable growth at 45°C; short rods and coryneforms; ribitol and glycerol teichoic acids	*L. casei, L. plantarum*	45–46
	L. curvatus	42–44
Heterofermentative:		
Produce about 50% lactic acid from glucose; produce CO_2 and ethanol; aldolase absent; phosphoketolase present; long and short rods; glycerol teichoic acid	*L. fermentum*	53
	L. brevis, L. buchneri	45
	L. kefir	41

19.26 Endospore-Forming Bacteria

The structure, mode of formation, and heat resistance of the bacterial endospore have been discussed in Section 3.12 and the process of spore formation itself was covered, as a model of cell differentiation, in Section 9.9. Several genera of endospore-forming bacteria have been recognized, distinguished on the basis of morphology, relationship to O_2, and energy metabolism (Table 19.37). The two genera most frequently studied are *Bacillus*, the species of which are aerobic or facultatively aerobic, and *Clostridium*, which contains the strictly anaerobic species. The genera *Bacillus* and *Clostridium* consist of Gram-positive or Gram-variable rods, which are usually motile, possessing peritrichous flagella. A major property of taxonomic value for distinguishing between species of *Bacillus* and *Clostridium* is the shape and position of endospores. These properties vary considerably among different species of endosporeformers (Figure 19.85), and is a useful starting point for beginning a taxonomic study of the group. Members of the genus *Bacillus* produce the enzymes catalase and superoxide dismutase. Clostridia do not produce catalase and produce only low levels of superoxide dismutase, and it is thought that one reason they are obligately anaerobic is that they have no way of getting rid of the toxic H_2O_2 and O_2^- produced from molecular oxygen (see Section 9.14).

Molecular taxonomic studies of the genus *Bacillus* have shown that it is a heterogeneous group and can hardly be considered an assemblage of closely related organisms. The DNA base composition of *Bacillus* species vary from about 30 to 70 percent GC, and studies on nucleic acid homologies by hybridization and genetic transformation also suggest considerable genetic heterogeneity. Although less genetic taxonomic work has been done with members of the genus *Clostridium*, data from DNA base compositions suggest more genetic homogeneity in this group than in bacilli; values from 23 to 54 percent GC have been reported, but most species cluster between 25 to 30 percent GC. Because of the complex series of enzymatic steps involved in sporulation, it seems reasonable to hypothesize that the ability to form endospores arose only once during evolution and that a primitive sporeformer was, by evolutionary divergence, the forerunner of the variety of spore-forming bacteria known today.

Even though they are not closely related genetically, all the spore-forming bacteria are *ecologically* related since they are found in nature primarily in soil. Even those species that are pathogenic to humans or animals are primarily saprophytic soil organisms, and infect hosts only incidentally. Spore formation should be advantageous for a soil microorganism because the soil is a highly variable environment. Although at some times nutrient supply is in excess, at other times it is deficient. Soil temperatures can be quite high in summer, especially at the surface. Thus a heat-resistant dormant structure should offer considerable survival value in nature. On the other hand, the ability to germinate and grow quickly when nutrients become available is also of value as it enables the organism to capitalize on a transitory food supply. The longevity of bacterial endospores is noteworthy. Records of spores surviving in the dormant state for over 50 years are well established, and isolations of thermophilic spore-forming bacilli from marine sediment cores nearly 6000 years old suggest that spores can exist for literally *thousands* of years under the appropriate conditions!

Bacillus Members of the genus *Bacillus* are easy to isolate from soil or air and are among the most common organisms to appear when soil samples are streaked on agar plates containing various nutrient media. Sporeformers can be selectively isolated from soil, food, or other material by exposing the sample to 80°C for 10 to 30 minutes, a treatment that effectively destroys vegetative cells while many of the spores present remain viable. When such pasteurized samples are streaked on plates and incubated aerobically, the colonies that develop are almost exclusively of the genus *Bacillus*. Bacilli usually grow well on synthetic media containing sugars, organic acids, alcohols, and so on, as sole carbon sources and ammonium as the sole nitrogen source; a few isolates have vitamin requirements. Many bacilli produce extracellular *hydrolytic enzymes* that break down polysaccharides, nucleic acids, and lipids, permitting the organisms to use these products as carbon sources and elec-

Table 19.37 The genera of endospore-forming bacteria

Characteristics	Genus	DNA (mole % GC)
Rods:		
Aerobic or facultative, catalase produced	*Bacillus*	32–69
Microaerophilic, no catalase	*Sporolactobacillus*	38–40
Anaerobic:		
Sulfate-reducing	*Desulfotomaculum*	37–50
Do not reduce sulfate, fermentative	*Clostridium*	24–54
Cocci (usually arranged in tetrads or packets), aerobic	*Sporosarcina*	40–42

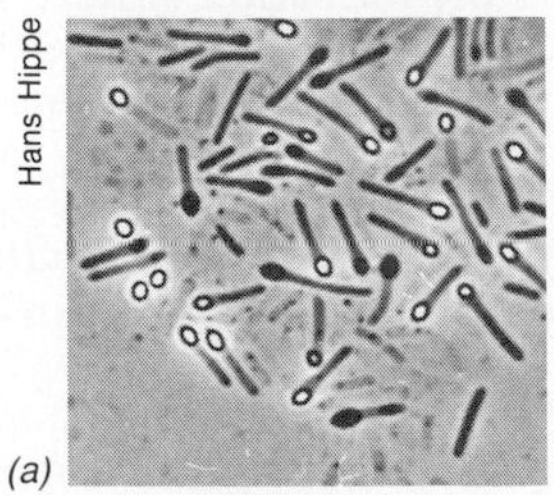
Hans Hippe
(a)

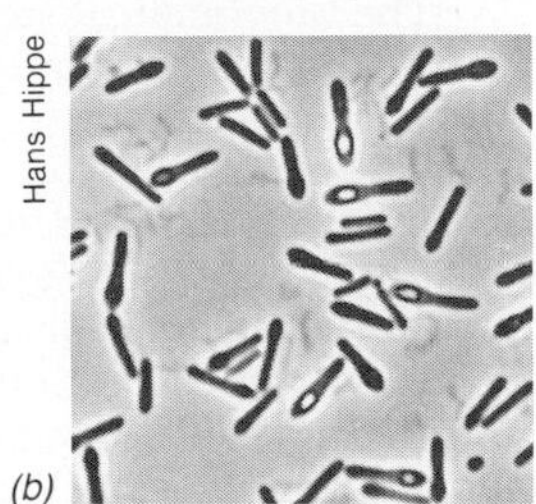
Hans Hippe
(b)

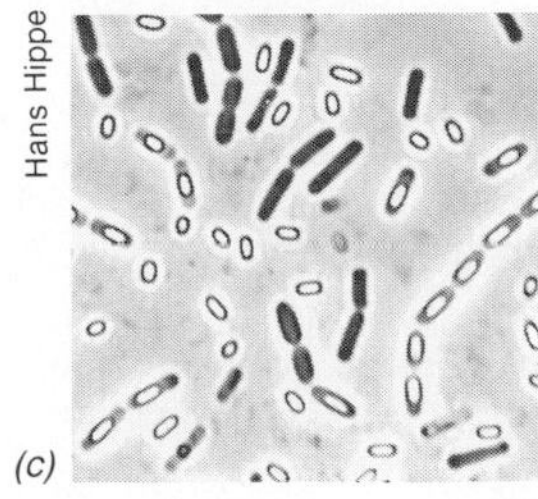
Hans Hippe
(c)

Figure 19.85 Phase contrast photomicrographs of various *Clostridium* species showing the different location of the endospore. (a) *C. cadaveris*, terminal spores. (b) *C. sporogenes*, subterminal spores. (c) *C. bifermentans*, central spores.

Table 19.38 Characteristics of representative species of the genus *Bacillus*

Characteristics	Species	DNA (mole % GC)
I. Spores oval or cylindrical, facultative aerobes, casein and starch hydrolyzed; sporangia not swollen, spore wall thin		
Thermophiles and acidophiles	*B. coagulans*	47
	B. acidocaldarius	60
Mesophiles	*B. licheniformis*	46
	B. cereus	35
	B. anthracis	33
	B. megaterium	37
	B. subtilis	43
Insect pathogen	*B. thuringiensis*	34
Sporangia distinctly swollen, spore wall thick		
Thermophile	*B. stearothermophilus*	52
Mesophiles	*B. polymyxa*	44
	B. macerans	52
	B. circulans	35
Insect Pathogens	*B. larvae*	—
	B. popilliae	41
II. Spores spherical, obligate aerobes, casein and starch not hydrolyzed		
Sporangia swollen	*B. sphaericus*	37
	B. pasteurii	38

tron donors. Many bacilli produce antibiotics, of which bacitracin, polymyxin, tyrocidin, gramicidin, and circulin are examples. In most cases antibiotic production seems to be related to the sporulation process, the antibiotic being released when the culture enters the stationary phase of growth and after it is committed to sporulation. An outline of the subdivision of the genus *Bacillus* is given in Table 19.38.

A number of bacilli, most notably *B. larvae, B. popillae* and *B. thuringiensis* are *insect pathogens*, and in recent years there has been considerable interest in these organisms because of their potential use in the biological control of insect infestations of plants. These insect pathogens form a crystalline protein during sporulation, called the **parasporal body**, which is deposited within the sporangium but outside the spore proper (Figure 19.86). These crystal-forming bacilli cause fatal diseases of moth larvae such as the silkworm, the cabbage worm, the tent caterpillar, and the gypsy moth, due to the action of this toxic substance.

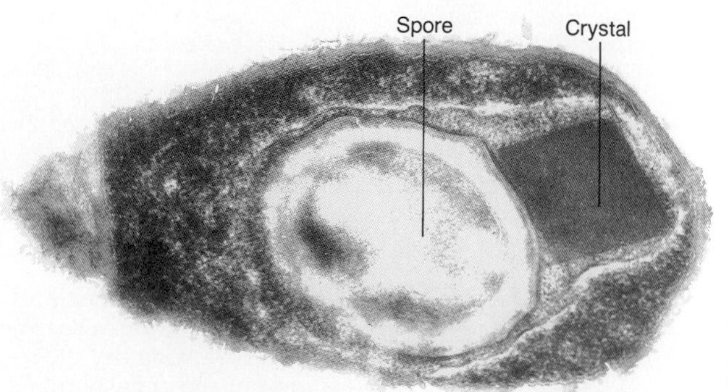

Figure 19.86 Formation of the toxic parasporal crystal in the insect pathogen *Bacillus thuringiensis*. Electron micrograph of a thin section.

Clostridium The clostridia lack a cytochrome system and a mechanism for electron-transport phosphorylation, and hence they obtain ATP *only* by substrate-level phosphorylation. A wide variety of anaerobic energy-yielding mechanisms are known in the clostridia (fermentative diversity is discussed in Section 16.18); indeed, the separation of the genus into subgroups is based primarily on these properties and on the nature of the electron donors used (Table 19.39).

A number of clostridia ferment sugars, producing as a major end product *butyric acid*. Some of these also produce *acetone* and *butanol*, and at one time the acetone-butanol fermentation by clostridia was of great industrial importance as it was the main commercial source of these products. Today, however, the chemical synthesis of acetone and butanol from petroleum products has mostly replaced the microbiological process. Some clostridia of the acetone-butanol type fix N_2; the most vigorous N_2 fixer is *C. pasteurianum*, which probably is responsible for most anaerobic nitrogen fixation in the soil. One group of clostridia ferments cellulose with the formation of acids and alcohols, and these are the main organisms decomposing cellulose anaerobically in soil. There is considerable industrial interest in the production of ethanol (an automotive fuel additive) by the clostridial fermentation of *cellulose*, and genetic studies are underway to increase the yield of ethanol and reduce the formation of acidic fermentation products, the goal being to use waste cellulose as a motor fuel.

The biochemical steps in the formation of butyric acid and butanol from sugars are well understood (Figure 19.87). Glucose is converted to pyruvate via the Embden-Meyerhof pathway, and pyruvate is split to acetyl-CoA, CO_2, and reducing equivalents (reduced ferredoxin) by a phosphoroclastic reaction (see Section 16.18). Acetyl-CoA is then reduced to fermentation products using the NADH derived from glycolytic reactions. The proportions of the various products are influenced by the duration and the conditions of the fermentation. During the early stages, butyric and

Table 19.39 Characteristics of some groups of the genus *Clostridium*

Key characteristics	Other characteristics	Species	DNA (mole % GC)
I. Ferment carbohydrates:			
Ferment cellulose	Fermentation products: acetic acid, lactic acid, succinic acid, ethanol, CO_2, H_2	*C. cellobioparum* *C. thermocellum*	28 38–39
Ferment sugars, starch, and pectin	Fermentation products: acetone, butanol, ethanol, isopropanol, butyric acid, acetic acid, propionic acid, lactic acid, CO_2, H_2; some fix N_2	*C. butyricum* *C. acetobutylicum* *C. pasteurianum* *C. perfringens*	27–28 28–29 26–28 24–27
Ferment sugars primarily to acetic acid	Total synthesis of acetate from CO_2; cytochromes present in some species	*C. aceticum* *C. thermoaceticum* *C. formicoaceticum*	33 54 34
Ferments sugars primarily to propionate	Fermentation products: propionate, isobutyrate, isovalerate	*C. propionicus*	—
II. Ferment proteins or amino acids	Fermentation products: acetic acid, fatty acids, NH_3, CO_2, sometimes H_2; some also ferment sugars to butyric and acetic acids; may produce exotoxins	*C. sporogenes* *C. tetani* *C. botulinum* *C. histolyticum* *C. tetanomorphum*	26 25-26 26-28 — —
III. Ferments carbohydrates or amino acids	Fermentation products from glucose: acetate, formate, small amounts of isobutyric and isovaleric acids	*C. bifermentans*	27
IV. Purine fermenters	Ferment uric acid and other purines, forming acetic acid, CO_2, NH_3	*C. acidurici*	28
V. Ethanol fermentation to fatty acids	Produces butyric acid, caproic acid, and H_2; requires acetate as electron acceptor; does not attack sugars, amino acids, or purines	*C. kluyveri*	30

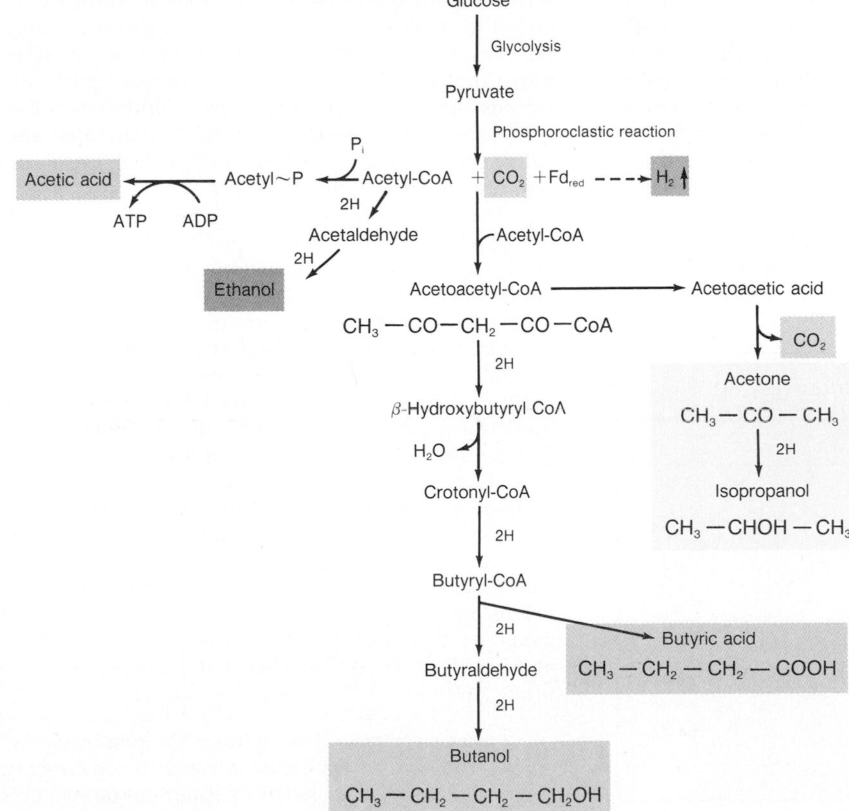

Figure 19.87 Pathway of formation of fermentation products from the butyric acid group of clostridia. The designation "2H" represents two electrons from one molecule of NADH.

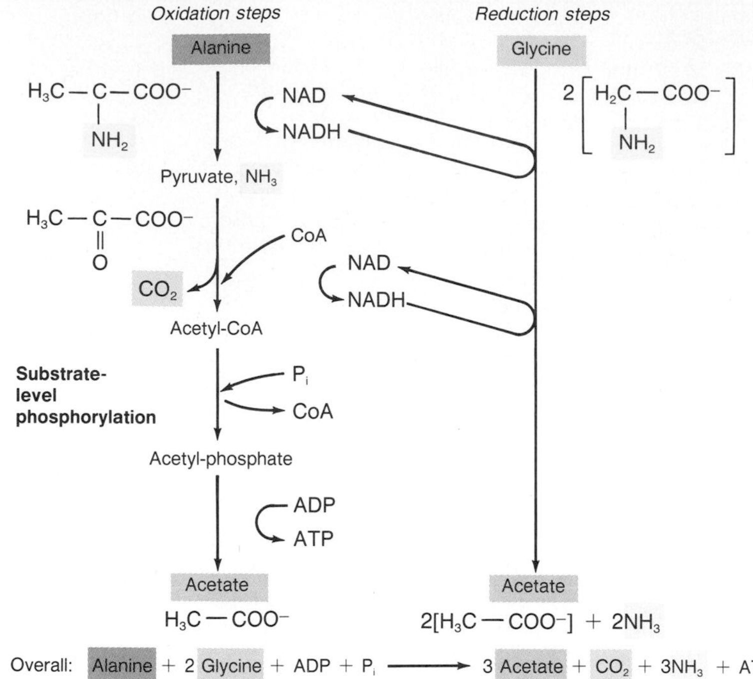

Figure 19.88 Coupled oxidation-reduction reaction (Stickland reaction) in *Clostridium sporogenes* between alanine and glycine.

acetic acids are the predominant products, but as the pH of the medium drops, synthesis of acids ceases and the neutral products acetone and butanol begin to accumulate. If the medium is kept alkaline with $CaCO_3$, very little of the neutral products are formed and the fermentation products consist of about 3 parts butyric and 1 part acetic acid. Studies of the enzymology of solvent production by clostridia indicate that acid production in some way triggers induction of enzymes involved in acetone and butanol synthesis.

Another group of clostridia obtain their energy by fermenting *amino acids*. Some strains do not ferment single amino acids, but when two separate amino acids are present in the medium, one functions as the electron *donor* and is *oxidized*, while the other acts as the electron *acceptor* and is *reduced*. This type of coupled decomposition is known as the **Stickland reaction**. For instance, *C. sporogenes* will catabolize a mixture of glycine and alanine, as outlined in Figure 19.88.

Various amino acids that can function as either oxidants or reductants in Stickland reactions are listed in Table 19.40.

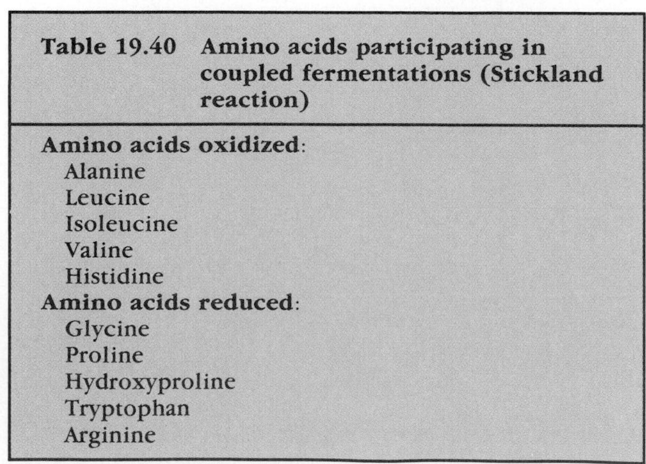
Table 19.40 Amino acids participating in coupled fermentations (Stickland reaction)

Amino acids oxidized:
Alanine
Leucine
Isoleucine
Valine
Histidine
Amino acids reduced:
Glycine
Proline
Hydroxyproline
Tryptophan
Arginine

The products of the oxidation are always NH_3, CO_2, and a carboxylic acid with one *less* carbon atom than the amino acid which is oxidized.

Some amino acids can be fermented singly, rather than in a Stickland-type reaction. These are alanine, cysteine, glutamate, glycine, histidine, serine, and threonine. It is usually found that each group of clostridia is specific in the kinds of substances it can ferment; usually either sugars or amino acids are utilized, although there are strains that can ferment both. Many of the products of amino acid fermentation by clostridia are foul-smelling substances, and the odor that results from putrefaction is a result mainly of clostridial action. In addition to butyric acid, other odoriferous compounds produced are isobutyric acid, isovaleric acid, caproic acid, hydrogen sulfide, methylmercaptan (from sulfur amino acids), cadaverine (from lysine), putrescine (from ornithine), and ammonia.

The main habitat of clostridia is the soil, where they live primarily in anaerobic "pockets," made anaerobic primarily by facultative organisms acting upon various organic compounds present. In addition, a number of clostridia have adapted to the anaerobic environment of the mammalian intestinal tract. Also, as was discussed in Section 11.10, several clostridia that live primarily in soil are capable of causing disease in humans under specialized conditions. Botulism is caused by *C. botulinum*, tetanus by *C. tetani*, and gas gangrene by *C. perfringens* and a number of other clostridia, both sugar and amino acid fermenters. These pathogenic clostridia seem in no way unusual metabolically, but are distinct in that they produce specific toxins or, in those causing gas gangrene, a group of toxins (Table 11.4). An unsolved ecological problem is what role these toxins play in the natural habitat of the organism. Many gas-gangrene clostridia also cause diseases in domestic animals, and botulism occurs in sheep and ducks, and a variety of other animals.

Sporosarcina The genus *Sporosarcina* is unique among endosporeformers because cells are *cocci* instead of rods. *Sporosarcina* consists of spherical to oval cells which divide in two or three perpendicular planes to form tetrads or packets of eight or more cells (see Figure 19.89). The organism is motile and strictly aerobic. Two species of *Sporosarcina* are known, *S. ureae* and *S. halophila*. The latter species is of marine origin and differs from *S. ureae* primarily in its requirement for sodium ions for growth. Analyses of cell walls have also shown that *S. ureae* and *S. halophila* differ in the nature of the diamino acid found in the crosslinking polypeptide of peptidoglycan; *S. ureae* contains lysine and *S. halophila* contains ornithine. The endospores of *Sporosarcina* are highly refractile, centrally located (see Figure 19.89), and contain dipicolinic acid, typical of endospores from other genera.

S. ureae can easily be enriched from soil by plating dilutions of a pasteurized soil sample on nutrient agar supplemented with 8 percent urea. Most soil bacteria are strongly inhibited by as little as 5 percent urea, however *S. ureae* actively decomposes urea to CO_2 and NH_3, and in so doing can dramatically raise the pH of unbuffered media (*S. ureae* is remarkably alkaline tolerant and will grow in media up to pH 10–11). *S. ureae* is common in soils, and studies of its distribution suggest that numbers of *S. ureae* are greatest in soils that receive inputs of urine (a source of urea), such as soils in which animals periodically urinate. Since many soil organisms are quite urea sensitive, these results suggest that *S. ureae* is ecologically important as a major urea degrader in nature.

Endospore formation The differences between the endospore and the vegetative cell are profound (see Table 9.3), and sporulation involves a very complex series of events. Bacterial sporulation does not occur when cells are dividing exponentially, but only when growth ceases owing to the exhaustion of an essential nutrient. For instance, if a culture growing on glucose as an electron donor exhausts the glucose in the medium, vegetative growth ceases and several hours later spores begin to appear. If more glucose is added to the culture just at the end of the growth period, sporulation is inhibited. Glucose probably prevents sporulation through *catabolite repression* (see Section 5.12), inhibiting the synthesis of the specific enzymes involved in forming the spore structures. Glucose is not the only substance repressing spore formation; many other electron donors also can do this. We thus see that growth and sporulation are opposing processes. It seems reasonable to assume that the elaborate control mechanisms in endospore-forming bacteria ensure that in nature sporulation will occur only when conditions are no longer favorable for growth.

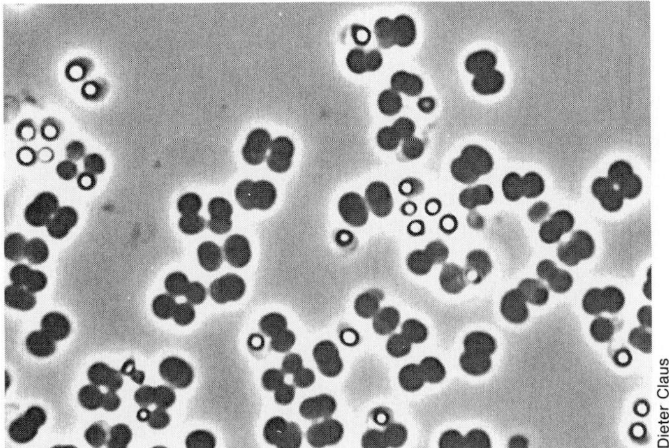

Figure 19.89 Phase-contrast photomicrograph of *Sporosarcina ureae*.

19.27 Mycoplasmas

Although seeming at first to be out of place in a discussion of Gram-positive bacteria because their lack of cell walls yields no reaction in the Gram stain, phylogenetically, mycoplasmas are at home with the Gram-positive bacteria because of their close evolutionary ties to the clostridia (see Section 18.5). The mycoplasmas are organisms without cell walls that do not revert to walled organisms. They are probably the smallest organisms capable of autonomous growth and are of special evolutionary interest because of their extremely simple cell structure.

The lack of cell walls in the mycoplasmas has been proved by electron microscopy and by chemical analysis, the latter showing that the key wall components, muramic acid and diaminopimelic acid, are missing. In Chapter 3 we discussed protoplasts and showed how these structures can be formed when cell-wall-digesting enzymes act on cells that are in an osmotically protected medium, and that when the osmotic stabilizer is removed, protoplasts take up water, swell, and burst. Mycoplasmas resemble protoplasts in their lack of a cell wall, but they are more resistant to osmotic lysis and are able to survive conditions under which protoplasts lyse. This ability to resist osmotic lysis is at least partially determined by the nature of the mycoplasma cell membrane, which is more stable than that of other procaryotes. In one group of mycoplasmas, the membrane contains **sterols** that seem to be responsible for stability, whereas in other mycoplasmas, carotenoids or other compounds may be involved instead. Those mycoplasmas possessing sterols in their membrane do not synthesize them but require them preformed in the culture medium. This sterol requirement is a current basis for separating the mycoplasmas into two groups (Table 19.41). Members of the genera *Mycoplasma, Ureaplasma, Spiroplasma* and *Anaeroplasma require* sterols, while *Acholeplasma* and *Thermoplasma do not. Thermoplasma* is phylogenetically unrelated to all other mycoplasmas because it is an archaebacterium. *Thermoplasma* is thermophilic and acidophilic, growing optimally at 55°C and pH 2; the properties of *Thermoplasma* are considered in more detail in Section 19.36.

Certain mycoplasmas contain compounds called **lipoglycans** (see Table 19.41). Lipoglycans are long chain heteropolysaccharides covalently linked to membrane lipids and embedded in the cytoplasmic membrane of many mycoplasmas. Lipoglycans resemble the lipopolysaccharides of Gram-negative bacteria (see Section 3.5) except that they lack the lipid A backbone and the phosphate typical of bacterial LPS. Lipoglycans presumably function to help stabilize the membrane, and have also been identified as facilitating attachment of mycoplasmas to cell surface receptors of animal cells. Like LPS, lipoglycans stimulate antibody production when injected into experimental animals.

Mycoplasma cells are usually fairly small, and they are highly pleomorphic, a consequence of their lack of rigidity. A single culture may exhibit small coccoid elements, larger, swollen forms, and filamentous forms of variable lengths, often highly branched (Figure 19.90). It is from the production of filamentous, funguslike forms that the name *Mycoplasma* (*myco* means "fungus") derives. A common growth form is seen in cultures that divide by budding;

19.41 Major characteristics of mycoplasmas

Genus	Properties	DNA (mole % GC)	Genome size	Presence of lipoglycans
I. Require sterols				
Mycoplasma	Many pathogenic; require sterols; facultative aerobes	23–40	5×10^8	+
Anaeroplasma	May or may not require sterols; obligate anaerobes; degrade starch, producing acetic, lactic, and formic acids plus ethanol and CO_2; inhibited by thallium acetate	29 (sterol-requiring strains) 40 (non-sterol-requiring strains)	−	+
Spiroplasma	Spiral to corkscrew-shaped cells; associated with various phytopathogenic (plant disease) conditions	25–31	10^9	−
Ureaplasma	Coccoid cell; occasional clusters and short chains; growth optimal at pH 6; strong urease reaction; associated with certain urinary tract infections in humans; inhibited by thallium acetate	27–30	5×10^8	−
II. Do not require sterols				
Acholeplasma	Facultative aerobes	27–36	10^9	+
Thermoplasma	Thermophilic, acidophilic archaebacterium found in heated coal refuse sites; no sterol requirement; DNA contains histones	46	10^9	+

division occurs with the cells remaining either directly attached or connected by thin hyphae (Figure 19.91).

The small coccoid elements (0.2 to 0.3 μm in size) are the smallest mycoplasma units capable of independent growth. Because of flexibility due to lack of a cell wall, mycoplasma cells pass through filters with pore sizes smaller than the true diameter of the cells, and this has led to erroneous estimates of the minimum cell size capable of growth. Cellular elements of diameters close to 0.1 μm exist in mycoplasma cultures, but these are not capable of growth. Even so, the minimum reproductive unit of 0.2 to 0.3 μm probably represents the smallest *free-living* cell. Additionally, the genome size of mycoplasmas is also smaller than that of most procaryotes, between 5×10^8 and 1×10^9 in molecular weight, which is comparable to that of the obligately parasitic chlamydia and rickettsia (see Sections 19.22 and 19.23), and about one-fifth to one-half that of *Escherichia coli*.

The mode of growth of mycoplasmas differs in liquid and agar cultures. On agar there is a tendency for the organisms to grow so that they become embedded in the medium, and the fibrous nature of the agar gel seems to affect the division process, perhaps by promoting separation of units from the growing mass. Colonies of mycoplasmas on agar exhibit a characteristic "fried-egg" appearance be-

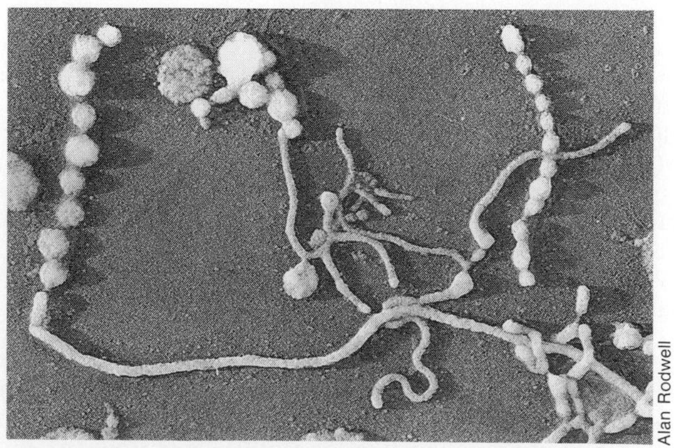

Figure 19.90 Electron micrograph of a metal-shadowed preparation of *Mycoplasma mycoides*. Note the coccoid and hyphalike elements.

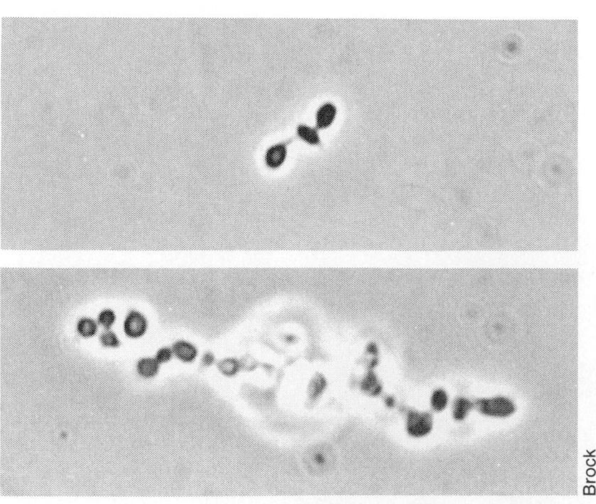

Figure 19.91 Photomicrograph by phase-contrast microscopy of a mycoplasma culture, showing typical cell arrangement.

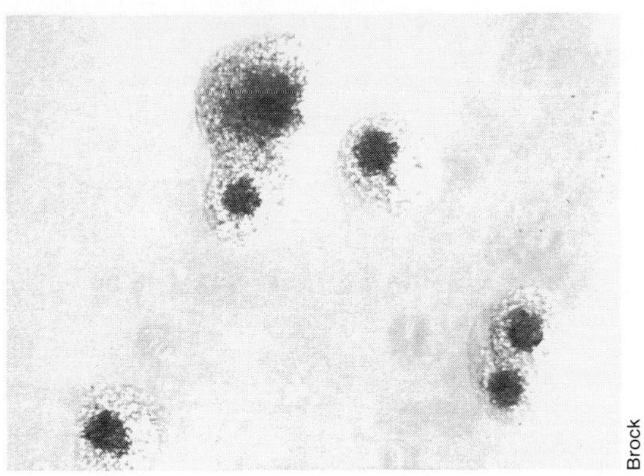

Figure 19.92 Typical "fried-egg" appearance of mycoplasma colonies on agar. The colonies are around 1 mm in diameter.

cause of the formation of a dense central core, which penetrates downward into the agar, surrounded by a circular spreading area that is lighter in color (Figure 19.92). Growth of mycoplasmas is not inhibited by penicillin, cycloserine, or other antibiotics that inhibit cell wall synthesis, but the organisms are as sensitive as other bacteria to antibiotics that act on targets other than the cell wall. Use is made of the natural penicillin resistance of mycoplasmas in preparing selective media for their isolation from natural materials. The culture media used for the growth of most mycoplasmas have usually been quite complex. Growth is poor or absent even in complex yeast extract-peptone-beef heart infusion media unless fresh serum or ascitic fluid is added. The main costituents provided by these two adjuncts are unsaturated fatty acids and sterols. Some mycoplasmas can be cultivated on relatively simple media, however, and synthetic media have been developed for some strains. Most mycoplasmas use carbohydrates as energy sources, and require a range of vitamins, amino acids, purines, and pyrimidines as growth factors. The energy metabolism of mycoplasmas is not unique. Some species are oxidative, possessing the cytochrome system and making ATP by electron-transport phosphorylation. Other species resemble the lactic acid bacteria in being strictly fermentative, producing energy by substrate-level phosphorylation and yielding lactic acid as the final product of sugar fermentation. Members of the genus *Anaeroplasma* are obligate anaerobes which ferment glucose or starch to a variety of acidic products.

The genus *Spiroplasma* consists of pleomorphic cells, spherical or slightly ovoid, which are often helical or spiral in shape (Figure 19.93). Although they lack a cell wall and flagella, they are motile by means of a rotary (screw) motion or a slow undulation. Intracellular fibrils that are thought to play a role in motility have been demonstrated. The organism has been isolated from ticks, the hemolymph and gut of insects, vascular plant fluids and insects that feed on fluids, and from the surfaces of flowers and other plant parts. *S. citri* has been isolated from the leaves of citrus plants, where it causes a disease called *citrus stubborn disease* and from corn plants suffering from *corn stunt disease*. A number of other mycoplasma-like bodies have been detected in diseased plants by electron microscopy, which indicates that a large group of plant-associated mycoplasmas may exist. Four species of *Spiroplasma* are recognized that cause a variety of animal diseases such as *honeybee spiroplasmosis, suckling mouse cataract disease*, and *lethargy disease* of the bettle *Melolontha*.

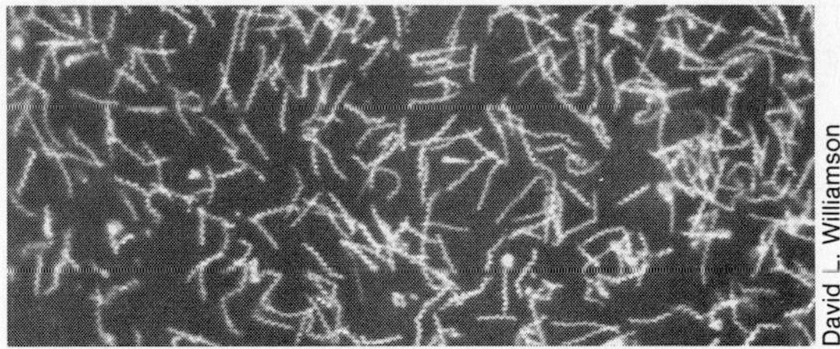

Figure 19.93 Dark field micrograph of the "sex ratio" spiroplasma removed from the hemolymph of the fly, *Drosophila pseudoobscura*. Female flies infected with the sex ratio spiroplasma bear only female progeny.

19.28 High GC Gram-Positive Bacteria: The "Actinomycetes"

An extremely large variety of bacteria falls under this heading, as evidenced by the entire volume (volume 4) of the new edition of *Bergey's Manual*, which is devoted to the filamentous actinomycetes, and major portions of volume 2 devoted to the rod-shaped relatives of this group. Phylogenetically, the actinomycetes form a subbranch of Gram-positive bacteria, distinct from the endosporeformers and Gram-positive cocci and asporogenous rods. Despite great morphological variability, the actinomycetes seem to form a tight phylogenetic unit and most representatives have GC ratios in the 60s and 70s mole percent. There are considerable difficulties in drawing clear-cut distinctions between various genera of actinomycetes. All of these organisms show a few common features: they are Gram-positive, rod-shaped to filamentous, and generally nonmotile in the vegetative phase (although motile stages are known). A continuum exists from simple rod-shaped organisms to rod-shaped organisms which occasionally grow in a filamentous manner, to strictly filamentous forms. Table 19.42 provides an overview of this group. In the following sections, we discuss some of the more interesting and important genera.

19.29 Coryneform Bacteria

The coryneform bacteria are Gram-positive, aerobic, nonmotile, rod-shaped organisms that have the characteristic of forming irregular-shaped, club-shaped, or V-shaped cell arrangements during normal growth. V-shaped cell groups arise as a result of a snapping movement that occurs just after cell division (called post-fission snapping movement or, simply, *snapping division*) (Figure 19.94). **Snapping division** has been shown to occur in one species because the cell wall consists of two layers; only the inner layer participates in cross-wall formation, so that after the cross-wall is formed, the two daughter cells remain attached by the outer layer of the cell wall. Localized rupture of this outer layer on one side results in a bending of the two cells away from the ruptured side (Figure 19.95), and thus development of V-shaped forms.

The main genera of coryneform bacteria are *Corynebacterium* and *Arthrobacter*. The genus *Corynebacterium* consists of an extremely diverse group of bacteria, including animal and plant pathogens as well as saprophytes. The genus *Arthrobacter*, consisting primarily of soil organisms, is distinguished from *Corynebacterium* on the basis of a cycle of development in *Arthrobacter* involving conversion

Table 19.42 Actinomycetes and related genera (all Gram-positive)

Major groups	DNA (mole % GC)
Coryneform group of bacteria: Rods, often club-shaped, morphologically variable; not acid-fast or filamentous; snapping cell division	
Corynebacterium: irregularly staining segments, sometimes granules; club-shaped swellings frequent; animal and plant pathogens, also soil saprophytes	51–65
Arthrobacter: coccus-rod morphogenesis; soil organisms	59–70
Cellulomonas: coryneform morphology; cellulose digested; facultative aerobe	71–73
Kurthia: rods with rounded ends occurring in chains; coccoid later	36–38
Propionic acid bacteria: anaerobic to aerotolerant; rods or filaments, branching	
Propionibacterium: nonmotile; anaerobic to aerotolerant; produce propionic acid and acetic acid; dairy products (Swiss cheese); skin, may be pathogenic	53–68
Eubacterium: obligate anaerobes; produce mixture of organic acids, including butyric, acetic, formic, and lactic; intestine, infections of soft tissue, soil; may be pathogenic; probably the predominant member of the intestinal flora	–
Actinomycetes: filamentous, often branching; highly diverse	
Group I. Actinomycetes: not acid-alcohol-fast; facultatively aerobic; mycelium not formed; branching filaments may be produced; rod, coccoid, or coryneform cells	
Actinomyces: anaerobic to facultatively aerobic; filamentous microcolony, but filaments transitory and fragment into coryneform cells; may be pathogenic for humans or animals; teeth	57–69
Bifidobacterium: smooth microcolony, no filaments; coryneform cells common; found in intestinal tract of breast-fed infants	55–67
Other genera: *Arachnia, Bacterionema, Rothia, Agromyces*	
Group II. Mycobacteria: acid-alcohol-fast, filaments transitory	
Mycobacterium: pathogens, saprophytes; obligate aerobes; lipid content of cells and cell walls high; waxes, mycolic acids; simple nutrition; growth slow; tuberculosis, leprosy, granulomas, avian tuberculosis; also soil organisms; hydrocarbon oxidizers	62–70
Group III. Nitrogen-fixing actinomycetes: nitrogen-fixing symbionts of plants; true mycelium produced	
Frankia: forms nodules of two types on various plant roots; probably microaerophilic; grows slowly; fixes N_2	–
Group IV. Actinoplanes: true mycelium produced; spores formed, borne inside sporangia	
Actinoplanes, Streptosporangium	69–71
Group V. Dermatophilus: mycelial filaments divide transversely, and in at least two longitudinal planes, to form masses of motile, coccoid elements; aerial mycelium absent; occasionally responsible for epidermal infections	
Dermatophilus, Geodermatophilus	–
Group VI. Nocardias: mycelial filaments commonly fragment to form coccoid or elongate elements; aerial spores occasionally produced; sometimes acid-alcohol-fast	
Nocardia: common soil organisms; obligate aerobes; many hydrocarbon utilizers	61–72
Rhodococcus: soil saprophytes, also common in gut of various insects	59–69
Group VII. Streptomycetes: mycelium remains intact, abundant aerial mycelium and long spore chains	
Streptomyces: Nearly 500 recognized species, many produce antibiotics	69–75
Other genera (differentiated morphologically): *Streptoverticillium, Sporichthya, Microcellobosporia, Kitasatoa, Chainia*	67–73
Group VIII. Micromonosporas: mycelium remains intact; spores formed singly, in pairs, or short chains; several thermophilic; saprophytes found in soil, rotting plant debris; one species produces endospores	
Micromonospora, Thermoactinomyces, Thermomonospora	54–79

from rod to sphere and back to rod again (Figure 19.96). However, some corynebacteria are pleomorphic and form coccoid elements during growth, so that the distinction between the two genera on the basis of life cycle is more quantitative than qualitative. The *Corynebacterium* cell frequently has a swollen end, so that it has a club-shaped appearance (hence the name of the genus: *koryne* is the Greek word for "club"), whereas *Arthrobacter* is less commonly club shaped.

Organisms of the genus *Arthrobacter* are among the most common of all soil bacteria. They are remarkably resistant to desiccation and starvation, despite the fact that they do not form spores or other resting cells. (The coccoid and rod-shaped forms seem equally resistant to desiccation and starvation.) Arthrobacters are a heterogeneous group that have considerable nutritional versatility, and strains have been isolated that decompose herbicides, caffeine, nicotine, phenol, and other unusual organic compounds.

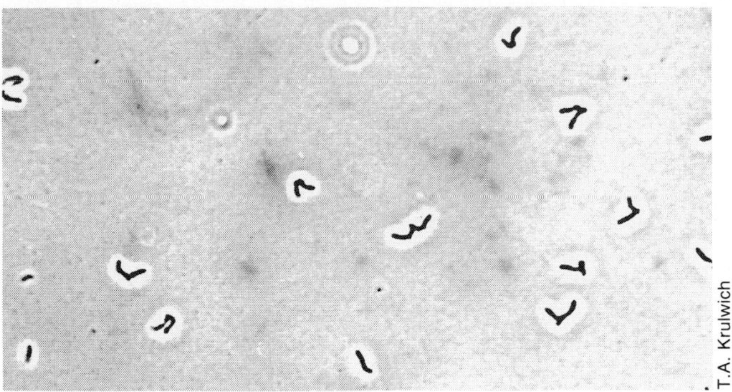

Figure 19.94 Photomicrograph of characteristic V-shaped cell groups in *Arthrobacter crystallopoietes*, resulting from snapping division.

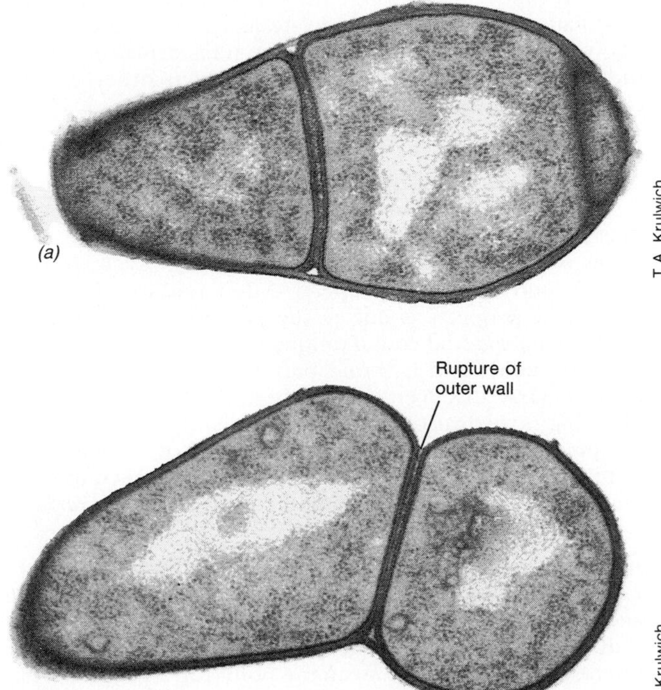

Figure 19.95 Electron micrograph of cell division in *Arthrobacter crystallopoietes*, illustrating how snapping division and V-shaped cell groups arise. (a) Before rupture of the outer cell-wall layer. (b) After rupture of the outer layer on one side.

19.30 Propionic Acid Bacteria

The propionic acid bacteria (genus *Propionibacterium*) were first discovered as inhabitants of Swiss (Emmentaler) cheese, where their fermentative production of CO_2 produces the characteristic holes; the presence of propionic acid is at least partly responsible for the unique flavor of the cheese. Although this acid is produced by some other bacteria, its production by the propionic acid bacteria is a distinguishing characteristic of the genus. The bacteria in this group are Gram-positive, pleomorphic, nonsporulating rods, nonmotile and anaerobic. They ferment lactic acid, carbohydrates, and polyhydroxy alcohols, producing propionic acid, succinic acid, acetic acid, and CO_2. Their nutritional requirements are complex, and they usually grow rather slowly. In some taxonomic schemes, the facultatively aerobic coryneforms are also classified as propionic acid bacteria.

The enzymatic reactions leading from glucose to propionic acid are of interest (Figure 19.97). The initial catabolism of glucose to pyruvate follows the Embden-Meyerhof pathway as in the lactic acid bacteria, but the NADH formed is reoxidized as one part of a cycle in which *propionic acid* is formed. Pyruvate accepts a carboxyl group from methylmalonyl-CoA by a transcarboxylase reaction, leading to the formation of oxalacetate and propionyl-CoA. The latter substance reacts with succinate in a step catalyzed by a CoA transferase, producing succinyl-CoA and propionate. The succinyl-CoA is then isomerized to methylmalonyl-CoA, and the cycle is complete (Figure 19.97). Reoxidation of NADH occurs in the steps between oxalacetate and succinate, and the oxidation-reduction balance is restored.

Most propionic acid bacteria also ferment lactate with the production of propionate, acetate, and CO_2. The anaerobic fermentation of lactic acid to propionate is of interest because lactic acid itself is an end product of fermentation for many bacteria. The propionic acid bacteria are thus able to obtain energy anaerobically from a substance that other bacteria are producing.

It is the fermentation of lactate to propionate that is important in Swiss cheese manufacture. The starter culture consists of a mixture of homofermentative streptococci and lactobacilli, plus propionic acid bacteria. The initial fermentation of lactose to lactic acid during formation of the curd is carried out by the homofermentative organisms. After the curd (protein and fat) has been drained, the propionic acid bacteria develop rapidly and usually reach numbers of 10^8 per gram by the time the cheese is 2 months old. Swiss cheese "eyes" are formed by the accumulation of CO_2, the gas diffusing through the curd and gathering at weak points.

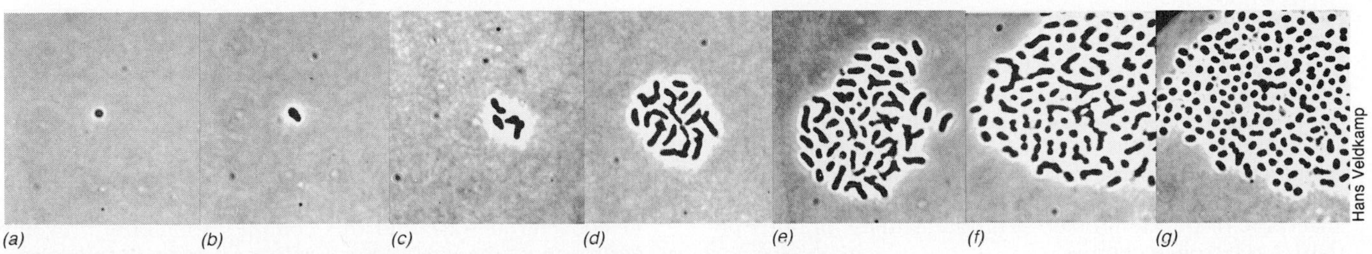

Figure 19.96 Stages in the life cycle of *Arthrobacter globiformis* as observed in slide culture; (a) single coccoid element; (b–e) conversion to rod and growth of microcolony consisting predominantly of rods; (f–g) conversion of rods to coccoid forms.

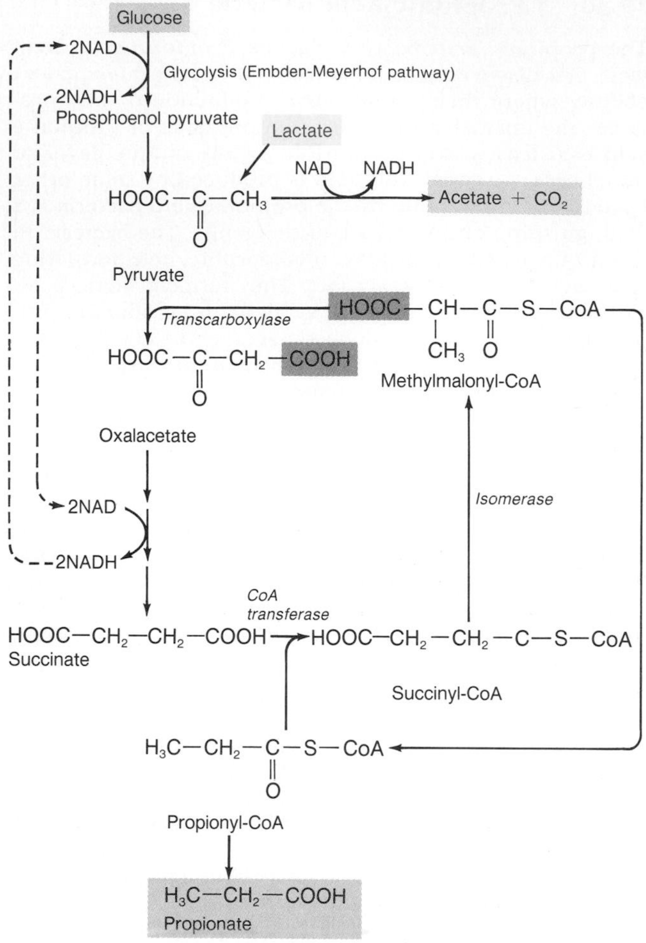

Figure 19.97 The formation of propionic acid by *Propionibacterium*. Either lactate, produced by the fermentative activities of other bacteria or glucose, can serve as starting substrate in the propionate fermentation.

Propionigenium A second bacterium, quite unrelated to *Propionibacterium*, also produces propionic acid fermentatively. *Propionigenium* is a Gram-negative, strictly anaerobic bacterium that ferments *succinate* to propionate and CO_2:

$$\text{Succinate} + H_2O \rightarrow \text{Propionate} + HCO_3^-$$
$$\Delta G^{0\prime} = -4.9 \text{ kcal/mole}$$

The free energy release of this transformation is the lowest energy yield known for any chemical transformation that supports growth of a bacterium. The succinate fermentation of *Propionigenium* releases less free energy than even the methanogenic fermentation of acetate (see Section 19.34), yet *Propionigenium* grows quite rapidly compared to aceticlastic methanogens. *Propionigenium* also grows on fumarate, malate, aspartate, oxaloacetate, and pyruvate as sole energy sources, but will not ferment sugars or carry out anaerobic respiration linked to nitrate, sulfate, or other potential electron acceptors.

The mechanism by which *Propionigenium* couples succinate fermentation to ATP synthesis is unknown, but it may be significant that all isolates of *Propionigenium*, whether from marine or freshwater mud, required *sodium* for growth. It is postulated that the fermentation of succinate (which essentially amounts to a decarboxylation of succinate) by *Propionigenium* is coupled to *proton gradient formation* in which sodium ions play an intimate role. There is some evidence that the decarboxylation of succinate by *Propionigenium* is linked to the *export* of sodium ions (Na^+) from the cell. If this occurs, then an *antiporter* (see Section 3.4) could exchange Na^+ for H^+, thus returning Na^+ to the *inside* of the cell and generating a proton gradient (outside acidic). The latter could then be used to drive ATP synthesis by normal chemiosmotic mechanisms (see Section 4.12 for discussion of chemiosmosis). However, regardless of the mechanics of succinate fermentation by *Propionigenium*, this organism undoubtedly plays an ecologically important role as a succinate degrader. Succinate is an important product of anaerobic degradation processes (see Section 19.20), being produced in fermentations and through the anaerobic reduction of fumarate. Since *Propionigenium* is limited to succinate, succinate derivatives, or pyruvate as energy sources, it is likely that it specializes in succinate transformations in nature and receives little competition from other heterotrophic anaerobes for its energy substrate.

19.31 Mycobacterium

The genus *Mycobacterium* consists of rod-shaped organisms, which at some stage of their growth cycle possess the distinctive staining property called **acid-alcohol fastness**. This property is due to the presence on the surface of the mycobacterial cell of unique lipid components called **mycolic acids** and is found only in the genus *Mycobacterium*. First discovered by Robert Koch during his pioneering investigations on tuberculosis, this unique staining property permitted the identification of the organism in tuberculous lesions; it has subsequently proved to be of great taxonomic use in defining the genus *Mycobacterium*.

Acid-alcohol fastness In the staining procedure (*Ziehl-Neelsen stain*) a mixture of the dye basic fuchsin and phenol is used in the primary staining procedure, the stain being driven into the cells by slow heating of the microscope slide to the steaming point for 2 to 3 minutes. The role of the phenol is to enhance penetration of the fuchsin into the lipids. After washing in distilled water, the preparation is decolorized with acid alcohol (3 percent HCl in 95 percent ethanol); the fuchsin dye is removed from other organisms, but is retained by the mycobacteria. After another wash in water, a final counterstain of methylene blue is used. Acid-alcohol-fast organisms on the final preparation appear *red* whereas the background and nonacid-alcohol-fast organisms appear *blue*.

As noted, the key component necessary for acid-alcohol fastness is a unique lipid fraction of mycobacterial cells called *mycolic acid*. Mycolic acid is actually a group of complex branched-chain hydroxy lipids with the overall structure shown in Figure 19.98*a*. The carboxylic acid group of the mycolic acid must be free (unesterified), and it reacts on a one-to-one basis with the fuchsin dye (Figure 19.98*b*). The mycolic acid is complexed to the peptidoglycan of the mycobacterial wall, and this complex somehow prevents approach of the acid-alcohol solvent during the decolorization step. It was also first demonstrated by Koch that disruption of cellular integrity destroys the acid-alcohol-fast property; thus cellular integrity is a necessary prerequisite of this property.

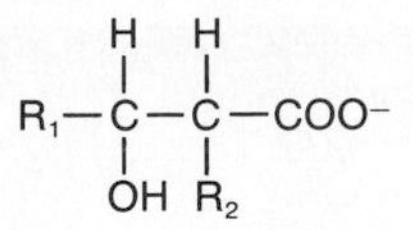

(a) Mycolic acid; R_1 and R_2 are long-chain aliphatic hydrocarbons

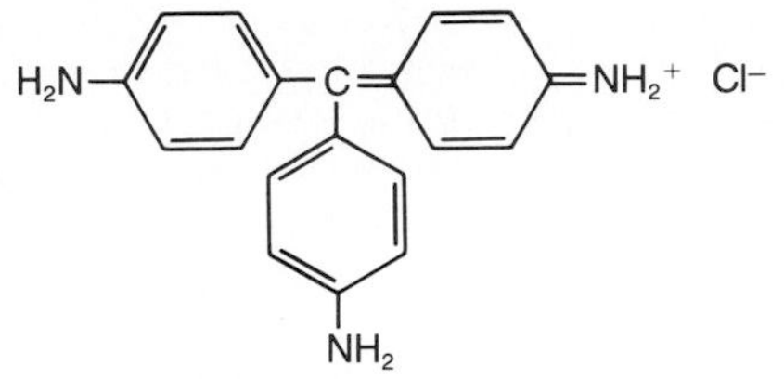

(b) Basic fuchsin

Figure 19.98 Structure of (a) mycolic acid and (b) basic fuchsin, the dye used in the acid-alcohol-fast stain. The fuchsin dye probably combines with the mycolic acid via ionic bonds between COO^- and NH_2^+.

The mycobacteria are not readily stained by the Gram method, because of the high surface lipid content, but if the lipoidal portion of the cell is removed with alkaline ethanol (1 percent KOH in absolute ethanol), the intact cell remaining is non-acid-alcohol fast but instead is Gram-positive. Thus *Mycobacterium* can be considered to be a Gram-positive bacterium. If the lipoidal portion is not removed, then the cells are resistant to decolorization by the Gram procedure even when stained with crystal violet alone (in the absence of iodine), whereas iodine is essential for the conventional Gram-staining procedure (see Figures 1.36 and 1.37).

Characteristics of mycobacteria Mycobacteria are generally rather pleomorphic, and may undergo branching or filamentous growth. However, in contrast to the actinomycetes, filaments of the mycobacteria become fragmented into rods or coccoid elements upon slight disturbance; a true mycelium is not formed. In general, mycobacteria can be separated into two major groups, *slow growers* and *fast growers* (Table 19.43). *M. tuberculosis* is a typical slow grower, and visible colonies are produced from dilute inoculum only after days to weeks of incubation. (The reason Koch was successful in first isolating *M. tuberculosis* was that he waited long enough after inoculating media.) When growing on solid media, mycobacteria generally form tight, compact, often wrinkled colonies, the organisms piling up in a mass rather than spreading out over the surface of the agar (Figure 19.99*a*). This formation is probably due to the high lipid content and hydrophobic nature of the cell surface. The characteristic slow growth of most mycobacteria is probably also due, at least in part, to the *hydrophobic* character of the cell surface, which renders the cells strongly impermeable to nutrients; species having less lipid grow considerably more rapidly.

For the most part, mycobacteria have relatively simple nutritional requirements. Growth often occurs in simple mineral salts medium with ammonium as nitrogen source and glycerol or acetate as sole carbon source and electron donor. Growth of *M. tuberculosis* is stimulated by lipids and fatty acids, and egg yolk (a good source of lipids) is often added to culture media to achieve more luxuriant growth. A glycerol-whole egg medium (Lowenstein-Jensen medium) is often used in primary isolation of *M. tuberculosis* from pathological materials. Perhaps because of the high lipid content of its cell walls, *M. tuberculosis* is able to resist such chemical agents as alkali or phenol for considerable periods of time, and this property is used in the selective isolation of the organism from sputum and other materials that are grossly contaminated. The sputum is first treated with 1 *N* NaOH for 30 minutes, then neutralized and streaked onto isolation medium.

A characteristic of many mycobacteria is their ability to form yellow carotenoid pigments (Figure 19.99*c*). Based on pigmentation, the mycobacteria can be divided into three groups: nonpigmented (including *M. tuberculosis, M. bovis*); forming pigment only when cultured in the light, a property called **photochromogenesis** (including *M. kansasii, M. marinum*); and forming pigment even when cultured in the dark, a property called **scotochromogenesis** (including *M. gordonae, M. paraffinicum*). The property of photochromogenesis is of some interest and has been extensively studied. This property is not unique to mycobacteria, as it also occurs in a number of fungi. Photoinduction of carotenoid formation involves short-wavelength (blue) light, and only occurs in the presence of O_2. The evidence indicates that the critical event in photoinduction is a light-catalyzed oxidation event, and it appears that one of the early enzymes in carotenoid biosynthesis is photoinduced. As with other carotenoid-containing bacteria, it has been suggested that carotenoids protect mycobacteria against oxidative damage involving singlet oxygen (see Section 9.14).

The cell walls of mycobacteria contain a peptidoglycan

Table 19.43 Some characteristics of representative mycobacteria

Species	Growth in 5% NaCl	Nitrate reduction	Growth at 45°	Human pathogen	Pigmentation
Slow growing species					
M. tuberculosis	−	+	−	+	None
M. avium	−	−	−	+	Old colonies pigmented
M. bovis	−	−	+	+	None
M. kansasii	−	+	−	+	Photochromogenic
Fast growing species					
M. smegmatis	+	+	+	−	None
M. phlei	+	+	+	−	Pigmented
M. chelonae	+	−	−	+	None
M. parafortuitum	+	+	−	−	Photochromogenic

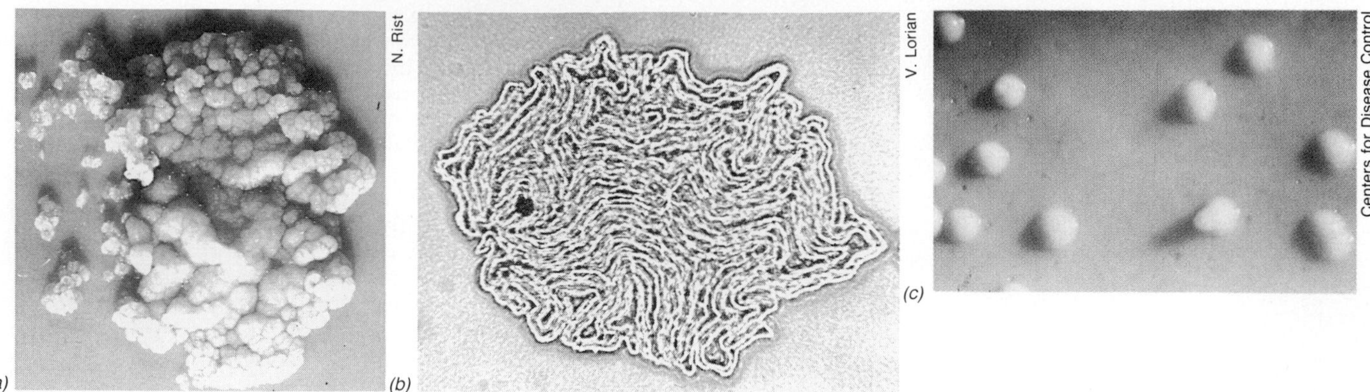

Figure 19.99 Characteristic colony morphology of mycobacteria. (a) *M. tuberculosis*, showing the compact, wrinkled appearance of the colony. The colony is about 7 mm in diameter. (b) A colony of virulent *M. tuberculosis* at an early stage, showing the characteristic cordlike growth. (c) Colonies of *M. avium* from a strain of this organism isolated as an opportunistic pathogen of an AIDS patient.

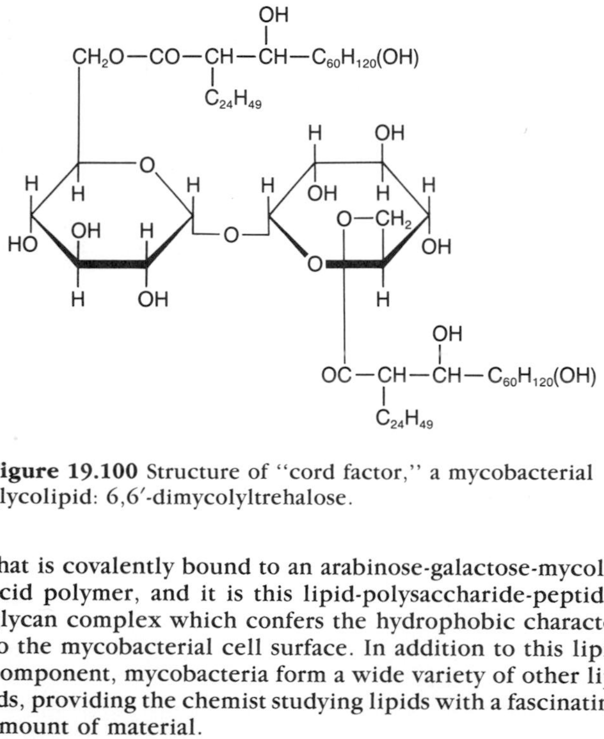

$CH_2O-CO-CH(C_{24}H_{49})-CH(OH)-C_{60}H_{120}(OH)$

$OC-CH(C_{24}H_{49})-CH(OH)-C_{60}H_{120}(OH)$

Figure 19.100 Structure of "cord factor," a mycobacterial glycolipid: 6,6′-dimycolyltrehalose.

that is covalently bound to an arabinose-galactose-mycolic acid polymer, and it is this lipid-polysaccharide-peptidoglycan complex which confers the hydrophobic character to the mycobacterial cell surface. In addition to this lipid component, mycobacteria form a wide variety of other lipids, providing the chemist studying lipids with a fascinating amount of material.

Virulence of *M. tuberculosis* cultures has been correlated with the formation of long cordlike structures (Figure 19.99) on agar or in liquid medium, due to side-to-side aggregation and intertwining of long chains of bacteria. Growth in cords reflects the presence on the cell surface of a characteristic lipid, the **cord factor**, which is a glycolipid (Figure 19.100). The pathogenesis of the disease tuberculosis is discussed in detail in Section 15.3.

19.32 Filamentous Actinomycetes

The actinomycetes are a large group of filamentous bacteria, usually Gram-positive, which form branching filaments. As a result of successful growth and branching, a ramifying network of filaments is formed, called a *mycelium* (Figure 19.101). Although it is of bacterial dimensions, the mycelium is in some ways analogous to the mycelium formed by the filamentous fungi. Most actinomycetes form spores; the manner of spore formation varies and is used in separating subgroups, as outlined in Table 19.42. The genus *Mycobacterium*, members of which often show a tendency to form branches, is also placed in the actinomycetes by some taxonomists. The DNA base compositions of all members of the actinomycetes fall within a relatively narrow range of 63 to 78 percent GC. Organisms at the upper end of this range have the highest GC percentage of any bacteria known. Phylogenetically, the filamentous actinomycetes form a coherent group; thus the mycelial spore-forming habit is of both phylogenetic as well as taxonomic importance. In the present discussion we concentrate on the genus *Streptomyces*.

Streptomyces *Streptomyces* is a genus represented by a large number of species and varieties. Over 500 species of *Streptomyces* are recognized by *Bergey's Manual*, although GC base ratios cluster tightly between 69–73 mol percent. *Streptomyces* filaments are usually 0.5 to 1.0 μm in diameter and of indefinite length, and often lack cross walls in the vegetative phase. Growth occurs at the tips of

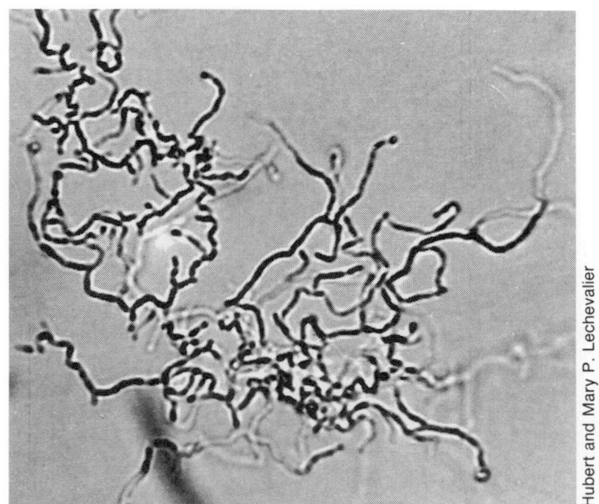

Figure 19.101 A young colony of an actinomycete, *Nocardia corallina*, showing typical filamentous cellular structure (mycelium).

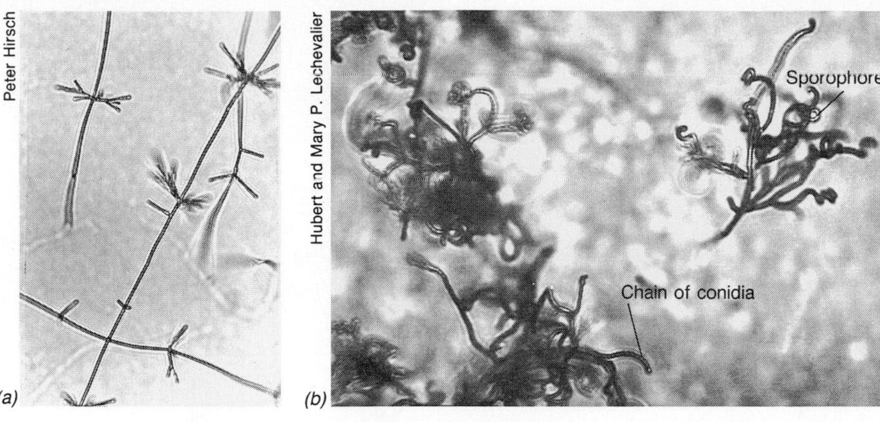

Figure 19.102 Photomicrographs of several spore-bearing structures of actinomycetes. (a) *Streptomyces*, a monoverticillate type. (b) *Streptomyces*, a spiral type.

the filaments and is often accompanied by branching so that the vegetative phase consists of a complex, tightly woven matrix, resulting in a compact convoluted colony. As the colony ages, characteristic aerial filaments called *sporophores* are formed, which project above the surface of the colony and give rise to spores (Figure 19.102). *Streptomyces* spores, usually called **conidia**, are not related in any way to the endospores of *Bacillus* and *Clostridium* since the streptomycete spores are produced simply by the formation of cross walls in the multinucleate sporophores followed by separation of the individual cells directly into spores (Figure 19.103). The surface of the conidial wall often has convoluted projections, the nature of which is characteristic of each species. Differences in shape and arrangement of aerial filaments and spore-bearing structures of various species are among the fundamental features used in separating the *Streptomyces* groups (Figure 19.104). The conidia and sporophores are often pigmented and contribute a characteristic color to the mature colony; in addition, pigments sometimes are produced by the substrate mycelium and contribute to the final color of the colony (Figure 19.105). The dusty appearance of the mature colony, its compact nature, and its color make detection of *Streptomyces* colonies on agar plates relatively easy.

Ecology and isolation of Streptomyces Although a few streptomycetes can be found in aquatic habitats, they are primarily *soil* organisms. In fact, the characteristic earthy odor of soil is caused by the production of a series of streptomycete metabolites called **geosmins**. These substances are sesquiterpenoid compounds, unsaturated ring compounds of carbon, oxygen, and hydrogen. The geosmin first discovered has the chemical name *trans*-1,10-dimethyl-*trans*-9-decalol. Geosmins are also produced by some cyanobacteria (see Section 19.2).

Alkaline and neutral soils are more favorable for the development of *Streptomyces* than are acid soils. Higher numbers of *Streptomyces* are usually found in well-drained soils (such as sandy loams, or soils covering limestone), and there is some evidence to suggest that *Streptomyces* require a lower water potential for growth than many other soil bacteria. Isolation of large numbers of *Streptomyces* from soil is relatively easy: a suspension of soil in sterile water is diluted and spread on selective agar medium, and the plates are incubated at 25°C. Media often selective for *Streptomyces* contain the usual assortment of inorganic salts to which starch, asparagine, or calcium malate is added as a carbon source and undigested casein or potassium nitrate as nitrogen source. After incubation for 5 to 7 days the

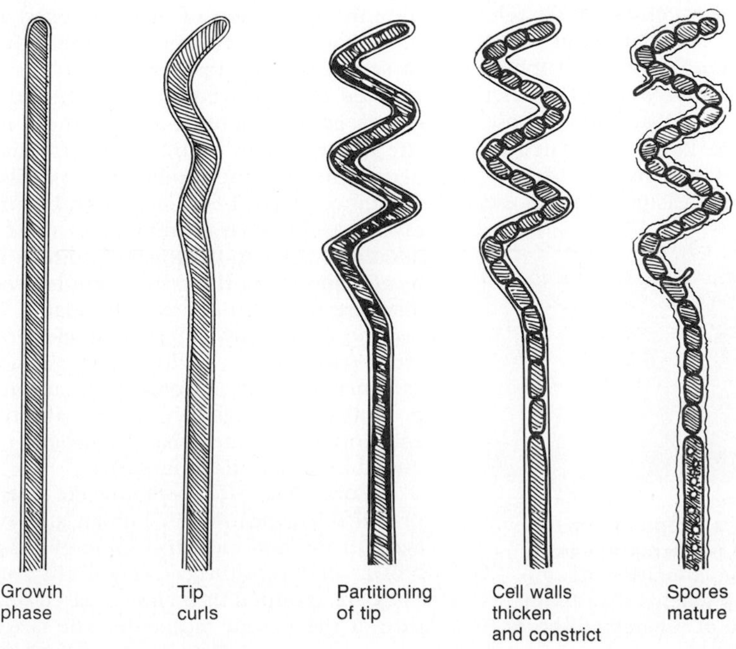

Figure 19.103 Diagram of stages in the conversion of a streptomycetes aerial hypha into spores (conidia).

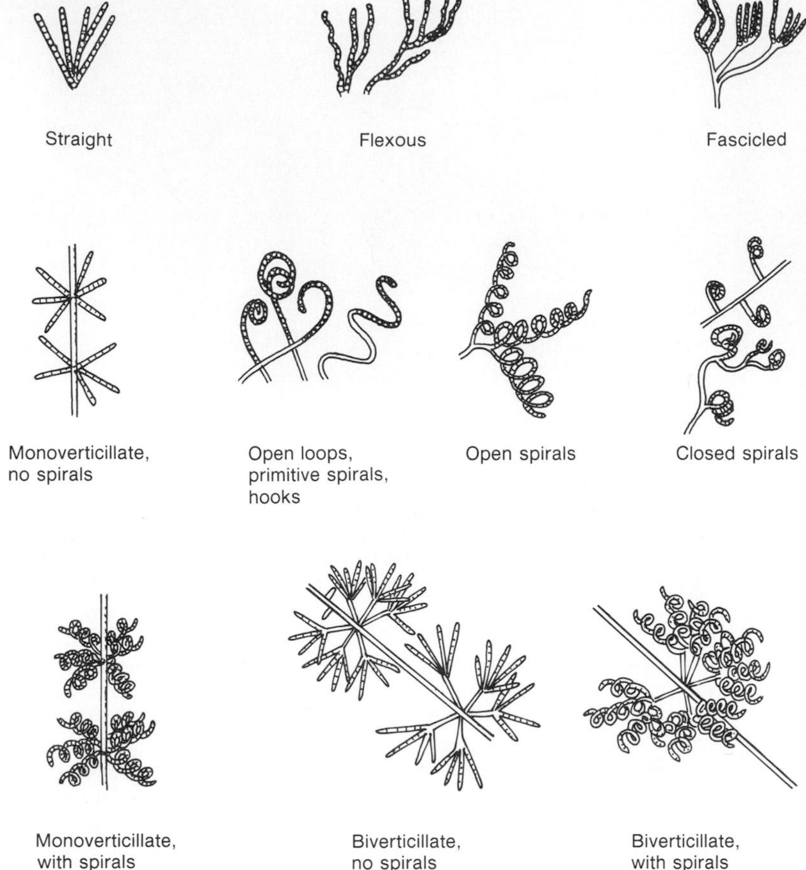

Figure 19.104 Various types of spore-bearing structures in the streptomycetes.

plates are examined for the presence of the characteristic *Streptomyces* colonies and spores of interesting colonies can be streaked and pure cultures isolated.

Nutritionally, the streptomycetes are quite versatile. Growth-factor requirements are rare, and a wide variety of carbon sources, such as sugars, alcohols, organic acids, amino acids, and some aromatic compounds, can be utilized. Most isolates produce extracellular hydrolytic enzymes that permit utilization of polysaccharides (starch, cellulose, hemicellulose), proteins, and fats, and some strains can use hydrocarbons, lignin, tannin, or even rubber. A single isolate may be able to break down over 50 distinct carbon sources. Streptomycetes are strict aerobes, whose growth in liquid culture is usually markedly stimulated by forced aeration. Sporulation usually takes place not in liquid culture but only when the organism is growing on the surface of agar or another solid substrate; it can occur, however, when organisms form a pellicle on the surface of an unshaken liquid culture.

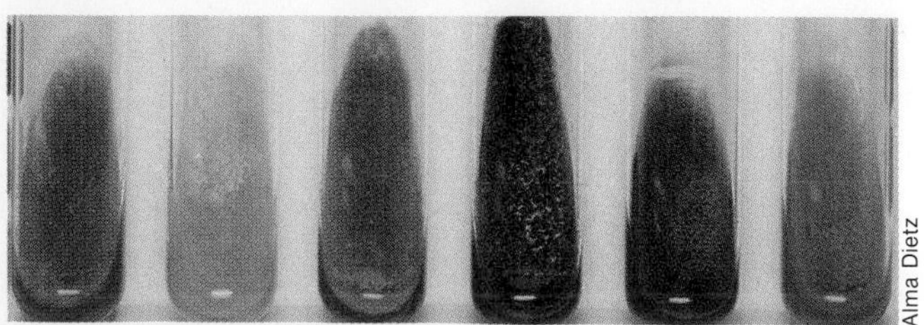

Figure 19.105 Typical appearance of a streptomycete growing on agar slants. Varying degrees of pigmentation are shown on different culture media. The coloration results from both the production of soluble pigments that diffuse into the agar and from the production of pigmented spores.

Antibiotics of Streptomyces Perhaps the most striking property of the streptomycetes is the extent to which they produce **antibiotics** (Table 19.44). Evidence for antibiotic production is often seen on the agar plates used in the initial isolation of *Streptomyces*: adjacent colonies of other bacteria show zones of inhibition (Figure 19.106). In some studies close to 50 percent of all *Streptomyces* isolated have proved to be antibiotic producers. Because of the great economic and medical importance of many streptomycete antibiotics, an enormous amount of work has been done on these producers. Over 500 distinct antibiotic substances have been shown to be produced by streptomycetes, and a large number of these have been studied chemically. Some organisms produce more than one antibiotic, and often the several kinds produced by one organism are not even chemically related. The same antibiotic may be formed by different species found in widely scattered parts of the world. A change in nutrition of the organism may result in a change in the nature of the antibiotic produced. The organisms are usually resistant to their own antibiotics, but they may be sensitive to antibiotics produced by other streptomycetes.

More than 50 streptomycete antibiotics have found practical application in human and veterinary medicine, agriculture, and industry. Some of the more common antibiotics of *Streptomyces* origin are listed in Table 19.44. They are grouped into classes based on the chemical structure of the parent molecule. The search for new strepto-

Table 19.44 Some common antibiotics synthesized by species of *Streptomyces*

Chemical class	Common name	Produced by	Active against*
Aminoglycosides	Streptomycin	*S. griseus*	Most Gram negatives
	Spectinomycin	*Streptomyces spp.*	*M. tuberculosis*, penicillinase-producing *N. gonorrhoeae*
	Neomycin	*S. fradiae*	Broad spectrum, usually used in topical applications due to toxicity
Tetracyclines	Tetracycline	*S. aureofaciens*	Broad spectrum, Gram positives, Gram negatives, rickettsias and chlamydias, *Mycoplasma*
	Chlortetracycline	*S. aureofaciens*	As for tetracycline
Macrolides	Erythromycin	*S. erythreus*	Most Gram positives, frequently used in place of penicillin, *Legionella*
	Clindamycin	*S. lincolnensis*	Effective against obligate anaerobes, especially *Bacteroides fragilis*
Polyenes	Nystatin	*S. noursei*	Fungi, especially *Candida* infections
	Amphocetin B	*S. nodosus*	Fungi
None	Chloramphenicol	*S. venezuelae*	Broad spectrum; drug of choice for typhoid fever

**Most antibiotics are effective against several different bacteria. The entries in this column refer to the most frequent clinical application of a given antibiotic.*

mycete antibiotics continues, since many infectious diseases are still not adequately controlled by existing antibiotics. Also, the development of antibiotic-resistant strains requires the continual discovery of new agents. We discussed the antibiotic industry in general and the role of *Streptomyces* in the commercial production of antibiotics in Section 10.6. Ironically, despite the extensive work on antibiotic-producing streptomycetes, and the fact that the antibiotic industry is a multi-billion dollar enterprise, the ecology of *Streptomyces* remains poorly understood.

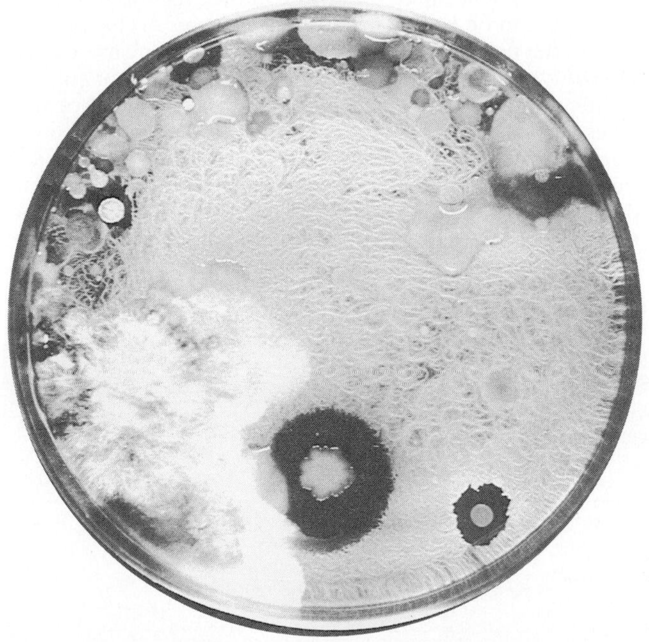

Eli Lilly & Co.

Figure 19.106 Antibiotic action of soil microorganisms on a crowded plate. The smaller colonies surrounded by inhibition zones are streptomycetes; the larger, spreading colonies are *Bacillus* sp.

19.33 Halophilic Archaebacteria

The rest of the organisms to be discussed in this chapter are members of the archaebacteria (see Section 18.6). We begin our discussion of the individual groups of archaebacteria with a consideration of the extreme halophiles. **Extremely halophilic archaebacteria** are a diverse group of procaryotes which inhabit highly saline environments such as solar salt evaporation ponds and natural lakes, or artificial saline habitats such as the surfaces of heavily salted foods like certain fish or meats. Such habitats are often called *hypersaline*. The term *extreme halophile* is used to indicate that these organisms are not only halophilic, but that their requirement for salt is *very high*, in some cases near that of saturation. A general definition of an extreme halophile is that the organism requires at least 1.5 M (8.8 percent) NaCl for growth and most require 3–4 M NaCl (17–23 percent) for optimal growth. Virtually all extreme halophiles can grow at 5.5 M NaCl (32 percent, the limits of saturation for this salt), although some species grow only very slowly at this salinity. Until recently only two genera of halophilic archaebacteria were formerly recognized, the rod-shaped *Halobacterium* and the coccus *Halococcus*. Further microbiological examination of highly saline habitats have yielded several new extreme halophiles; all of these organisms group phylogenetically within the archaebacteria (see Section 18.6). We begin our discussion of extremely halophilic archaebacteria by a consideration of the habitats of these unusual procaryotes, and proceed to discuss the taxonomy and physiology of the group. Finally, we consider a unique mechanism of light-induced ATP synthesis which does not involve chlorophyll pigments.

Saline environments Salty habitats are common throughout the world, but *extremely* saline habitats are rather rare. Most extremely saline environments are in hot, dry areas of the world, and such climatic conditions encourage evaporation and further concentration of the salts. Salt lakes can vary considerably in ionic composition. The

Figure 19.107 Hypersaline habitats. (a) Great Salt Lake, a lake in which the proportions of ions are similar to seawater. (b) General view near San Francisco Bay of a series of seawater evaporating ponds where solar salt is prepared. The red color is predominantly halobacteria.

predominant ions in a saline lake depend to a major extent on the surrounding topography, geology, and general climatic conditions. The Great Salt Lake in Utah (USA) (Figure 19.107*a*), for example, is essentially concentrated seawater because the relative proportions of the various ions resemble those of seawater, although the overall concentration of ions is much higher. Sodium is the predominant cation in Great Salt Lake while chloride is the predominant anion; significant levels of sulfate are also present at a slightly alkaline pH (Table 19.45). By contrast, another very saline basin, the Dead Sea, is relatively low in sodium but contains high levels of magnesium because of the abundance of magnesium minerals in the surrounding rocks (Table 19.45). The water chemistry of soda lakes resembles that of saline lakes such as Great Salt Lake, but because high levels of *carbonate* minerals are present in the surrounding rocks, the pH of soda lakes is quite high; pH values of 10–12 are not uncommon in these environments (Table 19.45).

Despite what may seem like rather harsh conditions, salt lakes can be highly productive systems. Archaebacteria are not the only microorganisms found. The eucaryotic alga *Dunaliella* (see Figure 1.27*c*) is the major, if not sole oxygenic phototroph in most salt lakes. In highly alkaline soda lakes where *Dunaliella* is absent, anoxygenic phototrophic purple bacteria of the genus *Ectothiorhodospira* (see Section 19.1) predominate. Because sulfate is generally abundant in salt lakes (Table 19.45), significant sulfate reduction can occur, leading to the formation of the sulfide needed for development of purple bacteria. At the high salt concentrations found in saline lakes, no higher plants can de-

Table 19.45 Ionic composition of some highly saline environments

Ion	Great Salt Lake	Concentration g/l Dead Sea	Typical soda lake	Seawater (for comparison)
Na^+	105	39	142	10.6
K^+	6.7	7.3	2.3	0.38
Mg^{2+}	11	41	<0.1	1.27
Ca^{2+}	0.3	17	<0.1	0.40
Cl^-	181	212	155	18.9
Br^-	0.2	5	—	0.065
SO_4^{2-}	27	0.5	23	2.65
HCO_3^-/CO_3^{2-}	0.7	0.2	67	0.14
pH	7.7	6.1	11	8.1

Table 19.46 Taxonomy of halophilic archaebacteria

Genus	DNA (mole % GC)	Habitat/Comments
Halobacterium		
H. halobium	66–68	Isolated from salted fish, hides, hypersaline lakes; related organisms, probably all the same species are *H. salinarium* and *H. cutirubrum*
H. volcanii	63	Dead Sea; requires high Mg^{2+}
H. mediterranii	60	Salterns; uses starch
H. saccharovorum	—	Salterns; uses sugars
H. vallismortis	—	Salt ponds, Death Valley, CA
Halococcus		
H. morrhuae	61–66	Salted fish
Natronobacterium		
N. gregoryi	65	All from highly saline soda lakes;
N. magadii	63	optimum pH for growth, 9.5
N. pharaonis	64	
Natronococcus		
N. occultus	64	Highly saline soda lake

velop, although in addition to microorganisms a few types of animals are capable of living (brine shrimp, brine flies). Organic matter originating from primary production by oxygenic or anoxygenic phototrophs then sets the stage for development of the halophilic archaebacteria, all of which are heterotrophic.

Marine salterns are also habitats for extremely halophilic bacteria. Marine salterns are small basins filled with seawater which are left to evaporate, yielding NaCl and other salts of commercial value (Figure 19.107*b*). As salterns approach the minimum salinity limits for extreme halophiles the waters turn a reddish purple in color indicative of the massive growth (called a *bloom*) of halophilic archaebacteria (the red coloration apparent in Figure 19.107*b* comes from carotenoids and other pigments of the halobacteria discussed below). Extreme halophiles have also been found in high salt foods such as certain sausages, marine fish, and salt pork. Other than causing some esthetic problems, growth of extreme halophiles in foods is of little consequence, since no extreme halophile has been shown to cause foodborne illness.

Taxonomy of halophilic archaebacteria Although many bacteria are known to tolerate fairly high salt concentrations and several organisms are mildly halophilic, all the heterotrophic bacteria of hypersaline habitats are archaebacteria. Table 19.46 lists the currently recognized species. Ribosomal RNA:DNA hybridization studies have defined four genera of extreme halophiles: *Halobacterium, Halococcus, Natronobacterium*, and *Natronococcus*. The halophilic archaebacteria are frequently referred to collectively as "halobacteria," probably because the genus *Halobacterium* was the first in this group to be described and is still the best studied representative of the group. As archaebacteria, halobacteria lack peptidoglycan in their cell walls, contain ether-linked lipids, and archaebacterial RNA polymerases. They also are insensitive to most antibiotics, and possess the other general attributes of representatives of this kingdom (see Section 18.8). *Natronobacterium* and *Natronococcus* differ from *Halobacterium* and *Halococcus* in being extremely *alkalinophilic* as well as halophilic. As befits their soda lake habitat (see Table 19.45), growth of natronobacteria is optimal at very low Mg^{2+} concentrations and high pH (9–11). Natronobacteria also contain unusual diether lipids not found in other extreme halophiles and cluster nicely as a phylogenetic group by RNA:DNA hybridization.

All halophilic archaebacteria are Gram *negative*, reproduce by binary fission, and do not form resting stages or spores. Most halobacteria are nonmotile, but a few strains are weakly motile by lophotrichous flagella. The genomic organization of *Halobacterium* and *Halococcus* is highly unusual in that large plasmids containing up to 25–30 percent of the total cellular DNA are present and the GC base ratio of these plasmids (57–60 percent GC) is significantly different from that of the chromosome (66–68 percent GC). Plasmids from extreme halophiles are among the largest naturally occurring plasmids known. In addition to the large amount of nonchromosomal DNA, the *Halobacterium* genome also contains considerable amounts of highly repetitive DNA, the function of which is unknown.

Physiology of halophilic archaebacteria All halophilic archaebacteria are heterotrophs and most species are obligate *aerobes*. Most halobacteria use amino acids or organic acids as energy sources and require a number of growth factors (mainly vitamins) for optimal growth. A few *Halobacterium* species will oxidize carbohydrates but this ability is relatively rare. Electron transport chains containing cytochromes *a, b*, and *c* are present in *Halobacterium* and energy is conserved during aerobic growth via a proton motive force arising from membrane-mediated chemiosmotic events. Some strains of *Halobacterium* have been shown to grow anaerobically. Anaerobic growth at the expense of sugar fermentation and by anaerobic respiration (see Section 16.13) linked to the reduction of nitrate, elemental sulfur, or thiosulfate has been demonstrated in certain strains. The finding that elemental sulfur serves as an electron acceptor for *Halobacterium* is especially interesting in light of the fact that both methanogenic bacteria (see Section 19.34) and the sulfur-dependent archaebacteria (see Section 19.35) can employ S^0 as an electron acceptor; sulfur reduction is thus a universal trait among archaebacteria and may represent one of the earliest types of anaerobic respiration to develop on Earth (see Sections 18.8 and 19.37).

All halophilic archaebacteria require large amounts of sodium for growth. In the case of *Halobacterium*, where detailed salinity studies have been performed, the requirement for Na^+ cannot be satisfied by replacement with another ion, even with the chemically related ion K^+. We learned in Section 9.13 that certain microorganisms can withstand the osmotic forces that accompany life in a high solute environment by accumulating organic compounds *intracellularly*; the latter compounds are generally referred to as **compatible solutes**. These compounds counteract the tendency of the cell to become dehydrated under conditions of high osmotic strength by placing the cell in positive water balance with its surroundings (see Section 9.13). Ironically, however, although *Halobacterium* only thrives in an osmotically stressful environment, it produces no *organic* compatible solute. Instead, cells of *Halobacterium* pump large amounts of K^+ from the environment into the cell such that the concentration of K^+ *inside* the cell is similar, if not even slighly greater than, the concentration of Na^+ *outside* the cell; thus *Halobacterium* employs an *inorganic ion* as its compatible solute (Table 19.47). No halophilic eubacteria, even those few eubacteria capable of growth at saturating salt, use K^+ as compatible solute. The use of a cation as a compatible solute is therefore a trademark of *archaebacterial* halophiles.

The cell wall of *Halobacterium* is stabilized by sodium ions; in low-sodium environments the cell wall breaks down. In electron micrographs of thin sections of *Halobacterium* (Figure 19.108) the organism appears similar to other Gram-negative bacteria in many respects, but the cell wall is quite different (compare Figure 19.108 with that of a Gram-negative eubacterium in Figure 3.22). Na^+ binds to the outer surface of the *Halobacterium* wall and is absolutely essential for maintaining cellular integrity; when insufficient Na^+ is present the cell wall breaks apart and the cell lyses. Peptidoglycan is absent in the cell wall of *Halobacterium* (as it is in the walls of all archaebacteria, see Section 3.6), and instead the cell wall is composed of *glycoprotein*. This protein has an exceptionally high content of the *acidic* (negatively charged) amino acids aspartate and glutamate. The negative charges contributed by the carboxyl groups of these amino acids in the cell wall glycoprotein are shielded by Na^+; when Na^+ is diluted away, the negatively charged parts of the proteins actively repel each other, leading to cell lysis.

Cytoplasmic proteins of *Halobacterium* are also highly acidic, but studies of several halobacterial enzymes have

Table 19.47 Levels of specific ions in the environment and in cells of *Halobacterium cutirubrum**		
Ion	**Concentration in medium (Molar)**	**Concentration in cells (Molar)**
Na^+	3.3	0.8
K^+	0.05	5.3
Mg^{2+}	0.13	0.12
Cl^-	3.3	3.3

**Data from Matheson, A. T., G. D. Sprott, I. J. McDonald, and H. Tessie. 1976. Can. J. Microbiol. 22:780–786.*

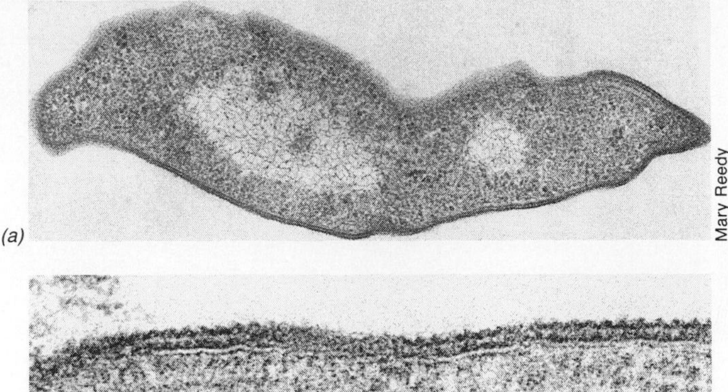

Figure 19.108 Electron micrograph of thin sections of the extreme halophile, *Halobacterium halobium*. (a) Longitudinal section. (b) High-magnification electron micrograph showing the regular structure of the cell wall.

shown that K^+, not Na^+, is required for activity. This, of course, is not surprising when it is recalled that K^+ is the predominate internal cation in cells of *Halobacterium* (Table 19.47). Besides a high acidic amino acid composition, halobacterial cytoplasmic proteins typically contain very low levels of *hydrophobic* amino acids. This phenomenon probably represents an evolutionary adaptation to the highly ionic cytoplasm of *Halobacterium*; in such an environment highly polar proteins would tend to remain in solution, whereas nonpolar molecules would tend to cluster and perhaps lose activity. The ribosomes of *Halobacterium* also require high K^+ levels for stability (ribosomes of nonhalophiles have no K^+ requirement). In summary, it appears that the halophilic archaebacteria are highly adapted, both internally and externally, to life in a highly ionic environment. Cellular components exposed to the external environment require high Na^+ for stability while internal components require high K^+. In no other group of bacteria do we find this unique requirement for specific cations in such high amounts.

Bacteriorhopsin and light-mediated ATP synthesis Certain species of extreme halophiles have the additional interesting property of showing a *light-mediated* synthesis of ATP that does not involve chlorophyll pigments. As discussed in Section 16.3, one of the key aspects of photosynthesis is *photophosphorylation*, the synthesis of ATP as a consequence of the light-dependent generation of a proton gradient across the photosynthetic membrane. We have seen the highly pigmented nature of the halophilic archaebacteria in Figure 19.107. Although lacking chlorophylls, under conditions of low aeration *Halobacterium halobium* and certain other extreme halophiles synthesize and insert into their membranes a protein called **bacteriorhodopsin**. Bacteriorhodopsin was named because of its functional similarity to the visual pigment of the eye, *rhodopsin*. Conjugated to bacteriorhodopsin is a molecule of *retinal*, a carotenoid-like molecule that can absorb light and catalyze the transfer of protons across the cytoplasmic membrane (Figure 19.109). Due to its retinal content, bacteriorhodopsin is purple in color, and cells of *Halobacterium* switched from growth under conditions of high aeration to oxygen-limiting conditions gradually change from

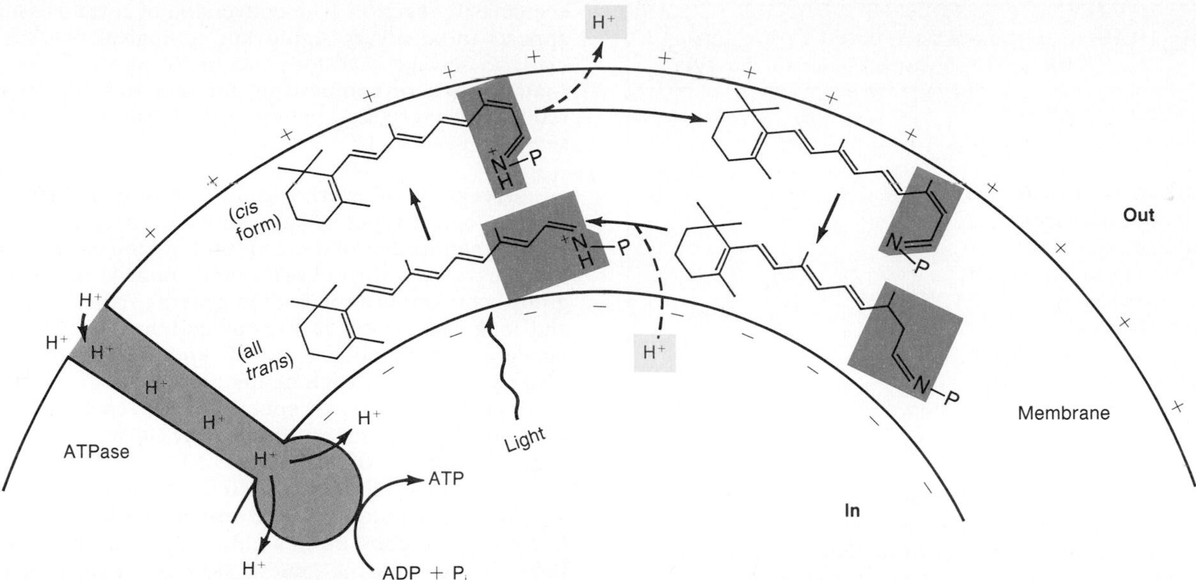

Figure 19.109 Model of the light-mediated bacteriorhodopsin proton pump of *Halobacterium*. The P stands for the protein to which the chromophore retinal is attached. *OUT* and *IN* designate opposite sides of the membrane.

an orange or red color (due to carotenoids which serve a photoprotective role) to more of a reddish purple in color because of the insertion into the cytoplasmic membrane of bacteriorhodopsin. Isolated purple membranes of *Halobacterium halobium* contain about 25 percent lipid and 75 percent protein, and the purple membrane appears at random in patches on the surface of the cytoplasmic membrane; this suggests that bacteriorhodopsin is wedged into preexisting membranes during the switch to low O_2.

Bacteriorhodopsin absorbs light strongly in the visible region of the spectrum at about 565 nm. The retinal chromophore of bacteriorhodopsin, which normally exists in an all ***trans*** configuration, is temporarily converted to the ***cis*** form following the absorption of light (Figure 19.109). This transformation results in the transfer of protons to the ***outside*** surface of the membrane (see Figure 19.109). The retinal molecule then returns to its more stable all ***trans*** isomer in the dark following the uptake of a proton from the cytoplasm, thus completing the cycle (Figure 19.109). As protons accumulate on the outer surface of the membrane, the proton electrochemical gradient (see Section 4.12) increases until the membrane is sufficiently "charged" to drive ATP synthesis through action of the membrane-bound ATPase (Figure 19.109).

Light-mediated ATP production in *H. halobium* has been shown to support slow growth of this organism anaerobically under nutritional conditions in which other energy-generating reactions do not occur, and light has been shown to maintain the viability of cultures of *Halobacterium* incubated anaerobically in the absence of organic energy sources. The light-stimulated proton pump of *H. halobium* also functions to pump Na^+ out of the cell by action of a Na^+/H^+ antiport system (see Section 3.4), and to drive the uptake of a variety of nutrients. In the latter connection the uptake of amino acids by *H. halobium* has been shown to be indirectly driven by light, because the transport of amino acids occurs with Na^+ uptake by an amino acid/Na^+ symporter (see Section 3.4). Continued uptake depends on the removal of Na^+ via the (light-driven) Na^+/H^+ antiporter.

Evolution of extreme halophiles Although they share virtually nothing with methanogenic bacteria from a *phenotypic* standpoint, the extreme halophiles are phylogenetically quite closely related to the methanogens (see Figure 18.12). This at first seems odd because most halophilic archaebacteria are obligate aerobes whereas methanogenic bacteria are all obligate anaerobes (see below). However, more detailed studies of halophilic archaebacteria have shown them to be a remarkably versatile lot, including many species capable of anaerobic growth by fermentation or anaerobic respiration. Indeed, even extremely halophilic and alkalinophilic methanogens have now been isolated, so that the connection between the two groups may not be as unusual as first thought. We will return to the theme of the evolution of archaebacteria in Section 19.37, after we have considered the remaining groups of archaebacteria.

19.34 Methane-Producing Bacteria: Methanogens

We described the overall process of methanogenesis in Section 16.16 and the ecology of methanogenesis in Section 17.12, and have noted that the biological production of methane is carried out by a unique group of procaryotes, the **methanogens**. Methane formation occurs only under strictly *anaerobic* conditions and thus methanogenesis is restricted to habitats which are highly anoxic (see Figure 17.23).

Nine substrates have been shown to be converted to methane by one or another methanogenic bacterium. Carbon dioxide, CO_2, is a nearly universal substrate for methanogens, the needed electrons usually being derived from H_2. When growing on $H_2 + CO_2$, the methanogens are *autotrophic*, with CO_2 serving as both carbon source and electron acceptor. In addition to CO_2, however, a variety of other compounds can be converted to methane by certain methanogenic species; these substrates are listed in Table 19.48. As described in Section 16.16, methane formation

Table 19.48 Substrates converted to methane by various methanogenic bacteria

CO_2-type substrates	
Carbon dioxide	CO_2
Formate	$HCOOH$
Carbon monoxide	CO
Methyl substrates	
Methanol	CH_3OH
Methylamine	$CH_3NH_3^+$
Dimethylamine	$(CH_3)_2NH_2^+$
Trimethylamine	$(CH_3)_3NH^+$
Acetoclastic substrate	
Acetate	CH_3COOH

from $H_2 + CO_2$ can be viewed as a type of *anaerobic respiration* in which CO_2 serves as the electron acceptor. However, biochemical studies of methanogenesis indicate that a conventional electron-transport system involving cytochromes and quinones is absent from methanogens grown on $H_2 + CO_2$. Thus, electron transport to CO_2 involves different electron carriers than that of other anaerobic respiratory processes such as nitrate and sulfate reduction. Although our understanding of methanogenesis is far from complete, the electron carriers involved in the reduction of CO_2 to methane are now fairly well understood. This process requires a host of specific coenzymes unique to methanogenic bacteria which serve as cofactors for enzymes which sequentially reduce various C-1 intermediates starting with CO_2 and eventually yielding CH_4.

Three *classes* of methanogenic substrates are known (see Table 19.48) and as expected, all are used with the release of free energy suitable for ATP synthesis. The first involves the use of *CO_2-type substrates:*

$$CO_2 + 4H_2 \rightarrow CH_4 + 2H_2O \qquad \Delta G^{0\prime}\ -32\ \text{kcal/reaction}$$

$$4HCOOH \rightarrow CH_4 + 3CO_2 + 2H_2O \qquad \Delta G^{0\prime}\ -67\ \text{kcal/reaction}$$

$$4CO + 2H_2O \rightarrow CH_4 + 3CO_2 \qquad \Delta G^{0\prime}\ -50\ \text{kcal/reaction}$$

The second class of reaction involves reduction of the *methyl group* of methyl-containing compounds to methane. In the case of methanol or methylamine the overall reaction to methane has the following stoichiometry:

$$4CH_3OH \rightarrow 3CH_4 + CO_2 + 2H_2O \qquad \Delta G^{0\prime}\ -77\ \text{kcal/reaction}$$

$$4CH_3NH_3Cl + 2H_2O \rightarrow 3CH_4 + CO_2 + 4NH_4Cl \qquad \Delta G^{0\prime}\ -55\ \text{kcal/reaction}$$

In these reactions some molecules of the substrate serve as an electron *donor* and are oxidized to CO_2, whereas other molecules are reduced and serve as electron *acceptor*. In growth on methyl compounds the reducing power for methanogenesis can also come from H_2. In the case of methanol the stoichiometry would be:

$$CH_3OH + H_2 \rightarrow CH_4 + H_2O \qquad \Delta G^{0\prime}\ -27\ \text{kcal/reaction}$$

The final methanogenic reaction is the **acetoclastic reaction**, the cleavage of acetate to CH_4 plus CO_2:

$$CH_3COOH \rightarrow CH_4 + CO_2 \qquad \Delta G^{0\prime}\ -31\ \text{kcal/mole}$$

Only two genera of methanogens, *Methanosarcina* and *Methanothrix*, have representatives able to carry out the acetoclastic reaction. The conversion of acetate to methane appears to be a very significant ecological process, especially in sewage digestors and in freshwater anoxic environments where competition for acetate between sulfate-reducing bacteria and methanogenic bacteria is not extensive (see Section 17.12).

Diversity of methanogenic bacteria A variety of morphological types of methanogenic bacteria have been isolated and studies of their physiology and molecular properties have served to classify methanogens into five major groups containing a total of ten genera (Table 19.49). Short and long rods, cocci in various cell arrangements, and filamentous methanogens are all known (Figures 19.110–19.112). Although each genus of methanogen is not morphologically unique, all species of a given genus are morphologically uniform and have DNA GC base ratios within about 10 percent of each other.

As discussed in Section 3.6, archaebacterial cell walls are chemically unique. **Pseudomurein**, the peptidoglycan-like molecule containing amino sugars and peptide cross-links, is found in the genera *Methanobacterium* (Figure 19.110), *Methanobrevibacter* (Figure 19.110), and *Methanothermus* (Figure 19.112). Most other methanogens have proteinaceous cell walls. Both Gram-positive and Gram-negative methanogens are known (see Table 19.49), thus the Gram stain is of little use in classifying these organisms. The current taxonomy of methanogenic bacteria leans heavily on molecular methods, in particular on 16S rRNA sequence comparisons (see Chapter 18 for a discussion of molecular phylogeny), and, to a lesser degree, on immunological methods. The creation of five major groups (taxonomically considered "families") has mainly been supported by ribosomal RNA analyses.

The *physiological* diversity of methanogens is rather limited. Unlike many bacteria, methanogens are restricted to a relatively limited diet of substrates (see Tables 19.48 and 19.49). When growing autotrophically (see below), CO_2 serves as carbon source, and those methanogens capable of metabolizing organic compounds to methane use intermediates generated during the catabolic process as biosynthetic precursors (see "Autotrophy in methanogens," below). Growth of virtually all methanogens is stimulated by acetate and the growth of some species is also stimulated by certain amino acids. For culture, many methanogens require complex additions such as yeast extract or casein digests, and some rumen methanogens require a mixture of branched chain fatty acids (see Section 17.13). Vitamins are required by some methanogens and the most commonly required vitamins are riboflavin, pantothenic acid, thiamin, biotin and *p*-aminobenzoate. All methanogens use NH_4^+ as a nitrogen source and a few species are known to fix molecular nitrogen (N_2 fixation, see Section 16.24). The trace metal *nickel* is required by all methanogens; it is a component of an important methanogenic coenzyme, *Factor F_{430}* (see below), and is also present in the enzymes *hydrogenase* and *carbon monoxide dehydrogenase* (see below). Iron and cobalt are also important trace metals for methanogens, and along with nickel, have been the only trace metals shown to be absolutely required for growth of these organisms.

Unique methanogenic coenzymes A number of coenzymes have been found in methanogens that are unique to this group of bacteria. Most of these coenzymes play important roles in the biochemistry of methanogenesis, and the structure and function(s) of most have now been elu-

Table 19.49 Characteristics of methanogenic bacteria

Genus	Morphology	Gram reaction	Number of species	Substrates for methanogenesis	DNA (mole % GC)
GROUP I					
Methanobacterium	Long rods	+ or −	3	$H_2 + CO_2$, formate	33–50
Methanobrevibacter	Short rods	+	3	$H_2 + CO_2$, formate	27–32
GROUP II					
Methanothermus	Rods	+	2	$H_2 + CO_2$	33
GROUP III					
Methanococcus	Irregular cocci	−	6	$H_2 + CO_2$, formate	31–40
GROUP IV					
Methanomicrobium	Short rods	−	2	$H_2 + CO_2$, formate	45–49
Methanogenium	Irregular cocci	−	4	$H_2 + CO_2$, formate	51–61
Methanospirillum	Spirilla	−	1	$H_2 + CO_2$, formate	46–50
GROUP V					
Methanosarcina	Large irregular cocci in packets	+	3	$H_2 + CO_2$ formate, methanol, methylamines, acetate	41–43
Methanococcoides	Irregular cocci	−	1	methanol, methylamines	42
Methanothrix	Long rods to filaments	−	2	Acetate	52

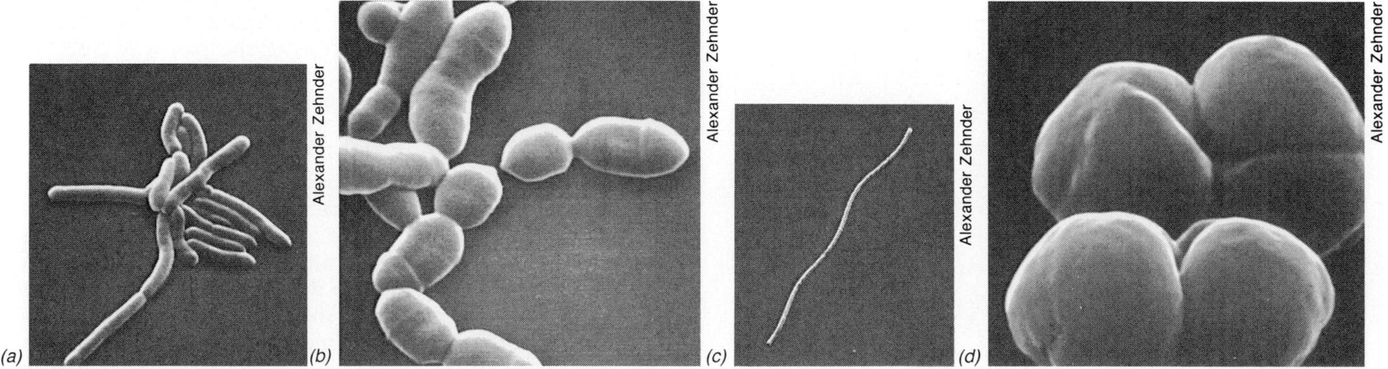

Figure 19.110 Scanning electron micrographs of whole cells of methanogenic bacteria, showing the considerable morphological diversity. (a) *Methanobrevibacter ruminantium*. (b) *Methanobacterium* strain AZ. (c) *Methanospirillum hungatii*. (d) *Methanosarcina barkeri*.

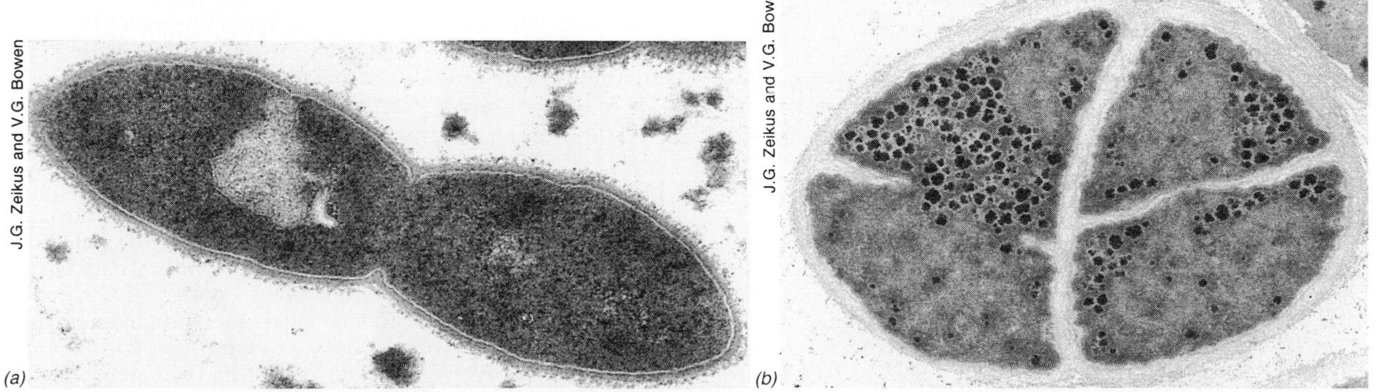

Figure 19.111 Transmission electron micrographs of methanogenic bacteria. (a) *Methanobrevibacter ruminantium*. (b) *Methanosarcina barkeri*, showing the thick cell wall and the manner of cell segmentation and cross-wall formation.

Helmut König and K.O. Stetter

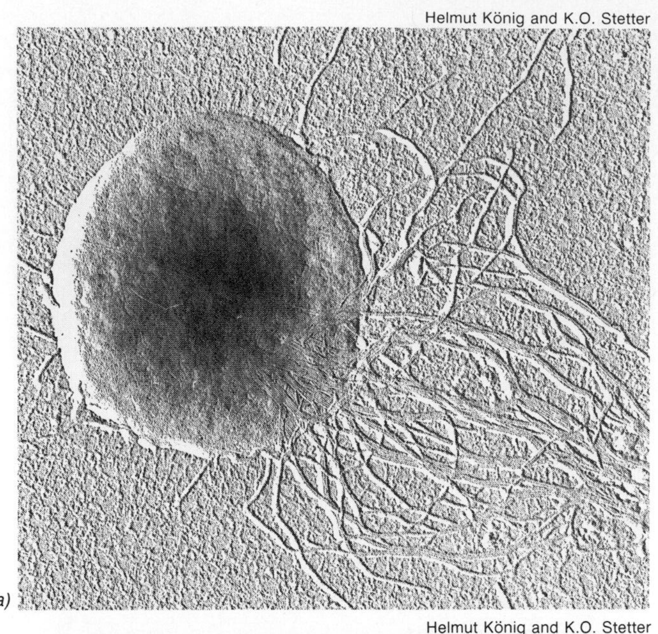

(a)

Helmut König and K.O. Stetter

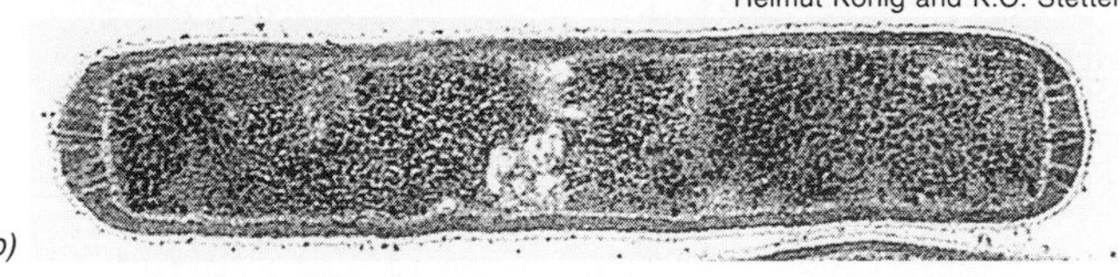

(b)

Stephen Zinder

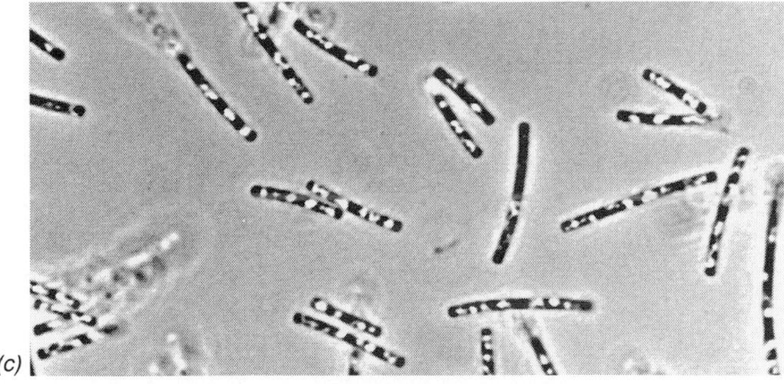

(c)

Figure 19.112 Thermophilic methanogens. (a) *Methanococcus jannaschii* (temperature optimum, 85°C), shadowed preparation electron micrograph. (b) *Methanothermus fervidus* (temperature optimum, 83°C), thin sectioned electron micrograph. (c) *Methanothrix* sp. (temperature optimum, 60°C), phase contrast micrograph.

cidated. Because of their importance in the energy yielding pathway, many of these coenzymes are present at far *higher* levels in methanogens than many of the common coenzymes such as NAD^+ or FMN are in other bacteria. In the next few sections we detail the structure and properties of each of these coenzymes as a prelude to our discussion of the biochemistry of methanogenesis.

Coenzyme F_{420} is a flavin derivative, structurally resembling the common flavin coenzyme, FMN (see Figure 4.14). The structure of F_{420} (Figure 19.113*a*) resembles that of FMN, but F_{420} lacks one of the nitrogen atoms of FMN in its middle ring and also lacks the methyl groups found on the benzene ring typical of true flavins (see Figure 4.14). F_{420} is a *two* electron carrier of rather low reduction potential ($E_0' = -0.37$ volts) and plays a physiological role in methanogens analogous to the role played by ferredoxin in other anaerobic bacteria. Coenzyme F_{420} interacts with a number of different enzymes in methanogens including hydrogenase and $NADP^+$ reductase. F_{420} also plays a major role in methanogenesis as the electron donor in two of the steps of CO_2 reduction (see below). The oxidized form of F_{420} absorbs light at 420 nm and fluoresces blue-green (Figure 19.114); upon reduction the coenzyme becomes colorless. The fluorescence of F_{420} is a useful tool for preliminary identification of an organism as a methanogen. Although important in methanogenesis, F_{420} may have other biological functions because cofactors similar to F_{420} have been detected in sulfate-reducing archaebacteria and in low levels in various eubacteria, including *Streptomyces* and some cyanobacteria.

Coenzyme F_{430} is a yellow, soluble, nickel-containing tetrapyrrole (Figure 19.113*b*) and plays an intimate role in the *terminal* step of methanogenesis as part of the methyl reductase system. Coenzyme F_{430} absorbs light strongly at 430 nm, but, unlike F_{420}, does not fluoresce. The nickel requirement for growth of methanogens reflects the abundance of F_{430} in the cells because careful measurements have shown that most of the nickel in cells of methanogens is associated with the F_{430}/methyl reductase system.

Methanofuran is a low-molecular-weight coenzyme that plays an important role in the *first* step of methanogenesis from CO_2 (see below). Methanofuran consists of a molecule of phenol, two glutamic acid molecules, an unusual long chain dicarboxyl fatty acid, and a furan ring (Figure 19.113*c*). CO_2 is reduced to the formyl level and bound by the amino side chain of the furan in the initial step of methanogenesis and is subsequently transferred to a second coenzyme in later steps of the pathway (see below).

Methanopterin is a methanogenic coenzyme containing a substituted pterin ring (Figure 19.113*d*). Methanopterin, which exhibits a bright blue fluorescence, was originally called Factor F_{342} because of its absorbance at 342 nm. Methanopterin resembles the vitamin *folic acid* (see Figure 9.39*c*) and serves as a C-1 carrier during the reduction of CO_2 to CH_4. The nitrogen atom highlighted in Figure 19.113*d* is the atom to which the C-1 intermediate binds. Methanopterin carries the C-1 unit during the majority of reductive steps in the methanogenic pathway, from the formyl (–CHO) level to the methyl (–CH_3) level (see below). *In vivo*, the reduced form of methanopterin, *tetrahydromethanopterin*, is thought to be the active form of the coenzyme.

Coenzyme M is involved in the *final step* in methane formation. Coenzyme M (Figure 19.113*e*), a very simple structure, has the chemical name 2-mercaptoethanesulfonic acid. The coenzyme is the carrier of the *methyl* group which is reduced to methane by the methyl reductase-F_{430} enzyme complex in the final step of methanogenesis:

$$CH_3\text{—S—CoM} + 2H \rightarrow HS\text{—CoM} + CH_4$$

Despite the structural simplicity of coenzyme M, it is highly specific in the methyl reductase reaction. A number of closely related analogs of coenzyme M have been found to be inactive (even the propane analog of coenzyme M, which differs by addition of only a single CH_2 group, is inactive). One compound which shows some activity with the methyl reductase enzyme is ethyl CoM, which is converted to ethane. However, since methanogenic bacteria do not normally form ethane, this reaction must be considered only a laboratory artifact. The rumen methanogen, *Methanobrevibacter ruminantium* has the interesting property of requiring coenzyme M as a growth factor. In this way coenzyme M can be considered to be a *vitamin*. Some methanogenic bacteria excrete coenzyme M, and this is appar-

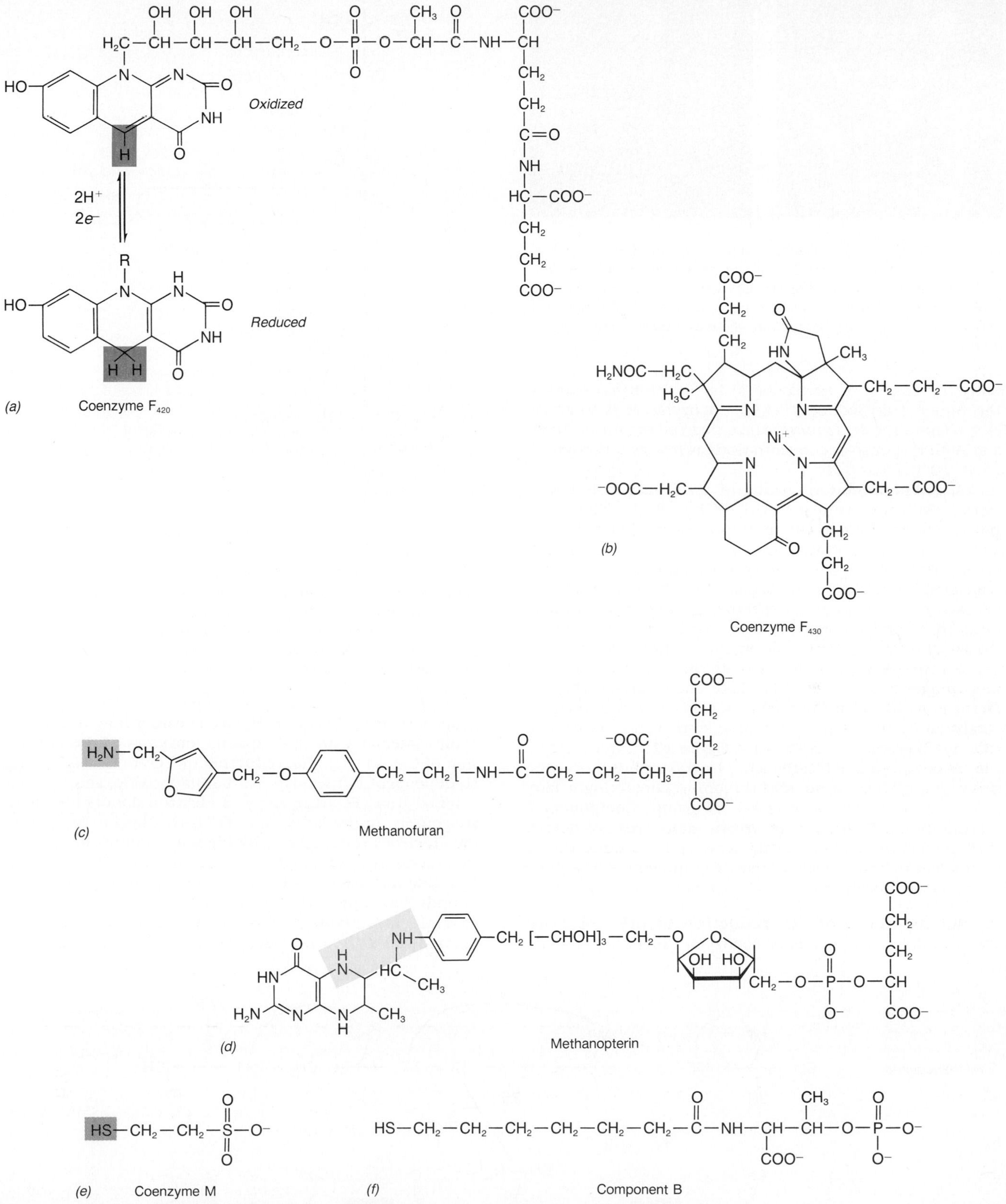

Oxidized

$2H^+$
$2e^-$

Reduced

(a) Coenzyme F_{420}

(b) Coenzyme F_{430}

(c) Methanofuran

(d) Methanopterin

(e) Coenzyme M

(f) Component B

Figure 19.113 Coenzymes unique to methanogenic bacteria. The atoms shaded in color are the sites of oxidation-reduction reactions (F_{420}) or the position to which the C-1 moiety is attached during the reduction of CO_2 to CH_4 (methanofuran, methanopterin, and coenzyme M).

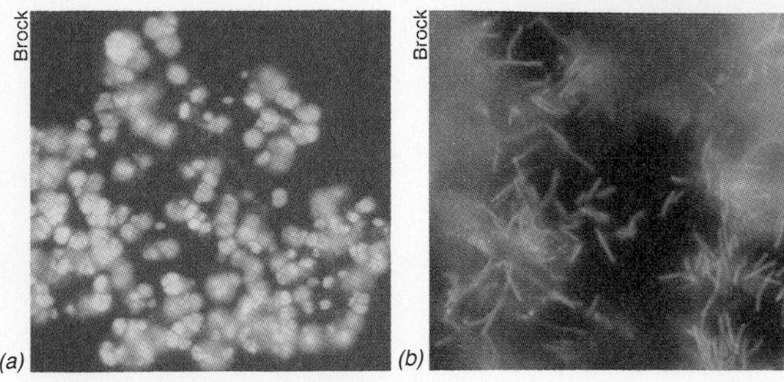

Figure 19.114 (a) Autofluorescence of the methanogen *Methanosarcina barkeri* due to the presence of the unique electron carrier F_{420}. The organisms were visualized with blue light in a fluorescence microscope. (b) F_{420} fluorescence in the methane bacterium *Methanobacterium formicicum*.

ently the source of coenzyme M for *M. ruminantium* in the rumen (see Section 17.13). Coenzyme M is so active as a vitamin for *M. ruminantium* that the organism shows a growth response at concentrations as low as 5 nanomolar (5×10^{-9} molar).

A potent inhibitory analog of coenzyme M is *bromoethanesulfonic acid*, Br—CH_2—CH_2—SO_3H. This compound causes 50 percent inhibition of methyl reductase activity at a concentration of 10^{-6} M, and it also inhibits growth of methanogenic bacteria. Because coenzyme M is restricted to methanogenic bacteria, the bromo analog can be used experimentally to specifically inhibit methanogenesis in natural environments in experimental studies of the anaerobic degradation of organic matter to methane.

Component B is the final unique coenzyme of the methanogens to be discussed. Like coenzyme M, this cofactor is involved in the *terminal* step of methanogenesis catalyzed by the methyl reductase system, but its precise role is still unknown. As shown in Figure 19.113*f*, the structure of component B is rather simple. It is a phosphorylated derivative of the amino acid threonine containing a fatty acid side chain with a terminal SH group. Component B resembles the vitamin *pantothenic acid* (part of acetyl-CoA, see Figure 4.13), and may serve as an electron *donor* to the methyl reductase system. Component B has been shown to definitely *not* be a C-1 carrier.

Biochemistry of CO_2 reduction to CH_4 Now that we have discussed the major coenzymes involved in methanogenesis from CO_2 plus H_2, we consider how these molecules interact with their specific enzymes in the conversion of CO_2 to CH_4. The reduction of CO_2 to CH_4 is usually H_2-dependent but formate, carbon monoxide, and even elemental iron (Fe^0) can serve as electron donors for methanogenesis. In the latter case, Fe^0 is oxidized to Fe^{2+}, with the electrons released combining with protons to form H_2 that serves as the immediate reductant in methanogenesis. In a few methanogens even certain simple *organic* compounds can supply the electrons for CO_2 reduction. For example, in *Methanospirillum* 2-propanol or 2-butanol can be oxidized to their keto derivatives, yielding electrons for

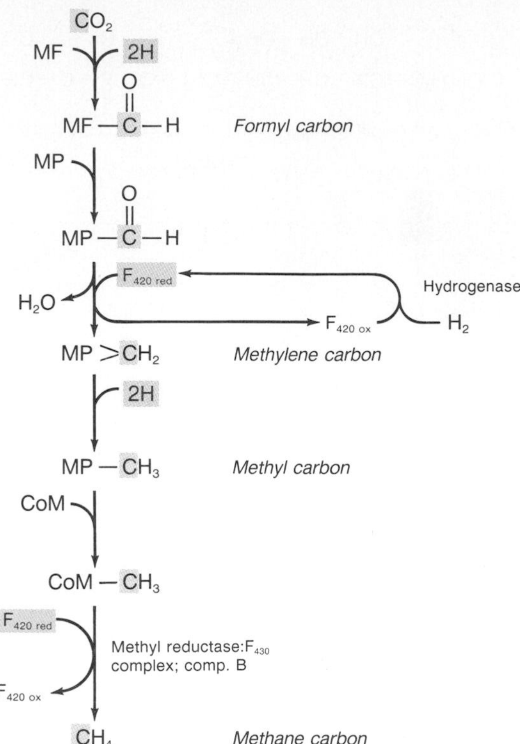

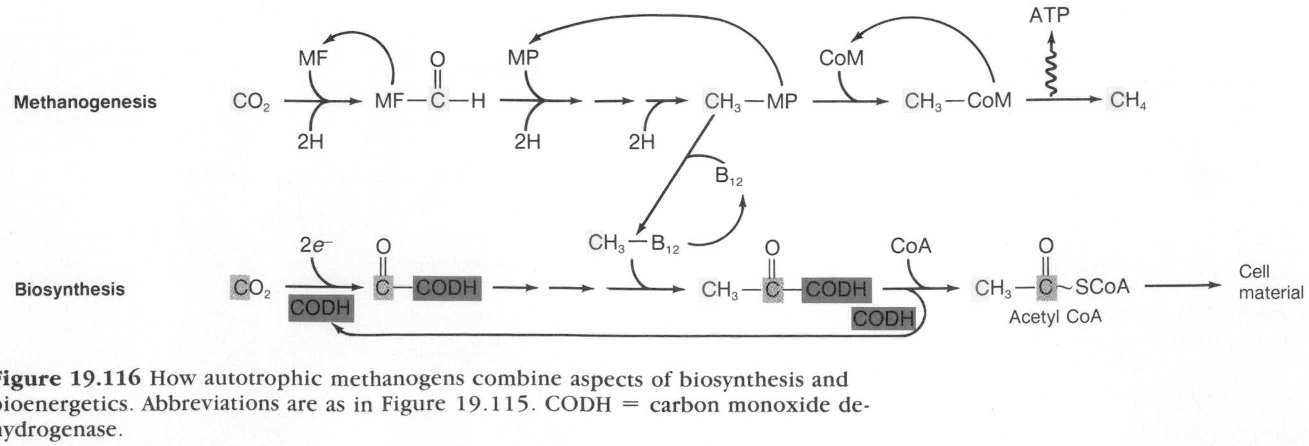

Figure 19.115 Pathway of methanogenesis from CO_2. Abbreviations: MF, methanofuran; MP, tetrahydromethanopterin; CoM, coenzyme M; Comp B, component B; F_{420}, coenzyme F_{420}; F_{430}, coenzyme F_{430}. The carbon atom reduced and the source of electrons are highlighted. See Figure 19.113 for the structures of the coenzymes.

Figure 19.116 How autotrophic methanogens combine aspects of biosynthesis and bioenergetics. Abbreviations are as in Figure 19.115. CODH = carbon monoxide dehydrogenase.

methanogenesis. But in general, the production of CH_4 from CO_2 is driven by molecular hydrogen.

The reduction of CO_2 to CH_4 occurs via several intermediates. The steps in CO_2 reduction, shown in Figure 19.115, are summarized below:

1. CO_2 is activated by methanofuran and subsequently reduced to the *formyl* level.
2. The formyl group is transferred from methanofuran to tetrahydromethanopterin (MP in Figure 19.115) and subsequently dehydrated and reduced in two separate steps to the *methylene* and *methyl* levels.
3. The methyl group is transferred from methanopterin to coenzyme M.
4. Methyl-coenzyme M is reduced to *methane* by the methyl reductase system in which F_{430}, component B, and F_{420} are involved.

In the steps of methanogenesis shown in Figure 19.115, the nature of the electron donor in several of the steps is unclear. Reduced F_{420} serves as electron donor in the reduction of the methenyl to the methylene group and in the reduction of CH_3-CoM to CH_4, but the nature of the direct electron donor in the remaining steps is unknown.

Autotrophy in methanogens It has been established that carbon dioxide is converted into organic form in methanogenic bacteria through the reactions of the **acetyl-CoA pathway** used also by the acetogens and sulfate-reducing bacteria. (See Section 19.9 for detailed discussion of this pathway.) However, unlike the other anaerobes which use this pathway, methanogens growing on $H_2 + CO_2$ integrate their biosynthetic and bioenergetic pathways because common intermediates are shared. This is possible because both the acetyl-CoA pathway and the methanogenic pathway lead to the production of CH_3 groups. As shown in Figure 19.116, autotrophically grown methanogens lack that part of the acetyl-CoA pathway leading to the production of methyl groups via tetrahydrofolate intermediates, and instead obtain methyl groups for the production of acetate for biosynthesis from the methanogenic pathway (compare Figure 19.116 with Figure 19.36). Specifically, methyl tetrahydromethanopterin donates methyl groups to a vitamin B_{12}-containing enzyme to yield CH_3-B_{12} (Figure 19.116). The CH_3 group is then transferred from CH_3-B_{12} to carbon monoxide dehydrogenase (which has previously reduced CO_2 to the level of CO), to eventually yield acetyl-CoA (Figure 19.116). Since, relative to methanogenesis, only a small amount of CO_2 is incorporated into cell material, this small drain on the methanogenic pathway is apparently of no consequence, and the merging of the two pathways probably effects an energy savings to the organism, since synthesis of additional enzymes of the acetyl-

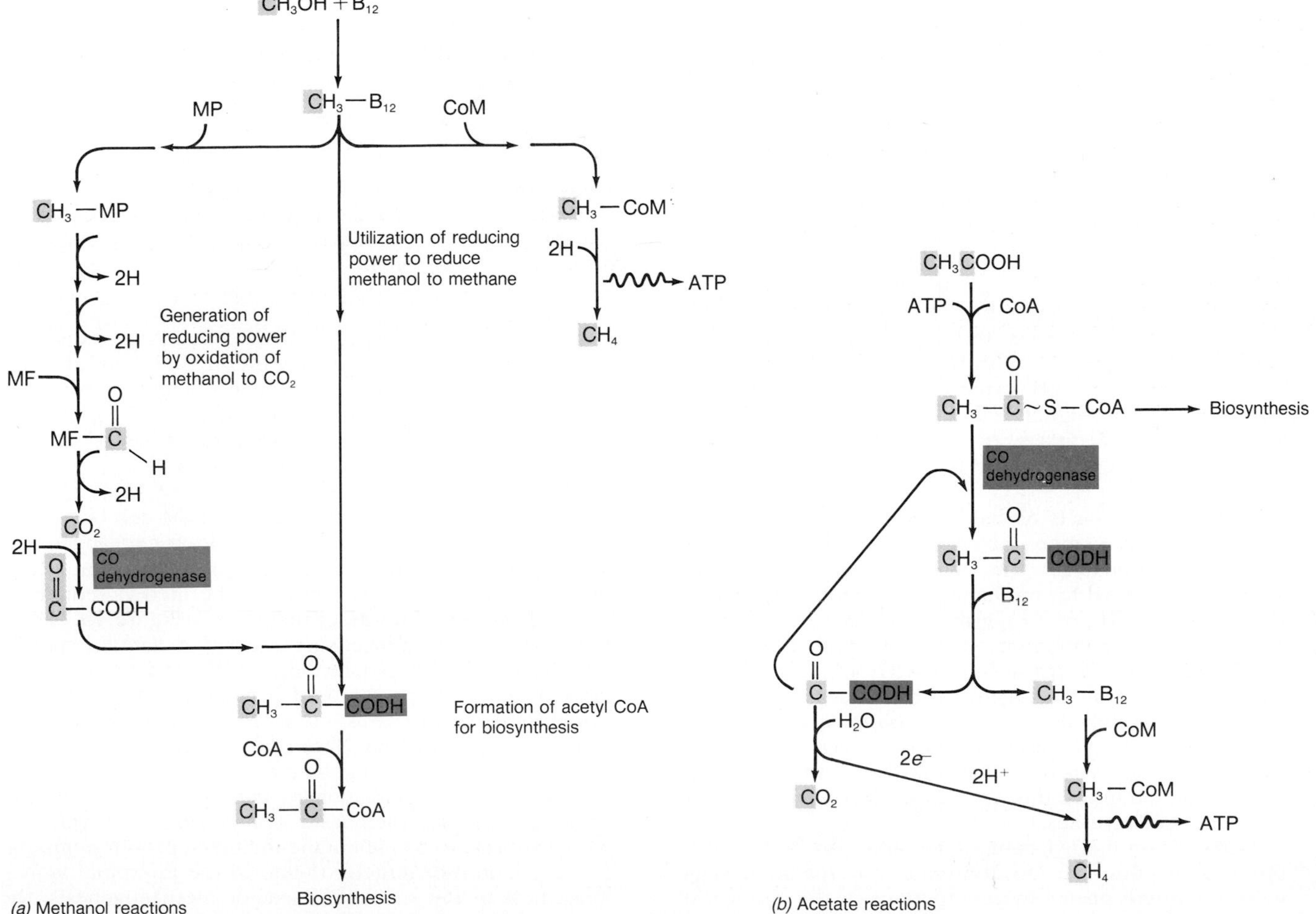

Figure 19.117 Utilization of reactions of the acetyl-CoA pathway during growth on methanol (a) or acetate (b) by methanogenic bacteria.

CoA pathway is not necessary in order to make the CH_3 group of acetate for biosynthesis.

Methanogenesis from methyl compounds and acetate Much less information is available concerning the pathway of methanogenesis from methyl compounds than from CO_2. However, reactions of the acetyl-CoA pathway discussed above do appear to be involved. Biochemical evidence suggests that methyl compounds such as methanol are catabolized by donating methyl groups to a vitamin B_{12} protein to form CH_3-B_{12} (Figure 19.117*a*). The latter donates the methyl group to CoM to give CH_3-CoM from which methane is obtained by reduction with electrons derived from oxidation of other molecules of methanol to CO_2 (Figure 19.117*a*). Carbon for biosynthesis originates by formation of CO from the oxidation of a methyl group from methanol and the combination of CO and a methyl group by carbon monoxide dehydrogenase to give acetate (Figure 19.117*a*).

Growth of acetoclastic methanogens is intimately tied to reactions of the acetyl-CoA pathway. In acetoclastic methanogens acetate is used directly for biosynthesis. For energy purposes acetate is also the energy source. Acetate is thought to be activated to acetyl-CoA which can interact with carbon monoxide dehydrogenase, following which the methyl group of acetate is transferred to the vitamin-B_{12} enzyme of the acetyl-CoA pathway to yield CH_3-B_{12} (Figure 19.117*b*). From here the methyl group is transferred to tetrahydromethanopterin and then to coenzyme M to yield CH_3-CoM. The latter is then reduced to CH_4 using electrons generated from the oxidation of CO to CO_2 by CO dehydrogenase (Figure 19.117*b*). Energy conservation presumably occurs in the acetoclastic reaction due to proton gradient formation during the methyl reductase step, as it does for $H_2 + CO_2$ or methylotrophically grown methanogens (see below).

Energetics of methanogenesis On theoretical grounds, ATP synthesis in methanogenic bacteria is thought to occur via electron-transport phosphorylation, since there is no known method for coupling substrate-level phosphorylation to methanogenesis from acetate, methanol, or H_2. Under standard conditions, the free energy change of the reduction of CO_2 to CH_4 with H_2 is -32 kcal/mole. However, concentrations of H_2 in methanogenic habitats are usually quite low, no higher than 1 micromolar, and because of the influence of concentration of reactants on free energy change (see Appendix 1), the free energy for the reaction forming CH_4 from $H_2 + CO_2$ by methanogenic bacteria in their natural habitat is probably much lower, about -15 kcal/mole. Thus no more than one ATP will be formed during CO_2 reduction to CH_4, and this agrees with molar growth yield data obtained for methanogenic bacteria growing on $H_2 + CO_2$. In $H_2 + CO_2$-grown methanogens no cytochromes or other common electron carriers are present. How then does ATP synthesis occur? It apppears that the terminal step of methanogenesis, the conversion of CH_3-CoM to CH_4, is linked to a *proton pump* that pumps H^+ to the outside of the cytoplasmic membrane (Figure 19.118*a*). Dissipation of the ensuing proton gradient by a membrane-bound proton-translocating ATPase (see Section 4.12) would then drive ATP synthesis.

Growth on methyl compounds may also be linked to this proton pump but an additional consideration is involved. Growth on methyl compounds in the absence of H_2 means that some of the substrate must be *oxidized* to CO_2 in order to generate the electrons needed for methyl *reduction* to methane. How does this occur? For growth of

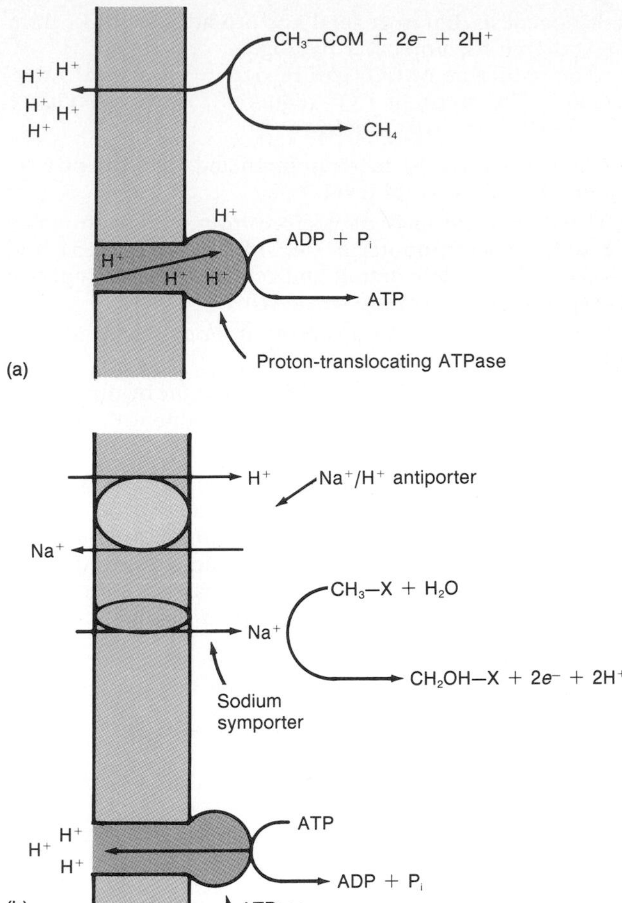

Figure 19.118 Energetics of methanogenesis. (a) Proposed mechanism of ATP production linked to the methyl reductase system. (b) Na^+/H^+ transport during methyl group oxidation. See text for details.

Methanosarcina barkeri on CH_3OH it has been found that *sodium* ions must be present in the medium, and it is hypothesized that the Na^+ requirement reflects the presence of a *sodium pump* in membranes of methanol-grown *M. barkeri*. It is thought that the oxidation of methyl groups, a thermodyamically *unfavorable* reaction, is driven by the influx into the cell of sodium ions (Figure 19.118*b*). Sodium is subsequently transported out of the cell by a membrane antiporter (see Section 3.4) which exchanges Na^+ for H^+ (Figure 19.118*b*) The original proton gradient would then be reestablished by action of the methyl reductase-coupled proton pump described above (Figure 19.118*a*). Methyl-grown methanogens are unusual in that they contain significant levels of cytochromes. Both cytochrome *b* and cytochrome *c* are present and it has been suggested that cytochromes may serve as initial electron *acceptors* during methyl group oxidation. Note that cytochromes are absent from $H_2 + CO_2$-grown methanogens.

Although the methanogens have given us many biochemical surprises, their basic mechanism of ATP synthesis, which appears to be membrane-mediated proton pumping, is a common bioenergetic theme in the microbial world. Regardless of the substrate used for methanogenesis, the *terminal* step of methanogenesis, driven by the methyl reductase system, seems to be the key to energy conservation in all cases.

19.35 Sulfur-Dependent Archaebacteria

A phylogenetically distinct branch of archaebacteria consists of organisms unified by their metabolic requirement for reduced sulfur compounds and their extremely thermophilic nature. This branch, referred to as the **sulfur-dependent archaebacteria**, contain representatives which are the most thermophilic of all known bacteria (see Section 9.11 for a discussion of thermophily). Besides their inherent evolutionary interest, sulfur-dependent archaebacteria have excited microbiologists interested in the molecular mechanisms of thermophily; several sulfur-dependent archaebacteria are capable of growth at temperatures above the normal boiling point of water! We begin with a brief overview of the group and proceed to a more detailed discussion of the major genera that have been characterized.

Overview of sulfur-dependent archaebacteria Sulfur-dependent archaebacteria have been variously called "extreme thermophiles," "thermoacidophiles," or "sulfur-dependent, extremely thermophilic bacteria." For simplicity, we will use the term *sulfur-dependent* archaebacteria. All sulfur-dependent archaebacteria have been isolated from geothermally heated soils or waters containing elemental sulfur. Elemental sulfur is formed from geothermal H_2S either by the spontaneous oxidation of H_2S with O_2 or by reaction of H_2S with SO_2 (the latter is a common component of volcanic gases). In terrestrial environments, sulfur-rich springs, mud pots, and soils may have temperatures up to 100°C and are generally mildly to extremely acidic due to production of sulfuric acid, H_2SO_4, from the biological oxidation of H_2S and S^0 (see Section 17.15). The term **solfatara** has been used to describe hot, sulfur-rich environments, and solfataric fields are found throughout the world (Figure 19.119*a*). Particularly extensive solfataras are found in Italy, Iceland, New Zealand, and Yellowstone National Park. Depending on the surrounding geology, solfataric environments may be either slightly alkaline or mildly acidic, pH 5–8, or extremely acidic, with pH values below 1 not uncommon. Sulfur-dependent archaebacteria have been obtained from both types of environments but the majority of these organisms inhabitat neutral or mildly acidic habitats. In addition to these natural habitats, thermophilic archaebacteria also thrive within artificial thermal habitats, in particular the boiling outflows of geothermal powerplants.

With only two known exceptions, sulfur-dependent archaebacteria are *obligate anaerobes*. Their energy-yielding metabolism is either heterotrophic or lithotrophic. The requirement for sulfur, nearly universal across the group, is based on the need for an electron *acceptor* to carry out anaerobic respiration. Elemental sulfur (S^0) is reduced to H_2S using electrons derived from the oxidation of organic or inorganic compounds. The energy-yielding reactions of some sulfur-metabolizing archaebacteria are shown in Table 19.50. Heterotrophic bacteria, such as those of the genera *Thermococcus* and *Thermoproteus*, oxidize a variety of organic compounds, in particular small peptides, glucose, and starch anaerobically in the presence of S^0 as electron acceptor. *Sulfolobus*, on the other hand, will utilize a variety of organic compounds (as well as elemental sulfur) as energy sources with O_2 as electron acceptor.

Many sulfur-dependent archaebacteria can grow *lithotrophically* with H_2 as energy source (Table 19.50). *Pyrodictium*, for example, grows strictly anaerobically in a

(a) Brock

(b) Brock

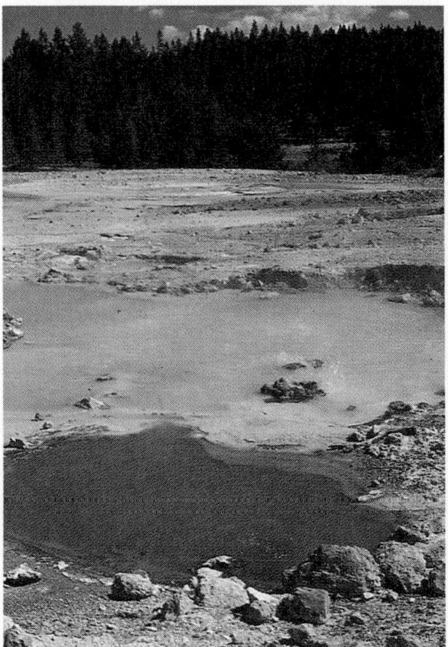
(c) Brock

Figure 19.119 Habitats of sulfur-dependent archaebacteria. (a) A typical solfatara in Yellowstone National Park. Steam rich in hydrogen sulfide rises to the surface of the earth. Because of the heat and acidity, higher forms of life do not develop. (b) Sulfur-rich hot spring, a habitat where *Sulfolobus* grows in profusion. (c) Iron-rich geothermal spring, another *Sulfolobus* habitat.

Table 19.50 Energy-yielding reactions of sulfur-dependent archaebacteria

Nutritional class	Energy-yielding reaction	Example
Heterotrophic	Organic compound + $S^0 \rightarrow H_2S + CO_2$	*Thermoproteus; Thermococcus; Desulfurococcus*
	Organic compound + $O_2 \rightarrow H_2O + CO_2$	*Sulfolobus*
	Organic compound $\rightarrow CO_2$ + fatty acids	*Staphylothermus*
Lithotrophic	$H_2 + S^0 \rightarrow H_2S$	*Acidianus; Pyrodictium; Thermoproteus*
	$2S + 3O_2 + 2H_2O \rightarrow 2H_2SO_4$	*Sulfolobus; Acidianus*
	$4FeS_2 + 15O_2 + 2H_2O \rightarrow 2Fe_2(SO_4)_3 + 2H_2SO_4$	*Sulfolobus*

mineral salts medium supplemented with H_2 and S^0 at temperatures up to 110°C. Growth of *Pyrodictium* is stimulated by the addition of organic compounds, but the latter cannot replace H_2, suggesting that this organism is an obligate lithotroph. Other lithotrophic growth modes among sulfur-dependent archaebacteria include the oxidation of elemental sulfur and ferrous iron (aerobically) by various *Sulfolobus* species (see Table 19.50). Thus, a variety of respiratory processes can be carried out by sulfur-dependent archaebacteria, but in all cases elemental sulfur plays a key role, either as an electron donor or an electron acceptor.

Extreme thermophiles from volcanic habitats As mentioned previously, volcanic habitats can have temperatures as high as 100°C and are thus suitable habitats for extremely thermophilic archaebacteria. The first such organism discovered, *Sulfolobus*, grows in sulfur-rich hot acid springs (Figure 19.119*b*) at temperatures up to 90°C and at pH values of 1–5. *Sulfolobus* (Figure 19.120*a*) is an obligate aerobe capable of oxidizing H_2S or S^0 to H_2SO_4 and fixing CO_2 as carbon source. *Sulfolobus* can also grow heterotrophically. *Sulfolobus* has a generally spherical shape and forms distinct lobes (Figure 19.120*a*). Cells adhere tightly to sulfur crystals where they can be visualized microscopically by use of fluorescent dyes (see Figure 16.19*b*). Besides an active *aerobic* metabolism, *Sulfolobus* is known to reduce Fe^{3+} to Fe^{2+} *anaerobically*. The ability of *Sulfolobus* to oxidize Fe^{2+} to Fe^{3+} aerobically (Figure 19.119*c*), however, has been used quite successfully in the high temperature leaching of iron and copper ores (see Section 17.17).

Although the mechanism of autotrophy in *Sulfolobus* has not been settled, it is clear that *Sulfolobus* does not use the Calvin cycle to fix CO_2 as do most autotrophs (see Section 16.6). And, although many anaerobic autotrophs, including the methanogens, fix CO_2 by way of the acetyl-CoA pathway (see Section 19.9), this pathway also does not seem to support autotrophy in *Sulfolobus*. Some evidence suggests that the reverse citric acid cycle is the mechanism by which CO_2 is fixed in *Sulfolobus*. This pathway, well established in green sulfur bacteria (see Section 19.1), has

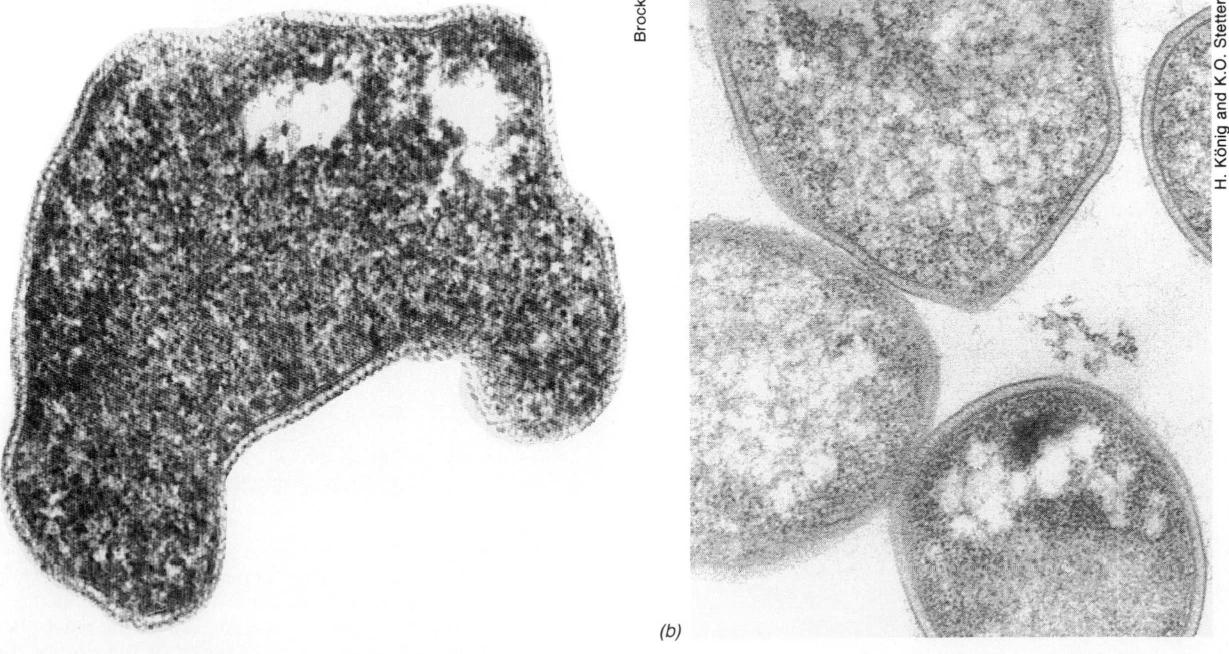

Figure 19.120 Acidophilic sulfur-dependent archaebacteria. (a) *Sulfolobus acidocaldarius*. Electron micrograph of a thin section. (b) *Acidianus infernus*. Electron micrograph of a thin section.

not been found in other organisms, but experiments using $^{14}CO_2$ point to this as the mechanism of autotrophy in *Sulfolobus*.

Recently a facultative aerobe resembling *Sulfolobus* has been isolated from acidic solfataric springs. This organism, named *Acidianus* (Figure 19.120*b*), differs from *Sulfolobus* primarily by virtue of its ability to grow *anaerobically*. Remarkably, *Acidianus* is able to use S^0 in both its aerobic and anaerobic metabolism. The generic name *Acidianus* is derived from the Latin terms *acidus* meaning sour, and Janus, the ancient Roman god with two faces looking in opposite directions. The latter term is particularly appropriate for describing the metabolism of *Acidianus*, because under *aerobic* conditions the organism uses S^0 as an electron *donor*, oxidizing S^0 to H_2SO_4. Anaerobically, *Acidianus* uses S^0 as an electron *acceptor* (with H_2 as electron *donor*) forming H_2S as the reduced product. Thus, the metabolic fate of S^0 in cultures of *Acidianus* depends upon the presence of O_2 and/or an electron donor.

Like *Sulfolobus, Acidianus* is roughly spherical in shape (Figure 19.120*b*). It grows at temperatures from about 65°C up to a maximum of 96°, with an optimum of about 90°C. Another property shared by *Sulfolobus* and *Acidianus* is an unusually low GC base ratio. The DNA of *Sulfolobus* is about 38 percent GC while that of *Acidianus* is even lower, about 31 percent. These low GC base ratios are intriguing when one considers the extremely thermophilic nature of these organisms; how do they prevent their DNA from melting? In the test tube DNA of 30–40 percent GC content would melt almost instantly at 90°C. Obviously extreme thermophiles have evolved some protective mechanisms to prevent DNA melting *in vivo* (see "Limits of microbial existence: temperature," below).

The genera *Thermoproteus* and *Thermofilum* consist of *rod-shaped* sulfur-dependent archaebacteria which inhabit neutral or slightly acidic hot springs. Cells of *Thermoproteus* are stiff rods about 0.5 μm in diameter and highly variable in length, ranging from short cells of 1–2 μm up to filaments of 70–80 μm in length (Figure 19.121*a*). Filaments of *Thermofilum* are thinner, some 0.17–0.35 μm in width with filament length ranging up to 100 μm (Figure 19.121*b*). Both *Thermoproteus* and *Thermofilum* are strict anaerobes which carry out a S^0-based anaerobic respiration. Unlike many other sulfur-dependent archaebacteria, the oxygen sensitivity of *Thermoproteus* and *Thermofilum* is extreme, comparable to that of the methanogens; thus elaborate precautions must be taken in their culture. Most *Thermoproteus* isolates can grow lithotrophically on H_2 or heterotrophically on complex carbon substrates such as yeast extract, small peptides, starch, glucose, ethanol, malate, fumarate, or formate. *Thermofilum* can be grown in either mixed culture with *Thermoproteus* or in pure culture only by the addition of a highly polar lipid fraction isolated from cells of *Thermoproteus*; in nature the two organisms probably coexist in close association. Both *Thermoproteus* and *Thermofilum* have similar GC base ratios (56–58 percent GC), but appear to be phylogenetically distinct by DNA:RNA hybridization analyses.

H. König and K.O. Stetter

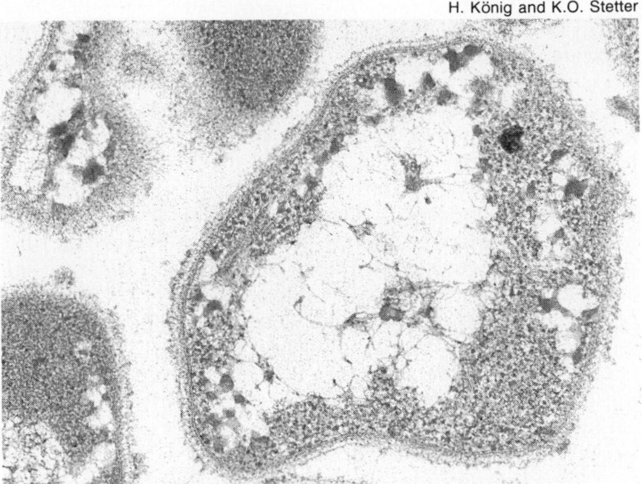

Figure 19.122 *Desulfurococcus*. Electron micrograph of a thin section of *D. saccharovorans*.

Desulfurococcus (Figure 19.122) is a spherical, obligately anaerobic S^0 respiring organism. *Desulfurococcus* grows best at near neutral pH and 80–90°C. The major features which differentiate *Desulfurococcus* from other sulfur-dependent archaebacteria are *genotypic* rather than phenotypic. DNA from *Desulfurococcus* contains 51 percent GC and hybridizes at only low levels with nucleic acid from *Thermoproteus*.

Extreme thermophiles from submarine volcanic areas We now turn our attention to *submarine* volcanic habitats where yet another phylogenetically distinct set of

H. König and K.O. Stetter

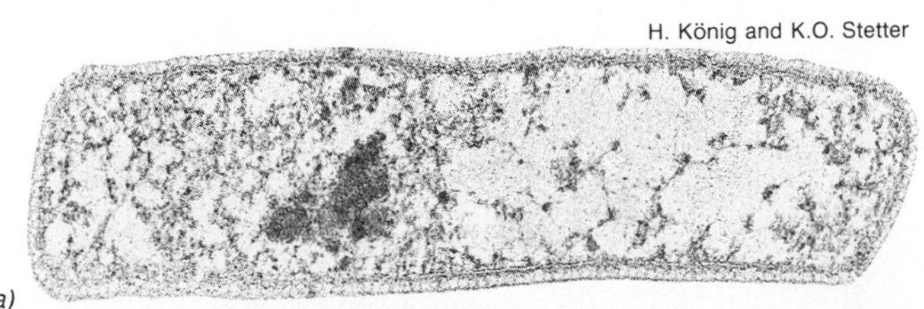

(a)

H. König and K.O. Stetter

(c)

(b)

H. König and K.O. Stetter

Figure 19.121 Rod-shaped sulfur-dependent archaebacteria from terrestrial volcanic habitats. (a) *Thermoproteus neutrophilus*. Electron micrograph of a thin section. (b) *Thermofilum librum*. Electron micrograph of shadowed cells. (c) *T. librum*. Electron micrograph of a thin section.

sulfur-dependent archaebacteria exist. Although these underwater microbial habitats are generally shallow ones, the pressure of even a few meters of water can raise the boiling point of water sufficiently to select for bacteria capable of growth in superheated liquids. We now consider these fascinating bacteria.

Geothermally heated sea floors exist in various locations around the world, and from a series of such habitats off the coast of Italy several genera of extremely thermophilic archaebacteria have been isolated. In submarine solfatara fields located in 2–10 meters of water off the coast of Vulcano, Italy, the sea floor consists of sandy sediments with cracks and holes from which geothermally heated water is emitted at temperatures up to 103°C. From such waters the following genera of extreme thermophiles have been isolated: *Pyrodictium, Thermococcus, Thermodiscus, Pyrococcus*, and *Staphylothermus*.

Pyrodictium is perhaps the most fascinating of all submarine volcanic extreme thermophiles because its growth temperature optimum, 105°C, is the highest of any known organism. Cells of *Pyrodictium* are irregularly disc and dish-shaped (Figure 19.123) and grow in culture as a mold-like layer upon sulfur crystals suspended in the medium. The cell mass consists of a huge network of fibers to which individual cells are attached (Figure 19.123). The fibers are hollow and consist of proteinaceous subunits arranged in a fashion similar to that of the flagellin protein of the bacterial flagellum (see Section 3.8). The filaments of *Pyrodictium* do not function in motility but apparently serve to *attach* the cells to a solid substratum. *Pyrodictium* is a strict anaerobe that grows lithotrophically at neutral pH on H_2 with S^0 as electron acceptor. Growth occurs between 82° and 110°C and is stimulated by the addition of organic compounds. The cell envelope of *Pyrodictium* consists of glycoprotein and lacks peptidoglycan (as do all archaebacterial cell walls, see Section 3.6). Phylogenetically, *Pyrodictium* shows affinities to *Sulfolobus* by both 16S and 5S ribosomal RNA sequencing. However, due to the very large number of posttranscriptionally modified bases in the ribosomal RNAs of *Pyrodictium*, the exact phylogenetic position of this organism is unclear. DNA from *Pyrodictium* shows a substantially higher base ratio (62 percent GC) than that of any of the sulfur-dependent archaebacteria from terrestrial habitats.

Thermodiscus resembles *Pyrodictium* morphologically but is *heterotrophic* rather than lithotrophic and is unable to grow above 100°C. *Thermodiscus* is an obligate heterotroph, with growth occurring on complex organic mixtures with S^0 as electron acceptor at temperatures between 75° and 98°C (optimum around 90°C). The DNA of *Thermodiscus* contains a significantly lower GC base ratio (49 percent) than that of *Pyrodictium*, but analyses of 16S rRNA sequences show that *Thermodiscus* is specifically related to *Pyrodictium*.

Thermococcus is a spherical sulfur-dependent archaebacterium indigenous to submarine thermal waters in various locations throughout the world. The spherical cells contain a tuft of polar flagella and are thus highly motile (Figure 19.124*a*). *Thermococcus* is an obligately anaerobic heterotroph which grows on proteins and other complex organic mixtures (including some sugars) with S^0 as electron acceptor. It is not as thermotolerant as *Pyrodictium*, the optimum growth temperature being only 88°C, with a temperature range from 70–95°C. *Thermococcus* is an especially interesting organism from a *phylogenetic* standpoint. As shown in Figure 18.12, 16S rRNA sequence comparisons show *Thermococcus* to be a member of a very

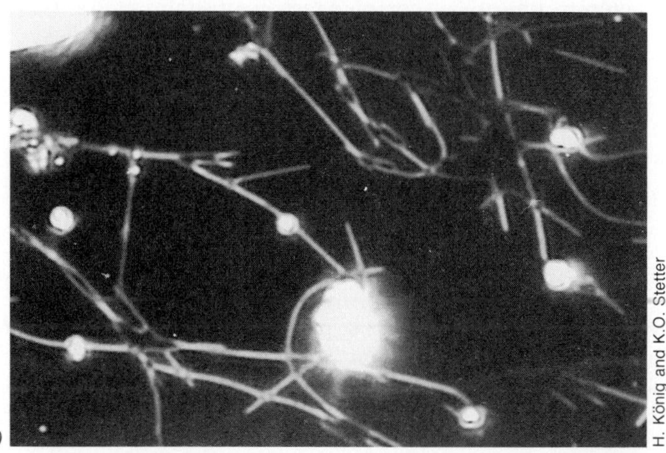
(a) H. König and K.O. Stetter

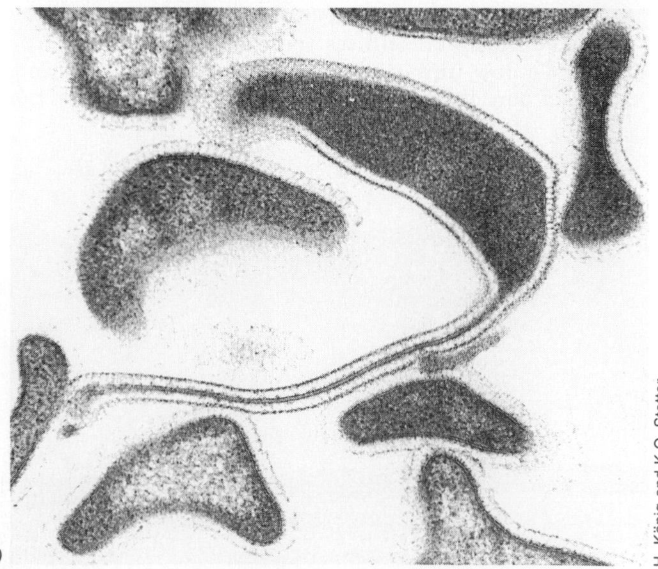
(b) H. König and K.O. Stetter

Figure 19.123 *Pyrodictium occultum*, the most thermophilic of all known bacteria. (a) Dark field micrograph. (b) Electron micrograph of a thin section.

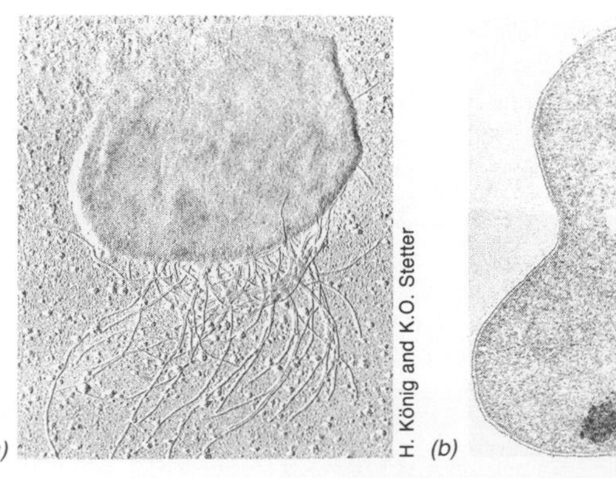
(a) H. König and K.O. Stetter (b) H. König and K.O. Stetter

Figure 19.124 Spherical-shaped sulfur-dependent archaebacteria from submarine volcanic areas. (a) *Thermococcus celer*. Electron micrograph of shadowed cells. (b) Dividing cell of *Pyrococcus furiosus*. Electron micrograph of thin section.

slowly evolving archaebacterial lineage. This means that *Thermococcus* has evolved the least (of known organisms) from the universal ancestor of all living organisms, and this implies that it should more closely resemble the ancestral phenotype than other extant organisms. Indeed, the fact that *Thermococcus* is an extremely thermophilic sulfur-respiring heterotroph fits nicely with the established ideas of what the chemical and physical conditions were like on Earth at the time life originated (see Sections 18.1 and 19.37).

An organism morphologically quite similar to *Thermococcus* is *Pyrococcus* (Figure 19.124*b*). *Pyrococcus* (the Latin derivation literally means "fireball") differs from *Thermococcus* primarily by its significantly different GC ratio (see Table 19.51) and its higher temperature requirements; *Pyrococcus* grows between 70° and 103°C with an optimum of 100°C. However, metabolically *Thermococcus* and *Pyrococcus* are quite similar: proteins, starch, or maltose are oxidized and S^0 is reduced to H_2S. Comparisons of 16S rRNA sequences (see Figure 18.12) show *Thermococcus* and *Pyrococcus* to be specifically related despite their significantly different GC base ratios and optimal growth temperatures (see Table 19.51).

The genus *Staphylothermus* is a morphologically unusual bacterium consisting of spherical cells about 1 μm in diameter which form aggregates of up to 100 cells resembling those of *Staphylococcus* (Figure 19.125). *Staphylothermus* is a strictly anaerobic extreme thermophile growing optimally at 92°C and capable of growth between 65° and 98°C. Although S^0 is required for growth, oxidation of complex organic compounds is not tightly coupled to S^0 reduction in *Staphylothermus* as it is in other sulfur-dependent archaebacteria, because fatty acids such as acetate and isovalerate are produced during growth by this organism, suggesting that some sort of fermentative metabolism is occurring. The GC base ratio of *Staphylothermus* is quite low, about 35 percent, and 16S rRNA sequence studies have not shown this organism to be closely related to any other sulfur-dependent species.

Isolates of *Staphylothermus* have been obtained from both a shallow marine hydrothermal vent and from samples taken from a "black smoker" at a depth of 2500 meters (see Section 17.9 for a discussion of black smokers). Since the temperature of the black smoker yielding the *Staphylothermus* isolate was some 330°C, it is assumed that the organism originated from thermally heated seawater and not from the superheated water itself (see "Limits of microbial existence: temperature," below). Nevertheless, *Staphylothermus* may be widely distributed near hot hydrothermal vents and may be an ecologically important consumer in these environments.

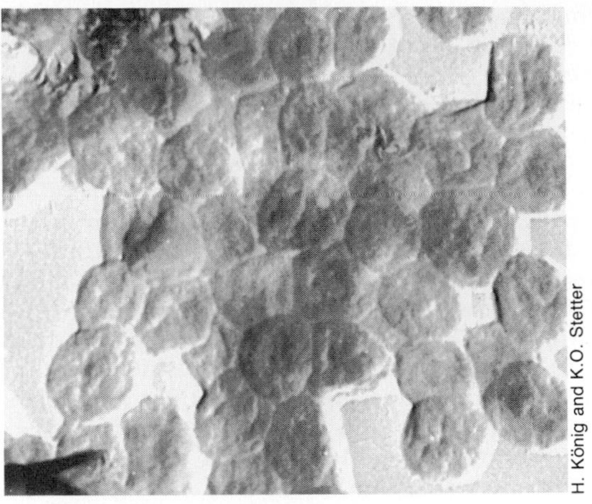

Figure 19.125 *Staphylothermus marinus*. Electron micrograph of shadowed cells.

Table 19.51 summarizes the major properties of both submarine and terrestrial sulfur-dependent archaebacteria.

Archaeoglobus In the preceding discussion we have emphasized that extremely thermophilic archaebacteria share a major metabolic theme: the use of S^0 as an electron acceptor for anaerobic growth. We have discussed earlier

Table 19.51 Properties of sulfur-dependent archaebacteria

Genus	Morphology	DNA (mole % GC)	Temperature °C			Optimum pH
			Minimum	Optimum	Maximum	
Terrestrial Volcanic Isolates						
Sulfolobus	Lobed-sphere	41	55	75–80	87	3
Acidianus	Sphere	31	60	85–90	95	2
Thermoproteus	Rod	56	60	88	96	6
Thermofilum	Rod	57	60	88	96	5.5
Desulfurococcus	Sphere	51	60	92	93	6
Submarine Volcanic Isolates						
Pyrodictium	Disc-shape with attached filaments	62	82	105	110	6
Pyrococcus	Sphere	38	70	100	103	6–8
Thermodiscus	Disc shaped	53	75	90	98	5.5
Staphylothermus	Spheres which aggregate into clumps	35	65	92	98	6–7
Thermococcus	Sphere	51	70	88	95	5.8

in this chapter other sulfur-reducing bacteria and the sulfate-reducing bacteria (see Section 19.8). Interestingly, however, all sulfur-dependent archaebacteria discussed thus far are unable to use *sulfate* as an electron acceptor. *Archaeoglobus*, however, is a truly sulfate-reducing extremely thermophilic archaebacterium. This organism couples the oxidation of H_2, lactate, pyruvate, glucose, or complex organic mixtures to the reduction of sulfate to sulfide (sulfite or thiosulfate are also reduced). Cells of *Archaeoglobus* are irregular motile spheres and cultures of this organism grow between 64° and 92° with an optimum of 83°C. *Archaeoglobus* is clearly an archaebacterium as adjudged from ribosomal RNA analyses, its lack of peptidoglycan, and its pattern of antibiotic resistance (see Section 18.8). However, one of the most remarkable aspects of this organism (besides its ability to reduce sulfate) is that it contains certain coenzymes previously found only in *methanogenic* archaebacteria. Specifically, coenzyme F_{420} and methanopterin, coenzymes intimately involved in biochemical reactions leading to the production of methane (see Section 19.34), are also present in *Archaeoglobus*. In fact, *Archaeoglobus* actually produces small amounts of methane during growth as a sulfate-reducing organism.

Archaeoglobus thus seems to share some metabolic features with methanogenic bacteria despite its major phenotypic property of reducing sulfate. A specific relationship between *Archaeoglobus* and methanogenic bacteria has also been detected by molecular sequencing methods. Analyses of 16S rRNA suggest that *Archaeoglobus* may be a "missing link" in the archaebacteria. The phylogenetic position of *Archaeoglobus* on the archaebacterial tree (see Figure 18.12) lies *between* the sulfur-dependent archaebacteria and the methanogenic bacteria. The addition of a unique metabolic phenotype to the archaebacterial kingdom has thus served to bridge two metabolic patterns, methanogenesis and S^0 reduction (whose previous connection was somewhat obscure), and has identified a type of transitional metabolism which may have been important in the metabolic diversification of archaebacteria.

Limits of microbial existence: temperature In the case of the sulfur-dependent archaebacteria we have seen growth at temperatures far higher than those supporting growth of the vast majority of known bacteria. What is the nature of this extreme heat tolerance and what are the temperature limits beyond which life is impossible? As we discussed in Section 9.11, the *macromolecules* of thermophiles and extreme thermophiles are stable to heat, and in the case of proteins, this stability resides in the unique folding of the molecules; the latter, of course, ultimately depends upon the *sequence* of amino acids in the protein. At the temperatures supporting growth of extreme thermophiles, however, a second consideration starts to become important: not only must *macromolecules* show stability, but small molecules, *monomers*, must also be unaffected by the destructive effects of heat. At temperatures above 100°C, a number of biologically significant monomers show some degree of heat lability. And, no matter how stable the macromolecules are, life is impossible in an environment in which the basic building blocks themselves are unstable. The thermal stability of biomolecules therefore may dictate the upper temperature for life, and already at temperatures as low as 100°C some important biomolecules are destroyed. For example, molecules such as ATP and NAD^+ hydrolyze quite rapidly at high temperature; the half-life of both of these molecules is less than 30 minutes at 100°C. However, temperatures up to 110°C are clearly still compatible with life because *Pyrodictium* grows at this temperature. Presumably thermolabile molecules can be resynthesized sufficiently quickly by *Pyrodictium* to allow growth of the organism at these temperatures.

Since life depends on liquid water, microbial habitats in excess of 100°C are limited to environments where water is under pressure, such as the sea floor. Indeed, all extremely thermophilic archaebacteria capable of growth at 100°C or higher seem restricted to these environments. As discussed in Section 17.9, extremely hot (but not boiling) water is emitted from deep-sea thermal vents (black smokers) at temperatures from 250° to 350°C. Will bacteria capable of growth at these temperatures ever be isolated? Probably not. Life in such environments seems impossible. Under black smoker conditions (250°C), macromolecules and simple organic molecules such as amino acids and nucleotides are spontaneously hydrolyzed so quickly that life *as we know it* could not occur. For example, the half life of DNA at 250°C is in the microsecond range and the half lives of proteins and ATP are less than a second. Hence, although the upper temperature for life has not yet been defined, the discovery of sulfur-dependent archaebacteria has shown that this limit is not lower than 110°C, and possibly even higher. Laboratory experiments on the heat stability of biomolecules predict that living processes could be maintained at temperatures as high as 150°C, but that above this temperature organisms would probably not be able to overcome the heat lability of the important biomolecules of life.

Because of their extreme thermophilicity, however, sulfur-dependent archaebacteria offer interesting model systems for studying the effect of heat on macromolecules. In particular, these organisms should yield important information about DNA-binding proteins necessary to prevent DNA from melting at their high growth temperatures (this is a particularly interesting question for those extreme thermophiles such as *Pyrococcus* whose DNA base ratio is very low, because the melting temperature of DNA is proportional to GC content, see Box, Chapter 5). Many interesting questions concerning the molecular mechanisms of thermophily can now be posed with experimental organisms like the sulfur-dependent archaebacteria available to do the critical experiments.

19.36 *Thermoplasma*

Thermoplasma is a cell wall-less archaebacterium that resembles the eubacterial mycoplasmas. *Thermoplasma* (Figure 19.126) is an acidophilic, aerobic heterotroph, and is mildly thermophilic. *Thermoplasma* grows optimally at 55°C and pH 2 in complex media supplemented with yeast extract. With one exception, all strains of *Thermoplasma* have been obtained from self-heated coal refuse piles (Figure 19.127). Coal refuse contains coal fragments, pyrite, and other organic materials extracted from coal, and when dumped into piles in coal mining operations tends to self heat by spontaneous combustion (Figure 19.127). This sets the stage for growth of *Thermoplasma* which apparently lives from organic compounds leached from the hot coal refuse. Because coal refuse piles are transitory environments, other potential habitats for *Thermoplasma* have been sought. Although acid hot springs would appear to be ideal habitats for this organism, only one isolate of *Thermoplasma* has ever been obtained from such an environment, suggesting that coal refuse piles, or perhaps coal itself, is the primary habitat of *Thermoplasma*.

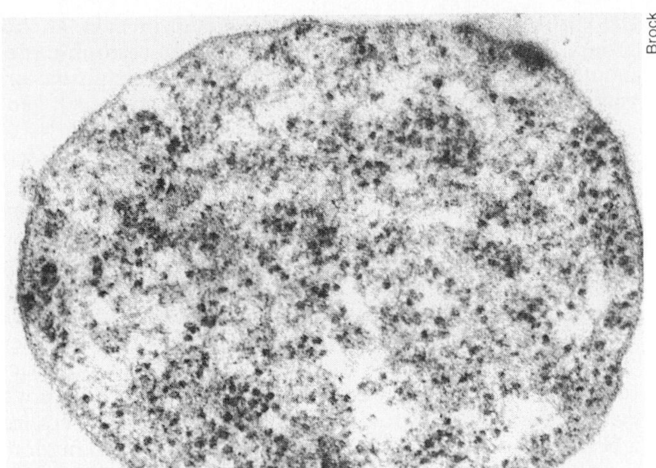

Figure 19.126 *Thermoplasma acidophilum*, an acidophilic, thermophilic, mycoplasma. Electron micrograph of thin section.

To survive the osmotic stresses of life without a cell wall and to withstand the dual environmental extremes of low pH and high temperature, *Thermoplasma* has evolved a cell membrane of chemically unique structure. The membrane contains lipopolysaccharide (referred to as *lipoglycan* in mycoplasmas, see Section 19.27) consisting of a *tetraether* lipid with mannose and glucose units (Figure 19.128). This molecule constitutes a major fraction of the total lipid composition of *Thermoplasma*. The membrane also contains glycoproteins but not sterols. Together these and other molecules render the *Thermoplasma* membrane stable to hot acid conditions.

The genome of *Thermoplasma* is of interest. *Thermoplasma* contains an extremely small genome, perhaps the smallest of all free-living bacteria. The DNA of *Thermoplasma* (molecular weight = 8×10^8) is surrounded by a highly basic DNA-binding protein that organizes the DNA into globular particles that resemble the nucleosomes of eucaryotic cells (see Section 3.16 for a discussion of the arrangement of DNA in eucaryotes). This protein strongly resembles the basic histone proteins of eucaryotic cells, and comparisons of amino acid sequences between the *Thermoplasma* protein and eucaryotic nuclear histones show significant sequence homology. Phylogenetically, *Thermoplasma* is clearly an archaebacterium, but by 16S rRNA sequence analyses fits nearer the methanogen/extreme halophile branch than to the sulfur-dependent archaebacteria (see Figure 18.12). The exact phylogenetic status of *Thermoplasma* is thus unclear.

19.37 Archaebacteria: Earliest Life Forms?

Now that we have considered the various archaebacterial groups, what can we say about their role in microbial evolution? Although metabolically diverse organisms, a common theme running through the archaebacterial kingdom is *adaptation to environmental extremes*. High salt, high temperature, strictly anaerobic conditions, these environmental obstacles have all been overcome by one or another archaebacterium. Recalling our description of the geochemical conditions prevalent on Earth at the time that life presumably arose (Section 18.1), do any particular groups of archaebacteria stand out as candidates for the earliest life forms? Recall that the surface of the early Earth was likely much *hotter* than it is today, perhaps as high as 100°C when living organisms first evolved. This implies that the earliest

(a)

(b)

Figure 19.127 A typical self-heating coal refuse pile.

[R]$_8$ Glu (α1 → 1) —O— (tetraether chain structure) —OH

R = Man (α1 → 2) Man (α1 → 4) Man (α1 → 3)

Figure 19.128 Structure of the tetraether lipoglycan of *Thermoplasma acidophilum*. Glu = glucose; Man = mannose.

life forms would have had to be thermophilic, or at least thermotolerant. As we now know, thermophily is common among archaebacteria; all sulfur-dependent archaebacteria are extreme thermophiles and several species of methanogens are also thermophilic (see Figure 19.112). Only the extreme halophiles lack what could be called truly thermophilic representatives. By contrast, most eubacteria and all eucaryotes are unable to grow at temperatures in excess of about 50°C.

Current phylogenetic evidence deduced from comparison of 16S rRNA sequences (see Chapter 18) suggest that archaebacteria are the *oldest* of the two procaryotic groups and that their rate of evolution has been substantially *slower* over time than that of either the eubacteria or the eucaryotes. This was shown in Figure 18.9 where it could be seen that the main branch leading to the archaebacterial kingdom is much *shorter* than those leading to either the eubacteria or the eucaryotes. It is not known why archaebacteria are the slowest evolving of the three kingdoms, but it may be related to their inhabiting extreme environments. For example, organisms living in thermal environments must maintain those genes which specify phenotypic characteristics adapting them to high temperatures; these genes cannot be significantly changed during evolution if the organism is to maintain itself in these environments. Indeed, by ribosomal RNA sequence criteria, organisms like the sulfur-dependent archaebacteria are likely to have been among the earliest life forms. The phenotypic properties of this group, thermophilicity, and anaerobic, heterotrophic metabolism, agree well with the phenotype of primitive organisms predicted from a consideration of early earth geochemical conditions. From sulfur-based respiration it is possible that the methanogenic subline of archaebacteria arose, perhaps via transitionary forms like *Archaeoglobus* (see above). Finally, extreme halophiles, thought to be the most derived (least ancient) of all archaebacteria, arose from methanogenic ancestors.

Because sulfur-dependent archaebacteria, particularly organisms like *Thermococcus*, play a central role in this evolutionary scenario, the question has arisen as to whether the universal ancestor of all cells was itself an archaebacterium. Unfortunately, we are not yet at a point where we can answer this question because to do so would require a better understanding of the true root of the three kingdom tree (Figure 18.9). Perhaps such information will emerge from additional molecular sequencing studies (see Chapter 18). At any rate, the discovery of archaebacteria as a third living kingdom has clearly given us a new picture of evolution in general and has forced a reevaluation of even the most basic of evolutionary questions. But most excitingly, with the help of molecular sequencing, evolutionary questions can now be approached *experimentally*. Therefore, it is hoped that in the next few years more precise answers to a variety of evolutionary questions will be forthcoming.

Supplementary Readings

Holt, J. G. (editor-in-chief). *Bergey's Manual of Systematic Bacteriology*. Volume 1, 1982. Gram-negative bacteria of medical or industrial importance. Volume 2, 1986. Gram-positive bacteria of medical or industrial importance. Volume 3, 1988. Other Gram-negative bacteria, cyanobacteria, archaebacteria. Volume 4, 1988. Other Gram-positive bacteria. This manual is the recognized authority on bacterial taxonomy. A good place to begin a literature survey on a specific bacterial group.

Schlegel, H. G. 1986. *General Microbiology*. 6th edition. Cambridge University Press, Cambridge,England. This English translation of a popular German textbook of microbiology contains a wealth of information on various bacterial groups.

Starr, M. P., H. Stolp, H. G. Trüper, A. Balows, and **H. G. Schlegel.** 1981. *The Prokaryotes—a handbook on habitats, isolation, and identification of bacteria*. Published in two volumes, 2284 pages, 169 chapters, Springer-Verlag, New York. The most complete reference on the characteristics of bacteria. Also includes formulation of various media and isolation procedures for virtually every procaryotic group known. Well illustrated.

Woese, C. R., and **Wolfe, R. S.** (eds.) 1985. Archaebacteria. Volume VIII of *The Bacteria—a treatise on structure and function*. Academic Press, New York. The most complete reference on all groups of archaebacteria. Well illustrated and covers both the biological and molecular aspects of archaebacteria.

Zehnder, A. J. B. (ed.) 1988. *Environmental Microbiology of Anaerobes*. A collection of chapters detailing the microbiology of the major groups of anaerobes: anoxygenic phototrophs, sulfate-reducers, methanogens, acetogens, denitrifyers, etc.

Appendix

1

Energy Calculations

Definitions

1. ΔG^0 = standard free energy change of the reaction, at 1 atm pressure and 1 M concentrations; ΔG = free energy change under the conditions used; $\Delta G^{0\prime}$ = free energy change under standard conditions at pH 7.
2. Calculation of ΔG^0 for a chemical reaction from the free energy of formation, G_f^0, of products and reactants:
   $\Delta G^0 = \Sigma\, \Delta G_f^0$ (products) $- \Sigma\, \Delta G_f^0$ (reactants)
   That is, sum the ΔG_f^0 of products, sum the ΔG_f^0 of reactants, and subtract the latter from the former.
3. For energy-yielding reactions involving H^+, converting from standard conditions (pH 0) to biochemical conditions (pH 7):
   $\Delta G^{0\prime} = \Delta G^0 + m\, \Delta G'_f\, (H^+)$
   where m is the net number of protons in the reaction (m is negative when more protons are consumed than formed), and $\Delta G'_f\, (H^+)$ is the free energy of formation of a proton at pH 7 = −9.55 kcal at 25°C.
4. Effect of concentrations on ΔG: with soluble substrates, the concentration ratios of products formed to exogenous substrates used are generally equal to or greater than 10^{-2} at the beginning of growth, and equal to or less than 10^{-2} at the end of growth. From the relation between ΔG and equilibrium constant (see item 8 below), it can be calculated that ΔG for the free energy yield in practical situations differs from free energy yield under standard conditions by at most 2.8 kcal, a rather small amount, so that for a first approximation, standard free energy yields can be used in most situations. However, with H_2 as a product, H_2-consuming bacteria present may keep the concentration of H_2 so low that the free energy yield is significantly affected. Thus in the fermentation of ethanol to acetate and H_2, the $\Delta G^{0\prime}$ at 1 atm H_2 is +2.3 kcal, but at 10^{-4} atm H_2 it is −8.6 kcal. With H_2-consuming bacteria present, therefore, the ethanol fermentation becomes useful. (See also item 9 below.)
5. Reduction potentials: by convention, electrode equations are written in the direction, oxidant + $ne^- \longrightarrow$ reductant (that is, as reductions), where n is the number of electrons transferred. The standard potential (E_0) of the hydrogen electrode, $2H^+ + 2e^- \longrightarrow H_2$ is set by definition at 0.0 V at 1.0 atm pressure of H_2 gas and 1.0 M (molar) H^+, at 25°C. E'_0 is the standard reduction potential at pH 7. See also Table A1.2.
6. Relation of free energy to reduction potential:
   $\Delta G^{0\prime} = -nF\, \Delta E'_0$
   where n is the number of electrons transferred, F is the faraday constant (23,062 cal), and $\Delta E'_0$ is the E'_0 of the electron-accepting couple minus the E'_0 of the electron-donating couple.
7. Equilibrium constant, K. For the generalized reaction
   $aA + bB \rightleftharpoons cC + dD$

$$K = \frac{[C]^c[D]^d}{[A]^a[B]^b}$$

   where A, B, C, and D represent reactants and products, a, b, c, and d represent number of molecules of each; and brackets indicate concentrations.

8. Relation of equilibrium constant, K, to free energy change. At constant temperature and pressure,
$$\Delta G = \Delta G^0 + RT\ln K$$
where R is a constant ($1.98\ cal \cdot mol^{-1} \cdot {}^oK^{-1}$) and T is the absolute temperature (°K).
9. Two substances can be coupled in a redox reaction even if the standard potentials are unfavorable, provided that the concentrations are appropriate. Assume that normally the reduced form of A would donate electrons to the oxidized form of B. However, if the concentration of the reduced form of A were low and the concentration of the reduced form of B were high, it would be possible for the reduced form of B to donate electrons to the oxidized form of A. Thus the redox couple would function in the reverse direction from that predicted from standard potentials. A practical example of this is the utilization of H^+ as an electron acceptor to produce H_2. Normally, H_2 production in fermentative bacteria is not extensive because H^+ is a poor electron acceptor; the midpoint potential of the $2H^+/H_2$ pair is −0.42 V. However, if the concentration of H_2 is kept low by continually removing it (a process done by methane bacteria, which use $H_2 + CO_2$ to produce methane, CH_4), the potential will be more positive and then H^+ will serve as a suitable electron acceptor.

Oxidation State or Number

1. The oxidation state of an element in an elementary substance (i.e., H_2, O_2) is zero.
2. The oxidation state of the ion of an element is equal to its charge (i.e., $Na^+ = +1$, $Fe^{3+} = +3$, $O^{2-} = -2$).
3. The sum of oxidation numbers of all atoms in a neutral molecule is zero. Thus H_2O is neutral because it has two H at +1 each and one O at −2.
4. In an ion, the sum of oxidation numbers of all atoms is equal to the charge on that ion. Thus, in the OH^- ion, $O(-2) + H(+) = -1$.
5. In compounds, the oxidation state of O is virtually always −2, and that of H is +1.
6. In simple carbon compounds, the oxidation state of C can be calculated by adding up the H and O atoms present and using the oxidation states of these elements as given in item 5, since in a neutral compound the sum of all oxidation numbers must be zero. Thus the oxidation state of carbon in methane, CH_4, is −4 (4H at +1 each = −4); in carbon dioxide, CO_2, the oxidation state of carbon is +4 (2O at −2 each = −4).
7. In organic compounds with more than one C atom, it may not be possible to assign a specific oxidation number to each C atom, but it is still useful to calculate the oxidation state of the compound as a whole. The same conventions are used. Thus the oxidation state of carbon

Table A1.1 Free energies of formation G'_f, (kcal/mole)

Carbon compound	Metal	Nonmetal	Nitrogen compound
CO, −32.78	Cu^+, +12.0	H_2, 0	N_2, 0
CO_2, −94.25	Cu^{2+}, +15.5	H^+, 0 at pH 0	NO, +20.7
CH_4, −12.13	CuS, −11.7	−1.36 per pH	NO_2, +12.4
H_2CO_3, −148.94	Fe^{2+}, −20.3	O_2, 0	NO_2^-, −8.9
HCO_3^-, −140.26	Fe^{3+}, −2.5	OH^-, −37.6 at pH 14	NO_3^-, −26.6
CO_3^{2-}, −126.17	FeS_2, −36.0	−47.2 at pH 7	NH_4^+, −19.0
Formate, −83.9	$FeSO_4$, −198	−56.7 at pH 0	N_2O, +24.9
Methanol, −41.9	PbS, −22.1	H_2O, −56.7	
Acetate, −88.29	Mn^{2+}, −54.4	H_2O_2, −32.05	
Alanine, −88.8	Mn^{3+}, −19.6	PO_4^{3-}, −245	
Aspartate, −167	MnO_4^{2-}, −120.9	Se^o, 0	
Butyrate, −84.3	MnO_2, −109	H_2Se, −18.4	
Citrate, −279	$MnSO_4$, −228	SeO_4^{2-}, −105	
Ethanol, −43.4	HgS, −11.7	S^o, 0	
Formaldehyde, −31.2	MoS_2, −53.8	SO_3^{2-}, −116	
Glucose, −219	ZnS, −47.4	SO_4^{2-}, −178	
Glutamate, −167		$S_2O_3^{2-}$, −122	
Glycerol, −117		H_2S, −6.5	
Lactate, −124		HS^-, +2.88	
Malate, −202		S^{2-}, +20.5	
Pyruvate, −113			
Succinate, −165			
Urea, −48.7			

Values for free energy of formation of various compounds can be found in Dean, J. A. 1973. Lange's Handbook of Chemistry, 11th ed, McGraw-Hill, New York; Garrels, R. M., and C. L. Christ. 1965. Solutions, Minerals, and Equilibria. Harper & Row, New York; Burton, K. 1957. Appendix in *Krebs, H. A., and H. L. Komberg, Energy Transformations in Living Matter, Ergebnisse der Physiologie. Springer-Verlag, Berlin; and Thauer, R. K., K. Jungermann, and K. Decker. 1977. Bacteriol. Rev. 41:100–180.*

in glucose, $C_6H_{12}O_6$, is zero (12H at +1 = 12; 6O at −2 = −12) and the oxidation state of carbon in ethanol, C_2H_6O, is −4 (6H at +1 = +6; one O at −2).

8. In all oxidation-reduction reactions there is a balance between the oxidized and reduced products. To calculate an oxidation-reduction balance, the number of molecules of each product is multiplied by its oxidation state. For instance, in calculating the oxidation-reduction balance for the alcoholic fermentation, there are two molecules of ethanol at −4 = −8 and two molecules of CO_2 at +4 = +8, so that the net balance is zero. When constructing model reactions, it is useful to calculate redox balances to be certain that the reaction is possible.

Calculating Free Energy Yields for Hypothetical Reactions

Energy yields can be calculated either from free energy of formations of the reactants and products or from differences in reduction potentials of electron-donating and electron-accepting partial reactions.

Calculations from free energy Free energies of formation are given in Table A1.1. The procedure to use for calculating energy yields of reactions follows.

1. *Balancing reactions* In all cases, it is essential to ascertain that the coupled oxidation-reaction is balanced. Balancing involves three things: (*a*) the total number of each kind of atom must be identical on both sides of the equation; (*b*) there must be an ionic balance, so that when positive and negative ions are added up on the right side of the equation, the total ionic charge (whether positive, negative, or neutral) exactly balances the ionic charge on the left side of the equation; and (*c*) there must be an oxidation-reduction balance, so that all of the electrons removed from one substance must be transferred to another substance. In general, when constructing balanced reactions, one proceeds in the reverse of the three steps listed above. Usually, if steps (*c*) and (*b*) have been properly handled, step (*a*) becomes correct automatically.
2. *Examples* (*a*) What is the balanced reaction for the oxidation of H_2S to SO_4^{2-} with O_2? First, decide how many electrons are involved in the oxidation of H_2S to SO_4^{2-}. This can be most easily calculated from the oxidation states of the compounds, using the rules given above. Since H has an oxidation state of +1, the oxidation state of S in H_2S is −2. Since O has an oxidation state of −2, the oxidation state of S in SO_4^{2-} is +6 (since it is an ion, using the rules given in items 4 and 5 above). Thus the oxidation of H_2S to SO_4^{2-} involves an eight-electron transfer (from −2 to +6). Since each O atom can accept two electrons (the oxidation state of O in O_2 is zero, but in H_2O is −2), this means that two molecules of molecular oxygen O_2, will be required to provide sufficient electron-accepting capacity. Thus, at this point, we know that the reaction requires one H_2S and $2O_2$ on the left side of the equation, and one SO_4^{2-} on the right side. To achieve an ionic balance, we must have two positive charges on the right side of the equation to balance the two negative charges of SO_4^{2-}. Thus two H^+ must be added to the right side of the equation, making the overall reaction

$$H_2S + 2O_2 \rightarrow SO_4^{2-} + 2H^+$$

By inspection, it can be seen that this equation is also balanced in terms of the total number of atoms of each kind on each side of the equation.

(*b*) What is the balanced reaction for the oxidation of H_2S to SO_4^{2-} with Fe^{3+} as electron acceptor? We have just ascertained that the oxidation of H_2S to SO_4^{2-} is an eight-electron transfer. Since the reduction of Fe^{3+} to Fe^{2+} is only a one-electron transfer, $8Fe^{3+}$ will be required. At this point, the reaction looks like this:

$$H_2S + 8Fe^{3+} \rightarrow 8Fe^{2+} + SO_4^{2-} \qquad \text{(not balanced)}$$

We note that the ionic balance is incorrect. We have 24 positive charges on the left and 14 positive charges on the right (16+ from Fe, 2− from sulfate). To equalize the

Table A1.2 Reduction potentials of some redox pairs of biochemical and microbiological importance

Redox pair	E'_0(V)
SO_4^{2-}/HSO_3^-	−0.52
CO_2/formate	−0.43
H^+/H_2	−0.41
Ferredoxin ox/red	−0.39
Flavodoxin ox/red*	−0.37
$S_2O_3^{2-}/HS^- + HSO_3^-$	−0.36
NAD/NADH	−0.32
Cytochrome c_3 ox/red	−0.29
CO_2/acetate$^-$	−0.29
S^0/HS^-	−0.27
SO_4^{2-}/HS^-	−0.25
CO_2/CH_4	−0.24
FAD/FADH	−0.22
$HSO_3^-/S_3O_6^{2-}$	−0.20
Acetaldehyde/ethanol	−0.20
Pyruvate$^-$/lactate$^-$	−0.19
FMN/FMNH	−0.19
Dihydroxyacetone phosphate/glycerolphosphate	−0.19
Flavodoxin ox/red*	−0.12
HSO_3^-/HS^-	−0.11
Menaquinone ox/red	−0.075
APS/AMP + HSO_3^-	−0.060
Rubredoxin ox/red	−0.057
Acrylyl-CoA/propionyl-CoA	−0.015
Glycine/acetate$^-$ + NH_4^+	−0.010
$S_4O_6^{2-}/S_2O_3^{2-}$	+0.025
Fumarate/succinate	+0.030
Cytochrome *b* ox/red	+0.030
Ubiquinone ox/red	+0.11
Cytochrome c_1 ox/red	+0.23
$S_2O_6^{2-}/S_2O_3^{2-} + HSO_3^-$	+0.23
NO_2^-/NO	+0.36
Cytochrome a_3 ox/red	+0.385
NO_3^-/NO_2^-	+0.43
Fe^{3+}/Fe^{2+}	+0.77
O_2/H_2O	+0.82
NO/N_2O	+1.18
N_2O/N_2	+1.36

Data from Thauer, R. K., K. Jungermann, and K. Decker, 1977. Bacteriol. Rev. 41:100–180.
**Separate potentials are given for each electron transfer in this potentially two-electron transfer.*

charges, we add $10H^+$ on the right. Now our equation looks like this:

$H_2S + 8Fe^{3+} \rightarrow 8Fe^{2+} + 10H^+ + SO_4^{2-}$
(not balanced)

To provide the necessary hydrogen for the H^+ and oxygen for the sulfate, we add $4H_2O$ to the left and find that the equation is now balanced:

$H_2S + 4H_2O + 8Fe^{3+} \rightarrow 8Fe^{2+} + 10H^+ + SO_4^{2-}$
(balanced)

In general, in microbiological reactions, ionic balance can be achieved by adding H^+ or OH^- to the left or right side of the equation, and since all reactions take place in an aqueous medium, H_2O molecules can be added where needed. Whether H^+ or OH^- is added will depend upon whether the reaction is taking place in acid or alkaline conditions.

3. *Calculation of energy yield for balanced equations from free energies of formation* Once an equation has been balanced, the free energy yield can be calculated by inserting the values for the free energy of formation of each reactant and product from Table A1.1, and using the formula in item 2 of the first section of this Appendix.

   For instance, for the equation

$$H_2S + 2O_2 \rightarrow SO_4^{2-} + 2H^+$$

G'_f values $\rightarrow$ $-6.5 + 0 - 177 + 2 \times -9.1$
(assuming pH 7)

$\Delta G^{0\prime} = -188.7$ kcal/mole

The values on the right are summed and subtracted from the values on the left, taking care to ensure that the signs are correct. From the data in Table A1.1, a wide variety of free energy yields for reactions of microbiological interest can be calculated.

Calculation of free energy yield from reduction potential Reduction potentials of some important redox pairs are given in Table A1.2. The amount of energy that can be released when two half-reactions are coupled can be calculated from the differences in reduction potentials of the two reactions and from the number of electrons transferred. The further apart the two half-reactions are, and the greater the number of electrons, the more energy released. The conversion of potential difference to free energy is given by the formula $\Delta G^0 = -nF\,\Delta E'_0$, where n is the number of electrons, F is the faraday constant (23 kcal) and $\Delta E'_0$ is the difference in potentials. Thus the $2H^+/H_2$ redox pair has a potential of -0.41 and the $\frac{1}{2}O_2/H_2O$ pair has a potential of $+0.82$, so that the potential difference is 1.23, which (since two electrons are involved) is equivalent to a free energy yield (ΔG^0) of -57 kcal/mole. On the other hand, the potential difference between the $2H^+/H_2$ and the NO_3^-/NO_2^- reactions is less, 0.84 V, which is equivalent to a free energy yield of -39 kcal/mole.

Because many biochemical reactions are two-electron transfers, it is often useful to give energy yields for two-electron reactions, even if more electrons are involved. Thus the SO_4^{2-}/H_2 redox pair involves eight electrons, and complete reduction of SO_4^{2-} with H_2 would require $4H_2$ (equivalent to eight electrons). From the reduction potential difference between $2H^+/H_2$ and SO_4^{2-}/H_2S (0.16 V), a free-energy yield of -29 kcal/mole is calculated, or -7 kcal/mole per two electrons. By convention, reduction potentials are given for conditions in which equal concentrations of oxidized and reduced forms are present. In actual practice, the concentrations of these two forms may be quite different, As discussed earlier in this appendix (item 9, first section), it is possible to couple half-reactions even if the potential difference is unfavorable, providing the concentrations of the reacting species are appropriate.

Appendix 2

The Mathematics of Growth and Chemostat Operation

Exponential Growth

Although many analyses of microbial growth can be done graphically, as discussed in Chapter 9, for some purposes it is convenient to use a differential equation that expresses the quantitative relationships of growth:

$$\frac{dX}{dt} = \mu X \tag{1}$$

where X may be cell number or some specific cellular component such as protein and μ is the *instantaneous growth-rate constant.* If this equation is integrated, we obtain a form which reflects the activities of a typical batch culture population:

$$\ln X = \ln X_0 + \mu(t) \tag{2}$$

where ln refers to the natural logarithm (logarithm base e), X_0 is cell number at time 0, X is cell number at time t, and t is elapsed time during which growth is measured. This equation fits experimental data from the exponential phase of bacterial cultures very well, such as those in Table 9.1. Taking the antilogarithm of each side gives

$$X = X_0 e^{\mu t} \tag{3}$$

This equation is useful because it allows prediction of population density at a future time by knowing the present value and μ, the growth-rate constant. As discussed in Section 9.1, an important constant parameter for an exponentially growing population is the doubling (or generation) time. Doubling of the population has occurred when $X/X_0 = 2$. Rearranging and substituting this value into equation 3 gives

$$2 = e^{\mu(t_{gen})} \tag{4}$$

Taking the natural logarithm of each side and rearranging gives

$$\mu = \frac{\ln 2}{t_{gen}} = \frac{0.693}{t_{gen}} \tag{5}$$

The generation time t_{gen} may be used to define another growth parameter, k, as follows:

$$k = \frac{1}{t_{gen}} \tag{6}$$

where k is the growth-rate constant for a batch culture. Combining Equations 5 and 6 shows that the two growth-rate constants, μ and k, are related:

$$\mu = 0.693k$$

It is important to understand that μ and k are both reflections of the same growth process of an exponentially increasing population. The difference between them may be seen in their derivation above: μ is the *instantaneous rate constant* and k is an *average* value for the population over a finite period of time. (See Table A2.1 for calculation of k values from experimental results.) This distinction is more than a mathematical point. As was emphasized in Chapter 9, microbial growth studies must deal with population phenomena, not the activities of individual cells. The constant k reflects this averaging assumption. However, the constant μ, being instantaneous, is a closer approximation of the rate at which individual activities are occurring. Further, the instantaneous constant μ allows us to consider bacterial growth dynamics in a theoretical framework separate from the traditional batch culture.

Table A2.1. Calculation of k values

Bacterial population densities are expressed in scientific notation with powers of 10, so Equation 3 may be converted to terms of logarithm base 10 and k substituted for the instantaneous constant μ:

$$k = \frac{\log_{10}X_t - \log_{10}X_0}{0.301t}$$

Example 1:

$X_0 = 1000\ (= 10^3)\ \log_{10}$ of $1000 = 3$

$X_t = 100{,}000\ (= 10^5)\ \log_{10}$ of $100{,}000 = 5$

$t = 4$ hours

$$k = \frac{5 - 3}{(0.301)4} = \frac{2}{1.204}$$

$k = 1.66$ doublings per hour

$t_{gen} = 0.60$ hour (36 minutes) for population to double

Example 2:

$X_0 = 1000\ (= 10^3)\ \log_{10}$ of $1000 = 3$

$X_t = 100{,}000{,}000\ (= 10^8)\ \log_{10}$ of $10^8 = 8$

$t = 120$ hours

$$k = \frac{8 - 3}{(0.301)120} = \frac{5}{36.12}$$

$k = 0.138$ doubling per hour

$t_{gen} = 7.2$ hours (430 minutes) for population to double

Mathematical Relationships of Chemostats

An especially important application of the instantaneous growth-rate constant μ is, of course, the chemostat, a culture device (Section 9.6) in which population size and growth rate may be maintained at constant values of the experimenter's choosing, over a wide range of values. As we saw in Section 9.5, the rate of bacterial growth is a function of nutrient concentration. This function represents a saturation process which may be described by the equation

$$\mu = \mu_{max}\frac{S}{K_s + S} \tag{8}$$

where μ_{max} is the growth rate at nutrient saturation, S is the concentration of nutrient,* and K_s is a saturation constant which is numerically equal to the nutrient concentration at which $\mu = \frac{1}{2}\mu_{max}$. Equation 8 is formally equivalent to the Michaelis–Menten equation used in enzyme kinetic analyses. Estimates of the unknown parameters μ_{max} and K_s can be made by plotting $1/\mu$ on the y axis versus $1/s$ on the x axis. A straight line will be obtained which intercepts the y axis, and the value at the intercept is $1/\mu_{max}$. The slope of the line is K_s/μ_{max}, and since μ_{max} is known, K_s can then be calculated. Values for K_s as a rule are very low. A few examples are: 2.1×10^{-5} M for glucose (*Escherichia coli*), 1.34×10^{-6} M for oxygen (*Candida utilis*), 3.5×10^{-8} M for phosphate (*Spirillum* spp.), and 4.7×10^{-13} M for thiamin (*Cryptococcus albidus*).

The key to the operation of the chemostat is that, in a steady-state population, the nutrient concentration is very low, usually very near 0. Therefore, each drop of fresh medium entering the chemostat supplies nutrient which is consumed almost instantaneously by the population and the experimenter has direct control over the growth rate: if nutrient is added more rapidly, μ increases; if it is added more slowly, μ decreases. In a chemostat the rate at which medium enters is equal to the rate at which spent medium and cells exit through the overflow. This rate is called the dilution rate, D, and is defined:

$$D = \frac{F}{V} \tag{9}$$

*It is usual to speak of a *limiting nutrient*, that is, some component of the growth medium (often the electron donor) which, if increased in concentration, causes an increase in growth rate. One must be careful with this terminology, however, for, as Section 9.5 showed, there is usually observed a saturation level for each nutrient. At saturation values the particular nutrient under consideration no longer limits growth and some other component or factor becomes limiting.

where F is the flow rate (in units of volume per time, usually ml/hour) and V is chemostat volume (in ml). Therefore, D has units of time^{-1} (usually hour^{-1}) as does μ. In fact, at steady state, $\mu = D$. That μ must equal D may be seen by consideration of Equations 8 and 9 and the discussion above. When S is low, μ is a direct function of S. But since nutrient is instantly consumed upon addition, the value of S at any moment is controlled by the rate of medium addition, that is, the dilution rate D. A further crucial consequence of the relation $\mu = D$ is that a chemostat is a *self-regulating* system within broad limits of D. Consider a steady-state chemostat functioning with $\mu = D$. If μ were to increase momentarily due to some variation in the culture, S would necessarily decline, as the increased growth caused increased nutrient consumption. The lower value of S would in turn cause μ to decrease back to its stable level. Conversely, if μ were to decrease momentarily, then S would increase as "extra" nutrient went unconsumed. This rise in S would lead to higher values of μ until steady state is again reached. This biological feedback system functions well in practice, as it does in theory. Chemostat populations may be maintained at constant growth rates (μ) for long periods of time: months or even years.

There is another important aspect to chemostat studies and that is population density. The relation between cell growth and nutrient consumption may be defined:

$$Y = \frac{dX}{dS} \tag{10}$$

where Y is the yield constant for the particular organism on the particular medium and dX/dS is the amount of cell increase per unit of substrate consumed. The yield constant is in units of

$$Y = \frac{\text{weight of bacteria formed}}{\text{weight of substrate consumed}}$$

and is a measure of the efficiency with which cells convert nutrient to more cell material. The composition of the medium in the reservoir is central to the successful steady-state operation of the chemostat. All components in this medium are present in excess, except one, the growth-limiting substrate. The symbol for the concentration of that substrate in the reservoir is S_R, whereas that for the same substrate in the culture vessel is S. Taking into account Equations 1, 8, and 10, the following considerations can be made.

When a chemostat is inoculated with a small number of bacteria, and the dilution rate does not exceed a certain critical value (D_c), the number of organisms will start to increase. The increase is given by

$$\frac{dX}{dt} = \text{growth} - \text{output or } \frac{dX}{dt} = \mu X - DX \quad (11)$$

since initially $\mu > D$, dX/dt is positive. However, as the population density increases, the concentration of the growth-limiting substrate decreases, causing a decrease of μ (Equation 8). As discussed above, if D is kept constant, a steady state will be reached in which $\mu = D$ and $dX/dt = 0$.

Steady states are possible only when the dilution rate does not exceed a critical value D_c. This value depends on the concentration of the growth-limiting substrate in the reservoir (S_R):

$$D_c = \mu_{max}\left(\frac{S_R}{K_s + S_R}\right) \quad (12)$$

Thus, when $S_R \gg K_s$ (which is usually the case), steady states can be obtained at growth rates close to μ_{max}. The change in S is given by

$$\frac{dS}{dt} = \text{input} - \text{output} - \text{consumption}$$

Since consumption is growth divided by yield constant (see Equation 10), we have

$$\frac{dS}{dt} = S_R D - sD - \frac{\mu X}{Y} \quad (13)$$

When a steady state is reached, $dS/dt = 0$.

From Equations 11 and 13, the steady-state values of organism concentration ($\bar{X}$) and growth-limiting substrate ($\bar{S}$) can be calculated:

$$\bar{X} = Y(S_R - S) \quad (14)$$

$$\bar{S} = K_s\left(\frac{D}{\mu_{max} - D}\right) \quad (15)$$

These equations have the constants, S_R, Y, K_s, and μ_{max}. Once their values are known, the steady-state values of $\bar{X}$ and $\bar{S}$ can be predicted for any dilution rate. As can be seen from Equation 15, the steady-state substrate concentration ($\bar{S}$) depends on the dilution rate (D) applied, and is independent of the substrate concentration in the reservoir (S_R). Thus D determines $\bar{S}$, and $\bar{S}$ determines $\bar{X}$. A specific example of the use of the equations to predict chemostat behavior can be obtained by study of Figure 9.9. In that figure, the constants are $\mu_{max} = 1.0$ hour^{-1}; $Y = 0.5$; $K_s = 0.2$ g/liter; $S_R = 10$ g/liter. With these constants, the curves in Figure 9.9 can be generated.

Experimental Uses of the Chemostat

One of the major theoretical advantages of a chemostat is that this device allows the experimenter to control growth rate (μ) and population density ($\bar{X}$) independently of each other. Over rather wide ranges (see above and Section 9.6) any desired value of μ can be obtained by alteration of D. In practice this means changing the flow rate, since volume is usually fixed (Equation 9). Similarly, the population density $\bar{X}$ may be determined by varying nutrient concentration in the reservoir (Equation 14). This independent control of these two crucial growth parameters is not possible with conventional batch cultures.

A practical advantage to the chemostat is that a population may be maintained in a desired growth condition (μ and $\bar{X}$) for long periods of time. Therefore, experiments can be planned in detail and performed whenever most convenient. Comparable studies with batch cultures are essentially impossible, since specific growth conditions are constantly changing.

Mathematics of the Stationary Phase

The differential equation (Equation 1) expressing the exponential growth phase can be modified to include the trend toward the stationary phase, which occurs at high population density. A second term is added to the equation that expresses the maximum population attainable for that organism under the environmental conditions specified, here called X_m. The equation then assumes a form often called the *logistics equation*:

$$\frac{dX}{dt} = \mu X - \frac{\mu}{X_m}X^2 \quad (16)$$

The second term in this equation essentially expresses self-crowding effects, such as nutrient depletion or inhibitor buildup. In integrated form, this equation will graph as a typical microbial growth curve, showing exponential and stationary phases. The length of the exponential phase will be determined by the size of X_m. When X is small, the second term will have little effect, whereas as X approaches X_m, the growth rate (dX/dt) will approach zero. The logistics equation is widely used by ecologists studying higher organisms to express the growth rate and carrying capacity of the habitat for animals and plants. The equation is equally applicable to microbial situations.

Bergey's Classification of Bacteria

Bergey's Manual of Systematic Bacteriology is the recognized authority on bacterial taxonomy. The *Manual* is divided into four volumes. Each volume is broken into several sections, with each section containing a number of related genera. In brief, the contents of each volume is as follows:

Volume I **Gram-negative bacteria of medical and commercial importance:** spirochetes, spiral and curved bacteria, Gram-negative aerobic and facultatively aerobic rods, Gram-negative obligate anaerobes, Gram-negative aerobic and anaerobic cocci, sulfate and sulfur-reducing bacteria, rickettsias and chlamydias, mycoplasmas.

Volume II **Gram-positive bacteria of medical and commercial importance:** Gram-positive cocci, Gram-positive endospore-forming and nonsporing rods, mycobacteria, non-filamentous actinomycetes.

Volume III **Remaining Gram-negative bacteria:** phototrophic, gliding, sheathed, budding and appendaged bacteria, cyanobacteria, lithotrophic bacteria, and the archaeobacteria.*

Volume IV **Filamentous actinomycetes and related bacteria.**

The detailed list of genera in *Bergey's Manual of Systematic Bacteriology* follows.

Volume I

SECTION 1
The Spirochetes
Order I: Spirochaetales
- Family I: Spirochaetaceae
  - Genus I: *Spirochaeta*
  - Genus II: *Cristispira*
  - Genus III: *Treponema*
  - Genus IV: *Borrelia*
- Family II: Leptospiraceae
  - Genus I: *Leptospira*

Other Organisms
- Hindgut Spirochetes of Termites and *Cryptocercus punctulatus*

SECTION 2
Aerobic/Microaerophilic, Motile, Helical/Vibrioid Gram-Negative Bacteria
- Genus: *Aquaspirillum*
- Genus: *Spirillum*
- Genus: *Azospirillum*
- Genus: *Oceanospirillum*
- Genus: *Campylobacter*
- Genus: *Bdellovibrio*
- Genus: *Vampirovibrio*

SECTION 3
Nonmotile (or Rarely Motile), Gram-Negative Curved Bacteria
- Family I: Spirosomaceae
  - Genus I: *Spirosoma*
  - Genus II: *Runella*
  - Genus III: *Flectobacillus*

Other Genera
Genus: *Microcyclus*
Genus: *Meniscus*
Genus: *Brachyarcus*
Genus: *Pelosigma*

SECTION 4
Gram-Negative Aerobic Rods and Cocci
Family I: Pseudomonadaceae
Genus I: *Pseudomonas*
Genus II: *Xanthomonas*
Genus III: *Frateuria*
Genus IV: *Zoogloea*
Family II: Azotobacteraceae
Genus I: *Azotobacter*
Genus II: *Azomonas*
Family III: Rhizobiaceae
Genus I: *Rhizobium*
Genus II: *Bradyrhizobium*
Genus III: *Agrobacterium*
Genus IV: *Phyllobacterium*
Family IV: Methylococcaceae
Genus I: *Methylococcus*
Genus II: *Methylomonas*
Family V: Halobacteriaceae
Genus I: *Halobacterium*
Genus II: *Halococcus*
Family VI: Acetobacteraceae
Genus I: *Acetobacter*
Genus II: *Gluconobacter*
Family VII: Legionellaceae
Genus I: *Legionella*
Family VIII: Neisseriaceae
Genus I: *Neisseria*
Genus II: *Moraxella*
Genus III: *Acinetobacter*
Genus IV: *Kingella*
Other Genera
Genus: *Beijerinckia*
Genus: *Derxia*
Genus: *Xanthobacter*
Genus: *Thermus*
Genus: *Thermomicrobium*
Genus: *Halomonas*
Genus: *Alteromonas*
Genus: *Flavobacterium*
Genus: *Alcaligenes*
Genus: *Serpens*
Genus: *Janthinobacterium*
Genus: *Brucella*
Genus: *Bordetella*
Genus: *Francisella*
Genus: *Paracoccus*
Genus: *Lampropedia*

SECTION 5
Facultatively Anaerobic Gram-Negative Rods
Family I: Enterobacteriaceae
Genus I: *Escherichia*
Genus II: *Shigella*
Genus III: *Salmonella*
Genus IV: *Citrobacter*
Genus V: *Klebsiella*
Genus VI: *Enterobacter*
Genus VII: *Erwinia*
Genus VIII: *Serratia*
Genus IX: *Hafnia*
Genus X: *Edwardsiella*
Genus XI: *Proteus*
Genus XII: *Providencia*
Genus XIII: *Morganella*
Genus XIV: *Yersina*
Other Genera of the Family *Enterobacteriaceae*
Genus: *Obesumbacterium*
Genus: *Xenorhabdus*
Genus: *Kluyvera*
Genus: *Rahnella*
Genus: *Cedecea*
Genus: *Tatumella*
Family II: Vibrionaceae
Genus I: *Vibrio*
Genus II: *Photobacterium*
Genus III: *Aeromonas*
Genus IV: *Plesiomonas*
Family III: Pasteurellaceae
Genus I: *Pasteurella*
Genus II: *Haemophilus*
Genus III: *Actinobacillus*
Other Genera
Genus: *Zymomonas*
Genus: *Chromobacterium*
Genus: *Cardiobacterium*
Genus: *Calymmatobacterium*
Genus: *Gardnerella*
Genus: *Eikenella*
Genus: *Streptobacillus*

SECTION 6
Anaerobic Gram-Negative Straight, Curved and Helical Rods
Family I: Bacteroidaceae
Genus I: *Bacteroides*
Genus II: *Fusobacterium*
Genus III: *Leptotrichia*
Genus IV: *Butyrivibrio*
Genus V: *Succinimonas*
Genus VI: *Succinivibrio*
Genus VII: *Anaerobiospirillum*
Genus VIII: *Wolinella*
Genus IX: *Selenomonas*
Genus X: *Anaerovibrio*
Genus XI: *Pectinatus*
Genus XII: *Acetivibrio*
Genus XIII: *Lachnospira*

SECTION 7
Dissimilatory Sulfate- or Sulfur-Reducing Bacteria
Genus: *Desulfuromonas*
Genus: *Desulfovibrio*
Genus: *Desulfomonas*
Genus: *Desulfococcus*
Genus: *Desulfobacter*
Genus: *Desulfobulbus*
Genus: *Desulfosarcina*

SECTION 8
Anaerobic Gram-Negative Cocci
Family I: Veillonellaceae
Genus I: *Veillonella*
Genus II: *Acidaminococcus*
Genus III: *Megasphaera*

SECTION 9
The Rickettsias and Chlamydias
Order I: Rickettsiales
Family I: Rickettsiaceae
Tribe I: *Rickettsieae*
Genus I: *Rickettsia*
Genus II: *Rochalimaea*
Genus III: *Coxiella*
Tribe II: *Ehrlichieae*
Genus IV: *Ehrlichia*
Genus V: *Cowdria*
Genus VI: *Neorickettsia*
Tribe III: *Wolbachieae*
Genus VII: *Wolbachia*
Genus VIII: *Rickettsiella*
Family II: Bartonellaceae
Genus I: *Bartonella*
Genus II: *Grahamella*
Family III: Anaplasmataceae
Genus I: *Anaplasma*
Genus II: *Aegyptianella*
Genus III: *Haemobartonella*
Genus IV: *Eperythrozoon*
Order II: Chlamydiales
Family I: Chlamydiaceae
Genus I: *Chlamydia*

SECTION 10
The Mycoplasmas
Order I: Mycoplasmatales
Family I: Mycoplasmataceae
Genus I: *Mycoplasma*
Genus II: *Ureaplasma*
Family II: Acholeplasmataceae
Genus I: *Acholeplasma*
Family III: Spiroplasmataceae
Genus I: *Spiroplasma*
Other Genera
Genus: *Anaeroplasma*
Genus: *Thermoplasma*
Mycoplasma-like Organisms of Plants and Invertebrates

SECTION 11
Endosymbionts
A: Endosymbionts of Protozoa
Endosymbionts of ciliates
Endosymbionts of flagellates
Endosymbionts of amoebas
Taxa of endosymbionts:
Genus I: *Holospora*
Genus II: *Caedibacter*
Genus III: *Pseudocaedibacter*
Genus IV: *Lyticum*
Genus V: *Tectibacter*
B: Endosymbionts of Insects
Blood-sucking insects
Plant sap-sucking insects
Cellulose and stored grain feeders
Insects feeding on complex diets
Taxon of endosymbionts:
Genus: *Blattabacterium*
C: Endosymbionts of Fungi and Invertebrates other than Arthropods
Fungi
Sponges
Coelenterates
Helminthes
Annelids
Marine worms and mollusks

Volume II

SECTION 12
Gram-Positive Cocci
Family I: Micrococcaceae
Genus I: *Micrococcus*
Genus II: *Stomatococcus*
Genus III: *Planococcus*
Genus IV: *Staphylococcus*

Family II: Deinococcaceae
Genus I: *Deinococcus*
Other Genera
Genus: *Streptococcus*
Pyogenic Hemolytic Streptococci
Oral Streptococci
Enterococci
Lactic Acid Streptococci
Anaerobic Streptococci
Other Streptococci
Genus: *Leuconostoc*
Genus: *Pediococcus*
Genus: *Aerococcus*
Genus: *Gemella*
Genus: *Peptococcus*
Genus: *Peptostreptococcus*
Genus: *Ruminococcus*
Genus: *Coprococcus*
Genus: *Sarcina*

SECTION 13
Endospore-forming Gram-Positive Rods and Cocci
Genus: *Bacillus*
Genus: *Sporolactobacillus*
Genus: *Clostridium*
Genus: *Desulfotomaculum*
Genus: *Sporosarcina*
Genus: *Oscillospira*

SECTION 14
Regular, Nonsporing, Gram-Positive Rods
Genus: *Lactobacillus*
Genus: *Listeria*
Genus: *Erysipelothrix*
Genus: *Brochothrix*
Genus: *Renibacterium*
Genus: *Kurthia*
Genus: *Caryophanon*

SECTION 15
Irregular, Nonsporing, Gram-Positive Rods
Genus: *Corynebacterium*
Plant Pathogenic Species of *Corynebacterium*
Genus: *Gardnerella*
Genus: *Arcanobacterium*
Genus: *Arthrobacter*
Genus: *Brevibacterium*
Genus: *Curtobacterium*
Genus: *Caseobacter*
Genus: *Microbacterium*
Genus: *Aureobacterium*
Genus: *Cellulomonas*
Genus: *Agromyces*
Genus: *Arachnia*
Genus: *Rothia*
Genus: *Propionibacterium*
Genus: *Eubacterium*
Genus: *Acetobacterium*
Genus: *Lachnospira*
Genus: *Butyrivibrio*
Genus: *Thermoanaerobacter*
Genus: *Actinomyces*
Genus: *Bifidobacterium*

SECTION 16
The Mycobacteria
Family: Mycobacteriaceae
Genus: *Mycobacterium*

SECTION 17
Nocardioforms
Genus: *Nocardia*
Genus: *Rhodococcus*
Genus: *Nocardioides*
Genus: *Pseudonocardia*
Genus: *Oerskovia*
Genus: *Saccharopolyspora*
Genus: *Micropolyspora*
Genus: *Promicromonospora*
Genus: *Intrasporangium*

Volume III

SECTION 18
Anoxygenic Photosynthetic Bacteria
Purple bacteria Family I: Chromatiaceae
Genus I: *Chromatium*
Genus II: *Thiocystis*
Genus III: *Thiospirillum*
Genus IV: *Thiocapsa*
Genus V: *Lamprobacter*
Genus VI: *Lamprocystis*
Genus VII: *Thiodictyon*
Genus VIII: *Amoebobacter*
Genus IX: *Thiopedia*
Family II: Ectothiorhodospiraceae
Genus: *Ectothiorhodospira*
Purple Nonsulfur Bacteria
Genus: *Rhodospirillum*
Genus: *Rhodopila*
Genus: *Rhodobacter*
Genus: *Rhodopseudomonas*
Genus: *Rhodomicrobium*
Genus: *Rhodocyclus*
Green Bacteria
Green Sulfur Bacteria
Genus: *Chlorobium*
Genus: *Prosthecochloris*
Genus: *Pelodictyon*
Genus: *Ancalochloris*
Genus: *Chloroherpeton*
Symbiotic Consortia
Multicellular, Filamentous, Green Bacteria
Genus: *Chloroflexus*
Genus: *Heliothrix*
Genus: *Oscillochloris*
Genus: *Chloronema*
Genera Incertae Sedis
Genus: *Heliobacterium*
Genus: *Erythrobacter*

SECTION 19
Oxygenic Photosynthetic Bacteria
A: Cyanobacteria
Family I: Chroococcaceae
Family II: Pleurocapsaceae
Family III: Oscillatoriaceae
Genus I: *Spirulina*
Genus II: *Arthrospira*
Genus III: *Oscillatoria*
Genus IV: *Lyngbya*
Genus V: *Pseudanabaena*
Genus VI: *Starria*
Family IV: Nostocaceae
Genus I: *Anabaena*
Genus II: *Aphanizomenon*
Genus III: *Nodularia*
Genus IV: *Cylindrospermum*
Genus V: *Nostoc*
Genus VI: *Scytonema*
Genus VII: *Calothrix*
Family V: Stigonemataceae
Genus I: *Chlorogloeopsis*
Genus II: *Fischerella*
Genus III: *Geitleria*
B: **Order: Prochlorales**
Family: Prochloraceae
Genus: *Prochloron*

SECTION 20
Aerobic Chemolithotrophic Bacteria and Associated Organisms
A: Nitrifiers
Family: Nitrobacteraceae
Genus I: *Nitrobacter*
Genus II: *Nitrospina*
Genus III: *Nitrococcus*
Genus IV: *Nitrosospira*
Genus V: *Nitrosomonas*
Genus VI: *Nitrosococcus*
Genus VII: *Nitrosospira*
Genus VIII: *Nitrosolobus*
Genus IX: *Nitrosovibrio*
B: Colorless Sulfur Bacteria
Genus: *Thiobacterium*
Genus: *Macromonas*
Genus: *Thiospira*
Genus: *Thiovulum*
Genus: *Thiobacillus*
Genus: *Thiomicrospira*
Genus: *Thiosphaera*
Genus: *Acidiphilium*
Genus: *Thermothrix*
C: Obligate Hydrogen Oxidizers
Genus: *Hydrogenobacter*
D: Iron and Manganese Oxidizing and/or Depositing Bacteria
Family: Siderocapsaceae
Genus I: *Siderocapsa*
Genus II: *Naumanniella*
Genus III: *Siderococcus*
Genus IV: *Ochrobium*
E. The Magnetotactic Bacteria
Genus: *Aquaspirillum*
Genus: *Bilophococcus*

SECTION 21
Budding and/or Appendaged Bacteria
A. Prosthecate Bacteria
1. Budding Bacteria
Genus: *Hyphomicrobium*
Genus: *Hyphomonas*
Genus: *Pedomicrobium*
Genus: *Ancalomicrobium*
Genus: *Prosthecomicrobium*
Genus: *Stella*
Genus: *Labrys*
2. Non-budding Bacteria
Genus: *Caulobacter*
Genus: *Asticcacaulis*
Genus: *Prosthecobacter*
Genus: *Thiodendron*
B. Non-prosthecate Bacteria
1. Budding Bacteria
Genus: *Planctomyces*
Genus: *Isosphaera*
Genus: *Blastobacter*
Genus: *Angulomicrobium*
Genus: *Gemmiger*
Genus: *Ensifer*

2. Non-budding Bacteria
Genus: *Gallionella*
Genus: *Nevskia*

C. Morphologically Unusual Budding Bacteria (involved in iron and manganese deposition)
Genus: *Seliberia*
Genus: *Metallogenium*
Genus: *Caulococcus*
Genus: *Kuznezovia*

D. Others
Spinate bacteria

SECTION 22
Sheathed Bacteria
Genus: *Sphaerotilus*
Genus: *Leptothrix*
Genus: *Haliscominobacter*
Genus: *Lieskeella*
Genus: *Phragmidiothrix*
Genus: *Crenothrix*
Genus: *Clonothrix*

SECTION 23
Non-Photosynthetic, Non-Fruiting, Gliding Bacteria
Order I: Cytophagales
Family: Cytophagaceae
Genus I: *Cytophaga*
Genus II: *Sporocytophaga*
Genus III: *Capnocytophaga*
Genus IV: *Flexithrix*
Genus V: *Flexibacter*
Genus VII: *Saprospira*
Genus VIII: *Herpetosiphon*
Order II: Lysobacterales
Family: Lysobacteriaceae
Genus: *Lysobacter*
Order III: Beggiatoales
Family: Beggiatoaceae
Genus I: *Beggiatoa*
Genus II: *Thioploca*
Genus III: *Thiospirillopsis*
Genus IV: *Thiothrix*
Others
Family I: Simonsiellaceae
Genus I: *Simonsiella*
Genus II: *Alysiella*
Family II: Pelonemataceae
Genus I: *Pelonema*
Genus II: *Peloploca*
Genus III: *Achroonema*
Genus IV: *Desmanthus*
Genera Incertae Sedis
Genus: *Chitinophaga*
Genus: *Microscilla*
Genus: *Herpetosiphon*
Genus: *"Toxothrix"*
Genus: *Leucothrix*
Genus: *Vitreoscilla*
Genus: *Desulfonema*
Genus: *Agitococcus*

SECTION 24
Gliding, Fruiting Bacteria
Order: Myxobacterales
Family I: Myxococcaceae
Genus: *Myxococcus*
Family II: Archangiaceae
Genus: *Archangium*
Family III: Cystobacteraceae
Genus I: *Cystobacter*
Genus II: *Melittangium*
Genus III: *Stigmatella*
Family IV: Polyangiaceae
Genus I: *Polyangium*
Genus II: *Nannocystis*
Genus III: *Chondromyces*

SECTION 25
Archaeobacteria*
Order I: Methanobacteriales
Family I: Methanobacteriaceae
Genus I: *Methanobacterium*
Genus II: *Methanobrevibacter*
Family II: Methanothermaceae
Genus: *Methanothermus*
Order II: Methanococcales
Family: Methanococcaceae
Genus: *Methanococcus*
Order III: Methanomicrobiales
Family I: Methanomicrobiaceae
Genus I: *Methanomicrobium*
Genus II: *Methanogenium*
Genus III: *Methanospirillum*
Family II: Methanosarcinaceae
Genus I: *Methanosarcina*
Genus II: *Methanococcoides*
Genus III: *Methanolobus*
Genus IV: *Methanothrix*
Genus V: *Halomethanococcus*
Family III: Methanoplanaceae
Genus: *Methanoplanus*
Others
Genus: *Methanosphaera*
Order IV: Thermoproteales
Family I: Thermoproteaceae
Genus I: *Thermoproteus*
Genus II: *Thermophilum*
Family II: Desulfurococcaceae
Genus I: *Desulfurococcus*
Genus II: *Thermococcus*
Genus III: *Thermodiscus*
Genus IV: *Pyrodictium*
Order V: Thermococcales
Family: Thermococcaceae
Genus I. *Thermococcus*
Genus II. *Pyrococcus*
Order VI: Sulfolobales
Family: Sulfolobaceae
Genus: *Sulfolobus*
Order VII: Thermoplasmales
Family: Thermoplasmaceae
Genus: *Thermoplasma*
Order VIII: Halobacteriales
Family: Halobacteriaceae
Genus I: *Halococcus*
Genus II: *Halobacterium*
Genus III: *Haloferax*
Genus IV: *Haloarcula*
Genus V: *Natronococcus*
Genus VI: *Natronobacterium*

*For the archaebacteria, the combining form archae*o* (from the Greek word *archaio*, meaning *ancient*) will be used in *Bergey's Manual* in order to fulfill certain rules of nomenclature. However, since the term *archaebacteria* is so firmly ingrained in the scientific literature, we will use this spelling throughout this book.

Volume IV

SECTION 26
Nocardioform Actinomycetes
Genus: *Nocardia*
Genus: *Rhodococcus*
Genus: *Nocardioides*
Genus: *Pseudonocardia*
Genus: *Oerskovia*
Genus: *Saccharopolyspora*
Genus: *Faenia (Micropolyspora)*
Genus: *Promicromonospora*
Genus: *Intrasporangium*
Genus: *Actinopolyspora*
Genus: *Saccharomonospora*

SECTION 27
Actinomycetes with Multi-Locular Sporangia
Genus: *Geodermatophilus*
Genus: *Dermatophilus*
Genus: *Frankia*

SECTION 28
Actinoplanetes
Genus: *Actinoplanes*
Genus: *Ampullariella*
Genus: *Pilimelia*
Genus: *Dactylosporangium*
Genus: *Micromonospora*

SECTION 29
Streptomyces and Related Genera
Genus: *Streptomyces*
Genus: *Streptoverticillium*
Genus: *Kineosporia*
Genus: *Sporichthya*

SECTION 30
Maduromycetes
Genus: *Actinomadura*
Genus: *Microbispora*
Genus: *Microtetraspora*
Genus: *Planobispora*
Genus: *Planomonospora*
Genus: *Spirillospora*
Genus: *Streptosporangium*

SECTION 31
Thermomonospora and Related Genera
Genus: *Thermomonospora*
Genus: *Actinosynnema*
Genus: *Nocardiopsis*
Genus: *Streptoalloteichus*

SECTION 32
Thermoactinomycetes
Genus: *Thermoactinomyces*

SECTION 33
Other Genera
Genus: *Glycomyces*
Genus: *Kibdelosporangium*
Genus: *Kitasatosporia*
Genus: *Saccharothrix*

Glossary

Only the major terms and concepts are included. If a term is not here, consult the index.

Acid-alcohol fastness A staining property of *Mycobacterium* species where cells stained with hot carbolfuschin will not decolorize with acid-alcohol.
Activation energy Energy needed to make substrate molecules more reactive; enzymes function by lowering activation energy.
Active immunity An immune state achieved by self-production of antibodies. Compare with passive immunity.
Aerobe An organism that grows in the presence of O_2; may be facultative or obligate.
Aerosol Suspension of particles in airborne water droplets.
Aerotolerant Of an anaerobe, not being inhibited by O_2.
Agglutination Reaction between antibody and cell-bound antigen resulting in clumping of the cells.
Allergy A harmful antigen-antibody reaction, usually caused by a foreign antigen in food, pollen, or chemicals; immediate-type hypersensitivity.
Allosteric Characteristic of some proteins, especially enzymes, in which a compound combines with a site on the protein other than the active site. The result is a change in conformation at the active site; useful in regulating activity.
Anabolism The biochemical processes involved in the synthesis of cell constituents from simpler molecules, usually requiring energy.
Anaerobe An organism that grows in the absence of oxygen.
Anaerobic respiration Use of an electron acceptor other than O_2 in an electron-transport oxidation. Most common anaerobic electron acceptors are nitrate, sulfate, and carbonate.
Anaphylatoxins The C3a and C5a fractions of complement which act to mimic some of the reactions of anaphylaxis.
Anaphylaxis (anaphylactic shock) A violent allergic reaction caused by an antigen-antibody reaction.
Anoxygenic photosynthesis Use of light energy to synthesize ATP by cyclic photophosphorylation without O_2 production in green and purple bacteria.
Antibiotic A chemical agent produced by one organism that is harmful to other organisms.
Antibody A protein present in serum or other body fluid that combines specifically with antigen.
Anticodon A sequence of three purine and pyrimidine bases in transfer RNA that is complementary to the codon in messenger RNA.
Antigen A substance, usually macromolecular, that induces specific antibody formation.
Antigenic determinants Portions of an antigen which elicit their own immune response.
Antimicrobial Harmful to microorganisms by either killing or inhibiting growth.
Antiparallel In reference to the structure of DNA, the orientation of the two strands of DNA (one strand runs $5' \rightarrow 3'$, the other $3' \rightarrow 5'$).
Antiseptic An agent that kills or inhibits growth, but is not harmful to human tissue.
Antiserum A serum containing antibodies.
Antitoxin An antibody active against a toxin.
Archaebacteria An evolutionarily distinct group of procaryotes, including the methanogenic, extremely halophilic, and sulfur-dependent bacteria.
Attenuation Selection from a pathogen of nonvirulent strains still capable of immunizing. Also: a process that plays a role in the regulation of enzymes involved in amino acid biosynthesis.
Autoimmunity Immune reactions of a host against its own self constituents.
Autolysis Spontaneous lysis.
Autotroph Organism able to utilize CO_2 as sole source of carbon.
Auxotroph A mutant that has a growth factor requirement. Contrast with a prototroph.
Axenic Pure, uncontaminated; an axenic culture is a pure culture.
Bacteremia The presence of bacteria in the blood.
Bactericidal Capable of killing bacteria.
Bacteriocin Agents produced by certain bacteria that inhibit or kill closely related species.
Bacteriophage A virus that infects bacteria.
Bacteriostatic Capable of inhibiting bacterial growth without killing.
Bacteroid A swollen, deformed *Rhizobium* cell, found in the root nodule.
Barophile An organism able to live optimally at high hydrostatic pressure.
Barotolerant An organism able to tolerate high hydrostatic pressure, although growing better at normal pressures.
Biosynthesis The production of needed cellular constituents from other (usually simpler) molecules.
Calorie A unit of heat or energy; that amount of heat required to raise the temperature of 1 gram of water by 1°C.
Capsid The protein coat of a virus.
Capsomere An individual protein subunit of the virus capsid.
Capsule A compact layer of polysaccharide exterior to the cell wall in some bacteria. See also *Glycocalyx* and *Slime layer*.

Carcinogen A substance that causes the initiation of tumor formation. Frequently a mutagen.
Carrier An individual that continually releases infective organisms but does not show symptoms of disease.
Catabolism The biochemical processes involved in the breakdown of organic compounds, usually leading to the production of energy.
Catabolite repression Repression of a variety of unrelated enzymes when cells are grown in a medium containing glucose.
Catalyst A substance that promotes a chemical reaction without itself being changed in the end.
Cell-mediated immunity An immune response generated by the activities of non-antibody-producing cells. Compare with *Humoral immunity*.
Chemiosmosis The use of ion gradients across membranes, especially proton gradients, to generate ATP. See *Proton-motive force*.
Chemoprophylaxis The use of chemicals or antibodies to prevent future infection or disease.
Chemostat A continuous culture device controlled by the concentration of limiting nutrient.
Chemotaxis Movement toward or away from a chemical.
Chemotherapy Treatment of infectious disease with chemicals or antibiotics.
Chromogenic Producing color; a chromogenic colony is a pigmented colony.
Chromosome The structure that contains the DNA in eucaryotes, usually complexed with histones.
Cilium Short, filamentous structure that beats with many others to make a cell move.
Cistron A gene which codes for a protein product.
Clone A population of cells all descended from a single cell; a number of copies of a DNA fragment obtained by allowing the inserted DNA fragment to be replicated by a phage or plasmid.
Cloning vector A DNA molecule that is able to bring about the replication of foreign DNA fragments.
Coccoid Sphere-shaped.
Coccus A spherical bacterium.
Codon A sequence of three purine and pyrimidine bases on DNA or messenger RNA that codes for a specific amino acid.
Coenzyme A low-molecular-weight chemical which participates in an enzymatic reaction by accepting and donating electrons or functional groups. Examples: NAD, FAD.
Coliforms Gram-negative, nonsporing, facultative rods that ferment lactose with gas formation within 48 hours at 35°C.
Colony A population of cells growing on solid medium, arising from a single cell.
Competence Ability to take up DNA and become genetically transformed.
Complement A complex of proteins in the blood serum that acts in concert with specific antibody in certain kinds of antigen-antibody reactions.
Complementary Nonidentical but related genetic structures that show precise base pairing.
Concatamer A DNA molecule consisting of two or more separate molecules linked end-to-end to form a long linear structure.
Conjugation In eucaryotes, the process by which haploid gametes fuse to form a diploid zygote; in procaryotes, transfer of genetic information from one cell to another by cell-to-cell contact.
Consortium A two-membered bacterial culture (or natural assemblage) in which each organism benefits from the other.
Contagious Of a disease, transmissible.
Cortex The region inside the spore coat of an endospore, around the core.
Covalent A nonionic chemical bond formed by a sharing of electrons between two atoms.
Crista Inner membrane in a mitochondrion, site of respiration.
Cryophile Psychrophile.
Culture A particular strain or kind of organism growing in a laboratory medium.
Cutaneous Relating to the skin.
Cyst A resting stage formed by some bacteria and protozoa in which the whole cell is surrounded by protective layer; not the same as spore.
Cytochrome Iron-containing porphyrin rings complexed with proteins, which act as electron carriers in the electron-transport system.
Cytoplasm Cellular contents inside the plasma membrane, excluding the nucleus.
Degeneracy In relation to the genetic code, the fact that more than one codon can code for the same amino acid.
Deletion A removal of a portion of a gene.
Denaturation Irreversible destruction of a macromolecule, as for example the destruction of a protein by heat.
Denitrification Conversion of nitrate into nitrogen gases under anaerobic conditions, resulting in loss of nitrogen from ecosystems.
Deoxyribonucleic acid (DNA) A polymer of nucleotides connected via a phosphatedeoxyribose sugar backbone; the genetic material of the cell.
Desiccation Drying.
Diauxic growth Growth occurring in two separate phases between which a temporary lag occurs.
Differentiation The modification of a cell in terms of structure and or function occurring during the course of development.
Diploid In eucaryotes, an organism or cell with two chromosome complements, one derived from each haploid gamete.
Disinfectant An agent that kills microorganisms, but may be harmful to human tissue.
Doubling time The time needed for a population to double. See also *Generation time*.
Electron acceptor A substance that accepts electrons during an oxidation-reduction reaction. An electron acceptor is an oxidant.
Electron donor A compound that donates electrons in an oxidation-reduction reaction. An electron donor is a reductant.
Electron-transport phosphorylation Synthesis of ATP involving a membrane-associated electron transport chain and the creation of a proton-motive force. Previously termed oxidative phosphorylation. See also *Chemiosmosis*.
Endemic A disease that is constantly present in low numbers in a population. Compare with *Epidemic*.
Endergonic A chemical reaction requiring input of energy to proceed.
Endocytosis A process in which a particle such as a virus is taken intact into an animal cell. Phagocytosis and pinocytosis are two kinds of endocytosis.
Endoplasmic reticulum An extensive array of internal membranes in eucaryotes.

Endospore A bacterial spore formed within the cell and extremely resistant to heat as well as to other harmful agents.
Endosymbiosis The hypothesis that mitochondria and chloroplasts are the descendants of ancient procaryotic organisms.
Endotoxin A toxin not released from the cell; bound to the cell surface or intracellular. Compare with *Exotoxin*.
Enteric Intestinal.
Enterotoxin A toxin affecting the intestine.
Enzyme A protein functioning as the catalyst of living organisms, which promotes specific reactions or groups of reactions.
Epidemic A disease occurring in an unusually high number of individuals in a community at the same time. Compare with *Endemic*.
Epidemiology The study of the incidence and prevalence of disease in populations.
Eubacteria All bacteria other than the archaebacteria.
Eucaryote A cell or organism having a true nucleus.
Exergonic reaction A chemical reaction that proceeds with the liberation of energy.
Exon The coding sequences in a split gene. Contrast with *Introns*, the intervening noncoding regions.
Exotoxin A toxin released extracellularly. Compare with *Endotoxin*.
Exponential phase A period during the growth cycle of a population in which growth increases at an exponential rate.
Expression The ability of a gene to function within a cell in such a way that the gene product is formed.
Expression vector Cloned DNA containing both the desired DNA and any necessary regulatory genes to allow the production of the desired gene product.
Facultative A qualifying adjective indicating that an organism is able to grow either in the presence or absence of an environmental factor (e.g., "facultative aerobe," "facultative psychrophile").
Feedback inhibition Inhibition by an end product of the biosynthetic pathway involved in its synthesis.
Fermentation Catabolic reactions producing ATP in which organic compounds serve as both primary electron donor and ultimate electron acceptor.
Fermentation (industrial) A large-scale microbial process.
Fermenter An organism which carries out the process of fermentation.
Fermentor A large growth vessel used to culture microorganisms on a large scale frequently for the production of some commercially valuable product.
Ferredoxin An electron carrier of low reduction potential; small protein containing iron.
Fever A rise of body temperature above the normal.
Filamentous In the form of very long rods, many times longer than wide.
Fimbria (plural fimbriae) Short filamentous structure on a bacterial cell; although flagella-like in structure, generally present in many copies and not involved in motility. Plays role in adherence to surfaces and in the formation of pellicles. See also *Pilus*.
Flagellum An organ of motility.
Flavoprotein A protein containing a derivative of riboflavin, which acts as electron carrier in the electron-transport system.
Fluorescent Having the ability to emit light of a certain wavelength when activated by light of another wavelength.
Fluorescent antibody Immunoglobulin molecule which has been coupled with a fluorescent dye so that it exhibits the property of fluorescence.
Frame shift Since the genetic code is read three bases at a time, if reading begins at either the second or third base of a codon, a faulty product usually results. This is called a frame shift (the reading frame refers to the pattern of reading).
Free energy Energy available to do useful work.
Fruiting body A macroscopic reproductive structure produced by some fungi (e.g., mushrooms) and some bacteria (e.g., myxobacteria). Fruiting bodies are distinct in size, shape, and coloration for each species.
Gametes In eucaryotes, haploid cells analogous to sperm and egg, which conjugate and form a diploid.
Gas vesicle A gas-filled structure in certain procaryotes that confers ability to float. Sometimes called *gas vacuole*.
Gene A unit of heredity; a segment of DNA specifying a particular protein or polypeptide chain.
Gene cloning The isolation of a desired gene from one organism and its incorporation into a suitable vector for the production of large amounts of the gene.
Generation time Time needed for a population to double. See also *Doubling time*.
Genetic map The physical arrangement and order of gene loci on the chromosome.
Genome The complete set of genes present in an organism.
Genotype The genetic complement of organisms. Compare with *Phenotype*.
Genus A group of related species.
Germicide A substance that inhibits or kills microorganisms.
Glycocalyx General term for polysaccharide components outside the bacterial cell wall. See also *Capsule* and *Slime layer*.
Glycolysis Reactions of the Embden-Meyerhof pathway in which glucose is oxidized to pyruvate.
Growth rate The rate at which growth occurs, usually expressed as the generation time.
Halophile Organism requiring salt (usually NaCl) for growth.
Haploid In eucaryotes, an organism or cell containing one chromosome complement and the same number of chromosomes as the gametes.
Hapten A substance not inducing antibody formation but able to combine with a specific antibody.
Hemagglutination Agglutination of red blood cells.
Hemolysins Bacterial toxins capable of lysing cells including red blood cells.
Hemolysis Lysis of red blood cells.
Heteroduplex A double-stranded DNA in which one strand is from one source and the other strand from another, usually related, source.
Heterofermentation Fermentation of glucose or other sugar to a mixture of products.
Heterotroph Organism obtaining energy and carbon from organic compounds.
Homofermentation Fermentation of glucose or other sugar leading to virtually a single product, lactic acid.
Homologous antigen An antigen reacting with the antibody it had induced.
Host An organism capable of supporting the growth of a virus or parasite.
Humoral immunity An immune response involving the activities of antibodies.

Hybridoma The fusion of a malignant cell with a single B-lymphocyte to produce a malignant lymphocyte producing monoclonal antibody.
Hydrolysis Breakdown of a polymer into smaller units, usually monomers, by addition of water; digestion.
Hydrophobic interactions Attractive forces between molecules due to the close positioning of nonhydrophilic portions of the two molecules.
Hydrothermal vents Warm or hot water emitting springs associated with crustal spreading centers on the sea floor.
Hypersensitivity An immune reaction, usually harmful to the animal, caused either by antigen-antibody reaction (see *Allergy*) or cellular-immune processes.
Icosahedron A geometrical shape occurring in many virus particles, with 20 triangular faces and 12 corners.
Immune Able to resist infectious disease.
Immunization Induction of specific immunity by injecting antigen or antibodies.
Immunogen An antigen that will react with an antibody or antibodies made against it.
Immunoglobulin Antibody.
Immunological Refers to processes of antibody formation and antigen-antibody reactions, whether or not immunity to infectious disease results.
In vivo In the body, in a living organism.
In vitro In glass, in culture.
Incidence In reference to disease transmission, the number of cases of the disease in a specific subset of the population.
Induced enzyme An enzyme subject to induction.
Induction The process by which an enzyme is synthesized in response to the presence of an external substance, the inducer.
Infection Growth of an organism within the body.
Inflammation Characteristic reaction to foreign particles and stimuli, resulting in redness, swelling, heat, and pain.
Inhibition Prevention of growth or function.
Inoculum Material used to initiate a microbial culture.
Insertion A genetic phenomenon in which a piece of DNA is inserted into the middle of a gene.
Insertion element Specific nucleotide sequences that are involved in the transfer and integration of pieces of DNA.
Integration The process by which a DNA molecule becomes incorporated into the genome.
Interferon A protein produced by cells as a result of virus infection which interferes with virus replication.
Interspecies hydrogen transfer The process in which organic matter is degraded anaerobically by the interaction of several groups of microbes in which H_2 production and H_2 consumption are intimately intertwined.
Intron The intervening noncoding regions in a split gene. Contrasted with *Exons*, the coding sequences.
Invasiveness Degree to which an organism is able to spread through the body from a focus of infection.
Ionophore A compound which can cause the leakage of ions across membranes.
Isotopes Different forms of the same element containing the same number of protons and electrons, but differing in the number of neutrons.
Lag phase The period after inoculation of a population before growth begins.
Latent virus A virus present in a cell, yet not causing any detectable effect.
Leukocidin A substance able to destroy phagocytes.
Leukocyte A white blood cell, usually a phagocyte.
Lipid Water-insoluble molecules important in structure of cell membrane and (in some organisms) cell wall. See also *Phospholipid*.
Lipopolysaccharide (LPS) Complex lipid structure containing unusual sugars and fatty acids found in many Gram-negative bacteria, and constituting the chemical structure of the outer layer.
Lithotroph An organism that can obtain its energy from oxidation of inorganic compounds.
Lophotrichous Having a tuft of polar flagella.
Luminescence Production of light.
Lymph A clear yellowish fluid found in the lymphatic vessels and which carries various white (but not red) blood cells.
Lymphocyte A white blood cell involved in antibody formation or cellular immune response.
Lymphokines Substances secreted from T-lymphocytes which stimulate the activity of macrophages.
Lysin An antibody that induces lysis.
Lysis Rupture of a cell, resulting in loss of cell contents.
Lysogeny The hereditary ability to produce virus. See also *Temperate virus*.
Lysosome A cell organelle containing digestive enzymes.
Macromolecule A large molecule formed from the connection of a number of small molecules. A polymer. Informational macromolecules play a role in transfer of genetic information.
Macrophage Large phagocytic cells involved in both phagocytosis and the antibody production process.
Magnetotaxis Directed movement of bacterial cells containing magnetite crystals by a magnetic field.
Major histocompatability complex A cluster of genes coding for key cell surface proteins important in cell-cell recognition processes.
Malignant In reference to a tumor, an infiltrating metastasizing growth no longer under normal growth control.
Mast cell Tissue cells adjoining blood vessels throughout the body, contain granules with inflammatory mediators.
Meiosis In eucaryotes, reduction division, the process by which the change from diploid to haploid occurs.
Membrane Any thin sheet or layer. See especially *Plasma membrane*.
Memory cell A differentiated B lymphocyte capable of rapid conversion to a plasma cell upon subsequent stimulation with antigen.
Mesophile Organism living in the temperature range around that of warm-blooded animals.
Messenger RNA (mRNA) An RNA molecule containing a base sequence complementary to DNA; directs the synthesis of protein.
Metabolism All biochemical reactions in a cell, both anabolic and catabolic.
Microaerophilic Requiring O_2 but at a level lower than atmospheric.
Micrometer One-millionth of a meter, or 10^{-6} m (abbreviated μm), the unit used for measuring microbes.
Microtubules Tubes that are the structural entity for eucaryotic flagella, have a role in maintaining cell shape, and function as mitotic spindle fibers.
Mitochondrion Eucaryotic organelle responsible for processes of respiration and electron-transport phosphorylation.
Mitosis A highly ordered process by which the nucleus divides in eucaryotes.

Mixotroph An organism able to assimilate organic compounds as carbon sources while using inorganic compounds as electron donors.
Monoclonal antibody A monospecific antibody.
Monocyte Circulating white blood cells which contain many lysosomes and can differentiate into macrophages.
Monomer A building block of a polymer.
Monotrichous The state of having a single polar flagellum.
Morbidity Incidence of disease in a population, including both fatal and nonfatal cases.
Mortality Incidence of death in a population.
Motility The property of movement of a cell under its own power.
Mutagen An agent that induces mutation, such as radiation or chemicals.
Mutant A strain differing from its parent because of mutation.
Mutation A sudden inheritable change in the phenotype of an organism.
Myeloma A malignant tumor of a plasma cell (antibody-producing cell).
Nitrification The conversion of ammonia to nitrate.
Nitrogen fixation Reduction of nitrogen gas to ammonia.
Nodule A tumorlike structure produced by the roots of symbiotic nitrogen-fixing plants. Contains the nitrogen-fixing microbial component of the symbiosis.
Nonsense mutation A mutation that changes a normal codon into one which does not code for an amino acid.
Nucleic acid A polymer of nucleotides. See *Deoxyribonucleic acid* and *Ribonucleic acid.*
Nucleoside A nucleotide minus phosphate.
Nucleotide A monomeric unit of nucleic acid, consisting of a sugar, phosphate, and nitrogenous base.
Nucleus A membrane-enclosed structure containing the genetic material (DNA) organized in chromosomes.
Nutrient A substance taken by a cell from its environment and used in catabolic or anabolic reactions.
Obligate A qualifying adjective referring to an environmental factor always required for growth (e.g., "obligate anaerobe").
Oligotrophic Describing a body of water in which nutrients are in low supply.
Open reading frame The entire length of a DNA molecule that starts with a start codon and ends with a stop codon.
Operator A specific region of the DNA at the initial end of the gene, where the repressor protein attaches and blocks mRNA synthesis.
Operon A cluster of genes whose expression is controlled by a single operator.
Opsonization Promotion of phagocytosis by a specific antibody in combination with complement.
Organelle A membrane-enclosed body specialized for carrying out certain functions.
Organotroph Heterotroph.
Osmosis Diffusion of water through a membrane from a region of low solute concentration to one of higher concentration.
Oxidation A process by which a compound gives up electrons, acting as an electron donor, and becomes oxidized.
Oxidation-reduction (redox) reaction A coupled pair of reactions, in which one compound becomes oxidized, while another becomes reduced and takes up the electrons released in the oxidation reaction.
Oxygenic photosynthesis Use of light energy to synthesize ATP and NADPH by noncyclic photophosphorylation with the production of O_2 from water.
Palindrome A nucleotide sequence on a DNA molecule in which the same sequence is found on each strand, but in the opposite direction, leading to the formation of a repetitious inversion.
Pandemic A worldwide epidemic.
Parasite An organism able to live on and cause damage to another organism.
Passive immunity Immunity resulting from transfer of antibodies from an immune to a nonimmune individual.
Pasteurization A process using mild heat to reduce the microbial level in heat-sensitive materials.
Pathogen An organism able to inflict damage on a host it infects.
Peptidoglycan The rigid layer of bacterial walls, a thin sheet composed of *N*-acetylglucosamine, *N*-acetylmuramic acid, and a few amino acids.
Periplasmic space The area between the plasma membrane and the cell wall, containing certain enzymes involved in nutrition.
Peritrichous flagellation Having flagella attached to many places on the cell surface.
Phagocyte A body cell able to ingest and digest foreign particles.
Phagocytosis Ingestion of particulate material such as bacteria by protozoa and phagocytic cells of higher organisms.
Phasmid A phage genome which contains the origin of replication of a plasmid.
Phenotype The characteristics of an organism observable by experimental means. Compare with *Genotype.*
Phospholipid Lipids containing a substituted phosphate group and two fatty acid chains on a glycerol backbone.
Photoautotroph An organism able to use light as its sole source of energy and CO_2 as sole carbon source.
Photoheterotroph An organism using light as a source of energy and organic materials as carbon source.
Photophosphorylation Synthesis of high-energy phosphate bonds as ATP, using light energy.
Photosynthesis Light-driven ATP synthesis. See also *Anoxygenic photosynthesis* and *Oxygenic photosynthesis.*
Phototaxis Movement toward light.
Phototroph An organism that obtains energy from light.
Phylogeny The ordering of species into higher taxa based on evolutionary relationships.
Phylum A taxonomic entity embodying a group of other taxonomic entities of lesser rank (e.g., the eubacteria comprise one phylum).
Pilus A fimbria-like structure that is present on fertile cells, both Hfr and F^+, and is involved in DNA transfer during conjugation. Sometimes called *Sex pilus.* See also *Fimbria.*
Pinocytosis In protozoa, the uptake of macromolecules into a cell by a drinking type of action.
Plaque A localized area of virus lysis on a lawn of cells.
Plasma The noncellular portion of blood.
Plasma cell A large differentiated B lymphocyte specializing in abundant (but short term) antibody production.
Plasma membrane The thin structure enclosing the cytoplasm, composed of phospholipid and protein, in a bimolecular leaflet structure.
Plasmid An extrachromosomal genetic element not essential for growth.

Platelet A noncellular disc-shaped structure containing protoplasm found in large numbers in blood and functioning in the blood clotting process.
Polar flagellation Condition of having flagella attached at one end or both ends of the cell.
Polyclonal antiserum A mixture of antibodies to a variety of antigens or to a variety of determinants on a single antigen.
Polymer A large molecule formed by polymerization of monomeric units.
Polymorphonuclear leukocyte (PMN) Small, actively motile white blood cells containing many lysosomes and specializing in phagocytosis.
Polypeptide A stretch of several amino acids linked together by peptide bonds.
Precipitation A reaction between antibody and soluble antigen resulting in a visible mass of antibody-antigen complexes.
Primer A short stretch of RNA or DNA used as a starting point for nucleic acid synthesis.
Probe A short stretch of nucleic acid complementary in base sequence to a nucleic acid from a given source, and used to selectively isolate the nucleic acid from a mixture of other nucleic acids.
Procaryote A cell or organism lacking a true nucleus, usually having its DNA in a single molecule.
Promoter The binding site for RNA polymerase, near the operator.
Prophage The state of a temperate virus when it is integrated into the host genome.
Prophylactic Treatment, usually immunologic, designed to protect an individual from a future attack by a pathogen.
Prostheca A cytoplasmic extrusion from a cell such as a bud, hypha, or stalk.
Prosthetic group The tightly bound, nonprotein portion of an enzyme; not the same as *Coenzyme*.
Protein A polymeric molecule consisting of one or more polypeptides.
Proton-motive force An energized state of a membrane created by expulsion of protons through action of an electron transport chain. See also *Chemiosmosis*.
Protoplasm The complete cellular contents, plasma membrane, cytoplasm, and nucleus; usually considered to be the living portion of the cell, thus excluding those layers peripheral to the plasma membrane.
Protoplast A cell from which the wall has been removed.
Prototroph The parent from which an auxotrophic mutant has been derived. Contrast with *Auxotroph*.
Provirus See *Prophage*.
Psychrophile An organism able to grow at low temperatures.
Pure culture An organism growing in the absence of all other organisms.
Pyogenic Pus-forming; causing abscesses.
Pyrogenic Fever-inducing.
Quarantine The limitation on the freedom of movement of an individual, to prevent spread of a disease to other members of a population.
Radioimmunoassay An immunological assay employing radioactive antibody for the detection of certain antigens in serum.
Reading-frame shift See *Frame shift*.
Recombinant DNA DNA molecules containing DNA from two or more sources.
Recombination Process by which genetic elements in two separate genomes are brought together in one unit.
Redox See *Oxidation-reduction reaction*.
Reduction A process by which a compound accepts electrons to become reduced.
Regulation Processes that control the rates of synthesis of proteins. Induction and repression are examples of regulation.
Replacement vector A cloning vector, such as a bacteriophage, in which most or all of the DNA of the vector can be replaced with foreign DNA.
Replication Conversion of one double-stranded DNA molecule into two identical double-stranded DNA molecules.
Repression The process by which the synthesis of an enzyme is inhibited by the presence of an external substance, the repressor.
Respiration Catabolic reactions producing ATP in which either organic or inorganic compounds are primary electron donors and inorganic compounds are ultimate electron acceptors.
Retrovirus An animal virus containing single-stranded RNA as its genetic material and which can produce a complementary DNA off of the RNA template by action of the enzyme, reverse transcriptase.
Ribonucleic acid (RNA) A polymer of nucleotides connected via a phosphate-ribose backbone, involved in protein synthesis.
Ribosome A cytoplasmic particle composed of RNA and protein, which is part of the protein-synthesizing machinery of the cell.
Secretion vector A DNA vector in which the protein product is both expressed and secreted (excreted) from the cell.
Septicemia Invasion of the bloodstream by microorganisms; bacteremia.
Serology The study of antigen-antibody reactions in vitro.
Serum Fluid portion of blood remaining after the blood cells and materials responsible for clotting are removed.
Shuttle vector A DNA vector which can replicate in two different organisms, used for moving DNA between unrelated organisms.
Site-directed mutagenesis The insertion of a different nucleotide at a specific site in a DNA molecule using synthetic DNA methodology.
Slime layer A diffuse layer of polysaccharide exterior to the cell wall in some bacteria. See also *Capsule* and *Glycocalyx*.
Species A collection of closely related strains.
Spheroplast A spherical, osmotically sensitive cell derived from a bacterium by loss of some but not all of the rigid wall layer. If all the rigid wall layer has been completely lost, the structure is called a *Protoplast*.
Spore A general term for resistant resting structures formed by many bacteria and fungi.
Stalk An elongate structure, either cellular or excreted, which anchors a cell to a surface.
Stationary phase The period during the growth cycle of a population in which growth ceases.
Sterile Free of living organisms.
Sterilization Treatment resulting in death of all living organisms and viruses in a material.
Strain A population of cells all descended from a single cell; a clone.
Substrate The compound undergoing reaction with an enzyme.

Substrate-level phosphorylation Synthesis of high-energy phosphate bonds through reaction of inorganic phosphate with an activated (usually) organic substrate.
Supercoil Highly twisted form of circular DNA.
Suppressor A mutation that restores wild-type phenotype without affecting the mutant gene, usually arising by mutation in another gene.
Symbiosis A relationship between two organisms.
Syntrophy A nutritional situation in which two or more organisms combine their metabolic capabilities to catabolize a substance not capable of being catabolized by either one alone.
Systemic Not localized in the body; an infection disseminated widely through the body is said to be systemic.
Taxis Movement toward or away from a stimulus.
Temperate virus A virus which upon infection of a host does not necessarily cause lysis but may become integrated into the host genetic material. See *Lysogeny*.
Thermophile An organism living at high temperature.
Tolerance In reference to immunology, the acquisition of nonresponsiveness to a molecule normally recognized by the immune system.
Toxigenicity The degree to which an organism is able to elicit toxic symptoms.
Toxin A microbial substance able to induce host damage.
Toxoid A toxin modified so that it is no longer toxic but is still able to induce antibody formation.
Transcription Synthesis of a messenger RNA molecule complementary to one of the two double-stranded DNA molecules.
Transduction Transfer of genetic information via a virus particle.
Transfer RNA (tRNA) A type of RNA involved in the translation process; each amino acid is combined with one or more specific transfer RNA molecules.
Transformation Transfer of genetic information via free DNA.
Translation The process during protein synthesis in which the genetic code in messenger RNA is translated into the polypeptide sequence in protein.
Transpeptidation The formation of peptide bonds between the short peptides present in the cell wall polymer, peptidoglycan.
Transposon A genetic element that can move from place to place; contains an insertion element at each end.
Transposon mutagenesis Insertion of a transposable element within a gene generally leads to inactivation of that gene, resulting in the appearance of a mutant phenotype.
Tricarboxylic acid cycle (citric acid cycle, Krebs cycle) A series of steps by which pyruvate is oxidized completely to CO_2, also forming NADH, which allows ATP production.
Vaccine Material used to induce antibody formation resulting in immunity.
Vacuole A small space in a cell containing fluid and surrounded by a membrane. In contrast to a vesicle, a vacuole is not rigid.
Vector An agent, usually an insect or other animal, able to carry pathogens from one host to another. Also: A genetic element able to incorporate DNA and cause it to be replicated in another cell.
Viable Alive; able to reproduce.
Viable count Measurement of the concentration of live cells in a microbial population.
Virion A virus particle; the virus nucleic acid surrounded by protein coat.
Viroid A small RNA molecule that has viruslike properties.
Virulence Degree of pathogenicity of a parasite.
Virus A genetic element containing either DNA or RNA that is able to alternate between intracellular and extracellular states, the latter being the infectious state.
Wobble In reference to the genetic code, the concept that base pairing between codon and anticodon is not as critical in the third position as in the first or the second.
Xenobiotic A completely synthetic chemical compound not naturally occurring on earth.
Xerophile An organism adapted to growth at very low water potentials.
Zoonoses Diseases primarily of animals which are occasionally transmitted to humans.
Zygote In eucaryotes, the single diploid cell resulting from the conjugation of two haploid gametes.

Photo Credits and Acknowledgments

CHAPTER 1 **1.2b** Herbert Voelz. **1.5** Richard Feldmann. **1.6a** Richard Feldmann. **1.11** Councilman Morgan. **1.12** Huntington Potter and David Dressler. **1.13** Arthur Kelman. **1.16a** Hans Hippe. **1.18** Carolina Biological Supply Co. **1.20a** Thomas D. Brock; **1.20b** Dennis Kunkel; **1.20c** Carolina Biological Supply Co. **1.24** Bob Harris. **1.27a,b** Carolina Biological Supply Co; **1.27c** Arthur M. Nonomura and Microbio Resources, Inc. **1.28** Kenneth B. Raper. **1.31b** D. Foster and the Woods Hole Oceanographic Institute. **1.32** NASA. **1.33** Carl Zeiss, Inc. **1.38** JEOL Company. **1.39a** Carolina Biological Supply Co; **1.39b** Thomas D. Brock; **1.39c,d** Norbert Pfennig.

CHAPTER 3 **3.2** D.E. Caldwell. **3.3** M. Kessel and Y. Cohen. **3.4a** Bryan Larsen; **3.4b** A. Umeda, et al. **3.5a** L.A. Mangels, et al.; **3.5b** A. Umeda and K. Amako. **3.7a** Walther Stoeckenius. **3.14** N.J. Lang. **3.22a** J.L. Pate; **3.22b** T.D. Brock and S.F. Conti; **3.22c,d** A. Umeda and K. Amako. **3.31** A. Umeda and K. Amako. **3.36** E. Frei and R.D. Preston. **3.38** E. Leifson. **3.40** R. Jarosch. **3.44a** Melvin S. Fuller. **3.46** Sydney Tamm. **3.47** M. Haberey. **3.50** D. Aswad and D. Koshland, Jr. **3.52** J.P. Duguid and J.F. Wilkinson. **3.53** C. Brinton, et al. **3.54a** Elliot Juni; **3.54b** Frank Dazzo and Richard Heinzen. **3.56a** Allan Konopka; **3.56b** A.E. Walsby. **3.57** A.E. Walsby. **3.59** H.S. Pankratz, T.C. Beaman, and Philipp Gerhardt. **3.62b,c** D.W. Fawcett **3.64** T. Slankis and S. Gibbs. **3.65** Stanley C. Holt. **3.68a** E. Guth, T. Hashimoto, and S.F. Conti; **3.68b** D.W. Fawcett.

CHAPTER 4 **4.3** Richard Feldmann. **4.15d** Richard Feldmann.

CHAPTER 5 **5.40a** R. Sivendra; **c** A.E. Walsby; **5.40b,d,e** Thomas D. Brock. **Methods Box** Page 160, 161. Pharmacia, Inc. Page 163. Schleicher and Schnell, Inc. Page 163. Carl Marrs and Martha Howe. Page 164. Madigan.

CHAPTER 6 **6.4a** J.T. Finch. **6.8a** W.F. Noyes; **6.8b** P.W. Choppin and W. Stoeckenius; **6.8c** M. Wurtz. **6.10** Paul Kaplan. **6.11** Paul Kaplan. **6.18** R.C. Valentine. **6.32** M. Wurtz. **6.39b** G. Chardonnet and S. Dales. **6.43a** Arthur J. Olson. **6.44** Erskine Caldwell. **6.46a** P.W. Choppin and W. Stoeckenius. **6.47** Heather Mayor. **6.48** Alexander Erb and Jerome Vinograd. **6.51** R.W. Horne. **6.52** Richard Feldmann.

CHAPTER 7 **7.4** Huntington Potter and David Dressler. **7.10** C. Brinton. **7.25** S.F. Conti and T.D. Brock.

CHAPTER 8 **8.3b** Frederick R. Blattner. **8.5a** William Reznikoff; **8.5b** James Shapiro.

CHAPTER 9 **9.16a** Katherine M. Brock; **9.16b** Thomas D. Brock. **9.17** Thomas D. Brock. **9.18** Thomas D. Brock. **9.22** Coy Laboratories, Inc. **9.31** B.A. Bridges. **9.32** Radiation Technologies, Inc. **9.34** Carlos Pedros-Alio and T.D. Brock. **9.37** Thomas D. Brock. **9.42** Thomas D. Brock.

CHAPTER 10 **10.4a** Queue Systems, Inc.; **10.4b,c** Novo Industri. **10.5** Novo Industri. **10.6** Elmer L. Gaden, Jr. **10.8b** Pfizer Inc. **10.24** The Christian Brothers Winery. **10.27a,b** American Mushroom Institute; **10.2c,d** Bob Harris.

CHAPTER 11 **11.4** T. Lie. **11.5** C. Lai, M.A. Listgarten, and B. Rosan. **11.6** I.L. Schechmeister and J. Bozzolla. **11.8** Dwayne Savage. **11.9** Dwayne Savage and R.V.H. Blumershine. **11.12** Harlan Sprague Dawley, Inc. **11.13** Walter Reed Army Institute of Research. **11.15a** E.T. Nelson, J.D. Clements, and R.A. Finkelstein; **11.15b** J.W. Costerton. **11.16** Leon J. Le Beau. **11.19** H.W. Smith and S. Halls. **11.24** Thomas D. Brock. **11.25** J.G. Hirsch. **11.27** C.L. Wisserman.

CHAPTER 12 **12.3a** Richard J. Feldmann; **12.3b** Craig Smith. **12.19** R.J. Poljak. **12.22** C. Weibull, W.D. Bickel, W.T. Hashius, K.C. Milner, and E. Ribi. **12.26** Wellcome Research Laboratories. **12.37** E. Munn. **in Box, page 455** Perkin-Elmer Cetus. **in Box, page 465** Keystone Diagnostics, Inc.

CHAPTER 13 **13.3a** Theodor Rosebury; **13.3b** Leon J. Le Beau. **13.5** Leon J. Le Beau. **13.6e** Leon J. Le Beau. **13.7a** William B. Cherry; **13.7b,c** Dharam Ablashi and Robert C. Gallo. **13.11b** Victor Tsang. **13.12** Lucy S. Tompkins.

CHAPTER 15 **15.3** Franklin H. Top, Communicable and Infectious Diseases, 6th ed. C.V. Mosby Co., St. Louis, MO, 1968. **15.4** Franklin H. Top, Communicable and Infectious Diseases, 6th ed. C.V. Mosby Co., St. Louis, MO, 1968. **15.5** Isaac Schechmeister. **15.6** Franklin H. Top, Communicable and Infectious Diseases, 6th ed. C.V. Mosby Co., St. Louis, MO. 1968. **15.7** Carey S. Callaway. **15.10** Aaron Friedman. **15.12** C.H. Binford **in** Top, F.H. and P.F. Wehrle, (ed.), Communicable and Infectious Diseases, 7th ed., C.V. Mosby Co., St. Louis, MO. 1972. **15.14** Irene T. Schulze. **15.19** Morris D. Cooper. **15.20** Theodor Rosebury. **15.21** S. Olansky and L.W. Shaffer, **in** Top, F.H. and P.F. Wehrle (ed.), Communicable and Infectious Diseases, 7th ed., C.V. Mosby Co, St. Louis, MO. 1972. **15.22** Gordon A. Tuffli. **15.24a-f,h** Centers for Disease Control; **15.24g** Jonathan W.M. Gold. **15.25** Centers for Disease Control. **15.26** Centers for Disease Control. **15.28** Willy Burgdorfer. **15.32a** D.E. Feely, S.L. Erlandsen, and D.G. Case; **15.32b** S.L. Erlandsen. **15.34a** Louisville Water Company; **15.34b,c** Metropolitan Water District of Southern California.

CHAPTER 16 **16.5a** Hilda Canter-Lund **16.7** Michael T. Madigan. **16.8** Norbert Pfennig. **16.12** Yehuda Cohen and Moshe Shilo. **16.14** Thomas Jensen. **16.16** Norbert Pfennig. **16.19** Thomas D. Brock. **16.21a** Bill Strode; **16.21b,c** Thomas D. Brock. **16.22** Thomas D. Brock. **16.35** B.V. Hofsten. **16.36** Katherine M. Brock. **16.37** Thomas D. Brock. **16.38** Thomas D. Brock.

CHAPTER 17 **17.1** Thomas D. Brock. **17.3** Thomas D. Brock. **17.7a** Thomas D. Brock. **17.14** T.R.G. Gray. **17.15** Jannasch, H.W., C.O. Wirsen and C.D. Taylor. 1982. Deep-Sea Bacteria: Isolation in the absence of decompression. Science 216: 1315-1317. **17.18a** Dudley Foster, Woods Hole Oceanographic Institution; **17.18b** James Agviar, Woods Hole Oceanographic Institution; **17.18c** Carl Wirsen, Woods Hole Oceanographic Institution. **17.19a** Colleen Cavanaugh. 1981. Science. 209:340-342; **17.19b** Colleen Cavanaugh. 1983. Nature. 302:58-61. **17.20a** Robert D. Ballard; **17.20b** Dudley Foster, Woods Hole Oceanographic Institution. **17.31** Thomas D. Brock. **17.33** Thomas D. Brock. **17.35** Thomas D. Brock. **17.38** Thomas D. Brock. **17.39** Standard Oil Company. **17.44** Thomas D. Brock. **17.45** Thomas D. Brock. **17.47** Jo Handelsman. **17.49** Wood, Keith V., D.W. Ow, M. DeLuca, J.R. deWet, D.R. Heliniski, and S.H. Howell. 1986. Science. 234:856-859. **17.50** Joe Burton. **17.51** Ben B. Bohlool, NifTAL Project, University of Hawaii. **17.52** Joe Burton. **17.54** F.B. Dazzo and W.J. Brill. 1977. Applied and Environmental Microbiology. 33:132-136. **17.55a** Ben B. Bohlool; **17.55b** Mark Dudley and Sharon Long. **17.56a** Phil Gates, Biological Photo Service; **17.56b** P.J. Dart and F.V. Mercer. **17.58** B. Dreyfus. **17.59** J-H. Becking. **17.60** J-H. Becking.

CHAPTER 18 **18.1** Malcolm Walter. **18.2** J.W. Schopf. 1968. Journal of Paleontology Pt. 1 42:651-688. **18.5** Carl Woese. **18.6** Carl Woese. **18.14** W. Zillig.

CHAPTER 19 **19.2b** Norbert Pfennig. **19.3a** Remsen, C.C., S.W. Watson, J.B. Waterbury, and H.G. Trüper. 1968. Journal of Bacteriology. 95:2374–2392; **19.3b** Jeffrey C. Burnham and S.F. Conti; **19.3c** G. Cohen-Bazire; **19.3d** F. Rudy Turner and Howard Gest. **19.4a–e** Norbert Pfennig; **19.4f** Peter Hirsch. **19.5a,b and d** Norbert Pfennig; **19.5c** Peter Hirsch; **19.5e** R. Scherrer. **19.6a and b** Norbert Pfennig; **19.6c** Madigan, M.T. and T.D. Brock. 1977. Journal of General Microbiology. 102:279–285; **19.6d,e** Gorlenko, V.M. and T.A.

Pivovarova. 1977. Bulletin of Academy of Sciences of U.S.S.R. Biological Series, No. 3, 396–409. **19.8** F. Rudy Turner and Howard Gest. **19.12** Douglas Caldwell. 1975. Canadian Journal of Microbiology. 21: 362–376. **19.13** Douglas Caldwell. **19.14** Kurt Hanselmann. **19.15** John Ormerod and the Norwegian Institute for Water Research. **19.17** Kurt Hanselmann. **19.18** Susan Barns and Norman Pace. **19.19** Edwards, M.R., D.S. Berns, W.D. Ghiorse, and S.C. Holt. 1968. Journal of Phycology. 4:283–298. **19.22** Kit W. Lee. **19.23** T. Burger-Wiersma. **19.24** S.W. Watson. 1971. International Journal of Systemic Bacteriology. 21: 254–270. **19.25** S.W. Watson. 1971. International Journal of Systematic Bacteriology. 21:254–270. **19.27** Jessup M. Shively. **19.29** Jessup M. Shively. **19.30** Frank Mayer. **19.31a** Smith, U. and D.W. Ribbons. 1970. Archiv Mikrobiologie. 74: 116–122; **19.31b** Smith, U., D.W. Ribbons, and D.S. Smith. 1970. Tissue and Cell. 2:513–520. **19.34** Charles R. Fisher. **19.35a and f** Norbert Pfennig; **19.35c–e** Fritz Widdel; **19.35b** N. Pfennig, F. Widdel, and H.G. Trüper. 1981. The dissimilatory sulfate-reducing bacteria. Pages 926–940. **In** Starr, M.P., H. Stolp, H.G. Trüper, A. Balows, and H.G. Schlegel,(eds.). The Prokaryotes—A Handbook of Habitats, Isolation, and Identification of Bacteria. Springer-Verlag, New York, N.Y. **19.37a** Pate, J.L. and E.J. Ordal. 1965. Journal of Cell Biology. 27:133–150; **19.37b** J.T. Staley. 1968. Journal of Bacteriology. 95:1921–1942; **19.37c** V.M. Gorlenko. **19.39a** Einar Leifson; **19.39b,c** Cohen-Bazire, G., R. Kunisawa, and J.S. Poindexter. 1966. Journal of General Microbiology. 42:301–308. **19.40** Bauld, J. and J.T. Staley. 1976. Journal of General Microbiology. 97:45–55. **19.42** W.C. Ghiorse. 1984. Annual Review of Microbiology. 38:515. **19.44** Peter Hirsch. **19.45** Conti, S.F. and P. Hirsch. 1965. Journal of Bacteriology. 89:503–512. **19.46a and b** Noel Krieg; **19.46c** Stanley Erlandsen; **19.46d** H.D. Raj. **19.47** Blakemore R.P. and R.B. Frankel. 1981. Scientific American. 245:58–65. **19.48a** Stolp, H. and H. Petzold. 1962. Phytopathologische Zeitschrift. 45:373; **19.48b,c** Burnham, J.C., T. Hashimoto, and S.F. Conti. 1968. Journal of Bacteriology. 96:1366–1381. **19.50a** Canale-Parola, E., S.C. Holt, and Z. Udris. 1967. Archiv Mikrobiologie. 59:41–48; **19.50b** Blakemore, R.P. and E. Canale-Parola. 1973. Archiv Mikrobiologie. 89:273–289. **19.51a** Joseph R. and E. Canale-Parola. 1972. Archiv Mikrobiologie. 81:146–168. **19.52** Ryter, A. and J. Pillot. 1965. Annale de l'Institut Pasteur 109:552–562. **19.53** Paster, B.J. and E. Canale-Parola. 1985. Applied Environmental Microbiology. 50:212–219. **19.54** Michael Richard. **19.55** Thomas D. Brock. **19.56** Michael Richard. **19.58** David White. **19.60a and b** Hans Reichenbach; **19.60c** Stephens, K., G.D. Hegeman, and D. White. 1982. Journal of Bacteriology. 149:739–747. **19.61** Grillone, P.L. and J. Pangborn. 1975. Journal of Bacteriology. 124:1558–1565. **19.62** Herbert Voelz. **19.63a** Thomas D. Brock; **19.63b and c** Hoeniger, J.F.M., H.D. Tauschel, and J.L. Stokes. 1973. Canadian Journal of Microbiology. 19:309–313 and plates I–VII. By permission of The National Research Council of Canada. **19.64** W.C. Ghiorse. 1984. Annual Review of Microbiology. 38:515. **19.65a** Arthur Kelman. **19.66** L.P. Lin and H.L. Sadoff. **19.67a and b** J.H. Becking. 1981. The family Azotobacteraceae. **In** Starr, M.P., H. Stolp, H.G. Trüper, A. Balows, and H.G Schlegel (eds.) 1981. The Prokaryotes—A Handbook on Habitats, Isolation, and Identification of Bacteria. Springer-Verlag, Berlin. **19.68a** Michael K. Ochman; **19.68b** J.H. Becking. 1981. The family Azotobacteriaceae. **In** Starr, M.P., H. Stolp, H.G. Trüper, A. Balows, and H.G. Schlegel (eds.) 1981. The Prokaryotes—A Handbook on Habitats, Isolation, and Identification of Bacteria. Springer-Verlag, Berlin. **19.72** Kenneth H. Nealson. **19.73a** Arthur Kelman. **19.76a** Willy Burgdorfer; **19.76b** Devauchelle, G., G. Meynadier, and C. Vago. 1972. Journal of Ultrastructure Research. 38:134–148. **19.77** Robert R. Friis. **19.79** T. Beveridge. 1980. Canadian Journal of Microbiology. 26:235–242. **19.80** A. Umeda, et al. 1980. Journal of Bacteriology. 141:838–844. **19.81** D. Moyles and R.G.E. Murray. **19.83a** Thomas D. Brock; **19.83b** Bryan Larsen, University of Iowa. **19.84** Otto Kandler. **19.85** Hans Hippe. **19.86** J.R. Norris. **19.89** Dieter Claus. **19.90** Alan Rodwell. **19.93** David L. Williamson. **19.94** Krulwich, T.A. and J.L. Pate. 1971. Journal of Bacteriology. 105:408–412. **19.95** Krulwich, T.A. and J.L. Pate. 1971. Journal of Bacteriology. 105:408–412. **19.96** Veldkamp, Hans, G. van den Berg, and L.P.T.M. Zevenhuizen. 1963. Antonie van Leeuwenhoek 29:35–51. **19.99a** N. Rist, Pasteur Institute, Paris; **19.99b** V. Lorian. 1968. Annual Review of Respiratory Diseases. 97:1133–1135; **c** Centers for Disease Control, Atlanta, GA. **19.101** Hubert and Mary P. Lechevalier. **19.102a** Peter Hirsch; **b** Hubert and Mary P. Lechevalier. **19.105** Alma Dietz. **19.106** Eli Lilly & Co. **19.107a** Thomas D. Brock. **19.108a,b** Mary Reedy. **19.110** Alexander Zehnder. **19.111** Zeikus, J.G. and V.G. Bowen. 1975. Comparative ultrastructure of methanogenic bacteria. Canadian Journal of Microbiology. 21:121–129. Reproduced by permission of the National Research Council of Canada. **19.112a and b** Helmut König and K.O. Stetter; **19.112c** Stephen Zinder. **19.120a** From Brock, T.D., K.M. Brock, R.T. Belly, and R.L. Weiss. 1972. Archiv Mikrobiologie. 84:54–68. **19.120b** H. König and K.O. Stetter. **19.121** H. König and K.O. Stetter. **19.122** H. König and K.O. Stetter. **19.123a** H. König and K.O. Stetter; **19.123b** Stetter, K.O, H. König, and E. Stackebrandt. 1983. Systematic and Applied Microbiology. 4:535–551. **19.124a** H. König and K.O. Stetter; **19.124b** Fiala, G. and K.O. Stetter. 1986. Archives of Microbiology. 145:56–61. **19.125** H. König and K.O. Stetter. **19.126** Thomas D. Brock. **19.127** Thomas D. Brock. **Table 19.17** Hans Reichenbach.

Index